OXFORD MEDICAL PUBLICATIONS

Iatrogenic Diseases

Iatrogenic Diseases

Third Edition

Edited by

P.F. D'ARCY, OBE, B PHARM, PH D, D SC, FPS, CCHEM, FRSC, FPSNI
Professor of Pharmacy, The Queen's University of Belfast, Formerly Professor of Pharmacology, and Dean, Faculty of Pharmacy, University of Khartoum

and

J.P. GRIFFIN, BSC, PH D, MB, BS, MRCP, FRC PATH
Formerly Professional Head of Medicines Division, Department of Health and Social Security. Currently Director, Association of the British Pharmaceutical Industry; Honorary Consultant Clinical Pharmacologist, Lister Hospital, Stevenage, Hertfordshire

OXFORD NEW YORK TOKYO

OXFORD UNIVERSITY PRESS

1986

Oxford University Press, Walton Street, Oxford OX2 6DP

Oxford New York Toronto
Delhi Bombay Calcutta Madras Karachi
Petaling Jaya Singapore Hong Kong Tokyo
Nairobi Dar es Salaam Cape Town
Melbourne Auckland

and associated companies in
Beirut Berlin Ibadan Nicosia

Oxford is a trade mark of Oxford University Press

British Library Cataloguing in Publication Data
Iatrogenic diseases.—3rd ed.—(Oxford medical
publications)
1. Iatrogenic diseases   2. Drugs—Toxicology
3. Drugs—Side effects
I. D'Arcy, P.F.   II. Griffin, J.P.
615'.7042   RC90
ISBN 0–19–261440–1

Library of Congress Cataloging in Publication Data
Main entry under title:
Iatrogenic diseases.
(Oxford medical publications)
Includes bibliographies and index.
1. Drugs—Side effects.   2. Iatrogenic diseases.
I. D'Arcy, P.F. (Patrick Francis)   II. Griffin, J.P.
(John Parry)   III. Series. [DNLM: 1. Drug Therapy—
adverse effects. 2. Iatrogenic Diseases. QZ 42 I114]
RM302.5.I28   1986   615'.7042   85–15593
ISBN 0–19–261440–1

Set by Colset Private Limited, Singapore
Printed in Great Britain by
St Edmundsbury Press
Bury St Edmunds, Suffolk

Foreword

by Professor Sir John Butterfield, OBE, MD, FRCP
Regius Professor of Physic, University of Cambridge

It is a privilege and a pleasure to introduce the new updated edition of this very successful reference work about iatrogenic diseases from medicines, one of the most prevalent conditions facing modern health services and occupying countless hospital beds all over the Western world.

Iatrogenic — springing from its two Greek roots, 'doctor-generated' — today carries with it something of the inevitability of Greek tragedy in which, if we are wise, patients will be cast as the heroes. The initial enemy is the old one of disease of one sort or another. But these diseases are now unexpectedly allied with the very effects — the adverse reactions, of today's pharmacological and pharmaceutical triumphs. So it all comes from the inevitable consequences of the ingenuity and enquiring minds of bioscientists and doctors questing for new and better ways to combat and alleviate, even prevent, the initial diseases.

The Fates, of course, know these therapeutic attempts are laudable but also that they are doomed to some degree of failure — recently less and less, one hopes, as more and more compounds are tested to find those with the fewest side-effects and adverse reactions. But the latter remain an ever present consideration, every time the physician begins to write a prescription. For all attempts to find the perfect remedy, no matter how nearly we reach it, will be frustrated to a greater or lesser extent by the innate feature of our patients as a whole, namely, individual variations without which, of course, it would not have been possible for us to evolve as a species at all.

Once one has variations between individuals in the amino-acid sequence of this enzyme or that cell surface protein or of a mucopolysaccharide body constituent, there must be variation in the tertiary structures thereof and therefore in the binding properties for passing drugs. The consequences are differences in the way individuals can handle, distribute, metabolize, and excrete the new medicines. This is the so-called trade off of the modern therapeutic revolution as between the improved therapeutic responses and the inevitable risks from the side-effects and adverse reactions and, of course, idiosyncrasies. This trade off has to be identified and tackled successively by the developer of the new medicine, by the statutory authorities responsible to society for the safety of medicines, and, ultimately, by a watchful medical profession ascertaining and reporting the adverse reactions as marketing of the new products extends further and further into the population. By the time products have found their way into this present book there will have been a powerful consensus of opinion all down the line supporting the case for the compounds' efficacy being acceptable in the face of any risks presently recognized and recorded.

One says risks presently recognized, because, as time passes, new drugs are increasingly used among elderly patients; however, with the passage of time any individual's body constituents at the molecular or cell surface level are also altered by life events on top of their genetic constitution — infections generate antibodies, cosmic radiation affects protein synthesis, contact with toxic elements at work or in the general environment in this industrial era, all these factors lead to more or less random variations in the make-up of the tissues and organ performances of elderly people.

Of course, by the time the increasing numbers of geriatric cases presenting themselves to all the specialist clinics today get there, doctors are faced also with the treatment and care of more than one organ deficiency. Consequently, the doctors have to grapple with additional complications when trying to plan treatment for elderly patients. Often they are dealing with multiple pathologies which incidentally make each elderly patient a much more fascinating and indeed often unique case disease mixture. For example the sequelae of an earlier renal-tract infection may affect the excretion of any medicines prescribed; ageing of the gastrointestinal tract and personal food fads may affect the absorption of medicines; metabolism of drugs by the liver may be affected not only by ageing changes in the hepatocytes, but by social problems such as bad drinking habits in earlier life. Furthermore, the vascular complications of previously unrecognized or untreated hypertension can disturb the pattern of cerebral circulation, making some parts of the brain less susceptible to circulating levels of drugs so larger doses may be needed to achieve the desired effects of compounds acting through the central nervous system. There seems little doubt that the reverse can obtain too in elderly patients, that is, lower doses may be desirable. And so on.

It is the interaction of all these and many more variables which explain the well-established fact the side-effects and adverse reactions are so much more prevalent in elderly patients presenting themselves in increasing proportions for therapy in hospitals. And this whole scene is even further compounded when the doctor, pre-registration house officer to consultant, has to treat several chronic diseases in the same patient, often during an added acute episode of, say, pneumonia. Thus we find these patients taking an antibiotic, an antihypertensive formulation, an anti-anginal compound, and maybe a diuretic or antidiabetic pill. This poses a therapeutic challenge in getting patients' co-operation taking pills after discharge from hospital. There are other thera-peutic problems which may lead to the prescription of incompatible drugs or drugs where side-effects may enhance dangerously the action of another compound in simultaneous use.

Thus the modern practitioner not only has to embrace much more fundamental knowledge than his forerunners — even his predecessors who trained 10 years ago — but he or she has to be at least as intrepid if not more so to start therapy. No wonder then that to avoid unnecessary suffering, to reduce the consequences of adverse reactions, to ensure that side-effects are recognized as such and not treated separately as additional diseases, easy and frequent access has to be made in the clinic and in the ward office to a compendium like the present volume.

We are all beginning to realize that the old advice, that to learn about and know thoroughly one compound, in each of the important categories of medicines meant that you could confidently practise, is no longer tenable. Referring physicians from outside hospitals and partners and trainees in general practice are all very likely these days to have different favourites from oneself. So the advice to all physicians today must be 'Don't feign knowledge of drugs you don't know about; if in any doubt at all, look them up'.

There is one last point which should be made for both the specialist and the generalist. It concerns the ever present need to think of the patient as a whole. Don't forget, for example, that the effects of drugs prescribed during lactation may have important consequences for the breast-fed baby.

A particular pleasure in presenting this book is to be able to say complimentary things about the authors, whom I have known for over a decade. Both are distinguished in their individual fields, but important too is the fact that they complement each other. Indeed one sees in this volume how the pharmacist and the physician interact fruitfully. They have also taken great advantage of their experiences at the centre of the web of pharmaceutical monitoring systems, so that they have had the opportunity of ensuring that this book is the most up-to-date compendium of iatrogenic medical effects. Thus, I am pleased to reiterate the reliability, the comprehensiveness, and the essential practicability of this volume. I wish its new edition every success and every success, too, to all the physicians who use it in their attempts to alleviate or prevent disease.

Preface to the third edition

The major causes of adverse reactions to medicines were categorized in the Preface to the first edition, and in general these have not altered. However, awareness of another major factor in the interaction between the patient and his or her medicine is becoming increasingly recognized: the contribution of the disease state in the manifestation of certain adverse reactions. The best known example of this is the characteristic ampicillin-induced rash in the patient with glandular fever. A more recently recognized example is the fact that patients with Hodgkin's disease are more likely to develop peripheral neuropathy during vincristine treatment than patients being treated with comparable doses for other malignancies. In the coming years it is to be hoped that greater emphasis will be placed on the identification of particular patient- and disease-related factors that may influence the reaction to therapeutic substances than to the over-hasty but popular and newsworthy condemnation of products that have produced adverse effects in a small number of patients. It is becoming increasingly necessary to determine the circumstances of an adverse event associated with exposure to a product as opposed to the more simple identification of association of an adverse reaction associated with a particular drug.

A classic example is the development of hepatorenal syndrome associated with benoxaprofen, where the mean duration of treatment of fatal cases was 8.5 ± 1.2 months and of non-fatal cases was 6.9 ± 1.9 months. The geographical distribution of the reports of benoxaprofen-associated hepatic and renal damage showed inequalities in distribution with over-representation of reports both by size of population and exact drug usage from Northern Ireland, Scotland, and England north of the Trent, with consequent under-representation in England south of the Trent and Wales. This is the converse of the reports of photosensitivity reactions associated with benoxaprofen which were relatively more common in the southern areas of Britain. These differences were statistically significant. For Wales and England south of the Trent, the number of photosensitivity reactions is larger than expectation on the basis of a proportionate share of the prescriptions ($p < 0.00002$) and the number of hepato-renal reactions, smaller ($p < 0.003$); for northern areas of Britain the converse applied.

It has therefore been postulated that patients in southern areas of the British Isles developed photosensitivity reactions and stopped taking benoxaprofen before serious damage to the liver and kidneys appeared.

The hepato-renal reactions to benoxaprofen therefore occurred predominantly in elderly female patients exposed to the substance for a considerable number of months and who lived in the northern areas of the British Isles, or who were housebound and did not get exposure to sunlight.

The public as well as the medical and pharmaceutical professions have become increasingly aware of the need to judge therapeutic substances, taking into account therapeutic benefit as well as potential risk. It has also become more generally recognized that the use of any therapeutic agent is inevitably attended by a small risk that the patient may react adversely to the prescribed agent. Risk–benefit assessment has therefore become a concept that is widely understood.

Within the last twelve months a further concept has been introduced and that is cost–benefit assessment of a medicine in purely economic terms. Attempts to quantitate the total economic benefit of treatment are becoming increasingly demanded. One of the more comprehensive studies in this area by Professor George Teeling Smith and Nicholas Wells has been summarized by them as follows:

In 1982, it is estimated that the pharmaceutical industry's medicines saved the NHS hospital service in England and Wales at least £1658 million. This saving is estimated *after* account has been taken of the cost of hospitalised cases of adverse drug reactions. In addition, the industry was responsible for the payment of an estimated £665 million in taxes to the Exchequer in England and Wales, including the taxes relating to pharmaceutical exports. These figures contrast with a total revenue for the industry from its sales to the National Health Service in England and Wales of £1225 million in 1982. In the latter year, it therefore appears that, overall, the pharmaceutical industry in England and Wales made a nett *contribution* to government finance of almost £1100 million. (Teeling-Smith, G., and Wells, N. 1985). The economic contribution of the industry in England and Wales. *Pharm. J.* **235**, 178–9.)

In a written Parliamentary reply on 13 November 1980

Sir George Young stated that in 1977, 129 366 patients were discharged from or died in hospitals in the United Kingdom due to adverse effects of medicinal agents. These figures from the Hospital Inpatient Enquiry for England and Wales includes iatrogenic adverse reactions to medicinal products as well as admission to intentional or accidental drug overdose and suicides. It is essential that this figure is put into perspective and this is demonstrated in Table 1.

Table 1   Discharges from or deaths in hospital attributable to the adverse effects of medicinal agents

Year	As percentage of all discharges and deaths	As percentage of all bed days
1977	2.43	0.45
1978	2.24	0.44
1979	2.04	0.40
1980	1.95	0.35
1981	1.92	0.37
1982	1.87	0.36
1983	1.63	0.30

Iatrogenic adverse reactions tend to occur at the two age extremes, i.e. in children and the elderly, whereas self-poisonings occur predominantly in females and in those aged 15–45 years and successful suicide in the older male sector of the population.

For 1983, adverse reactions to medicines due to all causes (i.e. iatrogenic adverse reactions and accidental and intentional overdose) resulted in 164 136 days of hospital bed occupancy out of a total of 55 245 770 days of hospital bed occupancy and thus represented a total cost of £13.31 million or 0.17 per cent of the cost of the hospital sector (source: Nicholas Wells, Office of Health Economics, London).

P. F. D'ARCY
Holywood,
Co. Down.
J. P. GRIFFIN
Digswell,
Herts.
November 1985

Preface to the first edition

The use of any therapeutic agent is inevitably attended by a small risk that the patient may react adversely to the prescribed agent. Adverse reactions can be caused by several different factors. In correlating the data for this book, it has been apparent that two factors are the major contribution to the manifestation of iatrogenic disease. These are the abnormal patient reaction to a drug, and the development of unexpected toxicity when several drugs are given in combination.

Predictable toxicity is the manifestation of secondary pharmacological actions; this is a hazard that can be well elucidated in the battery of tests to which the drug is subjected during its development stage. In such instances, assuming that the drug has a worthwhile place in therapy, the ratio of dosage of drug to produce the major effect, to dosage to produce a secondary (toxic) effect is the real factor which should be considered and not the 'built in' potentiality to the side-effect itself. Nevertheless, certain groups of patients exist who may be at risk from these predictable manifestations of toxicity. These patients are at peculiar risk because of a genetically determined defect of metabolism, or because their metabolism has been impaired by concomitant hepatic disease, or because their excretory function has been reduced by either liver or renal malfunction. In these instances, the drug or its metabolites may rapidly build up to toxic levels in the body, even at normal accepted therapeutic doses.

Intolerance is a lowered threshold to the normal pharmacological action of drugs. Individuals may vary widely from the well established norm in their reaction to drugs. The very old and the very young are liable to be more sensitive to drugs possibly because the metabolic and excretory mechanisms essential for the disposal of the drug are less efficient than in the adult. In addition, the reactions of the old or the young may also differ qualitatively from those of the adult.

Adverse reactions may follow the use of a drug, and these reactions may be unexpected, in that they are completely unrelated to the known toxicity of the drug. These reactions include hypersensitivity to the agent, in which the patient develops antibodies to the drug. The antigenic factor is usually a combination of drug with body protein. Skin rashes and eruptions are the most common symptoms of this type of allergic reaction, although haemolytic anaemia is not infrequent.

Idiosyncrasy involves a qualitatively abnormal response on the part of the patient to the drug; an example of this is drug-induced porphyria, in which a qualitatively abnormal response of porphyrin metabolism is induced by barbiturates in susceptible subjects. Similarly, mepacrine-induced haemolytic anaemia in glucose-6-phosphate dehydrogenase deficient subjects is an idiosyncratic reaction.

The role of polypharmacy in iatrogenic disease is not insignificant, since toxicities not shown by either drug singly may develop when used in combination. This ill-begotten offspring of medicine and pharmacy has been nurtured through the ages and William Withering wrote in 1785 that 'the ingenuity of man has ever been fond of exerting itself to varied forms and combinations of medicines'. This is equally true today and one objective of this book is to emphasize that the risk of untoward reaction bears a direct relation to the number of drugs prescribed at any one time.

The practice of treating trivial complaints by the simultaneous administration of a wide variety of drugs has been satirized by Moat (1969) in his article 'Life without Leeches' in the *Daily Telegraph Supplement* (255). He humorously described the treatment taken by a guest at his home:

> The other day a friend from the big world came to stay, and when I took him his breakfast I found him swallowing pills. He'd been depressed recently, which was why he took the yellow pills, antidepressants. The white pills were tranquillizers, and he took those because the yellow pills were inclined to agitate him. The mixture of white and yellow made him itch unbearably and affected his vision — hence the blue pills. He found this particular dosage of pills constipating, and for that he took a strong aperient. 'And just look at these', he said excitedly, waving at me a bottle of large multi-coloured pills. 'And what are those?' I asked. 'I don't know', he said, 'but I like to keep them till last'.
>
> I then asked him how he was feeling. He said he felt fine, except that the pills made him feel lethargic, which he personally found depressing.

Laurence, writing in the latest edition of his *Clinical Pharmacology*, expressed similar views when he said that 'habitual polypharmacy is sure to blur the outline of rational thought which should precede the use of any drug'. Unfortunately the danger of polypharmacy is somewhat concealed from the physician, since a patient, blissfully unaware of the hazards of drug interaction, may indulge in self-medication. For example, a prescribed monoamine oxidase inhibitor is taken together with a self-prescribed proprietary common cold remedy containing ephedrine, phenylephrine, or other sympathomimetic amine. A life-threatening hypertensive crisis may ensue.

In presenting the data in this book, it was felt that classification of drug-induced reactions into a systematic pathological approach would result in the most readable and useful presentation for the prescribing physician, for the student of medicine, for the pharmacist and the pharmacologist. We have attempted in this volume to produce an adjunct for the study of therapeutics and at the same time provide a reference book on the clinical aspects of drug toxicity.

Where we have referred to specific iatrogenic effects of drugs we have used the approved name rather than suggest that the reaction had been exhibited by any particular brand of drug. Nevertheless, to increase the usefulness of the book, we have given, in an appendix, the British, American and continental proprietary names of each drug, together with the approved name.

In conclusion, let all of us who contribute to the ultimate treatment of patients reflect on the prayer of Sir Robert Hutchinson which hangs on the wall of the Children's Ward at The London Hospital:

From inability to let well alone:
from too much zeal for the new
and contempt for what is old:
from putting knowledge before wisdom,
science before art, and cleverness before
common sense,
from treating patients as cases, and from
making the cure of the disease more
grievous than the endurance of the same,
Good Lord, deliver us.

P.F. D'ARCY
Pattishall

J.P. GRIFFIN
Welwyn

January 1971

Acknowledgements

The first edition of *Iatrogenic diseases* was almost totally the work of the editors, but in the second edition and *Iatrogenic diseases, Updates 1981, 1982*, and *1983* we have received the help of an increasing number of contributors. The third edition now has nearly 30 contributors and the amount of hard work that they have made to get their chapters to us on schedule is greatly appreciated, as is their liaison with each other to avoid much unnecessary duplication which has made our task as editors that much easier.

Various colleagues have supplied us with illustrations, authors and publishers have granted us permission to reproduce illustrations and tables, and, on several instances, sections of various chapters have been contributed by experts for our contributors. These are acknowledged by name in the text but our gratitude is expressed here.

Our publishers have been as always a great help to us. Especial thanks are due to those who have been most burdened with the production of a work of this magnitude and in this we include the secretarial help given by Miss Catherine Jarman, Miss Ruth Griffin, and Mrs Vera Markey. We should also like to thank the librarian of Medicines Division DHSS, Mrs Sheila Shrigley; the Medicines Division DHSS Information Section; Mrs Margaret Dow and Dr Cheryl Twomey, Dr Keith Lewis and the librarians of Queen's University, Belfast; and Miss D. Jones, the librarian of the Pharmaceutical Society of Great Britain.

Above all we are indebted to our wives for being so tolerant about the time given up to this task by ourselves and also for their many assistances both great and small rendered over the years.

Contents

Plates fall between pp. 240 and 241 of the text

xiii

Contributors

A.W. ASSCHER, BSC, MD, FRCP,
Professor of Renal Medicine and Honorary Director, Institute for Renal Diseases and Honorary Consultant Physician University Hospital of Wales and Cardiff Royal Infirmary.

J.B. BOURKE, MA, MB, B CHIR, FRCS,
Senior Lecturer in Surgery, University of Nottingham, University Hospital, Nottingham.

A.L. DIAMOND, LL.M.
Professor of Law and Director of the Institute of Advanced Legal Studies (University of London).

G.E. DIGGLE, MB, BS, AKC, DIP EL,
Senior Medical Officer, Medicines Division, Department of Health and Social Security, London.

N.M.G. DUKES, MD, LLD,
Professor of Drug Policy Science, University of Groningen, The Netherlands and Regional Officer for Pharmaceuticals and Drug Utilization, World Health Organization, European Office, Copenhagen.

A.P. FLETCHER, MB, BS, PH D,
Director of Medical and Regulatory Affairs, CTC International, Townfield House, Chelmsford CM1 1QL.

E.S. HARPUR, B SC, PH D, MPS,
Lecturer in Pharmacology, Department of Pharmacy, University of Aston in Birmingham.

LISA E. HILL, MD, MRCP, DCH,
Senior Medical Officer, Medicines Division, Department of Health and Social Security, London.

J.A. HOLGATE, MSC, MB, CH B, FI BIOL,
Principal Medical Officer, Medicines Division, Department of Health and Social Security, London.

D.J. KING, MD, MRC PSYCH, DPM,
Senior Lecturer in Clinical Neuropharmacology,
Department of Therapeutics and Pharmacology, The Queen's University of Belfast; Consultant Psychiatrist, Holywell Hospital, Antrim.

D.R. LAURENCE MD, FRCP,
Professor of Clinical Pharmacology, University College Hospital Medical School, London.

PATRICIA R. McELHATTON, MSC, PH D, MI BIOL,
Senior Research Scientific Officer, Department of Pharmacology, Guy's Hospital Medical School, London.

J.C. McELNAY, B SC, PH D, MPSNI,
Lecturer in Biopharmacy, Department of Pharmacy, The Queen's University of Belfast.

MARY G. McGEOWN, CBE, MD, PH D, D SC (HON.), FRCP ED.,
Consultant Nephrologist, Renal Unit, Belfast City Hospital; Honorary Reader in Nephrology, The Queen's University of Belfast.

A. McQUEEN, MD, CH B, MRC PATH,
Senior Lecturer (Pathology), Department of Dermatology, Anderson College, University of Glasgow; Honorary Consultant Pathologist, Western Infirmary, Glasgow.

R.D. MANN, MD, MRCGP, FCP,
Principal Medical Officer, Medicines Division, Department of Health and Social Security, London.

A.N. NICHOLSON, OBE, D SC, FRCP(Edin), FRC PATH.,
Consultant in Aviation Medicine, Royal Air Force Institute of Aviation Medicine, Farnborough, Hampshire.

LEO OFFERHAUS, MD,
Vice Chairman, Netherlands Board for Evaluation of Drugs, The Netherlands.

R.G. PENN, MD,
Principal Medical Officer, Medicines Division; Department of Health and Social Security, London.

JENS SCHOU, MD,
Professor of Pharmacology, University of
Copenhagen.

M.E. SCOTT, B SC, MD, FRCP, FRCPI,
Consultant Cardiologist, Cardiac Unit, Belfast City
Hospital, Belfast.

C.J. SPEIRS, MD, CH B, MRCP,
Department of Clinical Pharmacology, Royal
Postgraduate Medical School, London.

F. M. SULLIVAN, B SC,
Senior Lecturer, Department of Pharmacology, Guy's
Hospital Medical School, London.

SYLVIA M. WATKINS, MA, DM, FRCP,
Consultant Physician, Lister Hospital, Stevenage.

J.C.P. WEBER, MD,
Senior Medical Officer, Medicines Division,
Department of Health and Social Security, London.

J.R.B. WILLIAMS, MD, FRC PATH,
Consultant in Haematology, Lister Hospital,
Stevenage.

J.M.T. WILLOUGHBY, MA, DM, FRCP,
Consultant Physician, Lister Hospital, Stevenage.

SHIMONA YOSSELSON-SUPERSTINE, PHARM D,
Lecturer in Clinical Pharmacy, School of Pharmacy,
Hebrew University of Jerusalem.

Part I

1 Medicines and the media

J.P. GRIFFIN

In AD 43 Scribonius Largus accompanied the Emperor Claudius to Britain as his medical attendant and in the *Scribonius Largus Formulary* were 271 prescriptions. Girdwood (1981) wrote that in the *Edinburgh Royal Infirmary Pharmacopoeia* of 1935 he was able to identify 26 of these substances and in the 1942 edition of *Hale–White's Materia Medica* 35 of the substances known to Scribonius Largus were still cited. Conversely Girdwood pointed out that of the 1074 active drug substances incorporated in the 2032 products cited in the August 1980 issue of the *Monthly Index of Medical Specialities* (MIMS) 947 of these were not known in 1930.

From this very basic analysis it is plain that the pharmaceutical armamentarium has increased much further in the last 50 years than in the previous 1900 years. Such an advance in therapeutics has clearly brought a spate of new drug-induced diseases. Clearly for each new substance coming into therapeutic use an evaluation of its benefits and risks have to be made. If the benefits outweigh the risks, then the substance has a valuable place in therapeutics. However, 'our community is reluctant to accept that occasional risks are bound to accompany the benefits obtained from drugs' (Vere 1981). It is equally true to say that the medical and pharmaceutical professions, the media, and the public are unclear as to whether the nature of risk–benefit assessments once made stands for all time. A risk–benefit assessment can only be made for a particular point in time in the light of available alternative therapies. A drug which has an acceptable risk–benefit ratio at one point in time may have an adverse risk–benefit ratio as safer and more effective drugs become available. As an example in the 1950s and 1960s and even in the early-1970s the risk–benefit assessment of non-steroidal anti-inflammatory agents such as phenylbutazone and oxyphenbutazone were initially clearly and later marginally in favour of these two substances. However, as continuing reports of fatal blood dyscrasias accumulated and newer and sometimes safer non-steroidal anti-inflammatory drugs were introduced, the balance of risk–benefit assessment tipped the other way. In crude terms the drugs became obsolete.

Other examples of drugs becoming obsolete are the non-selective β-adrenergic stimulant drugs such as iso-prenaline and orciprenaline which have been superseded by newer selective β_2-stimulant drugs such as rimiterol, salbutamol, and terbutaline.

Medicines are news and the media focuses on three aspects of medicine:

1. Stories about wonder drugs that a regulatory authority refuses to license, thereby denying the public amazingly efficacious remedies.
2. Hazard stories relating to widely used and commercially successful products where the basic claim is that the regulatory authorities are failing to protect the public.
3. An outcry that when an obsolete drug is removed from the market, perhaps after a long and honourable history; the regulatory authority acted tardily in not dealing with the matter years earlier.

It does not seem to matter to the media that their criticisms are often self-contradictory. What does matter to the media is to have a story that can sell newsprint or increase television viewing of their programme. Unlike the pharmaceutical industry which has to obtain marketing authorization before it can sell its product, in our democratic society the journalist has to meet no peer review before he can peddle his wares. The public therefore believes the media to behave responsibly in their coverage both of the discovery of 'wonder cures' and of adverse reactions to drugs. This responsibility is sadly lacking in most areas of the media in recent years. One is left with the impression that the media are being hypocritical since they are using the same techniques to sell a story that they criticize the pharmaceutical industry for using to promote their products.

Wonder-drug stories

The essential ingredients for a 'wonder drug' is that it has to be 'magically efficacious' in common conditions that have a chronic clinical course which is either universally fatal or cannot be 'cured' by conventional medical science.

The classical areas for such wonder drugs to develop are in:

1. Cancer of all types;
2. Multiple sclerosis;
3. Rheumatoid arthritis;
4. Asthma;
5. Psoriasis.

There have been recent examples of all these areas being involved in wonder cures.

Laetrile in cancer

Laetrile (see also Chapters 12 and 38) has been in the forefront of news as a cure for cancer in the United States for some years, and more so as a result of the illness and subsequent death from cancer of the actor Steve McQueen who turned to this therapy in desperation. The danger of publicizing this form of treatment for this so-called anticancer drug is the utter immorality of generating public enthusiasm for a form of therapy which American experience has shown to be utterly ineffective and also dangerous.

Laetrile has not been licensed in either the United States by the Food and Drug Administration (FDA) or in the United Kingdom for *any* purpose.

There are two dangers inherent in the growth of demand for this treatment. Firstly, patients are seduced from conventional therapy, which, although not ideal, may hold some hope for them, to this useless drug. It is difficult to understand why the popularity of Laetrile as a cancer cure has survived so long in the face of lack of data on its efficacy and the undoubted evidence of its toxicity.

Laetrile is also known as vitamin B_{17} although it is not a vitamin, but was so designated by Dr Ernst Krebs who has been tireless in its promotion.

Laetrile is available in two forms. As an injection it is useless but safe. As a purified pellet, however, it can be lethal. If the stomach or faeces contain vegetable matter which is a source of beta-glucosidase, the patient may produce his own 'gas chamber' and die of prussic acid poisoning.

Laetrile in the form of amygdalin is also present in apricot and other fruit stones together with the enzyme beta-glucosidase. When the two are brought together by crushing or grinding, cyanide may be produced. Laetrile in this form can be found in health food stores. Much promotion of Laetrile in America therefore skirts around prohibitions which might involve its proponents in litigation by laying all the emphasis on the 'nutritional' aspect of the substance.

Warnings about the undesirability of Laetrile use and the problems associated with it have been covered extensively in the *FDA Bulletin* (Food and Drug Administration 1977) and also in a review by Dr R.G. Penn (1981).

It is interesting to note that after their initial enthusiasm for this supposed 'breakthrough' the media have since had cause to reflect; we now see reports of a rather different nature such as that which appeared in the *Daily Telegraph* reporting the results of American Government controlled studies (*Daily Telegraph* 1981). These showed that Laetrile had 'not produced any substantial benefit'. Moreover even the *Sun* whose high reputation for depth and accuracy can hardly be doubted has seen fit to publish a report of the same findings under the bold headline 'McQueen Drug may be a Killer' (MacMillan 1981) in which they refer to Laetrile as being 'worthless — and may actually kill' and quote the 'Research Head, Dr Vincent de Vita' as saying 'the hollow promise of this drug has led thousands of Americans away from potentially helpful therapy'.

A more serious statement concerning these studies which indicated lack of efficacy can be found in an editorial in *Science* (1981). The loyalty which the media has for this product can be likened to promotion from 'rags to riches' for the product followed by its public execution.

Oil of evening primrose in multiple sclerosis

Various attempts have been made to promote oil of evening primrose for a variety of indications ranging from the treatment and/or prevention of coronary artery disease to the 'cure' of multiple sclerosis. The claim to cure or even improve multiple sclerosis by inducing remission or preventing relapses was not substantiated by trials conducted by the Medical Research Council Unit at Newcastle-upon-Tyne.

'Wonder cures' for arthritis

Yucca

The next of these 'wonder' products is the mellifluously named 'yucca'. This is a hardy desert plant which is, in fact, widely available in this country as a house plant. The advocates of yucca extract claim that American Indians have used it for thousands of years as a general tonic and for the relief of pain. Needless to say, the evidence is scanty and in no way backs up the grandiose claims for its effectiveness. The evidence considered in America by the Arthritis Foundation, including a history of the uses of the plant, primitive Indian usage, and commercial use and the summary of a double-blind study involving 165 patients, was described as 'full of holes' so the claimed result cannot be said to have any scientific validity.

It is significant that yucca tablets are promoted as a 'food supplement', a technicality which we have seen applied in the case of Laetrile as a means of avoiding liabilities under medicines' legislation.

There is no evidence at the moment to suppose that yucca tablets are harmful, though there is equally no evidence that they are in any way effective in treating rheumatoid or osteoarthritis.

In 1982 this product was given the widest possible publicity on the BBC programme *Nationwide*. Although their objective was to show the activities of the producer and marketing agent in the UK in an unfavourable light, the result has been quite the opposite, namely, to increase the demand for this unproven remedy.

The green-lipped mussel

Another unproven remedy gaining popularity for the treatment of rheumatoid arthritis is extract of green-lipped mussel. This delectable creature is gathered in New Zealand and has been promoted as a 'natural cure' for all forms of arthritis. Some reputable rheumatologists have commenced studies more out of curiosity than conviction, though claims of significant efficacy have been made in a letter to the *Lancet* from Robin and Sheila Gibson of the Glasgow Homoeopathic Hospital (Gibson and Gibson 1981) in which they claim that experience of treating over 500 patients has led them to believe that green-lipped mussel extract is efficacious in the treatment of both rheumatoid and osteoarthritis. In the same letter, they dismiss the proposition that no component so far found in *Seatone* (the product in question) could account for any inflammatory activity as 'largely irrelevant'.

As a result of the publicity for this product demand has naturally increased, despite its at present unproven efficacy.

Tests are, however, being undertaken; the results of a small study carried out by Dr Huskisson and colleagues were reported in the *British Medical Journal* (Huskisson *et al*. 1981) in which the conclusion reached was that this product 'does not appear to be worthwhile except for the very considerable placebo effect that any new treatment has in rheumatoid arthritis'.

Dr Huskisson is reported to be undertaking a long-term study. It is doubtful, however, that, if the results underline the earlier conclusion, very much publicity will follow; there is no news value in an ineffective product which is not marketed (unless it has some disasterous side-effects).

Treatment for the relief of foot odour

If only to prove that the bounds of credulity never seem to be reached, it is worth mentioning the advertised 'cure' for foot odour involving the use of 'Formula 2000' which must be mixed with the urine of the sufferer in a bowl. The feet are then soaked in this mixture and dried; perhaps most alarming is the injunction of the firm's advertising material (in bold type) 'Do not wash feet for at least 48 hours'.

Dimethyl sulphoxide (DMSO)

Another case which has recently come to our attention is the use of dimethylsulphoxide (DMSO) in the treatment of arthritis. An article by Professor McIntyre and colleagues appeared in the *Lancet* on 8 November 1980 (Yellowless *et al*. 1980) relating the case of two elderly people who were given intravenous infusions of DMSO. These two patients (husband and wife) had been treated privately beginning in May 1979 for painful arthritic knees. They were given three daily intravenous infusions of 100 g of 20 per cent DMSO in dextrose solution, apparently without ill effect. However, after the second course of treatment in February 1980, the wife became drowsy, vomited blood, and was admitted to the Royal Free Hospital. Her husband was also admitted and after seven days both were discharged. Liver enzyme changes occurred in both cases. (Yellowless *et al*. 1980).

The use of DMSO in rheumatoid arthritis has not been licensed by the UK Licensing Authority, but it has been administered under Section 9 of the Medicines Act which permits the administration of medicinal products especially prepared for a named patient. In America the FDA have also not licensed DMSO for the treatment of arthritis; they have licensed it for a condition that they recognize, namely, interstitial cystitis, but British urologists do not recognize this diagnosis. However, following widespread publicity for this treatment on American television, the FDA issued a statement in August of 1983 expressing their concern at the increased use of DMSO in arthritis. They pointed out that there was no generally acceptable evidence that DMSO was either safe or effective in treatment for arthritis or bursitis. Also it had been tested for a plethora of disease conditions including genital herpes and mental retardation but no effectiveness had been proven.

The history of DMSO goes back many years into the distant past and in 1967 a whole issue of the *Annals of the New York Academy of Sciences* was devoted to DMSO (Leake 1967). Several of the papers published in that volume described both corneal and lenticular opacities developing in animals and patients treated with DMSO (see also Chapter 30).

At the present time, DMSO's sole justifiable place in therapeutics is an excipient to promote the percutaneous absorption of certain active ingredients, for example, idoxuridine.

'Shock-horror' stories

In the past years there have been a number of these which have been unjustified and have done untold harm, or have caused needless suffering to thousands of unfortunate patients or have caused drugs to be removed from the market groundlessly.

The classic ingredient of a 'shock-horror' story is almost the converse of a 'wonder drug' story, namely:

1. Drug damages unborn children or babies;
2. Drug caused cancer;
3. Wonder drug kills the chronically sick, which is a variety of the rags-to-riches story of a drug followed by its public execution. Examples of each of these 'shock-horror' stories will be cited.

Drugs damaging the unborn

Debendox

An example of character — assassination by the media of an innocent drug is *Debendox* (see also Chapter 22). Three times in as many years the Committee on Safety of Medicines (CSM) has carefully examined all the available data on the issue of whether this preparation for the treatment of morning sickness of pregnancy had produced on increased incidence of congenital abnormality in the offspring of mothers treated with the drug. The committee found no evidence to indicate that there was an increased risk of fetal damage associated with this product. However, the *Debenox*-Action-Group remained unpacified, and clearly only a total ban would satisfy them. The manufacturers in 1980 admitted a considerable fall in sales — almost to zero according to unofficial reports. No drug could be expected to survive such continued adverse publicity and the most optimistic marketing manager could never really have hoped for such a recovery in sales that would have allowed *Debendox* to recover its lead in the morning sickness remedy league.

Neither the CSM in the UK nor the FDA in the USA could find evidence to condemn *Debendox*; but in the USA in the courts uninformed lay juries with opinions coloured by press stories awarded damages in Orlando, Florida, against the manufacturer, these were subsequently squashed on appeal.

It was therefore with some relief that it was noted that Sir George Young (at that time Under Secretary of State for Health and Social Security) in a House of Commons adjournment debate on 24 April 1980 refused to align himself with those who 'suggest that the Health Minister of this country should act in response to scare mongering or the verdict of a lay jury in the United States, rather than the advice of an expert committee to align oneself with a movement which is at heart anti-science, anti-progress, anti-medicine and anti-the welfare of these people of this country' (Deitch 1980; *Hansard* 1980; *Lancet* 1980). The media campaign did not cease, and after several years which saw many legal suites in the United States against *Debendox*, the manufacturer withdrew the product from the market world-wide. The decision to withdraw the product was not made or any scientific grounds, but solely because the company's insurers had increased their premiums due to the law-suits, and it had became uneconomical to pay the excessive premiums.

Pertussis vaccine

A classic example of the shock-horror approach was the outcry regarding the safety or otherwise of pertussis vaccine which, it was alleged, produced brain damage in a large number of children vaccinated with triple vaccine (diphtheria, tetanus, and pertussis — DPT) (Gillie 1977) (see also Chapter 37).

If pertussis vaccine–induced brain damage does occur, it is nevertheless unjustifiable to base an incidence figure on the soft data available. The differing incidence figures produced by various sources have been quoted in *Whooping cough vaccination* (Department of Health and Social Security 1977); they vary from 1 in 300 000 children inoculated to 1 in 168 000 injections administered (Stewart 1977) which, bearing in mind that three injections of DPT are given per course of immunization per child, gives an incidence figure of about 1 in 56 000 children inoculated. It is a great pity that any such guesses at an incidence of pertussis vaccine–induced brain damage were ever made.

The furore generated by the media and politicians on this matter has been most detrimental to the immunization programme in the UK; moreover the uptake of all vaccines, not only pertussis, fell. This was followed by the most serious outbreak of pertussis in 20 years, which might never have occurred had the population's immunity to pertussis been maintained by an acceptable uptake of the vaccine. Dr A. Griffiths at a Medico Pharmaceutical Forum meeting on 'Medicine and the media' (see Cromie 1981) stated that this outbreak was of 100 000 cases of pertussis, of which 20 000 suffered bronchitis and 4000 suffered convulsions, most of which could have been prevented if pertussis vaccination had continued at the previous level. Griffiths' 1980 figures are for the whole of the UK; for England and Wales the figure for the epidemic was 66 000 cases (Geffen 1981).

As has been stated earlier, vaccination for all communicable diseases fell as a result of this irresponsible campaign, including polio vaccination and rubella vaccination. It is to be hoped that the DHSS campaign in 1983–84 for increasing rubella vaccination will rectify this state and that not only will the uptake in teenage girls return to its previous high level but that girls who in the past declined vaccination against rubella will avail themselves of this renewed offer.

Cancer horror stories

Valium

Yet another case of a media-inspired scare story is that of *Valium* and its suspected carcinogenic properties. On the basis of some studies performed by a British doctor working in Canada, which purported to show a link between *Valium* and cancer, a storm of suspicion and protest was whipped up against the drug. The doctor concerned had transplanted breast tumours into rats fed with diazepam (*Valium*) and observed that at low dosages the tumours grew faster. The effect, however, disappeared at high dosages.

The CSM reconsidered diazepam and concluded that there was no established link between diazepam and cancer; that it neither caused nor promoted the growth of cancer. Formal carcinogenicity studies conducted at an independent contract research centre were negative.

Here again was a case of a perfectly respectable, safe, and much-used drug, on the market for 18 years, suddenly held to be highly suspect on the strength of unsupported, questionable research findings blown up out of all proportion for their newsworthiness rather than for their scientific validity. It is hardly necessary to stress the anxiety caused by these unsubstantiated rumours to those thousands of people for whom *Valium* was a therapeutic agent.

Cimetidine

The drug cimetidine was granted a product licence by the CSM in 1976 after eight years of exhaustive research by Smith, Kline, and French. It was at the time regarded as a major breakthrough in the treatment of peptic ulcers (both gastric and duodenal) acting as an H_2-receptor antagonist. Since then it has been widely prescribed. Within a year of its launch, it had been given to 1 300 000 people, and by 1984 it had been taken by about 11 million people world-wide.

Case reports of gastric cancer arising during treatment with cimetidine have occurred as isolated cases in the medical literature, and have also been reported on the Adverse Drug Reaction reporting system of the CSM — the so-called yellow card system. These have, of course, been considered by the Adverse Reactions Subcommittee of the CSM. Of course, one interpretation of these incidents has been that they are caused by treatment with cimetidine, but without adequate investigation and diagnosis it is not possible to tell whether or not there was a pre-existing gastric cancer (see also Chapter 35). The short latency (a few months) between the start of cimetidine treatment and the appearance of the gastric cancer would tend to favour the view that there was such a pre-existing condition, since most chemically-induced cancers have a latency of many years.

There are, however, other aspects to this story since treatment with cimetidine reduces gastric acidity and allows an increased formation of nitrosamines, which are known carcinogens, from dietary nitrite and nitrates. This has been advanced as a possible mechanism of carcinogenesis.

Dr Reed of the Hammersmith Hospital presented a paper at the meeting of the World Congress of Gastroenterology in Hamburg in June 1980 (Reed *et al.* 1980) showing a correlation between the fall in gastric acidity and the formation of intragastric nitrosamine. Dr Reed's paper had not been published in full, but the possible link between the drug and cancer has been seized by the press and given considerable coverage, particularly in the *Sunday Times*. Whatever the merits of the scientific case, which we have yet fully to see, there seems little to be gained from premature publicity of a phenomenon so far only suspected.

Another aspect of the same story is the theoretical formation of nitroso-cimetidine *in vitro* which has not yet been demonstrated to occur *in vivo*.

Benoxaprofen and the pursuit of absolute safety

The biggest news feature on adverse drug reactions (ADRs) during 1982 was the suspension of the product licence for *Opren* (benoxaprofen), a non-steroidal, anti-inflammatory agent. The suspension of the product licence for benoxaprofen, initially for 90 days, by the United Kingdom Licensing Authority acting on the advice of the Committee on Safety of Medicines (CSM) took place on 3 August 1982. This was done on the basis of a wide spectrum of adverse drug reactions. Professor A. Goldberg, Chairman of the Committee on Safety of

Medicines, writing to the medical and pharmaceutical professions on 3 August 1982 stated:

The Committee on Safety of Medicines has received over 3500 reports of adverse reactions associated with this drug; included among these reports are 61 fatal cases, predominantly in the elderly. Having regard to these reports there is concern about the serious toxic effects of the drug on various organ systems, particularly the gastro-intestinal tract, the liver and bone marrow, in addition to the known effects on skin, eyes and nails.

When this announcement was made, benoxaprofen was already being marketed in 10 countries, including the United States, South Africa, West Germany, Switzerland, France, Denmark, and Spain. However, the Australian and New Zealand regulatory authorities had not approved the drug for marketing. Benoxaprofen was made available to UK general practitioners in autumn 1980 and was hailed by the manufacturers (Eli Lilly and their UK subsidiary, Dista Products) as a new anti-arthritic with disease-modifying properties, a claim based on animal studies. The drug was put on to the market with massive publicity on the radio and in newspapers, encouraging patients to believe it was a major advance in the treatment of arthritis and to ask their doctors specifically for *Opren*. It has been variously estimated that between 500 000 and 750 000 patients had received the drug in the UK before its withdrawal.

Gastrointestinal ulceration and haemorrhage, photosensitivity, and onycholysis were reported during the clinical trials of benoxaprofen (Mikulaschek 1980). The first reports of deaths associated with benoxaprofen came in April and May 1982 with accounts (Goudie *et al*. 1982; Taggart and Alderdice 1982) from Glasgow and Belfast of eight elderly women taking the drug who had died after developing cholestatic jaundice. Subsequently, several letters describing further cases were published in the correspondence columns of the *British Medical Journal* during June and July 1982 (Prescott *et al*. 1982; Fisher and McArthur 1982; Firth *et al*. 1982; Duthie *et al*. 1982).

In May 1982, at about the time these reports began to appear in the UK medical literature, the drug was launched in the United States as *Oraflex*. In June 1982, the Health Research Group, a US consumer organization, petitioned the US Department of Health and Human Services seeking an immediate ban on benoxaprofen because of the UK reports of liver damage. The FDA ordered a review of the benoxaprofen toxicity data but did not consider that drastic action was warranted. Meanwhile, in June 1982, Dista Products issued a 'Dear Doctor' letter indicating that the UK data sheet had been revised to include the warning that in patients over the age of 65 the daily dose should be halved. On 2 August 1982, the Danish regulatory authorities limited the pre-scribing of benoxaprofen mainly to hospitals. However, following further review of the available data, the British authorities decided to suspend the promotion and supply of benoxaprofen on 3 August 1982, only three months after the first report of liver damage associated with benoxaprofen appeared in the literature. Ironically, an article by Mikulaschek (1982) concluded that 'Studies with benoxaprofen in rheumatoid arthritis and osteo-arthritis, conducted in more than 2000 patients, continue to demonstrate its safety and effectiveness'.

The reaction of the medical profession to the suspension of the licence for this non-steroidal, anti-inflammatory drug varied from criticism for premature, precipitate, and unjustified withdrawal of the drug to criticism for undue delay in reacting to the reports of adverse reactions received. These extremes of view were reflected in the media. The CSM also came under attack because of a delay between the reporting of *Opren's* withdrawal in the news media and the doctors receiving the official notification. Professor Goldberg in his 3 August 1982 letter to doctors anticipated this occurrence and stated that this action was regrettable but necessary on grounds of safety. By mid-August 1982 the manufacturer of benoxaprofen had decided on a world-wide withdrawal of the drug from the market.

Subsequently, media and political scrutiny were brought to bear on the systems used currently within the UK for post-marketing surveillance of newly launched drugs. Proposals for new schemes of varying ambitious-ness and cost abounded. The statement of a European Workshop held in 1977 on 'Monitoring of drugs' which follows points out that absolute safety is unattainable and its pursuit may do more harm than good.

Medicines can never be entirely safe. Despite extensive testing and monitoring of medicines, unforeseen and unpredictable adverse reactions will continue to occur. The public needs to be aware that treatment with medicines always carries some risk. It is the duty of all concerned to maximise benefit and minimise risk. In the opinion of this group of European scientists it is now advisable to revise our methods of assessment of medicines. We must recognize that existing methods are unsatisfactory. We recommend more rational but less extensive laboratory studies without unnecessary multiplication of detailed clinical trials before registration. Instead we recommend much closer and more extensive surveillance of medicines after they are available for general prescription. Only by the careful study of medicines in everyday use can the greatest benefits be obtained from their administration, the untoward rare potential disaster recognised at the earliest possible moment, and the ill effects minimised. Absolute safety is unattainable and its pursuit, regardless of other considerations, is achieving more harm than good.

Another example of the effect of the benoxaprofen withdrawal was the development of a humorous but cynical approach to industrial pharmaceutical development and drug toxicity in the media, which is typified by the article by Miles Kington in *The Times* (London) of 6 October 1982.

A new miracle drug will be coming on the market next spring, called *Sufferin. It is claimed by its makers to be different from all drugs so far announced as new miracle drugs. Normally, even if a drug cures the condition it is treating, it also has unpleasant side-effects. *Sufferin is different. It only has side-effects and cures nothing . . .
'Yes, I'm very excited about the prospects of *Sufferin,' says chief chemist Louis Exocet. 'It's the very first time we have marketed a drug with an asterisk in front of the name. Previously, you know, we have had terrible trouble thinking up names which had not been registered by someone else. Now, by putting this little star in front, the name is bound to be different. This asterisk is truly the miracle ingredient!'
What about the drug itself? Is there really a market for a drug that cures nothing and only does you harm?
'That shows how little you know about the drug world,' says Exocet.
'People are already used to the idea. Millions of patients every day go to their doctor and say, "Doctor, that stuff you gave me, it hasn't cleared up my condition. But it's given me a funny rash." Well, the doctor is baffled. But with *Sufferin he can never be baffled, for that is the whole intention!'
'Also, it will be very good for the people who are malingerers and have nothing wrong really. The doctor has nothing to cure, and gives them *Sufferin, which cures nothing. But it also gives them some real symptoms, which subconsciously they were wanting all along.
'Above all, it is designed for the majority of ailments, which will go away anyway, whether people see a doctor or not. The doctor cannot cure those ailments, but he must give the patient some treatment, because that is the way the patient is comforted. So he gives him *Sufferin. *Sufferin gives him those side-effects. The doctor can cure the side-effects by telling him to stop taking *Sufferin.'
What exactly are the side-effects?
'Slight dizziness. A small rash. Blood-shot eyes. Nothing serious. There is also, though perhaps I should not mention it, a slight urge to take more *Sufferin.'
An addictive drug? Isn't that illegal?
'No more than alcohol.'
Finally, if it is possible to make a drug that has no cure, only side-effects, does this mean that one day there can be a drug which has a cure and no side-effects?
'My friend,' says Louis Exocet, 'you really know nothing about the drug world, do you?'

The benoxaprofen experience and the media reaction to it has reflected adversely on the public confidence in the pharmaceutical industry, the medical profession, and the drug regulatory authorities in the USA, the UK, West Germany, Denmark, and other countries who permitted the drug on to their national markets, but ought to reflect even more seriously on the media who hail trivial new drugs as major advances and raise popular expectation of great advances but are the first to bay for blood when the

expectations they themselves have created are shown to be a bubble.

A detailed investigation of the toxicity of benoxaprofen revealed some very interesting factors (see also Chapters 12, 30, 32, and 35). The reports of photosensitivity reactions were clearly correlated to the hours of sunlight and showed a seasonal distribution with a peak rate of reporting in June, July, and August 1981 and a second peak in June and July 1982. Photosensitivity reactions were commonest in the most southerly areas of the UK and this correlated with areas with the greatest amount of sunlight. Most patients who developed photosensitivity reactions stopped taking the drug after a relatively short exposure of several weeks. However, the cases of renal toxicity and hepatotoxicity occurred in patients who had been exposed to the drug for many months, for example, the fatal cases of hepatorenal syndrome occurred after a mean duration of exposure to benoxaprofen of over 10 months. The geographical distribution of cases of fatal hepatorenal syndrome were commonest in the northern areas of the UK and were most common in Northern Ireland where the mean daily hours of sunlight were lowest. These findings led us to conclude that prolonged exposure to benoxaprofen was necessary for the development of renal and hepatic damage and that prolonged exposure was more likely to occur in those areas of the UK where photosensitivity was not a limiting factor affecting the duration of therapy (Griffin 1984).

The worm that turns on motherhood

This story by Oliver Gillie in the *Sunday Times* of 13 February 1983 must rate as one of the best horror stories of the century. The report stated:

Doctors in the United States have discovered a parasitic worm in the human body which they believe is the cause of one of the commonest and most serious diseases of pregnancy. The disease, toxaemia, affects one in 10 pregnant women in Britain and in the most severe cases can kill.
The worm, discovered by doctors at Loyola University in Chicago, also appears to be an important cause of babies being born prematurely, small or in ill health. In addition, it may be a cause of sterility and high blood pressure.
The discovery will surprise the medical world because new human parasites (as distinct from the bacteria and viruses which cause many infections) are seldom discovered today. The worm is related to tapeworms, hookworms, and roundworms which cause disease mostly in tropical countries and in insanitary conditions. It has been named *Hydatoxic lualba*.
The organism was discovered by Judith Lueck, a scientist at Loyola University. Using a microscope, she observed something swimming among cells taken from a specimen of womb cancer. After many tests, she found a stain showing several worms up to

one sixteenth of an inch long. She concluded that the original smaller 'organism' which she had seen swimming was a sperm from the worm.

The worm, which exists in male, female, and immature forms was later found with its eggs in the blood and placentas of women suffering from various forms of womb cancer and from toxaemia of pregnancy. But Lueck and her collaborator, Dr Silvio Aladjem, were perplexed to find the worm in the blood of healthy women as well. To find out if it was important in causing toxaemia, they injected it into pregnant mice and beagles.

The pregnant beagles all went down with toxaemia. They suffered increased blood pressure, protein in their urine, and changes in their eyes and liver; their blood clotted abnormally easily. Half of the foetuses conceived by the beagles injected with the worms died in the womb and others were stunted in growth. But non-pregnant beagles injected with the worms remained perfectly healthy, as did pregnant beagles not injected. The same results were obtained with mice.

The experiments led Aladjem and Lueck to suggest that the worm multiplies in the bodies of some pregnant women to cause toxaemia. It may be that some women with a particular hereditary constitution are more vulnerable than others and that the general state of a woman's immune defences at the time of pregnancy are important. Women are more vulnerable in their first pregnancy, perhaps because the generally become immunised after the first exposure to an increase in numbers of the worm.

Oliver Gillie's story went far further than the original articles by Lueck *et al.* (1983) and Aladjem, Lueck, and Brewer (1983) in the *Am. J. Obstet. and Gynecol.* (Vol 145, pp. 15ff and pp. 27ff.) It also shows the danger of elaborating news stories on the basis of unconfirmed reports in the scientific and medical literature. In the *Lancet* of 21 May 1983 Gau *et al.* reported their investiga-

tions into the existence of this new helminth named by Lueck *Hydatoxic lualba* in the following words:

To investigate these findings we first looked at touch smears of placentas from normal pregnancies, using Lueck's staining technique which involved preliminary exposure of smears to concentrated sulphuric acid. All smears showed organisms identical to those described in the paper. Blood samples from non-pregnant, predominantly male subjects with no evidence of trophoblastic disease were also examined. All smears were positive. Specimens not subjected to sulphation were always negative.

Subsequently we looked at transverse sections of these 'worms' by both light and electron microscopy and found that they did not have any helminth structure but consisted simply of space surrounded by an a nuclear coagulum. Clearly these organisms are artefacts produced by the preliminary sulphation in Lueck's staining technique and cannot therefore be responsible for gestational, nor in fact for any other, pathological process.

It is of great interest to compare the pictures of this hypothetical worm *Hydratoxic lualba* in the papers of the scientific writers to the rather fanciful drawings in the *Sunday Times* (see Figs. 1.1 and 1.2). Perhaps it would be better if Oliver Gillie had entitled his article 'The Worm that Wasn't'.

The pill panic that never was

In October 1983 Prof Vessey and colleagues published a paper in the *Lancet* which presented data indicating that long-term use of oestrogen/progestogen-combined oral

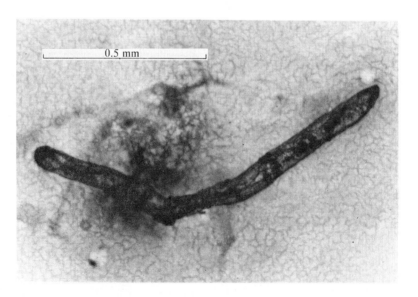

0.5 mm

Fig. 1.1 The 'worm' as photographed by Dr Gillian Gau.

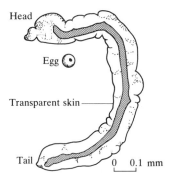

Fig. 1.2 The 'worm' as represented in the *Sunday Times*.

contraceptive preparations were associated with a greater risk in developing cervical carcinoma than users of intrauterine contraceptive devices. In the same issue of *Lancet*, Professor Pyke and his colleagues presented data linking an increased risk of breast cancer in women under 25 years who had used those combined oestrogen/progestogen pills with a high progestogen potency for prolonged periods, i.e. 5 years or more. The Committee on Safety of Medicines in a letter from the Chairman Sir Abraham Goldberg dated 20 October 1983 communicated to the medical profession in the following terms:

The Committee taking all considerations of oral contraceptive safety into account recommend that women should be prescribed a product with the lowest suitable content of both oestrogen and progestogen. The Committee is of the view that the potential risk does *not* require patients currently using products with a high progestogen potency to be prescribed other products immediately. Women should therefore be advised to complete the use of the contraceptive pills in their possession. In particular it is unwise and unnecessary to interrupt the current monthly cycle of treatment. Any woman who does wish to discontinue her present treatment in mid-cycle should be advised to adopt other methods of contraception immediately. Those women for whom a new combined oral contraceptive is prescribed with a low oestrogen and low progestogen content should be advised to use an additional contraceptive method for the first 14 days of their first cycle on the alternative products.

The media almost universally stated that these reports had caused panic. Gwen Farrow in *World Medicine* of 26 November 1983 reviewed the media reporting at length. 'It appeared that there was a nationwide panic among the $3\frac{1}{2}$ million women on the Pill. The *Sunday People* said women were 'stunned' and clinics were 'besieged'. The *Daily Mail* said 'doctors and clinics have been inundated with calls from thousands of worried women'. And the *Guardian*, which should have known better, said the Family Planning Association 'reported mounting panic among patients'.

Examining the facts

All this is odd. The FPA told me they reported no such thing. In Birmingham where the clinics take on 14 000 new patients annually, they received a total of 97 phone calls on the day after the *Sunday People* had said they were besieged, of which 10 led to appointments. The remaining 87 women were reassured. And Dr Nancy Loudon, Chairman of the National Association of Family Planning Doctors, reported from Glasgow: 'No panic'.

In clinics in Hampstead, there was a small increase in attendances and inquiries over the few days following the Pike and Vessey reports. The FPA itself, by the end of the week, had no evidence of panic and no one could find a besieged gynaecologist anywhere.

The *Sunday Mirror* managed to find a shock-horror angle all of its own. 'Pill Revolt by Wives', said its headline, followed by a story which began: 'Bedtime habits are set for a revolutionary change following new scares over the Pill'.

I fear the press did little better on the facts of the case. It is always dangerous when one of the quality Sunday papers sets out to get to the heart of something. The prestige of the paper and the very length of the treatment leads one to believe they have got it right. But the *Sunday Times*, despite the efforts of five researchers and two writers, managed to get it wrong. Its claim that 'it is easier to advise a low dosage pill than it is to get one' is nonsense. If the authors were suggesting doctors won't prescribe, they were wrong. And if they meant that the low-dosage pill is in short supply, they were still wrong. The *Sunday Times* article went on to say: 'If the more rigorous standard on oestrogen content favoured by the Family Planning Association is used, only one pill passes — Loestrin 20.' This the FPA quite naturally denies as a piece of misleading nonsense. In fact, *Loestrin 20* is not widely used, and carries unacceptable side-effects for very many women.

A representative of the *Star* attended a press conference at the FPA and reported the following day that FPA doctors had said the Pike and Vessey reports were probably wrong. The FPA doctors said no such thing. The *Daily Mirror* did no better. It said the Health Department had sent out new rules to doctors on prescribing 'only the lowest strength pill'. The department, of course, did nothing of the kind. And the good old *Guardian* dropped another clanger with the headline 'Demand for Pill Ban after Link with Cancers'. The story which followed reported no such demand, and rightly so, as there never was one.

The FPA press conference was interesting in its own right. A gentleman from the *London Standard* accused the distinguished panel of speakers of whitewashing the Pill. 'You've been doing it for years', he proclaimed. And an angry young man from the *Star* had devised a neat question designed to produce — if anyone was foolish enough to answer it — the perfect shock-horror headline: ''Would you agree that if these reports are right many women will get cancer who wouldn't have got it otherwise?'' ' (Farrow 1983)

What is clear is that the press cannot stand a complex story which lacks a clear-cut solution. It deals happily with disasters. It ignores good news (only three newpapers covered the news last year that the Pill protects against certain cancers). On this occasion the panic that it claimed to see never got beyond the pages of the newspapers themselves. One doctor who runs a large Family Planning Clinic in Manchester informed the author (JPG) that he had had four visits from women concerned about their oral contraceptive regime but six telephone calls from journalists. A letter in the *Lancet* from Jewell *et al.* (1983) confirms the lack of panic by the public.

Experience in a practice on the outskirts of Southampton with a list of 8200 patients indicated that women did not consult their doctors in a panic following the media coverage of the papers on oral contraception and cancer in the *Lancet* of October 22. In the two weeks beginning October 21 only 25 of the 500 women listed as receiving their oral contraception from us consulted specifically in response to the reports. Twenty others brought up the subject during a consultation for another matter. This comprised no more pill-related consultations than we would normally expect and was far fewer than we and others had been prepared for. Doctors and journalists may perhaps use the media more effectively for health education if they base their efforts on observed behaviour rather than theoretical expectation.

What was the motive of the media in going all out to generate panic? It can only be the need to sell a story — truth clearly does not feature!

The saga of proxicromil

When Fisons' drug *Proxicromil* was recently withdrawn before marketing, because of an undisclosed 'toxic event' in the animal-testing procedures, it was noticeable that maximum coverage was given to the potential benefit that could result from such a drug, the great cost of carrying out the exhaustive tests before it could be licensed, and so on, with the distinct implications that a valuable drug had been lost (and the company's shares devalued) because of over-rigidity and rigour of the licensing authority's standards.

This story was blown up to have all the advantages of wonder-drug story and the shock-horror story in one.

The nature of the animal toxicity was never disclosed in the press but it was enough to cause worry for those patients who had received the drug during its clinical trial. This story is an excellent example of: 'The truth that is told with bad intent beats all the lies you can invent.' (William Blake)

Sodium valproate

The Press Council severely censured the *Daily Mirror* following complaints that the newspaper irresponsibly published unbalanced reports with unjustified headlines about the anti-epileptic drug sodium valproate. The Press Council as reported in the *Daily Mirror* 26 March 1984 states 'The articles were unbalanced, themselves distressing to patients and potentially dangerous'. It stated that 'it was thoughtless of the newspaper to publish without including a warning to [epileptic] readers that it was dangerous to stop using the drug or to change the dose without consulting their own doctors'. 'The articles betrayed a lack of knowledge about epilepsy and of the importance of *Epilim* (sodium valproate) in treating it.'

Conclusion

Journalists would do well to remember the inscription on a thirteenth-century penbox from Anatolia exhibited in the Kuwait Museum which states in an Arabic inscription around the sides of the box

To its owner happiness and peace and long life in which no pigeon coos in a plaintive manner. Glory and continuance and veneration and appreciation and high rank and eminence and loftiness and praise and blessings.

But more telling is the comment on the lid of the box:

So do not write with your hand except that which will delight you if you see it on (the day of) judgement.

REFERENCES

ALADJEM, S., LUECK, J., AND BREWER, J.I. (1983). Experimental induction of a toxemia-like syndrome in the pregnant beagle. *Am. J. Obstet. Gynecol.* **145**, 27–37.

CROMIE, B.W. (1981) Medicine and the media. *J.R. Soc. Med.* **74**, 224–5.

DAILY TELEGRAPH (1981). 2 May 1981.

DEPARTMENT OF HEALTH AND SOCIAL SECURITY (1977). Whooping cough vaccine. Report of the evidence on whooping cough vaccination by the Joint Committee on Vaccination and Immunization, pp. 1–38. HMSO, London.

DEITCH, R. (1980). Adverse reactions to drugs. *Lancet* i, 1095–6.

DUTHIE, A., GLANFIELD, P., NICHOLLS, A., FREETH, M., MOORHEAD, P., AND TRIGER, D. (1982). Fatal cholestatic jaundice in elderly patients taking benoxaprofen. *Br. med. J.* **285**, 62.

FARROW, G. (1983). The panic that never was. *World Medicine* 26 November 1983.

FOOD AND DRUG ADMINISTRATION (1977). Update on Laetrile. *FDA Bull.* **7**, 2–4, 26–32.

FIRTH, H., WILCOCK, G.K., AND ESIRI, M. (1982). Side effects of benoxaprofen. *Br. Med. J.* **284**, 1784.

FISHER, B.M. AND MCARTHUR, J.D. (1982). Side effects of benoxaprofen. *Br. med. J.* **284**, 1783.

GAU, G.S., BHUNDIA, J., NAPIER, K., AND RYDER, T.A. (1983). The worm that wasn't. *Lancet* i, 1160–61.

GEFFEN, T. (1981). Vaccination against whooping cough. *Hlth Trends* **13**, 44–5.

GIBSON, R.G. AND GIBSON, S.L.M. (1981). Green-lipped mussel extract in arthritis, *Lancet* i, 439.

GILLIE, O. (1977). *Sunday Times*, 6 March 1977.

—— (1983). The worm that turns on motherhood. *Sunday Times* 13 February 1983, 7.

GIRDWOOD, R.H. (1981). Developments in therapeutics in the last fifty years. *Scott. med. J.* **26**, 53–8.

GOLDBERG, A. (1982). Benoxaprofen. Letter from chairman of the Committee on Safety of Medicines, 4 August 1982.

—— (1983). Use of combined oral contraceptive pills and cancer. Letter from chairman of Committee on Safety of Medicines, 20 October 1983.

GOUDIE, B.M., BIRNIE, G.F., WATKINSON, G., MACSWEEN, R.N.M., KISSEN, L.H., AND CUNNINGHAM, N.E. (1982). Jaundice associated with the use of benoxaprofen. *Lancet* i, 959.

GRIFFIN, J.P. (1984). The advantages and limitations of drug orientated schemes for monitoring ADRs — voluntary reporting. In *Monitoring for adverse drug reactions* (ed. S.R. Walker and A. Goldberg), pp. 21–30. MTP Press, Lancaster.

HANSARD (1980). Adjournment debate. House of Commons. 24 April 1980. [Also reported in *Pharm. J.* (1980) **224**, 505.]

HUSKISSON, E.C., SCOTT, J., AND BRYANS, R. (1981). Seatone is ineffective in rheumatoid arthritis. *Br. med. J.* **282**, 1358.

JEWELL, D., KINMOTH, A.L., METCALFE, C., FREER, C., FREEMAN, G., BAIN, J., AND BURKE, P. (1983). Publicity, the pill and cancer. *Lancet* ii, 1307.

KINGTON, M. (1982). *The Times* 6 October 1982.

LANCET (1980). Commentary from Westminster: adverse reactions to drugs. *Lancet* i, 1095.

—— (1983). Oral contraceptives and neoplasia. *Lancet* ii, 947–8.

LEAKE, C.D. (ED.) (1967). Biological actions of dimethylsulfoxide. *Ann. NY Acad. Sci.* **141**, 1–671.

LUECK, J., BREWER, J.I., ALADJEM, S., AND NOVOTNY, M. (1983). Observations of an organism found in patients with gestational trophoblastic disease and in patients with toxemia of pregnancy. *Am. J. Obstet. Gynecol.* **145**, 15–26.

MacMILLAN, S. *Sun* 2 May 1981.

MIKULASCHEK, W.M. (1980). Long-term safety of benoxaprofen. *J. Rheumatol.* **7**(Suppl. 6), 100–7.

—— (1982). An update on long-term efficacy and safety with benoxaprofen. *Eur. J. rheumatol. Inflamm.* **5**, 206–15.

PENN, R.G. (1981). Adverse reactions to herbal preparations. In D'Arcy, P.F., and Griffin, J.P. *Iatrogenic diseases, Update 1981*, pp. 204–17. Oxford University Press, Oxford.

PRESCOTT, L.F., LESLIE, P.J., AND PADFIELD, P. (1982). Side effects of benoxaprofen. *Br. med. J.* **284**, 1783.

PYKE, M.C., HENDERSON, B.E., KRAILO, M.D., DUKE, A., AND ROY, S. (1983). Breast cancer in young women and use of oral contraceptives. Possible modifying effect of formulation and age at use. *Lancet* ii, 926–9.

REED, P.L., SMITH, P.L.R., AND WALTERS, C.L. (1980). The influence of cimetidine treatment on gastric pH and nitrosamine concentration. *Eleventh Int. Cong. Gastroenterol.*, *Hamburg*, 8–13 June 1980.

REPORT OF A EUROPEAN WORKSHOP (1977). Towards a more rational regulation of the development of new medicines. *Eur. J. clin. Pharmacol.* **11**, 233–8.

SCIENCE (1981). Editorial: Laetrile brush fire is out, scientists hope. *Science* **212**, 758–9.

STEWART, G.T. (1977). Vaccination against whooping cough. Efficacy versus risks. *Lancet* i, 234.

STOFFER, S.S., SZPUNAR, W.E., COLEMAN, B., AND MALLOS, P. (1980). Advice from some health food stores. *J. Am. med. Ass.* **244**, 2045–6.

TAGGART, H.McA. AND ALDERDICE, J.M. (1982). Fatal cholestatic jaundice in elderly patients taking benoxaprofen. *Br. med. J.* **284**, 1372.

VERE, D.W. (1981). Controlled clinical trials: the current ethical debate. *J.R. Soc. Med.* **74**, 84–7.

VESSEY, M.P., LAWLESS, M., McPHERSON, K., AND YEATES, D. (1983). Neoplasia of the cervix, uteri and Contraception: a possible adverse effect of the pill. *Lancet* ii, 930–4.

YELLOWLEES, P., GREENFIELD, C., AND MC INTYRE, N. (1980). Dimethylsulphoxide-induced toxicity. *Lancet* ii, 1004–6.

2 Iatrogenic disease: an historical survey of adverse reactions before thalidomide

R.G. PENN

Introduction

An increasingly popular term is 'iatrogenic disease', which is defined by the *Concise Oxford Dictionary* as '[of disease] caused by process of diagnosis or treatment' and is derived from the Greek *iatros* (physician). The implication of the term is to blame the doctor for adverse reactions caused by treatment and logically this should include not only pharmaceutical agents but also the adverse effects of investigation and other medical procedures. Physicians have been concerned since the earliest times that no one else should treat the sick and that the prescription and dispensing of drugs, if not in their hands, was at least to be supervised by them. This concern has in the past probably owed more to the closed-shop principle rather than to humanitarian worries about the liability of the patient to adverse effects from medication. Doctors have never been eager to admit that medicines and treatment as used by them have any adverse effects, though they have endeavoured to give the impression that it might be a totally different matter if the patient were in the hands of someone else. There have been few warnings over the years that drugs have any actions other than the desired ones. This *laissez-faire* attitude is illustrated by the absence of any legislation in force in the UK to assess the safety of a drug until the passing of the Medicines Act in 1968 (with a few exceptions under the Therapeutic Substances Act of 1925). This is in sharp contrast to the plethora of laws over the preceding several centuries to control quality and who should be allowed to sell drugs.

Nevertheless, a few ancient writers did concern themselves, albeit briefly, with the fact that some medicines were not all they might be. It is of course impossible now to know the views of the ordinary practitioner of the time as sources of information are such a limiting factor. Future historians will have little difficulty in reconstructing the attitudes and problems of today due to the modern publications 'overkill', but for early historical times we are dependent on the odd comment or so made by no more than a handful of writers. This review will give a few illustrative examples of such writers up to modern times and then take a closer look at some late-nineteenth- and early-twentieth-century investigations into adverse reactions to drugs as these laid the foundations of the surveillance techniques of today.

Early concern with adverse drug reactions

Homer (*fl. c.* 700 BC) was certainly aware of the toxic nature of medicines and commented in the *Odyssey* (when speaking of Egypt), '. . . there the earth, the giver of grain, bears greatest store of drugs, many that are healing when mixed and many that are baneful . . . ' (*Odyssey*, IV.228). Homer also mentions in the *Odyssey* that Odysseus smeared the tips of his arrows with poison (*Odyssey*, I.251–67) and that the suitors importuning Penelope were worried that Odysseus would come back and put a deadly poison in their wine (*Odyssey*, II.325–30).

Hippocrates (*fl. c.* 400 BC) and without doubt the classic 'iatros' did give a few warnings about drugs: '. . . in acute diseases employ drugs very seldom and only at the beginning. Even then, never prescribe them until you have made a thorough examination of the patient . . .' (*Aphorisms*, I.24). This advice could hardly be bettered today. Hippocrates did warn specifically about hellebore which he said

' . . . is a dangerous drug for those with healthy flesh since in these it induces convulsions' (*Aphorisms*, IV.16). He further warned that these convulsions could be fatal (*Aphorisms*, V.1). Hippocrates also noted other adverse reactions '. . . when men died from excessive purgation following the administration of drugs, some vomited bile and some phlegm . . .' (*The nature of*

14

man, 6). Whether Hippocrates ever used the word 'iatrogenic' is open to question but doubtless he would have understood it.

Socrates (469–399 BC), who was a contemporary of Hippocrates, was sentenced to die by drinking hemlock and the central nervous effects of the potion were graphically described by Plato in his eye-witness account of the death scene. (*Phaedo*, 117–18).

There were the occasional further warnings about drugs in ancient times and Celsus (*fl.* AD 30) in 'On Medicine' (Book V.1) said

on the other hand Asclepiades dispensed with the use of these [medicines] for the most part, not without reason; and since nearly all medicaments harm the stomach and contain bad juices, he transferred his treatment rather to the management of the actual diet.

It was not only the ancient writers on medical topics who warned about the dangers of drugs as Ovid, somewhat earlier than Celsus (*circa* 1 BC), and admittedly in a rather dubious context said

. . . giving girls aphrodisiac drugs, too, is useless — and dangerous. Drugs can affect the brain, induce madness. Avoid all such nasty tricks. (*Ars Amatoria*, 2.104)

Over the next millennium and a half it is possible to pick out only the odd writer here and there who warned that drugs were not all they seemed. The early formularies were usually just that — descriptive lists of medicines with perhaps instructions for mixing and compounding them and were not meant to be textbooks of therapeutics. Indeed the first official pharmacopoeia in Britain, the *London Pharmacopoeia* of 1618 made its attitude perfectly clear in its preface. The efficacy and use of the medicines was not to be discussed at all in the *Pharmacopoeia* in case this information was seized upon and used by the quack and the itinerant drug peddler. The book was written for the already knowledgeable 'disciples of Apollo' and not for the common people.

The ancient formularies were often in use for many hundreds of years. That of Dioscorides, dating to the 1st century AD for example, was still in use in England late in the seventeenth century and the influence of Galen (*c.* AD 130–200) was to fossilize medicine for the next 1400 years. Contemporary with Galen was Aretaeus the Cappadocian whose works contained a good description of the effects of an overdose of atropine-like drugs. It was to Islamic physicians that the torch was then handed on, and Avicenna (*c.* AD 980–1037) gave a very good description of acute and chronic mercury poisoning. Haly Abbas, a Persian physician who died in AD 994 thought it important to search for new remedies but pointed out the need for caution and advised, among other things, that new remedies should first be tried in animals. He was a

little more adventurous than a somewhat earlier Islamic physician, Michael Mesue, who never employed a drug which had been in use for less than two centuries (Withington 1894).

It was Paracelsus (1493–1541) who became the figurehead of the new revolution which changed therapeutics and was to bring chemically derived mineral medicines into the apothecaries' shops and call into question the old Galenic remedies. Paracelsus may indeed be thought of as the founder of chemical pharmacology and the patron saint of the modern pharmaceutical company. It was he who in an impassioned defence of his new medicines (*The Third Defence*, concerning the description of the new receipts, *c.* 1537) claimed that all substances were poisonous and that it was only by paying attention to the dose that made it not so and that even food and drink taken in excess were 'poison'.

Attempts to regulate the administration of drugs centred in the sixteenth century, as in the twentieth, on new medication with officialdom generally on the side of the conservative Galenists. The story of antimony at this time is illustrative of the antagonism among the several schools of therapeutics. This antagonism was especially so in France where the Faculty of Medicine in Paris, which was intensely traditional and followed the Galenic school, in 1566 forbade the use of antimony which was often impure, mixed with arsenic, and had caused many fatalities. This is a very early example of withdrawal of a product from the market by official action due to severe adverse reactions. In 1658, however, the French King Louis XIV was attacked by what was probably typhoid in a campaign in Flanders. The galenicals of the royal physicians could do no good and a local physician was summoned to the king at Calais. After being purged 20 times with medicines containing one ounce of antimony, the king recovered. It is not surprising therefore that with such royal patronage, antimony was subsequently officially rehabilitated, and in 1666 the Faculty of Medicine voted in its favour.

In general, however, the governments of the time were content to pay attention only to quality and supply of medicines and to leave safety to take care of itself. France pursued a more forward policy and on 17 March 1781 a decree was passed by the Council of State in France 'Concerning the Discipline and Police of the three Corps of Medicine' which ruled that no patent medicine was to be sold without a licence which must specify the disease for which it was appropriate (Cumston 1919). The UK commitment to monitor adverse reactions to licensed products was foreshadowed in Article V of the French decree:

Art. V. His Majesty commands that his first Physician shall be bound to send two printed copies of each patent or grant, to the deans of the Faculties or Aggregations of Medicine, who shall take care to inform him exactly as to the success and ill consequences of the said remedies.

It is a matter for some dispute how successful the various French measures were as the passage of legislation does not necessarily cure the problem.

The eighteenth and nineteenth centuries

The eighteenth century saw the birth of real experimental pharmacology which was to change the face of medicine beyond recognition, and physicians began to warn against adverse reactions. William Withering (1741–99) published a textbook of botany in 1776, giving 'a botanical arrangement of all the vegetables naturally growing in Great Britain'. In the book he emphasized that valuable remedies could be obtained from many substances but some evoked rapid and intense effects and are called poisons and that these poisons in small doses were the best medicines though they were too dangerous in large doses. Withering is best known for his classic book *An account of the foxglove* which he published in 1785 and which gave a description of the toxic effects of digitalis which could stand today.

The foxglove when given in very large and quickly repeated doses occasions sickness, vomiting, purging, giddiness, confused vision, objects appearing green and yellow; increased secretion of urine, with frequent motions to part with it; and sometimes inability to retain it; slow pulse, even as slow as 35 in a minute, cold sweats, convulsions, syncope, death.

Withering was well aware of the need to titrate the dose to minimize the risk of these adverse reactions:

Let the medicine therefore be given in the doses and at the intervals mentioned above: let it be continued until it either acts on the kidneys, the stomach, the pulse or the bowels; let it be stopped upon the first appearance of any of these effects and I will maintain that the patient will not suffer from its exhibition nor the practitioner disappointed in any reasonable expectation.

At first during this period experimenters were concerned mainly with the overt toxicity of different substances and to find antidotes. Analysis of their mode of action was not possible in these early phases as this presupposes at least a rough idea of basic pharmacological and physiological changes. Nevertheless, owing to the pioneering efforts of workers like François Magendie (1783–1855), Claude Bernard (1813–78), Rudolf Buchheim (1820–79), and Oswald Schmiederberg (1838–1921), pharmacology and the investigation of drugs was made a separate and developing science.

Anaesthesia

One of the major therapeutic advances during this time was that of anaesthesia. The story has been repeatedly told of the use of ether by Crawford Long (1842), W.T.G. Morton (1846), and Robert Liston (1846), of nitrous oxide by Horace Wells (1845), and of chloroform by J.Y. Simpson (1847). The use of these powerful substances demonstrated dramatically that there was a critical relationship between dose and effect and that adverse reactions could occur and that patients could die as the result of anaesthesia. The profession became worried and realized that something had to be done.

In 1864 a committee appointed by the Royal Medical and Chirurgical Society (later to become the Royal Society of Medicine) enquired 'into the uses and the physiological, therapeutical, and toxical effects of chloroform'. Chloroform was first used as an anaesthetic in 1847 and as its use increased it was found that occasionally people died unexpectedly during anaesthesia. Not only did this committee, in a published report (1864) some 119 pages long, comment on animal experiments comparing chloroform with ether but also discussed in detail the obstetric and surgical use of chloroform. The report tabulates the details of 109 fatal cases, a further 9 incomplete reports of fatal cases, 4 'delayed' fatal cases, and 25 non-fatal cases who also had adverse reactions, and some cases where chloroform was used for suicidal purposes. The Committee found that chloroform depressed the action of the heart and frequently caused death by syncope in contrast with ether which had only a slight depressant effect on the heart.

In 1875 the British Medical Association, at its annual meeting in Edinburgh, appointed a committee to enquire into various anaesthetic agents. But in 1877 when the British Medical Association held its annual meeting in Manchester, Spencer Wells in his 'Address in Surgery' (Wells 1877) said:

I wish I could speak as confidently of the chemical composition of the fluid sold as bichloride of methylene as I can of its anaesthetic properties . . . I am sorry that some of the analytical chemists whom I have asked to clear up the question of its composition have not done so. It ought to be done, it can be done, and it must be done.
The Committee appointed in Edinburgh two years ago . . . but which has never met until this morning might very well undertake this task . . . in the words of the resolution; namely 'to inquire into and report upon the use in surgery of various anaesthetic agents and mixture of such agents and to collect and summarise the evidence of British practitioners in surgery and medicine as to the relative advantages of chloroform, ether, nitrous oxide gas and other agents and to carry on suitable experimental investigations' . . . I am not speaking too strongly if I say it is the duty of the Association at once, without any unnecessary delay, to satisfy the public that all that is possible is

being done to discover the means by which anaesthesia, effectual now, may be rendered safe for the future.

This excerpt gives some very interesting concepts as seen by an authoritative medical figure of his day and presented to the leaders of the profession. There is expressed a worry about the quality of some anaesthetic agents (i.e. the composition of bichloride of methylene), although Wells apparently had little doubt about its efficacy. The demand was forcibly put that it was the responsibility of the medical profession to assess the safety and efficacy of a group of potent and dangerous agents (not merely their *absolute* efficacy but *relative* also) and that this assessment should not only be clinical but be supported by experimentation. The onus was placed firmly on the professional users to *satisfy the public* as to the safety of anaesthesia. A 'sub-committee of research' was appointed with Dr McKendrick, Professor of Physiology at Glasgow, as chairman and convenor. The final report was published in 1880 (McKendrick *et al.* 1880). It is interesting to note that the McKendrick report quoted the earlier findings of the Royal Medical and Chirurgical Society (though only briefly) but gave the date of the report as 1874 (i.e. some 10 years later) — perhaps this was an unconscious slip of the BMA irked by the embryo Royal Society of Medicine.

The McKendrick report found that chloroform was hazardous because not only did it cause respiratory depression but even minor doses could cause cardiac arrest during the induction phase of anaesthesia. There were many doctors who disagreed with these findings and in 1889 E. Lawrie, Residency Surgeon at Hyderabad, carried out experiments on 128 pariah dogs and reported that chloroform may be given to dogs with perfect safety and that in no case did the heart become dangerously affected until after the breathing had stopped (Annotation, *Lancet* 1889). These results were hotly contested in the columns of the *Lancet* but were confirmed by the Second Hyderabad Commission of which Lawrie was the president, (Lawrie *et al.* 1890). The weight of clinical evidence, however, showed convincingly that chloroform could and did cause sudden cardiac arrest in man. Lawrie's work is one of the first major scientific demonstrations that toxicology testing in other species does not necessarily give the picture in man.

There were other investigations into chloroform and in 1893, for example, the *Lancet* sent out a questionnaire world-wide and showed that heart failure was the commonest cause of death with chloroform. The BMA set up a Commission on Anaesthesia in 1891 and a Special Chloroform Committee in 1901.

It was A. Goodman Levy who provided the key to the problem in 1911. He showed that cats under light chloroform anaesthesia could die from ventricular fibrillation especially if small doses of adrenaline were injected. Full chloroform anaesthesia did not produce these effects and neither did ether anaesthesia. Chloroform sensitizes the myocardium to the action of catecholamines, and this may produce cardiac arrhythmias, perhaps progressing to ventricular fibrillation especially during the induction stage of anaesthesia.

The twentieth century

Arsenobenzol compounds

More collaborative investigations into adverse effects of other drugs were carried out in later years, and in 1922, for example, the Medical Research Council reported the findings of an enquiry into epidemics of jaundice and hepatic necrosis following the use of organic arsenicals, such as salvarsan, to treat syphilis in soldiers returning from World War I (*Reports of the Salvarsan Committee 1922*).

The arsenobenzol compounds produced many important ill effects which included encephalitis haemorrhagica and acute exfoliative dermatitis. Jaundice also occurred, and the most puzzling was jaundice starting weeks or months after the termination of treatment and which commonly progressed to 'acute yellow atrophy of the liver'.

The committee was particularly concerned at the epidemic-like outbreaks of this late jaundice, classically described at the Cherry Hinton Military Hospital, Cambridge, in an outbreak which resulted in the deaths of 15 patients. In retrospect it is probable that these outbreaks were of serum hepatitis, transmitted via the intravenous infusion techniques used. The committee, whilst acknowledging the possibility of 'adventitious infection', tended to the conclusion that the ill effects were due to the inherent chemical nature of the arsenobenzol compounds and not to any particular specially toxic batches. They were influenced by the occurrence of 'toxic jaundice and acute yellow atrophy of the liver' during World War I among workers using tetrachlorethane ('dope' for aeroplane wings) and trinitrotoluene ('TNT' used in munitions).

The committee might possibly have been well advised to take closer note of the findings in their Report about the venereal disease centre at the Portobello Hospital, Dublin. Eight soldiers died of malaria between November and December 1917, though only one of these had been in a tropical country. The infection was undoubtedly transmitted from a carrier via the almost communal intravenous infusion apparatus used for the neoarsphenamine treatment. It gave the committee a clear pointer to a route which could transmit infection. The committee did make a plea about adverse reactions which foreshadows the future Yellow Card system.

The committee hope that a plain statement of the rare fatalities and other untoward effects known to occur after the use of arsenobenzol preparations may encourage the communication to the Ministry of Health of details concerning such accidents, for it is only in the light of such information that investigation and measures with regard to their prevention can be successfully undertaken. (*Salvarsan Committee Report*: Introduction)

During the first half of the twentieth century it became increasingly obvious that drugs were toxic and some were very toxic, but this knowledge came haphazardly and very slowly. There were voices raised that drugs must be assessed for their toxicity and in 1929 Chauncey B. Leake criticized the development of drugs based on 'empiric trials at the suggestion of interested commercial concerns', and emphasized the need for sound and unprejudiced scientific evidence:

Clinical trial of new chemical substances should be made only after critical disinterested pharmacologic study has estimated (a) the probable toxicity, (b) the type and mode of action, (c) the worthiness of application to human beings, and (d) the reasonableness of replacing existing drugs. There is no short cut from chemical laboratory to clinic, except one that passes too close to the morgue.

(Leake 1929)

Leake was a good example of a voice crying in the wilderness and further instances of drug toxicity continued (and continue) to be reported. A brief account of some of the classical cases will be given and these illustrate how long it takes for a particular toxicity to be recognized and how long it may take for this to be generally accepted by clinicians.

Amidopyrine (aminophenazone, aminopyrine)

Amidopyrine was introduced as an analgesic and antipyretic but it was not for nearly half a century that agranulocytosis, its major adverse reaction was recognized. Kracke and Parker (1934) in a comprehensive review discussed possible aetiologies of agranulocytosis of which they had collected nearly a thousand cases reported in the literature between 1922–1932. They commented on a series of 14 cases of agranulocytosis presented by Madison and Squier in 1933 in each of which there was a history of amidopyrine consumption. Kracke and Parker astutely observed in their series that agranulocytosis was more common among physicians, nurses, and their relatives than any other group (there was 40-fold greater incidence in physicians than in lawyers and 200-fold greater incidence in nurses than in female school teachers — an early example of a primitive case-control study). They also correlated these cases of agranulocytosis with the use of drugs containing a benzene ring, especially amidopyrine, and postulated that this drug was indiscriminately taken by the medical and allied professions.

Cinchophen

Cinchophen was introduced into medical practice in 1908 as an analgesic and was especially used to relieve the pain of rheumatoid arthritis and gout.

In 1923 Worster Drought described the case of a 59-year-old man with long-standing gout who took cinchophen (atophan). The patient started on a dose of $7\frac{1}{2}$ grains three times a day but discontinued the drug after 12 days when he developed urticaria (i. e. a total of 270 grains = 17.55 grammes). Three weeks later, the urticaria gone, he took one $7\frac{1}{2}$-grain dose only, developed severe urticaria and pruritus which persisted and ten days later jaundice appeared of a moderate degree with no enlarged or tender liver and which cleared up in several weeks. Other reports of severe liver toxicity followed but Wade and Beeley (1976) discuss an inquest in Birmingham in 1934 where several doctors said that they did not know that cinchophen could cause jaundice. Wade and Beeley comment that cinchophen is a good example of how long it can take before a serious adverse reaction is recognized and how slowly and incompletely the medical profession becomes aware of it.

Elixir of sulfanilamide–Massengill

It is, perhaps, an unfortunate truism that legislation often occurs after the event and this was particularly so in the USA with attempts to replace the unsatisfactory 1906 Pure Food and Drugs Act. These attempts were not successful until the passing of the Food, Drug and Cosmetic Act of 1938, following the *Elixir of Sulfanilamide-Massengill* disaster of 1937. The 1938 Act forbade the marketing of new drugs until these had been cleared by the Food and Drug Administration. At that time the extent of governmental authority provided by the 1906 Act and the Sherley Amendment of 1912 was very small. Manufacturers were entirely free to market a product provided it was not adulterated or mislabelled and no considerations were given to the safety and the efficacy of the product. During September and October 1937, at least 76 people died as a result of taking *Elixir of Sulfanilamide-Massengill* (Geiling and Cannon 1938). These 76 were confirmed deaths but it was alleged that the real death total was over 100. The Massengill Company was an old established firm who wanted to sell a liquid form of sulphanilamide, the first of the new sulpha drugs which it already formulated in tablets and capsules. The Elixir was essentially a 10-per cent solution of sulphanilamide in 72-per cent diethylene glycol with some flavouring and colourants. The firm could only be prosecuted for mislabelling the product as an elixir which applies to an alcoholic solution and not to diethylene glycol. No animal toxicity tests had been done on the preparation although there were already in the literature reports dealing with the toxicity of diethylene glycol (Oettingen and Jirouch 1931; Haag and Ambrose 1937).

Geiling and Cannon (1938) showed conclusively by animal toxicity tests and human investigation that the toxic ingredient in the Massengill elixir was the diethylene glycol and their report on behalf of the American Medical Association makes some very interesting demands to improve drug testing in animals. They said that the exact composition and method of manufacture should be given. The drug should be tested for toxicity at varying dosages in several species in both acute and chronic studies. They emphasized the need for careful and frequent observation of the animals during these studies and for thorough pathological examination of tissues. They suggested that the drug should also be tested in animals with experimental lesions of the excretory and detoxicating systems and that the absorption and excretion of the drug should be determined together with its concentration at different times in blood and tissues. They also recommended that possible interactions with foods and other drugs should be investigated and careful watch kept for any untoward reactions. Whilst conceding that many would consider these safeguards to be too rigid, Geiling and Cannon maintained that lives would be saved if such standards were put into effect and that the life and safety of the individual should not be subordinated to the competitive system of drug exploitation.

The toxicity of diethyleneglycol has again come to the fore with the discovery of Austrian wines contaminated with the substances, and warnings being issued of its toxicity (all media sources) during July 1985. Stocks of all Austrian wines were promptly withdrawn throughout the UK.

Stalinon

In 1957 a French pharmacist, the inventor of an oral preparation against boils called *Stalinon*, was sentenced to two years' imprisonment and heavily fined for not carrying out satisfactory pre-marketing safety tests (*British Medical Journal* 1958). *Stalinon* capsules contained 15 mg of diiodoethyltin and 100 mg of isolinoleic acid esters. Most people took them without untoward effects but a minority had signs of raised intracranial pressure such as headaches, confusion, and vomiting. It was alleged that *Stalinon* killed 102 people and permanently affected 100 more, some survivors having residual paraplegia. The enquiry established that the clinical trials had been performed with capsules supposedly containing 50 mg diiodoethyltin but which in reality contained only 3 mg owing to a dispensing error. The *Stalinon* capsules sold to the public therefore contained five times as much active material as was tested in the clinical situation and provided a graphic if unfortunate demonstration of the human toxicity of organic tin compounds. Later studies, both human and animal, showed intramyelinic vacuolation with astrocytic swelling but little or no evidence of neuronal degeneration and occurring almost exclusively in the central nervous system rather than peripheral nerves.

Phenacetin (acetophenetidin)

Phenacetin was first used in 1887 and is an effective analgesic and antipyretic. Whilst it is mainly de-ethylated to paracetamol in the body, the two substances should not be thought of as completely equivalent and phenacetin has metabolites such as *p*-phenetidine which paracetamol does not have. Unlike paracetamol, phenacetin can also produce psychological effects to cause relaxation and lessening of tension in a similar way to ethanol. These central nervous system effects are thought to be major factors contributing to the abuse potential of phenacetin-containing combinations which, if used for long periods, can lead to the development of analgesic nephropathy (see Chapter 34, this volume). In 1950 Zollinger and Spuhler implicated analgesics as a cause of chronic renal disease in Switzerland and this gave rise to intensive research in many countries. There was much controversy as to the existence of any such association especially as it was difficult to produce analogous damage in the laboratory animal. In 1963 Grimlund published a classical epidemiological study in the Swedish town of Huskvarna which did much to resolve the doubts. Huskvarna had some 13 000 inhabitants in 1963 of whom about 3000 were locally employed by the Husqvarna Company. These workers, mainly male, made a wide range of articles such as guns, sewing machines, and bicycles. In the influenza epidemic of 1918–1919 a local physician, Dr Hjorton, prescribed a compound mixture of 0.15 g caffeine and 0.5 g each of phenacetin and phenazone. *Hjorton's Powders* became very popular in the Husqvarna factory, being considered very beneficial and invigorating, and they were much used even when the takers were otherwise in good health. They were also used as presents, as bribes, at social gatherings, and were treated in much the same way as tobacco elsewhere. The taking of 10–12 powders a day was not uncommon. In the 1950s it was noticed that deaths from uraemia in Huskvarna were much in excess of those of nearby similar communities, being some 10 times those of Fagersta, a town of approximately the same size and character. There was also a marked preponderance of male deaths from uraemia unlike elsewhere in Sweden, where the sex ratio was approximately equal. A survey of the population at Huskvarna showed that the uraemic deaths and renal damage in living residents closely correlated with whether they had been long-term ingesters of the phenacetin-containing powders, and that many of the fatal cases of those with severely damaged kidneys had taken 10 kg or more of phenacetin over the years. In interviews, 189 out of 936 Husqvarna Company employees interviewed admitted abuse of the phenacetin powders and 34 per cent of these had renal damage of some degree compared with only 2.4 per cent in the 747 'non-takers'.

Phenacetin was removed from all OTC medicines in Sweden by governmental action in 1961. The number of deaths from erstwhile abusers in Huskvarna did not start to diminish until about 1968 — a time-lag to be expected from the natural history of the disease.

Since the paper by Grimlund a considerable international literature on analgesic nephropathy has built up but the general topic will not be considered further here. It is sufficient for the present purpose to say that analgesic nephropathy which is really a particular form of chronic renal papillary necrosis, although classically associated with long-term phenacetin usage, is also alleged to occur in the abuse of other similar analgesics. There are striking differences in incidence from one country to another and figures collected by Kincaid-Smith (1978) for the occurrence of renal papillary necrosis per 100 necropsies show an incidence in Australia varying between 3.7 to 21 per cent but only 0.2 per cent in the USA and 0.16 per cent in the UK.

In the UK in 1973 the Committee on Safety of Medicines drew attention to the general view of the medical profession in the country that phenacetin-containing products should be controlled because of the occurrence of analgesic nephropathy and carcinoma of the renal tract associated with its long-term use. As a result, a Statutory Order (The Medicines [Phenacetin Prohibition] Order 1974) was introduced to prohibit the sale, supply, or import of medicinal products consisting of phenacetin or containing more than 0.1 per cent phenacetin, although certain exemptions were made to permit the supply of phenacetin through a registered pharmacy on prescription by a doctor or dentist. It was hoped at the time of the 1974 Order that pharmaceutical companies would reformulate those products still left on the market

so as to exclude phenacetin. However, when the Committee on the Review of Medicines (CRM) considered phenacetin-containing products in 1977 as part of its analgesic review, it was clear that about 50 such products remained on the market, many having names which did not indicate the presence of phenacetin, e.g. Aspirin Compound Tablets, Compound Codeine Tablets. The CRM therefore recommended that on grounds of safety, phenacetin has no place in analgesic (including anti-inflammatory and antipyretic) therapy alone or in combination products. An Order was therefore made under Section 62 of the Medicines Act 1968 to prohibit the sale, supply or importation into the UK of medicinal products containing phenacetin and of phenacetin as a drug substance (The Medicines [Phenacetin Prohibition] Order 1979). The Order permits products containing less than 0.1 per cent phenacetin to be available without prescription; the main use of this small percentage of phenacetin is as a stabilizer in hydrogen peroxide. There is also an exemption for doctors who wish to prescribe phenacetin itself.

Thalidomide

The story of thalidomide is too well known to bear much repetition here but as it is the foundation stone on which the Medicines Act was built, it is relevant to summarize what happened. Thalidomide first went on sale in 1956 in West Germany as Contergan and enjoyed good sales both there and in other countries (in the UK in 1958 as *Distaval*) as a sleeping aid and as a treatment of vomiting in early pregnancy because of its prompt action, lack of hangover, and apparent safety. In 1960–61 it became evident that the prolonged use of thalidomide could cause hypothyroidism and peripheral neuritis and for this reason marketing approval was delayed in the United States which thus avoided the full horror of the story to come. In 1961 reports began to be made of a remarkable rise in West Germany since 1959 in the incidence of a peculiar malformation of the extremities of the newborn. This condition was characterized by defective long bones of the limbs which had normal to rudimentary hands or feet. Owing to its external resemblance to a seal's flipper, it was given the name of phocomelia. This condition had previously been very rare in West Germany but whereas no cases had been reported in the 10 years between 1949 and 1959, there were 477 cases in 1961 alone. It took several years for it to become clear epidemiologically that thalidomide was the cause of these abnormalities but by this time some thousands of babies had been born worldwide, suffering from severe skeletal deformities caused by the drug.

Although it had been known for many years that

chemical substances could act as teratogens in lower animals and in mammals, the medical world as a whole seemed to have paid little attention to the possibilities of danger to the developing human fetus. Most reputable manufacturers in the UK did actually carry out tests and clinical trials on their new drugs but were not legally required to do so. Certainly drug companies did not deem it necessary that new drugs should be tested for their hazards to the human embryo before marketing. Warnings against the use of drugs in pregnancy were, however, already being raised about this time but these were largely ignored. Until 1968, therefore and the passing of the Medicines Act, no statute in the UK required the pre-marketing approval of medicines on the grounds of their safety. Kenneth Robinson (later to become the Minister of Health under a Labour Government, October 1964–October 1968) observed in a parliamentary debate on 8 May 1963:

I come to my last main topic which is the control and safety of drugs. This is of course a subject which was thrust to the fore both in this House and in the public press a year or so ago as a result of the thalidomide tragedy. The House and the public suddenly woke up to the fact that any drug manufacturer could market any product, however inadequately tested, however dangerous, without having to satisfy any independent body as to its efficacy and safety and the public was almost uniquely unprotected in this respect.

(Robinson 1963)

The thalidomide disaster therefore caused widespread anxiety about the absence of safeguards to ensure that the harmful side-effects of drugs would be brought to the attention of physicians who could then weigh the risks against the expected benefits. It resulted, in Britain, in the passing of the Medicines Act 1968 which embraces all aspects of the control of medicines with reference to safety, quality, and efficacy. There is no legal definition in the Act of an adverse reaction, but the importance attached to the problem is shown by the Committee on Safety of Medicines being specifically charged with:

promoting the collection and investigation of information relating to adverse reactions, for the purpose of enabling such advice to be given.

REFERENCES

BRITISH MEDICAL JOURNAL (1958). 'Stalinon': a therapeutic disaster. *Br. med. J.* i, 515.

CELSUS (1935–38). *On medicine* [trans. W G Spencer]. The Loeb Classical Library, Harvard University Press. Heinemann, London.

CUMSTOM, C.G. (1919). The legal control of the sale of nostrums and poisons in France during the eighteenth century. *Ann. med. Hist.* 2, 396–9.

GEILING, E.M.K. AND CANNON, P.R. (1938). Pathologic effects of elixir of sulfanilamide (diethylene glycol) poisoning. *J. Am. med. Ass.* 111, 919–26.

GRIMLUND, K. (1963). Phenacetin and renal damage at a Swedish factory. *Acta med. scand.* Suppl. 405, 3–25.

HAAG, H.B. AND AMBROSE, A.M. (1937). Studies on the physiological effects of diethylene glycol. *J. Pharmacol. exp. Ther.* 59, 93.

HIPPOCRATES (1978). In *Hippocratic writings* (ed. G.E.R. Lloyd). Penguin Books, Harmondsworth, Middlesex.

HOMER (1938). *The Odyssey* (trans. A T Murray). The Loeb Classical Library, Harvard University Press. Heinemann, London.

KINCAID-SMITH, P. (1978). Drug-induced renal disease. *Practitioner* 220, 862–7.

KRACKE, R.R. AND PARKER, F.P. (1934). The etiology of granulopenia (agranulocytosis). *J. lab. clin. Med.* 119, 799–818.

LANCET (1889). Annotation. Hyderabad Medical School. Chloroform inhalation. *Lancet* i, 344.

LAWRIE, E., BRUNTON, T.L., BOMFORD, G., AND HAKIM, R.D. (1890). Report of the Second Hyderabad Chloroform Commission. *Lancet* i, 149–59.

LEAKE, C.D. (1929). The pharmacologic evaluation of new drugs. *J. Am. med. Ass.* 93, 1632–4.

McKENDRICK, J.G., COATS, J., AND NEWMAN, D. (1880). Report on the action of anaesthetics. *Br. med. J.* 2, 957–82.

OETTINGEN, VON, W.F. AND JIROUCH, E.A. (1931). The pharmacology of ethylene glycol and some of its derivatives in relation to their chemical constitution and physical chemical properties. *J. Pharmacol. exp. Ther.* 42, 355–72.

OVID (1982). *The erotic poems* (trans. Peter Green). Penguin Books, Harmondsworth, Middlesex.

REPORT OF THE COMMITTEE APPOINTED BY THE ROYAL MEDICAL AND CHIRURGICAL SOCIETY TO ENQUIRE INTO THE USES AND THE PHYSIOLOGICAL, THERAPEUTICAL AND TOXICAL EFFECTS OF CHLOROFORM (1864). *Medical-Chirurgical Transactions* 47, 323–442.

REPORTS OF THE SALVARSAN COMMITTEE (1922). II Toxic effects following the employment of arsenobenzol preparations. *Medical Research Council*, HMSO, London.

ROBINSON, K. *Parliamentary Debates.* House of Commons Official Report 677, 109, 8 May 1963, Vol. 439.

WADE, O.L. AND BEELEY. L. (1976). *Adverse reactions to drugs*, 2nd edn. Heinemann, London.

WELLS, S.T. (1877). Address in surgery. *Br. med. J.* 2, 174–8.

WITHINGTON, E.T. (1894). *Medical history from the earliest times*, 2nd reprint 1964. The Holland Press, London.

WORSTER-DROUGHT, C. (1923). Atophan poisoning. *Br. med. J.* 1, 148–9.

3 Two decades of drug-induced disasters

M.N.G. DUKES

Misadventure in Clayton Tunnel

On the sunny morning of Sunday, 25 August 1861, three trains left Brighton Railway Station bound for London. The first was an excursion from Portsmouth consisting of 16 wooden coaches, the second, a Brighton excursion of 17 such coaches, and the third, a well-filled regular train of 12 coaches. The line was worked on the time-interval system, which the Railway Company preferred because it was simple, cheap, and allowed frequent passage of trains, in this instance at intervals of five minutes. As the heavy trains lumbered up the gradient to the long tunnel at Clayton however, the intervals between them had lessened to four and three minutes, a risk inherent in the system.

As the first train approached the tunnel it should have triggered a signal to prevent others from following it, but the signal did not operate. The failure set off an alarm bell in the cabin, but the exhausted signalman who was on a continuous shift of 24 hours instead of the regulation 18 hours noticed it very late. He hurried out of his signal box waving a red flag. The driver of the second train glimpsed the flag as he passed, but his train carried brakes only on the locomotive, and it was deep in the tunnel before he could bring it to a standstill and throw it into a slow reverse. At that moment, the signalman received a telegraph message from the north end of the tunnel that a train had emerged. Believing that this must have been the second train, rather than the first, he unfurled his white flag and allowed the third of the trains to plunge into the tunnel. A few seconds and 250 yards later it crashed into the reversing Brighton excursion. In the mangled wooden wreckage, burned by hot coals and scalded by steam, 21 people died most horribly in the darkness. No fewer than 176 were seriously injured (Rolt 1982).

The Railway Inspectorate's extensive report on the Clayton Tunnel accident, and others like it, speeded the adoption of more adequate rules for safe railway working including the basic trio of interlocked signals, section-block working and continuous braking ('lock, block, and brake'). Such rules, adopted voluntarily or by decree, ultimately rendered the railway the safest form of mass transit ever developed.

There are some close and remarkable parallels between the history of railway development and regulation in the nineteenth century and the history of drug development and regulation in the twentieth; one day they should be examined in more detail. For the moment it is sufficient to notice that both have been influenced by misadventures which, like the accident in Clayton Tunnel, have often been attributable to a whole chain of faults — technical, human, commercial, and official. Whilst the history of accidents may be studied out of mere morbid curiosity or from a desire to seek out and punish the guilty party (if such a party exists), it can and should also be used to guide us towards principles which will lessen the risk of such things happening in the future.

The questions which need to be asked about any misadventure of this type are manifold. Could the event have been foreseen and prevented entirely? If not, how could it have been recognized and countered in sufficient time to prevent the worst from happening? Did its occurrence point to any general underlying risk which could manifest itself again?

When we look specifically at drug disasters there is good reason to limit our considerations to the events of the last 20 years. For one thing, it is the most recent events which in the changing world of drug research and development are likely to be most helpful in assessing the risks which we currently run. For another, limiting our review to the last two decades enables us to set aside at last the much-told tale of thalidomide, which is now nearly a quarter of a century old. The thalidomide tragedy looms so ponderously over the history of side-effects that any inclusion of it in a general discussion all too easily distorts the entire picture, causing other events to pale into insignificance and suggesting that since 1961 we have solved the worst of our problems. That is most certainly not true. The number of patients gravely injured or killed in epidemics of drug-induced disease since then is a vast multiple of the number of thalidomide victims; the range of injuries produced is also so wide that no simple solution to the problems seems likely to emerge.

All the events which will be summarized here have been reviewed before in these volumes or elsewhere, often at great length. The present purpose must be to seek to distil the essential elements from each of these stories and to learn from them what we can.

Clioquinol

The history of clioquinol is recounted in detail in Chapters 27 and 30, this volume; it will therefore suffice at this point to recall a few essential facts. Briefly, this halogenated hydroxyquinoline was first brought into use in internal medicine around 1930, largely for the prevention and treatment of diarrhoea; in due course it became extremely popular throughout the world, often being used for long periods of time. Scattered reports that the drug had certain neurotoxic effects appeared in the Argentine medical literature in 1935 and in a Swiss veterinary journal in 1965. Only in Japan did things take a quite different turn. In that country, from 1957 onwards, a series of eight regional epidemics of an unfamiliar neurological syndrome were reported. By 1963 it had become recognized that the eyes could also be involved and the term 'subacute myelo-opticoneuropathy' (SMON) was soon coined. By 1966 the total number of cases reported in Japan had risen to 1859 and in 1969 alone more than 2000 additional cases were reported. SMON was frequently severe; more than 90 per cent of the patients had paraesthesiae and dysaesthesiae of the lower extremities; total blindness occurred in 2.5 per cent. Nearly 4 per cent of patients suffered from proctoparalysis and paralysis of the bladder. After a period of violent dispute, it became quite clear that the vast majority of victims had been heavy users of clioquinol.

The paradoxes of the clioquinol story are many, but nothing is more puzzling than the fact that this condition occurred so much more frequently in Japan than elsewhere. A relatively small number of cases were described in Western Europe, and it is even more striking that the conditions occurred rarely or not at all in a number of South-East Asian countries where the consumption of the drug continued to be as intensive as in Japan, even after it had been removed from the market in the latter country. One hypothesis is that it might have potentiated the effect of certain encephalopathogenic metallic ions (particularly aluminium and bismuth, which were popular remedies in Japan and may often have been

taken simultaneously); this possibility is still being explored. It is, however, astonishing that at a world congress devoted to SMON in Kyoto in 1979 virtually no serious attention was paid to the need for an explanation of these geographical differences, an understanding of which could prove so vital to our recognition and even our anticipation of future complications of this type.

The clioquinol case provides a classic illustration of several other things as well, to begin with the fact that the incidence of an adverse reaction can change dramatically, for known or unknown reasons, at some point in time. There is no reason to think that these problems existed, even on a modest scale, in Japan in 1940 or 1950, despite the fact that the drug had already become very popular. Again, the remarkable geographical distribution of the complications which ensued underlines the need, often unrecognized, to monitor drugs in each area of the world; anyone who has had contact with medicine in the Far East or in Africa can confirm that there are numerous factors in indigenous medical practice and habits which might affect (and probably regularly do affect) the safety of modern synthetic drugs, rendering it dangerous to extrapolate from knowledge garnered only in Europe and North America. But the clioquinol episode also reminds us how early evidence could at one time be buried in inaccessible literature; not until Japanese researches looked very carefully and with hindsight for them was it realized that the Argentinian cases and the relevant Swiss veterinary literature even existed.

Practolol

The practolol case is equally well known and is reviewed in Chapters 16 and 30, this volume. This β-adrenergic blocker, the second to attain widespread importance after the introduction of propranolol, was marketed as early as 1970. In 1973 it was first reported to produce an eczematous or psoriasiform skin rash, with or without ocular discomfort. Within the next year a form of polyserositis was described, generally characterized by pleural and joint effusions, but which could also involve the peritoneal cavity. In 1974 and 1975 other reports delineated what is now recognized as the classic picture of practolol-induced sclerosing peritonitis. Shortly afterwards, the alarming picture of practolol-induced eye changes, comprising keratoconjunctivitis sicca, conjuctival scarring, fibrous metaplasia and shrinkage, even resulting in blindness, became clear. By October 1975 the practolol-induced 'oculomucocutaneous syndrome' had become recognized as a serious iatrogenic entity and the oral form of practolol was withdrawn.

Here one sees that even in an era of strict drug control a serious effect was only discovered relatively late; the process took some five years in all despite the gravity of the complication. There is no reason to think that there was (as there seems to have been with clioquinol) an actual increase in the frequency of the complication in course of time; early cases were apparently simply not recognized as being drug-induced. The history of practolol illustrates the difficulties which can arise when one is dealing with an adverse reaction which occurs only in a low percentage of all the cases treated, especially where there has been much co-medication. Again there appears to have been some geographical difference; both the absolute incidence of the practolol syndrome and the relative incidence of the individual symptoms — notably those involving the eyes, the skin, and the internal organs — were somewhat different in Great Britain and in the Netherlands, showing the need for a rapid and complete exchange of data on suspected adverse effects across national frontiers. Here, however, another technical bogey raised its head; in neither country did the health authorities possess detailed drug utilization data, and the question of the absolute incidence of the practolol syndrome and the exclusion of other causal drugs might have proven very difficult to solve had not the manufacturer — ICI — provided the missing data from its own records. In this and other respects one sees how valuable the co-operation of a conscientious company can be in establishing a link between a drug and an adverse effect at an early phase, however much anguish that may entail within the company's own ranks.

Metamizol

Metamizol is also known as novaminsulfon. Unfortunately it is also known by some 20 other names — generic, chemical and commercial — as well, and this is perhaps the first problem arose with respect to effective monitoring. Some of the confusion around its adverse effects has clearly been attributable to the failure of physicians, sometimes quite expert in this field, to recognize it under its various connotations (see also Chapters 12 and 32).

There is today no reasonable doubt that metamizol, by whatever name it be known, causes blood dyscrasias and in some countries the evidence on that score is impressive (Prescott 1980; Del Favero 1983). That conclusion can be reached on the totality of the material and despite the fact that the individual data are themselves frequently defective in detail; in the Netherlands, for example, it was possible to show not only that metamizol featured markedly in the reports of blood dyscrasias to the

national monitoring centre, but also that it was a common part of the drug history in patients admitted to hospital for such disorders. This is one of many instances where an attempt to discredit the proof of an adverse reaction by demanding that individual data all meet strict algorithmic standards can prove contrary to the public interest; a tower built of fragmented bricks may be as solid as any other.

Particularly remarkable in this case is the continuing insistence in many quarters that even this serious adverse effect is outweighed by the drug's therapeutic activity. Broadly classified as a non-steroidal anti-inflammatory agent, metamizol has been accorded the image of a non-addicting substitute for morphine. Whilst it has not been on sale in the UK for many years, there are many countries in which it is solemnly stated to be indispensable to the practice of medicine. Certainly it is an effective drug; indeed it works in colic, largely because in those countries where it is employed it is generally used in a fixed combination with a spasmolytic agent; but it is not by any means unique. In the years since metamizol was developed it has become evident that other non-steroidal anti-inflammatory agents too can be characterized as potent analgesics provided they are appropriately studied and given in adequate doses. In cases such as this — and the same problem can be said to have arisen with clioquinol — one must surely be wary of allowing well-documented evidence that an old drug is harmful to be outbalanced by considerations of therapeutic uniqueness unless these are equally well-documented.

Triazolam

Since triazolam is by now a widely used hypnotic drug which has gained a considerable popularity in many countries for its lack, in many users, of any troublesome hangover effect, it may appear surprising to encounter it in a chapter devoted to drug disasters (see also Chapter 28). Nevertheless it has a particular relevance, for it illustrates the fact that even drugs which are a blessing for some can, if they are handled the wrong way, prove a disaster to others.

The events which surrounded the withdrawal from the Dutch market of triazolam in 1979 were in part published at that time (Dukes 1980), yet the full story was never told. In retrospect, the truth behind it has become increasingly clear. If one simply goes by the view which was put around by the manufacturer and widely quoted in the press, triazolam was withdrawn because the Netherlands Committee for the Evaluation of Medicines capitulated to the pressures of a sensationalist press; the true story is somewhat different and deserves to be known for the general lessons which can be learnt from it.

Triazolam was introduced into the Netherlands in the course of 1978. During the first few months after its introduction, at which time it was still being used only on a relatively small scale, a handful of reports suggestive of excitation reactions reached the Adverse Reaction Monitoring Bureau near The Hague. They were too fragmentary to merit immediate action, but the Bureau was put on the alert. In the early summer of 1979, however, two things happened. In the first place, a considerable degree of advertising was released for the product, including certain press releases to the public media on the remarkable new hypnotic which could now be obtained from the physician on request; a considerable increase in sales appears to have ensued. In the second place a psychiatrist, also in The Hague, noted in his practice a number of cases in which the drug appeared to have induced something like an acute psychosis. In July 1979, the psychiatrist in question not only published his findings, but also, rather injudiciously, appeared on television to publicize them. A great deal of excitement in the media followed, and within only a few weeks the Adverse Reaction Monitoring Centre had received more than 1000 reports of similar reactions. Some of them were serious; a number of patients were reported to have commited suicide. In many of the graver cases, the picture was indeed one of acute psychic derangement. The drug's licence was suspended for further investigation, and it was this further investigation which led to the ultimate and final disappearance of triazolam from the Netherlands market.

The findings left the committee in no doubt whatsoever that, although a high proportion of the cases reported indeed might be attributed to suggestion induced by the media, there remained a substantial number which could not be explained away in this manner. In many of the patients concerned, other benzodiazepines had been well tolerated; in some there was a clear history of positive rechallenge; others clearly antedated the publicity in the mass media and could not be attributed to it. The psychotic reaction seemed to have occurred more particularly in individuals with a prior history of mental instability or neurasthenia. The reaction in many cases represented an accentuation of a pre-existing character trait; morose individuals became pathological depressed or suicidal; excitable persons became aggressive; individuals of a querulous nature became intolerable for their environment.

Looked at after a space of six years and in the light of the total benzodiazepine literature, before and since, the story of triazolam in the Netherlands is not only explicable but also significant and interesting. There can be no

reasonable doubt that the high doses studied and used at that time — up to 1.0 mg — created a situation which has been much less prone to occur in the countries where this hypnotic has since been sold; in that respect the story is reminiscent of that of the oral contraceptives which were initially developed and marketed at eight times the dose which was ultimately found to be sufficient. The nature of the triazolam reaction is however very clearly the same as that of the classic benzodiazepine withdrawal reaction which has been known since the introduction of chlordiazepoxide a quarter of a century ago. The essential difference is, apparently, that with an extremely short-acting benzodiazepine such as triazolam, having no active metabolites, the shock to the system induced by withdrawal can occur acutely, within a few hours, and will not be masked or alleviated by the slow decline of an active metabolite level in the blood. If the effect occurs during light sleep, nightmares may result; if it sets in after waking, it may suffice to derange an already uncertain mental equilibrium.

One lesson to be learned from the triazolam incident is quite certainly that matters like this should not forever remain locked within the files of a regulatory agency; it is regrettable that the Netherlands agency never saw fit to publish the factual basis of its decision. Had it done so, it would have contributed a great deal to the understanding of the benzodiazapine withdrawal phenomenon, particularly as regards short-acting compounds, a subject on which knowledge is now only very gradually emerging from other quarters.

A postscript to Holland's Halcion story was written early in 1985 when the Netherlands Council of State, ruling on an appeal by the manufacturer, upheld the Evaluation Committee's action with respect to the drug (Anon 1985).

Benoxaprofen

Benoxaprofen provides one of the most remarkable of stories, and is more fully discussed in Chapters 12, 30, 32, and 35. The incidents relating to it may be looked at specifically with regard to the compound itself (Del Favero 1983) or as part of a whole series of grave problems which have afflicted the field of non-steroidal anti-inflammatory therapy in the very recent past (Dukes 1984). Here we shall confine our attention to the drug itself.

Benoxaprofen was an antirheumatic compound; it was submitted to various drug regulatory agencies for licencing from about 1979 onwards. Some of them accepted it, some hesitated, others declined to license it, although the latter fact was not generally made known. The compound was fairly closely related to many earlier antirheumatic drugs and most of its effects appeared very similar; benoxaprofen did however appear to have a little less effect on prostaglandin synthetase than its predecessors and a little more on leukocyte migration. The theory was propagated that this might result in a lesser incidence of those adverse effects — notably gastric bleeding — which are thought to reflect prostaglandin synthetase inhibition. Early clinical work indeed suggested that benoxaprofen was well tolerated by the stomach, but this is quite a common early finding with most antirheumatic drugs of this class — the real gastric problems usually emerge only when the drug is used under field conditions on a large scale.

Once the problems associated with benoxaprofen began to emerge it was quickly apparent that they were both serious and unusual. Not only was the drug causing the usual pattern of gastric irritation, but it was also apparently inducing fatal hepatic reactions in elderly people and it was causing photosensitivity on a massive scale; onycholysis was occurring in a frequency of anything up to one in seven. Benoxaprofen was withdrawn from the world market in August 1982 and finally abandoned by its manufacturer, late in 1983, after it had additionally been found to be a liver carcinogen in the rat.

The case of benoxaprofen points to the suddenness with which a true epidemic of adverse effects can occur, breaking out in this instance mainly in the UK where the drug was first introduced. There is no reasonable doubt as to the reasons for this. Exploiting the potentially interesting but modest differences between the drug and its predecessors, the promotion was such as to imply that benoxaprofen was the sensation of the century. As a result, many patients who had been previously treated with other antirheumatic drugs were likely to be needlessly shifted to this one. The very large scale on which adverse reactions can occur under these conditions is evident; there must be many physicians who would be glad to see new drugs introduced with a greater degree of caution until they have proved their place in practice.

There are several other lessons to be learnt from this tragedy as well. Firstly, the need to study in the elderly drugs which are likely to be used in that age group, has also become evident with respect to other problem products such as *Osmosin* (discussion follows). Secondly, there is the need to study antirheumatic drugs, which are going to be used over very long periods, for more than a few months before they are marketed. This point has been made repeatedly by drug regulatory agencies and challenged just as repeatedly by many drug companies as imposing an impossible and unreasonable burden on them; when one, however, sees that the real trouble with benoxaprofen arose only after it had been

used for some nine months in the individual patient (Griffin 1984), one sees again the need for such long-term studies.

Zomepirac

It can be instructive at this point to look briefly at two other preparations in the series of non-steroidal anti-inflammatory agents, both of which are reviewed in more detail in Chapters 7 and 8 of this volume. The first of these is zomepirac. Like the benoxaprofen drama, that of zomepirac was limited largely to one country — in this instance, the United States. Zomepirac is basically a traditional anti-inflammatory analgesic agent. It is structurally almost a twin sister of tolmetin which has been sold for many years as a somewhat ordinary antirheumatic drug. Zomepirac was not however sold for this same purpose, but launched as a specific analgesic; the link to tolmetin must have been lost upon many a prescriber. Consequently, although physicians using tolmetin were no doubt aware that it was somewhat prone to produce anaphylactoid reactions, they had no reason to suspect that the same might happen with zomepirac, a drug which appeared from its presentation to be quite different. Small wonder, then, that they were taken entirely by surprise when it caused a large number of incidents of anaphylactic and allergic complications, some of them so serious that the drug was acutely withdrawn. Once again, as with benoxaprofen, though in a somewhat different manner, one must attribute the scope of the problems with zomepirac largely to the way in which it was introduced.

Osmosin (Indosmos, Osmogits)

The story of *Osmosin*, again told in more detail in Chapter 16 is not that of a new compound but of a new pharmaceutical form. A full explanation of the course of events relating to this product may now never be forthcoming, for the product is no longer with us. *Osmosin*, also known as *Indosmos*, was essentially a pharmaceutical form of indomethacin in which the drug, surrounded by a membrane with a single laser hole drilled in it, was released as the preparation passed through the gastrointestinal tract. Such a system delivers the drug more gradually than orthodox tablets and thus allows less frequent dosing.

The idea was interesting, and in theory the incidence of gastric adverse reactions might also have been reduced. There was some published or unpublished evidence that they were less frequent, but some other evidence that they were not. However that may be, some time after marketing the impression began to arise that the drug was perhaps more likely than the orthodox form of indomethacin to produce small intestinal perforation, a finding which was unique in the UK's *Adverse Drug Reaction Register*; there were also some reports of intestinal bleeding. On the basis of this evidence, and no doubt a lot more which seems to be buried in administrative files, the product was withdrawn world-wide. It subsequently became entirely clear that *Osmosin* had indeed produced intestinal perforations, some of them fatal, in a number of patients, especially elderly people; many of these patients had earlier tolerated plain indomethacin without difficulty. In several cases the empty *Osmosin* shells were found floating free in the peritoneal cavity. The new product would appear to have adhered to the intestinal mucosa or become lodged in diverticula where it exerted the erosive effect which indomethacin would normally have produced in the stomach with less serious consequences. Subsequent work showed that the coating of the tablet was 'sticky' under physiological conditions, so that it may have adhered to the mucosa; in addition, the excipient contained high levels of potassium, which itself is corrosive to the bowel.

The *Osmosin* story again illustrates several points of general importance. Why was this product, like benoxaprofen, introduced before it had been studied on a reasonable scale in the elderly population in which it was so likely to be used? It was, after all, a radically new principle, was presented as such, and thus deserved to be treated in a manner analogous to that which holds good for new compounds. No one can reasonably argue that a requirement for specific studies in the elderly would have been revolutionary; the World Health Organization has been very clear on the matter on more than one occasion (WHO 1981).

Once again, as with benoxaprofen, the introduction to the market was very emphatic, no doubt because of the competitive situation; the advertising certainly suggested a high degree of innovation, and was prone to induce a change in prescribing habits, even for patients who were being adequately treated with indomethacin or another antirheumatic drug. A more encouraging aspect of the *Osmosin* case is that the findings of the British control authorities, which were to precipitate the world-wide withdrawal of *Osmosin*, were not locked up in that country for a long time. There was rapid communication between agencies on the problem as it developed. What is striking is that although the British committee's journal *Current Problems* only made the case known to physicians in the UK around 15 August 1983, the *Osmosin* story was in the hands of governments around the world simultaneously and was much more widely disseminated

wherever the need existed. Rapid broadcasting of suspicions is not always a good thing, but it is clear that rapid distribution of information on possible serious adverse effects has become possible today in a manner which would have been unthinkable a decade ago; it may help us to cut short many such dramas in the future.

One other question which *Osmosin* may be said to raise relates to the volume of evidence on a serious, apparent, adverse effect of a new product which is needed before restrictive action is taken. By all accounts some eight months passed between the first marketing of this product and its withdrawal. Although old people can develop small-bowel perforations spontaneously, it is hardly common and it became apparent quite early that as an adverse reaction this was unique to *Osmosin*. In such cases a regulatory agency must and should have the courage to weigh the emergent evidence of special risk against the therapeutic advantages of the new product; where, as in this case, the latter are questionable in the extreme, there may be a good case for suspending the sales licence at a very early phase indeed whilst the matter is investigated further.

Discussion

The detailed evidence available from several of these cases reminds us, among other things, that the world community has allowed a situation to arise in which companies are tempted (and sometimes in fact obliged for the sake of their own survival) to act in a manner which is not in the best interests of society. Whilst some firms have done their best to accelerate a clarification of any suspicion which arises, others seem all too inclined to retard it. The latter reaction may be more prone to occur where large commercial interests are at stake and where extreme pressure is being exerted by a company's commercial management to gain a large market share rapidly with a new product; this very situation, however, brings with it a considerable risk that any major adverse reaction which does occur will suddenly involve a great many people and, where sales pressures are so extreme, a regulatory agency must be more than usually alert to every suspicion of emerging risk.

When one looks further into the way in which some of these major incidents have been dealt with, one must express some concern at the disproportion between the limited amount of scientific capacity often available in the public sector (i.e. within the drug control agencies of many countries) to investigate such matters and inform physicians about them impartially and the vast capacity which the drug industry maintains for examining the same material and presenting the sunnier side of the case if it wishes to do so. Some sectors of industry are, as implied earlier in this review, all too prone to seek to discredit evidence on the basis of the weakness of its individual components, thereby delaying the effect of cumulative proof. The demand for algorithms to assess the value of adverse reaction reports is surely being pressed too hard by people who would hope to see a great many inconvenient reports discredited in this way.

One last point raised by these events relates to the fate of what has sometimes been called lost and scattered evidence. Some of the material which one needs in order to assess adverse effects never gets in to the public domain at all. It is buried in the toxicological files of drug companies, or the archives of regulatory agencies, and it may even be scattered between the files of different products. Only occasionally is the medical scientist who is neither attached to a drug company nor bound to a regulatory agency able to obtain a clear overview and to piece together the fragments into a coherent hypothesis. Contacts between drug regulatory agencies may at least do something to generate a broader view and to mobilize some of these hidden data in the service of public health.

REFERENCES

ANON (1985). Handhaving besluit Halcion. *Pharmaceutisch. Weekblad*, **120**, 539

DEL FAVERO, A. (1983). Anti-inflammatory analgesics and drugs used in rheumatoid arthritis and gout. In *Side-effects of drugs annual* (ed. M.N.G. Dukes), Vol. 7, 104–25. Excerpta Medica, Amsterdam.

DUKES, M.N.G. (1980). The Van der Kroef syndrome. In *Side-effects of drugs annual* (ed. M.N.G. Dukes), Vol. 4, v–ix. Excerpta Medica, Amsterdam.

—— (1984). The seven pillars of foolishness. In *Side-effects of drugs annual* (ed. M.N.G. Dukes), Vol. 8, i–v. Elsevier Scientific Publishers, Amsterdam.

GRIFFIN, J.P. (1984). Spontaneous reporting. In *Monitoring for adverse drug reactions* (ed. S.R. Walker and A. Goldberg), MTP Press, Lancaster.

PRESCOTT, L.F. (1980). Anti-inflammatory analgesics and drugs used in rheumatoid arthritis and gout. In *Side-effects of drugs annual* (ed. M.N.G. Dukes), Vol. 4, 63–73. Excerpta Medica, Amsterdam.

ROLT, L.T.C. (1982). *Red for danger: A history of railway accidents and railway safety*. David & Charles, Newton Abbot.

WORLD HEALTH ORGANIZATION, REGIONAL OFFICE FOR EUROPE (1981). *The control of drugs for the elderly*. EURO Reports and Studies, 50. WHO, Copenhagen.

4 Epidemiological aspects of iatrogenic disease

P.F. D'ARCY

Incidence of drug-induced disease

Although many studies provide only relatively crude estimates of the epidemiology of adverse drug reactions (ADRs), these statistics do offer some insight into the magnitude of the problem. Opinions differ about how ADRs are defined, detected, and reported, and also about the manner in which the findings from one finite population are in time extrapolated to a national population. These discussions are, of course, valid but there is a danger in debating these, that one may lose sight of the basic problem that ADRs are sufficiently frequent and serious in their consequences to be considered an international health problem. It may therefore be more realistic to accept that the precision of the estimates regarding the incidence of drug-induced disease is less important than the acknowledgement that they do occur regularly. Certainly they are a continuing problem (Feldmann 1983).

It is not the intention of this chapter to present a catalogue of references and abstracts to all the epidemiological studies on ADRs; emphasis has been given instead to surveying the international scene and discussing and pinpointing specific problems of adverse drug effects, especially those relative to the incidence of ADRs in the elderly patient.

It was during the 10 years from 1964 to 1974 that much of the epidemiological basis of drug-induced disease was established and foremost among the studies were those done in Baltimore (Seidl *et al.* 1966*a,b*; Smith *et al.* 1966), in Montreal (Ogilvie and Ruedy 1967), and in Belfast (Hurwitz 1969*a,b*; Hurwitz and Wade 1969). Although these groups pioneered the hospital-based system of gaining epidemiological data on drug reactions, it is without doubt the Boston group which have made the biggest impact in this field. The Boston Collaborative Drug Surveillance (BCDS) Program has collected together similar quantitative information on consecutive patients admitted to medical wards. In 1968, the group published a report on the first 830 patients monitored in a chronic diseases hospital (Borda *et al.*

1968). There were collectively 7078 drug exposures and 405 reported adverse reactions, 22 per cent of which were thought to be due to a drug–drug interaction. In 1972, the group re-examined their data on 9900 monitored patients. There were 83 200 drug exposures and 3600 (36.4 per cent) reported adverse reactions. A total of 234 (6.9 per cent) of the ADRs were attributed by the attending physicians to a drug interaction. This was a considerably lower frequency than that which was previously reported; the change was probably due to the fact that all of the nine hospitals added to the Surveillance Program since the initial report were acute-diseases hospitals. In virtually all cases (230 out of 234), the reported drug–drug interactions resulted from cumulative pharmacological effects. In 1973, the group's total of monitored patients had reached 11 526; there were 103 770 drug exposures and the adverse reaction rate was 28.1 per cent (Miller 1973). In 1974, the patients numbered 19 000 with 171 000 drug exposures: the reaction rate had seemingly stabilized at 30 per cent (Jick 1974). It is interesting to note that during the six years or so of this study, awareness of the relatively high incidence of ADRs had not altered clinical practice in so far as multiple-drug therapy was concerned; patients at the beginning and at the end of the study received, on average, nine drugs during their stay in hospital. Table 4.1 summarizes some of the epidemiological data from the groups that have already been mentioned; additional data from other studies using a similar format of monitoring are also included.

Drug-related deaths

Members of the Boston group also reported on the 24 drug-related deaths in hospitalized medical patients (Porter and Jick 1977); this latter study monitored more than 26 000 acutely ill patients in seven countries and in this sense presented some insight at that time into the attendant hazards of drug therapy in the more advanced countries. Such data must not, however, be considered in isolation, or out of context of the diseases from which the

Table 4.1   Incidence of adverse drug reactions during hospital stay or as cause of hospital admission

Authors	Year of study and hospital	Duration of survey	Total patients in survey	Patients who developed reaction		Reaction rate (%)	Methods/Details
				In hospital	Requiring admission		
Schimmel (1964)	1960–61. Yale Univ. Service, New Haven, Conn. USA.	8 months	1 014	103	—	10.0	House-officer reports
MacDonald and Mackay (1964)	1962. Burlington, Vermont, USA.	1 year	9 557	98	—	1.0	Report cards and investigation
Seidl, et al. (1966b)	1964. Johns Hopkins, Baltimore, Md., USA.	3 months	714	97	—	13.6	Prospective surveillance
Reidenberg (1968)	1964–66. 5 Phila-delphia hospitals, USA.	2 years	86 100	772 reactions	—	0.9	Report cards
Smith et al. (1966)	1965. Johns Hopkins, Baltimore, Md., USA.	1 year	900	97	—	10.8	Prospective surveillance
Ogilvie and Ruedy (1967)	1965–66. Montreal General, Canada.	1 year	731	132	—	18.0	Report cards and investigations
Hoddinott et al. (1967)	1966. A Univ. hospital, London, Ontario, Canada.	59 days	104	16	—	15.0	Prospective surveillance
Borda et al. (1968)	1966–67. Lemuel Shattuck Hospital, Boston, USA.	11 months	830	405 reactions	—	35.0	Prospective surveillance (start of BCDS Program)*
Hurwitz and Wade (1969)	1965–66. Belfast City and Purdysburn hospitals, N. Ireland.	1 year	1 160	118	—	10.2	Prospective surveillance
Hurwitz (1969b)	1965–66. Belfast City and Purdysburn hospitals, N. Ireland.	1 year	1 268	—	37	2.9	Prospective surveillance
Gardner and Watson (1970)	1969. Shands Teaching Hospital, Univ. Florida, USA.	3 months	939	99	—	10.5	Pharmacist-based programme, prospective surveillance
Wang and Terry (1971)	1967–68. Wood Veterans Admin. Centre, Milwaukee, Wis., USA.	1 year	8 291	128	—	1.54	Nurse observer and clinical laboratory report
BCDS Program (1972)	1966–72. Boston group programme, USA.	6 years	9 900	3 600 reactions	—	36.4	Cumulative and ongoing data from surveillance programme*
Miller (1973)	1966–73. Boston group programme, USA, Canada, and Israel.	7 years	11 526	3 240	—	28.1	Cumulative and ongoing data from surveillance programme*
Jick (1974)	1966–74. Boston group programme, USA and elsewhere.	8 years / 8 years	19 000 / 7 017	— / —	— / 260 reactions	30.0 / 3.0	Cumulative and ongoing data from surveillance programme*

Table 4.1 *Continued*

Authors	Year of study and hospital	Duration of survey	Total patients in survey	Patients who developed reaction		Reaction rate (%)	Methods/Details
				In hospital	Requiring admission		
Caranasos *et al.* (1976)	1969–72. Univ. Florida Teaching Hospital, Gainesville, Fl., USA.	3 years	6 063	—	117	2.9	Pharmacist-based programme, prospective surveillance*
McKenney and Harrison (1976)	1974. Univ. Virginia Teaching Hospital, Richmond, Virginia, USA.	2 months	216	—	17	7.9	Pharmacist-based programme prospective surveillance
McKenzie *et al.* (1976)	Date not specified. Univ. Florida Shands Teaching Hospital, Florida, USA.	3 years	3 556	—	72 admissions (64 patients)	2.0	Prospective surveillance of paediatric medical admissions*
Levy *et al.* (1977)	1969–76, Boston group programme, Hadassah Univ. Hospital, Jerusalem, Israel.	4 years (1969–72) / 4 years (1973–6)	1 608 / 1 163	315 / 88	— / —	19.6 / 7.6	Cumulative and ongoing data from surveillance programme
Wiser *et al.* (1978)	1973–76. Loch Raven V.A. Hospital, Baltimore, Md., USA.	$2\frac{1}{2}$ years	43 cases of ADRs	—	16 preventable 7 questionably preventable 20 non-preventable	—	Prospective study to determine frequency of preventable ADRs
Hutcheon *et al.* (1978)	1973–76. Medical wards in Western Infirmary, and Stobhill General Hospital, Glasgow, Scotland.	3 years	2 580	—	85 normal drug dosage 66 drug overdosage	3.3 2.6	Boston Collaborative Drug Surveillance Program
Lawson *et al.* (1979)	Same study as above (Hutcheon *et al.*)	3 years	2 580	—	28 Life-threatening ADRs	1.1 of patients 0.2 of drug prescriptions	Same study as above; assessment of life-threatening reactions
Auzépy *et al.* (1979)	1968–77. Adult intensive care units in 9 University Hospitals in France	10 years	63 717	1 132 drug complications 252 patients died	—	1.8 of drugs	Retrospective study of severe drug complications
Falk (1979)	Coronary care unit, Brook General Hospital, London.	6 months	89 patients (65 treatments)	19 treatments	—	23.6 of treatments	Prospective study of adverse reactions to therapy in acute myocardial infarction patients
Hess *et al.* (1979)	Copenhagen Hospital, Herlev, Denmark	1 year	1 325 admissions 1 136 patients	235 ADRs in 210 patients	7.1% of ADRs caused admission	17.8 of admissions 1.86 of drugs	Prospective study in a department of internal medicine*

(*continued overleaf*)

Table 4.1   *Continued*

Authors	Year of study and hospital	Duration of survey	Total patients in survey	Patients who developed reaction		Reaction rate (%)	Methods/Details
				In hospital	Requiring admission		
Levy *et al.* (1979)	1969–76. Hadassah University Hospital, Jerusalem.	7 years	2 499	—	103 (5 fatalities, 11 life-threatening)	4.1	BCDS Program; survey of hospital admissions to general medical wards
Armstrong *et al.* (1980)	Nursing homes in Houston, Texas, USA.	—	11 173	potential for 362 clinically significant drug interactions in 298 patients	—	2.7 of population	Retrospective community pharmacy survey of prescription records
Ghose (1980)	1979. General medical unit, Cumberland Infirmary Carlisle, UK.	3 months	569 admissions	—	17 acute self-poisoning 15 other drug related problems	9.9 8.8	Prospective surveillance by consultants of primary cause of admission and stay in hospital
Harte and Timoney (1980)	1978. Five hospitals in Ireland	2–3 months	866	38	—	4.4	Retrospective study from paediatric prescription records*
Porter and Jick (1980)	1966–80 1973–80	14 years 7 years	5 311 patients on ampicillin 1 040 patients on amoxycillin	289 (rash) 70 (rash)	— —	5.4 6.7	BCDS Program; adverse skin reactions to two antibiotics
Shehadi and Toniolo (1980)	Data provided by radiologists in USA, Canada, Europe, and Australia	> 5 years	302 083 examinations	reaction rate varied from: 8.0% (cholangiography) to 2.06% (cerebral angiography) 18 fatal reactions overall.	—	5	Retrospective study on adverse reactions to intravascular contrast media; international survey under aegis of International Society of Radiology
Trunet *et al.* (1980)	1978–79. Multidisciplinary intensive care unit, Henri Mondor Hospital, Créteil, France	1 year	325	—	41 including 8 fatalities	12.6	Prospective study of admittances to multidisciplinary intensive care unit*
Williamson and Chopin (1980)	1975–76. 49 Geriatric Medicine Departments in England, Wales, and Scotland.	1 year	1 998	—	209 plus 39 noted at (but not cause of) admission	10.5	Prospective study by consultants of ADRs in geriatric patients admitted to hospital*
Bergman and Wiholm (1981)	Medical wards, Huddinge University Hospital, Sweden	$3\frac{1}{2}$ months	285 admissions	—	45 (ADR was main reason for 36; contributory reason for 9)	16	Prospective surveillnace of drug-related problems causing hospital admission
Ghodse *et al.*	1975–76. Accident	1 year	356 848	3 728	—	1.83	Prospective survey

(1981)	and Emergency Units in 7 hospitals in Greater London		incidents				of drug-related attendance to accident and emergency departments
Steel et al. (1981)	1979. University Hospital, Boston Medical Center, USA.	5 months	815	290 including 30 fatalities	—	36	Prospective study of iatrogenic illness on a general medical service*
Danielson et al. (1982)	1977–81. Surgical wards in five hospitals in USA, Scotland, and New Zealand	4 years	5 232 (46 868 drugs)	1 150 ADRs (62 major, 35 life-threatening)	—	2.2 of drugs	BCDS Program; intensive drug monitoring of surgical patients*
Levy et al. (1982)	1979–80. Hadassah University Hospital, Jerusalem	1 year	1 184	—	34	2.9	Prospective study by physician and clinical pharmacist to determine medical admission due to non-compliance with drug therapy
Yosselson-Superstine and Weiss (1982)	Hadassah-Hebrew Hospital, Jerusalem	7 months	906 paediatric admissions	—	160	17.7	Prospective survey by clinical pharmacists to estimate drug-related hospitalization of children*
Black and Somers (1984)	1979. General Medical Unit Royal Perth Hospital, Australia	1 year	481 admissions	—	13 definite or probable; 17 possible	6.2	Survey from in-patient discharge summaries to assess drug-related illness resulting in hospital admission
Black and Somers (1984)	1977. All hospital admissions, Royal Perth Hospital, Australia	1 year	37 318 admissions	—	245	0.66	Survey from Medical Records Department data to assess drug-related illness resulting in hospital admission

* See text of chapter for further details of the study

patients were suffering. In the Boston group study, most of the patients who died from drug therapy were suffering from severe, terminal illnesses such as cancer, leukaemia, pulmonary embolism, and cirrhosis. Viewed retrospectively (*British Medical Journal* 1977), only six out of 24 deaths occurring in 26 500 consecutive patients could have been prevented, and in only three cases did death result from treatment of patients who were otherwise only mildly unhealthy. The prevalence of *preventable* deaths in this group of medical in-patients was 1 per 10 000, and the drugs responsible were predominantly intravenous fluids and potassium chloride.

Similar conclusions were reached at that time on the results of other studies on adverse drug reactions and associated deaths. Irey (1976) from the US Armed Forces Institute of Pathology classified 827 autopsied cases of adverse drug reactions and found that only 25 were due to therapeutic errors that were unjustifiable and could have been prevented. Two hundred and twenty cases (26.6 per cent) were unexpected adverse drug reactions,

and although these reactions might not have been preventable, since they were not anticipated, the number could perhaps have been reduced by more careful selection and the use of anti-infective and anaesthetic agents since these accounted for more than half the deaths (*Journal of the American Medical Association* 1976).

Caranasos and his colleagues (1976) in Florida analysed 16 deaths from adverse drug reactions that occurred in 7423 hospitalized patients (0.22 per cent incidence). Eleven of the 16 patients were judged to be terminally ill and five seriously ill, and the therapy was a heroic effort to save life. The authors concluded from their study that deaths associated with therapeutic drugs are uncommon and usually occur in the terminally ill.

Armstrong et al. (1976) from the Boston group reported that drug-related deaths occurred in only two of a carefully defined group of surgical in-patients. In both these cases death was attributed to haemorrhage as a result of heparin administration to elderly women — a group known to be at a particularly high risk of toxicity from this anticoagulant (Jick *et al.* 1968). In a more recent study on the intensive drug monitoring of surgical patients, Danielson *et al.* (1982), also from the Boston group, monitored 5232 patients in selected wards in five hospitals in the United States, Scotland, and New Zealand from 1977 through 1981. Patients received on average nine drugs on the ward and adverse reactions were associated with 2.2 per cent of these drug orders. Of the 1150 drug-attributed adverse reactions, only 62 were considered to be 'major' by the attending physicians, and 35 (affecting 20 patients) were termed 'life-threatening'. There were, however, no drug-attributed deaths.

A number of countries have official ADR monitoring committees and their reports provide firm and substantiated data on fatal and non-fatal reactions. The Australian Adverse Reactions Advisory Committee (ADRAC) and the New Zealand Committee on Adverse Drug Reactions both publish annual reports: the ADRAC, for example, considered 2503 reports of suspected ADRs during 1980 and their report was published in May 1982 (ADRAC 1982). An analysis of the source of these reports showed that 1113 were made from hospitals and 791 from general practitioners. As in previous years, suspected ADRs experienced by women (1375 reports) were reported more frequently than those of men (945); a ratio of 1.46:1. The sex of the patient in 183 of the reports was not stated.

Five hundred and twenty-five separate drugs were listed as 'suspected' and the 10 most-reported drugs, which accounted for 27.2 per cent of the total reports were (in order of magnitude): co-trimoxazole, naproxen, sodium diatrizoate with meglumine diatrizoate (*Urovison*), amoxycillin, metoprolol, ampicillin, hydrochlorothiazide with amiloride hydrochloride (*Moduretic*), cimetidine, methyldopa, and erythromycin. Forty-one deaths were reported to the Committee in which the reporter suspected that concurrent therapy may have contributed to the outcome. Among the drugs so implicated were: azathioprine, allopurinol, sodium valproate, sodium diatrizoate alone or with meglumine diatrizoate, timolol maleate, halothane, naproxen, and warfarin.

The New Zealand report during the year April 1980 to March 1981 listed a total of 714 reports of ADRs (McQueen 1982); 27 of these were fatal. The youngest fatality was a 15-year-old-girl who received *Althesin*, alcuronium, nitrous oxide, and halothane during general anaesthesia (mastoiditis) and developed malignant hypothermia. The eldest patient, a woman of 84 years, developed a gastrointestinal haemorrhage after treatment with ibuprofen for osteoarthritis. When all reactions (fatal and non-fatal) were considered, non-steroidal, anti-inflammatory drugs headed the list as producing the largest category of reported reactions. There were 93 of these reports and many of the reactions were common to the whole group, being sequelae of prostaglandin synthesis inhibition. Of these, 23 involved peptic ulceration. Allergic skin reactions including the Stevens–Johnson syndrome and exfoliate dermatitis were reported with most of the drugs in the group. Other ADRs included fluid retention, bronchospasm, haematopoietic disorders, and CNS effects.

The Danish Board of Adverse Reactions to Drugs is also a fruitful source of data on ADRs; for example, Døssing and Andreasen (1982) described and analysed the drug-induced hepatotoxic reactions voluntarily reported to the Danish Board during the decade 1968–78. Halothane accounted for one-quarter of the total number (572) of reports of hepatotoxicity. Hepatotoxicity itself accounted for 6 per cent of the total number of ADRs reported. Thirteen per cent of the patients with halothane-induced hepatotoxicity died and the total frequency of fatal hepatotoxicity due to halothane and other drugs (e.g. oxyphenisatin, rifampicin, alpha-methyldopa, papaverine, phenytoin (all cytotoxic reactions) and chlorpromazine, phenylbutazone, and androgenic and anabolic steroids (all cholestatic reactions), represented 12 per cent of drug-related deaths in Denmark. This latter statistic is similar to figures reported in Britain (Read 1979), and in Sweden (Böttiger *et al.* 1979).

In Sweden, a nation-wide system for reporting ADRs has been functioning since 1965 and an analysis and report of the first 10 years of this system was made by Böttiger *et al.* (1979). There were 274 drug-induced deaths in Sweden during that 10-year period and the annual incidence has been remarkably constant with 25–30 reported cases per year. There was a marked increase with age in the incidence of fatal reactions, more so than for all drug reactions. Women were found to consume more drugs than men and get more reactions, but not more fatal reactions. Antimicrobial agents (antibiotics and sulphonamides) were responsible for 21 per cent of the fatal reactions, followed by oral hypoglycaemics (9 per cent), oral contraceptives (9 per cent) and anti-inflammatory agents (8 per cent). The blood and the bone marrow were the most susceptible organs, responsible for 40 per cent of the fatal reactions, followed by thromboembolism (10 per cent) and hepatocellular damage (9 per cent). Other Swedish studies of particular interest in the ADR field are those of Bergman and Wiholm (1981) on the association between hospital admissions and drug-related problems, and Gustafsson and Boëthius (1982) on the utilization of analgesics from 1970 to 1978; this latter study is particularly relevant within the Swedish drug scene since approximately one-tenth of reported ADRs in Sweden are due to analgesics. Neither of these studies, however, relates specifically to drug-related deaths so they will not be discussed further in this present context.

From Créteil in France, Trunet and his colleagues (1980) studied, prospectively, all patients admitted to a multidisciplinary intensive care unit to determine how many of their diseases were drug-induced and, of these, what number were potentially avoidable. Iatrogenic disease was responsible for 41 hospital admissions (12.6 per cent) of the 325 patients admitted in the course of one year. Many of these patients had concomitant serious illness. Nevertheless, 19 patients (46.3 per cent) were admitted with iatrogenic disease resulting from therapeutic or technical errors that were potentially avoidable. Iatrogenic disease was fatal in eight cases, life-threatening in 13, and moderate in 20.

The United States has always been a fruitful source of information on adverse drug events; Abramson and his colleagues (1980) from Pittsburg, for example, analysed 145 reports of adverse occurrences involving patients in a medical–surgical intensive care unit during a five-year period. They disclosed 92 instances of human error and 53 cases of equipment malfunction. Mortality for patients with an incident report filed during the ICU admission (41 per cent) was almost twice that for all ICU patients. Of the 92 instances of human error reported, 21 related to medicines and seven to intravenous fluids. An apparent clustering of incidence reports during July and August suggested an association with the annual summer influx of young physicians and nurses unfamiliar with the ICU setting.

In another American study, Steel et al. (1981) from Boston found that 36 per cent of 815 consecutive patients on a general medical service of a university hospital had an iatrogenic illness. In 9 per cent of all persons admitted, the incident was considered major in that it threatened life or produced considerable disability. In 2 per cent of the 815 patients, the iatrogenic illness was believed to have contributed to the death of the patient. Exposure to drugs was a particularly important factor in determining which patient had complications. Totally drugs accounted for 208 of the 497 complications reported; intravenous therapy accounted for a further 34 complications.

Since 1982, the epidemiological scene in Britain has been dominated by reports of adverse drug reactions and fatalities associated with a number of non-steroidal anti-inflammatory drugs. For example, benoxaprofen (*Opren, Oraflex*) was banned from clinical use in August 1982 on grounds of safety (Goldberg 1982). The Committee on Safety of Medicines (CSM) had received over 3500 reports of adverse reactions associated with this drug via the yellow-card reporting system. Included among these reports were 61 fatalities, predominantly among the elderly. With regard to these reports there was concern about the serious toxic effects of the drug on various organ systems, particularly the gastrointestinal tract, the liver, and the bone marrow, in addition to the known adverse effects on skin, eyes, and nails. The FDA in America had received reports of 550 adverse reactions and 12 fatal cases involving the drug.

The literature on benoxaprofen has become voluminous: original reports on its adverse reactions were published by Morgan and Behn (1981); Taylor et al. (1981); Dodd et al. (1981); Fenton et al. (1982); Halsey and Cardoe (1982); Hindson et al. (1982); Larkin et al. (1982); Ledermann et al. (1982); Marsden and Dahl (1982); Wilkins et al. (1982). There are reports on drug-induced jaundice in the elderly patient, sometimes with a fatal result by Firth et al. (1982); Fisher and McArthur (1982); Goudie et al. (1982); Prescott et al. (1982); and Taggart and Alderdice (1982).

In August 1983, the CSM issued a warning on *Osmosin* a controlled-release formulation of indomethacin (Committee on Safety of Medicines 1983), and the product licence for *Flosint*, a proprietary tablet preparation of indoprofen was suspended for a period of three months in December 1983 (Wills 1983). Added to this list of casualties was zomepirac (*Zomax*), which was temporarily withdrawn by the manufacturer because of reports of fatal anaphylaxis associated with its use (Paulus 1983). More recently, phenylbutazone and oxyphenbutazone, two of the longest time-serving members of the NSAIDs, have also come under the scrutiny of the CSM with the result that licences of products containing phenylbutazone have been varied to restrict the use of the drug to the treatment of ankylosing spondylitis and to limit its supply to hospitals only. The Committee has further recommended to the licensing authority that products containing oxyphenbutazone should have their licences revoked (this action has since been taken). This information was given to doctors and pharmacists in a letter from the Chairman of the Committee dated 7 March 1984 (Goldberg 1984).

The Committee had studied a total of 1693 reports of adverse reactions associated with the use of products containing phenylbutazone which had been received since 1964. Of these 445 were fatal; 207 of these were attributed to aplastic anaemia, 63 to various white blood cell disorders, and 19 to thrombocytopenia. The Committee examined this evidence in comparison with the relative risks of other drugs currently available in this class and concluded that the evidence of these risks outweighed the drug's benefits in the treatment of all

Table 4.2 Adverse reactions to nitrofurantoin as percentage of total reports received in the UK, Sweden, and The Netherlands (absolute number in parentheses)*

Group	UK (1964–80)		Sweden (1966–76)	The Netherlands (1975–80)	
Acute lung reactions	14.1 (64)		43.2 (398)	12.5 (11)	
Chronic lung reactions	2.0 (9)		5.3 (49)	3.4 (3)	
Allergic reactions	40.0 (182)		41.7 (384)	42.0 (37)	
Liver and gastrointestinal disturbances					
Liver damage	3.9			9.1	
Nausea and vomiting	11.4	16.0 (73)	5.4 (50)	4.5	13.6 (12)
Various	0.7				
Blood dyscrasias	2.4 (11)		2.2 (20)	6.8 (6)	
Peripheral neuropathy	14.1 (64)		2.2 (20)	9.1 (8)	
Various	11.4 (54)		Not known	12.5 (11)	

* Source: Penn and Griffin (1982).

conditions except ankylosing spondylitis, which represented less than 5 per cent of current usage of the drug.

Other information regarding the hazards of these two latter anti-inflammatory agents, particularly blood dyscrasias and fatalities, was summarized in the *Drug and Therapeutics Bulletin* (1984).

National variations in ADR reporting

There is evidence to suggest that any direct comparison of the range of ADRs to any drug cannot be made between countries, even those which at first sight seem similar. This is unfortunate since it prevents valid global assessments being made of the hazard of specific drugs. Whether such differences depend on the actual incidence of adverse reactions or whether they depend on the importance placed upon them by the doctor reporting them to the regulatory body is a matter of speculation.

An example of national variations in ADR reporting was given by Penn and Griffin (1982) who published the results of a study in which comparison was made between the reporting of adverse reactions to nitrofurantoin in Sweden, the Netherlands, and the UK. In all three countries nitrofurantoin is an approved drug for the treatment of urinary-tract infection.

The Swedish drug regulatory authorities received more reports of ADRs to nitrofurantoin than to any other drug, and the annual reporting rate is increasing despite a steady fall in the prescribing of nitrofurantoin. For example, in Sweden nitrofurantoin provided 1.5 per cent of the total number of ADRs during the period 1965–69, 7.9 per cent during 1970–74, and 7.2 per cent during 1975–79; during that time the ranking of nitrofurantoin as a cause of ADRs rose from tenth to second. In contrast, in the UK, the total number of adverse reactions to nitrofurantoin reported to the CSM via the

yellow-card system has been steadily falling for many years; e.g. 1.77 per cent of all ADRs in 1964 and 0.11 per cent in 1980. This decline in reports has been accompanied by a fall in the number of prescriptions.

Lung reactions and liver damage were more commonly reported in Sweden than in the UK where gastrointestinal symptoms and peripheral neuropathy were more common. Such figures as were available in the Netherlands showed an intermediate picture. The incidence of blood dyscrasias and allergic reactions were, however, similar in the three countries.

Table 4.2 shows the spectrum of adverse reactions to nitrofurantoin experienced in each of the three countries and compares the incidence reported in the UK (during 1964–80) with corresponding figures generated in Sweden and the Netherlands during the years 1966–76 and 1975–80, respectively.

The results of these comparisons indicate that not only is there a difference between the number of reports of ADRs to nitrofurantoin received by the drug regulatory authorities of the three countries, but the range of the adverse reactions reported is also different. For example, if the incidence of ADR reports on nitrofurantoin is related to the population of each country, then in Sweden there are 10.5 reports per million inhabitants per year, whilst in the UK the figure is about 0.5/million/year — an outstanding difference.

The reasons for such differences are not apparent and Penn and Griffin commented that it is disturbing to find that the range of side-effects to such a common drug can differ so much between countries with obviously close and similar ethnic, dietary, and economic backgrounds. They therefore posed two questions.

1. To what extent do ranges of ADRs differ from country to country with other drugs?

2. Do any such differences depend on the actual incidence of adverse reactions or on the importance placed upon them by the doctor reporting them to the regulatory body?

Factors that may influence the reporting rate are activities of the drug regulatory authorities in issuing warnings about the adverse reactions to a particular drug. Both the Swedish and the UK authorities issued warnings on nitrofurantoin, but their substance and emphasis differed and this may have introduced a bias. Interestingly, however, neither authority emphasized the problem of peripheral neuropathy, yet this has appeared to be the major difference in the profile of adverse reactions to the drug (e.g. 14.1 per cent of total reports in the UK, 9.1 per cent in the Netherlands, and 2.2 per cent in Sweden). Adverse reactions and interactions to nitrofurantoin have been extensively reviewed by D'Arcy (1985).

The last 5–10 years have seen the regulatory agencies and international organizations, concerned with drug safety and efficacy, tentatively feeling their way towards co-operation in the use of their knowledge, experience, and resources within formal institutional frameworks. These moves have indicated not only the potential benefits of closer co-operation but also a range of problems not immediately obvious. This national variation in ADR reporting is one such problem. Apart from nitrofurantoin, perhaps the most striking example of the effect of such variation is to be seen in the way in which clioquinol-associated subacute myelo-optic neuropathy (SMON) was almost entirely confined to Japan, despite an equally high consumption of clioquinol in other countries, e.g. India and the Philippines. Whether the differences in reaction were racial, environmental, or related to coincident disease has never been adequately determined despite the decision of the Tokyo Court incriminating clioquinol (see also Chapter 27). Further discussion on the advantages and problems of international co-operation in drug evaluation and marketing approval in Western Europe was presented by Griffin and Long (1982).

Medical incompetency: fact or fiction?

In January 1976, the *New York Times* published a five-part series on 'Medical incompetency' in the United States and said that, in addition to the 30 000 fatalities each year, 'perhaps ten times as many patients suffer life-threatening and sometimes permanent side-effects such as kidney failure, mental depression, internal bleeding, and loss of hearing or vision' as a result of wrong or unnecessary prescriptions. It is patently obvious that this conclusion was reached on the unjustifiable extrapolation of limited and selective data. The true picture that emerges on both sides of the Atlantic, in Europe and in Australia and New Zealand, is less serious than may be inferred from media reporting. Few patients appear to die¯ as a result of therapeutic endeavour, and most of those who do have been treated with powerful drugs to delay the progress of otherwise fatal disorders (*British Medical Journal* 1977). However, it is true, unfortunately, that there will always be an irreducible minimum number of people who 'get ill' from drugs (Ballin 1974).

Many cases of adverse drug reactions can be attributed to predisposing factors in the patient; individual patients may vary widely from the 'norm' in their reaction to drugs. A number of factors may interplay in the adverse reaction to medication; age, sex, genetic factors, and disease states are particularly relevant. The influence of genetic factors (see Chapter 5) and disease states (see especially Chapters 17 and 34) are discussed elsewhere in this book. This present chapter will therefore confine discussion on these ADR–predisposing factors to the effects of age (geriatrics and paediatrics) and also to differences in sex. Some specific discussion will also be given to ADRs in surgical patients, in community-care patients, and in dental patients.

Patient factors in iatrogenic disease

Age

Emphasis has been given in a number of reviews to patient factors that are known to influence the course of adverse drug reactions and interactions (Wallace and Watanabe 1977; Ouslander 1981; Braverman 1982; Greenblatt *et al.* 1982; Shaw 1982; Royal College of Physicians 1984); of these factors age has been pinpointed as a significant contributor to the outcome of adverse reactions (D'Arcy and McElnay 1983; Royal College of Physicians 1984). Senescence is frequently evoked to explain these unwanted drug effects, undoubtedly rightly — providing senescence is regarded as being accompanied, for example, by small, lean body mass, poor renal function, and impaired function of other organs, notably the liver. In elderly patients the reserve capacity of many organs may be considerably reduced, and because of this erosion there is a narrowing of the safety margin between the therapeutic and toxic dose of many drugs. As a result of this the elderly, as a group, get rather more than their fair share of drug-induced diseases; this especially applies to psychogeriatric patients, in whom the nature of the drugs, the number used concomitantly, and the long-term use presents peculiar hazards.

Sometimes adverse events may simply occur because the prescribed dose is too large for the weight of the patient; at other times the more subtle sequelae of senescence are responsible. Many adverse effects of drugs that may simply be a nuisance to the younger patients are much intensified in the elderly due to decreased metabolism and poor renal function. Indeed, these adverse effects of therapy may convert the elderly patient from a functional sentinent human being into a chair-fast incontinent wreck.

There is some urgency to investigate actively the effects of age on the epidemiology of ADRs since the proportion of the population aged over 65 years is growing in most countries of the Western world; in Britain, for example, the elderly represent only about 12 per cent of the population, but they are responsible for about 30 per cent of the National Health Service's expenditure on medicines (Crooks *et al.* 1975). In America, the elderly comprise 10 per cent of the population but receive more than 25 per cent of all prescribed medication (May *et al.* 1982). At present it is estimated that 40 per cent of physicians' office time and 33 per cent of hospital time is devoted to elderly patients; over the next four decades this is expected to rise to about 75 per cent of the total time of the health care services (Butler 1980). In France, elderly subjects receive as much as 25 to 30 per cent of all drug prescriptions (Ministère de la Santé, Ministère du Travail 1975). Correlating directly with these figures is the higher frequency of adverse drug reactions within the geriatric age group (Caird 1977; Braverman 1982; Greenblatt *et al.* 1982).

It is also relevant that in Britain and America it is projected that the elderly will comprise about 25–35 per cent of the population by the end of the century. Apart from other social problems that this partition may cause, health care needs are certain to increase and with this the need for more medication. The drug-related problems of the elderly today must therefore be solved and the lessons learnt must be programmed into future patterns of safe drug treatment of the elderly. Various articles have been published with the sole intent of making the prescribing of drug treatments for the elderly as safe as is therapeutically possible. Of several, the following are particularly worthy of study: Davidson (1971); Harman (1971); Hall (1972); Law and Chalmers (1976); Caird (1977); Ouslander (1981); Braverman (1982); Lamy (1982); May *et al.* (1982); Shaw (1982); Stewart *et al.* (1982); Campbell *et al.* (1983); Hyland (1983); Hallworth and Goldberg (1984); and Royal College of Physicians (1984).

Pharmacokinetics in the elderly

The rational use of drugs in the elderly can be exceptionally difficult at times because of the changes that commonly accompany ageing. These changes, which are both physiological and pathological, may lead to altered drug handling, especially renal, or to an altered response of a target organ to a given concentration of drug. Undoubtedly the elderly are at a substantially greater risk of experiencing adverse drug reactions than younger patients.

Examples that present a special hazard to the elderly when used alone or in combination with other agents are the antidepressants (tricyclic or tetracyclic compounds and monoamine oxidase inhibitors); the anti-inflammatory or antirheumatic agents (especially the non-steroidal); the cardiac glycosides; the oral diuretics that cause K^+ depletion; K^+ supplementation itself; the ulcer-healing drug carbenoxalone; barbiturates; oral hypoglycaemic drugs; tranquillizers, especially the phenothiazines; and lithium salts. This list is obviously not exhaustive but it will serve to illustrate the peculiar hazards of some drugs in this specific group of patients. These hazardous effects are summarized in Table 4.3. A leading article in the *British Medical Journal* (1978b) made much the same comment and agreed that much of this increased risk of medication in the elderly stems from relatively few drugs; in particular the leading article cited digoxin, benzodiazepines, phenothiazines, and hypotensive agents. Less frequently, it suggested, did problems arise from organ failure or from the cumulative effects of several drugs taken together.

Advancing age may grossly affect the circulating level of a drug, and elderly individuals often display much higher plasma concentrations of certain drugs than do younger subjects. Such changes obviously increase the risk of toxicity in this group of patients, who are often more susceptible to the effects of drugs due to other mechanisms. With relatively few exceptions (O'Malley *et al.* 1971; Crooks *et al.* 1976), most reports on the pharmacokinetic profile of drugs in the elderly relate to investigations on single drugs, usually given in single doses, in a small number of volunteer subjects. Collectively, however, they present a useful cross-section of drug problems in the elderly, which illustrates the extent to which advancing age can impair drug handling and impair the benefits of treatment. Such reports are far too numerous to be reviewed comprehensively here, however, brief details of drug handling in the elderly are given in Tables 4.4 and 4.5. Table 4.4. is an augmented form of that already published by Ritschel (1980) in his review of the disposition of drugs in geriatric patients,

Table 4.3   Drugs or drug combinations presenting a special hazard to the elderly patient (from D'Arcy and McElnay 1983)

Drug or drug combination	Hazardous effects
Antidepressants *Tricyclic compounds* e.g. amitriptyline, desipramine, dibenzepin, doxepin, imipramine, nortriptyline, opipramol, tofenacin, trimipramine, viloxazine, etc.	Hypotension, drowsiness, anticholinergic effects (e.g. dry mouth, constipation, urinary retention, blurred vision, etc.), interaction with drugs affecting adrenergic neurones (e.g. sympathomimetic amines, antihypertensive agents, MAO-inhibitors) and with anticholinergic agents, anticonvulsants, tranquillizers, neuroleptics, alcohol, other CNS depressants, and thyroid hormones.
MAO-inhibitors e.g. iproniazid, isocarboxazid, mebanazine, nialamide, phenelzine, tranylcypromine, etc.	MAO-inhibitors should only be used with caution in geriatric patients due to their potential for hazardous side-effects, and the danger of their interaction with other drugs (e.g. tricyclic antidepressants, phenobarbitone, carbamazepine, insulin, hypoglycaemic sulphonylureas, pethidine, sympathomimetic amines, etc.), and with tyramine-containing foods, and beverages containing tyramine, dopamine, or serotonin.
Antirheumatic (anti-inflammatory) agents e.g. corticosteroids, non-steroidal agents (salicylates, ibuprofen, indomethacin, etc.)	The incidence of gastric ulcer increases with age, and the whole spectrum of anti-inflammatory agents in clinical use will cause dyspepsia and precipitate or exacerbate gastric ulceration or bleeding.
Cardiac glycosides e.g. digitalis (the term is used collectively to denote digitalis and its glycosides of which digoxin is the one most commonly used in the clinic).	Overdosage with digitalis can cause practically every known arrhythmia in the elderly patient, especially atrial and nodal tachycardia and atrioventricular dissociation. Nausea and vomiting with marked slowing of the heart are less often seen in geriatrics. Increased heart rate with worsening failure may be due to overdosage. Mental confusion, mental depression, toxic psychosis, gynaecomastia, and often fatal acute abdominal syndrome resembling mesenteric artery occlusion occur. Combination of digitalis with K^+-depleting thiazide or other diuretics will precipitate digitalis toxicity as also will hypercalcaemia and hypothyroidism. There is evidence that many elderly patients on maintenance digoxin do not require such treatment.
Oral diuretics *Thiazide (benzothiadiazine) group* · e.g. bendrofluazide, chlorothiazide, cyclopenthiazide, hydrochlorothiazide, hydroflumethiazide, methyclothiazide, polythiazide, etc.	Geriatric patients are often K^+ depleted because of simple lack of intake, so that a serious K^+ deficiency may occur with prolonged and poorly supervised treatment. This commonly precipitates digitalis toxicity. Recommended K^+ supplements are: Potassium Chloride Effervescent Tablets BNF (*Sando K Tablets*) containing potassium 12 millimoles (12 mEq) and chloride 8 millimoles (8 mEq) in an effervescent base; Potassium Chloride Slow Tablets BNF (*Leo-K Tablets, Slow-K Tablets*) containing 8 millimoles (8 mEq) of each ion in a slow-release base. Combined tablets of thiazide diuretic and potassium chloride are not recommended because flexibility in dosage of each component is lost, and, for geriatric patients, the contained dose of potassium and chloride ions per dose of diuretic is not sufficient for the patient's requirements. Potassium supplementation has been associated with hyperkalaemia and other adverse reactions including intestinal ulceration. Thiazides in high dosage are potentially diabetogenic to latent or early diabetics.
Frusemide group (e.g. frusemide, bumetanide, mefruside). *Ethacrynic acid*	Intense and rapid diuresis of about 6 hours' duration occurs with these diuretics; they may act when the thiazides have failed but they have similar side-effects. Transitory or permanent deafness has been reported with high dosage of intravenous frusemide and oral or intravenous ethacrynic acid, especially in the presence of renal failure. Impairment of glucose tolerance has been reported after frusemide and reports conflict regarding ethacrynic acid.
Gastrointestinal drugs Carbenoxolone sodium	Although carbenoxolone promotes healing of the gastric ulcer, it must be regarded as a hazardous drug for the elderly patient or any patient with cardiac or renal

Table 4.3   *Continued*

Drug or drug combination	Hazardous effects
	disease. It promotes Na^+ retention and oedema and causes severe hypokalaemia and muscle paresis even in recommended doses. *Thiazide diuretics cannot be used to combat this oedema due to worsening of the hypokalaemia.*
Hypnotics (soporifics) Barbiturates	ACUTE EFFECTS: Hangover, confusion, increased nocturnal restlessness, excitement leading to delerium. CHRONIC EFFECTS: The clinical picture resembles arteriosclerotic dementia, i.e. intellectual impairment, slurred speech, unsteady gait. *Safe hypnotics are chloral hydrate, dichloralphenazone, or possibly one of the benzodiazepines (e.g. temazepam at half normal adult dose).*
Hypoglycaemic agents *Sulphonylureas* e.g. acetohexamide, tolbutamide, chlorpropamide, glibenclamide, tolazamide.	These oral hypoglycaemic agents cause prolonged and dangerous hypoglycaemia if potentiated by other drugs given in the treatment of concomitant disease, e.g. thiazide diuretics, sulphonamides, chloramphenicol, phenylbutazone, the coumarin anticoagulants, and phenelzine. An unpleasant disulfiram-like reaction may be produced by alcohol in a patient taking a sulphonylurea; the incidence of this effect increases with duration of drug treatment. β-blocking drugs may abolish the warning signs of hypoglycaemia that are dependent on the normal outpouring of adrenaline; these agents should be avoided by diabetic patients controlled on insulin or sulphonylureas.
Tranquillizers *Phenothiazines*	Induced parkinsonism, excess lethargy, skin photosensitivity reactions, cholestatic jaundice, impaired rate of liver function, hypotension, reduced thyroid function, and accidental hypothermia are more common adverse effects of phenothiazines in the elderly patient. Constipation may be made worse, and intestinal ulceration, obstruction, and gangrene of the large bowel have been reported. Phenothiazines potentiate the actions of hypnotics and opiates which may be advantageous in some instances.
Benzodiazepines e.g. chlordiazepoxide, diazepam, medazepam, etc.	In general the benzodiazepines have a remarkable safety record in regard to overdosage; they are unlikely to interact with co-prescribed drugs by liver enzyme induction since they seem to be poor (or not) enzyme inducers. Although they are relatively safer than the major tranquillizers (i.e. the phenothiazines), they commonly cause drowsiness, ataxia, and confusion in the elderly or debilitated patient. There is a risk of dependence of the barbiturate-alcohol type and withdrawal reactions, including convulsions, have been observed when a high dose schedule has been abruptly stopped. It is better not to give tranquillizers to patients suffering grief following bereavement because medication during the initial period of grief may produce delayed depression.
Drugs for urinary incontinence *Anticholinergic agents* e.g. propantheline, dicyclomine, emepronium, etc.	Incontinence of urine may sometimes respond to treatment with drugs which either control bladder function or eradicate urinary infection. Anticholinergic drugs, such as propantheline bromide, dicyclomine, and emepronium bromide sometimes help to reduce incontinence, but they are best avoided when glaucoma is present. They may cause troublesome constipation, and dryness of the mouth with oral and lingual ulceration.

Table 4.4   Change of drug parameters in the aged (from Ritschel 1980)

Drug	Biological half-life	Volume of distribution	Total clearance
Acetaminophen (paracetamol)	↑	U	↓
Acetanilide	↑	↓	U
Aminopyrine	↑	NI	NI
Amitriptyline	↑	↓	↓
Amobarbital	↑	U	↓
Ampicillin	↑	U	↓
Antipyrine (phenazone)	↑	↓	↓
Aspirin	NI	↑	↓
Carbenicillin	↑	↑	NI
Carbenoxolone	↑	↓	↓
Cefamandole	↑	↑	U
Cefazoline	↑	U	↓
Cefradin	↑	U	↓
Chlordiazepoxide	↑	↑	↓
Chlormethiazole	↑	↑	↓
Chlorthalidone	↑	U	↓
Cimetidine	↑	↓	↓
Cyclophosphamide	↑	↑	NI
Desipramine	↑	NI	NI
Desmethyldiazepam	↑	↑	↓
Diazepam	↑	↑	U or ↓
Digoxin	↑	↓	↓
Dihydrostreptomycin	↑	NI	NI
Doxycycline	↑	U	↓
Flurbiprofen	U	U	U
Gentamicin	↑	↓	↓
Imipramine	↑	NI	NI
Indomethacin	↑	NI	NI
Indoprofen	U	U	U
Kanamycin	↑	NI	↓
Levomepromazine	↑	U	↓
Lidocaine (lignocaine)	↑	↑	U
Lithium	NI	NI	↓
Lorazepam	U or ↑	↓	U or ↓
Methotrexate	↑	↓	↓
Metoprolol	↑	NI	NI
Morphine	↓	NI	NI
Netilmicin	↑	U	↓
Nitrazepam	↑	↑	U
Nortriptyline	↑	U	↓
Oxazepam	↑	↑	NI
Penicillin G	↑	NI	NI
Phenobarbital	↑	NI	NI
Phenylbutazone	↑	U	↓
Phenytoin	NI	NI	↑
Procaine penicillin	↑	NI	NI
Practolol	↑	↓	NI
Propicillin	↑	↓	NI
Propranolol	↑	↓	↓
Protriptyline	↑	↓	↓
Quinidine	↑	↓	↓
Spironolactone	↑	NI	NI
Sulbenicillin	↑	↑	NI
Sulfamethizole	↑	U or ↓	↓
Sulfisomidine (sulphasomidine)	↑	↑	↓
Tetracycline	↑	NI	NI
Theophylline	↑	U or ↓	↓
Thioridazine	↑	NI	NI
Tobramycin	↑	↓	↓
Tolbutamide	↑	↓	↓
Warfarin	↑	U	↓

↑ = increase; ↓ = decrease; U = unchanged; NI = no information available.

and it summarizes information on the biological half-life, the volume of distribution, and the total clearance of a number of drugs that are commonly prescribed for the elderly patient. Table 4.5 is reproduced from a review by Greenblatt *et al.* (1982) of drug disposition in old age.

Inappropriate and unnecessary drug treatment in the elderly

Inappropriate and unnecessary drug treatment is one of the prime causes of ADRs in the elderly; diuretics and psychotropic drugs have been especially incriminated.

Diuretics

Diuretics are one of the most frequently prescribed and ingested medicines among elderly patients, and it is significant in this respect that a leading article in the *British Medical Journal* (1978a) emphasized that many such patients who have diuretics prescribed for them do not need them and that their use is often inappropriate and potentially harmful. Many elderly patients thus needlessly experience the disadvantages of diuretic-induced urinary incontinence, acute retention, K^+ deficiency, and thus interference with social activities. Examples of enforced diuretic abuse are not difficult to find; for example, safe withdrawal of maintenance diuretic therapy was demonstrated by a double-blind controlled trial for 54 patients. Only eight required diuretic treatment to be resumed within 12 weeks (Burr *et al.* 1977). A further example is given by a study arranged by the British Geriatrics Society (Williamson and Chopin 1980); 1998 patients admitted to 49 geriatric medicine departments were studied for prescribing patterns and for adverse reactions to treatment; 81.3 per cent of these patients were on a prescribed drug at the time of admission. Diuretics were found to be the most widely prescribed drug (37.4 per cent of sample) and they caused

Table 4.5   Studies of the relation of age to the clearance of drugs cleared by hepatic biotransformation (from Greenblatt et al. 1982)

Drug or metabolite	Initial pathway of biotransformation*
Evidence suggesting age-related reduction in clearance	
Antipyrine † (phenazone)	Oxidation (OH, DA)
Diazepam †	Oxidation (DA)
Chlordiazepoxide	Oxidation (DA)
Desmethyldiazepam †	Oxidation (OH)
Desalkylflurazepam †	Oxidation (OH)
Clobazam †	Oxidation (DA)
Alprazolam †	Oxidation (OH)
Quinidine	Oxidation (OH)
Theophylline	Oxidation
Propranolol	Oxidation (OH)
Nortriptyline	Oxidation (OH)
Small or negligible age-related change in clearance	
Oxazepam	Glucuronidation
Lorazepam	Glucuronidation
Temazepam	Glucuronidation
Warfarin	Oxidation (OH)
Lidocaine (lignocaine)	Oxidation (DA)
Nitrazepam	Nitroreduction
Flunitrazepam	Oxidation (DA), nitroreduction
Isoniazid	Acetylation
Ethanol	Oxidation (alcohol dehydrogenase)
Metoprolol	Oxidation
Digitoxin	Oxidation
Prazosin	Oxidation
Data conflicting or not definitive	
Meperidine (pethidine)	Oxidation (DA)
Phenylbutazone	Oxidation (OH)
Phenytoin	Oxidation (OH)
Imipramine	Oxidation (OH, DA)
Amitriptyline	Oxidation (OH, DA)
Acetaminophen (paracetamol)	Glucuronidation, sulfation
Amobarbital	Oxidation (OH)

* OH denotes hydroxylation and DA dealkylation.
† Evidence suggests that the age-related reduction in clearance is greater in men than in women.

the largest number (60) of adverse reactions. Full recovery from these drug-induced reactions was noted in 73.3 per cent of patients when diuretics were stopped.

In America, Portnoi and Pawlson (1981) investigated the extent of diuretic abuse in a 160-patient nursing home. Twenty-seven of these elderly patients were receiving diuretics and the study determined their current need for the therapy and the effect of discontinuing it. Initial evaluation of the patients' medical records showed that in five cases there were no clinical notes to explain the reasons for diuretic therapy. In 12 cases there were physicians' notes giving the initial indications for the treatment, such as congestive heart failure or hypertension, but there was no record of these conditions ever having been re-evaluated. In the remaining 10 cases, there were nursing notes about ankle oedema at the time of the patient's admission to the nursing home but these were not followed by notes reflecting regular observations, despite continuous treatment with diuretics. The duration of therapy in the group varied from two months to at least five years. The total duration of such treatment could not be documented in most cases because of the virtual absence of good past medical records in the nursing home charts. Twenty-two of the 27 patients were receiving frusemide (furosemide), and the others a thiazide.

No signs of congestive heart failure or of elevated blood pressure were found in any of the 27 patients at the time of the initial evaluation by Portnoi and Pawlson. Thus it was concluded that discontinuation of the diuretic treatment could be attempted with safety, providing that there was careful observation of each patient's clinical status in the following few months. Three patients died during the six months of observation, two from pneumonia and one from myocardial infarction. The condition of the rest of the patients remained stable. There was no increase in blood pressure and no clinical data to suggest a recurrence of congestive heart failure.

Frusemide and the thiazides are potent diuretics and amongst the many adverse effects that have been reported commonly are hypokalaemia, hyperglycaemia, and hyperuricemia; in addition, postural hypotension, urinary incontinence, dehydration, mental confusion, fatigue, and weakness are common problems in elderly patients receiving such medication. Some of these adverse effects were seen in the 27 patients reported in the study and they subsided when diuretics were stopped.

A further insight into some of the problems commonly experienced by elderly patients taking diuretics was given by O'Malley and O'Brien (1980) in their review of the management of hypertension in the elderly. These reviewers emphasized the deterioration of glucose tolerance that can be associated with thiazides, and the modest increases in serum uric acid, although such thiazide-induced hyperuricaemia rarely leads to secondary gout. Nocturia is also a problem due to the long duration of effect of thiazides; on the other hand, frusemide is more likely to cause acute retention or urinary incontinence in susceptible individuals. The question of routine administration of K^+-sparing diuretics or K^+-supplementation with thiazides in the elderly is, according to these reviewers, a difficult one. The elderly are sensitive to the K^+-losing effects of thiazides, so K^+-supplements may be necessary. Unfortunately, the elderly are also prone to hyperkalaemia when K^+-supplementation is used, and they generally find effervescent potassium preparations unpalatable and have difficulty in swallowing large tablets containing potassium.

A further example of a hazard associated with diuretic treatment in the elderly is the report by Boulton and Hardisty (1982) of a life-threatening ventricular arrhythmia in a 70-year-old woman after inappropriate treatment with two diuretics.

Psychotropic drugs

A World Health Organization report on health care in the elderly (World Health Organisation 1981) concluded that elderly patients consume a lot of drugs, that polypharmacy seems to be the rule in acute hospital settings and in institutions, and that psychoactive drugs appear to be used where most of the mentally disabled are found.

Evidence strongly indicates that, in the elderly, drugs acting on the central nervous system (CNS) produce a relatively greater response than in the young for a given plasma concentration. This is made worse by the fact that there are very few drugs for which a specific geriatric dosage is recommended, and the dosage regimens for new drugs are still established on trial data obtained in younger individuals. Overdosage is a common occurrence in elderly patients: for example, Greenblatt and Allen (1978) (Boston Collaborative Drug Surveillance Program) expressed concern at the extensive prescribing of psychotropic drugs for elderly patients. They studied the potential hazards of nitrazepam therapy for insomnia in 2111 middle-aged to elderly patients (mean age 57 years) in hospital medical wards. Drowsiness or hangover were reported in 2.3 per cent of cases and nightmares, insomnia, and agitation in 0.7 per cent. CNS depression was significantly more frequent (11 per cent) in patients aged 80 years or over, but signs of CNS stimulation did not correlate with age. These investigators concluded that low doses of nitrazepam were safe for elderly patients but that these patients are readily susceptible to excessive CNS depression at high doses. They felt that there was little reason to exceed a dose of 5 mg in an elderly patient.

Chemical strait-jackets

It is perhaps pertinent at this stage to ask whether it is normal practice to keep the elderly in 'chemical strait-jackets' in some institutions? Certainly there are ample grounds on which to base such a belief. For example, Ray and others (1980) reviewed 'Medicaid' prescriptions for 5902 'continuous' nursing-home (NH) patients in Tennessee and compared them with those for matched ambulatory patients living in the community: 5739 patients (97 per cent) received 384 326 prescriptions (average 67 per person), while 4161 (71 per cent) in the community comparison group received 123 025 prescriptions (30 per person). Of the NH patients, 43 per cent received antipsychotic drug preparations, whereas these drugs were infrequently prescribed for the ambulatory community patients, and 34 per cent received drugs from two or more different categories and 1.6 per cent from four categories. The most common combination was an antipsychotic drug and a sedative-hypnotic (usually thioridazine and flurazepam), followed by a minor tranquillizer and a sedative-hypnotic (diazepam and chloral hydrate). The three most frequently prescribed antipsychotic drugs were thioridazine, chlorpromazine, and haloperidol.

Ray and his colleagues (1980) commented that the readiness of the practising physician to prescribe these drugs was not altogether surprising since there were no widely-accepted and clear guidelines for antipsychotic medication in the elderly patient. They urged that controlled studies be carried out in nursing homes to assess the efficacy of alternative methods of patient management that relied less on psychotropic drugs; many authorities already recommend that antipsychotic drugs should only be used to treat acute behavioural disorders in such patients. It is apparent, however, that such advice is generally unheeded or totally disregarded and one need look no further than the review by Salzman (1981) on psychotropic drug use and polypharmacy in a general hospital setting to see just how widespread is the use of drugs by hospital inpatients, young and old alike. Salzman (1981) assessed the usage of psychotropic drugs by all of 348 medical and surgical inpatients of a Boston teaching and referral hospital on a randomly chosen weekday. The psychiatric in-patient service, the specialized diabetic treatment unit, and the cardiac care unit were excluded from this survey to ensure that the patients surveyed were not receiving psychotropic drugs for psychiatric illness and were representative of average medical or surgical patients in a general hospital. Of all surveyed in-patients, 42.8 per cent were receiving at least one psychotropic medication. Sleep medications were the most frequently prescribed class of psychotropic drugs

(37.6 per cent) and flurazepam was the most commonly prescribed of all drugs (31.6 per cent). Phenothiazines and neuroleptics were given to control pain, agitation, or nausea, rather than psychosis, and in respect to the present context it was apparent that elderly patients formed the majority of patients receiving such medication (mean age 67.3 years). Antidepressants were prescribed without recorded justification in the medical record and if given for depression were underdosed; the mean age of the patients being so treated was 58.5 years. Diazepam was the most frequently prescribed anti-anxiety drug and the second most frequently prescribed (14 per cent) psychotropic drug. Polypharmacy was common, with the average patient receiving seven different drugs; indeed, when the most commonly prescribed drugs were ranked in order of prescription frequency, it was possible to draw up a hypothetical average list of drugs taken by a patient in that survey hospital, the list being: flurazepam, paracetamol (acetaminophen), milk of magnesia, an antibiotic, dioctyl sodium sulphosuccinate, diazepam, and dextropropoxyphene.

Only one of these drugs, the antibiotic, would be used to treat primary medical illness; all the other medicines would be used to treat secondary physical symptoms such as pain or constipation, symptoms resulting from the stress of being hospitalized (insomnia, anxiety), or adverse effects (e.g. constipation) resulting from the use of neuroleptics or antidepressants.

Salzman (1981) has drawn up an interesting comparison based on three authoritative studies on the percentage of hospital inpatients who receive at least one psychoactive drug; it is reproduced here as Table 4.6.

Clearly polypharmacy with antipsychotic and other drugs is a policy liable to evoke potentially hazardous interactions in the elderly patient and it should be borne in mind in this respect that the elderly patient is prone to suffer more serious sequelae of drug–drug interactions than the younger patient simply because he is more likely to be on a treatment regimen that requires careful stabilization (e.g. anticoagulant therapy, antidiabetic treatment, or antihypertensive care) and that, if unbalanced, can be very hazardous.

Table 4.6   Per cent of hospital inpatients who received at least one psychoactive drug (from Salzman 1981)

Drug	Salzman (1981) ($N = 348$)	Boston Collaborative Drug Survey (Miller and Greenblatt 1976) ($N = >20\ 000$)	Davidson et al. (1975) ($N = 1361$)
Sleep medications (total)	37.6	53.4	— [a]
Flurazepam	31.6	8.3	—
Chloral hydrate	2.0	32.0	—
Pentobarbital (pentobarbitone)	2.6	13.1	6.7
Secobarbital (quinalbarbitone)	1.1	10.0	—
Methyprylon	0.3	2.2	—
Major tranquillizers (total)	14.6	— [b]	16.0
Prochlorperazine	8.0	10.0	—
Haloperidol	2.3	—	1.0
Chlorpromazine	1.1	3.6	11.7
Thioridazine	1.1	—	5.5
Promethazine	0.9	3.8	—
Promazine	0.6	—	—
Trifluoperazine	0.3	—	—
Fluphenazine	0.3	—	—
Antidepressants (total)	4.9	— [b]	5.3
Amitriptyline	2.8	1.7	3.9
Imipramine	0.9	—	2.3
Doxepin	0.6	—	—
Amitriptyline-perphenazine	0.3	—	—
Minor tranquillizers (total)	19.8	—	7.9 [c]
Diazepam	14.0	19.8	7.1
Chlordiazepoxide	0.9	15.6	0.8
Hydroxyzine	4.9	0.9	—

[a] Data not provided in reference.

[b] Other neuroleptics such as haloperidol and other antidepressants such as imipramine were undoubtedly prescribed during the period of the Collaborative Survey. The published reference, however, does not include percentage of prescription of these and other drugs.

[c] Minor tranquillizers actually accounted for 72 per cent of the total prescriptions, but only 108 (7.9 per cent) different patients out of 1361 received these drugs.

Poor compliance with medication instructions

It must not be forgotten that in most Western countries the total number of drugs prescribed for the elderly is higher than that for younger age groups. It is against this background that the problem of poor compliance with medication instructions must be seen and appreciated. It is not only unfair but also grossly wasteful of resources to expect elderly and often confused patients to effect the final link in their medical treatment by complying with complicated schedules for several drugs. Lundin (1978) from the School of Nursing of the University of Minnesota has described encountering some of these problems during structured interviews with 50 persons aged over 65 years who were living independently in the community. Results showed that hazardous as well as wasteful practices were occurring in their medication-taking behaviour; 66 per cent of their medicines were being taken without adequate instructions and 25 per cent of the medicines were not being taken as labelled. This is by no means an isolated case; it is much more like the normal situation in America and in Europe. Indeed Atkinson and others (1978) considered the elderly as a special case in drug prescribing. Their recommendations for improving compliance include not using child-proof (which may also be 'Granny-proof') medicine containers; using medicine labels that can be read even with poor eyesight; issuing personal medical record cards; restricting prescribed drugs to the absolute minimum number essential; training patients in their drug-taking routine before discharge from hospital; and, most important, ensuring that the size, colour, and shape of the tablet dispensed by the hospital pharmacy is the same as that subsequently provided by the local community pharmacy (i.e. the same brand should be used). Such measures are common sense in pure pharmaceutical terms, but they can, however, be overlooked in determining more complicated (and more expensive) schemes to avoid unnecessary adverse drug reactions. Later studies by Lundin and her colleagues (1980) emphasized the need for education of the independent elderly patient in the responsible use of prescription medication.

Some insight into the elderly patient's thoughts about medicine taking and its potential dangers was given by the study of Gebhardt and her colleagues (1978) from the State University College at Cortland. A sample of non-institutionalized senior citizens was interviewed to gain their views on drug-related practices and selected drug issues. Data collected from these interviews revealed that the elderly had a number of potentially damaging beliefs about drug risks: only half the number of people interviewed thought, for example, that there were risks involved when taking prescribed medicines, and only one-third thought that personal knowledge about their medications would be beneficial, and they knew little about the possible hazards of drug interaction. There was a very positive opinion towards the use of OTC (over-the-counter) medicines, particularly laxatives and vitamin preparations; the belief that there was a need for a daily bowel movement was as common as the practice of using a laxative to keep bowel movements regular. The great majority of those interviewed believed that all older patients should take vitamins. About 60 per cent of the subjects said that they would discontinue a prescribed drug if it didn't seem to be working.

The frequency of drug use was high in this group; over 80 per cent were taking prescription drugs and only 7 per cent said they were not taking medication of any kind. They believed (mistakenly) that, as a group, they used less medicine than younger adults. None of these observations is surprising; it has perhaps not been stressed enough that the general attitude of elderly patients towards their medicines is largely flavoured and influenced by experience in the early life with less potent and less specific, or indeed less effective, medication. This may well indicate the need for careful and gentle re-education of the elderly community about the changes in the efficacy and attendant hazards of medicines now being prescribed.

Non-compliance may be intentional

It should not be assumed that non-compliance in the elderly patient is always accidental or circumstantial — it may well be intentional. For example, Cooper and others (1982) surveyed 11 non-institutionalized subjects over 60 years old (average age 70.5 years) to determine how they were taking prescription drugs and the reasons for taking them in any way other than prescribed. There were an average of 3.3 prescriptions per person and for 16 per cent of the drugs there was evidence of non-adherence to the instructions on the bottle label. Forty-three per cent of subjects did not adhere to at least one prescription and in 70 per cent of cases this was intentional. The reason for non-adherence by most (52 per cent) of these patients was that they did not believe that the drug was needed in the dose prescribed by the doctor. In only 15 per cent was it because of adverse effects and in 4 per cent of cases because patients thought they needed more than was prescribed. Non-adherence was unintentional in about 30 per cent of subjects, usually because of memory lapse (15 per cent).

Although geriatric patients present problems of forgetfulness, it is also clear from a survey by Chryssidis and others (1982) that these patients can equally present problems due to self-neglect. This survey from Australia is particularly interesting since, although carried out in suburban Adelaide, it involved 39 patients (mean age 70.5 years) from different ethnic backgrounds (Australian, English, German, Polish, Ukranian, Russian, Dutch, and Rumanian). Some patients displayed a lack of confidence in modern medications, and another thought that, once a patient became ill, medicines did not help the situation and furthermore believed that combinations of medicines 'poisoned the blood'; nothing could be done to convince this patient that his opinion was unfounded. Antibiotics were a particular problem and seven out of eight patients exhibited either high compliance or over-compliance. Amongst the latter patients there seemed to be motivation to complete short prescribed courses of treatment, and some wanted to do it more quickly than was expected, for example, one patient showed a compliance of 132.5 per cent. Another patient did not even commence a prescribed programme and a look in his medication cabinet showed that he had large stocks of other unused medications as well. Common beliefs were fear of habituation to drugs or a fear of poisoning, and many of the subjects receiving non-steroidal anti-inflammatory drugs said they preferred to endure a certain amount of pain rather than take a tablet, because they feared becoming 'immune' to the drug and having to suffer excruciating rheumatic or arthritic pain later. Table 4.7 summarizes the drug-compliance spectrum of these elderly patients.

ADRs in children

There are some notable gaps in adverse drug reaction monitoring, in particular there is considerable need for studies on the epidemiology of paediatric ADRs. Relatively few such exercises have been made, and these uniformly indicate that more detailed studies are required. Most of the studies reported to date have been conducted in hospitalized patients. For example, drug exposure amongst hospitalized children was examined by Mckenzie *et al.* (1973) in Gainesville, Florida, and also by the Boston Collaborative Drug Surveillance Program (1972). The Gainesville group monitored 658 consecutive paediatric admissions and the Boston group studied 361 hospitalized children. The results of these two studies were very similar; the average drug exposure rate was 4.3 drugs per patient during a mean hospital stay of nine days. The Boston group compared these results with those they obtained in 6312 adults among whom the mean hospital stay was 19 days and the mean drug exposure was 8.4 drugs per patient. When these figures were adjusted for duration of stay, the number of drugs per patient per hospital day were comparable: 0.44 in children versus 0.48 in adults, or about one new drug every other hospital day.

In the Boston study, 13 per cent of the children and about 20 per cent of the adults experienced an ADR during their period in hospital, the difference being due in part to the longer hospital stay for adults. The children experienced adverse reactions with 5.8 per cent of the drugs administered, and the adults with 4.8 per cent. Life-threatening reactions were sustained by 3.6 per cent of all hospitalized children (fatalities 0.55 per cent) and 4.7 per cent of the adults (fatalities 0.44 per cent). Similar reaction rates were observed by the Gainesville group; 10.6 per cent of the patients developed an ADR during hospitalization; 8.1 per cent had an ADR on admission; and 3.0 per cent were admitted as a direct result of drug toxicity. Twenty-two per cent of these reactions were life-threatening and two were fatal.

A later study by Mckenzie *et al.* (1976) reinforced these data by producing evidence from a three-year prospective epidemiologic surveillance for ADRs in 3566 paediatric medical admissions to Shands Teaching Hospital in the University of Florida. They identified 72 admissions (2.0 per cent) resulting from ADRs; 64 patients were involved. Adverse reactions were moderately severe in 56 per cent of the 72 admissions, severe or life-threatening in about 40 per cent, and contributed to death in four cases (5.5 per cent). Those who died were a 12-year-old boy, a 13-year-old girl, and a 14-year-old girl, all with acute lymphoblastic leukaemia, treated with different combinations of vincristine, prednisone, daunomycin (daunorubicin), 6-mercaptopurine, and L-asparaginase, and a 4-month-old boy with a history of heart failure, treated with digoxin.

The incidence of drug-induced admissions in this study was similar to the incidence in adults, but the types of drugs most often implicated (antineoplastics and corticosteroids) were different from those in adults (aspirin, digoxin, and warfarin). The fatality rate in this study cannot, however, be extrapolated to other hospitals as the number of acute leukaemia patients admitted to that hospital resulted in three of the four fatalities. The authors also found that the proportion of drug-reaction admissions was smaller up to 6 years of age than in patients over 6 years. They concluded, however, that this was probably due to the use of more toxic drugs to treat severe diseases for which the usual age of onset is 6–15 years.

Two further studies have given somewhat diverging views on the extent to which children are exposed to drugs and the degree to which their treatments are supervised or monitored. The first of these relates to hospital practice in Ireland; Harte and Timoney (1980) examined 866 paediatric prescriptions from five

Table 4.7 Summary of the drug compliance spectrum in elderly patients (from Chryssidis *et al.* 1982)

	Age and sex	Nationality*	Number in household	Number of medications	Mean compliance (per cent)
1	72 F	A	2	3	50
2	74 F	E	1	1	35.7
3	78 F	A	1	1	12
4	71 F	E	2	5	103.4
5	57 F	G	1	2	103.2
6	82 F	A	1	1	3.7
7	67 M	G	1	2	132.5
8	74 F	A	1	3	73.8
9	68 M	A	3	1	36.6
10	74 F	P	2	2	100
11	78 M	P	2	2	31.3
12	72 M	A	2	2	37.2
13	74 F	A	1	2	114.3
14	55 F	U	3	3	72
15	60 F	A	4	1	87
16	68 M	P	2	2	0
17	72 F	A	1	1	157
18	69 M	P	2	1	67
19	80 F	R	2	3	69.6
20	58 F	P	2	3	100
21	79 F	D	1	4	74.5
22	81 F	A	1	4	85.3
23	69 M	Ru	3	2	86
24	69 F	A	1	3	64
25	75 M	A	2	1	33
26	90 M	A	3	2	50
27	68 M	A	3	3	50
28	75 M	A	1	2	58.4
29	60 F	A	1	2	144
30	55 F	E	8	2	106.3
31	71 M	A	2	2	41.5
32	68 M	U	1	3	30
33	67 F	A	2	2	75.8
34	78 F	A	3	1	48.5
35	66 F	P	3	1	100
36	56 M	P	3	1	138.8
37	55 F	A	3	2	166.5
38	76 F	A	1	4	50
39	89 F	A	1	2	84
Mean 70.5 ± 1.4					73.7 ± 6.5

* Nationality: A = Australian; E = English; G = German; P = Polish; U = Ukranian; R = Russian; D = Dutch; Ru = Rumanian.

hospitals during a 2–3 month period. About 40 per cent of the patients were less than one year old. Antimicrobial agents dominated the list and accounted for over 50 per cent of the prescriptions. More than half of the antimicrobial agents prescribed were the penicillin-related antibiotics. The single most common antibiotic used was ampicillin which accounted for 11 per cent of the prescriptions surveyed. The use of amoxycillin was low; tetracycline was prescribed only once and chloramphenicol twice. Antibiotics in the cephalosporin group ranked as the ninth most frequently prescribed drug, and co-trimoxazole was widely used especially in younger children. The only non-antimicrobial drugs commonly prescribed were salbutamol, paracetamol, phenobarbitone, and thiopental; with one exception, thiopental was given rectally for basal anaesthesia.

Thirty-eight ADRs were recorded during this survey; eight occurred in children on ampicillin and symptoms included rash (three patients), diarrhoea (four patients), and thrush (one patient). Adverse reactions to phenobarbitone (five patients) were all described as gastrointestinal, a side-effect not commonly reported for this drug. Other drugs which were associated with more than a single adverse reaction included nitrofurantoin (three), ampicillin plus cloxacillin sodium (*Ampiclox*) (two), cytarabine (two), clindamycin (two), and erythromycin (two).

In the second paper (Catford 1980), paediatric prescriptions from a random sample of 72 general practitioners in Wessex were assessed during one month. Prescriptions for drugs known to be contraindicated for children, e.g. chloramphenicol, barbiturates, tetracyclines, and drugs affecting appetite were not encountered. Most prescriptions were for one drug only and about 1 per cent of prescriptions were questionable on the basis of current specialist teaching. However, 42 per cent of the doctors used drugs now considered hazardous or undesirable (usually for diarrhoea, vomiting or enuresis).

Catford concluded that prescribing for children appeared to be responsible in the main, but there was a lag in the availability or use of important information relevant to general practice.

Little has been published in non-European countries on ADRs in children and it is therefore particularly interesting to note the work of Yosselson-Superstine and Weiss (1982) who surveyed admissions to a paediatric ward in the Hadassah-Hebrew Hospital in Jerusalem during a seven-month period to determine drug-related hospitalization. Of the 906 admissions studied, 160 (17.7 per cent) were drug-related; of these 29 (3.2 per cent) were attributed to adverse reactions; 100 (11 per cent) were due to inappropriate drug therapy; and 31 (3.4 per cent) resulted from patient (or parent) non-compliance with medication instructions.

Antineoplastic agents caused the greatest number of ADRs leading to hospital admission. Corticosteroids and anticonvulsants were also responsible for hospitalization, far more than their percentage of use in the total admitted population might indicate. Antimicrobials caused 35.6 per cent of the reactions but they were also the most used group of drugs — 32.9 per cent of all admitted patients took antimicrobials prior to their admission. All medications causing ADRs were prescription drugs.

The appearance of a known side-effect was involved in 65.5 per cent of the total reactions: approximately 24 per cent were due to hypersensitivity reactions (allergy or idiosyncrasy) and about 7 per cent were due to drug interactions. Of all the reactions, 10.4 per cent were mild, 44.8 per cent were moderate, 41.4 per cent were severe, and one reaction (3.4 per cent) caused by cytotoxic therapy was fatal.

The 100 admissions that were a result of inappropriate therapy represented 165 deviations from correct and accepted treatment. In 36 cases, all of them infections, a drug which was not considered to be a drug of choice had been prescribed without a justifiable reason; in one-third of those cases it had occurred as a result of an incorrect diagnosis.

The most incorrectly prescribed group of drugs was the antimicrobials; they represented 146 errors or deviations (88.5 per cent of total deviations from recommended therapy). In 54 per cent of those cases ampicillin was prescribed. Ten errors (6 per cent) were found in anticonvulsant therapy, seven (4.2 per cent) in the prescribing of aminophylline, and one each with digoxin and insulin. Pneumonia had been diagnosed in 41 of the 100 hospitalizations due to inappropriate therapy. Inappropriate therapy was more frequent in the six-month to four-year-old age group than in other patients.

Of the 31 patients hospitalized as a result of non-compliance with medication instructions, 17 were admitted because of this reason and in the other 14 non-compliance was a contributory factor together with inappropriate therapy. Of this group of 31 children, 10 discontinued their medicines, 11 lowered the dosage, and six took medicines irregularly. The remaining four showed a mixed type of non-compliance. The reasons for poor compliance were commonly nausea and vomiting and forgetfulness or social reasons on the behalf of the parent. Two mothers blamed non-compliance on confusing directions for medicine usage, others thought that the child had recovered from the illness and therefore did not need any more medicine, some also thought that continuation of medication required more follow-up visits to the clinic which they considered to be a burden. Non-compliance (on part of patient or parent) was most pronounced in children between the ages of 6 and 12 months and in those over 10 years of age.

Interestingly, patients complied most poorly with antimicrobial agents (64.5 per cent), than with anticonvulsants (22.6 per cent), and digoxin (3.2 per cent). The high rate of failure to take anticonvulsants according to the doctors' instructions was most significant since this group of drugs represented only 2.5 per cent of the total number of drugs prescribed in the study. This seemed to be associated with denial of the disease or drowsiness or sedation caused by these drugs.

Sex differences in ADR incidence

Sex as well as age must be taken into account in studies of human drug metabolism (Proksch and Lamy 1977). For example the mean half-life of the 'marker' phenazone (antipyrine) has been shown to be 30 per cent longer in young males (mean age 27 years), but 78 per cent longer in elderly women (mean age 78 years) (O'Malley *et al.* 1971); it is also well known that agranulocytosis due to amidopyrine, phenylbutazone, and chloramphenicol occurs far more frequently in females than in males; the ratio is about 3:1 for these drugs. Pancytopenia due to chloramphenicol, is the result of bone-marrow damage through an antimetabolic effect and again females are more susceptible than males. The possibility that females, young girls in particular, may have an inferior defence mechanism against the toxic effects of chloramphenicol must be considered.

Limited studies have suggested that adverse drug reactions are more common in women. For example, Hurwitz (1969a), in her Belfast study of predisposing factors in adverse reactions to drugs, showed that more women than men developed such reactions. Chapman and Duggan (1969) compared patients with gastric and duodenal ulcer and found that a greater proportion of women 30 to 59 years of age with gastric ulcer used large quantities of aspirin. This effect was not evident among other women or among men. Stewart and Cluff (1974) reported on the gastrointestinal manifestations of adverse drug reactions in 3164 in-patients and 612 out-patients. In both groups a significantly higher predisposition to side-effects was observed in women. Caranasos and his colleagues (1974) investigated drug-induced illness leading to hospitalization; significantly more women than men were hospitalized for adverse drug reactions. The Boston group recorded rates of allergic skin reactions to commonly used drugs in 22 227 consecutively monitored medical in-patients (Arndt and Jick 1976). The rate of such reactions for women was about 50 per cent higher than that for men. In a prospective study in a teaching hospital (Ziegler spital, Bern) on 914 hospitalized patients, 171 had evidence of adverse drug reactions. Although almost an equal number of men and women were hospitalized, approximately two-thirds of the patients with reactions were women (Klein et al. 1976).

In Denmark, Hess and her colleagues (1979) examined medical records of admissions to a department of internal medicine for one year and registered drug consumption and ADRs immediately before and during hospitalization. They recorded 1325 admissions divided among 1136 patients. On average younger patients (15–50 years old) received three drugs, and older patients (51–70 + years old) five drugs during their stay in hospital. A total of 235 ADRs was recorded in 210 patients; of these 183 were classified as mild, 48 as severe, and in four cases ADRs were considered as a significant cause of death. Calculated as a percentage of admissions, 18 per cent drug reactions were found; 10.9 per cent appeared within two weeks before admission, and 7.1 per cent were the immediate cause of hospitalization. Of importance within the present context, a higher fre-

quency of ADRs were found in women as compared with men, and in older women as compared with younger. No age-related differences could be established, however, in the men. An overall incidence of ADRs of 1.8 per cent (calculated per given number of drugs) was found.

The Australian Adverse Drug Reactions Advisory Committee's report for 1980, which was published in May 1982 (ADRAC 1982), commented that of the 2503 reports of suspected ADRs, 1375 of these related to women. As in previous years the suspected ADRs experienced by women were reported more frequently than those in men; a ratio of 1.46:1.

Several reports in the literature (e.g. Stewart and Cluff 1971) indicated in the past that women use more medicines than men, and this characteristic was also found for the total drug usage (prescription and non-prescription) in the Dunedin Program in the United States; this programme was initiated in July 1975 to screen elderly subjects annually for previous undetected medical disorders (see Stewart et al. 1984). Dunedin is a retirement community on the midwest coast of Florida. May et al. (1982) in a survey of this community showed that elderly patients take a large number of over-the-counter medicines. They recorded prescription and non-prescription medication for 3192 Dunedin residents (2009 women and 1183 men) of which only 11.4 per cent were taking no medicines. Seventy per cent of women and 58 per cent of men were taking at least one non-prescription medicine regularly; the number of prescription drugs taken by these residents ranged from 0–9 for women, and 0–11 for men. Women on average were using 3.5 medicines as compared with 2.8 for men.

In the University of Chile drug surveillance programme (Domecq et al. 1980), it was also demonstrated that women suffer more ADRs than men. This programme monitored 1920 consecutive patients during their stay in hospital. Of the 941 women, 357 (37.9 per cent) suffered adverse reactions. Of the 979 men, 290 (29.6 per cent) suffered reactions and this difference was highly significant ($p < 0.0005$). A total of 1439 adverse reactions were recorded, 837 in women (43.4 per cent definite, 56.6 per cent probable) and 602 in men (52.1 per cent definite, 47.8 per cent probable). Only 4.3 per cent of the reactions in women were severe, as were only 5.8 per cent in men. Two patients died: 1 male cirrhotic with acute heart failure due to intravenous mannitol, and 1 woman with toxic epidermal necrolysis due to thiacetazone 150 mg plus isoniazid 300 mg. Ninety-three per cent of the reactions in women and 83 per cent in men were dose-related. There were sex differences in the manifestations of side-effects: gastrointestinal, 36.4 per cent in women, 21.4 per cent in men; allergic reactions, 9.7 per cent (skin 7.6 per cent) in women, 4.2 per cent (skin 2.5 per cent) in men. The higher rates for women were significant. However, men suffered more diuretic-induced metabolic/electrolyte disturbances than women (45.5 per cent vs. 25.1 per cent). Frusemide (furosemide), one of the most used drugs, caused the highest proportion of reactions 104/270 (38.5 per cent) in women, and 116/283 (41 per cent) in men. This latter difference in reaction rate was not significant.

Since women take more drugs than men, their increased incidence of ADRs may be due to their greater exposure to prescribed and non-prescribed medicines. However, there may be other contributory factors that have not yet been established. The result is clear as the preceding reports testify; however, the cause of this predisposition is uncertain.

In a very specialized context, the outcome of pregnancy among women in anaesthetic practice was investigated by Pharoah *et al.* (1977). A survey was made of the outcome of pregnancies of 5700 women doctors first registered in England and Wales in 1950 or later. Conceptions that occurred when the mother was in an anaesthetic appointment resulted in smaller babies, higher stillbirth rates, and more congenital malformations of the cardiovascular system than in the pregnancies of other women doctors. There was no significant difference in the spontaneous abortion rate between the two groups. A pronounced effect of age on this rate was evident among all groups examined.

Earlier reports from Copenhagen by Askrog and Harvald (1970), from California by Cohen *et al.* (1971, 1974), and Glasgow by Knill-Jones *et al.* (1972) suggested that there is a higher incidence of spontaneous abortion or premature labour in female anaesthetists who work during pregnancy than control groups. The incidence of spontaneous abortions in the Glasgow study was 18.2 per cent of children born to working female anaesthetists. Children born to working female anaesthetists had a congenital abnormality rate of 6.5 per cent compared with 2.5 per cent in female anaesthetists who did not work during pregnancy. The Glasgow study also indicated an incidence of sterility twice that of their control group. The Copenhagen study indicated an increased ratio of female to male births and suggested that this could be interpreted as indicating some degree of selective toxicity on the male fetus. This picture was further complicated by the study of Wyatt and Murray Wilson (1973) from the Sheffield region who demonstrated an excess of female children born to 157 male anaesthetists (56.8 per cent) compared with the men for the Sheffield region (46.6 per cent), and England and Wales as a whole (48.6 per cent), indicating a possible effect on either gametogenesis or on the viability of the gamete.

Based on the findings of these earlier studies, it is not surprising that Corbett *et al.* (1974) in America found a higher incidence of congenital malformations in the children of nurse anaesthetists. What is surprising, however, is that they found a threefold increase in the incidence of cancer in these nurses than in controls and a slight but not significant increase in cancer of the offspring.

Although these are very specialized examples of adverse effects of anaesthetic agents on the woman and her fetus, it does clearly illustrate one other facet of the specific adverse effect of drugs or environmental factors on women. If exposure to anaesthetic gases is regarded as an occupational hazard, then perhaps other occupations or other environmental contaminations should also be similarly investigated. Women routinely handling anticancer drugs, for example, may be at a similar risk (see Chapter 41).

ADRs in surgical patients

A number of drug and adverse reaction monitoring studies were done with multinational collaboration. One of these by Danielson *et al.* (1982) from the Boston Collaborative Drug Surveillance Program investigated drug monitoring of surgical patients and was carried out on selected general surgical wards in five hospitals in the United States, Scotland, and New Zealand during 1977 to 1981. The methods employed for the monitoring study were similar to those previously used by the BCDSP in investigating more than 40 000 hospitalized medical patients (Jick *et al.* 1970).

A total of 5232 surgical patients were monitored in this study; patients received an average of nine drugs (range 7 to 11 drug exposures per patient). Of the 46 868 drugs received on the wards, 1150 (2.5 per cent) were associated with an adverse reaction that, after clinical investigation, was thought to be definitely (30 per cent), probably (59 per cent), or possibly (12 per cent) drug-induced. Most of the adverse reactions were characterized as being of 'minor' or 'moderate' severity (95 per cent). There were 62 'major' adverse reactions (5 per cent) occurring in 42 patients. The 10 most frequently reported ADRs were: nausea/vomiting (23 per cent of all adverse reactions), constipation (6.3 per cent), diarrhoea (5.7 per cent), electrolyte imbalance (5.5 per cent), drowsiness (4.8 per cent), skin rash (4.6 per cent), disorientation (2.7 per cent), fever (2.3 per cent), hypotension (1.9 per cent), itchiness (1.6 per cent), vertigo (1.6 per cent), and other non-classified reactions (40 per cent).

There were 15 drugs in the list of highest reported ADRs; hydrochlorothiazide headed this list (17 per cent),

Table 4.8 Fifteen drugs* with highest reported adverse reaction rates in a multinational study on surgical patients (data modified from Danielson et al. 1982)

Drug	Number of exposures	Number of adverse reactions (per cent)
Hydrochlorothiazide	69	12 (17)
Whole blood*	151	22 (15)
Warfarin	60	7 (12)
Amoxicillin	87	10 (11)
Packed RBCs*	235	24 (10)
Dihydrocodeine	1084	108 (10)
Co-trimoxazole	251	23 (9)
Pentazocine	132	11 (8)
Ampicillin	410	30 (7)
Metronidazole	294	21 (7)
Lorazepam	57	4 (7)
Neomycin	120	8 (7)
Castor oil	94	6 (6)
Digoxin	238	15 (6)
Chlorpromazine	79	5 (6)

* Whole blood and packed RBCs were included by Danielson et al. (1982) under the general heading of drugs in their paper.

followed closely by whole blood (15 per cent), and warfarin (12 per cent). The placing of these other drugs in the ADR 'league table' are shown in Table 4.8 which is based on data given by Danielson et al. (1982).

Twenty patients suffered a total of 35 life-threatening reactions, but there were no drug-attributed deaths. Respiratory depression (five patients) was the most frequent life-threatening event, followed by pseudomembraneous colitis (three patients), anaphylactic reactions (three patients), pulmonary oedema (two patients), and cholestatic jaundice (two patients). The other life-threatening reactions were severe diarrhoea, hyponatraemia, thrombocytopenia, CNS depression, perforated ulcer, and dehydration/hypotension (one patient each).

Danielson and his colleagues emphasized that their current series of monitored patients were drawn from general surgical wards and therefore they did not include patients at high risk from surgical complications (e.g. cardiopulmonary or neurosurgery). Nevertheless, they were able to draw comparisons between the data collected on these surgical patients and data collected in other studies by the BCDSP on medical patients.

Surgical patients, for example, received proportionally more analgesics, intravenous fluids, and pre-operative medications when compared with medical in-patients previously monitored in many of the same hospitals (Jick et al. 1970). Since these three classes of medication tend to be comparatively non-toxic, it was not surprising that the overall rate of toxic reactions among surgical patients per drug received (2.5 per cent) was only half that for medical patients (5.1 per cent). The

adverse reaction rates, however, for particular drugs such as diuretics and antibiotics were similar for both groups of patients.

Anaesthetics and ancillary drugs used during anaesthesia (e.g. skeletal muscle relaxants) have been specifically pinpointed in other reports as being associated with disconcerting reactions. For example, the fifteenth report (1980) of the New Zealand Committee on Adverse Drug Reactions (McQueen 1981) listed that among non-fatal reactions, anaesthetics and relaxants produced known reactions. These included, out of a total of 863 ADR reports, three of jaundice associated with halothane and 10 instances of acute anaphylaxis with, respectively, Althesin (two), tubocurarine (two), pancuronium (one), and suxamethonium (five). In the succeeding report of the New Zealand committee (McQueen 1982), there were 714 adverse drug reactions listed of which 27 were fatal. Three of these fatalities were produced by general anaesthetics. A 15-year-old boy developed malignant hyperthermia after $2\frac{1}{2}$ hours of anaesthesia. It became known after the event that his cousin had also died from hyperthermia in the same way $4\frac{1}{2}$ years earlier. Hepatic necrosis after the giving of halothane was the cause of two deaths. General anaesthetics were also the subject of 18 reports of non-fatal reactions; eight of these involved halothane and six Althesin; among the latter were three cases of immediate hypersensitivity syndrome.

Muscle relaxants were associated with 17 cases of adverse reaction, including three fatalities and five of immediate hypersensitivity. Two of the cases of immediate hypersensitivity were with tubocurarine and three with suxamethonium. There were also three cases of suxamethonium apnoea. It is particularly relevant to this finding that Youngman et al. (1983) surveyed patients who suffered a life-threatening systemic reaction during general anaesthesia. The study was carried out in Auckland, New Zealand, and since 1977, a total of 158 patients who had had a severe systemic reaction were referred for skin testing. They were tested for each of the drugs used during their anaesthesia. Suxamethonium sensitivity was investigated in 85 of these patients, of whom 28 (three men, 25 women) had strong positive skin reactions to the drug; 11 of 37 patients tested for alcuronium sensitivity (two men, nine women), 15 of 54 tested for gallamine sensitivity (four men, 11 women), and 21 of 43 tested for tubocurarine sensitivity (two men, 19 women) were positive. Interestingly the female/male ratio was 8:1.

Death associated with anaesthesia was the subject of two studies. The first, from Finland by Hovi-Viander (1980), surveyed more than 300 000 anaesthetic inductions. Death during or within three days of the anaesthetic occurred in 626 patients (0.18 per cent). Abdominal operations produced the largest surgical category (39.2 per cent) and more than half the operations (62.4 per cent) were major surgery. There was also a high proportion of emergency procedures (56.5 per cent). The cause of death in most cases (71.9 per cent) was the primary disease, with many patients in poor condition or moribund before surgery. Death was attributed to surgery in 73 patients (11.7 per cent) and to the anaesthesia in 67 patients (10.7 per cent). In 24 of these 67 patients there was probably an overdose of anaesthetic in relation to their physical status and age.

In the second survey, Turnbull *et al.* (1980) reviewed 200 000 anaesthetic inductions at Vancouver General Hospital and found that death occurred within 48 hours of an anaesthetic in 423 patients (0.22 per cent). The initial state of 135 of these patients made it unlikely that they would survive surgery. Fifty-two of the patients who died were less than one month old and many of these suffered from congenital heart disease. The most common conditions in adult patients who died were cardiac disease with cardiopulmonary bypass, brain tumour or brain oedema, multiple injury, profound sepsis, and major vascular catastrophies.

ADRs in dental patients

The possibility of iatrogenic complications from routine use of drugs in patients also concerns the dental profession; in the USA, dentists write about 32 million prescriptions per year.

There has been virtually no documentation to determine the actual extent of the ADR problem in dentistry and apprehension is therefore largely speculative and is based on the experience in general medicine. The only article dealing with the potential for ADRs in the dental patient described a survey conducted by Suomi *et al.* (1975); they found that 33.6 per cent of a total of 529 dental patients reported previous drug reactions and sensitivities as being their most common medical concern. These data did not imply that dental drug reactions occurred to this extent, but rather that the possibility of recurrence was high enough to warrant consideration. This report and the absence of other reliable data stimulated the University of Southern California Dental School to conduct a 2-year study monitoring its patients' drug use in an effort to document the extent of potential dental drug reactions and to prevent their occurrence whenever possible (Oksas 1978).

The 2418 patients in this study received 2653 drugs and only 110 cases of possible reactions were encountered. These consisted of 50 reports of drug interactions, 29 reports of contraindication, and 31 actual cases of side-effects. All but the latter group of reactions were avoided by careful review of patients' medical data. Interestingly, 30.2 per cent of study patients had concurrent medical conditions which would contraindicate the use of some of the dental drugs; 21.7 per cent of the study patients were taking medically prescribed drugs on a routine basis which indicated the potential reservoir for dental/medical drug interactions. In addition, 23.4 per cent of the patients interviewed indicated prior drug-related reactions, many of which involved community used dental medication.

Oksas (1978) concluded that 4.24 per cent of the monitored dental prescriptions could result in drug reactions. He also commented that dental/medical drug interactions were the most prevalent type of potential ADRs found in the study; typically, these were CNS depressant synergy from narcotics. The dental prescribing of a drug that previously created untoward reactions was rare and aspirin accounted for most of such reactions. Gastrointestinal upset from analgesics and erythromycin therapy were common side-effects. The evaluation suggested that narcotic analgesics, erythromycin, and barbiturates carried a greater than average risk of involvement in dental drug reactions.

ADRs in community care patients

The importance of intensive surveillance as the means of obtaining a reliable estimate of the incidence of ADRs in the community was re-emphasized by Martys (1981) especially since the voluntary reporting by the patient is distinctly limited in this respect (Culliton and Waterfall 1980). Certainly Martys, a general practitioner in Derbyshire UK, had ample experience on which to base this emphasis. Relatively few studies concerned patients in the community, and one of the few was published by him (Martys 1979).

In his assessment of the incidence of ADRs in a general-practice setting he estimated that, of the 817 patients in his practice, 41 per cent were thought to have 'certainly' or 'probably' a reaction to the drug that was prescribed for them. Adverse effects on the gastrointestinal and central nervous systems (e.g. nausea, diarrhoea, dry mouth, drowsiness, headache, and dizziness) were the most frequently reported. Peak incidence was 1–3 days and 90 per cent of reactions had occurred by the fourth day of treatment.

Drugs acting on the CNS were responsible for a greater incidence of more severe reactions than drugs in other groups, and also for a greater overall incidence of ADRs of all grades of severity (51 per cent), followed closely by antihistamines (45 per cent), drugs acting on the cardio-respiratory centre (40 per cent), and antibiotics (35 per cent). Rash occurred in 3 per cent of patients and half the number of these were receiving or had received ampicillin. A further 3 per cent of patients complained of confusion or hallucination — reactions that caused immediate cessation of treatment. Pentazocine and dihydrocodeine were often associated with these latter symptoms.

At least one in six patients for whom a prescription was written took additional treatment on his/her own initiative; in some age groups the figure for self-medication was as high as one in four. More young adults and children took medication in addition to their prescribed treatment than did patients in other age groups.

In a follow-up paper, Martys (1981) commented that since his original study had indicated that the burden of drug-induced disease in the community might be larger than he had suggested, there was a considerable need for more information to be collected on a greater scale. A similar comment was expressed earlier by Cartwright (1979) from the Institute of Social Studies in Medical Care; he thought also that Martys' estimate of a 41 per cent frequency might be a serious underestimate since it was based on reactions in patients given single-drug treatments, whereas most patients on prescribed medication were known to take more than one drug concurrently. The previous discussion in this chapter on the incidence of drug-related hospital admissions, especially among the elderly, adds much weight to such observations and extrapolations.

Martys (1981) suggested that the problem of ADRs in community patients could be reduced if community doctors were made aware of the efficacy, adverse effects, and possible interactions of the drugs they employ. Certainly the pharmaceutical companies have a particular responsibility in this respect but it is not their sole responsibility. There is not the space in this current chapter to discuss the ways and means of keeping the general practitioner informed about drugs, there is space, however, to mention two reference articles which are highly relevant in this context. The first is by Strickland-Hodge and Jeqson (1980), and it deals with usage of information sources by general practitioners, and the second is a paper by Avorn *et al.* (1982) from the Department of Social Medicine and Health Policy at Harvard Medical School, Boston.

There is also some evidence to suggest that patients also should be more informed about drugs. George *et al.* (1983) from the Clinical Pharmacology Group at the University of Southampton issued leaflets containing information about medicines to 56 patients prescribed penicillin and 43 patients prescribed non-steroidal anti-inflammatory drugs. The patients were interviewed between four and 10 days later and their responses compared with those of 65 patients prescribed penicillin and 33 prescribed NSAIDs who did not receive a leaflet.

Patients who received a leaflet were more likely to be completely satisfied with their treatment and with the information that they had been given. They were also more likely to know the name of their medicine and were much more aware of potential unwanted effects. Although there was no evidence that knowledge increased the incidence of adverse effects, when these did occur they were more likely to be recognized as being due to the medicine.

Why are ADRs so often undetected initially?

With all the reports on ADRs that appear in the literature each year from official drug regulatory bodies and from investigating clinicians, it is difficult to understand why toxic reactions to individual drugs are so often undetected initially. Indeed this question has been posed by Vesell (1980) writing in the 'Sounding board' column of the *New England Journal of Medicine*. Apart from asking the question, he also suggested some answers which serve, at least in part, to explain this initial lag in detection. He suggested that the problem starts in the medical school pharmacology course where it is generally taught that, in almost all patients, each drug has a single and well-defined group of therapeutic and adverse effects. Therefore the clinical effects of differences among patients in drug response and toxicity are inadequately appreciated and taught.

Many patients do not respond to drugs according to textbook idealism; there are many factors that are now well established as affecting drug disposition and these may produce large variations in response in different patients. Genetic factors have a considerable influence on responses to drugs largely due to effects on hepatic metabolism. Exposure to environmental factors can alter the expression of the genetic factors involved. Disease states affecting the hormonal status or the functions of the gut, liver, heart, and kidneys can also alter the capacity of the patient to absorb, distribute, metabolize, and excrete drugs.

Age can also alter drug elimination and such changes in the elderly are mainly due to degenerative alterations and decreased physiological function in heart, liver, kidneys, and gut. Such changes occur at different rates in different subjects.

All these differences contribute to greater variations in response to a newly marketed drug than are found in clinical trials, in which the patients are usually selected for near-basal status with respect to these factors, unlike the patients treated in later clinical practice. Drugs which are mainly used in the very old are commonly tested on much younger populations and drugs considered safe or effective in younger adults may be neither in the very old. This highlights the importance of testing drugs in the elderly (Smith *et al.* 1983).

Undoubtedly Vesell is right in his explanations but it should also be appreciated that the 'numbers game' probably exerts influence as well. If an ADR is likely to occur in '*x*' per cent of patients, then there is no guarantee that this probability will be uniformly distributed among the finite and usually relatively small population that is involved in the typical clinical trial, especially if such patients are preselected to be of near-basal status. It is only when larger populations of patients are involved that the true extent of the *x* percentage of reactors is revealed. It may take a period of relatively extensive clinical use before the true extent of the ADR is revealed; this was so for benoxaprofen, chloramphenicol, halothane, practolol, ticrynafen (tienilic acid), phenytoin, the sulphonamides, and, of course, in a different context, the same is true for the hazards of cigarette-smoking.

However, an ADR has to be recognized and linked with specific drug therapy before it is detected; therefore perhaps, as Vesell has concluded, it all starts with the basic education of the clinician at medical school level; physicians must expect deviations from anticipated drug actions and be forewarned to deal with them by means of appropriate changes in dosage. Wider recognition of this fundamental therapeutic principle can lead to more diligent and efficient efforts to scrutinize the patient's responses and to individualize doses of drugs having low therapeutic indices. The advent of pharmacokinetics is a very useful tool in dealing with some of the problem cases, especially when consideration is given to whether or not there are factors present which may predispose the patient to the development of an adverse reaction to his treatment.

Counting the cost of ADRs

The present chapter has attempted to present some insight into the epidemiology of adverse drug reactions. Approximate and often localized figures have been placed against the incidence of adverse reactions and drug-induced fatalities. From these it is clear that adverse reactions are becoming an increasingly important problem in drug treatment. In Britain, perhaps 5 per cent of general hospital beds are occupied by patients

suffering from their treatment. In the United States, an estimate has been given of one in seven hospital beds taken up by patients under treatment for adverse reactions caused by drugs. The work of the Boston Collaborative Drug Surveillance group and smaller groups in other countries has gone some way forward to produce reliable data. Certainly their and other surveys have at least established the minimum extent of the problem.

Estimates of the cost of adverse drug reactions to the community are, however, largely undetermined; this is not entirely surprising since obviously estimates of cost must depend on reasonably comprehensive estimates of incidence. Some attempts have been made to identify such costs, although these also relate to a localized situation and cannot validly be extrapolated into a wider context.

A decade has passed since Mach (1975) discussed possible methods of counting the cost of adverse drug reactions; he gave emphasis especially to the direct and indirect costs sustained by the community; direct costs include cost of treatment, hospitalization, and prevention and detection of adverse reactions, while indirect costs include the lost contributions of the patient to the gross national product. The figures resulting from his calculations are equally valid today (allowing for inflationary increases) and they must be viewed within the American scene and allow for estimates of lost lifetime earnings, but they are nonetheless frighteningly high. For example, a man aged 45 years who died as a result of an adverse drug reaction during a 9-day stay in hospital represents a total cost of approximately $105 000; a 65-year-old woman, $31 300; and a 10-year-old boy, $126 300. Calculations for malformations, permanent disability, hospital admission, or extra stay in hospital ranged respectively from $76 700 (complete disability, boy under 1 year) to $200 (extra time in hospital, 5-year-old child).

In recent years, all over the world there has been an explosive growth in the cost of medical care; it must be clear therefore that avoidable drug reactions are an unnecessary and unacceptable drain on stretched medical resources. It may be difficult to express these in terms of precise monetary cost, but it is not so difficult to assess them in terms of morbidity and mortality. Surely this combination of costs is quite unacceptable.

Besides having the potential to cause morbidity and even mortality, ADRs result in loss of productive employment and earnings as well as significant increases in health care costs to the individual and to the state. Increased length of stay in hospital is but one of these additional costs; Spino *et al.* (1978) from Toronto Western Hospital have assessed the influence of ADRs on length of hospital stay in 204 patients who were receiving furosemide (frusemide). The mean duration of stay in patients with no ADRs was 13.5 days, and 27.2 days in patients experiencing one or more ADRs. Similar emphasis was given by Zook and Moore (1980).

The direct cost to society in the USA may currently be more than 5 billion dollars per year; it is a sobering thought that this figure approaches about half the total cost of prescription drugs produced and marketed annually in the United States.

Not as obvious, but of considerable concern, is the effect of ADRs on public confidence in the health care system in general and in prescription drugs in particular. Public fear of the worst possible effects of drug consumption can in itself be a major factor in patient non-compliance.

REFERENCES

ABRAMSON, N.S., WALD, K.S., GRENVIK, A.A.A., ROBINSON, D., AND SNYDER, J.V. (1980). Adverse occurrences in intensive care units. *J. Am. med. Ass.* **244**, 1582–4.

ADRAC (1982). Adverse Drug Reactions Advisory Committee Report for 1980. *Med. J. Austral.* **1**, 416–19.

ARMSTRONG, B., DINAN, B., AND JICK, H. (1976). Fatal drug reactions in patients admitted to surgical services. *Am. J. Surg.* **132**, 643–5.

ARMSTRONG, W.A. JR., DRIEVER, C.W., AND HAYS, R.L. (1980). Analysis of drug–drug interactions in a geriatric population. *Am. J. hosp. Pharm.* **37**, 385–7.

ARNDT, K.A. AND JICK, H. (1976). Rates of cutaneous reactions to drugs. A report from the Boston Collaborative Drug Surveillance Program. *J. Am. med. Ass.* **235**, 918–23.

ASKROG, V. AND HARVALD, B. (1970). Teratogen effekt of inhalations-anaestetika. *Nord. Med.* **83**, 498–500.

ATKINSON, L., GIBSON, I., AND ANDREWS, J. (1978). An investigation into the ability of elderly patients continuing to take prescribed drugs after discharge from hospital and recommendations concerning improving the situation. *Gerontology* **24**, 225–34.

AUZEPY, PH., DUROCHER, R., GAY, R., HAEGY, J.M., HARARI, A., MOTIN, A., PISOT, D., RIMAILHO, A., AND TREMOLIÈRES, F. (1979). Accidents médicamenteux graves chez l'adulte: incidence actuelle dans le recrutement des unités de réanimation. *Nouv. Presse Méd.* **8**, 1315–18.

AVORN, J., CHEN, M., AND HARTLEY, R. (1982). Scientific versus commercial sources of influence on the prescribing behavior of physicians. *Am. J. Med.* **73**, 4–8.

BALLIN, J.C. (1974). The ADR numbers game. *J. Am. med. Ass.* **229**, 1097–8.

BERGMAN, U. AND WIHOLM, B.E. (1981). Drug-related problems causing admission to a medical clinic. *Eur. J. clin. Pharmacol.* **20**, 193–200.

BLACK, A.J. AND SOMERS, K. (1984). Drug-related illness resulting in hospital admission. *J.R. Coll. Physcns, London* **18**, 40–1.

BOSTON COLLABORATIVE DRUG SURVEILLANCE PROGRAM (1968). (Borda, I.T., Slone, D., and Jick, H.). Assessment of adverse

reactions within a drug surveillance program. *J. Am. med. Ass.* **205**, 645–7.

—— (1972). Adverse drug interactions. *J. Am. med. Ass.* **220**, 1238–9.

BÖTTIGER, L.E., FURHOFF, A.K., AND HOLMBERG, L. (1979). Fatal reactions to drugs. A 10-year material from the Swedish. Adverse Drug Reaction Committee. *Acta. med. scand.* **205**, 451–6.

BOULTON, A.J.M. AND HARDISTY, C.A. (1982). Ventricular arrhythmias precipitated by treatment with non-thiazide diuretics. *Practitioner* **226**, 125, 128.

BRAVERMAN, A.M. (1982). Therapeutic considerations in prescribing for elderly patients. *Eur. J. rheumatol. Inflamm. Dis.* **5**, 289–93.

BRITISH MEDICAL JOURNAL (1977). Leading article: Deaths due to drug treatment. *Br. med. J.* **1**, 1492–3.

—— (1978a). Leading article: Diuretics in the elderly. *Br. med. J.* **1**, 1092–3.

—— (1978b). Leading article: Drugs in the elderly. *Br. med. J.* **1**, 1168.

BURR, M.L., KING, S., AND DAVIES, H.E.F. (1977). The effects of discontinuing long-term diuretic therapy in the elderly. *Age Ageing* **6**, 38–45.

BUTLER, R.N. (1980). The gray revolution and health. *Am. Pharm.* NS **20**, 9–13.

CAIRD, F.I. (1977). Prescribing for the elderly. *Br. J. hosp. Med.* **17**, 610–13.

CAMPBELL, A.J., McCOSH, L., AND REINKEN, J. (1983). Drugs taken by a population based sample of subjects 65 years and older in New Zealand. *NZ med. J.* **96**, 378–80.

CARANASOS, G.J., MAY, F.E., STEWART, R.B., AND CLUFF, L.E. (1976). Drug-associated deaths of medical in-patients. *Arch. intern. Med.* **136**, 872–5.

CARTWRIGHT, A. (1979). Adverse reactions to drugs in general practice. *Br. med. J.* **2**, 1437.

CATFORD, J.C. (1980). Quality of prescribing for children in general practice. *Br. med. J.* **280**, 1435–7.

CHAPMAN, B.L. AND DUGGAN, J.M. (1969). Aspirin and uncomplicated peptic ulcer. *Gut* **10**, 443–50.

CHRYSSIDIS, E., FREWIN, T.A., FREWIN, D.B., AND HOWARD, A.F. (1982). Drug compliance in the elderly. *Aust. J. hosp. Pharm.* **12**, 8–10.

COHEN, E.N., BELLVILLE, J.W., AND BROWN, B.W. (1971). Anesthesia, pregnancy and miscarriage: a study of operating-room nurses and anesthetists. *Anesthesiology* **35**, 343–7.

——, BROWN, B.W., BRUCE, D.L., CASCORBI, H.F., CORBETT, T.H., JONES, T.W., AND WHITCHER, C.E. (1974). Occupational disease among operating-room personnel. A National Study. Report of an ad hoc committee on the effect of trace anesthetics on the health of operating-room personnel. *Anesthesiology* **41**, 321–40.

COOPER, J.K., LOVE, D.W., AND RAFFOUL, P.R. (1982). Intentional prescription nonadherence (noncompliance) by the elderly. *J. Am. geriat. Soc.* **30**, 329–33.

COMMITTEE ON SAFETY OF MEDICINES (1983). *Osmosin* (controlled release indomethacin). *Current Problems*, no. 11. August 1983, Committee on Safety of Medicines, London.

CORBETT, T.H., CORNELL, R.G., LIEDING, K., AND ENDRES, J.L. (1974). Birth defects of children among Michigan nurse anesthetists. *Anesthesiology* **41**, 341–4.

CROOKS, J., O'MALLEY, K., AND STEVENSON, I.H. (1976).

Pharmacokinetics in the elderly. *Clin. Pharmacokinet.* **1**, 280–96.

——, SHEPHERD, A.H.M., AND STEVENSON, I.H. (1975). Drugs and the elderly: the nature of the problem. *Hlth. Bull.* **33**, 222–7.

CULLITON, B.J. AND WATERFALL, W.K. (1980). Post-marketing surveillance. *Br. med. J.* **280**, 1175–6.

DANIELSON, D.A., PORTER, J.B., DINAN, B.J., O'CONNOR, P.C., LAWSON, D.H., KELLAWAY, G.S.M., AND JICK, H. (1982). Drug monitoring of surgical patients. *J. Am. med. Ass.* **284**, 1482–5.

D'ARCY, P.F. (1985). Drug interactions and reactions update: nitrofurantoin. *Drug Intel. clin. Pharm.* **19**, 540–7.

—— AND McELNAY, J.C. (1983). Adverse drug reactions and the elderly patient. *Adv. Drug React. Ac. Pois. Rev.* **2**, 67–101.

DAVIDSON, J.R.T., RAFT, D., LEWIS, B.F., AND GEBHARDT, M. (1975). Psychotropic drugs on general medical and surgical wards of a teaching hospital. *Arch. gen. Psychiat.* **32**, 507–11.

DAVIDSON, W. (1971). Drug hazards in the elderly. *Br. J. hosp. med.* **6**, 83–95.

DODD, M.J., GRIFFITHS, I.D., HOWE, J.W., AND MITCHELL, K.W. (1981). Toxic optic neuropathy caused by benoxaprofen. *Br. med. J.* **283**, 193–4.

DOMECQ, C., NARANJO, C.A., RUIZ, I., AND BUSTO, U. (1980). Sex-related variations in the frequency and characteristics of adverse drug reactions. *Int. J. clin. Pharmacol. Ther. Toxicol.* **18**, 362–6.

DØSSING, M. AND ANDREASEN, P.B. (1982). Drug-induced liver disease in Denmark. An analysis of 572 cases of hepatotoxicity reported to the Danish Board of Adverse Reactions to Drugs. *Scand. J. Gastroenterol.* **17**, 205–11.

DRUG AND THERAPEUTICS BULLETIN (1984). Phenylbutazone and oxyphenbutazone: time to call a halt. *Drug Ther. Bull.* **22**, 5–6.

FALK, R.H. (1979). Adverse reactions to medication on a coronary care unit. *Postgrad. med. J.* **55**, 870–3.

FELDMANN, E.G. (1983). Editorial. Adverse drug reactions — a continuing problem. *J. Pharm. Sci.* **72**, 585.

FENTON, D.A., ENGLISH, J.S., AND WILKINSON, J.D. (1982). Reversal of male-pattern baldness, hypertrichosis, and accelerated hair and nail growth in patients receiving benoxaprofen. *Br. med. J.* **284**, 1228–9.

FIRTH, H., WILCOCK, G.K., AND ESIRI, M. (1982). Side effects of benoxaprofen. *Br. med. J.* **284**, 1784.

FISHER, B.M., AND MCARTHUR, J.D. (1982). Side effects of benoxaprofen. *Br. med. J.* **284**, 1783.

GARDNER, P. AND WATSON, L.J. (1970). Adverse drug reactions: a pharmacist-based monitoring system. *Clin. Pharmacol. Ther.* **11**, 802–7.

GEBHARDT, M.W., GOVERNALI, J.F., AND HART, E.J. (1978). Drug-related behavior, knowledge and misconceptions among a selected group of senior citizens. *J. Drug. Educ.* **8**, 85–92.

GEORGE, C.F., WATERS, W.E., AND NICHOLAS, J.A. (1983). Prescription information leaflets: a pilot study in general practice. *Br. med. J.* **287**, 1193–6.

GHODSE, A.H. AND OTHER PARTICIPANTS (1981). Drug-related problems in London accident and emergency departments. A twelve-month survey. *Lancet* **ii**, 859–62.

GHOSE, K. (1980). Hospital bed occupancy due to drug-related problems. *J.R. Soc. Med.* **73**, 853–6.

GOLDBERG, A. (1982). Letter to doctors and pharmacists. Opren — suspension of product licences. Committee on Safety of Medicines, London. 3 August 1982.

—— (1984). Letter to doctors and pharmacists. Phenylbutazone and oxyphenbutazone. Committee on Safety of Medicines, London. 7 March 1984.

GOUDIE, B.M., BIRNIE, G.G., WATKINSON, G., MACSWEEN, R.N.M., KISSEN, L.H., AND CUNNINGHAM, N.E. (1982). Jaundice associated with the use of benoxaprofen. *Lancet* **i**, 959.

GREENBLATT, D.J. AND ALLEN, M.D. (1978). Toxicity of nitrazepam in the elderly: a report from the Boston Collaborative Drug Surveillance Program. *Br. J. clin. Pharmacol.* **5**, 407–13.

——, SELLERS, E.M., AND SHADER, R.I. (1982). Drug disposition in old age. *New Engl. J. Med.* **306**, 1081–8.

GRIFFIN, J.P. AND LONG, J.R. (1982). International co-operation in drug evaluation and marketing approval in Western Europe. *Pharm. Int.* **3**, 153–8.

GUSTAFSSON, L.L. AND BOËTHIUS, G. (1982). Utilization of analgesics from 1970 to 1978: Prescription patterns in the County of Jämtland and in Sweden as a whole. *Acta med. scand.* **211**, 419–25.

HALL, M.A. (1972). Drugs and the elderly. *Adv. Drug React. Bull.* **35**, 108–11.

HALLWORTH, R.B. AND GOLDBERG, L.A. (1984). Geriatric patients' understanding of labelling of medicines. *Br. J. Pharmaceut. Pract.* **6**, 6–14(Part 1), 42–8(Part 2).

HALSEY, J.P. AND CARDOE, N. (1982). Benoxaprofen: side-effect profile in 300 patients. *Br. med. J.* **284**, 1365–8.

HARMAN, J.B. (1971). Prescribing for the elderly. *Prescribers' J.* **11**, 142–5.

HARTE, V.J. AND TIMONEY, R.F. (1980). Drug prescribing in paediatric medicine. *Irish med. J.* **73**, 157–61.

HESS, J., ANDERSEN, T., NIELSEN, I.K., CHRISTIANSEN, L.V., FOLKENBERG, F., ØSTERLIND, A.W., OLSEN, L., KAMPMANN, J.P., AND HANSEN, J.E.M. (1979). Drug consumption and adverse reactions in a department of internal medicine. *Ugeskr. Laeg.* **141**, 174–7.

HINDSON, C., DAYMOND, T., DIFFEY, B., AND LAWLOR, F. (1982). Side effects of benoxaprofen. *Br. med. J.* **284**, 1368–9.

HODDINOTT, B.C., GOWDEY, C.W., COULTER, W.K., AND PARKE, J.M. (1967). Drug reactions and errors in administration on a medical ward. *Can. med. Ass. J.* **97**, 1001–6.

HOVI-VIANDER, M. (1980). Death associated with anaesthesia in Finland. *Br. J. Anaesth.* **52**, 483–9.

HURWITZ, N. (1969a). Predisposing factors in adverse reactions to drugs. *Br. med. J.* **1**, 536–9.

—— (1969b). Admissions to hospitals due to drugs. *Br. med. J.* **1**, 539–40.

—— AND WADE, O.L. (1969). Intensive hospital monitoring of adverse reactions to drugs. *Br. med. J.* **1**, 531–6.

HUTCHEON, A.W., LAWSON, D.H., AND JICK, H. (1978). Hospital admissions due to adverse drug reactions. *J. clin. Pharmacol.* **3**, 219–24.

HYLAND, M. (1983). Prescribing for the elderly. *Irish med. J.* **76**, 209–12.

IREY, N.S. (1976). Adverse drug reactions and death. A review of 827 cases. *J. Am. med. Ass.* **236**, 575–8.

JOURNAL OF THE AMERICAN MEDICAL ASSOCIATION (1976). Editorial. Adverse drug reactions and associated deaths. *J. Am. med. Ass.* **236**, 592.

JICK, H. (1974). Drugs — remarkably nontoxic. *New Engl. J. Med.* **291**, 824–8.

——, MIETTINEN, O.S., SHAPIRO, S., LEWIS, G.P., SUSKIND, V., AND

SLONE, D. (1970). Comprehensive drug surveillance. *J. Am. med. Ass.* **213**, 1455-60.

——, SLONE, D., BORDA, I.T., AND SHAPIRO, S. (1968). Efficacy and toxicity of heparin in relation to age and sex. *New Engl. J. Med.* **279**, 284-6.

KLEIN, U., KLEIN, M., STURM, H., ROTHENBÜHLER, M., HUBER, R., STUCKI, P., GIKALOV, I., KELLER, M., AND HOIGNE, R. (1976). The frequency of adverse drug reactions as dependent upon age, sex and duration of hospitalization. *Int. J. clin. Pharmacol.* **13**, 187-95.

KNILL-JONES, R.P., RODRIGUES, L.V., MOIR, D.D., AND SPENCE, A.A. (1972). Anaesthetic practice and pregnancy: Controlled survey of women anaesthetists in the United Kingdom. *Lancet* **i**, 1326-8.

LAMY, P.P. (1982). The elderly, drugs and cost control. *Drug Intel. clin. Pharm.* **16**, 768-71.

LARKIN, J., PULLAR, T., AND MASON, D. (1982) Side effects of benoxaprofen. *Br. med. J.* **284**, 1784.

LAW, R. AND CHALMERS, C. (1976). Medicines and elderly people: a general practice survey. *Br. med. J.* **1**, 565-8.

LAWSON, D.H., HUTCHEON, A.W., AND JICK, H. (1979). Life threatening drug reactions amongst medical in-patients. *Scot. med. J.* **24**, 127-30.

LEDERMANN, J.A., HOFFMAN, B.I., AND CONTRERAS, M. (1982). Side effects of benoxaprofen. *Br. med. J.* **284**, 1784.

LEVY, M., KLETTER-HEMO, D., NIR, I., AND ELIAKIM, M. (1977). Drug utilization and adverse drug reactions in medical patients. Comparison of two periods, 1969-72 and 1973-76. *Israel J. med. Sci.* **13**, 1065-72.

——, LIPSHITZ, M., AND ELIAKIM, M. (1979). Hospital admissions due to adverse drug reactions. *Am. J. med. Sci.* **277**, 49-56.

——, MERMELSTEIN, L., AND HEMO, D. (1982). Medical admission due to noncompliance with drug therapy. *Int. J. clin. Pharmacol. Ther. Toxicol.* **20**, 600-4.

LUNDIN D.V. (1978). Medication taking behavior of the elderly: a pilot study. *Drug Intel. clin. Pharm.* **12**, 518-22.

——, EROS, P.A., MELLOH, J., AND SANDS, J.E. (1980). Education of independent elderly in the responsible use of prescription medications. *Drug Intel. clin. Pharm.* **14**, 335-42.

MACDONALD, M.G., AND MACKAY, B.R. (1964). Adverse drug reactions: Experience of Mary Fletcher Hospital during 1962. *J. Am. med. Ass.* **190**, 1071-4.

MACH, E.P. (1975). Counting the cost of adverse drug reactions. *Adv. Drug React. Bull.* **54**, 184-7.

MARSDEN, J.R. AND DAHL, M.G.C. (1982). Side effects of benoxaprofen. *Br. med. J.* **284**, 1782-3.

MARTYS, C.R. (1979). Adverse reactions to drugs in general practice. *Br. med. J.* **2**, 1194-7.

—— (1981). Monitoring adverse effects of drugs in the community. *Practitioner* **225**, 619-21.

MAY, F.E., STEWART, R.B., HALE, W.E., AND MARKS, R.G. (1982). Prescribed and nonprescribed drug use in an ambulatory elderly population. *South. med. J.* **75**, 522-8.

MCKENNEY, J.M. AND HARRISON, W.L. (1976). Drug-related hospital admissions. *Am. J. hosp. Pharm.* **33**, 792-5.

MCKENZIE, M.W., MARCHALL, G.L., NETZLOFF, M.L., AND CLUFF, L.E. (1976). Adverse drug reactions leading to hospitalization in children. *J. Pediat.* **89**, 487-90.

——, STEWART, R.B., WEISS, C.F., AND CLUFF, L.E. (1973). A pharmacist-based study of the epidemiology of adverse drug reactions in pediatric medicine patients. *Am. J. hosp. Pharm.* **30**, 898-93.

MCQUEEN, E.G. (1981). New Zealand committee on adverse drug reactions: fifteenth annual report 1980. *NZ med. J.* **93**, 194-8.

—— (1982). New Zealand Committee on adverse drug reactions: sixteenth annual report 1981. *NZ med. J.* **95**, 230-3.

MILLER, R.R. (1973). Drug surveillance utilizing epidemiologic methods: A report from the Boston Collaborative Drug Surveillance Program. *Am. J. hosp. Pharm.* **30**, 584-92.

—— AND GREENBLATT, D.J. (Eds.) (1976). *Drug effects in hospitalized patients.* Wiley and Sons, New York.

MINISTÈRE DE LA SANTE, MINISTÈRE DU TRAVAIL (Eds.) (1975). *Santé et Sécurité social.* La Documentation Française, Paris.

MORGAN, S.H. AND BEHN, A.R. (1981). Association between Stevens–Johnson syndrome and benoxaprofen. *Br. med. J.* **283**, 144.

OGILVIE, R.I. AND RUEDY, J. (1967). Adverse reactions during hospitalization. *Can. med. Ass. J.* **97**, 1445-50, 1450-57.

OKSAS, R.M. (1978). Epidemiologic study of potential adverse drug reactions in dentistry. *Oral Surg.* **45**, 707-13.

O'MALLEY, K., CROOKS, J., DUKE, E., AND STEVENSON, I.H. (1971). Effect of age and sex on human drug metabolism. *Br. med. J.* **3**, 607-9.

—— AND O'BRIEN, E. (1980). Management of hypertension in the elderly. *New Engl. J. Med.* **302**, 1397-1401.

OUSLANDER, J.G. (1981). Drug therapy in the elderly. *Ann. intern. Med.* **95**, 711-22.

PAULUS, H.E. (1983). FDA Arthritis Advisory Committee Meeting. *Arthritis Rheumatism* **26**, 1288-9.

PENN, R.G. AND GRIFFIN, J.P. (1982). Adverse reactions to nitro-furantoin in the United Kingdom, Sweden and Holland. *Br. med. J.* **284**, 1440-2.

PHAROAH, P.O.D., ELBERMAN, E., DOYLE, P., AND CHAMBERLAIN, G. (1977). Outcome of pregnancy among women in anaesthetic practice. *Lancet* **i**, 34-6.

PORTER, J. AND JICK, H. (1977). Drug-related deaths among medical inpatients. *J. Am. med. Ass.* **237**, 879-81.

—— AND —— (1980). Amoxicillin and ampicillin: rashes equally likely. *Lancet* **i**, 1037.

PORTNOI, V.A. AND PAWLSON, L.G. (1981). Abuse of diuretic therapy. *J. chron. Dis.* **34**, 363-5.

PRESCOTT, L.F., LESLIE, P.J., AND PADFIELD, P. (1982). Side effects of benoxaprofen. *Br. med. J.* **284**, 1783.

PROKSCH, R.A. AND LAMY, P.P. (1977). Sex variation and drug therapy. *Drug Intel. clin. Pharm.* **11**, 398-406.

RAY, W.A., FEDERSPIEL, C.F., AND SCHAFFNER, W. (1980). A study of antipsychotic drug use in nursing homes. Epidemiologic evidence suggesting misuse. *Am. J. publ. Hlth.* **70**, 485-91.

READ, A.E. (1979). *Liver and biliary disease.* W.B. Saunders Co, London.

REIDENBERG, M.M. (1968). Registry of adverse drug reactions. *J. Am. med. Ass.* **203**, 31-4.

RITSCHEL, W.A. (1980). Disposition of drugs in geriatric patients. *Pharm. Int.* **1**, 226-30.

ROYAL COLLEGE OF PHYSICIANS (1984). Medication for the elderly. A report of the Royal College of Physicians. *J. R. Coll. Physcns, London* **18**, 7-17.

SALZMAN, C. (1981). Psychotropic drug use and polypharmacy in a general hospital. *Gen. hosp. Psychiat.* **3**, 1-9.

SCHIMMEL, E.M. (1964). The hazards of hospitalization. *Ann. intern. Med.* **60**, 100-10.

SEIDL, L.G., FRIEND, D., AND SADUSK, J. (1966a). Meeting the problem. Panel discussion of experiences and problems involved in reporting adverse drug reactions. *J. Am. med. Ass.* **196**, 421–8.

——, THORNTON, G.F., SMITH, J.W., AND CLUFF, L.E. (1966*b*). Studies on the epidemiology of adverse drug reactions, **III**. Reactions in patients on a General Medical Service. *Bull. Johns Hopk. Hosp.* **119**, 299–315.

SHAW, P.G. (1982). Common pitfalls in geriatric drug prescribing. *Drugs* **23**, 324–8.

SHEHADI, W.H. AND TONIOLO, G. (1980). Adverse reactions to contrast media. A report from the Committee on Safety of Contrast Media of the International Society of Radiology. *Radiology* **137**, 299–302.

SMITH, C., EBRAHIM, S., AND ARIE, T. (1983). Drug trials, the 'elderly', and the very aged. *Lancet* **ii**, 1139.

SMITH, J.W., SEIDL, L.G., AND CLUFF, L.E. (1966). Studies on the epidemiology of adverse drug reactions, V. Clinical factors influencing susceptibility. *Ann. intern. Med.* **65**, 629–40.

SPINO, M., SELLERS, E.M., AND KAPLAN, H.L. (1978). Effect of adverse drug reactions on the length of hospitalization. *Am. J. hosp. Pharm.* **35**, 1060–4.

STEEL, K., GERTMAN, P.M., CRESCENZI, C., AND ANDERSON, J. (1981). Introgenic illness on a general medical service at a university hospital. *New Engl. J. Med.* **304**, 638–42.

STEWART, R.B. AND CLUFF, L.E. (1971). Studies on the epidemiology of adverse drug reactions, VI. Utilization and interactions of prescription and non-prescription drugs in outpatients. *Johns Hopkins med. J.* **129**, 319–31.

—— AND —— (1974). Gastrointestinal manifestations of adverse drug reactions. *Dig. Dis.* **19**, 1–7.

——, HALE, W.B., AND MARKS, R.G. (1982). Analgesic drug use in an ambulatory elderly population. *Drug Intel. clin. Pharm.* **16**, 833–6.

——, ——, AND MARKS, R.G. (1984). Drug use and adverse drug reactions in an ambulatory elderly population: A review of the Dunedin Program. *Pharm. Int.* **5**, 149–52.

STRICKLAND-HODGE, B. AND JEQSON, M.H. (1980). Usage of information sources by general practitioners. *J. R. Soc. Med.* **73**, 857–62.

SUOMI, J.D., HOROWITZ, H.S., AND BARBANO, J.P. (1975). Self-reported systemic conditions in an adult study population. *J. dent. Res.* **54**, 1092.

TAGGART, H.MC.A. AND ALDERDICE, J.M. (1982). Fatal cholestatic jaundice in elderly patients taking benoxaprofen. *Br. med. J.* **284**, 1372.

TAYLOR, A.E.M., GOFF, D., AND HINDSON, T.C. (1981). Association between Stevens-Johnson syndrome and benoxaprofen. *Br. med. J.* **282**, 1433.

TURNBULL, K.W., FANCOURT-SMITH, P.F., AND BANTING, G.C. (1980). Death within 48 hours of anaesthesia at the Vancouver General Hospital. *Can. anaesthetic Soc. J.* **27**, 159–63.

TRUNET, P., LE GALL, J.R., L'HOSTE, F., REGNIER, B., SAILLARD, Y., CARLET, J., AND RAPIN, M. (1980). The role of iatrogenic disease in admissions to intensive care. *J. Am. med. Ass.* **244**, 2617–20.

VESSELL, E.S. (1980). Why are toxic reactions to drugs so often undetected initially? *New Engl. J. Med.* **302**, 1027–8.

WALLACE, D.E. AND WATANABE, A.S. (1977). Drug effects in geriatric patients. *Drug Intel. clin. Pharm.* **11**, 597–603.

WANG, R.I.H. AND TERRY, L.C. (1971). Adverse drug reactions in a Veterans Administration Hospital. *J. clin. Pharmacol.* **11**, 14–18.

WILKINS, M.R., SCOTT, D.L., AND FARR, M. (1982). Side effects of benoxaprofen. *Br. med. J.* **284**, 1782–3.

WILLIAMSON, J. AND CHOPIN, J.M. (1980). Adverse reactions to prescribed drugs in the elderly: a multicentre investigation. *Age Ageing* **9**, 73–80.

WILLS, B.A. (1983). Letter. Flosint (indoprofen) — suspension of product licence. Department of Health and Social Security, London. 16 December 1983.

WISER, T.H., IRELAND, G.A., AND KUSHNER, H.A. (1978). Adverse drug reactions causing hospitalization. *Clin. Res.* **26**, 596A.

WORLD HEALTH ORGANIZATION (1981). Use of medicaments by the elderly. *Drugs* **22**, 279–94.

WYATT, R. AND MURRAY WILSON, A. (1973). Children of anaesthetists. *Br. med. J.* **1**, 675.

YOSSELSON-SUPERSTINE, S. AND WEISS, T. (1982). Drug related hospitalization in pediatric patients. *J. clin. hosp. Pharm.* **7**, 195–203.

YOUNGMAN, P.R., TAYLOR, K.M., AND WILSON, J.D. (1983). Anaphylactoid reactions to neuromuscular blocking agents: a commonly undiagnosed condition. *Lancet* **ii**, 597–9.

ZOOK, C.J. AND MOORE, F.D. (1980). High-cost users in medical care. *New Engl. J. Med.* **302**, 996–1002.

5 Pharmacogenetics and iatrogenic disease

J.P. GRIFFIN

In the first edition of this book in 1972 D'Arcy and Griffin wrote that in the consideration of adverse reactions to drugs three factors had to be borne in mind: firstly, the properties of the drug itself; secondly, the uniqueness of the patient being treated; and thirdly, the possibility of interaction between the drug and other drugs or environmental chemicals to which the patient is exposed.

Predictable toxicity is the manifestation of secondary pharmacological actions; this is a hazard that can be well elucidated in the battery of tests to which the drug is subjected during its development stage. In such instances, assuming that the drug has a worthwhile place in therapy, the ratio of dosage of drug to produce the major effect to dosage to produce a secondary (toxic) effect is the real factor which should be considered and not the 'built in' potentiality to the side-effect itself. Nevertheless, certain groups of patients exist who may be at risk from these predictable manifestations of toxicity. These patients are at peculiar risk because of a genetically determined defect of metabolism, or because their metabolism has been impaired by concomitant hepatic disease, or because their excretory function has been reduced by either liver or renal malfunction. In these instances, the drug or its metabolites may rapidly build up to toxic levels in the body, even at normal accepted therapeutic doses. A number of cases where the metabolism of the drug is delayed due to genetically determined enzyme deficiency and where, consequently, the primary pharmacological or secondary toxicological effects of the drug are enhanced can be cited: for example, plasma pseudocholinesterase deficiency or presence of atypical cholinesterases and slow acetylators of isoniazid.

Adverse reactions may follow the use of a drug, and these reactions may be unexpected, in that they are completely unrelated to the known toxicity of the drug. These reactions include hypersensitivity to the agent, in which the patient develops antibodies to the drug. The antigenic factor is usually a combination of drug with body protein. Skin rashes and eruptions are the most common symptoms of this type of allergic reaction, although haemolytic anaemia is not infrequent. Allergic reactions to drugs are not strictly speaking genetically determined diseases, but like many allergic conditions such as asthma and hayfever there may be a family history of similar allergic or idiosyncratic responses to drugs, for example, aspirin or tartrazine (see Chapter 7).

Idiosyncrasy involves a qualitatively abnormal response on the part of the patient to the drug; an example of this is drug-induced porphyria, in which a qualitatively abnormal response of porphyrin metabolism is induced by barbiturates in susceptible subjects (see Chapter 6). Similarly, mepacrine-induced haemolytic anaemia in glucose-6-phosphate dehydrogenase deficient subjects is an idiosyncratic reaction (see Chapter 12).

The term 'pharmacogenetic disorders' was originally coined by Vogel (1959) and was originally limited to hereditary disorders revealed solely by the use of drugs. It now embraces all genetic contributions to the considerable variation that exists in the interaction between man and the pharmacological agents that he uses. The term can therefore be taken to cover the adverse drug reactions that may occur when a particular agent is given to a patient with impaired renal function where this dysfunction is genetically determined, for example, renal polycystic disease.

The scope of this chapter, however, is limited to those pharmacogenetic disorders which present as medical emergencies of one form or other.

Pharmacogenetic disorders of neuromuscular response to therapeutic agents

Malignant hyperpyrexia

Malignant hyperpyrexia (hyperthermia) is a rare pharmacogenetic disease occurring both in man and in a strain of pig, the Landrace Strain. The mode of inheritance is autosomal dominant. In susceptible individuals any potent inhalational anaesthetic agent or any skeletal muscle relaxant causes fever, rigidity, hyperventilation, cyanosis hypoxia, respiratory and metabolic acidosis, hyperphosphataemia with a raised glucose. An initial hyperkalaemia and hypercalcaemia are followed by hypokalaemia and hypocalcaemia. Later there is elevation in the serum of enzymes of cardiac and skeletal muscle origin.

In the series reported by Britt and Kalow (1968) the mean maximum temperature was 108°F (42.2°C) and accompanying this there were signs of increased muscle metabolism, in particular, tachycardia, tachypnoea, sweating, and blotchy cyanosis. The mortality of the condition is high (60–70 per cent).

Malignant hyperpyrexia has been divided into two clinical categories, a rigid and a non-rigid type. Furniss (1970) showed that the age of the patient is significant when attempting to differentiate between the rigid and non-rigid conditions. Those reacting with hypertonus were usually under 20 years of age and the abnormality was probably hereditary, whilst most of the non-rigid cases occurred in patients over 20 years old and appeared to be sporadic.

The susceptibility to develop malignant hyperpyrexia is due to a genetic anomaly, and in the first family in which this was recognized this abnormality was clearly inherited as a dominant characteristic. Subsequently, further examples of the familial occurrence of malignant hyperpyrexia have been described and in a review of 115 cases in 1969, 43 of these had other members in the family who had also been affected. The inherited defect is associated with a mild or even subclinical myopathy, which can be detected in affected individuals and their relatives by finding raised serum-creatine-phosphokinase (CPK) levels. (Britt *et al.* 1969). Isaacs and Barlow (1970) investigated 99 members of a single family with a history of malignant hyperpyrexia. They found that the resting creatine-phosphokinase and -aldolase levels were high in many of these patients, and concluded that the high level of muscle enzyme in serum was evidence of a subclinical myopathy. They also suggested that anaesthetic agents such as halothane and neuromuscular blockers like suxamethonium damaged the abnormal muscle mechanisms and triggered off a fulminating hyperpyrexia, the rigidity following immediately on the administration of suxamethonium or alternatively developing progressively during the halothane anaesthetic. Suxamethonium and halothane most frequently are described as the causative agents, but two cases precipitated by enflurane have been described (Sutherland and Carter 1975; Carpopresso *et al.* 1975).

Denborough *et al.* (1970) described a patient, who had survived malignant hyperpyrexia, and three of his close relatives were found to have very high CPK levels.

Although the patient's muscles seemed normal, two of the three relatives had a mild but definite myopathy, affecting predominantly the lower muscles of the thigh. It seems that malignant hyperpyrexia develops in individuals with a myopathy which is inherited as an autosomal dominant, but which may be subclinical.

Denborough *et al.* (1973) described the electron microscopic structure of a biopsy of the rectus abdominus muscle of a 71-year-old aunt of the propositus in a family where there have been 10 deaths from malignant hyperpyrexia. This elderly lady had shown the features of a myopathy since childhood. The striking histological abnormality in this patient was the presence of cores in 55 per cent of her Type I muscle fibres. This histological appearance in a myopathy has been called central core disease (Shy and Magee 1956; Dubowitz and Pearse 1960). (Core-like structures have also been noted in the muscle fibres of Landrace pigs which show a susceptibility to develop malignant hyperpyrexia.) An *in-vitro* pharmacological study on muscle from this patient showed a markedly abnormal response to halothane. The correlation between the histological appearance of cores in muscle and the pharmacologically abnormal behaviour is still to be defined.

Investigations then turned towards the development of a diagnostic test, when Ellis *et al.* (1973) realized the necessity of screening relatives of patients who had developed malignant hyperpyrexia, so that appropriate precautions could be made if these people ever required an anaesthetic. They described a technique in which a biopsy of the *vastus medialis* muscle was removed under local anaesthesia. The motor point was identified by electrical stimulation and the specimens were taken to include the motor innervation. The muscle specimens were then set up in an isolated organ bath and exposed first to halothane and then to halothane plus suxamethonium. Four out of seven specimens taken developed contracture to halothane alone, and a further specimen developed a contracture to halothane in the presence of suxamethonium.

The diagnosis of patients prone to the non-rigid type of hyperpyrexia was somewhat simpler. These patients can be distinguished from normals, not on the basis of abnormal serum enzymes, as with the rigid type, but oddly enough on the failure of intra-arterial suxamethonium to produce a neuromuscular block. (See also Chapter 33.)

Drug-induced neuromuscular block due to suxamethonium

Suxamethonium (succinylcholine, diacetycholine) was introduced in 1949 by Bovet and his colleagues. It is a potent neuromuscular blocking agent of the depolarizing type with very brief duration. This is because suxamethonium is rapidly hydrolysed and inactivated by pseudocholinesterase present in the serum. It is used when only a brief period of muscle relaxation is needed or it may be given by intravenous drip to produce prolonged neuromuscular blockade.

Suxamethonium is a very useful muscle relaxant, although there have been reports since its early use of prolonged apnoea and muscular pain and stiffness in some cases. Bourne *et al.* (1952) and Evans *et al.* (1952, 1953) drew attention to the prolonged apnoea in man after suxamethonium in the presence of low plasma serum pseudocholinesterase. Lehmann and Ryan (1956) observed a familial incidence of low pseudocholinesterase in the absence of overt disease, and Lehmann and Simmons (1958) described suxamethonium apnoea in two brothers who had low plasma-pseudocholinesterase levels and both had, one of them repeatedly, a prolonged muscular paralysis following the injection of suxamethonium. Kaufman *et al.* (1960) recorded the first observation of suxamethonium apnoea in an infant; the family was studied in detail and there was a deficiency of pseudocholinesterase in one paternal and in one maternal grandparent. The parents had normal values for pseudocholinesterase, although these were in the lower end of the normal range.

Telfer *et al.* (1964) described seven patients who possessed an atypical pseudocholinesterase; six of these patients presented as prolonged response to suxamethonium. The familial incidence of the occurrence of an atypical cholinesterase was demonstrated in each of these cases. People who are abnormally sensitive to suxamethonium have been shown to have an atypical form of pseudocholinesterase in their serum (Kalow 1956, 1959). This atypical enzyme hydrolyses suxamethonium at a much slower rate than does the normal, or usual type of serum-cholinesterase, and consequently the apnoea which the drug induces is excessively prolonged.

Patients with an atypical form of pseudocholinesterase do not present any other recognizable abnormality, and they usually present as cases of prolonged apnoea following a single injection of suxamethonium. A technique was, however, developed to detect the presence of this atypical enzyme (Kalow and Genest 1957; Kalow and Staron 1957) by determining what is known as the 'dibucaine number'. This is the percentage inhibition of enzyme activity produced by the inhibitor dibucaine under certain standardized conditions. In the test, the activity of the enzyme (usual pseudocholinesterase, or atypical pseudocholinesterase) in the presence and in the absence of dibucaine (10^{-5}M) is measured by following the rate of hydrolysis of the substrate benzoylcholine

(5×10^{-5}M) spectrophotometrically at 240 μm. The reaction is carried out at pH 7.4 in phosphate buffer.

With this technique most people show a dibucaine number of about 80 (normal homozygotes), those with atypical pseudocholinesterase may be classified as dubucaine number 40–75 (heterozygotes) or dibucaine number 30 or below (abnormal homozygotes). The intermediate group (heterozygotes) have a mean dibucaine number of about 62 and are believed to synthesize both the 'usual' and 'atypical' forms of the enzyme. They do not generally respond to suxamethonium abnormally.

Evidence, either experimental or clinical, has been presented for each, or for combinations of, these four theories. Current opinion is that prolonged apnoea after suxamethonium is due to quantitatively or qualitatively deficient pseudocholinesterase. Qualitative defects of pseudocholinesterase activity are due to genetic abnormalities. Four allelic genes seem to control the inheritance of pseudocholinesterase; one normal, two atypical, and one silent allelic gene. They form pseudocholinesterase with varying activity.

Quantitative reduction of pseudocholinesterase occurs in patients who received organophosphorous compounds like the antineoplastics, cyclophophamide, and thiotepa (N, N', N"-triethylenethiophosphoramide). Ecothiopate iodide, an anticholinesterase miotic used in the treatment of glaucoma, also reduced pseudocholinesterase levels, and *Trasylol*, a polypeptide inactivator of kallikrein, obtained from animal sources, is also reported to cause prolonged apnoea (Chasapakis and Dimas 1966).

Prolonged apnoea after suxamethonium is still a serious complication and the treatment of prolonged apnoea depends mostly on good ventilation and maintaining physiological levels of pH, CO_2 and bicarbonate. Administration of whole blood or plasma has been recommended. When suxamethonium is present for a long time, or has been given in large doses, a dual neuromuscular block occurs in which the depolarizing block changes to a non-depolarizing type (i.e. curare-like); in this state edrophonium chloride or neostigmine methylsulphate may be helpful.

Drug-induced haemolytic anaemia in patients with inherited erythrocyte abnormalities

The lifespan of the erythrocyte in a normal subject varies between 110 and 125 days; in haemolytic anaemias the red cells are destroyed more rapidly and the average lifespan of the erythrocyte is correspondingly shorter. It has been known for many years that when certain ordinarily harmless drugs are administered to some individuals in ordinary therapeutic doses this is accompanied by rapid red-cell destruction and the development of acute haemolytic anaemia.

Glucose-6-phosphate dehydrogenase (G6PD) deficiency

The most typical clinical manifestation of glucose-6-phosphate dehydrogenase (G6PD) deficiency is a hae-molytic crisis precipitated by drugs (or by food or infections). Mild episodes may be recognized only by the incidental discovery of haemolytic anaemia in a patient receiving treatment for an unrelated condition. Neonatal hyperbilirubinaemia leading to kernicterus, occurs in G6PD-deficient infants in Africa, China, and the Mediterranean region after exposure of the child or mother to certain drugs or to naphtalene, and at times even in the absence of such exposure.

The initial work identifying the mechanism of certain drug-induced haemolytic anaemias with deficiency of the enzyme G6PD was conducted in the US Army Malaria Research Unit at Stateville Penitentiary, Illinois, where the mechanism of the haemolytic anaemia induced by the antimalarial drug primaquine was investigated. When a daily dose of primaquine (30 mg) was given to a sensitive individual, there was no untoward reaction for about three days. Then the urine began to darken, the haemo-globin level, red cell count, and haematocrit fell. In the more severe type of reaction immediate transfusion may become necessary. If, in spite of these reactions pri-maquine was continued, the haemolytic symptoms subsided after about a week. After about a month red-cell production had speeded up and the disease was 'self-limited'.

If primaquine was stopped for a month or two and then resumed, a new haemolytic episode similar to the initial one occurred. Once a 'self-limiting' stage had been reached, an increase in dose induced further haemolysis for a period until a new self-limiting phase was achieved. The explanation for the haemolysis followed by a recovery was demonstrated using red cell tagged with radioactive ^{59}Fe. By administering radioactive iron ^{59}Fe for a short period, newly produced red cells marked with this tracer were produced in susceptible individuals. A two-week course of primaquine given to these subjects produced the expected haemolytic episode but the young labelled red cells (8–21 days of age) were not destroyed. However, if primaquine was given later when the labelled red cells were 55 days old, all the labelled red cells were destroyed.

It was demonstrated that the haemolytic episodes only occurred in subjects whose whole red cells were deficient

Table 5.1   Incidence of glucose-6-phosphate dehydrogenase deficiency in Arab and African subjects (Vello and Ibrahim 1962)

Country	Incidence per cent
Congo	3–30
East Africa and Tanganyika	2–28
Gambia	15
Ghana	24
Nigeria	6–17
Northern Nigeria	21
Palestine (Arab)	3.4
South Africa (Bantu)	3
Sudan (Arab)	
adults	8.1
newborn (umbilical cord)	7.3

in the enzyme G6PD and that this deficiency in G6PD was more marked as the red blood cells aged. In those countries where malaria is endemic, prophylaxis of malaria and treatment of the condition are an every-day problem, and knowledge of the incidence of G6PD deficiency in these populations was determined in epi-demiological studies (Vello and Ibrahim 1962) and is shown in, Table 5.1. It was also found that G6PD defi-ciency existed in various forms and that a wide range of drugs could initiate haemolytic episodes in G6PD-defi-cient subjects (see Table 5.2) and that the severity of the haemolytic response was affected by the severity of the enzyme deficiency.

African type (A−)

The A− type of G6PD deficiency is characterized by mild enzyme deficiency (mean enzyme activity 8–20 per cent of normal) and high electrophoretic mobility. The youngest red cells have normal or almost normal enzyme levels, and red cells younger than 50 days have sufficient enzyme activity to be protected against damage by hae-molytic drugs. Thus, only the old cells are susceptible to destruction and, even if the offending drug continues to be administered, haemolysis will be self-limited. The risk of potentially fatal haemolysis is therefore less than with the Mediterranean type of enzyme deficiency and fewer drugs are potentially toxic (Table 5.2).

Mediterranean type

The Mediterranean type of G6PD deficiency is charac-terized by severe enzyme deficiency (0–4 per cent enzyme activity). Identification requires further enzyme charac-terization. Enzyme deficiency affects even the younger red cells, and haemolytic episodes are therefore not self-

Table 5.2 Drugs reported to induce haemolysis in subjects with G6PD deficiency

Drug	Haemolysis † Black subjects	Haemolysis † Caucasian subjects
Acetanilide	+ + + + +	
Dapsone	+ +	+ + +
Furazolidone	+ +	
Furaltadone	+ +	
Nitrofural	+ + + +	
Nitrofurantoin	+ +	+ +
Sulphanilamide	+ + +	
Sulphapyridine	+ + +	+ + +
Sulphacetamide	+ +	
Salazosulphapyridine	+ + +	
Sulphamethoxypyridazine	+ +	
Thiazosulphone	+ +	
Quindine		+ +
Primaquine	+ + +	+ + +
Pamaquine	+ + + +	
Pentaquine	+ + +	
Quinocide	+ + +	+ +
Naphthalene	+ + +	+ + +
Neoarsphenamine	+ +	
Phenylhydrazine	+ + +	
Toluidine blue	+ + + +	
Trinitrotoluene		+ + +

† Severity of the haemolytic event is indicated by the number of + on a five-point scale.

limited. Haemolysis is thus more severe and more often life-threatening, and more drugs are potentially harmful (Table 5.2).

Other common types of severe G6PD deficiency

The variants of G6PD deficiency that are common in East and South-East Asia (e.g. the variants Canton and Union) differ from the Mediterranean and A⁻ types. The changes in enzyme activity and the clinical implicants with regard to haemolysis have not yet been determined with these variants. It is likely that some at least may be as severe as the Mediterranean types of deficiency.

Much more work is required to characterize these G6PD types, and to assess the susceptibility to haemolysis of red cells containing them before their public health importance can be fully evaluated. Until more information becomes available, variants that have not yet been defined should be considered a potential risk if they are associated with an enzyme activity of less than 10 per cent. It should be stressed, however, that there is no complete correlation between enzyme activity, as estimated by standardized *in-vitro* techniques and clinical severity.

Glucose-6-phosphate dehydrogenase deficiency is known to be a sex-linked trait. In the male the gene is associated with the X chromosome derived from the mother. Males with G6PD deficiency have a single enzyme-deficient red-cell population and homozygous females also have a single red-cell population. Heterozygous females are more frequently encountered in a population than deficient males, and they have two red-cell populations, one normal and one deficient. The ratio of the two populations varies widely, with a mode of 50:50; in rare cases the proportion of abnormal cells may be as low as 1 per cent or as high as 99 per cent. Only abnormal cells are drug-sensitive, and in most heterozygous females drug-induced haemolysis is mild, since in the typical heterozygous female only half of the cells are enzyme-deficient. Only about one-third of all heterozygous females have a high enough proportion of abnormal cells to predispose them to clinically significant haemolysis.

Instances of haemolytic anaemia following administration of customary doses of these drugs are not limited to patients with inherited G6PD deficiency, but also applies to other inherited red-cell abnormalities. Haemolytic anaemia may develop on exposure to oxidant drugs in persons who, for example, inherit erythrocyte-glutathione (GSH)-reductase deficiency, erythrocyte-(GSH)-deficiency or the haemoglobinopathy associated with haemoglobin-Zurich, also haemoglobin-H disease and possibly other unstable haemoglobins may be implicated.

Haemolytic anaemia due to 7-gammaS globulin during administration of pyramidone has been reported.

Disturbances of porphyrin metabolism

The porphyrias are a group of disease mainly hereditary in origin, which have many different symptoms. In some the only problem is an undue sensitivity to sunlight — porphyria cutanea tarda (PCT); in others the normal life of the patient may be shattered by devastating attacks of abdominal pain, paralysis of limbs and profound mental upset — acute intermittent porphyria (AIP). These diseases have one thing in common — a marked disturbance of porphyrin metabolism, of which the striking clinical impression is often the passage of dark red urine (see Chapter 6).

Topical steroids, intraocular pressure, and glaucoma

Repeated topical application of glucocorticoids to the eye is followed by an increase in intraocular pressure

(Armaly 1968; Schwartz *et al.* 1972). The extent to which this happens depends upon the age of the subject and his genetic make-up. The European populations studied show a trimodal distribution: 66 per cent with low, 29 per cent with intermediate, and 5 per cent with high pressure changes in response to the drug. Family studies suggest that the response is influenced by two alleles P^L and P^H, with the three groups represented by the genotypes P^LP^L, P^LP^H, and P^HP^H respectively. The raised intraocular pressure is totally reversible by withdrawal of the drug.

The risk of developing open-angle hypertensive glaucoma 'spontaneously' is greatly increased in people with P^HP^H genetic make-up, and although the other genotypes are found in cases of glaucoma, the P^LP^L group is much smaller than would be expected for the normal population.

Fast and slow acetylation: example of genetic polymorphism

Many drugs are acetylated during their metabolism, for example hydrallazine, isoniazid, phenelzine, and sulphonamides, and people vary in their speed of acetylaton; this variation is genetically linked. The normal adult population may be divided into two approximately equal groups, the slow and the rapid acetylators, although there are racial differences in the incidence of slow acetylators in the population, for example 5 per cent of Eskimos are slow acetylators, 15 per cent of Chinese, 45 per cent of American blacks, 45 per cent of Europeans, 60 per cent of Indians, and 55–75 per cent of Jews.

This acetylation polymorphism is an excellent example of a pharmacogenetic phenomenon which is clinically relevant. Slow acetylators have higher plasma concentrations of the unchanged drug and a greater proportion of it in the urine than do rapid acetylators. They are more prone to experience severe side-effects during treatment with such drugs. On the other hand, the rapid acetylators often respond less favourably to treatment because they have lower blood levels of the drug and suffer from inadequate dosage.

One of the most important aspects of genetic variation in aceylator phenotype is associated with the use of the antitubercular drug, isoniazid.

Isoniazid

Isoniazid, a drug used in the treatment of tuberculosis, is usually taken by mouth and is completely and rapidly absorbed. Some patients were shown to achieve higher blood levels of isoniazid than others. It was shown that the differences in blood levels were due to differences in the speed of acetylation and inactivation of isoniazid. It was shown that biopsy specimens from the livers of 'rapid inactivators' of isoniazid had higher levels of hepatic N-acetyl transferase than biopsy specimens obtained from 'slow inactivators'. The levels of the N-acetyl transferase are genetically determined, and patients may be classified as 'slow inactivators' (autosomal homozygous recessive) or 'rapid inactivators' (heterozygous and homozygous dominants).

The relevance of the speed of acetylation of isoniazid in respect to therapy is twofold: firstly, to achieve adequate therapeutic levels of the drug to treat the tuberculosis in the rapid inactivator; secondly, to identify the slow inactivator who is more prone to the toxicity of isoniazid due to reduced ability to inactivate the drug. Slow inactivators are more prone to develop peripheral neuritis due to isoniazid toxicity.

Adams and White (1965) described a patient with encephalopathy following isoniazid; this was followed by a report of a further eight cases in a Tuberculosis Chemotherapy Centre at Madras (Devadatta 1965). Six of these eight patients developed convulsions of the grand mal type, the remaining two had symptoms of a toxic psychosis. Of these eight patients, six (five with convulsions and one with psychosis) also developed signs of peripheral neuropathy. All eight of these patients were slow inactivators of isoniazid. Evans *et al.* (1960) suggested that slow inactivators of isoniazid are more prone to develop neurological complications than rapid inactivators. This was well supported by Devadatta's (1965) experience of 43 cases of neurotoxicity caused by isoniazid where 36 patients (83.7 per cent) were classified as slow inactivators.

Obviously there was a need to develop techniques to pick out these patients who were at special risk in terms of drug toxicity. Rao and associates (1970) undertook sulphadimidine acetylation tests as a basis for classification of patients as slow or rapid inactivators of isoniazid. A total of 103 patients were examined in these tests; 52 have been previously classified as slow and 51 as rapid inactivators of isoniazid by a standard microbiological assay method. Each patient received sulphadimidine by mouth (44 mg/kg), and free and total sulphadimidine were estimated in blood and urine collected at six hours.

The results of these investigations suggested that a patient's status as an isoniazid acetylator can be predicted from an examination of his ability to acetylate sulphadimidine. For example, patients can be classified as slow inactivators of isoniazid if the proportion of acetylated sulphadimidine (total minus-free) is (a) less than 25 per cent in blood or (b) less than 70 per cent in urine. The sulphadimidine test was easy to perform and results could be determined on the same day as the test.

Work along similar lines to determine the acetylator phenotype was conducted by Schroder (1972) who administered sulphadimidine (10 mg/kg) to 50 healthy subjects, followed a week later by sulphapyridine (10 mg/kg), and after a further week, sulphasalazine (2–4 g). The proportion of subjects excreting less than 65 per cent of these sulphonamides were classified as slow inactivators. Complete uniformity in classification into slow or rapid acetylators was achieved with all three of the drugs. The technique used was reported to be rapid and urine from 15–20 patients could be screened per hour.

Other drugs

Slow acetylation has been implicated in adverse reactions to other drugs; for example, slow acetylators are more prone to experience severe side-effects during treatment with MAO-inhibtors, such as phenelzine.

Hydrallazine fell from favour temporarily as a hypotensive drug, partly on account of its liability to induce a systemic lupus reaction; one investigation of 12 such cases showed that all the patients were slow acetylators.

Slow acetylators on combined PAS and isoniazid therapy are extremely sensitive to phenytoin, and if treatment with these antitubercular drugs has to be given to such an epileptic then a reduced dosage of phenytoin will be required.

Oxidative genetic polymorphism

Mono-oxidation represents the single most important step in hepatic metabolism, taking place in the microsomal P-450 system. In recent years heterogeneity of this oxidative system has been described. The first to be documented related to the genetic polymorphism of debrisoquine hydroxylation. The phenotyping procedure consists of the administration of a single 10-mg dose of debrisoquine orally, followed by the collection of an 8-hour urine sample. In the urine, unchanged debrisoquine and its metabolite 4-OH debrisoquine are measured using a gas chromatographic method and the results expressed as a ratio of debrisoquine: 4-OH debrisoquine. The ratio is small in individuals capable of hydroxylating the drug efficiently and is large in poor metabolizers. In the general Caucasian population the distribution is bimodal with 90 per cent being extensive metabolizers and 10 per cent being poor metabolizers, the poor metabolizers being virtually unable to oxidize are forced to dispose of the compound which is therefore liable to accumulate.

Table 5.3 Adverse reactions due to poor microsomal oxidation

Adverse reaction	Drugs concerned
Lactic acidosis	Phenformin
	Metformin
Agranulocytosis	Carbimazole
	Phenylbutazone
	Chlorpromazine
	Nortriptiline
	Imipramine
	Thioridazine
	Captopril
Neuropathy or sensory disturbances	Perhexiline
	Phenytoin
Hepatic adenoma	Oral contraceptives
Cerebellar signs	Perhexiline
	Phenytoin
Cirrhosis	Perhexiline
Vitamin-D-deficiency-like state	Phenytoin
	Phenobarbitone
Folate-deficiency-like state	Phenytoin
	Phenobarbitone
Syncope	Prazosin
Malignant ventricular arrhythmias	Mexiletine
	Disopyramide
	Prenylamine
	Perhexiline
Bradycardia	Propranolol
	Metoprolol

* Source: Smith and Shah (1984, personal communication) and Smith and Idle (1981).

The same oxidative system as determined using debrisoquine as a metabolic probe is also known to control the metabolism of various other drugs and thus to have a bearing on a number of drug-induced disorders.

Table 5.3 shows those adverse reactions which are considered by Professor R.I. Smith and Rashmi Shah of St. Mary's Hospital Medical School to have a metabolic basis which are attributable to inter-individual differences in oxidation status.

It is known that in the UK some 8–10 per cent of individuals are poor hydroxylators of debrisoquine, and in Scandinavia some 2–3 per cent are poor hydroxylators. In recent years the heterogeneity of the microsomal cytochrome P-450 oxidative system has been further investigated. Three further mixed oxidase function defects which differ from the debrisoquine hydroxylation, poor and extensive hydroxylator phenotypes have been identified. These relate to the polymorphism in the oxidation of mephenytoin, carbocysteine, and tolbutamide; each of these systems have poor metabolizers and extensive metabolizers, with relative incidences of poor metabolizers at 6–10 per cent the population in Europe and North America.

The inheritance of 'poor metabolizer' status for debrisoquine, mephenytoin, carbocysteine, and tolbutamide is believed to be autosomal recessive and independently inheritable for each of the four systems.

RECOMMENDED FURTHER READING

PREISIG, R. (1983). Pharmacogenetics. *Pharm. Int.* **4**, 314–17.
WHO (1973). Pharmacogenetics. *Tech. Rep* Series no. 524. WHO, Geneva.

REFERENCES

ADAMS, P. AND WHITE, C. (1965). Isoniazid induced encephalopathy. *Lancet* **i**, 680–2.

ARMALY, M.F. (1968). Genetic factors related to glaucoma. *Ann. NY Acad. Sci.* **151**, 861–75.

BOURNE, J.G., COLLIER, H.O.J., AND SOMERS, G.F. (1952). Succinylcholine (succinoylcholine) muscle-relaxant of short action. *Lancet* **i**, 1225–9.

BRITT, B.A. AND KALOW, W. (1968). Hyperrigidity and hyperthermia associated with anaesthesia. *Ann. NY Acad. Sci.* **151**, 947–58.

——, LOCHER, W.G., AND KALOW, W. (1969). Hereditary aspects of malignant hyperthermia. *Can. anaesth. Soc. J.* **16**, 89–98.

CARPOPRESSO, P.R., GITTLEMAN, M.A., REILLY, D.J., AND PATERSON, L.J. (1975). Malignant hyperthermia associated with enflurane anaesthesia. *Arch. Surg.* **110**, 1491–3.

CHASAPAKIS, G. AND DIMAS G. (1966). Possible interaction between muscle relaxants and the kallikrein-trypsin inactivator 'Trasylol'. Report of three cases. *Br. J. Anaesth.* **38**, 838–9.

D'ARCY, P.F. AND GRIFFIN, J.P. (Eds.) (1972). *Iatrogenic disease*, 1st edn, Preface. Oxford University Press, Oxford. (See pp. ix–x of this edition.)

DENBOROUGH, M.A., DENNETT, X., AND ANDERSON, R.MCD. (1973). Central core disease and malignant hyperpyrexia. *Br. med. J.* **1**, 272–3.

——, EBELING, P., KING, J.O., AND ZAPF, P. (1970). Myopathy and malignant pyrexia. *Lancet* **i**, 1138–40.

DEVADATTA, S. (1965). Isoniazid-induced encephalopathy. *Lancet* **ii**, 440.

DUBOWITZ, V. AND PEARSE, A.G.E. (1960). Oxidative enzymes and phosphorylase in central-core disease of muscle. *Lancet* **ii**, 23–24.

ELLIS, F.R. (1973). Malignant hyperpyrexia. *Anaesthesia.* **28**, 245–52.

EVANS, D.A.P., MANLEY, K.A., AND MCKUSICK, V.A. (1960). Genetic control of isoniazid metabolism in man. *Br. med. J.* **2**, 485–91.

EVANS, F.T., GRAY, P.W.S., LEHMANN, H., AND SILK, E. (1952). Sensitivity to succinylcholine in relation to serum-cholinesterase. *Lancet* **i**, 1229–30.

——, ——, ——, AND —— (1953). Effects of pseudocholinesterase level on action of succinylcholine in man. *Br. med. J.* **1**, 136–8.

FURNISS, P. (1970). Hyperpyrexia during anaesthesia. *Br. med. J.* **4**, 745.

ISAACS, H. AND BARLOW, M.B. (1970) Malignant hyperpyrexia during anaesthesia: possible association with sub-clinical myopathy. *Br. med. J.* **1**, 275–77.

KALOW, W. (1956). Familial incidence of low pseudocholinesterase level. *Lancet* **ii**, 576–7.

—— (1959). *Ciba Foundation Symposium on the biochemistry of human genetics* (ed. G.E. Wolstenholme and M.C. O'Conner), p.39. CIBA, London.

——, BRITT, B.A., TERREAU, M.E., AND HAIST, C. (1970). Metabolic error of muscle metabolism after recovery from malignant hyperthermia. *Lancet* **ii**, 895–8.

—— AND GENEST, K. (1957). A method for the detection of atypical forms of human serum cholinesterase; determination of dibucaine numbers. *Can. J. Biochem.* **35**, 339–46.

—— AND STARON, N. (1957). On distribution of atypical forms of human serum cholinesterase, as indicated by dibucaine numbers. *Can. J. Biochem.* **35**, 1305–20.

KAUFMAN, L., LEHMANN, H., AND SILK, E. (1960). Suxamethonium apnoea in an infant: expression of familial pseudocholinesterase deficiency in three generations. *Br. med J.* **1**, 166–7.

LEHMANN, H. AND RYAN, E. (1956). Familial incidence of low pseudocholinesterase level. *Lancet* **ii**, 124.

—— AND SIMMONS, P.H. (1958). Sensitivity to suxamethonium: apnoea in two brothers. *Lancet* **ii**, 981–2.

RAO, K.V.N., MITHISON, D.A., NAIR, N.G.K., PREMA, K., AND TRIPATHY, S.P. (1970). Sulphadimidine acetylation test for classification of patients as slow or rapid inactivators of isoniazid. *Br. med J.* **3**, 495–7.

SCHRODER, H. AND EVANS, D.A. (1972). Acetylator phenotype and adverse effects of sulphasalazine in healthy subjects. *Gut* **13**, 278–84.

SCHWARTZ, J.T., REULING, F.H., FEINLIEB, M., GARRISON, R.J., AND COLLIE, D.J. (1972). Twin heritability study of the effect of corticosteroids on intraocular pressure. *J. med. Genet.* **9**, 137–43.

SHY, G.M. AND MAGEE, K.R. (1956). A new congenital non-progressive myopathy. *Brain* **79**, 610–21.

SMITH, R.L. AND IDLE, J.R. (1981). Genetic polymorphism in drug oxidation. In *Drug reactions and the liver* (ed. M. Davis, J.M. Tredger, and R. Williams), pp. 95–104. Pitman Medical, Bath.

SUTHERLAND, F.S. AND CARTER, J.R. (1975). Malignant hyperpyrexia during enflurane anaesthesia. *J. Tenn. med. Ass.* **68**, 785–6.

TELFER, A.B.M., MCDONALD, D.J.F., AND DINWOODIE, A.J. (1964). Familial sensitivity to suxamethonium due to atypical pseudocholinesterase. *Br. med. J.* **1**, 153–6.

VELLO, F. AND IBRAHIM, S.A. (1962). Erythrocyte glucose-6-phosphate deficiency in Khartoum. *Sudan med. J.* **1**, 136–7.

VOGEL, F. (1959). Modern problems of human genetics. *Ergebn. inn. Med. Kinderheil Kd.* **12**, 52–125.

6 Disorders of porphyrin metabolism

A.P. FLETCHER AND J.P. GRIFFIN

The porphyrins are highly coloured red substances which are precursors in the syntheses of haem, haemoglobin, myoglobin, and the cytochromes. The steps of the synthetic pathway are shown in Fig. 6.1. This synthesis can probably occur in cells generally but the most important organs in which it takes place are the bone marrow and liver.

The porphyrias are a group of diseases, mainly hereditary in origin, which have many different symptoms. In some the only problem is an undue sensitivity to sunlight; in others the normal life of the patient may be shattered by devastating attacks of abdominal pain, paralysis of limbs, and profound mental upset. These diseases have one thing in common — a marked disturbance of porphyrin metabolism, of which the striking clinical impression is often the passage of dark red urine. The porphyrias involve, as a primary or secondary biochemical abnormality, a fault in the synthesis of porphyrins with increased urinary and faecal excretion of metabolic intermediates.

Types of porphyria

The two commonest porphyrias in Britain are acute intermittent porphyria and porphyria cutanea tarda, one of the cutaneous hepatic porphyrias.

In acute intermittent porphyria the urine contains very high levels of δ-aminolaevulinic acid and porphobilinogen, both porphyrin precursors, while in porphyria cutanea tarda the urine and plasma contain uroporphyrin with no increase in the precursors. (The excess uroporphyrin found in the urine in acute intermittent porphyria is formed non-enzymically in the acid milieu of the urine.)

It is thought that the high circulating plasma levels of uroporphyrin in porphyria cutanea tarda contributes to the photosensitivity. Patients with acute intermittent porphyria have a striking increase in ALA-synthetase (now known as ALA-synthase), the rate-limiting enzyme in porphyrin synthesis, in their livers. There has been controversy as to the level of this enzyme in the livers of patients with porphyria cutanea tarda, but the overall evidence suggests that it is raised in this condition also. It

seemed likely that some additional biochemical abnormality was present in acute intermittent porphyria preventing the excessive formation of porphyrins in the liver, which always occurs in cutaneous porphyria. Work has been published which gives a rationale for this chemical finding in acute intermittent porphyria (Strand *et al.* 1970). Uroporphyrinogen-I synthetase (urosynthetase) and uroporphyrinogen-III co-synthetase are two enzymes which catalyse the conversion of four molecules of porphobilinogen to uroporphyrinogen III. In acute intermittent porphyria there is decreased activity of urosynthetase in the liver, and it has been suggested that the partial block at this level would not only interfere with the negative-feedback regulation of hepatic ALA-synthetase, leading to overproduction of ALA and porphobilinogen, but also would account for the normal or only slightly increased urinary and faecal porphyrins.

The cutaneous porphyrias

The first description of cutaneous porphyria (porphyria cutanea tarda) was made in Glasgow at the end of the nineteenth century in two brothers who were fishermen from Stornaway. These brothers were extremely sensitive to sunlight and each summer they described itching and burning sensations of the hands and face, followed by blistering. The urine was burgundy-red in colour and contained a great excess of porphyrins.

Skin photosensitivity and a marked increase in porphyrin formation by the liver can occur as a result of certain drugs (Table 6.1.) and other toxic substances. In South Africa many thousands of Bantus have this type of porphyria with evidence of hepatic dysfunction. The main precipitating factor seems to be the drinking of adulterated alcohol brewed and sold in urban areas. In 1956 many thousands of cases of cutaneous porphyria occurred in south-east Turkey, especially among children below the age of 15 years. This was caused by the distribution of seed-wheat dressed with the fungicide hexachlorobenzene, which was used for bread making. Many children died and had evidence of severe mutilation of exposed skin on the face, legs, and hands. They all had very large quantities of porphyrins in their urine and blood. The purely cutaneous form of the disease has a

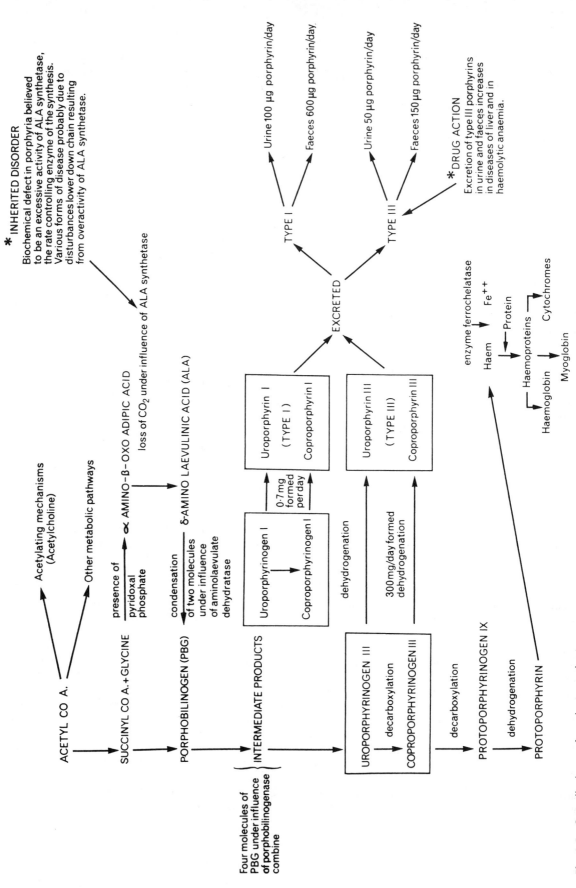

Fig. 6.1 Overall schema of porphyrin synthesis.

Table 6.1   Drugs classified as to their association with porphyria

Potentially porphyrogenic drugs	Drugs believed not to precipitate porphyria
Alcohol	Acetazolamide
Alphaxalone	Adrenaline
Aluminium	Alclofenac
2 allyloxy-3 methylbenzamide	*Amitriptyline*
Amidopyrine (A)	Aspirin
Aminoglutethemide	Atropine
Amitriptyline	
Amphetamines	B vitamins (except
Androgens (A, C)	pyridoxine)
Apronalide	Bethanidine
Avapyrazone	Biguanides
Azapropazone	Bromides
	Bumetanide
	Bupivacaine
Barbiturates (A, C)	Buprenorphine
Bemegride	
Busulphan	Cephalexin
	Cephalosporins
Carbromal	Chloral hydrate
Carbamazepine	*Chloramphenicol*
Chlorambucil	Chlormethiazole
Chloramphenicol	Chlorpheniramine
Chlordiazepoxide (A)	Chlorpromazine
Chlormezanone	*Chloroquine*
Chloroform	Chlorthiazides
Chloroquine (C)	Clobazam
Chlormethiazole	Clofibrate
Chlorpropamide (A, C)	*Clonazepam*
Cimetidine	Codeine
Clonazepam	Colchicine
Clonidine	Corticosteroids
Cocaine	Cyclizine
Colistin	Cyclopropane
Cyclophosphamide	
	Dexamethasone
Danazol	Diamorphine
Dapsone	*Diazepam*
Diazepam	Diazoxide
Dichloralphenazone (A)	Digitalis compounds
Diclofenac	Diphenhydramine
Dielthyhexyl phthalate	Dicoumarol anticoagulants
Diethylpropion	Disopyramide
Dimenhydrinate	Domperidone
	Droperidol
Enflurane	
Ergot preparations	
Erythromycin	EDTA
Ethchlorvynol	Ether (diethyl)
Ethinamate	
Ethosuximide	Fentanyl
Etomidate	Flurbiprofen
Eucalyptol	Fusidic acid
Flufenamic acid	Gentamicin
Flunitrazepam	Glipizide
Fluroxine	Guanethidine

Table 6.1   *Continued*

Potentially porphyrogenic drugs	Drugs believed not to precipitate porphyria
Frusemide	
	Heparin
Glutethimide	*Hydrallazine*
Gold preparations	Hyoscine
Griseofulvin (A)	
	Ibuprofen
Halothane	*Imipramine*
Hydantoins (phenytoin,	*Indomethacin*
ethotoin, mephentoin)	Insulin
Hydrallazine	
Hydrochlorthiazide	Ketamine
Hyoscine N-butyl bromide	Ketoprofen
Imipramine	Labetolol
Indomethacin	Lithium
	Lorazepam
Isoniazid	
Isopropylmeprobamate	Mandelamine
Lignocaine	Mecamylamine
	Meclozine
Mefenamic acid	*Mefenamic acid*
Mephenazine	Mersalyl
Meprobamate	Metformin
Mercury compounds	Methadone
Methoxyflurane	Methenamine mandelate
Methsuximide	Methylphenidate
Methyldopa	Morphine
Methyprylone	
Metoclopramide	
Metyrapone	Naproxen
Metronidazole	Neostigmine
	Nitrofurantoin
Nalidixic acid	Nitrous oxide
Nikethamide	Nortriptyline
Novobiocin	
Nitrazepam	*Oxazepam*
Nitrofurantoin	Oxpentifylline
Oral contraceptives	Paracetamol
Oxazolidinediones	*Paraldehyde*
(paramethadone and	Penicillamine
trimethadione)	Penicillins
Oestrogens (A, C)	*Pethidine*
Oxazepam	Phenformin
	Pheniramine
Pancuronium	Phenothiazines
Paraldehyde	(e.g. chlorpromazine)
Pargyline	Phenoperidine
Pentazocine	*Phenylbutazone*
Pentylenetetrazol	Prednisolone
Pethidine	Procaine
Phenazone	Prilocaine
Phenelzine	Promethazine
Phenoxybenzamine	Propantheline bromide
Phensuximide	*Propanidid*
Phenylbutazone	Primaquine
Phenylhydrazine	Propoxyphene
Primidone	Propranolol

Table 6.1 *Continued*

Potentially porphyrogenic drugs	Drugs believed not to precipitate porphyria
Probenecid	Prostigmine
Progestogens (A, C)	*Pyrimethamine*
Propanidid	
Pyrazinamide	Quinine
Pyrazolones (antipyrine,	
isopropylantipyrine, dipyrone,	Reserpine
sodium phenyl dimethyl	Resorcinol
pyrazolone)	*Rifampicin*
Pyridoxine	
Pyrimethamine	Streptomycin
	Succinylcholine
Ranitidine	
Rifampicin	*Tetracyclines*
	Tetraethylammonium
	bromide
Sodium valproate	Thiouracils
Spironolactone	Tricyclic antidepressants
Steroids	(amitriptyline)
Streptomycin	Trifluoperazine
Succinimides (ethosuximide,	Thiazides
methsuximide, phensuximide)	Tripelennamine
Sulphonal	Tubocurarine
Sulphonamides (A)	*Sodium valproate*
Sulphonylureas	
Sulthiame	Vitamin C
Tetracyclines	
Theophylline	
Tolazamide	
Tolbutamide (A, C)	
Tranylcypromine	
Trional	
Troxidone	
Valproic acid	
Viloxazine	
Xylocaine	

A, acute porphyria; C, cutaneous porphyria.
Substances marked in italics have been variously described as causing porphyria or being safe to administer to patients with porphyria (see text).

hereditary component but also occurs secondary to hepatocellular disease.

Acute porphyrias

The symptoms of acute porphyria are severe attacks of colicky abdominal pain, vomiting, and constipation. There may be weakness or even paralysis of the limbs and, more rarely, of the respiratory muscles. The psychiatric manifestations are most important and the patient may be wrongly diagnosed as having hysteria, psychoneurosis, paranoia, or schizophrenia. Charac-

teristically the urine darkens on standing and gives a positive Ehrlich test; it contains coproporphyrin and uroporphyrin in large quantities.

The most important cause of the disease is the hereditary factor since it is transmitted as a Mendelian dominant but with a varying degree of penetrance. The disease may be entirely symptomless or latent and provoked only by the administration of drugs such as barbiturates.

Mixed porphyrias

These are a group of porphyrias in which there are symptoms of skin photosensitivity, and also features of acute porphyria with pain, vomiting, constipation, paralysis, and psychiatric complications. This type of porphyria, also known as variegate porphyria, is most common in South Africa where there are about 10 000 cases, practically all of whom trace their descent to two Dutch settlers who married in the Cape in 1688. In the intervening period these porphyric families were just as large and prolific as other families, but it was the introduction of barbiturates at the beginning of the twentieth century that caused the disease to become much more lethal. In terms of practical therapy in patients with any form of porphyria, aspirin, methadone, pethidine, and morphine are acceptable analgesics but pentazocine and phenylbutazone should be avoided.

Increased porphyrin excretion caused by drugs in normal persons

This metabolic disturbance is like the experimental porphyrias induced by drugs in animals, and when the increased porphyrin formation is accompanied by clinical symptoms of photosensitivity it should be considered among the toxic varieties of the symptomatic hepatic porphyrias.

There is some divergence of opinion as to whether barbiturates and other drugs, known to precipitate attacks in patients with acute intermittent porphyria, can derange the liver porphyrin metabolism of normal individuals. However, several cases have been described in which a marked increase in porphyrin excretion and appearance of clinical symptoms of photosensitivity have been linked with the administration of certain drugs, in cases in which there was no obvious evidence of a genetic background of porphyria. Drugs that have been implicated in this disorder are: sulphonal (Harris 1898), phenobarbitone (Haxthausen 1927), tolbutamide (Rook and Champion 1960), chlorpropamide (Zarowitz and Newhouse 1965), and oestrogens (Becker 1965; Copeman *et al.* 1966; Felsher and Redeker 1966).

Most of these drugs can produce a condition of hepatic porphyria in animals or tissue culture of liver cells; this provides some support for the belief that they may also produce a porphyria-like syndrome in normal persons with the appearance of cutaneous symptoms in the most severe cases. However, it is a fact that, out of the very large number of patients receiving these drugs, only a very small proportion develop the porphyria-like syndrome; therefore the existence of some concurrent factor might be suspected, and the presence of a genetic disposition cannot be excluded. Indeed, there is no evidence that an acute attack of porphyria can be precipitated by drugs in the absence of a genetic disposition. In general, clinical experience indicates very strongly that symptoms of acute porphyria appear in those patients who are carriers of the genetic defect in a latent state.

A moderate increase in the excretion of porphyrins, or their precursors, without clinical symptoms is probably not uncommon among normal persons receiving certain drugs. For example, a small but significant increase in faecal and blood porphyrin levels was observed by Rimington et al. (1963) and Ziprkowski et al. (1966) in persons receiving the antifungal griseofulvin in the usual therapeutic doses, and a significantly increased excretion of urinary δ-aminolaevulinic acid was reported in a group of healthy females receiving oral contraceptives (Koskelo et al. 1966). Van Der Grient (1968) reviewed the literature on griseofulvin and porphyrin metabolism in humans and in animal studies, and concluded by recommending that griseofulvin be withheld from patients with a familial history of porphyria. With regard to oral contraceptives, it is interesting to note that Perlroth and co-workers (1965) showed that, in established acute intermittent porphyria, the administration of an oral contraceptive resulted in amelioration of the disease; obviously at present there is inadequate evidence to determine the true role of oral contraceptives in porphyrin metabolism.

Adverse effect of some drugs on the clinical and biochemical picture of hepatic porphyria

Non-barbiturate and barbiturate hypnotics

Hepatic porphyrias are inherited as Mendelian dominants and are characterized by excess production of porphyrin and related compounds by the liver. Acute intermittent porphyria is the most frequent, and the clinical features include a combination of abdominal colic, vomiting, and severe constipation, often with peripheral neuritis and psychological disturbance. Photosensitivity

is absent. The urine darkens on standing and gives a positive Ehrlich test; it contains coproporphyrin and uroporphyrin in large quantities. The liver synthesizes excess of the porphyrin precursors, δ-aminolaevulinic acid and porphobilinogen (Sherlock 1968).

It is now widely accepted that the administration of some drugs to patients with genetic hepatic porphyria of the acute intermittent or mixed types can worsen the course of the disease and also precipitate attacks in persons with latent porphyria. The worsening action involves not only the metabolic picture and excretion of pyrrole compounds, but also the clinical symptoms especially the nervous symptoms. This is a far more serious complication of therapeutic use of barbiturates and several other drugs than the induction of porphyria in normal persons. Patients with the genetic trait can be very sensitive to these drugs and, in a relatively small number of them, a single dose may be sufficient to change an asymptomatic abnormality of metabolism to a very serious clinical syndrome menacing life. Although an acute attack of the disease can be triggered by several other factors such as the ingestion of alcohol, dietary indiscretions (see MacAlpine et al. 1968), an acute infection, pregnancy, or menstruation, drug administration is probably the most important precipitating factor.

The causal relationship between some drugs and this disease has been known for many years. In 1889, Stockvis described the case of an elderly woman who, after taking the hypnotic sulphonal, passed dark red urine containing a pigment similar to haematoporphyrin and who subsequently died. One year later Harley (1890) reported the case of a patient, who, after ingestion of sulphonal exhibited many of the neurological features of what is now known as acute intermittent porphyria. Several other cases of acute porphyria were described in the following years and in some of these sulphonal or the related drugs tetronal and trional had been taken before the onset of symptoms.

The importance of barbiturates in relation to acute porphyria was also suspected fairly soon after the introduction of these agents in clinical therapy. Dobrschansky (1906) described a typical case of acute porphyria in a patient who had received prolonged treatment with barbitone, and in the following years several other groups reported deterioration in the clinical and metabolic state of porphyric patients after administration of barbiturates (Denny-Brown and Sciarra 1945; Eliaser and Kondo 1942; Prunty 1946; Whittaker and Whitehead 1956). From the study of a large number of patients in Sweden and Great Britain respectively, Waldenström (1937) and Goldberg (1959) were both convinced that barbiturates could precipitate attacks in persons with latent porphyria of the acute intermittent variety and that they seriously affected the prognosis of the disease. In South Africa, patients with mixed porphyria are extremely sensitive to barbiturates, and in these patients most, if not all, acute attacks appear to be precipitated by administration of either barbiturates or other drugs (Dean 1963). In patients with hereditary coproporphyria, acute attacks can also be precipitated by barbiturates and other drugs (Goldberg *et al.* 1967).

The onset of an acute attack, however, does not necessarily follow the administration of barbiturates, and Waldenström (1937, 1939), Schmid (1966), Eales (1966), and With (1965) had experience of patients with latent genetic porphyria of all three varieties, who were given barbiturates without ill effect. From a survey of all porphyric patients hospitalized in Seattle over an 11-year period and given barbiturate for general anaesthesia, Ward (1965) concluded that the precipitation of an acute attack was a very rare complication of barbiturate anaesthesia; some other authors (Günther 1922; Turner 1938; Discombe and D'Silva 1945) even failed to find any relation between barbiturates and acute porphyria. Several likely explanations can be offered for this discrepancy in findings. Firstly, some of these studies, in which an adverse effect of barbiturates was not noted or occurred only rarely, may have included patients with hepatic porphyrias of the aquired varieties, who are not expected to show any abnormal sensitivity to the barbiturates. This is likely to be the case, as suggested by Eales (1966), for the survey by Ward in which no discrimination appears to have been made between different types of hepatic porphyrias. A second very important reason is that there seems to be a considerable difference in sensitivity to barbiturates between different patients with latent porphyria; some develop very serious symptoms and even fatal paralysis after a single dose, others require prolonged administration of relatively large doses before showing symptoms (Goldberg 1959). Therefore the absence of symptoms of an acute attack after administration of single isolated doses of barbiturates to a patient, as was the case for the members of the family described by With (1965), does not necessarily mean that a patient can go on taking barbiturates with impunity. Finally the study of the literature suggests that there may also be a difference in sensitivity to barbiturates between the various genetic varieties of hepatic porphyria: South African mixed porphyria is probably the most sensitive, the acute intermittent variety less so, and hereditary coproporphyria even less.

More recently other drugs have been implicated in the precipitation of attacks of acute porphyria in patients who possess the genetic trait. Such drugs include sulphonamides, non-barbiturate sedatives, hypoglycaemics, anticonvulsants, tranquillizers, and griseofulvin. A list of these drugs which all resemble barbiturates in the production of acute porphyria is given in Table 6.1. In addition sex hormones and chloroquine have been reported to affect the condition of porphyric patients.

Sex hormones (oestrogens, progestogens, and androgens)

The relationship of sex hormones to porphyria is a diverse one; they can precipitate acute attacks of porphyria on the one hand, and on the other they can prevent the appearance of acute attacks in those patients in whom exacerbation of clinical symptoms occur regularly with menstruation.

Watson and associates (1962) described a female patient with a latent mixed hepatic porphyria who was being treated with stilboestrol for carcinoma. Her urinary porphobilinogen and uroporphyrin increased above their previously high levels and cutaneous manifestations developed, but no abdominal or nervous symptoms ensued.

There have been several reports of impaired liver function and jaundice after oral contraceptive therapy (Cohen *et al.* 1964); this could cause diversion of the porphyrins from the enteric route of excretion to the blood and urine and result in photosensitivity. In mixed porphyria photosensitivity is often associated with impairment of excretory liver function. It is not surprising therefore that oestrogens and progestogens administered together or separately, in oral contraceptives, have been reported to affect the biochemical picture and clinical state of patients with acute intermittent porphyria. Increased excretion of porphyrin precursors, and, less frequently, appearance of symptoms indicative of clinical exacerbation have been observed in patients of both sexes (Levit *et al.* 1957; Redeker 1963; Welland *et al.* 1964). An acute attack with peripheral paralysis in a patient with porphyria of the mixed type was reported by Nayrac and co-workers (1964). Dean (1963) reported increased blood and urinary porphyrin levels with the appearance of jaundice and photosensitivity in a patient with mixed porphyria who had been taking an oral contraceptive containing lynoestrenol and mestranol (*Lyndiol*).

In clear contrast to these findings, Haeger-Aronson (1963) described one case, and Perlroth and associates (1965) described two cases, where the exacerbation of acute intermittent porphyria occurring in association with menstruation was prevented by combined oestrogen and progestogen therapy. The fact that these hormones can exacerbate acute intermittent porphyria in some patients and ameliorate the condition in others, together with the fact that menstruation and pregnancy can adversely affect the course of the disease, would indicate most strongly that the hormonal balance of the patient is a prime factor in the pathogenesis of attacks of this disease.

It is not known however in what way the hormonal state is implicated or which hormone or endocrine system is primarily involved. In one of the patients described by Perlroth *et al.* (1965), complete control of the porphyric symptoms was also obtained by treatment with either an androgen or an oestrogen alone, as well as by administration of a combination of an oestrogen with a progestogen. Androgenic therapy was similarly effective in another patient described by Schmid (1966).

These observations suggest that a physiological response common to all three hormonal treatments may prevent acute attacks in these patients. Perlroth *et al.* (1965) suggested that inhibition of the secretion of pituitary gonadotrophins, with stabilization of endogenous steroid production at low level, may be the effective mechanism. This interpretation presupposes that a high level of endogenous sex hormones may be conducive to an attack of acute porphyria, whereas a high level of exogenous synthetic hormones may not; but this concept is not borne out by the finding of Redecker (1963) and Wetterberg (1964) of acute episodes after administration of synthetic sex hormones. Another possibility is that the recurring hormonal imbalance responsible for the periodic exacerbation of the clinical symptoms in these patients lies outside the gonadotrophins–ovary system and that large doses of sex hormones may affect other functions of the pituitary in addition to the secretion of gonadotropin.

The occurrence of porphyria cutanea tarda in patients receiving oestrogen therapy has been the subject of a number of reports (Roenigk and Gottlob 1970; Byrne *et al.* 1976; White 1977; Weimar *et al.* 1978). Most of these, Byrne *et al.* 1976, refer to male patients with a clinical diagnosis of carcinoma of the prostate but some involve women on oral contraceptives. These patients either present with, or go on to develop, the full clinical picture of porphyria cutanea tarda including photosensitivity, dermal blisters, and erosions, hypertrichosis, and hyperpigmentation of exposed skin. Diagnosis is confirmed by greatly raised levels of urinary coproporphyrin, uroporphyrin, ALA, and porphobilinogen. Hepatic enzyme (SGOT & AP) levels are often raised and biopsy reveals excess iron in hepatocytes and Kupffer cells.

Although there are now large numbers of reports in the literature where combined oestrogen–progesterone oral contraceptives have produced the neuropsychiatric and abdominal symptoms of acute intermittent porphyria together with the increased urinary excretion of porphyrins, in porphyria variegata oestrogens do not

apparently provoke acute attacks (Goldswain and Eales 1971). In variegata porphyria attacks provoked by barbiturates or sulphonamides are accompanied by changes in liver function. Five cases were reported in the literature (Eales 1966; Dean 1965; Baxter and Permowicz 1967; McKenzie and Acharya 1972; Fowler and Ward, 1975); in all of these five cases the characteristic skin lesions were produced, and four of these cases were jaundiced and in the other the liver function tests were abnormal. The enzyme changes and liver biopsy, where conducted, were indicative of a cholestatic jaundice, a known side-effect of oestrogen therapy. Fowler and Ward (1975) showed that at the height of the jaundice their patient showed a marked change in the pattern of porphyrin excretion., Normally coproporphyrin is preferentially, and protoporphyrin exclusively, excreted by the liver. In cholestasis there is a diversion of porphyrins from faecal to urinary excretion which reverses as the jaundice subsides. The impairment of biliary secretion may well result in high levels of porphyrin in skin tissue. This would account for the acute and striking photosensitivity which was an unusual feature of all these cases.

The porphyria variegata provoked by the contraceptive pill may therefore be manifest secondary to the drug-induced cholestatic jaundice.

Therapeutic implications

In both intermittent porphyria and in cutaneous porphyria the basic lesion is the striking increase in ALA-synthetase, and most of the drugs that can precipitate attacks of the porphyria are to a greater or lesser extent enzyme-inducers. When treatment is necessary, drugs which have no enzyme-inducing properties should be used. Those drugs which have been shown from experience to be safely used in patients with porphyria are relatively weak or ineffective enzyme-inducers (Eales 1971; Goldswain and Eales 1971; Beattie *et al.* 1973) and are shown in Table 6.1.

Symptomatic cutaneous porphyria and chloroquine

There have been a number of reports on the effect of the administration of the antimalarial chloroquine to porphyric patients (Linden *et al.* 1954; Marsden 1959; Cripps and Curtis 1962; Gertler 1962; Rimington 1964; Sweeney *et al.* 1965). Unlike the barbiturates and the other agents so far considered in relation to porphyria, chloroquine affects hepatic porphyria of the symptomatic variety (porphyria cutanea tarda) in which cutaneous symptoms are present without abdominal or neurological disturbances.

This purely cutaneous form of the disease is hereditary

and appears in middle age; it may occur secondary to hepatocellular disease. There is a well-established relationship between this condition and alcoholism and usually there is evidence of hepatic dysfunction and raised serum bilirubin levels. Liver biopsy sections show subacute hepatitis or cirrhosis. Uroporphyrin is increased in the liver. Exacerbation of symptoms has been connected with deterioration of liver function. At this time porphyrins which would be normally excreted into the bile may be directed via the kidneys to the urine. The tendency to porphyria is made manifest by coincident liver disease. In the healthy liver the porphyrin is excreted harmlessly into the bile; in the diseased or damaged liver it is retained in the blood. The porphyria itself may be hepatotoxic (Sherlock 1968). Chloroquine is concentrated to a remarkable degree in the liver; indeed this is the advantage for its use as a tissue amoebicide in the treatment of liver amoebiasis. It exerts no effective action on amoebae elsewhere and is quite valueless, for example, in the treatment of intestinal amoebiasis. It is not surprising therefore that with this specific concentration in the liver, chloroquine in the presence of porphyria cutanea tarda causes a hepatic reaction with transient fever, general malaise, tachycardia, a rise in serum transaminases, and hepatic necrosis. Hepatic uroporphyrin decreases and urinary excretion increases (Kofman et al. 1955). This effect does not recur and, on giving the chloroquine when the uroporphyrin excretion has considerably decreased, further administration of chloroquine does not cause any new increase in urinary uroporphyrin (Sweeney et al. 1965). Sweeney and co-workers (1965) were not able to state how long this refractory period to chloroquine effects lasted; however, Sherlock (1968) commented that this period may last for up to five months.

The nine cases reported by Sweeney et al. (1965) all had alcoholism in common as a predisposing factor of their symptomatic cutaneous porphyria and, although these patients had a severe reaction lasting three days following a 2–8-day latent period after chloroquine administration, they subsequently showed clinical improvement of their cutaneous porphyria. It is doubtful, however, whether the improvement could be related to the use of chloroquine and the authors were more of the opinion that a good diet, absence of alcohol and sunlight, and attention to their skin lesions were the cause of this. They considered that evidence of liver dysfunction contraindicated the use of chloroquine in symptomatic porphyria.

The massive uroporphyrinuria reported by Sweeney et al. (1965) was not accompanied by a significantly increased excretion of porphyrin precursors, and they suggested, for this reason, that the liver was the probable source of the urinary uroporphyrin and it was from that site that it was released during the reaction to chloroquine. A very careful study of the effects of chloroquine on liver porphyrin metabolism was made by Felsher and Redeker (1966) in five patients with symptomatic porphyria. A marked increase in urinary porphyrin excretion was observed after chloroquine administration and was accompanied by a 50 per cent reduction of uroporphyrin in the liver. This supported the view that chloroquine causes a release of uroporphyrin from the liver rather than an increased synthesis at that site. No effect on porphyrin and porphobilinogen excretion was observed when chloroquine was given to two patients with acute intermittent porphyria; chloroquine did not induce porphyria in liver cells cultured in vitro as do most of the agents responsible for precipitating attacks in patients with porphyria of the acute and mixed types. Chloroquine is known to form complexes with ferrihaemic acid and protoporphyrin in vitro, and Felsher and Redeker (1966) suggested that the formation of a similar complex might occur between chlorquine and liver uroporphyrin, leading to a large concentration of chloroquine in the liver cells, to liver damage, and, finally to leakage of the accumulated uroporphyrin. These investigators also reported a marked clinical and biochemical improvement in their patients once the initial transient reaction to chloroquine had subsided. No further symptoms of porphyria were seen in these patients for several months.

Apart from its use in the chemotherapy of malaria and amoebiasis, chloroquine is effective in relatively large doses in the treatment of discoid and systemic lupus erythematosus and in rheumatoid arthritis. This 'anti-inflammatory' use of chloroquine has become widespread in recent years and obviously the chance of chloroquine reaction in symptomatic cutaneous porphyria has also increased correspondingly, especially when coincident with alcoholism. Evidence of liver dysfunction must contraindicate the use of chloroquine with the possible exception of tissue amoebiasis of the liver, for which the drug is specifically indicated. There is a very close structural relationship between chloroquine and other members of the 4-aminoquinoline series of antimalarials (Fig. 6.2), and it is to be expected that similar effects might occur on porphyrin metabolism with amodiaquine and hydroxychloroquine.

Erythropoetic protoporphyria and photosensitizing drugs

A further type of abnormality of porphyrin metabolism, erythropoietic protoporphyria, has not so far been linked

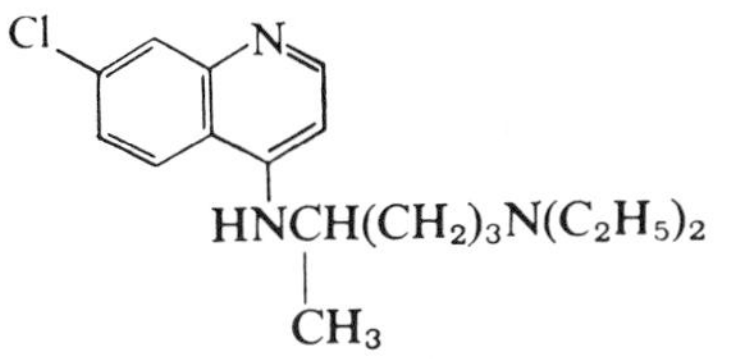

Fig. 6.2 Structural relationship within the 4-aminoquinolines.

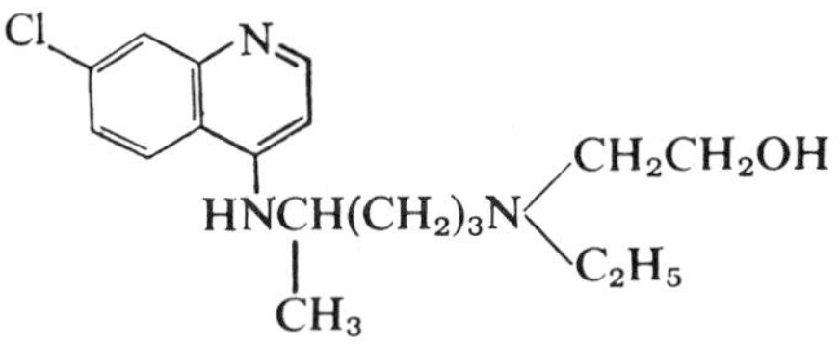

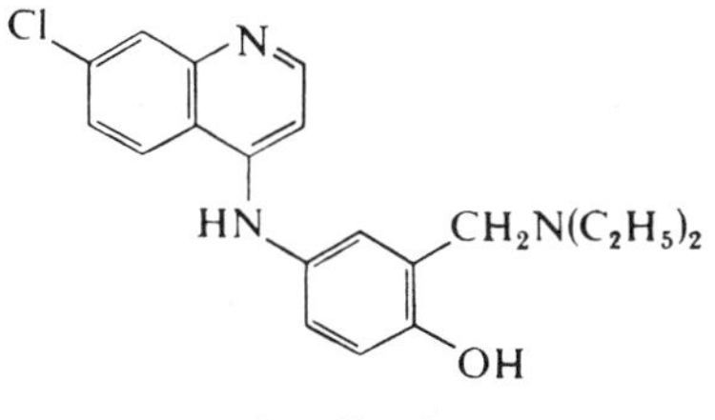

with drug administration. This may well be due to its rare occurrence. This condition is characterized by excess of erythrocyte protoporphyrin and the cardinal symptom is photosensitivity with erythematous papules and vesicles on face, ears, and extremities after exposure to sunlight. The urine is pink or reddish, and liver biopsies show characteristic focal, intrahepatic deposits of pigment containing protoporphyrin (Cripps and Scheuer 1965).

Although there have been no reports of drug precipitation or exacerbation of this condition, it is interesting to learn that when certain farm animals, notably sheep and pigs, fed on buckwheat (*Fagopyrum esculentum*) which contains phytoporphyrin, they may develop hypersensitivity and, if bare or slightly covered areas of the skin are then exposed to sunlight, vesiculation, ulceration, or even pyrexia and constitutional symptoms may ensue (Thorne 1966). This would suggest that dietary factors and drugs might possibly influence the course of erythropoietic protoporphyria; certainly it is within the realms of possibility. Any drug that produces photosensitivity would be expected to have an adverse effect on this condition. Unfortunately there are

a large number of drugs implicated in photosensitivity reactions (see Chapter 32), including certain tetracyclines, sulphonamides, sulphonylurea derivatives, phenothiazines, and thiazide diuretics. In addition, methoxsalen (8-methoxypsoralen), and trioxsalen (trimethylpsoralen), and other psoralens used in the treatment of vitiligo (leucoderma), produce intolerance to sunlight when taken orally or applied topically (D'Arcy 1965; 1966). The psoralens might also be expected to have a deleterious effect on the course of erythropoietic protoporphyria; indeed trioxsalen is specifically contraindicated in porphyria (Council on Drugs, 1966).

Other drugs and therapeutic procedures associated with the induction of porphyria

Although no particularly dramatic advances have been made in the field of drug-induced porphyria in the last two or three years the list of incriminated agents has continued to increase as the number of new chemical entities multiplies. The setting up of a National Register of porphyric patients in the UK is to be welcomed since it enables this section of the population to be identified and offers scope for patient education. There is also evidence that the recognition of the problems of these patients is being better understood, and reviews dealing with drug-precipitating factors have been published in many journals.

A selection of the numerous reports and reviews includes those of Larter (1978), adding ergot derivatives to the list of porphyrinogenic drugs; Millar (1980), dealing with the antituberculous drug rifampicin; Treece (1976), adding pyrazinamide; and Garcia-Merino and Lopez-Lozano (1980) adding sodium valproate. There have also been a number of publications from Eastern Europe (Kansky *et al.* 1978; Bielawska-Krasnowiecka and Jaremin 1979; Kostrzewska *et al.* 1979), in which the precipitating factors most commonly referred to are oral contraceptives, clomethiazole edisylate, an analgesic, probon, and industrial lead poisoning. A review of 153 acute attacks of porphyria in 138 patients of Afrikaner origin was published (Eales 1979) and is of general interest. A number of drugs are identified for which a temporal relationship could be shown between the use of the drug and an attack. These drugs were, in decreasing order of frequency: barbiturates, causing 81 attacks; analgesics, 16 acute attacks; sulphonamides, 16 acute episodes; non-barbiturate hypnotics, 15 acute episodes; anticonvulsants, alchohol, and oral conraceptives were each represented; ergot as a precipitating agent was also referred to.

In the same paper Eales suggested that the use of propranolol can give considerable symptomatic relief in acute drug-precipitated attacks, although there is of course no specific treatment. Propranolol does allay the anxiety and control the tachycardia and hypertension. Eales stated that he was loath to use haematin to abort the acute attack in view of the high incidence of renal failure in variegate porphyria and because haematin has been documented as being able to precipitate acute renal failure.

Comprehensive reviews of drug therapy associated with porphyria were published by Moore (1980) and Moore and Disler (1983). More than 100 drugs are listed in these publications as being associated with the induction of acute porphyria. A number of the same drugs are also listed as being safe for use in patients known to be susceptible to porphyria. The authors attributed this discrepancy to difficulties in interpreting clinical reports. The more important drugs about which this doubt exists are as follows: amitriptyline, chloramphenicol, chloroquine, chlormethiazole, clonazepam, diazepam, hydrallazine, imipramine, indomethacin, nitrofurantoin, oxazepam, paraldehyde, pethidine, phenylbutazone, propanidid, pyrimethamine, rifampicin, tetracyclines, and sodium valproate. It will be of great interest to follow future reports in the literature to see to which list each of these drugs should properly be assigned. Table 6.1 combines the drugs listed by Griffin (1979), Moore (1980), and Moore and Disler (1983).

Examination of the list of porphyrinogenic drugs reveals that a very wide diversity of chemical structure is involved. This prohibits any simple explanation in terms of the presence or absence of particular reactive groups. Moore (1980) postulated that the principal initiating mechanism is depletion of free haem, which in turn induces the enzyme δ-aminolaevulinic acid synthase, but it is not clear how the many different chemical structures of the drugs involved give rise to this single effect. It has been suggested that lipophilicity or the presence of allelic groups, the barbiturate nucleus, or certain steroid structures are implicated, which may be true in specific instances but is unlikely to provide a general mechanism for porphyrinogenicity.

Haemodialysis-associated porphyria

Porphyria cutanea tarda has generally been classified as an acquired disorder of porphyrin metabolism. More recently, it was suggested that it may be familial, probably transmitted as an autosomal dominant trait (Benedetto et al. 1978).

Skin lesions indistinguishable from porphyria cutanea tarda occur in about 1 per cent of patients on maintenance haemodialysis (Brivet et al. 1978), and usually there is no evidence of abnormal porphyrin metabolism, but two cases in which true porphyria cutanea tarda developed during haemodialysis were reported (Poh-Fitzpatrick et al. 1978); the primary defect was a decrease in uroporphyrinogen decarboxylase activity. Garcia Parilla et al. (1980) described a case of a 16-year old boy with chronic glomerulonephritis who developed hyperpigmentation, skin fragility, and bullae on his face and backs of hands after one year on chronic haemodialysis. This patient was recognized as having porphyria cutanea tarda, his uroporphyrin decarboxylase activity was found to be low (as was his mother's; but his father's, his sisters', and brothers' were all normal). The boy had been receiving 100 mg iron dextran monthly and had received 6 units of blood in the 12 months preceding the appearance of symptoms. His plasma ferritin was 878 μg/1 (normal range 10–250 μg/1).

Iron overload is common in patients undergoing haemodialysis and in this patient iron overload was certainly present. Chapman (1980) reported the cases of a mother and daughter of Turkish origin both with β-thalassaemia minor with iron overload who developed porphyria cutanea tarda. The serum ferritin levels in the mother were 1060 μg/1 and in the daughter 930 μg/1 (normal range 180 μg/1) and liver biopsy in both showed siderosis. Both patients had low uroporphyrin decarboxylase levels. It was suggested that the manifestation of porphyria cutanea tarda depended in these cases on the concordance of an enzyme defect, low uroporphyrin decarboxylase, and iron overload.

A further case which turned out to be a true porphyria cutanea tarda was reported by Topi et al. (1980). They also drew attention to the obscure aetiology of haemodialysis-associated porphyria and suggested photoallergic or phototoxic drug reactions, PVC intoxication, toxicity from plasticizer, aluminium overload, or hepatopathy as possible causes. It is of interest that the PVC plasticizer, diethylhexylphthalate, is on the list, as this is believed to accumulate in the liver and is known to be associated with marked peroxisome proliferation and hepatic enlargement. These two latter findings are interpreted as indicating serious hepatic malfunction and it seems possible that porphyric episodes associated with chronic haemodialysis may be explained in this way.

Rifampicin

The first case of rifampicin-induced porphyria was documented by Millar (1980) in a 55-year-old man who developed porphyria cutanea tarda when treated with rifampicin and isoniazid for a tuberculous psoas abscess.

Rechallenge with both drugs separately incriminated rifampicin.

Danazol

Exacerbations of acute porphyria associated with the administration of oestrogens is well known and exacerbations of acute intermittent porphyria during menstruation have been attributed to the endogenous hormonal changes occurring at that time. Danazol (17-α-pregn-4-en-20-yno(2,3-d)isoxazol-17-ol), which is a synthetic steroid with weak androgenic activity neither progestational nor oestrogenic but is antigonadotropic, was considered as a possible therapeutic agent in menses-associated porphyria. In practice, however, danazol appears to act as a porphyrinogenic drug and triggers attacks in susceptible women. In studies on the action of danazol on rat liver ALA-synthase (previously known as ALA-synthetase), Lamon *et al.* (1979) were unable to demonstrate any increase in enzyme activity. However, Hughes and Rifkind (1981) used chick embryo liver homogenates and cultured chick embryo hepatocyte, which they believed to be a more sensitive model system, to investigate the effect of danazol on ALA-synthase. They found a dose-related increase in protoporphyrin production in the cultured hepatocyte system and increased ALA-synthase activity in the intact chick embryo system. The increased ALA-synthase activity was less than that produced in control experiments by allylisopropylacetamide, which suggests that danazol is not a potent inducer. Addition of Ca Mg-EDTA, an iron chelator, to the cultured hepatocyte test system considerably enhanced danazol's stimulating effect on protoporphyrin production, presumably by preventing the conversion of porphyrins to haem.

Increased urinary excretion of 5 β-reduced steroid metabolites as compared to 5 α-reduction products due to a deficiency of hepatic 5 α-reductase is a possible aetiological factor in acute porphyria and is consistent with the fact that the 5 β-reduced metabolites are more powerful inducers of ALA-synthase than the 5 α-metabolites. On the assumption that the porphyrinogenic effect of danazol is mediated through a metabolite, Hughes and Rifkind (1981) postulated that danazol is more readily reduced to the 5 β-metabolite by chick embryo liver than by rat liver, which is supported by existing knowledge that rat liver metabolizes steroids predominantly by the 5 α-pathway whereas avian liver prefers the 5 β-pathway.

In the light of these studies there would seem to be some justification for the belief that the chick embryo liver system will prove to be a more sensitive and reliable screening test for porphyrinogenicity in drugs and chemicals.

Methods for screening drugs for porphyrinogenic potential

Certain forms of acute porphyria are probably the foremost examples of drug-induced disease. The list of drugs that are reported to have been associated with acute porphyria is so long and contains such broad therapeutic classes that there is scarcely any branch of medicine that is not affected. The possibility that attacks of acute porphyria may occur in association with hormones, antibiotics, anti-epileptic drugs, and cancer chemotherapeutic agents emphasizes the breadth of the problem. It is therefore of great importance that methods should be developed to detect new and existing drugs and chemicals that are porphyrinogenic so that they may be avoided by susceptible individuals. It is equally important that screening methods should be used to detect susceptible individuals. These people are usually relatives of patients known to have manifest or latent porphyria, although they may be totally unaware of their condition.

A major problem that still exists is the lack of knowledge of the underlying mechanisms of porphyrinogenesis by drugs and chemicals. It has frequently been stated that no chemical structure or grouping is common to all the substances known to be porphyrinogenic, an observation that would seem to be irrefutable when the list of those substance is considered. Nevertheless, the fact that they are all capable of causing acute episodes of porphyria strongly suggests that some common property exists. Many are known to be inducers of the enzyme aminolaevulinic acid synthase (ALA-synthase) and are inhibitors of cytochrome P-450 formation, although there are some porphyrinogenic substances which apparently do not exhibit either effect. Another hypothesis is that porphyrinogenesis is mediated through inhibition of hepatic ferrochelatase. This enzyme catalyses the insertion of iron into protoporphyrin IX, which results in decreased haem production, which in turn represses ALA-synthase.

In-vitro model systems which may give some indication of potential porphyrinogenicity have made use of rat or mouse hepatic cells or chick embryo livers or liver cells. *In-vivo* systems using rats dosed with substances such as hexachlorobenzene which induce a latent porphyria-like state have also been used. Unfortunately there is no general agreement on the validity of these models as predictors of porphyrinogenicity in drugs.

There is general agreement, however, that the major controlling influence on haem synthesis is the first, and rate-limiting enzyme, ALA-synthase. In the hepatic porphyrias the activity of ALA-synthase is greatly increased, resulting in excess production of δ-aminolaevulinate, which is known to inhibit γ-aminobutyrate release. Although haem is known to have negative feedback control on the synthesis of ALA-synthase, it is not known how porphyrinogenic agents interfere with this carefully controlled system. Although increased translation of ALA-synthase from polyadenylated RNA in chick embryo livers has been demonstrated (Brooker *et al.* 1980), it is difficult to believe that the great diversity of chemical structure that are known to induce porphyria all act through this single mechanism.

At least one substance (2-allyl-2-isopropylacetamide) is known to destroy the haem of cytochrome P-450 and this is the primary mechanism by which the substance induces ALA-synthase. A similar mechanism has also been suggested for the possible porphyrinogenic activity of contraceptive steroids. On the other hand, 2-isopropyl-2-propylacetamide, an analogue of 2-allyl-2-isopropylacetamide, and phenobarbitone have both been reported to either not affect or actually increase cytochrome P-450 haem. Nevertheless, in other experiments on chick embryo livers Lim *et al.* (1980) showed that a loss of cytochrome P-450 haem occurs over a short-time period (3 hours) even in the cases of phenobarbitone and 2-allyl-2-isopropylacetamide. It seems likely therefore that, initially at least, the induction of ALA-synthase in response to those two substances is also due to a destruction of cytochrome P-450 haem. Lim *et al.* (1980) proposed that induction of experimental porphyria is a result of:

1. Induction of ALA-synthase by destruction of cytochrome P-450 haem;
2. Maintenance of continued enzyme synthesis by inhibition of haem synthesis and/or continued removal of newly synthesized haem.

At the present time two systems are being investigated which make use of whole rats. In one method developed by Blekkenhorst *et al.* (1980), low doses of 3,5-diethoxy-carbonyl-1,4-dihydrocollidine are given to rats to induce a condition which resembles latent human variegate porphyria. If, subsequent to the induction of this latent porphyric state, a drug known to induce attacks of acute porphyria in humans is administered to the rats, a metabolic disorder similar to the human condition becomes manifest. Using this system Blekkenhorst *et al.* confirmed the porphyrinogenicity of phenobarbitone and showed that flunitrazepam and *Althesin* should also be regarded as porphyrinogenic and therefore used with caution in patients thought to be susceptible to acute porphyria.

In addition to inducing δ-aminolaevulinic acid synthase activity, 3,5-diethoxycarbonyl-1,4-dihydrocollidine has long been known to inhibit *in-vivo* hepatic ferrochelatase activity. Tephly *et al.* (1980) produced strong evidence to suggest that ferrochelatase inhibitor is a porphyrin-like substance derived from the catabolism of haem. The mechanism of ferrochelatase inhibition in this system has yet to be elucidated but it is clearly of great interest in understanding the cause of drug-induced attacks of porphyria.

Another animal model which may be of value in predicting which new drugs may be porphyrinogenic was described by Day *et al.* (1980). This makes use of the well-known porphyrinogenic agent hexachlorobenzene which produces a condition in rats that is closely similar to human symptomatic porphyria. This action may be partly explained by the ability of hexachlorobenzene to cause a deficiency of the enzyme uroporphyrin decarboxylase. Hexachlorobenzene is also capable of producing, in rats, an intermediate porphyric state similar to that produced by 3,5-diethoxycarbonyl-1,4-dihydrocollidine. The administration of phenylbutazone to rats in the hexachlorobenzene latent porphyric state induced an acute porphyric episode with similar characteristics to human acute porphyria. It remains for future research to determine the value of these two animal models as predictive methods for the study of potentially porphyrinogenic drugs.

In an experimental model using mouse hepatocytes, Cole *et al.* (1981) studied the effects of the two porphyrinogenic drugs, griseofulvin and 3,5-diethoxycarbonyl-1,4-dihydro-2,4,6-trimethylpyridine (DDC), on ferrochelatase activity. This enzyme is responsible for the insertion of iron into protoporphyrin IX and it is postulated that inhibition of its activity would lead to a decrease in available haem which would in turn release ALA-synthase from haem-mediated control. The activity of these two substances on avian model systems is contradictory. Although they are both porphyrinogenic, only DDC inhibits ferrochelatase activity in either chick embryo liver cells in culture or in intact 17-day-old chick embryos. In their study on mouse hepatocytes, Cole *et al.* (1981) demonstrated a similar difference between griseofulvin and DDC confirming that the two species behaved in the same way. The reasons for this difference are not clear but both substances are known to cause accumulation of an N-alkyl porphyrin, which directly inhibits ferrochelatase. The results of these experiments emphasize the difficulties involved in attempting to explain the porphyrinogenicity of certain drugs in terms of mechanisms worked out in model animal systems.

Although the presently available evidence is conflicting, the importance of haem as a controlling substance is supported by the fact that haematin, when administered to patients with porphyria, decreases ALA-synthase activity and is also effective in the treatment of the patients' conditions. A difficulty in further elucidating the role of haem in the control of ALA-synthase activity is the inability to measure the regulatory haem pool because it is so small. An indirect measure can be made, however, by estimating tryptophan pyrrolase activity. In a study on rat liver pyrrolase in animals pretreated with DDC, Badawy (1981) developed a screening system which distinguishes between a number of porphyrinogenic drugs (α-methyldopa, thiopentone, chlordiazepoxide, tolbutamid, and sedormid) and drugs believed to be non-porphyrinogenic (pheniramine, oxpentifylline, imipramine, ketamine, and mefenamic acid). Although these techniques have considerably advanced our knowledge of drug-induced porphyria, their general validity as predictors of porphyrinogenicity remains uncertain and further experience is required before recommendations for drug testing can be made.

RECOMMENDED FURTHER READING

BRITISH MEDICAL ASSOCIATION (1968). *Porphyria, a royal malady*. London.

DE MATTEIS, F. (1967). Disturbances of liver porphyrin metabolism caused by drugs. *Pharmacol. Rev.* 19, 523–9.

GOLDBERG, A. (1968). The porphyrias. In *Porphyria, a royal malady*, pp.66–68. British Medical Association, London.

—— AND RIMINGTON, C. (1962). *Diseases of porphyrin metabolism*. Thomas, Springfield, Ill.

MOORE, M.R., CAMPBELL, B.C., AND GOLDBERG, A. (1977). *Lead in environment and man* (ed. J.Lenihan and W.W. Fletcher). Vol. 6: *The chemical environment*, pp. 64–92. Blackie, Glasgow.

SCHMID, R. (1966). The porphyrias. In *The metabolic basis of inherited disease* (ed. J.B., Stanbury, J.B. Wyngaarden and D.S. Frederickson). McGraw-Hill, New York.

REFERENCES

BADAWY, A.A.B. (1981). Heme utilization by rat liver tryptophan pyrrolase as a screening test for exacerbation of hepatic porphyrias by drugs. *J. Pharmacol. Meth.* 6, 77–85.

BAXTER, D.L. AND PERMOWICZ, S.E. (1967). Variegate porphyria (mixed porphyrin). *Arch. Dermatol.* 96, 98–100.

BEATTIE, A.D., MOORE, M.R., GOLDBERG, A., AND WARD, R.L. (1973). Acute intermittent porphyria: Response of tachycardia and hypertension to propranolol. *Br. med. J.* 3, 257–60.

BECKER, F.T. (1965). Porphyria cutanea tarda induced by estrogens. *Arch. Dermatol.* 92, 252–6.

BENEDETTO, A.V., KRUSHNER, J.P., AND TAYLOR, J.S. (1978). Porphyria cutanea tarda in three generations of a single family. *New Engl. J. Med.* 298, 358–62.

BIELAWSKA-KRASNOWIECKA, G. AND JAREMIN, B. (1979). Przypadek Porfirii Skornej Poznej U Chorego Zawodowo Narazonego Na Dzialaine Olowiu I Etyliny. *Wiad. Lek.* 32, 199–202.

BLEKKENHORST, G.H., HARRISON, G.G., COOK, F.S., AND FAIES, I. (1980). Screening of certain anaesthetic agents for their ability to elicit porphyric phases in susceptible patients. *Br. J. Anaesth.* 52, 759–62.

BRIVET, F., DRUCKE, T., GUILLEMETTE, J., ZINGRAFF, J., AND CROSNIER, J. (1978). Porphyria cutanea tarda-like syndrome in haemodialysed patients. *Nephron* 20, 258–66.

BROOKER, J., MAY, B., AND ELLIOT, W. (1980). Synthesis of δ-aminlaevulinate synthase *in vitro* using hepatic mRNA from chick embryos with induced porphyrias. *Eur. J. Biochem.* 106, 17–24.

BYRNE, J.P.H., BOSS, J.M., AND DAWBER, R.P.R. (1976). Contraceptive pill-induced porphyria cutanea tarda presenting with oncholysis of the finger nails. *Postgrad. med. J.* 52, 535–8.

CHAPMAN, R.W.G. (1980). Prophyria cutanea tarda and beta thalassaemia minor with iron overload in mother and daughter. *Br. med. J.* 280, 1255.

COHEN, S.N., PHIFER, K.O., AND YIELDING, K.L. (1964). Complex formation between chloroquine and ferrihaemic acid *in vitro*, and its effect on the antimalarial action of chloroquine. *Nature*, London 202, 805–6.

COLE, S.P.C., MASSEY, T.E., MARKS, G.S., AND RACZ, W.J. (1981). Effects of porphyrin-inducing drugs on ferrochelatase activity in isolated mouse hepatocytes. *Can. J. Physiol. Pharmacol.* 59, 1155–8.

COPEMAN, P.W.M., CRIPPS, D.J., AND SUMMERLY, R. (1966). Cutaneous hepatic porphyria and oestrogens. *Br. med. J.* 1, 461–3.

COUNCIL ON DRUGS (1966). An agent for stimulating pigmentation or tolerance to sunlight: Trioxsalen (Trisoralen). *J. Am. med. Ass.* 197, 43.

CRIPPS, D.J. AND CURTIS, A.C. (1962). Toxic effect of chloroquine on porphyria hepatica. *Arch. Dermatol.* 86, 575–81.

—— AND SCHEUER, P.J. (1965) Hepatobiliary changes in erythropoietic protoporphyria. *Arch. Pathol.* 80, 500–8.

D'ARCY, P.F. (1965). The pharmacological basis of drug treatment in dermatology. *Pharm. J.* 194, 637–43.

—— (1966). The sun and the skin. *Pharm. J.* 196, 477–81.

DAY, R.S., BLEKKENHORST, G.H., AND EALES, I. (1980). Drug induced porphyric episodes in rats. *Int. J. Biochem.* 12, 1007–11.

DEAN, G. (1963). *The porphyrias. A story of inheritance and environment*. Pitman, London.

—— (1965). Oral contraceptives in porphyria variegata. *S. Afr. med. J.* 39, 278–80.

DENNY-BROWN, D. AND SCIARRA, D. (1945). Changes in nervous system in acute porphyria. *Brain* 68, 1–16.

DISCOMBE, G. AND D'SILVA, J.L. (1945). Acute idiopathic porphyria. *Br. med. J.* 2, 491–3.

DOBRSCHANSKY, M. (1906). Einiges über Malonal. *Wein. Med. Presse* 47, 2145–51.

EALES, L. (1966). Porphyria and thiopentone. *Anesthesiology* 27, 703–4.

—— (1971). The acute porphyria attack. III. Acute porphyria: the precipitating and aggravating factors. *S. Afr. med. J.* 45, 120–5.

—— (1979). Porphyria and the dangerous life-threatening drugs. *S. Afr. med. J.* 56, 914–17

ELIASER, M. AND KONDO, B.O. (1942). Electrocardiographic changes associated with acute porphyria. *Am. Heart J.* 24, 696–702.

FELSHER, B.F. AND REDEKER, A.G. (1966). Effect of chloroquine on hepatic uroporphyrin metabolism in patients with porphyria cutanea tarda. *Medicine, Baltimore* **45**, 575–83.

FOWLER, C.J. AND WARD, J.M. (1975). Porphyria variegata provoked by contraceptive pill. *Br. med. J.* **1**, 663–4.

GARCIA-MERINO, J.A. AND LOPEZ-LOZANO, J.J. (1980). Risks of valproate in porphyria. *Lancet* **ii**, 856.

GARCIA PARRILLA, J., ORTEGA, R., PENA, M.L., RODICIO, J.L., SALAMANCA, R.E., OLMES, A., AND ELDER, G.H. (1980). Porphyria cutanea tarda during maintenance haemodialysis. *Br. med. J.* **280**, 1358.

GERTLER, W. (1962). Latente Porphyria cutanea tarda manifestiert durch Resochintherapie bei Vitiligo. *Derm. Wschr.* **146**, 376–7.

GOLDBERG, A. (1959). Acute intermittent porphyria. A study of 50 cases. *Quart. J. Med.* **28**, 183–209.

——, RIMINGTON, C., AND LOCHHEAD, A.C. (1967). Hereditary coproporphyria. *Lancet* **i**, 632–6.

GOLDSWAIN, P.R. AND EALES, L. (1971). The oral contraceptive norinyl-1 in porphyria. *S. Afr. med. J.* **45**, 111–19.

GRIFFIN, J.P. (1979). Disturbances of porphyrin metabolism. In D'Arcy, P.F., and Griffin, J.P. *Iatrogenic diseases*, 2nd edn., pp. 201–10. Oxford University Press, Oxford.

GÜNTHER, H. (1922). Die Bedeutung der Hämatoporphyrine in Physiologie and Pathologie. *Ergebn allg. Path. Amat.* **20**, 608–764.

HAEGER-ARONSEN, B. (1963). Various types of porphyria in Sweden. *S. Afr. J. Lab. clin. Med.* **9**, 288–95.

HARLEY, V. (1890). Two fatal cases of an unusual form of nerve disturbance associated with dark-red urine, probably due to defective tissue oxidation. *Br. med. J.* **2**, 1169–70.

HARRIS, D.F. (1898). On the red ally of urohaematoporphyrin: A retrospect of twelve cases. *Br. med. J.* **1**, 361–2.

HAXTHAUSEN, H. (1927). Ein Fall von Hydroa aestivale ähnehndem Lichtausschlag bei einem Patienten mit Hämatoporphyrinurie, hervogerufen durch Luminal. *Derm. Wschr.* **84**, 827–9.

HUGHES, M.J. AND RIFKIND, A.B. (1981). Danazol, a new steroidal inducer of δ-aminolaevulinic acid synthetase. *J. clin. Endocrinol. Metab.* **52**, 549–52.

KANSKY, A., GREGORE, J., AND PAVLIC, M. (1978). Oral contraceptives and their influence on porphyrin concentrations in erythrocytes and urine. *Dermatologia, Basle* **157**, 181–5.

KOFMAN, S., JOHNSON, G.C., AND ZIMMERMAN, H.J. (1955). Apparent hepatic dysfunction in lupus erythematosus. *Arch. intern. Med.* **95**, 669–76.

KOSKELO, P., EISALO, A., AND TOIVONEN, I. (1966). Urinary excretion of porphyria precursors and coproporphyrin in healthy females on oral contraceptives. *Br. med. J.* **1**, 652–4.

KOSTRZEWSKA, E., GREGOR, A., AND TRACZYK, A. (1979). Ocena porfirynogennego dzialania wybranych lekow. *Polski Tygodnikhekarski* **34**, 817–19.

LAMON, J.M., FRYKHOLM, B.C., HERRERA, W., AND TSCHUDY, D.P. (1979). Danazol administration to females with menses-associated exacerbations of acute intermittent porphyria. *J. clin. Endocrinol. Metab.* **48**, 123–6.

LARTER, R.A. (1978). Anesthesia and the porphyrias. *J. Am. Ass. nurse Anesthet.* **46**, 271–81.

LEVIT, E.J., NODINE, J.H., AND PERLOFF, W.H. (1957). Progesterone-induced porphyria. *Am. J. Med.* **22**, 831–3.

LIM, L.K., SRIVASTAVA, G., BROOKER, J.D., MAY, B.K., AND ELLIOTT, W.H. (1980). Evidence that in chick embryos destruction of

hepatic microsomal cytochrome P-450 haem is a general mechanism of induction of δ-aminolaevulinate synthase by porphyria-causing drugs. *Biochem. J.* **190**, 519–26.

LINDEN, I.H., STEFFEN, C.G., NEWCOMER, V.D., AND CHAPMAN, M. (1954). Development of porphyria during chloroquine therapy for chronic discoid lupus erythematosus. *Calif. Med.* **81**, 235–8.

MACALPINE, I., HUNTER, R., AND RIMINGTON, C. (1968). Porphyria in the Royal Houses of Stuart, Hanover and Prussia. In *Porphyria, a royal malady*, pp. 1–16, 17–57. British Medical Association, London.

MAGNUS, I.A., CORBETT, M., AND HERXHEIMER, A. (1979). A national register for erythropoetic protoporphyria. *Br. med. J.* **2**, 508–9.

MARSDEN, C.W. (1959). Porphyria during chloroquine therapy. *Br. J. Dermatol.* **71**, 219–22.

McKENZIE, A.W. AND ACHARYA, U. (1972). The oral contraceptive and variegate porphyria. *Br. J. Dermatol.* **86**, 453–7.

MILLAR, J.W. (1980). Rifampicin-induced porphyria cutanea tarda. *Br. J. Dis. Chest* **74**, 405–8.

MOORE, M.R. (1980). International review of drugs in acute porphyria — 1980. *Inter. J. Biochem.* **12**, 1089–97.

—— AND DISLER, P.B. (1983). Drug-induction of the acute porphyrias. *Adv. Drug React. Ac. Pois. Ref.* **2**, 149–89.

NAYRAC, P., GRAUX, P., FOURLINNIE, J.C., AND PETIT, H. (1964). A propos de deux cas de porphyries (dont une forme mixte délenchée par l'association oestradiol-progestérone). *Lille méd.* **9**, 704–11.

PERLROTH, M.G., MARVER, H.S., AND TSCHUNDY, D.P. (1965). Oral contraceptive agents and the management of acute intermittent porphyria. *J. Am. med. Ass.* **194**, 1037–42.

POH-FITZPATRICK, M.B., BELLETT, N., DE LEO, V.A., GROSSMAN, M.E., AND BICKERS, D.R. (1978). Porphyria cutanea tarda in two patients treated with haemodialysis for chronic renal failure, *New Engl. J. Med.* **299**, 292–4.

PRUNTY, F.T.G. (1946). Acute porphyria: Investigations on pathology of porphyrins and identifications of excretion of uroporphyrin. I. *Archs intern. Med.* **77**, 623–42.

REDEKER, A. (1963). Conference discussion. *S. Afr. J. Lab. clin. Med.* **9**, 302–3.

RIMINGTON, C. (1964). Drug and enzyme interactions in the porphyrias. *Proc. R. Soc. Med.* **57**, 511–14.

——, MORGAN, P.N., NICHOLLS, K., EVERALL, J.D., AND DAVIS, R.R. (1963). Griseofulvin administration and porphyrin metabolism. A survey. *Lancet* **ii**, 315–22.

ROENIGK, H.H. AND GOTTLOB, M.E. (1970). Estrogen-induced porphyria cutanea tarda. *Arch. Dermatol.* **102**, 260–66.

ROOK, A. AND CHAMPION, R.H. (1960). Porphyria cutanea tarda and diabetes. *Br. med. J.* **1**, 860–1.

SCHMID, R. (1966). The porphyrias. In *The metabolic basis of inherited disease* (eds. J.B. Stanbury, J.B. Wyngaarden, and D.S. Fredrickson). McGraw-Hill, New York.

SHERLOCK, S. (1968). The liver in the porphyrias. In *Diseases of the liver and biliary system*, 4th edn. Blackwell, Oxford.

STOCKVIS, B.J. (1889). Over twee Zeldzame Kleurstoffen in urine van zieken. *Ned. T. Geneesk.* **25**, 409–17.

STRAND, L.J., FELSHER, B.F., REDEKER, A.G., AND MARVER, H.S. (1970). Enzymatic abnormality in heme biosynthesis in acute intermittent porphyria decreased hepatic conversion of porphobilinogen to porphyrins and increased δ-aminolevulinic acid synthetase activity. *Proc. natl Acad. Sci. USA* **67**, 1315-20.

SWEENEY, G.D., SAUNDERS, S.J., DOWDLE, E.B., AND EALES, L. (1965). Effects of chloroquine on patients with cutaneous porphyria of the 'Symptomatic' type. *Br. med. J.* **1**, 1281-5.

TEPHLY, T.R., GIBBS, A.H., INGALL, G., AND DE MATTEIS, F. (1980). Studies on the mechanism of experiments porphyric and ferrochelatase inhibition produced by 3,5-diethoxycarbonyl-1,4-dihydrocollidine. *Int. J. Biochem.* **12**, 993-8.

THORNE, N. (1966). Cosmetics and the dermatologist. The lucites and sunscreening preparations. *Br. J. clin. Pract.* **20**, 443-4, 447.

TOPI, G.C., GANDOLFO, L.D'A., DE COSTANZA, F., AND CANCARINI, G.C. (1980). Porphyria and pseudo-porphyria in haemodialyzed patients. *Int. J. Biochem.* **12**, 963-7.

TREECE, G.L., MAGNUSSEN, C.R., PATTERSON, J.R., AND TSCHUDY, D.P. (1976). Exacerbation of porphyria during treatment of pulmonary tuberculosis. *Am. Rev. resp. Dis.* **113**, 233.

TURNER, W.J. (1938). Studies on porphyria. III. Acute idiopathic porphyria. *Arch. intern. Med.* **61**, 762-73.

VAN DER GRIENT, A.J. (1968). Antifungal drugs. In *Side-effects of drugs*, Vol. VI. (ed. L. Meyler and A. Herxheimer), pp. 241-62, 315-19. Excerpta Medica Foundation, Amsterdam.

WALDENSTRÖM, J. (1937). Studien über die Porphyrie. *Acta. med. scand.* (Suppl.) **82**, 254.

—— (1939). Neurological symptoms caused by so-called acute porphyria. *Acta psychiat. scand.* **14**, 375-9.

WARD, R.J. (1965). Porphyria and its relation to anesthesia. *Anesthesiology* **26**, 212-15.

WATSON, C.J., RUNGE, W., AND BOSSENMAIER, I, (1962). Increased urinary porphobilinogen and uroporphyrin after administration of stilboestrol in a case of latent porphyria. *Metabolism* **11**, 1129-33.

WEIMAR, V.M., WEIMAR, G.W., AND CEILLEY, R.I. (1978). Estrogen-induced porphyria cutanea tarda complicating treatment of prostatic-carcinoma. *J. Urol.* **120**, 643-4.

WELLAND, F.H., HELLMAN, E.S., GADDIS, E.M., COLLINS, A., HUNTER, G.W.JR., AND TSCHUDY, D.P. (1964). Factors affecting the excretion of porphyrin precursors by patients with acute intermittent porphyria. I. The effect of diet. *Metabolism* **13**, 232-50.

WETTERBERG, L. (1964). Oral contraceptives and acute intermittent porphyria. *Lancet* **ii**, 1178-9.

WHITE, M.I. (1977). Porphyria cutanea tarda induced by oestrogen therapy. *Br. J. Urol.* **49**, 468.

WHITTAKER, S.R.F. AND WHITEHEAD, T.P. (1956). Acute and latent porphyria. *Lancet* **i**, 547-51.

WITH, T.K. (1965). Porphyria. *Lancet* **i**, 916-17.

ZAROWITZ, H. AND NEWHOUSE, S. (1965). Coproporphyrinuria with cutaneous reaction induced by chlorpropamide, *NY med. J.* **65**, 2385-7.

ZIPRKOWSKI, L., SVEINBERG, A., CRISPIN, M., KRAKOWSKI, A., AND SAIDMAN, J. (1966). The effect of griseofulvin in hereditary porphyria cutanea tarda. *Arch. Dermatol.* **93**, 21-7.

7 Drug-induced allergic and hypersensitivity reactions

J.P. GRIFFIN

Not every idiosyncratic reaction to a drug is due to allergic hypersensitivity. In the course of the years, various types of unexpected and unforeseeable drug-induced effects originally attributed to allergic mechanisms have been found to be due to biochemical abnormalities. These lead to a derangement of drug metabolism and thereby to an abnormal or exaggerated toxic response. True allergic and anaphylatic reactions to medicines do, however, quite commonly occur, and in a minority of cases they may be life-threatening. Large molecular drugs such as the polypeptide hormones can themselves act as complete antigens and, in exceptional cases, large-molecular contaminants or degradation products can lead to hypersensitivity reactions during medical treatment. Much more commonly, however, the drug involved in an allergic incident has only a low molecular weight and will only act as an antigen after it has become bound to a macromolecule. It is not clear whether the liability of certain persons to experience hypersensitivity reactions is due to the fact that they are more prone to form such immunogenic complexes or to the fact that their system responds more acutely to their presence. However, there can be no doubt that for many drugs a small fraction of the population is likely to respond adversely by exhibiting one or other of these reactions. It is often difficult to determine whether a 'hypersensitivity reaction' is occurring. If an adverse response is to be considered a hypersensitivity reaction it should meet most of the following criteria.

1. Sensitization to the drug substance may take some time to develop following either continuous or intermittent exposure.
2. After an allergic state has been established, an allergic reaction can be precipitated by minute amounts of the drug.
3. The reaction recurs upon repeated exposure to the allergen.
4. The reaction does not resemble the pharmacological action of the drug.
5. The symptoms are suggestive of known forms of allergic responses.

The frequency of allergic drug reactions is unknown but has been shown to comprise approximately 20–30 per cent of reported adverse drug reactions.

Hypersensitivity reactions were classified by Coombs and Gell (1975) into four main clinical types: Type I — anaphylactic; Type II — cytotoxic; Type III immune complex-mediated; and type IV — cell-mediated injury. Clinical hypersensitivity reactions to drugs may involve more than one type of reaction. Hypersensitivity reactions to drugs involve all of Types I, II, III, or IV.

Type I hypersensitivity reactions

These are of the 'immediate-type' and involve an antigen–antibody contact on the surface of the basophils and mast cells, as a result of which pharmacologically active mediators are released. It should be recalled that some of these mediators — histamine and prostaglandin E_2 — are themselves substances which are on occasion administered medicinally. A Type I reaction may be limited to cutaneous weals and flares, but it can also result in life-threatening systemic anaphylaxis, asthma, laryngeal angioneurotic oedema, or combinations of these. These symptoms and signs may occur within minutes of receiving the drug.

A few drugs release histamine from mast cells by a direct pharmacological action and in these cases there is no antibody–antigen reaction despite all the signs and symptoms of the effects of the chemical mediators being apparent. Drugs that release histamine in this way are tubocurarine, pethidine, morphine, and some antihistamines.

The acute anaphylactic reaction

In 1965, it was estimated that there might be as many as 300 fatal anaphylactic reactions annually to penicillin in the USA alone; no other country appears to have produced so high an estimate (Idse *et al.* 1968), but this grave

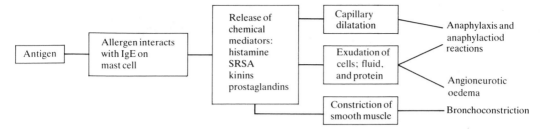

Fig. 7.1 Type I hypersensitivity reaction.

condition, commonly induced by a drug considered almost innocuous, is by no means rare. The Indonesian Committee for Adverse Reactions monitoring in the two-year period June 1981–May 1983 received 297 adverse reaction reports, 82 (27.6 per cent) of these were reports of anaphylactic shock with 24 fatalities. Seven of the fatalities were attributed to streptomycin, six to procaine penicillin, and five to penicillin.

The symptoms of anaphylactic shock are characteristic. Within some minutes of taking or absorbing the allergen, the susceptible individual develops peripheral circulatory collapse, severe bronchoconstriction, or both. His respiratory symptoms may not be dominant and they may indeed be absent altogether; subjectively the patient first experiences malaise with nausea, a metallic taste in the mouth, and commonly an urge to micturate or defaecate. In exceptional cases prolonged unconsciousness or vomiting may be major signs. Death may be due either to asphyxia or to cardiac arrhythmias with shock. Prompt treatment, commonly involving intensive care, is needed to help the patient through the emergency. Intramuscular injections of adrenaline (0.5 ml or 1 ml of a 1 : 1000 solution), repeated as often as needed at intervals of a few minutes, are always the first therapeutic measures to sustain a measurable blood pressure. Subcutaneous administration may be followed by retarded absorption as a result of impaired circulation in the small vessels. Intravenous injections of adrenaline, given very slowly, may sometimes be required to save the patient's life. Parenteral steroids in high doses should also be administered, eg. 'Dexamethasone Shock-Pack'. It is advisable, however, to supplement this primary therapy with a parenteral antihistamine (eg. chlorphenir-amine, 10 mg i.m.). Table 7.1 gives a list of the drugs most commonly reported to the Committee on Safety of Medicines (CSM) as being incriminated as precipitating anaphylactic or anaphylactoid reactions over the period 1964–83.

Anaphylactic shock due to acetylcysteine when given intravenously for the treatment of paracetamol

Table 7.1 Most common causes of drug-associated anaphylaxis or anaphylactoid reactions spontaneously reported in yellow cards to CSM Jan 1964–Dec 1983 (after Griffin 1983)

Suspect drug	Number of reactions reported	Deaths related
Althesin*	65	7
Dextran	43	9
Diatrizoic acid	40	11
Tetracosactrin	39	6
Specific desensitizing vaccine	36	7
Iron-dextran complex	35	2
Zomepirac	35	—
Grass pollen vaccine	30	1
Measles vaccine (live attenuated)	29	—
Sodium iothalamate	25	7
Diphtheria vaccine	23	—
Thiopentone	21	7
D/Vac pollen	21	—
Alcuronium	17	2
Pertussis vaccine	17	1
Tetanus vaccine	17	1
Co-trimoxazole	17	—
Penicillin NOS	16	15
Nalidixic acid	16	—
Acetylcysteine	16	—
Salmonella typhi vaccine	15	—
Suxamethonium	14	—
Lignocaine	13	2
Cephalexin	12	—
Propanidid*	10	—
Benzylpenicillin	10	3
BCG	10	—
Aprotinin	10	3
Dust mite extract	10	3

* *Althesin* and propanidid contain *Cremophor EL* as an excipient. Anaphylactoid reactions with these products are probably related to this solubilizing agent.
* Tartrazine used as a colouring agent can precipitate anaphylaxis.

poisoning was reported by Vale and Wheeler (1982). These authors also reviewed five other cases of allergic reactions to intravenous acetylcysteine which had been reported to the manufacturer: four of these five cases developed urticaria; three developed angioneurotic oedema affecting face and tongue; and in two cases there was severe bronchoconstriction. The CSM has received reports of severe bronchoconstriction to acetylcysteine when the substance has been given intravenously in the treatment of paracetamol overdose and when it has been given orally or by inhalation as a mucolytic.

Repeated use of cisplatin for bladder irrigation for the treatment of local malignant conditions was observed to cause anaphylactic shock in seven out of a series of 67 patients treated in this way by Denis (1983), despite the fact that it has been claimed that there is no systemic absorption of cisplatin when given intravascularly. Diclofenac-induced anaphylactic shock was described by Dux et al. (1983).

Drug-induced asthma

Of the various drugs capable of precipitating an asthmatic attack, including status asthmaticus, aspirin is the most notorious, although it is still not clear whether the problem is due to the compound itself or to combination with aspiryl anhydride, which is apparently highly immunogenic. The mechanism involved, for that matter, is also disputed; it is possible that selective inhibition of the synthesis of prostaglandin E is involved.

Some 25 per cent of asthmatic patients are intolerant to aspirin and many of these are also sensitive to other anti-inflammatory analgesics including some of quite different chemical structures. These include ibuprofen, indomethacin, fenoprofen, flufenamic acid, mefenamic acid, paracetamol, zomepirac, and phenylbutazone. Aspirin intolerance also occurs with benorylate, which is metabolized to aspirin, and with diflunisal, which is one of the latest salicylic acid derivatives.

A number of other classes of drugs have been shown to be capable of causing severe attacks of isolated bronchospasm, that is, bronchospasm not associated with a full anaphylactic syndrome. However, in various cases it is almost certain that the mechanism is not allergenic, for example, with β-adrenergic blocking agents in asthmatic patients. In other cases the allergic reaction may not be due to the drug substance but to an excipient or colorant such as tartrazine, or to cremophor in certain intravenous anaesthetic agents such as Althesin and propanidid or parenteral vitamin K.

Bronchospasm usually sets in within 15–30 minutes of taking the allergen and it may be extremely severe and resistant to treatment. A dose of 1 ml of Adrenaline

Table 7.2   Drugs most commonly reported to induce asthma or bronchospasm from spontaneous reports to CSM Jan 1964–Dec 1983 (after Griffin 1983)

Suspect drug	Number of reactions reported	Deaths related
Althesin*	71	5
Nadolol	57	—
Timolol**	40	4
Atenolol**	40	—
Ibuprofen	37	1
Indomethacin	26	—
Nitrofurantoin	25	—
Diatrizoic acid	24	—
Oxprenolol**	22	4
Sodium iothalamate	21	—
Dextran	19	—
Isoprenaline	18	10
Propranolol**	18	4
Naproxen	17	—
Ipratropium bromide	17	—
Benoxaprofen	16	—
Metoprolol tartrate**	15	—
Flurbiprofen	14	—
Salbutamol	13	9
D/Vac pollen	13	—
Labetalol**	13	3
Cimetidine	13	—
Ethinyloestradiol	12	—
Diclofenac sodium	12	—
Zomepirac	12	—
Co-trimoxazole	12	—

*   Althesin-induced bronchoconstriction is probably related to the Cremophor EL present as a solubilizing agent.
**   Bronchoconstriction associated with β-adrenergic blocking agents is non-allergenic.
Note: Tartrazine used as a colouring and aspirin widely incorporated in many proprietary medicines can cause severe asthma.

Injection BP, given subcutaneously is the immediate treatment of choice, followed as necessary by parenteral corticosteroids and either sympathomimetic or corticosteroid aerosols. A very few patients develop pneumonitis and pulmonary oedema, particularly if treatment is delayed.

Table 7.2 gives a list of the drugs most commonly reported as being associated with drug-induced bronchoconstriction or asthma over the period 1964–83.

Pulmonary alveolitis

Nitrofurantoin accounted for 921 adverse reaction reports to the Swedish Drug Regulatory Authorities in the period 1966–76, that is, 8 per cent of the total input to their adverse reaction register (Penn and Griffin 1982). Of these adverse reactions 48.5 per cent were pulmonary allergic reactions. In the UK adverse reactions to nitrofurantoin were less common and the 455 reports received formed only 0.5 per cent of the total input and of this input only 16.1 per cent were allergic pulmonary reactions (Penn and Griffin 1982). The reasons for this greater number of allergic pulmonary reactions in Sweden is unclear, but it might be related to the earlier use of nitrofurans as growth promotors in various sectors of the meat and poultry industry, resulting in a sensitized population.

Nitrofurantoin remained the commonest source of adverse reaction reports up to 1981 in Sweden. Allergic pulmonary reactions have been reported by Nader and Schillaci (1983) for naproxen, and by Akoun and Perrot (1983) for acebutolol.

Drug-induced angioneurotic oedema

Angioneurotic oedema presents as an emergency when the larynx is involved in a Type I reaction. This complication is much less common than the involvement of the face and the area of the skin where the allergen has been injected, but it should always be anticipated as a risk, since immediate tracheostomy may be required. Table 7.3 gives a list of the drugs most commonly reported to the CSM as being associated as precipitating angioneurotic oedema.

Drug-induced urticaria

The 25 drugs most commonly associated with urticarial eruptions based on spontaneous reports to the CSM are shown in Table 7.4.

Type II hypersensitivity reactions

These result from a complement-fixing reaction between antigen and antibody on the cell surface, leading to cytolysis. A possible consequence where the erythrocytes are involved is haemolysis, but white-cell lysis or platelet lysis can also occur. Combinations of these events may be observed and (in theory, at least) other types of cell in the body may be damaged in the same way.

Red cell immune disorders

Four different mechanisms have been described in drug-induced anaemia. Firstly, the drug or its metabolite acts as a hapten and binds with protein present on the red cell membrane to form a complete antigen. A specific antibody is then formed to the antigen attached to the red blood cell. This may cause a positive direct antiglobulin

Table 7.3 Most common causes of drug-induced angioneurotic oedema spontaneously reported on yellow cards to CSM Jan 1964–Dec 1983 (after Griffin 1983)

Suspect drug	Number of reactions reported	Deaths related
Co-trimoxazole	39	2
Naproxen	31	
Ibuprofen	19	
Diatrizoic acid	16	
Benoxaprofen	14	
Zomepirac	14	
Piroxicam	12	
Metronidazole	11	
Azapropazone	11	
Feprazone	11	
Fenbufen	11	
Indomethacin	10	
Ketoprofen	10	
Cimetidine	10	
Fenoprofen calcium	9	
Fenclofenac	9	
Diclofenac sodium	9	
Acetylsalicylic acid	8	1
Carbamazepine	8	
Ampicillin	8	1
Meprobamate	7	
Metoclopramide	7	
Phenylbutazone	7	
Paracetamol	7	
Trimethoprim	7	
Althesin	7	
Sulindac	7	
Flurbiprofen	7	
Diflunisal	7	

Note: 18 of the 29 drug substances most frequently reported to CSM as producing angioneurotic oedema are non-steroidal anti-inflammatory agents.

test (direct Coomb's Test). The antigen–antibody reaction may lead to haemolytic anaemia: ampicillin, carbenicillin, cephalosporins, methicillin, penicillin G, and streptomycin have all been described as being capable of inducing this type of reaction.

Secondly, immune complexes of drug- and anti-drug–antibody may be formed in the plasma, and this immune complex is non-specifically absorbed onto the red blood cell with subsequent activation of complement. Since the initial step is the formation of the drug immune complex, involvement of the red blood cell is secondary, and the red cell is an 'innocent-bystander' in the reaction. Chlorpromazine, chlorpropamide, hydrallazine, isoniazid, sulphonamides, streptomycin, stibophen, quinine, quinidine, phenocetin, dipyrone, para-aminosalicylic acid (PAS), rifampicin, tetracyclines, thiazides, and tolbutamide may cause this type of haemolytic anaemia.

Thirdly, drugs such as α-methyldopa, levodopa, mefe-

Table 7.4 Most common causes of drug-induced urticarial reactions reported to CSM Jan 1964–Dec 1983 (after Griffin 1983)

Suspect drug	Most important reaction	Deaths related
Fenbufen	199	—
Co-trimoxazole	194	—
Benoxaprofen	125	—
Ampicillin	110	—
Fenclofenac	104	—
Feprazone	98	—
Cimetidine	88	—
Alclofenac	64	—
Amoxycillin	55	—
Piroxicam	54	—
Nalidixic acid	53	—
Mefenamic acid	48	—
Paracetamol	45	—
Ethinyloestradiol	45	—
Diflunisal	45	—
Naproxen	42	—
Azapropazone	42	—
Zomepirac	42	—
Ibuprofen	39	—
Maprotiline	39	—
Indomethacin	37	—
Oxprenolol	35	—
Trimethoprim	35	—
Diatrizoic acid	35	—
Metronidazole	34	—

Note: 13 of the 25 drug substances most likely to cause urticarial reactions are non-steroidal anti-inflammatory agents.

namic acid, and vaccines such as typhoid, poliomyelitis and triple vaccine may trigger the formation of anti-red cell antibody. Approximately 10–20 per cent of patients receiving methyldopa develop a positive Coomb's test, but only 0.5–1.0 per cent develop a haemolytic anaemia. Approximately 70 per cent of reported drug-induced immune haemolytic anaemias are due to methyldopa.

Lastly, cephalosporins are known to chemically modify the red cell membrane and allow plasma proteins to attach through a non-immunological mechanism, this manifests itself as a positive Coomb's test, but only in rare cases has a haemolytic anaemia ensued.

White-cell immune disorders

The immunological basis of drug-induced agranulocytosis is not entirely clear, it is believed that the drug or metabolite attaches to the granulocyte as a hapten and forms an antigen which is then destroyed by the antibody and its complement. Amidopyrine is a classical example of a hapten–antibody reaction of this type and granulocytopoenia usually occurs 7–14 days after starting the drug if the drug is used continuously or more rapidly in a previously sensitized patient. Other drugs reported to cause this type of reaction are cephalothin, chlorpropamide, dipyrone, methimazole, phenylbutazone, sulphonamides, thiouracil, and tolbutamide.

Platelet-immune disorders

The drug or a metabolite acts as a hapten on the platelet, and lysis of the platelet follows in the presence of complement and antigen–antibody complexes. This results in thrombocytopoenic purpura. Drugs implicated in this type of reaction are acetazolamide, aspirin, antazoline, carbamazepine, cephalosporins, chlorthiazide, digitoxin, hydrochlorthiazide, imipramine, meprobamate, methyldopa, novobiocin, phenolphthalein, phenytoin, quinine, quinidine, rifampicin, spironolactone, stibophen, and thioguanine.

Vascular endothelium-immune disorder

A vascular endothelium disorder may be induced by Type II or Type III hypersensitivity reactions. The mechanism may involve an immune complex sensitization to vascular endothelium which results in the loss of integrity of small blood vessels, purpura, and other haemorrhagic manifestations; other variations may be arthropathy and renal lesions of proliferative glomerulonephritis. Chlorpromazine, iodides, penicillins, phenylbutazone, phenytoin, propylthiouracil, quinidine, and sulphonamides have been incriminated in various reports.

The treatment of most Type II hypersensitivity reactions involves the withdrawal of the offending drug,

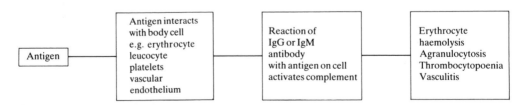

Fig. 7.2 Type II hypersensitivity reaction.

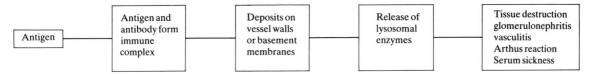

Fig. 7.3 Type III hypersensitivity reaction.

and the judicious use of corticosteroids may hasten recovery.

Type III hypersensitivity reactions

These are known as toxic immune complex reactions. These reactions occur when antibodies (usually IgM or IgG) react with tissue antigens, or when soluble antigen–antibody complexes are formed. These complexes deposit on tissue target cells. The complement is then activated and causes accumulation of neutrophilic leucocytes, which release lysosomal enzymes that cause tissue destruction. Diseases caused by this mechanism include glomerulonephritis, rheumatic (collagen) diseases, and various vasculitic skin eruptions. The predominant types of drug reactions that fall into this category are serum sickness (a systemic form) and arthus reaction (a localized form).

Drugs implicated in Type III hypersensitivity reactions are ACTH, barbiturtes, chloral, dextro-iron-complexes, penicillins, erythromycin, griseofulvin, heparin, hydrallazine, insulin, iodides, iodine contrast media, isoniazid, nitrofurantoin, phenolphthalein, phenylbutazone, phenytoin, procainamide, propylthiouracil, quinidine, quinine, salicylates, serums, streptomycin, sulphonamides, vaccines, and viomycin.

Drug-induced allergic injury to the kidneys

Although the kidneys, like the skin, are probably quite commonly involved in Type III allergic responses to drugs, severe complications are generally rare, except in patients treated with penicillamine. The nephrotic syndrome which has been described repeatedly in patients taking this compound is due to deposition of IgG on the basement membrane. Acute renal failure with oliguria can also very rarely be induced indirectly by allergic mechanisms, for example by para-aminosalicyclic acid (PAS). The primary process seems to be an immune haemolytic reaction; haemoglobinaemia, shock, and renal injury then ensue.

It must be emphasized that patients who have already experienced Type III renal damage should not undergo hyposensitization with the allergen concerned since progressive renal failure may well result. The same situation may arise in patients who are known to be hypersensitive if the original allergen continues to be given with immunosuppressants; the latter will suppress the acute allergic response but they will not protect the kidney from further injury.

Drug-induced lupus-erythematosus syndrome

For many years it has been recognized that a number of drugs can give rise to a syndrome in certain patients which has a very close resemblance to the systemic form of lupus erythematosus (SLE), and in some cases, it is virtually indistinguishable from it clinically. Most of the manifestations of SLE may be seen as adverse reactions to a number of drugs, but on the whole, renal involvement is not seen, and the severity of the disease is less in the drug-induced type. Joint pain, muscle pain and tenderness, a variable degree of lymphadenopathy and hepatosplenomegaly, lung involvement, pericarditis, Raynaud's phenomenon, and rash may all be seen, and pyrexial attacks may also occur. Laboratory investigations often give results typical of true SLE including the presence of LE cells, antinuclear antibodies, and antibodies to single-strand DNA. Antibodies to native DNA are very rarely found, and the DNA-binding capacity to date has been negative in the drug-associated SLE, in contradistinction to the finding in true SLE.

It is still a matter for speculation whether drug-induced SLE is an entity distinct from the spontaneously occurring type, or whether the administration of certain drugs leads to an exhibition of SLE by patients in whom the disease has been dormant, or who have a genetic predisposition to the disorder. It is known, however, that some cases may persist after discontinuation of the therapy responsible, but recovery is the more likely outcome when the drug is stopped. Occasionally, treatment of drug-induced SLE by the use of corticosteroids is necessary, but this must be regarded as the exception rather than the rule. Hydrallazine was the first drug to be recognized as having a potential for leading to the appearance of SLE, and this occurred in approximately 10 per cent of patients treated with the drug for hypertension on a long-term basis. An autoimmune basis was postulated and anti-bodies to hydrallazine could be demonstrated. The most frequently involved drugs are hydrallazine, isoniazid, phenytoin, and procainamide. Other important drugs include PAS, tetracyclines, streptomycin, sulphon-amides, methyldopa, methylthiouracil, D-penicillamine, and oral contraceptives. Reports of exacerbation of LE

Table 7.5   Drugs most commonly associated with LE syndrome based on spontaneous reports to the CSM Jan 1964–Dec 1983 (after Griffin 1983)

Suspect drug	Number of reactions reported	Deaths related
Hydrallazine	61	
Procainamide	22	
Practolol	18	
Labetalol	7	
Sulphasalazine	6	
Carbamazepine	4	
Ethinyloestradiol	4	
Methylopa	3	
Oxprenolol	3	
Co-trimoxazole	3	
Pheneturide	2	
Phenylbutazone	2	
Sodium aurothiomalate, other gold compounds	2	
Guanoxan	2	
Propranolol	2	
D-penicillamine	2	
Cimetidine	2	
Phenytoin	1	
Ethosuximide	1	
Fluphenazine	1	
Chlordiazepoxide	1	
Lithium carbonate	1	
Lignocaine	1	
Mestranol	1	
Propylthiouracil	1	
Tolbutamide	1	
Cyclopenthiazide	1	1
Ampicillin	1	
Streptomycin	1	
Oxytetracycline	1	
Sulphametopyrazine	1	
Isoniazid	1	
Nalidixic acid	1	
Yellow fever vaccine	1	
Tamoxifen	1	
Sodium valproate	1	
Levamisole	1	
Acebutolol	1	
Bromocriptine	1	
Atenolol	1	
Aminoglutethemide	1	
Insulin neutral Human (EMP)	1	1

by griseofulvin, rifampicin, and clofibrate have been published.

The drugs most commonly reported to the CSM as associated with LE syndrome, are shown in Table 7.5.

Drug-induced serum sickness

Intensive production of immune complexes following the administration of a drug can lead to the development of a severe reaction of the serum-sickness type. The name of this syndrome incorrectly suggests that it will result only from treatment with sera; in fact many drugs have now been implicated, including metronidazole (Weart and Hyman 1983), penicillins, and sulphonamides, aspirin and thiouracil. In some of those cases circulating anti-bodies have been unequivocally detected.

The timespan of a reaction is extraordinarily variable: occasional cases arise within a few hours of taking the offending drug, but a delay of some 7–10 days is more usual, and in a few instances symptoms appear much later, sometimes even after the drug has been withdrawn. The symptoms are likely to include pyrexia, painful swelling of the joints, lymphadenopathy, and urticaria. Two possible complications of the full-blown syndrome which may lead to emergency situations are myocarditis and acute intestinal nephritis.

Arthus reaction

The Arthus reaction is a localized serum-sickness reaction. It is localized because the sensitized person already has high antibody titres. As in serum sickness, antigen–antibody immune complex deposition occurs; it is, however, localized and may occur within 2–5 hours after injection of the antigen.

Probable clinical counterparts of the Arthus reaction include various hypersensitivity pneumonitis or extrinsic allergic alveolitis, for example, with nitrofurantoin.

Type IV reactions

These reactions are the 'delayed' type' and are a con-sequence of a direct reaction between an allergen and sensitized lymphocytes. Generally only cutaneous symp-toms result, but this type of mechanism has been held responsible for some serious types of drug reactions, including halothane-induced hepatotoxicity.

Most contact sensitivities are of this type. The mech-anism is the formation of a complete antigen in the skin by the binding of haptens to various proteins present in the skin; the SH and NH_2 groups of cysteine and cystine are known binding sites. The same allergen may bind

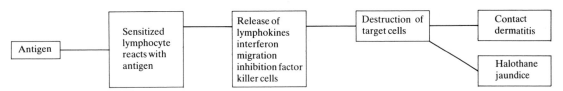

Fig. 7.4 Type IV hypersensitivity reaction.

with one or more skin protein; thus the clinical reaction may involve the combined response to binding of a hapten to multiple proteins. When the lymphocytes are exposed to the allergen, sensitization and proliferation of a clone of sensitized T-lymphocytes result and this is followed by the elicitation of the reaction involving release of lymphokines.

Clinically eczematous dermatitis elicited by contact sensitization is the most usual manifestation: acutely the reaction may be vesiculation, erythema, and oedema, but chronically exposed sensitized skin may show erythema, lichenification, and hyperpigmentation. The most common medicaments or ingredients of medicaments to cause contact sensitivity are benzyl alcohol, chromium compounds, ethylenediamine compounds, formaldehyde, lanolin, mercury derivatives such as merthiolate, mercurochrome, thiomersal, neomycin, nickel, para-aminobenzoic acid compounds, e.g. benzocaine, parabens, Peru balsam, and phenylenediamine compounds. Treatment is by identification and subsequent avoidance of the allergen; topical corticosteroids may be necessary. Topical antihistamines paradoxically may be potent sensitizers, particularly ethylenediamine-based molecules. Ethylenediamine hypersensitivity is a complex problem and multifactorial and will therefore be dealt with in some detail.

Ethylenediamine allergy

Allergic reactions to ethylenediamine were first observed in pharmacists who developed contact dermatitis of the hands and face while preparing aminophylline (theophylline–ethylenediamine) suppositories (Tas and Weissberg 1958; Baer *et al.* 1959). Similar reports of contact dermatitis due to ethylenediamine have been reported following industrial exposure in the rubber industry.

Ethylenediamine is used as an excipient in a number of dermatological creams, particularly antibacterial, antifungal creams, and some topical corticosteroid preparations, e.g. *Tri-Adcortyl* cream (triamcinolone acetonide 0.1 per cent, gramicidin 0.025 per cent, neomycin 0.25 per cent, and nystatin 100 000 units, parabens, and ethylenediamine). Papers by Burrey *et al.* (1973), White *et al.*

(1978), and Wright and Harman (1983) state that *Tri-Adcortyl* cream is the commonest cause of contact allergy to ethylenediamine.

Aminophylline is the ethylenediamine salt of theophylline and is used because its greater solubility enables it to be given parenterally as an injection or rectally as a suppository. In sensitized patients the parenteral administration of any ethylenediamine salts may cause rashes, fever, lymphadenopathy, and bronchospasm.

The recognition of ethylenediamine hypersensitivity may be difficult since it may present as a worsening of an existing dermatological condition which is being treated with a topical preparation containing ethylenediamine in its base.

The recognition of ethylenediamine-induced bronchoconstriction is comparatively easy in an industrial context (Tas and Weissberg 1958; Gelfand 1963; Lam and Chan-Yeung 1980). It is more difficult to identify in the context of the worsening bronchospasm in an asthmatic subject treated with parenteral aminophylline, which may be wrongly attributed to lack of efficacy, and the aminophylline injection may even be repeated. Elias and Levinson (1981) described an asthmatic patient with an aminophylline rash associated with worsening bronchospasm, and it is tempting to postulate that the worsening bronchoconstriction being recognized as an allergic reaction to ethylenediamine in this case was aided by the concurrent dermatitis. Hardy *et al.* (1983) described a case of a 61-year-old woman who developed an acute dermatitis after taking *Phyllocontin* (a slow-release oral formulation of aminophylline). Gibb and Thompson (1983) summarized data on 15 allergic skin reactions following parenteral aminophylline as three cases of acute dermatitis, six cases of generalized erythema, and six cases of exfoliactive dermatitis.

Immunological mechanism

There is considerable debate about the mechanism of aminophylline allergy. Hardy *et al.* (1983) state

The rash may take various forms but does not usually occur until 18 to 24 hours after exposure and thus is more characteristic of a cell mediated hypersensitivity than one involving IgE.

In other words, they believe that rashes are a Type IV reaction. Conversely Gibb and Thompson (1983) consider

The precise immunological mechanisms of aminophylline allergy are unknown, although in most cases positive reaction to skin patch tests with delayed response support a cell mediated mechanism. Patch tests have also given negative reactions; however, cases of generalised erythema, and clearly the few cases of acute urticaria are Type I immediate hypersensitivity reactions.

Ethylenediamine antihistamines and cross-sensitivity reactions

Antihistamines of the ethylenediamine type should not be used to treat any allergic reaction where there is a possibility of ethylenediamine sensitivity.

Cross-sensitivity between drugs used topically and related drugs used systemically is well recognized with the ethylenediamine group of antihistamines, e.g., mepyramine, tripelennamine. (White *et al.* 1978). White *et al.* also drew attention to the dangers of administering parenteral aminophylline to a patient with a contact allergy to ethylenediamine.

Burry (1978), Calnon (1975), and Wright and Harman (1983) all drew attention to the cross-sensitivity with piperazine in patients who have contact allergy to ethylenediamine. In cases described by Wright and Harman (1978) and Burry (1978) the contact sensitization was caused by ethylenediamine in *Tri-Adcortyl* cream (Squibb) or its Australian equivalent *Kenacomb* and subsequent reactions to piperazine. In the case of the 37-year-old man described by Wright and Harman the patient had a patchy red scaling rash affecting forearms, calves, hands, and face which had existed for 20 years and the topical preparation which he had used most frequently had been *Tri-Adcortyl*. On three occasions twice therapeutically and once as a test rechallenge, the patient received piperazine. In the test situation rechallenge, which was his third exposure to piperazine, 50 µg piperazine provoked a generalized macular papular rash within hours, and with shivering, anxiety, and a tachycardia of 140 beats/min.

To sum up, therefore, sensitization to ethylenediamine can occur as a result of exposure to the substance as an excipient in the base of creams used in topical medicines or cosmetics or from industrial exposure, or from parenteral aminophylline. Cross-sensitivity with ethylenediamine antihistamines, aminophylline, and piperazine can occur. See Fig. 7.5 for structural formulae of cross-reacting substances.

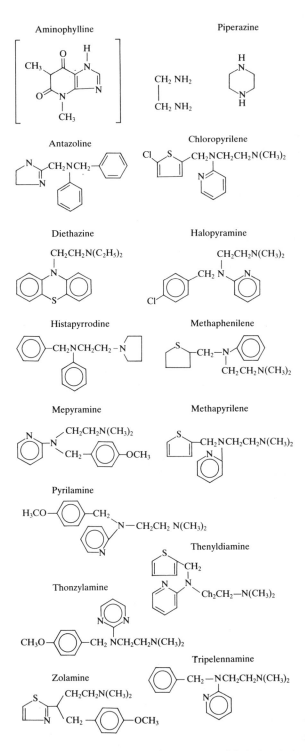

Fig. 7.5 Ethylenediamine antihistamines and their chemical structures.

Allergic reactions to insulin

The immunological responses to insulin are complex and are of varying types with different clinical presentations. All insulins are allergic to varying degrees and while IgE antibodies lead to insulin allergy, IgG and IgA antibodies have a neutralizing effect and, by binding a proportion of the injected insulin, diminish the biological effect and lead to immunological insulin resistance. Small amounts of IgG insulin–binding antibodies can be detected in most patients given exogenous insulin.

Significant amounts of insulin–neutralizing IgG or IgA antibodies are noted more often in patients treated with bovine insulin than those given porcine insulin or highly purified monocomponent or human enetically-engineered insulin. Fig. 7.6 gives a schematic representation of allergic reactions to insulin, taken from a review by Paterson *et al.* (1983).

Conclusion

Whatever the merits or shortcomings of the Coombs and Gell classification used in this chapter, it is important to recognize that in the individual case confronting the doctor, it may not be at all clear which type of hypersensitivity reaction is present or whether there are several reactions occurring simultaneously or indeed whether the reaction is due to hypersensitivity at all. Bizarre combinations of symptoms can occur, with only circumstantial evidence pointing to a possible drug association. For example, the role of allergic mechanisms in Lyell's syndrome and the Stevens–Johnson syndrome is still not clear; an acute vasculitis can proceed to vascular necrosis and exfoliative dermatitis to hear failure before one has had time to determine exactly what is happening and why. Toxic epidermal necrosis has been reported to be induced by sulphonamides and various non-steroidal anti-inflammatory agents such as benoxaprofen and indomethacin.

RECOMMENDED FURTHER READING

The reader is referred to standard works on adverse reactions to drugs and to recent review articles on allergic reactions to drugs.

COLLEGE OF PHARMACY PRACTICE (1983). Allergies to food and medicines. *Pharm. J.* **231**, 557–9.

DUKES, M.N.G. (Ed.) (1979). *Meyler's side-effects of drugs*, Vol. IX. Excerpta Medica, Amsterdam.

—— (Ed.) (1977, 1978, 1979, 1980, 1981, 1982). *Side-effects of drugs annual*, nos. 1–6. Excerpta Medica, Amsterdam.

—— (1980). Allergic and anaphylactic drug-induced emergencies. In *Drug-induced emergencies* (ed. P.F. D'Arcy and J.P. Griffin), pp. 91–101. John Wright & Sons, Bristol.

IDSØE, O., GUTHE, T., WILLCOX, R.R., AND DE WECK, A.L. (1968). Nature and extent of penicillin side-reactions with particular reference to fatalities from anaphylactic shock. *Bull. WHO* **38**, 159–88.

O'SULLIVAN, M. (1983). A case of toxic epidermal necrosis secondary to indomethacin. *Br. J. Rheumatol.* **22**, 47–9.

PARKER, C.W. (1982). Allergic reactions in man. *Pharmacol. Rev.* **34**, 85–104.

PATTERSON, R. (1982). Allergic reactions to drugs and biologic agents. *J. Am. med. Ass.* **248**, 2637–45.

WITTE, K.E. AND WEST, D.P. (1982). Immunology of adverse drug reactions. *Pharmacotherapy* **2**, 54–65.

REFERENCES

AKOUN, G.M. AND PERROT, J.Y. (1983). Acebutolol-induced hypersensitivity pneumonitis. *Br. med. J.* **286**, 266–7.

BAER, R.L., COHEN, H.J., AND NEIDORFF, A.H. (1959). Allergic eczematous sensitivity of aminophylline. *Arch. Dermatol.* **79**, 647–8.

BURRY, J.N. (1978). Ethylenediamine sensitivity with a systemic reaction to piperazine citrate. *Contact Dermatitis* **4**, 380.

——. KIRK, J., REID, J.G., AND TURNER, T. (1973). Environmental dermatitis: patch tests in 1000 cases of allergic contact dermatitis. *Med. J. Austral.* **2**, 681–5.

CALNON, C.D. (1975). Occupational piperazine dermatitis. *Contact Dermatitis* **1**, 126.

COOMBS, R.R.A. AND GELL, P.G.H. (1975). Classification of allergic reactions responsible for clinical hypersensitivity and disease. In *Clinical aspects of immunology* (ed. P.G.M. Gell, R.R.A. Coombs, and P.J. Lachmann), pp. 761–81. Blackwell Scientific, Oxford.

DENIS, L. (1983). Anaphylactic reactions to repeated intravesicular instillation with Cisplatin. *Lancet* **i**, 1378.

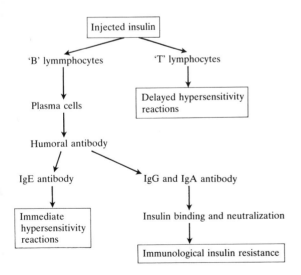

Fig. 7.6 Immunological reactions to insulin (after Patterson *et al.* 1983).

DUX, S., GROSLOP, I., GARTY, M., AND ROSENFELD, J.B. (1983). Anaphylactic shock induced by diclofenac. *Br. med. J.* **286**, 1861.

ELIAS, J.A. AND LEVENSON, A. (1981). Hypersensitivity reactions to ethylene diamine in aminophylline. *Am. Rev. resp. Dis.* **123** (5), 550–2.

GELFAND, H.H. (1963). Respiratory allergy due to chemical compounds encountered in the rubber, lacquer, shellac, and beauty industries. *J. Allergy* **34**, 374–81.

GIBB, W. AND THOMPSON, P.J. (1983). Allergy to aminophylline. *Br. med. J.* **287**, 501.

GRIFFIN, J.P. (1983). Drug-induced allergic and hypersensitivity reactions. *Practitioner* **227**, 1283–97.

HARDY, C., SCHOFIELD, O., AND GEORGE, C.F. (1983). Allergy to aminophylline. *Br. med. J.* **286**, 2051–2.

LAM, S. AND CHAN-YEUNG, M. (1980). Ethylenediamine-induced asthma. *Ann. Rev. resp. Dis.* **121** 151–5.

NADER, D.A. AND SCHILLACI, R.F. (1983). Pulmonary infiltrates with eosinophilia due to naproxen. *Chest* **83**, 280–2.

PATTERSON, K.R., PAICE, B.J., AND LAWSON, D.H. (1983). Undesired effects of insulin therapy. *Adv. Drug Reactions ac. Poison. Rev.* **2** , 219–34.

PENN, R.G. AND GRIFFIN, J.P. (1982). Adverse reactions to nitrofurantoin in the United Kingdom, Sweden, and Holland, *Br. med. J.* **284**, 1440–2.

TAS, J. AND WEISSBERG, D. (1958). Allergy to aminophylline. Report of a case. *Acta Allergologica* **12**, 39–42.

VALE, J.S. AND WHEELER, D.C. (1982). Anaphylactoid reactions to acetylcysteine. *Lancet* **ii**, 988.

WEART, C.W. AND HYMAN, L.C. (1983). Serum sickness associated with metronidazole. *South. med. J.* **76**, 410–11.

WHITE, M.I., DOUGLAS, W.S., AND MAIN, R.A. (1978). Contact dermatitis attributed to ethylenediamine. *Br. med. J.* **1**, 415–16.

WRIGHT, S. AND HARMAN, R.R.M. (1983). Ethylenediamine and piperazine sensitivity. *Br. med. J.* **287**, 463–4.

8 Prescription-related adverse reaction profiles and their use in risk–benefit analysis

C.J. SPEIRS

Introduction

There is an increasing awareness of the need to monitor adverse drug reactions (ADR) to newly marketed drugs as noted by the Working Party on Adverse Reactions of the Committee on Safety of Medicines (CSM) in its report of 1983. The pre-marketing clinical experience of a new drug, however, intensely studied, is limited in numbers to a few thousand patients. This number is miniscule in comparison to the number of patients who may be exposed to the drug within the first few years of marketing. Adverse reactions occurring at incidences well below those which could conceivably be detected by even the largest clinical studies may cause concern.

Serious reactions occurring with an incidence of 1: 10000 prescriptions are almost certain to be missed by pre-marketing studies and even those occurring at 1: 1000 prescriptions cannot be detected with reasonable certainty by pre-marketing clinical trials. Such incidences could lead to hundreds or even thousands of adverse reactions within a year in a single country of the size of the UK where, in many therapeutic categories, a new drug may reach annual sales figures of over one million prescriptions per year, within a year of marketing. The estimated number of prescriptions is used here as a denominator to determine 'incidence'. This in fact represents a considerable simplification of a complex problem. The incidence of an ADR may be 1: 10000 prescriptions overall but may vary considerably with age, many ADRs being more frequent in the elderly. Also incidence figures may be deceptive as, for example, with anaphylactic reactions which may go unnoticed initially with a zero incidence on first exposure and appear only after substantial numbers of patients have been exposed to the drug for the second time. The huge numbers of patients that need to be monitored in order to detect ADRs at such a low incidence was well-illustrated by Lasagna (1983), demonstrating the limitations of formal post-marketing studies.

The relatively large numbers of patients potentially at risk from undiscovered side-effects justifies considerable endeavour in their detection. A number of approaches to this problem have been suggested. In the UK specific suggestions for improving the speed of detection of drug ADRs were made following the discovery of the side-effect syndrome associated with practolol. These included suggestions for registered release (Dollery and Rawlins 1977) or controlled release of new drugs (Lawson and Henry 1977). Central to any effective cohort monitoring scheme is the need to link prescription records to patient records. Inman (1981) pioneered such a system in the UK. Cohort studies, such as these, offer one means of detecting ADRs which can establish valid associations between drugs and adverse reactions as well as an incidence of such adverse reactions. Other investigations such as control studies and studies of national morbidity and mortality data offer further approaches to the detection of ADRs but in the context of newly-marketed drugs they all suffer from practical shortcomings in the time taken and the cost and resources required.

Against this there is a need to identify any safety hazard on new drugs quickly as large numbers of potentially at-risk patients may be continually exposed to a potential hazardous drug. Two methods are available which allow regulatory authorities to obtain information both quickly and without imposing any large burden on resources and expenditure. The first is by monitoring literature reports, particularly those first reports linking a particular drug to a particular adverse reaction. The second is the accumulation and analysis of spontaneous reports of adverse reactions. Both methods are limited scientifically. In particular they are often insufficient to 'validate' the ADR or in other words they may fail to clearly establish the relationship between the drug and the ADR in question. Also a true incidence figure cannot be established by the study of such anecdotal reports alone. Nevertheless in practice much useful information may be gained from such data. Literature reports most often are the first alert for suspected adverse reactions but subsequent confirmation and validation of the ADR by literature reports may be slow as outlined by Venning (1983). National spontaneous reporting systems offer a complementary source of data on ADRs. The *Report of the Working Party on Adverse Reactions* (Committee on Safety of Medicines 1983) drew attention to three functions of the spontaneous voluntary reports (yellow card) data base held by the CSM:

1. To draw attention to previously unsuspected possible adverse reactions which can then be followed up;
2. To provide confirmatory evidence in the form of further reports of a particular problem following first alerts in the literature;
3. To assist in the assessment of the risk–benefit ratio of a drug compared with similar drugs.

It will be noted that these functions assume that such data are used to support, confirm, and assist in the interpretation of data available from other sources and to stimulate the generation of other data when required.

The use of the drug profile

At its simplest such a profile consists of adding together all the adverse reactions reported in relation to a particular drug and then dividing this total into the number of reactions to each particular system/organ/class as described in relation to antidepressants by Inman (1980) and in relation to non-steroidal anti-inflammatory drugs by Cuthbert (1977). Thus the proportion of reported adverse reactions to the various systems is obtained for each drug and can be displayed by histograms.

Figures 8.1–8.3 contrast the profiles obtained with anti-inflammatory drugs (Cuthbert 1977). It should be noted that the height of each column does not represent actual numbers of reported adverse reactions but only the proportions of those ADRs reported in association with that drug affecting the various system/organ/classes. There is obviously a considerable number of confounding factors which will be discussed later in this chapter which may distort such profiles. Nevertheless, the profound increase in the proportion of liver disease associated with ibufenac (Fig. 8.2) was sufficient to emphasize the risk associated with its administration. An analysis of reports showed that the drug was associated with cholestatic jaundice, and the drug was withdrawn in 1967 because of this toxicity. Similar profiles were constructed to show the increased proportion of haematopoietic disorders associated with phenylbutazone and

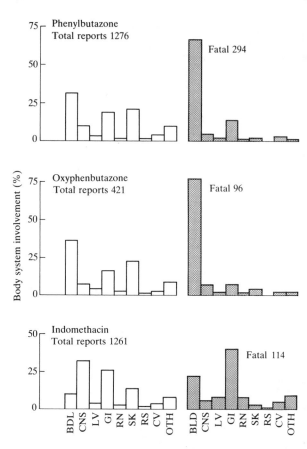

Fig. 8.1 Adverse reactions to phenylbutazone, oxyphenbutazone, and indomethacin reported to the CSM (1964 to January 1973). Analysis by body system involved and expressed as percentages for total and fatal reactions (from Ballantyne 1977). BLD, blood disorder; CNS, central nervous system; LV, liver toxicity; GI, gastrointestinal; RN, renal toxicity; SK, skin disorder; RS, respiratory disorder; CV, cardiovascular disorder; OTH, other.

oxyphenbutazone (Fig. 8.1). This sort of profile proved useful in stimulating further work on drug-induced blood dyscrasias (Inman 1977), resulting in warnings to the medical profession. Finally in 1984 the CSM proposed that the product licence for oxyphenbutazone be withdrawn and the licensed indications for the use of phenylbutazone be reduced to ankylosing spondylitis.

Thus drug profiles have been shown to be of assistance in drawing attention to specific problems as associated with individual drugs. Comparisons are limited to drugs of the same therapeutic class where the clinical usage of each drug is similar. This helps to limit the extent of a considerable number of biases which can exist in reporting ADRs. However the absence of any estimate of the

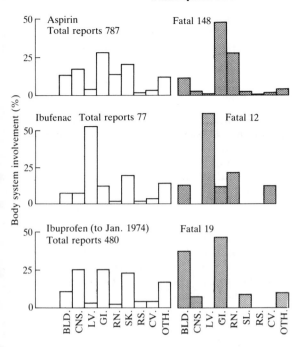

Fig. 8.2 Adverse reactions to aspirin, ibufenac, and ibuprofen, reported to the CSM (1964 to January 1973; ibuprofen to January 1974). Analysis by body system involved and expressed as percentages for total and fatal reactions (from Ballantyne 1977). BLD, blood disorder; CNS, central nervous system; LV, liver toxicity; GI, gastrointestinal; RN, renal toxicity; SK, skin disorder; RS, respiratory disorder; CV, cardiovascular disorder; OTH, other.

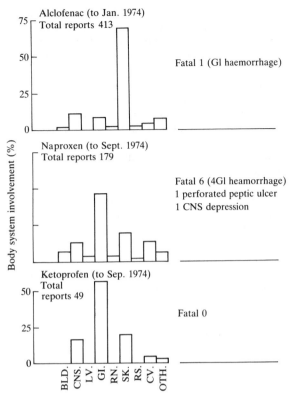

Fig. 8.3 Adverse reactions to alclofenac, naproxen, and ketoprofen reported to the CSM (alclofenac, 1964 to January 1974; naproxen and ketoprofen, 1964 to September 1974). Analysis by body system involved and expressed as percentage for total reactions (from Ballantyne 1977). BLD, blood disorder; CNS, central nervous system; LV, liver toxicity; GI, gastrointestinal; RN, renal toxicity; SK, skin disorder; RS, respiratory disorder; CV, cardiovascular disorder; OTH, other.

extent of use of the drugs compared limits considerably the value of the signals generated by these profiles. In order to increase their usefulness, the usage of the drugs has to be taken into account.

Advantages of prescription-related profiles

The comparison of simple proportional profiles cannot be relied on to highlight the greater toxicity of one drug as opposed to another. For example, with a drug such as benoxaprofen, the very large number of skin reactions reported in association with the drug tends to reduce the proportion of other more serious (hepatic and renal) reactions, which may therefore be masked by the sheer numbers of reports of skin rashes. If the incidence of adverse reactions reported is related to a denominator, such as prescription figures, then the resulting profile no longer masks such differences between drugs. It must be cautioned that what is derived is not a true incidence but a reported incidence of adverse reactions. Such prescript-

ion-related profiles may be employed usefully when these limitations and the need for appropriate further studies are realized.

The use of prescription data

The use of prescription data, as mentioned previously, is an arbitrary measurement of exposure to the drug. An example of an attempt to achieve a more accurate estimate of drug exposure is that of Sweden's Adverse Reaction Committee (SADRAC) which has adopted a *Defined Daily Dose* (DDD) for each drug (SADRAC 1983). Whatever the measurement adopted, quantification of total drug use has to be based on an estimation. Therefore differences in prescribing patterns may invalidate comparisons of estimated incidences of ADRs between drugs. Nevertheless such difficulties need not lead to large distortions, as comparisons are limited to drugs of the same therapeutic class, and their clinical use often follows similar patterns (indeed comparisons between drugs whose clinical usage varied substantially would be unhelpful in any case). Wherever possible, therefore, confirmatory data should be sought on patient ages, diagnostic indications, dose, and duration of treatment to ensure that approximately similar exposures are being measured for each drug. Such data may be obtained from both government and commercial sources in the UK.

Prescription data used have been made available from the DHSS Research and Statistics Division which analyses a 0.5 per cent sample of prescriptions from England and Wales and a 2.0 per cent sample from Scotland. These are prescriptions from general practitioners which account for about 85 per cent of total prescriptions. Data from commercial sources have allowed prescriptions for individual drugs of the same class to be compared to check any divergence in prescribing patterns, as between hospital and general practice prescribing, or in patient demography between drugs.

Prescription-related incidences of reported adverse drug reaction

Prescribing data, as described earlier, have been linked to data on adverse reactions reported to the CSM for non-steroidal anti-inflammatory drugs (NSAID). This has been plotted in two ways. Firstly, by plotting the incidence of total adverse reactions per million estimated prescriptions per year for all NSAID's currently marketed at the time of the survey (Figs. 8.4–8.7). Secondly, histograms of reported ADRs for the first year of marketing are shown in Fig. 8.8. The years used for the comparison are not calendar years. The reason for this is the rapid and profound decline in reporting rates following initial high rates when the drug is first marketed. This is so marked that comparisons of adverse reaction

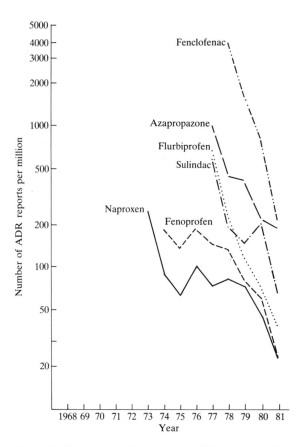

Fig. 8.4 Adverse reactions of selected drugs per million estimated prescriptions per year.

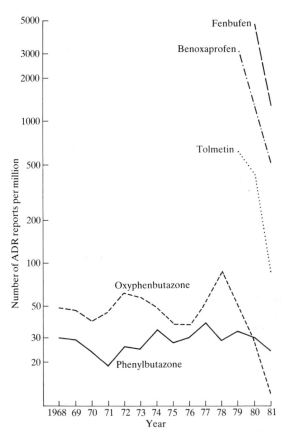

Fig. 8.5 Adverse reactions of selected drugs per million estimated prescriptions per year.

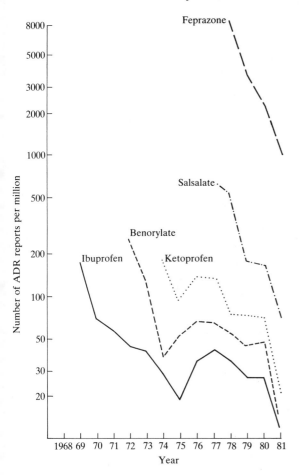

Fig. 8.6 Adverse reactions of selected drugs per million estimated prescriptions per year.

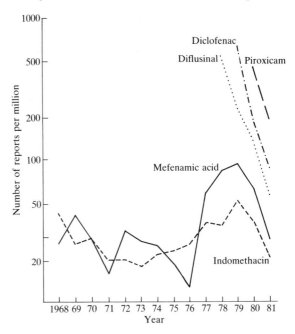

Fig. 8.7 Adverse reactions of selected drugs per million estimated prescriptions per year.

reporting rates of drugs must be controlled for the time elapsed since the marketing of each drug. Thus the first two years of marketing and adverse reaction reporting of drug A can be compared to the first two years of drug B, and so on. Therefore, in Figs. 8.3 and 8.4, yearly comparisons are made from the anniversary of marketing of each drug. (This is not true for the old established drugs, such as phenylbutazone and indomethacin, marketed before the introduction of the CSM-ADR monitoring scheme, for which calendar years are compared.)

When graphs comparing frequencies of ADR reports against time are plotted for each drug as in Figs. 8.4–8.7, the reported incidence of adverse reactions shows a marked uniformity of pattern between the different NSAIDs. The fall following initial high levels of reporting after marketing occurs in similar fashion with each drug. It can also be seen that NSAIDs which were avail-

able before the start of the CSM's adverse reaction reporting system (the yellow-card system was established in 1964) all have low rates of total reports per annum. The reason for this marked decline is an interesting phenomenon, confirming that doctors report adverse reactions more frequently on new drugs than on established drugs. Such a trend has been encouraged by the CSM for some years. (New drugs are marked with an inverted black triangle in the *British National Formulary* and *Data Sheet Compendium* in order to encourage more reporting.) This further encouraged reporting of adverse-reaction cases in order to obtain a safety profile on newly marketed drugs as speedily as possible (Committee on Safety of Medicines 1983). It is interesting that, although the total number of reports on established drugs falls, serious side-effects, particularly when well-described, continue to be relatively well reported. Thus large numbers of deaths from blood dyscrasias have continued to be reported annually in association with phenylbutazone and oxyphenbutazone and, even in 1977, years after the introduction of the drug in the UK Inman (1977) was able to conclude that probably about one in four of all deaths from blood dyscrasia related to these drugs had been reported to the CSM.

The profiles for each NSAID as established by numbers of ADR reports per million estimated prescriptions in its first year on the UK market are found in Fig.

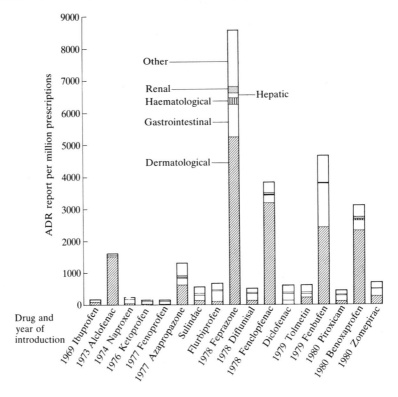

Fig. 8.8   Histograms of reported ADRs for first year of marketing for selected drugs.

8.8. All NSAIDs marketed since the CSM-ADR report-ing scheme was started (1964) and up until 1981 are included. Each histogram is divided to show the relative contribution made by dermal, gastrointestinal, hepatic, renal, haematological, and all other adverse reactions reported on each drug. The overall differences in reporting rates found in the first year of marketing are quite striking. It must first be said that a large number of confounding factors may significantly influence the totals and these are dealt with below. It should not be for-gotten that the figure represents reported rather than real incidences of ADRs. It is nevertheless a valuable starting-point for further study. Can reporting rates of over 40-fold difference be all caused by biases affecting doctors' reporting of adverse reactions and unrelated to the incidence of ADRs? Can two drugs (benoxaprofen and piroxicam) both launched with large sales within a year of one another have an over five fold difference in reporting rates of ADRs simply because of undetected factors influencing reporting rather than due to a real incidence of ADRs? Certainly on closer investigation the large proportion of reports, on adverse skin reactions with benoxaprofen, show a real difference. This is

revealed by examination of the yellow cards involved which shows benoxaprofen-induced photosensitivity to be largely responsible for the difference in reporting of dermatological ADRs. This is merely the simplest inves-tigation of such a profile. A more detailed appraisal of the information which may be provided must await a careful consideration of the biases which may influence the reporting of ADRs on individual drugs.

Confounding factors affecting the interpretation of reported rates of ADRs

These were discussed by Inman (1980) but it is important to consider all those likely to be relevant before attempt-ing any comparison between drugs.

1. The first in importance, and the most obvious con-founding factor, is publicity associating a given side-effect to a particular drug. Publicity may first arise in specialized journals but its effect on doctors' reporting of

ADRs increases as reporting spreads to the popular medical press. Undoubtedly the most potent stimulus to ADR reporting is widespread publicity in the lay press and media which can influence patients, or their relatives, to return to their doctors with problems which may or may not be related to the drug. Certainly this appears to increase doctors tendency to report not only the particular problem in question but also other events including some which are doubtfully associated with the drug. This leads to a great increase in reporting and much of this may be retrospective, leading to a further distortion of the reported rates. Thus the CSM continued to receive ADR reports on benoxaprofen well after the drug was withdrawn from the market, relating to events one or even two years previously.

2. The overall rates of reporting to the CSM have increased, particularly during the mid-1970s when the rate approximately doubled. The influence of this can be seen in Fig. 8.3 where there is a tendency for reporting rates for all but the newly introduced drugs (for which the rates of reporting are falling) to rise. Such changes in reporting rates will affect any comparison made between levels of reports received for, say, ibuprofen in 1969 (Fig. 8.2) and any drug marketed after the late-1970s and will be biased in favour of the former. Thus the calendar year of reports received must be taken into consideration.

3. The initial rate of sale has been thought to be a factor influencing reporting profiles, 'explosive marketing' being thought to be associated with a high reporting rate. An initially high sale will clearly affect the actual number of ADR reports associated with that drug. However it does not influence prescription-related levels of reporting. In these, the most heavily marketed drugs, benoxaprofen and piroxicam (Figs. 8.5 and 8.7), are associated with high and low levels of reporting. Conversely the least heavily marketed drugs, feprazone (Fig. 8.6) and tolmetin (Fig. 8.5), are also associated with widely disparate reporting rates. High initial sales levels do not, therefore, necessarily bias reporting rates but of course the actual numbers reported are much higher with widely prescribed drugs. Where the prescriptions of a drug are few and the ADR uncommon, the numbers reported will be very small. Using a prescription-related profile such small numbers are multiplied when expressed as an estimated incidence per million prescriptions. This is important since very small numbers may otherwise be missed. However, unless the nature of such data is clearly understood and the need for further investigation is recognized, extrapolation from such small numbers could be extremely misleading.

4. The different patterns of usage of drugs, as between hospitals and general practice, or between different diagnostic indications and age groups of patients, may influence both the incidence of ADRs due to a particular drug, and the incidence of reporting of ADRs. For example a drug used more in the elderly may cause a higher incidence of side-effects; those used more in hospitals may be subject to greater surveillance and there may be more thorough reporting of suspected ADRs. Although a difference between drugs in their usage need not necessarily lend a bias to the reported levels of ADRs, data on drug usage should be obtained to expose any serious variations in usage which may confound any comparison.

5. The numerous ADRs generated by post-marketing surveillance (PMS) studies are not routinely included in the CSM data base. However the more serious ADRs are included, and it appears that drugs undergoing intensive monitoring may have more reports entering the CSM data base as a result. One drug in particular, where such a study appeared to stimulate a number of adverse reaction, reports to the company which were subsequently included in the CSM's ADR data base, was fenbufen, as shown in Fig. 8.5. Over half of the total reports relating to this drug in its first year of marketing were received from the company whereas for all the other NSAIDs, the company reports constituted only a minority, usually around one-fifth of the total. Thus a simple comparison of total reported ADR's would appear to be biased against fenbufen and this has to be allowed for when comparisons are made.

6. The last point is that comparisons made between drugs should be of similar periods after marketing. The incidence of reporting of ADRs falls so markedly, as shown in Figs. 8.4–8.7, that to compare the levels of ADR's on two drugs in the same calendar year, where one had been marketed for four years or more, and the other for one year only, would give a reporting bias against the latter drug of up to 50-fold. The comparison in Fig. 8.8 is therefore for the first year of marketing only.

Between-drug comparison of profiles obtained on non-steroidal anti-inflammatory drugs (NSAID)

Estimated prescription-related side-effect profiles of all those NSAIDs (17) which were introduced on to the UK market following the introduction of the yellow-card ADR monitoring system in 1964 up until 1981, were prepared using the CSM data base, and the prescribing statistics prepared by the SR division of the DHSS as described earlier. The most notable finding is the nearly 50-fold variation between estimated reporting rates for the least frequently reported (ibuprofen: 177 reports per million prescriptions), and the most frequently reported (feprazone: 8557 per millon). The only obvious bias was that the general reporting rate, at the time ibuprofen was first marketed, was only about half that when feprazone was first marketed. This considerable difference in reporting rate is difficult to explain without postulating some increased risk from feprazone as perceived by reporting doctors. Indeed the same question arises with fenbufen, fenclofenac, benoxaprofen, alclofenac, and possibly azapropazone which were prominent in terms of overall reporting rates of ADRs in their first year of marketing. (As noted previously the high level associated with fenbufen was magnified by a level of company reports.) A further interesting feature is that all these drugs had a high incidence of reporting of skin rashes. Three — feprazone, benoxaprofen, and fenclofenac — appeared to have a higher proportion of the rarer, but often more serious, ADRs affecting the hepatic, renal, and haematological systems. Clearly the numbers of ADRs in these categories were too small to be necessarily related to drug toxicity, but could have arisen by chance. The correlation of a high incidence of skin rash and serious ADR reports affecting other systems was noted previously (Weber 1984).

Such a profile clearly projects two sorts of finding. The first consists of obviously heterogeneous data involving large numbers of reports, e.g. the totals received for each drug or for common causes such as dermatological reactions. The reasons for these increased numbers may be swiftly determined. In all the prominent drugs, skin reaction featured highly and the high incidence of skin rashes may be verified by reference to clinical trial data. The second is that uncommon ADRs comprise very small actual numbers of reports, particularly for infrequently prescribed drugs. The relative incidence of these, between drugs, may be compared but the numbers are too small to exclude the possibility of their being due only to chance. Even when a drug is heavily prescribed, it may take a considerable time before serious, but uncommon, adverse reactions have occurred often enough to appear prominently in the medical literature, or accumulate in significant numbers in the CSM's data base to attract attention, as found with benoxaprofen. Fig. 8.5 shows that the prescription-related data might have provided an early warning of problems concerning the renal and hepatic toxicity of benoxaprofen by the end of the first year of marketing. At the time, no such profile could be established because of the time required to obtain estimated prescription figures.

When a particular drug is associated with a high level of reporting of serious ADRs as compared to other drugs, the actual number of reports is usually small, and it is important that further investigation should be done. This should be aimed initially at verifying the data, that is ensuring the accuracy of the yellow-card report by reference to the original yellow-card report, and further enquiries as necessary. Validation of the causal role of the drug may be obtained in a variety of ways as discussed by Venning (1983b).

It is of interest that at the time of writing of the five drugs appearing as frequently associated with ADRs in this profile, alclofenac, benoxaprofen, feprazone and fenclofenac have been withdrawn from the market. The fifth drug, fenbufen, was comparatively frequently reported partly because of having the highest proportion of reports on the CSM register sent in by the pharmaceutical company marketing the drug (for the period of comparison).

The use of prescription-related profiles in assessment of risk–benefit analysis of drugs

While such profiles can act as a useful aid to the investigation of the adverse-reaction profile of a drug, they are only helpful when there exists a category of drugs which share a common mechanism of action and similar clinical applications.

It is axiomatic that these profiles deal only with evidence relating to safety in the risk–benefit analysis so that the benefits of all the drugs compared should be very similar. This requires a consensus of clinical judgement of those expert in the use of these drugs. Obviously a reasonable appreciation of their relative efficacy should be obtained before such an exercise is started.

Adverse-reaction profiles are increasingly being used as an aid to monitor the critical initial period of marketing of new drugs, where the exposure of the drug to patients escalates from a few thousand carefully selected and monitored patients on clinical trials, often to several millions, within one or two years. As adverse reactions begin to be reported in tens, twenties, hundreds, or even thousands, their significance can only be sensibly assessed by using estimates of the exposure of patients to the drug, and comparing the rates of reporting to experience with other drugs of the same class. To do this on a regular basis demands a rapid and reasonably accurate source of statistics on drug prescriptions or sales. Such data are becoming increasingly available and may be expressed in terms of prescriptions or daily doses.

However, when ADR prescription-related profiles are assembled it is important to remember:

1. That they represent reported rates, not true rates of occurrence of ADRs;

2. That the various factors which may distort the reported figures should be born in mind;

3. That the prescribing figures are estimates based on a sample, albeit appreciable.

Finally these profiles, properly used, serve to maximize the usefulness of data obtained from spontaneous reporting. They therefore assist in achieving the aims of the ADR reporting systems by providing important early warnings and hence a stimulus to further investigation, particularly early in the life of a new drug. In addition they serve to emphasize and confirm known or suspected adverse reactions.

REFERENCES

BALLANTYNE, B. (ED.) (1977). *Current approaches in toxicology.* John Wright and Sons, Bristol.

COMMITTEE ON SAFETY OF MEDICINES (1983). *Report of the Working Party on Adverse Reactions.* HMSO, London.

CUTHBERT, M.F. (1977). Adverse reactions to antirheumatic drugs: some correlations with animal toxicity studies. In *Current approaches in toxicology* (ed. B. Ballantyne), Chapter 19. John Wright and Sons, Bristol.

DOLLERY, C.T. AND RAWLINS, M.D. (1977). Monitoring adverse reactions to drugs. *Br. med. J.* 1, 96-7.

INMAN, W.H.W. (1977). Study of fatal bone marrow depression with special reference to phenylbutazone and oxyphenbutazone. *Br. med. J.* 1, 1500-5.

—— (Ed.) (1980). In *Monitoring for drug safety,* Chapter 1. MTP Press, Lancaster.

—(1981). Post marketing surveillance of adverse drug reactions in general practice. 11 Prescription event monitoring at Southampton University. *Br. med. J.* i, 1216-17.

LASAGNA, L. (1983). Discovering adverse reactions. *J. Am. med. Ass.* 249, 2224-5.

LAWSON. D.H. AND HENRY, D.A. (1977). Monitoring adverse reactions to new drugs: 'restricted release' or 'monitored release'? *Br. med. J.* 1, 691-2.

SWEDISH ADVERSE DRUG REACTION ADVISORY COMMITTEE (SADRAC) (1983). *SADRAC Bulletin,* no. 35 (ed. B.E. Wiholm). Swedish Adverse Drug Reaction Advisory Committee, Uppsala, Sweden.

VENNING, G.R. (1983*a*). Identification of adverse reactions to new drugs. II How were 18 important adverse reactions discovered and with what delays? *Br. med. J.* 1, 289-93.

—— (1983)*b*). Identification of adverse reactions to new drugs. IV Verification of suspected adverse reactions. *Br. med. J.* 1, 544-7.

WEBER, J.P.C. (1984). Side effects of anti-inflammatory/analgesic drugs. In *Advances in inflammation research,* Vol. 6 (ed. K.D. Rainsford and G. Velo), pp. 1-7. Raven Press, New York.

9 Mathematical models in adverse drug reaction assessment

J.C.P. WEBER

Assessment of a potential drug hazard, the possibility of which should be considered whenever a new chemical entity is marketed, depends on the evaluation of data from many sources. This chapter deals with mathematical methods, and although these are important, it should be understood that they represent only one facet of a many-sided situation.

We have seen, in the last chapter, how adverse drug reaction (ADR) profiles can identify target organs. We need to know, in addition, what I shall call the ADR performance of a drug, an inverse, perhaps, of its therapeutic performance. We need to know at what rate we are receiving reports and how this rate compares with that for other drugs of the same therapeutic group both in regard to total and to specific ADRs. We need to know whether the rate changes and in what way and in relation to what circumstances. We would like to know how patient-exposure to the drug influences reporting. Unfortunately, at present we are never able reliably to quantitate patient-exposure. Nevertheless, as I shall endeavour to show, we can make use of prescription data, which though not synonymous with patient-exposure, are clearly related.

Reporting rate

The study of adverse reactions, a developing art, is peculiarly beset with the shifting sands and pitfalls of soft data. It is often said that only 10 per cent of adverse reactions are reported. This statement is probably derived from Inman's (1977) paper on deaths due to aplastic anaemia in patients taking phenylbutazone or oxyphenbutazone in which he found that only 11 per cent of these fatalities had been reported to the Committee on Safety of Medicines (CSM). There is no doubt that, of the total number of ADRs actually occurring with a single drug, the proportion reported varies from time to time. The awareness of doctors may be stimulated by journal articles, by the Committee on Safety of Medicines Adverse Reaction Leaflets, and their publication *Current Problems*. Moreover, the number of doctors

reporting also fluctuates and it is probably as much or more this than the number of reports contributed by each individual doctor that is the main determinant of input. These factors are imponderable but the resultant input of ADR reports can be recorded in two simple ways as a regression on time.

Non-cumulative plot

The non-cumulative plot simply records adverse reactions, either total or specific to some organ, on a graph whose ordinate y represents the reactions and whose abscissa x, the units of time chosen. The points on the graph may merely be joined to give the type of plot shown in Fig. 9.1 which also shows a similar presentation of the prescriptions for a vasodilator drug, the appropriate ordinate for these being on the right-hand side of the graph. It is obvious that there were rather wild fluctuations in the total input of ADR reports per quarter and that such fluctuations were not accompanied by any similar changes in the prescribing of the drug. It is of interest that the peak of the graph occurs at a point about two years after marketing. A similar decline in the reporting of seven non-steroidal anti-inflammatory drugs at about two years of marketing has been demonstrated (Weber 1984) and seems to occur with many other drugs. The reason is not clear, although it must be remembered that doctors are bombarded with 20 or more new chemical entities yearly (in addition to new mixtures of existing drugs). It seems possible that this period may be the longest that an interest can be sustained in the reporting of adverse reactions to a single drug under these trying circumstances.

The graphic representation of the data can be taken a stage further by calculating the regression lines for the two quite separate ADR reporting trends. Regressions of this type are represented by the equation $y = mx + c$ where m is the regression coefficient and c the constant determined by the intercept of the regression line on y. Under these circumstances m represents the rate of increase of y with unit time, and the physicist would call it an acceleration.

This non-cumulative regression has certain disadvantages. First, the coefficient of correlation (r) may not be statistically significant if the points are extremely fluctuant, or if, as is seen in the second ADR slope in Fig. 9.1, there is no obvious increase in the rate of input with time. It should be recalled that what is being measured with this non-cumulative regression is an increase in rate per unit time or, as has already been said, an acceleration. An advantage of the non-cumulative regression, however, is that significant negative regression and correlation coefficients will indicate indisputably that input is decelerating. It is therefore, on occasion, a useful mathematical tool which should take its place with all the others available.

The equations derived from the data in Fig. 9.1 for the first eight and the last seven quarters are, respectively, $y = 82.6x - 57.6$ and $y = 2.5x + 16.7$. The respective correlation coefficients are 0.78 ($p = 0.01$) and 0.62 ($p = 0.09$). It comes as no surprise that the correlation coefficient for the second slope is not statistically significant, since there is clearly no acceleration in reporting during the period concerned.

The cumulative plot

When a similar regression is calculated on the basis of cumulative data on the y axis, m measures, as before, the increase of y with unit time. However, since the difference between each consecutive value on y is the input for the unit of time concerned, m gives an indication of rate, rather than rate of increase as with the non-cumulative plot. In physical terms we are therefore considering velocity instead of acceleration.

The data from Fig. 9.1 are plotted cumulatively in Fig. 9.2 and it is now obvious that there are three separate rates of input of ADR reports which are significantly different from each other. If there is doubt about the significance of a change in slope, the significance of the difference between the two regression coefficients can be calculated by standard statistical methods. It will be seen that the final rate of input is only 11 per cent of the rate for the quarters immediately preceding it. (46.9/419.6). Also shown on the same graph are the prescription totals, this time expressed cumulatively. As before, there is no obvious relationship between the slopes representing ADR input and those for prescription totals. A cumulative slope is always positive, even when the rate is falling, and this is a reason for expressing the data as both non-cumulative and cumulative plots.

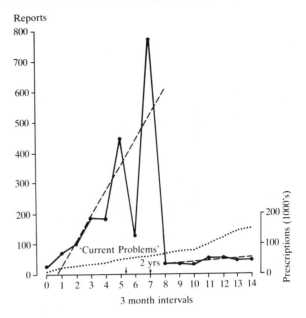

Fig. 9.1 Non-cumulative plots of ADR reports and prescriptions against time. Key: reports (————); calculated regression lines(- - - -); prescriptions (.........).

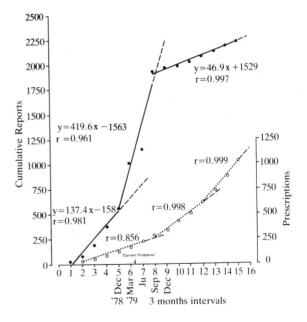

Fig. 9.2 Cumulative plots of ADR reports and prescriptions against time. Key: reports (————); prescriptions (.........).

Relationship between ADR input and prescribing

Thus far we have considered both the acceleration, if any, and velocity of ADR input, i.e. we have made measurements solely in relation to the passage of time. As has been said, the numerical exposure of patients cannot be adequately ascertained. The pool of prescriptions at any given point in time for any drug used in long-term treatment is made up of patients already on treatment together with new patients not previously treated. As we move to the next unit of time there will be a further change introduced by reduction due to patients stopping treatment either because of adverse reactions or because they no longer need the drug. (In long-term treatment the latter proportion is likely to be small.) A supposition quite often made is that if, for instance, a drug has produced 200 ADR reports when prescriptions are 200 000 yearly, then 1000 can be expected when prescriptions reach 1 000 000. The composite nature of the prescription pool already referred to and the complete absence of obvious correlation between ADR input and prescription trends render such an assumption both naïve and unpractical.

As will be seen later, it tends to overstate the expectation. If a prediction of this kind is necessary, the actual performance of the drug concerned must be used as a predictor. In order to do this it is clearly necessary to establish a predictable relationship between the trends for ADR input and prescribing. If the cumulative total of ADR be used on the y-axis, and the cumulative prescription total on the x-axis, the resultant graph is a curve which takes the form $y = m \log x + c$ as shown in Fig. 9.3. The example given is that of a non-steroidal anti-inflammatory drug but the same procedure has been followed with many other drugs and all adopt the same pattern unless something of community health significance occurs substantially to disrupt it, by producing an unexpected influx of retrospective reports. This curve can be used as a reasonably reliable predictor of adverse reaction totals, but curves are always more difficult to handle than linear graphs so that the next logical step is to modify the equation to $\log y = m\log x + c$ by using the logarithm of the cumulative totals instead of the totals themselves.

Figure 9.4 shows the original data used in Fig. 9.1 and 9.2 expressed in this form. The times corresponding to the various points on the slopes have been marked on this graph to facilitate comparison with Fig. 9.2. It will be seen that the reduction in reporting seen in Fig. 9.2 at the ninth quarter (December 1979), is faithfully represented. Also, comparison of the regression coefficients shows the same reduction to 11 per cent of the earlier input as that calculated from the coefficients of the non-log cumulative regression against time in Fig. 9.2. It is also evident that the first two distinct rates of input of ADR against time visible in Fig. 9.2 are not seen in Fig. 9.4. The interpretation of this is that the steep increase in ADR report input between December 1978 and September 1979, almost certainly brought about by an

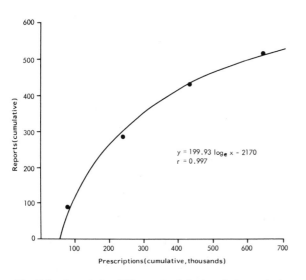

Fig. 9.3 Cumulative ADR reports plotted against cumulative prescription data.

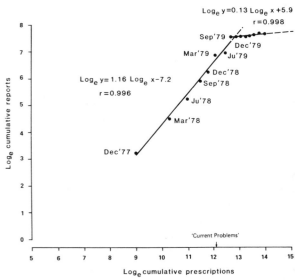

Fig. 9.4 Logarithm of cumulative reports against logarithm of cumulative prescriptions. (Same data as in Figs. 9.1 and 9.2.).

entry in *Current Problems* and by journal articles, was, nevertheless, not incompatible with the high level of prescribing. By September 1979, it was realized that an increase in angina pectoris, for which the drug was prescribed, was a possibility and that the drug had to be used carefully and selectively. Interest in the reporting of this possible adverse reaction subsequently declined and, as we have seen, the decline represents a real decrease in relation to the prescription totals.

Implications of the log relationship

The implications of the logarithmic relationship are probably best illustrated by the use of some examples.

Example 1

The ADR input rate suddenly falls producing a new cumulative linear regression against time. The question to be answered is whether or not this decline in input is due to reduced prescribing. Obviously, if the actual total of prescriptions is not falling the question can be very readily answered without the need for a logarithmic plot. If one were produced, it should show log correlation of ADR input with prescribing but with a markedly lower regression coefficient. This, of course, is the situation already presented in Fig. 9.4.

Example 2

If the report input is falling and prescriptions are also falling, it may be advantageous to know whether these two situations are mathematically related and therefore to be able to suggest the possibility that they may represent cause and effect.

Figure 9.5 shows the cumulative prescription totals plotted on a monthly basis for a non-steroidal anti-inflammatory drug. It will be seen that after September

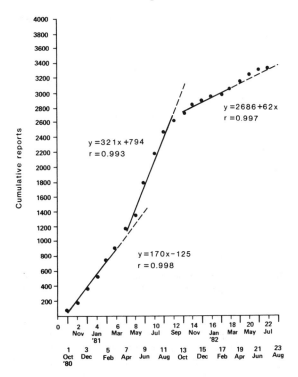

Fig. 9.6 Cumulative ADR reports for the same drug as in Fig. 9.5 plotted against time.

1981, the eleventh month of marketing, the rate of prescribing fell. The regression coefficient indicating 63 000 prescriptions per month is only 82 per cent of the previous coefficient indicating 76 700 prescriptions monthly. There was therefore a fall of 18 per cent in prescribing. Figure 9.6 presents ADR reports cumulatively on a monthly basis and shows a fall of 81 per cent for the same period of time considered in Fig. 9.5.

This seems to be a remarkable fall in ADR reports in the presence of an 18 per cent fall in the prescribing trend. The log linear graph is presented in Fig. 9.7 from which it will be seen that a single line adequately represents the situation from a mathematical point of view, the correlation coefficient being 0.996. The point representing September 1981 is the farthest from the regression line and is the eleventh in the series of 16 points shown. The point estimate of y for this month is, in logarithmic terms, 7.742 which corresponds to a total of 2303 adverse reactions. The observed number was 2610. It is possible to calculate the 95 per cent confidence interval for the estimate of y by standard methods (Snedecor and Cochran 1967). In this way, again in logarithmic terms, the interval is found to be 7.507–7.977 corresponding to 1820–2914 reports. The observed number of 2610 is thus well within this interval.

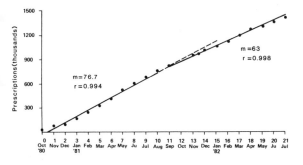

Fig. 9.5 Cumulative prescription data for a NSAID plotted against time.

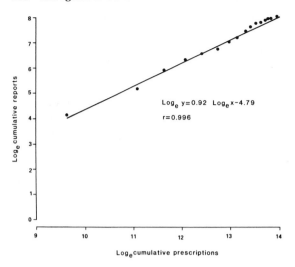

$\log_e y = 0.92 \ \log_e x - 4.79$

$r = 0.996$

Fig. 9.7 Logarithmic plot of cumulative reports against prescription data. (Same drug as in Figs. 9.5 and 9.6.)

The other neighbouring values to the month of September 1981 are also within the 95 per cent confidence interval and we can therefore be assured that the calculated slope shown in Fig. 9.7 reasonably represents the correlation of report input with prescribing. The trend is in contrast to that presented in Fig. 9.4 where report input fell markedly in relation to prescribing. The unchanged relationship between report input and prescribing, despite the fall in prescriptions shown in Fig. 9.5, leads to the conclusion that ADR reporting is likely to have fallen largely because prescribing has fallen, despite the apparent initial disparity between the two falls.

Example 3

It is often desirable to be able to estimate the likely cumulative total of ADR reports for a given drug when the prescription total shall have equalled that for another comparator drug of the same pharmacological group. As has been earlier pointed out, the use of simple comparative proportions for such an estimate leads to inaccuracies.

Let us assume that, using the data on which Fig. 9.5 and 9.6 are based, we note that in October 1981 a cumulative total of 2740 reports corresponded to a prescription total of 892 200. We shall use the drug concerned as its own comparator, the object of the exercise being to estimate how many ADR reports are likely to have been received when the prescription totals reach, say, 1 400 000. Since we are using an actual occurrence we shall use the prescription total most closely corresponding to this, which was that for June 1982, i.e. 1 392 500.

Now, if the estimate of reports be made by simple proportion, we have

$$\frac{2740}{892\,200} = \frac{\Sigma\,ADR}{1\,392\,500}$$

where $\Sigma\,ADR$ is the estimated total number of reports for the test drug. From this equation the $\Sigma\,ADR$ = 4276. The observed total of reports at June 1982 was in fact 3 290, so this calculation produces an overestimate of 30 per cent.

If the log linear slope shown in Fig. 9.7 be used instead, the estimate of y can be obtained by substitution into the equation:

$$\log_e \Sigma\,ADR = 0.918 \log_e 1\,392\,500 - 4.785$$

whence $\Sigma\,ADR$ = 3647, an overestimate of only 10 per cent.

It might be argued that the difference between the two estimates of some 629 reports is not of very great importance when total ADRs are being considered. Since the proportionate method invariably overestimates the total, it seems unlikely that the Pharmaceutical Industry would take this view. Moreover, when serious or fatal ADRs are under consideration, an overestimate becomes the more important since it may be a determinant of a regulatory decision.

Comparison of proportions and the inverted report/prescription ratio

It should not be concluded from the foregoing that comparison of ADR/prescription proportions is always to be avoided. It is important to use absolute totals when such comparisons are made, rather than any kind of extrapolation to a theoretical prescription total such as, say, 1 000 000. It is also important to use data derived from a period of time after marketing which is the same for the two or more drugs to be compared. For the reason given earlier in this chapter, this period of time can be about two years, though in practice comparisons may sometimes have to be made for a shorter period. The drugs to be compared must be of the same pharmacological group, in use for, broadly, the same clinical indications. The normal variate corrected for continuity provides an indication of statistically significant differences between the proportions compared, and allows them to be ranked. The writer has used such a method for comparison of nine non-steroidal anti-inflammatory drugs in regard both to total reports and to reports of five serious adverse reactions (Weber 1984) and it appears to provide consistently useful data. It should not be forgotten that such methods offer approximations to the probable truth and that all other available data must be used when a comparative assessment of a drug for regulatory purposes is being made.

Since it is undesirable to extrapolate to a fictitious number of prescriptions, it is better to use what can be called the inverse report/prescription (IRP) ratio, which is simply the total observed number of prescriptions related to one ADR report. When a large number of drugs of a particular therapeutic group is available for study, e.g. non-steroidal anti-inflammatory drugs or benzodiazepines, overall IRP ratios can be obtained for use as reference data for new members of the group. It is, for instance, unusual to obtain a sustained IRP ratio, for total reports, of less than 300, at least for non-steroidal anti-inflammatory drugs, and such a finding would indicate the need for detailed investigation. Similar data can be accumulated for specific ADRs, such as blood dyscrasias, hepatitis, and nephropathy.

Conclusion

Most of the relatively simple mathematical procedures described in this chapter can, with benefit, form part of the routine surveillance of medicines which are new chemical entities. Gross deviations of new members from the expected values should be interpreted as urgent warnings requiring detailed investigation using all available sources of data. The study of adverse drug reactions is a fascinating blend of clinical medicine, pharmacology, mathematics, and philosophy. Therein lie its weaknesses, its strength, and its interest to those who practise it.

REFERENCES

INMAN, W.H.W. (1977). Study of fatal bone marrow depression with special reference to phenylbutazone and oxyphenbutazone. *Br. med. J.* **1**, 1500–5.

SNEDECOR, G.W. AND COCHRAN, W.G. (1967). *Statistical methods*, 6th edn., p. 155. Iowa State Press, Ames, Iowa.

WEBER, J.C.P. (1984). Epidemiology of adverse reactions to non-steroidal anti-inflammatory drugs. In *Side-effects of anti-inflammatory/analgesic drugs*, Vol. 6 (ed. K.D. Rainsford and G. Velo), pp. 1–7. Raven Press, New York.

10 Strategy of risk–benefit analysis

A.W. ASSCHER

All things are poisonous for there is nothing without poisonous qualities. It is only the dose that makes things a poison.

(Paracelsus 1493–1541)

Half the modern drugs could well be thrown out of the window except that the birds might eat them.

(M.H. Fisher 1879).

Introduction

Adverse reactions to drugs are as old as medicine itself; this is why it is not difficult to find quotations which exhort therapeutic caution. The code of Hammurabi of 2200 BC dictated that doctors should lose their hands if they caused the death of their patients. I suspect the efficacy of this rule was absolute only so far as surgeons were concerned but less effective for physicians who may also harm by their words. Homer (950 BC) said of drugs that many were excellent but others were fatal and Rhazes (AD 860–932), the physician-in-chief of the Great Hospital in Baghdad, advised that prescriptions should always be kept simple. The Hippocratic policy of *primum non nocere* rightly continues to be emphasized in our undergraduate curriculum.

There are at least four reasons why analysis of the risks and benefits of medical treatment have assumed greater importance in recent years. Firstly, as the major killer diseases of the infectious type disappear from developed countries, modern treatment more often concerns marginal benefits. Compare the impact of smallpox and poliomyelitis vaccination with the dubious benefits of reducing raised serum lipids. The more marginal the gain, the more important to undertake risk–benefit analysis. Treatment is nowadays not infrequently offered to apparently healthy subjects detected during screening programmes which by themselves may not be devoid of danger. Potentially harmful investigation and treatment may therefore be undertaken on large numbers of subjects of whom only a few are likely to benefit. The second reason why risk–benefit analysis is now a matter of major concern to the medical profession is that the modern therapeutic armamentarium contains many bolder forms of intervention, both medical and surgical. Partly as the result of the wise use of these, the popu-

lation now exposed to these more dangerous treatments is older than it used to be. The age structure of our population is shifting to the side of greater longevity and it is well known that the dangers of all forms of treatment are magnified in the elderly because of altered pharmaco-dynamics and kinetics of drugs and lessened overall resistance to operative interventions (Hurwitz 1969).

The third and least palatable reason why doctors should concern themselves with risk–benefit judgements is the growth of medical litigation. Every decision in medicine involves a risk–benefit analysis which may have to stand up to critical examination by peers, patients, relatives, and the media as well as the courts. Lastly, and by no means least, objective analysis of risks and benefits plays a pivotal role in drug regulatory decisions. In view of the importance of the subject it is surprising how little attention is paid to it in the medical syllabus.

The difficulties of quantifying risks and benefits are enormous, as is the definition of 'acceptable' levels of risk. These problems should be a challenge rather than a reason for shying away from a subject of such importance. In this chapter I intend to identify some of the problems involved in risk–benefit analysis and to consider ways in which our decision-making processes could be improved and risks be minimized.

Components of risk

There is no hope of quantifying risks or benefits unless they are dissected into their components. The risks of any medical procedure are of two kinds, namely, risks to the individual and risks to society (*Journal of Medical Ethics* 1982). In turn individual risk can be viewed from three standpoints. These are: (1) the nature of the risk; (2) its likelihood; (3) the amount of harm it would produce. To illustrate the value of such a dissection of risk, let us take an injectable contraceptive as an example. The societal risks of releasing such a drug include its possible abuse as a means of limiting the fertility of certain population groups for political purposes or its administration to mentally handicapped girls who may not be able to give informed consent for its use. The societal benefits relate to curtailment of the population explosion with consequent saving of limited resources and benefit to the existing population. The individual risks might include menstrual irregularity, weight gain, mastalgia, depression, failure of return of fertility, vascular complications, hypertension, and the possiblity of increased risk of breast or other gynaecological cancers.

Advantages might include the facts that it is the most foolproof means of contraception and that the messiness or deliberate preparation for intercourse that accompanies the use of caps and spermicidal creams does not arise. Each one of these pros and cons can be put into its proper perspective by using analogies, e.g. the risk of death from taking an oral contraceptive for one year is less than that for a car passenger travelling 100 miles (Pochin 1982). Sometimes it is possible to assess the likelihood of a particular adverse effect from a knowledge of the genetic make-up of the patient, e.g. hydrallazine-induced systemic lupus erythematosus is significantly more common in females who are slow acetylators of the drug and whose HLA type is DR4 (Batchelor *et al.* 1980). The physician also has to evaluate from his knowledge of the patient how a particular adverse event would affect that patient. For example, a drug which produces alopecia would not be readily given to a young woman but might present less of a problem when prescribed for the author of this chapter!

From this brief account of the components of risk it is clear that all risks short of death are difficult to measure. The addition of the components of risk is like adding apples, motor cars, grapes, and houses, there does not appear to be a common denominator so what hope is there of ever drawing up an equation of risks and benefits? Distasteful as it might seem to doctors, it is sometimes possible to reduce the components of risk and benefit to a monetary value. For example, in the case of contraceptives the extent of financial stress on the family of an additional child, the risks of harmful effects on the wife and mother which may occur as the result of the need for an abortion if the drug were not given, and the consequences of time lost from work as a result of unwanted developments can all be expressed in monetary terms. Although such actuarial calculations are possible in a few instances, it is hardly likely to be sufficiently practical or precise for everyday risk–benefit judgement in clinical practice or for decision-making by drug regulatory authorities. Yet attempts to construct the equation are of value as they necessitate the enumeration of all components of risk and benefit and so serve to make judgement less biased and more explicit. In what follows I will concentrate on those aspects of individual risk (and benefit) which are predictable and/or quantifiable. However inadequate this may be in terms of assessing total risk (or benefit) it is the best we can do at the present time. It is perhaps a reflection of the extent of our concern about drug safety that much more has been written about the measurement and prediction of risks than those of benefits

Prediction and measurement of risk

Drugs may be divided into two main classes. Firstly, there are those used to supply something the body lacks such as vitamin B_{12} for pernicious anaemia or vitamin C for scurvy. Used in correct doses, for purely replacement purposes, such drugs seldom cause adverse reactions, given that they are sufficiently pure and bio-available. Secondly, there is the much larger group of drugs used for their pharmacological actions. Such drugs are bound to produce adverse reactions from time to time. Those who insist that nothing but the complete safety of drugs will suffice are demanding the impossible. If their demands were satisfied, therapeutic stagnation would result. As one director of a pharmaceutical company put it, 'If you do not want to take the risk of a drug, try the risk of disease'. A public that demands progress must be prepared for some risks. The job of the medical and allied professions and the drug regulatory authorities is to minimize them. Risk cannot be considered in isolation from the whole process of drug development because knowledge of adverse drug reactions evolves all the way along with the development of the drug and its subsequent marketing. The usual purpose of research in the drug industry is to discover drugs which are profitable. For this purpose to be achieved the drug must be both useful and safe. Adverse drug reactions are as important to the drug industry as they are to the patient and clinician. Ultimately the properties of a drug can only be determined by the clinician and his patients. The task of the industry is to predict the likely beneficial and adverse effects from animal experiments and to conduct closely monitored pharmacological studies on healthy subjects. The task of clinicians is to study the effect of the new drug in patients and to monitor its safety for the rest of the drug's existence. Hence close collaboration between clinicians, industry, and regulatory authorities is the key to smooth drug development. It has been said that sometimes too much time is wasted in animal studies before drugs are tested in humans but no clinician would contemplate use of a new drug in humans before the results of thorough preclinical tests are known. It must be stressed however that the real worth and dangers of any drug are not likely to be known until after extensive clinical use of that drug.

Preclinical testing

The development of a new drug is usually the work of a chemist; it is often based on his knowledge of the pharmacological activity of structurally related compounds. More rarely a completely novel compound is synthesized. After purification the new compound is given a preliminary pharmacological screen and, if this shows promising effects, it is subjected to more extensive pharmacokinetic and toxicological studies. Toxicity tests are of two types, namely, acute (single-dose) and chronic (repeat-dose) studies. The acute studies are aimed at the determination of the LD_{50} of the new compound, i.e. the dose required to kill half the animals. In such studies the mode of death of the animals is usually recorded, but no detailed post-mortem studies are done. LD_{50} studies have met with much criticism for they are wasteful of experimental animals and provide very limited information. After all it is at least as important to know how a drug kills as it is to know how much of it kills. Perhaps toxicity testing should be taken out of the hands of pathologists and regarded as an extension of physiological and pharmacological studies. I do not agree with those who seek to replace the LD_{50} with an *in-vitro* test that provides similar information. It is not worth replacing the LD_{50} for it is a useless, unstable measurement of toxicity. We really should design a new set of toxicity tests which utilize functional measurements in place of 'dead meat' pathology as end-points. This would not only meet the antivivisectionists part of the way but would also improve our knowledge of the toxic properties of new compounds. It might also serve to enhance the value of preclinical tests as predictors of human hazard.

One thing LD_{50} studies can be used for is to determine the therapeutic ratio of a new compound. This is defined as the lethal dose divided by the curative dose. The therapeutic index gives an indication of the safety margin of the drug. It is worth stressing that the industry does not usually determine the minimal curative dose of a drug. For example in the treatment of lower urinary-tract infection it is advised that trimethoprim should be used in a dose of 200 mg twice daily for five days and yet it has been established that a dose of 50 mg twice daily is just as effective as the higher dose (Trimethoprim Study Group 1981). Perhaps the industry should pay more attention to minimum effective dosage, for it would help to reduce risks as well as the expense of treatment.

Enormous difficulties are met in forecasting adverse effects that may occur or be recognized only long after the administration of the drug has been stopped. Such delayed adverse reactions may go undetected for many years. Concern for such delayed effects began with the thalidomide disaster, and other examples have come to light since.

Four such delayed types of adverse reaction can be distinguished, namely: carcinogenic, mutagenic, teratogenic, and effects on fertility. There is usually a fairly close association among these effects. Neither the results of bacteriological nor of animal tests for effects on cell division are necessarily applicable to humans and what is more the animal tests are of necessity carried out on limited numbers. Mathematical modelling techniques may be used to quantify risks from the limited preclinical toxicity data (see Chapter 9), but needless to say the mathematical treatment of such data can never rise above the original biological data. Nevertheless they do help in the objective assessment of risks and benefits as displayed by the preclinical tests.

With known chemical entities it may also be possible to limit the preclinical testing by the study of its quantitative structure activity relationships (QSAR). QSARs are mathematical models which relate biological activity to certain attributes of the molecular structure of the new drug. QSAR studies can only be used if the new drug is a member of a series of compounds which share a known molecular structure and mode of action. QSARs can also be used to design and later synthesize improved compounds of the series with greater activity and/or less toxicity. Several aspects of molecular structure are used as data base for QSAR analyses. These include molecular topology (so-called Free–Wilson analysis) and physicochemical properties (Hansch analysis). For instance the partition coefficients of the drugs in the series might be related to their activity/toxicity or the electron density plots displayed by computer graphics could be related to the toxic and/or wanted effects of the potential drugs. Again the conformation of the molecules could be related to their biological properties, e.g. the angles and distances between chemical groupings in the molecules. Computer analysis of these and other properties of the new molecule compared with the known characteristics of the existing substances of the series can thus be used to forecast the properties of the new compound and curtail animal experimentation.

What constitutes a safe drug is hard to define because it must be viewed in the light of intended use. When a drug is required as prophylaxis or to treat non-life-threatening disorders, the margin of safety (therapeutic ratio) should be very much greater than that demanded of a drug which can cause occasional cures of a disease which is generally speaking fatal. Thus drugs used in malarial prophylaxis should be extremely safe whereas the acceptable risk of a cytotoxic drug for treatment of cancer can be much higher. In this context the dictum that it may take a desperate remedy to treat a desperate illness applies.

Despite all the care taken over preclinical tests their value as predictors of human hazard is often in doubt. When toxicological tests show hazard in animals it is clearly unethical to proceed to human studies and in these cases it is therefore impossible to test the predictive value of the animal studies. The converse however is not true, i.e. substances which do not show toxicity in the animal models may still show toxicity in man. As Griffin (1983) pointed out this may be (1) because the syndrome of toxicity experienced in man does not occur in animal models (e.g. the practolol syndrome); or (2) because the appropriate animal model has not been devised or was not used; or (3) because one aspect of the drug's toxicity in animals may be a limiting factor, in that this aspect of toxicity appears first and limits the toxicity study in either duration or dosage so that another aspect of the drug's toxicity relevant to humans is not revealed. The latter is a common problem with non-steroidal anti-inflammatory drugs (NSAIDs) where the gastrointestinal side-effects 'overshadow' other adverse effects. On the whole animal toxicity studies are useful for predicting adverse reactions which are related to the primary pharmacological actions of the drug and for predicting those related to its direct toxic effects. They are not of much value in predicting subjective adverse reactions or those due to hypersensitivity reactions.

Clinical testing

Governmental agencies in a number of countries have developed guidelines for clinical testing. Initial safety testing is carried out in a few humans, usually healthy volunteers, and studies in larger numbers of healthy human subjects follow. Finally the drug is used in a clinical trial to prevent, combat, or control disease. Depending on the outcome, the clinical trial may be extended so that eventually the investigation and the findings as a whole comply with the drug laws and regulations and the drug is allowed for sale by the original manufacturers responsible for its development and investigation. It is to be noted that children, the aged, and pregnant women are generally excluded from the clinical trials of a new drug, and this may at a later stage be a source of unexpected adverse effects. A good example of this was the adverse effects of benoxaprofen encountered in the elderly. With the increasing age of our population it may become necessary to undertake testing of drugs in the geriatric age group prior to the granting of a product licence. Not infrequently various formulations of a single product are used during the course of the preclinical and clinical tests. This is done in order to select

the presentation which is most suitable for marketing. The final product is therefore unlikely to be identical with that with which much of the work relevant to toxicity and teratogenicity testing has been performed. Certain other reservations must also be made, shelf-life studies on the product for example are done under ideal conditions, somewhat removed from the actual conditions of a drug product in the field.

The foregoing is only a very rough outline of the process of drug development, but it may serve to remind the reader that limited information is available about a new drug by the time it receives its product licence. At the time the product licence is issued it is highly unlikely that the total spectrum of adverse and beneficial actions of a new drug is known. Take the example of a new drug which causes an adverse event in 1 in 10 000 patients treated. A trial involving 100 patients would only have a 1 per cent chance of revealing that adverse effect. Most drugs have not been given to more than a few hundred patients by the time they are marketed, and important adverse effects may therefore have been overlooked.

Post-marketing surveillance

Once a drug has been marketed the circumstances are no longer those of a controlled experiment. Many adverse drug reactions (ADRs) resemble naturally occurring events closely; it is essential therefore that all adverse events are reported in patients receiving a new drug. No scheme to achieve this objective is foolproof. In Britain as in many other developed countries, a voluntary reporting scheme (VRS) has been in operation since 1964, when the Committee on Safety of Medicines (CSM) was established; this is the so-called yellow-card system. More sophisticated methods of post-marketing surveillance have been suggested. These may supplement or even replace the voluntary reporting scheme as it exists today. Until this happens VRS continues to provide a valuable service in the detection of ADRs. VRS provides signals of drug safety problems which are generated by eyeballing the reports as they are received. They can also be used to investigate causality, to establish incidence, to facilitate a risk–benefit judgement and to inform prescribers and patients.

By itself VRS can only achieve the identification of safety problems but linked with other information its value can be extended. Although several other sources of information about ADRs exist, the yellow-card system is by far the most important source of information about adverse drug reactions; 76 per cent of all reports of suspected drug reactions are derived from the yellow cards (Inman 1980). Introduction of insert slips into GP prescription pads (FP10) has done much to encourage this method of reporting and yet there is still a vast amount of underreporting which relates to what Inman (1978) called the seven deadly sins. These are: (1) complacency, the mistaken belief that only safe drugs are allowed on the market; (2) fear of involvement in litigation, a lesser problem perhaps in the UK; (3) guilt because harm to the patient has been caused by the treatment the doctor has prescribed; (4) ambition to collect and publish a personal series of cases — a common human failing that may lead to serious delays in recognition of a hazard; (5) ignorance of the requirements for reporting, perhaps as a result of failure of communication between the reporting centre and the professions; (6) diffidence about reporting mere suspicions which might perhaps lead to ridicule; (7) lethargy — an amalgam of procrastination, lack of interest or time, inability to find a report card, and other excuses. Possibly the failure to report adverse reactions could be reduced by patient participation stimulated by the use of prescription information leaflets which are issued to patients and the yellow-card system might also be sharpened up by involvement of pharmacists and nurses in the ADR reporting. Even the simple step of providing a freephone facility for general practitioners to contact the regulatory authority might help to reduce underreporting. The more complete our adverse drug reaction register the more information can be obtained.

The type of information which may come to light includes the sex, race, and age incidence of adverse drug reactions, the occurrence of drug interactions, the identification of relatively more numerous adverse effects with one drug compared with another of its class, and the identification of an unusual or serious adverse effect. Although the number of patients treated with the new drug is often unknown, some estimates of incidence of the ADR may be derived from the knowledge of drug sales or from the number of prescriptions for the drug as recorded by the prescription pricing authority. Drug profiles which are dealt with in Chapter 8 may be constructed. These do not require a knowledge of drug usage since in this exercise the types of adverse drug reactions reported for different drugs of the same class are looked at. An example taken from Inman (1978) is shown in Fig. 10.1.

In-depth study of drug reaction reports may sometimes also enable the discovery of important associations between drug and adverse reactions. For instance the linkage of multiple exposure to halothane and hepatitis was a good example of what can be derived from an in-depth study of adverse drug reactions (Inman and Mushin 1978). In arriving at a decision as to whether or not causality exists between exposure to drugs and the observed adverse reaction, the criteria of Karch and Lasagna (1977) may be employed. These criteria are: (1)

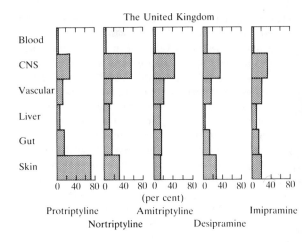

Fig. 10.1 Adverse-reaction profiles of tricyclic antidepressants based on 1109 reports. Note that skin reactions are more common with protriptyline than with other members of the group (from Inman 1978).

whether or not there is a definite temporal relationship between administration of the drug and emergence of the adverse reaction; (2) whether or not the ADR follows a known response pattern to the class of drug; (3) whether or not the condition improved when the drug was stopped; (4) whether the condition recurred on rechallenge; (5) whether or not the adverse effect could be reasonably explained in some other way. On the basis of this type of information the likelihood of drugs being responsible for adverse reaction may be graded into four categories, namely, definite, probable, possible, and doubtful.

Apart from underreporting another important defect of the yellow-card system is inadequate utilization of data. Venning (1983) claimed that the yellow-card system had made no contribution to 18 serious adverse reactions which had occurred since the thalidomide disaster. He stressed the value of anecdotal case reports as early warning signals followed by systematic disease- and/or case-orientated studies. Such studies he considered more sensitive indicators of adverse drug effects than voluntary reporting schemes. In defence of the yellow-card system Inman (1983) pointed out that only three of the drugs Venning had studied could have been tested by the yellow-card system. The ideal system for the identification of adverse drug effects would be if these adverse effects were to become part of medical record-linkage. Such a system would provide data on the incidence of reactions amongst drug users as well as the prevalence of drug usage among patients with diseases or syndromes that might be drug-induced. Apart from pilot studies such as that in Oxford (Skegg *et al.* 1981) this approach has not so far been applied on a large scale.

Various additional methods of post-marketing surveillance exist. These include surveillance by the pharmaceutical industry, prescription event monitoring, and studies by collaborative adverse reaction groups such as those established in Boston, Aberdeen, and Dundee. The development of local reporting schemes by yellow cards, the possible introduction of compulsory reporting schemes such as in Norway and Sweden, and the involvement of pharmacists as in the case of Australia are possible developments for the future. These and other new methods of post-marketing surveillance were reviewed by Wilson (1980).

Making the judgement

Good judgement requires a reliable and adequate data base, a knowledge of the consequences of not making any decision, a recognition of the frailties of human judgement, and a modicum of common sense.

In the foregoing sections we have discussed the ways in which the data base is constructed. Firstly, by dissecting the components of risk and benefit and then as is rarely possible, reducing them to a common denominator or at least making some attempt to quantify them. Even if quantitation is not possible, the mere process of dissection and subsequent recomposition of risks and benefits clarifies the issues prior to judgement. Three aspects of the data base have not so far been mentioned; firstly, that the description of risks and benefits should include areas of ignorance, for if these should be sufficiently large time for judgement may not be ripe. Secondly, that all risks and benefits should be enumerated in random order since there is a tendency for judgement to be biased by the order in which facts are presented (to the jury). Thirdly, it must be remembered that if a new drug is launched successfully that a large number of adverse effects are likely to be reported soon after the start of marketing. This clustering phenomenon may lead to exaggerated impression of hazard.

A knowledge of the consequences of inaction is vital to the decision-making process. What harm is likely if no treatment is given or if no product licence is granted? Let us turn to the case of *Depo-Provera*, the controversial long-acting intramuscular contraceptive. This drug would be indicated for women approaching the menopause, for women in special circumstances, such as air stewardesses and athletes or in other circumstances, e.g., for those women whose husbands are antagonistic to birth control or for some mentally retarded girls. In these minority groups intramuscular contraception might be

the preferred method. In numerical terms the consequences of not granting a licence for the drug are small but in the individual case non-availability of the drug could cause hardship. It follows that experts drawn from various disciplines should be involved in the affairs of drug regulatory authorities so that the needs of minority groups are given full consideration. The risk–benefit ratio should be considered separately for each population group.

There are three areas in which it is particularly important to understand the consequences of inaction. The first concerns the need to treat symptomless disorders such as mild hypertension or hyperlipidaemia. At what level of blood pressure and in which age, sex, and racial groups should raised blood pressure be treated? Should the criterion of success be reduction of blood pressure, increased longevity, reduction of heart attacks or strokes? Similarly, when one considers the problem of treating symptomless hyperlipidaemia, is success to be defined in biochemical terms or in terms of preventing heart attacks and/or strokes? The answers to these questions can only be obtained from long-term randomized controlled treatment trials. In situations where screening for symptomless disease and subsequent treatment is contemplated, it is as well to remember the criteria for successful screening procedures as laid down by Wilson and Jungner (1968). In simplified form these might be stated as follows: (1) the disease looked for must be a major health hazard; (2) there should be a latent phase of the disease which can be detected by a test acceptable to clinicians and the population alike; (3) the natural history of the symptomless disease and its response to and the safety of treatment should be known; and (4) the cost of case finding and treatment should be economically balanced against the expenditure on medical care as a whole. Few screening procedures satisfy all of these criteria. The treatment of mild hypertension, the treatment of symptomless bacteriuria in pregnancy, are examples of screening procedures where all these criteria have been satisfied. Before a drug is licensed for long-term treatment of symptomless abnormalities, it is essential that drug regulatory authorities are satisfied that the above criteria are met; otherwise, problems such as those encountered with the use of clofibrate in the treatment of symptomless hypercholesterolaemia for the prevention of ischaemic heart disease are likely to recur. In this instance it was found that although reduction of raised serum cholesterol with clofibrate reduced the incidence of ischaemic heart disease, long-term use of the drug caused an increased incidence of gallstones and was associated with an increased mortality (Report from Committee of Principal Investigators 1978).

The second situation in which it is important to

understand the consequence of inaction concerns the need to prevent recurrent disease. The prevention of recurrent myocardial infarction by anticoagulants, β-blockers or agents which reduce platelet adhesiveness is a good example. Here again our judgement of the risk–benefit ratio should be based on the outcome of ran-domized controlled trials, but even when the results of such trials do indicate benefit there remains the difficulty for the clinician of applying the results of such large-scale studies to the individual patient (Mitchell 1980).

The third instance where the consequences of inac-tivity need to be carefully weighed against those of inter-vention concerns prophylactic vaccination. The decision whether to recommend the use of a particular vaccine depends upon balancing the benefit likely to be conferred against the risks to the vaccinated individual. This balance is most in favour of immunization when the disease has a high incidence, morbidity, and mortality. Conversely, it is least in favour of immunization when the complications of the procedure are frequent and the disease is at low ebb. A very good illustration of this dilemma is to be found in the whooping cough immu-nization controversy which was the subject of detailed enquiry by the DHSS in 1977.

The frailties of human judgement are many. Judge-ment is an intuitive process; nevertheless, it is amenable to scientific study. Such scientific study of human judgement may pinpoint ways of improving judgement. Judgements are operations on information and it is necessary therefore to understand the human infor-mation processing system. One of the most important defects of the human information-processing system is that it is limited. This, as Hogarth (1980) pointed out, has

four major consequences. Firstly, we tend to select from all the information received, but in this process of selection anticipation plays an enormous part. By way of example, when the members of a regulatory authority are confonted with a vast amount of literature, no single member can usually digest all the information presented. Selection may therefore have to be practised. In this selection bias is inevitable, e.g. one member may only consider those items of information which had appeared in refereed journals, others might concentrate on their specialty interests such as societal risks, the issue of consent to treatment, or the question of carcinogenicity and effects on fertility. No single member can usually digest all of the evidence presented, no more than it is possible to appreciate more than about one-seventieth of what is present in one's visual field. This selection can introduce bias into judgement.

The second consequence of the limited information-processing capacity of the human brain is that information has to be processed sequentially. Sequential treatment of data also introduces bias. As any barrister or judge knows, the order in which data are presented to members of the jury can make a difference to their judge-ment. Take a simple example of two films, one showing a car speeding down a road, the other showing a man running very fast down the same road. If the runner were shown first and the car sequence second, it would seem that the car was chasing the runner. The opposite conclu-sion might be reached if the films were shown in reverse order. The third consequence of the limited information-processing capacity of the human brain is that it makes us resort to simpler strategies. We tend to see patterns where none exists. We tend to combine data where data should

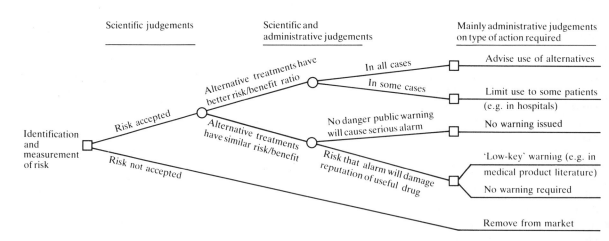

Fig. 10.2 Decision tree showing the interface between scientific and administrative judgements in drug monitoring (from Hogarth 1980).

not be combined. We tend to relate data to previous experiences which may or may not be correct.

The last and most serious consequence of our limited capacity for information-processing relates to the limitations of human memory. Memory changes depending on when and how reconstruction takes place. It would be wrong however to look upon humans as being simply inefficient computers. Humans have emotions, imagination; they have a tendency to question and may experiment and they may strive to go beyond the information they receive. These are properties which computers do not possess and which can lead to judgemental errors as well as illusions of judgemental ability. Several attempts have been made to improve human decision-making processes by means of decision analysis. A fundamental operational rule of decision analysis is to decompose a problem into manageable parts and then to recompose it. Pauker (1976) constructed a decision model concerning the desirability for medical versus surgical treatment of angina pectoris. It is a good example of the use of decision theory in medical judgement.

The underlying rationale of such a decision-theory approach to judgement is not only to dissect the problem in such a manner that all the elements which enter into the final decision are made explicit but also to express them in probabalistic terms. Attempts at quantification need not be very accurate since the importance of the quantitative elements in the judgement can be assessed at a later stage by a sensitivity analysis. This entails a retrospective study of how much change in any one parameter used to reach the judgement is needed to change that judgement. Usually enormous changes are needed, and this is why quantification of risk and benefits need not be too refined.

Drug monitoring is more complex than decisions such as the choice between medical and surgical treatment of angina pectoris as it involves different types of judgement by different parties, e.g. doctors, pharmacists, administrators, and sociologists. When people with such widely different interests face a common decision it is often assumed by the different parties that their varying interests imply different actions (Hogarth 1980). However this is not necessarily the case. Public decision-making can usually accommodate a wide range of apparently conflicting interests. However, unless alternative actions are explicitly evaluated against the interests of the different parties, advantage cannot be taken of the insensitivity implicit in the structure of many decisions. If real disagreement about actions exists, it is important that the source of the disagreement be identified. When many people are involved in a decision process, ineffective communication between groups is very common. For example, scientists who express

judgements at the first stage of drug licensing may find administrative judgements at subsequent stages to be unjustified. In reverse, administrators may fail to understand the basis of scientific judgement. Lack of understanding leads to lack of co-operation and as a consequence ineffective decision-making results. The steps in the decision-making process of a drug regulatory authority were schematically represented by Hogarth (1980) and are shown in Fig. 10.2. It is only by adopting such a logical approach that the intuitive element of human judgement can be minimized.

Minimizing the risks

Drugs are remarkably safe. Few patients would refuse an elective surgical operation with a risk of less than 1:10000. Yet for medicines much greater safety is demanded and achieved. From time to time it has been suggested that we need a Committee on Safety of Surgery more than we need a Committee on Safety of Medicines. Despite the relative safety of our drugs we do have an obligation to do all we can to make risks even less. Responsibility for this falls not only on the regulatory authorities but also on doctors, the pharmaceutical industry, pharmacists, the media, and patients themselves. There are several ways in which the risks of treatment can be reduced. Firstly, there is the role of education. On the one hand this education should be directed at good prescribing as well as spotting and reporting of adverse reactions. On the other, education should aim to enable the subject of adverse drug reactions to be seen in perspective so as to avoid alarmist attitudes. Too often do such attitudes lead to the early demise of potentially useful remedies. Drug regulatory authorities should be immune from political and public pressure and above all free from the pressures of action groups.

Following a decision to license a drug it is incumbent upon regulatory authorities to warn the medical profession and public of adverse events encountered with a new drug. If the signal is set off too early, it may prevent us every knowing whether the new drug was worthwhile. Various forms of feedback exist. These are: (1) the adverse reaction information service; (2) the register of adverse reactions; (3) the manufacturer's literature; (4) letters to medical journals; (5) 'Dear Doctor' letters; (6) scientific papers; (7) the *Current Problems* series; and (8) the adverse reaction series. In roughly ascending order these tend to have progressively greater impact on the media which may for better or sometimes for worse, act as an amplification loop. The regulatory authority has often been accused of providing inadequate feedback, which in turn may be part of the reason for the apathy amongst doctors to report adverse events. There is certainly truth in these accusations and the regulatory authority would do well to take heed of these accusations and improve its communications with the profession and public.

Concluding remarks

In this chapter I have outlined possible ways in which risk–benefit analysis in medical treatment can be made more objective and ways in which judgement of risks and benefits can be improved by creating an awareness of the frailties of human judgement. The safety of drugs is of concern to all. In clinical decision-making the main consideration is the welfare of the patient and at drug regulatory level the welfare of society must also be taken into account. Drug regulatory authorities should be free of political pressure just as pharmaceutical industries should not be pressurized to market their products hastily because of the relatively short life of their patents. The more marginal benefits of modern treatments demand detailed risk–benefit analysis of new drugs. All branches of the medical profession and industry should collaborate to achieve this so as to enable us to market the best possible medicines for use in the safest possible way to give maximum benefit to the patient.

Acknowledgement

I am most grateful to Dr J.P. Griffin, Director of the Association of the British Pharmaceutical Industry, for his constructive comments in the preparation of this chapter.

RECOMMENDED FURTHER READING

CAVALLA, J.F. (1981). Risk–benefit analysis in drug research. MTP Press, Lancaster.

GENT, M. AND SHIGEMATSU, I. (1978). Epidemiological issues in reported drug induced illnesses: SMON and other examples. McMaster University Library Press, Hamilton, Ontario.

GROSS, F.H. AND INMAN, W.H.W. (1977). *Drug monitoring.* Academic Press, London.

INMAN, W.H.W. (1980). *Monitoring for drug safety* MTP Press, Lancaster.

LUSTED, L.B. (1968). *Introduction to medical decision making.* Charles C. Thomas, Springfield, Illinois.

REFERENCES

BATCHELOR, J.R., WELSH, K.I., MANSILLA TINOCO, R., DOLLERY, C.T., HUGHES, G.R.V., BERNSTEIN, R., RYAN, P., NAIRN, P.F., ABER, G.M., BING, R.F., AND RUSSELL, G.I. (1980). Hydralazine-induced lupus erythematosus: influence of HLA-DR and sex on susceptibility. *Lancet* i, 1107–09.

GRIFFIN, J.P. (1983). Repeat-dose long-term toxicity studies. In *Discussion on the report of the FRAME Toxicity Committee* (ed. M. Balls, R.J. Ridell and A.N. Worden), pp. 98–107. Academic Press, London.

HOGARTH, R.M. (1980). Judgement, drug monitoring and decision aids. In *Monitoring for drug safety* (ed. W.H.W. Inman), pp. 439–75. MTP Press Ltd, Lancaster.

HURWITZ, N. (1969). Predisposing factors in adverse reactions to drugs. *Br. med. J.* 1, 536–9.

INMAN, W.H.W. (1978). Assessment of drug safety problems. In *Epidemiological issues in reported drug-induced illnesses S.M.O.N. and other examples* (ed. M. Gent and I. Shigematsu), pp. 17–24. McMaster University Library Press, Hamilton, Ontario.

—— AND MUSHIN, W.W. (1978). Jaundice after halothane; a further report. *Br. med. J.* 2, 1455–6.

—— (Ed.)(1980). The United Kingdom. In *Monitoring for drug safety*, pp. 9–47. MTP Press Ltd, Lancaster.

—— (1983). Adverse reactions to new drugs. *Br. med. J.* **286**, 719–20.

JOINT COMMITTEE ON VACCINATION AND IMMUNIZATION OF THE DHSS (1977). *Whooping cough vaccination. Review of evidence of whooping cough vaccination*, HMSO, London.

JOURNAL OF MEDICAL ETHICS (1982). Editoriali. *J. med. Ethics* **8**, 171–2.

KARCH, F.E. AND LASAGNA, L. (1977). Towards the operational identification of adverse drug reactions: *Clin. Pharmacol. Ther.* **21**, 247–54.

MITCHELL, J.R.A. (1980). Secondary prevention of myocardial infarction — the present state of the art. *Br. med. J.* **1**, 1128–30.

PAUKER, S.G. (1976). Coronary artery surgery: the use of decision analysis. *Ann. intern. Med.* **85**, 8–18.

POCHIN, E.E. (1982). Risk and medical ethics. *J. med. Ethics.* **8**, 180–4.

REPORT FROM THE COMMITTEE OF PRINCIPAL INVESTIGATORS (1978). A cooperative trial in the primary prevention of ischaemic heart disease using Clofibrate. *Br. med. J.* **40**, 1069–1118.

SKEGG, D.C.G., RICHARDS, S.M., AND DOLL, R. (1981). Record linkage for drug monitoring. *J. Epidemiol. Comm. Health.* **35**, 25–31.

TRIMETHOPRIM STUDY GROUP (1981). Comparison of trimethoprim at three dosage levels with co-trimoxazole in the treatment of acute symptomatic urinary tract infection in general practice. *Br. J. Antimicrob. Chemother.* **7**, 179–83.

VENNING, G.R. (1983). Identification of adverse reactions to new drugs. *Br. med. J.* **286**, 199–202; 289–92; 458–60; and 544–7.

WILSON, A.B. (1980). New surveillance schemes in the United Kingdom. In *Monitoring for drug safety* (ed. W.H.W. Inman), pp. 189–200. MTP Press Ltd, Lancaster.

WILSON, J.M.G. AND JUNGNER, G. (1968). *Principles and practice of screening for disease.* WHO Geneva.

11 Product liability in respect of drugs

A.L. DIAMOND and D.R. LAURENCE

English law has for many years contained principles which enable persons injured by defective goods or products to receive compensation in the form of damages — a money payment — from another person in certain circumstances. These principles are applicable to injuries caused by drugs as well as by any other product, though their application to drugs raises a number of special problems.

The law evolved in a pragmatic way, and followed two separate lines of development. One of these lines concentrated on the immediate relationship between seller and buyer, there being a contract between them. A 'contract' for this purpose can be quite informal: there is a contract of sale of goods between a retail pharmacist and a customer who buys a bottle of aspirin over the counter. The other line of development concentrated on a duty to be careful owed by one person to another, and does not depend on a direct contractual relationship.

Supply contracts

Where there is a contract to supply goods in return for a price, the law imposes obligations on the supplier in the form of terms of the contract. For example, where drugs are sold by a pharmacist for cash, whether or not on prescription (NHS prescriptions are dealt with separately later), there is a contract of sale of goods covered by the Sale of Goods Act 1979. In this contract the Act implies a number of 'conditions' relating to the goods sold, notably that the goods supplied will correspond with the description under which they were sold, that they will be of merchantable quality, and that they will be reasonably fit for the purpose for which they are bought.

These rules, now set out in the Act of Parliament, came into existence in the nineteenth century as a result of decisions by judges in particular cases which came before them. The policy of the law was set out in forthright terms by one of the judges, Chief Justice Best, in 1829.

It is the duty of the court in administering the law to lay down rules calculated to prevent fraud, to protect persons who are necessarily ignorant of the qualities of a commodity they purchase, and to make it the interest of manufacturers and those who sell to furnish the best article that can be supplied . . . . I wish to put the case on a broad principle. If a man sells an article he thereby warrants that it is merchantable — that is, fit for some purpose. If he sells it for a particular purpose he thereby warrants it fit for that purpose.

Not many cases arising from the sale of pharmaceutical products have come to court, but some important decisions have been made in cases involving food or drink. Thus in one case (*Frost* v. *Aylesbury Dairy Co.* in 1905), milk supplied by a dairy caused the death of Mrs Frost by typhoid. The dairy were sued for breach of the implied condition under the Sale of Goods Act 1893, then in force, that the milk would be reasonably fit for the purpose for which it was bought — that is, fit for human consumption. In its defence the dairy argued that it had taken every precaution possible in the then state of scientific knowledge to prevent infection, but the court said this was irrelevant. Under the Sale of Goods Act the dairy was deemed to promise that the milk was fit for human consumption, and it was not. Whether or not it was guilty of fault or negligence was beside the point. Liability of this kind, where fault is irrelevant, is sometimes known as 'strict liability'.

There is an important limitation on these liabilities under the Sale of Goods Act. Remedies under the Act are open only to the individual buyer who has entered into the contract of sale. Two cases where persons were scalded by defective hot-water bottles sold by pharmacists demonstrate this. In one case the injury was suffered by the buyer who had gone into the shop to buy the hot-water bottle. He was awarded damages under the Sale of Goods Act. In another case, a father bought the hot-water bottle but the injuries were suffered by his seven-year-old daughter. She had no claim under the Sale of Goods Act because she was not the buyer, and the father had no claim because he had not been injured. To obtain compensation, the daughter had to look for a different legal principle.

Negligence

This brings us to the other line of legal development, which does not depend on the existence of a contract. After some hesitation the courts eventually held, as recently as 1932, that a manufacturer owes to the ultimate consumer a duty to take reasonable care in making and putting up the product. Failure to take reasonable care is negligence, and the person injured can sue, contract or no contract. This was decided by the House of Lords in a famous case in which a decomposed snail found in a bottle of ginger beer was alleged to have made the consumer ill (*Donoghue* v. *Stevenson*, 1932). If the unfortunate lady had herself purchased the ginger beer she could have claimed from the seller under the Sale of Goods Act, and the seller would have been strictly liable — liable even though the bottle was opaque and had been sealed by the manufacturer. But since the bottle had been bought and paid for by a friend, she had no cliam against the seller (who had clearly not been negligent), and instead sued the manufacturer direct in a claim based on negligence.

Under this principle the seven-year-old girl scalded by the defective hot-water bottle succeeded when she sued the manufacturer. The manufacturer could not explain how the defect came about, and in the absence of any explanation the court concluded that an employee must have been negligent.

When the claim is based, not on a contract, but on an allegation of negligence, it is a defence to show that reasonable care was taken. The dairy whose milk was infected with typhoid bacteria would, it seems, have escaped liability if it had been sued for negligence rather than breach of the Sale of Goods Act. So the standards imposed under this head of the law are lower than those imposed under the sale contract. This head of the law makes it the interest of the manufacturer to take reasonable care, but not necessarily to make 'the best article that can be supplied'.

This principle of liability for negligence is not, of course limited to the manufacture and distribution of goods. It applies to every human activity. It is the principle that governs the legal liability of medical and other professional people who cause injury to their patients or clients: except in relation to the supply of goods, they are only held responsible if they have been at fault.

It is also the principle that governs the supply of medicines under the National Health Service. A private patient who buys a medicine on prescription from a pharmacist enters into a contract of sale of goods with the pharmacist. Thus the pharmacist is liable under the Sale of Goods Act if the medicine is not of merchantable quality (if, for example, there was a manufacturing defect and the medicine causes injury). But if the pharmacist dispenses the identical medicine under a national health prescription the Sale of Goods Act does not apply. There is no sale by the pharmacist to the patient and no contract between them. If the patient is injured as a result of a manufacturing defect, a claim for compensation must rest on the proof of negligence. Since the pharmacist is unlikely to have been negligent in these circumstances (unless he ought to have spotted the defect), any claim is likely to be brought against the manufacturer.

Strict liability

The distinctions between what has been called 'strict liability' under the Sale of Goods Act 1979 and liability for negligence have given rise to the anomalies in the law which were described above. These are the hard and fast line drawn by the law between the actual buyer (who takes the benefit of strict liability) and other persons, even members of his or her immediate family (negligence only); the distinction drawn between private patients or buyers of non-prescription medicines (strict liability) and recipients of NHS medicines (negligence); the fact that the retail pharmacist who enters into a contract with his customer is under a strict liability even for the contents of a sealed container which he cannot inspect while the manufacturer, who must bear some responsibility for the contents, is liable only for negligence. These anomalies (which do not exist in every legal system) have given rise to influential calls for reform of the law. The reform is a move towards the position that is to be found in French law under the Code Napoleon and in the law of most parts of the United States as a result of judicial decisions. This position is to hold the manufacturer under a strict liability to the ultimate consumer injured by the product.

This change in the law has been called for by a number of different bodies at home and abroad. In 1977 a report issued by the Law Commission for England and Wales and the Scottish Law Commission (two official bodies charged with the responsibility for eliminating anomalies from the law and for keeping it up to date) recommended that the reform should be made (Report on Liability for Defective Products). A year later the Royal Commission chaired by Lord Pearson on Civil Liability and Compensation for Personal Injury supported the recommendation. In Strasbourg, the Council of Europe adopted in 1977 the European Convention on Products Liability which sought to unify the laws of the different countries in Europe on the same lines (this Convention has not yet come into force). And in 1976 the Commission of the European Communities issued its proposal for a Directive on Liability for Defective Products which would require every member state of the European Economic Community (EEC) to bring its law into line with the basic principle of strict liability for the manufacturer of defective products.

This reform, if introduced into our law, would not be as revolutionary as it might be thought. Since 1893, when the first Sale of Goods Act was passed, manufacturers have been under a strict liability for defective products, for it must not be forgotten that manufacturers are sellers. If a buyer sues a retailer, the retailer in turn can sue his supplier and so on up the chain of distribution (for each seller was himself a buyer under an earlier contract). Ultimately, therefore, the strict liability would come to rest on the manufacturer. To this extent the reform is in part a procedural device to avoid expensive multiple lawsuits.

Various arguments have been adopted in support of the reform over and above the removal of the anomalies already described. Notably, it has been said that the question is essentially one of insurance. Not only is it cheaper for the manufacturer to insure against his possible liability (he is, if he is prudent, already insured against his liability for negligence) than for each consumer to take out an individual policy against the risk of being injured, but also the manufacturer can, by regarding the insurance premium as one of his production costs, spread the cost of insurance across all purchasers of his products. Moreover the liability creates an incentive to produce safer goods. There is also the moral argument adopted by some: that those who create a risk, and create it to make profit, have a special responsibility to bear any adverse consequences regardless of whether or not they are at fault.

The proposals for reform do not distinguish between manufacturing and design defects. A manufacturing defect affects individual articles or batches of articles when the act of manufacture has gone wrong. We are familiar with manufacturing defects when, for example, cars leaving the production line between certain dates, identified by their distinctive engine or chassis numbers, are recalled because of a failure to fix a particular component adequately so that there is, say, a weakness in the steering system. A design defect is when there is a flaw inherent in the product, affecting all items. Thalidomide has been said to be an example of a design defect, though to medical people the term 'design defect' will seem a curious way of describing the capacity of aspirin to cause gastric bleeding or of carbimazole to cause agranulocytosis. Whether these qualities would be regarded as defects at all is far from clear.

Are drugs different?

It has indeed been urged that drugs are so different from other products that strict liability should not be imposed on their manufacturers. All drugs are potentially dangerous, they are capable of being abused, and their defects cannot always be foreseen. Moreover, the imposition of a greater liability might inhibit research and the production of new and valuable remedies. This view has gained only limited support outside the pharmaceutical and medical professions. 'Drugs represent the class of product in respect of which there has been the greatest public pressure for surer compensation in cases of injury' stated the Pearson Royal Commission in its report in 1978, adding that this demand was now an international phenomenon.

No-fault schemes

However a number of countries, notably Sweden, have established schemes whereby individuals injured by medicines may obtain compensation, even though no one was at fault, without having to claim against a particular manufacturer. Such schemes are often (perhaps ambiguously) called 'no-fault schemes'. Although strict liability as described above operates whether or not anyone was at fault, the machinery involved contemplates a claim, ultimately backed up by a law-suit, by the individual injured party against the manufacturer who produced the product causing injury. A no-fault scheme involves the creation of a fund (whether administered by the government, by insurance companies, or by a group of manufacturers) against which claims may be made. The fund is administered independently of the particular manufacturer whose product was defective.

Vaccine damage

The United Kingdom does not have a general no-fault scheme for medicinal injury, but it has implemented a system of compensation in a single limited area, i.e. injury due to vaccines for certain listed diseases. The introduction of this limited scheme throws light on the public mind and on political pressures and processes. It provides lessons for those who are concerned to develop wider schemes. The vaccine scheme was introduced as a result of public pressure particularly related to brain damage from pertussis vaccine. The reasons include the fact that the government was promoting a policy of administering to healthy young children a product that might not only prevent them from contracting an unpleasant and dangerous disease but which was also intended to protect children in general. When parents discovered that this policy, promoted for the general good, could also result in injury to their own healthy children (regardless of what might happen to these children and others should they happen to contract pertussis), they reacted with emotion and quickly built up political pressure to achieve an Act of Parliament (the Vaccine Damage Payments Act 1979), which represents the only substantive recommendation of the Pearson Royal Commission to be implemented. It provides that some children injured by vaccines should be specially compensated by the government. Vaccine injury received massive publicity, and the rate of immunization fell dramatically. It was said that this resulted in an epidemic of pertussis that disabled and killed more children than would have been adversely affected had the vaccination programme been maintained (see Chapter 3).

The Act of 1979 has these important features:

1. Compensation, up to a maximum of £10 000, is payable only if the vaccine was used on a person aged under 18 years (except when vaccination was given at the time of an outbreak). Claims may only be made by children who survived over the age of two.
2. Disablement (judged by a tribunal) must be at least 80 per cent.
3. Causation must be shown on a balance of probabilities (this is the normal standard required in all civil litigation).
4. Payment is made by the government, not by the manufacturer (this is a no-fault scheme).

Although the scheme has resulted in compensation for some claimants, many claims have been rejected by reason of the 80 per cent disability rule or the failure to prove that vaccination caused the injury on a balance of probabilities. This has led to some bitterness, with accusations that the tribunals were insensitive to human suffering and were applying the rules too restrictively, coupled with demands that borderline cases be treated sympathetically — that is, that compensation should be awarded.

Compensation for all?

The truth is that wherever lines have to be drawn, some cases will fall on one side of the line and some cases on the other. The tribunals were applying the rules by which they were bound as best they could. Attempts to provide compensation for one class but not another are bound to cause distress to those whose claims are denied.

Consider four children. A has a 90 per cent disability caused by vaccination. B was vaccinated, suffered no ill effects as a result, but now has an identical 90 per cent disability not caused by the vaccine. C was not vaccinated, and subsequently became 90 per cent disabled by pertussis disease. D was born with a congenital 90 per cent disability. Of these four, only A will receive compensation (under the Vaccine Damage Payments Act). None of the others will qualify. Some people think there is an inherent unreason in this result: suffering is being relieved according to cause rather than according to need. How the loss was suffered, they would say, is past history; compensation should look to the future.

The decision to structure the fund in this way was a political decision, and the unequal treatment of like disabilities is a course that seems to be demanded by the public. If it is not possible to provide for all the sick and disabled at the high level that a warm heart desires, the argument goes, compensation must be rationed. In consequence, it becomes the aim of all potential claimants to seek to categorize themselves and their loved ones as specially deserving of a high level of resources, whilst at the same time expressing goodwill to those disabled by Act of God who do not qualify under any special programme and who will receive less.

The Pearson Royal Commission considered whether there should be established a no-fault scheme for all personal injuries and found against it on economic grounds. Whether their arguments were right or wrong, there seems to be no immediate prospect of government support for such a scheme in the face of the Royal Commission report.

Strict liability — the defences

There are two possible ways in which compensation for drug-induced injury might be handled in the future. One is under the existing or reformed law of product liability. The other, which we explore in more detail, is by the creation of a new no-fault scheme.

Where a claim for injury is based on negligence, and causation is proved on a balance of probabilities (as it always must be), the manufacturer may reply that the user was warned and knowingly took the risk on himself, or that the user himself behaved negligently. Whether a warning affords a defence to the manufacturer depends on many factors, including the prominence given to it, the way it was expressed, and in particular whether it enabled the user to look after his safety while using the product as it was intended to be used. It is self-evident that a manufacturer cannot protect himself against the consequences of his own negligence by a warning that 'The maker may have been negligent. This product may be lethal. Use it at your own risk.' Quite apart from warnings, however, the user's negligence does serve to reduce the manufacturer's liability. A court will apportion blame as between the parties and scale down the damages in proportion — 50/50 or 30/70 or as the case may be.

Under the regimes of strict liability envisaged by the reforms proposed, these defences would also be available. But a further defence, rejected by the law reformers, has been put forward by industry and espoused by the government. This is the defence that the manufacturer had developed the product to the highest current scientific standards and so should not be held responsible for what no one could prevent or foresee. This has become known as the 'state of the art' defence. It is a defence that (without acquiring any special name) has always been available in claims based on negligence, for negligence involves a failure to take reasonable care and the 'defence' simply negates negligence. In truth, the state of the art defence is one of the distinctions between strict liability and negligence liability. To speak of strict liability with a state of the art defence is a contradiction in terms: one is speaking of negligence liability. The case of the typhoid bacteria in milk described earlier is a good example: we can class the dairy's liability as strict liability rather than liability for negligence precisely because the defence of state of the art was raised and rejected.

Nevertheless, the EEC Directive (1985) to harmonize the laws of member states includes a defence that the state of scientific and technological knowledge did not permit the discovery of the defect.

A scheme for drug-induced injuries?

The state of the art defence is clearly one which would be of particular relevance to drug-induced injuries. Recent history has given us three examples, and indeed the demand for compensation for drug injury has stemmed chiefly from important accidents in which large numbers of people suffered major injuries or death due to a drug that has, as a consequence, been withdrawn. Thalidomide and clioquinol were given to people who had the expectation of a good life and this was shattered. Practolol was used in angina pectoris, a disease that generally occurs in later life, but its consequences were to put an end to the capacities for enjoyment of those affected.

All three episodes were marked by major campaigns of groups of patients against the pharmaceutical companies who made the drugs. Two companies delayed paying compensation while one (practolol) voluntarily set up an arrangement to compensate which, though criticized, was generally creditable. Of none of these companies could it be shown that it was negligent: indeed, the issue of negligence was not determined in the courts in any of these cases, and the issue of non-foreseeability would probably have negatived it.

In our view neither public demand nor public need is met by the expensive lottery of actions for negligence, and a regime of strict liability (so-called) with a state of the art defence would do no better. Something different is needed. It is clear from experience that serious injuries may occur, even in the absence of negligence. Among the possible reasons and circumstances are the following:

1. There may be a defect in the manufacturing process by which the drug was made; the product is not what was intended (manufacturing defect). This might or might not be a result of negligence.

2a. The drug has an inherent capacity to cause injury by its very nature; it continues in use because the overall benefits outweigh the inescapable risks ('design' defect).

2b. A new drug is found, after grant of a product licence, to have such serious adverse effects that its use is no longer acceptable and it is withdrawn ('design' defect).

3. The drug may be used unskilfully by the prescriber.

4. The patient may not follow the instructions given.

5. The patient may have an abnormality of his genetic constitution or disease which causes the drug to injure him.

6. A person may be totally unaware that he is taking a serious risk, whether it be so extremely remote that he is not told, for example, sudden death from penicillin, or because circumstances prevent it, as in some cases of serious illness, including cancer.

7. A person may know well he is taking a serious risk and deliberately accept it, or, having been told of future risk, may give priority to present gain or convenience.

Any workable and just scheme must take all these situations into account and do so without a lot of hair-splitting.

Compensation

As we have seen, compensation might be provided in essentially three different ways:

Negligence

In the absence of a contract this is the sole basis at present for obtaining compensation for injury due to drugs or medicines (except for the limited scheme for vaccine injuries). Negligence provides a totally inadequate basis for any general system of compensation for drug injury. Apart from difficulties of proof of negligence, expense, and delay, the fact is that most drug injuries which deserve compensation occur without fault.

Strict liability

As with negligence, this suffers from the disadvantage that, even if it were introduced into our law, a claim must be made against a named defendant and pursued through the courts with all the expense, delay, and general vexation that seem to be inseparable from civil litigation.

A no-fault scheme

In our view a no-fault scheme offers the best prospect of a workable system to provide compensation for drug injury. Since compensation is available from a central fund without proving fault, it avoids the confrontation and other disadvantages of litigation. Compensation is divorced from questions of fault and liability (though we would not exclude the possibility of the fund seeking reimbursement from a manufacturer where appropriate).

A no-fault scheme offers an administrative simplicity that litigation cannot provide. Causation must be proved, but once that is done compensation is payable.

Before describing how a no-fault scheme could be made to work we turn to the simple question why pharmaceutical companies should not be told by Parliament that they must compensate people injured by their products, and to get on with it.

The obligation of industry

When a drug is under development and goes into clinical trials sponsored by a company there is no doubt that the company should take responsibility for any injuries caused and, indeed, the Association of the British Pharmaceutical Industry (1983) has acknowledged this. We shall not further consider liability for injury prior to the award of a product licence (marketing).

But when a drug has a product licence, i.e. is marketed, it has passed through rigorous official regulatory review and will be used in circumstances and patients of infinite variety. There are occasions when it could be more accurate to say that injury is due to a defect in the patient rather than in the drug.

Patients take several drugs at a time, made by different companies; the drug causing the injury may not be iden-

tifiable, or identifiable as made by a particular company; the company may deny causation, perhaps especially where the alleged drug injury mimics spontaneous disease (e.g. thromboembolism and oral contraceptives), so that the claimant has to go to the courts which will also have great difficulty; the prescriber may have used the drug other than in strict accordance with the company's Data Sheet; the question of contributory negligence by the patient may cause the company to resist; the list of problems is endless. Despite the manifest and great difficulties, the interest of the pharmaceutical industry will be best served by the introduction of a scheme. Its reputation is seriously and continuously damaged by the absence of a scheme and it will remain under constant adverse criticism until this is remedied. The industry is known to be seriously concerned about its public image. It has expressed its willingness to discuss a scheme, but has also expressed its unwillingness to be left alone to find a scheme. We have sympathy with its position.

A no-fault scheme

There should be a no-fault scheme and it should not be run by the pharmaceutical industry. The government cannot stand aside; indeed we believe the prime reason the government seems to wish to stand aside is fear of the cost rather than because of any issue of principle.

A no-fault scheme could work as follows:

1. A central body (fund) should be set up.
2. The criteria for causation and for eligibility for compensation will be applied by a medical tribunal with legal and perhaps lay involvement.
3. If compensation is awarded then it can be paid at once, or at such time as the full extent of injury is determined.
4. If it seems appropriate, i.e. if negligence is suspected, then the central body will seek reimbursement from the body or person responsible (subrogation). But the injured person will not be involved. We regard the prevention of confrontations between patients and pharmaceutical companies as of paramount importance. The undoubted merits of this science-based industry will continue to be overshadowed by criticism as long as such confrontations continue. It is very much in the interest of industry that a scheme be implemented.
5. There would be no difficulty in awarding compensation even though there were doubts about the origin of a particular medicine or where a succession of different brands or generic preparations had been used.
6. Any scheme offering selective compensation for injury according to cause rather than according to need

must require that causation be proved according to the 'balance of probabilities'. This latter is the criterion for civil law rather than 'beyond reasonable doubt', which is the criterion of criminal law.

Inevitably there will be cases where the claimant will be totally convinced he or she was injured by a drug, and yet the medical tribunal will remain unpersuaded. There is no escape from this. Laxity or an excess of sympathy for a sad case will mean injustice and expenditure of scarce resources on people who are not intended to get special compensation and will deprive other disabled people. If society wants special schemes of compensation, this will have to be accepted, for not to do so will make nonsense of selective schemes and broaden them into one big no-fault scheme which society cannot afford (if it thought it could afford such a scheme it would presumably have implemented it before now and there would be no need for the present discussion).

7. The criteria generally proposed for judging eligibility for compensation are that the adverse event should carry a risk so remote that it would not ordinarily be taken into account when choosing the treatment, and the injury should be severe, e.g. blood dyscrasias, hepatic necrosis, epidermal necrolysis. Whether or not a patient was warned would not be relevant. Negligence by the patient would be relevant, at least to some degree.

Injuries that would not normally be compensatable include high-frequency serious effects, e.g. anticancer drugs, where a patient suffering from serious disease is taking serious risks to preserve life or escape serious disability.

Plainly there is room for much discussion, and there are plenty of areas of very real difficulty, e.g. where a drug increases the incidence of a spontaneous disease. But criteria such as the above have been in use in other countries, e.g. Sweden, for years, and we should profit by their experience.

8. It is necessary to draw substantial financial resources from somewhere. It is also necessary to get a proper balance between the interests of the individual, of industry, and of the professions.

In the Federal Republic of Germany manufacturers are required to be insured and have found the cost surprisingly low so far. In Sweden pharmaceutical companies have undertaken liability and cover it with insurance. Premiums are related to company turnover.

In the UK, where drug prices are negotiated with the government, it has been said by a Minister of Health that the cost of a compensation scheme sponsored by the industry would be taken into account in official price negotiations.

The possibility of a major catastrophe always remains and should this happen it would be impossible for the government to stand by when a well-constructed insurance scheme failed. We do not believe arguments over provision for a remote and enormous catastrophe should inhibit the introduction of a general scheme.

The climate of opinion is changing. The inevitability of some such scheme is increasingly being accepted. We believe discussions on financing a scheme should be opened between the principally interested parties with the aim of producing a scheme that is so obviously workable and in the public interest that the government will not wish to, indeed cannot, stand aside.

Conclusion

We believe we have said enough to show that, failing a comprehensive compensation scheme for all the misfortunes of life, a no-fault scheme for drug-induced injury is desirable, workable, and need not cost so much as to render serious discussion on implementation a waste of time. Schemes already operate in other countries and much can be learned from them. No doubt none of them is perfect, but they show that critics who spend their time on the easy art of proving that the difficulties are too great to contemplate action are wrong.

We consider that the principal interested parties should get together to propose a scheme which, after wide consultation, should be formally put forward. This should be done without waiting for the government to act, though ultimately government, and perhaps parliamentary, approval will be necessary.

REFERENCES

ASSOCIATION OF THE BRITISH PHARMACEUTICAL INDUSTRY (1983). *Guidelines: Clinical trials — compensation for medicine-induced injury. Brit. med. J.* **287**, 675.

THE LAW COMMISSION AND THE SCOTTISH LAW COMMISSION (1977). *Report on liability for defective products.* Cmnd. 6831, HMSO, London.

ROYAL COMMISSION ON CIVIL LIABILITY AND COMPENSATION FOR PERSONAL INJURY (1978) (Pearson Commission). Cmnd. 7054-1, HMSO, London.

Part II

12 Blood dyscrasias

J.R.B. WILLIAMS

Drug-induced suppression of one or more elements of haemopoiesis is the most serious haematological side-effect of drugs which in this chapter are classified according to the pharmacological activity involved. Antibody-linked cytopenias and haemolysis may lead to considerable distress and occasionally death and are discussed separately except where, as with antibiotics, the main · haematological effect is discussed elsewhere in the chapter. Reviews of haematological side-effects of drugs of all kinds have been published by Horler (1977) and Dawson (1979).

Agranulocytosis, thrombocytopenia, and aplastic anaemia

Neutropenia and thrombocytopenia are common unwanted effects of many drugs. They are usually due to either bone-marrow suppression or to the formation of antibodies against the drug or drug and cellular protein complex. Severe marrow suppression leading to pancytopenia or aplasia is fortunately infrequent but any drug that regularly causes neutropenia or thrombocytopenia may affect other cell lines of haemopoiesis and lead to complete marrow suppression.

Table 12.1 lists those drugs which may cause thrombocytopenia or neutropenia by bone-marrow suppression or by an immunological mechanism and which may cause pancytopenia or complete aplasia.

Bone-marrow suppression

Pathogenesis

Many drugs produce suppression of marrow activity or aplasia in a dose-related manner often, as with cytotoxics, limiting therapeutic dosage. Other drugs cause marrow suppression only rarely and this idiosyncratic response is independent of the dose. Many of the agents are used widely with safety and the sudden aplasia cannot be foretold by either the chemical structure of the drug or by any facet of the patient's clinical state. There is some indication however that abnormal metabolism may predispose to aplastic anaemia induced by phenylbutazone,

as in these patients oxidation of acetanilide is reduced.

In addition to this variation in liver enzyme activity, recent work has suggested that bone-marrow enzyme function may be important. In experimental studies it has been shown that aminopyrine N-demethylase, a cytochrome P-450 dependent enzyme in bone marrow, is increased by 3-methylchonanthrene, a known inducer of liver microsomal enzyme activity (Levere and Ibraham 1982). The affinity for the substrate in these experiments of the bone-marrow enzyme was lower than was the liver enzyme. As some chemicals such as benzene may be concentrated much more in the marrow than the liver, these differences in metabolism may be relevant to the haemopoietic toxicity of the compounds.

In humans variation in the ability of cells to detoxify drugs by P-450 dependent microsomal activity has been demonstrated (Gerson *et al.* 1983). An inherited defect in the metabolism of phenytoin and carbamazepine was shown using mouse hepatic microsomal preparations with the patients own lymphocytes *in vitro*. The results provide the first evidence for a role of arene oxide drug metabolites in the pathogenesis of aplastic anaemia, including the idiosyncratic form. More studies with other drugs are needed.

In-vitro culture of bone marrow may demonstrate a suppressive or stimulating effect on haemopoietic precursors of various drugs and the significance of the results is discussed in the relevant sections of this chapter. It can be shown moreover that decreased leucopoietic activity (CFU-C) may continue for years after an episode of drug-induced agranulocytosis in spite of normal peripheral blood and bone marrow. This raises the question of the presence of an underlying reduced myeloid proliferative potential which predisposes to the drug-induced neutropenia (Parmentier *et al.* 1978).

If *in-vitro* studies include the patients' serum with suitable controls it may be possible to show that a drug-dependent antibody acts against haemopoietic cells. This effect has been demonstrated with quinidine affecting both leucopoietic and erythropoietic cell lines (Kelton *et al.* 1979).

Benestad (1979) provided a review of drug mechanisms in marrow aplasia discussing the pathogenesis of chloramphenicol-, phenylbutazone-, gold-, and phenothia-

Table 12.1   Drugs associated with blood cytopenias and haemolysis*

Drug	Red-cell aplasia	Thrombocytopenia	Neutropenia	Pancytopenia	Haemolysis
Acetazolamide		+	+	+	
Ajmaline			+		
Allopurinol			+		
Alprenolol			+		
Amidopyrine		+	+ + +		+ +
Amiodarone	+				
Amphotericin B				+	
Amrinone		+ +			
Aprindine			+ +		
Antazoline		+ +			
L-Asparaginase		+ + +	+ + +	+ + +	+ +
Barbiturates		+		+	
Bendrofluazide		+			
Benoxaprofen			+		
Benzocaine					+ +
Captopril			+ +		+
Carbamazepine		+ +	+		
Carbimazole			+ +	+ +	
Carbutamide		+	+ + +	+	
Cephalosporins			+		+ +
Chloramphenicol		+	+ +	+ + +	
Chlordiazepoxide			+	+	
Chloroquine		+			
Chlorothiazides		+ +			
Chlorpropamide	+	+ +	+	+ +	+
Chlortetracycline				+	
Chlorthalidone			+		
Cimetidine		+	+ +	+	
Cincophen			+		
Clozapine			+ +		
Codeine		+			
Colchicine				+	
Cyclophosphamide		+ + +	+ + +	+ + +	+
Dapsone					+ + +
Desipramine		+ +			
Digitalis		+			
Digitoxin		+ +			
Diatrizoate					+ +
L-dopa					+ +
Erythromycin		+			
Ethacrynic acid			+		
Fenoprofen		+	+	+	
Feprazone		+			+
5-Fluorocytosine				+	
5-Fluorouracil		+ + +	+ + +	+ + +	+
Frusemide		+	+		
Gold salts	+	+ + +	+ + +	+ + +	
Heparin		+ +		+	
Hydroxychloroquine		+ +			
Ibuprofen			+		+
Imipramine			+ +		
Indomethacin		+	+ +	+	
Isoniazid		+		+	
Isosorbide dinitrate					+
Levamisole		+	+ + +		
Mebhydrolin napadisylate			+ +		
Mefenamic acid					+ + +

Table 12.1 *Continued*

Drug	Red-cell aplasia	Thrombocytopenia	Neutropenia	Pancytopenia	Haemolysis
Mepazine			+ +	+	
Meperidine		+			
Meprobamate		+	+	+	
Metiamide			+		
Methimazole			+ +		
α-Methyldopa		+ +			+ + +
Methotrexate		+ + +	+ + +	+ + +	+ +
Methylene blue					+
Metronidazole			+		
Mianserin			+ +		
Nalidixic acid					+
Naproxen				+	
Niflumic acid			+		
Nitrofurantoin			+ +		+
Novobiocin			+		
Oestrogens		+		+	
Oxyphenbutazone			+ +	+ + +	
Para-aminosalicylic acid		+ +			
Paracetamol		+	+		+ +
Paramethadione		+			
Penicillamine		+ +	+		
Penicillins		+	+ +	+	+ + +
Phenacetin		+	+ +		+ +
Phenazopyridine					+ + +
Phenindione			+		
Phenothiazines		+	+ +	+ + +	+
Phenylbutazone		+	+ +	+ + +	+
Phenytoin		+ +	+ +	+ +	+
Potassium iodide		+			
Potassium perchlorate				+ +	
Prednisone		+			
Primaquine					+ + +
Procainamide			+		
Procarbazine		+	+ +	+ +	+
Propylthiouracil		+	+ +	+	+
Pyramethamine			+		
Pyrazinamide		+		+ +	
Quinacrine				+ +	
Quinidine		+ + +	+		
Quinine		+ + +	+		
Reserpine		+			
Rifampicin		+ +	+		+ + +
Rondomycin				+	
Spironolactone			+		
Streptomycin		+		+	
Sulphamethoxazole with trimethoprim			+		
Sulindac	+	+	+	+	
Sulphonamides	+	+ +	+ +	+ +	+ +
Tetracyclines		+			+
Thenalidine			+		
Thiocyanate				+	
Thioridazine			+ +		
Thiosemicarbazine			+		
Tolbutamide		+ +	+	+ +	
Triamterene					+
Trimethadione		+	+	+ + +	

Table 12.1 *Continued*

Drug	Red-cell aplasia	Thrombocytopenia	Neutropenia	Pancytopenia	Haemolysis
Trinitroglycerine		+			
Valproate	+				
Vancomycin			+		

+ + + Indicates a substantial number of reports.
 + + Occasional reports.
  + Rare or single reports.
* Amplification of the side-effects and references may be found in the text.

zine-induced disease and discussed abnormalities of immunological and haemopoietic proliferative response, including the significance of *in-vitro* marrow-culture experiments.

Experimental aspects of aplastic anaemia were reviewed by Haak (1980) and Yunis and Salem (1980) gave a detailed review of drug and chemically-induced mitochondrial damage and sideroblastic changes, especially the effects of ethanol and isoniazid and their analogues. Another review was provided by Appelbaum and Fefer (1981).

Clinical reports

Two independent studies of drug-induced agranulocytosis in the Netherlands (Zwaan and Meyboom 1979) showed that in the years around 1974 approximately 50 per cent of the cases were due to pyrazolone anti-inflammatory drugs, especially noramidopyrine methane sulphonate. Between 19 and 15 per cent of other cases were due to antirheumatic drugs of other chemical series and a further 10 per cent were due to sulphonamides and related drugs.

In India a study of 60 patients with hypoplastic anaemia showed that 38 were associated with potentially myelotoxic drugs, especially chloramphenicol (Garewal *et al.* 1981). In Sweden 256 patients who developed agranulocytosis between 1973 and 1978 showed a change in the nature of the associated drugs from the latter half of the 1960s (Arneborn and Palmblad 1978, 1982). In the recent survey 51 had received cytotoxic drugs with the well-known drug-related effect but 84 were caused by other drugs. Of these the commonest were sulphonamides, antithyroid drugs, and thenalidine. Commenting on the report in the *Lancet* (1983) the reviewer stressed the problems in allotting responsibility for cytopenias or marrow suppression in most patients who have usually received more than one drug. The present author is strongly in favour of carefully watched rechallenge when possible, or if at all indicated for future use in the patient concerned. This should not be attempted after

resolution of an idiosyncratic aplasia but may be reasonably safe in mild thrombocytopenia or neutropenia of slow onset and probably dose-related.

Recent clinical reviews include one on agranulocytosis by Young and Vincent (1981) and on aplastic anaemia by Heimpel and Heit (1980).

Antibiotics and antibacterial agents

Chloramphenicol

Blood dyscrasias were reported even in the early days of clinical use of chloramphenicol and today the association between chloramphenicol dosage and serious blood dyscrasias such as aplastic anaemia, thrombocytopenia, or granulocytopenia is well recognized. These toxic effects may occur after the administration of small doses for short periods or may appear only after prolonged therapy. If the aplastic anaemia is detected early, and the antibiotic is discontinued, the bone marrow may recover, but often the disease ends fatally. Cases have been reported as occurring after the administration of only a few grams, often, tragically, for some trivial indication (*British Medical Journal* 1963). Aplastic anaemia from chloramphenicol or thiamphenicol is a conditional or idiosyncratic response (Benestad 1979) but bone-marrow suppression short of aplasia is dose-related.

Using *in-vitro* culture, Ratzan *et al.* (1974) showed that chloramphenicol is inhibitory in therapeutic concentrations but this effect is a dose-dependent reversible one rather than the irreversible lesion leading to aplastic anaemia (*British Medical Journal* 1980).

The toxic effects of chloramphenicol may occur at any age but are commonest in older children and more so in females than in males. Premature and newborn infants are particularly susceptible, and there have been several deaths reported (Hodgman 1961); a symptom-complex the 'grey syndrome' developed in these infants, all of whom received doses larger than 100 mg/kg of body weight daily.

The recommended dose in adults and children is

50–100 mg/kg daily giving plasma levels of 10–20 μg/ml. This dosage is also suitable for infants over 4 weeks in age, but from 2–4 weeks the dose should be 50 mg/kg and under two weeks 25 mg/kg per day to yield plasma concentrations of 15–25 mg controlled by monitoring as discussed by Mulhall *et al.* (1983). The symptoms of the 'grey syndrome' were described in detail by Ory and Yow (1963) and appeared after three or four days' treatment. Deaths occurred within 24–48 hours of onset of symptoms. Post-mortem studies did not reveal the cause of death and no characteristic changes were found. However, it is very likely that death was due to the abnormally progressive accumulation of chloramphenicol, and some of its degradation metabolites, since in the newborn infant the immature liver is deficient in conjugation mechanisms and there is a decrease in the renal tubular secretion of the chloramphenicol glucuronide.

Early in the development of chloramphenicol toxicity, vacuoles appear in the erythroid-cell and myeloid-cell precursors in the bone marrow (Rosenbach *et al.* 1960; Saidi and Wallerstein 1960; McCurdy 1961; Saidi *et al.* 1961). These lesions are reversible and their relation to later more serious bone-marrow depression is unknown.

Ingall and others (1965) observed vacuoles in erythroid-cell and myeloid-cell cytoplasm in the bone marrows of five out of eight children treated with short courses of chloramphenicol for a variety of infections. These changes were morphologically identical with vacuolization seen in phenylalanine deficiency. Treatment of these patients with 100 mg/kg/day of *laevo*-phenylalanine by mouth caused the vacuoles to disappear in two children after 48 and 96 hours, and in a third child the size and number of vacuoles were markedly diminished. All three children had continued to receive chloramphenicol in the same dosage.

Ingall *et al.* (1965) therefore suggested that chloramphenicol might act on the bone marrow by affecting phenylalanine metabolism. Chloramphenicol has also been shown to block the incorporation of alanine into protein moieties (Gale 1958); thus chloramphenicol could block phenylalanine utilization as a primary effect or secondarily as a result of interference with alanine metabolism.

The production of vacuoles in erythroblasts and myeloid precursors by more toxic chloramphenicol analogues suggests that the toxic effect is at stem-cell level. The block appears to be interference with messenger RNA during the earliest phases of protein synthesis (Weisberger 1969).

It has also been suggested that genetic factors might influence the susceptibility of patients to chloramphenico-induced aplastic anaemia. Nagao and Mauer (1969) reported the clinical courses of aplastic anaemia in identical twins. The first twin was treated with chloramphenicol for an upper respiratory-tract infection at the age of 5 months; at 6 months the symptoms recurred and treatment was continued. At 8 months, he was admitted to hospital with ecchymoses of 5 days' duration. On admission, his haemoglobin was 6.2 g/100 ml, his white-cell count was 6250/mm³, and the platelet count was 0. Aspirated bone marrow was hypoplastic and a diagnosis of aplastic anaemia due to chloramphenicol therapy was made.

The second twin suffered from the same respiratory-tract infections and also received chloramphenicol at the same time and doses as his twin. At the time of diagnosis of aplastic anaemia on his twin, he was admitted to hospital in the hope that he could be a source of bone-marrow transfusion. On admission, the second twin had a haemoglobin of 8.6 g/100 ml and a white-cell count of 14 500/mm³ with 19 per cent neutrophils and 79 per cent lymphocytes. The platelet count was 8500/mm³. The bone-marrow aspirate was cellular but contained 77 per cent lymphocytes. Gradually a pancytopenia developed and the bone marrow became hypoplastic. He was treated with packed red cells and platelet transfusions were given, and complete recovery ensued.

There have been other suggested mechanisms of how chloramphenicol induces blood dyscrasias. Inhibition of iron uptake has been suggested as also sensitization of bone marrow on the basis of allergy. These studies were well reviewed by Manten (1968).

The incidence of serious blood dyscrasias associated with chloramphenicol has been variously estimated as aplasia of bone marrow. 1 in 500 to 1 in 100 000 treated cases (*British Medical Journal* 1963), aplasia of bone marrow with consequent aplastic anaemia and agranulocytosis, 1 in 58 000 (Kähler 1965), and 1 in 75 000 (Girdwood 1964). The lethality of this irreversible effect has been estimated as 1 in 76 000 (Kahler 1965).

In the United Kingdom, the Committee on Safety of Medicines (CSM) announced that during the two-year period prior to 1967, it had received reports of 24 fatal cases of blood dyscrasias following the administration of chloramphenicol. These reports accounted for 80 per cent of all fatal cases of blood dyscrasias reported to have occurred in patients who received an antibiotic.

The CSM considered that chloramphenicol should never be used for the treatment of trivial infections and that it should not be used except when careful clinical assessment, usually supplemented by laboratory studies, indicated that no other antibiotic would suffice. The CSM recognized that chloramphenicol was a highly effective agent in typhoid fever and in *H. influenzae* meningitis, and in such cases they were of the opinion that the advantages of chloramphenicol greatly outweighed its hazards (*British Medical Journal* 1967).

Rhoades (1982) also stressed its value in typhoid and meningococcal meningitis and included severe infections due to *Bacteroides fragilis* and various rickettsiae among the indications. Since the CSM report was published there has been a considerable drop in the number of NHS prescriptions for chloramphenicol administered for systemic use (1968: 149000; 1969: 99000; and 1970: 58000), with the consequent reduction in incidence and deaths from aplastic anaemia. A similar reduction in the use of chloramphenicol occurred in Sweden (Bottiger and Westerholm 1974).

In a review of 641 cases of blood dyscrasias due to chloramphenicol Meyler *et al.* (1974) found that survival is more likely if the time between discontinuing the drug and development of pancytopenia is less than two months. Patients with a latent period of seven months or longer died.

Treatment with anabolic steroids and prednisolone with blood platelet concentrate and antibiotics when required may result in recovery of one-third of the patients (Hausman and Skrandies 1974) but the prognosis is greatly influenced by the response to platelet infusions. Five of 20 Israeli patients with aplastic anaemia induced by chloramphenicol developed acute leukaemia (Modan *et al.* 1975). This association has been reported by others and Brauer and Dameshek (1967) suggested that aplasia and acute leukaemia may represent a double response to marrow injury. They also comment on the occasional presentation of acute myeloblastic leukaemia by an aplastic phase. As chloramphenicol is much used in Israel the association of leukaemia and aplasia could perhaps be coincidental.

The intravenous use of the antibiotic is increasing but there is no clear indication as to whether the bone-marrow toxicity is less when given by this route (Feder *et al.* 1981). Ophthalmic administration of chloramphenicol occasionally causes aplastic anaemia and in wealthy countries alternatives are available (Dutro 1981). In the Middle East and Tropics however, minor conjunctival infections are frequent and the use of cheap chloramphenicol has made a great difference to the populations in these areas with little risk.

The combination of chloramphenicol with other drugs may increase the probability of aplastic anaemia occurring. Cimetidine may prove synergistic in this respect (Farber and Brody 1981), although in the patient they reported an effect from chlorpropamide or diazepoxide is not excluded.

Cephalosporins, penicillins

Agranulocytosis was reported as a complication of benzylpenicillin therapy as early as 1946 by Spain *et al.* but immune haemolysis is more frequent. Ley *et al.* (1959) first reported that penicillin may cause an immune haemolytic anaemia, and, during the next nine years, 13 further cases were recorded (Strumia and Raymond 1962; Beardwell 1964; Van Arsdel and Gilliland 1965; Clayton *et al.* 1965; Dawson and Segal 1966; Lai *et al.* 1966; Petz and Fudenberg 1966; Nesmith and Davis 1968; White *et al.* 1968). With the exception of two, these reports appeared in American literature. Reports of penicillin-induced haemolytic anaemia are fewer in the UK probably because lower doses of penicillin are used. It is possible, however, that immune haemolysis may go unrecognized, since high doses of penicillin are usually given to patients suffering from subacute bacterial endocarditis and septicaemia in whom increasing anaemia is part of the clinical picture.

White and colleagues (1968) analysed data from 14 cases. Nine patients received penicillin for subacute bacterial endocarditis, three had other infections, in one case the antibiotic was used as a post-operative cover, and in one case no diagnosis was given. Thirteen of the patients had been given 20 mega-units of penicillin per day at some stage of treatment and one patient had received 10 mega-units daily for 26 days. In six cases there was evidence of drug allergy before haemolytic anaemia developed; five patients developed a rash; and one had pyrexia and malaise.

In all cases, there was a positive direct antiglobulin reaction at the time the haemolysis was first diagnosed; this reaction was detected as early as two days and as late as 26 days after commencing penicillin therapy. Four patients who had initially received small doses of penicillin some days or weeks before larger doses were given, developed a sudden brisk haemolysis which contrasted with the slow onset seen with constant high dosage. The peripheral blood picture did not help diagnosis; spherocytosis was seen in one case and autoagglutination was not reported. There was an eosinophilia in peripheral blood or bone marrow in eight patients, and in one there was a leucopenia at the height of haemolysis, accompanied by a marked left shift in granulopoiesis.

No patient died as a result of uncontrolled haemolysis, although in one case the haemoglobin fell from 12.6 to 3 g/100 ml in 9 days accompanied by a reticulocytosis of 66 per cent. Six patients had an anaemia severe enough to require blood transfusion. In all cases haemolysis stopped when penicillin was withdrawn; corticosteroids were given to three patients without any apparent beneficial effects.

Penicillin-induced haemolytic anaemia is usually associated with a direct antiglobulin test and IgG antibodies and is dose-related (Erffmeyer 1981). At least two mechanisms are involved in the red-cell lysis: one is by cell-mediated cytotoxicity with direct contact with mixed mononuclear cells and the other by antibody-dependent mononuclear phagocytosis (Yust *et al.* 1982). Penicillin may rarely raise IgA, IgD, and IgM antibodies but so far there is only one report of haemolytic anaemia from IgM antibody formation (Bird *et al.* 1975) and none due to the other immunoglobulin classes. The β-lactam derivative involved is usually benzylpenicillin but induction by ticarcillin has been reported (Seldon *et al.* 1982). In this instance the antibody was IgG and the patient had previously been treated with and presumably immunologically primed by amoxycillin and benzylpenicillin. Treatment again with ticarcillin was forced by clinical requirements and this rechallenge led to recurrence of the anaemia.

Cross-reactivity with cephalosporins may occur with serum from patients developing Coomb's-positive haemolytic anaemia with penicillins (Nesmith and Davis 1968; Gralnick *et al.* 1967) and an immune-mediated haemolysis initiated by cephalothin has been reported by the same worker and his colleagues (Gralnick *et al.* 1971). The mechanism of haemolysis may be by adsorption of the drug on to the red-cell membrane, but in some instances it is due to the formation of immune complexes and an 'innocent bystander' effect (Duran-Suarez *et al.* 1981).

Penicillin-induced cytopenias

Penicillins may cause thrombocytopenia (Hsi *et al.* 1966), and this may occur with or without neutropenia. The latter is more frequent with the newer derivatives than with benzylpenicillin, notably with methicillin which may cause bone-marrow depression (Levitt *et al.* 1964; Godin *et al.* 1980). Cases of agranulocytosis have been reported, associated with treatment with doxacillin (Westerman *et al.* 1978; Krafft *et al.* 1978), oxacillin (Kahn 1978; Passoff and Sherry 1978), and nafcillin (Couchonnal *et al.* 1978; Carpenter 1980). Gosh (1979) briefly reviewed some of the reports of oxacillin-induced granulocytopenia and noted that high doses of antibiotic were administered in most cases.

Cloxacillin may cause side-effects of all kinds in 37 per cent of children (St. John and Prober 1981). In a series of 79 children treated with this drug only two developed transient neutropenia, but in 25 per cent eosinophilia occurred, mostly between the second and tenth day of treatment. In a prospective study neutropenia occurred in infants and children treated with nafcillin and oxacillin and eosinophilia developed in patients receiving methicillin and nafcillin (Nahata *et al.* 1982).

Bone-marrow cellularity varies from case to case. Leucopoiesis may be active at all stages but more often shows a reduction in numbers of the later phases of division with a maturation arrest, perhaps immune-mediated (Corbett *et al.* 1982). Bone-marrow suppression may extend further back in the sequence of differentiation as the patient reported by Krafft *et al.* (1978) showed an absence of myeloid forms except for myeloblasts and promyelocytes, presumably due to direct marrow suppression.

Neutropenia induced by benzylpenicillin is dose-related and may occur at 20 million units/day intravenously (Postelnick and Gaskins 1981). With piperacillin, dosages of 24 g daily may induce neutropenia (Wilson *et al.* 1979).

Proof of immune-mediation has been difficult but antineutrophil antibodies may be demonstrated by the incubation of normal neutrophils with antibody-sensitized neutrophils and measurement of the increased glucose oxidation induced. This measurement of serum opsonic activity was used by Weitzman *et al.* (1978) to show that neutropenia caused by nafcillin, dicloxacillin, oxacillin, cephalothin, and cefoxitin may be a drug-induced immune response. Opsonization related to the semi-synthetic penicillins was dependent on drug concentration but not on the presence of complement. The immunological response is however more complicated than this as ampicillin-induced neutropenia is less evidently dose-related. Two adult women treated with 2 g/day (approximately 30 mg/kg) developed neutropenia within a few days of starting treatment (Sidi *et al.* 1981). Recovery quickly followed discontinuation of the drug and in one patient migration-inhibition tests were positive in the presence of ampicillin and negative in the presence of the concomitantly administered α-methyldopa. Three children developed neutropenia during treatment with ampicillin in doses ranging from 100 to 400 mg/kg. All recovered on cessation of treatment with ampicillin, which was evidently the cause of neutropenia in two of them. A third also received chloramphenicol but recovery followed discontinuing ampicillin and the involvement of chloramphenicol, which could have been synergistic, is doubtful (Kumar and Kumar 1981).

Cephalothin occasionally causes neutropenia by an immune-response mechanism (Levin *et al.* 1971). It also induces neutropenia and thrombocytopenia; another case was reported by Naraqi and Raiser (1982). The circumstantial evidence implicated cephalothin but rechallenge failed to reproduce the haematological changes. Pancytopenia due to cephalothin may occur, and one report concerned a man treated for subacute bacterial endocarditis at a dose of 8 g daily (Tartas *et al.* 1981). Recovery commenced three days after discontinuing the drug and was complete a week later.

Neftel *et al.* (1981) showed that serum from affected patients modified lymphocyte culture indices and leucocyte/staphylococcal binding activity, although immunofluorescence and microcytoxicity tests failed to reveal antineutrophil antibodies. They suggested that penicillin combines with red-cell membranes with a hapten effect rather than an antibody/immune complex effect. Recently, however, Murphy *et al* (1983) demonstrated the presence of complement-fixing IgG antibody reacting with the patient's granulocytes and platelets in the presence of penicillin. They suggest that marrow hypoplasia in three of their patients was due to antibody suppression of penicillin-coated precursor cells.

Other antimicrobial agents

Vancomycin is a rare cause of neutropenia but West (1981, 1982) reported a further case of a 61-year-old man. He was treated for osteomyelitis with cefazolin followed by cephalothin. He was then given intravenous vancomycin and gentamicin and neutropenia occurred after the administration of 70 g vancomycin. He was also treated with heparin before the neutropenia occurred, but the timing of the drug administration suggests that the vancomycin was responsible. The bone marrow contained normal myeloid precursors and recovery was quick. Rapid recovery also occurred in the cases reported by Borland and Farrar (1979) and Kesarwala *et al.* (1981).

Thrombocytopenia has occasionally been reported as a complication of erythromycin, oxytetracycline, tetracycline, isoniazid, and pyrazinamide treatment; and neutropenia as a reaction to ristocetin (Gangarosa *et al.* 1958; Newton and Ward 1958). Rondomycin was reported as inducing temporary bone-marrow aplasia in a 20-month-old infant suffering from bronchitis, oedema, and an acute exudative erythema (Prusek *et al.* 1980). The patient's serum stimulated blast proliferation of autologous lymphocytes but not those of a normal infant and of an infant treated with tetracycline.

Sulphonamides and trimethoprim

Neutropenia occurring in association with sulphonamide drugs has been known since the early days of their use (Kracke 1938; Dolgopol and Hobart 1939). It then occurred particularly with sulphapyridine, mostly after substantial dosage, but now is seen during treatment with substituted sulphapyridines such as salicylazosulphapyridine (Ritz and Fisher 1960) or salicylsulphapyridine (Thirkettle *et al.* 1963).

In the UK, however, sulphamethoxazole used in conjunction with trimethoprim is now the commonest sulphonamide causing agranulocytosis. The bone-marrow suppression induced by this drug combination may extend to thrombocytopenia or pancytopenia (Tulloch 1976).

The pathogenesis of the neutropenia is not clear, although there have been suggestions that an immunological mechanism is involved (Evans and Ford 1958), especially as the onset of neutropenia may occur with skin rashes and other hypersensitivity phenomena (Rinkoff and Spring 1941) and is rapid in onset. Neutropenia of slow onset is commoner (delayed-onset neutropenia) and is presumably due to decreased cell production. The neutrophil concentration may stabilize at a relatively low level or even rise on continued treatment. The drug should be discontinued, however, if the neutrophil count is $1.5 \times 10^9/l$ or less for fear of progression to agranulocytosis.

Sulphonamides may also induce pure red-cell aplasia, and a female patient of Dunn and Kerr (1981) developed this complication during treatment with sulphasalazine. There was no evidence of haemolysis and recovery was rapid on discontinuing the drug.

Multiple haematological abnormalities may occur. Sulphasalazine-induced agranulocytosis with erythropoietic hypoplasia may accompany a considerable (80 per cent) plasma cell infiltration. These were abnormal and often immature but polyclonal in immunoglobulin production. Discontinuation of the sulphasalazine led to reduction of plasma cells to 10 per cent and reactive hyperplasia of erythropoiesis and myeloid division (Wheelan *et al.* 1982).

Sulphasalazine can also induce megaloblastic anaemia responsive to folate and in one instance accompanied by a complement-reacting positive direct antiglobulin test (Goldberg 1983). In this case sulphasalazine could be restarted with folate supplements without relapse of the anaemia.

An unusual combination of suppression of erythroid and megakaryocytic proliferation was reported by Davies and Palek (1980). A 78-year-old man was treated for granulomatous colitis with sulphasalazine, 2 g daily. After nine days he became anaemic. Hb 6.2 g dl⁻¹ and developed petechiae, platelets $33 \times 10^9\ l^{-1}$ with absence of erythroid precursors and megakaryocytes in the bone marrow. Leucopoiesis was almost normal. He died after seven months of supportive therapy including blood products, of pulmonary infiltration with neutropenia.

Further reports on bone-marrow suppression by cotrimoxazole provide further confirmation that previous megaloblastic anaemia or low folate levels from low intake or the increased demands of fever and infection may predispose to pancytopenia (Blackwell *et al.* 1978).

The pancytopenia in these cases was associated with severe megaloblastic change in the bone marrow, but reports of thrombocytopenia associated with trimethoprim–sulphamethoxazole to the Australian Adverse Drug Reaction Registry show that bone-marrow aplasia or reduction in megakaryocytes alone may occur (Dickson 1978). A prospective study of side-effects of trimethoprim–sulphamethoxazole in 50 children (Asmar et al. 1981) gave an incidence of 48 per cent developing haematological abnormalities; 34 per cent developed neutropenia, 12 per cent developed thrombocytopenia, and 6 per cent anaemia. Eosinophilia occurred in 14 per cent. These abnormalities quickly returned to normal when the treatment was discontinued but the authors pointed out the advisability of twice-weekly blood counts during trimethoprim–sulphamethoxazole treatment, which is certainly not current practice.

Megaloblastic anaemia may also occur as a result of treatment with these mixed agents but the patients usually have low folate reserves or abnormal metabolism beforehand. Magee et al. (1981) however reported an episode of acute megaloblastosis in a patient with abnormal folate metabolism evidently induced by the drug combination. A deoxyuridine suppression test indicated that deoxyuridine failed to inhibit the incorporation of ^{3}H-thymidine into DNA, and that this defect was corrected by the addition of folinic acid.

Direct inhibition of CFU-C activity has been shown to occur with trimethoprim and sulphamethoxazole (Golde et al. 1978). The inhibitive effect of these two drugs is reversed by folinic acid, suggesting that administration of folinic acid to patients may prevent the clinical haemopoietic toxic effects of trimethoprim–sulphamethoxazole, and showing that the investigation of possible therapeutic or prophylactic agents is possible by these in-vitro techniques.

Haematological toxicity with this combination is increased by accompanying immunosuppression with azathioprine (Bradley et al. 1980), especially if the co-trimoxazole is used in prophylaxis rather than in treatment of infection. These authors showed by in-vitro-culture that trimethoprim–sulphonamide potentiated the suppressive effect of 6-mercaptopurine. Immunosuppression by methotrexate with trimethoprim–sulphamethoxasole prophylaxis may also lead to pancytopenia with megaloblastosis (Kobrinsky and Ramsay 1981).

When co-trimoxazole is given to patients after bone-marrow transplantation for prophylaxis against infection with Pneumocystis carinii, there may be significant suppression of leucopoiesis (Schey and Kay 1984). The authors consider however that prophylaxis with this combination is important and should only be discontinued if there is dangerous neutropenia.

Rifampicin

Rifampicin causes a variety of haematological abnormalities of which haemolytic anaemia and thrombocytopenia are the more frequent, but neutropenia and eosinophilia occur occasionally. Blajchman and associates (1970) reported two cases of severe thrombocytopenia associated with rifampicin treatment. In the first case severe thrombocytopenia occurred during administration and readministration of the antibiotic. The patient's erythrocytes gave a positive direct antiglobulin test due to complement on the red cell surface: in the serum, complement-fixing antibodies were detected which were directed against the drug. In the initial administration of rifampicin, ethambutol was given as well but, as shown by the immunological studies, the latter was not incriminated in the thrombocytopenia.

Immunological studies showed antibodies, of both IgG and IgM type, capable of fixing complement to both normal and the patient's platelets, but only in the presence of rifampicin. In addition, the IgM type of antibody, but not the IgG, was capable of fixing complement to normal red cells; again only in the presence of the drug.

The second patient was on an intermittent high dose regimen for 4 months, preceded by 3 months of daily treatment. He did not receive ethambutol. The direct antiglobulin test was weakly positive, owing to complement on the red cell surface. Anti-rifampicin antibodies were found in the serum before a dose of rifampicin, and were not detectable 6 hours later when thrombocytopenia was observed.

A further case, reported by Devred et al. (1975), presenting with generalized purpura, responded quickly to discontinuation of the drug and steroid therapy and the platelet count returned to normal in 7 days. Hadfield (1980) reported the case of a 39-year-old man who presented with purpura after one dose of the drug following a four-month gap in daily therapy. Tests for rifampicin antibody were positive. Further clinical reports by Galietti et al. (1978) on thrombocytopenia stressed the importance of careful observation of patients receiving this drug. Rifampicin may cause severe haemolysis if given in high doses. The haemolysis is associated with a positive direct Coomb's test associated with fixation of complement to red cells (Lakshminarayan et al. 1973; Conen et al. 1979).

In a study of six patients with this complication it was shown that the IgG antibody concerned is specific for red cell antigen I (Duran-Suarez et al. 1981). Negative antiglobulin tests were found with cord red cells and adult Oi cells. This specificity suggests that in drug-induced haemolysis of the immune complex type, the erythrocyte cell membrane may play a more active part than has been thought previously.

There have been reports of transient leucopenia and of eosinophilia (Mungall and Standing 1978). The clinical significance of the eosinophilia is not known although the causal relationship is highly probable.

A detailed review of adverse reactions to rifampicin, including the haematological complications of thrombocytopenia and acute haemolytic anaemia is given by Girling (1977).

When aplastic anaemia occurs during modern combination therapy it may be difficult to decide on which drug was responsible. The deaths of three Nigerians from marrow aplasia during this treatment were probably attributable to the drugs which included thiacetazone and isoniazid, given to all three, and rifampicin, given to two of them (Williams *et al.* 1982).

Nalidixic acid

Nalidixic acid appears occasionally to induce autoimmune haemolytic anaemia (Gilbertson and Jones 1972). More often, however, haemolytic anaemic induced by this drug is in association with G6PD deficiency although a direct effect of nalidixic acid on red cells is not excluded in either adults (Alessio and Morselli 1972) or in infants (Belton and Jones 1965). The first case this affected was a newborn child who developed a haemolytic anaemia attributed to nalidixic acid in the breast-milk. The mother had received nalidixic acid 1 g q.d.s. and amylobarbitone sodium 65 mg t.d.s. from the ninth post-partum day. Lactation was established by the third day and was maintained; on the sixteenth day the infant appeared pale and jaundiced and haematological investigation indicated a drug-induced haemolytic anaemia; the direct Coomb's test was negative. On the same day, the infant was transfused with 150 ml of blood and bottle-feeding was started. The reticulocyte count gradually fell to 1 per cent in 10 days (previously 20 per cent of red cells) by which time no Heinz bodies could be seen (previously 50 per cent of the red cells contained one or more Heinz bodies).

Thereafter recovery was uneventful and 7 months after birth the child was perfectly normal. The concentration of nalidixic acid in breast-milk was calculated to be about 4 μg/ml, which, assuming a milk intake of 500 ml per day, would correspond to a dose of 2 mg per day of the drug, a very small amount to account for the haemolytic anaemia. The authors, however, suggested that reduced maternal urinary excretion of the drug due to reduced glomerular filtration associated with the raised blood urea may have raised the serum level in the mother and hence the amount of the drug present in the milk. How-

ever, other studies cited by Belton and Jones (1965) showed that patients with chronic renal failure treated with nalidixic acid show little difference in serum levels of the drug from those in normal subjects, although occasionally higher levels have been encountered.

A drug-linked immune response due to nalidixic acid can also cause severe haemolysis. Tafani *et al.* (1982) reported a fatal haemolytic episode associated with IgG and anti-complement antibodies and a direct antiglobulin test in a patient whose G6PD level was normal and in whom no other innate red-cell abnormality was demonstrated.

Antimalarials

Because of the high incidence of G6PD deficiency in races distributed in malarious areas, primaquine and the other 8-aminoquinoline antimalarials are common causes of haemolytic anaemia of this type. Many variants of this enzyme deficiency have been found, varying greatly in degree and therefore in the severity of drug-induced haemolysis. Details of these variants and their distribution are reviewed by Wintrobe *et al.* (1974).

When a daily dose of 30 mg of primaquine is given to sensitive individuals, there is no untoward reaction for about three days. Then the urine begins to darken, the haemoglobin level, red cell count and haematocrit fall, the reticulocyte count increases, Heinz bodies appear in the erythrocytes, and the patient feels ill (Beutler 1959). In the more severe type of reaction, haemolysis is almost explosive; there is, in addition, haemoglobinaemia and immediate transfusions are necessary (Hockwald *et al.* 1952). If, in spite of these reactions, primaquine therapy is continued, the haemolytic symptoms continue for about a week and then the patient slowly recovers. After about a month, during which time red cell production is speeded up to compensate for the haemolytic episode, the various blood parameters return to normal again. This 'self-limiting' nature of drug-induced haemolytic anaemia is characteristic of the disease (Dern *et al.* 1954). If therapy is stopped for a month or two and then resumed, a new haemolytic episode is observed similar in onset and in nature to the original episode. Once a self-limiting stage has been reached, an increase in the dose may induce further haemolysis for a period, until a new self-limiting phase is attained.

Heinz bodies (granular inclusions) are formed in the red cells of susceptible patients when treated with haemolytic-anaemia-producing drugs and this formation precedes by a day or two the haemolytic episode. Indeed recognition of Heinz bodies can be used to predict haemolysis but simple screening tests, often in kit form, are now widely available and patients who are likely to be susceptible may be tested before oxidant drugs are administered for the first time.

Substituted 2,4-diaminopyrimidines act as folate antagonists by interfering with the action of dihydro-reductase. Antimalarials of this series do not cause megaloblastosis in normal dosage but pyrimethamine may do so during the high dosage used in the treatment of polycythaemia vera (Girdwood, 1973).

Chloroquine

A 4-aminoquinoline used as an anti-inflammatory agent as well as an antimalarial may occasionally cause thrombocytopenia (Niewig et al. 1963), although its more toxic analogue quinacrine caused many cases of aplastic anaemia when extensively used as a malarial suppressive 40 years ago (Custer 1946).

Agranulocytosis may also occur in association with pyrimethamine when the drug is used as an antimalarial in conjunction with either dapsone (Friman et al. 1983; Whitehead and Geary 1983; Herbertson and Robson 1983) or sulphadoxine, (Olsen et al. 1982). The part played by the additional drug is not known, and dapsone in particular may be responsible for the suppression of leucopoiesis. When dapsone was used as a malarial prohylatic against Pl. falciparum in Vietnam, cases of agranulocytosis were frequent enough to lead to withdrawal of the drug for this purpose. The frequency is not enough to present a significant problem in the treatment of leprosy or dermatitis herpetiformis (Firkin and Mariani 1977; Wilson and Harris 1977). Maloprim, the combination of 12.5 mg pyrimethamine and 100 mg dapsone, may induce agranulocytosis if taken twice weekly but is safe at a dose of one tablet weekly. The medical profession and the general public should be aware of the risks of exceeding this dose (Bruce-Chwatt and Hutchinson 1984).

Metronidazole

Metronidazole may induce neutropenia or agranulocytosis (Taylor 1965; Levrat et al. 1978; Guerre et al. 1979; Dawson 1979; Smith 1980; and White et al. 1980). It occurs in 2–4 per cent of patients treated and may be dose-related. It has been suggested that the pathological process is immunological in nature but White et al. (1980) demonstrated marrow hypoplasia. It was further suggested that the combination of metronidazole with another drug that may cause neutropenia, such as azathioprine (McKendrick and Geddes 1979) or 5-fluorouracil may be unduly prone to produce this effect (Windle et al. 1979).

Other antimicrobial agents

The antiparasitic agent, mebendazole, used in high dosage (50 mg/kg daily to treat multiple abdominal cysts of hydatid disease caused moderate neutropenia with hypoplastic marrow (Miskovitz and Javitt 1980). The leucocyte count returned to normal within two weeks of discontinuing the drug, and the authors were of the opinion that the neutropenia was drug-induced. They commented on the increasing use of high-dose mebendazole and on the uncertain absorption affecting blood levels of the drug, with the implication that these may affect the haematological side-effects.

5-Fluorocytosine used as an antifungal agent may induce bone-marrow suppression, and aplastic anaemia caused by this drug was reported by Meyer and Axelrod (1974).

Salicylate and pyrazolone analgesics

Salicylates and allied analgesics

Cases of aplastic anaemia and other blood dyscrasias were associated with analgesic abuse reviewed by Prescott (1968). The extent of the problem was revealed by Murray et al. (1970) in a survey of psychiatric patients. Of 181 patients interviewed, 16 had consumed a total of more than 1 kg of analgesic; a further 26 patients admitted to a daily ingestion of analgesic during the previous six months. These data give an idea of the type of problem involved in this abuse, an abuse largely possible because of the ready availability of these analgesics.

Paracetamol (acetaminophen) may occasionally induce neutropenia or agranulocytosis. Reports concerning this drug include two patients who developed agranulocytosis following treatment with acetaminophen and recovered (Milman and Isager 1978) and a third (Evers and Knoop 1978) who required specific treatment with miconazole for Candida sepsis before remission of severe neutropenia. The drug also causes thrombocytopenia and haemolytic anaemia probably by an immune response to a metabolite of acetaminophen (Kornberg and Polliack 1978) and thrombocytopenia from this cause may be related to previous sensitization by aspirin (Scheinberg 1979). Further evidence of immune-mediation for paracetamol-induced thrombocytopenia was provided by a patient who ingested 1050 mg daily over two weeks. Improvement rapidly followed withdrawal of the drug, but accidental challenge resulted in relapse of purpura and thrombocytopenia. A test for migratory inhibition factor in the presence of paracetamol was positive, (Shoenfeld et al. 1980).

Paracetamol is extensively used in the UK and elsewhere without prescription and the rarity of reports of haematological complications when it is used in clinical dosage suggests that it is unusually safe from this point of view compared with other analgesics.

Leucopenia or agranulocytosis are readily produced by amidopyrine (Kracke 1938) and the more widely used derivative, dipyrone. Amidopyrine is no longer used in most countries but a report from Finland (Kantero *et al.* 1972) described 13 cases of agranulocytosis in children in which five were known to have been given amidopyrine suppositories, which were available without prescription. Several of the children included in this report had received many other drugs, some of which are known to cause agranulocytosis. A 2-year-old boy was given 13 drugs, including four different sulphonamides, two different barbiturates, amidopyrine, and aspirin, and a 9-year-old boy received chloramphenicol, amidopyrine, and diallylbarbituric acid with four other drugs.

These two children survived but the occurrence of agranulocytosis in them raises the question as to whether there was synergistic action in the pathogenesis of the neutropenia. It also underlines the danger of administering a large number of different drugs to the same patient whether prescribed or not. The sale of amidopyrine (dipyrone) continues in several parts of the world, although the drug has been withdrawn from distribution in most. Further cases of agranulocytosis were reported associated with administration of the drug in Mozambique (Epstein and Yudkin 1980) and Greece (Rees 1980). Levy (1980) however pointed out that the relationship between agranulocytosis and pyrazolone analgesic drugs is still uncertain as case-control studies have not been carried out and will be difficult to complete from an ethical point of view.

An international collaborative study group, co-ordinated in Israel but with a catchment population of 30 million people, from Europe, the United States, and South America, will study drug-induced agranulocytosis and may clarify this problem in three years time.

Amidopyrine and its close analogues produce neutropenia of rapid onset apparently by an immune mechanism, which has been demonstrated *in vitro*. Clinically the administration of a very small challenging dose of amidopyrine to a patient who had recovered from an attack of agranulocytosis was followed within 6–10 hours by disappearance of all the neutrophils from the blood (Dameshek and Colmes 1936). Furthermore it was shown that the transfusion of blood withdrawn from a sensitive patient 3 hours after administration of amidopyrine, produced in normal subjects significant granulocytopenia within 40 minutes, and recovery in 4 hours. The patient's serum was also shown to cause *in-vitro* lysis of granulocytes in the presence of the drug (Moeschlin and Wagner 1952). The possible immunological mechanisms

involved in this reaction were discussed in detail by Wintrobe *et al.* (1974).

Dipyrone may occasionally induce immune-mediated haemolytic anaemia. Ribera *et al.* (1981) reported the acute onset of haemolysis in a haemophiliac due to this drug. The antibody was shown to be IgG, complement-binding, and to induce haemolysis *in vitro* as well as giving positive direct and indirect antiglobulin tests.

Anti-inflammatory drugs and analgesics

Phenylbutazone and oxyphenbutazone

Phenylbutazone was a common cause of idiosyncratic drug-induced aplastic anaemia in the United Kingdom and was reported to the Committee of Safety of Medicines to have caused 173 cases from January 1964 to December 1973 (Cunningham *et al.* 1974). These authors compared the plasma clearance and half-lives of phenylbutazone and acetanilide in eight patients with phenylbutazone-induced aplastic anaemia, five with idiopathic marrow hypoplasia, and in normal volunteers. They showed that there was a correlation between the clearance of the two drugs and that there was a significant decrease in their clearance in patients with phenylbutazone-associated hypoplasia. They suggest that relatively poor paraoxidation of phenylbutazone may produce high blood concentrations and be a factor concerned in the aetiology of drug-associated aplastic anaemia. This mechanism cannot, however, be associated with aplastic anaemia resulting from the oxidation metabolite oxyphenbutazone which has also caused a substantial number of drug-induced cases of aplastic anaemia (Inman 1977). In Sweden between 1966 and 1970 this drug was the commonest cause of iatrogenic aplastic anaemia reported to the Swedish Adverse Drug Reaction Committee. In an *in-vitro* study using human granulocyte/monocyte cultures Smith *et al.* (1977) found that oxyphenbutazone was more toxic than phenylbutazone and that the metabolite γ-hydroxyphenylbutazone was inhibitory over a wider concentration and was inhibitory to the same degree, taking its plasma concentrations into account.

Phenylbutazone and its analogues are available only on prescription in most countries but aplastic anaemia occurred in a professional jockey who took phenylbutazone prepared for veterinary administration to horses (Ramsay and Golde 1976). It is also included with amidopyrine in 'Chinese herbal preparations' which have caused agranulocytosis (Ries and Sahud 1975).

Inman (1977) studied the causes of 269 deaths from aplastic anaemia and agranulocytosis reported in death certificates in the United Kingdom. Eighty-three deaths were probably caused by drugs of which 28 were caused by phenylbutazone and 11 by oxyphenbutazone-induced aplastic anaemia. Long-continued treatment with these drugs predisposes to the development of aplasia, especially in the elderly, but short courses given for the relief of traumatic pain or thrombophlebitis do not seem to carry much risk.

A short course of treatment in a young person however, may lead to neutropenia. A 32-year-old woman was treated with 400 mg phenylbutazone daily for a week each before and after laminectomy. Neutropenia followed, and this proved, unusually, to be cyclical in nature and to respond to high-dose prednisone (Rodgers and Shuman 1982). Spontaneous remission is also rare, occurring occasionally, but the effect of other therapy such as oxymethalone is not clear (Hashmi *et al.* 1980).

Indomethacin

Indomethacin occasionally produces mild to moderately severe anaemia, and neutropenia and thrombocytopenia have been observed (Prescott 1968). Aplastic anaemia was first reported by Canada and Burka (1968) in a 42-year old woman who was treated with indomethacin 75 to 100 mg daily. She developed a petechial rash with slight thrombocytopenia (platelets from 30–80 × 10⁹/l) with progressive anaemia and neutropenia. Bone-marrow examination showed complete aplasia, and the disease progressed to death in spite of treatment with prednisolone and testosterone. Fatal aplastic anaemia also followed the administration of allopurinol and indomethacin, again in usual dosage (Schattner *et al.* 1981). A positive lymphocyte transformation test in the presence of indomethacin *in vitro* and a negative one with allopurinol indicated that indomethacin was responsible.

Ibuprofen

An 82-year-old woman died of agranulocytosis after treatment with ibuprofen at 600 mg/day for 5 months (Gryfe and Rubenzahl 1976). She was also treated with frusemide, digoxin, and potassium chloride, but of these agents ibuprofen seemed to be the most likely cause. Autoimmune haemolytic anaemia is a very rare complication of treatment with ibuprofen. Korsager *et al.* (1981) studied the occurrence of haemolytic antibodies in 87 patients receiving long-term treatment with this drug. They found positive antiglobulin tests in eight patients with complement C3c and/or C3d on the red cells. None of these patients showed any evidence of haemolysis.

Benoxaprofen

Although cutaneous and hepatic side-effects of this drug were common and have forced its withdrawal from the UK market, haematological side-effects were not uncommon. The present author and his colleague (Essigman and Williams 1980) reported the onset of neutropenia during treatment with the drug and rapid improvement on its withdrawal. Hindson *et al.* (1982) referred to a further case. Allen and Littlewood (1982) commented on the effect of benoxaprofen in inhibition of the leucotactic agent, leucotriene B4, involved in the cyclo-oxygenase pathway of arachidonic acid to prostaglandins and related it to the good effect of the drug on some severe cases of psoriasis. They compared this effect of benoxaprofen with the lack of response of inflammatory joint disease to razoxane and hydroxyurea until neutropenia is induced. There is no evidence yet, however, that patients who develop neutropenia during benoxaprofen therapy show an exceptional improvement in their joint disease.

Sulindac

Sulindac has been reported to lead to marrow aplasia or erythropoietic suppression (Miller 1980; Bennett *et al.* 1980; Sanz *et al.* 1980), and there was report of sulindac-induced neutropenia with myelopoietic recovery in 10 days (Hynd *et al.* 1982). Another patient with sulindac-induced agranulocytosis studied by Romeril *et al.* (1981) showed increased bone-marrow leucopoiesis. CFU-C activity in *in-vitro* cultures of CFU-C activity was also increased during the neutropenic phase but not after clinical and haematological recovery. The cultures were not affected by the addition of sulindac or by recovery serum and complement, thus making either a direct effect of the drug or an immune-type pathogenesis unlikely. Sulindac-induced thrombocytopenia that was not confirmed by rechallenge was reported by Rosenbaum and O'Connor (1981).

Fenoprofen

Fatal agranulocytosis due to fenoprofen was reported by Simon and Kosmin (1978) and a further probable case with recovery by Treusch *et al.* (1979). Two fatal cases of aplastic anaemia associated with fenoprofen were reported (Ashraf *et al.* 1982). One presented after a dose of 1.8 g daily for 10 months and the other after 14 months of the same dose. In the second case, six months after discontinuing the drug the haemoglobin level and leucocyte count returned to normal but the platelet count remained at 50 × 10⁹ l⁻¹.

Acute thrombocytopenia induced by the same drug with rapid recovery on its withdrawal was reported by Simpson *et al.* (1978) and the unusual occurrence of red-cell aplasia possibly due to fenoprofen was observed in a 35-year-old Caucasian female with rheumatoid arthritis by Weinberger (1979). There was slow improvement on discontinuing the drug combined with treatment with high-dose prednisolone. The relationship between the administration of fenoprofen and the red-cell aplasia is not certain but no other cause for it was found.

Feprazone

A single case report of thrombocytopenia and haemolytic anaemia concerned a 21-year-old man who developed epistaxis 5 days and purpura 11 days after starting treatment with 400 mg daily. The bone marrow showed erythroid and megakaryocytic hyperplasia and this with a direct antiglobulin test of C3d specificity suggested that both the haemolysis and thrombocytopenia were part of a drug-linked immune response. Recovery was complete after treatment with prednisolone (Bell and Humphrey 1982).

Piroxicam

Two patients developed nephropathy with Henoch–Schönlein purpura during treatment with piroxicam. Renal function deteriorated in both patients but recovered on discontinuing the drug (Goebel and Mueller-Brodmann 1982). In one patient IgA immune complexes were demonstrated, and rechallenge with piroxicam again induced nephropathy and purpura.

Naproxen

Fatal aplastic anaemia followed intermittent treatment of recurrent back pain with naproxen over four years. Apart from a month's treatment with phenylbutazone 42 months before death, no other drugs were involved, but the occurrence of idiopathic aplastic anaemia could not be excluded (Arnold and Heimpel 1980).

Niflumic acid

There was a further report (Kallenberg and Roenhorst 1979) of agranulocytosis due to niflumic acid following the previous one of mild neutropenia by Telhag (1973).

Gold compounds

In the 1930–50 period gold was extensively used for the treatment of rheumatoid arthritis and other related disorders and was responsible for many cases of aplastic anaemia (Fitzpatrick and Schwartz 1948). After a period of relatively little use it was again more frequently prescribed but in lower dosage. In spite of this, further cases of bone-marrow depression have occurred (McCarty *et al.* 1962) and 14 cases of aplastic anaemia associated with gold therapy (sodium aurothiomalate) were reported to the CSM in a twenty-year period (Webber 1985). One patient with pure red cell aplasia after gold treatment was also reported (Reid and Patterson 1977).

Penicillamine

Thrombocytopenia is the usual haematological side-effect of penicillamine but agranulocytosis can occur. Ward and Weir (1981) reported a case occurring after seven weeks at the relatively low dose of 250 mg daily. Toxic epidermal necrolysis of staphylococcal origin accompanied the agranulocytosis but treatment with antibiotics and granulocyte infusions led to improvement in six days and recovery in eight weeks.

Previous treatment with gold resulting in toxic reactions may predispose to toxic reactions to penicillamine (Smith *et al.* 1982), and proteinuria or bone-marrow depression with gold may be followed by proteinuria or thrombocytopenia with penicillamine. This suggestion that bone-marrow suppression by penicillamine may be favoured by previous treatment with gold is supported by another care report (Aymard *et al.* 1980). On the other hand, Smith and Swinburn (1980) found no association between the incidence of side-effects to penicillamine and previous side-effects with gold; they also found that there is no increased risk for penicillamine if this drug is started within six months of stopping gold.

Sideroblastic anaemia followed complicated primary biliary cirrhosis treated with D-penicillamine 1 g daily for a year (Sullivan *et al.* 1981). The bone marrow contained ring sideroblasts which disappeared after discontinuing penicillamine and treatment with pyridoxine. The authors discussed the protective effect of pyridoxine in preventing penicillamine-induced weight loss in rats and suggested that in humans a weak subclinical antagonism between penicillamine and pyridoxine may have uncovered a latent pyridoxine deficiency in this patient.

Another unusual complication of penicillamine therapy was the demonstration of hypochromic anaemia and progressive hypocupraemia in two women (Cutolo 1982). The anaemia necessitated withdrawal of penicillamine after 13 and 19 months, respectively. Serum copper levels dropped from 25–28 μmol l^{-1} initially to 10–13 μmol l^{-1} at which time penicillamine was withdrawn. The authors suggested that, if serum copper levels drop to 13–14 μmol l^{-1} during penicillamine treatment, the drug should be discontinued.

Levamisole

Agranulocytosis due to levamisole was reported by Ruuskanen *et al.* (1976). The drug has been used to stimulate the immune response to tumour proliferation in lymphomata, in addition to its use in non-malignant immuno-deficiency states and its original use as an anthelminthic. Its use has increased and with it the incidence of side-effects. Schmidt and Mueller-Eckhardt (1977) reported three patients with levamisole-induced agranulocytosis in rheumatoid patients typed as HLA-B27. One of their patients also had thrombocytopenia and another patient with thrombocytopenia induced by levamisole was reported by El-Ghobarey and Capell (1977).

Symoens *et al.* (1978) reviewed the treatment of 6217 patients with levamisole. Of these patients 3900 were evaluated for side-effects of the drug and there were 267 reports of adverse reactions. Agranulocytosis or leucopenia occurred in 4.9 per cent of patients with rheumatic diseases of whom 81 per cent were female and these had an increased distribution of HLA-B27. In the various publications reviewed, from 0.2 to 1.0 per cent of patients suffering from other infections or inflammatory disorders developed neutropenia and 2.0 per cent of patients with malignant disease did so. Women treated with levamisole in addition to radiotherapy or chemotherapy for advanced carcinoma of the breast (Teerenhovi *et al.* 1978) developed agranulocytosis or severe granulocytopenia in 17 out of 174 cases. A higher proportion of 24 per cent of women with this disease receiving adjuvant chemotherapy of 5-fluorouracil, adriamycin, cyclophosphamide, and methotrexate with levamisole developed granulocytopenia (Vogel *et al.* 1978) and 27 per cent of women treated with the drug in addition to radiotherapy also developed severe leucopenia (Retsas *et al.* 1978).

The agranulocytosis is reversible on discontinuing the drug (Mielants and Veys 1978) and recovery may be heralded by an increase in the number of monocytes in the peripheral blood.

The occurrence of neutropenia in relation to levamisole seems to be potentiated by previous or simultaneous radiotherapy or cytotoxic treatment. There is evidence of reduction of leucopoietic activity in some patients affecting the later phases with survival of myeloblasts and promyclocytes. Reduced colony forming ability of bone-marrow cells may be found (Williams *et al.* 1978), although Schreml and Lohrmann (1979) found no difference in either the size of the granulocyte pool or CFU-C activity between their patients and controls. All their subjects, however, received adjuvant chemotherapy

for carcinoma of the breast with consequent reduction in the leucopoietic pool.

The effect of levamisole on granulocyte production may, however, be a dose-related one, as Mahmood and Robinson (1977) found that the drug stimulated release of colony stimulating factor in low concentrations but was inhibitory at higher ones.

Agranulocytosis, as apart from neutropenia, appears to be an idiosyncratic response and is probably due to an immunological reaction. Agglutinating antibodies acting on patient or donor neutrophils in the presence of levamisole have been reported (Symoens *et al.* 1978) but have not always been found in spite of severe agranulocytosis (Jensen *et al.* 1979; Heyns *et al.* 1979).

Investigation of patients with bladder cancer treated with levamisole showed that complement-dependent granulocytotoxic antibodies were formed in three patients with drug-induced neutropenia (Drew *et al.* 1980). The demonstration of this cytotoxicity did not require the presence of levamisole *in vitro* and affected only peripheral blood neutrophils. Red blood cells, T and B lymphocytes, monocytes, and colony-forming myeloid cells (CFU-C) were not affected. Using concentrations of levamisole of 0.1 to 10.0 μg ml^{-1} in agar cultures Hellman and Goldman (1980) found no effect on CFU-C activity. They were of the opinion that individual idiosyncrasy or hypersensitivity are responsible for the neutropenia but their findings do not exclude an immune response affecting mature neutrophils.

Further evidence of an immunological process was found by Thompson *et al.* (1980) who also demonstrated granulocytoxic antibodies and identified them as IgM. The thermal range and absorption characteristics may however differentiate the antibody in their patients from the auto-antibody reported by Drew *et al.* (1980). Thompson *et al.* suggested that, as normal myeloid colony-forming activity is not suppressed by serum from levamisole-induced neutropenic patients, the neutrophils of these patients may carry antigens not present on most individual's leucocytes.

It is now evident that the use of levamisole in patients whose immune response has been reduced by autoimmune disease, radiotherapy, or cytotoxic drugs is fraught with problems. The Danish Breast Cancer Co-operative Group (1980) found a frequency of 3.6 per cent agranulocytosis and 15–20 per cent leucopenia in patients treated with levamisole after mastectomy and local irradiation with frequent other side-effects, especially in postmenopausal women. Anthony (1980) suggested that this increased incidence of side-effects may be due to the frequency of autoimmune disease in older women.

The predisposition to neutropenia during treatment with levamisole does not appear to apply to HLA-B27 positive patients with malignant melanoma (Espinoza *et al.* 1979).

Cimetidine

Haematological complications associated with cimetidine have mostly occurred in patients who were already ill with renal or autoimmune disease or who were receiving drugs known to produce haematological side-effects. Neutropenia in a patient undergoing haemodialysis was considered to be cimetidine-induced (Ufberg *et al.* 1977), although the relationship was questioned (Lima 1978). Fatal agranulocytosis occurred in a 74-year-old uraemic man, 21 days after starting treatment with the drug (Crapper 1981). The dose administered was 1000 mg daily, and the author suggested that a lower dose is advisable in patients with renal impairment and that routine monitoring of the white-cell count is advisable. Commenting on this case report, Young and Vincent (1981) suggested that previously administered ampicillin may have been responsible.

The first fatal case of complete bone-marrow suppression was reported by Chang and Morrison (1979) but further reports by Elizaga *et al.* (1981) of agranulocytosis and by Tonkonow and Hoffman (1980) of aplasia probably implicate cimetidine but are not conclusive.

A further report by von Dölle and Sewing (1980) concerned a 57-year-old woman who developed agranulocytosis during treatment with carbimazole with full recovery. Subsequent treatment with cimetidine again produced severe neutropenia with recovery on discontinuing the drug. The relationship between the two episodes is not clear, but it seems likely that the patient's leucopoietic tissue was unusually sensitive to drug-induced suppression.

Confirmation by rechallenge was obtained in the case reported by Carloss *et al.* (1980), but the difficulty in deciding which drug or combination is shown in the report by Sazie and Jaffe (1980) in which cimetidine and phenytoin together appear to have caused granulocytopenia.

Marrow hypocellularity was present in a 43-year-old man treated with radiotherapy and chlorambucil for malignant lymphoma, followed by cimetidine for haematemesis and melaena (Posnett *et al.* 1979). The neutropenia occurred when the cimetidine dose was increased from 1200 mg to 2400 mg daily.

Rechallenge with cimetidine at the lower dose after marrow recovery produced no change in the neutrophil count. This occurrence suggests that the neutropenic effect of cimetidine may be dose-related. The authors discussed the possibility that cimetidine inhibits the effect of a natural H_2 agonist such as 4-methylhistamine

in triggering changes in the cell cycles of pluripotent haemopoietic cells. However, no inhibitors of CFU-C activity were found using cimetidine in *in-vitro* cultures (Johnson *et al.* 1977) .

Other patients who developed neutropenia during the administration of cimetidine have mostly had serious underlying disorders in addition to their gastrointestinal abnormality, and some of these have been of an autoimmune nature such as systemic lupus erythematosus (Littlejohn and Urowitz 1979), idiopathic thrombocytopenia (Klotz and Kay 1978), Behçet's disease (Druart *et al.* 1979), or in association with another drug known to cause neutropenia such as phenytoin (Al-Kawas *et al.* 1979). Cimetidine-associated thrombocytopenia was reported by Isaacs (1980) and both neutropenia and thrombocytopenia with haemolysis in a woman with cryptogenic cirrhosis by (Rate *et al.* 1979). The relationship between the autoimmune disorders and the pathogenesis of the cimetidine-induced haematological disorders is not known. A further case of cimetidine-associated thrombocytopenia was reported by Idvall (1979).

At high dose levels cimetidine inhibited granulopoiesis *in vitro* (Fitchen and Koeffler 1980) but this drug concentration was about 100 times the usual clinical peak drug level. At this therapeutic cimetidine concentration there was no demonstrable effect *in vitro*.

The side-effects of cimetidine were reviewed by McGuigan (1981) who concluded that the bone-marrow depression from this drug must be extremely rare and Richter *et al.* (1980) expressed doubts that some of the reported cases are certain in their pathogenesis. In some case reports the patients were treated with other drugs that are known to cause bone-marrow suppression or other clinical causes for neutropenia were present. It appears at present that cimetidine in clinical dosage may provoke bone-marrow suppression in highly susceptible individuals by an idiosyncratic reaction.

Ranitidine

There are two reports of haematological effects possibly related to this drug. In one, neutropenia followed 7 days after commencing treatment, but the clinical history was complicated, and cimetidine, spironolactone, and several antibiotics were administered previously. There was improvement in the two weeks after discontinuing the drug with regenerating myeloid tissue in the marrow (Harmon and Shuman 1984). In the other case, both neutropenia and thrombocytopenia followed 30 days treatment with ranitidine, also in a patient who had received cimetidine previously — the relationship to ranitidine was confirmed by rechallenge (Hervera *et al.* 1984).

These workers demonstrated *in vitro* CFU-GM inhibition by ranitidine at 300 μg per ml, but this concentration is much higher than that obtained clinically (peak 0.4 μg/ml). Inhibition of CFU-GM activity was also obtained by Aglietta *et al.* (1985), but again only at the high concentrations of 250–1000 μg/ml.

Antithyroid drugs

Propylthiouracil

The antithyroid drug propylthiouracil has commonly been reported to cause leucopenia during treatment and less commonly agranulocytosis (see Dalderup 1968), or acute thrombocytopenia (Fewell *et al.* 1950). However, two cases of propylthiouracil-induced aplastic anaemia were reported, the first by Martelo and associates (1967) and the second by Aksoy and Erdem (1968) from Istanbul.

The first case showed spontaneous clinical and haematological improvement after withdrawal of propylthiouracil but the second had more serious consequences. The patient, a 39-year-old man, was admitted to hospital with weakness, fatigue, palpitation, weight loss, pallor, epistaxis, and petechiae. Two months earlier he had been diagnosed as having Graves's disease and was treated with propylthiouracil. In a period of 3 weeks he received a fairly moderate course of treatment with 80×50 mg tablets of this drug. A week after propylthiouracil was stopped, the patient began to complain of epistaxis and weakness; blood transfusions were given without much success. Haematological and bone-marrow examinations led to a diagnosis of aplastic anaemia due to propylthiouracil. Aksoy and Erdem (1968) reported that the condition of this patient was still critical 2 months after the propylthiouracil had been stopped in spite of treatment with whole-blood transfusions (seven in 3 weeks), 250 mg of methyltestosterone weekly and dexamethasone, 6 mg per day. The clinical picture at that time was dominated by haemorrhagic manifestations (epistaxis, bleeding from the gums, and widespread purpuric spots); the red-cell count increased only to $2\,130\,000/mm^3$ ($1\,350\,000$ before treatment), the haemoglobin rose to 5.6 g/100 ml (3.5 before treatment), and platelets increased to $100\,000/mm^3$ ($10\,800$ before treatment).

The authors thought that the patient had some personal susceptibility to propylthiouracil which might possibly involve some essential metabolic process; they were not, however, prepared to suggest the nature of the metabolic process involved. An alternative possibility of an immune mechanism being involved was not supported by experimental evidence.

Neutropenia also occurs with methylthiouracil, methi-

mazole (Matsumoto *et al.* 1976), and carbimazole (Burrel *et al.* 1956; Tait 1957). The toxicity of carbimazole appears to be less than that of propylthiouracil but it is the most extensively used antithyroid drug in the United Kingdom and the occurrence of neutropenia is a regularly recurring clinical problem.

In the United States the close analogue methimazole is in general use and a review from Boston of 36 cases of agranulocytosis has provided new information on its haematological toxicity (Cooper *et al.* 1983). It is evident that agranulocytosis is likely to occur within two months of starting treatment, that it should be administered cautiously to the elderly, and that the risks are less at dosages less than 30 mg daily.

Lithium carbonate may cause goitre and hypothyroidism as a side-effect but lithium-induced thyrotoxicosis is a rarity. A patient with this disorder was treated with propythiouracil but developed agranulocytosis while still taking lithium. Lithium is known to increase the production of neutrophils but this effect was evidently inadequate to prevent the onset of agranulocytosis (Valenta *et al.* 1981).

Diuretics

Acetazolamide and methazolamide used in the treatment of glaucoma are known to induce aplastic anaemia or neutropenia. Two patients developed aplastic anaemia and one agranulocytosis in association with methazolamide (Werblin *et al.* 1979), although in each case other drugs known to cause bone-marrow suppression were also administered. The same authors (Werblin *et al.* 1980) reported two further patients with slight leucopenia whilst using both methazolamide and acetazolamide and another patient with acetazolamide-induced neutropenia was reported by Pearson *et al.* (1955).

The haematological side-effects of thiazide drugs were reviewed by Lundh and Hessegren (1979), Dargie and Dollery (1975), and Kutti and Weinfeld (1968). Chlorthiazide, hydrochlorothiazide, cyclopenthiazide, and bendrofluazide may all cause neutropenia or thrombocytopenia. Platelet counts of infants may be affected by the administration of thiazide diuretics to their mothers (Merenstein *et al.* 1970). Frusemide may cause neutropenia at high dosage (Wauters, 1975) but thrombocytopenia is a more regular if infrequent side-effect (Greenblatt *et al.* 1977; Lowe *et al.* 1979; Duncan *et al.* 1981). Duncan *et al.* (1981) demonstrated the presence of a frusemide-related antibody in their patient by the ^{51}Cr-release technique. Neutropenia also occurred during the administration of chlorthalidone (Klein 1963).

There is one report implicating spironolactone as a cause or agranulocytosis (Stricker and Oei 1984). This followed 5 weeks treatment with the drug, and as the bone marrow contained immature myeloid series cells, a sequestration effect, probably immune-mediated, seems likely. The relationship to spironolactone was confirmed by rechallenge.

Tranquillizers

Phenothiazines

Chlorpromazine may cause neutropenia, thrombocytopenia, or aplastic anaemia (Shelton *et al.* 1960). The onset of neutropenia is slow, three weeks of treatment usually being necessary. The bone marrow is hypocellular and chlorpromazine was shown to inhibit DNA and RNA incorporation into marrow cells *in vitro* in sensitive patients (Pisciotta 1969, 1971) and also in 75 per cent of random hospital patients. It has been suggested that the normal person who shows no ill effects from chlorpromazine administration overcomes delay in DNA synthesis because the proliferative potentiality of his marrow cells is sufficiently great to compensate for the drug-induced delay. However, the person who has a limited proliferative potential will develop neutropenia or other effects of marrow depression when treated with the drug. Occasionally, however, an immunological mechanism is implicated; for example Hoffman *et al.* (1963) demonstrated an antileucocyte antibody active against normal leucocytes in the presence of chlorpromazine but not without it.

Rapid neutropenia may follow overdosage and Burckhart *et al.* (1981) described this effect in a 5-year-old girl. The neutropenia occurred between 31 and 44 hours after ingestion and recovery was substantial by 79 hours. The authors suggest that phenothiazine-induced neutropenia may often be immune-mediated, although in their case at a dose level of 100–200 mg/kg direct suppression of leucopoiesis was likely.

Other phenothiazines, including promazine (Korst 1959) and phenothiazine itself (Fiore and Noonan 1959) may also cause agranulocytosis. Chlorpromazine has also been reported to cause haemolytic anaemia, probably by a direct toxic oxidant effect (How and Davidson 1977).

Benzodiazepines

A few patients treated with these compounds have developed neutropenia, including diazepam (Hollis 1969), nitrazepam, carbamazepine (Ganglberger 1968), and more recently chlordiazepoxide (Celada *et al.* 1977).

In a five-year period of reports to the CSM, chlordiazepoxide was associated with seven patients with thrombocytopenia, eleven with neutropenia of varying degree, and three with pancytopenia, including aplastic anaemia (Webber 1985). In 1970 there were 3.88 million prescriptions recorded for the drug in the NHS, so the risk of marrow depression is very small. The possibility of synergistic action of this widely used drug with another known to produce slow-onset neutropenia such as a sulphonamide or a phenothiazine should not be forgotten. The benzodiazapine, clozapine, was reported by Idänpään-Heikkilä *et al.* (1977), to have caused 17 cases of neutropenia or agranulocytosis amongst 3000 patients in six months. The authors suggest that there may be some inherited characteristic in the Finnish population predisposing to this tendency to agranulocytosis with the drug which is not so marked in other countries using the drug, although genetic studies gave no confirmation of this (Anderman and Griffith 1977; de la Chapelle *et al.* 1977). There were nine deaths in their series but eight of the patients also received phenothiazines, mostly before rather than during treatment so perhaps the phenothiazine sensitized an underlying hereditary tendency to the effect of clozapine (see above).

Flurazepam appears to have been the cause of agranulocytosis in a 45-year-old woman who was also taking carbamazepine. This drug had no effect on CFU-C in agar culture whereas flurazepam suppressed myeloid cells from the patient but not from normal marrow with or without the patient's serum (Hamaguchi *et al.* 1980).

Tricyclic and tetracyclic antidepressants

Thrombocytopenia may occur with imipramine (Goodman 1961) and with other tricyclics. Nixon (1972) reported thrombocytopenia induced by doxepin which recurred when the patient was treated with amitriptyline but not imipramine. He suggested that the difference in structure of the middle ring of imipramine may have altered the antigenicity. A therapeutic response to concomitant steroid was not excluded.

Agranulocytosis may also occur with tricyclics and has been reported as presenting after 4 weeks' treatment with desipramine (Hardin and Conrath 1982) and 7 weeks after starting dothiepin (Doery *et al.* 1982). The tetracyclic mianserin may induce agranulocytosis (Curson and Hale 1979) and this may lead to infections including those of the respiratory tract (Page 1982) or the intestinal tract with severe enterocolitis (Braye *et al.* 1982). Less severe neutropenia may also occur as a side-effect of this drug (McHarg and McHarg 1979).

Anticonvulsants and hypnotics

Aplastic anaemia has been caused by the relatively

little-used anticonvulsants, methylphenylethylhydantoin (Isaacson *et al.* 1956) and trimethadione, but occasional reports of aplastic anaemia associated with administration of phenytoin have been made to the CSM. Neutropenia and thrombocytopenia due to drugs of the hydantoin series have been reported (*Panel of Hematology, Council on Drugs, AMA,* Chicago 1964–7) and with various barbiturates including sodium amytal. Thrombocytopenia may also occur with butobarbitone (F. Young 1957), possibly with phenobarbitone (Boas and Erf 1936), and with sodium valproate (Winfield *et al.* 1976). It has also been reported with phenytoin, although in this instance a drug-linked antiplatelet antibody mechanism was demonstrated (Cimo *et al.* 1977).

Phenytoin and carbamazepine occasionally cause aplastic anaemia and a work of major importance by Gerson *et al.* (1983) provided a highly probable explanation of the mechanism, which may well apply to the suppressive effect of many drugs on the bone marrow. They demonstrated an inherited defect in metabolism using the patient's own lymphocytes as target cells. They suggested that arene oxides as intermediate metabolites although unstable, may affect lymphocyte blastogenesis *in vitro* and bone-marrow function *in vivo*.

The unusual occurrence of pure red-cell aplasia in a 9-year-old girl was attributed to sodium valproate given in a dose of 600 mg daily. Improvement followed discontinuation of the drug. Rechallenge with sodium valproate resulted in relapse of the anaemia with a reduction in bone-marrow red-cell precursors and recovery again on stopping the drug (MacDougall 1982).

Antidiabetic drugs

Tolbutamide, carbutamide, and chlorpropamide have occasionally caused neutropenia or agranulocytosis (Best 1963; *Panel on Haematology Reports AMA* , 1964, 1965, 1967; Stein *et al.* 1964; Kanefsky and Medoff 1980), and the chlorpropamide association with α-methyldopa has caused aplastic anaemia (McMurdoch *et al.* 1968). These drugs also may cause thrombocytopenia but by different mechanisms. Tolbutamide was reported by Schiff *et al.* (1970) to cause suppression of megakaryocyte activity in neonates whose mother received the drug, whereas chlorpropamide (Grace 1959) appeared to cause thrombocytopenia by an immunological process. The pathogenesis of carbutamide thrombocytopenia is not known.

Chlorpropamide has been reported to lead to pure red-cell aplasia (Gill *et al.* 1980), and Planas *et al.* (1980) reported another case where the patient received increasing doses, finally of 1000 mg/day. The bone marrow showed almost complete red-cell aplasia. A reticulocyte response occurred 5 days after discontinuing the drug,

and haemopoietic recovery was complete.

Haemolysis due to chlorpropamide was first only described by Logue *et al.* (1970). A second case reported by Saffouri *et al.* (1981) presented with acute onset only one week after starting chlorpropamide. Immunological studies gave a positive IgG direct antiglobulin and lysis of complement-sensitive PNH cells in the presence of chlorpropamide.

Drugs used in cardiovascular disorders

Amiodarone

Amiodarone is used increasingly to control recurrent cardiac arrhythmias. Side-effects including skin changes, corneal deposits, hypersensitivity pneumonitis, and fibrosing alveolitis have been reported; and one patient with the pulmonary changes also developed bone-marrow depression (Wright and Brackenridge 1982). This was relatively mild with slight anaemia, haemoglobin 9.2 g/dl^{-1} with normal white-cell and platelet counts. The bone marrow was hypoplastic and this continued until death, three months after discontinuing amiodarone, from cardiac disease.

Aprindine

Initial reports on aprindine-induced neutropenia (Bodenheimer and Samarel 1979; Köhler 1980; Opie 1980 *a,b*; Zipes *et al.* 1980) were difficult to interpret as other drugs were administered to the patients concerned and the pathogenesis was not clear. A further case report was published by Khan *et al.* (1983).

In-vitro studies show that aprindine inhibits incorporation of tritiated thymidine of unseparated murine and human marrow cells and inhibited CFU-C activity (Stryckmans *et al.* 1982). The toxic effects occur at aprindine concentrations close to the clinical therapeutic plasma levels. The authors also showed that the analogue moxaprindine has the same haematological toxicity but at higher relative concentrations.

A review of the pharmacology, clinical use, and toxicity of aprindine included comments on the occurrence of agranulocytosis (Stoel and Hagemeijer 1980). This usually occurs within the first six weeks of treatment and is reversible within 5–15 days of discontinuing the drug.

Captopril

Forslund *et al.* (1981) reported the occurrence of neutropenia in a young man treated with captopril, 75 mg daily. The drug was withdrawn and the leucocyte count improved. Low-dose captopril 12.5 mg every second day was tried in order to control the hypertension but again resulted in neutropenia (600×10^6 l and the drug was discontinued with improvement in a week. Bone-marrow examination showed toxic change and inhibition of myelopoiesis but the patient had been treated previously with a wide range of drugs including cyclophosphamide for two years which have influenced the leucopoietic response.

Agranulocytosis induced by captopril has been reported (Van Brummelen et al. 1980; Amman et al. 1980; Staessen et al. 1980; Elijovisch and Krakoff 1980; Walter et al. 1982). Edwards et al. (1981) reported successful reintroduction of captopril at 37.5 mg daily after sudden neutropenia (nadir 600×10^6 l during treatment at 300 mg daily. Another case was studied in detail (Staessen et al. 1981).

In vitro the drug did not inhibit CFU-C activity and the presence of circulating immune complexes suggested that the neutropenia is immunologically mediated. Rechallenge with a low dose may allow continuation of treatment with captopril without recurrence of neutropenia (Case et al. 1981).

Acute renal failure with autoimmune haemolytic anaemia, eosinophilia, and skin rash has been reported as probably captropril-induced by Luderer et al. (1981). The direct antiglobulin test was strongly positive with IgG antisera and weak with anti-complement antibody. The indirect antiglobulin test was positive with all erythrocytes except Rh-null cells. Withdrawal of captopril and treatment with prednisone led to improvement of skin lesions, renal failure, and the autoimmune haemolytic anaemia. Challenge with the other drugs the patient was receiving did not lead to relapse but rechallenge with captopril was understandably not attempted.

Reversible lymphadenopathy in two patients appears to have been induced by captopril (Åberg et al. 1981). Apart from slight eosinophilia in one patient the blood counts were normal.

Ajmaline and its derivatives

Ajmaline, although not widely used for the control of cardiac dysrhythmia has been associated with agranulocytosis (Spiel et al. 1977; Dupoirieux and Lobreau 1978; Bensaid et al. 1979). The analogue chloroacetylajmaline may also cause agranulocytosis and four cases are reported by Cassuto et al. (1980) who found the side-effect to be independent of dosage and, in the two cases examined, to have cellular bone marrow with reduced cellularity of later myeloid forms. These observations suggest that the response was immunological or allergic in nature, and this view was supported by the rapid recovery on discontinuing the drug.

Disopyramide

Agranulocytosis is a rare complication in the use of this drug (Conrad et al. 1978).

Amrinone

This new cardiac inotropic drug may cause thrombocytopenia (Rubin et al. 1979). In one patient 14 days treatment by 300 mg both intravenously and orally three times daily was followed by serum sickness and thrombocytopenia. Megakaryocytes in the bone marrow were increased in numbers and complement C3 levels were decreased, suggesting that the thrombocytopenia was immune-mediated. In a recent series of 12 patients treated with the drug, four developed thrombocytopenia and there was a progressive decline in platelet numbers in three others (Kinney et al. 1983). It is not yet known whether the thrombocytopenia is dose-related but it is evidently a frequent side-effect of the drug.

Hydrallazine

In a review of possible mechanisms of the hydrallazine-related lupus-like syndrome, Perry (1981) discussed the effect of the drug in binding pyridoxine. This occasionally leads to clinical depletion and peripheral neurological symptoms. Hydrallazine also binds iron, commonly causing a mild hypochromic anaemia. Severe anaemia may occur in patients with severe hydrallazine-induced lupus syndrome, and it is suggested that the DNA changes in this disorder may be related to the binding activity of the drug and DNA denaturation.

Alprenolol

This β-blocking agent may occasionally cause thrombocytopenia, and confirmation by rechallenge with the drug was reported by Magnusson and Rödjer (1980). In their case replacement of alprenolol by propanolol was successful, and thrombocytopenia did not recur.

Heparin

Thrombocytopenia induced by heparin has been shown to be due to an immunological response (Kapsch et al. 1979). The reaction may be due to IgG or IgM antibody and may be demonstrated by increased levels of platelet-bound immunoglobulin, by platelet aggregation in the presence of heparin, and by the formation of precipitin lines against a heterologous soluble platelet antigen by counter-current electrophoresis in agarose. The thrombocytopenia usually arises after seven days' treatment and may be suspected by increased heparin requirements associated with the release of platelet Factor 4 during *in-vivo* aggregation. In these circumstances heparin should be discontinued and alternative treatment instituted.

Kapsch *et al.* suggested oral anticoagulation and anti-platelet-aggregating agents and the present author has been forced to use streptokinase and ancrod in these circumstances. Continuing their previous studies with heparin, Eika *et al.* (1980) found a case each in 77 patients treated with two brands of hog-stomach heparin and 55 treated with beef-lung heparin.

A prospective study of the frequency of thrombocytopenia in patients treated with intravenous heparin of porcine gut origin indicated that with this preparation the incidence is low (Powers *et al.* 1979). Only four patients of 120 treated gave platelet counts below $150 \times 10^9 \, 1^{-1}$. In another prospective study five cases of thrombocytopenia developed in 43 patients randomized to receive either bovine 1 mg or porcine mucosal heparin (Ansell *et al.* 1980). Four of these patients were treated with bovine material and developed thrombocytopenia after 8–12 days. A lower incidence of five patients out of 200 treated with full intravenous doses of intestinal mucosal heparin developed thrombocytopenia (Olin and Graor 1981).

There is evidently much variation in the incidence of heparin-induced thrombocytopenia in different clinical units and the cause for this is obscure, perhaps depending on the source and brand of the preparation.

In-vitro experiments by Ansell *et al.* (1980) strongly suggested that heparin-induced thrombocytopenia is immune-mediated and stronger evidence of this was obtained by the isolation of a platelet-aggregating factor in the IgG fraction of patients sera (Chong *et al.* 1981). The antibody induced platelet aggregation and Factor 4 release and this effect could be inhibited by indomethacin, dipyridamole, and aspirin, which the authors suggest may have therapeutic implications. The platelets have impaired aggregation with collagen, ADP, arachnadonate and adrenaline (Chong *et al.* 1983).

In addition to thrombocytopenia heparin may prolong the bleeding time (Heiden *et al.* 1981). If aspirin is administered as well, the bleeding time is further prolonged, and this raises the question of bleeding in association with heparin therapy which may not necessarily be related to thrombocytopenia. It is, however, evident that during heparin therapy either thrombocytopenia or prolonged bleeding time or both together may lead to bleeding with otherwise normal dosage. The pathogenesis of heparin-induced thrombocytopenia was reviewed by Ansell and Deykin (1980). Occasionally disseminated intravascular coagulation with micro-angiopathic hae-molytic anaemia may accompany heparin-induced thrombocytopenia (Zalcberg *et al.* 1983).

Antihistamines

Agranulocytosis is still occasionally reported as occurring during antihistamine therapy. Hardin and Padilla (1978) reported a fatal case due to brompheniramine and 10 possible cases of neutropenia from antihistamines were observed in the Stockholm County Region during the years 1973–75. During this period thenalidine was the commonest cause of drug-induced neutropenia in Stockholm but the drug was withdrawn from use in Sweden in 1976 (Arneborn and Palmblad 1978, 1982). Between June 1973 and November 1981 there were 11 reports of neutropenia or agranulocytosis associated with mebhydrolin napadisylate to the Australian Drug Reactions Advisory Committee and of these seven were in patients in whom this drug was the only suspected one. In all cases recovery occurred or was in progress at the time of the report after discontinuing the drug (McEwen and Strickland 1982).

Other drugs

Quinine and quinidine

The compounds are well known to produce thrombocytopenia by an immune mechanism similar to that occurring with agranulocytosis from amidopyrine. Miescher and Miescher (1952) and Shulman (1964) suggested that the drug concerned forms a combination with a plasma protein or other molecule and that antibodies are formed to the complex. Reaction between them leads to an antigen–antibody complex which attaches to the platelet and damages it. Patients with quinine or quinidine-induced thrombocytopenia are extremely sensitive to the drug and traces of quinine found in tonic water may lead to haemorrhage. Purpura usually improves within a day or two of discontinuing the drug but the thrombocytopenia may be profound while it lasts. Van Leeuwen *et al.* (1982) studied the immunological reactions of sera from 14 patients with this disorder using normal platelets and those with Glanzmann's thrombasthenia and with Bernard–Soulier syndrome.

They showed that the antibody is IgG, always including subclass IgG_1 and sometimes IgG_3. IgM drug-dependent antibodies may occur. The IgG antibodies are complement-fixing and the studies provide further confirmation that in quinine/quinidine-related thrombocytopenia drug–antibody complexes are formed in the absence of platelets and then adhere to the Fc receptor on platelets. Platelets from patients with Glanzmann's disease react like normal platelets, but those from patients with Bernard–Soulier syndrome react less readily with quinine/quinidine-related antibody.

Agranulocytosis is occasionally caused by quinine but the pathogenesis is different from that causing thrombocytopenia. Sutherland *et al.* (1977) reported evidence of an inhibiting effect of quinine on their patient's marrow cells but not on normal marrow cells. Quinidine, however, had no effect on their patient's cells. Quinidine was, however, reported by Eisner *et al.* (1977) to have caused agranulocytosis by a drug-linked antileucocyte antibody reaction similar to the usual mechanism for quinidine-induced thrombocytopenia.

Thorotrast [^{232}thorium dioxide]

In Portugal, Da Silva Horta *et al.* (1965) checked records of 2377 individuals who had received injections of Thorotrast between 1930 and 1952. A total of 1107 cases were traced; of these 699 had died and 408 were still living: certified cause of death was obtained for the former group. Sixteen fatal blood dyscrasias had occurred; eight of these were leukaemias, six were aplastic anaemias (pancytopenia), and two were purpuras.

The authors considered that the toxicity of Thorotrast was such that its use was never justified in people with life expectation of more than two years. This agent is no longer used in the United Kingdom; it was banned in France in 1936 and in 1964 the United States Commissioner of Food and Drugs advised that it was unsafe for administration to man.

A further six cases were reported in 1977 by Johnson *et al.* in which Thorotrast was used for angiography or phlebography. Two patients developed acute leukaemia, one had marrow failure, and three had the haematological features of hypersplenism. The authors noted the diagnostic value of finding refractile granules of Thorotrast in macrophages in bone-marrow smears and confirmation of identity by the demonstration of α-particle tracks in autoradiographs.

Allopurinol

Further cases of agranulocytosis due to allopurinol occurred in Australia (Hawson and Bain 1980) and a mild neutropenia which recurred on challenge with a repeat course of the drug was reported from Canada (Rosenbloom and Gilbert 1981).

Cytotoxic drugs

Bone-marrow suppression is an expected and dose-related effect of most cytotoxic drugs occurring as part of their pharmacological effect directed primarily at leukaemic or other malignant tissue proliferation. We do not discuss this well-known effect but include unexpected haematological effects and comment on myelotoxicity of drugs in the stage of clinical trial. Cytotoxic drug-induced haemolysis is discussed later in this chapter.

VP-16-213, Etoposide

This drug is myelotoxic but not severely so at a dose of 200 mg a day orally for 5 days (G. Anderson *et al.* 1981). The addition of either methotrexate or hydroxydaunorubicin in pilot studies resulted in severe bone-marrow suppression. This has been explained as possibly due to protein-binding displacement by vincristine (Cantwell 1981) or by the timing of the 5-day oral cycle in suppressing marrow stem cells (G. Anderson *et al.* 1981).

PCNU

This new nitrosourea derivative is myelotoxic and this limits the maximum dose (Woolley 1981). Neutropenia occurs with a nadir at 20 days and thrombocytopenia with a nadir at 31 days with a cumulative effect from repeated doses. In its bone-marrow suppression it is more toxic than the equivalent dose of CCNU or methyl-CCNU.

Aziridinylbenzoquinone

This potentially cytotoxic drug has been administered to patients with malignant disease with partial improvement in some patients. The drug is myelotoxic at 6 mg/kg and this limits its clinical dosage (Bedikian *et al.* 1981; 1982).

Zinostatin

In combination with hydroxydaunorubicin in the treatment of hepatocellular cancer this cytotoxic proved to be severely myelotoxic. The use of the combination also proved to be more cardiotoxic than hydroxydaunorubicin alone and the use of these two drugs together should probably be avoided (Issell *et al.* 1981).

Dihydroxyanthracenedione

This drug has been used in the treatment of metastatic carcinoma of the breast. Some clinical improvement was obtained but dosage was limited to less than 4 mg/m^2 by granulocytopenia (Yap *et al.* 1981).

Laetrile

Agranulocytosis with the self-administration of this preparation resolved after its withdrawal. Repetition of the self-treatment resulted in relapse of the severe neutropenia, which again improved after discontinuing the material (Liegner *et al.* 1981). The authors suggest that contaminants or degradation products of amygdalin may be responsible for the suppression of leucopoiesis.

Other drugs

Neutropenia has been reported in the past to be induced by phenindione, ethacrynic acid, and rauwolfia and aplasia associated with bismuth, thiocyanate, colloidal silver and carbon tetrachloride (Wintrobe and Foerster 1974). Thrombocytopenia has been caused by digitoxin (Young *et al.* 1966) and methyldopa (Benraad and Schoemaker 1965) and in the past by organic arsenicals and antimony compounds.

Aluminium

Patients undergoing prolonged renal dialysis may develop encephalopathy with microcytic anaemia (Parkinson *et al.* 1981). This has been shown to be due to aluminium toxicity related to the oral intake. Administration of aluminium compounds may occasionally lead to these serious effects and patients in renal failure treated with aluminium hydroxide should have regular checks on their serum aluminium levels.

Drug-induced haemolytic anaemia

It has long been known that when certain ordinarily harmless drugs are administered to some individuals an acute haemolytic anaemia results. Cases of drug-induced haemolytic anaemia fall into several categories; some have been associated with drugs that have oxidant properties (Table 12.2), others depend on one of a number of immunological effects.

Antimalarials and glucose-6-phosphate dehydrogenase deficiency

Certain antipyretics, sulphonamides, primaquine, and other antimalarials of the 8-aminoquinoline group fall into the first category. Haemolysis can be induced in many individuals if very large doses are given, but in certain instances the administration of normal doses is accompanied by accelerated red cell destruction. Susceptibility to haemolysis by ordinary doses of these drugs

Table 12.2 Oxidant drugs associated with methaemoglobinaemia or haemolytic anaemia

Class of drug	Members
Antimalarials (8-aminoquinolines and 9-aminoquinolines)	primaquine pamaquine quinacrine pentaquine
Sulphonamides	sulphasalazine sulphafurazole sulphamethoxazole
Nitrofurans	nitrofurantoin nitrofurazone furazolidone
Sulphones	dapsone sulphoxone thiazosulphone solapsone
Probenecid	—
Vitamin K (water-soluble derivatives)	potassium menaphthosulphate menazodime menadiol sodium diphosphate chlormezanone
Cinchona alkaloids	quinine quinidine
Acetylated phenol antipyretics	paracetamol phenacetin aspirin amidopyrine
Pyridine derivatives	phenazopyridine
Local anaesthetics	benzocaine
Synthetic dyes	methylene blue
Radiographic contrast media	diatrizoate
Nitrate-containing compounds	isosorbide dinitrate

usually depends on a deficiency of the enzyme glucose-6-phosphate dehydrogenase (G6PD). Wintrobe *et al.* (1974) described the biochemical basis of the haemolytic anaemia that ensues. Wintrobe (1969) however, emphasized that instances of haemolytic anaemia following administration of customary doses of these drugs are not limited to patients with G6PD deficiency. Haemolytic anaemia may develop on exposure to oxidant drugs in persons who, for example, inherit erythrocyte glutathione (GSH) reductase deficiency, erythrocyte GSH deficiency, or a haemoglobinopathy such as Hb-H disease or an unstable haemoglobin such as Hb Zurich or Hb Köln.

The unstable haemoglobins are infrequent except in the Middle East and Eastern Asia where the occurrence of α-thalassaemia genes allow the presentation of Hb-H disease, but the unstable variants, whether hereditary or new mutations, may occur in any race. For example, the author has seen acute haemolytic anaemia induced in a 24-year-old Caucasian female who carried the unstable haemoglobin Hb Riverdale-Bronx by administration of a sulphonamide for a respiratory-tract infection.

In the hereditary enzyme deficiencies the abnormality appears to be an undue tendency for haemoglobin to be denatured due to a shortage of reduced glutathione, the formation of which is dependent on the provision of reduced nicotinamide adenine dinucleotide. The reactions involved are shown in Fig. 12.1. They are represented in the following formulae:

$$GSSG + reduced\ NADP \rightarrow 2GSH + NADP$$
$$\ldots Reaction\ (1)$$

Two dehydrogenases are known to function in erythrocytes and to produce reduced NADP, acting on the substrates glucose-6-phosphate (G6P) and 6-phosphogluconate (6PG) respectively. These reactions can be represented as follows:

$$G6P + NADP \rightarrow 6PG + reduced\ NADP$$
$$\ldots Reaction\ (2)$$
$$6PG + NADP \rightarrow ribulose\ 5\text{-}phosphate + CO_2 + reduced\ NADP$$
$$\ldots Reaction\ (3)$$

As may be seen from the formula, 6PG, the product of Reaction (2), is the substrate for Reaction (3), so that the latter reaction is dependent upon the former. These two reactions (2 and 3) direct carbohydrate metabolism along the oxidative-pentose-phosphate pathway, which is an alternative metabolic route to the anaerobic-glycolytic sequence.

In 1956, Carson and associate showed that whereas both Reaction (1), GSSG reductase, and Reaction (3), 6PG dehydrogenase, were about normal in erythrocytes from primaquine-sensitive individuals, there was a marked deficiency in the activity of Reaction (2), G6P dehydrogenase, in these cells.

The mechanism of haemolysis is still far from clear. The first detectable change even before haemolysis is a drop in the erythrocyte GSH level. Then Heinz bodies are formed, granular inclusions in the erythrocyte consist-

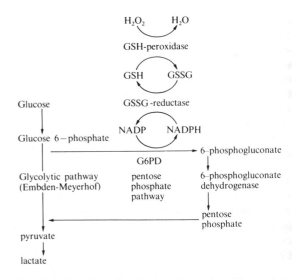

Fig. 12.1 Reactions involved in drug-induced haemolytic anaemia in G6PD-deficient subjects.

ing, it is thought, of denatured protein, probably altered haemoglobin. Normally this modified haemoglobin is reconverted back to its native form by reaction with GSH, which is itself oxidized to GSSG. Under normal conditions the GSSG will be then reduced again to GSH by Reaction (1), utilizing the reduced NADP produced in Reactions (2) and (3). However, in sensitive red cells there is a deficiency of G6PD and so Reaction (2) and hence Reaction (3) are blocked. No reduced NADP is forthcoming and consequently GSH cannot be regenerated in Reaction (1). When no free GSH remains in the erythrocyte, irreversible changes take place in the haemoglobin molecule leading to Heinz-body formation and eventually to cell destruction.

It has been suggested that the modified haemoglobin might be methaemoglobin which might then be converted irreversibly to sulphaemoglobin in deficient erythrocytes. There is a known enzymic reaction involving reduced NADP which converts methaemoglobin formed inside the red cell back to haemoglobin; this may be expressed as follows:

$$methaemoglobin + reduced\ NADP \rightarrow haemoglobin + NADP$$

However, a lack of reduced NADP would block this reaction and methaemoglobin would accumulate. There are several arguments against this theory; for example the reaction utilizes reduced NAD (nicotinamide adenine dinucleotide) as well as reduced NADP, and it cannot be presumed that the supply of reduced NAD, from glycolysis, is also depressed in mutant erythrocytes. A further pertinent fact is that methaemoglobinaemia and sulpha-emoglobinaemia occur clinically from time to time without being accompanied by gross haemolysis. The full story of the haemolysis is therefore clearly more complex than that outlined; indeed it may be quite different. Possibly, for example, both GSH depletion and G6PD deficiency are secondary to some unknown primary defect which itself leads to haemolysis.

Although infrequent, severe enzyme deficiencies may occur in Caucasians and have been reported from Finland (Vuopio *et al.* 1975) to South Africa (Cayanis *et al.* 1975). The latter authors reported a variant of G6PD deficiency whose propositus developed acute haemolytic anaemia following ingestion of a mixture of paracetamol and chlormezanone. Phenacetin has been reported to precipitate haemolysis in patients with G6PD deficiency (Woodbury and Fingl 1975) and other antipyretics, including amidopyrine, were possibly implicated in the severe haemolysis reported by Chan and Todd (1975). Their patients, however, suffered from viral hepatitis, which undoubtedly increases the probability of haemolysis in this enzyme deficiency (Pitcher and Williams 1963; Kattamis and Tjortjatou 1970).

In-vitro experiments show that pyruvate-kinase-deficient red cells from severely affected patients became depleted in ATP during incubation with salicylates (Glader 1976). There are no reports of haemolytic anaemia following administration of analgesics to patients with this and related enzyme deficiencies as yet but its occurrence in severely affected patients is quite likely.

Dapsone (diaminodiphenylsulphone)

Dapsone, a sulphone, was introduced in 1943 and is still the standard treatment for lepromatous or tuberculoid (neural) leprosy; its main advantage over newer agents such as thiambutosine and clofazimine is that it is very cheap, an important factor in the developing countries where it is mostly used.

Although dapsone has been definitely implicated in haemolytic anaemia, it occurs rarely and it is fair to comment that this has usually been associated with high dosage. Most leprologists keep the initial dosage low in the treatment of lepromatous leprosy, for example, 100 mg per week for the first month increasing stepwise over six months to 600 mg per week. This not only avoids the danger of haemolytic anaemia but also the erythema nodosum reaction (lepra reaction) on the skin when the patient becomes sensitive to the products of the mycobacteria that are being destroyed in his tissues. In the tuberculoid type of disease dosage should start at a low level and increase slowly to avoid reaction in, and destruction of, nerve tissue. In such cases a typical dosage

regime would be 50 mg of dapsone per week for the first month rising stepwise to 400 mg per week during the fifth month. There may also be frequent rest periods in both regimes.

Garrett and Corcos (1952), in a report on about 10 000 lepers receiving dapsone, did not mention the occurrence of haemolytic anaemia but Smith and Alexander (1959) noted Heinz bodies in the red cells of four patients but here the dosage of dapsone was 100–200 mg per day. Pengelly (1963) studied the survival of red cells labelled with ^{51}Cr in four patients on dapsone treatment for dermatitis herpetiformis. All four patients showed a reduced erythrocyte survival time on a dosage of 50–150 mg daily. In only two patients was there any previous suggestion of a haemolytic state. Glucose-6-phosphate dehydrogenase activity and glutathione stability was normal in all four cases. The author has also recently observed substantially reduced ^{51}Cr red cell survival time ($t_{1/2}$ = 15 days, normal 28–32) in a patient treated with dapsone, 100 mg daily, for dermatitis herpetiformis. The red-cell G6PD activity and haemoglobin types were normal but there was no anaemia in spite of a persistent reticulocytosis.

A normal volunteer was given ^{51}Cr-labelled red cells and the ^{51}Cr loss was followed by blood samples taken every 2 or 3 days. Dapsone was then given for a further 19 days and the total dosage was high 2050 mg. The ^{51}Cr curve remained within normal limits for the first 22 days and then fell below the normal range indicating that dapsone caused haemolysis of mature red cells. More recently there was a single case report (McConkey 1981) of haemolysis occurring during the treatment of rheumatoid arthritis with dapsone and other cases occurred in soldiers in Vietnam during treatment of *Plasmodium falciparum* malaria (Smithurst 1981). Prescriptions for dapsone may sometimes be dispensed with pyrimethamine without the knowledge of the prescriber. The two drugs are combined as a malarial prophylactic (*Maloprim*), but if used in the dapsone dosage required for the treatment of rheumatoid arthritis or other disorders megaloblastic anaemia may follow as a side-effect of pyrimethamine (Marks 1981).

Sulphonamides

Haemolytic anaemia due to sulphonamides without evidence of antibody formation, enzyme deficiency, or abnormal haemoglobin was reported by Shinton and Wilson (1960) and in association with an indirect antiglobulin reaction by Fishman *et al.* (1973). Abnormal red cells without frank haemolysis were observed in 35 of 50 patients receiving 2.5 g sulphasalazine daily as maintenance therapy in ulcerative colitis. A dose of 1.5 g/day appears to be a safer dose unless the patient is known to be a fast acetylator (Pounder *et al.* 1975).

Primaquine

Haemolysis in G6PD deficient patients treated with primaquine is well known, and usually relatively mild. In a series of 23 patients treated in Malaysia, seven haemolysed, five severely and required blood transfusion (Khoo 1981). Two of these severely affected patients developed renal failure.

Tetracycline

A further occurrence of the rare tetracycline-induced haemolytic anaemia (Mazza and Kryda 1980) differs from previous ones in that the process was oxidative in nature and not immune-mediated. Two separate courses of treatment with tetracycline both resulted in haemolysis with recovery on discontinuing the drug and tetracycline is therefore clearly identified as the causative agent.

Diatrizoate

This radiographic contrast agent caused acute haemolysis in a patient undergoing coronary angiography (Catterall *et al.* 1981) and another case was reported by Darr *et al.* (1981). In this instance the patient suffered from haemoglobin SC disease and was already predisposed to haemolysis, and the authors commented on the risk of diatrizoate in patients with SS or SC disease, especially with high doses of the contrast agent.

Phenytoin

Phenytoin caused intravascular haemolysis, possibly in association with halothane anaesthesia in five Japanese patients, one of whom developed marrow aplasia and another pure erythropoietic aplasia (Hotta *et al.* 1980). This most unusual occurrence raises questions as to the presence of degradation products in one of the drugs or to genetic predisposition in the patients affected.

Intralipid

Acute haemolysis followed the administration of 500 ml 10 per cent *Intralipid* in two hours instead of the usual eight hours (McGrath *et al.* 1982). Haptoglobin–haemoglobin complexes persisted for two weeks, possibly due to reticuloendothelial–system blockade by the intravenous lipid.

Isosorbide dinitrate

Acute haemolysis followed administration of this drug to two Iraqi Jews, both with Mediterranean type G6PD deficiency (Aderka *et al.* 1983). In one case the pathogenesis was confirmed by rechallenge with isosorbide-dinitrate.

Phenazopyridine

It is well known that methaemoglobinaemia and acute haemolysis follow an overdose of this drug or its administration to patients with renal failure. It also occurs readily in the presence of red-cell enzymatic defects or in patients with unstable haemoglobins. This is particularly so if other oxidant drugs such as sulphonamides or nitrofurantoin are given at the same time (Dickerman 1981; Jeffery *et al.* 1982).

Benzocaine

Methaemoglobinaemia has followed the administration of benzocaine as a local anaesthetic spray (Douglas and Fairbanks 1977). In their patients there were no underlying red-cell enzyme or haemoglobin abnormalities and the dose administered, although higher than recommended, was not exceptional. Similar cases were reported by Sandza *et al.* (1979); O'Donohue *et al.* (1980), and Olson and McEvoy (1981).

Benzocaine-containing teething medications can also cause methaemoglobinaemia in infants (Townes *et al.* 1977; McGuigan 1981), especially if the dose is excessively high (Potter and Hillman 1979). In neonates even insertion of a rectal probe lubricated with benzocaine containing material may result in methaemoglobinaemia (Sherman and Smith 1979).

Methlene blue

In most instances acute methaemoglobinaemia may be reversed by treatment with methylene blue, 250 mg/day in divided doses with ascorbic acid 440–500 mg/day. If there is a clinical urgency, methylene blue may be administered intravenously at 1–2 mg/kg. The effectiveness depends on the nature of the oxidant drug or chemical causing the methaemoglobinaemia and the biochemistry involved. With some substances, especially aniline, there may be no improvement with methylene blue and further administration of this compound may increase haemolysis (Harvey and Keitt 1983). Contrary to present standard therapeutics it is not advisable to repeat the dose of methylene blue, and if i.v. ascorbic acid failed, the present author would seriously consider exchange transfusion.

When used as a pigment for intra-amniotic injection to demonstrate premature rupture of the membranes, acute haemolysis may follow in the absence of abnormal red-cell enzymes. It is evident that methylene blue is a toxic agent, and its use in infants or in adults with abnormal red-cell stability should be avoided or used with care (Crooks 1982).

Drug-induced autoimmune haemolysis

These haemolytic anaemias are associated with a positive direct antiglobulin test (Coomb's) but several different mechanisms are involved. Classically these are divided into three (Garratty and Petz 1975) and the more important drugs involved in these processes are listed in Table 12.3. The pathogenesis of immune-mediated drug-induced haemolysis is complicated as the relationships between the drug, red cell membrane proteins, and antibody are still not clearly understood. Perhaps the most definite division is between the α-methyldopa type in which the antigen is an exposed red cell one and the others in which the specificity of the antibody is related to the chemical structure of the associated drug.

Haptene type

Penicillins. In the haptene type penicillins and some other drugs (Table 12.3) become attached to red cells and together stimulate antibody formation against the complex of red cell and drug. These antibodies are usually IgG in nature and red cells sensitized with the drug *in vivo* or *in vitro* react with the antibody as a Type II immune response. However, penicillin-induced haemolysis with disseminated intravascular coagulation and a negative direct antiglobulin reaction was reported by Brandslud *et al.* (1980) and haemolysis without either a direct antiglobulin test or disseminated intravascular coagulation by Spitzer (1981). Brandslud and his colleagues suggested that the reticuloendothelial system was blocked by immune complexes but there was no immunological confirmation of this and perhaps the same pathogenesis applies to Spitzer's case.

Insulin. Coomb's-positive haemolytic anaemia also occurred in a 9-year-old boy with insulin-resistant diabetes mellitus and circulating anti-insulin antibodies. A monoclonal IgG anti-insulin antibody coated the patient's red cells and reacted with exogenous insulin (Faulk *et al.* 1970).

Nomifensine. Nomifensine may cause haemolysis by an immune mechanism when taken in overdose (Prescott *et al.* 1980). After a probable dose of 2 g the patient developed renal failure and intravascular haemolysis. The direct antiglobulin test was positive with a high titre anti-IgG reaction but low serum antibody titre suggesting that the antibody was of high affinty to red cells.

Table 12.3 Reactions in three types of drug-induced haemolytic anaemias

Type of reaction	Role of drug or metabolite	Nature of attachment of antibody of red cell	Anti-globulin reaction	Mechanism of cell destruction	Drug
Haptene	Cell-bound haptene	To cell-bound drug	IgG	Agglutination	penicillin cephalosporins insulin
'Innocent bystander'	Antigen in circulating antigen–antibody complex	Adsorption as part of antigen–antibody–complement complex	Complement	Complement lysis	stibophen quinine quinidine phenacetin dipyrone PAS rifampicin
Direct autoimmune stimulation	Triggers formation of anti-red-cell antibody. No cross-reactivity with drug	To group-specific antigen site on red cell	IgG or IgM	Agglutination	α-methyldopa levodopa mefenamic acid immunization agent— typhoid, poliomyelitis, and triple vaccine

Source: Modified from Wintrobe *et al.* 1974.

Cisplatin. The occurrence of haemolysis as a side-effect of cyclophosphamide and procarbazine is reviewed by Lyman (1980) and new reports of immune-mediated haemolysis induced by cisplatin are published by Getaz *et al.* (1980) and Levi *et al.* (1981). The antibody as shown by the antiglobulin test may be specific in reaction with C3 and C3e (Levi *et al.* 1981), or with monospecific anti-IgG, as found by Getaz *et al.* who consider that the immune mechanisms similar to the classical penicillin-immune haemolysis characterized by strong binding of the drug to the red cell membrane. Hae-molysis with D.I.C. has also been reported as a side-effect of blood transfusion in patients treated with 5-fluouracil and mitomycin-C together (Jones *et al.* 1980).

Triamterene. Triamterene has also induced acute intra-vascular haemolysis with renal failure (Takahashi and Tsukuda 1979). An IgM λ antibody was demonstrated in the serum *in vitro* by haemagglutination inhibition and the serum haemolysed trypsinized red cells and PNH cells. The antibody cross-reacted with methotrexate which has a similar pteridine ring in its chemical structure.

Carbrital. Administration of *Carbrital* (carbromal and phenobarbitone) led to the development of a positive direct Coomb's test in three of 143 patients (Stefanini and Johnson 1970). *In-vitro* addition of patients' serum to red cells coated with *Carbromal* sensitized them to agglutinate with human antiglobin serum. Relatively low antibody titres were observed and there was no clinical evidence of haemolysis.

'Innocent bystander' type

In the 'innocent bystander' type, the red cell is injured during the interaction of an antibody with an extraneous antigen (i.e. drug). Stibophen, quinine, quinidine, and phenacetin produce this type of response as do other drugs listed in Table 12.3.

When the drug is oxidized metabolically it probably combines with some body protein, not red-cell protein. The drug-protein complex serves as an antigen and leads to the formation of an antibody. The antibody shows strong affinity for the red cell and, by binding to the cell, is able to activate the complement mechanism at the cell surface. This is detectable by the anticomplement anti-globulin reaction. Such cells, if heavily coated with complement, may be subjected to premature destruction in the reticuloendothelial system. The antigen–antibody complex appears to associate spontaneously from the damaged cell, and the IgG antiglobulin reaction is thus

negative. The 'innocent bystander' terminology is very appropriate since the immune reaction is directed not against autologous tissue components but against the drug, the blood cell is injured as an innocent bystander and is fundamentally a form of immune-complex disease (Type III immune response).

Haemolytic anaemia due to complement-fixing anti-bodies during administration of chlorpropamide was also reported by Logue *et al.* (1970) and a fatal haemoly-tic reaction to chlorpromazine seems to have had the same pathogenesis (Lindberg and Nordén 1961).

Immune haemolysis due to an IgM antibody followed the use of the new cytostatic drug 9-hydroxy-methyl-ellipticinium (Criel *et al.* 1980). Of eight patients treated with this drug three developed the drug-dependent anti-bodies; in two there was clinically evident haemolysis and one developed oliguric renal failure.

L-asparaginase may also induce immune haemolytic anaemia. The direct antiglobulin test is negative but the indirect is positive in the presence of the drug. The exact pathogenesis is not yet known (Cairo 1982).

Haemolysis as a complication of treatment with rifam-picin was considered earlier in this chapter.

Direct autoimmune stimulation (the α-methyldopa type)

Methyldopa. In 1965 Carstairs and colleagues found that three patients on methyldopa therapy had developed 'idiopathic' warm-antibody (IgG) autoimmune hae-molytic anaemia (Carstairs *et al.* 1966*b*). In January 1966 the CSM issued a warning about the association of methyldopa treatment and haemolytic anaemia based on nine cases reported by doctors or by the manufacturers of the drug (Cahal 1966). Subsequently some 30 cases were notified to the Committee.

Carstairs *et al.* (1966*a*) investigated this problem fur-ther and examined a consecutive series of hypertensive patients on methyldopa treatment who had no symptoms of haemolytic anaemia. Forty-one (20 per cent) of 202 unselected patients gave a positive direct antiglobulin test of the IgG type. In a control group of 76 hypertensive patients on other antihypertensive therapy, none had a positive reaction of the pure IgG type, although two had a positive test of the non-IgG type. No patient on methyl-dopa had overt haemolytic anaemia, but of the patients with a positive DAT six had raised reticulocyte counts and in a further six there was a possible slight depression of the platelet count. Red-cell survival was measured in four patients and in one there was a shortened survival time.

The incidence of a positive DAT was dose-dependent; 9 per cent of those taking 1 gram or less a day were affected and for doses of between 1 and 2 grams daily the figure was 19 per cent. Above 2 grams per day, 56 per cent had a positive DAT. The duration of treatment was also important, and in most cases the test became positive between 6 and 12 months after commencement of therapy.

Worlledge and her colleagues (1966) summarized the clinical, haematological, and serological data of the 30 cases who had developed autoimmune haemolytic anaemia whilst on methyldopa therapy, and had been reported to the CSM. Twenty-five of these cases were studied in detail; 24 had overt haemolytic anaemia. The patients had been treated for periods varying from 3 to 37 months, but 10 had been on treatment for 1 year or less: the majority of patients were receiving 1 gram or less of methyldopa per day.

The severity of the anaemia varied and was unrelated to the total dose of methyldopa. In all the anaemic patients the peripheral blood films showed spherocytosis and polychromasia; the direct antiglobulin reaction was positive in all cases and was entirely of the IgG (warm antibody) type. After discontinuing the drug the peripheral blood picture returns to normal in a few weeks followed by the autoantibodies but the DAT may take months or years to become negative (Worlledge 1969). The drug appears to cause an alteration in the red-cell antigens exposing or modifying the Rhesus specificity. This provokes antibody formation against normal red-cell antigens produced at the end of the life-span of the drug-modified red cells. This accounts for the delay in appearance of haemolytic anaemia after first taking the drug.

In detailed serological studies on 11 patients who developed antibodies after treatment with β-methyldopa (Lalezari et al. 1982) IgM antibodies with complement (C1q and sometimes C3 and C4) together with IgG were demonstrated or eluted from erythrocytes from haemolysing patients. The red cells and eluates from sensitized patients who were not haemolysing contained only IgG. Because of the non-agglutinating character of the IgM antibodies it appears that they are monomeric and that the haemolysis is mediated by the classic pathway of complement activation.

The possibility that a metabolite of methyldopa might be responsible for the haemolytic anaemia can be inferred from the work of Wurzel and Silverman (1966). Methyldopa, in a final concentration of 5 mg/ml, sensitized normal group O, Rh positive, donor red cells *in vitro* at room temperature. However, in a final concentration of one-tenth of that amount (0.5 mg/ml), seven out of 10 metabolites or closely related catecholamines sensitized the red cells more rapidly.

If in the absence of anaemia α-methyldopa treatment

is continued, the direct antiglobulin test usually remains positive. It has been suggested that the autoimmune phenomena induced by the drug are due to inhibition of suppressor lymphocyte activity (Kirtland et al. 1980). Occasionally on continued treatment the DAT may revert to negative (Habibi 1983) but whether this is due to recovery of suppressor cell activity or another mechanism is not known.

Severe haemolysis as a side-effect of this drug still occurs (Roy and Ghosh 1981), and alternative regimes are often now suitable for the treatment of hypertension. Immune thrombocytopenia may occur in association with α-methyldopa (Benraad and Schoenaker 1965) with rapid resolution on discontinuing the drug (Polk et al. 1982).

L–dopa. L–dopa readily induces a positive red cell direct antiglobulin reaction by the same pathogenesis as α-methyldopa (Henry et al. 1971; Joseph 1972). Haemolytic anaemia induced by this drug may also occur and is associated with a warm-reacting IgG anti-Rhesus antibody (Territo et al. 1973). Thrombocytopenia may also occur due to or stimulated by the drug (Wanamaker et al. 1976).

Mefenamic acid. Mefenamic acid was reported by Scott et al. (1968) to be associated with autoimmune haemolytic anaemia in three patients. In each case the autoimmune haemolytic anaemia was of the warm antibody IgG type, and the antibodies had some Rhesus specificity. The patients had received 1.5 g of mefenamic acid daily for between 1 and 2 years: all three patients recovered when the drug was withdrawn; two patients were given prednisone treatment (40 or 60 mg per day) and one received azathioprine 200 mg per day when the mefenamic acid was stopped. The direct anti-human globulin test remained positive for respectively 5, 9, and 3 months, and then became negative.

No inhibition of antibody activity was detected after pre-incubation of the patient's serum with mefenamic acid or its derivatives. Similarly, pre-incubation with normal cells with these compounds did not enhance the antibody activity. No Heinz bodies were found in normal red cells after 6 hours' incubation with mefenamic acid or its derivatives.

Chlorpromazine. This rarely causes haemolytic anaemia, usually of the α-methyldopa type, although a patient with a negative direct antiglobulin reaction was reported by Stein and Inwood (1980).

Methotrexate. Immune haemolytic anaemia induced by methotrexate is an unusual side-effect. A 53-year-old woman treated for metastatic breast carcinoma with 5-fluouracil and methotrexate developed immune haemolysis due to anti-IgG specificities anti-E and anti-C^w. The indirect antiglobulin test was only positive in the presence of methotrexate, a ^{51}Cr survival study with MTX treated autologuous red cells showed reduced survival, and a macrophage phagocyte uptake of methotrexate-treated cells was enhanced in the presence of patients' serum (Sacher *et al.* 1981).

A review of drug-induced haemolytic anaemia (Petz 1980) pays tribute to the memory of Dr Sheila Worlledge in recognition of her work in this field and especially of her classical studies in α-methyldopa-induced haemolysis. The review covers all aspects of pathogenesis and serology of the subject and lists 36 drugs which have been reported to cause a positive direct antiglobulin test with haemolytic anaemia.

Immune haemolysis of undetermined type

Haemolytic anaemia due to a 7S γ-globulin during administration of pyramidone was reported by Bernasconi *et al.* (1961). Acute autoimmune haemolytic anaemia was also reported by Zupańska *et al.* (1976) to follow immunization of children with typhoid, poliomyelitis, and triple vaccine. The antibodies concerned were of the warm-reacting type and similar antibodies were present in autoimmune haemolytic anaemia following acute infections in other children included in their series. The pathogenesis of the stimulation is not known but would seem likely to be the same following both natural and iatrogenic, bacterial or viral provocation.

An unusual combination of vasculititis with purpura, pulmonary cavitation, and anaemia occurred in two children receiving antithyroid treatment. They both received propylthiouracil and one was also treated with iodine. Although these are the only case reports of this combination, their severity justifies careful observation of children taking propylthiouracil for thyrotoxicosis (Cassorla *et al.* 1983). In only one case was the direct antiglobulin test positive, and the anaemia seems to have been of a mixed blood loss and haemolytic origin.

Ritodine hydrochloride may also induce haemolysis in conjunction with transient hepatitis, hypokalaemia, and leukaemoid reaction (Alcena 1982). The mechanism of haemolysis is not known.

Drug-induced megaloblastic anaemia

Cytotoxic drugs

Many cytotoxic drugs cause megaloblastic erythropoiesis as an expected part of their myelotoxicity, and this will not be discussed further here.

Anticonvulsant drugs

The development of megaloblastic anaemia in patients receiving anticonvulsant drugs, notably phenytoin, primidone and phenobarbitone is well recognized (Hawkins and Meynell 1954, 1958; Rhind and Varadi 1954; Ryan and Forshaw 1955; Fuld and Moorhouse 1956; Klipstein 1964; Reynolds *et al.* 1966*a*) and is particularly frequent when phenytoin and phenobarbitone are taken together. Methylphenobarbitone (Calvert *et al.* 1958) and a mixture of quinalbarbitone and amylobarbitone (Hobson *et al.* 1956) have also been reported to cause megaloblastic anaemia as has the anticonvulsant agent methoin.

Nitrofurantoin is chemically related to phenytoin (diphenylhydantoin) and has also been reported as causing megaloblastic anaemia (Bass 1963). The structure-activity and structure–toxicity relationship between these compounds is very clear; this is shown on Fig. 12.2.

All four compounds were thought to induce anaemia

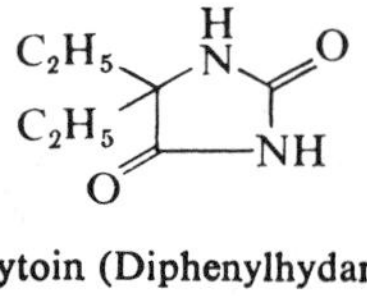

Phenytoin (Diphenylhydantoin)

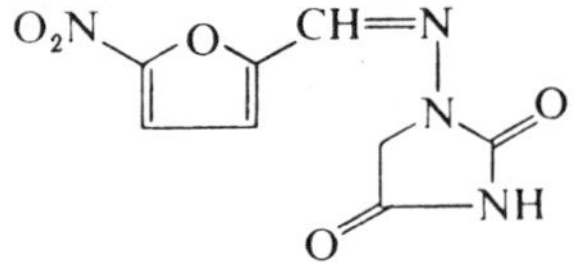

Nitrofurantoin

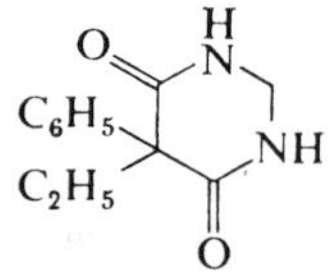

Primidone (Primaclone)

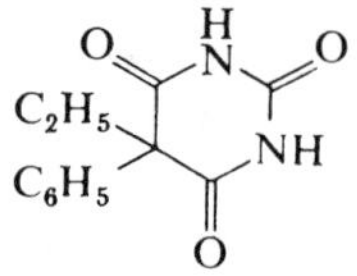

Phenobarbitone (Phenobarbital)

Fig. 12.2 Structural relationship between the anticonvulsant and other drugs known to cause megaloblastic anaemia.

by causing a disturbance in folic-acid metabolism. Hoff-brand and Necheles (1968) investigated the effect of phenytoin on folic-acid metabolism and showed that the drug inhibited folate absorption from folate polygluta-mates although there is some evidence against this (Baugh and Krumdieck 1969).

They also showed in studies *in vitro* that phenytoin inhibited the activity of folate conjugase from human jejunal mucosa. The important link between these two findings is that folate conjugase is responsible for the splitting of folate polyglutamates in the diet into simpler folate monoglutamates prior to absorption. Thus pheny-toin is thought to reduce the formation of folate mono-glutamates, which prevents the effective absorption of folate monoglutamate and thus leads to folate deficiency.

Rosenberg and collegues (1968) confirmed that pheny-toin inhibited the activity of folate conjugase *in vitro* and showed that, in patients, phenytoin reduced folate absorption from folate polyglutamates although the absorption of free folates was not affected.

Red-cell folate levels in 20 mentally retarded epileptics on anticonvulsant therapy and fed on known normal diets were significantly lower than in controls but there was no difference in haemoglobin levels or red-cell size or number (Weber *et al.* 1977). These authors suggest that there must be additional factors to cause symptoms and dietary deficiency seems to be likely in most instances (Flexner and Hartmann 1960; Reynolds *et al.* 1966*b*).

A different, or possibly additional, explanation for megaloblastic change during long-term administration of phenytoin and phenobarbitone was put forward by Wickramasinghe *et al.* (1976). They found that *in-vitro* culture of human bone-marrow cells with these drugs showed reduced ^{3}H-thymidine incorporation into DNA and increased ^{3}H-leucine incorporation into protein. Increased protein synthesis with decreased DNA syn-thesis would lead to uncoordinated cell development with megaloblastic dyshaemopoiesis. Taguchi *et al.* (1977) also found inhibition of ^{3}H-thymidine incorporation into DNA and suggested that failure of DNA synthesis could lead to death of these cells *in vivo* and an increased folate requirement due to increased activity of the folate coen-zyme activity involved in purine and pyrimidine synthe-sis. Treatment of anticonvulsant-induced megaloblastic anaemia with folic acid has been successful in some cases (Reynolds *et al.* 1966*b*).

Of interest in the general spectrum of phenytoin-induced iatrogenic effects is the well known unsightly manifestation of gum-hyperplasia seen in children under treatment for epilepsy with this anticonvulsant. The cause of this hyperplasia is unknown; there is no evidence to suggest that it is related to a folic-acid deficiency, although obviously with such cases a full investigation should be made of the haematological status of the child.

Folate antagonists and related substances

Pyrimethamine

The antimalarial agent pyrimethamine is well known to produce megaloblastic erythropoiesis, especially in high dosage. When it was used years ago in the treatment of polycythaemia vera at 25 mg/daily hypersegmented neu-trophils appeared in the peripheral blood in increased numbers after seven days and progressively increased thereafter and provided an early indication of megalo-blastic change (Chanarin 1964). Pre-existing folate defi-ciency increases the probability of this complication.

Trimethoprim

There are also reports of megaloblastic changes in the marrow following the administration of the structurally related antibacterial agent trimethoprim (Fig. 12.3)

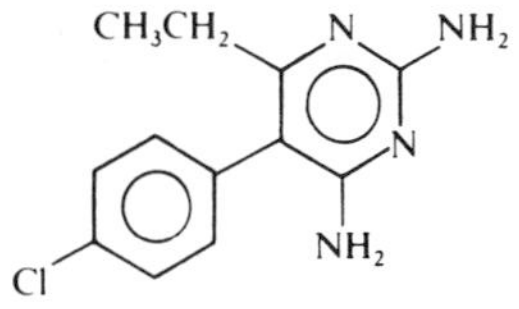

Pyrimethamine

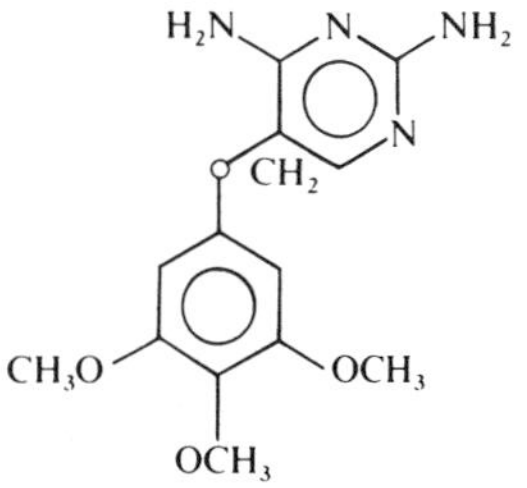

Trimethoprim

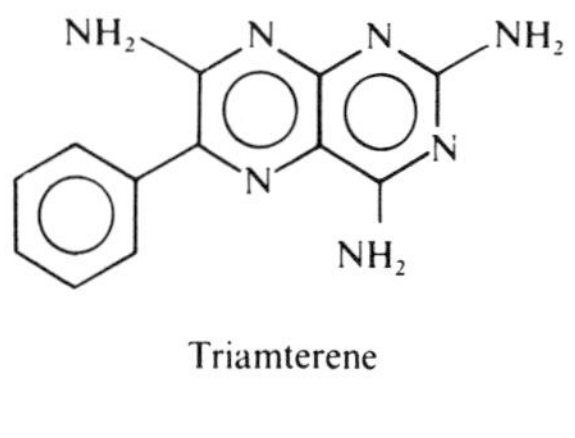

Triamterene

Fig. 12.3

(Kahn *et al.* 1968; Whitman 1969; Jewkes *et al.* 1970). Chanarin and England (1972) showed however that an underlying vitamin B_{12} or folate-deficiency anaemia may fail to respond to specific therapy if the patient is also being treated with co-trimoxazole. Withdrawal of this trimethoprim-sulphamethoxazole combination results in a reticulocyte response and rise in haemoglobin level. *In vitro*, the effect of trimethoprim is dose-related (Sive *et al.* 1972).

Other drugs of related molecular structure, including the diuretic triamterene (Fig. 12.3) could conceivably cause megaloblastic haemopoiesis if administered in sufficient dosage (Girdwood 1973) to patients who were already partially folate-depleted.

Proguanil

Patients with severe renal failure may also develop megaloblastic anaemia in association with proguanil (Boots *et al.* 1982).

Nitrous oxide

Pancytopenia with megaloblastic changes induced by nitrous oxide has been described over a long period (Lassen *et al.* 1956). The metabolic changes in DNA synthesis induced by this drug have been studied extensively by Chanarin and his colleagues and reviewed by them (Chanarin 1980; Chanarin *et al.* 1981). It is evident that several different metabolic functions are shut off, and that these effects of nitrous oxide give opportunities for detailed study of the relationships between cobalamins and folate metabolism in patients and experimental animals. In a study of nine surgical patients anaesthetized and then ventilated with nitrous oxide, it was found that a rise in numbers of hypersegmented neutrophils began on the fifth day, and megaloblastic changes were present in the bone marrow at 24 hours. The deoxyuridine-suppression test was abnormal in all cases and was corrected by vitamin B_{12} and by tetrahydrofolic acid and it 5-formyl derivative, but not by the 5-methyl one (Skacel *et al.* 1983).

The effects of nitrous oxide on haemopoiesis are summarized as:

1. Inactivation of methionine synthetase;
2. Curtailment of folate polyglutamate synthesis;
3. Reduction of methylation of deoxyuridine to deoxythymidine;
4. General folate deficiency because of impaired cellular uptake leading to elevated plasma folate levels and increased renal loss;
5. 5-methylfolate trapping is transient and does not

explain the changes following B_{12} inactivation by nitrous oxide by oxidation (Chanarin 1982).

Biguanides

Diabetic patients treated with metformin and phenformin may develop B_{12} malabsorption as shown by Schilling tests (Tomkin 1973). Following these previous studies on the inhibition of vitamin B_{12} absorption by metformin. Tomkin and his colleagues (Callaghan *et al.* 1980) reported frank megaloblastic anaemia in a woman after eight years' treatment with this drug. Absorption of radioactively-labelled vitamin B_{12} was reduced to 6.2 and 4.8 per cent of test doses without and with human intrinsic factor, respectively, and the serum B_{12} was low at 60 ng l^{-1} (normal range 150–900 ng l^{-1}). The anaemia responded to vitamin B_{12} treatment and the patient remained well for a further five years; maintained on chlorpropamide, metformin, and cyanocobalamin. Clinical presentation of vitamin B_{12} deficiency takes many years and patients treated with biguanides should have annual serum vitamin B_{12} assays.

Other drugs

Malabsorption of vitamin B_{12} leading to megaloblastic anaemia and reversible on ceasing to take the drug has been described in patients treated with colchicine (Webb *et al.* 1968), with neomycin (Jacobsen 1960), and slow-release potassium chloride (Salokannel *et al.* 1970). Malabsorption of vitamin B_{12} was shown to be the mechanism in the megaloblastic anaemias reported by Heinivaara and Palva (1965) during treatment of tuberculosis with *p*-aminosalicylic acid. Megaloblastic anaemia occurring as a response to treatment of tuberculosis is more often associated with isoniazid with or without PAS and is usually sideroblastic in type (Roberts *et al.* 1966).

Acyclovar was associated with megaloblastic haemopoiesis in three cases in whom the serum B_{12} and red cell folate were normal as was the dU-suppression tests. Interference with folate and B_{12} metabolism was thereby excluded, and the usually selective action of the drug in inhibiting viral DNA polymerase appears to have involved human DNA polymerase in these patients (Amos and Amess 1983).

Pelvic radiotherapy is an unusual iatrogenic cause for megaloblastic anaemia (C.G. Anderson *et al.* 1981). Nine years after radiotherapy for carcinoma of the cervix a 71-year-old woman developed anaemia due to iron and cobalamin deficiency due to ileal absorption deficiency. A resected length of ileum appeared thickened and histologically was oedematous but not fibrosed. Dawson in a review, (1979) discussed drug-induced megaloblastosis, commenting on the occurrence of vitamin B_{12} deficiency in patients ventilated with 50 per cent nitrous oxide after cardiac bypass surgery or in general intensive care. She also commented on the occurrence of acute megaloblastosis in patients receiving hyperalimentation with amino acids, sorbitol, and ethanol. Recent changes in this practice should avoid such complications.

Drug-induced myeloproliferative responses

Najman *et al.* (1980) described acute myelofibrosis following three years' treatment of multiple sclerosis with chlorambucil.

Acute myelofibrosis is also reported as a long-term effect of thorotrast administration, symptoms appearing 28 years later (Arnold and Oelbaum 1980). Progress of the disease was rapid and led to death from gram-negative septicaemia in eight months.

A myelodyspoietic syndrome with pancytopenia but increased bone-marrow cellularity with excess of blasts was associated with diethylstilboestrol therapy for carcinoma of the prostate (Anderson and Lynch 1980). The occurrence is probably a chance association but the authors comment on the profound effect of high-dose oestrogens on myelopoiesis in experimental animals in whom increased marrow cellularity may occur.

Cardiovascular prostheses

Haemolytic anaemia in which fragmented red cells were seen in the peripheral blood was reported as a complication of a Teflon cardiac repair by Sayed *et al.* (1961). Since then there have been many reports of significant or subclinical haemolysis following a wide range of intracardiac or vascular surgical procedures. Unsuccessful mitral valvoplasty (Ziperovich and Paley 1966) or mitral-valve replacement may lead to haemolysis (Marsh 1966) but haemolysis is much more frequent with aortic-valve prostheses especially when made of synthetic materials.

In most instances some form of haemodynamic defect is present resulting in marked turbulence which is probably responsible for the haemolysis (Sigler *et al.* 1963; Vanderbroucke *et al.* 1962). Whether haemolysis is severe or compensated, loss of iron in the urine may be substantial, and iron therapy may be required. Regular examination of the urine for haemosiderinuria will identify patients with cardiac prostheses who need iron maintenance therapy (Donnelly *et al.* 1972).

REFERENCES

ÅBERG, H., MÖRLIN, C., AND FRITHZ, G. (1981). Captopril-associated lymphadenopathy. *Br. med. J.* **283**, 1297-8.

ADERKA, D, GARFIWKEL, D., BOGRAD, H., FRIEDMAN, J. AND PINKHAS, J (1983). Isosorbide dinitrate-induced haenolysis in G6 P-D deficient subjects. *Acta. haemat.* **96**, 63-4.

AGLIETTA, M., STACCHINI, A., SANOVIO, F., AND PIACIBELLO, W. (1985). H_2-receptor antagonists and human granuloporesis. *Experientia* **41**, 375-6.

AKSOY, M. AND ERDEM, S. (1968). Aplastic anaemia after propylthiouracil. *Lancet* **i**, 1379.

ALCENA, V. (1982). Severe hemolytic anaemia, leukemoid reaction, acidosis, hypokalemia, and transient hepatitis associated with the administration of ritodrine hydrochloride (*Yutopar*). *Am. J. Obstet. Gynecol.* **144**, 852-4.

ALESSIO, L. AND MORSELLI, G. (1972). Occupational exposure to nalidixic acid. *Br. med. J.* **4**, 110-11.

AL-KAWAS, F.H., LENES, B.A., AND SACHER, R.A. (1979). Cimetidine and agranulocytosis. *Ann. intern. Med.* **90**, 992-3.

ALLEN, B.R. AND LITTLEWOOD, S.M. (1982). Side-effects of benoxaprofen. *Br. med. J.* **285**, 209.

AMMAN, F.W., BUHLER, F.R., BRONNER, F., RITZ, R., AND SPECK, B. (1980). Captopril-associated agranulocytosis. *Lancet* **i**, 150.

AMOS, R.J. AND AMESS, J.A.L. (1983). Megaloblastic haemopoiesis due to acyclovar. *Lancet* **i**, 242-3.

ANDERMAN, B. AND GRIFFITH, R.W. (1977). Clozapine-induced agranulocytosis: a situation report up to August, 1976. *Eur. J. clin. Pharmacol.* **11**, 199-201.

ANDERSON, A.L. AND LYNCH, E.C. (1980). Myelodyspoietic syndrome associated with diethylstilbestrol therapy. *Arch. intern. Med.* **140**, 976-7.

ANDERSON, C.G., WALTON, K.R., AND CHANARIN, I. (1981). Megaloblastic anaemia after pelvic radiotherapy for carcinoma of the cervix. *J. clin. Pathol.* **34**, 151-2.

ANDERSON, G. (1981). Oral VP-16-213 in advanced bronchogenic carcinoma. *Thorax* **36**, 719-20.

——, BOWYER, F., AND WILLIAMS, L. (1981). Oral VP-16-213 in advanced bronchogenic carcinoma and toxic effects when combined with methotrexate. *Thorax* **35**, 462-4.

ANSELL, J. AND DEYKIN, D. (1980). Heparin-induced thrombocytopenia and recurrent thromboembolism. *Am. J. Haematol.* **8**, 325-32.

——, SLEPCHUK, N., JR. KUMAR, R., LOPEZ, A., SOUTHARD, L., AND DEYKIN, D. (1980). Heparin-induced thrombocytopenia: a prospective study. *Thrombos. Haemostas.* **43**, 61-5.

ANTHONY, H.M. (1980). Adjuvant levamisole in breast cancer. *Lancet* **ii**, 1133.

APPELBAUM, F.R. AND FEFER, A. (1981). The pathogenesis of aplastic anaemia. *Sem. Haematol.* **18**, 241-57.

ARNEBORN, P. AND PALMBLAD, J. (1978). Drug-induced neutropenia in the Stockholm Region 1973-75. Frequency and causes. *Acta med. scand.* **204**, 283-6.

—— (1982). Drug-induced neutropenia — a survey for Stockholm 1973-78. *Acta med. scand.* **212**, 289-92.

ARNOLD, A.G. AND OELBAUM, M.H. (1980). Thorotrast administration followed by myelofibrosis. *Postgrad. med. J.* **56**, 124-7.

ARNOLD, R. AND HEIMPEL, H. (1980). Aplastic anaemia after naproxen. *Lancet* **i**, 321.

ASHRAF, M., PEARSON, R.M., AND WINFIELD, D.A. (1982). Aplastic anaemia associated with fenoprofen. *Br. med. J.* **284**, 1301-2.

ASMAR, B.I., MAQBOOL, S., AND DAJANI, A.S. (1981). Hematologic abnormalities after oral trimethoprim–sulfamethoxazole therapy in children. *Am. J. Dis. Childh.* **1351**, 1100–3.

AYMARD, J.P., BAILLE N., WITZ, F., COLOMB, J.N., AND LEDERLIN, P. (1980). Aplasia medullaire mortelle áprès prise de D-pénicillamine pour une polyarthrite rhumatoîde. *Nouv. Presse Méd.* **9**, 2255–6.

BASS, B.H. (1963). Megaloblastic anaemia due to nitrofurantoin. *Lancet* i, 530–1.

BAUGH, C.M. AND KRUMDIECK, C.L. (1969). Effects of phenytoin on folic-acid conjugases in man. *Lancet* ii, 19–21.

BEARDWELL, C.G. (1964). Acute haemolytic anaemia with antipenicillin antibodies complicating subacute bacterial endoearditis. *Proc. R. Soc. Med.* **57**, 332–3.

BEDIKIAN, A.Y., BODEY, G.P., BURGESS, M.A., AND FREIREICH, E.J. (1981). Phase 1 study of aziridinylbenzoquinone, (NSC 182986). *Cancer clin. Trials.* **4**, 459–63.

——, STROEHLEIN, J.R., KARLIN, D.A., KORINEK, J., AND BODEY, G.P. (1982). Phase 2 clinical evaluation of AZQ in colorectal cancer. *Am. J. clin. Oncol*(CCT) **5**, 535–7.

BELL, P.M. AND HUMPHREY, C.A. (1982). Thrombocytopenia and haemolytic anaemia due to feprazone. *Br. med. J.* **284**, 17.

BELTON, E.M. AND JONES, R.V. (1965). Haemolytic anaemia due to nalidixic acid. *Lancet* ii, 691.

BENESTAD, H.B. (1979). Drug mechanisms in marrow aplasia. In *Aplastic anaemia* (ed. G.C. Geary), pp. 26–41. Ballière Tindall, London.

BENNETT, L., SCHLOSSMAN, R., ROSENTHAL, J., BALZORA, J.D., AND ROSNER, F. (1980). Aplastic anaemia and sulindac. *Ann. intern. Med.* **92**, 874.

BENRAAD, A.H. AND SCHOENAKER, A.H. (1965). Thrombopenia after use of methyldopa. *Lancet* ii, 292.

BENSAID, J.J., DOUMEIX, J.J., AND GUALDE, N. (1979). Agranulocytose au cours d'un traitement par le chloroacétylajmaline. *Nouve Presse Méd.* **8**, 704.

BERNASCONI, C., BEDARIDA, G., POLLINI, G., AND SARIORI, S. (1961). Studio del meccanismo di emolisi in un caso di anemia emolitica acquista da piramidone. *Haematalogica, Pavia* **46**, 697–720.

BEST, W.R. (1963). Drug-associated blood dyscrasias. *J. Am. med. Ass.* **185**, 286–90.

BEUTLER, E. (1959). The hemolytic effect of primaquine and related compounds: a review. *Blood* **14**, 103–39.

BIRD, G.W.G., MCEVOY, M.W., AND WINGHAM, J. (1975). Acute haemolytic anaemia due to IgM penicillin-antibody in a 3-year-old child: A sequel to oral penicillin. *J. clin. Pathol.* **28**, 321–3.

BLACKWELL, E.A., HAWSON, G.A.T., LEER, J., AND BAIN, B. (1978). Acute pancytopenia due to megaloblastic arrest in association with co-trimoxazole. *Med. J. Austral.* **2**, 38–41.

BLAJCHMAN, M.A., LOWRY, R.C., PETTIT, J.E., AND STRADLING, P. (1970). Rifampicin-induced immune thrombocytopenia. *Br. med. J.* **3**, 24–6.

BOAS, E.P. AND ERF, L.A. (1936). Thrombocytopenic purpura following medication with sedormid and with phenobarbital. *NY State J. Med.* **36**, 491–4.

BODENHEIMER, H.C. AND SAMAREL, A.M. (1979). Agranulocytosis associated with aprindine therapy. *Arch. intern. Med.* **139**, 1181–2.

BOOTS, M., PHILLIPS, M., AND CURTIS, J.R. (1982). Megaloblastic anemia and pancytopenia due to proguanil in patients with chronic renal failure. *Clin. Nephrol.* **18**, 106–8.

BORLAND, C.D.R. AND FARRAR, W.E. (1979). Reversible neutropenia from vancomycin. *J. Am. med. Ass.* **242**, 2392–3.

BOTTIGER, L.E. AND WESTERHOLM, B. (1974). Drug-induced anaemia in Sweden with special reference to chloramphenicol. *Postgrad. med. J.* **50**, suppl. 5, 127–31.

BRADLEY, P.P., WARDEN, G.D., MAXWELL, J.G., AND ROTHSTEIN, G. (1980). Neutropenia and thrombocytopenia in renal allograft recipients treated with trimethoprim–sulphamethoxazole. *Ann. intern. Med.* **93**, 560–2.

BRANDSLUD, I., PETERSEN, P.H., STRUNGE, P., HOLE, P., AND WORTH, V. (1980). Haemolytic uraemic syndrome and accumulation of haemoglobin-haptoglobin complexes in plasma in serum sickness caused by penicillin drugs. *Haemostasis* **9**, 193–203.

BRAUER, M.J. AND DAMASHEK, W. (1967). Hypoplastic anaemia and myeloblastic leukaemia following chloramphenicol therapy. *New Engl. J. Med.* **277**, 1003–5.

BRAYE, S.G., COPPLESTONE, J.A., AND GARTELL, P.C. (1982). Neutropenic enterocolitis during mianserin-induced agranulocytosis. *Br. med. J.* **285**, 1117.

BRITISH MEDICAL JOURNAL (1963). Broad-spectrum antibiotics in today's drugs. *Br. med. J.* **1**, 1276–7.

—— (1967). Leading article: Toxicity of chloramphenicol. *Br. med. J.* **1**, 649.

—— (1980). Editorial: Laboratory studies in drug-induced pancytopenia. *Br. med. J.* **280**, 429–30.

BRUCE-CHWATT, L.J. AND HUTCHINSON, D.B.A. (1984). Agranulocytosis associated with Maloprim. *Br. med. J.* **288**, 65–6.

BURCKHART, G.J., SNIDOW, J., AND BRUCE, W. (1981). Neutropenia following acute chlorpromazine ingestion. *Clin. Toxicol.* **18**, 797–801.

BURREL, C.D., FRAZER, R., AND DONIACH, D. (1956). The low toxicity of carbimazole. *Lancet* i, 1453–6.

CAHAL, D.A. (1966). Methyldopa and haemolytic anaemia. *Lancet* i, 201.

CAIRO, M.S. (1982). Adverse reactions of L-asparaginase. *Am. J. Pediat: Hematol/Oncol.* **4**, 335–9.

CALLAGHAN, T.S., HADDEN, D.R., AND TOMKIN, G.H. (1980). Megaloblastic anaemia due to vitamin B_{12} malabsorption associated with long-term metformin treatment. *Br. med. J.* **281**, 1214–15.

CALVERT, R.J., HURWORTH, E., AND MACBEAN, A.L. (1958). Megaloblastic anaemia from methophenobarbital. *Blood* **13**, 894–8.

CANADA, A.T.JR. AND BURKA, E.R. (1968). Aplastic anaemia after indomethacin. *New Engl. J. Med.* **278**, 743–4.

CANTWELL, B. (1981). Oral VP-16-213 in advanced bronchogenic caricinoma. *Thorax* **36**, 719.

CARLOSS, H.W., TAVASSOLI, M., AND McMILLAN, R. (1980). Cimetidine-induced granulocytopenia. *Ann. intern. Med.* **93**, 57–8.

CARPENTER, J. (1980). Neutropenia induced by semisynthetic penicillin. *South. med. J.* **73**, 745–8.

CARSON, P.E., FLANAGAN, C.L., ICKES, C.E., AND ALVING, A.S. (1956). Enzymatic deficiency in primaquine-sensitive erythrocytes. *Science* **124**, 484–5.

CARSTAIRS, K.C., BRECKENRIDGE, A., DOLLERY, C.T., AND WORLLEDGE, S.M. (1966*a*). Incidence of a positive direct Coombs test in patients on α-methyldopa. *Lancet* ii, 133–5.

——, WORLLEDGE, S., DOLLERY, C.T., AND BRECKENRIDGE, A. (1966*b*). Methyldopa and haemolytic anaemia. *Lancet* i, 201.

CASE, D.B., WHITMAN, H.H., III, LARAGH, J.H., AND SPIERA, H. (1981). Successful low dose captopril rechallenge following drug-induced leucopenia. *Lancet* i, 1362–3.

CASSORLA, F.G., FINEGOLD, D.N., PARKS, J.S., TENORE, A. THAWERANI, H., AND BAKER, L. (1983). Vasculitis, pulmonary cavitation, and anaemia during antithyroid drug therapy. *Am. J. Dis. Childh.* 137, 118–22.

CASSUTO, J.P., CHICHMANIAN, R.M., DUJARDIN, P., AND AUDOLY, P. (1980). Caractéristiques des agranulocytoses à la chloroacétylajmaline. *Therapie* 35, 378–80.

CATTERALL, J.R., FERGUSON, R.J., AND MILLER, H.C. (1981). Intravascular haemolysis with acute renal failure after angiocardiography. *Br. med. J.* 282, 779–80.

CAYANIS, E., GOMPERTS, E.D., BALINSKY, D., DISLER, P., AND MYERS, A. (1975). G6PD Hillbrow: a new variant of glucose-6-phosphate dehydrogenase associated with drug-induced haemolytic anaemia. *Br. J. Haematol.* 30, 343–50.

CELADA, A., HERREROS, V., AND RUDOLF, H. (1977). Thrombocytopenic purpura during treatment with Librax, *Br. med. J.* 1, 268.

CHAN, T.K. AND TODD, D. (1975). Haemolysis complicating viral hepatitis in patients with glucose-6-phosphate dehydrogenase deficiency. *Br. med. J.* 1, 131.

CHANARIN, I. (1964). Studies in drug-induced megaloblastic anaemia. *Scand. J. Haematol.* 1, 280–8.

—— (1980). Cobalamins and nitrous oxide: a review. *J. clin. Pathol.* 33, 909–16.

—— (1982). The effects of nitrous oxide on cobalamins, folates, and on related events. *CRC Crit. Rev. Toxicol.* 10, 179–213.

——, DEACON, R., PERRY, J., AND LUMB, M. (1981). How vitamin B_{12} acts. *Br. J. Haematol.* 47, 487–91.

—— AND ENGLAND, J.M. (1972). Toxicity of trimethoprim–sulphanethoxazole in patients with megaloblastic haemopoiesis, *Br. med. J.* 1, 651–3.

CHANG, H.K. AND MORRISON, S.L. (1979). Bone marrow suppression associated with cimetidine. *Ann. intern. Med.* 91, 580.

CHONG, B.H., GRACE, C.S., AND ROZENBERG, M.C. (1981). Heparin-induced thrombocytopenia: effect of heparin platelet antibody on platelets. *Br. J. Haematol.* 49, 531–40.

——, ——, AND CASTALDI, P.A. (1983). Qualitative platelet defect in a case of heparin-induced thrombocytopenia. *Scand. J. Haematol.* 30, 427–9.

CIMO, P.L., PISCIOTTA, A.V., DESAI, R.G., PINO, J.L., AND ASTER, R.H. (1977). Detection of drug-dependent antibodies by the ^{51} Cr platelet lysis test: documentation of immune thrombocytopenia induced by diphenylhydantoin, diazepam and sulfisoxazole. *Am. J. Hematol.* 2, 65–72.

CLAYTON, E.M., ALISHULER, J., AND BOVE, J.R. (1965). Penicillin antibody as a cause of positive direct antiglobulin tests, *Am. J. clin. Pathol.* 44, 648–53.

CONEN, D., BLUMBERG, A., WEBER, S., AND SCHUBOTHIE, H. (1979). Hämolytische Krise und akutes Nierenversagen unter Rifampicin. *Schweiz med. Wochenschr.* 109, 558–62.

CONRAD, M.E., CUMBIE, W.G., THRASHER, D.R., AND CARPENTER, J.T. (1978). Agranulocytosis associated with disopyramide therapy. *J. Am. med. Ass.* 240, 1857–8.

COOPER, D.S., GOLDMINZ, D., LEVIN, A.A., LADENSON, P.W., DANIELS, G.H., MOLITCH, M.E., AND RIDGWAY, E.C. (1983). Agranulocytosis associated with antithyroid drugs. *Ann. intern. Med.* 98, 26–9.

CORBETT, G.M., PERRY, D.J., AND SHAW, T.R.D. (1982). Penicillin-induced leukopenia. *New Engl. J. Med.* 307, 642–3.

COUCHONNAL, G.J., HINTHORN, D.R., HODGES, G.R., AND LIU, C. (1978). Naficillin-associated granulocytopenia. *South. med. J.* 71, 1356–8.

CRAPPER, R.M. (1981). Fatal agranulocytosis attributable to cimetidine. *Med. J. Austral.* 2, 250–1.

CRIEL, A.M., HIDAJAT, M., CLARYSSE, A., AND VERWILGHEN, R.L. (1980). Drug dependent red cell antibodies and intravascular haemolysis occurring in patients treated with 9-hydroxy-methyl-ellipticinium. *Br. J. Haematol.* 46, 549–56.

CROOKS, J. (1982). Haemolytic jaundice in a neonate after intra-amniotic injection of methylene blue. *Arch. Dis. Childh.* 57, 872–86.

CUNNINGHAM, J.L., LEYLAND, M.J., DELAMORE, I.W., AND PRICE EVANS, D.A. (1974). Acetanilide oxidation in phenylbutazone-associated hypoplastic anaemia. *Br. med. J.* 2, 313.

CURSON, D.A. AND HALE, A.S. (1979). Mianserin and agranulocytosis. *Br. med. J.* 1, 378–9.

CUSTER, R.P. (1946). Aplastic anaemia in soldiers treated with atabrine (quinacrine). *Am. J. med. Sci. 1,* 212, 211–24.

CUTOLO, M., ACCARDO, S., CIMMINO, M.A., ROVETTA, G., BIANCHI, G., AND BIANCHI, V. (1982). Hypocupremia-related hypochromic anaemia during D-penicillamine treatment. *Arthritis Rheum.* 25, 119–20.

DALDERUP, C.B.M. (1968). Antithyroid drugs. In *Side-effects of drugs,* Vol. VI (ed. L. Meyler and A. Herxheimer). Excerpta Medica Foundation, Amsterdam.

DAMESHEK W. AND COLMES, A. (1936). The effect of drugs in the production of agranulocytosis with particular reference to amidopyrine hypersensitivity. *J. clin. Invest.* 15, 85–97.

DANISH BREAST CANCER CO-OPERATIVE GROUP (1980). Increased breast cancer recurrence rate after adjuvant therapy with levamisole. *Lancet* ii, 824–7.

DARGIE, H.J. AND DOLLERY, C.T. (1975). Adverse reactions to diuretic drugs. In *Meyler's side-effects of drugs,* Vol. VIII, (ed. M.N.G. Dukes), pp. 483–7. Excerpta Medica, Amsterdam.

DARR, M., HAMBURGER, S., KOPRIVICA, B., AND ELLERECK, E. (1981). Haemolytic anaemia associated with a radiopapque contrast agent in a patient with haemoglobin SC disease. *South. med. J.* 74, 1552.

DA SILVA HORTA, J., ABBATT, J.D., DA MOTTA, L.C., AND ROMZ, M.L. (1965). Malignancy and other late effects following administration of Thorotrast. *Lancet* ii, 201–5.

DAVIES, G.E. AND PALEK, J. (1980). Selective erythroid and megakaryocytic aplasia after sulfasalazine administration. *Arch. intern. Med.* 140, 1122.

DAWSON, A.A. (1979). Drug-induced diseases: drug-induced haematological disease. *Br. med. J.* 1, 1195–7.

DAWSON, R.B.JR. AND SEGAL, B.L. (1966). Penicillin-induced immunohaemolytic anaemia. *Arch. intern. Med.* 118, 575–9.

DE LA CHAPELLE, A., KARI, G., NORIMEN, M., AND HERNBERG, S. (1977). Clozapine-induced agranulocytosis. A genetic and epidemiologic study. *Hum. Genet.* 37, 183–94.

DERN, R.J., BEUTLER, E., AND ALVING, A.S. (1954). The hemolytic effect of primaquine, II. The natural course of the hemolytic anemia and the mechanism of its self-limited character. *J. Lab clin. Med.* 44, 171–6.

DEVRED, C., BERNADOU, A., AND BILSKI-PASQUIER, G. (1975). Purpura thrombopénique immuno-allergique du à la rifampicine. *Nouve Presse méd.* 4, 2042.

DICKERMAN, J.D. (1981). A familial hemolytic anemia associated with sulfa administration. *Hosp. Practice* 16, 41, 44, 47.

DICKSON, H.G. (1978). Trimethoprim–sulphamethoxazole and thrombocytopenia. *Med. J. Austral.* 2, 5–7.

DOERY, J.C.G., MEREDITH, H.A., AND MASHFORD, M.L. (1982). Agranulocytosis associated with dothiepin. *Med. J. Austral.* **2**, 389–90.

DOLGOPOL, V.B. AND HOBART, H.M. (1939). Granulocytopenia in sulfapyridine therapy. *J. Am. med. Ass.* **113**, 1012–17.

DONNELLY, R.J., RAHMAN, A.N., MANOHITHARAJAH, S.M., AND WATSON, D.A. (1972). Anaemia with artificial heart-valves. *Lancet* ii, 283.

DOUGLAS, W.W. AND FAIRBANKS, V.F. (1977). Methaemoglobinemia induced by a topical anaesthetic spray (Cetacaine). *Chest* **71**, 587–91.

DREW, S.I., CARTER, B.M., NATHASON, D.S., AND TERASAKI, P.I. (1980). Levamisole-associated neutropenia and auto-immune granulocytosis. *Ann. rheum. Dis.* **39**, 59–63.

DRUART, F., FROCRAIN, C., METOIS, P., MARTIN, J., AND MATUCHANSKY, C. (1979). Association of cimetidine and bone-marrow suppression in man. *Digest Dis. Sci.* **24**, 730–1.

DUNCAN, A., MOORE, S.B., AND BARKER, P. (1981). Thrombocytopenia caused by frusemide-induced platelet antibody. *Lancet* i, 1210.

DUNN, A.M. AND KERR, G.D. (1981). Pure red-cell aplasia associated with sulphasalazine. *Lancet* 2, 1288.

DUPOIRIEUX, J. AND LOBREAU, H. (1978). Agranulocytose suvenue au cours d'un traitment par chlorhydrate d'ajmaline. *Nouve Presse méd.* **7**, 1030.

DURAN-SUAREZ, J.R., MARTIN-VEGA, C., ARGELAGUES, E., MASSUET, L., RIBERA, A., AND TRIGINER, J. (1981). Red cell I antigen as immune complex receptor in drug-induced hemolytic anemias. *Vox Sang.* **41**, 313–15.

——, TRUJILLO, J., PRAT, L., AND MALDONADO, J. (1982). Anti-cephalosporins and immune complexes. *Transfusion* **22**, 541–2.

DUTRO, M.P. (1981). Chloramphenicol and aplastic anaemia. *Am. J. Ophthalmol.* **92**, 870–1.

EDWARDS, C.R.W., DRURY, P., PENKETH, A., AND DAMLUJI, S.A. (1981). Successful reintroduction of captopril following neutropenia. *Lancet* i, 723.

EIKA, C., GODAL, H.C., LAAKE, K., AND HAMBORG, T. (1980). Low incidence of thrombocytopenia during treatment with hog mucosa and beef lung heparin. *Scand. J. Haematol.* **25**, 19–24.

EISNER, E.V., CARR, R.M., AND MACKINNEY, A.A. (1977). Quinidine-induced agranulocytosis. *J. Am. med. Ass.* **238**, 884–6.

—— AND KASPER, K. (1972). Immune thrombocytopenia due to a metabolite of para-aminosalicylic acid. *Am. J. Med.* **53**, 790–6.

EL-GHOBAREY, A.F. AND CAPELL, H.A. (1977). Levamisole-induced thrombocytopenia. *Br. med J.* **2**, 555–6.

ELIJOVISCH, E. AND KRAKOFF, L.R. (1980). Captopril associated granulocytopenia in hypertension after renal transplantation. *Lancet* i, 927–8.

ELIZAGA, F.V., JIM, R.T.S., LAM, C.B.C., AND BERMANS, S.J. (1981). Haematological effects of cimetidine. *Ann. intern. Med.* **94**, 280.

EPSTEIN, P. AND YUDKIN, J.S. (1980). Agranulocytosis in Mozambique due to amidopyrine, a drug withdrawn in the West. *Lancet* ii, 254–5.

ERFFMEYER, J.E. (1981). Adverse reactions to penicillin. *Ann. Allergy.* **46**, 294–300.

ESPINOZA, L.R., DORVAL, G., AND OSTERLAND, C.K. (1979). Levamisole-induced adverse reactions and histocompatibility testing in malignant melanoma. *Tissue Antigens* **13**, 236–7.

ESSIGMAN, W.K. AND WILLIAMS, J.R.B. (1980). Transient neutropenia due to benoxaprofen. *Lancet* ii, 1383–4.

EVANS, R.S. AND FORD, W.P. (1958). Studies of the bone marrow in immunological granulocytopenia. *Arch. intern. Med.* **101**, 244–51.

EVERS, K.G. AND KNOOP, U.F. (1978). Miconazole treatment of candida sepsis in aminophenazone induced agranulocytosis. *Acta. paediat. belg.* **31**, 151–3.

FARBER, B.F. AND BRODY, J.P. (1981). Rapid development of aplastic anemia after intravenous chloramphenicol and cimetidine. *South. med. J.* **74**, 1257–8.

FAULK, W.P., TOMSOVIC, E.J., AND FUDENDER, G.H.H. (1970). Insulin resistance in juvenile diabetes mellitus. *Am. J. Med.* **49**, 133–9.

FEDER, H.M., OSIER, C., AND MADERAZO, E.G. (1981). Chloramphenicol: a review of its use in clinical practice. *Rev. infect. Dis.* **3**, 479–91.

FEWELL, R.A., ENGEL, E.F., AND ZIMMERMAN, S.L. (1950). Acute thrombocytopenic purpura associated with administration of propylthiouracil. *J. Am. med. Ass.* **143**, 891–2.

FIORE, J.M. AND NOONAN, F.M. (1959). Agranulocytosis due to mepazine (phenothiazine). *New Engl. J. Med.* **260**, 375–8.

FIRKIN, F.C. AND MARIANI, A.F. (1977). Agranulocytosis due to Dapsone. *Med. J. Austral.* **2**, 247–51.

FISHMAN, F.L., BARON, J.M., AND ORLINA, A. (1973). Non-oxidative haemolysis due to salicylazosulphapyridine: evidence for an immune mechanism. *Gastroenterology* **64**, 724.

FITCHEN, J.H. AND KOEFFLER, H.P. (1980). Cimetidine and granulopoiesis: bone marrow culture studies in normal man and patients with cimetidine-associated neutropenia. *Br. J. Haematol.* **46**, 361–6.

FITZPATRICK, W.J. AND SCHWARTZ, S.O. (1948). Aplastic anemia secondary to gold therapy. *Blood* **3**, 192–7.

FLEXNER, J.M. AND HARTMANN, R.C. (1960). Megaloblastic anaemia associated with anticonvulsant drugs. *Am. J. Med.* **28**, 386–96.

FORSLUND, T., BROGMÄSTARS, H., AND FYHRQUIST, F. (1981). Captopril-associated leucopenia confirmed by rechallenge in patients with renal failure. *Lancet* i, 166.

FRIMAN, G., NYSTRÖM-ROSANDER, C., JONSELL, G., BJORKMANN, A., LEKAS, G., AND SVENDSRUP, B. (1983). Agranulocytosis associated with malaria prophylaxis with Maloprim. *Br. med. J.* **286**, 1244–5.

FULD, H. AND MOORHOUSE, E.H. (1956). Observations on megaloblastic anaemias after primidone. *Br. med. J.* **1**, 1021–3.

GALE, E.F. (1958). Mode of action of chloramphenicol. In Ciba Foundation Symposium on *Amino acids and peptides with antimetabolic activity* (eds. G.E.W. Wolstenholme and C.M. O'Connor). London.

GALIETTI, E. POY, G., PASSERA, R., AND COSTA, M. (1978). La trombocitopenia da rifampicina. *Minerva Med.* **69**, 232–4.

GANGAROSA, E.J., LANDERMAN, N.S., ROSCH, P.J., AND HERNDON, E.G. JR. (1958). Hematologic complications arising during ristocetin therapy. *New Engl. J. Med.* **259**, 156–61.

GANGLBERGER, J.A. (1968). Erfahrungen mit dem psychotropen Antikonvulsivum Tegretol in der Neurochirurgie. *Wien med. Wochenschr.* **118**, 956–62.

GAREWAL, G., MOHANTY, D., AND DAS, K.C. (1981). A study of hypoplastic anaemia. *In d. J. med. Res.* **73**, 558–70.

GARRATTY, G. AND PETZ, L.D. (1975). Drug-induced haemolytic anaemia. *Am. J. med.* **58**, 398–407.

GARRETT, A.S. AND CORCOS, M.G. (1952). Dapsone treatment of leprosy. *Leprosy Rev. I* **23**, 106–8.

GERSON, W.T., FINE, D.G., SPIELBERG, S.P., AND SENSENBRENNER, L.L.

(1983). Anticonvulsant-induced aplastic anemia; increased susceptibility to toxic drug metabolites *in vitro*. *Blood* **61**, 889–93.

GETAZ, E.P., BECKLEY, S., FITZPATRICK, J., AND DOZIER, A. (1980). Cistplatin-induced haemolysis. *New Engl. J. Med.* **302**, 334–5.

GILBERTSON, C. AND JONES, R.D. (1972). Haemolytic anaemia with nalidixic acid. *Br. med. J.* **2**, 493.

GILL, M.J., RATLIFF, D.A., AND HARDING, L.K. (1980). Hypoglycemic coma, jaundice and pure r.b.c. aplasia following chlorpropamide therapy. *Ann. intern. Med.* **140**, 714–15.

GIRDWOOD, R.H. (1964). Drug-induced blood disorders. *Br. J. clin. Pract.* **18**, 701–7.

—— (1973). Drug-induced megaloblastic anaemia. In *Blood disorders due to drugs and other agents*. Excepta Medica, Amsterdam.

GIRLING, D.J. (1977). Adverse reactions to rifampicin in antituberculosis regimens. *J. antimicrob. Chem.* **3**, 115–32.

GLADER, B.E. (1976). Salicylate-induced injury of pyruvate-kinase-deficient erythrocytes. *New Engl. J. Med.* **294**, 916–18.

GODIN M., DESHAYES, P., DUCASTELLE, T., DELPECH, A., LELOËT, X., AND FILLASTRE, J.P. (1980). Agranulocytosis, haemorrhagic cystitis and acute interstitial nephritis during methicillin therapy. *J. Antimicrobiol. Chemother.* **6**, 296–7.

GOEBEL, K.M. AND MUELLER-BRODMANN, W. (1982). Reversible overt nephropathy with Henoch Schöenlein purpura due to piroxicam. *Br. med. J.* **284**, 311–12.

GOLDBERG, J. (1983). Sulfasalazine and folate deficiency. *J. Am. med. Ass.* **249**, 729.

GOLDE, D.W., BERSCH, N., AND QUAN, S.G. (1978). Trimethoprim and sulphamethoxazole inhibition of haematopoiesis *in vitro*. *Br. J. Haematol.* **40**, 363–7.

GOODMAN, H.L. (1961). Agranulocytosis associated with Tofranil. *Ann. intern. Med.* **55**, 321–3.

GOODMAN, L.S. AND GILMAN, A. (1975). *The pharmacological basis of therapeutics,* 5th edn. Macmillan, London.

GOSH, J.S. (1979). Oxacillin-induced granulocytopenia. *Acta. Haematol.* **61**, 59.

GRACE, W.J. (1959). Thrombocytopenia in a patient taking chlorpropamide. *New Engl. J. Med.* **260**, 711–2.

GRALNICK, H.R., MCGINNISS, M., ELTON, W., AND MCCURDY, P. (1971). Hemolytic anaemia associated with cephalothin. *J. Am. med. Ass.* **217**, 1193–7.

——, WRIGHT, L.D., AND MCGINNIS, M. (1967). Coomb's positive reactions associated with sodium cephalothin therapy. *J. Am. med. Ass.* **199**, 135–6.

GREENBLATT, D.J., DUHME, D.W., ALLEN, M.D. AND KOCH-WESER, J. (1977). Clinical toxicity of furosemide in hospitalized patients: a report from the Boston Collaborative Drug Surveillance Program. *Am. Heart J.* **94**, 6–13.

GRYFE, C.I. AND RUBENZAHL, S. (1976). Agranulocytosis and aplastic anaemia: possibly due to ibuprofen. *Can. med. Ass. J.* **114**, 877.

GUERRE, J., GAUDRIC, M., AND TUAL, J.L. (1979). Neutropenie preécoce au cours d'un traitement par le métronidazole. *Nouve Presse méd.* **8**, 699.

HAAK, H.L. (1980). Experimental drug-induced aplastic anaemia. *Clin. Haematol.* **9**, 621–39.

HABIBI, B. (1983). Disappearance of alpha-methyldopa induced red cell autoantibodies despite continuation of the drug. *Br. J. Haematol.* **54**, 493–5.

HADFIELD, J.W. (1980). Rifampicin-induced thrombocytopenia. *Postgrad. Med. J.* **56**, 59–60.

HAMAGUCHI, H., AMANO, M., SAKAMAKI, H., DAN, E., KURIYA, S., KAWADA, K., AND NOMURA, T. (1980). A case of agranulocytosis. Application of the *in vitro* cloning technique for identification of the offending drug. *Rinsho Ketsueki* **21**, 356–62.

HARDIN, A.S. AND PADILLA, E. (1978). Agranulocytosis during therapy with a brompheniramine-medication. *J. Ark. med. Soc.* **75**, 206–8.

HARDIN, T.C. AND CONRATH, F.C. (1982). Desipramine-induced agranulocytosis; a case report. *Drug Intel. clin. Pharm.* **16**, 62–3.

HARMAN, D.C. AND SHUMAN, R. (1984). Ranitidine. *New Eng. J. Med.* **310**, 1604.

HARVEY, J.W. AND KEITT, A.S. (1983). Studies of the efficacy and potential hazards of methylene blue therapy in aniline-induced methaemoglobinaemia. *Br. J. Haematol.* **54**, 29–41.

HASHMI, K.Z., MAKAR Y.F., AND ROUTIEDGE, R.C. (1980). Spontaneous remission in aplastic anaemia. *J. Pakistan med. Ass.* **30**, 185–6.

HAUSMAN, K. AND SKRANDIES, G. (1974). Aplastic anaemia following chloramphenicol therapy in Hamburg and surrounding districts. *Postgrad. med. J.* **50**, (Suppl. 5), 131–6.

HAWKINS, C.F. AND MEYNELL, M.J. (1954). Megaloblastic anaemia due to phenytoin sodium. *Lancet* **ii**, 373–8.

—— AND —— (1958). Macrocytosis and macrocytic anaemia caused by anticonvulsant drugs. *Quart. J. Med.* **27**, 45–63.

HAWSON, G.A.T. AND BAIN, B.J. (1980). Allopurinol and agranulocytosis. *Med. J. Austral.* **1**, 283–4.

HEIDEN, D., ROOVIEN, R., AND MIELKE, C.H. (1981). Heparin bleeding, platelet dysfunction, and aspirin. *J. Am. med. Ass.* **246**, 330–1.

HEINIVAARA, O. AND PALVA, I.P. (1965). Malabsorption and deficiency of vitamin B_{12} caused by treatment with paraminosalicylic acid. *Acta med. scand.* **177**, 337–41.

HEIMPEL, H. AND HEIT, W. (1980). Drug-induced aplastic anaemia: clinical aspects. *Clin. Haematol.* **9**, 641–62.

HELLMAN, A. AND GOLDMAN, J.M. (1980). Effect of levamisole on granulopoiesis in agar culture. *Biomedicine* **33**, 103–5.

HENRY, R.E., GOLDBERG, L.S., STURGEON, P., AND ANSEL, R.D. (1971). Serologic abnormalities associated with L-dopa therapy. *Vox Sang.* **20**, 306–16.

HERBERTSON, M. AND ROBSON, R.H. (1983). Agranulocytosis associated with Maloprim. *Br. med. J.* **286**, 1515.

HERRERA, A., SOLAL-CELIGNY, P., DRESCH, C., AND VALLIN, J. (1984). Ranitidine. *New Engl. J. Med.* **310**, 1604–5.

HEYNS, A.DU P., RETIEF, F.P., AND VORSTER, B.J. (1979). Agranulocytosis in an arthritic patient treated with levamisole. *S. Afr. med. J.* **55**, 177–9.

HINDSON, C., DAYMOND, T., DIFFEY, B., AND LAWLOR, F. (1982). Side-effects of benoxaprofen. *Br. med. J.* **284**, 1368–9.

HOBSON, Q.J.G., SELWYN, J.G., AND MOLLIN, D.L. (1956). Megaloblastic anaemia due to barbiturate. *Lancet* **ii**, 1079–81.

HOCKWALD, R.S., ARNOLD, J., CLAYMAN, C.B., AND ALVING, A.S. (1952). Status of primaquine, IV. Toxicity of primaquine in Negroes. *J. Am. med. Ass.* **149**, 1568–70.

HODGMAN, J.E. (1961). Chloramphenicol. *Pediat. Clin. N. Am.* **8**, 1027–42.

HOFFBRAND, A.V. AND NECHELES, T.F. (1968). Mechanism of folate deficiency in patients receiving phenytoin. *Lancet* **ii**, 528–30.

HOFFMAN, G.C., HEWLETT, J.S., AND GARZON, F.L. (1963). A drug-specific leuco-agglutinin in a fatal case of agranulocytosis due to chlorpromazine. *J. clin. Pathol.* **16**, 232–4.

HOLLIS, D.A. (1969). Diazepam: its scope in anaesthetic practice. *Proc. R. Soc. Med.* **62**, 806–7.

HORLER, A.R. (1977). Blood disorders. In *Textbook of adverse drug reactions* (ed. D.M. Davies), pp. 354–64. Oxford University Press.

HOTTA, T., HIRABAVASHI, N., KABAYASHI, T., AND NAKAMURA, S. (1980). Five cases showing haemoglobinuria and aplastic crisis following administration of diphenylhydantoin after craiotomy. *Rinsho Ketsueki* **21**, 536–43.

HOW, J. AND DAVIDSON, R.J.L. (1977). Chlorpromazine-induced haemolytic anaemia in anorexia nervosa. *Postgrad. med. J.* **53**, 278–9.

HSI, Y.-J., KUO, H.-Y., AND OUYANG, A. (1966). Thrombocytopenia following administration of penicillin. *Chinese med. J.* **85**, 249–51.

HYND, R.F., KLOFKORN, W.J., SHOLES, C.W., AND MOQUIN, R.B. (1982). Neutropenia and Pseudomonas septicaemia after sulindac therapy: case report. *Military Med.* **147**, 768–9.

IDÄNPÄÄN-HEIKKILÄ, J., ALHAVA, F., OLKINUORA, M., AND PALVA, L.P. (1977). Agranulocytosis during treatment with clozapine. *Europ. J. clin. Pharmacol.* **II**, 193–8.

IDVALL, J. (1979). Cimetidine-associated thrombocytopenia. *Lancet* **ii**, 159.

INGALL, D., SHERMAN, J.D., COCKBURN, F., AND KLEIN, R. (1965). Amelioration by ingestion of phenylalanine of toxic effects of chloramphenicol on bone marrow. *New Engl. J. Med.* **272**, 180–5.

INMAN, W.H.W. (1977). Study of fatal bone marrow depression with special reference to phenylbutazone and oxphenbutazone. *Br. med. J.* **1**, 1500–5.

ISAACS, A.J. (1980). Cimetidine and thrombocytopenia. *Br. med. J.* **280**, 294.

ISAACSON, S., GOLD, J.A., AND GINSBERG, V. (1956). Fatal aplastic anaemia after therapy with nuvarone (3-methyl-5-phenyl-hydantoin), *J. Am. med. Ass.* **160**, 1311–2.

ISSELL, B.F., GINSBERG, S.J., AND COMIS, R.L. (1981). Zinostatin and doxorubicin. *Cancer Clin. Trials* **4**, 323–6.

JACOBSON, E.D. (1960). An experimental malabsorption syndrome induced by neomycin. *Am. J. Med.* **28**, 524–33.

JEFFERY, W.H., ZELICOFF, A.P., AND HARDY, W.N. (1982). Acquired methemoglobinaemia and hemolytic anemia after usual doses of phenazopyridine. *Drug Intel. Clin. Pharm.* **16**, 157–9.

JENSEN, M.K., ERIKSEN, J., ANDERSEN, H., AND OSTERGAARD, P.A. (1979). Levamisole-induced granulocytopenia: cytologenetic and immunological studies. *Dan. med. Bull.* **26**, 14–17.

JEWKES, R.F., EDWARDS, M.S., AND GRANT, B.J.B. (1970). Haematological changes in a patient on long-term treatment with a tri-methoprim–sulphonamide combination. *Postgrad. med. J.* **46**, 723–6.

JOHNSON, N., MCL., BLACK, A.E., HUGHES, A.S.B., AND CLARKE, S.W. (1977). Leucopenia with cimetidine. *Lancet* **ii**, 1226.

JOHNSON, S.A.N., BATEMAN, C.J.T., BEARD, M.E.J., WHITEHOUSE, J.M.A., AND WATERS, A.H. (1977). Long-term haematological complications of Thorotrast. *Quart. med. J., New Series* **XLVI**, 259–71.

JONES, B.G., NEWMAN, C.E., FIELDING, J.W., HOWELL, A., AND BROOKES, V.S. (1980). Intravascular haemolysis and renal impairment after blood transfusion in two patients on long-term 5 fluorouracil and mitomycin-C *Lancet* **i**, 1275–7.

JOSEPH, C. (1972). Occurrence of positive Coomb's test in patients treated with levodopa. *New Engl. J. Med.* **286**, 1401–2.

KÄHLER, H.J. (1965). Chloramphenicol: Ein vielseitiges und zeitloses Antibiotikum. *Med. Klin.* **60**, 2005–11.

KAHN, J.B. (1978). Oxacillin-induced agranulocytosis. *J. Am. med. Ass.* **240**, 2632.

KAHN, S.B., FEIN, S.A., AND BRODSKY, I. (1968). Effects of trimethoprim on folate metabolism in man. *Clin. Pharmacol. Ther.* **9**, 550–60.

KALLENBERG, C.G.M. AND ROENHORST, H.W. (1979). Bijwerkingen van Geneesmiddelen: Agranulocytose door niflaminezuur (Inflaryl). *Ned. T. Geneesk.* **123**, 1207–8.

KANEFSKY, T.M. AND MEDOFF, S.J. (1980). Stevens-Johnson syndrome and neutropenia with chlorpropamide therapy. *Arch. intern. Med.* **140**, 1543.

KANTERO, I., MUSTALA, O., AND PALVA, I.P. (1972). Drug-induced agranulocytosis with special reference to aminophenazone. *Acta. med. scand.* **192**, 327–30.

KAPSCH, D.N., ADELSTEIN, E.H., RHODES, G.R., AND SILVER, D. (1979). Heparin-induced thrombocytopenia, thrombosis and haemorrhage. *Surgery* **86**, 148–55.

KATTAMIS, C.A. AND TJORTJATOU, F. (1970). The haemolytic process of viral hepatitis in children with normal or deficient glucose-6-phosphate dehydrogenase activity. *J. Pediat.* **77**, 422–30.

KELTON, J.G., HUANG, A.T., MOLD, N., LOGUE, G., AND ROSSE, W.F. (1979). The use of *in vitro* technics to study drug-induced pancytopenia. *New Engl. J. Med.* **301**, 621–4.

KESARWALA, H.H., RAHILL, W.J., AND ARARAM, N. (1981). Vancomycin induced neutropenia. *Lancet* **i**, 1423.

KHAN, A.H., CARLETON R.A., AND CHOWN, M. (1983). Aprindine therapy for refractory ventricular tachycardia. *Arch. intern. Med.* **143**, 229–32.

KHOO, K-K. (1981). The treatment of malaria by glucose-6-phosphate dehydrogenase deficient patients in Sabah. *Ann. trop. Med. Parasitol.* **75**, 591–5.

KINNEY, E.L., RALLARD, J.O., CARLIN, B., AND ZELIS, R. (1983). Amrinone-mediated thrombocytopenia Scand. J. Haematol. **31**, 376–80.

KIRTLAND, H.H., MOHLER, D.N., AND HORWITZ, D.A. (1980). Methyldopa inhibition of suppressor-lymphocyte function. *New Engl. J. Med.* **302**, 825–32.

KLEIN, M. (1963). Agranulocytosis secondary to chlorthalidone therapy. *J. Am. med. Ass.* **184**, 138–9.

KLIPSTEIN, F.A. (1964). Subnormal serum folate and macrocytosis associated with anticonvulsant drug therapy. *Blood* **23**, 68–86.

KLOTZ, S.A. AND KAY, B.F. (1978). Cimetidine and agranulocytosis. *Ann. intern. Med.* **88**, 579–80.

KOBRINSKY, N.L. AND RAMSAY, N.K.C. (1981). Acute megaloblastic anaemia induced by high-dose trimethoprim-methoxasole. *Ann. intern. Med.* **94**, 780–1.

KÖHLER, G.D.C. (1980). Antiarrhythmic agents and agranulocytosis. *Lancet* **i**, 1415–16.

KORNBERG, A. AND POLLIACK, A. (1978). Paracetamol-induced thrombocytopenia and haemolytic anaemia. *Lancet* **ii**, 1159.

KORSAGER, S. (1978). Haemolysis complicating ibuprofen treatment. *Br. med. J.* **1**, 79.

——, SØRENSEN, H., JENSEN, O.H., AND FALK, J.V. (1981). Antiglobulin-tests for detection of auto-immunohaemolytic anaemia during long-term treatment with ibuprofen. *Scand. J. Rheumatol.* **10**, 174–6.

KORST, D.R. (1959). Agranulocytosis caused by phenothiazine derivatives. *J. Am. med. Ass.* **170**, 2076–81.

KRACKE, R.R. (1938). Relation of drug therapy to neutropenic states. *J. Am. med. Ass.* **111**, 1255–9.

KRAFFT, T., PUGIN, P., AND MIESCHER, P.A. (1978). Agranulocytose et cloxacilline intraveineuse. *Schweiz. med. Wochenschr.* **108**, 1821–3.

KUMAR, K. AND KUMAR, A. (1981). Reversible neutropenia associated with ampicillin therapy in pediatric patients. *Drug Intel. clin. Pharm.* **15**, 802–6.

KUTTI, J. AND WEINFELD, A. (1968). The frequency of thrombocytopenia in patients with heart disease treated with oral diuretics. *Acta med. scand.* **183**, 245–50.

LAI, M., ROSNER, F., AND RITZ, N.F. (1966). Hemolytic anemia due to antibodies to penicillin. *J. Am. med. Ass.* **198**, 483–4.

LAKSHMINARAYAN, S., SAHN, S.A., AND HUDSON, L.D. (1973). Massive haemolysis caused by rifampicin. *Br. med. J.* **2**, 282.

LALEZARI, P., LOUIE, J.E., AND FADLALLAH, N. (1982). Serological profile of alpha-methyldopa-induced hemolytic anemia: correlation between cell-bound IgM and hemolysis. *Blood* **59**, 61–8.

LANCET (1983). Editorial: drug-induced neutropenia. *Lancet* i, 857–8.

LASSEN, H.C.A., NEUKIRCH F., AND KRISTENSEN, H.S. (1956). Treatment of tetanus; severe bone-marrow depression after prolonged nitrous-oxide anaesthesia. *Lancet* i, 527–30.

LEVERE, R.D. AND IBRAHAM, N.G. (1982). The bone marrow as a metabolic organ. *Am. J. Med.* **73**, 615–6.

LEVI, J.A., ARONEY, R.S., AND DALLEY, D.N. (1981). Haemolytic anaemia after cisplatin treatment. *Br. med. J.* **282**, 2003–4.

LEVIN, A.S., WEINER, R.S., FUDENBERG, H.H., SPATH, P., AND PETZ, L. (1971). Granulocytopenia caused by anti-cephalothin antibodies. *Clin. Res.* **19**, 424.

LEVITT, B.H., GOTTLIER, A.J., ROSENBERG, I.R., AND KLEIN, J.J. (1964). Bone-marrow depression due to methicillin, a semi-synthetic penicillin. *Clin. Pharmacol. Ther.* **5**, 301–6.

LEVRAT, R., JOLOT, A.Y., DURAND, D.V., AND MARTIN, D. (1978). Neutropenie prolongée au cours d'un traitement par le métronidazole. *Nouve Presse Méd.* **7**, 3053–4.

LEVY, M. (1980). Epidemiological evaluation of rare side-effects of mild analgesics. *Br. J. clin. Pharmacol.* **10**, 3955–95.

LEY, A.B., CAHAM, A., AND MAYER, K. (1959). In *Proceedings of the Seventh Congress of the International Society of Blood Transfusion*, Basel, p. 539.

LIEGNER, K.B., BECK, E.M., AND ROSENBERG, A. (1981). Laetrile-induced agranulocytosis. *J. Am. Med. Ass.* **246**, 2841–2.

LITTLEJOHN, G.O. AND UROWITZ, M.B. (1979). Cimetidine, lupus erythematosus and granulocytopenia. *Ann. intern. Med.* **91**, 317–8.

LIMA, M.S.S. (1978). Transient neutropenia and cimetidine. *Gastroenterology* **74**, 102–3.

LINDBERG, L.G. AND NORDEN, A. (1961). Severe haemolytic reaction to chlorpromazine. *Acta. med. scand.* **170**, 195–9.

LOGUE, G.L., BOYD., A.C., AND ROSSE, W.F. (1970). Chlorpropamide-induced immune haemolytic anaemia. *New Engl. J. Med.* **283**, 900–4.

LOWE, J., GRAY, J. HENRY, D.A., AND LAWSON, D.H. (1979). Adverse reactions to frusemide in hospital in patients. *Br. med. J.* **2**, 360–2.

LUDERER, J.R., SCHOOLWERTH, A.C., SINICROPE, R.A., BALLARD, J.O., LOOKINGBILL, D.P., AND HAYES, A.H. JR. ((1981). Acute renal failure, hemolytic anemia and skin rash associated with cap-

topril therapy. *Am. J. Med.* **71**, 493–6.

LUNDH, B., AND HESSEL GREN, K.H. (1979) Haematological side-effects from anti-hypertensive drugs. *Acta. med. scand.* **628**, 73–5.

LYMAN, G.H. (1980). Adjunctine agents in systemic chemotherapy. *Compr. Ther.* **6**, 16–25.

MACDOUGALL, L.G. (1982). Pure red cell aplasia associated with sodium valproate therapy. *J. Am. med. Ass.* **247**, 53–4.

MAGEE, F., O'SULLIVAN, H., AND MCCANN, S.R. (1981). Megaloblastosis and low-dose trimethoprim–sulphamethoxazole. *Ann. intern. Med.* **95**, 657.

MAGNUSSON, B. AND RÖDJER, S. (1980). Alprenolol-induced thrombocytopenia. *Acta med. scand.* **207**, 231–3.

MAHMOOD, T. AND ROBINSON, W.A. (1977). Effect of levamisole on human granulopoiesis *in vitro. Proc. Soc. exp. Biol. Med.* **156**, 359–64.

MANTEN, A. (1968). Antibiotic drugs. In *Side-effects of drugs*, Vol. VI (ed. L. Meyler and A. Herxheimer). Excerpta Medica Foundation, Amsterdam.

MARKS, J. (1981). Reactions to dapsone. *Lancet* ii, 585.

MAZZA, J.J. AND KRYDA, M.D. (1980). Tetracycline-induced hemolytic anemia. *J. Am. Acad. Dermatol.* **2**, 506–8.

MARSH, G.W. (1966). Mechanical haemolytic anaemia after mitral-valve replacement. *Br. med. J.* **2**, 31–2.

MARTELO, O.J., KATIMS, R.B., AND YUNIS, A.A. (1967). Bone-marrow aplasia following propylthiouracil therapy. Report of a case with complete recovery. *Arch. intern. Med.* **120**, 587–90.

MATSUMOTO, N., *ET AL.* (17 AUTHORS) (1976). Granulocytopenia associated with anti-thyroid drugs. *Jap. J. Clin. Haematol.* **171**, 82–7.

McCARTY, D.J., BRILL, J.M., AND HARROP, D. (1962). Aplastic anemia secondary to gold-salt therapy. *J. Am. med. ass.* **179**, 655–7.

McCONKEY, B. (1981). Adverse reactions to dapsone. *Lancet* ii, 525.

McCURDY, P.R. (1961). Chloramphenicol bone-marrow toxicity. *J. Am. med. Ass.* **178**, 588–93.

McEWEN, J. AND STRICKLAND, W.J. (1982). Mebhydrolin napadisylate: a possible cause of reversible agranulocytosis and neutropenia. *Med. J. Austral.* **2**, 523–5.

McGRATH, K.M., ZALCBERG, J.R., SLONIM, J., AND WILEY, J.S. (1982). Intralipid-induced haemolysis. *Br. J. Haematol.* **50**, 376–8.

McGUIGAN, J.E. (1981). A consideration of the adverse effects of cimetidine. *Gastroenterology* **80**, 181–2.

McGUIGAN, M.A. (1981). Benzocaine-induced methemoglobinemia. *Can. med. Ass. J.* **125**, 816.

McHARG, A.M. AND MCHARG, J.F. (1979). Leucopenia in association with mianserin therapy. *Br. med. J.* **1**, 623–4.

McKENDRICK, M.W. AND GEDDES, A.M. (1979). Neutropenia associated with metronidazole. *Br. med. J.* **2**, 795.

McMURDOCH, J.MCC., SPEIRS, C.F., AND MACE, M. (1968). Fatal marrow aplasia after chlorpropamide and methyldopa. *Lancet* i, 207.

MERENSTEIN, G.B., O'LOUGHLIN, E.P., AND PLUNKET, D.C. (1970). Effects of maternal thiazides on platelet counts of newborn infants. *J. Paediatr.* **76**, 766.

MEYER, R. AND AXELROD, J.L. (1974). Fatal aplastic anaemia resulting from flucytosine. *J. Am. med. Ass.* **228**, 1573.

MEYLER, L., POLAK, B.C.P., SCHUT, D., WESSELING, H., AND HERXHEIMER, A. (1974). Blood dyscrasias attributed to chloramphenicol. *Postgrad. med. J.* **50**, (Suppl. 5), 123–6.

MIELANTS, H. AND VEYS, E.M. (1978). A study of the haematological side effects of levamisole in rheumatoid arthritis with recommendations. *J. Rheumatol.* **5**, 77–83.

MIESCHER, P. AND MIESCHER, A. (1952). Die Sedormid-Anaphlyaxie. *Schweiz. med. Wochenschr.* **82**, 1279–82.

MILLER, J.L. (1980). Marrow aplasia and sulindac. *Ann. intern. Med.* **92**, 129.

MILMAN, N. AND ISAGER, H. (1978). Agranulocytose efter indtagelse at aminofenazonholdig udenlandsk hond kobsmedicin (Optalidon). *Ugeskr. Laeg.* **140**, 1168–9.

MISKOVITZ, P.F. AND JAVIT, N.B. (1980). Leucopenia associated with mebendazole therapy of hydatid disease. *Am. J. trop. Med. Hyg.* **29**, 1356–8.

MODAN, B., SEGAL, S., SHANI, M., AND SHEBA, C. (1975). Aplastic anaemia in Israel: evaluation of the etiological role of chloramphenicol on a community-wide basis. *Am. J. med. Sci.* **270**, 441–5.

MOESCHLIN, S. AND WAGNER, K. (1952). Agranulocytosis due to occurrence of leukocyte-agglutinins. *Acta. haematol., Basel* **8**, 29–41.

MUNGALL, I.P.F. AND STANDING, V.F. (1978). Eosinophilia caused by rifampicin. *Chest* **74**, 321–2.

MULHALL, A., DE LOUVOIS, J., AND HURLEY, R. (1983). Chloramphenicol toxicity in neonates; its incidence and prevention. *Br. med. J.* **287**, 1424–7.

MURPHY, M.F., RIORDAN T., MINCHINTON, R.M., CHAPMAN, J.F., AMESS, J.A.L., SHAW, E.J., AND WATERS, A.H. (1983). Demonstration of an immune-mediated mechanism of penicillin-induced neutropenia and thrombocytopenia. *Br. J. Haematol.* **55**, 155–60.

MURRAY, R.M., TIMBURY, G.C., AND LINTON, A.L. (1970). Analgesic abuse in psychiatric patients. *Lancet* **i**, 1303–5.

NAGAO, T. AND MAUER, A.M. (1969). Concordance for drug-induced aplastic anaemia in identical twins. *New Engl. J. Med.* **281**, 7–11.

NAHATA, M.C., DEBOLT, S.L., AND POWELL, D.A. (1982). Adverse effects of methicillin, nafcillin, and oxacillin in pediatric patients. *Dev. Pharmacol. Ther.* **4**, 117–23.

NAJMAN, A., GORIN, N.C., DUHAMEL, G., ROGER, M., AND DRY, J. (1980). Myélofibrose aiguë aprés traitement prolongé dùne sclerose en plaques par le chlorambucil. *Nouv. Presse Méd.* **9**, 1897.

NARAQI, S. AND RAISER, M. (1982). Nonrecurrence of cephalothin associated granulocytopenia and thrombocytopenia. *J. infect. Dis.* **145**, 281.

NEFTEL, K.A., WÄLTI, M., SPENGLER, H., VON FELTEN, A., WEITZMAN, S.A., BÜRGI, H., AND DE WECK, A.L. (1981). Neutropenia after penicillins: toxic or immune-mediated *Klin. Wochenschr.* **59**, 877–88.

NESMITH, L.W. AND DAVIS, J.W. (1968). Hemolytic anaemia caused by penicillin. *J. Am. med. Ass.* **203**, 27–30.

NEWTON, R.M. AND WARD, V.G. (1958). Leukopenia associated with ristocetin (spontin) administration. *J. Am. med. Ass.* **166**, 1956–9.

NIEWIG, H.O., BOUMA, K. DE VRIES, AND JANSZ, A. (1963). Haematological side-effects of some anti-rheumatic drugs. *Ann. rheum. Dis.* **22**, 440.

NIXON, D.D. (1972). Thrombocytopenia following doxepin treatment. *J. Am. med. Ass.* **220**, 418.

O'DONOHUE, W.J., MOSS, L.M., AND ANGELILLO, V.A. (1980). Acute methemoglobinemia induced by topical benzocaine and lidocaine. *Arch. intern. med.* **140**, 1508–9.

OLIN, J. AND GRAOR, R. (1981). Heparin-associated thrombocytopenia. *New Engl. J. Med.* **304**, 609.

OLSON, M.L. AND McEVOY, G.K. (1981). Methemoglobinemia induced by local anesthetics. *Am. J. hosp. Pharmacol.* **38**, 89–93.

OLSEN, V.V., LOFT, S., AND CHRISTENSEN, K.D. (1982). Serious reactions during malaria prophylaxis with pyrimethamine–sulfadoxine. *Lancet* **ii**, 994.

OPIE, L.H. (1980a). Antiarrhythmic agents. *Lancet* **i**, 861–68.

—— (1980b). Aprindine and agranulocytosis. *Lancet* **ii**, 689–90.

ORY, E.M. AND YOW, F.M. (1963). The use and abuse of the broad-spectrum antibiotics. *J. Am. med. Ass.* **185**, 273–9.

PAGE, C.E. (1982). Mianserin-induced agranulocytosis. *Br. med. J.* **284**, 1912–13.

PARKINSON, I.S., WARD, M.K., AND KERR, D.N.S. (1981). Dialysis encephalopathy, bone disease, and anaemia; the aluminium intoxication syndrome during regular haemodialysis. *J. clin. Pathol.* **34**, 1285–94.

PARMENTIER, C., TCHERNIA, G., SUBTIL, E., DIAKHATE, L., AND MORARDET, N. (1978). *In vitro* medullary granulocytic progenitor (CFU-C). Cultures from 6 cases of granulocytopenias. *Scand. J. Haematol.* **21**, 19–23.

PASSOFF, T.L. AND SHERRY, H.S. (1978). Oxacillin-induced neutropenia. *Clin. Orthop.* **135**, 69–70.

PEARSON, J.R., BINDER, C.L., AND NEBER, J. (1955). Agranulocytosis following diamox therapy. *J. Am. med. Ass.* **157**, 339–41.

PENGELLY, C.D.R. (1963). Dapsone induced haemolysis. *Br. med. J.* **2**, 662–4.

PERRY, H.M. (1981). Possible mechanisms of the hydralazine-related lupus-like syndrome. *Arthritis Rheumatol.* **24**, 1093–105.

PETZ, L.D. (1980). Drug-induced immune haemolytic anaemia. *Clinics Haematol.* **9**, 455–82.

—— AND FUDENBEG, F.D. (1966). Coomb's-positive hemolytic anemia caused by penicillin administration. *New Engl. J.* **274**, 171–8.

PISCIOTTA, A.V. (1969). Agranulocytosis induced by certain phenothiazine derivatives. *J. Am. med. Ass.* **208**, 1862–8.

PISCIOTTA, A.V. (1971). Drug-induced leukopenia and aplastic anaemia. *Clin. Pharmacol. Ther.* **12**, 13–43.

PITCHER, C.S. AND WILLIAMS, R. (1963). Reduced red-cell survival in jaundice and its relation to abnormal glutathione metabolism. *Clin. Sci.* **24**, 239–52.

PLANAS, A.T., KRANWINKEL, R.N., SOLETSKY, H.B., AND PEZZIMENTI, J.F. (1980). Chlorpropamide-induced pure RBC aplasia. *Arch. intern. Med.* **140**, 707–8.

POLK, O.D., KLETTER, G.G., SMITH, J., AND CASTRO, O. (1982). Impaired clot retraction in thrombocytopenia due to methyldopa. *South. med. J.* **75**, 374–5.

POSNETT, D.N., STEIN, R.S., GRABER, S.E., AND KRANTZ, S.B. (1979). Cimetidine-induced neutropenia. A possible dose-related phenomenon. *Arch. intern. Med.* **139**, 584–6.

POSTELNICK, M. AND GASKINS, J.D. (1981). Penicillin G-induced granulocytopenia. *Drug Intel. clin. Pharm.* **15**, 289.

POTTER, J.L. AND HILLMAN, J.V. (1979). Benzocaine-induced methemoglobinemia. *J. Am. Coll. emerg. Med.* **8**, 26–7.

POUNDER, R.E., CRAVEN, E.R., HENTHORN, J.S., AND BANNATYNE, J.M. (1975). Red-cell abormalties associated with sulphasalazine maintenance therapy for ulcerative colitis. *Gut* **16**, 181–5.

POWERS, P.J., CUTHBERT, D., AND HIRSH, J. (1979). Thrombocytopenia found uncommonly during heparin therapy. *J. Am. med. Ass,* **241**, 2396–7.

PRESCOTT, L.E. (1968). Antipyretic and analgesic drugs. In *Side-effects of drugs*. Vol. VI (ed. L. Meyler and A. Hexheimer). Excerpta Medica Foundation, Amsterdam.

—— ILLINGWORTH, R.N., CRITCHLEY, J.A.J.H., FRAZER, L., AND STIRLING, M.L. (1980). Acute haemolysis and renal failure after nomifensine overdosage. *Br. med. J.* **281**, 1392-3.

PRUSEK, W., NAWROCKA, E., KOZIEROWSKA, B., AND KOSINSKI, S. (1980). Exudative erythema with bone marrow aplasia after treatment with rondomycin. *Polsk. Tyg. Lek* **35**, 503-4.

RAMSAY, R. AND GOLDE, D.W. (1976). Aplastic anaemia from veterinary phenylbutazone. *J. Am. med. Ass.* **236**, 1049.

RATE, R., BONNELL, M., CHERVENAK, C., AND PAVINICH, G. (1979). Cimetidine and haematologic effects. *Ann. intern. Med.* **91**, 795.

RATZAN, R.J., MOORE, M.A.S., AND YUNIS, A.A. (1974). Effect of chloramphenicol and thiamphenicol on the *in vitro* colony-forming cell. *Blood* **43**, 363-9.

REES, J.K.H. (1980). Availability of amidopyrine preparations. *Lancet* **ii**, 581-2.

REID, G. AND PATTERSON, A.C. (1977). Pure red-cell aplasia after gold treatment. *Br. med. J.* **4**, 1457.

RETSAS, S., PHILLIPS, R.H., HANHAM, I.W.E., AND NEWTON, K.A. (1978). Agranulocytosis in breast cancer patients treated with levamisole. *Lancet* **ii**, 324-5.

REYNOLDS, E.H., HALLPIKE, J.E., PHILLIPS, B.M., AND MATTHEWS, D.M. (1966b). Reversible absorptive defects in anticonvulsant megaloblastic anaemia. *J. clin Pathol.* **18**, 593-8.

——, MILNER, G., MATTHEWS, D.M., AND CHANARIN, I. (1966a). Anticonvulsant therapy, megaloblastic haemopoiesis, and folic-acid metabolism. *Quart. J. Med.* **35**, 521-37.

RHIND, E.G. AND VARADI, S. (1954). Megaloblastic anaemia due to phenytoin sodium. *Lancet* **ii**, 921.

RHOADES, E.R. (1982). Seminar on antibiotics VIII, chloramphenicol. *J. Okla. State med. Ass.* **75**, 8-10.

RIBERA, A., MONASTERIO, G., ACEBEDO, G., TRIGINER, J., AND MARTIN, C. (1981). Dipyrone-induced haemolytic anaemia. *Vox. Sang.* **41**, 32-5.

RICHTER, J.E., GERHARDT, D.C., PASQUALE, D.N., AND CASTELL, D.O. (1980). Cimetidine and hematologic suppression: things are not always as they appear. *Dig. Dis. Sci.* **25**, 960-3.

RIES, C.A. AND SAHUD, M.A. (1975). Agranulocytosis caused by Chinese herbal medicines. *J. Am. med. Ass.* **231**, 352-5.

RINKOFF, S.S. AND SPRING, M. (1941). Toxic depression of the myeloid elements following therapy with the sulphonamides: report of 8 cases. *Ann. intern. Med.* **15**, 89-107.

RITZ, N.D. AND FISHER, M.J. (1960). Agranulocytosis due to administration of salicylazosulfapyridine (azulfidine). *J. Am. med. Ass.* **172**, 237-40.

ROBERTS, P.D., HOFFBRAND, A.V., AND MOLLIN, D.L. (1966). Iron and folate metabolism in tuberculosis. *Br. med. J.* **2**, 198-202.

RODGERS, G.M. AND SHUMAN, M.A. (1982). Acquired cyclic neutropenia: successful treatment with prednisone. *Am. J. Haematol.* **13**, 83-9.

ROMERIL, K.R., DUKE, D.S., AND HOLLINGS, P.E. (1981). Sulindac-induced agranulocytosis and bone-marrow culture. *Lancet* **ii**, 523.

ROSENBACH, L.M., CAVILES, A.P., AND MITUS, W.J. (1960). Chloramphenicol toxicity: Reversible vacuolization of erythroid cells. *New Engl. J. Med.* **263**, 724-8.

ROSENBAUM, J.T. AND O'CONNOR, M. (1981). Thrombocytopenia associated with sulindac. *Arthritis Rheumatol.* **24**, 753-4.

ROSENBERG, I.H., GODWIN, H.A., STREIFF, R.R., AND CASTLE, W.B. (1968). Impairment of intestinal deconjugation of dietary folate. *Lancet* **ii**, 530-2.

ROSENBLOOM, D. AND GILBERT, R. (1981). Reversible flu-like syndrome, leucopenia and thrombocytopenia induced by allopurinol. *Drug. Intel. clin. Pharm.* **15**, 286-7.

ROY, A. AND GHOSH, M.L. (1981). Coomb's positive haemolytic anaemia due to methyldopa. *Br. J. clin. Pract.* **35**, 54-5.

RUBIN, S.A., LEE, S., O'CONNOR, L., HUBENETTE, A., TOBER, J., AND SWAN, H.J. (1979). Thrombocytopenia and fever in a patient taking amrinone. *New Engl. J. Med.* **301**, 1185.

RUUSKANEN, O., REMES, M., MÄKELÄ, A.-L., ISOMÄKI, H., AND TOIVANEN, A. (1976). Levamisole and agranulocytosis. *Lancet* **ii**, 958-9.

RYAN, G.M.S. AND FORSHAW, J.W.B. (1955). Megaloblastic anaemia due to phenytoin sodium. *Br. med. J.* **2**, 242-3.

SACHER, R.A., WOOLLEY, P.V., PRIEGO, V.P., SCHANFIELD, M.S., AND BONNEM, E. (1981). Methotrexate-induced immune hemolytic anemia. *Transfusion* **21**, 625-6.

SAFFOURI, B., CHO, J.H., AND FELBER, N. (1981). Chlorpropamide-induced haemolytic anaemia. *Postgrad. med. J.* **57**, 44-5.

SAIDI, P. AND WALLERSTEIN, R.O. (1960). Effect of chloramphenicol on erythropoiesis. *Clin. Res.* **8**, 131.

——, ——, AND AGGELER, P.M. (1961). Effect of chloramphenicol on erythropoiesis. *J. Lab. clin. Med.* **75**, 247-56.

ST. JOHN, M.A. AND PROBER, C.G. (1981). Side-effects of cloxacillin in infants and chidren. *Can. med. Ass. J.* **125**, 458-60.

SALOKANNEL, S.J., PALVA, I.P., AND TAKKUNEN, J.T. (1970). Malabsorption of vitamin B_{12} during treatment with slow-release potassium chloride. *Acta med. scand.* **187**, 431-2.

SANDZA, J.G., ROBERTS, R.W., SHAW, R.C., AND CONNORS, J.P. (1979). Symptomatic methemoglobinemia with a commonly used topical anesthetic, Cetacaine. *Ann. Thoracic Surg.* **30**, 187-90.

SANZ, M.A., MARTINEZ, J.A., GOMIS, F., AND GARCIA-BORRAS, J.J. (1980). Sulindac-induced bone marrow toxicity. *Lancet* **ii**, 802-3.

SAYED, H.M., DACIE, J.V., HANDLEY, D.A., LEWIS, S.M., AND CLELAND, W.P. (1961). Haemolytic anaemia of mechanical origin after open-heart surgery. *Thorax* **16**, 356-60.

SAZIE, E. AND JAFFE, J.P. (1980). Severe granulocytopenia with cimetidine and phenytoin. *Ann. intern. Med.* **93**, 151-2.

SCHATTNER, A., SHTALRID, M., LEVY, R., AND BERREBI, A. (1981). Fatal aplastic anemia due to indomethacin-lymphocyte transformation tests *in vitro*. *Isr. J. med. Sci.* **17**, 433-6.

SCHEINBERG, I.H. (1979). Thrombocytopenic reaction to aspirin and acetaminophen. *New Engl. J. Med.* **300**, 678.

SCHEY, S.A. AND KAY, H.E.M. (1984). Myelosuppression complicating co-trimoxazole prophylaxis after bone-marrow transplantation. *Br. J. Haematol.* **56**, 179-80.

SCHIFF, D., ARANDA, J.V., AND STERN, L. (1970). Nconatal thrombocytopenia and congenital malformations associated with administration of tolbutamide to the mother. *J. Pediat.* **77**, 457-8.

SCHMIDT, K.L. AND MUELLER-ECKHARDT, C. (1977). Agranulocytosis, levamisole and HLA-B27. *Br. med. J.* **2**, 85.

SCHREML, W.P. AND LOHRMANN, H.P. (1979). No effects of levamisole on cytotoxic drug-induced changes of human granulopoiesis. *Blut* **38**, 331-6.

SCOTT, G.L., MYLES, A.B., AND BACON, P.A. (1968). Autoimmune haemolytic anaemia and mefanamic acid therapy. *Br. med. J.* **3**, 534-5.

SELDON, M.R., BAIN, B., JOHNSON, C.A., AND LENNOX, C.S. (1982). Ticarcillin-induced immune haemolytic anaemia. *Scand. J. Haematol.* **28**, 459–60.

SHELTON, J.G., KINGSTON, W.R., AND MCRAE, C. (1960). Aplastic anaemia and agranulocytosis following chlorpromazine therapy. *Med. J. Aust.* **1**, 130–1.

SHERMAN, J.M. AND SMITH, K. (1979). Methemoglobinemia owing to rectal probe lubrication. *Am. J. Dis. Childh.* **133**, 439.

SHINTON, N.K. AND WILSON, C. (1960). Autoimmune haemolytic anaemia due to phenacetin and p-amino-salicylic acid. *Br. med. J.* **1**, 226.

SHOENFELD, Y., SHAKLAI, M., LIVNI, E., AND PINKHAS, J. (1980). Thrombocytopenia from acetaminophen. *New Engl. J. Med.* **303**, 47.

SHULMAN, N.R. (1964). A mechanism of cell destruction in individuals sensitized to foreign antigens and its implications in auto-immunity. *Ann. intern. Med.* **50**, 506–21.

SIDI, Y., LIVNI, E., SOLOMON, F., AND PINKHAS, J. (1981). Ampicillin-induced neutropenia. *Haematologia* **66**, 216–19.

SIGLER, A.T., FORMAN, E.N., ZINKHAM, W.H., AND NEILL, CATHERINE A. (1963). Severe intravascular haemolysis following surgical repair of endocardial cushion defects. *Am. J. med.* **35**, 467–80.

SIMON, S.D. AND KOSMIN, M. (1978). Fenoprofen and agranulocytosis. *New Engl. J. Med.* **299**, 490.

SIMPSON, R.E., GOLDSTEIN, D.J., HJELTE, G.S., AND EVANS, E.R. (1978). Acute thrombocytopenia associated with fenoprofen. *New Engl. J. Med.* **298**, 629–30.

SIVE, J., GREEN, R., AND METZ, J. (1972). Effect of trimethoprim on folate-dependent DNA synthesis in human bone marrow. *J. clin Pathol.* **25**, 194–7.

SKACEL, P.O., HEWLETT, A.M., LEWIS, J.D., LUMB, M., NUNN, J.F., AND CHANARIN, I. (1983). Studies on the haemopoietic toxicity of nitrous oxide in man. *Br. J. Haematol.* **53**, 189–300.

SMITH, C.S., CHINN, S., AND WATTS, R.W.E. (1977). The sensitivity of human bone marrow granulocyte/monocyte precursor cells to phenylbutazone, oxyphenbutazone and gamma-hydroxphenylbutazone *in vitro* with observations on the bone marrow colony formation in phenylbutazone-induced granulocytopenia. *Biochem. Pharmacol.* **26**, 847–52.

SMITH, J.A. (1980). Neutropenia associated with metronidazole therapy. *Can. med. Ass. J.* **123**, 202.

SMITH, P.J. AND SWINBURN, W.R. (1980). Adverse reactions to D-penicillamine after gold toxicity. *Br. med. J.* **2**, 617.

——, ——, SWINSON, D.R., AND STEWART, I.M. (1982). Influence of previous gold toxicity on subsequent development of penicillamine toxicity. *Br. med. J.* **285**, 595–6.

SMITH, R.S. AND ALEXANDER, S. (1959). Heinz-body anaemia due to dapsone. *Br. med. J.* **1**, 625–7.

SMITHURST, B.A. (1981). Reactions to dapsone. *Lancet* **ii**, 585.

SPAIN, D.M. AND CLARK, T.B. (1946). A case of agranulocytosis occurring during the course of penicillin therapy. *Ann. intern. Med.* **25**, 732–3.

SPIEL, R., ENENKEL, W., AND FISCHER, M. (1977). Agranulocytosis after antiarrhythmic therapy with ajmalin. *Z. Kardiol.* **66**, 402–4.

SPITZER, T.R. (1981). Penicillin-induced haemolytic anaemia with negative direct antiglobulin test. *Lancet* **i**, 1361–2.

STAESSEN, J., BOOGAERTS, M., FAGARD, R., AND AMERY, A. (1981). Mechanism of captopril-induced agranulocytosis. *Acta clin. belg.* **36**, 87–90.

——, FAGARD, R., LIJNEN, P., AND AMERY, A. (1980). Captopril and agranulocytosis. *Lancet* **i**, 926–7.

STEFANINI, M. AND JOHNSON, N.L. (1970). Positive antihuman globulin test in patients receiving carbromal. *Am. J. med. Sc.* **359**, 49–55.

STEIN, J.H., HAMILTON, H.E., AND SHEETS, R.F. (1964). Agranulocytosis caused by chlorpropamide. *Arch. intern. Med.* **113**, 186–90.

STEIN, P.B. AND INWOOD, M.J. (1980). Hemolytic anemia associated with chlorpromazine therapy. *Can. J. Psychiat.* **25**, 659–61.

STOEL, I. AND HAGEMEIJER, F. (1980). Aprindine: a review. *Eur. Heart J.* **1**, 147–56.

STRICKER, B.H.Ch., AND OIE, T.T. (1984). Agramlocytosis caused by spironolactore, *Bir. med. J.,* **289**, 371.

STRUMIA, P.V. AND RAYMOND, F.D. (1962). Acquired haemolytic anaemia and antipenicillin antibody. *Arch. intern. Med.* **109**, 603–8.

STRYCKMANS, P.A., RONGE-COLLARD, E., DELFORGE, A., LAMBERT, M., AND SUCIU, S. (1982). Effect of two anti-arrhythmic drugs, aprindine and moxaprindine, on the replication capacity of murine and human haemopoietic cells. *Scand. J. Haematol.* **29**, 331–7.

SULLIVAN, A.L., BURAKOFF, R., AND WEINTRAUB, L.R. (1981). Sideroblastic anemia associated with penicillamine therapy. *Arch. intern. med.* **141**, 1713–4.

SUTHERLAND, R., VINCENT, P.C., RAIK, E., AND BURGESS, K. (1977). Quinine-induced agranulocytosis; toxic effect of quinine bisulphate on bone-marrow culture *in vitro*. *Br. med. J.* **1**, 605–7.

SYMOENS, J., VEYS, E., MIELANTS, M., AND PIENALS, R. (1978). Adverse reactions to levamisole. *Cancer treat. Rep.* **62**, 1721–30.

TAFANI, O., MAZZOLI, M., LANDINI, G., AND ALTERINI, B. (1982). Fatal acute immune haemolytic anaemia due to nalidixic acid. *Br. med. J.* **285**, 936–7.

TAGUCHI, H., LAUNDY, M., REID, C., REYNOLDS, E.H., AND CHANARIN, I. (1977). The effect of anticonvulsant drugs on thymidine and deoxyribosenucleic acid synthesis by human marrow cells. *Br. J. Haematol.* **36**, 181–7.

TAIT, G.P. (1957). Fatal agranulocytosis during carbimazole therapy. *Lancet* **i**, 303.

TAKAHASHI, H. AND TSUKADA, T. (1979). Triamterene-induced immune haemolytic anaemia with acute intravascular haemolysis and acute renal failure: *J. scand. Haematol.* **23**, 169–76.

TARTAS, N.E., BULLORSKY, E.O., JORGE, E.J.H., AND AVALOS, J.C.S. (1981). Pancytopenia induced by cephalothin. *J. Am. med. Ass.* **245**, 1148–9.

TAYLOR, J.A.T. (1965). Metronidazole and transient leucopenia. *J. Am med. Ass.* **194**, 1331.

TEERENHOVI, L., HEINONEN, E., GRÖHN, P., KLEFSTRÖM, P., MEHTONEN, M., AND THLIKAINEN, A. (1978). High frequency of agranulocytosis in breast-cancer patients treated with levamisole. *Lancet* **ii**, 151–2.

TELHAG, H. (1973). Niflumic acid in osteoarthrosis. Long-term tolerance. *Scand. J. Rheumat.* (Suppl.) 16–19.

TERRITO, M.C., PETERS, R.W., AND TANAKA, K.R. (1973). Autoimmune haemolytic anaemia due to levodopa therapy. *J. Am. med. Ass.* **226**, 1347–8.

THIRKETTLE, J.L., GOUGH, K.R., AND READ, A.E. (1963). Agranulocytosis associated with sulphasalazine (salazopyrin) therapy. *Lancet* **i**, 1395–7.

THOMPSON, J.S., HERBICK, J.M., KLASSEN, L.W., SEVERSON, C.D., OVERLIN, V.I., BLASCHKE, J.W., SILVERMAN, M.A., AND VOGEL, C.L. (1980). Studies on levamisole-induced agranulocytosis. *Blood* **56**, 388–96.

TOMKIN, G.H. (1973). Malabsorption of vitamin B_{12} in diabetic patients treated with phenformin: a comparison with metformin. *Br. med. J.* **2**, 673.

TONKONOW, B. AND HOFFMAN, R. (1980). Aplastic anemia and cimetidine. *Arch. intern. Med.* **140**, 1123–4.

TOWNES, P.L., GEERTSMA, M.A., AND WHITE, M.R. (1977). Benzocaine-induced methemoglobinemia. *Am. J. Dis. Childh.* **131**, 897–8.

TREUSCH, P.J., WOELKE, B.J., LEICHTMAN, D., AND TUCKER, R.A. (1979). Agranulocytosis associated with fenoprofen. *J. Am. med. Ass.* **241**, 2700–1.

TULLOCH, A.L. (1976). Pancytopenia in an infant associated with sulfamethoxazole trimethoprim therapy. *J. Pediat.* **88**, 499–500.

UFBERG, M.H., BROOKS, C.M., BOSANAC, P.R., AND KINTZEL, J.E. (1977). Transient neutropenia in a patient receiving cimetidine. *Gastroenterology* **73**, 635–8.

VALENTA, L.J., ELIAS, A.N., AND WEBER, D.J. (1981). Hyperthyroidism and propylthiouracil-induced agranulocytosis during chronic lithium carbonate therapy. *Am. J. Psychiat.* **138**, 1605–7.

VAN ARDSEL, P.P. JR. AND GILLILAND, B.C. (1965). Anaemia secondary to penicillin treatment. Studies on two patients with 'non-allergic' serum hemagglutins. *J. Lab. clin. Med.* **65**, 277–85.

VAN BRUMMELEN, P., WILLEMZE, R., TAN, W.D., AND THOMPSON. (1980). Captopril-associated agranulocytosis. *Lancet* i, 150.

VANDERBROUCKE, J., JOOSENS, J.V., AND VERWILGHEN, R. (1962). Intravascular haemolysis. *Br. med. J.* **1**, 1696–7.

VAN LEEUWEN, E.F., ENGELFRIET, C.P., KR. VON DEM BORNE, A.E.G. (1982). Studies on quinine-and quinidine-dependent antibodies against platelets and their reaction with platelets in the Bernard–Soulier syndrome. *Br. J. Haematol.* **51**, 551–60.

VOGEL, C.L., LIPSCOMB, D.L., SILVERMAN, M.A., KERNS, A.L., MANSELL, P.W., AND SUGARBAKER, E.V. (1978) Levamisole granulocytopenia in patients receiving adjuvant chemoimmunotherapy programme after surgery for breast carcinoma with axillary lymph node involvement. *Cancer Treat. Rep.* **62**, 1587–9.

VON DÖLLE, W. AND SEWING, K. FR. (1980). Agranulocytosis after cimetidine after agranulocytosis induced by carbimazole in a woman with cirrhosis of the liver, hypersplemism and hyperthyroidism, 2. *Gastroenterology* **18**, 81–2.

VUOPIO, P., HÄRKÖNEN, M., HELSKE, T., AND NÄVERI, H. (1975). Red-cell glucose-6-phosphate dehydrogenase deficiency in Finland. *Scand. J. Haematol.* **15**, 145–52.

WALTER, N.M.A., WHITWORTH, J.A., AND KINKAID-SMITH, P. (1982). Clinical experience with the angiotensin converting enzyme inhibitor captopril. *Clin. Exp. Pharmacol. Physiol.* **7**, 117–21.

WANAMAKER, W.M., WANAMAKER, S.J., CELLSIA, G.G., AND KOELLER, A.A. (1976). Thrombocytopenia associated with long-term levodopa therapy. *J. Am. med. Ass.* **235**, 2217–19.

WARD, K. AND WEIR, D.G. (1981). Life-threatening agranulocytosis and toxic epidermal necrolysis during low dose penicillamine therapy. *Ir. J. med. Sci.* **150**, 252–3.

WAUTERS, J.P. (1975). Unusual complication of high-dose frusemide. *Br. med. J.* **2**, 624.

WEBB, D.L., CHODOS, R.B., MAHAR, C.O., AND FALOON, W.W. (1968). Mechanism of vitamin B_{12} malabsorption in patients receiving colchicine. *New Engl. J. Med.* **279**, 845–50.

WEBBER, J.C.P (1985). Committee of Safety of Medicines, personal communication.

WEBER, T.H., KNUUTILA, S., TAMMISTO, P., AND TONTTI, K. (1977). Long-term use of phenytoin: effects on whole blood and red-cell folate and haematological parameters. *Scand. J. Haematol.* **18**, 81–85.

WEINBERGER, K.A. (1979). Fenoprofen and red cell aplasia. *J. Rheumatol.* **6**, 473–4.

WEISBERGER, A.S. (1969). Mechanisms of action of chloramphenicol. *J. Am. med. Ass.* **209**, 97–103.

WEITZMAN, S.A., STOSSEL, T.P., AND DESMOND, F.M. (1978). Drug-induced immunological neutropenia. *Lancet* i, 1068–72.

WERBLIN, T.P., POLLACK, L.P., AND LISS, R.A. (1979). Aplastic anemia and agranulocytosis in patients using methazolamide for glaucoma *J. Am. med. Ass.* **241**, 2817–18.

——, ——, AND —— (1980). Blood dyscrasias in patients using methazolamide (Neptazane) for glaucoma. *Ophthalmology (Rochester)* **87**, 350–4.

WEST, B.C. (1981). Vancomycin-induced neutropenia. *South. med. J.* **74**, 1255–6.

—— (1982). Vancomycin-induced neutropenia. *South. med. J.* **75**, 1576.

WESTERMAN, E.L., BRADSHAW, W., AND WILLIAMS, T.W. (1978). Agranulocytosis during therapy with orally administered doxacillin. *Am. J. clin. Pathol.* **69**, 559–60.

WHEELAN, K.R., COOPER, B., AND STONE, M.J. (1982). Multiple hematologic abnormalities associated with sulfasalazine, *Ann. intern. med.* **97**, 726–7.

WHITE, C.M., PRICE, J.J., AND HUNT, K.M. (1980). Bone marrow aplasia associated with metronidazole. *Br. med. J.* **1**, 647.

WHITE J.M., BROWN, D.L., HEPNER, G.W., AND WORLLEDGE, S.M. (1968). Penicillin-induced haemolytic anaemia. *Br. med. J.* **2**, 26–8.

WHITMAN, E.N. (1969). Effects in man of prolonged administration of trimethoprim and sulfisoxazole. *Postgrad. med. J.* **45**, (Suppl.), 46–51.

WHITEHEAD, S. AND GEARY, C.G. (1983). Agranulocytosis associated with Maloprim. *Br. med. J.* **286**, 1515.

WICKRAMASINGHE, S.N., SAUNDERS, J., AND WILLIAMS, GAIL (1976). Effects of anticonvulsant drugs on the synthesis of DNA and protein by human bone-marrow cells *in vitro*. *Scand. J. Haematol.* **17**, 312–16.

WILLIAMS, C.K.O., ADEROJU, E.A., ADENLE, A.D., SEKONI, G., AND ESAN, G.J.F. (1982). Aplastic anaemia associated with anti-tuberculosis chemotherapy. *Acta. Haematol.* **68**, 329–32.

WILLIAMS, G.T., JOHNSON, S.A.N., DIEPPE, P.A., AND HUSKISSON, E.S. (1978). Neutropenia during treatment of rheumatoid arthritis with levamisole. *Ann. Rheum. Dis.* **37**, 366–9.

WILSON, C., GREENHOOD, G., REMINGTON, J.S., AND VOSTI, K.L. (1979). Neutropenia after consecutive treatment courses with nafcillin and piperacillin. *Lancet* i, 1150.

WILSON, J.R. AND HARRIS, J.W. (1977). Hematologic side-effects of Dapsone. *Ohio State med. J.* **73**, 557–60.

WINDLE, R., MACPHERSON, S., AND BELL, P.R.E. (1979). Neutropenia associated with metronidazole. *Br. med. J.* ii, 1219.

WINFIELD, D.A., BENTON, P., AND ESPIR, M.L.E. (1976). Sodium valproate and thrombocytopenia. *Br. med. J.* **2**, 81.

WINTROBE, M.M. (1969). The therapeutic millennium and its

price: a view from the haematopoietic system. The Lilley Lecture, 1968. *J. R. Coll. Phycns., London,* **3**, 99–119.

——, LEE, G.R., BOGGS, D.R., BITHELL, T.C., ATHENS, J.W., AND FOERSTER, J. (1974). *Clinical haematology,* 7th edn. Lea and Febiger, Philadelphia.

WOODBURY, D.M. AND FINGL, E. (1975). Analgesic-antipyretics anti-inflammatory agents, and drugs employed in the therapy of gout. In *The pharmacological basis of therapeutics* (ed. C.S. Goodman and A. Gilman), p.345. Macmillan, London.

WOOLLEY, P.V., III, LUC, P. VAN T., RAHMAN, A., KORSMEYER, S.J., SMITH, F.P., AND SCHIEN, P.S. (1981). Phase 1 trial and clinical pharmacology of 1-(2-chloroethyl)-3-(2,6-dioxo-3-piperidyl)-1-nitrosurea. *Cancer Res.* **41**, 3896–900.

WORLLEDGE, S.M. (1969). Autoantibody formation associated with methyldopa (Aldomet) therapy. *Br. J. Haematol.* **16**, 5–8.

——, CARSTAIRS, K.C., AND DACIE, J.V. (1966). Autoimmune haemolytic anaemia associated with α-methyldopa therapy. *Lancet* **ii**, 135–9.

WRIGHT, A.J. AND BRACKENRIDGE, R.G. (1982). Pulmonary infiltration and bone marrow depression complicating treatment with amiodarone. *Br. med. J.* **284**, 303.

WURZEL, H.A. AND SILVERMAN, J.L. (1966). Methyldopa and haemolytic anaemia. *Lancet* **i**, 1158.

YAP, H-W., BLUMENSCHEIN, G.R., SCHELL, F.C., BUZDAR, A.V., VALDIVIESO, M. AND BODEY, G.P. (1981). Dihydroxyanthracene-dione: a promising new drug in the treatment of metastatic breast cancer. *Ann. intern. Med.* **95**, 694–7.

YOUNG, F. (1957). Severe post-operative thrombocytopenic purpura. *Br. med. J.* **2**, 919–20.

YOUNG, G.A.R. AND VINCENT, P.C. (1981). Cimetidine agranulocytosis. *Med. J. Austral.* **2**, 453.

YOUNG, R.C., NACHMAN, R.L., AND HOROWITZ, H.I. (1966). Thrombocytopenia due to digitoxin. *Am. J. Med.* **41**, 605–14.

YUNIS, A.A. AND SALEM, Z. (1980). Drug-induced mitochondrial damage and sideroblastic change. *Clin. Haematol.* **9**, 607–19.

YUST, I., FRISCH, B., AND GOLDSHER, N. (1982). Simultaneous detection of two mechanisms of immune destruction of penicillin-treated human red blood cells. *Am. J. Hematol.* **13**, 53–62.

ZALCBERG, J.R., MCGRATH, K., DAUER, R., AND WILEY, J.S. (1983). Heparin-induced thrombocytopenia with associated disseminated intravascular coagulation. *Br. J. Haematol.* **54**, 655–60.

ZIPEROVICH, S. AND PALEY, H.W. (1966). Severe mechanical haemolytic anaemia due to valvular heart disease without prosthesis. *Ann. intern. Med.* **65**, 342–6.

ZIPES, D.P., ELHARRAR, V., GILMOOR, R.E., HEGEL, J.F., AND PRYSTOWSKY, E.N. (1980). Studies with aprindine. *Am. Heart J.* **100**, 1055–62.

ZUPAŃSKA, B., LAWKOWICZ, W., GÓRSKA, B., KOZLOWSKA, J., OCHOCKA, M., ROKICKA-MILEWSKA, R., DERULSKA, D., AND CIEPIELEWSKA, D. (1976). Autoimmune haemolytic anaemia in children. *Br. J. Haematol.* **34**, 511–20.

ZWAAN, F.E. AND MEYBOOM, R.H.B. (1979). Causes and consequences of bone-marrow insufficiency in man. *Neth. J. Med.* **22**, 99–104.

13 Intravascular clotting

J.R.B. WILLIAMS

Normal haemostasis is maintained by a fine balance among platelet functional activity, coagulation, and fibrinolysis. Drugs may disturb any of these with resulting haemorrhage or thrombus formation. In large vessels the thromboembolic effects may be major and in arterioles the platelet and fibrin deposition may result in disseminated intravascular coagulation (DIC) with its serious clinical manifestations. In this Chapter the thromboembolic side-effects of oral contraceptives (OCs) are our main concern, but changes in haemostasis induced by other other drugs are also discussed.

Oral contraceptives

Haematological, biochemical, and physiological changes

Blood coagulation

Changes affecting chiefly the extrinsic coagulation cascade and antithrombin III activity are given in detail later and are shown diagrammatically (Fig. 13.1).

In a relatively small series of women Egeberg and Owren (1963) found an increase in factor VIII and VII

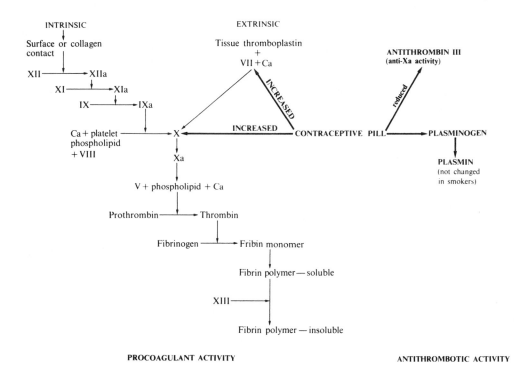

Fig. 13.1 Stages in process of blood coagulation, showing known influence of oral contraceptives in increasing levels of specific factors. International nomenclature of blood coagulation is shown together with synonyms (based on Bell *et al.* 1968).

during contraceptive medication. Rutherford *et al.* (1964) demonstrated increased levels of factor II, VII, IX, and X, and Poller and associates (1968) demonstrated an increase in factors VII and X in women taking oral contraceptives and showed that even with low-dose oral-contraceptive agents there was a significant rise in the level of both factors VII and X from the third month onwards. These changes did not appear to be dose-dependent. They also showed increases in the same factors, both involved in the extrinsic coagulation pathway in menopausal women receiving replacement therapy with equine (natural) oestrogens (Poller 1976; Poller *et al.* 1977).

Increased levels of factor VII and X were found by several other groups of workers including Meade *et al.* (1976) and Bonnar *et al.* (1976). The latter workers also found an increase in the intrinsic system factor IX in patients receiving the synthetic steroids mestranol–norethisterone but not in those taking conjugated equine oestrogens.

A rise in fibrinogen levels is usual, and this and factor X levels may be related to oestrogen dosage (Fawer *et al.* 1978). Changes in coagulation factor activity are less clearly associated with progestogen type or dose. It seems likely that only oestrogens affect factor VII levels but it may be that fibrinogen levels are influenced by the progestogen content of the contraceptive. The progestogen content of the combined oral contraceptives does not appear to influence venous thromboembolism, which is affected by the oestrogen dose and its influence on coagulation factor responses, especially factor VII (Meade 1981*a*). Reduction in coagulation-inhibitory activity of both antithrombin III (Greig and Notelowitz 1975) or more specifically anti-Xa (Kakkar *et al.* 1976; Meade *et al.* 1976; Sagar *et al.* 1976; Wessler *et al.* 1976) occurs.

Conard *et al.* (1979) confirmed the previously reported reduction in antithrombin III demonstrated by both the von Kaulla coagulation and the Mancini immunodiffusion methods in women taking both 50-μg and 30-μg oestrogen doses. No reduction in antithrombin III was found in users of the progestogen norgestrienone. Horváth *et al.* (1978) also confirmed the reduction in antithrombin III levels in most women using combined oestrogen/progestogen preparations. They found, however, that five of the 17 women they studied had raised antithrombin III levels and that they also had delayed bromsulphthalein plasma clearance, indicating subclinical cholestasis. The authors suggested that women with a predisposition to cholestasis induction by steroid-related drugs will develop high antithrombin III levels and that this association could be used as a screening test for either thrombotic or cholestatic tendency.

Abnormalities in haemostasis may also occur in workers employed in the manufacture of oral contraceptives (Poller *et al.* 1979).

Platelet function

Platelet aggregability by ADP is increased (Hampton 1970: Sanderson and Delamore 1973; von Kaulla 1976), and as shown in the Chandler's tube, by thrombin (Poller *et al.* 1977) but no change in platelet count or adhesiveness was found in the detailed study by Meade *et al.* (1976). Montanari *et al.* (1979) however found increased aggregation of platelets by ADP and Peters *et al.* (1979) demonstrated changes in the receptor sites on platelets for noradrenaline and 5-hydroxytryptamine. These changes were associated with increased platelet aggregation and provide a theoretical basis for the occurrence of a state of hypercoagulability in women taking oral contraceptives. Further confirmation of the possible importance of platelet function changes due to oral contraceptives was provided by the observations of Aranda *et al.* (1979) who found increased β-thromboglobulin levels in a study of 110 volunteers taking the pill compared with 35 normal controls of the same age. These increased levels clearly indicate increased platelet disintegration, presumably from aggregation.

The clinical importance of platelet functional changes was shown by a case report of Mazal (1978) who recorded increased spontaneous platelet aggregation in a 27-year-old woman who developed severe migraine while taking testranol 50 μg and norethisterone 1 mg for contraceptive purposes. Carotid arteriography showed widely distributed arterial and sinus occlusions, presumably due to multiple thrombus formation.

Fibrinolytic activity

Astedt *et al.* (1973) found a reduction in fibrinolytic activation in histological preparations of vein biopsies and both Brakeman (1968) and Hedlin (1975) found increased fibrinolytic activity in women taking oral contraceptives. Meade *et al.* (1976) also found increased fibrinolytic activity in non-smokers but found normal levels in women who smoke. These authors suggested that the increased fibrinolytic activity protects against the effects of decreased antithrombin III or anti-Xa activity and increased levels of factor X and VII. In smokers this protective effect is lost and women who smoke are therefore at greater risk from venous thromboembolism if they take oral contraceptives.

It was suggested by Alkjaersig *et al.* (1970) that 27 per cent of women on oral contraceptives had abnormal fibrinogen chromatographic patterns suggesting that repeated microscopic thromboses were occurring.

Analysis of nine patients developing reversible hypertension and/or impaired venous function while taking oestrogen-progestogen compounds showed evidence of microthrombi in glomerular capillaries or intrarenal arterioles in four of them (Boyd et al. 1975). These microangiopathic changes were associated with substantial rises in FDP levels and with thrombocytopenia.

Stadel (1981) in a review of oral contraceptives and cardiovascular disease recalled the observations of Alkjaersig et al. (1975) that the progestogen-only preparation, ethynodiol, may increase subclinical thrombosis as shown by chromatographic separation of fibrogen polymers. Demonstration of a rise in FDP fragment E may prove helpful (Gordon et al. 1977; Tso et al. 1980).

Plasma lipids

The metabolic effects of oral contraceptives on lipid metabolism were studied by Wynn and Niththyanthan (1981) in a group of 293 women wanting to start oral contraception and 536 women already taking oral contraceptives. Their main findings were that levenorgestrel-containing preparations and progestogen-only oral contraceptives significantly depressed HDL_2 cholesterol levels, and the ratio of HDL_2 to LDL. In view of the atherogenic protective effect of high-HDL levels, this demonstrable fall may be concerned with the arterial thrombotic side-effects.

A combination of norethisterone acetate with as little oestrogen as 20 μg ethinyloestradiol does not lower HDL or HDL_2 levels and the authors conclude that a combination of l mg norethisterone with 30 or 35 μg ethinyloestradiol should be satisfactory for most women. This view is supported by a follow-up report on the Framingham Study (Gordon et al. 1981) which confirms the inverse relation between HDL levels and ischaemic heart disease.

Haemopoiesis

Minor changes in the blood count induced by oral contraceptives probably do not influence coagulation. The mean red cell volume is increased by both smoking and oral contraceptives (Dodsworth et al. 1981), but other parameters are not influenced by oestrogens, although the neutrophil count is increased in smokers. The increase in red cell volume is probably related to minor degrees of folate deficiency. Both red cell and serum folate levels may be reduced, probably due to interference with absorption (Grace et al. 1982). Serum B_{12} levels however may be raised.

The folate studies were carried out in Boston, Massachussetts, but those by Dodsworth et al. (1981) were on residents in London. Further work by this group in the

United Kingdom on the haematological indices in four ethnic groups, white, black, Indian, and oriental showed no effect by oral contraceptives (Godsland et al. 1983). There were appreciable differences between the groups but these were determined by genetic factors, nutritional status, smoking frequency, and alcohol intake among other factors.

Vascular effects

Fawer et al. (1978) compared forearm blood flow, venous distensibility, and various coagulation factor levels in controls, pregnant women, and women taking both high progestogen and high oestrogen (sequential) contraceptives. They found an increase in venous distensibility in women taking the high progestogen: oestrogen ratio contraceptive but no change in blood flow in either contraceptive group.

They further suggested that the progestogen component increases venous capacitance and may induce venous stasis whereas the oestrogen component tends to affect coagulability. In a review of female hormones and vascular disease, Vessey (1980) commented on the need for more work on the coagulation and fibrinolytic systems and on changes in venous blood flow.

Interplay of physiological and metabolic factors

Thrombotic microangiopathy with thrombocytopenia induced by oestrogens was reported by Boyd et al. (1975). Another case with haemolytic anaemia and renal failure occurred in a 23-year-old multipara four months after starting a sequential oral contraceptive (Hauglustaine et al. 1981). Renal failure progressed requiring haemodialysis during which an oral progestogen contraceptive was used. Cadaver renal transplant was successful and the progestogen was later changed for a combined product containing ethinyloestradiol 50 μg and levonorgestrel 250 μg. Ten months later the haemolytic-uraemic syndrome returned but recovery followed transplant nephrectomy and a later second cadaver kidney transplant.

Ethnic differences are important as there is a different incidence of post-operative venous thrombosis in the Chinese (Tso et al. 1980). A detailed study of gynaecological patients in Hong Kong gave an overall incidence of 2.6 per cent positive ^{125}I fibrinogen leg scans in 154 patients. Women taking oral contraceptives and undergoing pelvic surgery for benign disease had 10.5 per cent positive scans and those treated for cervical carcinoma by Wertheim hysterectomy had 6.7 per cent.

Laboratory studies showed a rise in fibrinogen/fibrin degradation fragment E (FgE) after major surgery and this was increased by taking oral contraceptives, especially in those developing deep vein thrombosis. Plasmin levels were not affected by malignant disease or oral contraceptives but plasmin activator levels were decreased more in malignant disease than in benign cases and were not significantly changed in the oral contraceptive takers, at least in the first two post-operative days.

Antithrombin III levels were lower in three of the four patients developing deep vein thrombosis but did not fall in the other patients. This early decrease in plasmin activator levels was followed by increases in the antifibrinolytic activity of α_1 anti-trypsin and C$\bar{\text{i}}$ inhibitor, the greatest inhibitors being found in malignant cases. The rise in C$\bar{\text{i}}$ inhibitor levels was least in the oral contraceptive takers, and this with the higher FgE responses, indicating increased fibrinolytic activity suggests that the increased incidence of DVT in these patients is not likely to be due to decreased fibrinolysis. Tso and his colleagues suggested that measurement of FgE and antithrombin III in the post-operative period may prove useful in the diagnosis of deep-vein thrombosis. In the present author's experience both methods are useful in studying special clinical problems but their use in routine screening is unlikely to be widespread owing to the considerable commitment of laboratory time and cost.

The relationship between coagulation and platelet function change is complicated, as Bierenbaum et al. (1979) found an association between increased platelet aggregation with decreased high-density lipoprotein cholesterol. They considered that demonstration of these platelet and lipoprotein changes in a woman taking oral contraceptives are a definite contraindication to continuing them.

Another minor factor to be taken into account is the effect of high-dose ascorbic acid on the metabolism of ethinyloestradiol. Competitive inhibition of intestinal sulphation may lead to greater availability of oestrogen. Briggs (1981) showed that fibrinogen, factor VII, factor VIII, and caeuloplasmin levels are raised as are HDL-cholesterol and sex-hormone-binding globulin. Ascorbic acid may therefore induce biochemical changes which both predispose for and protect from cardiovascular side-effects of oral contraceptives. With the present vogue for self-administration of megadose vitamin C it may be necessary to take these metabolic effects into account in epidemiological, biochemical, or haematological studies concerned with oestrogens.

Meade et al. (1976) also suggested that the demonstration of increased coagulation factors, with low inhibitors of fibrinolytic activity, might select women who are unduly prone to thrombotic complications of oestrogen treatment. Laboratory supervision of this kind might perhaps allow women who have suffered from deep-vein thrombosis or pulmonary embolism as a complication of

trauma or surgical operation, and who do not smoke, to return cautiously to taking an oestrogen-progestogen contraceptive combination.

Reviewing the reports it is evident that it is not yet possible from laboratory and physiological measurements to predict the probability of venous or arterial thrombotic complications of the pill. It is apparent, however, that changes may be found which if severe enough may be assumed to make such an occurrence likely. Studies of coagulation changes should include assay of fibrinogen, factors VII, X, antithrombin III, and fibrinolytic activity. It is advisable to examine platelet aggregation, preferably spontaneous rather than after ADP or noradrenaline stimulation, and to measure fasting HDL-cholesterol levels. The value of limb plethysmography is as yet uncertain.

The combined effects of coagulation and platelet function may be assessed by thromboelastography, and Poller (1978) in a review article suggested that this may be the most useful single measurement, although in the author's laboratory assays of factor VII, factor X, antithrombin III, and fibrinolytic activity with HDL levels have proved useful.

Reviewing the various methods available, Mammen (1981) suggested that tests used should include as many aspects of coagulation and fibrinolytic function as possible, but it is still not clear which tests are best.

Clinical thrombotic effects — epidemiology

Oestrogen-induced thromboembolic disorders were reported by Daniel et al. (1967) during the suppression of lactation and by Bailar (1967) following treatment of prostatic cancer. Inman and Vessey (1968) found a highly significant correlation between pulmonary embolism and the use of oral contraceptives, and this with an association with deep-vein thrombosis was confirmed by Vessey and Doll (1968, 1969). Futher confirmation, this time including an association with cerebral thrombosis as well as venous thromboembolism, was provided by Vessey (1970) and an oestrogen dose-related effect was demonstrated by Inman et al. (1970). This was later confirmed by Bottiger et al. (1980).

Early reports of an increased risk of myocardial infarction by Oliver (1970) and Radford and Oliver (1973) were confirmed by Mann et al. (1975). Venous thrombosis at other sites occurring as a complication of the administration of oestrogen oral contraceptives includes five cases of superior mesenteric thrombosis reported by Rose (1972) and of intracranial thrombosis in two women by Fairburn (1973), and in one young woman earlier by Atkinson et al. (1970).

Since then there have been many case reports, small

Table 13.1   Major cohort studies of oral contraception (after Kols *et al.* 1982)

Study	Dates	Source of participants	Age	Groups of entry		% at entry
RCGP* UK	1968	General practice	15 +	OC users:	23 611	48
				Never users:	22 766	42
Oxford FPA UK	1968	Family planning clinics	25–39	OC users:	9653	48
				Diaphragm users:	4217	33
				IUD users:	3162	42
Walnut Creek US	1968–1977	Health insurance members	18–54	OC users:	6107	36
				Former users:	5547	36
				Never users:	6503	33

* Royal College of General Practitioners.

surveys, and reviews but the most helpful information has come from the three major surveys, two British and one American. The sizes and nature of the cohorts involved are given in Table 13.1.

Strong confirmation that oral contraceptives increase the mortality from diseases of the circulatory system was shown by the large prospective study carried out by the Royal College of General Practitioners (RCGP) in the United Kingdom (Beral and Kay 1977). In a further report the Royal College (RCGP 1978) analysed the frequency of venous thrombosis in relation to oral contraceptives. It showed an increased risk of four times for deep vein thrombosis and two and a half times for superficial thrombosis with an increase in superficial venous thrombosis in patients with varicose veins. There was a dose-related effect of oestrogens and progestogens on superficial venous thrombosis but no such relationship was found with regard to deep-vein thrombosis.

A recent report of the Royal College of General Practitioner's Oral Contraception Study (RCGP 1981) superseded the 1978 one. The analysis is of 249 deaths and shows from their data that the overall increased risk of death from vascular diseases for all ages and groups of users of the pill taken together compared with controls was 4.2. For the category of non-rheumatic heart disease and hypertension, ever-users had a relative risk of 5.6 compared with controls and ischaemic heart disease had a risk of 3.9. As in the previous report the risk of subarachnoid haemorrhage appears to be increased in ever-users (4.0). Considering the relative risk in current-users compared with former-users non-rheumatic heart disease and hypertension carried a ratio of 7.3 and 4.6 respectively, whereas in cerebrovascular disease the position was reversed (2.0 and 3.6 respectively). This reversal of risk may be due to cerebral vascular disease in some women resulting in the stopping of oral contraception and their subsequent grouping in the former-use category.

Cigarette-smoking apparently increases the probability of death from subarachnoid haemorrhage. In the study 71 per cent of the ever-users and 67 per cent of the controls who died of subarachnoid haemorrhage smoked compared with 48 per cent of all ever-users and 40 per cent of all controls.

The risk of death from cardiovascular disease is also increased by smoking. Taking all ages together the relative risk of oral contraceptive use in smokers is 5.1 compared with 3.2 in non-smokers and this risk has very greatly increased in women over 45 years in whom the relative risk was 18.1 compared with their non-smoking controls. From the way the data were collected, using smoking habits on entry to the study in 1968–69, it is probable that the statistics underestimated the enhanced risk due to smoking.

The study strongly suggests that the risk of mortality in pill-users increases with parity. As the authors stress, this observation is new and requires confirmation. The earlier suggestion that the duration of oral-contraceptive use increases the risk is no longer tenable, but doubt remains as to whether former use of the pill leads to a permanently increased risk. From the report it is implied but not proven that if a woman is normal in cardiovascular function and pathology when she discontinues the pill she is unlikely to carry an extra risk of disease in this system later. If hypertension or ischaemic heart disease have occurred, then the disease follows its usual course. The risk of cerebrovascular disease probably continues after stopping the pill.

The main risks of mortality from oral contraceptives are evidently from cardiac and arterial disease. Death from venous disorders, particularly pulmonary embolism, only account for 10 per cent of the increased risk. Taking the risks as a whole the excess mortality for women under 35 years of age is 1 per 77 000 women and could be due to chance. For women of 35–44 years the risk is 1 per 6700 in non-smokers and 1 in 2000 in smokers. The authors suggest that in non-smokers of this age group those who have no other risk factors associated with cardiovascular disease may find that the advantages of oral contraceptives outweigh the risks. At 45 years and above the risks are 1 in 2500 and 1 in 500 and therefore women over 45 years should not take oral contraceptives without very strong clinical indications.

Updating the report of the Oxford Family Planning Study (Vessey *et al.* 1977), Vessey *et al.* (1981) compared the RCGP study with their own follow-up of 17032 women studied by the Oxford Family Planning Association. Of these women 56 per cent were pill-users, 25 per cent were diaphragm-users, and 19 per cent used an intrauterine device. So far 89 deaths have occurred and 81 were available for study. The overall mortality rate was lower than the age-specific rates for England and Wales.

Various differences in the rates between the groups were observed, but only the increased death rate for non-rheumatic heart disease approached statistical significance. Their data support the view that oral contraceptives predispose to myocardial infarction, especially in women who smoke. Epidemiological studies usually indicate that the risk of myocardial infarction is largely confined to current users of oral contraceptives (Vessey 1980), but the present Oxford report supports the RCGP findings that women may have discontinued the pill because of cardiovascular disease and died later. There is no suggestion from the Oxford data that the pill is a cause of subarachnoid haemorrhage and there are several substantial differences in the observations of the two study groups. The reasons for this are not clear, but perhaps depend on the nature of the cohorts selected, general practitioners' patients on the one hand and Family Planning Association referrals on the other.

Progestogens also contribute to the increased risk of cardiovascular disease associated with oral contraceptives. Meade *et al.* (1980) studied nearly 2000 reports to the Committee on Safety of Medicines in the United Kingdom from 1964 to 1977. They found a significant positive association between the dose of norethisterone acetate and deaths from strokes and ischaemic heart disease, but not with hypertension or venous thrombosis. The higher dose of levonorgestral was associated with an excess of deaths from stroke and a possible excess from venous and nonvenous disease. Deaths from cardiovascular disease, venous and nonvenous, and nonfatal ischaemic heart disease were more frequent in users of oral contraceptives containing 50 μg of oestrogen than in those taking 30 μg, although as expected, the incidence of unwanted pregnancies was lower with the higher dose rate. The risk with either dosage is very small — between 0.1 and 0.2 per 100 woman-years. Meade *et al.* (1980)

however made a case for minimizing the dose of progestogen to reduce the chances of thromboembolism.

Reviews by Meade (1981*b*; 1982) include details of the work of his own team at Northwick Park Hospital, London. It is apparent that norgestrel may participate in the hypertension-inducing effect of 30 μg ethinyloestradiol-containing oral contraceptives. The effect of norethisterone in this respect may be dose-related and may be independent of the oestrogen content.

Reviewing previous studies Meade concludes that progestogens in combined oral contraceptives have an influence on the development of ischaemic heart disease and stroke. This effect may be mediated by changes in blood pressure or in blood lipid levels.

Epidemiological studies provide further evidence of the correlation of the incidence of hypertension and arterial disease (Oster *et al.* 1981) with the progestogen content of oral contraceptives and a follow-up report on the Framingham Study (Gordon *et al.* 1981) confirms the inverse relation between HDL levels and ischaemic heart disease.

A case-control study of deaths from subarachnoid haemorrhage in women aged 15–44 showed only a small excess of oral contraceptive use compared with healthy controls (Inman 1979). This increase was not statistically significant, and raises doubts as to the interpretation of the RCGP data. It is apparent that oral contraceptives may increase the blood pressure and lead to death of a few women from subarachnoid haemorrhage; hypertension *per se* is a much larger underlying risk.

Crawford *et al.* (1981) examined the rate of metabolism of oral contraceptives in smokers and non-smokers as a difference might explain the increased incidence of cardiovascular disease. They found however that plasma ethinyloestradiol and norgestrel clearances were similar in smokers and non-smokers.

The difficulties in interpreting the data from both case-control and cohort studies are discussed by MacRae (1980) in an analysis of the statistical and epidemiological problems involved. As he points out, the 'population mortality trends fail to show effects consistent with the magnitudes of risk postulated' in the published studies, and there must be differences in the populations studied, in the unavoidable bias involved in case-history-taking, and, perhaps, in the criteria for diagnosis of morbidity and cause of mortality.

The relationship between smoking and the cardiovascular side-effects of oral contraceptives were discussed in the *Updates* to the 2nd edition. These UK reports mostly indicate that smoking greatly enhances the risk of death in smokers compared with non-smokers taking oral contraceptives (Slone *et al.* 1981). A United States report from New York, Boston, and Delaware provided evidence that myocardial infarction, including surviving cases, is less frequent in smokers than non-smokers and that increased duration of oral-contraceptive use increases the risk of myocardial infarction. This also is in contradiction to the British reports. The explanation of the discrepancy is not obvious. Although at first sight the populations and living standards appear to be similar, the ethnic origins and dietary habits may be substantially different.

It is important for practitioners prescribing oral contraceptives to take serious account of risk factors in the medical history. Recent practice in this respect has probably led to a substantial decrease in mortality (Adam and Thorogood 1981). In this respect the effect of reducing the oestrogen content from 50 to 30 μg has been counteracted by the raising of the progestogen content. The interpretations of past and future work in this field are modified by a recent paper not specifically directed at the complications of oral contraceptives Vesey *et al.* (1982) examined the blood carboxyhaemoglobin and plasma thiocyanate levels in smokers and non-smokers. The found a partial correlation between the two levels after the number of cigarettes smoked had been allowed for ($r = 0.48$). Of the variation in carboxyhaemoglobin and thiocyanate concentrations 23 per cent was accounted for by the way the cigarettes were smoked and 21 per cent by the number smoked a day. The relationship between the levels reached an asymptote at rates above 25 cigarettes a day but the limits of the variation were wide — almost certainly determined by the smoker's inhalation and puffing habits. A few (2–4 per cent) of volunteers smoking 26–35 cigarettes a day appear to have the same risk as non-smokers, but on the other hand many smokers reached high carboxyhaemoglobin and thiocyanate levels on only 1–5 cigarettes a day.

The authors suggest that future epidemiological studies of smoking should include a biochemical parameter of smoke intake.

The alterations in normal physiology caused by cigarette-smoking and oral contraceptives were reviewed by McEwan (1979). Other reports vary considerably in the incidence and relationships of arterial and venous thrombosis, partly related to smoking but probably also related to other social patterns and to ethnic origin.

Petitti and Wingard (1978) and Petitti *et al.* (1979) found that smoking increased the risk of myocardial infarction, subarachnoid haemorrhage, other strokes, and venous thromboembolism whereas oral contraceptives were associated with an increased risk of subarachnoid haemorrhage and venous thromboembolism. Increased risk of myocardial infarction were found by Jick *et al.* (1978): Baudet *et al.* (1978); and Shapiro *et al.* (1979). The latter authors found an increased rate ratio overall for myocardial infarction for women who had used oral contraceptives in the previous month of 4.0 compared with those who had not. Women who smoked heavily and used oral contraceptives had an increased risk of 39.0 to 1 compared with those who did neither.

Pfeffer *et al.* (1978) found no association between the use of oestrogens and myocardial infarction in 220 women among a retired community population of 15 500 women. They commented on the low daily use of oestrogens and the short duration of the administration and the clinical and pathological relationship is clearly different from that arising in contraceptive oestrogen use.

An increased incidence of subarachnoid haemorrhage associated with the use of oestrogen contraceptive pills was reported previously and further reports were provided by Petitti and Wingerd (1978) who found an increased risk for non-smokers who used oral contraceptives of 6.5 times that of non-users. The risk in users who smoked was 22 times that of non-users. A continuing but reduced risk of 5.3 applied to past use of oral contraceptives.

It was suggested by Finn and St. Hill (1978), in comments on two women who developed subarachnoid haemorrhage while using oral contraceptives, that an immunological mechanism may initiate the haemorrhage. They observed depressed lymphocyte mitotic response to phytohaemagglutinin in women using oral contraceptives and considered that autoimmune arteritis should be looked for in various vascular diseases, presumably including complications of oestrogen/progestogen administration.

There have been other reports of arterial disease associated with oral contraceptives including ischaemic disease of the bowel (Ghahremani *et al.* 1977), in the vertebral, mesenteric, and renal arteries (Holt and Hollanders 1980) and in the arteries of the left arm in a male receiving oestrogen therapy (Danis *et al.* 1978).

A beneficial effect, probably partly due to the prevention of unwanted pregnancies has been the fall in maternal mortality rates in recent decades. The morbidity and mortality rate resulting from this oral contraceptive practice is not included in the maternity figures and it has been suggested (Beral 1979) that an alternative measure, the reproductive mortality rate, should be used, which includes complications of contraception as well as pregnancy and abortion.

Clinical dosage

The content of oestrogen and progestogen in preparations approved for sale in the United Kingdom are listed in the *British National Formulary* (1985), classified

according to drug concentrations. The chemical structures of the commoner agents are given in Fig. 13.2 which
shows the close relationships between oestrogens and
progestogens.

Corticosteroids

Treatment with corticosteroids and corticotrophin
(ACTH) may predispose to intravascular thrombosis and
thrombophlebitis especially if the duration of medication
is prolonged. Goodman *et al.* (1964) reported fatal pul-
monary-artery thrombosis in two children aged 4 years
and 5 years who had been treated for nephrosis and given
prednisone 25–80 mg daily and chlorothiazide for many
months.

A number of other investigators included thrombosis
and thrombophlebitis among the complications occurring with long-term treatment: for example, Turiaf and
colleagues (1962) reported seven cases of venous thrombosis, and two cases of cerebral vascular accident, among
268 patients on long-term cortisone therapy for asthma.
Saxena and Crawford (1965) treated 60 children with
nephrosis who were given corticosteroids or ACTH for 1
or 2 years; there were two cases of cerebrovascular
thrombosis. Bock (1966) reported on the side effects of
corticosteroid treatment in 412 patients with neurological disorders; patients (3 per cent) developed thrombosis and embolism. Huriez and Agache (1966) surveyed
2000 patients treated with ACTH and corticosteroids for
dermatological disorders; there was a 1 per cent incidence
of thrombosis with ACTH, 1 per cent with prednisolone,
and 0.2 per cent with cortisone. Cardiovascular accidents
occurred in 3 per cent of patients on ACTH, 3 per cent on
cortisone, 2 per cent on dexamethasone, and 0.4 per cent
on prednisone. Zuckner *et al.* (1967) listed side-effects of
intramuscular treatment with corticosteroid in 77
patients with rheumatoid arthritis over periods of treatment from 3 months to 5 years; three pateints suffered
thrombophlebitis. A Polish report (Florkiewicz and
Klamut 1966) described thrombophlebitis of the major
veins attributed to prednisone therapy in three cases.
Kornell (1966) reported on a patient with generalized
scleroderma in whom treatment with prednisolone coincided with an acceleration of involvement of the
myocardium.

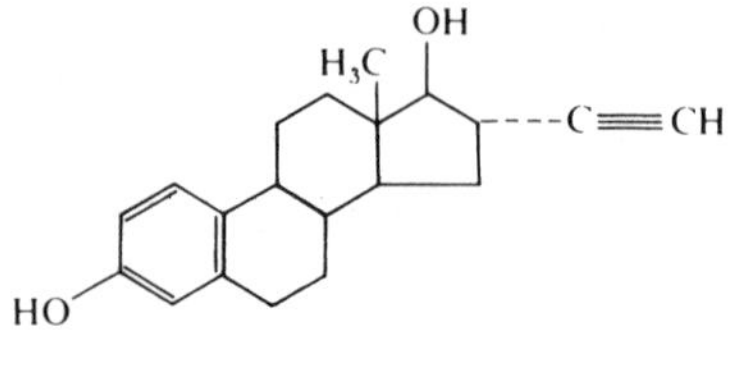

Fig. 13.2   Comparison of structural formulae of some
oestrogens and progestogens in combined oral contraceptive preparations.

OESTROGEN
Ethinyloestradiol (ethinylestradiol)

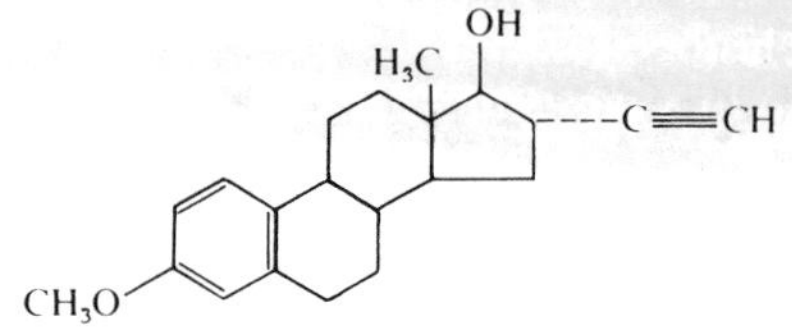

OESTROGEN
Mestranol

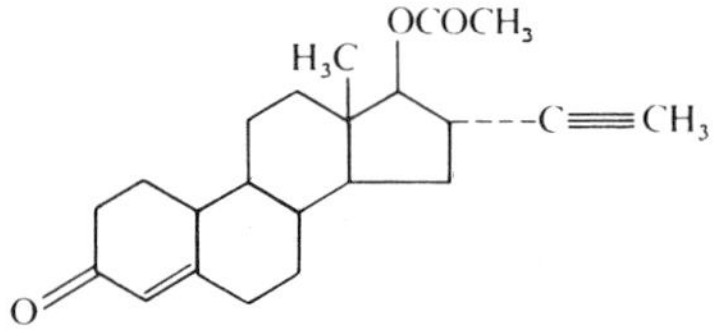

PROGESTOGEN
Norethisterone acetate (norethindrone acetate)

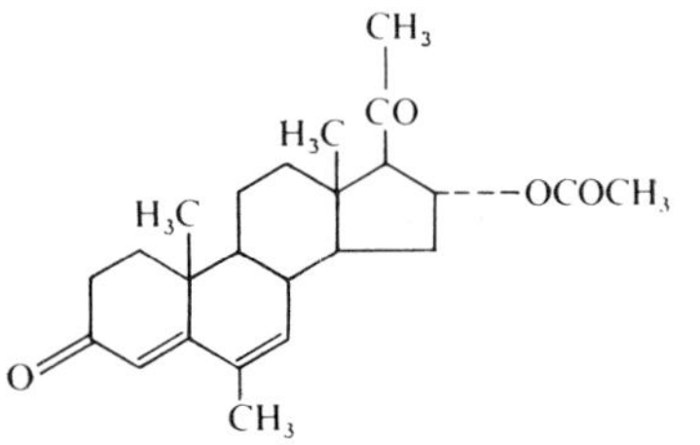

PROGESTOGEN
Megestrol acetate

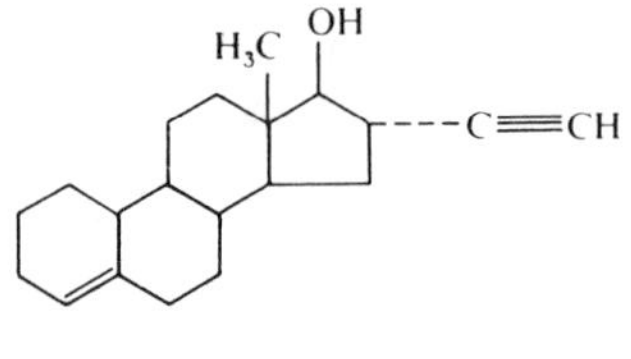

PROGESTOGEN
Lynoestrenol (lynestrenol)

Other drugs

Atenolol

Atenolol was reported as the cause of right renal artery thrombosis, resulting from a fall of blood pressure from 240/115 mmHg to 160/80 mmHg in a 70-year-old man (Shaw and Gopalka 1982). Only one dose of 100 mg was given but the blood urea was 19 mmol/l^{-1}, and the authors stress the need for care in the use of drugs in the elderly, especially in the presence of renal failure.

Anti-epileptic drugs

Sodium valproate may occasionally cause abnormalities of coagulation in conjunction with bone-marrow suppression and anti-platelet antibody formation (Smith and Boots 1980), although the formation of drug-induced anti-platelet antibody without DIC may occur in 14 per cent of patients receiving the drug (Sandler *et al.* 1979). The haematological effects of this drug are further reviewed in this volume (see Chapter 12) as haemolytic anaemia may also occur in conjuction with DIC. The same combination of haematological side-effects may occur after the administration of bromsulphthalein (Frick *et al.* 1979). Phenytoin (diphenylhydantoin) may cause purpura associated with DIC (Hanukoglu *et al.* 1980).

A 12-year-old boy with epilepsy had been treated for 6 years with clonazepam and 14 days before the onset of fever and purpura was treated also with phenytoin 300 mg daily. There was generalized lymphadenopathy accompanied by slight thrombocytopenia (platelets 90 × 10^9 1^{-1} with prolongation of the one-stage prothrombin time and partial thromboplastin time). An initial diagnosis of septicaemia was made with treatment with benzylpenicillin and gentamicin without improvement, which did occur on stopping phenytoin and combined antibiotics, and the septicaemia was never confirmed. It seems probable that the cutaneous vasculitis and DIC were related to the administration of phenytoin which was also responsible for the fever and lymphadenopathy.

Dopamine

Ischaemia producing gangrene of an extremity is a recognized complication of the use of dopamine. In a review of five patients in San Francisco who developed dopamine-induced gangrene, only one survived, and she required bilateral amputation. Three of them had laboratory evidence of disseminated intravascular coagulation, and this may be enhanced by the underlying shock for which the dopamine was used (Winkler and Trunkey 1981). The analogue dobutamine does not have this effect.

Ergot

Venous thrombosis is a rare iatrogenic effect of parenterally administered ergot preparations (Greene, 1959; Mintz *et al.* 1974), probably as a specific sensitivity. No abnormalities of platelets or coagulation factors have been found, but arterial thrombosis may follow arterial occlusion due to the drug (Spittell 1980).

β-Lactam antibiotics

Carbenicillin may produce a coagulation disorder in which inhibition of platelet aggregation and fibrin formation with an increase in antithrombin-III activity may lead to haemorrhage (Lurie *et al.* 1970); Lederer *et al.* 1973; Andrassy *et al.* 1976).

The latter authors report similar coagulation disorders after benzylpenicillin administration in uraemic patients and after high dosage in patients after cardiac surgery. There was prolongation of the bleeding time (> 30 min), appearing immediately after administration of penicillin, reduction of platelet aggregation induced by ristocetin and collagen, increase of antithrombin III, and inhibition of factor-Xa activity.

They showed that high doses of penicillin cause a latent coagulation disorder which may become apparent if superimposed on other haemostatic defects such as uraemia or surgery. They further suggested that the haemorrhagic diathesis of subacute bacterial endocarditis could stem from high-dose penicillin therapy.

Cazenave *et al.* (1977) suggested that benzylpenicillin, ampicillin, and oxacillin and also cephalothin may inhibit platelet function by coating the platelet surface and that the abnormalities of aggregation and the prolonged bleeding times result from this.

Moxalactam may induce a reversible bleeding state associated with a prolonged bleeding time with abnormal platelet function (Weitekamp and Aber 1983) and prolongation of the prothrombin time may also occur.

Mitomycin C and 5-fluorouracil

Acute renal toxicity with fibrin deposits in the kidneys and lungs occurred in three patients receiving long-term treatment with these two drugs. Two of the patients had received blood transfusions, and it is possible that these sensitized the patients in some way to the development of the haemolytic-uraemic syndrome (Crocker and Jones 1983).

Quinine

Intravascular coagulation has possibly also occurred as a

result of quinine hypersensitivity (Elliott and Trash 1979). A 41-year-old woman who was accustomed to drinking bitter lemon took 300 mg of quinine sulphate to relieve nocturnal calf cramps and developed abdominal pain and purpura. Investigations showed DIC with thrombocytopenia and attempts to demonstrate the usual immune mechanism for quinine-induced thrombo-cytopenia were unsuccessful. The authors think it is likely that the DIC was induced by quinine and suggest that this process may occur in other patients with quinine-associated purpura.

Prothrombin complex

Various preparations of prothrombin complex usually contain some activated factors, especially factor X and have been known to induce intravascular coagulation. Another report, in a 17-year-old haemophiliac with high titre factor-VIII inhibitors gives details of a good clinical response without improvement of blood clotting and with some laboratory evidence of intravascular coagulation (Fukui *et al.* 1981).

Psychotropic drugs

Treatment of depressed patients with psychotropic drugs of most pharmacological groups may be associated with pulmonary embolism or deep-vein thrombosis. These include various combinations of clomipramine, chlorimipramine, metapramine, lorazepam, chlorpromazine, lithium, chloral hydrate, and maprotiline (Ruh-Bernhardt *et al.* 1976).

A futher report on the formation of inhibitors to clotting factors induced by chlorpromazine provides a study of 75 schizophrenic patients (Zarrabi *et al.* 1979). Elevation of IgM levels and prolongation of the partial thromboplastin time was found in patients taking chlorpromazine, and there was correlation between the extent of these changes and the dose or duration of treatment with the drug. Attempts to identify the site of action of the coagulation inhibitor suggested that there was an effect on the contact activation phase. The pattern of inhibition was similar to that seen in some cases of coagulation-inhibitor formation in systemic lupus erythematosis.

Substitute heart valves

Weily *et al.* (1974) performed platelet studies in 55 patients after cardiac-valve replacement. Average platelet survival was normal in patients with homograft and Beall prostheses but was short in patients with old pros-

thetic mitral valves. Thromboembolism occurred in 62 per cent of patients with short survival but in only 9 per cent with normal platelet survival.

Reduction of platelet survival correlates well with thromboembolic episodes after valve replacement.

There are risks attached to prescribing oestrogens to patients with prosthetic heart valves; there are two reports of embolic phenomena in patients with these valves who were treated with conjugated equine oestrogens (Pitcher and Curry 1979; Ewing 1979).

Local thrombosis

Thromboses at the site of intravenous injections or infusion have been associated with many drugs. Among the fairly recently introduced drugs which regularly cause thrombosis without apparently leaking at the injection site are several of the major cytotoxic drugs, nitrogen mustard (Carpentieri *et al.* 1976), methotrexate, doxorubicin, vincristine, and bleomycin (Guthrie and Way 1974). Diazepam has also been reported to cause thrombosis (Padfield 1974) and most users have seen examples of this. Von Dardel (1976) administered diazepam dissolved in a lipid emulsion instead of propylene glycol or phenylcarbinol and found a greatly reduced incidence of local venous thrombosis. This liquid preparation is coming into greater use.

Accidental intra-arterial injection of diazepam may lead to severe damage to the limb, sometimes requiring amputation of a digit (Rees and Dormandy 1980). These authors stress the importance of injecting diazepam intravenously at a slow rate (less than 0.5 ml min^{-1}) so that an error in siting may be noticed. If intra-arterial injection is suspected and the cannula is in place, injection of papaverine or procaine may help, as may intravenous methyl prednisolone.

Intralipid may produce a hypersensitivity reaction and Kamath *et al.* (1981) reported the occurrence of urticaria and pruritus confirmed by challenge.

It is well known that thiopentone may cause thrombosis at the injection site. Other agents used in anaesthesia — methohexitone and propanidid — may also do so. Propanidid may in fact give extending thrombosis into the proximal veins (Conway *et al.* 1970; Hewitt *et al.* 1966; Thornton 1970). Local thrombosis is also a frequent complication of intravenous cephalosporin therapy (Siebert *et al.* 1976) but appears to be more severe with cephalothin compared with some other analogues.

Severe thrombophlebitis has also been reported as a complication of intravenous infusion of naftidrofuryl oxalate (*Praxilene*), due partly to the low pH of 3.4 of the solution but also to some factor in naftidrofuryl itself

(Woodhouse and Eadie 1977). The probability of thrombosis occurring increases with the duration of the infusion and those lasting less than two hours do not usually cause thrombophlebitis.

REFERENCES

ADAM, S.A. AND THOROGOOD, M. (1981). Oral contraception and myocardial infarction revisited: the effects of new preparations and prescribing patterns. *Br. J. Obstet. Gynaecol.* **88**, 838–45.

ALKJAERSIG, N., FLETCHER, A.P., AND BURSTEIN, R. (1970). Thromboembolism and oral contraceptive medication. *J. clin. Invest.* **49**, 3a (abstract 7).

——, ——, AND —— (1975). Association between oral contraceptive use and thromboembolism: a new approach to its investigation based on plasma fibrinogen chromatography. *Am. J. Obstet. Gynecol.* **122**, 199–211.

ANDRASSY, K., RITZ, E., HASPER, B., SCHERZ, M., WALTER, E., STORCH, H., AND VÖMEL, W. (1976). Penicillin-induced coagulation disorder. *Lancet* **ii**, 1039–41.

ARANDA, M., SAEZ, M., ABRIL, V., CARARACH, J., CASTELLS, E., AND CASTELLANOS, J.M. (1979). β-Thromboglobulin levels and oral contraception. *Lancet* **ii**, 308–9.

ASTEDT, B., ISACSON, S., NILSSON, I.M., AND PANDOLFI, M. (1973). Thrombosis and oral contraceptives: possible predisposition. *Br. med. J.* **2**, 631–4.

ATKINSON, E.A., FAIRBURN, B., AND HEATHFIELD, K.W.G. (1970). Intracranial venous thrombosis as complication of oral contraception. *Lancet* **i**, 914–5.

BAILAR, J.C. (1967). Thromboembolism and oestrogen therapy. *Lancet* **ii**, 560.

BAUDET, M., MAISONNEUVRE, D., NORMAND, J.P., BOSCHAT, J., AND MATHIVAT, A. (1978). Infarctus du myocarde et contraceptifs oraux. *Annls Méd. intern.* **129**, 456–62.

BELL, G.H., DAVIDSON, J.N., AND SCARBOROUGH, H. (1968). *Textbook of physiology and biochemistry,* 7th edn. Livingstone, Edinburgh.

BERAL, V. (1979). Reproductive mortality. *Br. med. J.* **2**, 632–4.

—— AND KAY, C.R. (1977). Mortality among oral-contraceptive users. Royal College of General Practitioners Oral Contraceptive Study. *Lancet* **ii**, 727–31.

BIERENBAUM, M.L., FLEISCHMAN, A.I., STIER, A., WATSON, P., SOMOL, H., NASO, A.M., AND BINDER, M. (1979). Increased platelet aggregation and decreased high-density lipoprotein cholesterol in women on oral contraceptives. *Am. J. Obstet. Gyneol.* **134**, 638–41.

BOCK, H.E. (1966). Kortikosteroidtherapie neurologischer Erkrankungen mit besonderer Berücksichtigung der Nebenwirkungen. *Ther. Umsch.* **23**, 112–19.

BONNAR, J., HADDON, M., HUNTER, D.H., RICHARDS, D.H., AND THORNTON, C. (1976). Coagulation system changes in postmenopausal women receiving oestrogen preparations. *Postgrad. med. J.* **52**, (suppl. 6), 30–4.

BÖTTIGER, L.E., BOMAN, G., EKLUND, G., AND WESTERHOLM, B. (1980). Oral contraceptives and thromboembolic disease, effects of lowering oestrogen content. *Lancet* **i**, 1097–101.

BOYD, W.N., BURDEN, R.P., AND ABER, G.M. (1975). Intrarenal vascular changes in patients receiving oestrogen-containing compounds — a clinical, histological and angiographic study.

Quart. J. Med. **175**, 415–31.

BRAKEMAN, P. (1968). In *Blood coagulation, thrombosis and female hormones* (ed. T.T. Astrup and I.S. Wright). James F. Mitchell Foundation, Washington.

BRIGGS, M.H. (1981). Megadose vitamin C and metabolic effects of the pill. *Br. med. J.* **283**, 1547.

BRITISH NATIONAL FORMULARY (1985). *Br. Nat. Formulary* **10**, 249–52.

CARPENTIERI, U., GUSTAVSON, L.P., LOCKHART, L.H., AND HAGGARD, M.E. (1976). Adverse reaction to nitrogen mustard therapy. *J. Pediat.* **88**, 1064.

CAZENAVE J.P., GUCCIONE, M.A., PACKHAM, M.A., AND MUSTARD, J.F. (1977). Effects of cephalothin and penicillin G on platelet function *in vitro. Br. J. Haematol.* **35**, 135–52.

CONARD, J., SAMAMA, M., HORELLOU, M.H., ZORN, J.R., AND NEAU, C. (1979). Antithrombin III and oral contraception with progestogen-only preparation. *Lancet* **ii**, 471.

CONWAY, C.M., AND ELLIS, D.B. (1970). Propanidid. *Br. J. Anaesth.* **42**, 249–54.

CRAWFORD, F.E., BACK, D.J., ORME, M.L.E., AND BRECKENRIDGE, A.M. (1981). Oral contraceptive steroid plasma concentrations in smokers and non-smokers. *Brit. med. J.* **282**, 1829–30.

CROCKER, J. AND JONES, E.L. (1983). Haemolytic-uraemic syndrome complicating long-term mitomycin C and 5-fluorouracil therapy for gastric carcinoma. *J. clin. Pathol.* **36**, 24–9.

DANIEL, D.G., CAMPBELL, H., AND TURNBULL, A.C. (1967). Puerperal thromboembolism and suppression of lactation. *Lancet* **ii**, 287–9.

DANIS, R.K., LASRY, G., MELLIERE, D., BEAUMONT, V., AND BOTTO, H. (1978). Thrombose aiguë artérielle sévère sous estrogénothérapie à forte dose. *Nouve. Presse méd.* **7**, 2396.

DARDEL, O. VON, MEBIUS, C., AND MOSSBERG, T. (1976). Diazepam in emulsion form for intravenous usage. *Acta anaesth. scand.* **20**, 221–4.

DODSWORTH, H.I., DEAN, A., AND BROOM, G. (1981). Effects of smoking and the pill on the blood count. *Br. J. Haematol.* **49**, 484–8.

EGEBERG, O. AND OWREN, P.A. (1963). Contraception and blood coagulability. *Br. med. J.* **1**, 220–1.

ELLIOTT, J.L. AND TRASH, D.B. (1979). Intravascular coagulation induced by quinine. *Scot. med. J.* **24**, 244–5.

EWING, A.Y. (1979). Emboli from prosthetic heart valve during postmenopausal oestrogen therapy. *Br. med. J.* **2**, 493.

FAIRBURN, B. (1973). Intracranial venous thrombosis complicating oral contraception: treatment by anti-coagulant drugs. *Br. med. J.* **1**, 647.

FAWER, R., DETTLING, A., WEIHS, D., WELTI, H., AND SCHELLING, J.L. (1978). Effect of the menstrual cycle, oral contraception and pregnancy on forearm blood flow, venous distensibility and clotting factors. *Eur. J. clin. Pharmacol.* **13**, 251–7.

FINN, R. AND ST. HILL, C.A. (1978). Oral contraceptives and subarachnoid haemorrhage. *Lancet* **ii**, 582.

FLORKIEWICZ, H. AND KLAMUT, M. (1966). Thrombophlebitis as a consequence of prednisone administration. *Pol. Tyg. lek.* **21**, 1932–4.

FRICK, G., BRUGMANN, E., AND FRICK, U. (1979). Disseminated intravascular coagulation after the bromsulphthalein test-indocyanin green test as an alternative test. *Mater. Med. Pol.* **11**, 260–4.

FUKUI, H., FUJIMORA, Y., TAKAHASHI, Y., MIKAMI, S., AND YOSHIOKA, A. (1981). Laboratory evidence of DIC under FEIBA treatment of a hèmophilic patient with intracranial bleeding and high titre factor VIII inhibitor. *Thrombosis Res.* **22**, 177–84.

GHAHREMANI, G.G., MEYERS, M.A., FARMAN, J., AND PORT, R.B. (1977). Ischaemic disease of the small bowel and colon associated with oral contraceptives. *Gastrointest. Radiol.* **2**, 221–8.

GODSLAND, I.F., SEED, M., SIMPSON, R., BROOM G., AND WYNN, V. (1983). Comparison of haematological indices between women of four ethnic groups and the effect of oral contraceptives. *J. clin. Path.* **36**, 184–90.

GOODMAN, N., GROSS, J., AND MENSCH, A. (1964). Pulmonary artery thrombosis: A complication occurring with prednisone and chlorothiazide therapy in two nephrotic patients. *Pediatrics* **34**, 861–8.

GORDON, T., KANNEL, W.B., CASTELLI, W.P., AND DAWBER, T.R. (1981). Lipoproteins, cardiovascular disease and death: the Framingham Study. *Arch. intern. Med.* **141**, 1128–31.

GORDON, Y.B., COOKE, E.D., BOWCOCK, S.A., RATKY, S.M., PILCHER, M.E., AND CHARD, T. (1977). Non-invasive screening for venous thromboembolic disease. *Br. J. Haematol.* **35**, 505–10.

GRACE, E., EMANS, J., AND DRUM, D.E. (1982). Clinical and laboratory observations: hematological abnormalities in adolescents who take oral contraceptive pills. *J. Pediat.* **101**, 771–4.

GREENE, R. (1959) Migraine, Part 1. *Br. med. J.,* **1**, 574–5.

GREIG, H.B.W. AND NOTELOWITZ, M. (1975). Natural oestrogens and antithrombin-III levels. *Lancet* **i**, 412–13.

GUTHRIE, D. AND WAY, S. (1974). Treatment of advanced carcinoma of the cervix with adriamycin and methotrexate combined. *Obstet. Gynec.* **44**, 586–9.

HAMPTON, J.B. (1970). Platelet abnormalities induced by the administration of oestrogens. *J. clin. Pathol.* **23**, (suppl. 3), 75–80.

HANUKOGLU, A., FRIED, D., AND GOTLIEB, A. (1980). Diphenyl-hydrantoin induced hypersensitivity reaction with an unusual purpuric skin leison. *Eur. J. Pediat.* **134**, 77–9.

HAUGLUSTAINE, D., VAN DAMME, B., VANRENTERGHEM, Y., AND MICHIELSEN, P. (1981). Recurrent hemolytic uremic sundrome during oral contraception. *Clin. Nephrol.* **15**, 148–53.

HEDLIN, A.N.C. (1975). The effect of oral contraceptive estrogen on blood coagulation and fibrinolysis. *Thrombos. Diathes. haemorrh. Stutt.* **33**, 370–9.

HEWITT, J.C., HAMILTON, R.C., O'DONNELL, J.F., AND DUNDEE, J.W. (1966). Clinical studies of induction agents XIV. A comparative study of venous complications following thiopentone, methohexitone and propanidid. *Br. J. Anaesth.* **38**, 115–18.

HOLT, P.M. AND HOLLANDERS, D. (1980). Massive arterial thrombosis and oral contraception. *Br. med. J.* **280**, 19–20.

HORVATH, T., NAGY, L., AND GOGL, A. (1978). Study of the activity of antithrombin III in latent cholestasis. *Acta med. Scient. Hung.* **35**, 105–13.

HURIEZ, C. AND AGACHE, P. (1966). La gamme des corticotropes en dermatologie. Indications tirées de 2000 cas traités. *Concours méd.* **88**, 4413–28.

INMAN, W.H.W. (1979). Oral contraceptives and fatal subarachnoid haemorrhage. *Br. med. J.* **2**, 1468–70.

—— AND VESSEY, M.P. (1968). Investigation of deaths from pulmonary, coronary and cerebral thrombosis and embolism in women of child-bearing age. *Br. med. J.* **2**, 193–9.

——,——, WESTERHOLM, B., AND ENGELUND, A. (1970). Thromboembolic disease and the steroidal content of oral contraceptives. A Report to the Committee on Safety of Drugs. *Br. med. J.* **2**, 203–9.

JICK, H., DINAN, B., HERMAN, R., AND ROTHMAN, K.J. (1978). Myocardial infarction and other vascular diseases in young women: role of oestrogens and other factors. *J. Am. med. Ass.* **240**, 2548–51.

KAKKAR, V.V., HIGGINS, A.F., SAGAR, S., AND DAY, T.K. (1976) Surgery, the contraceptive pill and postoperative deep vein thrombosis. *Br. J. Surg.* **62**, 162.

KAMATH, K.R., BERRY, A., AND CUMMINS, G. (1981). Acute hypersensitivity reaction to Intralipid. *New Engl. J. Med.* **304**, 360

KAULLA, K.N. VON (1976). Oestrogens and blood coagulation *Triangle* **15**, 9–17.

KOLS, A., RINEHART, W., PIOTROW, P.T., DOUCETTE, L., AND QUILLIN W.F. (1982). Oral contraceptives in the 1980s. *Population Rep* **10**, A191–222.

KORNELL, S. (1966). Adverse effect of corticosteroid therapy in a patient with progressive systemic sclerosis (generalized schler oderma). *Manitoba med. Rev.* **46**, 579–82.

LEDERER, D.A., DAVIES, T., CONNELL, G., DAVIES, J.A., AND MCNICOL G.P. (1973). The effect of carbenicillin on the haemostatic mechanism. *J. Pharm. Pharmacol.* **25**, 876–80.

LURIE, A., OGILVIE, M., TOWNSEND, D.R., GOLD C., MEYERS, A.M., ANI GOLDBERG, B. (1970). Carbenicillin-induced coagulopathy *Lancet* **i**, 1114–15.

MacRAE, K.D. (1980). Thrombosis and oral contraception. *Brit. J hosp. Med.* **24**, 438–42.

MAMMEN, E.F. (1981). Oral contraceptives and blood coagula tion: a critical review. *Am. J. Obstet. Gynecol.* **142**, 781–90

MANN, J.L., THOROGOOD, M., WATERS, W.E., AND POWELL, C. (1975) Oral contraceptives and myocardial infarction in youn, women: a further report. *Br. med. J.* **2**, 631–2.

MAZAL, S. (1978). Migraine attacks and increased platelet aggre gability induced by oral contraceptives. *Aust. NZ. J. Med.* **8** 646–8.

McEWAN, J. (1979). Smoking, age and the pill. *Int. J. gynaecol Obstet.* **16**, 529–34.

MEADE, T.W. (1981a). Oral contraceptives, clotting factors and thrombosis. *Am. J. Obstet. Gynecol.* **142**, 758–61.

—— (1981b). Effects of progestogens on the cardiovascula system. *Am. J. Obstet. Gynecol.* **142**, 776–80.

—— (1982). Recent developments on the association betwee oral contraceptives and cardiovascular disease. *Adv. Dru, React. accident Pois. Rev.* **1**, 243–54.

——, BROZOVIC, M., CHAKRABARTI, R., HOWARTH, D.J., NORTH W.R.S., AND STIRLING, Y. (1976). An epidemiological study o the haemostatic and other effects of oral contraceptives. *Br J. Haematol.* **34**, 353–64.

——, GREENBERG, G., AND THOMPSON, S.G. (1980). Progestogen and cardiovascular reactions associated with oral contracep tives and a comparison of the safety of 50- and 30-µg oestro gen preparations. *Brit. med. J.* **281**, 1157–61.

MINTZ, U., BAR-MEIR, S., AND DE VRIES, A. (1974). Ergotamine induced venous thrombosis. *Postgrad. med. J.* **50**, 244–6.

MONTANARI, C.M.D., VITTORIA, A., ROSSI, U., AND SALA, P. (1979) Human platelet aggregation curve and oral contraception *Acta haematol.* **61**, 230–2.

OLIVER, M.F. (1970). Oral contraceptives and myocardia infarction. *Br. med. J.* **2**, 210–13.

OSTER, P., ARAB, L., KOHLMEIER, M., MORDASINI, R., SCHELLENBERC B., AND SCHLIERF, G. (1981). Effect of estrogens and proges togens on lipid metabolism. *Am. J. Obstet. Gynecol.* **142** 773–5.

PADFIELD, A. (1974). Thrombosis following diazepam. *Br. J. Anaesth.* **46**, 413.

PETERS, J.R., ELLIOTT, J.M., AND GRAHAME-SMITH, D.G. (1979). Effect of oral contraceptives on platelet noradrenaline and 5-hydroxytryptamine receptors and aggregation. *Lancet* ii, 933–6.

PETITTI, D.B. AND WINGERD, J. (1978). The use of oral contraceptives cigarette smoking and risk of subarachnoid haemorrhage. *Lancet* ii, 234–6.

——, ——, PELLEGRIN, F., AND RAMCHARAN, S. (1979). Risk of vascular disease in women: smoking or contraceptives, noncontraceptive oestrogens, and other factors. *J. Am. med. Ass.* **242**, 1150–4.

PFEFFER, R.I., WHIPPLE, G.H., KUROSAKI, T.T., AND CHAPMAN, J.M. (1978). Coronary risk and oestrogen use in postmenopausal women. *Am. J. Epidemiol.* **107**, 479–98.

PITCHER, D. AND CURRY, P. (1979). Emboli from a prosthetic heart valve during postmenopausal oestrogen therapy. *Br. med. J.* **2**, 244–5.

POLLER, L. (1978). Oral contraceptives, blood clotting and thrombosis. *Br. med. Bull.* **34**, 151–6.

——, TABIOWO, A., AND THOMPSON, J.M. (1968). Effects of low-dose oral contraceptives on blood coagulation. *Br. med. J.* **2**, 218–19.

—— (1976). Natural oestrogen replacement therapy and blood clotting. *Postgrad. med. J.* **52**, 28–9.

——, THOMSON, J.M., AND COOPE, J. (1977). Conjugated equinine oestrogens and blood clotting: a follow-up report. *Br. med. J.* **1**, 935–6.

——, ——, OTRIDGE, B.W., YEE, K.F., AND LOGAN, S.H.M. (1979). Effects of manufacturing oral contraceptives on blood clotting. *Br. med. J.* **1**, 1761–2.

RADFORD, D.J. AND OLIVER, M.F. (1973). Oral contraceptives and myocardial infarction. *Br. med. J.* **2**, 428–30.

REES, M. AND DORMANDY, J. (1980). Accidental intra-arterial injection of diazepam. *Br. med. J.* **281**, 289–90.

ROSE, M.B. (1972). Superior mesenteric thrombosis and oral contraceptives. *Postgrad. med. J.* **48**, 430–3.

ROYAL COLLEGE OF GENERAL PRACTITIONERS' ORAL CONTRACEPTION STUDY (1978). Oral contraceptives, venous thrombosis, and varicose veins. *J. R. Coll. gen. Pract.* **28**, 393–9.

—— (1981). Further analyses of mortality in oral contraceptive users. *Lancet* i, 541–6.

RUH-BERNHARDT, D., FINANCE, F., ROHMER, F., AND SINGER, L. (1976). Incidence de la thérapeutique psychotrope sur la thrombogenèse et sur les fonctions plaquettaires. A propos de 4 cas d'accidents thromboemboliques survenus chez des malades traiteés par neuroleptiques et anti-dépresseurs. *L'Encephale* **2**, 239–55.

RUTHERFORD, R.N., HOUGIE, C., BANKES, A.L., AND COBURN, W.A. (1964). The effects of sex steroids and pregnancy on blood coagulation factors. Comparative study. *Obstet. Gynecol.* **24**, 886–92.

SAGAR, S., STAMATAKIS, J.D., THOMAS, D.P., AND KAKKAR, V.V. (1976). Oral contraceptives, antithrombin-III activity, and postoperative deep-vein thrombosis. *Lancet* i, 509–11.

SANDERSON, J.H. AND DELAMORE, I.W. (1973). Changes in platelet thrombotic tendency during oral contraception. *J. Obst. Gynaec. Brit. Comm.* **80**, 639–43.

SANDLER, R.M., BEVAN, P.C., ROBERTS, G.E., EMERSON, C., VOAK, D., DARNBOROUGH, J., AND HEELEY, A.F. (1979). Interaction between sodium valproate and platelets; a further study. *Br.*

med. J. **2**, 1476–7.

SAXENA, K.M. AND CRAWFORD, J.D. (1965). The treatment of nephrosis. *New Engl. J. Med.* **272**, 522–7.

SHAPIRO, S., SLONE, D., ROSENBURG, L., KAUFMAN, D.W., STOLLY, P.D., AND MIETTINEN, O.S. (1979). Oral contraceptive use in relation to myocardial infarction. *Lancet* i, 743–6.

SHAW, A.B. AND GOPALKA. S.K. (1982). Renal artery thrombosis caused by antihypertensive treatment. *Br. med. J.* **285**, 1617.

SIEBERT, W.T., WESTERMAN, E.L., SMILACK, J.D., BRADSHAW, M.W., AND WILLIAMS, T.W., JR. (1976). Comparison of thrombophlebitis associated with three cephalosporin antibiotics. *Antimicrob. agents and chemother.* **10**, 467–9.

SLONE, D., SHAPIRO, S., KAUFMAN, D.W., ROSENBERG, L., MIETHINEN, O.S., AND STOLLEY, P.D. (1981). Risk of myocardial infarction in relation to current and discontinued use of oral contraceptives. *New. Engl. J. Med.* **305**, 420–4.

SMITH, F.R. AND BOOTS, M. (1980). Sodium valproate and bone marrow suppression. *Ann. Neurol.* **8**, 197–9.

SPITTELL, J.A. (1980). Raynaud's phenomena and allied vasospastic disorders. In *Peripheral vascular diseases* (ed. Juergens, Spittell, and Fairbairn), 5th edn, pp. 573–5. Saunders, Philadelphia.

STADEL, B.V. (1981). Oral contraceptives and cardiovascular disease. *New Engl. J. Med.* **305**, 612–18.

THORNTON, J.A. (1970). Methohexitone and its application in dental anaesthesia. *Br. J. Anaesth.* **42**, 255–61.

TSO, S.C., WONG, V., CHAN, V., CHAN, T.K., MA, H.K., AND TODD, D. (1980). Deep vein thrombosis and changes in coagulation and fibrinolysis after gynaecological operations in Chinese — the effect of oral contraceptives and malignant disease. *Br. J. Haematol.* **46**, 603–12.

TURIAF, J., BASSET, G., GEORGES, R., JEANJEAN, Y., AND BATTESTI, J.P. (1962). Advantages, disadvantages and complications, metabolic changes and hormonosecretory disorders, caused by long-term cortisone therapy for asthma with continuous dyspnoea. *Rev. Tuberc., Paris* **26**, 1212–67.

VESEY, C.J., SALOOJEE, Y., COLE, P.V., AND RUSSELL, M.A.H. (1982). Blood carboxyhaemoglobin, plasma thiocyanate, and cigarette consumption: implications for epidemiological studies in smokers. *Br. med. J.* **284**, 1516–18.

VESSEY, M.P. (1970). Thrombosis aid the pill. *Prescribers' J.* **10**, 1–7

—— (1980). Female hormones and vascular disease — an epidemiological overview. *Br. J. Family Planning. Suppl.* **6**, 1–12.

—— AND DOLL, R. (1968). Investigation of relation between oral contraceptives and thromboembolic disease. *Br. med. J.* **2**, 199–205.

—— AND —— (1969). Investigation of relation between oral contraceptives and thromboembolic disease. A further report. *Br. med. J.* **2**, 651–7.

——, MCPHERSON, K., AND JOHNSON, B. (1977). Mortality among women participating in the Oxford Family Planning Association contraceptive study. *Lancet* ii, 731–3.

——, ——, AND YEATES, D. (1981). Mortality in oral contraceptive users. *Lancet* i, 549–50.

WEILY, H.S., STEELE, P.P., DAVIES, H., PAPAS, G., AND GENTON, E. (1974). Platelet survival in patients with substitute heart valves. *New Engl. J. Med.* **290**, 534–7.

WEITEKAMP, M.R. AND ABER, R.C. (1983). Prolonged bleeding times and bleeding diathesis associated with moxalactam administration. *J. Am. med. Ass.* **249**, 69–71.

WESSLER, S., GITEL, S.N., WAN, L.S., AND PASTERNACK, B.S. (1976). Estrogen-containing oral contraceptive agents: a basis for their thrombogenicity, *J. Am. med. Ass.* **236**, 2179–82.

WINKLER, M.J. AND TRUNKEY, D.D. (1981). Dopamine gangrene; association with disseminated intravascular coagulation. *Am. J. Surg.* **142**, 588–91.

WOODHOUSE, C.R.J. AND EADIE, D.G.A. (1977). Severe thrombophlebitis with Praxilene. *Br. med. J.* **1**, 1320.

WYNN, V. AND NITHTHYANTHAN, R. (1981). The effect of progestins in combined oral contraceptives on serum lipids with special reference to high-density lipoproteins. *Am. J. Obstet. Gynecol.* **142**, 766–72.

ZARRABI, M.H., ZUCKER, S., MILLER, F., DERMAN, R.M., ROMANO, G.S., HARTNETT, J.A., AND VARMA, A.O. (1979). Immunologic and coagulation disorders in chlorpromazine-treated patients. *Ann. intern. Med.* **91**, 194–9.

ZUCKNER, J., UDDIN, J., AND RAMSLEY, R.H. (1967). Prolonged effect from intramuscular corticosteroids. Triamcinolone acetonide in rheumatoid arthritis. *Acta rheum. scand.* **12**, 307–17.

14 Drug-induced cardiovascular disease

M.E. SCOTT

Introduction

Drug-induced cardiovascular disease may be considered under several headings. Firstly, we consider adverse cardiovascular reactions to drugs used to treat disorders of the heart and circulation, such as, heart failure, hypertension, or cardiac arrhythmias. With the increasing range of drugs available, problems of adverse drug interaction have become more numerous. Secondly, the cardiovascular system may be damaged by drugs used to treat non-cardiac disorders, such as, cancer, depression, or diabetes, or by hormones given to treat menstrual disorders or to prevent pregnancy. Thirdly, there may be damage to the heart and blood vessels from drugs aimed at preventing coronary heart disease.

Reactions to drugs used to treat cardiovascular disorders

Digitalis

The wider availability of serum digoxin assay has helped to clarify certain aspects of digitalis toxicity. The features of digitalis toxicity such as ventricular bigeminy, multiform ventricular extrasystoles, paroxysmal atrial tachycardia with block, ventricular tachycardia, and disturbances of conduction may all occur in other situations. In the absence of hypokalaemia, a low plasma digoxin level strongly suggests that an arrhythmia or conduction disturbance is not due to digitalis therapy (Opie 1980*a*). A slow heart rate is a poor indicator of digitalis toxicity (Williams *et al*. 1978). The use of digitalis after myocardial infarction can provoke sudden death. (Muller *et al*. 1983).

Special attention has been focused on the digitalis–quinidine interaction (Bigger 1979). When both drugs are administered together, the serum digoxin level rises substantially (Leahey *et al*. 1979). This is due to a decrease in the volume of distribution and clearance of digoxin (Schenck-Gustafsson and Dahlqvist 1980). Other mechanisms which contribute to the quinidine–digitalis interaction are reviewed by Bussey (1984). These include renal and non-renal clearance. Quinidine also increases the bio-availability of digoxin. When both drugs are indicated concurrently the dose of digoxin should be halved and serial measurements of serum digoxin made (Doering 1979). Conversely, when quinidine is omitted from such a regime, the dose of digoxin should be increased (Moench 1980). Verapamil also increases serum digoxin concentrations (Klein *et al*. 1982).

Moysey *et al*. (1981) observed that several patients receiving maintenance digoxin treatment abruptly developed clinical evidence of toxicity when amiodarone was administered in addition. Therefore, as in the case of quinidine therapy, it is recommended that the maintenance dose of digoxin be reduced when amiodarone is introduced. Despite this, intravenous amiodarone has been used successfully to treat a case of massive digoxin overdose which produced repeated ventricular fibrillation (Maheswaran *et al*. 1983). These authors suggested that its ability to reduce cardiac excitability at all levels probably contributed to the survival of their patient. Life-threatening digitalis intoxication is happily becoming less common, but still presents difficulties in management. In Boston, Smith and his colleagues (1982) developed purified digoxin-specific Fab. fragments of digoxin-specific antibodies obtained from sheep, and used them with considerable success to treat 26 patients with life-threatening digoxin or digitoxin toxicity. Cardiac rhythm disturbances and hyperkalaemia were reversed rapidly. There were no reported adverse reactions to the treatment. Treatment of digoxin overdose with antigen-specific fragments of digoxin-specific antibodies has also been used successfully in Britain (Rozkovec and Coltard 1982). Another case was described in detail by Harenberg *et al*. (1983). Recently the author used this treatment in the successful management of a man who had taken 12.5 mg of digoxin in a suicide bid, and whose plasma digoxin reached a peak of 19 μgl^{-1}. The problem of digitalis intoxication and its management was well reviewed by George (1983).

β-Adrenergic blocking drugs

After many millions of patient-treatment years, the established β-blocking drugs have emerged as being relatively safe. Though excessive bradycardia is often seen in patients responding adversely to normal or excessive dosage of β-blocking drugs, occasional examples of tachycardia have been recorded after excessive doses (Elonen et al. 1979).

The toxic effects of propranolol overdosage, including central nervous and cardiac effects, were well described by Buiumsohn et al. (1979). Occasional case reports linking specific β-blocking drugs to specific toxic reactions appear. Drug fever was described in a patient on oxprenolol (Hasegawa et al. 1980). A case of proximal myopathy was described associated with both sotalol and propranolol (Forfar et al. 1979).

In view of the spate of recent papers suggesting that β-blockers are beneficial in the management of patients with congestive cardiomyopathy, the report by Taylor and Silke (1981) suggests that great caution should be exercised. These authors found that in patients with heart failure on the basis of ischaemic myocardial damage, a small dose of oxprenolol further depressed left ventricular performance in all patients studied.

Although β-adrenergic blocking drugs reduce myocardial oxygen consumption in patients with coronary artery disease, propranolol has been reported to exacerbate coronary spasm in some patients with variant angina. A study by Kern et al. (1983) confirmed that propranolol can potentiate coronary vasoconstriction in some patients with coronary disease, possibly mediated by unopposed α-adrenergic vasomotor tone.

Anaphylaxis with cardiovascular collapse may be much more severe and much more resistant to treatment in patients receiving β-blocking drugs (Jacobs et al. 1981). Hannaway and Hopper (1983) described five illustrative cases and suggest that anaphylaxis potentiated by β-blockers is not an uncommon event.

With β-blocking drugs now used on a vast scale in the treatment of hypertension, two potential problems have continued to receive attention. The first is the use of β-blockers in pregnancy (Rubin 1981; Thorley et al. 1981; Caldroney et al. 1982). On the whole the β-blockers, especially propranolol, appear relatively safe in this condition, though often rather ineffective as the sole antihypertensive agent (Lubbe et al. 1982). However, in a report describing persistent bradycardia, hypotension, hypothermia, and poor peripheral perfusion in a newborn infant whose mother had been treated with atenolol, Woods and Morrell (1982) suggested that, if a cardioselective β-blocking drug is used to treat hypertension in pregnancy, a short-acting agent be chosen. The other anxiety concerns the rise in plasma lipids noted in patients given β-blocking drugs for the long-term management of mild hypertension. The general pattern in most studies is an increase in the concentration of very low-density lipoproteins and a decrease in the concentration of high-density lipoproteins during beta-blockade (Rossner 1982). These changes may be dose-related, as it has been shown that use of only 50 mg of atenolol daily did not produce these changes (Rossner and Weiner 1982).

Studies have shown that β-adrenergic blocking agents increase plasma triglyceride and urate levels, and reduce concentrations of high-density lipoprotein cholesterol (Leren et al. 1980). In order to determine if these effects persisted, Kristensen (1981) studied 47 patients with essential hypertension who had been treated with β-blockers for an average of 54 months. He compared them with 52 patients taking other antihypertensive medications. He found no significant differences in plasma triglyceride, total cholesterol, HDL-cholesterol, very low density lipoproteins, or urate in the two groups. It appears that the alteration in the plasma concentrations of these components is relatively short-lived, suggesting that β-blockers are unlikely to accelerate atherosclerosis, at least in this way.

Another concern related to the long-term use of β-blocking drugs has been that they might cause deterioration in glucose tolerance, and in this way promote atherosclerosis. Myers and Hope-Gill (1979) showed that infusion of propranolol reduces insulin response to a glucose load in normal subjects. However, a recent controlled study (Woods et al. 1981) has shown that neither propranolol nor the cardioselective β-blocker, metoprolol, had a significant effect on fasting plasma glucose, glucose tolerance, or insulin response. The β-blockers should therefore not produce long-term adverse effects on the cardiovascular system.

Several papers have appeared documenting the interaction between various β-blockers and cimetidine (Donovan et al. 1981; Kirch et al. 1981). It is now clear that extreme care needs to be taken when cimetidine is given simultaneously with propranolol or metoprolol, and probably with any β-blocking drug which is not almost solely excreted by the kidneys.

Interest has focused on metoprolol in relation to sclerosing obstructive peritonitis in patients on continuous ambulatory peritoneal dialysis, and Grefberg et al. (1983) encountered this complication in a patient treated with metoprolol and atenolol. It has been suggested that patients with hypertension and renal failure who develop sclerosing peritonitis while on a β-blocking drug, may already have had this condition prior to the exhibition of the drug, and may in fact have developed hypertension as a result of the sclerosing peritonitis (Pryor *et al.* 1983). The validity of this thesis was questioned by Bullimore (1983). Therefore it remains important for doctors using β-blockers to treat hypertension to be on the lookout for this serious development, which can in turn aggravate the hypertension.

Other antihypertensive drugs

Methyldopa

Methyldopa is losing ground to the β-adrenergic blocking drugs in the treatment of hypertension but is still prescribed (Opie 1984*a*). Toxic effects are few but serious (Opie 1980*b*). New adverse reactions to methyldopa continue to be described. Alfiro *et al.* (1981) described two cases of well-documented carotid sinus hypersensitivity producing syncopal attacks. In one of these rechallenge a month later reproduced the condition. A lupus-like syndrome was described by Dupont and Six (1982); withdrawal of the drug was followed by disappearance of the features of lupus, which included Raynaud's phenomenon, arthralgia, general weakness, morning stiffness, and a positive direct Coomb's test.

Hydrallazine

Hydrallazine continues to enjoy a revival of popularity in the treatment of systemic arterial hypertension, and is increasingly widely accepted as useful in the management of severe heart failure. The suspicion that tolerance to hydrallazine can develop and permit relapse of heart failure has been confirmed. The mechanism has been investigated in detail (Cogan *et al.* 1981; Packer *et al.* 1982*a*). Most such patients are already receiving digoxin. A factor contributing to tolerance may be a fall in plasma digoxin caused by increase in renal digoxin clearance, as a result of improved cardiovascular dynamics following hydrallazine therapy. A case of hypertension developing after the taking of hydrallazine has appeared, attributed to an increase in the pressure gradient across a stenosed renal artery (Webb and White, 1980).

Controversy has raged as to whether hydrallazine is useful and safe in the management of pulmonary arterial hypertension. Lupi-Herrera *et al.* (1982) documented beneficial effects in pulmonary hypertension of unknown cause, but Packer *et al.* (1982*b*) demonstrated deleterious effects. Several of their patients developed tachycardia and became symptomatically hypotensive within 24 hours of the initiation of treatment. They concluded that hydrallazine fails to produce consistent haemodynamic and clinical benefits in patients with primary and secondary pulmonary hypertension and that it frequently causes serious adverse reactions.

While hydrallazine in doses of 200 mg or less produces only a low incidence of side-effects (Chatterjee *et al.* 1978), in patients with slow-acetylator status and those with impaired renal function the blood concentrations will rise even with small doses. The author has personally observed a florid lupus erythematosus-like syndrome, with life-threatening pericardial effusion, develop in a hypertensive man with impaired renal function who was taking 150 mg hydrallazine daily for six years. The effusion and the other signs regressed completely on stopping hydrallazine and administering corticosteroids. The need for caution was also emphasized by Bing *et al.* (1980). These authors found that 3 per cent of patients developed a lupus-like syndrome while taking hydrallazine. All were slow acetylators and had not received more than 200 mg daily. Autoimmune phenomena were also more frequent in asymptomatic slow acetylators receiving 200 mg hydrallazine daily or less. Harland *et al.* (1980) suggested that a person may be a rapid acetylator of sulphadimidine and yet be a relatively slow acetylator of hydrallazine. They described a hydrallazine-induced lupus erythematosus-like syndrome in a patient of the rapid acetylator phenotype and suggested that a metabolic pathway other than acetylation may be concerned in the development of toxicity. The commonest skin manifestations are erythematous rashes, but two patients with ulcerating cutaneous vasculitis associated with hydrallazine treatment were described (Bernstein *et al.* 1980). The advent of endralazine, a new peripheral vasodilator, the antihypertensive effect of which appears to be uninfluenced by acetylator status is welcome (Holmes *et al.* 1983; Meredith *et al.* 1983).

Labetalol

The antihypertensive drug labetalol produces relatively few toxic effects. Some instances of muscle pain have been described, and a case of labetalol-induced toxic myopathy was reported by Teicher *et al.* (1981). This disappeared when the drug was withdrawn and reappeared on re-exposure to the drug. Labetalol also interacts with cimetidine, necessitating caution when the two drugs are used together (Daneshmend and Roberts 1981).

Labetalol, having both α- and β-adrenoceptor blocking actions can produce a hypertensive response in patients with phaeochromocytoma (Briggs *et al.* 1978). It was suggested (Reach *et al.* 1980) that this could be due to inadequate α-adrenergic blockade by too low a dose of the drug.

Indoramin

Indoramin is an α-adrenergic blocking agent which was

released recently in the United Kingdom for the treatment of hypertension. Reports of iatrogenic disorders associated with its use have appeared in the past three years. These have mainly related to the nervous system. In the first reported case of fatal self-poisoning with indoramin (Hunter 1982), death was due to severe depression of the central nervous system terminating with respiratory depression, bradycardia unresponsive to atropine and isoprenaline, and finally cardiac arrest. The fatal outcome may have been partly due to a very high concentration of plasma alcohol. There were also plasma levels of temazepam and oxazepam which were in the therapeutic range.

Captopril

Captopril, an angiotensin converting enzyme inhibitor has been available for clinical use for four years in the UK. It is proving very useful in the management of hypertension which fails to respond to the traditional first-line drugs. Dargie *et al.* (1983) listed many adverse effects which have been associated with its use. These include agranulocytosis, reversible lymphadenopathy, and pancytopaenia. In several of these reports the patients had renal insufficiency, indicating that in such patients frequent blood counts should be made during treatment with captopril. Immune complex glomerulopathy has also been described (Hoorntje *et al.* 1980).

A rise in plasma potassium occurs. This could produce cardiac dysrhythmias, especially in patients on digitalis for heart failure. The risk of hyperkalaemia is increased with concomitant administration of potassium supplements, or potassium-sparing agents, such as triamterene, amiloride, or spironolactone (Vidt *et al.* 1982; Textor *et al.* 1982).

An occasional problem with captopril is the occurrence of a severe first-dose effect when used to treat hypertension. Hodsman *et al.* (1983) reported that six out of 65 severely hypertensive patients developed symptoms of acute hypotension, including dizziness, stupor, dysphagia, and hemiparesis. They could not reliably predict a severe first-dose effect in individual patients, and recommend a close supervision of such patients for at least three hours after the first dose of captopril.

Diuretics

Thiazide

For many years thiazide diuretics have been used extensively in the treatment of patients with hypertension.They have been regarded as relatively free from adverse effects. However, the tendency is towards treating even mild hypertension with drugs. This results in the long-term treatment of very large number of younger people who are at relatively low risk from their medical condition. In this situation even mildly adverse consequences of treatment must be viewed with concern. There is growing disquiet about the long-term adverse metabolic effect of oral diuretic therapy (Murphy *et al.* 1982; Perez-Stable and Caralis 1983).

In the vast Multiple Risk Factor Intervention Trial (MRFIT), it appeared that thiazides may increase coronary mortality in hypertensive patients with an abnormal electrocardiogram (MRFIT Research Group 1982). Hypokalaemia, which tends to be intermittent, especially in hypertensive patients, is the best known effect of long-term thiazide therapy. The Veterans' Administration Cooperative Study Group on antihypertensive agents (1982) reported a correlation between a fall in serum potassium and ventricular ectopic activity. Because it may produce no symptoms until sufficiently severe to cause serious cardiac dysrhythmias, hypokalaemia must be sought biochemically. If acute infarction supervenes in its presence, the risk of death from ventricular fibrillation is doubled, perhaps because the release of adrenaline augments the adverse effect of hypokalaemia (Maclean and Tudhope 1983). Diuretics may also induce hypomagnesaemia (Swales 1983).

Other adverse effects of long-term thiazide diuretic therapy include raised blood sugar, uric acid, and possibly cholesterol, all of which may increase deaths from myocardial infarction (*Lancet* 1982a) Taken together, these adverse effects on the cardiovascular system signal the need for careful consideration of the risk–benefit ratio when prescribing them for mild hypertension. Possibly the calcium antagonists such as nifedipine may be preferable in those who can tolerate them (Murphy *et al.* 1983).

Frusemide

The loop diuretic frusemide has been implicated in increasing the incidence of patent ductus arteriosis in premature infants. (Green *et al.* 1983). The drug crosses the human placenta, and the plasma half-life is increased eightfold in neonates. It is also excreted in breast milk. Therefore Cohen (1983) recommended caution in the use of frusemide in mothers both before delivery and in the post-partum period.

Vasodilators for heart failure

Drugs such as prazosin, nitrates, nitroprusside, hydrallazine, and captopril, are being used increasingly in the management of acute and chronic heart failure (Wenting *et al.* 1983). When captopril is used in the treatment of heart failure, combined treatment with frusemide can result in hypotension, which could compromise coronary blood flow, particularly during exercise (Wenting *et al.* 1983). Because a fall in a plasma sodium has also been established (Nicholls *et al.* 1981), plasma sodium, as well as potassium concentration, should be monitored during captopril treatment for heart failure.

Enalapril maleate, another orally active converting enzyme-inhibitor is devoid of the sulphydryl group which is believed to cause many of the side-effects associated with captopril (Hodsman *et al.* 1982). The first reported case of suicidal overdosage with enalapril (300 mg) was described by Waeber *et al.* (1984). Oxazepam was also ingested. The patient became mildly hypotensive but this was well tolerated and was easily reversed by supportive infusion of fluid. Despite their drawbacks it is now established that the introduction of orally active converting enzyme-inhibitors constitutes a major advance in cardiac treatment (Hodsman and Robertson 1983; Braunwald and Colucci 1984).

The common adverse effects arising from the use of vasodilators in the management of heart failure include postural hypotension, syncope and tachycardia, but a major problem is the adverse and potentially dangerous effect of sudden cessation of treatment for any reason. Serious worsening of the heart failure was documented following sudden withdrawal of nitroprusside (Packer *et al.* 1979), hydrallazine (Black and Metha 1979), prazosin (Hanley *et al.* 1980), and captopril, (Dzau *et al.* 1980). The abrupt withdrawal of any drug acting on the cardiovascular system may produce a rebound if its duration of action is shorter than the readjustment of the homeostatic mechanisms activated (Gerber and Nies 1979). Evidence for such an adverse reaction has been well documented with clonidine and nitroglycerin, and it is likely that other drugs which alter cardiovascular haemodynamics or sympathetic function, and which have short durations of action will be found to produce a rebound after abrupt withdrawal.

Vasodilators such as glyceryl trinitrate are sometimes injected directly into the coronary arteries, usually to prevent or reverse coronary spasm. This practice has given rise to iatrogenic problems. Several cases of ventricular fibrillation have been encountered, including one of which the author has personal knowledge. Webb *et al.* (1983) noted that one formulation of trinitrate (*Tridil*) has a high potassium content and advised against its use by this route.

A case of marked hypersensitivity to prazosin was described (Ruzicka and Ring 1983). A prick test with prazosin led to a fall in blood pressure, dyspnoea, palpitation, and facial swelling, requiring treatment with corticosteroids and antihistamines. Two cases of priapism affect-

ing West Indians were described associated with prazosin therapy (Bhalla *et al.* 1979). A single report of hypothermia apparently associated with prazosin therapy appeared (de Leeuw and Birkenhäger 1980).

Oxpentifylline (*Trental*) is a vasodilator which is also used to reduce blood viscosity. Sznajder *et al.* (1984) described bradycardia and second-degree heart block in a patient taking the first documented massive overdose of the drug.

Calcium antagonists

The slow-channel calcium-blocking drugs include nifedipine, verapamil, prenylamine, diltiazem, indapamide, and perhexiline. The first to be widely used in this country was verapamil. This drug acquired a bad name for producing severe side-effect when occasional reports appeared of fatalities resulting when intravenous verapamil was given to patients with pre-existing atrioventricular disease, or who were taking β-blocking drugs (Sacks and Kennelly 1972). In a major study of the effects of giving oral verapamil to patients receiving high doses of β-blocking drugs, Packer *et al.* (1982c) concluded that combination therapy is potentially dangerous and should be used with caution in patients with angina pectoris. This supports the conclusion of Denis *et al.* (1977) who reported a group of patients receiving propranolol who developed circulatory collapse after intravenous verapamil. Another patient developed cardiogenic shock when propranolol was added to maintenance verapamil therapy. Kieval *et al.* (1982) added further support to the view that combining these two therapies is hazardous.

Since verapamil can shorten the refractory period of the accessory pathway in the Wolff–Parkinson–White syndrome, it too can cause an acceleration of the ventricular response during atrial fibrillation. Gulamhusein *et al.* (1982) found that, in four of eight patients, atrial fibrillation that was self-terminating before verapamil became sustained after the drug was given, and electrical conversion was required in two patients. Perrot *et al* (1984) described a case of sudden death occurring in a man with hypertrophic cardiomyopathy and atrial fibrillation which they felt could be attributed to verapamil. At the time of death the patient was wearing a Holter monitor which showed development of complete heart block followed by asystole. The potential of causing serious complications in patients with hypertrophic cardiomyopathy has been recognized before (Epstein and Rosing 1981).

Nifedipine has been used extensively in the treatment of angina pectoris. As Opie (1984b) indicated, very few life-threatening adverse reactions have been described with nifedipine. However, the drug has caused excess hypotension or cardiovascular depression in a few seriously ill patients. As the indications for its use widen to include heart failure, a note of caution must be sounded. Several reports of life-threatening adverse reactions to nifedipine have appeared. In two of six patients with severe heart failure treated by Brooks *et al.* (1981) there was a dangerous fall in cardiac output. Gillmer and Kark (1980) also reported a patient in whom pulmonary oedema was precipitated by nifedipine, indicating that caution should be exercised when using the drug in patients with poor left ventricular function. As with verapamil, this is particularly so when nifedipine is given along with β-blocking drugs (Anastassiades 1980; Opie and White 1980). Jee and Opie (1983) reported two cases in which the administration of nifedipine with prazosin caused an acute hypotensive response. They suggested that great care be taken when prescribing additive treatment with these two drugs.

Other calcium antagonists such as diltiazem, prenylamine, indapamide, and perhaps perhexiline probably share with nifedipine the potential to cause heart failure (Opie 1980*b*). Diltiazem has a significant negative chronotropic effect. This can lead to severe bradycardia. This drug also has a potent depressant effect on both impulse formation and atrioventricular conduction (Mitchell *et al.* 1982). Ishikawa *et al.* (1983) reported three patients, without symptoms of sinus-node dysfunction, who developed atrioventricular dissociation with diltiazem. One of these also developed severe sinus bradycardia and one sinus arrest.

The calcium antagonists have been particularly useful in treating patients with cerebral ischaemia. However, an occasional report has linked nifedipine with worsening of cerebral ischaemia (Nobile-Orazio and Sterzi 1981). These authors described two instances of cerebro-vascular accidents following closely on the administration of nifedipine. However, since both patients were elderly and had pre-existing factors for cerebral ischaemia, the cerebrovascular accidents were probably due to a non-specific reduction in cerebral blood flow in susceptible individuals, rather than a specific effect of the drug on the cerebral circulation. An unusual complication of nifedipine therapy, the Capgras syndrome, in which a loved one is believed to be an imposter, was described in an elderly man by Franklin *et al.* (1982). The author suggested that nifedipine-induced cerebral ischaemia may have played a role. However, there were other factors in the case, such as the patient's wife's rejection of him, which blur the causative role of nifedipine.

There is debate as to whether there is a calcium antagonist withdrawal syndrome similar to that which may follow abrupt withdrawal of β-blocking drugs (Nehring and Camm 1981). Subramanian *et al.* (1983) observed the effects of withdrawal of calcium antagonists in 143 patients using ambulatory ST segment monitoring with a frequency modulated tape recorder. They reported several apparent cases of sharply increased symptoms and signs of ischaemia. However, other workers have not observed this. In a placebo-controlled double-blind randomized cross-over study, Frishman *et al.* (1982) found no evidence of deterioration from placebo baseline levels following abrupt withdrawal of verapamil, whereas two patients had a severe exacerbation of angina following abrupt withdrawal of propranolol.

Anti-arrhythmic drugs

Since the second edition of *Iatrogenic diseases* a number of isolated case reports have added a little to our knowledge of the side-effects of these commonly administered drugs. Myocarditis resolving after discontinuation of procainamide was described by Myers *et al.* (1983). Although procainamide-induced pericarditis is well-known, myocarditis without pericarditis had not previously been reported. However, this case was complicated by the presence of an underlying cardiomyopathy which may have influenced the adverse reaction to the drug. A case of *torsades de pointes* produced by N-acetyl procainamide was reported (Olshansky *et al.* 1982). This drug, a metabolite of procainamide produced by the liver, has anti-arrhythmic potency independent of procainamide. Kluger *et al.* (1981) reported sudden deaths in four patients on N-acetyl procainamide, with drug plasma levels ranging from 15 to $66\mu g/ml$. It would appear that the plasma level of this drug can be critical in determining its usefulness or danger. Keidar *et al.* (1982) described sinoatrial arrest due to an injection of lignocaine in a patient with the sick-sinus syndrome who was already receiving amiodarone. It was probably the interaction between these two drugs, both of which may depress the sinus node, especially in patients with the sick-sinus syndrome, which caused the severe sinoatrial arrest. Since the second edition a new class I anti-arrhythmic drug flecainide acetate has been introduced. It is very potent and produces slight depression of myocardial performance in patients with coronary disease. Flecainide has been reported to enhance the degree of block in patients with underlying conduction disturbance, to abolish escape rhythms, and to increase pacing threshold, all potentially dangerous adverse effects. Therefore, it was concluded that flecainide acetate, while very effective, has rare but serious side-effects (*Lancet* 1984).

Amiodarone

Amiodarone has been available for many years in continental Europe and South America, but was only released on the British market for general use in 1981. It has valuable anti-arrhythmic properties associated with prolongation of the action potential duration and conduction velocity (Rowland and Krikler 1980; Ward *et al.* 1980). Amiodarone has been used frequently in the management of resistant supraventricular and ventricular dysrhythmias. Its therapeutic advantages are counteracted by adverse effects such as corneal deposits (Verin *et al.* 1971), cutaneous disorders, and altered thyroid function (McKenna *et al.* 1983; McGovern *et al.* 1983). Amiodarone can cause myocardial depression (Sicart *et al.* 1977).

Rees *et al.* (1981) emphasized that amiodarone, which is highly protein-bound, can interact with the anticoagulant warfarin. They described a patient in whom on two occasions the addition of amiodarone led to dangerous increases in the anticoagulant effects of warfarin treatment. This was confirmed in several other reports (Martinowitz *et al.* 1981; Serlin *et al.* 1981; Hamer *et al.* 1982). There is a high risk of haemorrhage and the effect can take several months to wear off.

Pneumonitis and pulmonary fibrosis are also well recognized complications of amiodarone treatment (Sobol and Rakita, 1982; McGovern *et al.* 1983; Forgoros *et al.* 1983). The dosage of amiodarone being received by many of these patients was greatly in excess of the level of 200–400 mg daily which is normally recommended.

In the management of cardiac arrhythmias a major cause of concern is the possibility that drug therapy may make the arrhythmia under treatment worse, or may provoke a much more dangerous arrhythmia. McComb *et al.* (1980) described amiodarone-induced ventricular fibrillation. In a few patients amiodarone increases the QT interval but this is sometimes difficult to detect because of the presence of U waves (McKenna *et al.* 1983). If it is considerable, this QT prolongation may indicate a risk of arrhythmogenicity and occasional but well-documented cases have been reported of patients who have developed *torsades de pointes* (Sclarovsky *et al.* 1983). *Torsades de pointes* has also been observed with quinidine, propafenone and mexiletine when given with amiodarone. Amiodarone has a marked effect on the kinetics of these and other commonly used anti-arrhythmic drugs such as digoxin, as well as having dynamic interactions with other drugs such as β-blockers and some calcium antagonists (Marcus 1983). Tartini *et al.* (1982) noted *torsades de pointes* in four patients given combined treatment with amiodarone/disopyramide, amiodarone/propafenone and amiodarone/mexiletine.

Velebit *et al.* (1982) noted a worsening of arrhythmia in 80 of 722 drug tests in 53 of 155 patients being treated for ventricular tachyarrhythmias. Such aggravation was noted with quinidine, procainamide, disopyramide, propranolol, metoprolol, pindolol, aprindine, mexiletine, and tocainide. Drug levels were always in the 'therapeutic' range. They conclude that a systematic approach to anti-arrhythmic drug testing is required before a patient is put on long term anti-arrhythmic therapy.

Ruskin *et al.* (1983) used electrophysiologic techniques to evaluate six patients who survived cardiac arrest occurring outside a hospital. All had been taking anti-arrhythmic drugs and had been without clinical evidence of drug toxicity at the time of the cardiac arrest. The anti-arrhythmic drugs involved were quinidine, disopyramide, quinidine/procainamide, and quinidine/disopyramide. When these were withdrawn, no arrhythmia could be induced by programmed cardiac stimulation; but when they were rechallenged with the same drugs, ventricular tachycardia could be initiated in four patients and high-grade atrioventricular block in another one. These observations confirm that anti-arrhythmic drugs may contribute to the occurrence of cardiac arrest in some patients.

Inotropic agents

Adrenaline

The ability of catecholamines to cause myocardial damage is well established. These hormones have been implicated in the myocardial necrosis occurring in some patients with phaeochromocytoma. A recent case is reported of severe myocardial ischaemia induced by intravenous adrenaline given for a suspected anaphylactic reaction (Horak *et al.* 1983).

Amrinone and milrinone

Amrinone is a bipyridine derivative which has positive inotropic properties unrelated to sympathomimetic or cardiac glycoside activity. It also has vasodilator properties. Therefore, large doses can reduce preload excessively and lower cardiac output, as occurred in two cases described by Wilmshurst *et al.* (1983). It often produces potentially serious side-effects, especially thrombocytopaenia (Wilmshurst and Webb-Peploe 1983). Milrinone, a derivitive of amrinone with nearly 20 times the inotropic potency has not produced these adverse effects in preliminary testing and shows promise for long-term treatment (Baim *et al.* 1983).

Adverse reactions to drugs used to treat non-cardiac conditions

Antineoplastic drugs

Anthracyclines

Anthracycline toxicity can present in several ways. Acute effects include various arrhythmias, mainly supraventricular, and effects on myocardial contractility and coronary vascular resistance. Subacute effects include transient left ventricular dysfunction and a rare pericarditis–myocarditis syndrome (Bristow *et al*. 1978). It is the chronic cardiotoxicity of doxorubicin which causes by far the greatest concern, since all patients given the drug develop some degree of cardiomyopathy (Bristow 1980). All the commonly used anthracyclines produce cardiomyopathic effects involving individual myocardial cells. In its most advanced form anthracycline toxicity presents with a typical clinical picture of congestive cardiac failure. This can develop rapidly and has led to the setting of an empiric upper dose limit of 550 mg m^{-2}. However, individual sensitivity to the drugs varies greatly (McKillop *et al*. 1983). In some patients and especially in children this dose may be too high.

In a major study of 389 children treated with doxorubicin, Goorin *et al*. (1981) reported 15 cases of congestive heart failure. Twelve survived, but six of the nine who did not die of their malignancies had persistently abnormal left ventricular function as measured by the echocardiogram. The toxicity of doxorubicin is discussed in a major review article by Young *et al*. (1981).

The author observed two 13-year-old patients who died of intractable heart failure after receiving doses of doxorubicin of 535 and 540 mg m^{-2}, respectively. Both had been receiving methotrexate and had a single dose of vincristin. A third patient, a child of seven years, also developed heart failure after receiving 400 mg m^{-2}. However, this was detected at an early stage as a result of a programme of intensive echocardiographic monitoring, instituted after the two fatalities. Happily, the child responded to treatment with digoxin and diuretics. Though the echocardiogram still shows reduced left ventricular function, the patient has no overt symptoms or signs of heart failure three years after the last dose of doxorubicin. It may be that in children the safe upper dosage level should be revised downwards, probably to below 400 mg m^{-2}, especially when doxorubicin is used in conjuction with other antimitotic agents.

The usefulness of measuring serial radionuclide ejection fractions in the assessment of doxorubicin cardiotoxicity was established recently in patients with normal and abnormal baseline resting left ventricular performance (McKillop *et al*. 1983; Choi *et al*. 1983). The sensitivity of the technique is improved by exercise. Using this technique it has been confirmed that some patients will develop congestive heart failure at doses lower than those usually considered safe.

Efforts are being made to discover ways of protecting patients from the cardiotoxicity of anthracyclines without impairing their effectiveness as antitumour agents. It was shown (Bristow *et al*. 1979) that combined histaminic and adrenergic blockade prevents nearly all the cardiomyopathic effects of adriamycin in the rabbit model. Such blockades do not affect the antitumour response in mice. Vitamin E has also conferred some protection on mice. Digoxin has been claimed to be valuable in patients treated with adriamycin, but perhaps it simply masks the adverse changes (Whittaker and al-Ismail 1984). Ultimately it may be possible to manipulate the structure–activity relationships of the anthracyclines, producing an active analogue devoid of cardiotoxicity, but at present all patients should be monitored for early signs of cardiac damage.

Vinblastine

Reports linking vinblastine therapy, with or without associated radiotherapy, and myocardial ischaemia and infarction have appeared. Harris and Wong (1981) described two such cases, but both had several preexisting risk factors for coronary artery disease and the evidence that the vinblastine was responsible for the attacks is poor. However, Lejonc *et al*. (1980) described myocardial infarction following vinblastine in a patient without major coronary risk factors. It would seem prudent to do at least a resting electrocardiogram in all patients, including children, before commencing vinblastine.

Antidepressant drugs

Tricyclic drugs

The adverse cardiovascular reactions to the tricyclic antidepressants must be considered in relation to the dangers of alternative methods of treatment, such as electroconvulsive therapy. The relative dangers of different methods of treating depression are still debated. Potentially dangerous swings in heart rate and blood pressure produced by modified electroconvulsive therapy have been documented by the author (Scott 1970). The risk of producing cardiac arrhythmias or conduction disturbances with the antidepressant drugs, especially when taken in overdose, has received much attention in recent years. All drugs used to treat depressive illness and notably the tricyclic group have some associated cardiovascular hazards. The elderly, and those with cardiovascular disease are at particular risk. In such patients it has been suggested that treatment with newer drugs such as mianserin or nomifensine is preferable (*Drug and Therapeutics Bulletin* 1979). In a major review of poisoning with tricyclic and related antidepressants, Starkey and Lawson (1980) found that the danger of toxic effects has probably been overestimated, at least in older children and adults. In 316 admissions over a 12-year period, they found severe cardiac arrhythmias or conduction disturbances in only 1.5 per cent of the cases. Their population of patients may represent less seriously ill patients than those reported from intensive care units today.

Maprotiline

Maprotiline hydrochloride is less sedative than amitryptyline, but does not appear to have fewer adverse cardiovascular effects. Knudsen and Heath (1984) describe the cardiovascular reactions in 43 consecutive episodes of maprotiline overdosage admitted to an intensive care unit. Eighteen had sinus tachycardia, 11 first-degree atrioventricular block, and 28 an increased QRS interval. Supraventricular or ventricular arrhythmias occurred in nine patients. Hypotension occurred in eight patients, and cardiac arrest occurred in three of these. The authors concluded that the cardiotoxicity of maprotiline is equal to, if not greater than, that of conventional tricyclic antidepressants.

Carbamazine

Cardiac complications, in the form of severe myocardial depression with bradycardia, conduction defects, hypotension, and oliguria, have been reported in a patient with carbamazine intoxication (Leslie *et al.* 1983). There appears to be great individual variation in the response to this drug's cardiac effects, as depressed atrioventricular conduction with complete heart block has occurred with the therapeutic use of carbamazine (Hamilton 1978). In addition to the cardiac hazards of antidepressants taken in overdose, there are hazards associated with drugs used in the management of such cases. Pentel and Peterson (1980) described asystole complicating physostigmine treatment of tricyclic antidepressant overdose. Another drug used by anaesthetists, the neuromuscular blocking agent atracurium, has given rise to bradycardia in four cases described by Carter (1983) who recommends that atropine should be considered as part of the premedication of patients receiving the drug.

H$_2$-receptor antagonists

Renewed concern regarding possible adverse effects of the H$_2$-receptor antagonists has arisen (*Lancet* 1982*b*).

This has been stimulated in part by the recent release of the histamine-receptor antagonist ranitidine. It has been suggested that it has a cholinergic-like effect (Bertaccini and Coruzzi 1981), and severe bradycardia following its use has been described (Camarri *et al* 1982; Shah 1982). However, the bradycardia responded to administration of atropine and may not have been the result of a cholinergic action of ranitidine (Jack *et al.* 1982). In a major review of ranitidine (Zeldis *et al.* 1983) it was suggested that the occasional occurrence of bradycardia may have resulted from a rise in plasma histamine which has been shown to occur after rapid infusions of ranitidine or cimetidine. Ranitidine was felt to be preferable to cimetidine in patients taking drugs such as warfarin, lignocaine, or propranolol, the metabolism of which is affected by cimetidine (Jack *et al.* 1983).

Tolbutamide and insulin

The claim by the University Group Diabetes Programme (UGDP) (1970) that oral hypoglycaemic agents cause premature death from cardiovascular disease remains a subject for debate. The UGDP Group further reported evidence that insulin treatment was of no value in preventing cardiovascular deaths in maturity-onset diabetics (Knatterud *et al.* 1978). However, Kilo *et al.* (1980) have presented a further detailed analysis of the figures published by the UGDP and have pointed out that there were some anomalies which undermined the validity of their conclusion. There was a notable discrepancy in the sex ratio of cardiovascular death rates in placebo-treated patients. Fewer than one-fifth as many of the placebo-treated women died of cardiovascular disease as men. The cardiovascular death rate in placebo-treated women, who constituted 70 per cent of the UGDP subjects studied, was spuriously low giving the false impression of increased cardiovascular deaths in all other treatment groups. They also pointed out that all excess cardiovascular deaths in the tolbutamide-treated group were in subjects whose blood glucose values were poorly controlled during the study.

Kilo *et al.* (1980) also noted that in the insulin versus placebo study, the control subjects and the subjects receiving the two insulin regimes differed significantly in terms of risk factors such as age, severity of diabetes, and diastolic blood pressure. When these factors were standardized by excluding the high-risk subjects from the analysis, the results favoured the two insulin groups. These controversies are still imperfectly resolved; while therapeutic caution remains in order, it does not seem proven that the risks of using oral hypoglycaemic drugs or insulin in the treatment of maturity-onset diabetes outweigh the advantages.

Corticosteroids

Corticosteroid drugs are often life-saving in severe asthmatic attacks, but no drug is entirely free from adverse effects. Ward and Pritchard (1980) described a severe bronchial reaction and cardiopulmonary arrest following an intravenous injection of hydrocortisone in a non-atopic, non-aspirin sensitive asthmatic. Pulmonary oedema due to left-sided heart failure has been reported in patients receiving both corticosteroids and β-sympathomimetic drugs (Elliot *et al*. 1978).

Salbutamol

A case of acute heart failure developing in a hypertensive woman on α-methyldopa who was given salbutamol for premature labour has appeared (Whitehead *et al*. 1980). The suspected mechanism was via an induced metabolic acidosis and hypokalaemia. Together they may have depressed myocardial function at a time when the circulating blood volume was high owing to prior treatment with α-methyldopa.

Ergotamine tartrate

Ergotamine preparations are used extensively in the treatment of migraine headaches. Several reports of the development of multiple arterial stenoses, with documented reversal after discontinuation of ergotamine have appeared (Greene *et al*. 1977). Corrocher *et al*. (1984) reported another case in which the arterial narrowing did not disappear completely after discontinuing the drug.

Ergonovine is sometimes administered to see whether coronary spasm is precipitated, during electrocardiographic monitoring, thallium-201 imaging, or coronary angiography. Occasionally, complete occlusion of a coronary artery results, which is usually reversed rapidly by nitroglycerine. A case of irreversible coronary artery occlusion was reported by Crevey *et al*. (1981). The patient developed an acute myocardial infarct but survived. Other reported complications of ergonovine administration include short runs of ventricular tachycardia, ventricular ectopics and bradycardia requiring temporary pacing. In view of these adverse effects, the author feels that the indications for this provocative test are rare.

Prothrombin-complex concentrate

Several cases of acute myocardial infarction occurring in a teen-aged haemophiliac with factor VIII inhibition who had received large amounts of prothrombin-complex concentrate were reported (Fuerth and Mahrer 1981; Agrawal *et al*. 1981; Gruppo *et al*. 1983). All three patients had inhibitors, and the dosage of prothrombin-complex concentrates greatly exceeded the generally recommended amount. There is also a report of acute myocardial infarction following prolonged intensive treatment with an activated prothrombin-complex concentrate (Schimpf *et al*. 1982). As there are no reports of myocardial infarction occurring in a patient receiving the standard dosage, multiple dosage should be avoided, as should concurrent use of fibrinolytic inhibitors such as epsilon-aminocaproic acid (Lusher 1984).

Oral contraceptives and oestrogens

In the Oxford Family Planning Association contraceptive study (Vessey *et al*. 1977) nine deaths from cardiovascular causes were observed among women in the oral contraceptive entry group observed for 49 681 woman-years, while no such deaths were observed among women not on the pill, observed for 39 146 woman-years. The Royal College of General Practitioners (1977) reported a relationship between norethisterone acetate in the pill and hypertension, and later with superficial venous thrombosis (1978). The doses of both the progesterone and the oestrogen components in the pill influence the risk. In a major report (Meade *et al*. 1980), a significant positive association was found between the dose of norethisterone acetate in different pills and deaths from stroke and ischaemic heart disease. There were no associations between the dose of norethisterone acetate nor of levonorgesterol with hypertension of venous thrombosis. Preparations with 50μg of oestrogen were associated with significantly more reports of death and ischaemic heart disease than those with only 30μg. Further evidence has accumulated confirming a relation between the use of oestrogen–progestogen oral contraceptives and the risk of cardiovascular problems such as venous thromboembolism, stroke, and acute myocardial infarction (Vessey 1980). (See also Chapter 13.)

Venning (1983) invited 20 physicians to compile lists of the 10 most important adverse reactions to drugs since thalidomide. The oculomucocutaneous syndrome caused by pracolol was agreed to be the most important, but thromboembolism due to oral contraceptives came a close second. Between 1964 and 1980, 268 fatal cases of pulmonary embolism and 136 cases of fatal myocardial infarction were attributed to oral contraceptives. The rate per million general practitioner prescriptions was 4.3. Recent reports indicate that some excess risk of cardiovascular disease persists in women after they discontinued the use of oral contraceptives (Slone *et al*. 1981;

Stadel 1982). Slone *et al.* (1981) reported a case-control study of 556 women admitted to hospital with first attacks of myocardial infarction. They confirmed that the risk of myocardial infarction among current users of oral contraceptives aged 25–49 years is between three and four times greater than that among comparable women who had never used oral contraceptives. The risk fell to normal after stopping the pill unless it had been used for five years or more. The risk of myocardial infarction among long-term past users aged 40–49 years was approximately twice that among comparable women who had never used oral contraceptives, and this risk appears to persist for up to 10 years after the drugs have been discontinued. Prolonged use of these drugs, especially in smokers over the age of 35 is associated with an unacceptably high death rate from cardiovascular disease. Further research is needed to enable this very effective form of contraception to be used with the absolute minimum risk of damage to the cardiovascular system. Happily, in a case-controlled pilot study, deaths from acute myocardial infarction and subarachnoid haemorrhage did not appear to be significantly commoner in middle-aged women receiving sex hormone replacement (Adam *et al.* 1981).

The use of oestrogens in the treatment of prostatic carcinomas can be harmful to the cardiovascular system (Heritier and Hessler 1978). These authors found a very high death rate from cardiovascular causes among oestrogen-treated patients who had previous evidence of cardiovascular disease. Again the risk seemed related to the dose of oestrogen used.

Drugs used to prevent coronary disease

Primary prevention of coronary disease which deals generally with individuals at low risk, differs from the treatment of diseases where some risk associated with treatment must be accepted (Rose and Shipley 1980). The past few years have seen the publication of results of several important studies designed to evaluate various approaches to the primary and secondary prevention of coronary heart disease. These have included trials of a variety of drugs believed likely to alter favourably plasma lipids or platelet function. On the whole, the results of these trials have been disappointing. They have clearly indicated the importance of documenting the potential hazards of such drug treatment. Indeed, all events occurring in drug trials need to be recorded (Simpson *et al.* 1980). In the major WHO multicentre clofibrate study (Committee of Principal Investigators 1978), there was a higher incidence of cholelithiasis in the group taking the drug, with more cholecystectomies and inevitably some associated operative deaths. The suspicion that other bowel disorders such as colonic neoplasms may have been commoner in the treated group has been raised (Committee of Principal Investigators 1980).

Some epidemiological studies have suggested an inverse relation between serum cholesterol and mortality from cancer. Experimental work suggests that the relationship may actually be between low retinoid levels (which are associated with low cholesterol levels) and cancer (Marenah *et al.* 1983). However, while enthusiasm for mass drug treatment of populations with higher than average plasma levels is not justified, several recent reports indicate that in men with markedly raised serum cholesterol, the incidence of coronary heart disease is reduced when serum cholesterol is reduced by cholestyramine therapy plus dietary measures (Lipid Research Clinics Coronary Primary Prevention Trials Results 1984). Furthermore, there was no indication of serious toxic effects resulting from long-term cholestyramine therapy (Levy 1983).

With regard to prevention of re-infarction, reports on trials of aspirin (Elwood and Sweetman 1979), sulphinpyrazone (Anturane Reinfarction Trial Research Group 1980), and dipyridamole (Aspirin Myocardial Infarction Study Group 1980) have shown, on balance, small but not statistically significant benefits. The adverse effects of treatment especially with aspirin have been significantly greater than in the control groups (Calliton and Waterfall 1980). In the aspirin myocardial infarction, 23.7 per cent of subjects receiving aspirin reported symptoms suggestive of peptic ulcer, gastritis, or gastric mucosal erosion, compared with 14.9 per cent of subjects receiving placebo. Thus the risk of causing iatrogenic disease to the cardiovascular or any other system, demands from the physician even more caution where mass medication is proposed than is already customary when prescribing for individual patients (Oliver 1982).

REFERENCES

ADAM, S., WILLIAMS, V., AND VESSEY, M.P. (1981). Cardiovascular disease and hormone replacement treatment: pilot case-control study. *Br. med. J.* **282**, 1277–8.

AGRAWAL, B.L., ZELKOWITZ, L., AND HLETKO, P. (1981). Acute myocardial infarction in a young hemophiliac patient with factor IX concentrate and epsilon-aminocaproic acid. *J. Pediat.* **98**, 931–3.

ALFIRO, P.A., THANAVARO, S., KLEIGER, R.E., TEFFREN, B.G., ARONSON, T.A., AND RUFFY, R. (1981). Alpha-methyldopa and carotid sinus hypersensitivity. *New Engl. J. Med.* **305**, 344.

ANASTASSIADES, C.J. (1980). Nifedipine and beta-blocker drugs.

Br. med. J. **281**, 1251–2.

ANTURANE REINFARCTION TRIAL RESEARCH GROUP (1980). Sulfin-pyrazone in the prevention of sudden death after myocardial infarction. *New Engl. J. Med.* **302**, 250–6.

ASPIRIN MYOCARDIAL INFARCTION STUDY GROUP (1980). A randomized controlled trial of aspirin in persons after myocardial infarction. *New Engl. J. Med.* **302**, 357–61.

BAIM D.S., McDOWELL, A.V., CHERNILES, J., MONRAD, E.S., PARKER, J.A., EDELSON, J., BRAUNWALD, E., AND GROSSMAN, W. (1983). Evaluation of a new bipyridine inotropic agent — milrinone — in patients with severe congestive heart failure. *New Engl. J. Med.* **309**, 748–56.

BERNSTEIN, R.M., EGERTON-VERNON, J., AND WEBSTER, J. (1980). Hydralazine-induced cutaneous vasculitis. *Br. med. J.* **280**, 156–7.

BERTACCINI, G. AND CORUZZI, G. (1981). Cholinergic-like effect of the new histamine H_2-receptor antagonist ranitidine. *Agents Actions* **12**, 168–71.

BHALLA, A.K., HOFFBRAND, B.I., AND PHATAK, P.S. (1979). Prazosin and priapism. *Br. med. J.* **2**, 1039.

BIGGER, J.P. JR. (1979). The quinidine–digoxin interaction: what we do not know about it? *New Engl. J. Med.* **301**, 779–81.

BING, R.F., RUSSELL, G.I., THURSTON, H., AND SWALES, J.D. (1980). Hydralazine in hypertension: Is there a safe dose? *Br. med. J.* **281**, 353–4.

BLACK, J.R. AND MEHTA, J. (1979). Precipitation of heart failure following sudden withdrawal of hydralazine. *Chest* **75**, 724–5.

BRAUNWALD, E. AND COLUCCI, W.S. (1984). Vasodilator therapy of heart failure: Has the promissory note been paid? *New Engl. J. Med.* **310**, 459–61.

BRIGGS, R.S.J., BIRTWELL, A.J., AND POHL, J.E.F. (1978). Hypertensive response to labetalol in phaeochromocytoma. *Lancet* **i**, 1045–6.

BRISTOW, M.R., ED. (1980). *Drug-induced heart disease.* Elsevier/North Holland Biomedical Press, Amsterdam.

——, BILLINGHAM, M.E., MINOBE, W.A., MASEK, M.A., AND DANIELS, J.R. (1979). Demonstration that adriamycin cardiotoxicity is mediated by vasoactive amines. *J. Molec. Cell. Cardiol.* **11** (Suppl. 1), 10.

——, THOMPSON, P.D., MARTIN, R.P., MASON, J.W., BILLINGHAM, M.E., AND HARRISON, D.C. (1978). Early anthracycline cardiotoxicity. *Am. J. Med.* **65**, 823–32.

BROOKS, N., CATTELL, M., PIDGEON, J., AND BALCON, R. (1981). Unpredictable response to nifedipine in severe cardiac failure. *Br. med. J.* **281**, 1324.

BUIUMSOHN, A., EISENBERG, E.S., JACOB, H., ROSEN, H., BOCK, J., AND FRISHMAN, W.H. (1979). Seizures and intraventricular conduction defect in propranolol poisoning. *Ann. intern. Med.* **91**, 860–62.

BULLIMORE, D.W. (1983). Do beta-adrenoceptor blocking drugs cause retroperitoneal fibrosis? *Br. med. J.* **288**, 719–20.

BUSSEY, H.I. (1984). Update on the influence of quinidine and other agents on digitalis glycosides. *Am. Heart J.* **107**, 143–6.

CALDRONEY, R.D., UELAND, K., AND RUBIN, P.C. (1982). Beta-blockers in pregnancy. *New Engl. J. Med.* **306**, 810.

CALLITON, B.J. AND WATERFALL, W.K. (1980). Studies of myocardial infarction *Br. med. J.* **280**, 1370–1.

CAMARRI, E., CHIRONE, E., FANTERIA, G., AND ZOCCI, M. (1982). Ranitidine-induced bradycardia. *Lancet* **ii**, 160.

CARTER, M.L. (1983). Bradycardia after the use of atracurium. *Br. med. J.* **287**, 247–48.

CHATERJEE, K., MASSIE, B., RUBIN, S., GELBERG, H., BRUNDAGE, B.H., AND PORTS, E.A. (1978). Long-term out-patient vasodilator therapy of congestive heart failure. *Am. J. Med.* **65**, 134–45.

CHOI, B.W., BERGER, H.J., SCHWARTZ, P.E., ALEXANDER, J., WACKERS, F.J., GOTTSCHALK, A., AND ZARET B.L. (1983). Serial radionuclide assessment of doxorubicin cardiotoxicity in cancer patients with abnormal baseline resting left ventricular performance. *Am. Heart J.* **106**, 638–43.

COGAN, J.J., HUMPHREYS, M.H., CARLSON, C.J., BENOWITZ, N.L., AND RAPAPORT, E. (1981). Acute vasodilator therapy increases renal clearance of digoxin in patients with congestive cardiac failure. *Circulation* **64**, 973–3.

COHEN, J.I. (1983). Promotion of patent ductus arteriosis by furosemide. *N. Engl. J. Med.* **309**, 432.

COMMITTEE OF PRINCIPAL INVESTIGATORS (1978). A cooperative trial in the primary prevention of ischaemic heart disease using clofibrate. *Br. Heart J.* **40**, 1069–118.

—— (1980). Report of a W.H.O. cooperative trial on primary prevention of ischaemic heart disease using clofibrate to lower serum cholesterol mortality: follow up. *Lancet* **ii**, 379–84.

CORROCHER, R., BRUGNARA, C., MASO, R., TADDEI, G., MEZZELANI, P., AND DE SANDRE, G. (1984). Multiple arterial stenoses in chronic ergot toxicity. *New Engl. J. Med.* **310**: 261.

CREVEY, B.J., OWEN, S.F., AND PITT, B. (1981). Irreversible coronary occlusion related to administration of ergonovine. *Circulation* **64**, 853–4.

DANESHMEND, T.K. AND ROBERTS, C.J.C. (1981). Cimetidine and bio-availability of labetalol. *Lancet* **i**, 505.

D'ARCY, P.F. AND GRIFFIN, J.P. (Eds.) (1979). *Iatrogenic diseases,* 2nd edn. Oxford University Press, Oxford.

DARGIE, H.J., BALL, S.G., ATKINSON, A.B., AND ROBERTSON, J.I.S. (1983). Converting enzyme inhibitors in hypertension and heart failure. *Br. Heart J.* **49**, 305–8.

DE LEEUW, P.W. AND BIRKENHÄGER, W.H. (1980). Hypothermia: a possible side-effect of prazosin. *Br. med. J.* **281**, 1181.

DENIS, B., PELLET, J., MACHECOURT, J., AND MARTIN-NOEL, P. (1977). Verapamil and beta blockade: a dangerous combination. *Nouv. Presse Méd.* **6**, 2075.

DOERING, W. (1979). Quinidine digoxin interaction. *New Engl. J. Med.* **301**, 400–4.

DONOVAN, M.A., HEAGERTY, A.M., POHL, J., CASTLEDEN, M., AND POHL, J.E.F. (1981). Cimetidine and bioavailability of propranolol. *Lancet* **i**, 164.

DRUG AND THERAPEUTICS BULLETIN (1979). Cardiac effects of antidepressive drugs. *Drug Ther. Bull.* **17**, 13–14.

DUPONT, A. AND SIX, R. (1982). Lupus-like syndrome induced by methyldopa. *Br. med. J.* **285**, 693–4.

DZAU, V.J., COLUCCI, M.D., WILLIAMS, G.H., CURFMAN, G., MEGGS, M.D., AND HOLLENBERG, N.K. (1980). Sustained effectiveness of converting enzyme inhibition in patients with severe congestive heart failure. *New Engl. J. Med.* **302**, 1373–9.

ELLIOTT, H.R., ABDULLA, U., AND HAYES, P.J. (1978). Pulmonary oedema associated with ritodrine infusion and betamethasone administration in premature labour. *Br. med. J.* **2**, 799–800.

ELONEN, N., NEUVONEN, P.J., TARASSANEN, L., AND KALA, R. (1979). Sotalol intoxication with prolonged QR interval and severe tachyarrhythmias. *Br. med. J.* **1**, 1184.

ELWOOD, P.C. AND SWEETNAM, P.M. (1979). Aspirin and secondary mortality after myocardial infarction. *Lancet* **ii**, 1313–15.

EPSTEIN, S.E. AND ROSING, D.R. (1981). Verapamil: its potential for causing serious complications in patients with hypertrophic cardiomyopathy *Circulation* **64**, 437–41.

FORFAR, J.C., BROWN, G.J., AND CULL, R.E. (1979). Proximal myopathy during beta-blockade. *Br. med. J.* **2**, 1331-2.

FORGOROS, R.N., ANDERSON, K.P., WINKLE, R.A., SWERDLOW, L.D., AND MASON, J.W. (1983). Amiodarone: clinical efficacy and toxicity in 96 patients with recurrent drug-refractory arrhythmias. *Circulation* **68**, 88-94.

FRANKLIN, G.S., BROWN, J.W., AND FREEDMAN, M.L. (1982). Paired Capgras syndrome and nifedipine. *Lancet* **ii**, 321.

FRISHMAN, H., KLEIN, N., STROM, J., COHEN, M.N., SHAMOON, H., WILLENS, H., KLEIN, P., ROTH, S., IORIO, L., LEJEMTEL, T., POLLACK, S., AND SONNENBLICK, E.H. (1982). Comparative effects of abrupt withdrawal of propranolol and verapamil in angina pectoris. *Am. J. Cardiol.* **50**, 1191-5.

FUERTH, J.H. AND MAHRER, P. (1981). Myocardial infarction after factor 1X therapy. *J. Am. med. Ass.* **245**, 1455-6.

GEORGE, C.F. (1983). Digitalis intoxication: a new approach to an old problem. *Br. med. J.* **286**, 1533-4.

GERBER, G.J., AND NIES, A.S. (1979). Abrupt withdrawal of cardiovascular drugs. *New Engl. J. Med.* **301**, 1234-5.

GILLMER, D.J. AND KARK, P. (1980). Pulmonary oedema precipitated by nifedipine, *Br. med. J.* **280**, 1420-1.

GOORIN, A.M., BOROW, K.M., GOLDMAN, D., WILLIAMS, R.G., HENDERSON, I.C., SALLEN, S.E., COHEN, H., AND JAFFE, N. (1981). Congestive heart failure due to adriamycin cardiotoxicity: its natural history in children. *Cancer* **47**, 2810-16.

GREEN, T.P., THOMPSON, T.R., JOHNSTON, D.E., AND LOCK, J.E. (1983). Furosemide promotes patent ductus arteriosus in premature infants with the respiratory distress sydrome. *New Engl. J. Med.* **308**, 743-8.

GREENE, F.L., ARIYAN, S., AND STANSEL, H.C., JR. (1977). Mesenteric and peripheral vascular ischaemia secondary to ergotism. *Surgery* **81**, 176-9.

GREFBERG, N., NILSSON, P., AND ANDRÉEN, T. (1983). Sclerosing obstructive peritonitis, beta-blockers, and continuous ambulatory peritoneal dialysis. *Lancet* **ii**, 733-4.

GRUPPO, R.A., BOVE, K.E., AND DONALDSON, V.H. (1983). Fatal myocardial necrosis associated with prothrombin-complex concentrate therapy in hemophilia A. *New Engl. J. Med.* **309**, 242-3.

GULAMHUSEIN, S., KO, P., CARRUTHERS, S.G., AND KLEIN, G.J. (1982). Acceleration of the ventricular response during atrial fibrillation in the Wolff-Parkinson White Syndrome after verapamil. *Circulation* **65**, 348-51.

HAMER, A., PETER, T., MANDEL, W.J., SCHEINMAN, M.M., AND WEISS, D. (1982). The potentiation of warfarin anticoagulation by amiodarone. *Circulation* **65**, 1025-29.

HAMILTON, D.V. (1978). Carbamazepine and heart block. *Lancet* **i**, 1365.

HANLEY, S.P., COWLEY, A., AND HAMPTON, J.R. (1980). Danger of withdrawal of vasodilator therapy in patients with chronic heart failure. *Lancet* **i**, 735-6.

HANNAWAY, P.J. AND HOPPER, G.D.K. (1983). Severe anaphylaxis and drug-induced beta-blockade. *New Engl. J. Med.* **308**, 1536.

HARENBERG, J., WAHL, P., STAIGER, CH., AND SMOLARZ, A. (1983). Treatment of digitalis intoxication with digoxin-specific Fab antibody fragments. *New Engl. J. Med.* **309**, 245-6.

HARLAND, S.J., FACCHINI, V., AND TIMBRELL, J.A. (1980). Hydrazine-induced lupus erythematosus-like syndrome in a patient of the rapid acetylator phenotype. *Br. med. J.* **281**, 273-4.

HARRIS, A.L. AND WONG, C. (1981). Myocardial ischaemia, radio-

therapy and vinblastine. *Lancet* **i**, 787.

HASEGAWA, K., SAKAMOTO, Y., MITSUNAGA, K., AND KASHIWAGI, H. (1980). Drug fever due to oxprenolol. *Br. med. J.* **2**, 27-8.

HERITIER, P. AND HESSLER, D. (1978). Les complications cardiovasculaires du traitement oestrogenique de la prostate. *Ther Unsch.* **35**, 841-4.

HODSMAN, G.P., BROWN, J.J., DAVIES, D.L., FRASER, R., LEVER, A.F., MORTON, J.J., MURRAY, G.D., AND ROBERTSON, J.I.S. (1982). Converting-enzyme inhibitor enalapril (MK421) in treatment of hypertension with renal artery stenosis. *Br. med. J.* **285**, 1697-99.

——, ISLES, C.G., MURRAY, G.D., USHERWOOD, T.P., WEBB, D.J., AND ROBERTSON, J.I.S. (1983). Factors related to first dose hypotensive effect of captopril: prediction and treatment. *Br. med. J.* **286**, 832-4.

—— AND ROBERTSON, J.I.S. (1983). Captopril: five years on. *Br. med. J.* **287**, 851-2.

HOLMES, D.G., BOGERS, W.A.J.L., WIDEROE, T.E., HUUNAN-SEPPALA, A., AND WIDEROE, B. (1983). Endralazine, a new peripheral vasodilator: absence of effect of acetylator status on antihypertensive effect. *Lancet* **i**, 670-1.

HOORNTJE, S.J., KALLENBERG, C.G.M., WEENING, J.J., DONKER, A.J.M., THE, T.H., AND HOEDMAEKER, P.J. (1980). Immune complex glomerulopathy in patients treated with captopril. *Lancet* **ii**, 1297.

HORAK, A., RAINE, R., OPIE, L.H., AND LLOYD, E.A. (1983). Severe myocardial ischaemia induced by adrenaline. *Br. med. J.* **286**, 519.

HUNTER, R. (1982). Death due to overdose of indoramin. *Br. med. J.* **285**, 1011.

ISHIKAWA, T., IMAMURA, T., KOIWAYA, Y., AND TANAKA, K.(1983). Atrioventricular dissociation and sinus arrest induced by oral diltiazem. *New Engl. J. Med.* **309**, 1124-5.

JACK, D., MITCHARD, M., AND SMITH, R.N. (1983). Influence of ranitidine on plasma metoprolol concentrations. *Br. med. J.* **287**, 1218.

——, RICHARDS, D.A., AND GRANATA, F. (1982). Side-effects of ranitidine. *Lancet* **ii**, 264-5.

JACOBS, R.L., RAKE, G.W., JR., FOURNIER, D.C., CHILTON, R.J., CULVER, W.G., AND BECKMAN, C.H. (1981). Potentiated anaphylaxis in patients with drug-induced beta-adrenergic blockade. *J. Allergy clin. Immunol.* **68**, 125-7.

JEE, L.D. AND OPIE, L.H. (1983). Acute hypotensive response to nifedipine added to prazosin in treatment of hypertension. *Br. med. J.* **287**, 1514.

KEIDAR, S., GRENADIER, E., AND PALANT, A. (1982). Sinoatrial arrest due to lidocaine injection in sick sinus syndrome during amiodarone administration. *Am. Heart J.* **104**, 1384-5.

KERN, M.J., GANZ, P., HOROWITZ, J.D., GASPAR, J., BARRY, W.H., LORELL, B.H., GROSSMAN, W., AND MUDGE, G.H. JR. (1983). Potentiation of coronary vasoconstriction by beta-adrenergic blockade in patients with coronary artery disease. *Circulation* **67**, 1178-85.

KIEVAL, J., KIRSTEN, E.B., KESSLER, K.M., MALLON, S.M., AND MYERBURG, R.J. (1982). The effects of intravenous verapamil on haemodynamic status of patients with coronary artery disease receiving propranolol. *Circulation* **65**, 653-9.

KILO, C., MILLER, J.P., AND WILLIAMSON, J.R. (1980). The achilles heel of the University Group Diabetes Programme. *J. Am. med. Ass.* **243**, 450-7.

KIRCH, W., KOHLER, J., SPAHN, H., AND MUTSCHLER, E. (1981). Interaction of cimetidine with metoprolol, propranolol or atenolol. *Lancet* ii, 531–2.

KLEIN, H.O., LANG, R., WEISS, E., DI SEGNI, E., LIBHABER, C., GUERRERO, J., AND KAPLINSKY, E. (1982). The influence of verapamil on serum digoxin concentration. *Circulation* **65**, 998–1003.

KLUGER, J., LEECH, S., REIDENBERG, M.M., LLOYD V., AND DRAYER, D.F. (1981). Long-term anti-arrhythmic therapy with acetylprocainamide. *Am. J. Cardiol.* **49**, 1124–32.

KNATTERUD, G.L., KLIMT, C.R., LEVIN, M.E., JACOBSON, M.E., AND GOLDNER, M.G. (1978). Effects of hypoglycaemic agents on vascular complicatións in patients with adult-onset diabetes: VII Mortality and selected nonfatal events with insulin treatment. *J. Am. med. Ass.* **240**, 37–42.

KNUDSEN, K. AND HEATH, A. (1984). Effects of self-poisoning with maprotiline *Br. med. J.* **288**, 601–3.

KRISTENSEN, B. (1981). Effect of long-term treatment with beta-blocking drugs on plasma lipids and lipoproteins. *Br. med. J.* **281**, 191.

LANCET (1982*a*). Diuretic or beta-blocker as first-line treatment for mild hypertension? *Lancet* ii, 1136–7.

—— (1928*b*). Cardiovascular histamine H_2 receptors. *Lancet* ii, 421–2.

—— (1984). Flecainide acetate. *Lancet* i, 85–6.

LEAHEY, E.B. JR., REIFFEL, J.A., HISSENBUTEL, R.H., DRUSIN, R.E., LOVEJOY, W.P., AND BIGGER, J.T. (1979). Enhanced cardiac effect of digoxin during quinidine treatment. *Arch. intern. Med.* **139**, 519–21.

LEJONC, J.L., VERNANT, J.P., MAZQUIN, I., AND CASTAINGE, A. (1980). Myocardial infarction following vinblastine treatment. *Lancet* ii, 692.

LEREN, P., HELGELAND, A., HOLME, I., FOSS, P.O., HJERMANN, I., AND LUND-LARSEN, P.G. (1980). Effect of propranolol and prazosin on blood lipids. The Oslo Study. *Lancet* ii, 4–6.

LESLIE P.J., HEYWORTH, R., AND PRESCOTT, L.F. (1983). Cardiac complications of carbamazepine intoxication: treatment by haemoperfusion. *Br. med. J.* **286**, 1018.

LEVY, R.I. (1983). The influence of cholestyramine-induced lipid changes on coronary artery disease progression: The NHLBI Type II coronary intervention study. *Circulation* **68** (suppl. 3) 188–91.

LIPID RESEARCH CLINICS CORONARY PRIMARY PREVENTION TRIAL RESULTS (1984). The relationship of reduction in incidence of coronary heart disease to cholesterol lowering. *J. Am. med. Ass.* **251**, 365–74.

LUBBE, W.F., HODGE, J.V., AND KELLAWAY, G.S.M. (1982). Antihypertensive treatment and fetal welfare in essential hypertension in pregnancy. *NZ med. J.* **95**, 1–5.

LUPI-HERRERA, E., SANDOVAL, J., SEOANE, M., AND BIALOSTOZKY, D. (1982). The role of hydralazine therapy for pulmonary arterial hypertension of unknown cause. *Circulation* **65**, 645–50.

LUSHER, J.M. (1984). Myocardial necrosis after therapy with prothrombin-complex concentrate. *New Engl. J. Med.* **310**, 464.

MACLEAN, D. AND TUDHOPE, G.R. (1983). Modern diuretic treatment *Br. med. J.* **i**, 1419–22.

MAHESWARAN, R., BRAMBLE, M.G., AND HARDISTY, C.A. (1983). Massive digoxin overdose: successful treatment with intravenous amiodarone. *Br. med. J.* **287**, 392–3.

MARCUS, F.I. (1983). Drug interactions with amiodarone. *Am. Heart J.* **106**, 924–9.

MARENAH, C.B., LEWIS, B., HASSALL, D., LA VILLE, A., CORTESE, C.,

MITCHELL, W.D., BRUCKDORFER, K.R., SLAVIN, B., MILLER, N.E., TURNER, P.R., AND HEDUAN, E. (1983). Hypocholesterolaemia and non-cardiovascular disease: metabolic studies on subjects with low plasma cholesterol concentrations. *Br. med. J.* **286**, 1603–6.

MARTINOWITZ, V., RABINOVICI, J., GOLDFARB, D., MANY, A., AND BANK, H. (1981). Interaction between warfarin sodium and amiodarone. *New Engl. J. Med.* **304**, 671–2.

McCOMB, J.M., LOGAN, K.R., KHAN, M.M., GEDDES, J.S., AND ADGEY, A.A.J. (1980). Amiodarone-induced ventricular fibrillation. *Eur. J. Cardiol.* **11**, 381.

McGOVERN, B., GARAN, H., KELLY, E., AND RUSKIN, J.N. (1983). Adverse reactions during treatment with amiodarone hydrochloride. *Br. med. J.* **287**, 175–80.

McKENNA, W.J., ROWLAND, E., AND KRIKLER, D.M. (1983). Amiodarone: the experience of the past decade. *Br. med. J.* **287**, 1654–6.

McKILLOP, J.H., BRISTOW, M.R., GORIS, M.L., BILLINGHAM, M.E., AND BOCKEMUEHL, K. (1983). Sensitivity and specificity of radionuclide ejection fractions in doxorubicin cardiotoxicity. *Am. Heart J.* **106**, 1048–56.

MEADE, T.W., GREENBERG, G., AND THOMPSON, S.C. (1980). Progestogens and cardiovascular reactions associated with oral contraceptives and a comparison of the safety of 50 ànd 30 μg oestrogen preparations. *Br. med. J.* **1**, 1157–61.

MEREDITH, P.A., ELLIOT, H.L., McSHARRY, D.R., KELMAN, A.W., AND REID, J.L. (1983). The pharmacokinetics of endralazine in essential hypertensives and in normotensive subjects. *Br. J. Clin. Pharmacol.* **16**, 27–32.

MITCHELL, L.B., JUTZY, K.R., LEWIS, S.J., SCHROEDER, J.S., AND MASON, J.W. (1982). Intracardiac electrophysiologic study of intravenous diltiazem and combined diltiazem–digoxin in patients. *Am. Heart J.* **103**, 57–66.

MOENCH, T.R. (1980). The quinidine–digoxin interaction. *New Engl. J. Med.* **302**, 864.

MOYSEY, J.O., JAGGARAO, N.S.V., GRUNDY, E.N., AND CHAMBERLAIN, D.A. (1981). Amiodarone increases plasma digoxin concentrations. Br. med. J. **282**, 272.

MULLER, J., TURL, Z., AND STONE, P. (1983). Does digoxin therapy increase mortality following myocardial infarction? *Circulation* **68** (Suppl. 3) 368–72.

MULTIPLE RISK FACTOR INTERVENTION TRIAL (MRFIT) Research Group (1982). Risk factor changes and mortality results. *J. Am. med. Ass.* **248**, 1465–77.

MURPHY, M.B., LEWIS, P.J., KOHNER, E.M., SCHUMER, B., AND DOLLERY, C.T. (1982). Glucose tolerance in hypertensive patients; a 14-year follow-up. *Lancet* ii, 1293–5.

——, SCRIVEN, A.J.I., AND DOLLERY, C.T. (1983). Role of nifedipine in treatment of hypertension. *Br. med. J.* **287**, 257–9.

MYERS, M.G. AND HOPE-GILL, H.F. (1979). The effect of d- and dl-propranolol on glucose-stimulated insulin release. *Clin. Pharmacol. Ther.* **25**, 303–8.

——, O'CONNELL, J.B., AND SUBRAMANIAN, R. (1983). Myocarditis resolving after discontinuation of procainamide. *Int. J. Cardiol.* **4**, 322–4.

NEHRING, J. AND CAMM, A.J. (1983). Calcium antagonist withdrawal syndrome. *Br. med. J.* **286**, 1057.

NICHOLLS, M.G., ESPINER, E.A., IKRAM, H., AND MASLOWSKI, A.N. (1981). Hyponatraemia in congestive heart failure during treatment with captopril. *Br. med. J.* **281**, 909.

NOBILE-ORAZIO, E. AND STERZI, R. (1981). Cerebral ischaemia after nifedipine treatment. *Br. med. J.* **283**, 948.

OLIVER, M.F. (1982). Risks of correcting the risks of coronary disease and stroke with drugs. *New Engl. J. Med.* **306**, 297–8.

OLSHANSKY, B., MARTINS, J., AND HUNT, S. (1982). N-Acetyl procainamide causing torsades de pointes. *Am. J. Cardiol.* **50**, 1439–41.

OPIE, L.H. (1980*a*). Drugs and the heart V. Digitalis and sympathomimetic stimulants. *Lancet* **i**, 912–18.

—— (1980*b*). Calcium antagonists. *Lancet* **i**, 806–9.

—— (1984*a*). Drugs and the heart four years on. *Lancet* **i**, 496–501.

—— (1984*b*). Calcium antagonists. Mechanisms, therapeutic indications and reservations: a review. *Quart. J. med.* **209**, 1–16.

—— AND WHITE, D.A. (1980). Adverse interaction between nifedipine and beta-blockade. *Br. med. J.* **281**, 1462.

PACKER, M., GREENBERG, B., MASSIE, B., AND DASH, H. (1982*b*). Deleterious effects of hydralazine in patients with pulmonary hypertension. *New Engl. J. Med.* **306**, 1326–31.

——, MEDINA, N., YUSHAK, M., AND GORLIN, R. (1982*a*). Haemodynamic characterisation of tolerance to long-term hydralazine therapy in severe chronic heart disease. *New Engl. J. Med.* **306**, 57–62.

——, MELLER, J., MEDINA, R.N., GORLIN, M., AND HERMAN, M. (1979). Clinical and hemodynamic deterioration following abrupt withdrawal of nitroprusside in severe heart failure. *Am. J. Cardiol.* **32**, 428.

——, ——, —— YUSHAK, M., SMITH, H., HOLT, J., GUERERRO, J., TODD, G.D., McALLISTER, R.G., JR., AND GORLIN, R. (1982*c*). Hemodynamic consequences of combined beta-adrenergic and slow calcium channel blockade in man. *Circulation* **65**, 660–6.

PENTEL, P. AND PETERSON, C.D. (1980). Asystole complicating physostigmine treatment of tricyclic antidepressant overdosage. *Ann. Emerg. Med.* **9**, 588–90.

PEREZ-STABLE, E. AND CARALIS, P.V. (1983). Thiazide-induced disturbances in carbohydrate, lipid, and potassium metabolism. *Am. Heart J.* **106**, 245–51.

PERROT, B., DANCHIN, N., AND TERRIER DE LA CHAISE, A. (1984). Verapamil: a cause of sudden death in a patient with hypertrophic cardiomyopathy. *Br. Heart J.* **51**, 352–4.

PRYOR, J.P., CASTLE, W.M., DUKES, D.C., SMITH, J.C., WATSON, M.E., AND WILLIAMS, J.L. (1983). Do beta-adrenoceptor drugs cause retroperitoneal fibrosis? *Br. med. J.* **2**, 639.

REACH, G., THIBONNIER, M., CHEVILLARD, C., CORVOL, P., AND MILLIEZ, P. (1980). Effect of labetalol on blood pressure and plasma catecholamine concentration in patients with phaeochromocytoma. *Br. med. J.* **280**, 1300–1.

REES, A., DALAL, J.J., REID, P.G., HENDERSON, A.H., AND LEWIS, M.J. (1981). Dangers of amiodarone and anticoagulant treatment. *Br. med. J.* **282**, 1756–7.

ROSE, G. AND SHIPLEY, M.J. (1980). Plasma lipids and mortality: a source of error. *Lancet* **i**, 523–6.

ROSSNER, S. (1982). Serum lipoproteins and ischaemic vascular disease: on the interpretation of serum lipid versus serum lipoprotein concentrations. *J. Cardiovasc. Pharmacol.* **4** (S), 201–5.

—— AND WEINER, L. (1982). Atenolol and timolol. *New Engl. J. Med.* **306**, 123.

ROWLAND, E. AND KRIKLER, D.M. (1980). Electrophysiological assessment of amiodarone in the treatment of resistant supraventricular arrhythmias. *Br. Heart J.* **44**, 82–90.

ROYAL COLLEGE OF GENERAL PRACTITIONERS (1977). Effect on hypertension and benign breast disease of progestogen component in combined oral contraceptives. *Lancet* **i**, 624.

—— (1978). Oral contraceptives, venous thrombosis and varicose veins. *J. R. Coll. gen. Pract.* **28**, 393–9.

ROZKOVEC, A. AND COLTARD, J. (1982). Treatment of digoxin overdose with antigen-specific fragment of digoxin-specific antibodies. *Br. med. J.* **285**, 1315–16.

RUBIN, P.C. (1981). Beta-blockers in pregnancy. *New Engl. J. Med.* **305**, 1323–6.

RUSKIN, J.N., McGOVERN, B., GARAN, H., DI MARCO, J.P., AND KELLY, E. (1983). Antiarrhythmic drugs: a possible cause of out-of-hospital cardiac arrest. *New Engl. J. Med.* **309**, 1302–5.

RUZICKA, T. AND RING, J. (1983). Hypersensitivity to prazosin. *Lancet* **i**, 473–4.

SACKS, H. AND KENNELLY, B.M. (1972). Verapamil in cardiac arrhythmias. *Br. med. J.* **2**, 716.

SCHENCK-GUSTAFSSON, K. AND DAHLQVIST, R. (1980). Pharmacokinetic evaluation of the quinidine-digoxin interaction. *8th Eur. Congr. Cardiol. Abstr.* **1518**, 126.

SCHIMPF, K.L., ZELTSCH, C.H., AND ZELTSCH, P. (1982). Myocardial infarction complicating activated prothrombin-complex concentrate substitution in patient with haemophilia A. *Lancet* **ii**, 1043.

SCLAROVSKY, S., LEWIN, R.F., KRACOFF, O., STRASBERG, B., ARDITTI, A., AND AGMON, J. (1983). Amiodarone-induced polymorphous ventricular tachycardia. *Am. Heart J.* **105**, 6–12.

SCOTT, M.E. (1970). A haemodynamic assessment of certain therapeutic procedures. M.D. Thesis, Queen's University of Belfast, Addendum pp. 225–56.

SERLIN, M.J., SIBEON, R.G., AND GREEN, G.J. (1981). Dangers of amiodarone and anticoagulant treatment. *Br. med. J.* **283**, 58.

SHAH, R.R. (1982). Histamine H_2-antagonists and the heart. *Lancet* **ii**, 1821–2.

SICART, M., BESSE, P., CHOUSSAT, A., AND BRICAUD, H. (1977). Action haemodynamique de l'amiodarone intra-veineuse chez l'homme. *Arch. Mal. Coeur Vaiss.* **70**, 219–27.

SIMPSON, R.J., TIPLADY, B., AND SKEGG, D.C.G. (1980). Event recording in a clinical trial of a new medicine. *Br. med. J.* **280**, 1133–4.

SLONE, D., SHAPIRO, S., KAUFMAN, D.W., ROSENBERG, L., MIETTINEN, O.S., AND STOLLEY, P.D. (1981). Risk of myocardial infarction in relation to current and discontinued use of oral contraceptives. *New Engl. J. Med.* **305**, 420–4.

SMITH, T.W., BUTLER, V.P., HABER, E., FOZZARD, H., MARCUS, F.I., BREMNER, W.F., SCHULMAN, I.C., AND PHILLIPS, A. (1982). Treatment of life-threatening digitalis intoxication with digoxin-specific fab antibody fragments. *New Engl. J. Med.* **307**, 1357–62.

SOBOL, M. AND RAKITA, L. (1982). Pneumonitis and pulmonary fibrosis associated with amiodarone treatment: a possible complication of a new antiarrhythmic drug. *Circulation* **65**, 819–22.

STADEL, B.V. (1982). Oral contraceptives and cardiovascular disease. *New Engl. J. Med.* **305**, 672–7.

STARKEY, I.R. AND LAWSON, A.A.H. (1980). Poisoning with tricyclic and related antidepressants — a ten-year review. *Quart. J. Med.* **49**, 33–49.

SUBRAMANIAN, V.B., BOWLES, M.J., KHURMI, N.S., DAVIES, A.B., O'HARA, M.J., AND RAFTERY, E.B. (1983). Calcium antagonist withdrawal syndrome: objective demonstration with frequency-modulated ambulatory ST-segment monitoring. *Br. med. J.* **286**, 520–1.

SWALES, J.D. (1983). Magnesium deficiency and diuretics. *Br. med. J.* **2**, 1377.

SZNAJDER, I., BENTUR, Y., AND TAITELMAN, U. (1984). First and second degree atrioventricular block in oxpentifylline overdose. *Br. med. J.* **288**, 26.

TARTINI, R., KAPPENBERGER, L., STEINBRUNN, W., AND MEYER, U.A. (1982). Dangerous interaction between amiodarone and quinidine. *Lancet* i, 1327–9.

TAYLOR, S.H. AND SILKE, B. (1981). Haemodynamic effect of beta-blockade in ischaemic heart failure. *Lancet* ii, 835–7.

TEICHER, A., ROSENTHAL, T., KISSIN, E., AND SAROVA, I. (1981). Labetalol-induced toxic myopathy. *Br. med. J.* **282**, 1824–5.

TEXTOR, S.C., BRAVO, E.L., FOUAD, F.M., AND TARAZI, R.C. (1982). Hyperkalemia in azotemic patients during angiotensin-converting enzyme inhibition and aldosterone reduction with captopril. *Am. J. Med.* **73**, 719–25.

THORLEY, K.J., McANISH, J., AND CRUIKSHANK, J.M. (1981). Atenolol in the treatment of pregnancy-induced hypertension. *Br. J. clin. Pharmacol.* **12**, 725–30.

UNIVERSITY GROUP DIABETES PROGRAMME (1970). A study of the effects of hypoglycaemic agents on vascular complications of adult-onset diabetes. Sections I and II. *Diabetes* **19** (Suppl.), 747–830.

VELEBIT, V., PODRID, P., LOWN, B., COHEN, B.H., AND GRABOYS, T.B. (1982). Aggravation and provocation of ventricular arrhythmias by antiarrhythmic drugs. *Circulation* **65**, 886–92.

VENNING, G.R. (1983). Identification of adverse reactions to new drugs. 1. What have been the important adverse reactions since thalidomide? *Br. med. J.* **286**, 199–202.

VERIN, P., GENDRE, P., BARCHEWITZ, G., LAURENT-BRONCHAT, G., YACOUBI, M., AND MORAN, S. (1971). Thesaurismose cornéenne par amiodarone. *Arch. Opthalmol.* **31**, 581–96.

VESSEY, M.P. (1980). Female hormones and vascular disease. Epidemiological overview. *Br. J. fam. Planning* (Suppl.) **6**, 1–12.

——, McPHERSON, K., AND JOHNSON, B. (1977). Mortality among women participating in the Oxford Planning Association Contraceptive Study. *Lancet* ii, 731–3.

VETERANS' ADMINISTRATION CO-OPERATIVE STUDY GROUP ON ANTI-HYPERTENSIVE AGENTS (1982). Comparison of propranolol and hydrochlorothiazide for the initial treatment of hypertension. II Results of long-term therapy. *J. Am. med. Ass.* **248**, 2004–11.

VIDT, D.G., BRAVO, E.L., AND FOUAD, F.M. (1982). Captopril. *New Engl. J. Med.* **306**, 214–19.

WAEBER, B., NUSSBERGER, J., AND BRUNNER, H.R. (1984). Self-poisoning with enalapril. *Br. med. J.* **288**, 287–8.

WARD, D.E., CAMM, A.J., AND SPURRELL, R.H.J. (1980). Clinical anti-arrhythmic effect of amiodarone in patients with resistant paroxysmal tachycardia. *Br. Heart J.* **44**, 91–5.

WARD, K. AND PRITCHARD, J.S. (1980). Severe bronchial reaction and cardiopulmonary arrest following intravenous injection of hydrocortisone in a non-atopic non-aspirin sensitive asthmatic. *Irish J. med. Sci.* **149**, 124–6.

WEBB, D.B. AND WHITE, J.P. (1980). Hypertension after taking hydralazine *Br. med. J.* **280**, 1582.

WEBB, S.C., CANEPA-ANSON, R., RICKARDS, A.F., AND POOLE-WILSON, P.A. (1983). High potassium concentration in a parenteral preparation of glyceryl trinitrate — need for caution if given by intracoronary injection. *Br. Heart J.* **50**, 395–6.

WENTING, G.J., MANIN'TVELD, A.J., WOITTIEZ, A.J., BOOMSMA, F., LAIRD-MEETER, K., SIMOONS, M.L., HUGENHOLTZ, P.G., AND SCHALEKAMP, M.A.D.H. (1983). Effects of captopril in acute and chronic heart failure — correlations with plasma levels of noradrenaline, renin, and aldosterone. *Br. Heart J.* **49**, 65–76.

WHITEHEAD, M.I., MANDER, A.M., HERTOGS, K., WILLIAMS, R.M., AND PETTINGALE, K.W. (1980). Acute congestive cardiac failure in a hypertensive woman receiving salbutamol for premature labour. *Br. med. J.* **280**, 1221–2.

WHITTAKER, J.A. AND AL-ISMAIL, S.A.D. (1984). Effect of digoxin and vitamin E in preventing cardiac damage caused by doxorubicin in acute myeloid leukaemia. *Br. med. J.* **288**, 283–4.

WILLIAMS, P., ARONSON, J., AND SLEIGHT, P. (1978). Is a slow pulse rate a reliable sign of digitalis toxicity? *Lancet* ii, 1340–2.

WILMSHURST, P.T., THOMPSON, D.S., JENKINS, B.S., COLTART, D.J., AND WEBB-PEPLOE, M.M. (1983). Haemodynamic effects of intravenous amrinone in patients with impaired left ventricular function. *Br. Heart J.* **49**, 77–82.

——, —— AND WEBB-PEPLOE, M.M. (1983). Side-effects of amrinone therapy. *Br. Heart J.* **49**, 447–51.

WOODS, D.L. AND MORRELL, D.F. (1982). Atenolol: side-effects in a newborn infant. *Br. med. J.* **285**, 691–2.

WOODS, K.L. WRIGHT, A.D., KENDALL, M.J., AND BLACK, E. (1981). Lack of effect of propranolol and metoprolol on glucose tolerance in maturity-onset diabetics. *Br. med. J.* **281**, 1321.

YOUNG, R.C., OZOLS, R.F., AND MYERS, C.E. (1981). Anthracycline anti-neoplastic drugs. *New Engl. J. Med.* **305**, 139–53.

ZELDIS, J.B., FRIEDMAN, L.S., AND ISSELBACHER, K.J. (1983). Ranitidine: a new H_2-receptor antagonist. *New Engl. J. Med.* **309**, 1368–73.

15 Iatrogenic lung disease

LISA E. HILL

A variety of drugs may provoke adverse drug reactions which may closely resemble respiratory disease due to other causes. The lungs fulfil many functions in addition to their more obvious function related to alveolar oxygen–carbon dioxide exchange. It has become apparent that highly perfused pulmonary tissue can play an important role in the fate of drugs and for this reason the lungs may be specifically involved as a site of drug toxicity.

The activity of the lungs as a metabolic organ is illustrated by their ability to inactivate active substances which occur naturally in the circulation, e.g. noradrenaline, 5-hydroxytryptamine, and the E and F prostaglandins (Vane 1968). In addition, the lungs possess saturable carrier systems which can actively extract drugs; this particularly applies to basic amines such as propranolol, methadone, and imipramine (*British Medical Journal* 1976).

In brief, the lungs can exert a buffering effect when a drug passes through the pulmonary circulation. In this situation, both drug uptake and considerable mixing occur which may effectively protect the heart, the brain, and in particular the coronary arteries from high concentrations in the blood of the drug when an injection is made into a peripheral vein. It is clear that this function may be interfered with in pulmonary disease associated with shunting, or when several drugs compete for uptake sites. In addition, the specific binding of drugs and their active metabolites by the lungs may lead to direct toxic effects. Probably the best example of this mechanism is in paraquat poisoning where the pulmonary damage is associated with an active transport process by which the drug is accumulated (Rose *et al.* 1974). Some other commonly prescribed drugs such as chlorphentermine and the tricyclic antidepressants are also localized in the lungs in animal studies but the significance of these observations to the clinical situation is not known.

Obstructive airways disease

Bronchoconstriction resulting in airflow obstruction is the most common response of the bronchial tree to noxious stimuli. It may be caused by a wide variety of factors, including ingested or inhaled drugs, and may be part of more generalized reactions, as in anaphylaxis. Asthma, or reversible airways obstruction, is the most commonly recognized form and may be due to an immediate allergic reaction (Type I), involving such mediators as histamine, prostglandins, or kinins. Thus any drug whose pharmacological action affects these mediators, or causes mast-cell degranulation, may cause acute bronchoconstriction. Other allergic reactions, such as Type III, where drugs may contribute to the formation of immune complexes, or Type IV reactions, which are cell mediated, may also induce asthma, although this more usually part of a more general or more extensive pulmonary pathological process. See also Chapter 7 for a more exhaustive description of drug-induced hypersensitivity reactions.

Acute allergic asthma

Antibiotics

Antimicrobial agents, penicillin, and cephalosporin derivatives, spiramycin, polymyxin, and, less commonly, tetracyclines, can all evoke acute allergic bronchoconstriction, even in non-asthmatics, and irrespective of the route of administration. Asthma following inhaled antibiotics, often occurring in industrial situations, has been well documented with challenge tests (Coutts *et al.* 1981; Wilson 1981), which were highly specific for the antibiotic involved (Paggiaro *et al.* 1979). The reactions usually occur after previous sensitization and may be severe and accompanied by generalized reactions such as fever, skin lesions, and eosinophilia, or anaphylaxis (Idsøe *et al.* 1968). The mechanism is in many cases antibody-dependent (Klaus and Fellner 1973; Levine and Zolov 1969). In addition basic polypeptides, such as polymyxin B, can cause mast-cell degranulation with histamine release (Wilson 1981).

Aspirin and other prostaglandin synthetase inhibitors

Aspirin is the commonest cause of drug-induced asthma. It has been estimated that some 20–25 cent cent of asthmatics are intolerant to aspirin, although its possible role in precipitating or aggravating asthma is often over-looked (Phills and Perelmutter 1974). In aspirin-sensitive patients bronchospasm usually develops 15-30 minutes after ingestion and may often be severe, prolonged, and occasionally fatal. Angioneurotic oedema and urticaria may accompany aspirin-induced asthma attacks.

Samter and Beers (1967, 1968) studied 182 aspirin-sensitive subjects and found a low incidence of atopy. Few patients had positive skin tests to common allergens, and there was no correlation between exposure to allergens and symptoms. In these subjects, there were no adverse reactions with sodium salicylate, salicylic esters, choline salicylate, thioaspirin, or paracetamol. There-fore, intolerance to aspirin is not intolerance to sali-cylates generally.

Similar reactions with acute airways obstruction occur with all types of prostaglandin inhibitors such as non-ste-roidal anti-inflammatory drugs (NSAIDs) in aspirin-sensitive patients, and have been reliably documented with provocation tests (Salberg and Simon 1980; Szczeklik 1980). Since some of the reactions have been near-fatal (Cohen *et al.* 1982), the wisdom of challenge must be questioned.

The common mechanism for these reactions is considered to be an imbalance of the arachidonic acid metabolism. Normally there is a balanced state in the lung between prostaglandin E_1 (PGE_1), which is a power-ful inhibitor of histamine release via cyclic AMP, and $PGF_{2\alpha}$ which is thought to effect the release of slow-reacting substance-of-anaphylaxis (SRS-A) and hista-mine. Both prostaglandins are derived from arachidonic acid, which is released from cell membranes by a number of stimuli — allergic, inflammatory, and immunological. Non-steroidal anti-inflammatory agents, aspirin, and the pyrazolones inhibit only the cyclo-oxygenase but not the lipoxygenation pathway, which results in reduced PGE_1 with unopposed $PGF_{2\alpha}$ production and thus SRS-A and histamine release. This explains the mechanism of the provocation of asthma by these drugs. It is, however, not clear why only some asthmatics are effected (*Lancet* 1981).

Other, chemically related compounds may have similar effects, such as for instance zomepirac, a new analgesic, chemically related to the non-steroidal anti-inflammatory drugs. A near fatal attack of broncho-spasm was described in an aspirin-sensitive asthmatic following a single tablet of zomepirac (Ross *et al.* 1982). The importance of the reaction to this particular com-pound lies in the fact that it may not be recognized as being related to the non-steroidal anti-inflammatory drugs, since it is promoted as an analgesic, in conditions other than musculoskeletal disorders.

Contrast media

Iodinated contrast media can induce acute bron-chospasm, which may be part of a general hypersensi-tivity reaction. As part of a six-year survey of the safety of contrast media, the International Society of Radio-logy examined a total of 4120 verified case reports of adverse reactions (Shehadi 1982). One hundred and fifty-six non-asthmatic patients experienced bronchospasms (4 per cent), but there was a much higher incidence in asthmatics, namely 14.93 per cent. Twelve of the total experienced pulmonary oedema (0.3 per cent). In only three did pulmonary oedema occur alone; in the remainder it was part of a more general life-threatening reaction. Two of the 12 also had bronchospasm, eight had dyspnoea, two also had laryngeal oedema, and two had hypotension (more than one symptom per patient). An interesting finding of the survey was that although patients might have life-threatening reactions at one exa-mination, it did not follow that a similar reaction would occur the next time. Similarly, the pre-examination skin test was found to be of no value in predicting whether a reaction would occur, but the test itself produced severe or life-threatening reactions (nine patients).

Alcohol

In a recent survey, Ayres and Clark (1983) found that in a population of 168 patients with asthma, 32.1 per cent reported that one, or more types of alcoholic drink made their asthma worse, the principal offenders being wines, beer, and whisky. On the other hand 23.2 per cent reported that alcohol improved their asthma. Alcohol has been shown to cause bronchoconstriction and this has usually been considered due to the presence of congeners (e.g. sodium metabisulphate) (Malish 1982). However, Gong *et al.* (1981) in an exhaustive inves-tigation of a Chinese-American with asthma showed that challenge with ethanol administered by different routes (oral, inhalation, or intravenous) reproducibly resulted in lowering of airway conductance, forced vital capacity (FVC), and forced expiratory volume in 1 second (FEV_1) with an increase in airways resistance. This effect was not prevented by pretreatment with sodium cromoglycate or isoprenaline, but atropine had a partial inhibitory effect. Similarly, pretreatment with oral chlorphe-niramine partially prevented the fall in airway con-ductance, whereas oral cyproheptadine had no effect. The authors suggest that alcohol may cause several mediators to act in concert, and that possibly there might be a release of the slow-reacting-substance-of-anaphy-laxis (SRS-A) for which they did not however test. It is not known at present whether the genetic disposition to ethanol-induced alcohol flushing in Oriental subjects constitutes a particular population prone to alcohol-induced bronchospasm.

In a study of 291 diabetics (Leslie *et al.* 1980), 191 reported facial flushing and 12 became dyspnoeic on exposure to chlorpropamide and alcohol. Five of the dyspnoeic patients developed wheezing and abnormal respiratory function tests indicative of bronchial asthma. In this study the asthmatic reaction was prevented by disodium cromoglycate and by naloxone. The authors concluded that asthma induced by chlorpropamide and alcohol may be mediated by endogenous peptides. The importance of these findings lies in the fact that many 'cold' remedies and cough mixtures are alcohol-based, and might therefore constitute a hazard to asthmatics, at least in certain subpopulations.

Tartrazine

Tartrazine, is an orange-yellow dye, is one of the colouring substances permitted under the EEC regulations for use as colourant in food, soft drinks, and drugs. The prevalence of sensitivity is not known but it has been suggested to be 0.01 per cent (*Drug and Therapeutics Bulletin* 1980).

Cross-reactivity with aspirin has been reported (Samter and Beers 1967) and tartrazine sensitivity is usually seen in aspirin-sensitive patients but sensitivity to tartrazine alone can occur (Zlotlow and Settipane 1977). Cross-sensitivity has been described to other azo dyes and to benzoates used as preservatives. Probably between 8 and 20 per cent of those sensitive to aspirin are also sensitive to tartrazine and reactions to both agents are thought to be non-allergic in nature.

The diagnosis of tartrazine sensitivity can be difficult since no simple test exists, and it can only be established from the history, elimination of tartrazine from the diet, and subsequent challenge. Asthmatics with aspirin-sensitivity need evaluation to exclude tartrazine intolerance especially since many cough mixtures and antihistamine preparations contain tartrazine as a colourant.

If tartrazine-sensitivity is established it is important to check that none of the foods, beverages, or drugs contain tartrazine and a tartrazine-free diet be prescribed. A list of drugs containing tartrazine is given in Table 15.1. Unfortunately, the presence of tartrazine in a product does not have to be declared by the present UK legislation though for drugs this is required in the USA (*Food and Drug Bulletin*, 1979). As can be seen from this list, tartrazine is present in a number of products which might be used by asthmatics. There is a movement now to bring in legislation to bar tartrazine from medicaments likely to be used by asthmatics or introduce a labelling requirement.

Cremophor EL

Severe, occasionally fatal, bronchospasm has been associated with the intravenous steroidal anaesthetic agent *Althesin*, which contains alphaloxone and alphadolone (Clarke *et al.* 1975; Evans and Keogh 1977; Dye and Watkins 1980). The reaction consists of bronchospasm and/or hypotension, which may be preceded by purplish generalized flushing of the skin. The solvent *Cremophor EL* (polyoxyethylated castor oil) was suspected by

---

Table 15.1   Drugs containing tartrazine (reproduced from *Drug and Therapeutics Bulletin* 1980)

The amounts of tartrazine per dose range from less than 1 μg to 2 mg or more

Antihistamines	*Actifed* syrup; *Atarax* 25 mg; *Dimotane* tablets; *Fenostil Retard*; Optimine* syrup; *Piriton**
Analgesics	*Cafergot* tablets; *Depronal SA; Fortagesic; Fortral; Hypon; Migraleve* (yellow tablets only); *Napsalgesic; Paragesic*
Antibiotics	*Achromycin; Aureomycin; Bactrim Drapsules* and dispersible tablets; *Broxil* tablets and capsules; *Ceporex* capsules*; *Crystapen V; Imperacin; Kantrex* capsules; *Ledermycin; Megaclor*; Rondomycin; Terramycin; Tetrex; Totomycin* mixture
Anticonvulsants	*Mysoline and phenytoin*
Antihypertensives	*Aldomet* 250 mg and 500 mg
Bronchodilators	*Amesec* capsules; *Expansyl Spansules*
Diuretics	*Lasix* 500 mg; *Lasikal; Lasix* + K
Iron preparations	*Fe-Cap; Fe-Cap C; Fe-Cap* folic; *Kelfolate*
Oral contraceptives	*Minovlar*; Minovlar ED**
Psychotropic drugs	*Atensine; Bellergal* and *Benergal Retard; Merital; Nembutal; Sparine; Tryptizol* 25 mg; *Valium* 5 mg
Miscellaneous	*Cedilanid; Deseril; Parlodel; Sanomigran*
BPC preparations	codeine linctus; diabetic linctus; diamorphine linctus; ephedrine elixir; isoniazid elixir; methadone linctus; neomycin elixir; noscapine linctus; paediatric compound tolu linctus; phenobarbitone elixir; pholcodine linctus; piperazine citrate elixir; strong pholcodine linctus

* These products are about to be replaced by a tartrazine-free version.

Notcutt (1973) when a reaction to *Althesin* occurred, 2 weeks after exposure to propanidid, which also contains cremophor. It has been shown that the reaction is immunologically based, with positive skin tests to cremophor, and involving IgG antibodies (Moneret-Vautrin *et al.* 1983). Both *Althesin* and propranidid have been withdrawn from the market by their manufacturers.

Other less frequent causes of allergic asthma

α-Methyldopa

Asthma from the inhalation of α-methyldopa powder was demonstrated in a woman who was handling the powder in a pharmaceutical factory (Harries *et al.* 1979). The patient started work in 1973 as an analytical chemist and developed intermittent asthma over several months. She was atopic and antibodies to α-methyldopa could not be demonstrated. The diagnosis was by a provocation test.

Chloramine

Bourne *et al.* (1979) described asthmatic symptoms which developed in seven brewery workers using chloramine as a sterilizing agent. The subjects showed positive weal and flare reactions to skin-prick tests and the symptoms did not recur when the men were removed from the areas where chloramine was handled. It is shown that in addition to causing irritant effects, the inhalation of dry or liquid aerosols of chloramine may cause sensitization with the development of allergic asthma on re-exposure.

Metabisulphite

Sulphites are added to certain infusion fluids in order to prevent decomposition and discolorization during sterilization. Severe asthmatic attacks have been reported following the use of IV fluids containing sulphites (Bassler and Heidenreich 1984). Inhalation challenge resulted in bronchoconstriction in asthmatics, even if they were not sensitized to sulphite, but not in non-asthmatic controls (Koepke *et al.* 1985). It is suggested that the sulphite content of infusion fluids should be declared.

Carbamazepine

The anti-epileptic carbamazepine may cause asthma as part of a more generalized sensitivity reaction with fever and eosinophilia, the syndrome resolving with withdrawal of the drug (Lewis and Rosenbloom 1982).

Disopyramide, tubocurarine, and alcuronium

Single reports of acute allergic reactions with bronchoconstriction have implicated the anti-arrhythmic disopyramide (Porterfield *et al.* 1980) and the muscle relaxants tubocurarine and alcuronium. On the latter cases skin tests were positive to both these drugs (Yeung *et al.* 1979).

Haemodialysis fluid

Similarly the acetate ion in the perfusate of *haemodialysis fluid* caused asthma, eosinophilia, and raised IgE with each perfusion, and inhalation challenge to acetate was positive. A change from acetate to bicarbonate resulted in abolition of all symptoms (Ei *et al.* 1979).

Paradoxical asthma due to bronchodilators

Paradoxical effects may be produced by bronchodilators drugs, or drugs which are intended to prevent mast-cell degranulation, which contrary to expectation may have a bronchoconstrictive effect.

Isoprenaline

The earliest reports of β-agonists such as isoprenaline inducing bronchospasm were by Keighley (1966) who described three patients with a family history of atopy who developed severe bronchoconstriction with isoprenalin aerosols, and who responded positively to rechallenge.

Terbutaline sulphate

Drexel *et al.* (1982) reported an attack of severe bronchospasm following five hours after the ingestion of terbutaline sulphate (2.5 mg) in an asthmatic. Because it was considered impossible that such a reaction could occur, therapy was resumed, resulting in a near-fatal attack of asthma, with, in addition, inspiratory stridor, cyanosis, and respiratory depression. The authors considered this a severe hypersensitivity reaction, but comment on the long delay (4–5 hours) between the ingestion of the drug and the reaction, suggestive of a metabolite of terbutaline being responsible.

Ipatropium bromide, deptropine citrate, and atropine methonitrate

Anticholinergic agents such as ipatropium bromide are also capable of inducing bronchoconstriction. Kennedy (1965) noted that out of a group of 20 asthmatic patients, the bronchoconstriction worsened in four patients after the inhalation of deptropine citrate, and in one patient after atropine methonitrate. The latter patient also developed severe asthma 30 minutes after inhalation of isoprenaline combined with deptropine citrate.

Connolly (1982) reported three atopic patients with reversible airways obstruction in whom ipatropium bromide produced an immediate, progressive fall in peak expiratory flow rate, which returned to pre-ipatropium levels when the drug was stopped. The one patient who was challenged with ipatropium, showed a positive test. It should be noted that one of the patients was receiving nebulized ipatropium, which excludes any effect due to the Freon propellant.

The author postulated that the effect might have been due to increased viscosity of bronchial secretions. However, Jolobe (1982) considered that there might be a fundamental change in the reactivity of the bronchial smooth muscle to anticholinergic drugs in atopic asthmatics, and that the action of the competitive antagonist attached to the receptor may then produce potentiating effects on acetylcholine instead of the usual inhibition.

Freon propellants

The role of the Freon propellants in provoking bronchial reactions was investigated by Bryant and Pepys (1976) who found that, challenge with a canister containing placebo and the propellants produced a fall of 20 per cent in FEV_1 which could be prevented by sodium cromoglycate or reversed by salbutamol. However, the inhalation of the propellant gases (as compressed gas) did not produce any change; furthermore, the gases only produced minimal histamine release from the patients' white blood cells. They suggested that the only other agents to which exposure may have occurred are the aluminium canister and the extractions from rubber components of the valves.

Other anti-asthma agents

Rarely hypersensitivity reactions occur to the mast-cell stabilizer sodium cromoglycate (Price 1982) with worsening of asthma, although in some cases bronchoconstriction may be due simply to mechanical irritation.

Acetyl-cysteine, used as a mucolytic by the oral and inhalational, and in paracetamol poisoning by the intravenous route, may induce severe hypersensitivity reaction with airways obstruction. Reports of asthma have appeared in the literature and in the adverse reactions register of the Committee on Safety of Medicines, irrespective of the route of administration (CSM, personal communication, 1984, Ho and Beilin 1983).

Allergic, or idiosyncratic reactions to intravenous or inhaled hydrocortisone have been reported in aspirin-sensitive asthmatics (Partridge and Gibson 1978) as well as in non-atopic, non-aspirin-sensitive asthmatics (Ward and Prichard 1980).

Changes in drug-induced mortality from asthma

The controversial rise in death rate occurring amongst asthmatics in Great Britain in the 1960s, while commonly attributed to the use of aerosols or propellants containing sympathomimetic amines (notably isoprenaline), has never been satisfactorily explained or the mechanism established. Although sympathomimetic amines may have had a direct contributory role, it may well be that a misplaced confidence in aerosols, for patients who in fact required corticosteroids, may have been a major factor (Inman 1974).

Wilson *et al.* (1981) reported an apparent increase in young people dying suddenly from acute asthma in New Zealand and attributed this to a change in prescribing habits for asthma with an increase in the use of oral theophylline drugs, especially slow-release formulations in place of inhaled steroids and cromoglycate. The authors suggest that there may be an additive toxic effect with inhaled β-agonists in high doses resulting in cardiac arrest.

Their findings have been widely criticized on a number of points:

1 No data were given on whether the change in prescribing habits mentioned was real or assumed, nor are there indications whether the change from inhaled steroids and cromoglycate was due to the fact that the latter were ineffective (Grant 1981; Beaglehole *et al.* 1981).

2 No indication was given as to how well controlled or supervised these patients were (Grant 1981).

3 The methodology of arriving at their figures (Beaglehole *et al.* 1981; Whittington 1981).

While it is agreed that self-administration of either theophylline or inhaled β-agonists may lead to overdose, all authors commenting on Wilson's findings, suggest that undertreatment and poor medical supervision are more likely to lead to sudden death in asthma. Only a properly constructed study can solve the problem as to whether increased toxicity really does exist, since experience and figures from other countries do not support these findings (Lambert 1981).

Acute non-allergic asthma

Drug-induced asthma may result from the normal pharmacological action of drugs, such as, for instance, the exacerbation of airways obstruction by drugs which antagonize the adrenergic tone of the bronchi, e.g. β-adrenergic blocking agents.

β-Adrenergic blocking agents

In asthma dependence on β-adrenergic tone may be critical, and β-adrenergic blocking agents, by their pharmacological action may therefore be expected to worsen asthma and to potentiate the effects of acetylcholine, histamine, and 5-hydroxytryptamine. Chronic bronchitics with reduced respiratory reserve may respond in a similar manner, with a fall in forced expiratory volume at 1 second (FEV_1) and forced vital capacity (FVC) (Sinclair 1979). It has furthermore been shown that the degree of bronchoconstriction is directly related to the potency of the β-adrenergic blocking agent (Ind and Barnes, personal communication, 1983). It might therefore be expected that cardioselective β_1-blocking agents are less likely to induce bronchoconstriction; however it has been shown that every class of blocking agent may worsen or precipitate asthma, and near-fatal and fatal results have been reported following the use of these drugs (Lewis 1981; Raine et al. 1981; Harries 1981).

One blocking agent which has been found to produce less severe reactions, is labetalol, which combines α- and β-blocker elements (Adam et al. 1982). Reports are now appearing that propranolol may induce severe bronchospasm even in non-asthmatics; in most of these cases the drug has been given for 1–5 months before symptoms appeared (Clark et al. 1982). Popio et al. (1983) suggested that those patients with chronic obstructive pulmonary disease, who are at risk from propranolol, may be identified beforehand by their response to the inhalation of carbachol.

Topical β-adrenergic blocking agents

Timolol maleate ophthalmic solution, for use in ocular hypertension, was first marketed in 1979, its advantage over existing ocular hypotensive agents lying in its action being free from concomitant change in pupil size and diminution of vision which occur with other drugs. Furthermore, it seemed not to have systemic effects such as change in heart rate and blood pressure. However, within the first year of marketing reports appeared in the literature of individual cases, with previously well controlled or asymptomatic asthma, who developed acute worsening of their airways obstruction following its use.

The connection between timolol ophthalmic drops and increased airways resistance has now been well documented by several authors, who measured pulmonary function before and after topical timolol. Challenge with as little as a single drop of 0.25 per cent timolol intra-ocularly was shown to produce a drop of 20 per cent in forced expiratory volume at one minute (FEV_1) within 15 minutes of administration and an increase in airways resistance (S_{raw}) of 40 per cent (Holtman et al. 1980). Administration of pilocarpine did not produce these changes. A dose relationship of the effect was demonstrated by Charan and Lakshminarayan (1980) showing a fall in FEV_1 of 20 per cent after one drop, and of 47 per cent if two drops of timolol were administered. It has further been shown that patients with 'chronic bronchitis' with irreversible airways obstruction may be adversely affected (Schoene et al. 1981). Since many of the latter patients may be elderly and therefore may be at risk from glaucoma, it has been suggested that, before instituting ocular timolol therapy, a spirometric test following a small challenge should be undertaken.

Prostaglandins

Prostaglandin F_2 is used in midtrimester abortion and for the induction of labour. It is a known bronchoconstrictor and prostaglandin F_2-induced bronchoconstriction was seen in 1:17 patients given the drug intravenously, in 1:22 cases of intrauterine application, and in 5:41 cases of intra-amniotic injection (Wislicki 1982).

Interstitial pulmonary disease and pulmonary fibrosis

Interstital lung disease or alveolitis (sometimes termed 'pneumonitis') is the term used for the non-specific morphologic reaction of the lung to injury. It has two distinct phases — an acute exudative phase (which is seen in ARDS) characterized firstly by destruction of Type I alveolar cells with intra-alveolar inflammatory changes and the formation of hyaline membranes, followed by a proliferative phase with Type II cells repopulating the denuded basement membrane with thickening and organization of the inflammatory exudate, resulting in incipient pulmonary fibrosis. Pulmonary function tests show diminished gas transfer with a restrictive defect, and sometimes an element of obstruction. Drug-induced injury of this kind may be identified in some cases as due to direct toxicity, or, to immunological damage (Types III and IV), but in the majority of cases the mechanism is not clear.

Anti-arrhythmic agents

Amiodarone

The first reports of interstitial alveolitis following the use of amiodarone appeared in 1981, soon after the introduction of the drug to the market. Typically increasing dyspnoea supervened 1–4 months after starting treatment. X-ray showed diffuse infiltration; pulmonary function tests revealed a restrictive defect and severe impairment of gas transfer. Histology showed thickened alveolar septa, with prominent fibroblasts and collagen, the pneumocytes lining the alveolar spaces were hyperplastic, atypical, and multinucleate (Rotmensch et al. 1980; Heger et al. 1981).

On discontinuing the drug and, with the use of prednisone, there is usually clinical improvement, accompanied by resolution of pulmonary infiltrates and return to normal of pulmonary function tests (Riley et al. 1982 Wright and Brackenridge 1982). However, Sobol and Rakita (1982) found pulmonary fibrosis in four of the six patients who had histological examinations, and McGovern et al. (1983) reported that in two of their four patients with alveolitis, lung function tests had not returned to normal by the end of six months following cessation of amiodarone treatment.

From all the published papers there appears to be a correlation between the use of steroids and recovery. The possibility that the reaction to amiodarone is dose-dependent was raised by Marchlinski et al. (1982). They treated 23 patients with amiodarone, but found that only four who had doses greater than 400–800 mg developed pulmonary complications. A possible immunological mechanism was postulated by Suarez et al. (1983) who found C_3 deposition on the alveolar septa in their case of amiodarone-dependent pneumonitis.

Tocainide

Tocainide, also a recently introduced anti-arrhythmic drug with pharmacological properties similar to those of lidocaine, has been reported to produce a similar picture to that of amiodarone, with a similar time interval after the start of administration, and, like amiodarone resolving after the drug was withdrawn (Perlow et al. 1981; Braude et al. 1982).

Cytotoxic agents and radiotherapy

Lung damage due to cytotoxic chemotherapy was first reported in 1961 (Oliner 1961) and recently other cytotoxic drugs have been associated with lung damage, although most of the drugs had been used for some time before their toxic effects on the lungs was recognized. In order to establish a causative role of cytotoxic drugs in causing lung damage it is necessary to show: (i) a history of drug exposure; (ii) the presence of lung damage; and (iii) the absence of other causes of lung damage.

Clinical symptoms usually develop slowly with increasing dyspnoea and cough with signs of basal crepitations — a picture which must be differentiated from chest infection. Radiologically there is pulmonary infiltration, which may mimic secondary tumours, and pulmonary functions show a restrictive defect with diminished gas transfer and hypoxaemia. Histologically the alveoli are filled with proteinaceous fluid, with desquamation of pneumocytes, alveolar epithelial dyplasia, thickening of septa, and interstitial fibroblastic proliferation (Weiss and Muggia 1980). The incidence of this type of reaction has been estimated to be in the region of 3–10 per cent for bleomycin, the drug most likely to produce pulmonary changes (Blum et al. 1973).

Weiss and Thrush (1982), in a major review of chemotherapeutic agents which cause pulmonary toxicity have ranked these agents into different categories with regard to pulmonary toxicity, namely:

High risk	bleomycin, busulfan, carmustine
Medium-to-low-risk	chlorambucil, cyclophosphamide, cytarabine, melphalam, methotrexate, mitomycin
Very low risk	azathioprine, 6-mercaptopurine, procarbazine, semustine, teniposide, uracil mustard, zinostatin

A single report has described the acute appearance of pulmonary hyaline-membrane disease in a 20-year-old patient with cerebellar medulloblastoma with the anti-cancer agent podophyllotoxin, but its pulmonary toxicity has not yet been quantified (Commers and Foley 1979). There is also now good evidence that the total dose of some cytostatics may be related to the development and eventual prognosis of the pulmonary damage. These include cyclophosphamide (less than 10 g total having a good prognosis), carmustine (with a possible 50 per cent risk of lung disease at total doses of 1500 mg/m^2) (Batist and Andrews 1981). The route of administration appears to be immaterial. For instance, methotrexate administered intrathecally may produce lung damage which appears to be reversible with steroids and is thought to be a hypersensitivity reaction (Weiss and Muggia 1980). Furthermore, the administration of certain cytotoxic agents together, such as for instance cyclophosphamide and carmustine, or the addition of high oxygen concentration in the inspired air to a cytostatic regimen, all increase the likelihood of pulmonary damage (Weiss and Muggia 1980).

Injury to the lungs, with pneumonitis, fibrosis, or effusion has for many years been a recognized complication or sequel of radiation therapy to the lungs, oesophagus, and mediastinum, the severity and reversibility of the damage depending on the source, dose, and rate employed (Chacko 1981).

It has now been clearly shown that there is enhanced pulmonary toxicity from the simultaneous administration of radiation and cytotoxic agents, irrespective of whether the radiation is given concomitantly or sequentially. It should be noted that the pulmonary damage may not be apparent for some time after treatment and pulmonary damage has now been reported after a latent interval in children receiving combination chemotherapy for childhood leukaemia or lymphomatous conditions. Two of these cases were fatal, one after a two-year remission (Muller *et al*. 1979). An even longer latent interval of 4-6 years between cytotoxic chemotherapy (cyclophosphamide) and the appearance of pulmonary fibrosis has been described suggesting that the damaging effect may continue after cessation of therapy (Weiss and Muggia 1980).

As the survival in some paediatric cancers improves, the recognition of the early signs of pulmonary complications of cytotoxic chemotherapy becomes increasingly important.

Finally, it should not be forgotten that cytotoxic drugs may be employed in disease other than malignancy. Burke *et al*. (1982) report two fatal cases of pulmonary fibrosis, resistant to steroids, which followed the use of cyclophosphamide for glomerulonephritis. It should also be borne in mind that where cytotoxic agents are employed for the treatment of renal disease, renal failure may increase their toxic effects (Perry *et al*. 1982).

Gold

Podell *et al*. (1980) report a case of widespread interstitial fibrosis and alveolar infiltrates which occurred in a woman after a total dose of 395 mg sodium aurothiomalate by injection. Treatment with oral methylprednisolone resulted in resolution of dyspnoea and radiographic appearances.

A second case is described by Smith and Ball (1980) in a woman who developed a pruritic macular rash on her trunk and face and mouth ulcers after a total dose of 480 mg sodium aurothiomalate by injection. A week after the injection she developed cough and dyspnoea on exertion. Chest radiography showed diffuse interstitial infiltration

and the diagnosis was confirmed histologically. Prednisone treatment alleviated cough and dyspneoa and the chest radiograph was normal within two months.

Two similar cases of interstitial pulmonary fibrosis associated with similar doses of gold by injection were described by Terho *et al*. (1979). Gold was stopped in both cases. In the first case, a 61-year-old man, steroids resulted in resolution within five months. However, in the second case, a 57-year-old woman, treatment with prednisolone for a year with the addition of azathioprine failed to result in improvement of her symptoms or in the radiographic appearances.

There is now some evidence that the reaction has an immunological basis. McCormick *et al*. (1980) recorded two women on gold therapy who developed progressive dyspnoea and pulmonary infiltrates with a restrictive respiratory defect accompanied by reduced gas transfer, in whom assays of cellular immunologic reactivity to gold salt were performed. The patients' lymphocytes were shown to react normally to stimulation with PHA (Phytohaemagglutinin), but were not stimulated to transform by the addition of gold salts. They did however produce lymphokine, as demonstrated by the appearance of migration inhibition factor and macrophage chemotactic factor in the supernatant. Normal subjects, patients with rheumatoid arthritis, and patients with rheumatoid arthritis on gold therapy, but not sensitive to gold, were used as controls. None of these control groups showed lymphokine activity. The authors postulate that gold, like beryllium (a well-documented hypersensitivity-inducing metal), binds with serum proteins to make complete antigens.

Similar *in-vitro* evidence of a hypersensitivity reaction was described by Alcalay *et al*. (1979), although the picture here was complicated by a more general immunological reaction to gold.

Nitrofurantoin

The acute reaction to nitrofurantoin with eosinophilia, chills, fever, and cough is described under Pulmonary eosinophilia, p. 210. There are, in addition, more insidious cases, especially after prolonged treatment, and in the elderly where a direct toxic effect has been suggested (Holmberg *et al*. 1980), and where the association with treatment may be missed with resulting pulmonary fibrosis. Although rarer than in adults, cases have also been reported in children. Rantala *et al*. (1979) reported a case in a child treated with co-trimoxazole and nitrofurantoin for a urinary-tract infection over a two-year period. Dyspnoea developed insidiously and chest radiograph showed diffuse intense interstitial fibrosis and a small pneumothorax. Resolution occurred on withdrawal of nitrofurantoin and treatment with prednisone.

In Sweden there has been a rising incidence of adverse reactions to nitrofurantoin in the decade 1966–76, and pulmonary reactions, mainly acute, constituted about half of these. During this period of increase in reactions, the sales figures of nitrofurantoin declined (Holmberg and Boman 1981). Penn and Griffin (1982), comparing the incidence of these reactions in three countries, found that the number of reactions to nitrofurantoin have declined (as have sales) in the UK and Holland. The reported pulmonary reactions between 1964 and 1980 in the UK were 73 (64 acute; 9 chronic) compared with 447 (398 chronic; 49 acute) in Sweden. The authors discussed the possible factors which might account for these differences, such as reporting differences, but so far an explanation has not been found.

Penicillamine

Penicillamine is known for many adverse effects, and as early as 1968 Walshe described acute respiratory distress together with urticarial reaction on two successive occasions, following the attempted initiation of penicillamine therapy in a boy with Wilson's disease.

Since that time several isolated reports of obstructive lung disease in association with penicillamine therapy have appeared in the literature, but the causal relationship has only recently been suspected, probably because in many cases the patients suffered from rheumatoid arthritis, where diffuse interstitial lung disease may occur as a part of the disease process.

In all cases there was a time relationship with gradual onset after some months' treatment with penicillamine possibly due to a cumulative effect (Scott et al. 1981), although more acute reactions have also occurred. In all, progressive dyspnoea and cough developed before radiological changes of interstitial fibrosis appeared. A severe fixed obstructive defect, accompanied by some restrictive element and diminished gas transfer, was demonstrated, in many cases irreversible even after stopping the drug and instituting treatment with steroids, although in the case described by Matloff and Kaplan (1980) plasmapheresis and immunosuppressants produced dramatic improvement with gradual resolution of the pulmonary changes.

A hypersensitivity or immune-dependent reaction has been suspected, and a Goodpasture-like syndrome has been described in a patient treated with penicillamine for biliary cirrhosis, although the typical renal lesion with deposition of IgG against the glomerular membrane was absent (Matloff and Kaplan 1980). Penicillamine anti-

bodies have been described, both in penicillin-sensitive and normal subjects (Assem and Vickers 1974).

The histopathology of the lung showed diffuse bronchiolitis with narrowing and obliteration of the bronchioles, infiltration with inflammatory cells, mainly mononuclear. While the process was focused in the small airways, there was a tendency to spread out into the surrounding lung parenchyma (Murphy et al. 1981).

While in each individual case there is no absolute proof, since confirmatory challenge might have precipitated acute respiratory failure, reports on some 30 patients treated with penicillamine for a variety of conditions who have developed pulmonary complications make a strong presumptive case for this adverse reaction.

Phenytoin

Acute pulmonary adverse effects of phenytoin and the related drug mesantoin were described by several authors in a total of four cases (Fruchter and Laptook 1981; Michael and Rudin 1981; Goffman et al. 1982). In all patients symptoms of a drug reaction appeared within three to six weeks of starting the drug with fever, rash, lymphadenopathy, and cough. The chest X-ray was initially normal in two, but diffuse pulmonary infiltrates appeared within a week of the start of symptoms and, in the one patient where this was tested, pulmonary functions showed reduced forced vital capacity and expiratory flow, also diminished diffusing capacity. Eosinophilia was present in two. All recovered after discontinuing phenytoin, two with additional steroid therapy.

Sulphasalazine

A number of individual cases link sulphasalazine, the accepted therapeutic agent for ulcerative colitis, with lung damage. The daily dosage of sulphasalazine ranged from 1.5 to 6 g and the period of exposure from 1.5 to 7 months. In all patients symptoms supervened after some weeks' treatment with salazopyrin, showing pulmonary infiltrates on radiogram, a restrictive functional defect, and, on biopsy, interstitial proliferative alveolitis. The damage may be reversible or may progress to pulmonary fibrosis (Matek et al. 1980; Williams et al. 1982). Pulmonary and systemic eosinophilia were also described.

Methysergide/ergotamine/bromocriptine

Pleuropulmonary fibrosis has been recognized as a rare adverse effect of methysergide therapy for nearly 20 years (Graham 1968). Similar changes have more recently been reported with the chemically related drug ergotamine, a potent drug in the treatment of migraine. The patient, a man in his early 50s had been using cafergot suppositories for nearly 10 years (together with other prophylactic drugs, but never methysergide) before developing dyspnoea and fever. On investigation he was found to have pleural thickening and also a tumour-like mass between the middle and lower lobe. Biopsy showed non-specific pleural and pulmonary interstitial fibrosis. On stopping ergotamine, he improved clinically as did his pulmonary function tests and his chest radiograph. (Taal *et al.* 1983).

Bromocriptine, an ergot derivative, has similarly been shown to cause pleuropulmonary fibrosis. Eight patients with parkinsonism developed a uniform picture of increasing dyspnoea, with fibrosis of the pleura and lung tissue which could partly be reversed after treatment with corticosteroids. The duration of treatment with bromocriptine had varied from 15 days to 3 years before symptoms developed (Vergeret *et al.* 1984).

Acebutolol and propranolol

Single reports have associated hypersensitivity pneumonitis with acebutolol and also propranolol (Akoun *et al.* 1983; Thompson and Grennan 1983). In the former case challenge confirmed the association with the typical clinical X-ray and pulmonary function features.

Antazoline

Palissa *et al.* (1979) describes a case of antazoline-induced acute interstitial pneumonitis and rash which was confirmed by a provocation test. A 35-year-old woman was treated with antazoline hydrochloride on two occasions 200-400 mg daily after which she developed dyspnoea, cough, fever, and an exanthematous rash. There was no eosinophilia and immunoglobulins were normal. Vital capacity and transfer factor were reduced. The symptoms and abnormal pulmonary function tests recurred on challenge with antazoline 50 mg, on two occasions some seven hours after the second dose.

Oxygen

Prolonged ventilation with high concentrations of oxygen may result in pulmonary disease, the histological features of which are typical of the following stages of interstitial pulmonary damage: (1)an early exudative phase characterized by congestion, alveolar oedema, intra-alveolar haemorrhage, and fibrin exudate, with the formation of prominent hyaline membrane without an associated inflammatory component, and (2) a late proliferative phase showing alveolar and interlobular septal oedema and fibroblastic proliferation, with early fibrosis and prominent hyperphasia of the alveolar lining cells (Nash *et al.* 1967). Oxygen should not be withheld from patients who need it for fear of possible toxic effect on the lungs, but the concentration should be reduced to 40 per cent or less as soon as measurements of arterial blood gases show that this can be done safely.

Pulmonary eosinophilia

Drug-induced pulmonary eosinophilia is a hypersensitivity reaction, acute or subacute, characterized by increasing dyspnoea, or the adult respiratory distress syndrome, diffuse pulmonary infiltration, restrictive/obstructive ventilatory defects with peripheral and pulmonary eosinophilia.

The earliest and most frequent reports relate to penicillin and its derivatives (Poe *et al.* 1980). However a wide variety of drugs are capable of causing this reaction. Repeated administration of nitrofurantoin was reported to result in a rise in peripheral eosinophils from 650 to 1625 mm^3 within 24 hours, accompanying pulmonary infiltration and a small pleural effusion (Israel and Diamond 1962; Holmberg *et al.* 1980). Phenytoin, in addition to causing interstitial pneumonitis, may result in pulmonary eosinophilia as part of a hypersensitivity reaction (Michael and Rudin 1981) and cessation of therapy resulted in resolution of signs and symptoms.

Carbamazepine (*Tegretol*), used in the treatment of trigeminal neuralgia and epilepsy has been reported rarely to give rise to hypersensitivity reactions in the lung, with airways obstruction, diminished gas transfer, and, on biopsy, eosinophilic interstitial changes. In two cases challenge with the drug confirmed the relationship (Schmidt and Brugger, 1980; Lee *et al.* 1981). Similarly, naproxen (Nader and Schillaci 1983), acebutolol (Thompson and Grennan 1983), clofibrate (Hendricksen and Simpson 1982) and desipramine (Mutnick and Schneiweiss 1982), gold (Scott *et al* 1981), and sulphasalazine (Williams *et al.* 1982) have been shown to cause pulmonary, and sometimes peripheral, eosinophilia with pulmonary infiltration.

Pulmonary oedema

Non-cardiogenic pulmonary oedema caused by drug ingestion is thought to be due to changes in membrane permeability or increases in pulmonary microvascular pressure, such as is thought to occur with neurogenic pulmonary oedema due to massive adrenergic discharges

(Staub 1981). The changes in permeability caused by drugs are possibly due to direct damage or immune mechanisms.

Hydrochlorthiazide

Acute pulmonary oedema following the use of hydrochlorthiazide is a rare occurrence and appears to be a true idiosyncratic reaction to the drug.

In the few cases reported the reaction occurred suddenly and dramatically, sometimes without previous exposure to the drug, although in the majority milder reactions to hydrochlorthiazide had previously occurred.

Acute dyspnoea developed within 20-60 minutes after ingestion of the drug with clinical and radiological signs of pulmonary oedema, a fall in arterial oxygen gradient which took several weeks to revert to normal, and was considered to be proportional to the severity of the reaction. Patients who were challenged had further severe episodes confirming the relation. Immunological investigations in one case showed normal B and T lymphocytes and complement levels, though very high concentrations of hydrochlorthiazide showed a very small degree of migration inhibition, which is considered only doubtfully related to the reaction (Bell and Lippmann 1979; Dorn and Walker 1981).

Beta-mimetic agents in pregnancy

Sympathomimetic drugs, such as ritodrine, terbutaline, or fenoterol are given by intravenous infusion in the prevention of preterm labour, and beta- or dexamethasone may be given concurrently by intramuscular injection in order to prevent hyaline membrane disease in the newborn. In all (probably a total of 10) reported cases except one, the patients were previously healthy young women with no history of pulmonary or heart disease. Only one patient had mild atrial disease. In all, acute pulmonary oedema occurred within 24 to 96 hours after the start of treatment (Jacobs *et al.* 1980) and responded well to diuretics and digoxin. Philipsen *et al.* (1981) in a prospective study in 23 patients with premature uterine contractions examined the effect of ritodrine when given in saline or isotonic glucose solution. There was a ritodrine dose-related, statistically significant fall in haemoglobin, haematocrit, and serum albumin similar with both infusin media. Serum renin and aldosterone levels were significantly elevated at six hours but then declined. Seven patients in the saline group, but none in the glucose group, developed pulmonary congestion, the saline group retaining fluid.

The results confirm that the beta-agonist-related pulmonary oedema is likely to be dependent on a significant increase in plasma volume superimposed on the already enlarged plasma volume of pregnancy and is quantitatively related to the amount of ritodrine administered. The mechanism is probably due to ritodrine-stimulated antidiuretic hormone (although this latter was not estimated) and the renin-aldosterone system.

Since the patients treated with ritodrine in saline showed a positive fluid balance, whereas those receiving glucose were normal, one cannot recommend treating pregnant women in premature labour with ritodrine in isotonic saline, and it is suggested that it is advisable to use the minimum possible effective dose of ritodrine in a restricted amount of glucose. All authors agree that tachycardia and twin pregnancy (with its further increase of plasma volume) are predisposing factors in the development of beta-agonist-related pulmonary oedema.

It now seems unlikely that increased pulmonary arterial pressure plays a part. Benedetti *et al.* (1982) measured pulmonary artery pressure in a patient with ritodrine-induced pulmonary oedema and found it to be normal and similar to post-partum values, with normal pulmonary vascular resistance, but with augmented systemic flow.

Isoxsuprine, a vasodilator with beta-adrenergic stimulating effects, also inhibits uterine contraction and has similarly been used in arresting premature labour. Reports have appeared of pulmonary oedema following the use of intravenous isoxsuprine in premature labour (Nagey and Crenshaw 1982; Guernsey *et al.* 1981). The mechanism is considered to be the same for other β-agonists.

Contrast media

Acute haemorrhagic pulmonary oedema with clinical rales and frothy haemorrhagic sputum, and typical radiological changes of pulmonary oedema, has occurred in non-asthmatic healthy young people following the intravenous injection of meglumine diatrizoate and sodium iothalamate. The mechanism is not explained but two theories are put forward: a local non-antibody-medicated immunological reaction, leading to increased capillary permeability, or an osmotic mechanism due to the hypertonicity of the agents (Chamberlain *et al.* 1979; Greganti and Flowers 1979).

Salicylates

In a retrospective study of 11 consecutive patients with salicylate intoxication, Walters *et al.* (1983) found pulmonary oedema to be present in 35 per cent. They identified the following risk factors: age over 30, cigarette smoking, chronic salicylate ingestion — a component of metabolic acidosis, and the presence of neurological symptoms on admission. They considered the aetiology to be multifactorial, but to centre around altered vascular permeability in the lungs.

Other drugs

Rare or isolated reports of drug-induced pulmonary oedema have implicated heparin (Ahmed and Nussbaum, 1981), methadone (Zyroff *et al.* 1974), haloperidol (Mahutte *et al.* 1982), methylmethacrylate (Safwat and Dror 1982), and nifedipine (Gillmer and Kark 1980).

In addition, fulminant pulmonary oedema has been reported following the administration of protamine after cardio pulmonary bypass in open heart surgery (Just-Viera *et al.* 1984), and its needless administration should therefore be discouraged.

Adult respiratory distress syndrome

The adult respiratory distress syndrome (ARDS) is characterized by marked respiratory distress, diminished pulmonary compliance with reduced PaO_2, reduced oxygen transfer in spite of ventilatory assistance, and sometimes an element of obstruction. Radiologically there is diffuse pulmonary infiltration. Histologically there is hyperaemia, oedema, and hyaline membrane formation. The syndrome arises within 1–96 hours after a pulmonary insult and is refractory to usual methods of therapy, requiring positive-end expiratory pressure ventilation and steroids.

Lidocaine

Lidocaine has repeatedly been reported to have caused ARDS. It occurred twice in a 57-year-old woman; the first time after topical application, when 5 ml of a 4 per cent solution was used to anaesthetize her oropharynx, and then again after 0.5 ml of a 1 per cent solution was administered subcutaneously (Howard *et al.* 1982). A similar, but fatal case was reported by Promisloff and DuPont (1983) after topical application and by Woelke and Tucker (1983) after subcutaneous infiltration.

Salicylates

Salicylates are a recognized cause of ARDS and cases have been reported with chronic salicylate ingestion in both adults and children and in salicylate overdose (Anderson and Refstad 1978; Kahn and Blum 1979;

Heffner and Sahn 1981). Leatherman and Drage (1982) discussed the difficulties of treatment and the pitfalls of diagnosis. The mechanism is not known but pulmonary capillary permeability is increased as in salicylate-induced pulmonary oedema.

Methotrexate

Bernstein *et al.* (1982) reported the rapid onset of respiratory distress in two children following the injection of methotrexate into the cerebrospinal fluid (intraventricular and lumbar routes respectively). In both children, symptoms and signs of pulmonary oedema supervened within 6 to 24 hours of the dose, in one case with fatal outcome. Histology showed alveolar damage with hyaline membrane formation.

The authors reviewed three similar cases reported in the literature, but consider that in two of them pulmonary oedema was part of a more general hypersensitivity reaction. They suggest a direct toxic effect of methotrexate on the hypothalamus with the release of a massive centrally mediated sympathetic discharge, leading to fluid shifts in the lung. This theory, however, does not explain the case in the literature which they considered similar to theirs, but in whom the drug was administered orally.

Barbiturates

Although the use of barbiturates is generally declining, there are a number of situations where they are routinely used. For instance in anaesthesia and neurosurgery, high-dose artificial barbiturate coma is used to preserve brain function during ischaemia due to vascular accidents, or to decrease intracranial pressure due to trauma. The aim is to give high doses of thiopental in order to keep the EEG in a state of burst suppression activity. Patients are mechanically ventilated. Ducati *et al.* (1981) undertook a prospective study of 26 comatose patients of whom 13 were treated with artificial barbiturate coma; the remainder were given only routine intensive care. Patients with lung disease, shock, and traumatic or infective chest lesions were excluded.

They found respiratory complications much more common among patients receiving barbiturates: 10 out of 13 on barbiturates had the adult respiratory distress syndrome (abnormal gas analysis, abnormal chest X-ray, decreasing chest compliance) whereas only two of the 13 on intensive care alone had the syndrome. Blood gases returned to normal rapidly once barbiturate therapy was stopped, the chest X-ray took 3-4 days to do so. No relation was found between the total dose of barbiturate given and the severity of the syndrome. The authors postulate a direct toxic effect by barbiturates either on the alveolar membrane, or on the myocardium. The latter is known to cause a negative inotropic effect, resulting in pulmonary hypertension. The latter hypothesis would fit in with the report by Wilhelms et al. (1981) of 'fluid' lung (pulmonary oedema) in barbiturate abusers.

Amphotericin-B and leucocyte transfusion

Therapeutic leucocyte transfusions are widely used in the supportive care of patients with marrow aplasia and profound neutropenia, and amphotericin-B is added where sepsis is suspected.

Two cases of adult respiratory distress syndrome, with new diffuse interstitial pulmonary infiltrates on X-ray, were described during, or shortly after, an infusion of amphotericin-B in leukaemia patients who were receiving daily leucocyte transfusion (Wright et al. 1981).

Wright et al. also reviewed seven years' records of leucocyte transfusion recipients (from a clinical trial of leucocyte transfusion) and found respiratory deterioration to be 10 times more common in patients who had amphotericin added to their antibiotic regimen than in patients who received leucocyte transfusion alone. The control group of patients, who had no leucocyte transfusion, but of whom five received amphotericin-B did not exhibit pulmonary complications. Wright et al. (1981) further noted that in seven of eight patients who were on amphotericin before starting leucocyte transfusion, pulmonary complications did not develop, and they postulated that possibly endotoxaemia may play a part in the mechanism. This theory appears to be supported by reports from other authors who, however, have not examined the possible factors in quite the same way (De Gregorio et al. 1981).

Cytosine arabinoside

In a review of 181 autopsies on patients who had died of leukaemia, a highly significant increase was found in the incidence of unexplained massive, or moderate, pulmonary oedema in patients who had been treated with cytosine arabinoside 1-30 days prior to death. These changes were not found in patients untreated, or treated more than 30 days before death. There was no dose effect.

The oedema was both interstitial and intra-alveolar and highly proteinaceous, a feature of ARDS. Clinically respiratory failure occurred in 76 per cent of these patients, 33 per cent suffered respiratory distress during the administration of the drug, and 43 per cent did so within 27 days of discontinuing (Haupt et al. 1981).

Cyclosporin

Administration of cyclosporin into a central vein in organ transplantation was shown, in 2 out of 12 patients, to cause ARDS within 5 days. This was considered to be due to the high concentration of cyclosporin reaching the pulmonary circulation and not part of a generalized 'capillary leak' syndrome (Powell-Jackson et al. 1984).

Diazoxide

Diazoxide is a potent, rapidly acting smooth-muscle relaxant which is occasionally used as a tocolytic. A single case of severe ARDS with severe hypoxaemia, unresponsive to high levels of inspired oxygen, normal pulmonary capillary pressure was reported by Di Orio and Brauner (1982). Although no cases have been reported of ARDS with this agent when used as a hypotensive, it is important to bear the possibility in mind.

Pulmonary hypertension

Fenfluramine

Two cases of pulmonary hypertension due to the prolonged use of fenfluramine for weight reduction were reported (35 weeks and 18 months followed by a further 40 weeks, respectively). Both patients experienced increasing breathlessness, palpitations, and one also had exertional chest pain. Electrocardiogram showed corpulmonale. Raised pulmonary artery pressure was present with no evidence of other cardiovascular disease. When fenfluramine was stopped, symptoms disappeared over a period of weeks in both cases. In one case, fenfluramine was reinstituted after re-gain of weight, and once again the symptoms of exercise dyspnoea reappeared; they had gone within two days of stopping the drug (Douglas et al. 1981).

Hydrallazine

In certain circumstances hydrallazine may cause pulmonary hypertension. Kronzon et al. (1982), having read reports of favourable results with the use of oral hydrallazine in the treatment of patients with primary pulmonary hypertension, employed hydrallazine (intravenously in one case, orally in the other) in two patients with primary pulmonary hypertension. In both cases this was followed by shortness of breath (within 10 minutes after the intravenous dose; after the second day of treatment after the oral dose) signs of right ventricular failure, and a rapid decline in clinical status. Cardiac catheterization showed a marked further increase of pulmonary artery pressure by a mean of 16 mmHg in both cases, in excess of the initial hypertensive value, in addition with increased right ventricular and right atrial pressures. The authors suggested that the cases which fail to show a response to vasodilators may have advanced disease, with the pulmonary vasculature unresponsive to vasodilators — the systemic vasodilation leading to increased cardiac output and increased pulmonary artery pressure.

Prostaglandin synthetase inhibitors

Indomethacin and other prostaglandin synthetase inhibitors are gaining usage in the treatment of premature labour. A complication of this therapy is an increase in pulmonary arterial pressure associated with this procedure.

Pulmonary hypertension associated with the inhibition of prostaglandin synthetase in infants has previously been described with indomethacin (Zuckerman *et al.* 1974; Manchester *et al.* 1976; Csaba *et al.* 1978). Levin *et al.* (1978) also reported two deaths in which they found an increase in pulmonary arterial smooth muscle and a decrease in the number of pulmonary vessels.

Wilkinson *et al.* (1979) described persistent pulmonary hypertension in three premature infants born to mothers treated with naproxen, another potent inhibitor of prostaglandin synthetase. Naproxen was given in an attempt to delay parturition. The ductus arteriosus closed and severe hypoxaemia, abnormalities in blood coagulation, renal function, and bilirubin metabolism were also associated, and one infant died.

Radiotherapy

Pulmonary hypertension has been reported as long as 14 years after irradiation of a child for neuroblastoma. The girl when aged seven months had an operation for removal of the tumour, followed by irradiation of the mediastinum and the left chest. At the age of 14, following 18 months increasing dyspnoea, she was found to have an underdeveloped left chest with a restrictive pulmonary defect and diminished gas transfer. Her pulmonary artery pressure was 70/30 mm Hg (systemic arterial pressure 95/70 mmHg) and her left pulmonary artery was found to be hypoplastic on angiography (Butler *et al.* 1981).

Pulmonary vascular disease

Anticoagulants

Intrapulmonary bleeding due to anticoagulant therapy

has only rarely been reported and is usually accompanied by frank haemoptysis. In the absence of the latter it may present diagnostic difficulties.

Granthill *et al.* (1981) described two patients on long-term anticoagulant therapy (one for previous myocardial infarction, one for previous arterial embolus) who developed a severe haemorrhagic syndrome (melena and/or epistaxis and anaemia followed within 24-48 hours by severe respiratory distress syndrome with increasing dyspnoea, severe hypoxaemia, but without major haemoptysis). Radiology showed diffuse bilateral miliary opacities. The diagnosis was made on the finding of large numbers of alveolar macrophages and histiocytes loaded with haemosiderin.

Chakraborty and Dreisin (1982) reporting a further case underlined the diagnostic difficulties: the patient, on anticoagulants for four years, presented with mild respiratory distress, a massive pleural effusion, and well-defined rounded opacity at both bases. There was no evidence of any overt or occult haemorrhagic episodes, but his prothrombin time was greatly prolonged. The pleural fluid was found to be grossly haemorrhagic and a wedge resection of one of the masses showed a large intrapulmonary haematoma, the alveoli containing haemosiderin-laden macrophages. No other pathology was found, and after re-establishing anticoagulation control the patient improved over the following year and his X-ray showed marked clearing.

Isolated reports have implicated nitrofurantoin toxicity with fatal pulmonary haemorrhage (Averbuch and Yungbluth 1980) and reversible pulmonary vasculitis with haemoptysis after propylthiouracil therapy (Cassorla *et al.* 1983).

Oral contraceptives

It seems appropriate in this section to mention the well-established relationship between the dose of oestrogen in combined oestrogenprogestogen oral contraceptives and the incidence of thromboembolic disease (Inman *et al.* 1970) since patients may not uncommonly present with a pulmonary embolism. Hence the occurrence of pleuritic pain in a patient on oral contraceptives, with or without haemoptysis, may well be the first indication of this diagnosis.

Pulmonary infections

Corticosteroids

Bacterial pneumonias and pulmonary tuberculosis are well recognized as complications of corticosteroid therapy. In this respect an additional complication has appeared in association with the increasing use of steroid aerosols in the treatment of asthma. In a survey of 936 patients, 5.5 per cent of those receiving beclomethasone diproprionate by aerosol were found to have clinical oro-pharyngeal candidiasis (Milne and Crompton, 1974). The incidence was considerably higher than in patients receiving oral prednisolone.

Immunosuppression

The immunosuppressive effects of cytostatic agents, and the increasing use of these drugs may lay the patient open to opportunistic infections of the lung with tuberculosis, fungi (such as *Aspergillus*), actinomycetes (such as *Nocardia*), protozoa (such as *Pneumocystis carinii*), and viruses (such as cytomegalovirus) which may be mistaken for the direct adverse effect of the drug on the lung (Altus and Andrew 1982). Suppurative pneumonia and the formation of lung abscess are also common in 'main-line' drug addicts (Briggs *et al*. 1967).

Pleuropulmonary reactions

β-blockers

Pleural fibrosis has been associated with oxprenolol (Page 1979) and acebutolol (Wood *et al*. 1982). In both cases there was dense fibrous tissue covering the pleura patchily, and, in the second case, granulomata in the lung subpleurally. Withdrawal of acebutolol resulted in recovery.

Ergotamine and methysergide

Pleuro-pulmonary fibrosis may also be related to methysergide therapy (Graham 1968; Roberts and Breckenridge 1975). In reviewing the literature on methy-sergide-induced retroperitoneal fibrosis, Graham found numerous examples of pleuro-pulmonary episodes which had occurred in these patients but which had been over-shadowed by the genitourinary problem.

The striking features of methysergide-induced pul-monary fibrosis have been the repeated episodes of pleuritic chest pain often accompanied by fever and pleural effusion, the presence of friction rubs being occa-sionally audible to the patient himself, and the chronic dense pleural thickening tending to cause gross limitation of motion of the chest. The X-ray findings show a large, often tumour-shaped, fibrotic lesion in the posterior chest which on biopsy shows inflammatory fibrosis encasing loculated fluid. There has been a tendency for these abnormalities to increase if methysergide therapy continues and for them slowly to improve when the drug

is withdrawn. The chemically related drug, ergotamine, has been reported to induce similar lesions (Taal *et al*. 1983).

Dantrolene

Dantrolene sodium (*Dantrium*) is a central-acting muscle relaxant employed in the treatment of patients with spastic neurological disorders. Adverse effects such as drowsiness, fatigue, nausea, intestinal atony, and potentially fatal hepatitis have previously been recorded. Petusevosky *et al*. (1979) described chronic pleural effusion in three patients, one of whom had an associated pericarditis. A fourth patient had both pleural and peri-cardial effusions. The pleural fluid was sterile, there was no pulmonary involvement and pleural fluid and peri-pheral blood eosinophilia occurred in all cases. The authors concluded that although a causal relationship remains unproved, these reports warrant careful observation in those receiving long-term dantrolene therapy.

Methotrexate

Urban *et al*. (1983) related that 18 of their 210 patients receiving high-dose methotrexate (8.5 per cent) developed pleuritic pain accompanied by thickening of the pleura, the onset of the pleurisy occurring only after 3–4 courses of methotrexate, with a transient and benign course.

Blood products

Blood products, e.g. Factor VIII (antihaemophilic globulin, AHG) from US sources has been reported to be contaminated with AIDS virus (human T-cell lympho-tropic virus type III, HTLV-III) which has led to hae-mophilic patients developing AIDS, which may manifest itself with pulmonary infection. Such cases have been reported in both the UK and USA, the commonest pul-monary infection being with *Pneumocystis carinii* (Seale 1985).

Barotrauma/pneumothorax

Barotrauma associated with artificial ventilation

Artificial ventilation with intermittent positive pressure is well known at times to result in barotraumas such as pneumothorax, pneumomediastinum, or subcutaneous emphysema. A less obvious complication, that of bron-chiolectasis, has now been described, in particular with maximum levels of positive end expiratory pressure (Slavin *et al.* 1982). In a retrospective study of 11 consecutive necropsies in patients who had died in the intensive therapy unit after artificial ventilation, the seven patients who had been treated with positive-end expiratory pressure all showed signs of bronchiolectasis, whereas those who had standard artificial ventilation did not. There was, furthermore, a correlation between the maximum level of positive end expiratory pressure, the duration of its application, and the severity of the condition. No patient had clinical or radiological signs of bronchiolectasis during life, but three had other signs of pulmonary barotrauma, due in two to the injury causing their admission but unexplained in one.

Histological changes ranged from slight dilatation of the terminal and respiratory bronchioles, in relation to the adjacent alveoli, to severe bronchiolectasis with interstitial emphysema; the more severe changes could be recognized with the naked eye, in the cut specimen, giving a 'honeycombed' appearance to the lung.

There were some confounding factors present, in particular the fact that positive-end respiratory pressure artificial ventilation is required in those patients where pulmonary function is already compromised in a life-threatening condition, making analysis of the effects difficult. However, the data make an association likely.

The authors recommend that high levels of positive end expiratory pressure should be used 'with great caution if at all'.

Pneumothorax as a complication of chemotherapy

Wilson *et al.* (1982) describe the fatal outcome from pulmonary fibrosis and recurrent spontaneous pneumothorax following treatment with BCNU (a total of 1400 mg m^{-2}) for a frontoparietal astrocytoma. Each attack was more serious, resulting finally in a severe airleak which could not be controlled even by oversewing at thoracotomy. At autopsy the lung showed diffuse fibrosis, mainly of the alveolar septa and blood vessels, and the alveolar lining cells were desquamated and showed mucinous metaplasia. The astrocytoma had been eliminated. The authors draw attention to the high doses of BCNU employed for glioma patients and the greater risk of developing pulmonary fibrosis. Lote *et al.* (1981) described two cases of recurrent pneumothorax in patients treated with combination chemotherapy for malignant disease (adriamycin, cyclophosphamide, and actinomycin). In both cases pneumothoraces first appeared within 3–6 weeks of starting chemotherapy and recurrence was linked to re-starting therapy. The authors suggested that cell necrosis (possibly due to fibrosis or

malignancy) and defective repair processes may result in bronchopleural fistulae.

Drug-induced pulmonary neoplasms

Barbiturates and lung cancer

In the Kaiser Permanente Medical Care Program (Friedman 1981) looking for possible carcinogenic effects of drugs in humans, there was a significant association between lung cancer and three barbiturates (pentobarbitone, phenobarbitone, and secobarbitone). Individual drugs also showed a weaker association with cancer of the ovary (secobarbitone) and thyroid (pentobarbitone).

The association with lung cancer involved 10 874 patients and remained unchanged when the figures were subjected to further analysis to allow for time lag and smoking. There were however features which put this relation in doubt, namely, there was no association with a specific histological type, nor was there any association with duration or intensity of drug use. The author concludes that in view of the uncertainties surrounding these findings, the link between lung cancer and barbiturates should be regarded as a hypothesis worthy of further investigation.

Isoniazid in pregnancy

There have been a number of studies which have examined the possible link between isoniazid and cancer, but three well-controlled large investigations have failed to find any relation. However, suspicion has been raised anew by a single report: a nine-year-old boy, with no history of any exposure to asbestos, developed malignant mesothelioma of the pleura; his mother had been treated with isoniazid during the second and third trimester of pregnancy, because of a positive tuberculin test. Note is taken of the report because

1 Isoniazid in pregnant mice produces lung tumours;
2 Mesothelioma is rare in childhood;
3 The long latency in this report: in an earlier negative review of children whose mothers took isoniazid in childhood, the longest median follow-up was eight years and, certainly so far as asbestos is concerned, the latent period from exposure is very prolonged (Tuman *et al.* 1980).

Respiratory depression

The central respiratory depression caused by anaesthetic agents, hypnotics, and strong analgesics, and the respiratory paralsysis caused by muscle relaxants is too well known to require description. There are however a few points worth mentioning.

Muscle relaxants

The concomitant use of antibiotics, including the polymyxins, tetracyclines, and the aminoglycosides with the non-depolarizing neuromuscular blocking agents (tubocurarine, gallamine, pancuronium) has occasionally resulted in prolonged muscular paralysis with apnoea.

In addition, acute post-operative respiratory failure, occurring 20-90 minutes after the apparent antagonism of the relaxant drug with neostigmine, occurred in four patients with renal failure (Miller and Cullen 1976). It appears likely that reduced renal clearance of muscle relaxants in patients with impaired renal function is responsible. Finally, the neuromuscular blocking action of suxamethonium has recently been shown dramatically prolonged in patients under treatment with lithium carbonate (Hill *et al.* 1976).

Phenothiazines

Phenothiazines may give rise to adverse pulmonary reactions by two separate mechanisms.

Central effect

The central anticholinergic syndrome was described by Zadrobilek and Draxler (1981) in a girl who had been premedicated with chlorprothixene (plus atropine and pethidine). The syndrome consists of apnoea, with or without loss of consciousness, and was reversible with physostigmine. The authors dismissed the possibility of potentiation of effects by atropine, and consider the reaction due to idiosyncratic hypersensitivity to phenothiazines. It is possible that the sudden infant death syndrome (SIDS) associated with phenothiazine syrups given for upper respiratory infections, could be due to the same cause. The association of phenothiazines with SIDS was investigated by Kahn and Blum (1982) in a prospective study comparing 52 SIDS victims and 36 'near miss' infants with a matched group of 175 infants admitted for an all-night polysomnographic recording at the request of their parents or paediatrician. (83 were children of mothers who had lost infants to SIDS). The incidence of upper respiratory-tract infection was similar in all three groups, but the use of phenothiazine syrups was significantly higher in the SIDS victims and the 'near miss' group, than in the control subjects (23 per cent of SIDS victims, 22 per cent of 'near miss', but only 2 per

cent of the controls). This relationship held true when related to the number with nasopharyngitis in each group who had phenothiazines (71 per cent SIDS victims, 57 per cent 'near miss' infants, and 8 per cent of controls). The central depressant effect of phenothiazines as well as hypersensitivity could be additional factors in increasing apnoea, and disturbances of the natural awakening process might also contribute.

Extrapyramidal effect

Tardive dyskinesia is a well-known adverse effect of neuroleptic therapy. Faheem *et al.* (1982) reviewed the literature with regard to respiratory difficulties in patients who had, or later developed, tardive dyskinesia, possibly due to involvement of the respiratory muscles. In addition, they described a patient on long-term treatment with chlorpromazine who developed isolated respiratory dyskinesia, with difficulty in breathing in the absence of demonstrable pulmonary functional abnormality. He also had loss of control of laryngeal movement. On stopping all neuroleptic therapy his symptoms worsened, necessitating re-introduction of mesoridazine. The 'malignant' syndrome of neuroleptics, described by Destee *et al.* (1981), with acute respiratory distress in the presence of extrapyramidal symptoms, would seem to be a more acute form of the extrapyramidal effects of phenothiazines, although a central component cannot be excluded.

Pulmonary manifestations of drug-induced generalized disease

Drug-induced generalized disease such as the LE syndrome and polyarteritis may involve the lungs and may present with pleurisy, pneumonitis or pulmonary oedema, transient infiltration or infarction. Drugs such as hydrallazine, gold salts, penicillins, hydantoins, or sulphonamides may be responsible. The possibility that drug-induced pulmonary disease may be part of a more generalized pathological process should therefore always be borne in mind.

REFERENCES

ADAM, W.W., MEAGHER, E.J., AND BARTER, C.E. (1982). Labetalol, beta-blockers and deterioration of chronic airway obstruction. Clin. Exp. Hypertension Theory Practice **A4** (8), 1419–28.
AHMED, S.S. AND NUSSBAUM, M. (1981). Development of pulmonary oedema related to heparin administration. *J. clin. Pharmacol.* **21**, 126–8.

AKOUN, G.M., HERMAN. D.P., MAYAUD, C.M., AND PERROT, J.Y. (1983). Acebutolol-induced hypersensitivity pneumonitis. *Brit. med. J.* **286**, 266–7.

ALCALAY, M., TOUCHARD, G., PATTE, P., BABIN, PH., REBOUX, J.F., THOMAS, PH., PATTE, D., AND BONTOUX, D. (1979). Nephropathie, pneumopathie et hepatopathie des sels d'or de survenue simultanée avec étude ultrastructurale des lesions pulmonaires et renales. *Rev. Rhumatisme* **46**, 498–8.

ALTUS. P. AND ANDREW, C.C. (1982). Miliary tuberculosis concurrent with busulfan lung. *South. med. J.* **75**, 755–7.

ANDERSON, R. AND REFSTAD, S. (1978). Adult respiratory distress syndrome, precipitated by massive salicylate poisoning. *Intens. Care Med.* **4**, 211.

ASSEM, E.S.K. AND VICKERS, M.R. (1974). Immunological response to penicillin allergic patients and normal subjects. *Postgrad. med. J.* **50** (Suppl. 2.), 65–70.

AVERBUCH, S.D. AND YUNGBLUTH, P. (1980). Fatal pulmonary haemorrhage due to nitrofurantoin. *Arch. intern. Med.* **140**, 271–3.

AYRES, J.G. AND CLARK, T.J.H. (1983). Alcoholic drinks and asthma: a survey. *Br. J. Dis. Chest.* **77**, 370–5.

BATIST, G. AND ANDREWS, J.L., (1981). Pulmonary toxicity of antineoplastic drugs. *J. Am. med. Ass.* **246**, 1449-53.

BASSLER, K.H., AND HEIDENREICH, O., (1984). Problems with the addition of sulfite to infusion solutions. *Infusionsther. klin. Esnaha.* **11**, 31–4.

BEAGLEHOLE, R., HARRIS, E.A., AND REA. H.H. (1981). Has the change to beta-agonists combined with oral theophylline increased cases of fatal asthma? *Lancet* **ii**, 38.

BELL, R.T. AND LIPPMANN, M. (1979). Hydrochlorthiazide-induced pulmonary oedema. *Arch. intern. Med.* **139**, 817–19.

BENEDETTI, T.J., HARGROVE, J.C. AND ROSENE, K.A. (1982). Maternal pulmonary edema during premature labour inhibition. *Obstet. Gynaecol.* **59**, 33S-7S.

BERNSTEIN, M.L., SOBEL, D.B., AND WIMMER, R.S. (1982). Noncardiogenic pulmonary oedema following injection of methotrexate into the cerebrospinal fluid. *Cancer* **50**, 866-8.

BLUM, R.H., CARTER, S.K., AND AGRE, K. (1973). A clinical review of bleomycin — a new antineoplastic agent. *Cancer* **31**, 903–14.

BOURNE, A.S., FIINDT, M.L.H., AND MILES WALKER, J. (1979). Asthma due to the industrial use of chloramine. *Br. med. J.* **2**, 10–12.

BRAUDE, A.C., DOWNAR, E., CHAMBERLAIN, D.W., AND ROBUCK, A.S. (1982). Tocainide-associated interstitial pneumonitis. *Thorax* **37**, 309–10.

BRIGGS, J.H., MC KERRON, C.G., SOUHAMI, R.L., TAYLOR, D.S.E., AND ANDREWS, H. (1967). Severe systemic infections complicating 'mainline' heroin addiction. *Lancet* **ii**, 1227–31.

BRITISH MEDICAL JOURNAL (1976). Leading article: Drug reactions and the lung. *Br. med. J.* **2**, 1030.

BRYANT, D.H. AND PEPYS, J. (1976). Bronchial reactions to aerosol inhalant vehicle. *Br. med. J.* **1**, 1319–20.

BURKE, D.A., STODDART, J.C., WARD, M.K., AND SIMPSON, C.G.B. (1982). Fatal pulmonary fibrosis occurring during treatment with cyclophosphamide. *Br. med. J.* **285**, 696.

BUTLER, P., CHAHAL, P., HUDSON, N.M., AND HUBNER, P.J.B. (1981). Pulmonary hypertension after lung irradiation in infancy. *Br. med. J.* **283**, 1365.

CASSORLA, F.G., FINEGOLD, D.N., PARKS, J.S., TENORE, A., THAWERANI, H., AND BAKER, L. (1983). Vasculitis, pulmonary cavitation and anaemia during antithyroid drug therapy. *Am. J. Dis. Child.* **137**, 118-22.

CHACKO, D.C. (1981). Considerations in the diagnosis of radiation injury. *J. Am. med. Ass.* **245**, 1255–8.

CHAKRABORTY, A.K. AND DREISEN, R.B. (1982) Pulmonary haematoma secondary to anticoagulant therapy. *Ann. intern. Med.* **96**, 67–9.

CHAMBERLIN, W.H., STOCKMAN, G.D., AND WRAY, N.P. (1979). Shock and non-cardiogenic pulmonary oedema following meglumine diatrizoate for intravenous pyelography. *Am. J. Med.* **67**, 684.

CHARAN, N.B. AND LAKSHMINARAYAN, S. (1980). Pulmonary effects of topical timolol. *Arch. intern. Med.* **140**, 843–4.

CLARK, B.G., ARAKI, M., AND RAWLINGS, J.L. (1982). Propranolol-induced dyspnoea in a non-asthmatic male. *Drug Intel. Clin. Pharm.* **16**, 776–7.

CLARKE, R.S.J., DUNDEE, J.W., GARRETT, R.T., McARDLE, G.K., AND SUTTON, J.A. (1975). Adverse reactions to intravenous anaesthetics. *Brit. J. Anaesth.* **47**, 575–85.

COHEN, R.D., BATEMAN, E.D., AND POTGIETER, P.D. (1982). Near fatal bronchospasm in an asthmatic patient following ingestion of flurbiprofen. *S. Afr. med. J.* **61**, 803.

COMMERS, J.R. AND FOLEY, J.F. (1979). Pulmonary hyaline membrane disease, occurring in the course of VM-26 therapy. *Cancer Treat. Rep.* **63**, 2093–5.

CONNOLLY, C.K. (1982). Adverse reaction to ipatropium bromide. *Br. med. J.* **285**, 934–5.

COUTTS, I.I., DALLY, M.B., NEWMAN TAYLOR, A.J., PICKERING, C.A., AND HORSFIELD, N. (1981). Asthma in workers manufacturing cephalosporins. *Br. med. J.* **283**, 950.

CSABA, I.F., SULYOK, E., AND ERTI. T. (1978). Relationship of maternal treatment with indomethacin to persistent fetal circulation syndrome. *J. Paediat.* **92**, 484.

DESTEE, A., PETIT, H., AND WAROT, M. (1981). Le syndrome malin des neuroleptiques. *Nouv. Presse méd.* **10**, 178.

DI ORIO, J. AND BRAUNER, R.E. (1982). Adult respiratory distress syndrome occurring after therapy with diazoxide and beta-methasone for premature labour, a case report. *R.I. med. J.* **65**, 275–7.

DORN, M.R. AND WALKER, B.K. (1981). Non-cardiogenic pulmonary oedema due to hydrochlorthiazide: positive re-challenge. *Chest* **79**, 482.

DOUGLAS, J.G., MUNRO, J.F., KITCHIN, A.H., MUIR, A.L., AND PROUDFOOT, A.T. (1981). Pulmonary hypertension and fenfluramine. *Br. med. J.* **283**, 881–3.

DREXEL, H., REGELE, M., AND LÄNGE, U. (1982). Successful treatment of terbutaline-induced bronchospasm with orciprenaline sulphate *Lancet* **ii**, 446.

DRUG AND THERAPEUTICS BULLETIN (1980). Tartrazine: a yellow hazard. *Drug ther. Bull.* **18**, 53–5.

DUCATI, A., SIGNORINI, G., MELI, M., LOBASCIO, A.E., MASSEI, R., AND INFUSO, L. (1981). Respiratory complications during artificial barbiturate coma. *J. Neurosurg. Sci.* **25**, 27–34.

DYE, D. AND WATKINS, J. (1980). Suspected anaphylactoid reaction to Cremophor EL. *Br. med. J.* **1**, 1353.

EI, K., HANAI, I., HORIUCHI, T., HANAI, J., GOTOH, H., HIRASAWA, Y., GEYJO, F., AND AIZAWA, Y. (1979). Haemodialysis-associated asthma in a renal failure patient. *Nephron.* **25**, 247–8.

EVANS, J.M. AND KEOGH, J.A.M. (1977). Adverse reactions to intravenous anaesthetic agents. *Br. med. J.* **2**, 735–6.

FAHEEM, A.D., BRIGHTWELL, D.R., BURTON, G.C., AND STRUSS, A. (1982). Respiratory dyskinesia and dysarthria from prolonged neuroleptic use: tardive dyskinesia? *Am. J. Psychiat.* **139**, 517–18.

FOOD AND DRUG BULLETIN (1979). Yellow no. 5 (tartrazine), labelling on drugs to be required. *Food Drug Bull.* **9**, 18.

FRIEDMAN, G.D. (1981). Barbiturates and lung cancer in humans. *J. nat. Cancer Inst.* **67**, 291–5.

FRUCHTER, L. AND LAPTOOK, A. (1981). Diphenylhydantoin hypersensitivity reaction associated with interstitial pulmonary infiltrates and hypereosinophilia. *Ann. Allergy (US)* **47**, 453–5.

GILLMER, D.J. AND KARK, P. (1980). Pulmonary oedema precipitated by nifedipine. *Br. med. J.* **280**, 1420–1.

GOFFMAN, T.E., DICICCO, B.S., AND TSOU, E. (1982). Mesantoin and pulmonary changes. *Ann. intern. Med.* **96**, 254–5.

GONG, H., TASHKIN, D.P., AND CALVARESE, B.M. (1981). Alcohol induced bronchospasm in an asthmatic patient. *Chest* **80**, 167–73.

GRAHAM, J.R. (1968). Fibrosis associated with methysergide therapy, in *Drug-induced diseases*, Vol. 3 (ed. L. Meyler and H.M. Peck), pp. 249–69. Excerpta Medica Foundation, Amsterdam.

GRANT, I.W.B. (1981). Has the change to beta-agonists combined with oral theophylline increased cases of fatal asthma? *Lancet* **ii**, 36–7

GRANTHILL, C., COLAVOLPE, C., HOUVENAEGHEL, M., AND FRANCOIS, G. (1981). Hemorragies occultes intrapulmonaires par accidentes des anticoagulants. *Ann. anaesth. Franc.* **1**, 53–6.

GREGANTI, M.A. AND FLOWERS, W.M. (1979). Acute pulmonary oedema after the intravenous administration of contrast media. *Radiology* **132**, 583–5.

DE GREGORIO, M.W., LEE, W.P., AND RIES, C.A. (1981). Letter to the editor. Relation of treatment with amphotericin-B, leucocytes or both to pulmonary deterioration. *New Engl. J. Med.* **305**, 585.

GUERNSEY, B.G., VILLARREAL, Y., SNYDER, M.D., AND GABERT, H.A. (1981). Pulmonary edema associated with the use of beta mimetic agents in preterm labour. *Am. J. hosp. Pharm.* **38**, 1942–8.

HARRIES, A.D. (1981). Beta blockade in asthma. *Br. med. J.* **282**, 1321.

HARRIES, M.G., NEWMAN TAYLOR, A., WOODEN, J., AND MACAUSLAN, A. (1979). Bronchial asthma due to alpha-methyl-dopa. *Br. med. J.* **1**, 1461.

HAUPT, H.M., HUTCHINS, G.M., AND MOORE, G.W. (1981). Ara-C lung. Non-cardiogenic pulmonary oedema complicating cytosine arabinoside therapy of leukaemia. *Am. J. Med.* **70**, 256–61.

HEFFNER, J.E. AND SAHN, S.A. (1981). Salicylate-induced pulmonary edema. *Ann. intern. Med.* **45**, 405.

HEGER, J.J., PRYSTOWSKY, E.N., JACKMAN, W.M., NACCARELLI, G.V., WARFEL, K., RINKENBERGER, R.L., AND ZIPES, D.P. (1981). Clinical efficacy and electrophysiology during long term therapy for recurrent ventricular tachycardia or ventricular fibrillation. *New Engl. J. Med.* **305**, 539–45.

HENDRICKSEN, R.M. AND SIMPSON, F. (1982). Clofibrate and eosinophilic pneumonia. *J. Am. med. Ass.* **247**, 3082.

HILL, G.E., WONG, K.C., AND HODGES, M.R. (1976). Potentiation of succinylcholine neuromuscular blockade by lithium carbonate. *Anaesthesiology* **44**, 439–42.

HO, S.W-C. AND BEILIN, L.J. (1983). Asthma associated with acetyl cysteine infusion and paracetamol poisoning: a report of two cases. *Br. med. J.* **287**, 87.

HOLMBERG, L. AND BOMAN, G. (1981). Pulmonary reactions to nitrofurantoin. *Eur. J. resp. Dis.* **62**, 180–9.

——, ——, BOTTIGER, L.E., ERIKSSON, B., SPROSS, R., AND WESSLING, A. (1980). Adverse reactions to nitrofurantoin. *Am. J. Med.* **69**, 733–8.

HOLTMANN, H.W., HOLLE, J.P., AND GIANZER, K. (1980). Alteration of bronchial flow resistance due to timolol 0.25 per cent in bronchial asthma. *Klin. Monatsbl. Augenheilk.* **176**, 441–4.

HOWARD, J.J., MOHSENIER, Z., AND SIMONS, S.M. (1982). Adult respiratory distress syndrome following administration of lidocaine. *Chest* **81**, 644–5.

IDSØE, O., GUTHE, T., WILCOX, R.R., AND DE WECK, A.L. (1968). Nature and extent of penicillin side reactions with particular reference to fatalities from anaphylactic shock. *Bull. Wld. Hlth. Org.* **38**, 159–88.

INMAN, W.H.W. (1974). Recognition of unwanted drug effects with special reference to pressurised bronchodilator aerosols. In *Evalution of bronchodilator drugs, ARC Symposium* (ed. D.M. Burley, S.W. Clarke, M.F. Cuthbert, J.W. Paterson, and J.H. Shelley), pp. 191–200. Trust for Education and Research in Therapeutics, London.

——, VESSEY, M.P., WESTERHOLM, B., AND ENGLELUND, A. (1970). Thromboembolc disease and the steroidal content of oral contraceptives. A report to the Committee on Safety of Drugs. *Br. med. J.* **2**, 203–9.

ISRAEL, H.L., AND DIAMOND, P. (1962). Recurrent pulmonary infilration and pleural effusion due to nitrofurantoin sensitivity. *New Engl. J. Med.* **266**, 1024–6.

JACOBS, M.M., KNIGHT, A.B., AND ARIAS, F. (1980). Maternal pulmonary oedema resulting from betamimetic and glucocorticoid therapy. *Obstet. Gynaecol.* **56**, 56–9.

JOLOBE, O.M.P. (1982). Adverse reactions to ipatropium bromide. *Br. med. J.* **285**, 1425–6.

JUST-VIERA, J.O., FISHER, C.R., GAGO, D. AND MORRIS, J.D. (1984). Acute reaction to protamine. Its importance to surgeons. *Am. Surg.* **50**, 52–60

KAHN, A. AND BLUM, D. (1979). Fatal respiratory distress syndrome and salicylate intoxication in a 2-year old. *Lancet* **ii**, 1130.

—— AND —— (1982). Phenothiazines and sudden infant death. *Pediatrics* **70**, 75–8.

KEIGHLEY, J.F. (1966). Iatrogenic asthma associated with adrenergic aerosols. *Ann. intern. Med.* **65**, 985–95.

KENNEDY, M.C.S. (1965). Bronchodilator action of deptropine citrate with and without isoprenaline by inhalation. *Br. med. J.* **2**, 916–17.

KLAUS, M.V. AND FELLNER, M.J. (1973). Penicilloyl-specific antibodies in man. Analysis in 592 individuals from the newborn to old age. *J. Gerontol.* **28**, 312–16.

KOEPKE, J.W., STAUDENMAYER, H., AND SELNER, J.C. (1985). Inhaled metabisulfute sensitivity. *Am. Allergy* **54**, 213–5.

KRONZON, I., COHEN, M., AND WINER, H.E. (1982). Adverse effect of hydralazine in patients with primary pulmonary hypertension. *J. Am. med. Ass.* **247**, 3112–14.

LAMBERT, P.M. (1981). Oral theophylline and fatal asthma. *Lancet* **ii**, 200–1.

LANCET (1981). Leading article. Arachidonic acid, analgesics, and asthma. *Lancet* **ii**, 1266.

LEATHERMAN, J.W., AND DRAGE, C.W. (1982). Adult respiratory distress syndrome due to salicylate intoxication. *Minn. Med.* **65**, 677–8.

LEE, T., COCHRANE, G.M., AND AMLOT, P. (1981). Pulmonary eosinophilia and asthma associated with carbamazepine. *Br. med. J.* **282**, 400.

LESLIE, R.D.G., BELLAMY, D., AND PYKE, D.A. (1980). Asthma induced by enkephalin. *Br. med. J.* **280**, 16–18.

LEVIN, D.L., FIXLER, D.E., MORRISS, F.C., AND TYSON, J. (1978). Morphological analysis of the pulmonary vascular bed in infants exposed in utero to prostaglandin synthetase inhibitors. *J. Paediat.* **92**, 478–83.

LEVINE, B.B. AND ZOLOV, D.M. (1969). Prediction of penicillin allergy by immunological tests. *J. Allergy* **43**, 231–44.

LEWIS, G.R.J. (1981). Beta-blockers and airways disease. *NZ med. J.* **93**, 93.

LEWIS I.J. AND ROSENBLOOM, L. (1982). Glandular fever-like syndrome, pulmonary eosinophilia and asthma associated with carbamazepine. *Postgrad. med. J.* **58**, 100–1.

LOTE, K., DAHL, O., AND VIGANDER, T. (1981). Pneumothorax during combination chemotherapy. *Cancer* **47**, 1743–5.

McCORMICK, J., COLE, S., LAHIRIR, B., KNAUFT, F., COHEN, S., AND YOSHIDA, T. (1980). Pneumonitis caused by gold salt therapy: Evidence for the role of cell mediated immunity in its pathogenesis. *Am. Rev. resp. Dis.* **122**, 145–52.

McGOVERN, B., GARAN, H., KELLY, E., AND RUSKIN, J.N. (1983). Adverse reactions during treatment with amiodarone hydrochloride. *Br. med. J.* **287**, 175–80.

MAHUTTE, C.K., NAKASATO, S.K., AND LIGHT, R.W. (1982). Haloperidol and sudden death due to pulmonary edema. *Arch. intern. Med.* **142**, 1951–2.

MALISH, D.M. (1982). Alcohol-induced bronchospasm. *Chest* **81**, 3.

MANCHESTER, D., MARGOLIS, H.S., AND SHELDON, R.E. (1976). Possible association between maternal indomethacin therapy and primary pulmonary hypertension in the newborn. *Am. J. Obst. Gynecol.* **126**, 467–9.

MARCHLINSKI, F.C., GANSLER, T.S., WAXMAN, H.L., AND JOSEPHSON, N.E. (1982). Amiodarone pulmonary toxicity. *Ann. intern. Med.* **97**, 839–45.

MATEK, W., ROSCH, W., AND BECKER, V. (1980). Pathologische Lungen und Leberveränderungen nach Therapie einer Colitis ulcerosa. *Forstschr. Med.* **98**, 491–6.

MATLOFF, D.S., AND KAPLAN, M.M. (1980). D-penicillamine induced Goodpasture-like syndrome in primary biliary cirrhosis — successful treatment with plasmapheresis and immunosuppressive. *Gastroenterol.* **78**, 1046–9.

MICHAEL, J.R. AND RUDIN, M.L. (1981). Acute pulmonary disease caused by phenytoin. *Ann. intern. Med.* **95**, 452–4.

MILLER, R.D. AND CULLEN, D.J. (1976). Renal failure and post-operative respiratory failure: recurarisation? *Br. J. Anaesth.* **48**, 253–6.

MILNE, L.J.R. AND CROMPTON, G.K. (1974). Beclomethasone dipropionate and oropharyngeal candidiasis. *Br. med. J.* **3**, 797–8.

MONERET-VAUTRIN, D.A., LAXENAIRE, M.C., AND VIRBY-BABEL, F. (1983). Anaphylaxis caused by anti-cremophor EL, IgG, STS antibodies in a case of reaction to Althesin. *Br. J. Anaesth.* **55**, 469–71.

MULLER, K.M., MENNE, R., HUTHER, W., AND BROBE, H. (1979). Fatal pneumopathy after cytostatic treatment for leukaemia in children. *J. Cancer Res. clin. Oncol.* **94**, 287–94.

MURPHY, K.C., ATKINS, C.J., OFFER, R.C., HOGG, J.C., AND SREIN, H.B. (1981). Obliterative bronchiolitis in two rheumatoid arthritis patients treated with penicillamine. *Arthritis Rheumatism* **24**, 557–60.

MUTNICK, A. AND SCHNEIWEISS, F. (1982). Desipramine-induced pulmonary interstitial eosinophilia. *Drug Intel. clin. Pharm.* **16**, 966–7.

NADER, D.A. AND SCHILLACI, R.F. (1983). Pulmonary infiltrates with eosinophilia. *Chest* **83**, 280–2.

NAGEY, D.A. AND CRENSHAW, M.C. (1982). Pulmonary complications of isoxsuprine therapy in the gravida. *Obstet. Gynaecol.* **59**, (suppl. 6), 38S–42S.

NASH, G., BLENNERHASSETT, J.B., AND PONTOPPIDAN, H.C. (1967). Pulmonary lesions associated with oxygen therapy and artificial ventilation. *New Engl. J. Med.* **276**, 368–74.

NOTCUTT, W.G. (1973). Adverse reactions to Althesin. *Anaesthesia* **28**, 673.

OLINER, H., SWARTZ, R., RUBIO, F., AND DAMESHEK, W. (1961). Interstitial pulmonary fibrosis following busulphan therapy. *Am. J. Med.* **31**, 134–9.

PAGE, R.L. (1979). Progressive pleural thickening during oxprenolol therapy. *Br. J. Dis. Chest* **73**, 195–9.

PAGGIARO, P.L., LOY, A.M., AND TOMA, G. (1979). Bronchial asthma and dermatitis due to spiramycin in a chick breeder. *Clin. Allergy* **9**, 571–4.

PALISSA, A., GUARDIA, J., BOFILL, J.M., AND BACARDI, R. (1979). Antazoline-induced allergic pneumonitis. *Br. med. J.* **2**, 1328.

PARTRIDGE, M.R., AND GIBSON, G.J. (1978). Adverse bronchial reactions to intravenous hydrocortisone in two aspirin-sensitive asthmatic patients. *Br. med. J.* **1**, 1521–2.

PENN, R.G. AND GRIFFIN, J.P. (1982). Adverse reactions to nitrofurantoin in the United Kingdom, Sweden and Holland. *Br. med. J.* **284**, 1440–2.

PERLOW, G.M., BIMAL, P.J. PAUKER, S.G., ZARREN, H.S. WISTRAN, D.C., AND EPSTEIN, R.L. (1981). Tocainide-associated interstitial pneumonitis. *Ann. intern. Med.* **94**, 489–90.

PERRY, D.J., WEISS, R.B., AND TAYLOR, H.G. (1982). Enhanced bleomycin toxicity during acute renal failure. *Cancer treat. Rep (US)* **66**, 592–3.

PETUSEVOSKY, M.L., FALING, L.J., ROCKLIN, R.E., SNIDER, G.L., MERLISS, A.D., MOSES, J.M., AND DORMAN, S.A. (1979). Pleuropericardial reaction to treatment with dantrolene. *J. Am. med. Ass.* **242**, 2772–4.

PHILIPSEN, T., ERIKSEN, D.S., AND LYNGGÅND, F. (1981). Pulmonary edema following ritodrine-saline infusion in premature labour. *Obstet. Gynaecol.* **58**, 304–8.

PHILLS, J.A. AND PERELMUTTER, L. (1974). IgE-mediated and Non-IgE-mediated allergic-type reactions to aspirin. *Acta allerg., Kbh.* **29**, 474–90.

PODELL, T.E., KLINENBERG, J.R., KRAMER, L.S., AND BROWN, H.V. (1980). Pulmonary toxicity with gold therapy. *Arthritis Rheumat.* **23**, 347–50.

POE, R.H., CONDEMI, J.J., WEINSTEIN, S.S., AND SHUSTER, R.J. (1980). Adult respiratory distress syndrome related to ampicillin sensitivity. *Chest* **77**, 449–51.

POPIO, K.A., JACKSON, D.H., UTELL, M.J., SWINBURNE, A.J., AND HYDE, R.W. (1983). Inhalation challenge with carbachol and isoproterenol to predict bronchospastic response to propranolol in COPD. *Chest* **83**, 175–9

PORTERFIELD, J.G., ANTMAN, E.M., AND LOWN, B. (1980). Respiratory difficulty after use of disopyramide. *New Engl. J. Med.* **303**, 584.

POWELL-JACKSON, P.R., CARMICHAEL, F.J., CALNE, R.Y., AND WILLIAMS, R. (1984). Adult respiratory distress syndrome and convulsions associated with administration cyclosporine in liver transplant recipients. *Transplantation* **38**, 341–3.

PRICE, H.V. (1982). Asthma attacks precipitated by disodium cromoglycate in a boy with α_1 anti-trypsin deficiency. *Lancet.* **ii**, 606–7.

PROMISLOFF, R.A., AND DUPONT, D.C. (1983). Death from ARDS and cardiovascular collapse following lidocaine adminstrtion. *Chest* **83**, 585.

RAINE, J.M., PALAZZO, M.G., KERR, J.H., AND SLEIGHT, P. (1981). Near fatal bronchospasm after oral nadolol in a young asthmatic and reponse to ventilation with halothane. *Br. med. J.* **1**, 548–9.

RANTALA, H., KIRVELA, O., AND ANTTOLAINEN, I. (1979). Nitrofurantoin lung in a child. *Lancet* **ii**, 798–9.

RILEY, S.A., WILLIAMS, S.E., AND COOKE, N.J. (1982). Alveolitis after treatment with amiodarone. *Br. med. J.* **284**, 161–2.

ROBERTS, J.B. AND BRECKENRIDGE, A.M. (1975). Drugs affecting autonomic function. In *Meyler's Side-effects of drugs,* Vol 8 (ed. M.N.G. Dukes), Chapter 13. Excerpta Medica Foundation, Amsterdam.

ROSE, M.S., SMITH, L.L., AND WYATT, I. (1974). Evidence for energy-dependent accumulation of paraquat into rat lung. *Nature, London* **252**, 314–15.

ROSS, S.R., FRIEDMAN, C.J., AND LESNEFSK, E.J. (1982). Near fatal bronchospasm induced by zomepirac sodium. *Am. Allergy* **48**, 233–4.

ROTMENSCH, H.H., LIRON, M., TUPILSKI, M., AND LANIADO, S. (1980). Possible association of pneumonitis with Admiodarone therapy. *Am. Heart J.* **100**, 412.

SAFWAT, A.M. AND DROR, A. (1982). Pulmonary capillary leak associated with methylmethacrylate during general anaesthesia. *Clin. Orthop.* **168**, 59–63.

SALBERG, D.J. AND SIMON. M.R. (1980). Severe asthma induced by naproxen, a case report and review of the literature. *Ann. Allergy* **45**, 372–5.

SAMTER, M. AND BEERS, R.F. JR (1967). Concerning the nature of intolerance to aspirin. *J. Allergy* **40**, 281–93.

—— AND —— (1968). Intolerance to aspirin. Clinical studies and consideration of its pathogenesis. *Ann. intern. Med.* **68**, 975–83

SCHMIDT, MIAND BRUGGER, E. (1980). Ein Fall von Carbamazepin-induzierter interstitieller Pneumonie. *Med. Klin.* **75**, 29–31.

SCHOENE, R.B., MARTIN, T.R., CHARAN, N.B., AND FRENCH, C.L. (1981). Timolol-induced bronchospasm in asthmatic bronchitis. *J. Am. med. Ass.* **245**, 1460–1.

SCOTT, D.L., BRADBY, G.V.H., AITMAN, T.J., ZAPHIROPOULOS, G.C., AND HAWKINS, C.F. (1981). Relationship of gold and penicillamine therapy to diffuse interstitial lung disease. *Ann. rheum. Dis.* **40**, 136–41.

SEALE, J. (1985). AIDS virus infection: prognosis and transmission. *J.R. Soc. Med.* **78**, 613–15.

SHEHADI, W.H. (1982). Contrast media adverse reactions: occurrence, recurrence and distribution patterns. *Diagnostic Radiol.* **143**, 11–17.

SINCLAIR, D.J.M. (1979). Comparison of effects of propranolol and metoprolol on airways obstruction in chronic bronchitis. *Br. med. J.* **1**, 168.

SLAVIN, G., NUNN, J.F. CROW, J. AND DORE, C.J. (1982). Bronchiolectasis–a complication of artificial ventilation. *Br. med. J.* **285**, 931–4.

SMITH, W. AND BALL, G.V. (1980). Lung injury due to gold treatment. *Arthritis Rheumat.* **23**, 351–4.

SOBOL, S.M. AND RAKITA, L. (1982). Pneumitis and pulmonary fibrosis associated with amiodarone treatment: a possible complication of a new antiarrhythmic drug. *Circulation* **65**, 819–24.

STAUB, N.C. (1981). Pulmonary edema due to increased microvascular permeability. *Ann. Rev. Med.* **32**, 291.

SUAREZ, L.D., PODEROSO, J.J., ELSNER, B., BUNSTER, A.H., ESTEVA, H., AND BELOTTI, M. (1983). Subacute pneumopathy during amiodarone therapy. *Chest* **83**, 566–8.

SZCZEKLIK, A. (1980). Analgesics, allergy and asthma. *Br. J. clin. Pharmacol.* **10**, 401S–405S.

TAAL, B.G., SPIERINGS, E.L.H., AND HILVERING, C. (1983). Pleuropulmonary fibrosis associated with chronic and excessive intake of erogtamine. *Thorax* **38**, 396–8.

TERHO, E.O., TORKKO, M., AND VALTA, R. (1979). Pulmonary damage associated with gold therapy. *Scand. J. resp. Dis.* **60**, 345–9.

THOMPSON, R.N. AND GRENNAN, D.M. (1983). Acebutolol induced hypersensitivity pneumonitis. *Br. med. J.* **286**, 894.

TUMAN, K.J., CHILCOTE, R.R., BERKOW, R.I., AND MOOHR, J.W. (1980). Mesothelioma in child with prenatal exposure to isoniazid. *Lancet* **ii**, 362.

URBAN, C., NIRENBERG, A., CAPARROS, B., ANAC, S., CACAVIO, A., AND ROSEN, G. (1983). Chemical pleuritis as the cause of acute chest pain following high-dose methotrexate treatment. *Cancer* **51**, 34–7.

VANE, J.R. (1968). The release and assay of hormones in the circulation. In *The scientific basis of medicine annual reviews,* Chapter 19. British Postgraduate Medical Federation. Athlone Press, London.

VERGERET, J., BARAT, M., TAYTARD, A., BELLVERT, P., DOMBLIDES, P., DOUVIERT, J.J., AND FREOUR, P. (1984). Pleuropulmonary fibrous and bromocriptine. *Sem. Hôp. Paris* **60**, 741–4.

WALSHE, J.M. (1968). Toxic reactions to penicillamine in patients with Wilson's disease. *Postgrad. med. J.* **44** (Suppl.), 6–8.

WALTERS, J.S., WOODRING, J.H., STELLING, C.B., AND ROSENBAUM, H.D. (1983). Salicylate induced pulmonary oedema. *Radiology* **146**, 289–93.

WARD, K. AND PRICHARD, J.S. (1980). Severe bronchial reaction and cardiopulmonary arrest following intravenous injection of hydrocortisone in a non-atopic, non-aspirin sensitive asthmatic. *Ir. J. med. Sci.* **149**, 124–6.

WEISS, R.B. AND MUGGIA, F.M. (1980). Cytotoxic drug-induced lung disease: Update 1980. *Am. J. Med.* **68**, 259–66.

—— AND THRUSH, D.M. (1982). A review of the pulmonary toxicity of cancer chemotherapeutic agents. *Oncol. Nurs. Forum (US)* **9**, 16–21.

WHITTINGTON, J.R. (1981). Has the change to beta agonists combined with oral theophylline increased cases of fatal asthma? *Lancet* **ii**, 37–8.

WILHELMS, E., FLACHSBART, F., BECKER, H., AND POSER, W. (1981). Fluid lung bei Missbrauch von Barbiuraten und barbiturähnlichen Substanzen. *Prax. Pneumo.* **35**, 371–3.

WILKINSON, A.R., AYNSLEY-GREEN, A., AND MITCHELL, M.D. (1979). Persistent pulmonary hypertension and abnormal prostaglandin E levels in preterm infants after maternal treatment with naproxen. *Archs. Dis. Childh.* **54**. 942–5.

WILLIAMS, T., EIDUS, L., AND THOMAS, P. (1982). Fibrosing alveolitis, bronchiolitis obliterans and sulfasalazine therapy. *Chest* **81**, 766–8.

WILSON, F.E. (1981). Acute respiratory failure secondary to polymyxin-B inhalation. *Chest* **79**, 237–9.

WILSON, J.D., SUTHERLAND, D.C., AND THOMAS, A.C. (1981). Has the change to beta-agonists combined with oral theophylline increased cases of fatal asthma? *Lancet* **i**, 1235–7.

WILSON, K.S., BRIGDEN, M.L., ALEXANDER, S., AND WORTH, A. (1982). Fatal pneumothorax in BCNU lung. *Med. pediat. Oncol.* **10**, 195–9.

WISLICKI, L. (1982). Systemic adverse reactions to prostaglandin F₂. *Int. J. Biol. Res. Pregnancy* **3**, 158–60.

WOELKE, B.J., AND TUCKER, R.A. (1983). Adult respiratory distress syndrome following lignocaine. *Chest* **83**, 933.

WOOD, G.M., BOLTON, R.P., MUERS, M.F., AND LOSOWSKY, M.S. (1982). Pleurisy and pulmonary granulomas after treatment with acebutolol. *Br. med. J.* **285**, 936.

WRIGHT, A.J. AND BRACKENRIDGE, R.G. (1982). Pulmonary infiltration and bone marrow depression complicating treatment with amiodarone. *Br. med. J.* **284**, 1303.

WRIGHT, D.G., ROBICHAUD, K.J., PIZZO, P.A., AND DEISSEROTH, A.B. (1981). Lethal pulmonary reactions associated with the combined use of amphotericin-B and leucocyte transfusions. *New Engl. J. Med.* **304**, 1185–9.

YEUNG, M.G., NG, L.Y., AND KOO, M.W.L. (1979). Severe bronchospasm in an asthmatic patient following alcuronium and D-tubocurarine. *Anaesthes. intens. Care* **7**, 62–4.

ZADROBILEK, E. AND DRAXLER, V. (1981). Chloprothixen-induziertes zentral anticholinerges Syndrom. *Anaesthetist* **30**, 307–8.

ZLOTLOW, M.J. AND SETTIPANE, G.A. (1977). Allergic potential of food additives: a report of a case of tartrazine sensitivity without aspirin intolerance. *Am. J. clin. Nutr.* **30**, 1023–5.

ZUCKERMAN, H., REISS, U., AND RUBINSTEIN, I. (1974). Inhibition of human premature labour by indomethacin. *Obstet. Gynec.* **44**, 787–92.

ZYROFF, J., SLOVIS, T.L., AND NAGLER, J. (1974). Pulmonary oedema induced by oral methadone. *Radiol.* **112**, 567–8.

16 The alimentary system

J.M.T. WILLOUGHBY

Introduction

Symptoms of gastrointestinal intolerance have at some time followed the ingestion of almost every medication prescribed. The syndrome of retrosternal or abdominal discomfort, bloating, flatulence, borborygmi, nausea, vomiting, colicky pains, and diarrhoea occurs with varying degrees of completeness and severity as a reaction which, because of the differences in susceptibility between different subjects, must be regarded as a defence mechanism tailored to each individual's physiological constitution. The adage 'one man's meat is another man's poison' tells us as much about drugs as about food and drink. The important characteristics of this type of reaction are: that it is chiefly determined by host factors; that it is non-specific once initiated; that it may involve any part of the alimentary tract; and that removal of the stimulus will usually lead to complete recovery. This account of iatrogenic conditions affecting the gastrointestinal tract will be confined largely to those that are specific in respect either of a particular treatment or of the part affected, or both, and in which mere withdrawal of the offending agent may not ensure recovery. Untoward effects of diagnostic instrumentation, medication, and surgical therapeutic procedures employed exclusively in the management of gastrointestinal disorders will not be discussed.

Drug-induced oral ulceration

Most cases are due to retention in the mouth of drugs known to be ulcerogenic elsewhere in the alimentary tract, e.g. aspirin (Van Wyk 1967; Glick *et al.* 1974), emepronium bromide (Strouthidis *et al.* 1972), gold salts (Gibbons 1979), and potassium chloride (see p. 239). However the more common mechanical action of ill-fitting dentures may also be potentiated by drugs reaching the mouth via the circulation, as in the case attributed to indomethacin treatment by Guggenheimer and Ismail (1975).

The occurrence of buccal erosions clinically and histologically typical of lichen planus in a man with rheumatoid arthritis who had been taking indomethacin for 10 months led Hamburger and Potts (1983) to investigate the association of this reaction with non-steroidal anti-inflammatory drugs (NSAIDs) generally. Not only indomethacin in suppository form but also oral preparations of ibuprofen, fenclofenac, flurbiprofen, and diflunisal (though not aspirin) were found to cause an identical eruption in this patient. Reviewing the records of 75 patients treated at Birmingham Dental School for oral lichen planus they observed that 20 had been taking NSAIDs at the time. Of these seven had recovered when the drug was withdrawn, including two who then relapsed on challenge.

Drug sensitization of the buccal mucosa analogous to contact dermatitis is apparently rare, and perhaps surprisingly so. Muniz and Berghman (1978) reported erosive stomatitis induced by lithium carbonate in a schizophrenic who had been taking this medication without untoward effects for four years. When, after healing, lithium was given again to control deterioration in the patient's behaviour, the ulcers recurred within 48 hours.

Altered taste sensation

Penicillamine

Impairment of taste sensation is a common effect of penicillamine treatment. It was first described in a patient with Wilson's disease as loss of taste for sweetness and salt (Sternlieb and Scheinberg 1964). In the experience of Jaffe (1968) such patients had been far less frequently affected than those with rheumatoid arthritis, in whom incidence of taste (but not olfactory) loss was at least 25 per cent. However, lower dosage schedules adopted in recent years for maintenance treatment with penicillamine may well have reduced the incidence of hypogeusia from this cause. Onset is usually within the first six weeks of treatment, and normal taste returns within six months, even when the drug is continued. Keiser *et al.* (1968) reported that recovery did not occur in their patients with scleroderma and cystinuria until treatment was stopped, but this exceptional finding may have been a consequence of doses approaching — and in some cases exceeding — 2 g daily.

The action of penicillamine here may depend on chelation of Cu^{2+} or Zn^{2+} ions. Henkin (1976) favoured binding of Cu^{2+} because treatment with copper sulphate had reversed hypogeusia in patients with low serum levels and a high urine output of this ion. This would also explain the relative immunity from hypogeusia of patients with Wilson's disease. Lyle (1974) was not convinced that either copper or zinc supplements had assisted the resolution of penicillamine hypogeusia and pointed out that dimercaprol, which forms stable complexes with both ions, had been given for long periods without impairing taste sensation. He suggested instead that penicillamine might act, like certain other amino acids, by inhibiting afferent impulses from the taste buds.

Griseofulvin

A more complex disorder of taste sensation was described by Fogan (1971) in a patient treated for several weeks with griseofulvin at 750 mg daily. Not only was there a general loss of taste acuity but also a tendency for all foods to taste the same. Sensation returned to normal a week after the treatment had been stopped. The author drew attention to the chemical similarity between griseofulvin and penicillamine, but in an appended comment Henkin speculated that the free electrons of the unsaturated keto group in the griseofulvin molecule might form covalent bonds with a protein component of the taste-bud membrane and so alter both its conformation and its function.

Thiocarbamides

Erikssen et al. (1975) reported hypogeusia in two patients, one on thiamazole and the other on carbamizole. In one this persisted for a year after discontinuation of treatment. The effect was ascribed to a common thiol group chelating heavy-metal ions, though without direct evidence.

Levodopa

In a review of the first 100 cases of Parkinson's disease treated with levodopa in his unit Barbeau (1970) reported that 22 patients had developed a change in taste sensation after two months on the drug. This consisted of an added metallic or garlic-like taste which did not usually persist. Among 514 patients receiving a lower dose of levodopa in combination with a decarboxylase inhibitor only 4.5 per cent reported abnormalities in taste (Siegfried and Zumstein 1971). In most this consisted of a disagreeable taste

difficult to describe, which might last for nine months. From its limitation to the anterior two-thirds of the tongue the authors postulated a corda tympani lesion as the cause.

Lithium carbonate

Within days of starting a course of lithium carbonate the patient of Duffield (1973) noted a strong and unpleasant flavour superimposed upon the flavours of butter, cream, cheese, or celery. Two days without treatment sufficed to abolish this. From their study of 450 patients Himmelhoch and Hanin (1974) calculated that one in 20 experienced the flavour with dairy products and even more with celery. Experiments showing that rat brain concentrated lithium carbonate in the olfactory bulb suggested a central site of action.

One case was described in which the apparently trivial disability of ageusia for salt due to lithium carbonate seriously compromised management of the mania for which this had been prescribed (Bressler 1980). The patient's natural response of increasing his salt intake led to a relapse associated with low blood levels of lithium, probably the result of more rapid urinary excretion.

Captopril

An angiotensin-converting enzyme (ACE)-inhibitor given for hypertension, captopril, causes both hypogeusia (Vlasses and Ferguson 1979; Atkinson et al. 1980; White et al. 1980) and dysgeusia. Two of the three patients who developed hypogeusia after treatment with captopril by McNeil et al. (1979) also suffered a persistent salty taste in the mouth, which proved to be the most prominent adverse reaction among the 16 patients treated. Of the 37 patients in the last three series cited above, six (16 per cent) experienced some form of taste disturbance. Only in the two cases of Atkinson et al. (1980) did taste sensation return to normal — after two and five weeks respectively — in the course of continued captopril treatment. Should the relatively low dosage of 25 mg t.d.s. now considered adequate for most cases of hypertension be widely adopted the frequency of this and other side-effects is likely to decrease.

Captopril binds to the Zn^{2+} ion in ACE, but the failure of oral zinc supplements to reverse the disturbance when given during treatment to two of their patients inclined McNeil et al. (1979) to doubt the relevance of this reaction. Contrary evidence comes from a study of enalapril, which is an ACE-inhibitor lacking the sulphydryl group of captopril and apparently also its characteristic side-effects (Gavras et al. 1981; Hodsman et al. 1983).

Metronidazole

Metronidazole is well established as a treatment for protozoal infestation and anaerobic bacterial infection, and also as a prophylactic in bowel surgery. In neither is it customarily taken for other than short periods, which may explain why the first account of its effect on taste sensation came so long after its introduction. Following a preliminary report two years earlier Brandt et al. (1982) described the use of metronidazole in extended courses to control the symptoms of Crohn's disease, and mentioned that 24 out of 26 patients had complained of a metallic taste while taking it. Several other cases were notified to the reviewer in personal communications, so it may be supposed that this effect is under-reported rather than uncommon.

Other drugs

Other drugs capable of inducing abnormalities in taste sensation include flurazepam, which may have a bitter or metallic after-taste; dipyridamole, which has caused dysgeusia with nausea; metformin; and benoxaprofen (Willoughby 1983).

Chronic haemodialysis

Without commenting on its prevalence among dialysis patients in general Atkin-Thor et al. (1978) studied dysgeusia in 20 such patients who had complained that protein foods in particular had become either bitter or tasteless. All except one had some degree of hypogeusia on formal testing, and in 12 this was severe. In 18 patients, hair Zn^{2+} levels were below the lower normal limit and after six weeks of zinc sulphate treatment at 220 mg per day significant increases in both Zn^{2+} levels and taste acuity were detectable. Similar results were reported by Mahajan et al. (1980).

Irradiation

Some loss of taste sensation occurs, as might be expected, after irradiation of tumours affecting the mouth or surrounding structures (Kolmus and Farnsworth 1959; McCarthy-Leventhal 1959). Formal testing has shown that impairment is measurable some three weeks after the beginning of therapy and is greater for bitter and salt qualities than for sweet, perhaps because the last is served by more taste buds (Mossman and Henkin 1978). Dysgeusia was complained of by most affected patients. Abnormalities persisted for at least a year in some patients, but could be ameliorated by zinc supplements. Observations of taste disturbance in patients with

carcinoma of the larynx are complicated by the fact that the tumour itself may induce abnormalities which can then be reversed by successful treatment (Kashima and Kalinowski 1979).

Iatrogenic lesions of the salivary glands

Iodine-containing compounds

Painful enlargement of one or both parotid glands has occurred as a rare complication of both oral and parenteral iodide administration (Davidson et al. 1974). It was first observed to follow intravenous urography with iodine-containing contrast medium (Sussman and Miller 1956), and most subsequent cases have arisen under similar conditions. A typical history is that described by Imbur and Bourne (1972) in a 64-year-old man who underwent intravenous urography with a preparation which included a mixture of sodium and meglumine diatrizoates. Three hours later there was severe, painful, bilateral swelling of the parotid and submandibular glands, which had improved by the next morning and resolved within 48 hours. After an interval of 3 days a different iodine-containing medium was injected under antihistamine and corticosteroid cover. Once again the swelling occurred after 3 hours, but on this occasion to a lesser degree. This study clearly implicated iodine as the cause of the swelling, and suggested inflammation of allergic origin as the pathological process. Harden (1968), however, delivered a carefully quantified oral challenge to a patient who had first developed submandibular adenitis while taking a cough mixture containing iodide, and observed that clinical relapse became evident at approximately the same level of plasma iodide as that obtaining in the original attack. Since far higher iodide levels have been recorded in the absence of salivary gland damage (Nakadar and Harris-Jones 1971), it may be that this effect is indeed dose-dependent but confined to individuals in whom normal constraints on glandular concentration of the ion are lacking.

Phenylbutazone and oxyphenbutazone

Sialadenitis on a different time-scale but otherwise similar has been reported in patients receiving phenylbutazone (Cohen and Banks 1966; Murray-Bruce 1966; Garfunkel et al. 1974) and oxyphenbutazone (Cardoe 1964; Stahl 1966; Gross 1969). The patient of Cohen and Banks (1966) was a 44-year-old man who presented with a fever and symmetrical enlargement of the parotid glands. Complement-fixation tests for mumps S and V antibodies were negative. The patient made a spontaneous recovery, but contracted an identical brief illness the following year. On each occasion he had been prescribed phenylbutazone two days previously for tennis elbow, and the onset of resolution could be related to his stopping the drug. Murray-Bruce (1966) described a case where phenylbutazone had been given by intramuscular injection to a merchant seaman for gout. The patient developed parotid swelling which lasted for a month until his treatment was changed to sulphinpyrazone. About three years later the parotid swelling recurred, and persisted without pain, after another course of phenylbutazone. A sialogram was performed on this occasion, and proved normal. In the case reported by Garfunkel *et al.* (1974) painful swelling of both parotid and submandibular glands was accompanied by a febrile, myalgic response to phenylbutazone. Although the glands were indurated and tender no lymphadenopathy could be detected. Improvement was rapid during a course of erythromycin and continued more slowly over a 3-month period during which function, as judged by handling of ^{99m}Tc-pertechnate, returned to normal. Gross (1969) described a case of febrile parotitis caused by oxyphenbutazone in a 57-year-old woman. On the third day of treatment the patient developed a temperature of 40°C and painless parotid swelling. Serological tests for mumps were negative.

Most patients have responded with a dramatic drop in fever to corticosteroids or to simple withdrawal of the drug; and since a second episode of sialadenitis associated with phenylbutazone therapy tends to be worse than the first it seems probable that the underlying mechanism is some form of hypersensitivity reaction.

Adrenergic agonists and antagonists

Isoprenaline was reported to cause a painless swelling of the parotid glands in the absence of fever and malaise (Borsanyi and Blanchard 1961).

In cases ascribed to treatment with bretylium tosylate and guanethidine sulphate the swelling was accompanied by pain. Mårdh *et al.* (1974) published the histories of two patients who took methyldopa intermittently and developed swelling of the submandibular and parotid glands within 48 hours of ingesting a tablet. A third patient received methyldopa for one month, but it was not until 11 days after the drug's withdrawal that unilateral parotid enlargement was observed. In all these cases sialadenitis was accompanied by fever, but tests for both viral and bacterial infection proved negative. The fact that features suggestive of lupus erythematosus were also present in one patient raised the possibility of an immunological basis for this as for other forms of drug-induced sialadenitis.

Nitrofurantoin

In a survey of adverse reactions to nitrofurantoin reported from three European countries, Penn and Griffin (1982) included three cases of painless bilateral salivary gland enlargement, all in British women who were taking no other drug at the time. Pellinen and Kalske (1982) described a case from Finland in which painful bilateral swelling of the parotid glands with fever recurred twice on challenge with the drug. The initial reaction in this woman, as in two Dutch women similarly affected (Meyboom *et al.* 1982), occurred within a few days of starting nitrofurantoin treatment. An immune mechanism is suggested by the fact that, in each of the patients treated more than once, a single 50-mg tablet sufficed to cause swelling on the second occasion.

Busulphan

A chronic form of sialadenitis was responsible for dryness of the mouth and dysphagia in a man who had been taking busulphan for chronic lymphocytic leukaemia over a nine-year period (Morales Polanco *et al.* 1977). Microscopic damage to the salivary glands, studied at autopsy, included atypia of the duct epithelial cells, which had undergone nuclear changes, and periductal fibrosis accompanying infiltration of the stroma with lymphocytes and plasma cells.

Instrumentation

Acute, painless, transient swelling of the salivary glands may follow endotracheal anaesthesia (Attas *et al.* 1968; Matsuki *et al.* 1975) or upper gastrointestinal endoscopy (Slaughter and Boyce 1969; Slaughter 1975; Gordon 1976). Incidence is believed to be about 0.2 per cent in anaesthesia and from 0.05 to 0.1 per cent in endoscopy. Of five cases encountered in one year by Matsuki *et al.* (1975), two involved enlargement of both parotid glands, two of the right parotid only, and one of the left submandibular. A soft swelling of the glands might be noted at any time from the operation itself to 24 hours afterwards, gradually diminishing over the next 8–60 hours. At no stage have patients experienced appreciable discomfort, nor has any treatment been required. There is some suggestion that endoscopy is more likely to affect the submandibular than the parotid glands, although in other respects the clinical features of the swelling thus induced are identical to those described after anaesthesia.

From the consistency of the gland and the finding of

Table 16.1 Drugs known to cause oesophageal ulceration

Drug	References
Potassium chloride	Pemberton 1970; Whitney and Croxon 1972; McCall, 1975
Emepronium bromide	Habeshaw and Bennett 1972; Kavin 1977; Pilbrant, 1977
Doxycycline*	Bokey and Hugh 1975; Schneider 1977; Carlborg et al. 1978
Tetracycline	Crowson et al. 1976; Channer and Hollanders 1981
Clindamycin	Sutton and Gosnold 1977
Co-trimoxazole	Bjarnason and Björnsson 1981
5-fluorouracil	Pannuti et al. 1973
Indomethacin	Gardies et al. 1978; Agdal 1979; Bataille et al. 1982
Aspirin-containing mixtures	Carlborg et al. 1978; Williams 1979
Phenylbutazone-containing mixtures	Lennert and Kootz 1967; Juncosa 1970
Theophylline	Enzenauer et al. 1984; Stoller 1985
Alprenolol	Carlborg et al. 1979
Quinidine	Bohane et al. 1978; Teplick et al. 1980; Mason and O'Meara 1981
Proguanil	Evans 1981
Naftidrofuryl	McCloy and Kane 1981
Chlormethiazole	Rohner et al. 1982
Iron	Abbarah et al. 1976; Carlborg et al. 1978; Kobler et al. 1979
Ascorbic acid	Walta et al. 1976
Pantogar (Vitamin B complex, para-aminobenzoic acid, etc.)	Kobler et al. 1979

* Now available as dispersable tablet, *Vibramycin D.*

Slaughter and Boyce (1969) that sialography could reproduce the typical features of the condition it appears probable that the swelling is caused by retained secretions. Inflammatory change was seemingly absent in the one biopsy examination recorded (Matsuki et al. 1975). Most observers have agreed that a combination of ductal occlusion and mechanically stimulated oversecretion is needed to induce clinically evident swelling, but it remains to be determined whether the duct is damaged by direct pressure or by the violence of the patient's coughing and retching as intubation proceeds.

Drug-induced oesophagitis

Reports of dysphagia and a localized area of oesophagitis have been linked with delay in the passage of numerous capsules and tablets at a point where any corrosive property they possess is concentrated to an extent unrecognized by conventional tests of toxicity. The list given in Table 16.1, with a selection of references, omits few of these, but it should be noted that the first three alone, potassium chloride, emepronium bromide, and the tetracycline derivative, doxycycline, account for most of the episodes on record. The vulnerability of the oesophagus

to damage by drugs in common use had hardly been appreciated before the report of Pemberton (1970) drew attention to the inflammatory potential of enteric-coated potassium in patients with cardiomegaly, which offered a convenient mechanical explanation for the hold-up so essential to this effect. However, many of the patients receiving tetracyclines for infection or emepronium bromide for bladder hypotonia have been young and generally healthy individuals in whom no simple predisposing pathology was evident.

As accounts of patently drug-induced oesophagitis accumulated, certain common factors in cases due to a wide variety of compounds emerged and were repeatedly emphasized. Enough instances occurred in the elderly to suggest that subclinical dystonia of the oesophagus, which increases in incidence with age, might be important. Moreover, the fact that a 15-year-old girl whose lesion had been caused by emepronium bromide was found to have intra-oesophageal pressure changes indicative of absent peristalsis even after healing (Collins et al. 1979) identified smooth muscle dysfunction as a factor to be reckoned with at any age. On many occasions the first intimation of trouble has been a retrosternal pain severe enough to wake the patient in the middle of the night. These patients are usually found to have taken their medication just before retiring and without water to wash it down. Evans and Roberts (1976) found that 57 out of 98 barium sulphate tablets the size of a regular aspirin were retained in the oesophagus for longer than 5 minutes, and in this light it is easy to visualize how the loss of gravitational effect in the recumbent position, together with the loss of peristaltic activity considered normal in such circumstances, might allow time for a capsule or tablet to discharge most of its contents over a very short segment of oesophageal mucosa.

Channer and Virjee (1982) arrived at a quantitative estimate of the contribution both gravity and peristalsis make to the passage of a standard gelatin capsule through the oesophagus. Fifty patients shown to have normal motility in a barium swallow examination were given capsules filled with barium sulphate to take either standing or lying and with either 15 ml or 60 ml water. All capsules swallowed with 60 ml water while standing entered the stomach within 5 seconds. Both posture and water volume had highly significant effects on transit ($p < 0.001$), but a greater effect of posture was suggested by the finding that, at the 10 min stage, three times as many capsules remained after 60 ml water drunk supine than after 15 ml drunk standing. Delayed transit of over 20 seconds was the only abnormality in 20 per cent of patients, all known to have gastro-oesophageal reflux. Transit was arrested long enough for the capsule to disintegrate in a further 52 per cent, and in a majority of these patients the site of arrest was just above the lower oesophageal sphincter. Disintegration occurred between 2 and 3 minutes after the capsule had been ingested, leaving a period of 7–8 minutes before all barium had been cleared from the site. This group included eight of the 10 patients with dysphagia for food, and 14 of the 22 who complained of tablets sticking, compared with five in the 'delayed transit' group. Despite these relationships only three of 26 (11.5 per cent) patients with arrested transit had been aware of a trial capsule lodging in the oesophagus.

Perhaps the main question left unanswered by this study and another published simultaneously (Hey et al. 1982) is why the area of oesophagitis is usually much higher than the site at which most of the trial capsules had lodged. The lower site favoured by these experiments may well, however, be relevant to an investigation which showed that patients with gastro-oesophageal reflux are likely to be at particular risk of forming a stricture if they also take potentially irritative oral medication. Heller et al. (1982) found that six out of 76 patients coming to Eder–Puestow dilatation for benign oesophageal stricture had been receiving either emepronium bromide or potassium preparations. Of the remaining 70 patients 22 had been taking a NSAID, a proportion significantly higher than the 10 so treated out of 70 control patients

without stricture who had been matched with them for sex and age.

The chief symptom of drug-induced oesophagitis is severe constant retrosternal pain which is aggravated by eating and drinking. Such pain is promptly investigated, but if a barium swallow is performed the result is likely to be negative. Endoscopy, however, invariably shows a narrow zone of inflammation which occupies the mid-oesophagus in four out of five cases. One or more discrete ulcers are usually seen, with or without minimal stricture formation. The consequences are seldom serious, though deaths from haemorrhage have followed ulceration due to potassium chloride (Whitney and Croxon 1972; McCall 1975) and indomethacin (Agdal 1979). As a rule spontaneous improvement begins as soon as the source of the pain is identified and the offending medication stopped. Resolution both of symptoms and the endoscopic damage is complete within two weeks in most cases. Occasionally there remains a stricture narrow enough to cause residual dysphagia and to require dilatation.

Drug-induced disorders of oesophageal motility

Carbachol and rupture of the oesophagus

Cochrane (1973) described a case in which the injection of carbachol for urinary retention in a 58-year-old man was followed by haematemesis and oesophageal rupture, presumably a direct result of its cholinergic action, from which the patient died.

Chlormethiazole and oesophageal dystonia

Dewis et al. (1982) reported the case of a woman aged 50 who had received diazepam and chlormethiazole (dose 2–4 g daily) for 10 years in the treatment of a personality disorder. She first complained of dysphagia one year after starting treatment, but it was only after seven years that she consented to investigation. Barium swallow examination showed a delay in filling of the oesophagus, and manometry a motility disorder characterized by spontaneous non-propagated contractions throughout its length. Endoscopic appearances were normal. Six weeks after withdrawal of chlormethiazole alone clinical improvement had been accompanied by a return almost to normal of the motility pattern.

Chlormethiazole is understood to act chiefly on central mechanisms at various levels, but has not hitherto been known to affect the control of deglutition.

Anti-inflammatory drugs and the upper gastrointestinal tract

Anti-inflammatory drugs are associated with two forms of gastrointestinal toxicity which should be clearly distinguished. The first is a non-specific dyspepsia which in a given patient may occur with one agent only, while even others of the same chemical class are perfectly tolerated. The second is a tendency to damage the mucosa, chiefly that of the stomach, and thus to constitute a factor in ulceration of the upper gastrointestinal tract. It is not difficult to imagine how anti-inflammatory activity, being directed against the excessive manifestation of a natural defensive response elsewhere, may harm a mucosa already vulnerable to its own secretions, but the mechanism of non-specific dyspepsia is as obscure in this as in other instances.

Aspirin

Aspirin had been in common use for over half a century before Douthwaite and Lintott (1938) established its relation to gastric bleeding by gastroscopic observation of hyperaemia and submucosal haemorrhage in 13 of 16 subjects who had taken it just before the examination. The clinical importance of this effect was soon confirmed (Hurst and Lintott 1939) when a patient presenting with haematemesis and melaena was seen to be bleeding from sites at which fragments of aspirin tablets were still in contact with the gastric mucosa. Muir and Cossar (1955) not only witnessed bleeding after ingestion of aspirin at gastroscopy but also noted that 54 of 166 patients admitted with haemorrhage had taken salicylates within the preceding 6 hours, and singled out 21 in whom the circumstances suggested aspirin to be a major factor. Alvarez and Summerskill (1958) found that 49 of 103 patients with upper gastrointestinal bleeding had taken salicylates within 24 hours of admission, as against only seven of 103 matched controls. The apparent success of these and other early epidemiological surveys in identifying aspirin as an important cause of acute upper gastrointestinal haemorrhage did not, however, stand up to detailed examination of the methods they employed (Langman 1970; Cooke 1973; Rees and Turnberg 1980). Pitfalls, such as the difficulty of obtaining an accurate estimate of aspirin ingestion from the patient's history, and the inadequacy of other in-patients as controls for those admitted with haemorrhage, were shown to have resulted in such wide quantitative variations in the findings of different series as to throw serious doubt upon the association. Furthermore, despite good endoscopic and histological evidence for erosive gastritis as a consequence of aspirin ingestion, it has not yet been possible to define the lesion specific to aspirin which would enable a firm conclusion to be reached. It also remains to be explained why bleeding is so rare an accompaniment of aspirin overdosage.

It must be conceded nevertheless that later studies designed to circumvent some of the epidemiological problems have tended to support a role for aspirin in acute haemorrhage, albeit far weaker than that previously envisaged. A report from the Boston Collaborative Drug Surveillance Program (BCDSP) (Levy 1974) established a significant relationship between severe haemorrhage and regular aspirin medication in patients without evidence of duodenal ulceration ($p < 0.01$) and also in those with a newly-diagnosed benign gastric ulcer ($p < 0.05$). The suggestion here that aspirin is especially liable to cause bleeding from acute gastric lesions recalls the finding of Allibone and Flint (1958) that among patients admitted for haemorrhage only those with acute lesions had taken significantly more aspirin than controls during the previous two weeks. In a similar survey by Valman et al. (1968), 80 per cent of patients with acute gastric lesions but only 50 per cent of those bleeding from chronic peptic ulcer had taken aspirin within the preceding week.

Langman's (1970) concern that the influence of aspirin on acute bleeding might have been exaggerated by the failure of other investigators to distinguish between patients who took aspirin for early symptoms of haemorrhage and those in whom bleeding followed aspirin ingestion led to a study in which enquiry was also made as to the consumption of paracetamol by patients and controls (Coggon et al. 1982). Paracetamol causes negligible microscopic damage to the gastric mucosa (Ivey et al. 1978; Hoftiezer et al. 1982), and in the BCDSP has not been associated with bleeding (Jick 1981). However, Coggon et al. (1982) found that patients more frequently admitted to recent paracetamol ingestion than did their matched controls from the general population. This increase fell short of the increase in numbers taking aspirin, and the authors took the difference to indicate a genuine but small effect of aspirin, such that perhaps one-third of cases of bleeding in patients taking this drug had occurred as a result of their doing so.

The probability that aspirin will precipitate haemorrhage in any one individual may be greatly increased by contributory factors. The combination with alcohol is particularly dangerous (Goulston and Cooke 1968; Needham et al. 1971; DeSchepper et al. 1978), for alcohol has at least three separate actions: it enhances the solubility of aspirin, has a direct toxic effect of its own on the gastric mucosa, and stimulates acid secretion. Pre-existing chronic gastritis may render the mucosa more vulnerable (Winawer *et al.* 1971), although achlorhydria is protective (St.John and McDermott 1970). There is evidence from both human and animal studies that vitamin C deficiency can prolong the haemorrhage caused by aspirin (Russell *et al.* 1968; Russell and Goldberg 1968). Bile is a well-recognized gastric irritant, and the mechanism by which this may aggravate the effect of aspirin is suggested by a finding of Cochran *et al.* (1975) that taurocholic acid lowers the transmucosal potential difference, implying pathological back-diffusion of H$^+$ ions into the epithelium. A correlation has been found between bile reflux and bleeding in patients receiving aspirin (Capron *et al.* 1977).

Less controversy attends the induction by aspirin of chronic occult blood loss. This was first demonstrated by Stubbé (1958) who investigated 140 patients with rheumatic conditions and 40 normal subjects using the benzidine test, and found blood in the stools of 70 per cent of those taking aspirin. By tagging patients' erythrocytes with ^{51}Cr Scott *et al.* (1961) were able not only to confirm the incidence of occult bleeding but also to quantify it. In 74 patient-trials 70 per cent of patients bled a mean 4.9 ml each day, the highest mean daily loss in a single trial amounting to 18 ml, and the maximum lost by any patient in a 24-hour period to 106 ml. No increase in bleeding followed increase of dosage from 2 to 6 g per day. Taking aspirin with meals did not consistently decrease blood loss either in this trial or in that of Pierson *et al.* (1961), which gave generally similar results with the same technique.

Since these early studies so much reliance has come to be placed on ^{51}Cr excretion as a measure of blood loss from the stomach or duodenum that some examination of its validity is in order. Although it cannot identify the site of bleeding, the combined evidence against any erosive action of systemic aspirin on gastrointestinal mucosae and in favour of both decreased toxicity and increased absorption of intraluminal aspirin at higher pH levels (Cooke and Goulston 1969; Welch *et al.* 1978; Khoury *et al.* 1979; Hunt and Fisher, 1980) argues strongly against the occurrence of any appreciable damage beyond the duodenum. However, Rees and Turnberg (1980) drew attention to studies showing that ^{51}Cr is excreted in the bile and, as aspirin is known to stimulate bile secretion (Cooper and Williamson 1982), there is urgent need for a quantitative estimate of the error thus introduced. Most investigations depending on measurement of faecal ^{51}Cr loss as an index of occult haemorrhage cover no more than a few days' aspirin ingestion. The true clinical significance of such short-term studies was called into question by a preliminary report of MacKercher *et al.* (1977*b*) pointing out that the gastric mucosa of patients with rheumatoid arthritis who

had been taking up to 5.2 g aspirin daily for at least six months was more resistant to damage by a single dose of aspirin than that of normal volunteers. Graham *et al.* (1983) dosed 12 normal males with 2.6 g aspirin daily for a week, and found gastroscopic damage to be most severe at three days, after which healing took place despite continued treatment.

Attempts have been made to diminish the toxicity of aspirin by presenting it in special formulations. Tablets designed to disperse more rapidly than plain aspirin, whether entitled 'soluble' or in effervescent form, were found by Scott *et al.* (1961) to cause no less blood loss. Buffered aspirin has been observed endoscopically to damage both gastric and duodenal mucosae as severely as plain aspirin after a day's (Hoftiezer *et al.* 1982) or a week's (Lanza *et al.* 1980) administration to healthy subjects. Silvoso *et al.* (1979) noted a greater incidence of gastric ulceration in patients with rheumatic diseases who were on long-term treatment with the buffered version. Endoscopic damage to the stomach has been substantially less with enteric-coated (EC) aspirin, given either to arthritic patients or for one week to normal volunteers (Silvoso *et al.* 1979; Lanza *et al.* 1980; Hoftiezer *et al.* 1980). However, there is evidence that the duodenum may become as severely affected by EC as by plain aspirin in the course of time (Lockard *et al.* 1980; Hoftiezer *et al.* 1980). ^{51}Cr excretion studies show a significant increase in blood loss when either an EC tablet or a capsule containing EC microspheres is taken, but generally this is much less than that which occurs with plain aspirin (Scott *et al.* 1961; Frenkel *et al.* 1968; Mielants *et al.* 1981; Portek *et al.* 1981). A sustained-release preparation in which the aspirin is bonded to a matrix relatively insoluble in an acid medium has also been associated with significant reduction in bleeding (Treadwell *et al.* 1973). Measurement of occult blood loss in two studies comparing plain with micro-encapsulated aspirin revealed in each case significant bleeding, but a lesser effect of the microencapsulated version (Brandslund *et al.* 1979; Dybdahl *et al.* 1980).

It was commented by both Lanza *et al.* (1980) and Silvoso *et al.* (1979) that despite their differing mucosal effects the plain and EC aspirins gave similar blood salicylate levels; while in the study of Dybdahl *et al.* (1980) there was no correlation between blood loss and salicylate level. Ivey *et al.* (1980) infused aspirin intravenously for two hour periods in normal subjects to give them blood levels of salicylate in the therapeutic range for rheumatic diseases, and found no change in the potential difference or in light and scanning electron microscopic appearances of the gastric mucosa. Such results would seem to confirm the conclusions of most earlier investigators that the action of aspirin on the gastric mucosa is a local one, independent of changes in bleeding time.

Another means of reducing occult blood loss while retaining the anti-inflammatory effect of aspirin is to inactivate it by chemical combination with a second molecule until it reaches the site of absorption in the small intestine. Benorylate, an ester of aspirin and paracetamol, has been shown to cause less than 50 per cent of the blood loss attributable to soluble aspirin in equivalent dosage (Croft *et al.* 1972). Salsalate, an esterification product of two molecules of salicylic acid, was compared with plain and EC aspirin in a single dose of 3 g taken by 12 healthy subjects, and was found to give similar serum levels of salicylate (Mielants *et al.* 1981). In another part of this study 42 patients admitted for orthopaedic surgery were treated with salsalate for one week, and only two showed an increase in gastrointestinal blood loss, a proportion significantly lower than those observed in groups receiving soluble, EC or intravenous aspirin in comparable doses. The difluorinated salicylate, diflunisal, has been associated with major gastrointestinal haemorrhage in several single-case reports, but at 500 g daily in normal subjects DeSchepper *et al.* (1978) found it to cause no increase in ^{51}Cr excretion, even when taken with alcohol in amounts sufficient to potentiate the effect of aspirin. Given for a week at double this dosage to patients with osteoarthritis, diflunisal proved comparable to paracetamol in its effect on ^{51}Cr excretion (Tringham *et al.* 1980). Choline magnesium trisalicylate at a dosage giving similar serum levels of salicylate to aspirin at 3 g daily produced far less change in the gastric mucosa among 10 normal subjects who received each for 5 days in a cross-over study (Kilander and Dotevall 1983).

In attempting to assess the clinical relevance of such indirect tests of occult blood loss as that which measures faecal ^{51}Cr excretion, the best guide we have is the incidence of iron-deficiency anaemia in patients who take aspirin regularly. Summerskill and Alvarez (1958) were the first to document the association in two patients. Among 52 of their patients in whom aspirin caused bleeding, Scott *et al.* (1961) identified six with an iron-deficiency anaemia they considered mainly due to prolonged aspirin consumption. Baragar and Duthie (1960), however, compared a group of 75 rheumatoid patients who had received at least 2.6 g of aspirin daily with 31 who had taken it only sporadically over a period of 6 years and found that mean haemoglobin had risen slightly in each group. Stubbé (1961) used aspirin challenge to help incriminate this drug as the principal cause of anaemia in 14 patients who took it regularly and whose benzidine tests reverted to normal as soon as he withdrew it. Bannerman *et al.* (1964) investigated 68

patients with iron deficiency and persistent occult bleeding from the gastrointestinal tract. In only 17 were they able to trace the cause, and in 12 of these it proved to be aspirin ingestion. Heggarty (1974) found that withdrawal of high-dose aspirin treatment was the only measure needed to prevent recurrence of iron-deficiency anaemia in five children with occult blood loss.

Knowledge of the microscopic changes induced in the gastric mucosa by contact with aspirin depends chiefly on specimens obtained from experimental animals. However Croft and Wood (1967) ingeniously prepared human material by giving five patients admitted for elective partial gastrectomy 3.4 g soluble aspirin daily for 2–5 days, and compared the mucosal appearances in their resected specimens with those in specimens from five other patients who had not received aspirin. Erosions, one to five in number, were found (mainly around the lesser curve) in all of the patients who had taken aspirin but in only one, a case of carcinoma, among those who had not. Frank haemorrhage was seen in some lesions, and the mucosa stripped more readily in specimens processed after aspirin treatment. Microscopically the erosions commonly penetrated the muscularis mucosae and were surrounded by zones of haemorrhage.

In another series of three patients given 3.0 g aspirin daily for up to five days before gastrectomy, Hahn *et al.* (1975) observed by light microscopy cytolysis, with or without desquamation, in many surface mucous cells. Under the electron microscope they saw pyknosis of nuclei, clumping of chromatin, and lysis of cytoplasmic organelles. All these appearances could be reproduced closely in the gastric mucosa of guinea pigs by feeding them aspirin for four days. The only change peculiar to human mucosa was the appearance in parietal cells of secondary lysosomes containing various inclusions which could have been products of cell degeneration. Rainsford (1975a) also identified changes in parietal cells as the first to occur in apparently normal tissue outside an erosion, the greatest damage being to the mitochondria. Aspirin causes mucosal erosions in many species, usually in the body of the stomach and affecting first the mucous cells. Swelling of surface mucous cells was shown by scanning electron microscopy to be one of the earliest changes in the cat (Frenning and Öbrink 1971). In the mouse ultrastructural damage to mucous cells was found 3 minutes after administration of 20 mmol aspirin (Hingson and Ito 1971). Using a special technique to study wider areas of mucosa Yeomans *et al.* (1973) showed that rats given aspirin in a dosage equivalent to three to four tablets in the average human develop extensive superficial necrosis after 30 minutes. Most of this area is re-epithelialized after 4 hours, but some deep erosions remain and take up to one week to heal.

If aspirin taken regularly can cause such sustained damage to the upper gastrointestinal mucosa as to induce iron deficiency, the question must be asked whether it has any role in the genesis of chronic peptic ulcer. There is little to associate it with duodenal ulcer, but various lines of evidence (St. John 1975; Duggan 1976) combine to suggest that it may contribute to chronic gastric ulceration. Kiser (1963) reported five cases in which gastric ulcer followed a high intake of aspirin, healing rapidly after the drug had been stopped. The Boston study (Levy 1974) revealed an association between benign gastric ulcer and regular ingestion of aspirin. In New South Wales, Billington (1960, 1965) investigated a sudden increase in the incidence of gastric ulcer among women of a certain age-group, and found that excessive aspirin consumption was another characteristic of this cohort. The same pattern has since been recognized elsewhere in Australia. Further investigation of the relationship between aspirin ingestion and chronic peptic ulcer in Sydney has supported earlier work elsewhere in showing no effect on susceptibility to duodenal ulceration (Piper *et al.* 1981). Surprisingly, however, although the group of females with gastric ulcer was again found to include more individuals with a history of heavy aspirin consumption than that of controls matched for age and social class ($p < 0.01$), the association of gastric ulcer with paracetamol conumption was even stronger ($p < 0.001$). If this finding throws doubt on the causative role of aspirin even in gastric ulcer, it is at least consistent with the further observation in the same patients that, provided healing was complete by the time they left hospital, those who then resumed analgesics proved no more susceptible to recurrence over the next 4 years than those who abstained (Piper *et al.* 1977).

Among those investigators who agree that gastric ulcer is unduly prevalent in patients with rheumatoid arthritis, there is nevertheless controversy as to whether this may not be a property of the disease rather than its treatment and, if it is a property of the treatment, whether a combination of corticosteroid and aspirin may not be needed to precipitate the ulcer. Two studies have suggested that gastric ulcers in patients with a history of regular aspirin ingestion are more likely to occur in the antrum than in the generally favoured area high on the lesser curve or posterior wall (Cameron 1973; McDonald 1973). Chronic ulcers are not readily inducible in animals other than by excessively high, regular dosage with aspirin. In one such model the rat was found to develop ulcers in an otherwise normal antral mucosa (St. John *et al.* 1973).

Acetic acid derivatives

Indomethacin

Early trials of indomethacin yielded an alarming rate of peptic ulceration, but as the ulcerogenic effect of this drug is dose-related (Rothermich 1966) and as the dosage now commonly employed is far lower, its use should at present carry less hazard. Lövgren and Allander (1964) reported a series of 18 patients receiving 200–300 mg indomethacin daily. Five developed gastro-duodenal ulcers and two of these died, one after massive haemorrhage, the other after resection for multiple gastric ulcers. Three of the five had a previous history of peptic ulcer, but barium-meal examinations were negative at the start of treatment. Among 11 of 234 (4.7 per cent) patients who developed ulcers or gastrointestinal haemorrhage while taking indomethacin, Rothermich (1966) found four with a past history of peptic ulceration. Taylor *et al.* (1968) described 10 cases of gastric ulcer in patients given indomethacin, three of whom were receiving suppositories only. In favour of an association with indomethacin were the following facts: (i) only one of the patients had a past history of gastric ulcer; (ii) only two were on other potentially ulcerogenic drugs (one ACTH, one aspirin); and (iii) six of the ulcers were in the antrum. It is of interest that one of these prepyloric ulcers healed while the patient continued indomethacin at a reduced dosage. Sun *et al.* (1974) prescribed indomethacin for 14 patients with rheumatoid arthritis: one developed a gastric ulcer and three developed duodenal ulcers. This experience in a series which was characterized by an unusually high ulcer incidence in most of the regimes studied offers a marked contrast to the figure of 5.5 per cent for peptic ulceration obtained by Cooke (1973) in a review of five other series covering 634 patients on indomethacin treatment.

Wanka *et al.* (1964) found that indomethacin increased faecal loss of ^{51}Cr-tagged erythrocytes in five of 13 patients, but to a lesser extent than aspirin. Even with the high dosage of 200 mg per day the patients of Beirne *et al.* (1974) lost a mean of 2.1 ml blood per 24 hours when taking indomethacin as against 6.1 ml on aspirin. In comparing the effects of indomethacin and aspirin on Heidenhain pouches in the dog, Chvasta and Cooke (1972) noted that the latter was four times more likely to cause mucosal bleeding. In man the haemorrhagic lesions seen on gastroscopy after a short course of indomethacin at 100 mg daily were only slightly fewer and less severe than those caused by aspirin at 3.6 g daily (Lanza *et al.* 1975).

Two recent reports have implicated indomethacin, given to neonates as a medical means of closing a patent ductus arteriosus, in perforation of the stomach, though both the infants described had nasogastric tubes, and in one the stomach had been inadvertently distended with oxygen (Gray and Pemberton 1980; Campbell *et al.* 1981). Since, moreover, there was no evidence of inflammation in the first of these cases the relevance of indomethacin to either must be seriously questioned. Nasogastric tubes in three out of 39 infants treated by Yanagi *et al.* (1981) yielded blood, but, whatever the cause of this haemorrhage, it proved insufficient to affect haemoglobin level, and was in no case associated with radiological evidence of ulceration. However, one severe bleed probably due to oral indomethacin has been reported (Kennedy *et al.* 1982).

An attempt to reduce the toxicity of indomethacin by presenting it in a rigid case from which it was leached at 7 mg per hour by osmosis (*Osmosin*) does not seem to have affected the frequency of haemorrhagic episodes, but to have extended its action to more distal segments of bowel (Day 1983). Indomethacin is probably safest when given in a suppository containing 100 mg (Holt and Hawkins 1965; Woolf 1965; Keenan *et al.* 1984).

Sulindac

Sulindac is closely related to indomethacin. Bianchi Porro *et al.* (1977) compared gastroscopic appearances in six rheumatoid subjects who had received aspirin (3.6–4.8 g daily) for 6–8 weeks with those in patients who had received sulindac (300–400 mg daily). Five of the six patients on aspirin showed gross changes (peptic ulcer in two cases), whereas only one of those on sulindac had developed slight oedema and erythema of the mucosa. Its effect on gastrointestinal blood loss when taken by 10 subjects at two daily dosages (240 and 400 mg) has been compared with that of aspirin at 4.8 g per day and placebo by the ^{51}Cr technique (Cohen 1976). For sulindac in either dose the mean of measurements made on the eighth and fifteenth days of treatment was at 0.90 ml no greater than that for placebo, whereas aspirin was associated with a mean loss of 3.65 ml. The blood recovered from gastric washings in healthy volunteers given sulindac at 400 mg daily exceeded the amount following ingestion of placebo on no more than one-third of study days (Hunt *et al.* 1983). The relative safety of sulindac indicated by this and earlier findings is attributed by the authors to its poor solubility — and therefore incomplete absorption — in an acid medium.

Attempts to compare the gastric toxicity of NSAIDs by histological studies are made infrequently, perhaps because sampling error is difficult to avoid. This might explain why McIntyre *et al.* (1981) found that sulindac at 400 mg caused a greater increase in the microscopic criteria for acute gastritis than indomethacin at 75 mg daily.

Tolmetin

An erosive effect of tolmetin on gastric mucosa was demonstrated in the rat by Mann (1977). The damage, both visible and microscopic, caused in human gastric mucosa by tolmetin given to normal volunteers for a week at 2 g daily was found by Lanza *et al.* (1978) to be intermediate in severity between the effect of aspirin at 3.6 g and that of ibuprofen at 2.4 g. The chemically similar compound, zomepirac, prescribed for 2 weeks at 300 mg daily, resulted in somewhat less blood loss than aspirin at 3.9 g (Johnson 1980), but in a long-term treatment study the same dose was associated with no less clinically apparent bleeding or peptic ulceration than aspirin at 3 g daily (Honig 1980).

Isoxepac

Svendsen *et al.* (1981) found that isoxepac given for one week at a daily dosage of 600 mg to 12 healthy volunteers in a cross-over study caused minimal gastric damage, and in this respect was comparable to indomethacin at 150 mg, while haemorrhagic lesions occurred exclusively in a third period during which the same subjects received aspirin at 3 g.

Fenclofenac

Fenclofenac at dosages up to 1200 mg daily has been observed to cause less blood loss than diclofenac at 150 mg (Uthgenannt and Letzel 1981), and in one endoscopic study appeared to *diminish* the number of gastric erosions among 10 patients with osteoarthritis after 8 weeks' treatment (Hradsky 1981).

Suprofen

Although a potent inhibitor of prostaglandin synthesis, suprofen is considered more of an analgesic than an anti-inflammatory agent. A cross-over trial in 20 normal subjects has shown 800 mg daily to cause significantly less occult bleeding than aspirin at 2.6 g (Arnold and Berger 1983). The long-term clinical safety of suprofen at this dosage has yet to be established, but trivial gastrointestinal toxicity has been encountered in patients receiving 200 mg or 400 mg daily over periods of at least 3 months (Yeadon *et al.* 1983).

Propionic acid derivatives

All drugs of this class are liable to provoke dyspeptic

symptoms, and several representatives have been shown to cause gastric mucosal damage at recommended dosage levels (Caruso and Bianchi Porro 1980). Microscopic study has confirmed endoscopic impressions that the lesions are inflammatory in nature, and bleeding detectable by the ^{51}Cr technique may ensue. It is generally agreed, however, that in comparable antirheumatic dosage the propionic acid derivatives are less corrosive than aspirin.

Fenbufen is a propionic acid derivative described as a 'pro-drug' because it has no anti-inflammatory activity until metabolized by the liver to biphenylacetic acid. An endoscopic study has shown it to cause somewhat less visible damage to the gastric mucosa after a week's treatment at 1000 mg daily than either naproxen at 750 mg or indomethacin at 150 mg (Lanza and Nelson 1982). Both in acute dosage (Collins *et al.* 1983) and at 900 mg taken daily for a month (Lussier *et al.* 1982*b*) it causes less gastrointestinal blood loss than aspirin.

Ibuprofen, which is probably both less active and less irritant than any of its analogues, causes less haemorrhagic lesions in the gastric mucosa at 1.6 g daily (Lanza *et al.* 1975, 1978), but is still capable of inducing blood loss when given in therapeutically adequate doses (Thompson and Anderson 1970; Schmid and Culic 1976; Bianchi Porro *et al.* 1977a). In a cross-over trial against aspirin at 3 g daily ketoprofen 100 mg was found to cause less severe endoscopic damage (Rahbek 1976), and under similar conditions 150 mg daily for one week gave comparable occult bleeding to that obtained with naproxen at 500 mg (Magnusson *et al.* 1977). This in turn has been shown to induce gastric lesions intermediate in severity between those given by two dosage levels of ibuprofen (Lanza *et al.* 1978). Substituting naproxen at 500 mg for aspirin at doses sufficient to give guaiac-positive bleeding in normal subjects Arsenault *et al.* (1975) found that bleeding decreased to placebo levels over the next two weeks. Loebl *et al.* (1977) added successively aspirin and fenoprofen to paracetamol in the treatment of 14 patients with rheumatoid arthritis. Gastroscopy after 2 weeks on the trial drug showed that six patients had developed antral ulceration during the period of aspirin treatment but only one during the fenoprofen period. Mean 24-hour blood loss measured by ^{51}Cr excretion amounted to 0.8 ml with paracetamol alone, 2.2 ml when this was supplemented with fenoprofen, and 5.0 ml when aspirin was given. In comparative studies against aspirin at 2.1 g and phenylbutazone at 600 mg daily, flurbiprofen at 300 mg proved least corrosive to the gastric mucosa, and, although one subject on flurbiprofen bled appreciably, mean occult blood loss in this group was no greater than the placebo level (Vakil *et al.* 1977*a,b*). Tiaprofenic acid gave little evidence of ulcerogenicity in one

long-term clinical trial (Camp 1981). Oxaprozin at 1200 mg daily was shown to give less severe gastric erosion (Lanza *et al.* 1981) and less occult blood loss (Lussier *et al.* 1982*a*) than aspirin at 3.9 g.

Benoxaprofen is a relatively weak inhibitor of prostaglandin synthesis, and has emerged from comparative studies as causing less gastric damage than other NSAIDs (Ridolfo 1980; Yeung Laiwah *et al.* 1981). Moreover surveys of its side-effect profile after extended use in nearly 2000 patients have included no cases of major haemorrhage (Mikulaschek 1980; Halsey and Cardoe 1982). Other serious side-effects in elderly patients have, however, prompted its withdrawal; and a report of bleeding from duodenal ulcers in three elderly women at intervals of 1 to six weeks after starting benoxaprofen suggests that this age-group may also be especially vulnerable to its action on the gastro-duodenal mucosa (Stewart 1982).

An early study of ^{51}Cr excretion after treatment with indoprofen at 300 or 600 mg daily equated its effect with that of ibuprofen (Bianchi Porro *et al.* 1977a). Among 4042 patients with osteoarthritis who received indoprofen orally at a dosage of 400–800 mg daily for a month in a multinational study (Bruni *et al.* 1982) only eight (0.2 per cent) had clinically severe bleeding, although nearly 10 per cent reported some abdominal pain. Experience with general use of this drug in the UK has resulted in more reports to the Committee on Safety of Medicines (CSM) associating it with serious gastrointestinal bleeding and perforation than might have been expected from surveys elsewhere, and the licence for its sale in this country has therefore been suspended.

Pyrazolones

Phenylbutazone, like indomethacin, is a common cause of non-specific dyspepsia and early acquired a bad name for peptic ulceration because of the large doses originally prescribed. In a study of patients with rheumatoid arthritis Kern *et al.* (1957) treated 100 with phenylbutazone, most receiving 400 mg a day. During courses of up to 37 months, seven developed new ulcers (four of these were duodenal), which represented a higher incidence than that found in the groups taking corticosteroids or aspirin at the same clinic. Sperling (1969) followed a series of 562 patients on maintenance doses of either phenylbutazone or oxyphenbutazone and found only three in whom dyspepsia could be ascribed to an ulcer. Scott *et al.* (1961) detected an increase in faecal excretion of ^{51}Cr in two of 22 patients taking phenylbutazone, but one of these had just completed a course of aspirin. No occult blood loss could be demonstrated with this technique by MacFarlane *et al.* (1976) even though damage to the gastric mucosa was evident in seven of 13 healthy subjects who received a short course of phenyl-butazone at 400 mg a day and were then examined by gastroscopy. Two of these subjects displayed widespread submucosal haemorrhages, but in general the damage was less severe than that caused by indomethacin.

Feprazone is a compound of phenylbutazone and a terpenyl group which is claimed to be comparable with phenylbutazone in activity and better tolerated by the gastrointestinal tract (Fletcher *et al.* 1975). It does not appear to cause occult blood loss (MacFarlane *et al.* 1976). Both these drugs potentiate the activity of warfarin but feprazone to a lesser extent than phenyl-butazone (Chierichetti *et al.* 1975).

Fluproquazone given at 300 mg daily for one week to 12 healthy volunteers induced only two erosions in one of them, whereas 11 of the same group developed a total of 80 erosions after aspirin at 3 g daily (Gillberg *et al.* 1981). The same study found no significant increase in excretion of ^{51}Cr among subjects receiving fluproquazone.

Fenamates

Mefenamic acid and flufenamic acid share with other anti-inflammatory agents the property of inhibiting cyclo-oxygenase, but differ in causing diarrhoea more than dyspepsia. Peptic ulceration is not associated with either of these drugs taken alone, although Prescott (1968) cited one case of acute gastric ulcer in which mefe-namic acid was taken with corticosteroids and another case where haematemesis without radiological evidence of gastric ulcer occurred during mefenamic acid treat-ment. Slight though significant occult blood loss has also been demonstrated in patients receiving mefenamic acid (Skyring and Bhanthumnavin 1967).

In an endoscopic cross-over comparison of tolfenamic acid at 600 mg with aspirin at 3 g daily Axelsson *et al.* (1977) found that only one of 10 normal volunteers suffered an increase of gastritis over a week's treatment, whereas only two escaped inflammatory change during the aspirin period. Various forms of gastrointestinal intolerance were displayed by a few individuals among 41 patients completing a 6-month trial of tolfenamic acid, but haemorrhage was not reported (Sørensen and Chris-tiansen 1977). As with other fenamates diarrhoea was not uncommon, and proved severe enough in three patients to require their withdrawal from the trial.

Piroxicam

This popular agent has been associated with upper gastrointestinal bleeding in the experience of most endo-scopy units. A series of five cases presented to one hospital within a period of 9 weeks (Emery and Grahame 1982).

Mechanisms of gastric damage by NSAIDs

In man the gastric effects of aspirin are mediated locally from within the lumen since intravenous administration does not increase occult blood loss even though it is as effective as oral dosage in prolonging the bleeding time (Cooke and Goulston 1969; Leonards and Levy 1970). Aspirin in the plain form tends to reach the gastric mucosa as large particles, 'soluble' aspirin in fine dispersion, yet there is no conclusive evidence that one is more damaging than the other. The mechanism of its action remains obscure despite much detailed investi-gation, chiefly in animals (Rainsford 1975*b*). A fact not disputed is that an acid medium greatly potentiates the destructive action of aspirin on the mucosa, probably first by maintaining it in the non-ionized state necessary to rapid absorption and then by increasing the number of H^+ ions available for back-diffusion into the epithelial cell. This would explain why aspirin appears to cause far less damage when given with antacids or in an alkaline form, as in effervescent tablets containing bicarbonate (Pierson *et al.* 1961; Thorsen *et al.* 1968), and why patients with achlorhydria are relatively immune from its action (St. John and McDermott 1970). The converse effects of salicylate on acid secretion are at least two in number and oppose each other. Johnson and Overholt (1967) have demonstrated liberation of histamine, which presumably augments acid production, and Anderson (1964), a local depressive action. Rainsford (1975*b*) analysed the complex series of effects ascribed to aspirin into 'primary' (or immediate) and 'secondary' (or longer-term). The primary include:

1. Sloughing of mucus by the acetic and salicylic acids into which it is hydrolyzed at the epithelial surface;
2. Denaturation and killing of surface cells;
3. Breaking down the mucosal barrier (Davenport 1967) so that H^+ ions may diffuse back and assist in cell destruction.

The secondary series involve numerous enzyme-dependent reactions. In the gastric mucosa of the bull-frog aspirin has been shown to depress mitochondrial oxidative phosphorylation (Spenney and Bhown 1977), upon which depends a variety of synthetic and transport activities. Here too salicylate appears to be the active moiety, although the finding common to rat and man that choline magnesium trisalicylate causes less damage than aspirin in doses giving equal absorption of salicylate has been taken to imply that the acetyl grouping may have more clinical relevance (Danesh *et al.* 1983; Kilander and Dotevall 1983). That aspirin may affect another fundamental step in the production of energy was shown by Kuo and Shanbour (1976) who observed depletion of ATP when aspirin was instilled into the mucosal surface of the canine stomach. They attributed to this an early inhibition of mucosal ion transport which was followed by increased permeability.

The mucous layer has long been accepted as important in protecting the gastric mucosa from secreted acid, and defects in its production have been shown to follow the administration of either indomethacin or phenylbutazone (Menguy and Desballiets 1967a,b; Parke et al. 1975). Azuumi et al. (1980) found that aspirin given to rats substantially reduced one of three glycoprotein peaks on gel chromatography of samples from this layer before any macroscopic evidence of mucosal damage could be detected. Williams and Turnberg (1981) threw further light on the chemical 'structure' of the mucous layer, when serial measurements made with an antimony micro-electrode passed through this towards the mucosa of rabbit stomach in vitro showed a linear increase in pH which became maximal at the epithelial surface. The authors interpreted this result as supporting an earlier postulate (Heatley 1959) that bicarbonate has a part to play in gastric mucosal protection. Further work on the same model (Rees et al. 1983) established that both aspirin and indomethacin inhibited secretion of bicarbonate in the fundus, and that in the case of indomethacin only this effect could be blocked by pretreating the mucosa with the synthetic prostaglandin (PG), 16,16 dimethyl PGE_2. The same PG had earlier been shown to stimulate bicarbonate production in the amphibian fundus (Garner et al. 1979), and here also to protect against the inhibitory action of indomethacin, implying a basic mechanism which might well be common to all species.

However, the findings of Rees et al. imply too that one NSAID may differ from another in the way it causes damage. A further example of this is furnished by the contrasting effects of aspirin and indomethacin on transmucosal potential difference (PD). Compared under identical conditions either in the gastric (Murray et al. 1974) or in the buccal (Huston 1978) mucosa of man aspirin consistently reproduces the fall in PD which prompted Davenport's theory of the 'mucosal barrier', while indomethacin does not. These distinctions made, it must also be recognized that most unwanted effects of NSAIDs on the gastro-duodenal mucosa can be linked to the opposing protective effect of PGs. Müller et al. (1981) have shown in healthy volunteers given aspirin that the gastric PD can be maintained by simultaneous administration of 16,16 dimethyl PGE_2. As yet only the 'epidermal growth factor' of Konturek et al. (1981), a polypeptide synthesized by the salivary and duodenal glands of the rat, can be claimed to represent an entirely independent protective system.

The efficacy of PGs in preventing gastric ulceration was first demonstrated by Robert et al. (1968) who found that PGE_1 administered to rats by subcutaneous infusion inhibited the ulcerogenic effect both of ligating the pylorus and of prednisolone injection. Later both PGE_2 and certain synthetic analogues were shown in the rat to protect the gastric mucosa against indomethacin (Lippmann 1974). Although some PGs are known to inhibit gastric acid secretion (Robert et al. 1967), others without this property offer comparable protection against indomethacin (Robert 1976; Cohen and Pollett 1976), aspirin (Cohen and Pollett 1976; Cohen 1981), and flurbiprofen (Robert 1976).

Reviewing all the evidence against an antisecretory mechanism as the means by which PGs protect the gastric mucosa Bennett et al. (1973) pointed out that indomethacin too could decrease pentagastrin-stimulated acid secretion in man. Alternatives proposed by Robert et al. (1967) included the vasodilator effect of PGs familiar in other tissues, a mechanism favoured by Bennett and colleagues who had found that indomethacin reduced gastric mucosal blood flow in the dog. However the pathological characteristics of ulcers induced by NSAIDs are not such as to suggest an ischaemic aetiology, and in the rat Fang et al. (1977) were unable to reproduce the effect of indomethacin by arterial ligation or the administration of vasopressin. Such considerations led to the concept of 'cytoprotection' (Robert 1976) in which PGs were seen as maintaining the integrity of epithelial cells, perhaps by acting as trophic hormones. Chaudhury and Jacobson (1978) found that indomethacin, like aspirin, inhibited active transport of sodium in the canine mucosa and proposed that, as 16,16 dimethyl PGE_2 both opposed this action and increased the mucosal content of cyclic AMP, the role of natural PGs might be one of sustaining the sodium pump. Support for cytoprotection has come from the finding that gastric epithelial cells from the rat, cultured in monolayers as a means of excluding circulatory factors, may be protected from aspirin damage by 16,16 dimethyl PGE_2 (Terano et al. 1984). However, in this model neither stimulation of cyclic AMP nor addition of an active analogue could reproduce the protective effect.

Prophylaxis against damage by NSAIDs

Failure to define the physiological role of PGs in the upper gastrointestinal tract has not prevented preliminary studies of their efficacy as pharmacological agents designed to eliminate the hazards of anti-inflammatory treatment in chronic rheumatic disease, although long-term trials in patients remain to be described. Cohen (1978) reported that PGE_2 at 4 mg daily by mouth abolished the ^{51}Cr loss induced in normal volunteers by aspirin at 2.4 g, but feared that the dose employed might prove unsafe in pregnant women. Yik *et al.* (1982) found that 2 mg daily sufficed to inhibit completely the bleeding due to a similar dose of aspirin, and singled out freedom from diarrhoea as the principal benefit of keeping dosage of PGE_2 below the 4 mg level. Kollberg *et al.* (1981) added either PGE_2 at 1 mg daily or 15-(R)-15 methyl PGE_2 at 120 μg daily to indomethacin given to arthritic patients at 150 mg daily for short periods, and both significantly reduced ^{51}Cr loss without diminishing the therapeutic response. Such results should encourage manufacturers to test formulations in which PG and NSAIDs are combined.

Any of the modern anti-ulcer agents might be expected to offer some protection against NSAID-induced damage, but perhaps chiefly the histamine H_2-receptor antagonists because of their favourable effect on intra-luminal acidity. Some attempt has been made to answer two questions: (1) Do these drugs prevent the primary erosive action of NSAIDs, and (2) can they render anti-inflammatory treatment safe for patients with chronic peptic ulcers? Rats dosed acutely with aspirin, naproxen, or tolmetin developed less gastric erosions if metiamide or cimetidine had been given first (Mann 1977; Mann and Sachdev 1977). In acute human studies MacKercher *et al.* (1977*a*) measured aspirin-induced damage to the gastric mucosa through the concomitant fall in PD, and found this to be prevented by pretreatment with cimetidine. Konturek *et al.* (1983) were similarly able to prevent gastric bleeding due to aspirin, given in four doses of 0.5 g during a single day, by pretreatment with 100 mg ranitidine 15 minutes before each dose.

Answers to the second question have been somewhat inconclusive. Croker *et al.* (1980) observed healing of ulcers in 15 out of 21 patients with rheumatoid or osteoarthritis treated with cimetidine during 3 weeks of anti-inflammatory therapy. Moreover seven out of eight patients maintained on cimetidine for between 6 and 12 months thereafter were shown endoscopically to be free of recurrence at the end of the study period despite continuation of their drugs at unchanged or increased dosage. The limitations of this small open study were conceded by the authors, but no comparable double-blind trial has been published. O'Laughlin *et al.* (1982) identified gastric ulcers as 'aspirin-associated' in 18 patients with rheumatic diseases, and on a double-blind basis compared the healing in groups given cimetidine and placebo respectively. Again numbers were small, and whether because of this or because antacids were given to patients on placebo an apparent advantage of cimetidine

proved statistically non-significant. Perhaps more relevant to rheumatological practice was the finding of Welch *et al.* (1978) that cimetidine reduced occult blood loss in patients receiving long-term aspirin treatment. Although it is reasonable to ascribe any protective effect of H_2 receptor antagonists to their anti-secretory action, the possibility of an alternative mechanism in man was raised by Guth *et al.* (1979) who demonstrated in the pylorus-ligated rat an independent cytoprotective action of cimetidine.

Clinically effective protection without acid suppression has been demonstrated with sucralfate, an aluminium salt of sucrose sulphate which binds to exposed proteins in the ulcer base (Tesler and Lim 1981).

Corticosteroids and the stomach

In an exhaustive search of the literature from 1950 to 1975 Conn and Blitzer (1976) collected 50 controlled studies of corticosteroid treatment (32 being double-blind) which together provided data on 6102 cases. Analysis of the figures for both prevalence and serious complications of peptic ulcer showed no significant difference between patients receiving corticosteroids and their controls. The occurrence of peptic ulcer in no more than 1 per cent of either group is at first sight suspiciously low, but the authors point out that patients with previous ulceration were specifically excluded from most trials, and demonstrate how the frequency of ulcers increases in both groups — but with statistical significance only in corticosteroid-treated patients ($p < 0.05$) — as the period of observation is extended. Contrary to the findings of earlier investigators (Kern *et al.* 1957; Kammerer *et al.* 1958), Conn and Blitzer have not been able to show any relationship between the daily dosage of corticosteroid and ulcer incidence. Although a total dose of corticosteroid equivalent to more than 1000 mg prednisone is associated with a highly significant increase in frequency of ulceration among the treated patients ($p < 0.001$), the same significance attaches to a parallel increase in frequency among their controls. Little doubt can remain from this survey and the reviews of other authorities (Cushman 1970; Cooke 1973) that in most clinical situations treatment with corticosteroids has no influence on the development of peptic ulcer. However, as Conn and Blitzer concede, some degree of fallibility is unavoidable when studies are undertaken in a clinical context, and even without this source of error figures such as theirs could conceal a genuine causal association between corticosteroid therapy and peptic ulcer in exceptional circumstances.

Perhaps this is why a more recent study on the same lines (Messer *et al.* 1983) has shown the same trend towards a higher incidence of ulceration in patients receiving corticosteroids, for whom overall the risk ratio was 2.3. Here the difference was significant, even though ulcers were detected in only 55 of 3064 (1.8 per cent) patients on corticosteroids and 23 of 2897 (0.8 per cent) on alternative treatment.

A more liberal use was made of corticosteroids in the 1950s and 1960s than in the present decade for treating rheumatoid arthritis, and it is notable that most claims for an ulcerogenic effect of these compounds relate to the management of rheumatoid patients in that era. Kammerer *et al.* (1958) examined all 117 patients of their series radiologically, whether or not they had complained of dyspepsia, and found that 31 per cent of those given corticosteroids developed peptic ulcer as against 9 per cent of those not so treated. Similar figures emerged from many other studies of the period, and even from the more recent account of Sun *et al.* (1974) who subjected all 140 of their rheumatoid patients to barium-meal examination, revealing ulcers in 40 per cent of those on corticosteroids and in none of a group receiving gold salts. In a series of 2114 patients with rheumatoid arthritis treated at the Mayo Clinic, Bowen *et al.* (1960) were unable to detect any such effect of corticosteroids: the incidence of peptic ulcer being 8.1 per cent in patients on other regimes, 7.5 per cent overall among those on corticosteroids, and 8.2 per cent in patients displaying 'hypercortisonism'. What they did observe was a higher ratio of gastric to duodenal ulcers in the corticosteroid group; and in rheumatoid patients generally an incidence of peptic ulcer three to four times higher than that found in other patients attending the Clinic. Atwater *et al.* (1965) incorporated both a clinical study and an autopsy series in their investigation of patients with rheumatoid arthritis. Each suggested that the disease doubled the risk of peptic ulcer while treatment with corticosteroids introduced no further risk. Nine of the 10 duodenal ulcers found at autopsy had been diagnosed before corticosteroids were prescribed, whereas all four gastric ulcers had followed this treatment. Spiro and Milles (1960) were satisfied that gastric ulcers could be caused by corticosteroids and pointed to the unusual frequency of antral involvement in patients so treated, not only as favouring this interpretation but also as implying a relationship between these and 'stress' ulcers.

Evidence against corticosteroids being in themselves a significant cause of peptic ulcer comes from experience of their use in other chronic diseases, such as bronchial asthma (Duprez and Simons 1963), ulcerative colitis (Kirsner *et al.* 1959; Truelove and Witt 1955), and Crohn's disease (Jones and Lennard-Jones 1966;

Sparberg and Kirsner 1966), which has not resulted in any apparent excess of peptic ulcers.

If it is possible to reconcile the mass of conflicting results in this field and attempt a brief summary, one might venture the opinion that corticosteroids are not intrinsically ulcerogenic but that they may cooperate with some factor in the disease for which they are being given to induce gastric ulcer in a few susceptible subjects.

Unlike aspirin, corticosteroids do not break down the gastric mucosal barrier to increase ionic fluxes and decrease the transmucosal potential difference (Chvasta and Cooke 1972; Murray *et al.* 1974; Ivey *et al.* 1975) or inhibit prostaglandin synthesis to any appreciable degree (Vane 1971). Although both prednisolone (Aubrey and Burns 1973) and hydrocortisone (Watson *et al.* 1973) potentiate the gastric acid response to food in dogs, even this effect may be negligible in man. Dreiling and Janowitz (1957) were unable to detect any increase in acid output by 55 patients after 25–40 mg ACTH, 25–85 mg hydrocortisone, or 50 mg prednisolone; Kammerer *et al.* (1958) a slight and probably insignificant increase in six of 14 normal volunteers receiving either hydrocortisone or prednisone. Both ACTH (Hirschowitz *et al.* 1955) and cortisone (Menguy and Masters 1963) reduce secretion of mucus, and parathyroid extract has been shown to protect the gastric mucosa of the rat from damage by corticosteroids through its action in stimulating mucus production (Menguy and Masters 1965).

Miscellaneous drug-induced gastric ulceration

High concentrations of KCl may ulcerate the gastric as well as other mucosal surfaces of the alimentary tract (see pp. 239–40). Spironolactone has twice been incriminated as a principal cause of gastric ulceration (Brown *et al.* 1971; MacKay and Stevenson 1977). The second instance concerned a woman with alcoholic cirrhosis — but no varices — who presented with haematemesis when receiving spironolactone, 600 mg daily, as her sole medication. At endoscopy two bleeding acute ulcers, each 2 cm in diameter, were found on the greater curvature of the stomach. Both had healed two weeks after withdrawal of treatment. Goodman (1977) speculated that the property of spironolactone displayed in these cases might have been identical with that known to inhibit healing of gastric ulcer by carbenoxolone sodium.

Another diuretic, acetazolamide, caused anorexia and abdominal discomfort in a 65-year-old man who had been treated with this drug over a period of 2 months for glaucoma. Gastroscopy revealed multiple erosions throughout the stomach. The tablet was stopped, the symptoms regressed, and the stomach was found to be normal on repeat examination 26 days later (Herman *et al.* 1980).

A 15-year-old girl given zinc sulphate 220 mg b.d. for acne suffered epigastric discomfort after taking each capsule. On the day she fainted and passed melaena stools she was found to be tender epigastrically and to have a haemoglobin of 5.4 g/dl. Endoscopy showed a patchy gastritis with a resolving haemorrhagic erosion. No further dyspepsia or bleeding followed withdrawal of treatment (Moore 1978).

Massive gastric ulceration followed proximal displacement of a catheter in the hepatic artery delivering 5-fluorouracil for dissolution of hepatic metastases in eight cases described by Narsete *et al.* (1977).

Potassium chloride ulceration

Ulceration of the small intestine by enteric-coated tablets of potassium chloride (EC KCl) was the subject of two almost simultaneous 'first' reports, one from Sweden (Lindholmer *et al.* 1964), the other from the USA (Baker *et al.* 1964). The latter described 12 patients treated surgically at one hospital during a period of 15 months; 11 presenting with small-bowel obstruction, three with perforation, and one with gastrointestinal haemorrhage. All had cardiovascular disease, and 11 had been receiving EC KCl either in combination with or as a supplement to hydrochlorothiazide. Duration of treatment varied from 8 days to 33 months. In one patient an undigested tablet of KCl was found at the site of perforation. The lesions consisted of a marked stenosis 8–25 mm in length and, where perforation had occurred, this was always just proximal to the stenosis. Ulcers were small and punched-out in apparently normal mucosa. Microscopically there was acute and chronic inflammation of the mucosa, fibrosis of the submucosa, and little or no damage to deeper layers except in cases of perforation. Lindholmer *et al.* (1964) encountered four patients requiring operation for small-bowel obstruction due to stricture during the months of April and May 1964. In a retrospective survey of similar cases treated in Stockholm they found a total of 20 patients who had been receiving thiazide diuretics at the time, but were uncertain whether to blame the diuretic or the potassium most of these were also taking. Later one of this group, Räf (1967), wrote a monograph which remains today the most comprehensive account of small-intestinal ulceration due to EC KCl.

Alarmed by these early reports the manufacturers instigated a survey of small-bowel ulceration in selected hospitals throughout the world, which they completed a month after the paper of Baker *et al.* (1964) had appeared, and published the following year (Lawrason *et al.* 1965). A total of 395 patients was collected and of these 275 (57 per cent) had received either a diuretic or KCl or the two in combination. Most were over 50 years of age and had cardiovascular disease. These data were supplemented by the results of animal experiments at the laboratories concerned, showing that ulceration could readily be induced in monkeys (though not in dogs) by as little as 100 mg KCl. If given in liquid form or in tablets that disintegrated rapidly it would ulcerate the stomach, if EC its effect would be seen in the small intestine. EC preparations of hydrochlorothiazide, cyclothiazide, or bendrofluazide alone were innocuous. Evidence for a specific local effect of K^+ ions was provided by the finding that EC sodium chloride induced diarrhoea but never ulceration. In man, unlike the monkey, gastric ulceration by EC KCl has been reported, with accompanying pain or haemorrhage (Jacobs and Pringot 1973).

In consequence of these and other studies alternative slow-release preparations of KCl were developed and have proved in practice to be far safer. It is therefore surprising that the manufacture of tablets containing KCl in an enteric coating still continues, and that four of these are currently licensed in UK. Ball (1976) reported from the Isle of Skye (population 7000) no fewer than four cases of small-intestinal ulceration occurring within two years in patients taking *Salupres*, a combination of hydrochlorothiazide, reserpine, and KCl, now withdrawn. Two other cases, one involving a death after operation for a perforated ileal ulcer, were subsequently recorded (Makey 1977).

The slow-release preparation generally favoured is one in which crystals of KCl are embedded in wax and the resulting matrix is coated with sugar. Its use is so extensive that clinicians should be aware of the instances in which this tablet too has caused ulceration of the alimentary tract. McAvoy (1974) described the development of multiple punched-out, painless ulcers on the tongue and buccal mucosa of a patient who sucked these tablets instead of swallowing them. A further series of reports concerned ulceration and stricture of the oesophagus, just below the carina, arising within a few days of cardiac-valve surgery (Pemberton 1970; Whitney and Croxon 1972; Howie and Strachan 1975; Peters 1976). This is due perhaps to some slight displacement of the oesophagus at operation which, with the temporary loss of motility that must result from handling of the tissues and the proximity of enlarged heart chambers, could well delay the passage of a tablet long enough to permit local accumulation of K^+ ions in dangerous concentrations. McCall (1975) reported a case in which hold-up occurred in the absence of preceding surgery. His patient, who had aneurysmal dilatation of the left atrium, died of massive haematemesis from an aortic fistula. Sumithran *et al.* (1979) described a fistula between the oesophagus and left atrium arising in similar circumstances and likewise fatal.

Symptomatic ulceration of the stomach has rarely been attributed to slow-release KCl (McMahon and Akdamar 1976). However, in a prospective gastroscopic study the same wax-matrix tablet given to healthy volunteers for a week at 96 mmol KCl daily caused visible damage to the gastric antrum or some other part of the upper gastrointestinal tract in over half the subjects (McMahon *et al.* 1982). By contrast, a micro-encapsulated preparation in identical dosage proved innocuous unless accompanied by an anti-cholinergic drug to delay gastric emptying. A similar difference was observed by Barkin *et al.* (1983), but its clinical relevance will remain in question until larger numbers have been studied, for gastroscopic appearances failed to correlate with symptoms and in no subject was occult bleeding detected. The relative safety of wax-matrix KCl is attested by the fact that no excess of gastrointestinal bleeding was found among over 900 in-patient users monitored by the BCDSP (Aselton and Jick 1983).

Ulceration, stricture, and perforation of the small bowel, identical to that found with EC KCl, has also occurred (Heffernan and Murphy 1975; Dayer *et al.* 1977; Flintholm *et al*, 1981; Lofgren *et al.* 1982; Weiss *et al.* 1977), and in one case was shown to be associated with appreciable slowing of small intestinal transit (Farquharson-Roberts *et al.* 1975). It would thus seem wise to employ parenteral or effervescent preparations of KCl whenever there is reason to suspect the presence of stasis anywhere in the alimentary tract.

Miscellaneous drug-induced ulceration of the lower gastrointestinal tract

Discrete inflammatory lesions of the jejunum or ileum are rare. Their clinical manifestations depend usually on an associated stricture or perforation, and in the absence of specific underlying pathology the typical microscopic picture is frequently such as to suggest ischaemia as the initiating process. Isolated perforations of the colon occur more frequently, but almost always in elderly subjects whose circulation may be compromised by atherosclerosis and in whom the presence of thin-walled diverticula at a site of heavy bacterial colonization should make for additional susceptibility. Until the advent of EC KCl, it was exceptional for suspicion to fall on any drug as a cause of ulceration in the small intestine, although there was already good circumstantial evidence

for corticosteroid treatment as a factor in colonic perforation (Sautter and Ziffren 1959). Subsequent reports of focal inflammation in the lower gastrointestinal tract may at times have fixed too readily on current medication as its cause, but in some instances the association is strong, some are supported by parallel accounts in the literature, some implicate drugs known to be ulcerogenic in the upper gastrointestinal tract, and in others anatomical features, such as a Meckel's diverticulum, or established pathology, such as Crohn's disease, may accentuate the ulcerogenic potential of drugs that are safe for most patients. Many cases involve subjects with rheumatoid arthritis, perhaps necessarily because of the treatment this requires, but it may be that here too there is some predisposition in the disease itself. It is in the light of such considerations that the following drugs are regarded as capable of damaging the lower gastrointestinal tract.

NSAIDs

One instance of fulminating ileocolitis was reported in a young woman who had received a few days of phenylbutazone to a total dose of 4 g (Bárta *et al.* 1965). The authors suspected this episode might have represented activation of a subclinical ulcerative colitis. Ulceration localized to the sigmoid colon manifested as rectal bleeding in a 70-year-old man every time phenylbutazone was represcribed after a 9-year course at 600 mg daily (Bravo and Lowman 1968). Oxyphenbutazone was the only treatment given to relieve post-episiotomy pain in a 32-year-old woman who, after 5 days and a total dose of 4 g, developed peritonitis due to perforation of a small caecal ulcer (Debenham 1966). This was not related to a diverticulum, and the surrounding mucosa was free of inflammation. Neoptolemos and Locke (1983) have described recurrent small-bowel obstruction in a 73-year-old woman who had been taking phenylbutazone for 3 years and who at laparotomy was found to have multiple strictures with ulceration over a short segment of proximal ileum.

A suggestive association between oral indomethacin and small-bowel ulceration was described in several cases (Shack, 1966; Sturgess and Krone 1973). However, the paucity of such episodes in relation to the frequency with which indomethacin has been prescribed over a 20-year period should be regarded as indicating a high level of resistance to damage by the conventional capsule of indomethacin throughout the lower gastrointestinal tract. In the course of 3 months between August and November 1983, the CSM received reports of 11 cases in which ulcers or perforation of the intestine had been associated with indomethacin treatment; no fewer than

Plates

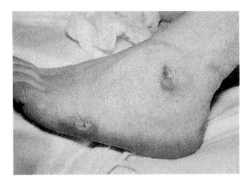

a Bullae due to barbiturate overdosage

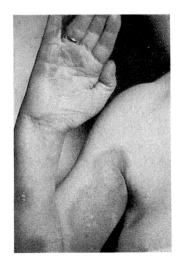

b Eruption induced by phenolphthalein

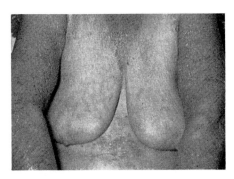

c Sulphonamide-induced photosensitivity

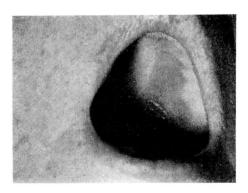

d Mepacrine pigmentation in mouth

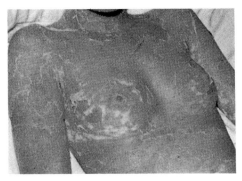

e Toxic epidermal necrolysis due to sulphonamides

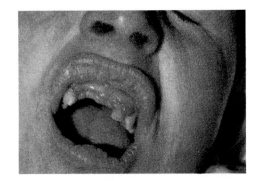

f Gingival hyperplasia due to diphenylhydantoin (phenytoin)

Plate 1 Drug-induced skin reactions. **a–f** by courtesy of Dr. A. McQueen

a Practolol-induced eruption

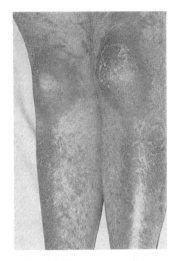

b Practolol-induced eruption

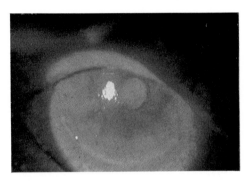

c Practolol eye: filamentary keratitis

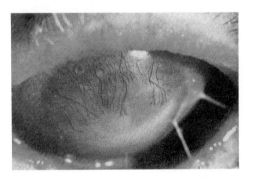

d Practolol eye: corneal vascularization and scarring, excess mucus present (therapeutic contact lens *in situ*)

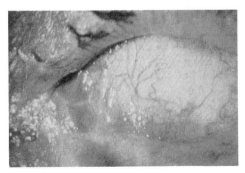

e Practolol eye: fornix obliteration and scarring; loss of structure of inner canthus

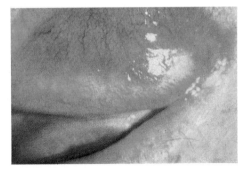

f Practolol eye: upper tarsal subconjunctival scarring

Plate 2 Practolol-induced oculomucocutaneous syndrome. **a** and **b** by courtesy of Dr. A. McQueen; **c–f** by courtesy of Mr. Peter Wright, Moorfields Eye Hospital, London.

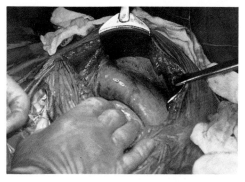

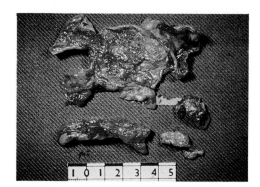

a Practolol-induced sclerosing peritonitis   **b** Practolol-induced sclerosing peritonitis

Plate 3 a and **b** Practolol-induced sclerosing peritonitis (**a** by courtesy of Dr. R. P. H. Thompson and Professor W. I. Cranston, St. Thomas's Hospital, London; **b** by courtesy of Professor A. E. Reed, Bristol Royal Infirmary).

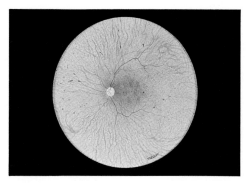

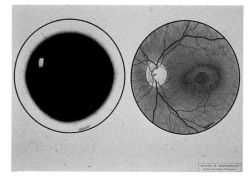

c Chloroquine retinopathy   **d** Cornea and retina after chloroquine treatment

Plate 3 c and **d** Drug-induced disorders of the eye, by courtesy of Dr. P. Hansell, Director of A-V Communications, Institute of Ophthalmology, University of London

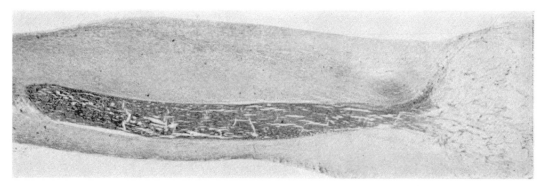

Plate 3 e Methysergide-induced thickening and fibrosis of mitral valve. P.M. specimen, H & E stain, by courtesy of Dr. Klaus Misch, The Lister Hospital, Stevenage, Hitchin

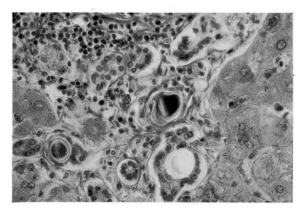

Plate 4a

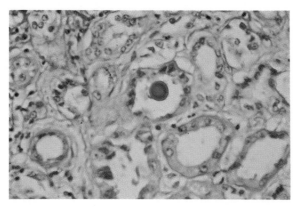

Plate 4b

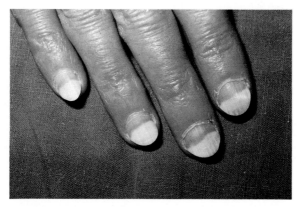

Plate 4c

Plate 4 Benoxaprofen-induced concretions in liver, **a** and kidney, **b** (by courtesy of Professor R. N. M. MacSween, Western Infirmary, Glasgow). Benoxaprofen-induced oncholysis, **c** (by courtesy of Professor R. N. M. MacSween and Dr. J. English).

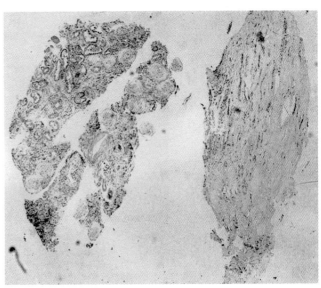

a (above) Analgesic papillary necrosis. Post-mortem specimen from a female patient aged 79 years; 8 Compound Codeine Tablets per day had been taken for many years for rheumatism and she was known to have suffered from analgesic nephropathy. The kidney shows complete loss of the renal papilla. Courtesy of Dr. D. J. Pollock, Institute of Pathology. The London Hospital

b (right) Analgesic papillary necrosis. Two pieces of renal biopsy from a male patient aged 45 years who had previously had a colectomy for ulcerative colitis. This patient had controlled his iliostomy by taking Compound Codeine Tablets for many years. The biopsy specimens show interstitial fibrosis and glomerular hyalinization. Courtesy of Dr. D. J. Pollock, Institute of Pathology, The London Hospital

Plate 5

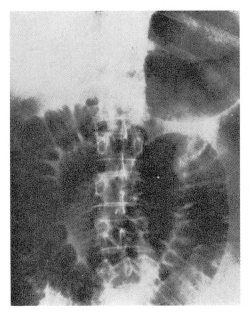

a 'Thick bowel' sign with dilatation of proximal small bowel, in plain abdominal radiograph

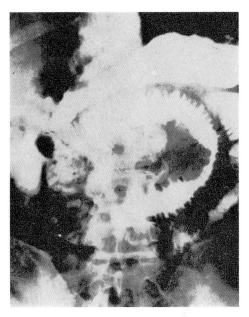

b 'Picket fence' sign in small bowel meal

Plate 6 a and **b** Anticoagulant haematoma. Sears, A. D. *et al.* (1964) J. A. M. Roentg. **91,** 808–13

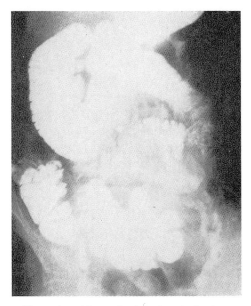

c Shortening of ileum simulates haustration in small bowel meal

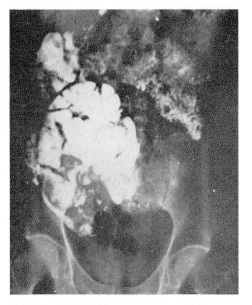

d Barium-filled ileal loops compressed into an oval cocoon by a radiolucent band which represents thickened peritoneum

Plate 6 c and **d** Sclerosing peritonitis. Marshall, A. J. *et al.* (1977) *Quart. J. Med.* **46,** 135–49

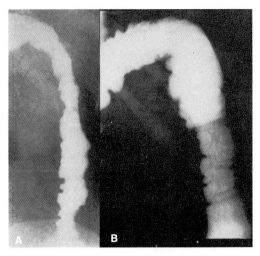

b Pseudomembranous colitis. Autopsy specimen with plaques 2–6 mm in diameter. Stanley, R. J. *et al.* (1974) *Radiology*, **111,** 519–24

a Ischaemic colitis in the splenic flexure attributed to oral contraceptives. A. Stricture and thumb-printing in the first barium enema film. B. One month later the segment has widened and sacculation has developed. Cotton, P.B. and Lea Thomas, M. (1971) *Brit, med. J.* **3,** 27–8.

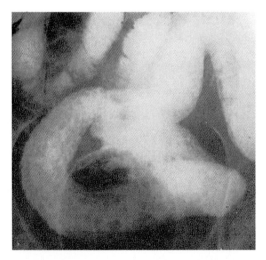

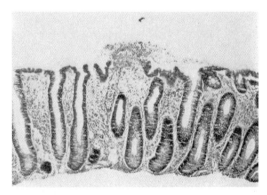

d Pseudomembranous colitis. Rectal biopsy showing a small spray of fibrin, polymorphs, and epithelial debris arising from the superficial interglandular area in close proximity to a dilated capillary (type 1 lesion). Price, A.B. and Davies, D.R. (1977). *J. clin. Path,* **30,** 1–12

c Pseudomembranous colitis. In the barium enema film plaques appear as multiple rounded filling defects. (Stanley, R.J. *et al.* (1974) *Radiology,* **111,** 519–24)

Plate 7

six of these having a fatal outcome. This dramatic change coincided with the introduction of *Osmosin*, a preparation of indomethacin in which the drug is leached gradually by osmosis from a non-absorbable capsule. To drive the process 158 mg potassium bicarbonate is added, and this could increase the erosive effect of indomethacin accumulating in high concentration wherever the capsule might lodge. An elderly man was changed from conventional to slow-release indomethacin after taking the former without ill effect for 10 years, and within 3 weeks had died of peritonitis from multiple perforating ulcers of ileum and colon. In this case capsules were found adjacent to some of the ulcers at autopsy, while in a 79-year-old woman, who was known to have diverticulosis coli, laparotomy revealed one capsule within a perforation of the sigmoid colon and five nearby in the pouch of Douglas (Day 1983). *Osmosin* has since lost its product licence. Designed to release indomethacin at a single site, the suppository containing 100 mg rarely causes local irritation, although two cases of clinically significant rectal bleeding have been associated with its use (Walls *et al.* 1968; Levy and Gaspar 1975).

Both naproxen and aspirin have been implicated, without very firm evidence, in ulcerative lesions of the lower gastrointestinal tract (Schwartz 1981; Shaunak and O'Donohue 1983), and provocation testing has shown aspirin to be capable of inducing acute inflammatory change at various sites, including the distal large bowel, in one idiosyncratic patient (Pearson *et al.* 1983).

Diarrhoea is a well-recognized side-effect of mefenamic acid treatment, and steatorrhoea has also been demonstrated in this connection (Marks and Gleeson 1975). It was not until recently, however, that a possible basis for these reactions was found in the form of an enterocoiitis severe enough in one case to cause bloody diarrhoea, and in both the cases described recurring within hours of challenge (Hall *et al.* 1983). Rectal, with or without ileal, inflammation was found in three of four cases in another series related to mefenamic acid treatment, and a latent period of at least 3 months was noted as being a potential obstacle to diagnosis (Phillips *et al.* 1983).

Penicillamine

A 61-year-old woman with rheumatoid arthritis developed acute proctosigmoiditis after many years of treatment with indomethacin suppositories (Hickling and Fuller 1979). She had, however, recently started treatment with D-penicillamine in addition. Both drugs were stopped because of the proctitis and only the D-penicillamine was re-prescribed when this remitted. In 11 days there was a relapse of bloody diarrhoea, which

again remitted when the D-penicillamine was finally withdrawn.

Gold salts

Since the first case of enterocolitis due to chrysotherapy was described by Goldhammer (1935) only 17 have been recorded, one-third of them fatal (Nagler and Paget 1983). Several preparations have been incriminated, and the total dosage of gold at onset of symptoms has varied widely between approximate limits of 50 and 250 mg. The typical presentation is with profuse watery diarrhoea, in which blood may appear, inconstant but sometimes severe abdominal pain, vomiting, and pyrexia. A skin rash and neutropenia may also occur. Most cases have involved a colitis, readily appreciable at sigmoidoscopy, and histologically non-specific. Affected small bowel has been characterized either by the radiological appearances of generalized mucosal oedema with a malabsorption syndrome (Kaplinsky *et al.* 1973; Siegman-Igra *et al.* 1976) or by irregular stricture of the distal ileum simulating Crohn's disease (Martin *et al.* 1981). This appearance, and for some patients a history of transient diarrhoea or rectal bleeding before treatment, may indicate pre-existing inflammatory bowel disease in which the extra-intestinal feature of a polyarthritis was initially more apparent than the intestinal disturbance itself. Management consists of withdrawing gold therapy and instituting general supportive measures. By these means it should be possible to avoid the septicaemic shock from which some patients have died, although there is as yet no specific protection against the massive necrosis of bowel sometimes found at autopsy. One patient survived colectomy following perforation due to haemorrhagic necrosis (Fam *et al.* 1980).

The fact that all cases of gold enterocolitis have been associated with relatively low dosage and the resemblance of other toxic effects of gold to immune complex disease together suggest an extension of this mechanism to bowel (Stein and Urowitz 1976). In the case of Szpak *et al.* (1979) the small intestinal biopsy showed subtotal villous atrophy, and the patient's lymphocytes reacted specifically with sodium aurothiomalate *in vitro*.

Corticosteroids

Perforation of colonic diverticula in patients receiving corticosteroids has been reported chiefly after renal transplantation, but the first case on record occurred in a patient given cortisone for rheumatoid arthritis (Sautter and Ziffren 1959). The clinical characteristics of corticosteroid-induced diverticular perforation, as encountered in 11 cases, were compared by Carter and Shorb (1971) with those of perforation from other causes, and included a more acute presentation, a higher incidence of free perforation, and over double the mortality. ReMine and McIlrath (1980) showed that mortality from intestinal perforation in patients treated with prednisone rises sharply when the daily dose exceeds 20 mg.

Oral contraceptives

In a report of severe and mostly acute abdominal symptoms supervening in five previously healthy young women whose only medication was an oral contraceptive, Morowitz and Epstein (1973) included two cases of typical ischaemic disease but also two of distal (ulcerative) colitis and one with the characteristics of Crohn's ileocolitis. Although these three cases met the recognized criteria for a diagnosis of chronic inflammatory bowel disease they did not respond to appropriate medication until the contraceptive pill had been withdrawn. Bonfils *et al.* (1977) compiled a total of six case histories from French centres in which the common features were non-bloody diarrhoea and a total colitis. Four of the patients were young women on the contraceptive pill. All six women receiving oestrogen, progestogen, or tablets combining the two, in an American series (Tedesco *et al.* 1982*b*) suffered bloody diarrhoea, with the radiological picture of superficial total or subtotal colitis. They were found to have rectal sparing at colonoscopy, and where ulcers were seen in the oedematous proximal mucosa they were so discrete as to mimic Crohn's disease. However in no case was granulomatous change or radiological involvement of the small bowel encountered. Three cases of inflammatory bowel disease (two accompanied by abnormalities of the liver) were described by Camilleri *et al.* (1981) in similar subjects. One had radiologically typical ileitis and rectal biopsy changes which included a non-caseating granuloma, another a colitis with features of Crohn's disease. The latter resolved when the contraceptive pill was withdrawn.

Although the authors of these reports were willing to consider the association between oral contraception and inflammatory bowel disease coincidental in at least some of their patients, the overall picture — and particularly the consistent trend towards recovery when the Pill is stopped — is such as to suggest a causal relationship. There is some evidence that chronic inflammatory bowel disease is aetiologically heterogeneous, and there are as yet no grounds for discounting the possibility that ischaemia may act as a contributory factor.

Iron

The irritant properties of iron have proved insufficient to damage the intestine except when a tablet has lodged within a diverticulum. Alaily (1974) reported gangrene of a Meckel's diverticulum which contained a *Ferro-Gradumet* tablet, and Ingoldby (1977), a fatal case of peritonitis following perforation of a jejunal diverticulum by the same preparation of ferrous sulphate in a non-absorbable matrix.

Ethacrynic acid

Data from the BCDSP published by Slone *et al.* (1969) indicated that 20 per cent of 157 patients treated with intravenous ethacrynic acid had developed gastrointestinal bleeding. This finding seemed to gain support from a second analysis (Jick and Porter 1978) which identified ethacrynic acid as the drug most frequently associated with severe bleeding among the 16 646 consecutive medical in-patients who had been monitored up to that time. Meanwhile, however, a sample of the cases first reported had been re-examined (Wilkinson *et al.* 1969), and it was found that in a substantial proportion bleeding had already occurred before the administration of ethacrynic acid. Whether or not such an approach misrepresents routine clinical monitoring as exemplified by the BCDSP, the chance that this undoubted relationship could be causative seems remote, involving as it does only the intravenous preparation of ethacrynic acid and having as yet received no hint of confirmation from other sources.

Digitalis

A single report of 11 cases in which congestive heart failure was accompanied by massive venous gangrene of bowel identified treatment — and in most cases overdosage — with digitalis compounds as the significant common factor (Gazes *et al.* 1961). The authors offered no rationale for their choice and, as the circumstances of this series could hardly be reproduced in an age of effective diuretic treatment, the role of digitalis as a precipitant of acute intestinal necrosis is likely to remain conjectural.

Methyldopa

The US Bureau of Drugs investigated 11 cases in which methyldopa was provisionally held to be the cause of an acute colitis, and was satisfied as to the significance of the association in six (Graham *et al.* 1981). The first case recorded was that of Bonkowsky and Brisbane (1976) in a

55-year-old man who developed malaise, fever, chills, and loose bloody stools one week after starting methyldopa. Other changes included a macular skin eruption and tender hepatomegaly. Rectal biopsies showed crypt abscesses, epithelial damage, hyperaemia, and oedema of the lamina propria, and a patchy inflammatory infiltrate containing polymorphs, eosinophils, and plasma cells. This patient was one of three in the total series subjected to challenge with methyldopa. Like the others he relapsed, developing bloody diarrhoea, chills, and fever within hours of taking a single tablet.

Clofazimine

Karat (1975) described two cases of ileitis in patients with leprosy who had been receiving clofazimine for 6 and 18 months, respectively. Dosage had been 300 mg daily for the first 3 months, reducing then to 100 mg daily. The diagnosis was made radiologically, but at least one of these patients also underwent laparotomy. The terminal 6 inches of ileum was thickened and oedematous, and a few mesenteric nodes were seen to be enlarged. Biopsies of bowel and lymph node showed multiple foreign-body granulomata containg crystals of clofazimine. In these and most subsequent cases the chief symptom was colicky abdominal pain, although nausea, vomiting, and diarrhoea have also been described. Onset is usually gradual, but the patient of Jagadeesan et al. (1975) presented with an acute abdomen. At laparotomy multiple longitudinal hyperaemic streaks in the terminal 14 inches of ileum were found to correspond with Peyer's patches. Microscopically these showed massive infiltration with eosinophils, plasma cells, and lymphocyctes. Nodular thickening of the upper ileum, reported by Mason et al. (1977) in a more chronic case, was associated histologically with eosinophilic and histiocytic infiltration of the submucosa, and thus probably represented a later stage of the same lesion. This patient was helped by prednisone in high dosage, but others recovered spontaneously within weeks of clofazimine withdrawal. Resolution of a lesion characterized by multiple polypoid filling defects in the ileum was followed by serial barium studies in the case of de Bergeyck et al. (1980). Little change was seen 2 months after clofazimine had been stopped, but appearances had returned to normal by 14 months.

Corn starch

A case of haemorrhagic proctitis arose 3 months after colectomy for diverticulitis in a woman of 79 (Gill and Piris 1978). Mucosal biopsies revealed a granulomatous reaction containing aggregated starch granules. A similar reaction to skin-testing with corn starch suggested hypersensitivity to this as the cause, and the inflammation responded well to prednisolone retention enemas.

Drug-induced malabsorption

Although many drugs are known to interfere with intestinal absorption relatively few have been thoroughly investigated from this viewpoint. The proportion of patients receiving such drugs whose health is seriously compromised by the resulting malabsorption must be small, but practitioners should know those that have been incriminated so as to avoid prescriptions which may accentuate pre-existing malabsorption or contribute to malnutrition of whatever cause. Mechanisms by which drugs are known to affect absorption have been placed by Longstreth and Newcomer (1975) in four categories:

1 A direct toxic effect causing morphological changes in the mucosa of the small intestine;
2 Inhibition of mucosal enzymes with or without morphological evidence of mucosal damage;
3 Binding and precipitation of micellar components, such as bile acids and fatty acids;
4 Alteration of the physicochemical state of another drug or dietary ion.

Laxatives

Liquid paraffin has long been recognized as an inhibitor of lipid absorption (Food Standards Committee Report 1962). Magnesium sulphate too causes mild steatorrhoea (Race et al. 1970). Systematic study of malabsorption due to laxatives has been largely confined to the chemical stimulants of peristalsis and would suggest that even these achieve a clinically significant effect only when taken regularly and in such excess as to induce the syndrome of 'laxative abuse' (Heizer et al. 1968; Cummings et al. 1974). Steatorrhoea, osteomalacia, and malabsorption of sugars have been described; but unexplained diarrhoea, a characteristic pattern of metabolic abnormalities, and radiological dilatation of the bowel remain the important diagnostic features. If phenolphthalein is the cause its presence in the stool may readily be established by the pink colour developing when alkali is added. There is no apparent mucosal damage (Heizer et al. 1968). Phenolphthalein has been shown to inhibit uptake of glucose (Hand et al. 1966), while oxyphenisatin in the rat leads also to a glucose leak from the mucosa, indicating a block in active transport as well as uptake (Hart and McColl 1968).

Anion-binding resins

Van Itallie *et al*. (1961) were unable to demonstrate in any of five patients treated with cholestyramine for the pruritus of primary biliary cirrhosis the steatorrhoea that might have been expected to follow its action in sequestering bile salts. Cholestyramine has, however, been shown to impair fat absorption in normal subjects (Zurier *et al*. 1965) and has doubled faecal fat excretion in patients receiving it for diarrhoea due to ileal resection (Hofmann and Poley 1969). West and Lloyd (1975) studied 18 children with familial hypercholesterolaemia on long-term treatment with massive doses of cholestyramine. Their principal finding was one of marked folate deficiency, with a mean fall in mean serum levels from 7.7 to 4.4 ng/ml and a proportionate deficit in red cell folate, which could be more than compensated by replacement with 5 mg folic acid daily. The polyglutamates which represent 80 per cent of dietary folate are anionic and thus predictably bound by cholestyramine. Five out of seven children with a normal fat intake developed steatorrhoea, considered to arise partly as a result of reduced micelle formation but also from direct binding of fatty acids by cholestyramine. Prothrombin times were unchanged, but serum levels of vitamins A and E, together with that of inorganic phosphate, fell toward the lower limit of normal. Elsewhere hypoprothrombinaemia with haemorrhage has been reported (Gross and Brotman 1970). Polidexide, an anion-exchange gel based on dextran, was shown to cause transient mild malabsorption of vitamin B_{12} but to have no effect on the absorption of fats, glucose, or iron (Ritland *et al*. 1975).

Neomycin

Of all antibiotics neomycin has been the most closely studied as a cause of malabsorption, presumably because, being poorly absorbed itself and prescribed solely to kill intestinal bacteria, it may be supposed to remain longer in contact with a greater area of mucosa than any other commonly used antibacterial agent. The first account of a sprue-like syndrome in patients receiving neomycin was that of Faloon *et al*. in 1958. The same workers found they could induce a reversible malabsorption of fat, cholesterol, carotene, iron, vitamin B_{12}, xylose, glucose, and nitrogen with oral neomycin (Jacobson *et al*. 1960*a*). Hvidt and Kjeldsen (1963) showed that a dose as small as 3 g daily could cause steatorrhoea of up to 20 g fat per 24 hours in normal subjects, with reversion to baseline levels within a week of its discontinuation. At least two mechanisms are involved in the impairment of lipid absorption by neomycin.

Through its cationic amino groups it binds the anions of micelles in the intestinal lumen and precipitates these so that their component molecules of cholesterol, monoglyceride, fatty acid, and fat-soluble vitamins cannot be taken up by the epithelial cells (Thompson *et al*. 1971). It also causes extensive microscopic damage in the mucosa itself, the first changes being detectable within six hours of the ingestion of a 2-g dose (Keusch al. 1970). Within a month of starting neomycin at 12 g daily, 11 of 12 patients studied by Jacobson *et al*. (1960*b*) yielded jejunal biopsies showing clubbed villi and oedema of the lamina propria, with vascular dilatation and a heavy infiltrate of lymphocytes and plasma cells. Dobbins *et al*. (1968) confirmed their findings, noting also the presence of macrophages containing a characteristic pigment. Ultrastructural abnormalities included ballooning and fragmentation of the microvilli, nuclear changes, and swelling of both mitochondria and the endoplasmic reticulum.

Kendall (1971) investigated the absorption of xylose after acute dosage with neomycin, finding that a single dose of 2 g can reduce this by 30 per cent. Glucose absorption is impaired to a similar extent. In the case of disaccharides the effect of non-specific damage by neomycin is enhanced by its inhibition of enzymic activity (Paes *et al*. 1967).

Tetracycline

Following an observation that transient steatorrhoea occurred in some patients who had been given tetracycline to label the bone mineralization front preparatory to bone biopsy, Mitchell *et al* (1982) undertook a formal study of fat absorption in 27 patients given either a single dose of 2 g tetracycline or 600 mg demethyl-chlortetracycline daily in divided doses for 2–4 days. Faecal collections started between one and 6 days after dosing with tetracycline showed that six patients excreted at least 10 g fat per 24 hours on a dietary intake of 50–70 g, with two others giving borderline values. Jejunal biopsy was normal in all cases. Although all the patients studied were under investigation for osteoporosis or osteomalacia, those with steatorrhoea could be shown to have regained normal fat absorption at varying intervals after stopping tetracycline.

Sulphasalazine

On the basis of a finding in their laboratory that sulphasalazine inhibited folate conjugase Halsted *et al*. (1981) compared the folate status of seven patients receiving 2–4 g sulphasalazine daily with that of 11 not so treated. The former had significantly lower serum folate levels. When sulphasalazine was perfused into the jejunum at concentrations greater than 1 mmol l^{-1} they noted not only impaired hydrolysis of polyglutamate but also intraluminal accumulation of monoglutamate, implying two separate mechanisms by which folic acid absorption could be decreased by this drug. Swinson et al. (1981) confirmed the tendency to folate malabsorption in 80 patients treated with sulphasalazine but found only two in whom haematological consequences of folate deficiency could be attributed to the drug alone.

Colchicine

Use of colchicine in the treatment of gout is now infrequent, but in the high doses needed for controlling an acute attack this drug consistently induced both diarrhoea and malabsorption of xylose, fat, and vitamin B$_{12}$ (Webb et al. 1968; Race et al. 1970). This effect could be obtained as well with parenteral as with oral administration of colchicine (Faloon and Chodos 1969). High doses, especially if given parenterally, when the drug is excreted through the intestinal mucosa, lead to mitotic arrest in the crypt cells of both rats (Hawkins et al. 1965) and man (Stemmermann and Hayashi 1971). More sustained administration of colchicine in man has been associated with anything from the mildest of changes to villous atrophy, abnormalities of epithelial-cell nuclei, and chronic inflammatory-cell infiltration of the lamina propria (Hawkins et al. 1965; Race et al. 1970). Mucosal enzymes depressed by colchicine include the disaccharidases (Race et al. 1970) and a number of dehydrogenases (Luketic et al. 1964).

Earlier acute measurements which allowed an estimate of the malabsorption induced by colchicine when this was given for gout can now be supplemented by data on its chronic effect in patients treated prophylactically for familial Mediterranean fever. Ehrenfeld et al. (1982) selected for their study 12 patients who had been receiving colchicine for at least 3 years. They found that three had mild steatorrhoea, with faecal excretion of 7.5–9.9 g fat daily. These and one other patient displayed also the decreased Na-K-ATPase activity already well documented in acute animal experiments (Rachmilewitz et al. 1978). It was perhaps surprising that jejunal histology proved to be normal in all these patients.

Methotrexate

In the rat methotrexate causes mitotic arrest of crypt cells, from which recovery is rapid after a single dose (Loehry and Creamer 1969). Striking morphological change may follow in the human jejunal mucosa (Trier 1962a,b), and a study of children treated with methotrexate for acute lymphoblastic leukaemia demonstrated malabsorption of xylose, the severity of which correlated with the cumulative dose of the drug (Craft et al. 1977). In another series malabsorption of xylose failed to correlate with either total dosage of methotrexate or duration of treatment, and xylose absorption could not be used to predict absorption of the methotrexate itself (Pinkerton et al. 1981). Comparison between studies in this field is made difficult by differences in the regimes employed, especially as most involve simultaneous administration of other cytotoxic drugs.

Although most forms of cytotoxic treatment cause symptoms of gastrointestinal intolerance, none apart from methotrexate has been consistently associated with either morphological or functional disturbance of the human small bowel (Shaw et al. 1979; Mitchell and Schein 1982).

Biguanides

The effect of phenformin, metformin, and buformin in lowering blood glucose levels has been attributed in part to malabsorption of glucose but in part also to enhancement of the action of insulin (Butterfield et al. 1961) and to anorexia (Stowers and Bewsher 1969). Malabsorption is implied by the finding that phenformin depresses the hyperglycaemic response to an oral but not to an intravenous load of glucose (Czyzyk et al. 1968). When Kruger et al. (1970) pre-treated rats with phenformin they found glucose transport in everted small intestinal sacs to be markedly decreased. In rats also Lorch (1971) was able to show that the hypoglycaemic effect of different biguanides correlated with their potency as inhibitors of glucose transport. Perfusion studies in man (Arvanitakis et al. 1973) demonstrated that phenformin may reduce glucose absorption by 50 per cent. Xylose absorption is impaired by both phenformin (Stowers and Bewsher 1969) and metformin (Berchthold et al. 1969).

Evidence has been adduced for the malabsorption of several substances other than sugars in animal or human subjects receiving biguanides, but only in the case of vitamin B$_{12}$ does the disturbance appear to affect patients on therapeutic doses. In a comparison of diabetics taking metformin with those on chlorpropamide Tomkin et al. (1971) found that 30 per cent of the former had abnormal Schilling tests and that four of these patients also had low serum levels of vitamin B$_{12}$. In a later study of 24 patients who had been on phenformin for at least 3 years, 11 (46 per cent) had depressed vitamin B$_{12}$ absorption (Tomkin 1973). Serum levels of the vitamin were lower in the patients receiving metformin, and the author speculated that this might reflect a greater degree of malabsorption due simply to the fact that the recommended dose of this drug was 10 times greater than that of phenformin. Appropriately enough the first case of megaloblastic anaemia attributable to biguanide medication occurred in a woman who had been treated for 8 years with metformin (Callaghan *et al*. 1980).

Arvanitakis *et al*. (1973) examined jejunal biopsies from their subjects on phenformin and were unable to detect any change in light microscopic appearances. Under the electron microscope, however, there was complete disappearance of matrix granules from the mitochondria of epithelial cells which, from the work of others with metabolic inhibitors, they took to be a sign of depressed energy metabolism.

Para-aminosalicylic acid (PAS)

Studies by Tygstrup *et al*. (1959, 1961) on tuberculous patients indicated that PAS lowered the plasma cholesterol and caused malabsorption of radioactive triolein from the gut. Levine (1968) investigated the effect of PAS on fat absorption in seven healthy males. During the 4-week study period, acid PAS, recrystallized in the presence of ascorbic acid (PAS-C) to minimize gastrointestinal symptoms and to give higher blood levels of PAS, was administered in a dosage of 6 g — therapeutically equivalent to 12 g PAS — daily, and, after 10 days without treatment, was given again in double dosage. A significant effect was seen with the higher dose only, which induced moderate steatorrhoea in all subjects, increasing the mean daily output of fat in the faeces from 5.4 to 11.3 g. No damage could be detected in jejunal biopsies under light microscopy. The hypocholesterolaemia observed by Tygstrup's group (1959) and by many others since cannot be correlated with either steatorrhoea (Levine 1968) or triglyceride absorption (Samuel *et al*. 1965) and may be a consequence of the folic acid deficiency frequently seen in patients taking PAS (Longstreth and Newcomer 1975).

In many individual cases of steatorrhoea attributable to PAS treatment there has been concomitant malabsorption of other substances , including xylose, folic acid, and vitamin B_{12} (Akhtar *et al*. 1968; Coltart 1969). Heinivaara and Palva (1965) performed Schilling tests before and after institution of PAS treatment. Impairment of vitamin B_{12} absorption was detectable within 2–6 weeks, and tests reverted to normal within 2 weeks of the drug's being discontinued. Subsequently the same group suggested that PAS, because of its structural similarity to folic acid, might interfere with the absorption of vitamin B_{12} by competitive inhibition (Palva *et al*. 1966). This hypothesis derives some support from the fact that both the production of intrinsic factor and its binding to vitamin B_{12} remain normal in patients taking PAS (Palva

and Heinivaara 1966; Halsted and McIntyre 1972), but is weakened by the failure of attempts by other investigators to correct the absorptive defect with folic acid supplements (Halsted and McIntyre 1972; Toskes and Deren 1972).

NSAIDs

Kendall *et al*. (1971) compared the effects of oral aspirin (1.2 g) and of indomethacin (100 mg) given as a single dose either together with oral xylose or 30 minutes before intravenous xylose. Both drugs significantly decreased urinary xylose excretion: indomethacin only when the sugar was given orally, implying that it had decreased absorption, but aspirin to an equal extent with either route of administration, implying a parenteral effect. Since there had been no change in the microscopic appearance of the jejunal mucosa after indomethacin the authors put forward its known action of accelerating gastric emptying as a possible mechanism in this instance. Also considered was its ability to uncouple oxidative phosphorylation, a property which it shares with most of the commonly used NSAIDs and which might therefore also explain the finding of Saleh *et al*. (1969) that salicylates and phenylbutazone inhibit glucose absorption in everted small-intestinal sacs of the rat, as well as the observation that aspirin may decrease jejunal glucose absorption by as much as 50 per cent in human subjects (Arvanitakis *et al*. 1977). In the latter study an almost equivalent depression by aspirin of mucosal ATP concentration was demonstrated.

A case of steatorrhoea in the presence of normal xylose absorption was associated with mefenamic acid treatment by Marks and Gleeson (1975).

Allopurinol

An 80-year-old Israeli farmer with gout, who had not received colchicine, complained of diarrhoea and was found to be anaemic one year after starting allopurinol (Chen *et al*. 1982). In time he came to associate bouts of diarrhoea and weight loss with his (intermittent) allopurinol treatment, and after 3 years presented with general malnutrition, malabsorption of iron, vitamin B_{12} and xylose, and a faecal fat content of 75 g per 24 hours. Barium follow-through showed dilated loops of small bowel, and jejunal biopsy appearances were consistent with coeliac disease. Although he was at first treated with gluten withdrawal, it later became apparent that this patient could remain well and with normal absorptive function on an unrestricted diet, provided allopurinol was discontinued.

Methyldopa

In a single case, carefully observed, both damage to the jejunal mucosa, in the form of partial villous atrophy with granulomatous inflammation, and malabsorption of xylose and vitamin B_{12} were ascribed to methyldopa treatment (Schneerson and Gazzard 1977).

Anticonvulsants and folic acid

Badenoch (1954) described two cases of megaloblastic anaemia in epileptics taking phenytoin and phenobarbitone, one of whom responded to treatment with folic acid. Later the same year Hawkins and Meynell (1954) reported another case in which phenytoin had been the only anticonvulsant prescribed. Hobson *et al.* (1956) specifically investigated folic acid absorption in their patient whose megaloblastic anaemia had followed heavy dosage with a widely used combination of amylobarbitone sodium and quinalbarbitone sodium and found no deficit. Evidence was subsequently obtained for an effect of phenytoin in depressing the activity of the intestinal conjugase thought to be necessary for the release of folic acid from dietary polyglutamates (Hoffbrand and Necheles 1968); but the quantitative significance of this and other intestinal actions of the barbiturate group remains controversial and most authorities now prefer to follow Maxwell *et al.* (1972) in accepting hepatic microsomal enzyme induction, which accelerates the degradation of folic acid, as the principal cause of its deficiency in patients taking these drugs.

Drug interactions in the gastrointestinal tract

Although it has as yet had little impact on clinical practice, there is already an extensive literature on the effects which drugs simultaneously ingested may have upon one another at any stage up to and including the process of absorption (see also Chapter 39). Few of the investigations published cover more than two selected drugs, yet in the average physician's practice there are many patients on three or more. Some idea of the task confronting workers in this field may be gained from the finding in a study of 277 patients attending an anticoagulant clinic that no less than 117 were receiving at least two drugs other than the anticoagulant itself and 26 a total of six or more (Williams *et al.* 1976).

Pancreatitis

A critical review of the English-language literature on drug-induced pancreatitis (Mallory and Kern 1980) classified treatments according to the strength of their relationship to acute attacks. When, as in most instances, the aetiology is seemingly multifactorial the role of any single drug may be hard to determine unless relapse occurs on challenge or support is provided by parallel cases on record. Sifting the cumulative evidence has helped to distinguish a group of drugs with which the risk is relatively high, but no real advance can be made until more is known of the mechanisms involved in pathogenesis. Here experimental work is likely to play the greatest part, although detailed observation of the changes attending clinical acute pancreatitis in individual patients has already suggested that some treatments may act indirectly by inducing a metabolic disturbance, such as hyperlipidaemia (Davidoff *et al.* 1973; Call *et al.* 1977) or hypercalcaemia (Manson 1974; Izsak *et al.* 1980), of which pancreatitis is a known complication. Meanwhile the possibility that current medication may be relevant to any given attack of acute pancreatitis should be considered with particular care whenever cholelithiasis or alcoholism, the commonest of associated pathologies, are absent, as in 203 (34.5 per cent) of the 590 cases collected by Trapnell (1972). It is therefore especially disappointing that, ignoring the growth of the literature on iatrogenic pancreatitis since that date, the compilers of an equally large but far more recent series (Corfield *et al.* 1985) failed to ascertain drug histories, even in the 23 per cent of cases for which they could find no cause.

Diuretics and sulphonamides

Diuretics were first implicated when Johnston and Cornish (1959) described acute pancreatitis in three patients who had been receiving chlorothiazide for 3–4 months. In one hyperglycaemia also occurred and this, like the pancreatitis, regressed after chlorothiazide had been withdrawn. These authors did not challenge patients with the drug to confirm its relationship with the pancreatitis, but chose instead two other means of exploring the association (Cornish *et al.* 1961). Firstly, they made serial measurements of serum amylase in 20 patients starting treatment with chlorothiazide, and found that it rose to as much as twice the original value in 10 of them, decreasing once again when the drug was stopped. No patient developed symptoms suggestive of pancreatitis. Secondly, they treated mice for 1–6 months with doses of chlorothiazide equivalent to 1 g daily for a 70-kg human, killing 50 animals each month. Twenty-two (7 per cent) of the mice contracted an acute pancreatitis, which was severe in seven animals. In a majority this could be appreciated only as microscopic inflammation of the parenchyma, but in some animals the pancreas was enlarged and hyperaemic. Wenger and Gross (1964) reported three cases of fatal necrotizing pancreatitis in patients who had been taking hydrochlorothiazide for several months, one of whom had in the past been alcoholic. Chlorthalidone, a sulphonamide derivative of phthalimidine rather than of benzothiadiazine, was associated with haemorrhagic pancreatitis in a patient known to suffer from severe atherosclerosis (Jones and Caldwell 1962), Although differing markedly in age from most others in this category, the first patient whose pancreatitis was ascribed to frusemide treatment, a 23-year-old alcoholic with malignant hypertension, must at least have shared with them a severe degree of degenerative vascular disease (Wilson et al. 1967). Since ischaemia has been suspected to account for many cases of 'idiopathic' acute pancreatitis in the elderly (Trapnell 1972) some caution must be exercised in ascribing a causative role to the diuretic in any arteriopathic patient whose medication happens to include one of these. Frusemide was, however, clearly implicated by challenge in a patient who developed a recurrence of both typical epigastric pain and hyperamylasaemia 5 days after the drug had been re-prescribed, and who returned to normal in both respects 36 hours after its final withdrawal (Jones and Oelbaum 1975).

An autopsy study of patients who had died from acute pancreatitis revealed hyperplasia of the parathyroid glands only in those who had been taking thiazide diuretics (Pickleman et al. 1979). Since none had had hypercalcaemia while alive, the parathyroid effect was thought to be a secondary phenomenon causally unrelated to the pancreatitis. That diuretic action as such may be important is suggested by the findings of Bourke et al. (1978) who took drug histories from 100 patients in their first attack of acute pancreatitis, each matched for age and sex with controls who had been admitted to hospital with acute abdominal pain and a normal serum amylase. The difference between the pancreatitis patients and controls was almost entirely accounted for by takers of cyclopenthiazide with potassium chloride and of frusemide, especially the former.

A report of acute pancreatitis closely related to ingestion of both sulphasalazine and phthalysulphathiazole in a single patient (Block et al. 1970) lends further support to the existence of a specific effect, either toxic or allergic, of the sulphonamide grouping.

Oral contraceptives

Bank and Marks (1970) reported the cases of two women who developed severe abdominal pain and vomiting in the first and second cycles respectively of treatment with oral contraceptive tablets combining an oestrogen with a progestogen. Pancreatitis was confirmed in both at laparotomy. A similar clinical pattern, including relapse after inadvertent challenge with the Pill in one instance, was seen in the two cases of Davidoff et al. (1973); and in each of these studies a clue to the mechanism involved was given by the lipaemic appearance of serum taken for laboratory tests. It had been appreciated first by Klatskin and Gordon (1952) that pancreatitis could occur as a complication of hyperlipidaemia, and Havel (1969) had documented the association of pancreatitis with familial Type I hyperlipidaemia, suggesting that the necrosis might follow micro-infarction due to embolism of the smallest vessels by chylomicrons. Bank and Marks (1970) found that one of their patients remained hyperlipidaemic after stopping the Pill, although less so than while she was taking it, and that two of her brothers also had hyperlipidaemia. In Case 1 of Davidoff et al. (1973) the pancreatitis regressed when fat restriction was imposed and relapsed as soon as the patient resumed a normal diet. Both this woman and her mother had a Type IV electrophoretic pattern. The second patient displayed a heavy pre-β band which disappeared on carbohydrate restriction. Since the chylomicron count was negligible in both their patients these authors discounted the concept of microembolism and postulated that the pancreatitis must be related simply to the quantitative increase in triglyceride levels. Glueck et al. (1972) revived an earlier speculation that exposure of excess triglycerides to lipase in the pancreatic circulation might lead to a release of free fatty acids in such concentration as to damage the capillary walls, and thus initiate a process of self-destruction through ischaemia. That some other mechanism may operate in certain cases is suggested by the report of Mungall and Hague (1975) whose patient had normal levels of cholesterol and triglyceride in the serum while receiving oestradiol with megestrol acetate, yet died of pancreatic necrosis.

Call et al. (1977) described a case of pancreatitis in a woman who developed Type V hyperlipoproteinaemia when treated with oestrogen (Premarin) for menopausal symptoms. She was, however, also receiving frusemide, and when challenged subsequently by each of these drugs in turn, experienced pain with a raised serum amylase only in response to the frusemide. It may be speculated that the lipid disorder had in some way 'primed' the pancreas for the action of frusemide.

Corticosteroids

Zion et al. (1955) reported a fatal case of acute-on-chronic pancreatitis in a woman with scleroderma who had been treated for five months with cortisone. Cortisone had also been the principal medication of two children who developed acute pancreatitis (Baar and Wolf 1957), and in another child ACTH therapy had preceded the pancreatitis (Marczynska-Robowska 1957). Carone and Liebow (1957) studied the pancreas at autopsy in 54 patients who had received ACTH or corticosteroids and compared the appearances with those in an equal number of age-matched patients who had not been treated with corticosteroids. Sixteen of the test group showed histological evidence of either acute pancreatitis or fat necrosis, or both; the lesion being severe in three, moderately severe in seven, and mild in six. Gross pathology was present in six cases although none of these patients had complained of symptoms suggesting pancreatitis. Among the controls there were two cases of acute pancreatitis, one in association with polyarteritis nodosa. Microscopically the acini were dilated, with flattening of their lining cells, in 59 per cent of corticosteroid-treated patients. Of the controls only 24 per cent showed this change, and then to a lesser degree. Such appearances in the pancreas had hitherto been associated with severe infection elsewhere, which was evident in 39 of the test group but only 19 of the controls. Focal inflammation with a mixed leucocytic infiltrate was present, mainly in the interstitial rather than the parenchymal tissue. There were no important vascular abnormalities.

Carone and Liebow (1957) noted that pancreatitis developing in rabbits given corticosteroids by Lazarus and Bencosme (1956) and by Stumpf et al. (1956) had shown similar microscopic features. The latter had also been able to demonstrate that acinar dilatation under these conditions was associated with increased viscosity of the pancreatic juice, a finding later confirmed in dogs by Nelp et al. (1961). In favour of a role for corticosteroids in the aetiology of clinically significant acute pancreatitis is the fact that many of the patients developing this while on corticosteroids have been children, in whom it is otherwise rare (Riemenschneider et al. 1966). However, Steinberg and Lewis (1981) drew attention to the conflict between findings that point to drugs of this class as causative and those that demonstrate their efficacy in its treatment, judging that the latter carry more conviction.

Among possible contributary factors must be considered the disease for which the corticosteroids have been prescribed, and other drugs used in its treatment. Most patients have been debilitated by potentially lethal disorders, such as the disseminated lupus erythematosus (DLE) present in four of the six children described by Riemenschneider et al. (1966). This particular association is lent additional significance by a report of acute pancreatitis occurring on two separate occasions as part of a drug-induced DLE syndrome in a patient at first treated with procainamide and then taking it again inadvertently (Falko and Thomas 1975). Kaplan and Dreiling (1977) reviewed cases in which corticosteroids had been implicated in the aetiology of acute pancreatitis, and concluded that in many instances the underlying disease or other drugs might have been more to blame. Synergy between drugs is suggested by the same group's finding that hydrocortisone given with azathioprine altered exocrine pancreatic function in dogs; when azathioprine was used alone it had had no such effect (Dreiling and Nacchiero 1978).

A dangerous feature of acute pancreatitis associated with corticosteroid treatment is the likely delay in diagnosis occasioned by the ability of these drugs to suppress symptoms and by their reputation as a frequent cause of upper abdominal pain with connotations far less sinister.

Cytotoxic drugs

Nogueira and Freedman (1972) reported the case of a woman receiving prednisone for Crohn's disease who developed severe upper abdominal pain with nausea and a serum amylase level of 410 Somogyi units (SU) 3 weeks after azathioprine had been added to her regime. Symptoms regressed when the prednisone was given in higher dosage and the azathioprine discontinued. Subsequent challenge with azathioprine led to a recurrence of pain and increased the serum amylase level to 1026 SU. A similar case was described by Kawanishi et al. (1973) who commented on the rapidity with which symptoms recurred after azathioprine challenge, suggesting an immune mechanism. Their patient was also receiving sulphasalazine, which failed to provoke symptoms on challenge. Azathioprine used as the sole anti-inflammatory medication for Crohn's disease in a controlled trial has been associated with six cases of acute pancreatitis among 113 patients treated (Singleton et al. 1979).

A mild attack of pancreatitis occurred after several days' treatment with 6-mercaptopurine in one patient among 58 who were receiving this drug for Crohn's disease (Present et al. 1980). Acute pancreatitis attributable to azathioprine or its metabolite, 6-mercaptopurine, is very rare outside the USA, but one case has been reported in the UK involving azathioprine treatment of a patient with rheumatoid arthritis (Pozniak et al. 1981).

Using the dog pancreas as an experimental model Broe and Cameron (1983) found that azathioprine, added to the perfused blood at up to 5 mg/kg of the animal's weight, had no effect on the gross appearance of the pancreas or on serum amylase levels, but doubled the volume of its overall secretion and bicarbonate output, while profoundly depressing trypsin output.

Of 43 children receiving L-asparaginase for acute leukaemia, seven (16.2 per cent) showed evidence of acute pancreatitis at various stages up to 10 weeks after cessation of therapy (Weetman and Baehner, 1974). Onset of pancreatitis from this cause may be insidious clinically, and any clinical suspicion may then be lulled by a relative hypoamylasaemia also due to L-asparaginase. Such were the circumstances in which the diagnosis was missed until autopsy in the case described by McLean et al. (1982). In one series of 303 therapeutic trials of L-asparaginase, there were no episodes of overt pancreatitis, but autopsy revealed foci of pancreatic inflammation and necrosis in six out of 37 patients who had died within 2 weeks of completing a course (Oettegen et al. 1970). Pancreatitis as a side-effect of treatment with L-asparaginase is less likely to occur with a continuous low-dose regimen than with high doses given at longer intervals (Jaffe et al. 1972).

For some years L-asparaginase stood alone among the drugs used in acute leukaemia as a cause of acute pancreatitis, but now others are implicated with varying degrees of certainty. Altman et al. (1982) reported the case of a 14-year-old boy with acute lymphocyctic leukaemia who had suffered non-specific abdominal pains during a course of L-asparaginase and then, while receiving cytosine arabinoside some 4 months after his last dose of L-asparaginase, developed a well-authenticated attack of pancreatitis. At least one further attack followed when the treatment was resumed. Murphey and Josephs (1981) diagnosed acute pancreatitis in two leukaemic patients who had received pentamidine isethionate in daily dosages of 220 mg and 240 mg respectively for intercurrent *Pneumocystis carinii* pneumonia. In one of these cases the clinical picture was highly suggestive, but laboratory support for the diagnosis was confined to an increase in the amylase–creatinine clearance ratio; while in the other a mildly increased level of serum amylase, recurrent on challenge, constituted the only evidence.

Phenformin

Graeber et al. (1976) described the cases of two elderly diabetic patients receiving phenformin in whom acute pancreatitis followed the rebound alkalosis resulting from treatment of life-threatening lactic acidosis with intravenous sodium bicarbonate. Nothing in their analysis of these cases or in their review of several others previously reported gave any hint as to what the mechanism might have been.

Valproic acid

Two cases of acute pancreatitis in children receiving valproic acid, one authenticated by a recurrence of typical symptoms and hyperamylasaemia after challenge (Batalden et al. 1979) and the other by laparotomy findings (Camfield et al. 1979), were the subjects of almost simultaneous 'first reports'. From these, together with another probable case described by Camfield et al. and a fourth recorded by Sasaki et al. (1980), it would seem a dosage of at least 25 mg per kg weight per day is needed to induce pancreatitis, and that an induction period of several months is the rule. Despite this unusually slow evolution, the fact that other children may take up to 65 mg per kg body weight of valproic acid daily without developing even biochemical evidence of pancreatitis, together with the more rapid onset of a second attack at a lower challenge dose, suggest that the causative mechanism may involve some form of hypersensitivity.

Cimetidine

The original report of a case in which cimetidine was blamed for precipitating acute pancreatitis (Arnold et al. 1978) provoked a reply (Dammann and Augustin 1978) which gave details of its use in the treatment of 11 patients with alcoholic pancreatitis and concluded that at least it had not impaired the effect of the other measures employed. However, Meshkinpour et al (1979) undertook a formal trial of cimetidine against placebo in acute pancreatitis and found that hyperamylasaemia persisted slightly longer in the cimetidine group. Wilkinson et al. (1981) reported two cases in which patients who developed acute pancreatitis while taking cimeditine, were later challenged with the drug, and relapsed within 24 hours.

The study of Broe et al. (1982), in which 52 of 116 patients presenting with pancreatitis received cimetidine at 1200 mg daily as part of the treatment, while 64 did not, was unable to distinguish between the subsequent course of the disease in the two groups, suggesting that any damage caused by cimetidine may represent an idiosyncratic response confined to a very small proportion of the patients who take it.

Tetracycline

Until recently the only good evidence for acute pancreatitis as a complication of tetracycline treatment related to its association with fatty liver, usually in pregnant women. Elmore and Rogge (1981) have now described two attacks of clinical pancreatitis occurring without liver disease in a 13-year-old girl given tetracycline at 1–2 g daily for severe cystic acne. On the second occasion ultrasonic scanning showed the head of the pancreas to be swollen.

Sulindac

This drug was associated with acute pancreatitis in three cases (Goldstein *et al.* 1980; Siefkin 1980; Lilly 1981). That of Siefkin involved a patient who relapsed within 3 weeks of starting a second course and recovered only when sulindac was again withdrawn. The patient described by Lilly had been taking both hydrochlorothiazide and an oestrogen for some years before sulindac was prescribed at 200 mg b.d. Typical acute pancreatitis followed 3 weeks later, and twice again on challenge. Although the patient remained tolerant to her other drugs some form of synergy may well have been responsible for this case.

Methyldopa

The relationship between methyldopa treatment and acute pancreatitis in a single case was carefully documented by Rominger *et al.* (1978) who were able to demonstrate a marked rise in temperature and serum levels of both amylase and lipase within 12 hours of challenge. Subsequent reports incriminated methyldopa as the cause of pancreatitis in a patient with membranous glomerulonephritis (Warren *et al.* 1980), who might therefore have been exposed to unusually high levels of the drug, a patient shown to have gallstones, and a third in whom no other contributory factor could be identified (Van der Heide *et al.* 1981) The two latter relapsed clinically within hours of challenge, when their serum amylase levels again rose sharply.

One case of chronic pancreatitis was reasonably ascribed to methyldopa (Ramsay *et al.* 1982).

Miscellaneous precipitants

Although as yet, it would appear, recorded at no more than one centre apiece, relationships between acute pancreatitis and the following treatments have been satisfactorily established: clonidine (Amery *et al.* 1973), mefenamic acid (Van Walraven *et al.* 1982), and nitrofurantoin (Nelis 1983).

Three cases of acute pancreatitis occurring within hours of translumbar aortography were reported by Imrie *et al.* (1977), but two at least were in subjects rendered susceptible by biliary disease and alcoholism respectively, and neither a suggestive clinical picture nor a rise in serum amylase was found in any of 50 further patients screened prospectively. In two cases described by Larsen *et al.* (1962) haemorrhagic pancreatitis was attributed to inadvertent overdosage with warfarin, but each patient was also receiving a thiazide diuretic.

Parenteral nutrition

When Manson (1974) treated a patient with parenteral nutrition for enterocutaneous fistula there developed a hypercalcaemia which went unnoticed for several weeks until she complained of severe upper abdominal pain. At that time serum calcium was reported as 4.0 mmol l^{-1} and serum amylase as 948 SU. A laparotomy revealed fat necrosis of the pancreas, and clinical recovery followed adjustment of the serum calcium level. This effect of hypercalcaemia was presumed to derive from a known action of Ca^{2+} ions in accelerating the activation of trypsinogen by trypsin.

Izsak *et al.* (1980) described simultaneous hypercalcaemia and hyperamylasaemia in six further patients receiving total parenteral nutrition (TPN). Gallstones, alcohol, and likely drugs were excluded as contributory factors in these cases. Three of the patients had developed hyperlipidaemia during lipid infusions, but their pancreatitis recurred after serum lipid levels had returned to normal. The cause of the hypercalcaemia remained obscure in some cases, but solutions containing vitamin D were incriminated in others. Acute pancreatitis is a particularly serious complication for patients who are already so ill as to require TPN, and two of these patients died from it.

Surgical procedures

Indications that cardiac surgery might be complicated more frequently than other types of surgery by acute pancreatitis were first published by three separate centres in 1968. In a later series eight patients were found to have pancreatitis at autopsy and 18 had serum amylase levels of greater than 150 SU at some time during the first 5 post-operative days (Panebianco *et al.* 1970). In order to evaluate the role of cardiac bypass Hennings and Jacobson (1974) compared serum amylase levels in patients who had required bypass for cardiac surgery, finding values of 400–3300 SU in nine out of 12 cases, with those in patients undergoing pulmonary surgery without bypass, which remained normal in all 22 cases. Ferrier (1976) studied the pancreas in 182 autopsies following cardiac surgery and found pancreatitis in 29 cases, an incidence of 16 per cent. Mild and moderate forms were seen as a peripheral necrosis spreading outwards to involve surrounding fat. Severe inflammation destroyed the basic glandular morphology, causing haemorrhagic necrosis as the most striking feature. Because the earliest parenchymal damage occurred peripherally Ferrier (1976) favoured ischaemia as the primary mechanism of this pancreatitis. Nine cases gave microscopic evidence of thromboembolism, but whether this had originated with platelet thrombi forming on the valves or apparatus, disseminated intravascular coagulation, haemoconcentration, splanchnic vasoconstriction, or a multiplicity of such factors could not be determined.

Sorrell *et al.* (1975) pioneered a new end-to-end (10″–10″) jejuno-ileal bypass operation in the treatment of refractory obesity. Of their first 16 patients three developed acute pancreatitis at intervals of 1–6 months post-operatively. In all these patients the gall-bladder had been apparently normal at the time of operation, and one had had a normal cholecystogram; yet when pancreatitis was diagnosed all had non-functioning gall-bladders, two of which contained multiple calculi. It was speculated that the disturbance of enterohepatic bile-salt circulation created by this operation had rendered the bile lithogenic, but experiments with a similar operation in dogs reduced the bile-salt pool without causing pancreatitis.

Surgery has also been implicated as a cause of acute pancreatitis in patients undergoing renal transplantation. This association is discussed below.

Because emergency laparotomy must compound the dangers of acute pancreatitis, the most important aspect of management in all such cases as those described above is an awareness of iatrogenic pancreatitis that will suggest the diagnosis. Estimation of serum amylase should then confirm it, so that the appropriate conservative measures may be instituted without delay.

Chronic haemodialysis and renal transplantation

In order to assess the extent to which treatment of chronic renal failure may damage the alimentary tract some knowledge is required of abnormalities that may be attributed to the uraemic state itself. Now that metabolic disturbance can be minimized by dietary measures alone few patients remain truly untreated, and it is only from a study of the older literature that the entire spectrum of digestive disease among such patients may be appreciated (Jaffé and Laing 1934; Thoroughman and Peace 1959). Since, however, the value of this knowledge depends solely on the diminishing number of patients exposed to the full consequences of chronic renal failure it should suffice to record here that in uraemia the principal lesion is a form of ischaemic enterocolitis traceable to vascular degeneration and having its most severe effect in the colon, probably because of the potential for secondary bacterial infection at this site, whereas the greatest danger for a treated patient lies in upper gastrointestinal ulceration.

Peptic ulcer

Understanding of the mechanisms involved in the much increased risk of peptic ulcer which seems to follow the institution of chronic haemodialysis or a transplant operation remains fragmentary, although advances have been made in recent years. Certainly it is now recognized that the concept of chronic renal failure as depressing gastric acid secretion (Lieber and Lefèvre 1959; Gingell *et al.* 1968) and of treatment as enhancing it (Venkataswaran *et al.* 1972) is an over-simplification.

Fillastre *et al.* (1965) studied 33 patients with end-stage renal failure whose mean total acid response to histamine stimulation proved to be significantly higher than that of a normal control group. Thirty-four of the 56 patients tested with pentagastrin by Gordon *et al.* (1972) gave a response within the normal range, and in females mean peak acid output was nearly double that for healthy controls. The only patient whose response changed after dialysis showed a decrease to normal from high values. McConnell *et al.* (1975) detected gross impairment of acid secretion in only 10 out of 25 patients before haemodialysis, and noted that only four of these swung over to a state of excess secretion after treatment. Serum gastrin levels were higher than normal both before and after haemodialysis, and they believed that this must induce a tendency to increased acid secretion in many patients with chronic renal failure, the effect of which, however, might be largely masked before treatment by a factor in uraemia toxic to gastric mucosa. An alternative masking mechanism postulated by Lieber and Lefèvre (1959) is the direct neutralization of gastric acid by excess ammonia in the uraemic stomach. Hällgren *et al.* (1979) found among patients receiving haemodialysis a similar variation of stimulated acid response to that observed by McConnell *et al.* (1975), but with seemingly opposite changes in serum gastrin levels; for the patients in whom these proved to be highest were achlorhydric, and 75 per cent of the patients in whom they were normal or only moderately elevated produced an excess of acid. I.L. Taylor *et al.* (1980) discussed the discrepancies in gastrin data, stressing the relevance of molecular size to activity and clearance and concluding that changes in gastrin secretion are unlikely to be responsible for the increased incidence of duodenal ulcer in chronic renal failure. Doherty and McGeown (1977) reported that mean gastric acidity was slightly, though non-significantly, higher in patients with chronic renal failure than in normal controls, but that the distribution of individual values was also much wider than normal, giving substantial numbers of both hyper- and hyposecretors. After haemodialysis the means of both basal and peak acid output increased significantly (Doherty 1978). High overnight secretion of acid seems to occur both before and after haemodialysis (Dekkers *et al.* 1971; Shepherd *et al.* 1973).

Most workers have found the acid response to be potentiated by renal transplantation, and there is evidence that the transition from haemodialysis may be accompanied by an increased incidence of peptic ulceration (Gingell *et al.* 1968; Chisholm *et al.* 1977). In Belfast, however, neither basal nor peak acid output rose further after transplantation, and there was in fact a significant fall in peak values with lapse of time after operation (Doherty 1978).

Both endoscopically and histologically the abnormalities in the gastroduodenal mucosa which follow transplantation closely resemble those to be found in patients on chronic haemodialysis (Franzin *et al.* 1982*a*, *b*; Milito *et al.* 1983). Hyperplasia of parietal cells is accompanied by hypertrophy and distal spread of the acid-secreting mucosa, with a high incidence of inflammatory change, especially in the duodenum. A lesion as yet described only in patients receiving haemodialysis is focal angiodysplasia of the gastric submucosa (Cunningham 1981). Before as well as after transplantation the frequency of ulceration correlates well with hypersecretion (Gingell *et al.* 1968; Dekkers *et al.* 1972; Milito *et al.* 1983).

Peptic ulceration in recipients of renal allografts tends to be multifocal (Blohmé 1975) and to present with massive haemorrhage demanding immediate surgical intervention. Despite early and often heroic surgery few of these patients recover. The majority develop their ulcers in the course of rejection (Aldrete *et al.* 1975; Berg *et al.* 1975; Meech *et al.* 1979) and therefore at a time of particularly intensive immunosuppression, which not only adds defective healing to a severe metabolic disadvantage but may also be responsible for the ulceration itself. In children the incidence of peptic ulcer is probably no less but its prognosis is far better (Schnyder *et al.* 1979).

Upper gastrointestinal haemorrhage after transplantation is most likely to occur in patients with a past history of gastric or duodenal inflammatory disease. Analysing a series of 620 renal allografts performed in Gothenberg, Blohmé (1975) found that gastroduodenal complications developed in 8 per cent of patients who had no such history, in 19 per cent of those who had had a peptic ulcer, and in 48 per cent of those with a history of 'gastritis'. Eleven patients among 109 studied by Owens *et al.* (1976) had had previous peptic ulceration and seven of these suffered a recurrence after transplantation, while the remaining 96 patients included only 10 destined to develop peptic ulcer post-operatively ($p < 0.0002$).

Some authorities have concluded that prophylactic surgery should be performed in patients selected for transplantation who are known to have had a peptic ulcer (Spanos *et al.* 1975; Owens *et al.* 1977). In 19 of 30 such patients subjected to vagotomy with antrectomy or pyloroplasty by Spanos *et al.* (1975) there were only three whose transplantation was complicated by peptic ulceration, whereas bleeding occurred either before or after transplantation in nine of the remaining 11. Many units, however, have opted for conservative measures in the hope that peptic ulcers will be among the lesions affected by changing patterns of immunosuppressive therapy (discussed later) and on the basis of the success claimed for prophylaxis with H_2-receptor antagonists in most trials to date. Ulcers encountered during haemodialysis have been cured with low doses of cimetidine (Jones *et al.* 1979). Doherty and McGeown (1978) described a programme of routine upper gastrointestinal assessment before transplantation which revealed evidence of chronic peptic ulcer in 34 out of 131 (26 per cent) patients. Only one out of 13 such patients given either conventional antacids or cimetidine developed bleeding after transplantation. An alternative to this selective approach giving equally good results is routine prophylaxis with cimetidine, which at King's College Hospital enabled 93 consecutive transplants to be performed without one case of haemorrhage (Jones *et al.* 1978; Rudge *et al.* 1979). At a Belgian centre mortality from peptic ulcer had been decreased by prophylactic surgery but seemingly abolished by the use of cimetidine (Lerut *et al.* 1980). Elsewhere results have been less dramatic, although in one large series cimetidine appeared to have reduced the need for blood transfusion and surgery and also the mortality among patients who presented with bleeding (Garvin *et al.* 1982). In contrast to the apparent efficacy of cimetidine reported in open trials, a double-blind comparison of this drug with placebo (Schiessel *et al.* 1981) was unable to distinguish between the treatment groups in respect of either peptic ulceration or upper gastrointestinal bleeding. Another study likewise failed to show any advantage of cimetidine over other antacids in the prevention of ulcers but, curiously, a slight enhancement of graft survival in patients receiving cimetidine (Stuart *et al.* 1981).

Thus a high incidence of peptic ulceration is recognized universally as the commonest and perhaps the most dangerous of gastrointestinal sequelae to renal transplantation. Conflicting data on changes in acid secretion that may follow the onset of uraemia or its treatment suggest that among patients with chronic renal failure, as in less selected populations, an excess of acid is only one factor to be considered and probably relevant only to duodenal ulcer. The additional stress of major surgery might be expected to increase the tendency to post-operative ulceration in any group of patients sharing a chronic metabolic disturbance, and the special features of acute peptic ulceration in this group engender a suspicion that the high rate of cytomegalovirus (CMV) infection associated with conventional immunosuppressive therapy may be responsible for a further significant proportion of cases.

Oesophagitis

Few series contain examples of oesophagitis following renal transplantation, but in that of Owens et al. (1976) this entity accounted for over one-fifth of all side-effects involving the alimentary tract. Although reflux of excess gastric acid must play a part, fungal infection encouraged by immunosuppression is probably more important, and it is notable that Hadjiyannakis et al. (1971) made this diagnosis in all their cases of oesophagitis. Jones et al. (1982) were unable to eradicate oesophageal candidiasis in two diabetic patients whose infection became disseminated after the treatment of rejection episodes, and proved fatal in both instances.

Colonic ulceration and perforation

A discrete form of colonic ulceration, probably identical to that recorded in the early literature on gastrointestinal consequences of largely chronic renal failure, is restricted to patients receiving long-term haemodialysis (Mills et al. 1981; Huded et al. 1982). It occurs chiefly in the right colon, may be single or multiple, presents with pain and bloody diarrhoea, and is histologically non-specific. Intestinal perforation following renal transplantation is seen most frequently in the colon, usually in diverticula where some predisposing infection is to be expected and where the wall lacks its muscle coat. Of 13 colonic complications encountered by Bernstein et al. (1973) no less than five were of this nature. The mortality of the consequent peritonitis has been particularly high in these patients, most of whom have been under treatment for graft rejection. Powis et al. (1972) lost all five of their patients, in whom it appeared that the perforation had complicated clinically silent ischaemic colitis. Occlusion of the mural vessels with atherosclerosis, fibrinoid necrosis, or thrombosis was, however, inconstant in this series and totally absent in three cases of bowel perforation and two of extensive necrosis reported by Demling et al. (1975), who therefore favoured immunosuppression as the principal causative factor. These authors ascribed the survival of two of their patients to the fact that corticosteroids were stopped as soon as the diagnosis had been made. The detailed analysis of Thompson et al. (1975) provided further evidence to incriminate cortico-

steroids specifically in the aetiology of post-transplant perforation. Since perforation of colonic diverticula is as avoidable as it is serious some have advocated routine barium-enema examination in all patients over 40 years old for whom renal transplantation is contemplated, and, where feasible, prior resection of areas of diverticulosis (Bernstein et al. 1973; Hognestad and Flatmark 1976).

Occasional cases of colitis have been mostly of the pseudomembranous variety arising in the course of graft rejection. One patient described by Aldrete et al. (1975) recovered after the graft had been resected and his health had been restored by haemodialysis. Other cases reported by Penn et al. (1970) and by Bernstein et al. (1973) were associated with systemic sepsis. Perloff et al. (1976) published four case histories of their own and reviewed 11 others drawn from five major series covering 1156 transplantations. All were characterized by deteriorating renal function, and the mortality was 90 per cent. In view of the pathological appearances antibiotics were among the aetiological factors discussed, but the authors were satisfied that ischaemia was the essential cause in every instance.

Margolis et al. (1977) identified ischaemia as the cause of nearly half the major gastrointestinal complications they had met, and argued that in their cases hypotension during the surgery itself or during post-operative haemodialysis was a more likely precipitant than any of the factors more commonly incriminated. A special hazard favouring colonic perforation in older patients, presumably by an ischaemic mechanism, is corticosteroid-induced diabetes (Bailey et al. 1971).

Intestinal obstruction

A few patients have required laparotomy for intestinal obstruction, which has proved to be due either to massive right-sided faecal impaction (Penn et al. 1970; Bernstein et al. 1973) or to adhesions (Aldrete et al. 1975). Some degree of intestinal hypotonicity is thought to result from the uraemic state (Adams et al. 1982) and, among other contributory factors, the trauma of a major operation and the constipating effect of the antacids these patients sometimes routinely receive would seem the most important.

Pancreatitis

Since the first report of acute pancreatitis following renal transplantation by Starzl (1964) most centres have recorded several instances. Johnson and Nabseth (1970) established the worldwide incidence of pancreatitis in allograft recipients at 2 per cent, and found this complication to be fatal in half the cases. In some individual series the incidence has been substantially higher: the figure of 6 per cent at the University of Utah (Renning *et al.* 1972) being attributed partly to surgical trauma at nephrectomy, and that of 5.6 per cent at Denver (Penn *et al.* 1972) to severe associated disease such as septicaemia and hepatitis. So many of the factors discussed above as contributing to the aetiology of acute pancreatitis are present in allografted patients that it is perhaps surprising not to find it reported more frequently. Hill *et al.* (1967) noted that many such patients were discovered at autopsy to have a non-haemorrhagic pancreatitis of which clinical features had been absent. As with other gastrointestinal complications acute pancreatitis appears to have a predilection for the patient whose graft is either infected or undergoing rejection.

Thus high doses of corticosteroids and azathioprine are almost always being taken; and other factors considered relevant to individual cases include the hypercalcaemia of post-transplant parathyroid hyperplasia (Tilney *et al.* 1966; Penn *et al.* 1972; Fernandez and Rosenberg 1976), the virus of hepatitis (Penn *et al.* 1972; Fernandez and Rosenberg 1976), cytomegalovirus (CMV) (Tilney *et al.* 1966; Van Geertruyden and Toussaint 1967; Parham 1981), ischaemia from hypertensive or autoimmune arteriolar disease (Tilney *et al.* 1966; Woods *et al.* 1972), uraemia (Van Geertruyden and Toussaint 1967), alcoholism (Woods *et al.* 1972), and gallstones (Penn *et al.* 1972).

Role of viral infection

The relatively late emergence of infection as the most potent threat to success in renal transplantation may be due partly to an early concentration of research upon the effort to prevent rejection, but owes much also to increasing awareness that the presence of viral inclusion bodies in recipient tissue might be relevant to the outcome. Both clinical and serological criteria of infection by CMV, the virus most frequently identified, appear uncertain (Neild and Southee 1979), but the epidemiological evidence for transmission via the graft (Fiala *et al.* 1975; Betts 1982) has received support from identification of the same strain in both recipients of a single donor's kidneys (Wertheim *et al.* 1983). The fact that one of these patients responded with complement-fixing antibodies and remained well, whereas the other's fatal infection had become clinically severe before any titre could be detected, would seem to bear out the postulate of Simmons *et al.* (1977) that it is the individual's ability to mount an immune response which determines the significance of CMV infection. The stomach is a site where the severity of CMV infection following renal

transplantation may vary from multiple ulceration causing potentially lethal haemorrhage (Diethelm *et al.* 1976) to the appearance of inclusion bodies in milder forms of gastritis, correlating wth post-operative seroconversion (Franzin *et al.* 1981). CMV has also been cultured from ulcers of the colon, perforating in at least one case (Aldrete *et al.* 1975; Lerut *et al.* 1980).

Lymphoma

Lymphomata ascribed to transplantation and immuno-suppression may occur in the bowel either as an isolated lesion or as part of a multisystem process. Described variously as 'plasmacytoma', 'reticulum-cell sarcoma', 'large-cell sarcoma' or 'immunoblastic sarcoma', they probably represent a single entity which may arise in lymphoid tissue at any point in the gastrointestinal tract. The association of the first reported case with hepatitis (Kuster *et al.* 1972), three others with CMV infection (Pinkus *et al.* 1974; Calne *et al.* 1979), and a fourth with herpes-type virus particles (Jamieson *et al.* 1981) would be consistent with induction by viruses. Moreover, the relatively brief latent period of a few months between operation and onset of a febrile illness, with weight loss, anaemia and perhaps frank haemorrhage, may suggest infection, and the correct diagnosis has rarely been made in life. Haemorrhagic gastric erosions in the case of Hara *et al.* (1979), recalling those in which others have found CMV inclusion bodies, proved at autopsy to be representative of a multifocal infiltration in which ileum, caecum, and ascending colon were also involved. Stomach with duodenum (Calne *et al.* 1979; Jamieson *et al.* 1981), jejunum (Calne *et al.* 1979; Kašliková *et al.* 1981), ileum (Kuster *et al.* 1972) and colon (Pinkus *et al.* 1974; Coggon *et al.* 1981) have all been recorded as sites of more localized disease. A report from Starzl *et al.* (1984) offers hope that a proportion of such 'lymphomata' may regress without rejection of the graft when immunosuppressive treatment is withdrawn.

Changes in immunosuppressive policy

The intimate relationship between prophylaxis against renal transplant rejection and infective complications, which now appear to account for most of the morbidity and mortality of transplantation, has served to broaden the appeal of schemes for reducing the dosage of immunosuppressive drugs. The first published attempt to eliminate gastrointestinal complications of renal transplantation by manipulating immunosuppresive therapy was that of Siegel *et al.* (1972). Their aim was to give the corticosteroid intermittently, and of the first 28 cases started on one daily dose of prednisolone, 18 were successfully converted to alternate-day dosage at an early stage. After 823 patient-months there had been 23 rejection reactions, none of which had proved irreversible, and no gastrointestinal complications. McGeown and colleagues (1980) in Belfast instituted low prednisolone dosage as long ago as 1968, between which year and the writing of their report 151 transplants were followed from the first postoperative day with oral prednisolone at 20 mg together with azathioprine at 3 mg per kg body weight. Of 42 grafts lost only 17 failed through rejection, and infection was a contributory factor in only four of the 23 deaths. Gastrointestinal morbidity comprised: one death from gastric haemorrhage, one further haemorrhage severe enough to require transfusion, and non-fatal perforation of a duodenal ulcer in two patients.

Other centres have compared the outcome of existing high-dose regimens with that of reducing the corticosteroid component. In Birmingham Buckels *et al*. (1981) adopted McGeown's scheme, and in a year's trial of this against a regimen in which the starting dose of prednisolone was, at 75 mg daily, nearly four times as great, found no difference in graft survival and only one-fifth the mortality ($p < 0.025$). The incidence of non-renal complications in this study was not stated. Morris *et al*. (1982) did not abandon high-dose intravenous methylprednisolone, either routinely at the end of the first postoperative week or in the treatment of rejection. Even though their lower dose of prednisolone allowed a much higher cumulative intake than in Belfast or Birmingham, early infections were reduced by over 50 per cent. CMV infection was implicated in all of the high-dose but none of the low-dose deaths, and the combined incidence of peptic ulcer and colitis was double in the high-dose group. Another low-dose regimen adopted at Guy's Hospital (Papadakis *et al*. 1983) led to a decrease in infections other than those of the urinary and respiratory tracts, and it may be considered of special relevance to the alimentary tract that despite the relatively small numbers studied infections with CMV were significantly reduced ($p < 0.05$). Inflammatory lesions of the upper gastrointestinal tract, though few, numbered more in the low-dose group. Four intestinal perforations occurred among the 34 high-dose patients but none among the 33 on low dosage. In a retrospective survey of 367 renal allografts performed at Duke University, North Carolina, between 1965 and 1978 it was recorded that HLA identity of donor and recipient had been achieved 45 times (Meyers *et al*. 1979). Late gastrointestinal complications had occurred in 11 of the 32 patients who were given long-term maintenance treatment with corticosteroids and in only one of the 13 from whom this was withheld.

The preliminary results of a European multicentre trial in which 117 patients were treated with cyclosporin A as sole immunosuppressive for periods up to 11 months showed that none had developed gastrointestinal side-effects (European Multicentre Trial 1982). Two gastric ulcers, one duodenal ulcer, and one case of severe duodenitis among 115 controls on prednisolone with azathioprine were attended with serious consequences, including the death of one patient.

Irradiation injury

The clinico-pathological syndrome resulting from damage to the intestine by therapeutic irradiation has remained surprisingly constant in character over many years of technical innovation and appears to be affected little by the variations in approach required for the treatment of a diversity of intra-abdominal tumours. Most cases have followed radiotherapy, external or local, or both, to carcinoma of the uterine cervix; but other malignancies treated include carcinomata of the endometrium, ovary, testis, prostate, bladder, kidney, rectum, and stomach, together with lymphomata involving the intra-abdominal lymph nodes. In earlier days the chances of gastrointestinal symptoms developing after irradiation were increased by imperfect control of the dosage and its distribution, but now by the greater energy of the radiation used and by the longer periods of survival associated with improved techniques.

Damage to the bowel during therapy or immediately afterwards is a common occurrence. Of 44 children given whole-abdominal irradiation, mainly for lymphoma or renal tumours, 31 (70.5 per cent) developed vomiting and diarrhoea at the time (Donaldson *et al*. 1975). All of the nine patients treated by Wolloch *et al*. (1973), comprising eight with uterine carcinoma and one with hypernephroma, suffered from diarrhoea, tenesmus, and rectal bleeding, which settled spontaneously after completion of treatment. This account and most others follow closely that of Walsh (1897), who reported the first case of radiation diarrhoea in an X-ray worker a mere 18 months after the publication of Roentgen's discovery. Cases of diarrhoea persisting beyond this stage or of an acute necrotizing enteritis which may develop during treatment (Duncan and Leonard 1965) are rare.

Incidence

The incidence of chronic irradiation injury is low. Strickland (1954) believed that late symptoms arose in 2.4 per cent of patients who had received pelvic irradiation, and Schmitz *et al*. (1974) made an estimate of some 2.0 per cent in cases of uterine carcinoma. Much higher figures are given by others, such as DeCosse *et al*. (1969) who found an incidence of 11.6 per cent in patients treated for carcinoma of the cervix during a 15-year period at the University of Cleveland, and Palmer and Bush (1976) who noted late symptoms in 10 per cent of their patients in this category. Even in a later series of 400 patients treated for uterine carcinoma between 1974 and 1978 as many as 7 per cent developed serious intestinal injury (Cochrane et al. 1981). Although the incidence of small-bowel damage was estimated at less than 0.5 per cent overall at one centre (Poddar et al. 1982), another has claimed that a dosage of 5000 cGy or more brings 5 per cent involvement of small bowel (Morgenstern et al. 1977). Because small intestine injury may be clinically silent it is possible that the more marked discrepancies between series as to its frequency result more from differences in the diligence with which it has been sought than from other variables, such as dosage. Among the more recent surveys a slight predominance of small over large intestinal lesions is the rule (Cochrane et al. 1981; Kwitko et al. 1982).

Dosage and predisposing factors

A broad correlation between the severity and extent of bowel damage and the dose of radiation has been observed (Schmitz et al. 1974), but investigators have more often been impressed with the low dosage sufficient to induce damage in some patients and with the contrasting resistance offered by others to exceptionally high doses. The consequent speculation as to predisposing factors has ranged widely. Most authors have recorded how many of their patients had undergone previous abdominal surgery or had a history of pelvic inflammatory disease, either of which could produce adhesions to tether segments of normally mobile bowel within the field of irradiation. Graham and Villalba (1963) stated that all six of their patients with ileitis gave such a history. Loiudice et al. (1977) found that 75 per cent of patients with radiation enteropathy had had prior surgery as against 10 per cent of matched irradiated controls. Nussbaum et al. (1981) noted in their series a particularly strong correlation between recent surgery and irradiation injury. Adjuvant chemotherapy lowers the threshold for damage to the small bowel (Scarpello and Sladen 1977; Kwitko et al. 1982) as well as the upper alimentary tract. DeCosse et al. (1969) obtained an association with hysterectomy ($p < 0.025$) but none with pelvic inflammatory disease. They were more interested in the high prevalence of hypertension among their patients and by an almost equally significant association of diabetes mellitus and cardiovascular disease with late irradiation damage, since both the pathology and the natural history of the disease are such as to implicate ischaemia in its

pathogenesis. Of nine patients who presented 5 or more years after their course of radiotherapy, six were subject to advancing cardiovascular disease, and in one patient the symptoms of ileitis could be activated by stopping her digitalis and relieved by restarting it. In a case described by Wellwood and Jackson (1973) rectovaginal fistula developed simultaneously with manifestations of severe aortoiliac atherosclerosis. Potish et al. (1979) studied nine variables which they considered of possible relevance and discovered that most of the predictive power lay in the trio: number of previous laparotomies, thin physique, and hypertension.

Generalizing from the literature Schmitz et al. (1974) stated that if the dose of radiation is less than 4000 cGy, divided into several well-spaced treatments, the only damge to be expected is a mild inflammation with oedema, increased mucus production, and perhaps superficial ulceration, all of which should regress on completion of treatment. In doses of more than 4500 cGy given over short periods, there is a danger not only of more severe acute inflammation but also of chronic damage. The cases of proctosigmoiditis studied by Gilinsky et al. (1983) had received median summated doses in the range 5600–7200 cGy. Absolute dosage may not be as important as the mode of its administration, for the use of intracavitary sources has been associated with a higher incidence of both enteritis (Morgenstern et al. 1977) and intestinal lesions in general (Russell and Welch 1979).

It is generally agreed that the small bowel is more vulnerable to this type of injury than the large, perhaps because the turnover rate of its epithelial cells is only half as rapid (Tankel et al. 1965). Symptoms, however, are most commonly those of a proctocolitis, since the rectum, and to a lesser extent the sigmoid colon, is fixed at a site close to the majority of tumours treated; while the small bowel, being both mobile and highly motile, rarely presents any one segment within the zone of high dosage long enough to incur serious damage. From a study of 94 patients coming to laparotomy, Palmer and Bush (1976) were nevertheless able to demonstrate at least a mild mural fibrosis of the ileum in most of the patients who had presented with proctocolitis.

Pathology

The earliest gross change detectable in irradiated bowel at laparotomy is a degree of induration together with areas of serosal telangiectasia to which thin, filmy adhesions may be attached. The wall then becomes thickened, segments of stricture and dilatation develop, and the serosa takes on an opaque white appearance, alternating with areas of more marked telangiectasia. The mesentery becomes thicker and shorter, adhesions more dense. At any stage ischaemia may render the intestinal wall friable or frankly necrotic, so that perforation may occur or a fistula be formed if the inflamed surface is already adherent to another hollow viscus. The mucosa is oedematous, and later becomes fixed to the submucosa. Any ulceration may be discrete, when the intervening mucosa is relatively normal, or confluent, when such mucosa as may remain is heaped up a manner reminiscent of pseudopolyposis. Telangiectasia in the mucosa has been identified as the cause of severe chronic ileal bleeding (Taverner *et al.* 1982).

The kinetics and morphology of the microscopic changes of irradiation enterocolitis were established 44 years ago by the classic studies of Warren and Friedman (1942). They described oedema and deposition of fibrin in all layers of the wall, succeeded by hyaline change in the collagen bundles, and some increase of fibrous tissue. The characteristic hyaline thickening was most prominent in the submucosa and subserosa, and from the latter would spread to involve mesenteric tissue. The fibroblasts themselves were atypical, having abnormal branching processes and nuclei with clumped chromatin and one or more nucleoli. In the walls of arteries oedema would again be accompanied by hyaline change which, with hypertrophy of the muscle coats, tended to obliterate the lumen. Obliterative change was usually even more marked in the veins, while dilatation of lymphatics and veins provided the most striking vascular feature of the reaction, presumably being responsible for the chylous ascites occasionally found in these patients (Murray and Massey 1974; Hurst and Edwards 1979). The epithelial lining away from ulcerated areas showed ballooning of mucous cells and distortion of glands, with or without nuclear atypia. Later authors have described pyloric metaplasia, varying degrees of round-cell infiltration of the lamina propria, and depletion of lymphoid tissue. In the small intestine the villi become blunted, with disorganization of the epithelium and distortion of crypts.

Wiernik (1966) took jejunal biopsies after single doses of 800 cGy, noting a decrease in mitotic figures within 24 hours and subsequent destruction of all crypt cells. Over the next 11 days the villi became progressively more stunted as cells sloughed off from their tips into the lumen without being replaced by fresh cells from the crypts. If the intervals between treatments were not sufficient to allow complete recovery some broadening of the villi was still detectable over three months after radiotherapy had been completed. Trier and Browning (1966) obtained similar results, adding a detailed account of the inflammatory-cell exudate and including a description of ultrastructural changes in the nuclei, cytoplasmic organelles, and microvilli of the epithelial cells.

Clinical features of irradiation enterocolitis

Even in patients destined to suffer disabling symptoms at a later stage the acute phase is usually followed by an interval of apparently normal health, which may last as little as 2 months or as long as 30 years (DeCosse *et al.* 1969). Some patients then present with rectal bleeding only, and in most of these spontaneous recovery will follow (Gilinsky *et al.* 1983). In others there is a recurrence of diarrhoea with or without blood and sometimes with the characteristics of steatorrhoea, accompanied by varying degrees of abdominal distension, colicky pains, and weight loss. Occasionally the predominant symptom is constipation rather than diarrhoea, but as a rule this develops later, giving way before long to symptoms of partial or total obstruction in either the large or small bowel, or both. In some instances blood supply to a segment of bowel falls below the critical level, leading to a necrosis which, if localized, will result in fistulation or perforation and if more extensive may be followed by gangrene. Typical sigmoidoscopic appearances include a generalized proctitis (with a large shallow ulcer on the anterior wall of the rectum adjacent to the uterine cervix in cases where this was the target of therapy), and around the rectosigmoid junction, a stricture which may be impassable.

On barium-enema examination the colon shows a general narrowing, loss of haustration, irregularity, and stiffening, with or without short strictures. These appearances are non-specific, but given the patient's history the only important differential diagnosis is that of tumour spread, and there is a real danger that active treatment may be abandoned in a patient who remained fit for, say, three years after treatment unless irradiation colitis is considered and its evolutionary pattern appreciated. Small-bowel meal shows variable dilatation which may be the predominant change or may alternate with irregular segmental narrowing. Fistulae may be demonstrated by either type of contrast study. A full account of the radiological picture associated with irradiation injury is given by Mason *et al.* (1970).

The value of colonoscopy in resolving diagnostic problems that arise in proven irradiation colitis is demonstrated by the results of Reichelderfer and Morrissey (1980) who examined 13 patients over a period of eight years. The indication in each case was either haemorrhage or development of a radiological stricture, or both, between six months and 14 years after radiotherapy. Difficulties caused by fixation or narrowing of the bowel could sometimes be overcome by using a paediatric gastroscope (*sic*). Among eight cases examined for 'stricture', two showed no stricture (suggesting spasm as the cause of the radiological narrowing), two suspected of having a new primary carcinoma and one of a metastasis from the irradiated tumour were cleared of malignancy, and one had normal mucosa at the site of the stricture. In this case the report was one of probable extrinsic tumour compressing the bowel, a diagnosis later confirmed at laparotomy.

Malnutrition and malabsorption

Some patients become seriously malnourished as a result of anorexia, vomiting, diarrhoea, and loss of protein (with or without blood) from the inflamed mucosal surface. In addition damage to the small bowel may cause clinically significant malabsorption. Transient and mostly slight steatorrhoea has been reported as occurring in the early phase (Reeves *et al.* 1959; Tarpila, 1971), but malabsorption sufficient to affect the patient's health rarely becomes manifest within three months of treatment. Greenberger and Isselbacher (1964) found grossly defective vitamin B_{12} absorption and faecal fat excretion increased to a mean of 58 g daily in a patient who had been exposed to excessive irradiation several years previously. Although the authors favoured an exudative mechanism related to the characteristic histological abnormality of mucosal lymphangiectasia, the absence of hypoproteinaemia and the fact that the latest in a series of ileal resections had rid this patient of all macroscopic disease pointed rather to a loss of absorptive capacity, resulting chiefly from surgery, as the cause of malnutrition in her case. The possibility of exudation was, however, raised again in Case 1 of Tankel *et al.* (1965) when faecal fat output increased progressively in the presence of an almost normal mucosa until it outstripped the patient's intake. Wellwood and Jackson (1973) who reported steatorrhoea in six of their 38 cases, vitamin B_{12} malabsorption in five, and hypocalcaemia in two, were surprised by the degree of malabsorption in patients who were found at laparotomy to have little gross involvement of the small bowel. Failure of small intestinal function to correlate with the severity of microscopic change has also been noted (Tankel *et al.* 1965; Trier and Browning 1966; Tarpila 1971). Among possible explanations for these inconsistencies two seem worthy of serious consideration. Firstly, chronic obstruction of small intestine, frequently at multiple points of stricture or adhesion, has been a prominent feature in most cases, creating the perfect milieu for bacterial overgrowth. Although cultures of luminal contents from the small bowel of such patients can be difficult, success in identifying a pathogen has been reported (Conklin and

Anuras 1981), enabling diarrhoea to be controlled with appropriate antibiotic treatment. Secondly, the triad of mesenteric sclerosis, invasion of the submucosa with hyaline connective tissue, and mucosal lymphangiectasia would provide the basis for a diminished clearance and perhaps also secretion of triglycerides and long-chain fatty acids, and it would be of interest to know whether steatorrhoea due to irradiation could be influenced by feeding fat in the form of short- or medium-chain compounds.

The enteropathy of pelvic irradiation may be clinically silent and therefore a potential cause of nutritional deficiencies in subjects who consider themselves cured. In most such cases the pattern of malabsorption is such as to suggest selective damage to the terminal ileum. McBrien (1973) studied 14 consecutive patients who had had up to 3600 cGy cobalt teletherapy for carcinoma of the bladder and were returning for check cystoscopy at a mean 2 years after treatment. Eight (57 per cent) had depressed vitamin B_{12} absorption in the presence of intrinsic factor. Among 17 women who had been treated for genital cancer few complained of diarrhoea but all except one gave abnormal results in the ^{14}C-cholylglycine breath test, indicating impaired bile salt absorption (Newman *et al.* 1973). In a similar study Andersson *et al.* (1978) found that among 20 patients with diarrhoea following pelvic irradiation only seven had steatorrhoea, but all gave evidence of bile salt malabsorption in the breath test.

Treatment

Acute enterocolitis occurring in the course of radiotherapy may respond to anti-emetics and antidiarrhoeals alone. For more intractable cases Wolloch *et al.* (1973) found sedation, a low-residue diet, and hydrocortisone retention enemas effective. Mennie *et al.* (1975) obtained a significant reduction of stool frequency in 11 of 14 patients by using aspirin in an alkaline effervescent preparation (*Alka-Seltzer*) on the basis that the diarrhoea might be prostaglandin-dependent.

Most patients developing chronic damage have required surgery at some time, usually to relieve obstruction; but results are as a rule poor because the technical difficulty of handling friable tissues bound together with adhesions leads to a high incidence of serious post-operative complications in patients who are already debilitated by their disease. In the series of Wellwood and Jackson (1973) 14 of 38 (37 per cent) patients died, some after multiple operations, and this experience was by no means atypical (Graham and Villalba 1963; DeCosse *et al.* 1969). Russell and Welch (1979) encountered major complications in 65 per cent of their operations, half being due to anastomotic leakage. Even a later report, Kwitko *et al.* (1982), features a death rate of 32 per cent among patients operated for enteritis, most fatalities following breakdown of anastomoses. With such experiences in mind Cochrane *et al.* (1981) recommended wide resection, and perhaps employing study of frozen sections to establish the microscopic normality of tissues at the suture line. They preferred bypass to resection wherever the damage had been so great as to make the latter technically difficult, and practised temporary exteriorization to assist the healing of doubtful anastomoses.

Success with the limited surgery required for procto-sigmoiditis seems equally elusive. No less than 79 per cent of operations performed at Groote Schuur between 1974 and 1981 gave rise to complications (Gilinsky *et al.* 1983). One approach to this problem is subtotal resection of the rectum with sleeve anastomosis, a procedure which Cooke and De Moor (1981) found had left 65 per cent of their patients incontinent one year post-operatively. Although others have fared better (Localio *et al.* 1969), it would seem advisable to limit operative manipulation to the most conservative procedure possible in each individual case (Deveney *et al.* 1976; Swan *et al.* 1976; Poddar *et al.* 1982). A lesson of the utmost importance, which seems to have been relearned at the expense of much morbidity and even mortality at many separate centres, is that the pathological process responsible for delayed irradiation damage to the intestine is essentially progressive and will always tend to attack previously uninvolved bowel, however wide the margin of 'normal' tissue resected at the last operation.

For improved management of this intractable condition recourse will be necessary to more positive medical measures, and it may be that recent advances in parenteral nutrition and elemental diets have already begun to change its prognosis. Even without these, remarkable results have been reported from some centres. Faced with the extensive small-intestinal involvement that might be expected from whole abdominal irradiation in children, Donaldson *et al.* (1975) treated all five of their patients with a low-fat, low-residue diet entirely free of gluten, lactose, and cow's-milk protein. Not only was there rapid relief of symptoms, but in all cases a regression of radiological abnormalities was observed, and after gradual relaxation of the diet the children had resumed normal growth on normal food. Goldstein *et al.* (1976) have achieved comparable success in four adults treated with sulphasalazine, oral prednisone (in one case), and a bland diet. Bosaeus *et al.* (1979) detected malabsorption of bile salts in nine patients complaining of watery diarrhoea. After their daily fat intake had been limited to 40 g for periods of at

least 3 months, eight reported a normal bowel habit, though the ninth required additional treatment with cholestyramine. In a similar series described by van Blankenstein (1981) 15 women who were well, apart from painless, urgent diarrhoea with incontinence after pelvic irradiation, found their symptoms alleviated by cholestyramine at 1–8 g daily. Patients at the other end of the clinical spectrum of irradiation enteritis are unable to assimilate food taken by mouth. The 10 patients of Miller *et al.* (1979) included three who had derived no benefit from prednisone, with or without sulphasalazine. Excluding from the total two who died of recurrent carcinoma, two of the remaining eight died during the course of home parenteral nutrition (HPN), and to date the others had required continuous treatment of up to 52 months' duration. In these there had been partial healing of some fistulae and substantial nutritional improvement, amounting to a mean weight gain of 8.7 kg and an increase of 4.4 g 1^{-1} in serum albumin. Lavery *et al.* (1980) regarded five of their patients as being totally dependent on parenteral nutrition for survival, and assessed their progress after 6–30 months' HPN. Only three had survived; one having died from recurrent cancer and another from a pharmaceutical error. Those remaining had gained a mean 13 kg in weight and were free of technical problems relating to their mode of nutrition. There is, of course, scope for combining parenteral nutrition with corticosteroid treatment, exemplified by the greater success of parenteral than of enteral feeding as an adjunct to methylprednisolone in the series of Loiudice and Lang (1983). Where stricture formation favours overgrowth of intestinal pathogens, as in the case of Conklin and Anuras (1981) which yielded a culture of *Gaffkya anaerobia*, an antibiotic may alter the course of the disease.

For patients with distal proctocolitis of mild-to-moderate severity corticosteroid retention enemas frequently suffice to control symptoms.

Prophylaxis

In order to minimize small-bowel injury Green *et al.* (1975) suggested that those most at risk (women, thin, and elderly subjects) should receive doses only just within the therapeutic range to a modified field while lying in the prone position. Later the same group (Green *et al.* 1978) reported the results of routine small-bowel contrast radiology before treatment in 74 patients treated between 1975 and 1977. Small intestine was seen at the upper level of the carcinoma in 19 patients and overlapping it in 10. They claimed to have avoided injury in all these patients by using the precautions listed earlier and by attempting to reduce the size of the tumour with hormones or whole pelvic irradiation given before the definitive course of radiotherapy. Such a claim would seem premature after so short a lapse of time, and the results of longer-term studies are still awaited. A form of medical prophylaxis advocated by Morgenstern *et al.* (1977) involves giving anti-inflammatory drugs, such as corticosteroids and sulphasalazine, with an elemental diet, during the course of radiotherapy.

Carcinoma

There have been several reports of carcinoma arising in irradiated colon and rectum (Slaughter and Southwick 1957; DeCosse *et al.* 1969; MacMahon and Rowe 1971). Localio *et al.* (1969) described a patient who developed a first carcinoma in the caecum and, some years after this had been resected, a second in the sigmoid colon, both at sites of severe radiation damage. The six cases of primary adenocarcinoma of rectum or rectosigmoid occurring among 913 patients given radiotherapy for carcinoma of the cervix by MacMahon and Rowe (1971) followed both early and delayed proctitis, with latencies of 14 months to 25 years before presentation of the tumour. Castro *et al.* (1973) summarized the epidemiological evidence for a causative association in 26 patients, adding the observation that colloid carcinomas, which comprised no more than 10 per cent of the total series of large-bowel cancers from their institution, were found in 58 per cent of irradiated patients. Sandler and Sandler (1983) have calculated the risk ratio for carcinoma in such patients as lying betwen 2.0 and 3.6.

Injury to the upper gastrointestinal tract

Lesions of the upper gastrointestinal tract have attracted less attention but can be equally disabling. Damage to the mouth and salivary glands may occur when tumours of the tonsils or nasopharynx are treated with radiotherapy. Xerostomia is a distressing result of irreversible injury to the salivary glands. Loss of protective IgA encourages infection, especially with *Candida albicans*, and permits overgrowth of the bacteria responsible for dental caries. The health of the teeth can, however, be preserved by daily application of sodium fluoride gel (Dreizen *et al.* 1977). Ulceration of the buccal mucosa usually resolves within a few weeks, and loss of taste sensation within 4 months. Zinc sulphate may be given either as treatment for hypogeusia or in a prophylactic role. Schneider *et al.* (1977) found that of 1922 patients followed up for a mean 29.9 years, 19 had developed benign tumours of the salivary glands (as against less than one expected) and eight cancers (as against 0.2 expected). The latent period between irradiation and diagnosis of the tumour varied

from 7 to 32 years, and after the first 15 years incidence became constant. It has been estimated that carcinoma develops in the parotid glands of 8 per cent of patients irradiated at this level (Walker *et al.* 1981).

In a series of 20 patients irradiated for bronchial carcinoma all developed oesophagitis (Seaman and Ackerman 1957). Eleven experienced nothing more than a substernal burning sensation from the third week of treatment until several weeks after its completion. In four patients this was more severe, and was accompanied by dysphagia for up to three months. In the remaining five patients the inflammation caused fibrous strictures requiring dilatation. On barium swallow examination fine ulceration was seen with or without later appearance of a smooth stricture. Microscopically there was little evidence of vascular damage, but a transient loss of epithelium followed by gross thickening of the submucosa and degeneration of smooth muscle fibres. The authors noted that most cases of severe oesophagitis were in patients treated early in their experience of the betatron, when the doses given were more than sufficient for therapeutic success. A study (published 15 years later) of 95 patients given mediastinal irradiation for lymphoma showed that only 30 (32 per cent) had developed oesophagitis, and that only two of these had been severely affected (Marks *et al.* 1974). Refinement of technique enabled Carmel and Kaplan (1976) to claim that no case of oesophagitis had occurred among patients with Hodgkin's disease treated at Stanford over a 20-year period. Pearson (1969) was able to find no more than 20 long-term survivors in a series of 99 treated for squamous carcinoma of the oesophagus. Of these 11 had significant stricture formation, causing anxiety over possible tumour recurrence as well as obstructive symptoms.

Most strictures respond to mechanical dilatation alone, but additional infiltration of the mucosa with hydrocortisone has been recommended for those that are refractory (Nelson *et al.* 1979). From experiments in opossums given massive gamma irradiation of the oesophagus it seems that indomethacin might prove effective in human prophylaxis by suppressing prostaglandin synthesis (Northway *et al.* 1980).

The oesophagus may also be affected indirectly by radiotherapy to the head and neck. The patient of Thorpe *et al.* (1982) received 3680 cGy for a nasopharangeal lymphosarcoma, and 40 years later was found to have developed achalasia in the presence of vagal denervation.

Not uncommonly, combined chemotherapy is given as well as irradiation in the treatment of small-cell carcinoma of the bronchus. In such cases severe oesophageal damage may be seen with X-ray dosage one-half (or less) the 6000 cGy stated by Seaman and Ackerman

(1957) to be the maximum tolerated (Greco *et al*. 1976; Chabora *et al*. 1977). To apportion blame among individual drugs cannot be easy, but there is some evidence that doxorubicin (*Adriamycin*) (Horwich *et al*. 1975; Greco *et al*. 1976) is especially liable to potentiate irradiation damage. *Adriamycin* has been shown to increase cell-killing by X-rays *in vitro* (Watring *et al*. 1974). It is thought to interfere with cellular repair by binding DNA.

Short-term studies in 13 patients who had received pelvic irradiation for uterine carcinoma and four who had had para-aortic irradiation for testicular carcinoma detected no impairment of biliary or pancreatic secretion after courses of up to 5000 cGy given over 5 weeks (Becciolini *et al*. 1979). There are, however, cases in which chronic pancreatitis can reasonably be ascribed to radiotherapy given years before onset of symptoms. In the patient of Burbige *et al*. (1977) steatorrhoea and weight loss began 3 years after he had received 6120 cGy for bladder carcinoma. Small intestinal studies were normal, but endoscopic retrograde pancreatography (ERP) showed dilatation of the main pancreatic duct behind a segment of stenosis. Mitchell *et al*. (1979) described two patients who had been given 3650 and 4500 cGy, respectively, to the para-aortic nodes for seminoma, with latent periods of 15 and 24 years before onset of symptoms suggesting chronic pancreatitis. The diagnosis was made by ERP in one and at laparotomy in the other. Of the three patients above only one is recorded as having a second condition which might have contributed to the aetiology, namely gallstones, but as he suffered also from bilateral radiation nephritis there can be little doubt that his fibrosing pancreatitis had the same cause.

Ileus and obstruction

Drugs with which ileus has been associated include many prescribed for their action on nervous mechanisms and a few known for their neurotoxicity. Although some, such as the opiates, work primarily by stimulating the circular muscle of the intestine and others, such as the phenothiazines and tricyclic antidepressants by decreasing muscular tone, the initial effect is nearly always a loss of propulsive activity inducing constipation. Because this effect is so familiar and the next stage, paralytic ileus, so uncommon clinicians may well miss the transition, which has been known to occur quite abruptly (Burkitt and Sutcliffe 1961; Milner and Hills 1966; Toghill and Burke 1970), and thus needlessly endanger the lives of their patients.

Ganglion-blocking agents

Among the earliest cases of drug-induced ileus many were due to the ganglion-blocking agents which for long constituted the only effective means of reducing hypertension but now are used only for emergency treatment. This group includes hexamethonium, pentamethonium, pentolinium tartrate, mecamylamine, and pempidine. Goldstone (1952) described ileus in a middle-aged hypertensive treated with pentamethonium. At laparotomy the whole of the small bowel was grossly distended, although the colon was normal in calibre. When the gut was emptied of gas by puncture it did not contract but merely became flaccid. The patient died 48 hours after operation. A further 12 cases of paralytic ileus caused by ganglion-blocking drugs were described by Munster and Milton (1961).

Psychotropic and antiparkinsonian drugs

Much reliance is placed by psychiatrists on three classes of drugs, the tricyclic antidepressants, the phenothiazines, and an antiparkinsonian group, which share atropine-like activity and therefore a propensity for inducing constipation. The fact that their therapeutic effects differ increases the chance that member drugs of these classes will be prescribed in combination; a special case being the need for antiparkinsonian cover in all schizophrenics on long-term phenothiazine maintenance. Eight cases of ileus were described by Warnes *et al*. (1967) in psychiatric patients taking two or more of these drugs together. In three cases death from circulatory collapse followed shortly after diagnosis. The patient of Giordano *et al*. (1975), a 19-year-old who was receiving chlorpromazine 1 g, trifluoperazine 40 mg, and benztropine maleate 3 mg daily, developed sudden crampy abdominal pains after 6 weeks' treatment, and within a further 8 hours had become grossly distended. At laparotomy the entire bowel was dilated, with a greenish-yellow discoloration, and free fluid was found. Despite decompression and intensive therapy the patient died 5 hours post-operatively. *Cl. perfringens* was cultured from the peritoneal fluid. The authors postulated a sequence in which stretching of the bowel wall by dilatation had compromised venous return and had thus caused ischaemia of the mucosa, followed by ulceration, bacterial invasion, and incipient necrosis of bowel, together with a fatal Gram-negative septicaemia. Such a train of events, although highly speculative even in this case, would provide an explanation for the rapidity with which shock has supervened in many instances of drug-induced ileus.

Use of a single psychotropic drug has also on occasion been associated with paralytic ileus. Cases attributable to amitriptyline (Burkitt and Sutcliffe 1961; Milner and Buckler 1964; Clarke 1971), nortriptyline (Milner and Hills 1966), chlorpromazine (Davis and Nusbaum 1973), and orphenadrine (Daggett and Ibrahim 1976) have all been described. Davis and Nusbaum (1973) reported the unusual circumstance of colonic ileus occurring without prior constipation in a patient whose dose of chlorpromazine had been increased from 200 mg to 600 mg daily a week before onset. Haloperidol is a butyrophenone derivative used for the treatment of psychosis and noted for the weakness of its anticholinergic properties. However a case was recorded (Maltbie *et al.* 1981) in which a 52-year-old woman with paranoid schizophrenia developed constipation during the first 7 weeks of treatment with haloperidol at a mean daily dosage of 41.4 mg. At the end of this period abdominal distension with mild tenderness, diminished bowel sounds, and multiple fluid levels on an erect plain film of the abdomen indicated the onset of ileus, which responded well to conservative measures.

Methadone

Although opiates must often have been suspected of contributing to paralytic ileus this association was poorly documented until high doses of methadone came to be employed in the substitution therapy of heroin addiction. Spira *et al.* (1975) described gross faecal impaction leading to obstruction in five cases. In some of these abdominal rigidity due to heroin withdrawal combined with distension and loss of bowel sounds to give the false impression that perforation had occurred. These patients recovered, but a fatal case was later reported by the same group (Rubenstein and Wolff 1976).

Dantrolene

Dantrolene is prescribed for relief of spasticity, and its principal action is therefore on striated muscle. Among 75 patients receivng this drug in the recommended dosage of 50-100 mg q.d.s. Shaivitz (1974) found three who developed ileus as a sequel to severe constipation.

Clonidine

Clonidine in doses adequate to control systemic hypertension not infrequently causes constipation. Bear and Steer (1976) reported a case of intractable hypertension in a patient who had undergone renal transplantation and was taking methyldopa, hydrallazine, and propranolol as well as clonidine in a dosage of 2 mg daily. Constipation was followed by the development of an ileus which responded to withdrawal of the clonidine alone.

Vincristine

A few cases of ileus have complicated the neuropathy caused by treatment of malignant disease with vincristine. In a series of 19 patients studied by Martin and Compston (1963) there were 10 who developed constipation, and in one of these ileus was subsequently diagnosed. Toghill and Burke (1970) reported fatal ileus in a patient who had received a single 6-mg dose of vincristine.

Colonic pseudo-obstruction (Ogilvie's syndrome)

To other categories of iatrogenic ileus should be added the acute colonic pseudo-obstruction also called 'Ogilvie's syndrome' after the author of the first published report (Ogilvie 1948). A review by Nanni *et al.* (1982) covering 351 cases shows that most have been due to surgery, trauma, or sepsis, and gives a breakdown into individual aetiologies. Abdominal and pelvic surgery has been responsible for 12.2 per cent of cases, caesarian section for 7.7 per cent and orthopaedic surgery for 5.4 per cent. A cause only recently described (Lopez *et al.* 1981) is the combination of external irradiation for carcinoma of the cervix uteri with subsequent implantation of radium locally, which obtained in three cases at a single centre.

Characteristic features include fever, distension, abdominal pain, vomiting, and gaseous dilatation of the colon which is usually maximal on the right side. Since the advent of fibreoptic endoscopy colonoscopic decompression has been the treatment of choice (Kukora and Dent 1977; Robbins *et al.* 1982).

Mechanical obstruction

Obstruction of the bowel during aluminium hydroxide treatment is not necessarily the consequence of constipation induced over a long period, as in patients with chronic renal failure. The series of Korenman *et al.* (1978) included five cases of acute renal failure in which bezoars of *Amphogel* came to occupy segments as long as the colon itself, and in one instance the entire bowel. An important result was patchy necrosis of the wall, from which perforation might follow. Daily doses of *Amphogel* quoted by these authors were in the range 120–600 ml.

Emergency surgery was required to relieve acute obstruction caused by the accumulation of non-absorbable residues from *Ferrograd-Folic* tablets proximal to the narrowed terminal ileum of a patient with Crohn's disease (Shaffer *et al.* 1980). The authors named 18 other preparations available in UK which incorporated a non-absorbable matrix, suggesting that all were contraindicated in patients with intestinal strictures.

Bran is enthusiastically recommended by most clinicians as a cheap, safe, and effective means of softening the stool. This too, however, is susceptible to overdosage, as found by Kang and Doe (1979) whose patient was given bran to supplement other treatment for constipation due to antidepressive therapy. This woman was so impressed by the efficacy of bran that she had increased her intake to between 160 and 200 g a day for 6 months before presenting with nausea, abdominal cramps, distension, and incontinence of faeces. She recovered after 'enormous quantities'of unaltered bran had been evacuated manually. Sandeman *et al.* (1980) described oesophageal obstruction in an 80-year-old woman due to a bolus of the popular gum laxative, *Normacol Standard*, which had become wedged at the level of the aortic arch and subsequently came to fill the entire lower two-thirds of the oesophagus. The main local change was an acute oesophagitis, but the patient died from aspiration pneumonia directly attributable to the obstruction. The authors of this report suggested that instructions for the administration of hygroscopic gum laxatives should emphasize the importance of taking adequate quantities of water *with* the preparation, since water drunk afterwards to assist passage of the bolus is likely to aggravate the obstruction. Elderly patients in particular should be advised against taking such preparations immediately before retiring.

The presence of gastroparesis in some poorly-controlled diabetics predisposes to bezoar formation. Among three cases reported by Canivet *et al.* (1980) was one of a patient for whom a fibre-rich diet had been prescribed. The current trend towards using fibre as an adjuvant to insulin treatment for unstable diabetes could lead to more such cases unless the danger is foreseen and a check is made on the gastric motility of individual patients for whom this regimen is intended.

Haemorrhage due to anticoagulants

Gross haemorrhage into the alimentary tract of patients receiving anticoagulants is rare unless prothrombin activity has been allowed to fall well below the therapeutic range, yet the frequency of severe bleeding from this cause is higher in the gut than at any other site (Sixty Plus Reinfarction Study Research Group 1982).

The first two cases recorded were in patients taking dicoumarol, and the descriptions given by Berman and Mainella (1952) of their clinical course and pathology differ in no important respect from most subsequent accounts. Both patients were middle-aged and had been anticoagulated for myocardial infarction. Onset of bleeding was signalled in each case by the development of cramping abdominal pains, accompanied in one by bloody diarrhoea. There followed progressive distension and tenderness of the abdomen with vomiting, pyrexia, and signs of intestinal obstruction. Each patient had entered a state of shock at the time of laparotomy, and in each the peritoneal cavity contained free blood. The first had a retroperitoneal haematoma, with petechial haemorrhages throughout the serosal surface of the bowel and gangrene of the left side of the colon. In the other was found, somewhat more typically, a black sausage-like thickening of 10 inches of the distal jejunum with marked distension of proximal small bowel and many haemorrhagic areas in the serosa on either side. Grossly the resected specimen in this case showed occlusion of the lumen by intramural haematoma, and microscopically complete preservation of the mucosa down to the muscularis, with extensive haemorrhage in the submucosa.

There is no evidence that any one oral anticoagulant is more likely than others to cause intramural haematoma of the bowel: in one series of seven cases, three were due to phenindione, two to dicoumarol, and one each to nicoumalone and ethyl biscoumacetate (Beamish and McCreath 1961). Warfarin occurs most frequently in the literature because it has been the oral preparation of choice for many years, and heparin least frequently, perhaps because it is rare for symptoms to arise within 3 weeks of initial anticoagulation. Modern requirements for long-term heparin administration, as in haemodialysis and parenteral nutrition left to the patient at home, may lead to a change here. Male patients outnumber female in a ratio of 2:1. As the principal lesion is a localized haematoma symptoms are predominantly — sometimes exclusively — those of mechanical obstruction rather than of the ileus that might follow more diffuse involvement. Overt haemorrhage, indicating substantial blood loss into the lumen, is an unusual feature and presumably follows intramural haematoma formation (Birns *et al.* 1979). The jejunum was the site of over half the lesions reported (Guivarc'h *et al.* 1979), the ileum is next most commonly affected, followed by the duodenum, and least commonly the colon and oesophagus (Gabriele and Conte 1964; Andress 1971; Snyder *et al.* 1973). Haemoperitoneum is the rule, varying in quantity from a few millilitres to several litres. The mucosa, although it is dissected off underlying coats by the haematoma, invariably appears intact, and necrosis of the bowel was only once reported. Thus despite the alarming appearance it so often presents, the affected segment in cases of anticoagulant haematoma is usually viable and responsive to medical management. The dangers surgery holds for a patient seriously ill with uncontrolled anticoagulation superimposed on occlusive vascular disease make it imperative, where possible, to achieve the diagnosis without operating.

The features of intestinal obstruction in a patient known to be on anticoagulant therapy and, in most cases, showing a prolonged prothrombin time should suggest the diagnosis strongly enough to commit the clinician to a supportive regime of blood transfusion and fluid and electrolyte replacement, with or without injections of vitamin K, while requesting a plain radiograph. The presence of multiple fluid levels in the erect film and of dilated bowel loops in the supine will confirm obstruction, but search should also be made for the hazy appearance that indicates free fluid and for the 'thick bowel sign' of Sears et al. (1964). This consists of 'thin curvilinear, continuous gas shadows delineating a narrowed bowel lumen which is coarsely irregular from distorted mucosal folds', and enabled its discoverers, having observed it in one case diagnosed by other means, to make the diagnosis with confidence from plain radiographs in three subsequent patients (see Plate 6a). If uncertainty remains and the patient is fit for contrast radiology, a small-bowel meal should be undertaken. The appearances given by a traumatic haematoma in the small bowel were described as resembling a 'coiled spring' by Felson and Levin (1954), but it was not until seven years later that Culver et al. (1961) found the description equally applicable to a case of anticoagulant haematoma. Almost simultaneously Senturia et al. (1961) reported how they had been able to make a preoperative diagnosis in two patients from films showing 'a rigid segment with narrowed lumen from which spike-like projections extended out to the normal width of the bowel', and Wiot et al. (1961) displayed examples of an identical phenomenon in two cases of duodenal haematoma, coining for it the term 'picket-fence sign', which has since become standard (see Plate 6b).

Most patients undergo rapid improvement on conservative therapy (Silbert et al. 1962; Beamish and McCreath 1967; Stanton et al. 1974). If none is seen within 48–72 hours a laparotomy will be required to exclude another lesion or the rare development of necrosis in the involved segment (Herbert 1968; Zer et al. 1972). Clinical and radiological recovery is usually complete within 3 weeks. Whether or not anticoagulants should then be represcribed must depend on the individual circumstances of each case, but, in general, patients have not been subject to recurrence of intestinal haematoma if adequately controlled.

Ischaemic enterocolitis

Oral contraceptives

It may never be possible to prove a causal relationship between treatment with oral-contraceptive agents and ischaemic lesions of the bowel in women because challenge is ethically indefensible. Yet the number of reports now testifying to an association between these drugs (known for their thrombogenic potency) and acute mesenteric vascular occlusion in healthy adults of the group least likely to develop atherosclerosis leaves little room for doubt as to its nature. In the first recorded case a 37-year-old woman receiving norethynodrel and ethinyloestradiol-3-methyl ether presented with the picture of peritonitis and was found by Reed and Coon (1963) to have infarcted nearly 3 m of small intestine. Autopsy confirmed thrombosis of the superior mesenteric vein as the immediate cause but revealed no pathology in the vein itself, or elsewhere, to explain the thrombosis. Milne and Thomas (1976) reported a case differing clinically in that no more than 15 cm of ileum was affected, but in which thrombus was extruded from a mesenteric vein at operation, and microscopy of the resected specimen showed venous thrombi undergoing organization and recanalization. In most published cases the lesion has not been severe enough to warrant laparotomy, and the diagnosis has been inferred from the history and the irregular narrowing, 'thumb-printing' and stiffness seen in barium contrast radiographs of small bowel (Nothmann et al. 1973), proximal (Kilpatrick et al. 1968; Prust and Kumar 1976) or distal (Cotton and Lea Thomas 1971) colon (see Plate 7a).

At the first recorded colonoscopic examination in a case of colitis associated with oral-contraceptive medication (Bernadino and Lawson 1976) a group of small, discrete ulcers was seen at the splenic flexure. In a recent study of five patients with more severe symptoms, including pain and bloody diarrhoea, colonoscopy identified two principal patterns: discrete ulceration with normal intervening mucosa, and a confluent erythema with friability (Tedesco et al. 1982b)

Apart from the bleeding characteristic of large-bowel ischaemia the most notable feature of all but the worst cases has been the transience of the illness. Complete radiological recovery of the small bowel was demonstrable at 2 weeks in the case of Nothmann et al. (1973), while a patient with haemorrhagic colitis described by Prust and Kumar (1976) became symptom-free in 36 hours and within a week showed normal appearances on repeat barium-enema examination.

Vasopressin

In view of its world-wide use for many years as a means of acutely lowering portal venous pressure by constriction of the splanchnic arteries in patients bleeding from gastro-oesophageal varices, it is perhaps surprising that there are so few reports of ischaemic injury to the bowel from vasopressin infusion. Conn *et al.* (1972) described a case of extensive and fatal small intestinal necrosis following intra-arterial infusion of vasopressin in a man of 52 with a history of stroke and angiographically-demonstrated superior mesenteric artery stenosis. In recent years it has been found equally effective to use a peripheral vein for the infusion, and there can be little doubt that this is a safer route. In one such case, however, Lambert *et al.* (1982) have recorded the passage of bright red blood per rectum by a man of 54 who had received vasopressin for 6 hours, at a maximum rate of 0.4 u per minute, 48 hours previously. Colonoscopy revealed a 5-cm length of inflammation and submucosal ecchymoses at the rectosigmoid junction, and biopsies from this point were reported as showing acute haemorrhagic necrosis consistent with ischaemic colitis. The bleeding stopped spontaneously after two days, and repeat colonoscopy 18 days later was normal. In this patient also angiography showed diffuse atherosclerotic change which could have contributed to the ischaemia.

Ergot

Ischaemic disease of both small (Sayfan *et al.* 1977) and large (Stillman *et al.* 1977) bowel has been described in patients taking ergot. The patient of Stillman *et al.* had spontaneously increased her dosage of ergotamine tartrate from 1.3 to 3.9–5.2 mg daily during the three days before admission, and was also receiving ethinyloestradiol 0.05 mg daily.

Abdominal aortic surgery

Damage to the distal colon following surgical repair of the abdominal aorta and its branches received little attention in the literature until Smith and Szilagyi (1960) examined the results of 120 aneurysmectomies they had performed. Eleven led to colitis varying in severity from transient mucosal ulceration to massive gangrene of the left colon. Three patients developed bloody diarrhoea within 48 hours of operation. In them symptoms were over in a few days, and barium enema showed no residual change thereafter. Of six who suffered more severe diarrhoea, which started up to two weeks post-operatively, several died with fulminating proctosigmoiditis and others were left with a chronic stricture. One patient died

on the seventh post-operative day after interference with an anomalous blood supply, and another shortly after the procedure from necrosis of the left colon caused by multiple atheromatous embolism of the left colic and inferior mesenteric arteries. Johnson and Nabseth (1974) recorded nine personal cases of ischaemic colitis and one of enteritis following 187 operations on the abdominal arteries, including 103 aneurysmectomies. Six of the nine patients with colitis died from this complication. In a review covering 20 reports and over 6300 cases of abdominal aortic surgery they found 99 patients had developed ischaemic colitis (an incidence of approximately 1.5 per cent), with a mortality of 40 per cent. Colitis was more likely to occur after aneurysmectomy than after reconstructive procedures, and the most frequent cause seemed to be ligation of the inferior mesenteric artery. Hagihara *et al.* (1979) performed post-operative colonoscopy in 133 out of 180 patients who underwent abdominal aortic surgery at Lexington between 1974 and 1976. Enabled thus to detect the mildest cases of distal colitis they claimed an incidence of nearly 13 per cent. Again operation for aneurysm provided most of the cases, and in this instance all seven deaths.

Ernst (1983) recommended thorough pre-operative work-up of patients destined for abdominal aortic reconstruction, listing the radiological and other indices which in his experience offer the best means of assessing the risk of post-operative intestinal ischaemia in each individual patient.

Mechanical injuries of the alimentary tract

Orthopaedic devices and procedures

Fixation of the spine by braces and casts is well recognized as a cause of upper gastrointestinal symptoms. Oesophageal pathology due to reflux of acid and small intestinal contents was not, however, reported until Gryboski *et al.* (1978) described oesophagitis in three children with thoracolumbosacral braces and one who had been placed in a body cast after surgery for kyphoscoliosis. A 3-year-old became anaemic from occult bleeding and a 17-year-old suffered massive haematemesis. Two patients underwent manometry, which showed intragastric pressure to exceed intraoesophageal only when the brace was on.

The term 'body-cast syndrome' is commonly taken to refer to the consequences of gastric dilatation, with or without obstruction of the duodenum when this becomes compressed in its third part by the superior mesenteric artery. The first case, described by Willett (1878) in a boy of 17 for whom he had fashioned a plaster-of-paris jacket in the treatment of kyphosis, was characterized, like others since, by protracted vomiting which failed to relent when the appliance was removed, and by a fatal outcome. Death is presumably due to dehydration and electrolyte disturbance in most instances, though ischaemic necrosis of the stomach has been described (Letac *et al.* 1971). In the case reported by Buzzard (1880) the jejunum had rotated on the ligament of Treitz, perhaps as a direct result of the pressure exerted by the cast, and through failing to return to its correct position had caused the obstruction from which all else followed. This patient had developed paraplegia from Pott's disease, and so here there might well have been a degree of neurogenic ileus as well. The role of compression by the superior mesenteric artery is taken for granted by most authors, but does not seem to have been demonstrated at autopsy. Dorph (1950) described a case, clinically typical, in which the stomach and duodenum were grossly dilated, with a sudden transition at the ligament of Treitz to normal appearances in the jejunum. No evidence of organic obstruction was found. Attention has been drawn to the association of the cast syndrome with asthenic build (Almgren and Juhl 1977), a recent growth spurt in children (Wayne and Burrington 1972), and recent weight loss in adults (Wayne *et al.* 1971), any of which could in theory narrow the angle between the superior mesenteric artery and the aorta, so that gravity and the pressure of a cast might more easily make a pincer of the arteries to obstruct the duodenum. Mansberger *et al.* (1968) showed in an angiographic study that this angle was narrower in patients with the 'compression syndrome' than in normal controls.

The first principle in management is awareness of the complication, for its consequences may become irreversible with surprising rapidity. The cast or traction must be removed, and the patient nursed prone or in the left lateral position until the symptoms have settled. Gastric suction may be helpful, and losses of fluid and electrolyte must be made good by intravenous infusion. On occasion surgery will be required (Evarts *et al.* 1971; Wayne *et al.* 1971; Altman and Puranik 1973).

Haimovic *et al.* (1978) reported the development of colonic ileus in three patients undergoing laminectomy. All presented with vague abdominal discomfort and pyrexia, with tympanitic bowel sounds. One had a lymphoma and another hypertensive heart disease with renal insufficiency. The authors contrasted this condition with the ileus of spinal trauma which is characterized by atony of the entire bowel, and speculated that enhanced sympathetic activity could be involved here.

Three isolated cases of intestinal injury due to hip surgery were reported. Puranen and Koivisto (1978)

noted penetration of the acetabulum by a 130° blade plate which, after 8 months, caused three perforations in the ileum and two in the bladder. Foster and Bourke (1978) blamed heat from the cement used in fixing the acetabular component of a hip prosthesis for development of an omental band which obstructed the sigmoid colon 6 years post-operatively. In a similar case described by King *et al.* (1983) an 84-year-old woman developed small-bowel obstruction 10 days after insertion of a Charnley hip prosthesis. Medial extrusion of the bone cement had brought it into contact with a loop of ileum which became fixed by the heat of polymerization, enabling a volvulus to form. Septicaemic shock ensued, and the patient died after 120 cm of infarcted ileum had been resected.

Gynaecological devices and procedures

Intrauterine contraceptive devices occasionally become displaced, but resulting bowel injury is rare. In the patient of D'Amico and Israel (1977), who presented with fever, abdominal distension, and rebound tenderness 2 months after her Lippe's loop had been noted to lie within the peritoneal cavity, the loop was found in a large abscess around a perforated distal ileum. A two-month history of colic, pyrexia, and vaginal discharge was given by a woman whose Dalkon Shield had entered the transverse colon, with consequent formation of multiple actinomycotic abscesses (Wagner *et al.* 1979). In each of these cases coincident pelvic inflammatory disease may have influenced the outcome. Copper devices are intrinsically irritant, and in two cases described by Watney (1978) a Copper 7 had become embedded in the intestinal wall as well as provoking a local peritoneal reaction. Recurrent abdominal pain and fever led to the discovery of a Copper 7 within an inflammatory mass formed by the caecum and appendix of the second case reported by Key and Kreutner (1980). Their first patient, however, was discovered at laparotomy to have the greater part of a Tatum-T device within the jejunal lumen over a year after its insertion, during which period she had complained of no symptoms. Another Copper 7 was recovered from the sigmoid mucosa of a woman whose only symptom had been prolapse of the threads per anum after defaecation (Beard 1981). In case 1 of Smith *et al.* (1983) a lost thread led to fatal strangulation of the ileum.

Amniocentesis exposes the fetus to various hazards. In the case reported by Rickwood (1977) an infant was born with a paraumbilical fistula discharging meconium. When explored the fistula was found to be related to an ileal atresia. In the infant described by Swift *et al.* (1979) obstruction of a different type followed the creation by amniocentesis of a defect in the upper abdominal wall, through which a knuckle of strangulated small intestine projected at birth. Adhesions were present around the neck of the hernia.

Caecal volvulus is a rare complication of caesarian section which has occurred twice at a single centre (Alinovi *et al.* 1980). This report ascribes the mishap to inadvertent displacement of a caecum which had been incompletely fixed during embryogenesis, and recommends that the position and state of the caecum should be specifically checked before abdominal closure.

Laparoscopic sterilization is very rarely complicated by perforation of an intestinal loop as a consequence of thermal injury. Such accidents usually affect the small bowel and from the figures of two large series may be expected to follow 0.05–0.15 per cent of such operations (Loffer and Pent 1975; Maudsley and Qizilbash 1979).

Peritoneal dialysis

Perforation of a viscus may arise during peritoneal dialysis as a hazard easily explained and perhaps surprisingly uncommon (Simkin and Wright 1968; Rigolosi *et al.* 1969; Dunea 1971). It is most likely to occur when the bowel is fixed by adhesions, dilated, or oedematous as a result of disease. Among the 443 catheter insertions of Simkin and Wright 6 (1.4 per cent) were thus complicated, and in a review of 490 dialyses between 1965 and 1971 Roxe *et al.* (1976) noted an incidence of 1 per cent. The other chief intra-abdominal complication of peritoneal dialysis, peritonitis, was diagnosed with the same frequency in this series, but had been encountered in up to 10 per cent of procedures elsewhere (Dunea 1971).

Peritonitis is a relatively common complication of continuous ambulatory peritoneal dialysis (CAPD). Perhaps its main danger is that it may conceal intestinal perforation, a circumstance which may well have contributed to the death of the patient described by Watson and Thompson (1980). More frequently, however, perforation of the bowel in CAPD would seem to be a benign condition which can be managed conservatively. About 6000 dialyses, with mean duration of 9 months, had been undertaken by Rubin *et al.* (1976) when they reported six perforations in five patients. Dialysis had been continued during treatment with intraperitoneal and systemic antibiotics, avoidance of 'dwell time' being the only concession considered necessary to healing.

Cardiac pacemakers

An 86-year-old man developed abdominal cramps, fever, and vomiting several years after insertion of a cardiac pacemaker between deep fascia and peritoneum in the left lower quadrant (Matern *et al.* 1977). He became shocked, and at laparotomy both leads were found to be intraperitoneal, giving rise to a thick mesenteric band which had obstructed and infarcted 170 cm of small bowel. Theuer (1979) reported the case of a woman who developed chronic nausea and vomiting, associated with a severe but localized gastritis, following invagination of the stomach by a pacemaker which had gradually slipped downwards from a high epigastric site in the abdominal wall.

Antibiotic-associated colitis

Descriptions of a pseudomembranous colitis (PMC) appeared in textbooks as early as the mid nineteenth century (Newman 1956), and in time this change came to be known as a rare complication of many serious disorders, especially the dysenteries, uraemia, and obstruction of the large bowel by carcinoma, as well as a possible sequel to abdominal surgery. An apparent upsurge in its incidence was noted by pathologists in the United States shortly after it had become clear to clinicians in the late 1940s that their enthusiastic use of the first broad-spectrum antibiotic, chlortetracycline, had unleashed a potent new cause of iatrogenic diarrhoea (Reiner *et al.* 1952). Among the patients developing antibiotic-associated colitis were a number who had been given high doses to sterilize the bowel in preparation for surgery, and for all the earliest observers it was a fundamental characteristic of this condition that it involved widespread mucosal inflammation in the absence of any known pathogen (Reiner *et al.* 1952; Klotz *et al.* 1953).

Confusion arose when the stools of some patients with PMC ascribed to antibiotics yielded a heavy growth of *Staph. aureus* (Newman 1965; Hartmann and Angevine 1956). It seems that in a proportion of these cases the lesion was indeed PMC and the growth of *Staph. aureus* an incidental occurrence of no pathogenic significance, as in certain cases of ulcerative colitis successfully treated with corticosteroids (Valberg and Truelove 1960). Finland *et al.* (1954) suspected that overgrowth of *Staph. aureus* was encouraged by particularly high doses of a broad-spectrum antibiotic, and found that this organism could sometimes be eliminated merely by withdrawing the drug. A majority of patients, however, had developed frank staphylococcal infection which, because it affected the small bowel at least as severely as the colon (Newman 1956), was more properly designated 'staphylococcal enterocolitis'. This fulminant condition, pursuing a more rapid course than PMC and frequently ending in death from septicaemia, is now rare; perhaps in part because the use of oral tetracycline for pre-operative bowel preparation has been discontinued, in part because the epidemic phage type 80/81 *Staph. aureus* is no longer seen (Keusch and Present 1976).

Aetiology

Evidence for a bacterial enterotoxin as the cause of antibiotic-associated colitis was first put forward by Larson *et al.* (1977). A toxin later extracted by Larson and Price (1977) from the faeces of nine patients with PMC and two with antibiotic-associated non-specific colitis, and also independently by Rifkin *et al.* (1977) from the faeces of two further patients with PMC, was shown to be neutralized by *Cl. sordellii* antitoxin. Against a role of *Cl. sordellii*, however, were the facts that this organism had never been cultured from the faeces of patients with antibiotic-associated colitis, that it had failed to cause intestinal lesions in hamsters, and that cell-free extracts from its culture were not cytotoxic *in vitro* (Bartlett *et al.* 1978). Instead Bartlett and his co-workers in Boston obtained cultures of *Cl. difficile* which produced a toxin that not only caused enterocolitis in hamsters but was also neutralized by *Cl. sordellii* antitoxin. Simultaneous work in Birmingham (George *et al.* 1978) and in Los Angeles (George, Sutter *et al.* 1978) provided confirmation from two independent sources that a toxin cytopathic in tissue culture and produced solely by strains of *Cl. difficile* could be isolated from the faeces of every patient tested. Faecal extracts from hamsters with clindamycin-associated colitis have induced haemorrhagic ileocaecitis in others of the species, and are neutralized by a multivalent clostridial antitoxin (Rifkin *et al.* 1978). Hamsters may be protected from this experimental lesion by vancomycin (Bartlett *et al.* 1977), an antibiotic which has been found to be effective against seven strains of *Cl. difficile* (George *et al.* 1978) and which, when administered orally in a dosage of 2 g daily to patients with antibiotic-associated colitis, leads not only to rapid clinical recovery but also to disappearance of toxin from the stools (Rifkin *et al.* 1977; Tedesco *et al.* 1978). There is now, therefore, no doubt that treatment with certain antibiotics encourages the growth of *Cl. difficile*, which in turn secretes one or more toxins capable of inducing a colitis frequently characterized by pseudomembrane formation. The paradox by which a given antibiotic may act as either cause or cure for this colitis was elegantly resolved by the animal model of O'Connor *et al.* (1981). Syrian hamsters were infected with isolates of *Cl. difficile* previously shown to be highly sensitive *in vitro* to rifampicin, which duly gave them some protection against the resulting colitis. After 6 months of these experiments, however, many animals died in a sudden epidemic of caecitis caused by the same *Cl. difficile*, which had now become resistant.

Antimicrobial regimens

Cases of colitis displaying a notable uniformity in both clinical and pathological features have been associated on occasion with the use of many different antibiotics and antimicrobials, including chlortetracycline (Reiner *et al.* 1952; Klotz *et al.* 1953), oxytetracycline (Hartmann and Angevine 1956), chloramphenicol (Reiner *et al.* 1952), ampicillin (Schapiro and Newman 1973; Keating *et al.* 1974; Christie and Ament, 1975; Berkowitz *et al.* 1976), cloxacillin (Goodman *et al.* 1977), oxacillin (Friedman *et al.* 1980), erythromycin, neomycin, co-trimoxazole (Slagle and Boggs 1976), the cephalosporins (Slagle and Boggs 1976; Tures *et al.* 1976), rifampicin (Fournier *et al.* 1980; Borriello *et al.* 1980), and metronidazole (Saginur *et al.* 1980; Thomson *et al.* 1981), as well as occurring more consistently with lincomycin and clindamycin. In numerous other instances the colitis has been preceded by treatment with more than one antibiotic, as in the series of Clark *et al.* (1976) who encountered no case of colitis among 33 patients given lincomycin alone but six among 25 patients who had received both this and another antibiotic.

Our present understanding of antibiotic-associated colitis as a product of selective bacterial overgrowth in the intestinal lumen suggests a need for fundamental revision of the principles governing pre-operative antibiotic prophylaxis. Keighley *et al.* (1979) reported that six out of seven cases of PMC among 93 patients receiving prophylaxis for colorectal surgery with metronidazole and kanamycin, given either orally or parenterally, were from the group of 47 on oral treatment. Among the less common modes of antibiotic administration, whole-gut irrigation with neomycin and erythromycin base in preparation for large-bowel surgery (Weidema *et al.* 1980), local application of clindamycin solution to the face for acne (Milstone *et al.* 1981), and intraperitoneal instillation of cephalothin (Coleman *et al.* 1981) or cefuroxime (Gokal *et al.* 1982) for peritonitis arising during continuous ambulatory peritoneal dialysis have all been associated with *Cl. difficile* colitis.

Incidence of diarrhoea and colitis

Diarrhoea is a familiar accompaniment of antibiotic therapy, and whether or not this reflects a state of predisposition to overgrowth with *Cl. difficile* there can be little doubt that its cause most commonly lies elsewhere. However, the features of a severe colitis following lincomycin therapy have been described both as 'non-specific' (Kaplan and Weinstein 1968; Manashil and Kern 1973) and as showing the typical picture of PMC (Benner and Tellman 1970). In a prospective study of cases related to clindamycin treatment Tedesco *et al.* (1974) noted only a difference of degree, in that symptoms were milder and the inflammation was less intense in those patients whose colitis was not of the pseudomembranous type. Dispute as to the pathological spectrum of *Cl. difficile* colitis may perhaps also be due in part to the fact that the lesion of PMC is patchy and may be missed if examination is confined to sigmoidoscopy (Goodacre *et al.* 1977). Two of the most complete pathological studies of PMC yet conducted (Reiner *et al.* 1952; Price and Davies 1977) gave virtually identical descriptions of the earliest specific histological lesion, which is far too minute to be apparent clinically, and even lesions that are obvious to the naked eye may be misinterpreted as 'non-specific' unless subjected also to microscopy (Burbige and Milligan 1975).

The incidence of iatrogenic PMC is unknown. Some idea of the relative diarrhoeogenic potencies of those antibiotics most frequently associated with PMC may be gained from a survey of 1158 orthopaedic patients receiving prophylaxis at Guy's Hospital (Beavis *et al.* 1976); lincomycin inducing diarrhoea in 22.2 per cent, clindamycin in 15.3 per cent, and ampicillin in 9.6 per cent of patients, for whom each respectively had been prescribed. During the study period only three cases of PMC were documented, two after lincomycin and one after clindamycin treatment. Of 200 patients in St. Louis treated with clindamycin by Tedesco *et al.* (1974) 21 per cent developed diarrhoea, 10 per cent PMC. Swartzberg *et al.* (1976) could not calculate the incidence of PMC among 1000 patients receiving clindamycin at Palo Alto, but obtained a figure of no more than 6.6 per cent for diarrhoea and noted in addition that the diarrhoea in their patients was markedly less severe than that of the St. Louis series. Variation proved to be temporal as well as geographical, for in the year of this study no case of PMC was verified at Palo Alto, although 14 had been diagnosed the year before. A similar phenomenon was reported from Michael Reese Hospital in Chicago where during the first 20 months of its use clindamycin was associated with not one case of PMC, yet appeared to induce nine over the next two months (Kabins and Spira 1975).

The discovery that PMC was a consequence of infection with *Cl. difficile* at once explained such clustering, and outbreaks of *Cl. difficile* colitis among hospital patients have now been frequent enough to establish the importance of cross-infection in humans, although the precise means by which this occurs has yet to be discovered (Greenfield *et al.* 1981; Kim *et al.* 1981; Rogers *et al.* 1981). Because the frequency of *Cl. difficile* colitis is likely to vary from time to time according to variations in the extent of cross-infection, studies designed to establish the proportion of cases of antibiotic-associated diarrhoea attributable to *Cl. difficile* cannot be expected to yield absolute figures. Lishman *et al.* (1981) tested the stools of 53 patients who had developed diarrhoea after a course of antibiotic and found 10 (19 per cent) positive for *Cl. difficile* toxin. However, four out of 53 (7.5 per cent) control patients without diarrhoea gave stools in which toxin was present at comparable titres. At another centre Viscidi *et al.* (1981) obtained very different findings, in that only one out of 56 (2 per cent) patients without diarrhoea gave stools positive for toxin as against 130 out of 215 (60 per cent) patients with diarrhoea. Such contrasting results tend not only to confirm the supposition that the role of *Cl. difficile* in antibiotic-associated diarrhoea is inconstant, but also to indicate that its toxin will not invariably cause diarrhoea.

As yet no strain of *Cl. difficile* responsible for an outbreak of colitis has been traced to its source. Borriello *et al.* (1983) noted that several large wild and domestic animals were known to carry *Cl. difficile*, and gave figures for their own survey of household pets in which 11 out of 52 (21 per cent) dogs and six out of 20 (30 per cent) cats harboured the organism. None of these had apparent gastrointestinal disease. Pathology of the genitourinary tract is associated with a relatively high rate of colonization by *Cl. difficile*, and it may be significant that dysuria and sterile pyuria were among the symptoms suffered in a recent American outbreak of *Cl. difficile* colitis (McKinley *et al.* 1982). Whatever their origin it seems that reservoirs of infection are readily established in the hospital environment, although not always at an identifiable site. In a cluster of 10 cases studied by Pierce *et al.* (1982) at one hospital during 1979 no fewer than five patients had been admitted within a 14-day period, yet no home contact, nurse, food-handler, or item of equipment used in the processing or serving of food yielded a growth of *Cl. difficile*, and the only link between patients (apart from treatment with antibiotics) was the finding that eight had undergone some gastrointestinal procedure during this period. A point of epidemiological interest was that as many as 36 per cent of controls for these patients, selected from the same wards, proved to be carriers of *Cl. difficile*, whereas carriage rates among healthy adults are mostly one tenth of this figure, and even a population selected for stool culture has given only 8 per cent positive specimens (Nash *et al.* 1982). Normal neonates have a particularly high carriage rate for both *Cl. difficile* and its cytotoxin. Richardson *et al.* (1983) showed that colonization occurs during the first 4 weeks of life, and in the series of Lishman *et al.* (1983) isolation rates did not differ in subgroups with and without bowel disturbance.

Two patients with necrotizing enterocolitis yielded consistently negative stools.

Clinical features and diagnosis

Patients developing antibiotic-associated colitis typically describe a sudden onset of watery diarrhoea during the course of treatment or up to 2 weeks after its completion. This is accompanied by diffuse abdominal aching or colicky pain, often with tenesmus and distension. As the condition progresses the patient becomes febrile, and blood may appear in the stools. When admitted to hospital even those patients without predisposing pathology are ill and dehydrated, with a clinical picture resembling severe infective or ulcerative colitis. Like these, antibiotic-associated colitis may be complicated by toxic megacolon and its attendant mortality (Slater 1982). Occasionally diarrhoea is absent or inconspicuous in relation to the pain and distension, so that the case may present as an abdominal emergency (Tedesco et al. 1975). Sigmoidoscopic appearances are variable. The rectal mucosa may show a non-specific granularity and friability or be covered with pseudomembranous plaques 0.2–2.0 cm in diameter, creamy-white and sometimes tinged with green or grey (see Plate 7b). Apart from a hyperaemic rim to the plaques and oedema elsewhere the exposed mucosa may seem normal. Very small plaques are readily concealed by the tenacious film of mucus often present, and in such instances an erroneous diagnosis will be made unless the mucosa is biopsied.

Barium studies in PMC usually show diffuse involvement of the colon with a lesion identified as ulceration by some authors but convincingly shown in the illustrations of Stanley et al. (1974) to represent coating of pseudomembranous plaques (see Plate 7c.) In the knowledge that even severe PMC may be patchy or confined to the proximal colon it is logical that negative findings at rigid sigmoidoscopy should be regarded as an indication for fibreoptic examination when the clinical picture is otherwise typical. Early authors specified colonoscopy (Seppälä 1978; Tedesco 1979), but more recently Tedesco et al. (1982a) reported seeing pseudomembranes in 20 out of 22 patients with the flexible sigmoidoscope alone.

Treatment

If the antibiotic is continued the lesion tends to progress, but the patient's condition will not necessarily be improved by withdrawal of the drug alone. In many instances bed-rest and attention to fluid and electrolyte balance are the only additional measures required (Benner and Tellman 1970; Le Frock et al. 1975).

Although few clinicians now would wish to withhold antimicrobial treatment from even the mildest cases, variable success was achieved with other agents before the infective origin of antibiotic-associated colitis had been established. The status of antidiarrhoeal medication is frankly controversial. Pittman (1975) found that all those patients in whom colitis was most severe gave a history of constipation or of treatment with diphenoxylate with atropine (Lomotil) or both, whereas Tedesco (1976) used Lomotil routinely in his scheme of treatment and claimed success in all but two of 47 cases. Striking improvement was attributed by several authors to their use of cholestyramine at the low dosage of 2–4 g daily (Burbige and Milligan 1975; Sinatra et al. 1976; Tures et al. 1976). In the case of Tures et al. (1976) PMC had been aggravated by numerous courses of treatment with cephalosporins; yet when the patient died of myocardial infarction 3 weeks after institution of cholestyramine his colon was found to be normal. This effect of cholestyramine may be explained by the finding of Chang et al. (1978) that it binds the toxin of Cl. difficile in vitro.

In selecting vancomycin as one of two antibiotics with which to protect hamsters against the colitis inducible with clindamycin Bartlett et al. (1977) had in mind the efficacy of this poorly-absorbed compound in human staphylococcal enterocolitis. It was then promptly adopted for the treatment of colitis due to Cl. difficile (Rifkin et al. 1977; Modigliani and Delchier 1978; Tedesco et al. 1978), and has since remained the standard against which other antimicrobials are judged. Most authorities agree that nothing is to be gained by giving more than 125 mg 6-hourly for an initial course of 5 days, orally if possible. However from the experience of vancomycin as an agent capable of precipitating PMC in hamsters Larson et al. (1978) predicted the likelihood of relapse in patients receiving this treatment, an event which has since become familiar (George et al. 1979; Finch et al. 1979; Walters et al. 1983). Some patients respond to one or more further courses of vancomycin alone, but it may be better to institute combined therapy. In a series of 11 patients Tedesco (1982) was able to secure a final remission by adding colestipol to adsorb and inactivate the cytotoxin while the antibiotic dosage was tapered over a period of some 3 weeks. Of alternative antibiotics tetracycline has been successful in cases of antibiotic-associated diarrhoea without colitis (DeJesus and Peternel 1978), and bacitracin has effected a cure both in new cases and in those that had proved refractory to vancomycin (Chang et al. 1980). The efficacy of metronidazole is now well documented (Trinh Dinh et al. 1978; Matuchansky et al. 1978; Pashby et al. 1979), although resistance to this too has been described (Mogg *et al.* 1979). In a direct comparison between oral vancomycin at 2 g daily and metronidazole at 1 g daily for 10 days, Teasley *et al.* (1983) found the drugs to be equivalent in efficacy and relapse rate, but preferred metronidazole because of its lower cost. The price advantage of metronidazole would have remained substantial even if this American trial had been made with the much lower dose of vancomycin usually employed, a point to which attention had already been drawn in the UK (Bolton 1980).

Antitoxin has prevented clindamycin-induced colitis in hamsters when given up to 24 hours after administration of the toxin (Allo *et al.* 1979), but its value at the relatively late stage of diagnosis in human disease has yet to be determined.

Although effective treatment may be expected to restore a normal bowel habit within two weeks, some macroscopic evidence of mucosal inflammation is usually detectable for at least two months in patients with disease so well established.

In a small proportion of fulminant cases colectomy may prove a life-saving measure (Scott *et al.* 1973; Levine *et al.* 1976; Goodacre *et al.* 1977; Eriksson *et al.* 1982), although even the development of toxic megacolon is not necessarily an indication for surgery. In a survey of the literature Cone and Wetzel (1982) found that operative treatment had fared no better than conservative, with an overall mortality of 33 per cent. Vancomycin alone sufficed to control both the colitis and its complication in the case of Templeton (1983). The first patient to develop toxic megacolon in antibiotic-associated PMC (Brown *et al.* 1968) recovered after intensive intravenous replacement and a combination of corticosteroids with methicillin, nystatin, and *Lactobacillus casei*. This and the recent success of a retention enema prepared from faeces in eliminating *Cl. difficile* after repeated courses of vancomycin had failed (Schwan *et al.* 1983) suggest that in some instances replacement of the normal flora might obviate the need for antibiotic treatment of antibiotic-associated colitis.

Pathology

Gross pathological examination of resected bowel in PMC confirms the impression given by radiological studies that the lesion is effectively limited to the colon, although congestion and oedema of the terminal ileum may be present. The earliest microscopic change is a general increase in epithelial mucus production, leading to ballooning of individual cells, some of which burst and discharge their contents into the cyrpts. Being unduly tenacious this mucus may then accumulate and distort the glandular pattern. An early pseudomembrane is formed at this stage, consisting mainly of mucus and polymorphonuclear leucocytes. Death of epithelial cells is first seen to occur in minute form on the 'interglandular summits'. Typically a few adjacent cells desquamate and their basement membrane ruptures. As the site is infiltrated by ploymorphonuclear leucocytes a fibrinous exudate develops, and these elements, together with epithelial debris, erupt into the lumen as through the summit of a volcano to form the stalk of the definitive pseudomembrane [see Plate 6(d)]. Although necrosis then extends outwards from this point to involve surrounding glands, with a corresponding increase in the surface area of the pseudomembrane, the attachment of the pseudomembrane to the mucosa remains relatively narrow. Unless, therefore, special care is taken with the presentation of biopsies for histological study their two main components readily become separated, allowing the true nature of the lesion to be missed. This applies particularly to early cases, where the lesion may not be large enough to appear in more than a few of the sections made. The principal inflammatory exudate in the lamina propria consists of lymphocytes, plasma cells, and macrophages, and extends with diminishing intensity into the submucosa. Here oedema and hyperaemia are the main characteristics, with subsequent endothelial proliferation and, in the severest lesions only, capillary and arteriolar microthrombi. It is emphasized in all the key pathological texts that thrombosis of small vessels is a late manifestation and therefore almost certainly a complication rather than a cause of mucosal necrosis in this condition. (Reiner *et al.* 1952; Goulston and McGovern 1965; Price and Davies 1977).

New pathological studies of antibiotic-associated colitis in the hamster suggest that the primary site of damage is the microvillous membrane. Different patterns of pathological change correlate well with the three clinical patterns recognized in man: diarrhoea without obvious colitis; PMC; and areas of extensive sloughing in which pseudomembranes are replaced by a uniformly haemorrhagic surface (Humphrey *et al.* 1979b; Price *et al.* 1979).

Pathogenic toxins

The cytotoxin of *Cl. difficile* incriminated in most studies to date has yet to be chemically defined. However, after partial purification of the material obtained from hamsters with antibiotic-associated colitis Humphrey *et al.* (1979a) estimated its molecular weight at about 107 000 daltons. It was heat-labile and readily bound by cholestyramine *in vitro*. One explanation for cases of antibiotic-associated colitis in which *Cl. difficile* but not its cytotoxin is isolated could be that this organism elaborates more than one toxin capable of inducing colitis

(Nash *et al.* 1982). In the USA N.S. Taylor *et al.* (1980) discovered a distinct enterotoxin which was similar to the cytotoxin in molecular weight but had only a fraction of its cytopathic effect and was best assayed by the rabbit ileal loop test. The existence and properties of the new toxin were soon confirmed by Burdon *et al.* (1981) in Birmingham. Relevance of this enterotoxin to human disease was implied by the finding of N.S. Taylor *et al.* (1980) that, when injected intracaecally into hamsters, it caused a fatal caecitis, though a crude extract which probably contained both cytotoxin and enterotoxin was shown by Hughes *et al.* (1983) to induce Cl$^-$ secretion by rabbit ileum *in vitro* through a Ca^{2+}-dependent mechanism involving no microscopic damage to the mucosa. Failure of Riley *et al.* (1983) to stimulate secretion in either cat jejunum or the suckling mouse assay with any of 30 *Cl. difficile* strains isolated from adults with diarrhoea suggests a marked interspecies variability in the response to *Cl. difficile* toxins which must in particular throw serious doubt on the role of the enterotoxin in human disease.

Where neither *Cl. difficile* nor its toxin can be identified in the stools of patients with antibiotic-associated colitis the possibility that another member of the genus *Clostridium* is the cause should be considered. Reference is made in the following section to *Cl. butyricum* as an opportunist pathogen in Crohn's disease (Graham 1982), while Lamont *et al.* (1979) have identified in rabbits given clindamycin a toxin of 45 000 daltons which was specifically neutralized by antiserum to *Cl. perfringens* type E.

Variants of antibiotic-associated colitis

Cl. difficile in inflammatory bowel disease

The concept of intercurrent infection as a cause of exacerbations in chronic inflammatory bowel disease is now well recognized. When in such cases the infecting organism is *Cl. difficile* the role of antibiotic therapy must be considered doubtful, though Pokorney and Nichols (1981) had some evidence that in their patient sulphasalazine contributed to the episode. In the series of Lamont and Trnka (1980) four out of five patients with ulcerative colitis were on sulphasalazine, but two were able to continue this without ill effect after the toxin had been eliminated. Among 109 patients with either ulcerative colitis or Crohn's disease Greenfield *et al.* (1983) found no relationship between sulphasalazine treatment and the isolation of either *Cl. difficile* or its toxin from faeces, though treatment with antibiotics during the previous month was associated with an isolation rate

more than double that for the group as a whole ($p < 0.01$). Bolton *et al.* (1980) screened 56 patients with diarrhoea for the presence of *Cl. difficile*. Of the nine giving a positive test only one had been receiving antibiotics, and five of the remainder were patients with inflammatory bowel disease on corticosteroids. Conversely, antibiotic-associated clostridial infection may be the means by which a chronic colitis of idiopathic type is brought to light. When a 57-year-old man developed colitis with bloody diarrhoea during a course of erythromycin and gave a 12-year history of minor bowel irregularity, there must have been a suspicion of underlying chronic inflammation, presumably strengthened by the development of toxic megacolon (Graham 1982). The patient came to colectomy, when the full histological picture of Crohn's disease was found in the resected specimen. It so happened that the infecting organism in this case was *Cl. butyricum*, but whichever species had been cultured the question could be asked whether the need for colectomy might have been avoided if corticosteroids had been given in addition to vancomycin.

Neutropenic enterocolitis

When patients under treatment for malignant disease develop *Cl. difficile* colitis the cause may be wholly the antibiotics they so frequently require, but other factors must be considered. Of the two cases reported by Cudmore *et al.* (1982) one had received only 5-fluorouracil; the other had received antibiotics during a phase of neutropenia, which has also been a feature of atypical cases seen in the Cambridge Leukaemia Unit (Rampling *et al.* 1982). Some of these had no pseudomembranes or even diarrhoea, but presented with jaundice and ascites. The mucosal surface of one resected colon was oedematous, with focal haemorrhages but no ulcers or exudate. Patchy necrosis was found in the deeper layers of the caecum, where the serosa was oedematous and covered with a fine layer of fibrin. Although others identify this form of colitis with the necrotizing enterocolitis that is recognized as a complication of leukaemia with agranulocytosis (Couzigou *et al.* 1982) both groups would seem to agree that the special features are primarily the result of failure to localize infection in the absence of neutrophils and that vancomycin is indicated as strongly here as in classical *Cl. difficile* colitis. The prime significance of neutropenia in this enterocolitis, which is characteristically confined to the proximal colon and distal ileum, is supported by a case in which it followed agranulocytosis due to mianserin treatment (Braye *et al.* 1982). *Cl. difficile* and its toxin were not sought here, but an infiltrate of Gram-positive bacilli was seen in the sub-mucosa of the resected terminal ileum.

Ampicillin colitis

Discussion at an early workshop on antibiotic-associated colitis (Keusch and Present 1976) touched on an apparently distinct form of colitis, characterized by bloody diarrhoea, absence of pseudomembranes, and relatively mild inflammation in patients who had been taking ampicillin. Subsequently Toffler *et al.* (1978) described five cases of an acute transient colitis following treatment with ampicillin in one, amoxycillin in another, and ampicillin together with oral penicillin in the remainder. All presented with abdominal cramps and bloody diarrhoea. Sigmoidoscopic abnormality was minimal, but in three patients who underwent barium enema examination there were gross radiological changes resembling those of acute ischaemia. Features included 'thumb-printing', spasm responsive to glucagon, diffuse thickening of mucosal folds, nodularity, and punctate ulceration. Such appearances were largely confined to the right colon, as were the sharply-demarcated haemorrhagic and spastic lesions seen at colonoscopy by Sakurai *et al.* (1979) in eight patients who had likewise developed sudden, severe abdominal cramps and bloody diarrhoea after ampicillin treatment. In this series only one patient had pseudomembranes, and microscopic evidence of inflammation was slight. Without any specific treatment the patients of both these series recovered clinically in a matter of three days. Sakurai and colleagues found a normal mucosa in all six of the patients they subjected to repeat endoscopy within 4–12 days of the first examination. Four cases of a similar transient colitis seen in the UK (Freeman and Low 1981) comprised two attributable to amoxycillin and two to phenoxymethyl penicillin. All the patients were young females, of whom two were also taking a contraceptive pill. In one of these, however, the Pill was resumed after the attack without provoking a recurrence. No pathogen has yet been grown from the stool of patients with this transient haemorrhagic colitis, and failure to culture *Cl. difficile* in particular is a consistent feature (McKinley and Toffler 1980; Freeman and Low 1981; Gould *et al.* 1982).

Sclerosing peritonitis

Practolol

The gastrointestinal side-effects of practolol, a β-adrenergic receptor antagonist, include symptoms of acute intolerance, pneumatosis coli in a single published case (Thein and Asquith 1977), and, most importantly, sclerosing peritonitis. Following descriptions of an early eczematous or psoriasiform skin rash, with or without ocular discomfort, in patients receiving practolol, Raftery and Denman (1973) reported the onset of a syndrome resembling disseminated lupus erythematosus after 6 months' treatment in three cases. It was not until the next year that it became clear that this polyserositis, which was characterized by pleural and joint effusions, could also involve the peritoneal cavity. Independent reports by Brown *et al.* (1974) and Windsor *et al.* (1975) described a new entity, sclerosing (or fibrinous) peritonitis, in four patients who had been taking practolol for 2 years. The epidemiological, clinical, and pathological picture of sclerosing peritonitis given by these early accounts was quite distinctive, and has since been closely reproduced in many individual case reports. Drawing on their unequalled experience of 16 patients with sclerosing peritonitis the Bristol group published a description of this disorder that is likely to be definitive (Marshall *et al.* 1977), since the oral preparation of practolol for maintenance therapy is no longer available.

All the patients known to Marshall *et al.* (1977) had been subject earlier to other side-effects of practolol. Eleven had had dry eyes, with or without corneal ulceration, and nine a skin eruption which, in all except one, had been accompanied by ocular symptoms. Three had had pleural involvement, and two parenchymatous lung disease, progressing in one case to chronic fibrosing alveolitis. No patient had been receiving practolol for less than 15 months at onset of abdominal symptoms and all save two had been treated for more than two years. Confusion as to the cause of the peritonitis had arisen in two cases where practolol had been discontinued eight months previously. However, latent periods of even greater length have been recorded by other authors, amounting to a year in at least two cases (Halley and Goodman 1975; Kristensen *et al.* 1975), 18 months in another (Allan and Cade 1975), and 22 months in a fourth (Trudinger and Fitchett 1976).

The most usual history has been one in which the patient noted gradual onset of abdominal distension and discomfort or intermittent colicky pains, unrelated to meals, over a period of weeks or months. Nausea has been common, leading frequently to vomiting, which may be copious and such as to suggest pyloric stenosis. Weight loss of at least 5 kg is not uncommon. In most cases the patient is admitted to hospital with the symptoms and signs of small-bowel obstruction, occasionally acute in onset (Allan and Cade 1975). Twelve of the 16 patients in the Bristol series presented with an abdominal mass, five of these being in the suprapubic region where, as in several cases reported from elsewhere

(Kristensen *et al.* 1975; Trudinger and Fitchett 1976), they simulated genital tumours.

Adequate barium studies have seldom been performed, but Marshall *et al.* (1977) were able to identify characteristic small-intestinal changes in eight follow-through series. Common to all were the dual features of dilatation and sacculation, giving the ileum a superficial resemblance to haustrated colon [see Plate 5(c)]. The extent of the sacculation correlated well with shortening of the bowel found at laparotomy. In six cases the loops of small intestine were well separated, and this effect occurring with sacculation in a film illustrated by Cook and Foy (1976) enabled these authors to predict shortening pre-operatively. On occasion several loops were observed by the Bristol group to lie enclosed within a narrow curvilinear translucency, another effect explicable on the basis of laparotomy findings [see Plate 5d]. Fluoroscopy showed the bowel to be markedly fixed, and transit of barium was correspondingly slow in nearly all subjects. Small-bowel function has rarely been tested in patients with sclerosing peritonitis since by the time most present both the clinical picture and the appearances in plain abdominal radiographs indicate early surgery for relief of obstruction. Two patients at Bristol had steatorrhoea associated with overgrowth of *E. coli* and *Bacteroides* in the jejunum.

Findings at laparotomy have been strikingly uniform in all reports. The small bowel is encased in a sheath of greyish, opaque fibrous tissue so as to form a bag or cocoon filled with a mass of intestinal loops that are shortened, dilated, and kinked but otherwise normal. The mesentery may be unaffected (Allan and Cade 1975) or covered to a variable extent with the same membrane. Retroperitoneal tissues are invariably spared and the colon almost always so. As a rule the fibrous covering is readily removed by incising it along the antimesenteric border of the bowel, making further perpendicular incisions towards the root of the mesentery if necessary, and then peeling it off by blunt dissection (Bendtzen and Søborg 1975; Cook and Foy 1976; Trudinger and Fitchett 1976). In some cases vascular adhesions to a thickened parietal peritoneum render complete removal of the abnormal tissue impracticable, so that multiple operations may be needed (Marshall *et al.* 1977) and the chances of a fatal outcome to intra-abdominal complications are substantially increased (Halley and Goodman 1975; Marshall *et al.* 1977). Even without such difficulties post-operative complications are frequent in these patients, since all have advanced cardiovascular disease and many are elderly.

The principal microscopic feature in all 11 specimens of the Bristol series was a coat of laminated fibrous tissue 0.5–4.0 mm thick deposited immediately beneath the mesothelium and infiltrated with eosinophilic fibrinous material, which in three cases penetrated the mesothelium to form macroscopic nodules on its surface. A cellular infiltrate, consisting chiefly of mononuclear cells, was found in relation to blood vessels of the underlying adipose layer and extending sparsely into the fibrous tissue itself. The nature of this inflammatory reaction has varied considerably, for in the case of Allan and Cade (1975) heavy lymphocytic infiltration was described, while Kristensen *et al.* (1975) noted similar appearances in one of their cases and in another no inflammation whatever. The superficial fibrinous exudate was prominent in the specimen examined by Trudinger and Fitchett (1976), and was associated with polymorphonuclear infiltration. Vascularity, although likewise variable, can be intense in both fibrous and adipose layers. No instance has yet been reported of this process invading the intestinal wall.

The mechanism by which practolol induces sclerosing peritonitis is unknown. One pharmacological effect proposed to explain the DLE-like syndrome is inactivation by practolol of the suppressor T-lymphocytes which normally prevent elaboration of auto-antibodies (Raftery and Denman 1973). Another would depend on its action of interfering with endogenous catecholamines in respect of their ability to control the release of lysosomes from phagocytic cells and thus prevent inappropriate inflammatory reactions (Marshall *et al.* 1977). β-adrenergic activity also suppresses the production of collagen by fibroblasts, and in cultures of such cells obtained from human lung and skin propranolol was shown to inhibit the suppression exerted by isoprenaline (Moss *et al.* 1979). In favour of a pharmacological effect is the apparent relationship between total dose of practolol and peritonitis, as well as the failure by Raftery and Denman (1973) to detect any immune reactivity to practolol in their patients: against it the extreme rarity of sclerosing peritonitis attributable to other β-adrenergic blocking agents. There are reasons for doubting any direct association between the DLE-like syndrome and sclerosing peritonitis, not least among them the absence of this phenomenon from both spontaneous and drug-induced DLE, although each would seem to depend in part upon the dose of practolol received. There is no evidence from HLA studies that patients who contract sclerosing peritonitis do so because they inherit an abnormal immune constitution (Dick *et al.* 1978).

Other β-blocking agents

Gurry *et al.* (1975) were disturbed that their case of sclerosing peritonitis followed 6 months' treatment with propranolol in a patient who had previously received practolol. The epidemiological evidence now available is adequate to exonerate propranolol in this instance, but subsequent events have justified the concern felt by Marshall *et al.* (1977) over the results of their decision to study asymptomatic patients on β-blocking drugs, which included dilatation and sacculation of the small bowel in two patients taking propranolol and three taking oxprenolol. One patient on 120 mg propranolol daily underwent a second small-bowel meal examination 6 months after the first, when marked progression of the earlier changes were demonstrated, and it was noted that during the intervening period he had developed abdominal pain and nausea. Withdrawal of the β-blocking agent was associated with some regression of the radiological abnormalities in this and five other cases.

Since this study two cases of sclerosing peritonitis caused by propranolol have been reported. A 43-year-old woman noted symptoms a mere 2 months after propranolol had been prescribed at 80 mg daily for angina pectoris (Harty 1978). Lower abdominal pain, distension, and tenesmus were followed by ocular and skin lesions. Polyserositis became apparent when she developed a pleural effusion and ascites. At laparotomy the mesentery was found to be shortened and continuous with a dense layer of fibrous tissue that encased the small bowel, loops of which were grossly distended. Microscopy of the peritoneum showed a diffuse proliferation of fibrous tissue and a cellular infiltrate of lymphocytes and polymorphs. This case is notable not only as the first in which a typical sclerosing peritonitis had followed the use of a β-blocking agent other than practolol but also for the exceptionally short duration of treatment required to induce the syndrome of which this formed a part. The patient of Ahmad (1981) was a 56-year-old man who had been taking this drug for an unspecified period at 320 mg daily when he underwent coronary arterial bypass. During convalescence he suffered a profuse rectal haemorrhage, and at laparotomy multiple adhesions were found. Those in the right upper quadrant were especially dense, fixing the organs there to an extent that effectively precluded their mobilization. The impression gained clinically was that this fibrotic process had caused the haemorrhage by inducing local ischaemia in the right colon. A resected segment from that side showed extensive ulceration, with a zone of fibrinoid necrosis underlain by granulation tissue, containing a mixed inflammatory exudate and proliferating fibroblasts and capillaries. The inflammation and fibrosis spread through the muscular wall to involve serosa and mesentery. In the adjacent ileum only the serosa was substantially involved. Before ultimately recovering the patient developed ileus, an extensive left pleural effusion, and acute renal failure.

Clark and Terris (1983) reported the case of a 63-year-old woman who presented with a five-day history of colicky central abdominal pain, distension, vomiting, and absolute constipation after receiving metoprolol 200 mg and cyclopenthiazide 250 mg daily for 3 years in a hypotensive regimen. The only previous abdominal surgery had been an appendicectomy, and the pathological features found at laparotomy on this occasion were typical of sclerosing peritonitis.

The rarity of sclerosing peritonitis among the millions of patients who have taken propranolol does not, as claimed elsewhere (Marigold *et al.* 1982), invalidate the association provided that the evidence for this is strong enough in the cases reported. It does, however, suggest that some form of exceptional predisposition must have obtained in these cases. Now that metopolol too has been implicated, reassurance about the safety of β-blocking agents other than practolol seems more than ever inappropriate, and informed vigilance the approach to be advised. The danger of rarity in the association between a treatment and one of its more serious adverse effects is that the possibility of its applying in a particular case may be ignored, so that the patient is subjected to an unnecessary risk of severe ill-health or even death.

Continuous ambulatory peritoneal dialysis (CAPD)

A form of sclerosing peritonitis has been described in patients who have developed chronic symptoms after months to years of CAPD (Gandhi *et al.* 1980; Bradley *et al.* 1983). This is almost always preceded by recurrent bacterial or 'aseptic' peritonitis, and the symptoms are more those that would be expected of a simple chronic peritonitis than those of the intestinal obstruction which prompts patients with classical sclerosing peritonitis to seek medical advice. At laparotomy the principal finding is a generalized thickening of the peritoneum. Possible causes apart from repeated acute peritonitis, which does not necessarily induce chronic structural changes (Sorkin *et al.* 1982), include the relatively low pH or other potentially irritative chemical qualities of some commercially prepared dialysis fluids. Oreopoulos *et al.* (1983) commented that fluids containing acetate seemed to be associated with a higher incidence of peritoneal fibrosis than those containing lactate. However, several atypical pathogens have been incriminated in recurrent peritonitis related to CAPD, and failure to eradicate these — or more familiar organisms insensitive to the antibiotics used — may predispose to fibrosis from continuous low-grade inflammation (Eisenberg *et al.* 1983). Finally, it has been noted by most authors that some of their patients have been on long-term treatment with β-blocking agents during the period in which their peritoneal fibrosis developed.

Acknowledgments

My thanks for help in collecting the data upon which this chapter is based are due to all members of the library staff at Lister Hospital, BMA House, and RN Hospital Haslar, but especially to Mrs Sheila Lindsell and Mrs Sally Knight. I am grateful also to Mrs Wendy Miller for the patience she has displayed in typing the manuscript.

RECOMMENDED FURTHER READING

COOKE, A.R. (1976). Drugs and gastric damage. *Drugs* 11, 36–44.

HART, F.D. (1975). The new antirheumatic drugs. *Drugs* 9, 321–5.

KENDALL, M.J. AND CHAN, K. (1973). Drug-induced malabsorption. *Xenobiotica* 3, 727–44.

KIKENDALL, J.W., FRIEDMAN, A.C., OYEWOLE, M.A., FLEISCHER, D., AND JOHNSON, L.F. (1983). Pill-induced esophageal injury: case reports and review of the medical literature. *Dig. Dis. Sci.* 28, 174–82.

KINSELLA, T.J. AND BLOOMER, W.D. (1980). Tolerance of the intestine to radiation therapy. *Surg. Gynecol. Obstet.* 151, 273–84.

MILLER, T.A. AND JACOBSON, E.D. (1979). Gastrointestinal cytoprotection by prostaglandins. *Gut* 20, 75–87.

REFERENCES

ABBARAH, T.R., FREDELL, J.E., AND ELLENZ, G.B. (1976). Ulceration by oral ferrous sulfate. *J. Am. med. Ass.* 236, 2320.

ADAMS, P.L., RUTSKY, E.A., ROSTAND, S.G., AND HAN, S.Y. (1982). Lower gastrointestinal tract dysfunction in patients receiving long-term hemodialysis. *Arch. intern. Med.* 142, 303–6.

AGDAL, N. (1979). Medicininducerede esophagusskader. En oversigt samt et tilfaelde af indometacinfremkaldt ulceration med dodelig udgang. *Ugeskr. Laeg.* 141, 3019–21.

AHMAD, S. (1981). Sclerosing peritonitis and propranolol. *Chest* 79, 361–2.

AKHTAR, A.J., CROMPTON, G.K., AND SCHONELL, M.E. (1968). Para-aminosalicylic acid as a cause of intestinal malabsorption. *Tubercle* 49, 328–31.

ALAILY, A.B. (1974). Gangrene of Meckel's diverticulum in pregnancy due to iron tablet. *Br. med. J.* 1, 103.

ALDRETE, J.S., STERLING, W.A., HATHAWAY, B.M., MORGAN, J.M., AND DIETHELM, A.G. (1975). Gastrointestinal and hepatic complications affecting patients with renal allografts. *Am. J. Surg.* 129, 115–24.

ALINOVI, V., HERZBERG, F.P., YANNOPOULOS, D., AND VETERE, P.F. (1980). Cecal volvulus following cesarean section. *Obstet. Gynecol.* 55, 131–4.

ALLAN, D. AND CADE, D. (1975). Delayed fibrinous peritonitis after practolol treatment. *Br. med. J.* 4, 40.

ALLIBONE, A. AND FLINT, F.J. (1958). Gastrointestinal haemorrhage and salicylates. *Lancet* ii, 1121.

ALLO, M., SILVA, J., FEKETY, R., RIFKIN, G.D., AND WASKIN, H. (1979). Prevention of clindamycin-induced colitis in hamsters by *Clostridium sordellii* antitoxin. *Gastroenterology* 76, 351–5.

ALMGREN, B. AND JUHL, M. (1977). Superior mesenteric artery syn-

drome complicating treatment with balanced traction. A case report. *Acta sorthop. cand.* 48, 25–8.

ALTMAN, A.J., DINNDORF, P., AND QUINN, J.J. (1982). Acute pancreatitis in association with cytosine arabinoside therapy. *Cancer* 49, 1348–56.

ALTMAN, D.H. AND PURANIK, S.R. (1973). Superior mesenteric artery syndrome in children. *Am. J. Roentg.* 118, 104–8.

ALVAREZ, A.S. AND SUMMERSKILL, W.H.J. (1958). Gastrointestinal haemorrhage and salicylates. *Lancet* ii, 920–5.

AMERY, A., VANDENBROUCKE, J., DESBUQUOIT, J.L., AND DE GROOTE, J. (1973). Pancreatitis during clonidine treatment. *Tijdschr. Gastroent.* 16, 179–85.

ANDERSON, K.W. (1964). A study of the gastric lesions induced by aspirin in laboratory animals. *Arch. int. Pharmacodyn. Thér.* 152, 379–91.

ANDERSSON, H., BOSAEUS, I., AND NYSTRÖM, C. (1978). Bile salt malabsorption in the radiation syndrome. *Acta radiol. (ther.)* 17, 312–18.

ANDRESS, M. (1971). Submucosal haematoma of the oesophagus due to anticoagulant therapy. Report of a case. *Acta radiol. Diagn.* 11, 216–19.

ARNOLD, F., DOYLE, P.J., AND BELL, G. (1978). Acute pancreatitis in a patient treated with cimetidine. *Lancet.* i, 382–3.

ARNOLD, J.D. AND BERGER, A.E. (1983). Comparison of fecal blood loss after use of aspirin and suprofen. *Pharmacology* 27 (suppl. 1), 14–22.

ARSENAULT, A., VARADY, J., LeBEL, E., AND LUSSIER, A. (1975). Effect of naproxen on gastrointestinal microbleeding following acetylsalicylate medication. *J. clin. Pharmacol.* 15, 340–6.

ARVANITAKIS, C., CHEN, G.-H., FOLSCROFT, J., AND GREENBERGER, N.J. (1977). Effect of aspirin on intestinal absorption of glucose, sodium and water in man. *Gut* 18, 187–90.

——, LORENZSONN, V., AND OLSEN, W.A. (1973). Phenformin-induced alterations of small-intestinal function and mitochondrial structure in man. *J. Lab. clin. Med.* 82, 195–200.

ASELTON, P.J. AND JICK, H. (1983). Short-term follow-up study of wax matrix potassium chloride in relation to gastrointestinal bleeding. *Lancet* i, 184.

ATKINSON, A.B., BROWN, J.J., LEVER, A.F., AND ROBERTSON, J.I.S. (1980). Combined treatment of severe intractable hypertension with captopril and diuretic. *Lancet* ii, 105–8.

ATKIN-THOR, E., GODDARD, B.W., O'NION, J., STEPHEN, R.L., AND KOLFF, W.J. (1978). Hypogeusia and zinc depletion in chronic dialysis patients. *Am. J. clin. Nutr.* 31, 1948–51.

ATTAS, M., SABAWALA, P.B., AND KEATS, A.S. (1968). Acute transient sialadenopathy during induction of anesthesia. *Anesthesiology* 29, 1050–2.

ATWATER, E.C., MONGAN, E.S., WIECHE, D.R., AND JACOX, R.F. (1965). Peptic ulcer and rheumatoid arthritis. *Arch. intern. Med.* 115, 184–9.

AUBREY, D.A. AND BURNS, G.P. (1973). Effect of topical prednisolone and acetic acid on the antral phase of gastric secretion. *Am. J. Surg.* 125, 676–80.

AXELSSON, C.K., CHRISTIANSEN, L.V., JOHANSSEN, AA., AND EJBY POULSEN, P. (1977). Comparative effects of tolfenamic acid and acetylsalicylic acid on human gastric mucosa. A double-blind crossover trial employing gastroscopy, external gastro-camera and multiple biopsies. *Scand. J. Rheumatol.* 6, 23–7.

AZUUMI, Y., OHARA, S., ISHIHARA, K., OKABE, H., AND HOTTA, K. (1980). Correlation of quantitative changes of gastric mucosal glycoproteins with aspirin-induced gastric damage in rats. *Gut* **21**, 533–6.

BAAR, H.S. AND WOLFF, O.H. (1957). Pancreatic necrosis in cortisone-treated children. *Lancet.* **i**, 812–15.

BADENOCH, J. (1954). The use of labelled vitamin B_{12} and gastric biopsy in the investigation of anaemia. *Proc. Roy Soc. Med.* **47**, 426–7.

BAILEY, G.L., MOCELIN, A.J., GRIFFITHS, H.J.L., HAMPERS, C.L., AND MERRILL, J.P. (1971). Renal homotransplantation in the 50–80 year age group. *Proc. Dial. Transpl. Forum* **1**, 1–5.

BAIRD, G.M. AND DOSSETOR, J.F.B. (1981). Methotrexate enteropathy. *Lancet.* **i**, 164.

BAKER, D.R., SCHRADER, W.H., AND HITCHCOCK, C.R. (1964). Small-bowel ulceration apparently associated with thiazide and potassium therapy. *J. Am. med. Ass.* **190**, 586–90.

BALL, J.R. (1976). Potassium strictures of the upper alimentary tract. *Lancet.* **i**, 495–6.

BANK, S. AND MARKS, I.N. (1970). Hyperlipaemic pancreatitis and the Pill. *Postgrad. med. J.* **46**, 576–8.

BANNERMAN, R.M., BEVERIDGE, B.R., AND WITTS, L.J. (1964). Anaemia associated with unexplained occult blood loss. *Br. med. J.* **1**, 1417–19.

BARAGAR, F.D. AND DUTHIE, J.J. (1960). Importance of aspirin as a cause of anaemia and peptic ulcer in rheumatoid arthritis. *Br. med. J.* **1**, 1106–8.

BARBEAU, A. (1970). L-DOPA therapy, past, present and future. *Ariz. Med.* **27**, 1–4.

BARBIER, P., PRINGOT, J., HEIMANN, R., FIASSE, R., AND JACOBS, E. (1976). Digestive lesions induced by kalium chloride. *Acta gastro-enter Belg.* **39**, 261–74.

BARKIN, J.S., HARARY, A.M., SHAMBLEN, C.E., AND LASSETER, K.C. (1983). Potassium chloride and gastrointestinal injury. *Ann. intern. Med.* **98**, 261–2.

BÁRTA, K., SEIDLOVÁ, V., AND BENÝ ŠEK, L. (1965). Prudká ileokolitida po fenylbutazonu. *Vnitřní lékařstvi* **11**, 326–30.

BARTLETT, J.G., ONDERDONK, A.B., AND CISNEROS, R.L. (1977). Clindamycin-associated colitis in hamsters: protection with vancomycin. *Gastroenterology* **73**, 772–6.

——, CHANG, T., AND ONDERDONK, A.B. (1978). Will the real *Clostridium* species responsible for antibiotic-associated colitis please step forward? *Lancet.* **i**, 338.

BATAILLE, C., SOUMAGNE, D., LOLY, J., AND BRASSINNE, A. (1982). Esophageal ulceration due to indomethacin. *Digestion* **24**, 66–8.

BATALDEN, P.B., VAN DYNE, B.J., AND CLOYD, J. (1979). Pancreatitis associated with valproic acid therapy. *Pediatrics* **64**, 520–2.

BATES, B. (1965). Granulomatous peritonitis secondary to corn starch. *Ann. intern. Med.* **62**, 335–47.

BEAMISH, R.E., AND MCCREATH, N.D. (1961). Intestinal obstruction complicating anticoagulant therapy. *Lancet.* **ii**, 390–2.

BEAR, R. AND STEER, K. (1976). Pseudo-obstruction due to clonidine. *Br. med. J.* **1**, 197.

BEARD, R.J. (1981). Unusual presentation of translocated intrauterine contraceptive device. *Lancet.* **i**, 837.

BEAVIS, J.P., PARSONS, R.L., AND SALFIELD, J. (1976). Colitis and diarrhoea: a problem with antibiotic therapy. *Br. J. Surg.* **63**, 299–304.

BECCIOLINI, A., CIONINI, L., CAPPELLINI, M., AND ATZENI, G. (1979). Biliary and pancreatic secretion in abdominal irradiation. *Acta radiol. Oncol. Radiat. Phys. Biol.* **18**, 145–54.

BEIRNE, J.A., BIANCHINE, J.R., JOHNSON, P.C., AND WORTHAM, G.F.

(1974). Gastrointestinal blood loss caused by tolmetin, aspirin, and indomethacin. *Clin. Pharmac. Ther.* **16**, 821–5.

BENDTZEN, K. AND S.OBORG, M. (1975). Sclerosing peritonitis and practolol. *Lancet.* **i**, 629.

BENNER, E.J. AND TELLMAN, W.H. (1970). Pseudomembranous colitis as a sequel to oral lincomycin therapy. *Am. J., Gastroent.* **54**, 55–8.

BENNETT, A., STAMFORD, I.F., AND UNGER, W.G. (1973). Prostaglandin E_2 and gastric acid secretion in man. *J. Physiol.* **229**, 349–60.

BERCHTHOLD, P., BOLLI, P., ARBENZ, V., AND KEISER, G. (1969). Disturbances of intestinal absorption following metformin treatment. *Diabetologia* **5**, 405–12.

BERG, B., GROTH, C.-G., MAGNUSSON, G., LUNDGREN, G., AND RINGDÉN, O. (1975). Gastrointestinal complications in 248 kidney transplant recipients. *Scand. J. Urol. Nephrol.* **29** (Suppl.) 19–20.

BERKOWITZ, D., BEZAHLER, G., AND BRANDT, L.J. (1976). Ampicillin-associated colitis. *Am. J. Gastroent.* **66**, 362–5.

BERMAN, H. AND MAINELLA, F.S. (1952). Toxic results of anticoagulant therapy. *NY. St. J. Med.* **52**, 725–7.

BERNADINO, M.E. AND LAWSON, T.L. (1976). Discrete colonic ulcers associated with oral contraceptives. *Am. J. dig. Dist.* **21**, 503–6.

BERNSTEIN, W.C., NIVATVONGS, S., AND TALLENT, M.B. (1973). Colonic and rectal complications of kidney transplantation in man. *Dis. Colon Rectum.* **16**, 255–63.

BETTS, R.F. (1982). Cytomegalovirus infection in transplant patients. *Prog. med. Virol.* **28**, 44–64.

BIANCHI PORRO, G., CORVI, G., FUCCELLA, L.M., GOLDANIGA, G.C., AND VALZELLI, G. (1977). Gastrointestinal blood loss during administration of indoprofen, aspirin and ibuprofen. *J. int. med. Res.* **5**, 155–60.

——, PETRILLO, M., CARUSO, E., AND FUMAGALLI, M. (1977). Sulindac and gastric mucosa. *Lancet.* **i**, 1152–3.

BILLINGTON, B.P. (1960). Gastric ulcer: age, sex, and a curious retrogression. *Australas. Ann. Med.* **9**, 111–21.

—— (1965). Observations from New South Wales on the changing incidence of gastric ulcer in Australia. *Gut* **6**, 121–33.

BIRNS, M.T., KATON, R.M., AND KELLER, F. (1979). Intramural hematoma of the small intestine presenting with major upper gastrointestinal hemorrhage. Case report and review of the literature. *Gastronterology* **77**, 1094–1100.

BJARNASON, I., AND BJÖRNSSON, S. (1981). Oesophageal ulcers. An adverse reaction to co-trimoxazole. *Acta med. scand.* **209**, 431–2.

BLOCK, M.B., GENANT, H.K., AND KIRSNER, J.B. (1970). Pancreatitis as an adverse reaction to salicylazosulfapyridine. *New Engl. J. Med.* **282**, 380–2.

BLOHMÉ, I. (1975). Gastro-duodenal bleeding after renal transplantation. *Scand. J. Urol. Nephrol.* **29** (Suppl.), 21–3.

BOHANE, T.D., PERRAULT, J., AND FOWLER, R.S. (1978). Oesophagitis and oesophageal obstruction from quinidine tablets in association with left atrial enlargement. A case report. *Aust. paediat. J.* **14**, 191–2.

BOKEY, L. AND HUGH, T.B. (1975). Oesophageal ulceration associated with doxycycline therapy. *Med. J. Aust.* **1**, 236–7.

BOLTON, R.P. (1980). Vancomycin dose for pseudomembranous colitis. *Lancet.* **ii**, 428.

—— AND READ, A.E. (1982). *Clostridium difficile* in toxic megacolon complicating acute inflammatory bowel disease. *Br. med. J.* **285**, 475-6.

——, SHERRIFF, R.J., AND READ, A.E. (1980). *Clostridium difficile* associated diarrhoea: a role in inflammatory bowel disease? *Lancet.* **i**, 383-4.

BONFILS, S., HERVOIR, P., GIRODET, J., LE QUINTREC, Y., BADER, J.P., AND GASTARD, J. (1977). Acute spontaneously recovering ulcerating colitis (ARUC). Report of 6 cases. *Am. J. dig. Dis.* **22**, 429-36.

BONKOWSKY, H.L. AND BRISBANE, J. (1976). Colitis and hepatitis caused by methyldopa. *J. Am. med. Ass.* **236**, 1602-3.

BORRIELLO, S.P., HONOUR, P., TURNER, T., AND BARCLAY, F. (1983). Household pets as a potential reservoir for *Clostridium difficile* A infection. *J. clin. Path.* **36**, 84-87.

——, JONES, R.H., AND PHILLIPS, I. (1980). Rifampicin-associated pseudomembranous colitis. *Br. med. J.* **281**, 1180-1.

BORSANYI, S.J. AND BLANCHARD, C.L. (1961). Asymptomatic enlargement of the parotid glands due to the use of isoproterenol. *Maryland med. J.* **10**, 572-3.

BOSAEUS, I., ANDERSSON, H., AND NYSTRÖM, C. (1979). Effect of a low-fat diet on bile salt excretion and diarrhoea in the gastrointestinal radiation syndrome. *Acta radiol. Oncol. Radiat. Phys. Biol.* **18**, 460-4.

BOURKE, J.B., McILLMURRAY, M.B., MEAD, G.M., AND LANGMAN, M.J.S. (1978). Drug-associated primary acute pancreatitis. *Lancet.* **i**, 706-8.

BOWEN, R., MAYNE, J.G., CAIN, J.C., AND BARTHOLOMEW, L.G. (1960). Peptic ulcer in rheumatoid arthritis and relationship to steroid treatment. *Proc. Staff Meet. Mayo Clin.* **35**, 537-44.

BRADLEY, J.A., McWHINNIE, D.L., HAMILTON, D.N.H., STARNES, F., MacPHERSON, S.G., SEYWRIGHT, M., BRIGGS, J.D., AND JUNOR, B.J. (1983). Sclerosing obstructive peritonitis after continuous peritoneal dialysis. *Lancet.* **ii**, 113-14.

BRANDSLUND, I., RASK, H., AND KLITGAARD, N.A. (1979). Gastrointestinal blood loss caused by controlled-release and conventional acetylsalicylic acid tablets. *Scand. J. Rheumatol.* **8**, 209-15.

BRANDT, L.J., BERNSTEIN, L.H., BOLEY, S.J., AND FRANK, M.S. (1982). Metronidazole therapy for perineal Crohn's disease: a follow-up study. *Gastroenterology.* **83**, 383-7.

BRAVO, A.J. AND LOWMAN, R.M. (1968). Benign ulcer of the sigmoid colon: an unusual lesion that can simulate carcinoma. *Radiology* **90**, 113-15.

BRAYE, S.G., COPPLESTONE, J.A., AND GARTELL, P.C. (1982). Neutropenic enterocolitis during mianserin-induced agranulocytosis. *Br. med. J.* **285**, 1117.

BRESSLER, B. (1980). An unusual side-effect of lithium. *Psychosomatics,* **21**, 688-9.

BROE, P.J. AND CAMERON, J.L. (1983). Azathioprine and acute pancreatitis: studies with an isolated perfused canine pancreas. *J. surg. Res.* **34**, 159-63.

——, ZINNER, M.J., AND CAMERON, J.L. (1982). A clinical trial of cimetidine in acute pancreatitis. *Surg. Gynecol Obstet.* **154**, 13-16.

BROWN, C.H., FERRANTE, W.A., AND DAVIS, JR. W.D. (1968). Toxic dilatation of the colon complicating pseudomembranous enterocolitis. *Am. J. dig. Dis.* **13**, 813-21.

BROWN, J.J., FERRISS, J.B., FRASER, R., LEVER, A.F., AND ROBERTSON, J.I.S. (1971). Spirnolactone in the treatment of hypertension with aldosterone excess. In *The medical uses of spirono-lactone* (ed. G.M. Wilson), pp. 27-36. Excerpta Medica, Amsterdam.

BROWN, P., BADDELEY, H., READ, A.E., DAVIES, J.D., AND McGARRY, J. (1974). Sclerosing peritonitis, an unusual reaction to a β-adrenergic blocking drug (practolol) *Lancet.* **ii**, 1477-81.

BRUNI, G., LAVEZZARI, M., PERBELLINI, A., BATTAGLIA, A., AND EMANUELI, A. (1982). Adverse reactions to indoprofen: a survey based on a total of 6764 patients. *J. int. med. Res.* **10**, 306-24.

BUCKELS, J.A.C., MACKINTOSH, P., AND BARNES, A.D. (1981). Controlled trial of low versus high dose oral steroid therapy in 100 cadaveric renal transplants. *Proc. Eur. Dial. Transplant. Ass.* **18**, 394-9.

BURBIGE, E.J., AND MILLIGAN, F.D. (1975). Pseudomembranous colitis: association with antibiotics and therapy with cholestyramine. *J. Am. med. Ass.* **231**, 1157-8.

——, TARDER, G.L., AND BELBER, J.P. (1977). Malabsorption following radiation therapy. *Am. J. Gastroent.* **67**, 589-92.

BURDON, D.W., THOMPSON, H., CANDY, D.C.A., KEARNS. M., LEES, D., AND STEPHEN, J. (1981). Enterotoxin (s) of *Clostridium difficile.* *Lancet.* **ii**, 258-9.

BURKITT, E.A. AND SUTCLIFFE, C.K. (1961). Paralytic ileus after amitriptyline ('Tryptizol'). *Br. med. J.* **2**, 1648-9.

BUTTERFIELD, W.J.H., FRY, I.K., AND WHICHELOW, M.J. (1961). The hypoglycaemic action of phenformin: studies in diabetics after short-term therapy. *Lancet.* **ii**, 563-7.

BUZZARD, T. (1880). A case of paraplegia from Pott's disease, treatment by Sayre's jacket, intestinal obstruction, death from a kink in the duodenum. *Trans. clin. Soc., Lond.* **13**, 157-66.

CALL, T., MALARKEY, W.B., AND THOMAS, F.B. (1977). Acute pancreatitis secondary to furosemide with associated hyperlipidemia. *Am. J. dig. Dis.* **22**, 835-8.

CALLAGHAN, T.S., HADDEN, D.R., AND TOMKIN, G.H. (1980). Megaloblastic anaemia due to vitamin B_{12} malabsorption associated with long-term metformin treatment. *Br. med. J.* **280**, 1214-15.

CALNE, R.Y., ROLLES, K., WHITE, D.J.G., THIRU, S., EVANS, D.B., McMASTER, P., DUNN, D.C., CRADDOCK, G.N., HENDERSON, R.G., AZIZ, S., AND LEWIS, P. (1979). Cyclosporin A intially as the only immunosuppressant in 34 recipients of cadaveric organs: 32 kidneys, 2 pancreases, and 2 livers. *Lancet.* **ii**, 1033-6.

CAMERON, A.J. (1973). Aspirin intake in patients with peptic ulcer. *Gastroenterology* **64**, 705.

CAMFIELD, P.R., BAGNELL, P., CAMFIELD, C.S., AND TIBBLES J.A.R. (1979). Pancreatitis due to valproic acid. *Lancet.* **i**, 1198-9.

CAMILLERI, M., SCHAFIER, K., CHADWICK, V.S., HODGSON, H.J., AND WEINBREN. K. (1981). Periportal sinusoidal dilation, inflammatory bowel disease, and the contraceptive pill. *Gastroenterology* **80**, 810-15.

CAMP, A.V. (1981). Tiaprofenic acid in the treatment of rheumatoid arthritis. *Rheumatol. Rehabil.* **20** 181-3.

CAMPBELL, A.N., BEASLEY, J.R., AND KENNA, A.P. (1981). Indomethacin and gastric perforation in a neonate. *Lancet.* **i**, 1110-11.

CANIVET, B., CREISSON, G., FREYCHET, P., AND DAGEVILIE, X. (1980). Fibre, diabetes, and risk of bezoar. *Lancet.* **ii**, 862.

CANTER, J.W., AND SHORB, P.E. (1971). Acute perforation of colonic diverticula associated with prolonged adreocorticosteroid therapy. *Am. J. Surg.* **121**, 46-51.

CAPRON, J.-P., DUPAS, J.-L., JOLY, J.-P., AND LORRIAUX, A. (1977).

Duodenogastric reflux of bile and aspirin-induced gastric damage in man. *Lancet*. **i**, 601–2.

CARDOE, N. (1964). The place of oxyphenbutazone in the treatment of rheumatoid arthritis and allied conditions. *Med. J. Aust.* **2**, 986–8.

CARLBORG, B., DENSERT, O., AND KUMLIEN, A. (1979). Esofagusskador av alprenolol. *Läkartidningen* **76**, 2706–8.

——, KUMLIEN, A., AND OISSON, H. (1978). Medikamentella esofagusstrikturer. *Läkartidningen* **75**, 4609–11.

CARMEL, R.J. AND KAPLAN, H.S. (1976). Mantle irradiation in Hodgkin's disease. An analysis of technique, tumor eradication, and complications. *Cancer, NY*. **37**, 2813–25.

CARMICHAEL, H.A., NELSON, L.M., AND RUSSELL, R.I. (1978). Cimetidine and prostaglandin: evidence for different modes of action on the rat gastric mucosa. *Gastroenterology* **74**, 1229–32.

CARONE, F.A., AND LIEBOW, A.A. (1957). Acute pancreatic lesions in patients treated with ACTH and adrenal corticoids. *New Engl. J. Med.* **257**, 690–7.

CARUSO, I. AND BIANCHI PORRO, G. (1980). Gastroscopic evaluation of anti-inflammatory agents. *Br. med. J.* **280**, 75–8.

CASTRO, E.B., ROSEN, P.P., AND QUAN, S.H.Q. (1973). Carcinoma of large intestine in patients irradiated for carcinoma of cervix and uterus. *Cancer NY*. **31**, 45–52.

CHABORA, B.M., HOPFAN, S., AND WITTES, R. (1977). Esophageal complications in the treatment of oat cell carcinoma with combined irradiation and chemotherapy. *Radiology* **123**, 185–7.

CHANG, T.-W., GORBACH, S.L., BARTLETT, J.G., AND SAGINUR, R. (1980). Bacitracin treatment of antibiotic-associated colitis and diarrhea caused by *Clostridium difficile* toxin. *Gastroenterology* **78**, 1584–6.

——, ONDERDONK, A.B., AND BARTLETT, J.G. (1978). Anion-exchange resins in antibiotic-associated colitis. *Lancet*. **ii**, 258–9.

CHANNER, K.S. AND HOLLANDERS, D. (1981). Tetracycline-induced oesophageal ulceration. *Br. med. J.* **282**, 1359–60.

—— AND VIRJEE, J. (1982). Effect of posture and drink volume on the swallowing of capsules. *Br. med. J.* **285**, 1702.

CHAUDHURY, T.K. AND JACOBSON, E.D. (1978). Prostaglandin cytoprotection of gastric mucosa. *Gastroenterology* **74**, 59–63.

CHEN, B., SHAPIRA, J., RAVID, M., AND LANG, R. (1982). Steatorrhoea induced by allopurinol. *Br. med. J.* **284**, 1914.

CHIERICHETTI, S., BIANCHI, G., AND CERRI, B. (1975). Comparison of feprazone and phenylbutazone interaction with warfarin in man. *Curr. ther. Res.* **18**, 568–72.

CHISHOLM, G.D., MEE, A.D., WILLIAMS, G., CASTRO, J.E., AND BARON, J.H. (1977). Peptic ulceration, gastric secretion, and renal transplantation. *Br. med. J.* **1**, 1630–3.

CHRISTIE, D.L. AND AMENT, M.E. (1975). Ampicillin-associated colitis. *J. Pediat.* **87**, 657–8.

CHVASTA, T.E. AND COOKE, A.R. (1972). The effect of several ulcerogenic drugs on the canine gastric mucosal barrier. *J. Lab. clin. Med.* **79**, 302–15.

CLARK, C.E., THOMPSON, H., McLEISH, A.R., POWIS, S.J.A., DORRICOTT, N.J., AND ALEXANDER-WILLIAMS, J. (1976). Pseudomembranous colitis following prophylactic antibiotics in bowel surgery. *J. amtimicrob. Chemother.* **2**, 167–73.

CLARK, C.V. AND TERRIS, R. (1983). Sclerosing peritonitis associated with metoprolol. *Lancet*. **i**, 937.

CLARK, I.M.C. (1971). Adynamic ileus and amitriptyline. *Br. med. J.* **2**, 531.

COCHRAN, K.M., MacKENZIE, J.F., AND RUSSELL, R.I. (1975). Role of taurocholic acid in production of gastric mucosal damage after ingestion of aspirin. *Br. med. J.* **1**, 183–5.

COCHRANE, J.P.S., YARNOLD, J.R., AND SLACK, W.W. (1981). The surgical treatment of radiation injuries after radiotherapy for uterine carcinoma. *Br. J. Surg.* **68**, 25–28.

COCHRANE, P. (1973). Spontaneous oesophageal rupture after carbachol therapy. *Br. med. J.* **1**, 463–4.

COGGON, D., LANGMAN, M.J.S., AND SPIEGELHALTER, D. (1982). Aspirin, paracetamol and haematemesis and melaena. *Gut* **23**, 340–4.

——, ROSE, D.H., AND ANSELL, I.D. (1981). A large bowel lymphoma complicating renal transplantation. *Br. J. Radiol.* **54**, 418–20.

COHEN, A. (1976). Intestinal blood loss after a new anti-inflammatory drug, sulindac. *Clin. Pharmac. Ther.* **20**, 238–40.

COHEN, L. AND BANKS, P. (1966). Salivary gland enlargement and phenylbutazone. *Br. med. J.* **1**, 1420.

COHEN, M. (1978). Mucosal cytoprotection by prostaglandin E_2. *Lancet*. **ii**, 1253–4.

——(1981). Prevention of aspirin-induced fall in gastric potential difference with prostaglandins. *Lancet*. **i**, 785.

——, CHEUNG, G., AND LYSTER, D.M. (1980). Prevention of aspirin-induced faecal blood loss by prostaglandin E_2. *Gut* **21**, 602–6.

—— AND POLLETT, J.M. (1976). Prostaglandin in E_2 prevents aspirin and indomethacin damage to human gastric mucosa. *Surg. Forum.* **27**, 400–2.

COLEMAN, D.L., JUERGENSEN, P.H., BRAND, M.H., AND FINKELSTEIN, F.O. (1981). Antibiotic associated diarrhoea during administration of intraperitoneal cephalothin. *Lancet*. **i**, 1004.

COLLINS, A.J., NOTARIANNI, L.J., AND DIXON, A. St. J. (1983). Acute gastric microbleeding after aspirin ingestion, compared with a similar dose of fenbufen by measurement of haemoglobin in gastric aspirates. *J. Pharm. Pharmacol.* **35**, 610–12.

COLLINS, F.J., MATTHEWS, H.R., BAKER, S.E., AND STRAKOVA, J.M. (1979). Drug-induced oesophageal injury. *Br. med. J.* **1**, 1673–6.

COLTART, D.J. (1969). Malabsorption induced by para-aminosalicylate. *Br. med. J.* **1**, 825–6.

CONE, J.B. AND WETZEL, W. (1982). Toxic megacolon secondary to pseudomembranous colitis. *Dis. Colon Rectum*, **25**, 478–82.

CONKLIN, J.L. AND ANURAS, S. (1981). Radiation-induced recurrent intestinal pseudo-obstruction. *Am. J. Gastroent.* **75**, 440–4.

CONN, H.O. AND BLITZER, B.L. (1976). Nonassociation of adrenocorticosteroid therapy and peptic ulcer. *New Engl. J. Med.* **294**, 473–9.

——, RAMSBY, G.R., AND STORER, E.H. (1972). Selective intraarterial vasopressin in the treatment of upper gastrointestinal hemorrhage. *Gastroenterology* **63**, 634–45.

COOK, A.I.M. AND FOY, P. (1976). Sclerosing peritonitis and practolol therapy. *Ann. R. Coll. Surg.* **58**, 473–5.

COOKE, A.R. (1973). Drugs and peptic ulceration. In *Gastrointestinal disease*. (ed. M.H. Sleisinger and J.S. Fordtran), Chapter 52. W.B. Saunders Co., Philadelphia.

—— AND GOULSTON, K. (1969). Failure of intravenous aspirin to increase gastrointestinal blood loss. *Br. med. J.* **3**, 330–2.

COOKE, S.A.R. AND DE MOOR, N.G. (1981). The surgical treatment of the radiation-damaged rectum. *Br. J. Surg.* **68**, 488–92.

COOPER, M.J. AND WILLIAMSON, R.C.N. (1982). Effect of sodium and acetylsalicylates on bile flow in man. *Gut* **23**, A881–2.

CORFIELD, A.P., COOPER, M.J., AND WILLIAMSON, R.C.N. (1985). Acute pancreatitis: a lethal disease of increasing incidence. *Gut* 26, 724–9.

CORNISH, A.L., McCLELLAN, J.T., AND JOHNSTON, D.H. (1961). Effects of chlorothiazide on the pancreas. *New Engl. J. Med.* 265, 673–5.

COTTON, P.B. AND LEA THOMAS, M. (1971). Ischaemic colitis and the contraceptive pill. *Br. med. J.* 3, 27–8.

COUZIGOU, P., REIFFERS, J., LEVÊQUE, A.–M., BROUSTET, A., AND BÉRAND, C. (1982). Les entérocolites aiguës nécrosantes des agranulocytoses sont des formes majeures de colites pseudomembraneuses. *Gastroenterol. clin. biol.* 6, 48–51.

CRAFT, A.W., KAY, H.E.M., LAWSON, D.N., AND McELWAIN, T.J. (1977). Methotrexate-induced malabsorption in children with acute lymphoblastic leukaemia. *Br. med. J.* 2, 1511–12.

CROFT, D.N. AND WOOD, P.H.N. (1967). Gastric mucosa and susceptibility to occult gastrointestinal bleeding caused by aspirin. *Br. med. J.* 1, 137–41.

—— CUDDIGAN J.H.P., AND SWEETLAND, C. (1972). Gastric bleeding and benorylate, a new aspirin. *Br. med. J.* 3, 545–7.

CROKER, J.R., COTTON, P.B., BOYLE, A.C., AND KINSELLA, P. (1980). Cimetidine for peptic ulcer in patients with arthritis. *Ann. rheum. Dis.* 39, 275–8.

CROWSON, T.D., HEAD, L.H., AND FERRANTE, W.A. (1976). Esophageal ulcers associated with tetracycline therapy. *J. Am. med. Ass.* 235, 2747–8.

CUDMORE, M.A., SILVA, J., FEKETY, R., LIEPMAN, M.K., AND KIM, K-H. (1982). *Clostridium difficile* colitis associated with cancer chemotherapy. *Arch. intern. Med.* 142, 333–5.

CULVER, G.J., PIRSON, H.S., MILCH, E., BERMAN, L., AND ABRANTES, F.J. (1961). Intramural hematoma of the jejunum: a case report. *Radiology* 76, 785–9.

CUMMINGS, J.H., SLADEN, G.E., JAMES, O.F.W., SARNER, M., AND MISIEWICZ, J.J. (1974). Laxative-induced diarrhoea: a continuing clinical problem. *Br. med. J.* 1, 537–41.

CUNNINGHAM, J.T. (1981). Gastric telangiectasias in chronic hemodialysis patients: a report of six cases. *Gastroenterology* 81, 1131–3.

CUSHMAN, P. (1970). Glucocorticoids and the gastrointestinal tract: current status. *Gut* 11, 534–9.

CZYZYK, A., TAWECKI, J., SADOWSKI, J., POWIKOWSKA, I., AND SZCZEPANIK, Z. (1968). Effect of biguanides on intestinal absorption of glucose. *Diabetes* 17, 492–8.

DAGGETT, P. AND IBRAHIM, S.Z. (1976). Intestinal obstruction complicating orphenadrine treatment. *Br. med. J.* 1, 21–2.

D'AMICO, J. AND ISRAEL, R. (1977). Bowel obstruction and perforation with an intraperitoneal loop intrauterine contraceptive device. *Am. J. Obstet. Gynec.* 129, 461–2.

DAMMANN, H.G. AND AUGUSTIN, H.J. (1978). Cimetidine and acute pancreatitis. *Lancet.* i, 666.

DANESH, B.J.Z., NELSON, L.M., MORGAN, R.J., ORJIOKE, C., DOCHERTY, C., AND RUSSELL, R.I. (1983). Does the acetyl linkage in acetylsalicylic acid (ASA) contribute to the gastric mucosal damaging effect of this drug? *Gut,* 24, A1001.

DAVENPORT, H.W. (1967). Salicylate damage to the gastric mucosal barrier. *New Engl. J. Med.* 276, 1307–12.

DAVIDOFF, F., TISHLER, S., AND ROSOFF, C. (1973). Marked hyperlipidemia and pancreatitis associated with oral contraceptive therapy. *New Engl. J. Med.* 289, 552–5.

DAVIDSON, D.C., FORD, J.A., AND FOX, E.G. (1974). Iodide sialadenitis in childhood. *Arch. Dis. Childh.* 49, 67–8.

DAVIS, J.T., AND NUSBAUM, M. (1973). Chlorpromazine therapy and functional large-bowel obstruction. *Am. J. Gastroenterol.* 60, 635–9.

DAY, T.K. (1983). Intestinal perforation associated with osmotic slow release indomethacin capsules. *Br. med. J.* 287, 1671–2.

DAYER, P., CECH, P., AND COURVOISIER, B. (1977). Chlorure de potassium et ulcères non spécifiques de l'intestin grêle. *Schweiz. med. Wochenschr.* 107, 379–80.

DEBENHAM, G.P. (1966). Ulcer of the cecum during oxyphenbutazone (Tandearil) therapy. *Can. med. Ass. J.* 94, 1182–4.

DE BERGEYCK, E., JANSSENS, P.G., AND DE MUYNCK, A. (1980). Radiological abnormalities of the ileum associated with the use of clofazimine (Lamprene; B663) in the treatment of skin ulceration due to *Mycobacterium ulcerans. Lepr. Rev.* 51, 221–8.

DeCOSSE, J.J., RHODES, R.S., WENTZ, W.B., REAGAN, J.W., DWORKEN, H.J., AND HOLDEN, W.D. (1969). The natural history and management of radiation injury of the gastrointestinal tract. *Ann. Surg.* 170, 369–84.

DEJESUS, R., AND PETERNEL, W.W. (1978). Antibiotic-associated diarrhea treated with oral tetracycline. *Gastroenterology* 74, 818–20.

DEKKERS, C.P.M., ENDEMAN, J.H., POEN, H., JESSURUN, R.E.M., AND TEN THIJE, O.J. (1972). Upper gastrointestinal complications and gastric secretion studies in advanced renal insufficiency. *Arch. Franc. Mal. Appar. dig.* 61, 273c.

DEMLING, R.H., SALVATIERRA, O., AND BELZER, F.O. (1975). Intestinal necrosis and perforation after renal transplantation. *Archs. Surg.* 110, 251–3.

DeSCHEPPER, P.J., TJANDRAMAGA, T.B., DeROO, M., VERHAEST, L., DAURIO, C., STEELMAN, S.L., AND TEMPERO, K.F. (1978). Gastrointestinal blood loss after diflunisal and after aspirin: effect of ethanol. *Clin. Pharmacol. Ther.* 23, 669–76.

DEVENEY, C.W., LEWIS, F.R., AND SCHROCK, T.R. (1976). Surgical management of radiation injury of the small and large intestine. *Dis. Colon Rectum* 19, 25–9.

DEWIS, P., LOCAL, F., ANDERSON, D.C., AND BANCEWICZ, J. (1982). Reversible oesophageal dysphagia and long-term ingestion of chlormethiazole. *Br. med. J.* 284, 705–6.

DICK, H.M., WRIGHT, P., CHAPMAN, C.M., ZACHARIAS, F.J., AND NICHOLLS, J.T. (1978). Adverse reactions to practolol: some observations on the possible relevance to immune mechanisms. *Allergy* 33, 71–5.

DIETHELM, A.G., GORE, I., CH'IEN, L.T., STERLING, W.A., AND MORGAN, J.M. (1976). Gastrointestinal hemorrhage socondary to cytomegalovirus after renal transplantation: a case report and review of the problem. *Am. J. Surg.* 131, 371–4.

DOBBINS, W.O., HERRERO, B.A., AND MANSBACH, C.M. (1968). Morphologic alterations associated with neomycin-induced malabsorption. *Am J. med. Sci.* 255, 63–77.

DOHERTHY, C.C. (1978). Studies on gastric acid secretion in chronic renal failure. *Irish J. med. Sci.* 147, 376–7.

—— AND McGEOWN, M.G. (1977). Peptic ulceration, gastric secretion, and renal transplantation. *Br. med. J.* 2, 188.

—— AND —— (1978). Prevention of upper gastrointestinal complications after kidney transplantation. *Proc. Eur. Dial. Transplant. Ass.* 15, 361–71.

DONALDSON, S.S., JUNDT, S., RICOUR, C., SARRAZIN, D., LEMERLE, J., AND SCHWEISGUTH, O. (1975). Radiation enteritis in children: a retrospective review, clinicopathologic correlation, and dietary management. *Cancer* 35, 1167–8.

DORPH, M.H. (1950). The cast syndrome. Review of the literature and report of a case. *New Engl. J. Med.* **243**, 440–2.

DOUTHWAITE, A.H. AND LINTOTT, G.A.M. (1938). Gastroscopic observation of effect of aspirin and certain other substances on the stomach. *Lancet.* **ii**, 1222–5.

DREILING, D.A. AND JANOWITZ, H.D. (1957). Effects of ACTH and adrenal steroid hormones on gastric secretion. *Clin. Res. Proc.* **5**, 110.

—— AND NACCHIERO, M. (1978). The effect of imuran on pancreatic secretion: a preliminary report. *Am. J. Gastroenterol.* **69**, 491–3.

DREIZEN, S., DALY, T.E., DRANE, J.B., AND BROWN, L.R. (1977). Oral complications of cancer radiotherapy. *Postgrad. Med.* **61**, 85–92.

DUFFIELD, J.E. (1973). Side effects of lithium carbonate. *Br. med. J.* **1**, 491.

DUGGAN, J.M. (1976). Aspirin in chronic gastric ulcer: an Australian experience. *Gut.* **17**, 378–84.

—— (1981). The pathogenesis of the aspirin-related gastric lesion. *J.R. Coll. Physcns. London.* **15**, 117–18.

DUNCAN, W. AND LEONARD, J.C. (1965). The malabsorption syndrome following radiotherapy. *Quart. J. Med.* **34**, 319–29.

DUNEA, G. (1971). Peritoneal dialysis and hemodialysis. *Med. Clin. N. Amer.* **55**, 155–75.

DUPREZ, A. AND SIMONS, M. (1963). A propos de quelques drames abdominaux au cours de traitements à la cortisone. *Acta gastro-enter. Belg.* **26**, 609–20.

DYBDAHL, J.H., DAAE, L.N.W., LARSEN, S., EKELI, H., FRISLID, K., WIIK, I., AND AASTAD, L. (1980). Acetylsalicylic acid-induced gastrointestinal bleeding determined by a ^{51}Cr method on a day-to-day basis. *Scand. J. Gastroenterol.* **15**, 887–95.

EHRENFELD, M., LEVY, M., SHARON, P., RACHMILEWITZ, D., AND ELIAKIM, M. (1982). Gastrointestinal effects of long-term colchicine therapy in patients with recurrent polyserositis (familial Mediterranean fever). *Dig. Dis. Sci.* **27**, 723–7.

EISENBERG, E.S., ALPERT, B.E., WEISS, R.A., MITTMAN, N., AND SOEIRO, R. (1983). Rhodotorula rubra peritonitis in patients undergoing continuous ambulatory peritoneal dialysis. *Am. J. Med.* **75**, 349–52.

ELMORE, F. AND ROGGE, J.D. (1981). Tetracycline-induced pancreatitis. *Gastroenterology* **81**, 1134–6.

EMERY, P. AND GRAHAME, R. (1982). Gastrointestinal blood loss and piroxicam. *Lancet.* **i**, 1302–3.

ENZENAUER, R.W., BASS, J.W., AND McDONNELL, J.T. (1984) Esophageal ulceration associated with oral theophylline. *New Engl. J. Med.* **310**, 261.

ERIKSSEN, J., SEEGAARD, E., AND NAESS, K. (1975). Side-effects of thiocarbamides. *Lancet.* **i**, 231–2.

ERIKSSON, B., ÖHMAN, U., SCHMIDT, D., AND NORLANDER, A. (1982). Severe antibiotic-associated colitis treated by colectomy. Report of two cases. *Acta chir. scand.* **148**, 629–31.

ERNST, C.B. (1983). Prevention of intestinal ischemia following abdominal aortic reconstruction. *Surgery.* **93**, 102–6.

EUROPEAN MULTICENTRE TRIAL. (1982). Cyclosporin A as sole immunosuppressive agent in recipients of kidney allografts from cadaver donors. Preliminary results of a European Multicentre Trial. *Lancet.* **ii**, 57–60.

EVANS, K.T. (1981). Tablets and dysphagia. British Society of Gastroenterology, Autumn Meeting, Exeter.

—— AND ROBERTS, G.M. (1976). Where do all the tablets go? *Lancet.* **ii**, 1237–9.

EVARTS, C.M., WINTER, R.B., AND HALL, J.E. (1971). Vascular compression of the duodenum associated with the treatment of scoliosis. Review of the literature and report of eighteen cases. *J. Bone Jt. Surg.* **A53**, 431–44.

FALKO, J.M. AND THOMAS, F.B. (1975). Acute pancreatitis due to procainamide-induced lupus erythematosus. *Ann. intern. Med.* **83**, 832–3.

FALOON, W.W. AND CHODOS, R.B. (1969). Vitamin B_{12} absorption studies using colchicine, neomycin and continuous ^{57}Co-B_{12} administration, *Gastroenterology* **56**, 1251.

——, FISHER, C.J., AND DUGGAN, K.C. (1958). Occurrence of a sprue-like syndrome during neomycin therapy. *J. clin. Invest.* **37**, 893.

FAM, A.G., PATON, T.W., SHAMESS, C.J., AND LEWIS, A.J. (1980). Fulminant colitis complicating gold therapy. *J. Rheumatol.* **7**, 479–85.

FANG, W.-F., BROUGHTON, A., AND JACOBSON, E.D. (1977). Indomethacin-induced intestinal inflammation. *Am. J. dig. Dis.* **22**, 749–59.

FARQUHARSON-ROBERTS, M.A., GIDDINGS, A.E.B., AND NUNN, A.J. (1975). Perforation of small bowel due to slow-release potassium chloride. (Slow-K). *Br. med. J.* **3**, 206.

FELSON, B. AND LEVIN, E.J. (1954). Intramural hematoma of the duodenum: diagnostic roentgen sign. *Radiology* **63**, 823–31.

FERNANDEZ, J.A., AND ROSENBERG, J.C. (1976). Post-transplantation pancreatitis. *Surg. Gynecol Obstet.* **143**, 795–8.

FERRIER, H. (1976). Pancreatitis after cardiac surgery: a morphologic study. *Am. J. Surg.* **131**, 684–8.

FIALA, M., PAYNE, J.E., BERNE, T.V., MOORE, T.C., HENLE, W., MONTGOMERIE, J.Z., CHATTERJEE, S.N., AND GUZE, L.B. (1975). Epidemiology of cytomegalovirus infection after transplantation and immunosuppression. *J. infect. Dis.* **132**, 421–33.

FILLASTRE, J.-P., BLAISE, P., ARDAILLOU, R., AND RICHET, G. (1965). Sécrétion gastrique des urémies chroniques. *Rev. fr. Études clin. biol.* **10**, 180–90.

FINCH, R.G., McKIM THOMAS, H.J., LEWIS, M.J., SLACK, R.C.B., AND GEORGE, R.H. (1979). Relapse of pseudomembraneous colitis after vancomycin therapy. *Lancet.* **ii**, 1076–7.

FINLAND, M., GRIGSBY, M.E., AND HAIGHT, T.H. (1954). Efficacy and toxicity of oxytetracycline (terramycin) and chlortetracycline (aureomycin): with special reference to use of doses of 250 mg every four to six hours and to occurrence of staphylococcic diarrhea. *Arch. intern. Med.* **93**, 23–43.

FLETCHER, M.R., LOEBL, W., AND SCOTT, J.T. (1975). Feprazone a new anti-inflammatory agent: studies of potency and gastrointestinal tolerance. *Ann. rheum. Dis.* **34**, 190–4.

FLINTHOLM, J., ANDERSEN, F.H., AND PEDERSEN, N.O. (1981). Kaliuminduceret tyndtarmsperforation. *Ugeskr. Laeg.* **143**, 550–1.

FOGAN, L. (1971). Grisoefulvin and dysgeusia: implications? *Ann. intern. Med.* **74**, 795–6.

FOOD STANDARDS COMMITTEE REPORT ON MINERAL OIL IN FOOD (1962). HMSO, London.

FOSTER, G.E. AND BOURKE, J.B. (1978). Acetabular cement causing intestinal obstruction. *Lancet.* **ii**, 267.

FOURNIER, G., ORGIAZZI, J. LENOIR, B., AND DECHAVANNE, M. (1980). Pseudomembranous colitis probably due to rifampicin. *Lancet.* **i**, 101.

FRANZIN, G., MUOLO, A., AND GRIMINELLI, T. (1981). Cytomegalovirus inclusions in the gastrointestinal mucosa of patients after renal transplantation. *Gut* **22**, 698–701.

——, MUSOLA, R., AND MENCARELLI, R. (1982*a*). Morphological changes of the gastroduodenal mucosa in regular dialysis uraemic patients. *Histopathology*. **6**, 429–37.

——, ——, AND —— (1982*b*). Changes in the mucosa of the stomach and duodenum during immunosuppressive therapy after renal transplantation. *Histopathology*, **6**, 439–49.

FREEMAN, A.H. AND LOW, F.M. (1981). Acute transient colitis due to ampicillin. British Society of Gastroenterology, Autumn Meeting, Exeter.

FRENKEL, E.P., McCALL, M.S., DOUGLASS, C.C., AND EISENBERG, S. (1968). Fecal blood loss following aspirin and coated aspirin microspherule administration. *J. clin. Pharmacol, and J. New Drugs* **8**, 347–51.

FRENNING, B. AND ÖBRINK, K.J. (1971). The effects of acetic and acetylsalicylic acids on the appearance of the gastric mucosal surface epithelium in the scanning electron microscope. *Scand. J. Gastroenterol.* **6**, 605–12.

FRIEDMAN, R.J., MAYER, I.E., GALAMBOS, J.T., AND HERSH, T. (1980). Oxacllin-induced pseudomembranous colitis. *Am. J. Gastroenterol* **73**, 445–7.

GABRIELE, O.F. AND CONTE, M. (1964). Spontaneous intramural hemorrhage of the colon. Associated with anticoagulants. *Arch Surg.* **89**, 522–6.

GANDHI, V.C., HUMAYUN, H.M., ING, T.S., DAUGIRDAS, J.T., JABLOKOW, V.R., IWATSUKI, S., GEIS, W.P., AND HANO, J.E. (1980). Sclerotic thickening of the peritoneal membrane in maintenance peritoneal dialysis patients. *Arch. intern. Med.* **140**, 1201–3.

GARDIES, A., GEVAUDAN, J., LE ROUX, C., CORNET, C., WARNET-DUBOSCQ, J., AND VIGUIE, R. (1978). Ulcère iatrogène de l'oesophage. *Nouv. Presse méd.* **7**, 1032.

GARFUNKEL, A.A., ROLLER, N.W., NICHOLS, C., AND SHIP, I.I. (1974). Phenylbutazone-induced sialadenitis. *Oral Surg.* **38**, 223–6.

GARNER, A., FLEMSTRÖM, G., AND HEYLINGS, J.R. (1979). Effects of anti-inflammatory agents and prostaglandins on acid and bicarbonate secretions in the amphibian-isolated gastric mucosa. *Gastroenterology* **77**, 451–7.

GARVIN, P.J., CARNEY, K., CASTANEDA, M., AND CODD, J.E. (1982). Peptic ulcer disease following transplantation: the role of cimetidine. *Am. J. Surg.* **144**, 545–8.

GAVRAS, H., BIOLLAZ, J., WAEBER, B., BRUNNER, H.R., GAVRAS, I., AND DAVIES, R.O. (1981). Antihypertensive effect of the new oral angiotensin converting enzyme inhibitor 'MK-421'. *Lancet* **ii**, 543–6.

GAZES, P.C., HOLMES, C.R., MOSELEY, V., AND PRATT-THOMAS, H.R. (1961). Acute hemorrhage and necrosis of the intestines associated with digitalization. *Circulation* **23**, 358–64.

GEORGE, R.H. (1981). Metronidazole and antibiotic-associated colitis. *Br. med. J.* **283**, 1468.

——, SYMONDS, J.M., DIMOCK, F., BROWN, J.D., ARABI, Y., SHINAGAWA, N., KEIGHLEY, M.R.B., ALEXANDER-WILLIAMS, J., AND BURDON, D.W. (1978). Identification of *Clostridium difficile* as a cause of pseudomembranous colitis. *Br. med. J.* **1**, 695.

——, YOUNGS, D.J., JOHNSON, E.M., AND BURDON, D.W. (1978). Anion-exchange resins in pseudomembranous colitis. *Lancet* **ii**, 624.

GEORGE, W.L., SUTTER, V.L., GOLDSTEIN, E.J.C., LUDWIG, S.L., AND FINEGOLD, S.M. (1978). Aetiology of antimicrobial-agent-associated colitis. *Lancet* **i**, 802–3.

——, VOLPICELLI, N.A., STINER, D.B., RICHMAN, D.D., LIECHTY, E.J., MOK, H.Y.I., ROLFE, R.D., AND FINEGOLD, S.M. (1979). Relapse of pseudomembranous colitis after vancomycin therapy. *New Engl. J. Med.* **301**, 414–5.

GIBBONS, R.B. (1979). Complications of chrysotherapy: a review of recent studies. *Arch. intern. Med.* **139**, 343–6.

GILINSKY, N.H., BURNS, D.G., BARBEZAT, G.O., LEVIN, W., MYERS, H.S., AND MARKS, I.N. (1983). The natural history of radiation-induced proctosigmoiditis: an analysis of 88 patients. *Quart. J. Med.* **52**, 40–53.

GILL, P.G. AND PIRIS, J. (1978). Proctitis caused by cornstarch glove powder: report of a case. *Dis. Colon Rectum* **21**, 207–8.

GILLBERG, R., KORSAN-BENGSTEN, K., MAGNUSSON, B., AND NYBERG, G. (1981). Gastrointestinal blood loss, gastroscopy and coagulation factors in normal volunteers during administration of acetylsalicyclic acid and fluproquazone. *Scand J. Rheumatol.* **10**, 342–6.

GINGELL, J.C., BURNS, G.P., AND CHISHOLM, G.D. (1968). Gastric acid secretion in chronic uraemia and after renal transplantation. *Br. med. J.* **4**, 424–6.

GIORDANO, J., HUANG, A., AND CANTER, J.W. (1975). Fatal paralytic ileus complicating phenothiazine therapy. *South. med. J.* **68**, 351–3.

GLICK, G.L., CHAFFEE, R.B., SALKIN, L.M., AND VANDERSALL, D.C. (1974). Oral mucosal chemical lesions associated with acetylsalicylic acid. *NY St. dent. J.* **40**, 475–8.

GLUECK, C.J., SCHEEL, D., FISHBACK, J., AND STEINER, P. (1972). Estrogen-induced pancreatitis in patients with previously covert familial type V hyperlipoproteinemia. *Metabolism* **21**, 657–66.

GOKAL, R., RAMOS, J.M., FRANCIS, D.M.A., FERNER, R.E., GOODSHIP, T.H.J., PROUD, G., BINT, A.J., WARD, M.K., AND KERR, D.N.S. (1982). Peritonitis in continuous ambulatory peritoneal dialysis. Laboratory and clinical studies. *Lancet* **ii**, 1388–91.

GOLDHAMMER, S. (1935). Ein Fall von tödlicher Solganalvergiftung. *Medsche. Klin.* **31**, 645–7.

GOLDSTEIN, F., KHOURY, J., AND THORNTON, J.J. (1976). Treatment of chronic radiation enteritis and colitis with salicylazosulfapyridine and systemic corticosteroids: a pilot study. *Am. J. Gastroenterol.* **65**, 201–8.

GOLDSTEIN, J., LASKIN, D.A., AND GINSBERG, G.H. (1980). Sulindac associated with pancreatitis . *Ann. intern. Med.* **93**, 151.

GOLDSTONE, B. (1952). Death associated with methonium treatment. *S. Afr. med. J.* **26**, 552–4.

GOODACRE, R.L., HAMILTON, J.D., MULLENS, J.E., AND QIZILBASH, A. (1977). Persistence of proctitis in 2 cases of clindamycin-associated colitis. *Gastroenterology* **72**, 149–52.

GOODMAN, M.J. (1977). Gastric ulceration induced by spironolactone. *Lancet* **i**, 752.

——, KENT, P.W., AND TRUELOVE, S.C. (1977). Glucosamine synthetase activity of the colonic mucosa in membranous colitis. *Gut* **18**, 229–31.

GORDON, E.M., JOHNSON, A.G., AND WILLIAMS, G. (1972). Gastric assessment of prospective renal-transplant patients. *Lancet* **i**, 226–9.

GORDON, M.J. (1976). Transient submandibular swelling following esophagogastroduodenoscopy. *Am. J. dig. Dis.* **21**, 507–8.

GOULD, P.C., KHAWAJA, F.I., AND ROSENTHAL, W.S. (1982). Antibiotic-associated hemorrhagic colitis. *Am. J. Gastroenterol.* **77**, 491–3.

GOULSTON, K. AND COOKE, A.R. (1968). Alcohol, aspirin, and gastrointestinal bleeding. *Br. med. J.* **4**, 664–5.

GOULSTON, S.J.M. AND McGOVERN, V.J. (1965). Pseudomembranous colitis. *Gut* **6**, 207–12.

GRAEBER, G.M., MARMOR, B.M., HENDEL, R.C., AND GREGG, R.O. (1976). Pancreatitis and severe metabolic abnormalities due to phenformin therapy. *Arch Surg.* **111**, 1014–16.

GRAHAM, C.F., GALLAGHER, K., AND JONES, J.K. (1981). Acute colitis with methyldopa. *New Engl. J. Med.* **304**, 1044–5.

GRAHAM, D.Y., SMITH, J.L., AND DOBBS, S.M. (1983). Gastric adaptation occurs with aspirin administration in man. *Dig. Dis. Sci.* **28**, 1–6.

GRAHAM, J.B. AND VILLALBA, R.J. (1963). Damage to the small intestine by radiotherapy. *Surg. Gynecol. Obstet.* **116**, 665–8.

GRAHAM, J.R. (1982). *Clostridium difficile* in toxic megacolon. *Br. med. J.* **285**, 889.

GRANT, J.B.F., DAVIES, J.D., AND JONES, J.V. (1976a). Allergic starch peritonitis in the guinea-pig. *Br. J. Surg.* **63**, 867–9.

——, ——, ESPINER, H.J., AND ELTRINGHAM, W.K. (1976b). The immunogenicity of starch glove powder and talc. *Br. J. Surg.* **63**, 864–6.

GRAY, P.H. AND PEMBERTON, P.J. (1980). Gastric perforation associated with indomethacin therapy in a pre-term infant. *Aust. paediat. J.* **16**, 65–6.

GRECO, F.A., BRERETON, H.D., KENT, H., ZIMBLER, H., MERRILL J., AND JOHNSON, R.E. (1976). Adriamycin and enhanced radiation reaction in normal esophagus and skin. *Ann intern. Med.* **85**, 294–8.

GREEN, N., IBA, G., AND SMITH, W.R. (1975). Measures to minimize small intestine injury in the irradiated pelvis. *Cancer, NY* **35**, 1633–40.

——, MELBYE, R.W., IBA, G., AND KUSSIN, L. (1978). Radiation therapy for carcinoma of the prostate. The experience with small intestine injury *Int. J. Radiat. Oncol. Biol. Phys.* **4**, 1049–53.

GREENBERGER, N.J. AND ISSELBACHER, K.J. (1964). Malabsorption following radiation injury to the gastrointestinal tract. *Am. J. Med.* **36**, 450–6.

GREENFIELD, C., AGUILAR RAMIREZ, J.R., POUNDER, R.E., WILLIAMS, T., DANVERS, M., MARPER, S.R., AND NOONE, P. (1983). *Clostridium difficile* and inflammatory bowel disease. *Gut* **24**, 713–17.

——, BURROUGHS, A., SZAWATHOWSKI, M., BASS, N., NOONE, P., AND POUNDER, R. (1981). Is pseudomembranous colitis infectious? *Lancet* **i**, 371–2.

GROSS, L. (1969). Oxyphenylbutazone-induced parotitis. *Ann. intern. Med.* **70**, 1229–30.

—— AND BROTMAN, M. (1970). Hypoprothrombinemia and hemorrhage associated with cholestyramine therapy. *Ann. intern. Med.* **72**, 95–6.

GRYBOSKI, J.D., KOCOSHIS, S.A., SEASHORE, J.H. GUDJONSSON, B., AND DRENNAN, J.C. (1978). 'Body-brace' oesophagitis, a complication of kyphoscoliosis therapy. *Lancet* **ii**, 449–51.

GUGGENHEIMER, J. AND ISMAIL, Y.H. (1975). Oral ulcerations associated with indomethacin therapy: report of three cases. *J. Am. dent. Ass.* **90**, 632–4.

GUIVARC'H, M., GEAU-BRISSONNIÈRE, O., AND ROULLET-AUDY, J.C. (1979). A propos d'une série récente de dix hématomes du grêle sous anticoagulants. *Chirurgie, Paris*, **105**, 524–34.

GURRY, J.F., CUNNINGHAM, I.G.E., AND BROOKE, B.N. (1975). β-blockers and fibrinous peritonitis. *Br. med. J.* **2**, 498.

GUTH, P.H., AURES, D., AND PAULSEN, G. (1979). Topical aspirin plus HCl gastric lesions in the rat. Cytoprotective effect of prostaglandin, cimetidine and probanthine. *Gastroenterology* **76**, 88–93.

HABESHAW, T. AND BENNETT, J.R. (1972). Ulceration of mouth due to emepronium bromide. *Lancet* **ii**, 1422.

HADJIYANNAKIS, E.J., SMELLIE, W.A.B., EVANS, B.D., AND CALNE, R.Y. (1971). Gastrointestinal complications after renal transplantation. *Lancet* **ii**, 781–5.

HAGIHARA, P.F., ERNST, C.B., AND GRIFFEN, W.O. (1979). Incidence of ischemic colitis following abdominal aortic reconstruction. *Surg. Gynecol. Obstet.* **149**, 571–3.

HAHN, K.-J., KRISCHKOFSKI, D., WEBER, E., AND MORGENSTERN, E. (1975). Morphology of gastrointestinal effects of aspirin. *Clin. Pharmacol. Ther.* **17**, 330–8.

HAIMOVIC, I.C., ARBIT, E., AND POSNER, J.B. (1978). Colonic ileus complicating laminectomy. *Neurosurgery* **3**, 369–72.

HALL, R.I., PETTY, A.H., COBDEN, I., AND LENDRUM, R. (1983). Enteritis and colitis associated with mefenamic acid. *Br. med. J.* **287**, 1182.

HALLEY, W. AND GOODMAN, J.D.S. (1975). Practolol and sclerosing peritonitis *Br. med. J.* **2**, 337–8.

HÄLLGREN, R., LANDELIUS, J., FJELLSTRÖM, K.E., AND LUNDQVIST, G. (1979). Gastric acid secretion in uraemia and circulating levels of gastrin, somatostatin, and pancreatic polypeptide. *Gut* **20**, 763–8.

HALSEY, J.P. AND CARDOE, N. (1982). Benoxaprofen: side-effect profile in 300 patients. *Br. med. J.* **284**, 1365–8.

HALSTED, C.H., GANDHI, G., AND TAMURA, T. (1981). Sulfasalazine inhibits the absorption of folates in ulcerative colitis. *New Engl. J. Med.* **305**, 1513–17.

—— AND McINTYRE, P.A. (1972). Intestinal malabsorption caused by aminosalicylic acid therapy. *Arch. intern. Med.* **130**, 935–9.

HAMBURGER, J. AND POTTS, A.J.C. (1983). Non-steroidal anti-inflammatory drugs and oral lichenoid reactions. *Br. med. J.* **287**, 1258.

HAND, D.W., SANFORD, P.A., AND SMYTH, D.H. (1966). Polyphenolic compounds and intestinal transfer. *Nature, London*. **209**, 618.

HARA, H., YAMANE, T., AND YAMASHITA, K. (1979). Extramedullary plasmacytoma of the gastrointestinal tract in a renal transplant recipient. *Acta pathol. Jpn.* **29**, 661–8.

HARDEN, R. McG. (1968). Submandibular adenitis due to iodide administration. *Br. med. J.* **1**, 160–1.

HART, S.L. AND McCOLL, I. (1968). The effect of the laxative oxyphenisatin on intestinal glucose absorption in the rat and man. *Br. J. Pharmacol. Chemother.* **32**, 683–6.

HARTMANN, H.A. AND ANGEVINE, D.M. (1956). Pseudomembranous colitis complicating prolonged antibiotic therapy. *Am. J. med. Sci.* **232**, 667–73.

HARTY, R.F. (1978). Sclerosing peritonitis and propranolol. *Arch intern. Med.* **138**, 1424–6.

HAVEL. R.J. (1969). Pathogenesis, differentiation and management of hypertriglyceridemia. *Adv. intern. Med.* **15**, 117–54.

HAWKINS, C.F., ELLIS, H.A., AND RAWSON, A. (1965). Malignant gout with tophaceous small intestine and megaloblastic anaemia. *Am. rheum. Dis.* **24**, 224–33.

—— AND MEYNELL, J.M. (1954). Megaloblastic anaemia due to phenytoin sodium. *Lancet* **ii**, 737–8.

HEATLEY, N.G. (1959). Mucosubstance as a barrier to diffusion. *Gastroenterology* **37**, 313–17.

HEFFERNAN, S.J. AND MURPHY, J.J. (1975). Ulceration of small intestine and slow-release potassium tablets. *Br. med. J.* **2**, 746.

HEGGARTY, H. (1974). Aspirin and anaemia in childhood. *Br. med. J.* **1**, 491-2.

HEINIVAARA, O. AND PALVA, I.P. (1965). Malabsorption and deficiency of vitamin B_{12} caused by treatment with para-aminosalicyclic acid. *Acta med. scand.* **177**, 337-41.

HEIZER, W.D., WARSHAW, A.L., WALDMANN, T.A., AND LASTER, L. (1968). Protein-losing gastroenteropathy and malabsorption associated with factitious diarrhea. *Ann. intern. Med.* **68**, 839-52.

HELLER, S.R., FELLOWS, I.W., OGILVIE, A.L., AND ATKINSON, M. (1982). Non-steroidal anti-inflammatory drugs and benign oesophageal stricure. *Br. med. J.* **285**, 167-8.

HENKIN, R.I. (1976). Taste dysfunction and penicillamine. *J. Am. med. Ass.* **236**, 250-1.

HENNINGS, B. AND JACOBSON, G. (1974). Postoperative amylase excretion: a study following thoracic surgery with and without extracorporeal circulation. *Ann. clin. Res.* **6**, 215-22.

HERBERT, D.C. (1968). Anticoagulant therapy and the acute abdomen. *Br. J. Surg.* **55**, 353-7.

HERMAN, J., NASSAR, F., AND DALLI, N. (1980). Endoscopically proved erosive gastritis due to acetazolamide. *Isr. J. med. Sci.* **16**, 866-7.

HEY, H., JØRGENSEN, F., SØRENSEN, K., HASSELBALCH, H., AND WAMBERG, T. (1982). Oesophageal transit of six commonly used tablets and capsules. *Br. med. J.* **285**, 1717-19.

HICKLING, P. AND FULLER, J. (1979). Penicillamine causing acute colitis. *Br. med. J.* **2**, 367.

HILL, R.B., DAHRLING, B.E., STARZL, T.E., AND RIFKIND, D. (1967). Death after transplantation: an analysis of 60 cases. *Am. J. Med.* **42**, 327-34.

HIMMELHOCH, J.M. AND HANIN, I. (1974). Side effects of lithium carbonate. *Br. med. J.* **4**, 233.

HINGSON, D.J. AND ITO, J. (1971). Effect of aspirin and related compounds on the fine structure of mouse gastric mucosa. *Gastroenterology* **61**, 156-77.

HIRSCHOWITZ, B.I., STREETEN, D.H.P., POLLARD, H.M., AND BOLDT, H.A. (1955). Role of gastric secretions in activation of peptic ulcers by corticotropin (ACTH) *J. Am. med. Ass.* **158**, 27-32.

HOBSON, Q.J.G., SELWYN J.G., AND MOLLIN, D.L. (1956). Megaloblastic anaemia due to barbiturates. *Lancet* **ii**, 1079-81.

HODSMAN, G.P., BROWN, J.J., CUMMING, A.M.M., DAVIES, D.L., EAST, B.W., LEVER, A.F., MORTON, J.J., MURRAY, G.D., ROBERTSON, I., AND ROBERTSON, J.I.S. (1983). Enalapril in the treatment of hypertension with renal artery stenosis. *Br. med. J.* **287**, 1413-17.

HOFFBRAND, A.V. AND NECHELES, T.F. (1968). Mechanism of folate deficiency in patients receiving phenytoin. *Lancet* **ii**, 528-30.

HOFMANN, A.F. AND POLEY, J.R. (1969). Cholestyramine treatment of diarrhea associated with ileal resection. *New Engl. J. Med.* **281**, 397-402.

HOFTIEZER, J.W., O'LAUGHLIN, J.C., AND IVEY, K.J. (1982). Effects of 24 hours of aspirin, Bufferin, paracetamol, and placebo on normal human gastroduodenal mucosa. *Gut* **23**, 692-7.

——, SILVOSO, G.R., BURKS, M., AND IVEY, K.J. (1980). Comparison of the effects of regular and enteric-coated aspirin on gastroduodenal mucosa of man. *Lancet* **ii**, 609-12.

HOGNESTAD, J. AND FLATMARK, A. (1976). Colon perforation in renal transplant patients. *Scand. J. Gastroenterol.* **11**, 289-92.

HOLGATE, S.T., WHEELER, J.H., AND BLISS, B.P. (1973). Star peritonitis: an immunological study. *Ann. R. Coll. Surg.* **52**, 182-8.

HOLT, L.P.J. AND HAWKINS, C.F. (1965). Indomethacin: studies of absorption and of the use of indomethacin suppositories. *Br. med. J.* **1**, 1354-56.

HONIG, S. (1980). Preliminary report: long-term safety of zomepirac. *J. clin. Pharmacol.* **20**, 392-6.

HOON, J.R. (1974). Bleeding gastritis induced by long-term release aspirin. *J. Am. med. Ass.* **229**, 841-2.

HORWICH, A., LOKICH, J.J., AND BLOOMER, W.D. (1975). Doxorubicin, radiotherapy, and oesophageal stricture. *Lancet* **ii**, 561-2.

HOWIE, A.D. AND STRACHAN, R.W. (1975). Slow-release potassium chloride treatment. *Br. med. J.* **2**, 176.

HRADSKY, M. (1981). Endoscopic evaluation of gastric tolerance to fenclofenac. *Upsala. J. med. Sci.* **86**, 99-103.

HUDED, F.V., POSNER, G.L., AND TICK, R. (1982). Nonspecific ulcer of the colon in a chronic hemodialysis patient. *Am. J. Gastroenterol* **77**, 913-16.

HUGHES, S., WARHURST, G., TURNBERG, L.A., HIGGS, N.B., GIUGLIANO, L.G., AND DRASAR, B.S. (1983). *Clostridium difficile* toxin-induced intestinal secretion in rabbit ileum *in vitro*. *Gut* **24**, 94-8.

HUMPHREY, C.D., CONDON, C.W., CANTEY, J.R., AND PITTMAN, F.E. (1979*a*). Partial purification of a toxin found in hamsters with antibiotic-associated colitis. Reversible binding of the toxin by cholestyramine. *Gastroenterology* **76**, 468-76.

——, LUSHBAUGH, W.B., CONDON, C.W., PITTMAN, J.C., AND PITTMAN, F.E. (1979*b*). Light and electron microscopic studies of antibiotic associated colitis in the hamster. *Gut* **20**, 6-15.

HUNT, J.N. AND FISHER, M.A. (1980). Aspirin-induced gastric bleeding stops despite rising plasma salicylate. *Dig. Dis. Sci* **25**, 135-9.

——, SMITH, J.L., AND JIANG, C.L. (1983). Gastric bleeding and gastric secretion with sulindac and naproxen. *Dig. Dis. Sci.* **28**, 169-73.

HURST, A. AND LINTOTT, G.A.M. (1939). Aspirin as a cause of haematemesis: a clinical and gastroscopic study. *Guy's Hosp. Rep.* **89**, 173-6.

HURST, P.A. AND EDWARDS, J.M. (1979). Chylous ascites and obstructive lymphoedema of the small bowel following abdominal radiotherapy. *Br. J. Surg.* **66**, 780-1.

HUSTON, G.J. (1978). The effects of aspirin, ethanol, indomethacin and 9-fludrocortisone on buccal mucosal potential difference. *Br. J. clin. Pharmacol.* **5**, 155-60.

HVIDT, S. AND KJELDSEN, K. (1963). Malabsorption induced by small doses of neomycin sulphate. *Acta med. scand.* **173**, 699-705.

IGNATIUS, J.A. AND HARTMANN, W.H. (1972). The glove starch peritonitis syndrome. *Ann. Surg.* **175**, 388-97.

IMBUR, D.J. AND BOURNE, R.B. (1972). Iodide mumps following excretory urography. *J. Urol.* **108**, 629-30.

IMRIE, C.W., GOLDRING, J., POLLOCK, J.G., AND WATT, J.K. (1977). Acute pancreatitis after translumbar aortography. *Br. med. J.* **2**, 681.

INGOLDBY, C.J.H. (1977). Perforated jejunal diverticulum due to local iron toxicity. *Br. med. J.* **1**, 949-50.

IVEY, K.J., PAONE, D.B., AND KRAUSE, W.J. (1980). Acute effect of systemic aspirin on gastric mucosa in man. *Dig. Dis. Sci.* **25**, 97-9.

——, PARSONS, C., AND WEATHERBY, R. (1975). Effect of prednisolone and salicylic acid on ionic fluxes across the human stomach. *Aust. NZ J. Med.* **5**, 408–12.

——, SILVOSO, G.R., AND KRAUSE, W.J. (1978). Effect of paracetamol on gastric mucosa. *Br. med. J.* **1**, 1586–8.

IZSAK, E.M., SHIKE, M., ROULET, M., AND JEEJEEBHOY, K.N. (1980). Pancreatitis in association with hypercalcemia in patients receiving total parenteral nutrition. *Gastroenterology* **79**, 555–8.

JACOBS, E. AND PRINGOT, J. (1973). Gastric ulcers due to the intake of potassium chloride. *Am. J. dig. Dis.* **18**, 289–94.

JACOBSON, E.D., CHODOS, R.B., AND FALOON, W.W. (1960a). An experimental malabsorption syndrome induced by neomycin, *Am. J. Med.* **28**, 524–33.

——, PRIOR, J.T., AND FALOON, W.W. (1960b). Malabsorption syndrome induced by neomycin: morphologic alterations in the jejunal mucosa. *J. Lab. clin. Med.* **56**, 245–50.

JAFFE, I.A. (1968). Effects of penicillamine on the kidney and on taste. *Postgrad. med. J. (Suppl.)* **44**, 15–18.

——, TRAGGIS, D., DAS, L., KIM, B.S., WON, H., HANN, L., MOLONEY, W.C., AND DOHLWITZ, A. (1972). Comparison of daily and twice-weekly schedule of L-asparaginase in childhood leukemia. *Pediatrics* **49**, 590–5.

JAFFE, R.H. AND LAING, D.R. (1934). Changes of the digestive tract in uremia: a pathologic anatomic study. *Arch. intern. Med.* **53**. 851–64.

JAGADEESAN, K., VISWESWARAN, M.K., AND HARIHARA IYER, K. (1975). Acute abdomen in a patient treated with Lamprene. *Int. Surg.* **60**, 208–10.

JAIN, R. AND RAMANAN, S.V. (1978). Iatrogenic pancreatitis. A fatal complication in the induction therapy for acute lymphocytic leukemia. *Arch intern. Med.* **138**, 1726.

JAMIESON, N.V., THIRU, S., CALNE, R.Y., AND EVANS, D.B. (1981). Gastric lymphomas arising in two patients with renal allografts. *Transplantation* **31**, 224–5.

JICK, H. (1981). Effects of aspirin and acetaminophen in gastrointestinal hemorrhage. Results from the Boston Collaborative Drug Surveillance Program. *Arch. intern. Med.* **141**, 316–21.

—— AND PORTER, J. (1978). Drug-induced gastrointestinal bleeding. *Lancet* **ii**, 87–9.

JOHANSSON, C., KOLLBERG, B., NORDEMAR, R., AND BERGSTRÖM S. (1979). Mucosal protection by prostaglandin E$_2$. *Lancet* **i**, 317.

JOHNSON, L.R. AND OVERHOLT, B.F. (1967). Release of histamine into gastric venous blood following injury by acetic or salicylic acid. *Gastroenterology* **52**, 505–9.

JOHNSON, P.C. (1980). A comparison of the effects of zomepirac and aspirin on fecal blood loss. *J. clin. Pharmacol.* **20**, 401–5.

JOHNSON, W.C. AND NABSETH, D.C. (1970). Pancreatitis in renal transplantation. *Ann. Surg.* **171**, 309–14.

—— AND —— (1974). Visceral infarction following aortic surgery. *Ann. Surg.* **180**, 312–18.

JOHNSTON, D.H. AND CORNISH, A.L. (1959). Acute pancreatitis in patients receiving chlorothiazide. *J. Am. med. Ass.* **170**, 2054–6.

JONES, J.H. AND LENNARD-JONES, J.E. (1966). Corticosteroids and corticotrophin in the treatment of Crohn's disease. *Gut* **7**, 181–7.

JONES, J.M., GLASS, N.R., AND BELZER, F.O. (1982). Fatal *candida* esophagitis in two diabetics after renal transplantation. *Arch. Surg.* **117**, 499–501.

JONES, M.F. AND CALDWELL, J.R. (1962). Acute hemorrhagic pancreatitis associated with administration of chlorthalidone:

report of a case. *New Engl. J. Med.* **267**, 1029–31.

JONES, P.E. AND OELBAUM, M.H. (1975). Frusemide-induced pancreatitis. *Br. med. J.* **1**, 133–4.

JONES, R.H., LEWIN, M.R., AND PARSONS, V. (1979). Therapeutic effect of cimetidine in patients undergoing haemodialysis. *Br. med. J.* **1**, 650–2.

——, RUDGE, C.J., BEWICK, M., PARSONS, V., AND WESTON, M.J. (1978). Cimetidine: prophylaxis against upper gastrointestinal haemorrhage after renal transplantation. *Br. med. J.* **1**, 398–400.

JUNCOSA, L. (1970). Ulcus péptico yatrógeno del esófago. *Revta esp. Enferm. Apar. dig.* **30**, 457–8.

KABINS, S.A. AND SPIRA, T.J. (1975). Outbreak of clindamycin-associated colitis. *Ann. Intern. Med.* **83** 830.

KAMMERER, W.H., FREIBERGER, R.H., AND RIVELIS, A.L. (1958). Peptic ulcer in rheumatoid patients on corticosteroid therapy: a clinical experimental and radiologic study. *Arthritis Rheum.* **1**, 122–41.

KANG, J.Y. AND DOE, W.F. (1979). Unprocessed bran causing intestinal obstruction *Br. med. J.* **1**, 1249–50.

KAPLAN, K. AND WEINSTEIN, L. (1968). Lincomycin *Pediat. Clins. N. Am.* **15**, 131–9.

KAPLAN, M.H. AND DREILING, M.A. (1977). Steroids revisited II. Was cortisone responsible for the pancreatitis? *Am J. Gastroenterol* **67**, 141–7.

KAPLINSKY, N., PRAS, M., AND FRANKL, O. (1973). Severe enterocolitis complicating chrysotherapy. *Ann rheum. Dis* **32**, 574–7.

KARAT, A.B.A. (1975). Long-term follow-up of clofazimine (Lamprene) in the management of reactive phases of leprosy. *Lepr. Rev. Suppl.* **46**, 105–9.

KASHIMA, H.K. AND KALINOWSKI, B. (1979) Taste impairment following laryngectomy. *Ear Nose Throat J.* **58**, 88–92.

KASLIKOVA, J., KOČANDRLE, V., ZÁSTAVA. V., JIRKA, J., SKÁLA, I., AND PIRK, F. (1981). Mutiple immunoblastic sarcoma of the small intestine following renal transplantation. *Transplantation* **31**, 481–2.

KAVIN, H. (1977). Oseophageal ulceration due to emepronium bromide. *Lancet* **i**, 424–5.

KAWANISHI, H., RUDOLPH, E., AND BULL, F.E. (1973). Azathioprine-induced acute pancreatitis. *New Engl. J. Med.* **289**, 357.

KEATING, J.P., FRANK, A.L., BARTON, L.L., AND TEDESCO, F.J. (1974). Pseudomembranous colitis associated with ampicillin therapy. *Am. J. Dis. Childh* **128**, 369–70.

KEENAN, D.J.M., CAVE, K., LANGDON, L., AND LEA, R.E. (1984). Rectal indomethacin for control of postoperative pain. *Br. med. J.* **288**, 240.

KEIGHLEY, M.R.B., ARABI, Y., ALEXANDER-WILLIAMS, J., YOUNGS, D.W., AND BURDON, D.W. (1979). Comparison between systemic and oral antimicrobial prophylaxis in colorectal surgery. *Lancet* **i**, 894–7.

KEISER, H.R., HENKIN, R.I., AND BARTTER, F.C. (1968). Loss of taste during therapy with penicillamine. *J. Am. med. Ass.* **203**, 381–3.

KENDALL, M.J. (1971). M.D. Thesis, University of Birmingham.

——, NUTTER, S., AND HAWKINS, C.F. (1971). Xylose test: effect of aspirin and indomethacin. *Br. med. J.* **1**, 533–6.

KENNEDY, J.D., JONES, R.C.M., HUDSON, S.A., AND CHOUHAN, U.M. (1982). Patent ductus arteriosus in premature babies *Br. med. J.* **284**, 115.

KERN, F., CLARK, G.M., AND LUKENS, J.G. (1957). Peptic ulceration occurring during therapy for rheumatoid arthritis. *Gastroenterology* **33**, 25–33.

KEUSCH, G.T. AND PRESENT, D.H. (1976). Summary of a workshop on clindamycin colitis. *J. infect. Dis.* **133**, 578–87.

——, TRONCALE, F.J., AND PLAUT, A.G. (1970). Neomycin-induced malabsorption in a tropical population. *Gastroenterology* **58**, 197–202.

KEY, T.C. AND KREUTNER, A.K. (1980). Gastrointestinal complications of modern intrauterine devices. *Obstet. Gynecol.* **55**, 239–44.

KHOURY, W., GERACI, K., ASKARI, A., AND JOHNSON, M. (1979). The effect of cimetidine on aspirin absorption. *Gastroenterology* **76**, 1169.

KILANDER, A. AND DOTEVALL, G. (1983). Endoscopic evaluation of the comparative effects of acetylsalicylic acid and choline magnesium trisalicylate on human gastric and duodenal mucosa. *Br. J. Rheumatol* **22**, 36–40.

KILPATRICK, Z.M., SILVERMAN, J.F., BETANCOURT, E., FARMAN, J., AND LAWSON, J.P. (1968). Vascular occlusion of the colon and oral contraceptives: possible relation. *New Engl. J. Med.* **278**, 438–40.

KIM, K.H., FEKETY, R., BATTS, D.H., BROWN, D., CUDMORE, M., SILVA, J., AND WATERS, D. (1981). Isolation of *Clostridium difficile* from the environment and contacts of patients with antibiotic-associated colitis. *J. infect. Dis.* **143**, 42–50.

KING, P.M., CRAWSHAW, H.M., AND McLEAN ROSS, A.H. (1983). Small bowel volvulus following total hip replacement. *Br. J. Surg.* **70**, 100.

KIRSNER, J.B., PALMER, W.L., SPENCER, J.A., BICKS, R.O., AND JOHNSON, C.F. (1959). Corticotrophin (ACTH) and adrenal steroids in management of ulcerative colitis: observations in 240 patients. *Ann. intern. Med.* **50**, 891–925.

KISER, J.R. (1963). Chronic gastric ulcer associated with aspirin ingestion: a report of 5 cases and review of the literature. *Am. J. dig. Dis.* **8**, 856–9.

KLATSKIN, G. AND GORDON, M. (1952). Relationship between relapsing pancreatitis and essential hyperlipemia. *Am. J. Med.* **12**, 3–23.

KLOTZ, A.P., PALMER, W.L., AND KIRSNER, J.B. (1953). Aureomycin proctitis and colitis: a report of five cases. *Gastroenterology* **25**, 44–7.

KOBLER, E., NÜESCH, H.J., BÜHLER, H., JENNY, S., AND DEYHLE, P. (1979). Medikamentös bedingte ösophagusulzera. *Schweiz. med. Wochenschr.* **109**, 1180–2.

KOLLBERG, B., NORDEMAR, R., AND JOHANSSON, C. (1981). Gastrointestinal protection by low-dose oral prostaglandin E_2 in rheumatic diseases. *Scand. J. Gastroenterol.* **16**, 1005–8.

KOLMUS, H. AND FARNSWORTH, D. (1959). Impairment and recovery of taste following irradiation of the oropharynx. *J. Laryng. Otol.* **73**, 180–2.

KONTUREK, S.J., BRZOZOWSKI, T., PIASTUCKI, I., DEMBINSKI, A., RADECKI, T., DEMBINSKA-KIEC, A., ZMUDA, A., AND GREGORY, H. (1981). Role of mucosal prostaglandins and DNA synthesis in gastric cytoprotection by luminal epidermal growth factor. *Gut* **22**, 927–32.

——, KWIECIEN, N., OBTULOWICZ, W., POLANSKI, M., KOPP, B., AND OLEKSY, J. (1983). Comparison of prostaglandin E_2 and ranitidine in prevention of gastric bleeding by aspirin in man. *Gut* **24**, 89–93.

——, OBTULOWICZ, W., SITO, E., OLESKY, J., WILKUN, S., AND KIEC-DEMBINSKA, A. (1980). Distribution of prostaglandins in gastric and duodenal mucosa of healthy subjects and duodenal ulcer patients: effects of aspirin and paracetamol. *Gut* **22**, 283–9.

KORENMAN, M.D., STUBBS, M.B. AND FISH, J.C. (1978). Intestinal obstruction from medication bezoars. *J. Am. med. Ass.* **240**, 54–5.

KRISTENSEN, K., SAND-KRISTENSEN, J., AND THORBORG, J.B. (1975). Practolol and sclerosing peritonitis. *Lancet* i, 741–2.

KRUGER, F.A., ALTSCHULD, R.A., HOLLOBAUGH, S.L., AND JEWETT, B. (1970). Studies on the site and mechanism of action of phenformin. II. Phenformin inhibition of glucose transport by rat intestine. *Diabetes* **19**, 50–2.

KUKORA, J.S. AND DENT, T.L. (1977). Colonoscopic decompression of massive nonobstructive cecal dilation. *Arch. Surg.* **112**, 512–17.

KUO, Y-J. AND SHANBOUR, L.L. (1976). Mechanism of action of aspirin on canine gastric mucosa. *Am. J. Physiol.* **230**, 762–7.

KUSTER, G., WOODS, J.E., ANDERSON, C.F., WEILAND, L.H., AND WILKOWSKE, C.J. (1972). Plasma cell lymphoma after renal transplantation. *Am. J. Surg.* **123**, 585–7.

KWITKO, A.O., PIETERSE, A.S., HECKER, R., ROWLAND, R., AND WIGG, D.R. (1982). Chronic radiation injury to the intestine: a clinicopathological study. *Aust. NZ J. Med.* **12**, 272–7.

LAMBERT, M., DE PEYER, R., AND MULLER, A.F. (1982). Reversible ischemic colitis after intravenous vasopressin therapy. *J. Am. med. Ass.* **247**, 666–7.

LAMONT, J.T., SONNENBLICK, E.B., AND ROTHMAN, S. (1979). Role of clostridial toxin in the pathogenesis of clindamycin colitis in rabbits. *Gastroenterology* **76**, 356–61.

—— AND TRNKA, Y.M. (1980). Therapeutic implications of *Clostridium difficile* toxin during relapse of inflammatory bowel disease. *Lancet* i, 381–3.

LANGMAN, M.J.S. (1970). Epidemiological evidence for the association of aspirin and acute gastrointestinal bleeding. *Gut* **11**, 627–34.

——, (1977). The role of analgesic antirheumatic drugs in precipitating acute upper gastrointestinal bleeding. *Proc. R. Soc. Med.* **70** (Suppl. 7), 16–17.

——, AND COOKE, A.R. (1976). Gastric and duodenal ulcer and their associated diseases. *Lancet* i, 680–3.

LANZA, F.L., HUBSHER, J.A., AND WALKER, B.R. (1981). Gastroscopic evaluation of the effect of aspirin and oxaprozin on the gastric mucosa. *J. clin. Pharmacol.* **21**, 157–61.

——, AND NELSON, R.S. (1982). Endoscopic comparison of the effects of fenbufen, naproxen, indomethacin, and placebo on gastric mucosa. *Study published by Lederle Laboratories.*

——, ROYER, G., AND NELSON, R (1975). An endoscopic evaluation of the effects of non-steroidal anti-inflammatory drugs on the gastric mucosa. *Gastrointest. Endosc.* **21**, 103–5.

——, ——, AND —— (1978). Gastroscopic evaluation of the effects of Motrin, Indocin, aspirin, Naprosyn, Tolectin, and placebo on gastric mucosa of normal volunteers. *Arthritis Rheum.* **21**, 572.

——, —— AND —— (1980). Endoscopic evaluation of the effects of aspirin, buffered aspirin, and enteric-coated aspirin on gastric and duodenal mucosa. *New Engl. J. Med.* **303**, 136–8.

LARSEN, R.R., SAVOYER, R.B., SAWYER, K.C., AND McCURDY, R.E. (1962). Hemorrhagic pancreatitis complicating anticoagulant therapy *NY St. J. Med.* **62**, 2397–9.

LARSON, H.E., LEVI, A.J., AND BORRIELLO, S.P. (1978). Vancomycin for pseudomembranous colitis. *Lancet* ii, 48.

——, AND PRICE, A.B. (1977). Pseudomembranous colitis: presence of clostridial toxin. *Lancet* ii, 1312–14.

——, PARRY, J.V., PRICE, A.B., DAVIES, D.R., DOLBY, J., AND

TYRRELL, D.A.J. (1977). Undescribed toxin in pseudomembranous colitis. *Br. med. J.* **1**, 1246–8.

LAVERY, I.C., STEIGER, E., AND FAZIO, V.W. (1980). Home parenteral nutrition in management of patients with severe radiation enteritis. *Dis. Colon Rectum* **23**, 91–3.

LAWRASON, F.D., ALPERT, E., MOHR, F.L., AND McMAHON, F.G. (1965). Ulcerative-obstructive lesions of the small intestine. *J. Am. med. Ass.* **191**, 641–4.

LAZARUS, S.S. AND BENCOSME, S.A. (1956). Development and regression of cortisone-induced lesions in rabbit pancreas. *Am. J. clin. Pathol.* **26**, 1146–56.

LEE, C.M., COLLINS, W.T., AND LARGEN, T.L. (1952). A reappraisal of absorbable glove powder. *Surg. Gynecol. Obstet.* **95**, 725–37.

LEFROCK, J.L., KLAINER, A.S., CHEN, S., GAINER, R.B., OMAR, M., AND ANDERSON, W. (1975). The spectrum of colitis associated with lincomycin and clindamycin therapy. *J. infect. Dis.* **131** (Suppl.), 108–15.

LENNERT, K.A. AND KOOTZ, F. (1967). Nil nocere! Arzneimittelbedingte dünndarmulzera. *Münch. med. Wochenschr.* **109**, 2058–62.

LEONARDS, J.R. AND LEVY, G. (1970). Aspirin-induced occult gastrointestinal blood loss: local versus systemic effects. *J. pharm. Sci.* **59**, 1511–13.

LERUT, J., LERUT, T., GRUWEZ, J.A., MICHIELSEN, P., AND VAN RENTERGHEM, I. (1980). Surgical gastro-intestinal complications in 277 renal tranplantations. *Acta. chir. belg.* **79**, 383–9.

LETAC, R., DELANNE, A., BAROUK, L., BONDONNY, J.-M., AND SAINT-SUPÉRY, G. (1971). Deux cas de nécrose de l'estomac avec perforation au cours de dilatation aiguë chez des porteurs de corsets plâtrés. *Ann. Chir. infant.* **12**, 101–6.

LEVINE, B., PESKIN, G.W., AND SAIK, R.P. (1976). Drug-induced colitis as a surgical disease. *Arch. Surg.* **111**, 987–9.

LEVINE, R.A. (1968). Steatorrhoea induced by para-aminosalicylic acid. *Ann. intern. Med.* **68**, 1265–70.

LEVY, M. (1974). Aspirin use in patients with major upper gastrointestinal bleeding and peptic ulcer disease: a report from the Boston Collaborative Drug Surveillance Program, Boston University Medical Center. *New Engl. J. Med.* **290**, 1158–62.

LEVY, N. AND GASPAR, E. (1975). Rectal bleeding and indomethacin suppositories. *Lancet* **i**, 577.

LIEBER, C.S. AND LEFÈVRE, A. (1959). Ammonia as a source of gastric hypoacidity in patients with uremia. *J. clin. Invest.* **38**, 1271–7.

LILLY, E.L. (1981). Pancreatitis after administration of sulindac. *J. Am. med. Ass.* **246**, 2680.

LINDHOLMER, B., NYMAN, E., AND RÄF, L. (1964). Nonspecific stenosing ulceration of the small bowel: preliminary report. *Acta. chir. scand.* **128**, 310–11.

LIPPMANN, W. (1974). Inhibition of indomethacin-induced gastric ulceration in the rat by perorally-administered synthetic and natural prostaglandin analogues. *Prostaglandins* **7**, 1–10.

LISHMAN, A.H., AL-JUMAILI, I.J., AND RECORD, C.O. (1981). Spectrum of antibiotic-associated diarrhoea. *Gut* **22**, 34–7.

——, SHIBLEY, T., AL-JUMAILI, A.J., AND RECORD, C.O. (1983). *Clostridium difficile* and its toxin in neonates. *Gut* **24**, A501.

LOCALIO, S.A., STONE, A., AND FRIEDMAN, M. (1969). Surgical aspects of radiation enteritis. *Surg. Gynecol. Obstet.* **129**, 1163–72.

LOCKARD, O.O., IVEY, K.J., BUTT, J.H., SILVOSO, G.R., SISK, C., AND HOLT, S. (1980). The prevalence of duodenal lesions in patients with rheumatic diseases on chronic aspirin therapy. *Gastrointest. Endosc.* **26**, 5–7.

LOEBL, D.H., CRAIG, R.M., CULIC, D.D., RIDOLFO, A.S., FALK, J., AND SCHMID, F.R. (1977). Gastrointestinal blood loss: effect of aspirin, fenoprofen, and acetaminophen in rheumatoid arthritis as determined by sequential gastroscopy and radioactive fecal markers. *J. Am. med. Ass.* **237**, 976–81.

LOEHRY, C.A. AND CREAMER, B. (1969). Three-dimensional structure of the rat small intestinal mucosa related to mucosal dynamics. Part I. Mucosal structure and dynamics in the rat after the administration of methotrexate. *Gut* **10**, 112–16.

LOFFER, F.D. AND PENT, D. (1975). Indications, contraindications and complications of laparoscopy. *Obstet. Gynecol. Surv.* **30**, 407–27.

LOFGREN, R.P., ROTHE, P.R., AND CARLSON, G.J. (1982). Jejunal perforation associated with slow-release potassium chloride therapy. *South med. J.* **75**, 1154–5.

LOIUDICE, T., BAXTER, D., AND BALINT, J. (1977). Effects of abdominal surgery on the development of radiation enteropathy. *Gastroenterology* **73**, 1093–7.

—— AND LANG, J.A. (1983). Treatment of radiation enteritis: a comparison study. *Am. J. Gastroenterol.* **78**, 481–7.

LONGSTRETH, G.F. AND NEWCOMER, A.D. (1975). Drug-induced malabsorption. *Mayo Clin. Proc.* **50**, 284–93.

LOPEZ, M.J., MEMULA, N., DOSS, L.L., AND JOHNSTON, W.D. (1981). Pseudo-obstruction of the colon during pelvic radiotherapy. *Dis. Colon Rectum* **24**, 201–4.

LORCH, E. (1971). Inhibition of intestinal absorption and improvement of oral glucose tolerance by biguanides in the normal and in the streptozotocin-diabetic rat. *Diabetologia* **7**, 195–203.

LÖVGREN, O. AND ALLANDER, E. (1964). Side-effects of indomethacin. *Br. med. J.* **1**, 118.

LUKETIC, G.C., MYREN, J., SACHS, G., AND HIRSCHOWITZ, B.I. (1964). Effect of therapeutic doses of colchicine on oxidative enzymes in the intestine. *Nature, London* **202**, 608–9.

LUSSIER, A., LeBEL, E., AND TÉTREAULT, L. (1982a). Gastrointestinal blood loss of oxaprozin and aspirin with placebo control. *J. clin. Pharmacol.* **22**, 173–8.

——, TETREAULT, L., AND LeBEL, E. (1982b). Comparison of the gastrointestinal microbleedings caused by aspirin, fenbufen and placebo. *Study published by Lederle Laboratories.*

LYLE, W.H. (1974). Penicillamine and zinc. *Lancet* **ii**, 1140.

MacFARLANE, J.D., MAISEY, M.N., GRAHAME, R., AND BURRY, H.C. (1976). Gastrointestinal blood loss on phenylbutazone and feprazone. *Rheumatol. and Rehabil.* **15**, 108–11.

MACKAY, A. AND STEVENSON, R.D. (1977). Gastric ulceration indued by spironolactone. *Lancet* **i**, 481.

MacKERCHER, P.A., IVEY, K.J., BASKIN, W.N., AND KRAUSE, W.J. (1977a). Protective effect of cimetidine on aspirin-induced gastric mucosal damage. *Ann. intern. Med.* **87**, 676–9.

——, ——, ——, ——, TATUM, W., AND MORRIS, J. (1977b). Comparison of effect of aspirin on gastric mucosa of chronic rheumatoid arthritics with normal subjects. *Gastroenterology* **72**, 1092.

MacMAHON, C.E. AND ROWE, J.W. (1971). Rectal reaction following radiation therapy of cervical carcinoma: particular reference to subsequent occurrence of rectal carcinoma. *Ann. Surg.* **173**, 264–9.

MAGGS, R.L., AND REINUS, F.Z. (1959). Peritonitis, caused by surgical glove starch powder, treated with steroids. *Am. J. Surg.* **98**, 111-15.

MAGNUSSON, B., SÖLVELL, L., AND ARVIDSSON, B. (1977). A comparative study of gastrointestinal bleeding during administration of ketoprofen and naproxen. *Scand. J. Rheumatol.* **6**, 62-4.

MAHAJAN, S.K., PRASAD, A.S., LAMBUJON, J., ABBASI, A.A., BRIGGS, W.A., AND McDONALD, F.D. (1980). Improvement of uremic hypogeusia by zinc: a double-blind study. *Am. J. clin. Nutr.* **33**, 1517-21.

MAKEY, D.A. (1977). Why prescribe enteric-coated potassium? *Lancet* **i**, 704.

MALLORY, A. AND KERN, F. (1980). Drug-induced pancreatitis: a critical review. *Gastroenterology* **78**, 813-20.

MALTBIE, A.A., VARIO, I.G., AND THOMAS, N.U. (1981). Ileus complicating therapy. *Psychosomatics* **22**, 158-9.

MANASHIL, G.B. AND KERN, J.A. (1973). Nonspecific colitis following oral lincomycin therapy. *Am. J. Gastroenterol.* **60**, 394-9.

MANN, N.S. (1977). Naproxen-and tolectin-induced acute erosive gastritis: its prevention by metiamide and cimetidine. *Gastroenterology* **72**, 1096.

—— AND SACHDEV, A.J. (1977). Aspirin- and bile-induced acute erosive gastritis: its prevention by metiamide therapy. *Arch. pathol. lab. Med.* **101**, 206-7.

MANSBERGER, A.R., HEARN, J.B., BYERS, R.M., FLEISIG, N., AND BUXTON, R.W. (1968). Vascular compression of the duodenum. Emphasis on accurate diagnosis. *Am. J. Surg.* **115**, 89-96.

MANSON, R.R. (1974). Acute pancreatitis secondary to iatrogenic hypercalcemia: implications of hyperalimentation. *Arch. Surg.* **108**, 213-15.

MARCZYNSKA-ROBOWSKA, M. (1957). Pancreatic necrosis in a case of Still's disease. *Lancet* **i**, 815-16.

MÅRDH, P.A., BELFRAGE, I., AND NAVARSTEN, E. (1974). Sialadenitis following treatment with α-methyldopa. *Acta med. scand.* **195**, 333-5.

MARGOLIS, D.M., ETHEREDGE, E.E., GARZA-GARZA, R., HRUSKA, K., AND ANDERSON, C.B. (1977). Ischemic bowel disease following bilateral nephrectomy or renal transplant. *Surgery* **82**, 667-73.

MARIGOLD, J.H., POUNDER, R.E., PEMBERTON, J., AND THOMPSON, R.P.H. (1982). Propranolol, oxprenolol, and sclerosing peritonitis. *Br. med. J.* **284**, 870.

MARKS, J.E., MORAN, E.M., GRIEM, M.L., AND ULTMANN, J.E. (1974). Extended mantle radiotherapy in Hodgkin's disease and malignant lymphoma. *Am. J. Roentg.* **121**, 772-88.

MARKS, J.S. AND GLEESON, M.H. (1975). Steatorrhoea complicating therapy with mefenamic acid. *Br. med. J.* **4**, 442.

MARSHALL, A.J., BADDELEY, H., BARRITT, D.W., DAVIES, J.D., LEE, R.E.J., LOW-BEER, T.S., AND READ, A.E. (1977). Practolol peritonitis: a study of 16 cases and a survey of small-bowel function in patients taking β-adrenergic blockers. *Quart. J. Med.* **46**, 135-49.

MARTIN, D.M., GOLDMAN, J.A., GILLIAM, J., AND NASRALLAH, S.M. (1981). Gold-induced eosinophilic enterocolitis: response to oral cromolyn sodium. *Gastroenterology* **81**, 1567-70.

MARTIN, J. AND COMPSTON, N. (1963). Vincristine sulphate in the treatment of lymphoma and leukaemia. *Lancet* **ii**, 1080-3.

MASON, G.H., ELLIS-REGLER, R.B., AND ARTHUR, J.F. (1977). Clofazimine and eosinophilic enteritis. *Lepr. Rev.* **48**, 175-80.

MASON, G.R., DIETRICH, P., FRIEDLAND, G.W., AND HANKS, G.E. (1970). The radiological findings in radiation-induced enteritis and colitis: a review of 30 cases. *Clin. Radiol.* **21**, 232-47.

MASON, S.J. AND O'MEARA, T.F. (1981). Drug-induced esophagitis. *J. clin. Gastroenterol.* **3**, 115-20.

MATERN, W.E., JAFFE, M.S., AND TOWBIN, R. (1977). Unusual complication from a cardiac pacemaker. *J. Am. med. Ass.* **238**, 969.

MATSUKI, A., WAKAYAMA, S., AND OYAMA, T. (1975). Acute transient swelling of the salivary glands during and following endotracheal anaesthesia. *Anaesthesist* **24**, 125-8.

MATUCHANSKY, C., ARIES, J., AND MAIRE, R. (1978). Metronidazole for antibiotic-associated pseudomembranous colitis. *Lancet* **ii**, 580-1.

MAUDSLEY, R.F. AND QIZILBASH, A.H. (1979). Thermal injury to the bowel as a complication of laparascopic sterilization. *Can. J. Surg.* **22**, 232-4.

MAXWELL, J.D., HUNTER, J., STEWART, D.A., ARDEMAN, S., AND WILLIAMS, R. (1972). Folate deficiency after anticonvulsant drugs: an effect of hepatic enzyme induction? *Br. med. J.* **1**, 297-9.

McADAMS, G.B. (1956). Granulomata caused by absorbable starch glove powder. *Surgery* **39**, 329-36.

McAVOY, B.R. (1974). Mouth ulceration and slow-release potassium tablets. *Br. med. J.* **4**, 164-5.

McBRIEN, M.P. (1973). Vitamin B_{12} malabsorption after cobalt teletherapy for carcinoma of the bladder. *Br. med. J.* **1**, 648-50.

McCALL, A.J. (1975). Slow-K ulceration of oesophagus with aneurysmal left atrium. *Br. med. J.* **3**, 230-1.

McCARTHY-LEVENTHAL, E. (1959). Post-irradiation mouth blindness. *Lancet* **ii**, 1138-9.

McCLOY, E.C. AND KANE, S. (1981). Drug-induced oesophageal ulceration. *Br. med. J.* **282**, 1703.

McCONNELL, J.B., STEWART, W.K., THJODLEIFSSON, B., AND WORMSLEY, K.G. (1975). Gastric function in chronic renal failure. Effects of maintenance haemodialysis. *Lancet* **ii**, 1121-3.

McDONALD, W.C. (1973). Correlation of mucosal histology and aspirin intake in chronic gastric ulcer. *Gastroenterology* **65**, 381-9.

McGEOWN, M.G., DOUGLAS, J.F., BROWN, W.A., DONALDSON, R.A. KENNEDY, J.A., LOUGHRIDGE, W.G., MEHTA, A., NELSON, S.D. DOHERTY, C.C., JOHNSTONE, R., TODD, G., AND HILL, C.M. (1980). Advantages of low dose steroid from the day after renal transplantation. *Transplantation* **29**, 287-9.

McINTYRE, R.L.E., IRANI, M.S., AND PIRIS, J. (1981). Histological study of the effects of three anti-inflammatory preparations on the gastric mucosa. *J. clin. Pathol.* **34**, 836-42.

McKINLEY, M. AND TOFFLER, R.B. (1980). Antibiotic-associated hemorrhagic colitis. *Dig. Dis. Sci.* **25**, 812-3.

McKINLEY, M.J., TRONCALE, F., SANGREE, M.H., SCHOLHAMER, C., AND BRAND, M. (1982). Antibiotic-associated colitis: clinical and epidemiological features. *Am. J. Gastroenterol.* **77**, 77-81.

McLEAN, R., MARTIN, S., AND LAM-PO-TANG, P.R.L. (1982). Fatal case of L-asparaginase induced pancreatitis. *Lancet* **ii**, 1401-2.

McMAHON, F.G. AND AKDAMAR, K. (1976). Gastric ulceration after 'Slow-K'. *New Engl. J. Med.* **295**, 733-4.

——, RYAN, J.R., AKDAMAR, K., AND ERTAN, A. (1982). Upper gastrointestinal lesions after potassium chloride supplements: a controlled clinical trial. *Lancet* **ii**, 1059-61.

McNEIL, J.J., ANDERSON, A., CHRISTOPHIDIS, N., JARROTT, B., AND LOUIS, W.J. (1979). Taste loss associated with oral captopril treatment. *Br. med. J.* **2**, 1555-6.

MEECH, P.R., HARDIE, I.R., HARTLEY, L.C.J., STRONG, R.W., AND CLUNIE, G.J.A. (1979). Gastrointestinal complications following renal transplantation. *Aust. NZ J. Surg.* **49**, 621–5.

MENGUY, R. AND DESBAILLETS, L. (1967*a*). Role of inhibition of gastric mucous secretion in the phenomenon of gastric mucosal injury by indomethacin. *Am. J. dig. Dis.* **12**, 862–6.

—— —— (1967*b*). Influence of phenylbutazone on gastric secretion of mucus. *Proc. Soc. exp. Biol. Med.* **125**, 1108–11.

—— AND MASTERS, Y.F. (1963). Effect of cortisone on mucoprotein secretion by gastric antrum of dogs: pathogenesis of steroid ulcer. *Surgery* **54**, 19–27.

—— —— (1965). Influence of parathyroid extract on gastric mucosal content of mucus. *Gastroenterology* **48**, 342–9.

MENNIE, A.T., DALLEY, V.M., DINNEEN, L.C., AND COLLIER, H.O.J. (1975). Treatment of radiation-induced gastrointestinal distress with acetylsalicylate. *Lancet* **ii**, 942–3.

MESHKINPOUR, H., MOLINARI, M.D., GARDNER, L., BERK, J.E., AND HOEHLER, F.K. (1979). Cimetidine in the treatment of acute alcoholic pancreatitis. A randomized, double-blind study. *Gastroenterology* **77**, 687–90.

MESSER, J., REITMAN, D., SACKS, H.S., SMITH, H., AND CHALMERS, T.C. (1983). Association of adrenocorticosteroid therapy and peptic ulcer diseases. *New Engl. J. Med.* **309**, 21–4.

MEYBOOM, R.H.B., VAN GENT, A., AND ZINKSTOK, D.J. (1982). Nitrofurantoin-induced parotitis. *Br. med. J.* **285**, 1049.

MEYERS, W.C., HARRIS, N., STEIN, S., BROOKS, M., JONES, R.S., THOMPSON, W.M., STICKEL, D.L., AND SEIGLER, H.F. (1979). Alimentary tract complications after renal transplantation. *Ann Surg.* **190**, 535–42.

MIELANTS, H., VEYS, E.M., VERBRUGGEN, G., AND SCHELSTRAETE, K.. (1981). Comparison of serum salicylate levels and gastrointestinal blood loss between salsalate (Disalcid) and other forms of salicylates. *Scand, J. Rheumatol.* **10**, 169–73.

MIKULASCHEK, W.M. (1980). Long-term safety of benoxaprofen *J. Rheumatol.* **7**, (Suppl. 6), 100–7.

MILITO, G., TACCONE-GALLUCCI, M., BRANCALEONE, C., NARDI, F., FILINGERI, V., CEXA, D., AND CASCIANI, C.U. (1983). Assessment of the upper gastrointestinal tract in hemodialysis patients awaiting renal transplantation. *Am. J. Gastroenterol.* **78**, 328–31.

MILLER, D.G., IVEY, M., AND YOUNG, J. (1979). Home parenteral nutrition in treatment of severe radiation enteritis. *Ann. intern Med.* **91**, 858–60.

MILLS, B., ZUCKERMAN, G., AND SICARD, G. (1981). Discrete colon ulcers as a cause of lower gastrointestinal bleeding and perforation in end-stage renal disease. *Surgery* **89**, 548–52.

MILNE, P.Y. AND THOMAS, R.J.S. (1976). Mesenteric venous thrombosis associated with oral contraceptives: a case report. *Aust. NZ. J. Surg.* **46**, 134–6.

MILNER, G. AND BUCKLER, E.G. (1964). Adynamic ileus and amitriptyline. *Med. J. Aust.* **1**, 921–2.

—— AND HILLS, N.F. (1966). Adynamic ileus and nortriptyline *Br. med. J.* **1**, 841–2.

MILSTONE, E.B., McDONALD, A.J., AND SCHOLHAMER, C.F. (1981). Pseudomembranous colitis after topical application of clindamycin *Arch Dermatol.* **117**, 154–5.

MITCHELL, C.J., SIMPSON, F.G., DAVISON, A.M., AND LOSOWSKY, M.S. (1979). Radiation pancreatitis: a clinical entity? *Digestion* **19**, 134–6.

MITCHELL, E.P. AND SCHEIN, P.S. (1982). Gastrointestinal toxicity of chemotherapeutic agents. *Sem. Oncol.* **9**, 52–64.

MITCHELL, T.H., STAMP, T.C.B., AND JENKINS, M.V. (1982). Steatorrhoea after tetracycline. *Br. med. J.* **285**, 780.

MODIGLIANI, R. AND DELCHIER, J.C. (1978). Vancomycin for antibiotic-induced colitis. *Lancet* **i**, 97–8.

MOGG, G.A.G., BURDON, D.W., AND KEIGHLEY, M. (1979). Oral metronidazole in *Clostridium difficile* colitis. *Br. med. J.* **2**, 335.

MOORE, R. (1978). Bleeding gastric erosion after oral zinc sulphate. *Br. med. J.* **1**, 754.

MORALES POLANCO, M.R., BUTRON, L., ECHEGOYEN, G., AND PIZZUTO CHAVEZ, J. (1977). Sialoadenitis crónica. Manifestacion tóxica del empleo prolongado de busulfán en el tratamiento de la leucemia granulocítica crónica. *Sangre (Barc.)* **22**, 243–51.

MORGENSTERN, L., THOMPSON, R., AND FRIEDMAN, N.B. (1977). The modern enigma of radiation enteropathy: sequelae and solutions. *Am. J. Surg.* **134**, 166–72.

MOROWITZ, D.A. AND EPSTEIN, B.H. (1973). Spectrum of bowel disease associated with use of oral contraceptives. *Med. Ann. District Columbia* **42**, 6–10.

MORRIS, P.J., CHAN, L., FRENCH, M.E., AND TING, A. (1982). Low dose oral prednisolone in renal transplantation. *Lancet* **i**, 525–7.

MOSS, J., BERG, R.A., BAUM, B.J., AND CRYSTAL, R.G. (1979). *In vitro* model for fibrosis induced by β-adrenergic blockers: propranolol inhibits β-adrenergic suppression of collagen production by human fibroblasts. *Clin. Res.* **27**, 445A.

MOSSMAN, K.L. AND HENKIN, R.I. (1978). Radiation-induced changes in taste acuity in cancer patients. *Int. J. Radiat. Oncol Biol. Phys.* **4**, 663–70.

MUIR, A. AND COSSAR, I.A. (1955). Aspirin and ulcer. *Br. med. J.* **2**, 7–12.

MÜLLER, P., FISCHER, N., DAMMANN, H.G., KATHER, H., AND SIMON, B. (1981). Simultaneous addition of 16, 16-dimethyl-prostaglandin E_2 prevents aspirin and bile salt damage to human gastric mucosa. *Z. Gastroent.* **19**, 373–6.

MUNGALL, I.P.F. AND HAGUE, R.V. (1975). Pancreatitis and the Pill. *Postgrad. med. J.* **51**, 855–7.

MUNIZ, C.E. AND BERGHMAN, D.H. (1978). Contact stomatitis and lithium carbonate tablets. *J. Am. med. Ass.* **239**, 2759.

MUNSTER, A. AND MILTON, G.W. (1961). Paralytic ileus due to ganglion-blocking agents. *Med. J. Aust.* **2**, 210–13.

MURPHEY, S.A. AND JOSEPHS, A.S. (1981). Acute pancreatitis associated with pentamidine therapy. *Arch. intern. Med.* **141**, 56–8.

MURRAY, H.S., STROTTMAN, M.P., AND COOKE, A.R. (1974). Effect of several drugs on gastric potential difference in man. *Br. med. J.* **1**, 19–21.

MURRAY, J.M. AND MASSEY, F.M. (1974). Chylous ascites after radiation therapy for ovarian carcinoma. *Obstet. Gynecol.* **44**, 749–51.

MURRAY-BRUCE, D.J. (1966). Salivary gland enlargement and phenylbutazone. *Br. med. J.* **1**, 1599–600.

NAGLER, J. AND PAGET, S.A. (1983). Nonexudative diarrhoea after gold salt therapy: case report and review of the literature. *Am. J. Gastroenterol.* **78**, 12–14.

NAKADAR, A.S. AND HARRIS-JONES, J.N. (1971). Sialadenitis after intravenous pyelography. *Br. med. J.* **3**, 351–2.

NANNI, G., GARBINI, A., LUCHETTI, P., NANNI, G., RONCONI, P., AND CASTAGNETO, M. (1982). Ogilvie's syndrome (acute colonic pseudo-obstruction): review of the literature (October 1948 to March 1980) and report of four additional cases. *Dis. Colon Rectum* **25**, 157–66.

NARSETE, T., ANSFIELD, F., WIRTANEN, G., RAMIREZ, G., WOLBERG, W., AND JARRETT, F. (1977). Gastric ulceration in patients receiving intrahepatic infusion of 5-fluorouracil. *Ann. Surg.* **186**, 734–6.

NASH, J.Q., CHATTOPADHYAY, B., HONEYCOMBE, J., AND TABAQCHALI, S. (1982). *Clostridium difficile* and cytotoxin in routine faecal specimens. *J. clin. Pathol.* **35**, 561–5.

NEEDHAM, C.D., KYLE, J., JONES, P.F., JOHNSTON, S.J., AND KERRIDGE, D.F. (1971). Aspirin and alcohol in gastrointestinal haemorrhage. *Gut* **12**, 819–21.

NEILD, G. AND SOUTHEE, T. (1979). Cytomegalovirus infection and renal transplantation. *Lancet* i, 604.

NELIS, G.F. (1983). Nitrofurantoin-induced pancreatitis: report of a case. *Gastroenterology* **84**, 1032–4.

NELP, W.B., BANWELL, J.G., AND HENDRIX, T.R. (1961). Pancreatic function and the viscosity of pancreatic juice before and during cortisone administration. *Bull. Johns Hopkins. Hosp.* **109**, 292–301.

NELSON, R.S., HERNANDEZ, A.J., GOLDSTEIN, H.M., AND SACA, A. (1979). Treatment of irradiation esophagitis. Value of hydrocortisone injection. *Am. J. Gastroenterol.* **71**, 17–23.

NEOPTOLEMOS, J.P. AND LOCKE, T.J. (1983). Recurrent small bowel obstruction associated with phenylbutazone. *Br. J. Surg.* **70**, 244–5.

NEWMAN, A., KATSARIS, J., BLENDIS, L.M., CHARLESWORTH, M., AND WALTER, L.H. (1973). Small intestinal injury in women who have had pelvic radiotherapy. *Lancet* ii, 1471–3.

NEWMAN, C.R. (1956). Pseudomembranous enterocolitis and antibiotics. *Ann. intern. Med.* **45**, 409–44.

NICHOLLS, J.G. (1971). Starch granulomatosis of the peritoneum. *Br. med. J.* **4**, 426.

NOGUEIRA, J.R., AND FREEDMAN, M.A. (1972). Acute pancreatitis as a complication of Imuran therapy in regional enteritis. *Gastroenterology* **62**, 1040–1.

NORTHWAY, M.G., LIBSHITZ, H.I., OSBORNE, B.M., FELDMAN, M.S., MAMEL, J.J., WEST, J.H., AND SZWARC, I.A. (1980). Radiation esophagitis in the oppossum: radioprotection with indomethacin. *Gastroenterology* **78**, 883–92.

NOTHMANN, B.J., CHITTINAND, S., AND SCHUSTER, M.M. (1973). Reversible mesenteric vascular occlusion associated with oral contraceptives. *Am. J. dig. Dis.* **18**, 361–8.

NUSSBAUM, H., GILBERT, H., KAGAN, A.R., CHAN, P., WOLLIN, M., WINKLEY, J., RAO, A., KWAN, D., AND HINTZ, B. (1981). Guidelines for radiation injury to bowel and bladder from external irradiation alone. *Cancer clin. Trials.* **4**, 295–9.

O'CONNOR, R.P., SILVA, J., AND FEKETY, R. (1981). Rifampicin and antibiotic-associated colitis. *Lancet* i, 499.

OETTGEN, H.F., STEPHENSON, P.A., SCHWARTZ, M.K., LEEPER, R.D., TALLAL, L., TAN, C.C., CLARKSON, B.D., GOLBEY, R.B., KRAKOFF, I.H., KARNOFSKY, D.A., MURPHY, M.L., AND BURCHENAL, J.H. (1970). Toxicity of *E. coli* L-asparaginase in man. *Cancer NY* **25**, 253–78.

OGILVIE, H. (1948). Large-intestine colic due to sympathetic deprivation: a new clinical syndrome. *Br. med. J.* **2**, 671–3.

O'LAUGHLIN, J.C., SILVOSO, G.K., AND IVEY, K.J. (1982). Resistance to medical therapy of gastric ulcers in rheumatic disease patients taking aspirin. A double-blind study with cimetidine and follow-up. *Dig. Dis. Sci.* **27**, 976–80.

OREOPOULOS, D.G., KHANNA, R., AND WU, G. (1983). Sclerosing obstructive peritonitis after CAPD. *Lancet* ii, 409.

OWENS, M.L., PASSARO, E., WILSON, S.E., AND GORDON, H.E. (1977). Treatment of peptic ulcer disease in the renal transplant patient. *Ann. Surg.* **186**, 17–21.

——, WILSON, S.E., SALTZMAN, R., AND GORDON, E. (1976). Gastrointestinal complications after renal transplantation: predictive factors and morbidity. *Arch. Surg.* **111**, 467–71.

PAES, I.C., SEARL, P., RUBERT, M.W., AND FALOON, W.W. (1967). Intestinal lactase deficiency and saccharide malabsorption during oral neomycin administration. *Gastroenterology* **53**, 49–58.

PAINE, C.G. AND SMITH, P. (1957). Starch granulomata. *J. clin. Pathol.* **10**, 51–5.

PALMER, J.A. AND BUSH, R.S. (1976). Radiation injuries to the bowel associated with the treatment of carcinoma of the cervix. *Surgery* **80**, 458–64.

PALVA, I.P. AND HEINIVAARA, O. (1966). Drug-induced malabsorption of vitamin B_{12}: *in vitro* studies using the dialysis technique. *Scand. J. Haematol.* **3**, 33–7.

—— AND MATTILA, M. (1966). Drug-induced malabsorption of vitamin B_{12}. III. Interference of PAS and folic acid in the absorption of vitamin B_{12}. *Scand. J. Haematol.* **3**, 149–53.

PANEBIANCO, A.C., SCOTT, S.M., DART, C.H., TAKARO, T., AND ECHEGARY, H.M. (1970). Acute pancreatitis following extracorporeal circulation. *Ann. thorac. Surg.* **9**, 562–8.

PANNUTI, F., MARTONI, A., POLLUTRI, E., CAMERA, P., AND CASTELLARI, S. (1973). Studio dei principali parametri di biotollerabilità clinica ed ematochimica al 5-fluorouracile somministrato per os. *Gazz. med. ital Milano* **132**, 303–15.

PAPADAKIS, J., BROWN, C.B., CAMERON, J.S., ADU, D., BEWICK, M., DONAGHEY, R., OGG, C.S., RUDGE, C., WILLIAMS, D.G., AND TAUBE, D. (1983). High verses 'low' dose corticosteroids in recipients of cadaveric kidneys: prospective controlled trial. *Br. med. J.* **286**, 1097–1100.

PARHAM, D.M. (1981). Post-transplantation pancreatitis associated with cytomegalovirus (report of a case). *Hum. Pathol.* **12**, 663–5.

PARKE, D.V., LINDUP, W.E., SHILLINGFORD, J.S., AND SMITH, M.J. (1975). The effects of ulcerogenic and ulcer-healing drugs on gastric mucus. *Gut* **16**, 396.

PARRY, D.J. AND WOOD, P.H.N. (1967). Relationship between aspirin-taking and gastroduodenal haemorrhage. *Gut* **8**, 301–7.

PASHBY, N.L., BOLTON, R.P., AND SHERRIFF, R.J. (1979). Oral metronidazole in *Clostridium difficile* colitis *Br. med. J.* **1**, 1605.

PEARSON, D.J., STONES, N.A., AND BENTLEY, S.J. (1983). Proctocolitis induced by salicylate and associated with asthma and recurrent nasal polyps. *Br. med. J.* **287**, 1675.

PEARSON, J.G. (1969). The value of radiotherapy in the management of esophageal cancer. *Am. J. Roentg.* **105**, 500–13.

PELLINEN, T.J. AND KALSKE, J. (1982). Nitrofurantoin-induced parototis *Br. med. J.* **285**, 344.

PEMBERTON, J. (1970). Oesophageal obstruction and ulceration caused by oral potassium therapy. *Br. Heart. J.* **32**, 267–8.

PENN, I., BRETTSCHNEIDER, L., SIMPSON, K., MARTIN, A., AND STARZL, T.E. (1970). Major colonic problems in human homotransplant recipients. *Arch. Surg.* **100**, 61–5.

——, DURST, A.L., MACHADO, M., HALGRIMSON, C.G., BOOTH, A.S., PUTMAN, C.W., GROTH, C.G., AND STARZL, T.E. (1972). Acute pancreatitis and hyperamylasemia in renal homograft recipients. *Arch. Surg.* **105**, 167–72.

——, GROTH, C.G., BRETTSCHNEIDER, L., MARTIN, A.J., MARCHIORO, T.L., AND STARZL, T.E. (1968). Surgically correctable intraabdominal complications before and after renal homotransplantation. *Ann. Surg.* **168**, 865–70.

PENN, R.G. AND GRIFFIN, J.P. (1982). Adverse reactions to nitrofurantoin in the United Kingdom, Sweden, and Holland. *Br. med. J.* **284**, 1440–2.

PENNY, W.J., RHODES, J., AND THOMSON, W. (1981). Protection against aspirin-induced blood loss in man: assessment of a new mucolytic agent. *Br. J. clin. Pharmacol.* **11**, 626–9.

PERLOFF, L.J., CHON, H., PETRELLA, E.J., GROSSMAN, R.A., AND BARKER, C.F. (1976). Acute colitis in the renal allograft recipient *Ann. Surg.* **183**, 77–83.

PERPER, J.A., PIDLAON, A., AND FISHER, R.S. (1971). Granulomatous peritonitis induced by rice-starch glove powder. A clinical and experimental study. *Am. J. Surg.* **122**, 812–17.

PETERS, J.L. (1976). Benign oesophageal stricture following oral potassium chloride therapy. *Br. J. Surg.* **63**, 698–9.

PHILLIPS, M.S., FEHILLY, B., STEWART, S., AND DRONFIELD, M.W. (1983). Enteritis and colitis associated with mefenamic acid. *Br. med. J.* **287**, 1626.

PICKLEMAN, J., STRAUS, F.H., AND PALOYAN, E. (1979). Pancreatitis associated with thiazide administration. A role for the parathyroid glands? *Arch Surg.* **114**, 1013–16.

PIERCE, P.F., WILSON, R., SILVA, J., GARAGUSI, V.F., RIFKIN, G.D., FEKETY, R., NUNEZ-MONTIEL, O., DOWELL, V.R., AND HUGHES, J.M. (1982). Antibiotic-associated pseudomembranous colitis: an epidemiologic investigation of a cluster of cases. *J. infect. Dis.* **145**, 269–74.

PIERSON, R.N., HOLT, P.R., WATSON, R.M., AND KEATING, R.P. (1961). Aspirin and gastrointestinal bleeding: chromate[51] blood loss studies. *Am. J. Med.* **31**, 259–65.

PILBRANT, Å. (1977). Ulceration due to emepronium bromide tablets. *Lancet* **i**, 749.

PINKERTON, C.R., BRIDGES, J.M., AND WELSHMAN, S.G. (1981). Enterotoxic effect of methotrexate: does it influence the drug's absorption in children with acute lymphoblastic leukaemia? *Br. med. J.* **282**, 1276–7.

PINKUS, G.S., WILSON, R.E., AND CORSON, J.M. (1974). Reticulum cell sarcoma of the colon following renal transplantation. *Cancer, NY* **34**, 2103–8.

PIPER, D.W., GREIG, M., LANDECKER, K.D., SHINNERS, J., WALLER, S., AND CANALESE, J. (1977). Analgesic intake and chronic gastric ulcer, acute upper gastrointestinal haemorrhage, personality traits and social class. *Proc. R. Soc. Med.* **70**, (Suppl. 7), 11–15.

——, McINTOSH, J.H., ARIOTTI, D.E., FENTON, B.H., AND MacLENNAN, R. (1981). Analgesic ingestion and chronic peptic ulcer. *Gastroenterology* **80**, 427–32.

PITTMAN, F.E. (1975). Lomotil and antibiotic colitis. *Ann. intern. Med.* **83**, 124–5.

PODDAR, P.K., BAUER, J., GELERNT, I., SALKY, B., AND KREEL, I. (1982). Radiation injury to small intestine. *Mt. Sinai J. Med.* **49**, 144–9.

POKORNEY, B.H. AND NICHOLS, T.W. (1981). Pseudomembranous colitis. A complication of sulfasalazine therapy in a patient with Crohn's colitis. *Am. J. Gastroenterol.* **76**, 374–6.

PORTEK, I., GRAHAM, G., AND FLEMING, A. (1981). Enteric-coated pelletized aspirin. Gastrointestinal blood loss and bioavailability. *Med. J. Aust.* **2**, 39–40.

POTISH, R.A., JONES, T.K., AND LEVITT, S.H. (1979). Factors predisposing to radiation-related small-bowel damage. *Radiology* **132**, 479–82.

POWIS, S.J.A., BARNES, A.D., DAWSON-EDWARDS, P., AND THOMPSON, H. (1972). Ileocolonic problems after cadaveric renal transplantation. *Br. med. J.* **1**, 99–101.

POZNIAK, A.L., AHERN, M., AND BLAKE, D.R. (1981). Azathioprine-induced shock. *Br med. J.* **283**, 1548.

PRESCOTT, L.F. (1968). Antipyretic analgesic drugs. In *Sideeffects of drugs*, Vol. VI (ed. L. Meyer and A. Herxheimer), p. 132. Excerpta Medica Foundation, Amsterdam.

PRESENT, D.H., KORELITZ, B.I., WISCH, N., GLASS, J.L., SACHAR, D.B., AND PASTERNACK, B.S. (1980). Treatment of Crohn's disease with 6-mercaptopurine. A long-term, randomized, double-blind study. *New Engl. J. Med.* **302**, 981–7.

PRICE, A.B. AND DAVIES, D.R. (1977). Pseudomembranous colitis *J. clin. Pathol.* **30**, 1–12.

——, LARSON, H.E., AND CROW, J. (1979). Morphology of experimental antibiotic-associated enterocolitis in the hamster: a model for human pseudomembranous colitis and antibiotic-associated diarrhoea. *Gut.* **20**, 467–75.

PRUST, F.W. AND KUMAR, G.K. (1976). Massive colonic bleeding and oral contraceptive 'pills'. *Am. J. Obstet. Gynecol.* **125**, 695–8.

PURANEN, J. AND KOIVISTO, E. (1978). Perforation of the urinary bladder and small intestine caused by a trochanteric plate. *Acta orthop. scand.* **49**, 65–7.

RACE, T.F., PAES, I.C., AND FALOON, W.W. (1970). Intestinal malabsorption induced by oral colchicine: comparison with neomycin and cathartic agents. *Am. J. med. Sci.* **259**, 32–41.

RACHMILEWITZ, D., FOGEL, R., AND KARMELI, F. (1978). Effect of colchicine and vinblastine on rat intestinal water transport and Na-K-ATPase activity. *Gut* **19**, 759–64.

RÄF, L.E. (1967). Enteric-coated potassium chloride tablets and ulcer of the small intestine. *Acta chir. scand. Suppl.* 374.

RAFTERY, E.B. AND DENMAN, A.M. (1973). Systemic lupus erythematosus syndrome induced by practolol. *Br. med. J.* **2**, 452–5.

RAHBEK, I. (1976). Gastroscopic evaluation of the effect of a new anti-rheumatic compound ketoprofen (19.583 R.P.), on the human gastric mucosa. A double-blind cross-over trial against acetylsalicylic acid. *Scand. J. Rheumatol.* (Suppl. 14), 63–72.

RAINSFORD, K.D. (1975a). Electronmicroscopic observations on the effect of orally administered aspirin and aspirin-bicarbonate mixtures on the development of gastric mucosal damage in the rat. *Gut* **16**, 514–27.

—— (1975b). The biochemical pathology of aspirin-induced gastric damage. *Agents Actions* **5**, 326–44.

—— (1977). The comparative gastric ulcerogenic activities of non-steroid anti-inflammatory drugs. *Agents Actions* **7**, 573–7.

RAMPLING, A., WARREN, R.E., BERRY, P.J., SWIRSKY, D., HOGGARTH, C.E., AND BEVAN, P.C. (1982). Atypical *Clostridium difficile* colitis in neutropenic patients. *Lancet* **ii**, 162–3.

RAMSAY, L.E., WAKEFIELD, V.A., AND HARRIS, E.E. (1982). Methyldopa-induced chronic pancreatitis. *Practitioner* **226**, 1166–9.

REED, D.L. AND COON, W.W. (1963). Thromboembolism in patients receiving progestational drugs. *New Engl. J. Med.* **269**, 622–4.

REES, W.D.W., GIBBONS, L.C., AND TURNBERG, L.A. (1983). Effects of non-steroidal anti-inflammatory drugs and prostaglandins on alkali secretion by rabbit gastric fundus *in vitro. Gut.* **24**, 784–9.

—— AND TURNBERG, L.A. (1980). Reappraisal of the effects of aspirin on the stomach. *Lancet* **ii**, 410–13.

REEVES, R.J., SANDERS, A.P., ISLEY, J.K., SHARPE, K.W., AND BAYLIN, G.J. (1959). Fat absorption from the human gastro-intestinal tract in patients undergoing radiation therapy. *Radiology* **73**, 398–401.

REICHELDERFER, M. AND MORRISSEY, J.F. (1980). Colonoscopy in radiation colitis. *Gastrointest. Endosc.* **26**, 41–3.

REINER, L., SCHLESINGER, M.J., AND MILLER, G.M. (1952). Pseudomembranous colitis following aureomycin and chloramphenicol. *Arch Pathol.* **54**, 39–67.

ReMINE, S.G. AND McILRATH, D.C. (1980). Bowel perforation in steroid-treated patients. *Ann. Surg.* **192**, 581–6.

RENNING, J.A., WARDEN, G.D., STEVENS, L.E., AND REEMTSMA, K. (1972). Pancreatitis after renal transplantation. *Am. J. Surg.* **123**, 293–6.

RICHARDSON, S.A., ALCOCK, P.A., AND GRAY, J. (1983). *Clostridium difficile* and its toxin in healthy neonates. *Br. med. J.* **287**, 878.

RICKWOOD, A.M.K. (1977). A case of ileal atresia and ileocutaneous fistula caused by amniocentesis. *J. Pediat.* **91**, 312.

RIDOLFO, A.S., CRABTREE, R.E., JOHNSON, D.W., AND ROCKHOLD, F.W. (1980). Gastrointestinal microbleeding: comparisons between benoxaprofen and other nonsteroidal anti-inflammatory agents. *J. Rheumatol.* **7**, (Suppl. 6), 36–47.

RIEMENSCHNEIDER, T.A., WILSON, J.F., AND VERNIER, R.L. (1966). Glucocorticoid-induced pancreatitis in children. *Pediatrics* **41**, 428–37.

RIFKIN, G.D., FEKETY, F.R., SILVA, J., AND SACK, R.B. (1977). Antibiotic-induced colitis: implication of a toxin neutralised by *Clostridium sordellii* antitoxin. *Lancet* **ii**, 1103–6.

——, SILVA, J., AND FEKETY, R. (1978). Gastrointestinal and systemic toxicity of fecal extracts from hamsters with clindamycin-associated colitis. *Gastroenterology* **74**, 52–7.

RIGOLOSI, R.S., MAHER, J.F., AND SCHREINER, G.E. (1969). Intestinal perforation during peritoneal dialysis. *Ann. intern. Med.* **70**, 1013–15.

RILEY, T.V., BOWMAN, R.A., ROBINSON, J., AND BURKE, V. (1983). Clostridium difficile and its toxins. *Lancet* **i**, 1386.

RITLAND, S., FAUSA, O., GJONE, E., BLOMHOFF, J.P., SKREDE, S., AND LÅNNER, A. (1975). Effect of treatment with a bile-sequestering agent (Secholex) on intestinal absorption, duodenal bile acids, and plasma lipids. *Scand. J. Gastroenterol.* **10**, 791–800.

ROBBINS, R.D., SCHOEN, R., SOHN, N., AND WEINSTEIN, M.A. (1982). Colonic decompression of massive cecal dilatation (Ogilvie's syndrome) secondary to cesarian section. *Am. J. Gastroenterol.* **77**, 231–2.

ROBERT, A. (1976). Antisecretory, antiulcer, cytoprotective and diarrheogenic properties of prostaglandins. *Adv. Prostaglandin Thromboxane Res.* **2**, 507–20.

——, NEZAMIS, J.E., AND PHILLIPS, J.P. (1968). Effect of prostaglandin E_1 on gastric secretion and ulcer formation in the rat. *Gastroenterology* **55**, 481–7.

ROGERS, T.R., PETROU, M., LUCAS, C., CHUNG, J.T.N., BARRETT, A.J., BORRIELLO, S.P., AND HONOUR, P. (1981). Spread of *Clostridium difficile* among patients receiving non-absorbable antibiotics for gut decontamination. *Br. med. J.* **282**, 408–9.

ROHNER, H.G., BERGES, W., AND WEINBECK, M. (1982). Clomethiazol tablets induce ulcers in the esophagus. *Z. Gastroent.* **20**, 469–73.

ROMINGER, J.M., GUTIERREZ, J.G., CURTIS, D., AND CHEY, W.Y. (1978). Methyldopa-induced pancreatitis. *Am. J. dig. Dis.* **23**, 756–8.

ROTHERMICH, N.O. (1966). An extended study of indomethacin. *J. Am. med. Ass.* **195**, 1102–6.

ROXE, D.M., ARGY, W.P., FROST, B., KERWIN, J., AND SCHREINER, G.E. (1976). Complications of peritoneal dialysis. *South med. J.* **69**, 584–7.

RUBENSTEIN, R.B. AND WOLFF, W.I. (1976). Methadone ileus syndrome: report of a fatal case. *Dis. Colon Rectum* **19**, 357–9.

RUBIN, J., OREOPOULOS, D.G., AND LIO, T.T. (1976). Management of peritonitis and bowel perforation during chronic peritoneal dialysis *Nephron* **16**, 220–5.

RUDGE, C.J., JONES, R.H., BEWICK, M., WESTON, M.J., AND PARSONS, V. (1979). Peptic ulcer after renal transplantation. *Lancet* **i**, 562.

RUSSELL, J.C. AND WELCH, J.P. (1979). Operative management of radiation injuries of the intestinal tract. *Am. J. Surg.* **137**, 433–42.

RUSSELL, R.I. AND GOLDBERG, A. (1968). Effect of aspirin on the gastric mucosa of guinea pigs on a scorbutogenic diet. *Lancet* **ii**, 606–8.

——, WILLIAMSON, J.M., GOLDBERG, A., AND WARES, E. (1968). Ascorbic-acid levels in leucocytes of patients with gastrointestinal haemorrhage. *Lancet* **ii**, 603–6.

RUTLIN, E., BERSTAD, A., AND REFSUM, N. (1977). Gastric mucosal damage caused by plain and microencapsulated acetylsalicylic acid tablets in healthy subjects: a gastrocamera study. *Scand. J. Gastroenterol.* **12**, 989–92.

SAGINUR, R., HAWLEY, C.R., AND BARTLETT, J.G. (1980). Colitis associated with metronidazole therapy. *J. infect. Dis.* **141**, 772–4.

St JOHN, D.J.B. (1975). Gastric mucosal damage by aspirin. *C.R.C. crit. Rev. Toxicol.* **3**, 317–44.

—— AND McDERMOTT, F.T. (1970). Influence of achlorhydria on aspirin-induced occult gastrointestinal blood loss: studies in Addisonian pernicious anaemia. *Br med. J.* **2**, 450–2.

——, YEOMANS, N.D., AND DE BOER, W.G.R.M. (1973). Chronic gastric ulcer induced by aspirin: an experimental model. *Gastroenterology* **65**, 634–41.

SAKURAI, Y., TSUCHIYA, H., IKEGAMI, F., FUNATOMI, T., TAKASU, S., AND UCHIKOSHI, T. (1979). Acute right-sided hemorrhagic colitis associated with oral administration of ampicillin. *Dig. Dis. Sci.* **24**, 910–15.

SALEH, S., KHAYYAL, M.T., EL-MASRI, A.M., AND GHAZAL, A.M. (1969). Effect of acetylsalicylic acid and phenylbutazone on glucose absorption *in vitro*. *Metabolism* **18**, 599–605.

SAMUEL, P., PRICE, H.C., AND SCHALCHI, O.B. (1965). Serum cholesterol reduction by para-aminosalicylates in man: 1^{131} triolein absorption studies. *Proc. Soc. exp. Biol. Med.* **118**, 654–8.

SANDEMAN, D.R., CLEMENTS, M.R., AND PERRINS, E.J. (1980). Oesophageal obstruction due to hygroscopic gum laxative. *Lancet* **i**, 364–5.

SANDLER, R.S. AND SANDLER, D.P. (1983). Radiation-induced cancers of the colon and rectum: assessing the risk. *Gastroenterology* **84**, 51–7.

SASAKI, M., TONODA, S., AOKI, Y., AND KATSUMI, M. (1980). Pancreatitis due to valproic acid. *Lancet* **i**, 1196.

SAUTTER, R.D. AND ZIFFREN, S.E. (1959). Adrenocortical steroid therapy resulting in unusual gastrointestinal complications. *Arch. Surg.* **79**, 346–56.

SAXÉN, L., KASSINEN, A., AND SAXÉN, E. (1963). Peritoneal foreign-body reaction caused by condom emulsion. *Lancet* **i**, 1295–6.

SAYFAN, J., ADAM, Y., AND SIGAL, B. (1977). Ergot-induced small bowel perforations. *Harefuah* **93**, 197–8.

SCARPELLO, J.H.B. AND SLADEN, G.E. (1977). Malabsorption in relation to abdominal irradiation and quadruple chemotherapy for lymphosarcoma. *Postgrad. med. J.* **53**, 219–21.

SCHAPIRO, R.L. AND NEWMAN, A. (1973). Acute enterocolitis: a complication of antibiotic therapy. *Radiology* **108**, 263–8.

SCHIESSEL, R., STARLINGER, M., WOLF, A., PINGGERA, W., ZAZGORNIK, J., SCHMIDT, P., WAGNER, O., SCHWARZ, S., AND PIZA, F. (1981). Failure of cimetidine to prevent gastroduodenal ulceration and bleeding after renal transplantation. *Surgery* **90**, 456–8.

SCHMID, F.R. AND CULIC, D.D. (1976). Anti-inflammatory drugs and gastrointestinal bleeding: a comparison of aspirin and ibuprofen. *J. clin. Pharmacol. and J. New Drugs.* **16**, 418–25.

SCHMITZ, R.L., CHAO, J.-H., AND BARTOLOME, J.S. (1974). Intestinal injuries incidental to irradiation of carcinoma of the cervix of the uterus. *Surg. Gynecol. Obstet.* **138**, 29–32.

SCHNEERSON, J.M. AND GAZZARD, B.G. (1977). Reversible malabsorption caused by methyldopa. *Br. med. J.* **2**, 1456–7.

SCHNEIDER, A.B., FAVUS, M.J., STRACHURA, M.E., ARNOLD, M.J., AND FROHMAN, L.A. (1977). Salivary gland neoplasms as a late consequence of head and neck irradiation. *Ann. intern. Med.* **87**, 160–4.

SCHNEIDER, R. (1977). Doxycycline esophageal ulcers. *Am. J. dig. Dis.* **22**, 805–7.

SCHNYDER, P.A., BRASCH, R.S., AND SALVATIERRA, O. (1979). Gastrointestinal complications of renal transplantation in children. *Radiology* **130**, 361–6.

SCHWAN, A., SJÖLIN, S., AND TROTTESTAM, U. (1983). Relapsing *Clostridium difficile* enterocolitis cured by rectal infusion of homologous faeces. *Lancet* ii, 845.

SCHWARTZ, H.A. (1981). Lower gastrointestinal side-effects of nonsteroidal antiinflammatory drugs. *J. Rheumatol.* **8**, 952–4.

SCOTT, A.J., NICHOLSON, G.I., AND KERR, A.R. (1973). Lincomycin as a cause of pseudomembranous colitis. *Lancet* ii, 1232–4.

SCOTT, J.T., PORTER, I.H., LEWIS, S.M., AND DIXON, A. St. J. (1961). Studies of gastrointestinal bleeding caused by corticosteroids, salicylates, and other analgesics. *Quart. J. Med.* **30**, 167–88.

SEAMAN, W.B. AND ACKERMAN, L.V. (1957). The effect of radiation on the esophagus. A clinical and histologic study of the effect produced by the betatron. *Radiology* **68**, 534–40.

SEARS, A.D., HAWKINS, J., KILGORE, B.B., AND MILLER, J.E. (1964). Plain roentgenographic findings in drug induced intramural hematoma of the small bowel. *Am. J. Roentg.* **91**, 808–13.

SEINO, S., SEINO, Y., MATSUKURA, S., KÚRAHACHI, H., IKEDA, M., YAWATA, M., AND IMURA, H. (1978). Effect of glucocorticoids on gastrin secretion in man. *Gut* **19**, 10–13.

SENTURIA, H.R., SUSMAN, N., AND SHYKEN, H. (1961). The roentgen appearance of spontaneous intramural hemorrhage of the small intestine associated with anticoagulant therapy. *Am. J. Roentg.* **86**, 62–9.

SEPPÄLÄ, K. (1978). Colonoscopy in the diagnosis of pseudomembranous colitis. *Lancet,* ii, 435.

SHACK, M.E. (1966). Drug-induced ulceration and perforation of the intestine. *Ariz. Med.* **23**, 517–23.

SHAFFER, J.L., HIGHAM, C., AND TURNBERG, L.A. (1980). Hazards of slow-release preparations in patients with bowel strictures. *Lancet* ii, 487.

SHAIVITZ, S.A. (1974). Dantrolene. *J. Am. med. Ass.* **229**, 1282–3.

SHAUNAK, S. AND O'DONOHUE, J. (1983). Massive haematemesis and melaena from a Meckel's diverticulum. *Postgrad. med. J.* **59**, 786–7.

SHAW, M.T., SPECTOR, M.H., AND LADMAN, A.J. (1979). Effects of cancer, radiotherapy and cytotoxic drugs on intestinal structure and function. *Cancer treat. Rev.* **6**, 141–51.

SHEPHERD, A.M.M., STEWART, W.K., AND WORMSLEY, K.G. (1973). Peptic ulceration in chronic renal failure. *Lancet* i, 1357–9.

SIEFKIN, A.D. (1980). Sulindac and pancreatitis. *Ann. intern. Med.* **93**, 932.

SIEGEL, R.R., LUKE, R.G., AND HELIEBUSCH, A.A. (1972). Reduction of toxicity of corticosteroid therapy after renal transplantation. *Am. J. Med.* **53**, 159–69.

SIEGFRIED, J. AND ZUMSTEIN, H. (1971). Changes in taste under L-DOPA therapy. *J. Neurol.* **200**, 345–8.

SIEGMAN-IGRA, Y., YARON, M., SILETZKI, M., SCHUJMAN, E., AND GILAT, T. (1976). Colitis and death following gold therapy. *Rheumatol. Rehabil.* **15**, 245–7.

SIEMSSON, O.J., ANDERSEN, J.T., MEYHOFF, H.H., NORDLING, J., AND WALTER, S. (1981). Effect of emepronium bromide on lower oesophageal sphincter. *Br. med. J.* **282**, 1928–9.

SILBERT, B., FIGIEL, L.S., AND FIGIEL, S.J. (1962). Intramural jejunal hematomas secondary to anti-coagulant therapy. *Am. J. dig. Dis.* **7**, 892–9.

SILVOSO, G.R., IVEY, K.J., BUTT, J.H., LOCKARD, O.O., HOLT, S.D., SISK, C., BASKIN, W.N., MacKERCHER, P.A., AND HEWETT, J. (1979). Incidence of gastric lesions in patients with rheumatic disease on chronic aspirin therapy. *Ann. intern. Med.* **91**, 517–20.

SIMKIN, E.P. AND WRIGHT, F.K. (1968). Perforating injuries of the bowel complicating peritoneal catheter insertion. *Lancet* i, 64–6.

SIMMONS, R.L., MATAS, A.J., RATTAZZI, L.C., BALFOUR, H.H., HOWARD, R.J., AND NAJARIAN, J.S. (1977). Clinical characteristics of the lethal cytomegalovirus infection following renal transplantation. *Surgery* **82**, 537–46.

SINATRA, F., BUNTAIN, W.L., MITCHELL, C.H., AND SUNSHINE, P. (1976). Cholestyramine treatment of pseudomembranous colitis. *J. Pediat.* **88**, 304–6.

SINGLETON, J.W., LAW, D.H., KELLEY, M.L., MEKHJIAN, H.S., AND STURDEVANT, R.A.L. (1979). National Cooperative Crohn's Disease Study: adverse reactions to study drugs. *Gastroenterology* **77**, 870–82.

SIXTY PLUS REINFARCTION STUDY RESEARCH GROUP (1982). Risks of long-term oral anticoagulant therapy in elderly patients after myocardial infarction. *Lancet* i, 64–8.

SKYRING, A. AND BHANTHUMNAVIN, K. (1967). Gastrointestinal bleeding: a comparison of mefenamic acid and aspirin. *Med. J. Aust.* **1**, 601–3.

SLAGLE, G.W. AND BOGGS, H.W. (1976). Drug-induced pseudomembranous enterocolitis: a new etiologic agent *Dis. Colon Rectum* **19**, 253–5.

SLATER, D. (1982). *Clostridium difficile* in toxic megacolon. *Br. med. J.* **285**, 888–9.

SLAUGHTER, D.P. AND SOUTHWICK, H.W. (1957). Mucosal carcinoma as a result of irradiation. *Arch. Surg.* **74**, 420–9.

SLAUGHTER, R.L. (1975). Parotid gland swelling developing during peroral endoscopy. *Gastrointest. Endosc.* **22**, 38–9.

—— AND BOYCE, H.W. (1969). Submaxillary salivary gland swelling developing during peroral endoscopy. *Gastroenterology* **57**, 83–8.

SLONE, D., JICK, H., LEWIS, G.P., SHAPIRO, S., AND MIETTINEN, O.S. (1969). Intravenously given ethacrynic acid and gastrointestinal bleeding: a finding resulting from comprehensive drug surveillance. *J. Am. med. Ass.* **209**, 1668–71.

SMITH, P.A., ELLIS, C.J., SPARKS, R.A., AND GUILLEBAUD, J. (1983). Deaths associated with intrauterine contraceptive devices in the United Kingdom between 1973 and 1983. *Br med. J.* **287**, 1537–8.

SMITH, R.F. AND SZILAGYI, D.E. (1960). Ischemia of the colon as a compication in the surgery of the abdominal aorta. *Arch. Surg.* **80**, 806–21.

SNEIERSON, H. AND WOO, Z.P. (1955). Starch powder granuloma: a report of two cases. *Ann. Surg.* **142**, 1045–50.

SNYDER, N., PATTERSON, M., AND HUGHES, W.S. (1973). Esophageal hematoma. *South. med. J.* **66**, 1079–80.

SØRENSEN, K. AND CHRISTIANSEN, L.V. (1977). Long-term therapy with tolfenamic acid pINN. A clinical and toxicological study with special reference to clinical and chemical laboratory parameters. *Scand. J. Rheumatol.* **Suppl. 20.**

SORKIN, M.I., LUGER, A.M., PROWANT, B., KENNEDY, J., MOORE, H., AND NOLPH, K.D. (1982). Histological and functional characteristics of the peritoneal membrane of a diabetic patient after 34 months of CAPD. *Periton Dial. Bull.* **2**, 24–7.

SORRELL, V.F., KNIGHT, D.H., AND BURCHER, S.K. (1975). Pancreatitis following intestinal bypass for obesity. *Aust. NZ. J. Surg.* **45**, 163–7.

SPANOS, P.K., SIMMONS, R.L., AND NAJARIAN, J.S. (1974). Peptic ulcer disease in the transplant recipient. *Arch Surg.* **109**, 193–7.

SPARBERG, M. AND KIRSNER, J.B. (1966). Long-term corticosteroid therapy for regional enteritis: an analysis of 58 courses in 54 patients. *Am. J. dig. Dis.* **11**, 865–80.

SPENNEY, J.G. AND BHOWN, M. (1977). Effect of acetylsalicyclic acid on gastric mucosa. II. Mucosal ATP and phosphocreatine content, and salicylate effects on mitochondrial metabolism. *Gastroenterology* **73**, 995–9.

SPERLING, I.L. (1969). Adverse reactions with long-term use of phenylbutazone and oxyphenbutazone. *Lancet* **ii**, 535–7.

SPIRA, I.A., RUBENSTEIN, R., WOLFF, D., AND WOLFF, W.I. (1975). Fecal impaction following methadone ingestion simulating acute intestinal obstruction. *Ann. Surg.* **181**, 15–19.

SPIRO, H.M. AND MILLES, S.S. (1960). Clinical and physiologic implications of the steroid-induced peptic ulcer. *New Engl. J. Med.* **263**, 286–94.

STAHL, R. (1966). Parotitis in Tandearil treatments. *Nord. Med.* **75**, 170.

STANLEY, R.J., MELSON, G.L., AND TEDESCO, F.J. (1974). The spectrum of radiographic findings in antibiotic-related pseudomembranous colitis. *Radiology* **111**, 519–24.

STANTON, P.E., WILSON, J.P., LAMIS, P.A., AND LETTON, A.H. (1974). Acute abdominal conditions induced by anticoagulant therapy. *Am. Surg.* **40**, 1–14.

STARZL, T.E. (1964). *Experience in renal transplantation.* W.B. Saunders Co, Philadelphia.

——, NALESNIK, M.A., PORTER, K.A., HO, M., IWATSUKI, S., GRIFFITH, B.P., ROSENTHAL, J.T., HAKALA, T.R., SHAW Jr., B.W., HARDESTY, R.L., ATCHISON, R.W., JAFFE, R., AND BAHNSON, H.T. (1984). Reversibility of lymphomas and lymphoproliferative lesions developing under cyclosporin-steroid therapy. *Lancet* **i**, 583–7.

STEIN, H.B. AND UROWITZ, M.B. (1976). Gold-induced enterocolitis. Case report and literature review. *J. Rheumatol.* **3**, 21–6.

STEINBERG, W.M. AND LEWIS, J.H. (1981). Steroid-induced pancreatitis: does it really exist? *Gastroenterology* **81**, 799–808.

STEMMERMANN, G.N. AND HAYASHI, T. (1971). Colchicine intoxication: a reappraisal of its pathology based on a study of three fatal cases. *Hum. Pathol.* **2**, 321–32.

STERNLIEB, I. AND SCHEINBERG, I.H. (1964). Penicillamine therapy for hepatolenticular degeneration. *J. Am. med. Ass.* **189** 748–54.

STEWART, I. (1982). Gastrointestinal haemorrhage and benoxaprofen. *Br. med. J.* **284**, 163–4.

STILLMAN, A.E., WEINBERG, M., MAST, W.C., AND PALPANT, S. (1977). Ischemic bowel disease attributable to ergot. *Gastroenterology* **72**, 1336–7.

STOLLER, J.L. (1985). Oesophageal ulceration and theophylline. *Lancet* **ii**, 328–9.

STOPA, E.G., O'BRIEN, R., AND KATZ, M. (1979). Effect of colchicine on guinea pig intrinsic factor-B_{12} receptor. *Gastroenterology* **76**, 309–14.

STOWERS, J.M. AND BEWSHER, P.D. (1969). Studies on the mechanism of weight reduction by phenformin. *Postgrad. med. J.* **45**, (May Suppl.), 13–16.

STRICKLAND, P. (1954). Damage to the rectum in the radium treatment of carcinoma of the cervix. *Br. J. Radiol.* **27**, 630–4.

STROUTHIDIS, T.M., MANKIKAR, G.D., AND IRVINE, R.E. (1972). Ulceration of the mouth due to emepronium bromide. *Lancet* **i**, 72–3.

STUART, F.P., RECKARD, C.R., SCHULAK, J.A., AND KETEL, B.L. (1981). Gastroduodenal complications in kidney transplant recipients. *Ann. Surg.* **194**, 339–42.

STUBBÉ, L. TH. F.L. (1958). Occult blood in faeces after administration of aspirin. *Br. med. J.* **2**, 1062–6.

—— (1961). Ijzergebrek-anemie en het gebruik van acetosal. *Ned Tijdschr. Geneesk.* **105**, 1673–8.

STUMPF, H.H., WILENS, S.L., AND SOMOZA, C. (1956). Pancreatic lesions and peripancreatic fat necrosis in cortisone-treated rabbits. *Lab Invest.* **5**, 224–35.

STURGES, H.F. AND KRONE, C.L. (1973). Ulceration and stricture of the jejunum in a patient on long-term indomethacin therapy: report of a case. *Am. J. Gastroenterol.* **59**, 162–9.

SUGARBAKER, P.H., Mc REYNOLDS, R.A., AND BROOKS, J.R. (1974). Glove starch granulomatous disease. An unsolved surgical problem. *Am. J. Surg.* **128**, 3–7.

SUMITHRAN, E., LIMM, K.H., AND CHIAM, H.L. (1979). Atrio-oesophageal fistula complicating mitral valve disease. *Br. med. J.* **2**, 1552–3.

SUMMERSKILL, W.H.J. AND ALVAREZ, A.S. (1958). Salicylate anaemia *Lancet* **ii**, 925–8.

SUN, D.C.H., ROTH, S.H., MITCHELL, C.S., AND ENGLUND, D.W. (1974). Upper gastrointestinal disease in rheumatoid arthritis. *Am. J. dig. Dis.* **19**, 405–10.

SUSSMAN, R.M. AND MILLER, J. (1956). 'Iodide mumps' after intravenous urography. *New Engl. J. Med.* **255**, 433–4.

SUTTON, D.R. AND GOSNOLD, J.K. (1977). Oesophageal ulceration due to clindamycin. *Br. med. J.* **1**, 1598.

SVENDSEN, L.B., HANSEN, O.H., AND JOHANSEN, Aa (1981). A comparison of the effects of HP 549 (isoxepac), indomethacin and acetylsalicyclic acid (aspirin) on gastric mucosa in man. *Scand. J. Rheumatol.* **10**, 186–8.

SWAN, R.W., FOWLER, W.C., AND BORONOW, R.C. (1976). Surgical management of radiation injury to the small intestine. *Surg. Gynecol. Obstet.* **142**, 325–7.

SWARTZBERG, J.E., MARESCA, R.M., AND REMINGTON, J.S. (1976). Gastrointestinal side-effects associated with clindamycin: 1000 consecutive patients. *Arch. intern. Med.* **136**, 876–9.

SWIFT, P.G.F., DRISCOLL, I.B., AND VOWLES, K.D.J. (1979). Neonatal small-bowel obstruction associated with amniocentesis. *Br. med. J.* **1**, 720.

SWINSON, C.M., PERRY, J., LUMB, M., AND LEVI, A.J. (1981). Role of sulphasalazine in the aetiology of folate deficiency in ulcerative colitis. *Gut* **22**, 456–61.

SZPAK, M.W., JOHNSON, R.C., BRADY, C.E., AND BOSWELL, R.N. (1979). Gold (Au) induced enterocolitis. *Gastroenterology* **76**, 1257.

TAFT. D.A., LASERSOHN, J.T., AND HILL, L.D. (1970). Glove starch granulomatous peritonitis. *Am. J. Surg.* **120**, 231–6.

TANAKA, M., UCHIYAMA, M., AND IKEDA, S. (1977). Duodenal mucosal damage associated with chronic use of antiinflammatory drugs. *Endoscopy* **9**, 136–9.

TANKEL, H.I., CLARK, D.H., AND LEE, F.D. (1965). Radiation enteritis with malabsorption. *Gut* **6**, 560–9.

TARPILA, S. (1971). Morphological and functional response of human small intestine to ionizing radiation. *Scand. J. Gastroenterol.* **6** (Suppl.), 9–48.

TAVERNER, D., TALBOT, I.C., CARR-LOCKE, D.L., AND WICKS, A.C.B. (1982). Massive bleeding from the ileum: a late complication of pelvic radiotherapy. *Am. J. Gastroenterol.* **77**, 29–31.

TAYLOR, I.L., SELLS, R.A., McCONNELL, R.B., AND DOCKRAY, G.J. (1980). Serum gastrin in patients with chronic renal failure. *Gut* **21**, 1062–7.

TAYLOR, N.S., THORNE, G.M., AND BARTLETT, J.G. (1980). Separation of an enterotoxin from the cytotoxin of *Clostridium difficile*. *Clin. Res.* **28**, 285A.

TAYLOR, R.T., HUSKISSON, E.C., WHITEHOUSE, G.H., DUDLEY HART, F., AND TRAPNELL, D.H. (1968). Gastric ulceration during indomethacin therapy. *Br. med. J.* **4**, 734–7.

TEASLEY, D.G., OLSON, M.M., GEBHARD, R.L., GERDING, D.N., PETERSON, L.R., SCHWARTZ, M.J., AND LEE, Jr, J.T. (1983). Prospective randomized trial of metronidazole versus vancomycin for Clostridium-difficile-associated diarrhoea and colitis. *Lancet* **ii**, 1043–6.

TEDESCO, F.J. (1976). Clindamycin-associated colitis: review of the clinical spectrum of 47 cases. *Am. J. dig. Dis.* **21**, 26–32.

—— (1979). Antibiotic associated colitis with negative proctosigmoidoscopy examination. *Gastroenterology* **77**, 295–7.

—— (1982). Treatment of recurrent antibiotic- associated pseudomembranous colitis. *Am. J. Gastroenterol.* **77**, 220–1.

——, ANDERSON, C.B., AND BALLINGER, W.F. (1975). Drug-induced colitis mimicking an acute surgical condition of the abdomen. *Arch. Surg.* **110**, 481–4.

——, BARTON, R.W., AND ALPERS, D.H. (1974). Clindamycin-associated colitis: a prospective study. *Ann. intern. Med.* **81**, 429–33.

——, CORLESS, J.K., AND BROWNSTEIN, R.E. (1982a). Rectal sparing in antibiotic-associated pseudomembranous colitis: a prospective study. Gastroenterology **83**, 1259–60.

——, MARKHAM, R., GURWITH, M., CHRISTINE, D., AND BARTLETT, J.G. (1978). Oral vancomycin for antibiotic-associated pseudomembranous colitis. *Lancet* **ii**, 226–8.

——, VOLPICELLI, N.A., AND MOORE, F.S. (1982b). Estrogen- and progesterone-associated colitis: a disorder with clinical and endoscopic features mimicking Crohn's colitis. *Gastrointest. Endosc.* **28**, 247–9.

TEMPLETON, J.L. (1983). Toxic megacolon complicating pseudomembranous colitis. *Br. J. Surg.* **70**, 48.

TEPLICK, J.G., TEPLICK, S.K., OMINSKY, S.H., AND HASKIN, M.E. (1980). Esophagitis caused by oral medication. *Radiology* **134**, 23–5.

TERANO, A., MACH, T., STACHURA, J., TARNAWSKI, A., AND IVEY, K.J. (1984). Effect of 16, 16-dimethyl prostaglandin E_2 on aspirin-induced damage to rat gastric epithelial cells in tissue culture. *Gut* **25**, 19–25.

TESLER, M.A. AND LIM, E.S. (1981). Protection of gastric mucosa by sucralfate from aspirin-induced erosions. *J. clin. Gastroenterol.* **3** (Suppl. 2), 175–9.

THEIN, S.L. AND ASQUITH, P. (1977). Pneumatosis coli: complication of practolol. *Br. med. J.* **1**, 268.

THEUER, D. (1979). Erosive gastritis caused by peripatetic pacemaker. *Lancet* **ii**, 309–10.

THOMPSON, G.R., BARROWMAN, J., GUTIERREZ, L., AND DOWLING. R.H. (1971). Action of neomycin on the intraluminal phase of lipid absorption. *J. clin. Invest.* **50**, 319–23.

THOMPSON, M. AND ANDERSON, M. (1970). Studies of gastrointestinal blood loss during ibuprofen therapy. *Rheumatol. phys. Med.* **10**, (Suppl.), 104–7.

THOMPSON, W.M., SEIGLER, H.F., AND RICE, R.P. (1975). Ileocolonic perforation: a complication following renal transplantation. *Am. J. Roentg.* **125**, 723–30.

THOMSON, G., CLARK, A.H., HARE, K., AND SPILG, W.G.S. (1981). Pseudomembranous colitis after treatment with metronidazole. *Br. med. J.* **282**, 864–5.

THOROUGHMAN, J.C. AND PEACE, R.J. (1959). Abdominal surgical emergencies caused by uremic enterocolitis: report of 12 cases. *Am. Surg.* **25**, 533–9.

THORPE, J.A.C., OAKLAND, C., ADAMS, I.P., AND MATTHEWS, H.R. (1982). Irradiation-induced motor disorder of the oesophagus. *Gut* **23**, 710–11.

THORSEN, W.B., WESTERN, D., TANAKA, Y., AND MORRISSEY, J.F. (1968). Aspirin injury to the gastric mucosa: gastrocamera observations of the effect of pH. *Arch. intern. Med.* **121**, 499–506.

TILNEY N.L., COLLINS, J.J., AND WILSON, R.E. (1966). Hemorrhagic pancreatitis: a fatal complication of renal transplantation. *New Engl. J. Med.* **274**, 1051–7.

TINKER, M.A., TEICHER, I., AND BURDMAN, D. (1977). Cellulose granulomas and their relationship to intestinal obstruction. *Am. J. Surg.* **133**, 134–9.

TOFFLER, R.B., PINGOUD, E.G., AND BURRELL, M.I. (1978). Acute colitis related to penicillin and penicillin derivatives. *Lancet* **ii**, 707–9.

TOGHILL, P.J. AND BURKE, J.D. (1970). Death from paralytic ileus following vincristine therapy. *Postgrad. med. J.* **46**, 330–1.

TOMKIN, G.H. (1973). Malabsorption of vitamin B_{12} in diabetic patients treated with phenformin: a comparison with metformin. *Br. med. J.* **3**, 673–5.

——, HADDEN, D.R., WEAVER, J.A., AND MONTGOMERY, D.A.D. (1971). Vitamin B_{12} status of patients on long-term metformin. *Br. med. J.* **2**, 685–7.

TOSKES, P.P. AND DEREN, J.J. (1972). Selective inhibition of vitamin B_{12} absorption by para-aminosalicylic acid. *Gastroenterology* **62**, 1232–7.

TRAPNELL, J. (1972). The natural history and management of acute pancreatitis. *Clin. Gastroenterol.* **1**, 147–66.

TREADWELL, B.L.J., CARROLL, D.G., AND POMARE, E.W. (1973). Gastrointestinal blood loss with sustained-release aspirin. *NZ med. J.* **78**, 435–7.

TRIER, J.S. (1962a). Morphologic alterations induced by methotrexate in the mucosa of human proximal intestine. I. Serial observations by light microscopy. *Gastroenterology* **42**, 295–305.

—— (1962*b*). Morphologic alterations induced by methotrexate in the mucosa of human proximal intestine. II. Electron microscopic observations. *Gastroenterology* **43**, 407–24.

—— AND BROWNING, T.H. (1966). Morphologic response of the mucosa of human small intestine to X-ray exposure. *J. clin. Invest.* **45**, 194–204.

TRINGHAM, V.M., YOUNG, J.H., AND COCHRANE, P. (1980). Aspirin, paracetamol, diflunisal and gastrointestinal blood loss. *Eur. J. Rheumatol. Inflamm. Dis.* **3**, 175–9.

TRINH DINH, H., KERNBAUM, S., AND FROTTIER, J. (1978). Treatment of antibiotic-associated colitis by metronidazole. *Lancet* **i**, 338–9.

TRUDINGER, B.J. AND FITCHETT, D.H. (1976). Practolol peritonitis presenting as a pelvic mass. *Br. J. Obstet. Gynaecol.* **83**, 326–8.

TRUELOVE, S.C. AND WITTS, L.J. (1955). Cortisone in ulcerative colitis: final report on therapeutic trial. *Br. med. J.* **2**, 1041–8.

TURES, J.F., TOWNSEND, W.F., AND ROSE, H.D. (1976). Cèphalosporin-associated pseudomembranous colitis. *J. Am. med. Ass.* **236**, 948–9.

TYGSTRUP, N., WINKLER, K., AND WARBURG, E. (1959). Effect of *p*-aminosalicylic acid on serum cholesterol. *Lancet*, **i**, 503.

——, ——, AND JORGENSEN, K. (1961). Treatment of hypercholesterolaemia with para-aminosalicylic acid. *Ugeskr. Laeg.* **123**, 255.

UTHGENANNT, H., AND LETZEL, H. (1981). Gastrointestinal blood loss in volunteers following fenclofenac and diclofenac. *Br. J. clin. Pract.* **35**, 229–32.

VAKIL, B.J., KULKARNI, R.D., KULKARNI, V.N., MEHTA, D.J., CHARPURE, M.B., AND PISPATI, P.K. (1977*a*). Estimation of gastro-intestinal blood loss in volunteers treated with non-steroidal anti-inflammatory agents. *Curr. med. Res. Opin.* **5**, 32–7.

——, SHAH, P.N., DALAL, N.J., WAGHOLIKER, U.N., AND PISPATI, P.K. (1977*b*). Endoscopic study of gastro-intestinal injury with non-steroidal anti-inflammatory drugs. *Curr. med. Res. Opin.* **5**, 38–42.

VALBERG, L.S. AND TRUELOVE, S.C. (1960). Noninfective pseudo-membranous colitis following antibiotic therapy. *Am. J. dig. Dis.* **5**, 728–38.

VALMAN, H.B., PARRY, D.J., AND COGHILL, N.F. (1968). Lesions associated with gastroduodenal haemorrhage in relation to aspirin intake. *Br. med. J.* **4**, 661–3.

VAN BLANKENSTEIN, M. (1981). Diarrhoea and bile acid malabsorption in mild radiation enteropathy. *Gut* **22**, A433.

VAN DER HEIDE, H., TEN HAAFT, M.A., AND STRICKER, B.H.CH. (1981). Pancreatitis caused by methyldopa. *Br. med. J.* **282**, 1930–1.

VANE, J.R. (1971). Inhibition of prostaglandin synthesis as a mechanism of action for aspirin-like drugs. *Nature New Biol.* **231**, 232–5.

VAN GEERTRUYDEN, J. AND TOUSSAINT, C. (1967). Pancréatite aigüé après transplantation rénale. *Acta chir. Belg.* **3**, 271–80.

VAN ITALLIE, T.B., HASHIM, S.A., CRAMPTON, R.S., AND TENNENT, D.M. (1961). The treatment of pruritus and hypercholesteremia of primary biliary cirrhosis with cholestyramine. *New Engl. J. Med.* **265**, 469–74.

VAN WALRAVEN, A.A., EDELS, M., AND FONG, S. (1982). Pancreatitis caused by mefenamic acid. *Can. med. Ass. J.* **126**, 894.

VAN WYK, C.W. (1967). The oral lesion caused by aspirin: a cli-nicopathological study. *J. dent. Ass. S. Afr.* **22**, 1–7.

VENKATESWARAN, P.S., JEFFERS, A., AND HOCKEN, A.G. (1972). Gastric acid secretion in chronic renal failure. *Br. med. J.* **2**, 22–3.

VISCIDI, R., WILLEY, S., AND BARTLETT, J.G. (1981). Isolation rates and toxigenic potential of *Clostridium difficile* isolates from various patient populations. *Gastroenterology* **81**, 5–9.

VLASSES, P.H. AND FERGUSON, R.K. (1979). Temporary ageusia related to captopril. *Lancet* **i**, 526.

WAGNER, M., KISELOW, M.C., GOODMAN, J.J., BIEVER, P., AND GILL, L. (1979). The relationship of the intrauterine device, actinomy-cosis infection and bowel abscesses. *Wisc. med. J.* **78**, 23–6.

WALKER, M.J., CHAUDHURI, P.K., WOOD, D.C., AND DAS GUPTA, T.K. (1981). Radiation-induced parotid cancer. *Arch. Surg.* **116**, 329–31.

WALLS, J., BELL, D., AND SCHORR, W. (1968). Rectal bleeding and indomethacin. *Br. med. J.* **2**, 52.

WALSH , D. (1897). Deep tissue traumatism from roentgen ray exposure. *Br. med. J.* **2**, 272–3.

WALTA, D.C., GIDDENS, J.D., JOHNSON, L.F., KELLEY, J.L., AND WAUGH, D.F. (1976). Localized proximal esophagitis secondary to ascorbic acid ingestion and esophageal motor disorder. *Gastroenterology* **70**, 766–9.

WALTERS, B.A.J., ROBERTS, R., STAFFORD, R., AND SENEVIRATNE, E. (1983). Relapse of antibiotic associated colitis: endogenous persistence of *Clostridium difficile* during vancomycin therapy. *Gut* **24**, 206–12.

WANKA, J., JONES, L.I., WOOD, P.H.N., AND DIXON, A.St. J. (1964). Indomethacin in rheumatic diseases: a controlled clinical trial. *Ann. rheum. Dis.* **23**, 218–25.

WARNES, H., LEHMANN, H.E., AND BAN, T.A. (1967). Adynamic ileus during psychoactive medication: a report of three fatal and five severe cases. *Can. med. Ass. J.* **96**, 1112–13.

WARREN, S. AND FRIEDMAN, N.B. (1942). Pathology and pathologic diagnosis of radiation lesions in the gastrointestinal tract. *Am. J. Pathol.* **18**, 499–507.

WARREN, S.E., MITAS, J.A., AND SWERDLIN, A.H.R. (1980). Pancreatitis due to methyldopa: case report. *Milit. Med.* **145**, 399–400.

WARSHAW, A.L. (1972). Diagnosis of starch peritonitis by para-centesis. *Lancet* **ii**, 1054–6.

WATNEY, P.J.M. (1978). Copper intrauterine devices and the small intestine. *Br. med. J.* **2**, 255–6.

WATRING, W.G., BYFIELD, J.E., LAGASSE, L.D., LEE, Y.D., JUILLARD, G., JACOBS, M., AND SMITH, M.L. (1974). Combination adria-mycin and radiation therapy in gynecologic cancers. *Gynecol. Oncol.* **2**, 518–26.

WATSON, L.C., REEDER, D.D., AND THOMPSON, J.C. (1973). Effect of hydrocortisone on gastric secretion and serum gastrin in dogs. *Surg. Forum* **24**, 354–6.

—— AND THOMPSON, J.C. (1980). Erosion of the colon by a long-dwelling peritoneal dialysis catheter. *J. Am. med. Ass.* **243**, 2156–7.

WAYNE, E., MILLER, R.E., AND EISEMAN, B. (1971). Duodenal obstruction by the superior mesenteric artery in bedridden combat casualties. *Ann. Surg.* **174**, 339–45.

—— AND BURRINGTON, J.D. (1972). Duodenal obstruction by the superior mesenteric artery in children. *Surgery* **72**, 762–8.

WEBB, D.I., CHODOS, R.B., MAHAR, C.Q., AND FALOON, W.W. (1968). Mechanism of vitamin B_{12} malabsorption in patients receiv-ing colchicine. *New Engl. J. Med.* **279**, 845–50.

WEETMAN, R.M. AND BAEHNER, R.L. (1974). Latent onset of clinical pancreatitis in children receiving L-asparaginase therapy. *Cancer* **34**, 780–5.

WEIDEMA, W.F., VON MEYENFELDT, M.F., SOETERS, P.B., WESDORP, R.I.C., AND GREEP, J.M. (1980). Pseudomembranous colitis after whole gut irrigation with neomycin and erythromycin base. *Br. J. Surg.* **67**, 895–6.

WEISS, S.M., RUTENBERG, H.L., PASKIN, D.L., AND ZAREN, H.A. (1977). Gut lesions due to slow-release KCl tablets. *New Engl. J. Med.* **296**, 111–12.

WELCH, R.W., BENTCH, H.L., AND HARRIS, S.C. (1978). Reduction of aspirin-induced gastrointestinal bleeding with cimetidine. *Gastroenterology* **74**, 459–63.

WELLWOOD, J.M. AND JACKSON, B.T. (1973). The intestinal complications of radiotherapy. *Br. J. Surg.* **60**, 814–18.

WENGER, J. AND GROSS, P.R. (1964). Acute pancreatitis related to hydrochlorothiazide therapy. *Gastroenterology* **46**, 768.

WERTHEIM, P., BUURMAN, C., GEELEN, J., AND van der NOORDAA, J. (1983). Transmission of cytomegalovirus by renal allograft demonstrated by restriction enzyme analysis. *Lancet* **i**, 980–1.

WEST, R.J. AND LLOYD, J.K. (1975). The effect of cholestyramine on intestinal absorption. *Gut* **16**, 93–8.

WHITE, N.J., RAJAGOPALAN, B., YAHAYA, H., AND LEDINGHAM, J.G.G. (1980). Captopril and frusemide in severe drug-resistant hypertension. *Lancet* **ii**, 108–10.

WHITNEY, B. AND CROXON, R. (1972). Dysphagia caused by cardiac enlargement. *Clin Radiol.* **23**, 147–52.

WIERNIK, G. (1966). Radiation damage and repair in human jejunal mucosa. *J. Path. Bact.* **91**, 389–93.

WILKINSON, M.L., O'DRISCOLL, R., AND KIERNAN, T.J. (1981). Cimetidine and pancreatitis. *Lancet* **i**, 610–11.

WILKINSON, W.H., CIMINERA, J.L., AND SIMPKINS, G.T. (1969). Intravenously given ethacrynic acid and gastrointestinal bleeding. *J. Am. med. Ass.* **210**, 347.

WILLETT, A. (1878). Fatal vomiting, following the application of the plaster-of-Paris bandage in case of spinal curvature. *St. Bart's Hosp. Rep.* **14**, 333–5.

WILLIAMS, J.G. (1979). Drug-induced oesophageal injury. *Br. med. J.* **2**, 273.

WILLIAMS, J.R.B., GRIFFIN, J.P., AND PARKINS, A. (1976). Effect of concomitantly administered drugs on the control of long-term anticoagulant therapy. *Quart. J. Med.* **45**, 63–73.

WILLIAMS, S.E. AND TURNBERG, L.A. (1981). Demonstration of a pH gradient across mucus adherent to rabbit gastric mucosa: evidence for a 'mucus-bicarbonate' barrier. *Gut* **22**, 94–6.

WILLOUGHBY, J.M.T. (1983). Drug-induced abnormalities of taste

sensation. *Adverse Drug React. Bull.* No. 100.

WILSON, A.E., MEHRA, S.K., GOMERSALL, C.R., AND DAVIES, D.M. (1967). Acute pancreatitis associated with frusemide therapy. *Lancet* **i**, 105.

WINAWER, S.J., BEJAR, J., McRAY, R.S., AND ZAMCHECK, N. (1971). Hemorrhagic gastritis: importance of associated chronic gastritis. *Arch. intern. Med.* **127**, 129–31.

WINDSOR, W.O., KURREIN, I., AND DYER, N.H. (1975). Fibrinous peritonitis: a complication of practolol therapy. *Br. med. J.* **2**, 68.

WIOT, J.F., WEINSTEIN, A.S., AND FELSON, B. (1961). Duodenal hematoma induced by coumarin. *Am. J. Roenig* **86**, 70–5.

WOLLOCH, Y., CHAIMOFF, C., AND DINTSMAN, M. (1973). Late complications of radiation-induced damage to the gastro-intestinal tract. *Am. J. Proctol.* **24**, 473–80.

WOODS, J.E., ANDERSON, C.F., FROHNERT, P.P., AND PETRIE, C.R. (1972). Pancreatitis in renal allografted patients. *Mayo Clin. Proc.* **47**, 193–5.

WOOLF, D.L. (1965). Indomethacin suppositories. *Br. med. J.* **1**, 1497.

YANAGI, R.M., WILSON, A., NEWFELD, E.A., AZIZ, K.U., AND HUNT, C.E. (1981). Indomethacin treatment for symptomatic patent ductus arteriosus: a double-blind control study. *Pediatrics* **67**, 647–52.

YEADON, A., RAINA, M., GARDNER, M.C., MILAK, D.M., AND SMITH, K.E. (1983). Suprofen. An overview of long-term safety. *Pharmacology* **27**, (Suppl. 1), 87–94.

YEOMANS, N.D., St JOHN, D.J.B., AND DE BOER, W.R.G.M. (1973). Regeneration of gastric mucosa after aspirin-induced injury in the rat. *Am. J. dig. Dis.* **18**, 773–80.

YEUNG LAIWAH, A.C., HILDITCH, T.E., HORTON, P.W., AND HUNTER, J.A. (1981). Antiprostaglandin synthetase activity of non-steroidal anti-inflammatory drugs and gastrointestinal micro-bleeding: a comparison of flurbiprofen with benoxaprofen. *Ann. rheum. Dis.* **40**, 455–61.

YIK, K.Y., DREIDGER, A.A., AND WATSON, W.C. (1982). Prostaglandin E$_2$ tablets prevent aspirin-induced blood loss in man. *Dig. Dis. Sci.* **27**, 972–5.

ZER, M., CHAIMOFF, CH., AND DINTSMAN, M. (1972). Anticoagulant ileus with intestinal necrosis. *Isr. J. med. Sci.* **8**, 154–7.

ZION, M.M., GOLDBERG, B., AND SUZMAN, M.M. (1955). Corticotrophin and cortisone in treatment of scleroderma. *Quart. J. Med.* **24**, 215–17.

ZURIER, R.B., HASHIM, S.A., AND VAN ITALLIE, T.B. (1965). Effect of medium-chain triglyceride on cholestyramine-induced steatorrhea in man. *Gastroenterology* **49**, 490–5.

17 Drug-induced hepatic dysfunction

J.B. BOURKE

Zimmerman (1978) pointed out that adverse reactions to drugs account for only a small fraction of overt liver disease. Only 2 per cent of patients admitted to general hospitals in Boston (Koff *et al.* 1970) and Copenhagen (Bjornboe *et al.* 1967) were considered to have drug-induced jaundice. However, in various surveys Zimmerman pointed out that, among the causes of massive hepatic necrosis, drug-induced injury accounts for between 20–30 per cent of cases. The seemingly paradoxical disparity between the relatively small proportion of all cases of jaundice attributable to drugs and the major importance of drug-induced injury as a cause of acute hepatic failure finds ready explanation in the gravity of drug-induced hepatocellular damage. The case–fatality rate ranges from 10 to 50 per cent, thus the severity of drug-induced hepatocellular injury converts a low absolute incidence to a major cause of serious hepatic necrosis.

Zimmerman (1978) supported this statement with estimated mortality rates for drug-induced acute hepatocellular injury, for example, cinchophen 50 per cent, halothane 50 per cent, iproniazid 15 per cent, phenytoin 40 per cent, isoniazid 10 per cent and methyldopa 10 per cent. The mortality rates for drug-induced cholestatic jaundice have estimated mortalities around 1 per cent or less.

Several hundred substances have been reported to cause liver damage (Ludwig and Alexelsen 1983). The degree of certainty of cause and effect with the agents cited varies from being well documented to solitary case reports in the literature, these latter are nevertheless listed for completeness (see Table 17.1).

Drug-induced jaundice

Oxytocin and neonatal jaundice

A retrospective study by Friedman *et al.* (1978) of 12 461 single births confirmed an association between maternal oxytocin infusion and neonatal jaundice. The effect of oxytocin on jaundice was independent of gestational age at birth, sex, race, epidural anaesthesia, method of delivery, and birth weight, each of which was significantly associated with neonatal jaundice. The effect of oxytocin was, however, small, producing a calculated mean increase in peak plasma bilirubin concentrations of 8.6 μmol/l (0.5 mg/100 ml) at one week of age: this excess was independent of sex and less than the effect of the baby being born one week earlier.

Oxytocics

Neonatal jaundice was studied in 739 infants who were delivered vaginally in the vertex position without major complication. Labour was induced or stimulated after randomization to one of three oxytocics (prostaglandin E_2 orally, oxytocin intravenously, or demoxytocin buccally). Lange *et al.* (1982) showed that gestational age had a highly significant influence on the risk of jaundice (serum bilirubin > 205 μmol/l). An apparent influence of three oxytocics was not significant, although they may have had a slight effect. However, any such effect could be a consequence of the infants of the mothers who received oxytocics being less mature than those whose mothers did not receive them. The duration of labour and maternal age had no effect on risk of jaundice. Lange *et al.* (1982) concluded that neonatal jaundice after induced and stimulated labour seems to be primarily associated with fetal maturity and that any pharmacological side-effects, if any, of oxytocics are of no importance.

Salicylates and Reye's syndrome

Reye's syndrome is an acute systemic disorder characterized by an encephalopathy with cerebral accumulation lacking cellular infiltrate and a striking accumulation of fat within the liver (Reye *et al.* 1963). Reye's syndrome is typically seen in a child who is recovering from a mild respiratory-tract infection, chicken pox, or other viral illness. The case fatality rate was 23 per cent (Centre for Disease Control 1980), and there are no long-term hepatic effects in survivors. Many exogenous factors have been charged with a possible role in Reye's syndrome such as insecticide-related chemicals and aflatoxin. Salicylates are often implicated and Partin *et al.* (1982) found salicylates in 74 per cent of patients with biopsy-proven Reye's syndrome of all grades of

Table 17.1 Drugs causing jaundice

Drugs causing hepatocellular damage		Drugs causing cholestatic jaundice	
Acetaminophen	Iproniazid	Acetohexamide	Methyltestosterone
Acetohexamide	Isocarboxazide	Ajmaline	Nitrofurantoin
Amitriptyline	Isoniazid	Anabolic steroids	Norethandrolone
Amodiaquine	Lergotrile mesylate	(C-17 alkylated)	Norethindrone
Antimonials	Mepacrine (quinacrine)	Arsenicals (organic)	Norethisterone
L-Asparginase	Mercaptopurine	Azathioprine	Norethynodrel
Azacytidine	Metahexamide	Busulphan	Oxacillin
Azepinamide	Methotrexate	Carbimazole	Para-aminobenzyl-
Azathioprine	Methoxyflurane	Chlorothiazides	caffeine
Azauridine	Methyldopa	Chlorpromazine	Para-aminosalicylic acid
Benziodarone	Mithramycin	Chlorpropamide	Penicillamine
Bleomycin	Mitomycin	Diflunisal	Phenindione
Calvacin	Nitrofurantoin	Ectylurea	Phenothiazines
Carbutamide	Oleandomycin	Griseofluvin	Probenecid
Chlorambucil	Oxacillin	Ibufenac	Prochlorperazine
Chloramphenicol	Oxyphenybutazone	Mepazine	Sulphadiazine
Chloroform	Oxyphenisatin	Methimazole	Sulphanilamide
Chlorpropamide	Papaverine	Methylandrostanalone	Thiazides
Chromomycin	Para-aminosalicylic acid (PAS)	Methyldopa	Thiouracil
Cinchophen	Paracetamol	Methylestrenolone	Tolazamide
Colchicine	Phenazopyridine		Tolbutamide
Clindamycin	Phenylbutazone		
Cyclophosphamide	Phenytoin	**Drugs causing a mixed lesion of cholestasis and**	
Cyproheptadine	Procainamide	**hepatocellular damage**	
Dantrolene sodium	Propylthiouracil	Acetohexamide	Gold salts
Disulfiram	Puromycin	Azathioprine	Hydrazine-type
Emetine	Pyridinolcarbamate	Busulphan	MAOIs
Ethacrynic acid	Pyrrolidizine	Carbarsone	Idoxuridine
Ethionamide	Quinidine	Carbutamide	Iprindole
Ethrane	Rifampicin	Chlorambucil	Isoniazid
Fenclozic acid	Salazopyrin	Chlorpromazine	Novobiocin
Ferrous sulphate	Sulphonamides	Chloramphenicol	Para-aminosalicylic
Fluroxene	Sulphones	Dantrolene sodium	acid
Griseofulvin	Tannic acid	Diethylstilboestrol	Phenylhydrazide
Halothane	Tetracycline	Dinitrophenol	Primaquine
Hycanthone	(degradation products)	Erythromycin estolate	Sulphonamides
Hydroxystilbamides	Thiosemicarbazone		Thiouracil
Ibufenac	Urethane		
Ibuprofen	Valproic acid	**Hepatic tumours (benign and or malignant)**	
Indomethacin	Vitamin A	Androgenic anabolic steroids	Phenelzine
Iprindole	Zoxazolamine	Oral contraceptives	Thorotrast

severity. The salicylate concentration in Reye's syndrome is rarely in the toxic range but the salicylate threshold may be lowered by some genetic factor of the preceding illness, which is one of the features of Reye's syndrome.

Valproic acid

Abnormal liver function tests in patients treated for epilepsy with sodium valproate or valproic acid was reported by Sussman and McLain (1979) in nine patients aged 12 years to 30 years without evidence of overt clinical signs of liver dysfunction. The abnormal values

returned promptly to normal when sodium valproate was discontinued or the dose reduced.

Jeavons (1980) stated that in various series the incidence of raised transaminase levels in patients treated varied from 0 to 18 per cent. An editorial in the (1980) *Lancet* stated that the incidence of hepatic dysfunction in patients on sodium valproate has been variously reported as from around 3 per cent to as high as 44 per cent. In a study from the United States (Coulter and Allen 1980), only one of the 44 children with raised transaminase levels became unwell and in all 44 the biochemical abnormalities resolved completely when the dose of sodium valproate was reduced by 25–50 per cent. Thirty-two patients subsequently were able to tolerate their previous drug dosage without a further rise in SGOT and only six had abnormal liver function tests for a second time. Thus, if all patients in whom liver function tests became abnormal were to stop taking sodium valproate, then a large number would be deprived of the drug unnecessarily. In the UK Jeavons (1980) treated over 500 patients with sodium valproate without seeing overt liver disease although, in a small series of 109 patients, three patients did show raised transaminase levels. Jeavons concluded that routine monitoring of liver function tests is not of much value in predicting which patients will develop clinical liver damage.

Severe hepatotoxicity associated with sodium valproate treatment of epileptics has been reported, and fatal cases were reported by Suchy *et al.* (1979), Gerber *et al.* (1979), Addison and Gordon (1980), Jacobi *et al.* (1980), Le Bihan *et al.* (1980), and Ware and Millward-Sandler (1980) in which histopathology of the liver has been conducted on one or more of the fatal cases reported (see Table 17.1). The liver injury is usually reported as centrilobular necrosis with or without bile-duct proliferation.

It is important to distinguish between the hyperammonaemia associated with liver damage and the association between hyperammonaemia without liver damage and sodium valproate therapy which has been reported in children (Coulter and Allen 1980; Coulter 1980; Sills *et al.* 1980). In one of the two cases reported by Sills *et al.* death occurred and at autopsy the liver was normal.

The attention of the medical professions in Australia (Australian Drug Evaluation Committee 1980) and in the United Kingdom (July 1981) have been drawn to this problem. The Committee on Safety of Medicines (CSM) issued a warning in *Current Problems* (1981) on the problem of liver damage associated with sodium valproate stating, 'it is important to be aware that sudden failure of seizure control with malaise, anorexia and vomiting may indicate incipient liver damage at a time when liver function tests may be within the normal range'. The CSM also pointed out that hyperammonaemia is a recognized consequence of liver failure, but some reports indicate that it may occur, in those patients receiving sodium valproate, in the absence of liver damage. In such patients it has been suggested that there may be derangement of propionic acid metabolism secondary to metabolic effects of sodium valproate. The symptoms include anorexia, vomiting, and ataxia and there may be an encephalopathy with increasing clouding of consciousness.

The overall position of valproate-associated liver damage world-wide was summarized up to July 1981 by Gosling (personal communication to J.P. Griffin).

Forty-eight fatal cases of liver damage associated with sodium valproate were known to the manufacturers. Four cases had occurred in Australia, seven cases in the UK, 24 cases in the USA, four cases in West Germany, three cases in France, two cases in Belgium, and one case each in Canada, India, Italy, and Mexico. The median age was 7.0 years with a range of 2½ months to 34 years. The sex distribution was 26 males and 22 females. The dosage of sodium valproate used in these patients ranged from 67 to 3900 mg/day (media 625 mg/day). The duration of therapy prior to death had ranged from 3 to 180 days (the median duration was 60 days). In 41 cases the patients were receiving other anticonvulsant drugs and in only seven cases was sodium valproate the only form of anticonvulsant therapy administered.

The *FDA Drug Bulletin* (July 1981) refers to 43 deaths from liver failure of whom only six received sodium valproate as sole therapy for epilepsy. The manufacturers estimate that in July 1981 there were 600 000 patients world-wide receiving sodium valproate for control of their epilepsy.

Fatal liver disease seems, therefore, to be a rare complication of what is otherwise a fairly non-toxic and very useful drug, but clinicians should always suspect hepatotoxicity when a patient becomes unwell or control of fits deteriorates in a patient on valproate.

The mechanism of valproate-associated liver damage is unknown, but valproic acid is structurally similar to other short-chain fatty acids such as 4-pentenoic acid, propionate, and butyrate, which have been used as experimental hepatotoxins. There are also similarities with Jamaican vomiting sickness which is due to hypoglycin found in ackee fruit (a fact commented upon by Suchy *et al.* 1979; Addison and Gordon 1980: *Lancet* 1980). Vomiting encephalopathy, and fatty degeneration of the liver and hyperammonaemia are features of Jamaican vomiting sickness.

Sugimoto *et al.* (1983) described a Reye-like syndrome in a patient being treated with valproic acid. Hyperammonaemia and severe liver damage as well as diffuse small droplets in the liver biopsy material were demonstrated.

Direct hepatotoxicity caused by drugs

Halothane

Evidence for a direct hepatotoxic action of halothane has been derived from several experimental studies in which animals have been exposed to prolonged periods of sub-anaesthetic doses of the agent. It has, thus, been possible to produce hepatic damage in rats, mice, guinea pigs, and dogs, although in each instance this has been mild. In humans minor abnormalities in liver function with elevations of transaminases and bromsulphthalein retention have been shown to occur more frequently after anaesthesia with halothane than with other agents.

Dundee and colleagues (McIlroy *et al.* 1979; Fee *et al.* 1979) described in their first paper the methodology of a large prospective study on the influence of repeated anaesthetics on liver function. The most suitable patients were considered to be those presenting for endoscopic examination of the bladder and urethra, for urethral dilatation, and for cervical implantation of radium. 63 patients received two or more administrations of halothane, and 66 received two or more administrations of enflurane, both drugs being given with nitrous oxide. Of these patients, only 30 were exposed to a second anaesthetic within 4 weeks of their first exposure. Moreover, the average age of the patients was over 60 years and about 70 per cent had carcinoma. The study cannot, therefore, be said to meet the optimistic description of the authors of 'a large prospective study'. These authors used six tests of liver enzyme changes and eosinophil counts which are probably not the best indices of liver damage. Despite these criticisms the study did show a greater frequency of increased enzymatic changes following repeated administrations of halothane than following enflurane and the average alanine amino-transferase and gamma-glutamyl-transpeptidase levels were increased to a greater degree following halothane than enflurane. This may indicate that halothane is more hepatotoxic than enflurane but it should be remembered that since the selection of the subjects was largely of patients with bladder carcinoma, it is likely that these patients had received anaesthetics prior to those given in the course of study, and it is likely that these anaesthetics were of halothane rather than enflurane. Prior sensitization to halothane was, therefore, more likely than to enflurane.

In the study of Dundee and his colleagues (McIlroy *et al.* 1979; Fee *et al.* 1979) the high incidence of liver enzyme changes contrasted with the relatively low incidence of florid hepatic necrosis estimated to be between 1 in 22 000 and 1 in 100 000 of halothane anaesthetics given, and this is compatible with a mild hepatotoxic effect of halothane. This has been attributed to the formation of reactive metabolites which are capable of binding to hepatocyte macromolecules. The overall biotransformation of halothane appears to be affected by genetic factors and environmental factors such as exposure to hepatic microsomal enzyme-inducing substances. The degree of obesity seems to be relevant in that obese subjects who are particularly prone to develop halothane hepatitis show greater rises of plasma inorganic fluoride concentration. To explain the florid heptatocellular necrosis which occasionally follows multiple exposures to halothane, a sensitization to halothane-altered liver constituents, leading to immune-mediated destruction of hepatocytes has been postulated (Vergani *et al.* 1978; Davis *et al.* 1979).

It is possible that only those individuals with a predisposition to organ-specific auto-immunity will be at risk, and this may be governed by genetic factors, as appears to be the case for patients with chronic active hepatitis. In keeping with this is the observation that a high proportion of patients with halothane hepatitis showed elevated titres of thyroid autoantibodies, which persisted even after hepatocellular damage had resolved (Walton *et al.* 1976). The rarity of severe hepatocellular necrosis, estimated as occurring after between 1 in 22 000 (Mushin *et al.* 1971) and 1 in 35 000 (Strunin 1977) halothane anaesthetics, could be explained by the necessity for such defects in immune regulation to be superimposed on abnormalities in the metabolism of the drug via hepatotoxic derivatives before this complication can develop.

The effects of halothane may be different in children. Wark (1983) reported on 165 400 anaesthetics given at the hospital for sick children, Great Ormond Street, during the 23-year period, 1957–79. Almost all of the patients would have been exposed to halothane. Seventy-four patients became jaundiced for the first time in the post-operative period. Halothane-associated hepatitis was excluded as the cause of post-operative jaundice in all but two. In these two patients in whom the diagnosis of halothane-associated hepatitis was possible the hepatic illness was mild and both patients made an uneventful recovery. In this study the risk of a patient becoming jaundiced due to halothane-associated hepatitis was less than 1 in 82 000. Thus, halothane can be used in children whenever it is warranted and can be used with safety.

Paracetamol

In large doses, in man and animals, paracetamol produces centrilobular necrosis of liver parenchymal cells. In overdosed patients nausea and vomiting may occur within a few hours to be followed by upper abdominal pain and hepatic tenderness. Biochemical evidence of liver damage becomes apparent 12–26 hours after ingestion but maximum abnormalites of liver function tests may be delayed for at least three days. In severely poisoned patients who recover, the plasma aspartate (AST) and alanine aminotransferase (ALT) may rise to 10 000 units/l or more, the bilirubin to 70–100 mmol/l (4–6 mg/100 ml) and the prothrombin ratio to 2–3. Regeneration is usually rapid, and liver function tests return to normal with 7–14 days. With more severe intoxication fulminant hepatic failure develops at three to six days with deepening jaundice, coma, hyperventilation, acidosis, hypoglycaemia, cerebral oedema, renal failure, disseminated intravascular coagulation, and haemorrhage. Liver failure is treated conventionally but the prognosis is very poor. Acute renal tubular necrosis may occur in the absence of hepatic failure.

Glutathione

Therapeutic doses of paracetamol do not cause liver damage because the small amounts of the highly reactive metabolite formed are rapidly inactivated by preferential conjugation with hepatic glutathione and subsequently excreted in the urine as cysteine and mercapturic acid conjugates. However, hepatic glutathione is rapidly depleted by large doses of paracetamol, and when stores are reduced to less than 30 per cent of normal the excess metabolite is free to alkylate vital cell constituents, causing necrosis. Glutathione also inhibits the covalent binding of paracetamol to microsomal protein.

Glutathione does not enter the liver from the blood stream but its precursors, the sulphydrylamino acids, do. Treatment of paracetamol overdosage has, therefore, been based on the administration of sulphydryl compounds such as cysteamine, L-methionine, and N-acetylcysteine.

N-acetylcysteine

Prescott *et al.* (1979) treated 100 cases of severe paracetamol poisoning with intravenous N-acetylcysteine (acetylcysteine). There was virtually complete protection against liver damage in 40 patients treated within 8 hours after ingestion (mean maximum serum alanine transaminase activity 271 U/l). Only 1 out of 62 patients treated within 10 hours developed severe liver damage compared with 33 out of 57 patients (58 per cent) studied retrospectively who received supportive treatment alone. Early treatment with acetylcysteine also prevented renal impairment and death. The critical ingestion treatment interval for complete protection against severe liver damage was 8 hours. Efficacy diminished progressively thereafter, and treatment after 15 hours was ineffective.

Prescott *et al.* (1979) recommended the following treatment regime: on admission blood should be taken for emergency estimation of plasma paracetamol concentration (a specific method should be used). Gastric aspiration and lavage should be performed on patients admitted within four hours of ingestion and of all those in coma. Decision to treat with N-acetylcysteine should be on the basis of plasma paracetamol levels and time from ingestion. The treatment line joins plots of 200

mg/l at 4 hours after ingestion and 30 mg/l at 15 hours on semilog paper. If the plasma concentration is above this time/concentration line, then Prescott (1979) recommended N-acetylcysteine be administered with an initial dose of 150 mg/kg diluted in 5 per cent dextrose given intravenously over a 15-minute period, followed by 50 mg/kg in 500 ml of 5 per cent dextrose infused over 4 hours. A further 100 mg/kg in 1 litre of 5 per cent dextrose should be given over the next 16 hours. The quantity of intravenous fluid used in children should be modified to take into account age and weight.

N-acetylcysteine is commercially available as a 20 per cent solution in 10 ml ampoules (*Parvolex*).

N-acetylcysteine is usually free of side-effects but an anaphylactoid reaction in one patient was reported (Walton *et al.* 1979). Gilligan *et al.* (1980) *considered that all patients with suspected paracetamol overdose should be given i.v. N-acetylcysteine immediately.* Their reasons for this advice are that to be effective N-acetylcysteine must be given within no more than 15 hours, and the sooner the better, after paracetamol ingestion; side-effects of treatment are few. The decision whether or not to treat on the basis of whether the blood level is above a certain line relating blood level with time after ingestion relies on the patient's estimate of time of ingestion, and this is notoriously inaccurate. If this advice is followed, emergency measurement of blood levels of paracetamol (which may take several hours during which time therapy is withheld) may logically be a thing of the past.

Methione

An alternative is oral methionine in a dose of 2.5 g every four hours up to a total of 10 g. However, this regimen is not always effective, and the oral route is unreliable because nausea and vomiting usually develop within a few hours of ingestion of a hepatotoxic dose of paracetamol. Methionine is contraindicated in severe liver disease and may contribute to encephalopathy in paracetamol poisoning.

Cysteamine

Cysteamine invariably causes unpleasant side-effects and need no longer be used.

Other drugs

Other supportive therapy for paracetamol poisoning consists of maintenance intravenous fluids and the cautious use of anti-emetics in patients with persistent nausea and vomiting. In patients admitted too late for effective treatment where severe liver damage can be anticipated, vitamin K_1 and fresh frozen plasma or clotting factor concentrate should be given if the pro-thrombin time ratio exceeds 3.0. Hepatic failure is treated conventionally.

Prescott (1979) stated that in his experience, patients who are alcoholics or have been on drugs which cause microsomal enzyme induction such as barbiturates and anticonvulsants, are particularly susceptible to the hepatotoxic effects of paracetamol overdose.

Prescott's experience indicates that children under the age of 10 years are much less susceptible to liver damage by large doses of paracetamol. Meredith *et al.* (1978) reviewed 116 cases of paracetamol overdose in children from the records of the National Poisons Information Service and except for one, there were no hepatic complications. The exception was a child who had hepatic encephalopathy following methionine given at a later state than recommended, which may have contributed to the encephalopathy. The patients seemed to fall into two groups: children aged 1 to 4 years, who had an equal male: female ratio, and ingested small quantities of paracetamol with minimum damage; and children aged 11 to 13 years with a female: male ratio of 3:1 who ingested larger quantities of paracetamol with a greater likelihood of liver damage. The peak incidence of all forms of poisoning in young children is known to occur at 1 to 4 years, and this probably represents true accidental poisoning; in contrast, the high incidence of paracetamol poisoning at 11 to 13 years is likely to be due to self-poisoning. From these findings it is clear that small children rarely require treatment with a specific antidote, though older children of about 11–13 years of age may require active measures.

Cytotoxic agents

Methotrexate

Methotrexate used in the treatment of various neoplastic conditions and psoriasis is associated with elevated serum alkaline phosphatase and transaminase levels. Vaughan *et al.* (1979) in 24 patients treated for carcinoma of the breast with cyclophosphamide, methotrexate, and 5-fluorouracil, and Parker *et al.* (1980) in 36 children with acute lymphoblastic leukaemia found that the rises in alkaline phosphatase and transaminase levels were too variable to be used firstly as evidence of metastatic breast disease or, secondly, as indications of drug-induced liver damage in the children.

Methotrexate is also widely used in the treatment of psoriasis and the study by Zachariae *et al.* (1980) is valuable in giving some indication of the incidence of liver damage in these patients.

Seven hundred and sixty-four liver biopsies were performed in 328 psoriatics on treatment with methotrexate or being considered for systemic treatment either with methotrexate or with psoralens and long-wave ultraviolet light. The diagnosis of cirrhosis was established histologically in 21 patients. Two patients had cirrhosis in their pre-methotrexate biopsy and were not given methotrexate. The remainder all showed no signs of cirrhosis or fibrosis in their pre-methotrexate biopsy. The difference between the post-methotrexate biopsies and the pre-methotrexate biopsies was highly significant. Among 39 patients treated for more than 5 years, 10 developed cirrhosis (25.6 per cent). Almost all patients were on a divided-dose intermittent-oral-dosage schedule. The cumulative dose of methotrexate, when cirrhosis was first found, ranged from 590 to 8105 mg, with an average dosage of 2200 mg. Other factors contributing to cirrhosis in this study seem to be previous treatment with arsenic, a previous intake of alcohol, and lowered renal function.

Data on later biopsies from 14 patients of whom 11 continued to receive methotrexate, it seemed that methotrexate-induced liver cirrhosis is not of a very aggressive nature.

Indicine N-oxide

Indicine N-oxide, a pyrrolizidine alkaloid was given to a five-year-old boy with refractory acute myelocytic leukaemia. Three days after receiving the drug the boy developed acute hepatic failure and died nine days later. Massive hepatic necrosis was present at post-mortem. This may have been caused by Indicine N-oxide (Cook *et al.* 1983).

Hydrallazine and dihydrallazine

Hydrallazine has been widely used for treating hypertension. Hepatotoxicity in the form of acute hepatic necrosis, which occurred two days after commencing treatment with hydrallazine 50 mg daily, was first reported by Bartoli *et al.* (1979). The association was supported by biopsy and rechallenge with the drug. Barnett *et al.* (1980) reported a case of a 37-year-old woman who was known to be a rapid acetylator and was treated with 25 mg hydrallazine twice daily. Within two weeks the patient was complaining of malaise and generally feeling unwell. Hydrallazine was stopped but after three months of attempting to control her hypertension by other means hydrallazine was reintroduced. Within three days of restarting hydrallazine the liver function tests were grossly abnormal and indicated a mixed hepatocellular and cholestatic dysfunction. Liver function tests remained grossly abnormal for a further two weeks but were normal after five weeks.

Hydrazine has been found in the urine of patients treated with hydrallazine (Timbrell and Harland 1979), and this is thought to be the mechanism of toxicity, hydrazine being well known to cause this type of liver damage.

Dihydrallazine

Dihydrallazine is an analogue of hydrallazine which is widely used in Europe. Pariente *et al.* (1983) reported a patient with drug-induced hepatitis which quickly improved when dihydrallazine treatment was stopped and recurred when it was resumed. Hepatitis was severe with hepatic encephalopathy and prolonged pro-thrombin time. There was centrizonal and bridging necrosis. Fibrotic sequelae were seen on repeat liver biopsy.

Tannic acid

Tannic acid caused liver damage when applied to burns due to its systemic absorption, and mild liver damage has been reported after barium enemas to which tannic acid had been added to improve coating of the mucosa. The drug is itself hepatotoxic, without undergoing metabolic transformation to an active molecule (Baker and Handler 1943; McAlister *et al.* 1963; Keeling and Thompson 1979).

Tetracycline

Tetracyclines depress liver metabolism but in the usual therapeutic dose are quite safe. Their use in the last trimester of pregnancy has, however, been associated with acute fatty liver. Wenk *et al.* (1981) indicated that this condition is more common than usually realized and described two patients who developed tetracycline-induced fatty liver in the immediate puerperium, in both cases with a fatal outcome. The first patient, a 19-year-old, was treated with tetracycline for a suspected urinary-tract infection. Caesarian section was performed and, following operation, the patient developed marked epigastric pain and tachypnoea with anxiety and vomited coffee-ground material. Her pulse rate increased to 140/min, followed by cyanosis and shock. Vasopressor and transfusion therapy were started in conjunction with tetracycline given intravenously. Ascites reappeared, shock persisted, and she died. Microscopic examination of the liver at autopsy showed extensive diffuse fatty change manifested as pronounced smooth, yellow discoloration in general and non-coalescent microva-

cuolization of the liver cells located centrally in the lobules. The second patient, a 21-year-old with no history of tetracycline therapy underwent caesarian section. Her post-partum period was filled with many complications. Her condition gradually deteriorated, ending in death 14 days after hospitalization. Among other findings autopsy showed morphological and histological changes consistent with fulminant fatty liver of pregnancy. Tissue extraction and chemical analysis demonstrated qualitative evidence of tetracycline.

Drug-induced hepatitis

Captopril

Captopril has only been marketed for the treatment of hypertension for a short period of time, but it is already revealed that side-effects include liver and renal toxicity and blood dyscrasias. Vandenburg *et al.* (1980) described the case of a 64-year-old man who became jaundiced on captopril therapy; investigation revealed that the patient's liver disorder was primarily hepatocellular damage with secondary cholestatic elements.

Mild increases in liver enzyme level have been recorded in some patients taking captopril (Elmenhary *et al.* 1981). Vandenburg *et al.* (1982) reported a patient with hepatitis, and a further patient was reported by Zimran *et al.* (1983). Twenty-six days after starting captopril a 74-year-old woman became jaundiced with raised bilirubin and alkaline phosphatase and without concomittant raised enzymes; serum and urine amylase were also raised. Her jaundice increased and itch developed. After exclusion of extrahepatic biliary obstruction, the possibility of drug jaundice was considered. Captopril was stopped and the jaundice resolved in a week, and alkaline phosphatase and amylase activities returned to normal within two weeks. At no time did she complain of abdominal pain or show signs of acute pancreatitis. In view of her age and rapid improvement after cessation, rechallenge and biopsy were not performed. The clinical and laboratory picture were more in keeping with a drug-induced rather than a viral hepatitis.

Carbamazepine

Three male patients with no history of liver disease developed a febrile illness suggestive of biliary-tract infection within four weeks of starting carbamazepine therapy (Levy *et al.* 1981). The dose of carbamazepine in each patient was 800, 1200, and 200 mg daily, respectively. Liver function values were all mildly elevated. Within three days of carbamazepine withdrawal the fever and clinical symptoms disappeared. Laboratory values quickly returned to normal except for alkaline phosphatase levels which remained elevated for four months. In one patient, carbamazepine was restarted when symptoms of *tic douloureux* flared up. Within 10 hours, fever and a pruritic rash appeared. The drug was withdrawn. All three liver biopsies showed granulomatous hepatitis, with acute cholangitis in two patients.

Clorazepate

Clorazepate is a tranquillizer related to the benzodiazepine group. Parker (1979) described a case of jaundice and hepatic necrosis associated with clorazepate in a 27-year-old male being treated for depression. Other cases resembling this were reported to the CSM in patients receiving clorazepate, but cause and effect has not been established.

Cyclofenil

Cyclofenil is a non-steroidal drug with a simultaneous effect on ovulation. Olsson *et al.* 1983 reported hepatic reactions in 30 patients taking cyclofenil and the liver damage was probably related to a metabolic idiosyncrasy and was reversible in all patients on stopping the drug. It is clear that 80 per cent of the hepatic reactions to cyclofenil occurred one to four months after starting therapy.

Cyclofenil has been used in Sweden, UK, France, Italy, Japan, and Germany and the only country to report hepatic reactions is Sweden where the incidence of liver damage is 1.3 per cent (Olsson *et al.* 1983). Increased aminotransferase values have recently been reported from Sweden in patients with systemic sclerosis treated with cyclofenil (Blom-Bulow *et al.* 1981). This may be a geographical difference in susceptibility to the drug which has been observed in oestrogen-induced liver disease, which is more common in Scandinavia (Simmon 1978).

Cyclosporin A

Sixty-six patients receiving 67 renal transplants were treated with cyclosporin A at 17.5 mg kg^{-1} on the day of operation and for the next eight weeks, after which the dose was lowered to the 10 mg kg^{-1} range (Klintmalm *et al.* 1981). Most patients also received prednisone 200 mg on the day of operation, tapered thereafter by 40 mg daily to a final maintenance dose of 20 mg daily. Six patients were hepatitis-B, surface antigen positive preoperatively and one had cirrhosis. Thirteen of the 66 patients developed hepatotoxicity (defined as a serum bilirubin greater than 34.2 μmol l^{-1}). This occurred in 11 of the patients within two weeks to two months of the operation. When cyclosporin A dosage was lowered to about 10 mg kg^{-1} daily in these 11 patients, bilirubin levels returned to normal. Three patients developed late toxicity at 7–13 months post-operatively and were changed from cyclosporin A to azathioprine, after which liver function returned to normal.

Diclofenac

Diclofenac, a derivative of phenylacetic acid, is a non-steroidal, anti-inflammatory agent. Some patients have developed liver function test abnormalities during treatment (Ciccolunghi *et al.* 1978) but acute hepatitis has not been previously reported. Dunk *et al.* (1982) reported a 52-year-old man who developed jaundice after five months of taking diclofenac. On withdrawal of the drug, the jaundice resolved and on rechallenge it recurred. Liver biopsy showed a moderately severe acute hepatitis with canalicular cholestasis, Kupffer-cell proliferation, and plasma cells and eosinophils in the inflammatory infiltrate. The patient's liver function returned to normal within six weeks of finally stopping diclofenac. Dunk *et al.* (1982) concluded that their patient had an idiosyncratic drug-induced hepatitis of the 'metabolic aberration' type (Zimmerman 1981).

Feprazone

Two patients developed jaundice within a month of starting feprazone, a non-steroidal, anti-inflammatory agent (Wiggins and Scott 1981). They were a 74-year-old man treated for arthralgia and a 36-year-old woman with low back pain. The man had also been taking präxilene, 300 mg daily for peripheral vascular disease and phenobarbitone 60 mg daily for epilepsy for one year. The only abnormal findings in both patients were liver function tests. Tests for hepatitis-B surface antigen were negative in both. Liver biopsy showed granulomatous reaction in the first case and hepatitic changes in the second, similar to changes reported with phenylbutazone (structurally related to feprazone). Clinical and biochemical changes resolved in both cases after stopping feprazone. The man is still taking his other drugs and has been well with normal liver function tests for two years.

Floctafenine

A 53-year-old alcoholic with cirrhosis, chronic pancreatitis, and duodenal ulcer was treated unsuccessfully with floctafenine (600 mg daily) for relief of radiating epigastric pain (Pasqua *et al.* 1982). The dose was increased to 1200 mg daily and 24 hours later the patient entered a deep coma (grade IV) and all oral drugs were stopped, including his previous medications which consisted of spironolactone (200 mg daily), frusemide (furosemide; 25 mg every other day), and cimetidine (400 mg at night). However, cimetidine (800 mg daily) was continued i.v. along with lactulose enemas and the patient awoke over the next 24 hours. Spontaneous worsening of liver function as a cause of encephalopathy seems unlikely as the alcoholic hepatitis was improving at the time of coma. The close temporal relation between floctafenine administration and coma indicates a casual link.

Fusidic acid

Humble *et al.* (1980), Vickers and Menday (1980), and Talbot and Beeley (1980) all reported jaundice developing in patients treated with intravenous and oral fusidic acid. Talbot and Beeley (1980) drew attention to the fact that, at the time of writing their letter, of 116 adverse reaction reports to the CSM related to fusidic acid, 44 were reports of jaundice.

Ketoconazole

Clinical research with ketoconazole started in 1977 and marketing began in March 1981. It was used for treating various fungal infections and a single dose of 200 mg ketoconazole is effective and safe. In July 1980 one patient was reported to have had jaundice and hepatitis during ketoconazole treatment (Peterson *et al.* 1980). Heiberg and Svejgaard (1981) reported another patient, and three more were reported in 1981 (Macnair *et al.* 1981).

By March 1982 Janssen Pharmaceuticals (Janssen and Symoens 1982) had received 31 world-wide reports of icteric signs or anicteric symptoms while 300 000 patients were estimated to have taken the drug. Only one patient died from diffuse hepatic necrosis. Symptomatic reactions usually occur during the first few months of treatment and as early as one or two weeks after treatment commences. They subside when treatment is discontinued. However, one symptomatic patient recovered while continuing treatment (Firebrace 1981).

Up to December 1982 1.3 million patients had received ketoconazole (Boughton 1983). By January 1983 96 cases of symptomatic liver damage had been reported and three had died. Many of these patients with liver damage may have received griseofulvin therapy without response and then been given ketoconazole. It is known that griseofulvin may cause intrahepatic cholestasis (Chiprut *et al.* 1976). Boughton (1983) therefore recommended that there should be a one-month gap between cessation of griseofulvin therapy and the start of ketoconazole

therapy and in this way hepatic reactions may be avoided.

Mistletoe hepatitis

Mistletoe-attributed hepatitis has been reported in a 49-year-old woman who had taken a proprietary herbal preparation containing motherwort, kelp, wild lettuce, skullcap, and mistletoe (Harvey and Colin-Jones 1981). While there is little doubt that the herbal preparation used induced the hepatitis, the attribution to the mistletoe is circumstantial.

Naproxen

Victorino *et al.* (1980) describe a case of jaundice and hepatitis associated with naproxen. Liver biopsy revealed eosinophilic infiltration and steatosis and was compatible with drug-induced acute hepatitis. Naproxen-associated jaundice and hepatitis had been described previously (Bass 1974; Law and Knight 1976) but liver biopsies had not been performed. Diagnosis hinged on the clinical picture and liver function tests of raised transaminases and alkaline phosphatase.

Nifedipine

Rotmensch *et al.* (1980) described the case of a 69-year-old man treated with nifedipine for severe angina. Ten days after starting the drug he developed anorexia, nausea, fever, and jaundice. Laboratory investigations were compatible with toxic hepatitis. Recovery occurred on withdrawal of the drug; on nifedipine rechallenge four months after recovery, fever, raised alkaline phosphatase, and raised IgG levels were noted. Immunological studies indicated an allergic hepatitis with lymphocyte sensitization as the mechanism of damage.

Papaverine and verapamil

In a study by Pathy and Reynolds (1980) a double-blind placebo-controlled trial of sustained-release papaverine against placebo in elderly patients was undertaken to assess its effect on intellectual impairment.

Six of 14 patients receiving a sustained-release papaverine preparation developed abnormal liver function tests, raised alkaline phosphatase and transaminases. One patient had jaundice and another abnormal liver histology on a biopsy specimen. This is only the third report of impaired liver function following the administration of papaverine — a drug which has been used in clinical medicine for over 100 years.

Brodsky *et al.* (1981) described hepatotoxicity associated with verapamil, a papaverine derivative, in a 55-year-old male being given the drug for angina. Serum enzymes and bilirubin level returned to normal after drug withdrawal. On rechallenge on two occasions enzyme changes and bilirubin increases returned.

Perhexiline maleate

In an attempt to assess the incidence of perhexiline maleate–associated liver damage Poupon *et al.* (1980) conducted a study on 46 patients treated with the drug at 100–400 mg/day for their angina. Hepatomegaly was found in six patients; abnormal bromsulphthalein clearance was found in 74 per cent of cases and raised transaminases in 30 per cent, other liver function tests were abnormal in 10 per cent of patients.

Liver biopsy was undertaken in 11 of the patients with abnormal bromsulphthalein clearance. Diffuse micro- and macro-vacuolar fatty change and hepatocellular hyperplasia were found in all specimens. Histochemical studies revealed abnormal amounts of phospholipids and gangliosides in some of the specimens. Hepatocellular necrosis was present in four of the eleven biopsies.

Atkinson *et al.* (1980) described a single case of papilloedema and hepatic dysfunction associated with perhexiline maleate therapy. The patient presented with right-sided incoordination and on examination had marked papilloedema of the right eye (the other had been enucleated some years previously). Serum transaminases and lactic dehydrogenase levels were markedly raised. The drug was withdrawn and both papilloedema, ataxia, and abnormal liver function tests showed resolution over the next six weeks.

Hay and Gwynne (1983) reported two patients who developed cirrhosis following therapy with perhexiline maleate. Liver failure and polyneuropathy caused death in one patient who had received 300 mg daily for three years. Cirrhosis was an unexpected finding in the other patients whose perhexiline maleate dose was 200 mg daily for five years. Hay and Gwynne (1983) recommended that perhexiline maleate should be prescribed cautiously and discontinued if liver function tests become abnormal.

Pieterne *et al.* (1983) noted that this potent anti-anginal drug may cause an alcoholic-type hepatitis and cirrhosis. They reported a patient who developed cirrhosis after sixteen months of perhexiline maleate therapy.

Roberts *et al.* (1981) described a patient who developed hepatic failure after uncomplicated orthopaedic surgery and died. He had been taking perhexiline for 18 months before operation; after jaundice developed, a liver biopsy was performed. This showed the changes normally associated with acute or chronic alcoholic liver disease. The established cirrhosis was probably related to his heavy alcohol intake which had ceased several months ago while the more acute degenerative changes resulted form perhexiline maleate.

D-pencillamine

Rosenbaum *et al.* (1980) described two patients with rheumatoid arthritis who developed evidence of heptatoxicity while on D-penicillamine. One patient developed fever and jaundice; both patients showed raised alkaline phosphatase and transaminase levels. Both patients recovered after withdrawal of the drug. Hepatotoxicity, though rare, should be added to the spectrum of D-penicillamine toxicity.

Seibold *et al.* (1981) reported the case of a 26-year-old woman who had been treated with corticosteroids for four years for symptomatic control of systemic lupus erythematosis. Because of persistent joint pain despite various non-steroidal anti-inflammatory agents, it was decided to start her on a course of penicillamine. Ten days after starting D-penicillamine at 250 mg daily the patient developed fever of 102°F (38.9°C), chills, and a non-pruritic erythematous rash. D-penicillamine was discontinued and all symptoms abated within 48 hours. Five days later D-penicillamine therapy was restarted. Four hours after taking a single capsule fever, arthritis and rash recurred and diarrhoea was noted. D-penicillamine was discontinued and all symptoms diminished. It was noted that the patient was jaundiced. Serum glutamic oxaloacetic transaminase and serum glutamic pyruvic transaminase were abnormally elevated. The liver function abnormalites resolved slowly over seven weeks, although later a transient rise in alkaline phosphatase level was noted. It was considered that this patient had developed drug-induced cholestasis.

Propylthioracil

Weiss *et al.* (1980) described two cases of propylthiouracil-induced hepatic damage in two women being treated for thyrotoxicosis. In one case in a 28-year-old woman acute epigastric pain, nausea, and vomiting were the presenting symptoms, and abnormal liver function tests indicative of hepatocellular damage were found: raised SGOT and bilirubin levels. That propylthiouracil was the causative agent was proven by rechallenge. In the second case a liver biopsy of a 54-year-old woman indicated chronic active hepatitis with spontaneous remission on ceasing treatment.

Drug-induced intrahepatic cholestasis

Ajmaline

Ajmaline is a Rauwolfia alkaloid used in parts of continental Europe for the treatment of cardiac arrhythmias.

There have been several reports of cholestatic jaundice in patients treated with ajmaline; seven cases of jaundice probably caused by ajmaline have been reported to the Swedish Adverse Drug Reaction Committee. Jaundice has usually appeared within 1–3 weeks of commencing the drug, but occasionally the duration of treatment has been longer. Fever has occurred in some patients and eosinophilia has been reported. Liver biopsy has usually shown biliary stasis. The jaundice has resolved within 2 weeks to 4 months of withdrawing the drug in all reported patients except one; that patient had also received methyltestosterone and ethinyloestradiol and jaundice persisted for more than two years.

Azathioprine

Cholestatic jaundice has been described in patients receiving azathioprine following renal transplantation (Sparberg et al. 1969; Davis and Williams 1977). Allergic reactions to azathioprine, consisting of rashes, pyrexia, nausea, vomiting, and anorexia, are occasional complications of immunosuppressant therapy.

Davis et al. (1980) described a case of chronic active hepatitis and cholestasis in a 55-year-old woman who also experienced hypersensitivity reactions. Davis et al. (1980) postulated that these adverse reactions are due to two different portions of the azathioprine molecule. The 6-mercaptopurine portion being the factor responsible for the cholestasis and the imidazole side-chain for the allergic and hypersensitivity reactions. It has long been known that 6-mercaptopurine can cause cholestatic jaundice (McIlvanie and MacCarthy 1959).

Benoxaprofen

On 3 August 1982 Sir Abraham Goldberg, the Chairman of the CSM announced that the Licensing Authority had suspended the benoxaprofen product licence immediately on safety grounds; the CSM had received over 3500 reports of adverse reactions which included 61 fatalities, mainly in the elderly. Benoxaprofen had already been known to have side-effects on skin (photosensitivity) and nails (onycholysis) but other organs such as the gut, the liver, and the bone marrow were now implicated (see also Chapter 4).

Benoxaprofen was reported as being effective in rheu-

matoid arthritis (Huskisson 1979; Bacon et al. 1980), osteoarthritis (Alarcon-Segovia 1980; Blechman 1980), and ankylosing spondylitis (Bird et al. 1980). It was favoured as it could be given once daily since it had a long half-life (Berry et al. 1980). It is known that it has a much longer half-life in very old patients (111 hours at mean age 82) than in younger patients (29 hours at mean age 41) (Hamdy et al. 1982).

In May 1982, Taggart and Alderdice reported five cases of cholestatic jaundice with fatal outcome in elderly women taking benoxaprofen. All were 80 years or older and all died within three weeks of the onset of jaundice. The typical necropsy findings were of intrahepatic cholestasis and a mild inflammatory reaction in the portal triads. Taggart and Alderdice comment that the liver injury was of the canalicular cholestatic type of drug-induced jaundice in which there is no hepatocellular disease and little portal inflammation (Zimmerman 1981). Goudie et al. (1982) reported the occurrence of jaundice in three patients on benoxaprofen. All were over 70 years old. Two recovered when benoxaprofen was stopped and one died of bronchopneumonia.

Prescott et al. (1982) reported two patients taking benoxaprofen who developed cholestatic jaundice and renal failure; one died. Liver biopsies showed intrahepatic cholestasis. They noted that the duration of treatment with benoxaprofen was similar to that reported by Taggart and Alderdice (1982). Prescott et al. pointed out that benoxaprofen is partly excreted in the urine as the glucuronide conjugate (Nash et al. 1980). The acidic, anti-inflammatory drugs are nephrotoxic under certain conditions (Prescott 1982) and they suggested that the prolonged plasma half-life may be further prolonged in the presence of renal damage and cholestasis. Fisher and McArthur (1982) reported a further death due to severe cholestatic jaundice in a middle-age man taking benoxaprofen. Firth et al. (1982) reported a further fatality with gross intrahepatic cholestasis.

A survey of the cases of hepato-renal syndrome associated with benoxaprofen reported to the CSM showed that the fatal cases had received the drug for an average of 8.5 months, and the non-fatal cases for an average of 6.9 months. Two of the fatal cases had in fact been receiving benoxaprofen for over 2 years.

The geographical distribution of reports of hepatic and renal adverse reactions showed inequalities in distribution with over-representation of reports both by size of population and exact drug usage from Northern Ireland, Scotland, and England north of the Trent, with consequent under-representation in England south of the Trent and Wales. This is the converse of the reports of photosensitivity reactions which were relatively more common in the southern areas of the British Isles, these differences were statistically significant. For Wales and England, south of the Trent, the number of photosensitivity reactions is larger than expectation on the basis of a proportionate share of the prescriptions ($p < 0.00002$) and the number of hepato-renal reactions smaller ($p = 0.003$). For Scotland, Northern Ireland, and England north of the Trent, the converse situation, of course, applies (Table 17.2). It is therefore tempting to postulate that patients in the southern areas of the British Isles developed photosensitivity reactions and stopped taking the drug before serious damage to liver and kidneys occurred. Anecdotal reports tend to support this view and several cases of photosensitivity reactions have been reported where the drug was stopped and then restarted in the autumn in patients who then subsequently developed liver or kidney damage (Griffin 1984).

Within two years of marketing in a blaze of publicity, benoxaprofen had its product licence suspended in August, 1982 and was also withdrawn at the same time in the United States of America. It has subsequently been abandoned by its makers.

Table 17.2 Photosensitivity and hepatorenal reactions and prescriptions of benoxaprofen: geographical distribution (Griffin 1984)

	Benoxaprofen prescriptions number (per cent)	Photosensitivity reactions number (per cent)	Hepatorenal reactions number (per cent)
Wales and south of Trent	829 400 (55)	967 (60)	8 (27.0)
Scotland, N. Ireland, N. of England and Trent	673 400 (45)	630 (40)	22 (73)

Cefaclor

Cefaclor, a cephalosporin antibiotic was reported to have caused a cholestatic jaundice in a child (Bosio 1983).

Cloxacillin

Enat *et al.* (1980) reported severe intrahepatic cholestasis which occurred in a 69-year-old female patient who had taken nitrofurantoin, ampicillin, and cloxacillin within the previous four weeks. As only nitrofurantoin was known to cause cholestasis, the reaction was attributed to nitrofurantoin. She was given cloxacillin again two years later. The cholestasis reappeared at once. A macrophage inhibition factor test confirmed that cloxacillin was the offending drug.

Co-trimoxazole

Ogilvie and Toghill (1980) described the case of a patient who developed cholestatic jaundice following therapy with co-trimoxazole. Inadvertent rechallenge with the drug resulted in further clinical and biochemical deterioration but complete recovery ensued on stopping this drug.

Abi-Mansur *et al.* (1981) reported on a 52-year-old woman who developed severe cholestasis seven days after the administration of trimethoprim–sulphamethoxazole. Laparotomy was undertaken and extrahepatic obstruction was excluded. A wedge liver biopsy showed marked pericentral cholestasis. The symptoms subsided and liver function returned to normal after stopping the drug. On detailed questioning the patient gave a previous history of urticarial reaction following trimethoprim–sulphamethoxazole and the authors, therefore, suggested that the jaundice is a hypersensitivity-type reaction. Nair *et al.* (1980) reported on a further patient with trimethoprim–sulphamethoxazole–induced cholestasis.

Cyproheptadine

Cyproheptadine, a potent histamine and serotonin antagonist, has a similar chemical structure to the phenothiazines. Henry *et al.* (1978) reported a single case of cholestatic jaundice in a patient on cyproheptadine, whose clinical features, biochemical abnormalites, and histological appearance of liver biopsy specimen, and slow resolution of the jaundice suggested a drug-induced cholestasis of the phenothiazine type.

Diflunisal

A single case of cholestatic jaundice associated with diflunisal was reported by Warren (1978): an earlier report by Kaklamanis *et al.* (1978) reported a single case of increased liver transaminases associated with the drug.

Disopyramide

Gottschall *et al.* (1970), Riccioni *et al.* (1977), Meinertz *et al.* (1977), Craxi *et al.* (1980), and Edmonds and Hayler (1980) all described one or more cases of intrahepatic cholestasis associated with disopyramide. Two of the cases reported by Craxi *et al.* (1980) were subjected to laparotomy because of continuing jaundice despite withdrawal of the drug. The biliary tree was normal in these cases and jaundice took four months and seven months, respectively, to clear.

Disopyramide may cause a cholestatic jaundice which develops during the first week of treatment. There is prompt clinical resolution on stopping the drug and laboratory abnormalities may persist for 18 months (Bakris *et al.* 1983). The authors found 21 cases including their patient in the literature since 1969. There were no deaths. Liver biopsy was performed in only one patient and this showed a cholestatic jaundice. In the unbiopsied patients there can be either cholestatic or hepatocellular features. (Bakris *et al.* 1983).

Erythromycin estolate and erythromycin ethyl succinate

Keefe *et al.* (1982) reported two patients who had hepatotoxicity with erythromycin estolate. Thirteen and 15 years later, respectively, they received erythromycin ethyl succinate and had a further hepatotoxic reaction. These two patients demonstrate erythromycin cross-sensitivity and these preparations should not be used in patients with previous erythromycin-associated liver injury.

The prescription-event monitoring technique was used by Inman and Rawson (1983) to determine if erythromycin estolate was a more frequent cause of jaundice than erythromycin stearate or tetracycline. 12 208 patients for whom 5343 doctors had prescribed one of the three drugs were studied. There were 16 reports of jaundice of which four were due to gallstones, three to cancer, six to viral hepatitis, and only three were possibly related to an antibiotic. All three had received erythromycin stearate. No case was attributable to the estolate, which had previously been suspected of being a more frequent cause of jaundice.

Glibenclamide

A 61-year old man presented with hypoglycaemic coma and was noticed to be slightly jaundiced and to have cutaneous bullae. After reversal of his hypoglycaemic coma he was found to have a mildly raised bilirubin level and also asparate and alanine aminotransferase levels. His alkaline phosphatase level was greatly raised. Glibenclamide was stopped and 10 days later a liver biopsy was normal. His bullae dried up over a month, his enzymes and bilirubin returned to normal in days, and his alkaline phosphatase returned to normal in weeks. Wong Paitoon *et al.* (1981) noted that this combination of reversible intrahepatic cholestasis and cutaneous bullae due to glibenclamide had not previously been recorded. In this man the probable diagnosis of drug-induced intrahepatic cholestasis was determined by exclusion of extrahepatic biliary obstruction, viral hepatitis, and primary liver disease. The cholestasis cleared within five weeks of glibenclamide withdrawal whilst other drug therapy

continued. Rechallenge with glibenclamide was not considered as the patient no longer needed oral hypoglycaemic agents.

Haloperidol

Haloperidol is an effective antipsychotic agent which is widely used in Europe and the USA (Dincsory and Saelinger 1982). Dincsory and Saelinger (1982) reported a 15-year-old patient with chronic cholestatic liver disease associated with haloperidol therapy. They noted that the clinical and morphological features and the close temporal relationship between medication and onset of illness in this patient led them to infer that haloperidol was responsible for the chronic cholestasis reaction on a hypersensitivity basis. Twenty-eight months after stopping haloperidol the patient was asymptomatic but still had mildly raised alkaline phosphatase, transaminase and-γ-glutamyl transpeptidase.

Phenothiazines

Jaundice with the clinical and biochemical features of biliary obstruction occurs as a rare complication following the therapeutic use of some drugs, including a number of phenothiazines. The mechanism of this jaundice is stasis of bile in the canaliculi of the centrilobular zones of the liver lobules. Chlorpromazine jaundice will be considered as an example of this type of phenothiazine reaction. Other drugs producing a similar effect will then be referred to more briefly.

Many reports have appeared of jaundice occurring during treatment with chloropromazine. Probably about 1 per cent of the patients who take the drug for more than 2 weeks develop jaundice. In most cases, jaundice occurred 7–36 days after treatment started, although occasionally it first appeared after administration of the drug had been discontinued for as long as 18 days. However, the appearance of jaundice has been unrelated to total dose consumed, and it has been described after amounts as small as 75 mg. Jaundice has usually been preceded by a prodromal stage of fever, abdominal discomfort, vomiting, or diarrhoea for a few days. An enlarged, tender liver has been recorded in some patients, and splenomegaly was occasionally found. The clinical and biochemical findings closely mimicked those of extrahepatic biliary obstruction. Stools were pale and urine contained bile pigment. Serum bilirubin and alkaline-phosphatase levels were raised; cephalin flocculation and thymol turbidity have almost always been normal; prothrombin time was normal; and eosinophilia in the blood was sometimes found early in the course of the jaundice.

On withdrawal of chlorpromazine, the jaundice has usually subsided gradually during the succeeding 4 weeks, with ultimate complete recovery. There have been several cases, however, in which jaundice was very prolonged, even up to 3 years, and in 10 per cent of cases jaundice lasted for more than 3 months. On the other hand, in some patients, the jaundice faded even though the drug was still being taken regularly. In few cases, in spite of apparent clinical recovery, histological changes in the liver have persisted throughout follow-up. Occasional cases of xanthomatous biliary cirrhosis following chlorpromazine administration have been reported.

It has been suggested that corticosteroids or corticotrophin hastened the disappearance of the jaundice, although it is doubtful whether this treatment greatly modified the progress of the disorder. Knowledge of a previous history of chlorpromazine ingestion should save many unnecessary laparotomies, but in cases in which the jaundice persists for a prolonged period after stopping the drug, the possibility of obstruction to the extrahepatic bile duct occurring in a patient who just happened to have had chlorpromazine must, of course, be considered. Prolonged jaundice usually merits surgical exploration of the bile ducts with cholangiography performed at the time of operation to exclude obstruction to the main bile ducts.

Many liver biopsies have now been performed in this type of jaundice and there has been a general uniformity of findings; the most characteristic features were the biliary thrombi found within the distended canaliculi of the centrilobular zones, with little or no evidence of hepatic-cell necrosis.

The mechanism by which chlorpromazine produces canalicular stasis of bile is still obscure. It has been suggested, although without proof, that the bile stasis is the result of increased viscosity in the bile canaliculi, or possibly of a defect in the normal process of hydration of bile. However, it has not been possible to reproduce this effect of chlorpromazine in animals. The lack of correlation with total dosage, the time-relationship between start of therapy and onset of jaundice, and the presence of eosinophilia have suggested that a sensitization phenomenon is more likely. This has been supported by reports of an acceleration reaction of the same type produced by a small dose of chlorpromazine in patients who have recovered from chlorpromazine jaundice. On the other hand, the response to the 'challenging' dose has not always occurred, and the occasional disappearance of jaundice while the drug was continued is difficult to understand.

An unexplained pyrexia during chlorpromazine therapy should suggest the possibility of incipient obstructive jaundice. In some milder examples of this drug reaction, jaundice may not develop, and fever, increased serum alkaline phosphatase, and eosinophilia may be the only indications. Chlorpromazine should probably be avoided in patients with previous liver disease, in malnourished patients, or when other potentially hepatotoxic agents are being used. Clearly, the further administration of chlorpromazine to a patient who has previously become jaundiced following that drug would be most unwise.

Jaundice appears to occur less often following the use of some of the newer phenothiazine drugs than with chlorpromazine. There have been reports of occasional cases of jaundice after prochlorperazine, pecazine, trifluoperazine, and promazine, but it is not at present possible to estimate its frequency with these drugs.

There is no method of testing in which patients chlorpromazine jaundice is likely to develop, indeed cutaneous tests for sensitivity are thought to be valueless. No cross-sensitivity with promazine has been demonstrated, so possibly other phenothiazine derivatives might be substituted for chlorpromazine after development of sensitivity.

In case where chlorpromazine jaundice occurred in a 59-year-old man the liver biopsy showed the typical picture of cholestatic jaundice, the centrilobular bile canaliculi being distended with bile plugs. A repeat liver biopsy performed 8 months after the initial presentation showed no bile stasis but chronic inflammatory reaction in the portal tracts was now much more intense with destruction of adjacent liver parenchyma; a diagnosis of active chronic hepatitis initiated by chlorpromazine was made.

Sulindac

Sulindac is a recently introduced non-steroidal, anti-inflammatory drug. Anderson (1979), Wolfe (1979), and Dhand et al. (1981) reported single case reports of cholestatic jaunclice . Whittaker et al. (1982) reported two further case reports. In one patient both the serum transaminase and alkaline phosphatase were moderately raised. Jaundice had developed 10 days after starting oral sulindac. A month after stopping sulindac, the patient's liver function test had returned to normal, and a month later the patient was rechallenged with sulindac, 200 mg twice daily. Within 24 hours of starting the drug again, the patient complained of fever and malaise and SGOT was raised from 18 I.U./l before challenge to 62 I.U./l the morning after challenge. Within two days of discontinuing the drug she felt well, and one month later liver function was normal. This patient's liver biopsy showed a normal liver parenchyma except for prominence of ceroid pigment within Kupffer cells in the centrilobular zone III areas. The portal tracts were described as unremarkable.

The second patient of Whittaker *et al.* (1982) showed evidence of hepato-cellular damage largely within the centrilobular areas (zone III). The portal tracts were normal. This female patient became anicteric seven months after taking sulindac and her liver function enzymes had returned to normal.

Kaul *et al.* (1981) described a 12-year-old girl with acute hepatitis about six weeks after starting sulindac for rheumatoid arthritis. The drug was stopped and symptoms were reversed. No biopsy is reported but symptoms recurred with challenge.

Five patients have now been described who developed reversible hepatic dysfunction with sulindac (Anderson 1979; Wolfe 1979; Dhand *et al.* 1981; Whittaker *et al.* 1982). One patient was described with a predominantly cholestatic picture (Whittaker *et al.* 1982). The mechanism of hepatotoxicity is unknown.

Thiabendazole

A family of five — two parents and three daughters, — were given thiabendazole for pin-worm infestation (Fink *et al.* 1979). The mother and one daughter who were infested developed keratoconjunctivitis, xerostomia, and cholangiostatic jaundice. It was suggested by these authors that thiabendazole may have acted as a hapten and by binding to the body protein induced the production of auto-antibodies which may have acted against the biliary epithelium, the salivary duct epithelium, and the lacrimal gland ducts.

Thiabendazole is a relatively safe and effective agent with a wide range of activity against nematodes which infest the gastrointestinal tract. Rex *et al.* (1983) described a 55-year-old man who developed prolonged jaundice and sicca complex after a course of thiabendazole therapy. ERCP showed a normal pancreatic and biliary tree. Liver biopsy was consistent with a drug-induced cholestasis. The patient recovered from his jaundice but his sicca complex persisted one year after stopping thiabendazole therapy. Rex *et al.* (1983) also noted that there were eight other cases in the literature and that the majority of those submitted to liver biopsy showed cholestatic features. They concluded that thiabendazole occasionally causes a syndrome of severe prolonged sicca complex with a drug-induced cholestasis.

Trazodone

Trazodone is a recently introduced antidepressant. There have been two reports of liver disease with this drug. Chu

et al. (1983) reported a 63-year-old man in whom trazodone dose was gradually increased to 500 mg daily over a 10-day period. Three weeks later trazodone was withdrawn due to lack of therapeutic response. Liver enzyme levels were raised compared with normal pre-treatment values. A liver biopsy showed mild portal expansion with moderate eosinophils, some mononuclear and polymorphonuclear leucocytes, scattered foci of Kupffer cells, and acidophil bodies. The patient also had a skin rash. The rash and the liver function tests returned to normal four weeks after drug withdrawal.

Sheikh and Nies (1983) reported a 71-year-old lady who started trazodone, 50 mg daily. Two weeks later she complained of malaise and jaundice. Her liver enzymes were raised and a liver biopsy showed canalicular cholestasis with numerous acidophil bodies. There was Kupffer-cell hyperplasia and portal inflammation with lymphocytes. Recovery occurred eight weeks after withdrawal of the drug.

Drug-induced hepatic neoplasia

Angiosarcoma of the liver and hepatocellular carcinoma

Angiosarcoma of the liver is rare, it has been associated with thorium dioxide (*Thorotrast*), organic arsenic, and vinyl chloride monomer. Daneshmend *et al.* (1979) reported a case of angiosarcoma of the liver associated with phenelzine therapy for a period of six years. Toth (1976) also pointed out that phenelzine causes angiosarcoma of the liver in mice.

Falk *et al.* (1979) reported the results of a retrospective epidemiological study of deaths from hepatic angiosarcoma in the United States during the 10-year period 1964–74. There were 168 such cases, of which 37 (22 per cent) were associated with previously known causes (vinyl chloride, *Thorotrast*, and inorganic arsenic) and four (3.1 per cent) of the remaining 131 cases with the use of androgenic–anabolic steroids. It was suggested that the long-term use of androgenic–anabolic steroids is the fourth cause of hepatic angiosarcoma, the majority of cases still being of unknown aetiology.

Evans *et al.* (1983) described the liver changes seen in five vinyl chloride workers. The five livers showed angioformative and hepatocellular growth disturbance in varying proportions. Angiosarcoma was present in four and liver cell hyperplasia, in all. Hyperplasia nodules were present in three and hepatocellular carcinoma was found in two. In one the transition from hyperplastic nodule to hepatocellular carcinoma was demonstrated. The relationship between these changes and vinyl chloride exposure is documented, and it was shown that they are causally related.

Hepatocellular adenoma

In a case-control study of hepatocellular adenoma conducted by the Centre of Disease Control and the Armed Forces Institute of Pathology (Rooks *et al.* 1979), 79 women with hepatocellular adenoma and 220 age- and neighbourhood-matched controls were interviewed. Information was obtained on nine additional patients with hepatic adenoma who died. The mean age at diagnosis for the 88 women was 30.4 years; the most important presenting signs were intraperitoneal haemorrhage in 29 patients (34 per cent) and bleeding into tumour in 13 patients (15 per cent). In the patients who did not present with haemorrhage, 20 sought attention because of awareness of an abdominal mass, and eight sought attention because of pain.

Rooks *et al.* (1979) confirmed the association of hepatocellular carcinoma with oral contraceptive use; only six out of 88 patients had not taken oral contraceptives. The risk of hepatocellular adenoma increased with duration of oral contraceptive use, and high oestrogen content, and increasing age over 30 years further increased the risk. Long-term users of oral contraceptives had an estimated annual incidence of hepatocellular adenoma of 3 to 4 per 100 000.

Neuberger *et al.* (1980) described a series of 10 women with oral contraceptive–associated liver tumours, seven of which were found to be hepatocellular carcinoma and three of which were benign tumours.

Mahboubi *et al.* (1981) reviewed the 266 cases in the literature of hepatic neoplasms that have been reported in the world literature from 1972 to 1978 in young women using oral contraceptive therapy. They concluded that, although most commonly benign, these vascular tumours rupture and cause severe haemorrhage in 30 per cent of cases and had a consequent mortality rate of over 10 per cent. In their own study they collected a series of 25 benign liver tumours in young Nebraska women taking oral contraceptives. The age range of these women ranged from 19 to 43 years and the duration of oral contraceptive use and type of tumour are shown in Table 17.3.

Mahboubi *et al.* estimated that these 25 cases of liver tumours associated with oral contraceptive use had occurred in a potential population of oral contraceptive users of 74 000 women within Nebraska.

Oral contraceptive-induced tumours have usually been reported as benign vascular tumours but Mahboubi *et al.* (1981) included four cases of hepatocellular carcinoma. Tesluk and Lawrie (1981) record the transformation of a hepatic adenoma to hepatocellular carcinoma in a 34-year-old woman who had been taking oral contraceptives for five years. Monroe *et al.* (1981) document the case of a 32-year-old woman with an eight-year history of oral contraceptive use who developed a hepatic angiosarcoma. These cases and others in the literature indicate that both benign and malignant tumours of the liver can arise associated with oral contraceptive use and that benign tumours can undergo malignant change.

It should be remembered that, in animal carcinogenicity studies in rodents using combined oestrogen–progestogen preparations, the liver was the site of neoplastic change with several combinations (Committee of Safety of Medicines 1972).

Shar and Kew (1982) reported on a 23-year-old woman who died two years after presenting with a haemoperitoneum from rupture of a hepatocellular carcinoma. This patient has been taking an oral contraceptive for periods of 1 year and 10 months separated by a pregnancy. Histologically the lesion was a moderately differentiated hepatocellular carcinoma. They reviewed the literature and recorded 23 other patients who presented with hepatocellular carcinoma while taking an oestrogen and progesterone or oestrogen-alone oral contraceptive preparation. This number can be expanded to

Table 17.3 **Histological diagnoses in 25 female patients with hepatic tumours associated with oral contraceptives (after Mahboubi *et al.* 1981)**

Years of use	Hepato-cellular adenoma	Hepato-cellular carcinoma	Focal nodular hyperplasia	Ruptured vascular lesion	Hamartoma
0–1	1	1		0	0
1–2	1	0	1	1	0
2–3	1	2		0	0
3–4	2	0	1	0	0
4–5	2	0		0	1
5–6	3	1	1	1	0
6–7	1	0	2	0	0
7 +	0	0	2	0	0
Total (per cent)	11 (44)	4 (16)	7 (28)	2 (8)	1 (4)

87 known cases if the American College of Surgeons Commission of Cancer (Vana and Murphy 1979) cases and the University of Louisville Registry cases (Christopherson *et al.* 1978) are included. They noted that hepatocellular carcinoma can arise in the liver parenchyma *de novo* or arise in an adenoma. Oral contraceptives had been used for from 6 to 12 years before diagnosis and the average time was 65 months.

Since the initial report (Baum *et al.* 1973) there have been a growing number of reports of benign hepatocellular neoplasia in women receiving oral contraceptives. Several described the pathological picture of peliosis hepatitis either alone or concurrently with adenomas and focal nodular hyperplasia. Schonberg (1983) described another example and reviewed the previous reports. The term 'peliosis hepatitis' means blood-filled cysts of the liver and was first described by Yanoff and Rawson (1964). Liver palpation in patients using oral contraceptives is important (Schonberg 1983). Schonberg concluded that peliosis hepatitis may be reversible when oral contraceptives are discontinued. Peliosis hepatitis has also been described after androgen therapy (Burger and Marcuse 1952).

Westaby *et al.* (1983) describe three patients with androgen-related primary hepatic tumours in non-Fanconi syndrome patients who had taken androgens for 10–17 years. They concluded that androgens had to be taken over long periods of time before tumours developed in non-Fanconi syndrome patients. Two of their patients presented with haemoperitoneum at laparotomy. The other patient histologically had a hepatic adenoma. Following androgen withdrawal there has been no tumour progression or metastasis. However, the tumour has not regressed. Westaby *et al.* (1983) noted that haemoperitoneum is the cause of mortality and morbidity. Patients with androgenic and anabolic steroid-related hepatic tumours had taken 17-alkylated compounds. The non-17-alkylated steroids have been suggested as a safe alternative but their period of assessment is short and they should be carefully monitored.

Cholangiocarcinoma

Littlewood *et al.* (1980) raise the possibility that a case of cholangiocarcinoma in a 21-year-old woman taking oral contraceptives could be causally related. This tumour has its peak incidence in the sixth decade and is exceptionally rare in the young. Klatskin (1977) in his review of 250 liver tumours associated with oral contraceptives had a single case of cholangiocarcinoma which he considered to be causally related.

Haemangioendothelial sarcoma of the liver in women has been reported in association with oral contraceptive agents containing oestrogen (see *Iatrogenic disease*, 2nd edn, p. 354). Horn *et al.* (1980) reported the case of an 81-year-old man treated for nine years with stilboestrol, 5 mg twice daily, for presumed — but unproven — carcinoma of the prostate (no biopsy was taken at time of commencing treatment). A wedge biopsy of the liver was performed but four months after discharge from hospital the patient was readmitted to hospital with ascites, dyspnoea, and peripheral oedema and died shortly after. Autopsy confirmed the biopsy report of haemangioendothelial sarcoma: however, no evidence to support the original diagnosis of carcinoma of the prostate was found. Hoch-Ligeti (1978) also reported a single case of angiosarcoma of the liver associated with diethylstilboestrol.

Liver damage caused by herbal products

Pyrrolidizine alkaloids

Self-medication has increased in popularity and the current interest in 'alternative medicine' has brought 'health' foods and herbal medicines into fashion. Neither patient or doctor may be aware of their constituents, let alone the toxicity of these substances taken as medicines.

Various herbal remedies contain alkaloids such as pyrrolidizine, which may cause severe liver damage and death. The genera in which these alkaloids occur include *Senecio* (of which the English ragwort is a species). *Crotalaria*, and *Heliotropium* — all found all over the world. Pyrrolidizine alkaloids have caused liver disease in, for example, Jamaica, South Africa, Israel. Egypt, and India. The plants are ingested as a herbal infusion and also as a food (*ackee*). Such liver damage was first reported from Jamaica, where occlusive disease of the small branches of the hepatic veins was endemic. This was linked with drinking bush-tea — infusions are made of any available herbs (over 299 species are known to be used). *Crotalaria* species are not normally used to make a beverage because of their bitterness, but they are often used for their alleged medicinal properties. A painful enlargement of the liver (without much jaundice) results: this may be followed by hepatic failure and death, non-portal cirrhosis, or complete recovery. There is a time-lag between ingestion of these alkaloids and the onset of symptoms — as long as three months in one fatal case in an Indian epidemic of *Heliotropium* toxicity (Datta *et al.* 1978).

Hepatotoxicity of other herbals

Liver injury has also resulted from eating *Vicia* faba

beans, cereals contaminated with *Pencillium* and groundnuts with *Aspergillus* flavus. The fungi *Amanita phalloides* and *Helvella* have also been the cause of hepatotoxicity in 'natural remedies'.

Pennyroyal oil has been reported to be associated with massive hepatic necrosis when taken in large doses as an abortifacient. This is thought to be due to oil which contains 85 per cent pulagone giving rise to an epoxide which is detoxified by glutathione; when glutathione supply is exhausted the hepatic damage is produced by this metabolite. Sullivan *et al.* (1979) by analogy with a similar situation with paracetamol-induced liver damage recommend the use of acetylcysteine as an antidote.

Mistletoe is one of the oldest and most widely used herbal remedies and has been advocated for infertility, asthma, epilepsy, and as an aphrodisiac. Harvey and Colin-Jones (1981) described a case of mistletoe-induced hepatitis in a 49-year-old woman: and demonstrated this by rechallenge.

Budd-Chiari syndrome

Oral contraceptives

The Budd–Chiari syndrome in women taking oral contraceptives continues to occur and be reported. Tsung *et al.* (1980) reported a case in a 21-year-old woman and reviewed 18 other cases reported in the literature in which 12 of these women died. Death occurred between 38 days and 1½ years after the onset of symptoms.

Lewis *et al.* (1983) collected 47 cases of Budd–Chiari syndrome associated with oral contraceptives: 29 from literature and 18 additional cases by questionnaire. They compared medical and surgical management and concluded that a satisfactory response may accompany either medical or surgical management of patients. Those with severe occlusive disease may benefit most from surgical decompression of the hepatic veins.

Dacarbazine

Dacarbazine is used in the treatment of malignant melanoma. Greenstone *et al.* (1981) described two cases of death following acute liver failure due to intrahepatic vascular occlusion and referred to five other fatal cases of death from intravascular occlusion due to dacarbazine treatment. The histology of one case of overt Budd–Chiari syndrome due to dacarbazine therapy had been examined by one of the authors in which the vascular occlusion had been in the hepatic vein.

Dacarbazine (DTIC)

Dancygier *et al.* (1982) reported on the use of dacarbazine in 100 patients with malignant melanoma. The morphological spectrum includes thrombosis of terminal hepatic veins with extended centrilobular necrosis infiltrated with eosinophils, subtotal lobular necrosis without inflammation, granulomatous hepatitis, and acute toxic hepatitis. In clinically mild courses of jaundice there was loss of ribosomes from the rough endoplasmic reticulum and dilatation of the vesicles of the smooth muscle endoplasmic verticulum. In the clinically severe cases, electron microscopy showed rupture of the nuclear membrane, fragmentation of nucleoli, cytoplasmic vacuole formation, and bleeds of the cell membrane. There was also dilatation of the bile canaliculi with stunting of electron dense fibrillary material.

The data of Dancygier *et al.* (1983) has only been published in abstract form and they conclude that their 'morphological investigations indicate that dacarbazine may exert toxic effect on the microfilamentous cytoskeleton of the hepatocytes.

Dancygier *et al.* (1983) reported a 61-year-old man who developed signs of hepatic failure during the second treatment cycle with dacarbazine for malignant melanoma. Light microscopy showed extensive centrilobular liver necrosis. Terminal hepatic venules did not show vasculitis or thrombosis and there was a lack of inflammatory infiltration. The patient received 250 mg methylprednisolone intravenously at the onset of symptoms and was discharged 12 days after the peak rise of transaminases with normal liver parameters.

Single agent chemotherapy after surgical treatment for malignant melanoma with dacarbazine was used by Feaux de Lacroix *et al.* (1983). They reported four deaths. During the second therapy cycle liver failure due to massive hepatic necrosis occurred with widespread thrombosis of hepatic veins. Budd–Chiari syndrome caused by an allergic thrombophlebitis with secondary liver cell necrosis seems the probable explanation.

As dacarbazine leads to remission in only about 20 per cent of patients with malignant melanoma its therapeutic application may have to be re-evaluated.

RECOMMENDED FURTHER READING

KEELING, P.W.N. AND THOMPSON, R.P.H. (1979). [*See References.*]
LUDWIG, J. AND ALEXELSEN, R. (1983). Drug effects on the liver. An updated compliation of drugs and drug-related hepatic diseases. *Dig. Dis. Sci.* **28**, 651–66.
ZAFRANI, E.S., VON PINAUDEAU, Y., AND DHUMEAUX, D. (1983). Drug-induced vascular lesions of the liver. *Arch. intern. Med.* **143**, 495–502.
ZIMMERMAN, H.J. (1978). [*See References.*]

REFERENCES

ABI-MANSUR, P., ARDACA, M.C., ALLAM, C., AND SHAMMA, M. (1981). Trimethoprim–sulfamethoxazole-induced cholestasis. *Am. J. Gastroenterol.* **76**, 356–9.

ADDISON, G.M. AND GORDON, N.S. (1980). Sodium valproate and acute hepatic failure. *Dev. Med. Child Neurol.* **22**, 248–9.

ALCARCON-SEGOVIA, D. (1980). Long-term treatment of symptomatic osteoarthritis with benoxaprofen. Double-blind comparison with aspirin and ibuprofen. *J. Rheumatol.* **7** (Suppl. 6), 89–99.

ANDERSON, R.J. (1979). Severe reaction associated with sulindac administration. *New Engl. J. Med.* **300**, 735–6.

ATKINSON, A.B., MACREAVEY, D., AND TROPE, E. (1980). Papilloedema and hepatic dysfunction apparently induced by perhexiline maleate. *Br. Heart J.* **43**, 490–1.

AUSTRALIAN DRUG EVALUATION COMMITTEE (1980). Sodium valproate, Letter to medical profession. November 1980.

BACON, P.A., DAVIES, J., AND RING, F.J. (1980). Benoxaprofen dose range studies using quantative thermography. *J. Rheumatol.* **7** (suppl. 6), 48–53.

BAKER, R.D. AND HANDLER, P. (1943). Animal experiments with tannic acid suggested by the tannic acid treatment of burns. *Ann. Surg.* **118**, 417.

BAKRIS, G.L., CROSS, P.D., AND HAMMARSTEN, (1983). Disopyramide-associated liver dysfunction. *Mayo Clin. Proc.* **58**, 265–7.

BARNETT, D.B., HUDSON, S.A., AND GOLICHTLY, P.W. (1980). Hydrallazine-induced hepatitis. *Br. med. J.* 1165–6.

BARTOLI, E., MASSARELLI, G., SOLIMAS, A., FAEDDA, R., AND CHIANDUSSI, L. (1979). Acute hepatitis with bridging necrosis due to hydrallazine intake. *Arch. intern. Med.* **139**, 698–9.

BASS, B.H. (1974). Jaundice associated with naproxen. *Lancet* **i**, 998.

BAUM, J.K., BOOKSTEIN, J.J., HOLTZ, F., AND KLEIN, E.W. (1973). Possible association between benign hepatomas and oral contraceptives. *Lancet* **ii**, 926–9.

BERRY, H., BLOOM, B., AND HAMILTON, E.B.D. (1980). Dose-range studies of benoxaprofen compared with placebo in patients with severe active rheumatoid arthritis. *J. Rheumatol.* **7** (Suppl. 6), 54–9.

BIRD, H.A., RHIND, B.M., PICKUP, M.E., AND WRIGHT, V. (1980). A comparative study of benoxaprofen and indomethacin in ankylosing spondylitis. *J. Rheumatol.* **7** (Suppl. 6) 139–42.

BJORNBOE, M., IVERSON, O., AND OLSEN, S. (1967). Infective hepatitis and toxic jaundice in a municipal hospital during a five-year period. *Acta med. scand.* **182**, 1.

BLECHMAN, W.J. (1980). Crossover comparison of benoxaprofen and naproxen in osteoarthritis. *J. Rheumatol.* **7** (Suppl. 6), 132–8.

BLOM-BULOW, B., OBERG, K., WOLLHEIM, A., PERSSON. B., JONSON, B., MALMBERG, P., BOSTROM, M., AND HERBAI, G. (1981). Gyclofenil versus placebo in progressive systemic placebo in progressive systemic sclerosis. *Acta med. scand.* **210**, 419–28.

BOSIO, M. (1983). Cholestatic jaundice and haematuria due to hypersensitivity to cefaclor in a child. *J. Toxicol. Clin. Toxical.* **20**, 79–84.

BOUGHTON K. (1983). Ketoconazole and hepatic reactions. *South Afr. med. J.* **63**, 955.

BRITISH MEDICAL JOURNAL (1979). Editorial: Liver injury, drugs and popular poisons. *Br. med. J.* **1**, 574–5.

BRODSKY, S.L., CUTLER, B.S., WEINER, D.A., AND KLEIN, M.D. (1981). Hepatotoxicity due to treatment with verapamil. *Ann. intern. Med.* **94**, 490–1.

BURGER, R.A. AND MARCUSE, P.M. (1952). Peliosis hepatitis. Report of a case. *Am. J. clin. Pathol.* **22**, 569–72.

CENTRE FOR DISEASE CONTROL. (1980). Follow up on Reye's Syndrome. *US Morbid, Mortal. Wkly. Rep.* **29**, 321–2.

CHIPRUT, R.O., VITERI, A., JAMROG, C., AND DYCK, W.P. (1976). Intrahepatic cholestasis after griseofulvin administration. *Gastroenterology* **70**, 1141–3.

CHRISTOPHERSON, W.M., MAYS, E.T., AND BARROWS, G. (1978). Hepatocellular carcinoma in young women on oral contraceptives. *Lancet* **ii**, 38–5.

CHU, A.G., GUNSOLLY, B.L., SUMMERS, R.W., ALEXANDER, B., McCHESNEY, C., AND TANNA, V.L. (1983). Trazodone and liver toxicity. *Ann. intern. Med.* **99**, 128–9.

CICCOLUNGHI, S.N., CHAUDRI, H.A., SCHUBIGER, B.I., AND REDDROP, R. (1978). Report on a long-term tolerability study of up to two years with dicloflenic sodium (*Valtarol*). *Scand. J. Rheumatol.* **22**, 86–96, *Suppl 22*.

COMMITTEE ON SAFETY OF MEDICINES (1972). *Carcinogenicity tests of oral contraceptives. A report.* HMSO, London.

—— (1981). *Current Problems* No. 6, July.

COOK, B.A., SINNHUBER, J.R., THOMAS, P.J., OLSON, T.A., SILVERMAN, T.A., JONES, R., WHITEHEAD, V.M., AND RUYMANN, F.B. (1983). Hepatic failure secondary to indicine N-oxide toxicity. *Cancer* **52**, 61–3.

COULTER, D.L. (1980). Valproate hyperammonaemia and hyperglycinaemia. *Lancet* **ii**, 260.

—— AND ALLEN, R.J. (1980). Secondary hyperammonaemia a possible mechanism for valproate encephalopathy. *Lancet* **i**, 1310–11.

CRAXI, A., GATTO, G., MARINGHINI, A., ORSINI, S., PINZELLO, G., AND PAGLIARO, L. (1980). Disopyramide and cholestasis. *Ann. intern. Med.* **93**, 150–1.

DANCYGIER, H., RUNNE, U., LEUSCHNER, U., MILBRADT, R., AND CLASSEN, M. (1983). Dacarbazine (DTIC)-induced human liver damage: light and electron microscopy findings. *Hepatogastroenterology* **30**, 93–5.

——, ——, WACKER, D., HAUK, H., LEUSCHNER, U., AND CLASSEN, M. (1982). Dacarbazine (DTIC)-induced human liver injury. *Gut.* **23**, A447.

DANESHMEND, T.K., SCOTT, G.L., AND BRADFIELD, J.W.B. (1979). Angiosarcoma of liver associated with phenelzine. *Br. med. J.* **1**, 1679.

DATTA, D.V., KHUROO, M.S., MATTOCKS. A.R., AIKAT, B.K., AND CHHUTANI, P.N. (1978). Herbal medicines and veno-occlusive disease in India. *Postgrad. med. J.* **54**, 511–15.

DAVIS, M. AND WILLIAMS, R. (1977). Drugs and the liver. In *Textbook of adverse drug reactions* (ed. D.S. Davies). Oxford University Press, Oxford.

——, EDDLESTON, A.L.W.F., AND WILLIAMS, R. (1980). Hypersensitivity and jaundice due to azathioprine. *Postgrad. med. J.* **56**, 274–5.

——, VERGANI, D., EDDLESTON, A.L.W.F., AND WILLIAMS, R. (1979). Halothane hepatitis–toxicity and immunity. In *Immune reactions in liver disease* (ed. A.L.W.F. Eddleston, J.C.P. Weber, and R. Williams), pp. 235–46. Pitman Medical London.

DHAND, A.K., LA BRECQUE, D.R., AND METZGER, J. (1981). Sulindac (*Clinoral*) hepatitis. *Gastroenterology* **80**, 585–6.

DINCSORY, H.P. AND SAELINGER, D.A. (1982). Halperidol-induced chronic cholestatic liver disease. *Gastroenterology* **83**, 694–700.

DONAT, J.F., BOCCHINI, J.A., GONZALEZ, E., AND SCHWENDIMANN, R.N. (1979). Valproic acid and fatal hepatitis. *Neurology* **29**, 273–4.

DUNK, A.A., WALT, R.P., JENKINS, W.J., AND SHERLOCK, S.S. (1982). Diclofenac hepatitis B. *Br. med. J.* **1**, 1605–6.

EDMONDS, M.E. AND HAYLER, A.M. (1980). Letter to editor (no title). *Eur. J. clin. Pharmacol.* **18**, 285–6.

ELMENHARY, M.M., SHAKER, A., RAMADAN, M., HAMZA, S., AND TARDOS, S.S. (1981). Control of essential hypertension by captopril, an angiotensin-converting inhibitor. *Br. J. clin. Pharmacol.* **11**, 469–75.

ENAT, R., POLLOCK, S., BEN-ARIEH, Y., AND BARZILAI, D. (1980). Cholestatic jaundice caused by cloxacillin: macrophage inhibition factor test in preventing rechallenge with hepatotoxic drugs. *Br. med. J.* **281**, 982–3.

EVANS, D.M., WILLIAMS, W.J., AND KING, I.T. (1983). Angiosarcoma and hepatocellular carcinoma in vinyl chloride workers. *Histopathology* **7**, 377–88.

FALK, H., THOMAS, L.B., POPPER, H., AND ISHAK, K. G. (1979). Hepatic angiosarcoma associated with androgenic-anabolic steroids. *Lancet* **ii**, 1120–23.

FEAUX DE LACROIX, W., RUNNE, V., HAUK, H., DOEPFMER, K., GROTH, W., AND WACKER, D. (1983). Acute liver dystrophy with thrombosis of hepatic veins: a fatal complication of dacarbazine treatment. *Cancer treat. Rep.* **67**, 779–84.

FEE, J.P.H., BLACK, G.W., DUNDEE, J.W., McILROY, P.D.A., JOHNSTON, H.M.L., JOHNSTON, S.B., BLACK, I.H.C., McNEILL, H.G., NEILL, D.W., DOGGART, J.R., MERRETT, J.D., McDONALD, J.R., BRADLEY, D.S.G., HAIRE, M., AND McMILLAN, S.A. (1979). A prospective study of liver enzyme and other changes following repeat administration of halothane and enflurane. *Br. J. Anaesth.* **51**, 1133–41.

FINK, A., MACKAY, C.J., AND CUTLER, S.S. (1979). Sicca complex and cholangiostatic jaundice in two members of a family, probably caused by thiabendazole. *Ophth.* **86**, 1892–6.

FIREBRACE, D.A.J. (1981). Hepatitis and ketoconazole therapy. *Br. med. J.* **283**, 1058–9.

FIRTH, H., WILOCK, G.K., AND ESIRI, M. (1982). Side-effects of benoxaprofen. *Br. med. J.* **284**, 1784.

FISHER, B.M. AND McARTHUR, J.D. (1982). Side-effects of benoxaprofen. *Br. med. J.* **284**, 1783.

FRIEDMAN, L., LEWIS, P.J., CLIFTON, P., AND BULPITT, C.J. (1978). Factors influencing the incidence of neonatal jaundice. *Br. med. J.* **1**, 1235–7.

GERBER, N., DICKINSON, R.G., HARLAND, R.C., LYNN, R.K., HOUGHTON, D., ANTONIAS, J., AND SCHIMSCHOCK, J.C. (1979). Reye-like syndrome associated with valproic acid therapy. *J. Paediat.* **95**, 142–4.

GILLIGAN, J.E., KEMP, R., AND PHILLIPS P.J. (1980). Paracetamol concentrations, hepatotoxicity and antidotes. *Br. med. J.* **280**, 114.

GOTTSCHALL, C.A.M., CUNHA, J.P., MILLER, V., AND TEIXEIRA, J.C. (1970). Treatment of arrhythmias with disopyramide. *Rev. Ass. med Rio Grande do Sul* **14**, 195.

GOUDIE, B.M., BIRNIE, G.F., WATKINSON, G., MACSWEEN, R.N.M., KISSON, L.H., AND CUNNINGHAM, N.E. (1982). Jaundice associated with the use of benoxaprofen. *Lancet* **i**, 959.

GREENSTONE, M.A., DOWD, R.M., MIKHAILIDS, D.P., AND SCHEUER, P.J. (1981). Hepatic vascular lesions associated with dacarbazine treatment. *Br. med. J.* **282**, 1744–5.

GRIFFIN, J.P. (1984). Spontaneous reporting. In *Monitoring for adverse reactions* (ed. S.R. Walker and A. Goldberg). MTP Press, Lancaster.

HAMDY, R.C., MURNANE, B., PERERA, N., WOODCOCK, K., AND KOCH, I.M. (1982). The pharmacokinetics of benoxaprofen in elderly subjects. *Eur. J. Rheumat. Inflammation* **5**, 69–76.

HARVEY, J. AND COLIN-JONES, D.G. (1981). Mistletoe hepatitis. *Br. med. J.* **282**, 186–7.

HAY, D.R., AND GWYNNE, J.F. (1983). Cirrhosis of the liver following therapy with perhexiline maleate. *NZ. med. J.* **96**, 202–4.

HEIBERG, J.S. AND SVEJGAARD, E. (1981). Toxic hepatitis during ketoconazole treatment. *Br. med. J.* **2**, 825.

HENRY, D.A., LOWE, J.M., AND DONNELLY, T. (1978). Jaundice during cyproheptadine treatment. *Br. med. J.* **1**, 753.

HOCH-LIGETI, C. (1978). Angiosarcoma of the liver associated with diethylstilboestrol. *J. Am. med. Ass.* **240**, 1510–11.

HORN, J.M., PIROLA, R.C., AND CROUCH, R.L. (1980). Haemangio-endothelial sarcoma of the liver associated with long-term oestrogen therapy in man. *Dig. Dis. Sci.* **25**, 879–83.

HUMBLE, M.W., EYKYN, S.J., AND PHILLIPS, I. (1980). Staphylococcal bacteraemia fusidic acid and jaundice. *Br. med. J.* **280**, 1495–8.

HUSKINSSON, E.C. (1979). Clinical studies with benoxaprofen. *Eur. J. Rheumatol. Inflammation* **1**, 29–36.

INMAN, W.H. AND RAWSON, N.S. (1983). Erythromycin estolate and jaundice. *Br. med. J.* **286**, 1954–55.

JACOBI, G., THORBECK, R., RITZ, A., JANSSEN, W., AND SCHMIDTS, H.L. (1980). Fatal hepatotoxicity in child on phenobarbitone and sodium valproate. *Lancet* **i**, 712–13.

JANSSEN, J. AND SYMOENS, J. (1982). *Hepatic reactions during ketoconazole treatment.* Janssen Pharmaceutica, Beerse, Belgium.

JEAVONS, P.M. (1980). Sodium valproate and acute hepatic failure. *Dev. Med. Child. Neurol.* **22**, 547–8.

KAKLAMANIS, P.H., SFIKAKIS, P., DEMETRIADES, P., TSACHALOS, P., THOUAS, B., AND SEITANIDES, B. (1978). Double-blind 12-week comparison trial of diflunisal and acetylsalicylic acid in the control of pain of osteoarthritis of the hip and/or knee. *Clin. Ther.* **1** (Suppl. A), 20–4.

KAUL, A., REDDY, J.C., FAGMAN, E., AND SMITH, G.F. (1981). Hepatitis associated with the use of sulindac in a child. *J. Paediat.* **99**, 650–1.

KEEFE, E.B., REIS, T.C., AND BERLAND, J.E. (1982). Hepatotoxicity to both erythromycin and erythromycin ethylsuccinate. *Dig. Dis. Sci.* **27**, 701–4.

KEELING, P.W.N. AND THOMPSON, R.P.H. (1979). Drug-induced liver disease. *Br. med. J.* **1**, 990–3.

KLATSKIN, G. (1977). Hepatic tumours: possible relationship to use of oral contraceptives. *Gastroenterology* **73**, 386–94.

KLINTMALM, G.B.G., IWATSUKI, S., AND STARZL, T.E. (1981). Cyclosporin A hepatotoxicity in 66 renal allograft recipients. *Transplantation* **32**, 488–9.

KOFF, R.S., GARDNER, R., HARINASUTA, U., AND PIHL, C.O. (1970). Profile of hyperbilirubinemia in three hospital populations. *Clin. Res.* **18**, 680.

LANCET (1980). Editorial: Sodium valproate and the liver. *Lancet* **ii**, 1119–20.

LANGE, A.P., SECHER, N.J., WESTERGAARD, J.G., AND SKOVGARD, I.

(1982). Neonatal jaundice after labour induced of stimulated by prostaglandin E_2 or oxytocin. *Lancet* i, 991–4.

LAW, I.P. AND KNIGHT, H. (1976). Jaundice associated with naproxen. *New Engl. J. Med.* **295**, 1202.

LE BIHAN, G., BOURREILLE, J., SAMPSON, M., LEROY, J., SZEKELY, A.M., AND COQUEREL, A. (1980). Fatal hepatic failure and sodium valproate. *Lancet* ii, 1298–9.

LEVY, M., GOODMAN, M.W., VAN DYNE, B.J., AND SUMNER, H.W. (1981). Granulomatous hepatitis secondary to carbamazepine. *Ann. intern. Med.* **95**, 64–5.

LEWIS, J.H., TICE, H.L., AND ZIMMERMAN, H.J. (1983). Budd–Chiari syndrome associated with oral contraceptive steroids. Review of treatment of 47 cases. *Dig. Dis. Sci.* **28**, 673–83.

LITTLEWOOD, E.R., HARRISON, I.G., MURRAY-LYON, I.M., AND PARADINAS, F.J. (1980). Cholangiosarcoma and oral contraceptives. *Lancet* i, 310–11.

MACNAIR, A.L., GASCOIGNE, E., AND HEAP, J. (1981). Hepatitis and ketoconazole therapy. *Br. med. J.* 1058.

MAHBOUBI, E.O., SAYED, G.M., SCHENKEN, J.D., AND SHUBIK, P. (1981). Benign hepatic tumours and steroid hormones in Nebraska. *Nebr. med. J.* **66**, 14–19.

MATHIS, R.K., SIBLEY, R.K., AND SHARP, H.L. (1979). Valproic acid and liver toxicity. *Gastroenterology.* **17**, A25.

McALISTER, W.H., ANDERSON, M.S., BLOOMBERG, G.R., AND MARGULIS, A. (1963). Lethal effects of tannic acid in the barium enema; Report of three fatalities and experimental studies. *Radiology* **80**, 765.

McILROY, P.D.A., FEE, J.P.H., DUNDEE, J.W., BLACK, G.W., DOGGART, J.R., JOHNSTONE, H.M.L., HAIRE, M., AND NEILL, D.W. (1979). Methodology of a prospective study of changes in liver enzyme concentration following repeat anaesthetics. *Br. J. Anaesth.* **51**, 1125–32.

McILVANIE, S.K. AND MacCARTHY, J.D. (1959). Hepatitis in association with 6-mercaptoparine therapy. *Blood* **14**, 80–90.

MEINERTZ, T., LANGER, K.H., KASPER, W., AND JUST, H. (1977). Disopyramide-induced intrahepatic cholestasis. *Lancet* ii, 828–9.

MEREDITH, T.J., NEWMAN, B., AND GOULDING, R. (1978). Paracetamol poisoning in children. *Br. med. J.* **2**, 478–9.

MONROE, P.S., RIDDELL, R.H., SIEGLER, M., AND BAKER, A.L. (1981). Hepatic angiosarcoma: possible relationship to long-term oral contraceptive ingestion. *J. Am. med. Ass.* **246**, 64–5.

MUSHIN, W.W., ROSEN, M., AND JONES. E.V. (1971). Posthalothane jaundice on relation to previous administration of halothane. *Br. med. J.* **3**, 18.

NAIR, S.S., KAPLAN, J.M., AND LEVINE, M. (1980). Trimethoprim–sulphamethoxazole–induced cholestasis. *Ann. intern. Med.* **92**, 511–12.

NASH, J.F., CARMICHAEL, R.H., RIDOLFO, A.S. AND SPRADLIN, C.T. (1980). Pharmacokinetic studies of benoxaprofen after therapeutic doses with a review of related pharmacokinetic and metabolic studies. *J. Rheumatol.* **7** (Suppl. 6), 12–19.

NEUBERGER, J., PORTMANN, B., NUNNERLEY, H.B., LAWS, J.W., DAVIS, M., AND WILLIAMS R. (1980). Oral contraceptive associated liver tumours: occurrence of malignancy and difficulties in diagnosis. *Lancet* ii, 273–6.

OGILVIE, A.L. AND TOGHILL, P.J. (1980). Cholestatic jaundice due to co-trimoxazole. *Postgrad. med. J.* **56**, 202–4.

OLSSON, R., TYLLSTROM, J., AND ZETTERGREN, L. (1983). Hepatic reactions to cyclofenil. *Gut* **24**, 260–3.

PARIENTE, E.A., PESSAYRE, D., BERNUAV, J., DEGOTT, C., AND BENHAMOV, J.P. (1983). Dihydralazine hepatitis: report of a case and review of the literature. *Digestion* **27**, 47–52.

PARKER, D., BATE, C.M., CROFT, A.W., GRAHAM-POLE, J., MALPAS, J.S., AND STANSFELD, A.G. (1980). Liver damage in children with acute leukaemia and non-Hodgkin's lymphoma on oral maintenance therapy. *Cancer Chemother. Pharmacol.* **4**, 121–7.

PARKER, J.L.W. (1979). Potassium clorazepate (Tranxene)-induced jaundice. *Postgrad. med. J.* **55**, 908–10.

PARTIN, J.S., PARTIN, J.C., SCHUBERT, W.K., AND HAMMOND, J.G. (1982). Serum salicylate concentration in Reye's syndrome. *Lancet* i, 191–4.

PASQUA, P., CRAXI, A., AND PAGLIARA, L. (1982). Floctafenine and coma in cirrhosis. *Ann. intern. Med.* **96**, 253.

PATHY, M.S. AND REYNOLDS, A.J. (1980). Papaverine and hepatotoxicity. *Postgrad. med. J.* **56**, 488–90.

PETERSON, E.A., DAVID, W.A., AND KIRKPATRICK, H. (1980). Treatment of chronic mucocutaneous candidiasis with ketoconazole. *Ann. intern. Med.* **93**, 791–5.

PIETERSE, A.S., ROWLANDER, R., AND DUNN, D. (1983). Perhexiline maleate induce cirrhosis. *Pathology Australia* **15**, 201–3.

POUPON, R., ROSENSZTAJN, L., PRUDHOMME DE SAINT-MAUR, P., LAGERON, A., GOMBEAU, T., AND DARRIS, F. (1980). Perhexiline maleate-associated hepatic injury prevalance and characteristics. *Digestion* **20**, 145–50.

PRESCOTT, L.F. (1979). Poisoning with salicylates, paracetamol and other analgesics. *Prescribers' J.* **19**, 169–75.

——, ILLINGWORTH, R.N., CRITCHLEY, J.A.J.H., STEWART, M.J., ADAM R.D., AND PROUDFOOT, A.T. (1979). Intravenous N-acetylcysteine: the treatment of choice for paracetamol poisoning. *Br. med. J.* **2**, 1097–100.

—— (1982). Analgesic nephropathy: a reassessment of the role of phenacetin and other analgesics. *Drugs* **23**, 75–149.

——, LESLIE, P.J., AND PADFIELD, P. (1982). Side-effects of benoxaprofen. *Br. med. J.* **284**, 1783.

REX, D., LUMENG, L., EBLE, J., AND REX, L. (1983). Intrahepatic cholestasis and sicca complex after thiabendazole. Report of a case and review of the literature. *Gastroenterology* **85**, 18–21.

REYE, R.D.K., MORGAN, G., AND BARAL, J. (1963). Encephalopathy and fatty degenerative of the viscera: a disease entity in childhood. *Lancet* ii, 749–52.

RICCIONI, N., BOZZI, L., SUSINI, N., AND RONI, P. (1977). Disopyramide-induced intrahepatic cholestasis. *Lancet* ii, 1362.

ROBERTS, R.K., COHN, D., PETROFF, V., AND SENEVIRATNE.,B. (1981). Liver disease induced by perhexiline maleate. *Med. J. Aust.* **2**, 553–4.

ROOKS, J.B., ORY, H.W., ISHAK, K.G., STRAUSS, L.T., GREENSPAN, J.R., HILL, A.P., AND TYLER, C.W. (1979). Epidemiology of hepatocellular adenoma. The role of oral contraceptive use. *J. Am. med. Ass.* **242**, 644–8.

ROSENBAUM, J., KATZ, W.A., AND SCHUMACHER, H.R. (1980). Hepatotoxicity associated with use of D-penicillamine in rheumatoid arthritis. *Ann. Rheum. Dis.* **39**, 152–4.

ROTMENSCH, H., ROTH, A., LIRON, M., RUBINSTEIN, A., GEFEL, A., AND LIVNI, E. (1980). Lymphocyte sensitisation in nifedipine-induced hepatitis. *Br. med. J.* **281**, 976–7.

SCHONBERG, L.A. (1982). Peliosis hepatitis and oral contraceptives. A case report. *J. reprod. Med.* **27**, 753–6.

SEIBOLD, J.R., LYNCH, C.J., AND MEDSGER, T.A. (1981). Cholestasis associated with D-penicillamine. Case report and review of literature. *Arth. Rheum.* **24**, 554–6.

SHAR, S.R. AND KEW, M.C. (1982). Oral contraceptives and hepatocellular carcinoma. *Cancer* **49**, 407–10.

SHEIKH, H.H. AND NIES, A.S. (1983). Trazodone and intrahepatic cholestasis. *Ann. intern. Med.* **99**, 572.

SILLS, J.A., TREFOR-JONES, R.H., AND TAYLOR, W.H. (1980). Valproate hyperammonaemia and hyperglycinaemia. *Lancet* **ii**, 261–2.

SIMMON, F.R. (1978). Effects of estrogen on the liver. *Gastroenterology* **75**, 512–54.

SPARBERG, M., SIMON, N., AND DEL GRECO, F. (1969).Intrahepatic cholestasis due to azathioprine. *Gastroenterology*. **57**, 439.

STRUNIN, L. (1977). In *Side-effects of drugs annual* (ed. N.M.G. Dukes), pp. 186–98. Excerpta Medica, Amsterdam.

SUCHY, F.J., BALISTRERI, W.F., BUCHINO, J., SONDHEIMER, J., BATES, S.R., KEARNS, L.G., STULL, J.D., AND BOVE, K.E. (1979). Acute hepatic failure associated with the use of sodium valproate. *New Engl. J. Med.* **300**, 962–6.

SUGIMOTO, T., NISHIDA, N., YASUHARA, A., ONE, A., SAKANE, Y., AND MATSUMURA, T. (1983). Reye-like syndrome associated with valproic acid. *Brain Dev.* **5**, 334–7.

SULLIVAN, J.B., RUMACK, B.H., THOMAS, H., PETERSON, R.G., AND BRYSON, P. (1979). Pennyroyal oil poisoning and hepatotoxicity. *J. Am. med. Ass.* **242**, 2873–4.

SUSSMAN, N.M. AND McLAIN, L.W. (1979). A direct hepatoxic effect of valproic acid. *J. Am. med. Ass.* **242**, 1173–4.

TAGGART, H.M. AND ALDERDICE, J.M. (1982). Fatal cholestatic jaundice in elderly patients taking benoxaprofen. *Br. med. J.* **284**, 1372.

TALBOT, J. AND BEELEY, L. (1980). Fusidic acid and jaundice. *Br. med. J.* **281**, 308.

TESLUK, H. AND LAWRIE, J. (1981). Hepatocellular adenoma: its transformation to carcinoma in a user of oral contraceptives. *Arch. Pathol. Lab. Med.* **105**, 296–9.

TIMBRELL, J.A. AND HARLAND, S.J. (1979). Identification and quantitation of hydrazine in the urine of patients treated with hydrallazine. *Clin. Pharmacol. Ther.* **26**, 81–8.

TOTH, B. (1976). Tumorigenicity of beta-phenylethylhydrazine sulfate in mice. *Cancer Res.* **36**, 917–20.

TSUNG, S.H., HAN, D., LOH, W.P., AND LIN, J.I. (1980). Budd–Chiari syndrome in women taking oral contraceptives. *Ann. Clin. Lab. Sci.* **10**, 518–22.

VANA, J. AND MURPHY, G.P. (1979). Primary malignant liver tumours: association with oral contraceptives. *NY State J. Med.* **79**, 321–5.

VANDENBURG, M., PARFREY, P., WRIGHT, P., AND LAZDA, E. (1980). Hepatitis associated with captopril. *Br. J. clin. Pharmacol.* **11**, 105–6.

——,——,——,——. (1982). Hepatitis associated with captopril treatment. *Br. J. Clin Pharmacol.* **11**, 105–106.

VAUGHAN, W.P., WILCOX, P.M., ANDERSON, P.O., ETTINGER, D.S., AND ABELOFF, M.D. (1979). Hepatic toxicity of adjuvant chemotherapy for carcinoma of the breast. *Med. paediat. Oncol.* **7**, 351–9.

VERGANI, D., TSANTOULAS, D., EDDLESTON, A.L.W.F., DAVIS, M., AND WILLIAMS, R. (1978). Sensitisation to halothane-altered liver

components in severe hepatic necrosis after halothane anaesthesia. *Lancet* **ii**, 801.

VICKERS, C.F.H. AND MENDAY, A.P. (1980). Fusidic acid and jaundice. *Br. med. J.* **281**, 308.

VICTORINO, R.M.M., SILVEIRA, J.C.B., BAPTISTA, A., AND DE MOURA, M.C. (1980). Jaundice associated with naproxen. *Postgrad. med J.* **56**, 368–70.

WALTON, B., SIMPSON, B.R., STRUNIN, L., DONIACH, D., PERRIN, J., AND APPLEYARD, A.J. (1976). Unexplained hepatitis following halothane. *Br. med. J.* **1**, 1171.

WALTON, N.G., MANN, T.A.N., AND SHAW, K.M. (1979). Anaphylactoid reaction to N-acetylcysteine. *Lancet* **ii**, 1298.

WARE, S. AND MILLWARD-SANDLER, G.H. (1980). Acute liver disease associated with sodium valproate. *Lancet* **ii**, 1110–13.

WARK, H.J. (1983). Postoperative jaundice in children. The influence of halothane. *Anaesthesia* **38**, 237–42.

WARREN, N.S. (1978). Diflusinal-induced cholestatic jaundice. *Br. med. J.* **2**, 736–7.

WEISS, M., HASSIN, D., AND BAK, H. (1980). Prophylthiouracil-induced hepatic damage. *Arch. intern. Med.* **140**, 1184–5.

WENK, R.E., GEBHARDT, F.D., BHAGAVAN, B.S., LUSTGARTEN, J.A., AND MCCARTHY, E.F. (1981). Tetracyline-associated fatty liver of pregnancy, including possible pregnancy risk after chronic dermatological use of tetracycline. *J. reprod. Med.* **26**, 135–41.

WESTABY, D., PORTMANN, B., AND WILLIAMS, R. (1983). Androgen-related primary hepatic tumours in non-Fanconi patients. *Cancer*, United States **51**, 1947–52.

WHITTAKER, S.J., AMAR, J.N., WANLESS, I.R., AND HEATHCOTE, J. (1982). Sulindac hepatotoxicity. *Gut* **23**, 875–7.

WIGGINS, J. AND SCOTT, D.L. (1981). Hepatic injury following feprazone therapy. *Rheumatol. Rehabil.* **20**, 44–5.

WOLFE, P.B. (1979). Sulindac and jaundice. *Ann. intern Med.* **91**, 656.

WONG PAITOON, V., MILLS P.R., RUSSELL, R.I., AND PATRICK, R.S. (1981). Intrahepatic cholestasis and cutaneous bullae associated with glibenclamide therapy. *Postgrad. med. J.* **57**, 244–6.

YANOFF, M. AND RAWSON, A.J. (1964). Peliosis from cavernous haemangioma. *Arch. Pathol.* **77**, 159–62.

YOUNG, R.S.K., BERGMAN, I., AND GANG, D.L. (1980). Fatal Reye-like syndrome associated with valproic acid. *Ann. Neurol.* **7**, 389.

ZACHARIAE, H., KRAGBALLE, K. AND SOGAARD, H. (1980). Methotrexate-induced liver cirrhosis. *Br. J. Dermatol.* **102**, 407–12.

ZIMMERMAN, H.J. (1978). Drug-induced liver disease. In *Hepatotoxicity: the adverse effects of drugs and other chemicals on the liver*, pp. 349–69. Appleton-Century-Crofts, New York.

ZIMMERMAN, H.J. (1981). Drug hepatotoxicity: Spectrum of clinical lesions. In *Drug reactions and the liver* (ed. M. Davis, J.M. Tredger, and R. Wilkins), pp. 35–53. Pitman Medical, London.

ZIMRAN, A., ABRAHAM, A.S., AND HERSHKO, C. (1983). Reversible cholestatic jaundice and hyperamylasaemia associated with captopril treatment. *Br. med. J.* **287**, 1676.

18 Drug-induced disorders of carbohydrate and fat metabolism

J.P. GRIFFIN

Drug-induced hyperglycaemia

A large number of drugs in various therapeutic classes have been reported to induce hyperglycaemia with or without ketosis.

Anti-neoplastic drugs

Asparginase (colopase)

Hyperglycaemia, diabetic glucose tolerance curves, decreased serum immunoreactive insulin, and clinical diabetes without ketosis attributable to asparginase have been described by several authors. Two cases of diabetic ketoacidosis were also reported (Gillette *et al.* 1972; Aussanaire *et al.* 1972). Pui *et al.* (1981) set out to determine the risk factors for the development of hyperglycaemia in patients treated with asparginase (colopase). A retrospective study of 421 children with leukaemia given colopase (10 000 units/m/dose, on three different dosage schedules), prednisone (40 mg/m daily), and vincristine (1.5 mg/m dose weekly, plus daunorubicin in some) as remission induction therapy, identified several risk factors for hyperglycaemia in these children. Forty-one children (9.7 per cent) developed hyperglycaemia, 39 of them within one week of the first dose of colopase. Ages of the 421 children studied ranged from 3 months to 20 years (mean 6.8 years) but children over 10 years old were more at risk than younger children. A family history of diabetes mellitus also increased the risk; hyperglycaemia occurred in 19 children (20 per cent) of those with such a history but in only 20 (6.3 per cent) of the remainder. Obesity and Down's syndrome were also positively associated with development of hyperglycaemia. In patients with more than one risk factor the frequency of hyperglycaemia increased sharply. Hyperglycaemia resolved in all patients; in 32, hyperglycaemia resolved despite continued treatment. Close monitoring for glycosuria is recommended during treatment with colopase in patients with the risk factors shown by this study.

Cyclophosphamide

Development of diabetes mellitus was described by Pengelly (1965) in three women who were being treated with cyclophosphamide for carcinoma of the breast.

Antitubercular drugs

Isoniazid

Glycosuria and hyperglycaemia during treatment with isoniazid have been reported (Dickson 1962). In a group of 50 patients receiving isoniazid, three developed glycosuria and two of these presented all the classical characteristics of true diabetes. The urine of the third reverted to normal on the cessation of isoniazid therapy. There was no pre-existent diabetes in any of these cases.

Rifampicin

An interesting study from Japan by Takasu *et al.* (1982) demonstrated early-phase hyperglycaemia, associated with increased rates of insulin and C-peptide secretion after oral administration of 100 g glucose, which was observed among patients with pulmonary tuberculosis who were taking rifampicin. This early-phase hyperglycaemia appeared shortly after rifampicin was started and it disappeared completely a few days after rifampicin was discontinued. No difference in oral glucose tolerance was noted between healthy normal subjects and patients with pulmonary tuberculosis who were not taking any medication.

Antituberculous drugs other than rifampicin did not induce early-phase hyperglycaemia. Because intravenous glucose tolerance was normal in patients treated with rifampicin, it is suggested that rifampicin produces an early-phase hyperglycaemia possibly by augmenting intestinal absorption of glucose. It is doubtful whether these changes have any pathological significance.

Diuretics

The fact that diuretics could disturb carbohydrate metabolism and induce hyperglycaemia was recognized soon after the introduction of chlorothiazide, and this effect has been noted in almost every diuretic introduced subsequently. The hyperglycaemia so caused can be severe and hyperosmolar non-ketotic diabetic coma has been reported in patients treated with the following agents: diazoxide, chlorothiazide, hydrochlorothiazide, bendrofluazide, chlorthalidone, and frusemide (Dargie and Dollery, 1975).

Thiazides (benzothiadiazines)

The possibility that thiazide diuretics can induce a diabetes-like state was first noted by Finnerty (1959) and Wilkins (1959), who reported that hyperglycaemia and glycosuria developed in patients taking such drugs. Other reports soon appeared relating the aggravation of known diabetes, or the uncovering of 'latent diabetes' to the administration of thiazide compounds. Zatuchni and Kordasz (1961) reported the effects of a single dose of chlorothiazide (1 g) or trichlormethiazide (8–16 mg) on the two-hour post-prandial blood sugar level in 25 patients hospitalized for reasons other than diabetes. Fourteen (56 per cent) of the 25 patients had abnormal (above 120 mg/100 ml) two-hour blood glucose levels after the thiazide dosage. However, it must be noted that 10 (40 per cent) had abnormally high levels even prior to administration of the diuretic, and that six of these worsened after the single-dosage treatment. Only four out of the 25 patients showed an apparent alteration from normal to abnormal blood glucose levels.

Further investigations on this effect were carried out by Goldner and colleagues (1960), who also studied a group of hospitalized patients, 20 of whom were diabetic and 20 of whom were non-diabetic with no family history of the disease. The drugs administered in this study were chlorothiazide (500–1000 mg/day), hydrochlorothiazide (100–300 mg/day), and dihydroflumethiazide (20 mg/day). Each drug was given for at least 5 days. None of the non-diabetic patients, and only six of the diabetic patients, showed an increase in fasting blood-sugar level following treatment, and such effects were only seen at the higher levels of dosage.

In a similar study, Runyan (1962) found that benzthiazide in large doses (50 mg/t.i.d.) for one week resulted in a 45 per cent increase in fasting blood sugar level of patients with mild uncomplicated diabetes. When dosage was reduced to 50 mg daily the increase in fasting blood-sugar level was only 14 per cent. Shapiro and colleagues (1961) studied the effects of thiazides on car-

bohydrate metabolism in middle-aged and elderly hypertensive patients and showed that chlorothiazide (1 g/day) for two weeks raised the fasting blood sugar level in patients who were judged as 'potential diabetics', due to family history of diabetes or to previous abnormal glucose-tolerance tests, but did not affect non-diabetic control subjects.

The effect of long-term administration of thiazide diuretics on carbohydrate metabolism has also been studied by a number of clinicians. Roediger and associates (1964) compared the effects of 3 months chlorothiazide therapy (1 g/day) with 3 months bendroflumethiazide treatment (10 mg/day) in 35 patients under treatment for hypertension and diabetes mellitus. Fasting blood-sugar levels were raised to comparable levels (147–160 mg/100 ml) with either thiazide. Wolff et al. (1963) focused attention on the long-term effects of benzothiadiazine on hypertensive patients. They reported a mean fasting blood glucose level of 88.4 mg/100 ml in patients receiving placebo dosage and a value of 97 mg/100 ml in those patients on the thiazide diuretic. Five patients developed frank diabetes mellitus whilst on thiazide therapy, and, in two of these patients, a return to normal glucose levels followed after treatment was discontinued. The remaining three patients were not obese, had no family history of diabetes, and, prior to the thiazine treatment, had normal carbohydrate tolerance. Apparently these patients suffered permanent damage in their ability to handle carbohydrate.

Brown and Brown (1967) studied 11 normal men with negative family histories of diabetes who received 50 mg hydrochlorothiazide daily for 5 months. At the end of that time one man showed diminished ability to utilize carbohydrates. That individual, plus four others, continued taking the drug for a total of 14 months, at which time three of the five displayed an abnormal oral glucose-tolerance test. Thus it would appear that thiazide-induced carbohydrate intolerance is not limited to persons with other disease states or with a family history of diabetes but can occur in normal individuals.

The effect of thiazide compounds on oral glucose tolerance in pregnancy was investigated by Sugar (1961) who reported aggravation of known diabetes during pregnancy by thiazide diuretics. Ketoacidosis occurred in two pregnant patients, one receiving chlorothiazide (50 mg/day) and the other benzhydroflumethiazide (5 mg/day). In contrast to this Esbenshade and Smith (1965) did not observe any increase in the two-hour post-prandial glucose level in 23 patients in the third trimester who had received hydrochlorothiazide 50 mg twice per day for 7 days in spite of the fact that four of these patients had a family history of diabetes mellitus.

Lewis et al. (1976) reported the results of a study in

which glucose-tolerance tests were performed in 137 patients before diuretic therapy and after 1 year of continuous treatment with one of four diuretics, clorexolone, chlorothiazide, chlorthalidone, or frusemide. Six years later these 137 patients were followed up and asked to attend for a further glucose-tolerance test. Sixty-seven patients were studied; 51 had been on continuous diuretic therapy for the 6 years plus (mean duration of treatment 80 ± 1 month), and 41 of these patients were on bendrofluazide. Sixteen patients had discontinued diuretic treatment after a mean of 42 ± 2 months.

In both groups of patients the glucose-tolerance test was normal after 1 year's treatment, but in those on continuous treatment for 6 years there had been a significant deterioration in glucose tolerance. In those patients in whom diuretic therapy had been discontinued after about three and a half years the glucose tolerance tested 6 years after the commencement of diuretic therapy had not deteriorated.

These workers concluded that the magnitude of the diabetogenic effect was not large and in 80 months treatment none of the 51 patients had developed clinical diabetes but 11 patients had developed a diabetic glucose-tolerance curve. It would appear that these results indicate that length of treatment may be significant in the manifestation of the diabetogenic effect of thiazide diuretics.

In a survey by Amery et al. (1978) 119 elderly, hypertensive patients were followed-up for 1 year and 48 for 2 years in a double-blind, randomized, controlled trial in which they received either placebo or 25–50 mg hydrochlorothiazide and 50–100 mg of triamterene daily. Half of the active treatment group also received 250 mg to 2 g methyldopa daily. After 2 years the active treatment group had an average increase in fasting blood-sugar of 9.6 mg/dl compared with an average fall of 3.1 mg in the placebo group ($p < 0.001$).

Blood-glucose rose by an average of 26.6 mg/dl in the active group when determined 1 hour after 50 g oral glucose and decreased by an average of 5.3 mg/dl in patients who had been on placebo for two years ($p < 0.05$). The hyperglycaemic effect of diuretics appeared to be related to potassium loss since, in both groups, impairment of glucose tolerance was most marked in those in whom the serum-potassium decreased over the 2 years.

The mechanism of thiazide-induced diabetes would appear to be complex, and there is evidence for both a pancreatic and a peripheral mechanism. Patients who develop moderately severe diabetes while on diuretics have a low plasma insulin concentration. The plasma insulin has been shown to rise in such cases after the withdrawal of the diuretic (Samaan et al. 1963). Intravenous

hydrochlorothiazide or frusemide lowers the serum insulin while having little effect on blood glucose (Walkhaus 1967). This effect appears to be due to an action on the release of insulin from the pancreas. A peripheral action is indicated by the fact that chlorothiazide directly inhibits the uptake of glucose into rat adipose tissue incubated in vitro (Weller and Borondy 1967).

Marks et al. (1981) reported on a study of 40 patients who had been treated with hydrochlorothiazide for the control of their hypertension for a minimum period of 10 years. A glucose-tolerance test (100 g glucose) was conducted on each of these patients, and an abnormal glucose tolerance test was defined as a fasting level in excess of 6.1 mmol/l or a 2-hour value above 6.6 mmol/l. Thirty-six patients had completely normal glucose-tolerance curves and four had diabetic-type curves. Three other patients were obese with hyperinsulinaemia, having insulin values three times normal. Those four patients with abnormal glucose-tolerance curves all had a family history of diabetes and in two cases were obese and had cardiac failure. Marks et al. concluded that patients taking thiazides are at particular risk of developing diabetes if they have a family history of diabetes or were obese, or had cardiac failure. Marks et al.'s conclusions are completely in accord with a much earlier study by Wolff et al. (1963). Wolff and his colleagues took into account the patients' family history. Of the 45 patients studied, 11 were found to have a diabetic type of glucose-tolerance test; of these, five had a family history of diabetes mellitus. Of the remaining six patients, four were obese, with mean weight of 175 kg. Thus, only two out of 45, or 8.9 per cent, developed diabetes without a recognized predisposition. Many patients were in cardiac failure at some time and required a multiplicity of other drugs in addition to their diuretic therapy.

In a study by Dollery and colleagues (Murphy et al. 1982) at the Hammersmith Hospital, glucose tolerance was studied in a prospective investigation of 34 hypertensive patients treated with oral thiazide diuretics for 14 years without any interruption of their treatment. Standard oral glucose tolerance tests were carried out before treatment and after 1, 6, and 14 years. Mean fasting blood glucose increased from 4.7 ± 0.6 to 6.0 ± 2.3 mmol/l and the 2-h value rose from 5.5 ± 1.9 to 8.0 ± 4.7 mmol/l after 14 years. Withdrawal of thiazide therapy for 7 months in 10 of the patients resulted in mean reductions of 10 per cent in fasting blood glucose and 25 per cent in the 2-h value. Murphy et al. (1982) concluded that 'it is possible that the long-term use of thiazides, while reducing mortality from cerebrovascular disease and renal impairment, could be increasing the incidence of coronary vascular disease. The development of glucose intolerance during long-term thiazide treatment suggests that we may be substituting one long-term cardiovascular risk factor for another'. The basis for this conclusion was that in the Whitehall study, a long-term prospective study of the health of a large number of civil servants, Fuller *et al.* (1980) reported a higher incidence of coronary-artery disease in subjects with symptomless glucose intolerance than in subjects with normal glucose tolerance. In another 10-year prospective study, Jarrett *et al.* (1982) found a higher coronary-heart-disease mortality rate in patients with glucose intolerance than in normal controls.

Diazoxide

Diazoxide was the outcome of research to separate the antihypertensive and diuretic properties of clorothiazide, to which it has a close structural resemblance. There are, however, important differences in the pharmacology of chlorothiazide and diazoxide. Diazoxide has no diuretic properties and produces a sharp and immediate fall in blood pressure when given intravenously, which chlorothiazide does not.

Diazoxide causes a predictable hyperglycaemia when administered to animals or man and it has therefore proved easier to study the mechanism of the diabetogenic effect with it than with the thiazide diuretics. It may be unwise to assume that the diabetogenic action of diazoxide and the thiazide diuretics are the same because their pharmacological actions are different.

Diazoxide is a potent diabetogenic agent in humans and its effect is increased by combining it with a thiazide diuretic. Dollery *et al.* (1962) reported two patients who were treated with diazoxide 400 mg daily for hypertension. As the patients became oedematous, hydrochlorothiazide was added to treatment in a dose of 50 mg daily. In each case after about four weeks' treatment the patients became diabetic with high fasting blood sugar levels. One of these patients was known to have a normal glucose-tolerance test before diazoxide was given and the glucose-tolerance test returned to normal 17 days after stopping the drugs. The plasma insulin-like activity, assayed by the rat epididymal-fat-pad technique, was reduced during the diabetic period.

Okun *et al.* (1963) used diazoxide and trichlormethiazide together to treat hypertension and found that approximately half the patients developed hyperglycaemia. Subsequently, it was shown that this effect could be reproduced in animals. Kyam and Stanton (1964) gave diazoxide, 100–400 mg/kg to fasting rats. A rise in blood glucose followed, which could be antagonized by administration of tolbutamide. Serum insulin-like activity in rats receiving diazoxide was not significantly

lower than in control rats, though a marked hyperglycaemia had been produced. A β-adrenergic blocking drug (MJ 1999) was capable of modifying the hyperglycaemia produced by diazoxide. Potassium deficiency in the rats strikingly exaggerated the hyperglycaemic action of diazoxide, while potassium supplements (KCl 750 mg/kg) reduced it.

Tabachnick *et al.* (1964) carried out an extensive investigation into the pharmacological basis of the hyperglycaemia produced by diazoxide. They produced hyperglycaemia in de-pancreatized dogs, alloxan-treated mice, propylthiouracil-treated mice, and nephrectomized mice. These observations suggested that the main locus of action was not in the pancreas, thyroid, or kidneys. Diazoxide hyperglycaemia was partly suppressed by adrenalectomy, hypophysectomy, or administration of isopropyl-methoxamine, a specific inhibitor of the metabolic effects of adrenaline. However, some doubt is cast upon the importance of the participation of the sympathetic nervous system in this effect, since reserpine-treated mice respond to diazoxide with hyperglycaemia. These previous observations suggest that diazoxide diabetes is produced by an extra-pancreatic action.

Seltzer and Allen (1965) studied eight normal subjects given 600 mg of diazoxide and 8 mg of trichlormethiazide daily for 5 days. Blood glucose and plasma insulin (immuno-assay) were determined after oral and intravenous glucose loads at the end of the 5-day period on the drugs and again 2 weeks later. Diazoxide raised the fasting, 60- and 120-minute blood sugars and reduced the rise in plasma insulin produced by the glucose load. The ratios of total insulin response to total glycaemia stimulus were 1.81 in controls and 0.4 during diazoxide. In the intravenous glucose-tolerance tests plasma insulin was maximal at 5 minutes in controls and then paralleled the falling blood sugar. Diazoxide almost completely inhibited insulin release under these circumstances. Seltzer and Allen (1965) concluded that diazoxide diabetes stems from the same functional defects which typify spontaneous diabetes. There is initially a sluggish response to acute stimulus and proportionately less insulin production per unit of secretory stimulus. Since the abnormalities produced in normal controls could rapidly be reversed by withdrawal of the drugs they suggested that this might prove a useful experimental model for further study of human diabetes.

Diazoxide has been used to treat children with leucine-sensitive hypoglycaemia and patients with insulin-producing tumours (Unger 1966). Drash and Wolff (1964) found that diazoxide in doses of 12 mg/kg per day elevated the blood glucose response to oral glucose in a 4-year-old boy with leucine-sensitive hypoglycaemia.

Samols and Marks (1966) reported similar observations in two children with leucine-sensitive hypoglycaemia. It is of interest that this unwanted and potentially dangerous hyperglycaemic effect of diazoxide should have been put to good use in treating patients with intractable hypoglycaemic attacks.

There is very strong evidence that diazoxide has both a peripheral and pancreatic action. Its ability to produce hyperglycaemia in de-pancreatized or alloxan-diabetic animal proves the peripheral action, while the alterations in plasma insulin produced in intact animals, or slices of pancreas, suggest an effect on insulin formation or release. Howell and Taylor (1966) showed that diazoxide inhibits insulin secretion from the rabbit pancreas *in vitro* in response to glucose.

Smith *et al.* (1982) reported the development of neonatal hyperglycaemia after prolonged maternal treatment with diazoxide. The report of Smith *et al.* is the first report of transplacental diabetogenesis with this agent.

Frusemide

Impairment of glucose tolerance also occurs with frusemide, an oral diuretic which is structurally dissimilar to the benzothiadiazines (thiazides). Frusemide is a member of a group of monosulphamyl diuretics, which also includes guinethazone, chlorthalidone and clorexolone, while ethacrynic acid is a related desulphamyl compound.

Konigstein (1965) showed that five of 11 non-diabetic obese subjects showed deterioration of glucose tolerance following an oral dosage of 75 mg frusemide daily for 5 days, although two diabetic subjects who showed deterioration of glucose tolerance on thiazide therapy showed no impairment while taking frusemide 20–25 mg daily for many months. Toivonen and Mustala (1966) described deterioration of glucose tolerance in one patient treated with frusemide 80 mg daily for 4 weeks; the glucose tolerance reverted to normal after cessation of therapy. Walkhaus (1967) demonstrated a fall in serum insulin levels following intravenous frusemide.

Ethacrynic acid

Ethacrynic acid is a related desulphamyl compound, and although ethacrynic acid has been implicated in impaired glucose tolerance, the evidence for this is somewhat controversial. There are conflicting opinions as to whether or not this diuretic is diabetogenic.

Jones and Pickens (1967) described a single case of a non-diabetic woman whose glucose tolerance had become impaired after thiazide therapy, but had returned to normal after treatment ceased. Diuretic treatment was then recommended with ethacrynic acid and her glucose tolerance again became impaired; diuretic treatment was again discontinued with a return to a normal glucose-tolerance curve. Jones and Pickens (1967) suggested that any patient, who has shown glycosuria and diminished glucose tolerance on thiazide diuretics, should be carefully monitored when placed on an ethacrynic-acid diuretic regime.

Feldman and Diamond (1967) studied 15 patients, some of whom had normal glucose tolerance and others who had mild or severe diabetes. Glucose tolerance studies were done prior to starting treatment with ethacrynic acid and were repeated after 1 month and 3 months of continuous modication. Only one patient, a non-diabetic, showed any impairment of glucose tolerance. Glucose tolerance reverted to normal in this patient after withdrawal of the diuretic.

Lebacq and Marcq (1967) studied 24 normal subjects and 15 diabetics during ethacrynic acid treatment. Carbohydrate intolerance was induced or worsened by the diuretic in eight of the 24 normal subjects and eight of the 15 diabetics. Serum amylase levels were unchanged during treatment. Ethacrynic acid resulted in lower insulin levels in the fasting state and reduced insulin response to glucose load.

A study by Dige-Petersen (1966) indicated that ethacrynic acid did not adversely affect glucose metabolism but the balance of evidence is that it does.

Mefruside

Mefruside is a relatively new oral sulphonamide diuretic with a structural similarity to frusemide. Mefruside has been shown to cause hyperglycaemia in animals (Wales *et al.* 1968).

Metolazone

Metolazone is a quinethazone derivative and has been relatively recently introduced; it has been shown to cause mild impairment of glucose tolerance in some patients (Bennett and Porter 1973).

Spironolactone

Jariwalla *et al.* (1981) described the case history of a 57-year-old insulin-dependent diabetic man with alcoholic cirrhosis who following a minor surgical procedure developed ketoacidosis and life-threatening hyperkalaemia which was attributed to spironolactone.

Tienilic acid

Tienilic acid was withdrawn from clinical use because of reported hepatotoxicity. During a clinical trial of tienilic acid (Lund-Johansen 1981) one of 14 patients in the study developed diabetes which persisted several weeks after the withdrawal of the drug.

Triamterene

In-vitro studies have indicated that lowered potassium levels may impair the pancreatic insulin release. A study does exist that suggests that potassium administration can reverse diuretic-induced hyperglycaemia (Rapoport and Hurd 1964); this has not been substantiated.

Triameterene, which is a potassium-sparing diuretic, can also induce hyperglycaemia (Wales *et al.* 1968) and studies of glucose tolerance during states of potassium depletion have shown no abnormality (Kaess *et al.* 1971).

Hyperosmolar non-ketotic diabetic syndrome caused by diuretics

Fonseca and Phear (1982) reviewed the case notes of 11 consecutive patients admitted with the hyperosmolar non-ketotic diabetic syndrome to determine how often it was precipitated by diuretic treatment. The criteria used for diagnosis were severe hyperglycaemia (blood glucose greater than 30 mmol l^{-1} (541 mg/100 ml), absence of ketoacidosis, dehydration, variable neurological signs, including depressed sensorium or frank coma, and serum osmolarity above 340 mOsm/kg (calculated using the formula; osmolarity (mOsm/kg) = 2 × (serum sodium + potassium) + blood glucose + blood urea). The results of their study are shown in Table 18.1. Eight of the 11 patients were receiving diuretic therapy. Two patients died, one from a myocardial infarction. Of the survivors, four required insulin for control of diabetes and two were controlled on diet alone, the rest being controlled with oral hypoglycaemics in modest dosage. Two of the four patients needing insulin did not develop the syndrome while on diuretics (cases 2 and 3). None of the patients were given diuretics after recovery and hyperosmolar non-ketotic diabetes did not recur. (Eight patients were followed up for at least a year.)

The syndrome is a well-known complication of thiazide diuretics, but it was initially thought that frusemide only caused the syndrome when given in enormous doses. In Fonseca and Phear's series, three cases occurred with doses of 40–80 mg frusemide per day.

Bumetanide and indapamide have not previously been reported to cause hyperosmolar non-ketotic diabetic syndrome.

Beta-blockers have also been reported by Podolsky (1970) as precipitating the hyperosmolar non-ketotic diabetic syndrome and three of Fonseca and Phear's series were receiving a beta-blocker plus a diuretic.

Nardone and Bouma (1979) drew attention to the risk of diabetic coma developing in patients receiving thiazides and propranolol therapy and document two such cases in men aged 52 and 55 years, respectively. The younger man had developed diabetic ketoacidosis, the other had diabetic hyperosmolar coma and was similar to the cases described by Fonseca and Phear.

The patients reported by Fonseca and Phear were all elderly and these investigators advise that 'antihypertensive treatment with diuretics should be started carefully and a close watch kept on these patients over the next few months for the development of diabetes. Early intervention at the onset of symptoms may prevent the patient from reaching the late stage of hyperosmolar coma with its associated high mortality'. To which warning should be added the advice — do not be too hasty in adding a beta-blocker to this regime.

The mortality of hyperosmolar non-ketotic diabetic syndrome is high in all series.

Anticonvulsants — phenytoin

Millichap (1965) first showed that diphenylhydantoin and other anticonvulsants could cause hyperglycaemia. Belton *et al.* (1965) confirmed, in rabbit experiments, that diphenylhydantoin and phenobarbitone at anticonvulsant doses caused a significant rise in blood sugar.

Millichap (1969) records that, in view of the hyperglycaemic action of diphenylhydantoin, this drug could be used to control the hypoglycaemic seizures seen in leucine-sensitive hypoglycaemia. Over a 12-month period, diphenylhydantoin successfully maintained a normal fasting blood-sugar level in a child with leucine-sensitive hypoglycaemia and controlled the child's convulsions. The abnormal hypoglycaemic response to leucine was not prevented in a subsequent sensitivity test, but the child did not convulse.

The mechanism of the hyperglycaemic action of diphenylhydantoin has been variously suggested as being due to a direct action of diphenylhydantoin on the hypothalamus (Belton *et al.* 1965) to stimulation of the pituitary and the adrenal cortex (Woodbury 1952), or to an impairment of the insulin response to carbohydrate, i.e. causing a relative hypoinsulinaemia (Peters and Samaan 1969).

Table 18.1 Details of 11 patients who developed hyperosmolar non-ketotic diabetes after diuretic treatment (after Fonseca and Phear 1982)

Case no.	Previous diabetes	Age (years) sex	Drugs	Duration of treatment	Blood glucose (mmol/l)	Na (mmol/l)	K (mmol/l)	Blood urea (mmol/l)	HCO_3 (mmol/l)	Osmolarity	Outcome
1	No	60/F	Moduretic 2 tablets daily	3 wk	3.2	156	6	50	24	406	Recovered; diabetes controlled on diet
2	No	71/F	—	—	75	136	5	43	24	400	Recovered; diabetes controlled on insulin
3	No	64/M	—	—	39	156	4	22	27	381	Recovered; diabetes controlled on insulin
4	No	75/F	Indapamide 2.5 mg	3 mnth	42	147	4.3	24	22	369	Recovered; diabetes controlled by diet
5	No	68/M	Propranolol 80 mg qds; bumetanide	1 y	115	143	5	17	22	428	Recovered; controlled on insulin and propranolol
6	No	79/F	Cyclopenthiazide 0.5; oxprenolol 80 mg	2 mnth	34	144	4.5	19	25	350	Recovered; controlled on metformin and labetalol
7	Yes (on diet)	70/F	Prednisolone 20 mg	2 wk	45	144	5	18	24	361	Recovered; controlled on phenformin
8	Yes (chlorpropamide)	75/F	Frusemide 80 mg	1 wk	36	142	5	15	22	345	Died of myocardial infarction
9	Yes (on diet)	67/F	Acetazolamide 250 mg	10 d	45	147	4	18	22	365	Recovered; controlled on chlorpropamide
10	Yes (on diet)	75/F	Frusemide 40 mg	10 d	43	153	5	21	27	380	Recovered; controlled on insulin
11	Yes (metformin)	70/F	Frusemide 40 mg; oxprenolol 80 mg	6 mnth	37	170	5.5	31	23	419	Died

Conversion: SI to traditional units — Blood glucose: 1 mmol/l $\approx$ 18 mg/100 ml. Na, K, HCO_2: 1 mmol/l = 1 mEq/l. Blood urea: 1 mmol/l $\approx$ 6 mg/100 ml.

In contrast there is no evidence that prolonged administration of phenytoin leads to any increase in the incidence of diabetes in epileptics, although few systemic studies have been carried out. In a group of children on chronic diphenylhydantoin therapy, no significant differences were found in glucose tolerance compared with normal children of similar age. Also, in a study on normal adults not taking any other drugs, no significant deterioration in oral glucose tolerance was observed after a standard dose of phenytoin had been taken for 10 days, but insulin response showed a significant reduction which, although it varied between subjects, directly correlated with the serum concentration of the drug. Both phenytoin and diazoxide (a non-diuretic thiazide) have similar effects on the β-cells. *In-vitro* studies have confirmed that both these drugs can inhibit pancreatic insulin release; with diazoxide this inhibition can be reversed by administration of tolbutamide, whereas with phenytoin it is not reversible — suggesting that this drug interferes with insulin synthesis as well as with insulin release (Levin *et al.* 1972).

Following observations that some patients on phenytoin had hypercholesterolaemia, the effect of the drug on plasma lipids was studied in normal volunteers. Six volunteers were given 300 mg of phenytoin daily for 12 weeks or placebo for 12 weeks and then crossed over to the alternative preparation. Blood lipids were measured weekly. In another study five volunteers received placebo for 1 week and then received 300 mg phenytoin daily for 4 weeks, and blood lipids were monitored twice weekly. Both these studies showed that phenytoin caused increases in levels of both cholesterol and plasma trigylcerides (Jubitz *et al.* 1974).

A case of diabetic ketoacidosis in a 64-year-old black woman with maturity-onset diabetes receiving phenytoin for a seizure disorder was reported by Carter *et al.* (1981). The woman was admitted to the hospital with a one-day history of polyuria and polydipsia. For the 10 months before admission, her diabetes was controlled with isophane insulin suspension 27 units daily. She also took phenytoin 100 mg orally three times a day. This was prescribed approximately six weeks earlier for right-sided focal seizures that were detected by electro-encephalogram during a previous hospitalization for non-ketotic hyperosmolar coma. No other medications were taken. The patient was treated with intravenous fluids and intermittent doses of i.v. insulin. Her condition rapidly improved and insulin zinc suspension 35 units daily was prescribed on discharge. Phenytoin was discontinued because the seizure disorder was considered secondary to the previous episode of hyperosmolar coma. Phenytoin-induced hyperglycaemia is not uncommon, but ketosis associated with phenytoin is much rarer.

Hyperglycaemia induced by steroidal hormones

Corticosteroids

Most patients who develop steroid diabetes have a predisposition to diabetes mellitus as shown by a positive family history of the disease. It is rare for a steroid-induced diabetic to develop ketosis, but this has been described by Blereau and Weingarten (1964). In this case, a 29-year-old man was given corticosteroid therapy for pulmonary sarcoidosis; his glucose-tolerance test before steroid therapy was normal. He developed severe hyperglycaemia (520 mg 100 ml), acidosis, and ketosis. After withdrawal of steroids his glucose tolerance test returned to normal. Schubert and Schulte (1963) reported a series of 214 patients treated with corticosteroids for periods in excess of 3 days: 14 per cent developed steroid diabetes. Disturbances of carbohydrate metabolism were observed earlier in patients on high doses of steroids or who had liver disease which retarded the elimination of the steroid as a glucuronide. Matsunaga *et al.* (1963) found that of 235 patients treated on steroid therapy 5.5 per cent developed diabetes. No patients on corticotrophin or cortisone developed steroid diabetes although 7 per cent of those patients treated with prednisolone, 23 per cent of those treated with paramethasone, and 20 per cent of those treated with betamethasone developed diabetes.

Systemic administration or topical application of corticosteroids to a diabetic under good control may disturb carbohydrate balance and the requirements for insulin or oral hypoglycaemic agents may increase. Kershbaum (1963) reported four cases of diabetes treated with topical fluocinolone acetonide for skin conditions, who experienced an increase in insulin or tolbutamide requirements after commencing this therapy.

A further complication of steroid therapy in diabetic patients is the development of ketonuria without accompanying glycosuria. This was reported by Madison (1964) who observed this in four diabetic patients who were treated with topical triamcinolone acetonide for necrobiosis lipoidica. No explanation for this phenomenon is known.

Gomez and Frost (1976) described the case of a 30-year-old female psoriatic patient who had an abnormal glucose-tolerance test without other evidence of diabetes who then developed post-prandial hyperglycaemia and glycosuria during a period of topical administration of halcinonide cream (0.1 per cent) under occlusion. A similar case in a psoriatic 47-year-old man with an initially abnormal glucose-tolerance curve who developed clinical diabetes when treated with topical betamethasone valerate cream (0.1 per cent) was described by the same authors.

Dumler *et al.* (1982) studied the development of steroid-induced diabetes in 143 renal allograft recipients; steroid-induced diabetes developed in 9.8 per cent of patients. However, in blacks its incidence was significantly higher than in whites (17.3 vs. 5.5 per cent, respectively, $p < 0.01$). The development of steroid-induced diabetes was not associated with a higher frequency of HLA-BB or HLA-Bw15 in either race; in black graft recipients, HLA-B14 was significantly more frequent ($p < 0.001$) among those who developed steroid-induced diabetes, than in insulin-dependent diabetic (Type I) and non-diabetic recipients. The clinical course of patients with steroid-induced diabetes has been similar to that of non-insulin-dependent diabetics (Type II).

Oral contraceptives

Impairment of glucose tolerance in women taking oral contraceptives has been recognized since the drugs were introduced (Waine *et al.* 1963) and has been confirmed by some (Peterson *et al.* 1966; Wynn and Doar 1966) but debated by others (Taylor and Kass 1968; Clinch *et al.* 1969). Descriptions of the severity of the impairment and its frequency have varied in the reports and evidence has been presented that the impairment was transient and that it reverted to pre-treatment levels while contraceptives were continued or when they were discontinued.

For the purpose of this review it is convenient to classify reports on impairment of carbohydrate metabolism due to oral-contraceptive agents into three categories, studies in diabetic women, studies in women with potential diabetes, and studies in normal subjects.

There are many early reports suggesting that oral contraceptives adversely affected carbohydrate tolerance and caused a change in diabetic control. Usually there was an increased insulin demand and the previously stable diabetic became unstable. An illustrative case is one that has been described by Paros (1964). A patient with previously stable diabetes was given norethynodrel with ethinyloestradiol; after the third day of medication, glycosuria, polyuria, and polydipsia reappeared and insulin demand increased. Even after discontinuation of the contraceptive it was nearly 2 months before the diabetes was stable again.

With regard to the second category, Paros (1964) reported on 12 patients who had a positive family history of diabetes, of whom seven showed increased blood-sugar levels after using oral contraceptives.

In one study, Szabo *et al.* (1970) drew attention to the unwise complacency surrounding the use of these agents in women with potential diabetes. The inserts packaged with oral contraceptive agents advise caution when taken by women with manifest diabetes, but give no guidance regarding their use in those with potential diabetes. In their study Szabo *et al.* (1970) performed oral glucose-tolerance tests periodically on 15 women who had shown abnormal tolerance to glucose in the third trimester of pregnancy, but who had normal tolerance after delivery. Subsequently five of these women received a combination of norethindrone (1 mg) and mestranol (0.05 mg) as an oral contraceptive cyclically. The other 10 women practised other methods of contraception and served as controls to the study.

Abnormal glucose tolerance recurred in all five women while they were taking the contraceptive agent. When therapy ceased the glucose-tolerance test remained diabetic in three cases; it temporarily improved but became abnormal again in the fourth woman. The fifth woman was lost to follow-up observation after discontinuing contraceptive pills. Three of the 10 controls showed deterioration of the glucose-tolerance test. The authors concluded that screening tests for diabetes should be performed early during contraceptive treatment, and that women with latent diabetes should be excluded from hormonal birth control.

It is, however, in the third category that the full effect of the contraceptive agent on an otherwise normal carbohydrate tolerance is shown and in this respect the studies of Gershberg *et al.* (1964), Clinch *et al.* (1969), and Wynn and Doar (1969) are illustrative of the problems created.

Gershberg and colleagues (1964) investigated 59 patients of whom 51 had used norethynodrel with mestranol for two years or longer. Blood sugar levels were measured fasting, and 1 and 2 hours after administration of glucose. The fasting level was outside normal limits in six of these patients, and the 2-hour blood sugar level was raised above normal limits in 27 of these women.

Clinch *et al.* (1969) studied the effect of oral-contraceptive medication in 42 women. An intravenous glucose-tolerance test was used in preference to the oral test. Blood glucose levels were higher than normal in 32 of the 42 women when they were on contraceptive medication.

Wynn and Doar (1969) made longitudinal studies of plasma glucose, non-esterified fatty acids, insulin, and blood pyruvate levels during oral and intravenous glucose-tolerance tests in women treated with combined oral-contraceptive preparations.

Group A: consisted of 91 women tested before and during oral-contraceptive medication. In terms of total area between the plasma curve and the abscissa, oral and intravenous glucose tolerance deteriorated during therapy in respectively 78 and 70 per cent of these women. Thirteen per cent of women in this group developed chemical diabetes mellitus as a result of medication.

Group B: consisted of 38 women tested during contraceptive therapy and again after discontinuing medication. An improvement occurred in the oral glucose tolerance in 90 per cent and in the intravenous glucose tolerance in 85 per cent after medication was discontinued.

Group C: consisted of 22 women whose glucose-tolerance tests were similar to those measured in Group B, while on oral contraceptives. These women were re-tested while continuing on contraceptives and there was no change in the oral or intravenous glucose tolerance.

Plasma levels of non-esterified fatty acids were unchanged before and after oral or intravenous glucose-tolerance tests in Groups A and B. As a result of medication, the fasting blood pyruvate rose in Group A. The mean fasting insulin levels were unchanged in Groups A and B on or off therapy.

The conclusion drawn from these studies by Wynn and Doar (1969) was that the impaired glucose tolerance was steroid diabetes, caused by elevated plasma cortisol levels secondary to the oestrogen component of the oral contraceptive.

Oestrogen–progestogen oral contraceptives, especially of the combined type, offer almost complete protection against pregnancy. All other methods are substantially less effective and, in assessing the significance of the iatrogenic risks of oral contraception, due consideration must be given to the dangers of unwanted pregnancy, especially since pregnancy itself can worsen a diabetic or potential diabetic state.

It is understandable, although somewhat disappointing, that medical opinion, whilst emphasizing the association between the use of oral contraceptives and thromboembolic disorders (e.g. Vessey 1970), has not also given some authoritative warning to doctors that some of their patients may also be at risk from a breakdown of carbohydrate homeostasis, especially if they are controlled diabetics or, perhaps more important, if they are undiagnosed potential diabetics. Certainly a routine investigation of carbohydrate-metabolism status would be a wise and sensible precaution, if not in all patients, then certainly in those who might be suspected through familial history or *avoirdupois* to have an impaired or a 'strained' carbohydrate tolerance. The danger of precipitating permanent diabetes has, we believe, not been properly appreciated, and it is perhaps towards gaining informative data on this aspect that long-term studies should be initiated.

Since these earlier publications, the situation has changed considerably in respect to the association between oral contraceptives and the development of diabetes mellitus. Two major prospective long-term studies, which both began in 1968, were conducted in the United Kingdom by the Royal College of General Practitioners and the Oxford Family Planning Association. They surveyed 67990 and 54900 women-years of oral contraceptive use, respectively, and both studies included appropriate control groups. An association between the development of diabetes mellitus and oral contraceptive use was not demonstrated (Wingrave *et al.* 1979). These findings might reflect the reduced diabetogenic potential of the lower-dose oestrogen contraceptive pills introduced at about the time these studies commenced compared with the original high-dose oestrogen preparations which were in use when many of the reports of oral contraceptive–induced diabetes mellitus were made between 1963 and 1969.

Ginseng

Ginseng has been reported to cause hyperglycaemia and this is undoubtedly due to its high content of oestrogenic substances (see Chapter 38).

Hyperglycaemia induced by miscellaneous drugs

β-Endorphin

Van Loon and Appel (1981) in animal studies demonstrated that synthetic human β-endorphin increased plasma glucose concentration when administered intracisternally in a chronically cannulated, conscious, unrestrained, adult male by prior systemic administration of naloxone, supporting mediation of the effect at opioid receptors in brain. Adrenal denervation blocked the beta-endorphin-induced increase in plasma glucose, supporting a thesis that this effect is mediated at least in part by increased epinephrine secretion. The hyperglycaemic response to intracerebral β-endorphin was also blocked by either intracerebral hemicholinium-3 or somatostatin, supporting both a cholinergic link and a somatostatin neuron in the brain mechanism regulating endorphin-induced stimulation of sympathetic outflow.

Studies in man of β-endorphin have been limited but a hyperglycaemic effect has been reported and a mechanism for the reaction has been advanced.

Cimetidine

Pomare (1978) described deterioration in glucose handling and hyperosmolar non-ketotic diabetic coma after intravenous cimetidine. Reddy (1981) described a 58-year-old male patient who developed polydipsia, polyuria, and glycosuria while on cimetidine therapy. The patient was found to have marked hyperglycaemia. Carbohydrate metabolism returned to normal within four weeks of stopping cimetidine. A second case of hyperglycaemia and renal failure attributed to cimetidine by Reddy is not convincing since the renal failure was more likely to be due to the concomitantly administered gentamicin. The patient, a 69-year-old Vietnamese woman, died and the death was undoubtedly iatrogenic but not attributable to cimetidine.

Lithium salts

Hyperglycaemia has been described in patients receiving lithium therapy. A rise in blood glucose has been demonstrated to occur within 30 min of ingesting a loading dose of a lithium salt. Crammer (1982) drew attention to the fact that obesity is a major side-effect of lithium treatment and that obesity may be a major contributing factor in the development of diabetes during long-term lithium therapy.

Indomethacin

Tkach (1982) published a report of a computer-assisted survey of the literature related to indomethacin. He found that in an earlier survey of a similar type covering 13 studies and 649 patients no cases of hyperglycaemia due to indomethacin therapy had been recorded; however in three further studies covering 169 patients, three patients developed hyperglycaemia during indomethacin therapy.

Lomotil overdose

Modi (1981) reported a single case of *Lomotil* (diphenoxylate and atropine) — poisoning in an 11-month-old boy. On arrival at hospital, the baby was comatose with shallow respiration, constricted pupils, and peripheral cyanosis. On admission, blood glucose was high on two consecutive *Dextrostix* estimations, greater than 13.9 mmol ℓ^{-1} (250 mg/100 ml) and was determined in the laboratory as 30.8 mmol l^{-1} (555 mg/100 ml). In view of the history of *Lomotil* ingestion, the baby was treated with naloxone and three hours later the blood glucose was 5.0 mmol l^{-1} (90 mg/100 ml). The baby made an uneventful recovery and neither glycosuria nor hyperglycaemia were detected on any subsequent occasion.

Malathion poisoning

Ramu *et al.* (1973) reported four cases of hyperglycaemia and glucosuria in children treated with malathion shampoos for head lice.

Two cases of malathion poisoning in which hyperglycaemia was a major complicating factor were reported by Mellor *et al.* (1981). The patients were an 81-year-old woman and her 39-year-old son. The method of exposure to the malathion was not detailed, but two other members of the family were less severely affected than the two patients described, who were both admitted to hospital in coma with miosis, muscle fasciculation, metabolic acidosis, and hyperglycaemia. The family cat and dog died during the exposure to malathion. The level of hyperglycaemia in the 39-year-old man on admission was 29.0 mmol l^{-1} (522 mg/100 ml) and, in the 81-year-old woman, 25.5 mmol l^{-1} (460 mg/100 ml).

Organophosphorus poisoning

Hyperglycaemia has been described as a complication of organophosphorus poisoning. For example, in a series of 105 African cases of organophosphorus poisoning, 44 per cent had depressed levels of consciousness, 42 per cent had muscle fasciculation; 14 per cent had glycosuria, and 7 per cent had hyperglycaemia (Hayes *et al.* 1978).

Mellor *et al.* (1981) concluded that the presence of non-ketotic hyperglycaemic coma with metabolic acidosis should suggest, in addition to diabetes mellitus, organophosphorus poisoning, which will be characteristically associated with miosis and fasciculation.

Oxymetholone

Woodward *et al.* (1981) determined immunoreactive insulin and plasma glucose during a glucose-tolerance test in seven patients with aplastic anaemia of whom six were receiving oxymetholone therapy. All the patients receiving oxymetholone therapy showed impaired glucose tolerance and an excessive immunoreactive insulin response to glucose load. The single patient not receiving oxymetholone had a normal response to glucose load.

These results were interpreted to indicate that oxymetholone caused insulin-resistant hyperglycaemia. In the six patients studied the results were not correlated with changes in growth hormone, cortisol, or glucagon.

Propranolol plus chlorpropamide

Hyperglycaemia was reported by Holt and Gaskins (1981) in a 59-year-old male with pancreatitis who was being treated with propranolol 120 mg daily for his hypertension and chlorpropamide 400 mg daily for his diabetes mellitus. Prior to receiving chlorpropamide, his diabetes was controlled by 5 units daily of regular insulin whilst receiving propranolol 60 mg twice daily. Two days after beginning chlorpropamide, at a time when his pancreatitis was resolving, his blood sugar increased from 120 to 285 mg per cent, which may have been due to the concomitant propranolol-blocking insulin release.

Sympathomimetic amines

The sympathomimetic amines fenoterol, ritodrine, salbutamol, and terbutaline given orally or by intravenous infusion in the suppression of premature labour may cause hyperglycaemia and ketosis in both normal and diabetic women. The combination of any of these sympathomimetic agents with a corticosteroid such as β-methasone or dexamethasone significantly increases the risk of hyperglycaemia and ketosis. (Benedetti 1983; Borberg *et al.* 1978; Hertz *et al.* 1978). Hertz *et al.* (1978) found that the daily insulin requirement of the pregnant diabetic mother increased by an average of 18 per cent when ritodrine alone was given and 44 per cent when ritodrine was given with a corticosteroid. Mordes *et al.* (1982) described two further cases of ritodrine-induced ketoacidosis when used to delay labour in 20-year-old and 30-year-old diabetic women. Terbutaline when used to delay the onset of labour has also been shown to induce hyperglycaemia, hyperinsulinaemia, hyper-lactacidaemia, and hyperkalaemia (Cotton *et al.* 1981). Terbutaline was also shown to induce hypoerglycaemia in the fetus or neonate when given to the mother to suppress labour (Svenningsen 1982); however, there were no adverse long-term sequelae.

Other drugs

Other agents that have been reported to have caused hyperglycaemia on a limited number of occasions, although cause and effect has not in most cases been established, are chenodeoxycholic acid, clofibrate, colchicine, danazol, nicotinic acid, nifedipine, praziquantel, somatostatin, theophylline, and salmon calcitonin.

Parenteral nutrition in infants

In a study conducted in Athens by Anagnostakis and Matsaniotis (1982) on 36 very low-birth-weight infants, i.e. less than 1200 g at birth, admitted to their unit over an 18-month period, seven were fed orally while the remaining 29 received, during the first 4 days of life, parenteral glucose infusion either as a supplement to oral feedings (*n:* 11) or as the only source of fluids (*n:* 18). Among these 29 infants 21 (72 per cent) manifested hyperglycaemia (blood glucose, $>$, 125 mg/dl). On the contrary none of the seven infants receiving oral feedings exclusively manifested hyperglycaemia. Hyperglycaemia was related to high rates of glucose infusion (>0.4 g/kg/hr). These data attest to the fragile nature of glucose metabolism in infants of very low birth-weight, and excess parenteral glucose should be avoided.

Drug-induced hypoglycaemia

The commonest cause of drug-induced hypoglycaemia is treatment of diabetic patients with insulin or oral hypoglycaemia agents; in this context the older sulphonylurea agents, tolbutamide and chlorpropamide, and the newer agents, glibenclamide, gliclazide, and glipizide, have all been incriminated. The biguanide hypoglycaemics, phenformin, buformin, and metformin, have also produced hypoglycaemic episodes in diabetics. The biguanide derivative pentamidine isethionate used as an antiprotozoal agent has also produced hypoglycaemic coma.

Rosenbaum and Rosenbaum (1982) described a case of extreme sensitivity to chlorpropamide in a male patient with coeliac disease. The coeliac disease presenting as severe hypoglycaemia in a previously stable diabetic patient with the severe diarrhoea clearly being a destabilizing factor.

Anti-neoplastic agents

Hypoglycaemia associated with extra pancreatic neoplasms is well recognized and has been described in association with hepatomas and adrenocortical carcinomas. Walden *et al.* (1979) described two cases, one with leukaemia and the other with a trophoblastic testicular teratoma in whom hypoglycaemia occurred as a terminal event. Neither of these tumours is known to be associated with the development of hypoglycaemia and Walden *et al.* considered that the hypoglycaemia could have been related to the cytotoxic therapy given to these patients. Marks *et al.* (1974) had earlier drawn attention to hypoglycaemia developing in patients receiving alkylating agents, nitrogen mustard, and cyclophosphamide-combination therapy.

Aspirin

In high doses salicylates may lower blood sugar. Salicylate intoxication in children may be accompanied by hypoglycaemia and this may be the cause of death.

β-Adrenergic blocking drugs

Kotler *et al.* (1966) noted that propranolol precipitated hypoglycaemia in an insulin-dependent diabetic, but did not change his total insulin requirements. Abramson *et al.* (1966) suggested that propranolol increases sensitivity to insulin in normal subjects by damping down the rebound of blood glucose level after its initial fall, while Sussman *et al.* (1967) suggested that propranolol could be a potent stimulator of insulin secretion and that the observed hypoglycaemia could have been due to raised circulating insulin levels.

Divitiss *et al.* (1968) showed that blockade of β-adrenergic receptors by propranolol inhibited, within certain limits, the usual response of blood glucose to tolbutamide during the tolbutamide-load test. They thought that the antagonistic effect of β-receptor blockade took place at the level of insulin release from the granules in the β-cells of the islets of Langerhans, and that insulin was released from β-cell granules only in the presence of β-adrenergic activity. An alternative hypothesis was that decreased splanchnic blood-flow due to blockade of β-adrenergic receptors reduced access to the pancreas of the injected tolbutamide.

Whatever the mechanism of action of propranolol, Reveno and Rosenbaum (1968) thought that its effect might be of value in the management of the labile diabetic. The object of their studies was to flatten the unpredictable hyperglycaemic peaks and, if possible, to reduce the variable insulin requirement. Oral propranolol in varying dosage and timing was tried in four patients treated with calorie diet restriction, isophane (NPH) insulin, and supplementary crystalline insulin, treatment which had either failed to achieve good control of persistent hyperglycaemia, or had poorly controlled glycosuria and hyperglycaemic attacks.

The combined propranolol, diet, and insulin regime produced a severe hypoglycaemic reaction in one patient which persisted even though the dosage of insulin and propranolol was reduced. In a second patient, a severe hypoglycaemic reaction also occurred and this ceased when propranolol dosage was reduced, although persistent glycosuria and raised morning blood-sugar levels remained. In the third patient, an improved status resulted from the combined treatment which continued for over 6 months, although heavy proteinuria continued and fresh retinal haemorrhages recurred at intervals. In the final case, the insulin requirement was reduced by the combined treatment, although fasting blood sugar levels remained high and glycosuria was unaltered. The previous episodes of hypoglycaemia which occurred on diet and insulin alone did not recur.

It may be assumed that in the labile diabetic the β-adrenergic blockade of propranolol, by opposing catecholamine action, might enhance hypoglycaemia in sufficient degrees to overcome the hyperglycaemia resulting from the multiple factors at play in this disease. This assumption, as stated by Reveno and Rosenbaum (1968), would necessarily ascribe a dominant role to the catecholamines and a weak compensatory or defensive reaction to these multiple factors. It is more likely, however, that there is already a trend towards hypoglycaemia when propranolol is effective and that the induced hypoglycaemia is the result of β-blockade interfering with the attainment of normoglycaemia.

Reveno and Rosenbaum (1968) concluded that propranolol has no practical value in the management of the brittle diabetic.

Another mechanism by which β-adrenergic blockade may produce hypoglycaemia is by blocking the release of glucose from the breakdown of liver glycogen.

The most important action of propranolol and other β-adrenergic blocking drugs with respect to the diabetic patient is the loss of those signs of hypoglycaemia mediated by adrenergic mechanisms, notably shaking and sweating by which diabetics and their friends and relatives recognize the signs of a hypoglycaemic attack (Griffin and D'Arcy 1984).

Most cases of hypoglycaemia associated with propranolol and other β-adrenergic blocking drugs have been in diabetic adults. A few reports of propranolol-induced hypoglycaemia in non-diabetics have been documented both in children and in adults. Kallen *et al.* (1980) drew attention to the production of hypoglycaemic coma with a blood glucose of 38 mg/100 ml in a 14-month-old boy who was being treated with propranolol for hypertension associated with coartation of the aorta. Treatment with 10 mg propranolol every six hours had continued a month prior to the child being found unresponsive in his cot one morning. After receiving glucose he recovered promptly, and no further episodes of hypoglycaemia occurred following discontinuation of propranolol.

The authors reviewed several other cases of hypoglycaemia associated with propranolol therapy in children and concluded that a prolonged period without food, for example 12–20 hours, was a common factor in these cases. It is recommended that prolonged periods without food are avoided in children receiving propranolol. Belton *et al.* (1980) also recorded a case of hypoglycaemic coma in a 12-year-old girl treated with propranolol for hypertension associated with a prolonged fast following admission to hospital after an epileptic seizure.

Belton *et al.* (1980) also recorded a case of hypo-glycaemic coma in a 22-year-old man treated with pro-pranolol for hypertension. In this patient the coma was attributed to the additive hypoglycaemic effects of β-blockade and alcohol. Weiner *et al.* (1982) reported the development of severe hypoglycaemia in a 62-year-old Arab woman who had no evidence of diabetes mellitus. The woman had been admitted to a coronary care unit with chest pain and ECG changes of cardiac ischaemia. She was started on propranolol 20 mg four times per day. On the second day the patient became hypoglycaemic with a blood sugar 37/100 ml. On the third day in hospital she was inadvertently given 20 mg propranolol and again became hypoglycaemic (blood glucose 29 mg/100 ml). A glucose tolerance test done on the tenth day in hospital was normal. A further challenge with pro-pranolol on the seventh and tenth day with 40 mg and 120 mg propranolol did not induce hypoglycaemia.

This report of propranolol-induced hypoglycaemia in a non-diabetic patient shortly after a myocardial ischaemic attack by Weiner *et al.* (1982) is rather reminiscent of a report by Wray and Sutcliffe (1972). Wray and Sutcliffe's non-diabetic patient was subsequently fasted and hypoglycaemia with low insulin levels developed following a test dose of propranolol.

The possibility of hypoglycaemia, which may closely mimic cardiogenic shock, should be considered when beta-blockers are given to severe cardiac patients who undergo dramatic haemodynamic events. In these cases, a prompt and accurate diagnosis of hypoglycaemia and its relatively simple treatment may induce a striking improvement in the patient's condition.

Disopyramide

Disopyramide is an anti-arrhythmic drug of fairly recent introduction. The first case of disopyramide-induced hypoglycaemia was described by the Japanese workers Taketa and Yamamoto (1980) in a 41-year-old male diabetic controlled with insulin (50 units/day) who developed hypoglycaemia after starting disopyramide (200 mg/day) to control his tachycardia. Lower levels of insulin were required and the daily insulin requirement fell to 20 units/day. After disopyramide was withdrawn, insulin requirements rose to their original level.

Goldberg *et al.* (1980) described an 88-year-old patient who developed severe morning hypoglycaemia (29 mg/100 ml) on disopyramide. Blood sugar recovered after withdrawal of the drug, but rechallenge with diso-pyramide produced hypoglycaemia three hours after ingestion.

Quevedo *et al.* (1981) described two further cases of hypoglycaemia associated with disopyramide in a 72-year-old woman and a 72-year-old man. Inadvertent reintroduction of disopyramide in the male patient caused a second episode of hypoglycaemia.

Quevedo *et al.* (1981) comment that neither of their two patients nor that of Goldberg *et al.* (1980) were well nourished and that these patients were all elderly; one patient had definite impairment of renal function. Clinicians should therefore be alert to the risk of fasting hypo-glycaemia as a complication of disopyramide therapy in elderly, underweight patients, particularly if renal function is impaired.

Nappa *et al.* (1983) described two further cases of severe hypoglycaemia induced by disopyramide. The first was a 74-year-old man who had received a single dose of 450 mg disopyramide orally and became unconscious 1½ hours after ingestion. His blood pressure was stabilized with metaraminol but 7 hours later he again became hypertensive and his blood glucose was 9 mg/100 ml. Repeated attempts to stabilize blood pressure and to restore blood glucose levels were unsuccessful and the patient died on the third day. The second patient was a 58-year-old man who was prescribed 200 mg disopyramide 6-hourly. On the tenth day the patient had a cardiac arrest and his blood glucose immediately after resuscitation was 4 mg/100 ml.

Hartigan (1983) reported the case of an 86-year-old woman who had been prescribed 100 mg disopyramide four times per day for 1 year, who developed hypogly-caemia while in hospital with pneumonia. Disopyramide was suspected. The patient was given a single 400-mg dose of disopyramide and blood sugars were measured hourly for 4 hours. Marked hypoglycaemia was induced by this dose of disopyramide at 2, 3, and 4 hours after ingestion.

Parenteral nutrition combined with haemodialysis

Miller *et al.* (1982) described the case of a 24-year-old woman who had sustained serious injuries in a road traffic accident, required renal dialysis daily, and was fed intravenously with a solution containing 25 per cent dextrose. Subsequently insulin had to be added to the parenteral fluid to maintain blood glucose concentrations at physiological values. On one occasion parenteral feeding was continued until dialysis was started; she became comatose and the plasma glucose concentration was found to be 1 mmol l⁻¹ (18 mg/100 ml). She responded rapidly to a 50-ml intravenous bolus of 50 per cent dextrose.

When parenteral feeding and dialysis are used simultaneously, glucose passes across the semi-permeable membrane from the blood to the dialysate so that hypoglycaemia may occur. Insulin added to the parenteral fluid further decreases blood glucose concentrations. Stopping parenteral feeding 40–45 minutes before dialysis is started eliminates this danger of hypoglycaemia.

Pentoxifylline

Pentoxifylline, a methyl xanthine derivative, is claimed to improve peripheral blood flow by reducing whole blood viscosity. The Data Sheet on the product in the United Kingdom draws attention to the fact that the drug increases insulin release from the pancreas and as a result some diabetic patients may need less insulin.

Perhexiline

Perhexiline maleate may rarely induce profound hypoglycaemia but the mechanism has not been elucidated (Roger *et al.* 1975).

Beta-adrenergic blocking agents are probably the second commonest cause of hypoglycaemic attacks. In this context perhexiline, which is widely used in the treatment of angina, is also well documented to cause hypoglycaemia. There may be peculiar risks if both beta-adrenergic blocking drugs and perhexiline are used simultaneously in the same patient (Bourmayan *et al.* 1978).

Streptozotocin

Streptozotocin is isolated from *Streptomyces achromogenes*. It was used by Murray-Lyon *et al.* (1968) to treat islet-cell tumours of the pancreas. A variety of toxic effects have been described including hypoglycaemia associated with hyperinsulinaemia, severe renal toxicity with proteinuria, and nephrogenic diabetes insipidus. Lactic acidosis has also been described as a complication of therapy (Narins *et al.* 1973). An acute toxic effect of streptozotocin that can be life-threatening is acute hypoglycaemia caused by enormous rises in the level of circulating insulin.

Sulphonamides

Arem *et al.* (1983) described severe hypoglycaemia with inappropriately elevated insulin levels in a 65-year-old woman with chronic renal failure who was taking two tablets of sulphamethoxazole and trimethoprim twice a day for a urinary-tract infection. Hypoglycaemia was readily corrected with intravenous glucose and did not recur after discontinuation of the sulphonamide. Insulin and glucose determinations during a 48-hour fast while the patient was rechallenged with this compound, as compared with those obtained during a 72-hour fast performed 12 days after discontinuation of the therapy, suggest that the hypoglycaemic episode was related to hyperinsulinaemia, probably induced by the sulphonamide. Other factors, including congestive heart failure, growth hormone deficiency, and hypoalaninaemia might have contributed to the development of hypoglycaemia in this patient.

Drug interactions affecting absorption of insulin from injection sites

Madsbad *et al.* (1980) showed that diabetic smokers need on average 15-20 per cent more insulin than non-smokers, and this percentage increases to 30 per cent in heavy smokers. The degree of glycaemic control in insulin-dependent diabetic patients is influenced by the rate of insulin absorption from subcutaneous tissue to blood. Kolendorf *et al.* (1979) showed a correlation between adipose tissue blood flow and insulin disappearance from subcutaneous tissue.

Smoking increases the concentration of catecholamines in the blood (Madsbad *et al.* 1980). In normal subjects, inhaling cigarettes produces a transient rise in blood pressure and pulse rate and a fall in skin temperature. This peripheral vasoconstriction lasts from several minutes up to one hour, and habitual smokers do not show tolerance to smoking by a decrease in these effects. The duration of this peripheral vasoconstriction caused by smoking was considered to have an effect on insulin absorbption. Klemp *et al.* (1982) tested this hypothesis in nine ketosis-prone insulin-dependent diabetic patients (six women and three men), aged 22 to 62 years (mean 35 $\pm$ S.D. 13 years), who were selected for the study after informed consent was obtained. All were known to have had diabetes for 3 to 17 years (mean 8 $\pm$ S.D. 6 years). Their daily insulin dose varied from 32 to 46 U (mean 38 $\pm$ 6 U). All were within standard weight $\pm$ 10 per cent, and none had signs or symptoms of neuropathy or nephropathy. The patients were all habitual smokers smoking 5-40 cigarettes daily (mean 13 $\pm$ S.D. 11).

The study began in the morning after the subjects had abstained from smoking overnight. The patients had their normal breakfast and their normal morning dose of intermediate-acting insulin. After 30 minutes' rest in the supine position, 8 U (0.2 ml) of rapid-acting ^{125}I-labelled insulin (1.2 uCi/40 U *Actrapid*) was injected subcutaneously. The disappearance rate of the tracer was measured starting 60 minutes after the ^{125}I-insulin injection. Ninety minutes after the ^{125}I-insulin injection the patients were asked to smoke a cigarette (*Prince*, filter-tipped, 815 mg tobacco, 2.0 mg nicotine) and inhale with their habitual depth and frequency, usually about once a minute. The smoking was finished after eight minutes, and the disappearance rate of the tracer was followed for an additional 60 minutes. Room temperature was kept constant ($22 \pm$ S.D. $0.5°C$) during the experiments. To prevent ^{125}I-accumulating in the thyroid, the patients were given 100 mg potassium iodide for five days after the insulin injection. Student's *t*-test for paired samples and linear regression analysis were used to analyse the data.

For all nine subjects, mean half-time was 158 $\pm$ S.E.M. 22 minutes in the first 30-minute period before smoking, 336 $\pm$ S.E.M. 97 minutes during smoking ($p < 0.05$, $n = 8$), and 207 $\pm$ S.E.M. 29 minutes in the first 30-minute period after smoking ($p < 0.05$). In the last 30 minutes the mean half-time was 155 $\pm$ S.E.M. 16 minutes. These half-lives correspond to an average decrease of 113 per cent in insulin absorption during smoking and a 31 per cent decrease in insulin absorption in the first 30 minutes after smoking.

Klemp *et al.* (1982) clearly showed that smoking has a direct influence on the rate of disappearance of ^{125}I-insulin from the injection sites. The duration of impaired absorption is associated with the duration of peripheral vasoconstriction caused, by smoking.

The effect of smoking as a vasoconstrictor, together with a direct metabolic effect associated with cate-cholamine release caused by smoking is considered to account for the increased insulin requirement in diabetic heavy smokers.

Adrenergic agonists, such as isoprenaline, dopamine, adrenaline, salbutamol, and dobutamine, interfere with glucose tolerance in diabetics resulting in hyperglycaemia and in some cases ketoacidosis (Thomas *et al.* 1977; Leslie and Coat 1977; Goldberg *et al.* 1975; Wood *et al.* 1981). Use of these adrenergic agonists in diabetics may be associated with increased insulin requirements.

Drug-induced lactic acidosis

Biguanides

Overproduction of lactic acid or failure of removal by the liver and kidney can result in metabolic acidosis. The best-known form of clinical lactic acidosis is that associated with shock. In this condition, Type A, insufficient oxygen is available to the peripheral tissues to oxidize pyruvate derived from glycolysis, and lactate is formed.

Another form of lactic acidosis, Type B, has been recognized and occurs without clinical evidence of poor perfusion or arterial desaturation. Except in rare hereditary forms, this type always arises in association with some precipitating factor, and the most important drug-induced causes are biguanide antidiabetic agents and over-rapid infusion of fructose-containing solutions. The main clinical features of Type B lactic acidosis are rapidity of onset, striking hyperventilation, drowsiness, and coma. Blood pressure is normal, peripheral circulation is good, and there is no cyanosis. Blood lactate varies according to the severity of the condition but may reach 20-30 mmol/l. The pathogenesis of Type B lactic acidosis is comparatively poorly understood since the crucial observations to determine the relative contributions of lactate overproduction and under-utilization have not been made. In Type B lactic acidosis it is a common event for shock to supervene after a few hours of adequate circulation. Furthermore severe acidosis itself inhibits hepatic removal of lactate from the blood.

About half the recorded cases of Type B lactic acidosis have occurred in patients receiving phenformin and there is no doubt that phenformin is a direct cause of lactic acidosis (Wise *et al.* 1976). Type B lactic acidosis not due to phenformin or other iatrogenic factor may occur in the course of Gram-negative septicaemia (Tranquada *et al.* 1966), pulmonary embolism (Huckabee 1961), and cerebral thrombosis (Dossester *et al.* 1965).

The mortality of Type B lactic acidosis is very high; for non-phenformin-associated cases the mortality is about 80 per cent. In phenformin-induced lactic acidosis the mortality is about 50 per cent if treatment commences before cardiovascular collapse, and 65 per cent or more thereafter.

Johnson and Waterhouse (1968) demonstrated that phenformin increased and prolonged the rise in lactate levels in diabetic subjects given glucose or alcohol. Varma *et al.* (1972) studied 12 diabetic patients; each had a diabetic glucose-tolerance test at the start of treatment. They were stabilized on phenformin 150 mg/day and a daily diet of 150 g carbohydrate. In the 12 patients the average fasting lactate level was 15.4 mg/100 ml. After intravenous glucose there was a rise in lactate levels to a maximum at 60 minutes with an average level of 19.6 mg/100 ml. Pyruvate levels did not rise correspondingly. In four patients in whom the phenformin was discontinued the mean fasting lactate was 6.5 mg/100 ml and in these there was no significant rise in lactate following intravenous glucose.

Cohen *et al.* (1973) reviewed published literature and drew attention to the fact that of 56 documented cases of phenformin-induced lactic acidosis in only one was the blood urea normal. A further study by Conlay and Loewenstein (1976) revealed that all cases with severe lactic acidosis admitted to a teaching hospital during a 17-month period were taking phenformin hydrochloride. Prerenal azotaemia was present at the time of admission in all but one patient, and in all survivors at the time of discharge renal function was normal. Bengtsson *et al.* (1972) also described the return of renal function to normal in seven patients who had shown abnormal urea and creatinine serum levels on admission with phenformin-induced lactic acidosis. Conlay and Loewenstein (1976) pointed out that in these patients it may be that they were azotaemic prior to the onset of lactic acidosis; it is equally possible that the lactic acidosis developed first and contributed to the development of prerenal azotaemia. There are three mechanisms by which this could happen. The first is the direct depressant effect of acidosis on the myocardium which would decrease cardiac output and thus decrease renal perfusion. The second is dehydration consequent on nausea and vomiting induced by acidosis. The third mechanism is that there might be an osmotic diuresis induced by lactate excretion leading to hypovolaemia.

Cohen *et al.* (1973) also pointed out that of 52 cases reported in the literature surveyed by them that abnormal renal function was present in 36 cases, hypotension and dehydration in 24 cases, infection in 13 patients, cirrhosis in four patients, and ethanol ingestion was excessive in four. These workers did not think however that lactic acidosis could have resulted purely from such long-standing conditions as cirrhosis or renal failure in these diabetics. They were influenced to this conclusion by the fact that in 24 out of 34 patients, in whom the duration of treatment was recorded, lactic acidosis occurred within two months of the start of treatment and in 17 of these lactic acidosis occurred within two weeks.

Phenformin has been used in association with an anabolic steroid such as ethyloestrenol to reduce plasma fibrinogen, platelet stickiness, and serum cholesterol (Fearnley *et al.* 1971). A patient with oliguric renal failure complicating accelerated hypertension was given phenformin and ethyloestrenol as fibrinolytic therapy. On the fifth day of this regime a profound and fatal metabolic acidosis developed despite good control of uraemia by peritoneal dialysis.

It was incorrectly stated by Jackson (1967) that with metformin (Fig. 18.1) the problem of metabolic acidosis never arises. A case of fatal metabolic acidosis with metformin was reported by Lebacq and Tirzmalis (1972) which followed a similar course to phenformin-induced lactic acidosis.

Gale and Tattersall (1976) stated that, if a biguanide is indicated, metformin should be used since it is less likely to precipitate lactic acidosis. In France, where metformin is used by three times as many patients as phenformin,

there were, in 1975, 68 cases of lactic acidosis associated with phenformin and four cases associated with metformin therapy.

During 1965 and 1977, 51 cases of lactic acidosis and four cases of unspecified acidosis were reported to the Swedish Adverse Drug Reaction Committee (Bergman *et al.* 1978). One case of lactic acidosis was due to metformin and was not fatal; the remaining 54 cases were caused by phenformin and 21 cases were fatal.

In normal subjects both biguanides decrease lactate clearance (increase the lactate half-life) (Phillips *et al.* 1977). (See Table 18.2.)

Table 18.2   Lactate half-life in minutes in normal and diabetic subjects on biguanides (after Phillips *et al.* 1977)

	Lactate half-life in minutes	
Treatment	Normal subjects	Diabetic patients
Control	9	—
Metformin	15	39
Phenformin	19	68

Evidence indicates that the precipitation of lactic acidosis by phenformin is a dose-dependent or body-fluid concentration-dependent effect (Cohen *et al.* 1973) and Conlay and Loewenstein (1976) showed that in one patient in whom serum phenformin was measured during a lactic acidotic coma this was found to be between six to nine times the usual therapeutic serum level. These observations tie in with the observations that the inhibition of gluconeogenesis from lactate in the isolated perfused rate or guinea-pig liver is concentration-dependent (Altschuld and Kruger 1968; Woods 1971). If the mechanism by which lactate half-life is prolonged is by the inhibition of reconversion of lactate to glucose then the greater effect of phenformin and metformin may be explained by the hepatic cumulation of phenformin which does not occur with metformin. Hall *et al.* (1968) using ^{14}C-labelled phenformin found that rat liver cumulated ^{14}C-

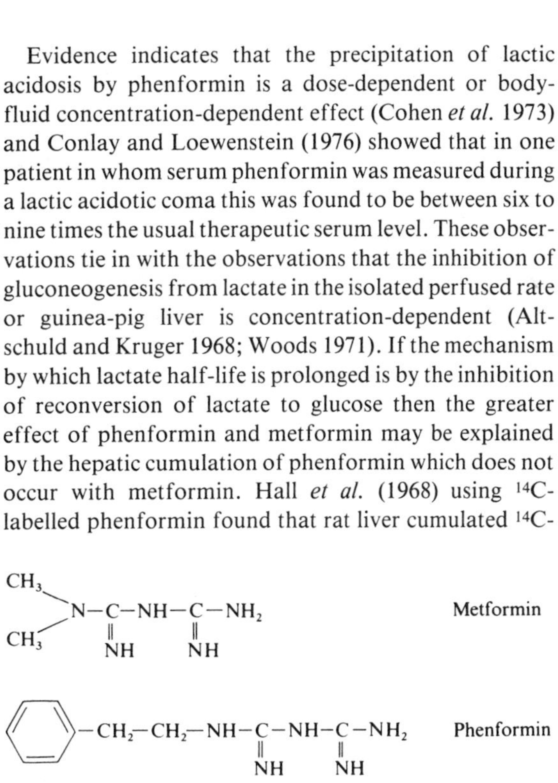

Fig. 18.1   Comparative chemical structures of metformin and phenformin.

labelled material. In a patient with fatal lactic acidosis the liver concentration of phenformin was 20 times that in the blood.

The fact that metformin is less likely to cause lactic acidosis than phenformin is demonstrated by the fact that Bertrand and Duwoos (1969) and Tomkins *et al.* (1972) documented cases of patients who developed lactic acidosis on phenformin and who after recovery were treated with metformin with no subsequent recurrence of lactic acidosis. However, despite metformin's lesser propensity compared with phenformin to cause lactic acidosis it should not be regarded as free of this unwanted side-effect. With both these biguanides, alcohol is likely to precipitate lactic acidosis.

In an extensive review of the world literature on biguanide-induced lactic acidosis by Luft *et al.* (1978), 429 cases were identified, but in 99 cases, clinical, laboratory, or treatment data could not be associated with specific patients and 330 cases were left for analysis. The average age of patients developing lactic acidosis was 64 years and the youngest was 13 years old and the eldest 90 years old; 32.7 per cent were male and 67.3 per cent female.

In 281 cases the biguanide incriminated was phenformin; in 30 cases buformin, and in 12 cases metformin were the causative agents. In four cases phenformin and metformin had been administered simultaneously to the patient (in three cases the biguanide involved was not specified).

At the time of development of lactic acidosis, 214 patients had other illnesses, predominantly cardiovascular or renal or both.

The concomitant use of other drugs has been known to precipitate lactic acidosis in biguanide-treated patients. In 330 cases examined 14 patients were receiving antibiotics (tetracycline 5, gentamicin 4, ampicillin 3, cephalothin 2, choramphenicol 2); 20 patients were digitalized, 32 patients were on diuretics.

In 17 cases the patients were alcoholic or had taken large amounts of alcohol immediately prior to the onset of lactic acidosis.

Symptoms of lactic acidosis

In only 195 cases were these recorded in sufficient detail for analysis (see Table 18.3).

In 45 per cent of patients the diagnosis was made within 24 hours of the onset of symptoms and 73 per cent were diagnosed within 72 hours. The duration of treatment for 64 per cent was less than 24 hours and in 88 per cent the outcome of the illness was determined within 72 hours. The overall duration of the illness was therefore very short.

Table 18.3   Symptoms complained of in 195 patients with lactic acidosis (after Luft *et al.* 1978)

Symptoms	Number of patients
Vomiting	100
Somnolence	98
Nausea	31
Epigastric pain	69
Loss of appetite	52
Hyperpnoea	50
Lethargy	29
Diarrhoea	27
Thirst	8

The overall mortality was 50.2 per cent; between 20-40 years the mortality was 25 per cent; between 40-60 years the mortality was 40.3 per cent; and over 60 years was 54.9 per cent.

Intravenous xylitol

Xylitol, a regular intermediate in carbohydrate metabolism, has been used as a calorie source in parenteral nutrition. Xylitol has the potential to induce lactic acidosis and osmotic diuresis, and has also been implicated in a more obscure, equally dangerous, and highly characteristic metabolic disturbance. The adverse effects have been recorded within 24 hours of the infusion being started, and consist of oliguria, uraemia, raised uric-acid levels, increase in bilirubin and serum transaminases, and sodium-oxalate crystal deposition. This syndrome may be fatal; and following a number of deaths in Australia, the Australian Drug Evaluation Committee withdrew xylitol from clinical use in 1970 (Thomas *et al.* 1972a, b) and other regulatory authorities have followed this move. The Australian Drug Evaluation Committee expressed reservations about the use of laevulose and sorbitol in intravenous infusions in the following terms:

'The main advantages claimed for the use of these substances rather than glucose are invalid. Glucose, apart from hyperglycaemia and hyperosmolarity, has no side-effects and is cheaper. It also poses no danger to those few people with hereditary fructose intolerance (laevulose can be fatal to infants with this rare condition). The Australian Drug Evaluation Committee has decided that for these reasons there is no justification for using solutions of laevulose and/or sorbitol on their own.'

'Combination proprietary preparations containing laevulose or sorbitol (for example, with amino acids and ethanol) are designed to provide as complete a nutrition source as possible and may be administered via a superficial vein. Glucose cannot be used in the same container in place of laevulose or sorbitol because it reacts with other components. Because it is irritant, glucose must be administered via a deep-vein catheter. Expertise in the technique of deep-venous catheterization is not universal, and so occasions may arise when these compound preparations will be useful (probably in emergency and isolated conditions). However, the Australian Drug Evaluation Committee has limited the allowable carbohydrate in the form of laevulose and/or sorbitol to a total of 5 per cent w/v, a level at which harmful side-effects are unlikely to occur.' (Kearney 1976).

Intravenous fructose

Over-rapid infusion of fructose has been reported to cause lactic acidosis (Craig and Crane 1971; Woods and Alberti 1972).

Methanol and ethanol

Other causes of lactic acidosis that have been reported are ethanol intoxication and methanol poisoning. Oliva (1970) reported a severe case of lactic acidosis due to ethanol ingestion; the importance of ethanol lies in its ability to inhibit hepatic lactate uptake (Krebs et al. 1969). The presence of ethanol may further increase the blood lactate concentration caused by other agents; for example, Johnson and Waterhouse (1968) demonstrated this effect in phenformin-treated patients given a lactate load.

Three case reports of unusual causes of lactic acidosis have appeared in the literature, none of which were fatal.

Terbutaline

Fahlen and Lapidus (1980) reported the case of a 28-year-old woman who became comatose and developed hypokalaemia and lactic acidosis after an overdose of 225 mg terbutaline, 1750 mg clomipramine, 150 mg oxazepam, and 5 g chloralhydrate. The patient recovered. Lactic acidosis has been reported with epinephrine intoxication (Kolendorf and Moller-Broch 1974).

Diazepam

Kapoor et al. (1981) described the case of a 90-year-old woman who had tetanus and was treated with 3000 units tetanus immune globulin, tetracycline, heparin, and diazepam 5 mg intravenously every four hours. Lactic acidosis was diagnosed on the fourth and fifth day of treatment. Diazepam dosage was reduced and the lactate level declined dramatically and blood pH became normal. On the seventh day when lactate levels were

normal, diazepam was again given intravenously but lactic acidosis again developed. Lactate levels fell when diazepam was replaced with metocurine, and there was no further lactic acidosis.

Dithiazanine iodide (Telmid)

Dithiazanine iodide is a cyanide dye which was used for treatment of intestinal worm infections in humans until its withdrawal from the market some years ago. Mather et al. (1967) described a patient who developed lactic acidosis after absorbing dithiazanine iodide from the gastrointestinal tract. This case is of interest because there was recovery in this patient following treatment with methylene blue, and it serves to show a connection between lactic acidosis and a drug shown experimentally to increase hepatic glycolysis.

Drug-induced disorders of lipid metabolism

β-adrenergic-blocking drugs

Durrington and Cairns (1982) reported the case of a 64-year-old man with a 20-year-old history of peripheral and coronary arterial disease who developed acute pancreatitis while on atenolol 100 mg daily. The pancreatitis was considered to be associated with impaired triglyceride clearance consequent to inhibition of lipoprotein lipase by β-blockers. Triglyceride clearance, measured by intravenous administration of a 10 per cent lipid solution (1 mg kg^{-1}) showed that with β-blockade, the elimination half-life of triglyceride was increased from a normal 16.5 min to 24.5 and 27 min, with metoprolol and atenolol, respectively. Recurrent angina necessitated administration of atenolol (100 mg daily) and metoprolol (100 mg b.i.d.) for a period of two weeks. Although lipoprotein lipase was said to remove triglycerides from both chylomicrons and low-density lipoproteins, the change in value of chylomicrons before and after treatment with β-blockers was not mentioned.

Leven et al. (1980) reported the results of a study to compare the effects of propranolol and prazosin, two drugs commonly used in the treatment of hypertension, itself a risk factor in coronary artery disease.

In 23 hypertensive men, aged 47–55, propranolol reduced serum high-density-lipoprotein (HDL) cholesterol by 13 per cent, reduced the ratio of HDL to low-density-lipoprotein (LDL) + very-low-density-lipoprotein (VLDL) cholesterol by 15 per cent, increased total triglycerides by 24 per cent, and increased serum uric acid by 10 per cent. Prazosin reduced total serum cholesterol by 9 per cent, LDL + VLDL cholesterol by 10 per cent, and total triglycerides by 16 per cent. These changes are statistically highly significant. On combined treatment with propranolol and prazosin, HDL cholesterol was still significantly reduced but changes in other blood lipids were small and insignificant.

Propranolol reduced HDL cholesterol and in several studies a low HDL cholesterol has been found to be associated with an increased incidence of coronary heart disease. The clinical importance of these differences between the two anti-hypertensive drugs is uncertain and it is also uncertain whether the effects seen with propranolol can be extrapolated to other β-adrenergic blocking drugs. Timolol is now being promoted for the secondary prevention of myocardial infarction following a large-scale prospective study in Norway, which showed a statistical benefit to timolol-treated patients. Similar indications for atenolol and metoprolol have also been approved in the U.K. It is uncertain whether these effects of β-adrenergic blocking drugs on HDL and on secondary prevention of myocardial infarction are general to this class of drug, since the two effects at first sight appear contradictory.

Clofibrate-paradoxical hypercholesterolaemia

Increases in serum cholesterol levels after a few months on clofibrate were reported in four patients with primary biliary cirrhosis. In three patients the cholesterol levels doubled and the xanthomas became worse (Schaffner 1969).

In some patients with Fredrickson Type IV hyperlipidaemia lowering of VLD lipoprotein levels has been associated with a rise in LD lipoproteins during clofibrate therapy (Strisower et al. 1968).

Drug-induced increases in plasma cholesteral and plasma triglyceride level with oral contraceptives, injectable contraceptives and hormone replacement therapy

Combination oestrogen–progestogen products

Wide acceptance of oral-contraceptive therapy has been associated with an increasing recognition of diverse metabolic side-effects. However, particular concern has arisen following reports of increased levels of plasma triglycerides in women receiving this form of contraception (Aurell et al. 1966; Wynn et al. 1966, 1969). The increase in triglyceride levels in these women was such that the sex differential between men and women of similar ages was reduced. These observations have raised the possibility that the relative immunity from atherosclerotic vascular disease enjoyed by premenopausal women may be severely jeopardized.

A detailed investigation of this problem was carried out by Hazzard et al. (1969) using 10 young women of 20-33 years in good health, with normal menses, and who were non-obese, within the 95 percentile of the ideal Metropolitan Life Assurance standards. Identical studies were performed before and after 14 days' treatment with ethinyloestradiol 0.05 mg and medroxyprogesterone acetate 10 mg daily. During this treatment, none of the 10 subjects showed a net gain in weight, and the mean plasma triglyceride level rose from 45 mg/100 ml to 64 mg/100 ml ($p < 0.005$): the post-hepatic lipolytic activity fell from 0.373 μEq FFA/ml/min to 0.199 μEq FFA/ml/min ($p < 0.001$): the mean basal immunoreactive insulin levels increased in eight of 10 subjects, the mean increase for the group being from 10.5 μV/ml before treatment of 14.5 μV/ml during treatment ($p < 0.05$). The one subject who showed no increase in plasma triglyceride during the initial 14-day period and who was subsequently followed-up for three complete cycles, showed a rise in plasma triglyceride level of nearly 100 per cent during the second cycle.

It would appear that the rise in plasma triglyceride level might be due to increased endogenous (hepatic) triglyceride synthesis, which would be supported by the increase in immunoreactive insulin levels, and/or to decreased triglyceride removal as suggested by the fall in post-hepatic lipolytic activity.

Wynn et al. (1969) made a longitudinal study of the effects of oral contraceptives on fasting serum lipid and low-density-lipoprotein levels in two groups of women. Group A consisted of 116 women tested before and during contraceptive therapy. Group B consisted of 48 women initially tested during therapy and tested again after it had been discontinued. In both groups, while on contraceptive therapy, there was a higher mean serum triglyceride and cholesterol level. Elevated mean fasting S_f0-12, S_f20-100, and S_f100-400 serum lipoprotein levels were found during therapy, but S_f12-20 lipoprotein and chylomicron triglyceride levels were unchanged. Serum triglyceride levels increased in 95 per cent of Group A women during therapy and decreased in 88 per cent of Group B women after therapy was discontinued. No relationships was found between the nature of the oestrogen–progesterone combination, day of treatment cycle, duration of therapy, degree of obesity, parity, or family history of diabetes, or of oral glucose-tolerance abnormality.

The clinical significance of these findings is unknown but they cannot be ignored, since they raise the possibility of irreversible structural changes such as atherosclerosis after 10 or 20 years on oral contraceptives. Both raised serum lipid levels and chemical diabetes are associated with an increased incidence of occlusive vascular disease. The risks of venous thrombosis, pulmonary embolism, and cerebral thrombosis have already been shown to be raised in women taking oral contraceptives. In view of these doubts, the wisdom of administering such compounds to healthy women for many years was seriously considered. (*Lancet* 1969).

It is considered that patients having a serum cholesterol higher than 300 mg/100 ml and/or plasma triglycerides higher than 200 mg/100 ml should not receive treatment with oral contraceptives (De Gennes *et al.* 1973). Fredrickson Type IV, V, and possibly Type I hyperlipidaemia increase in severity during treatment with oral-contraceptive agents (Turpin and Menage 1973). In this context it may be relevant to comment that oestrogen-induced pancreatitis has been observed predominantly in patients with pre-existing hypertriglyceridaemia (Glueck *et al.* 1972).

The discovery that the high-density lipoprotein (HDL) fraction is an anti-risk factor for coronary heart disease has modified the lipid theory of atherogenesis. High coronary risk is associated with high levels of cholesterol-rich, low-density lipoprotein (VLDL) and with a low level of HDL. The effect of drugs on lipid metabolism therefore has taken on new significance.

Wallace *et al.* (1979) reported the results of a study of women attending 10 North American lipid research clinics. Plasma total cholesterol, triglyceride, low-density (LDL), very-low-density (VLDL), and high-density (HDL) lipoprotein levels in those taking oral contraceptives (OC) and in those taking oestrogens for menopausal symptoms were compared with those in women not taking gonadal hormones, after adjustment for age, educational attainment, and body-mass index. Oral contraceptives and oestrogen-users were leaner than non-users. Compared with controls, oral contraceptive-users showed increased cholesterol, triglyceride, and LDL VLDL cholesterol levels, but HDL cholesterol levels were similar. Cholesterol, triglyceride, and HDL and VLDL cholesterol levels were positively associated with the quantity of the oestrogen component of the oral contraceptive preparations. Compared with non-users, menopausal oestrogen-users had slightly lower cholesterol and triglyceride levels, significant decreases in LDL and VLDL cholesterol, and a significant increase in HDL cholesterol.

Metenolone

Garbrecht *et al.* (1981) described 10 cases of hyperlipidaemia, Fredrickson Type II, in a series of 28 post-meno-

pausal women treated with metenolone enanthate; in a further two patients a Fredrickson Type IIb hyperlipidaemia was induced. The hyperlipidaemia regressed in each case on ceasing therapy with metenolone.

Medroxyprogesterone acetate (Depo-Provera)

In the Netherlands, Kremer *et al.* (1980, 1981) found that in 23 women who received a contraceptive injection of 150 mg of depot-medroxyprogesterone acetate (*Depo-Provera*) intramuscularly every 12 weeks for at least one year, the mean serum high-density-lipoprotein (HDL)-cholesterol concentration was distinctly lower than in 23 women matched for age, body weight, and alcohol consumption and using intrauterine devices for contraception. The high serum levels of medroxyprogesterone during the first few weeks after intramuscular injection and the lower levels at 12 weeks after injection were both associated with the same low levels of HDL-cholesterol. Kremer *et al.* considered that the fall in serum HDL-cholesterol was not a direct effect of the drug but rather an indirect effect on the ovaries affecting the production of endogenous oestrogen.

For this reason the depot injections of both medroxyprogesterone acetate (*Depo-Provera*) and norethisterone enanthate do not necessarily have the advantage of reduced risk of coronary arterial disease or thromboembolic phenomena claimed; in fact, Kremer *et al.* go as far as to conclude that it would be advisable to avoid prescribing medroxyprogesterone acetate to women with increased risk of atherosclerosis.

Briggs and Briggs (1977) studied the changes in plasma HDL-cholesterol concentrations in four groups of non-smoking women: (1) 96 untreated controls; (2) 71 volunteers receiving either norethisterone acetate or ethinyl oestradiol; (3) 409 women taking oral contraceptives; and (4) 40 woman receiving oestrogens or progestogens for gynaecological indications, including nine women who had been receiving Depo-medroxyprogesterone acetate (50 mg weekly) for at least 2 months. There was a statistically significant reduction (approximately 15 per cent less than control levels) in HDL-cholesterol levels in women receiving Depo-medroxyprogesterone acetate. The authors conclude that, because HDL-cholesterol concentration is thought to be negatively correlated to cardiovascular disease risks, products which induce a large decrease are to be avoided, although their discussion is confined to the use of oral contraceptives.

Norethisterone enanthate

In a UK study on 75 women given 12-24 consecutive intragluteal injections of norethisterone enanthate (200 mg) every 56 days, Howard et al. (1982) demonstrated reductions in high-density-lipoprotein levels, when compared with 21 age-matched controls who were not using any form of steroidal contraception. Reduction in serum HDL-cholesterol levels was not correlated with either the serum norethisterone concentrations or the length of use of norethisterone enanthate nor was it affected by obesity or smoking.

Another investigation on norethisterone enanthate was carried out by Fotherby et al. (1982) but as they are also the authors of the study described above, it is possible that the same women may have been observed in both studies. Fotherby et al. found that the HDL-cholesterol level in 74 blood samples from 61 women who had received norethisterone enanthate (200 mg) for between 2 and 4½ years was significantly reduced (mean decrease of approximately 20 per cent compared to control levels). In a study on four women, most of the decrease in HDL-cholesterol levels occurred with the first injection of norethisterone enanthate.

An inverse relation has been reported between serum HDL-cholesterol and coronary heart disease. Whether drug-induced reduction in this possibly protective lipid fraction could in the long term lead to an increased incidence of atheroma and myocardial infarction is not known. However, analysis of the Framingham prospective study, which now has data on 24 years of follow-up of atherosclerotic disease, still attaches the greatest prognostic importance to the level of total serum cholesterol which was not increased in the studies by Howard et al. (1982) and Fotherby et al. (1982).

Miscellaneous drugs inducing hyperlipidaemia

Etretinate and isotretinoin hypertriglyceridaemia

Gollnick and Orfanos (1981) reported that about 25 per cent of patients receiving etretinate (Tigason) for the treatment of acne show a reversible rise in plasma triglycerides. The significance of these changes in patients with pre-existing hyperlipidaemia, diabetes, obesity, or liver damage may be greater than in otherwise normal patients. Similar changes have been recorded for isotretinoin, another vitamin A analogue used in the treatment of acne.

Phenytoin-induced hypercholesterolaemia

Following clinical observations that some patients on phenytoin developed hypercholesterolaemia, studies to investigate this were conducted on two groups of volunteers who were dosed with 300 mg phenytoin per day for 4

weeks or 12 weeks. These studies showed that phenytoin could cause hypercholesterolaemia (Jubiz et al. 1974).

Vitamin-E-induced hypercholesterolaemia

Only a few adverse reactions on vitamin E have been reported, but in a study on 52 persons taking 300 mg vitamin E daily there was a mean rise of 74 mg 100ml in plasma cholesterol levels (Dahl 1974).

Insulin-induced lipoatrophy

Lipoatrophy occurs in about 10 per cent of patients receiving conventional insulins (Wright et al. 1979). Highly purified porcine insulin is not generally thought to cause this complication and is normally advocated for its treatment. Lipoatrophy due to monocomponent porcine insulin has been reported by Hanai et al. (1976) in Japan and Jones et al. (1981).

The cause of insulin-induced lipoatrophy is unknown, but an immune pathogenesis has been suggested to explain its appearance in patients treated with conventional insulins. The appearance of lipoatrophy with monocomponent insulins is not consistent with this hypothesis.

Other metabolic disorders

Hyperammonaemia and sodium valproate

Sodium valproate (sodium dipropylacetate) hepatotoxicity is a well established clinical entity, and appears to be distinct from the rarer occurence of hyperammonaemia (Coulter and Allen 1980; Sills et al. 1980). It is known that sodium valproate may produce metabolic upset by interference with propionic acid metabolism causing secondary hyperammonaemia. This may manifest itself clinically as vomiting ataxia and increasing clouding of consciousness. Should these symptoms occur, sodium valproate should be discontinued.

Hyperglycinaemia and sodium valproate

Administration of sodium valproate to rats reduces the activity of the glycine cleavage enzyme system in the liver (Kochi et al. 1979) and the same enzyme defect was reported in ketotic hyperglycaemia (Tada et al. 1974). Jaeken et al. (1977) reported that in a series of 14 children, all over nine months of age, on sodium valproate (10.75 mg/kg/day) the mean urine glycine was 7355 ± 949 μmol/g creatinine, while in 52 control children, it was 1867 ± 134 μmol/g creatinine. In 13 children on sodium valproate the serum glycine was 418 ± 8 μmol. Both these biochemical changes reported by

Jaeken *et al.* (1977) were highly significant statistically; it, however, remains to be established whether this mechanism is clinically relevant in the development of sodium valproate-associated encephalopathy.

Simila *et al.* (1979) described two pregnancies in which the mothers had been given sodium valproate throughout the pregnancy. The two neonates were both found to have hyperglycinaemia. Both these infants showed normal psychomotor development, and there was no evidence that moderate rises in glycine concentration impaired these two children's CNS development.

Effect of propranolol on arterial ammonia in patients with liver cirrhosis

In a study by Van Buuren *et al.* (1982), propranolol (20 mg four times a day) was given to patients with liver cirrhosis or fatty infiltration of the liver. In six patients with cirrhosis and a stable arterial plasma ammonia concentration before treatment, blood ammonia was increased significantly on day-3 of propranolol treatment. Arterial plasma ammonia concentration was still high on day-6 of propranolol. Individual percentage change in arterial ammonia ranged from 8 to 66 per cent. After propranolol had been discontinued, ammonia concentration returned to pretreatment concentrations in 3 to 6 days. In three patients with fatty livers and normal pretreatment ammonia concentrations, no change was detected in arterial plasma ammonia while they were on propranolol.

Most of the ammonia generated in the bowel is absorbed and first-pass elimination in the normal liver is high. Hepatic clearance depends on hepatic blood flow and the most likely explanation for the effect of propranolol described by Van Buuren *et al.* is porto-systemic shunting of blood away from the liver.

Drug-induced vitamin deficiency

Vitamin deficiency may be induced by a variety of drugs. Some of these are dealt with more extensively in other sections of these volumes but are summarized briefly below.

Folate deficiency

This may be associated with phenytoin therapy for epilepsy and with folate antagonists used in cancer chemotherapy, for example, methotrexate, but also occurs with other cytotoxic agents, for example, the purine and pyrimidine analogues mercaptopurine, azathioprine, 5-fluorouracil, cytosine arabinoside, and thioguanine.

Folate deficiency with associated megaloblastic anaemia has been reported in association with sulphonamide combinations with trimethoprim.

Vitamin B$_{12}$ deficiency

This is due to drug-induced malabsorption and has been reported in association with paraaminosalicylic acid therapy, colchicine, neomycin, slow-release potassium preparations, and the biguanides, metformin and phenformin.

Pyridoxine deficiency

This is associated with oral-contraceptive therapy has been reported to result in depression which can be relieved by pyridoxine supplements. L-dopa also causes pyridoxine deficiency but pyridoxine supplements reduce the efficacy of this anti-parkinsonian agent.

Nicotinic acid deficiency

Following the use of isoniazid in the treatment of tuberculosis, nicotinic acid deficiency has been reported to result in pellegra (McConnell and Cheetham 1952; Cohen *et al.* 1974; Bender *et al.* 1979; Meyrick-Thompson *et al.* 1981).

Vitamin D deficiency

This deficiency and subsequent osteoporosis may follow the long-term use of enzyme-inducing agents, for example, phenobarbitone, phenytoin, and glutethimide.

Vitamin K deficiency

This can be induced by any oral antibiotic that destroys the indigenous gut flora. The resulting vitamin K deficiency may be sufficient to disrupt a previously stabilized anti-coagulant regime.

REFERENCES

ABRAMSON E.A., ARKY, R.A., AND WOEBER, K.A., (1966). Effects of propranolol on the hormonal and metabolic responses to insulin-induced hypoglycaemia *Lancet* ii. 1386-8.

ALTSCHULD, R.A. AND KRUGER, F.A. (1968). Inhibition of hepatic gluconeogenesis in guinea pig by phenformin. *Ann. NY Acad. Sci.* **148**, 612-22.

AMERY, A., BERTHAUX, P., BULPITT, C.E., DERUYTTE, M., DE SCHAEPDRYVER, A., DOLLERY, C., FAGARD, R., FORETTE, F., HELLEMANS, J., LUND-JOHANSEN, P., MUTSERS, A., AND TUO MILEHTO, J. (1978). Glucose intolerance during diuretic therapy. Results of trial by the European Working Party on Hypertension in the elderly. *Lancet* i, 681-3.

ANAGNOSTAKIS, D. AND MATSANIOTIS, N. (1982). Hyperglycaemia in very low birth weight infants: a common iatrogenic problem. *Pädiatrie Pädologie* **17**, 585–90.

AREM, R., GARBER, A.J., AND FIELD, J.B. (1983). Sulphonamide-induced hypoglycaemia in chronic renal failure. *Arch intern.Med.* **143**, 827–9.

AURELL, M., CRAMER, K., AND RYBO, G. (1966). Serum lipids and lipoproteins during long-term administration of an oral contraceptive. *Lancet* **i**, 291–3.

AUSSANAIRRE, M., CABANES, J., CHAUSSAIN, J.L., AND MARIA, J. (1972). Diabète aigu acidocétosique chez une enfant traitée par la L'asparginase. *Arch, franc, Pédiat.* **29**, 527.

BELTON, N.R., ETHERIDGE, J.E.JR., AND MILLICHAP, J.G. (1965). Effects of convulsions and anticonvulsants on blood sugar in rabbits. *Epilepsia, Boston* **6**, 243–9.

BELTON, P., CARMODY, M., DONOHOE, J., AND O'DWYER, W.F. (1980). Propranolol associated hypoglycaemia in non-diabetics. *J. Irish med. Sci.* **73**, 173.

BENDER, W., GREIL, W., RUTHER, E., AND SCHNELLE, K. (1979). Effects of the beta-adrenoreceptor blocking agent sotalol on CNS: sleep, EEG and psychophysiological parameters. *J. clin. Pharmacol.* **19**, 505–12.

BENEDETTI, T.J. (1983). Maternal complications of parenteral B-sympathomimetic therapy for premature labor. *Amr. J. Obstet. Gynecol.* **145** 1–6.

BENGTSSON, K., KARLBERG, B., AND LINDGREN, S. (1972). Lactic acidosis in phenformin-treated diabetics. *Acta med. scand.* **191**, 203–8.

BENNETT, W.M. AND PORTER, G.A. (1973). Efficacy and safety of metolazone in renal failure and nephrotic syndrome, *J. clin. Pharmacol.* **13**. 357–64.

BERGMAN, U., BOMAN, G., AND WTHOLM, B-E. (1978). Epidemiology of adverse drug reactions to phenformin and metformin. *Br. med. J.* **2**, 464–6.

BERTRAND, C.M. AND DUWOOS, H. (1969). Syndrome musculaire a Teffort revelateur d'un debut d'intoxication par l'acide lactique endogene consecutif a l'administration de phenformin. *Ann. Endocrin., Paris* **30**, 570–5.

BLEREAU, R.P., BLEREAU, R.P. AND WEINGARTEN, C.M. AND WEINGARTEN, C.M. (1964). Diabetic acidosis secondary to steroid therapy. *New Engl. J. Med.* **271**, 836.

BORBERG, C., GILLMER, M.D.G., BEARD, R.W., AND OAKLEY, N.W. (1978). Metabolic effects of beta sympathomimetic drugs and dexamethasone in normal diabetic pregnancy. *Br. J. Obstet. Gynaecel.* **85**, 184.

BOURMAYAN, C., FOURNIER, C., DUDOGNON, P., AND GERBAUX, A. (1978). Hypoglycaemic au cours d'un traitment par le maleate de perhexiline et le pindolol. *Concours med.* **100**, 3071.

BRIGGS, M.H. AND BRIGGS, M. (1979). Plasma lipoprotein changes during oral contraception. *Curr. Med. Res. Opin.* **6**, 249–54.

BROWN, W.J.JR AND BROWN, F.K. (1967). Thiazide-induced alteration of carbohydrate tolerance in normal men. *Curr. ther. Res.* **9**, 200–7.

CARTER, B.L., SMALL, R.E., MANDEL, M.D., AND STARKMAN, M.T. (1981). Phenytoin induced hyperglycaemia. *Am. J. hosp. Pharm.* **38**, 1508–12.

CLINCH, J., TURNBULL, A.C., AND KHOSLA, T. (1969). Effect of oral contraceptives on glucose tolerance. *Lancet* **i**, 857–8.

COHEN, L.K., GEORGE, W., AND SMITH, R. (1974). Isoniazid induced acne and pellegra. *Arch. Dermatol* **109**, 377–81.

COHEN, R.D., WARD, J.D., BRAIN, A.J.S., MURRAY, C.R. SAVEGE, T.M., AND LES R.A. (1973). The relation between phenformin and lactic acidosis. *Diabetologica* **9**, 43–6.

CONLAY, L.A. AND LOEWENSTEIN, J.E. (1976). Phenformin and lactic acidosis. *J. Am. med. Ass.* **235**, 1575–8.

COTTON, D.B., STRASSNER, H.T., LIPSON, L.G., AND GOLDSTEIN, D.A. (1981). The effects of terbutaline on acid base, serum electrolytes, and glucose remostasis during the management of preterm labor. *Am. J. Obstet Gynecol.* **141**, 617–24.

CRAIG, G.M. AND CRANE, C.W. (1971). Lactic acidosis complicating liver failure after intravenous fructose. *Br. med. J.* **4**, 211–12.

COULTER, D.C. AND ALLEN, R.J. (1980). Secondary hyperammonaemia: a possible mechanism for valproate encephalopathy. *Lancet* **i**, 1310–11.

CRAMMER, J.L. (1982). Lithium diabetes and obesity. *Agressologica* **23B**, 99–101.

DAHL, S. (1974). Vitamin E in clinical medicine. *Lancet* **i**, 465.

DARGIE, H.J. AND DOLLERY, C.T. (1975). Adverse reactions to diuretic drugs. In *Side-effects of drugs*. Vol. VIII (ed. M.N.G. Dukes). Excerpta Medica, Amsterdam.

DEGENNES, J.L., TURPIN, G., AND KARTANEGARA, S. (1973). Hyperlipidaemias and thromboembolic complications in the course of contraceptive treatment with estroprogestatives. *Gynecologie* **24**, 405.

DICKSON, L. (1962). Glycosuria and diabetes mellitus following INAH therapy. *Med. J. Aust.* **49**, 325–6.

DIGE-PETERSEN, H. (1966). Ethacrynic acid and carbohydrate metabolism. *Nord. Med.* **75**, 123–5.

DIVITISS, O.DE, GIORDANO, E., GALO, B., AND JACONO, A. (1968). Tolbutamide and propranolol. *Lancet* **i**, 749.

DOLLERY, C.T., PENTECOST, B.L., AND SAMAAN, N.A. (1962). Drug-induced diabetes. *Lancet* **ii**, 735–6.

DOSSETOR, J.B., ZBOROWSKI, D., DIXON, H.B., AND PARE, J.A.P. (1965). Hyperlactatemia due to hyperventilation: Use of CO_2 inhalation. *Ann. NY Acad. Sci.* **119**, 1153–64.

DRASH, A. AND WOLFF, F. (1964). Drug therapy in leucine-sensitive hypoglycemia. *Metabolism* **13**, 487–92.

DUMLER, F., HAYASHI, H., HUNTER, J., AND LEVIN, N.W. (1982). Racial difference in the incidence of steroid diabetes in rena transplant patients. *Henry Ford Hosp. med. J.* **30**, 14–16.

DURRINGTON, P.N. AND CAIRNS, S.A. (1982). Acute pancreatitis: a complication of beta-blockade. *Br. med. J.* **284**, 1016.

ESBENSHADE, J.H.JR AND SMITH, R.J. (1965). Thiazides an pregnancy: a study of carbohydrate tolerance. *Am. J. Obstet Gynecol* **92**, 270–1.

FAHLEN, M. AND LAPIDUS, L. (1980). Lactic acidosis and beta adrenergic agents. *Lancet* **ii**. 390.

FEARNLEY, G.R., CHAKRABARTI, AND EVANS, J.F. (1971). Mode o action of phenformin plus ethyloestrenol on fibrinolysis *Lancet* **i**, 723–5.

FELDMAN, E. AND DIAMOND, S. (1967). Ethacrynic acid. A non diabetogenic diuretic. *Dis. Chest.* **51**, 282–7.

FINNERTY, F.A. (1959). Discussions of special problems o therapy. In *Hypertension* (ed. J.H. Mayer). W.B. Saunders Philadelphia.

FONSECA, V. AND PHEAR, D.N. (1982). Hyperosmolar non-ketoti diabetic syndrome precipitated by treatment with diuretics *Br. med. J.* **284**, 36–7.

FOTHERBY, K., TRAYNER, I., HOWARD, G., HAMANI, A., AND ELDER M.G. (1982). Effect of injectable norethisterone œnanthat (Norigest) on blood lipid levels. *Contraception* **25**, 435–46.

FULLER, J.H., SHIPLEY, M.J., ROSE, G., JARRETT, R.J., AND KEEN, I (1980). Coronary heart disease risk and impaired glucos tolerance. The Whitehall Study. *Lancet* **i**, 1373–6.

GALE, E.A.M. AND TATTERSALL, R.B. (1976). Can phenformin-induced lactic acidosis be prevented? *Br. med. J.* **2**, 972–5.

GARBRECHT, M., LEHMANN, U., O'BRIEN, S., STOLZENBACK, G., AND MULLERLEILE, U. (1981). Hyperlipoproteinamie bei additiver Behandlung des metastasierenden Manna Karrzinoms mit Metenolononanthat. *Dtsch. med. Wschr.* **106**, 400–3.

GERSHBERG, H., JAVIER, Z., AND HULSE, M. (1964). Glucose tolerance in women receiving an ovulatory suppressant. *Daibetes* **13**, 378–82.

GILLETTE, P.C., HILL, L., STARLING, K.A., AND FERNBACK, D.J. (1972). Transient diabetes mellitus secondary to L-asparginase therapy in acute leukaemia. *J. Paediat.* **8**, 109–11.

GLUECK, C.J., FORD, S., AND FALLATT, R. (1972). Lipids and lipases in normal women on progestational oral contraceptive. *Clin. Res.* **20**, 426.

GOLDBERG, I.J., BROWN, L.K., AND RAYFIELD, E.J. (1980). Disopyramide-induced hypoglycaemia. *Am. J. Med.* **69** 463–6.

GOLDBERG, R., JOFFE, B.I., BERSOLN, L., VAN AS, M., KRUT, L., AND SEFTEL, H.C. (1975). Metabolic responses to selective β-adrenergic stimulation in man. *Postgrad. med. J.* **51**, 53–8.

GOLDNER, M.G., ZAROWITZ, H., AND AKGUN, S. (1960). Hyperglycemia and glycosuria due to thiazide derivatives administered in diabetes mellitus. *New Engl. J. Med.* **262**, 403–5.

GOLLNICK, H. AND ORFANOS, C.E. (1981). Das verhalten von trigly-ceriden und cholesterin im serum bei hautkranken unter oraler behandlung mit aromatschen retinoid. *Z. Hautkr.* **56**, 1183–96.

GOMEZ, E.C. AND FROST, P. (1976). Induction of glycosuria and hyperglycaemia by topical corticosteroid therapy. *Arch. Dermatol* **112**, 1559–62.

GRIFFIN, J.P. AND D'ARCY, P.F. (1984). *Manual of adverse drug interactions, 3rd edn.* J. Wright, Bristol.

HALL, H., RAMACHANDER, G., AND GLASSMAN, J.M. (1968). Tissue distribution and excretion of Phenformin in normal and diabetic animals. *Ann. NY Acad. Sci.* **148**, 601–11.

HANAI, N., KATO, N., KAWASAKI, M., *et al.* (1976). Lipoatrophy due to monocomponent insulin J. *Jap. Diabetic Soc.* **19**, 427–8.

HARTIGAN, J.D. (1983). Severe hypoglycaemia complicating disopyramide *(Norpace)* therapy. A case report. *Nebraska med. J.* **68**, 36–8.

HAYES, M.M., VAN DER WESTHUIZEN, N.G., AND GELFOND, M. (1978). Organophoshorus poisoning in Rhodesia. A study of the clinical features and management in 105 patients. *S. Afr. med. J.* **54**, 230–4.

HAZZARD, W.R., SPIGER, M.J., BAGDADE, J.D., AND BIERMAN, E.L. (1969). Studies on the mechanism of increased plasma trigly-ceride levels induced by oral contraceptives. *New Engl. J. Med.* **280**, 471–4.

HERTZ, J., LARSEN, P., AND PEDERSEN, L.M. (1978). The diabetogenic effect of ritodrine. Severe hyperglycaemia and keto-acidosis in five pregnant diabetics and in one non diabetic during treatment with ritodrine or ritodrine combined with betamethasone. *Vgeskr. Laeg.* **140**, 223–6.

HOLT, R.J. AND GASKINS, J.D. (1981). Hyperglycaemia associated with propranolol and chlorpropamide. *Drug/Intel. clin. Pharm.* **15**, 599–600.

HOWARD, G., BLAIR, M., FOTHERBY, K., TRAYNER, I., HAMAWI, A., AND ELDER, M.G. (1982). Some metabolic effects of long term use of the injectable contraceptive norethisterone oenanthate. *Lancet* **i**, 423–4.

HOWELL, S.L. AND TAYLOR, K.W. (1966). Effects of diazoxide on insulin secretion *in vitro. Lancet* **i**, 128–9.

HUCKABEE, W.E. (1961). Abnormal resting blood lactate. The significance of hyperlacticocidaemia in hospitalized patients. *Am. J. Med.* **30**, 833–9.

JACKSON, W.P.U. (1967). The oral antidiabetic biguanides. *S Afr. G.P. Review* **1**, 1–4.

JAEKEN, J., CORBEEL, L., CASAER, P., CARCHON, H., EGGERMONT, E., AND EECKELS, R. (1977). Dipropylacetate (valproate) and glycine metabolism. *Lancet* **ii**, 617.

JARIWALLA, A.G., JONES, C.R., LEVER, A., AND HALL, R. (1981). Spironolactone and diabetic ketosis. *Post-grad. med. J.* **57**, 573–4.

JARRETT, R.J., MCCARTNEY, P., AND VEEN, H. (1982). The Bedford Survey. Ten-year mortality rates in newly diagnosed diabetics and normoglycaemic controls and risk indices for coronary artery disease in borderline diabetics. *Diabetologica* **22**, 79–84.

JOHNSON, H.K. AND WATERHOUSE, C. (1968). The relationship of alcohol and hyperlactatemia in diabetic subjects treated with phenformin. *Am. J. Med.* **45**, 98–104.

JONES, G.R., STATHAM, B., OWENS, D.R., JONES, M.K., AND HAYES, T.M. (1981). Lipoatrophy and monocomponent porcine insulin. *Br. med. J.* **282**, 190.

JONES, I.G. AND PICKENS, P.T. (1967). Diabetes mellitus following oral diuretics. *Practitioner* **199**, 209–10.

JUBIZ, W., WILSON, D., AND ANSTALL, H.B. (1974). Lipoprotein abnormalities induced by diphenylhydantoin (Dilantin) administration. *Clin. Red.* **22**, 192A.

KAESS, H., SCHLIERF, G., EHLERS, W., VON MIKULICZ-RADECKI, J.G., KASSENSTEIN, P., WALTER, K., BRECK, W., AND HENGSTMANN, J. (1967). The carbohydrate metabolism of normal subjects during potassium depletion. *Diabetologica* **7**, 82–6.

KALLEN, R.J., MAHLER, J.H., AND LIN, H.L. (1980). Hypoglycaemia — a complication of treatment of hypertension with propranolol. *Clin. Paediat.* **19**, 567–8.

KAPOOR, W., CAREY, P., AND KARPF, M. (1981). Induction of lactic acidosis with intravenous diazepam in a patient with tetanus. *Arch. intern. Med.* **141**, 944–5.

KEARNEY, J.J. (1976). Limits on infusions of laevulose and sorbitol: a statement prepared and published at the request of the Australian Drug Evaluation Committee. *Med. J. Aust.* **1**, 582.

KERSHBAUM, A. (1963). Diabetogenic effect of fluorine-containing steroids. *Br. med. J.* **2**, 253.

KLEMP, P., STABERG, B., MADSBAD, S., AND KOLENDORF, K. (1982). Smoking reduces insulin absorption from subcutaneous tissue. *Br. med. J.* **284**, 237.

KOCHI, H., HAYASAKA, K., HIRAGA, K., AND KIKUCHI, G. (1979). Reduction of the level of the glycine cleavage system in rat liver resulting from the administration of dipropylacetate: an experimental approach to hyperglycaemia. *Arch. Biochem. Biophys.* **198** 589–97.

KOLENDORF, K., BOJSEN, J., AND NIELSEN, S.L. (1979). Adipose tissue blood flow and insulin disappearance from subcutaneous tissue. *Clin. Pharmacol. Ther.* **25**, 598–604.

—— AND MÖLLER-BROCH, B. (1974). Lactic acidosis in epinephrine poisoning. *Acta med. scand.* **196**, 465–6.

KONIGSTEIN, R.P. (1965). Die Anwendung von Furosemid (Lasix) in der Geriatrie. Mit einem Beitrag zur sogenannten diabetogenene wirkung der Salidiuretika. *Wien. Klin. Wschr.* **77**, 93–7.

KOTLER, M.N., BERMAN, L., AND RUBENSTEIN, A.H. (1966). Hypoglycaemia precipitated by propranolol. *Lancet* ii, 1389–90.

KREBS, H.A., FREELAND, R.A., HEMS, R., AND STUBBS, M. (1969). Inhibition of hepatic gluconeogenesis by ethanol. *Biochem. J.* **112**, 117-24.

KREMER, J., DEBRUUN, H.W., AND HINDRIKS, F.R. (1980). Serum high density lipoprotein cholesterol levels in women using a contraceptive injection of depot-medroxyprogesterone acetate. *Contraception* **22**. 359–67.

——, ——, AND —— (1981). Prikpil, serum HDL-cholesterol en hart infarct. *Ned. T. Geneesk* **125**, 1418–21.

KVAM, D.C. AND STANTON, H.C. (1964). Studies on diazoxide hyperglycemia. *Diabetes* **13**, 639–44.

LANCET (1969). Leading article: Metabolic effects of oral contraceptives. *Lancet* ii, 783–4.

LAVELLE, K.J., ATKINSON, K.F., AND STUART, A.K. (1976). Hyperlactaemia and haemolysis in G6PD deficiency after nitrofurantoin ingestion. *Am. J. med. Sci.* **272**, 201.

LEBACQ, E.C. AND MARCQ, M. (1967). A study of the mechanism of ethacrynic-acid-induced hyperglycemia. *Rev. franc. Étud. clin. biol.* **12**, 160–2.

LESLIE, D. AND COATS, P.M. (1977). Salbutamol-induced diabetic ketoacidosis. *Br. med. J.* ii, 768.

LEVEN, P., FOSS, P.O., HELGELAND, A., HJERMANN, I., HOLME, I., AND LUND-LARSEN, P.G. (1980). Effect of propranolol and prazosin on blood lipids. (The Oslo Study). *Lancet* ii, 4–6.

LEVIN, S.R., CRODSKY, G.M., HAGURA, R., AND SMITH, D.F. (1972). Comparisons of the inhibitory effects of diphenylhydantoin and diazoxide upon insulin secretion from the isolated perfused pancreas. *Diabetes* **21**, 856–62.

LEWIS, P.J., KOHNER, E.M., PETRIE, A., AND DOLLERY, C.T. (1976). Deterioration of glucose tolerance in hypertensive patients on prolonged diuretic treatment. *Lancet* i, 564–6.

LUFT, D., SCHMULLING, R.M., AND EGGSTEIN, M. (1978). Lactic acidosis in biguanide-treated diabetics. *Diabetologia* **14**, 75–89.

LUND-JOHANSEN, P. (1981). Haemodynamic and metabolic long-term effects of tielinic acid in essential hypertension *Acta.med.scand. (Suppl)* **646**, 106–14.

MADISON, J.F. (1964). Ketonuria after local steroids in necrobiosis lipoidica. *Arch. Dermatol.* **90**, 477–8.

MADSBAD, S., McNAIR, P., CHRISTENSEN, M.S., CHRISTIANSEN, C., FABER, O.K., BINDER, C., AND TRANSBØL, I. (1980). Influence of smoking on insulin requirement and metabolic status in diabetes mellitus. *Diabetes Care* **3**, 41–3.

MARKS, L.J., STEINKE, J, POLOSKY, S., AND EGDAHL, R.H. (1974). Hypoglycaemia associated with neoplasia. *NY Acad.Sci.* **230**, 147–160.

MARKS, P., NIMALASURIYA, A., AND ANDERSON, J. (1981). The glucose tolerance test in hypertensive patients treated long-term with thiazide diuretics. *Practitioner* **225**, 392–3.

MATHER, B.J., GINN, H.E., AND MERCIER, R.K. (1967). Lactic acidosis in dithiazanine intoxication. *Ann. intern. Med.* **66**, 1060.

MATSUNAGA, F., KUBO, A., AND KATAKURA, G. (1963). Steroid diabetes, diabetes mellitus appearing during treatment with pituitary-adrenal hormones. *J. Ther., Tokyo* **45**, 1988–95.

McCONNELL, R.B. AND CHEETHAM, H.D. (1952). Acute pellagra during isoniazid therapy. *Lancet* ii, 959–60.

MELLOR, D., FRASER, I., AND KRYGER, M. (1981). Hyperglycaemia in anticholinesterase poisoning. *Can. med. ass. J.* **124**, 745–8.

MEYRICK – THOMSON, R.H., ROWLAND-PAYNE, C.M.E., AND BLACK, M.M. (1981). Isoniazid induced pellegra. *Br. med. J.* **283**, 287–9.

MILLER, J.D.B., BROOM, J., AND SMITH, G. (1982). Severe hypoglycaemia due to combined use of parenteral nutrition and renal dialysis. *Br. med. J.* **285**, 9–10.

MILLICHAP, J.G. (1965). Anticonvulsant drugs. In *Physiological pharmacology*, Vol. II (ed. W.S. Root and F.S. Hofmann). Academic Press, New York. (1969). Hyperglycemic effect of diphenylhydantoin. *New Engl. J. Med.* **281**, 447.

MODI, N. (1981). Hyperglycaemia in Lomotil poisoning. *Arch. Dis. Childh.* **56**, 157.

MORDES, D., KREUTNER, K., METZGER, W., AND COLWELL, J.A. (1982). Dangers of intravenous ritodrine in diabetic patients. *J. Am. med. Ass.* **248**, 973–5.

MURPHY, M.B., LEWIS, P.J., KOHNER, E., SCHUMER, B., AND DOLLERY, C.T. (1982). Glucose intolerance in hypertensive patients treated with diuretics: a fourteen year follow-up, *Lancet* **2**, 1293–5.

MURRAY-LYON, I.M., EDDLESTON, A.L.W.F., WILLIAMS, R., BROWN, R., HOGBIN, B.M., BENNETT, A., EDWARDS, J.C., AND TAYLOR, K.W. (1968). Treatment of multiple-hormone-producing malignant islet-cell tumour with streptozotocin. *Lancet* ii, 895–8.

NAPPA, J.M., DHANANI, S., LOVEJOY, J.R., AND VANDER ARK C. (1983). Severe hypoglycaemia associated with disopyramide. *Western J. Med.* **138**, 95–7.

NARDONE, D.A. AND BOUMA, D.J. (1979). Hyperglycaemia and diabetic coma: possible relationship to diuretic — propranolol therapy. *S. med. J.* **72**, 1607–8.

NARINS, R.G., BLUMENTHAL, S.A., FRASER, D.W., TIZIANELLO, A., AND SOLOW, J. (1973). Streptozotocin-induced lactic acidosis. A case report and some experimental observations. *Am. J. med. Sci.* **265**, 455–61.

OKUN, R., RUSSELL, R.P., AND WILSON, W.R. (1963). Use of diazoxide with trichlormethiazide for hypertension. *Archs intern. Med.* **112**, 882–8.

OLIVA, P.B. (1970). Lactic acidosis. *Am. J. Med.* **48**, 209–25.

PAROS, N.L. (1964). Side-effect of oral contraceptives. *Br. med. J.* **1**, 630.

PENGELLY, C.R. (1965). Diabetes mellitus and cyclophosphamide. *Br. med. J.* **1**, 1312–13.

PETERS, H.B. AND SAMAAN, N.A. (1969). Hyperglycemia with relative hypoinsulinemia in diphenylhydantoin toxicity. *New Engl. J. Med.* **281**, 91–2.

PETERSON, W.F., STEEL, M.W.JR, AND COYNE, R.V. (1966). Analysis of the effect of ovulatory suppressants on glucose tolerance. *Am. J. Obstet. Gynecol.* **95**. 484–8.

PHILLIPS, P.J., THOMAS, D.W., AND HARDING, P.E. (1977). Biguanides and lactic acidosis. *Br. med. J.* **1**, 234.

PODOLSKY, S. (1970). A possible role for propranolol in recurrent non-ketotic hyperosmolar diabetic coma. *Diabetes* **19**, 398.

POMARE, E.W. (1978). Hyperosmolar non-ketotic diabetes and cimetidine. *Lancet* i, 1202.

PUI, C.H., BURGHEN, G.A., BOWMAN, W.P., AND AUR. R.J. (1981). Risk factors for hyperglycaemia in children with leukaemia receiving L-asparginase and prednisolone. *J. Pediat.* **99**, 46–50.

QUEVEDO, S.F., KRAUSS, D.S., CHAZAN, J.A., CRISAFULLI, F.S., AND KAHN, C.B. (1981). Fasting hypoglycaemia secondary to disopyramide therapy. *J. Am. med. Ass.* **245**, 2424.

RAMU, A., SLONIN, E.A., AND EGAL, F. (1973). Hyperglycaemia in acute malathion poisoning. *Isr. J. med. Sci.* **9**, 631.

RAPOPORT, M.I. AND HURD, H.F. (1964). Thiazide-induced glucose intolerance treated with potassium. *Arch. intern. Med.* **113**, 405-8.

REDDY, J. (1981). Hyperglycaemia and renal failure related to the use of ritodrine. *NZ med J.* **93**, 354-5.

REVENO, W.S. AND ROSENBAUM, H. (1968) Propranolol and hypoglycaemia. *Lancet* i, 920.

ROEDIGER, P.M., ALDEN, J., BEARDWOOD, D., AND HUTCHISON, J.C. (1964). Benzothiadiazines and diabetes mellitus. II. Comparison of hyperglycemia patients with diabetes mellitus. *Curr. ther. Res.* **6**, 670-6.

ROGER, P., NOGUE, F., RAGNAND, J.M., MANCIET, G., AND DOUMAX, V. (1975). Hypoglycaemia apres maléate de perhexiline. *Nouv, Presse med.* **4**, 2663.

ROSENBAUM, L.H. AND ROSENBAUM, H. (1982). Chlorpropamide-induced hypoglycaemia. *J. Am. med. Ass.* **247**, 813-16.

RUNYAN, J.W.JR. (1962). Influence of thiazide diuretics on carbohydrate metabolism in patient with mild diabetes. *New Engl. J. Med.* **267**, 541-3.

SAMAAN, N.A., DOLLERY, C.T., AND FRASER, R. (1963). Diabetogenic action of benzothiadiazines. *Lancet* ii, 1244-6.

SAMOLS, E. AND MARKS, V. (1966). The treatment of hypoglycaemia with diazoxide. *Proc R. Soc. Med.* **59**, 811-14.

SCHAFFNER, F. (1969). Paradoxical elevation of serum cholesterol by clofibrate in patients with primary biliary cirrhosis. *Gastroenterology* **57**, 253-5.

SCHUBERT, G.E. AND SCHULTE, H.D. (1963). Contribution to the clinical picture of steroid diabetes. *Dtsch. med. Wschr.* **88**, 1175-88.

SELTZER, H.S. AND ALLEN, E.W. (1965). Inhibition of insulin secretion in diazoxide diabetes. *Diabetes* **14**, 439.

SHAPIRO, A.P., BENEDEK, T.G., AND SMALL, J.L. (1961). Effects of thiazides on carbohydrate metabolism in patients with hypertension. *New Engl. J. Med.* **265**, 1028-33.

SILLS, J.A., JONES, R.N.T., AND TAYLOR, W.H. (1980). Valproate, hyperammonaemia, and hyperglycinaemia. *Lancet* ii, 260-1.

SIMILA, S, von WENDT, L, HARTIKAINEN-SORRI, A.L., KÄÄPÄ, P. AND SAUKKONEN, A.L. (1979). Sodium valproate, pregnancy, and neonatal hyperglycinaemia. *Arch. Dis. Childh.* **54**, 985-6.

SMITH, M.J., AYNSLEY-GREEN, A., AND REDMAN, L.W. (1982). Neonatal hyperglycaemia after prolonged maternal treatment with diazoxide. *Br. med. J.* **284**, 1234.

STRISOWER, E.H., ADAMSON, G., AND STRISOWER, B. (1968). Treatment of hyperlipidaemias. *Am. J. Med.* **45**, 488-501.

SUGAR, S.N.J. (1961). Diabetic acidosis during chlorothiazide therapy. *J. Am. med. Ass.* **175**, 618-19.

SUSSMAN, K.E., STJERNHOLM, M.R., AND VAUGHAN, G.D. (1967). Propranolol and hypoglycaemia. *Lancet* i, 626.

SVENNINGSEN, N.W. (1982). Follow-up studies on pre-term infants after maternal β-receptor agonist treatment. *Acta-obstet. gynecol. scand Suppl.* **108**, 67-70.

SZABO, A.J., COLE, H.S., AND GRIMALDI, R.D. (1970). Glucose tolerance in gestational diabetic women during and after treatment with a combination-type oral contraceptive. *New Engl. J. Med.* **282**, 646-50.

TABACHNIK, I.I., GULBENKIAN, A., AND SEIDMAN, F. (1964). The effect of a benzothiadiazine, diazoxide, on carbohydrate metabolism. *Diabetes* **13**, 408-18.

TADA, K., CORBEEL, L., EECKELS, R., AND EGGERMONT, E. (1974). A block in glycine cleavage reaction as a common mechanism in ketotic and non-ketotic hyperglycinaemia. *Pediat. Res.* **8**, 721-3.

TAKASU, N., YAMADA, T., MIURA, H., SAKAMOTO, S., KORENAGA, M.,

NAKAJIMA, K., AND KANSYAMA, M. (1982). Rifampicin-induced early phase hyperglycaemia in humans. *Am. Rev. resp. Dis.* **125**, 23-7.

TAKETA, K. AND YAMAMOTO, Y. (1980). Hypoglycaemic effect of disopyramide in a case of diabetes mellitus under insulin treatment. *Acta med. Okayama* **34**, 289-92.

TAYLOR, M.B. AND KASS, M.B. (1968). Effect of oral contraceptives on glucose metabolism. *Am. J. Obstet. Gynecol.* **102**, 1035-8.

THOMAS, D.J.B., GILL, B., STUBBS, AND W.A., BROWN, P. (1977). Salbutamol-induced diabetic ketoacidosis. *Br. med. J.* ii, 438.

THOMAS, D.W. GILLIGAN, J.E., EDWARDS, J.B., AND EDWARDS, R.G. (1972). Lactic acidosis and osmotic diuresis produced by Xylitol infusion. *Med. J. Austr.* **1**, 1246-8.

——, EDWARDS, J.B., GILLIGAN, J.E., LAWRENCE, J.R., AND EDWARDS, R.G. (1972). Complications following intravenous administration of solutions containing xylitol. *Med. J. Aust.* **1**, 1238-46.

TKACH, J.R. (1982). Indomethacin induced hyperglycaemia in psoriatic arthritis *J. Am. Acad. Dermatol.* **7**, 802.

TOIVONEN, S. AND MUSTALA, O. (1966). Diabetogenic action of frusemide. *Br. med. J.* **1**, 920-1

TOMKINS, A.M., JONES, R., AND BLOOM, A. (1972). Lactic acidosis occurring during phenformin therapy. *Postgrad. med. J.* **48**, 386-7.

TRANQUADA, R.E., GRANT, W.J., AND PETERSON, C.R. (1966). Lactic acidosis. *Arch. intern. Med.* **117**, 192-202.

TURPIN, G. HÔP. AND MENAGE, J.J. (1973). Study of metabolism of circulating lipids during oral contraception. *Sem. Hop., Paris* **49**, 2833.

UNGER, R.H. (1966). Treatment of hypoglycaemia associated with pancreatic and extrapancreatic tumours. *Mod. Treatm.* **3**, 386-98.

VAN BUUREN, H.R., VAN DER VELDEN, P.C., KOOREVAR, G., AND SILBERBUSCH, J. (1982). Propranolol increases arterial ammonia in liver cirrhosis. *Lancet* ii, 951-2.

VAN LOON, G.H. AND APPEL, N.M. (1981). β-endorphin induced hyperglycaemia is mediated by increased central sympathetic outflow to adrenal medulla. *Brain Res.* **204**, 236-41.

VARMA, S.K., HEANEY, S.J. WHITE, W.G., AND WALKER, R.S. (1972). Hyperlacticacidaemia in phenformin-treated diabetics. *Br. med. J.* **1**, 205-6.

VESSEY, M.P. (1970). Thrombosis and the Pill. *Prescribers J.* **10**, 1-7.

WAINE, H., FRIEDEN, E.H., CAPLAN, H.I., AND COLE, T. (1963). Metabolic effects of Enovid in rheumatic patients. *Arthr. Rheum.* **6**, 796.

WALDEN, P.A.M., DENT, J., AND BAGSHAWE, K.D. (1979). Fatal hypoglycaemia after Cytotoxic Chemotherapy. *Cancer* **44**, 2029-31.

WALES, J.K., GRANT, A., AND WOLFF, F.W. (1968). Studies on the hyperglyaemic effect of non-thiazide diuretics. *J. Pharmacol. exp. Ther.* **159**, 229-35.

WALLACE, R.B., HOOVER, J., BARRETT-CONNER, E., RIFKIND, B.M., HUNNINGLAKE, D.B., MACKENTHUN, A., AND HEISS, G. (1979). Altered plasma lipid and lipoprotein levels associated with oral contraceptive and oestrogen use. Report of the Medications Working Group of the Lipid Research Clinics Program. *Lancet* ii 111-14.

WALKHAUS, W. (1967). Zur diabetogenen Wirkung von Saluretika. *Wien. Klin. Wschr.* **14**, 256.

WEINER, P., PELLED B., ALSTER, R. AND PLAVNICK, L. (1982). Propranolol-induced hypoglycaemia in a nondiabetic patient during acute coronary insufficiency. *Isr. J. med. Sci.* **18** 725–6.

WELLER, J.M. AND BORONDY, M. (1967). Inhibitory effect of chlorothiazide *in vitro* on glucose metabolism of adipose tissue. *Proc. Soc. Exp. Biol. Med.* **124**, 220–3.

WINGRAVE, S.J., KAY, C.R., AND VESSEY, M.P. (1979). Oral contraceptives and diabetes mellitus. *Br. med. J.* **1**, 23.

WILKINS, R.W. (1959). New drugs for the treatment of hypertension. *Ann. intern. Med.* **50**, 1–10.

WISE, P.H., CHAPMAN, M., THOMAS, D.W., CLARKSON, A.R., HARDING, P.E., AND EDWARDS, J.B. (1976). Phenformin and lactic acidosis. *Br. med. J.* **1**, 70.

WOLFF, F.W. AND PARMLEY, W.W. (1963). Aetiological factors in benzothiadiazine hyperglycaemia. *Lancet* **ii**, 69.

——, ——, WHITE, K., AND OKUN, R. (1963). Drug-induced diabetes; diabetogenic activity of long-term administration of benzothiadiazines. *J. Am. med. Ass.* **185**, 568–74.

WOOD, S.M., MILNE, J.R., EVANS, S.F., AND RODGERS, P. (1981). Effect of dobutamine on insulin requirements in diabetics. *Br. med. J.* **282**, 946–7.

WOODBURY, D.M. (1952). Effects of chronic administration of anticonvulsant drugs alone and in combination with desoxy-corticosterone on electroshock seizure threshold and tissue electrolytes. *J. Pharmacol. exp. Ther.* **105**, 46–57.

WOODS, H.F. (1971). Some aspects of lactic acidosis. *Br. J. hosp. Med.* **11**, 668–76.

—— AND ALBERTI, K.G.M.M. (1972). Dangers of intravenous fructose. *Lancet* **ii**, 1354–7.

WOODWARD, T.L., BURGHEN, G.A., KITABCHI, A.E., AND WILIMAS, J.A. (1981). Glucose intolerance and insulin resistance in aplastic anaemia treated with oxymethalone. *J. Endocrinol Metab.* **53**, 905–8.

WRAY, R. AND SUTCLIFFE, S.B.J. (1972). Propranolol-induced hypoglycaemia and myocardial infarction. *Br. med. J.* **2** 592.

WYNN, V. AND DOAR, J.W.H. (1966). Some effects of oral contraceptives on carbohydrate metabolism. *Lancet* **ii** 715–19.

—— AND —— (1969). Some effects of oral contraceptives on carbohydrate metabolism. *Lancet* **ii**, 761–6.

——, ——, AND MILLS, G.L. (1966). Some effects of oral contraceptives on serum-lipid and lipoprotein levels. *Lancet* **ii**, 720–3.

——, ——, ——AND STOKES, T. (1969). Fasting serum triglyceride, cholesterol and lipoprotein levels during oral contraceptive therapy. *Lancet* **ii**, 756–60.

ZATUCHNI, J. AND KORDASZ, F. (1961). The diabetogenic effect of thiazide diuretics. *Am. J. Cardiol* **7**, 565–7.

19 Drug-induced disorders of mineral metabolism

L. OFFERHAUS

Mineral and fluid balance disturbances are quite common in clinical medicine. Drugs are only relatively rarely responsible for such changes, but they may aggravate pre-existing disorders of absorption, fluxes, and excretion. As all body mineral and fluid pools are in a continuous dynamic state, only minor changes in regulatory mechanisms or specific fluxes may over longer periods cause major balance shifts (Brass and Thompson 1982; Oh and Carroll 1983). The following review will discuss the interaction of a number of drugs with such pathophysiological mechanisms and the resulting changes in the balance of water, sodium, potassium, magnesium, calcium, and related ions.

Altered sodium levels

Hyponatraemia

Hyponatraemia is not an unusual finding in critical care medicine. Osmolarity may occasionally be normal (hyperlipaemia, hyperproteinaemia), but usually it is lowered, either from a loss of sodium ions, water retention, or both, causing cellular overhydration. Drug-induced hyponatraemia may either be caused by diuretic therapy in the presence of volume depletion, reduced glomerular filtration, and normal levels of antidiuretic hormone (i.e. the kidney is unable to correct the excessive loss of sodium, as in salt-wasting renal disease), or by the syndrome of inappropriate ADH secretion (SIADH): water retention despite hypoosmolarity and normal fluid balance. The syndrome was first described in the presence of certain tumours, pulmonary and central nervous system disorders, glucocorticosteroid deficiency, and hypothyroidism. Despite continuing renal sodium losses in iso- or even hypoosmolar urine, arterial fluid volume and renal function are normal. A number of different drugs are now known to cause SIADH and hyponatraemia either by the enhancement of the effect of endogenous ADH or by a direct renal effect. Certain other drugs may cause simultaneous retention of water and sodium ions through a direct renal tubular effect: The kidney retains water and sodium, and, though arterial fluid volume is increased, serum osmolarity remains within the normal range (Oh and Carroll 1983).

Drug-induced syndrome of inappropriate ADH secretion

Chlorpropamide and other sulphonylurea derivatives

Since it became known that chlorpropamide may cause SIADH and hyponatraemia, and may therefore also be useful in the treatment of ADH deficiency (diabetes insipidus), a considerable number of cases of hyponatraemia, sometimes with overt signs of water intoxication (headache, drowsiness, weakness, disorientation, stupor, and eventually coma) have been described in the literature. The drug has no direct antidiuretic effect, but its action on the renal tubules depends on the presence of endogenous ADH. Most reported cases were elderly patients with decreased renal function on long-term treatment. The coexistence of congestive heart failure seems to predispose to the condition, as well as (relative) overdosage and/or elevated serum levels. The incidence in a diabetic population varies between 4 and 16 per cent. (Brass and Thompson 1982). Though isolated cases of SIADH and hyponatraemia due to tolbutamide have been described, the incidence of this complication in diabetic patients treated with this drug seems to be quite rare. It is not improbable that the difference with chlorpropamide is explained by the extremely long half-life and duration of action (and the resulting accumulation) of the latter drug. Of the other sulphonylurea derivatives thus far only glipizide has been implicated in the development of SIADH and hyponatraemia (Ducobu and Dupont 1981).

Carbamazepine

Whether carbamazepine has an antidiuretic action on the renal tubules of its own in the absence of endogenous ADH or not has remained a controversial problem (Meinders *et al.* 1974). The drug has been effectively used in the treatment of diabetes insipidus. Hyponatraemia, though definitely rare, has been observed during treatment with large doses (700–1800 mg/day) and/or elevated serum concentrations of the drug. The antidiuretic effect may be masked by concurrent administration of phenytoin, which inhibits ADH secretion (Sordillo *et al.* 1978).

Other drugs implicated in development of SIADH

A number of other drugs have occasionally caused SIADH-like syndromes, dilutional hyponatraemia, and/or clinical signs of water intoxication. With the possible exception of vincristine and oxytocin, the mechanism by which the drug causes such symptoms has rarely been elucidated. In none of the other cases reported in the literature did hyponatraemia or SIADH lead to clinical complications. Vincristine seems to cause SIADH through a direct neurotoxic action on the neurohypophysis (Rosenthal and Kaufman 1974); the development of SIADH seems to run parallel to other symptoms of neurotoxicity. The toxicity seems to be dose-related and has been confirmed histologically in animal experiments. Severe acute water intoxication has been described in parturient women receiving large intravenous doses of oxytocin together with an abundant intravenous fluid supply. Fortunately the number of such reports has declined since 1970, as it became clear that such toxicity can be avoided by a more careful regulation of fluid intake and monitoring of venous pressure and urine flow. A single report has mentioned water retention and hyponatraemia due to cyclophosphamide. In isolated cases similar symptoms during treatment with clofibrate, tiotixene, thioridazine, amitryptiline, tranylcypromine, biguanides, paracetamol (large single doses), diazoxide, isoprenaline (Brass and Thompson 1982; Griffin 1982), clonidine (Burrows and Gribbin 1979), ibuprofen (Blum and Aviram 1980), and even with chlorthalidone and related thiazide diuretics (Hamburger *et al.* 1981) have been mentioned. Complicating factors, such as congestive heart failure, often confound the assessment of the causality (Nicholls *et al.* 1980), though with some drugs a positive rechallenge has been definite proof of the direct link.

Use of drugs in the maintenance treatment of diabetes insipidus

If reduction of water intake is not feasible or impossible some drugs can be advantageously used in the treatment of chronic hyponatraemia, whatever its cause. Both lithium and the tetracycline analogue demeclocycline, which reduce renal cyclic AMP production and may also antagonize its action, have been used in small doses to that purpose, but the application is limited by the potential nephrotoxicity of the drugs. The same effect may be obtained with the 'loop' diuretics, furosemide and ethacrynic acid (Oh and Carroll 1983).

Drug-induced sodium loss as a cause of hyponatraemia

Excessive treatment with thiazide diuretics, especially if complicated by hypokalaemia and volume contraction, may occasionally result in hyponatraemia with high urinary osmolality and high urinary sodium concentrations despite lowered serum sodium levels. A particularly dangerous situation, especially in elderly patients, seems to exist if thiazide diuretics are combined with the potassium-sparing agent amiloride, which drug has a profound influence of its own on sodium fluxes across biological membranes in animal models (Tarssanen *et al.* 1980; Griffing *et al.* 1983).

Hypernatraemia

Hypernatraemia may either be caused by relative water loss or by a sodium excess. It is always characterized by an increase in plasma osmolality and cellular dehydration. Most drug-induced forms of hypernatraemia are caused by excessive renal water loss, e.g. renal diabetes insipidus, occasionally as a result of more or less permanent renal damage, or, more rarely, pituitary diabetes insipidus (central inhibition of ADH secretion). Excessive sodium (re-) absorption has incidentally been reported as a result of drug treatment. The renal effects of some drugs on water reabsorption have been effectively used in the treatment of SIADH.

Lithium

Renal diabetes insipidus caused by long-term treatment with lithium salts seems to be far more common than is usually realized. Urinary concentration disorders have been reported in 16 to 96 per cent (average 37 per cent) of all lithium-treated patients, many of whom manifest with all the clinical characteristics of renal diabetes insipidus (Hansen 1981; De Paulo *et al.* 1981). Though both the clinical picture and the microscopical changes are usually reversible after stopping lithium administration, continued treatment may lead in approximately 14 per cent of all such patients to a persisting irreversible interstitial nephropathy with decreased glomerular filtration rate (Hestbech *et al.* 1977; De Paulo *et al.* 1981), histologically characterized by interstitial cortical fibrosis, dilated tubules with microcyst formation, swelling of tubular mitochondria, and nuclear pycnosis. The severity of these changes depends on the duration of treatment and the total dose of lithium (Hansen *et al.* 1979; Aurell *et al.* 1981). The response to exogenous vasopressin is abolished. The most obvious clinical symptoms are polydipsia and polyuria, and occasionally severe dehydration and life-threatening hypernatraemia have been observed.

Other causes of drug-induced diabetes insipidus

The tetracycline derivative demeclocycline has been implicated in a number of cases of mild nephrogenic diabetes insipidus, mainly manifesting itself as a decreased urinary concentrating ability, unresponsive to exogenous ADH administration. The renal tubular effect is probably caused by interference with the function of cyclic AMP. The changes are reversible on discontinuing the drug. The anaesthetic agent methoxyflurane is metabolized to oxalic acid; deposition of oxalic acid crystals in renal tissue, particularly in patients with advanced renal failure, may occasionally lead to renal damage and the clinical symptoms of nephrogenic diabetes insipidus with polyuria and hypernatraemia. However, the exact mechanism is not quite clear, because fluoride ions may also be implicated. Some cases of nephrogenic diabetes insipidus have been described after overdosages of propoxyphene and after therapeutic doses of the fungicide antibiotic amphotericin B, but this seems to be an extremely rare complication.

The only drug — except alcohol — which has been reported to cause inhibition of ADH secretion at the level of the pituitary is phenytoin; this property has also been successfully used in the treatment of SIADH, and it may mask the symptoms of carbamazepine-induced SIADH.

Other causes of drug-induced hypernatraemia

Hypertonic oral or intravenous sodium salts (mainly chloride, bicarbonate, citrate, and lactate) may occasionally cause transient hypernatraemia, particularly in the rehydration of infantile diarrhoea, but it is rarely life-threatening. In some reports the osmotic action of lactulose syrup in patients with hepatic pre-coma seems to have contributed to a rise of serum sodium levels. The mineralocorticoid action of high glucocorticosteroid doses may rarely cause hypernatraemia, which can be antagonized by the aldosterone antagonist spironolactone.

Drug-induced sodium and water retention

Well-known causes of renal retention of sodium ions together with the appropriate quantity of fluid are the corticosteroids, drugs with pronounced mineralocorticoid action such as carbenoxolone, the oral-contraceptive steroids, and a number of non-steroidal anti-inflammatory agents. The water- and salt-retaining properties of the corticosteroids are to some degree dependent on the mineralocorticoid action of these drugs, being more pronounced with the 'first-generation' drugs than with newer drugs like dexamethasone. Fluid and sodium retention always go together with a loss of potassium ions. Since the introduction of a great number of 'low-oestrogen' contraceptive pills, fluid retention, weight-gain and hypertension have become far less important than some years ago. Though oedema, blood pressure increase, and weight-gain used to be frequent complications of treatment with some of the older non-steroidal anti-inflammatory drugs such as phenylbutazone and its main metabolite oxyphenbutazone, these complications are much less of a problem with the second- and third-generation antirheumatic drugs, with the possible exception of indomethacin, which may also completely antagonize both the diuretic and the anti-hypertensive effects of the thiazide diuretics and the 'loop' diuretics bumetanide and furosemide, possibly through its effect on renal prostaglandin synthetase. Since the introduction of the histamine H_2-blockers the anti-ulcer drug carbenoxolone, which combines a specific action on gastric ulcer healing with a strong — and sometimes life-threatening — mineralocorticoid effect, has lost most of its importance, and it is to be hoped that it will slowly pass into oblivion.

Altered potassium levels

Hypokalaemia

Hypokalaemia is defined as a reduction in the plasma potassium concentration and may therefore not necessarily reflect a decrease of total body potassium stores. A low plasma potassium may either result from a renal or intestinal loss of potassium ions, a reduced potassium intake or a shift of potassium ions into the cell, as in the correction of acidosis or during the concurrent administration of glucose and insulin, for instance, in the treatment of diabetic coma.

Drug-induced changes in plasma-potassium levels are mainly caused by increased renal excretion of potassium ions through a number of different mechanisms. Reports on other causes are mainly anecdotal: Massive laxative abuse may rarely cause profound hypokalaemia and dehydration with resultant renal tubular damage. Such 'factitious' hypokalaemia ('pseudo-Bartter's syndrome') has also been observed as a consequence of abuse or misuse of diuretics (Spratt and Pont 1982; Offerhaus

1984). Moderate decreases of serum potassium values, caused by a glucose-transport-linked intracellular shift have been reported during intravenous infusion of β-adrenergic agonists (salbutamol, isoxsuprine, ritrodrine, and fenoterol) in the treatment of asthma or premature labour.

Diuretic-induced renal potassium loss

Though the thiazide diuretics have been much maligned as a frequent cause of life-threatening hypokalaemia (Perez-Stable and Caralis 1983), normal long-term thera-peutic use of thiazide diuretics in hypertensive patients does not result in a depletion of total body potassium (Kassirer and Harrington 1977; Sandor *et al.* 1982). In a recent review of 17 publications on the occurrence of ventricular arrhythmias possibly induced by hypokalaemia, only one fulfilled all criteria for a definite causal relationship, i.e. adequate documentation of a temporal relationship between serum potassium and the arrhythmia, an adequate monitoring system, exclusion of other possible causes, presence of controls, and resolution of the arrhythmia after restoring serum potassium to normal (Harrington *et al.* 1982). The same equivocal results were obtained from the analysis of the data from a large-scale trial of thiazide diuretics in mild-to-moderate hypertension. No unique or unambiguous link could be made between the occurrence of ventricular arrhythmias and low plasma potassium concentrations (MRC Working Party 1983). Nevertheless depletion of body potassium stores might occur, and might have serious clinical consequences, in some patients with potassium wasting, in patients with congestive heart failure or uncompensated hepatic cirrhosis and secondary hyperaldosteronism, and in patients on concomitant digitalis therapy. All benzothiadiazine — and related oral diuretics — including clorthalidone (Ram *et al.* 1981; Falch and Schreiner 1981), indapamide (De Ortiz *et al.* 1983), and xipamide (Boulton and Hardisty 1982) may occasionally cause life-threatening hypokalaemia, but the risk seems to be equal or even greater with the use of the potent loop diuretics, particularly furosemide.

Other drugs causing increased renal potassium loss

Transient tubular damage caused by some nephrotoxic antibiotics (aminoglycosides, particularly gentamicin, outdated tetracycline, neomycin absorbed from large burnt areas, and amphotericin B) may occasionally cause increased renal loss of potassium ions, often together with magnesium and calcium. However, the evidence is equivocal and the mechanism is not fully elucidated. In most reports high doses were used for long periods

without the currently usual pharmacokinetic monitoring of serum levels. Large tubular loads of anionic drugs such as massive doses of penicillin or synthetic penicillin derivatives may need potassium ions to be excreted, thus causing profound losses of potassium ions. Particularly, the use of carbenicillin in haematological patients has repeatedly been implicated as a cause of severe hypokalaemia. Levodopa may occasionally cause hypokalaemia through a still undefined mineralocorticoid action which can be blocked by methyldopa. Moreover, all drugs with mineralocorticoid action, such as deoxycorticosterone, fludrocortisone, most glucocorticosteroids and carbenoxolone (as well as the parent compound, liquorice) may cause hypokalaemia. Rarer causes include lithium nephropathy and rapid infusion of the ganglion-blocking drug trimetaphan in anaesthesia.

Hyperkalaemia

A considerable number of drugs may increase plasma-potassium concentrations, either by a shift of potassium ions from the cells into the extracellular space, or by a reduction of renal potassium excretion. Shifts over cellular membranes are mainly caused by drugs which affect one of several cellular ion transport mechanisms, or, if potassium ions are leaking through damaged cellular membranes into the extracellular fluid tubular retention of potassium ions may either be caused by a direct effect on the tubular epithelium or via inhibition of aldosterone or angiotensin secretion. Moreover, excessive oral intake of potassium salts — either as a pharmaceutical preparation for the correction or prophylaxis of hypokalaemia, or as a salt substitute, may cause hyperkalaemia if renal potassium output does not keep pace with input (renal insufficiency, simultaneous use of so-called potassium-sparing diuretics).

Drug-induced hyperkalaemia caused by cellular ion shifts

A massive potassium outflux from muscle cells may be caused by the depolarizing agent suxamethonium, which may be particularly dangerous in patients with an added risk of high plasma potassium levels, as in severe trauma, extensive burns, and/or renal damage. The effect is apparently dose-related and can be partially prevented by the simultaneous administration of d-tubocurarine. Other depolarizing drugs are not known to carry this particular risk. Massive tumours responding rapidly to cytostatics, such as Burkitt's lymphoma and lymphosarcoma treated with cyclophosphamide, may release enough potassium ions to markedly raise plasma potassium concentration. Temporary hyperkalaemia during infusion of the amino-acid arginine in the treatment of metabolic alkalosis may also be life-threatening in patients with impaired renal and/or hepatic function. As transmembrane transport of potassium ions is partially regulated by adrenergic mechanisms, and β-adrenergic agonists are known to lower plasma potassium, it is not unexpected that β-adrenoceptor blocking drugs may cause an outward shift of potassium ions from the cell into the extracellular space. This effect should allegedly be an advantage for the combined use of β-blockers and thiazide diuretics, though it cannot possibly prevent intracellular potassium depletion. Excessive hyperkalaemia during β-blockade has thus far only been observed after cardiac bypass surgery (Brass and Thompson 1982).

Drug-induced hyperkalaemia due to renal mechanisms

Life-threatening hyperkalaemia has mainly been observed as a consequence of the use of the potassium-retaining drug triamterene as sole agent, of combination preparations of thiazide diuretics, and one of these so-called potassium-sparing diuretics, i.e. triamterene, amiloride, and the aldosterone antagonist spironolactone, or with accidental combinations of these drugs with oral potassium salts. In a hospital population monitored by the Boston Collaborative Drug Surveillance Program 4.1 per cent of 2081 patients receiving potassium supplements and 7.3 per cent of 783 patients receiving spironolactone had plasma potassium levels over 6.0 mmol/ (Lawson *et al.* 1982). Moderate hyperkalaemia may be caused by long-term, high-dose intravenous administration of heparin, which has aldosterone-inhibiting properties. The angiotensin-converting enzyme inhibitors captopril and enalapril may occasionally raise plasma potassium concentrations, particularly in patients with poor renal function; this property has been successfully used to combat hypokalaemia induced by primary or secondary hyperaldosteronism (Griffin 1982; Griffing *et al.* 1983).

Hyperkalaemia due to drug-induced hyporeninaemic hypoaldosteronism

Non-steroidal anti-inflammatory drugs have been used in treating prostaglandin-induced hypokalaemia in Bartter's syndrome. However, for the same reason they may cause hyperkalaemia in patients with previously normal serum potassium levels, particularly in patients with pre-existing renal disease. The clinical picture is characterized by hypervolaemia, hyperchloraemic acidosis, moderate-to-severe renal insufficiency, and low plasma aldosterone and renin concentrations. The condition is reversible on discontinuing the offending drug. In appro-

ximately 25 per cent of the published cases hyperkalaemia was symptomatic or even life-threatening. Indomethacin has been reported to cause hyporeninaemic hypoaldosteronaemic hyperkalaemia even in patients with previously normal or only slightly disturbed renal function (Findling *et al.* 1980; Goldszer *et al.* 1981). A similar case due to the administration of oxyphenbutazone to a gouty patient with an unilaterally normally functioning kidney has been published (Merkt 1983). A rechallenge with the same drug was positive. Both plasma aldosterone and plasma renin activity were extremely low and unresponsive to standing upright or to diuretic treatment. These abnormal findings resolved quickly and spontaneously after stopping oxyphenbutazone. Piroxicam has also been reported to cause severe hyperkalaemia (Miller *et al.* 1984). NSAID-induced hyperkalaemia seems to be far more common than usually assumed (Zimran *et al.* 1985).

Hyperkalaemia induced by renal potassium overloading

Mention has already been made of the risk of causing dangerous hyperkalaemia by the administration of oral potassium supplements (Wilson and Farndon 1982), particularly in combination with potassium-sparing agents (Lawson *et al.* 1982). It is little known that some drugs, such as penicillin and methicillin, may contain enough potassium to raise plasma potassium concentrations to dangerous levels, particularly in patients with impaired renal function. Plasma potassium should be very closely monitored in such patients, particularly in those suffering from salt-loosing nephritis.

Altered magnesium levels

Drug-induced magnesium depletion

It has been known since a long time that chronic treatment with thiazide diuretics may, besides a loss of potassium ions, also cause a depletion of body stores of magnesium (Offerhaus 1984). Interest in the consequences of low serum magnesium levels has recently been revived by the observation that adequate magnesium suppletion in magnesium-deficient patients on long-term diuretic treatment facilitates correction of intracellular potassium (Swales 1982). The correction of magnesium depletion may help to restore responsiveness to diuretics and to correct hypokalaemia, particularly, in elderly patients with severe congestive heart failure with diuretic-induced hypokalaemia and secondary hyperaldosteronism. In a selected group of patients with diminished magnesium intake (magnesium-deficient diet, soft water supply, chronic alcoholism) favourable experience has been gained by correcting these mineral deficiencies with intramuscular magnesium sulphate, a diet rich in magnesium (bananas) or spironolactone (Sheehan and White 1982). The effect of other potassium-sparing agents (triamterene and amiloride) on magnesium metabolism has not been properly established. Hypomagnesaemia in hypertensive patients on long-term diuretic treatment — cannot be corrected with amiloride and clinical symptoms are exceptions rather than the rule (Webster and Dickner 1984). 'Concomitant use of thiazide diuretics and sodium cellulose phosphate in the prevention of renal stone formation may cause profound magnesium depletion (Johansson et al. 1983).

Drug-induced magnesium excess

Isolated reports have pointed to the risk of excessive magnesium absorption in renal patients taking large doses of magnesium-containing antacids. Though under normal circumstances magnesium ions are extremely poorly absorbed, there are exceptions — particularly during concomitant treatment with vitamin D analogues for renal osteodystrophy — in which magnesium poisoning may occur, manifesting itself as lethargy, coma, circulatory collapse, and eventually respiratory paralysis and death. Preferably such patients should therefore not be treated with magnesium-containing antacids.

Drug-induced disorders of calcium metabolism

Diuretic-induced disorders

Thiazides and most other related diuretics — furosemide (Chandler 1977), chlorthalidone, indapamide, and tienilic acid (Bloch et al. 1981) — inhibit tubular secretion of calcium ions, a property which has been successfully used in preventing stone formation in idiopathic hypercalcinuria. However, such treatment has also been reported to cause transient hypercalcaemia (Offerhaus 1984) and increased mineral content of bone, which might be an added advantage in patients with osteoporosis and postmenopausal hypertensive women (Wasnich et al. 1983). However, calcium stone formation cannot be prevented unless overt hypercalcinuria is present; this treatment seems useless in the average patient with recurrent calcium stones. Inclusion criteria should therefore be very strict in order to achieve success. Though raised serum calcium concentrations have been mentioned in a great number of reports, full data on the degree of

volume depletion, serum albumin concentration, and the concentration of ionized calcium in the serum are rarely available, so these observations should be interpreted with utmost caution. However, persistent hypercalcaemia after discontinuation of thiazide treatment warrants an intensive search for parathyroid adenomas. On the other hand, furosemide causes transient hypercalcinuria, an effect which has been used in the treatment of acute hypercalcaemia.

Other drug-induced disorders of calcium metabolism

Hypercalcaemia

Natural or semisynthetic hormones and vitamins with a specific action on calcium metabolism such as calcitonin, parathyroid hormone, glucagon, and glucocorticoids may as a matter of course cause profound changes in calcium absorption, release, or excretion, particularly if no clear deficiency exists or if doses go beyond the deficiency. A particularly notorious example is hypervitaminosis D, which is often accidental (cumulation of vitamin D analogues from different medical and nonmedical sources), and often leaves permanent renal, dental, bone and other damage. Even the properly controlled use of certain shorter-acting vitamin D analogues in renal osteodystrophy such as dihydrotachysterol and other 1-alpha-HO analogues does carry a certain risk of life-threatening hypercalcaemia. Similar problems have been described in the treatment of post-thyroidectomy hypocalcaemia with dihydrotachysterol (Griffin 1982). Hypercalcaemia was in the past a well-known and dreaded complication of the treatment of disseminated bone metastases of mammary carcinoma with oestrogens and anabolic steroids. Since 1978 a considerable number of similar cases have been described during treatment with the oestrogen-receptor antagonist tamoxifen. Though hypervitaminosis A may be complicated by hypercalcaemia, normal prophylactic or therapeutic dosages of this vitamin have not been implicated in hypercalcaemia. However, a recent report suggests that vitamin A analogues, such as isotretinoin given in accepted therapeutic dosages, may occasionally cause hypercalcaemia (Valentic et al. 1983). An unusual case of hypercalcaemia resulting from massive absorption of calcium ions from resorbable haemostatic calcium acetate–containing compresses applied to a liver tear was described by Texier et al. (1982). Particularly in anuric patients or patients with haemorrhagic shock, it is clear that omitting to count the number of such compresses used during an operation and not monitoring the serum calcium concentration could be very hazardous. It is, however, questionable whether such compresses should be used at all.

Hypocalcaemia

Hypocalcaemia may be caused by hypercalcinuria as a result of renal tubular defects due to a number of cytostatic drugs, i.e. mithramycin, doxorubicin hydrochloride, cytarabine, and cisplatinum, often in combination with excessive magnesium losses. Such deficiencies should be looked for and corrected. Another well-known cause of hypocalcaemia is the administration of large oral or rectal doses of phosphates. The recent resurgence of interest in the use of intravenous sodium edetate infusions as an 'alternative' treatment of atherosclerosis opens the possibility of life-threatening complications due to hypocalcaemia and renal damage due to insoluble tubular deposits of calcium edetate complex.

Almost all anticonvulsant drugs and a number of hynotics with enzyme-inducing properties (methaqualone, glutethimide) have been implicated in the development of hypocalcaemia and resultant osteomalacia; though the mechanism is still far from clear, the resultant changes are very similar to those caused by the diphosphonates (etidronate, clodronate) in the treatment of Paget's disease of the bone and bone metastases (Brass and Thompson 1982). Excessive intestinal loss of calcium ions due to laxative (phenolphthalein) abuse has occasionally been reported to cause hypocalcaemia and osteomalacia (Frame *et al.* 1971).

Other drug-induced osteoarthropathies

A number of drugs may cause bone and joint disorders which are partly or indirectly related to disorders of calcium metabolism. Arthritis and arthralgia may be part of the clinical picture of drug-induced SLE (systemic lupus erythematosus) syndrome, i.e. due to hydrallazine and procainamide, or they may be due to serum-sickness type reactions caused by drug hypersensitivity. Some drugs may cause joint pain without demonstrable cause, i.e., some tuberculostatic drugs such as isoniazid, pyrazinamide, and ethionamide, particularly if given in combination; some of the older 'high-oestrogen' contraceptive pills, sulphonamides, and corticosteroids which produce peripheral manifestations of a generalized polyarteritis and some vaccines (Hart 1974). More recent, but isolated, observations mention the possibility that similar syndromes may be caused by nalidixic acid (Griffin 1979). It seems also likely that cimetidine may cause arthritis with joint effusion with or without muscle pain (*Current Problems* 1981), and that the new tetracyclic antidepressant mianserin may cause a diffuse polyarthritis, particularly of the small peripheral joints (hands, feet, knees and elbows), possibly due to an allergic reaction to the drug (*Current Problems* 1982).

Clinical syndromes such as glucocorticoid-induced osteoporosis and dexamethasone-induced avascular necrosis of the hip used to be rather common (Hart 1974), but due to the increased awareness of these hazardous complications of long-term and/or high-dose corticosteroids the incidence has fortunately decreased (Hart 1982). Some drugs (heparin, penicillamine, methotrexate — in children) if used in high doses during long periods, may exceptionally cause *osteoporosis*; the direct link with calcium metabolism is not always clear, because only the bone matrix may be affected, as in penicillamine-induced osteopathy (decreased synthesis of hydroxyproline?). Some drugs which are useful in the treatment of osteoporosis, such as fluoride and the diphosphonates (clodronate sodium, etidronate, and similar drugs) if used incautiously or as the sole means of treatment may actually cause osteomalacia (Boyce *et al.* 1982; Nagant de Deuxchaisnes *et al.* 1982). This paradoxical outcome is not always preventable by addition of calcium and/or vitamin D (Compston *et al.* 1980). Osteomalacia and ricketts, particularly in institutionalized children with lowered physical activity and exposure to sunlight, are known to result from anticonvulsant treatment, particularly if the two potent hepatic drug-metabolizing enzyme-inducers — phenytoin and phenobarbital — are given together (Hahn 1976). This complication can be prevented and, if necessary, treated with adequate vitamin D supplementation. However, other well-known enzyme-inducing agents, such as rifampicin in combination with isoniazid, used in the long-term treatment of tuberculosis in Indian patients, apparently do not cause either hypocalcaemia or osteomalacia (Perry *et al.* 1982).

Other iatrogenic disorders of mineral metabolism

Aluminium

In 1964 Lotz *et al.* reported that the use of excessive doses of aluminium preparations as phosphate binders in renal bone disease could be the cause of severe bone pain due to osteomalacia not responsive to vitamin D and analogues, and since 1970 it has been known that these symptoms are associated with elevated serum and bone aluminium concentrations (Berlyne *et al.* 1970). However, not until 1976 (Alfrey *et al.*) was the association made with the clinical picture of haemodialysis encephalopathy, manifesting itself as progressive dementia. Because bone tissue avidly binds aluminium ions, these conditions may only very slowly improve during haemodialysis with deionized dialysate or after transplantation. The growing number of such cases has led to the awareness of tap water as an additional source of aluminium ions in haemodialysis fluids, and steps have been taken on an international level to prevent this hazard: Aluminium concentrations in the dialysate over 20 μg/l are hazardous, and, if spontaneous aluminium levels in dialysate water exceed this value, the source should be treated by reverse osmosis or deionization. Overt aluminium intoxication has been successfully treated with deferoxamine (Milliner *et al.* 1984; Mudde *et al.* 1985). However, haemodialysis encephalopathy has not only been traced to this cause, but occasionally also to the use of aluminium-containing antacids in renal osteodystrophy (Kaye *et al.* 1983; Andreoli *et al.* 1984), though the benefit of such treatment is generally felt to outweigh the risks (Griffin 1983).

Recently a number of cases of aluminium intoxication due to the use of contaminated albumin solutions have been described (Milliner *et al.* 1985; Loeliger *et al.* 1985).

RECOMMENDED FURTHER READING

ANDREOLI, S.P., BERGSTEIN, J.M., SHERRARD, D.J. (1984). Aluminium intoxication from aluminium-containing phosphate binders in children with azotemia not undergoing dialysis. *New Engl. J. Med.* **310**, 1079–83.

BRASS, E.P., AND THOMPSON, W.L. (1982). Drug-induced electrolyte abnormalities. *Drugs* **24**, 207–28.

OFFERHAUS, L. (1984). Diuretics. In *Meyler's side effects of drugs*, 10th edn. (ed. N.M.G. Dukes and J. Elis, pp. 369–85. Excerpta Medica, Amsterdam.

OH, M.S. AND CARROL, H.J. (1983). Electrolyte disorders. In *The pharmacologic approach to the critically ill patient* (ed. B. Chernow, and C.R. Lake), pp. 715–27. Williams & Wilkins, Baltimore.

REFERENCES

ALFREY, A.C., LEGENDRE, G.R., AND KAEHNY, W.D. (1976). The dialysis encephalopathy syndrome: possible aluminium intoxication. *New Engl. J. Med.* **294**, 184–8.

AURELL, M., SVALANDER, C., WALLIN, L., AND ALLING, C. (1981). Renal function and biopsy findings in patients on long-term lithium treatment. *Kidney Int.* **20**, 663–70.

BERLYNE, G.M., BEN-ARI, J., AND PEST, D. (1970). Hyperaluminaemia from aluminium resin in renal failure. *Lancet* ii, 494–6.

BLOCH, R., STEIMER, C., WELSCH, M., AND SCHWARTZ, J. (1981). L'effet hypocalciurique de l'hydrochlorothiazide, de la chlorthalidone, de l'indapamide et de l'acide tienilique. *Therapie* **36**, 567–74.

BLUM, M. AND AVIRAM, A. (1980). Ibuprofen-induced hyponatremia. *Rheumatol. Rehabil.* **19**, 258–60.

BOULTON, A.J.M. AND HARDISTY, C.A. (1982). Ventricular arrhythmias precipitated by treatment with non-thiazide diuretics. *Practitioner* **226**, 125–8.

BOYCE, B.F., SMITH, L., FOGELMAN, I., RALSTON, S., AND BOYLE, I.T. (1982). Diphosphonates and inhibition of bone mineralisation. *Lancet* i 964–5.

BURROWS, A.W. AND GRIBBIN, B. (1979). Clonidine-induced dilutional hyponatremia. *Postgrad. med. J.* **55**, 42–5.

CHANDLER, P.T. (1977). Increased serum calcium levels induced by furosemide. *South. med. J.* **70**, 571–4.

COMPSTON, J.E., CHADHA, S., AND MERRETT, A.L. (1980). Osteomalacia developing during treatment of osteoporosis with sodium fluoride and vitamin D. *Br. med. J.* **II**, 910–11.

CURRENT PROBLEMS (1981). Cimetidine and arthropathy. *Current Problems*, no. 7.

—— (1982). Mianserin (Bolvidon, Norval) and arthropathy. *Current Problems*, no. 8.

DUCOBU, J. AND DUPONT, P. (1981). Glipizide-induced hyponatremia. *Therapie* **36**, 597–9.

FALCH, D.K. AND SCHREINER, A.M. (1981). Changes in urinary electrolytes versus serum electrolytes during treatment of primary hypertension with chlorthalidone alone and in combination with spironolactone. *Acta med. scand.* **209**, 111–14.

FINDLING, J.W., BECKSTROM, D., RAWSTHORNE, L., KOZIN,F., AND ITSKOVITZ, H. (1980). Indomethacin-induced hyperkalemia in three patients with gouty arthritis. *J. Am. med. Ass.* **244**, 1127–28.

FRAME, B., GUING, H.L., FROST, H.M., AND REYNOLDS, W.A. (1971). Osteomalacia induced by laxative (phenolphthalein) ingestion. *Arch. intern. Med.* **128**, 794–6.

GOLDSZER, R.C., COODLEY, E.L., ROSNER, M.J., *et al.* (1981). Hyperkalemia associated with indomethacin. *Arch. intern. Med.* **141**, 802–4.

GRIFFIN, J.P. (1979). Drug-induced disorders of mineral metabolism. In D'Arcy, P.F. and Griffin, J.P. *Iatrogenic diseases*, 2d. edn. pp. 226–38. Oxford University Press, Oxford.

GRIFFING, G.T., SINDLER, B.H., AURECCHIA, S.A., AND MELBY, J.C. (1983). Reversal of diuretic-induced secondary hyperaldosteronism and hypokalemia by enalapril (MK-421): A new angiotensin-converting enzyme inhibitor. *Metabolism* **32**, 711–16.

HAHN, T.J. (1976). Bone complications of anticonvulsants. *Drugs* **12**, 201–24.

HAMBURGER, S., KOPRIVICA, B., ELLERBECK, E., AND CORINSKY, J.O. (1981). Thiazide-induced syndrome of inappropriate secretion of antidiuretic hormone — time course of resolution. *J. Am. med. Ass.* **246**, 1235–6.

HANSEN, H.E. (1981). Renal toxicity of lithium. *Drugs* **22**, 461–76.

——, HESTBECH, J., SORENSEN, J.L., AND AURELL, M. (1979). Chronic interstitial nephropathy in patients on long-term lithium treatment. *Quart. J. Med.* **48**, 577–91.

HARRINGTON, J.T., ISNER, J.M. AND KASSIRER, J.P. (1982). Our national obsession with potassium. *Am. J. Med.* **73**, 155–9.

HART, F.D. (1974). Drug-induced arthritis. *Curr. Med. Res. Opin.* **2**, 505–12.

—— (1982). Systemic corticosteroids and corticotrophin. In *Drug treatment of the rheumatic diseases* (ed. F. D. Hart), pp. 92–103. ADIS Press, New York.

HESTBECH, J., HANSEN, H.E., AMDISEN, A., AND AURELL, M. (1977). Chronic renal lesions following long-term treatment with lithium. *Kidney Int.* **12**, 205–13.

JOHANSSON, G., BACKMAN, U., DANIELSON, B.G., FELLSTRÖM, B., LJUNGHALL, S., AND WIKSTRÖM, B., (1983). Magnesium metabolism during treatment with benroflumethiazide or sodium cellulose phosphate in renal stone disease. *Magn. Bull.* **1**, 4–7.

KASSIRER, J.P. AND HARRINGTON, J.T. (1977). Diuretics and potassium metabolism: a reassessment of the need, effectiveness and safety of potassium therapy. *Kidney Int.* **11**, 505–11.

KAYE, M. (1983). Oral aluminium toxicity in a non-dialysed patient with renal failure. *Clin. Nephrol.* **20**, 208–11.

LAWSON, D.H., O'CONNOR, P.C., AND JICK, H. (1982). Drug attributed alterations in potassium handling in congestive cardiac failure. *Eur. J. clin. Pharmacol.* **23**, 21–6.

LOELIGER, E.A. AND WOLFF, F.A. DE (1985). Aluminium contamination of albumin replacement solutions. *New Engl. J. Med.* **312**, 1389–90.

LOTZ, M., NEY, R., AND BARTTER, F.C. (1964). Osteomalacia and debility resulting from phosphorus depletion. *Transa. Ass. Am. Physcns.* **77**, 281–5.

Medical Research Council Working Party on mild to moderate hypertension (1983). Ventricular extrasystoles during thiazide treatment: substudy of MRC mild hypertension trial. *Br. med. J.* **2**, 1249–53.

MERKT, J. (1983). Oxyphenbutazon-induzierte Hyperkaliaemie beim selektiven Hyperaldosteronismus des Erwachsenen. *Deutsch. Med. Wochenschr.* **108**, 1535–6.

MEINDERS, A.E., CEJKA, V., AND ROBERTSON, G.L. (1974). The antidiuretic action of carbamazepine in man. *Clin. Sci. Molec. Med.* **47**, 289–99.

MILLER, K.P., LAZAR, E.J., AND FOTINO, S. (1984). Severe hyperkalaemia during piroxicam therapy *Arch. intern. Med.* **144**, 2414–5.

MILLINER, D.S., NEBEKER, H.G., OTT, S.M., ANDRESS, D.L., SHERRARD, D.J., ALFREY, A.C., SLATOPOLSKY, E.A., AND COBURN, J.W. (1984). Use of the deferoxamine infusion test in the diagnosis of aluminium-related osteodystrophy. *Ann. intern. Med.* **101**, 775–80.

MILLINER, D.S., SHINABERGER, J.H., SHUMAN, P., AND COBURN, J.W. (1985). Inadvertent aluminum administration during plasma exchange due to aluminum contamination of albumin-replacement solutions. *New Engl. J. Med.* **312**, 165–6.

MUDDE, A.H., ROODVOETS, A.P., AND GRONINGEN, K.VAN (1985). Hypercalcaemic osteomalacia and encephalopathy due to aluminium intoxication in haemodialysis patients. *Neth. J. Med.* **28**, 2–5.

NAGANT DE DEUXCHAISNES, C., ROMBOUTS-LINDEMANS, C., HUAUX, J.P., AND DEVOGELAER, J.P. (1982). Diphosphonates and inhibition of bone mineralisation. *Lancet* **ii**, 607–8.

NICHOLLS, M.G., ESPINER, E.A., IKRAM, H., AND MASLOWSKI, A.H. (1980). Hyponatraemia in congestive heart failure during treatment with captopril. *Br. med. J.* **2**, 909.

ORTIZ, H.DE, DE QUATTRO, E., STEPHANIAN, E., AND DE QUATTRO, V. (1983). Long-term effectiveness of indapamide in hypertension: neural, renin and metabolic responses. *Clin. Exp. Hypertension*, **A5**(5), 665–72.

PAULO, DE, J.R., CORREA, E.I., AND SAPIR, D.G. (1981). Renal toxicity of lithium and its implications. *Johns Hopkins med.J.* **149**, 15–21.

PEREZ-STABLE, E. AND CARALIS, P.V. (1983). Thiazide-induced disturbances in carbohydrate, lipid, and potassium metabolism. *Am. Heart J.* **106**, 245–51.

PERRY, W., EROOGA, M.A., BROWN, J., JENKINS, M.V., SETCHELL, K.D.R., AND STAMP, T.C.B. (1982). Calcium metabolism during rifampicin and isoniazid therapy for tuberculosis. *J.R. Soc. med.* **75**, 533–6.

RAM, C.V.S., GARRETT, B.N., AND KAPLAN, N.M. (1981). Moderate sodium restriction and various diuretics in the treatment of hypertension — effects of potassium wastage and blood pressure control. *Arch. intern. Med.* **141**, 1015–19.

ROSENTHAL,S. AND KAUFMAN,S. (1974). Vincristine neurotoxicity. *Ann. intern. Med.* **80**, 733–737.

SANDOR, F.F., PICKENS, P.T., AND CRALLAN, J. (1982). Variations of plasma potassium concentrations during long-term treatment of hypertension with diuretics without potassium supplements. *Br. med. J.* **1**, 711–12.

SHEEHAN, J. AND WHITE, A. (1982). Diuretic-induced hypomagnesaemia. *Br. med. J.* **2**, 1157–60.

SORDILLO, P., SAGRANSKY, D.M., MERCADO, R., AND MICHELIS, R.F. (1978). Carbamazepine-induced syndromes of inappropriate antidiuretic hormone secretion. *Arch. intern. Med.* **138**, 299–301.

SPRATT, D.I. AND PONT, A. (1982). The clinical features of covert diuretic use. *West. J. Med.* **137**, 331–5.

SWALES, J.D. (1982). Magnesium deficiency and diuretics. *Br. med. J.* **2**, 1377–78.

TARSSANEN, L., HUIKKO, M., AND ROSSI, M. (1980). Amiloride-induced hyponatremia. *Acta med. scand.* **208**, 491–4.

TEXIER, D., CHEVALLIER, P., PERROTIN, D., AND GUILMOT, J.-L. (1982). Hypercalcaemia associated with resorbable haemostatic compresses. *Lancet* **i**, 688–9.

VALENTIC, J.P., ELIAS, A.N., AND WEINSTEIN, G.D. (1983). Hypercalcemia associated with oral isotretinoin in the treatment of severe acne. *J. Am. med. Ass.* **250**, 1899–900.

WASNICH, R.D., BENFANTE, R.J., YANO, K., HEILBRUN, L., AND VOGEL, J.M. (1983). Thiazide effect on the mineral content of bone. *New Engl. J. Med.* **309**, 344–6.

WEBSTER, P.D., AND DYCKNER, T. (1984). Problems with potassium and magnesium in diuretic-treated patients. *Acta pharmacol. toxicol.* **54**, Suppl. 1, 59–65.

WILSON, R.G. AND FARNDON, J.R. (1982). Hyperkalaemic cardiac arrhythmia caused by potassium citrate. *Br.med.J.* **1**, 197.

ZIMRAN, A., KRAMER, M., PLASKIN, M., AND HERSHKO, C. (1985). Incidence of hyperkalaemia induced by indomethacin in a hospital population. *Br. med. J.* **291**, 107–8.

20 Drug-induced endocrine dysfunction

A. P. FLETCHER

Adrenal dysfunction

Two categories of toxic effects are observed in the therapeutic use of adrenocorticosteroids. Firstly, there are those resulting from continued use of doses in excess of normal physiological requirements, and, secondly, there are those resulting from withdrawal of the extended use of higher doses.

Corticosteroid-induced Cushing's disease

The principal complications resulting from prolonged therapy with corticosteroids are:

1. Fluid and electrolyte disturbances.
2. Hyperglycaemia and glycosuria.
3. Peptic ulceration which may bleed or perforate.
4. Osteoporosis with vertebral compression fractures.
5. Myopathy characterized by weakness of the proximal musculature of arms and legs and associated shoulder and pelvic musculature.
6. Psychoses, which may take various forms, e.g. nervousness, insomnia, changes in mood, and psychotic changes of manic depressive or schizophrenic types; suicidal tendencies are not uncommon.
7. Hypercoagulability of blood with thromboembolic episodes.
8. Cushingoid appearance, consisting of 'moon-face', 'buffalo-hump', supraclavicular fat pads, central obesity, striae, acne, and hirsutism.

Nielsen *et al.* (1963) described the side-effects in a series of 50 patients aged 30 to 50 years treated with corticosteroids for prolonged periods of time, up to 12 years in some cases. The overall duration of corticosteroid exposure in this series of rheumatoid arthritic patients was 300 patient-years. The total incidence of major side-effects was 24 per cent with mental changes; 18 per cent with peptic ulceration; 6 per cent with vertebral fractures; and an overall increase in serious infections.

Topical steroid preparations

Cushing's syndrome associated with the topical application of steroid preparations has been described previously and, in one case of a child on long-term betamethasone, death from Addisonian crisis was reported (Staughton and August 1975; Sneddon 1970). A further fatality was reported by Nathan and Rose (1979) involving a 60-year-old woman with an 8-year history of submammary intertrigo. This was treated with large amounts of clobetasol propionate and betamethasone valerate which had been repeatedly applied to ever-increasing areas of her trunk. The patient presented with typical Cushingoid appearance, paper-thin skin, and widespread bruising. She also had hypokalaemia, glycosuria, and raised blood glucose. Topical steroids were withdrawn and decreasing doses of oral hydrocortisone substituted. The patient resisted mobilization and was put on anticoagulant therapy with heparin because of the risk of thrombosis. In spite of this she became anuric and showed signs of venous obstruction in both legs. The possibility of Addisonian crisis was excluded and systemic steroids were increased to combat the stress but this supportive treatment was unable to prevent the death of the patient. Autopsy showed extensive venous obstruction, and gross renal atrophy. The pituitary showed Crooke's hyaline change of the basophils thought to be due to long-term exogenous steroids. This case is of interest in that it shows that excessive use of topical steroids may result in death from causes other than Addisonian crisis.

Cushing's syndrome was also reported by Ortega and Grande (1979) following excessive use of dexamethasone nasal spray. The patient, a 45-year-old woman, presented with weight gain, facial changes, acne, dry coarse skin, hirsuitism, and ecchymoses. Chronic rhinitis had been treated with dexamethasone nasal spray for the previous 26 months. The patient had increased the dose until she was using as much as two containers (4.5 mg/container) per day. Clinical examination showed the typical features of Cushing's syndrome and biochemical examination revealed a slightly abnormal glucose-tolerance test, decreased 17-hydroxycorticosteroids, 17-ketosteroids, and serum cortisol. ACTH stimulation showed diminished adrenal response. A diagnosis of iatrogenic Cushing's syndrome was made and the patient treated with decreasing doses of prednisone. Twelve weeks later the prednisone was stopped, the patient having made a complete recovery.

Another rather similar report by Heroman *et al.* (1980) described Cushingoid changes in a four-year-old child given dexamethasone nose drops at a dose of 1.2 mg per day for eight weeks. During that time the child gained 2.0 kg in body weight, developed a 'moon-shaped' face, protruberant abdomen, and hoarseness. Stimulation by exogenous ACTH at the time the patient was taken off dexamethasone revealed adrenal suppression. Decreasing doses of oral hydrocortisone succeeded in returning the patient to normal appearance and an adequate cortisol response to ACTH.

It is clear that a variety of topical steroid preparations are capable of producing adrenal suppression, sometimes at unexpectedly low dosage. In the case described by Heroman *et al.* (1980) it is possible that use of a solution rather than an aerosol may have encouraged overdosage and unintentionally high absorption.

Corticosteroid-induced adrenal hypofunction

Prolonged administration of steroids reduces the weight of the adrenal glands and causes adrenal atrophy. The decrease in adrenal mass is accompanied by a loss of physiological sensitivity to pituitary corticotrophin (ACTH), which can be readily shown by a diminished output of adrenal steroids as compared with a normal gland after a standard dose of ACTH. This adrenal suppression is due to the administered steroids inhibiting the release of ACTH from the adenohypophysis, and is in keeping with the concept of a feedback homeostatic mechanism in which the circulating corticosteroid level regulates the rate of ACTH release from the pituitary (Sayers 1950).

All corticosteroids cause pituitary inhibition of corticotrophin secretion; on a weight-for-weight basis some have a greater inhibitory action than others, but since the more potent ones are usually given in smaller dosage than the less potent parent steroids, cortisone and cortisol, the resultant degree of pituitary inhibition is about the same.

Adrenal weight reduction and histological changes are only a rough guide to the degree of suppression, which can be assessed more accurately by studying the response of the cortex to stimulation with a standard dose of corticotrophin and measuring the output of adrenocortical steroids. The degree of adrenocortical suppression may

be more related to the duration of corticosteroid treatment than to the dosage. There are, however, wide individual variations in the response of patients' adrenals to corticosteroid dosage, and this probably explains why some patients previously treated with steroids react badly to surgery or other stress while others do not. In these circumstances the patient is at special risk and an adrenal-insufficiency state can be caused by, for example, an omitted dose, a dose reduced below the physiological requirement, or the exposure of the patient to a stressful episode when the cortex is unable to supply the endogenous cortisol required of it.

Surgical and anaesthetic trauma are perhaps the best examples of controlled stressful episodes, and historically the first report came in 1952 when a 34-year-old man, who had been treated for 18 months with cortisone acetate for rheumatoid arthritis, had a cup arthroplasty on his hip. He developed irreversible shock and died (Fraser *et al.* 1952). Attention was drawn immediately to the fact that this man had died as a result of adrenocortical failure due to adrenal suppression by exogenous corticosteroids. Nevertheless, there were many subsequent reports in the early literature of severe shock developing not only in patients having corticosteroid therapy at the time of surgery, but also in those who had stopped treatment weeks or months before (Lewis *et al.* 1953; Salassa *et al.* 1953; Downs and Cooper 1955; Harnagel and Kramer 1955; Kittredge 1955; Hayes 1956; Hayes and Kushlan 1956; Allanby 1957; Plumer and Armstrong 1957; Slaney and Brooke, 1957; Schneewind and Cole 1959).

In humans there is little information on how long adrenal suppression persists after stopping treatment with corticosteroids, and such data as there are indicate a wide individual variation. Some patients may respond normally to corticotrophin within 4 days of stopping treatment, whereas Christy *et al.* (1956) found in an 11-year-old child, who had had 75 mg of cortisone daily for a year, that 20 units of corticotrophin gel per day for 20 days failed to restore normal adrenal responsiveness.

Clinical experience also shows that there may be a very long delay before normal responsiveness returns. Patients who have stopped steroid treatment for as long as 4½ to 24 months may develop irreversible shock after even minor surgical procedures, and this had led to the general adoption of special pre-operative preparation for any patient who had had steroid treatment.

Corticosteroids can cross the placental barrier and cause fetal adrenal failure. Bongiovanni and McPadden (1960) reviewed 260 pregnancies in which corticosteroids had been administered to the mothers. One case of adrenocortical failure was found; the mother had received 5000 mg of cortisone during pregnancy. The baby was in a state of collapse for the first 72 hours after birth but eventually recovered. A second case was reported by Oppenheimer (1964); a male infant weighing 1.49 kg was born after 6 months' gestation. The mother suffered from Boeck's sarcoidosis and had been treated with high doses of prednisolone. The infant died shortly after birth. At autopsy, the adrenals were small; the outer zone showed necrosis, haemorrhage, and cyst formation. In this infant the development and function of the adrenals had been arrested by the large doses of steroids given to the mother. After the child's birth, deprivation of the circulating maternal steroids precipitated adrenal insufficiency because of the hypoplastic cortex.

Cyproterone acetate

The treatment of precocious puberty with cyproterone acetate has been reported as causing adrenal suppression by Savage and Swift (1981). They concluded that adrenal hypofunction is caused by the drug's glucocorticoid-like properties. It is recommended that children receiving cyproterone acetate should carry information to warn doctors that cover might be required in case of illness or anaesthetic.

Etomidate

Etomidate is a fairly recently introduced intravenous anaesthetic agent which may be used for both induction and maintenance of anaesthesia. Chemically etomidate is an ethyl-phenylethyl imidazole carboxylate sulphate and so is distinct from the commonly used intravenous agents. During the last year a report from Ledingham and Watt (1983) drew attention to the possibility that etomidate was associated with an increase in mortality in multiply-injured patients treated in intensive care.

A survey of 428 patients, covering an 11-year period (1969–80), suggested that treatment with opiates and etomidate was associated with a statistically significant increase in mortality when compared with another group treated with opiates and benzodiazepines. Criticism of this report (Savege, 1983; Doenicke 1983) on the grounds that it was retrospective, uncontrolled, and did not take into account other variables has cast some doubt on the conclusions of Ledingham and Watt (1983) that increased mortality is due to suppression of adrenocortical function. Notwithstanding these criticisms there seems to be fairly convincing confirmatory evidence that infusion of etomidate is associated with suppression of plasma cortisol levels and negative *Synacthen* tests (Klausen *et al*. 1983; Allolio *et al*. 1983). It should be pointed out that the most serious problems seem to be encountered when etomidate is used on a chronic basis

for the sedation of critically ill patients over a period of several weeks. Such use is outside the licensed indications in the UK and promotion for purposes other than induction and maintenance of surgical anaesthesia is illegal.

Industrial hazard

Although adrenal suppression due to therapeutically administered corticosteroids is well documented, the danger of this occurring in those who manufacture the drugs is less well known. A man who had worked for 16 years in the manufacture of a potent corticosteroid was found to be suffering from chronic adrenocortical insufficiency attributed to chronic absorption of the glucocorticoid. Eleven other symptom-free workers were therefore screened. Two of these workers, like the first patient, gave grossly abnormal responses to the *Synacthen* (tetracosactrin) test (Newton *et al*. 1978). All workers manufacturing potent steroids should therefore be screened regularly by measurement of their plasma cortisol concentrations and should be moved regularly to processing other drugs.

Drug-induced thyroid dysfunction

Iodine

Iodine is one of the oldest remedies for disorders of the thyroid gland, being used in the treatment of endemic goitre and of thyrotoxicosis (Lugol's iodine). An unexpected and paradoxical complication of iodine treatment is hyperthyroidism. This was first described for six patients by Coindet (1821) as follows:

I have observed goitrous patients who have been greatly affected by the treatment: acceleration of the pulse, palpitation, dry frequent cough, insomnia, rapid emaciation, loss of strength, in others only swelling of the legs or tremor or painful hardening of the goitre, sometimes a shrinkage of the breasts, remarkable and sustained increase in appetite.

This clear description of iodine-induced hyperthyroidism appeared only nine years after the element was discovered and some time before the classic descriptions of thyrotoxicosis by Parry (1825), Graves (1835), and von Basedow (1840). Later workers confirmed the existence of the phenomenon, which they called Jod–Basedow.

The term 'Jod–Basedow' fell into disuse over the years and some doubted the existence of iodine-induced thyrotoxicosis, but Means *et al*. (1963) again drew attention to the fact that administration of iodine in the small amount of a few hundred milligrams per day could induce persistent thyrotoxicosis notably in subjects with non-toxic goitre.

During the 1920s in the United States iodinization of salt was introduced and was followed with an increase in the incidence of thyrotoxicosis (Kimball 1925; Jackson 1925; McClure 1927). Van Leeuwen (1954) described an increase in the incidence of thyrotoxicosis in Holland following addition of small quantities of iodine to bread. Connolly *et al.* (1970) reported that following the iodinization of bread in Tasmania the number of cases of thyrotoxicosis seen in the two clinics on the island doubled.

Vagenakis *et al.* (1972) reported the development of thyrotoxicosis in four out of eight patients with non-toxic goitre given potassium iodide following thyroid scanning with radioactive iodine. All the patients either had a non-toxic goitre before the appearance of thyrotoxicosis or had lived in endemic goitre areas. One suggestion has been that in iodine-deficient areas the thyroid gland is likely to be under considerable stimulation by thyrotrophin (thyroid-stimulating hormone, TSH) in an attempt to compensate for iodine depletion. When large amounts of iodine are then given they are incorporated into thyroxine and triiodothyronine under the influence of the high TSH levels and produce thyrotoxicosis. Although this may well happen to some extent, it would not account for the prolonged duration of thyrotoxicosis in many of the reported cases.

The persistence of thyroid over-activity implies that the abnormality is more than a transient overshoot. A second possibility is that the nodular goitres contained one or more autonomous adenomata (which are know to be more common in endemic goitre areas) and that these adenomata, because of concomitant iodine deficiency, could not secrete enough thyroid hormone to produce clinical thyrotoxicosis until the iodine deficiency was suddenly corrected. Similarly, administration of iodine might unmask a diffuse thyroid overactivity of the Graves' disease type, which had previously been inapparent because of iodine deficiency. It certainly seems probable that some underlying abnormality of the thyroid must be present, for the large majority of people given extra iodine do not develop thyrotoxicosis.

There certainly seems to be some evidence for a difference in the biochemistry of thyrotoxicosis in iodine-deficiency areas. Hollander *et al.* (1972) found that in most cases the laboratory confirmation of hyperthyroidism most commonly rests on the findings of raised serum levels of thyroxine (T4). Puzzled by the occasional case of what seemed to be typical hyperthyroidism in the absence of increased serum total and free T4 levels, they investigated the possibility that excess thyroidal secretion of triiodothyonine (T3) rather than T4 was a fault in such instances. In 1968 they identified a patient who was clearly hyperthyroid with normal serum T4 levels but raised T3 concentrations by gas–liquid chromatography.

These workers have shown that the incidence of T3 thyrotoxicosis in New York among patients with hyperthyroidism is about 4 per cent of the total. In Chile the incidence of thyrotoxicosis with normal T4 levels but raised T3 levels was 56 (12.5 per cent) of 449 patients. It is suggested by these workers that T3 thyrotoxicosis is higher in areas of iodine deficiency.

In a recent review Fradkin and Wolff (1983) gave a valuable account of iodine-induced thyrotoxicosis classifying patients into: (1) those from endemic goitre areas; (2) those with previous non-endemic goitre; (3) those without previous thyroid disease and (4) those with previous hyperthyroidism or with Graves' disease. Although the occurrence of thyrotoxicosis in association with increased iodide intake is well documented, they quote the fact that 10^8 tablets of potassium iodide are prescribed annually in the USA but only a handful of cases are actually reported.

Ingestion of iodine salts is also occasionally associated with a hypothyroid state with the appearance of myxoedema and the development of non-toxic goitre. The development of goitre is not uncommon (Begg and Hall 1963; Frey 1964; Helgason 1964; Horden *et al.* 1964). Several cases of potassium iodide-induced goitres with or without hypothyroidism have been reported after prolonged ingestion of iodine-containing drugs. In particular the use of iodine-containing mixtures in chronic bronchitis and asthma has been incriminated. Frey (1964) described five cases in adults where the dosage totalled 300 g of potassium iodide per year for one patient and more than 500 g per year for the other four. Discontinuation of iodide administration was followed by a return of the thyroid condition to normal within one or two months, followed by a short period of slight thyroid hyperactivity. There was no relationship between the size of the thyroid gland and the severity of hypothyroidism.

In a survey of the incidence on non-toxic goitre it was found that in a group of 24 males with non-toxic goitre there were three drug-induced cases. All three patients had taken iodine-containing preparations for asthma over a period of years. One of the patients had, however, also taken the antithyroid drug carbimazole (Horden *et al.* 1964).

In one case the occurrence of a small thyroid adenoma was noted after continuous administration of iodide over a period of 3 years; the nodule disappeared completely following withdrawal of iodine (Siegal 1964). Twenty-three asthmatic patients who had taken powders containing iodopyrine with or without additional iodide in the form of mixtures or elixirs, were found by Begg and Hall (1963) to have goitre and hypothyroidism either alone or in combination. Iodine medication had continued regularly from 1 to 22 years, and the thyroid disorder became evident after periods varying from 6 months to 20 years, most usually after 3 to 8 years of treatment. Several cases recovered when iodine medication was withdrawn and others improved with the aid of thyroid hormone.

Miscellaneous problems

Two interesting cases of iodine thyrotoxicosis were reported by Skare and Frey (1980) in which the causative agent was apparently powdered sea kelp. Both were euthyroid prior to reportedly sporadic but long-term ingestion of the sea kelp. The condition improved after cessation of the excess iodine intake. A case of drug-induced primary hypothyroidism and hyperprolactinaemia was reported by Kable (1981) in a female patient suffering a schizo-affective disorder. Drug treatment with thioridazine and lithium carbonate was associated with raised serum thyroxine and there was bilateral galactorrhoea. The primary hypothyroidism was attributed to lithium therapy and the hyperprolactinaemia to thioridazine.

Other therapeutic agents associated with thyroid dysfunction

Numerous drugs have been reported as being associated with thyroid dysfunction, both hyper- and hypo-activity occurring in different patients. The drugs so involved can be classified into those containing and those not containing iodine.

Iodine-containing agents

Amiodarone

Amiodarone is a benzofuran derivative with a thyroxine-like structure. The half-life of the drug in the body is between 28 and 100 days, and it is known to accumulate in muscle and adipose tissue during the first month of intake. The regular dose of 400 mg daily yields about 9 mg of iodine per day which is 100 times the normal dietary intake. In some cases the drug, or the effects of the iodine, is known to be active 11 months after it was discontinued. Amiodarone causes a significant increase in serum thyroxine and a fall in the level of T3. In normal subjects thyroid hormone metabolism may be affected by reducing de-iodination of T4 to T3 and inducing the preferential production of 3, 5, 3-triiodothyronine (reverse T_3). In earlier reports of the effect of amio-

darone on thyroid function tests, no signs of hyperthyroidism or hypothyroidism were observed but since then cases of both abnormal states have been reported.

In one of the cases described by Keidar *et al.* (1980) a clue to the appearance of iatrogenic hyperthyroidism was the recurrence of paroxysmal atrial fibrillation. It would seem that recurrence of sinus or atrial tachycardia or atrial fibrillation in a patient on amiodarone should alert the physician to the possibility of drug-induced hyperthyroidism.

Amiodarone has now been in clinical use for more than 15 years and its demonstration as a highly effective anti-arrhythmic agent has led to its increased popularity. Although it has been available for a good number of years, it is only fairly recently that more detailed information on its pharmacokinetic behaviour in humans has been published (Holt *et al.* 1983). These studies have shown that amiodarone has a very large volume of distribution and a comparatively low total clearance. The drug has an exceptionally long half-life of elimination (52.6 ± 23.7 for the parent compound and 61.2 ± 31.1 days for the desethyl metabolite). The mean half-life is observed to be much longer in patients than in volunteer subjects, which could be explained by time-dependent differences between single and repeated dosing. Oral bioavailability is low when it is considered that amiodarone is very lipid soluble. This is thought to be due to incomplete absorption across the intestinal mucosa rather than to first-pass metabolism. Tissue concentrations have been found to be high and are relatively much higher than in plasma.

The relationship between dosage and adverse effects has been studied by Heger *et al.* (1983), showing that higher maintenance doses predisposed to pulmonary toxicity but the absence of thyroid dysfunction in this group of 53 patients precluded the demonstration of a dose relationship for that adverse effect. In another study on 140 patients, Harris *et al.* (1983) observed four cases of thyroid dysfunction (two hyper- and two hypo-thyroidism) and five other cases of thyroid-related biochemical abnormalities.

The occurrence of these adverse effects was not related to either duration or level of dosage. It would seem from these various studies that, although pharmacokinetic data is in reasonable agreement with observed plasma and tissue levels of parent compound and metabolite and that these levels in turn correlate with the occurrence of dermal, pulmonary, and neurological side-effects, no such relationships exist with associated thyroid dysfunction.

Low T3 syndrome. Patients chronically treated with amiodarone undergo a very significant iodine burden as shown by the urinary excretion of iodine during treatment and by the fact that large amounts of organic iodine remain present in the blood and continue to be excreted through the kidneys many months after the drug has been withdrawn. Treated patients were discovered to develop either hyperthyroidism or hypothyroidism during treatment with amiodarone or after its withdrawal (Massin *et al.* 1971; Jonckheer *et al.* 1973; 1976; Savoie *et al.* 1975; Pucciarelli 1978). Peculiar clinical features were found in these patients, chiefly characterized by the paucity of clinical signs of hyperthyroidism and the fact that apparently, in nearly half of them, no underlying thyroid disease could be demonstrated, Jonckheer *et al.* (1976) considered that because of the β-adrenergic blocking effect of amiodarone, hyperthyroidism in patients treated with this drug may be clinically masked.

In a study on 62 patients, Jonckheer *et al.* (1978) demonstrated that in patients chronically treated with amiodarone the peripheral metabolism of T4 was altered. In the 48 patients remaining euthyroid while on amiodarone or after its withdrawal, but still under its influence as shown by iodine overload, a 'low T3' syndrome was observed. This state is characterized by a high total T4, a low free T4, a normal T3 resin uptake, a low total T3, a normal free T3, and a high reverse T3, and a relative TSH unresponsiveness to TRH. In this group of 62 patients, 12 were thyrotoxic and two were hypothyroid with the appropriate changes in laboratory parameters.

Burger *et al.* (1976) had made essentially similar observations finding that this drug induced a decrease in serum T3 whereas serum T4 and reverse T3 increased. Burger *et al.* suggested that amiodarone changes thyroid-hormone metabolism, possibly by reducing deiodination of T4 to T3 and inducing a preferential production of reverse T3. One of the normal participants in the experiment, who took 400 mg amiodarone for 28 days, had overt thyrotoxicosis a few months after the experiment was over.

In a more recent study on 100 patients treated for between 6 weeks and 8 years Jaggarao *et al.* (1982) reported one case of thyrotoxicosis, 10 cases of latent or overt hypothyroidism, and 25 cases that remained clinically euthyroid but manifested raised free thyroxine levels. Total triiodothyronine levels were normal in 19 cases and decreased in one. TRH levels were abnormal in four of the 13 cases in which it was measured. The authors believe that there is an inhibition of peripheral thyroxine to triiodothyronine conversion with diversion to reversed triiodothyronine.

Singh and Nademanee (1983) pointed out that cardiac arrhythmias are extremely uncommon in patients with hypothyroid states and that the effects of amiodarone on the metabolism and action of thyroid hormones is there-fore of great interest to cardiologists and endocrinologists. They also raise the question of whether the action of amiodarone on the thyroid is due to a specific effect of the drug or whether it is simply due to iodine overload. It is not impossible that the antiarrhythmic action of amiodarone is mediated, at least in part, by a selective inhibition of thyroxine action on the heart. The overall effects of the administration of 400–600 mg/day of amiodarone would seem to be to increase serum thyroxine (T4), to decrease slightly triiodothyronine (T3) and to increase reverse T3. Since the administration of inorganic iodine does not produce the same spectrum of biochemical changes, it may be concluded that the effects of amiodarone on thyroid function are due to a specific action of the drug rather than simple iodine overload. It is postulated that amiodarone reduces deiodination of T4 to T3 and also induces a preferential production of reverse T3.

Recovery of normal thyroid indices after drug withdrawal is slow, which is consistent with the very long half-life of elimination. In particular reverse T3 levels were still significantly raised 6 weeks after drug withdrawal. Although patients on amiodarone therapy have hyperthyroxinaemia they do not usually develop clinical hyperthyroidism but do exhibit a paradoxical bradycardia. Nevertheless it is clear that a small proportion of patients do develop hyper- or hypothyroidism in an unpredictable way after prolonged amiodarone treatment. The suggestion is that, because the adverse effect of clinically apparent thyroid dysfunction is (1) not dosage-related, (2) is rare, and (3) has a variable geographical incidence, it is related to an idiosyncratic response to iodine in susceptible individuals. This conclusion is in contrast to the explanations given for purely biochemical abnormalities. The frequency of hyperthyroidism would seem to be 1–5 per cent and hypothyroidism 1–2 per cent.

Other drugs causing low T3-syndrome

Very few drugs apart from amiodarone have been shown to produce low T3-syndrome. Iopanoic acid (Burgi *et al.* 1976) and dexamethasone (De Groot and Hoye 1976) have been so incriminated.

Radiographic contrast agents

Many of these agents have been implicated in the development of hyperthyroid states, and the present situation has recently been briefly reviewed by Fradkin and Wolff (1983). The iodine content of these agents is 1–4 g for cholecystography and 10 g for urography. Each dose may contain as much as 5 mg of free iodine as a contaminant and further iodide becomes available through deiodination of the parent compound. This may result in a 200-fold elevation of blood iodide for several days. Cholecystographic agents have also been shown to inhibit conversion of T4 to T3 with subsequent increase in serum levels of T4, reverse T3 and TSH. When the very large number of radiocontrast investigations are taken into account, the adverse effect of hyperthyroidism is certainly a very infrequent event.

Grubeck-Loebenstein et al. (1983) described changes of thyroid function in 20 patients who received amidotrizoic acid (*Urographin*) during the course of coronary angiographic examination. They found an increase in serum thyroxine, T3 and plasma thyrotropin, both in the basal state and after TRH stimulation. It is of interest that reverse T3 was not found to be raised and that the biochemical changes were not related to the amount of contrast medium administered.

Congenital goitre and hypothyroidism

Maternal ingestion of iodide has been implicated in the causation of congenital goitre since Parmelee and associates (1940) reported three cases of newborn infants with goitre. All three mothers had taken iodine-containing preparations throughout pregnancy; none of the mothers had goitre. Carswell et al. (1970) reviewed the literature on iodide goitre in adults and children and added eight cases of their own of congenital goitre and hypothyroidism due to maternal ingestion of iodide. These eight cases were seen in the Glasgow area in the previous 14 years. There were four deaths, two of which were due to unrelated causes. Two of the survivors were mentally retarded, due presumably to fetal hypothyroidism.

These authors strongly recommended that iodide-containing preparations should not be used during pregnancy and they should cease to be available without prescription. As a consequence, steps have been taken in the United Kingdom to control the sale and supply of iodine-containing products, without a prescription, to those products containing less than a daily dose of 10 mg iodine (Table 20.1). In the past the major source of concern was maternal self-medication with iodine-containing antiasthmatic preparations, and this problem was reviewed by D'Arcy and Griffin (1972) and Carswell et al. (1970). It has, therefore, been particularly disquieting to find that case reports of congenital goitre and hypothyroidism due to maternal use of iodine-containing asthma preparations continue to appear in various parts of the world, e.g. the United States (Melvin et al. 1978), Australia (Penfold et al. 1978), and France (Crepin et al. 1978). There is a very strong case that iodine-containing asthma preparations should carry clear warnings that they should not be taken during pregnancy. In fact, since these products are of very doubtful efficacy, a case could be made for the removal from the market of all proprietary preparations of this type and the deletion from official monographs of iodine-containing cough and asthma preparations which contain unacceptable levels of iodine for administration during pregnancy. Doctors should be aware of these risks and take the necessary steps to ensure that their asthmatic and bronchitic patients are warned of these dangers, and that their pregnant patients are especially questioned regarding such self-medication and advised accordingly.

Another disquieting aspect of iodine-induced neonatal hypothyroidism is that which results from the cutaneous application of topical formulations of iodide-containing preparations. A recent case was reported by Chabrolle et al. (1978). Hypothyroidism with moderate goitre was discovered in a newborn presenting with repeated apnoea,

Table 20.1 The following formulary products contain iodine or iodides, only aqueous solution of iodine is BP 1980. Unfortunately there is no way of legally deleting out-of-date monographs, and they persist

	Formulary	Dose	Daily dose of I_2 or KI
Clioquinol (substance only)	BP 1980	750–1500 mg (daily)	311–622 mg I_2
Di-iodohydroxyquinoline tablets	BP 1973	1–2 g (daily)	639–1278 mg I_2
Aqueous solution of iodine (Lugol's solution)	BP 1980	1 ml (daily)	= 130 mg total I_2
Belladonna and ephedrine mixture paediatric	BPC 1973	1 yr 5 ml	50 mg KI
		1–5 yr 10 ml	100 mg KI
Lobelia and stramonium compound mixture	BPC 1973	10 ml	200 mg KI
Potassium iodide mixture ammoniated	BPC 1973 (BNF 1971)	10–20 ml	150–300 mg KI
Stramonium and potassium iodide mixture	BPC 1973 (BNF 1971)	10 ml	200 mg KI
Stramonium and potassium iodide mixture pro infant	BPC 1954 (NF 1954)	4–8 ml 3 years	73–146 mg KI
Stramonium and ephedrine mixture pro infant	BPC 1949 (NF 1949)	4–8 ml	73–146 mg KI
Creosote and potassium iodide mixture	BPC 1949	15–30 ml	343.5–687 mg KI

indicating a severe risk of sudden death. The mother was given no iodine-rich drug. However, the child was submitted to several large applications of iodized alcohol to the skin for numerous blood samplings. Treatment with thyroid extract resulted in the disappearance of the respiratory pauses and return to euthyroidism within 3 months.

Povidone iodine

Hyperthyroidism was also reported by Jacobson *et al.* (1981), following intravaginal iodine administration. The patient, who was undergoing elective vaginal hysterectomy, was prepared preoperatively with undiluted intravaginal providone iodine. Ten per cent povidone iodine (14.5 ml) had also been similarly administered four days and one day preoperatively. The immediate post-operative period was uneventful but on the eighth day the patient noticed fatigue, excessive sweating, heat intolerance, and anxiety. At 17 days post-operation the pulse rate was 140 per minute, the thyroid was diffusely enlarged to about three times the normal size, the skin was warm and moist, reflexes were brisk, and a fine hand tremor was present. Thyroidal radioactive iodine uptake was decreased and both total serum iodine and inorganic iodine levels were markedly raised. Over the next 34 days the size of the thyroid gland decreased to normal dimensions and the patient became clinically and biochemically euthyroid.

Felsol powders

A case of iatrogenic dwarfism was described by Wilkinson *et al.* (1972). The patient was an asthmatic child and had taken *Felsol* powders since the age of 7 years. (Each *Felsol* powder contains phenazone 768 mg, phenacetin 97 mg, caffeine 100 mg, iodine 12 mg, extract of mistletoe 2 mg.) At the age of 11 years he was noted to have a goitre but was euthyroid. At the age of 24 years he was a dwarf of 124 cm height and 26.8 kg weight, noticeably hypothyroid with a large goitre, and pubertal changes had not occurred. *Felsol* was stopped and within 5 months thyroid function had returned to normal. Between the ages of 24 and 28 puberty occurred and quite amazingly an increase in height of 23 cm occurred.

Administration of T3 or T4

Simple obesity is sometimes treated by the administration of either T3 or T4 and it was pointed out by Dornhorst *et al.* (1981) that this may be associated with the development of a persistent hypothyroid state after treatment is discontinued. Five obese patients treated with

thyroid hormone for between one and nine years became hypothyroid. All patients were clinically euthyroid prior to treatment and in the four cases in which thyroid function tests were done they were found to be normal. After stopping thyroid hormone all patients became clinically and biochemically hypothyroid within three months. At the time hypothyroidism was detected, four out of five cases were shown to have low-titre thyroid microsomal antibodies. In four cases thyroid hormone administration had to be restarted but, in the fifth, thyroid function spontaneously improved over a period of nine months. The administration of thyroxine to euthyroid patients usually only causes temporary suppression of thyroid function but in the four cases reported it was permanent. The authors comment on the unexpectedly high incidence of microsomal antibodies, pointing out that they are only present in 9–16 per cent of the general population and that this correlates with lymphocytic infiltration and small thyroid gland size post-mortem. It is suggested that subclinical autoimmune thyroiditis may make an individual more susceptible to persistent thyroid suppression by treatment with thyroid hormone.

Therapeutic agents not containing iodine

Lithium

Lithium administration may be followed, sometimes within a few weeks, by a decrease in the amount of thyroxine secreted by the thyroid (Sedvall *et al.* 1968; Shopsin 1970). A few patients develop goitre, usually as a 'late' side-effect occurring after months or years of lithium therapy. There have been few reports of overt hypothyroidism. Wiggers (1968) reported the occurrence of clinical and biochemical myxoedema in a patient five months after beginning lithium. Shopsin *et al.* (1969) described the case of a middle-aged woman who was found to be hypothyroid with a goitre after taking lithium for two years; investigation suggested that she suffered from an underlying chronic thyroiditis. Rogers and Whybrow (1971) described two cases of clinical hypothyroidism, one developing after 20 months and the other after three months of lithium therapy. Both patients, however, had received a variety of other psychotropic drugs, and in the second case symptoms suggestive of hypothyroidism were noted before treatment with lithium was begun.

Candy (1972) described the case of a depressed 48-year-old woman who became severely myxoedematous within seven weeks of starting lithium therapy. The thyroid was diffusely enlarged and soft, her voice was hoarse, there was swelling of the face, fingers, and ankles, and she complained of drowsiness and sensitivity to cold. Within two months of stopping the lithium she was clinically euthyroid, and thyroid function tests had returned to normal within four months of stopping treatment.

The precise mechanism of goitre formation is obscure, but there is some evidence that it results from impaired synthesis of hormone in the gland. While stimulation of thyroid adenyl cyclase by TSH is inhibited by lithium, this is not the full explanation of the effects, since the action of cyclic adenosine monophosphate (AMP) on the thyroid is also blocked. The goitre seldom needs treatment, but if it is large, or if hypothyroidism occurs, there is a good response to thyroxine.

Perrild et al. (1978) described permanent hypothyroidism is a 56-year-old man a 54-year-old woman after prolonged lithium therapy.

Several epidemiological studies have been conducted to determine the prevalence of lithium-induced hypothyroidism. Piziak et al. (1978) conducted a study which involved a survey of 2590 patients treated by 70 clinicians. The major finding was that lithium-related hypothyroidism was not a common problem. Goitre occurred in 0.3 per cent of patients treated and clinical hypothyroidism in 1–2 per cent.

In a study by Cho et al. (1979) 195 patients on lithium therapy were studied, seven patients, all women, developed hypothyroidism while taking lithium therapy. Further investigation revealed that five of these seven lithium-induced hypothyroid patients were 'rapid cyclers' with respect to their manic-depressive psychosis. Only 16 of the 115 women attending the clinic were 'rapid cyclers', so the incidence of lithium-induced hypothyroidism was significantly higher among rapid-cycling women than non-rapid-cycling women (31.2 per cent vs. 2.1 per cent $X^2 = 7.64, p < 0.01$).

A third study to look into the incidence of lithium-induced hypothyroidism was conducted by Transbol et al. (1978) and involved 86 patients on lithium therapy and appropriate controls. Hypothyroidism was diagnosed on the basis of raised TSH levels in 20 patients, all were female and only one was under 40 years of age. This prevalence of cases of hypothyroidism on lithium therapy was 19 cases out of 56 female patient over 40 years of age (34 per cent). It would appear that a subgroup of patients most liable to develop hypothyroidism can be defined. These are 'rapid-cyclers' who are female and over 40 years of age.

While the most usual effect of lithium on the thyroid would appear to be to induce hypothyroidism, a single case of what appears to be lithium-induced thyrotoxicosis in a 22-year-old woman was reported by Reus et al. (1979).

In a long-term study of thyroid function during lithium therapy Lazarus et al. (1981) evaluated 73 patients both clinically and biochemically. Goitre was found in 37 per cent, exophthalmos in 23 per cent, positive thyroid auto-antibodies in 24 per cent, and abnormal TRH in 49 per cent. A high prevalence of abnormal thyroid-releasing hormone tests (49.3 per cent) was also noted. They concluded that a direct action of lithium on the hypothalamic-pituitary axis was unlikely to be the case but that the effect was more probably due to antibody-mediated damage to the thyroid.

Thalidomide

In 1958 Murdoch and Campbell studied the effects of thalidomide (2-phthalimidoglutarimide) on ^{131}I-uptake by the thyroid gland in nine euthyroid patients. The results from this investigation indicated that thalidomide had mild but definite antithyroid action. These workers did not advance any hypothesis to explain the mode of action of thalidomide's antithyroid activity but did issue a clear warning that it would be unjustified to use this drug as a long-term sedative hypnotic. Despite this relatively early work, thalidomide was used in the very indications that these workers had warned against. Alexander (1961) described a case of myxoedema following long-term therapy with thalidomide, and Simpson (1962) described a further two cases of myxoedema induced by thalidomide; Lillicrap (1962) reported another case.

In the cases reported, the onset of myxoedema occurred within two or three months of commencing therapy and was described as of acute onset. Lillicrap's description is very detailed, two months after starting therapy the patient, a 49-year-old woman, complained of lethargy, intolerance of cold, rough dry skin, and deepening of the voice. The patient was started on thyroid but continued to take thalidomide. One year later she stopped taking thyroid and her symptoms returned, and a month after stopping the thyroid the thalidomide was also stopped; it was then that her thyroid function began to improve and eventually returned to normal. This patient also developed the typical thalidomide-induced peripheral neuritis after six months on the drug; these symptoms did not disappear when thalidomide was stopped. The thalidomide story is now very well known; but this aspect of thalidomide iatrogenesis is worthy of mention since if any series of cases were needed to illustrate its potential harmful effects certainly these showed that the potential spectrum of thalidomide-induced disease was highlighted some four years before the appearance of the first reported teratogenic effect. Thalidomide is currently used therapeutically in the treatment of Lepra reactions and Behçets disease.

Salicylates

Thyroid function may be depressed by salicylates; this had been demonstrated both in animal studies and in patients. Hetzel and associates (1963) showed that salicylates caused a fall in plasma-bound iodine while Myhill and Hales (1963) showed that there was a fall in plasma-bound iodine in six euthyroid and eight hyperthyroid subjects receiving 6.5 g of calcium salicylate for four days.

The literature on this effect of salicylates was reviewed by Prescott (1968). The general picture is that salicylates lower the plasma protein-bound iodine and decrease the level of circulating pituitary TSH. This depression in TSH release corresponds with a rise in free circulating thyroxine which is caused by the salicylates displacing thyroxine from thyroid-binding pre-albumin. The increased level of circulating free thyroxine is thought to act on the pituitary-thyroid feedback to reduce the output of TSH (Good *et al*, 1965; Marshall *et al*. 1965).

Of all the salicylates used in therapy, the dosage of para-aminosalicylic acid in tuberculosis is the greatest. It is therefore not surprising that Marchese *et al*. (1963) reported goitre with myxoedema as a rare complication of the therapy. Generally, however, the effects of salicylates on thyroid function would seem to be of minor clinical importance, although results of thyroid-function tests may be influenced by a concomitant salicylate ingestion.

Phenylbutazone

The administration of phenylbutazone may inhibit the uptake of iodine by the thyroid. The development of a non-toxic goitre in a man with psoriatic arthritis was reported by Benedek (1962). The goitre was reversible and related to phenylbutazone therapy. It did not occur while he received either phenylbutazone or vitamin A separately, but occurred during each of two periods in which both drugs were administered. No effect on thyroid size and function was seen in two other men who received the same drugs in the same combination for a similar length of time. Therefore, it seemed likely that not only the combination of these drugs, each of which is known to cause hypothyroidism, but an idiosyncrasy was also involved in the above patient.

There have, however, been other cases in which phenylbutazone has been implicated in a thyroid-inhibiting action without involvement of vitamin A. These cases were reviewed by Prescott (1968) and it would seem that the goitrogenic effects of phenylbutazone are more marked in patients with pre-existing thyroid enlargement.

Animal studies have confirmed the thyroid-inhibiting action of phenylbutazone and it is interesting to note that the lethal toxicity of phenylbutazone in these studies was reduced by about one-half by simultaneous administration of triiodothyronine (Eger and Fernholz 1965), this may, however, have been due to increased metabolism of the phenylbutazone rather than to any reduction of toxicity due to thyroid inhibition.

Sulphonylurea hypoglycaemic agents

In 1957 Mamou described a single case of myxoedema following 30 days treatment of maturity-onset diabetes with carbutamide. Brown and Solomon (1958) described a fall of ^{131}I-uptake and protein-bound iodine levels during carbutamide therapy, and Seegers *et al*. (1957) found that in elderly diabetics treated with carbutamide 2 g daily there was a slight decrease in protein-bound iodine and basal metabolic rate over a period of 20 weeks.

Creutzfeldt and Soeling (1960) described a further two cases of carbutamide-induced hypothyroidism with an onset within a few months of commencing carbutamide therapy at 2 g daily. In these patients the protein-bound iodine, ^{131}I-uptake, and the basal metabolic rates were affected, but these effects diminished with continuing therapy. Carbutamide is no longer used as an oral hypoglycaemic, but both tolbutamide and chlorpropamide have hypothyroid effects. Skinner (1959) noted a fall in protein-bound iodine in 23 diabetics treated on chlorpropamide, in three of whom the protein-bound iodine had fallen to hypothyroid levels in 19 weeks.

Hypothyroidism has also been described following tolbutamide therapy at 2 g daily by Burda (1965). In the patient described by Burda there was a low protein-bound iodine ^{131}I-uptake. Tolbutamide was discontinued and the thyroid stimulated with TSH, and thyroid function improved. When thyroid function had recovered tolbutamide was restarted, and 27 months later the patient was again noted to be hypothyroid. TSH was given after stopping tolbutamide and thyroid function recovered. Thereafter the patient's diabetes was controlled on insulin. Schless (1966) described a male diabetic patient who became hypothyroid during treatment with tolbutamide 500 mg/day. The patient was changed to phenformin and the thyroid function recovered.

The largest documented study is that of Hunton *et al*. (1965) from Sheffield who studied 220 diabetics treated with sulphonylurea hypoglycaemic drugs and 229 diabetics treated with diet alone (113 patients), or insulin (93 patients), or biguanides alone (23 patients). Hypothyroidism was found in 30 patients on sulphonylureas com-

pared with eight in the control group. In the sulphonylurea group eight were on tolbutamide and 22 on chlorpropamide therapy. In one patient on chlorpropamide
therapy a rise in protein-bound iodine was measured on
withdrawing chlorpropamide with a subsequent fall on
restarting the drug. In 12 patients in whom sulphonylureas were stopped an increase in protein-bound iodine
to within normal limits occurred without any additional
therapy.

Hunton *et al.* (1965) demonstrated that the incidence
of hypothyroidism was related to the duration of therapy
ranging from 10 per cent after 6–8 months' treatment to
as much as 25 per cent after 6-years' treatment. The incidence of goitre was 14 per cent in the sulphonylurea-
treated group and 16 per cent in the controls.

The mechanism of sulphonylurea hypothyroidism was
never clarified but some reports suggested that pituitary
TSH release was affected rather than a block in thyroxine
synthesis. It is of considerable interest that reports of sulphonylurea hypothyroidism have not appeared in the
literature for some 15 years and clinically it does not
appear to be a problem; yet for 10 years from the
mid-1950's to the mid-1960s' reports were common in the
literature. In a study by Balodimos *et al.* (1968) an evaluation of post-mortem findings on a group of 55
diabetic patients treated with sulphonylurea compounds
— predominately tolbutamide for from 24 to 112 months
— showed no increase in the incidence of thyroid pathology. (This study did show an increase in the incidence of
islet-cell tumours compared with the general population
— three tumours in 55 patients — but this had not been
confirmed by other workers.)

Resorcinol

Resorcinol is widely used in a number of dermatological
preparations, and may be used as a keratolytic agent in
lotions, ointments, and pastes. Excessive absorption
through the skin can result in tinnitus, tachycardia, and
methaemoglobinaemia. Bull and Fraser (1950) described
three patients who developed myxoedema associated
with a resorcinol-containing preparation applied to
varicose ulcers. The thyroid gland in these cases was
described as soft and diffusely enlarged. On withdrawal
of the resorcinol the myxoedema recovered and thyroid
function tests gradually returned to normal. Since this
first association of resorcinol treatment with myxoedema
several other reports have confirmed this finding.

In patients treated with resorcinol-containing dermatological preparations it is commonly found that the
serum protein-bound iodine is decreased and the 24-hour
uptake of ^{131}I into the gland is reduced.

Antithyroid drugs

Propylthiouracil   Propylthiouracil is an important drug
in the treatment of hyperthyroidism during pregnancy. It
is known that propylthiouracil crosses the placenta and
at high doses may cause foetal goitre and hypothyroidism. It is generally considered that propylthiouracil
in low doses of 300 mg or less daily satisfactorily controls
maternal hyperthyroidism and is not believed to cause
hypothyroidism in the infant. Neonatal goitres do,
however, occur on maternal doses of 100 to 200 mg propylthiouracil daily, although it is not known whether the
fetal thyroid stimulation is due to fetal TSH or maternal
thyroid-stimulating globulin. Cheron *et al.* (1981)
noticed a goitre and hypothyroidism in a neonate born to
a mother who had been receiving 400 mg propylthiouracil daily, and as a consequence of this observation
initiated a prospective study of neonatal thyroid function
in 11 cases. The mothers were receiving 100–200 mg propylthiouracil daily at term and their serum T4, TSH, and
rT3 concentrations were normal for pregnant women.
All neonates were normal birthweight and none had
goitre. Mean cord-serum T4 concentrations were lower
than controls as were mean cord-serum free-T4 indices.
The serum T4 concentrations rose during the first three
neonatal days but were still significantly lower than the
controls. Mean cord-serum TSH levels in the infants of
treated mothers were normal but after one day this rose
to 2.5 times normal. After three days serum TSH
returned to normal values. Serum T3 concentrations in
the infants of treated mothers did not differ from
controls, neither did cord-serum or T3 levels. However,
the latter parameter was open to doubts because of different assay methods.

Treatment of the mother was aimed at maintaining a
euthyroid or mildly hyperthyroid state. Biochemical
measurements showed that this was achieved but even
on this minimal dosage regime neonatal hypothyroxin-
aemia was present. Cheron *et al.* (1981) were not able to
determine from their results whether transient hypothy-
roxinaemia has longer-term effects on growth or
development neither could they exclude the possibility of
adverse effects on neural tissue.

Solomon (1981) pointed out that Cheron *et al.* (1981)
significantly extended previous observations on the treatment of maternal hyperthyroidism with propylthiouracil. Although propylthiouracil is able to cross the
placenta, maternal T2 and T3 cannot, so overt fetal
hypothyroidism and goitre may exist even in the presence
of maternal hyperthyroidism. It is recommended that the
dosage of propylthiouracil be individually adjusted to
maintain the patient's serum T4 and T3 at slightly above
the upper limits of normal. Since mild maternal hyper-

thyroidism appears to be well tolerated and maternal hypothyroidism poorly tolerated by the fetus, the maintenance of a mildly hyperthyroid state in the mother would seem to be an acceptable objective of therapy. Solomon (1981) further drew attention to the much more serious consequence of fetal hyperthyroidism and advocates the measurement of maternal serum long-acting thyroid stimulator (TSI) late in pregnancy in women with Graves' disease. Since propylthiouracil reaches the fetal thyroid it should be used in low doses if high TSI levels are present.

Other antithyroid drugs   Antithyroid drugs can cross the placenta and impair thyroid function in the fetus. Neonatal goitres in infants born to mothers who have taken antithyroid drugs during pregnancy for thyrotoxicosis probably result from increased secretion of fetal thyroid-stimulating hormone (TSH) in response to the depressed fetal thyroid hormone output induced by the placental transfer of these drugs. Increased TSH levels in the cord blood have been reported in infants who have drug-induced goitres. However, little is known about thyroid hormone or TSH levels in the neonatal period in infants who are exposed to antithyroid drugs *in utero* but have no goitre or clinical evidence of hypothyroidism. Low *et al.* (1978) produced information on two babies born apparently euthyroid following treatment of their mothers for thyrotoxicosis with carbimazole, and in both these babies cord blood TSH was raised 6 h and 24 h after delivery but had returned to normal by 72 h. Serum thyroxine T4 and triiodothyronine T3 were low but recovered by 72 h. These data were consistent with a state of relative hypothyroidism despite the babies apparent euthyroid state.

Fenclofenac

The anti-inflammatory drug fenclofenac has been shown to interfere with thyroid function tests (Isaacs and Monk 1980; Ratcliffe *et al.* 1980). This is attributed to the ability of fenclofenac to displace T4 and T3 from their binding sites on serum proteins. Euthyroid patients on fenclofenac were found to have total thyroxine indices well down in the hypothyroid region whereas free T4 values tended to remain in the euthyroid range.

Humphrey *et al.* (1980) studied the effects of fenclofenac on thyroid function test in 10 male volunteers to whom the drug was given at a dose of 600 mg twice a day for 28 days. The mean concentration of total T4 fell markedly to reach a plateau after 10 days; however the concentration of free T4 showed much less fall over the same period of time. Mean serum TSH fell, reaching its lowest value on day-4, but rose again to pre-treatment

values by day-14. TRH tests indicated a suppression of mean maximal TSH increment at day-7 but this recovered to normal by day-28. All indices had returned to normal 28 days after cessation of fenclofenac administration.

Midgely and Wilkins (1980) developed a new technique of measuring free T4 using a ^{125}I-labelled derivative of T4 that cross-reacts with a T4-specific anti-serum but not binding to thyroxine-binding serum proteins. They examined nine blood samples from patients on fenclofenac and confirmed earlier findings that T3-uptake and total T4 values are markedly altered and that FT1 suggested gross hypothyroidism. Before fenclofenac treatment, seven of the patients were euthyroid and had TSH values in the normal range but the other two were mildly hypothyroid with high TSH values. When on fenclofenac the TSH values indicated mild hypothyroidism. FT1 values were the same for euthyroid and hypothyroid groups, demonstrating that it is a poor indicator of thyroid function in patients on fenclofenac. However, the two groups could be distinguished on the basis of free T4-and TSH-measurements, suggesting that those two parameters are of value in patients on fenclofenac.

Danazol

The effect of danazol, a pregnanediene compound used in the treatment of endometriosis, on thyroid function tests has been described by Wortsman *et al.* (1979), using danazol at a dose of 400–800 mg daily on 19 patients with endometriosis. They observed that it was commonly associated with a decrease in the thyroxine T4 index, although thyrotropin levels remained unaltered. In a few cases a marked decrease in the T4 index was observed, but in all cases values returned to normal when danazol was discontinued. Danazol is known to block the ovulatory peak of the gonadotrophins and may therefore interfere with hypothalamic-pituitary control of thyroid-stimulating hormone. Wortsman *et al.* (1979) also suggested that reversible hypothyroidism may be the cause of the increase in serum creatine phosphokinase during treatment with danazol.

Barbieri (1980) questioned the suggestion of Wortsman *et al* (1979) that danazol might cause a hypothyroid state by pointing out that danazol lowers thyroid-binding globulin capacity, lowers total T4, but produces no change in free T4, and therefore does not alter TSH. Barbieri (1980) postulated that this may be explained by the androgenic effects of danazol, causing a decreased production of thyroid-binding globulin by the liver or alternatively by direct binding of danazol to thyroid-binding globulin.

In a later communication Wortsman and Hirschowitz

(1980) maintained their view that danazol is associated with hypothyroidism since at least five of their patients had low free T4 levels. They suggest the controversy may be resolved by correlating high serum creatine phosphokinase levels with low free T4 by demonstrating that serum creatine phosphokinase level returns to normal if the hypothyroid state is reversed by thyroid replacement therapy.

Anticonvulsant-induced increase in thyroid hormone metabolism

Yeo *et al.* (1978) investigated serum total and free thyroid hormone concentrations in 42 patients with epilepsy taking anticonvulsants (phenytoin, phenobarbitone, and carbamazepine either singly or in a combination). There was a significant reduction in total thyroxine (TT4), free thyroxine (FT4), and free triiodothyronine (TFT3) in the treated group compared with controls. Free hormone concentrations were lower than total hormone concentrations, suggesting that increased clearance of thyroid hormones occurs in patients receiving anticonvulsants. In a more recent paper (Aanderud and Strandjord 1980) two cases of overt clinical hypothyroidism were reported. Both patients were on long-term epileptic therapy, one with phenytoin and carbamazepine and the other with carbamazepine alone. The clinical diagnosis was confirmed by laboratory tests, and both patients returned to the euthyroid state after withdrawal of the drugs.

Two cases of autoimmune thyroiditis associated with anticonvulsant therapy were reported by Nishiyama *et al.* (1983). Both were young girls (15 and 12 years of age) and both presented with goitrous hypothyroidism and elevated anti-DNA and antithyroid antibodies. The 15-year-old girl was on treatment with phenytoin, phenobarbital, and ethosuximide and the 12-year-old was on phenytoin and phenobarbital. In the first case ethosuximide was discontinued and replaced by sodium valproate, other anticovulsants being continued, and in the second case the phenobarbital was discontinued and the phenytoin dose was reduced from 50 to 25 mg per day. These measures were effective after about one year in returning the patients to normality in respect of their thyroid abnormalities. The authors regard these cases as supporting evidence for a direct relationship between anticonvulsant therapy and the development of anti-thyroid antibodies.

Oestrogens

The oestrogenic component of the oral contraceptives increases the level of thyroxine-binding globulin. The increased plasma binding alters the results of some of the thyroid-function tests. Protein-bound iodine, butanol-extractable iodine, and thyroxine (T4) levels are elevated, and the corresponding triiodothyronine (T3) levels are decreased. The basal metabolic rate, cholesterol, ^{131}I-uptake, and unbound thyroxine levels are unchanged, which suggests that thyroid function itself is unaltered.

Diabetes insipidus

Lithium

Polyuria and polydipsia can be induced in rats by toxic doses of lithium. The rats passed dilute urine and appeared to be unable to concentrate their urine even after injection of vasopressin. These effects are most easily produced in salt-deprived animals. Death if it occurs is due to oliguric renal failure.

Some patients may develop a disturbing degree of polyuria and polydipsia even when treated with conventional doses of lithium, and it seems probable that many others have these symptoms in a milder degree. Angrist *et al.* (1970) and Ramsey *et al.* (1972) reported on five patients who developed persistent polyuria, beginning between two weeks and seven months after starting long-term lithium carbonate therapy in conventional doses, and who had safe plasma-lithium concentrations. No patient had clinical or conventional biochemical evidence of pre-existing renal disease or electrolyte imbalance. Neither water deprivation nor administration of a hypertonic saline caused these patients to concentrate their urine or reduce the urine flow, despite haemoconcentration and loss of weight, indicating a diagnosis of diabetes insipidus. The urine did not become concentrated after administration of vasopressin, and it was concluded that the patients had acquired nephrogenic diabetes insipidus; the condition was reversible when the lithium was discontinued.

The mechanism of this interference with the action of vasopressin is not known but it has been suggested that it is similar to that produced experimentally by magnesium, manganese, and caesium. In some of these instances the toxic agent inhibits the activation of cyclic AMP by vasopressin, and this is probably true for lithium. Unlike other forms of nephrogenic diabetes insipidus, it is unsafe to treat the patient by salt depletion and thiazide diuretics, because salt depletion will enhance the toxicity of lithium due to increase in the plasma level which should be kept in range of 0.7 to 1.3 mEq/l.

The symptoms of lithium toxicity vary from mild symptoms such as sleepiness, vertigo, and slurred speech to more severe symptoms of muscle hypotonia, hyperextension of the limbs, epileptiform attacks and coma.

Drugs affecting hypothalamic–pituitary–gonadal function

Oral contraceptives

Under the influence of oral contraceptives, the normal ovulatory cycle is changed to a non-ovulatory one, vaginal bleeding following the discontinuation of oral-contraceptive medication for several days between each cycle is, in fact, withdrawal bleeding and not a normal menstruation.

Plate (1968) reported on the histology of ovarian tissue from 11 women who had used ovulation inhibitors for contraception during 4–30 cycles; all the women were under 40 years of age and all had had regular menstruation before they started to use ovulation inhibitors. The specimens of ovarian tissue were obtained at operation in nine cases and post-mortem in two cases. The ovulation inhibitors used and the dosages per tablet were:

Lyndiol	5 mg lynoestrenol and 0.15 mg mestranol
Lyndiol 2.5	2.5 mg lynoestrenol and 0.075 mg mestranol
Gynovlar 21	3 mg norethisterone acetate and 0.05 mg ethinyloestradiol.
Anovlar	4 mg norethisterone acetate and 0.05 mg ethinyloestradiol.
Planovin	4 mg megestrol acetate and 0.05 mg ethinyloestradiol.
Orthonovin	2 mg norethisterone and 0.1 mg mestranol

No corpora lutea were observed in any of the ovaries, but follicle growth or follicle maturation was present in nine women, hypothecosis in eight women, peri- or intra-follicular bleeding in four women, and follicle cysts in 10 women.

The most conspicuous change was fibrosis, which occurred in seven women, and a thickened *tunica albuginea* in six women. As a result of this fibrosis, the follicles were sometimes embedded completely in connective tissue. These changes were most marked in women who had been on oral contraceptive therapy for some considerable time.

Plate (1968) also reviewed the literature on fibrotic changes associated with oral-contraceptive medication. Ryan and colleagues (1964) found a 'focal cortical condensation of stroma in some degree' in about 50 per cent of 18 women investigated; O'Neil (1965) observed a 'definite Stein–Leventhal syndrome' in two women, and Graudenz and Beirao de Almeida (1965) found moderate fibrosis, especially in the cortex, even after short-term

use of ovulation inhibitors. In one of these patients the enlarged ovaries were reminiscent of the Stein–Leventhal syndrome. Other investigators found thickening of the *tunica albuginea* (Diddle *et al.* 1966; Starup 1967), while Van Roy (1966) described fibrotic granulations in the cortex. These changes were not, however, observed by all investigators; for example Linthorst (1966) did not find any thickening of the *tunica albuginea* in ovaries of patients who had been on ovulation inhibitors for a long time. In contrast others (Zussman *et al.* 1967) found increased fibrosis in 10 patients who had been on oral contraceptives for only 1–3 cycles.

To discover whether fibrosis also occurred in the ovaries when ovulation was inhibited during pregnancy, Plate (1968) examined the ovarian tissue of six women, five with full-term pregnancy and one who was 7 months pregnant. There was slight fibrosis in four cases but there was no embedding of the follicles in fibrous tissue. No fibrosis was evident in the other two cases. None of the fibrosis seen in these cases was comparable with the degree of fibrosis seen after the use of oral contraceptives.

Plate (1968) also assumed that since, in the majority of women, regular menstruation returned after oral contraceptives were stopped, the fibrotic changes in the ovaries were reversible. To test this assumption he examined ovarian tissue from six patients who had ceased using oral contraceptives. The fibrosis of the stroma soon disappeared (within 2–5 cycles), and thickening of the *tunica albuginea* disappeared a little later. He found that the length of time for which oral contraceptives had been taken was of no great importance in this respect.

Having established these points, Plate (1968) then questioned whether fibrosis persists in cases in which amenorrhoea or anovulatory haemorrhage occurs after oral contraceptives have been stopped. Whitelaw *et al.* (1966) reported personal communications from a number of gynaecologists who observed amenorrhoea for up to 18 months after withdrawal of oral contraceptives. They observed that the relative frequency of anovulatory cycles and amenorrhoea after stopping oral-contraceptives medication was much higher than had been suspected. The findings of other groups (referenced by Plate 1968) were in general agreement. There was, however, little evidence in the literature to assist in determining what changes had taken place in the ovaries of women with such amenorrhoea or anovular bleeding. One such report by Lorrain (1966) cited a 22-year-old woman who was amenorrhoeic for three months after using *Enovid* for three months. Her ovaries were enlarged and contained follicle cysts with hyperthecosis, but no thickening of the *tunica albuginea*.

Plate (1968) summarized his study by concluding that,

although in most cases the fibrotic ovarian changes induced by oral-contraceptive medication were reversible, it would be unwise to assume that this was always the case. He therefore advocated that oral contraceptives should be withheld after specific time (he suggested a year) and resumed only after a few ovulations had occurred. Shearman (1971) described 69 cases of secondary amenorrhoea of more than 12 months' duration after treatment with oral contraceptives. Two patients appeared to have undergone a premature menopause, and in the remainder the level of disturbance was hypothalamic. Galactorrhoea was the only other abnormal physical finding and was present in 11 women. Sixteen of 36 patients conceived after treatment with clomiphene citrate or human gonadotrophins, 19 improved spontaneously, and 25 continued to have amenorrhoea despite treatment.

The incidence of secondary amenorrhoea after discontinuing oestrogen–progestogen oral contraceptives was evaluated in a survey conducted by Steele et al. (1973). Out of 210 women seen at the Middlesex Hospital with secondary amenorrhoea the 63 who developed it after stopping oral contraceptives were fully investigated. Five had organic disease sufficient to account for the amenorrhoea (one had severe diabetes, one a pituitary tumour, and three premature ovarian failure); two patients had galactorrhoea (one of who also had a pituitary tumour); two had anorexia nervosa. Of the 63 women 40 (63 per cent) had suffered from amenorrhoea or prolonged or irregular menstrual cycles before taking the pill, and this suggested that combined oestrogen–progestogen oral contraceptives should be used with caution for women with irregular menstruation.

Nineteen patients wished to become pregnant and 12 did so after treatment with clomiphene or gonadotrophins.

In the same paper Steele et al. reported the time at which 204 women recorded their first menstrual cycle after discontinuing the pill. All except 74 patients had a menstrual cycle within 5 weeks of stopping, and all except five had a cycle within three months, and only one had amenorrhoea lasting longer than six months. These results indicate that the incidence of secondary amenorrhoea after stopping oral contraceptives is relatively low.

Hancock et al. (1976) reported a series of 64 patients with ovulatory dysfunction after stopping oral contraceptives that were presented at their Leeds Unit from 1968 to 1974. Forty-nine women had had amenorrhoea for at least 12 months and 15 were menstruating fewer than five times per year. Thirty of these patients had had regular cycles before commencing oral-contraceptive therapy. In this group of patients with previously normal cycles 22 patients were classified with respect to their ideal weight compared with actual weight. Only two of these were above their ideal weight, and in seven the actual body-weight was less than their ideal body-weight by 10 per cent or more. Hancock et al. considered that women of low body weight may be at particular risk of developing post-Pill amenorrhoea.

The most obvious explanation for this is perhaps on the basis of a dose–weight ratio, Hancock et al. considered that support for this may be derived from yearly incidence of patients seen with 'post-Pill' amenorrhoea. In their series there was a steady rise from five cases in 1968 to 18 cases in 1971 and a fall to only four cases in 1974. In 1970 the Committee on Safety Medicines (CSM) issued a warning about the increased risk of thromboembolic disease associated with oral-contraceptive agents with oestrogen contents of above $50\mu g$/day. This warning was followed by the general use of low-dose oestrogen combinations. The fall in the incidence of post-Pill amenorrhoea may be related to the reduction in oestrogen content of the preparations now available.

In the most recent study by Vessey et al. (1978) the fertility of both nulligravid and parous women who stopped taking oral contraceptives was initially impaired in comparison with that of women who stopped using other methods of contraception. But the effect of oral contraceptives on fertility had become negligible by 42 months after cessation of contraception in nulligravidae and by 30 months on multiparae. Impairment seemed to be independent of the length of use of oral contraceptives. These results suggest that, although women may have temporary impairment of fertility after discontinuing oral contraception, they are unlikely to become permanently sterile through taking the pill.

L-dopa–post-menopausal vaginal bleeding

Post-menopausal bleeding is a comparatively rare complication of L-dopa therapy in post-menopausal women treated for parkinsonism. Krause-Larsen and Garde (1971) described two cases in women both aged 59 years who had reached the menopause 10 and 11 years before therapy with L-dopa commenced. In both patients vaginal bleeding started within a few days of starting therapy, and in each patient bleeding was provoked three times and disappeared when L-dopa was stopped. These authors had come across this symptom in two patients in a series of 23 post-menopausal women treated.

Ansel (1970) described a similar effect earlier, and Sandler (1971) stated that this side-effect had been predicted on theoretical grounds even earlier on the basis that L-dopa administration to the female rat, like that of progesterone, appears to provoke a specific increase in uterine monoamine oxidase (MAO) activity: pronounced increases in MAO activity have been noted in human endometrium in the late secretory phase of the menstrual cycle, a possible but underdetermined factor in the onset of menstruation.

Virilizing effects of anabolic steroids

The various anabolic steroids have virilizing effects to a greater or lesser degree. These effects may disturb the menstrual pattern, may increase the libido in both sexes, and may cause failue of spermatogenesis. Boys may show precocious sexual development and in girls the clitoris may hypertrophy. Virilization of the voice is a troublesome early symptom reported after treatment with anabolic agents (De Lange and Doorenbos 1968).

Hypergonadotrophic hypogonadism

Association of cimetidine treatment with transitory alopecia and hypergonatotrophic hypogonadism was reported by Vircburger *et al.* (1981). The patient, who was a television reporter, suffered loss of hair from the head of sufficient extent to require the wearing of a wig. Loss of beard and body hair was also noticed. Plasma testosterone levels were shown to be low and LH and FSH levels were increased. Cimetidine treatment had been given at a dose of 1000 mg/day for six weeks. About three months after treatment was stopped regeneration of body hair was observed and over the next five months the patient gradually returned to normal both with respect to hair distribution and hormone levels.

Neuroleptic agents

Although menstrual irregularity in women and diminished potency and libido in men are well recognized side-effects of neuroleptic agents and although such symptoms point to an effect on the pituitary–gonadal system, at the doses used clinically, minimal or no effect on LH levels have been detected. Effects on testosterone levels have been variable, depending upon the actual drug used, and possibly suggest that thioridazine is the neuroleptic most frequently associated with sexual dysfunction.

In a study on 42 men with clinically stable schizophrenia Brown *et al.* (1981) compared effects of thioridazine, chlorpromazine hydrochloride, trifluoperazine hydrochloride, fluphenazine hydrochloride, perphenazine, and haloperidol on testosterone, LH, and prolactin levels. The patients treated with thioridazine had lower testosterone and LH levels than patients on any of the other drugs. The authors suggest that the low LH levels seen in the patients on thioridazine may support the view that the testosterone-lowering effects of this drug may be mediated by LH. They pointed out that, if thioridazine were an inhibitor of the synthesis and secretion of testosterone, then a compensating rise of LH would be expected rather than the fall which is observed. It would seem that thioridazine in therapeutic doses has a greater effect on the pituitary–gonadal axis than other neuroleptic agents. The reasons for this apparent drug specificity are not clear but the absence of a correlation among LH and testosterone levels and prolactin levels make it unlikely that the effects are mediated by the effects of prolactin on the pituitary. However, the previously demonstrated fact that thioridazine produces serum neuroleptic levels many times higher than the other drugs was confirmed in this study, and it seems likely that high peripheral neuroleptic levels may have a greater effect on the pituitary, which lies outside the blood–brain barrier.

Although testosterone and LH levels are consistently lowered in patients on thioridazine treatment, they do remain within the normal range. The clinically observed adverse effects such as loss of libido and erectile dysfunction are therefore not readily explained unless the patients concerned were in the low normal range before treatment.

Loperamide

This relatively new drug is a peripheral opiate agonist used in the treatment of diarrhoea. Although it is a powerful inhibitor of gastric motility, it does not exert central opiate activity. This would appear to be due to its poor penetration of the blood–brain barrier because in brain homogenates it has high affinity for opiate receptors. In a study on 19 healthy male volunteers Caldara *et al.* (1981) found that although loperamide caused an increase in plasma glucose and serum free fatty acid levels and a lowering of serum immunoreactive insulin and C-peptide there was no effect on pituitary hormone secretion even after high doses. This is in contrast to effects observed on centrally acting opiate agonists and supports the concept that loperamide's lack of central activity is due to its failure to cross the blood–brain barrier.

Aminoglutethimide

This agent, which is a potent inhibitor of adrenal steroid biosynthesis, is used in the treatment of metastatic breast cancer. It is well known that recovery of hypothalamic-pituitary-adrenal function after chronic administration of pharmacological doses of corticosteroids is very prolonged and is a serious and troublesome adverse effect. What little evidence there is on the time-course of recovery of hypothalamic–pituitary–adrenal function after chronic inhibition of steroidogenesis by aminoglutethimide indicates a similar prolonged course to that seen after suppression by exogenous steroids. In a study reported by Worgul *et al.* (1981) a series of 10 patients with metastatic breast cancer were treated with a combination of aminoglutethimide and sufficient hydrocortisone to prevent symptoms of adrenal insufficiency and to prevent a reflex increase in ACTH secretion. It is of considerable interest that abrupt discontinuation of treatment in these patients produced no signs or symptoms of adrenal insufficiency even though some patients had been on treatment for up to five years. It seems likely that prevention of the reflex increase in ACTH secretion by concomitant hydrocortisone administration limits the marked adrenal hypertrophy and lipid accumulation that it known to occur after the administration of aminoglutethimide alone. It is postulated that this preserves basal cortisol levels at normal values so that cessation of treatment is not accompanied by adverse effects.

Combination chemotherapy of childhood malignancies

In recent years advances in the treatment of certain malignancies of childhood, notably some leukaemias and Hodgkin's disease have met with considerable success and in an increasing number of cases has resulted in permanent cure. In these circumstances adverse effects which might have been acceptable if an increase in life expectancy of a few years at most were all that could be expected are now becoming a matter of concern. The most important of these must be the possibility that one or other of the chemotherapeutic agents used or irradiation may be the cause of a second malignant disease in later life. It is beyond the scope of this chapter to look into this particular serious possibility but the adverse effects of potent cytotoxic agents and irradiation do involve several aspects of the endocrine system. Although signs and symptoms of gonadal dysfunction and decrease in growth rate in children and young adults treated with combined chemotherapy protocols with or without irradiation are frequent and well documented, there is still lack of agreement on the causative factors and the mechanisms involved. There is no doubt that many of the young patients in these programmes fail to achieve the stature of their healthy school friends and there is no doubt that many of the males will have small testicular size and are eventually oligospermic or azoospermic. Although the female reproductive system appears to be more resistant than the male, ovarian abnormalities and menstrual dysfunction have been observed. In spite of general agreement on the occur-

rence of these adverse effects, there is less agreement on the precise causative factors. Thus opinions vary on the relative importance of chemotherapy or irradiation in causing growth retardation, and there is similar disagreement on the roles of tubular or Leydig-cell testicular damage in male patients.

In a report on 19 boys treated with combination chemotherapy (mechlorethamine, vincristine, procarbazine, and prednisone) for Hodgkin's disease Sherins *et al.* (1978) found a high incidence of gynaecomastia which they interpreted as indicating Leydig-cell dysfunction. Gynaecomastia occurred in nine out of the 19 patients and only in the pubertal subjects. The six prepubertal subjects showed no signs of gynaecomastia. Histological examination of biopsies of the testes in six pubertal boys revealed germinal aplasia in all of them. The authors consider that mechlorethemine (nitrogen mustard) and procarbazine are the two components of the combination responsible for testicular damage. Low serum testosterone levels and increased serum LH levels indicate that the Leydig cells are susceptible to damage.

In a later series of 17 young patients (15 males and two females) Whitehead *et al.* (1982) saw only one case of gynaecomastia, although severe testicular damage appeared to be very common. All 17 patients except one had been previously exposed to a course of combination chemotherapy (mustine, vincristine, procarbazine, and prednisolone), and 15 had also received neck or mantle irradiation with doses from 2500–3000 cGy. The two girls in the study had apparently normal menarches and one had a regular ovulatory menstrual cycle. The authors conclude that gonadal function in the male is more vulnerable than in the female, although the prepubertal ovary is not immune from damage. Severe tubular damage was attributed to mustine and procarbazine in the combination. Whitehead *et al.* (1982) were unable to explain the big discrepancy in the incidence of gynaecomastia in their study when compared to the study of Sherins *et al.* (1978).

A very similar situation is found in children treated with combination chemotherapy for acute lymphoblastic leukaemia. Testicular histology was studied by Lendon *et al.* (1978) in 44 boys whose treatment protocols included vincristine, prednisolone, 6-mercaptopurine and methotrexate. A number of other drugs were also used, including asparaginase, cyclophosphamide, cytosine arabinoside, and doxorubicin. The tubular fertility index was calculated on each biopsy specimen and was found to be significantly related to three variables: (a) the total dose of cyclophosphamide; (b) whether the total dose of cytosine arabinoside exceeded $1g/m^2$, and (c) the length of time between cessation of therapy and biopsy. The authors found encouragement in the third related variable which suggested that tubular fertility index improved with increasing time after cessation of treatment.

In a later paper Shalet *et al.* (1981) investigated the same group of patients to see if the stage of puberty was related to later gonadal dysfunction after combination chemotherapy. They found that there were more frequent abnormalities of FSH secretion in the pubertal boys as compared to the prepubertal ones, but that this was not explained by the more severe tubular damage in the former group. The authors believe that in the prepubertal boys gonadotrophin levels are controlled by small amounts of androgen produced in the interstitial tissues and that inhibin, the postulated FSH-suppressing hormone which acts at the hypothalamic–pituitary level, plays only a minor role. They conclude that the combination chemotherapy protocol used in their study produced minimal Leydig-cell damage, as shown by the response to HCG stimulation, but it did cause frequent and severe tubular damage. From the clinical standpoint these boys would be expected to mature normally through puberty but would end up either oligospermic or azoospermic.

It is generally felt that the female gonadal system is considerably less vulnerable to chemotherapy than the male system, and most available evidence would seem to support that view. In a study of ovarian development in leukaemic children Himelstein-Braw *et al.* (1978) examined specimens of ovary from 31 children dying of leukaemia. Ovaries from 28 normal children dying from accidents served as a control. The leukaemic children had been treated with a variety of cytotoxic agents in combination with steroids. Methotrexate and 6-mercaptopurine were used as maintenance therapy but in addition cytosine arabinoside, L-asparaginase, and cyclophosphamide were used. Six of the patients also received cerebrospinal irradiation. All the ovaries from normal children showed active follicle growth and none were quiescent. Most of the ovaries of leukaemic children were abnormal, although leukaemic infiltration occurred in only one case. There was no follicle growth in 22 per cent of cases, but the presence of 'scars' of large follicles suggested that follicular growth and atresia had occurred previously. The number and size of large antral follicles were reduced significantly in all cases. It was not clear whether follicular growth was inhibited by the disease itself or by the chemotherapy. All ovaries of children treated for more than two months showed inhibition of follicle growth. More prolonged treatment caused a reduction in the number of small oocytes and also affected the ovarian stroma. In view of the known adverse effects of cyclophosphamide and other cytotoxic agents on gonadal tissue, the authors concluded that the

abnormalities seen in the ovaries of treated leukaemic children were probably due to the drugs used.

It is also well known that children treated for malignant disease by chemotherapy and irradiation fail to achieve the same stature as normal children of the same age. In a study on 95 children over a period of 3–4 years Griffin and Wadsworth (1980) showed that radiotherapy to the central nervous system is associated with inhibition of growth. They are of the view that this inhibition of growth coincides with a transient depression of growth hormone levels due to radiation damage to the pituitary region. The reduction in growth in this group was not severe and appeared only to affect the first year after presentation and was normal thereafter. Measurement of growth hormone levels showed a blunting of peak levels after irradiation of the central nervous system. At the present time there is no general agreement on the level of growth hormone necessary to maintain normal growth, but an arbitrary criterion of growth hormone deficiency is taken as an inability to produce a serum growth hormone level of 20 mU/l after adequate provocation.

Essentially similar findings were reported by Shalet *et al.* (1979) in a study of 26 children with acute lymphoblastic leukaemia all of whom had been treated with cranial irradiation. Although mean standing heights was significantly less than normal, only a small minority of the children were clinically growth-hormone-deficient as measured by the response to insulin induced hypoglycaemia and an arginine test. However, whereas Griffin and Wadsworth (1980) attribute diminished stature in these children to the effects of irradiation, Shalet and Price (1981) believed that the combination of cytotoxic agents and steroids is the causative factor. There would seem to be insufficient evidence available at the present time to distinguish between these two hypotheses.

Anti-epileptic drugs

Diminished libido and reduced fertility are known to occur in male epileptic patients on chronic anti-epileptic therapy. Some evidence is available to show that this may be due to a reduction in free testosterone in these patients. Plasma concentrations of sex-hormone-binding globulin are raised, total testosterone is either normal or high, and gonadotrophin levels are often increased. In the group of 37 male patients studied by Dana-Haeri *et al.* (1982) sex-hormone-binding globulin was significantly raised (49.0 nmol/l compared with 23.3 nmol/l in controls) but calculated free testosterone was significantly reduced. The authors concluded that plasma-total testosterone in epileptic patients is not a satisfactory index of androgenic activity. They postulate that the reduction in plasma-free testosterone concentration may be due to a combination of enhanced metabolism from hepatic microsomal enzyme-induction and increased levels of sex-hormone-binding globulin. However the possibility that central control of gonadal function is disturbed cannot be neglected.

Drug-induced parathyroid dysfunction

Lithium

In a survey of 96 manic-depressive patients studied while on long-term lithium therapy, Christiansen et al. (1978) found statistically significant elevations of serum immunoreactive parathyroid hormone ($p < 0.001$) as well as the protein-corrected levels of serum calcium ($p < 0.001$) and serum magnesium ($p < 0.001$), thus indicating a state of 'primary' hyperparathyroidism. The patients as a group had normophosphataemia and normophosphatasia, supporting the impression of a rather mild state of biochemical hyperparathyroidism.

The administration of lithium is known to be associated with an elevation of serum-calcium levels and a lowering of urinary calcium excretion. In addition serum phosphate may be lowered. Davis et al. (1981) investigated 19 psychiatric out-patients, all diagnosed as having manic-depressive psychosis. Treatment was with lithium carbonate, 0.6–1.8 g/day for periods varying from one month to three years. Serum samples were assayed for parathyroid hormone, calcium, phosphate, total protein, globulin, magnesium, lithium, creatinine, and liver enzymes. Controls consisted of 15 healthy adults. All patients had normal renal and hepatic function. Mean serum-calcium and magnesium levels showed no differences from the controls but mean serum-phosphate levels were significantly lowered. There was also a positive correlation between serum-calcium and lithium levels and serum-phosphate, magnesium, or parthyroid hormone values. Nevertheless, parathyroid hormone levels were significantly higher in lithium-treated subjects than in controls. It is concluded that lithium can produce a drug-dependent, reversible, biochemical hyperparathyroidism. However, the complications of primary hyperparathyroidism, nephrolithiasis, and osteitis fibrosa, have not been observed in patients on long-term lithium therapy. Although the patients reported by Davis et al. (1981) did not have hypocalcaemia, it is possible that lithium may cause a functional hypocalcaemia since lithium is known to antagonize a number of calcium-dependent processes. This means that in spite of normal serum-calcium levels there may be, in the presence of lithium, a deficiency of available calcium for biological activity. Nevertheless, the mechanism of parathyroid hormone elevation in patients on lithium therapy is not known, although it could be indicative of early renal failure.

A much more acute response in an elderly female patient (aged 70 years) suffering from manic-depressive psychosis was described by Rothman (1982). Although the patient had either recently received or was still receiving several other drugs, including dipyridamole, flurazepam, hydrochlorthiazide, metoprolol, methyldopa, thiothixene, and haloperidol, the author attributed the patient's spectrum of adverse side-effects to increasing doses of lithium. Lithium intoxication was diagnosed on the basis of tremor, ataxia, incontinence, and delirium. In this case serum calcium levels began to rise on the first day after the initiation of lithium therapy and the parathyroid hormone level was raised after only 10 days. Serum calcium levels became normal 2 days after cessation of lithium therapy and parathyroid hormone levels were normal after 8 days. Although there were several confusing factors in this case, it does indicate that hyperparathyroidism may become manifest within a few days of initiating lithium therapy.

In a careful study on a 46-year-old male patient who manifested a hyperparathyroid state whilst on chronic lithium therapy, Shen and Sherrard (1982) concluded that in lithium-induced hyperparathyroidism there is a reset of the 'set-point' or 'calciostat'. They based this conclusion on the observation that the regression line, obtained when serum immunoreactive parathyroid hormone is plotted against serum calcium, is shifted to the right implying that it necessitates higher serum calcium concentrations to inhibit parathyroid hormone secretion than is found in normal individuals. The authors are of the view that this explanation for raised parathyroid hormone levels taken together with the obvious lack of complications and the prompt resolution of hypercalcaemia on withdrawal of lithium are contradictions to surgery in these cases.

[131]I-induced hypoparathyroidism

Jialal et al. (1980) report a case presenting with both clinical and biochemical evidence of hyperthyroidism. Treatment with [131]I (5 mCi) was given orally and six months later the patient returned complaining of restlessness, generalized stiffness of the muscles, and episodes of muscular cramps. Clinically, she still appeared mildly thyrotoxic but biochemically she proved to be euthyroid. However, examination of the nervous system revealed brisk bilateral tendon reflexes, positive Chvostek's reflex, and Trousseau's sign. The patient was hypocalcaemic and there was no measurable parathyroid hormone in the blood. Immunological tests were unable to demonstrate antibodies to parietal cells, mitochondria, smooth muscle, or adrenals, and tests for antinuclear factor were negative. Cases of ^{131}I-induced hypoparathyroidism are very rare but have been reported. The present case is of interest in that an auto-immune aetiology could be excluded, and there was no reason to believe that the patient was in a hypopara-thyroid state before treatment. It would seem to be a reasonable precaution to monitor serum-calcium levels for at least a year after ^{131}I treatment.

Aluminium overload

Some recent studies have suggested that parathyroid function may be affected by increased levels of aluminium. There are two situations in which patients ingest large amounts of aluminium, one being chronic peptic ulceration and the other chronic haemodialysis. It was suggested by Alfrey et al. (1976) that the dialysis–encephalopathy syndrome is caused by aluminium intoxication, and it is also known that hypocalcaemia, which is present in long-term dialysis patients, is a stimulus for parathyroid hormone secretion. Recent work by Cann et al. (1979) has shown that ingested aluminium may be preferentially localized in certain tissues, in particular, the parathyroid glands. Tissue was obtained from patients undergoing surgery for parathyroidism and from autopsies. Four of the 19 patients studied had taken aluminium-containing drugs and the parathyroid glands in those cases contained markedly higher concentrations of aluminium. The authors claimed that there was a linear relationship between dietary intake and concentration in the parathyroid glands. It is also pointed out that the intake of aluminium in those patients was far below that used for phosphate-binding therapy in haemodialysis. The possibility of hyperparathyroidism should be considered in patients on long-term aluminium-containing therapy. It should be noted that the clinical presentation of aluminium overload in patients on continuous ambulatory, peritoneal dialysis resembles hyperparathyroidism in that headaches and abdominal cramping pains are predominant features.

Oral contraceptives

The administration of oestrogen is known to suppress bone turnover and lowering of serum phosphate has been reported during treatment with oestrogen and oral contraceptives. It is also known that in patients with hypoparathyroidism oestrogen treatment may induce hypocalcaemia. The case of a 41-year-old woman who had no relevant previous medical problems who developed secondary hyperparathyroidism whilst on a norethindrone/mestranol oral contraceptive was reported by Moses and Notman (1982). Extensive laboratory investigations on the patient indicated mild hypocalcaemia, hypophosphataemia, increased parathyroid hormone levels, and increased urinary excretion of cAMP. The administration of 50 000 units of vitamin D daily resulted in the calcium and phosphate levels returning to normal. When all treatment was withdrawn those levels decreased slightly but remained within the normal range. When the same oral contraceptive was recommenced the same picture of secondary hyperparathyroidism was produced. The authors concluded that in this case the administration of the oral contraceptive pill was the causative factor.

General

It is known that plasma antidiuretic hormone (ADH) levels are increased in response to nausea caused by administration of apomorphine. There are no significant changes in plasma osmolarity or mean arterial pressure so it is thought that the ADH-lowering effect of apomorphine may be attributed to its dopaminergic agonist properties. It is possible however that the effect may be caused by a response to nausea and/or emesis rather than a direct central effect of apomorphine.

In a study on chemotherapy-induced nausea and/or emesis in humans Fisher et al. (1982) observed similar increases in ADH levels. Patients on cisplatin were excluded because of the likelihood of serum osmolarity changes in response to mannitol diuresis and intravenous fluid replacement. However patients treated with dactinomycin, doxorubicin, bleomycin, cyclophosphamide, methotrexate, streptozocin, vincristine, and vinblastine were included. Since these chemotherapeutic agents are known not to enter the central nervous system it seemed unlikely that the ADH elevations were caused by a direct central action of the drug or drugs. From the fact that patients receiving one or several of these agents, who did not suffer nausea and/or emesis, did not manifest raised ADH levels supports the hypothesis that it is the induction of nausea or emesis that is the triggering factor. This observation may be of considerable importance in the management of these patients and further studies to determine the interrelationship between nausea, ADH release, and plasma osmolarity are indicated.

The significance of possible ADH elevations associated with other drugs should not be overlooked. Nausea and/or vomiting and very common adverse effects of drugs are often considered to be little more than an unpleasant but trivial consequence of treatment.

Alterations of ADH levels, particularly in old, debilitated, or seriously ill patients may require more serious consideration.

REFERENCES

AANDERUD, S. AND STRANDJORD, R.E. (1980). Hypothyroidism induced by anti-epileptic therapy. *Acta neurol. scand.* **61**, 330–3.

ALEXANDER, I.R.W. (1961). Acute myxoedema, *Br. med. J.* **2**, 1434.

ALFREY, A.C., LE GENDRE, G.R., AND KAEHNY, W.D. (1976). The dialysis encephalopathy syndrome. *New Engl., J. Med.* **294**, 184–8.

ALLANBY, K.D. (1957). Deaths associated with steroid hormone therapy. *Lancet* **i**, 1104–10.

ALLOLIO, B., STUTTMANN, R., FISCHER, H., LEONHARD, W., AND WINKELMANN, W. (1983). Long-term etomidate and adrenocortical suppression. *Lancet* **ii**, 626.

ANGRIST, B.M., GERSHON, S., LEVITAN, S.J., AND BLUMBERG, A.G. (1970). Lithium-induced diabetes insipidus-like syndrome. *Comprehens. Psychiat* **11**, 141–6.

ANSEL, R.D. (1970). In *L-dopa in parkinsonism* (ed. A Barbeau and F.H. McDowell), p. 317. F.A. Davis, Philadelphia.

BALODIMOS, M.O., GRAHAM, C.A., MARBLE, A., AND KRALL, L.P. (1968). Acetohexamide in the therapy of diabetes mellitus. *Metabolism* **17**, 669–80.

BARBIERI, R.L. (1980). Danazol and thyroid function. *Ann. intern. Med.* **92**, 133.

BASEDOW VON, C.A. (1840). Exophthalmos durch Hypertrophie des Zellgewebes in der Augenhohle. *Wochenschr. des HeilKunde.* **13**, 197–204, 220–8.

BEGG, T.B., AND HALL, R. (1963). Iodide goitre and hypothyroidism. *Quart. J. Med.* **32**, 351–62.

BENEDEK, T.G. (1962). Goitre formation as a result of phenylbutazone and vitamin A. *J. clin. Endocrinol.* **22**, 959–62.

BONGIOVANNI, A.M., AND McPADDEN, A.J. (1960). Steroids during pregnancy and possible fetal consequences. *Fertil. Steril.* **11**, 181–6.

BROWN, J. AND SOLOMON, D.H. (1958). Mechanism of antithyroid effects of a sulfonylurea in the rat. *Endocrinology* **63**, 473–80.

BROWN, W.A., LAUGHREN, T.P., AND WILLIAMS, B. (1981). Differential effects of neuroleptic agents on the pituitary-gonadal axis in men. *Arch. gen. Psychiat.* **38**, 1270–72.

BULL, G.M. AND RUSSELL FRASER, T. (1950). Myxoedema from resorcinol ointment applied to leg ulcers. *Lancet* **i**, 851–5.

BURDA, C.D. (1965). Sulphonylurea hypothyroidism in diabetics. *Lancet* **ii**, 1016–17.

BURGER, A., DINICHERT, D., NICOD, P. JENNY, M., LEMARCHAND-BERAUD, T., AND VALLOTON, M.B. (1976). Effect of amiodarone on serum tridiodothyronine, reverse triiodothyronine, thyroxin, and thyrotropin. A drug influencing peripheral metabolism of thyroid hormones. *J. clin. Invest* **58**, 255–9.

BURGI, H., WIMPFHEIMER, C., BURGER, A., ZAUNBAUER, W., ROSLER, H., AND LEMARCHAND-BERAUD, T. (1976). Changes of circulating thyroxine-triiodothyronine and reverse T3 after radiographic contrast agents. *J. clin. Endocr. Metab.* **43**, 1203–10.

CALDARA, R., TESTORI, G.P., PERRARI, C., ROMUSSI, M., RAMPINI, P., BORZIO, M. AND BARBIERI, C. (1981). Effect of loperamide, a peripheral opiate agonist, on circulating glucose, free fatty acids, insulin, C-peptide and pituitary hormones in healthy man. *Eur. J. clin. Pharmacol.* **21**, 185–8.

CANDY, J. (1972). Severe hypothyroidism — an early complication of lithium therapy. Br. med. J. **3**, 277.

CANN, C.E., PRUSSIN, S.G., AND GORDON, G.S. (1979). Aluminium uptake by the parathyroid glands. *J. clin. Endocrinol. Metab.* **49**, 543–5.

CARSWELL, F., KERR, M.M., AND HUTCHINSON, J.H. (1970). Congenital goitre and hypothyroidism produced by maternal ingestion of iodides. *Lancet* **i**, 1241–3.

CHABROLLE, J.-P., MONOD, N., PLOUIN, P., LELOI'H., DE MONTIS, G., AND ROSSIER, A. (1978). Surcharge iodee post-natale avec hypothyroidie et pauses respiratoires. Danger de l'application cutanee de produits iodes. *Arch Fr. Pédiat.* **35**, 432–7.

CHERON, R.G., KAPLAN, M.M., LARSEN, P.R., SELEKOW, H.A. AND CRIGLER, J.F. (1981). Neonatal thyroid function after propylthiouracil therapy for maternal Graves' disease. *New Engl. J. Med.* **304**, 525–8.

CHO, J.T., BONE, S., DUNNER, D.L., COLT, E., AND FIEVE, R.R. (1979). The effect of lithium treatment on thyroid function in patients with primary affective disorder. *Am. J. Psychiat.* **136**, 115–16.

CHRISTIANSEN, C., BAASTRUP, P.C., LINDGREEN, P., AND TRANSBOL, I. (1978). Endocrine effects of lithium. II. Primary hyperparathyroidism. *Acta Endocr., Copenh.* **88**, 528–34.

CHRISTY, N.P., WALLACE, E.Z., AND JAILER, J.W. (1956). Comparative effects of prednisone and cortisone in suppressing the response of the adrenal cortex to ACTH. *J. clin. Endocrinol.* **16**, 1059–74.

COINDET, J.R. (1821). Nouvelles recherches sur les effets de l'iode et sur les precautions à suivre dans le traitement du goitre par un nouveau remède. *Ann. Chim. Phys.* **16**, 252–66.·

CONNOLLY, R.J., VIDOR, G.I., AND STEWART, J.C. (1970). Increase in thyrotoxicosis in endemic goitre area after iodation of bread. *Lancet* **i**, 500–2.

CREPIN, G., DELAHOUSSE, G., DECOCQ, J., DELCROIX, M., CAQUANT, F., QUERLEU, D., AND TALEB, L. (1978). Dangers des medicaments iodes chez la femme enceinte. *Phlebologie* **31**, 279–85.

CREUTZFELDT, W. AND SOELING, H.D. (1960). Oral diabetes therapy and its experimental principles. *Ergebn. Med. Kinderheilk.* **15**, 1–213.

DANA-HAERI, J., OXLEY, J. AND RICHENS, A. (1982). Reduction in free testosterone by antiepileptic drugs. *Brit. Med. J.* **284**, 85–6.

D'ARCY, P.E. AND GRIFFIN, J.P. (Eds.) (1972). *Iatrogenic diseases*, 1st edn., pp. 109–10. Oxford University Press. London.

DAVIS, B.M., PFEFFERBAUM, A., KRUITZIK, S., AND DAVIS, K.L. (1981). Lithium's effect on parathyroid hormone. *Am. J. Psychiat.* **138**, 489–92.

DE GROOT, L.J. AND HOYE, K. (1976). Dexamethasone suppression of serum T3 and T4. *J. clin. Endocr. Metab.* **42**, 976–8.

DE LANGE, W.E. AND DOORENBOS, H. (1968). Hormones and synthetic substitutes. In *Side-effects of Drugs*, Vol. VI, (ed. L. Meyler and A. Herxheimer), pp. 383–421. Excerpta Medica Foundation, Amsterdam.

DIDDLE, A.W., WATTS, G.E., GARDNER, W.H., AND WILLIAMSON, P.J.

(1966). Oral contraceptive medication. A prolonged experience. *Am. J. Obstet. Gynec* **95**, 489-95.

DOENICKE, A., (1983). Etomidate. *Lancet* **ii**, 168.

DORNHORST, A., DAVIE, M.W.J., FAIRNEY, A. AND WYNN, V. (1981). Possible iatrogenic hypothyroidism. *Lancet* **i**, 52.

DOWNS, J.W. AND COOPER, W.G. (1955). Surgical complications resulting from ACTH and cortisone medication. *Am Surg.* **21**, 141-6.

EGER, W. AND FERNHOLZ, J. (1965). Uber den thyreostatischen Effekt des Phenylbutazon und des Oxyphenbutazon unter dem Einfluss des Trijodthyronin. *Med. Pharmacol. exp., Basel* **13**, 17-23.

FELLOWS, I.W., BYRNE, A.J. AND ALLISON, S.P. (1983). Adrenocortical suppression with etomidate. *Lancet* **ii**, 54-55.

FISHER, R.D., RENTSCHLER, R.E., NELSON, J.C., CODFREY, T.E., AND WILBUR, D.W. (1982). Elevation of plasma antidiuretic hormone (ADH) with chemotherapy-induced emesis in man. *Cancer treat. Rep.* **66**, 25-9.

FRADKIN, J.E. AND WOLFF, J. (1983). Iodide-induced thyrotoxicosis. *Medicine* **62**, 1-20.

FRASER, C.G., PREUSS, F.S. AND BIGFORD, W.D. (1952). Adrenal atrophy and irreversible shock associated with cortisone therapy. *J. Am. med. Ass.* **149**, 1542-3.

FREY, H. (1964). Hypofunction of the thyroid gland due to prolonged and excessive intake of potassium iodide. *Acta endocr., Kbh.* **47**, 105-20.

GOOD, B.F., HETZEL, B.S., AND HOGG, B.M. (1965). Studies of the control of thyroid function in rats: effects of salicylate and related drugs. *Endocrinology* **77**, 674-82.

GRAUDENZ, M.G. AND BEIRAO DE ALMEIDA, A. (1965) Noresteroides anti-concepcionais. Estudo experimental baseado na histologia do utero, ovario e embriao. *Rev. Ginec. Obstet., Rio de J.* **116**, 108-27.

GRAVES, R.J. (1835) Clinical Lectures. *London Med. Surg. J.* **7**, 516.

GRIFFIN, N.K. AND WADSWORTH, J. (1980). Effect of treatment of malignant disease on growth in children. *Arch. Dis. Childh.* **55**, 600-3.

GRUBECK-LOEBENSTEIN, B., KRONIK, G., MOSSLACHER, H., AND WALDHAUSL, W. (1983). The effect of iodine-containing contrast medium on thyroid function of patients undergoing coronary angiography. *Exp. Clin. Endocrinol.* **81**, 59-64.

HANCOCK, K.W., SCOTT, J.S., PANIGRAHI, N.M., AND STITCH, S.R. (1976). Significance of low body weight in ovulatory dysfunction after stopping oral contraceptives. *Br. med. J.* **2**, 399-401.

HARNAGEL, E.E. AND KRAMER, W.G. (1955). Severe adrenocortical insufficiency following joint manipulation. *J. Am. med. Ass.* **158**, 1518-19.

HARRIS. L., McKENNA, W.J., ROWLAND, E. AND KRIKLER, D. M. (1983). Side effects and possible contraindications of amiodarone use. *Am. Heart J.* **106**, 916-23.

HAYES, M.A. (1956). Surgical treatment as complicated by prior adrenocortical steroid therapy. *Surgery* **40**, 945-50.

—— AND KUSHLAN, S.D. (1956), Influence of hormonal therapy for ulcerative colitis upon course of surgical treatment. *Gasteroenterology* **30**, 75-84.

HEGER, J.J., PRYSTOWKSY, E.N. AND ZIPES, D.P. (1983). Relationships between amiodarone dosage, drug concentrations and adverse side-effects. *Am. Heart J.* **106**, 931-5.

HELGASON, T. (1964). Iodides, goitre, and myxoedema in chronic

respiratory disorders. *Br. J. Dis. Chest* **58**, 73-7.

HEROMAN, W.M. BYBEE, D.E., CARDIN, M.J., BASS, J.W., AND JOHNSONBAUGH, R.E. (1980). Adrenal suppression and cugshingoid changes secondary to dexamethasone nose drops, *J. Paediat.* **96**, 500-1.

HETZEL, B.S., GOOD, B.F., AND WELLBY, M.L. (1963). Salicylate action and thyroidal autonomy in hyperthyroidism. *Lancet* **ii**, 93-4.

HIMELSTEIN-BRAW, R., PETERS, H., AND FABER, M. (1976). Morphological study of the ovaries of leukaemic children. *Br. J. Cancer.* **38**, 82-7.

HOLLANDER, S., STEVENSON, C., MITSUMA, T., PINEDA, G., SHENKMAN, L., AND SILVA, E. (1972). T3 thyrotoxicosis in an iodine-deficient area. *Lancet* **ii**, 1276.

HOLT, D.W., TUCKER, G.T., JACKSON, P.R., AND STOREY, G.C.A. (1983). Amiodarone pharmacokinetics. *Am. Heart J.* **106**, 840-7.

HORDEN, R.M., ALEXANDER, W.D., AND HARRISON, M.T. (1964). Non-toxic goitre in males. *Br. med. J.* **1**, 1419-21.

HUMPHREY, M.J., CAPPER, S.J., AND KURTZ, A.B. (1980). Fenclofenac and thyroid hormone concentrations. *Lancet* **i**, 487-8.

HUNTON, R.B., WELLS, M.V., AND SKIPPER, E.W. (1965). Hypothyroidism in diabetics treated with sulphonylurea. *Lancet* **ii**, 449-51.

ISAACS, A.J., AND MONK, B.E. (1980). Fenclofenac interferes with thyroid-function tests. *Lancet* **i**, 267-8.

JACKSON, A.S. (1925). Iodine hyperthyroidism: an analysis of fifty cases. *Boston med. surg. J.* **193**, 1138-40.

JACOBSON, J.M., HANKINS, G.V., MURRAY, J.M., AND YOUNG, R.L. (1981). Self-limited hyperthyroidism following intravaginal iodine administration. *Am. J. Obstet. Gynecol.* **140**, 472-3.

JAGGARAO, N.S.V., SHELDON, J., GRUNDY, E.N., VINCENT, R., AND CHAMBERLAIN, D.A. (1982). The effects of amiodarone on thyroid function *Postgrad. med. J.* **58**, 693-6.

JIALAL, I., PILLARY, N.I., AND ASMAL, A.C. (1980). Radio-iodine induced hypoparathyroidism. *S. Afr. med. J.* **58**, 939.

JONCKHEER, M.H., BLOCKX, P., KAIVERS, R., AND WYFFELS, G. (1973). Hyperthyroidism as a possible complication of the treatment of ischemic heart disease with amiodarone. *Acta card. bel.* **28**, 192-200.

——, ——, BLOCKX, P., AND BERNARD, R. (1976). Amiodarone et function thyroidienne. *Arch. mal. Coeur Vaiss.* **12**, 1315-19.

——, ——, BROECKAERT, I., CORNETTE, C., AND BECKERS, C. (1978). Low T3 syndrome in patients chronically treated with an iodine-containing drug. Amiodarone. *Clin. Endocrinol.* **9**, 27-35.

KABLE, W.T. (1981). Drug induced primary hypothyroidism and hyperprolactinaemia. *Fertil. Steril.* **35**, 483-4.

KEIDAR, S., GRENADIER, E., AND PALANT, A. (1980). Amiodarone induced thyrotoxicosis: four cases and a review of the literature. *Postgrad. med. J.* **56**, 356-8.

KIMBALL, O.P. (1925). Induced hyperthyroidism. *J. Am. med. Ass.* **85**, 1709.

KITTREDGE, W.E. (1955). Potential hazards of cortisone in treatment of prostatic cancer: report of a fatal case. *J. Urol.* **73**, 585-90.

KLAUSEN, N.O., MOELGAARD, J., FERGUSON, A.H., KAALUND JENSEN, J., LARSEN, C., AND PAABY, P. (1983). Negative Synacthen test during etomidate infusion. *Lancet* **ii**, 848.

KRAUSE-LARSEN, C., AND GARDE, K. (1971). Post-menopausal bleeding: another side-effect of L-dopa. *Lancet* **i**, 707-8.

LAZARUS, J.H., JOHN, R., BENNIE, E.H., CHALMERS, R.J., AND CROCKETT, G. (1981) Lithium therapy and thyroid function: a long term study. *Psychol. Med.* **II**, 85–92.

LEDINGHAM, I. McA. AND WATT, I. (1983). Influence of sedation on mortality in critically ill multiple trauma patients. *Lancet* i, 1270.

LENDON, M., HANN, I.M., PALMER, M.K., SHALET, S.M., AND MORRIS-JONES, P.H. (1978). Testicular histology after combination chemotherapy in childhood for acute lymphoblastic leukaemia. *Lancet* ii, 439–41.

LEWIS, L., ROBINSON, R.F., YEE, J., HACKER, L.A., AND EISEN, G. (1953). Fatal adrenal cortical insufficiency precipitated by surgery during prolonged continuous cortisone treatment. *Ann. intern. Med.* **39**, 116–26.

LILLICRAP, D.A. (1962). Myxoedema after thalidomide *(Distaval).* *Br. med. J.* **1**, 477.

LINTHORSRT, G. (1966). The effects of oral contraceptives on the ovary. In *Social and medical aspects of oral contraception.* Round-table Conference, Scheveningen, Netherlands (ed. M.N.G. Dukes), pp. 95–7. Excerpta Medica Foundation, Amsterdam.

LORRAIN, J. (1966). Ovaries polykystiques ßa la suite d'un traitement au noréthynodrel. *Un. méd. Can.* **95**, 1053.

LOW, L.C.K. RATCLIFFE, W.A., AND ALEXANDER, W.D. (1978). Intrauterine hypothyroidism due to antithyroid-drug therapy for thyrotoxicosis during pregnancy. *Lancet* ii, 370–1.

MAMOU, H. (1957). Myxoedème après traitements par les sulfamides antidiabétiques. (Myxedema after antidiabetic sulfonamide therapy). *Sem. Hôp., Paris* 33, 1044–5.

MARCHESE, M.J., BERTORELLO, M.C., AND CARDOZO, E. (1963). Mixedema en el curso de tratamiento antituberculoso. *Rev. Asoc. med. argent.* 77, 542–4.

MARSHALL, J.S., LEVY, R.P., AND LEONARDS, J.R. (1965). The acute effect of salicylate administration on human thyroxine transport. *J. Lab. clin. Med.* **66**, 1001.

MASSIN, J.P., THOMOPOULOS, P., KARAM, J., AND SAVOIE, J.C. (1971). Le resque thyroidien d'un nouveau coronaro-dilatateur iode: l'amiodarone (Cordarone). *Annls Endocr.* 32, 438–48.

McCLURE, R.D. (1927). Experiences with thyroid problems in Detroit, *Ann. Surg.* 85, 333–78.

MEANS, J.H., DE GROOT, L.J., AND STANBURY, J.B. (1963). *The thyroid and its diseases.* McGraw Hill, New York.

MELVIN, G.R., ACETO, T., BARLOW, J., MUNSON, D., AND WIERDA, D. (1978). Iatrogenic congenital goitre and hypothyroidism with respiratory distress in a newborn. *S. Dakota med. J.* October 15–19.

MIDGLEY, J.E.M. AND WILKINS, T.A. (1980). Hypothyroidism in patients on fenclofenac. *Lancet* ii, 704.

MOSES, A.M. AND NOTMAN, D.D. (1982). Secondary hyperparathyroidism caused by oral contraceptives. *Arch intern. Med.* **142**, 128–9.

MURDOCH, J. McC. AND CAMPBELL, G.D. (1958). Antithyroid activity of N-phthalyl glutamic acid imide (K17). *Br. med. J.* **1**, 84–5.

MYHILL, J. AND HALES, I.B. (1963). Salicylate action and thyroidal autonomy in hyperthyroidism. *Lancet* i, 802–5.

NATHAN A.W. AND ROSE, G.L. (1979). Fatal iatrogenic Cushing's syndrome. *Lancet* i, 207.

NEWTON R.W., BROWNING, M.C.K. IQBAL, J., PIERCY, N., AND ADAMSON, D.G. (1978). Adrenocortical suppression in workers manufacturing synthetic glucocorticoids. *Br. med. J.* **1**, 73–4.

NIELSON, J.B., BRIVSHOLM, A., FISCHER, F., AND BRØCHNER-MORTENSEN, K. (1963). Long-term treatment with corticosteroids in rheumatoid arthritis. *Acta Med. scand.* **173**, 177–83.

NISHIYAMA, S., MATSUKURA, M., FUJIMOTO, S., AND MATSUDA, I. (1983). Reports of two cases of autoimmune thyroiditis while receiving anticonvulsant therapy. *Eur. J. Pediat.* **140**, 116–7.

O'NEIL, R. (1965). Discussion. In *Recent advances in ovarian and synthetic steroids* (ed. R.P. Shearman), p. 24. Globe Commercial Ltd, Sydney.

OPPENHEIMER, E.H. (1964). Lesions in the adrenals of an infant following maternal corticosteroid therapy. *Bull. Johns Hpk. Hosp.* **114**, 146–51.

ORTEGA, L.D. AND GRANDE, R.G. (1979). Cushing's syndrome due to abuse of dexamethasone nasal spray. *Lancet* ii, 96.

PARMELEE, A.H., ALLEN, E. STEIN, I.F. AND BUXBAUM, H. (1940). Three cases of congenital goitre. *Am. J. Obstet. Gynecol* 40, 145–7.

PARRY, C.H. (1825). Collections from the unpublished medical writings of Dr. C.H. Parry. Underwood, London, 1895.

PENFOLD, J.L., PEARSON, C.C. SAVAGE, J.P., AND MORRIS, L.L. (1978). Iodide-induced goitre and hypothyroidism in infancy and childhood. *Aust. Paediat. J.* **14**, 69–73.

PERRILD, H., MADSEN, S.N., AND HANSEN, J.E.M. (1978). Irreversible myxoedema after lithium carbonate *Br. med. J.* **1**, 1108–9.

PIZIAK, V., SELLMAN, J.E., AND OTHMER, E. (1978). Lithium and hypothyroidism. *J. clin. Psychiat.* **39**, 709–11.

PLATE, W.P. (1968). Ovarian changes after long-term oral contraception. In *Drug-induced diseases,* Vol. III (ed L. Meyler, and H.M. Peck), pp. 235–8. Excerpta Medica Foundation, Amsterdam.

PLUMER, J.N. AND ARMSTRONG, R.S. (1957). Adrenocortical failure following long-term steroid therapy. *Ariz. Med.* **14**, 202–5.

PRESCOTT, L.F. (1968). Antipyretic analgesic drugs. In *Side-effects of drugs ,* Vol. VI eds. L. Meyler and A. Herxheimer, pp. 101–39. Excerpta Medica Foundation, Amsterdam.

PUCCIARELLI, G. (1978). Amiodarone e tiroide: a proposito di due casi di ipotiroidism. *Clin. ter* 84, 81–4.

RAMSEY, T.A., MENDELS, J., STOKES, J.W., AND FITZGERALD, R.G. (1972). Lithium carbonate and kidney function: a failure in renal concentrating ability. *J. Am. med. Ass.* 219, 1446–9.

RATCLIFFE, W.A., HAZELTON, R.A. AND THOMPSON, J.A. (1980). Effect of fenclofenac on thyroid-function tests. *Lancet* i, 432.

REUS, V.L., GOLD, P., AND POST, R. (1979). Lithium-induced thyrotoxicosis. *Am. J. Psychiat.* **136**, 724–5.

ROGERS, M.P. AND WHYBROW, P.C. (1971). Clinical hypothyroidism occurring during lithium treatment: two case histories and a review of thyroid function in 19 patients. *Am. J. Psychiat.* **128**, 158–63.

ROTHMAN, M. (1982). Acute hyperparathyroidism in a patient after initiation of lithium therapy. *Am. J. Psychiat.* **139**, 362–3.

RYAN, G.M., CRAIG, J., AND REID, D.E. (1964). Histology of the uterus and ovaries after long-term cyclic norethynodrel therapy. *Am. J. Obstet. Gynecol.* **90**, 715–25.

SALASSA, R.M., BENNETT, W.A., KEATING, F.R., AND SPRAGUE, R.G. (1953). Post-operative adrenal cortical insufficiency. *J. Am. med. Ass.* **152**, 1509–15.

SANDLER, M. (1971). Gynaecological side-effects of levodopa therapy. *Lancet* i, 807.

SAVAGE, D.C.L. AND SWIFT, P.G. (1981). Effect of cyproterone acetate on adrenocortical function in children with precocious puberty. *Arch. Dis. Childh.* **56**, 218–22.

SAVEGE, T.M. (1983). Etomidate and adrenocortical function. *Lancet* i, 1434.

SAVOIE, J.C., MASSIN, J.P., THOMOPOULOS, P., AND LEGER, F. (1975). Iodine-induced hypertoxicity in apparently normal thyroid glands. *J. clin. Endocrinol,* **41**, 685–91.

SAYERS, G. (1950). The adrenal cortex and homeostasis. *Physiol. Rev.* **30**, 241–320.

SCHNEEWIND, J.H. AND COLE, W.H. (1959). Steroid therapy in surgical patients. *J. Am. med. Ass.* **170**, 1411–21.

SCHLESS, G.L. (1966). Hypothyroidism secondary to tolbutamide, *J. Am. med. Ass.* **195**, 1073.

SEDVALL, G., JONSSON, P., PETTERSSON, U., AND LEVIN, K. (1968). Effects of lithium salts on plasma protein-bound iodine and uptake of iodine-131 in thyroid gland of man and rat. *Life Sci.* **7**, 1257–64.

SEEGERS, W., McGAVACK, T.H., HARR, H., ERK, V.O., AND SPELLEN, B. (1957). Influence of the arylsulfonylureas on thyroid function in older diabetic men and women. *J. Am. geriat. Soc.* **5**, 739–46.

SHALET, S.M. AND PRICE, D.A. (1981). Effect of treatment of malignant disease on growth in children. *Arch. Dis. Childh.* **56**, 235–6.

——, ——, BEARDWELL, C.G., MORRIS-JONES, P.H., AND PEARSON, D. (1979). Normal growth despite abnormalities of growth hormone secretion in children treated for acute leukaemia. *J. Pediat.* **94**, 719–22.

——, HANN, I.M., LENDON, M., MORRIS-JONES, P.H. AND BEARDWELL, C.G. (1981). Testicular function after combination chemotherapy in childhood for acute lymphoblastic leukaemia. *Arch. Dis. Childh.* **56**, 275–8.

SHEARMAN, R.P. (1971). Prolonged secondary amenorrhoea after oral contraceptive therapy. *Lancet* ii, 64–6.

SHEN, F. AND SHERRARD, D.J. (1982). Lithium-induced hyperparathyroidism: an alteration of the 'set-point'. *Ann. intern Med.* **96**, 63–4.

SHERINS, R.J., OLWENY, C.L.M., AND ZIEGLER, J.L. (1978). Gynecomastia and gonadal dysfunction in adolescent boys treated with combination chemotherapy for Hodgkin's disease. *New Engl. J. Med.* **299**, 12–16.

SHOPSIN, B. (1970). Effects of lithium on thyroid function. A review. *Dis. nerv. Syst.* **31**, 237–44.

——, BLUM, M., AND GERSHON, S. (1969). Lithium-induced thyroid disturbance: case report and review. *Comprehensive Psychiat.* **10**, 215–23.

SIEGAL, S. (1964). The asthma-suppressive action of potassium iodide. *J. Allergy* **35**, 252–70.

SIMPSON, J.A. (1962). Myxoedema after thalidomide (Distaval). *Br. med. J.* **1**, 55.

SINGH, B.N. AND NADEMANEE, K. (1983). Amiodarone and thyroid function: clinical implications during antiarrhythmic therapy. *Am. Heart J.* **106**, 857–69.

SKARE, S. AND FREY, H.M.W. (1980). Iodine-induced thyrotoxicosis in apparently normal thyroid glands. *Acta Endocrinol.* **94**, 332–6.

SKINNER, N.S., HAYES, R.L., AND HILL, S.R. (1959). Studies on the use of chlorpropamide in patients with diabetes mellitus. *Ann. N.Y. Acad. Sci* **74**, 830–44.

SLANEY, G. AND BROOKE, B.N. (1957). Post-operative collapse due to adrenal insufficiency following cortisone therapy. *Lancet* i, 1!67–70.

SNEDDON, M.I.B. (1970). Les effects nocifs des steroîde en application locale. *Bull. Soc. Fr. derm. Syph,* **77**, 670–2.

SOLOMON, D.H. (1981). Pregnancy and PTU. *New Engl. J. Med.* **304**, 538–9.

STARUP J. (1967). The effects of gestagen and oestrogen treatment on the development of ovarian follicles, laboratory observations. *Acta obstet. gynec. scand.* **46**, Suppl. 9, 15.

STAUGHTON, R.C.D. AND AUGUST, P.J. (1975). Cushing's syndrome and pituitary-adrenal suppression due to clobetasol propionate. *Br. med. J.* **2**, 419–21.

STEELE, S.J., MASON, B., AND BRETT, A. (1973). Amenorrhoea after discontinuing combined oestrogen-progestogen oral contraceptives. *Br. med. J.* **4**, 343–5.

TRANSBOL., I., CHRISTIANSEN, C., AND BASSTRUP, P.C. (1978). Endocrine effects of lithium. I. Hypothyroidism its prevalence in long-term treated patients. *Acta endocrinal Copenh.* **87**, 759–67.

VAGENAKIS, A.G., WANG, C., BURGER, A., MALOOF, F., BRAVERMAN, L.E., AND INGBAR, S.H. (1972). Iodide-induced thyroxicosis in Boston. *New Engl. J. Med.* **287**, 523–7.

VAN LEEUWEN, E. (1954). Een Vam Van Genuine Hyperthyreose (M. Basedow Van Gejodeered Brood). *Ned. Tijdschr. Geneesk.* **98**, 81–9.

VAN ROY, M. (1966). Cited by Plate, W.P. (1968).

VESSEY, M.P., WRIGHT, N.H., McPHERSON, K., AND WIGGENS, P. (1978). Fertility after stopping different methods of contraception. *Br. med. J.* **1**, 265–7.

VIRCBURGER, M.I., PRELEVIC, G.M., BURKIE, S., ANDREJEVIE, M.M., AND PERIC, L.J.A. (1981). Transitory alopecia and hypergonadotrophic hypogonadism during cimetidine treatment. *Lancet* i, 1160–1.

WHITEHEAD, E., SHALET, S.M., MORRIS JONES, P.H., BEARDWELL, C.G., AND DEAKIN, D.P. (1982). Gonadal function after combination chemotherapy for Hodgkin's disease in childhood. *Arch. Dis. Childh.* **47**, 287–91.

WHITELAW, M.J., NOLA, V.F., AND KALMAN, C.F. (1966). Irregular menses, amenorrhoea and infertility following synthetic progestational agents. *J. Am. med. Ass.* **195**, 780–2.

WIGGERS, S. (1968). Lithiumpovirkning af glandula thyreoidea. *Ugeskrift for Laeger* **130**, 1523–5.

WILKINSON, R., ANDERSON, M., AND SMART, G.A. (1972). Growth-hormone deficiency in iatrogenic hypothyroidism. *Br. med. J.* **2**, 87–8.

WORGUL, T.J., KENDALL, J. AND SANTEN, R.J. (1981). Recovery of hypothalamic-pituitary-adrenal function after long-term suppression by aminoglutethimide and hydrocortisone. *J. clin. Endocrinol. Metab.* **53**, 879–82.

WORTSMAN, J., AND HIRSCHOWITZ, J.S. (1980). Danazol and thyroid function: in comment. *Ann. intern. Med.* **92**, 133–4.

——, —— AND SOLER, N. (1979). Danazol and thyroid function. *Ann intern. Med* **91**, 321.

YEO, P.P.B., BATES, D., HOWE, J.G., RATCLIFF, W.A., SCHARDT, C.W., HEATH, A., AND EVERED, D.C. (1978). Anticonvulsants and thyroid function. *Br. med. J.* **1**, 1581–3.

ZUSSMAN, W.V., FORBES, D.A., AND CARPENTER, R.J.JR. (1967). Overian morphology following cyclic norethindrone mestranol therapy. *Am. J. Obstet. Gynecol.* **99**, 99–105.

21 Drug-induced diseases of the breast

A. P. FLETCHER

Mammotropic action of drugs

Drugs having a mammotropic action can be divided into two main types. Firstly, there are some drugs which fairly commonly cause breast hypertrophy in men, notably digitalis, spironolactone, ethionamide, and griseofulvin. This particular action of these drugs is rarely noticed or complained of in women patients, especially since none of these agents has ever been reported as inducing lactation. Drugs of this type are believed to exert their mammotropic action by mimicking oestrogen or progesterone at peripheral receptor sites.

The second group of drugs causes gynaecomastia in male patients and in women induces both mammary hypertrophy and galactorrhoea usually associated with amenorrhoea. This type of action is exhibited by the rauwolfia alkaloids such as reserpine; all the phenothiazine derivatives such as chlorpromazine, prochlorperazine, perphenazine, trifluoperazine, thioridazine, methopromazine, and aminopromazine; members of the thioxanthene series such as flupenthixol decanoate; chlorprothixene; butyrophenone agents such as haloperidol; benzodiazepine derivatives such as chlordiazepoxide; tricyclic antidepressants such as imipramine; and the antihypertensive agent α-methyldopa.

In this second group of drugs the mammotropic effect appears to be correlated with increased secretion of prolactin. Most of the compounds in this group have adrenolytic activity and it is considered that these agents affect the hypothalamo–pituitary link. Surgical disruption of this link in premenopausal women has been shown to cause amenorrhoea and lactation (Ehni and Eckles 1959) and it is possible that these drugs produce a pharmacological interruption of this link.

Hypopituitarism may be associated with galactorrhoea and raised plasma prolactin levels (Forsyth *et al.* 1971). In these cases, presumably, a lesion in the hypothalamus or interruption of the pituitary-stalk portal capillary system prevents the supply of the hypothalamic hypophysiotropic factors to the pituitary. Thus prolactin-inhibitory factors fail to reach the pituitary so that the normal tonal suppression of prolactin secretion is removed resulting in high circulating prolactin levels and galactorrhoea.

Shearman (1971) described galactorrhoea as being common in women suffering from secondary amenorrhoea after discontinuing oral-contraceptive therapy. In Shearman's series, 11 out of 69 patients had copious galactorrhoea; Friedman and Goldfien (1969) described galactorrhoea as being present in nine of 21 similar patients.

Group 1

Digitalis

A relatively large number of cases of gynaecomastia have occurred in men on digitalis therapy, and in the majority of cases reported, therapy had been given in the form of digitalis leaf rather than digoxin or digitoxin (Conn 1964). Cases of mammary adipose and glandular hypertrophy in women were only reported on three occasions (Calov and Whyte 1954; Bloch 1961; Capeller *et al.* 1959). In these three instances, the patients were postmenopausal and were receiving digitalis leaf.

It is postulated that digitalis, or one of its metabolic products, has sufficient structural similarity to oestrogen (Fig. 21.1) for oestrogen-like effects to be produced under appropriate conditions. It is unlikely that this condition would be recognized in a plethora of endogenous oestrogen; hence it is only recognized when it occurs in men or postmenopausal women. It is not surprising therefore that only one case of mammotropic activity due to digitalis has been described in a premenopausal woman (Wolf 1964).

Since oestrogen and digitalis are both metabolized and conjugated in the liver, it has been suggested that a second condition necessary for the mammotropic action of digitalis to become apparent is a disturbance of hepatic metabolism, possibly attributable to the concurrent circulatory disturbance of congestive heart failure. However, limited studies in men with digitalis-induced gynaecomastia have failed to reveal any unusual

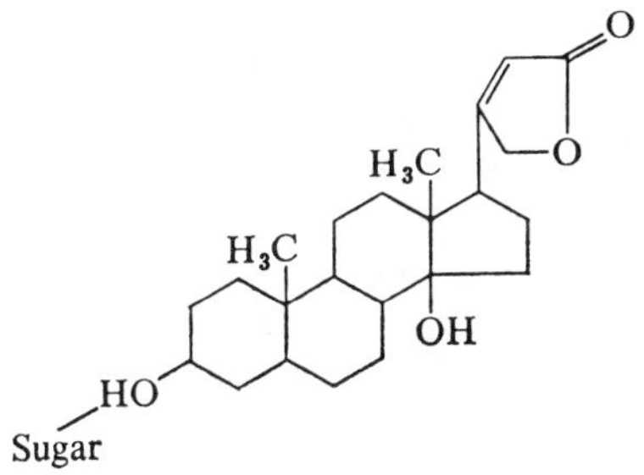

Basic structure of digitalis glycoside

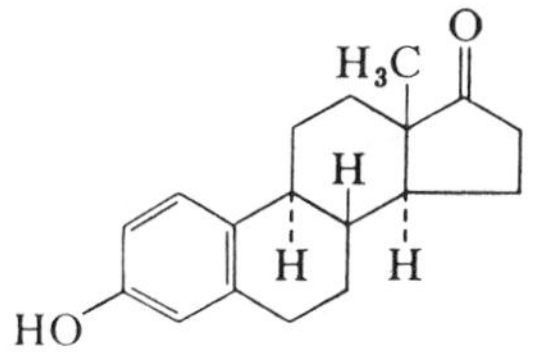

Oestrone

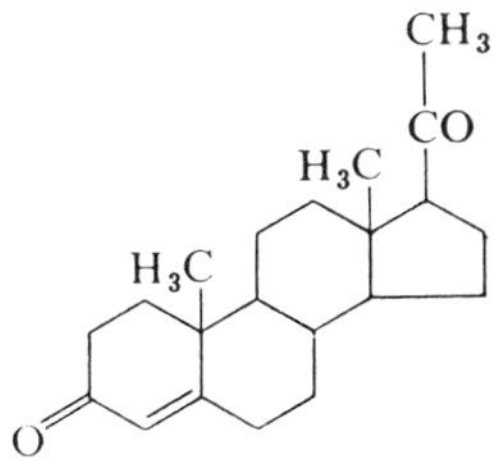

Progesterone

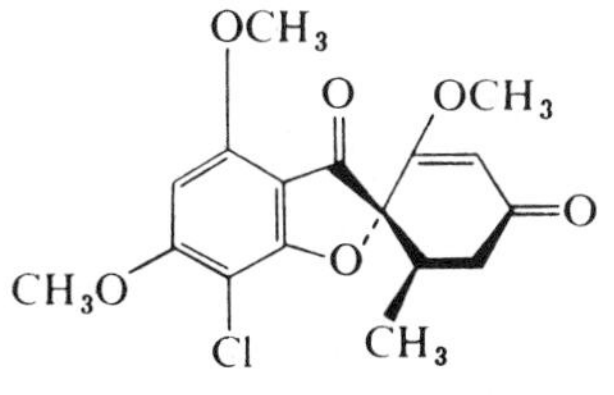

Griseofulvin

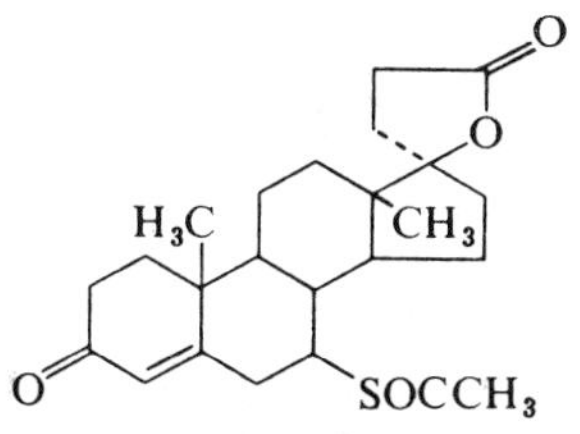

Spironolactone

Fig. 21.1 Structural similarity between some mammotropic drugs and oestrogen or progesterone.

or unexpected alteration of liver-function tests. In a report prepared for the US National Cancer Institute (Stenkvist *et al.* 1979), it was noted that breast cancer patients on cardiac glycosides (usually digoxin) had tumours with a tumour-cell population that was built up of cells that were smaller and more uniform in morphology, density, and size than in those patients with breast cancer not on digitalis. It was also noted that the tumour volume was smaller at diagnosis, and that, two years after diagnosis of breast cancer, distant spread was less common in patients on digitalis treatment — 2/33 compared with 28/146 not on digitalis (p = 0.07). This effect of digitalis on breast cancer was considered to be due to an interaction with oestrogen receptors in the tumour cells. Cove and Barker (1979) investigated the suggestion that digoxin interferes with oestrogen receptor (ER) on cytosols from 11 primary human-breast carcinomas containing ER (mean 143 fmol ml^{-1} cytosol, range 46–304); the *in-vitro* binding of H-oestradiol 0.9 nmol l^{-1} was unaffected by 10 μmol l^{-1} digoxin. In the cytosol from two of these tumours, lower concentrations of digoxin (100, 10, 1.0, and 0.1 mmol l^{-1}, 1.0 μmol l^{-1}) also failed to compete with the binding of oestrogen to ER. Progesterone receptor (PgR) is another possible site of action of digoxin, especially since progesterone has been reported to bind to cardiac digoxin receptor. However, in cytosol from each of seven primary breast tumours containing PgR (mean 327 fmol ml^{-1} cytosol, range 92–740), digoxin at two concentrations (1.0 nmol l^{-1} and 10 μmol l^{-1}) did not interfere with the binding of the synthetic progesterone analogue R5020).

The higher concentrations of digoxin used were considerably greater than the therapeutic range (0.65–2.60 nmol l^{-1}) and it is therefore unlikely that digoxin has any direct action on the binding of oestradiol or progesterone to their cytosol receptors in breast cancer in *in vivo*. These results do not exclude the possibility that digoxin exerts other direct effects on breast cancer or that one of the other cardiac glycosides competes for ER or PgR.

Spironolactone

Spironolactone was first incriminated as a cause of gynaecomastia by Smith (1962) and this was quickly followed by other reports of the same complication (Restifo and Farmer 1962; Williams 1962; Mann 1963; Sussman 1963).

Clarke (1965) reported on a series of 12 patients treated with spironolactone, four of the seven male patients and none of the five women complained of hypertrophy of the breast. The high incidence of gynaecomastia in this group of men indicates that it may be a common complication of spironolactone therapy. However, three of the four men with gynaecomastia had been treated with digitalis in addition to spironolactone, and digitalis may have potentiated a mammotropic action of spironolactone. Nevertheless, spironolactone has caused gynaecomastia without the synergistic action of digitalis. The mammotropic action of spironolactone is thought to depend upon its structural similarity (Fig. 21.1) to progestational hormones (Sussman 1963).

Bridgeman and Buckler (1974) described two cases of spironolactone-induced gynaecomastia in male patients. In both these men the 24-hour urinary excretion of 17-hydroxycorticosteroids and 17-ketosteroids was reduced and it was also found that urinary oestrone and luteinizing hormone were depressed. All these parameters increased when spironolactone was discontinued. A study by Miyatake et al. (1978) indicated that long-term spironolactone treatment can increase the serum levels of oestrone and oestradiol in hypertensive men followed by the development of gynaecomastia. Thus, the elevation in circulating oestrogens could well explain the oestrogenic side-effects of spironolactone treatment. These authors presented evidence that although spirono-lactone and its metabolite canrenone can interfere with the radioimmunoassay for progestogen, aldosterone, and 11-deoxycorticosteroids, it did not interfere with their techniques for measuring oestrone and oestradiol. Serum testosterone, LH, and prolactin levels were not affected. The exact mechanism of serum oestrogen elevation is obscure in this study, however, it may be due to increased peripheral conversion rate of androstenedione to oestrone, and then to oestradiol. This hypothesis is consistent with observations by Rose et al. (1977) who reported that increased blood oestradiol levels were primarily due to a significant increase in the rate of peripheral conversion of testosterone into oestradiol in their spironolactone-treated patients.

Oestrogens

The human breast seems to be uniquely sensitive to ovarian steroids. Woman is the only primate in which the gross morphological breast development is completed at puberty; in all other primates the mammary gland develops only as a consequence of the hormonal changes of pregnancy. Because of this sensitivity women are aware of breast changes during the normal menstrual cycle, with a feeling of fullness and a tingling sensation preceding menstruation. Women taking oral contraceptives seem to experience similar cyclic breast symptoms.

Milligan et al. (1975) measured the volume of the left and right breasts daily in four nulliparous woman during normal menstrual cycles and after the use of oral contraceptives. A glass mixing bowl 7 inches (17.8 cm) in diameter standing inside a container on the floor was filled to the brim with water. The woman, kneeling on the floor, lowered one breast into the bowl, thus displacing water into the surrounding container; the volume of water displaced was measured. Variability due to postural changes was controlled by marking positions for the container, hands, knees, and elbows on a sheet of plastic. Each woman made three consecutive measurements on each breast every day at the same time, using water of about the same temperature. These observations provided data on three complete normal cycles and six complete contraceptive-controlled cycles. The mean total change in volume/breast throughout the cycle was 100 ml under natural conditions and 66 ml on oral contraceptives.

Oestrogen/progestogen oral-contraceptive preparations therefore appear to reduce the monthly cyclical changes in breast volume. There are no comparable data on the effects of depot injections of progestational substances such as Depo-Provera which are administered once every 3-4 months.

Oestrogen effects on the mammary gland outside the normal circumstances of therapeutics are illustrated by the following reports. Fara et al. (1979) reported an epidemic of breast enlargement in girls and boys attending a school in Milan, first noted in November 1977, was followed up until the end of 1978: 213 boys aged 3-14 years and 110 girls aged 3-7 years were studied; control children attending five other schools were also examined. In total of 1647 boys and 476 girls were examined. Breast enlargement was significantly more common in boys (29.0 per cent) and girls (21.6 per cent) aged 3-5 years, boys (58.0 per cent) aged 6-10, and girls (67.1 per cent) aged 6-7 years from the school in Milan, than in age- and sex-matched children at control schools. Breast enlargement was not pronounced and disappeared within eight months. Hormonal determinations were within normal limits except for 17β-oestradiol which was slightly raised. Although oestrogen contamination was not detected when samples of school meals were tested, an uncontrolled supply of poultry and beef was suspected as being the cause of this outbreak. In Italy the meat of young animals is preferred and despite regulations oestrogens may be fed to farm animals to accelerate their weight gain. Weight for weight, meat consumption in children is likely to be higher than in adults.

On 16 November 1978, many of the British national papers (The Times, the Guardian, Daily Express, Daily Mirror, Daily Mail) ran stories that male rapists, child molesters, transsexuals, and other sex offenders in various prisons including Dartmoor and Grendon Underwood had been given the oestrogen, oestradiol, as an implant. About 70 of these prisoners developed breasts as a side-effect of this attempt to control their sexual urges. All had had to undergo mastectomy, and claims that disfiguring scars had resulted from this surgery were made.

Di Raimondo *et al.* (1980) described the case of a 70-year-old man who developed marked gynaecomastia shortly after his wife started to use an oestrogen-containing vaginal cream as a lubricant before intercourse. It is, however, hard to imagine the man getting sufficient oestrogen by this means!

Ginseng

Ginseng (*Panax ginseng*) contains small quantities of oestrone, oestradiol, and oestriol in the root. Palmer *et al.* (1978) described ginseng-induced swelling of the breasts with diffuse nodularity and tenderness.

Clomiphene

Clomiphene citrate has been found to be quite effective in the treatment of subfertile men with idiopathic oligospermia and more recently in the treatment of men with varicoceles. In general, the drug is well tolerated by men with minimal (if any) side-effects such as slight weight gain and occasional transient visual disturbances. Check *et al.* (1978) reported the first case of clomiphene-induced gynaecomastia in a 32-year-old man being treated for primary infertility with the drug.

In this patient, elevated serum oestrogens were detected and these fell to within the normal range 4 months after stopping the clomiphene. Other tests showed that follicle-stimulating hormone (FSH), luteinizing hormone (LH), and testosterone levels were within the normal range.

Ethionamide

Gynaecomastia was observed in 13 patients out of a total of 446 tuberculous patients treated with this drug (Gernez-Rieux *et al.* 1963).

Griseofulvin

Durand *et al.* (1964) described four patients in whom treatment with griseofulvin resulted in gynaecomastia, hyperpigmentation of the breast areolae and of the external genitalia, hypertrophy of the clitoris, and vaginal discharge. An oestrogenic activity of griseofulvin was postulated.

Group 2

Reserpine

Gynaecomastia was reported by Wilkins (1954) to have developed in male patients following administration of rauwolfia alkaloids, and Khazan *et al.* (1962) described five cases of galactorrhoea in a series of 43 female patients treated with reserpine alone and 30 women treated with reserpine and chlorpromazine.

In animal studies Khazan *et al.* (1962) found that reserpine in adequate doses could induce lactation in adult female rats; in these animals the morphological and histological changes in ovaries and uteri were similar to those encountered in normally lactating rats.

Phenothiazine derivatives

Most of the phenothiazines used clinically have been reported as causing engorgement of the female breast with secretion of either milk or colostrum. The incidence of this mammotropic effect and secretion of colostrum occurs in some 10–15 per cent of patients receiving phenothiazine derivatives (Wright 1955; Robinson 1957). Robinson found that all the patients who lactated were below 43 years of age and were taking a dose of chlorpromazine in excess of 300–400 mg/daily. Khazan *et al.* (1962) found that 33 women showed galactorrhoea in a series of 650 cases treated with chlorpromazine.

Khazan *et al.* (1962) studied the mammotropic action of phenothiazine derivatives on adult female rats 2–3 months old and weighing 120–150 g. Each drug was injected subcutaneously into 10 female rats for seven days. The relative effectiveness of the drugs was classified according to a mammotropic index ranging from 1 to 5. The most influential drugs in producing mammary hypertrophy (Grade 5) were trifluoperazine, prochlorperazine, perphenzine, and triflupromazine. Chlorpromazine and chlordiazepoxide were moderately effective (Grades 3 and 4). L-mepromazine, methopromazine, and iminopromazine showed less effect (Grades 1 and 2).

Trifluoperazine was the most effective phenothiazine in producing mammary hypertrophy in laboratory animals; this has also been borne out in clinical experience. Khazan *et al.* (1962) described the following case.

A 20 year-old unmarried woman suffering from schizophrenia was given 20 mg/day Stelazine (trifluoperazine) for a period of 60 days. Prior to treatment she had regular menstrual cycles with no evidence of any endocrine disorders. It was noted that Stelazine caused enlargement of the breasts and very copious milk secretion, flowing almost freely from the breasts at the slightest pressure. The effect was much more intensive than that seen in all other cases treated with chlorpromazine or reserpine, and lactation ceased when the drug was withdrawn.

Complications of the mammotropic actions of phenothiazines can occur. Whiffen (1963) described a case of a 26-year-old woman treated with chlorpromazine 300–400 mg daily for almost five years, who developed a non-tender mass in the left breast. On frozen section the lesion appeared to be a lipogranuloma, and on permanent section an increased number of ducts and acini were noted, some of which were slightly dilated and mildly inflamed. One segment of duct had lost its epithelial lining and there was a heavy infiltrate of lymphocytes, histiocytes, and lipophagocytes. The appearance was of chronic mastitis with galactocoele formation. The excess mammary secretions had extravasated from the duct in this area and formed a lipogranuloma.

Thioridazine (and the thioxanthene, chlorprothixene)

These agents have also been reported as producing gynaecomastia in men and lactation and amenorrhoea in women (Khazan et al. 1962).

Imipramine

Galactorrhoea, associated with swelling of the mammary glands, developed in a 34-year-old woman after six months of treatment with 75–100 mg imipramine daily. The lactation ceased on withdrawal of the drug, and recurred when therapy was recommended. (Khazan et al. 1962).

Amoxapine

Galactorrhoea and hyperprolactinaemia associated with amoxapine therapy in a 29-year-old woman with primary depression was reported by Gelenberg et al. (1979). Amoxapine was gradually increased to 300 mg by the fifth day of therapy; galactorrhoea started on day 14 of treatment. After three months amoxapine therapy was discontinued. Depressive symptoms returned and amoxapine was resumed and galactorrhoea and oligomenorrhoea returned. A further attempt to discontinue amoxapine after seven months of treatment was accompanied by the same course of events. Prolactin levels were measured on seven occasions and found to be increased during amoxapine therapy and to fall on withdrawal of amoxapine to normal levels.

Benzodiazepines

Lactation due to chlordiazepoxide was described by

Lampe (1967) but no reports of a mammotropic effect of diazepam in males was made until 1979 when Moerck and Magelund reported the case of a 55-year-old man who abused diazepam (taking 100 mg/daily) and who developed bilateral gynaecomastia. At this time serum prolactin levels were raised but there were no changes in other endocrine parameters; thyroid hormones, LH, FSH, testosterone, urinary 17-ketosteroids, chorionic gonadotrophin, and oestrogens were within normal levels. After the gynaecomastia had resolved following withdrawal of diazepam the patient was rechallenged with 20 mg diazepam and no rise in serum prolactin was observed.

Diazepam

A case of diazepam-induced gynaecomastia in which a raised serum prolactin level was the only hormonal abnormality was referred to in Update 1981. A further five cases have now been reported by Bergman et al. (1981) who demonstrated raised serum oestradiol levels in all patients while on the drug. The gynaecomastia diminished gradually after withdrawal of diazepam. Thyroid disease, HCG-producing tumours, liver disease, and hypogonadism were excluded by the appropriate laboratory tests. The authors postulated enhanced conversion of testosterone to oestradiol, increased sex-hormone-binding globulin, or decreased metabolism or excretion of oestrogen as possible mechanisms.

Amphetamines and diethylpropion

The amphetamines were first described as a cause of gynaecomastia by Tooley and Lack (1949), and Bridgeman and Buckler (1974) described gynaecomastia due to diethylpropion. Both amphetamines and diethylpropion exert their appetite-suppressant activity through the hypothalamus.

Bridgeman and Buckler (1974) measured the 24-hour urinary 17-oxosteroids, 17-oxogenic steroids, oestrone, and luteinizing hormone, all of which were increased in the diethylpropion-treated men with gynaecomastia. These levels fell to within normal limits when the diethylpropion was discontinued. These workers contrasted the effects of diethylpropion and spironolactone, both of which produce gynaecomastia, as typical examples of the two different modes of mammotropic action of drugs. Diethylpropion increased 17-oxosteroids, oxogenic steroids, oestrone, and luteinizing hormone in 24-hour urinary collections; spironolactone caused a fall below normal levels in all these determinations.

Methyldopa

Pettinger *et al.* (1963) described a series of 15 female patients who were treated for hypertension with α-methyldopa, five of who commenced lactation. These women were aged 33–47 years; four were premenopausal and one had undergone hysterectomy and oophorectomy. Relationship of the medication to lactation was verified when cessation of lactation occurred within three weeks in two subjects in whom the drug was discontinued; resumption of methyldopa therapy resulted in recurrence of lactation. Changing patients over to guanethidine (another sympatholytic drug) did not support a continuation of methyldopa-induced lactation.

Steiner *et al.* (1976) demonstrated that single doses of methyldopa of 750 or 1000 mg given to hypertensive patients caused a significant increase in serum prolactin levels within four to six hours after drug administration. Long-term methyldopa treatment was associated with threefold to fourfold increases in basal prolactin levels compared with those in normal subjects. In patients treated with methyldopa for two to three weeks the growth-hormone response to insulin-induced hypoglycaemia was significantly greater than that measured in normal subjects or untreated hypertensive patients. In patients treated with methyldopa for prolonged periods (mean in excess of 12 months) the growth-hormone response to insulin-induced hypoglycaemia was indistinguishable from that seen in normal subjects.

The significance of methyldopa effects on growth-hormone release is unknown.

Lactation following therapeutic abortion with prostaglandins

Sato *et al.* (1974) showed that intravenous administration of prostaglandins ($PGF_2\alpha$, PGE_1, and PGE_2) caused elevation in blood prolactin in castrated male rats. Hafs (1975) showed that in 16 heifers the plasma prolactin averaged 26 ng/ml^{-1} before intramuscular injection of $PGF_2\alpha$ (60 mg), increased to 81 ng/ml^{-1} within 10 min, and 104 ng/ml^{-1} at 1 hour, 72 ng/mg^{-1} at 2 hours, and fell to normal at 4 hours. Hafs (1975) in reviewing the action of prostaglandins on the pituitary–hypothalamic axis considered that the hypothalamus is the principal site of action of prostaglandins in causing prolactin and ACTH release. The release of growth hormone in response to prostaglandins probably represents a direct action on the pituitary. Whether prostaglandins modulate the release of TSH is not determined, but there is some evidence that PGE_1 can increase TSH release from rat pituitaries *in vitro*.

Prostaglandins E_1 and E_2 given by intraventricular injection into the third ventricle caused release of luteinizing hormone and follicle-stimulating hormone but caused no effect on these hormones on direct injection into the anterior pituitary.

Smith *et al.* (1972) drew attention to the fact that, although lactation rarely occurs in women following spontaneous or induced termination of pregnancy before the fourth to fifth month, they had observed that a large percentage of women commenced to lactate within three days of abortion induced in early pregnancy with intrauterine infusion of prostaglandin $F_2\alpha$. In a prospective study in 80 women having their pregnancy terminated by intrauterine infusion of prostaglandin $F_2\alpha$ compared with a similar number having their prgnancy terminated by surgical means. The numbers of women lactating in the prostaglandin $F_2\alpha$-terminated group ranged from 70 to 84.2 per cent depending on the stage of pregnancy; in the surgically terminated group the number lactating ranged from 0 to 25 per cent.

This apparently clear-cut association of lactation and exposure to prostaglandin $F_2\alpha$ is ruined by the fact that, although both groups of women had pethidine as an analgesic, the women who were receiving intra-amniotic prostaglandin $F_2\alpha$ received metoclopramide (*Maxolon*) for the control of nausea and vomiting. Metoclopramide has some structural relationship to the phenothiazines and can therefore be expected to induce lactation by release of prolactin. In this study by Smith *et al.* (1972), therefore, no clear answer can be given as to whether this increase in laction was due to prostaglandin $F_2\alpha$ or due to the metoclopramide given to control the nausea, vomiting, and diarrhoea which occur as side-effects of prostaglandin $F_2\alpha$.

Cannabis

Olusi (1980) described three cases of gynaecomastia in chronic cannabis smokers in all of whom the serum prolactin levels were markedly raised. The question of whether the gynaecomastia associated with cannabis was due to prolactin release or to the known oestrogenic effects of tretrahydrocannabinol was discussed, but a prolactin-release mechanism was favoured.

Cimetidine

Since Hall (1976) reported the development of gynaecomastia in two patients treated with cimetidine, various other cases of gynaecomastia and galactorrhoea have been reported (Bateson *et al.* 1977; Delle Fave *et al.* 1977*a*; Sharpe and Hawkins 1977; Bezuidenhout 1978). An extensive study of 25 patients was reported by Spence and Celestin (1979). Gynaecomastia occurred unilaterally or bilaterally in five out of 25 male duodenal-ulcer patients after more than four months treatment with cimetidine 1.6 g daily. All elected to continue treatment to 12 months and their breast enlargement regressed rapidly and disappeared after stopping treatment. During treatment all patients were found to have normal concentrations of plasma testosterone and oestradiol, and serum prolactin was normal in the two patients measured. Excision biopsy of the subareolar tissue in one patient revealed histology typical of the florid state of gynaecomastia. Blockade of androgen-responsive receptors in the target organ was considered to be the most likely mechanism involved.

Other studies (Delle Fave *et al.* 1977*b*; Bateson *et al.* 1977), however, found raised prolactin levels associated with gynaecomastia. Burland *et al.* (1978) demonstrated increases in prolactin levels after 400 mg cimetidine was given intraveneously to volunteers.

The mechanism of cimetidine-induced gynaecomastia is open to debate and the choice lies between a peripheral anti-androgen effect and increased prolactin release.

Alkylating agents

The treatment of various malignancies with alkylating agents is well known to be associated with adverse effects on the seminiferous tubules with consequent oligo- or azoospermia, reduction in testicular size, and elevated follicle-stimulating hormone. Friedman and Plymate (1980) report three cases of adult males treated with alkylating agents — two with cyclophosphamide and one with galacticol.

Three different malignancies were involved, one patient having a hypernephroma, one malignant lymphoma, and the other lymphosarcoma. All presented with decreased beard growth. Clinical examination revealed bilateral gynaecomastia and soft, small testes. Biochemically all had elevated FSH and LH concentrations, low-to-normal testosterone levels, and no detectable beta-subunit of human chorionic gonadotrophin. The authors point out that seminiferous-tubule dysfunction is sometimes associated with a state of compensated Leydig-cell failure which is characterized by elevated LH levels, low testosterone levels, and an exaggerated response of serum LH to luteinizing hormone-releasing hormone. The three cases they reported conformed to this picture and in addition had gynaecomastia. The cause of Leydig-cell dysfunction in these patients is not known but in rats with seminiferous-tubule dysfunction ultrastructural changes in Leydig-cells have been observed. Treatment with testosterone oenanthate in oil suppressed LH levels in all patients and led to satis-

factory resolution of gynaecomastia and an improvement in the other adverse effects.

D-penicillamine

Therapy with D-penicillamine is known to be occasionally associated with breast enlargement although it is not understood why this occurs. Massive mammary hyperplasia was reported to have occurred in a 39-year-old woman whose bust size increased from 36 inches to 48 inches. Bilateral total mastectomy was carried out and 18 pounds of breast tissue was removed (Desai 1973). Another case of penicillamine-induced breast enlargement which was successfully treated with danazol was described by Taylor *et al.* (1981). The patient, a 41-year-old woman suffering from rheumatoid arthritis, developed massive, tender bilateral breast enlargement after approximately two years of therapy with D-penicillamine. Two courses of treatment with danazol achieved a satisfactory reduction in breast size.

Other drugs reported to cause gynaecomastia

A variety of other drugs have been reported to cause gynaecomastia; in many cases it is not known which is the mechanism of this action, in others it is clear. The following are agents reported to cause gynaecomastia but which have not been discussed above.

Androgens	Heroin
BCNU	Isoniazid
Busulphan (*Myleran*)	*o, p*-DDD
Cannabis	Phenelzine (*Nardil*)
Chlortetracycline	Stilboestrol
Clonidine	Vincristine (*Oncovin*)
Diethylstilboestrol (DES)	Vitamin D_2
Human chorionic gonadotrophin	Hydrazine derivatives

Drugs affecting prolactin levels in blood

Over the past few years the significance of altered prolactin levels, which may be associated with a variety of endocrine signs and symptoms, has been debated in the literature. At the present time it would seem to be generally agreed that sustained hyperprolactinaemia is a causative factor in hypogonadism, a condition which may in turn be associated with gynaecomastia, galactorrhoea, menstrual irregularities, infertility, and oligospermia. In a series of cases studied at the Upstate

Medical Center, Syracuse, New York, Badawy *et al.* (1980) found that two-thirds of the patients with hyperprolactinaemia had discernable pituitary tumours. It was suggested (Boyar *et al.* 1974) that this may be a consequence of lactotrope hypertrophy and hyperplasia, possibly stimulated by drugs or other extrinsic agents. There is no doubt that an increasing number of drugs are known to be associated with hyperprolactinaemia. In view of the possibly serious consequences of this effect these drugs should be kept under careful surveillance.

Sodium valproate

Although the precise causes and effects of hyperprolactinaemia are still largely obscure, a number of common factors are beginning to emerge. In an interesting paper by Melis *et al.* (1982) the central role of gamma-aminobutyric acid (GABA) in the control of prolactin secretion was considered in some detail. Their hypothesis that endogenous GABA exerts an inhibitory effect on prolactin secretion was based on observations that the anti-epileptic drug, sodium valproate, after acute oral administration lowers plasma prolactin levels in humans and that ethanolamine-O-sulphate, a substance known to stimulate the GABAergic system, produces similar effects. Both substances are inhibitors of the enzyme GABA-transaminase and it is thought that, since GABA receptors are present in the anterior pituitary, the inhibition of prolaction secretion may be due to a direct effect of GABA on pituitary cells. However it seems that *in-vitro* GABA exerts a lesser inhibitory effect than dopamine and that *in-vivo* GABA manifests its inhibitory effect by blunting stimulated prolactin release. It is suggested that GABA inhibition of prolactin secretion could be masked by strong dopaminergic control in resting conditions, GABA effects becoming apparent after pharmacological raising of prolactin levels or stimulation of the GABAergic system. Earlier reports that GABA-agonist drugs produce a rise in plasma prolactin levels are explained in terms of pharmacological interactions with other neurotransmitter systems known to be affected by exogenous GABAergic stimulation.

Melis *et al.* (1982) also showed that, in hyperprolactinaemic subjects with no radiological evidence of pituitary tumours, stimulation of the GABAergic system by sodium valproate lowers plasma prolactin concentrations. In patients with evidence of pituitary adenomas, no changes in prolactin secretion were observed. Although these conclusions are largely speculative, they are consistent with many of the observations reported to date. The failure of sodium valproate to affect the increased prolactin secretion caused by prolactinomas may be explained by a defect of the GABAergic system in the presence of such tumours. Alternatively, GABA receptors at the pituitary level may be altered.

Neuroleptic agents

Changes in the tuberoinfundibular dopamine system during chronic administration of neuroleptic agents were suggested by Brown and Laughren (1981) as an explanation of the apparent tolerance developed to the prolactin-elevating effects of those agents. It seems that, immediately after the initiation of neuroleptic treatment (thioridazine, 150–300 mg/daily; chlorpromazine, 200–450 mg/daily; haloperidol, 15 mg/daily; trifluoperazine, 5 mg/daily), serum prolactin levels rise markedly for approximately one week after which a steady fall occurs over several months. Although there is apparently a transient rise about eight weeks after initiation of treatment the fall continues until a stable level, approaching that of the drug-free period, is reached that may be maintained for years. This behaviour suggests that a state of tolerance to the prolactin-lowering effect of neuroleptics is developed during chronic administration.

Brown and Laughren (1981) raised the possibility that the steady decrease in prolactin levels during the first few months of treatment may be due to depletion of lactotropic cells. However, they pointed out that this is unlikely in the light of the work of Naber *et al.* (1979) who showed that the normal prolactin levels achieved during the chronic administration of neuroleptic agents are markedly increased by giving TRH. Thus it seems unlikely that a depletion of the prolactin pool occurs in response to neuroleptic drugs. There seems little doubt, however, that the changes in prolactin levels during chronic treatment with neuroleptic agents are mediated through changes in the dopamine system. These changes may be due to either altered dopamine synthesis or changes in dopamine receptor sensitivity. In addition there is evidence (Grudelsky 1981) that prolactin may contribute to the control of its own secretion through a positive feedback action on the release of dopamine from tuberoinfundibular neurons.

Cimetidine

The treatment of peptic ulceration by cimetidine, a histamine H_2-receptor antagonist, has been reported as being associated with a number of adverse effects. These include breast pain, gynaecomastia, galactorrhoea, oligospermia, and impotence, all effects known to be associated with hyperprolactinaemia. However reports of prolactin levels in patients treated with cimetidine have been contradictory, some showing no change and others showing a rise. In a study on 20 men and five women patients Krawiec et al. (1981) attempted to resolve the problem. Plasma prolactin levels were found to rise in one-half of the patients, but the upper limit of normal was exceeded in only one patient. Mean prolactin levels were not statistically different from normal after six weeks of treatment on 1000 mg/daily. One of the female patients in this study had initial hyperprolacti-naemia which returned to normal after cimetidine was withdrawn. Neither this patient nor the one male subject who developed postcimetidine hyperprolactinaemia showed any side-effects. In the series of six patients reported by Delle Fave et al. (1977b), treated at a dose of 1600 mg/daily, all showed hyperprolactinaemia and two developed gynaecomastia. It is thus apparent that raised prolactin levels and the presence or absence of gynaeco-mastia are not well correlated in patients treated with this H_2-receptor antagonist. It is known that cimetidine inhibits the binding of androgens to their receptors, interacts with oestrogen receptors, and induces oestra-diol and progesterone receptors; it may be that the variable adverse-effect response pattern to cimetidine treatment is explicable in terms of differing sensitivity at the hormone-receptor sites.

In relation to this it is of interest that Beck (1981), in a study on boys with marked gynaecomastia, was unable to demonstrate raised prolactin levels. Although hyperpro-lactinaemia is associated with galactorrhoea it is apparently not involved in prepubertal gynaecomastia which does appear to be correlated with raised oestradiol levels.

Ranitidine

Ranitidine is an H_2-receptor antagonist similar in action to cimetidine but four to 10 times more potent on a molar basis in inhibiting gastric-acid secretion in man. Although the intravenous administration of ranitidine stimulates prolactin secretion, standard oral treatment does not. Studying a group of 19 males and six females set against a control group of 20 males and nine females, Robins and McFadyen (1981) demonstrated that oral ranitidine treatment (300 mg/daily) does not cause hyperprolactinaemia. In-vitro experiments have shown that ranitidine does not have dopaminergic effects and does not alter prolactin levels through action on the pituitary (Yeo et al. 1980). In view of the fact that oxmetidine hydrochloride, a substance very similar to cimetidine, failed to raise serum prolactin in healthy volunteers, Sharpe et al. (1980) have suggested that cimetidine's hyperprolactinaemic effect is due to some particular feature of the cimetidine molecule.

Oral contraceptives

A possible association between the use of the contraceptive pill and hyperprolactinaemia has been suggested in recent years. The overall picture, although somewhat confused, does suggest that the oestrogen content of the combined oral contraceptive may be responsible for hyperprolactinaemia in some women users through its known stimulatory effect on the lactotropes. In a study on patients who were being evaluated for infertility and/or menstrual irregularities Badawy et al. (1981) attempted to clarify the situation. The series involved 123 patients presenting with menstrual irregularities and/or infertility starting within one year after termination of oral contraceptive use. Patients with abnormal thyroid function, previous treatment with phenothiazines, reserpine or morphine, menstrual irregularities prior to oral contraceptive use, or a history of trauma or chest surgery were excluded from the study. In this highly selected group the presence of hyperprolactinaemia was apparently correlated with oral-contraceptive use of more than one year, particularly if use started before the age of 25. The long-term significance of hyperprolactinaemia in these women is not known. In view of the poor correlation between hyperprolactinaemia and other adverse effects (e.g. gynaecomastia), caution should be observed in extrapolating these findings beyond the selected group studied by Badawy et al. (1981).

Radioiodine, propylthiouracil, and sub-total thyroidectomy

The association of hyperprolatinaemia and/or galactor-rhoea with thyroid failure is well known clinically but, as in the previously quoted examples in this chapter, the precise factors involved are not clear. It has been suggested that the hyperprolactinaemia associated with hypothyroidism may be a result of decreased hypotha-lamic dopamine secretion rather than of an increase in TRH secretion. In a study on spontaneous and iatrogenic hypothyroidism, Contreras et al. (1981) examined 34 patients (27 female) with iatrogenic hypothyroidism. Hyperprolactinaemia was much more common in spon-taneous (88.2 per cent) than in iatrogenic (31 per cent) hypothyroidism. This difference might well be due to a much longer duration of the disease before diagnosis in the spontaneous cases than in the iatrogenic ones (66.3 $\pm$ 10.8 vs. 6.7 $\pm$ 1.8 months). The prolactin response to metoclopramide, a dopamine-blocking agent, which was studied in 13 of the women patients with spontaneous disease was inversely correlated with basal prolactin levels. It was suggested that this would support the hypothesis that a decreased hypothalamic dopamine content might result in a greatly reduced TRH effect upon pituitary lactotropes which would be observed as a decreased response to dopamine receptor blocking agents. The study also indicated that male patients with spontaneous disease had significantly lower prolactin levels than comparable female patients.

Labetolol

Labetolol is an antihypertensive drug possessing both alpha- and beta-adrenoceptor blocking properties. An interesting feature of the drug is its ability to raise serum prolactin levels after intravenous administration (Barbieri *et al.* 1981). The effect is most marked in women patients but is also observed to a lesser extent in males. A similar effect has also been observed in other antihypertensive drugs such as reserpine and methyldopa. It is unlikely that lowering of the blood pressure is the causative factor because the same effect is not seen with clonidine or prazosin. Since the dopaminergic system is generally regarded as a major factor in prolactin control, an effect of labetolol on this system must be considered possible although at the present time this has not been demonstrated.

Buprenorphine

Opiate agonists and antagonists are known to affect levels of plasma-luteinizing hormone (LH) and prolactin both in animals and man although the effects of opiate antagonists on prolactin are very variable. The morphine-like drug, buprenorphine, has mixed agonist–antagonist activity and is 25–40 times more potent than morphine itself. It seems likely that species differences, diurnal effects, and degree of stress may contribute to the sometimes variable effects that have been reported.

Interest in buprenorphine centres on its actions both as an opiate agonist and antagonist. In a study on male heroin addicts (Mendelson *et al.* 1982), buprenorphine was shown to produce a signficiant rise in prolactin levels in all subjects. Moreover the prolactin levels remained raised throughout maintenance therapy on buprenorphine. It is also of interest that LH levels were significantly decreased over the same trial period. The authors are careful to point out that findings in chronic heroin addicts may not be generally applicable even though drug-free baseline plasma prolactin and LH levels were well within normal limits for adult males.

Since opiate agonist drugs have been convincingly shown to decrease plasma LH levels it would seem that this mixed agonist–antagonist is acting predominantly as an agonist on the LH system. This is supported by the observation that buprenorphine produces a greater suppression of LH than 40 mg/daily i.v. of heroin and is consistent with the fact that it is known to be 24–40 times more potent than morphine.

Similarly, buprenorphine stimulation of prolactin levels is consistent with a predominantly agonist action although the relatively small increase in plasma levels in comparison with the larger effects of lower doses of less potent opiate agonists suggests that antagonist mechanisms are also active.

Metoclopramide, sulpiride, and domperidone

These three related gastrointestinally active compounds were all reported as being associated with raised serum prolactin level (Aono *et al.* 1978; Cann *et al.* 1983) which in many cases is also associated with galactorrhoea and mastalgia. The prolactin-raising effects of these drugs are not surprising as they are believed to act by antagonizing dopamine in the medium eminence, which is outside the blood–brain barrier. A total of 18 women were studied by Aono *et al.* (1978), 13 being treated with sulpiride and five with metoclopramide, all 18 of whom manifested galactorrhoea and/or menstrual disorders. Elevated prolactin levels were observed in seven patients (five on sulpiride therapy and two on metoclopramide) out of the nine patients who were still on drug treatment during the course of the study. Biochemical abnormalities tended to correlate with severity of clinical signs and symptoms.

In the series of 30 patients (18 female, two male) treated with domperidone by Cann *et al.* (1983), serum prolactin levels were found to be raised above normal limits in 13 of the 18 patients in which they were measured. Clinical adverse effects were reported in seven of these 18 patients, five in the raised serum prolactin group and two in the normal group. The most frequent adverse effect complained of was constipation (6 out of 7 patients) and the authors postulate that constipation in some way sensitizes the patient to the action of prolactin or sex steroids. They point out that the incidence of mastalgia and galactorrhoea seen in their series is only likely to be a therapeutic problem if the drug is used on a chronic basis. If its use is confined to short-term treatment of severe acute nausea and vomiting, as is permitted in the current data sheet, then adverse effects are unlikely.

Müller *et al.* (1983) recently reviewed prolactin-lowering and -releasing substances giving a useful classification of these agents.

Prolactin-lowering drugs

1. Direct-acting dopamine receptor agonists
2. Indirect-acting dopamine receptor agonists
3. Drugs which impair serotoninergic neurotransmission
4. γ-Aminobutyric acid-mimetic drugs.

Prolactin-elevating drugs

1. Dopamine receptor antagonists.
2. Drugs otherwise capable of impairing central nervous system dopamine function
3. Drugs enhancing serotoninergic neurotransmission
4. Blockers of serotonin reuptake
5. H_1-receptor agonists
6. H_2-receptor antagonists

The first group of prolactin-lowering drugs was discussed in some detail by Müller *et al.* (1983) as it contains the important ergot derivatives bromocriptine, lergotrile, lysuride, pergolide, and metergoline. These substances inhibit prolactin secretion in all vertebrates tested so far and do not require the presence of an intact hypothalamo–pituitary connection in order to do so. These drugs are powerful suppressors of physiological, pathological, and iatrogenic lactation and in addition restore regular menstruation, libido, and potency when these are impaired by hyperprolactinaemia.

Examples of drugs falling into the second group of prolactin-lowering agents are amphetamine, nomifensine, methylphenidate, amineptine, mazindol, and diclofensine. These substances would seem to enhance the availability of dopamine at its receptors either by inducing dopamine release or by blocking dopamine uptake processes. In contrast to the direct-acting dopamine receptor agonists, the second group rely upon the integrity of the connection between the hypothalamus and the anterior pituitary. Well-known examples of the first group of prolactin-elevating drugs are psychotropic agents of the phenothiazine, butyrophenone, thioxanthene, benzodiazepine, and substituted benzamide groups. Therapeutic agents in these groups have the property of blocking dopamine receptors and thus remove or reduce the inhibiting effect of dopamine on prolactin secretion. In their review Müller *et al.* (1983) referred to the possibility that long-term antipsychotic treatment with dopamine-receptor blocking agents may be associated with the development of tolerance to their prolactin-elevating effects. They postulated that this is due to a decrease in the amount of drug bound to pituitary receptors. Other drugs causing elevation of serum prolactin levels through effects on catecholamine neurotransmission are the tyrosinehydroxylase inhibitors (e.g. α-methyl-*p*-tyrosine), diethyldithiocarbonate, α-

methyldopa, and the L-aromatic amino acid decarboxylase inhibitors (e.g. carbidopa and benserazide).

The situation with respect to drugs which may cause elevated prolactin secretion through their effects on serotoninergic function is decidedly confused. Fenfluramine, an appetite suppressant, which is an indirect-acting serotonin agonist has been shown to cause elevated prolactin levels in rats but there seem to be no similar reports relating to its use in man. Although the serotonin-uptake inhibitor clomipramine has been shown to raise prolactin levels in man, the similarly acting compounds clovoxamine and and fluvoxamine have so far only been shown to be prolactin-elevating in rats.

The situation with respect to GABA-mimetic drugs is somewhat similar to that existing for the serotoninergic agents in that their mode of action is still poorly understood. Sodium valproate, a blocker of GABA-transaminase, induces a lowering of prolactin levels both in normo- and hyperprolactinaemic women with no evidence of pituitary tumours but does not do so in patients with prolactinomas. It is thought that this effect is mediated through inhibition of GABA degradation leading to increased GABA concentrations in the hypothalamus. This is then released through the hypophyseal portal vessels to reach GABA receptors in the pituitary.

The wide variety of commonly used drugs that may raise or lower prolactin levels emphasizes the growing importance of this adverse effect. Physicians should maintain vigilance as previous experience with other adverse effects associated with a broad range of different agents suggest that the list of offending drugs tends to increase with time.

Drug-induced neoplastic disease of the breast

It is beyond the scope of this chapter to present a comprehensive review of the huge volume of literature that exists on this subject. The major concern arises from the widespread use of the combined oestrogen/progestogen contraceptive pill but also to a lesser extent from the use of substances with oestrogenic and/or progestational activity in a variety of pre- and postmenopausal conditions. Since both classes of substance have physiological effects on the breast, it is reasonable to enquire whether or not this activity may be associated with the development of benign or malignant breast neoplasms.

Oral contraceptives have been available to women in the UK since 1960 so we are now entering a period when a small number of women exist who may have been taking them continuously or intermittently for more than 20 years. In these circumstances, even without sophisticated epidemiological studies, it is encouraging to know that no obvious association with breast neoplasms has been observed. Unfortunately, from the epidemiologist's point of view, there have been numerous changes in dosage and formulation during those 24 years which seriously compromise all the studies which have so far been completed. In the early years pills containing 50 μg or more of the oestrogen component were those most commonly in use. Since that time the possible association of adverse effects, particularly those affecting the cardiovascular system, with high oestrogen content has led to the development of contraceptives with a markedly lower dosage of that component. It is not surprising that over the years argument has continued over the relative contribution made by oestrogen and progestogen to the occurrence of adverse effects. The tendency has been to lump together all oral contraceptives whether of high or low oestrogen or progestogen content and to treat them as though they were all the same. This highly dubious habit has greatly confused what was already a very unclear situation and it seems that many more years of careful epidemiological study are needed before any definitive conclusions can be drawn.

Malignant breast disease is unfortunately a condition with very numerous predisposing factors such as age, parity, age at first pregnancy, menstrual history, presence or absence of pre-existing benign breast disease, etc. The latter would seem to be a particular problem as benign neoplasms are fairly clearly negatively associated with oral contraceptive use whereas certain types are probably associated with a greater incidence of malignant disease.

A comprehensive review of the vast literature on this subject is beyond the scope of the present chapter but a few brief comments on the most recent views will help to put the situation in perspective.

Pike *et al.* (1983) pointed to a possible association between the use of high-progestogen content pills and breast cancer in women aged 37 and less at diagnosis. Although the numbers involved were small the authors attribute excess risk to those combination products which contain high-progestogen potency. This attribution is weakened by the use of a potency scale that relies on a delay-of-menses test that is of dubious relevance to progestogen activity on the breast. The delay-of-menses test relies on endometrial haemostatic effects, whereas possible carcinogenicity in the breast is related to interactions between progestogen and specific receptor sites. Some explanation also has to be given for the fact that medroxyprogesterone acetate, a pure progestogen given unopposed by oestrogen, is apparently associated with a lower than expected incidence of breast cancer.

The final report of the extensive study conducted by Sir Richard Doll and Prof. M. Vessey (Vessey *et al.* 1983) gives a useful brief review of papers published up to 1983. These authors found no evidence of a causal relationship between oral contraceptive use and breast cancer in 1176 case-controlled patients. Even though deficiencies in the data are recognized, they concluded that the results of their studies give cause for cautious optimism.

Unopposed oestrogen therapy and breast cancer

Although the major concern has been the combined oestrogen/progestogen contraceptive pill, epidemiological studies have also been conducted on patients taking oestrogen alone. As in the case of the oral contraceptives, the evidence available is somewhat confusing in that some studies have suggested a positive relationship whereas others have been negative.

In 1950 a trial was set up to evaluate the effects of large doses of stilboestrol and ethisterone on the rates of fetal loss in pregnant diabetic women. Eighty women were allocated at random to receive the hormonal treatment and 76 to receive inactive tablets of identical appearance. Beral and Colwell (1980) reported the result of a follow-up of the participants in this study. Twenty-seven years later, information was obtained on about 97 per cent of the women, all but four being traced. All respondents were unaware who had received hormones. The overall mortality was 4.5 times that of women of comparable age in England and Wales, most deaths being from complications of diabetes. More tumours, mainly benign, of the reproductive tract were reported in the hormone-exposed than the nonhormone-exposed group 14 (18 per cent) and two (3 per cent), respectively. Four cases of malignant breast disease were reported in the hormone-exposed women and none in the non-exposed.

These findings support other evidence linking oestrogen treatment and breast cancer and suggest that the latent period before the tumour becomes clinically apparent may be 15 years or longer. A study published by Gambrell *et al.* (1980) gave results which led to an opposite conclusion being drawn. During the six-year period 1972–77, 123 postmenopausal women with breast cancer either had the disease diagnosed at Wilford Hall USAF Medical Centre or were referred there for therapy. Their ages ranged from 33 to 90 (mean, 56.6 years). Of these women, 64.2 per cent had never taken hormones, 25.2 per cent were oestrogen users, 4.9 per cent had a history of hormone usage, and one patient was using oestrogen vaginal cream. In a subgroup of 27 clinic patients (1975–77 period) during 14 548 patient years of observation, breast cancer was diagnosed for an overall incidence of 185.6: 100 000 women/year. Among the 27 patients, the annual incidence of breast cancer was highest in the untreated group at 410.5:100 000 women. In comparison, the incidence in the oestrogen users was 137.7:100 000 women — a significant difference ($p < 0.01$). The incidence in oestrogen–progestogen users was 155.6:100 000 a difference which was also statistically significant ($p < 0.05$). There was no significant difference in the incidence of breast cancer between the oestrogen users and the oestrogen–progestogen users. These data indicate that oestrogen therapy decreases the risk of breast cancer and that, unlike the situation with adenocarcinoma of the endometrium, progestogens do not offer additional protection from breast cancer.

One should be cautious in accepting the figures given by Gambrell et al. (1980) for the incidence of cancer of the breast per 100 000 women per year based on a sample of only 27 cases of breast cancer diagnosed in 1975–77 out of a total of 14 548 women-years for all women on the base.

On balance it would seem reasonable to conclude that an increased risk of breast tumours following oestrogen therapy is not great, if indeed it exists at all, at the dose levels used for hormone replacement therapy.

However, no discussion on the role of oestrogens and breast cancer would be complete without the reminder that men given oestrogens have an increased risk of developing the disease, men with breast cancer excrete an abnormally high concentration of urinary oestrogens (Dao et al. 1973), and breast cancer is abnormally prevalent in men with Klinefelter's syndrome (Dodge et al. 1969).

Diazepam and breast cancer

An editorial on the general subject of 'Mind and cancer' appeared in the *Lancet* on 31 March 1979. A letter (Horrobin et al. 1979), written in response to the editorial, raised the possibility that diazepam might function as a promoter in breast carcinoma. A number of individual scientific observations have been adduced in support of this hypothesis but the chain of reasoning is incomplete and a number of essential logical connectives (e.g. the necessary demonstration that high prostaglandin levels are causes, rather than effects, of cancer) cannot be found.

Raised prostaglandin (PG) levels have been reported in a number of neoplastic conditions (Voekel et al. 1975; Jaffee and Condon 1976; Tashjian et al. 1977; Bennet et al. 1977), and it has been suggested (Karmali et al. 1979) that this could arise from abnormal regulation of PG synthesis. A further speculation is that the abnormal

regulation might result from a reduction in the activity of thromboxane A2 (TXA2), a substance supposedly involved in the normal inhibitory regulation of PG synthesis. Following this hypothetical line of thought, it has been theorized (Karmali et al. 1978) that substances which interfere with the actions (or synthesis) of TXA2 might somehow affect the rate of tumour growth, and it has been postulated that diazepam might act as a competitive TXA2-inhibitor.

The suggestion that diazepam might accelerate the growth of human mammary carcinomata (Horrobin et al. 1979) evidently has its roots in these ideas and in the finding that the growth rate of transplanted neoplastic tissue was increased in diazepam-treated female rats (Karmali et al. 1978). In this study, 10 rats received 2.5 mg/kg diazepam after transplantation of R-3230-AC tumour tissue. There were 10 control animals. At four weeks the tumours were reported to be significantly ($p < 0.001$) heavier in dosed animals. There is no reference to techniques used for randomization, blind dosing, blind dissection, etc. in this work. When the experiment was repeated, with seven rats per group, a similar result was reported (Karmali et al. 1979). Subsequent work, using an 'ascites' tumour (Karmali et al. 1980) and a promoter assay (Trosko and Horrobin 1980; Boyland 1981) was adduced in support of the hypothesis that diazepam accelerates cancer growth.

A conventional rat tumourigenicity study (Jackson and Harris 1981), using adequate numbers of animals, failed to show that diazepam either increased the incidence or decreased the latency of tumours. If diazepam were a promoter as defined (Weinstein and Wigler 1977), an increase in incidence would have been expected. This orthodox study showed no such effect.

It is clear that a satisfactory answer to the hypothesis that diazepam accelerates human breast cancer is required from an adequate analysis of retrospective human information. A small study has been reported (Stoll 1976), but this was not designed to answer the question addressed here; its aim was to establish whether distress (indicated by 'drug' use) over the years preceding the diagnosis of breast cancer was related to the degree of advancement of the disease at its presentation. Several tentative hypotheses could arise from this limited study: that developing but undiagnosed cancer causes anxiety; that anxiety accelerates cancer development; that various psychotropic agents accelerate cancer development. A preliminary screening study (Friedman and Ury, 1980) of 95 drugs in 143 000 patients for four years found that there were 13 000 users of diazepam: 324 cases of cancer (all types were observed, 334.6 being expected). In the Boston case control study (Boston Collaborative Drug Surveillance Programme 1973) there were suggestions that an association might exist between breast cancer and rauwolfia, but no association with diazepam was found, although relatively few patients had used the compound for long periods. An American group at the National Cancer Institute published a report (Kleinerman *et al.* 1981) of a case-control study which provides no evidence that diazepam accelerates the development of breast cancer. In this study both cases and controls were selected from subjects identified in a multicentre breast-cancer screening programme involving 280 000 participants. It can be stated with considerable certainty that there is no firm scientific basis for alarmist articles that have appeared in the lay press following Horrobin's publication (*News of the World*, 28 January 1980; *Daily Telegraph*, 2 January 1981; *Financial Times*, 2 January 1981; *The Times*, 2 January 1981).

It is reassuring that a major epidemiological study recently reported has failed to show a relationship between diazepam use and breast cancer. In a study by the Boston Drug Surveilliance Programme (Kaufman *et al.* 1982) the possible association between breast cancer and diazepam was evaluated on a case-control basis in 1236 women with breast cancer and 728 control subjects with other malignancies. Compared to women who never used diazepam, the relative risk for women who used the drug at least four days per week for at least six months was estimated to be 0.9, with 95 per cent confidence limits of 0.5 and 1.6. There was no apparent association for recent use, or for use in the distant past, although confidence intervals were fairly wide in these categories. The results were not explained by various potential confounding factors, including the major risk factors for breast cancer.

Reserpine therapy and breast cancer

The Boston Collaborative Drug Surveillance Programme (1974) demonstrated an association between cancer of the breast and reserpine therapy. It appeared that the risk of breast cancer was threefold greater in women exposed to reserpine compared with women not exposed. Because this association was made known to two other groups of clinical epidemiologists, two other retrospective surveys were undertaken. Armstrong *et al.* (1974) in the United Kingdom and Heinonen *et al.* (1974) in Finland both found that there was an increased risk of developing breast cancer in women receiving rauwolfia alkaloids.

Mack *et al.* (1975), O'Fallon *et al.* (1975), and Laska *et al.* (1975) also conducted surveys, which however did not support the association between breast cancer and treatment with rauwolfia alkaloids such as reserpine. A further and more carefully controlled study by Armstrong *et al.* (1976) confirmed the association between rauwolfia alkaloids and breast cancer. In Armstrong *et al.*'s survey, three groups of women were selected from a sample of death certificates that had been coded by the Office of Population Censuses and Surveys for all conditions mentioned on them.

1. All women with reference to both breast cancer and hypertension;
2. All women with reference to hypertension and other cancers;
3. A group of women with reference to hypertension without cancer, selected to match the breast-cancer patients with respect to five features.

When women with breast cancer were compared both with the women with other cancers and the women without cancer, there was a positive association between breast cancer and the use of rauwolfia alkaloids, although neither the difference in frequency nor in duration observed was statistically significant. The association appeared to be strongest for use near the time of diagnosis of the cancer. This agrees with nearly all other data and would be expected if rauwolfia derivatives promoted the development of breast cancer from previously initiated cells.

Armstrong *et al.* (1976) pointed out that the risk of developing carcinoma of the breast did not increase with duration of antihypertensive therapy with rauwolfia alkaloids, but that the association was greatest with administration near the time of diagnosis. If the data of Laska *et al.* (1975) are re-examined, analysis shows a relative risk of 1.6 with the use of rauwolfia derivatives in the first year before diagnosis, increasing to 4.0 if use in the first and second year before diagnosis is considered. Use of rauwolfia alkaloids more than two years before diagnosis was not associated with increased risk.

Attempts to explain this association of breast cancer with reserpine therapy have centred on the prolactin-releasing properties of the drug. If Armstrong *et al.* (1976) are correct, this mechanism of action of rauwolfia activating the development of breast cancer from previously initiated cells would be in agreement with the warnings of Ward (1974) and Thorner *et al.* (1974) that drugs which raise circulating prolactin levels should not be given to patients with breast cancer.

In practical therapeutics therefore the lesson to be learnt is to avoid prolactin-releasing agents such as those referred to in a previous section of this chapter in women with breast carcinoma.

Alcohol and breast cancer

A study from the Boston Drug Surveillance Programme

Group (Rosenberg *et al.* 1982) indicated that there was a relation between breast cancer and alcoholic-beverage consumption which was evaluated in a case-control study of 1152 women with breast cancer and two groups of control women — 519 with endometrial or ovarian cancer and 2702 with non-malignant disorders. The relative risk estimate of breast cancer, with allowance for all potential distorting factors, for women who had ever drunk alcoholic beverages relative to those who had never drunk was 1.4 (95 per cent confidence interval, 1.0–2.0) when the comparison group was the group with endometrial or ovarian cancer and 1.9 (1.5–2.4) when the controls who had non-malignant disorders were the comparison group. The association was evident for beer, wine, and spirits.

The suggestion that alcohol might be a carcinogen had been raised earlier by the USA's Third National Cancer Survey. Alcohol ingestion was associated with a higher occurrence of cancers of the breast, thyroid, and malignant melanoma. Williams (1976) suggested that a unifying hypothesis to explain these apparently diverse associations might be advanced on the basis that alcohol might stimulate anterior pituitary secretion of prolactin, thyroid-stimulating hormone, and melanocyte-stimulating hormone. Under the trophic effects of these hormones, Williams postulated that the target organs exhibited increased mitotic activity and increase susceptibility to malignancy.

Rosenberg *et al.* (1982) noted an earlier report (Breslow and Enstrom 1974) of a correlation between alcohol consumption (especially beer) and breast cancer mortality in 41 states of the USA. However, Rosenberg *et al.* (1982) were cautious in their interpretation of the findings and indicated that one possible explanation for this association is that alcoholic consumption may be related to certain dietary habits which are more directly related to the increased risk.

International differences in alcohol intake, fat consumption, and breast cancer may indeed be more significant than alcohol intake. La Vecchia *et al.* (1982) made this point very clearly.

Data on total per caput alcohol consumption are readily available for fourteen countries and, if anything, show a negative relation with breast cancer mortality ($r = 0.52$, $p = 0.054$). The same countries illustrate the well known and very strong, positive correlation between total fat consumption and breast cancer ($r = 0.86$, $p < 0.001$), and the partial correlation coefficient for alcohol and breast cancer after correcting for fat consumption remains slightly negative ($r = 0.34$, $p = 0.26$).

(See also Table 21.1.) International correlation studies have serious limitations especially where estimates of consumption refer to the entire population and not only

Table 21.1   Total per caput alcohol consumption, total fat intake, and age-standardized death from breast cancer in 14 countries (after La Vecchia 1982)

Country	Per caput alcohol consumption (litres of absolute alcohol 1966 or 1967)*	Total fat intake (1964–6)†	Age-standardized death rate for breast cancer (per 100 000 women 1966–7)‡
France	24.66	136.6	16.98
Italy	18.00	85.6	16.09
Portugal	17.57	71.7	11.93
Austria	14.47	119.1	17.47
West Germany	13.63	133.8	17.65
Australia	10.71	129.2	19.15
Czechoslovakia	10.27	94.6	14.89
Canada	8.95	141.4	23.47
Belgium	8.42	138.9	20.97
UK	7.66	141.9	24.40
Ireland	7.64	135.8	21.03
Denmark	7.50	157.0	24.62
Netherlands	6.19	153.3	26.45
Finland	2.16	114.4	14.65

Data from: *Smith, R. (1981) *Br. med. J.* **283**, 895–98 (derived from the 1968 annual report of the Dutch Distillers' Association); † food balance sheets, 1964–6 average (Food and Agriculture Organization 1971); ‡ Logan, W.P.D. (1975) *WHO Chron.* **29**, 462–71.

to women at risk. However, the typical pattern of alcohol consumption (mainly wine) in the countries with the highest alcohol intake does not suggest a remarkable disproportion between the sexes. Furthermore, the strongest previously reported association with breast cancer was with beer, which is traditionally drunk by men. It is difficult to reconcile the apparently negative correlation between total alcohol intake and breast cancer with a direct carcinogenic action for alcohol *per se*. However, in the USA, alcohol consumption increases with affluence, and it might be that within that country, certain dietary or general life-style habits associated with greater affluence are more directly responsible for the increased risk.

Mammography

Simon (1977) and Bailar (1977) drew attention to the fact that the investigative procedure of mammography may of itself cause breast cancer. Bailar cautioned that in women under 60–65 years mammography should only be undertaken if there is specific reason for concern or where mammography may give results which will affect the management of the case. Whereas early diagnosis of breast cancer is desirable, we must be sure that efforts to stop the disease do not lead to greater problems than they are intended to prevent.

Ahmed and Steel (1962) described mammary carcinoma occurring 30 years after *Thorotrast* mammography conducted in 1939.

REFERENCES

AHMED, M.Y., AND STEELE, H.D. (1972). Breast carcinoma 30 years after *Thorotrast* mammography. *Can. J. Surg.* **15**, 45–9.

AONO, T., SHIOJI, T., KINUGASA, T., ONISHI, T., AND KURACHI, K. (1978). Clinical and endocrinological analyses of patients with galactorrhoea and menstrual disorders due to sulpiride or metoclopramide. *J. Clin. Endocrinol. Metab.* **47**, 675–80.

ARMSTRONG, B., SKEGG, D., WHITE, G., AND DOLL, R. (1976). Rauwolfia derivatives and breast cancer in hypertensive women. *Lancet* ii, 8–12.

——, STEVENS, N., AND DOLL. R. (1974). Retrospective study of the association between use of rauwolfia derivatives and breast cancer in women. *Lancet* ii, 672–5.

BADAWY, S.Z.A., NUSBAUM, M.L., AND OMAR, M. (1980). Hypothalamic pituitary evaluation in patients with galactorrhoea–amenorrhoea and hyperprolactinaemia. *Obstet. Gynecol.* **55**, 1.

——, REBSCHER, F., KOHN, L., WOLFE, H., OATES, R.P., AND MOSES, A. (1981). The relationship between oral contraceptive use and subsequent development of hyperprolactinaemia. *Fertil. Steril.* **36**, 464–7.

BAILAR, J.C. (1977). Mammography — a time for caution. *J. Am. med. Ass.* **237**, 997–8.

BARBIERI, C., FERRARI, C., CALDARA, R., CROSSIGNANI, R.M., AND BERTAZZONI, A. (1981). Endocrine and metabolic effects of labetolol in man. *J. cardiovasc. Pharmacol.* **3**, 981–91.

BATESON, M.C., BROWNING, M.C.K., AND MACONNACHIE, A. (1977). Galactorrhoea with cimetidine. *Lancet* ii. 247–8.

BECK, W. (1981). Normoprolactinaemia in boys with marked gynaecomastia. *Eur. J. Pediat.* **137**, 41–4.

BENNETT, A., CHARLIER, E.M., MCDONALD, A.M., SIMPSON, J.S., STANFORD, I.F., AND ZEBBO, T. (1977). Prostaglandins and breast cancer. *Lancet* ii, 624–6.

BERAL, V. AND COLWELL, L. (1980). Randomised trial of high doses of stilboestrol and ethisterone in pregnancy: long term follow up of mothers. *Br. med. J.* **281**, 1098–1101.

BERGMAN, D., FUTTERWEIT, W., SEGAL, R., AND SIROTA, D. (1981). Increased oestradiol in diazepam related gynaecomastia. *Lancet* ii, 1225–6.

BEZUIDENHOUT, D.J.J. (1978). Ginekomastia tydens simetidin — terapie. *S. Afr. med. J.* **54**, 263.

BLOCH, K. (1961). On pathogenesis of breast hypertrophy from digitalis therapy. *Z. Kreisl-Forsch* **50**, 591–5.

BOSTON COLLABORATIVE DRUG SURVEILLANCE PROGRAM (1973). Oral contraceptives and venous thromboembolic disease, surgically confirmed gall-bladder disease and breast tumours. *Lancet* i, 1339–404.

—— (1974). Reserpine and breast cancer. *Lancet* ii, 669–71.

BOYAR, R.M., KAPEN, S., FINKELSTEIN, J.W., PERLOW, M. SASSIN, J.F., FUKUSHIMA, D.K., WEITZMAN, E.D., AND HELLMAN, L. (1974). Hypothalamic-pituitary function in diverse hyperprolactin-aemic states. *J. clin. Invest.* **53**, 1588–8.

BOYLAND, E. (1981). Diazepam and tumour promotion. *Lancet* i, 455.

BRESLOW, N.E. AND ENSTROM, J.E. (1974). Geographic correlations between mortality rates and alcohol tobacco consumption in the United States. *J. nat. Cancer. Inst.* **53**, 631–4.

BRIDGEMAN, J.F. AND BUCKLER, J.M.H. (1974). Drug-induced gynaecomastia. *Br. med. J.* **3** 520–1.

BROWN, W.A. AND LAUGHREN, T.P. (1981). Tolerance to the prolactin-elevating effects of neuroleptics. *Psychiat. Res.* **5**, 317–22.

BURLAND, W.L., GLEADLE, R.I., LEE, R.M., ROWLEY-JONES, D., AND GROOM, G.V. (1978). Cimetidine and serum prolactin. *Br. med. J.* **1**, 717.

CALOV, W.L., AND WHYTE, H.M. (1954). Oedema and mammary hypertrophy: toxic effects of digitalis leaf. *Med. J. Aust.* **1**, 556–7.

CANN, P.A., READ, N.W., AND HOLDSWORTH, C.D. (1983). Galactorrhoea as side effect of domperidone. *Br. Med. J.* **286**, 1395–6.

CAPELLER, D. VON, COPELAND, G.D., AND STERN, T.N. (1959). Digitalis intoxication: a clinical report of 148 cases. *Ann. intern. Med.* **50**, 869–78.

CHECK, J.H., MURDOCK, M.G., CARO, J.F., AND HERMEL, M.B. (1978). Cystic gynaecomastia in a male treated with clomiphene citrate. *Fertil. Steril.* **30**, 713–15.

CLARKE, E. (1965). Spironolactone therapy and gynecomastia. *J. Am. med. Ass.* **193**, 163–4.

CONN, H.L.JR. (1964). Digitalis therapy and gynecomastia. *J. Am. med. Ass.* **190**, 1018–19.

CONTRERAS, P., GENERINI, G., MICHELSEN, H., PUMARINO, H., AND CAMPINO, C. (1981). Hyperprolactinaemia and galactorrhoea: spontaneous versus iatrogenic hypothyroidism. *J. clin. Endocrinol. Metab.* **53**, 1036–9.

COVE, D.H., AND BARKER, G.A. (1979). Digoxin and hormone receptors. *Lancet*, i, 204.

DAO, T.L., MORREAL, C., AND NEMOTO, T. (1973). Urinary estrogen excretion in men with breast cancer. *New Engl. J. Med.* **289**, 138–40.

DELLE FAVE, G.F., TAMBURRANO, G., DE MAGISTRIS, L., NATOLI, C., SANTORO, M.L., CARRATU, R., AND TORSOLI, A. (1977a). Gynaecomastia with cimetidine. *Lancet* i, 1319.

——, ——, ——, ——, ——, ——, AND —— (1977b). Variations in serum prolactin following cimetidine treatment for peptic ucler disease. *Rend. Gastroent.* **9**, 142–3.

DESAI, S.N. (1973). Sudden gigantism of the breasts: drug induced? *Br. J. Plastic Surg.* **26**, 371–2.

DI RAIMONDO, C.V., ROACH, A.C., AND MEADOR, C.K. (1980). Gynecomastia from exposure to vaginal estrogen cream. *New Engl. J. Med.* **302**, 1089–90.

DODGE, O.G., JACKSON, A.W., AND MULDAL, S. (1969). Breast cancer and interstitial-cell tumour in a patient with Klinefelter's syndrome. *Cancer* **24**, 1027–32.

DURAND, P., BORRONE, C., SCARABICCHI, S., AND RAZZI, A. (1964). Hyperpigmentation of breast areolae and external genitals with gynaecomastia following griseofulvin treatment. *Minerva, Med.* **55**, 2422–5.

EHNI, G., AND ECKLES, N.E. (1959). Interruption of the pituitary stalk in the patient with mammary cancer. *J. Neurosurg.* **16**, 628–52.

FARA, G.M., CORVO, G., DEL BERNUZZI, S., BIGATELLO, A., DI PIETRO, C., SCAGLIONI, S., AND CHIUMELLO, G. (1979). Epidemic of breast enlargement in an Italian school. *Lancet* ii, 295–7.

FORSYTH, I., BESSER, G.M., EDWARDS, C.R.W., FRANCIS, L., AND MYRES, R.P. (1971). Plasma prolactin activity in inappropriate lactation. *Br. med. J.* **3**, 225–7.

FRIEDMAN, G.D., AND URY, H.K. (1980). Initial screening for carcinogenicity of commonly used drugs. *J. Nat. Cancer Inst. (US)* **65**, 723–33.

FRIEDMAN, N.M., AND PLYMATE, S.R. (1980). Leydig cell dysfunction and gynaecomastia in adult males treated with alkylating agents. *Clin. Endocr.* **12**, 553–6.

FRIEDMAN, S., AND GOLDFIEN, A. (1969). Amenorrhea and galactorrhea following oral contraceptive therapy. *J. Am. Med. Ass.* **210**, 1888–91.

GAMBRELL, D., MASSEY, F.M., CASTENADA, T.A., AND BODDIE, A.W. (1980). Estrogen therapy and breast cancer in post-menopausal women. *J. Am. geriat. Soc.* **28**, 251–7.

GELENBERG, A.J., COOPER, D.S., DOLLER, J., AND MALOOF, F. (1979). Galactorrhoea and hyperprolactinaemia associated with Amoxapine therapy. *J. Am. med. Ass.* **242**, 1900–1.

GERNEZ-RIEUX, C., TACQUET, A., AND MACQUET, V. (1963). Perfusions with ethionamide in the treatment of pulmonary tuberculosis. *G. ital. Chemioter.* **10**, 87–98.

GRUDELSKY, G.A. (1981). Tuberoinfundibular dopamine neurons and the regulation of prolactin secretion. *Psychoneuroendocrinology* **6**, 3–16.

HALL, W.H. (1976). Breast changes in males on cimetidine. *New Engl. J. Med.* **295**, 841.

HAFS, H.D. (1975). Prostaglandins and the control of anterior pituitary hormone secretion. In *Hypothalamic hormones*, ed. M. Motta, P.G. Crosignani, L. Martini. Proceedings of Serons Symposia no. 6 Academic Press, New York.

HEINONEN, O.P., SHAPIRO, S., TUOMINEN, L., AND TURUNEN, M.I. (1974). Reserpine use in relation to breast cancer. *Lancet* **ii**, 675–7.

HORROBIN, D.F., GHAYUR, T., AND KARMALI, R.A. (1979). Mind and cancer. *Lancet* **i**, 978.

JACKSON, M.R., AND HARRIS, P.A. (1981). Absence of effect of diazepam on tumours. *Lancet* **i**, 104.

JAFFEE, B.M., AND CONDON, S. (1976). Prostaglandins E and F in endocrine diarrheagenic syndromes. *Ann. Surg.* **184**, 516–24.

KARMALI, R.A., HORROBIN, D.F. GHAYUR, T., MANKU, M.S., CUNNANE, S.C., MORGAN, R.O., ALLY, A.I., KARMAZYN, M., AND OKA, M. (1978). Influence of agents which modulate thromboxane A2 synthesis or action on R3230AC mammary carinoma. *Cancer Lett.* **5** 205–8.

——, VOLKMAN, A., MUSE, P., AND LOUIS, T.M. (1979). The influence of diazepam administration in rats bearing the R3230AC mammary carcinoma. *Prostaglandins Med.* **3**, 193–8.

——, ——, SPIVEY, W., MUSE, P., AND LOUIS, T.M. (1980). Intrarenal growth of the Walker 256 tumour and renal vein concentrations of PGE_2, $PGF_{2\alpha}$ and TXB_2; Effects of diazepam. *Prostaglandins Med.* **4**, 239–46.

KAUFMAN, D.W., SHAPIRO, S., SLONE, D., ROSENBERG, L., HELMRICH, S.P., MIETTINEN, O.S., STOLLEY, P.D., LEVEY, M., AND SCHOTTENFELD, D. (1982). Diazepam and the risk of breast cancer. *Lancet* **i**, 537–9.

KHAZAN, N., PRIMO, C., DANON, A., ASSAEL, M., SULMAN, F.G., AND WINNIK, H.Z. (1962). The mammotropic effect of tranquilizing drugs. *Arch. int. Pharmacodyn.* **141**, 291–305.

KLEINERMAN, R.A., RINTON, L.A., HOOVER, R., AND FRAUMENI, J.F., JR. (1981). Diazepam and breast cancer. *Lancet* **i**, 1153.

KRAWIEC, J., KRUGLIAK, P., LEVY, J., LAMORECHT, S.A., AND ODES, H.S. (1981). Effect of oral cimetidine on plasma prolactin. *Isr.* J. med. Sci. **17**, 1183–4.

LAMPE, W.T. (1967). Lactation due to psychotropic agents. *Metabolism* **16**, 257–8.

LASKA, E.M., SEIGEL, C., MEISNER, M., FISCHER, S., AND WANDERLING, J. (1975). Matched-pairs study of reserpine use and breast cancer. *Lancet* **ii**, 296–300.

LA VECCHIA, C., FRANCESCHI, S., AND CUZICK, J. (1982). Alcohol and breast cancer *Lancet* **i**, 621.

MACK, T.M., HENDERSON, B.E., GERKINS, V.R., ARTHUR, B., BAPTISTA, J., AND PIKE, M.C. (1975). Reserpine and breast cancer in a retirement community. *New Engl. J. Med.* **292**, 1366–71.

MANN, V.M. (1963). Gynecomastia during therapy with spironolactone. *J. Am. med. Ass.* **184**, 778–80.

MELIS, G.B., PAOLETTI, A.M. MAIS, V., MASTRAPASQUA, N.M., STRIGINI, F., FRUZZETTI, F., GUARNIERI, G., GAMBACCIANI, M., AND FIORETTI, P. (1982). The effects of the GABAergic drug, sodium valproate, on prolactin secretion in normal and prolactinaemic subjects. *J. clin. Endocrinol. Metab.* **54**, 485–9.

MENDELSON, J.H., ELLINGBOE, J., MELLO, N.K., AND KUEHNLE, J. (1982). Buprenorphine effects on plasma luteinizing hormone and prolactin in male heroin addicts. *J. Pharmacol. exp. Ther.* **220**, 252–5.

MILLIGAN, D., DRIFE, J.O., AND SHORT, R.V. (1975). Changes in breast volume during normal menstrual cycle and after oral contraceptives. *Br. med. J.* **4**, 494–6.

MIYATAKE, A., NOMA, K., NAKAO, R., MORIMOTO, Y., AND YAMAMURA, Y. (1978). Increased serum oestrone and oestradiol following spironolactone administration in hypertensive men. *Clin. Endocr.* **9**, 523–33.

MOERCK, H.J., AND MAGELUND, G. (1979). Gynaecomastia and diazepam abuse. *Lancet* **i**, 1344–5.

MÜLLER, E.E., LOCATELLI, V., CELLA, S., PEÑALVA, A., NOVELLI, A., AND COCCHI, D. (1983). Prolactin-lowering and -releasing drugs: mechanisms of action and therapeutic applications *Drugs* **25**, 399–432.

NABER, H., FISCHER, H., AND ACKENHEIL, M. (1979). Effect of long term neuroleptic treatment on dopamine tuberoinfundibular system: development of tolerance? *Commun. Psychopharmacol.* **3**, 59.

O'FALLON, W.M., LABARTHE, D.R., AND KURLAND, L.T. (1975). Rauwolfia derivatives and breast cancer. *Lancet* **ii**, 292–6.

OLUSI, S.O. (1980). Hyperprolactinaemia in patients with suspected cannabis induced gynaecomastia. *Br. med. J.* **280**, 255.

PALMER, B.V., MONTGOMERY, A.C.V., AND MONTEIRO, J.C.M.P. (1978). Ginseng and mastalgia. *Br. med. J.* **1**, 1284.

PETTINGER, W.A., HORWITZ, D., AND SJOERDSMA, A. (1963). Lactation due to methyldopa. *Br. med. J.* **1**, 1460.

PIKE, M.C., HENDERSON, B.E., KRAILO, M.D., DUKE, A. AND ROY, S. (1983). Breast cancer in young women and use of oral contraceptives: possible modifying effect of formulation and age at use. *Lancet* **ii**, 926–30.

RESTIFO, R.A., AND FARMER, T.A. (1962). Spirono lactones and gynaecomastia. *Lancet* **ii**, 1280.

ROBINS, A.H. AND McFADYEN, M.L. (1981). Effect of the new H_2-receptor antagonist ranitidine on plasma prolactin levels in duodenal ulcer patients. *J. Pharm. Pharmacol.* **33**, 615–16.

ROBINSON, B. (1957). Breast changes in male and female with chlorpromazine and reserpine therapy. *Med. J. Aust.* **2**, 239.

ROSE, L.I., UNDERWOOD, R.H., NEWMARK, S.R., KISCH, E.S., AND WILLIAMS, G.H. (1977). Pathophysiology of spironolactone-induced gynecomastia. *Ann. intern. Med.* **87**, 398–403.

ROSENBERG, L., SLONE, D., SHAPIRO, S., KAUFMAN, D.W., HELMRICH, S.P., MIETTINEN, O.S., STOLLEY, P.D., LEVY, M., ROSENSHEIN, N.E., SCHOTTENFELD, D., AND ENGLE, R.L. (1982). Breast cancer and alcoholic-beverage consumption. *Lancet* **i**, 267–71.

SATO, T., HIRONO, M., IEEAKA, I., SHIGESHIRO, T., AND TAYA, K. (1974). Action of prostaglandin (PG) on FSH, LH and prolactin release in rats. *Folia Endocrinol (Jpn).* **50**, 661.

SHARPE, P.C., AND HAWKINS, B.W. (1977). Efficacy and safety of cimetidine. Long-term treatment with cimetidine. In *Cimetidine: Proc. 2nd Int. Symp. Histamine H$_2$-receptor Antagonists* (ed. W.L. Burland and M.A. Simpkins) pp. 358–66. Excerpta Medica. Amsterdam.

——, MELVIN, M.A., MILLS, J.G., BURLAND, W.L., AND GROOM, G.V. (1980). in *Further experience with H$_2$-receptor antagonists in peptic ulcer disease* (ed. A. TORSOLI, P.E. LUCCHELLI, AND R.W. BRIMBLECOMBE) pp. 366–70. Excerpta Medica. Amsterdam.

SHEARMAN, R.P. (1971). Prolonged secondary amenorrhoea after oral contraceptive therapy. *Lancet* **ii**, 64–6.

SIMON, N. (1977). Breast cancer induced by radiation. *J. Am. med. Ass.* **237**, 789–90.

SMITH, I.D., SHEARMAN, R.P., AND KORDA, A.R. (1972). Lactation following therapeutic abortion with prostaglandin F$_2\alpha$. *Nature, London* **240**, 411–12.

SMITH, W.G. (1962). Spironolactone and gynaecomastia. *Lancet* **ii**, 886.

SPENCE, R.W. AND CELESTIN, L.E. (1979). Gynaecomastia associated with cimetidine. *Gut* **20**, 154–7.

STEINER, J., CASSAR, J., MASHITER, K., DAWES, I., AND RUSSELL FRASER, T. (1976). Effects of methyldopa on prolactin and growth hormone. *Br. med. J.* **1**, 1186–8.

STENKVIST, B., BENGTSSON, E., ERIKSSON, O., HOLMQUIST, J., NORDIN, B., WESTMAN-NAESER, S., AND EKLUND, G. (1979). Cardiac glycosides and breast cancer. *Lancet* **i**, 563.

STOLL, B.A. (1976). Psychosomatic factors and tumour growth. In: *Risk factors in breast cancer* (ed. B.A. STOLL), pp. 193–203. William Heinemann Books, London.

SUSSMAN, R.M. (1963). Spironolactone and gynaecomastia. *Lancet* **i**, 58.

TASHJIAN, A.H., VOEKEL, E.F., AND LEVINE, L. (1977). Plasma concentrations of 13, 14-hydro-15-keto-prostaglandin E2 in rabbits bearing the VX2 carcinoma: Effects of hydrocortisone and indomethacin. *Prostaglandins* **14**, 309–17.

TAYLOR, P.J., CUMMING, D.C., AND CORENBLUM, B. (1981). Successful treatment of D-penicillamine-induced breast gigantism with danazol. *Br. med. J.* **282**, 362–3.

THORNER, M.O., VOLANS, G., BESSER, G.M., AND MCNEILLY, A.S. (1974). Antiemetics, prolactin and breast cancer. *Br. med. J.* **3** 467.

TOOLEY, P.H., AND LACK, C. (1949). Gynaecomastia during treatment with amphetamine. *Lancet* **i**, 650–1.

TROSKO, J.E. AND HORROBIN, D.F. (1980). The activity of diazepam in a Chinese homster V79 lung cell assay for tumour, IRCS. *J. med. Sci.* **8**, 887.

VESSEY, M., BARON, J., DOLL, R., McPHERSON, K., AND YEATES, D. (1983). Oral contraceptives and breast cancer: final report of an epidemiological study. *Br. J. Cancer* **47**, 455–62.

VOEKEL, E.F., TASHKIAN, A.H., FRANKLIN, B., WASSERMAN, E., AND LEVINE L. (1975). Hypercalcemia and tumour prostaglandins in the VX2 carcinoma model in the rabbit. *Metabolism* **24**, 973–86.

WARD, H.W.C. (1974). Metaclopramide and prolactin. *Br. med. J.* **3**, 169.

WEINSTEIN, I.B. AND WIGLER, M. (1977). Cell culture studies provide the information on tumour promoters. *Nature* **270**, 659–60.

WHIFFEN, J.D. (1963). Unusual surgical consideration of phenothiazine therapy. *Am. J. Surg.* **106**, 991–2.

WILKINS, R.W. (1954). Clinical usage of rauwolfia alkaloids, including reserpine (Serpasil). *Ann. NY Acad. Sci.* **59**, 36–44.

WILLIAMS, E. (1962). Spironolactone and gynaecomastia. *Lancet* **ii**, 1113.

WILLIAMS, R.R. (1976). Breast and thyroid cancer and malignant melanoma promoted by alcohol-induced pituitary secretion of prolactin, TSH and MSH. *Lancet* **ii**, 996–9.

WOLF, H.L. (1964). Digitalis therapy and gynecomastia. *J. Am. med. Ass.* **190**, 1018.

WRIGHT, V.K. (1955). Complications of chlorpromazine treatment. *Dis. nerv. Syst.* **16**, 114.

YEO, T., DELITALA, G., BESSER, G.M., AND EDWARDS, C.R. (1980). The effects of ranitidine on pituitary hormone secretion in vitro. *Br. J. Pharmacol.* **10**, 171–3.

22 The teratogenic and other toxic effects of drugs on reproduction

F.M. SULLIVAN AND PATRICIA R. Mc ELHATTON

Introduction

It has been known for over 50 years that chemicals can induce congenital malformations in animals. In the early-1950s it was even known in man that drugs like aminopterin, if used unsuccessfully to induce abortion, could cause severe central nervous system malformations of the fetus, and that the synthetic progestogens, if used in an attempt to prevent abortion, could cause masculinization of female fetuses. However, it was not until thalidomide was revealed as the cause of disastrous malformations of babies in 1961 that teratogenicity was regarded as a real drug hazard for man. Up to that time, toxicological assessment of new drugs was relatively primitive and in most countries, including the UK, was not even required legally. Immediately after the thalidomide experience, however, public pressure led to the rapid setting-up of government committees to approve all new drugs before marketing was allowed. Among the pre-clinical tests which now have to be carried out on all new drugs there is a requirement in almost every country of the world for some tests of teratogenicity or reproductive toxicity.

Reproductive toxicity

Teratogenicity

Originally defined as the production of monsters, and currently meaning the production of structural abnormalities, teratogenicity is only one part of a much wider spectrum of toxic effects which drugs can induce in relation to reproduction. Drugs can increase or decrease fertility in both males and females. During pregnancy, apart from the induction of congenital malformations, drugs can affect behavioural development of the offspring. This was first defined by Werboff and Gottlieb (1963) *as behavioural teratology* meaning the ability of drugs given during pregnancy to affect the behaviour or functional adaptation of the offspring to its environment. Public interest was first aroused by the observations of Werboff and his colleagues that the administration of the then widely-used tranquillizer meprobamate to pregnant rats could affect the learning capacity of the offspring. Since then, a wide range of, mainly centrally acting, drugs have been shown to affect learning ability of offspring in animals (see review by Barlow and Sullivan 1975). There is also evidence in man that smoking during the later stages of pregnancy may affect the behaviour and intelligence of the children even up to 11 years of age (Butler and Goldstein 1973)

Transplacental carcinogenicity

This is another recently described phenomenon which was first described in animals and is now known to be a hazard for man. It was first shown that small doses of carcinogens, too low to affect the mother, if given in the late stages of pregnancy in rats could result in 100 per cent of the offspring developing cancer after a latency of one year or more which is well into 'middle age' for a rat (see review by Tomatis 1974). Subsequent work showed that the sensitive time for cancer induction was in very late pregnancy and in the neonatal period and that a very wide range of carcinogens could produce this effect.

Herbst and his co-workers (Herbst *et al.* 1971, 1974, 1975) found that the use of the synthetic oestrogen, stilboestrol, during pregnancy could result in the induction of a rare type of vaginal cancer in the daughters of such women, which did not manifest itself until they reached the age of 15–20 years. These observations have since been confirmed and extended to show that the male offspring are also affected with testicular and genital malformations and the production of abnormal spermatozoa. Other drugs suspected of being involved in human transplacental carcinogenesis are the folic-acid antagonists aminopterin and methotrexate, phenytoin, chloramphenicol, and immunosuppressive drugs (Tomatis 1974), and, although a link has been established between maternal influenza or chickenpox infection in pregnancy and childhood cancer, it is still not clear whether the infections or the drugs used in their treatment are the causative factors.

Mutagenesis

Mutagenesis is another possible reproductive hazard which has been induced, so far only in animals, by drugs. There are two major types of mutagenic effect which can be induced by drugs:

1. In somatic cells, which can result in cancer induction;
2. In germ cells, which can result in inherited defects, for example, reduced fertility leading after two or three generations to complete infertility.

There are many different types of mutation which can be produced by drugs ranging from dominant lethal mutations which result in prompt death of the offspring, to recessive point mutations with very low lethality, and which could spread widely in the community before there was any awareness of their presence.

There has been considerable research recently on the correlation between mutagenicity and carcinogenicity in attempts to find short-term mutagenic tests which could indicate the possible carcinogenic hazard of chemicals including drugs. The insidious action of such substances necessitates a constant awareness of their potential risks. Reports on the mutagenic potential of over 7000 chemicals have been collected by the Environmental Mutagen Information Centre (see Malling and Wassom 1977). Since cells undergoing active division are most susceptible to mutagens, the hazards for men, with large numbers of dividing cells in the testis, are probably much greater than for women. Thus the effects of drugs taken by *men* on the subsequent development of their offspring must always be borne in mind.

Animal tests for teratogenicity

A number of good books have been written describing the methodology and interpretation of animal studies. Several hundreds of drugs have been tested for teratogenicity in animals and details of many of these were catalogued by Shepard (1976). Because of the diversity of test procedures, species of animal, doses, routes, etc. used in these tests it is not really of much interest or value for the non-specialist to read through such a list. Furthermore, by careful selection of drug doses, times, and routes of administration it is possible to produce some congenital abnormalities or fetal death with almost any drug, so that the mere fact that a drug has produced some malformations in an animal test is no indication that it is hazardous for use during pregnancy in man. For this reason the effects of drugs in animals will not be reviewed except to amplify human data if necessary.

However, for completeness, a brief review of some of the more important aspects of animal teratology tests will be given.

Species

The three most commonly used species for teratogenicity testing for new drug studies for regulatory authorities are the mouse, the rat, and the rabbit. This is probably for two reasons. Mice and rats have been used by reproductive endocrinologists for very many years so that we have good background information on the reproductive processes in these species. Furthermore they are easy to breed with a short gestation period of 20–22 days and have large litters, commonly of 10–15 pups. Thus a large amount of information can be had at a relatively low cost. Rabbits are probably used as a second species to satisfy the requirements of most regulatory agencies which ask for tests in a rodent (mice or rats) and a non-rodent (usually rabbit), which may react differently; for example, it was possible to demonstrate thalidomide-induced abnormalities in rabbits but not in rodents. Various other species — hamsters, ferrets, pigs, cats, dogs, and chickens — have also been used sporadically but do not seem to have striking advantages over the more common ones. Primates are sometimes used, especially to check on equivocal results obtained in rodents or rabbits. Not many laboratories are equipped for primate studies and using wild-caught monkeys with little background information could be misleading. Commonly six to 15 monkeys are used for a test, and while positive results could obviously give cause for concern, negative results in 15 animals would give little assurance on a statistical basis. One small primate, the marmoset, has been bred in laboratories for teratological work but insufficient data is available to assess its value. It is interesting from the point of view of assessment of drug safety that although the human and rat are both very susceptible to the teratogenic action of folic-acid antagonists, attempts to produce malformations with such substances in subhuman primates have so far been unsuccessful.

Effects of increasing doses

In principle, if high enough doses of drugs are used in animal studies then either the mother will be affected first, showing signs of toxicity at doses which leave the fetus unaffected, or else the fetus will be affected first. The latter is usually the case and fetal death occurs before maternal death. If a drug has teratogenic potential, then this is usually seen with doses just sublethal to the fetus. It is thus important in teratological studies that a range of

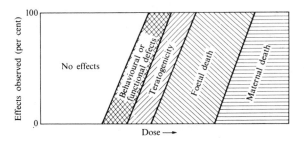

Fig. 22.1 Relationship between dose and effects observed with typical drugs. At any given dose an overlap may exist with some fetal death, some teratogenic effects, and some functional defects. It is important with individual drugs to define the threshold doses for each effect. For some embryotoxic drugs, no significant teratogenic range may exist.

doses is used sufficient to establish whether a teratogenic range actually exists and if so what the lower threshold is. The relationship between dose and effects is shown diagrammatically in Fig. 22.1 for a typical drug, though with some compounds no significant teratogenic range may be present. In general, the wider the teratogenic range and the more distant it is from the maternal toxic range, the more hazardous the drug must be considered for man.

Relevance of animal studies for man

From what has been said above it is clear that wide inter- and even intra-species differences exist in response to drug treatment. The mere isolated observation that a drug has teratogenic potential in an animal species does not imply that it will also have such an effect in man. However, drugs, which are teratogenic in several species and over a wide range of doses must be regarded as having teratogenic potential for man and it is important that monitoring is carried out to assess the possible hazard for man. As will appear in the drug review which follows in this chapter, there is remarkably little information on the safety or otherwise for use in pregnancy of the vast majority of drugs in clinical use today.

Legislative requirements

From 1964, all new drugs submitted for registration in the UK had to be tested for teratogenic potential in two animal species, usually the rat or mouse and the rabbit. This involved administration of the test drug during the period of embryogenesis and subsequent examination of the fetuses for structural malformations. Some other

countries, notably the USA, had rather wider requirements involving drug administration to male as well as female animals, and with some of the offspring allowed to be reared to look for postnatal effects. Since 1975, the UK requirements have been similarly widened and now consist of three parts:

1. A fertility test, usually carried out on rats or mice in which males are treated for 60–80 days before mating to cover all stages of spermatogenesis, and are then mated with females treated throughout the whole of pregnancy. By careful examination of some of the females killed during pregnancy, this part of the test will give information on adverse effects on male or female fertility and on mutagenic potency as evidenced by an increase in dominant lethal mutations, as well as effects on fetal development.

2. The second part is a classical teratological test in two species with pregnant females treated during the period of embryogenesis.

3. In the third part, females are treated during the last third of gestation (not covered in 2 above), through parturition, and up to weaning. This tests for effects on fetal and uterine growth, parturition, and postnatal development and lactation.

Some of the offspring from parts 1 and 3 are allowed to grow up so that late effects on behaviour, auditory and visual function, and reproductive capacity may be assessed (for details see Griffin 1977).

It is hoped that these newer requirements, which are broadly similar to those in the USA, Canada, Japan, and some other European countries will provide some safeguards against the introduction of drugs with potent reproductive toxic effects. However, the species differences described above in response to teratogenic effects must still mean that careful scrutiny of all new drugs must be carried out during the early clincial period to ensure that any adverse reproductive effects in man will be detected as soon as possible.

Principles of teratology as applied to man

Among the many lessons which have been learned from the thalidomide experience in man and subsequently from animal experiments, certain basic principles have emerged which are clinically important for physicians using drugs in women of child-bearing age. Some factors which may be considered are stage of pregnancy, dose and duration of drug administration, placental transfer, whether to avoid all drugs in pregnancy.

Stage of pregnancy

Most physicians are now aware of the concept of 'critical

periods' of developmental sensitivity to drugs. In general, the organs which are developing at any stage of embryogenesis are the ones most sensitive to the teratogenic effects of drugs given at that time. Thus the grossest congenital malformations are most likely to be induced during organogenesis, e.g. anencephaly or spina bifida induced by drugs interfering with closure of the anterior or posterior neuropore, or phocomelia or amelia by interference with limb-bud formation, etc. As a result of the experiences with thalidomide and rubella many physicians believe that only the first three months of pregnancy are hazardous from the point of view of drug-induced effects on the fetus. This is not the case, however. During the later stages of fetogenesis, histological, functional, and biochemical development are at their most active stages and drugs may interfere with these. Brain development in particular is most rapid in man towards the end of pregnancy and during the first two years postnatally, so that although drugs may induce anencephaly by an action in the first month of gestation, other drugs may cause mental retardation by an action during the latter months of pregnancy. As mentioned above for animal experiments, the sensitive period for transplacental cancer induction is the very end of pregnancy. Thus *all* stages of pregnancy must be regarded as potentially hazardous for the fetus and *not just the first trimester*.

Drug dosage

As described above for animal experiments, increasing doses of drugs during pregnancy may lead progressively to retardation of growth and functional development, malformation, death, and abortion. At any given dose, the effect will vary, not only with the stage of pregnancy, but also with the genotype of the individual and with other environmental (e.g. drugs or stress) factors operating at the same time. The majority of malformations are probably multi-functionally induced, and although with thalidomide almost every women exposed at the appropriate time had a malformed child irrespective of dose, one must expect that with most drugs, only a percentage of exposed women will have affected children and that this will be dose-related. This makes detection of teratogens difficult, but because of this as in all of therapeutics, the least effective dose of drug should be used.

Pharmacokinetics and placental transfer

Although the concept of a placental barrier is an old one which seemed justified by acute experiments in animals and in man, it is now clear that most substances which are free in the maternal plasma will cross to the fetus (Nau and Liddiard 1978). The extent of transfer will depend on molecular size, charge, lipid-solubility, and the other general factors governing the transfer of drugs across membranes. However, a very important factor is time, and in the situation where a mother is chronically dosed with a drug, there will be sufficient time for considerable drug transfer to occur into the fetus even for highly charged molecules like hexamethonium. This was well demonstrated by the neonatal mortality following hexamethomium use in the treatment of hypertension in late pregnancy, with resultant paralytic ileus in the fetus and neonate (Morris 1953).

Furthermore, some teratogens may act on the mother's endocrine or metabolic status or affect the utero-placental blood flow and transfer function and so cause abnormalities without even crossing the placenta (Sullivan 1972). Thus, with rare exceptions, all drugs should be considered as potentially hazardous despite any statements regarding placental transfer.

Should all drugs be avoided in pregnancy

Since malformations may be induced during the first two or three weeks of gestation when the women may not even be aware that she is pregnant, this question could more correctly be referred to all women of child-bearing age. In general, most studies on drug usage during pregnancy suggest that drugs are not a major cause of malformations in man. It must also be remembered that severe maternal disorders such as hyperemesis, hypertension, or epileptic fit, may be more dangerous for the fetus than the drugs used in their treatment. Thus a conscious choice should be made where the information exists to balance the advantages of drug therapy against the possible disadvantages. As in most things, extreme views either way may be equally bad.

Evidence on the safety or otherwise of specific drug classes

It is unfortunately true that there is very little evidence on the safety or otherwise of the majority of drugs in use today. This applies even to those drugs which have been in use for decades and this lack of evidence should not be taken to indicate safety in use. For example, stilboestrol has been in use for 40 years and has only recently been shown to present a carcinogenic and teratogenic hazard if used by pregnant women. On the other hand, anecdotal accounts of individual malformed children born to women who had taken a drug cannot indicate a hazard since 3–6 per cent of all pregnancies result in malformed or malfunctioning children, and coincidences with drug administration are common. Where good evidence exists, this is mentioned below, but failure to mention specific drugs must not be taken to imply safety, but merely lack of evidence one way or the other.

Alcohol

The fetal alcohol syndrome

During the last few years there have been an increasing number of reports of children born to alcoholic mothers with multiple defects known as the fetal alcohol syndrome (FAS). Although there is evidence dating back to biblical times and from Greek mythology indicating that alcohol can cause congenital malformations, it is only comparatively recently that attempts have been made to investigate this scientifically. Although Sullivan (1899) conducted the first studies on the effects of alcohol in pregnancy showing a 2.5-fold increase in infant mortality and stillbirths in the offspring of alcoholic mothers, it was not until the late-1960s that data describing a cluster of defects known as FAS was published (Lemoine *et al.* 1968). To date 245 cases of FAS have been reported (Streissguth *et al.* 1980*a*). FAS is characterized by four general categories of defects, namely, central nervous system (CNS) dysfunction, growth deficiencies, facial abnormalities, and a variety of major and minor structural malformations (Jones and Smith 1975; Streissguth *et al.* 1980*b*).

The term CNS dysfunction has been used to describe a whole range of defects which may involve intellectual impairment, structural malformations of the brain, and behavioural changes. Many of these children have microcephaly and abnormally shaped heads, sometimes associated with mild-to-moderate mental retardation. IQ studies in Germany, France, Sweden, and the USA show similar results with the average IQ of FAS children about 68, considered to be mild retardation, but the range is large (Streissguth *et al.* 1980*b*). Others have been described as having poor muscle tone and co-ordination, or as being irritable and hyperactive.

Growth deficiencies are often apparent at birth. These infants are lighter in weight and shorter in length than expected and often have a smaller head circumference. Although chronic alcoholics often suffer from malnutrition and other diseases it is unlikely that FAS is a result of malnutrition for several reasons. Many of the mothers are upper social class and well nourished. The length of the babies is affected more than the weight. Furthermore, the growth deficiencies persist into later life (Streissguth *et al.* 1978*a*), and the children do not seem to catch up either in body weight or stature with normal children of the same age. A common complaint

from parents of children diagnosed FAS is that their children are 'skinny' and they 'cannot fatten them up'; in other words, there is failure to thrive. No similar syndrome is seen in the infants of malnourished women, and moreover, similar effects can be produced in animals with alcohol alone.

The facial abnormalities are probably the most difficult to recognize from the point of view of the nonexpert. In general, the face is broad and flat with a short upturned nose. The eyes are often small and deeply set with short palpebral fissures. Occasionally there are more obvious structural malformations. There is also an altered spatial relationship between the eyes, nose, ears, and mouth. Micrognathia is often present and sometimes the ears are low set and abnormal in shape.

Some of the children have other defects of varying severity. Major defects of the heart, kidneys, and brain have been reported. Clarren *et al.* (1978) examined two brains from FAS infants and found leptomeningeal neuroglial heterotopias. Less severe defects involving the genital organs, club foot, dislocated hip, and cleft palate have been also reported. However, both the incidence and the range of these malformation is very variable.

FAS has been recognized in France, Germany, Sweden, Ireland, South Africa, Australia, and parts of the USA. In the Soviet Union, there have been reports of neurological impairment in children born to alcoholic mothers. Although the frequency of FAS is not known in Canada, the health authorities do acknowledge its existence and are attempting to collect and assess information concerning the use of alcohol in pregnancy. Similarly in the UK, the incidence of FAS is not known, in fact, we cannot say with certainty whether or not the syndrome exists in the UK. The annual alcohol consumption per capita for persons aged 15 years and over in the UK has increased from 5.2 l absolute alcohol in 1950 to 9.7 l absolute alcohol in 1976. Beer drinking has increased moderately, but the overall increase is due to wines and certain spirits. Furthermore, the trend is for young people and women to drink more (Faculty of Community Medicine 1981). As far as we are aware there are no retrospective or prospective studies in progress in the UK to examine the effects of alcohol in pregnancy. However, there are some physicians who do not believe that the syndrome is due to alcohol, but is caused probably by vitamin and nutritional deficiencies, smoking, and or hereditary factors. This may be so, but in studies where these other factors have been analysed, alcohol still comes out as the dominant factor (Rosett *et al.* 1976; Streissguth *et al.* 1978*a*; Kaminski *et al.* 1978; Little 1977).

In this situation, the development of an animal model, in which alcohol could be given under controlled conditions would be of value in determining the degree to which alcohol might be implicated as a teratogen. Although this approach seems simple enough, the technical difficulties which have emerged concerning the pharmacokinetics, route of administration, dose, and duration of administration have made this a somewhat formidable task. Experiments in rats, mice, rabbits, dogs, guinea pigs, and the monkey have shown various effects on the nutritional status of the dam and fetus, on the endocrine status, brain biochemistry, and behavioural changes in the offspring as well as structural malformations. Some of these effects are similar to those seen in the FAS; however, there is no completely satisfactory animal model which has all these features. One of the most interesting findings was the possibility that it was not alcohol *per se* which was the teratogen, but possibly one of its metabolites, acetaldehyde (O'Shea and Kaufman 1979). This could have important clinical implications especially for pregnant women who are undergoing alcohol aversion therapy with drugs such as disulfiram. Flynn *et al.* (1981) measured the zinc status of 25 pregnant alcoholic women and the same number of controls and they found that the manifestations of fetal morphogensis seemed to be related to the zinc status. Jameson (1976) had shown previously that zinc deficiency was associated with chronic alcoholism. In a series of 234 pregnancies, he found that those women with low zinc levels had low-birth-weight babies. Eight of these babies were malformed and each was from the lowest zinc-level group. Furthermore, Jameson stated that the defects in these infants showed similarities to those described in FAS. However, at present, because of the diverse biological and pharmacological effects of alcohol, its exact mode of action is difficult to define.

A question often asked is whether FAS only occurs in babies born to alcoholic mothers. Full FAS has invariably occurred only in children whose mothers were diagnosed alcoholic or who drank heavily during their pregnancy (Rosett *et al.* 1976; Kaminski *et al.* 1978). Attempts have been made to define what is meant by heavy drinking and it is now generally agreed that the chronic consumption of 89 ml pure ethanol per day (equivalent to six spirit drinks) constitutes a major risk to the fetus. Lower levels of consumption, i.e. what is often termed social drinking, or the less frequent use of alcohol, carries an unknown risk and is associated with less severely affected children. In a study of social drinkers which was controlled for other factors such as smoking, an incidence of this syndrome of 11 per cent has been reported for women consuming 25–50 g of alcohol per day, and 19 per cent for women consuming 50 g or more (Hanson *et al.* 1977, 1978). However, no absolutely safe level of alcohol has been established. Majewski *et al.*

(1976) concluded that the critical issue was not the exact amount of alcohol consumed but the chronicity of the alcoholism. Reports on both retrospective and prospective studies have confirmed the adverse effects of excessive alcohol consumption (Ouellette *et al.* 1977; Olegard *et al.* 1979) and demonstrated some effects on postnatal development at the slightly more moderate level of four drinks per day (? social drinking) (Streissguth *et al.* 1980*a*). Approximately 1 in 300 babies show some evidence of alcohol-induced damage, about half of which had full FAS (Olegard *et al.* 1979). There is inadequate evidence as yet to define the most sensitive period of pregnancy, and it would seem that the fetus is at risk at all stages of pregnancy. The elimination of alcohol in newborn infants is much slower than in adults and, in babies with FAS, alcohol withdrawal symptoms have been reported to occur up to 6–12 hours after delivery.

An interesting brief review by Rosett (1980) cautioned physicians against overstating the risk since less than 1 oz of alcohol per day has not been demonstrated to produce any defects. However, the brain pathology described by Clarren *et al.* (1978) and Peiffer *et al.* (1979) suggested that the mothers were not heavy drinkers and that the occasional session of binge-drinking might be as harmful as persistent heavy drinking. Induction of spontaneous abortion due to alcohol consumption in the second trimester has also been reported. Two studies showed that there is an increased risk, of about twofold, in women consuming 1 oz or more of alcohol daily. Furthermore, in one of the studies, there was still a significant increase in those consuming 1 ounce of alcohol twice per week or more. These effects were independent of age, race, parity, previous abortions, or smoking habits (Harlap and Shiono 1980; Kline *et al.* 1980).

As we do not know what amount of alcohol predisposes the fetus to abnormal development, it preempts the question will it help to change drinking habits once it is known that a woman is pregnant. Olegard *et al.* (1979) in a combined retrospective and prospective study in Sweden showed that, if mothers who were heavy drinkers stopped or limited their drinking to less than two drinks per day during early pregnancy, they had normal babies. On the other hand, if alcohol consumption was not reduced until the fifth or sixth month of pregnancy, the babies did not have full FAS, but they were lighter in weight and shorter in length than the reference population. In the group of women who continued to drink heavily during their pregnancy, their children had various features of FAS. It is not known whether it is persistent heavy drinking or the occasional binge-drinking during pregnancy which produces the most severe effects. In fact, the wide range of defects seen in the FAS may well be due to the different amounts of alcohol consumed at different stages of pregnancy. Little *et al.* (1980) examined three groups of 50 infants: those born to mothers who reported total abstinence during pregnancy, but had a history of prior alcoholism; those born to mothers who drank heavily throughout their pregnancy; and those born to nondrinking controls (mean daily intake < 0.1 ounce ethanol). There were significant differences in mean birth weights in those who drank heavily during pregnancy; their babies on average weighed 493 g less than the controls ($p < 0.001$). Those who had a history of alcoholism but did not drink during the pregnancy, had babies that weighed 258 g less than the controls ($p < 0.05$). No other factors, other than alcohol, could be linked directly to the decrease in birth weight. This suggests that a maternal history of alcoholism may predispose the fetus to additional risks even if the mother was sober during that particular pregnancy.

There have been several follow-up studies on children diagnosed as FAS. Olegard *et al.* (1979) showed that small size at birth was correlated with reduced mental performance later in life. Amongst the children examined, 58 per cent had IQ scores less than 85 and 19 per cent had scores less than 70. Furthermore, 8 per cent had cerebral palsy; this is an incidence of 1/5000, i.e. every sixth case of cerebral palsy is related to the alcohol consumption of the mother. Sokol *et al.* (1980) in a prospective cohort study of 12 127 pregnancies throughout a 52-month period showed that 204 (1.7 per cent) of pregnant mothers had abused alcohol. Although the authors fail to define 'alcohol abuse' they claim that when other factors such as age, parity, etc. were taken into account as many as 50 per cent of the infants of alcohol abusers were at risk because of alcohol abuse. There was a significant increase in babies with FAS (2.5 per cent) compared with none in the controls and an 11.9 per cent incidence of genitourinary abnormalities. However, neonatal deaths and perinatal mortality were not significantly different from the control population. They also reported an increase in intrauterine growth retardation of 1.8-fold associated with smoking alone, 2.4-fold with alcohol alone, and 3.9-fold with cigarette smoking and alcohol together. Shaywitz *et al.* (1980) examined 87 children, 6½–18½ years old, referred to the Learning Disorders Unit at Yale New Haven Hospital, for prenatal exposure to alcohol. Of these 15 of 87 were born to mothers who drank heavily in pregnancy. There were 11 males and four females in this group and all growth measurements were affected. These children had a continuum of dysmorphic features of FAS with an inverse relationship noted between the age at

presentation and the intensity of the dysmorphic features. What is particularly interesting about these children is that all had average intelligence, the IQ range being 82–113, yet they experience persistent academic failure, partially attributable to problems of hyperactivity and attention regulation. Of approximately 245 reports of children diagnosed as having FAS, 126 have undergone standardized IQ testing. The IQ scores in 85 per cent of these children were less than 70 (Streissguth *et al.* 1980b).

Sandor *et al.* (1981) examined 76 FAS children for cardiac malformations. The age range in these patients was birth to 18 years old. Of these 43 were males and 33 females of whom seven were Caucasian and the other 69 were North American Indian. There were 31 patients (41 per cent) with cardiac lesions of whom 15 (26 per cent) had ventricular septal defects. Another 12 patients (16 per cent) had functional murmurs. In addition there were six patients with variable cardiac defects. It is interesting that the incidence of cardiac defects is higher in the natives of British Columbia than in the rest of the population, which may help to explain the increase in cardiac defects in this study compared with other studies, although Olegard *et al.* (1979) did report an increase in cardiac defects. Bartolome *et al.* (1981) reported the results of experiments in rats exposed *in utero* to ethanol during the second half of pregnancy in which there was acceleration of maturation of cardiac sympathetic function in the offspring, an alteration which may lead to abnormal ECG and disturbed maturation of the myocardium. These authors suggest that this might be a factor in the abnormal ECG of babies born to alcoholic mothers.

Having described the potential risk to the fetus reported to be associated with *in utero* exposure to alcohol, we are left with the problem of whether or not this can be prevented. The developing brain seems to be the most susceptible organ as far as the effects of alcohol are concerned (Olegard *et al.* 1979). Sobriety even after a bout of heavy drinking does not always protect the developing infant from brain damage, nor from prenatal growth deficiency (Little *et al.* 1980). Sobriety may prevent the expression of some of the physical features of the FAS, and thus any mild mental retardation or brain dysfunction may not be observed until the child is several years old. Although two drinks per day have not been associated with malformations, some degree of brain damage cannot be entirely discounted. At present there is no known safe limit of alcohol consumption during pregnancy. Therefore, if FAS is to be prevented, it is important that women of child-bearing age should be educated about the risks of alcohol for the fetus so that, if necessary, they can adjust their drinking habits before they become pregnant rather than during their pregnancy.

Anaesthetics

General anaesthetics

General anaesthetics are one of the groups of drugs about which there is increasing evidence of reproductive hazard, not only for women but also for men. Studies from Russia, Denmark, USA and the UK have all suggested that female anaesthetists and operating-theatre personnel have higher abortion rates than expected. The American Society of Anesthesiologists after a study of 49 585 exposed operating-room staff compared with 23 911 unexposed controls (Ad Hoc Committee 1974) concluded that there was approximately a twofold increase in spontaneous abortions and in congenital malformations in exposed females, and a 25 per cent increase in congenital malformations in the offspring of exposed males. There was a similar 1.3–2.0-fold increase in the incidence of cancer in exposed females. Subsequent studies have confirmed these observations. Pharoah *et al.* (1977) in a survey of 5700 women doctors in the UK found that conception which occurred when the mother was in anaesthetic practice resulted in smaller babies, higher stillbirth rates, and more congenital malformations of the cardiovascular system than pregnancies of other women doctors. Corbett *et al.* (1974) in the USA also found a higher incidence of congenital malformations in the children of nurse anaesthetists. They also found a threefold increase in the incidence of cancer in these nurses than in controls and a slight but not significant increase in cancer of the offspring. The possible transplacental carcinogenic effects of anaesthetics were discussed by Corbett (1976) who demonstrated the possibility of such an effect in animals exposed to isoflurane. At present there is insufficient evidence to come to any conclusions regarding such hazards in humans, but the chemical similarity of the known human carcinogen vinyl chloride to anaesthetics like trichlorethylene would suggest that such information is badly needed (Fig. 22.2).

Possible effects on male anaesthetists have also been suggested by the observation of decreased fertility and an increase in spontaneous abortion in the wives of male anaesthetists (Askrog and Harvald 1972). More carefully controlled studies, however, have not confirmed these observations (Knill-Jones *et al.* 1972), though these latter studies did find a significant increase (of about 30 per cent) in the incidence of 'minor' congenital malformations in the children of exposed fathers. They also found a twofold increase in the rate of spontaneous abortion if

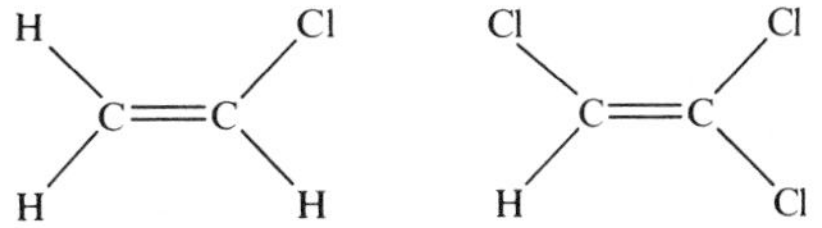

Vinyl chloride   Trichloroethylene

Fig. 22.2

the wife was exposed to anaesthetics.

Further evidence on the hazards of anaesthetic practice has been provided by a retrospective survey of all anaesthetists working in the West Midlands area of England (Tomlin 1979). By means of a questionnaire and follow-up visits Tomlin obtained a 92.4 per cent response from 340 anaesthetists, including 83 women, working in the region. He had a small control group of 60 anaesthetists who had either had a family before starting practice or had been away from anaesthetic practice for more than one year, as well as comparison with other control groups from the published literature. He demonstrated an increased risk of abortion, especially in exposed females, with an overall rate of 18.2 per cent compared with about 10 per cent in controls. Malformations, especially of the CNS and musculoskeletal system, were also about twice as high (9.4 per cent) as expected and with girl babies being significantly ($p < 0.05$) more affected than boys. The author, an obvious enthusiast for his cause, was criticized by a number of other workers (see Vowles *et al.* 1979) for choice of control data and the lack of rigid statistical criteria in analysis of his data. However, his survey did show that 40 per cent of anaesthetizing areas still had no ventilation and in many others it was extremely primitive, despite the warnings and recommendations by the DHSS for reduction of pollution in operating areas.

A report of a survey based on registry data of infants born to women working in operating theatres in Sweden compared with those working in any type of medical work has been published (Ericson and Källen 1979). They found no difference in birth weight, perinatal deaths, or malformation rates. Because of the brevity of the report it is difficult to assess its validity, but it is surprising that no serious malformations at all were observed in 226 pregnancies in operating-theatre personnel.

The actual cause of the adverse effects on pregnancy in operating-theatre personnel has not been determined. Animal studies have shown that both inhalation and injectable anaesthetics may cause congenital malformations, fetal death, and mutagenic effects, and it has also been suggested that metabolites or breakdown products of anaesthetics may be responsible. However, the cause of the effects observed in humans may be related more to the stress of working in an operating-theatre environment than to any individual anaesthetic. However, health authorities both in the USA and in the UK have issued warnings about the hazards to pregnant women of working with anaesthetics and have suggested that pollution in theatres should be reduced.

Very little information exists as to the hazard for patients undergoing operative procedures involving anaesthesia. It is not known if any particular anaesthetic or stage of pregnancy is especially hazardous. It has been suggested, however, that women of child-bearing age should have elective procedures done during the first half of the menstrual cycle to avoid the risk of accidental exposure of the early embryo to anaesthetics at a stage when the woman may not be aware that she is pregnant.

Local anaesthetics

Although not of teratological interest, the increasing use of local anaesthetics to produce regional analgesia during delivery has resulted in a number of publications on the placental transfer and fetal toxicity of these compounds. As would be expected local anaesthetics cross the placenta easily and may produce pharmacological and toxic effects on the fetus at plasma levels similar or rather less than those causing effects in the mother (Pedersen *et al.* 1978). There is good evidence that the fetus and neonate can metabolize local anaesthetics and the high levels in the neonate following epidural anaesthesia are slowly metabolized and excreted in the urine in the first few days of life (Kuhnert *et al.* 1979). Among the toxic effects reported following local analgesia in labour are fetal bradycardia, respiratory difficulties, seizures, and hyperirritability (Hill and Stern 1979).

A large co-operative study carried out at St. Mary's Hospital, London compared the effects on the fetus and newborn of intramuscular pethidine, epidural bupivacaine, or no drugs (Lieberman *et al.* 1979). These were studied prospectively in a group of 145 normal healthy uncomplicated pregnancies. No direct effects were observed on the fetus, but 12 of 51 infants in the pethidine group were given oxygen and three required intermittent positive-pressure respiration (IPPR); six of 59 in the bupivacaine group were given oxygen, and two required IPPR compared with none needing resuscitation in the 35 untreated controls. The one-minute Apgar scores were significantly depressed in the pethidine group but were normal at 5 minutes. Measurement of blood levels of the drugs showed that they freely crossed the placenta and were only slowly eliminated by the newborns with half-lives of pethidine and bupivacaine of 22.7 h and 14.0 h compared with 3.2 h and 1.25 h for

adults. Minimal effects on the subsequent behaviour of the neonates were observed but the authors stress that these conclusions only apply to healthy uncomplicated deliveries. In fact, in two subsequent papers by the same authors (unpublished, personal communication) the behaviour of the infants was further analysed by the Brazelton Neonatal Behavioural Assessment Scale during the first six weeks of life, in relation to the blood levels and total exposure to pethidine and bupivacaine at delivery. Clear drug-related effects on alertness, attention, social interactions, and numerous subtle effects were observed persisting throughout the six-week period, especially in the bupivacaine group. Obviously, other long-term follow-up studies with these and other local anaesthetics are required.

Anti-asthmatics

The effects of asthma on pregnancy and vice versa have been extensively reviewed and there seems to be a small increase in morbidity for both the mother and fetus with some reduction in birth weight. Whether the effects on the fetus are due to the disease state of the mother with hypoxaemia or due to the drugs used cannot be defined. However, the majority of authors feel that drug therapy for asthma need not be altered due to pregnancy (Greenberger and Patterson 1978; Weinstein *et al.* 1979; De Swiet 1979; Turner *et al.* 1980; Hernandez *et al.* 1980; Pratt 1981).

Sympathomimetics

These drugs have been used extensively in late pregnancy to suppress uterine activity in premature onset of labour. Buphenine, feneterol, hexoprenaline, isoprenaline, isoxuprine, orciprenaline, ritodrine, salbutamol, and terbutaline have all been used with varying degrees of success. The ratio of cardiovascular stimulation to uterine relaxation, i.e. their relative $\beta_1 : \beta_2$ activity is an important factor since increased heart rate, tremor, and anxiety are limiting factors in their use. In a comparison of isoprenaline, orciprenaline, ritodrine, and terbutaline it was shown that isoprenaline had the worst therapeutic ratio as a uterine relaxant and terbutaline the best. The use and safety of these drugs in late pregnancy was well reviewed by Crowley (1983) but is not relevant to their potential for teratogenicity when used for asthma treatment in early pregnancy.

Adrenaline

There is little information on the safety of adrenaline in

pregnancy, but in the Boston Collaborative Project (Heinonen *et al.* 1977*b*) a small but significant increase in malformations was observed in a group of 189 babies whose mothers had been exposed to adrenaline in early pregnancy. Since adrenaline tends to be used only in severe asthma attacks this may as easily reflect an effect of hypoxaemia as a direct action of the drug.

Isoprenaline

There is very little information on the safety of isoprenaline in human pregnancy. It is not teratogenic if administered to animals in early pregnancy. No significant increase in malformations was observed in the Heinonen *et al.* (1977*b*) study, but the number of exposed mothers was too small to be of significance.

Ephedrine

No increase in malformations was observed in the Heinonen *et al.* (1977*b*) study in 373 babies whose mothers had been exposed to ephedrine in early pregnancy.

Phenylephrine

No increase in malformations was observed in the Heinonen *et al* (1977*b*) study in 1249 babies whose mothers were exposed to phenylephrine in early pregnancy.

Phenylpropanolamine

A small but significant increase in malformations was observed in the Heinonen *et al.* (1977*b*) study in 726 babies whose mothers had been exposed to phenylpropanolamine in early pregnancy.

Salbutamol, terbutaline

There is no evidence about the safety of these drugs in early pregnancy in humans, though they do not cause malformations if given to animals. They have both been extensively used in late pregnancy to cause uterine relaxation without any important adverse effect on the fetus.

Xanthines

Theophylline and aminophylline are both safe for use during pregnancy. Heinonen *et al.* (1977*b*) found no increase in malformations in 117 babies whose mothers took theophylline or in 76 babies whose mothers took aminophylline in early pregnancy. The wide clinical use of these substances in pregnancy without reports of malformations supports their safety for use. The extensive data on the safety of caffeine is reviewed separately in this chapter.

Corticosteroids

The systemic use of steroids is referred to elsewhere in this chapter. In general it is felt that the use of steroids when necessary for asthma treatment in pregnancy does not carry any excessive risk for the fetus (Turner *et al.* 1980). Where use has been prolonged, however, the mother may need additional support during labour. Often the use of systemic steroids can be reduced or even replaced by topical steroid aerosols which are probably safer.

Beclomethasone dipropionate

The topical use of steroids by inhalation is widely practised now in the treatment of asthma. Very little of the drug is absorbed so that adrenal suppression is minimal and the chance of an effect on the fetus is very unlikely. In one study of 45 pregnancies in which beclomethasone aerosol was used 4–16 times per day (average dose 336 μg) along with prednisone when necessary, no increase in malformations was observed (Greenberger and Patterson 1983). This supports the earlier observation of Morrow-Brown *et al.* (1977) who reported 20 normal babies following beclomethasone use in 20 pregnancies.

Betamethasone valerate

There is no specific information on the use of this drug in pregnancy but it is unlikely to differ from beclomethasone above.

Sodium cromoglycate *(Intal)*

This drug is virtually non-toxic when given by inhalation. Less than 10 per cent of a dose is absorbed into the circulation. The drug is not teratogenic in animals even after high parenteral doses. In a review of cromoglycate, Kingsley and Cox (1978) state without any details or reference that it has been used in 150 pregnancies in the first trimester without any obvious increase in birth abnormalities. Wilson (1982) reported on 10 year's use of cromoglycate in Sri Lanka in 296 women during the whole of pregnancy. He found four diverse malformations only (1.35 per cent) in the total births. Although there were no control data available, this would not suggest any significant teratogenic risk from cromoglycate.

Adrenocorticosteroids

The majority of the synthetic and naturally occurring corticosteroids are teratogenic in animals. The commonest defect produced is cleft palate in mice and rabbits, but the incidence is very dependent on the genetic constitution, with 0-100 per cent of offspring being affected. The evidence in man suggests that corticosteroids have a relatively low teratogenic potential though there may be a higher-than-normal incidence of stillbirths. Although there are many individual case-reports linking congenital malformations to the use of corticosteroids in pregnancy, several large surveys on well over 1000 women have not confirmed these (Schardein 1985). On the other hand, although the incidence of congenital malformations may not be increased, in one series of 34 pregnancies in which the mothers were given prednisolone during pregnancy there were eight stillbirths (two with anencephaly), nine babies with placental insufficiency requiring special care, and one respiratory distress syndrome; i.e. a total of 18 of 34 babies dead or at considerable risk compared with four of 34 control babies from mothers with similar general diseases but without steroid treatment (Warrell and Taylor 1968). Occasional cases of adrenal insufficiency have also been reported in infants, but no sign of adverse effects on the reproductive organs of male infants (Zondek and Zondek 1974).

Anticoagulants

There is an increasing body of evidence that the coumarin-type of oral anticoagulants, especially warfarin, are teratogenic in man. The warfarin embryopathy which is a syndrome consisting of nasal hypoplasia, stippled epiphyses sometimes associated with optic atrophy, mental retardation, and various other defects has been described by several workers (Pettifor and Benson 1975; Warkany 1975; Pauli and Hall 1979; Hall *et al.* 1980).

Until quite recently it was thought that heparin, which, because of its molecular size, hardly crosses the placenta at all, might be a safe alternative. However, reviews by Pauli and Hall (1979) and Hall *et al.* (1980) indicated that this may not be true. Hall *et al.* (1980) collected data (1945–78) on 423 pregnancies in which coumarin derivatives were used and a further 135 published reports on the use of heparin during pregnancy.

In pregnancies where oral anticoagulants were used they stated that, as well as the haemorrhagic complications associated with coumarins, only two-thirds of the pregnancies will result in an apparently normal live-born infant. One in six will have either CNS defects and/or the warfarin embryopathy and one in six will end in abortion or stillbirth. The results with heparin use are similar (see below).

The warfarin embryopathy seems to be associated with exposure to coumarins during the first trimester (weeks 6–9).

CNS abnormalities and eye defects were originally thought to be part of the warfarin embryopathy, but it now seems more likely that they result from exposure to coumarins during the second and/or third trimester.

Hall *et al.* (1980) reviewed in detail data on 24 children with the warfarin embryopathy. Follow-up of 17 of these children indicated that some did exhibit catch-up growth both in weight and height. However, the nose may show no apparent catch-up growth and in severely affected cases it often remains small and sunken into the face. Of the 24 children who were assessed, five died (21 per cent), four (17 per cent) had scoliosis, three (12 per cent) were blind and had developmental delays, three (12 per cent) were deaf, two (8 per cent) had congenital heart defects, and one (4 per cent) had seizures. Overall, 12 of the 24 had no serious debility. Although the calcific stippling which is incorporated into the epiphyses does not seem to cause asymmetric growth, it may account for the scoliosis observed in four of the children.

Because of the diversity of tests used to assess development only an estimate of overall development is reported. Adequate information has been collected on 16 children, five of whom (31 per cent) showed significantly retarded development, three being mildly retarded; the other two were moderately-to-severely retarded. All of the children who were mentally retarded were exposed to coumarins in the last two trimesters as well as in the first trimester.

Thirteen cases of CNS abnormalities were described of which five also had the warfarin embryopathy (exposed in first trimester), six had structural malformations, seven had eye anomalies, and two had encephaloceles. No consistent pattern of CNS malformation was observed and there was no correlation between the type of abnormality produced and the time of exposure to coumarins, other than exposure in the second or third trimester. Sequelae from CNS abnormalities have been more debilitating than those of the warfarin embryopathy. None of these children were normal at follow-up; all were mentally retarded and 50 per cent of them were blind. In addition four (31 per cent) were spastic, three (23 per cent) had seizures, one (8 per cent) was deaf, one (8 per cent) had scoliosis, one (8 per cent) exhibited growth failure, and three (23 per cent) of the children died.

Hall and co-workers have proposed a mechanism of action for the production of the warfarin embryopathy. They think that a molecular mechanism similar to the effect of coumarin derivatives on vitamin K-dependent clotting factors may be involved. Coumarin derivatives inhibit vitamin K-dependent post-transitional carboxylation of coagulant proteins. The special proteins or osteo-calcins which may well control calcification during embryogenesis are also vitamin K-dependent and therefore, if coumarins inhibit such proteins during a critical period of ossification, this could result in the occurrence of stippled calcification, nasal hypoplasia, vertebral abnormalities, and shortening of extremities.

Similar data were described in a review by Stevenson et al. (1980) in which they described a child with the warfarin embryopathy, bilateral optic atrophy plus other defects; these have also been described by Baillie et al. (1980).

Kort and Cassel (1981) reported on 40 patients with cardiac disease who received warfarin during pregnancy. There was a high fetal mortality rate of 12.5 per cent; five fetuses died in utero; and three were macerated when detected at 38–40 weeks gestation. Congenital abnormalities were present in three of 36 live-births; one bilateral congenital dislocation of the hip, one cleft palate and harelip, and one meningomyelocele. The mothers of these three infants had received warfarin in the first trimester. It is interesting that in this study no cases of congenital epiphyseal stippling were observed.

Heparin was thought to confer a theoretical advantage over the coumarins, because it did not cross the placenta. However, from the data presented on 135 pregnancies in which herapin was used, this would not seem to be correct (Hall et al. 1980). There is a high maternal complication rate, 14 of 135 (10 per cent) had severe haemorrhage and three (2 per cent) died. Whether this is related to the condition for which heparin is given is not certain. But, even excluding the maternal morbidity and mortality, only two-thirds of the pregnancies will result in a normal outcome, one in eight pregnancies will end in a stillbirth and one in five pregnancies will end in premature delivery. Thus the overall picture when heparin is used during pregnancy is little better than that seen for the coumarin derivatives. A comparison of the effects of heparin and coumarin on fetal outcome is summarized in Table 22.1.

It is unknown whether heparin carries an increased risk at any particular stage of pregnancy. There are no reports of studies on the mechanism of action of heparin in pregnancy. However, Hall and co-workers (1980) suggested that, as heparin does not cross the placenta but

Table 22.1 Summary of the effects on in utero exposure to anticoagulant drugs*

Effects on the fetus	Warfarin (total n = 418)	Heparin (total n = 135)
Live-born — normal	293	86
Live-born — with problems	57	30
Haemorrhage	7	0
Warfarin embryopathy	11	0
CNS abnormalities	6	0
CNS abnormalities + warfarin embryopathy	5	0
Premature, but normal	8	19
Premature, died	4	10
Other problems (unspecified)	12	1
Spontaneous abortion	36	2
Stillborn — no haemorrhage reported	25	17
Stillborn — with haemorrhage	7	0

* Compiled from data presented by Hall et al. (1980).

is a strong chelating agent, it may act indirectly by depleting calcium or other cations in the mother and subsequently in the developing fetus. Hall and co-workers recommended pre-conceptional counselling and in most cases the prevention of pregnancy. Furthermore, if a woman should become pregnant or require anticoagulation during pregnancy, they feel that therapeutic abortion should be offered.

However, these views are not supported by other workers, and the use of heparin is recommended as the drug of choice in pregnancy (Cohlan 1980; Globus 1980; Walters and Humphrey 1980). This decision seems to be based on the theory that heparin does not cross the placenta and therefore cannot produce fetal side-effects, a theory which cannot be accepted. Although theoretically heparin could be considered as the drug of choice for pregnant women because it has not been associated with fetal anticoagulation or an increase in the incidence of congenital malformations, nevertheless, it has not been shown to be a safe alternative because of the high incidence of maternal complications and fetal and neonatal losses. It was also found that the anticoagulant effect of heparin is markedly decreased in late pregnancy, thus leading to an increase in heparin dose requirements (Whitfield et al. 1983).

Thus it would seem that anticoagulants should be given during pregnancy only if absolutely necessary. Both coumarin derivatives and heparin have risks associated with their use. Most current data indicates that only two-thirds of the babies will be normal if exposed to either drug, and the use of heparin carries a greater risk for the mother.

Antidepressants

These drugs have little teratogenic activity in animals though malformations have been produced both with the monoamine oxidase inhibitors and the tricyclics.

Tricyclic antidepressants

The only drug of this group which has been studied in any detail in pregnancy is imipramine. McBride (1972) first reported three cases of limb reduction deformities related to imipramine and amitryptiline usage. However, other workers in studies covering about 697 pregnancies have failed to confirm such an association (Kuenssberg and Knox 1973; Crombie *et al.* 1970; Idänpään-Heikkila and Saxen 1973; Wilson and Fraser 1977). Although the data suggest there is no association of tricyclics and congenital malformations, the effects observed in the neonate are more readily acknowledged. The neonate metabolizes the tricyclic antidepressants very slowly and both anticholinergic and sympathomimetic effects such as irritability, muscle tremors, and tachycardia have been observed.

Tetracyclic antidepressants

There is no evidence either way as regards human teratogenicity for the newer antidepressants such as maprotiline. Their safety for use in human pregnancy has yet to be established. Although there is no published information on mianserin in human pregnancy, the manufacturers reported that when used in very high doses in animals it was not teratogenic. Furthermore, they know of five pregnancies in which therapeutic doses of mianserin were taken and all the babies were normal (personal communication, Organon).

Monoamine oxidase inhibitors

There is no evidence either way to implicate this group of drugs, e.g. tranylcypromine, isocarboxazid, phenelzine, as teratogens in human pregnancy although phenelzine has been reported to cause malformations in some species of animals. Kullander and Källen (1976a) reported on a prospective study on the use of psycho-

pharmaca (actual drugs used not specified but they comprised barbiturates, tranquillizers, and anti-depressants) during 6376 pregnancies and the correlation with certain socio-medical variables and pregnancy outcome. An association between drug use and later miscarriage or legal abortion was demonstrated which may reflect an increased use of these drugs in unwanted pregnancies. No association between use of these drugs and congenital malformations or stillbirths was found. Although pregnancies in which these drugs were used during the first trimester tended to be of longer duration, there were no effects on the birth weights of full-term infants.

In conclusion, based on the negative data associating the tricyclic antidepressants with congenital malformations in human pregnancies, it would seem that the risk to the fetus, if any would be small. The monoamine oxidase inhibitors, because of their inherent toxicity should be used only when absolutely necessary.

Anti-emetics and antihistamines

Meclozine, cyclizine, and chlorcyclizine

It is convenient to consider these groups of drugs together as the same substances may be used for both purposes. Obviously the anti-emetics are a popular group of drugs for use in early pregnancy to combat early-morning sickness. Shortly after the effects of thalidomide were discovered, the anti-emetic meclozine either alone or in combination with pyridoxine as *Ancoloxin* was suspected of being teratogenic in man. Suspicion likewise fell on other piperazine antihistamines like cyclizine and chlorcyclizine which were shown to be teratogenic in animals. A number of large prospective and retrospective studies failed to confirm this suspicion (Stalsberg 1963; Yerushalmy and Milkovitch 1965). A committee was set up by the FDA in America to review the evidence up to 1965 (Sadusk and Palmisano 1965). They concluded, after reviewing a very large amount of evidence on over 30 000 subjects, that there was still not enough data to rule out unequivocally a cause-and-effect relationship in a very small fraction of human congenital defects. Taking into account the doubtful evidence of efficacy of the drugs in any case, they recommended that a warning should be put on the labels of the drugs advising against their use in pregnancy without medical advice. Lenz (1966) reviewed the pooled evidence of 15 studies and found that 12 children of a total of 3333 mothers who had taken meclozine during the first trimester of pregnancy had cleft lip and/or cleft palate which he stated to be two to three times the expected number. Since similar facial defects had been produced in animal studies, he felt that the drug must be suspected of a slight teratogenic effect. McBride (1969) in a survey of women who had had babies with cleft palate found that a higher-than-expected number had taken cyclizine, but he also found a higher (and similar) incidence of cleft palate in women with nausea and vomiting in pregnancy who did not have treatment. He suggested that the defects may be related to some underlying pathology of the pregnancy rather than to specific drugs. No association of congenital malformations with meclozine treatment was found in the large retrospective survey of Nelson and Forfar (1971) or with cyclizine treatment in a Finnish study (Saxen 1974). In an analysis of the data from the Boston Collaborative study in which 1014 women took meclozine during the first four months of pregnancy, no overall increase in malformations was observed though there was a small but significant (threefold) increase in ocular defects (Shapiro et al. 1978).

Dimenhydrinate, diphenhydramine

No evidence of a teratogenic effect of dimenhydrinate was found in one study of 327 pregnancies (McColl 1963). Diphenhydramine was associated with a higher than expected incidence of facial clefts in the large Finnish study mentioned above (Saxen 1974).

Phenothiazines

The use of phenothiazines was not associated with an increased risk of congenital malformations, perinatal mortality, low birth weight, or poor postnatal development when assessed in 1309 women who took these drugs during the first four months of pregnancy. Of these, 403 were heavily exposed (Slone et al. 1977). Analysis for effects of individual drugs was not however performed. Ananth (1975) in a review of the published studies on the phenothiazines, including chlorpromazine and trifluoperazine, concluded that there was no evidence that they were teratogenic. However, Kullander and Källen (1976b) in a prospective study of anti-emetics in 6376 pregnancies in Sweden found a positive association between the use of anti-emetics and congenital malformations. This was primarily due to a threefold higher than expected incidence of congenital dislocation of the hip in users of drugs, particularly promethazine.

Debendox

One of the antinauseants, Debendox (Bendectin, Merbental, Lenotan) has been the subject of much controversy as regards its safety for use in pregnancy. As a consequence of this, it has now been withdrawn from the market by the manufacturers (June 1983) because of media pressure and the high cost of liability insurance. However, drug regulatory authorities both in the USA and UK have reiterated their assurances that Debendox is not a teratogen in man.

Debendox was first marketed in 1956 as a three-ingredient tablet containing dicyclomine hydrochloride, doxylamine succinate, and pyridoxine. In 1976, it was reformulated in the USA and marketed without dicyclomine hydrochloride. Some other countries adopted the reformulated tablet, but in the UK until late-1982 it was still manufactured as the original formula containing the three substances. Although Debendox was at one time available over the counter in the UK, the majority of women obtained it on prescription.

Two large surveys demonstrated the safety of a combination of dicyclomine hydrochloride and doxylamine succinate (the major components of Bendectin, Debendox, Merbental, and Lenotan) in pregnancy, Smithells and Sheppard (1978) found no increase in congenital malformations in 2298 patients taking Bendectin during pregnancy when compared with a control population. Furthermore, there was no evidence of any difference in malformation rates of these offspring where the mother had taken the drug during the first 10 weeks of pregnancy (1372 cases) compared with those taking the drug in later pregnancy. This lack of adverse effect is supported by analysis of the Boston Collaborative Perinatal Project data by Shapiro et al. (1977) for doxylamine and dicyclomine separately. No increase in congenital malformations perinatal mortality, or reduced birth weight was observed nor was there any effect on IQ at 4 years of age of about 900 exposed offspring.

Following the recent legal case in the USA concerning the possible teratogenicity of Bendectin (Debendox) there has been much activity reassessing the available data on this drug. A report (Journal of the American Medical Association 1979) from the Center for Disease Control, Atlanta reviewed their data on 1183 malformed children and found no evidence that Bendectin increased the incidence of limb, cardiac, CNS, or palatal defects. Reviewing the data from the Boston Collaborative Perinatal Project in which 5401 of 50 282 women took an antinauseant or antihistamine in the first four months of pregnancy, the evidence suggests that Bendectin was in fact the safest of the antinauseants for which substantial data existed (Heinonen et al. 1977b).

Several other retrospective studies (Harron et al. 1980; Jick et al. 1981a; Mitchell et al. 1981; Cordero et al. 1981; Correy and Newman 1981; Clarke and Clayton 1981) and prospective studies (Gibson *et al.* 1981; Fleming *et al.* 1981; Morelock *et al.* 1982) were also published indicating that *Debendox* does not significantly increase the incidence of congenital malformations. However, several research groups have indicated that there may be a slight increase in specific types of malformations and have expressed concern that the methods of analysis used in these surveys may not be sensitive enough to detect small increases in the incidence of malformations. Jick *et al.* (1981*a*) in Seattle reported no overall increase in malformations in a large population of women receiving *Bendectin*. However, he found a weak association between *Bendectin* use (two-ingredient tablet) and gastrointestinal atresia. The increased rate, however, was not significant, and there was no relationship with the number of prescriptions issued, so a causal relationship could not be established. Similarly, Cordero *et al.* (1981) in Atlanta published data indicating no overall increase in major malformations but did find an association between *Bendectin* use and amniotic bands, a subgroup of limb reduction deformities. Furthermore, they found a small increased risk of encephalocele with the two-ingredient tablet but not the three-ingredient tablet, and of oesophageal atresia with the three-ingredient tablet only. However, the authors do not consider these results to constitute a substantial risk, and, if the risk is causally related to *Bendectin*, it would be extremely low, totalling 1.5 per 1000 for the three defects mentioned. (See also MacMahon 1981.)

In the prospective study by Gibson *et al.* (1981) in Adelaide comparing 1817 *Debendox* users with 5771 non-users there was no evidence that *Debendox* increased either the overall malformation rate or the incidence of skeletal and cardiovascular defects. However, there were two unexpected findings, namely, a small increase in genital-tract anomalies in male babies and there was a synergistic effect of tobacco smoking and the use of *Debendox* in early pregnancy. The authors found these results difficult to explain. In addition to a small number of case reports of malformations associated with *Bendectin* and *Debendox* (Patterson 1977; Donnai and Harris 1978), there have been two studies associating the use of *Debendox* with congenital heart defects (Rothman *et al.* 1979) and cleft lip with or without cleft palate (Golding and Baldwin 1980).

It was also suggested that some of the malformations observed in the babies of women who took *Debendox* resemble those of the Poland anomaly (Mellor 1978). However, this was disputed by David (1982) who examined 78 cases of Poland anomaly and found no association with the use of *Debendox*.

It is estimated that over 30 million pregnant women have taken *Bendectin/Debendox* and there seems to be no increase in the overall malformation rate. Nevertheless, despite these reassurances, there is still public concern about its safety and parents have formed action groups to investigate matters further. In the USA where about 25 per cent of pregnant women took *Bendectin*, the FDA reviewed the evidence regarding its safety in 1980 (Kolata 1980). Experts studied data from both animal and epidemiological studies from which they concluded that there was no demonstrable relationship between the drug and birth defects. Nevertheless, they did recognize that, because of the impossibility of providing absolute safety, there was a 'residual uncertainty' about its safety in pregnancy. Despite these findings, law suits are still being filed against the manufacturers.

Because recent publicity has indicated that there is new evidence, based on animal studies concerning the safety of *Bendectin* in pregnancy, the FDA have agreed to examine this data and reassess the situation (Kolata 1982). There is one unpublished rat study from West Berlin, showing a slight increase in diaphragmatic hernia when very high doses of *Bendectin* were used, but this was not a dose-related effect. A summary statement of the raw data has been sent to the FDA, and both Roll and the FDA are to repeat this study. There are also two unpublished primate studies, one by Hendrickx, from the Primate Research Center in California, the other by McClure from the Yerkes primate research centre.

Hendrickx treated 12 cynomologus monkeys with 10–12 times the human dose of *Bendectin* throughout organogenesis. Two of the monkeys aborted, and three went to term and produced normal offspring. The other seven monkeys were delivered two months before term and, of these, four offspring had intraventricular septal defects. However, due to lack of control data the significance of this is uncertain since time of septal closure is not precisely known. However, because of the concern that *Debendox* might increase the incidence of CHD in humans (Rothman *et al.* 1979), further investigations have been initiated. A 2½-year double-blind study is now in progress in which four groups of monkeys (20 per group) are being treated with three different doses of *Bendectin*. The study of McClure in nine rhesus monkeys, showed that *Bendectin* did not cause malformations in the offspring.

As a result of these studies, the FDA requested that researchers involved in epidemiological studies, look for an association between *Bendectin*, diaphragmatic hernias, and cardiac defects. So far, none has been reported. Although *Bendectin* has been withdrawn, it is still very much under investigation.

Conclusion

There seems no doubt that anti-emetic drugs have been vastly overprescribed, e.g. in 1978 in the USA there were 3 million prescriptions for *Bendectin*, i.e. one for every pregnant woman, but where severe nausea and vomiting exists the use of anti-emetics is justified.

Because of the difficulty of excluding a small risk of teratogenicity when any drug is used in pregnancy, it is recommended that before antinauseants are used, dietary measures should be tried as far as possible. However, if treatment with an antinauseant is necessary, then, in the absence of *Debendox* which is no longer available, the best choice may be *Ancoloxin* or a phenothiazine other than promethazine.

Anti-epileptic drugs

In the last 20 years there have been a number of epidemiological surveys in various parts of the world reporting an increase in the incidence of congenital malformations in the children born to epileptic women, though it is difficult to determine whether it is the disease itself or the drugs used in its treatment which are responsible for the increased malformation rate. However, in most of the surveys where it has been measured the incidence of malformations is higher in epileptics receiving drug treatment during the relevant pregnancy than in those receiving no drug treatment. Another important factor to be considered is the involvement of folate. It has been suggested that since anticonvulsant drugs lower serum folate levels it may be the folate deficiency which is responsible for the malformations observed.

Earlier studies of anti-epileptic drugs (1960–80)

Evidence implicating anticonvulsant drugs

Janz and Fuchs (1964) reported that in the Netherlands where cleft lip with or without cleft palate constitutes 10 per cent of all malformations there was a fivefold increase in this defect in babies born to epileptic women. This was followed by a report (Elshove and Van Eck 1971) of an eightfold increase in all congenital malformations in babies exposed to anticonvulsant drugs *in utero* among these the incidence of cleft lip with or without cleft palate was 29 times higher than was expected. In England, Meadow (1968, 1970) also observed an increased malformation rate in the babies of epileptic women, and launched an appeal to physicians to report any similar observations. In the USA, German *et*

al. (1970) suggested that the oxazolidines paramethadione and trimethadione were responsible for the malformations and retarded intra-uterine growth which they had observed in babies born to mothers suffering from *petit mal.* However, the authors did not observe an overall increase in the incidence of congenital malformations in babies born to epileptic mothers in New York.

Since then several surveys on congenital malformations and epilepsy have been published. Speidel and Meadow (1972) reported the results of a survey in Leeds on the outcome of 427 pregnancies in 186 epileptic women which showed a two- to threefold increase in the incidence of congenital malformations. There was also a twofold increase in perinatal mortality and a seven- to eightfold increase in the number of mentally subnormal children. The drugs implicated in this survey were phenobarbitone, phenytoin, primidone, troxidone, pheneturide, methylphenobarbitone, and diazepam. In a small study at St. Thomas's Hospital Maternity Unit, London (South 1972) the incidence of cleft lip with or without cleft palate was shown to be 70 times greater in a group of 31 women taking anti-epileptic drugs compared with those not receiving drug treatment.

A seven-year survey of congenitally malformed babies in Cardiff (Lowe 1973) showed that the malformation rate among the total birth population was 2.7 per cent, whereas in epileptic mothers treated with anticonvulsants during the first trimester, the malformation rate was 6.7 per cent. Millar and Nevin (1973) carried out a similar study in Northern Ireland in which they observed a sevenfold increase in the incidence of cleft lip with or without cleft palate. This correlated well with the findings of the US Air Force Hospital study 1965–71 (Niswander and Wertelecki 1973) in which there was a five- to sixfold increase in cleft lip with or without cleft palate and a twofold increase in cardiac defects and club foot. Specific skeletal abnormalities have also been observed in babies born to mothers who had taken phenytoin during pregnancy (Loughnan *et al.* 1973). These abnormalities included hypoplasia and irregular ossification of the phalanges which produces short, narrow, and misshapen ends to the fingers and toes. In most cases the fingers were more severely affected than the toes. Similar findings associating phenytoin with skeletal defects were reported by Barry and Danks (1974) in Australia. They observed seven children within a year with hypoplasia of the digits, cleft lip with or without cleft palate, and diaphragmatic hernia associated with the maternal use of phenytoin.

Fedrick (1973) published a survey by the Oxford Record Linkage Group which reported the outcome of 233 babies born to 168 epileptic women. This study showed that there was a significant increase in malformations in the epileptic group compared with the controls (13.8 and 5.6 per cent respectively, $p < 0.0005$). Fedrick also suggested that there was a dose-related effect with phenobarbitone but not with phenytoin, although phenytoin was more likely to cause malformations than phenobarbitone. However, if the two drugs were given together, the teratogenic effects may be potentiated, which emphasizes the need for assessing the possibilities of interactions between anti-epileptic drugs. Primidone also was implicated as a possible teratogen in this survey as it was associated with malformations in six out of 24 babies.

Hill (1973a), in an editorial article on teratogenesis and anti-epileptic drugs, evaluated 22 publications describing the frequency with which congenital malformations occurred in 2283 children exposed *in utero* to anticonvulsant drugs. He reported that the total abnormalities included cleft lip with or without cleft palate (5.0 per cent), skeletal anomalies (1.9 per cent), congenital heart disease (1.4 per cent), CNS defects (1.2 per cent), anomalies of the gastrointestinal tract (1.1 per cent), facial and ear abnormalities (1.0 per cent), mental retardation (0.7 per cent), genito-urinary anomalies (0.6 per cent), plus other isolated anomalies. One of the most frequently occurring defects was cleft lip with or without cleft palate, the incidence of which was 10 per 1000 compared with 1.5 per 1000 in the control population. Furthermore Hill stated that babies born to epileptic women not on drug therapy had a similar malformation rate (29.5 per 1000) to the control population (24.5 per 1000). This suggests that it may be the drugs rather than the epilepsy which are involved in causing the malformations. In a follow-up study (Hill 1973b) of children born to epileptic mothers, it was shown that there was a persistent adverse effect on performance in the Gessel Developmental Schedules carried out at 9, 18, and 24 months of age in the babies exposed to anticonvulsants *in utero*. These effects on physical and mental development observed by Hill (1973b) correlate well with the findings of other workers (Speidel and Meadow 1972; Bjerkedal and Bahna 1973; Annegers *et al.* 1974; Illingworth 1976).

It is a commonly-held belief that drugs given to pregnant women will only affect fetal development if they are given during the first trimester. Thus, the suggestion that drugs can influence mental development and subsequent behaviour of the offspring is an important concept. Brain development continues throughout the whole of pregnancy, particularly in the third trimester and into the neonatal period. The fetus may be exposed to anticonvulsant drugs for nine months *in utero* and then may also be exposed to these drugs via the breast-milk for several months more, and may thus receive doses of drugs which it cannot metabolize effi-ciently because of the lack of the necessary enzymes. This covers the time of maximum brain growth and thus it is a very real possibility that these drugs can interfere with mental development and behaviour.

The possibility that not all anti-epileptic drugs are equally teratogenic is suggested by the work of Starreveld-Zimmerman *et al.* (1973). Whilst not being able to implicate any particular anticonvulsant as a teratogen, they did observe that in 50 babies born to epileptic mothers, who in addition to other anticonvulsants also received carbamazepine, there were no malformations. This group did not differ in any other characteristic from the control sample. Furthermore, in a more recent report from the Netherlands (Starreveld-Zimmerman *et al.* 1975) phenytoin, primidone, the barbiturates, ethosuximide, and trimethadione have all been implicated as teratogens, but not carbamazepine or the benzodiazepines, e.g. nitrazepam, diazepam, and chlordiazepoxide. This statement agrees with the previous findings of this research group (Starreveld-Zimmerman *et al.* 1973) but is contrary to the observations of several other workers (Millar and Nevin 1973; Barry and Danks 1974; Speidel and Meadow 1974), who have presented positive evidence implicating carbamazepine and diazepam. Starreveld-Zimmerman *et al.* (1975) reported that 10 per cent of the pregnancies in which phenobarbitone and phenytoin were given alone or in combination with other anticonvulsants result in the birth of congenitally malformed children. However, in 50 pregnancies where carbamazepine alone was given during the first trimester, no malformations occurred. These authors suggested that a shift in anti-epileptic medication towards carbamazepine might be beneficial for women of child-bearing age. Monson *et al.* (1973) investigated a cohort of 50 897 pregnancies and the highest malformation rate (61 per 1000) was seen in the 98 children exposed to phenytoin during early pregnancy, compared with 25 per 1000 in the control children of non-epileptic women. Speidel and Meadow (1974) reviewed the clinical data and estimated that the risk of an epileptic woman having a malformed child is one in 10, but nevertheless they felt that anticonvulsant therapy should not be altered merely because of the pregnancy.

Annegers *et al.* (1974) in the USA studied 284 births to 138 women with epilepsy and amongst the 141 children born to mothers taking one or more anticonvulsant drugs, 10 had serious malformations. The malformations included five babies with congenital heart disease, two with cleft palate, and one baby with both heart and palatal defects. Amongst the 56 children born to epileptics not on drug therapy in the first trimester, there was only one malformation, agenesis of the testis. There were no other differences as regards age, parity, or type of seizure. This particular study is of importance in that it did take into account the types of seizures experienced by the epileptic group on drug therapy compared with their matched controls. Annegers *et al.* (1974) also reported that there were no malformations amongst the 26 babies born to mothers whose epilepsy was in remission or amongst the 61 babies from mothers whose pregnancy pre-dated the epilepsy. The authors could not say with certainty which drug was producing the teratogenic effects, but the results of the survey indicated that there was just as much chance of having a malformed child after barbiturate therapy as after phenytoin therapy. This differs from the results published by other authors (Loughnan *et al.* 1973; Barry and Danks 1974; Fedrick 1973) which have tended to implicate phenytoin as a more potent teratogen than phenobarbitone.

The effects of anticonvulsant medication in the father were also studied. In one survey (Annegers *et al.* 1974) no correlation was found between treating the father and malformation of the offspring. However, in another study (Shapiro *et al.* 1976*b*) it was shown that having an epileptic father produced a malformation rate in the children intermediate to that of having an epileptic mother and the non-epileptic control group.

Smithells (1976) reviewed the evidence to date and reported a 5.4 per cent incidence of malformation in 2403 children born to epileptic women as compared with an incidence of 2.2–3.5 per cent in six control populations, i.e. about a twofold increase in incidence. Annegers *et al.* (1978) extended their earlier studies using the medical records linkage system of the Rochester Project to study births from 1922 to 1976 to women with epilepsy or whose husbands had epilepsy. They found no excess in malformations among 215 offspring in which the birth preceded the onset of epilepsy, or was after remission, or where no medication for epilepsy was taken. However, there was a high incidence (19/177 or 10.7 per cent) of major malformations where the mother took anti-epileptics during the pregnancy. The rate was not higher when the husband had epilepsy (9/234 or 3.8 per cent).

Evidence suggesting that factors other than anticonvulsants are implicated in teratogenesis

Several workers (Fraser and MacGillivray 1969; Staples 1972; Livingstone *et al.* 1973; Kuenssberg and Knox 1973; Janz 1975; Shapiro *et al.* 1976*b*) have disputed the implications that anticonvulsant drugs are teratogenic in the human. One of the major problems in assessment is the choice of a control group. Even if one compares treated epileptics with non-treated epileptics, those on drug therapy may have a more severe form of epilepsy

than those receiving no medication. A study in Scotland during 1965–67 of over 15 000 pregnancies showed that of 30 cases of cleft lip with or without cleft palate none was born to epileptic mothers whether on anticonvulsant therapy or not (Kuenssberg and Knox 1973). However, other types of malformations were reported (five cases in all, which constituted 10 per cent of the total malformations), such as hydrocephalus and mild hypospadiasis, associated with two or more of the following drugs; phenobarbitone, phenytoin, ethosuximide, and primidone. Nevertheless, Kuenssberg and Knox concluded from this data that there was no good evidence of a causal relationship between anticonvulsant drugs and the malformations observed. A brief report (Livingstone *et al.* 1973) of 100 pregnant epileptic women at Baltimore City Hospital indicated that there was no increase in the incidence of congenital malformations. Janz (1975) in a review article suggested that the evidence in both animals and man that phenytoin increases the incidence of oro-facial clefts is highly suspect, especially as no dose-related effects have been obtained in many instances. Shapiro *et al.* (1976*b*) presented the results of two surveys, one in Finland and one in the USA, which seemed to indicate that the fetal damage previously attributed to phenobarbitone and phenytoin plus other anticonvulsant drugs might be due to the epilepsy itself, i.e. it was not the drugs, but the epilepsy, in either parent, which was related to an increased risk of having a malformed child.

Hanson and co-workers (Hanson *et al.* 1976) also analysed the same data from the same collaborative perinatal project as Shapiro *et al.* (1976*b*), but their conclusions were different. A particular pattern of malformations was found which has been called the fetal-hydantoin syndrome (Hanson and Smith 1975), plus the possibility of impaired intellectual ability, with about one-third of fetuses exposed to phenytoin being affected. Other workers (Visser *et al.* 1976) also disputed the results presented by Shapiro *et al.* (1976*b*) and stressed that most epileptics patients are on multiple-drug therapy and this was not taken into account when the USA and Finnish data were assessed.

Recent studies

Recently a number of comprehensive reviews have been published surveying almost every aspect of anti-epileptic drug treatment during pregnancy (Janz 1982*a, b*; Annegers *et al.* 1983; Canger *et al.* 1983; Friis 1983, Hiilesmaa *et al.* 1983).

An important collaborative multi-centre study from Japan was published recently (Nakane *et al.* 1980). Eleven centres participated in this study from 1974 to

1977 and analysed both retrospective and prospective data on 657 epileptic women on anticonvulsant therapy and another 162 women not on therapy. In the anticonvulsant group, 73 per cent of the women had live babies, 14 per cent had miscarriages or stillbirths, and 13 per cent had induced abortions. In contrast, in the nonmedicated group, 80 per cent had live babies and only 4 per cent had miscarriages or stillbirths. In this series there were 638 live births, 63 (9.9 per cent) of which were malformed; 55 (11.5 per cent) from the anticonvulsant-treated group compared with 3 (2.3 per cent) in the nonmedicated group which falls within the expected range. Furthermore, as the number of anticonvulsant drugs used in combination therapy increased in the first trimester, so did the incidence of congenital malformations. The incidence of malformations was particularly high when three or more drugs were used. If three drugs were used there was a twofold increase (11 per cent), and with the use of four drugs a fourfold increase (23 per cent) when compared with the incidence of malformations (5.5 per cent) in those receiving two drugs. Only 15 per cent of the patients in this study were on monotherapy.

In this study the incidence of cleft lip $\pm$ cleft palate (3.14 per cent) and cardiovascular defects (2.95 per cent) was higher in the anticonvulsant group than in the nonmedicated group. The normal incidence of these defects in Japan is 0.27 and 0.15 per cent, respectively. Overall there was a fivefold increase in malformations in the anticonvulsant group ($p < 0.01$). There was no relationship between malformations and maternal age, disease, or duration of treatment. However, in the anticonvulsant group there was an association for the following factors: occurrence of partial seizures during pregnancy ($p < 0.10$), malformation in the mother ($p < 0.001$), previous pregnancies ending in stillbirths or miscarriages ($p < 0.05$), and abnormal delivery ($p < 0.05$). In the non-medicated group there was an association for two factors only; epilepsy due to an organic cause ($p < 0.05$) and a history of anticonvulsant therapy, especially of more than 5 years duration ($p < 0.05$). The malformation rate was highest (12.7 per cent) amongst the treated group who had seizures during pregnancy compared with (3.3 per cent) in the nonmedicated group. Analysis of the data on the safety of individual anticonvulsant drugs indicated that the highest risk was associated with the use of trimethadione.

Metabolism and breast milk

Carbamazepine is similar to phenytoin in that the concentration of the drug in maternal and neonatal plasma is about the same. The half-life of the drug in the neonate is very similar to that of adults after receiving multiple doses of the drug which indicates that the neonate has an increased capacity for metabolizing the drug through transplacental autoinduction. Enzyme induction also occurs when the neonate has been exposed *in utero* to phenobarbitone, the half-life being shorter than in infants not previously exposed to the drug. Thus high concentrations at birth seem to be associated with rapid metabolism. In addition, phenytoin, the barbiturates, and carbamazepine are secreted into breast milk. Because of the acidic nature of the hydantoins and barbiturates the concentration of the drugs in breast milk is significantly below that of the plasma due to the pH difference between milk and plasma. For carbamazepine the concentration in the milk is about 60 per cent of that of maternal plasma. However, when assessed on a mg/kg/day basis the dose to the baby is about one-fiftieth of the therapeutic dose and would seem negligible. The amount of diazepam which is present in breast milk is about 10 per cent of the maternal plasma concentration, but because of the very slow metabolism and clearance of the drug by the baby, mothers taking daily doses greater than 10 mg may be advised not to breast-feed. It is important to emphasize that both the fetus and neonate do receive considerable amounts of the drugs given to the mother, and as brain development continues throughout the whole of pregnancy and well into the neonatal period, chronic exposure to the fetus and neonate to these drugs may well influence mental development and subsequent postnatal behaviour.

Serum folate concentrations

Hiilesmaa *et al.* (1983) reported on the serum folate concentrations, blood counts, and anti-epileptic drug concentrations in 133 pregnancies from 125 pregnant epileptic women. Twelve women had no drugs, 20 had carbamazepine only, 53 had phenytoin only, and 48 had combination therapy. Amongst the 133 pregnancies there was an inverse correlation between serum folate concentrations and those of phenytoin and phenobarbitone. The number of epileptic seizures during pregnancy showed no association with serum folate concentrations. No cases of maternal tissue folate deficiency or fetal damage attributable to low maternal serum folate were observed. However, there were nine infants with structural malformations, 10 infants had one or more features of the fetal hydantoin syndrome (FHS), and one had both a structural defect and FHS. No cases of cleft palate or neural-tube defects were observed. All cases of the FHS were considered to be mild. There were also five perinatal deaths. It is interesting that the serum folate concentrations for infants in the perinatal death group and with congenital malformations, including FHS, were similar to those for the normal babies. These authors recommended that a low dose (100–1000 μg daily) of folate supplement should be sufficient for pregnant women with epilepsy despite the antifolate action of the anti-epileptic drugs. Furthermore, folate concentrations in pregnant women with high serum concentrations of phenobarbitone or phenytoin should be carefully monitored. This is a particularly interesting study, because the relationship between folate, anti-epileptic drugs, and the production of congenital malformations has been the subject of much debate. Both animal and human studies on these interrelationships have produced conflicting data and it is still not possible to say whether the administration of folate is really beneficial and it certainly cannot be assumed that excess folate is entirely harmless (McElhatton 1977).

Individual drugs

Phenytoin

As well as the evidence mentioned above implicating phenytoin as a teratogen, further evidence on the adverse mental effects on the offspring when exposed to phenytoin during pregnancy were published briefly by Hill (1979) and Smith (1979), the latter author recommending strongly that phenytoin should not be used in pregnancy, because of the high incidence of both physical and mental handicap produced by that drug. However, this view was not shared by the Committee on Drugs of the American Academy of Pediatrics (1979) who merely recommended that women who require medication for epilepsy should be advised that they have a 90% chance of having a normal child but that the risk of congenital malformations and mental retardation is two to three times greater than average. . . . They did not recommend switching from phenytoin to other anticonvulsants. In another set of recommendations (Montouris et al. 1979) physicians are recommended to avoid trimethadione if possible but otherwise not to alter drug therapy, other than to control carefully the blood levels which may vary because of the metabolic changes associated with pregnancy.

Dieterich et al. (1980) published some interesting data on 37 children born to treated mothers and 22 children born to treated fathers. In the group of children born to treated mothers, nine of 37 had the complete fetal hydantoin syndrome (FHS), although two of the mothers were not receiving hydantoins. Another 15 children had one or more signs of the embryopathy. Thus in all, 24 of the 37 children (64.8 per cent) were affected, which is a

much higher figure than has been quoted in other studies. There was no relationship between the type of epilepsy and the severity of the malformations. In the group of 22 children born to fathers taking anticonvulsant drugs, none had any features of FHS. Birth weight and length were normal. The actual height and weight were below the expected values for children of epileptic mothers, but normal for children of epileptic fathers. The head circumference was normal in both groups.

The authors concluded that a characteristic pattern of acrofacial anomalies occurs with in-utero exposure to various anticonvulsant drugs, not just phenytoin, and suggest that the term 'anticonvulsant embryopathy' should be used rather than FHS. Furthermore, the facial clefts and other minor anomalies which occur more frequently in the children of both epileptic mothers and fathers are not part of this embryopathy, but are related to parental epilepsy rather than to prenatal exposure to anticonvulsant drugs.

There have been numerous reports on the adverse effects of phenytoin on the fetus, but its mechanism of action is unknown. Hall et al. (1980) put forward a theory to explain the acrofacial anomalies observed in FHS. Phenytoin interferes with coagulation factors possibly by competitive inhibition of vitamin K. Vitamin K is also involved in post-translational modification of proteins by formation of γ-carboxyglutamyl residues. This type of modification is necessary for many proteins to be able to bind calcium. If phenytoin depletes vitamin K and thus interferes with the calcium-binding capacity of certain proteins at a critical stage of embryogenesis, it may well interfere with the ossification processes and thus produce the digital defects seen in FHS.

Apart from the acrofacial anomalies and retarded mental and motor development associated with phenytoin, there is new evidence to suggest that it may also be associated with the formation of neuroblastomas (Seeler et al. 1979; Smith 1980). In a review of FHS, Smith reported at least four children with FHS who have also developed a neuroblastoma. In 1979, Seeler et al. were aware of three cases of neuroblastoma and they calculated that three such cases would only appear by chance, once in 60 years. This is particularly interesting because phenytoin is metabolized to an epoxide, and it is known that some carcinogenic substances such as alkylating agents have strained ring structures similar to the epoxides. It was previously suggested that anticonvulsants have carcinogenic activity in humans (Cincinnati 1970) but this was disputed by other workers (Clemmesen et al. 1974) who produced data from the Danish Cancer Registry on 9136 epileptic patients in whom there was no increase in the incidence of cancer. Nevertheless, it raises the possibility that, since the fetus is often more sensitive to carcinogens than adults, phenytoin and or its epoxide may be acting as a transplacental carcinogen.

Dam and Dam (1980) in a review of epilepsy in pregnancy discussed the placental transfer of various anti-epileptic drugs and their concentration in the neonate. Phenytoin is present in equal concentrations in maternal plasma and cord plasma at birth, and clearance from the neonate is not achieved until 4–5 days postpartum. Furthermore, there is considerable variation in the half-life of the drug during the first three days of life (60 h) compared with when the baby is one week old (about 7½ h). In premature babies this is even more variable, being in the range 75 ± 65 h.

Primidone

There have been several case reports involving primidone as a teratogen (Rudd and Freedom 1979; Myhre and Williams 1981) and the term primidone embryopathy is now also being used. Pharmacokinetic studies on the placental transfer of primidone and its metabolites and primidone plus other anticonvulsant drugs have also been reported (Nau et al. 1980).

Carbamazepine

Hiilesmaa et al. (1981) followed up 133 epileptics on various anticonvulsant drugs and 143 controls. Seventy-seven of the epileptic group were on monotherapy, the remainder were on combination therapy. It was found that carbamazepine alone, or combination therapy including phenobarbitone, was associated with retardation of fetal head growth. The mean head circumference for babies from mothers on carbamazepine alone was 7 mm less than the controls, and for those on phenobarbitone or primidone combination therapy it was 6 mm less. There was no catch-up in growth by 18 months of age. Growth variables did not correlate with the maternal drug concentrations, seizures, or the type of epilepsy. In the epileptic group of 133 mothers, there were 143 babies and of these six were malformed. Two babies had hypospadias, three had congenital dislocation of the hip, and one baby had an inguinal hernia. Unfortunately, the control data for malformations was omitted. Hiilesmaa et al. concluded that as the reduced head circumference occurred only with certain drugs, it is indicative of a drug-related effect rather than of the presence of maternal epilepsy.

It is interesting that phenytoin alone was not associated with a small head circumference, but phenytoin plus phenobarbitone was. Fedrick (1973) also reported that there were dose-related effects with phenobarbitone but not with phenytoin, but if the two drugs were given together the teratogenic effects were potentiated. However, although the head circumference in the carbamazepine group was less than that of the controls, it still fell within the normal range and there was no correlation between head circumference and intelligence. The authors suggested that it might be indicative of some subtle effect on fetal brain development and as such should be distinguished from a 'genetically' small head.

Sodium valproate *(Epilim)*

Sodium valproate has been known to be teratogenic in rodents and rabbits since its introduction on the market. It produced a dose-related spectrum of malformations including renal, palatal, vertebral, and neural-tube anomalies. The relevance of this for man, however, was unknown. Studies in animals have shown that valproate readily binds to zinc and it has been suggested that this may account for its teratogenic action (Hurd et al. 1983).

Sodium valproate readily crosses the human placenta and the concentration in fetal blood may be two to three times higher than in maternal blood (Kaneko et al. 1983).

From 1979 a number of individual case reports were published of malformed children born after exposure to valproate alone (Alexander 1979; Dalens et al. 1980) or in combination with other anti-epileptics (Dickinson and Gerber 1980; Hiilesmaa et al. 1981; Gomez 1981). No clear pattern emerged however until Robert and Guibaud (1982) reported results from the monitoring system for the Rhone–Alpes region of France, where there are about 72 000 births each year. Between August 1979 and August 1982, 72 cases of lumbrasacral neural-tube defects alone or associated with other malformations were recorded. Nine of the infants were born to epileptic mothers who had taken sodium valproate during pregnancy. This was a very much greater incidence than would be expected by chance. There were no cases associating sodium valproate with anencephaly among the 17 cases reported for that period. In five cases, sodium valproate was used alone; it was used in combination with phenobarbitone in three cases, and with clonazepam in one case. In seven of the cases the doses of sodium valproate used were high, in excess of 1 g daily. This study is part of the International Clearing House for Birth Defects Monitoring System. An association of valproate use with spina bifida was also found in Italy but not in other countries in the programme. This, however, was probably because of the much lower use of valproate in these other countries (Bjerkedal et al. 1982). The data from France and Italy have been updated by Robert (1983a, b) and show that of 71 French women with convulsive disorders who delivered malformed babies, 10 mothers, of whom nine had taken sodium valproate, had babies with spina bifida. Since about 35 per cent of epileptic women in that region of France were given valproate, the excess incidence in the spina bifida group is very highly significant. From the data presented it is impossible to calculate the precise risk of spina bifida in patients taking valproate but a value of about 1 per cent is reasonable.

Reports on the use of *Epilim* in pregnancy were collected by the manufacturer (Sanofi UK Ltd, personal communication, 1983). Of 52 pregnancies in which *Epilim* was used alone, eight resulted in malformed babies, three of whom had neural-tube defects. Of 71 pregnancies in which *Epilim* was used in combination with other anticonvulsants, 22 were abnormal of whom seven had neural-tube defects. Since doctors are more likely to report abnormal outcomes than normal ones, the true incidence of defects cannot be derived from this data. However, to have 10 babies with spina bifida, meningocele, or hydrocephaly out of a total of 30 abnormal babies is at least twice as high as expected. Cardiovascular and digital defects were also observed.

Trimethadione

As mentioned above, the possibility that trimethadione was a teratogen was reported by German *et al.* (1970). Since then the oxazolidines have been implicated in a number of cases, and a syndrome of malformations — the trimethadione syndrome — similar to, but not identical with the fetal-hydantoin syndrome has been observed. The trimethadione syndrome can be distinguished from the hydantoin syndrome by such anomalies as V-shaped eyebrows and low set ears with anteriorly folded helix (Feldman *et al.* 1977; Goldman and Yaffe 1978).

Summary

In summary, there is still no clear evidence that any one anticonvulsant drug is safer than the others. However, several authorities recommend the use of a single-drug regime if at all possible. Trimethadione seems to be contraindicated and ethosuximide is to be preferred for treatment of petit mal epilepsy. For other forms of epilepsy some authorities recomend low doses of phenytoin its use remains controversial. It has been suggested that carbamazepine should be tried first rather than phenytoin or phenobarbitone in women of child-bearing age. However, from the data now available there seems to be some degree of risk attached to all of the anti-epileptic drugs. In general, therefore, whichever drug is used to treat epilepsy, it should be given at the lowest possible effective dose. There is no good evidence that medication should be withdrawn in pregnancy; in fact fetal hypoxia due to maternal fits may be more damaging than the drugs themselves. However, as the metabolism of anti-epileptic drugs may change unpredictably during pregnancy, monitoring of maternal blood levels is advised. There certainly seems to be increased risk associated with multiple-drug therapy, so that, where possible, the use of monotherapy in women of reproductive capacity is to be recommended.

The preference of the patient also has to be taken into account. In one study (Philbert and Pedersen 1983) of 83 epileptic women of child-bearing age being treated with sodium valproate, half of the women intending to get pregnant did not wish to change to a different drug after counselling because they found valproate so satisfactory. In this case, where the potential malformation was mainly spina bifida, the offer of amniocentesis for AFP analysis with the offer of abortion if positive was preferred to changing to another drug.

There is a slightly increased risk of oral-contraceptive failure in women on anti-epileptics (Coulam and Annegers 1979; *British Medical Journal* 1980) so that it may be necessary to consider the use of additional contraceptive methods if pregnancy is absolutely to be avoided.

Antihypertensive drugs

Until recently there has been very little information concerning the adverse effects of drug therapy for hypertension on the well-being of the fetus and neonate. One major difficulty in assessing the effects of drugs on the fetus is that the presence of severe hypertension is itself a high risk factor for the fetus and more so if renal damage is also involved. Thus single case reports of outcomes of such pregnancies cannot give much indication of the toxicity of drugs. The pathophysiology of toxaemia and hypertension in pregnancy with emphasis on the role of various prostaglandins was reviewed by Ferris (1983). Breart *et al.* (1982) reported on a study of 2 997 women who at their first visit to the antenatal clinic had no signs of hypertension. These patients were assessed throughout their pregnancies and from this data Breart and co-workers produced several tables of various risk factors, and showed that maternal hypertension developing between 28 and 35 weeks is a major risk factor for fetal growth retardation.

The usual principles involved in the treatment of hypertension in middle or late life do not necessarily apply to the management of young women during pregnancy. Although moderately high blood pressure

(140/90 – 170/110) may predispose to aortic dissection, and bleeding from cerebral aneurysms and myocardial infarction, these complications are relatively rare during pregnancy and do not necessarily require treatment. But, if the rise in blood pressure is secondary to pre-eclampsia, treatment to lower the blood pressure may be required but may have inherent risks for the fetus. Furthermore, if hypertension was present prior to pregnancy, the development of pre-eclampsia increases the risk to the fetus about fivefold. It is thought that the dangers of chronic hypertension for the fetus arise entirely from the superimposed pre-eclampsia, without which the perinatal mortality is thought to be no greater than average. In fact, Leather et al. (1968) demonstrated that, once pre-eclampsia had developed, treatment confers no benefit on the fetus, although it may be essential for maternal well-being.

These views are supported by data published by Suonio (1982) who studied placental blood flow in late pregnancy (28–41 weeks) using a radionuclide method in three groups of pregnant patients. The first group of 120 women were normotensives, the second group of 103 women were pre-eclamptic, and the third group of 93 women had chronic hypertension. He found that in the pre-eclamptic group the placental blood flow was inversely correlated with the severity of the disease and was inadequate in about one-third of cases. In the chronic hypertensive group the mean placental blood flow was reduced even in those with mild labile disease but was adequate in about 90 per cent of cases and hypotensive therapy was not necessary for these and could be disadvantageous. The β-mimetic ritodrine could improve placental blood flow in both pre-eclamptic and chronic hypertension without causing any change in the maternal blood pressure. The fetal outcome of 157 hypertensive pregnancies, the majority with pre-eclampsia and/or nephrosclerosis, whose underlying disease had been established by renal biopsy, was reported by Lin et al. (1982). These authors concluded that gestational hypertension alone had relatively little influence on the outcome of pregnancy, but the underlying disease process was of most importance. A combination of hypertension and proteinuria was associated with high perinatal mortality which was worst in the multiparous pre-eclamptic women in whom the perinatal mortality was 31 per cent. These adverse effects of severe maternal hypertension were supported by a study of hypertensive pregnancies by Brazy et al. (1982). Sibai et al. (1983) studied 211 consecutive pregnancies complicated by mild maternal chronic hypertension in which antihypertensive therapy was discontinued at the first prenatal visit. Only 13 per cent of the women required antihypertensive therapy later in pregnancy.

They found that discontinuance of therapy did not adversely affect the antepartum course or perinatal outcome. In fact in the absence of pre-eclampsia, the perinatal mortality of these patients approached that of the general population and therefore suggested that factors other than an increase in blood pressure might be responsible for poor obstetric outcome in the pre-eclamptic women.

Thus there is somewhat conflicting evidence on the fetal outcome of pregnancy-associated hypertension, and the decision whether or not to treat is controversial. In mild chronic hypertension the prognosis for the fetus is unlikely to be altered by hypotensive therapy. In pre-eclampsia, the more severe the disease the worse the outcome is for the fetus in terms of perinatal mortality and morbidity and the welfare of the mother is the determining factor in deciding on treatment.

BETA-BLOCKERS

Propranolol

Two studies were published which suggest that propranolol may be damaging to the fetus. Lieberman et al. (1978) compared the outcome of nine pregnancies in which propranolol was used (with or without other drugs), with the outcome in 15 pregnancies treated over the same period with similar drugs but without propranolol, all for the treatment of severe hypertension. Seven of the nine propranolol-treated pregnancies and five of the 15 treated with other drugs ended in fetal or neonatal death. The propranolol babies were also significantly lower in weight. Careful analysis of the patients did not suggest the propranolol-treated mothers were different from the others in terms of maximum diastolic pressure, blood urea, proteinuria, parity, etc. Significantly more of them (all nine), however, had been treated with diuretics than in the other group (6/15, $p < 0.004$). The perinatal mortality among babies whose mothers had received diuretics was significantly ($p < 0.005$) higher than when they had not. A multiple regression analysis, however, suggested that the probability of fetal or neonatal death was significantly higher ($p < 0.05$) when the mother had been treated with propranolol. In reviewing the possible mechanisms of action of propranolol, the authors suggest that, while β-blockade may be unimportant in unstressed or only mildly stressed fetuses, it may be fatal under extreme conditions of placental insufficiency.

Another report on the use of propranolol for a variety of conditions in 12 pregnancies was studied. The most significant effect was intrauterine growth retardation with seven of the 12 babies at or below the tenth centile for head circumference and five of the 12 for body weight (Pruyn *et al.* 1979). However, in a series of nine high-risk pregnancies with hypertension, all treated with propranolol, a reasonably satisfactory outcome was achieved, but with very careful control being kept over the mothers (Tcherdakoff *et al.* 1978). In conclusion it would seem wise to avoid the use of propranolol in pregnancy unless other therapy fails, and when used, very careful monitoring of the mother and fetus seems essential.

Atenolol

Liedholm (1983) reported on a retrospective analysis of all hypertensive pregnancies seen in Lund, Sweden for 1980. From a total of 3077 pregnancies, 113 (3.7 per cent) were hypertensive, among which were six sets of twins, 119 babies in all. Twenty-two (19 per cent) of these women received no treatment, 53 (59 per cent) received a β_1 blocker, atenolol, and the remaining 35 (39 per cent) received atenolol plus hydrallazine. Nine women were on treatment when they became pregnant and with one exception all the women remained on treatment to term. Drug treatment resulted in a significant decrease in blood pressure. There was no maternal mortality and no eclampsia, in fact there were very few maternal side-effects. However, the perinatal mortality rate was somewhat high, 2.5 per cent (3/119), all of whom were stillborn. This is approximately three times higher than the norm for Sweden, but is very similar to the figures quoted by other workers. Eleven of the babies (9.5 per cent) were small for gestational age, four of whom had neonatal hypoglycaemia. There were three cases of respiratory distress syndrome and five others with a mild respiratory problem of which five were ascribed to either low Apgar scores or premature delivery and another two were small for gestational age. No other risks to the neonates were observed.

Liedholm concluded that atenolol can be used either alone or with hydrallazine at all stages of pregnancy for both pre-existing and pregnancy-induced hypertension with little adverse effect on either the mother or the fetus. Rubin *et al.* (1983) report similar findings in a placebo-controlled, double-blind study in 120 women with mild-to-moderate pregnancy-associated hypertension. Atenolol once daily caused a reduction in blood pressure, delayed the onset of proteinuria, and resulted in fewer hospital admissions. Loss of blood pressure control leading to withdrawal from the study occurred more often in the placebo group, and these babies had a higher morbidity. Intrauterine growth retardation (IUGR), neonatal hypoglycaemia, and hyperbilirubinaemia occurred with the same frequency in the

atenolol group as in the placebo group. Neonatal bradycardia occurred more frequently following atenolol, but the systolic blood pressure of these babies was the same in both groups. Respiratory distress syndrome occurred in the placebo group only. Both neonatal weight and maturity were the same for both groups. The authors concluded that atenolol is effective in lowering the maternal blood pressure without adversely affecting the baby. Although the results of this study sound very encouraging, the report is purely descriptive and no data are presented and thus a critical assessment of the study is not possible.

Acebutalol, pindolol, and atenolol

Dubois *et al.* (1983*a, b*) reported on a study of 121 patients aged 20–44 years old who were considered to be high-risk pregnancies. In all 125 babies were born, four had two successive pregnancies, and there was one set of twins. Three β-blockers were used, acebutalol (56 women), pindolol (38 women), and atenolol (31 women). In 110 cases β-blocker therapy alone was used, but in the remaining 15 cases other drugs hydrallazine, (12); methyldopa, (2); and clonidine, (1) were added in the ninth month. No placebo group was included because of ethical constraints. Seventy of these patients (56 per cent) had a previous history of hypertension. Treatment was started when the DBP was 90 mmHg, and this regimen was successful in controlling the blood pressure in 95 per cent of cases. Maternal tolerance to the drugs was excellent and did not require interruption and no untoward uterine contractions were observed (a purported theoretical disadvantage with the use of β-blockers) in pregnancy. Three groups of newborns were defined; those weighing less than 2.5 kg; those weighing between 2.5–2.8 kg; and those weighing more than 2.8 kg. The distribution of these babies according to drug exposure is shown in Table 22.2. The incidence of low birth weights (<2.5 kg) being highest in the atenolol group.

Table 22.2 Effects of β-blockers on birth weight

| Birth weight (kg) n = total number | No. cases (per cent) for each drug | | |
	Acebutalol (n = 56)	Pindolol (n = 38)	Atenolol (n = 31)
< 2.5 n = 20	7 (12.5)	3 (7.9)	10 (32.3)
2.5 – 2.8 n = 15	5 (8.9)	6 (15.8)	4 (12.9)
> 2.8 n = 90	44 (78.5)	29 (76.3)	17 (54.8)

There was no evidence of low Apgar scores,

bradycardia, or hypotension in the neonates. Four of the infants (3.2 per cent) had congenital malformations, three of whom had been exposed *in utero* to pindolol and one to acebutalol. However, in all cases the drugs had been administered after 24 weeks of gestation and therefore the malformations were unlikely to be drug-induced.

Dubois and co-workers recommended that atenolol should not be used during early pregnancy, but acebutalol and pindolol can be used. Furthermore, they recommended that β-blocker therapy should be maintained up to delivery to prevent rebound hypertension.

Labetalol

Labetalol has been reported to be embryo-lethal rather than teratogenic when given in high doses to rats and rabbits. It is known to cross the placenta in humans, and it has been used successfully to treat hypertension arising in the second and third trimesters of pregnancy. No adverse effects on the fetus or neonate have been directly attributed to its use, but its safety for use in pregnancy has yet to be established. Animal evidence showed that labetalol can bind to melanin in fetal eyes, but the clinical significance of this is unknown. Opthalmological examination in 15 infants born to mothers treated with labetalol during the third trimester showed no retinal abnormalities at birth (Batagol 1983).

Oxprenolol

In a comparison of the use of oxprenolol or methyldopa (with or without hydrallazine) in 43 moderately severe hypertensives (Gallery *et al.* 1979), the outcome following oxprenolol was satisfactory and superior to methyldopa.

Metoprolol

Its safety for use in pregnancy has not been established.

Timolol

Its safety for use in pregnancy has not been established.

Methyldopa

Methyldopa is probably one of the most commonly used hypotensive drugs given in pregnancy. It does cross the placenta, and its use must be balanced against its possible adverse effects on the fetus. There have been two recent studies on the use of methyldopa during pregnancy. One was a prospective study in Oxford which began in 1970 (Ounsted *et al.* 1980) in which 353 infants were examined. Women with mild hypertension prior to 28-weeks gestation were randomly allocated to one of two groups. The first group had hypotensive therapy, mainly methyldopa, until delivery. There were 100 infants born to mothers in this group. In the second group no specific treatment was given; mothers in this group gave birth to 102 infants. A third group was introduced for comparison which was a random sample of the hospital population; this group had 151 infants. There was one neonatal death from a major congenital malformation (unspecified) from the group that received no treatment, but no other data on congenital malformations were given.

At birth there was no association between intrauterine growth retardation and the severity of maternal hypertension. There were no differences amongst the three groups in the incidence of moderate or severe pre-eclampsia or in smoking habits. The children of the hypertensive mothers were in general slightly retarded in a wide range of developmental parameters studied, compared with the normal control group. However, in all parameters the children from the treated mothers were less affected than those of the untreated mothers. The effects tended to decrease with age up to four years and seemed to be due to a general delay in development rather than a big effect in a few children. There was a significant excess of newborn infants, particularly boys, in the treated hypertensive group with head circumferences below the mean for their gestational age when compared with either the untreated hypertensive group or the randomly selected group. No relationship was established between head circumference and the total amount or duration of treatment with methyldopa, nor did it affect the overall developmental scores. It is interesting that the differences in head circumference were confined to babies whose mothers began treatment during 16–20 weeks of gestation. The differences in head size between these groups persisted until two months of age, but after corrections were made for birth weight, gestation, and sex, they were no longer apparent at 6 months and 12 months of age. In both the treated and untreated groups there was an excess of infants with delayed fine motor function at six months of age. Analysis of perinatal factors and those relating to the childs growth, development, and neurological status at 12 months showed no significant differences. On average, the children from the randomly selected non-hypertensive group were more advanced when assessed by a global score of development.

There were 275 children available for follow-up at four years of age, 86 from the treated group, 82 from the untreated group, and 107 from the randomly selected group. The distribution of total and sector scores was normal within each group. Some of the children born to women with hypertension were mildly retarded in certain aspects of development at four years of age. Although induction of labour, fetal distress, and delivery by Caesarian section occurred more often in the hypertensive groups, these perinatal factors made a negligible contribution to the total overall development scores at four years of age. These children are being reassessed at seven years of age. The authors conclude that maternal hypertension is associated with slight developmental delay in early childhood but that treatment with methyldopa reduced this effect. Therefore, it would seem relatively safe to use methyldopa during pregnancy if the mother's condition requires it.

The results of the 7½-year follow-up study on 56 of the children whose mothers had superimposed pre-eclampsia compared with 176 children whose mothers had hypertension alone, have now been published (Ounsted *et al.* 1983). Perinatal mortality in the hypertension alone group was similar to that for the hospital population in general at the time of their birth. However, in the group that developed pre-eclampsia it was significantly higher. At 7½ years of age, there were no differences in frequency of health, handicap, sight and hearing problems, weight, height, head circumference, and standing or supine blood pressure between the groups. Children in the pre-eclampsia group had slightly higher mean scores for six aspects of intellectual development; and in one of these, 'perceptual matching', the difference was significant after adjustment had been made for other confounding variables. Thus, the authors concluded that pre-eclampsia superimposed on hypertension does not increase the likelihood of impaired growth and development among those children who survive the perinatal period.

The second study on methyldopa was designed to see if the systolic blood pressure was lower in infants of the same birth weight in mothers who were or were not treated with hypotensive drugs (Whitelaw 1981). Indications for treatment were either long-standing hypertension before pregnancy which had been treated with hypotensives or a diastolic pressure consistently higher than 95 mmHg before 28 weeks gestation.

Over a 12-month period 24 mothers were given 500 mg-2 g of methyldopa daily and were delivered after 37 weeks gestation. The systolic blood pressure was measured in these 24 babies at 12, 36, 60, 84, and 108 hours after delivery and were compared with 50 randomly selected full-term infants. There were no significant differences between the groups for birth weight or gestational age, but there were differences in the systolic blood pressure. In the babies exposed *in utero* to methyldopa, the resting systolic blood pressure was 4.5 mmHg lower than in the controls on day 1 post-partum and 4.3 mmHg lower on day 2. However, there were no significant differences on days, 3, 4, and 5 postpartum. None of the babies in the methyldopa group had a heart rate less than 120 beats per minute. There were three cases of fetal distress during labour, but the condition of all the babies at birth and their Apgar scores were satisfactory. It is interesting that the reduction in systolic blood pressure of 4–5 mmHg in the first two days of life is comparable to that observed in normotensive adults. Furthermore, the duration of the reduced systolic pressure was consistent with the slow elimination of the drug by the neonate. Normal BP was achieved by about three days of age by which time the babies' plasma levels of methyldopa were low.

Whitelaw concludes that this mild reduction in BP should not seriously compromise babies and there was no evidence that their cardiovascular system could not respond to stress.

DIURETICS

Diuretics, i.e. thiazides and diazoxide, are best avoided in pregnancy as they have undesirable effects on both the mother and fetus (Wood and Blainey 1981). Apart from maternal choleostasis and pancreatitis they may also cause a reduction in plasma volume and thereby lower placental perfusion and thus compromise the fetus (Walters and Humphrey 1980; Batagol 1983). Neonatal thrombocytopenia has also been reported (Bynum 1977).

Diazoxide

Animal studies have shown that diazoxide crosses the placenta and can cause diabetes mellitus in the fetus almost immediately after the mother receives the drug (reviewed by Batagol 1983). Batagol reported briefly on the use of diazoxide and chlormethiazole for pre-eclamptic toxaemia which was associated with profound hypotonia, decreased ventilatory effort, and aponea in the neonates. Two of the five infants who required ventilation assistance died. These effects have not been reported with the use of chlormethiazole alone.

Frusemide

There is no evidence either way regarding the teratogenicity of frusemide. The same precautions should be observed as for the thiazides.

Carbonic anhydrase inhibitors

These are not in general use as diuretics. Although they have been reported to cause malformations in animals there is no evidence regarding their teratogenicity in man.

Potassium-sparing diuretics *(Spironolactone, triamterene, amiloride)*

There is no published evidence regarding the teratogenicity of the drugs in humans.

Combination therapy

A study of 58 patients with chronic mild hypertension, their BP being 140/90 mmHg prior to pregnancy but with no target organ damage, who were treated with combination therapy was described by Arias and Zamora (1979). In this study 29 patients received hypotensive therapy throughout the whole of pregnancy and 29 received no treatment; no placebos were used. The treated group were assigned arbitarily to three different types of therapy; 11 women received methyldopa plus hydrochlorothiazide, 10 received hydrallazine plus hydrochlorothiazide, and eight received all three drugs. The dose of hydrochlorothiazide was kept constant, 50 mg daily, but the dose of the other drugs were varied in order to keep the diastolic pressure below 90 mmHg. Untreated patients whose hypertension was aggravated during pregnancy were treated before delivery, but for analysis purposes were regarded as untreated. Thirteen of the 29 (41.3 per cent) of the untreated patients developed pregnancy-aggravated hypertension, i.e. a diastolic pressure $>$ 100 mmHg after 28 weeks, compared with 4 of 29 (13.7 per cent) in the treated group ($p < 0.05$). These four treated patients all delivered compromised infants whereas only five of 13 untreated patients had a poor fetal outcome ($p = 0.05$) The authors suggest that this may be indicative of haemodynamic alteration caused by diuretic hypotensive drugs which may be aggravated by maternal vasoconstriction. There were no significant differences for gestational age, birth weight, intrauterine growth retardation, or fetal distress between the treated and control groups. No congenital malformations were observed and no fetal deaths apart from one neonatal death in the untreated group.

It is interesting that although there was a higher incidence of pregnancy-aggravated hypertension in the untreated group, there were no significant differences in fetal outcome.

Conclusion

In conclusion it would seem that the diuretics are contraindicated in the management of hypertension in pregnancy because they may reduce plasma volume and thereby lower placental perfusion which may compromise the fetus. There is contradictory evidence concerning the use of β-blockers. Both propranolol and atenolol have been associated with intrauterine growth retardation and an increase in perinatal mortality but results with other β-blockers have been very satisfactory. There is no evidence either way for some of the newer drugs such as labetalol, metoprolol, and timolol. Their use in pregnancy remains controversial. The majority of studies recommend methyldopa (with or without hydrallazine or clonidine) as the drugs of choice for the treatment of hypertension in pregnancy. The 7½-year follow-up study of infants exposed *in utero* to methyldopa has been very encouraging and would support these recommendations.

Caffeine

As caffeine is a common constituent of coffee, tea, cocoa, chocolate, cola-type drinks, and some analgesics and 'cold cures', recent reports concerning its reproductive toxicity have provoked considerable interest in the lay and medical press. There is no doubt that, when used in very high doses, caffeine can produce mutagenic effects *in vitro* and teratogenic effects in animals. However, the important question is what the relevance of this is for man. At a recent symposium on caffeine in Athens (*Food Chemical News* 1982), it was stated that the evidence available from animal and epidemiological studies, together with a long history of human use indicated that it was highly unlikely that caffeine was carcinogenic, mutagenic, or teratogenic at the concentrations occurring in humans as a result of normal consumption of caffeine-containing beverages.

In the USA, largely as a result of consumer-group concern over animal experiments, the Food and Drug Administration (FDA) in autumn 1980, removed caffeine from their list of G.R.A.S. (generally recognized as safe) substances and issued a number of recommendations that pregnant women should avoid caffeine-containing beverages or, at least, reduce their consumption to less than five cups of tea or coffee a day. There have been many animal and some human studies carried out. The present discussion will be limited mainly to the human data.

Studies, most of which have been retrospective, have not indicated that caffeine is a teratogen for man

(Crainicianu 1969; Heinonen *et al.* 1977*b*; Weathersbee *et al.* 1977; Streissguth *et al.* 1978*b*; Borlée *et al.* 1978; Rosenberg *et al.* 1982; Linn *et al.* 1982). Apart from the inherent problems of retrospective surveys, it has not been possible to distinguish the effects of caffeine *per se* from the effects of the caffeine-containing beverages. Other confounding factors, such as smoking and alcohol consumption which are known to correlate with caffeine consumption and are also known to affect fetal development, have usually not been adequately controlled. Although five of the six retrospective studies reported no increase in the incidence of congenital malformations, there are considerable differences in the other observations reported.

Weathersbee *et al.* (1977) carried out a survey of 489 households, 75 per cent of which were Mormon in which alcohol, caffeine, and nicotine were proscribed. The results of this survey showed that high caffeine consumption was associated with an increase in abortions and perinatal mortality, but not congenital malformations. The authors suggested that caffeine intakes in excess of 600 mg daily, i.e. equivalent to eight cups of coffee daily, may predispose women to reproductive problems.

Heinonen *et al.* (1977*b*) in the Collaborative Perinatal Project found no evidence of teratogenicity amongst high caffeine consumers, they estimated the standardized risk in the offspring of 12 700 mothers who took caffeine in pregnancy to be 0.99. In a survey of 1439 babies whose mothers had been exposed to caffeine during pregnancy there was a twofold increase in abortion rates and a threefold increase in breech presentations in women who had ingested more than six cups of coffee per day (Streissguth *et al.* 1978*b*). Furthermore, the women who were heavy coffee users had a history of significantly more spontaneous abortions in previous pregnancies. However, there was no increase in the incidence of congenital malformations. Twenty of the babies from women consuming more than 444 mg/day caffeine were rated as abnormal due to a significant increase in cyanosis and acrocyanosis ($p < 0.025$). The Brazelton neonatal assessment was performed on 417 one-day-old babies born to mothers in the high-caffeine group. This showed that these babies had lower activity scores, poorer muscle tone, and more motor immaturity. All of the above observations were independent of smoking, alcohol consumption, sex, parity, examination age, or the examiners used, but the effects were slight.

Borlée *et al.* (1978) studied 202 malformed children and 175 normal control children in Belgium from 1972 to 1974. No definite association between coffee consumption, smoking, drug use, spontaneous abortion, or length of gestation was found. However, there was a slight increase (44 malformed *vs.* 21 controls, $p < 0.05$) in the number of mothers of malformed babies compared with the control mothers who drank more than eight cups of coffee per day. Few details were given in this study, but the authors felt that the data presented did not show a definite association between congenital malformations and coffee consumption.

More recently Rosenberg *et al.* (1982) of the Boston drug epidemiology unit, reported the data obtained on a case control study of 2030 malformed babies. Six selected birth defects were evaluated in relation to maternal ingestion of caffeine from tea, coffee, or cola. The defects studied were; inguinal hernia (380); cleft lip with or without cleft palate (299); cardiac defects (277); pyloric stenosis (194); isolated cleft palate (120); and neural-tube defects (101). These were compared with 712 infants with other types of malformations.

There were no noticeable differences between the distributions of case and control mothers as to the type or amount of beverage consumed. Overall, 75 per cent of mothers drank tea and about 50 per cent drank coffee and cola drinks. Decaffeinated coffee was taken by 18 per cent of the mothers, and these were similarly distributed amongst case and control populations. Only 21 mothers (1 per cent) were estimated to have taken as much as 1000 mg caffeine daily, whereas 218 mothers (11 per cent) had consumed about 400 mg caffeine, equivalent to at least four cups of coffee daily. These authors observed no significant effects due to caffeine consumption and concluded that it was not a major teratogen with regard to the six defects evaluated. Furthermore, when caffeine consumption was assessed in the controls, there was no consistent evidence of association for any specific defect with caffeine. The authors estimated that, in this study, over 85 per cent of the daily dietary intake of caffeine came from coffee and tea and during pregnancy women seemed to be more moderate drinkers of these beverages. As only 11 per cent of the mothers drank four or more cups of coffee daily, birth defects could not be evaluated in relation to very high levels of caffeine consumption. As most of this data was collected before the hypothesis that caffeine caused birth defects became popular and the question on beverage consumption was only one amongst a large number of items on which information was elicited, the authors feel that their data are fairly free from recall bias. Unfortunately no data were published on either alcohol consumption or smoking habits.

Linn *et al.* (1982) analysed the medical records and interview data on 12 205 women (non-diabetic and non-asthmatic) to evaluate coffee consumption and adverse pregnancy outcome. Coffee and tea consumption, but not other sources of caffeine such as soft drinks or drugs, were assessed in the first trimester, and confounding factors such as smoking and alcohol were assessed in each trimester. Previous medical and obstetric histories were also taken into account. About 34 per cent drank no tea or coffee during the first trimester; 57 per cent did not drink coffee. Only 5 per cent drank four or more cups of coffee/day and less than 1 per cent drank more than seven cups of coffee per day.

The results of this study showed that there was no excess of malformations among heavy coffee drinkers. In the group consuming no tea/coffee, 9.2 per cent of the women had malformed babies compared with 7.4 per cent in the heavy coffee consumers group. Most of the malformations were minor. Major malformations were reported in 2.5 per cent of babies from the no tea/coffee group compared with 2.0 per cent in the high coffee group. The study is large enough to have an 85 per cent chance of detecting a true twofold increase in major malformations and a 93 per cent chance of detecting a 1.5 fold increase in all malformations ($p < 0.05$), but the power to detect small increases in specific defects was low. Low birth weight and short gestation occurred more often among the offspring of women who drank four or more cups of coffee per day but, after analysis to control for confounding factors, it could be shown that this effect was almost entirely due to smoking and there was no relationship between low birth weight or short gestation and heavy coffee consumption independent of smoking. Thus it would seem that smoking rather than coffee consumption influenced the length of gestation and birth weight.

More heavy coffee drinkers had previous poor obstetric outcomes such as a stillbirth or a spontaneous abortion, and a positive relationship was also found between high coffee consumption and premature rupture of the membranes and breech presentation, with about a 1.5-fold increase in incidence. This supports the data reported by Streissguth et al. (1978b),

The study by Linn et al. has been the subject of some criticism mainly relating to the fact that the analysis was based on cups of coffee consumed per day and that they did not have information on the consumption of other sources of caffeine such as soft drinks, chocolate, or caffeine-containing drugs. This could have reduced the power of the study (Bracken et al. 1982; Luke 1982; Steinmann 1982). However, coffee's lack of a teratogenic effect is supported by Kurppa et al. (1982, 1983) from Finland. In the Finnish study, which is still in progress, information has been collected on 706 mothers of malformed children and an equal number of matched controls. The malformations studied were CNS (112) structural-skeletal malformations (210) cardiovascular (143) and oral clefts (241). Analysis of the results gave no

indication that coffee intake was associated with an increase in congenital malformations.

Although many of the criticisms of the Linn et al. study are valid they merely emphasize the inherent difficulties in the design and interpretation of such retrospective studies on caffeine consumption. The confounding effects of alcohol consumption and smoking which are known to relate to coffee intake may be very large. Van den Berg (1977) from the Californian Child Health and Development Study 1959–66 also reported that the apparent association between coffee consumption and prematurity was in fact, largely due to confounding factors such as smoking. However, Hogue (1981) opposed these conclusions. She re-analysed Van den Berg's data and showed that there was a statistically significant relative risk (1.24) of low birth weight amongst consumers of seven or more cups of coffee daily, even after controlling for smoking.

The only prospective study published is that of Mau and Netter (1974). They studied 5200 maternity cases and found no association of congenital malformations with coffee consumption. However, they did report a higher incidence of low birth weight (<2.5 kg) babies born to women who were heavy coffee drinkers.

Mulvihill (1973) ascribed the absence of congenital malformations in humans to the rapid metabolism of caffeine, less than 1 per cent being excreted unchanged. More recent studies have indicated that the metabolism of caffeine alters during pregnancy (Aldridge et al. 1981; Miller and Harris 1981; Knutti et al. 1981) and is greatly influenced by smoking. Pregnancy prolongs the plasma half-life of caffeine from six hours in the first trimester to between 10 and 18 hours in the second and third trimesters. By one week postpartum, values have fallen to non-pregnant values of five to six hours. In women who smoke, the plasma half-life of caffeine is decreased by about 50 per cent to 3.5 hours. In a small study in pregnant women, Kirkinen et al. (1983) showed that intervillous placental blood flow was slightly reduced 30 minutes after drinking two cups of coffee. In seven hypertensive women however, placental blood flow was already lower than in this normotensive women and the coffee consumption did not reduce this further.

The overall conclusions to be drawn from the epidemiological studies are that there is some evidence for an increase in spontaneous abortions, breech presentations, and low birth weights, in relation to high caffeine consumption, but there is no evidence of an increase in the incidence of congenital malformations. An important consideration in these studies in the known correlation between high coffee consumption, alcohol intake, and smoking. Since both alcohol and smoking are known to affect fetal development, birth weight, neonatal mortality, and postnatal development, assessment of any effect of caffeine, must involve careful control of these confounding variables. The studies described above have not adequately controlled for these factors. Furthermore, it is not clear from these studies whether effects observed are related to the total caffeine intake or more specifically related to the coffee, tea, etc. which are sources not only of caffeine but of many other chemicals also.

None of the studies described have shown a causal relationship between congenital malformations and the consumption of caffeinated beverages. Thus it would seem that drinking normal amounts of these beverages during pregnancy holds no major teratogenic hazard for the fetus.

Chemotherapy

ANTIBIOTICS

Antibiotics are among the most commonly prescribed drugs during pregnancy and the perinatal period. As most antibiotics readily cross the placenta and are present in breast milk, it is most important that consideration is given to the possibility of adverse effects in the fetus and neonate as well as in the mother.

Although pharmacokinetic data on placental transfer of antibiotics in humans is very limited, there is evidence that the data obtained in non-pregnant subjects cannot be directly extrapolated to the pregnant subject (Schwarz 1981; Landers et al. 1983; Ledward 1983). In a comprenhensive review of antibiotic use during pregnancy and in the postpartum period Landers et al. (1983) emphasized that the physiological changes which occur during pregnancy, i.e. alterations in plasma protein concentrations, increased cardiac output and renal flow, and an increase in gut motility during late pregnancy, may alter the pharmacokinetics of the particular drug which may either enhance its toxicity, or cause a decrease in blood levels of the drug, thus giving inadequate protection to both the mother and fetus. It has been estimated that the overall percentage decrease of antibiotic serum concentration in pregnancy is in the range 10–50 per cent compared with that in the nonpregnant woman. Thus it is of extreme importance that once a decision has been taken to give an antibiotic an adequate dose of the drug is used.

Although there is good evidence that almost all antimicrobial drugs cross the placenta (Pomerance and Yaffe 1973), in general they do not present any serious teratological hazard. However, suspicions were first aroused by the report that of 85 women treated in the first trimester with antibiotics, 12 delivered malformed

infants (Carter and ...
studies, however, failed ...
(Nelson and Forfar 1971).

Recently there have been ...
published on the toxic effects of a ...
and recommendations as to the drug ...
bacterial infections (Schwarz 1981; S ...
Louvois 1983; Batagol 1983; Ledward 19 ...
al. 1983). Predictions concerning safety can be ...
a certain amount of confidence for some ...
antibiotics such as the penicillins which have been ...
for many years, but with some of the newer antibiot ...
very little is known about their safety in pregnancy.

In the following section, each of the classes of antimicrobial drugs will be discussed with regard to their safety for use in pregnancy.

Penicillins

Penicillins such as benzyl penicillin, amoxycillin, ampicillin, and flucloxacillin cross the placenta readily, but seem to pose very little risk to the fetus. Although penicillin usage in the first trimester has been associated with an increase in cleft lip and palate (Saxen 1975); other reports have not confirmed this. The penicillins have been extensively used in pregnancy and it would seem that those in current use are not hazardous to the fetus or neonate. If maternal hypersensitivity is a problem, cephalosporins may be used instead, although care must be exercised if penicillin hypersensitivity is very serious; because approximately 10 per cent of people allergic to penicillin will be allergic to cephalosporins also. The newer penicillins such as mezlocillin, azlocillin, and piperacillin are reported to have a wider range of activity, but they have not been in use long enough to assess any possible hazards for the fetus.

Cephalosporins

The cephalosporins such as cephaloridine, cephadrine, cefuroxime, cefaclor, and cefoxitin, are broad-spectrum antibiotics with pharmacological activity similar to the penicillins. Cephalosporins cross the placenta and reach sufficiently high concentrations in the amniotic fluid and fetal blood to inhibit both the gram-positive and some of the gram-negative organisms commonly found in intrauterine infections (Stewart et al. 1973). The cephalosporins have been extensively used in pregnancy and they seem to be relatively safe as judged by the lack of reports associating their use with congenital malformations or other toxic effects. Animal studies have shown a lack of teratogenicity with most of the cephalosporins, but some of the newer members of the

or, when
nic effect
is to their
eem that
more on a
her than
erapeutic
, if high
e fetus,
such as
drugs of

treating fetal infections such as syphilis (Philipson *et al.* 1973). Erythromycin itself does not seem to cause any particular hazard to the fetus. However, erythromycin estolate should be avoided as it causes hepatotoxicity in the mother; treatment for 10–14 days for instance can give rise to choleostatic jaundice and thus may have indirect toxic effects on the fetus.

Aminoglycosides

The aminoglycosides such as neomycin, gentamicin, kanamycin, streptomycin, tobramycin, and amikacin are bactericidal and have been associated with some degree of ototoxicity and nephrotoxicity. Although placental transfer of these drugs is poor, there is evidence that they cause eighth-nerve damage and hearing loss in the offspring. Some damage was reported in 11 per cent of children whose mothers were treated with streptomycin, and furthermore the effect was not dose-related (Ganguin and Rempt 1970). However, not all investigations have confirmed this (see pp. 726–8 for further consideration). Nevertheless, the use of aminoglycosides in pregnancy is best avoided. However, they may be necessary for the treatment of severe genital-tract infections, in which case careful monitoring of the plasma levels is to be recommended. Furthermore, it has been recommended that, if the aminoglycosides are used in the third trimester, especially if the baby is premature and requires an incubator, auditory function should be carefully monitored up to school age (Batagol 1983). Gentamicin has been used in labour and so far has not been associated with adverse effects. On the other hand, amikacin has been used so infrequently that no recommendations can be made regarding its safety for use in pregnancy.

Sulphonamides

The use of sulphonamides such as sulphamethoxazole, sulphamethizole, sulphadiazine, sulphadimethoxine, and sulphametopyridazine has increased as a result of increasing bacterial resistance to antibiotics which are in general more active and less toxic. The sulphonamides are transferred rapidly to the fetus via the placenta, and they have been associated with a small increase in congenital malformations by Nelson and Forfar (1971) but not by Richards (1969) or Saxen (1975). Their use in late pregnancy has been associated with neonatal jaundice and kernicterus. The sulphonamides displace bilirubin from binding sites on the plasma proteins and compete with it for conjugation with glucuronic acid. This results in high levels of bilirubin and allows it to diffuse into the fetal brain to cause damage (Boggs *et al.*

acycline,
pectrum
antibiotics whose usefulness has decreased as a result of bacterial resistance. However they do remain the drugs of choice for infections caused by mycoplasma, rickettsia, and chlamydia and in severe acne vulgaris.

Tetracyclines if given during pregnancy have been shown repeatedly to affect the developing teeth of the offspring. If given after the twelfth week of gestation, hypoplasia and yellow/brown discoloration of the deciduous teeth may occur, and in late pregnancy the permanent teeth may be affected (Weyman 1965). Oxytetracycline which produces a creamy discoloration may be the best choice if use of such a drug is essential in pregnancy. Incorporation into fetal bones with a reversible growth retardation also occurs (Greene 1976). Tetracyclines have also been implicated as a cause of cataracts though the evidence is slight (Harley *et al.* 1964).

There is also an increased risk of maternal toxicity such as fatty necrosis of the liver, pancreatitis, and liver damage associated with tetracycline use in pregnancy. Another important factor is that in recent years tetracyclines have become very popular for treating acne vulgaris and in such cases a woman may be exposed to tetracyclines before she realizes she is pregnant. Thus considering the possible toxic effects on both mother and fetus, tetracyclines are best avoided in pregnancy and should be used cautiously in women of child-bearing age.

Marcrolides

Erythromycin has a similar though not identical antibacterial spectrum to that of the penicillins, and may be used in cases of penicillin hypersensitivity. Although erythromycin crosses the placenta, relatively low levels are achieved in the fetal circulation when therapeutic doses are given to the mother. Thus it is of limited value in

1967). The risks seem to be greater in pre-term babies with immature liver and kidney function. There is also a risk of neonatal methaemoglobinaemia if the drug is administered in late pregnancy. Thus the sulphonamides are not recommended for use in late pregnancy, i.e. after the 38th week or if premature onset of labour is suspected. They are also contraindicated in subjects with G6PD-deficiency.

Co-trimoxazole

Co-trimoxazole is a mixture of five parts sulphamethoxazole and one part trimethoprim and is available under the trade names of *Bactrin, Fectrim,* and *Septrin.* Although there is no evidence either way as regards its toxic effects on the fetus, it is generally considered to be contraindicated in the first trimester. Trimethoprim is a folate antagonist, and low folate levels had been associated with an increased incidence of congenital malformations. Sulphamethoxazole is a sulphonamide, of which the properties and the adverse effects on the fetus and neonate have been discussed above. Thus it would seem prudent to avoid its use in pregnancy especially if the folate status of the mother is in doubt, if there is a malabsorption syndrome, or if she is receiving anticonvulsant drugs. If the use of co-trimoxazole is considered to be essential, the co-administration of folic acid supplements (5–10 mg daily) is recommended. It is interesting that trimethoprim has been reported to affect sperm production. A drop in sperm count has been reported in patients who had a course of co-trimoxazole (Murdia *et al.* 1978), but this has yet to be confirmed by other studies. It has been suggested that trimethoprim, by inhibiting folate reductase, may deprive the rapidly dividing spermatogenic cells of active folate.

Other antibiotics:

Clindamycin and Lincomycin

Their safety for use in pregnancy has not been established but they do cross the placenta readily.

Novobiocin

Its safety in pregnancy has not been established. This drug has limited use owing to the frequency of adverse effects and the development of bacterial resistance. It is known that novobiocin crosses the placenta and gains access to the fetus. If used shortly before delivery, it may predispose to neonatal jaundice and thrombocytopenia (Ledward 1983). Thus the risk–benefit ratio must be very carefully assessed prior to its use in the pregnant patient.

Chloramphenicol

When administered in the first and second trimesters it has been associated with maternal toxicity, such as aplastic anaemia, rather than fetal toxicity. Its use in the third trimester could be associated with the development of the 'Grey baby syndrome' which may occur following treatment of the neonate, although there are no reports to substantiate this. However, it is known that the fetus/neonate lacks the necessary enzymes to conjugate the drug which leads to a prolongation of the serum half-life and a concomitant increase in toxicity. Thus its use throughout the whole of pregnancy is best avoided where possible.

Nitrofurantoin

Nitrofurantoin has been used for over 25 years and no adverse effects on fetal development have been reported. In a retrospective study of 101 pregnant women treated with nitrofurantoin, 200–400 mg daily, and the same number of controls, no adverse effects were reported (Perry *et al.* 1967). However, if nitrofurantoin is used near term it may cause neonatal haemolysis in G6PD-deficient babies. It is contraindicated in subjects with renal impairment, and for this reason it is not recommended for use in the neonate or in premature infants.

Nalidixic acid

Nalidixic acid has been reported to produce teratogenic effects in rats and rabbits. However, it has been used to treat urinary-tract infections during pregnancy and there have been no reports on adverse effects in the fetus (Kucers and Bennett 1979). The drug is concentrated in the urine, with only low levels in the serum.

Amphotericin B

The safety in use of amphotericin B during pregnancy has yet to be established. One report has indicated that, when used to treat serious fungal infections, amphotericin B was not associated with any adverse effects on the fetus (Philpot and Lo 1972).

Summary

In conclusion, the majority of the commonly used antibiotics seem to be safe for use during pregnancy provided that the usual precautions are taken that also apply in the nonpregnant patient. Changes in metabolism do occur during pregnancy so that the pharmacokinetics of the antibiotic may differ during pregnancy requiring serum levels to be carried out to ensure adequate control of infections. The safety of many of the newer antibiotics has not been established in pregnancy so that use of the traditional drugs is preferred where these are adequate for control of infection. Use of the penicillins and cephalosporins seems safe. Antibiotics to be avoided during pregnancy are the tetracyclines and aminoglycosides because of known fetal toxicity, and it is best to avoid the use of sulphonamides, co-trimoxazole, and chloramphenicol if other antibiotics would be equally effective.

Antituberculosis drugs

The introduction of antituberculosis drugs such as streptomycin, isoniazid, paraaminosalicylic acid, ethambutol, ethionamide, and rifampicin has improved the prognosis for both mother and fetus. Although the administration of drugs to pregnant women is not to be recommended routinely, antituberculosis drugs are amongst the exceptions, since pregnancy can aggravate active TB and thus endanger maternal and fetal well-being. However, the use of these drugs during pregnancy has presented new problems, namely, are any or all of these drugs teratogenic; are any particular drug combinations safer than others; is drug administration safer at any particular stage of pregnancy. Many physicians are so concerned about the possible adverse effects of these drugs and/or the disease on the fetus that they still advocate induced abortion, although this view need no longer be held. Data on the teratogenic effects of some of the antituberculosis drugs in animals, and more recently in man, has been accumulating. The opinions of Pomerance and Yaffe (1973) and Wilson and Fraser (1977) indicating no increase in the incidence of congenital malformations associated with streptomycin, isoniazid, and *p*-aminosalicylic acid were mentioned briefly in the original chapter. Subsequent workers do not necessarily agree with these findings. There have been several reports (Bobrowitz 1974; Rocker 1977; *Lancet* 1980*b*; Good *et al.* 1981) and reviews (Warkany 1979; Snider *et al.* 1980) on most of the antituberculosis drugs in use today. Where possible each antituberculosis drug will be evaluated individually for its effects in pregnancy, although most have been assessed as part of combination therapy.

There have been several reports on the effects of ethambutol (EMB) in pregnancy (Place 1964; Bobrowitz 1974; Lewit *et al.* 1974; Warkany 1979; Snider *et al.* 1980). In some of the studies EMB was used as part of combination therapy. Bobrowitz (1974) examined 42 pregnancies from 1964 to 1973 in 38 women who had

chronic pulmonary TB treated with ethambutol. In a follow-up of these children, the oldest being about nine years of age, he found that eight of the 42 children had a variety of different minor anomalies which were not thought to be related to treatment. His overall conclusion was that EMB had no teratological effect and was safe for use in pregnancy. Although a loss of vision often accompanied the use of this drug in the early days of its use, several reports (Place 1964; Lewit *et al.* 1974; Brock and Roach 1981) found no ocular defects in the offspring of mothers exposed to EMB either in the first trimester or throughout pregnancy. Snider *et al.* (1980) reviewed reports of 655 pregnancies from 650 women who had been treated with EMB usually in combination with INH. There were 592 normal infants, and 14 abnormal infants with a variety of defects. None of these were ocular defects which supports the above observations. There were also 26 premature births and five still births reported.

The use of streptomycin (SM) and dihydrostreptomycin (DHSM) was reviewed by Warkany (1979). Placental transfer has been established in animals, but on the whole the fetal concentrations were lower than those of the mother. Ototoxicity is known to occur in adult patients and there are conflicting reports as to how often eighth-nerve damage occurs in the offspring. By 1976 there were about 30 reports of children with vestibular or hearing deficiencies whose mothers had taken SM or DHSM during their pregnancies. There is no clear indication that exposure in the first trimester is more or less dangerous than during later stages of pregnancy. Although full adult development of the inner ear is said to be complete by midterm, even the mature adult ear can be damaged by SM or DHSM, An editorial in the *Lancet* (1980*b*) also warns that SM may cause deafness. Snider *et al.* (1980) reported a high incidence of ototoxicity in fetuses exposed to SM which seemed to be unrelated to the time of exposure. They quote that one in six fetuses develop some hearing or vestibular defect, and, furthermore, a substantial number of them were deaf. In this respect Snider and co-workers regarded SM as potentially dangerous throughout the whole of pregnancy which confirms the overall conclusions of Warkany (1979).

There are only a few reports associated with the use of isoniazid (INH) in pregnancy. There is one report by Monnet *et al.* (1967) which described five children with severe encephalopathies. These children were exposed to INH at different times during pregnancy; only two of whom were exposed in the first trimester. However, four out of five of them had severe psychomotor retardation and convulsions. The other child had cerebral hemiplegia. The authors speculated that it might be a hypovitaminosis of vitamin B_6 which might be the teratogenic factor.

Snider *et al.* (1980) reviewed 1302 reports of women exposed to INH with a total of 1480 pregnancies. Of these, over 400 were known to have taken INH during the first four months of pregnancy. The most common dose used was 300 mg daily and was often accompanied (1417 cases) by ethambutol. Over 95 per cent of the pregnancies resulted in normal babies and only 16 of the infants had malformations.

Ethionamide (ETA) has been known to cause abortions and malformations in rodents and rabbits. However, there are contradictory reports concerning its use in human pregnancy. Potworowska *et al.* (1966) in Poland reported malformations in seven of 23 infants exposed *in utero* to ETA. On the other hand, Schardein (1985) cites several studies, comprising a total of 70 cases in all in which no relationship between malformations and ETA exposure could be established.

Rifampicin (RMP) has been the subject of many reports in recent years. Jentgens (1975) described 82 pregnancies in which various combination therapies including RMP and EMB were used. There was no increase in the overall malformation rate, and he was strongly against termination on the grounds of risk from chemotherapy alone. Warkany (1979) states that, although there is good evidence in rodents that RMP is teratogenic, there is no evidence of teratogenicity in children exposed to moderate therapeutic doses. He cites (without giving references) 34 cases of normal children who had been exposed to RMP *in utero* with normal fetal development. Snider and co-workers (1980) collected data on 446 pregnancies exposed to RMP and the outcomes were 386 normal-term infants, seven spontaneous abortions, two premature births, nine stillbirths, and 14 abnormal offspring. The 14 abnormal children had a variety of structural defects including three limb reduction defects. The overall incidence of malformations with rifampicin was 3.4 per cent which is only slightly above the incidence expected in the general population. The overall increased risk is therefore small with this drug, but seems greater than with INH or ethambutol. This drug is also known to cause hepatotoxicity in adults, menstrual irregularities, amenorrhoea, and loss of oral contraceptive efficacy.

Warkany (1979) thought that the risk of therapy during pregnancy is reduced if therapy is given after the first trimester and doses are kept to a minimum. However, as adults are also prone to eighth-nerve damage when treated with SM, the safety of the fetus cannot be guaranteed at any stage of pregnancy. Snider *et al.* (1980) on the basis of an extensive review of the literature, however, recommend the following drug regime for pregnant tuberculous patients.

INH which has a high efficacy and patient acceptability seems to be the safest drug which has had the most extensive clinical use, with 95 per cent of pregnancies exposed to INH giving rise to normal infants, and only 1.09 per cent with malformations. They recommend the use of INH with ethambutol if the disease is not extensive. The overall malformation risk with ethambutol being 2.19 per cent in 655 pregnancies reviewed. If a third drug is required, rifampicin should be used. Streptomycin should be avoided if possible because of the risk of eighth-nerve damage, but can be used if RMP is contraindicated. They do not recommend advising therapeutic abortion for pregnant women who are taking first-line antituberculous therapy.

Metronidazole

Robinson and Mirchandani (1965) reported on a study in which 300 women were treated with metronidazole, usually in late pregnancy, and there were no adverse effects on the fetuses. Since then there have been several reports on the use of metronidazole in pregnancy. Morgan (1978, 1979) reported the results of a study of 597 pregnant women treated with oral metronidazole for trichomonas infection. There was another group of 283 pregnant women in whom the infection was left untreated. The majority of women were diagnosed at the first examination, but treatment was often deferred until the second trimester. The treated group received either a 7- or 10-day course of oral metronidazole 200 mg three times per day. Of these 62 were treated during the first trimester, 284 during the second trimester, and 251 during the third trimester.

There were no differences in the incidence of either premature births, small-for-dates babies, stillbirths, or major or minor congenital malformations in the treated compared with the untreated group. However, Morgan recommended that if possible treatment should be deferred until the second trimester.

The safety of the drug for long-term use has been questioned because it was found to be mutagenic in bacteria and carcinogenic in rodents. The significance of this for man, however, is uncertain (*British Medical Journal* 1980b). Long-term therapy (more than three months) in adults is associated with peripheral neuropathies, which are usually reversible on cessation of treatment.

Guilhou and Meynadier (1980) felt that these were sufficient reasons to avoid its use in pregnancy. The authors regarded it as relatively safe for the short-term treatment of rosacea in pregnancy, but as the long-term effects are unknown, they suggested limiting treatment to three months.

On the other hand Burton and Saihan (1980) stated that careful monitoring of the infants in several studies over the last 20 years has shown no increase in the incidence of congenital malformations, abortions, or stillbirths even when the drug is used during the first trimester. Biggs and Allan (1981) in a general review of medication in pregnancy positively recommended the use of metronidazole for trichomonas infections though other authors (Rao and Arulappu 1981; Landers *et al.* 1983) felt it should be reserved for treatment of serious infections only, and, if possible, it should be avoided in the first trimester.

Likewise Robbie and Sweet (1983) in a very comprehensive review of metronidazole use in obstetrics and gynaecology concluded that during pregnancy and lactation, metronidazole should be avoided if at all possible. Although they quoted no good evidence for metronidazole being teratogenic in humans, they recommended that, if it is indicated in pregnancy, its use should be limited to the second and third trimester only. As metronidazole concentrations in serum and milk are comparable, if it is warranted during lactation, they recommend a single dose of 2 g, with lactation interrupted for 24–48 hours since the neonate has immature development of hepatic and renal enzymes and drug metabolism and elimination may occur more slowly.

The increase in the incidence of cancer in rodents treated with metronidazole has caused some concern. There have been two retrospective studies in humans: one involved 2460 subjects treated with metronidazole from 1969 to 1973 (Friedman 1980); the other involved 771 women treated from 1960 to 1969 (Beard *et al.* 1979), and no increase in cancer due to metronidazole was observed. In a critical review of the experimental and clinical data on the mutagenicity, carcinogenicity, and teratogenicity of metronidazole, Roe (1983) concluded that it was essentially free from carcinogenic and teratogenic risk. He did not, however, consider the possibility of a transplacental carcinogenic effect. Since the fetus may be more susceptible to carcinogens than adults, caution should be used in prescribing this drug for long periods in pregnant women.

In conclusion, although there seems to be no increase in congenital malformations associated with metronidazole use, nevertheless it is desirable where possible to defer treatment until after the first trimester. If there is a possibility of transplacental carcinogensis, then it should be prescribed for pregnant women only if absolutely necessary and long-term follow-up of the infants is desirable.

Antimalarial drugs

The incidence of malaria world-wide has continued to change in recent years and has worsened in a number of countries, particularly in South-East Asia. In many countries the resurgence of malaria has been attributed to vector resistance to prophylactic drugs such as chloroquine. This widespread development of drug resistance causes many problems regarding the choice of an appropriate drug for use in a given area, and these problems were reviewed in a report of a meeting of the Ross Institute (1981). The Ross Institute recommended the use of six main drugs which span the range of antimalarial activity; the 4-aminoquinolines such as chloroquine and amodiaquine; drugs with antifolate activity, such as pyrimethamine and proguanil, and drugs which are a combination of pyrimethamine plus dapsone (*Maloprim*) or pyrimethamine plus sulpha-doxine (*Fansidar*).

Although it is generally accepted that visitors to malarious areas should take prophylaxis regularly while in that area and for 4–6 weeks afterwards, the situation as regards treating pregnant women and infants is not so clear because of the possible risk to the fetus from the prophylactic drugs. Pregnancy interferes with the immune processes of malaria, a disease which itself alters immune reactivity. Mothers in highly endemic areas may lose some of their immunity and thus suffer more severe attacks of malaria (Bruce-Chwatt *et al.* 1981; Bruce-Chwatt 1983).

The placenta often becomes infected, particularly with *P. falciparum*, which can cause intrauterine growth retardation and failure to thrive. Neonatal mortality is much greater in this group of children. Occasionally, malaria may be transmitted from mother to child via the placenta, but it is quite a rare event and it is thought to occur more commonly in untreated, non-immune women when compared with women in highly endemic areas and with high immunity (Logie and McGregor 1970; Hindi and Azimi 1980; Bruce-Chwatt *et al.* 1981). Another risk of malaria is severe anemia of pregnancy, probably due to a combination of nutritional and parasitological causes, in highly endemic areas which often results in maternal death. Thus withholding malarial prophylaxis from pregnant women can have profound effects on the mother and fetus which may be fatal. On the other hand, most of the commonly used antimalarial drugs have been reported to produce teratogenic effects in animals and there have been a number of case reports in humans suggesting a causal relationship. Several reviews on malaria prophylaxis during pregnancy have been published (Bruce-Chwatt *et al.* 1981; Trussel and Beeley 1981; Ross Institute 1981;

Batagol 1983; Bruce-Chwatt 1983) and, without exception, these authors recommended that malaria prophylaxis should be given, because malaria is a greater hazard than the drugs to both the mother and the fetus. The choice of malarial prophylaxis in pregnancy has however been the subject of much debate, and each of the six commonly used drugs will be discussed individually.

Chloroquine

Chloroquine, which is neurotoxic in adults was suspected of teratogenicity following the case of one woman who produced two deaf children while on this drug, but four normal children when not on the drug (Smith 1966). There have been several case reports associating the use of chloroquine with a variety of malformations including optic and otic defects (reviewed by Schardein 1985). However, these malformations have followed the use of much higher doses of chloroquine than those used for malaria prophylaxis. Where the parasite is sensitive to chloroquine it remains the drug of choice. The risk of treatment would seem relatively small in comparison to those of untreated malaria.

Amodiaquine

The risk to the fetus is thought to be similar to that quoted for chloroquine.

Pyrimethamine

Pyrimethamine is teratogenic in animals when given in high doses (reviewed by Schardein 1985). There are a few case reports associating pyrimethamine with congenital malformations. This drug has been used for several years and there is no good evidence that there is an increased risk of malformations in humans when it is used for prophylaxis (Ross Institute 1981; Batagol 1983). However, as pyrimethamine is a folic acid antagonist, its use in the first trimester of pregnancy is best avoided if possible. When it is used, folate supplements should be given. Pyrimethamine, which is tasteless and needs to be given only once weekly for prophylaxis, may be the drug of choice if vomiting of pregnancy is a problem.

Proguanil

There are no data which specifically implicate proguanil as a teratogen in humans. However, as it is a folate antagonist, the same precautions as described above for pyrimethamine should be observed.

Maloprim (pyrimethamine plus dapsone)

Both pyrimethamine alone and dapsone alone have been used in higher doses than those used in *Maloprim* for many years without evidence of a teratogenic effect. Because of the teratogenic effects of large doses of pyrimethamine in animals, the use of Maloprim in pregnancy is not recommended by several national authorities or by the WHO. However, where the risk of chloroquine-resistant malaria is high, the Ross Institute (1981) recommended that *Maloprim* may be used together with 10–15 mg folic or folinic acid.

Fansidar (pyrimethamine plus sulphadoxine)

Very little is known about the effects of *Fansidar* in human pregnancy. Animal studies have shown that doses of *Fansidar* equivalent to half the human therapeutic dose, given to rats during organogenesis produce teratogenic effects, fetal loss, and a decrease in maternal weight gain. Pyrimethamine has been discussed previously, and there is no evidence that sulphadoxine alone has toxic effects in pregnancy when given in therapeutic doses. As so little is known about the side-effects of this drug, it is best avoided in pregnancy if at all possible, its use being strictly limited to instances where there is a high risk of exposure to chloroquine-resistant malaria. Because of the risk of kernicterus in the newborn following sulphonamide use, *Fansidar* is best avoided in the third trimester.

Quinine

Although the newer antimalarial drugs are considered to be safer than quinine for use in pregnancy, there are instances where it may be necessary to use quinine. In situtations where the malarial parasite has become resistant to the newer drugs, quinine may be the only effective drug to use. Quinine has often been used to induce abortions and this has resulted in some cases in the development of congenital malformations. Tanimura and Lee (1972) cited 21 cases of teratogenicity related to quinine use in humans between 1949 and 1972. However, the doses used were very high, between 1 and 4 g and were taken to produce abortion. There are other examples in the literature with a wide range of malformations reported, but no consistent syndrome, although several cases of ototoxicity and retinal damage have been reported (reviewed by Schardein 1985). On the other hand there is no indication that the drug is teratogenic in normal therapeutic doses used for malaria treatment (Lenz 1966; Heinonen *et al.* 1977b; Batagol 1983; Bruce-Chwatt 1983).

Primaquine

Primaquine, is best avoided in the third trimester of pregnancy because of the possibility of neonatal haemolysis, methaemoglobinaemia, and kernicterus. It is also contraindicated in women with glucose-6-phosphate dehydrogenase deficiency.

Recommended scheme for choosing malarial prophylaxis for the pregnant women

Up-to-date information on the predominant type of malaria and whether drug resistance is a problem in the particular area to which the person intends to travel must be obtained. Advice on this together with recommendations on the most suitable drugs for that area can be obtained in the UK from the Bureau of Hygiene and Tropical Medicine or from British Airways.

Having obtained this information, it is necessary to assess the risks associated with the particular drugs recommended against the clinical status of the mother, e.g. G6PD-deficiency, folate deficiency malnutrition, or sulphonamide sensitivity and known toxic effects on the fetus. The best possible advice to the pregnant woman is not to visit areas where malaria is prevalent if it can be avoided. If this is not possible, Table 22.3 gives what is

thought to be a suitable regimen for malarial prophylaxis in the pregnant woman.

Cancer chemotherapeutics

The anticancer drugs which are specifically designed to inhibit cell division include examples of known teratogens in man. Practically all types of anticancer drugs have been shown to be teratogenic in most species including man, i.e. alkylating agents, antimetabolites, antitumour antibiotics, and steroids. However, not every woman treated with large doses of antitumour drugs will have a malformed child. There are numerous cases on record where a woman has had more than one pregnancy while on a prolonged course of chemotherapy resulting in both normal and abnormal (or aborted) fetuses. In a review of the use of anticancer drugs in pregnancy (Gililland and Weinstein 1983) the overall malformation rate, considering all drugs used, was about 14 per cent which illustrates the high proportion of normal babies born despite intensive treatment covering all stages of gestation. Catanzarite and McHargue (1983) on the basis of recent literature suggested that a fetal survival rate of 80–95 per cent may be expected in pregnancy compli-

Table 22.3   Regime for malarial prophylaxis in pregnancy

Drugs in order of preference	Comments
Chloroquine	The drug of choice: major concern is the increased number of areas where the parasite is chloroquine-resistant (often resistant to pyrimethamine and proguanil as well).
Amodiaquine	Similar to chloroquine — possible alternative if there is chloroquine resistance.
Proguanil	Resistance — often associated with chloroquine resistance. Folic acid antagonist — best avoided in first trimester. Always give folate supplements.
Pyrimethamine	Folic acid antagonist. Avoid in first trimester if possible. Always give folate supplements. Resistance — often associated with resistance to chloroquine and proguanil.
Maloprim (pyrimethamine plus dapsone)	Give folic acid supplements (10 mg). Use only if risk of malaria is high and if there is chloroquine resistance.
Fansidar (pyrimethamine plus sulphadoxine)	Teratogenic in animals at doses less than human therapeutic dose. Little information as regards human pregnancy. Avoid use in late pregnancy. Restrict use to situations where other prophylactics are ineffective.
Primaquine	Resistance — often associated with chloroquine resistance. Best avoided in third trimester, particularly last week of pregnancy — risk of neonatal kernicterus haemolysis and methaemoglobinaemia. Contraindicated in G6PD deficiency.
Quinine	Reports of teratogenicity in animals and humans with high doses. Restrict use to situations where there is chloroquine resistance and other prophylactics are ineffective.

cated by nonlymphocytic leukaemia. Few specific patterns of malformations have been described following chemotherapy, but almost every organ system may be affected depending on the time of administration. The literature was reviewed by Schardein (1985) and more recently by Gililland and Weinstein (1983). The latter authors attempted to catalogue the drugs used, the time of administration, and the subsequent outcome of pregnancy. Most of the evidence they presented is based on individual case reports, sometimes backed up by animal data. Although the significance of individual case reports is difficult to assess in terms of overall risk, they are useful in order to pick up any pattern of defects which may occur for future risk assessment.

Alkylating agents

Schardein (1985) attempted to assess the teratogenic risk of some of the alkylating agents based on the ratio of malformed to normal babies reported in the literature.

Busulphan	1:9
Chlorambucil	1:2
Cyclophosphamide	1:6
Nitrogen mustard	1:3
TEM	0 (seven reported normal pregnancies)

Since physicians are less likely to report normal than abnormal outcomes of pregnancy, it is probable that these estimates err on the high side. Gililland and Weinstein (1983) reported four congenital abnormalities and 26 normal babies who had been exposed to busulphan.

Antitumour antibiotics

The antitumour antibiotics, such as actinomycin C and D, chromomycin A3, daunomycin, azastreptonigrin, puromycin, sarcomycin, streptonigrin, but not streptozotocin, are very teratogenic in animals but have not been reported to cause malformations in man. Negative results were reported in three women treated with actinomycin C, in six women on actinomycin D seven women on daunorubicin, seven women on doxorubicin, and seven women on adriamycin (Schardein 1985). One pregnancy in which bleomycin was used ended in the birth of a normal child (Gililland and Weinstein 1983).

Antimetabolites

The antimetabolites which act by interfering with the actions of normal metabolites like folic acid, nicotinamide, purines, and pyrimidines are teratogenic in animals and man (reviewed by Schardein 1985). Aminopterin, a folic-acid antagonist, was one of the first drugs to be formally tried as an abortifacient in women (Thiersch 1956; Goetsch 1962) and, of 42 women treated, only 29 aborted and three live congenitally malformed babies were delivered in the survivors (plus two dead malformed abortuses). A number of other malformed children were born following treatment with aminopterin or methotrexate (Shaw and Rees 1980; Gililland and Weinstein 1983). A wide variety of defects were observed depending again on the time of administration but malformations of the head such as anencephaly, hydrocephalus, wide-spaced eyes, and ear defects have been commonly reported (Milunsky et al. 1968). Overall, aminopterin seems to be more teratogenic and fetotoxic than methotrexate where the risk seems to be about 1 in 20. Another antimetabolite, 6-azauridine, has also been implicated as a teratogen (Vojta and Jirasek 1966). The only other antimetabolite about which there is a reasonable amount of information is 6-mercaptopurine, with over 24 cases of normal outcome of pregnancy being reported (Schardein 1985) though malformations following its use have been reported. There have also been occasional case reports associating the use of 5-fluorouracil (Stephens et al. 1980), cytarabine (Wagner et al. 1980), and combination therapy with cytarabine and thioguanine (Schafer 1981) with congenital malformations.

Vinca alkaloids

Although the vinca alkaloids such as vincristine and vinblastine are very powerfully teratogenic and fetotoxic in animals there is insufficient evidence to assess the risk for man. However, it is encouraging that normal pregnancies have resulted following the use of these drugs. Gililland and Weinstein (1983) report on 25 pregnancies where the fetus was exposed to the vinca alkaloids often in combination with other anticancer drugs, and of these two babies were malformed.

Immunosuppressants

Although little information is available as regards immunosuppressant drugs in pregnancy, from data on 100 women treated with azathioprine or cyclophosphamide it would seem that there is a high incidence of intra-uterine growth retardation (Scott 1977). However, there does not seem to be an increase in reports of congenital malformations in infants exposed to azathioprine even during the first trimester (Gililland and Weinstein 1983).

Other cyctotoxic drugs

There is little information regarding the safety in use during human pregnancy of urea derivatives, nitroso-compounds, colchicine, and demecolcine. All of these have been shown to be teratogenic and fetotoxic in animals. Both normal and abnormal outcomes have been reported in the few pregnancies where the mother has been treated with procarbazine or colchicine. However, negative results have been reported so far when triethe-lenemelamine (four cases) and urethane (seven cases) have been used (Gililland and Weinstein 1983).

There has been one large study published by Blatt *et al.* (1980) describing a survey of 488 patients treated with cancer chemotherapy at the National Cancer Institute, USA. This group consisted of 212 females and 236 males, of whom 23 of the females had 30 pregnancies and seven males fathered 12 pregnancies. The diagnoses for the patients were variable, but all had received moderate-to-high doses of combination chemotherapy. In the female group, 19 of the 23 women (83 per cent) conceived (1 month to 9 years) after therapy was stopped. Of the remaining four women, one conceived and was treated during the first two months of pregnancy, one started treatment in the second trimester, and the other restarted treatment in the third trimester. In the group of seven males, two were receiving therapy at the time of conception.

Within the whole group there were 10 elective abortions, two spontaneous abortions, and 28 singleton full-term births in which all the babies had normal birth weights. The spouses of two of the treated males were still pregnant. Four of the 28 infants had defects. One child had a congenital dislocation of the hip, but this was probably not related to treatment as chemotherapy had stopped six years prior to the pregnancy. Another child, whose mother had been on chemotherapy had a history of febrile seizures, but was otherwise normal at 5½ years of age. There was one child with bilateral pilonidal dimples but no spina bifida, whose father was on chemotherapy; this child was otherwise normal at three months. The fourth child, whose father was off therapy for 4½ years prior to conception, was born with a capillary nevus, but was otherwise normal at 15 months old.

At follow-up examination the children ranged in age from a few days old to 12 years, and, as far as it was possible to ascertain, their growth and development as well as school performance seemed normal. These authors concluded that chemotherapy given prior to conception, or after the first trimester, or given to males prior to conception, does not necessarily jeopardize the infant, and in the majority of cases a normal outcome

may be expected. However, caution is necessary in the interpretation of this study since the number of subjects is small.

Studies of this kind, including follow-up, are very important, because now that various forms of cancer can be controlled, the likelihood is that more women will want to become pregnant.

The above study emphasizes the point made earlier that despite the teratogenicity and mutagenicity of such drugs in animal models, not every woman who is on or has been on therapy will have a poor pregnancy outcome. This information is important when counselling such patients who ask for advice.

Conclusion

Large doses of the cytotoxic drugs are teratogenic in animals and there are reports implicating the majority of them as teratogens in humans also. Nevertheless, the overall malformation rate in women treated with anticancer drugs during pregnancy is probably less than 15 per cent. This means that even when treatment is carried out during the first trimester, the chances of having a normal baby are still better than six to one. When treatment is carried out late in gestation the risk of anaemia or leukopenia should be considered in the newborn.

Cough suppressants

Cough suppressants containing morphine derivatives or codeine would be expected to present the same hazards as the use of these drugs as analgesics (see above). Iodide-containing cough suppressants have been implicated in teratogensis in man. There are a number of individual reports of women taking iodide in cough mixtures (or for hypothyroidism) who have given birth to children with goitres. This may be severe enough to kill the child (Louw 1963; Iancu *et al.* 1974).

Hallucinogens, etc.

Lysergic acid diethylamide (LSD), mescaline, mari-juana, methltryptamine, and tetrahydrocannabinol have all been shown to be teratogenic in animals but there is little evidence regarding their effects in man. LSD has been suspected of being teratogenic and the cases reported were reviewed by Long (1972). Of 161 children born to mothers who took LSD, 140 were normal. Of the 21 abnormal children, 14 had limb defects of some sort, including five with congenital absence of hands, fingers, or toes. Thus, although the overall incidence of abnormalities is not very high, the incidence of limb defects is sufficient to raise a high degree of suspicion. There are no good data on the effects in humans of the other drugs.

A review of the effects of marijuana in pregnancy concludes that, although large doses in animals may cause fetal death, the production of malformations is relatively unusual except when very high doses are given. There are marked species differences and malformations have not been produced in primate studies. Human data are inadequate for assessment but one study in Canada of 291 pregnant women did not find an effect of cannabis use on birth weight, head circumference, or length. The number of heavy users, however, are very small (Abel 1983).

Hexachlorophane

Hexachlorophane (HCP) is classified as an antiseptic and skin disinfectant, and as such is used in creams, ointments, lotions, and talcs. It was first marketed in the 1940s but it was not until the 1970s that it was discovered that repeated use could cause toxicity problems.

Hexachlorophane can penetrate unbroken skin and lead to significant blood concentrations of the drug. Smaller quantities were found to accumulate in adipose tissue, brain, and liver as well as other organs (Food and Drugs Administration 1978; Halling 1979). The half-life of hexachlorophane is about 19 hours and it is rapidly excreted after administration, but with persistent use complete elimination may take 4–7 days. About 84 per cent is excreted unchanged in the faeces and urine within 96 hours of a topical application.

In a study in the USA in which 75 volunteers washed their hands four times a day with 5 ml of a 3 per cent hexachlorophane solution, the mean blood level attained after 21 days was $0.21\ \mu g\ ml^{-1}$. On the tenth day of a similar study on eight men and eight women the mean blood level was $0.37\ \mu g\ ml^{-1}$. However, a number of the women consistently had much higher values than this (Lockhart 1972).

Local application of hexachlorophane to skin can produce vacuolation and oedema in the CNS, status spongiosus in the heavily myelinated long tracts of the brainstem, and intramyelinic vacuolation of the interperiod line. Similar neuropathies have been found in animal studies. Multiple defects have been reported in rabbit and rodent studies and optic-nerve damage and sometimes blindness have occurred in sheep. Moreover, the blood concentrations of hexachlorophane in these animals were very similar to those reported for man.

An *FDA Drug Bulletin* published in September 1978 issued an interim precaution to surgeons and nurses along with other health care personnel, especially if they were pregnant, to avoid hexachlorophane scrubs.

This warning was based partly on the data from France where infants who had been exposed accidentally to high (6 per cent) concentrations of hexachlorophane in talcum powder had suffered brain damage. This outbreak of accidental hexachlorophane poisoning in France in 1972–73 affected 204 children of whom 36 died. The source of the poisoning was a baby talcum powder which would normally have contained no hexachlorophane, but which, because of a manufacturing error, containing 6.3 per cent hexachlorophane. The syndrome consisted of ulcerative skin lesions, fever, vomiting, diarrhoea, muscular weakness or paralysis, encephalopathy, seizures, and coma. Post-mortem examination of the 36 infants, mostly after exhumation, showed that hexachlorophane was widely distributed and concentrated in brain, liver, kidney, lung, and skin. A single blood sample obtained from a dying child had a concentration of $1.15\ \mu g\ ml^{-1}$ hexachlorophane. Light microscopy of samples of brain and spinal cord obtained from 12 children immediately after death showed vacuolation of the white matter, similar to the 'status spongiosus' lesions described in rats treated with hexachlorophane. Studies are still in progress to investigate any long-term or delayed health effects in the infants who survived the epidemic (Martin-Bouyer et al. 1982). Studies in rhesus monkeys also reported the occurrence of brain lesions associated with the topical application of hexachlorophane.

In 1972 the FDA banned all non-prescription use of the drug. Hexachlorophane was available as a surgical scrub to health care personnel only. In the UK several proprietary baby lotions and ointments were removed from the market, and re-marketed at a later date, minus the hexachlorophane. The FDA recommend a maximum concentration of 0.1 per cent as a preservative in cosmetics. Safety tests are now compulsory on all products containing 0.1–0.7 per cent HCP. Products containing more than 0.7 per cent HCP are available on prescription only. It is recommended for washing neonates only if infection is present and there is no other satisfactory treatment.

Hexachlorophane dusting talcs and lotions such as *pHiso-MED* and *Ster-Zac DC* are still available in the UK but most of the antiseptic lotions now contain chlorohexidene rather than hexachlorophane.

There have been two studies on the use of hexachlorophane in Swedish hospitals by Halling (1979) and Baltazar et al. (1979). Halling reported the findings of a retrospective study carried out in six hospitals for chronic diseases in Sweden. Severe congenital malformations occurred more frequently in neonates born to mothers who used hexachlorophane soaps 10–60 times per day in handwashing and as handcreams, during the first trimester. According to these figures, 25 of 460 infants had severe malformations in the hexachlorophane group compared with none of 233 control infants born to mothers similarly employed, but who did not use hexachlorophane-containing products. The incidence of minor anomalies was also higher in the hexachlorophane group than in the controls. Other factors, such as age, infections, drug usage, or smoking habits, did not seem to differ between the two groups.

In two of the hospitals during the years 1970–76, six of 65 neonates were severely malformed (anal atresia oesophageal altresia, pulmonary hypoplasia, renal polycystic disease, microphthalmia, and cleft lip and palate). Their mothers had used the soap *Sanitval* (0.5 per cent HCP) or *pHisohex* (3 per cent HCP). There were also several other infants with minor anomalies such as club foot and cardiac anomalies. None of the 68 control babies had major malformations.

One other hospital in the group assessed the data (1969–75) on the nursing staff who washed their hands 14–67 times per day with a liquid soap containing 1 per cent HCP. During 1969, a handcream (0.5 per cent HCP) was also issued to them. These mothers were exposed at least during the first trimester and gave birth to 82 infants (1970–75) of whom 14 had malformations. Four of these infants had very severe cardiac defects. None of the 46 controls was malformed. In the other three hospitals, examination of hexachlorophane-exposed staff showed that 15 children with major malformations were born from a total of 313 deliveries.

This paper was severely criticized by Källen (1978) who described in detail the method of selection of cases used by Halling. He stated that she selected three clusters of malformations picked up from a much larger National Study where no overall effect was observed. The selection of controls were also biased and this is supported by the lack of any serious malformations in the 233 controls (seven expected, $p < 0.01$). Doubts on the validity of the study were also cast by Janerich (1979a). The full report of the Swedish national study mentioned above was published by Baltazar *et al.* (1979). They carried out two types of study. In the first, 1500 women who worked in 31 hospitals for chronic diseases and who gave birth during 1965–75 were studied. In the second study, all women working in medical occupations in Sweden and who delivered babies during 1973–75 were investigated. This comprised a cohort of 30 048 babies which represented 9.3 per cent of all infants delivered in Sweden during these years. Neither study revealed any

overall increase in malformation rate though the presence of the clusters reported by Halling was detected. There was no evidence, however, that this related to the use of hexachlorophane. The malformation rate and perinatal death rate in hospitals using hexachlorophane did not differ from those in hospitals not using hexachlorophane.

In conclusion it seems likely that the observation of teratogenic risk associated with the use of hexachlorophane reported by Halling (1979) was due to selection bias including a number of clusters of malformations of unknown aetiology. The much larger studies by Baltazar *et al.* (1979) did not show any increased risk of malformations associated with hexachlorophane use. However, the severe neurotoxicity of hexachlorophane in babies in the French disaster would caution against the use of this substance in pregnancy and in the neonatal period, since marked dermal absorption does occur and mild neurological damage would not have been detected in the Baltazar *et al.* study.

Hypnotics and sedatives

Most of the anxiolytics will induce sleep when given in large doses at night and most hypnotics will sedate if given in divided doses during the day.

Benzodiazepines

The benzodiazepines such as nitrazepam, temazepam, flurazepam, and triazolam are ones commonly used. There is no evidence that these are any better or any worse than diazepam when used during pregnancy (see 'Minor tranquillizers').

Barbiturates

Phenobarbitone has already been discussed in the section on anti-epileptic drugs. There are epidemiological studies implicating the shorter-acting barbiturates, such as amylobarbitone, as teratogenic if taken during the first trimester but the evidence is not strong (Nelson and Forfar 1971). There have also been occasional case reports associating several drugs in this group, i.e. barbitone, secobarbitone, and heptobarbitone, with congenital malformations (Schardein 1985). More recently there has been growing concern regarding the effects of barbiturates used in late pregnancy and prior to term. Respiratory depression, sedation, coagulation defects, and withdrawal symptoms have been reported in neonates. Some of these symptoms may persist for several months into postnatal life and thus the possibility of effects on brain development and subsequent postnatal behaviour cannot be ignored.

Non-barbiturates:

Thalidomide

It is not necessary to review again the effects of thalidomide in any detail. It was a mild hypnotic of very low acute toxicity in animals, yet proved to have disastrous effects on the human fetus. It is not known precisely what the incidence of defects was, but almost 100 per cent of women who took thalidomide during the sensitive period, i.e. days 21–36 of gestation, produced malformed infants. The malformations consisted of short or absent arms or legs as the most obvious defect, but deformities of the ears, heart, gastrointestinal tract, and of the renal and urogenital systems were also produced. In about three-quarters of the cases, only the arms were affected. The total number of children affected was estimated by Lenz (1966) at about 7000–8000 mostly in W. Germany, Britain, and Japan. Despite a considerable amount of work, its mechanism of action as a teratogen is still unknown.

Glutethimide

Glutethimide because of its close structural similarity to thalidomide has been studied in some detail, but does not seem to be teratogenic in animals (Bennet 1962). However, withdrawal symptoms in the neonate from this drug have been reported (Batagol 1983).

Chloral hydrate

The little evidence available so far, seems to indicate that chloral hydrate is safe (Nelson and Forfar 1971). Although large doses of chloral hydrate have been reported to cause fetal death, there is no evidence to indicate that it was teratogenic (Batagol 1983).

Conclusions

There seems to be good evidence that barbiturates taken during the first trimester can produce a small increase (two-threefold at most) in the incidence of congenital malformations. Thus in real terms this still means that the baby has at least a 90 per cent chance of being normal. The risk of malformations with benzodiazepines, if any, would seem to be low. Thalidomide has been removed from general use; however, it is used in special circumstances, and thus its teratogenic effects must be

borne in mind when considering its use in women of child-bearing age. Glutethimide and chloralhydrate would seem not to pose any particular hazard to the fetus as regards structural malformations.

However, with all of the drugs which affect the CNS, the late effects on the fetus and neonate such as respiratory depression, hypotonia, and effects on brain development and postnatal behaviour, must be carefully considered.

Hypoglycaemics

Diabetes is associated with increased risks to the fetus and good control of the pregnant diabetic is essential. In general, insulin seems to be the drug of choice in the pregnant diabetic though even with good control the incidence of defective children may still be higher than normal. A number of reports associating congenital malformations with the use of the oral hypoglycaemics have been published. Overall an incidence of about 5 per cent of malformations in the offspring of women on sulphonylureas would imply an approximate twofold increase in malformations. No clear syndrome of defects has been observed but since these drugs are potent teratogens in animals, it would seem wise to avoid their use in man if possible. Some more recent studies, however, (Sutherland et al. 1973; Jackson 1974) have suggested that chlorpropamide in doses up to 200–250 mg per day is harmless to the fetus. This, however, confines the use of the drug to the milder diabetic.

Although not a hypoglycaemic drug, saccharin is widely used by diabetics as well as by many normal women as an aid to weight control. Following the Canadian Government report that saccharin was carcinogenic in the offspring of rats treated with very high doses, a prospective study was set up to investigate whether saccharin use was associated with an increase in abortion rates in women. This might be expected if saccharin was mutagenic or teratogenic. A series of 574 consecutive spontaneous abortions was compared with 320 pregnant controls matched for a number of variables such as age, race, marital status, educational level, previous abortions, and smoking. The use of sugar substitutes was compared between the two groups and no significant association was found between spontaneous abortion and use of sugar substitutes (Kline et al. 1978).

Major analgesics

The major or narcotic analgesics can be divided into two main groups; firstly those which are used to treat mild-to-moderate pain such as codeine, dextropropoxyphene, and pentazocine; secondly those used to treat moderate-to-severe pain, such as morphine, heroin, and methadone. On the whole the risk of addiction with drugs in the first group is low, although there is an increasing problem of drug abuse with dextropropoxyphene combinations. The second group of drugs frequently produce tolerance and dependence with repeated use.

Drugs for mild-to-moderate pain relief:

Codeine

Codeine is prescribed usually in combination with other drugs for a variety of complaints but mainly for the relief of mild-to-moderate pain or as an antitussive. Many of these preparations can be bought over the counter. Despite the fact that codeine has been extensively used, there are very few data available on the effects of codeine in either animal or human pregnancy.

Saxen (1975) published the results of a survey on human congenital malformations from the Finnish Register of Congenital Malformations 1967–71. He reported that codeine had been taken by three times as many women in the study group during the first trimester when compared with the control group. Among the total malformations observed, cleft lip with or without cleft palate had the highest incidence (8.6 per cent).

A neonatal withdrawal syndrome was also reported in a baby whose mother had taken large doses of codeine and aspirin throughout pregnancy (Van-Leeuwen et al. 1965). The symptoms were relieved by the administration of codeine to the baby. Four other cases were cited by Batagol (1983) — two cases where codeine had been taken as an antitussive for 10 days to three weeks prior to delivery and two other cases of neonatal withdrawal symptoms in infants of non-addicted mothers.

Because there is so little experimental or clinical data available it is not possible to assess the teratogenic potential of codeine. Caution must be used in assessing the data from the Finnish study because codeine is rarely used alone but is normally combined with other analgesics and may be used for a variety of conditions which may themselves affect pregnancy.

Dextropropoxyphene (propoxyphene)

Propoxyphene is a synthetic derivative of morphine and bears structural similarities to methadone. It is often used in combination with paracetamol (*Distalgesic*) to treat moderate-to-severe pain. Despite its widespread use, there is very little information available regarding its

effects in pregnancy. Equivocal results have been reported from teratogenicity studies in animals. There have been occasional case reports associating the use of dextropropoxyphene with congenital malformations (Barrow and Suder 1971; Douglas Ringrose 1972; Schardein 1985; Golden et al. 1982). However, the evidence is equivocal, as in most of the above cases, other drugs had been taken as well.

Pentazocine

According to the manufacturers' reports, there is no evidence from studies in rats and rabbits that pentazocine has any adverse effects on fertility, length of gestation, litter size, or on fetal development (reviewed by Schardein 1985). Occasional case reports have indicated that infants exposed *in utero* to pentazocine suffer from withdrawal symptoms (Batagol 1983). There have been reports of two normal children being born to pentazocine-treated mothers (Kopelman 1975; Kwan 1970), but no large studies on the use of this drug in pregnancy have been reported.

Drugs used for moderate-to-severe pain

Morphine, methadone, and derivatives are non-teratogenic in most animal tests, though they have been shown to be teratogenic in hamsters (Geber and Schramm 1975). One of the difficulties in interpreting animal data is that these drugs depress respiration and food intake, both of which are known to cause malformations in animals. However, it does seem that prenatal exposure to narcotic drugs such as methadone can produce adverse effects on postnatal development. Walz et al. (1983) recently reported on a multigeneration study in which rats of both sexes were treated with methadone for 60 days prior to mating. There was a reduction in body weight, delayed development of reflexes, and altered behaviour patterns in the offspring. Furthermore, breeding offspring (F1) of which both parents had methadone gave F2 generation offspring for which neonatal mortality was significantly increased. However, no methadone-related effects were seen in the F3 generation. Szeto (1983) also reported effects of methadone in fetal lambs.

Other than isolated case reports, which cannot be evaluated, there is no good evidence that narcotic analgesics are teratogenic in man (reviewed by Schardein 1985; Batagol 1983). A number of studies have shown no increase in either total or specific defects in offspring of women taking heroin, with or without methadone (Blinick et al. 1973; Lipsitz and Blatman 1973). However, women taking such drugs near term are very liable to induce an addictive state with a pronounced withdrawal syndrome in the baby after delivery. There may be low birth weight and severe respiratory distress leading to a high neonatal mortality, though with adequate care this is not inevitable (Harper *et al.* 1974). Unfortunately there is now evidence available that prenatal exposure to these narcotic drugs may produce effects on postnatal behaviour and subsequent development (Hutchings 1982; Johnson and Rosen 1982; Lifschitz *et al.* 1983).

Hutchings (1982) in a comprehensive review of methadone and heroin during pregnancy discussed the evidence for behavioural effects in both human and animal offspring. He concluded that clinical and animal studies showed that prenatal exposure to opiates produced effects that occur in two phases. Firstly, there is an early acute phase, which can be quite prolonged lasting for 12 weeks to six months in humans, during which there is a 'neonatal abstinence syndrome', CNS arousal, hyperactivity, disturbed sleep, and increased lability of state. It is thought that these protracted symptoms may result from the slow clearance of the drug from the neonatal tissues. The second phase of the syndrome, seems to be less well understood, but, the limited clinical evidence suggests that exposure to heroin, particularly as part of poly-drug abuse, can result in impaired organizational and perceptual abilities, poor self-adjustment, and heightened activity in situations requiring motor inhibition. Although studies of preschool children exposed to methadone *in utero* showed no ill effects on intellectual and cognitive function, they did reveal hyperactivity, impaired motor inhibition, impulsive behaviour, and brief attention span. It is interesting that many of these behavioural patterns have been picked up in behaviour screens in animals also.

Johnson and Rosen (1982) studied the effects of prenatal methadone exposure in early infancy as part of an ongoing longitudinal study. Forty-one children born to methadone-maintained mothers and 23 children from matched backgrounds, but with a negative history of maternal drug abuse were evaluated at six months of age. Physical and neurological examinations as well as a battery of behavioural assessment tests were carried out on each of these children. The data collected from these children at six months of age showed no significant differences between methadone-exposed and control infants as regards suspect abnormal neurological signs, visual habituation, mental development, motor development, or object permanence. However, suspect abnormal signs were significantly correlated with low Bayley Scales for methadone infants which suggests that those experiencing difficulties do so in multiple areas. Unfortunately there are no clear indications as to which

factors predispose these infants to such developmental disorders. The only significant factor related to developmental outcome was gender — there were significantly more low scores among methadone-exposed males than females. The authors concluded that, although they could not define a 'fetal methadone syndrome', there was a higher incidence of low developmental scores amongst the methadone group, particularly in males at six months of age which they felt warrants further follow-up.

Lifschitz *et al.* (1983) reported on fetal and postnatal growth of children born to narcotic-dependent women delivered at Houston Public Hospital from 1974 to 1977. They studied the effects of heroin and methadone on birth length and three-year stature of children of 22 untreated heroin addicts, 21 women receiving methadone maintenance therapy (95 per cent of whom were poly-drug users), and a drug-free comparison group of 28. The mean birth lengths of both groups of drug-exposed infants were significantly lower than that of the comparison group; however, after correction for various factors such as sex, race, prenatal care, obstetric history, maternal education, and smoking, the group means were similar.

At three years of age the mean height was comparable for all three groups. However, when adjusted for birth length, parental height, and smoking, the methadone group was significantly shorter than that of the children exposed *in utero* to heroin; the control group assumed an intermediate position. The authors concluded that the influence of intra-uterine exposure to narcotics on fetal growth cannot be separated from the influences of other factors closely associated with the life-style of women who abuse drugs.

This is a particularly important consideration as drug abuse is on the increase in UK.

Minor analgesics and non-steroidal anti-inflammatory drugs

Minor analgesics used extensively for the relief of mild-to-moderate pain, such as aspirin, paracetamol, codeine, etc., have been on the market for several decades. However, very little is known about the effects on fetal development and the safety for use in pregnancy of some of these compounds. It is only in recent years that the differences in metabolism in the pregnant and nonpregnant subject, and in the fetus and neonate have been fully appreciated. This has led to the publication of numerous reviews on the use of these drugs in pregnancy, labour, and in the postpartum period (Schardein 1985;

Rudolph 1981; Collins 1981; Batagol 1983; Fidler and Ellis 1983).

The introduction of the non-steroidal anti-inflammatory drugs such as naproxen and indomethacin has posed new problems thought to be associated with their proposed mechanism of action as prostaglandin synthesis inhibitors (Nuki 1983). There are some animal data available on the reproductive toxicity of certain of these newer drugs, but in most cases their safety for use in human pregnancy has not yet been established. However, in a comprehensive review of rheumatoid arthritis in pregnancy, Bulmash (1979) concluded that none of the newer anti-inflammatory drugs should be regarded as safe and that they should be avoided, and that no new drugs should be started during pregnancy. Furthermore, patients already taking such drugs should have the dosage reduced or stopped if possible and be maintained on aspirin alone. He recommends that, despite the potential fetal hazards associated with aspirin, it is the safest drug available to treat arthritis in pregnancy.

SALICYLIC ACIDS

There is a voluminous literature showing that salicylates are teratogenic in a variety of animal species including mice, rats, guinea-pigs, cats, and monkeys, but less so in the rabbit. The doses used in all of these experiments were very high, however, so the relevance to man is doubtful. There is, however, a good deal of evidence regarding the use of these drugs in man. With the possible exception of vitamins, analgesics are the drugs most commonly used by pregnant women. In developed countries the incidence of usage is remarkably similar. Bleyer *et al.* (1970) in a carefully controlled study showed that in their patients in the USA aspirin was the most commonly taken drug (average 8.7 drugs per woman) and was used by 69 per cent of the patients. Hill *et al.* (1972) and Hill (1973*a*) showed that in Texas women took an average of 13 drugs during pregnancy with analgesics being the most common and used by 64 per cent. Forfar and Nelson (1973) showed that in Scotland the average number of drugs prescribed for pregnant women was four, and after iron preparations, analgesics topped the list with 63 per cent of women using them (three-quarters of them being self-prescribed). Aspirin was used by 54 per cent of the mothers, and they state that '2 in every 3 mothers take aspirin in full dosage for 6 weeks during pregnancy.' In a survey of 4000 Australian patients (Collins and Turner 1973), it was shown that the Mediterranean migrants (i.e. one-third of the out-patient population) did not take salicylate regularly, but of the 2600 Australian-born women, 44 per cent did take

aspirin regularly. These women could be divided into two groups, the constant takers (i.e. every day) and the intermittent takers (i.e. at least once a week). A large study in Boston, USA (Stone *et al.* 1976) on 50 282 women showed that 32 164 of them (64 per cent) took aspirin at some time during their pregnancy, 29.6 per cent of them taking it during the first four months.

With such widespread use of aspirin and with the known problems of underreporting of self-medicated drugs and the wide variety of unrecognized preparations containing aspirin, it is difficult to assess by normal epidemiological methods the possible hazards of analgesics in pregnancy, though they must obviously be fairly small. Furthermore, analgesics are always taken to relieve some underlying problem be it headache, stress, or influenza and these confounding variables must always be borne in mind when assessing the epidemiological findings.

Aspirin

Since 1970 there have been numerous surveys on the teratogenic effects of aspirin on human pregnancy. An editorial article in the *Medical Journal of Australia* (1971) reported a significant increase in the consumption of aspirin taken by mothers of children with congenital malformations during the first 28 days of pregnancy. A retrospective study carried out by Nelson and Forfar (1971) in Edinburgh showed that a high proportion of mothers took aspirin during the first 56 days of pregnancy. There was a significant difference for aspirin consumption in the first 28 days: eight out of 458 mothers of malformed children were aspirin-takers compared with three out of 911 in the controls ($p < 0.05$). The average dose of aspirin in the study group was just over half that of the controls. The abnormalities in these eight babies included achondroplasia, hydrocephalus, congenital heart disease, mongolism, congenital dislocation of the hip, hydrocele, talipes, and papilloma of the forehead. Richards (1969, 1972) reported the results of a retrospective survey in South Wales in which it was found that salicylates had been taken by 22.3 per cent of the mothers with malformed babies, but by only 14.4 per cent of the controls, during the first trimester. Moreover, with the exception of the cardiovascular system, in each affected system there was a higher proportion of salicylate-takers in the malformed group than in the controls. These effects were particularly marked if salicylates were taken during the first trimester as opposed to the second and third trimesters. The main defects observed were anencephaly, spina bifida, cleft lip with or without cleft palate, pyloric stenosis, talipes, and gastrointestinal tract abnormalities. However, a check with the

Royal College of General Practitioners regarding drugs on prescription showed that there was no excess of prescription for 'aspirin-type' drugs given to the mothers of malformed children. Thus, Richards, concluded that aspirin may be a weak teratogen in man. Sever (1972) raised the question of whether it was aspirin itself or the condition for which aspirin was taken that was responsible for the malformations. He reported that salicylates taken to treat influenza rather than the virus *per se* may be responsible for the increase in malformations after an influenza epidemic, as an increase in drug usage (drug sales) correspond with the time of the epidemic. He too reported a significantly higher usage of aspirin in the mothers of malformed children compared with their matched controls. Sever concluded that population differences in drug usage or differences in teratogenic sensitivity may account for the conflicting data. McNiel (1973) reported eight cases of congenital malformations associated with the maternal use of aspirin; however, in each of these cases other drugs had been taken and therefore one cannot discount the fact that these other drugs themselves, or in combination, may have been responsible for the defects observed.

Data compiled from the Finnish Register of Congenital Malformations (1967–71) indicates that the consumption of salicylates during the first trimester was nearly three times more frequent in mothers of children with oro-facial clefts than by the control mothers ($p <$ 0.001). The rate was highest for cleft lip ($\pm$ palate) in that it reached about four times the control values. Differences were also noted between the control and study groups during the second and third trimesters, but they were not so marked as in the first trimester. Karkinen-Jaaskelainen and Saxen (1974) also examined 80 mothers with children with CNS defects who had had influenza in the first trimester and compared them with matched control groups. Of the group with influenza plus defects, 52 per cent took drugs (43 per cent took analgesics, mainly aspirin (34 per cent), amidopyrine, and phenacetin); of the control with CNS defects but without influenza, 28 per cent took drugs (16 per cent analgesics); and, in the absolute controls, 11–17 per cent took drugs. They concluded that it was impossible to separate possible drug effects from the influenzal effects.

A large prospective epidemiological study of the Boston Collaborative Peri-natal Project (Slone *et al.* 1976) on a cohort of 50 282 pregnancies in the USA of whom 64 per cent took aspirin at some time during the pregnancy, showed that the malformation rates in the babies were similar in all the three groups of:

1. 35 418 women who did not take aspirin during the first four months of pregnancy;

2. the 9736 women with 'intermediate' exposure ($<$8 days' use);

3. the 5128 women with 'heavy' exposure ($>$8 days' use).

They concluded that the study gave no evidence that aspirin ingestion during pregnancy was associated with an increase in malformations. Furthermore, they stated that the confidence limits of the analysis are such that it was unlikely that any substantial teratogenic effect would have been missed.

There are numerous individual reports of cases in which aspirin has been implicated as a possible teratogen but with a background congenital malformation rate of 2–3 per cent these claims cannot be validated, and there are also many cases of women taking large doses of aspirin throughout pregnancy without ill-effects on the baby. In fact, it was suggested (Ben Ismail *et al.* 1975) that aspirin may be safer, for women requiring anticoagulant therapy, than coumarins and more convenient than heparin.

Although the majority of the evidence on aspirin suggests that it is not teratogenic if used in normal dosage, there is very clear evidence that it may produce other toxic effects on both the mother and the baby if used in late pregnancy.

One of the most important deleterious effects of aspirin taken in late pregnancy relates to an association between aspirin and an inhibition of platelet function. In the non-pregnant human it has been shown that aspirin alters platelet function by loss of secondary aggregation in response to ADP or adrenaline and inhibition of aggregation by collagen (Al-Mondhiry *et al.* 1970). Following on their investigation of drug usage in late pregnancy (Bleyer *et al.* 1970) and the observations of Palmisano and Cassiday (1969) that in 272 consecutive newborns screened for plasma salicylate, 10 per cent had levels greater than 1 mg/100 ml, Bleyer and Breckenridge (1970) examined the effect of aspirin on the baby. They compared 14 babies exposed to aspirin in the last week of pregnancy with 17 not exposed. They found two adverse drug-related reactions with significant differences between the groups. First, there was platelet dysfunction with inhibition of collagen-induced aggregation, and secondly diminished factor XII (Hageman factor). Three of the aspirin babies had bleeding episodes compared with one in the control group. They suggested that the use of aspirin should be restricted in the last month of pregnancy. Several other workers have since reported similar findings. In a study on seven infants whose mothers had taken unspecified doses of aspirin during pregnancy, cord blood was collected at delivery and tests for platelet function were performed. These tests showed that the platelets of newborn infants are less reactive than those of the adult to all of the tested aggregating agents (Casteels-Van Daele *et al.* 1972). The platelet-release reaction was also impaired in the newborn whose mothers had taken aspirin just a few days prior to delivery. Studies on the prenatal effects of aspirin have shown that there was a decrease in platelet aggregation and dimished Hageman factor (XII) activity in the infants whose mothers had taken aspirin during the last week of pregnancy (the dose in some cases being as little as one tablet — 5 grains, 325 mg). An impaired ability to detoxify and excrete salicylates and a tendency of haemorrhage was also indicated in the neonate (Pochedly and Ente 1972). Brown (1974) showed that all of 10 babies whose mothers had taken aspirin in the last week of pregnancy (and eight of the mothers) had impaired platelet aggregation, and she showed that neonatal platelets are 10 times more sensitive to aspirin than adult platelets.

The results of another survey on the use of drugs and the production of congenital malformations implicates aspirin as a causative agent in the occurrence of haemorrhages just before term, at delivery, and neonatally. There is also the possibility of severe haemorrhaging at delivery if the mother has taken salicylate just prior to this (Schenkel and Vorherr 1974).

Rumack *et al.* (1981) reported on a prospective study of 108 infants, born at 34 weeks gestation or earlier or weighing 1500 g or less, carried out to determine the incidence of intracranial haemorrhage and the multiple risk factors involved in causing or aggravating this condition in premature infants. On day two postpartum the mothers were asked whether they took aspirin or paracetamol in the last week of pregnancy. The infants were scanned between days 3 and 7 postpartum. Intracranial haemorrhages developed in 53/108 infants (49 per cent) overall and in 31/71 (44 per cent) of infants whose mothers had not taken either aspirin or paracetamol. The number of intracranial haemorrhages in the aspirin-exposed group (12/17, 71 per cent) was significantly greater than that seen in the controls or the control and paracetamol groups ($p < 0.05$). The incidence of haemorrhages in the paracetamol group was not significantly different from the controls. Thus the authors concluded that the use of aspirin in the last three months of pregnancy is highly questionable and probably inappropriate.

The results of an Australian survey of 144 mothers, who had taken salicylates during pregnancy for varying times and had both cord and maternal blood samples taken at delivery have been published (Turner and Collins 1975). These results showed that the babies were of low birth weight (after correction for smoking and

birth order) and, moreover, it was shown that the birth weight decreased with increasing length of time of salicylate medication. Some of them had features of dysmaturity such as cracked skin, and a lack of subcutaneous fat. There were no serious haematological features; however, some did develop jaundice of a transient nature. Turner and Collins concluded from this study that chronic salicylate ingestion is associated with an increase in perinatal mortality and a decrease in uterine growth, but is not involved in teratogenesis. This study was criticized by Shapiro *et al.* (1976a) on the basis of the small number of subjects and the inclusion of data on fetal wastage from previous pregnancies in order to increase the sample size. Furthermore, the amount of analgesic taken by Australian women seems to be excessively high compared with most other populations. The most recent study by the Boston Collaborative Project on 41 337 women (Shapiro *et al.* 1976a) found no evidence that aspirin, as taken during pregnancy in the United States, was a cause of stillbirth, neonatal mortality, or reduced birth weight.

A number of studies have shown that aspirin administration in late pregnancy will delay the onset of parturition in rats. There is also evidence that in humans aspirin interferes with labour and parturition. A retrospective survey of 103 patients with arthritis or collagen disease taking high daily doses (3.25 g = 50 grains) of aspirin for at least the last six months of gestation were compared with diseased patients without aspirin, and also with controls (Lewis and Schulman 1973). They showed a significant increase (11 days) in the average length of gestation, in the frequency of post-maturity (> 15 days), and mean duration of spontaneous labour (70 per cent). The infant birth weight was not significantly lower but there was more blood loss during labour in the aspirin-treated group than in either of the control groups ($p < 0.025$). These authors suggest that these effects are related to the inhibition of prostaglandin synthesis, though Horan (1974) suggests that the effects may be due to a decreased rate of deciduoma growth during the early weeks of pregnancy. Aspirin has also been shown to cause a marked prolongation of the induction-abortion interval in induced abortions (Waltman *et al.* 1973; Waltman and Tricomi 1974). However, Smith *et al.* (1975) reported that although aspirin applied to isolated human pregnant myometrial strips stimulated by prostaglandin $F_{2\alpha}$ caused a dose-dependent inhibition of the force of contraction it had no significant effects on the induction-abortion interval in patients undergoing $PGF_{2\alpha}$-induced abortion in the second trimester. Nevertheless, it has been reported that in 12 women due for second trimester abortion, injections of arachidonate into the amniotic sac resulted in abortion in all cases, whereas two other women who had ingested aspirin prior to the arachidonate injection did not abort. In fact aspirin and aspirin-type drugs have been successful in the treatment of threatened premature labour (Mosler 1975). This was attributed to a decrease in uterine activity due to inhibition of prostaglandin synthesis and release. Other workers, however, feel that aspirin should not be given deliberately during pregnancy and especially in late pregnancy (Waltman and Tricomi 1974). On the other hand, Brenner et al. (1975) found that a mixture of drugs containing a total of 4 g aspirin helped to reduce the side-effects of abortion induced with 15-(S)-15-methyl PGE_2 and did not interfere with its effectiveness.

PROPIONIC ACID DERIVATIVES:

Naproxen

Its safety in human pregnancy has not been established. Reproductive studies in which rats and rabbits were exposed to toxic doses of naproxen throughout organogenesis produced an increase in the number of dead and macerated fetuses and delayed ossification. Furthermore, when naproxen was given in late gestation, parturition was delayed and littering down was more difficult in the rabbit (reviewed by Batagol 1983). There have been occasional reports of persistent pulmonary hypertension presenting during the first few hours of life, when naproxen has been given to delay parturition (Wilkinson et al. 1979; Wilkinson 1980). A high level of naproxen was found in the cord blood at birth in three infants, thus confirming that naproxen crosses the placenta. It has been suggested that it inhibits prostaglandin synthesis in the fetus, probably PGE with subsequent constriction of the ductus arteriosus and pulmonary arterial hypertension. The underlying pathology is not clear, but may relate to functional changes in the reactivity of the pulmonary arterioles. Since the outcome is often fatal, the risks of this type of therapy seem unwarranted.

Fenbufen

Reproduction studies in animals have shown that it is a prostaglandin inhibitor. Its safety in human pregnancy has not been established.

Ibuprofen

Studies in rodents and rabbits have not shown teratogenic effects. Ibuprofen does however, delay parturition and closure of the ductus arteriosus in animals.

It safety in human pregnancy has not been established.

Fenopropen calcium

Animal studies have shown no evidence of teratogenicity or embryo-toxicity. However, there was prolongation of gestation in 50 per cent of the animals which was attributed to inhibition of uterine prostaglandin synthesis. An increase in abortions was observed in rabbits treated throughout organogenesis with doses of fenopropen equivalent to the human therapeutic dose.

Its safety for use in human pregnancy has not been established.

Carprofen

Teratology studies in the rat in which carprofen 2–20 mg/kg daily was given orally throughout organogenesis, had no adverse effects on fetal development as judged by the incidence of congenital malformation, resorptions, litter size, or fetal weight, despite some maternal toxicity in the highest dose group (McClain and Hoar 1980). Fertility and peri-postnatal studies showed a slight increase in the incidence of dead pups at birth, and increase in resorption rates and prolonged gestation.

No data are available on the toxicity of this drug in human pregnancy.

ACETIC ACID DERIVATIVES:

Indomethacin

Indomethacin is considered to be a potent inhibitor of prostaglandin synthesis, probably PGE. The use of indomethacin during pregnancy has been the subject of much concern and has raised serious ethical issues. There is good evidence that some of the prostaglandins such as PGE_2, PGI_2, and PGD_2 are involved in regulating the fetal and neonatal circulation. In particular, PGE_2 seems to be involved in maintenance of ductus arteriosus patency during fetal life. However, the role of PGI_2 and PGD_2 in mediating the fall in pulmonary vascular resistance following the first breath is equivocal.

Two major problems have arisen concerning the use of indomethacin: firstly, its use in late pregnancy for the treatment of onset of premature labour; and, secondly, its use in the neonatal period of management of persistent patent ductus arteriosus. The use of indomethacin (and to a lesser extent aspirin and naproxen) has been associated with persistent pulmonary hypertension presenting during the first few hours of life. This is thought to be due to the inhibition of prostaglandin synthesis, probably PGE, in the fetus with subsequent constriction of the ductus arteriosus and pulmonary arterial hypertension (Levin et al. 1978; Csaba et al. 1978;

Rubaltelli *et al.* 1979; Wilkinson *et al.* 1979). The under-lying pathology is not clear but may relate to functional changes in the reactivity of the pulmonary arterioles. Since the outcome is often fatal, this type of therapy seems unwarranted. Intrauterine congestive heart failure is a common pathological mechanism for oedema forma-tion and it is often associated with premature closure of the ductus arteriosus. There has been a recent case report in which indomethacin was given to arrest premature labour and the mother subsequently gave birth to twins, one of whom had cardiac failure and the other hydrops fetalis (Mogilner *et al.* 1982). The clinical picture strongly supports a connection between hydrops fetalis and cardiac failure due to premature closure of the ductus arteriosus by indomethacin.

The problems involved in the use of indomethacin to treat patent ductus arteriosus in the neonate were reported by Strauss *et al.* (1982) and Gersony *et al.* (1983). Strauss *et al.* (1982) reported on two large premature newborn infants with respiratory distress syndrome and ductus-dependent congenital heart disease. Pharmacological closure of the ductus with indomethacin was attempted without knowledge of the presence of congenital heart disease. In one of the infants, ductual closure had to be reversed with alpro-stadil (PGEL) prior to surgery when congenital heart disease was diagnosed. Thus in both of these cases the use of indomethacin seemed to unmask clinical signs of potentially life-threatening cardiac disease. This empha-sizes the need to be aware of the possibility of congenital heart defects when treating the pre-term infant with respiratory distress syndrome and persistent fetal circula-tion, before using indomethacin to induce closure of the ductus arteriosus. Gersony *et al.* (1983) reported on the results of a national collaborative study in which the role of indomethacin in the management of patent ductus arteriosus (PDA) was evaluated. In a two-year period, 1978–80, 3559 newborn infants with a birth weight of less than 1750 g were studied at 13 centres, and of these 421 met the criteria for 'haemodynamically significant' PDA. Three treatment strategies were compared: (1) indomethacin as part of initial medical therapy (fluid restriction, diuretics, possibly digoxin) with surgical closure if needed; (2) indomethacin only when usual medical therapy failed, with surgical closure if needed; and (3) no indomethacin, but surgical closure of PDA infants in whom usual medical therapy was unsuccessful. Both the efficacy and toxicity associated with these three treatment regimes were also reported. Sixteen infants had to be excluded during the course of the trial because they were found to have congenital anomalies, mainly VSD. Thus the results dealing with closure and re-opening were based on 405 infants only. After 48 h treat-ment, 79 per cent of the infants given indomethacin no longer met the criteria for haemodynamically significant PDA, compared with 28 per cent of those who received placebo. The opening rates among 135 infants given indomethacin was 26 per cent, but most of these even-tually closed without surgical intervention. Thus the ductus arteriosus was permanently closed in 79 per cent of the infants who received indomethacin as part of the initial therapy without need for back-up therapy compared with only 35 per cent in the medical therapy alone group, the final closure ratio being 2.3:1 which is highly significant ($p < 0.001$). Furthermore, these closure rates were not related to birth weight, gestational age, gender, or race. Indomethacin as a back-up to usual medical therapy resulted in similar closure rates. In order to assess the overall effects the three treatment strategies were compared. There were no significant differences in mortality rates. Infants given indomethacin only if usual therapy failed (strategy 2), had a lower incidence of bleeding than those to whom it had been given as part of the initial therapy (strategy 1). Furthermore, infants on strategy 2, also had lower rates of pneumothorax and retrolental fibroplasia than those who did not receive indomethacin (strategy 3).

The authors concluded that the treatment of choice in small premature infants with haemodynamically sig-nificant PDA is to use indomethacin only after an appropriate course of usual medical therapy fails. It is expected that one-third of the infants would respond to the usual medical therapy alone and thus not be exposed to the additional problems associated with indome-thacin. Indomethacin given at such times seems to be equally effective in closing the ductus. Whether early treatment of asymptomatic babies is warranted remains speculative.

Sulindac

Studies in rats and mice showed no adverse effects on fertility or evidence of teratogenicity. A range of doses 20–60 mg/kg approximately 3–10 times the human thera-peutic dose have been administered to rabbits during organogenesis. No adverse effects were reported at the lower doses, but there were teratogenic effects at the highest dose. Delayed parturition was also observed in the mouse. Its safety in human pregnancy has not been established.

FENAMATES:

Flufenamic acid

Safety in pregnancy has not been established.

Mefenamic acid

Safety in pregnancy has not been established.

PYRAZOLONES:

Phenylbutazone

There have been no specific reports associating the use of phenylbutazone with congenital malformations. However, phenylbutazone does have transient antithyroid effects and appears to inhibit the uptake of iodine. Goitre and hypothyroidism have been reported following long-term use of the drug (cited by Batagol 1983). Other potentially toxic effects of phenylbutazone on the haematopoietic system are also factors for consideration before using the drug in pregnancy.

Paracetamol and phenacetin

Although paracetamol has been used extensively as an analgesic over the last 30 years, very little is known about its effects in pregnancy. The College of General Practitioners survey failed to show any evidence of a teratogenic effect of paracetamol and in fact the incidence of congenital malformations following prescriptions for paracetamol and phenacetin was significantly lower than expected (Crombie et al. 1970).

There have been occasional case reports associating the use of paracetamol with congenital malformations. These were reviewed by Schardein (1985), who cited four case reports of malformations associated with its use in the first trimester. In addition, he reviewed another seven case reports of malformations associated with paracetamol in combination with other drugs, particularly aspirin. Paracetamol has been shown to cross the placenta in humans (Levy et al. 1975). However, paracetamol has been implicated as hazardous particularly in the production of effects on the fetal kidney from the second month of gestation onwards (Schenkel and Vorherr 1974; Char et al. 1975; Golden et al. 1982). Fetal anaemia and methaemoglobinaemia, together with liver toxicity and jaundice, have also been associated with maternal consumption of paracetamol. These views were supported by Collins (1981) in a review on the effects of paracetamol in pregnancy.

Phenacetin when taken by pregnant women is metabolized to paracetamol which is transferred to the fetus (Levy et al. 1975). The use of this drug is no longer recommended.

In conclusion, from the clinical data available especially the College of General Practitioners report, it would seem that paracetamol is free from any significant teratogenic effects if used during pregnancy. However,

the fetus does seem susceptible to the same degree of hepatoxic and nephrotoxic effects as the adult. Thus, although occasional use of paracetamol during pregnancy may be safe, prolonged consumption of high doses may be hazardous to the fetus.

Major tranquillizers

The major tranquillizers, phenothiazines and haloperidol, have antifertility and embryo-lethal effects in animals, but have little teratogenic activity in general.

Phenothiazines

There is inadequate evidence to assess the risks in humans. There are numerous individual case reports concerning the teratogenicity of chlorpromazine, trifluoperazine, prochlorperazine, perphenazine, promazine, thioridazine, and acepromazine (Schardein 1985), but, in view of the normal background incidence of malformations, one or two individual cases can never prove a causal relationship. Women on such drugs are usually also exposed to several other drugs at the same time. A number of small studies have failed to show any teratogenicity; in 52 women treated with chlorpromazine and in 56 women treated with perphenazine (Schardein 1985) no ill-effects were found.

The results of a large prospective French study on 12 764 women were published and did show a significant increase in congenital malformations in children whose mothers took phenothiazines during the first trimester. A variety of malformations was produced and was significantly increased only for the three-carbon side-chain phenothiazines (acetylpromazine, chlorpromazine, methotrimeprazine, oxomemazine) and not for the two-carbon analogues (promethazine, propiomazine) or the piperidine or piperazine side-chain phenothiazines (Rumeau-Rouquette et al. 1977). However, in the Boston prospective cohort study of 50 282 pregnancies, the malformation rates were similar in the 1309 offspring of mothers who had taken phenothiazines during the first 4 months of pregnancy to those in 48 973 children whose mothers had not taken these drugs. Furthermore there were no differences in perinatal mortality, mean birth weights or IQ scores measured at 4 years of age. However, there was a suspicion of an association between phenothiazine exposure and cardiovascular malformation, but this was not statistically significant (Slone et al. 1977).

Although there is thus conflicting evidence concerning the association of phenothiazines and congenital malformations, the overall risk is small. There does seem to be more agreement concerning the possible effects on the neonate. It is not unreasonable to expect that the phenothiazines may have pharmacological effects on the fetus–neonate similar to those on the mother. Although the phenothiazines have been shown to induce fetal liver enzymes, they are not particularly associated with the occurrence of fetal–neonatal jaundice, as they are in the adult (Batagol 1983).

Infants born to mothers receiving phenothiazines may exhibit extrapyramidal effects which begin within the first 24 hours postpartum and can often persist for anything from several weeks up to nine months if left untreated (Beeley 1981b; Batagol 1983).

There have also been occasional reports of the phenothiazines producing adrenergic blockade giving rise to problems of thermoregulation in the neonate. The anticholinergic effects of these drugs have also been implicated in two neonates with functional intestinal obstruction born to schizophrenic women treated with phenothiazines plus antiparkinson drugs (Falterman and Richardson 1980).

Fluphenazine

Batagol (1983) cited the results of a retrospective study in 394 women who took either fluphenazine or placebo for nausea or vomiting during pregnancy. No adverse effects were observed as regards spontaneous abortion, perinatal mortality, premature births, or twinning. Congenital malformations occurred in 2.7 per cent of 226 live and stillborn infants in the fluphenazine group, compared with 3.5 per cent of 143 live and stillborn deliveries in the matched placebo group. The incidence of malformations in either group is not significantly different from the spontaneous incidence for the general population. There have been occasional case reports of extrapyramidal effects in the neonate associated with prenatal exposure to fluphenazine.

Prochlorperazine

Although this drug is reported to cause malformations in rodents, its association with congenital deformities in human pregnancies is equivocal (Rumeau-Rouquette et al. 1977; Heinonen et al. 1977b; Schardein 1985).

Trifluoperazine

As with prochlorperazine, there are equivocal reports of malformations in rodents, but no specific associations with abnormalities in human pregnancies; nevertheless, the manufacturers recommend that it not be used in pregnancy unless absolutely necessary.

Butyrophenones

Haloperidol has also been implicated in several individual case reports of malformed children. However, a retrospective study of 98 women treated with haloperidol for hyperemesis gravidarum failed to confirm this (Van Waes and Van der Velde 1969). Batagol (1983) cited a study by Hansen and Oakley in Atlanta in which the association between haloperidol and limb defects was reviewed for the period 1970–73. The overall conclusion was that there was no causal relationship; however, the study was small, involving only 86 infants, and no other details are given in order to facilitate critical assessment of the study.

Lithium

Lithium carbonate, used in manic-depressive psychoses, was known to have some teratogenic potential in animals before it was used in man.

A registry was therefore set up to record the outcome of pregnancy in women receiving lithium treatment during early pregnancy. Recent evidence also suggests that lithium treatment in late pregnancy can have adverse effects on the fetus and neonate. The first report on 118 babies included five stillbirths and seven neonatal mortalities; a total of nine (7.6 per cent) babies had malformations (Schou et al. 1973). Further cases have been added to this total, 187 children being registered up to 1977 (Beeley 1981b). Twenty of these children had congenital malformations, 15 of whom had cardiac defects, including Ebstein's anomaly. Since this anomaly has a spontaneous incidence of only one in 20 000 births, it seems very likely that lithium does present a significant risk of teratogenesis.

The use of lithium in late pregnancy needs to be carefully monitored as the amounts of lithium required to provide adequate control vary considerably during pregnancy. Inadequate control, either insufficient or excessive amounts of lithium, can prove hazardous to both the mother and the fetus. Lithium toxicity in the neonate frequently produces cyanosis and hypotonia. There are also occasional reports of neonatal goitre, transient hypothyroidism, and nephrogenic diabetes insipidus (Beeley 1981b).

Conclusions

In conclusion, almost all of the major tranquillizers have been shown to cause both structural malformations and behavioural effects, i.e. to affect postnatal growth and development, in animal studies. There are conflicting data as to whether these drugs produce congenital malformations in humans but the overall risk is small. Long-term follow-up studies of children treated with these drugs are urgently required. In the case of lithium, where there has been follow-up, both structural and functional defects have been reported in the fetus and neonate. In view of this, it is strongly recommended that lithium not be used during pregnancy, unless all other forms of therapy have proved to be inadequate.

Minor tranquillizers

The benzodiazepines (anxiolytics) are among the most widely prescribed drugs in pregnancy, with 30–40 per cent of all pregnant women being given anti-anxiety drugs at some stage of pregnancy. Although many new anti-anxiety drugs have been introduced during the last few years, there has been remarkably little effort to establish their safety for use in human pregnancy. There is growing concern about the addictive properties and the withdrawal symptoms induced by these so-called 'minor' tranquillizers with long-term use. As these drugs are likely to have pharmacological effects on the fetus and neonate similar to those in the adult, it is of the utmost importance that these facts are carefully considered before prescribing these drugs to women of child-bearing age.

Diazepam (Valium)

Safra and Oakley (1975) in a study of 278 malformed children found that diazepam ingestion was four times more common in the mothers of children with cleft lip and/or palate than among mothers of children with other defects. Data from the Finnish register of congenital malformations supported these findings (Saxen and Saxen 1975).

The side-effects of diazepam on the neonate have been well documented. (Cree *et al.* 1973; Haram 1977; Gillberg 1977; Woods and Malan 1978; Rowlatt 1978). These effects on the neonate are often referred to as 'the floppy infant syndrome'.

Cree *et al.* (1973) reported that a total dose of diazepam of 30 mg or less during the 15 hours prior to onset of labour had little effect on the state of the infant. However, larger doses were associated with low Apgar scores at birth, apnoeic spells, hypotonia, reluctance to feed, and impaired metabolic responses to cold stress. Furthermore, significant concentrations of both diazepam and one of its active metabolites, methyldiazepam, were still detectable in some babies for up to eight days postpartum. This correlates well with the pharmacokinetic and placental transfer data of diazepam and its disposition in the neonate reported by Mandelli *et al.* (1975).

Roche, the manufacturers of *Valium* are carrying out a follow-up study from birth to 12 years, in children exposed to diazepam *in utero* but no results are available yet (personal communication).

Lorazepam

There is little evidence whether the newer anti-anxiety drugs such as clorazepate, medazepam, or lorazepam are any better or worse than diazepam. However, there is information available on the use of some of these drugs during late-pregnancy and in labour.

A recent study by Whitelaw *et al.* (1981) reported the effects of lorazepam administered to 51 mothers who delivered 53 babies, who were followed up until five days postpartum. Two groups of hypertensive mothers, none of whom developed eclampsia, were studied. One group of 35 with mild hypertension were treated with oral lorazepam 2–7.5 mg per day, and the second group of 16 with fulminating hypertension were treated with intravenous infusion of lorazepam as required, and were also given hydrallazine i.v. Eleven of the 16 were delivered by Caesarian section.

The maternal plasma lorazepam concentrations were higher than the cord plasma levels. Cord plasma concentrations of 45 μg l^{-1} and above were associated with 75 per cent of the infants who required ventilation at birth. The neonate conjugates lorazepam to its inactive glucuronide very slowly; this is excreted in the urine over a period of about seven days. The amount of lorazepam secreted into breast milk is insignificant.

It is interesting that full-term neonates whose mothers had received oral lorazepam had no major complications at delivery other than a slight delay in establishing feeding. Furthermore, this was associated in seven of 29 mothers with high doses of lorazepam, 5.5 mg daily for 22 days. Those mothers who received lorazepam i.v. had infants with significantly lower Apgar scores, hypothermia, need for ventilation, and poor sucking ability. The preterm babies whose mothers had received lorazepam by either route had similar problems. Infants that had been exposed to lorazepam had lower Apgar scores than others receiving diazepam, but the diazepam babies were heavier and more mature.

The authors recommended that the use of lorazepam prior to 37 weeks or when given intravenously for the treatment of severe pre-eclampsia, should be restricted to hospitals with facilities for neonatal intensive care.

Clorazepate

There have been occasional case reports associating clorazepate with congenital malformations. However, a causal relationship is difficult to prove. Its safety in pregnancy has not been established.

Medazepam

Medazepam (*Nobrium*) in rabbits had no adverse effects on fertility and was not teratogenic, although at very high doses it caused maternal sedation. There was a high incidence of neonatal mortality within the first few days, attributed to the dams not feeding the pups. Multigeneration studies in rats, mice, and rabbits indicate no problems when medazepam is used in therapeutic doses. There is no available data on medazepam in human pregnancies, but it is thought to act in a similar way to diazepam, and thus there is a possibility of a $1\frac{1}{2}$-fold increase in the incidence of oro-facial clefts (personal communication, Roche).

Meprobamate and chlordiazepoxide

A carefully controlled prospective study showed that meprobamate use in early pregnancy is associated with a four- to fivefold increase in the expected incidence of malformations. A wide variety of defects was observed, particularly cardiac defects (Milkovich and Van den Berg 1974). The same study also revealed a fourfold increase in incidence of malformations following the use of chlordiazepoxide, again with a variety of defects. These results were not supported by a large collaborative study in the USA (Hartz *et al.* 1975) where no significant effects of meprobamate or chlordiazepoxide could be found.

Conclusions

In conclusion, there is some evidence that the benzodiazepines such as diazepam are associated with a small increase in the incidence of cleft lip and/or palate, when used in the first trimester. The use of the benzodiazepines in late pregnancy also may lead to problems in that they produce neonatal hypotonia, apnoeic spells, low Apgar scores, impaired metabolic responses to cold stress, and poor sucking ability. Furthermore these drugs are present in breast milk, but whether they reach sufficiently high levels to produce sedation in the infant is still a contentious subject.

Overall, if benzodiazepines are really essential during pregnancy, it is strongly recommended that the shorter-acting ones should be used and that their use be discontinued, if possible, well prior to delivery. The evidence concerning the teratogenicity of meprobamate and chlordiazepoxide remains equivocal.

Nutritional supplements

Optimal fetal growth and intrauterine development require a constant supply of nutrients from the mother to the fetus. Although the average diet in developed countries is thought to contain an adequate supply of vitamins and minerals to maintain health and cope with the extra demands of pregnancy, it has become common practice to give iron and possibly vitamins routinely throughout pregnancy.

Controversy exists as to whether these supplements are really necessary as both nutrient deficiences (folic acid) and excesses (vitamin A) have been reported to increase the incidence of congenital malformations. The risks and benefits of such supplements have been extensively reviewed by Moghissi (1981).

It does not seem out of place in a chapter on drug-induced teratogenesis to conclude with a report of a paper on drug prevention of teratogenesis.

In 1980 Smithells and co-workers reported some initial data on the possible prevention of neural-tube defects (NTD) by treatment of the mother around the time of conception with a multi-vitamin and iron preparation, *Pregnavite Forte F*. Since then additional information has become available which is very encouraging. In a multicentre trial (Smithells *et al.* 1981*a*), women with a history of one or more children with a neural-tube defect and who faced a 5 per cent risk of recurrence were given three multi-vitamin tablets per day from at least 28 days before conception until at least the second missed period, i.e. after the time of neural-tube closure. Those who complied with the above conditions were classed as fully supplemented (FS). Those who conceived within 28 days of taking the pills or took them after conception but before the estimated time of neural tube closure, or had missed taking the pills for more than one day were classed as partially supplemented (PS). The babies born to these women were compared with babies from 300 mothers who were unsupplemented (US). Because of the ethical constraints placed upon the experimenters the control population was not treated in a double-blind manner with placebo, so that some controversy followed the publication of the results. However, the results were extremely encouraging and have now been supported by the preliminary data published from the second cohort study by the some group of workers (Smithells *et al.* 1981*b*). The same protocol was used as in the first study and those interim results are from pregnancies that have ended, or those in which the results of alpha fetoprotein (AFP) analyses are known.

	1 previous NTD		2 previous NTD		Combined	
	FS	US	FS	US	FS	US
No NTD	175	172	25	16	200	188
NTD	1	9	1	1	2	10
Totals	176	181	26	17	202	198

If these results are added to those of the first cohort, then the recurrence of NTD in the vitamin-supplemented group is 3/397 (0.76 per cent) compared with 23/493 (4.67 per cent) ($p < 0.0003$). The authors concluded that there were no consistent effects of maternal age on the incidence or recurrence rate of NTD, but there was an effect of social class. There was a widely observed social-class gradient in the incidence of NTD and in this study more social class III-V mothers occurred in the US group which does suggest a nutritional contribution to the causation of NTD which may be overcome by vitamin supplements. The authors accept the criticism that the mothers were recruited in different ways. The supplemented mothers were not pregnant when recruited to the study and had received counselling after the birth of their previous child with a NTD, whereas, the unsupplemented mothers had received no counselling and were usually pregnant when recruited to the study. Thus in the FS group a history of NTD in the immediately previous pregnancy was more common (79 per cent) than in the unsupplemented group (60 per cent). The corollary is that the incidence of spontaneous abortion, i.e. 9 and 17 per cent, respectively, and all other outcomes are under represented in the FS group. Among the women whose previous pregnancy did not result in NTD the proportion of immediately preceding pregnancies ending in spontaneous abortions was very similar, 42 per cent in the FS group and 40 per cent in the US group. As it is well known that there is an increased frequency of abortion prior to the NTD births, the possibility of a causal relationship has been the subject of some debate. However, Smithells and his co-workers feel it is unlikely that the deficit of earlier spontaneous abortions in their supplemented group could explain the decrease in NTD recurrence rates.

The results for a second cohort of 254 mothers with a history of previous neural-tube defect have been published (Smithells *et al.* 1983). In the second cohort there were two NTD recurrences (0.9 per cent of 234 infants/fetuses examined), which is significantly fewer than the 11 NTD recurrences (5.1 per cent of 215 infants/fetuses examined) born to 219 unsupplemented mothers in the same centre over the same period.

Since both cohorts were managed according to the same protocol and, as the second cohort followed imme-

diately after the first, the authors have combined the data obtained. Thus the overall recurrence rate of NTD is 0.7 per cent for 454 FS mothers compared with 4.7 per cent for the 519 US mothers. The recurrence rates after one previous NTD were 0.5 per cent for FS mothers and 4.2 per cent for the US mothers. Furthermore, after two or more previous NTDs, the recurrence rates were 2.3 per cent for FS mothers and 9.6 per cent for US mothers. There were no recurrences amongst the offspring of a further 114 mothers whose duration of supplementation fell short of the full regimen, i.e. the partially supplemented (PS) group.

The authors took into account factors such as maternal age, social class, previous reproductive history, and pregnancy outcome when analysing these results. They concluded that some of these factors contributed substantially to the differences in recurrence rates between the FS and US groups.

There is no way of actually testing the validity of this at the moment because of the lack of a placebo group. However, the British Medical Research Council (MRC) early in 1983, agreed to a controlled study being carried out to see if vitamin supplementation did have an anti-teratogenic effect in mothers at risk.

Nevertheless, this is still quite a controversial subject, the practicality and ethics of which have been seriously criticized (Leck 1983).

A brief report was published by Tolarova (1982) from Czechoslovakia on the use of periconceptual vitamin and folic acid supplementation to prevent the recurrence of cleft lip. Tolarova carried out a prospective study in women who had one previous child with unilateral harelip with or without cleft palate and no other family history of facial clefts. These women who were selected from the register of oro-facial clefts in Bohemia 1970–75, were asked to take a multi-vitamin preparation, *Spofavit*, three times daily plus 10 mg folic acid for at least three months prior to conception to the end of the first trimester. Eighty women agreed to take the vitamin supplements and 212 refused, and these were used as controls. No substantial differences in social class were noted. In the supplemented pregnancies the recurrence rate was 1/85 (1.2 per cent) compared with controls 15/212 (7.1 per cent ($p = 0.023$). This frequency of clefts in the controls was much higher than the risk calculated from 2487 sibships (4.1 per cent for unilateral or bilateral cleft lip with or without cleft palate) in previous studies. Again, this seems highly suggestive that vitamin supplementation may have a protective effect; however, the author does acknowledge the need to carry out larger studies to investigate this. Although controversy exists as to the value of giving nutritional supplements *per se* during pregnancy, there is considerable disagreement among those who do believe in supplementation as to which vitamin preparations should be given. This point is well illustrated by the debate surrounding the study by Laurence *et al.* (1981) who carried out a randomized, controlled, double-blind trial in South Wales to prevent the recurrence of NTD in women who had one previously affected child. This data was collected from the malformation register, paediatric records, Local authority records, and other sources between 1954 and 1969.

Sixty women were asked to take 2 mg folic acid twice daily and 51 women given a placebo starting from the time that contraceptive measures were stopped. Only 44 of the 60 women complied with the instructions. The women were revisited at around 6 weeks, 6 months, and at the end of pregnancy; blood folate levels were monitored between 6 and 9 weeks of pregnancy; and the diets were assessed as good, fair, or inadequate. In the vitamin-supplemented group, if the blood folate levels at 6–9 weeks were less than $10 \, \mu g \, l^{-1}$, they were regarded as non-compliers; in the placebo group all folate levels were below $12 \, \mu g \, l^{-1}$. There were no recurrences amongst compliant mothers (0/44) but there were two NTDs in the 16 non-compliant mothers. In the placebo group there were four NTDs out of 51. Thus there were no recurrences in the supplemented group compared with 6/67 in the non-compliers-plus-placebo group ($p = 0.04$). All of the NTD babies occurred in the women taking an inadequate diet. The authors conclude that women on poor diets can decrease their chances of having babies with NTD by either improving their diets or taking folate supplements.

These authors were highly critical of the preliminary data published by Smithells *et al.* (1981*a*) in that they felt that *Pregnavite Forte F* was a 'blunderbuss preparation' and some of its constituents might actually be hazardous to the fetus. They felt that folic acid supplements are safer and cheaper. This has lead to a great deal of lively correspondence between the two groups of workers.

Some of the major criticisms of the Laurence study are that the numbers used were small and they tested compliers vs. non-compliers vs. placebo. It is possible that the non-compliers and placebo group are two different populations (Mamtani and Watkins 1981). If the non-compliers were included with the treatment group, then the effect of good diet remains highly significant but that of folate does not. Smithells *et al.* (1981*c*) questioned the statement that the trial was double-blind since on the basis of the folate measurements taken between 6 and 9 weeks of pregnancy the mothers were classed as compliers or non-compliers; thus the code was broken. Laurence and Campbell (1981) refuted this by emphasizing that the code was not broken until the end of the study. Smithells *et al.* (1981*c*) reported that they had found low levels of ascorbic acid, riboflavin, and folic acid in mothers with NTD babies, so although folic acid tablets were cheaper it was felt that *Pregnavite Forte F* was more likely to be effective. They also point out that the non-compliers did, in fact, comply in part, and therefore the details of the two babies with NTD were crucial in order to determine the minimum effective period of supplementation. Furthermore, they feel that there is no way of knowing that the placebo group did not receive vitamin supplements which are readily bought over the counter.

From these studies it would seem that multi-vitamin supplements during pregnancy do reduce the recurrence risk of NTD. The use of such supplements would also seem warranted when it is known that the diet is poor or when drugs are being used which may alter the nutritional status of the mother and thus alter the supply of micro-nutrients to the developing fetus. Hopefully the data obtained from the MRC double-blind trial will help to define more clearly the role of vitamins in the prevention of neural-tube defects.

Organic solvent exposure and abuse: glue-sniffing, etc.

During the last decade there has been a marked increase in glue-sniffing, and since 1970 glue-sniffing has been directly responsible for at least 60 deaths in the UK. In some areas in the UK glue-sniffing or solvent abuse has reached almost epidemic proportions, particularly amongst adolescents. A recent figure quoted was that glue-sniffers were inhaling anything from one small tube of glue to 4 pints of glue per day. Most or the glues used contain either toluene or benzene as solvents, both of which have a high affinity for adipose tissues, particularly in the CNS, from which it is released slowly.

The animal data available on the effects of toluene in pregnancy show that it is not teratogenic but there is a decrease in fetal weight, growth retardation, and embrolethality at high concentrations. However, the relevance of this for pregnant women is difficult to assess. Barlow and Sullivan (1982) in a review of reproductive hazards to women at work, described data obtained on women working in glue and solvent factories who had been exposed to high solvent levels, especially toluene. In these studies no teratogenic effects could be confirmed and there were no obvious effects on fertility, but there was a twofold increase in low birth weights of the offspring, a result which was also found in animal studies.

There was one case-referent study in Finland which began in 1976 and is still in progress (Holmberg 1979; Holmberg and Nurminen 1980). They report data from

1976 to 1978 in which 120 mothers of children with congenital central-nervous-system defects and their matched paired controls were assessed for exposure to noxious agents. It was found that significantly more case mothers than controls had been exposed to organic solvents during the first trimester ($p < 0.01$). Anencephaly was the most common defect which occurred in five of the 14 children of solvent-exposed parents. No other consistent pattern of drug use or related factors could be identified.

Although this study has been criticized as regards to the lack of adequate control groups and analysis of social factors (Sheikh 1979), the results demonstrate the need for further investigation.

Holmberg et al. (1982) reported the results of a cumulative case-referent study of selected exposures during pregnancy among 388 mothers of children born with oral clefts in Finland from December 1977. Significantly more case mothers (14) than referent mothers (4) had been exposed to organic solvents during the first trimester ($p < 0.05$). Nine children of exposed case mothers had cleft palate, one of these children had multiple defects, and five had cleft lip with or without cleft palate (two with additional malformations). Eight of the case mothers had been exposed to organic solvents at work, and the other six mothers had been exposed to solvents at home. The corresponding figures for the referent mothers were two and two, respectively. A variety of solvents were implicated but no specific one stood out.

A report by King et al. (1981) from Glasgow described encephalopathies in children who were glue-sniffers. This study analysed data obtained from 19 children aged 8–14 years who were admitted to hospital (1974–80) with acute encephalopathy due to toluene intoxication. The medical histories of these children describe CNS disturbances such as euphoria, hallucinations, ataxia, convulsions, coma, and behavioural changes associated with diplopia. There was complete recovery in 13 of these children, the remaining five children still having psychological impairment and personality changes. One child still had persistent cerebellar ataxia one year after the acute attack despite the absence of further exposure.

Thus exposure to solvents may occur both at work and at home, as well as from the increase in solvent abuse, particularly amongst adolescents. There is very little data available on the effects of these substances in human pregnancy. However, as solvents such as toluene and benzene cross the placenta, the fetus as well as the mother will be exposed to the adverse effects of these substances. As brain development continues throughout the whole of pregnancy, particularly in the third trimester, and into the neonatal period, there is a very real possibility that these solvents may affect the developing brain and the subsequent mental development and behaviour of the offspring.

There are now a few case reports which tend to support this theory. One case of a baby with features very similar to those of the fetal alcohol syndrome was reported where the baby's mother was a chronic solvent abuser, primarily of toluene (Toutant and Lippmann 1979). In the same year, Hunter et al. (1979) reported data concerning the appearance of a syndrome of multiple anomalies consisting mainly of mental retardation, abnormal muscle tone, and poor postnatal growth at Shamattawa from 1973 to 1977. Gasoline-sniffing and alcohol abuse were known to be common problems in this American-Indian community.

Streicher et al. (1981) reported on the clinical and laboratory findings in 25 adults aged 18–40 years who were hospitalized for problems related to paint-sniffing. Three women continued to sniff paint throughout pregnancy. All had liveborn infants, one of whom had cerebellar dysfunction, one of whom was apparently normal, and the third of whom was too young for assessment.

In conclusion, there are a few epidemiological studies indicating that solvent exposure during pregnancy, particularly in the work place, may cause an increased incidence of congenital malformations or low birth weight in the offspring. Case reports of malformations associated with solvent abuse such as glue-sniffing are as yet inconclusive. However, it must be borne in mind that the pharmacological and toxicological effects which these substances produce in adults may also occur in the developing fetus.

Ovulation inducers

Bromocriptine

The dopaminergic agonist bromocriptine (Parlodel) is increasingly being used to treat women with infertility due to hyperprolactinaemia. Since treatment is usually continued for some time after conception occurs, knowledge about the possible hazard of this drug for the fetus is important.

The Drug Monitoring Centre of Sandoz reported (Turkalj et al. 1982) on the outcome of 1410 pregnancies in 1335 women who had received bromocriptine therapy. In the majority of cases (82 per cent) bromocriptine was given for conditions such as amenorrhoea, galactorrhoea, or luteal insufficiency, whereas in 256 pregnancies (18 per cent) pituitary tumours, including acromegaly were reported as the primary diagnosis. The age of the patients ranged from 18 to 42 years, and the doses of bromocriptine used were 1–40 mg. It was possible to ascertain the duration of fetal exposure in 1278 pregnancies (90.6 per cent), but in most cases bromocriptine was taken in the first eight weeks after conception.

These pregnancies resulted in 197 (14 per cent) early terminations — this included 25 (1.8 per cent) induced abortions and 1213 (86 per cent) live births. There were no significant differences in the incidence rate of spontaneous abortions (11.1 per cent), ectopic pregnancies (0.9 per cent), and major (1.0 per cent) or minor (2.5 per cent) malformations from those quoted for normal populations. However, the incidence of twin pregnancies (1.8 per cent) was slightly, although, not significantly higher if correction is made for concomitant therapy with other agents inducing ovulation such as clomiphene and HCG. The birth weights of both singleton and twin births were comparable with those in normal populations.

Congenital abnormalities were detected at birth in 43 (3.5 per cent) of 1241 live and stillborn infants. The pattern of the organ systems affected as well as the incidence rate of particular types of anomalies were similar to those quoted by Heinonen *et al.* (1977*b*) in data from the Boston Collaborative Perinatal Study. Furthermore these malformations were not related to either the dose used or the duration of exposure to bromocriptine.

The authors concluded that the use of bromocriptine in the treatment of infertile women is not associated with an increased risk of abortion, mutliple pregnancies, or congenital abnormalities. Nevertheless, as the risk of abortion is not increased by interruption of treatment it is still recommended that bromocriptine therapy be stopped as soon as pregnancy is confirmed, unless there are definite indications for continued use.

These results were supported by data published by Bergh *et al.* (1982), who reported the clinical course and outcome of 19 bromocriptine-induced pregnancies in 14 women with previously untreated large prolactinomas. These women, aged 25–37 years, were infertile with amenorrhoea of six months to 15 years duration.

Two women showed signs of postpartum tumour enlargement, but only one of them developed visual field disturbances at 30 weeks of pregnancy, which after treatment, went uneventfully to term. This is in agreement with the data published by Nillius (1980).

The other 12 women had 17 uneventful pregnancies with no pituitary tumour complications. The authors conclude that in properly investigated patients with prolactinomas the risk of serious, irreversible tumour complications during bromocriptine-induced pregnancies is very small.

An editorial in the *Lancet* (1982) discussed the management of infertile hyperprolactinaemic women. Although bromocriptine rapidly restores fertility, there is a chance of precipitating tumour growth. However, the frequency of serious complications of this type seems to be low. A study by Nillius (1980) was cited, in which only nine complications, seven of which were minor, due to tumour expansion occurred in 162 pregnancies. It was emphasized that, even when complications do arise, the tumour can be safely decompressed during pregnancy either by bromocriptine treatment or by operation, neither of which endangers the mother or child.

In conclusion, although there are obvious difficulties in obtaining a completely satisfactory control group for such a population, comparison with various published control series strongly suggests that bromocriptine therapy before and during early pregnancy is not associated with an increased risk of spontaneous abortions, multiple pregnancies, or congenital malformations. Furthermore, preliminary data from the ongoing study at the Drug Monitoring Centre of Sandoz do not suggest any adverse effects on postnatal development either.

Clomiphene

Induction of ovulation in infertile women with clomiphene was first reported by Greenblatt *et al.* (1961). Since then there have been numerous reports describing its use and comparing its efficacy with other agents such as gonadotrophins in promoting fertility. Although it has become fairly well established that clomiphene is very efficient at inducing ovulation, concern has been expressed as regards the outcome of some of the pregnancies in relation to abortion rate, multiple pregnancies, and congenital malformations.

Dyson and Kohler (1973) reported two cases of anencephaly associated with the maternal use of clomiphene, although one of the mothers did take promethazine, an antinauseant, as well. The authors, while acknowledging that abnormal development of the CNS and treatment with ovulation stimulants might be purely fortuitous, appealed for information on similar cases. In response to this request several workers (Sandler 1973; Barrett and Hakim 1973; Field and Kerr 1974) reported the occurrence of anencephaly, four cases in all, associated with the use of clomiphene. However, this association was disputed by James (1973), who, using data from the British Perinatal Mortality survey, hypothesized that couples who produce anencephalic infants may be subfertile. Thus, one might expect anencephalic births to be associated with, but not necessarily caused by, drugs used to treat infertility.

In addition, malformations other than neural-tube defects were reported also. Berman (1975) reported one case of a baby with multiple skeletal malformations born to a lady who had taken clomiphene and dimenhydrinate. One case of cystic dysplasia of the ovaries in the newborn was reported by Ford and Little (1981), where the mother had been treated with clomiphene immediately before conception. Ovarian enlargement in the mother, but not in the fetus, as a result of clomiphene therapy, has been well documented. Ford and Little concluded that, although no definite causal relationship could be established, clomiphene is excreted slowly and may still be present six weeks after its administration, and thus could affect ovarian development.

Congenital retinal aplasia in a baby whose mother had required three months treatment with clomiphene to achieve the pregnancy was reported by Laing *et al.* (1981). The pregnancy, delivery, and routine examination of the baby at birth were normal. The baby, at five months of age was referred to a neurologist and was diagnosed as having retinal aplasia. Follow-up of the child at 11 months of age showed that its vision in the dark had improved suggesting that he may have had scotoma similar to that recorded in adults treated with clomiphene. In contrast to these single case reports however, several larger studies have not suggested that clomiphene treatment is related to an increase in congenital malformations.

Kurachi *et al.* (1983) studied the effects of clomiphene citrate on the incidence of congenital malformations in newborn infants from 1034 pregnancies after ovulation induction recorded at nine university hospitals in Japan from 1976 to 1980. In all, 935 infants were born of whom 21 (2.3 per cent) showed visible malformations. This was not significantly different from the incidence of 1.7 per cent reported in 30033 infants after spontaneous ovulation. Furthermore the types of malformations observed were similar in both groups. The authors concluded that these malformations were not related to maternal age nor to the dose of clomiphene used. However, from the data presented there would seem to be a small dose-related trend. These data supported the results of an earlier study by MacGregor *et al.* (1968) which reported a malformation rate of 1.8 per cent in 2000 pregnancies achieved with clomid. Harlap (1976) reported similar findings on a study of 225 infants born after clomiphene therapy.

Hack *et al.* (1972) at Tel-Hashomer Government Hospital reported data on 96 pregnancies from 86 patients induced with clomiphene from 1966 to 1970. Although 37 of the women had 44 previous pregnancies, fetal wastage had been high (70.5 per cent) producing only 13 live infants prior to clomiphene therapy. Twelve patients received prednisone in addition to clomiphene. After clomiphene treatment all the pregnancies were normal, except for a high incidence of toxaemia in 16 women (16.7 per cent). There were 88 singleton pregnancies and eight (9.4 per cent) twin pregnancies. The gestational age and birth weight were normal. The total fetal and neonatal loss was 6.7 per cent, the highest losses, 25 per cent, occurring in the twin births, probably due to prematurity. Two infants were malformed at birth and another two had suspected dislocation of the hip. No malformations occurred in the infants exposed to prednisone. The physical development of these children was followed up to one year of age. Two infants developed pyloric stenosis in the first two months of life, but apart from this, all other children were normal when assessed for head circumference and height at 6 and 12 months.

However, somewhat conflicting evidence was presented by Ahlgren *et al.* (1976) who reported a slight though not significant increase in the incidence of severely malformed infants in pregnancies occurring after clomiphene therapy. They reported on the outcome of 159 pregnancies following clomiphene treatment in Sweden during 1967–74, and found that eight of 148 babies had major malformations. This incidence of 5.4 per cent was not significantly higher than the expected rate of 3.2 per cent. The authors, however, felt that the results were suggestive of an increase especially since three of the babies had the same defect of club foot.

Correy *et al.* (1982) reported the results of 156 pregnancies induced with clomiphene at two Tasmanian hospitals. An increase in multiple pregnancies and a small increase in perinatal mortality was found, but there was no increase in abortion rate. Five of 137 infants had congenital malformations. Two of these had undescended testes, one had hydrocephalus, one had an imperforate anus, and one had multiple defects. This 3.6 per cent incidence of malformations was 2.5 times higher than the state incidence reported for Tasmania.

Although the majority of these studies have shown either no or only a slight increase in major malformations, a number of other effects have been reported. An increase in multiple pregnancies, particularly twins, is consistently reported. A study by Lee *et al.* (1982) reported on 24 chronic anovulatory women who became pregnant after treatment with clomiphene. A total of 25 live babies were born including two sets of twins. One pregnancy ended in spontaneous abortion, and three in premature deliveries. Two of the babies had malformations, one with cryptorchidism and the other with hypogammaglobulinaemia. There were no differences in birth weight compared with a matched control group. Thus, overall there was no adverse effects of clomiphene seen on fetal growth or malformation rate, but there was an increase in twin births.

Seki *et al.* (1983) reported on the outcome of 103/121 pregnancies induced by clomid. The total abortion rate was 18.5 per cent. There were 80 full-term babies, three premature deliveries, 19 spontaneous abortions, and one termination. None of the 83 live babies had congenital malformations at birth. However, at follow-up three children with abnormalities were found, one dislocated hip, one inguinal hernia, and one testicular hydrocele and inguinal hernia. Seven of the children suffered from various diseases during development of which two, Kawasake's disease and infantile autism, were worthy of note.

Other ovulation stimulants

Several studies also report the outcome of pregnancies achieved with other ovulation stimulants, such as PHMG, HCG, and PMS, with or without clomiphene.

Senez and Gillet (1978) studied 53 consecutive cases treated with clomiphene and HCG. Treatment reduced the abortion rate from 43 to 13 per cent. There were 48 liveborn infants two of whom had congenital malformations. One infant had spina bifida and hydrocephalus; the other had absence of the upper limb, possibly due to amniotic bands. There was also an increase in the incidence of twin births and prematurity, probably as a result of this. The authors concluded that there was no overall increase in the incidence of malformations or in abortion rate.

Kurachi *et al.* (1983) in the study mentioned above compared the results of 1034 pregnancies induced with clomiphene with 186 pregnancies induced with HMG–HCG treatment. There was a higher incidence of abortions (19.4 per cent), and premature births (13.4 per cent) in the HMG–HCG group compared with those (14.2 and 5.9 per cent, respectively) in the clomiphene-treated group. The incidence of malformations was lower (1.9 per cent), but not significantly different from that (2.2 per cent) in the clomiphene-treated group, neither of which were significantly different from the incidence of malformations (1.7 per cent) in the group of 30 033 infants born after spontaneous ovulation. The numbers in this study (HMG–HCG) are relatively small and thus the authors regarded the findings as tentative.

Seki *et al.* (1983) attempted to study the outcome of 271 pregnancies achieved with various methods of ovulation induction; i.e. gonadotrophins-HMG, HCG, PMS, clomiphene, clomiphene plus gonadotrophin, sexovid (*cyclofenil*), and epimastrol. The overall abortion rate was 18.4 per cent and the incidence of multiple pregnancies was high (20 per cent) after HMG. The male-to-female ratio was within normal limits and the overall incidence of birth defects was 1.8 per cent. The body weights and lengths of the neonates were similar to those for the general population of Japan, although the weights of those in the gonadotrophin groups were slightly lower.

In an attempt to follow up the later development of these children, questionnaires were sent to 141 mothers of whom 113 replied. No particular tendency was observed in the malformations and/or abnormalities found at the time of follow-up ($<$ 1 y to 9 y) or in the diseases suffered during the course of development. Body heights and weights at the time of follow-up fitted well the physical development curves for children throughout Japan.

Conclusion

In conclusion the majority of large studies on the use of bromocriptine or clomiphene for ovulation induction show no significant increase in the incidence of congenital malformations even though treatment is often continued into early gestation. Although it is difficult to define a control group for such a population of women, there would not appear to be any marked increase in spontaneous abortion rate either. However, a consistent finding with clomiphene therapy is an increase in multiple births though this is usually only in the form of twin pregnancies.

Sex hormones

Introduction

Since the 1950s, concern has been expressed that progestational agents could cause masculinization of female fetuses. Furthermore, this clinical data was supported by the results from animal studies. The genital anomalies in humans were considered to be mild and were usually reversible (Smithells 1965).

This concern about the relationship between sex hormones and congenital malformations progressed from progestational agents used as hormone support therapy (HST), to hormone pregnancy tests (HPT), and oral contraceptives (OC), many of which have a synthetic progestogen as one of their constituents.

Several workers have reported an association between progestogens such as medroxyprogesterone acetate (MPA), e.g. *Provera, Depot-Provera* (Burstein and Wasserman 1964; Limbeck *et al.* 1969) or hydroxyprogesterone caproate, e.g. *Primolut Depot* (Dillon 1970), and congenital malformations other than genital anomalies.

Gal *et al.* (1967) were the first to report an increased incidence of neural-tube defects in the offspring of women who had used HPTs. Since then a number of publications concerning the possible association between HPTs and non-genital malformations have appeared.

A variety of defects have been reported to be increased after maternal sex hormone therapy during pregnancy, e.g. cardiac defects, especially transposition of the great vessels (TGV) (Levy *et al.* 1973), spina bifida (Dillion 1970), limb reduction defects (Janerich *et al.* 1973), and cleft palate (Brogan 1975). Other workers however have failed to confirm the above studies (Mulvihill *et al.* 1974; Laurence *et al.* 1971; Spira *et al.* 1972; Oakley *et al.* 1973; Haller 1974).

Nora and Nora (1973 *a, b*) drew attention to an increase in malformations of the skeletal, gastrointestinal, renal, and cardiovascular systems to which they gave the acronym VACTERL (Vertebral, Anal, Cardiac, Tracheo-oesophageal, Renal, and Limb defects). Since then a number of workers have published data supporting an association between oestrogen and progestogen intake during pregnancy and the VACTERL syndrome in the offspring (Kaufman 1973; Balci *et al.* 1973; Oakley *et al.* 1973). The evidence presented is often difficult to assess because it is not always clear which hormones were used or for what purpose they were used (i.e. HST, OC, or HPT). This raises the possibility that the underlying maternal pathology could be a causal factor in the production of malformed offspring rather than the hormones *per se*. This poses several important questions: namely, are those hormones maintaining an already defective fetus, or, if it is the hormones that are responsible for the malformations, is it because the dose is too high and it is acting as a toxin, or is the dose too low and thus inadequate to maintain a good enough pregnancy to sustain normal fetal development. Thus the interpretation of these studies is difficult, particularly when trying to distinguish studies which fail to show an effect from those that show there is no effect.

Some of the data on the use of hormones in pregnancy were reviewed in the 2nd edition (pp. 452–3). Since then new data have been published which have caused public concern and in some instances have lead to litigation proceedings. Although there is considerable confusion in the literature as to the particular value of these hormones and their subsequent effects on the fetus, an attempt will be made to review the data in two sections: firstly, hormones used in pregnancy tests or as support therapy and, secondly, those used in oral contraceptives.

Hormone pregnancy tests

There are several studies expressing a diversity of opinion as to whether or not HPTs are a causal factor in the production of congenital malformations. However, very few studies actually state which HPT was used and whether its sole use was for pregnancy diagnosis or, whether, it may have been misused as an abortifacient or for other purposes. The first suspicion that hormones in early pregnancy might be teratogenic came from the retrospective studies by Gal (1972; Gal *et al.* 1967) who found a higher usage of HPT by mothers of children with CNS malformations than by a control population. However, it has been stated that the affected children came from a wide area of south-east England but the controls did not, so that variations in prescribing habits were not accounted for. Thus this study can only be suggestive of an association. Her data was supported by the observations of Dillon (1970, 1976) but this was only a small study of women with hormone support therapy (HST) and poor pregnancy histories, and was uncontrolled so that it is of limited predictive value. Gal's study was also supported by Janerich *et al.* (1974, 1977) who reported increased hormone use in mothers and children with limb deformities and also in mothers of children with heart defects. Nora and Nora (1973a, b, 1976, 1978) and Nora *et al.* (1976, 1978) also reported excess hormone use in mothers of children with heart defects but the studies of both of these groups are subject to the same criticism as above about the difficulties of choosing appropriate control groups. Furthermore, the levels of statistical significance attained in some of these studies are very unimpressive making a causal link between the malformations and hormone exposure seem rather unlikely. An important study is that by Greenberg *et al.* (1977) using the data base of the OPCS in Britain.

This study did control for the general practitioner by choosing controls from the same practice as the index case but thereby lost control over other important factors. This study did show a significant but small excess of hormone use in the mothers of malformed children but no specific malformation pattern was observed.

Against these studies are three other retrospective studies which suggest that hormones are not teratogenic. Burstein and Wasserman (1964) studied 174 women exposed to medroxyprogesterone acetate in the first trimester of pregnancy had found only two malformed children, one with a heart defect and one with a genital defect. This seems a surprisingly low malformation rate and, as they were primarily interested in genital defects, it seems possible that other types of defect were missed. Smithells and Chin (1964) reported only three malformed children from 189 mothers exposed to HPT. This again seems a small number of defects but would at least suggest a low order of teratogenicity if indeed it existed at all. Laurence *et al.* (1971) criticized the report by Gal and reported no significant increase in HPT usage by 271 mothers of spina bifida children compared with controls.

From the retrospective studies, therefore, one is unable to come to any conclusion except that, if there is a risk from HPTs, it is likely to be small. Because of the acknowledged deficiences of the retrospective studies, a number of prospective studies have been carried out.

Only three studies have suggested a positive correlation between consumption of sex hormones and an increase in malformations and none of these are on the authors' own admission very convincing. Heinonen *et al.* (1977*b*) using the data of the Boston Collaborative Study showed a twofold increased risk of heart defects in women using oestrogens plus progestogens ($p < 0.05$) but not for either drug alone. Because of the low level of significance the authors state that further studies are needed. One difficulty with the Boston data base is that it was not originally devised for detecting teratogens and different workers using the same data base have in the past arrived at different conclusions regarding the teratogenicity of drugs (anti-epileptics, for example).

Harlap *et al.* (1975) using data from Jerusalem suggested an association of increased malformations with hormone use. However, data on drug intake were not routinely recorded in this survey so the reliability of the drug data is suspect.

Nora and Nora (1976, 1978) following their retrospective surveys attempted to perform a prospective study. They reported a slight but not significant increase in heart defects, but owing to the rapid decline in use of hormones in pregnancy felt it was unlikely that they would be able to complete a satisfactory study. They feel that their original risk estimates may have been too high and are unlikely to be substantiated.

Against these three studies are at least four studies suggesting no increased risk from hormones in pregnancy. Spira *et al.* (1972) and Goujard and Rumeau-Roquette (1977) presumably reviewing data on up to 12764 women reported no increase in malformation rates in hormone-exposed women compared with non-exposed. Many subgroups of hormones were looked at with no signs of significant effects.

Haller (1974) studied 3588 women in Göttingen and found no difference in malformation rates in women exposed to HPT or HST in pregnancy compared with non-exposed women. Kullander and Källen (1976*c*) in a Swedish study on 5753 births found no difference in hormone use between mothers of malformed children and normal children. In 156 women subjected to HPT, the malformation rate was not different from those not so exposed. The report of the Deutsche Forschungsgemeinschaft (1977) on 7120 pregnancies reported no significant correlation between hormone use and malformations. Little actual data is reported however so the significance of the work cannot be assessed.

There are several other studies in the literature but none which shed any further light on this problem. The overall conclusion must be that most prospective studies fail to show any increased risk from hormone exposure in early pregnancy. This may mean *either* that there is no risk, *or* that the risk is too small to be detected by the studies carried out.

These views were supported by Wilson and Brent (1981) is an extensive review of the possible teratogenicity of female sex hormones.

An important study from Japan by Matsunaga and Shiota (1979) investigated the relationships between maternal genital bleeding, hormone support therapy (HST), and external malformations in the fetus with special reference to the critical period of organogenesis. Data were collected on 667 undamaged embryos from induced abortions whose mothers had genital bleeding during early pregnancy and some of whom had also received hormone treatment. In addition 90 embryos with polydactyly and 38 embryos with limb reduction deformities (LRD), were also studied. The evidence presented indicated that for major malformations, e.g. CNS anomalies, cleft lip, polydactyly, and LRD, the maternal genital bleeding was not the cause but a consequence of the conception of an abnormal embryo. Analysis of the timing of hormone therapy compared with the critical period for the induction of these defects did not suggest that hormone treatment was implicated in causing the defects. Furthermore, it would seem that sex hormones are not capable of salvaging malformed fetuses. However, the proportion of embryos found dead *in utero*, whether externally normal or abnormal, was higher in the group treated with hormones than in the untreated group. Matsunaga and Shiota indicated that these findings are probably due to selection of embryos with potentially lethal conditions resulting from the practice of giving HST to those patients who are considered to be at highest risk. This interpretation of the data was criticized by Janerich (1979*b*) who suggested that HST caused the death of a larger proportion of externally normal embryos, and furthermore, that this threefold relative risk of fetal death from HST could hide or prevent recognition of a large relative risk of defective fetuses when ascertainment occurs at term Oakley (1979) supported the conclusions of the original authors and quotes the data presented in a study by Colvin *et al.* (1950). In this study the outcome of 1570 consecutive cases of threatened abortion was followed. Seventy per cent of the fetuses went to term, 28 per cent underwent spontaneous abortion, and 2.1 per cent ended in premature labour. All of the abortuses were systematically examined and it was found that 72 per cent resulted from blighted ova, 13.6 per cent had fetuses with developmental defects or pathology of the placenta and/or membranes, and in the remaining 11.3 per cent no cause was found. Thus in this group of 1570 untreated cases of threatened abortion, only 62 (3.9 per cent) could possibly have been prevented by HST.

After reviewing all of the above data what conclusions can be reached regarding the hazards of exogenous sex hormones in pregnancy, other than their effects on genital development? It is superfluous to state that drugs should not be given during pregnancy unless really required and that, if there is even the slightest suspicion about the safety of particular drugs, then this must apply even more forcibly. It is generally agreed that hormone pregnancy tests are quite unnecessary now that urine tests are so simple and satisfactory, and even in the absence of doubts about the safety of hormones in early pregnancy, the decision to ban hormone pregnancy tests was a correct one. However, it is important nevertheless to examine the evidence about the hazards of hormones in pregnancy to try to assess how convincing it is, and whether the evidence of hazard is greater and more convincing than the evidence for no hazard.

Care must be taken not to reach conclusions solely on the basis of statistical analysis of numbers of malformed children in exposed as compared with non-exposed women. Certain groups, for example those receiving hormone support therapy, are obviously being treated with hormones in an attempt to maintain a pregnancy which is already at risk. It is well known that the prognosis for such pregnancies is poor so that one must be careful not to ascribe the poor outcome to the use of hormones. The correct method to establish the teratogenicity of hormones in such women would be to compare two comparable groups differing only in that one group was treated with hormones and the other was not. Such studies have not been carried out and indeed would be very difficult to carry out, both on ethical grounds and also because it would require that examination should be carried out of all aborted fetuses as well as liveborns to ensure that the hormones had not merely prevented abortion of malformed fetuses.

A major confusing factor in retrospective studies relates to the choice of suitable control populations. It is well known, for example, that many maternal factors correlate with increased malformation rates, such as age, social class, parity, previous reproductive history, etc. It is also well known that there are strong geographical influences on malformation rates, e.g. a high incidence of anencephaly and spina bifida in parts of Wales compared with south-east England. What is not so generally known is that drug-prescribing habits also show marked geographical variations, often of a very localized nature depending on many factors. Thus if studies are carried out in centres for children with specific malformations, e.g. CNS or heart defects, then the children may be drawn from a wide area and unless the control normal children are taken from the same areas, perhaps even from the same general practitioners, associations may be found which are not in any way causally related. This could lead to erroneous conclusions about the hazards of drugs. This is one of the many reasons why prospective studies are preferred to retrospective studies for assessment of potential drug hazard.

The majority of evidence indicates that there is a lack of causal relationship between sex hormones administered in pregnancy and non-genital malformations in the offspring. If there is an increased risk of any particular non-genital malformation, the risk must be small and may not necessarily be causal but may be related to the underlying maternal pathology or to the presence of a pre-existing malformation.

Oral contraceptives

There are conflicting reports on the teratogenic risks associated with the use of oral contraceptives (OCs) just prior to, or in early pregnancy. It has been estimated that in Europe and the USA up to 5 per cent of women will be taking OCs in early pregnancy and some 25–35 per cent in the three months prior to conception. Thus the ovum and fetus may be exposed to sex hormones prior to conception and during embryogenesis when the major organ systems are developing.

Although the majority of studies have shown that there is no overall increase in congenital malformations attributable to OC usage (RCGP Survey 1976; Ortiz-Perez *et al.* 1979; Alberman *et al.* 1980; Janerich *et al.* 1980; Smithells 1981; Savolainen *et al.* 1981), some workers have reported a small increase in specific malformations (Harlap *et al.* 1975; Harlap and Eldor 1980; Kasan and Andrews 1980; Klinger and Glasser 1981). Because of the widespread use of OCs, these reports of adverse effects in pregnancy warrant careful consideration. In addition to those studies discussed in the original chapter (*Iatrogenic diseases*, 2nd edn, pp. 452–3) some of the newer studies were reviewed by Darling and Hawkins (1981).

There are five areas of major concern: namely, the production of chromosome abnormalities, particularly triploidy; a shift in the sex distribution of the offspring; an increase in spontaneous abortions; a higher incidence of ectopic pregnancies; and an increased risk of congenital malformations such as cardiac defects, neural-tube defects, and limb reduction deformities.

There is some evidence that chromosome anomalies such as triploidy and polyploidy as well as breakages occur more frequently in the pregnancies of previous OC users (Carr 1967, 1970; Klinger *et al.* 1970; Littlefield *et al.* 1975). However, these studies were carried out on spontaneous abortuses between 1969 and 1973 when the levels of sex hormones in OCs were much higher than they are now. Furthermore, the degree of statistical significance reached with such small sample sizes is at best, marginal. Other studies by Lauritsen (1975), Alberman *et al.* (1980), and Janerich *et al.* (1980) have failed to confirm these effects. Thus, there appears to be no clear causal relationship between chromosome damage and the use of OCs and it must be concluded that if a risk exists it must be low.

Several studies have indicated a shift in sex ratio amongst offspring of previous OC users or a preponderance of males in malformed offspring (RCGP Survey 1976; Harlap and Davies 1978; Janerich *et al.* 1980; Harlap and Eldor 1980). The numbers involved in these studies were small, however, and larger studies have failed to confirm these findings in either abortuses or neonates (Rothman and Liess 1976; Klinger and Glasser 1981; the review by Smithells 1981).

There is very little evidence that the incidence of spontaneous abortion is higher in ex-pill users. The RCGP survey (1976) indicated that there was a higher abortion rate, but this was based on a very small sample size. Furthermore, when the data was re-analysed it was found that this increase was due to induced abortions rather than spontaneous abortions. Alberman *et al.* (1980) showed that there was a small excess of spontaneous abortions which occurred after OC use, but there was no evidence of a substantial excess or deficit which might have explained previous reports of excess chromosomal anomalies in spontaneous abortuses after OC use. They concluded that a poorer outcome was just slightly more common in pregnancies occurring after OC use particularly those occurring within a month or after a year of cessation. Thus it would seem that the length of time taken to conceive may be one of the factors influencing pregnancy outcome.

Moreover, women who become pregnant whilst taking OCs presumably do not wish to become pregnant at that time. The possibility of attempts at procuring an abortion must be taken into consideration, so that malformed fetuses resulting from failed abortions may be ascribed to contraceptive drugs.

There have been several case reports indicating a possible association between oral contraceptives and ectopic pregnancies (Bonnar 1974; Weiss *et al.* 1976). However, the sample sizes were very small and other workers have failed to confirm the above findings (Hawkins 1974; Berger *et al.* 1976). The study by Berger *et al.* (1976) reported data on 869 young women who were followed up retrospectively between 1961 and 1970. Amongst these women, 463 chose to use OCs whilst 406 did not. In the OC-users group 2/666 pregnancies were ectopic (3/1000) compared with 3/533 (5.6/1000) in the non-users group. This difference was not significant and

the authors felt that there was no evidence to implicate OCs or a temporary hormone imbalance in either of the ectopic pregnancies which occurred in the OC users. At present there seems to be no clear causal relationship between OC use and ectopic pregnancies.

There is still concern that exposure to OCs in early pregnancy may lead to congenital malformations in the offspring. In addition to the studies reviewed in *Iatrogenic diseases*, 2nd edn, pp. 452–3 (*Levy et al.* 1973; Nora and Nora 1973*a, b*, 1974, 1975; Janerich *et al.* 1973; Harlap *et al.* 1975; Heinonen *et al.* 1977*a*; Schardein 1985; Goujard and Rumeau-Rouquette 1977), there have been three subsequent reports indicating an adverse pregnancy outcome.

Janerich *et al.* (1980) compared the history of OC use in 715 women who gave birth to malformed infants with those of matched controls (715) with normal children. The case mothers were found to have used OCs after the last menstrual period or just before conception slightly more often than the controls but this difference was not significant. There was no difference in OC use overall in the mothers of malformed children compared with the mothers of normal children. When a subgroup of malformed children were isolated with one or more of the structural abnormalities of the VACTERL syndrome, OC use was approximately twofold greater ($p < 0.01$) in their mothers than in the controls. Cytogenetic abnormalities and hypospadias were not associated with OC use, at or after the time of conception. As previously mentioned they found a predominance of males (13 of 16; 81 per cent) amongst the malformed babies. The authors concluded that the association between birth defects and OCs is not large, but the possibility of a causal relationship cannot be excluded. These findings were not supported by Harlap and Eldor (1980) who found no increase in congenital malformations in 108 infants conceived while the mothers were taking oral contraceptives. They reported a significantly large number of twins and an excess of perinatal mortality however.

Kasan and Andrews (1980) studied 10479 singleton births in South Glamorgan in 1974–6 of which 27.3 per cent of infants were born to women who had used OCs in the three months prior to their last menstrual period or during early pregnancy. There were significant differences in the two populations of 'users' versus 'non-users' in age distribution, parity, social class, and smoking, habits. There was no difference in the overall malformation rate in the two groups or in the incidence of any specific malformation except neural tube defects. There were significantly more infants with neural-tube defects (18 of 2859; 0.63 per cent) amongst OC users compared with 19 of 7620 (0.25 per cent $p < 0.01$) in non-users.

Although the authors acknowledge that South Wales has a relatively high incidence of NTD which may partly explain the present findings, it would have been expected that non-users and users would have been equally affected and in fact, in their study the overall incidence of NTD was lower than the national average. A prospective study of OC use in relation to subsequent outcome of pregnancy was reported by Savolainen *et al.* (1981) using the Finnish register of congenital malformations which is based on a matched-pair approach. They found no indication of any increase in total congenital malformations or of any specific defect in relation to OC use in approximately 700 matched pairs.

Because of the inherent difficulties of obtaining and assessing data on the use of OCs in pregnancy it is not possible to exclude completely the possibility of teratogenic effects of poor fetal outcome. However, it would seem that any risk of malformation is small and would not necessarily constitute grounds for termination.

Because of the possibility of increased sensitivity of the fetus to circulating sex hormones, it may be advisable to delay conception for 3 to 4 months after stopping OC use so that these hormones are cleared from the maternal circulation.

Thyroid drugs

Thyroid hormones

Thyroid-stimulating hormones, thyroxine (T4) and triiodothyronine (T3), have all been shown to be teratogenic if given in large doses to animals. However, in humans, when used in the usual doses for the treatment of hypothyroidism, T3 and T4 have been shown to cross the placenta only very poorly. Turnover of these substances is not altered in the mother during pregnancy so there is no need to alter the maternal treatment because of the pregnancy (Burr 1981).

Antithyroid drugs

The antithyroid drugs cross the placenta readily and can inhibit thyroid function in the fetus. The most obvious effect of this is to cause a compensatory thyroid hypertrophy and goitre which can be large enough to compress the fetal trachea and cause neonatal asphyxial death. Over 400 cases of such goitres have been recorded following iodide exposure in pregnancy, and drugs, such as cough medicine containing iodides as well as iodide itself and radio-iodine, should be avoided during pregnancy.

Apart from the production of goitres, teratogenic effects have not been clearly demonstrated following use of other antithyroid agents. Scalp defects have been reported following methimazole therapy (Milham and Elledge 1972) and limb defects and cataracts after carbimazole therapy (McCarroll *et al.* 1976), but the relationship of these defects with the treatment is uncertain (Ramsay 1983). Both propylthiouracil (Burrow 1978) and carbimazole (Sugrue and Drury 1980) have been used during pregnancy with good results both for the mother and fetus, and long-term follow-up of the children up to 13 years of age has not shown any long-term disadvantage for the children in terms of physical or mental development (McCarroll *et al.* 1976). Some authors recommend stopping treatment during the last month of pregnancy to reduce the risk of neonatal goitre, and because the disease is often much milder at the end of gestation.

All of the antithyroid drugs pass into the milk but not equally. The proportion of the adult dose consumed by a breast-fed baby has been calculated to be about 0.07 per cent for propylthiouracil, 0.5 per cent for carbimazole, and 10 per cent for methimazole (Tagler and Lindström 1980). Thus, if a mother wishes to breast-feed, it would be advisable to use propylthiouracil, if possible, to avoid adverse effects on the child.

The use of β-blockers for more than a short period of up to seven days is not recommended.

Women with thyrotoxicosis may also be at risk of producing thyrotoxic babies due to the transplacental passage of the immunoglobulins, TSI and TSI-P, and this risk can be tested for prenatally by analysis of maternal serum (Dirmikis and Munro 1975).

Vaginal spermicides

During the last 30 years there has been a shift in interest from barrier methods of contraception to more sophisticated methods such as IUDs and the oral contraceptive pill. However, in the past few years concern has been expressed about the adverse effects of oral contraceptives on women's health and on pregnancy outcome. Because of this many people have returned to the use of barrier methods such as diaphragms, condoms, and spermicides (Kleinman 1980). In general the spermicides in common use are phenoxypolyethoxyethanol derivatives which are formulated as pessaries, pastes, creams, gels, foams, and soluble films. The major ones used in the UK contain nonoxynol-9, di-isobutyl-phenoxypolyethoxyethanol and octoxynol.

However, the possibility has now been raised that the use of spermicides may be associated with the presence of congenital malformations as well as an increase in spontaneous abortions. In 1977 Smith *et al.* in a small case-control study reported an association of limb reduction deformities (LRD) with the use of spermicides; since then there have been a number of studies investigating the association between the use of spermicides and congenital malformations. Two of these studies indicated a higher incidence of malformations and spontaneous abortions (Jick *et al.* 1981*b*, 1982) and another indicated a higher incidence of Down's syndrome (Rothman 1982), among the offspring of ex-spermicide users. However, data from the Boston Collaborative Perinatal Project (Shapiro *et al.* 1982), the Oxford Family Planning Association Contraceptive Study (OFPACS) (Huggins *et al.* 1982), as well as data published by Polednak *et al.* (1982), Mills *et al.* (1982), and Bracken and Vita (1983) showed no increase in malformations.

The data provided by Jick *et al.* (1981*b*) was obtained from the Boston Collaborative Drug Surveillance Programme. They studied 763 live infants born to white women who had obtained vaginal spermicides during the 10 months prior to conception (computer file check of prescriptions). In this study group the prevalence of major malformation was 2.2 per cent (17/763) compared with 1.0 per cent (39/3902) in the controls. Four major categories of malformations were shown to be increased: limb reduction deformities with a prevalence rate (PR) of 3.9 compared with 0.3 in the controls; neoplasms, nesidioblastosis and medulloblastoma, were seen in two babies (PR = 2.6) compared with none in the controls. In fact, these neoplasms are so rare that neither have been seen in the 20 000 babies born at Group Health since 1972. The prevalence of chromosome abnormalities was 3.9 in the study group (three babies with Down's syndrome) but only 0.3 in the controls (one baby with trisomy 18). Hypospadias was seen in two children (PR = 2.6) in the study group but was absent in the controls.

There was also a 1.8-fold increase in spontaneous abortions which required hospitalization in spermicide users when compared with non-users. Recalculation of the prevalence ratios with age stratification did not affect the results. The authors however were very cautious in the interpretation of this study. Since there was absence of a single well-defined syndrome and lack of information on actual use and time of use of the spermicides in relation to conception as well as the possibility of other confounding factors involved in the choice of this type of contraception, they felt that the results should be regarded as tentative. This report was criticized by Oakley (1982) as regards its design and the conclusions drawn from the data presented. He pointed out that the incidence of major malformations in the control group of only 1 per cent was about half the expected incidence and furthermore it is likely that vaginal spermicides would have been purchased over the counter and not on prescription and this group may well be included in the controls. Shapiro *et al.* (1982) also commented on the fact that no cases of Down's syndrome were observed in the 3902 babies of non-users whereas about four would be expected.

A more recent paper by Jick *et al.* (1982) reported the incidence of miscarriages amongst women who used vaginal spermicides (20 per cent used products containing nonoxynol-9 and 80 per cent used products containing octoxynol), oral contraceptives, or neither of these methods. Amongst the 813 women who obtained vaginal spermicides within 48 weeks of the estimated date of fertilization, 47 (5.8 per cent) had an early miscarriage whereas, in the oral contraceptive group 35/1127 subjects, (3.1 per cent) miscarried. The miscarriage rate in the group of women using neither method (140/4231 = 3.3 per cent) was very similar to that of the oral contraceptive group. The authors estimated the risk ratio of spermicide users versus non-spermicide users to be 1.8 (90 per cent confidence interval 1.4, 2.3). Furthermore, the association seemed to be strongest amongst women who obtained a spermicide within 12 weeks of the expected date of fertilization. Examination of the abortuses revealed that the association with spermicides was strongest among those where an abnormal fetus was present, but there was no increase in any particular type of malformation.

Rothman (1982) published data collected from a case-control study of 460 babies with congenital heart disease (CHD) who were born during 1973–75 in Massachusetts. The controls (1500) were obtained from a random sample of Massachusetts birth certificates over the same period and the majority of the cases were obtained from the New England Regional Infant Cardiac Programme. After each year of the study, questionnaires were sent to the mothers and from this data it was found that 16 of 390 children with CHD had also been diagnosed as having Down's syndrome. The contraceptive histories of these mothers showed that 4 of 16 (25 per cent) had been spermicide users (PR = 3.6), compared with 109 of 1254 (8.7) of the controls or 40 of 374 (10.7 per cent) with CHD only. Furthermore, it was independent of maternal age. Rothman considers his data to be free from recall-bias because there was no suspicion of a link between spermicide use and Down's syndrome when the data was collected. Although the number of cases was small, he felt that there was a tentative link between spermicide use and at least one type of Down's syndrome which supports the findings of Jick *et al.* (1981*b*). However, the same criticisms apply to this study as applied to the major study of Jick *et al.* (1981*b*), in which Rothman was a co-author, in that there is no information on maternal obstetric histories, other drug treatment, or time of fetal exposure to the spermicides.

In 1977 data was published from the Boston Collaborative Perinatal Project indicating that local contraceptives were not associated with malformations (Heinonen *et al.* 1977*b*). However, because of the contradictory evidence presented by Jick *et al.* (1981*b*) and Rothman (1982) this data was re-analysed in greater detail (Shapiro *et al.* 1982).

Since the spermicides in common use in 1958–65 when the data were collected were nonoxynol-9 and octoxynol (those most commonly used in the Jick study), which are still in use, and phenylmercuric acetate, which is no longer in use, the results with the latter have been evaluated separately.

In the total cohort study of 50 282 pregnancies, 462 women had used vaginal spermicides (other than phenylmercuric acetate) actually during the first trimester of pregnancy and of these, 438 (95 per cent) had also used them in the month prior to their last menstrual period (LMP). So, in this study spermicide users are quite well defined — i.e. exposure was limited to within 16 weeks of the LMP, which covers the period of embryogenesis and as 95 per cent of the users had used the spermicides in the month prior to the LMP this would cover the time when sperm or ova could have been damaged during fertilization.

Overall, 31 of 462 (6.7 per cent) babies exposed during the first trimester were malformed compared with 3217 of 49 820 (6.5 per cent) in the non-exposed group. Major malformations were observed in 2.2 per cent exposed and 2.8 per cent controls. The estimated rate ratio for major malformations was 0.9. Minor malformations were observed in 2.8 per cent exposed and 1.8 per cent unexposed giving an adjusted risk ratio of 1.2 (95 per cent confidence limits 0.7–2.2). In the group of 889 mothers who used phenylmercuric acetate spermicides during the first trimester, 47 (5.3 per cent) of the babies were malformed compared with 3201 of 49 393 (6.4 per cent) of the controls. Overall no excess of LRD, neoplasms, hypospadias, or Down's syndrome, or any other specific malformations were observed in this study. Shapiro *et al.* (1982) concluded that non-mercurial spermicides do not increase the overall malformation rate, and even with the phenylmercuric acetate spermicides, which are no longer in use, an increase in major malformations of more than 1–2-fold can be ruled out. Their study was not large enough however to exclude the possibility that certain specific malformations may be increased.

Huggins *et al.* (1982) published data from the Oxford Family Planning Association Contraceptive Study (OFPACS). This covered a population of 17 032 participants over the period 1968–74. The data in the OFPACS is limited to the main method of contraception used at any one time, so it cannot distinguish diaphragm and condom users who also use spermicides from those who do not. However, the response to a questionnaire about spermicide use showed that almost 100 per cent of users of diaphragms always used a spermicide. They analysed their data in three sections: singleton planned pregnancies, singleton unplanned pregnancies, and multiple pregnancies; within each group the outcome of pregnancy in condom users and diaphragm users was compared with other groups using oral contraceptives, IUDs, or some other or no method. Therefore this study does have the advantage that it can distinguish between women who had stopped using contraceptives — planned pregnancies — from those who experienced contraceptive failure.

Analysis of the data from 5729 singleton planned pregnancies showed no significant increase in spontaneous abortions or in malformations. Birth weight and sex ratio were unaffected. In the group of 1552 singleton unplanned pregnancies there was no evidence that women who became pregnant while using a diaphragm or condom had an increase of spontaneous abortions. Birth weight and sex ratio were unaffected. However, the occurrence of congenital malformations was difficult to interpret because of the small numbers.

In the group of 81 women who had multiple pregnancies (65 planned and 16 unplanned), there was no association between the use of spermicides and multiple pregnancies. Of the planned pregnancies, five resulted in a total of six babies with birth defects; four of these mothers were from the condom group. These four mothers were followed up: three of the four had not used spermicides; the other had discontinued using spermicides for more than a year prior to conception. So in this particular group there seems to be no association between spermicide use and malformations.

The conclusions drawn from this study by Huggins *et al.* (1982) indicated that there were no adverse effects on spontaneous abortion rate, birth weight, or sex ratio; however, there was some concern as regards congenital malformations. The authors commented that, although none of the infants had malignant neoplasms, each of the other three anomalies specifically mentioned by Jick *et al.* as possibly being associated with spermicide use (limb reduction defects, hypospadias, and Down's syndrome) did occur more commonly among infants born to mothers who had ever used diaphragms than amongst infants born to mothers who had not. Among the unplanned pregnancies, the apparent aggregation of a number of serious birth defects in the diaphragm group was also worrying. While neither set of findings is by any means conclusive, it seems the data on malformations presented in this study do perhaps add some weight to the views put forward by Jick *et al.*

From these findings they drew the following conclusions. Spermicide use may possibly have some small adverse effect on the risk of congenital malformation, especially among infants conceived as a result of contraceptive failure. No evidence was found for any other adverse effect on the outcome of pregnancy, and the data provides strong evidence against the hypothesis that spermicide use has any appreciable effect on the risk of spontaneous abortion.

The overall lack of teratogenic effects of non-hormonal contraceptives was supported by a study by Bracken and Vita (1983). These authors reported the results of a case-control study (1427 cases and 3001 controls) at five Connecticut hospitals in which they found little relationship between delivery of a malformed infant and the use by the mother of non-hormonal contraceptive methods at conception. The analysis was focused mainly on three contraceptive methods, rhythm, spermicides, and IUD. There was a lack of association of malformations with the use of any of these methods.

In a detailed analysis of specific malformations and contraceptive use, two types of malformations were associated with the rhythm method, i.e. cleft lip and palate and congenital hydrocele, and one group, i.e. multiple anomalies, was associated with IUD use. However, all of these are very difficult to evaluate as the numbers involved are small.

In an abstract report of a large study on 3146 women using spermicides compared with 13 148 using other methods of birth control, no evidence was found that spermicide exposure before or after the last menstrual period was teratogenic (Mills *et al.* 1982).

Polednak *et al.* (1982) published data from two studies designed to investigate the possible effect of maternal spermicide (mostly nonoxynol-9) use on the development of the fetus.

The first was a cohort study consisting of 302 women who used spermicides and 716 women who used no contraceptive methods in the year prior to pregnancy resulting in the live birth of a normal child. There was no evidence that spermicide use prior to the last menstrual period (LMP) had any effect on either mean birth weight or on the proportion of lower birth weights. However, the mean birth weight of female infants was significantly lower in the post-LMP spermicide users than in pre-LMP only spermicide users or no contraceptive users.

The second was a case-control study of 715 births with selected birth defects, i.e. LRD, hypospadias, anencephaly, spina bifida, Down's syndrome, CVD, and multiple defects, and 715 controls, matched for maternal age and race. The data showed that no significant increase in relative risk was associated with maternal spermicide use either prior to or after the last menstrual period. The relative risks for post-LMP spermicide use were greater than 1.00 for hypospadias (eight cases, two controls from 99 pairs), and for LRD (six cases, three controls from 108 pairs), but neither of these differences was significant.

Apart from the inherent problems associated with any retrospective studies, these studies present additional problems inasmuch that each has defined spermicide exposure differently and it is not always apparent whether the pregnancies were associated with contraceptive failure and there is no indication of maternal health or obstetric histories. Thus interpretation of the data on congenital malformations is very difficult.

Several mechanisms of action have been proposed: namely, damage to a genetic locus which could give rise to the split hand or foot type deformities; damage during the first or second meiotic division giving rise to chromosome non-disjunction which might be associated with Down's syndrome; or direct damage to the developing embryo during early pregnancy (LRD, hypospadias) or late pregnancy (neoplasms).

In fact there seems to be little evidence that spermicides cause genetic damage. Although this was proposed by Jick *et al.* (1981*b*) as a mechanism of action in LRD, it remains highly speculative and controversial (Shapiro *et al.* 1982; Oakley 1982). The evidence put forward indicating damage during meiosis seems to be based on one study only, and this has yet to be published in full. This preliminary report by Warburton and his co-workers quoted by Shapiro *et al.* (1982) presented data on four aborti in which there was meiotic disruption giving rise to three cases of tetraploidy and one triploidy, whose mothers had used vaginal spermicides from the time of conception and from this inferred an association between vaginal spermicides and chromosome abnormalities. However, it was pointed out by other workers (Shapiro *et al.* 1982) that the mechanism for tetraploidy is likely to be mitotic non-disjunction at the first cleavage and therefore the findings do not justify an inference of a possible effect on the risk of trisomy as in Down's syndrome or other chromosomal abnormalities occurring during meiosis. In fact the results of the Boston Collaborative Perinatal Project (Heinonen *et al.* 1977*b*; Shapiro *et al.* 1982) and the OFPACS (Huggins *et al.* 1982) in which data was presented from women who were exposed prior to and during their pregnancies would support this.

Thus it would seem likely that exposure to vaginal spermicides has little or no effect on the incidence of spontaneous abortions or overall congenital malformation rate. There is some indication that there may be a small increase in certain malformations such as NTD, hypospadias, LRD, and Down's syndrome. However, as no definite syndrome can be defined and as in most cases the time of exposure is not known, it is not possible to infer a causal relationship. Much larger studies would be required to identify an increase in specific malformations.

Miscellaneous drugs

Isotretinoin *(Roaccutane)*

This drug which is a retinoid and related to vitamin A is very effective when used orally for the treatment of very severe acne. It is one of the few drugs which are known to be potent teratogens in humans. Rosa (1983) reported on pregnancy outcome following the use of this drug in the USA. At that time only 14 people were known to have been exposed to the drug during pregnancy and 13 of these had had an adverse outcome. Eight ended in spontaneous abortions and five in babies with birth defects (two hydrocephalus, two hydrocephalus with cardiovascular defect and small or absent ears or eyes, one with small ears and heart murmur). The majority of the cases were identified prospectively and only one exposed patient is known to have had a normal pregnancy outcome. More recently, Stern *et al.* (1984) and Rosa *et al.* (1985) have reported on 34 malformed babies with the characteristic features of brain, cardioaortic, microtia, facial palsy, micrognathia, cleft palate and/or thymic aplasia associated with isotretinoin use. In addition, 30 spontaneous abortions were reported. Exposures were usually in the first 4 weeks post-conception. It is emphasized that this drug must not be used in women who may become pregnant. It is to be used only when it is known that the patient is taking effective contraceptive measures. Many of the exposures which have occurred have been related to failures of contraceptive methods and barrier methods may not be adequate. The possibility of interactions between antibiotics such as tetracycline and oral contraceptives should also be considered. Because of the persistence of the drug and metabolites in the body, it is recommended that pregnancy should be deferred for at least one month after cessation of treatment.

Etretinate *(Tigason)*, tretinoin *(Retin A)*

These drugs are also retinoids related to isotretinoin. Etretinate is used systemically for the treatment of psoriasis and tretinoin is used topically for the treatment of acne. There is no good evidence regarding their safety in human pregnancy, however in animal studies both of them show adverse effects on fertility and both are more potent teratogens than isotretinoin in rodents and rabbits (Kamm 1982). Happle *et al.* (1984) reported that 6 of 19 women who had used etretinate around the time of conception had babies or aborted fetuses with severe craniofacial or skeletal malformations. Rosa *et al.* (1985) reported on 8 malformed babies and one stillbirth following etretinate use. Use of this drug should therefore also be avoided in pregnancy, and because of its persistence in the body it is also advisable to avoid pregnancy for some months after cessation of treatment.

Penicillamine

This drug is used as a chelating agent in the treatment of Wilson's disease, in severe rheumatoid arthritis, and in cystinuria. There are two case reports of babies born with a similar type of connective-tissue disorder following treatment of the mothers throughout pregnancy. One received 2 g daily for cystinuria (Mjolnerod *et al.* 1971) and the other 900 mg daily for arthritis (Solomon *et al.* 1977). In both cases the babies died. However, in a series of 27 women treated with penicillamine for arthritis or cystinuria (Lyle 1978) and in a series of 29 pregnancies in which it was used for Wilson's disease (Scheinberg and Sternlieb 1975) no increase in malformations was observed. In a review of all the published reports of penicillamine use in pregnancy, Endres (1981) found a total of 87 pregnancies with only two malformations (mentioned above) present in the babies. It may be that the dosage is an important factor. In Wilson's disease the penicillamine is largely chelated by the high copper levels and is rapidly excreted thus having some protective effect against placental transfer. Overall, the risk from penicillamine seems low but efforts should be made to keep the dose as low as possible, and its use in arthritis should be avoided if possible.

Sulphasalazine

Mogadam *et al.* (1980) studied the outcome of pregnancy in patients with ulcerative colitis and Crohn's disease. They reported briefly on 531 pregnancies in patients of whom 287 were treated with either or both corticosteroids and sulphasalazine and 244 received neither drug. The frequency of underweight babies, spontaneous abortions, prematurity, stillbirths, and severe jaundice was significantly less in all of the patients with inflammatory bowel disease than in the general population (possibly due to better prenatal care) and there was no increase in congenital abnormalities in either the treated or untreated group. Fahrlander (1980) also did not report any increase in malformations associated with sulphasalazine use. However, one child with fatal cardiovascular defects and stillborn twins both with genitourinary malformations were detected by the Tasmanian Birth Notification Scheme (Newman and Correy 1983). They also referred to another case of a child with hydrocephaly and cleft lip and palate, all following sulphasalazine therapy. As no pattern of defects was obvious in these individual case reports and as the much larger studies referred to above did not indicate any excess risk, it would seem safe to use this drug when indicated.

REFERENCES

ABEL, E.L. (1983). *Marihuana, tobacco, alcohol and reproduction*, pp. 31–42. CRC Press, Florida.

AD HOC COMMITTEE (1974). Occupational disease among operating-room personnel: a national study. *Anesthesiology* **41**, 321–40.

AHLGREN, M., KÄLLEN, B., AND RANNEVICK, G. (1976). Outcome of pregnancy after clomiphene therapy. *Acta obstet. gynec scand.* **55**, 371.

ALBERMAN E., PHAROAH, P., CHAMBERLAIN, G., ROMAN, E., AND EVANS, S. (1980). Outcome of pregnancies following the use of oral contraceptives. *Int. J. Epidemiol,* **9**, 207–13.

ALDRIDGE, A., BAILEY, J., AND NEIMES, A.H. (1981). The disposition of caffeine during and after pregnancy. *Seminars Perinatol.* **5**, 310–14.

ALEXANDER, F.W. (1979). Sodium valproate and pregnancy. *Arch. dis. Child.* **54**, 240–5.

AL-MONDHIRY, H., MARCUS, A.J., AND SPAET, T.H. (1970). On the mechanism of platelet function inhibition by acetyl salicylic acid. *Proc. Soc. exp. Biol. Med.* **133**, 632–6.

ANANTH, J. (1975). Congenital malformations with psychopharmacologic agents. *Comprehens. Psychiat.* **16**, 437–45.

ANNEGERS, J.F., ELVEBACK, L.R., HAUSER, W.A., AND KURLAND, L.T. (1974). Do anticonvulsants have a teratogenic effect? *Arch. Neurol.* **31**, 364–73.

——, HAUSER, W.A., ELVEBACK, L.R., ANDERSON, V.E., AND KURLAND, L.T. (1978). Congenital malformations and seizure disorders in the offspring of parents with epilepsy. *Int. J. Epidemiol* **7**, 241–7.

——, KURLAND, L.T. AND HAUSER, W.A. (1983). Teratogenicity of anticonvulsant drugs. In Epilepsy (ed. AA. Ward, J.K. Penry, and D. Purpura), pp. 239–48. Raven Press, New York.

ARIAS, F. AND ZAMORA, J. (1979). Antihypertensive treatment and pregnancy outcome in patients with chronic mild hypertension. *Obstet. Gynaecol.* **53**, 489–94.

ASKROG, V. AND HARVALD, B. (1972). Teratogenic effects of inhalation anaesthetics. *Nord. Med.* **83**, 498–500.

BAILLIE, M., ALLEN, E.D., AND ELKINGTON, A.R. (1980). The congenital warfarin syndrome: a case report. *Br. J. Ophthalmol.* **64**, 633–5.

BALCI, S., SAY, B., PIRNAR, T., AND HICSONMEZ, A (1973). Birth defects and oral contraceptives. *Lancet* **ii**, 1098.

BALTAZAR, B., ERICSON, A., AND KÄLLEN, B. (1979). Delivery outcome in women employed in medical occupation in Sweden. *J. occup. Med.* **21**, 543–8.

BARLOW, S.M. AND SULLIVAN, F.M. (1975). Behavioural teratology. In *Teratology, trends and applications* (ed. C.L. Berry and D.E. Poswillo), pp. 103–20. Springer-Verlag, Berlin.

—— AND —— (1982). *Reproductive hazards of industrial chemicals; and evaluation of animal and human data.* Academic Press, New York.

BARRETT, C. AND HAKIM, C. (1973). Anencephaly, ovulation stimulation, subfertility, and illegitimacy. *Lancet* **ii**, 916–17.

BARROW, M.V. AND SOUDER, D.E. (1971). Propoxyphene and congenital malformation. *J. Am. med. Ass.* **217**, 1551–2.

BARRY, J.E. AND DANKS, D.M. (1974). Anticonvulsants and congenital abnormalities. *Lancet* **ii**, 48–9.

BARTOLOME, J.V. SCHANBERG, S.M. AND SLOTKIN, T.A. (1981). Premature development of cardiac sympathetic neurotransmission in the fetal alcohol syndrome. *Life Sci.* **28**, 571–6.

BATAGOL, R. (1983). *The Royal Women's Hospital reference guide on drugs and pregnancy incorporating the 1983 update to the 3. A.W. reference guide on drugs and pregnancy.* Melbourne, Australia.

BEARD, M.C., NOLLER, K.L., O'FALLON, W.M., KURLAND, L.T. AND DOCKERTY, M.B. (1979). Lack of evidence for cancer due to metronidazole. *New Engl. J. Med.* **301**, 519–22.

BEELEY, L. (1981*a*). Adverse effects of drugs in the first trimester of pregnancy. In *Clinics in obstetrics and gynaecology* (ed. S.M. Wood and L. Beeley), Vol. 8, no. 2, pp. 261–74. W.B. Saunders. London.

—— (1981*b*). Adverse effects of drugs in later pregnancy. In *Clinics in obstetrics and gynaecology* W.B. Saunders, London. (ed. S.M. Wood and L. Beeley), Vol. 8, no. 2, pp. 275–90.

BEN ISMAIL, M., GHARIANI, M., AND JAIS, J.M. (1975). Pregnancy, heart valve prostheses and antiplatelet aggregation treatment. *Nouv. Presse Méd.* **4**, 746.

BENNETT, J.S. (1962). Note on glutethimide. *Can. med. Ass.* **87**, 571.

BERGER, G.S., TAYLOR, R.N. AND TRELOAR, A.E. (1976). Ectopic pregnancy and the pill. *Lancet* **ii**, 961.

BERGH, T., NILLIUS, S.J., ENOKOSSON, P., LARSSON, S.-G., AND WIDE, L. (1982). Bromocriptine induced pregnancies in women with large prolactinomas. *Clin. Endocrinol.* **17**, 625–31.

BERMAN, P. (1975). Congenital abnormalities associated with maternal clomiphene ingestion. *Lancet* **ii**, 878.

BIGGS, J.S.G. AND ALLAN, J.A. (1981). Medication and pregnancy. *Drugs* **21**, 69–75.

BJERKEDAL, T. AND BAHNA, S.L. (1973). The course and outcome of pregnancy in women with epilepsy. *Acta obstet. gynaecol. scand.* **52**, 245–8.

——, GIEZEL, A., GOUJARD, J. KALLEN, B., MASTROIACOVA, P., NEVIN, N., OAKLEY, G., AND ROBERT, E. (1982). Valproic acid and spina bifida. *Lancet* **ii**, 1096.

BLATT, J., MULVIHILL, J.J. ZIEGLER, J.L., YOUNG, R.C., AND POPLACK, D.G. (1980). Pregnancy outcome following cancer chemotherapy. *Am. J. Med.* **69**, 828–32.

BLEYER, W.A. AND BRECKENRIDGE, R.T. (1970). Studies on the detection of adverse drug reactions in the newborn. II. The effects of prenatal aspirin on the newborn hemostasis. *J. Am. med. Ass.* **213**, 2049–53.

——, AU, W.Y.W., LANGE, W.A., AND RAISZ, L.G. (1970). Studies on the detection of adverse drug reaction in the newborn. I. Fetal exposure to maternal medication. *J. Am. med. Ass.* **213**, 2046–8.

BLINICK, G., JEREZ, E., AND WALLACH, R.C. (1973). Methadone maintenance, pregnancy and progeny. *J. Am. med. Ass.* **225**, 477–9.

BOBROWITZ, L.D. (1974). Ethambutol in pregnancy. *Chest* **66**, 20–4.

BOGGS, T.R., HARDY, J.G., AND FRAZIER, T.M. (1967). Correlation of neonatal serum total bilirubin concentrations and developmental status at age eight months. *J. Pediat.* **71**, 553.

BONNAR, J. (1974). Progestagen-only contraception and tubal pregnancies. *Lancet* **i**, 170–1.

BORLÉE, I., LECHAT, M.E. BOUKAERT, A., AND MISSON, C. (1978). Le café, facteur de risque pendant la grossess? *Louvain Med.* **97**, 279–84.

BRACKEN, M.B. BRYCE-BUCHANAN, C., SILTEN, R., AND SRISUPHAN, W. (1982). Coffee consumption during pregnancy. *New Engl. J. Med.* **306**, 1548–9.

—— AND VITA, K. (1983). Frequency of non-hormonal contraception around conception and association with congenital malformations in offspring. *Am. J. Epidemiol.* **117**, 281–91.

BRAZY, J.E., GRIMM, J.K. AND LITTLE, V.A. (1982). Neonatal manifestations of severe maternal hypertension occurring before the thirty-sixth week of pregnancy. *J. Pediat.* **100**, 265–71.

BREART, G. RABARISON, Y., PLOUIN, P.F., SUREALL, C., AND RUMEAU-ROUQUETTE, C. (1982). Risk of fetal growth retardation as a result of maternal hypertension. *Dev. Pharmacol. Ther.* **4**, (Suppl. 1), 116–23.

BRENNER, W.E., DINGFELDER, J.R., AND STAUROVSKY, L.G. (1975). The efficacy and safety of intra-muscularly administered 15s. 15 methyl prostaglandin E-2 methyl ester for induction of artificial abortion. *Am. J. Obstet. Gynec.* **123**, 19–31.

BRITISH MEDICAL JOURNAL (1980a). Drug interaction with oral contraceptive steroids. *Br. med. J.* **281**, 93–4.

—— (1980b). Editorial: Drug innovation in bureaucracy. *Br. med. J.* **280**, 1484.

—— (1981). Editorial: Teratogenic risk of antiepileptic drugs. *Br. med. J.* **283**, 515–16.

BROCK, P.G. AND ROACH, M. (1981). Antituberculous drugs in pregnancy. *Lancet* **i**, 43.

BROGAN, W.F. (1975). Cleft lip and palate and pregnancy tests. *Med. J. Aust.* **1**, 44.

BROWN, A.K. (1974). The susceptibility of the fetus and child to chemical pollutants. Special susceptibility of the fetal and neonatal hematopoietic system to chemical pollutants including drugs administered to the mother. *Pediatrics* **53**, 816–17.

BRUCE-CHWATT, L.J. (1983). Malaria and pregnancy. *Br. med. J.* **286**, 1457–8.

——, BLACK, R.H., CANFIELD, C.J., CLYDE, D.F., PETERS, W., AND WERNSDORFER, W.H. (1981). *Chemotherapy of malaria*, 2nd ed. WHO, Geneva, Switzerland.

BULMASH, J.M. (1979). Rheumatoid arthritis and pregnancy. *Obstet. Gynecol. Ann.* **8**, 223–76.

BURR, W.A. (1981). Thyroid disease. In *Clinics in Obstetrics and Gynecology* (ed. S.M. Wood and L. Beeley), Vol. 8, no. 2, pp. 341–51. W.B. Saunders, London.

BURROW, G.N. (1978). Maternal-fetal considerations in hyperthyroidism. *Clins Endocr. Metab.* **7**, 115–25.

BURSTEIN, R. AND WASSERMAN, H.C. (1964). The effect of Provera on the fetus. *Obstet. Gynecol.* **23**, 931–4.

BURTON, J.L. AND SAIHAN, E.M. (1980). Reply on the use of metronidazole for rosacea in pregnancy. *Br. J. Dermatol.* **103**, 586.

BUTLER, N.R. AND GOLDSTEIN, H. (1973). Smoking in pregnancy and subsequent child development. *Br. med. J.* **4**, 573–5.

BYNUM, T.E. (1977). Hepatic and gastrointestinal disorders in pregnancy. *Med. Clin. N. Am.* **61**, 129–38.

CANGER, R., CORNAGGIA, C.M., AND BIANCHI, M. (1983). Epilepsy and pregnancy *Prog. clin. biol. Res.* **124**, 353–60.

CARR, D.H. (1967). Chromosomes after oral contraceptives. *Lancet* **ii**, 830–1.

CARR, D.H. (1970). Chromosome studies in selected spontaneous abortions. I. Conception after oral contraceptives. *Can. Med. Ass. J.* **103**, 343–8.

CARTER, M.P. AND WILSON, F. (1965). Antibiotics in early pregnancy and congenital malformations. *Dev. Med. Child Neurol.* **7**, 353–9.

CASTEELS-VAN DAELE, M., JAEKEN, J., EGGERMOUNT, E., GOETANODE, G., AND VERMIJLEN, J. (1972). More on the effects of antenatally administered aspirin on the aggregation of platelets of neonates. *J. Pediat.* **80**, 685–6.

CATANZARITE, V.A. AND MCHARGUE, A. (1983). Leukemia and pregnancy. *Am. J. Obstet. Gynecol.* **145**, 384.

CHAR, V.C., CHANDRA, R., FLETCHER, A.B., AND AVERY, G.B. (1975). Polyhydramnios and neonate renal failure — a possible association with maternal acetaminophen ingestion. *J. Pediatr.* **86**, 638–9.

CINCINNATI, J.J.A. (1970). Malignant lymphoma associated with hydantoin drugs. *Arch. Neurol.* **22**, 450–4.

CLARKE, M. AND CLAYTON, D.G (1981). Safety of Debendox. *Lancet* **i**, 659–60.

CLARREN, S.K., ALVORD, E.C., SUMI, S.M., STREISSGUTH, A.P., AND SMITH, D.W., (1978). Brain malformations related to prenatal exposure to ethanol. *J. Pediatr.* **92**, 64–7.

CLEMMESEN, J., FUGLSANG-FREDERIKSEN, V., AND PLUM, C.M. (1974). Are anticonvulsants oncogenic? *Lancet* **i**, 705–7.

COHLAN, S.Q., (1980). Drugs and pregnancy. In *Progress in clinical and biological research*, Vol. 44, pp. 77–96. Liss, New York.

COLLINS, E. (1981). Maternal and fetal effects of acetaminophen and salicylates in pregnancy. *Obstet. Gynecol.* **58**, 57S–62S.

—— AND TURNER, G. (1973). Salicylates and pregnancy. *Lancet* **ii**, 1494.

COLVIN, E.D., BARTHOLOMEW, R.A., GRIMES, W.H., AND FISH, J.S. (1950). Salvage possibilities in threatened abortion. *Am. J. Obstet. Gynecol.* **59**, 1208–22.

COMMITTEE ON DRUGS, AMERICAN ACADEMY OF PEDIATRICS (1979). Anticonvulsants and pregnancy. *Pediatrics* **63**, 331–3.

CORBETT, T.H. (1976). Cancer and congenital anomalies associated with anaesthetics. *Ann. NY Acad. Sci.* **271**, 58–66.

——, CORNELL, R.G., LIEDING, K., AND ENDRES, J.L. (1974). Birth defects of children among Michigan nurse anesthetists. *Anesthesiology* **41**, 341–4.

CORDERO, J.F., OAKLEY, G.P., GREENBERG, F., AND JAMES, L.M. (1981). Is Bendectin a teratogen? *J. Am. med. Ass.* **245**, 2307–10.

CORREY, J.F., MARSDEN, D.E., AND SCHOKMAN, F.C.M. (1982). The outcome of pregnancy resulting from clomiphene-induced ovulation. *Aust. NZ J. Obstet. Gynaecol.* **22**, 18–21.

—— AND NEWMAN, N.M. (1981). Debendox and limb reduction deformities. *Med. J. Aust.* **1**, 417–18.

COULAM, C.B. AND ANNEGERS, J.F. (1979). Do anticonvulsants reduce the efficacy of oral contraceptives? *Epilepsia* **20**, 519–26.

CRAINICIANU, A. (1969). Pharmacodynamic action of caffeine: inference in clinical obstetrics. *Rev. fr. Gynecol. Obstet.* **64**, 415–20.

CREE, J.E. MEYER, K., AND HAILEY, D.M. (1978). Diazepam in labour. Its metabolism and effect on the clinical condition and thermogenesis of the newborn. *Br. med. J.* **4**, 251–5.

CROMBIE, D.L., PINSENT, R.J.F.H., SLATER, B.C., FLEMING, D., AND

CROSS, K.W. (1970). Teratogenic drugs. R.C.G.P. Survey, *Br. med. J.* 4, 178–9.

CROWLEY, P. (1983). Premature labour. In *Drugs and pregnancy* (ed. D.F. Hawkins) pp. 155–83. Churchill Livingstone, Edinburgh.

CSABA, I.F., SULYOK, E., AND ERTL, T. (1978). Relationship of maternal treatment with indomethacin to persistence of fetal circulation syndrome. *J. Pediat,* 92, 484.

DALENS, B., RAYNAUD, E.J., AND GAULME, J. (1980). Teratogenicity of valproic acid. *J. Pediatr.* 97, 332–3.

DAM, M. AND DAM. A.Y. (1980). Epilepsy in pregnancy. In *The treatment of epilepsy* (ed. J.H. Tyrer), pp. 323–47. MTP Press Ltd, Lancaster, England.

DARLING, M.R. AND HAWKINS, D.G. (1981). Sex hormones in pregnancy. In *Clinics obstetrics and gynaecology* (ed. S.M. Wood and L. Beeley), Vol. 8, no. 2, pp. 405–19. W.B. Saunders. London.

DAVID. T.J. (1982). Debendox does not cause Poland anomaly. *Arch. Dis. Childh.* 57, 479–80.

DE LOUVOIS, J. (1983). Antibiotics and antimicrobial chemotherapy in pregnancy. In *Clinical pharmacology in obstetrics* (ed. P. Lewis), pp. 57–71. Wright PSG, Bristol, London, Boston.

DE SWIET, M. (1979). Respiratory disease in pregnancy. *Postgrad. med. J.* 55, 325–8.

DEUTSCHE FORSCHUNGSGEMEINSCHAFT. (1977). *Schwangerschaftsverlauf und Kindesentwicklung.* Published by Harold Boldt.

DICKINSON, R.G. AND GERBER, N. (1980). Teratogenicity of valproic acid. *J. Pediatr.* 97, 333.

DIETERICH, E., STEVELING, A., LUKAS, A., SEYFEDDINIPUR, N., AND SPRANGER, J. (1980). Congenital anomalies in children of epileptic mothers and fathers. *Neuropediatrics* 11, 274–83.

DILLON, S. (1970). Progestogen therapy in early pregnancy and associated congenital defects. *Practitioner* 205, 80–4.

—— (1976). Congenital malformations and hormones in pregnancy. *Br. med. J.* 2, 1146.

DIRMIKIS, S.M. AND MUNRO, D.S. (1975). Placental transmission of thyroid-stimulating immunoglobulins. *Br. med. J.* 2, 665–6.

DONNAI, D. AND HARRIS, R. (1978). Unusual fetal malformations after antiemetics in early pregnancy. *Br. med. J.* 1, 691–2.

DOUGLAS RINGROSE, C.A. (1972). The hazard of neutrotropic drugs in the fertile years. *Can. med. Ass. J.* 106, 1058.

DUBOIS, D., PETITCOLAS, J., TEMPERVILLE, B., KLEPPER, A., AND CATHERINE, PH. (1983a). Beta blocker therapy in 125 cases of hypertension during pregnancy. *Clin. exp. Hypertension* B2 (1), 41–59.

——, ——, ——, AND —— (1983b). Treatment with atenolol of hypertension in pregnancy. *Drugs* 25, (Suppl. 2), 215–18.

DYSON, J.L. AND KOHLER, H.G. (1973). Anencephaly and ovulation stimulation. *Lancet* i, 1256–7.

ELSHOVE, J. AND VAN ECK. J.H.M. (1971). Congenital malformations particularly cleft lip with or without cleft palate in children of epileptic mothers. *Ned. Tijdsch. Geneesk.* 115, 1371.

ENDRES, W. (1981). D-penicillamine in pregnancy — to ban or not to ban? *Klin. Wochenschr.* 59, 535–7.

ERICSON, A. AND KÄLLEN, B. (1979). Survey of infants born in 1973 or 1975 to Swedish women working in operating rooms during their pregnancies. *Anesth. Analg.* 58, 302–5.

FACULTY OF COMMUNITY MEDICINE (1981). Report: A recommendation for prevention of alcohol related disorders.

FAHRLANDER, H. (1980). Salazosulfapyridin im der Schwangerschaft. *Dtsch. Med. Wochenschr.* 105, 1729–31.

FALTERMAN, C.G. AND RICHARDSON, C.J. (1980). Small left colon syndrome associated with maternal ingestion of psychotropic drugs. *J. Pediatr.* 97, 308–10.

FDA DRUG BULLETIN (1978). Hexachlorophene — interim caution regarding use in pregnancy. *FDA Drug Bull.* 8, 26–7.

FEDRICK, J. (1973). Epilepsy and pregnancy: A report from the Oxford Record Linkage Study. *Br. med. J.* 2, 442–8.

FELDMAN, G.C. WEAVER, D.D., AND LOVRIEN, E.W. (1977). The fetal trimethadione syndrome. *Am. J. Dis. Childh.* 131, 1389–92.

FERRIS, T.F. (1983). The pathophysiology of toxaemia and hypertension during pregnancy. *Drugs* 25, (Suppl. 2), 198–205.

FIDLER, J. AND ELLIS, C. (1983). Analgesia in pregnancy. In *Clinical Pharmacology in obstetrics* (ed. P. Lewis), pp. 49–56. Wright PSG, London.

FIELD, B. AND KERR, C. (1974). Ovulation stimulation and defects of neural-tube closure. *Lancet* ii, 1511.

FLEMING, D.M., KNOX, J.D.F., AND CROMBIE, D.L. (1981). Debendox in early pregnancy and fetal malformation. *Br. med. J.* 283, 99–101.

FLYNN, A., MARTIER, S.S., SOKOL, R.J., MILLER, S.I., GOLDEN, N.L., AND DEL VILLANO, B.C. (1981). Zinc status of pregnant alcoholic women: A detriment of fetal outcome. *Lancet* i, 572–5.

FOOD CHEMICAL NEWS (1982). Editorial. *Food chem. News* 24, (33), 32–34.

FORD, W.D.A. AND LITTLE, K.E.T. (1981). Fetal ovarian dysplasia possibly associated with clomiphene. *Lancet* ii, 1107.

FORFAR, J.O., AND NELSON, M.M. (1973). Epidemiology of drugs taken by pregnant women: Drugs that may affect the fetus adversely. *Clin. Pharmacol Ther.* 14, 632–42.

FRASER, W.I. AND MAC GILLIVRAY, R.C. (1969). Anticonvulsant drugs and congenital abnormalities. *Lancet* i, 56.

FRIEDMAN, G.D. (1980). Cancer after metronidazole. *New Engl. J. Med.* 302, 519–20.

FRIIS, M.L. (1983). Antiepileptic drugs and teratogenesis. *Acta neurol. scand. Suppl.* 94, 39–43.

GAL, I. (1972). Risks and benefits of the use of hormonal pregnancy test tablets. *Nature* 240, 241–2.

——, KIRMAN, B., AND STERN, J. (1967). Hormonal pregnancy tests and congenital malformations. *Nature* 216, 83.

GALLERY, E.D.M., SAUNDERS, D.M., HUNYOR, S.N., AND GYÖRY, A.Z. (1979). Randomised comparison of methyldopa and oxprenolol for treatment of hypertension in pregnancy. *Br. med. J.* 1, 1591–4.

GANGUIN, G. AND REMPT, E. (1970). Streptomycin in pregnancy and its effect on the hearing of the children. *Z. Laryngol. Rhinol. Otol.* 49, 496.

GEBER, W.F. AND SCHRAMM, L.C. (1975). Congenital malformations of the central nervous system produced by narcotic analgesics in the hamster. *Am. J. Obstet. Gynecol.* 123, 705–13.

GERMAN, J., EHLERS, K.H., KOWAL, A., DE GEORGE, F.V., ENGLE, M.A., AND PASSARGE, E. (1970). Possible teratogenicity of trimethadione and paramethadione. *Lancet* ii, 261.

GERSONY, W.M., PECKHAM, G.J., CURTIS ELLISON, R., MIETTINEN, O.S., AND NADAS, A.S. (1983). Effects of indomethacin in premature infants with patent ductus arteriosus: Results of a national collaborative study. *J. Pediat.* 102, 895–906.

GIBSON, G.T., COLLEY, D.P., MCMICHAEL, A.J., AND HARTSHORNE, J.M. (1981). Congenital anomalies in relation to the use of doxylamine/dicyclomine and other antenatal factors. *Med. J. Aust.* 1, 410–14.

GILILLAND, J. AND WEINSTEIN, L. (1983). The effects of cancer chemotherapeutic agents on the developing fetus. *Obstet. Gynecol. Surv.* **38**, 6–13.

GILLBERG, C. (1977). Floppy Infant Syndrome and maternal diazepam. *Lancet* **ii**, 244.

GLOBUS, M.D. (1980). Teratology for the obstetrician. Current status. *Obstet. Gynaecol.* **55**, 269–77.

GOETSCH, C. (1962). An evaluation of aminopterin as an abortifacient. *Am. J. Obstet. Gynecol.* **83**, 1474–7.

GOLDEN, N.L., KING, K.C., AND SOKOL, R.J. (1982). Propoxyphene and acetominophen. *Pediatrics* **21**, 752–4.

GOLDING, J. AND BALDWIN, J. (1980). Clefts of lip and palate and maternal drug consumption. I. antenauseants. Read before the FDA Fertility and Maternal Health Drugs and Advisory Committee. Washington. DC September 1980.

GOLDMAN, A.S. AND YAFFE, S.J. (1978). Fetal trimethadione syndrome. *Teratology* **17**, 103–6.

GOMEZ, M.R. (1981). Possible teratogenicity of valproic acid. *J. Pediatr.* **98**, 505–9.

GOOD, J.T., ISEMAN, M.D., DAVIDSON, P.T., LAKSHMINARAYAN, S., AND SAHN, S.A. (1981). Tuberculosis in association with pregnancy. *Am. J. Obstet. Gynaecol.* **140**, 492–8.

GOUJARD, J. AND RUMEAU-ROUQUETTE, C. (1977). First trimester exposure to progestogen/oestrogen and congenital malformations. *Lancet* **i**, 482–3.

GREENBERG, G., INMAN. W.H.W., WEATHERALL, J.A.C., ADELSTEIN, A.M. AND HASKEY. J.C. (1977). Maternal drug histories and congenital abnormalities. *Br. med. J.* **2**, 853–6.

GREENBERGER, P. AND PATTERSON, R. (1978). Safety of therapy for allergic symptoms during pregnancy. *Ann. intern. med.* **89**, 234–7.

—— AND —— (1983). Beclomethasone diproprionate for severe asthma during pregnancy. *Ann. Intern. Med.* **98**, 478–80.

GREENBLATT, R.B., BARFIELD, W.E., JUNGEK, E.C., AND RAY, A.W. (1961). Induction of ovulation with MRL/41. *J. Am. med. Ass.* **178**, 101–4.

GREENE, G.R. (1976). Tetracycline in pregnancy. *New Engl. J. Med.* **295**, 512–13.

GRIFFIN, J.P. (1977). Current problems and current requirements in reproduction studies. In *Current approaches in toxicology* (ed. B. Ballantyne), pp. 41–53. Wright, Bristol.

GUILHOU, J.J. AND MEYNADIER, J. (1980). Rosacea, metronidazole and pregnancy. *Br. J. Dermatol.* **103**, 585–6.

HACK, M., BRISH, M., SERR, D.M., INSLER, V., SALOMY, M., AND LUNENFELD, B. (1972). Outcome of pregnancy after induced ovulation. *J. Am. med. Ass.* **220**, 1329–33.

HALL, J.G., PAULI, R.M., AND WILSON, K.M. (1980). Maternal and fetal sequelae of anticoagulation during pregnancy. *Am. J. Med.* **68**, 122–40.

HALLER, J. (1974). Hormone therapy during pregnancy. *Deutsches Arzteblatt* **14**, 1013–15.

HALLING, H. (1979). Suspected link between exposure to hexachlorophene and malformed infants. *Ann. NY Acad. Sci.* **320**, 426–35.

HANSON, J.W. AND SMITH, D.W. (1975). The fetal hydantoin syndrome. *J. Pediat.* **87**, 285–90.

——, MYRIANTHOPOULOS, N.C., HARVEY, M.A., AND SMITH, D.W. (1976). Risks to the offspring of women treated with hydantoin anticonvulsants, with emphasis on the fetal hydantoin syndrome. *J. Pediat.* **89**, 662–8.

——, STREISSGUTH, A.P., AND SMITH, D.W. (1977). The effects of moderate alcohol consumption during pregnancy on fetal growth and morphogenesis. Fifth International Conference on Birth Defects. *Excerpta Medica Int. Cong. Series no.* **426**.

——, ——, AND —— (1978). The effects of moderate alcohol consumption during pregnancy on fetal growth and morphogenesis. *J. Pediat.* **92**, 457–60.

HAPPLE, R., TRAUPE, H., BOUNAMEAUX, Y., AND FISCH, T. (1984). Teratogene Wirkung von Etretinat beim Menschen. *Dtsch. med. Wschr.* **109**, 1476–80.

HARAM, K. (1977). Floppy Infant Syndrome and maternal diazepam. *Lancet* **ii**, 612–13.

HARLAP, S. (1976). Ovulation induction and congenital malformations. *Lancet* **ii**, 961.

——, AND DAVIES, A.M. (1978). The pill and births. The Jerusalem Study. Bethesda, M.D., Center for Population Research, NICHD.

—— AND ELDOR, J. (1980). Births following oral contraceptive failures. *Obstet. Gynecol.* **55**, 447–52.

——, PRYWES, R., AND DAVIES, A.M. (1975). Birth defects and oestrogens and progesterones in pregnancy. *Lancet* **i**, 682–3.

—— AND SHIONO, P.H. (1980). Alcohol, smoking and the incidence of spontaneous abortions in the first and second trimester. *Lancet* **ii**, 173–6.

HARLEY, J.D., FARRAR, J.F., GRAY, J.B., AND DUNLOP, I.C. (1964). Aromatic drugs and congenital cataracts. *Lancet* **i**, 472.

HARPER, R.G., SOLISH, G.I., PUROW, H.M., SANG, E., AND PANEPINTO, W.C. (1974). The effect of a methadone treatment programme upon pregnant heroin addicts and their newborn infants. *Paediatrics* **54**, 300–5.

HARRON, D.W.G., GRIFFITHS, K., AND SHANKS, R.C. (1980). Debendox and congenital malformations in Northern Ireland. *Br. med. J.* **281**, 1379–80.

HARTZ, S.C., HEINONEN, O.P., SHAPIRO, S., SISKIND, V., AND SLONE, D. (1975). Antenatal exposure to meprobamate and chlordiazepoxide in relation to malformations, mental development and childhood mortality. *New Engl. J. Med.* **292**, 726–8.

HAWKINS, D.F. (1974). Progestogen only contraception and tubal pregnancies. *Br. med. J.* **1**, 387.

HEINONEN, O.P., SLONE, D., MONSON, R.R., HOOK, E.B., AND SHAPIRO, S. (1977a). Cardiovascular birth defects and antenatal exposure to female sex hormones. *New Engl. J. Med.* **296**, 67–70.

——, ——, AND SHAPIRO, S. (1977b). *Birth defects and drugs in pregnancy*. Publ. Sciences Group Inc, Littleton, Massachusetts.

HERBST, A.L., ULFEDER, H., AND POSKANZER, D.C. (1971). Adenocarcinoma of the vagina: association of maternal stilboestrol therapy with tumour appearance in young women. *New Engl. J. Med.* **284**, 878–81.

——, ROBBOY, S.J., SCULLY, R.E., AND POSKANZER, D.C. (1974). Clear-cell adenocarcinoma of the vagina and cervix in girls: Analysis of 170 registry cases. *Am. J. Obstet. Gynecol.* **119**, 713–24.

——, POSKANZER, D.C., ROBBOY, S.J., FRIEDLANDER, L., AND SCULLY, R.E. (1975). Prenatal exposure to stilboestrol. A prospective comparison of exposed female offspring with unexposed controls. *New Engl. J. Med.* **292**, 334–9.

HERNANDEZ, E., ANGELL, C.S., AND JOHNSON, J.W.C. (1980). Asthma in pregnancy: Current concepts. *Obstet. Gynecol.* **55**, 739–43.

HIILESMAA, V.K., TERAMO, K., GRANSTRÖM, M.L., AND BARDY, A.H. (1981). Fetal head growth retardation associated with maternal antiepileptic drugs. *Lancet* **ii**, 165–7.

——, ——, ——, AND —— (1983). Serum folate concentrations during pregnancy in women with epilepsy: relation to antiepileptic drug concentrations, number of seizures, and fetal outcome. *Br. med. J.* **287**, 577–9.

HILL, R.M. (1973*a*). Teratogenesis and antiepileptic drugs. *New Engl. J. Med.* **289**, 1089–90.

—— (1973*b*). Drugs ingested by pregnant women. *Clin. Pharmacol. Ther.* **14**, 654–9.

——, HORNING, M., AND HORNING, E.C. (1972). Pattern of drug ingestion in gravid females. *Clin. Res.* **20**, 92.

HILL, R. (1979). Anticonvulsant medication, *Am. J. Dis. Childh.* **133**, 449–50.

HILL, R.B. AND STERN, L. (1979). Drugs in pregnancy: effects on the fetus and newborn. *Drugs* **17**, 182–97.

HINDI, R.D. AND AZIMI, P.H. (1980). Congenital malaria due to plasmodium falciparum. *Pediatrics* **66**, 977–9.

HIRSCH, H.A. (1971). The use of cephalosporin antibiotics in pregnant women *Postgrad. med. J.* **47** (suppl.), 90–3.

HOGUE, C.J. (1981). Coffee in pregnancy. *Lancet* **i**, 554.

HOLMBERG, P.C. (1979). Central nervous system defects in children born to mothers exposed to organic solvents during pregnancy. *Lancet* **ii**, 177–9.

——, HERNBERG, S., KURPPA, K, RANTALA, K., AND RIALA, R. (1982). Oral clefts and organic solvent exposure during pregnancy. *Int. Arch. occup. environ. Hlth* **50**, 371–6.

—— AND NURMINEN, M. (1980). Congenital defects of the central nervous system and occupational factors during pregnancy. A case-referent study. *Am. J. indust. Med.* **i**, 167–76.

HORAN, A.H. (1974). Aspirin, prostaglandin and gestation. *Lancet* **i**, 31.

HUGGINS, G., VESSEY, M., FLAVEL, R., YEATES, D., AND MCPHERSON, K. (1982). Vaginal spermicides and outcome of pregnancy: Findings of a large cohort study. *Contraception* **25**, 219–30.

HUNTER, A.G.W., THOMPSON, D., AND EVANS, J.A. (1979). Is there a fetal gasoline syndrome? *Teratology* **20**, 75–80.

HURD, R.W., WILDER, B.J., AND VAN RINSVELT, H.A. (1983). Valproate, birth defects and zinc. *Lancet* **i**, 181.

HUTCHINGS, D.E. (1982). Methadone and heroin during pregnancy: a review of behavioural effects in human and animal offspring. *Neurobehav. Toxicol. Teratol.* **4**, 429–34.

IANCU, H., BOYANOWER, Y., AND LAURIAN, N. (1974). Congenital goitre due to maternal ingestion of iodide. *Am. J. Dis. Child.* **128**, 528–30.

IDÄNPÄÄN-HEIKKILA, J. AND SAXEN, L. (1973). Possible teratogenicity of imipramine/chloropyramine. *Lancet* **ii**, 282–4.

ILLINGWORTH, R.S. (1976). Factors that influence fetal development. *Mimms Magazine* (Jan.), 53–6.

JACKSON, W.P.U. (1974). Chlorpropamide in diabetic pregnancy. *Lancet* **ii**, 843.

JAMES, W.H. (1973). Anencephaly, ovulation stimulation, subfertility and illegitimacy. *Lancet* **ii**, 916.

JAMESON, S. (1976). Effects of zinc deficiency in human reproduction. *Acta med. scand.* (Suppl.) **593**, 1–89.

JANERICH, D.T. (1979*a*). Environmental causes of birth defects: the hexachloraphane issue. *J. Am. med. Ass.* **241**, 830–1.

—— (1979*b*). Supportive hormone therapy and birth defects. *Teratology* **20**, 483–4.

——, DUGAN, J.M., STANDFAST, S.J., AND STRITE, L (1977). Congenital heart disease and prenatal exposure to exogenous sex hormones. *Br. med. J.* **1**, 1058–60.

——, PIPER, J.M., AND GLEBATIS, D.M. (1973). Hormones and limb-reduction deformities. *Lancet* **ii**, 96.

——, ——, AND —— (1974). Oral contraceptives and congenital limb-reduction defects. *New Engl. J. Med.* **291**, 697–700.

——, ——, AND —— (1980). Oral contraceptive use and birth defects. *Am. J. Epidemiol.* **112**, 73–9.

JANZ, D. (1975). The teratogenic risk of anti-epileptic drugs. *Epilepsia* **16**, 159–69.

—— (1982*a*). On major malformations and minor anomalies in the offspring of parents with epilepsy: review of the literature. In *Epilepsy, pregnancy and the child* (ed. D. Janz, A. Richens, L. Bossi, H. Helge, and D. Schmidt), pp. 211–22. Raven Press, New York.

—— (1982*b*). Antiepileptic drugs and pregnancy: altered utilization patterns and teratogenesis. *Epilepsia* **23** (suppl. 1), 553–63.

—— AND FUCHS, U. (1964). Sind Antiepileptische Medikamenta Während der schwangerschaft Wädlich? *Deutsche Med. Wschr.* **89**, 241–3.

JENTGENS, H. (1975). Antituberkulotische Therapie mit Ethambutol und rifampicin (RIF) in der Schwangerschaft. Prax. Pnemol. **30**, 42–5.

JICK, H., HOLMES, L.B. HUNTER, J.R., MADSEN, S., AND STERGACHIS, A. (1981*a*). First trimester drug use and congenital disorders. *J. Am. med. Ass.* **246**, 343–6.

——, SHIOTA, K., SHEPARD, T.H., HUNTER, J.R., STERGACHIS, A., MADSEN, S., AND PORTER, J.B. (1982). Vaginal spermicides and miscarriage seen primarily in the emergency room. *Teratogenesis, Carcinogenesis, Mutagenesis* **2**, 205–10.

——, WALKER, A.M., ROTHMAN, K.J., HUNTER, J.R., HOLMES, L.B., WATKINS, R.N., D'EWART, D.C., DANFORD, A., AND MADSEN, S. (1981*b*). Vaginal spermicides and congenital disorders. *J. Am. med. Ass.* **245**, 1329–32.

JOHNSON, H.L. AND ROSEN, T.S. (1982). Prenatal methadone exposure: effects on behaviour in infancy. *Pediat. Pharmacol.* **2**, 113–20.

JONES, K.L. AND SMITH, D.W. (1975). The fetal alcohol syndrome. *Teratology* **12**, 1–10.

JOURNAL OF THE AMERICAN MEDICAL ASSOCIATION (1979). C.D.C. Study: no evidence for teratogenicity of Bendectin. *J. Am. med. Ass.* **242**, 2518.

KÄLLEN, B. (1978). Hexachlorophane teratogenicity in humans disputed *J. Am. med. Ass.* **240**, 1585–6.

KAMINSKI, M., RUMEAU-ROUQUETTE, C., AND SCHWARTZ, D. (1978). Alcohol consumption in pregnant women and the outcome of pregnancy. *Alcoholism* **2**, 155–63.

KAMM, J.J. (1982). Toxicology, carcinogenicity and teratogenicity of some orally administered retinoids. *J. Am. Acad. Dermatol.* **6**, 652–9.

KANEKO, S., OTANI, K., FUKUSHIMA, Y., SATO, T., NOMURA, Y., AND OGAWA, Y. (1983). Transplacental passage and half life of sodium valproate in infants born to epileptic mothers. *Br. J. clin. Pharmacol.* **15**, 503–5.

KARKINEN-JAASKELAINEN, M. AND SAXEN, L. (1974). Maternal influenza, drug consumption and congenital defects of the central nervous system. *Am. J. Obstet. Gynecol.* **118**, 815–18.

KASAN, P.N. AND ANDREWS, J. (1980). Oral contraception and congenital abnormalities. *Br. J. Obstet. Gynaecol.* **87**, 545–51.

KAUFMAN, R.I. (1973). Birth defects and oral contraceptives. *Lancet* **i**, 1396.

KING, M.D., DAY, R.E., OLIVER, J.S., LUSH, M., AND WATSON, J.M. (1981). Solvent encephalopathy, *Br. med. J.* **283**, 663–5.

KINGSLEY, P.J. AND COX, P.S.G. (1978). Cromolyn sodium (sodium cromoglycate) and drugs with similar activities. *Allergy Principles Practice* **1**, 481–98.

KIRKINEN, P., JOUPPILA, P., KOIVULA, A., VUORI, J., AND PUUKKA, M. (1983). The effect of caffeine on placental and fetal blood flow in human pregnancy. *Am. J. Obstet. Gynecol.* **147**, 939–42.

KLEINMAN, R.L. (1980). *I.P.P.F. family planning handbook for doctors*. London.

KLINE, J., SHROUT, P., STEIN, Z., SUSSER, M., AND WARBURTON, D. (1980). Drinking during pregnancy and spontaneous abortion. *Lancet* ii, 176–80.

——, STEIN, Z.A., SUSSER, M., AND WARBURTON, D. (1978). Spontaneous abortion and the use of sugar substitutes (saccharin). *Am. J. Obstet. Gynec.* **130**, 708–11.

KLINGER, H.P. AND GLASSER, M. (1981). Contraceptives and the conceptus. II. Sex of the fetus and neonate after oral contraceptive use. *Contraception* **23**, 367–74.

——, ——, AND KAVA, H.W. (1970). Contraceptives and the conceptus. I. Chromosome abnormalities of the fetus and neonate related to maternal contraceptive history. *Obstet. Gynecol.* **48**, 40–8.

KNILL-JONES, R.P., RODRIGUES, L.V., MOIR, D., AND SPENCER, A.A. (1972). Anaesthetic practice and pregnancy: controlled survey of women anaesthetists in the United Kingdom. *Lancet* i, 1326–8.

KNUTTI, R., ROTHWELL, H., AND SCHALATTE, C. (1981). Effect of pregnancy on the pharmacokinetics of caffeine. *Eur. J. clin. Pharmacol.* **21**, 121–6.

KOLATA, G.B. (1980). How safe is Bendectin? *Science* **210**, 518–19.

—— (1982). F.D.A. to re-examine Bendectin data. *Science* **217**, 335.

KOPELMAN, A.E. (1975). Fetal addiction to pentazocine. *Pediatrics* **55**, 888–9.

KORT, H.I. AND CASSEL, G.A. (1981). An appraisal of warfarin therapy during pregnancy. *S. Afr. med. J.* **60**, 578–9.

KUCERS, A. AND BENNETT, N. MCK. (1979). *The use of antibiotics — a comprehensive review with clinical emphasis*, 3rd ed. Heinemann, London.

KUENSSBERG, E.V. AND KNOX, J.D.E. (1973). Teratogenic·effect of anticonvulsants. *Lancet* i, 198.

KUHNERT, B.R., KNAPP, D.R., KUHNERT, P.M., AND PROCHASKA, A.L. (1979). Maternal, fetal and neonatal metabolism of lidocaine. *Clin. Pharmacol. Ther.* **26**, 213–20.

KULLANDER, S. AND KÄLLEN, B. (1976a). A prospective study of drugs and pregnancy. I. Psychopharmaca. *Acta obstet. gynecol. scand.* **55**, 25–33.

—— AND —— (1976b). A prospective study of drugs and pregnancy. II. Antiemetic drugs. *Acta obstet. gynecol. scand.* **55**, 105–11.

—— AND —— (1976c). A prospective study of drugs and pregnancy. *Acta-obstet. gynecol. scand.* **55**, 221–4.

KURACHI, K., AONO, T., MINAGAWA, J., AND MIYAKE, A. (1983). Congenital malformations of newborn infants after clomiphene induced ovulation. *Fertil. Steril.* **40**, 187–9.

KURPPA, K., HOLMBERG, P.C., KUOSMA, E., AND SAXEN, L. (1982). Coffee consumption during pregnancy. *New Engl. J. Med.* **306**, 1548.

——, ——, ——, AND —— (1983). Coffee consumption during pregnancy and selected congenital malformations. A nationwide case-control study. *Am. J. publ. Hlth* **73**, 1397–9.

KWAN, V.W. (1970). Pentazocine in pregnancy. *J. Am. med. Ass.* **211**, 1544.

LAING, I.A., STEER, C.R., DUDGEON, J., AND BROWN, J.K. (1981). Clomiphene and congenital retinopathy. *Lancet* ii, 1107–8.

LANCET (1980a). Vitamins, neural-tube defects and ethics committees. *Lancet* i, 1061–2.

—— (1980b). Editorial: Antituberculous drugs in pregnancy. *Lancet* ii, 1285–6.

—— (1982). Editorial: Prolactinomas: bromocriptine rules O.K.? *Lancet* i, 430–1.

LANDERS, D.V., GREEN, J.R. & SWEET, R.L. (1983). Antibiotic use during pregnancy and the post-partum period. *Clin. Obstet. Gynecol.* **26**, 391–406.

LAURENCE, K.M. AND CAMPBELL, H. (1981). Trial of folate treatment to prevent the recurrence of neural tube defect. *Br. med. J.* **282**, 2131.

——, JAMES, N., MILLER, M.H., TENNANT, G.B., AND CAMPBELL, H. (1981). Double-blind randomised controlled trial of folate treatment before conception to prevent recurrence of neural tube defects. *Br. med. J.* **282**, 1509–11.

LAURENCE, M., MILLER, M., VOWLES, M., EVANS, K., AND CARTER, C. (1971). Hormonal pregnancy tests and neural tube malformations. *Nature* **233**, 495–6.

LAURITSEN, J.G. (1975). The significance of oral contraceptives in causing chromosome anomalies in spontaneous abortions. *Acta obstet. gynecol. scand.* **54**, 261–4.

LEATHER, H.M., HUMPHREYS, D.M., BAKER, P., AND CHADD, M.A. (1968). A controlled trial of hypotensive agents in hypertension in pregnancy. *Lancet* ii, 488–90.

LECK, I. (1983). Spina bifida and anencephaly: fewer patients, more problems. *Br. med. J.* **286**, 1679–80.

LEDWARD, R.S. (1983). Antimicrobial drugs in pregnancy. In *Drugs and pregnancy* (ed. D.F. Hawkins), pp. 102–15. Churchill Livingstone London.

LEE, F., NELSON, N., FAIMAN, C., CHOI, N-W., AND REYES, F.I. (1982). Low dose corticoid therapy for anovulation: Effect on fetal weight. *Obstet. Gynecol.* **60**, 314–17.

LEMOINE, P., HAROUSSEAU, H., BORTEYRU, J-P., AND MENUET, J.C., (1968). Les enfants des parents alcooliques anomalies observees. A propos de 127 cas. *Quest Medicale* **21**, 476–82.

LENZ, W. (1966). Malformations caused by drugs in pregnancy. *Am. J. Dis. Childh.* **112**, 99–106.

LEVIN, D.L., FIXLER, D.E., MORRISS, F.C., AND TYSON, J. (1978). Morphological analysis of the pulmonary vascular bed in infants exposed to in utero to prostaglandin synthetase inhibitors. *J. Pediat.* **92**, 478–83.

LEVY, E.P. COHEN, A., AND FRASER, F.C. (1973). Hormone treatment during pregnancy and congenital heart defects. *Lancet* i, 611.

LEVY, G., GARRETTSON, L.K. AND SODA, D.M. (1975). Evidence of placental transfer of acetominophen. *Pediatrics* **55**, 895.

LEWIS, R.B. AND SCHULMAN, J.D. (1973). Influence of acetylsalicylic acid, an inhibitor of prostaglandin synthesis, on the duration of human gestation and labour. *Lancet* ii, 1159–61.

LEWIT, J., NEBEL, L., TERRACINA, S., AND KARMAN, S. (1974). Ethambutol in pregnancy: observations on embryogenesis. *Chest* **66**, 25–6.

LIEBERMAN, B.A., ROSENBLATT, D.B., BELSEY, E., PACKER, M., REDSHAW, M., MILLS, M., CALDWELL, J., NOTARIANNI, L., SMITH, R.L., WILLIAMS, M., AND BEARD, R.W. (1979). The effects of maternally administered pethidine or epidural bupivacaine on the fetus and newborn. *Br. J. Obstet. Gynec.* **86**, 598–606.

——, STIRRAT, G.M., COHEN, S.L., BEARD, R.W., PINKER, G.D., AND BELSEY, E. (1978). The possible adverse effect of propranolol on the fetus in pregnancies complicated by severe hypertension. *Br. J. Obstet. Gynaec.* **85**, 678–83.

LIEDHOLM, H. (1983). Atenolol in the treatment of hypertension of pregnancy. *Drugs* **25** (Suppl. 2), 206–11.

LIFSCHITZ, M.H., WILSON, G.S., O'BRIEN SMITH, E., AND DESMOND, M.M. (1983). Fetal and postnatal growth of children born to narcotic dependent women. *J. Pediat.* **102**, 686–91.

LIMBECK, G.A., RUVALCABA, R.A., AND KELLEY, V.C. (1969). Simulated congenital adrenal hyperplasia in a male neonate associated with medroxyprogesterone therapy during pregnancy. *Am. J. Obstet. Gynecol.* **103**, 1169–70.

LIN, C-C., LINDHEIMER, M.D., RIVER, P., AND MOAWAD, A.H. (1982). Fetal outcome in hypertensive disorders of pregnancy. *Am. J. Obstet. Gynecol.* **142**, 255–260.

LINN, S., SCHOENBAUM, S.C., MONSON, R.R., ROSNER, B., STUBBLEFIELD, P.G., AND RYAN, K.J. (1982). No association between coffee consumption and adverse outcome of pregnancy. *New Engl. J. Med.* **306**, 141–5.

LIPSITZ, P.J. AND BLATMAN, S. (1973). The early neonatal period of 100 live-borns of mothers on methadone. *Paediat. Res.* **7**, 404.

LITTLE, R.E. (1977). Moderate alcohol use during pregnancy and decreased infant birth weight. *Am. J. publ. Hlth* **67**, 1154–6.

——, STREISSGUTH, A.P., BARR, H.M. AND HEMAN, C.S. (1980). Decreased birth weight in infants of alcoholic women who abstained during pregnancy, *J. Pediatr.* **96**, 974–7.

LITTLEFIELD, L.G., LEVER, W.E., MILLER, F.L., AND GOH, K-O (1975). Chromosome breakage studies in lymphocytes from normal women and women taking oral contraceptives. *Am. J. Obstet. Gynecol.* **121**, 976.

LIVINGSTONE, S., BERMAN, W., AND PAULI, L.L. (1973). Maternal epilepsy and abnormalities of the fetus and newborn. *Lancet* **ii**, 1265.

LOCKHART, J.D. (1972). Hexachlorophene and the food and drug administration. *J. clin. Pharmacol.* **9**, 445–50. ·

LOGIE, D.E. AND McGREGOR, I.A. (1970). Acute malaria in newborn infants. *Br. med. J.* **3**, 404–5.

LONG, S.Y. (1972). Does LSD induce chromosomal damage and malformations? *Teratology* **6**, 75–90.

LOUGHNAN, P.M., GOLD, H., AND VANCE, J.C. (1973). Phenytoin teratogenicity in man *Lancet* **i**, 70–2.

LOUW, J.H. (1963). Congenital goitre. A review with a report of three cases of suffocative goitre in the newborn. *S. Afr. med. J.* **37**, 976–83.

LOWE, C.R. (1973). Congenital malformations among infants born to epileptic women. *Lancet* **i**, 9–10.

LUKE, B. (1982). Coffee consumption during pregnancy. *New Engl. J. Med.* **306**, 1549.

LYLE, W.H. (1978). Penicillamine in pregnancy. *Lancet* **i**, 606–7.

MacGREGOR, A.H., JOHNSON, J.E., AND BUNDE, C.A. (1968). Further clinical experience with clomiphene citrate. *Fertil. Steril.* **19**, 616–22.

MacMAHON, B. (1981). More on Bendectin. *J. Am. med. Ass.* **246**, 371–2.

MAJEWSKI, F., BIERICH, J.R., LOSER, H., MICHAELIS, R., LEIBER, S., AND BETTECKEN, F. (1976). Zur klinik und pathogenese der alkohol embryopathie. Bericht uber 68 Fälle. *Muench med. Wochenschr.* **118**, 1635–42.

MALLING, H.V. AND WASSOM, J.S. (1977). Action of mutagenic agents. In *Handbook of teratology*, Vol. 1 (ed. J.G. Wilson and F. Clarke Fraser), pp. 99–152. Plenum Press, New York.

MAMTANI, R., AND WATKINS, S.J. (1981). Trial of folate treatment to prevent recurrence of neural tube defect. *Br. med. J.* **282**, 2056–7.

MANDELLI, M., MORSELLI, P.L. NORDIO, S., PARDI, G., PRINCIPI, N., SERENI, F., AND TOGNONI, G. (1975). Placental transfer of diazepam and its disposition in the newborn. *Clin. Pharmacol. Ther.* **17**, 564–72.

MARTIN-BOUYER, G., LEBRETON, R., TOGA, M., STOLLEY, P.D., AND LOCKHART, J. (1982). Outbreak of accidental poisoning in France. *Lancet* **i**, 91–5.

MATSUNAGA, E., AND SHIOTA, K. (1979). Threatened abortion, hormone therapy and malformed embryos. *Teratology* **20**, 469–80.

MAU, G. AND NETTER, P. (1974). Are coffee and alcohol consumption risk factors in pregnancy? *Geburtsh, U. Frauenheilk.* **34**, 1018–22.

McBRIDE, W.G. (1969). An aetiological study of drug ingestion by women who gave birth to babies with cleft palate. *Aust. NZ Obstet. Gynec.* **9**, 103–4.

—— (1972). The teratogenic effects of imipramine. *Teratology* **5**, 262.

McCARROLL, A.M., HUTCHINSON, M., MCAULEY, R., AND MONTGOMERY, D.A. (1976). Long-term assessment of children exposed *in utero* to carbimazole, *Arch. Dis. Childh.* **51**, 532–6.

McCLAIN, R.M. AND HOAR, R.M. (1980). Reproduction studies with Carprofen, a non-steroidal anti-inflammatory agent in rats. *Tox. appl. Pharmacol.* **56**, 376–82.

McCOLL, J.D. (1963). Dimenhydrinate in pregnancy. *Can. med. Ass. J.* **88**, 861.

McCORMACK, W.M., GEORGE, H., DONNER, A., KODGIS, L.F., ALPERT, S., LOWE, E.W., AND KASS, E.H. (1977). Hepatotoxicity of erythromycin estolate during pregnancy. *Antimicrob. Agents Chemother.* **12**, 630–2.

McELHATTON, P.R. (1977). Effects on fetal development of anti-epileptic drugs administered during pregnancy in the mouse. PhD Thesis, University of London 1977, pp. 26–33.

McNIEL, J.R. (1973). The possible teratogenic effect of salicylates on the developing fetus. Brief summaries of eight suggestive cases. *Clin. Pediat.* **12**, 347–50.

MEADOW, S.R. (1968). Letter. *Lancet* **ii**, 1296.

—— (1970). Congenital abnormalities and anticonvulsant drugs. *Proc. Roy. Soc. Med.* **63**, 48–9.

MEDICAL JOURNAL OF AUSTRALIA (1971). Editorial: The analgesic powder. *Med. J. Aust.* **2**, 689–90.

MELLOR, S. (1978). Fetal malformation after Debendox treatment in early pregnancy. *Br. med. J.* **1**, 1055.

MILHAM, S. AND ELLEDGE, W. (1972). Maternal methimazole and congenital defects in children. *Teratology* **5**, 125.

MILKOVICH. L. AND VAN DEN BERG, B.J. (1974). Effects of prenatal meprobamate and chlordiazepoxide hydrochloride on human embryonic and fetal development. *New Engl. J. Med.* **291**, 1268–71.

MILLAR, J.H.D. AND NEVIN, N.C. (1973). Congenital malformations and anticonvulsant drugs. *Lancet* **1**, 328.

MILLER, S.A. AND HARRIS, J.E. (1981). *Drugs in our food supply: Caffeine and other substances in beverages.* FDA publication. Washington, DC.

MILLS, J., HARLEY, E., REED, G., AND BERENDES, H. (1982). Are spermicides teratogenic? *Am. J. Epidemiol.* **116**, 584.

MILUNSKY, A., GRAEF, J.W., AND GAYNOR, M.P. (1968). Methotrexate-induced congenital malformations. *J. Pediat.* **72**, 790–5.

MITCHELL, A.A., ROSENBERG, L., SHAPIRO, S., AND SLONE, D. (1981). Birth defects related to Bendectin use in pregnancy. *J. Am. med. Ass.* **245**, 2311–14.

MJOLNEROD, O.K., RASMUSSEN, K., DOMMERUD, S.A., AND GJERULDSEN, S.T. (1971). Congenital connective tissue defect probably due to D-penicillamine treatment in pregnancy. *Lancet* **i**, 673–5.

MOGADAM, M., DOBBINS, W.D., AND KORELITZ, B.I. (1980). The safety of corticosteroids and sulphsalazine in pregnancy associated with inflammatory bowel disease. *Gastroenterology* **78**, 1224.

MOGHISSI, K.S. (1981). Risks and benefits of nutritional supplements during pregnancy. *Obstet. Gynecol.* **58** (Suppl.), 685–785.

MOGILNER, B.M., ASHKENAZY, M., BORENSTEIN, R., AND LANCET, M. (1982). Hydrops fetalis caused by maternal indomethacin treatment. *Acta obstet. gynecol. scand.* **61**, 183–5.

MONNET, P., KALB, J.C., AND PUJOL, M. (1967). De l'influence nocive de l'isoniazide sur le produit de conception. *Lyon Med.* **218**, 431–55.

MONSON, R.R., ROSENBERG, L., HARTZ, S.C., SHAPIRO, S., HEINONEN, O.P., AND SLONE, D. (1973). Diphenylhydantoin and selected congenital malformations. *New Engl. J. Med.* **289**, 1049–52.

MONTOURIS, G.D., FENICHEL, G.M., AND MCLAIN, L.W. (1979). The pregnant epileptic. A review and recommendations. *Archs Neurol.* **36**, 601–3.

MORELOCK, S., HINGSON, R., KAYNE, H., DOOLING, E., ZUCKERMAN, B., DAY, N., ALPERT, J.J., AND FLOWERDEW, G. (1982). Bendectin and fetal development. *J. Obstet. Gynecol.* **142**, 209–13.

MORGAN, I. (1978). Metronidazole treatment in pregnancy. *Int. J. Gynaecol. Obstet.* **15**, 501–2.

—— (1979). Metronidazole treatment in pregnancy. In *Metronidazole, Proceedings*, Geneva, Switzerland, April 25–27 (ed. I. Phillips and J. Collier), pp. 237–41. Academic Press, New York.

MORRIS, N. (1953). Hexamethonium compounds in the treatment of pre-eclampsia and essential hypertension during pregnancy. *Lancet* **i**, 322–4.

MORROW-BROWN, H. STOREY, G., AND JACKSON, F.A. (1977). Beclomethasone dipropionate aerosol in long term treatment of perennial and seasonal asthma in children and adults — a report of five and one-half years experience in 600 asthmatic patients. *Br. J. clin. Pharmacol.* **4**, 259S–267S.

MOSLER, K.H. (1975). The treatment of threatened premature labour by tocolytics Ca^{++} antagonists and anti-inflammatory drugs. *Arzneim Forsch.* **25**, 263–6.

MULVIHILL, J.J. (1973). Caffeine as a teratogen and mutagen. *Teratology* **8**, 69–72.

——, MULVIHILL, C.G., AND NEILL, C.A. (1974). Congenital heart defects and prenatal sex hormones. *Lancet* **i**, 1168.

MURDIA, A., MATHUR, V., KOTHARI, L.K., AND SINGH, K.P. (1978). Sulphatrimethoprim combinations and male fertility. *Lancet* **ii**, 375–6.

MYHRE, S.A. AND WILLIAMS, R. (1981). Teratogenic effects associated with maternal primidone therapy. *J. Pediat.* **99**, 160–62.

NAKANE, Y., OKUMA, T., TAKAHASHI, R., SATO, Y., WADA, T., SATO, T., FUKUSHIMA, Y., KAMASHIRO, H., ONO, T., TAKAHASHI, T., AOKI, Y., KAZAMATSURI, H., INAMI, M., KOMAI, S., SEINO, M., MIYAKOSHI, M., TANIMURA, T., HAZAMA, H., KAWAHARA, R., OTSUKI, S., HOSOKAWA, K., INANAGA, K., NAKAZAWA, J., AND YAMAMOTO, Y. (1980). Multi-institutional study on the teratogenicity and

fetal toxicity of anti epileptic drugs. A report of a collaborative study group in Japan. *Epilepsia* **21**, 663–80.

NAU, H. AND LIDDIARD, C. (1978). Placental transfer of drugs during early human pregnancy. In *Role of pharmacokinetics in prenatal and perinatal toxicology*, Part IV. (ed. D. Neubert, H.J. Marker, H. Nail, and J. Langman), pp. 465–82. Georg Thieme, Stuttgart.

NAU, H., RATING, D., HAUSER, I., JAGER, E., KOCH, S., AND HELGE, H. (1980). Placental transfer and pharmacockinetics of primidone and its metabolites phenobarbital, PEMA and hydroxyphenobarbital in neonates and infants of epileptic mothers. *Eur. J. clin. Pharmacol.* **18**, 31–42.

NELSON, M.M. AND FORFAR, J.O. (1971). Association between drugs administered during pregnancy and congenital abnormalities of the fetus. *Br. med. J.* **1**, 523–7.

NEWMAN, N.M. AND CORREY, J.F. (1983). Possible teratogenicity of sulphasalazine. *Med. J. Aust.* **1**, 528–9.

NILLIUS, S.J. (1980). Medical therapy of prolactin-secreting pituitary tumours. In *Progress in reproductive biology*, Vol. 6 (ed. O.P. Hubinont), pp. 194–221. S. Karger, Basle.

NISWANDER, J.D. AND WERTELECKI, W. (1973). Congenital malformation among offspring of epileptic women. *Lancet* **i**, 1062.

NORA, A.H. AND NORA, J.J. (1978). Maternal exposure to exogenous progestogens/estrogens as a potential cause of birth defects. *Adv. Planned Parenthood* **12**, 156–69.

——, ——, BLU, J., INGRAM, J., FOUNTAIN, A., PETERSON, M., LORTSCHER, R.E., AND KIMBERLING, W.J. (1978). Exogenous progestogen and oestrogen implicated in birth defect. *J. Am. med. Ass.* **240**, 837–43.

——, ——, FERINCHIEF, A.G., INGRAM, J.W., FOUNTAIN, A.K., AND PETERSON, M.J. (1976). Congenital abnormalities and first trimester exposure to progestogen/oestrogen. *Lancet* **i**, 313–14.

NORA, J.J. AND NORA, A.H. (1973*a*). Birth defects and oral contraceptives. *Lancet* **i**, 941–2.

—— AND —— (1973*b*). Preliminary evidence for a possible association between oral contraceptives and birth defects. *Teratology* **7**, A24.

—— AND —— (1974). Can the pill cause birth defects? *New Engl. J. Med.* **291**, 731–2.

—— AND —— (1975). A syndrome of multiple congenital anomalies associated with teratogenic exposure. *Archs environ. Hlth* **30**, 17–21.

—— AND —— (1976). Prospective study of infants born to mothers receiving exogenous hormones in pregnancy. *Teratology* **13**, 32A.

NUKI, G. (1983). Non-steroidal analgesic and anti-inflammatory agents *Br. med. J.* **287**, 39–43.

OAKLEY, G.P. (1979). Threatened abortion, hormone therapy and malformed embryos., *Teratology* **20**, 481–2.

—— (1982). Spermicides and birth defects. *J. Am. med. Ass.* **247**, 2405.

——, FLYNT, J. W., AND FALEK, A. (1973). Hormonal pregnancy tests and congenital malformation. *Lancet* **ii**, 256.

OLEGARD, R., SABEL, K-G., ARONSSON, M., SANDIN, B., JOHANSSON, P.R., CARLESSON, C., KYLLERMAN, M., IVERSEN, K., AND HRBEK, A., (1979). Effects on the child of alcohol abuse during pregnancy. *Acta paediatr. scand.* Suppl. **275**, 112–21.

ORTIZ-PEREZ, H.E., FUERTES DE LA HABA, A., BANGDIWALA, I.S., AND ROURE, C.A. (1979). Abnormalities among offspring of oral and non-oral contraceptive users. *Am. J. Obstet. Gynecol.* **134**, 512–17.

O'SHEA, K.S. AND KAUFMAN, M.H. (1979). The teratogenic effect of acetaldehyde: implications for the study of the fetal alcohol syndrome. *J. Anat.* **128**, 65–76.

OUELLETTE, E.M., ROSETT, H.L., ROSMAN, N.P., AND WEINER, L. (1977). Adverse effects on the offspring of maternal alcohol abuse during pregnancy. *New Engl. J. Med.* **297**, 528–30.

OUNSTEAD, M., COCKBURN, J., MOAR, V.A., AND REDMAN, C.W.G. (1983). Maternal hypertension with superimposed pre-eclampsia: effects on child development at $7\frac{1}{2}$ years. *Br. J. Obstet. Gynec.* **90**, 644–9.

OUNSTED, M.K., MOAR, V.A., GOOD, F.J., AND REDMAN, C.W.G. (1980). Hypertension during pregnancy with or without specific treatment — the children at the age of 4 years. *Br. J. Obstet. Gynaecol.* **87**, 19–24.

PALMISANO, P.A. AND CASSIDAY, G. (1969). Salicylate exposure in the perinate. *J. Am. med. Ass.* **209**, 556–8.

PATTERSON, D.C. (1977). Congenital deformities association with Bendectin. *Can. med. Ass. J.* **116**, 1348.

PAULI, R.M. AND HALL, J.G. (1979). Warfarin embryopathy. *Lancet* ii, 144.

PEDERSEN, H., MORISHIMA, H.O., AND FINSTER, M. (1978). Uptake and effects of local anaesthetics in mother and fetus. *Int. Anesthesiol. Clins* **16**, 73–89.

PEIFFER, J., MAJEWSKI, F., FISCHBACH, H., BIERICH, J.R., AND VOLE, B. (1979). Alcohol embryopathy — neuropathology of three children and three fetuses. *J. neurol. Sci.* **41**, 125–39.

PERRY, J.E., TONEY, J.D., AND LE BLANC, A.L. (1967). Effect of nitrofurantoin on the human fetus. *Tex. Rep. Biol, Med.* **25**, 270.

PETTIFOR, J.M., AND BENSON, R. (1975). Congenital malformations associated with the administration of oral anticoagulants during pregnancy. *J. Pediat.* **86**, 459–62.

PHAROAH, P.O.H., ALBERMAN, E., DOYLE, P., AND CHAMBERLAIN, G. (1977). Outcome of pregnancy among women in anaesthetic practice. *Lancet* i, 34–6.

PHILBERT, A. AND PEDERSEN, B. (1983). Treatment of epilepsy in women of child bearing age: Patients opinion of teratogenic potential of valproate. *Acta neurol. scand.* Suppl. **94**, 35–8.

PHILIPSON, A., SABATH, L.D. AND CHARLES, D. (1973). Transplacental passage of erythromycin and clindomycin. *New Engl. J. Med.* **288**, 1219–21.

PHILPOT, C.R. AND LO, D. (1972). Cryptococcal meningitis in pregnancy. *Med. J. Aust.* **2**, 1005–8.

PLACE, V.A. (1964). Ethambutol administration during pregnancy: A case report. *J. New Drugs* **4**, 206–8.

POCHEDLY, C. AND ENTE, G. (1972). Adverse hematologic effects of drugs. *Pediat. Clins. N. Am.* **19**, 1095–111.

POLEDNAK, A.P., JANERICH, D.T., AND GLEBATIS, D.M. (1982). Birth weight and birth defects in relation to maternal spermicide use. *Teratology* **26**, 27–38.

POMERANCE, J.J. AND YAFFE, S.J. (1973). Maternal medication and its effect on the fetus. *Curr. Prob. Paediat.* **4**, 1–61.

POTWOROWSKA, M., SIANOZECKA, E., AND SZUFLADOWICZ, M. (1966). Treatment with ethionamide in pregnancy. *Gruzlica* **34**, 341–7.

PRATT, W.R. (1981). Allergic diseases in pregnancy and breast feeding. *Ann. Allergy* **47**, 355–60.

PRUYN, S.C., PHELAN, J.P., AND BUCHANAN, G.C. (1979). Long term propranolol therapy in pregnancy: maternal and fetal outcome. *Am. J. Obstet. Gynaecol.* **135**, 485–9.

RAMSAY, I. (1983). Drug treatment of thyroid and adrenal disease during pregnancy. In *Clinical pharmacology in obstetrics* (ed.

P. Lewis), pp. 232–50. Wright PSG, Bristol.

RAO, J.M. AND ARULAPPU, R. (1981). Drug use in pregnancy: how to avoid problems. *Drugs* **22**, 409–444.

RICHARDS, I.D.G. (1969). Congenital malformations and environmental influences in pregnancy. *Br. J. prev. soc. Med.* **23**, 218–25.

—— (1972). A retrospective enquiry into possible teratogenic effects of drugs in pregnancy. *Adv. exp. Med. Biol.* **27**, 441–55.

ROBBIE, M.O. AND SWEET, R.L. (1983). Metronidazole use in obstetrics and gynecology. *Am. J. Obstet. Gynecol.* **145**, 865–81.

ROBERT, E. (1983a). Valproic acid in pregnancy — Association with spina bifida: a preliminary report. *Clin. Pediatr.* **22**, 336.

—— (1983b). Valproate: A new cause of birth defects — reports from Italy and follow-up from France. *Morbidity Mortality Wkly Rep.* Aug. 26th, pp. 438–9.

—— AND GUIBAUD, P. (1982). Maternal valproic acid and neural tube defects. *Lancet* ii, 937.

ROBINSON, S.C. AND MIRCHANDANI, G. (1965). Trichomonas vaginalis. Further observations on metromidazole. (Flagyl) including infant follow-up. *Am. J. Obstet. Gynecol.* **93**, 502.

ROCKER, I. (1977). Rifampicin in early pregnancy. *Lancet* ii, 48.

ROE, J.C. (1983). Toxicological evaluation of metronidazole with particular reference to carcinogenic, mutagenic, and teratogenic potential. *Surgery* **93**, 158–64.

ROSA, F.W. (1983). Teratogenicity of isotretinoin. *Lancet* ii, 513.

ROSA, F., WILK, A., AND KELSEY, F. (1985). *Human Retinoid Teratogenicity. Proc. 13th Conf. of the European Teratology Society*, September 16–20, Rostock-Warnemünde, GDR.

ROSENBERG, L., MITCHELL, A.A., SHAPIRO, S., AND SLONE, D. (1982). Selected birth defects in relation to caffeine-containing beverages. *J. Am. med. Ass.* **247**, 1429–32.

ROSETT, H.I., (1980). Guest editorial: A clinical perspective of the fetal alcohol syndrome. *Alcoholism: clin. exp. Res.* **4**, 119–22.

——, OUELLETTE, E.M., AND WEINER, L. (1976). A pilot prospective study of the fetal alcohol syndrome at the Boston City Hospital. Part I. Maternal drinking. *Ann. NY Acad. Sci.* **273**, 118–22.

ROSS INSTITUTE (1981). Malaria prevention in travellers from the United Kingdom. *Br. med. J.* **283**, 214–18.

ROTHMAN, K.J. (1982). Spermicide use and Down's syndrome. *Am. J. public Hlth* **72**, 399–401.

——, FYLER, D.C., GOLDBLATT, A., AND KREIDBERG, M.B. (1979). Exogenous hormones and other drug exposures of children with congenital heart disease. *Am. J. Epidemiol.* **109**, 433–9.

——, AND LIESS, J. (1976). Gender of offspring after oral contraceptive use. *New Engl. J. Med.* **295**, 859–61.

ROWLATT, R.J. (1978). Effect of maternal diazepam on the newborn. *Br. med. J.* **1**, 985.

ROYAL COLLEGE OF GENERAL PRACTITIONER ORAL CONTRACEPTIVE STUDY (RCGP SURVEY) (1976). The outcome of pregnancy in former oral contraceptive users. *Br. J. Obstet. Gynaecol.* **83**, 608–16.

RUBALTELLI, F.F., CHIOZZA, M.L., ZANARDO, V., AND CANTARUTTI, F. (1979). Effect on neonate of maternal treatment with indomethacin, *J. Pediat.* **94**, 161.

RUBIN, P.C., BUTTERS, L., LOW, R.A., CLARK, D.C., AND REID, J.L. (1983). Atenolol in the management of hypertension during pregnancy *Drugs* **25** (Suppl. 2), 212–14.

RUDD, N.L., AND FREEDOM, R.M. (1979). A possible primidone

embryopathy, *J. Pediat.* **94**, 835-7.

RUDOLPH, A.M. (1981). Effects of aspirin and acetaminophen in pregnancy and in the newborn. *Arch intern. Med.* **141**, 358-63.

RUMACK, C.M., GUGGENHEIM, M.A., RUMACK, B.H., PETERSON, R.G., JOHNSON, M.L., AND BRAITHWAITE, W.R. (1981). Neonatal intracranial haemorrhage and maternal use of aspirin. *Obstetrics Gynecology* **58**, 52S–56S.

RUMEAU-ROQUETTE, C., GOUJARD, J., AND HULL, G. (1977). Possible teratogenic effect of phenothiazines in human beings. *Teratology* **15**, 57-64.

SADUSK, J.F. AND PALMISANO, P.A. (1965). Teratogenic effect of meclizine cyclizine and chlorcyclizine. *J. Am. med. Ass.* **194**, 987-9.

SAFRA, M.J. AND OAKLEY, G.P. (1975). Association between cleft lip with or without cleft palate and prenatal exposure to diazepam. *Lancet* **ii**, 478-80.

SANDLER, B. (1973). Anencephaly and ovulation stimulation. *Lancet* **ii**, 379.

SANDOR, G.S., SMITH, D.F., AND MAC LEOD, P.M. (1981). Cardiac malformations in the fetal alcohol syndrome. *J. Pediat.* **98**, 771-3.

SAXEN, I. (1975). Associations between oral clefts and drugs taken during pregnancy. *Int. J. Epidemiol.* **4**, 37-44.

—— AND SAXEN, L. (1975). Association between maternal intake of diazepam and oral clefts. *Lancet* **ii**, 498.

SAXEN, L. (1974). Cleft palate and maternal diphenhydramine intake. *Lancet* **i**, 407-8.

SAVOLAINEN, E., SAKSELA, E., AND SAXEN, L. (1981). Teratogenic hazards of oral contraceptives analysed in a national malformation register. *Am. J. Obstet. Gynecol.* **140**, 521-4.

SCHAFER, A.I. (1981). Teratogenic effects of antileukemic chemotherapy. *Arch. intern. Med.* **141**, 514-15.

SCHARDEIN, J.L. (1985). *Chemically induced birth defects.* Marcel Dekker Inc. New York and Basel.

SCHEINBERG, I.H. AND STERNLIEB, I. (1975). Pregnancy in penicillamine treated patients with Wilson's disease. *New Engl. J. Med.* **293**, 1300-2.

SCHENKEL, B. AND VORHERR, H. (1974). Non-prescription drugs during pregnancy. Potential teratogenic and toxic effects on embryo and fetus. *J. reprod. Med.* **12**, 27-45.

SCHOU, M., GOLDFIELD, M.D., WEINSTEIN, M.R., AND VILLENEUVE, A. (1973). Lithium and pregnancy. I. Report from the Register of Lithium Babies. *Br. med. J.* **2**, 135-6.

SCHWARZ, R.H. (1981). Considerations of antibiotic therapy during pregnancy. *Obstetrics Gynecology* **58** (suppl.), 95S–97S.

SCOTT, J.R. (1977). Fetal growth retardation associated with maternal administration of immuno-suppressive drugs. *Am. J. Obstet. Gynecol.* **128**, 668-76.

SEELER, R.A., ISRAEL, J.N., ROYAL, J.E., KAYE, C.I., RAO, S., AND ABULABAN, M. (1979). Ganglioneuroblastoma and fetal hydantoin alcohol syndromes. *Pediatrics* **63**, 524-7.

SEKI, K., SEKI, M., AND KATO, K. (1983). Outcome of pregnancy and follow-up of children conceived by ovulation induction. *Asia-Oceania J. Obstet. Gynaecol.* **9**, 59-69.

SENEZ, PH. AND GILLET, J.Y. (1978). Avenir et qualite des grossesses obtenues par l'association citrate de clomifene et gonadotrophine chorionique *J. Gyn. Obst. Biol. Repr.* **7**, 987-90.

SEVER, L.E. (1972). Influenza and congenital malformations of the central nervous system. *Lancet* **i**, 910-11.

SHAPIRO, S., HEINONEN, O.P., SISKIND, V., KAUFMAN, D.W., MONSON, R.R., AND SLONE, D. (1977). Antenatal exposure to doxylamine succinate and dicyclomine hydrochloride (Bendectin) in relation to congenital malformation, perinatal mortality rate, birth weight and intelligence quotient score. *Am. J. Obstet. Gynecol.* **128**, 480-5.

——, KAUFMAN, D.W., ROSENBERG, L., SLONE, D., MONSON, R.R., SISKIND, V., AND HEINONEN, O.P. (1978). Meclizine in pregnancy in relation to congenital malformations. *Br. med. J.* **1**, 483.

——, SLONE, D., HARTZ, S.C., ROSENBERG, L., SISKIND, V., MONSON, R.R., MITCHELL, A.A., HEINONEN, O.P., IDANPÄÄN-HEIKKILÄ, J., HÄRÖ, S., AND SAXEN, L. (1976b). Anticonvulsants and perinatal epilepsy in the development of birth defects. *Lancet* **i**, 272.

——, ——, HEINONEN, O.P., KAUFMAN, D.W., ROSENBERG, L., MITCHELL, A.A., AND HELMRICH, S.P. (1982). Birth defects and vaginal spermicides. *J. Am. med. Ass.* **247**, 2381-4.

——, SISKIND, V., MONSON, R.R., HEINONEN, O.P., KAUFMAN, D.W., AND SLONE, D. (1976a). Perinatal mortality and birth weight in relation to aspirin taken during pregnancy. *Lancet* **i**, 1375-6.

SHAW, E.B., AND REES, E.L. (1980). Fetal damage due to aminopterin ingestion. Follow up at $17\frac{1}{2}$ years of age. *Am. J. Dis. Childh.* **134**, 1172-3.

SHAYWITZ, S.E., COHEN, D.J., AND SHAYWITZ, B.A. (1980). Behaviour and learning difficulties in children of normal intelligence born to alcoholic mothers. *J. Pediat.* **96**, 978-82.

SHEIKH, K. (1979). Teratogenic effects of organic solvents. *Lancet* **ii**, 963.

SHEPARD, T.H. (1976). *Catalog of teratogenic agents*, 2nd edn. Johns Hopkins University Press, Baltimore.

SIBAI, B.M., ABDELLA, T.N., AND ANDERSON, G.D. (1983). Pregnancy outcome in 211 patients with mild chronic hypertension. *Obstet. Gynecol.* **61**, 571-6.

SLONE, D., SISKIND, V., HEINONEN, O.P., MONSON, R.R., KAUFMAN, D.W., AND SHAPIRO, S. (1976). Aspirin and congenital malformations. *Lancet* **i**, 1373-5.

——, ——, ——, ——, ——, AND —— (1977). Antenatal exposure to the phenothiazines in relation to congenital malformations, perinatal mortality rate, birth weight and intelligence quotient score. *Am. J. Obstet. Gynecol.* **128**, 486-8.

SMITH, D.W. (1966). Dysmorphology (teratology). *J. Pediat.* **69**, 1150-69.

—— (1979). Anticonvulsant medication. *Am. J. Dis. Childh.* **133**, 450-1.

—— (1980). Hydantoin effects on the fetus. In *Phenytoin induced teratology and gingival pathology* (ed. T.M. Hassell, M.C. Johnston, and K.H. Dudley), pp. 35-40. Raven Press, New York.

SMITH, E.S.O., DAFVE, C.S., MILLER, J.R., AND BANNISTER, P. (1977). An epidemiological study of congenital reduction deformities of the limbs. *Br. J. prev. soc. Med.* **31**, 39-41.

SMITH, I.D., TEMPLE, D.M., AND SHEARMAN, R.P. (1975). The antagonism by anti-inflammatory analgesics of prostaglandin F2 alpha-induced contractions of human and rabbit myometrium *in vitro*. *Prostaglandins* **10**, 41-57.

SMITHELLS, R.W. (1965). The problem of teratogenicity. *Practitioner* **194**, 104-10.

—— (1976). Environmental teratogens of man. *Br. med. Bull.* **32**, 27-33.

—— (1981). Oral contraceptives and birth defects. *Develop. Med. Child. Neurol.* **23**, 369–83.

—— AND CHIN, E.R. (1964). Meclozine and fetal malformations. A prospective study. *Br. med. J.* **1**, 217–18.

——, NEVIN, C., SELLER, M.J., SHEPPARD, S., HARRIS, R., READ, A.P., FIELDING, D.W., WALKER, S., SCHORAN, C.J., AND WILD, J. (1983). Further experience of vitamin supplementation for prevention of neural tube defect recurrences. *Lancet* i, 1027–31.

—— AND SHEPPARD, S. (1978). Teratogenicity testing in humans; a method demonstrating safety of Bendectin. *Teratology* **17**, 31–6.

——, ——, SCHORAH, C.J., NEVIN, N.C., AND SELLER, M.J. (1981*a*). Trial of folate treatment to prevent recurrence of neural tube defects. *Br. med. J.* **282**, 1793.

——, ——, ——, SELLER, M.J., NEVIN, N.C., HARRIS, R., READ, A.P., AND FIELDING, D.W. (1981*b*). Apparent prevention of neural tube defects by periconceptional vitamin supplementation. *Arch. Dis. Childh.* **56**, 911–18.

——, ——, ——, ——, ——, ——, ——, ——, AND WALKER, S. (1981*c*). Vitamin supplementation and neural tube defects. *Lancet* ii, 1425.

——, SCHORAH, C.J., SELLER, M.J., NEVIN, N.C., HARRIS, R., READ, A.P., AND FIELDING, D.W. (1980). Possible prevention of neural tube defects by periconceptional vitamin supplementation. *Lancet* i, 339–40.

SNIDER, D.E., LAYDE, P.M., JOHNSON, M.W., AND LYLE, M.A. (1980). Treatment of tuberculosis during pregnancy. *Am. Rev. resp. Dis.* **122**, 65–79.

SOKOL, R.J., MILLER, S.I., AND REED, G. (1980). Alcohol abuse during pregnancy: an epidemiologic study. *Alcoholism: clin. exp. Res.* **4**, 135–45.

SOLOMON, L., ABRAMS, G., DINNER, M., AND BERMAN, L. (1977). Neonatal abnormalities associated with D-penicillamine treatment during pregnancy. *New. Engl. J. Med.* **296**, 54–5.

SOUTH, J. (1972). Teratogenic effects of anticonvulsants. *Lancet* ii, 1154.

SPEIDEL, B.D. AND MEADOW, S.R. (1972). Maternal epilepsy and abnormalities of the foetus and newborn. *Lancet* ii, 839–43.

—— AND —— (1974). Epilepsy, anticonvulsants and congenital malformations. *Drugs* **8**, 354–65.

SPIRA, W., GOUJARD, J., HUEL, G., AND RUMEAU-ROUQUETTE, C. (1972). Investigation into the teratogenic action of sex hormones first results of an epidemiological survey involving 20,000 women. *Revue Medicine* **41**, 2683–94.

STALSBERG, H. (1963). Antiemetics and congenital malformations — meclazine, cyclizine and chlorcyclizine. *Tidsskr. Nor. Laegeforen.* **85**, 1840–1.

STAPLES, R.E. (1972). Teratology. In *Anti-epileptic drugs* (ed. D.M. Woodbury, J.K. Penry, and R.P. Schmidt). Raven Press, New York.

STEINMANN, W.C. (1982). Caffeine consumption during pregnancy. *New Engl. J. Med.* **306**, 1549.

STEPHENS, J.D., GLOBUS, M.S., MILLER, T.R., WILBER, R.R., AND EPSTEIN, C.J. (1980). Multiple congenital anomalies in a fetus exposed to 5-fluorouracil during the first trimester. *Am. J. Obstet. Gynecol.* **137**, 747–9.

STERN, R.S., ROSA, F., AND BAUM, C. (1984). Isotretinoin and pregnancy. *J. Am. Acad. Dermatol.* **10**, 851–4.

STEVENSON, R.E., BURTON, O.M., FERLAUTO, G.J., AND TAYLOR, H.A. (1980). Hazards of oral anticoagulants during pregnancy. *J. Am. med. Ass.* **243**, 1549–51.

STEWART, K.S. (1981). Bacterial infections. In *Clinics in obstetrics and gynaecology* (ed. S.M. Wood and L. Beeley), Vol. 8, no. 2, pp. 315–32. W.B. Saunders, London.

——, SHAFI, M., ANDREWS, J., AND WILLIAMS, J.D. (1973). Distribution of parenteral ampicillin and cephalosporins in late pregnancy. *J. Obstet. Gynaec. Br. Cmwlth* **80**, 902–8.

STRAUSS, A., MODANLOU, H.D., GYEPES, M., AND WITTNER, R. (1982). Congenital heart disease and respiratory distress syndrome. *Am. J. Dis. Childh.* **136**, 934–6.

STREICHER, H.Z., GABOW, P.A., MOSS, A.H., KONO, D., AND KAEHNY, W.D. (1981). Syndromes of toluene sniffing in adults. *Ann. intern. Med.* **94**, 758–62.

STREISSGUTH, A.P., BARR, H.M., MARTIN, D.C., AND HERMAN, C.S. (1980*a*). Effects of maternal alcohol, nicotine and caffeine use during pregnancy on infant mental and motor development at eight months. *Alcoholism: clin. exp. Res.* **4**, 152–64.

——, HERMAN, C.S., AND SMITH, D.W. (1978*a*). Intelligence behaviour and dysmorphogenesis in the fetal alcohol syndrome: a report on 20 patients. *J. Pediat.* **92**, 363–7.

——, LANDESMAN-DWYER, S., MARTIN, J.C., AND SMITH, D.W. (1980*b*). Teratogenic effects of alcohol in humans and laboratory animals. *Science* **209**, 353–61.

——, MARTIN, D.C., AND BARR, H.M. (1978*b*). Caffeine effects on pregnancy outcome. First International Caffeine Committee Workshop Hawaii, 8–10 Nov., 1978.

STARREVELD-ZIMMERMAN, A.A.E., VAN DER KOLK, W.J., ELSHOVE, J., AND MEINARDI, H. (1975). Teratogenicity of anti-epileptic drugs. *Clin. Neurol. Neurosurg.* **77**, 81–95.

——, ——, MEINARDI, H., AND ELSHOVE, J. (1973). Are anticonvulsants teratogenic? *Lancet* ii, 48–9.

SUGRUE, D. AND DRURY, M.I. (1980). Hyperthyroidism complicating pregnancy: Results of treatment by antithyroid drugs in 77 pregnancies. *Br. J. Obstet. Gynecol.* **87**, 970–5.

SULLIVAN, F.M. (1972). Mechanisms of teratogenesis. In *Adverse drug reactions* (ed. D.J. Richards, and R.K. Rondell), pp. 19–25. Churchill-Livingstone, Edinburgh.

SULLIVAN, W.C. (1899). A note on the influence of maternal inebriety on the offspring. *J. ment. Sci.* **45**, 489–503.

SUONIO, S. (1982). Maternal placental blood flow in hypertensive pregnancy and the effect of Beta Mimetic drugs on it. *Dev. Pharmacol. Ther.* **4** (Suppl. 1), 99–108.

SUTHERLAND, H.W., STOWERS, J.M., CORMACK, J.D., AND BEWSHER, P.D. (1973). Evaluation of chlorpropamide in chemical diabetes diagnosed during pregnancy. *Br. med. J.* **3**, 9–13.

SZETO, H.H. (1983). Effects of narcotic drugs on fetal behavioural activity: acute methadone exposure. *Am. J. Obstet. Gynecol.* **146**, 211–16.

TANIMURA, T., AND LEE, S. (1972). Discussions on the suspected teratogenicity of quinine in humans. *Teratology* **6**, 122.

TCHERDAKOFF, P.H., COLLIARD, M., BERRARD, E., KREFT, C., DUPAY, A., AND BERNAILLE, J.M. (1978). Propranolol in hypertension during pregnancy. *Br. med. J.* **2**, 670.

TEGLER, L. AND LINDSTRÖM, B. (1980). Antithyroid drugs in milk. *Lancet* ii, 591.

THIERSCH, J.B. (1956). The control of reproduction in rats with the aid of antimetabolites and early experiments with antimetabolites as abortifacient agents in man. *Acta Endocrinol.* **23**, 37–45.

TOLAROVA, M. (1982). Periconceptional supplementation with vitamins and folic acid to prevent recurrence of cleft lip. *Lancet* ii, 217.

TOMATIS, L. (1974). Role of prenatal events in determining cancer risks. In *Modern trends in toxicology no. 2* (ed. E. Boyland

and R. Goulding), pp. 163–78. Butterworths, London.

TOMLIN, P.J. (1979). Health problems of anaesthetists and their families in the West Midlands. *Br. med. J.* **1**, 779–84.

TOUTANT, C. AND LIPPMANN, S. (1979). Fetal solvents syndrome. *Lancet* **i**, 1356.

TRUSSEL, R.R. AND BEELEY, L. (1981). Infestations. In *Clinics in obstetrics and gynaecology* (ed. S.M. Wood and L. Beeley), Vol. 8, no. 2, pp. 333–40. W.B. Saunders, London.

TURKALJ, I., BRAUN, P., AND KRUPP, P. (1982). Surveillance of bromocriptine in pregnancy. *J. Am. med. Ass.* **247**, 1589–91.

TURNER, E.S., GREENBERGER, P.A., AND PATTERSON, R. (1980). Management of the pregnant asthmatic patient. *Ann. interna. Med.* **93**, 905–18.

TURNER, G. AND COLLINS, E. (1975). Fetal effect of regular salicylate ingestion in pregnancy. *Lancet* **ii**, 338–9.

VAN DEN BERG, B.J. (1977). Epidemiological observations of prematurity: effects of tobacco, coffee and alcohol. In *The epidemiology of prematurity* (ed. D.M. Reed and F.J. Stanley), pp. 157–76.

VAN LEEUWEN, G., GUTHERIE, R., AND STANGE, F. (1965). Narcotic withdrawal reaction in a newborn infant due to codeine. *Pediatrics* **36**, 635–6.

VAN WAES, A. AND VAN DER VELDE, E. (1969). Safety evaluation of haloperidol in the treatment of hyperemesis gravidarum. *J. clin. Pharmacol.* **9**, 224–7.

VISSER, G.H.A., HUISJES, H.J., AND ELSHOVE, J. (1976). Anticonvulsants and fetal malformations. *Lancet* **i**, 970.

VOJTA, M., AND JIRASEK, J. (1966). 6-azauridine induced changes of the trophoblast in early human pregnancy. *Clin. Pharmacol. Ther.* **7**, 162–5.

VOWLES, M., PETHYBRIDGE, R.J., AND BRIMBLECOMBE, F. (1979). Letter. *Br. med. J.* **1**, 1079.

WAGNER, V.M., HILL, J.S., WEAVER, D., AND BAEHNER, R.L. (1980). Congenital abnormalities in baby born to cytarabine treated mother. *Lancet* **ii**, 98–9.

WALTERS, W.A.W. AND HUMPHREY, M.D. (1980). Common medical disorders in pregnancy and their treatment. *Drugs* **19**, 455–63.

WALTMAN, R., AND TRICOMI, V. (1974). Non-steroid anti-inflammatory drugs and pregnancy. *Prostaglandins* **8**, 397–400.

——, ——, AND PALAV, A. (1973). Aspirin and indomethacin effect on instillation abortion time of mid-trimester hypertonic-saline-induced abortion. *Prostaglandins* **3**, 47–58.

WALZ, M.A., DAVIS, W.M., AND PACE, H.B. (1983). Parental methadone treatment: A multigenerational study of development and behaviour in offspring. *Dev. Pharmacol. Ther.* **6**, 125–37.

WARKANY, J. (1975). A warfarin embryopathy? *Am. J. Dis. Childh.* **129**, 287–8.

—— (1979). Antituberculous drugs. *Teratology* **20**, 133–7.

WARRELL, D.W., AND TAYLOR, R. (1968). Outcome for the fetus of mothers receiving prednisolone during pregnancy. *Lancet* **i**, 117–18.

WEATHERSBEE, P.S., OLSEN, L.K., AND LODGE, J.R. (1977). Caffeine and pregnancy — a retrospective survey. *Postgrad. Med.* **62**, 64–9.

WEINSTEIN, A.M., DUBIN, D.B., PODLESKI, W.K., SPECTOR, S.L., AND FARR, R.S. (1979). Asthma and pregnancy. *J. Am. med. Ass.* **241**, 1161–5.

WEISS, D.B., ABOULAFIA, Y., AND MILEWIDSKY, A. (1976). Ectopic pregnancy and the pill. *Lancet* **ii**, 196.

WERBOFF, J., AND GOTTLIEB, J.S. (1963). Drugs in pregnancy: behavioural teratology. *Obstet. Gynecol. Surv.* **18**, 420–3.

WEYMAN, J. (1965). The clinical appearances of tetracycline staining of the teeth. *Br. dent. J.* **118**, 289–91.

WHITELAW, A. (1981). Maternal methyl dopa treatment and neonatal blood pressure. *Br. med. J.* **283**, 471.

WHITELAW, A.G.L., CUMMINGS, A.J., AND MC FADYEN, I.R. (1981). Effect of maternal lorazepam on the neonate. *Br. med. J.* **282**, 1106–8.

WHITFIELD, L.R., LELE, A.S., AND LEVY, G. (1983). Effect of pregnancy on the relationship between concentration and anticoagulant action of heparin. *Clin. Pharmacol. Ther.* **34**, 23–8.

WILKINSON, A.R. (1980). Naproxen levels in pre-term infants after maternal treatment. *Lancet* **ii**, 591–2.

——, AYNSLEY-GREEN, A., AND MITCHELL, M.D. (1979). Persistent pulmonary hypertension and abnormal prostagladin E. levels in preterm infants after maternal treatment with naproxen. *Arch. Dis. Childh.* **54**, 942–5.

WILSON, J. (1982). Use of sodium cromoglycate during pregnancy. *J. Pharm. Med.* **8** (2) suppl., 45–51.

WILSON, J.G., AND BRENT, R.L. (1981). Are female sex hormones teratogenic? *Am. J. Obstet. Gynecol.* **141**, 567–80.

—— AND FRASER, F.C. (1977). *Handbook of teratology*, Vol. 1. Plenum Press, New York and London.

WOOD, S.M. AND BLAINEY, J.D. (1981). Hypertension and renal disease. In *Clinics in obstetrics and gynaecology*, (ed. S.M. Wood, and L. Beeley), Vol. 8, no. 2, pp. 439–53. W.B. Saunders, London.

WOODS, D.L. AND MALAN, A.F. (1978). Side effects of maternal diazepam on the newborn infant. *S. Afr. med. J.* **54**, 636.

YERUSHALMY, H. AND MILKOVITCH, L. (1965). Evaluation of the teratogenic effect of meclizine in man. *Am. J. Obstet. Gynecol*, **93**, 553–62.

ZONDEK, L.H. AND ZONDEK, T. (1974). Observations on the maturational appearance of the reproductive organs of male infants after treatment with ACTH and corticosteroids. *Halv. Pediat. Acta* **29**, 173–9.

23 Drugs in milk

P.F. D'ARCY and J.C. McELNAY

Published information on drugs excreted in human milk, with few exceptions, is replete with isolated or anecdotal reports, and with inaccurate, premature, misleading, and undocumented conclusions. It is clear that most drug substances are excreted into human milk, and although some early reports of finding no drugs in milk have been given importance, it must be clearly understood that such conclusions are only as good as the analytical techniques used and in many cases it is now certain that there were analytical deficiencies. However, it is not just the presence of a drug in milk that is critical, it is the concentration in which it appears and the volume of milk ingested that is important. When the concentration of drug in milk is an appreciable percentage of the maternal plasma level of drug, the infant may well experience the full pharmacological and toxicological spectrum of that agent. Even when the drug is present in milk in low concentration, when compared with the maternal plasma concentration of drug, suckling may still not be safe because the volume of milk ingested and the half-life of the drug in the infant may be factors that predispose towards hazard. Some drugs, particularly antibiotics, may act as sensitizing agents at low concentration in the child without producing any adverse effect in the mother.

A pharmacokinetic approach

Recently some attention has been given in the literature to pharmacokinetic studies on drugs in the plasma and milk of nursing mothers. This is a welcome sign since it indicates a trend away from the isolated or anecdotal reports of earlier publications. Among such work that of Wilson *et al.* (1980) did much to advocate and advance a scientific approach to the whole question of drugs in breast milk. In their comprehensive review, Wilson and his colleagues described in fine detail the principles, pharmacokinetics, and projected consequences of drug excretion in human breast milk.

Fundamental principles of breast-milk excretion have been used to construct a pharmacokinetic approach useful for the study of most drugs. An infant-modulated three-compartment open model has been prepared for drug distribution and elimination in the breast-feeding

woman. Milk/plasma concentration ratios have been projected on the basis of pH partitioning and, while some studies have confirmed these projections, others have clearly demonstrated the need to consider additional factors such as lipid solubility and protein-binding characteristics of a drug in milk.

Such human pharmacokinetic studies, however, to be effective need firm data from conceptually valid and controlled studies on medicines consumed or administered during lactation. In spite of current interest in such studies, such data are conspicuously lacking for many of the drugs in common clinical use. This lack of good information very much impedes authoritative recommendations on caution, contraindication, or relative safety of drugs during lactation. Advice given on the potential risk to the infant therefore remains largely speculative.

Fortunately, relatively few drugs are definitely contraindicated in breast-feeding; there is, however, a much larger list of drugs in the 'use with caution in breast-feeding' category. It is in this latter group that uncertainty and speculation arise since advice given is largely based on inadequate reports often made with inconclusive recommendations on safety or hazard to the infant. Proscriptive or permissive advice is not possible and the drugs therefore remain in the limbo of a cautionary group.

Contraindication, caution, or 'safe' drugs in breast-feeding

Table 23.1 lists those drugs which are

1. Contraindicated in breast feeding;
2. Those which should only be used with caution by the nursing mother.

Table 23.2 shows those drugs which **appear** to be safe in breast-feeding.

Information in the latter category is largely based on work by the Mersey Regional Drug Information Service (1981) in Britain which has done a great deal to provide this type of information to the practitioner. This regional

Table 23.1 Drugs and breast-feeding: contraindications and cautions

Contraindicated in breast-feeding
(stop drug or stop breast-feeding)

Androgens
Antithyroid drugs (e.g. carbimazole, iodine, methimazole, propylthiouracil, thiouracil)
Atropine
Bromides
Cancer chemotherapy (e.g. cyclophosphamide, fluorouracil, methotrexate, etc.)
Carisoprodol
Chloramphenicol
Chlorthalidone
Chlortrianisene
Ergot alkaloids (e.g. ergotamine)
Ethisterone
Heroin
Iodine, iodides, and radioactive iodine
Indomethacin
Lithium
Methadone
Metronidazole (oral dosage)
Medroxyprogesterone (injectable contraceptive)
Nalidixic acid
Nitrofurantoin (in G6PD-deficient infants)
Phenindione and other indanedione anticoagulants
Povidone–iodine vaginal gel
Radio-pharmaceuticals
Reserpine
Sulphonamides (including co-trimoxazole; in G6PD-deficient infants)
Tetracyclines

Use with caution in breast-feeding
(monitor infant regularly; avoid drug if possible)

Amantadine
Aminoglycoside antibiotics (e.g. gentamicin, kanamycin, neomycin, tobramycin, streptomycin, etc.)
Aminophylline/theophylline
Antihistamines (e.g. clemastine)
Aspirin in high doses taken chronically
Barbiturates and non-barbiturates hypnotics (e.g. chloral derivatives, dichloral phenazone) and sedatives
Benzodiazepines (clorazepate, chlordiazepoxide, diazepam, lorazepam, lormetazepam, nitrazepam, oxazepam, etc. particularly at high dosage)
Butorphanol
Cis (Z)-flupenthixol
Carbamazepine
Beta-blockers (e.g. atenolol, metoprolol, mepindolol, nadolol, oxprenolol, propranolol, sotalol, timolol, and ophthalmic timolol)

Cefoxitin
Ceftazidime
Chlorpromazine and other phenothiazine tranquillizers (high-dose)
Cimetidine
Clindamycin
Clonidine
Corticosteroids (e.g. prednisone, prednisolone)
Co-trimoxazole
Dapsone
Dextropropoxyphene (propoxyphene)
Diphenoxylate
Disopyramide
Domperidone
Dyphylline
Ethosuximide
Glibenclamide
Haloperidol
Ibuprofen
Isoniazid
Lincomycin
Loperamide
Meprobamate
Methyldopa
Metoclopramide
Metrizamide
Naproxen
Novobiocin
Oestrogens
Oral anticoagulants (excluding warfarin and acenocoumarol, but including other coumarins)
Oral contraceptives (combined and progestogen-only types. Note: medroxyprogesterone injection is contraindicated)
Oral hypoglycaemic agents
Orciprenaline
Pentazocine
Phenobarbitone
Phenytoin sodium (diphenylhydantoin)
Primidone
Procainamide
Purgatives (senna and other anthraquinone derivatives, e.g. danthron, also phenolphthalein)
Pyridostigmine
Sisomicin
Sodium valproate
Sulphasalazine
Sulpiride
Thiazide and some other diuretics (e.g. bendrofluazide, frusemide, but not chlorthalidone)
Tolbutamide
Vitamin D (high doses, i.e. therapeutic use)

Note: if the nursing mother has impaired renal function, then drugs excreted mainly by the kidneys may appear in higher concentration in her milk.

Table 23.2   Drugs which with available evidence APPEAR to be safe in breast-feeling. (Modified from Mersey Regional Information Service, 1981)

Acenocoumarol (short-term treatment)
Acetaminophen (paracetamol)
Amitriptyline (and other tricyclic antidepressants)
Aspirin (low-dose, occasional use)
Baclofen
Captopril
Cephalosporins (e.g. cefadroxil, cefotaxime,
  cephalexin, cephalothin, cephapirin)
Codeine (in cough and analgesic preparations)
Digoxin
Erythromycin
Ethambutol
Flufenamic acid
Folic acid
Heparin
Hydrallazine
Imipramine (and other tricyclic antidepressants)
Insulin
Iron (ferrous sulphate)
Mefenamic acid
Mexiletine
Nitrofurantoin (in absence of G6PD-deficiency)
Pencillins (e.g. ampicillin, amoxycillin,
  penicillin V)
Phenothiazines (low-dose)
Sodium cromoglycate
Terbutaline
Verapamil
Vitamin A and D (low-dose, i.e. prophylactic use)
Vitamin C
Warfarin

service has published a drug information letter, *Drugs and breast feeding* which lists, among cautions and contraindications, 27 drugs or classes of drugs which appear to be safe, although the general caution of avoiding the taking of any drug during breast-feeding is emphasized. Apart from the drugs listed by the regional service, other drugs have been added which, from literature reports, appear to present little hazard to the suckling infant. It must be emphasized, however, that 'appears to be safe' is exactly what it means and absolute safety cannot be assured or assumed.

Milk-to-plasma drug concentration ratios

Table 23.3 presents some data collated from the literature on human milk/plasma (M/P) drug concentration ratios which provide some basic information on which the risk to the suckling infant can be predicted. This must, however, only be regarded as a guide and nothing

Table 23.3   Human maternal milk-to-plasma ratio (M/P) for various drugs. (Based on data cited by Vorherr (1974 *a, b*), Lien (1979), Wilson *et al.* (1980), Gardner and Rayburn (1982), Chaplin *et al.* (1982), and papers reviewed in this current chapter: lowest and highest recorded ratios are tabulated)

Drug/Ratio

Acetaminophen
  (paracetamol) 0.76–1.42
Alcohol 1.0
Amitriptyline 0.11–0.17
Amoxycillin 0.04
Ampicillin 0.03–0.3
Aspirin 0.6–1.0
Atenolol 1.3–6.8
Caffeine 0.61–0.87
Captopril 0.007
Carbamazepine 0.4–0.7
Carisoprodol 2.0–4.0
Cefadroxil 0.02
Cefazolin 0.02–0.07
Cefotaxime 0.16
Cephalexin 0.14
Cephalothin 0.5
Cephapirin 0.48
Chloral hydrate 0.5
Chloramphenicol 0.05–0.6
Chlorazepate 0.13–0.3
Chlorpromazine 0.3–0.5
Chlortetracycline 0.25–0.5
Chlorthalidone 0.03
Cimetidine 3.5–11.76
Codeine 2.2–2.31
Cis(Z)-flupenthixol 1.3
Colistin sulphate 0.17–0.18
Cycloserine 0.3–1.18
Demeclocycline 0.33–0.5
Desmethyldiazepam 0.1
Dextropropoxyphene 0.5
Diazepam 0.01–0.02
Dicoumarol 0.01–0.02
Digoxin 0.45–1.0
Disopyramide 0.9
Doxepin 0.33
Doxycycline 0.32–0.37
Dromperidone 0.25
Dyphylline 2.08
Erythromycin 0.5–3.0
Ethosuximide 0.8
Ethylbiscoumacetate 0.01
Folic acid 0.02
Flufenamic acid < 0.01
Haloperidol 0.59–0.69
Imipramine 0.08–0.5
[131]Iodine 65
Isoniazid 1.0
Kanamycin 0.05–1.0
Levonorgestrel (formerly
known as *d*-norgestrel)
  0.15–0.20
Lincomycin 0.13–2.25

Table 23.3 *Continued*

Drug/Ratio	
Lithium salts 0.25–0.77	Propranolol
Lormetazepam 0.06	< 0.001–2.0
Medroxyprogesterone 1.0	Propylthiouracil 0.13–0.47
Mepindolol 0.4	Pyrazolone 1.0
Meprobamate 2–4	Pyridostigmine 0.66
Methacycline 0.5	Pyrimethamine 0.2–0.43
Methadone 0.8	Quinine sulphate 0.14
Methenamine hippurate 1.08	Ranitidine 2–6.7
Methimazole 0.98–1.16	Rifampicin 0.2–0.6
Methotrexate 0.08–1.0	Sodium valproate 0.05–0.1
Metoclopramide 1.83	Sotalol 2–8
Metoprolol 2.6–3.7	Spironolactone 0.8
Metronidazole 0.45–1.8	Streptomycin 0.5–1.0
Mexiletine 0.78–2.0	Sulphacetamide 0.08
Morphine 1.07–4.14	Sulphadiazine 0.21
Nadolol 2.7–4.6	Sulphadimidine 0.51
Nalidixic acid 0.08–0.13	Sulphafurazole 1.0
Nortriptyline 1.0	Sulphanilamide 1.0
Novobiocin 0.1–0.25	Sulphapyridine 0.6–1.0
Oxprenolol 0.29	Sulphasalazine 0.33
Penicillin G 0.02–0.2	Sulphathiazole 0.3–0.5
Phenacetin 0.39–0.61	*Sulphetrone* 6.4
Phenindione 0.012–0.06	Terbutaline 1.4–2.9
Phenobarbitone 0.17–0.46	Tetracycline 0.2–1.4
Pentothal 1.0	Theobromine 0.82
Phenylbutazone 0.1–0.3	Theophylline 0.63–0.87
Phenytoin 0.13–2.0	Thiouracil 3.0
Piroxicam 0.009–0.014	Timolol 0.8
Prednisolone 0.15	Tolbutamide 0.09–0.4
Prednisone 0.67	Trimethoprim > 1.0
Primidone 0.81	Verapamil 0.23
Procainamide 4.3	

more, since the data have been gained from vastly different types of studies, some involving acute drug dosage, others chronic, often at single-unit dosage levels with milk obtained under varying times and conditions of sampling. Nor indeed can it be assumed that the analytical methods employed have equatable degrees of specificity or sensitivity.

Avery (1979) advocated that, as a general rule, special concern should be given to those drugs which exhibit an M/P ratio > 1. In the absence of other guidelines or definite evidence to the contrary, this is sensible advice.

A Third World problem

Mothers in Third World countries frequently breast-feed their babies for periods in excess of two years. As compared with former years, it is now more likely that the nursing mother may also be receiving prophylactic medication or drug treatment for a tropical disease. This is a matter for concern since, apart from dapsone (Table

23.1, 'cautionary category') and *Sulphetrone* and quinine sulphate (Table 23.3), there is almost a total lack of information on breast-milk excretion and M/P ratios for antimalarial drug combinations, for drugs against amoebiasis, schistosomiasis, leishmaniasis, tryanosomiasis, or indeed against any of the other infections or infestations which are endemic or prevalent in developing countries. With such an extended pattern of breast-feeding, it may well be that the suckling child could ingest significant amounts of relatively toxic drugs over a lengthy period.

Restrictions on breast-feeding would not be feasible nor indeed desirable in such cases because prolonged breast-feeding is of critical importance in helping to lower existing high infant morbidity and mortality. However, knowledge of drug excretion in milk and particularly of M/P drug concentration ratios might enable maternal drug dosage schedules to be adjusted to reduce the hazard to the suckling child. Research is badly needed on the breast-milk excretion of these drugs and it is obvious that such research can only be done in the Third World countries where the cases are available.

Milk banks also have drug problems

The excretion of drugs in breast milk has become not only the concern of nursing mothers but also of those involved in the collection of human breast milk for milk banks. Cash and Giacoia (1981) have reported useful guidelines for donor recruitment and screening. Prospective donors are excluded if they are heavy smokers > 20 cigarettes per day), drink alcoholic or caffeine beverages regularly, or receive any medications except an occasional aspirin. Milk is not accepted if the donor or her infant are ill.

The rest of this chapter will discuss drugs for which relevant information is available. For convenience the drug groups discussed are presented in alphabetical order. The reader is reminded that case-report data are best treated as early warning systems which require substantiation.

Analgesics

Aspirin

Jamali and Keshavarz (1981), using a sensitive HPLC method, found that, following oral dosage of 500, 1000, and 1500 mg of acetylsalicylic acid to six healthy nursing mothers, salicylate appeared in milk not later than one hour post-dosage and that it reached maximum concentrations (5.8, 16.0, and 38.7 μg ml^{-1}, respectively) in 2–6 hours. Additional work by these investigators suggested that maternal plasma concentrations of salicylate were more than ninefold that of corresponding levels in breast milk.

Much higher concentrations of salicylate in milk were found in studies by Berlin et al. (1980) who demonstrated milk salicylate peak concentrations of 173–483 μg ml^{-1} after administration of a single oral dose of 650 mg acetylsalicylic acid to 10 nursing mothers. In contrast to both these findings, Erickson and Oppenheim (1979) failed to find detectable levels (50 μg ml^{-1}) of salicylate in one case after the mother had taken 4 g of acetylsalicylic acid daily for four days.

Studies by Findlay et al. (1981), in which single oral doses of a compound analgesic (total 454 mg aspirin) were given to two lactating women, showed that salicylate penetrated poorly into milk with peak concentrations of only 1.12–1.69 μg ml^{-1}, whereas corresponding peak plasma figures were 33–43.4 μg ml^{-1}.

These diverging results typify some of the problems inherent in studying the excretion of drugs in milk. Different analytical procedures, different sampling systems, and different dosage protocols add to the confusion in the literature. In these examples, the varied formulations

of acetylsalicylic (including combination preparations), and possibly their varied bioavailability, could account for some of this conflicting evidence.

It is difficult from the four studies that have been described to draw any firm conclusions about the hazard to the nursing infant of maternal dosage with acetylsalicylic acid. The general view, however, as expressed by the Mersey Regional Drug Information Service (1981) is that the occasional use of low doses appears to be safe, but if high doses of acetylsalicylic acid are used for prolonged periods then the drug should be put into the 'cautionary' category, and the suckling child should be monitored for rash and platelet dysfunction.

A case report by Clark and Wilson (1981) reinforces this caution. The report describes a 16-day-old girl who was admitted to hospital with metabolic acidosis. The child improved after the correction of the acidosis; however, a serum salicylate level on her third hospital day was still 24 mg dl^{-1}. The presence of salicylate metabolites in her urine were demonstrated by gas chromatography. After excluding other possibilities for salicylate ingestion by the infant the authors suggested that the drug may have been derived from her mother's milk. The mother had been taking 600 mg of aspirin every four hours for arthritis and had been nursing the infant until the day of hospitalization. A further case report (Bailey et al. 1982) described the examination of salicylate content of a nursing mother taking chronic therapeutic doses of aspirin. Based on the drug content in this mother's milk, it was estimated that more than 25 litres of milk at its peak drug concentration would have to be consumed by the infant to provide the salicylate content of one aspirin tablet. Although these anecdotal reports add little to the scientific literature on the subject they do, together with other reports on salicylates (e.g. Berlin et al. 1980), indicate the need for caution.

Butorphanol

This new potent synthetic agonist–antagonist analgesic has been detected in the milk of lactating women following oral and intramuscular administration. Six women were studied for each administration route. The amount of the drug detected was very small and therefore the potential for a neonatal effect from administration of the drug to the nursing mother would appear minimal (Pittman et al. 1980). The drug has initially, however, been placed in the 'caution' section of Table 23.1 until more data are available.

Dextropropoxyphene (propoxyphene)

Dextropropoxyphene is a commonly used analgesic

either taken alone or more commonly in combination with paracetamol. This analgesic is related in chemical structure to methadone and, after overdosage, symptoms are similar to those of morphine poisoning except that convulsions are a prominent feature.

There is current concern on the hazard presented when dextropropoxyphene formulations are taken with alcohol (Carson and Carson 1977; Whittington 1977; Young and Lawson 1980; Whittington and Barclay 1981) and indeed on the increased use of this combination in deaths from overdose (Locket 1980). Although early studies showed that dextropropoxyphene did not present problems when used in labour, there is now one report of possible neonatal dextropropoxyphene withdrawal syndrome in a baby whose mother had taken one to four 64-mg capsules daily for nearly four years (Tyson 1974).

Catz and Giacoia (1972) reported that, following a suicide attempt with dextropropoxyphene by a nursing mother, the quantity of the drug in breast milk was half the concentration in the mother's plasma. This would suggest that an infant weighing 5–7 kg could ingest up to 1 mg of the analgesic per day, if the mother was taking the maximum recommended dose. Ananth (1978) in considering this possibility thought that the 1 mg/day intake by the infant was a significant amount. Caution should therefore be emphasized if the nursing mother is taking this drug; if possible it should be stopped during lactation.

Narcotic analgesics

Babies born to women using heroin or methadone have shown signs of withdrawal (Tylden 1973), which can be severe or even fatal. Opiate withdrawal may first become manifest as fetal distress in labour — 3 hours or so after the time of the last dose taken by the mother. A baby seemingly normal at birth may pass into a withdrawal state 3–4 hours after the last maternal dose of heroin, or as long as three days after the last dose of methadone.

There is some disagreement as to the advisability of a methadone-maintained mother breast-feeding her child. Smialek and his associates (1977) from Wayne County, USA, instructed narcotic-addicted mothers to bottle-feed their infants; that this is good advice is illustrated by one of their cases in which one infant received methadone through breast milk and died. The mother was a former heroin addict and had received maintenance doses of methodone during and after pregnancy. The child was five weeks old and at autopsy the blood methadone level was found to be 0.04 mg dl^{-1}.

Considering the risk of alternatives to the methadone treatment of heroin addiction (Zinberg, 1977), this drug is likely to remain a therapeutic agent for years to come:

it is therefore essential to contraindicate breast-feeding if the mother is receiving such treatment. Another valid reason to contraindicate breast-feeding is that the addicted mothers tend to treat their minor mood swings with large doses of other drugs, for example diazepam, phenobarbitone, tricyclic antidepressants, or hypnotics in the post-partum period when craving becomes intense. The amount of these other drugs excreted in their milk can cause dangerous coma in the infant.

Against this advice, a review article from the American Academy of Clinical Toxicology (Thoman 1978) suggested that only small amounts of methadone are found in breast milk and that breast-feeding should be permitted. However, on grounds of safety, contraindication of breast-feeding is advised until such time that other evidence emerges to suggest that this is an unnecessary restriction.

Paracetamol (acetaminophen)

Although paracetamol is a widely used analgesic and antipyretic, its disposition in human milk was not reported until a study by Berlin et al. (1980) produced evidence that maternal ingestion of this drug in effective analgesic doses does not present a risk to the nursing infant.

Twelve nursing mothers were given a single 650-mg oral dose of paracetamol; simultaneous saliva and milk samples were collected from the subjects at intervals after the dosing and in two subjects plasma samples were also obtained at several times during the first six hours.

Paracetamol appeared in maternal saliva and milk and peak levels (10–15 μg ml^{-1}) were achieved by 1–2 hours post-dosage. Saliva/milk ratios during the elimination phase ranged from 0.7 to 1.1 with most of the values between 0.9 and 1.0. In the two subjects studied, saliva/plasma ratios were 0.9 to 1.0. No adverse effects, especially behavioural changes in the infants, were noted.

Assuming that each infant took 90 ml of milk at 3, 6, and 9 hours after maternal dosage with the analgesic, the authors calculated that the amount of paracetamol available for ingestion would range from 0.28 to 1.51 mg (mean $\pm$ SE: 0.88 $\pm$ 0.31 mg). This represented from 0.04 to 0.23 per cent (mean: 0.14 $\pm$ 0.04 per cent) of the maternal dose of the analgesic. They concluded that it is possible for the nursing mothers to take an effective analgesic/antipyretic dosage of paracetamol with very minimal possible drug exposure to the nursing infant.

It is on this evidence that this analgesic has been placed in the 'appears to be safe' category (Table 23.2).

Evidence confirming this decision was given by Findlay et al. (1981), who studied paracetamol as a metabolite of phenacetin dosage to nursing mothers. Paracetamol was found at higher concentrations than the parent phenacetin in both maternal milk and plasma; but it was not thought that infants would be at risk from these concentrations in milk. However, these investigators commented that some drugs including paracetamol are cleared significantly more slowly by the neonate than by the mother; thus there could be hazard to the infant if excessive doses were taken continously by the mother.

Bitzén et al. (1981) monitored maternal milk and plasma levels of paracetamol in three lactating women after a single oral dose of 500 mg. Paracetamol concentrations were consistently lower in milk than in plasma, with a mean M/P AUC ratio of 0.76. The half-lives of paracetamol in plasma and milk were almost identical with an overall mean of 2.7 hours. As less than 1 per cent of the maternal dose would be present in 1 litre of milk (i.e. 29.1 μmol), the authors concluded that breast-feeding need not be discontinued or interrupted due to paracetamol treatment in conventional dosage.

Pentazocine

The Mersey Regional Information Service (1981) puts pentazocine into the 'cautionary' category. It recommends that the suckling infant be monitored for drowsiness. No other new information is available.

Antibiotics and antibacterial agents

Amantadine

Vomiting, urinary retention, and skin rashes have been reported in infants fed by mothers taking amantadine (Knowles 1972; O'Brien 1974). This antiviral and antiparkinsonian drug has been placed in the 'caution' category (Table 23.1).

Aminoglycoside antibiotics

Streptomycin was reported to be excreted in breast milk in concentrations of 1–3 mg/100 ml after therapeutic doses to the mother (Vorherr 1974b) and to give a maternal milk/plasma ratio of 0.5–1.0 (Wilson et al. 1980). Previous reviewers (for example, O'Brien 1974) suggested that the risks of harmful effects on the child (including ototoxicity) outweigh the benefits of nursing. However, streptomycin is poorly absorbed from the infant's gut and outright contraindication may not be justified. Breast-feeding should, however, be avoided if there is maternal renal insufficiency since milk levels of streptomycin can then increase some 25-fold (Knowles 1973; Chaplin et al. 1982).

Not a great deal is known about the distribution or hazard of other aminoglycoside antibiotics in breast-milk. Amikacin was not detected in milk after a single i.m. dose of 200 mg (Yuaso 1974); gentamicin levels in milk were 15.7 μg/100 ml one hour after dosage of 80 mg i.m. (Ho 1970); kanamycin appeared in milk (18.4 μg/100 ml) after a 1 g i.m. dose, and approximately 0.05 per cent of the administered drug appeared in the milk per day (M/P ratio 0.4–1.0) (Chyo et al. 1962; Gardner and Rayburn 1982). Tobramycin gives concentrations in milk ranging from 0 to 60 μg/100 ml and persisting for up to eight hours after maternal dosage of 80 mg i.m. (Takase et al. 1975; Lien 1979).

The best available advice about these other aminoglycosides and breast-feeding is that they are probably safe, but like streptomycin they may be dramatically increased in concentration in milk if there is maternal kidney impairment; under such circumstances breast-feeding is contraindicated.

Chloramphenicol

Chloramphenicol blood levels are not usually sufficient to produce the grey baby syndrome but they may harm the bone marrow (Anon 1974; Levin 1975). One study has reported a number of adverse effects including: refusal of breast, falling asleep during feeding, and vomiting after the feed (Havelka and Frankova 1972). Breast-feeding is therefore contraindicated during maternal chloramphenicol treatment (Anon 1974; Havelka and Frankova 1972; Vorherr 1974 a, b).

Clindamycin, erythromycin, and lincomycin

Concentrations of clindamycin and lincomycin were found to be insignificant in breast milk and were without effect on the infant (White 1978). The same type of assessment was also made for erythromycin (White 1978) although earlier studies listed observed milk/plasma ratios varying from 0.5 (Knowles 1965) to 2.5–3.0 (Vorherr 1974 b).

The Mersey Regional Drug Information Service (1981) listed both clindamycin and lincomycin in their 'caution' category and recommended that the infant be monitored for diarrhoea; this is not surprising in view of the association of these two antibiotics with colitis. A similar category has been given for both antibiotics in this present review (Table 23.1).

A Swedish paper (Stéen and Rane 1982) helped to clarify the situation concerning the excretion of clindamycin in breast milk. In a study of five women who were given 150 mg three times daily for at least a week, the milk concentrations of clindamycin ranged from about

1/10 to several times the concentration in plasma. The actual amounts secreted were small; however, the lack of knowledge about the disposition and effects of clindamycin in newborn infants is a strong argument against nursing during clindamycin therapy. A case report has in fact implicated clindamycin in toxicity in a child who had two grossly bloodly stools after the mother had received clindamycin for presumed endometrial infection. Although this case did not prove a causal relationship between clindamycin intake by the mother and the baby's bloody diarrhoea, its reporter suggested a striking similarity between this case and clindamycin enterocolitis (Mann 1980).

Erythromycin was placed in the 'appears safe' category by the Mersey Regional Drug Information Service (1981); the same category has been given in this text (Table 23.2).

Dapsone

This is the standard treatment for leprosy and unavoidably it is often taken by lactating mothers who are afflicted. Little can be done about this since often there are no alternative drugs available and it would be totally impractible to contraindicate breast feeding. None the less, a case report by Sanders *et al.* (1982) of an infant developing a mild haemolytic anaemia during breast-feeding from a dapsone-treated mother presents evidence that this drug should, when possible, only be used with extreme caution in mothers who intend to feed their infants. Earlier reports indicated hypermelanosis, raised serum bilirubin concentration, and raised reticulocyte count in the suckling infants (Hocking 1968; Maegraith and Giles 1971).

Metronidazole

Oral dosage with metronidazole is not normally recommended during lactation. This drug has become a commonly used antimicrobial agent in the treatment of puerperal infections and is found in breast milk in concentrations equivalent to those in maternal serum (Gray *et al.* 1961; Study Group 1978; Erickson *et al.* 1981). Wilson *et al.* (1980) commented that metronidazole is generally contraindicated in the nursing mother, although an earlier review by Thoman (1978) suggested that there was no hazard to the infant but cautioned against the drug's use in high dosage in the nursing mother.

In spite of these controversial views, the general view is that if it is necessary to use metronidazole to control anaerobic infections then the drug should be used but breast-feeding should stop. Trichomonas vaginitis is best treated by local preparations and, since systemic absorption is minimal, this presents no undue hazard in breast-feeding.

Much useful information on the comparative levels of metronidazole in mothers and infants during therapeutic administration for puerperal infection in lactating mothers was generated by Heisterberg and Branebjerg (1983) in Copenhagen. Fifteen mothers and their babies were included in this survey; four of the mothers received 400 mg of metronidazole thrice daily, and 11 mothers 200 mg thrice daily during days 0 to 22 after parturition. The results of this survey is summarized in Tables 23.4 (a) and (b).

Maternal milk/plasma ratios approached one and baby/mother plama ratio was 0.13 to 0.17; infant total clearance of metronidazole was calculated to be about 60 per cent of maternal clearance by body weight and 24 per cent by body surface, independent of dosage, while maximum infant metronidazole intake was estimated to be 3.0 mg/kg/day assuming a daily intake of 500-ml milk.

On the basis of these data, Heisterberg and Branebjerg (1983) recommended avoidance of simultaneous breast-feeding and metronidazole therapy until possible harmful long-term effects on the neonates are known.

Some work has been done on alternative forms of maternal dosage and it was suggested by Moore and Collier (1979) that metronidazole administered to the mother by rectum may not present such a potential hazard to the suckling child. They reported the use of metronidazole rectal suppositories in the prevention of infection following caesarian section. However, metronidazole administered by rectum for the treatment of anaerobic infection will have to be absorbed systemically if it is to be effective, therefore to differentiate between the effective use of oral or rectal metronidazole does not seem to be particularly meaningful.

Withholding breast-feeding for specific periods after maternal metronidazole dosage has also been advocated. For example, Erickson *et al.* (1981) assayed breast milk concentrations of the drug in three women who had been treated with a single 2-g dose for trichomoniasis. Highest concentrations of metronidazole were found within 2 to 4 hours after dosage (2 hours level: 45.8 μg ml[-1]), and they declined over the next 12 to 24 hours (24–48 hour level: 3.5 μg ml[-1]).

This report therefore suggests that, if breast-feeding is withheld for 12 to 24 hours after dosage, the infant will be exposed to a much reduced amount of metronidazole. This exposure would be approximately 9.8 mg over 48 hours, if breast-feeding were withheld for 12 hours, and 3.5 mg over 48 hours if withheld for 24 hours.

Since the single dose of 2-g metronidazole has been

Table 23.4(a)   Concentrations of metronidazole (M) in μg/ml in mothers and infants. Time after dosing/feed in minutes (from Heisterberg and Branebjerg 1983).

Patient	M (mg/day)	Mother Plasma						Milk		Infant Plasma	
		Time	Conc.	Time	Conc.	Time	Conc.	Time	Conc.	Time	Conc.
1	600	120	8.8	270	7.8			270	6.2	120	1.4
2	600	140	3.3	120	1.7			120	1.6	*	—
3	1 200	120	7.8	240	8.5	120	15.5	120	12.9	45	4.9
4	600	120	1.9	0	4.6	30	5.2	30	5.3	60	0.3
5	600	130	6.6	240	5.0			240	5.0	120	0.6
6	600 (twins)	120	4.5	240	3.5	120	5.7	120	5.5	60 / 60	0.6 / 0.5
7	600	120	8.3	240	7.9	120	11.6	120	12.2	60	0.6
8	600	120	1.0	240	1.0	150	3.8	150	4.4	60	0.7
9	600	120	3.4	30	5.6			30	7.0	60	1.1
10	600	120	4.2	240	4.2	180	5.5	180	5.8	60	1.0
11	600	105	5.0	240	3.9	60	6.1	60	3.8	60	0.4
12	1 200	140	3.7	135	13.6			135	13.1	60	0.6
13	600	140	4.3	20	5.3			20	5.8	60	1.2
14	1 200	120	13.5	180	11.7	240	9.6	180	11.6	120	2.3
15	1 200										

* Sample too small to analyse

Table 23.4(b)   Concentrations of metronidazole in breast milk and infant and maternal plasma in μg/ml and milk/plasma and infant/mother plasma ratios (from Heisterberg and Branebjerg 1983).

Metronidazole (mg/day)	Number patients	Concentrations of metronidazole in					
		Mothers				Infants	
		Plasma		Milk		Plasma	
		Mean	Range	Mean	Range	Mean	Range
600	11	5.0	1.0–11.6	5.7	1.6–12.2	0.8	0.3–1.4
1 200	4	12.5	3.7–17.9	14.4	11.6–18.0	2.4	0.6–4.9

Metronidazole (mg/day)	Number patients	Mothers milk/plasma		Baby/mother plasma	
		Mean	Range	Mean	Range
600	11	0.99	0.62–1.25	0.13	0.05–0.23
1 200	4	0.98	0.83–1.13	0.17	0.04–0.32

shown to be as effective in treating *Trichomonas vaginalis* infection as former regimens of 250 mg thrice daily for seven days, the single dose with cessation of breast-feeding offers a possible regimen of choice for the treatment of trichomoniasis in lactating women.

It is worthy of note, however, with all these various recommendations that the presence of metronidazole in milk has not actually been shown to present any hazard to the suckling infant. The strictures that have been recommended are all based on the theoretical possibility that milk levels of the drug which equate with maternal plasma levels may produce harmful effects. In this respect it is significant to note that no adverse reactions to the drug were seen in mothers or infants in the study by Heisterberg and Branebjerg (1983), nor in any other milk study so far reported. The concern seems to depend on an early report from the American FDA which indicated possible low risk of carcinogenesis (Food and Drugs Administration 1976) and on other reports in the literature of mutagenic effects in some strains of bacteria, increased chromosome aberrations in patients after prolonged high dosage, and congenital abnormalities in infants exposed to metronidazole in the first trimester of pregnancy (see summary in Martindale 1982).

Nalidixic acid, nitrofurantoin

These antibacterial agents were included by the Mersey Regional Drug Information Service (1981) in their information letter on drugs and breast feeding.

Nalidixic acid was placed in their 'avoid' category probably in view of one report of haemolytic anaemia in the breast-fed child; the mother was also taking amylobarbitone (Belton and Jones 1965).

Nitrofurantoin has also been placed in the 'avoid' category for G6PD-deficient infants, but in the 'appears safe' category for others.

Sulphonamides are in the 'avoid' category for G6PD-deficient infants, and in the 'caution' category for other infants since they may cause haemolytic anaemia. The Regional Service recommends monitoring the infant for jaundice. One case of haemolytic anaemia was reported by White (1978).

Similar categories for these drugs are given in this present review except that sulphonamides have been retained in their 'contraindicated' group (Table 23.1) due, not only to the potential risk of haemolytic anaemia, but also due to the risk of hyperbilirubinaemic encephalopathy (Anderson 1977; Wilson et al. 1980); in addition prolonged exposure to sulphonamides in milk may cause bacterial resistance (Vorherr 1974b).

Novobiocin

Possible kernicterus was reported in the neonate (Knobden et al. 1973), but the antibiotic has been used to treat infections in infants without untoward effects (Knowles 1965; O'Brien 1974). It is placed in the 'cautionary' category in Table 23.1.

Penicillins and cephalosporins

The concentrations of amoxycillin and five cephalosporins in breast milk were studied by Kafetzis et al. (1981) in 42 lactating mothers. Each mother received one single dose of 1 g of either an orally or intravenously administered antibiotic. Amoxycillin, cephalexin, and cefadroxil were given orally and peak milk concentrations averaged 0.81 ± 0.33 (SD) μg ml^{-1} at 5 hours, 0.50 ± 0.23 μg ml^{-1} at 4 hours, and 1.64 ± 0.73 μg ml^{-1} at 6 hours, respectively. Cephalothin, cephapirin, and cefotaxime were given i.v. as a bolus injection, and peak milk concentrations at 2 hours averaged 0.47 ± 0.14 μg ml^{-1}, 0.43 ± 0.16 μg ml^{-1}, and 0.32 ± 0.09 μg ml^{-1}, respectively.

Milk/serum ratios for each antibiotic showed maximum values of: 0.04 (amoxycillin), 0.14 (cephalexin), 0.02 (cefadroxil), 0.50 (cephalothin), 0.48 (cephapirin), and 0.16 (cefotaxime).

The authors of this report were somewhat vague in interpreting their findings. However, the milk/serum ratios are low for each antibiotic and it is unlikely that side-effects other than sensitization or allergic reactions would occur.

The excretion of cefoxitin, a new semisynthetic cephamycin derivative, in breast milk was investigated by Dubois et al. (1981). Mammary excretion of the drug was examined in 16 nursing mothers. No cefoxitin was detectable in any of the milk samples from 30 minutes to 24 hours after a 1 g intramuscular injection. The authors were careful to point out, however, that their findings did not exclude the possibility of a cefoxitin passage into maternal milk under other conditions, namely when higher doses of cefoxitin were used or after repeated or intravenous injections of cefoxitin. Further work is therefore required to clarify the situation before meaningful recommendations can be given as to the safety of this particular drug in the nursing mother. This cephalosporin has therefore been temporarily placed in the 'cautionary' category of Table 23.1.

Ceftazidime, another new cephalosporin, has a relatively long half-life in the body and this could account for its transfer into milk. However, Blanco et al. (1983) showed that after oral dosage (2 g t.i.d. for five days) drug concentrations in milk stay relatively constant at about 4–5 μg ml^{-1}. There is thus no evidence of progressive accumulation probably because 80 per cent of the drug is excreted unchanged 8 hours after a dose. It has been temporarily placed in the 'cautionary' category of Table 23.1 until more information is available.

Earlier studies on the excretion of penicillins and cephalosporins in milk were reviewed by Wilson et al. (1980); they also show that these antibiotics attain relatively low concentrations in milk, the milk/plasma ratios being in the order of 0 to 0.2.

The Mersey Regional Drug Information Service (1981) listed cephalosporins (e.g. cephalexin) and penicillins (e.g. ampicillin, amoxycillin, penicillin V) in their 'appears safe' category. On the basis of this collective evidence and opinion, the penicillins and cephalosporins (with the exception of cefoxitin and ceftazidime) are placed in the 'appears to be safe' category (Table 23.2).

Povidone-iodine

Postellon and Aronow (1982) reported a case in which a breast-fed infant's serum and urine iodine levels were grossly elevated. The 7½-month-old child's mother had been using a povidone–iodine vaginal gel (Betadine) once daily for six days without douching. Inorganic iodine levels of breast milk were almost 25 times that formed in maternal serum. This report is a further illustration that iodine is contraindicated in nursing mothers and that vaginal iodine preparations cannot be considered benign medications.

Sisomicin

The pharmacokinetics of this antibiotic were examined in 131 obstetrics patients, 11 of whom were breast-feeding mothers (Von Kobyletzki *et al.* 1979). The drug was demonstrated in breast milk in small amounts. When medication was continuous the amount excreted reached 100 μg by the third day. When the medication was discontinued, it disappeared rapidly from the mother's milk (mostly within 24 hours). Until more is known about the disposition of sisomicin in children after ingestion of the drug from breast milk, caution should be exercised during its use in nursing mothers.

Tetracyclines

Tetracyclines may cause mottling of the teeth if absorbed by the infant, but are probably bound by calcium in milk which retards absorption (*British Medical Journal,* 1969; Catz and Giacoia 1972; Knowles 1972); nevertheless it has been recommended by some authorities (Catz and Giacoia 1972; Vorherr 1974 *b*) that tetracyclines should not be given to nursing mothers.

Anticoagulants

Oral anticoagulants are secreted somewhat unpredictably in milk, and bleeding episodes have occurred in infants after surgery or trauma. These drugs should be used with caution (Anderson 1977); some authorities consider breast-feeding to be contraindicated (Catz and Giacoia 1972; Knowles 1974), although warfarin appears to be safe (Baty *et al.* 1976; Orme *et al.* 1977, 1979) (see also below). The indanediones (e.g. phenindione) should be avoided during lactation since they may cause prothrombin deficiency and anaemia (Takyi 1970; Anon 1974; O'Brien 1974). One case was reported in which the suckling infant developed incisional and scrotal haemorrhage after inguinal herniotomy (Eckstein and Jack 1970). Heparin does not pass into milk.

Acenocoumarol

In a study of 20 patients who received acenocoumarol post partum (19 prophylactically after Caesarean section and one therapeutically for deep vein thrombosis) no drug could be detected in breast milk from any mother after at least five days of treatment (Houwert-de Jong *et al.* 1981). The authors suggested that these results indi-

cated that there was no need to advise against breast feeding of full-term healthy infants by mothers taking acenocoumarol for a short time.

Warfarin

Orme and his colleagues (1977, 1979) reported that nursing mothers taking warfarin may safely breast-feed their infants. Warfarin concentrations in the plasma and breast-milk of 13 mothers were measured; less than 0.08 μmol of warfarin per litre of breast milk was found in each instance at up to 10 days after delivery. Seven of the mothers were breast-feeding their infants; in none of them was warfarin detected in the plasma. Furthermore, in three infants studied, the British Corrected Ratio (BCR) was appreciably less than the maternal value. The range of BCRs in the three infants (1.1–1.8) was well within the expected range for newborn infants who had not been given vitamin K at birth. The plasma warfarin concentrations in the 13 mothers (1.6–8.5 μmol l^{-1}; 0.5–2.6 μg ml^{-1}) were within the range of patients with a prothrombin time within the therapeutic range.

The authors warned and emphasized that their results applied only to warfarin and that their conclusions should not be extrapolated to other coumarin anticoagulants or to phenindione.

Orme and his colleagues did not measure the plasma or breast-milk concentrations of warfarin metabolites since they argued that the anticoagulant effect of warfarin alcohols (the only active metabolites) is some hundred-fold less than warfarin and therefore the possible presence of these substances would not appear to be therapeutically important. They suggested also that, on a physicochemical basis, it would seem unlikely that warfarin alcohols would pass into breast milk if warfarin itself did not. They suggested that the early reports of warfarin being present in breast milk utilized non-specific warfarin assay methods which monitored inactive warfarin metabolites in the milk (Orme *et al.* 1979).

Confirmation of this opinion was given by McKenna *et al.* (1983) who reported their findings in two mother-infant pairs studied for any evidence of a reduction in the one-stage Quick prothrombin activity or factor II or VII/X activities in the infants while being breast-fed by mothers who were receiving therapeutic doses of warfarin sodium. They showed that there was neither an immediate nor a delayed biologic effect on coagulation tests. They therefore also recommended that mothers receiving warfarin sodium be allowed to breast-feed their infants.

Warfarin is therefore placed in the 'appears to be safe' category (Table 23.2).

Anticonvulsants

There have been relatively few reports about the concentration of anticonvulsants in human milk and therefore a study from Japan by Kaneko *et al.* (1979) is noteworthy. These workers investigated the relative concentrations of diphenylhydantoin (phenytoin), phenobarbitone, primidone, carbamazepine, and ethosuximide in maternal milk and serum.

The study was carried out on nine patients with epilepsy who had received several anticonvulsant drugs; milk and serum samples were collected simultaneously from the day 3 to 32 after delivery. Women who took their medication irregularly before and after their delivery, or those who were given prolactin were also included in the study. Serum and milk levels of the drugs were measured using gas–liquid chromatography and enzyme immunoassay.

Table 23.5 shows the mean values of, and the relationship between, the serum and milk concentrations of the five anticonvulsants and the range of concentrations in which they were found. Serum levels of drugs were somewhat lower than therapeutic levels reported in the literature (see, for example, Penry and Newmark 1979); this was especially so for phenytoin, primidone, and ethosuximide and may reflect poor compliance with medication instruction during pregnancy and the puerperium or an effect of pregnancy on drug metabolism or distribution.

Kaneko and his colleagues did not make any firm recommendations regarding the safety or otherwise of allowing their patients to breast-feed their babies. Specific comment was made, however, that phenobarbitone reached a concentration of 33 μg ml^{-1} in milk of one of the mothers; this would represent a drug intake of about 1.6 mg daily in the feed, and in that case they recommended bottle-feeding.

Previous advice given in the 2nd edition of this book (D'Arcy 1979) dealt only with three anticonvulsant drugs: phenytoin sodium, primidone, and phenobarbitone, and it was recommended that caution be exercised in that the infant should be monitored regularly for signs of drug intoxication if it were breast-fed. This advice still stands; the Japanese study, although interesting, is probably too small to be useful as a sample; furthermore ethnic differences in enzyme induction may present a major problem in extrapolating data from Japanese studies to the general situation.

There is also one additional consideration concerning phenytoin resulting from the Japanese study: there is often decreased absorption of phenytoin during the last trimester of pregnancy and blood concentrations of the anticonvulsant may fall below the therapeutic range (5–20 μg ml^{-1}); there is also an increased volume of distribution of the drug and haemodilution. Increased dosage of phenytoin may therefore be required to maintain anti-epileptic protection. Increased levels of the drug may be expected in milk unless, after delivery, the dosage of phenytoin is appropriately reduced.

Phenobarbitone is excreted in breast milk in sufficient quantity to induce the infant's hepatic enzymes. Infants may become sleepy and one death has been reported (Juul 1976). However, other studies which have been summarized by Wilson *et al.* (1980) are also relevant; they indicate that barbiturate levels in milk may be as low as 1.5 per cent and in normal anti-epileptic doses in the mother this should have no clinical effect on the infant except in cases of undue sensitivity to the drug.

With phenytoin sodium there is also a chance for enzyme inhibition in the nursing infant. As with phenobarbitone, this anticonvulsant appears in milk in concentrations such that the infant would receive more than 1 per cent of the total daily maternal dose (Vorherr 1974 *a, b*). Wilson's survey of the literature does not, however, provide any firm evidence for contraindicating phenytoin or phenobarbitone during lactation or vice versa. Nor indeed is there any such evidence for contraindication in the study by Stéen *et al.* (1982) who investigated the quantitative transfer of phenytoin (dose range 200–400 mg daily) into breast milk in six epileptic mothers who were nursing their infants (1–3 months old).

Mean maternal plasma phenytoin-levels ranged from

Table 23.5 Concentrations of anticonvulsant drugs in human serum and milk samples*

Drug	Mean serum concn (μg ml^{-1} ± s.d.)	Mean milk concn (μg ml^{-1} ± s.d.)	Milk/serum concn %
Phenytoin	4.5 ± 1.4 (2.1–5.7)	0.8 ± 0.3 (0.5–1.4)	18
Phenobarbitone	19.3 ± 14.5 (2.5–42)	10.4 ± 10.8 (0.5–33)	46
Primidone	2.8 ± 2.5 (0.8–15.7)	2.3 ± 2.2 (0.5–6.7)	81
Carbamazepine	4.3 ± 1.7 (3.2–6.2)	1.9 ± 1.6 (0.8–3.8)	40
Ethosuximide	29.3 ± 8.0 (18–39)	21.3 ± 2.8 (18–24)	79

Figures in parentheses are range of values.

* Source: modified from Kaneko *et al.* (1979).

12.8 to 78.5 μmol l^{-1} in plasma and 0.8 to 11.7 μmol l^{-1} in milk (Table 23.6). The mean milk/plasma concentration was 0.13. Phenytoin plasma levels were detectable in two infants at 0.72 and 0.46 μmol l^{-1}. The calculated daily phenytoin dose to the infant (0.03–0.47 mg kg^{-1}) was well below the therapeutic dose for infants (10 mg kg^{-1}) and no infant in this series showed any signs of toxicity due to phenytoin.

On the basis of their experiences Stéen and his colleagues do not contraindicate breast-feeding because of phenytoin treatment, nor do they think that there is any need to time nursing periods since plasma and milk levels are fairly constant.

Rane and Tunell (1981) studied maternal and infant's plasma concentrations of ethosuximide as well as the milk concentration in one epileptic mother and her child during a period of 4.5 months after delivery. The milk concentration of the anticonvulsant was similar to that in the maternal plasma on the third day after delivery. During the following two months the average milk/maternal plasma concentrations ratio was 0.8. These data are further evidence that ethosuximide, together with the other anticonvulsant drugs carbamazepine, phenobarbitone, phenytoin, primidone, and sodium valproate, should be used with caution during breast-feeding (Table 23.1) and that, if used, the infant should be monitored regularly.

The Mersey Regional Drug Information Service (1981) adopted a similar classification for barbiturates (high-dose) and carbamazepine and cautioned that infants should be monitored for sedation or drowsiness and poor feeding. Primidone, phenytoin, and ethosuximide are not included in their list, but sodium valproate is, under the 'caution' classification, with the advice that the infant be monitored for liver and platelet dysfunction. Sodium valproate is therefore included on the 'cautionary' list in Table 23.1.

An excellent and comprehensive research review by Nau et al. (1982) gave relevant pharmacokinetic data on anticonvulsant drugs widely used during pregnancy and the neonatal period. This review gives some direct answers to some long-standing questions about the pre- and postnatal toxicology of this class of compounds.

Antidiabetic agents

Tolbutamide was found in low concentration (0.3–1.8 mg/100 ml) in the milk of two of five subjects who had taken a single dose of 500 mg. These levels represent 9–40 per cent of maternal serum levels of the drug. Effects of ingestion of the drug in milk were not studied (Moiel and Ryan 1967). This single and dated study has been cited erroneously in subsequent reviews to contraindicate tolbutamide in breast-feeding; there is insufficient data to support such a recommendation although breast-fed infants of mothers taking all oral hypoglycaemic agents should be carefully monitored.

It has not been established whether glibenclamide is transferred to human milk. However, other sulphonylureas have been found in milk and there is no evidence to suggest that glibenclamide differs from the group in this respect (ABPI 1983–84), and therefore all oral hypoglycaemics are placed in the 'cautionary' category of Table 23.1.

Table 23.6   Plasma and milk concentrations of phenytoin and 4-OH-phenytoin for mother–infant pairs*

| | Phenytoin | | | | | 4-OH-Phenytoin | | |
| | Maternal mean plasma concentration | AUC in plasma | Milk mean concentration | Milk/plasma concentration | Infant mean plasma concentration | Maternal mean plasma concentration[a] | Milk mean concentration[a] | Infant mean plasma concentration |
Pair	(μmol l^{-1})	(μmol l^{-1} × h)	(μmol l^{-1})	ratio	(μmol l^{-1})	(μmol l^{-1})	(μmol l^{-1})	(μmol l^{-1})
1	78.5	430	11.7	0.15	0.72	1.0/11.6	0.4/1.0	0.4
2	13.2	91	1.3	0.09	NM	0.7/7.5	0.9/0.7	—
3	64.5	379	8.9	0.14	NM	—	—	—
4	43.2	305	7.7	0.18	NM	—	—	—
5	46.9	338	8.0	0.17	0.46	—	—	—
6	12.8	58	0.8	0.06	NM	—	—	—

[a] Unconjugated to conjugated metabolite ratio.

AUC, area under the plasma concentration versus time curve; NM, not measurable.

* Source: Stéen et al. (1982).

Antithyroid agents

Antithyroid drugs are usually contraindicated during breast-feeding due to the presumed risk of potential antithyroid effects on the infants. Thiouracil, for example, is known to concentrate in breast milk and is contraindicated in breast feeding as also are carbimazole and methimazole (Wilson *et al.* 1980).

Carbimazole and methimazole

Carbimazole has been associated with danger of hypothyroidism or goitre in suckling infants and its use is contraindicated in the nursing mother (Takyi 1970; O'Brien 1974; APBI 1983–84). Methimazole is the active metabolite of carbimazole (rapid and complete biotransformation occurs), and it is of interest to note that Johansen *et al.* (1982) have reported on the excretion of methimazole in human milk.

Methimazole was monitored in the blood and milk of five lacating women after oral administration of carbimazole (40 mg). The mean milk/serum ratio was 0.98, and 0.14 per cent of the dose administered was excreted in milk over the eight-hour period following dosing. Although no specific recommendation was given by the authors concerning methimazole, it is still maintained within the contraindicated table (Table 23.1), until further information on the effects of antithyroid drugs on the developing child are known.

Thiouracil and propylthiouracil

Thiouracil was found to be concentrated in breast milk in a 1944 study in two lactating women (Williams *et al.* 1944), and because of danger of hypothyroidism or goitre in the infant, and in rare cases, agranulocytosis, it is contraindicated in breast-feeding (Williams *et al.* 1944; Knowles 1965). Propylthiouracil has also been contraindicated in breast-feeding (Williams *et al.* 1944) but more recent evidence has suggested that propylthiouracil may not suffer from this restriction.

Nine lactating mothers were given propylthiouracil, 400 mg by mouth (Kampmann *et al.* 1980); at 1.5 hours the mean serum concentration of the drug reached 7.7 μg ml^{-1}, falling to 3.9 μg ml^{-1} at 4 hours. The concentration of propylthiouracil in milk was 0.7 μg ml^{-1} at 1.5 hours and 0.5 μg ml^{-1} at 4 hours. The mean volume of milk collected over 4 hours was 184 ml and the mean total of drug excreted into milk was 99 μg (range 27–308 μg) or 0.025 per cent (range 0.007–0.077 per cent) of the ingested dose. One infant was followed-up for five months while the mother received 200–300 mg propyl-

thiouracil daily; there were no changes in the infant in serum thyroxine, tri-iodothyronine, thyroid stimulating hormone, and T_3-resin-uptake test throughout that time.

Kampmann *et al.* (1980) suggested that lactating mothers on propylthiouracil may continue breast-feeding if they wish, though close supervision of the infant is advised. Similar advice was given by Thoman (1978) on the basis of a literature review.

Nonetheless, potential hazard is so evident that on balance it is still recommended in this present review that these drugs should not be given to mothers who are breast-feeding or alternatively that breast-feeding should cease (Table 23.1). It would be wise to err on the side of safety and advise the restrictions at least until additional evidence is forthcoming to allow another assessment of the situation.

Cancer chemotherapy

Breast-feeding is usually contraindicated in any woman receiving antineoplastic or immunosuppressant drugs (Catz and Giacoia 1972; O'Brien 1974; Vorherr 1974*b*; Savage 1976; Lewis 1978; Beeley 1981; Chaplin *et al.* 1982).

Cardiovascular drugs

Beta-blockers

Relatively little is known about the excretion of beta-blocking drugs and other antihypertensive agents into breast milk. New evidence has suggested, however, that atenolol, metoprolol, mepindolol, nadolol, oxprenolol, propranolol, sotalol, and timolol should be placed in the 'cautionary' category.

Atenolol and metoprolol

Liedholm *et al.* (1981) assessed the passage from serum to milk of atenolol in seven hypertensive, lactating women, and that of metoprolol in three healthy women who agreed to take this beta-blocker at cessation of lactation. For both drugs, the concentration in milk was higher than in maternal serum at every time studied and the resulting AUC (area under the concentration–time curve) values were 1.5–6.8 times (atenolol) and 2.6–3.7 times (metoprolol) greater in milk than in maternal serum. They calculated, however, that the dose to the infant following a meal at the time of maximum maternal drug concentration would not exceed 0.13 mg atenolol or 0.05 mg metoprolol. None of the atenolol-exposed infants in this study showed any symptoms of beta-blockade.

Liedholm *et al.* (1981) commented therefore that it was unlikely that, unless renal (atenolol) or hepatic (metoprolol) function in the infant was pronouncedly impaired, breast-feeding need be interrupted due to maternal medication with normal doses of either drug.

Sandstrom (1980) also examined metoprolol and he found that this beta-blocker attained levels in breast milk that were on average 3.5 times higher than those of maternal plasma. He calculated that an infant consuming 1 litre of breast milk per day would receive about 2 per cent of the normal daily dose of the drug for a hypertensive adult. Such an amount was considered to be unlikely to induce adverse reactions in the normal breast-fed child.

Thorley and McAinsh (1983) compared the breast-milk elimination of atenolol (with low lipid solubility and protein binding) 100 mg daily with that of propranolol (with high lipid solubility and protein binding) 40 mg b.i.d. in 10 women in the puerperium.

The pH of the milk samples had a mean of 7.54. Individual drug concentrations in milk varied between 380 and 1040 ng ml^{-1} for atenolol and 14 and 36 ng ml^{-1} for propranolol. The mean milk-to-plasma ratios were 1.3:1 for atenolol and 2.0:1 for propranolol. There was no sign of clinical β-blockade in the babies.

A neonatal milk intake of 500 ml daily would give an infant about 0.3 mg of atenolol and 0.01 mg of propranolol daily. These authors concluded that both drugs may be used safely in lactating women but the choice of drug should be made on the basis of clinical observation.

Mepindolol

Mepindolol was also shown to be excreted in breast milk in five mothers (Krause *et al.* 1982) after one and five daily doses of mepindolol sulphate (20 mg). The average plasma/milk drug concentration ratio was 2.6 $\pm$ 1.6. Except for one baby, plasma levels were always below the limit of detection. The authors, however, concluded that, due to the high potency of the drug, discontinuation either of drug treatment or breast feeding is recommended at least in those patients receiving high doses of the beta-blocker. No drug-related effects were found in the babies in this study.

Nadolol

Devlin *et al.* (1981) collected simultaneous serum and milk samples over a 10-day period from 12 normotensive, lactating subjects who took 80-mg nadolol daily for five days. Steady-state serum concentrations of the drug were attained in three days; milk concentrations of nadolol were much higher than maternal serum concen-

trations by day 3 and throughout the remainder of the study. The mean ($\pm$ SE) steady-state levels of nadolol in milk (356.9 $\pm$ 40.4 ng ml^{-1}) were almost five times higher than corresponding levels in maternal serum (77.3 $\pm$ 6.9 ng ml^{-1}).

The authors of this report estimated that a 5-kg nursing infant would consume about 2–7 per cent of the daily adult therapeutic dose of nadolol; they therefore recommended that caution should be exercised in its use in lactating patients.

Oxprenolol

Fidler *et al.* (1983) measured the excretion of oxprenolol in breast milk of nine hypertensive postpartum patients receiving oral oxprenolol 80 mg twice daily. The mean milk-to-maternal plasma ratio of oxprenolol was 0.29 ($\pm$ 0.14, SD), which indicated that the total quantity of drug that would be ingested by a fully breast-fed infant was unlikely to be therapeutically significant.

Propranolol

Reports on the partitioning of propranolol into breast milk gave varying figures. Bauer *et al.* (1979), for example, found a milk/plasma ratio of between 0.4 and 0.7 when propranolol was taken at a dosage of up to 240 mg daily by the mother. From such data they concluded that milk concentrations of the drug were sufficiently low that the nursing infant would consume only 1 per cent of the therapeutic dose for children. Similar conclusions were reached by Lewis and her colleagues (1981). However, Smith *et al.* (1983) found milk/plasma ratios of 0.33 to 1.65 in three lactating women who were receiving chronic propranolol treatment for hypertension and who were each in the first week of breast-feeding. The half-life of elimination of propranolol from breast milk was 6.5 $\pm$ 3.4 hours (mean $\pm$ SD) which was significantly longer than the half-life of elimination of propranolol from plasma (2.6 $\pm$ 1.2 hours (mean $\pm$ SD)).

Based on their data, Smith and his colleagues calculated that the suckling infant would ingest less than 0.1 per cent of the maternal dose of propranolol and therefore would be unlikely to be at risk from the effects of this beta-blocker during continued nursing.

Further support for these views was given by two case reports, one by Taylor and Turner (1981) and another by Brodrick (1982) of mothers who received 40 mg daily and 40 mg twice daily of propranolol, respectively. These reports indicated minimal transfer into breast milk. Taylor and Turner (1981) estimated a daily intake in breast milk of propranolol of about 3 μg. Based on these findings, propranolol is unlikely to give rise to overt pharmacological effects in the child; however, Brodrick (1982) felt that it would be wise to monitor serum glucose levels in newborn infants; this caution refected the opinions of earlier workers who advocated regular examination of the suckling infant for signs of beta-blockade (e.g. bronchospasm. hypotension, congestive cardiac failure, and hypoglycaemia) if the nursing mother was being treated with propranolol (Levitan and Manion 1973; Fidler 1974; Karlberg *et al.* 1974; Anderson and Salter 1976; Savage 1976).

Sotalol

In one study carried out in Belfast (O'Hare *et al.* 1980) the concentration of sotalol was measured in samples of maternal plasma and breast milk from five nursing mothers. High sotalol concentrations were found in breast milk (mean plasma: milk ratio was 1:5.4). After calculating the dose that an infant would ingest in breast milk, it was found that this overlapped on a mg/kg/day basis with the therapeutic dose in adults. These data would therefore indicate that beta-blockade could occur in the infant; however, the authors of the report were unable to demonstrate this in the infants in their study.

With all these beta-blockers, however, caution in the use should be exercised when the mother is breast-feeding; the infant should be monitored for signs of beta-blockade, notably for bradycardia and hypoglycaemia. All these beta-blockers have therefore been placed in a 'cautionary' category (Table 23.1).

Timolol

Two recent publications have provided some data on the excretion of timolol in breast milk. In the first of these Fidler *et al.* (1983) measured maternal plasma and milk levels of timolol during a dosage regimen of 5 mg thrice daily by mouth in nine hypertensive postpartum patients. The mean milk-to-plasma ratio was 0.80 ($\pm$ 0.21 SD). The authors concluded that the total quantity of drug ingested by a fully breast-fed infant was unlikely to be therapeutically significant.

In the second report, Lustgarten and Podos (1983) described how a 34-year-old nursing mother with elevated intraocular pressure voluntarily and without their knowledge applied 0.5 per cent timolol maleate twice daily to her right eye. A milk sample 1.5 hours after administration of drug showed a much higher level of timolol (5.6 ng ml^{-1}) than a plasma sample drawn at the same time (0.93 ng ml^{-1}). The patient's milk sample 12 hours after her last timolol dose contained 0.5 ng ml^{-1} of timolol.

These clinicians warned that ophthalmic timolol should be used with caution by nursing mothers and only when it was absolutely necessary. If such therapy is warranted, the infant should be observed carefully for signs of beta-blockade. They were guided into giving this advice by an earlier report in which an 18-month-old child suffered several spells of apnoea when treated with a 0.25 per cent ophthalmic solution of timolol (Williams and Ginther 1982). They calculated that the blood level of timolol in the latter case was only about four times higher than the potential maximal dose that could be ingested by a suckling infant.

Captopril

New evidence has also been documented for the anti-hypertensive agent, captopril; this was studied by Devlin and Fleiss (1981). They examined blood and milk concentrations of captopril in 12 lactating, normotensive subjects after the administration of 100-mg captopril tablets on a three times daily regimen for a total of 73 days. Peak blood concentrations of captopril after the seventh tablet were 713.1 $\pm$ 140.6 ng ml^{-1} (mean $\pm$ SE) as compared with the average milk concentration of 4.7 $\pm$ 0.7 ng ml^{-1}. Time to peak blood concentration averaged 1.1 $\pm$ 0.2 hours, while that of milk was 3.8 $\pm$ 0.6 hours.

Assuming a milk production of 1 litre per day, and assuming a concentration of 4.7 ng ml^{-1} free captopril in milk, the amount of free captopril in the daily maternal milk (4.7 μg) would be a miniscule amount (0.002 per cent) of the 300-mg oral daily adult dose of the drug. On a weight basis, a 5-kg infant would consume a maximum of 0.9 μg kg^{-1} free captopril in a 24-hour period. It is on this evidence that captopril has been included in the 'appears to be safe' category (Table 23.2).

Clonidine

No problems are anticipated when the nursing mother takes *Dixarit* (25 μg clonidine hydrochloride per tablet) for the prophylaxis of migraine or recurrent vascular headache (Boehringer Ingelheim 1976). However, the suckling infant should be monitored carefully when higher doses of this drug are taken by the mother as, for example, with the antihypertensive formulations, *Catapres* (0.1 or 0.3 mg clonidine hydrochloride per tablet) or *Catapres Perlongets* (0.25 mg clonidine hydro-chloride per sustained-release capsule) (Boehringer Ingelheim 1976).

Digoxin

Levy *et al.* (1977), from Israel, cited five women with

rheumatic heart disease who were on long-term digoxin therapy and who wanted to nurse their babies. Samples of milk and maternal blood were taken at the same time at least six hours after the last digoxin dose and drug concentrations were measured by radioimmunoassay.

The results of these measurements are summarized in Table 23.7 and show, in most cases, that the digoxin concentrations for serum and milk were similar; in no case was the digoxin concentration in milk higher than that in the serum.

Levy and his colleagues commented, however, that, owing to the large volume of distribution of digoxin in the body, total daily excretion of digoxin in mothers with therapeutic serum digoxin levels would not exceed 1 to 2 μg; this, in their view, was not a sufficient amount to affect the child. A similar view was expressed by Wilson et al. (1980).

In a more recent study, Reinhardt et al. (1982) examined the transfer of digoxin to milk in 11 nursing mothers after intravenous or oral administration of a single dose of 0.5 to 0.75 mg of digoxin. Their data indicated a milk-to-serum drug ratio of 0.6 to 0.7. Using kinetic analysis it was estimated that, even in the case of long half-lives, only about 3 per cent of the therapeutic drug levels would be reached in the child. This supports the general view of safety for suckling infants whose mothers are treated with appropriate doses of digoxin (Table 23.2).

Disopyramide

A case report on a West Indian woman who was breast-feeding while receiving disopyramide (200 mg three times daily) detailed the presence of disopyramide in breast milk in a similar concentration to plasma. The milk/plasma ratio was 0.9 ± 0.17 (SD) in paired samples taken on the fifth to eight day of treatment. The estimated dose ingested by the infant was less than 2 mg/kg/day. The milk/plasma ratio of the active N-monodesalkyl metabolite was 5.6 ± 2.9 (SD) and was found in milk in concentrations similar to those of

Table 23.7   Digoxin concentrations in serum and milk

Case no.	Digoxin concentration (ng ml^{-1})	
	Serum	Milk
1	0.6	< 0.5
2	1.1	0.5
3	0.7	0.7
4	0.7	0.5
5	1.2	1.0

Source: Levy et al. (1977).

disopyramide. Parent drug or metabolite were not detected in the infant's serum and no adverse effects were noted in the infant. The authors felt, however, that close observation of the baby and measurement of both disopyramide and its active metabolite in breast milk and infant plasma was prudent pending further investigation (Barnett et al. 1982). Due to this latter recommendation disopyramide is provisionally placed in the caution section of Table 23.1.

Hydrallazine

Liedholm et al. (1982) in a study involving six mothers treated with hydrallazine and atenolol for hypertension of pregnancy examined the concentrations of hyrallazine in the breast milk of one patient. They calculated that the dose per milk feed of 75 ml would not exceed 0.013 mg and therefore that breast-feeding would not result in a clinically relevant concentration in the infant. This drug has initially been placed in the 'appears to be safe' table (Table 23.2); however, further studies involving more mothers are required before final recommendations can be made.

Methyldopa

Methyldopa appears in breast milk and the use of the drug in women who are nursing their newborn infant requires that anticipated benefits be weighed against possible risks (ABPI 1983–84). It is therefore placed in the 'use with caution' category (Table 23.1).

Mexiletine

Timmis et al. (1980) measured two paired milk and plasma samples and found mexiletine milk/plasma concentration ratios of 2.0 and 1.1, which is in good agreement with a later study by Lewis et al. (1981). The latter workers obtained 12 paired samples of breast milk and blood (collected 2–5 days postpartum), from a 30-year-old woman who was on 8-hourly propranolol (20 mg) and mexiletine (200 mg).

Propranolol was detectable in nine of the samples of blood (maximum concentration 4.7 ng ml^{-1}) and in only four of the milk samples (maximum concentration 5 ng ml^{-1}).

Mexiletine was measurable in all samples, and the milk/plasma ratio varied between 0.78 and 1.89 (mean 1.45). However, the authors commented that the large volume of distribution of mexiletine made it unlikely that the small dose received from the milk would be detrimental to the health of the infant. Assuming an average daily intake of 0.5 litres of milk and a maternal plasma mexiletine concentration of 2.0 μg ml^{-1}, they calculated that it was unlikely that the infant would receive more than 1.25 mg of mexiletine in any 24-hour period.

It is on the basis of this evidence that mexiletine is placed in Table 23.2 in the 'appears to be safe' category.

Procainamide

Pittard and Glazier (1983) reported on procainamide excretion in the milk of a 30-year-old mother who was prescribed the drug in a pregnancy complicated by premature ventricular contractions. She was given procainamide 375 mg orally four times a day and this was later increased to 500 mg q.i.d. The raised dosage was continued postpartum while she was breast-feeding her baby. Mean ($\pm$ SD) concentrations of procainamide (1.1 $\pm$ 0.6 mg l^{-1}) and its primary active metabolite N-acetyl-procainamide (1.6 $\pm$ 0.8 mg l^{-1}) were consistently lower in maternal serum than in simultaneously collected milk specimens. Mean ($\pm$ SD) milk to serum for PA and NAPA was 4.3 $\pm$ 2.4 and 3.8 $\pm$ 2.4, respectively.

Despite these M:P ratios, which suggested accumulation of the drug and its metabolite in milk, the authors estimated that amounts of drug and metabolite available to the nursing infant were small and that maximal exposure would not be expected to yield clinically significant plasma concentrations in the infant (< 1 mg l^{-1} total active drug). Indeed no adverse effects were reported on this breast-fed child, and no special precautions are listed by the manufacturer in breast-feeding. Procainamide is listed provisionally in the 'cautionary' category of Table 23.1.

Reserpine

Lethargy, diarrhoea, and sufficient nasal congestion to interfere with the infant's respiration were reported in suckling infants whose mothers were treated with reserpine (Knobden et al. 1973; Anon 1974; Vorherr 1974b). Reserpine may also produce galactorrhoea (O'Brien 1974). It is therefore placed in the 'contraindicated' category in Table 23.1.

Verapamil

The concentration of this anti-arrhythmic agent in milk and serum was determined in samples taken within the same hour, 3–5 days after delivery from a 25-year-old woman treated with verapamil (240 mg daily) for paroxysmal supraventricular tachycardia (Andersen 1983). The average concentration of verapamil in milk was found to be 23 per cent of that in serum, and the total excretion in milk was less than 0.01 per cent of the administered dose. The serum concentration of verapamil in the child was 2.1 ng ml^{-1} during the treatment, and verapamil could not be detected (concentration less than 1 ng ml^{-1}) in serum from the child 38 hours after the last maternal verapamil dose.

Anderson concluded that this presence of verapamil in milk did not offer any other risk to a normal child than the always existing possibility of allergy. This drug is therefore provisionally placed in the 'appears to be safe' category in Table 23.2.

Corticosteroids

Theoretically corticosteroids, taken by the mother while breast-feeding, could suppress the infant's growth and interfere with endogenous corticosteroid synthesis or produce other unwanted effects in the infant related to the pharmacological and toxicological spectrum of these drugs. No adverse effects have been reported but some authorities have advised that breast-feeding be stopped if the mother is taking large doses of corticosteroids (Catz and Giacoia 1972; Anon 1974; Knowles 1974; Vorherr, 1974 b). However, some early work reviewed by Wilson et al. (1980) suggested that a very small dose of steroid is delivered to the infant via milk when the mother receives conventional doses of prednisone or prednisolone. Evidence supporting this view was given by Sagraves et al. (1981) in their case study of a 26-year-old breast-feeding mother who was taking Deltasone (prednisone) 20 mg q.d. for systemic lupus erythematosus.

Milk, serum, and urine samples were collected 3 weeks postpartum at intervals after the morning dose, and analysed by HPLC for prednisone and prednisolone. Radioimmunoassay was also utilized for prednisolone detection.

Prednisone was detected in milk within 45 minutes, with its metabolite, prednisolone, being noted 2 hours after dosage; highest concentrations of both steroids were noted in milk 2 hours after dosage, with a prednisone concentration of 69.3 ng ml^{-1} and a milk/plasma ratio of 0.67, while the prednisolone concentration was 32.7 ng ml^{-1} with a ratio of 0.15.

Average milk concentrations over the eight-hour collection period were 41.5 ng ml^{-1} for prednisone and 14.5 ng ml^{-1} for prednisolone. The authors did not comment on the safety or otherwise of these steroid contents in milk to the suckling infant, nor indeed did these studies give any information on the long-term effects of such exposure.

In the absence of further definitive evidence, corticosteroids have been retained in the 'cautionary' category (Table 23.1); the Mersey Regional Drug Information

Service (1981) gave a similar recommendation for high dosage of corticosteroids (e.g. > 10 mg prednisolone daily), and advised that the infant be monitored for adrenal suppression. They suggested for asthmatic patients, that the use of aerosol formulations (e.g. *Becotide, Bextasol*) may avoid any possible problems.

Diuretics

Hydrochlorothiazide and other thiazides

Diuretics can cause a decrease in milk production, and have at times been used to suppress lactation in place of oestrogens (Healy 1961; Catz and Giacoia 1972; Knowles 1972). Apparently no adverse effects have been reported in the infant due to diuretics in milk (Catz and Giacoia 1972), but the possibility of idiosyncratic reaction to sulphonamide-derivative diuretics (e.g. thiazides) should be borne in mind (Knobden *et al.* 1973), although it is doubtful whether significant amounts of thiazide diuretics appear in milk. Werthmann and Krees (1972), for example, demonstrated that, in 11 lactating women who were each given a single 500-mg dose of chlorothiazide, the concentrations of the diuretic in milk samples taken one, two, and three hours after the dose were all < 1000 ng ml^{-1}. They concluded that the amount of chlorothiazide that could be transferred from a lactating mother to her breast-fed infant was negligible.

With chlorothiazide and hydrochlorothiazide the manufacturer's recommendation is that the patient should stop nursing if the use of either drug is deemed to be essential (ABPI 1983–84). Miller *et al.* (1982) do not agree with this recommendation; they showed, in a study on hydrochlorothiazide disposition in a mother and her breast-fed infant, that therapeutic doses of hydrochlorothiazide are not excreted into milk at concentrations great enough to produce any significant intake in the suckling child. In their study there was a mean milk hydrochlorothiazide concentration of about 80 ng ml^{-1}. Blood specimens from the baby showed no detectable levels of the diuretic (detection limit 20 ng ml^{-1}), and serum electrolyte concentrations, blood glucose concentrations, and blood urea nitrogen levels were normal in the child.

In the light of these two studies it is difficult to support the manufacturer's recommendation to contraindicate breast-feeding; these diuretics are therefore placed in a 'use with caution' category in Table 23.1.

Chlorthalidone

In nine women given this diuretic (50 mg/day) for control of oedema in toxaemia of pregnancy, the drug was found in amniotic fluid, in maternal and cord blood at delivery, and in milk 3 days after delivery (Mulley *et al.* 1978).

Concentrations of chlorthalidone in maternal blood were consistently higher than in other fluids. With one exception, drug concentrations in cord blood were 13 to 16 per cent of maternal blood concentrations at delivery. Concentrations in amniotic fluid and milk were similar and were consistently lower than those in cord blood. The authors of the report calculated that a 3.5-kg baby would have about 250 μg of chlorthalidone in its blood at birth and, if breast-fed, would receive an additional 180 μg daily. They suggested that intake of this diuretic could cause haemoconcentration in the child and therefore recommended stopping breast-feeding for a few days postpartum. In respect of this advice it is particularly important to bear in mind the long half-life of chlorthalidone in adults (about 60 hours).

This diuretic is placed in the 'contraindicated' category in Table 23.1.

Dopamine antagonists

Dopamine antagonists are used to augment lactation; they owe this action to the stimulation of prolactin release from the pituitary. Three drugs are currently in use: domperidone, metoclopramide, and sulpiride, and current studies have largely been directed towards establishing their safety to the breast-fed child.

Domperidone

Domperidone raises the mean serum prolactin level in normal women from 8.1 to 110.9 ng ml^{-1} after one 20-mg dose (Brouwers *et al.* 1980). In a pilot study to assess safety for the suckling infant, Hofmeyr and van Iddekinge (1983) showed that domperidone was secreted in considerably smaller amounts in breast milk relative to the therapeutic dose than were either metoclopramide or sulpiride (Table 23.8). It is provisionally placed in the 'cautionary' category of Table 23.1.

Metoclopramide

Kauppila *et al.* (1984) reported that stimulation of pituitary prolactin release to improve milk secretion was accomplished by the administration of 10-mg metoclopramide t.i.d. for two weeks to 18 mothers in early and late puerperium.

Concentrations of metoclopramide were constantly higher in breast milk than in the plasma; peak maternal plasma levels occurred 2–3 hours after the administration of the drug followed by peak milk concentration. Plasma

Table 23.8 Dopamine antagonist secretion in human breast milk after oral administration (from Hofmeyr and van Iddekinge 1983)

Drug*	Mean levels (ng/ml)	
	Serum	Breast milk
Metoclopramide	68.5 (n = 10)	125.7 (n = 10)
Sulpiride		970 (n = 20)
Domperidone	10.3 (n = 5)	2.6 (n = 30)

* Metoclopramide 10 mg, with sampling 2 h after a single dose. Sulpiride 50 mg twice daily for 7 days with sampling 2 h after a morning dose. Domperidone 10 mg 8-hourly for 4 days with serum sampling 1.75–3 h after a dose and sampling of all milk.

and milk concentrations increased on the second and third treatment days and not significantly thereafter. The plasma half-life of the drug varied between 2 and 5 hours.

Although the estimated exposure of the new-born to metoclopramide was about 5 per cent of the recommended daily dose for children, the safety of this level must not be accepted unquestioningly as elimination of metoclopramide may be slower in the newborn child due to immaturity. Metoclopramide is placed in the 'cautionary' category of Table 23.1.

Sulpiride

An initial suggestion of infant safety during the use of sulpiride for inadequate lactation in nursing mothers was provided by Ylikorkala *et al.* (1982). Three mothers out of a total of 14 patients taking sulpiride complained of mild side-effects (headache and tiredness); however, no noenatal side-effects were reported. This anecdotal evidence, however, requires support from studies detailing the drug's excretion in milk. It is provisionally placed in the 'cautionary' category of Table 23.1.

Gastrointestinal medications

Cimetidine

A report from West Germany by Somogyi and Gugler (1979) presented the results of a study of cimetidine excretion into the milk of nursing mothers following single and then chronic dosing. The subject was a 25-year-old nursing mother who had been breast-feeding for six months; she was receiving cimetidine for a radiologically proven duodenal ulcer.

Cimetidine was assayed in plasma and milk by an established HPLC method. In all milk samples cimetidine was found in concentrations higher than the corresponding maternal plasma levels. In the single-dose study, drug concentrations in plasma were highest at 1 hour after dosing with cimetidine (400 mg), but in milk the

peak concentration was later, occurring between 1 and 3 hours after dosing. During a total collection time of 10 hours, 325 μg of cimetidine was excreted into 130 ml of milk; milk-to-plasma ratio levels (about 3.5) were twice theoretical values (1.6) calculated using a general pH partition equation.

Following chronic dosing with cimetidine (200 mg × 3/day, plus 400 mg at night), pronounced excretion of cimetidine in milk was observed; milk concentrations of the drug at the three points of minimum plasma levels were exceedingly high and similar to each other (Table 23.9). Whereas plasma concentrations of drug varied by a factor of 2.0, milk concentrations differed by only 1.2. Milk-to-plasma ratios of cimetidine were larger (4.6 to 11.76) than those observed in the single-dose study.

Somogyi and Gugler concluded that cimetidine appeared to equilibrate slowly with milk and, in the pseudo-distribution equilibrium state, the milk-to-plasma ratio of drug concentration was twice that predicted. However, when the drug was allowed to accumulate, as in the chronic dosing study, plasma concentrations of the drug were in the expected range but milk concentrations of cimetidine remained elevated and constant which was suggestive of an active transport mechanism. They calculated that the maximum amount of cimetidine an infant could ingest, assuming an intake of 1 l of milk per day and being fed at the time of peak drug concentration in milk, would be about 6 mg (i.e. 1.5 mg/kg daily) and therefore cautioned against nursing mothers taking cimetidine.

Diphenoxylate and loperamide

Wilson *et al.* (1980) suggested that gastrointestinal drugs such as diphenoxylate, loperamide, and cimetidine should be placed in a 'cautionary' category. Cimetidine has already been placed in such a list (Table 23.1). The evidence against diphenoxylate is entirely based on pharmacological prediction, but is none the less useful because of this. It is a potent antidiarrhoeal agent which

Table 23.9 Plasma and milk concentrations following chronic cimetidine dosing*

Dose (mg)	Time since last dose (h)	Plasma** concn (μg ml^{-1})	Milk concn (μg ml^{-1})	Milk/plasma concn ratio
400	10	0.51	6.00	11.76
200	5.3	0.78	5.80	7.44
200	4.2	1.06	4.88	4.60

* Source: Somogyi and Gugler (1979).
** Plasma concentrations of cimetidine represent the minimum levels before the morning, midday, and late afternoon doses, respectively.

apparently acts upon opiate receptors to produce its effect; it is said to be almost devoid of central-morphine-like actions, except in exceptionally high doses. The manufacturers of diphenoxylate warn of its presence in human milk.

Diphenoxylate is marketed in combination with a subtherapeutic dose of atropine (25 μg per adult unit dose) to discourage abuse; this may be significant in the present context since atropine is a contraindicated substance during lactation (Table 23.1). This view is still controversial (Wilson *et al.* 1980), although intoxication (mydriasis, tachycardia, constipation) was reported in infants sensitive to atropine (Knowles 1965; O'Brien 1974), and atropine may inhibit lactation.

Even less evidence is available, factual or predictive, about loperamide. It is thought to have less central action than diphenoxylate but it is not known whether it appears in breast milk in sufficient quantities to affect the nursing infant.

Non-steroidal, anti-inflammatory agents

Indomethacin

Manufacturer's literature warns that the safety of indomethacin for use during pregnancy or lactation has not been established (ABPI 1983–84) and advice to clinicians world-wide has consistently contraindicated its use at these times (Fairhead 1978). A single case of convulsions in a breast-fed infant after maternal indomethacin dosage was reported from Sweden by Eeg-Oloffson and colleagues (1978) and adds some new information to a rather sparse bibliography.

Because of severe pain from symphyseolysis, the mother was given indomethacin from the fourth-day after delivery in doses increasing to 200 mg daily. On the seventh day dosage was stopped due to complaints of severe headache and the mother was discharged from hospital with her healthy child. Two hours after returning home the child was found with general seizures lasting about 5 minutes; he was admitted to hospital and the next day suffered another general seizure episode lasting about 2 minutes. Examination revealed a normal boy with a normal neurological state.

Extrapolation of data via a simulated pharmacokinetic computer program suggested that the maternal plasma concentration of indomethacin was 1–4 μg ml^{-1}. Assuming that the milk concentration of the drug was similar to that of the maternal plasma during the steady state, and assuming that the child had a milk intake of 500 ml daily, the infant would have received 0.5 to 2.0 mg of indomethacin per 24-hour period (i.e. 0.1–0.5 mg/kg).

Glucuronidation is known to be an important mechanism for the elimination of this drug and it is known also that this capacity in a one-week-old child is limited. The drug would therefore be likely to accumulate in the child and, since indomethacin is known to lower the convulsive threshold, the authors of the report suggested that indomethacin ingestion in milk might be harmful to the newborn who already had a tendency to convulsions. They emphasize warning against the use of indomethacin as an analgesic in nursing mothers.

Ibuprofen and naproxen

Townsend *et al.* (1982) were unable to find ibuprofen in the breast milk of 12 mothers who ingested one 400-mg ibuprofen (*Motrin*) tablet every six hours over a 24-hour period for relief of post-Caesarean-section pain. The detection limit of their assay was 1 μg ml^{-1}. In their review, however, Chaplin *et al.* (1982) commented, from a personal communication with the Boots Company, that a single patient showed a concentration of 50 μg/100 ml three hours after a maternal dose of 400 mg three times daily. They recommended that caution should be exercised with the use of this drug in nursing mothers. Until more data are available on chronic ibuprofen dosage, it has also been placed within the cautionary category.

Naproxen levels in one nursing mother's milk (single case study) was found to be maximal four hours after dosing on a 375 mg b.i.d. dosing schedule (Jamali *et al.* 1982). The range of concentrations found as 176–237 μg/100 ml. Total cumulative naproxen excreted in the urine of the infant constituted 0.26 per cent of that of the mother during a dosing interval. Although the authors of this report felt that the extent of naproxen ingestion from breast milk was unlikely to cause significant consequences in the suckling infant, naproxen has been placed together with ibuprofen in the cautionary section of Table 23.1 until more data are available.

Oral contraceptives and sex hormones

Oral and injectable contraceptives

Oestrogens, oestrogen/progestogen and oestrogen/androgen combinations are used in high doses to suppress lactation. In the relatively small doses used in oral contraceptive formulations they seem to have little effect on milk flow in most women (Gambrell 1970; Vorherr 1973). In some women, however, contraceptive doses can suppress lactation sufficiently to stop breast-feeding (Gambrell 1970), especially when lactation is not well established (*British Medical Journal* 1970). Progestogen-only contraceptives (oral and depot forms) do not appear to adversely affect milk flow (Karim *et al.* 1971; Sammour *et al.* 1973), although decrease in fat, protein, and material content of milk has been reported (Kader *et al.* 1969; Toaff *et al.* 1969; Sammour *et al.* 1973) (see also later in text). Two cases have been reported of oral-contraceptive-induced breast enlargement in male infants (Curtis 1964; Marriq and Oddo 1974) and proliferation of the vaginal epithelium in female infants (Lauritzen 1967) attributable to oral contraceptives, although a clear cause-and-effect relationship has not been firmly established (*British Medical Journal* 1970). In one study (Wong and Wood 1971), but not in another (Gould *et al.* 1974), there was a correlation between prior contraceptive use and breast-milk jaundice.

Although most physicians generally discourage use of oestrogens during lactation, contraceptive efficacy and safety can still be achieved by the use of progestogens alone in a low-dose regimen. Obviously, the contraceptive should have minimal effects on lactation and preferably should not be excreted into milk.

A number of studies have investigated both these aspects; for example Mettler *et al.* (1977), in Germany, showed that lynoestrenol 0.5 mg daily appeared to be a satisfactory postpartum contraceptive in 167 women followed-up for 8–32 weeks. Milk yield was not significantly different in 67 women when compared with 20 untreated mothers during a seven-day observation period postpartum.

Nilsson *et al.* (1977) from Sweden measured levonorgestrel (formerly named *d*-norgestrel) concentrations by radioimmunoassay in plasma and milk of 15 lactating women who took three different oral contraceptives containing different amounts of the progestogen 2 months postpartum. The levonorgestrel plasma-to-milk ratio was about 100 to 15, with about 1 per cent of the given dose transferred with 600 ml of milk per day (0.3 and 0.15 μg of levonorgestrel respectively). No levonorgestrel was detected in the milk of women taking 30 μg per day.

In a later study, Nilsson and Nygren (1980) calculated that with a minipill of 30 μg of levonorgestrel daily to the mother the fully breast-fed infant will receive an amount of approximately 1 μg per month; this corresponds to 1 minipill per 2–3 years. Considering this extremely low amount of progestogen and the absence of reported progestogenic side-effects in neonates, they saw no reason to contraindicate the use of low-dose progestogens during lactation, if non-hormonal methods were unacceptable.

Although there has been little new evidence on the use of hormonal contraceptives during lactation, the debate still continues (American Academy of Pediatrics 1981).

Kincl (1980) listed side-effects of oestrogens given to rodents during lactation and to women during pregnancy — which is entirely another problem. His arguments are largely based upon circumstantial evidence; nonetheless, he believes that this is sufficient to contraindicate the use of oral contraceptives, and he states 'and indeed any other drugs taken for convenience' in women breast-feeding their children (one might well disagree with the term convenience!!).

Nilsson and Nygren (1980) largely refuted this viewpoint, although they agree that oestrogens should not be used during lactation and indeed that they are not needed. Contraception, in their view can be achieved by the use of gestagens alone, and in a low-dose regimen that presents no hazard to the suckling infant.

The only relatively new evidence of any significance is that produced by Lonnerdal *et al.* (1980). Total nitrogen, non-protein nitrogen, lactose, and the individual milk proteins, lactoferrin and α-lactalbumin, and albumin were analysed in 21 lactating mothers before and after introduction of oral contraceptives.

The four oral contraceptives used were different combinations of levonorgestrel, megestrol acetate, and ethinyl oestradiol; 24-hour milk volumes were registered by weighing the infants before and after each feed.

The mean milk volumes of the mothers remained within normal limits before and after starting and no significant effects were observed in milk levels of α-lactalbumin, protein or non-protein nitrogen at any time. Introduction of oral contraceptives (OCs) did, however, significantly decrease albumin and lactoferrin levels in milk as compared to controls. Three of the mothers attributed a positive effect of oral contraceptives on milk yield, but 12 thought that the OCs suppressed lactation.

The authors of this report suggested that possibly the use of oral contraceptives during lactation should be limited. It is doubtful, however, whether the decrease in lactoferrin levels in milk induced by OCs is of any great concern since, in earlier studies Lonnerdal *et al.* (1976*a*, *b*) showed that marked interindividual variation is seen for the milk content of both α-lactalbumin and lactoferrin in normal lactation.

In an earlier study Toddywalla *et al.* (1977) from India also showed that contraceptives differed in their effects on milk composition. They studied the effects of oral and injectable contraceptives on milk quality in 36 postpartum women and demonstrated that, apart from a significant increase in protein content and a slight increase in quantity, the milk of mothers on a three-monthly depot medroxyprogesterone acetate (150 mg) was similar to that of a control group of women using conventional non-steroidal contraceptives. Women receiving 300 mg of the depot injection, six-monthly, increased the quantity of their milk significantly but the protein, fat, and calcium content of the milk was reduced. A low-dose progesterone group on norethisterone (0.35 mg) showed a significant decrease in the fat and calcium content of their milk.

Other workers investigated the concentration of medroxyprogesterone acetate in milk and some concern has been expressed at their findings. For example, Saxena *et al.* (1977) from London showed that, after an injection of 150 mg of medroxyprogesterone, the drug levels in breast milk were similar to maternal plasma levels in seven lactating women. The ratio was almost 1:1 throughout the study period of up to 87 days. In contrast, in women who took oral preparations of the progestogen-only-type (norethisterone, 350 mg, five cases) or of the combined-type (*d*-norgestrel (150 mg) plus ethinyl oestradiol (30 μg, two cases), the levels of steroids in milk were about one-tenth or less of those found in maternal plasma.

On the basis of these various and international studies, it would seem that the use of oral contraceptives by nursing mothers, progestogen-only or combined type, does not grossly affect milk composition nor do they appear in any appreciable amount in milk. This cannot, however, be said for injections of medroxyprogesterone which can patently tamper with the composition of breast milk and can appear in milk in concentrations equivalent to those of the maternal plasma. This high level must give cause for concern and must indicate that, in such cases, continuation of breast-feeding cannot be recommended. Certainly the long-term effects on the infants who ingest medroxyprogesterone is far from certain and it would be wise to err on the side of safety in the absence of any indication of long-term safety. In this context the practice of routinely injecting the mother with depot-medroxyprogesterone at the time of Rubella vaccination would contraindicate the continuation of breast-feeding.

It is a pity that the whole question of contraception during breast-feeding is still confused. Earlier authoritative advice was generally against prescribing oral contraceptives to nursing mothers (Spellacy 1972); it may well be that progestogen-only or combined-type oral contraceptives only enter into milk in low and insignificant concentrations but more evidence is needed to confirm this before general statements of safety can be made. At best, such agents should only be used with caution in the nursing mother. Therefore in the absence of any other definitive information, the combined and progestogen-only types of OCs are retained in the 'cautionary' category (Table 23.1). The Mersey Regional Drug Information Service (1981) does not include oral contraceptives in their pamphlet on drugs and breast-feeding. They do, however, place low-dose ($< 50 \mu$g) oestrogens in their 'caution' category and they advise monitoring for decreased milk production. High-dose oestrogens ($> 50 \mu$g) are listed in the 'avoid' category since long-term safety is in doubt and they may inhibit lactation; breast enlargement has been reported in male infants, and proliferation of vaginal epithelia in female nurslings. The earlier literature references to these latter effects were cited by Lien (1979) in his review and survey into the excretion of drugs in milk.

The injectable depot medroxyprogesterone acetate, which is not licensed in the USA, and approved only for short-term use in Britain, has been approved as a contraceptive in 78 countries around the world (Senanayaka 1980) including 12 in Europe of which Sweden is one of the latest to give such approval (National Board of Health and Welfare, Sweden 1981; Scrip 1981).

There is inadequate information presently available on possible effects on the nursing infant of this contraceptive in breast milk and it is generally contraindicated in breast-feeding (Table 23.1) yet an estimated one and a half million women, the majority of whom live in Third World countries, choose it in preference to other forms of contraception. One must assume that most of these women also breast-feed their children for relatively long periods.

Depo-Provera is culturally accepted in Third World countries by women: current evidence and opinion suggests that it interferes less with lactation than do combined oral contraceptives; nor apparently does it seem to have any adverse effects on the suckling child. It is also of importance that the injection can be kept secret by the women whose husbands oppose birth control; under similar circumstances supplies of the 'pill' are difficult to hide and compliance with dosage instructions are generally poor.

Other hormones

All androgens (e.g. testosterone, methyltestosterone, and fluoxymesterone) should be avoided during breast-feeding due to possible virilization of the suckling child (Martindale 1982); fluoxymesterone has been used to suppress lactation (Arena 1966). Ethisterone (ethinyltestosterone) is a progestogen with oestrogenic and androgenic properties; it should be avoided during breast feeding (ABPI 1977); it may cause skeletal advancement of the infant (O'Brien 1974). Chlortrianisene, a synthetic non-steroidal oestrogen, has been used to suppress lactation. All these hormonal agents are placed in the 'contraindicated' category in Table 23.1.

Psychoactive drugs

For convenience, sedative, hypnotic, tranquillizer, and antidepressant drugs are grouped together in this text under the umbrella of psychoactive drugs. Anticonvulsant drugs are, however, placed in a separate section.

Barbiturates and non-barbiturate hypnotics

Barbiturates

Barbiturates stimulate the metabolism of other drugs in the mother and also endogenous steroids in the infant, even if low levels pass into the milk (Knowles 1972; Anon 1974; Vorherr 1974*b*). Shorter-acting barbiturates appear to be relatively preferable to longer-acting barbiturates because lower concentrations appear in the milk (Matranga 1970; Horning *et al.* 1975). Large single doses of barbiturate seem to have more potential for causing drowsiness in the infant than do multiple small doses (Tyson *et al.* 1938).

Bromides

Bromides should be avoided during breast-feeding; they are obsolete and it has long been known that they may cause skin rash and drowsiness in the suckling infant (Kwit and Hatcher 1935; Tyson *et al.* 1938; Yeung 1950; Illingworth 1953). *Note: some non-prescription sleeping aids may contain bromides.*

Chloral hydrate and derivatives

Chloral hydrate and dichloralphenazone were both reported to cause drowsiness in the infant when taken by the nursing mother (Bernstine *et al.* 1956; Lacey, 1971).

Benzodiazepines

Diazepam is excreted into breast milk directly or as its active metabolites, N-desmethyldiazepam and oxazepam. The relative proportions probably vary with their degrees of protein-binding. Chlordiazepoxide, oxazepam, and clorazepate (a pro-drug for N-desmethyldiazepam) are probably excreted in breast milk but adequate documentation is lacking (Wilson *et al.* 1980). An early study suggested avoidance of diazepam during breast-feeding due to weight loss and lethargy of the suckling infant and possible hyperbilirubinaemia (Patrick *et al.* 1972).

Two later studies involved lorazepam and lormetazepam. One mother who was breast-feeding her child required 2.5 mg lorazepam b.i.d. for five days after delivery. On the fifth day her milk contained 12 μg l^{-1} free and 35 μg l^{-1} conjugated drug (Whitelaw *et al.* 1981). Based on this information, the authors felt that the maximum amounts that a child could absorb would be pharmacologically insignificant.

In a further study (Hümpel *et al.* 1982) involving lormetazepam ingestion in five mothers, the milk concentration of lormetazepam was below 0.2 ng ml^{-1}, the detection limit of the assay. However, an unknown compound, possibly a metabolite, was detected. The plasma-to-milk ratio for the glucuronide of the drug was 0.04. The quantity of free and conjugated active ingredient transferred to the children via breast milk was calculated to be at most 100 ng kg^{-1}, corresponding to 0.35 per cent of the maternal dose; this was regarded as tolerable.

Potassium clorazepate is a benzodiazepine with actions and uses similar to those of diazepam. The drug and its metabolites are excreted into milk in small amounts (ABPI 1983–84). Recommendations for use during lactation are as for other benzodiazepines.

It would seem therefore that low doses of benzodiazepines appear safe in breast-feeding, and indeed this is the category in which these agents in low dosage are placed by the Mersey Regional Drug Information Service (1981). However, since the agents are demonstrable in breast milk, and, since monitoring for sedation, poor suckling, and jaundice presents few problems, benzodiazepines are retained in the 'cautionary' category (Table 23.1) at this time. Such monitoring is particularly advisable if the mother is taking high doses of any of the benzodiazepines.

Chlorpromazine and other tranquillizers

Chlorpromazine

Chlorpromazine was detected in all milk samples taken from four nursing mothers who received the drug during the puerperium for control of psychiatric symptoms (Wiles *et al.* 1978). Concentrations of drug ranged from 7 to 98 ng ml^{-1} compared with plasma levels of 16 to 52 ng ml^{-1}. In two mothers, plasma concentrations were lower than the milk concentration of chlorpromazine. Metabolites of chlorpromazine were also present in milk.

One of two breast-fed babies in this study showed no obvious adverse effect to the presence of drug in milk (7 ng ml^{-1}) but the other was drowsy and lethargic (milk concentration of drug: 92 ng ml^{-1}). The authors of this report caution in allowing nursing mothers receiving chlorpromazine and related drugs to breast-feed their babies, especially during the first few weeks after delivery.

Although this advice is clear and conforms with the

findings of the study, the literature on the safety of the phenothiazines in milk still remains confused. In a comprehensive review on the side-effects to the neonate from psychotropic agents excreted in milk, Ananth (1978) gave conflicting advice to that of Wiles *et al.* (1978). He accepted that various phenothiazines including chlorpromazine, trifluoperazine, prochlorperazine, thioridazine, mesoridazine, and piperacetazine were excreted in milk but commented that no significant neonatal effects has been observed. From a review of the literature available at that time he concluded that the quantities of the phenothiazines excreted in milk were probably harmless to the neonate. The review by Wilson *et al.* (1980) added further confusion: haloperidol appears in milk in similar concentration to that in maternal blood but they stated that chlorpromazine may transfer to milk although several studies have been unable to detect its presence in milk.

The situation is therefore uncertain; Uhlif and Ryznar (1973) reported that the daily maternal dose of chlorpromazine must be 200 mg or more for the drug to enter milk. It seems therefore, from the evidence currently available and especially from the cases of Wiles *et al.* (1978), that the potential hazard to the neonate of phenothiazine in milk depends on the concentration at which it is present, the volume of milk ingested, and also on the individual susceptibility of the infant to the drug. Until more reports emerge, it would be wise to adopt caution (Table 23.1) and recommend avoiding these drugs during breast-feeding. However, if both phenothiazine and breast-feeding are necessary, then these drugs should only be used if mother and baby are kept under supervision by doctor or nurse.

Cis(Z)-flupenthixol

The concentrations in milk of the neuroleptic drug cis(Z)-flupenthixol were about 30 per cent higher than in serum in five mothers treated either intramuscularly or orally at the time of giving birth (Kirk and Jørgensen 1980). The authors felt that the amounts of drug administered to the neonate in milk were of no importance unless the neonate differed considerably from the adult in its sensitivity to the drug or its ability to metabolize it. Such information is not known, therefore the drug has initially been placed in the cautionary section of Table 23.1.

Haloperidol

Two case reports have discussed the excretion of haloperidol in breast milk. Levels of haloperidol in breast milk in the first study (Stewart *et al.* 1980) were 5 ng ml^{-1} after an average dose of 30 mg daily. In the second study,

Whalley *et al.* (1981) found levels of up to 23.5 ng ml^{-1} at a dosage level of 5 mg b.i.d. In this latter study the infant was apparently not sedated; it fed well and continued to thrive. Until more data are available, and especially since adverse behavioural effects may occur in the young of nursing animals given haloperidol (Lundborg 1972), this drug is placed in the 'cautionary' category (Table 23.1).

Lithium

The classification of lithium into a contraindicated category (Wilson *et al.* 1980; Mersey Regional Drug Information Service 1981) (Table 23.1) seems to depend on a single case reported by Tunnessen and Hertz (1972). The mother took 600–1200 mg of lithium carbonate daily and within a few hours after parturition the baby became cyanotic, hypnotic, and hypothermic. After warming in an incubator, the cyanosis disappeared and muscle tone improved. Breast-feeding was started and the infant remained hypotonic during the next three days. Serum lithium concentrations were 1.5 mEq l^{-1} in the mother and 0.6 mEq l^{-1} in the baby at five days of age. Breast milk content was 0.6 mEq l^{-1}. On stopping breast-feeding, the infant's serum lithium concentration fell to 0.21 mEq l^{-1} by seven days of age.

Earlier data was presented by Sykes and Quarrie (1976), in which the milk/plasma ratio of lithium concentrations in one mother, who received 800 mg of lithium per day, decreased from about 0.77 to 0.25 from day 28 to day 42. No adverse effects on the infant were reported.

Tricyclic antidepressants

Serum and milk concentrations were similar for amitriptyline and its active metabolite, nortriptyline, in a mother who was receiving sustained-release amitriptyline (Brixen-Rasmussen *et al.* 1982). The transfer of drug in milk to the child was calculated to be about 1 per cent of the dose given to the mother. No active drug was detected in the baby's serum and there was no evidence of drug-related adverse effects. This confirms the finding of an earlier report by Bader and Newman (1980). Tricyclic antidepressants are thus maintained in the 'appear to be safe' category of Table 23.2.

Radio-pharmaceuticals

Available information suggests that breast-feeding should be stopped when radio-pharmaceuticals are administered to the lactating mother. Some advice is available on when it is safe to resume breast-feeding; this is, however, based on the opinion of the author or authors of specific reports and it has no official sanction *per se*. A summary of the available evidence and advice is given in Table 23.10.

Respiratory drugs

Aminophylline and theophylline

Wilson *et al.* (1980) reviewed literature on the pharmacokinetics of theophylline in plasma and breast milk and on adverse reactions in the suckling infant. Theophylline and aminophylline have been found in milk, apparently enough to cause irritability, fretfulness, and insomnia in the suckling infant. They warn that the asthmatic patient who requires theophylline preparations for the control of her asthma and who wishes to nurse her infant must be careful of the additive effects of co-consumed caffeine in coffee, tea, or cola soft-drinks, or theobromine in chocolate. These xanthines could augment the concentration of theophylline or aminophylline in breast milk to the extent of provoking adverse CNS-stimulant effects in the infant.

Aminophylline and theophylline are therefore placed in the 'cautionary' category in Table 23.1. The Mersey Regional Drug Information Service (1981) adopted a similar category and cautioned that the infant should be monitored for irritability. Findlay *et al.* (1981) studied the disposition of some analgesic drugs and caffeine in breast milk and plasma of two lactating mothers; they concluded that infants would not be at risk from the effects of caffeine ingested via milk providing the mothers took recommended doses. They also commented, however, that detailed studies of caffeine concentrations in milk of mothers who ingest large quantities of caffeine-containing beverages would be of great interest.

Dyphylline

This theophylline derivative was determined in the breast milk of a population of 20 lactating mothers (Jarboe *et al.* 1981). The drug was administered as a single deep intragluteal injection at a dose of 5 mg kg^{-1}. The ratio of dyphylline distribution between milk and serum was 2.08 ± 0.52; this ratio is about three times that of theophylline. Although the dose likely to be received by a child, based on these results, is unlikely to produce any pharmacological action, dyphylline has been placed in a 'cautionary' category in Table 23.1 until more is known about the pharmacokinetic handling of this drug in children.

Table 23.10 Information and advice on breast-feeding during maternal dosage with radio-pharmaceuticals

Preparation	Advice	Reference
^{67}Gallium citrate	Breast-feeding contraindicated	Larson and Schall (1971)
^{125}I-labelled albumin	Breast-feeding should be stopped for at least 10 days after dosage	Bland *et al.* (1969)
^{131}Iodine	Breast-feeding contraindicated after large therapeutic doses, and should be withheld for a minimum of 24 hours after smaller diagnostic doses. Iodine appears to be concentrated in milk.	Neurenberger and Lipscomb (1952) Miller and Weetch (1955) Weaver *et al.* (1960)
^{131}I-hippurate	Stop breast-feeding for 24 hours after dosage.	Schwartz *et al.* (1968)
^{133}I-labelled macro-aggregated albumin	Stop breast-feeding for 10–12 days after dosage	Wyburn (1973)
^{99m}TcO$_4$	Preparation appears to be concentrated in milk although there is much variation; stop breast-feeding for 32–72 hours after dosage	Spencer *et al.* (1970) Vagenakis *et al.* (1971) Wyburn (1973)
^{99m}TcO$_4$-labelled macroaggregated albumin	It appears to be safe to resume breast-feeding after 24 hours.	Berke *et al.* (1973)

Orciprenaline

One case of respiratory failure was reported in a breast-fed infant whose mother had taken a combination product of orciprenaline and bromhexine (Korsner 1976). The manufacturers do not recommend any special precautions during lactation.

Potassium or sodium iodide

Iodine or iodide salts (expectorants) appear to be concentrated in milk (Miller and Weetch 1955) and should be avoided during breast-feeding especially since there are suitable alternative cough medicines. They present a danger of hypothyroidism, goitre, and skin rashes to the breast-fed child (Anon 1970; Catz and Giacoia 1972; Anon 1974).

Note: iodides are present in some non-prescription cough and expectorant mixtures.

Terbutaline

Boréus *et al.* (1982) reported two cases in which breast milk and blood were collected from two nursing mothers who were treated with oral terbutaline (2.5 mg t.i.d.). The maximum recorded milk-to-plasma drug concentration ratio was 2.9. No symptoms of β-adrenoceptor stimulation were found at routine clinical examination and both children developed normally. The authors felt it was unlikely that the small amounts of terbutaline ingested via milk by infants at this maternal dosage range would lead to plasma levels of pharmacological significance.

Two similar cases were reported by Lönnerholm and Lindström (1982) and these workers also found very small amounts of terbutaline in breast milk and no accumulation of the drug in the infants. They concluded that breast-feeding need not be interrupted due to maternal medication with terbutaline. This drug is therefore placed in the 'appears to be safe' category (Table 23.2).

Vitamins

Vitamin B_1 (thiamine)

Thiamine should only be given to mothers in conditions in which there is a known or assumed deficiency; severely thiamine-deficient mothers should not breast-feed because thiamine deficiency causes excretion of toxic methylglyoxal into milk (Fehily 1944).

Vitamin D

In early studies the vitamin D concentration in human milk was reported to be very low. It is now clear, however, that these early determinations were made on the lipid fraction of the milk, and the aqueous phase was discarded. It is now known that most of the vitamin D in human milk is present as the water-soluble conjugate of vitamin D with sulphate.

Lakdawala and Widdowson (1977) showed, in a study involving 22 women, that the vitamin D sulphate level in milk collected 3–8 days postpartum was 1.0 to 7.78 μg dl^{-1}. In another group of 14 lactating mothers the vitamin D sulphate concentration in milk averaged 0.91 μg dl^{-1} during 4–6 weeks postpartum.

It is interesting to note in this respect that a concentration of 0.9 μg dl^{-1} in human milk is about the same as the concentration of vitamin D stated to be present in milk formulae fortified with vitamin D and diluted according to the manufacturer's instructions (Department of Health and Social Security 1974).

The recommended intake of vitamin D for a lactating women in 10 μg daily (Department of Health and Social Security 1969) and in full lactation the mother might be supplying her infant with approximately this amount. The mother may lose some vitamin D in her urine and she will derive an unknown and variable amount from the action of sunlight on her skin. It may therefore be necessary to consider the vitamin D requirements of the mother who is breast-feeding her baby and supplement this when necessary.

The recommended intake of vitamin D for infants is generally agreed to be 400 I.U. or 10 μg daily (Department of Health and Social Security 1969) and, in view of the findings of Lakdawala and Widdowson (1977), it is very relevant to go back over some earlier publications and consider the advice that was given. For example, Goldberg (1972) suggested that the ingestion by the mother of large pharmacological amounts of vitamin D for hypoparathyroidism might lead to undesirable amounts of vitamin D metabolites (especially 25-hydroxy vitamin D) in breast milk. He urged caution in allowing breast-feeding in such unusual circumstances. Obviously this caution is still important and should be re-emphasized. It is large pharmacological doses of vitamin D (e.g. 50 000 I.U.) which have caused idiopathic hypercalcaemia of infancy.

In the light of the evidence presented by Lakdawala and Widdowson (1977), the requirement for vitamin D of the mother who is breast-feeding her baby should be reconsidered and the mother advised to take vitamin D supplements, if these are necessary.

Miscellaneous drugs

Baclofen

This drug, which is used in the treatment of multiple sclerosis and other spastic conditions, has recently been examined for its distribution in breast milk in a single patient (Eriksson and Swahn 1981). The total amount excreted in milk during 24 hours was about one-thousandth of the ingested maternal dose and the authors felt that the concentration of the drug likely to appear in the newborn's serum would not reach toxic levels.

Carisoprodol

This centrally acting muscle relaxant should not be administered to nursing mothers; the suckling infant may be exposed to adverse effects ranging from CNS depression to gastrointestinal upsets. This drug is excreted in milk in concentrations 2–4 times that in maternal plasma (O'Brien 1974).

Clemastine

A case report (Kok et al. 1982) described a previously well child who became drowsy, irritable, and refused to feed soon after her mother started taking clemastine (an antihistamine) 1 mg twice daily for the treatment of catarrh. As the symptoms cleared when the drug was stopped, these authors suggested that clemastine is contraindicated in breast-feeding and recommended careful monitoring of other antihistamine use by nursing mothers. Chaplin et al. (1982) also documented a case of infant irritability after a mother took a long-acting oral antihistamine decongestant preparation (6 mg dexbrompheniramine maleate and 120 mg ephedrine sulphate twice daily). Clemastine and other antihistamines are therefore placed in the 'cautionary' category of Table 23.1. This reclassification is thought advisable since some antihistamines are combined with other drugs in formulated preparations.

Ergot preparations

Ergotamine tartrate (e.g. with caffeine in *Cafergot*) should be avoided during lactation. Vomiting, weak pulse, unstable blood pressure, and diarrhoea were reported in the suckling infant (Fomina 1934; Knowles 1965; Shane and Naftolin 1974) and repetitive doses may inhibit lactation (Shane and Naftolin 1974). The manufacturers contraindicate the preparation during breast-feeding (ABPI 1983–84).

Sulphasalazine

Sulphasalazine (SASP) is widely used as maintenance treatment for ulcerative colitis. After oral dosage SASP is cleard by gut bacteria into its two constituents, 5 aminosalicylate and sulphapyridine (SP). Since many of the patients are women in the reproductive age, there is the question of the extent to which this drug and its metabolites reach the fetus and the degree to which they appear in milk if treatment is continued during lactation. Khan and Truelove (1979) investigated both these potential hazards and reported the extensive use of the drug during pregnancy without effect on its course or danger to the fetus.

With respect to the excretion of SASP in milk, they investigated three mothers receiving the drug, 0.5 mg four times daily. The drug and its metabolites passed into milk and 1 week after delivery the concentration of SASP in milk was about 30 per cent of the level in maternal serum. The mean total concentration of the major metabolite, sulphapyridine (SP), in milk exceeded 50 per cent of the level in serum, and metabolites of sulphapyridine itself were present also in milk in a concentration roughly equivalent to their level in the serum.

These investigators concluded, since the concentrations of sulphasalazine and its major metabolites in milk were much lower than those of maternal serum, that they were unlikely to cause harmful side-effects in the suckling infant.

Berlin and Yaffe (1979, 1980) reported sulphasalazine (SASP) disposition in breast milk in one nursing mother taking the drug for ulcerative colitis (500 mg by mouth every 6 hours). Plasma SASP levels were between 9.4 and 15.4 μg ml^{-1}. Maternal sulphapyridine (SP) levels were 12–18 μg ml^{-1}, and total 'sulpha' compounds were 18–22 μg ml^{-1}.

Sulphasalazine was not found in any of 44 milk samples taken over a two-month period, but sulphapyridine appeared in breast milk 8 hours after dosing and remained constant at levels of 3.5 μg ml^{-1}; total 'sulpha' compounds in milk were 4–7 μg ml^{-1}. Milk/plasma ratio for SP was 0.60.

Assuming a continual milk level for SP of 5 μg ml^{-1}, the 24-hour dose available to the infant was 4 mg or about 0.32 per cent of the maternal oral dose of SP in the form of SASP. Levels of SP and metabolites in random urine samples from the infant were 2–3 μg ml^{-1} or an estimated 24-hour urinary excretion of 1.4–2.1 mg total sulpha compounds.

These observations on one patient are interesting, firstly, unlike Khan and Truelove (1979), these latter workers were not able to detect sulphasalazine in milk — only its metabolite; secondly, they showed that the absorption of the main metabolite, sulphapyridine, in human milk did occur in the nursing infant. Of importance also they reported that no adverse effects were observed in this infant. This drug is provisionally placed in the 'cautionary' section of Table 23.1.

Acknowledgement

The authors wish to thank the Trent and West Midlands (England) Drug Information Centres for kindly providing some of the information used in this chapter.

RECOMMENDED FURTHER READING

FINDLAY, J.W.A. (1983). The distribution of some commonly used drugs in human breast milk. *Drug Metab. Rev.* **14**, 653–84.

RAYBURN, W.F. AND ZUSPAN, F.P. (EDS.) (1982). *Drug therapy in obstetrics and gynecology*. Appleton-Century-Crofts, Norwalk, Connecticut.

WILSON, J.T. (1983). Determinants and consequences of drug excretion in breast milk. *Drug Metab. Rev.* **14**, 619–52.

REFERENCES

ABPI (1977). Ethisterone, p. 61. In *ABPI data sheet compendium 1977*. Datapharm Publications Ltd, London.

—— (1983–84). Potassium clorazepate (*Tranxene*), p. 198; glibenclamide (*Daonil*) pp. 484–5; methyldopa (*Aldomet*), pp. 760–1; hydrochlorothiazide (*Hydrosaluric*), pp. 805–6; chlorothiazide (*Saluric*), pp. 815–16; carbimazol (*Neo-Mercazole*), pp. 871–2; indomethacin (*Indocid*) pp. 838–9; ergotamine (*Cafergot*), p. 1328; ergotamine (*Migril*), p. 1395. In *ABPI data sheet compendium 1983–84*. Datapharm Publications Ltd, London.

AMERICAN ACADEMY OF PEDIATRICS (1981). Committee on Drugs: Breast-feeding and contraception. *Pediatrics* **68**, 138–40.

ANANTH, J. (1978). Side effects in the neonate from psychotropic agents excreted through breast-feeding. *Am. J. Psychiat.* **135**, 801–5.

ANDERSEN, H.J. (1983). Excretion of verapamil in human milk. *Eur. J. clin. Pharmacol.* **25**, 279–80.

ANDERSON, P.O. (1977). Drugs and breast-feeding — a review. *Drug Intel. clin. Pharm.* **11**, 208–23.

—— AND SALTER, F.T. (1976). Propranolol therapy during pregnancy and lactation. *Am. J. Cardiol.* **37**, 325.

ANON (1970). Drug-induced goitres in fetus, and in children and in adults. *Med. Lett. Drugs Ther.* **12**, 61–2.

—— (1974). Drugs in breast milk. *Med. Lett. Drugs Ther.* **16**, 25–7.

ARENA, J.M. (1966). Drugs and breast-feeding. *Clin. Pediat.* **5**, 472.

AVERY, G.S. (ED.) (1979). *Drug treatment*. Adis Press, Sydney; Publishing Sciences Group, Acton, Mass; Churchill-Livingstone, Edinburgh. [Cited by Wilson, J.T. *et al.* (1980).]

BADER, T.F. AND NEWMAN, K. (1980). Amitriptyline in human breast milk and the nursing infant's serum. *Am. J. Psychiat.* **137**, 855–6.

BAILEY, D.N., WEIBERT, R.T., NAYLOR, A.J., AND SHAW, R.F. (1982). A study of salicylate and caffeine excretion in the breast milk of two nursing mothers. *J. anal. Toxicol.* **6**, 64–9.

BARNETT, D.B., HUDSON, S.A., AND McBURNEY, A. (1982). Disopyramide and its N-monodesalkyl metabolite in breast milk. *Br. J. clin. Pharmacol.* **14**, 310–12.

BATY, J.D., BRECKENRIDGE, A., LEWIS, P.J., ORME, M., SERLIN, M.J., AND SIBEON, R.G. (1976). May mothers taking warfarin breast feed their infants? *Br. J. clin. Pharmacol.* **3**, 969P.

BAUER, J.H., PAPE, B., ZAJICEK, J., AND GROSHONG, T. (1979). Propranolol in human plasma and breast milk. *Am. J. Cardiol.* **43**, 860–2.

BEELEY, L. (1981). Drugs and breast-feeding. *Clin. Obstet. Gynaecol.* **8**, 291–5.

BELTON, E.M. AND JONES, R.V. (1965). Haemolitic anaemia due to nalidixic acid. *Lancet* **ii**, 691.

BERKE, R.A., HOOPS, E.C., KEREIAKES, J.C., AND SAENGER, E.L. (1973). Radiation dose to breast-feeding child after mother has 99mTc-MAA lung scan. *J. Nucl. Med.* **14**, 51–2.

BERLIN, C.M., PASCUZZI, M.J., AND YAFFE, S.J. (1980). Excretion of salicylate in human milk. *Clin. Pharmacol. Ther.* **27**, 245–6.

—— AND YAFFE, S.J. (1979). Disposition of sulfasalazine (*Azulfidine*) in human breast milk, plasma and saliva. *Pediat. Res.* **13**, 396.

—— AND —— (1980). Disposition of salicylazosulfapyridine (*Azulfidine*) and metabolites in human breast milk. *Dev. Pharmacol. Ther.* **1**, 31–9.

BERNSTINE, J.B., MEYER, A.E., AND BERNSTINE, R.L. (1956). Maternal blood and breast milk estimation following the administration of chloral hydrate in the puerperium. *J. Obstet. Gynaecol. Br. Emp.* **63**, 228–31.

BITZÉN, P.-O., GUSTAFSSON, B., JOSTELL, K.G., MELANDER, A., AND WÅHLIN-BOLL, E. (1981). Excretion of paracetamol in human breast milk. *Eur. J. clin. Pharmacol.* **20**, 123–5.

BLANCO, J.D., JORGENSEN, J.H., CASTANEDA, Y.S., AND CRAWFORD, S.A. (1983). Ceftazidime levels in human breast milk. *Antimicrob. Agents Chemother.* **23**, 479–80.

BLAND, E.P., DOCKER, M.I., SELWYN CRAWFORD, J., AND FARR, R.F. (1969). Radioactive iodine uptake by the thyroid of breast fed infants after maternal blood-volume measurements. *Lancet* **ii**, 1039–41.

BOEHRINGER INGELHEIM (1976). via West Midlands Drug Information Service.

BORÉUS, L.O., DE CHÂTEAU, P., LINBERG, C., AND NYBERG, L. (1982). Terbutaline in breast milk. *Br. J. clin. Pharmacol.* **13**, 731–2.

BRITISH MEDICAL JOURNAL (1969). Tetracycline in breast milk. *Br. med. J.* **4**, 791.

—— (1970). Contraceptive steroids in breast milk. *Br. med. J.* **4**, 731–2.

BRIXEN-RASMUSSEN, L., HALGRENER, J., AND JØRGENSON, A. (1982). Amitriptyline and nortriptyline excretion in human breast milk. *Psychopharmacology* **76**, 94–5.

BRODRICK, A. (1982). Propranolol in breast milk. An investigation by high performance liquid chromatography. *Br. J. Pharmaceut. Pract.* **3**, 23–4.

BROUWERS, J.R.B.J., ASSIES, J., WIERSINGA, W.M., HUIZING, G., AND TYTGAT, G.N. (1980). Plasma prolactin levels after acute and subchronic oral administration of domperidone and of metoclopramide: a cross-over study in healthy volunteers. *Clin. Endocrinol.* **12**, 435–40.

CARSON, D.J.L. AND CARSON, E.D. (1977). Fatal dextropropoxyphene poisoning in Northern Ireland. *Lancet* **i**, 894–7.

CASH, J.K. AND GIACOIA, D. (1981). Organising and operation of a human milk bank. *J. obstet., gynecol., neonatal Nursing* **10**, 434–8.

CATZ, C.E., AND GIACOIA, G.P. (1972). Drugs and breast milk. *Pediat. Clins. N. Am.* **19**, 151–66.

CHAPLIN, S., SANDERS, G.L., AND SMITH, J.M. (1982). Drug excretion in human breast milk. *Adv. Drug React. Acute Pois. Rev.* **1**, 255–87.

CHYO, N., SUNADA, H., AND NOHARA, S, (1962). Clinical studies of kanamycin applied in the field of obstetrics and gynecology. *Asian med. J.* **5**, 293–303.

CLARK, J.H. AND WILSON, W.G. (1981). A 16-day-old breast-fed infant with metabolic acidosis caused by salicylate. *Clin. Pediat.* **20**, 53–4.

CURTIS, E.M. (1964). Oral-contraceptive feminization of a normal male infant: Report of a case. *Obstet. Gynecol.* **23**, 295–6.

D'ARCY, P.F. (1979). Drugs in milk. In D'Arcy, P.F. and Griffin, J.P. *Iatrogenic diseases*, 2nd edn., pp. 425–35. Oxford University Press, Oxford.

DEPARTMENT OF HEALTH AND SOCIAL SECURITY (1969). Rep. publ. Hlth. med. Subj., No. 120, London.

—— (1974). Rep. Hlth. social Subj., No. 9. London.

DEVLIN, R.G., DUCHIN, K.L., AND FLEISS, P.M. (1981). Nadolol in human serum and breast milk. *Br. J. Clin. Pharmacol.* **12**, 393–6.

—— AND FLEISS, P.M. (1981). Captopril in human blood and breast milk. *J. clin. Pharmacol.* **21**, 110–13.

DUBOIS, M., DELAPIERRE, D., CHANTEUX, L., DEMONTY, J., LAMBOTTE, R., KRAMP, R., AND DRESSE, A. (1981). A study of the transplacental transfer and the mammary excretion of cefoxitin in humans. *J. clin. Pharmacol.* **21**, 477–83.

ECKSTEIN, H.B. AND JACK, B. (1970). Breast-feeding and anticoagulant therapy. *Lancet* **i**, 672–3.

EEG-OLOFFSON, O., MALMROS, I., ELWIN, C.-E., AND STÉEN, B. (1978). Convulsions in a breast-fed infant after maternal indomethacin. *Lancet* **ii**, 215.

ERICKSON, S.H. AND OPPENHEIM, G.L. (1979). Aspirin in breast milk. *J. fam. Pract.* **8**, 189–90.

——, ——, AND SMITH, G.H. (1981). Metronidazole in breast milk. *Obstet. Gynecol.* **57**, 48–50.

ERIKSSON, G., AND SWAHN, C.-G. (1981). Concentrations of baclofen in serum and breast milk from a lactating woman. *Scand. J. clin. Lab. Invest.* **41**, 185–7.

FAIRHEAD, F.W. (1978). Convulsions in a breast-fed infant after maternal indomethacin. *Lancet* **ii**, 576.

FOOD AND DRUGS ADMINISTRATION (1976). Metronidazole box warning. *FDA Drug Bull.* **6**, 22–3.

FEHILY, L. (1944). Human milk intoxication due to B$_1$ avitaminosis. *Br. med. J.* **2**, 590–2.

FIDLER, G.I. (1974). Propranolol and pregnancy. *Lancet* **ii**, 722–3.

FIDSLER, J., SMITH, V., AND DE SWIET, M. (1983). Excretion of oxprenolol and timolol in breast milk. *Br. J. Obstet. Gynaecol.* **90**, 961–5.

FINDLAY, J.W.A., DE ANGELIS, R.L., KEARNEY, M.F., WELCH, R.M., AND FINDLAY, J.M. (1981). Analgesic drugs in breast milk and plasma. *Clin. Pharmacol. Ther.* **29**, 625–33.

FOMINA, P.I. (1934). Untersuchungen über den Übergang des aktiven Agens des Mutterkorns in die Milch sillender Mütter. *Arch. Gynaek.* **157**, 275.

GAMBRELL, R.D. (1970). Immediate postpartum oral contraception. *Obstet Gynecol.* **36**, 101–6.

GARDNER, D.K. AND RAYBURN, W.F. (1982). Drugs in breast milk. In *Drug therapy in obstetrics and gynecology* (ed. W.F. Rayburn, and F.P. Zuspan), pp. 175–96. Appleton-Century-Crofts, Norwalk, Connecticut.

GOLDBERG, L.D. (1972). Transmission of a vitamin-D metabolite in breast milk. *Lancet* **ii**, 1258–9.

GOULD, S.R., MOUNTROSE, U., BROWN, D.J., WHITEHOUSE, W.L., AND BARNARDO, D.E. (1974). Influence of previous oral contraception and maternal oxytocin infusion on neonatal jaundice. *Br. med. J.* **3**, 228–30.

GRAY, M.S., KANE, P.O., AND SQUIRES, S. (1961). Further observations on metronidazole (Flagyl). *Br. J. Vener. Dis.* **37**, 278–9.

HAVELKA, J. AND FRANKOVA, A. (1972). A study of side effects of maternal chloramphenicol therapy in newborns. *Cesk. Pediat.* **21**, 31–3.

HEALY, M. (1961). Suppressing lactation with oral diuretics. *Lancet* **i**, 1353–4.

HEISTERBERG, L. AND BRANEBJERG, P.E. (1983). Blood and milk concentrations of metronidazole in mothers and infants. *J. perinat. Med.* **11**, 114–20.

HO, T. (1970). Studies on the absorption and excretion of gentamicin in newborn infants. *Jpn J. Antibiot.* **23**, 298–311.

HOCKING, D.R. (1968). Neonatal haemolytic disease due to dapsone. *Med. J. Aust.* **1**, 1130–1.

HOFMEYR, G.J. AND VAN IDDEKINGE, B. (1983). Domperidone and lactation. *Lancet* **i**, 647.

HORNING, M.G. STILLWELL, W.G., NOWLIN, J., LERTRATANANGKOON, K., STILLWELL, R.N., AND HILL R.M. (1975). Identification and quantification of drugs and drug metabolites in human breast milk using GC-MS-COM methods. *Mod. Probl. Paediat.* **15**, 73–9.

HOUWERT -DE JONG, M., GERARDS, L.J., TETTEROO-TEMPELMAN, C.A.M., AND DE WOLF, F.A. (1981). May mothers taking acenocoumarol breast feed their infants? *Eur. J. clin. Pharmacol.* **21**, 61–4.

HÜMPEL, M., STOPPELLI, I., MILIA, S., AND RAINER, E. (1982). Pharmacokinetics and biotransformation of the new benzodiazepine, lormetrazepam, in man. *Eur. J. clin. Pharmacol.* **21**, 421–5.

ILLINGWORTH, R.S. (1953). Abnormal substances excreted in human milk. *Practitioner* **171**, 533–8.

JAMALI, F., AND KESHAVARZ, E. (1981). Salicylate excretion in breast milk. *Int. J. Pharmaceutics* **8**, 285–90.

——, TAM, Y.K., AND STEVENS, R.D. (1982). Naproxen excretion in breast milk and its uptake by suckling infant. *Drug Intel. clin. Pharm.* **16**, (abstract 32), 475.

JARBOE, C.H., COOK, L.N., MALESIC, I., AND FLEISCHAKER, J. (1981). Dyphylline elimination kinetics in lactating women: blood to milk transfer. *J. clin. Pharmacol.* **21**, 405–10.

JOHANSEN, K., ANDERSEN, N., KAMPMANN, J.P., MØLHOLM HANSEN, J., AND MORTENSEN, H.B. (1982). Excretion of methimazole in human milk. *Eur. J. clin. Pharmacol.* **23**, 339–41.

JUUL, S. (1976). Fenemalforgitting via modermaelken? *Uneskr. Laeg.* **131**, 2257–8.

KADER, M.M.A., HAY, A.A. EL-SAFOURI, S., AZIZ, M.T.A., EL-DIN, J.S., KAMAL, I., HEFNAWI, F., GHONEIM, M., TALAAT, M., YOUNIS, N., TAGUI, A., AND ABDALLA, M. (1969). Clinical, biochemical, and experimental studies on lactation III. Biochemical changes induced in human milk by gestagens. *Am. J. Obstet. Gynecol.* **105**, 978–85.

KAFETZIS, D.A., SIAFAS, C.A., GEORGAKOPOULOS, P.A., AND PAPADA-

TOS, C.J. (1981). Passage of cephalosporins and amoxicillin into the breast milk. *Acta paediat. scand.* **70**, 285–8.

KAMPMANN, J.P., JOHANSEN, K., MØLHOLM HANSEN, J., AND HELWEG, J. (1980). Excretion of propylthiouracil in human milk. Revision of a dogma. *Lancet* i, 736–8.

KANEKO, S., SATO, T., AND SUZUKI, K. (1979). The levels of anticonvulsants in breast milk. *Br. J. clin. Pharmacol.* **7**, 624–6.

KARIM, M., AMMAR. R., EL MAHGOUB, S., EL GANZOURY, B., FIKRI, F., AND ABDOU, I. (1971). Injected progestogen and lactation. *Br. med. J.* **1**, 200–3.

KARLBERG, B., LUNDBERG, D., AND ÅBERG, H. (1974). Excretion of propranolol in human breast milk. *Acta pharmacol. toxicol.* **34**, 222–4.

KAUPPILA, A., ARVELA, P., KOIVISTO, S., YLIKORKALA, O., AND PELKONEN, O. (1984). Metoclopramide and breast feeding: transfer into milk and the newborn. *Eur. J. clin. Pharmacol.* **25**, 819–23.

KHAN, A.K.A. AND TRUELOVE, S.C. (1979). Placental and mammary transfer of sulphasalazine. *Br. med. J.* **2**, 1553.

KINCL, F.A. (1980). Debate on the use of hormonal contraceptives during lactation. *Res. Reprod.* **12**, 1.

KIRK, L. AND JØRGENSEN, A. (1980). Concentration of cis(Z)-flupenthixol in maternal serum, amniotic fluid, umbilical cord serum and milk. *Psychopharmacol.* **72**, 107–8.

KNOBDEN, J.E., ANDERSON, P.O., AND WATANABE, A.S. (1973). *Handbook of clinical drug data.* 3rd edn., pp. 89–118. Drug Intl. Pub., Hamilton, Illinois.

KNOWLES. J.A. (1965). Excretion of drugs in milk, a review. *J. Pediat.* **66**, 1068–82.

—— (1972). Drugs in milk. *Pediat. Currents, Ross Laboratories* **21**, 28–32.

—— (1973). Effects on the infant of drug therapy in nursing mothers. *Drug Ther.* **3**, 57–65.

—— (1974). Breast milk. A source of more than nutrition for the neonate. *Clin. Toxicol.* **7**, 69–82.

KOK, T.H.H.G., TAITZ, L.S., BENNETT, M.J., AND HOLT, D.W. (1982). Drowsiness due to clemastine transmitted in breast milk. *Lancet* i, 914–15.

KORSNER, A. (1976). Excretion of drugs in milk. *Pharm. J.* **217**, 293.

KRAUSE, W., STOPPELLI, I., MILIA, S., AND RAINER, E. (1982). Transfer of mepindolol to newborns by breast-feeding mothers after single and repeated daily doses. *Eur. J. clin. Pharmacol.* **22**, 53–5.

KWIT, N.T. AND HATCHER, R.A. (1935). Excretion of drugs in milk. *Am. J. Dis. Child.* **49**, 900–4.

LACEY, J.H. (1971). Dichloralphenazone and breast milk. *Br. med. J.* **4**, 684.

LAKDAWALA, D.R. AND WIDDOWSON, E.M. (1977). Vitamin-D in human milk. *Lancet* i, 167–8.

LARSON, S.M., AND SCHALL, G.L. (1971). Gallium 67 concentration in human breast milk. *J. Am. med. Ass.* **218**, 257.

LAURITZEN, C. (1967). On endocrine effects of oral contraceptives. *Acta endocrinol.* **124** (Suppl.), 87–100.

LEVIN, R.H. (1975). Teratogenicity and drug excretion in breast milk (Maternogenecity). In *Clinical pharmacology and therapeutics* (ed. E.T. Herfindal and J.L. Hirschman), pp. 23–44. Williams Wilkins Co., Baltimore, Maryland.

LEVITAN, A.A. AND MANION, J.C. (1973). Propranolol therapy during pregnancy and lactation. *Am. J. Cardiol.* **32**, 247.

LEVY, M., GRANT, L., AND LAUFER, N. (1977). Excretion of drugs in human milk. *New Engl. J. Med.* **297**, 789.

LEWIS, A.M., PATEL, L., JOHNSON, A., AND TURNER, P. (1981). Mexiletine in human blood and breast milk. *Postgrad. med. J.* **57**, 546–7.

LEWIS, P. (1978). Drugs in breast milk. *J. matern. child Hlth.* (April 1978), 128–32.

LIEDHOLM, H., MELANDER, A., BITZÉN, P.-O., LÖNNERHOLM, G., MATTIASSON, I., AND NILSSON, B. (1981). Accumulation of atenolol and metoprolol in human breast milk. *Eur. J. clin. Pharmacol.* **20**, 229–31.

——, WÅLIN-BOLL, E., HANSON, A., INGEMARSSON, I., AND MELANDER, A. (1982). Transplacental passage and breast milk concentration of hydralazine. *Eur. J. clin. Pharmacol.* **21**, 417–19.

LIEN, E.J. (1979). The excretion of drugs in milk: a survey. *J. clin. Pharmacol.* **4**, 133–44.

LOCKET, S. (1980). Acute poisoning caused by drugs in excessive dosage. In *Drug-induced emergencies* (ed. P.F. D'Arcy and J.P. Griffin), pp. 198–231. Wright, Bristol.

LONNERDAL, B., FORSUM, E., AND HAMBRAEUS, L. (1976*a*). The protein content of human milk. *Proc. 10th Int. Cong. Nutrition*, Kyoto, Japan, p. 698. Victroy-sha Press.

——, ——, AND —— (1976*b*). A longitudinal study of the protein, nitrogen and lactose contents of human milk from Swedish well-nourished mothers. *Am. J. clin. Nutr.* **29**, 1127–33.

——, ——, AND —— (1980). Effect of oral contraceptives on composition and volume of breast milk. *Am. J. Clin. Nutr.* **33**, 816–24.

LÖNNERHOLM, G. AND LINDSTRÖM, B. (1982). Terbutaline excretion into breast milk. *Br. J. clin. Pharmacol.* **13**, 729–30.

LUNDBORG, P. (1972). Abnormal ontogeny in young rabbits after chronic administration of haloperidol to the nursing mothers. *Brain Res.* **44**, 684–7.

LUSTGARTEN, J.S. AND PODOS, S.M. (1983). Topical timolol and the nursing mother. *Arch. Ophthalmol.* **101**, 1381–2.

MAEGRAITH, B.G. AND GILES, H.M. (1971). *Management and treatment of tropical diseases*, p. 658. Blackwell, Oxford.

MANN, C. (1980). Clindamycin and breast-feeding. *Pediatrics* **66**, 1030.

MARRIQ, P. AND ODDO, G. (1974). La gynécomastie induite chez le nouveau-né par le lait maternel? *Nouv. Presse Méd.* **3**, 2579.

MARTINDALE (1982). Metronidazole. In *The extra pharmacopoeia*, 28th edn. (ed. J.E.F. Reynolds), pp. 968–73. Pharmaceutical Press, London.

MATRANGA, A. (1970). Drugs excreted in breast milk. *Drug Information Service Newsletter*, Alta Bates Community Hospital **2**, 22–30.

McKENNA, R., COLE, E.R., AND VASAN, U. (1983). Is warfarin contraindicated in the lactating mother? *J. Pediatr.* **103**, 325–7.

MERSEY REGIONAL DRUG INFORMATION SERVICE (1981). *Drugs and breast feeding. Drug Information Letter,* No. 35, February 1981.

METTLER, L., MÜLLER, M., DITTMAR, F.W., AND SEMM, K. (1977). Postpartum contraception: effect of lynoestrenol during the lactating period. *Münch. med. Wschr.* **119**, 853–6.

MILLER, H. AND WEETCH, R.S. (1955). The excretion of radioactive iodine in human milk. *Lancet* ii, 1013.

MILLER, M.E., COHN, R.D., AND BURGHART, P.H. (1982). Hydrochlorothiazide disposition in a mother and her breast-fed infant. *J. Pediat.* **101**, 789–91.

MOIEL, R.H. AND RYAN, J.R. (1967). Tolbutamide (*Orinase*) in human breast milk. *Clin. Pediat.* **8**, 840.

MOORE, B. AND COLLIER, J. (1979). Drugs and breast-feeding. *Br. med. J.* **2**, 211.

MULLEY, B.A., PARR, G.D., PAU, W.K., RYE, R.M., MOULD, J.J., AND SIDDLE, N.C. (1978). Placental transfer of chlorthalidone and its elimination in maternal milk. *Eur. J. clin. Pharmacol.* **13**, 129–31.

NATIONAL BOARD OF HEALTH AND WELFARE, SWEDEN (1981). Press release from the Pharmaco-therapeutic Unit, S.L.A., 11th September 1981.

NAU, H., KUHNZ, W., EGGER, H.-J., RATING, D., AND HELGE, H. (1982). Research review. Anticonvulsants during pregnancy and lactation. Transplacental, maternal and neonatal pharmacolinetics. *Clin. Pharmacokinet.* **7**, 508–43.

NILSSON, S. AND NYGREN, K.-G. (1980). Debate on the use of hormonal contraceptives during lactation. *Res. Reprod.* **12**, 1–2.

——, ——, AND JOHANSSON, E.D.B. (1977). *d*-Norgestrel concentrations in maternal plasma, milk and child plasma during administration of oral contraceptives to nursing women. *Am. J. Obstet. Gynecol.* **129**, 178–84.

NURNBURGER, C.E. AND LIPSCOMB, A. (1952). Transmission of radioiodine (I131) to infants through human milk. *J. Am. med. Ass.* **150**, 1398–400.

O'BRIEN, T.E. (1974). Excretion of drugs in human milk. *Am. J. hosp. Pharm.* **31**, 844–54.

O'HARE, M.F., MURNAGHAN, G.A., RUSSELL, C.J., LEAHEY, W.J., AND VARMA, M.P.S. (1980). Sotalol as a hypotensive agent in pregnancy. *Br. J. Obstet. Gynaecol.* **87**, 814–20.

ORME, M.L'E., LEWIS, P.J., SWIET, M. DE, SERLIN, M.J., SIBEON, R., AND BATY, J.D. (1977). May mothers given warfarin breast-feed their infants? *Br. med. J.* **1**, 1564–5.

——, SERLIN, M.J., AND BRECKENRIDGE, A. (1979). Thromboembolism in pregnancy. *Br. med. J.* **2**, 333.

PATRICK, M.J., TILSTONE, W.J., AND REAVEY, P. (1972). Diazepam and breast-feeding. *Lancet* i, 542–3.

PENRY, J.K. AND NEWMARK, M.E. (1979). The use of antiepileptic drugs. *Ann. intern. Med.* **90**, 207–18.

PITTARD, W.B. III AND GLAZIER, H. (1983). Procainamide excretion in human milk. *J. Pediat.* **102**, 631–3.

PITTMAN, K.A., SMYTH, R.D., LOSADA, M., ZIGHELBOIM, I., MADUSKA, A.I., AND SUNSHINE, A. (1980). Human perinatal distribution of butorphanol. *Am. J. Obstet. Gynecol.* **138**, 797–800.

POSTELLON, D.C. AND ARONOW, R. (1982). Iodine in mothers milk. *J. Am. med. Ass.* **247**, 463.

RANE, A. AND TUNELL, R. (1981). Ethosuximide in human milk and in plasma of a mother and her nursed infant. *Br. J. clin. Pharmacol.* **12**, 855–8.

REINHARDT, D., RICHTER, O., GENZ, T., AND POTTHOFF, S. (1982). Kinetics of the translactal passage of digoxin from breast feeding mothers to their infants. *Eur. J. Pediat.* **138**, 49–52.

SAGRAVES, R., KAISER, D., AND SHARPE, G.L. (1981). Prednisone and prednisolone concentrations in the milk of a lactating mother. *Drug Intel. clin. Pharm.* **15**, 484.

SAMMOUR, M.B., RAMADAN, M.E., AND SALAH, M. (1973). Effect of chlormadinone on the composition of human milk. *Fertil. Steril.* **24**, 301–4.

SANDERS, S.W., ZONE, J.J., FOULTZ, R.L., TOLMAN, K.G., AND ROLLINS, D.E. (1982). Hemolytic anemia induced by dapsone transmitted through breast milk, *Ann. intern. Med.* **96**, 465–6.

SANDSTROM, B. (1980). Metoprolol excretion into breast milk. *Br. J. clin. Pharmacol.* **9**, 518–19.

SAVAGE, R.L. (1976). Drugs and breast-milk. *Adv. Drug React. Bull.* no. 61, 212–15.

SAXENA, B.N. SHRIMANKER, K., AND GRUDZINSKAS, J.G. (1977). Levels of contraceptive steroids in breast milk and plasma of lactating women. *Contraception* **16**, 605–13.

SCHWARTZ, K.-D., POTSCHWADEK, B., AND SCHOLZ, B. (1968). The excretion of I131 in breast milk in isotope nephrography performed with I131 hippurate post partum. *Rad. Biol. Ther.* **9**, 259–62.

SCRIP (1981). Sweden strongly approves *Depo-Provera* as a contraceptive. *Scrip* No. 629, 28th September 1981, p.3.

SENANAYAKA, P. (1980). International Planned Parenthood Federation: Testimony before Subcommittee on International Economic Policy and Trade, Committee on Foreign Affairs, US House of Representatives, Washington, DC, 9th September 1980.

SHANE, J.M. AND NAFTOLIN, F. (1974). Effect of ergonovine maleate on puerperal prolactin. *Am. J. Obstet. Gynecol.* **120**, 129–31.

SMIALEK, J.E., MONFORTE, J.R., ARONOW, R., AND SPITZ, W.U. (1977). Methadone deaths in children. A continuing problem. *J. Am. med. Ass.* **238**, 2516–17.

SMITH, M.T., LIVINGSTONE, I., HOOPER, W.D., EADIE, M.J., AND TRIGGS, E.J. (1983). Propranolol, propranolol glucuronide, and naphthoxylactic acid in breast milk and plasma. *Therap. Drug Monit.* **5**, 87–93.

SOMOGYI, A. AND GUGLER, R. (1979). Cimetidine excretion into breast milk. *Br. J. clin. Pharmacol.* **7**, 627–8.

SPELLACY, W.N. (1972). Oral contraceptives contraindicated for nursing mothers. *J. Am. med. Ass.* **221**, 1415.

SPENCER, R.P., CORNELIUS, E.A., AND KASE, N.G. (1970). Breast secretion of 99mTc in the amenorrhea–galactorrhea syndrome. *J. nucl. Med.* **11**, 467.

STÉEN, B. AND RANE, A. (1982). Clindamycin passage into human milk. *Br. J. clin. Pharmacol.* **13**, 661–4.

——, —— LÖNNERHOLM, G., FALK, O., ELWIN, C.-E., AND SJÖQVIST, F. (1982). Phenytoin excretion in human breast milk and plasma levels in nursed infants. *Therap. Drug Mon.* **4**, 331–4.

STEWART, R.B., KARAS, B., AND SPRINGER, P.K. (1980). Haloperidol excretion in human milk. *Am. J. Psychiat.* **137**, 849–50.

STUDY GROUP (1978). An evaluation of metronidazole in the prophylaxis of anaerobic infections in obstetrical patients. *J. Antimicrob. Chemother.* **4**, (Suppl. C), 55–62.

SYKES, P.A. AND QUARRIE, J. (1976). Lithium carbonate and breast-feeding. *Br. med. J.* **2**, 1299.

TAKASE, Z., SHIRAFUJI, H., UCHIDA, M., AND KANEMITSU, M. (1975). Laboratory and clinical studies on tobramycin in the field of obstetrics and gynaecology. *Chemotherapy.* **23**, 1399–402.

TAKYI, B.E. (1970). Excretion of drugs in human milk. *J. hosp. Pharm.* **28**, 317–26.

TAYLOR, E.A. AND TURNER, P. (1981). Antihypertensive therapy with propranolol during pregnancy and lactation. *Postgrad. med. J.* **57**, 427–30.

THOMAN, M. (1978). Editorial: News from the American Academy of Clinical Toxicology. Breast-feeding and drugs. *Vet. human Toxicol.* **20**, 246–75.

THORLEY, K.J. AND McAINSH, J. (1983). Levels of the beta-blockers atenolol and propranolol in the breast milk of women treated for hypertension in pregnancy. *Biopharmaceut. Drug Dispos.* **4**, 299–301.

TIMMIS, A.D., JACKSON, G., AND HOLT, D.W. (1980). Mexiletine for the control of ventricular dysrhythmias in pregnancy. *Lancet* ii, 647–8.

TOAFF, R., ASHKENAZI, H., SCHWARTZ, A., AND HERZBERG, M.

(1969). Effects of estrogen and progestagen on the composition of human milk. *J. Reprod. Fertil.* **19**, 475-82.

TODDYWALLA, V.S., JOSHI, L., AND VIRKAR, K. (1977). Effect of contraceptive steroids on human lactation. *Am. J. Obstet. Gynecol.* **127**, 245-9.

TOWNSEND, R.J., BENEDETTI, T., ERICKSON, S., GILLESPIE, W.R., AND ALBERT, K.S. (1982). A study to evaluate the passage of ibuprofen into breast milk. *Drug Intel. clin. Pharm.* **16**, 482 (abstract 72).

TUNNESSEN, W.W. AND HERTZ, G.C. (1972). Toxic effects of lithium in newborn infants: A commentary. *J. Pediat.* **81**, 804-7.

TYLDEN, E., (1973). The effect of maternal drug abuse on the foetus and infant. *Adv. Drug React. Bull.*, No. 38, 120-3.

TYSON, H.K. (1974). Neonatal withdrawal symptoms associated with maternal use of propoxyphene hydrochloride (*Darvon*). *J. Pediat.* **85**, 684-5.

TYSON, R.M., SHRADER, E.A., AND PERLMAN, H.H. (1938). Drugs transmitted through breast milk. Part II. Barbiturates. *J. Pediat.* **13**, 86-90.

UHLIF, F. AND RYZNAR, J. (1973). Appearance of chlorpromazine in mother's milk. *Activ. Nerv. Sup.* **15**, 106.

VAGENAKIS, A.G., ABREAU, C., AND BRAVERMAN, L. (1971). Duration of radioactivity in the milk of a nursing mother following ^{99m}Tc administration. *J. Nucl. Med.* **12**, 188.

VON KOBYLETZKI, D., SCHMITZ, M.S.G., GILLISEN, A.P.J., AND SCHEER, M. (1979). Pharmacokinetic studies with sisomicin in obstetrics. *Infection* **7**, (Suppl. 3), S276.

VORHERR, H. (1973). Contraception after abortion and post partum. *Am. J. Obstet. Gynecol.* **117**, 1002-25.

—— (1974a). *The breast*. Academic Press, New York.

—— (1974b). Drug excretion in breast milk. *Postgrad. med. J.* **56**, 97-104.

WEAVER, J.C., KAMM, M.L., AND DOBSON, R.L. (1960). Excretion of radioiodine in human milk. *J. Am. med. Ass.* **173**, 872.

WERTHMANN, M.W. JR. AND KREES, S.V. (1972). Excretion of chlorothiazide in human breast milk. *J. Pediat.* **81**, 781-3.

WHALLEY, L.J., BLAIN, P.G., AND PRIME, J.K. (1981). Haloperidol secreted in breast milk. *Br. med. J.* **282**, 1746-7.

WHITE, M. (1978). *Breast-feeding and drugs in human milk.* pp. 1-31. La Leche League, Franklin Park, Illinois.

WHITELAW, A.G.L., CUMMINGS, A.J., AND McFADYEN (1981). Effect of maternal lorazepam on the neonate. *Br. med. J.* **282**, 1106-8.

WHITTINGTON, R.M. (1977). Dextropropoxyphene (*Distalgesic*) overdosage in the West Midlands. *Br. med. J.* **2**, 172-3.

—— AND BARCLAY, A.D. (1981). The epidemiology of dextropropoxyphene (Distalgesic) overdose fatalities in Birmingham and the West Midlands. *J. clin. hosp. Pharm.* **6**, 251-7.

WILES, D.H., ORR, M.W., AND KOLAKOWSKA, T. (1978). Chlorpromazine levels in plasma and milk of nursing mothers. *Br. J. clin. Pharmacol.* **5**, 272-3.

WILLIAMS, R.H., KAY, G.A., AND JANDORF, B.J. (1944). Thiouracil. Its absorption, distribution and excretion. *J. clin. Invest.* **23**, 613-27.

WILLIAMS, T. AND GINTHER, T. (1982). Hazard of ophthalmic timolol. *New Engl. J. Med.* **306**, 1485-6.

WILSON, J.T., BROWN, R.D., CHEREK, D.R., DAILEY, J.W., HILMAN, B., JOBE, P.C., MANNO, B.R., MANNO, J.E., REDETZKI, H.M., AND STEWART, J.J. (1980). Drug excretion in human breast milk: principles, pharmacokinetics and projected consequences. *Clin. Pharmacokinetics* **5**, 1-66.

WONG, Y.K. AND WOOD, B.S.B. (1971). Breast-milk jaundice and oral contraceptives. *Br. med. J.* **4**, 403-4.

WYBURN, J.R. (1973). Human breast milk excretion of radionuclides following administration of radiopharmaceuticals. *J. nucl. Med.* **14**, 115-17.

YEUNG, G.T.C. (1950). Skin eruption in new born due to bromism derived from mother's milk. *Br. med. J.* **1**, 769.

YLIKORKALA, O., KAUPPILA, A., KIVINEN, S., AND VIINIKKA, L. (1982). Sulpiride improves inadequate lactation. *Br. med. J.* **285**, 249-51.

YOUNG, R.J. AND LAWSON, A.A.H. (1980). Distalgesic poisoning — cause for concern. *Br. med. J.* **1**, 1045-7.

YUASO, M. (1974). A study of amikacin in obstetrics and gynaecology. *Jpn J. Antibiotics.* **27**, 371-81.

ZINBERG, N.E. (1977). The crisis in methadone maintenance. *New Engl. J. Med.* **296**, 1000-2.

24 Drug-induced sexual dysfunction

J.P. GRIFFIN

Drug-induced impotence and ejaculation failure

In a survey conducted at the Minneapolis Veterans Administration Medical Centre (Slag *et al.* 1983), 1180 men in the medical out-patient clinic were interviewed to determine the presence or absence of impotence: 401 (34 per cent) men admitted to impotence; these men had a mean age of 59.4 years. A comprehensive evaluation of the problem was made for each of the 188 men with impotence who requested help. The cause of the impotence was diagnosed as being due to medication in 25 per cent; psychogenic, 14 per cent; neurological, 7 per cent; urological, 6 per cent; primary hypogonadism, 10 per cent; secondary hypogonadism 9 per cent; diabetes mellitus, 9 per cent; hypothyroidism, 5 per cent; hyperthyroidism, 1 per cent; hyperprolactinaemia, 4 per cent; miscellaneous and unknown causes, 11 per cent. In 7 per cent of these cases there was concurrent alcoholism.

These workers concluded that erectile dysfunction is common in middle-aged men and is frequently overlooked as a problem. The commonest single cause (25 per cent) of this distressing condition is iatrogenic.

Slag *et al.* (1983) report that the commonest medications implicated were diuretics, antihypertensives, and vasodilators. These patients usually gave a history of normal sexual function before the use of the medication. In many instances the patients stopped taking the medication without telling their physicians why and were labelled non-compliant, because they were too embarrassed to tell their doctor the real reason for their failure to take the medicine, i.e. erectile dysfunction.

Patients are reluctant to volunteer information about their sexual difficulties and male patients are particularly loath to admit that they are impotent. Doctors should be aware of this and not assume that a problem not spontaneously reported does not exist. The truth may be elicited more easily from the partner than the patient. A further confounding factor is that any impairment of sexual function may be due to the influence either of drugs (see Table 24.1), or to the influence of disease on the patients at risk. Psychological factors may also be important in certain conditions, for example, in men with previous myocardial infarction normal sexual activity is not resumed within a 12-month period in a large percentage of cases. This is partly due to the patient's own fear, but in other cases the patient's spouse believes the excitement may be detrimental to the patient's health.

In this rather complex area reliable data on the incidence of drug-induced impotence has been obtained in only a small number of well-designed studies (Bulpitt and Dollery 1973; Bulpitt *et al.* 1974); nevertheless, large numbers of case reports of drug-induced impotence and/or ejaculation failure have appeared in the literature. Any attempt to classify such problems on the basis of the pharmacology of the incriminated drug is difficult but certain generalizations can be made.

Drugs which interfere with autonomic nervous function

Failure to achieve or sustain penile erection is known as erectile impotence. Parasympathetic, cholinergic fibres and adequate blood supply are the critical factors at his stage. The next phase is contraction of the vasa deferentia, prostate and seminal vesicles (the internal genitalia) which is under sympathetic adrenergic control and propels the semen into the bulbar urethra; this process is known as emission. It leads immediately and inevitably to ejaculation, the final stage, in which the bladder-neck is tightly closed and rhythmic contraction of the bulbar muscles (striated) results in expulsion of the seminal fluid. Both parasympathetic and somatic motor nerves are involved at this stage. Because they are so closely related, emission and ejaculation can be considered together and failure of this stage is usually referred to as ejaculatory failure. Ejaculation itself may be triggered even in the absence of a proper emission phase and the result is a 'dry orgasm'. Another cause of an apparently dry orgasm is failure to close off the bladder-neck adequately, which results in retrograde ejaculation of the sperm into the bladder.

Adrenergic neurone blocking drugs

Almost all members of this group that are given to ambulant patients on a maintenance dose for periods of

Table 24.1 **Drugs reported to cause impotence and/or ejaculation failure**

1. Antihypertensive drugs (22)

   (a) Adrenergic neurone-blocking drugs
       Bethanidine sulphate (1)
       Clonidine hydrochloride (2, 5, 9)
       Debrisoquine sulphate (3)
       Guanethidine sulphate (1)
       Guanoclor sulphate
       Guanoxan sulphate

   (b) Ganglion-blocking drugs (5)
       Hexamethonium bromide
       Mecamylamine hydrochloride (4)
       Pempidine tartrate
       Pentolinium tartrate

   (c) Enzyme inhibitors
       Methyldopa (8)

   (d) Rauwolfia alkaloids (5)

   (e) Beta-adrenergic blocking agents
       Propranolol (11)

   (f) Other antihypertensive agents
       Prazosin (10)
       Hydrallazine

2. Antidysrrhythmic agents

   Disopyramide (6, 12)
   Perhexiline (13, 16)

3. Cholesterol-lowering agents

   Clofibrate (14)
   Probucol (15)

4. Antidepressants

   (a) Tricyclic antidepressants, e.g. imipramine,
       amoxapin
   (b) Monoamine oxidase inhibitors (MAOI) derived
       from hydrazine, e.g. phenelzine sulphate (17)
                                  iproniazid phosphate (18)

5. Drugs producing gynaecomastia and impotence

   Amphetamines (23)
   Cannabis (23, 25)
   Cimetidine (26, 27)
   Diethylpropion (21, 23)
   Ethionamide (23)
   Fenfluramine (21, 23)
   Haloperidol (20)
   Imipramine
   Mesoridazine (7, 19, 23)
   Phenothiazine derivatives (23)
   Reserpine (6, 23)
   Spironolactone (23)
   Thioridazine (7, 19, 23)

6. Miscellaneous drugs

   Alcohol
   Baclofen
   Liquorice and liquorice derivatives (24)
   Thiabendazole

References:
(1) Bulpitt and Dollery (1973); Bulpitt *et al*. (1974).
(2) ABPI (1979–80), *Catapres*, pp. 149–50.
(3) ABPI (1979–80), *Declinax*, p. 849.
(4) ABPI (1979–80), *Inversine*, pp. 657–8.
(5) Klein (1972).
(6) The *Lancet* (1979).
(7) Kotin *et al*. (1976).
(8) Alexander and Evans (1975).
(9) *Medical Letter on Drugs and Therapeutics* (1977).
(10) Tester-Dalderup (1979).
(11) Warren and Warren (1977).
(12) McHaffie *et al*. (1977).
(13) Howard and Rees (1976).
(14) Schneider and Kaffarnik (1975).
(15) Dukes (1978).
(16) ABPI (1979–80), *Pexid*, p. 682.
(17) ABPI (1979–80), *Nardil*, p. 1062.
(18) ABPI (1979–80), *Marsilid*, p. 862.
(19) Duskalov (1969).
(20) Council of Drugs (1968).
(21) Connell (1977).
(22) Hamilton and Mahapatra (1972).
(23) Griffin (1979).
(24) Langman (1972).
(25) Connell (1972).
(26) Peden *et al*. (1979).
(27) Jensen *et al*. (1983).

Table 24.2 Drugs reported to cause impotence or ejaculation failure presented in descending order of importance. Ranking is based on total reports to the Committee on Safety of Medicines with adjustment for number of years on the UK market and usage based on prescription figures. (With acknowledgements to Dr J.C.P. Weber)

Impotence	Ejaculation failure
Mazindol	Labetolol
Perhexilene	Clomimpramine
Labetolol	Cimetidine
Cimetidine	Propranolol
Atenolol	
Disopyramide	
Metoprolol	
Clofibrate	
Fenfluramine	
Guanethidine	
Hydroflumethiazide	
Clonidine	
Oxprenolol	
Chlorthalidone	
Propranolol	
Methyldopa	

months have been reported to cause both impotence and ejaculation failure, for example, bethanidine, clonidine, debrisoquine, guanethidine, guanoclor, and guanoxan.

Ganglion-blocking drugs

Ganglion-blocking drugs such as hexamethonium, mecamylamine, pempidine, and pentolinium have all been reported to produce impotence.

Drugs with anticholinergic properties

Theoretically any drug with anticholinergic properties may affect sexual function. Tricyclic antidepressants have anticholinergic properties and affect sexual function, but again the extent of the problem is difficult to assess since depression itself may affect sexual ability. Imipramine which is a potent anticholinergic agent is also a prolactin- and gonadotrophin-releasing agent and may cause impotence by hormonal mechanisms.

Disopyramide which is used as an anti-arrhythmic agent is also a potent anticholinergic and was reported by McHaffie *et al.* (1977) to produce impotence.

Alpha-adrenergic blocking drugs

The alpha-adrenergic blocking drug, indoramin is marketed for use as an antihypertensive agent. In a study by Gould *et al.* (1981) 27 patients with essential hypertension were entered in a double-blind trial; seven

patients withdrew due to extreme lethargy (indoramin has potent antihistamine effects) and, of the remaining 20, 13 were male. Four of these 13 patients reported failure of ejaculation while on indoramin.

Beta-adrenergic blocking drugs

Propranolol is the only drug in this class reported to cause impotence (Warren and Warren 1977), other than timolol given as eye drops, *vide infra.*

Drugs producing gynaecomastia and impotence

Prolactin- and gonadotrophin-releasing agents that produce both gynaecomastia and impotence

In this category there are a considerable number of drugs with a widely differing primary pharmacology, for example, amphetamine, chlorprothixene, cimetidine, cyproterone, diethyl-propion, fenfluramine, haloperidol, imipramine, methyldopa, phenothiazines, reserpine, and thioridazine and its active metabolite mesoridazine. In addition, the deliberate induction of high prolactin levels in a number of normal male volunteers using metoclopramide (Falaschi *et al.* 1978) was reported to result in loss of both libido and spontaneous erections; changes in seminal composition were also described in some cases.

Cimetidine was incriminated by Peden *et al.* (1979) in causing loss of libido and failure to achieve erection within a few weeks of commencing therapy. Such, however, is the reluctance to report such a complication that of three patients in Peden's series, one waited 7 months and another 11 months before complaining of their problem, Peden *et al.* (1979) also documented that 23 cases of impotence associated with cimetidine therapy had been reported to the Committee on Safety of Medicines; all these men were over 37 years old. Various mechanisms for this effect of cimetidine on male sexual function have been suggested. The most credible is that cimetidine induces the release of prolactin and gonadotrophins and such events are known to be associated with sexual dysfunction in the male. Cimetidine also exerts an anti-androgenic effect in animals (Leslie and Walker 1977), and this could also be a factor, but how important male sex hormones are in these mechanisms is still unclear since normal erections may occur in castrated men. Adaikan and Karim (1979) pointed out that the smooth muscle of the erectile corpora cavernosum of the human penis *in vitro* was either contracted or relaxed by histamine. The former effect was abolished by mepyramine (a histamine H_1-receptor antagonist), which also potentiates the relaxant effect of histamine on this tissue. The relaxant effect of histamine on the human penis is abolished by burimamide, which is chemically related to cimetidine and is known to antagonize actions of histamine mediated through H_2-receptors. Thus cimetidine, by blocking H_2-receptors on the body of the penis, may prevent erection.

Some indication of the incidence of impotence induced by psychotropic agents was given by Kotin *et al.* (1976). In one group of 57 patients on thioridazine, 34 had difficulties with ejaculation and retrograde ejaculation was prominent. In another series of 60 patients taking other major tranquillizers, sexual problems were reported in 15 cases, usually difficulties with erection. There were no cases of retrograde ejaculation.

Drugs which produce gynaecomastia mimicking oestrogen or progestogen at peripheral sites and have also been reported to cause impotence

Among these are stilboestrol, spironolactone, and eithionamide. Men working in the production of oral contraceptive agents may also be at risk. Beetz and Schiller (1969) described the circumstances in which five out of 11 men closely involved in the production of a chlormadinone/mestranol oral contraceptive preparation developed painful gynaecomastia, loss of libido, and depressed spermatogenesis.

Drugs which may produce gynaecomastia and impotence by various other mechanisms

Examples of such agents are cannabis, morphine derivatives, hydrazine derivatives such as isoniazid, cytotoxic agents including BCNU, busulphan and vincristine, and the MAOI antidepressants phenelzine and iproniazid. In this connection it should be noted that tranylcypromine sulphate, which is a non-hydrazine MAOI, has not been reported to cause gynaecomastia or impotence.

Miscellaneous drugs

A considerable number of drugs have been reported to cause impotence and or ejaculation failure for which no obvious mechanism can be discerned. This group contains the cholesterol-lowering agents, clofibrate and probucol; the antidysrrhythmic agent, perhexiline; liquorice and liquorice derivatives; and baclofen and thiabendazole.

Impotence in male patients with cardiovascular disease

One confounding factor is that any impairment of sexual function may be due either to drugs or to the influence of disease on the patients at risk. Even untreated, 17 per cent of hypertensives complained of impotence (erectile failure) when surveyed by questionnaire as compared with 7 per cent of a normal population of similar age. If indeed hypertension does cause erectile failure the mechanism is obscure. Yet another puzzling finding is that individual patients may experience problems from one drug known to impair sexual function yet not from another with the same propensity.

The group of drugs most likely to affect sexual function are those that act on the sympathetic and parasympathetic systems. Antihypertensive and psychoactive drugs are the two largest categories. Of patients on treatment for hypertension, as many as 43 per cent complained of impotence. Bulpitt *et al.* (1974), in a study in which failure of ejaculation affected 26 per cent of the treated male hypertensives, found no such problems in their normal controls. Thus, while hypertension itself may cause erectile failure, antihypertensive treatment makes the impairment of sexual function worse. Bulpitt found that guanethidine and bethanidine (causing erectile failure in 54 and 68 per cent of patients, respectively) were substantially worse than other drugs, for which the mean incidence of erectile failure was 36 per cent. Guanethidine and bethanidine also caused relatively more failures of ejaculation. No class of antihypertensive drug seems free from an effect on sexual function, however, and case reports testify to the occurrence of erectile failure with methyldopa and clonidine, while diuretics and beta-adrenoceptor blocking agents also cause problems.

The observation that almost any form of antihypertensive therapy appears to cause an increase in the incidence of impotence is further illustrated by the study conducted by the American Veterans Administration of the treatment of hypertension in relatively young patients (Perry 1978). He reported that in two groups of young men with mild hypertension treated with either placebo or chlorthalidone ± 0.25 mg reserpine for 2 years, eight of the 504 patients on placebo reported impotence compared with 61 of 508 patients on active treatment.

The anti-arrhythmic agents perhexiline and disopyramide have also been reported to cause impotence in the patients treated for dysrrhythmias with these drugs, it is difficult to separate the effect of the drug from the effect of the disease state and the associated psychological condition. In the case of impotence associated with disopyramide it would be reasonable to attribute this effect to the anticholinergic effect of the drug, since many other drugs with anticholinergic effects have been reported to cause impotence, e.g. imipramine.

The lipid-lowering agents clofibrate and probucol which have also been reported to cause impotence are likely to be used in the same subpopulation as that in which antihypertensive and anti-arrhythmic drugs are used. Schneider and Kaffarnik (1975) reported three cases of impotence in a series of about 100 patients treated with clofibrate. All three patients had Type IV hyperlipidaemia and two had suffered myocardial infarctions. One patient continued on clofibrate therapy and remained impotent; the other two recovered within 3–4 weeks of discontinuing clofibrate.

The view has been taken that the medical condition, i.e. hypertension or myocardial ischaemia, may be as important a factor as the therapy in producing sexual dysfunction.

Impotence in male patients with glaucoma

However, McMahon et al. (1979) described impotence developing in two male patients treated with the selective β-adrenergic blocking drug, timolol, in ophthalmic drops for their glaucoma. When the timolol eye drops were discontinued, one noted a return of his sexual function while the other noted no change. In McMahon et al.'s cases there was no hypertension to confuse the situation and it must be concluded that the β-adrenergic blocking drug was inducing impotence as a direct pharmacological effect with little psychosomatic overlay.

Glaucoma patients themselves may also have a high incidence of lack of libido and Wallace et al. (1979) described decreased libido in 34 male and five female patients with glaucoma treated with carbonic anhydrase inhibitors acetazolamide, ethoxazolamide. The male patients in this series ranged from 19 to 72 years and the females from 27 to 47 years. In a number of these patients there was rechallenge, and reduced libido and/or impotence recurred on restoring carbonic anhydrase therapy.

Antidepressant drugs

Both depression and antidepressant therapy have been reported to be associated with reduced libido and impotence. However, there are good pharmacological mechanisms to explain the production of impotence and ejaculatory failure with most antidepressant drugs.

Two reports of painful ejaculation with amoxapine have appeared in the literature. Schwarcz (1982) reported the case of a 33-year-old man who had suffered from depression which was unresponsive to drug treatment for 10 years, who was given amoxapine 50 mg three times daily after a recurrence of depression. The next day he noted the onset of painful ejaculation but as his depression was relieved, he continued to take amoxapine for a further seven days. Within one week of stopping the drug, his ejaculation became normal but his depression

increased so amoxapine 50 mg three times daily was reinstituted. Although his depression was again relieved, the ejaculatory disturbance recurred and was not relieved by halving the total daily dose to 75 mg. Amoxapine was therefore discontinued. The second report by Kulik and Wilbur (1982) concerned a 29-year-old man with a two-year history of anxiety and depression who was given 75 mg amoxaphine at bedtime. Three days later he had a painful spasmodic ejaculation. During the five weeks he took amoxapine, painful ejaculation occurred on 15 occasions. The drug was stopped and three to four days later his ejaculatory function was normal. Maprotiline 75 mg at bedtime was substituted for amoxapine and the problem did not recur.

Non-steroidal, anti-inflammatory agent

Indomethacin

Del Favero (1981) reported the case of a 23-year-old man who became impotent while on an eight-day course of indomethacin. The subject had not had previous problems of this nature.

Naproxen

Hazelman et al. (1979) reported a case of impotence associated with naproxen therapy; Wei and Hood (1980) described the case of a 66-year-old male who developed ejaculatory dysfunction while on a short course of naproxen. Ejaculatory function returned to normal within one week of stopping naproxen, but after reinstitution of the drug he was again unable to ejaculate.

The mechanism of action for this side-effect of non-steroidal, anti-inflammatory agents has been attributed to effects on prostaglandin synthesis. Similar reports were made to the Committee on Safety of Medicines relating to the following non-steroidal, anti-inflammatory agents:

Benoxaprofen, flurbiprofen, and sulindac — three reports each;
Phenylbutazone, ketoprofen, mefanamic acid, and indomethacin — two reports each.

If this is a real association of an adverse reaction with this class of drugs, it is comparatively rare in view of their widespread use.

Drugs causing male infertility

Anti-leprosy agents—dapsone

Dapsone was reported to cause infertility in two male

patients with recovery of spermatogenesis after stopping the drug (Grieve 1979).

Antibiotics

Timmermans (1974) demonstrated that gentamicin, oxytetracycline, spiramycin, and framycetin impaired spermatogenesis in the albino rat. Timmermans examined human testis biopsies in patients treated with gentamicin prior to prostatic surgery and demonstrated interrupted spermatogenesis prior to meiosis.

Cyproterone acetate

Cyproterone is used to control libido in severe hypersexuality and or sexual deviation in the male; it acts as an anti-androgen by blocking androgen receptors. While this drug produces loss of libido and impotence it cannot be classified as a side-effect since this is the therapeutic objective; it does, however, also produce gynaecomastia.

Cyproterone causes reduction of sperm count and volume of the ejaculate and infertility is usual. There may be azoospermia after eight weeks and also slight atrophy of the seminiferous tubules. Spermatogenesis usually recovers three to five months after stopping this agent but it may be delayed up to 20 months. There is evidence that abnormal sperms which might give rise to malformed embryos are produced during such treatment (Androcur Data Sheet).

Pesticides

Whorton *et al.* (1977) reported a number of cases of infertility discovered among men working in a Californian pesticide factory. The suspected cause was exposure to the chemical 1,2-dibromo-3-chloropopane (DBCP). The major effects, seen in 14 of 25 non-vasectomized men, were azoospermia or oligospermia and raised serum-levels of follicle-stimulating and luteinizing hormones. No other major abnormalities were detected, and testosterone levels were normal. Although a quantitative estimation of exposure could not be obtained, the observed effects appeared to be related to duration of exposure to the pesticide,

Cytotoxic agents

After modern chemotherapy programmes fertility in the male is usually so impaired that procreation is impossible; reports of fertility after treatment are therefore noteworthy. One leukaemic male treated for 18 months with methotrexate and 6-mercaptopurine, fathered a child while on drugs and against medical advice. The mother after an uneventful pregnancy produced a normal child (Kroner and Tschumi 1977). Lillyman (1979) described a man with acute lymphoblastic leukaemia treated for two years with cytarabine but no cyclophosphamide who sired a normal daughter three years after stopping therapy. Hinkes and Plotkin (1973) recorded a similar case. In contrast, two cases were recorded by Russell *et al.* (1976) in which both fathers had received cytosine arabinoside, thioguanine, and daunorubicin; both of their children had major abnormalities, namely Fallot's tetralogy and anencephaly, respectively. Both fathers were just past the intensive period of therapy at the time of conception when genetic damage would be expected to be at its maximum.

The results of the most comprehensive survey on the effect of cyclical combination chemotherapy on male gonadal function were reported by Chapman *et al.* (1979c). The effect of cyclical chemotherapy on fertility and gonadal function was investigated in 74 male patients who had been treated for advanced Hodgkin's disease. All patients were azoospermic after therapy, and, with a median follow-up period of 27 months (range 1–62 months), only four patients had regained spermatogenesis. Testicular biopsy showed an absence of germinal epithelium without other gross architectural changes. Despite this high degree of infertility, 60 per cent of patients were practising contraception. A decline in libido and sexual performance with frequent long periods of sexual inactivity was noted by most men during therapy. Although some recovery was apparent once therapy was stopped, this was incomplete in approximately half the number of patients. Follicle-stimulating hormone levels were consistently raised after therapy at all periods of study. Median luteinizing-hormone levels were at or just above, upper limit of normal and median testosterone levels were normal. Increased prolactin levels were seen in 42 per cent of patients, of whom about a half had an identifiable cause for hyperprolactinaemia. Return of spermatogenesis could not be predicted by serial hormone assessment. Because of the guaranteed infertility and the low frequency and unpredictability of recovery of spermatogenesis, sperm storage should be available for male patients undergoing cytotoxic therapy, since most of these patients may enjoy prolonged survival, hormone-replacement therapy will usually be unnecessary. However, the probability of major changes in libido and sexual performance should be discussed with patients so that additional stress can be avoided. Contraceptive advice should be available to those who require it.

Nitrofurantoin

Yunda and Kushniruk (1974*a, b*) and Yunda *et al.* (1974) in a series of studies in male albino rats indicated that nitrofurantoin affected spermatogenesis in the late stages, and caused a reduction in the nucleic acid content in the spermatocyte and spermatid series. The spermatozoa formed had reduced motility. Full recovery of the process of spermatogenesis occurred on withdrawal of nitrofurantoin. These workers claimed that the toxic effect of nitrofurantoin on the testis in the rat could be prevented by cystine and vitamin C simultaneously with the nitrofurantoin.

In man, Nelson and Bunge (1957) reported a transitory reduction in sperm counts due to maturation arrest if the nitrofurantoin dosage exceeded 10 mg/kg/day in about a third of men so treated. Iunda and Kushniruk (1975) found a depressant effect of nitrofurantoin on spermatocyte count, sperm motility, and ejaculation volume.

Smoking

Evans *et al.* (1981) took sperm samples from a carefully matched group of 43 cigarette-smokers and 43 non-smokers attending an infertility clinic. The smokers were found to have a greater percentage of abnormal sperm forms that the nonsmokers. It was concluded that the sperm abnormalities in smokers may reflect genetic damage to these cells as a consequence of exposure to smoke products.

Sulphonamide-trimethoprim combinations

Murdia *et al.* (1978) reported decreases in sperm count in 14 male patients out of 40 male patients treated with a sulphonamide-trimethoprim combination product co-trimoxazole) for 7–10 days; the sperm counts were taken before and one month after co-trimoxazole treatment. The fall in count varied from 6.7 to 88 per cent and in all but four patients the fall was over 50 per cent of the initial count. Guillebaude (1978) pointed out that in 42 per cent of Murdia *et al.*'s patients, a rise in sperm count was noted. It was also pointed out that the fall occurred within one month of treatment and in the human male spermatogenesis takes 70–80 days and that any effect on spermatogenesis would not be expected to be apparent in less than this time. Murdia *et al.*'s report is therefore unlikely to indicate any real risk.

Sulphasalazine (Salazopyrine)

Levi *et al.* (1979) reported infertility associated with oligospermia in four patients, all immigrants from Asia, or of Asian origin, whose ulcerative colitis was controlled with sulphasalazine. No other cause for infertility was found in any of them nor in their wives. Findings of semen analyses rapidly improved in all patients on withdrawal of sulphasalazine which resulted in four pregnancies in three of the wives. Reintroduction of sulphasalazine was followed by rapid deterioration in the semen of two patients.

Toth (1979*a*) reported six cases of male infertility associated with sulphasalazine therapy, and Toth (1979*b*) reported that a further four cases had been seen. Out of the 10 treated patients, six managed to stay off the drug, and all six wives reported pregnancies. It takes about three months for complete regeneration of the spermatogenesis within the testes. The effect seems to be completely reversible. A characteristic cell type was present in all the cases examined — namely a ballooned enlarged spermatozoon with a pale-staining head structure. The toxic effect of the drug may be due to a breakdown product structurally related to salicylic acid and one which blocks prostaglandin synthesis.

Two studies into the effect of sulphasalazine on male fertility were conducted — one by a group of workers in London (Toovey *et al.* 1981) and the second by a group in Glasgow (Birnie *et al.* 1981). In London 18 of the 28 patients had gross abnormalities of their semen but no endocrine abnormality: the semen improved two months after stopping sulphasalazine and 10 pregnancies occurred. In Glasgow 18 or the 21 patients had abnormal semen and 15 had oligospermia. One patient discontinued treatment and had a normal semen-analysis result two months later. The mechanism of the side-effect is not clear, but it seems that either sulphasalazine or a metabolite is toxic to developing spermatozoa. However, some men remain fertile while having treatment. In Glasgow five patients had become fathers while taking sulphasalazine; one baby was stillborn but the others were normal.

Tobias *et al.* (1982) reported the case of a 39-year-old white male with a 15-year history of ulcerative colitis who had been treated for one year with sulphasalazine. He was investigated because of infertility following 18 months marriage and was found to have severe oligospermia and low serum testosterone. Three months after stopping sulphasalazine the testosterone level returned to normal and semen showed sperm counts compatible with fertility. His wife's pregnancy was confirmed five months after stopping sulphasalazine.

Drugs affecting libido

Old herbals are full of recipes for aphrodisiacs which

were usually based on the doctrine of signatures, i.e. the physical appearance of the substance being used resembled the sexual organs. On this basis the roots of mandrake and ginseng for their phallic resemblance and oysters because of their resemblance to the vulva, were used as aphrodisiacs.

In modern pharmacology Buffum (1982) identified the following drug substances which have been used to enhance male sexual function: levodopa, volatile nitrites, bromcriptine, zinc salts, parachlorphenylalinine, L-tryptophan, yohimbe, pheromones, clomiphene, luteinizing hormone (LH), and naloxone. None of these substances have been spectacularly successful.

Endralazine

Zacest and Reece (1983) in a study of endralazine, an antihypertensive agent, in 10 healthy male volunteers aged 20–34 gave intravenous infusion of 0.05 mg/kg infused over five minutes. Four volunteers experienced sexual arousal with penile erections, three others experienced psychological sexual arousal, and three experienced no such effects. In the seven volunteers reporting sexual arousal the effect occurred within one hour of administration.

Flufenazine

The endocrine side-effects of fluphenazine are usually manifest by impotence in men and increased libido in women. Two case reports by Gomez (1981) are at variance with this general picture.

A 25-year-old man with six-year history of schizophrenia was treated with fluphenazine 25 mg i.m. every three weeks. As he improved the dose was decreased to 12.5 mg. Six weeks after the decrease, his 30-year-old sister reported that 'he was horny all the time' and was masturbating and frequently propositioning the neighbourhood girls. He was receiving no other medication. Laboratory studies showed no changes except increased prolactin (16.5 ng/ml) and luteinizing hormone (62.5 mIU/ml). Fluphenazine was withdrawn but four weeks later he was hospitalized after a relapse.

A 28-year-old man with a four-year history of chronic schizophrenia, had been in satisfactory remission for about one year, on fluphenazine 12.5 mg i.m. every three weeks. His wife reported that he was demanding sex every day and when she refused he went to pornographic movies and engaged in sexual activity with prostitutes.

Laboratory tests were once again negative except for increased prolactin (16.7 ng/ml) and luteinizing hormone (40 mIU/ml). He was switched to a dihydro-indolone compound and his hypersexuality decreased moderately.

Effect of drugs on spermatozoal motility

Since the development of a transmembrane migration method, based on the ability of progressive forward moving sperms to move through 5 μm pores of a nucleopore membrane by Hong et al. (1981a) it has been possible to measure objectively the dose–response relationship between drugs and sperm motility.

Turner and his co-workers (Hong et al. 1981a, b, 1982) showed that drugs which belong to different therapeutic classes but possess a common local anaesthetic property, such as propranolol, D-propranolol, lignocaine, procaine, tetracaine, and chlorpromazine, all inhibited human sperm motility with similar dose–response curves. Their sperm-immobilizing potencies correlated with their local anaesthetic potencies measured on nerve preparations.

The same group of workers in a study of five different beta-adrenoceptor blocking drugs, which had different anaesthetic potencies as well as varied lipid solubilities, also showed that the sperm-immobilizing effect was related to local anaesthetic activity. The concentrations of these drugs which were required to inhibit human sperm motility fell in the millimolar range. Although information concerning the presence and quantity of these drugs in the genital tract is not available, they are unlikely to reach such high concentrations in the genital tract after systemic administration. For propranolol the concentration required to immobilize sperm is more than 1000 times higher than its therapeutic serum concentration. Thus the application of any of these drugs as a systemic contraceptive is not promising, and their direct adverse effect on sperm motility as a cause of drug-induced infertility cannot be substantiated.

Caffeine was demonstrated in this in-vitro model (Hong et al. 1981b) to antagonize the inhibitory effect of procaine and propranolol on sperm motility. Caffeine itself increases sperm motility in vitro. It is unknown whether caffeine has any significant effect on sperm motility in vivo, but caffeine ingestion by both male or female partner could theoretically increase sperm motility in the vagina.

Turner and his co-workers (Verma et al. 1983) demonstrated that indoramin, a selective α-adrenoceptor blocking agent, also inhibits sperm motility in vitro.

On the basis of these studies clinical trials have been initiated to investigate an intravaginal preparation of D-propranolol as a contraceptive agent in several centres. A report by Zipper *et al.* (1983) from Chile showed that the efficacy and tolerability of 80-mg propranolol tablets as a vaginal contraceptive were studied in 198 fertile women for 11 months. The calculated one-year life table pregnancy rate was 3.4/100 women and the Pearl index was 3.9/100 women years. No major adverse effects were encountered.

The findings suggested that propranolol is an effective vaginal contraceptive whose failure rate compares favourably with that of other methods of contraception.

Reduction in serum and salivary testosterone

Ketoconazole

De Felice *et al.* (1981) described three cases of gynaeco-mastia associated with exposure to the antifungal agent, ketoconazole (an imidazole derivative). Other reports have referred to gynaecomastia, loss of libido, and impotence associated with ketoconazole. Work by Schurmeyer and Neilschlag (1982) indicated that these effects may be associated with a ketoconazole-induced reduction of plasma and salivary testosterone. The West German doctors reported a study on five healthy male volunteers who received a single oral dose of 400 mg ketoconazole. Both serum and saliva testosterone concentrations fell sharply, the lowest values being obtained four to six hours after drug ingestion, when serum and saliva levels had fallen to 30 and 34 per cent of basal levels, respectively, well within the range of values found in hypogonadal men. Pharmacokinetic studies revealed a reciprocal relationship between testosterone levels and ketoconazole levels.

Pointing out that the salivary testosterone concentration is good index of serum free testosterone, the West German researches conclude that the parallel decrease in total serum testosterone is accompanied by a decrease in available testosterone at the target organs. Investigations of long-term effects of ketoconazole treatment on hormonal indices and fertility should be carried out. The effects of ketoconazole are therefore that of temporary chemical castration.

Nitroimidazoles

In the search for male antifertility agents, the nitroimidazole derivatives have been regarded as a particularly potent group of compounds for development. Administered to immature male rates for 14 or 30 days, five nitroimidazole compounds were effective in producing marked testicular weight reduction when administered in doses of 100 mg/kg body weight. In histological section

of these testes the oldest type of all present was the pachytene spermatocyte; these were significantly reduced in numbers and appeared to be undergoing degeneration. Many tubules contained only spermatogonia and Sertoli cells or even Sertoli cells alone. Leydig cells were morphologically unchanged. The mode of action of the nitroimidazoles appears to differ from that of the imidazole derivative ketoconazole.

Niridazole

Niridazole (*Ambilhar*) is 1-(5 nitro-2 thiazolyl)-2-oxo-tetra hydroimidazole and is an effective antibilharzial agent. Because of the antifertility effect of niridazole in animals on the meiotic stage of spermatogenesis, the effect of niridazole in human males was studied (El Beheiry *et al.* (1982) by semen analyses performed twice, one week apart, and by testicular biopsy in 20 male bilharzial patients. Niridazole was found to induce defective spematogenesis in the form of spermatocyte arrest or germinal cell hypoplasia. The effect was transient, with recovery occurring within three months of ceasing therapy. Use of niridazole should be embarked upon with caution in male patients who are subfertile.

Drug-induced priapism

Priapism, persistent painful erection of the penis, is rare but has many causes including drugs, particularly the phenothiazines (Meiraz and Fishelovitch 1969), chlorpromazine, and thioridazine (Dorman and Schmidt 1976). How phenothiazines cause priapism is not clear but it has been attributed to their ability to block alpha-adrenergic receptors. Parasympathetic dominance seems to encourage erection, and sympathetic blockade seems to inhibit ejaculation and detumescence.

Priapism has been described in patients taking guanethidine, and the direct acting vasodilator hydrallazine (Rubin 1968). Two cases of priapism in West Indian patients taking prazosin for hypertension were described by Bhalla *et al.* (1979). The mechanism of production of the priapism was attributed to the alpha-adrenergic blocking action of prazosin.

Law *et al.* (1980) described priapism in a 25-year-old male on labetolol and attributed this to its alpha-adrenergic blocking effect.

Drug-induced Peyronie's disease

Peyronie's disease is the formation of fibrous tissue, usually as plaques, in the erectile tissue of the corpora cavernosa with limitation of erectile function on the affected side which usually prevents intercourse. Kristensen (1979) described a case of Peyronie's disease occurring in an 18-year-old man who was being treated for severe hypertension with the alpha- and beta-adrenergic blocking agent labetolol. They also reviewed the literature reports of 146 patients with Peyronie's disease; 19 had been treated with a beta-adrenergic blocker (12 propranolol, six practolol, and one with both). Peyronie's disease has also been described with metoprolol.

Drug-induced changes in female libido

While it has been known for many years that drugs may affect male libido or produce impotence, the problems of drug-induced impairment of female sexuality have been largely overlooked. Drug-induced reduction in female libido has been described in association with oral contraceptives, tranquillizers, and more recently chemotherapeutic agents.

Cytotoxic-induced ovarian failure

It has been recognized that some alkylating agents administered year after year in chronic leukaemia (busulphan, chlorambucil) may eventually cause sterility in men. But the medical profession were unaware that in women the weekly or monthly pulses of drugs used in Hodgkin's disease destroyed libido, ruining personal relations and disrupting families — this in addition to causing sterility. An unbelievable amount of human misery has been silently borne by the patients for three decades.

The problem of ovarian failure in women treated with cytotoxic drugs for Hodgkin's disease was revealed by a study from St. Bartholomews Hospital, London by Chapman and her colleagues (1979a). Forty-one women who had been treated with chemotherapy for Hodgkin's disease between 1969 and 1977 were evaluated for ovarian function. All but one of the patients had received standard MVPP therapy, and the remaining patient had received one course of MVPP and nine courses of chlorambucil-VPP. Of these, 21 women had received additional MVPP as maintenance therapy or had received other agents (doxorubicin, bleomycin, cyclophosphamide, and chlorambucil alone or in combination) because the disease had relapsed after MVPP. At the beginning of the study, seven women still received chemotherapy, and 34 had finished therapy from two to 62 months previously. All patients were further observed during a 16-month period.

Histories and pretreatment ovarian biopsy specimens indicated normal fertility before therapy, thus implying no adverse effect of Hodgkin's therapy by menstrual history, serial basal body temperatures, and hormonal levels. Each case was assigned to one of three categories primary ovarian failure (failed ovary), irregular ovarian activity (failing ovary), and normal cyclic ovarian activity (functioning ovary). After therapy, 20 of 41 patients (49 per cent) were categorized as failed, 14 (34 per cent) as failing, and only seven (17 per cent) as functioning in 16 months of further observation, progressive loss of ovarian function occurred that was clearly age-related but not statistically dose-related. Induction of premature, irreversible menopause presented a need for effective hormonal replacement and patient counselling. The fact that these women experienced considerable distress as a result of ovarian failure came to light as a result of the letter written by a 21-year-old woman with Hodgkin's disease to Dr Ramona Chapman complaining of no periods for three months, hot flushes, irritability, and loss of libido and consequent sexual problems. To investigate the extent of this problem in women treated, a questionnaire was submitted to the same 41 patients evaluated over a 16-month period in the previous study (Chapman et al. (1979a) and the results reported in a separate paper (Chapman et al. 1979b). Each patient's sexual history was recorded in a questionnaire completed by the patient and her physician in conference. She was asked to recall her behaviour before, during, and after therapy under the following headings: (a) libido and sexual function; (b) menstrual function; (c) symptoms such as flushes, irritability, poor sleep pattern, dry vagina, painful intercourse, and decreased libido; (d) separation and divorce; (e) other relationships.

When the results of the questionnaires were analysed the following results were obtained.

(a) Libido and sexual function

Of 37 women, 34 (92 per cent) stated that their libido was strong or moderate before therapy. During therapy 13 women (35 per cent) stated their libido was none for the first time, and after completion of therapy (median time after therapy 36 months), 27 women (73 per cent) stated they had mild or no libido. When pressed, some women who claimed mild libido after therapy in fact had none, in that they never would have desired sexual activity themselves, but engaged in it only to please their partner; they experienced no arousal in sexual foreplay.

(b) Menstrual function

Menstrual changes reflected those of libido ratings. Whereas 82 per cent had regular menses before therapy, amenorrhea became a major factor during therapy (47 per cent) and by the time that the majority of women had discontinued therapy for approximately three years, only 12 per cent had regular menses, while those groups with irregular menses (26 per cent) and amenorrhea (62 per cent) had increased proportionately.

Symptoms

A constellation of symptoms, including hot flushes, irritability, poor sleep pattern, dry vagina, painful intercourse, decreased libido, and a poor self-image, was present in part or entirely in women with chemical castration and amenorrhea.

(d) Separation and divorce

Except in two cases, disintegration of relationship occurred only in women of 31 years or younger; 33 women were at risk as a partner in a couple. Ten couples experienced stress strong enough to cause separation or divorce (30 per cent). From the years 1969 to 1977, the cumulative divorce rate in Great Britain in women aged 25 to 30 years was 8 per cent (0.9 per cent yearly). Thus, the couple in this study suffered a severance rate nearly four times that of the general population.

(e) Other relationship

Loss of friendships, rows in the family, and irritability with children, or loss of jobs were common. Women younger than 40 years of age at the onset of ovarian dysfunction, suffered these symptoms more intensively and more frequently than older women.

Hormone replacement was given to women with failed ovarian function using the following regime medroxy-progesterone acetate 5 mg orally for the first seven days of each calendar month, and ethinyl oestradiol, 30 to 50 μg daily from day 8 to the end of each calendar month.

Hormonal replacement resulted in a dramatic relief of symptoms, especially loss of hot flushes and irritability and return of normal libido. Although hormone replacement was remarkable only one of the 10 broken couples reunited; all the other men were so bitter that they refused to see the investigators.

Other drugs causing diminished female libido

In view of the serious effects of drug-induced loss of libido on a woman's life, drugs with this action should be used with a degree of reluctance and only after explaining that this side-effect may occur and ensuring that the woman and her partner are aware of its nature.

The effects of oral contraceptive steroids on sexual function in females may also be influenced by the psychological effects of using a reliable contraceptive method; libido may increase when the risk of unwanted pregnancy is removed. Oral contraceptives can induce depression, but diminished sexual response can occur in the absence of clinical features of depression. Adams and his colleagues (1978) studied the effect of contraceptives on the mid-cycle peak in female-initiated sexual activity, which is probably hormone-dependent. This peak was abolished by the oral contraceptive but not by other contraceptive techniques.

Cyproterone acetate is used as a contraceptive in West Germany, but is not licensed for use in the United Kingdom. In one study of 500 cycles in 602 women, loss of libido occurred in 10 per cent of patients on the preparation. This loss of libido appeared to be more common with cyproterone acetate than with danazol. This latter agent, an inhibitor of pituitary gonadotrophins, is used in the treatment of endometriosis and occasionally in the treatment of fibrocystic disease. Danazol is commonly reported to cause loss of libido (Westerholm 1977).

Fenfluramine, an antiobesity agent, when used in high doses of 240 mg/day produces loss of libido as its main side-effect. At this dosage 85 per cent of women treated may suffer loss of libido (Connell 1977).

Methimazole, an antithyroid drug, was reported by Hempel and Hester (1976) to cause reversible loss of libido in a 34-year-old hyperthyroid woman during two consecutive courses of treatment with the drug. Other drugs reported to cause loss of libido are the tranquilizers sulpiride and lenperone, and the antimigraine drug pizotifen.

Substances reported to cause increased female libido

It is unusual for women to complain of increased libido or for it to be regarded as an adverse effect of therapy by the patient herself or her partner. In fact the search for an effective and reliable aphrodisiac has been sought since time immemorial; mandrake root has been reputed to have this effect and was known by Jacob's wife Leah (*Genesis* 30, verses 15 *et seq.*). Various preparations have been popularly acclaimed to act as aphrodisiacs, such as cantharides (Spanish Fly) and various strychnine-containing formulations. Ginseng has been promoted as an aphrodisiac but there is little evidence to support such a claim. Ginseng does contain small quantities of oestrone, oestradiol, and eostriol in the root and Palmer *et al.* (1978) described ginseng-induced swelling of the breasts with diffuse nodularity and tenderness and this may be relevant in this present context. Friesen (1976) reported that about 5 per cent of female patients treated with mazindol have reported an aphrodisiac effect. One woman aged 40 years had such strong craving for sexual intercourse, that her husband who worked away from home during the week, had to change his job to be home each night.

Drugs of abuse are generally believed to decrease libido. In former heroin addicts the dose of methadone required to control withdrawal effects is inversely correlated with sexual activity of both males and females. Heroin use is itself associated with decline in sexual function. However, a significant proportion of female alcoholics believe that alcohol increases sexual desire and enjoyment.

Inhaled nitrites have been abused for the purpose of increasing sexual enjoyment, particularly by male homosexuals, and various pharmacological actions, apart from vasodilation, have been postulated to explain the alleged stimulant effects.

A number of drugs, including vitamins, are reputed to increase sexual desire and performance, but the benefits may well be more imaginary than real. In a study of the effect of vitamin E therapy there was no significant effect on sexual arousal or behaviour, but subjects taking the vitamin had significantly more nonsexual adverse effects (Herold et al. 1979). Expectation of effect is clearly the explanation, and this is strikingly illustrated by one report from Australia in which a patient attributed her change in libido to an allergy to the tartrazine content of the tablet coating of an oral contraceptive steroid (Williams 1979).

Phenelzine-induced orgasmic failure

Phenelzine, a MAO inhibitor used to treat a women of 24 years of age for agrophobia, was initially given at a dose of 45 mg daily with some improvement. When the daily dose of phenelzine was increased to 60 mg daily almost complete relief was obtained but was accompanied by failure to achieve an orgasm. The ability to have an orgasm returned when the dose was lowered. These dose-related effects were repeated on several occasions over the succeeding months, orgasmic inhibition and improvement of agrophobia occurring at the higher dosage of phenelzine (Barton 1979). (Phenelzine was also reported to cause ejaculatory failure in two men.) Further cases of phenelzine orgasmic failure in young sexually active women were described in three cases by Moss (1983) and one case by Pohl (1983).

Degan (1982) incriminated trifluoperazine and fluphenazine decanoate as causes of anorgasmia in young women. Anorgasmia associated with thioridazine was described in a 26-year-old woman by Shen and Park

(1982). Two cases of clomipramine-induced anorgasmia were described by Quirk and Einarson (1982).

Drugs affecting female fertility

In many reviews of drug-induced sexual dysfunction, a pessimistic note has been sounded with respect to a woman's fertility following 'pill-induced amenorrhoea' or following cytotoxic therapy for malignant disease. The studies cited below give a more optimistic picture.

Oral contraceptives

In a study by Hull et al. (1981) of women with secondary amenorrhoea after oral contraceptive use in 48 women in whom primary ovulation failure was excluded, comparison was made with 47 women whose secondary amenorrhoea did not occur following oral contraceptive use. Both groups of women were subjected to treatment aimed at inducing ovulation; for example, patients with hyperprolactinaemia received bromcriptine; others where appropriate were treated with clomiphene. In patients with 'post-pill' amenorrhoea, 91 per cent became pregnant within 12 months and 98 per cent within 24 months; this did not differ from the conception rate in patients whose amenorrhoea was not due to oral contraceptive therapy.

Cytotoxic drugs

Rustin et al. (1981) analysed data from 611 women treated with cytotoxic drugs for their chorioncarcinoma. There were 104 deaths from the initial disease, 83 questionnaires were not returned, and 38 patients from overseas were not traceable. In all, obstetric histories were obtained for 375 women, of whom 177 had become pregnant since finishing chemotherapy. There were a total of 315 pregnancies resulting in 245 live births from 167 women and 76 miscarriages or terminations. Only five women who wished to become pregnant had failed to do so. All women had received methotrexate and 40 women who had become pregnant had also received cyclophosphamide and actinomycin.

REFERENCES

ABPI 1985-6). *ABPI data sheet compendium 1985-6*. Datapharm Publications Ltd, London.
ADAIKAN, P.G. AND KARIM, S.M.M. (1979). Male sexual dysfunction during treatment with cimetidine. *Br. med. J.* 1, 1282-3.
ADAMS, D.B., GOLD, A.R., AND BURT, A.D. (1978). Rise in female-ini-

tiated sexual activity at ovulation and its supression by oral contraceptives. *New Engl. J. Med.* **299**, 1145–50.

ALEXANDER, W.D. AND EVANS J.I. (1975). Side effects of methyl-dopa *Br. med. J.* **2**, 501.

BARTOS, J.L. (1979). Orgasmic inhibition in a young woman. *Am. J. Psychiat.* **136**, 1616.

BATEMAN, D.N. (1980). Drugs and sexual dysfunction. *Adverse Drug React. Bull.* **85**, 308–11.

BEETZ, D. AND SCHILLER, F. (1969). Andrologische Verande-rungen bei Beschaftigen der Producktion Oraler Kontra-zeptive, *Z. ges. llyg.* **15**, 924.

BHALLA, A.K. HOFFBRAND, B.I. PHATAK, P.S., AND REUBEN, S.R. (1979). Prazosin and priapism. *Br. Med. J.* **2**, 1039.

BIRNIE, G.G. McLEOD, T.I.F., AND WATKINSON, G. (1981). Incidence of sulphasalazine induced male infertility, Gut **22**, 542–5.

BUFFUM, J. (1982). Pharmacosexology: the effects of drugs on sexual function. A review. *J. psychoactive Drugs* **14**, 5–44.

BULPITT, C.J. AND DOLLERY, C.T. (1973). Side effects of hypoten-sive agents evaluated by a self-administered questionnaire, *Br. med. J.* **3**, 485–90.

——, ——, AND CARNE, S. (1974). A symptom questionnaire for hypertensive patients. *J. chron. Dis.* **27**, 309–23.

CHAPMAN, K.M. SUTCLIFFE, S.B., AND MALPAS, J.S. (1979*a*). Cyto-toxic-induced ovarian failure in women with Hodgkin's disease. I. Hormone failure. *J. Am. med. Ass.* **242**, 1878–81.

——, ——, AND —— (1979*b*). Cytotoxic-induced ovarian failure in Hodgkin's Disease 2. Effects on sexual function *J. Am. med. Ass.* **242**, 1882–5.

——, ——, REES, L.H., EDWARDS, C.K.W., AND MALPAS, J.S. (1979*c*). Cyclical combination chemotherapy and gonadal function. Retrospective study in males. *Lancet* **i**, 285–9.

CONNELL, P.H. (1972). Hallucinogens. In *Side effects of drugs,* Vol. 7 (ed. L. Meyler and A. Herxheimer), p. 38. Excerpta Medica, Amsterdam.

—— (1977). Hallucinogens. In *Side effects of drugs annual.* Vol. 1 (ed. N.M.G. Dukes), p. 4. Excerpta Medica, Amsterdam.

COUNCIL OF DRUGS (1968). Evaluation of a new anti psychotic agent Haloperidol, *J. Am. med. Ass.* **205**, 577.

DE FELICE, R., JOHNSON, D.G. AND GALGIANI, J.N. (1981). Gyneco-mastia with ketoconazole. *Antimicrob. Agents Chemother.* **19**, 1073–4.

DEL FAVERO, A. (1981). Anti-inflammatory analgesics and drugs used in the treatment of rheumatism and gout. In *Side effects of drugs annual* no. 5 (ed. M.N.G. Dukes), pp. 88–117. Excerpta Medica, Amsterdam.

DEGAN, K. (1982). Sexual dysfunction in women using major tranquillisers. *Psychosomatics* **23**, 959–61.

DORMAN, B.W. AND SCHMIDT, J.D. (1976). Association of priapism in phenothiazine therapy *J. Urol.* **116**, 51–3.

DUKES, N.M.G. (1978). Drugs affecting lipid metabolism. In *Side effects of drugs annual.* Vol. 3 (ed. N.M.G. Dukes), p. 360. Excerpta Medica. Amsterdam.

DUSKALOV, D. (1969). Behandlung des stotterns mid dem neuen midium tranquilizer Mesoridazin. *Praxis* **58**, 630.

EL-BEHEIRY, A.H., KAMEL, M.N., AND GAD. A. (1982). Niridazole and fertility in bilharzial men. *Arch. Androl.* **8**, 297–300.

EVANS, H.J., FLETCHER, J., TORRANCE, M., AND HARGREAVE, T.B. (1981). Sperm abnormalities and cigarette smoking. *Lancet* **i**, 627–9.

FALASCHI, P., FRAJESE, G., SCIARRA, F., ROCCA, A., AND CONTI, C. (1978). Influence of hyperprolactinaemia due to metoclopra-mide on gonadal function in men. *Clin. Endocrinol.* **8**, 427–33.

FRIESEN, L.V.C. (1976). Aphrodisia with mazindol. *Lancet* **ii**, 974.

GOMEZ, E.A. (1981). Hypersexuality in men receiving fluphena-zine deconoate. *Am. J. Psychiat.* **138**, 1263.

GOULD, B.A., MARM, S., DAVIES, A.B., AND RAFTERY, E.B. (1981). Failure of ejaculation with indoramin. *Br. med. J.* **282**, 1796.

GRIEVE, J. (1979). Male infertility due to sulphasalzine. *Lancet* **ii**, 464.

GRIFFIN, J.P. (1979). Endocrine dysfunction. In D'Arcy, P.F. and Griffin, J.P. *latrogenic diseases.* 2nd edn., pp. 239–51. Oxford University Press.

GUILLEBAUD, J. (1978). Sulpha-trimethoprim combinations and male infertility. *Lancet* **ii**, 523.

HAMILTON, M. AND MAHAPATRA, S.B. (1972). Antidepressive drugs. In *Side effects of drugs,* No. 7 (ed. L. Meyler and A. Herxheimer), p. 22. Excerpta Medica, Amsterdam.

HAZELMAN, B.L., MOWAT, A.G., STURGE, R.A., HALL, M. SEIFERT, M.H., LIYANGE, S.P., MATTHEWS, J.A., JENNER, J.R., AND ENGLER, C. (1979). A long term trial of Naproxen in the treatment of rheumatoid arthritis ankylosing spondylitis and osteo-arthritis. *Eur. J. Rheumatol. Inflam.* **2**, 56–64.

HEMPEL, R.D. AND HESTER, R. (1976). Libido verlust unter methi-mazol. *Dt. Gesundheit. Wes.* **31**, 2157.

HEROLD, E., MOTTIN, J., AND SABRY, Z. (1979). Effect of vitamin E on human sexual functioning. *Arch. sex Behav.* **8**, 397–403.

HINKES, E. AND PLOTKIN, D. (1973). Reversible drug induced ste-rility in a patient with acute leukaemia. *J. Am. med. Ass.* **223**, 1490–1.

HONG, Y.C., CHAPUT DE SAINTONGE, D.M., AND TURNER, P. (1981*a*). A simple method to measure drug effects on sperm motility. *Br. J. clin. Pharmacol.* **11**, 385–87.

——, —— AND —— (1981*b*). The inhibitory action of procaine, (+)-propanolol, and (±) propanolol on human sperm moti-lity antagonism by caffeine. *Br. J. clin. Pharmacol.* **12**, 751–3.

—— AND TURNER, P. (1982). Drugs and sperm. *Br. med. J.* **284**, 1194.

HOWARD, D.J. AND REES, I.R. (1976). Long-term perhexiline maleate and liver dysfunction. *Br. med. J.* **1**, 133.

HULL, M.G.R., BROMHAM, D.R., SAVAGE, P.E., AND JACKSON, J.A.M. (1981). Normal fertility in women with post-pill ame-norrhoea. *Lancet* **i**, 1329–32.

IUNDA. I.F. AND KUSHNIRUK, Y.I. (1975). Functional state of the testis after the use of certain antibiotic and nitrofuran prepa-rations. *Antibiotiki* **9**, 843.

JENSEN, R.T., COLLEN, M.J., PANDOL, S.J., ALLENDE, H.D., RAUFMAN, J.P., BISSONNETTE, B.M., DUNCAN, W.C., DURGIN, P.L., GILLIN, J.C., AND GARDNER, J.D. (1983). Cimetidine-induced impotence and breast changes in patients with gastric hypersecretory states. *New Engl. J. Med.* **308**, 883–7.

KLEIN, F. (1972). Hypotensive drugs. In *Side effects of drugs,* Vol. 7 (ed. L. Meyler and A. Herxheimer), pp. 297–306. Excerpta Medica. Amsterdam.

KOTIN, J., WILBERT, D.E., VERBURG, D., AND SOLDINGER, S.M. (1976). Thioridazine and sexual dysfunction. *Am. J. Psychiat.* **133**, 82–5.

KRISTENSEN, R.O. (1979). Labetalol-induced Peyronie's disease case-report. *Acta med. Scand.* **206**, 511–12.

KRONER, T.H. AND TACHUMI, A. (1977). Conception of normal child during chemotherapy of acute lymphoblastic leukaemia in the father. *Br. med. J.* **1**, 1322–3.

KULIK, F.A. AND WILBUR, R. (1982). Case report of painful ejaculation as a side effect of Amoxapine. *Am. J. Psychiat.* **139**, 234-5.

LANCET (1979). Editorial: Drugs and male sexual function *Lancet* **ii**, 883-4.

LANGMAN, M.J.S (1972). Gastrointestinal drugs. In *side effects of drugs*, Vo. 7 (ed. L. Meyler and A. Herxheimer), p. 501. Excerpta of Medica, Amsterdam.

LAW, M.R., COPELAND, R.F.P., ARMITSTEAD, J.G., AND GABRIEL, (1980). Labetalol and Priapism. *Br. med. J.* **280**, 115.

LESLIE, G.B. AND WALKER, T.P. (1977). Toxicological profile of cimetidine. In *Cimetidine* (ed. W.L. Burland and M.A. Simkins), pp. 24-33. Excerpta Medica, Amsterdam.

LEVI, A.J., FISHER, A.M., HUGHES, L. AND HENDRY, W.F. (1979). Male infertility due to sulphasalazine. *Lancet* **i**, 276-8.

LILLYMAN, J.S. (1979). Male fertility after successful chemotherapy for lymphoblastic leukaemia. *Lancet* **ii**, 1125.

McHAFFIE, D.J. GUZ, A., AND JOHNSTON, A. (1977). Impotence in patients on disopyramide. *Lancet* **i**, 859.

McMAHON, C.D., SHAFFER, R.N., HOSKINS, H.D., AND HENDERSON, J. (1979). Adverse effects experienced by patients taking timolol. *Am. J. Ophthalmol.* **88**, 736-8.

MEDICAL LETTER ON DRUGS AND THERAPEUTICS (1977). *Med. Lett. Drugs Therapeutics* **19**, (20).

MEIRAZ, D. AND FISHELOVITCH, J. (1969). Priapism and largactil medication in Israel. *Med. J. Sci.* **5**, 1254.

MOSS, H.B. (1983). More cases of anorgasmia after MAOI treatment. *Am. J. Psychiat.* **140**, 266.

MURDIA, A., MATHUR, V., KOTHARI, L.K., AND SINGH, K.P. (1978). Sulpha-trimethoprim combinations and male fertility. *Lancet* **ii**, 375-6.

NELSON, W.O. AND BUNGE, R.G. (1957). The effect of therapeutic dosages of nitrofurantoin upon spermatogenesis in man. *J. Urol.* **77**, 275.

PALMER, B.V., MONTGOMERY, A.C.V., AND MONTEIRO, J.C.M.P. (1978). Ginseng and mastalgia. *Br med. J.* **1**, 1284.

PEDEN, N.R., CARGILL, J.M., BROWNING, M.C.K., SAUNDERS, J.H.B., AND WORMSLEY, K.G. (1979). Male sexual dysfunction during treatment with cimetidine. *Br. med. J.* **1**, 659.

PERRY, H.M.JR (1978). Veterans Administration cooperative studies of hypertension. *Angiology* **29**, 804-16.

POHL, R. (1983). Anorgasmia caused by MAOIs. *Am. J. Psychiat.* **140**, 510.

QUIRK, K.C. AND EINARSON, T.R. (1982). Sexual dysfunction and clomipramine. *Can. J. Psychiat.* **27**, 228-31.

RUBIN, S.O. (1968). Priapism as a probable sequel to medication. *Scand. J. Urol. Nephrol.* **2**, 81-5.

RUSSELL, J.A., POWLES, R.L., AND OLIVER, R.T.D. (1976). Conception and congenital abnormalities after chemotherapy of acute myelogenous leukaemia in two men. *Br. med. J.* **1**, 1508.

RUSTIN G.J.S., BAGSHAWE, K.D., NEWLANDS, E.S., AND BEGENT, R.H.J. (1981). Cytotoxic drugs and sterility. *Lancet* **i**, 1316.

SCHNEIDER, J. AND KAFFARNIK, H (1975). Impotence in patients treated with clofibrate. *Atherosclerosis* **21**, 455.

SCHURMEYER, T. AND NEILSCHLAG, E. (1982). Ketoconazole-induced drop in serum and saliva testosterone. *Lancet* **ii**, 1098.

SCHWARCZ, G. (1982). Case report of inhibition of ejaculation and retrograde ejaculation as side effects of Amoxapine. *Am. J. Psychiat.* **139**, 233-4.

SHEN, W.W. AND PARK, S. (1982). Thioridazine-induced inhibition of female orgasm. *Psychiat. J. Univ. Ottawa* **7**, 249-51.

SLAG, M.F., MORLEY, J.E., ELSON, M.K., TRENCE, D.L., NELSON, C.J., NELSON, A.E., KINLAW, W.B., BEYER, H.S., NUTTALL, F.Q., AND SHAFER, R.B. (1983). Impotence in medical clinic outpatients. *J. Am. med. Ass.* **249**, 1736-40.

TESTER-DALDERUP, L.B.M. (1979). Hypotensive drugs in *Side effects of drugs annual*, Vol 3 (ed N.M.G. ed. Dukes), p.194. Excerpta Medica. Amsterdam.

TIMMERMANS, L. (1974). Influence of antibiotics on spermatogenesis. *J. Urol.* **112**, 348-9.

TOBIAS, R., SAPIRE, K.E., COETZEE, T., AND MARKS, I.N. (1982). Male infertility due to sulphasalazine. *Postgrad. med. J.* **58**, 102-3.

TOOVEY, S., HUDSON, E., HENDRY, W.F., AND LEVI, A.J. (1981). Sulphasalazine and male infertility: reversibility and a possible mechanism. *Gut* **22**, 445-51.

TOTH, A. (1979a). Reversible toxic effect of salicylazosulfapyridine on semen quality. *Fertil. Steril.* **31**, 538.

—— (1979b). Male infertility due to sulphasalazine. *Lancet* **ii**, 904.

VERMA, R.B., ABRAMS, S.M.L., AND TURNER, P. (1983). Effect of indoramin on sperm motility. *Br. J. clin. Pharmacol.* **15**, 127-8.

WALLACE, T.R., FRAUNFELDER, F.T., PETURSSON, G.J., AND EPSTEIN, D.L. (1979). Decreased libido side effect of carbonic anhydrase inhibitor. *Ann. Ophthalmol.*, 1563-6.

WARREN, S.C. AND WARREN, S.G. (1977). Propranolol and sexual impotence. *Ann. intern. Med.* **86**, 112.

WEI, N. AND HOOD, J.C. (1980). Naproxen and ejaculatory dysfunction. *Ann. intern. Med.* **93**, 933.

WESTERHOLM, B. (1977). Sex hormones anabolic agents and related drugs. In *Side effects of drugs annual*. Vol. 1 (ed. N.M.G. Dukes), pp. 292-311. Excerpta Medica, Amsterdam.

WHORTON, D., KRAUSS, R.M., MARSHALL, S., AND MILBY, T.H. (1977). Infertility in male persticide workers. *Lancet* **ii**, 1259-61.

WILLIAMS, W. (1979). Drug/food allergy and sexual responsiveness. *Med. J. Aust.* **1**, 281.

YUNDA, I.F. AND KUSHNIRUK, Y.I. (1974a). Effect of nitrofurantoin preparations on spermatogenesis. *Bull. exp. Biol. Med.* **77**, 534-6.

—— AND —— (1974b). The effect of nitrofuran preparations on spermatogenesis. *Byull. eksper. Biol. Med. (USSR)* **77**, 68-70.

——, MELNIK, A., AND KUSHNIIRUK, Y.I. (1974). Experimental study of gonadotoxic effects of nitrofurans and its prevention. *Int. Urol. Nephrol., Budapest* **6**, 125-35.

ZACEST, R., AND REECE, P.A. (1983). Endralazine and sexual arousal. *Lancet* **i**, 1221.

ZIPPER, J., WHEELER, R.G., POTTS, D.M., AND RIVERA, M. (1983). Propranolol as a novel, effective spermicide, preliminary findings. *Br. med. J.* **287**, 1245-6.

25 Drug-induced disorders of muscular function

JENS SCHOU

Adverse drug effects on muscular function may be due to either a direct toxic effect, which can be elicited through a number of different mechanisms, or they may be caused indirectly through involvement of the motor-nerve function of the skeletal muscles. The possibility of a drug-induced disorder should be considered in any patient with neuromuscular symptoms, since early withdrawal of a drug provoking such a condition is often followed by prompt recovery.

The drug-induced effects on muscular function nearly exclusively hit (injure) the voluntary skeletal muscles; very little is known about toxic injuries to the involuntary smooth muscles. The voluntary skeletal muscles represent a significant amount of the total body weight constituting about 40 per cent of the total body mass. On the other hand the smooth muscles are of very limited quantity, but functionally they represent an important part of the autonomously innervated involuntary organ functions as they include the muscles in the gastro-intestinal and urinary system and also in the arteries and veins and are thus of utmost importance to the cardiovascular homeostasis.

Smooth muscle and cardiac muscle disorders

Smooth-muscle function is affected by all autonomic drugs as targets for drug effects. However, the adverse reactions described are pharmacological dose-related effects and shall not be treated here.

However, the cardiac muscle is worth mentioning as a target for hazardous effects of drugs. The main risk to life posed by many drugs taken in overdose is caused by their affect on cardiac-muscle function. An example of this is the tricyclic antidepressant, imipramine. The tricyclic drugs have a quinidine-like effect on the heart which may be traced even at higher therapeutic levels by changes in the electrocardiogram. This cardiotoxicity can be described as an impairment of the conduction system with stimulation of ectopic centres in the cardiac muscle. In acute poisoning atrioventricular block may be

seen and death from heart-failure may result. Drug-related myocarditis was reviewed by Fenoglio *et al.* (1981). Similar impairment of heart function is also seen as a result of tablet automatism leading to overdoses of propoxyphene. This drug may also lead to quinidine-like cardiotoxicity. Self-poisoning can result in cardiac failure through mechnism similar to that described for tricyclic antidepressant drugs.

In addition to propoxyphene, its primary metabolite, norpropoxyphene, shows similar toxicity (Lund-Jacobsen 1978). In this respect propoxyphene and norpropoxyphene are equipotent, but only the parent compound has analgesic properties. The pharmacokinetics of propoxyphene in humans is well known (Wolen *et al.* 1971; Gram *et al.* 1979; Inturrisi *et al.* 1982). It is, after both single and repeated doses, characterized by a very significant first-pass metabolism. During a prolonged dose schedule a steady state was reached after 2–3 weeks when the plasma concentration of the toxic metabolite norpropoxyphene was about five times the concentration of the analgesic drug propoxyphene due to first-pass metabolism and hepatic biotransformation. As the metabolite is cardiotoxic and does not depress the respiratory centre as does propoxyphene, a rather low overdose in a patient on long-term treatment with propoxyphene may lead to fatal cardiotoxicity, since the chronic overdose is built upon the accumulated metabolite toxicity (Way and Schou 1979).

Disorders of skeletal muscle function

Several reviews on adverse drug effects on muscles were published which focus selectively on skeletal disorders (Mastaglia 1982; Neuhaus 1978; Mastaglia and Argov 1980, 1981*a, b*; Argov and Mastaglia 1979).

Drug-induced disorders of muscular function result either from direct effects on the structural and functional integrity of the skeletal muscles, or they may be secondary to a disturbance of the motor innervation of the muscles. Targets for the latter neuromuscular effects can be found in the central nervous system, e.g. the parkinson-like syndrome during therapy with neuroleptic drugs, or they can be elicited at different levels of the neuromuscular impulse transmission before or at the neuromuscular junction. Here there are four possible sites of drug action:

1. Presynaptic inhibition of the propagation of the action potential, similar to the action of local anaesthetics;
2. Impaired formation and/or release of acetylcholine, perhaps due to interference with calcium ion movement through the nerve terminal and the mitochondrial membrane;
3. Blocking of the acetylcholine receptors;
4. Failure of acetylcholine to interfere with the muscle action membrane potential at the end-plate membrane.

Examples of drugs causing disorders through acting at the different sites in the neuromuscular system are given in Table 25.1.

Impairment of the normal functional structure of muscular tissue can either occur as a result of drug administration by intramuscular injections (focal myopathies) or may result from the adverse effects of systemic drugs leading to diffuse, generalized myopathies. The extreme case is necrotizing myopathy, for example, acute rhabdomyolysis (Lane and Mastaglia 1978) in which the muscles may be swollen and in which there may even be a virtually complete paralysis and myoglobinuria which can lead to renal failure.

Focal myopathies

Local muscle tissue damage after intramuscular injections of drug solutions for parenteral use was first described for local anaesthetics (Brun 1959) and for broad-spectrum antibiotics (Hanson 1961). It was later described for many different therapeutic groups and agents including chemotherapeutic (Cioc and Schilling 1965; Rasmussen and Svendsen 1976), chlorpromazine (Bree *et al.* 1971), and digoxin (Steiness *et al.* 1974) to mention but a few significant examples. Even the vehicle

alone may lead to damage (Pizzolato and Mannheimer 1961). It should be noted that the drug injections may lead to local histological changes; pain at injection is a warning sign. Addition of hyaluronidase and local anaesthetics to injection solutions only obscures this physiological warning, and a rise in serum enzymes, such as the serum creatinine phosphokinases, can be taken as an indication of the quantity of muscle damage (Sidell *et al.* 1974; Steiness *et al.* 1978).

Repeated injections over several months or years may produce severe contractures (Blain 1984) due to muscular fibroses. This emphasizes that the oral route for drug administration should be used whenever feasible, intramuscular injections being an alternative which should be used only when oral medication is not practicable.

It should be emphasized, however, that while damage can be easily demonstrated in animal experiments, very little documentation is presented as to the morphological implications for human muscles of the clinical situation with intramuscular medication. However, evidence was recently published concerning the retention of the oily vehicle of hormonal preparations which, years earlier, had been injected into the gluteal muscle in the treatment of climacteric disorders (Harrestrup Andersen 1983). In the injection zone the fibrotic remnants could be demonstrated by X-ray examination, and they were thereafter removed surgically. This relieved three patients from the myalgia that had brought them to hospital. In experiments on dogs, microembolism in the lung was demonstrated after intramuscular application of two different oily vehicles, sesame oil and *Viscoleo* (Svendsen and Aaes-Jørgensen 1979).

Systemic myopathies

Drug-induced systemic myopathies were subgrouped by Mastaglia (1982) into necrotizing myopathies, myotonias, drug-induced lipidosis of muscles, malignant hyperpyrexia, corticosteroid myopathy, β-adrenoreceptor-blocker and hypokalaemic myopathy, and inflammatory myopathies. This scheme shall be followed

Table 25.1   Examples of drugs causing disorders of skeletal muscular function with sites of action at different levels of the neuromuscular system

| Central nervous system | Presynaptic | Impaired acetylcholine | | Receptor blockade | Interference with action potential |
		Formation	Release		
Phenothiazine Neuroleptic drugs	Propranolol Chloroquine Lincomycin	Hemicholinium Diethylamino ethanol	Procainamide Lithium Aminoglycoside antibiotics	Phenytoin Penicillamine Chlorpromazine Trimethadione	Quinine Imipramine

here in general, except that malignant hyperpyrexia is extensively described in Chapter 33 of this volume.

Necrotizing myopathies

Necrotizing myopathies are initially characterized by muscle pain, tenderness, and weakness and specifically by a marked elevation of serum creatinine phosphokinase level as an indication of destruction of muscular tissue. The muscles may be swollen and in severe cases there may be muscular paralysis and myoglobinuria, which may lead to acute renal failure. Such cases have been referred to as 'acute rhabdomyolysis' (Lane and Mastaglia 1978), and such necrotizing myopathies have been associated with clofibrate and aminocaproic acid (EACA) and also with chronic alcoholism (Perkoff *et al.* (1966), heroin (Richter *et al.* 1971), and phencyclidine addiction (Cogen *et al.* 1978).

Milder cases of such systemic myopathies have been associated with injections, especially leading to fibrous myopathies and contractures which are more severe than if they were only provoked by injections (see later for pentazocine-induced fibrous myopathies).

Clofibrate-induced myopathy

Myopathy associated with clofibrate therapy is rare and was first reported by Langer and Levy (1968); other reports by Smals *et al.* (1977) and Abourizk *et al.* (1979) confirmed this association.

Some days after commencement of clofibrate therapy the patient complains of muscular aches and pains typically in the shoulders and calves; serum creatine phosphokinase is usually elevated; electromyographic changes are typical of a myopathy. In one case (Abourizk *et al.* 1979) the muscle was examined histologically, including electron microscopy; the myopathy included breakdown of contractile material, deranged mitochondria, dilated sarcoplasmic reticulum profiles, accumulation of membrane–bound dense bodies, discontinuities in the sarcolemma, and thickening of the capillary basement membrane. Macrophages invaded the damaged tissue.

Langer and Levy (1968) surveyed 60 patients with hyperlipoproteinaemia treated with clofibrate; five cases of elevated serum transaminase and creatine phosphokinase were observed. In two, severe myalgia, stiffness, weakness, and malaise coincident with drug administration developed. Cessation of therapy resulted in prompt resolution of symptoms. An identical clinical syndrome reappeared in one of these patients on rechallenge with clofibrate. Three patients had asymptomatic elevations in transaminase and creatine phospho-

kinase; in one case these abnormalities were clearly dose-related. Usually, the abnormalities persisted throughout the period of therapy. In some cases, however, transaminase became normal but creatine phosphokinase remained elevated while treatment continued. Since creatine phosphokinase is absent from liver, these findings suggest that skeletal muscle is the major source of the elevated serum enzymes and that periodic examinations for evidence of muscle tenderness and dysfunction and frequent determinations of transaminase and creatine phosphokinase should be performed on all patients receiving clofibrate who experience myalgia.

Katsilambros *et al.* (1972) described the case of a 36-year old diabetic woman who had diabetic reinopathy and nephropathy. After a myocardial infarction she was found to have hypercholestrolaemia and hypertriglyceridaemia and was treated with clofibrate. After 13 months treatment with clofibrate 500 mg four times per day she was admitted to hospital following a six-month history of general muscular aches, mainly in the thighs, and progessing to complete inability to walk. The main neurological findings were muscle weakness and tenderness, lack of tendon reflexes, and impaired pinprick and vibration appreciation. The serum enzymes were remarkably increased (namely the SGOT, CPK, and LDH); a bilateral quadriceps biopsy showed single-fibre atrophy.

Despite the presence of diabetes and renal damage it was considered that clofibrate could be a causative factor. Clofibrate was discontinued, elevated serum enymes fell within two days and were normal 15 days after stopping clofibrate. There was, however, no clinical improvement in muscle power.

Schneider (1980; Schneider *et al.* 1980) followed-up 31 patients (20 female, 11 male) taking clofibrate for 3–5 years who had intermittent episodes of elevated creatine phosphokinase (CPK) but without symptoms of acute muscular syndrome or chronic cellular damage; five of these, all male, had persistently raised creatine phosphokinase levels. Levy (1972), on the other hand, described myopathy after less than five weeks treatment with clofibrate in two subjects.

It is clear that subclinical myopathy in patients treated with clofibrate is much commoner than overt myalgia, and muscle weakness is in turn less common than myalgia.

Oliver and his colleagues (Smith *et al.* 1970) found no changes in serum creatine phosphokinase, serum transaminase, or alkaline phosphatase in a double-blind study on 211 men with cholesterol levels in the upper third of the normal range (110 were treated with clofibrate and 101 received identical capsules containing olive oil). These workers also reported no instance of myalgia or muscle stiffness in 452 men treated with clofibrate for one year.

Clofibrate-induced myopathy was described in four normolipidaemic children being treated with the drug for diabetes insipidus (Cario *et al.* 1979). In these patients creatine phosphokinase and transaminases were raised and EMG changes were also noted. On the basis of the occurrence of myopathy and increased creatine phospho-kinase activity and EMG changes, two of these authors took clofibric acid themselves. In both test persons sub-clinical myopathy was produced. After stopping the drug, transitional hypertriglyceridaemia occurred.

Myopathy associated with epsilon aminocaproic acid (EACA)

Cases of proximal muscular pain and tenderness asso-ciated with the use of EACA as a fibrinolytic agent were reported by several authors. Six cases of fully developed myopathy were also reported (Korsan-Bengsten *et al.* 1969; Bennett 1972; MacKay *et al.* 1978; Lane *et al.* 1979). The two cases described by Lane *et al.* illustrate the pattern of the syndrome. Muscle pain and tenderness are usually the earliest symptoms, rapidly followed by weakness with widespread involvement of proximal and axial muscles, the neck muscles being unaffected.

Myoglobinuria may occur, together with marked ele-vation of serum enzyme levels reflecting the severity of the myopathy. These cases also illustrate the reversibility of the myopathy on withdrawing the drug.

The fact that symptoms only appear after 4–6 weeks of treatment with daily doses of 18–30 g suggests a cumula-tive dose-related effect. However, many patients have taken up to 32 g daily for long periods without developing a myopathy (Korsan-Bengsten *et al.* 1969), raising the possibility of an idosyncratic reaction. All patients reported have elevated serum creatine phospho-kinase (SCPK). The mechanism of the myopathy is unknown but MacKay *et al.* (1978) postulated throm-bosis in skeletal muscle microvasculature. A case of severe myopathy and acute renal failure in a 20-year-old nurse treated with aminocaproic acid following a subara-chnoid haemorrhage was described by Biswas *et al.* (1980). Myalgia, haematuria, and proteinuria were noted after six weeks treatment with 30-mg aminocaproic acid/day; full recovery took three months after ceasing treatment.

Acute necrotizing myopathy continues to be reported in patients receiving aminocaproic acid (Biswas *et al.* 1980; Swash 1980). The pathogenesis of this disorder remains uncertain but it was suggested by Kennard *et al.* (1980) that microvascular thrombosis is probably the casual mechanism, while Vanneste and Wijngaarden

(1982) proposed muscular vasculitis or an auto-immune reaction proposed as the cause.

Pentazocine-induced fibrous myopathy

De Latour and Halliday (1978) described two cases of fibrous myopathy developing in patients into whom pen-tazocine had been injected. The fibrous myopathy and contractures that developed in these patients were greater than would have been expected from the injection pro-cedure alone. There is some debate as to whether the pen-tazocine myopathy also occurs in non-injected muscles in animal studies.

Phencyclidine-induced myoglobinuria

Phencyclidine was first used in the United States as an anaesthetic agent in the late-1950s under the trade name *Sernyl* and was eventually withdrawn from human cli-nical use by the manufacturer because of postanaesthetic excitement, visual disturbances, and delirium. Since 1967 it has been commercially available in the United States as an anaesthetic or tranquillizing agent for veterinary use.

It has, however, also come to the fore as a drug of abuse under the street names of 'Peace Pill'. 'PCP', 'Angel Dust', 'Hog', 'Rocket Fuel', 'Mist', and 'Monkey tranquilliser' and the material is very versatile in that it can be smoked, inhaled as snuff, taken orally, or injected.

Rhabdomyolysis with renal failure has been described following phencyclidine abuse (Hoogwerf *et al.* 1979; Barton *et al.* 1980; Patel *et al.* 1980). The patients were, if conscious, restless and incoherent with general muscle pain; others were admitted comatose. Oliguria, myoglo-binuria, proteinuria, and casts were almost universally detected in the 11 cases described in these three papers. Raised plasma myoglobin, CPK levels were also noted. Renal failure was severe enough to warrant haemo-dialysis in five of these patients.

Other drugs rarely associated with acute rhabdomyo-lysis include methadone (Penn *et al.* 1972), barbiturates (Clark and Sumerling 1966), meprobamate, and dia-zepam (Wattel *et al.* 1978). Muscle-fibre necrosis has also been related to emetine, vincristine, and to severe hypokalaemia.

Myotonias

Hereditary myotonias, such as myotonia congenita or dystrophobic myotonia, may be exacerbated and pre-viously undetected myotonias may be unmasked by a number of drugs, e.g. depolarizing muscle relaxants (Mitchell *et al.* 1978). This may lead to problems with tracheal intubation and ventilation during general anaesthesia; this problem does not occur with non-depolarizing neuromuscular blocking drugs (e.g. tubocurarine).

Nifedipine

Keider *et al.* (1982) described three cases of severe muscle spasms in both legs and hands in men, aged 56, 67, and 73, respectively, who were being treated with nifedipine for angina. The muscle spasms were severe and painful enough to discontinue treatment.

Miscellaneous

Myotonias were also found to be activated by the β-adrenergic blockers, propranolol (Blessing and Walsh 1977) and pindolol (Ricker *et al.* 1978), and the β_2-adrenergic agonist fenoterol in some patients. Muscle pains, cramps, spasms, and weakness occurring as side-effects during treatment with diuretics, such as frusemide, are possibly due to induction of myotonia (Bretag *et al.* 1980).

Drug-induced lipidosis

Drenckhahn and Lüllman-Rauch (1979) listed a number of drugs that can cause myopathy characterized by lysosomal degeneration of muscle fibres in man (chloroquine, vincristine, emetine, amiodarone, and perhexiline) and in experiments on animals (quinacrine, plasmocid, imipramine, iprindole, and chlorphentermine). These drugs are both water- and lipid-soluble and, being poorly ionized at the pH of plasma, get easy access to the cells through the plasma membrane. By their high affinity to polar lipids they get absorbed to intracellular membranes where they form non-digestible drug-lipid complexes. Eventually they accumulate within the lysosomes in the form of lamellated membranous and crystalloid structures (Lüllman *et al.* 1978). This mechanism may characterize a number of the below-mentioned reactions.

Carnitine-induced myasthenia

Carnitine plays a role in the transport of long-chain fatty acids into mitochondria, and carnitine deficiency has been demonstrated in patients with lipid storage myopathy. Carnitine deficiency has been demonstrated to occur during chronic haemodialysis of patients in renal failure.

Battistella *et al.* (1978), Bohmer *et al.* (1978), and Maebashi *et al.* (1978) all used carnitine to correct carnitine deficiency in patients on haemodialysis and also to lower plasma triglyceride levels. Bazzato *et al.* (1979) and De Grandis *et al.* (1980) described a myasthenia-like syndrome developing in some patients treated with carnitine in this manner. The mechanism of this effect was attributed to the known action of carnitine and hemicholinium in inhibiting the uptake of choline into neuronal tissue (Cornford *et al.* 1978). Such an antagonism on the terminal branches of the motoneurone would result in a reduction in synthesis of acetylcholine at the motor endplate.

In the series of 20 haemodialysis patients treated with carnitine by De Grandis *et al.* (1980), four developed severe weakness of the lower cranial and girdle muscles. The onset of weakness commenced about 30 days after the commencement of treatment with carnitine. The weakness presented as profound dysphagia, difficulties in swallowing and chewing. With exercise the strength of the proximal muscles of the pectoral and pelvic girdles waned. However, ptosis and weakness of ocular muscles was not observed.

Edrophonium increased muscle performance in these patients. Neurophysiological examination showed an impairment of neuromuscular transmission with short-term reduction of evoked responses to repetitive stimulation and the presence of post-activation exhaustion phenomena.

All clinical and neurophysiological changes associated with carnitine therapy disappeared between four and five days after stopping carnitine administration.

Amiodarone-induced neuromyopathy

Meier *et al.* (1979) described a single case in which marked distal motor and sensory impairment with distal muscular atrophy were observed clinically and confirmed by nerve and muscle biopsy in an 80-year-old man treated with amiodarone.

It was suggested that lipid storage in nerve and muscle tissue of this patient might have been related to the accumulation of the drug metabolites in these tissues. Iodine content of nerve and muscle were increased 40 times.

Emetine-induced myopathy

Emetine is highly active against extraintestinal amoebiasis and against the symptoms of amoebic dysentery. Thus emetine has remained the basis of treatment of the most important forms of amoebic disease in spite of some disadvantages. While the importance of its toxicity may have been exaggerated, it is certainly true that this has precluded its administration in the high doses necessary to treat severe infections and that, if it were better tolerated, emetine would be a more acceptable treatment, especially for use in ambulatory patients.

Prolonged administration and large doses of emetine may produce degenerative changes in the heart, skeletal muscles, gastrointestinal tract, liver, and kidneys. Cardiovascular effects are the most important toxic manifestations of emetine treatment and include tachycardia, electrocardiographic abnormalities, pericarditis, and hypotension. Excretion and detoxification of emetine is slow and therefore accumulation is a considerable factor in its toxicity.

Reports of muscular weakness and peripheral neuritis are less common than reports of cardiovascular disturbances. Four cases were reported by Keng and Swee (1966). The first, a patient suffering from amoebic hepatitis, developed bilateral foot-drop and absent knee and ankle jerks after two emetine injections. Chloroquine therapy was substituted and the patient was discharged three months later with normal muscle and reflex function.

The second patient complained of extreme weakness and paraesthesia of both legs after receiving 350 mg of emetine. His ECG records indicated toxic myocarditis as well. Leg movement returned within nine days and some six weeks later the patient was able to walk without support; at the same time the ECG returned to normal.

The third patient received combination therapy of chloroquine, oxytetracycline, and a total dose of 780 mg of emetine. The patient had head-drop, the facial muscles and the sternomastoids were weak, the palate was paralysed, and there was pronounced weakness in both limbs. The ECG showed inversion of T-waves.

The fourth case had received a total of 650 mg of emetine together with chloroquine. Eighteen days later this patient experienced aching pain in both thighs, associated with tiredness and weakness of both legs, and at a later time, of the hands also. His ECG showed low voltage, flattening and inversion of the T-waves in all leads. Deltoid-muscle biopsy suggested a primary myopathy.

The search for new less toxic ameobicides received a considerable stimulus when work aimed at the elucidation of the chemical structure of natural emetine opened up the way to the synthesis of new emetine derivatives. This work by Brossi et al. (1959) in the Roche Laboratories in Switzerland and England led to dehydroemetine (Dehydroemetine) and at about the same time this compound was also synthesized at Glaxo Laboratories (Mebadin).

Experimental chemotherapy and toxicology showed that dehydroemetine possessed equal or slightly superior amoebicidal activity to emetine, that it was eliminated more than twice as rapidly as emetine, and that it was half as toxic as natural emetine.

In the clinic, dehydroemetine has been better tolerated than emetine by injection and also by mouth as dehydroemetine resinate. Some patients complained of weakness but this was not severe. Cardiovascular disturbances were evident but treatment with dehydroemetine was far safer than with emetine.

The derivative was effective in both hepatic amoebiasis and amoebic dysentery giving a cure rate of about 95 per cent in both conditions (Blanc et al. 1961a, b; Armengaud et al. 1962; Blanc and Nosny 1962; Powell et al. 1962; Blanc 1963; Sarin 1963; Sardesai et al. 1963; Chhabra et al. 1964; Dempsey and Salem 1966; Shroff and Bodiwala 1966; Patel and Mehta 1967).

Chloroquine neuromyopathy

Chloroquine has proved to be an effective antimalarial agent. In the treatment of malaria it is used for relatively short periods of time and in the prophylaxis of the disease the dosage is usually low (300 mg chloroquine base). Complications from its antimalarial use are therefore rare; however, chloroquine is an effective anti-inflammatory agent and larger doses extending up to or beyond 600 mg of base daily are used in the treatment of rheumatoid arthrities or discoid lupus erythematosus. It is with these larger doses that toxic side-effects occur.

Neuromyopathy due to chloroquine is a rare side-effect although there have been casual references to muscle weakness in the literature. However, a number of clear-cut cases of chloroquine-induced myopathy or neuromyopathy have been described in several papers, for example: Loftus (1963); Whisnant et al. (1963); Begg and Simpson (1964): British Medical Journal (1964); Bureau et al. (1965); Blom and Lundberg (1965); Blomberg (1965); Journe et al. (1965); Lenoir et al. (1965); Renier (1965); Sanghvi and Mathur (1965); Bonard (1966); Eadie and Ferrier (1966); Millingen and Suerth (1966); Smith and O'Grady (1966).

Neuromyopathy usually appeared only in patients receiving 500 mg or more chloroquine daily. Whisnant and colleagues described four patients; in these cases symptoms started with muscular weakness in the lower limbs, usually arising in the proximal muscles and progressing slowly to the upper limbs. Sometimes the trunk, neck, and facial muscles were eventually affected. As a rule tendon reflexes were reduced or absent without any sensory disturbances.

In reviewing these and other cases, all patients had clinical similarities, but differed in detail. Most patients were on long-term treatment but symptoms were slow to develop, although in one case symptoms were apparent within a few weeks of the start of treatment. Neuropathic and myopathic abnormalities were demonstrated by electromyography and muscle biopsy. Histological studies showed focal necrosis and fibrosis of muscle; some 50 per cent of the sarcoplasmic material was found to have been replaced by vacuoles. Some patients also suffered from blurring of vision due to corneal opacity without retinopathy. Usually a gradual return to normal muscle function followed withdrawal of the chloroquine, although in some cases deep reflexes were still diminished, and in others some muscle weakness still remained. According to some reports, regression of the condition after chloroquine withdrawal is favourably influenced by corticosteroids.

The mechanism of these effects is not clear; individual susceptibility to the pathological reactions seems probable since there is no well-established correlation between dose and duration and development of a vacuolar myopathic pattern. The vacuolar changes are accompanied by glycogen accumulation and it was suggested that chloroquine may act on the muscle by inhibiting enzymes involved in glycogen breakdown in muscle tissue. It was shown that chloroquine interferes with hexokinase and the flavine adenine nucleotide enzymes; also an increase of the SGO and SGP transaminases was observed (Schindel 1968).

Loss of motor units and a rather low speed of motor conduction suggest neurogenic injury. The absence of reflexes, even when atrophy and paresis are not very pronounced, is also significant although this could be due to muscle damage affecting either the muscle-spindle receptors or their motor-nerve end-plate innervation by γ-fibres. The whole clinical picture, however, suggests neuromyopathy rather than pure myopathy. Clinically the condition is rather like that of steroid myopathy but differs clearly from nerve injuries caused by, for example, thalidomide, nitrofurantoin, or isoniazid (Schindel 1968).

An interesting experimental tool was produced by Smith and O'Grady (1966) who induced chloroquine myopathy in rats and rabbits. Large doses of chloroquine were required, 800–1050 mg/kg, although only for a short period. Histologically there was necrosis of both cardiac and skeletal muscle fibres. In muscle the main damage was in 'red' fibres and was thought to be due to binding of chloroquine by myohaemoglobin. Histological changes were not observed in the peripheral nervous system although there was damage to spindle-muscle fibres. These changes were thought to account for diminished tendon reflexes.

Hughes et al. (1970) described two cases of chloroquine myopathy; in one case the chloroquine had been given for rheumatoid arthritis and in the other case for

sarcoidosis. Skeletal and cardiac muscles were severely affected and one of these patients died of heart failure. At post mortem and on histological examination there was evidence of cardiomyopathy and in the skeletal muscle there was striking vacuolar degeneration of 50 per cent of muscle fibres. Histochemical examination showed that the granular (Type I) muscle fibres were preferentially affected. Electron microscopy showed the ultrastructural detail of the extensive degeneration of the muscle fibres which contain large numbers of myelin figures thought to be a type of lysosome. There were mitochondrial changes which appear to precede the formation of the myelin figures.

In a review of patients with chloroquine myopathy it was noted that some of these patients were rheumatoid-arthritic and had received simultaneous steroid therapy and it is important to realize that both agents can cause a myopathy and a correct assessment of the causative agent is important. In these circumstances it is best to remove the chloroquine first since recovery is usually rapid and may be beneficially influenced by steroid therapy.

Karstarp et al. (1973) described a case of chloroquine neuromyopathy in a young woman of Swedish origin who developed marked wasting of thigh muscles and loss of ankle and knee jerks while in Kenya and receiving antimalarial prophylaxis. This patient showed high levels of chloroquine in the presence of normal liver and renal function (levels of 80 μg/100 ml compared with 20–40 μg/100 ml in most patients in a steady state after four weeks).

Corticosteroid myopathy

A well-known complication to corticosteroid therapy is the development of muscle weakness and atrophy of muscles. Mastaglia (1982) suggested that this was the most common form of drug-induced myopathy in clinical practice. This complication, which is like the muscular weakness found in Cushing's syndrome, is particularly likely to occur during systemic treatment with 9α-fluorinated steroids, such as triamcinolone (Lancet 1965).

Dubois (1963) reported on side-effects experienced by 31 patients with systemic lupus erythematosus whilst on paramethasone (6α-fluoro-16α-methylprednisolone). One patient developed muscle weakness, and vacuolar degeneration was noted on muscle biopsy in another patient.

Braun et al. (1965a) described six patients with rheumatoid arthritis who developed a myopathy whilst on long-term corticosteroid treatment. There was a progressive loss of muscular strength; in all cases the pelvic region was affected; in addition four patients showed affections of the humeral-scapular region and one patient had affected calf muscles. These changes were symmetrical and were not accompanied by pain or cramp. Four cases showed electromyogram changes and muscle biopsy disclosed degenerative changes including granules, hyaline degeneration, vacuolar degeneration, and necroses.

Lerique and Chaumont (1965) described six cases of patients who developed myopathy while on treatment with triamcinolone or methylprenisolone; their electromyographs returned to normal three or four months after the steroid treatment was stopped. Coomes (1965) reported 34 patients on corticosteroids who showed mild Cushingoid signs; electromyogram reaction was normal, but in the low range. In 17 more severe cases the electromyogram reaction was well below the normal range.

Other reports of corticosteroid-induced myopathy were given by Braun *et al.* (1965*b*), Coste *et al.* (1965*a, b*), and Raffi *et al.* (1966). Features of the cases are similar: there is a delay in return to normal muscle function after stopping the steroid treatment, and there is no clear relationship between the severity of the myopathy and the duration of treatment. There is a suggestion that the incidence of myopathy is greatest in patients being treated with 9α-fluorinated corticosteroids. Hypokalaemia is a probable contributory cause of this steroid-induced myopathy and, in this respect it is significant that deoxycorticosterone, one of the most potent of the mineralocorticoids, also produces muscle weakness associated with hypokalaemia. Powell (1969) described myopathy with hypokalaemia in patients with ulcerative colitis treated with steroids.

Danazol-induced myalgia

Danazol, a synthetic steroid that inhibits the release of pituitary gonadotrophins, is used in the treatment of endometriosis and is effective in preventing attacks of angioneurotic oedema. Spaulding (1979) described the case of a 40-year-old woman treated with danazol for this indication who developed myalgia with increased serum creatine phosphokinase. On ceasing treatment with danazol the muscle pain resolved and the raised creatine phosphokinase fell.

Muscle pains and cramps have been reported in about 4 per cent of all patients treated with danazol (Spooner 1977; Young and Blackmore 1977) in two series of 704 and 452 patients, respectively.

Myopathy associated with β-adrenoceptor blocking agents

Patients treated with β-blockers commonly complain of weakness and muscle fatigue, particularly in the lower limbs (Stone 1979). This effect is seen with both non-selective and cardioselective blockers. It is not clear if it is due to a direct effect of the drugs on the muscles or whether the symptoms are due to the cardiovascular effects (Grimby and Smith 1978).

Forfar *et al.* (1979) described a simple case of a 68-year-old woman with angina who was treated with sotalol 80 mg/day. After six months treatment on sotalol she developed proximal muscle weakness. Treatment was changed to propranolol but weakness was unchanged. After some months on propranolol her serum creatine phosphokinase (SCPK) was found to be raised and isoenzyme studies showed it to be of muscle origin. EMG changes in the quadriceps were consistent with a mild myopathy.

Propranolol was discontinued and proximal muscle weakness rapidly resolved and SCPK concentrations returned to normal. After a period of two months off beta-adrenergic blocking drugs, propranolol was restarted and after four months weakness had returned. SCPK was again raised, and EMG changes again showed myopathic changes. Propranolol was again discontinued with a rapid return of muscle power and the patient was normal in this respect at six months follow up. Angina was controlled with nitrites and perhexiline.

After the report of Forfar *et al.* (1979), two other reports followed in which a rather different but probably much more important problem was reported. Uusitupa *et al.* (1980) reported thyrotoxicosis developing in three male patients during treatment with the beta-adrenergic blocking drugs timolol, pindolol, and propranolol. In all these patients the manifestations of thyrotoxicosis were obscured by the beta-adrenergic agents and the presenting features in all patients were wasting of the proximal muscles of the extremities; muscle power was only slightly impaired.

Little attention has been paid to the development of thyrotoxicosis during beta-blockade. This coincidence is probably rather common, because both treatment with beta-blocking agents and thyrotoxicosis are common clinical conditions. Beta-blocking agents mask many of the most typical clinical signs of thyrotoxicosis, such as tachycardia, tremor, and nervousness, without affecting the underlying metabolic disorder, except the deiodination of thyroxine to triiodothyronine. On the other hand, the development of thyrotoxic myopathy is not prevented by beta-blockade. Thus thyrotoxicosis presenting with atypical clinical symptoms and signs is probably the most common aetiological factor to be considered in proximal myopathy developing during beta-blockade.

Ramsey (1980) concurred with Uusitupa *et al.* (1980) on the importance of beta-adrenergic agents in masking thyrotoxicosis and emphasized the importance of considering thyrotoxicosis in any patient with proximal muscle myopathy receiving beta-blockers.

Labetalol-induced myopathy

Muscle pain associated with labetalol treatment of hypertension was reported by Andersson *et al.* (1976), Bolli and Wool-Manning (1977), and Jennings and Parsons (1976) but these were comparatively rare occurrences. Teicher *et al.* (1981) described the first case of toxic myopathy associated with labetalol treatment of hypertension. A 27-year-old male patient with essential hypertension was treated with labetalol 600 mg/day. While on this treatment the man began to complain of muscle pain, especially in the legs. Neurological examination did not show any abnormality: the muscle strength was normal, there was no wasting, and reflexes were normal. The activities of muscle enzymes in the blood-creatine phosphokinase, lactic dehydrogenase, and aldolase — were persistently high. Triiodothyronine and thyroxine concentrations were normal. On electrodiagnostic study the motor-nerve conduction velocity from ulnar, peroneal, and posterior tibial nerves on both sides was normal. The electromyogram findings were compatible with myositis, and a biopsy specimen was taken from the deltoid muscle. Light microscopy did not show any histological changes, and the histochemical examination (adenosinetriphosphatase succinic dehydrogenase staining) showed a normal typing of muscle fibres. Electronmicroscopy study showed many vacoules of various sizes in the sarcoplasm under the sarcolemma, without staining and without connection to any structure in the cell. These findings confirmed nonspecific toxic myopathy. The labetalol was stopped for 10 days, during which time the muscle pains disappeared and the enzyme activities returned to normal. A few days later, with the informed consent of the patient, labetalol was restarted. The muscle pain reappeared and the serum muscle enzyme activity rose. Labetalol was stopped and atenolol 100 mg twice daily was prescribed to control blood pressure: the response was satisfactory and no adverse reaction occurred.

This report of toxic myopathy to labetalol differs from the myopathy reported with beta-adrenergic blocking agents, where these agents had masked a toxic myopathy due to thyrotoxicosis.

Inflammatory and autoimmune myopathies

Inflammatory and autoimmune processes may be involved in the myopathy elicited during D-penicillamine therapy, which has been associated with the onset of polymyosites or dermatomyositis in a number of cases.

Penicillamine-induced myositis

Morgan *et al.* (1981) described the case of a 50-year-old man with a 10-year history of rheumatoid arthritis who developed right foot-drop and was admitted to hospital with a diagnosis of rheumatoid vasculitis. His history included allergy to penicillin. Based on the results of clincial laboratory and neurologic tests, the patient was given 250 mg penicillamine, 50 mg pyridoxine, and 60 mg prednisone daily. The dosage of penicillamine was increased to 250 mg twice daily and to 250 mg thrice daily 3 and 6 weeks later, respectively. Five days after penicillamine was given in a total daily dose of 750 mg the patient developed dysphagia for solid and liquid foods, bulbar weakness, and an absent gag reflex. Enzyme studies showed abnormally high creatine phosphokinase and serum glutamic oxaloacetic transainase values. Penicillamine therapy was stopped, and treatment with 100-mg intravenous methylprednisolone was begun. Electromyography demonstrated profuse spontaneous fibrillation in muscles of the proximal extremities and patches of 'myopathic' motor units on a background of giant units. Muscle biopsy confirmed myositis. With prednisone therapy, his condition gradually improved over the next six months. Morgan *et al.* (1981) raised the issue as to whether penicillamine induces myositis or causes exacerbation of previously existing myositis in rheumatoid arthritis. Another interesting point raised by these authors concerns the possible relationship of penicillin allergy to the complications of penicillamine therapy.

This report of penicillamine-induced myositis appears to be a completely distinct condition from the more common penicillamine-induced myasthenia gravis which is discussed later in this chapter.

Drug-induced lupus (LE) syndrome

A drug-induced lupus-erythematosus-like syndrome reported in patients treated with a number of drugs including procainamide (Blomgren *et al.* 1972), hydrallazine, phenytoin and mesantoin (Mastaglia and Argov 1981*b*) and levodopa (Wolf *et al.* 1976) shows an interstitial form of myositis characterized by muscle pain, stiffness, and mild weakness. It seems possible that the drug-induced inflammatory myopathy is a result of a loss of tolerance to muscle antigens due to a disturbance of the immunoregulatory mechanism as suggested by Mastaglia and Argov (1981*b*).

Other drugs associated with myopathy

Scattered information in the literature connects a number of various drugs to the occurrence of myopathies of relatively unspecified nature. A number of such reports shall be mentioned.

Cimetidine-induced myopathy

Walls *et al.* (1980) described two cases of motor neuropathy possibly associated with cimetidine therapy, both patients were asthmatic. Feest and Read (1980) described a single case of a 71-year-old man who developed weakness of both the pelvic and pectoral muscles after one year's therapy of cimetidine at a dose of 1 g/day. The cimetidine was stopped and over one year a remarkable recovery was noted. The only abnormal result on investigation was the EMG from both the deltoids and quadriceps which showed a low voltage myopathic pattern with no fibrillation.

Myalgia associated with cimetidine has been recognized from the time of introduction of the drug (Burland 1978).

Polymyositis associated with cimetidine was described by Matthiesen (1979).

Tetracycline

A 15-year-old girl, treated for acne with 250 mg tetracycline twice daily, after two months found she had difficulty climbing stairs. Electromyographic studies indicated a mild myopathic disease of uncertain cause with superimposed fibrillation. A biopsy of the left quadriceps showed randomized atrophy of muscle fibres. Recovery took place on drug withdrawal (Sinclair and Phillips 1982).

Rifampicin-induced myopathy

Jenkins and Emerson (1981) described the case of a woman diagnosed as having active pulmonary tuberculosis who was treated for 16 months with isoniazid and rifampicin, supplemented for the first three months by daily injections of streptomycin. Six years later reactivation of the tuberculosis was diagnosed after chest radiography revealed fresh cavitation and the sputum yielded acidfast bacilli and Lowenstein-Jensen cultures yielded colonies of myobacterium. Treatment was started with streptomycin, rifampicin, isoniazid, and ethambutol. After four weeks she was readmitted, complaining of increasing proximal muscle weakness in the arms and legs and inability to stand unaided. There was no paraesthesia, anaesthesia, or muscle pain. Facial and ocular muscles were not affected, and she did not have dysphagia or abnormal fatigability to suggest myasthenia gravis.

Physical examination showed no abnormality apart from appreciable proximal muscle weakness, which was greater in the legs than the arms and was accompanied by symmetrical proximal muscle wasting. No rash, muscle tenderness, or fasciculation was present. Normal values were obtained for full blood count; urea and electrolyte, serum calcium, and plasma protein concentrations; creatine phosphokinase activity; and thyroid function tests. Erythrocyte sedimentation rate was 48 mm in the first hour. A *Tensilon* test showed no response in any muscle group.

All drugs were stopped and within 24 hours muscle power had increased dramatically. An electromyogram performed six days after the drugs were stopped showed no abnormality. Eight days after the drugs were stopped muscle power had returned to normal. Isoniazid was reintroduced at full dosage without any problems. Rifampicin was then added at full dosage and within 24 hours severe muscle weakness recurred. An electromyogram again showed no abnormality, but a muscle biopsy specimen showed predominant Type-II muscle-fibre atrophy associated with sarcolemmal nuclei proliferation and increase in connective and fatty tissues, consistent with a drug effect. Rifampicin was withdrawn and within five days power returned to normal. Ethambutol was subsequently reintroduced without any further problem. Rifampicin antibodies were absent six weeks after the drug was stopped.

Mercaptopropionyl glycine

Hales *et al.* (1982) described a single case of myopathy in a 38-year-old woman being treated for cystinuria with mercaptopropionyl glycine. Mansell (1982) did not consider that the cause -and- effect relationship between the myopathy and mercaptopropionyl glycine in the case reported by Hales *et al.* was established and was very critical of their report.

Lithium-induced myopathy

Julien *et al.* (1979) described a case of a 62-year-old woman who was treated with lithium for manic depressive psychosis. In an episode of acute lithium intoxication she developed myopathy and cerebellar ataxia; the muscle weakness persisted after the acute episode but made a gradual recovery over 11 months. EMG changes during this period were consistent with myopathy. Investigation of muscle tissue was possible after some time due to the patient's suicide; the deltoid muscle showed some non-specific myopathic changes. It was concluded that this was a toxic myopathy associated with acute lithium intoxication.

Amiodarone

Meier *et al.* (1979) described a single case of neuro-myopathy occurring in association with chronic (8 years) amiodarone therapy. Electromyography revealed denervation potentials and severe loss of motor units in the M. extensor digitorum brevis and in the M. tibialis anterior. Biopsy indicated both neurogenic and myogenic changes and physical and chemical analysis of the muscle revealed the content of iodine to be 40 times normal value.

Ipecacuanha

Brotman *et al.* (1981) described a case of myopathy developing in an 18-year-old girl with anorexia with episodes of bulimia (voracious eating) who began to use ipecacuanha syrup in 85-ml doses to induce vomiting. Bennett *et al.* (1982) described a similar case of toxic myopathy in a 19-year-old girl with anorexia who had been using ipecacuanha syrup to induce vomiting over a period of two years. Biopsy of muscle indicated changes similar to those seen in dermatomyositis. In both these cases strength returned after cessation of ipecacuanha abuse. Although neither of these cases were strictly speaking iatrogenic diseases, they were clearly drug-induced.

Drug-induced disorders of neuromuscular transmission

The principles of the drug-induced disorders of the motor-nerve system to the voluntary muscles were described in the beginning of this chapter. Clinically post-operative respiratory depression and the unmasking or aggravation of myasthenia gravis occur most frequently.

Post-operative respiratory depression may either result from prolonged sucinylcholine action due to a genetic insufficiency in pseudocholinesterase activity in plasma, or may be due to interaction of certain antibiotics and other drugs having neuromuscular blocking qualities with suxamethonium or other muscle relaxants used during anaesthesia.

Drug-induced neuropathies may also present themselves as muscle weakness and some drugs may cause a neuropathy, a myopathy, or a myasthenic syndrome. There is inevitably some overlap in the handling of this subject on a purely systematic basis. Table 25.2 gives a classification on a clinicopathological basis and may be compared to Table 25.1.

Prolonged neuromuscular blockade

The drug-induced modification of ionic conductance at the neuromuscular function was extensively reviewed (Lambert *et al.* 1983), and an excellent survey of drug-induced neurological-disorders was published by Lane and Routledge (1983).

Prolonged suxamethonium block due to genetic failure in pseudocholinesterase

Suxamethonium (succinylcholine, diacetylcholine) was introduced in 1949 by Bovet and his colleagues. It is a potent neuromuscular blocking agent of the depolarizing type with very brief duration. This is because suxamethonium is rapidly hydrolysed and inactivated by pseudocholinesterase present in the serum. It is used when only a brief period of muscle relaxation is needed or it may be given by intravenous drip to produce prolonged neuromuscular blockade. Suxamethonium is a very useful muscle relaxant although there have been reports since its early use of prolonged apnoea and muscular pain and stiffness in some cases.

Bourne *et al.* (1952) and Evans *et al.* (1952, 1953) drew attention to the prolonged apnoea in man after suxamethonium in the presence of low plasma serum pseudocholinesterase. Lehmann and Ryan (1956) observed a familial incidence of low pseudocholinesterase in the absence of overt disease, and Lehmann and Simmons (1958) described suxamethonium apnoea in two brothers who had low plasma pseudocholinesterase levels and who both had, one of them repeatedly, a prolonged muscular paralysis following the injection of suxamethonium. Kaufman *et al.* (1960) recorded the first observation of suxamethonium apnoea in an infant; the family was studied in detail and there was a deficiency of pseudocholinesterase in one paternal and in one maternal grandparent. The parents had normal values for pseudocholinesterase although these were in the lower end of the normal range. Parbrook and Pierce (1960) compared the incidence of muscular pain and stiffness after the use of suxamethonium and suxethonium for dental in-patient anaesthesia and found no significant difference in the incidence or character of the postoperative pain or stiffness. Telfer *et al.* (1964) described the case histories of seven patients who possessed an atypical form of pseudocholinesterase; six of these patients presented as pro-

Table 25.2   Drugs causing neuromuscular dysfunction

Type of disorder	Drugs
Pharmacogenetic disorders	
(a) Malignant hyperpyrexia (see Chapter 33).	General anaesthetics, particularly halothane and enflurane, morphine, pethidine, promethazine, thiopentone, trimethazine. MAOI/tricyclic antidepressant combination
(b) Abnormal suxamethonium neuromuscular block in patients with deficient or abnormal pseudochlinesterases	Suxamethonium
Antibiotic-induced neuromuscular blockade	
(a) Curare-like block	Amikacin, dihydrostreptomycin, framycetin, gentamicin, neomycin paramomycin, sisomycin, streptomycin, tobramycin, viomycin
(b) Depolarizing block	Colistin, kanamycin, polymyxin B
(c) Uncharacterized	Clindamycin, lincomycin
Myalgia	Clofibrate, danazol
Myasthenia gravis-like syndrome	(a) aminoglycoside antibiotics and tetracycline with cause worsening of pre-existing myasthenia gravis
	(b) agents producing myasthenia in normal patients, carnitine, gentamicin, penicillamine
Myopathy	Aminocaproic acid, beta-adrenergic blocking drugs, cimetidine, clofibrate, corticosteroids, emetine, penicillamine, perhexiline, rifampicin
Neuromyopathy	Amiodarone, chloroquine, cimetidine, dapsone, perhexiline
Rhabdomyolysis	Phencyclidine, succinylcholine

longed response to suxamethonium. The familial incidence of the occurrence of an atypical cholinesterase was demonstrated in each of these cases.

This prolonged peripheral neuromuscular block commonly presenting as prolonged apnoea has been attributed over the years to one or more of the following mechanisms.

Succinylmonocholine and choline block. Suxamethonium (succinylcholine chloride or bromide) is normally rapidly hydrolysed by pseudocholinesterase to succinylmonocholine and choline. The succinylmonocholine in turn is hydrolysed to succinic acid and choline (Fig. 25.1). Succinylmonocholine and choline each have a weak depolarizing action; however, effective concentrations of these two metabolites are unlikely to accumulate unless massive doses of suxamethonium have been given, e.g. 1.5–2.0 g (Wylie and Churchill-Davidson 1960).

Dual block. After an intravenous injection of suxamethonium, certain subjects develop a non-depolarizing block at the motor end-plate following the original depolarizing block. This is usually seen after multiple doses of suxamethonium, although occasionally, it may appear even after a single injection. Patients with myasthenia gravis are particularly prone to develop this type of response (Churchill-Davidson 1955). This type of block is temporarily improved by edrophonium and is reversed by neostigmine.

Deficiency of pseudocholinesterase. Plasma pseudocholinesterase is responsible for the rapid destruction of suxamethonium, so that any reduction in pseudocholinesterase activity or plasma level of the enzyme will prolong the duration of the neuromuscular blockade. Production of pseudocholinesterase is depressed in certain pathological conditions including liver disease, severe malnutrition, and hyperproteinaemia other than

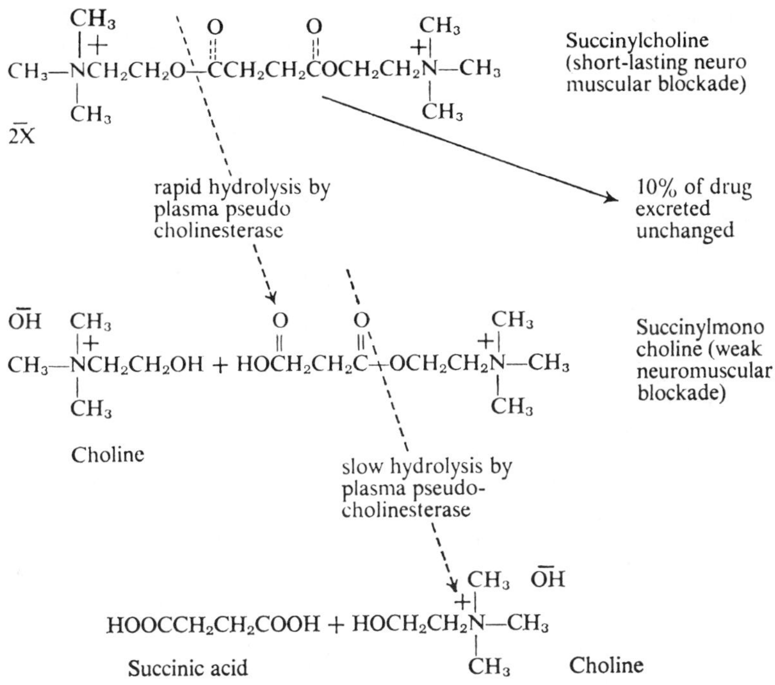

Fig. 25.1 Metabolic fate of succinylcholine (suxamethonium).

that due to renal disease. The action of the enzyme is inhibited by organophosphorous compounds such as are used in nerve-gases and insecticides. A commoner cause of a transient inactivation of pseudocholinesterase is the excessive administration by the anaesthetist or surgeon of drugs with anticholinesterase activity such as cocaine, procaine, and lignocaine. Some ganglionic blocking agents, trimetaphan camsylate and phenactropinium chloride (phenacylhomatropinium chloride), are also powerful cholinesterase inhibitors (Lehmann and Simmons 1958).

There may also be a genetically determined deficiency of the enzyme (Allot and Thompson 1956; Lehmann and Ryan 1956; Kalow and Staron 1957; Lehmann and Simmons 1958). Complete absence of pseudocholinesterase has been reported (Hart and Mitchell 1962).

Presence of an atypical form of pseudocholinesterase. Some people who are abnormally sensitive to suxamethonium have been shown to have an atypical form of pseudocholinesterase in their serum (Kalow 1956, 1959). This atypical enzyme hydrolyses suxamethonium at a much slower rate than does the normal, or usual type of serum cholinesterase, and consequently the apnoea which the drug induces is excessively prolonged.

The genetics of the condition were reviewed by Lehmann and Liddell (1962).

Patients with an atypical form of pseudocholinesterase do not present any other recognizable abnormality, and they usually present as cases of prolonged apnoea following a single injection of suxamethonium. A technique has, however, been developed to detect the presence of this atypical enzyme (Kalow and Genest 1957; Kalow and Staron 1957) by determining what is known as the 'dibucaine number'. This is the percentage inhibition of enzyme activity produced by the inhibitor dibucaine under certain standardized conditions. In the test, the activity of the enzyme (usual pseudocholinesterase, or atypical pseudocholinesterase) in the presence and in the absence of dibucaine (10^{-5} M) is measured by following the rate of hydrolysis of the substrate benzoylcholine (5×10^{-5} M) spectrophotometrically at 240 nm. The reaction is carried out at pH 7.4 in phosphate buffer.

With this technique most people show a dibucaine number of about 80 (normal homozygotes); those with atypical pseudocholinesterase may be classified as dibucaine number 40–75 (heterozygotes) or dibucaine number 30 or below (abnormal homozygotes). The intermediate group (heterozygotes) have a mean dibucaine number of about 62 and are believed to synthesize both the 'usual' and 'atypical' forms of the enzyme. They do not generally show an abnormal response to suxamethonium.

Evidence, either experimental or clinical, has been presented for each, or for combinations of, these four theories. Current opinion is that prolonged apnoea after suxamethonium is due to quantitatively or qualitatively deficient pseudocholinesterase (Lealock *et al.* 1966; Mone and Mathie 1967). Qualitative defects of pseudocholinesterase activity are due to genetic abnormalities. Four allelic genes seem to control the inheritance of pseudocholinesterase: one normal, two atypical, and one silent allelic gene. They form pseudocholinesterase with varying activity. Quantitative reduction of pseudocholinesterase occurs in patients who receive organophosphorous compounds like the antineoplastics, cyclophosphamide, and thiotepa (N, N′, N″-triethylene-thiophosphoramide). Ecothiopate iodide, an anticholinesterase miotic used in the treatment of glaucoma, also reduced pseudocholinesterase levels (McGavie 1965), and *Trasylol*, a polypeptide inactivator of kallikrein, obtained from animal sources, is also reported to cause prolonged apnoea (Chasapakis and Dimas 1966).

Prolonged apnoea after suxamethonium is still a serious complication and many authors have reported their cases (see Van Dijl 1968). The treatment of prolonged apnoea depends mostly on good ventilation and maintaining physiological levels of pH, CO_2 and bicarbonate. Administration of whole blood or plasma has been recommended. When suxamethonium is present for a long time, or has been given in large doses, a dual neuromuscular block occurs in which the depolarizing block changes to a non-depolarizing type (i.e. curare-like); in this state edrophonium chloride or neostigmine methylsulphate may be helpful.

The muscle pain that occurs quite frequently after the use of suxamethonium can be so severe as to immobilize the patient and even impair ventilation. Prevention or reduction of this postoperative pain were the goals of several investigators; their studies and investigations were reviewed by Van Dijl (1968).

Gallamine, injected (5 mg) just before suxamethonium, reduced the degree and incidence of pain; a prophylactic dose of 3 mg D-tubocurarine completely prevented the pain although with this a higher dose of suxamethonium was required to produce muscle relaxation; slow infusion of suxamethonium in place of intermittent injection of doses is claimed to prevent muscle pain, and diazepam reduced the percentage of postoperative muscle pains to a low figure.

Pretreatment with vecuronium was also reported to be a useful prophylactic drug against post-suxamethonium muscle pain (Ferres *et al.* 1983), even better than gallamine, tubocurarine, or pancuronium.

Abnormal response to suxamethonium due to non-genetic patient factors

It is established that suxamethonium may liberate potassium from skeletal muscles. An excessive rise in extracellular potassium levels, which may lead to cardiac arrest, frequently occurs in patients with traumatic or burn injuries to muscle, in patients with myopathies, and in patients with renal insufficiency. This subject was reviewed by Agoston (1975) who found that a considerable amount of literature on this reaction had appeared in the preceding five years. Several authors had reported hyperkalaemia (up to 13.6 mEq/l) associated with cardiac arrest following suxamethonium administration to a variety of conditions varying from patients with tetanus, burns, trauma to muscle, pseudohypertrophic (Duchenne-type) muscular dystrophy), and other myopathies. Where trauma or burns had been involved, the patients seemed most susceptible to the hyperkalaemic action of suxamethonium between 20 and 60 days after receiving the trauma.

Prolonged apnoea after suxamethonium may be seen in patients with renal failure. Hill *et al.* (1976) described apnoea lasting four hours in a manic-depressive patient taking lithium carbonate (1.5 g/day) following a single injection of suxamethonium. Studies in dogs led these workers to conclude that this was a true potentiation of suxamethonium by lithium.

Also di-tubocurarine is potentiated by lithium chloride in cats (Basuray and Harris 1977). It was reported that the local instillation of organophosphate anticholinesterase drugs for glaucoma into the eye has been followed by prolonged suxamethonium effect (Koenig and Ohrloff 1981; Packman *et al.* 1978; Marco and Randels 1979).

Antibiotic-induced neuromuscular blockade

Aminoglycoside antibiotics. The neuromuscular blocking action of neomycin sulphate was first reported by Pridgen in 1956, who noted respiratory depression following the instillation of this antibiotic into the peritoneal cavity. Engel and Denson (1957) reported respiratory depression in four infants who had received 1 g neomycin intraperitoneally. Other reports soon followed (e.g. Pittinger and Long 1958; Pittinger *et al.* 1958), and the treatment of this condition with cholinesterase inhibitors and/or calcium was advocated. This complication was also seen with oral and intrapleural administration of neomycin. It is more apt to occur when the

Table 25.3   Antibiotics implicated in producing neuromuscular blockade

Antibiotic	Source	Chemical class	Neuromuscular block	Grading of potency of neuromuscular block
Neomycin	*Streptomyces fradiae*	Aminoglycoside	Curare-like	+ + +
Streptomycin sulphate	*Streptomyces griseus*	Aminoglycoside	Curare-like	+ +
Dihydrostreptomycin sulphate	*Streptomyces griseus*	Aminoglycoside	Curare-like	+
Gentamicin sulphate	*Micromonospora purpurea*	Aminoglycoside	Curare-like	+ +
Kanamycin sulphate	*Streptomyces kana-myceticus*	Aminoglycoside	Depolarizing block	+ + +
Viomycin sulphate	*Streptomyces floridae, S. punaceus, S. vinaceus*	Aminoglycoside	Curare-like	+
Tobramycin	*Streptomyces tenebrarius*	Aminoglycoside	Curare-like	+ +
Framycetin sulphate	*Streptomyces fradiae, S. decarius*	Aminoglycoside	Curare-like	+ + +
Paromomycin (*Aminosidin*) sulphate	*Streptomyces rimosus*	Aminoglycoside	Curare-like	+ + +
Amikacin sulphate	*Semisynthetic*	Aminoglycoside	Curare-like	+ +
Sisomicin	*Micromonospora inyoensis*	Aminoglycoside	Curare-like	+
Gramicidin	*Bacillus brevis*	Polypeptide		
Polymyxin B sulphate	*Bacillus polymyxa*	Polypeptide	Depolarizing block	
Colistin sulphate Colistin sulphomethate sodium	*Bacillus species*	Polypeptide	Depolarizing block	
Clindamycin hydrochloride Clindamycin phosphate	*Streptomyces lincolnensis*	Lincomycin	Neuromuscular block (? type)	
Lincomycin Lincomycin hydrochloride	*Streptomyces lincolnensis*	Lincomycin	Neuromuscular block (? type)	
Tetracyclines		Tetracyline	Potentiation of neuromuscular blocking agents also produces increase in neuromuscular weakness in myasthenia patients	
Erythromycin	*Streptomyces erythreus*	Macrolide	Has been reported to produce myasthenia-like syndrome in normal adults and deterioration in myasthenic patients	

antibiotic is used in excessive amounts or when it is used in conjunction with ether anaesthesia or with di-tubocurarine.

Other antibiotics have also been implicated in neuro-muscular blocking activity (Table 25.3); Timmerman and associates (1959) screened a number of antibiotics on the sciatic nerve-gastrocnemius muscle preparation of the anaesthetized rabbit and showed that streptomycin sulphate, dihydrostreptomycin sulphate, polymyxin B sulphate, kanamycin sulphate, and neomycin sulphate produced neuromuscular blockade in the rabbit at various dosage levels. The blocking action of these five antibiotics, with the exception of polymyxin B sulphate, was enhanced by the simultaneous administration of d-tubocurarine chloride. Neostigmine methylsulphate antagonized the neuromuscular blockade produced by neomycin, streptomycin, and dihydrostreptomycin but tended to enhance the neuromuscular blocking action of polymyxin and kanamycin. Other antibiotics tested, tetracycline hydrochloride, ristocetin, oleandomycin, erythromycin, bacitracin, and penicillin G, showed no neuromuscular blocking activity.

Other investigator showed that viomycin and colistin have weak neuromuscular-blocking properties; that due to viomycin can be reversed by anticholinesterase whereas that produced by colistin, at least under clinical conditions, is not atagonized by neostigmine (Barlow and Groesbeek 1966; Pohlmann 1966), nor counteracted by the cholinesterase-inhibitor edrophonium.

These results would suggest that neomycin, viomycin, streptomycin, and dihydrostreptomycin have a curare-like neuromuscular-blocking action, whereas polymyxin B, colistin, and kanamycin have a depolarizing neuro-muscular-blocking action. Neomycin, viomycin, strepto-mycin, and dihydrostreptomycin have a common amino-glycoside structure and are all derived from species of actinomycetes (Table 25.3); it is not surprising therefore that they all share neuromuscular-blocking activity of the same type. Kanamycin, however, would seem to be the odd one out since it is also an aminoglycoside but has a depolarizing, not a curare-like action.

Streptomycin and dihydrostreptomycin have pro-duced neuromuscular blockade when used intraperi-toneally, and patients on prolonged intramuscular therapy of 1 g streptomycin per day have developed muscular weakness, fatigability, and blurred vision, though secondary to this effect. Continued therapy has been made possible by the concomitant administration of oral neostigmine. Hokkanen (1964) reported that patients with myasthenia gravis were unusually sensitive to the blocking effect of streptomycin. Kanamycin also produced respiratory depression when given intra-venously or intraperitoneally in two patients. This effect responded favourably to neostigmine in one instance and calcium gluconate in another.

All members of the aminoglycoside group have the ability to a greater or lesser extent to produce neuromus-cular blockade. The most potent aminoglycosides in this respect would appear to be neomycin/framycetin/paro-momycin and aminosidin. Of the newer aminoglycoside antibiotics, gentamicin shown in rats was to be only half as potent as neomycin in producing neuromuscular blockade. Tobramycin also appears from animal studies to be less potent than neomycin in this respect.

Polypeptide antibiotics.   Polymyxin B is one of a family of polypeptide antibiotics, the polymyxins, derived from bacterial species (Table 25.3); they form a separate group of antibiotics with characteristic microbiological, phar-macological, and toxicological features. Polymyxin E is identical with colistin; it is therefore not surprising that colistin and polymyxin B have neuromuscular-blocking properties of the same type. The polypeptide antibiotics, polymyxin A, polymyxin B, and colistin, seem to have an unusually high potential for producing neurological side-effects, and many reports of the neurotoxicity of parenteral colistin or polymyxin B have been published (see Manten 1968). The symptoms of intoxication vary from minor neurological dysfunctions, such as circum-oral tingling, numbness, and tingling in the limbs, to severe episodes such as convulsions, apnoea, and ultimately death. These polypeptides are all nephrotoxic and initially circum-oral numbness, ataxia, and bizarre behaviour were noted to occur in patients with impaired renal function where significant accumulation of the drug was thought to occur. Some patients have pro-gressed to generalized muscular weakness and apnoea. However, occasionally marked neuromuscular blockade occurs even with normal 'therapeutic' blood levels of colistin methanesulphonate resulting from doses of the order of 75 mg twice daily or every 12 hours (Gold and Richardson 1965, 1966). In these patients there was normal renal function and no obvious predisposing factor; premonitory signs were minimal or absent. The block was only partially reversed by edrophonium (10 mg i.v.); limb weakness improved but not weakness of facial muscles. After colistin was withdrawn, the myasthenia disappeared.

It was suggested that the nephrotoxic and neurotoxic actions of colistin are due to drug-induced potassium-losing nephropathy and hypokalaemia. However, in a study of drug-induced decrease of plasma potassium and sodium in relation to coexistent nervous disturbance in three patients, the disturbances did not resemble those observed in either azotaemic or hypokalaemic patients.

A direct action of colistin on the nervous system was therefore suggested (Baron *et al.* 1967).

Other antibiotics. Some investigators reported a slight neuromuscular blocking effect for tetracycline and chloramphenicol. However, most studies have been negative in this regard, and for the present at least, tetracycline and chloramphenicol may be considered free of this hazard.

The key to the successful treatment of antibiotic-induced neuromuscular blockade, especially of respiratory muscles, is early recognition of the occurrence. Once it is established that a patient is developing neuromuscular blockade, immediate supportive therapy should be commenced. Since the greatest danger is respiratory depression, anoxia, and cardiac arrest, the patient's respiratory function must be closely monitored, and he should be intubated and ventilated at the first signs of distress. If adequate ventilation has been assured, or if muscular weakness is only minimal, the physician may attempt to hasten recovery by the following methods. Calcium gluconate, 10 to 200 mg, should be given by slow intravenous infusion to minimize its cardiac effect. If improvement does not occur, edrophonium, 10 mg, can be given intravenously over one minute carefully observing the patient for signs of improvement over the next 10 minutes. If a favourable effect is noted, neostigmine, 1–2 mg, intramuscularly every four hours should be given as the action of edrophonium is fleeting. Atropine 0.4 to 0.6 mg, may be given with the neostigmine if one wishes to minimize the muscarinic effects of salivation, increased bronchial secretions, and gastrointestinal cramps of the latter medication (Creese 1969, personal communication).

Lincomycin and clindamycin are also reported to possess ability to enhance the action of neuromuscular-blocking agents (data sheets 1972 for Lincocin and Dalacin C. Upjohn). In addition neomycin was reported to interact with the neuromuscular blockade of pancuronium and tubocurarine (Giala *et al.* 1982).

Lippmann *et al.* (1982) studied the neuromuscular-blocking effects of tobramycin, gentamicin, and cefazolin in 40 patients undergoing extensive surgery. In recommended single doses, these drugs lack clinical neuromuscular-blocking and subclinical relaxant-potentiating effects. Gentamicin potentiates the magnesium-induced neuromuscular weakness in neonate babies whose mothers are treated with $MgSO_4$ for pre-eclampsia (L'Hommedieu *et al.* 1983*a, b*).

Drug-induced myasthenia gravis and similar conditions

Myasthenic symptoms may occur during treatment with

several drugs of very different character. Argov and Mastaglia (1979) subdivided these disorders into two groups, the one being characterized by the unmasking or aggravation of myasthenia gravis, the other as the drug-induced myasthenic syndrome. In the first group quinidine, propranolol, and lithium are given as examples while penicillamine, oxprenolol, and phenytoin are examples of drugs in the second group. This subdivision seems a little artificial, since there is considerable overlap.

Penicillamine-induced myasthenia gravis and other penicilamine-induced neuromuscular disorders

A syndrome clinically indistinguishable from classical myasthenia gravis may develop in patients treated with D-penicillamine. Except for two patients treated with D-penicillamine for Wilson's disease all the reports in the literature have related to patients receiving the drug for the treatment of rheumatoid arthritis.

The incidence of spontaneously occurring myasthenia gravis in patients with rheumatoid arthritis was estimated as 1 in 400 cases (Sundstrom and Schuna 1979) but Masters *et al.* (1977) found antibodies to skeletal muscle in 11 out of 56 rheumatoid patients treated with D-penicillamine. Trnavsky and Zbojanova (1979) found one case of D-penicillamine-induced overt myasthenia in 23 cases of rheumatoid arthritis treated by them. Many case reports of D-penicillamine-induced myasthenia have been published since the first reports appeared in 1974 (Gordon and Burnside 1977; Masters *et al.* 1977; Francois *et al.* 1978; Ongerboer de Visser and van Soesbergen 1978; Schumm and Stohr 1978; Weiss *et al.* 1978; Froelich *et al.* 1979; Sundstrom and Schuna 1979; Trnavsky and Zbojanova 1979). Despite these and other reports and reviews (Argov and Mastaglia 1979) no clear indication of the incidence of this serious complication has emerged but it is clear that immunological changes are much commoner than the overt manifestations of the disorder.

The finding of elevated levels of antibody to the neuromuscular junction acetylcholine receptor in 80 per cent of cases of overt D-penicillamine-induced myasthenia suggests that an immunological mechanism similar to that of classical myasthenia is involved. A gradual decline in antibody titres and clinical improvement on withdrawal of the drug has been observed in most cases where the antibody tries have been measured sequentially. Penicillamine is also known to produce other auto-immune disorders such as a lupus-like syndrome (Jaffe 1968), Goodpastures syndrome (Sternleib *et al.* 1975), and pemphigus (Hewitt *et al.* 1975).

Treatment of D-penicillamine-induced myasthenia

has, where necessary, ranged from short-term treatment with anticholinesterase agents followed by spontaneous recovery to thymectomy or thymic irradiation.

An association of D-penicillamine-induced myasthenia with the histocompatability antigens HLA-AI and HLA-B3, and B8 or both has been described and this is typical of classical myasthenia gravis (Russell and Lindstrom 1977; Seitz et al. 1976), but Keesey and Novom (1979) described a 52-year-old woman with rheumatoid arthritis who developed myasthenia when treated with penicillamine and showed typical anti-human acetylcholine receptor antibodies but her HLA antigens were A3, A11, B35, C4 DRwl, and DRw8; thus she possessed none of the HLA antigens commonly associated with classical myasthenia gravis, namely HLA-B8 and HLA-DRw3.

Miehlke (1974) Broll (1974) and Ott (1974) each reported that they had observed muscle weakness and myasthenia signs in a total of four patients with rheumatoid arthritis who had been treated with penicillamine. Camus (1974) described similar features in a patient with systemic lupus erythematosus being treated with penicillamine. Bucknall et al. (1975) described four patients with rheumatoid arthritis who developed myasthenia gravis after taking penicillamine. In one patient withdrawal of the drug was followed by spontaneous remission of the myasthenia, and in two the dose of anticholinesterase was subsequently reduced. In the fourth patient continuing penicillamine treatment was associated with increasingly severe myasthenic features, but on withdrawal of the drug these resolved. Other reports followed these, e.g. Yates (1975). Bucknall (1977) reviewed the 21 patients reported in the literature and added nine new cases: 18 cases were female and three were male; all suffered from rheumatoid arthritis. Sixteen patients showed improvement on anticholinesterase agents, 13 patients made a full recovery after withdrawal of penicillamine, but six required a period of time on anticholinesterase drugs. Four patients were continuing to receive these agents, four patients were uncontrolled by anticholinesterase, two had thymectomy, one had thymic irradiation, and one died. Myasthenia associated with penicillamine was not reported in patients with Wilson's disease treated with the drug.

Reports of myasthenia gravis associated with the use of D-penicillamine in the treatment of rheumatoid arthritis continue to appear in the literature; there are still no reports of myasthenia gravis complicating the use of D-penicillamine in Wilson's disease other than those of Czlonkowska (1975) and Masters et al. (1977).

The incidence of myasthenia gravis in rheumatoid-arthritic patients treated with D-penicillamine received some quantification as a result of the papers of Dawkins

et al. (1981) from Western Australia and Wysocka et al. (1981) from Poland.

Dawkins et al. (1981) made some generalizations based on their clinical experience on the incidence of myasthenia gravis in D-penicillamine-treated rheumatoid patients. In their series of 500 patients they only identified one case of overt myasthenia gravis, however, they found that antistriational antibodies occurred in about 20 per cent of D-penicillamine-treated rheumatoid patients. In a series of 11 patients with a high titre of antistriational antibodies using electromyography and edrophonium, three cases with possible early myasthenia were detected. (One of these 11 patients with antistriational antibodies but without myasthenia was receiving D-penicillamine for Wilson's disease.) On the basis of these findings, Dawkins et al. estimated the incidence of myasthenia gravis to range from 1 to 6 per cent of rheumatoid-arthritic patients treated with D-penicillamine. Dawkins et al. concluded that the induction of antistriational antibodies must raise the possibility that D-penicillamine could lead to the induction of a thymoma.

Wysocka et al. (1981) reported on 310 patients treated between 1971 and 1979 with D-penicillamine for rheumatoid arthritis. In this time two cases of myasthenia gravis developed, one in a 41-year-old man given the drug for eight months and the other in a 50-year-old woman given the drug for four months. In both patients, pneumomediastinography revealed enlargement of the thymus. Thymectomy followed by histological examination in both cases showed the existence of germinal centres and Hassall's bodies in the typical glandular texture. Follow-up examination three years later confirmed complete recovery in the male patient and a marked improvement in the female patient.

Fawcett et al. (1982) carried out extensive immunological and electrophysiological studies on a patient with rheumatoid arthritis who developed myasthenic symptoms after 10 months of therapy with D-penicillamine.

Ocular myasthenia was the presenting symptom in those few cases of myasthenia gravis associated with D-penicillamine use in Wilson's disease (Czlonkowska 1975; Masters et al. 1977). Kimbrough et al. (1981) described diplopia in a 58-year-old woman receiving D-penicillamine for severe rheumatoid arthritis. Her only neurological disorder was extra-ocular muscle imbalance. Diplopia disappeared after edrophonium chloride (Tensilon), 10 mg, was given intravenously. It was concluded that the patient had a D-penicillamine-induced myasthenia and D-penicillamine was discontinued and *Mestinon* instituted. Diplopia resolved within one month; *Mestinon* was discontinued.

Many individual case reports have also appeared in the literature but the report by Naidoo (1981) of a 36-year-old Indian lady who developed myasthenia gravis with symptoms of diplopia, drooping eyelids, fatigue, and difficulty of mastication while on D-penicillamine for rheumatoid arthritis is important as it describes a case that failed to resolve after withdrawal of the drug and after a one-year follow-up she still required neostigmine 60 mg/day.

In addition to myasthenia gravis, penicillamine has been implicated in other neuromuscular disorders including polymyositis. Reeback et al. (1979) reported an additional neuromuscular complication of penicillamine therapy, namely neuromyotonia. The neuromyotonia developed in a 61-year-old woman with seropositive rheumatoid arthritis. On examination, she had difficulty in relaxing muscles after contraction. For example, her fingers after clenching remained flexed for up to 30 seconds before the forearm flexors suddenly relaxed. Relaxation could be induced by gently extending the fingers. She had similar prolonged contraction in her periorbital facial muscles with difficulty in opening her eyes after blinking. This was mistaken at first for ptosis. Extra-ocular movements were normal. There was no myasthenic weakness or percussion myotonia; sensory examination was normal; ankle jerks were diminished; joint deformities were only slight. Concentric needle electromyography (CMG) in the small hand muscles, biceps, and deltoids showed prominent spontaneous discharges. These consisted of one or more motor units firing repetitively at 10 to 30 s. Electrode movement induced bursts of motor unit activity lasting for between 21 and 26 seconds. After brief voluntary muscular contraction, similar activity was seen, continuing for the same time. Rare pseudomyotonic discharges were seen. Motor-unit action potentials during voluntary activity were normal, and the interference pattern was full. Single-fibre EMG showed no jitter or blocking, and the fibre density was normal. Nerve conduction velocity measurements were normal. A needle biopsy of the right quadriceps muscle was normal; the serum creatine phosphokinase (SCPK) concentration was also normal.

Neuromyotonia was diagnosed and phenytoin 300 mg daily was given. The patient noted improvement within three days. EMG studies six weeks later showed much reduction in the neuromyotonia. The penicillamine was stopped and the neuromyotonia disappeared. It did not recur when phenytoin was also discontinued.

Chloroquine

Schumm et al. (1981) described myasthenic reaction with partial neuromuscular block in the electromyogram and increased antibodies against acetycholine-receptor protein which developed during chloroquine admi-

nistration over two months in a 52-year-old man known for eight years to have rheumatoid arthritis. When the drug was discontinued and pyridostigmine administration begun, myasthenia improved within six weeks and had completely disappeared after three months. During the same period abnormal neuromuscular transmission regressed. Also, the significantly increased antibodies against acetylcholine-receptor protein became normal. It remains undecided whether this was a drug-induced myasthenia gravis or only a latent myasthenia manifested by the drug.

Other drugs causing clinical deterioration in myasthenia gravis

Aminoglycoside antibiotics can produce neuromuscular blockade or can potentiate the neuromuscular blocking effect of succinylcholine. All these agents can cause increasing weakness and deterioration of clinical state in myasthenic patients. Other antibiotics, oxytetracycline (Wullen et al. 1967), and rolitetracycline (Gibbels 1967), have also been reported to aggravate myasthenia gravis.

Martens and Ansink (1979) described a rather different condition, which occurred in a 71-year-old man treated with 120 polymethylacrylate gentamicin-containing beads, each containing 7.5 mg gentamicin for chronic osteomyelitis of the tibia following a compound fracture. After 12 days all but 24 of the beads were removed.

Two months after insertion of these beads the patient was readmitted to hospital; general health was good and the sinus was almost healed. However, the patient complained that exercise provoked within a few minutes progressive symptoms such as bulbar dysarthria, ptosis on the left side, bilateral paresis of the external rectus muscles, difficulty with swallowing and chewing, weakness of the upper extremities, and a feeling of general fatigue. These symptoms improved with rest.

Shortly after admission the condition worsened and the patient had difficulty in breathing. Anticholinesterase therapy was introduced and the patient improved. Neurological investigation indicated a combination of polyneuropathy and myasthenia.

At a further operation 23 beads of the remaining 24 were removed, following which the nerve conduction velocities improved and the myasthenia improved sufficiently for the anticholinesterase therapy to be discontinued.

Most of the antibiotic drugs listed in Table 25.3 which have the potential to produce neuromuscular blockade of themselves or to potentiate neuromuscular-blocking agents may cause an increased weakness in patients with myasthenia gravis.

A myasthenic syndrome was reported in some patients taking phenytoin, and exacerbation of existing myasthenia gravis by phenytoin was also described (Norris *et al.* 1964; Brumalik and Jacobs 1974). So and Penry (1981) reviewed the effects of phenytoin on peripheral nerves and the neuromuscular junction.

Lithium carbonate was reported to unmask myasthenia gravis (Neil *et al.* 1976). Lithium carbonate also prolongs the neuromuscular-blocking effects of curare and succinylcholine. Granacher (1977) observed the occurrence of myasthenia gravis in a patient with recurrent episodes of mania during treatment with lithium carbonate.

Quinidine was reported to aggravate or unmask myasthenia gravis (Weisman 1949; Kornfeld *et al.* 1976); its L-isomer quinine was at one time used as a provocative test in the diagnosis of myasthenia gravis. Procainamide may also aggravate myasthenia gravis (Kornfeld *et al.* 1976; Drachman and Skom 1965). The beta-adrenergic blocking agents propranolol, oxprenolol, practolol, or timolol were reported to induce a myasthenic syndrome or to unmask myasthenia gravis (Herishanu and Rosenberg 1975; Hughes and Zacharias 1976; Shaivitz 1979).

Other agents reported as rare causes of exacerbation of myasthenia gravis are corticosteroids and ACTH, replacement doses of thyroid hormones (Drachman 1962), oral contraceptive agents (Bickerstaff 1975), and dapsone (*Avlosulfon*) (Rosén and Sörnäs 1982).

Drug-induced neonatal myasthenia

Buckley *et al.* (1968) described a male infant born of a myasthenic patient who was clinically normal until 24 hours after delivery, when he became lethargic, had only a faint cry, and had weak sucking and grasping reflexes. Several cyanotic and apnoeic attacks occurred and the baby was clinically myasthenic.

A normal electromyogram had been obtained at 12 hours after delivery but at 24 hours the pattern was characteristic of myasthenia. The diagnosis was confirmed by the improvement in clinical condition, and by an increase in neuromuscular transmission measured by electromyogram after an injection of neostigmine (0.1 mg i.m.). The baby was treated for 12 days with intramuscular injection of neostigmine in progressively decreasing dosage, and at the end of this period anticholinesterase therapy was stopped without the reappearance of any myasthenic symptoms.

During pregnancy, electromyogram tests indicated that the mother was overtreated with pyridostigmine and measurements with ^{14}C-labelled pyridostigmine showed that she excreted this drug more slowly than other myasthenic patients.

The diagnosis of drug-induced neonatal myasthenia due to placental transfer of pyridotigmine was made on the basis of the low level of plasma cholinesterase in maternal and cord blood, the onset of symptoms 24 hours after birth, and the decreasing requirements for neostigmine.

Miscellaneous drug-induced neuromuscular effects

Ryan *et al.* (1971) investigated the frequency of myoglobinaemia after a single intravenous dose of succinylcholine in 40 children and 30 adults. Myoglobin was measured in serial serum samples by a specific immune-precipitin technique sensitive to 4 μg/ml. Myoglobinaemia developed in one adult and 16 children after the intravenous administration of succinylcholine but not in 12 children after intramuscular injection. These findings point to a difference in muscle responsiveness in children to this depolarizing muscle-relaxant. This response diminishes with the onset of puberty. The development of myoglobinaemia after the administration of succinylcholine, although it did not appear to affect recovery from surgery, accentuates the need for another type of short-acting muscle-relaxant for use in children.

A 39-year-old man with ischaemic heart disease developed clinical myotonia while taking propranolol. The myotonia disappeared when administration of the drug ceased. The patient appeared to have dystrophia myotonica which had not been evident before propranolol therapy (Blessing and Walsh 1977).

Perhexiline maleate was reported to cause neuropathy (L'Hermitte *et al.* 1976; Howard and Rees 1976); also a single case of myopathy was documented (Tomlinson and Rosenthal 1977). The myopathy was reported to have occurred in a 67-year-old woman treated for angina pectoris with perhexiline maleate and to have presented within three weeks of commencing treatment with myopathy affecting the muscles of the shoulders and hips. The myopathy resolved within two months of stopping treatment.

REFERENCES

ABOURIZK, N., KHALIL, B.A., BAHUTH, N., AND AFIFI, A.K. (1979). Clofibrate-induced muscular syndrome. *J. neurol. Sci.* **42**, 1–9.

AGOSTON, S. (1975). Muscle relaxants. In *Side-effects of drugs*, Vol. VIII (ed. M.N.G. Dukes.) pp. 262–303. Excerpta Medica, Amsterdam.

ALLOT, E.N. AND THOMPSON, J.C. (1956). The familial incidence of low pseudocholinesterase level. *Lancet* ii, 517.

ANDERSSON, O., BERGLUND, G., AND HANSSON, L. (1976). Antihypertensive action, time of onset and effects on carbohydrate metabolism of labetalol. *Br. J. clin. Pharmacol.* **3**, (suppl. 3) 757–61.

ARGOV. Z. AND MASTAGLIA, F.L. (1979). Disorders of neuromuscular transmission caused by drugs. *New Engl. J. Med.* **301**, 409–13.

ARMENGAUD, M., BOURGOIN. J.J., AND GUÉRIN, M. (1962). L'amibiase aiguë de l'Africain en mileiu hospitalier (à propos de 153 observations). *Bull. Soc. méd. Afr. noire, Langue franc.* **7**, 783–93.

BARLOW, M.B. AND GROESBEEK, A. (1966). Apparent potentiation of neuromuscular block by antibiotics. *S. Afr. med. J.* **40**, 135–6.

BARON, F., JOINVILLE, R., PIQUET, R., DEJOUR, B., AND PRESSARD, P. (1967). Accidents due à la colistine. *Quest méd.* **20**, 148.

BARTON, C.H., STERLING, M.L., AND VAZIRI, N.D. (1980). Rhadbomyolysis and acute renal failure associated with phencyclidine intoxication. *Arch. intern. Med.* **40**, 568–9.

BASURAY, B.N. AND HARRIS, C.A. (1977). Potentiation of di-tubocurarine (d-Tc) neuromuscular blockade in cats by lithium chloride. *Eur. J. Pharmacol.* **45**, 79–82.

BATTISTELLA, P.A., ANGELINI, C., VERGANI, L., BERTOLI, M., AND LORENZI, S. (1978). Carnitine deficiency induced during haemodialysis. *Lancet* **i**, 939.

BAZZATO, G., MEZZINA, C., CIMON, M., AND GUARIERI, G. (1979). Myasthenia-like syndrome associated with carnitine in patients on long-term haemodialysis. *Lancet* **i**, 1041–2.

BEGG, T.B. AND SIMPSON, J.A. (1964). Chloroquine neuromyopathy. *Br. med. J.* **1**, 770.

BENNETT, J.R. (1972). Myopathy from E-aminocaproic acid: a second case. *Postgrad. med. J.* **48**, 440.

BENNETT, H.S., SPIRO, A.J., POLLACK, M.A., AND ZUCKER, P. (1982). Ipecac-induced myopathy simulating dermatomyositis. *Neurology* **32**, 91–4.

BICKERSTAFF, E.R. (1975). *Neurological complications of oral contraceptives*. Oxford University Press, London.

BISWAS, C.K., MILLINGAN D.A.R., AGTE, S.D., KENWARD, D.H., AND TILLEY, P.J.B. (1980). Acute renal failure and myopathy after treatment with aminocaproic acid. *Br. med. J.* **281**, 115–16.

BLAIN, P.G. (1984). Adverse effects of drugs on skeletal muscle. *Adverse Drug React. Bull.* February 384–7.

BLANC, F. (1963). Reflexions sur l'amibiase colique et son traitement. *Rev. Prat., France* **13**, 2861–70.

—— AND NOSNY, Y. (1962). Un amoebicide synthétique diffusible: la 2-déhydro-émétine. *Marseille-med.* **99**, 153–60.

——, ——, ARMENGAUD, M., SANKALE, M., MARTIN, M., AND CHARMOT, G. (1961a). Un amoebicide synthétique susceptible de remplacer l'émétine: la 2-déhydro-émétine. *Presse méd.* **69**, 1548–50.

——, ——, ——, ——, AND NOSNY, P. (1961b). La 2-déhydro-émétine dans le traitement de l'amibiase. *Bull. Sac. Path. exot.* **54**, 29–39.

BLESSING, W.W. AND WALSH, J.C. (1977). Myotonia precipitated by propranolol therapy. *Lancet* **i**, 73–4.

BLOM, S. AND LUNDBERG, P.O. (1965). Reversible myopathy in chloroquine treatment. *Acta med. scand.* **177**, 685–8.

BLOMBERG, L.H. (1965). Dystrophia myotonica probably caused by chloroquine. *Acta neurol. scand.* **41**, (Suppl. 13), 647–51.

BLOMGREN, S.E., CONDEMI, J.J., AND VAUGHAN, J.H. (1972). Procainamide-induced lupus erythematosus. *Am. J. Med.*. **52**, 338–48.

BOHMER, T., BERGREN, H., AND EIKLID, K. (1978). Carnitine deficiency induced during intermittent haemodialysis for renal failure. *Lancet* **i**, 126–8.

BOLLI, P. AND WOOL-MANNING, H.J. (1977). Treatment of hypertension with Labetalol. *New Zealand med. J.* **86**, 557–63.

BONARD, E.C. (1966). Neuropathie due à la chloroquine. *Schweiz. med. Wschr.* **96**, 1103–5.

BOURNE, J.G., COLLIER, H.O.J., AND SOMERS, G.F. (1952). Succinylcholine (succinoylcholine) muscle-relaxant of short action. *Lancet* **i**, 1225–9.

BRAUN, S., AUREL, M., COSTE, FE., AND DELBARRE, F. (1965a). La myopathie des corticoides: *Sem. Hôp., Paris* **41**, 1717–35.

——, COSTE, F., AND AUREL, M. (1965b). Les myopathies de corticoides. *Rev. Rhum.* **32**, 561–5.

BREE, M.M., COHEN, B.J., AND ABRAMS, G.D. (1971). Injection lesions following intramuscular administration of chlorpromazine in rabbits. *J. Am. vet. med. Ass.* **159**, 1598–1602.

BRETAG, A.H., DAWE, S.R., KERR, D.I.B., AND MOSKWA, A.G. (1980). Myotonia as a side effect of diuretic action. *Br. J. Pharmacol.* **71**, 467–71.

BRITISH MEDICAL JOURNAL (1964). Leading article: Chloroquine neuromyopathy. *Br. med. J.* **1**, 452.

BROLL, H. (1974). In *Penicillamine: an international symposium*. Royal Society of Medicine, London. Quoted by Bucknall (1977).

BROSSI, A., BAUMANN, M., CHOPPARD-DIT-JEAN, L.H., WURSCH, J., SCHNEIDER, F., AND SCHNIDER, O. (1959). Syntheseversuche in der Emetin-Reihe 4. Mitt: Racemisches 2-Dehydoemetin. *Helv. chim. Acta* **42**, 772–8.

BROTMAN, M.C., FORBATH, N., GARFINKEL, P.E., AND HUMPHREY, J.G. (1981). Myopathy due to ipecac syrup poisoning in a patient with anorexia nervous. *Can. med. Ass. J.* **125**, 453–4.

BRUMLIK, J. AND JACOBS, R.S. (1974). Myasthenia gravis associated with diphenylhydantoin therapy for epilepsy. *Can. J. neurol. Sci.* **1**, 127–9.

BRUN, A. (1959). Effect of procaine carbocaine and xylocaine on cutaneous muscle in rabbits and mice. *Acta anaesth. scand.* **3**, 59–73.

BUCKLEY, G.A., ROBERTS, D.V., ROBERTS, J.B., THOMAS, B.H., AND WILSON, A. (1968). Drug-induced neonatal myasthenia. *Br. J. Pharmacol.* **34**, 203P–204P.

BUCKNALL, R.C. (1977). Myasthenia associated with D-penicillamine therapy in rheumatoid arthritis. In *Penicillamine at 21: its place in therapeutics now.* ed. W.H. Lyle, and R.L. Kleinman *Proc. R. Soc. Med.* **70**, (Suppl. 3), 114–17.

——, DIXON, A. ST. J., GLICK, E.N. WOODLAND, J., AND ZUISHI, D.W. (1975). Myaesthenia gravis associated with penicillamine treatment for rheumatoid arthritis. *Br. med. J.* **1**, 600–2.

BUREAU, Y., BARRIERE, H., LITOUX, P., AND BUREAU, M. (1965). Les accidents neuromusculaires de la chloroquine. *J. méd. Nantes* **5**, 107.

BURLAND, W. (1978). Evidence for the safety of cimetidine in the treatment of peptic ulcer disease. In *Proceedings of the Third Symposium on H2 receptor antagonists* (ed. W. Creutzfeld), pp. 233–58. Excerpta Medica. Amsterdam.

CAMUS, J.P. (1974). In *Penicillamine: an international symposium*. Royal Society of Medicine, London. Quoted by Bucknall (1977).

CARIO, W.R., HOFFMAN, J., HUBSCHUMANN, K., KUNZE, D., AND REGLING,G. (1979). Myopathy induced by clofibrate treatment in normolipidaemic patients. *Acta paediat. Acad. Sci. Hung.* **20**, 1–9.

CHASAPAKIS, G. AND DIMAS, C. (1966). Possible interaction

between muscle relaxants and the kallikrein-trypsin inactivator *Trasylol. Br. J. Anaesth.* **38**, 838–9.

CHHABRA, R.H., BAMJI, D.D., AND DESAI, M.M. (1964). Clinical trials of injection of dehydroemetine in amoebiasis. *Curr. med. Pract.* **8**, 114–16.

CHURCHILL-DAVIDSON, H.C. (1955). Abnormal response to muscle relaxants. *Proc. R. Soc. Med.* **48**, 621–4.

CIOC, M. AND SHILLING, B. (1965). Changes in muscular tissue produced by injection of some known drugs. *Toxicol. appl. Pharmacol.* **7**, 179–89.

CLARK, J.G. AND SUMERLING, M.D. (1966). Muscle necrosis and calcification in acute renal failure due to barbiturate intoxication *Br. Med. J.* **2**, 214–15.

COGEN, F.C., RIGG, G., SIMMONS, J.L., AND DOMINO, E.F. (1978). Phencyclidine-associated acute rhabdomyolysis. *Ann. intern. Med.* **88**, 210–12.

COOMES, E.N. (1965). Corticosteroid. *Ann, rheum. Dis.* **24**, 465–72.

CORNFORD, E.M., BRAUN, L.D., AND OLDENDORF, W.H. (1978). Carrier mediated blood brain barrier transport of choline and certain choline analogs. *J. Neurochem.* **30**, 299–308.

COSTE, F., DELBARREBRAUN, S., RONDEAU, P., BEDORSEAU, P., AND AUREL, M. (1965*a*). La myopathie des corticoides. *Presse méd.* **73**, 2636.

——, ——, ——, ——, ——, AND —— (1965*b*). La myopathie des corticoides. *Lyon méd.* **214**, 91.

CZLONKOWSKA, A. (1975). Myasthenia syndrome during penicillamine treatment. *Br. med. J.* **2**, 726.

DAWKINS, R.L., GARLEPP, M.J., McDONALD, B.L., WILLIAMSON, J., ZILKO, P.J., AND CARRANO, J. (1981). Myasthenia gravis and D-penicillamine. *J. Rheumatol.* **8**, (Suppl. 7) 169–72.

DEMPSEY, J.J. AND SALEM, H.H. (1966). An enzymatic electrocardiographic study on toxicity of dehydroemetine. *Br. Heart J.* **28**, 505–11.

DE GRANDIS, D., MEZZINA, C., FIASCHI, A., PINELLI, P., BAZZAIO, G., AND MORACHIELLO, M. (1980). Myasthenia due to carnitine. *J. neurol. Sci.* **46**, 365–71.

DE LATOUR, B.J. AND HALLIDAY, W.R. (1978). Pentazocine fibrous myopathy. A report of two cases and literature review. *Arch. Phys. med. Rehabil.* **59**, 394–6.

DRACHMAN, D.B. (1962). Myasthenia gravis and the thyroid gland. *New Engl. J. Med.* **266**, 330–3.

——, AND SKOM, J.H. (1965). Procainamide — a hazard in myasthenia gravis. *Arch. Neurol.* **13**, 316–20.

DRENCKHAHN, D. AND LÜLLMANN-RAUCH, R. (1979). Experimental myopathy induced by amphiphilic cationic compounds including several psychotropic drugs. *Neuroscience* **4**, 549–62.

DUBOIS, E.L. (1963). Paramethasone in the treatment of systemic lupus erythematosus. Analysis of results in 51 patients with emphasis on single daily oral doses. *J. Am. med. Ass.* **184**, 463–9.

EADIE, M.J. AND FERRIER, T.M. (1966). Chloroquine myopathy *J. Neurol. Neurosurg. Psychiat.* **29**, 331–7.

ELLIS, F.G. (1962). Acute polyneuritis after nitrofurantoin therapy. *Lancet* ii, 1136–8.

ENGEL, H.L. AND DENSON, J.S. (1957). Respiratory depression due to neomycin. *Surgery* **42**, 862–4.

EVANS, F.T., GRAY, P.W.S., LEHNANN, H., AND SILK, E. (1952). Sensitivity to succinylcholine in relation to serum cholinesterase. *Lancet* i, 1229–30.

——, ——, ——, AND —— (1953). Effects of pseudocholinesterase level on action of succinylcholine in man. *Br. med. J.* **1**, 136–8.

FAWCETT, P.R.W., McLACHLAN, S.M., NICHOLSON, L.V.B., AND MASTAGLIA, F.L. (1982). D-penicillamine-associated myasthenia gravis: immunological and electrophysiological studies. *Muscle Nerve* **5**, 328–34.

FEEST, T.G. AND READ, D.J. (1980). Myopathy associated with cimetidine. *Br. med. J.* **281**, 1284–5.

FENOGLIO, J.J., McALLISTER, H.A., AND MULLICK, F.G. (1981). Drug related myocarditis. *Human pathol.* **12**, 900–7.

FERRES. C.J., MIRAKHUR, R.K., CRAIG, H.J.L. BROWNE, E.S., AND CLARKE, R.S.J. (1983). Pretreatment with vecuronium as a prophylatic against post-suxamethonium muscle pain. *Br. J. Anaesthiol.* **55**, 735–41.

FORFAR, J.C., BROWN, G.J., AND CULL, R.E. (1979). Proximal myopathy during beta-blockade. *Br. med. J.* **2**, 1331–2.

FRANCOIS, J., VERBRAEKEN, H., GABRIEL, P., AND WILEE, C. (1978). Syndrome myasthenique apres traitment peroral a la penicillamine. *Ophthalmol., Basel* **177**, 88–91.

FROELICH, C.J., HASHIMOTO, F., SEARLES, R.P., AND BANKHURST, A.D. (1979). D-penicillamine-induced myasthenia gravis in rheumatoid arthritis. *J. Rheumat.* **6**, 237–9.

GIALA, M. SAREYIANNIS, C., CORTSARIS, N., PARADELIS, A., AND LAPPAS, D.G. (1982). Possible interaction of pancuronium and tubocurarine with oral neomycin. *Anaesthesia, England* **37**, 776.

GIBBELS, E. (1967). Weitere Beobachtungen zur Nebenwirkung intravenoser Reverin-Gaben bei Myasthenia gravis pseudoparalytica. *Dt. med. Wochschr.* **92**, 1153–4.

GOLD, G.N. AND RICHARDSON, A.P. (1965). Myasthenic reaction to colistinmethate. *J. Am. med. Ass.* **194**, 1151–2.

—— AND —— (1966). An unusual case of neuromuscular blockade seen with therapeutic blood levels of colistin methanesulphonate (Coly-mycin). *Am. J. Med.* **41**, 316–21.

GORDON, R.A. AND BURNSIDE, J.W., (1977). D-penicillamine induced myasthenia gravis in rheumatoid arthritis. *Ann. intern. Med.* **87**, 578.

GRAM, L.F., SCHOU, J., WAY, W.L., HELTBERG, J., AND BODIN, N.O. (1979). d-Propoxyphene kinetics after single oral and intravenous doses in man. *Clin. Pharmacol. Ther.* **26**, 473–82.

GRANACHER, R.P. (1977). Neuromuscular problems associated with lithium. *Am. J. Psychiat.* **134**, 6.

GRIMBY, G. AND SMITH, U. (1978). Beta-blockade and muscle function. *Lancet* ii, 1318–19.

HALES, D.S.M. SCOTT, R. AND LEWI, H.J.E. (1982). Myopathy due to mercaptopropionyl glycine. *Br. med. J.* **285**, 939.

HANSON, D.J. (1961). Local toxic effects of broad-spectrum antibiotics following injections. *Antibiot. Chemother.* **11**, 390–404.

HARRESTRUP ANDERSEN, A. (1983). Intramuskulære injektioner [Intramuscular injections]. *Ugeskr. Læger* **145**, 182–3.

HART, S.M. AND MITCHELL, J.V. (1962). Suxamethonium in the absence of pseudocholinesterase. *Br. J. Anaesth..* **34**, 207–9.

HERISHANU, Y. AND ROSENBERG, P. (1975). β-Blockers and myasthenia gravis. *Ann. intern. Med.* **83**, 834–5.

HEWITT, J., BENVENISTE, M., AND LESSONA-LEIBOWITCH, M. (1975). Pemphigus induced by D-penicillamine. *Br. med. J.* **3**, 371.

HILL, G.E., WONG, K.C., AND HODGES, M.R. (1976). Potentiation of succinylcholine neuromuscular blockade by lithium carbonate. *Anaesthesiology* **44**, 439–42.

HOKKANEN, E. (1964). The aggravating effect of some antibiotics on the neuromuscular blockage in myasthenia gravis. *Acta neurol. scand.* **40**, 346–52.

HOOGWERF, B., KERN., J., BULLOCK, M., AND COMPTY, C.M. (1979). Pencyclidine induced rhabdomyolysis and acute renal failure. *Clin. Toxicol.* **14**, 47–53.

HOWARD, D.J. AND RUSSELL REES, J. (1976). Long-term perhexiline maleate and liver function. *Br. med. J.* **1**, 133.

HUGHES, J.T., ESIRI, M., OXBURY, J.M., AND WHITTY, C.W. (1970). Chloroquine myopathy. *Quart. J. Med.* **157**, 85–93.

HUGHES, R.O. AND ZACHARIAS, F.J. (1976). Myasthenic syndrome during treatment with practolol. *Br. med. J.* **1**, 460–1.

INTURRISI, C.E., COLBURN, W.A., VEREBY, K., DAYTON, H.E., WOODY, G.E., AND O'BRIEN, C.P. (1982). Propoxyphene and norpropoxyphene kinetics after single and repeated doses of propoxyphene. *Clin. Pharmacol. Ther.* **31**, 157–67.

JAFFE, I.A. (1968). Effects of penicillamine on the kidney and taste. *Postgrad. med. J.* Suppl. **23**, 15–18.

JENKINS, P. AND EMERSON, P.A. (1981). Myopathy induced by refampicin. *Br. med. J.* **283**, 105–106.

JENNINGS, K. AND PARSONS, V. (1976). A study of labetalol in patients of European West Indian and West African Origin. *Br. J. clin. Pharmacol.* **3**, (suppl. 3) 773–5.

JOURNE, P., LENOIR, P., MOREL, H., TURPING J., DUPONT, B., AND BOUREL, M. (1965). Une observation de myopathie vacuolaire au cours d'un traitement par la chloroquine. *Quest méd.* **18**, 990.

JULIEN, J., VALLAT, J.M., LAGUENY, A., AND VITAL, C. (1979). Myopathy and cerebellar syndrome during acute poisoning with lithium carbonate. *Muscle Nerve* May, June, 240.

KALOW, W. (1956). Familial incidence of low pseudocholinesterase level. *Lancet* ii, 576–7.

—— (1959). In Ciba Foundation Symposium on the *Biochemistry of human genetics* (ed. G.E. Wolstenholme and M.C. O'Conner), pp. 39. Churchill-Livingstone, London.

—— AND GENEST, K. (1957). A method for the detection of atypical forms of human serum cholinesterase; determination of dibucaine numbers. *Can. J. Biochem.* **35**, 339–46.

—— AND STARON, N. (1957). On distribution of atypical forms of human serum cholinesterase, as indicated by dibucaine numbers. *Can. J. Biochem.* **35**, 1305–20.

KARSTARP, A., FERNGREN, H., LUNDBERGH, P., AND LYING-TURNELL, U. (1973). Neuromyopathy during malaria suppression with chloroquine. *Br. med. J.* **4**, 736.

KATSILAMBROS, N., BRAATEN, J., FERGUSON, B.D., AND BRADLEY, R.F. (1972). Muscular syndrome after clofibrate. *New Engl. J. Med.* **286**, 1110–11.

KAUFMAN, L., LEHMANN, H., AND SILK, E. (1960). Suxamethonium apnoea in an infant: expression of familial pseudocholinesterase deficiency in three generations. *Br. med. J.* **1**, 166–7.

KEESEY, J. AND NOVOM, S. (1979). HLA antigens in penicillamine induced myasthenia gravis. *Neurology. Minneapolic* **29**, 528–9.

KEIDAR, S., BINENBOIM, C., AND PALANT, A. (1982). Muscle cramps during treatment with nifedipine. *Br. med. J.* **285**, 1241–2.

KENG, C.B. AND SWEE, Y.O. (1966). Neuromuscular manifestations of emetine toxicity. *Singapore med. J.* **7**, 156–63.

KENNARD, C., SWASH, M., AND HENSON, R.A. (1980). Myopathy due to aminocaproic acid. *Muscle Nerve* **3**, 202–6.

KIMBROUGH, R.L., MEWIS, L., AND STEWARD, R.H. (1981). D-penicillamine and the ocular myasthenic syndrome. *Ann. Ophthalmol.* **13**, 1171–8.

KOENIG, A. AND OHRLOFF, C. (1981). Beeinflussung der neuromuskulären Übertragung durch lokale Applikation von Isoptomax®-Augentropfen. *Klin. Mbl. Augenheilk.* **179**, 109–12.

KORNFELD, P., HOROWITZ, S.H., AND GENKINS, G. (1976). Myasthenia gravis unmasked by antiarrhythmic agents. *Mt Sinai J. Med. NY* **43**, 10–14.

KORSAN-BENGSTEN, K., YSANDER, L., BLOHME, G., AND TIBBLIN, E. (1969). Extensive muscle necrosis after long term treatment with aminocaproic acid (EACA) in a case of hereditary periodic oedema. *Acta med. scand.* **185**, 341.

LAMBERT, J.J. DURANT, N.N., AND HENDERSON, E.G. (1983). Drug-induced modification of ionic conductance at the neuromuscular junction. *Ann. Rev. Pharmacol. Toxicol.* **23**, 505–39.

LANCET (1965). Corticosteroid myopathy, annotation. *Lancet* ii, 1118.

LANE, R.J.M. AND MASTAGLIA, F.L. (1978). Drug-induced myopathies in man. *Lancet* ii, 562–6.

—— AND ROUTLEDGE, P.A. (1983). Drug-induced neurological disorders. *Drugs* **26**, 124–47.

—— McLELLAND, NJO, MARTIN, A.M., AND MASTAGLIA, F.L. (1979). Epsilon amino caproic acid (EACA) myopathy *Postgrad. med. J.* **55**, 282–5.

LANGER, T., AND LEVY, R.I. (1968). Acute muscular syndrome associated with administration of clofibrate. *New Engl. J. Med.* **279**, 856–8.

LEALOCK, A.M., CAMPBELL, D.J., AND McINTYRE, J.W.R. (1966). A clinical and biochemical approach to cholinesterase problems in anaesthesia. *Can. Anaesth. Soc. J.* **13**, 550–6.

LEHMANN, H. AND LIDDELL, J. (1962). In *Modern trends in anaesthesia*, Series II (ed. F.T. Evans and T.C. Gray). Butterworths, London.

—— AND RYAN, E. (1956). Familial incidence of low pseudocholinesterase level. *Lancet* ii, 124.

—— AND SIMMONS, P.H. (1958). Sensitivity to suxamethonium: apnoea in two brothers. *Lancet* ii, 981–2.

LENOIR, P., JOURNE, P., URVOY, M., AND BOUREL, M. (1965). Les complications musculaires et oculaires des traitements par la chloroquine. *Méd. et Hyg., Genéve* **714**, 1223–5.

LERIQUE, J. AND CHAUMOUNT, P. (1965). Exploration electrolytique au cours des syndromes cortisoniques. *Rev. neurol.* **112**, 342.

L'HERMITTE, F., FARDEAU, M., CHEDON, F., AND MALLECOURT, (1976). Polyneuropathy after perhexiline maleate therapy. *Br. med. J.* **1**, 1256.

L'HOMMEDIEU, C.S., HUBERT, P.A., AND RASACH, D.K. (1983a). Potentiation of magnesium-induced neuromuscular weakness by gentamicin. *Critical Care Med.* **11**, 55–6.

——, NICHOLAS, D., ARMES, D.A., JONES, P., NELSON, T., AND PICKERING, L.K. (1983b). Potentiation of magnesium sulfate-induced neuromuscular weakness by gentamicin, tobramycin, and anukacin. *J. Pediat.* **102**, 629–31.

LIPPMANN, M.Y., AU, E., AND LEE, C. (1982). Neuromuscular blocking effects of tobramycin, gentamycin, and cefazolin. *Anesth. Analg.* **61**, 767–70.

LOFTUS, L.R. (1963). Peripheral neuropathy following chloroquine therapy. *Can. med. Ass. J.* **80**, 407–12.

LUND-JACOBSEN, H. (1978). Cardiorespiratory toxicity of propoxyphene and norpropoxyphene in conscious rabbits. *Acta Pharmacol. Toxicol.* **42**, 171–8.

LÜLLMANN, H., LÜLLMANN-RAUCH, R., AND WASSERMANN, O. (1978). Lipidosis induced by amphiphilic cationic drugs. *Biochem. Pharmacol.* **27**, 1103–8.

MacKAY, A.R., LOI SANG, U., AND WEINSTEIN, P.R. (1978). Myopathy associated with epsilon-amino-caproic acid EACA therapy. *J. Neurosurg.* **49**, 597–601.

MAEBASHI, M., KAWAMURA, N., SATO, M., IMAMURA,A., AND YOSHI-NAGA, K. (1978). Lipid lowering effect of carnitine in patients with type IV hyperlipoproteinaemia. *Lancet* ii, 805–7.

MANSELL, M.A. (1982). Myopathy due to mercaptopropionyl glycine. *Br. med. J.* **285**, 1356–7.

MANTEN, A. (1968). Antibiotic drugs. In *Side-effects of drugs*, Vol. VI (ed. L. Meyer and A. Herxheimer). Excerpta Medica Foundation, Amsterdam.

MARCO, L.A. AND RANDELS, P.M. (1979). Succinylcholine drug interaction during electroconvulsive therapy. *Biol. Psychiat.* **14**, 433–55.

MARTENS, E.I.F. AND ANSINK, B.J.J. (1979). A myasthenia like syndrome and polyneuropathy complications of gentamicin therapy. *Clin. Neurol. Neurosurg.* **81**, 241–6.

MASTAGLIA, F.L. (1982). Adverse effects of drugs on muscles. *Drugs* 24, 304–21.

—— AND ARGOV, Z. (1980). Immunologically mediated drug-induced neuromuscular disorders. In *Pseudo-allergic reactions* (ed. Schlumberger *et al.*), pp. 1–24. Karger, Basle.

—— AND —— (1981*a*). Drug-induced neuromuscular disorder in man. In *Disorders of voluntary muscle* (ed. J.N. Walton), pp. 873–906. Churchill Livingstone, Edinburgh.

—— AND —— (1981*b*). Drug-induced myopathies and disorders of neuromuscular transmission. In *Advances in neurotoxicology (ed. Manzo et al.*), pp. 319–28. Pergamon Press, Oxford.

MASTERS, C.L., DAWKINS, R.L. ZILKO, P.J., SIMPSON, J.A., LEEDMAN, R.J., AND LINDSTROE, J. (1977). Penicillamine-associated myasthenia gravis, anti-acetylcholine receptor and anti-striational antibodies. *Am. J. Med.* **63**, 689–94.

MATTHIESEN. J. (1979). Polymyositis so en mulig birvirkning af cimetidin behandling. *Ugeskr, Laeg.* 141, 2762.

McGAVIE, D.D.M. (1965). Depressed levels of serum pseudocholinesterase with ecothiopate-iodide eyedrops. *Lancet* ii, 272–3.

MEIER, C., KAUER, B., MULLER, U., AND LUDIN, H.P. (1979). Neuromyopathy during chronic amiodarone therapy. *J. Neurol.* **220**, 231–9.

MEIHLKE, K. (1974). In *Penicillamine: an international symposium.* Royal Society of Medicine, London. Quoted by Bucknall (1977).

MILLINGEN, K.S. AND SUERTH, E. (1966). Peripheral neuromyopathy following chloroquine therapy. *Med. J. Aust.* **1**, 840–1.

MITCHELL, M.M., ALI, H.H., AND SAVARESE, J.J. (1978). Myotonia and neuromuscular blocking agents. *Anaesthesiology* 49, 44–8.

MONE, J.G. AND MATHIE, W.E. (1967). Qualitative and quantitative defects of pseudocholinesterase activity. *Anaesthesia* 22, 55–68.

MORGAN, G.J., McGUIRE, J.L., AND OCHOA, J. (1981). Penicillamine-induced myositis in rheumatoid arthritis. *Muscle Nerve* 4, 137–40.

NAIDOO, P.D. (1981). Penicillamine-induced myasthenia gravis in a patient with rheumatoid arthritis. *S. Afri. med. J.* **60**, 478.

NEIL, J.F., HIMMELHOCH, J.M., AND LICATA, S.M. (1976). Emergence of myasthenia gravis during treatment with lithium carbonate. *Arch. gen. Psychiat.* 33, 1090–2.

NEUHAUS, G.A. (1978). Muskelschwäche als Leit- oder Warn-symptom unter Medikamentösen Terapie und bei Intoxi-kationen. *Verh. Dtsch. Ges. Inn. Med.* **84**, 852–62.

NORRIS. F.H. JR. COLELLA, J., AND McFARLIN. D. (1964). Effect of diphenylhydantoin on neuromuscular synapse. *Neurology, Minneapolis* **14**, 869–76.

ONGERBOER DE VISSER, B.W. AND VAN SOESBERGEN, R.M. (1978). Myasthenia gravis als complicatie van D-pencillamine bij reumatoide arthritis. *Ned. T. Geneesk.* **122**, 2055–8.

OTT, V.R. (1974). In *Penicillamine: an international symposium.* Royal Society of Medicine, London. Quoted by Bucknall (1977).

PACKMAN, P.M., MEYER, D.A., AND VERDUN, R.M. (1978). Hazards of succinylcholine administration during electrotherapy. *Arch. gen. Psychiat.* **35**, 1137–41.

PARBROOK. G.D. AND PIERCE, G.F.M. (1960). Comparison of post-operative pain and stiffness after the use of suxamethonium and suxethonium compounds. *Br. med. J.* **2**, 579–80.

PATEL, J.C. AND MEHTA, A.B. (1967). Oral dehydroemetine in intestinal and extra-intestinal amoebiasis. *Ind. J. med. Sci.* **21**, 1–5.

PATEL, R., DAS, M., PALAZZOLO, M., AND ANSARI BAKASUB RAMANIAM, S. (1980). Myoglobinuric acute renal failure in phencyclidine overdose. Report on eight cases. *Ann. emerg. Med.* **9**, 549–53.

PENN, A.S., ROWLAND, L.P., AND FRASER, D.W. (1972). Drugs, coma and myoglobinuria. *Arch. Neurol.* **26**, 336–44.

PERKOFF, G.T., HARDY, P., AND VELEZ-GARCIA, E. (1966). Reversible acute muscle syndrome in chronic alcoholism. *New Engl. J. Med.* **274**, 1277–85.

PITTINGER, C.B. AND LONG, J.P. (1958). Neuromuscular blocking action of neomycin sulphate. *Antibiot. Chemother.* **8**, 198–203.

——, ——, AND MILLER, J.R. (1958). The neuromuscular blocking action of neomycin: a concern of the anaesthesiologist. *Curr. Res. Anesth.* **37**, 276–82.

PIZZOLATO, P. AND MANNHEIMER, W. (1961). Histopathological effects of local anaesthetic drugs and related substances. Charles C. Thomas, Springfield, Illinois.

POHLMANN, G. (1966). Respiratory arrest associated with intravenous administration of polymyxin B sulphate. *J. Am. med. Ass.* **196**, 181–3.

POWELL, J.R. (1969). Steroid and hypokalaemic myopathy after corticosteroids in ulcerative colitis. Am. J. Gastro-enterol. **52**, 425–32.

POWELL, S.J., McLEOD, I.N., WILMOT, A.J., AND ELSDON-DEW, R. (1962). Dehydroemetine in amebic dysentery and amebic liver abscess. *Am. J. trop. Med. Hyg.* **11**, 607–9.

PRIDGEN, J.E. (1956). Respiratory arrest thought to be due to intraperitoneal neomycin. *Surgery* 40, 571–4.

RAFFI, A., OPPERMAN, H., PAGEAUT, G., PEQUENGOT, J., AND LONGCHAMP, D. (1966). Trois cas de myopathie stéroidienne. *Pédiatrie* 21, 608–10.

RAMSEY, I. (1980). Beta blockade, myopathy, and thyrotoxicosis. *Br. med. J.* **280**, 718.

RASMUSSEN, F. AND SVENDSEN, P. (1976). Tissue damage and concentrations at the injection site after intramuscular injection of chemotherapeutics and vehicles in swine. *Res. vet. Sci.* **20**, 50–60.

REEBACK, J., BENTON, S., SWASH, M., AND SCHWARTZ, M.S. (1979). Penicillamine-induced neuromyotonia. *Br. med. J.* **1**, 1464–5.

RENIER, J. Cl. (1965). Une observation de neuromyopathie due à l'hydroxychloroquine. *Rev. Rhum.* **32**, 681–2.

RICHTER, R.W., CHALLENOR, Y.B., PEARSON, J., KAGEN, L.J., HAMILTON, L.L., AND RAMSEY, W.H. (1971). Acute myoglobinuria associated with heroin addiction. *J. Am. med. Ass.* **216**, 1172–6.

RICKER, K., HAAS, A., AND GLÖTZNER, F. (1978). Fenotferol precipitating myotonia in a minimally affected case of recessive myotonia congenita. *J. Neurol.* **219**, 279–82.

ROSÉN, I. AND SÖRNÄS, R. (1982). Peripheral motor neuropathy caused by excessive intake of dapsone (*Avlosulfon*). *Arch. Psychiatr. Nervenkr.* **232**, 63–9.

RUSSELL, A.S. AND LINDSTROM, J.M. (1977). Penicillamine-induced myasthenia gravis associated with antibodies to acetylcholine receptor. *Neurology. Minneapolis* **28**, 847–9.

RYAN, J.F., KAGAN, L.J., AND HYMAN, A.I. (1971). Myoglobinaemia after a single dose of succinyl choline. *New Engl. J. Med.* **285**, 824–7.

SANGHVI, L.M. AND MATHUR, B.B. (1965). Electrocardiagram after chloroquine and emetine. *Circulation* **32**, 281–9.

SARDESAI, H.V., SULE, C.R., AND GAVANKAR, S.S. (1963). Clinical trials with dehydroemetine (Ro 1–9334) in acute amoebiasis. *Indian J. med. Sci.* **17**, 334–5.

SARIN, B.P. (1963). Dehydroemetine in the treatment of intestinal amoebiasis in paediatric practice. *Burma med. J.* **11**, 233–5.

SCHINDEL, L. (1968). Antiprotozoal drugs. In *Side-effects of drugs*, Vol. VI (ed. L. Meyler and A. Herxheimer). Excerpta Medica Foundation. Amsterdam.

SCHNEIDER, J. (1980). Creatine Kinase in hyperlipoproteinemic patients treated with clofibrate. *Artery* **8**, 164–70.

—— MÜHLFELLER, G., AND KAFFARNIK, H. (1980). Drug safety in clofibrate treatment. In *Proc. VII Int. Symp. Drugs affecting lipid Metabolism*, p. 125. Milan.

SCHUMM, F. AND STOHR, M. (1978). Myasthene Syndrome unter Penicillamine-therapie. *Klin. Wschr.* **56**, 139–44.

——, WIETHOLTER, H., AND FATEH-MOGHADAM, A. (1981). Myasthenie-syndrom unter Cholorquin-therapie. *Dtsch. med. Wschr.* **106**, 1745–7.

SEITZ, D., HOPE, H.C. JANZEN, R.W.C. AND MEYER, W. (1976). Penicillamin-induzierte Myasthenie bei chronischer Polyarthritis. *Dt. med. Wschr.* **101**, 1153–8.

SHAIVITZ, S.A. (1979). Timolol and myasthenia gravis. *J.Am. med. Ass.* **242**, 1611–12.

SHROFF, K.R. AND BODIWALA, N.K. (1966). Clinical trial of dehydroemetine in amoebiasis. *Curr. med. Pract.* **10**, 233–8.

SIDELL, F.R., CULVER, D.L., AND KAMINSKIS, A. (1974). Serum creatinine phosphokinase activity after intramuscular injection. *J. Am. med. Ass.* **229**, 1894–97.

SINCLAIR, D. AND PHILLIPS, C. (1982). Transient myopathy apparently due to tetracyline. *New Engl. J. Med.* **307**, 821–2.

SMALS, A., BEEX, L., AND KLOPPENBORG, P. (1977). Clofibrate-induced muscle damage with myoglobinuria and cardiomyopathy. *New Engl. J. Med.* **296**, 942.

SMITH, A.F., MacFIE, W.G., AND OLIVER, M.F. (1970). Clofibrate, enzymes and muscle pain. *Br. med. J.* **2**, 86–8.

SMITH, B. AND O'GRADY, F. (1966). Experimental chloroquine myopathy. *J. Neurol. Neurosurg. Psychiat.* **29**, 255–8.

SO, E.L. AND PENRY, J.K. (1981). Adverse effects of phenytoin on peripheral nerves and neuromuscular junction: A review. *Epilepsia* **22**, 467–73.

SPAULDING, W.B. (1979). Myalgia and elevated creatine phosphokinase with danazol in hereditory angioneurotic oedema. *Ann. intern. Med.* **90**, 854.

SPOONER, J.P. (1977). Classification of side effects of danazol therapy. *J. int. med. Res.* **5** (Suppl. 3), 15–17.

STEINESS, E., RASMUSSEN, F., SVENDSEN, O., AND NIELSEN, P. (1978). A comparative study of serum creatinine phosphokinase (CPK) activity in rabbits, pigs and humans after intramuscular injection of local damaging drugs. *Acta Pharmacol. Toxicol.* **42**, 357–64.

——, SVENDSEN, O., AND RASMUSSEN, F. (1974). Plasma digoxin after parenteral administration. Local reactions after intramuscular administration. *Clin. Pharmacol. Ther.* **16**, 430–4.

STERNLIEB, I., BENNET, B., AND SCHEINBER, I.H. (1975). D-penicillamine-induced Goodpasture's syndrome in Wilson's disease. *Ann. intern. Med.* **82**, 673–4.

STONE, R. (1979). Proximal myopathy during beta-blockade. *Br. med. J.* **2**, 1583.

SUNDSTROM, W.R. AND SCHUNA, A.A. (1979). Penicillamine-induced myasthenia gravis. *Arthritis Rheum.* **22**, 197–8.

SVENDSEN, O. AND AAES-JØRGENSEN, X. (1979). Studies on the fate of vegetable oil after intramuscular injection into experimental animals. *Acta Pharmacol. Toxicol.* **45**, 352–78.

SWASH, M. (1980). Aminocaproic acid myopathy. *Br. med. J.* **281**, 454.

TEICHER, A., ROSENTHAL, T., KISSEN, F., AND SAROVA, I. (1981). Labetalol induced toxic myopathy. *Br. med. J.* **282**, 1824–5.

TELFER, A.B.M. MACDONALD. D.J.Z., AND DINWOODIE, A.J. (1964). Familial sensitivity to suxamethonium due to atypical pseudo-cholinesterase. *Br. med. J.* **1**, 153–6.

TIMMERMAN, J.C., LONG, J.P., AND PITTINGER, C.B. (1959). Neuromuscular blocking properties of various antibiotic agents. *Toxicol. appl. Pharmacol.* **1**, 299–304.

TOMLINSON, I.W. AND ROSENTHAL, F.D. (1977). Proximal myopathy after perhexiline maleate treatment. *Br. med. J.* **2**, 1319–20.

TRNAVSKY, K. AND ZBOJANOVA, M. (1979). Nezadouci-ucinky lecby progresivni Polyartrididy D Penicillaminem Fysiatr. *Rheumat. Vestn.* **57**, 87–91.

UUSITUPA, M., ARO, A., KORHONEN, T., AND JUKKA, E. (1980). Beta-blockade, myopathy and thyrotoxicosis. *Br. med. J.* **1**, 183.

VAN DIJL, W. (1968). Muscle-relaxant drugs. In *Side-effects of drugs*, Vol. VI (ed. L. Meyler and A. Herxheimer). Excerpta Medica Foundation, Amsterdam.

VANNESTE, J.A.L. AND VAN WIJNGAARDEN, G.K. (1982). Epsilon-aminocaproic acid myopathy. *Eur. Neurol.* **21**, 242–8.

WALLS, T.J., PEARCE, S.J., AND VENABLES, G.S. (1980). Motor neuropathy associated with cimetidine. *Br. med. J.* **281**, 974.

WATTEL, F., CHOPIN, C., DUROCHER, A., AND BERZIN, B. (1978). Rhabdomyolysis in acute intoxications. *Nouv. Presse méd.* **7**, 2553–60.

WAY, W.L. AND SCHOU, J. (1979). Entrance into brain of dextropropoxyphene and the toxic metabolie norpropoxyphene. *Arch. Toxicol.* (Suppl. 2), 367–70.

WEISMAN, S.J. (1949). Masked myasthenia gravis. *J. Am. med. Ass.* **141**, 917–18.

WEISS, A.S., MARKENSON, J.A., WEISS, M.S., AND KAMMERER, W.H. (1978). Toxicity of D-penicillamine in rheumatoid arthritis. A report of 63 patients including 2 with aplastic anemia and 1 with nephrotic syndrome. *Am. J. Med.* **64**, 114–20.

WHISNANT, J.P., ESPINOSA, R.E., KIERLAND, R.R., AND LAMBERT, E.H. (1963). Chloroquine neuromyopathy. *Proc. Mayo Clin.* **38**, 501–13.

WOLEN, R.L., GRUBER, C.M., KIPLINGER, G.F., AND SCHOLZ, N.E. (1971). Concentrations of propoxyphene in human plasma

following oral intramuscular, and intravenous administration. *Toxicol. appl. Pharmacol.* **19**, 480–92.

WOLF, S., GOLDBERG, L.S., AND VERITY, A. (1976). Neuromyopathy
and periarteriolitis in a patient receiving levodopa. *Arch.
intern. Med.* **136**, 1055–7.

WÜLLEN, F., KAST, G., AND BRUCK, A. (1967). Über Nebenwirkungen bei Tetracyclinverabreichung an myastheniker. *Dt.
Med. Wschr.* **92**, 667–9.

WYLIE, W.D. AND CHURCHILL-DAVIDSON, H.C. (1900). *A practice of
anaesthesia*, pp. 587. Lloyd-Luke, London.

WYSOCKA, K., FABIAN, F., AND LISTEWNIK, M. (1960). Myasthenia
gravis as a complication of D-penicillamine therapy in rhema-
toid arthritis. *Rhematologie* **40**, 135–7.

YATES, A. (1975). Myaesthenia associated with penicillamine
treatment. *Br. med. J.* **1**, 600–2.

YOUNG, M.D. AND BLACKMORE, W.P. (1977). The use of danazol in
the management of endometriosis. *J. int. med. Res.* **5**, (Suppl.
3), 86–91.

26 Drug-induced peripheral neuropathies

SYLVIA M. WATKINS

Nature scarcely seems capable of giving us any but quite short illnesses. But medicine has developed the art of prolonging them.

Marcel Proust: *A la recherche du temps perdu.*

Drug-induced peripheral neuropathy is a fairly common side-effect of drugs. This chapter will include motor, sensory, autonomic, and mixed peripheral neuropathies induced by drugs, as well as optic neuritis and the Guillain–Barré syndrome.

The clinical pictures seen in this context are extremely variable, and may be predominantly or exclusively motor or sensory and may be acute, subacute, or chronic. Sensory neuropathy may consist of only mild peripheral numbness, dysaesthesiae, or paraesthesiae with or without some subjective impairment of light touch and pin-prick sensation, usually in a symmetrical 'stocking and glove' distribution; there may be painful extremities, and tendon reflexes are usually diminished or absent. At the other extreme there may be widespread and profound impairment of all modalities of sensation including proprioception, resulting in ataxia.

Motor neuropathies result in flaccid weakness, wasting, and often tenderness of the muscles, most marked distally. The legs are usually affected more severely than the arms. In extreme cases there may be profound paresis or even total paralysis, which is life-threatening if it affects the respiratory muscles.

Autonomic neuropathy may include any combination of postural hypotension, bowel and bladder disturbances, impotence, abnormal sweating and cardiac dysrhythmias.

The Guillain–Barré syndrome includes a generalized polyneuritis, usually predominantly motor in character, associated with a very high protein in the cerebrospinal fluid (up to 300 mg per cent) without an increase in cells — the 'albumino-cytologic dissociation' (Guillain *et al.* 1916). In this syndrome, the proximal muscles may be affected as severely as the distal ones. Cranial nerves may be affected, resulting in ocular, facial, swallowing, and speech difficulties, while respiratory muscle failure is likely to be fatal if artificial ventilation is not instituted early enough.

The mechanisms for the development of drug-induced peripheral neuropathy are variable and not fully understood. Some cytotoxic drugs (e.g. vincristine and colchicine) appear to affect the function of neurotubules and axonal transport. Inhibition of vitamin B metabolism may be the mechanism of action of certain drugs including isoniazid, ethionamide, and perhaps D-penicillamine (which have an anti-pyridoxine effect) and chloramphenicol which can produce vitamine B_{12} deficiency. Nitrofurantoin probably causes the neuropathy by inhibiting the metabolism of pyruvate. Perhexiline, amiodarone, and hexachlorophene act by causing segmental demyelination, possibly by interfering with glycolipid metabolism. Peripheral sensory neurones are particularly sensitive to the direct toxic effects of doxorubicin and megadose pyridoxine. Most of the other drugs causing peripheral neuropathy have a direct toxic effect on the axon (Schlaepfer 1971; Bradley *et al.* 1970; Spencer and Schaumburg 1980; Lane and Routledge 1983).

Drug-induced peripheral neuropathies in some cases are dose-related and therefore to some extent predictable. With most drugs, however, the side-effects occur sporadically and therefore largely unpredictably, though certain predisposing factors are recognized:

1. Genetic, e.g. the Japanese are extremely susceptible to clioquinol neuropathy, but the Indians are relatively resistant to it.

2. Metabolic, e.g. slow acetylators are very susceptible to isoniazid neuropathy. Perhexiline neuropathy is more likely to occur in individuals with impaired ability to oxidize certain drugs. This type of metabolic anomaly is probably gentically controlled.

3. Renal, e.g. poor renal function predisposes to colistin, nitrofurantoin, and ethambutol neuropathy, presumably due to failure of excretion of the drug or its metabolites resulting in accumulation to toxic levels.

4. Nutritional: patients with diabetes mellitus, alcoholism, malignant disease, malnutrition, and/or vitamin deficiencies are more susceptible than others to drug-induced neuropathy.

5. Underlying pathology, e.g. lymphoma patients are more susceptible than patients with other types of malignancy to the neurotoxic effects of vincristine.

Another unresolved problem is the mechanism whereby drugs have toxic effects in some nerves but not in others: for example, clioquinol affects the optic but not the peripheral nerves. Nor is it clear why sensory or motor nerves are preferentially affected by various drugs.

The drugs which have been reported to cause peripheral neuropathy when used in normal doses are listed in Table 26.1, and will be discussed individually under the headings of their therapeutic groups.

Antibiotics and antibacterial drugs

Antituberculous drugs

Isoniazid

This drug has been widely used for decades in the management of tuberculosis. It is almost always given in combination with other drugs, usually rifampicin and ethambutol, or streptomycin and para-amino salicylic acid (PAS). Peripheral neuropathy occurs relatively commonly. Central nervous system complications (convulsions and psychological changes) are described elsewhere in this book.

The overall incidence of isoniazid-induced neuropathy is about 2 per cent (Goldman and Braman 1972). The neuropathy is to some extent dose-related and is more liable to occur when the dose exceeds 300 mg daily. It is also commoner in 'slow acetylators' (Evans et al. 1960); indeed, Devadatta (1965) found that over 80 per cent of isoniazid-induced neuropathy occurred in slow acetylators. Rao et al. (1970) undertook sulphadimidine acetylation tests for classification of patients as slow or rapid inactivators of isoniazid. A total of 103 patients were studied, 52 of whom had been classified as slow and 51 as rapid inactivators of isoniazid by a standard microbiological assay method. Each patient received an oral dose of sulphadimidine, 44 mg/kg body weight. Blood and urine were collected after six hours, and the free and total sulphadimidine were measured. The results showed that an individual must be classified as a slow acetylator if the proportion of acetylated sulphadimidine (total minus free) is less than 25 per cent in the blood or less than 70 per cent in the urine. The results were found to be unaltered, even if the urine was stored at room temperature for up to one week. The test is easy to perform, and thus gives a simple, reliable, and cheap method for identifying those patients who may be at risk of developing isoniazid peripheral neuropathy.

Isoniazid competitively inhibits the action of pyridoxine whose metabolites play essential roles in the metabolism of protein, carbohydrate, and fatty acids (Snider 1980). Daily administration of 6 mg of pyridoxine prevented the development of peripheral neuritis in the Madras study (Tuberculosis Chemotherapy Centre, Madras 1963). Slow acetylators probably need a higher dose. In the treatment of established isoniazid peripheral neuropathy, pyridoxine alone is usually not adequate to reverse the neuropathy while isoniazid therapy continues, and therefore the drug should be discontinued if neuropathic sympoms occur.

Malnutrition and other situations in which there is even a mild pyridoxine deficiency, such as pregnancy, chronic liver disease, alcoholism, epileptics on anticonvulsants, uraemia, and old age, also appear to predispose to isoniazid-induced nerve damage, which may, however, be prevented by adding a large dose (50–100 mg daily) of pyridoxine hydrochloride to the regime (Snider 1980; Girling, 1982; Yoshikawa and Nagami 1982). Diabetes mellitus may also predispose to the development of isoniazid neuropathy.

Experimental work has shown that, in rats at least, isoniazid affects motor nerve fibres more than sensory, and that the degenerative neuropathy caused by this drug is essentially a multifocal axonal lesion, rather than a lesion resulting from cell-body damage (Jacobs et al. 1979a,b). The mechanism of isoniazid neuropathy is uncertain. In rats, the drug causes an increase in leucin and a decrease in fucose incorporation into myelin. This biochemical change occurs at the same time as myelin ovoid formation in the nerve fibres, and is proportional to the degree of nerve degeneration (Colip et al. 1981).

Clinically, isoniazid neuropathy usually affects the lower limbs first, with numbness and paraesthesiae and later, muscle atrophy, weakness, and ataxia. The development of optic neuropathy is rare, but it can be induced by isoniazid, and is also associated with pyridoxine deficiency. (Liebold 1971).

Bahemuka et al. (1982) demonstrated slowing of nerve conduction velocity in isoniazid-treated patients without clinical evidence of neuropathy; they suggested that serial electrophysiological studies should be performed regularly to detect evidence of peripheral nerve damage.

Ethambutol

Ethambutol causes optic neuritis (Kuming and Braude 1979) or retrobulbar neuritis (Narang and Varma 1979; Johnston and Audet 1978). Peripheral neuritis is much less common, but in one report it was observed that peripheral neuropathy may precede the optic complications

Table 26.1 **Drugs reported to cause peripheral neuropathy and/or optic neuritis (when used at standard doses)**

Antibiotics and antibacterial drugs	Antimalarial, antihelminthic antiprotozoal and antifilarial drugs	Antineoplastic drugs and radiosensitizers	Drugs used in cardiovascular disorders	Anticonvulsant drugs	Antirheumatic drugs
Amphotericin-B	Chloroquine	Amsacrine	Amiodarone	Carbamazepine	Chloroquine
Chloramphenicol	Clioquinol	BCNU	?Captopril	Phenobarbitone	Gold Salts
Colistin	Emetine	Chlorambucil	Clofibrate	Phenytoin	Iboprufen
Demethylchlor-	Halquinol	Cis-platinum (DDP)	Dichloroacetate	Primidone	Indomethacin
tetracycline	Pentamidine	Cytosine arabinoside	Digitalis	Sulthiame	Penicillamine
Ethambutol	Stilbamidine	Desmethyl-	Disopyramide		Phenylbutazone
Ethionamide	Suramin	misonidazole	Hydrallazine		
Gentamicin	Tetramisole	Doxorubicin	Perhexiline		
Isoniazid	Thiabendazole	Etoposide	maleate		
Metronidazole		5-fluorouracil	Phenytoin		
Nalidixic acid		Hexamethylmelamine	Quinidine		
Nitrofurantoin		Methylglyoxal	Thiazide		
Penicillin		Misonidazole	diuretics		
Streptomycin		Nitrogen mustard			
Sulphonamides		Procarbazine			
Sulphones		Vinblastine			
		Vincristine			
		Vindesine			
		(Laetrile)			

Hypnotic and psychotropic drugs	Drugs used in endocrine disorders	Drugs used in gastrointestinal disorders	Vaccines and antitoxins	Miscellaneous	Drugs causing neuropathy when given in excessive doses
Amitriptyline	Chlorpropamide	?Cimetidine	Influenza vaccine	Arsenicals	Amitriptyline
Chlorprothixine	Propylthiouracil	Trithiozine	Rabies vaccine	Azathioprine	Chloroquine
Diazepam	Tolbutamide		Tetanus toxoid	Calcium carbamide	Colchicine
Imipramine	Contraceptive pill		Tetanuse antitoxin	Danazol	Dapsone
Lithium				Dimethylsulfoxide	Digoxin
Methaqualone				plus sulindac	Disulfiram
Phenelzine				Disulfiram	Ergotamine
Thalidomide				Methysergide	Heroin
Zimeldine				Podophyllum	Hexachlorophene
				Quinine	Lithium
				Thalidomide	Nitrous oxide
				Zinc pyrithione	Pyridoxine
					Vincristine

by some months, thus providing a warning of impending visual toxicity (Nair *et al.* 1980).

Although ethambutol rarely causes symptomatic neuropathy, Takeuchi *et al* (1980a) found that 10 per cent of patients taking this drug showed abnormalities of sensory nerve action potentials and/or reduction of mixed nerve conduction velocity. Most of the affected individuals were elderly, and/or had received high doses of the drug. Ethambutol neuropathy is commner in patients over 45 (Takeuchi *et al.* 1980b) or in the presence of renal failure (Aelony and Locks 1980), although individual susceptibility may play a role (Forester 1980). It appears that the dose of ethambutol taken after the onset of visual symptoms is important in determing the severity of the neurological symptoms. Sensory disturbances are more conspicuous than motor, and electrical studies show a neurogenic pattern with decrease in the motor unit, with usually normal motor conduction velocity, but reduced sensory conduction velocity, and unrecordable sensory nerve action potentials. (Takeuchi *et al.* 1980*b*).

Experimental work in rats has shown that the main pathological change is axonal degeneration; those animals which had taken only small doses, but for long periods also showed some axonal regeneration (Matsuoka *et al.* 1981).

Streptomycin

Streptomycin's most infamous neurotoxic side-effect is ototoxicity. Optic neuritis may also occur, and rarely also peripheral neuritis (Wilson *et al.* 1979; Janssen 1960).

Ethionamide

Peripheral neuropathy, deafness, visual disturbance and convulsions have all been reported as occasional side-effects of ethionamide therapy for tuberculosis (Data Sheet). The drug may also cause mental changes. An anti-pyridoxine effect is probably the mechanism for the neuropathy (Lane and Routledge 1983). Therefore ethionamide treatment should be supplemented by vitamin B complex, especially if patients who have a basically poor diet or need to receive large doses of the drug.

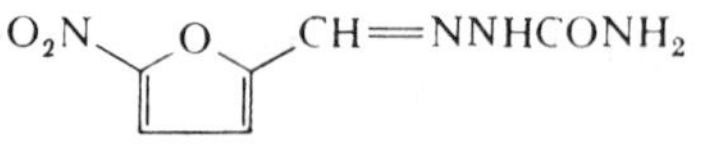

Nitrofurazone

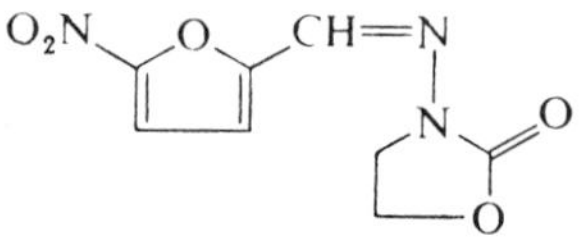

Furazolidone

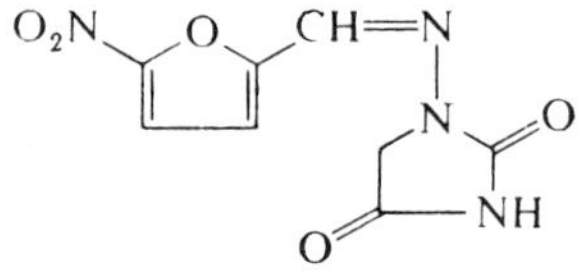

Nitrofurantoin

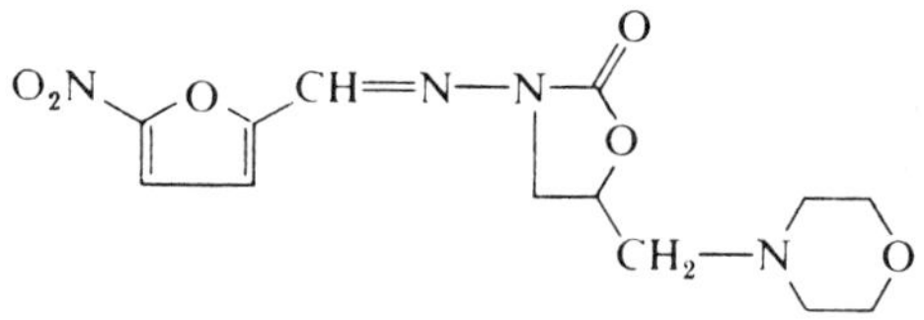

Furaltadone
(furmethonol)
(nitrofurmethone)
Withdrawn from clinical use

Fig. 26.1 Structural formulae of furan derivatives in current or previous clinical use.

Other antibiotics and antibacterial drugs

Nitrofurantoin

This commonly used antibacterial drug is one of a group of chemically related furan derivatives (Fig. 26.1), all of which have serious side-effects. Furaltadone was withdrawn 20 years ago on account of its neurotoxicity. Nitrofurazone which is used only topically may cause hypersensitivity reactions. Furazolidone has many non neurological side-effects, but there is also a single report of deafness and tinnitus possibly due to the drug (Van der Grient 1968).

Nitrofurantoin, which has now been in use for over 30 years, is particularly useful in urinary-tract infections because it is readily absorbed, and about 40 per cent of it is excreted in the urine. It has an antibacterial spectrum similar to that of chloramphenicol.

Gastro-intestinal side-effects (anorexia, nausea, vomiting, and diarrhoea) are common; rashes, haematological problems (haemolytic anaemias, folate-deficiency anaemia, eosinophilia, leucopenia), and pulmonary infiltration have been reported. However, peripheral neuropathy seems to be fairly common, and there have been reports of nitrofurantoin neuropathy from all over the world (Larsen and Bertelsen 1956; Olivarius 1956; Palmlov and Tunevall 1956; Falk *et al.* 1957; Briand and Tygstrup 1959; Hafstrom 1959; Collings 1960; Ellis 1962; Loughbridge 1962; Martin *et al.* 1962;

Uesu 1962; Willett 1963; Lhermitte *et al.* 1963; Heffelfinger and Allen 1964; Roelsen 1964; Rubinstein 1964,1968; Beverungen *et al.* 1965; Vickers 1965; Herndon and Fox 1966; Morris 1966). The reported incidence of nitrofurantoin neuropathy varied from 1.6 per cent in patients treated with doses of 4 mg/kg daily, to 26 per cent in patients receiving more than 7 mg/kg daily in a Japanese study (Kumamoto *et al.* 1972; Holmberg *et al.* 1980). Penn and Griffin (1982) noted considerable variation in the reported incidence of nitrofurantoin neuropathy in three different countries (United Kingdom, Sweden, and Holland).

Nitrofurantoin neuropathy is related to the blood levels of the drug and is thus to some extent dose-related. Not surprisingly, impaired renal function predisposes to neuropathy as the blood levels may readily reach toxic concentrations. (Loughbridge, 1962); Lhermitte *et al.* 1963; Beverungen *et al.* 1965). Behar *et al.* (1965) showed that normal patients on a dose of 300 mg of nitrofurantoin orally per day, had blood levels of 1.8–2.2 μg/ml; however, in patients with a blood urea above 45 mg/100 ml (7.5 mmol/l) the blood level of the drug rose to 5.1–6.5μg/ml. Nitrofurantoin should not be given to the elderly, nor to patients with poor renal function. Not only are such people likely to get toxic effects, but in addition, the drug will probably be inactive, as the kidneys will be unable to concentrate it sufficiently to give bacteriostatic levels in the urine (Hubman and Bremer 1965). Nor should the drug be given to pregnant women, as it does cross into the fetal circulation, at least in animals. Fetal kidneys do not excrete nitrofurantoin (Buzard and Conklin 1964) and hence the risk of toxic effects on the fetus would be considerable.

Pathologically, nitrofurantoin produces severe axonal degeneration, with interstitial oedema, but no inflammatory cells (Collings 1960; Loughbridge 1962; Herndon and Fox 1966; Morris 1966; Schaumburg and Spencer 1979; Yiannikas *et al.* 1981). Swelling of the anterior horn cells of the spinal cord, with eccentrically placed nuclei and severe chromatolysis have also been observed (Herndon and Fox 1966). Behar *et al.* (1965) produced experimental neuropathy in animals by giving very high doses of nitrofurantoin over a short period. The treated animals showed progressively decreasing conduction velocity and axonal dystrophy.

The pathogenesis of nitrofurantoin neuropathy is uncertain. It may be mediated by a reversible inhibition of the conversion of pyruvate and coenzyme A to acetyl CoA (Loughbridge 1962).

Clinically, nitrofurantoin neuropathy consists of a mixed sensorimotor picture, with distal numbness and paraesthesiae, more marked in the legs than in the arms. Later there may be wasting and flaccid weakness. The symptoms and signs usually start distally, and spread to involve proximal areas. All modalities of sensation may be affected, though joint position sense is usually spared. Tendon reflexes are absent or diminished, plantar responses flexor. Electrophysiological changes may be detected before the onset of symptoms (Lindholm 1967).

If the drug is discontinued when only sensory symptoms are present, the neuropathy is usually reversible and sensation recovers fairly quickly. However, once motor symptoms have occurred, the prognosis is poor and recovery is liable to be slow and incomplete (Toole and Parrish 1973). In all patients it is important to watch out carefully for symptoms indicative of peripheral neuropathy and, if these develop, the drug should be discontinued immediately.

Metronidazole

Metronidazole may cause convulsions and also a sensory neuropathy, sometimes associated with autonomic features. The neuropathy may be severe and the incidence seems to be related to the total cumulative dose (Coxon and Pallis 1976; Karlsson and Hamlyn 1977, Bradley *et al.* 1977; Frytak *et al.* 1978; Said *et al.* 1978; You and Grilliat 1981). Nerve biopsy showed some segmental loss of myelinated fibres and axonal degeneration of the remaining fibres (Bradley *et al.* 1977). The evolution of metronidazole neuropathy is unpredictable, and in severe cases recovery may be incomplete (Coxon and Pallis 1976; Hishon and Pilling 1977; Karlsson and Hamlyn 1977). Large doses of the drug should be restricted to the shortest possible time.

Other antibiotics

Gentamicin may rarely produce a myasthenia-like syndrome and polyneuropathy. Martens and Ansink (1979) reported a single case of this syndrome in a 70-year-old man treated with local application of gentamicin for chronic osteomyelitis, even though the blood levels of the drug were normal at the time.

Watson *et al.* (1981) reported a case of tetany induced by aminoglycosides: both gentamicin and amikacin may cause urinary loss of electrolytes, resulting in hypokalaemia, hypomagnesaemia, and hypocalcaemia, and hence tetany.

Chloramphenicol, especially if given for prolonged periods, may cause peripheral (axonal) neuropathy or optic neuritis, though neither of these complications is common (Joy *et al.* 1960; Murayama *et al.* 1973; Inoué *et al.* 1973; Asbury and Johnson 1978; Schaumburg and Spencer 1979; Wilson *et al.* 1979). This drug may cause vitamin B_{12} deficiency (Lane and Routledge 1983).

Other antibiotics reported as occasionally causing paraesthesiae or peripheral neuropathy include amphotericin B, colistin, nalidixic acid, demethylchlortetracycline, and sulphamethoxypyridazine (Davis 1968; Cohen 1970). Penicillin was reported as causing a mononeuritis or radiculopathy (Kolb and Gray 1946; Massey *et al.* 1979).

Sulphones

Sulphones, especially dapsone (diaminodiphenylsulphone), are used to treat leprosy and dermatitis herpetiformis (and also less commonly erythema elevatum diunitum and acne). Dapsone neuropathy, which appears to be dose-related, is predominantly motor involving progressive wasting and weakness, and sometimes associated with mild sensory changes (Wyatt and Stevens 1972; Castelein and de Weerd 1974; Gutmann *et al.* 1976). It appears six weeks to five years after starting treatment, usually at high doses (200–500 mg/day) (Revuz and Hornac 1981). The higher the dose, the earlier the symptoms appear.

Although dapsone-induced neuropathy is partly dose-related, there does also appear to be a dose-independent susceptibility to dapsone neuropathy, probably related to acetylator status, with slow acetylators being more liable to neurotoxicity (Gutmann *et al.* 1976; Rosén and Sörnäs 1982). Electrical studies indicate that the basic mechanism is a distal axonal degeneration of the motor neurones with a dying-back phenomenon. Sensory conduction studies have repeatedly proved normal (Koller *et al.* 1977; Helander and Partanen 1978; Rosén and Sörnäs 1982).

Leprosy affects the peripheral nerves, and this results in some of the most disabling and disfiguring forms of the disease. Clearly, then, one must be extremely careful when using a potentially neurotoxic drug like dapsone in the treatment of the disease, since the drug may precipitate acute or chronic nerve lesions (Warren 1980). It is possible that the high incidence of neuropathy in lepers may be partly related to the administration of excessive doses of dapsone (Warren 1980).

Peripheral neuropathy was also described in a man being treated with sulphoxone sodium for dermatitis herpetiformis. This patient had a predominantly sensory neuropathy, involving paraesthesiae and sensory impairment in both thumbs (Volden 1977).

Sulphone neuropathy is reversible and, if the drug is discontinued or the dose reduced, the symptoms and signs gradually improve. Once a patient has had sulphone neuropathy, he is liable to develop symptoms rapidly if the drug is reintroduced.

Antimalarial drugs

Chloroquine

Chloroquine is used in the treatment and prophylaxis of malaria, and also in the management of rheumatoid arthritis and systemic lupus erythematosus. In the past it has also been used for sarcoidosis and glomerulonephritis. It may cause a neuromyopathy, usually after prolonged treatment at high doses (e.g. 500 mg daily or more) such as are used in the management of systemic lupus erythematosus or rheumatoid arthritis (Whisnant *et al.* 1963; Begg and Simpson 1964; Garcin *et al.* 1965; Hicklin 1968; Hughes *et al.* 1971). In such cases the interpretation may be difficult, as the neuromyopathy could be related to the underlying disease, rather than to the treatment. However, it occurs also at doses used in malaria prophylaxis (varying between 600 mg weekly and as much as 200 mg daily) (Karstorp *et al.* 1973; Lhermitte *et al.* 1977). Chloroquine is absorbed rapidly and completely, but excretion is very slow, so that the drug may be present in appreciable quantities for months after stopping treatment.

The clinical picture is mainly of weakness and severe wasting, mainly proximally, starting in the legs, later spreading to the shoulder girdle. Reflexes are lost early. Sensory symptoms, if they occur at all, are mild and peripheral. There may also be pseudomyasthenic symptoms, with dysarthria, dysphonia, dysphagia, and generalized fatigability. Serum transaminases may be elevated. Electrical studies have shown features of neuropathy, myopathy, or both, and in one case there was evidence of a neuromuscular block. (Whisnant *et al.* 1963; Begg and Simpson 1964; Lhermitte *et al.* 1977). Histology of the muscles shows neurogenic atrophy of fibres, and autophagic vacuolation without inflammation or cellular infiltration. Recovery after withdrawing the drug is usually complete within a few months, although areflexia may be permanent.

Antihelminthic, antiprotozoal, and antifilarial drugs

There are occasional reports of neurotoxicity due to drugs in this group. A single dose of thiabendazole caused a Guillain–Barré syndrome in a four-year-old child (Katznelson and Gross 1978) and tetramisole was reported as the cause of a transient optic neuritis (Bomb *et al.* 1979). Optic atrophy following suramin treatment for ocular onchocerciasis is fairly common: possible mechanisms include inflammation of the optic nerve due to slow death of the microfilariae; a direct toxic effect of the drug; or the result of an antigen–antibody reaction (Thylefors and Rolland 1979). Emetine may cause muscle stiffness, myalgia, weakness of the limbs and facial muscles. Paraesthesiae are rare. The lesions may be neuropathic, myopathic, or they may affect the neuro-muscular junction (Manigand 1982).

Stilbamidine and pentamidine may produce a peripheral neuropathy, mainly sensory and autonomic. It is most likely to occur if the drugs are given for long periods. Hydroxystilbamidine may cause trigeminal neuritis, usually presenting with bilateral facial paraesthesiae, and sensory symptoms may also affect the neck, arms, and trunk (Gehin *et al.* 1981; Manigand 1982).

Hydroxyquinolines

The syndrome of subacute myelo-optic neuropathy (SMON) caused by clioquinol is described elsewhere in this book. Peripheral neuropathy and optic atrophy was described after the administration of halquinol (Hansson and Herxheimer 1981).

Antineoplastic drugs and radiosensitizers

Kaplan and Wiernik (1982) published an excellent review on the neurotoxicity of antineoplastic drugs. The vinca alkaloids are probably the commonest offenders in this field, although the wider use of cis-dichlorodiammine-platinum (II) (DDP) is reflected in the increasing incidence of neuropathy in patients undergoing cancer chemotherapy.

Vinca alkaloids

Infusions made from the flowers of Vinca rosea (*Catharanthus roseus*) have long had the reputation in folklore of controlling diabetes mellitus. Scientific investigation of this property showed that there was no significant effect on carbohydrate metabolism, but the side-effect of severe bone-marrow depression was almost universal. This finding led to the separation and identification of many vinca alkaloids, including several with powerful oncolytic activity. Two such alkaloids, vincristine and vinblastine, have been in use for two decades, and a third one, vindesine, has been introduced relatively recently. These drugs are useful in a large variety of malignant diseases including Hodgkin's and non-Hodgkin lymphomas, acute leukaemias, various childhood tumours, carcinomas of breast and bronchus, germinal-cell and other urogenital tumours, and a number of other cancers. The dosage of these drugs is limited by side-

effects, particularly vincristine-induced neurotoxicity and vinblastine-induced bone-narrow suppression, though all three drugs have neurotoxic side-effects to some extent.

Vincristine

Although peripheral neuropathy is the commonest and most serious dose-limiting side-effect of vincristine, other toxic effects have been reported. Significant haemotological depression is rare (Casey *et al.* 1973; Roeser *et al.* 1975) and usually occurs only when high doses are given (Martin and Compston 1963; Shaw and Bruner 1964; Haggard *et al.* 1968; Holland *et al.* 1973; Yap *et al.* 1979). Central nervous system disturbances, including irritability, depression, hallucinations, convulsions, and cerebellar dysfunction, have been described clinically and histologically in vincristine-treated patients, though in many cases the relationship between the drug and the cerebral symptoms is uncertain (Evans *et al.* 1963; Karon *et al.* 1966; Mathé *et al.* 1966; Sandler *et al.* 1969; Hardisty *et al.* 1969; Wisniewski *et al.* 1970; Holland *et al.* 1973; Rosenthal and Kaufman 1974; Jürgensson and Pichler 1976).

Partial or complete alopecia occurs frequently when doses are high (Karon *et al.* 1962; Martin and Compston 1963; Selawry and Hananian 1963; Holland *et al.* 1973) but is far less common using standard doses. The hair usually regrows after stopping treatment. Less common side-effects include anorexia, nausea and vomiting, rashes, dysuria, haemorrhagic cystitis, thrombophlebitis at the site of injection, and possibly heart damage (Lassman *et al.* 1965; Shaw and Bruner 1964; Heyn *et al.* 1966; Desai *et al.* 1970; Holland *et al.* 1973; Roeser *et al.* 1975).

The incidence of vincristine-induced neuropathy

The reported incidence of vincristine-induced neurotoxicity varies according to the criteria used to define 'neuropathy' and also according to the doses used. Table 26.2 shows the findings recorded in 13 different papers.

With a few exceptions, depressed tendon reflexes were found very commonly (up to 100 per cent). Paraesthesiae occurred less often, but were the commonest symptomatic manifestation of neuropathy affecting up to 90 per cent of the patients. On the other hand, constipation and muscle weakness were found in 7–50 per cent and 14–36 per cent of patients, respectively. Other symptoms and signs of neurotoxicity were in general less common. The frequency of stopping treatment or modifying dosage on account of neuropathy was also extremely variable: for example, Davis *et al.*(1974) reduced the dose or stopped

Table 26.2   Incidence of vincristine neurotoxicity according to various authors

	Depressed tendon reflexes, %	Numbness and/or paraesthesiae, %	Constipation, %	Muscle weakness, %	'Neurotoxicity' (varying criteria), %	Neuropathy sufficient to stop or modify dose, %
Bohannon et al. (1963)	75	90	50	—	—	50
Davis et al. (1974)	—	—	—	—	96	96
Edelstyn et al. (1975)	—	—	—	—	20	—
Gubisch et al. (1963)	14	81	19	14	100	33
Haggard et al. (1968)	12	5	7	18	57	10
Heyn et al. (1966)	33	11	32	36	—	—
Holland et al. (1973)	'nearly universal'	57	33	23	—	—
Karon et al. (1966)	—	—	—	—	35	—
Martin and Compston (1963)	89	58	47	21	—	—
Roeser et al. (1975)	—	—	—	—	—	44
Sandler et al. (1969)	100	46	—	34	—	—
Shaw and Bruner (1964)	—	—	—	—	100	—
Watkins and Griffin (1978)						
overall incidence	'nearly universal'	30	—	5	—	30
non-lymphoma patients	—	—	—	—	—	14
lymphoma patients	—	—	—	—	—	61
Whitelaw et al. (1963)	82	70	36	26	—	18

the drug in the event of depressed tendon reflexes or paraesthesiae, resulting in modified treatment in 96 per cent of patients. Others, however, stopped the vincristine only if the symptoms were 'severe', resulting in a change of dose in a much smaller proportion of patients (Whitelaw et al. 1963; Gubisch et al. 1963; Haggard et al. 1968).

Most authors are agreed that vincristine neuropathy is dose-dependent, and that the incidence rises with the individual doses and the duration of therapy (Karon et al. 1962; Evans et al. 1963: Whitelaw et al. 1963: Shaw and Bruner 1964; Hildebrand and Coërs 1965; Karon et al. 1966; Sandler et al. 1969; Hardisty et al. 1969; Desai et al. 1970; Bradley et al. 1970; Heiss et al. 1978).

Casey et al.(1973) observed that a high dose over a short period of time caused more neurotoxicity than the same total amount given in smaller doses over a longer period of time.

Dosage and duration of treatment are not the only factors influencing the development of neurotoxicity. Severe neuropathy has been observed after relatively small doses (O'Callaghan and Ekert 1976). A fulminating vincristine neuropathy after a single dose may represent an idiosyncratic reaction (Mubashir and Bart 1972). In general there appears to be variation in individual susceptibilities to both the therapeutic and toxic effects of vincristine (Rosenthal and Kaufman 1974).

Pre-existing disturbances of liver function render patients more susceptible to vincristine neuropathy. This

is probably due to a combination of impaired hepatic inactivation and delayed biliary excretion of the drug, resulting in higher blood levels and hence greater toxicity (Bohannon et al. 1963; Shaw and Bruner 1964; Tobin and Sandler 1968; Sandler et al. 1969).

Patients with advanced lymphomas (both Hodgkin's and non-Hodgkin's) are more susceptible to symptomatic vincristine neurotoxicity (61 per cent) than other cases, including leukaemia and breast-cancer patients (14 per cent) (Watkins and Griffin 1978).

The possibility that elderly patients may be more susceptible to vincristine toxicity has been suggested (Bradley et al. 1970; Hancock and Naysmith 1975) but not all workers would agree with this (Watkins and Griffin 1978). However, in general children do appear to tolerate the drug better than adults.

Physical activity appears to reduce the severity of vincristine neuropathy (Sakamoto 1974). It is difficult to assess the influence of pre-existing neurological disease on the development of vincristine neurotoxicity, and so far only occasional case reports are available. One patient wth Charcot–Marie–Tooth disease developed severe neuropathy after a small dose of the drug (Weiden and Wright 1972). On the other hand, a patient with pre-existing diabetic sensory neuropathy did not appear to be particularly susceptible (Casey et al. 1973). Six patients with paraparesis or parkinsonism were given vincristine in an attempt to reduce rigidity or spasticity; neurological improvement was minimal, but they did not appear to be unduly susceptible to neurotoxicity (Tobin and Sandler 1968).

Another possible potentiating factor in the development of the neuropathy is the combination of vincristine with other potentially neurotoxic drugs such as DDP or other neurotoxic cytotoxic drugs, or antibiotics such as isoniazid or metronidazole (Hildebrand and Kenis 1971; Hanefeld and Riehm 1980). Caution is recommended when using such drugs in combination. Thant *et al.* (1982) found clinical and pathological evidence that vincristine neuropathy may be enhanced by etoposide. Microscopy of such patients' nerves showed peripheral axonal and myelin degeneration similar to that seen in vincristine neuropathy; however, electron microscopy revealed electron-opaque granular degeneration of the myelin lamellae of affected nerve fibres, an appearance which is unusual in straightforward vincristine neuropathy (see below).

Vincristine neuropathy appears to be augmented by irradiation of the peripheral nerves (Cassady *et al.* 1980). In this context it is interesting to note that radiation neurotoxicity involves demyelination, whereas vincristine produces axonal degeneration and neurofilament abnormalities.

In summary, mild degrees of vincristine neuropathy are common; the more severe cases occur less frequently and appear to be dose- and time-related in most cases. Hepatic dysfunction, advanced lymphomas, other neurotoxic drugs, and physical inactivity may contribute to the severity of the neuropathy.

Clinical features of vincristine neurotoxicity

The clinical manifestations of vincristine neurotoxicity have been well reviewed elsewhere (Sandler *et al.* 1969; Bradley *et al.* 1970; *Lancet* 1973; Casey *et al.* 1973; Rosenthal and Kaufman 1974; Weiss *et al.* 1974). The changes are usually bilateral and symmetrical, the earliest and commonest finding being depression of the Achilles-tendon reflexes, followed by other reflexes. Impairment of tendon reflexes which is often unaccompanied by symptoms, is maximal 17 days after the administration of a single dose (Sandler *et al.* 1969).

Paraesthesiae, described as tingling, prickling or burning without numbness, are also common, usually affecting the hands before the feet. Sensory impairment in the finger-tips and toes, less often extending as far as the wrist or ankle, may occur, but extensive objective anaesthesia is uncommon at currently acceptable doses (Casey *et al.* 1973; Holland *et al.* 1973). Perception of vibration, pin-prick, touch, joint-position sense, and two-point discrimination is usually normal or only minimally impaired, even in the presence of an otherwise

severe neuropathy (Sandler *et al.* 1969; Casey *et al.* 1973; Mubashir and Bart 1972). In general, sensory symptoms are considerably worse than the signs.

Severe jaw pain, usually recurring after every injection of vincristine, is an extremely unpleasant but relatively uncommon side-effect of the drug (Evans *et al.* 1963; Windmiller *et al.* 1966: Haggard *et al.* 1968; Hardisty *et al.* 1969: Sandler *et al.* 1969; Holland *et al.* 1973). It may affect the maxilla, mandible, or throat. Other pains, occurring even less commonly, are deep, boring bone pains in axial or appendicular skeleton (Holland *et al.* 1973), unspecified 'limb pains' (Evans *et al.* 1963; Hardisty *et al.* 1969) 'neuritic pains' (Selawry and Hananian 1963), and myalgia (Bohannon *et al.* 1963: Sandler *et al.* 1969: Bradley *et al.* 1970). This latter symptom may be attributed either to muscle damage, or to intramuscular sensory nerve damage (Bradley *et al.* 1970).

Muscle weakness is more disabling than sensory symptoms, but fortunately less common (Karon *et al.* 1962; Selawry and Hananian 1963; Haggard *et al.* 1968; Sandler *et al.* 1969; Hardisty *et al.* 1969; Desai *et al.* 1970; Mubashir and Bart 1972; Holland *et al.* 1973). Weakness of the grip and small muscles of the hand results in considerable clumsiness, and further disability is caused by weakness of wrist extension, dorsiflexion and eversion of the feet, resulting in wrist-drop, foot-drop, and a stepping gait. Quadriceps weakness produces difficulty in climbing stairs. In general, proximal weakness is much less marked than distal, and legs are more severely affected than arms. Severe flaccid quadriparesis is rare, and may be the result of an idiosyncratic reaction to the drug (Mubashir and Bart 1972). Mononeuropathy has occasionally been described, but is an uncommon manifestation of vincristine toxicity (Desai *et al.* 1970; Levitt and Prager 1975). Atrophy of the interosseous, thenar, and hypothenar mucles may occur as a late manifestation of motor neuropathy (Selawry and Hananian 1963; Holland *et al.* 1973). An unusual presentation is of severe paralysis from vincristine-induced neuropathy affecting all four limbs, but with only mild sensory changes (Mueller and Flaherty 1978).

Cranial nerves may also be reversibly affected by vincristine. There have been reports of blindness associated with optic neuropathy, diplopia or blurred vision usually accompanied by ophthalmoplegia, ptosis, trigeminal or facial-nerve palsies, acoustic-nerve involvement, laryngeal-nerve palsies, and dysphagia with normal oesophagoscopy and barium swallow (Karon *et al.* 1962; Bohannon *et al.* 1963; Selawry and Hananian 1963; Albert *et al.* 1967; Sandler *et al.* 1969; Bradley *et al.* 1970; Holland *et al.* 1973; Haggard *et al.* 1968; Sanderson *et al.* 1976; Whittaker and Griffith 1977; Chisholm and Curry 1978; Norton and Stockman 1979;

Mahajan *et al.* 1981; Jean *et al.* 1981; Awidi 1980; Delaney 1982). Obviously in patients with malignant disease, other causes for cranial-nerve palsies should be sought before assuming that they are due to vincristine neurotoxicity.

The autonomic system may also be involved in vincristine neurotoxicity. Constipation, usually associated with abdominal pain, is reported in up to 50 per cent of patients (Table 26.2), and has been attributed to autonomic neuropathy (Selawry and Hananian 1963; Gubisch *et al.* 1963; Whitelaw *et al.* 1963; Martin and Compston 1963; Bohannon *et al.* 1963; Karon *et al.* 1966; Heyn *et al.* 1966; Haggard *et al.* 1968; Sandler *et al.* 1969; Bradley *et al.* 1970; Holland *et al.* 1973). It is usually possible to control this distressing symptom by means of laxatives and enemata, but sometimes adynamic ileus (occasionally fatal) may develop (Bohannon *et al.* 1963; Sandler *et al.* 1969; Toghill and Burke 1970; Hancock and Naysmith 1975).

Micturition disturbances vary from difficulty in initiating micturition to complete retention with bladder atony (Sandler *et al.* 1969; Bradley *et al.* 1970; Gottlieb and Cuttner 1971; Holland *et al.* 1973; Hancock and Naysmith 1975).

An interesting case of bowel and bladder atony, associated with a severe sensorimotor neuropathy in the legs, mimicking spinal-cord compression was described by Raphaelson *et al.* (1983).

The incidence of impotence appears to depend as much as anything on how frequently the right questions are asked (Holland *et al.* 1973).

Postural hypotension is a rare occurrence during vincristine therapy (McLeod and Penny 1969; Carmichael *et al.* 1970; Hancock and Naysmith 1975; di Bella 1980). Two such cases showed abnormal Valsalva responses, typical of autonomic neuropathy (Hancock and Naysmith 1975) and another study gave biochemical, physiological, and pharmacological evidence suggesting that the lesions are in the adrenergic post-ganglionic sympathetic nerves (Carmichael *et al.* 1970).

Another uncommon manifestation of vincristine neuropathy is hyponatraemia, attributed to inappropriate secretion of antidiuretic hormone (Slater *et al.* 1969; Meriwether, 1971; Robertson *et al.* 1973; Čáp *et al.* 1975; O'Callaghan and Ekert 1976; Lorini *et al.* 1977; Tzortzatou *et al.* 1979). This syndrome includes abdominal distension, drowsiness, convulsions, and coma, and is reversed on stopping the drug. Robertson *et al.* speculated that this syndrome may occur as a result of changes in osmoregulatory function caused by autonomic neuropathy which produces a 'resetting of osmoreceptor sensivity'. On the other hand, Slater *et al.*(1969) suggested that it may be due to a direct neurotoxic effect of vincristine on central nervous system sites of antidiuretic hormone formation and/or storage.

Another unusual vincristine-induced neurotoxic problem is an extrapyramidal syndrome associated with peripheral neuropathy (Jean *et al.* 1981).

There is a remarkable range of neuropathic disorders which have been ascribed to vincristine therapy. Most of the serious ones are however uncommon in practice, provided that therapy is discontinued when significant symptomatic neurotoxicity is established.

Electrophysiology of vincristine neuropathy

Detailed electrophysiological studies have been undertaken by several workers. It is generally agreed that the electrical findings indicated axonal degeneration with secondary chronic degenerative changes in the muscles (McLeod and Penny 1969; Bradley *et al.* 1970; Casey *et al.* 1973).

The evidence for axonal degeneration comes from nerve-conduction studies. Motor nerve-conduction velocity is normal or only slightly reduced (Hildebrand and Coërs 1965; Tobin and Sandler 1968; McLeod and Penny 1969; Bradley *et al.* 1970; Casey *et al.* 1973). The absence of gross slowing indicates axonal degeneration rather than segmental demyelination. Occasionally a complete nerve-conduction block may occur (Hildebrand and Coërs 1965; Bradley *et al.* 1970). Reduction in sensory action potential is common and may precede clinical evidence of neuropathy; some impairment of sensory conduction and increased distal sensory latency may occur (McLeod and Penny 1969; Bradley *et al.* 1970; Casey *et al.* 1973).

Mamoli *et al.*(1980) showed that the decrease in nerve conduction velocity and nerve action potential is dose-dependent, and that sensory fibres are damaged earlier and more severely than motor ones. They suggested that, if vincristine is discontinued or the dose halved when the action potential amplitude in extensor digitorum brevis falls below 1 mV, severe, irreversible polyneuropathy may be avoided.

Electromyographic abnormalities are consistent with chronic denervation without significant re-innervation. The findings include fibrillation potentials and positive sharp waves at rest; reduction of the amplitude of the maximum voluntary interference pattern, and an increase in the proportion of polyphasic action potentials (Bradley *et al.* 1970).

Preservation of the H reflex in spite of the loss of deep tendon reflexes may be found in the early stages of vincristine neuropathy (Sébille and Parlier 1981). Sandler *et al.* (1969) felt that this proved an intact reflex arc from the site of stimulation of the afferent nerve, via the spinal cord and efferent nerve to the muscles; they concluded that the earliest toxic effect of vincristine must be on the muscle spindle, damaging either the gamma innervation, or the distal portion of the afferent fibres conveying impulses from the spindle. McLeod and Penny (1969) on the other hand felt that this phenomenon represented only lower sensitivity of a slightly damaged nerve to impulses arising from the muscle spindle, compared with electrically stimulated impulses.

There is electrical evidence for a dying-back phenomenon, with loss of the H reflex later in the course of the disease, and greater impairment of the sensory action potential distally than proximally (Casey *et al.* 1973; Guiheneuc *et al.* 1980; Sébille and Parlier 1981) Métral *et al.* (1968) showed that repetitive nerve stimulation in vincristine neuropathy may result in progressive diminution of amplitude of response of a myasthenic type.

Histology of vincristine neuropathy

Histological examinations of nerve biopsy and autopsy specimens have confirmed that the main lesion is axonal degeneration. Both large and small myelinated fibres are affected by active Wallerian degeneration with swollen myelin sheaths and secondary demyelination; some axons are completely disrupted. Non-myelinated fibres may also be affected. A few fibres show signs of remyelination; others show abnormal variability in the relationship between nerve-fibre diameter and internodal length (Gottschalk *et al.* 1968; McLeod and Penny 1969; Bradley *et al.* 1970; Vallat *et al.* 1973). McLeod and Penny (1969) did find a few fibres with segmental demyelination, but this has not been confirmed by other workers.

Electron microscopy studies also show axonal degeneration, with myelin-sheath disruption, and aggregates of malaligned neurofilaments down the length of the axon, especially proximally, and in the cell body. Occasional damaged axons with degenerate mitochondria and myelin figures within autophagic vacuoles have been seen inside intact myelin sheaths, and non-myelinated fibres disrupted by large vacuoles have been observed. Blood vessels and perineural fibroblasts look normal. (Bradley *et al.* 1970; Wisniewski *et al.* 1970).

Muscle histology shows denervation atrophy in the distal skeletal muscles, and a toxic necrotizing myopathy with segmental necrosis and phagocytosis proximally (Bradley *et al.* 1970). Excessive variation in the size of the motor end-plates has been observed (Hildebrand and Coërs 1965). On electron microscopy Vallat *et al.* (1973) reported that most fibres were normal, but Bradley *et al.* found extensive changes including disruption of myofibrillary pattern, loss of myofibrils and of Z-lines.

Shelanski and Wisniewski (1969) found changes in the central nervous system, but Bradley *et al.*(1970) found no abnormalities in the spinal cord and dorsal root ganglia. Sanderson *et al.*(1976) found loss of retinal ganglion cells and atrophy of the corresponding optic nerve fibres in a man who developed optic neuritis on vincristine.

Studies in animals show changes similar to those observed in man, with Wallerian degeneration of the axons and demyelination (Uy *et al.* 1967; Gottschalk *et al.* 1968; Schlaepfer 1971; Folk *et al.* 1974; Cho *et al.* 1983). Muscle changes seem to be more prominent in animals than in man (Anderson *et al.* 1967; Casey *et al.* 1973).

Pathogenesis of vincristine neuropathy

Vincristine exerts a toxic action on motor and sensory axons, but sensory nerves are affected earlier and more severely than motor nerves (Caccia *et al.* 1977). The vinca alkaloids act on the mitotic spindle and inhibit RNA synthesis. This explains the oncolytic action of the drug, but little is known about the pathogenesis of the toxic neuropathy. Schlaepfer (1971) suggested that the vincristine causes disruption of neurotubules resulting in accumulation of neurofilaments; Bradley *et al.* (1970) speculated that disruption of the neurotubules with impairment of rapid axoplasmic transport mechanisms may be caused by vincristine, the 'dying-back' neuropathy possibly resulting from damage to the slower components of axoplasmic transport. McLeod and Penny (1969) felt that the drug may interfere with Schwann-cell as well as nerve-cell metabolism.

Treatment and prognosis

The only known treatment for established vincristine neuropathy is to reduce the dose or withdraw the drug altogether. This results in partial or complete recovery in most cases, though this may take weeks or months (Bohannon *et al.* 1963; Gubisch *et al.* 1963; Whitelaw *et al.* 1963; Albert *et al.* 1967; Sandler *et al.* 1969; Weiss *et al.* 1974; Casey *et al.* 1973; Rosenthal and Kaufman 1974; Hancock and Naysmith 1975). The paraesthesiae usually recover before the motor and sensory deficits, while reflexes return slowly if at all (Bohannon *et al.* 1963; Rosenthal and Kaufman 1974).

Awidi's patient (1980) emphasizes the danger of continuing treatment in the presence of neurological symptoms: the patient noticed blurred vision after a total dose of 10 mg, optic atrophy and blindness after 12 mg, and a mixed peripheral neuropathy including cranial-nerve involvement after 16 mg. The author concluded that regular visual field examination and fundoscopy should be undertaken in all patients receiving the drug.

Once a vincristine neuropathy has developed, neither vitamin B_{12} nor thiamine nor edrophonium chloride have any effect on the neuropathic symptoms or signs (Karon *et al*. 1962; Gubisch *et al*. 1963).

Neuropathy is a common and occasionally serious side-effect of vincristine therapy. In spite of this problem the drug is nevertheless so useful in the management of a large variety of malignant disorders that the risk of such toxicity is acceptable, and should not be regarded as a contraindication to using the drug (Warot *et al*. 1965; Casey *et al*. 1973). However, patients on vincristine therapy should be observed carefully for the onset of neuropathy, and the drug should be discontinued if there is evidence of progressive neurotoxicity. Minor non-progressive sensory symptoms are however acceptable, and are a small price to pay if the drug is proving efficacious.

Vinblastine

Reversible neuropathy affecting peripheral, cranial (especially laryngeal), and autonomic nerves, as well as central nervous system disturbances, has been reported after vinblastine therapy in man (Hertz *et al*. 1960; Frei *et al*. 1961; Warot *et al*. 1965; Gottschalk *et al*. 1968; Brook and Schreiber 1971; Whittaker and Griffith 1977). It has also been described in rodents (Németh *et al*. 1970). Sensory and motor symptoms with areflexia may occur, including a purely motor involvement which is rare in vincristine neuropathy (Armstrong *et al*. 1962). Encephalomyelopathy in children has also been reported. (Wisniewski *et al*. 1970).

The reason for the comparative rarity of vinblastine neurotoxicity is that, unlike vincristine, vinblastine therapy usually causes dose-limiting bone-marrow suppression much earlier than neuropathy, which consequently is rare in clinical practice, though readily demonstrable in animals.

Vindesine

Vindesine also causes peripheral neuropathy. The toxicity is dose-related, usually appearing after two to three courses of treatment at a dose of three to four mg per square metre every two weeks. The neurotoxicity is commoner in patients who have previously received treatment with other vinca alkaloids. Paraesthesiae, constipation, and weakness each appeared in approximately one-third of the patients of Valdivieso *et al*. (1981*a*,*b*). Myalgia and transient paralytic ileus were less common. Electrical studies revealed varying degrees of

delayed peripheral nerve conduction. Gralla *et al*. (1979) found a reversible peripheral neuropathy (usually mild) in all the 52 lung cancer patients treated with three to four mg per square metre per week. Bedikian *et al*. (1980) reported a 35 per cent incidence of neuropathy in patients receiving two or more courses of treatment at a dose of four mg per square metre every two weeks. Walker *et al*. (1982) using similar doses found mild peripheral neuropathy (not requiring dose reduction) in 79 per cent and constipation in 32 per cent. Combined DDP, bleomycin, and vindesine produced some degree of reversible peripheral neuropathy in all the patients treated by Itri *et al*. (1983), but in only one of the 52 was it severe, which is surprising considering that two of the three drugs used are neurotoxic.

It is widely believed that vindesine is less neurotoxic than vincristine, but a recent study by Jewkes *et al*. (1983) comparing weekly vincristine (1.4 mg per square metre) and vindesine (3 mg per square metre) in inoperable small-cell bronchial carcinoma showed a higher incidence of severe neurotoxicity in the patients treated with vindesine (nine out of 35) than in those who received vincristine (four out of 28), in spite of using standard doses of each. Thus the relative neurotoxicity of the two drugs remains to be established with certainty, but will no doubt become apparent with the increasing use of vindesine.

Cis-dichlorodiammineplatinum (II) (DDP)

In the few years that DDP has been in general use there have been several reports of mixed sensorimotor neuropathy which may or may not be reversible after stopping the drug. (Kedar *et al*. 1978; Hadley and Herr 1979; Arnold and Williams 1979; Reinstein *et al*. 1980; Hurley *et al*. 1982). Retrobulbar neuritis has also been reported (Ostrow *et al*. 1978; Becher *et al*. 1980), and so has anosmia (Gastaut *et al*. 1982).

DDP also causes dose-related renal damage which may cause hypomagnesaemia and hypocalcaemia, which in turn may result in tetany (Hayes *et al*. 1979; Stuart-Harris *et al*. 1980). It is important to monitor magnesium and calcium levels in patients on DDP, and to give replacement therapy as required. Ototoxicity is common with DDP (about 30 per cent in patients receiving 50 mg per square metre or more (Kovach *et al*. 1973; Hurley *et al*. 1982). This problem is reviewed elsewhere in this book.

The combination of DDP, vindesine, and bleomycin caused mild peripheral neuropathy in all of the 52 patients treated by Itri *et al*. (1983), but in only one case was it severe (see above).

Both electrical studies and nerve biopsies have shown that the neurotoxic effect of DDP results in axonal degeneration (Gastaut *et al.* 1982).

Misonidazole

Misonidazole is a hypoxic radiosensitizer whose use is unfortunately limited by neurotoxic side-effects: convulsions at higher doses and peripheral neuropathy, which is common and may be severe, at lower doses (Saunders *et al.* 1978; Kogelnik *et al.* 1978). The drug causes some degree of peripheral neuropathy in approximately half the patients who receive it, and almost one in 10 suffer from central nervous system damage and/or ototoxicity (Phillips *et al.* 1981). The neurotoxicity is dose-related, and may be avoided by limiting the total dose, giving longer overall treatment times, and monitoring plasma levels (Dische *et al.* 1978a, b, 1979; Wasserman *et al.* 1979). By increasing the overall treatment time, the total dose of the drug may be gradually increased without risking severe side-effects (Kogelnik 1980). In one study, neurotoxicity was associated with elevated plasma concentrations on the day after treatment, and prolonged plasma half-life of the drug (Walker and Strike 1980). Some protection against neuropathy may be achieved by the concurrent administration of steroids and also by phenytoin which appears to shorten the plasma half-life of the drug (Walker and Strike 1980; Wasserman *et al.* 1980).

The results of EMG studies by Mamoli *et al.* (1979) indicated a primary axonal neuropathy with greater changes in the sensory than in the motor nerves. There was only slight reduction in motor nerve conduction velocity, but increase in the distal latency occurred early in the course of the neuropathy, and this may be a useful indicator of subclinical neurotoxicity. The electrophysiological changes reverted to normal within six months of stopping treatment. Nerve biopsy has shown residual axonal degeneration, segmental demyelination, and remyelination, affecting both large and small myelinated nerve fibres (Urtasun *et al.* 1978).

Clearly, some means of predicting the development of neurotoxicity in individual patients would make it safer and easier to use the drug. Schwade *et al.* (1982) have produced pharmocokinetic models to predict the probability of developing peripheral neuropathy. Experimental work in rats showed that analysis of nerve trains evoked response may facilitate a diagnosis of neurotoxicity before the onset of overt clinical signs (Edwards *et al.* 1982). This technique might prove useful in evaluating patients receiving this drug.

Desmethylmisonidazole

Desmethylmisonidazole which is related chemically and therapeutically to misonidazole has an incidence of associated peripheral neuropathy similar to that found after misonidazole therapy (Dische *et al.* 1981).

Hexamethylmelamine

Hexamethylmelamine may produce a reversible sensorimotor neuropathy and/or depression, sleep disturbances, and various extrapyramidal syndromes (Stolinksy *et al.* 1972; Kaplan and Wiernik 1982). It proved neurotoxic in nearly half the patients with advanced ovarian carcinoma who received it as a single agent in the series of Wharton *et al.* (1979). Stolinsky *et al.* (1972) also reported neuropathy in patients who received hexamethylmelamine in combination with DTIC.

There have been reports of patients getting relief from their symptoms by taking pyridoxine, but it is not clear whether the vitamin is really protective against the neurotoxic effects of hexamethylmelamine (Wharton *et al.* 1979; Kaplan and Wiernik 1982).

Nitrogen mustard

The neurotoxicity and ototoxicity of nitrogen mustard used in high doses is the main dose-limiting factor. Most of the patients treated in this way had headaches, tinnitus, and nerve deafness which was often irreversible (Conrad and Crosby 1960; Glode 1979). Westbury (1962) reported on eight patients who received intra-arterial perfusions of nitrogen mustard; three of them developed mononeuritis or radiculitis.

Other cytotoxic drugs

Occasional case reports have appeared concerning patients who developed optic or peripheral neuritis whilst receiving cancer chemotherapy. These include a case of mononeuritis multiplex in a patient receiving oral 5-fluorouracil and total hepatic irradiation (Langley *et al.* 1978), two patients with acute myeloblastic leukaemia who developed peripheral neuropathy whilst receiving cytosine arabinoside (Russell and Powles 1974), a man with chronic lymphatic leukaemia who was afflicted by a severe sensorimotor neuropathy whilst taking chlorambucil (Sandler and Gonsalkorale 1977), a case of optic neuroretinitis in a patient on BCNU and procarbazine (Lennan and Taylor 1978), and sensory neuropathy in three patients receiving the new cytotoxic drug 4'-(9-acridnyl-amino)-methanesulphon-m-anisidide (amsacrine) (Van Echo *et al.* 1979). There have been occasional reports of peripheral neuropathy in patients receiving procarbazine, doxorubicin, etoposide, and methylglyoxal bis-dihydrochloride. In most of these cases it is impossible to prove beyond all doubt that the suspect drug was really responsible for the neuropathy, but circumstantial evidence is provided by the fact that it usually resolved on stopping the drug concerned, and in some instances relapsed when the drug was restarted (Brunner and Young 1965; Samuels *et al.* 1967; Falkson *et al.* 1975; Schaumburg and Spencer 1979; Spaulding *et al.* 1982).

Laetrile (amygdalin)

A worrying report appeared recently concerning a woman of 67 who was treated with laetrile at a dose containing 25 to 75 mg of cyanide daily (Kalyanaraman *et al.* 1983). She developed a neuromyopathy associated with raised levels of thiocyanate and cyanide in the blood and urine. The symptoms and signs improved when the drug was withdrawn. Sural-nerve biopsy showed demyelination and axonal degeneration; muscle biopsy showed deveration and myopathy with Type II atrophy. The appearances were similar to those seen in ataxic neuropathy due to high dietary cyanide.

This report is worrying because laetrile has no proven beneficial effect in cancer (*British Medical Journal* 1977a) and therefore it is indefensible to expose patients to the possibility of such serious side-effects. It is quite another matter taking such risks when using drugs of proven efficacy.

Drugs used in cardiovascular disorders

Perhexiline maleate

In 1976 Lhermitte *et al.* described peripheral neuropathy induced by perhexiline maleate. In July 1977 the Committee on Safety of Medicines issued a notice in its Adverse Reaction Series drawing attention to the development of peripheral neuropathy in patients being treated with this drug for angina. Since then there have been many reports of perhexiline-induced neuropathy, which is usually of mixed sensorimotor type, with or without autonomic or cranial neuropathy (Howard and Russell Rees 1976; Fraser and Miller 1978; Said 1978). The Guillain–Barré syndrome has also been described (Wallace 1979; Heathfield and Carabott 1982). An unusual presentation was that of an elderly patient who developed parkinsonism and fasciculation in the hands after taking the drug for five months; there were severe electrophysiological disturbances (Gordon and Gordon 1981). Convulsions have been reported, and cerebellar

disturbances have also been described in association with perhexiline-induced neuropathy (Turpin *et al.* 1983). The neuropathy is usually reversible after stopping the drug especially in the early stages, though recovery may be slow (Wijesekera *et al.* 1980; Gordon and Gordon 1981). The Guillain–Barré syndrome may be irreversible (Wallace 1979). Hepatotoxicity due to perhexiline maleate may also be a serious problem. It is discussed elsewhere in this book.

The reported incidence of perhexiline-induced neuropathy is variable. Reversible mild paraesthesiae were described in seven of 363 patients (Lockhart and Masheter 1976) but Sébille and Rozenstajn (1978), using neurophysiological studies, reckoned that some degree of neuropathy occurred in two-thirds of their patients.

Perhexiline-induced neuropathy is to some extent dose-related. In occasional patients it may develop as early as three weeks after starting treatment. Pagès *et al.* (1980) reported a patient who was trouble-free for a year while taking 300–400 mg per day of the drug, but who, when the dose was increased to 600 mg, developed a neuropathy which improved when the dose was reduced again.

Twenty per cent of individuals are slow metabolizers of perhexiline maleate, although this phenomenon is not found in all patients with perhexiline-induced neuropathy (Jallon *et al.* 1978). Shah *et al.* (1982), testing the ability of patients to oxidate a single oral dose of debrisoquine, showed that perhexiline-induced neuropathy may be related to poor oxidative metabolism which is independent of hepatic function, concurrent drug administration, tobacco, or alcohol. Presumably individuals with poor oxidative metabolism readily accumulate perhexiline maleate to toxic levels. The ability to oxidize certain drugs is genetically controlled, and hence the susceptibility to perhexiline-induced neuropathy may also be genetically determined. The authors suggested that their technique could be used as an 'oxidation phenotyping procedure' to identify those at risk of developing perhexiline neuropathy. More recently they have described a patient in whom they anticipated neuropathy through this test; by means of electrophysiological testing they demonstrated demyelination in the absence of symptoms. The electrical abnormalities reverted to normal 16 weeks after stopping the drug (Shah *et al.* 1983).

Perhexiline seems to have a direct toxic action on the nerves (Pagès *et al.* 1980) and is an unusual form of drug-induced neuropathy in that it involves the proximal portions of peripheral nerves (i.e. radiculoneuropathy) and is associated with increased protein in the cerebrospinal fluid indicating demyelination.

Electrical studies suggest a mainly demyelinating

process with slowed nerve conduction and prolonged distal latency (Bouche *et al.* 1979; Turpin *et al.* 1983). The presence of active denervation indicates a poor prognosis, and in such cases there is often only slight improvement after withdrawing the drug (Bouche *et al.* 1979).

Histology shows reduction of myelinated nerve fibres, frequent segmental demyelination, severe Wallerian degeneration, and polymorphic inclusions in various cells, especially in the Schwann cells. These inclusions contain lipid complexes, phospholipids, gangliosides, and cholesterol. The nerves contain increased levels of gangliosides. The abnormal phospholipid turnover may be related to a fall in sphingomyelinase which can be demonstrated experimentally (Said 1978; Pollet *et al.* 1979; Pagès *et al.* 1980; Hauw *et al.* 1981; Turpin *et al.* 1983). Clinical symptoms occur only when a large number of fibres have been lost. The toxicity of perhexiline maleate is greatest in those cells which have a high lipid turnover (notably Schwann cells and liver cells) and Hauw *et al.* (1981) felt that perhexiline-induced neuropathy should be regarded as a drug-induced phospholipidosis. Turpin *et al.* (1983) regarded it as a thesaurosis (i.e. a condition caused by the storing up of abnormal substances).

In view of the sometimes severe nature and occasional irreversibility of perhexiline-induced neuropathy as well as the hepatotoxicity, the drug should be avoided if possible; in any case there are many other effective anti-anginal drugs available. If perhexiline must be used, the patients should be watched carefully for the earliest signs of neurological or hepatic toxicity; any slowing of nerve conduction velocity or symptoms of peripheral neuropathy or the development of abnormal liver function tests should be regarded as indications for stopping the drug.

Amiodarone

Amiodarone is an anti-arrhythmic drug with many common side-effects which limit its usefulness. Sensorimotor neuropathy may occur and may be severe, especially in patients receiving high doses. The neuropathy is usually reversible if the dose is reduced or the drug discontinued. The reported incidence is variable, and may be as high as 10 per cent, though in some series there has been a high incidence (up to 20 per cent) of proximal muscle weakness and fatigue, without actual neuropathy (Ahmad 1979; Meier *et al.* 1979; Heger *et al.* 1981; Larre *et al.* 1981; Waxman *et al.* 1982; Lubbe and Mercer 1982; Martinez-Arizala *et al.* 1983; Harris *et al.* 1983; Patte *et al.* 1983).

Electrical studies have shown abnormal latency and reduced nerve conduction velocity (Heger *et al.* 1981;

Harris *et al.* 1983). Histological findings include segmental demyelination, with loss of large myelinated fibres and reduction of small myelinated and unmyelinated fibres (Meier *et al.* 1979; Ahmad 1979), and various inclusions in the Schwann cells, endothelial cells, pericytes, and keratinocytes. The inclusions are probably intralysosomal lipid complexes. The appearances are similar to those seen in perhexiline-induced neuropathy (Larre *et al.* 1981; Harris *et al.* 1983).

Hydrallazine

Hydrallazine neuropathy is uncommon but is dose-related and is more likely to occur in slow acetylators. Most of the reported cases occured at high doses, often when the drug was used as a sole agent in the management of hypertension. Now that it is more commonly used at lower doses (around 100–200 mg daily) and in combination with other drugs, neurotoxicity occurs rarely (Burley and Steen 1979). The pathogenesis is probably mediated by the formation of a pyridoxal-hydrallazine complex (Tsujimoto *et al.* 1981).

Other drugs used in cardiovascular disorders

Case reports have appeared of peripheral neuropathy in a patient on disopyramide (Dawkins and Gibson 1978) and also in a patient on high doses of phenytoin used as an anti-arrhythmic agent (Tindall and Willerson 1978). Phenytoin-induced neuropathy is discussed in further detail in the section on anticonvulsant drugs.

Propranolol may cause paraethesiae (Davies 1968). Optic neuritis may be caused by digitalis (Wilson *et al.* 1979) or quinidine (Naranjo *et al.* 1981).

Reversible distal peripheral neuropathy associated with proximal myopathy has been reported in a patient receiving clofibrate while on regular haemodialysis. The renal failure and the haemodialysis may have contributed to the problem; however, the drug had a very long half-life in this patient and the neuropathy resolved when it was withdrawn (Gabriel and Pearce 1976). Another patient with nephrotic syndrome developed myopathy and transient mononeuropathy while on clofibrate (Pokroy *et al.* 1977). Another cholesterol-lowering drug which has been tried in hypercholesterolaemia is dichloroacetate: unfortunately it too may cause sensorimotor neuropathy, which limits its usefulness (Stacpoole *et al.* 1979; Moore *et al.* 1979).

An interesting report by Miller and Moses (1978) described a man taking the thiazide diuretic chlorothiazide. After seven months on the drug he developed external ophthalmoplegia and a diabetic glucose tolerance curve, both of which resolved when the drug was withdrawn. The authors concluded that patients on continuous thiazide treatment may not only develop impaired glucose tolerance (or worsening of pre-existing diabetes) but may also suffer from the same complications as do diabetics.

There is a report of two cases with a mixed peripheral neuropathy, one of whom also had a Guillain–Barré syndrome; both these patients were taking both captopril and cimetidine. It is uncertain which of the drugs was responsible for the neuropathy, or whether it was due to the combination of the two (Atkinson *et al.* 1980).

Anticonvulsant drugs

Phenytoin

Phenytoin's most serious side-effect is a cerebellar syndrome, which occurs mainly when the drug is given in toxic doses (Selhorst *et al.* 1972). However, phenytoin may also cause peripheral neuropathy, which may or may not be reversible. Prolonged treatment with phenytoin may cause loss of ankle tendon reflexes, and there may be only electrophysiological evidence of peripheral neuropathy, or neuropathic symptoms with or without cerebellar disturbances.

Swift *et al.* (1981) found no relationship between clinical and electrical abnormalities on the one hand, and blood levels of the drugs or duration of treatment with either phenytoin or phenobarbitone on the other. However, others have found that the risk of peripheral neuropathy increases with dose levels, duration of treatment, and blood levels of phenytoin (Lusins and Jutkowitz 1972; So and Penry 1981; Lovelace 1982). According to Tindall and Willerson (1978) symptomatic peripheral neuropathy is commonest in patients who have taken the drug for over 10 years, but it is uncommon at normal therapeutic serum levels, except in the presence of chronic renal failure, diabetes mellitus, alcoholism, or nutritional deficiency. Mochizuki *et al.* (1981) found a significant correlation between total dosage and duration of phenytoin treatment and the reduction of motor nerve conduction velocity; they concluded that long-term phenytoin treatment can cause latent impairment of peripheral nerve function in children who have no clinical evidence of peripheral neuropathy.

Other anticonvulsant drugs

Although phenytoin is the drug most often mentioned in the context of anticonvulsant-induced neuropathy, the available evidence suggests that other anticonvulsants including phenobarbitone, primidone, and carbamazepine are just as likely to have neurotoxic side-effects (Swift *et al.* 1981). Carbamazepine neuropathy was also demonstrated in rats (Moglia *et al.* 1983). However, the exact incidence of neuropathy for different anticonvulsant drugs is difficult to determine, because so many patients receive combined therapy. Martinez-Figueroa *et al.* (1980) produced evidence that neuropathy in patients on anticonvulsants is related to folate deficiency: giving folate intramuscularly reversed the abnormalities in motor and sensory distal latencies.

Sulthiame may cause peripheral and perioral paraesthesiae (Manigand 1982).

Antirheumatic drugs

Gold salts

Gold-induced neuropathy is well recognized. Dick and Raman (1982) reported a severe polyneuropathy of the Guillain–Barré type in a woman who received gold treatment for rheumatoid arthritis. She made a full recovery after stopping the gold and receiving treatment with methylprednisolone and dimercaprol. A detailed study of three cases of gold neuropathy in patients with rheumatoid arthritis was undertaken by Katrak *et al.* (1980). They all had acute, symmetrical progressive polyneuropathy, and tended to improve (after initial deterioration) when the chrysotherapy was stopped. Histology showed axonal degeneration with segmental remyelination. Animal studies indicated that the severity of the neuropathy was dose-related. It is uncertain whether gold has a direct neurotoxic action, or whether the toxicity is mediated by immunological mechanisms (Dick and Raman 1982).

D-penicillamine

Penicillamine-induced peripheral, cranial or optic neuropathy is rare, and is indeed less common than the myasthenic syndrome (Wilson *et al.* 1979; Meyboom 1977; Argov and Mastalgia 1979). The commonest presentations are ptosis and oculomotor palsies. Mononeuritis multiplex and/or peripheral neuropathy are extremely rare. In a case of chronic rheumatoid arthritis treated with D-penicillamine, the patient developed mononeuritis multiplex, peripheral neuropathy, and oculomotor palsies; electrical studies showed prolonged distal latency, and the nerve biopsy showed axonal degenerative neuropathy and angiitis. The patient also had a raised alkaline phosphatase, albuminuria, elevated titres of antinuclear factor, and anti-DNA antibodies. When the drug was withdrawn, the symptoms, signs, electrical abnormalities, and antibodies all regressed rapidly (Mayr *et al.* 1983).

Indomethacin

Eade *et al.* (1975) described a patient who was treated for generalized arthritic symptoms and stiffness with a variety of anti-inflammatory drugs and analgesics, before finally taking indomethacin, 75 mg and later 150 mg daily. Five months later he developed paraesthesiae and progressive weakness. Nerve conduction studies revealed slowing of motor nerve conduction with normal sensory latency. The symptoms, signs, and nerve conduction abnormalities regressed within weeks of stopping the drug. There have been occasional reports of similar cases to the Committee on Safety of Medicines: all had reversible distal paraethesiae and weakness.

Other antirheumatic drugs

Ibuprofen may cause optic neuritis (Wilson *et al.* 1979); phenylbutazone may cause peripheral neuropathy (Manigand 1982); and chloroquine may give rise to neuromyopathy (Hicklin 1968) and to an irreversible ototoxicity (Manigand 1982). Chloroquine is discussed in detail above under the heading of Antimalarial drugs. Colchicine occasionally causes neuropathy at normal doses (Prescott 1975); the rarely reported cases of peripheral neuropathy and myopathy are more often the result of accidental or deliberate massive overdoses.

Hypnotic and psychotropic drugs

The tricyclic antidepressants are chemically related to the phenothiazines (Fig. 26.2). These groups of psychotropic drugs share many side-effects, some of which are potentially serious (Connell 1968). The tricyclic antidepressants have a blocking effect on the parasympathetic nervous system, and there are frequent reports of dry mouth, tachycardia, disturbances of accommodation, constipation, sweating, and impaired sexual function, particularly failure of ejaculation.

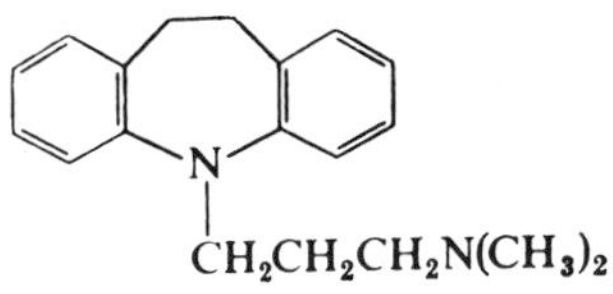

Imipramine

Freyhan (1959) described paraesthesiae in 5 per cent of his patients on imipramine. Tremors, mainly affecting the hands and tongue, are common and 14 per cent of patients treated with this drug in Sharp's series (1960) developed twitching and myoclonic episodes. Grand mal epilepsy may occur if the dose exceeds 300 mg daily. Motor neuropathy has also been described in a few cases, the youngest of whom was aged 57 (Collier and Martin 1960; Saavedra *et al.* 1960; Miller 1963). Motor neuropathy has occurred at doses ranging fom 75 to 300 mg

$CH_2CH_2CH_2N(CH_3)_2$

Imipramine

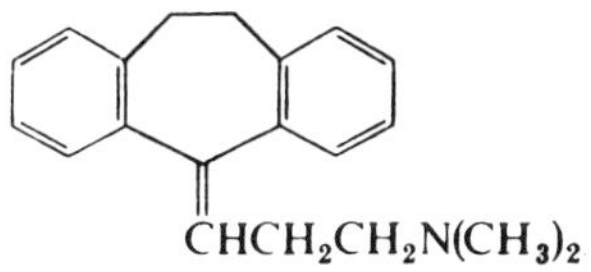

$CHCH_2CH_2N(CH_3)_2$

Amitriptyline

$CH_2CH_2CH_2N(CH_3)_2$

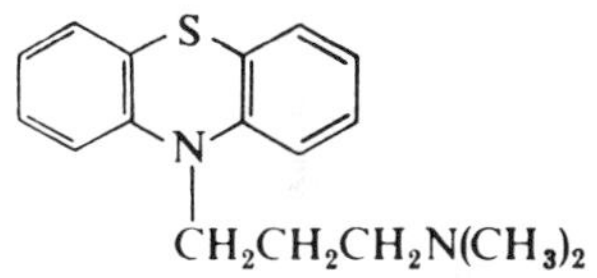

Chlorpromazine, a typical phenothiazine

Fig. 26.2 Structural similarity between imipramine, amitriptyline, and the phenothiazines.

daily. Symptoms, which generally developed within days or weeks of starting treatment, consisted of either peroneal mononeuritis (unilateral or bilateral) or a distal flaccid paresis. The patients with mononeuritis improved only slightly if at all after the drug was discontinued. However, the patients with distal motor neuropathy recovered on withdrawal of the drug.

Amitriptyline

Amitriptyline may cause numbness, burning, and tingling affecting the mouth, face, and distal parts of the limbs (Isaac and Carlish 1963; Davies 1968). A recent case report (Meadows *et al.* 1982) described a 39-year-old woman who had been taking amitriptyline for three years and was then given lithium. She developed peripheral paraesthesiae and numbness, starting in the feet and spreading up one leg. The lithium was stopped, and she was given pyridoxine 50 mg daily, resulting in improvement in her symptoms. Subsequently the lithium was reintroduced without causing a recurrence, and various sequential changes in her treatment established that it was the amitriptyline, rather than the lithium, which was responsible for the neuropathy. However, the depression was severe and required continued treatment with amitriptyline (at a dose of 200 mg daily) which was given together with pyridoxine hydrochloride. She increased the dose of the latter gradually, and found that she could eliminate the paraesthesiae by taking up to one gram of pyridoxine daily. Scarcely surprisingly, she had high plasma levels of pyridoxal phosphate. Amitriptyline is metabolized (at a rate which varies from one individual to another) to desmethylnortriptyline. This can combine with pyridoxal phosphate, and it is possible that the neuropathy is related to a reduction in the remaining available pyridoxal phosphate, hence the improvement in the neuropathy when taking massive doses of pyridoxine supplements.

Lithium

Lithium has many serious side-effects affecting the central nervous system, kidneys, thyroid, and other organs, which are described elsewhere in this book. At toxic blood levels there may be a motor neuropathy (Heim *et al*. 1981). Moreover, Girke *et al*. (1975) described abnormalities in nerve conduction velocity in individuals on lithium, but without symptoms of neuropathy.

Diazepam

Intravenous diazepam frequently causes local thrombophlebitis, and there has also been a report of neuropathy or neuralgia developing near the site of the injection (Donaldson and Gibson 1979).

Chlorprothixine

Danner (1981) described neuropathy in two patients who were on long-term treatment with the tranquillizer chlorprothixine. The clinical picture was unusual in that the patients had evidence of a mononeuropathy as well as polyneuropathy. Electrical studies indicated demyelinating polyneuropathy and axonal degeneration. Both patients improved when the drug was discontinued.

Other psychotropic drugs

The antidepressant drug zimeldine was withdrawn in 1983 on account of serious side-effects including neuropathies and the Guillain–Barré syndrome (Committee on Safety of Medicines 1983). The monoamine oxidase inhibitor phenelzine and the hypnotic methaqualone have been blamed for peripheral neuropathy (Davies 1968; Markes and Sloggen 1976). The hypnotic thalidomide is no longer available and need not be discussed here (see *Iatrogenic diseases*, 2nd edn., pp. 301–2). However, it is still occasionally used in the treatment lepra reactions, Behçet's disease, and of discoid lupus erythematosus, and a recent report of thalidomide-induced neuropathy

in such patients is described in the 'miscellaneous' section of this chapter (p.572) (Ludolph and Matz 1982).

Drugs used in endocrine disorders

There are reports of paraesthesiae in patients taking propylthiouracil (Frawley and Koepf 1950) and of optic neuritis in patients on chlorpropamide (Wilson *et al*. 1979). Ellenberg (1959) described peripheral neuropathy in diabetics treated with tolbutamide, but he felt that the drug had precipitated diabetic neuropathy rather than causing direct toxic effects.

A recent paper described investigations on the effect of prolonged insulin-induced hypoglycaemia in rats (Sidenius and Jakobsen 1983). The rats developed weakness, and the investigations showed reduced amplitude of the evoked muscle action potential, distal axonal degeneration, and perikaryal alterations of lower motor neurones. Peripheral neuropathy has not been described as a side-effect of insulin, but it may occur in patients with insulinomas, which may be related to the fact that peripheral-nerve metabolism requires adequate supplies of glucose. The authors speculate as to whether insulin-induced hypoglycaemia might contribute to peripheral-nerve disorders in long-standing diabetics.

Oral contraceptives

Carpal tunnel syndrome with sensory and motor changes affecting the median nerve in the hand is fairly common in patients on the contraceptive pill, especially if they are prone to fluid retention (Sabour and Fadel 1970). The condition is fully reversible, but liable to recur if the pill is resumed. The phenomenon is thought to be due to the retention of fluid causing oedema and hence compression of the median nerve within the carpal tunnel.

There is a solitary report of a 23-year-old woman on an oral contraceptive pill developing optic neuritis which regressed when the pill was withdrawn (German 1979).

Drugs used in gastrointestinal disorders

Trithiozine is available in some countries as a gastric antisecretory and anti-ulcer agent. Crespi *et al*. (1981) reported three cases of predominantly sensory neuropathy. Histology showed Wallerian degeneration of the myelinated fibres, especially those of large calibre. Electrical studies showed reduced motor or sensory conduction velocity in two of the three patients.

There is a report of mixed peripheral neuropathy in two patients, one of whom had a Guillain–Barré syndrome; the patients were both taking cimetidine and captopril, and it is uncertain which of the drugs was responsible for the neuropathy, or whether it was due to the combination of the two (Atkinson *et al.* 1980).

Cimetidine was also suspected by Walls *et al.* (1980) as the cause of a motor neuropathy in two patients on many different drugs. However, in spite of these occasional reports, it remains unproven that cimetidine can cause peripheral neuropathy.

Vaccines and antitoxins

There are reports of peripheral neuropathy and/or Guillain–Barré syndrome after various viral vaccinations. However, the Guillain–Barré syndrome maybe associated with many different viral infections including chickenpox, glandular fever, influenza A, parainfluenza 3, herpes, mumps, and tick-borne encephalitis. Comparing post-viral and post-vaccination Guillain–Barré syndromes there is remarkable similarity in the clinical features, and in the timing of the onset of the neurological problems in relation to the original viral illness or to the vaccination. The Guillain–Barré syndrome is probably due to an immunological reaction to the original virus, whether living or dead.

Influenza vaccine

The Guillain–Barré syndrome and/or peripheral neuropathy is a rare complication which may result from any of the inactivated influenza vaccines (Meyer *et al.* 1978; Marks and Halpin 1980). However, there was a severe outbreak of the Guillain–Barré syndrome in the USA in 1976–7, during an immunization programme using killed influenze vaccine (A/New Jersey 76)for the prevention of the so-called 'swine flu' (*British Medical Journal* 1977b; *Morbidity and Mortality Weekly Report* 1977). Although at the time there were many cases of Guillain–Barré syndrome in unvaccinated people, analysis of the figures leaves no doubt that the incidence was far higher in vaccinated than in unvaccinated individuals; the risk factor varied from state to state from a minimum of 5.1 to a maximum of 38 (mean 7.5). The neurological problems occurred one to four weeks after vaccination in three-quarters of the patients, with only a few cases being recorded after less than one week or more than four. The comment of a leader-writer at the time was that this episode was another example of an attempt by society to protect its members from an unpleasant illness (mortal to those who are already old, infirm or have a chronic respiratory illness) running into unexpected, serious side effects (*British Medical Journal* 1977b).

Other viral vaccines

Antirabies vaccine is known to cause neuroparalytic episodes, including motor and sensory neuropathy, encephalomyelitis, and less commonly Guillain–Barré syndrome (Sharp and McDonald 1967; *British Medical Journal* 1977b). Reducing the dose of the vaccine may decrease the incidence of such accidents, but does not eliminate them; a man of 71 developed fatal Guillain–Barré syndrome after receiving reduced doses of antirabies vaccine. (Vergara *et al.* 1979). The newer forms of antirabies vaccine appear to be much less toxic than those which were in use until recently.

Smallpox vaccination may cause serious encephalomyelitis but not peripheral neuropathy (Herrlich 1959).

Tetanus toxoid

Administration of tentanus toxoid may produce poly- or mononeuritis, cranial nerve palsies, polyradiculitis, or polyneuromyeloencephalopathy (Wirth 1965; Blumstein and Kreithen 1966; Eichler and Neundörfer 1969; Harrer *et al.* 1971; Gersbach and Waridel 1976; Schlenska 1977; Baust *et al.* 1979; Holliday and Bauer 1983).

Neurotoxicity is commoner in those who have had frequent booster doses of the toxoid. For example, a patient developed rapidly-spreading urticaria around the injection site, starting a few hours after a booster dose, followed eight days later by paralysis of the left vocal cord. He gradually improved over the next two months following treatment with steroids, antihistamines, vitamin B complex, and local electrotherapy (Eichler and Neundörfer 1969). Another patient developed dysphagia 10 days after a booster dose of tetanus toxoid, and the following day he had paralysis of accommodation. Wirth *et al.* (1965) described a reversible eighth-nerve deafness following a tetanus toxoid booster vaccination. Holliday and Bauer (1983) described a patient who developed a cauda equina syndrome and later neuropathy in the upper limbs, starting three days after the administration of tetanus and diphtheria toxoids. Diphtheria vaccine is not known to cause neuropathy, and the authors therefore assumed that it was due to the tetanus toxoid.

Such neuropathies are usually reversible, but the longer the interval between vaccination and the onset of the neurological symptoms, the slower the rate of recovery.

Tetanus antitoxin

There have been two cases reported in which tetanus anti-toxin resulted in serum sickness associated with neuralgia and sensorimotor neuropathy (Woolling and Rushton 1950).

Miscellaneous

Disulfiram

Disulfiram (*Antabuse*), used for treating alcoholics, has produced occasional cases of distal sensorimotor neuro-pathy or mononeuritis, sometimes associated with ence-phalopathy (Moddel *et al.* 1978; Norris 1979; Watson *et al.* 1980; Lemperière *et al.* 1982; Marra 1981; Mokri *et al.* 1981; Nukada and Pollock 1981; Ansbacher *et al.* 1982). The commonest presentation of disulfiram neuropathy is that of a mixed sensory and motor peripheral neuro-pathy. It is probably dose-related, and may be aggra-vated by the administration of chloral (Nukada and Pollock 1981). Norris (1979) described two obese patients with a peroneal mononeuropathy; he thought that it was probably due to a combination of disulfiram therapy, acute nutritional imbalance, prior alcoholism, and vulnerability of the peroneal nerve on crossing the legs.

Electrical studies show slowing of nerve conduction velocity, and denervation of distal muscles, suggesting a dying-back axonal degeneration, which has been confirmed by biopsy (Watson *et al.* 1980; Marra 1981; Olney and Miller 1980).

Moddel *et al.* (1978) in electron microscopy studies, found degenerating myelinated fibres, with dissolution of microtubules and neurofilaments, and axonal disin-tegration in the most severely affected fibres.

An occasional diagnostic problem arises from the fact that disulfiram neuropathy is difficult to distinguish from alcoholic neuropathy, which shows similar clinical, electrical, and histological features (Watson *et al.* 1980).

Disulfiram-induced neuropathy is usually reversible when the drug is discontinued. However, Worner (1982) described a patient who developed disulfiram-induced neuropathy on a dose of 250 mg daily: he then had par-aesthesiae and sensory impairment, but a normal EMG. The dose was reduced to 125 mg daily, and pyridoxine, thiamine, and multivitamin tablets were added. The neuropathic symptoms and signs regressed, in spite of continuing the disulfiram, albeit at a lower dose.

The precise mechanism of disulfiram neurotoxicity is not clear. The drug is metabolized to carbon disulphide, which, in animals at least, is neurotoxic (Ansbacher *et al.* 1982). An alternative explanation is that disulfiram inhibits the enzyme aldehyde dehydrogenase, resulting in accumulation of excess acetaldehyde in the presence of ethanol; however, toxicity may occur even without a drug–alcohol interaction (Marra 1981).

Arsenicals

Arsenic toxicity affects most systems including the skin, gastrointestinal tract, liver, kidneys, blood, and nervous system. Polyneuropathy (and other complications) induced by therapeutic arsenicals have virtually dis-appeared since the use of these drugs has almost stopped. However, rare cases are still seen of sensorimotor neuro-pathy or optic neuritis (usually of rapid onset but with slow recovery) (Wilson *et al.* 1979; Manigand 1982). Arsenicals are now administered mainly topically, for example in Fowler's solution. Robinson (1975) reported a case of polyneuropathy due to caustic arsenical paste applied by a lay 'cancer curer' to a facial ulcer (which at post-mortem examination was found to be benign). The patient's severe sensorimotor neuropathy was associated with high blood levels of arsenic. He had chronic bron-chitis, and in spite of treatment with dimercaprol, he died. This is another worrying story of dangerous 'alter-native treatment' for cancer, in this case administered without even the precaution of a diagnosis.

Thalidomide

Thalidomide-induced neuropathy was discussed fully in *Iatrogenic diseases*, 2nd edn., pp. 301–2. The drug was withdrawn when the disastrous side-effects and terato-genic problems were recognized. It is no longer generally available, so recent reports of thalidomide-induced neuropathy are few and far between. However, the drug is still used occasionally in the management of discoid lupus erythematosus. Ludolph and Matz (1982) described 26 such patients, who were closely monitored both clinically and electrophysiologically. Many had paraesthesiae, and sensory impairment, with or without weakness. Electrical studies showed reduced motor and sensory conduction velocity, and prolonged distal motor latency, and in most cases the electromyogram was abnormal. However, some asymptomatic patients had similar abnormalities, though in general, the worse the symptoms, the greater the electrophysiological abnorma-lities. There was no relationship between the severity of the neuropathy, and the total dose of thalidomide received. It is uncertain whether the neurotoxicity is mediated primarily via demyelination or axonal degene-ration, though the latter seems to be more probable.

Podophyllum

Podophyllum is a preparation made from either American mandrake or May-apple (*Podophyllum peltatum*) or Indian mandrake (*Podophyllum emodi*). Podophyllum resin is made from the dried rhizome or roots: Indian mandrake resin contains 40 per cent podophyllotoxin, and American mandrake resin 10 per cent. A variety of mandrake preparations have been in use for over a century as laxatives, and indeed are still occasionaly used, in spite of the toxicity. Podophyllum derivatives are also sometimes taken as a 'liver tonic', anthelminthic agent, or as a herbal slimming pill: there is no evidence of any efficacy in any of these fields, and therefore this type of use of podophyllum derivatives should be discouraged. Podophyllum is also used as an abortifacient (Clark and Parsonage 1957) which is equally unwise.

The topical application of an alcoholic extract of podophyllum resin to condylomata acuminata and vulval warts is an effective, if potentially dangerous, therapeutic procedure, originally introduced by Kaplan in 1942. Trouble also occasionally arises when patients accidentally ingest resin extract, intended for topical application.

Podophyllum is antimitotic, and podophyllotoxin derivatives have been investigated in the field of cancer chemotherapy. Epidophyllotoxin (etoposide, VP16-213) is being increasingly used in the management of various forms of malignant disease, and it too has occasional neurotoxic side-effects (see above).

Toxic effects following oral use or topical application include peripheral neuropathy, acute encephalopathy, bone-marrow suppression, coughing, hepatic dysfunction, nausea, vomiting, and profuse watery diarrhoea with loss of electrolytes. Podophyllum derivatives are possibly teratogenic and fetotoxic (Ward *et al.* 1954; Clark and Parsonage 1957; Balucani and Zellers 1964; Schirren 1966; Chamberlain *et al.* 1972; Rate *et al.* 1979); Filley *et al.* 1982).

The neurotoxic side-effects of podophyllum comprise a sensorimotor neuropathy, which may be only partly reversible, and also encephalopathy causing ataxia, psychosis, confusion, sometimes followd by coma and occasionally death.

There are a number of case reports involving neurotoxicity following ingestion of the drug, the earliest such report being made in 1890 by Dudley: he described a woman who took five grains of podophyllum resin by mouth; she developed vomiting, became hypotensive, then comatose, and finally died. Clark and Parsonage (1957) described a girl who took 2.8 g of the resin in an attempt to procure an abortion; she became unconscious within 24 hours, but when she regained consciousness she had evidence of peripheral neuropathy as well as encephalopathy.

Other reports have described the neurotoxic effects of topical application of podophyllum resin to the vulva, prepuce, and perianal regions for condylomata acuminata or vulval warts (Ward *et al.*1954; Schirren 1966; Chamberlain *et al.* 1972; Rate *et al.* 1979; Filley *et al.* 1982). Two of these five patients lost consciouness and one died. All developed motor and sensory neuropathy, with numbness, paraesthesiae, and muscle weakness, which in some cases was profound. Diplopia, respiratory difficulty, hypokalaemia, and normal cerebrospinal fluid were documented in the case of Chamberlain *et al.* (1972): this girl was pregnant, and was delivered of a macerated, stillborn infant, 10 days after the application of podophyllum resin. In the fatal case described by Ward *et al.* (1954) there was a post-mortem examination which revealed a rarified cortex, an increase in glial cells, and neuronophagia.

Rate *et al.* (1979) noted that their patient had failed to wash off the podophyllum resin after application; they commented that there is only a very narrow therapeutic margin between the effective dose and the toxic dose.

Other drugs

One case of peripheral neuropathy in an alcoholic being treated with calcium carbamide (*Abstem*) was reported (Reilly 1976). Optic neuritis has been described after the use of quinine (Wilson *et al.* 1979), and bilateral carpal tunnel syndrome in a patient on danazol (Gray 1978). Paraesthesiae may occur during ergotamine or methysergide treatments, probably mediated by vasoconstriction (Davies 1968). The mixed peripheral neuropathy or Guiilain–Barré syndrome in patients on a combination of cimetidine and captopril, has already been mentioned (Atkinson *et al.* 1980).

Farthing *et al.* (1980) described a case of reversible polyneuritis associated with other features of azathioprine sensitivity reaction (a clinical picture resembling serum sickness). They therefore suggested that extra care is required when using azathioprine in the management of immune-mediated neuropathies, myopathies, myasthenia, polyarteritis nodosa, and multiple sclerosis.

One curious report concerns a man who used *Head and Shoulders* shampoo, which contains 2 per cent *zinc pyrithione*, which is absorbed through the skin. This was thought to be the cause of his peripheral neuropathy, which resolved when he changed to a different shampoo (Beck 1978).

Another strange story concerns a man of 62 who was taking sulindac for degenerative arthritis. After six trouble-free months on the drug, he applied 90 per cent dimethylsulphoxide topically to his arms and legs, after which he developed a profound sensorimotor neuropathy. Electrical studies indicated demyelination and axonal neuropathy. He recovered gradually but incompletely (Reinstein *et al.* 1982).

Overdoses

The toxicity of drugs taken in excessive doses may be very different from that found when they are given in therapeutic doses. However, it is useful to recognize such types of toxic effects.

Neurotoxicity due to abuse of nitrous oxide appears to be an occupational hazard of dentists, while others suffer from prolonged exposure to the gas when working in poorly ventilated surgeries. One case report describes a patient with severe burns using *Entonox* (50 per cent nitrous oxide in oxygen) as required, for pain relief. He had consumed a total of 40 000 litres in three months when he developed neuropathy (Hayden *et al.* 1983).

Individuals exposed to excessive quantities of nitrous oxide may develop a syndrome reminiscent of subacute combined degeneration of the spinal cord, with sensorimotor neuropathy, and evidence of involvement of the posterior and lateral columns of the spinal cord. The patients present with a variety of neurological symptoms including intellectual impairment, spastic paraparesis and muscle weakness, paraesthesiae, impairment of all modalities of sensation, ataxia, loss of balance, and evidence of autonomic neuropathy including diarrhoea, impotence, and sphincter disturbances. Some patients have a positive Lhermitte's sign. The victims do not always recover fully when nitrous oxide is withdrawn. The cerebrospinal fluid is normal. Electrical studies have shown slowing of nerve conduction and are suggestive of axonal neuropathy. Macrocytosis and low serum vitamin B_{12} levels have been reported, and it has been suggested that nitrous oxide may interfere with the action of vitamin B_{12} in the nervous system (Adornato 1978; Layzer 1978; Layzer *et al.* 1978; Sahenk *et al.* 1978; Hayden *et al.* 1983; Blanco and Peters 1983).

Lithium intoxication may occur either from deliberate overdosage or from excessive therapeutic doses. There was a report of acute generalized sensorimotor peripheral neuropathy with cranial neuritis after an overdose (Brust *et al.* 1979). Heim *et al.* (1981) described a thyrotoxic woman of 48 who was treated with a variety of drugs including lithium in gradually increasing doses.

When the blood lithium level reached 3 mEql^{-1}, she had a sudden onset of vomiting, dysarthria, hyperkinesis, and myoclonus; this unusual clinical picture was associated with motor neuropathy and loss of reflexes, but without loss of sensation. The neuropathy improved, but the extrapyramidal symptoms and signs were permanent, in spite of stopping the drug. Pamphlett and Mackenzie (1982) described a man who developed neurotoxicity when the dose was increased from 1500 to 1800 mg per day, resulting in high blood levels of lithium: he developed confusion and weakness prior to becoming comatose. He had evidence of severe sensorimotor neuropathy and central nervous system disturbance associated with cerebral atrophy. Nerve conduction studies showed no motor or sensory responses in the legs, and little in the arms. Nerve biopsy revealed axonal degeneration. The authors commented that lithium neurotoxicity may be commoner and more serious than has been supposed until now. Again, there is a very narrow therapeutic margin with this drug.

Dapsone may be neurotoxic at therapeutic doses (see above): it is scarcely surprising to find that it is very toxic at high doses. Homeida *et al.* (1980) described a case of predominantly motor neuropathy and optic atrophy in a young man who had received dapsone in very high doses for a short period. The presence of the neuropathy led to an erroneous diagnosis of leprosy. Another patient with neuropathy on a high dose of dapsone (800 mg daily) was a slow acetylator, which may have contributed to the development of the motor neuropathy (Rosen and Sörnäs 1982). Electrophysiological studies in this patient showed low-amplitude muscle responses, prolonged distal latencies, and reduced motor conduction velocities. Sensory neurography was normal. The patient made a full clinical recovery, with electrophysiological improvement, after the dose of dapsone was reduced to 300 mg per week.

Another unusual case of severe neurotoxicity was reported by Graveleau *et al.* (1980): a young woman taking disulfiram already had some symptoms of neuropathy. She inadvertently increased the dose for a few days, and then developed quadriplegia plus cranial neuropathy.

Vincristine is commonly neurotoxic at standard doses and profoundly so when large doses are given. (Reports have recorded accidental overdoses as high as 32 mg.) Vincristine overdosage may result in a neuropathy affecting sensory, motor, cranial, and autonomic nerves, as well as causing confusion, hypertension, inappropriate secretion of antidiuretic hormone, and sometimes death (Berenson 1971; Kaufman *et al.* 1976; Fernandez *et al.* 1981). A fatal myeloencephalopathy occurred after inadvertent intrathecal administration of vincristine

(Slyter *et al.* 1980). Post-mortem examination showed that the neurones were swollen by aggregates of neuro-filaments, similar to those described in experimental models of vincristine neurotoxicity.

Amitriptyline overdose may cause coma, but patients who are semiconscious or only in light coma may have total external ophthalmoplegia, including absent oculo-cephalic and oculovestibular reflexes. Pupil reactions are usually present though they may be sluggish, and corneal reflexes are preserved. Occasional sharp conjugate downward movements of the eyes have also been described. The ophthalmoplegia is rapidly reversible after the administration of intravenous physostigmine, which also produces improvement in the level of consciousness. These observations have diagnostic importance: comatose patients with absent reflex ocular movements are usually assumed to be in deep coma; patients with amitriptyline overdose may have the ophthalmoplegia long before they are in deep coma. Furthermore, the differential diagnosis of amitriptyline overdosage from a brainstem lesion may be difficult, especially if there is no history available: the external ophthalmoplegia in a conscious patient, and the rapid reversibility after physostigmine would favour the diagnosis of amitriptyline overdosage (Smith 1979; Spector and Schnapper 1981; Delaney and Light 1981; Beal 1982).

Carbamazepine overdosage may produce a picture similar to that caused by amitriptyline in large doses, as described above. The essential features are external ophthalmoplegia with intact corneal reflexes and reactive pupils (Mullally 1982; Noda and Umezaki 1982). The diagnostic implications described in the section on amitriptyline overdosage apply also to patients who have taken excessive amounts of carbamazepine.

Colchicine at high doses causes peripheral neuritis and myopathy. The drug interferes with neuromuscular transmission (Prescott 1975).

In spite of the widely-held belief that water-soluble vitamins are totally safe, peripheral neuropathy caused by pyridoxine abuse has recently been reported (Schaumburg *et al.* 1983). Megadose pyridoxine has been prescribed for oedema (including premenstrual oedema), as a dietary supplement, for 'body building', carpal tunnel syndrome, schizophrenia, autism, and in one case it was recommended by an 'orthomolecular psychiatrist'. Patients complained of unsteadiness and had impairment of all modalities of sensation in the hands, feet, and lips. They had a positive Lhermitte's sign and pseudoathetosis. These patients had very high blood levels of pyridoxine. Electrical studies showed normal or minimal slowing of motor nerve conduction, and normal electromyogram patterns, but no sensory nerve action poten-

tials could be elicited. Sural nerve biopsy showed widespread non-specific axonal degeneration, affecting both large and small myelinated fibres. Rudman and Williams (1983) believe that high levels of biologically inactive pyridoxine in the peripheral nervous system may competitively inhibit pyridoxal phosphate, which normally acts as a coenzyme in the synthesis of sphingomyelin. The resulting deficiency of available pyridoxal phosphate may thus result in peripheral neuropathy. The brain and spinal cord are protected as little pyridoxine crosses the blood–brain barrier.

Lhermitte *et al.* (1977) reported a single case of a 26-year-old who had received prophylactic chloroquine 100–200 mg daily, but later took 600 mg per day for two months. He developed a motor neuropathy in the legs and a generalized myasthenic syndrome. Muscle biopsy revealed a vacuolar myopathy.

Bilateral optic atrophy was described in a woman who was drinking 10–15 ml of hexachlorophene daily, as well as using it as a topical application, for 10 months. There was no evidence of other neurotoxicity. (Slamovits *et al.* 1980). Experimental work in rats has produced a possible explanation of hexachlorophene neuropathy: the drug causes nerve oedema within the myelin sheath; the resulting increase in endoneural pressure causes reduction in nerve blood flow, and hence excerbation of the neuropathy (Myers *et al.* 1982).

Massive, acute overdoses of colchicine may rarely cause serious peripheral neuropathy (Manigand 1982).

A typical trigeminal neuralgia was described after toxic doses of digoxin (Bernat and Sullivan 1979).

A woman taking grossly excessive doses of ergotamine (up to 12 mg per day and 60 mg per week) developed ischaemia of the arm and a median-nerve neuropathy, probably due to ischaemia of the nerve (Fitzgerald 1978). The neuropathy resolved after stopping the ergotamine.

A case of polyradiculoneuropathy after intravenous heroin (after four years' abstinence from the drug) was reported (Loizou and Boddie 1978).

The neurotoxic effects of prolonged alcohol abuse, including peripheral and optic neuropathy, are multifactorial in origin. They are well known and need not be discussed further.

Misadventures

Occasionally there are accidental, unforeseen, and unfortunate consequences of using drugs in accepted ways and at standard therapeutic doses. These may be classified as misadventures.

For example, patients receiving anticoagulants have

developed femoral neuropathy. This problem is more likely to arise if the anticoagulant treatment is uncontrolled, and therefore the dose excessive. Haemorrhage into the iliac muscle, tracking down into the inguinal region, causes compression of the femoral nerve. Clinically there is evidence of motor and sensory abnormalities in the distribution of the femoral nerve, plus a mass (and later haematoma) in the iliac fossa, and sometimes associated tachycardia, hypotension, and leucocytosis (Wells and Templeton 1977; Cranberg 1979; Manigand 1982). A similar mishap was reported in a patient on systemic streptokinase therapy (Ganel *et al.* 1979).

Comparable anticoagulant mishaps inducing mononeuropathies have been described affecting the sciatic, lateral popliteal, radial, and median nerves (Manigand 1982).

Trauma to the sciatic nerve due to careless intramuscular injections still occur from time to time. A recent report concerned painful neuropathy following intramuscular quinine injections (Bourrel and Souvestre 1982). The patients had wasting and weakness, sometimes with sensory impairment, and some developed neuropathic feet with perforating ulcers.

Neurolytic lumbar sympathetic blockade, using alcohol or phenol in water, for relief of pain in patients with occlusive vascular disease caused severe temporary neuralgia in the distribution of L.1 in up to 40 per cent of cases, and in some it lasted for over one week (Cousins *et al.* 1979). These workers also reported minor sensory impairment in the L.1 to L.4 territory, and slight weakness at the hip in about 5 per cent of cases.

Retrobulbar alcohol injections for the relief of pain in blind, painful eyes, sometimes produces external ophthalmoplegia, probably due to accidental infiltration of the motor nerves (Olurin and Osuntokun 1978). Irrigation of the lacrimal duct (after probing the canaliculus) with a 20 per cent chloramphenicol solution resulted in optic atrophy and permanent loss of vision in one patient: it appeared that a false passage was created, allowing the solution to gain access to the orbital tissues; the ensuing orbital oedema probably resulted in retinal-artery occlusion (Rothkoff *et al.* 1979).

An unusual case of aseptic meningitis, arachnoiditis, communicating hydrocephalus, and Guillain–Barré syndrome was described after metrizamide lumbar myelography (Kelley *et al.* 1980). This type of complication was commoner with earlier myelographic dyes.

A curious case of acrodynia including sensory neuropathy due to mercury poisoning has been reported in a patient who had received gamma globulin injections for 15 years on account of agammaglobulinaemia. The mercury poisoning was due to the bacteriostatic agent merthiolate, which is present in commercial gamma globulin (Matheson *et al.* 1980).

Future

Drug-induced peripheral neuropathies will continue to occur, since many useful drugs must be used in spite of their known neurotoxicty. The important thing is to be aware of potential neurotoxic side-effects, to watch carefully for early sumptoms and signs, and, if appropriate, to withdraw the drug before the neuropathy is irreversible.

In some instances electrical testing may reveal early neuropathy in drug-treated patients without neurological symptoms (Girke *et al.* 1975; Lindholm 1967; Sébille and Rozenstajn 1978; Takeuchi *et al.* 1980a; Mochizuki *et al.* 1981; Bahemuka *et al.* 1982; Shah *et al.* 1983). Arezzo *et al.* (1983) described a simple battery-powered device which can measure vibration sensation in the fingertips; such a tool might be useful in the detection of subclinical neuropathy. It might sometimes be appropriate to introduce such a screening device to monitor patients being treated with certain potentially neurotoxic drugs.

Finally, there is some exciting experimental work being done in Italy and elsewhere on the therapeutic use of gangliosides in peripheral neuropathy: rats with carbon disulphide neuropathy, treated with high doses of gangliosides, showed better morphological regeneration of their peripheral nerves than controls or than rats treated with low-dose gangliosides or with vitamins. However, there was no difference clinically or electrophysiologically between the different groups (Maroni *et al.* 1981). Some patients with diabetic or alcoholic neuropathy treated with gangliosides showed electrophysiological improvements (Bassi *et al.* 1982).

Gangliosides or other compounds may ultimately prove useful in the management of drug-induced and other forms of peripheral neuropathy. However, in the absence of any definitive therapeutic measures for neuropathy, it is as well to remember that prevention is better than cure.

REFERENCES

ADORNATO, B.T. (1978). Nitrous oxide and vitamin B_{12}. *Lancet* **ii**, 1318.

AELONY, Y. AND LOCKS, M.O. (1980). Peripheral neuropathy associated with ethambutol. *Chest* **78**, 898.

AHMAD, S. (1979). Amiodarone. *Arch. intern. Med.* **139**, 1319.

ALBERT, D.M. WONG, V.G., AND HENDERSON ,E.S. (1967). Ocular complications of vincristine therapy. *Arch. Ophthalm.* **78**, 709–13.

ANDERSON, P.J., SONG, S.K., AND SLOTWINE, P. (1967). The fine structure of spheromembranous degeneration of skeletal muscle produced by vincristine. *J. Neuropathol. exp. Neurol.* **26**, 15–24.

ANSBACHER, L.E., BOSCH, E.P., AND CANCILLA, P.A. (1982). Disulfiram neuropathy: neurofilamentous distal axonopathy. *Neurology* **32**, 424–8.

AREZZO, J.C., SCHAUMBURG, H.H., AND PETERSEN, C.A. (1983). Rapid screening for peripheral neuropathy: a field study with the Optacon. *Neurosci. behav. Physiol.* **33**, 626–9.

ARGOV, Z. AND MASTALGIA, F.L. (1979). Drug-induced peripheral neuropathies. *Br. med. J.* **1**, 663–6.

ARMSTRONG, J.G., DYKE, R.W., AND FOUTS, P.J. (1962). Initial clinical experience with leucocrystine, a new alkaloid from *Vinca rosea. Proc. Am. Ass. Cancer Res.* **3**, 301.

ARNOLD, A.M. AND WILLIAMS, C.J. (1979). Drug-induced peripheral neuropathies. *Br. Med. J.* **1**, 955.

ASBURY, A.K. AND JOHNSON, P.C. (1978). Pathology of peripheral nerve. In *Major problems in pathology*, Vol. 9. (ed. J.L. Bennington), p. 73. W.B. Saunders, Philadelphia.

ATKINSON, A.B., BROWN, J.J., LEVER, A.F., McAREAVEY, D, ROBERTSON, J.I.S., BEHAN, P.O., MELVILLE, I.D., AND WEIR, A. I. (1980). Neurological dysfunction in two patients receiving captopril and cimetidine. *Lancet* ii, 36–7.

AWIDI, A.S. (1980). Blindness and vincristine. *Ann. intern. Med.* **93**, 781.

BAHEMUKA, M., KIOY, P.M., WANJAMA, M., AND WAINAINA, C.M. (1982). Electrophysiological evidence of peripheral nerve damage in patients treated with thiazine. *East Afr. med. J.* **59**, 798–803.

BALUCANI, M. AND ZELLERS, D. D. (1964). Podophyllum resin poisoning with complete recovery. *J. Am. med. Ass.* **189**, 639–40.

BASSI, S., ALBIZZATI, M.G., CALLONI, E., AND FRATTOLA, L. (1982). Electromyographic study of diabetic and alcoholic polyneuropathic patients treated with gangliosides. *Muscle Nerve* **5**, 351–6.

BAUST, W., MYER, D., AND WACHSMUTH, W. (1979). Peripheral neuropathy after administration of tetanus toxoid. *J. Neurol.* **222**, 131–3.

BEAL, M.F. (1982). Amitriptyline ophthalmoplegia. *Neurology* **32**, 1409.

BECHER, R., SCHÜTT, P., OSIEKA, R., AND SCHMIDT, C.G. (1980). Peripheral neuropathy and ophthalmologic toxicity after treatment with cis-dichlorodiamminoplatinum (II). *Cancer Res. clin. Oncol.* **96**, 219–21.

BECK, J.E. (1978). Zinc pyrithione and peripheral neuritis. *Lancet* i, 444.

BEDIKIAN, A.Y. VALDIVIESO, M., MAROUN, J., GUTTERMAN, J.U., HERSH, E.M., AND BODEY, G.P. (1980). Evaluation of vindesine and MER in colorectal cancer. *Cancer* **46**, 463–7.

BEGG, T.B. AND SIMPSON, J.A. (1964). Chloroquine myopathy. *Br. med. J.* **1**, 770.

BEHAR, A., RACHMILEWITZ, E., RAHAMIMOFF, R., AND DENMAN, M. (1965). Experimental nitrofurantoin polyneuropathy in rats: early histological and electrophysiological alterations in peripheral nerves. *Arch. Neurol., Chicago.* **13**, 160–3.

BERENSON, M.P. (1971). Recovery after inadvertent massive overdosage of vincristine. *Cancer Chemother. Rep.* **55**, 525–6.

BERNAT, J.L. AND SULLIVAN, J.K. (1979). Trigeminal neuralgia from digitalis intoxication. *J. Am. med. Ass.* **241**, 164–5.

BEVERUNGEN, W., FRITZ, K.W., AND ROSS, J. (1965). Polyneuritis nach Behandlung mit Nitrofurantoin bei einem Niereninsuffi-

zienz-Kranken. *Münch. med. Wschr.* **107**, 953–5.

BLANCO, G. AND PETERS, H.A. (1983). Myeloneuropathy and macrocytosis associated with nitrous oxide abuse. *Arch. Neurol.* **40**, 416–18.

BLUMSTEIN, G.I. AND KREITHEN, H. (1966). Peripheral neuropathy following tetanus toxoid administration. *J. Am. med. Ass.* **198**, 1030–1.

BOHANNON, R.A., MILLER, D.G., AND DIAMOND, M.D.(1963). Vincristine in the treatment of lymphomas. *Cancer Res.* **23**, 613–21.

BOMB, B.S., BEDI, H.K., MODI, A.P., AND SAHAI, R.N. (1979). Transient optic neuritis following tetramisole. *Trans. R. Soc. trop. Med. Hyg.* **73**, 110.

BOUCHE, P., BOUSSER, M.G., PEYTOUR, M.A., AND CATHALA, H.P. (1979). Perhexiline maleate and peripheral neuropathy. *Neurology, Minneapolis* **29**, 739–43.

BOURREL, P. AND SOUVESTRE, R. (1982). Traumatologie nerveuse particulière: les lésions du nerf sciatique par injections intrafessieres de quinine. *Med. Trop.* **42**, 209–13.

BRADLEY, W.G., KARLSSON, I.J., AND RASSOL, C.G. (1977). Metronidazole neuropathy: *Br. med. J.* **2**, 610–11.

——, LASSMAN, L.P., PEARCE, G.W., AND WALTON, J.N. (1970). The neuromyopathy of vincristine in man. Clinical electrophysiological and pathological studies. *J. Neurol. Sci.* **10**, 107–32.

BRIAND, P. AND TYGSTRUP, I. (1959). Polyneuritis following nitrofurantoin (*Furadantin*) treatment. *Ugeskr. Laegr.* **121**, 664–6.

BRITISH MEDICAL JOURNAL (1977a). Leading article: Laetrile: quacks and freedom. *Br. med. J.* **1**, 3–4.

—— (1977b). Leading article: Guillain–Barré syndrome and influenza vaccine. *Br. med. J.* **1**, 1373–4.

BROOK, J., AND SCHREIBER, W. (1971). Vocal cord paralysis: a toxic reaction to vincristine therapy. *Cancer Chemother. Rep.* **55**, 591–3.

BRUNNER, K.W. AND YOUNG, C.W. (1965). A methylhydrazine derivative in Hodgkin's disease and other malignant neoplasms: therapeutic and toxic effects studied in 51 patients. *Ann. intern. Med.* **63**, 69–86.

BRUST, J.C.M., HAMMER, J.S., CHALLENOR, Y., HEALTON, E.B., AND LESSER, R.P. (1979). Acute generalized polyneuropathy accompanying lithium poisoning. *Ann. Neurol.* **6**, 360–2.

BURLEY, D. AND STEEN, J. (1979). Drug-induced peripheral neuropathies. *Br. med. J.* **1**, 1082.

BUZARD, J.A. AND CONKLIN, J.D. (1964). Placental transfer of nitrofurantoin and furaladone. *Am. J. Physiol.* **206**, 189–92.

CACCIA, M.R., COMOTTI, B., UBIALI, E., AND LUCCHETTI, A. (1977). Vincristine polyneuropathy in man. *J. Neurol.* **216**, 21–6.

ČÁP, J., MISIKOVÁ, Z., AND KOVAC, R. (1975). Neurotoxicity of vincristine. *Cesk. Pediatr.* **30**, 565–6.

CARMICHAEL, S.M., EAGLETON, L., AYERS, C.R., AND MOHLER, D. (1970). Orthostatic hypotension during vincristine therapy. *Arch. intern. Med.* **126**, 290–3.

CASEY, E.B., JELLIFE, A.M., LE QUESNE, P.M., AND MILLETT, Y.C. (1973). Vincristine neuropathy. Clinical and electrophysiological observations. *Brain* **96**, 69–86.

CASSADY, J.R., TONNESSEN, G.L., WOLFE, L.C., AND SALLAN, S.E. (1980). Augmentation of vincristine neurotoxicity by irradiation of peripheral nerve. *Cancer treat. Rep.* **64**, 963–5.

CASTELEIN, S. AND DE WEERD, C.J. (1974). Neurotoxische eigenschappen van Dapson. *Ned.T. Geneesk* **118**, 342–4.

CHAMBERLAIN, M.J., REYNOLDS, A.L., AND YEOMAN, W.B. (1972). Toxic effects of podophyllum application in pregnancy. *Br. med. J.* **3**, 391–2.

CHISHOLM, B.C., AND CURRY, S.B. (1978). Vincristine-induced dysphagia. *Southern med. J.* **71**, 1364–5.

CHO, E.S., LOWNDES, H.E., AND GOLDSTEIN, B.D. (1983). Neurotoxicology of vincristine in the cat. Morphological study. *Arch. Toxicol.* **52**, 83–90.

CLARK, A.N.G. AND PARSONAGE, M.J. (1957). A case of podophyllum poisoning with involvement of the nervous system. *Br. med. J.* **2**, 1155–7.

COHEN, M.M. (1970). Toxic neuropathy. In *Handbook of clinical neurology,* Vol. 7 (ed. P.J. Vinken and G.W. Bruyn), pp. 510–6. North Holland/Elsevier, Amsterdam.

COLIP, M.P., BAUGHMAN, S., AND PETERSON, R.G. (1981). Alteration of fucose/leucine incorporation into PNS myelin by isoniazid neuropathy. *J. Neurobiol.* **12**, 193–7.

COLLIER, G. AND MARTIN, A. (1960). Side-effects of *Tofranil.* General review of three cases of polyneuritis of the lower extremities. *Ann. med.-psychol.* **118**, 719–38.

COLLINGS, H. (1960). Polyneuropathy associated with nitrofurantoin therapy. *Arch. Neurol., Chicago* **3**, 656–60.

COMMITTEE ON SAFETY OF MEDICINES (1983). *Current problems,* no. 11.

CONNELL, P.H. (1968). Central nervous system stimulant and antidepressant drugs. In *Side-effects of drugs,* Vol. VI (ed. L. Meyler and A. Herxheimer), pp. 1–50. Excerpta Medica, Amsterdam.

CONRAD, M.E. AND CROSBY, W.H. (1960). Massive nitrogen mustard therapy in Hodgkin's disease with protection of bone marrow by tourniquets. *Blood* **16**, 1089–103.

COUSINS, M.J., REEVE, T.S., GLYNN, C.J., WALSH, J.A., AND CHERRY, D.A. (1979). Neurolytic lumbar sympathetic blockade: duration of denervation and relief of rest pain. *Anaesth. intens. Care* **7**, 121–35.

COXON, A. AND PALLIS, C.A. (1976). Metronidazole neuropathy. *J. Neurol. Neurosurg. Psychiat.* **39**, 403–5.

CRANBERG, L. (1979). Femoral neuropathy from iliac hematoma: report of a case. *Neurol., Minneapolis,* **29**, 1071–2.

CRESPI, V., PETRUCCIOLI PIZZINI, M.G., TREDICI, G., BEVILACQUA, L., BOGLIUN, G., AND MANDELLI, A. (1981). Trithiozine polyneuropathy: clinical, neurophysiological and histopathological study of three cases. *J. neurol. Sci.* **2**, 291–5.

DANNER, R. (1981). Periphere Polyneuropathie bei Chlorprothixen-Therapie — Eine klinisch-neurophysiologische Dokumentation *Fortschr. Neurol. Psychiat.* **49**, 455–9.

DAVIES, D.M. (1968). Drug-induced paraesthesiae. *Adv. Drug React. Bull.* **9**, 19.

DAVIS, H.L.JR., RAMIREZ, G., ELLERBY, R.A., AND ANSFIELD, J. (1974). Five-drug therapy in advanced breast cancer: factors influencing toxicity and response. *Cancer* **34**, 239–45.

DAWKINS, K.D., AND GIBSON, J. (1978). Peripheral neuropathy with disopyramide. *Lancet* **i**, 329.

DELANEY, P. (1982). Vincristine induced laryngeal nerve paralysis. *Neurology, NY* **32**, 1285–8.

—— AND LIGHT, R. (1981). Gaze paresis in amitriptyline overdose. *Ann. Neurol.* **9**, 513.

DESAI, D.V., ESDINLI, E.Z., AND STUTZMAN, L. (1970). Vincristine therapy of lymphomas and chronic lymphocytic leukaemia. *Cancer* **26**, 352–9.

DEVADATTA, S. (1965). Isoniazid-induced encephalopathy. *Lancet* **ii**, 440.

DI BELLA, N.J. (1980). Vincristine-induced orthostatic hypotension: a prospective clinical study. *Cancer treat. Rep.* **64**, 359–60.

DICK, D.J. AND RAMAN, D. (1982). The Guillain-Barré syndrome following gold therapy. *Scand. J. Rheumatol.* **11**, 119–20.

DISCHE, S., SAUNDERS, M.I., ANDERSON., P., URTASUN, R.C., KARCHER, K.H. KOGELNIK, H.D., BLEEHEN, N., PHILLIPS, T.L., AND WASSERMAN, T.H. (1978a). The neurotoxicity of misonidazole: pooling of data from 5 centres. *Br. J. Radiol.* **51**, 1023–4.

——, ——, AND FLOCKHART, I.R. (1978b). The optimum regime for the administration of misonidazole and the establishment of multicentre clinical trails. *Br. J. Cancer* **37** (Suppl.3), 318–21.

——, ——, LEE, M.E., AND ANDERSON, P. (1979). Misonidazole — a drug for trial in radiotherapy and oncology. *Int. J. Radiat. Oncol. biol. Phys.* **5**, 851–60.

——, ——, AND STRATFORD, M.R.L. (1981). Neurotoxicity with desmethylmisonidazole. *Br. J. Radiol.* **54**, 156–7.

DONALDSON, D. AND GIBSON, G. (1979). Local complications with intravenous diazepam. *Can. dent. Ass. J.* **45**, 337–41.

DUDLEY, W.H. (1890). Fatal podophyllin poisoning. *Med. Rec.* **37**, 409.

EADE, O.E., ACHESON, E.D., CUTHBERT, M.F., AND HAWKES, C.H. (1975). Peripheral neuropathy and indomethacin. *Br. med. J.* **2**, 66–7.

EDELSTYN, G.A., BATES, T.D., BRINKLEY, D., MACRAE, K.D., SPITTLE, M.F., AND WHEELER, T. (1975). Comparison of 5-day, 1-day, and 2-day cyclical combination chemotherapy in advanced breast cancer. *Lancet* **ii**, 209–11.

EDWARDS, M.S., BOLGER, C.A., LEVIN, V.A., PHILLIPS, T.L., AND JEWETT, D.L. (1982). Evaluation of misonidazole peripheral neurotoxicity in rats by analysis of nerve trains evoked response. *Int. J. Radiat. Oncol. biol. Phys.* **8**, 69–74.

EICHLER, W. AND NEUNDÖRFER, B. (1969). Recurrenslähmung nach Tetanustoxoid-Auffrischimpfung (mit allergischer Lokalreaktion). *Münch. med. Wschr.* **111**, 1692–5.

ELLENBERG, M. (1959). Diabetic neuropathy precipitated by diabetic control with tolbutamide. *J. Am. med. Ass.* **169**, 1753–7.

ELLIS, F.J. (1962). Acute polyneuritis after nitrofurantoin. *Lancet* **ii**, 1136–8.

EVANS, A.E., FARBER, S., BRUNET, S., AND MARIANO, P.J. (1963). Vincristine in the treatment of acute leukaemia in children. *Cancer* **16**, 1302–6.

EVANS, D.A.P.,MANLEY.K.A., AND MCKUSICK, V.A. (1960). Genetic control of isoniazid metabolism in man. *Br. med. J.* **2**, 485–91.

FALK, G., REIS, G. VON, AND GEERSTEEGH-LIND, A. (1957). Polyneurit vid furadantin-behandling. *Opusc. med., Stockh.* **2**, 130–4.

FALKSON, G., VAN DYK, J.J., VAN EDEN, E.B., VAN DER MERWE, A.M., VAN DEN BERGH, J.A., AND FALKSON, H.C. (1975). A clinical trial of the oral form of 4-demethyl-epidophyllotoxin-β-D ethylidene glucoside (NSC 141540) VP. 16-213. *Cancer, NY* **35**, 1141–4.

FARTHING, M.J.G., COXON, A.Y., AND SHEAFF, P.C. (1980). Polyneuritis associated with azathioprine sensitivity reaction. *Br. med. J.* **1**, 367.

FERNANDEZ, M.N., CABRERA, R., SANJUAN, I., BARBOLLA, L., AND ZABALA, P. (1981). Hipertension arterial, gran mal, disquinesia intestinal, neuropatia sensitivomotriz e hiponatremia como consecuencias de sobredosificacion aguda de vincristina. *Sangre* **26**, 209–15.

FILLEY, C.M, GRAFF-RADFORD, N.R., LACY, J.R., HEITNER, M.A., AND EARNEST, M.P. (1982). Neurologic manifestations of podophyllin toxicity. *Neurology, NY* **32**, 308–11.

FITZGERALD, B. (1978). Saint Anthony's fire or carpal tunnel syndrome? A case of iatrogenic ergotism. *Hand* **10**, 82–6.

FOLK, R.M., PETERS, A.C., PATKOV, K.L., AND SWENBERG, J.A. (1974), Vincristine: a retrospective toxicological evaluation in monkeys and dogs using weekly intravenous injections for six weeks. *Cancer Chemother. Rep.* (Suppl.) **5**, 17–23.

FORESTER, D. (1980). Toxic effects of ethambutol. *Chest* **78**, 496–7.

FRASER, D.M. AND MILLER, H.C. (1978). Perhexilene-induced neuropathy. *Br. med. J.* **1**, 858–9.

FRAWLEY, T.F. AND KOEPF, G.F. (1950). Neurotoxicity due to thiouracil and thiourea derivatives. *J. clin. Endocrinol* **10**, 623–9.

FREI, E.III, FRANZINO, A., SCHNIDER, B.I., COSTA, G., COLSKY, J., BRINDLEY, C.O., HORSLEY, H., HOLLAND, J.F., GOLD, G.L., AND JONSSON, U. (1961). Clinical studies of vinblastine. *Cancer Chemother. Rep.* **12**, 125–9.

FREYHAN, F.A. (1959). Clinical effectiveness of *Trofanil* in the treatment of depressive psychoses. *Can. psychiat. Ass. J.* **4**, (Suppl.), S86–S99.

FRYTAK, S., MOERTEL, C.G., CHILDS, D.S., AND ALBERS, J.W. (1978). Neurologic toxicity associated with high dose metronidazole therapy. *Ann. intern. Med.* **88**, 361–2.

GABRIEL, R. AND PEARCE, J.M.S. (1976). Clofibrate-induced myopathy and neuropathy. *Lancet* **ii**, 906.

GANEL, A., ZWEIG, A., AND ADAR, R. (1979). Femoral neuropathy complicating systemic streptokinase treatment for ergotism. *Angiology* **30**, 192–4.

GARCIN, R., RONDOT, P., AND FARDEAU, M. (1965). Sur les accidents neuromusculaires et en particulier sur une myopathie vacuolaire observés au cours d'un traitement prolongé par la chloroquine. Amélioration rapide aprés arrêt du médicament. *Rev. Neurol.* **113**, 650.

GEHIN, P., BRICHET, B., VESPIGNANI, H., BARROCHE, G., AND WEBER, M. (1981). La neuropathie sensitive isolée du trijumeau. Discussion du role possible d'un facteur toxique. *Rev. Otoneuroophtalmol.* **53**, 193–8.

GERMAN, G.A. (1979). B-scan ultrasonography in the diagnosis of oral-contraceptive related optic neuritis. *J. Am. optometric Ass.* **50**, 243–4.

GERSBACH, P. AND WARIDEL, D. (1976). Paralysis aprés prévention antitétanique. *Schweiz. med. Wschr.* **106**, 150–3.

GASTAUT, J.L., PELLISSIER, J.F., JEAN, P, TUBIANA, N., AND CARCASSONNE, Y. (1982). Neuropathie périphérique au cisplatine: une observation. *Nouv. Presse Méd.* **11**, 1113–17.

GILES, H.MC.C. (1963). Imipramine poisoning in childhood. *Br. med. J.* **2**, 844–6.

GIRKE, W., KREBS, F.A. AND MÜLLER-OERLINGHAUSEN, B. (1975). Effects of lithium on electromyographic recordings in man: studies in manic depressive patients and normal volunteers. *Int. Pharmocospsychiat.* **10**, 24–36.

GIRLING, D.J. (1982). Adverse effects of antituberculous drugs. *Drugs* **23**, 56–74.

GLODE, L.M. (1979). Dose-limiting extramedullary toxicity of high dose chemotherapy *Exp. Hematol.* **7**, (Suppl. 5), 265–78.

GOLDMAN, A.L. AND BRAMAN, S.S. (1972). Isoniazid: a review with emphasis on adverse effects. *Chest* **62**, 71–7.

GORDON, M. AND GORDON, A.S. (1981). Perhexiline maleate as a cause of reversible Parkinsonism and peripheral neuropathy. *J. Am. geriat. Soc.* **29**, 259–62.

GOTTLIEB, R.J. AND CUTTNER, J. (1971). Vincristine-induced bladder atony. *Cancer* **23**, 674–5.

GOTTSCHALK, P.G., DYCK, P.J., AND KIELY, J.M. (1968). Vinca alkaloid neuropathy: nerve biopsy studies in rats and man. *Neurology, Minneapolis* **18**, 875–82.

GRALLA, R.F., RAPHAEL, B.G., GOLBEY, R.B., AND YOUNG, C.W. (1979). Phase II evaluation of vindesine in patients with non-small cell carcinoma of the lung. *Cancer treat. Rep.* **63**, 1343–6.

GRAVELEAU, J., ECOFFET, M., AND VILLARD, A. (1980). Les neuropathies dues au disulfirame. *Nouv. Presse Méd.* **9**, 2905–7.

GRAY, R.G. (1978). Bilateral carpal tunnel syndrome and arthritis associated with danazol administration. *Arthritis Rheumat.* **21**, 493–4.

GUBISCH, N.N., NORENA, D., PERLIA, C.P., AND TAYLOR, S.G. III (1963). Experience with vincristine in solid tumours. *Cancer Chemother. Rep.* **32**, 19–22.

GUTHENEUC, P., GINET, J., GROLEAU, J.Y., AND ROJOUAN, J. (1980). Early phase of vincristine neuropathy in man. *J. neurol. Sci.* **45**, 355–66.

GUILLAIN, G., BARRÉ, J.A., AND STROHL, A. (1916). Sur un syndrome de radiculonévrite avec hyperalbuminose du liquide céphale-rachidien sans réaction cellulaire. Remarqes sur les caractères cliniques et graphiques des réflexes tendineux. *Bull.Soc. méd. Hôp., Paris* **40**, 1462.

GUTMANN, L., MARTIN, J.C., AND WELTO, W. (1976). Dapsone motor neuropathy. An axonal disease. *Neurology, NY* **26**, 514–16.

HADLEY, D. AND HERR, H.W. (1979). Peripheral neuropathy associated with cis-dichlorodiamminoplatinum (II) treatment, *Cancer* **44**, 2026–8.

HAFSTROM, T. (1959). Prognosen vid furadantin-polyneurit. *Opusc. med., Stockholm.* **4**, 17–23.

HAGGARD, M.E., FERNBACH, D.J., HOLCOMB, T.M., SUTOW, W.W., VIETTI, T.J., AND WINDMILL, J. (1968). Vincristine in acute leukaemia of childhood. *Cancer* **22**, 438–44.

HANCOCK, B.W. AND NAYSMITH,A. (1975). Vincristine-induced autonomic neuropathy. *Br. Med. J.* **3**, 207.

HANEFELD, F. AND RIEHM, H. (1980). Therapy of acute lymphoblastic leukaemia in childhood: effects on the nervous system. *Neuropaediatrie* **11**, 3–16.

HANSSON, O. AND HERXHEIMER, A. (1981). Neuropathy and optic atrophy associated with halquinol. *Lancet* **i**, 450.

HARDISTY, R.M., McELWAIN, T.J., AND DARBY, C.W. (1969). Vincristine and prednisone for the induction of remissions in acute childhood leukaemia. *Br. med. J.* **2**, 662–5.

HARRER, G., MELNIZKY, U., AND WENDT, H. (1971). Akkomodationsparese und Schlucklähmung nach Tetanus Toxoid Auffrischimpfung. *Wien. med. Wschr.* **121**, 296–7.

HARRIS, L., McKENNA, W.J., ROWLAND, E., HOLT, D.W., STOREY, G.C. AND KRIKLER, D.M. (1983). Side effects of long-term amiodarone therapy. *Circulation* **67**, 45–51.

HAUW, J.J., MUSSINI, J.M., BOUTRY,J.M., ESCOUROLLE, R., POLLET, S., ALBOUZ, S., HARPIN, M.L., AND BAUMANN, N. (1981). Perhexiline maleate induced lipoidosis in human peripheral nerve and tissue culture: ultrastructural and biochemical changes. *Clin. Toxicol.* **18**, 1405–9.

HAYDEN, P.J., HARTEMINK, R.J., AND NICHOLSON, G.A. (1983). Myeloneuropathy due to nitrous oxide. *Burns incl. therm. Inj.* **9**, 267–70.

HAYES, F.A., GREEN, A.A., SENZER, N., AND PRATT, C.B. (1979). Tetany: a complication of cis-dichlorodiammineplatinum (II) therapy. *Cancer treat. Rep* **63**, 547–8.

HEATHFIELD, K.W.G. AND CARABOTT, F. (1982). Adverse effects of perhexiline. *Lancet* **ii**, 507–8.

HEFFELFINGER, J.C. AND ALLEN, R.J. (1964). Neurotoxicity with nitrofurantoin. A case report. *J. Pediat.* **65**, 611–12.

HEGER, J.J., PRYSTOWSKY, E.N., JACKMAN, W.M., NACCARELLI, G.V., WARFEL, K.A., RINKENBERGER, R.L., AND ZIPES, D.P. (1981). Amiodarone: clinical efficacy and electrophysiology during long-term therapy for recurrent ventricular tachycardia or ventricular fibrillation. *New Eng. J. Med.* **305**, 539–45.

HEIM, J., PINEL, J.F., ALLANNIC, H., FERRAND, P., SABOURAUD, O., AND LORCY, Y. (1981). Sequelles d'une intoxication par le lithium lors du traitement d'une cardiothyréose. *Sem Hôp. Paris* **57**, 1349–52.

HEISS, W.D., TURNHEIM, M., AND MAMOLI, F. (1978). Combination chemotherapy of malignant glioma. Effect of postoperative treatment with CCNU, vincristine, amethopterine and procarbazine. *Eur. J. Cancer* **14**, 1191–202.

HELANDER, J. AND PARTANEN, J. (1978). Dapsone-induced distal axonal degeneration of the motor neurons. *Dermatologica* **156**, 321–4.

HERNDON, R.F. AND FOX, G.E. (1966). Polyneuropathy due to nitrofurantoin. *Illinois med. J.* **129**, 164–6.

HERRLICH, A. (1959). Über Vakzineantigen: Versuch einer Prophylaxe neuraler Impfschaden. *Münch. med. Wschr.* **101**, 12–14.

HERTZ, R., LIPSETT, M.B., AND MOY, R.M. (1960). Effect of vinca leukoblastine on metastatic choriocarcinoma and related trophoblastic tumours in women. *Cancer Res.* **20**, 1050–3.

HEYN, R.M., BEATT, E.C., HAMMOND, D., LOUIS, J., PIERCE, M., MURPHY, M.L., AND SEVERO, N. (1966). Vincristine in the treatment of acute leukaemia in children. *Pediatrics* **38**, 82–91.

HICKLIN, J.A. (1968). Chloroquine neuromyopathy. *Ann. Phys. Med.* **9**, 189–92.

HILDEBRAND, J. AND COËRS, C. (1965). Étude clinique, histologique et électrophysiologique des neuropathies associées au traitement par la vincristine. *Eur. J. Cancer* **1**, 51–8.

—— AND KENIS, Y. (1971). Additive toxicity of vincristine and other drugs for the peripheral nervous systems. Three case reports. *Acta neurol. belg.* **71**, 486–91.

HISHON, S. AND PILLING, J. (1977). Metronidazole neuropathy. *Br. Med. J.* **2**, 832.

HOLLAND, J.F., SCHARLAU, C., GAILANI, S., KRANT, M.J., OLSON, K.B., HORTON, J., SCHNIDER, B.I., LYNCH, J.J., OWENS, A., CARBONE, P.P., COLSKY, J., GROB, D., MILLER, S.P., AND HALL, T.C. (1973). Vincristine treatment of advanced cancer: a co-operative study of 392 cases. *Cancer Res.* **33**, 1258–64.

HOLLIDAY, P.L. AND BAUER, R.B. (1983). Polyradiculoneuritis secondary to immunisation with tetanus and diphtherta toxoids. *Arch. Neurol.* **40**, 56–7.

HOLMBERG, L., BOMAN, G., BÖTTIGER, L.E., ERIKSSON, B., SPROSS, R., AND WESSLING, A. (1980). Adverse reactions to nitrofurantoin. Analysis of 921 reports. *Am. J. Med.* **69**, 733–8.

HOMEIDA, M., BABIKR, A., AND DANESHMEND, T.K. (1980). Dapsone-induced optic atrophy and motor neuropathy. *Br. med. J.* **2**, 1180.

HOWARD, D.J. AND RUSSELL, REES, J. (1976). Long-term perhexiline maleate and liver function. *Br. med. J.* **1**, 133.

HUBMANN, R. AND BREMER, G. (1965). Die Ausscheidung von Furadantin bei manifester Niereninsuffizienz. *Med. Welt. Stuttgart.* **19**, 1039–49.

HUGHES, J.T., ESIRI, M., OXBURY, J.M., AND WHITTY, C.W.M. (1971). Chloroquine myopathy. *Quart. J. Med.* **40**, 85–93.

HURLEY, M., STEPHENS, R.L., DAVIDNER, M.L., AND MAGRINA, J.F. (1982). Vulvar carcinoma. A case of DDP-induced neurotoxicity in chemotherapy. *J. Kansas med. Soc.* **83**, 239–40.

INOUÉ, K., MANNEN, T., TSUKAGOSHI, A., AND TOYOKURA, Y. (1973). Case of chloramphenicol opticoneuropathy. *Clin. Neurol.* **13**, 128–38.

ISAACS, A.D. AND CARLISH, S. (1963). Peripheral neuropathy after amitriptyline. *Br. med. J.* **1**, 1739.

ITRI, L.M., GRALLA, R.J., KELSEN, B.P., CHAPMAN, R.A., CASPER, E.S., BRAUN, D.W. JR., HOWARD, J.E., GOLBEY, R., AND HEELAN, R.T. (1983). Cisplatin, vindesine and bleomycin (CVB) combination chemotherapy of advanced non-small cell lung cancer. *Cancer* **51**, 1050–5.

JACOBS, J.M., MILLER, R.H., AND CAVANAGH, J.B. (1979*a*). The distribution of degenerative changes in INH neuropathy. *Acta neuropathol., Berlin.* **48**, 1–9.

——, ——, WHITTLE, A., AND CAVANAGH, J.B. (1979*b*). Studies on the early changes in acute isoniazid neuropathy in the rat. *Acta neuropathol., Berlin.* **47**, 85–92.

JALLON, P., LOISEAU, P., ORGOGOZO, J.M., AND SINGLAS, E. (1978). Polyneuropathy with normal metabolism of perhexiline maleate. *Ann. Neurol.* **4**, 385–6.

JANSSEN, P.J. (1960). Peripheral neuritis due to streptomycin. *Am. Rev. resp. Dis.* **81**, 726–8.

JEAN, P., FAVRE, R. GASTAUT, J.L., DODEMANT, P., RODOR, F., AND JOUGLARD, J. (1981). Syndrome extrapyramidal et vincristine. *Thérapie* **36**, 343–6.

JEWKES, J., HARPER, P.G., TOBIAS, J.S., GEDDES, D.M., SOUHAMI, R.L., AND SPIRO, S.G. (1983). Comparison of vincristine and vindesine in the treatment of inoperable non-small cell bronchial carcinoma. *Cancer treat. Rep.* **67**, 1119–21.

JOHNSTON, R.F. AND AUDET, P.R. (1978). Antituberculous chemotherapy. *Am. fam. Physn.* **17**, 136–9.

JOY, R.J., SCALETTAR, R., AND SODEE, D.M. (1960). Optic and peripheral neuritis, probable effect of prolonged chloramphenicol therapy. *J. Am. med. Ass.* **173**, 1731–4.

JÜRGENSSEN, O.A, AND PICHLER, E. (1976). Zerebellare Vincristine-Toxizität. *Klin. Pädiatr.* **188**, 455–8.

KALYANARAMAN, U.P., KALYANARAMAN, K., CULLINAN, S.A., AND McLEAN, J.M. (1983). Neuromyopathy of cyanide intoxication due to "laetrile" (amygdalin). A clinicopathologic study. *Cancer* **51**, 2126–33.

KAPLAN, I.W. (1942). Condylomata acuminata. *New Orleans M. & S. J.* **94**, 388–90.

KAPLAN. R.S. AND WIERNIK, P.H. (1982). Neurotoxicity of antineoplastic drugs. *Semin. Oncol.* **9**, 103–30.

KARLSSON, I.J. AND HAMLYN, A.L. (1977). Metronidazole neuropathy. *Br. med. J.* **2**, 832.

KARON, M., FREIREICH, E.J., AND FREI, E. III. (1962). A preliminary report on vincristine sulphate. A new active agent for the treatment of acute leukaemia. *Pediatrics* **30**, 791–6.

——, ——, ——, TAYLOR, R., WOLMAN, I.J., DJERASSI, I., LEE, S.L., SOWITSKY, A., HANANIAN, J., SELAWRY, O., JAMES, D.J.R., GEORGE, P., PATTERSON, R.B., BURGERT, O., HAURANI, F.I., OBERFIELD, R.A., MACY, C.T., HOOGSTRATEN, B., AND BLOM, J. (1966). The role of vincristine in the treatment of childhood acute leukaemia. *Clin. Pharmacol. Ther.* **7**, 332–9.

KARSTORP, A., FERNGREN, R. , AND LUNDBERG, P. (1973). Neuromyopathy during malaria suppression with chloroquine. *Br. med. J.* **4**, 736.

KATRAK, S.M., POLLOCK, M., O'BRIEN, C.P., NUKADA, H., ALLPRESS, S., CALDER, C., PALMER, D.G., GRENNAN, D.M., McCORMACK, P.I., AND LAURENT, M.R. (1980). Clinical and morphological features of gold neuropathy. *Brain* **103**, 671–93.

KATZNELSON, D. AND GROSS, S. (1978). Guillain-Barré syndrome following the use of thiabendazole. *Acta paediatr. scand.* **67**, 791–2.

KAUFMAN, P.A., KUNG, F.H., KOENING, H.M., AND GIAMMONA, S.T. (1976). Overdosage with vincristine. *J. Paediat.* **89**, 671–4.

KEDAR, A., COHEN, M.E., AND FREEMAN, A.I. (1978). Peripheral neuropathy as a complication of cisdichlorodiammineplatinum (II) treatment: a case report. *Cancer treat. Rep.* **62**, 819–21.

KELLEY, R.E., DAROFF, R.B., SHEREMATA, W.A., AND McCORMICK, J.R. (1980). Unusual effects of metrimazide lumbar myelography. Constellation of aseptic meningitis, arachnoiditis, communicating hydrocephalus and Guillain-Barré syndrome. *Arch. Neurol.* **37**, 588–9.

KOGELNIK, H.D. (1980). Clinical experience with misonidazole. High dose fractions versus daily low doses. *Cancer clin. Trials* **3**, 179–86.

——, MEYER, H.J., JENTZSCH, K., SZEPESI, T., KARCHER, K.H. MAIDA, L., MAMOLI, B., WESSELY, P., AND ZAUNBAUER, P. (1978). Further clinical experiences of a phase I study with the hypoxic cell radiosensitizer misonidazole. *Br. J. Cancer.* **37**, (Suppl. 3), 281–5.

KOLB, L.C. AND GRAY, S.J. (1946). Peripheral neuritis as a complication of penicillin therapy. *J. Am. med. Ass.* **132**, 323–6.

KOLLER, W.C., GEHLMANN, L.K., MALKINSON, F.D., AND DAVIS, F.A. (1977). Dapsone-induced peripheral neuropathy. *Arch. Neurol.* **34**, 644–6.

KOVACH, J.S., MOERTEL, C.G., SCHUTT, A.J., REITEMEIER, R.G., AND HAHN, R.G. (1973). Phase II study of cis-diamminedichloroplatinum (NSC-119875) in advanced carcinoma of the large bowel. *Cancer Chemother. Rep.* **57**, 357–9.

KUMAMOTO, Y., TAMIYA, T., ONISHI, S., KODAMA, N., AND SASAKI, T. (1972). Gastrointestinal tolerance to nitrofurantoin macrocrystals and microcrystals. *Chemother., Tokyo* **20**, 588–91.

KUMING, B.S. AND BRAUDE, L. (1979). Anterior optic neuritis caused by ethambutol toxicity. *S. Afr. med. J.* **55**, 4.

LANCET (1973). Leading article : Neurotoxicity of vincristine. *Lancet* **i**, 980–1.

LANE, R.J. AND ROUTLEDGE, P.A. (1983). Drug-induced neurological disorders. *Drugs, Australia* **26**, 124–47.

LANGLEY, J.R., ROSATO, R.E., AND EL-MAHDI, A. (1978). A primary malignant hemangioendothelioma of the liver: survival following non-operative treatment. *J. Surg. Oncol.* **10**, 533–41.

LARRE, P., COQUET, M., AND MAUPETIT, J. (1981). Neuropathie à l'amiodarone. *Nouv. Presse Méd.* **10**, 2750.

LARSEN, H.W. AND BERTELSEN, S. (1956). Neurologic symptoms during treatment with furadantins. *Ugeskr. Laeg.* **118**, 751–3.

LASSMAN, L.P., PEARCE, G.W., AND GONG, J. (1965). Sensitivity of intracranial glioma to vincristine sulphate. *Lancet* **i**, 296–8.

LAYZER, R.B. (1978). Myeloneuropathy after prolonged exposure to nitrous oxide. *Lancet* **ii**, 1227.

——, FISHMAN, R.A., AND SCHAFER, J.A. (1978). Neuropathy following abuse of nitrous oxide. *Neurol., Minneapolis.* **28**, 504–6.

LEMPERIÈRE, T., ADES, J., AND HARDY, P. (1982). Complications neuropsychiques des traitements au Disulfiram. A propos d'une observation. *Ann. med. Psychol.* **146**, 123–9.

LENNAN, R.M. AND TAYLOR, B.R. (1978). Optineuroretinitis in association with BCNU and procarbazine therapy. *Med. Pediatr. Oncol.* **4**, 43–8.

LEVITT, L.P. AND PRAGER, D. (1975). Mononeuropathy due to

vincristine toxicity. *Neurology, Minneapolis.* **25**, 894–5.

LHERMITTE, F., FARDEAU, M., CHEDRU, F., AND MALLECOURT, J. (1976). Polyneuropathy after perhexiline maleate. *Br. med. J.* **1**, 1256.

——, FRITEL, D., CAMBIER, J., MARTEAU, R., GAUTIER, J-C., AND NOCTON, F. (1963). Polynévrites au cours de traitement par la nitrofurantoïne. *Presse méd.* **71**, 767–70.

——, MARTEAU, R., CHEDRU, F., MALLECOURT, J. ESTRADE, G., GODET-GUILLAIN, J., AND CHEVALLAY, M. (1977). Neuromyopathie à la chloroquine. *Nouv. Presse. Méd.* **6**, 3205–7.

LIEBOLD, J.E. (1971). Drugs having a toxic effect on the optic nerve. *Int. Ophthalm. Clin.* **11**, 137.

LINDHOLM, T. (1967). Electromyographic changes after nitrofurantoin (*Furadantin*) therapy in non-uraemic patients. *Neurology, Minneapolis.* **17**, 1017–20.

LOCKHART, J.D.F. AND MASHETER, H.C. (1976). Report of a multicentre monitored release study of perhexiline maleate in the prevention of angina pectoris. *Br. J. clin. Practice* **30**, 172–7.

LOIZOU, L.A. AND BODDIE, H.G. (1978). Polyradiculoneuropathy associated with heroin abuse. *J. Neurol. Neurosurg. Psychiat.* **41**, 855–7.

LORINI, R., LARIZZA, D., GARIBALDI, E., AND POLITO, E. (1977). Ileoparalitico, convulsioni e ipoatremia in corso di trattamento con vincristina. *Minerva Pediatrica* **29**, 1733–8.

LOUGHBRIDGE, L.W. (1962). Peripheral neuropathy due to nitrofurantoin. *Lancet* **ii**, 1133–5.

LOVELACE, R.E. (1982). Peripheral neuropathy in epileptics. *Neurology, NY* **32**, 457.

LUBBE, W.F. AND MERCER, C.J. (1982). Amiodarone: its side effects, adverse reactions and dosage schedules. *NZ med. J.* **95**, 502–4.

LUDOLPH, A. AND MATZ, D.R. (1982). Elektrophysiologische Veränderungen bei Thalidomid-Neuropathie unter Behandlung des Lupus erythematodes discoides. *EEG-EMG* **13**, 167–70.

LUSINS, J.O. AND JUTKOWITZ, R. (1972). Residual cerebellar system dysfunction and peripheral neuropathy after diphenlhydantoin therapy. *Mt. Sinai J. Med.* **39**, 617–21.

MAHAJAN, S.L., IKEDA, Y., MYERS, T.J., AND BALDINI, M.G. (1981). Acute acoustic nerve palsy associated with vincristine therapy. *Cancer* **47**, 2404–6.

MAMOLI, B., WESSELY, P., KOGELNIK, H.D, MÜLLER, M., AND RATHKOLB, O. (1979). Electroneurographic investigations of misonidazole polyneuropathy. *Eur. Neurol.* **18**, 405–14.

——, HEISS, W.D., PODREKA, I., AND TURNHEIM, M. (1980). Elektroneurographische Untersuchungen bei der Vincristin-Polyneuropathie. *EEG-EMG* **11**, 21–7.

MANIGAND, G. (1982). Accidents neurologiques des médicaments. Neuropathies périphériques. Accidents ototoxiques. Blocs neuromusculaires. *Thérapie.* **37**, 225–48.

MARKES, P. AND SLOGGEN, J. (1976). Peripheral neuropathy caused by methaqualone. *Am. J. med. Sci.* **272**, 323–4.

MARKS, J.S. AND HALPIN, T.J. (1980). Guillain-Barré syndrome in recipients of A/New Jersey influenza vaccine. *J. Am. med. Ass.* **243**, 2490–4.

MARONI, M., COLOMBI, A., GILIOLI, R., ROTA, E., DE PASCHALE, G., CASTANO, P., FOA, V., AND DUCA, G. (1981). Effects of ganglioside therapy on experimental CS2 neuropathy. *Clin. Toxicol.* **18**, 1475–84.

MARRA, T.R. (1981). Disulfiram neuropathy. *Wisconsin med. J.* **80**, 29–30.

MARTENS, E.I.F. AND ANSINK, B.J.J. (1979). A myasthenia-like syndrome and polyneuropathy: complications of gentamycin therapy. *Clin. Neurol. Neurosurg.* **81**, 241–6.

MARTIN, J.AND COMPSTON, N. (1963). Vincristine sulphate in the treatment of lymphoma and leukaemia. *Lancet* **ii**, 1080–3.

MARTIN, W.J., CORBIN, K.B., AND UTZ, D.C. (1962). Paraesthesias during treatment with nitrofurantoin: report of case. *Proc. Mayo Clin.* **37**, 288–92.

MARTINEZ-ARIZALA, A., SOBOL, S.M., McCARTY, G.E., NICHOLS, B.R., AND RAKITA, L. (1983). Amiodarone neuropathy. *Neurosci. Behav. Physiol.* **33**,643–5.

MARTINEZ-FIGUEROA, A, JOHNSON, R.H., LAMBIE, D.S., AND SHAKIR, R.A. (1980). The role of folate deficiency in the development of peripheral neuropathy caused by anticonvulsants. *J. neurol. Sci.* **48**, 315–23.

MASSEY, E.W., PLEET, A.B., AND BRANNON, W.L. (1979). Penicillin neuropathy. *Anaesth. Analges.* **58**, 63–4.

MATHE, G., SCHNEIDER, M., SCHWARZENBERG, L., AMIEL, J.L., CATTAN, A., SCHLUMBERGER, J.R., AND GRACIA, O. (1966). The alkaloids of *Vinca rosea (vincaleucoblastine, leucocristine, leucosine)* in the treatment of leukaemia and of haemosarcoma. *Excerpt Medica Int. Congr. Ser.* **106**, 97.

MATHESON, D.S., CLARKSON, T.W., AND GELFAUD, E.W. (1980). Mercury toxicity (acrodynia) induced by long-term injection of gammaglobulin. *J. Pediat.* **97**, 153–5.

MATSUOKA, Y., TAKAYANAGI, T., AND SOBUE, I. (1981). Experimental ethambutol neuropathy in rats. *J. neurol. Sci.* **51**, 89–99.

MAYR, N., GRANINGER, W., AND WESSELY, P. (1983). Polyneuropathie bei chronischer Polyarthritis unter D-Penicillamin medikamentös induziert. *Wien. Klin. Wschr.* **95**, 86–8.

McLEOD, J.G. AND PENNY, R. (1969). Vincristine neuropathy: an electrophysiological and histological study. *J. Neurol. Neurosurg. Psychiat.* **32**, 297–304.

MEADOWS, G.G., BUFF, M.R., AND FREDRICKS. S. (1982). Amitriptyline-related peripheral neuropathy relieved during pyridoxine hydrochloride administration. *Drug Intel. clin. Pharm.* **16**, 876–7.

MEIER, C., KAUER, B., MÜLLER,U., AND LUDIN, H.P. (1979). Neuromyopathy during chronic amiodarone treatment. *J. Neurol.* **220**, 231–9.

MERIWETHER, W.D. (1971). Vincristine toxicity with hyponatremia and hypochloremia in an adult. *Oncology* **25**, 234–8.

METRAL, S., TCHERNIA, G., AND TRACH, T.-L. (1968). Effets des stimulations répétitives au cours des neuromyopathies induites par la vincristine: observations preliminaires. *Rev. Neurol.* **123**, 275–9.

MEYBOOM, R.H.B. (1977), Heavy metal antagonists. In *Side effects of drugs annual.* (ed. M.N.G. Dukes), Vol. 1, pp. 190–3. Excerpta Medica, Amsterdam.

MEYER, H.M., HOPPS, H.L., PARKMAN, P.D., AND ENNIS, F.A. (1978). Review of existing vaccines for influenza. *Am. J. clin. Pathol.* **70**, 146–52.

MILLER, M. (1963). Neuropathy, agranulocytosis and hepatotoxicity following imipramine therapy. *Am. J. Psychiat.* **120**, 185–6.

MILLER, N.R. AND MOSES, H. (1978). Transient oculomotor nerve palsy. Association with thiazide-induced glucose intolerance. *J. Am. med. Ass.* **240**, 1887–8.

MOCHIZUKI, Y., SUYEHIRO, T., TANIZAWA, A., OHKUBO, H., AND MOTOMURA, T. (1981). Peripheral neuropathy in children on long-term phenytoin therapy. *Brain Dev.* **3**, 375–83.

MODDEL,G., BILBAO, J.M., PAYNE, D., AND ASHBY, P. (1978). Disul-

firam neuropathy. *Arch. Neurol.* **35**, 658–60.

MOGLIA, A., ARRIGO, A., ZANDRINI, C., SCELSI, R., MUTANI, R., MONACO, F., FASSIO, F., AND TRACCIS, M.S. (1983). Peripheral nerve function in rats chronically treated with carbamazepine. *Farmaco (Prat.)* **38**, 200–4.

MOKRI, B., OHNISHI, A., AND DYCK, P.J. (1981). Dusulfiram neuropathy. *Neurology, NY* **31**, 730–5.

MOORE, G.W., SWIFT, L.L., RABINOWITZ, D., CROFFORD, O.B., OATES, J.A., AND STACPOOLE, P.W. (1979). Reduction of serum cholesterol in two patients with homozygous familial hypercholesterolemia by dichloroacetate. *Atherosclerosis* **33**, 285–93.

MORBIDITY AND MORTALITY WEEKLY REPORT (1977). (14 January 1977) Follow-up on Guillain-Barré syndrome-United States. United States Department of Health Education and Welfare Public Health Service/Center for Disease Control, Atlanta, Georgia.

MORRIS, J.S. (1966). Nitrofurantoin and peripheral neuropathy with megaloblastic anaemia. *J. Neurol. Neurosurg. Psychiat.* **29**, 224–8.

MUBASHIR, B.A., AND BART, J.B. (1972). Vincristine neurotoxicity. *New Engl. J. Med.* **287**, 517.

MUELLER, J.M. AND FLAHERTY, J. (1978). Vincristine-induced quadriparesis. *South med. J.* **71**, 1310–11.

MULLALLY, W.J. (1982). Carbamazepine-induced ophthalmoplegia. *Arch. Neurol.* **39**, 64.

MURAYAMA, E., MIYAKAWA, T., SUMIYOSHI, S., DESIMARU, M., AND SUGITA, K. (1973). Retrobulbar optic neuritis and polyneuritis due to prolonged chloramphenicol therapy. *Clin. Neurol.* **13**, 213–20.

MYERS, R.R., MIZISIN, A.P., POWELL, H.C., AND LAMPERT, P.W. (1982). Reduced nerve blood flow in hexachlorophene neuropathy. Relationship to elevated endoneurial fluid pressure. *J. Neuropathol, exp. Neurol.* **41**, 391–9.

NAIR, N.S., LEBRUN, M., AND KASS, I. (1980). Peripheral neuropathy associated with ethambutol. *Chest* **77**, 98–100.

NARANG, R.K. AND VARMA, B.M.D. (1979). Ocular toxicity of ethambutol (a clinical study), *Indian J. Ophthalmol.* **27**, 37–40.

NARANJO, C.A., FORNAZZARI, L., AND SELLERS, E.M. (1981). Clinical detection and assessment of drug induced neurotoxicity. *Prog. Neuropsychopharmacol.* **5**, 427–34.

NEMETH, L., SOMFAI, S., GAL, F., AND KELLNER, B. (1970). Comparative studies concerning the tumour inhibition and the toxicity of vinblastine and vincristine. *Neoplasia* **17**, 345–7.

NODA, S. AND UMEZAKI, H. (1982). Carbamazepine-induced ophthalmoplegia. *Neurology, NY* **32**, 1320.

NORRIS, F.H. (1979). Neuropathy associated with disulfiram administration. *Arch. Neurol.* **36**, 386–7.

NORTON, S.W., AND STOCKMAN, J.A. (1979). Unilateral optic neuropathy following vincristine chemotherapy. *J. Pediatr. Ophthalmol. Strabismus* **16**, 190–3.

NUKADA, H. AND POLLOCK, M. (1981). Disulfiram neuropathy — a morphometric study of sural nerve. *J. neurol. Sci.* **51**, 51–67.

O'CALLAGHAN, M.J. AND EKERT, M. (1976). Vincristine toxicity unrelated to dose. *Arch. Dis. Childh.* **51**, 289–92.

OLIVARIUS, H.DE F. (1956). Polyneuropathy during treatment with nitrofurantoin. *Ugeskr. Laeg.* **118**, 753–5.

OLNEY, R.K. AND MILLER, R.G. (1980). Peripheral neuropathy associated with disulfiram administration. *Muscle Nerve* **3**, 172–5.

OLURIN, O. AND OSUNTOKUN, O. (1978). Complications of retrobulbar alcohol injections. *Ann. Ophthalmol.* **10**, 474–6.

OSTROW, S., HAHN, D., WIERNIK, P.H., AND RICHARDS, R.D. (1978).

Ophthalmologic toxicity after cis-dichlorodiammineplatinum (II) therapy. *Cancer treat. Rep.* 62, 1591–4.

PAGÈS, M., MARTY-DOUBLE, CH., MARTY, R., FUENTES, C., AND PAGÈS, A.M. (1980). Polyradiculonevrite au maleate de perhexiline: etude quantimtrique et ultrastructurale. *Arch. d'Anat. Cytol. Pathol.* 28, 50–3.

PALMLOV, A. AND TUNEVALL, G. (1956). Furadantin as a urinary tract antiseptic. *Svensha Lak. Tidn.* 53, 186–75.

PAMPHLETT, R.S., AND MACKENZIE, R.A. (1982). Severe peripheral neuropathy due to lithium intoxication. *J. Neurol. Neurosurg. Psychiat.* 45, 656.

PATTE, D., LÉGER, F.A., SAVOIE, J.C., MÉNAGE, J.J., SAMSON, Y., NIVET, M., AND GOULON, M. (1983). Thyréotoxicose, puis hypothyroïde par surcharge iodée (amiodarone), associées à une neuropathie. *Ann. Med. Interne.* 134, 31–4.

PENN, R.G. AND GRIFFIN, J.P. (1982). Adverse reactions to nitrofurantoin in the United Kingdom, Sweden and Holland. *Br. med. J.* 1, 1440–2.

PHILLIPS, T.L., WASSERMAN, T.H., JOHNSON, R.J., LEVIN, V.A., AND VAN RAALTE, G. (1981). Final report on the United States Phase I clinical trial of the hypoxic cell radiosensitiser misonidazole. *Cancer* 48, 1697–1704.

POKROY, N., RESS, S., AND GREGORY, M.C. (1977). Clofibrate-induced complications in renal disease. *S. Afr. med. J.* 52, 806–8.

POLLET, S., HAUW, J.J., TURPIN, J.C., LE SAUX, F., ESCOUROLLE, R., AND HAUMANN, N. (1979). Analysis of the major lipid classes in human peripheral nerve biopsies. *J. neurol. Sci.* 41, 199–206.

PRESCOTT, L.F. (1975). Anti-inflammatory analgesics and drugs used in the treatment of rheumatoid arthritis and gout. In *Meyler's side effects of drugs* , Vol. 8 (ed. M.N.G. Dukes), pp. 207–40. Excerpta Medica, Amsterdam.

RAO, K.V.N., MITCHISON, D.A., NAIR, N.G.K., PREMA, K., AND TRIPATHY, S.P. (1970). Sulphadimidine acetylation test for classification of patients as slow or rapid inactivators of isoniazid. *Br. Med. J.* 3, 495–7.

RAPHAELSON, M.I., STEVENS, J.C., AND NEWMAN, R.P. (1983). Vincristine neuropathy with bowel and bladder atony mimicking spinal cord compression. *Cancer treat. Rep.* 67, 604–5.

RATE, R.G., LECHE, J., AND CHERVENAK, C. (1979). Podophyllin toxicity. *Ann. intern. Med.* 90, 723.

REILLY, T.M. (1976). Peripheral neuropathy associated with citrated calcium carbamide. *Lancet.* i, 911–12.

REINSTEIN, L., MAHON, R. JR. AND RUSSO, G.L. (1982). Peripheral neuropathy after concomitant dimethy sulfoxide use and sulindac therapy. *Arch. Phys. Med. Rehabil.* 63, 581–4.

——, OSTROW, S.S., AND WIERNIK, P.H. (1980). Peripheral neuropathy after cis-platinum (II) (DDP) therapy. *Arch. Phys. Med. Rehabil.* 61, 280–2.

REVUZ, J. AND HORNAC, P. (1981). Effets secondaires des traitements par les sulfones. *Larc med.* 1, 68–70.

ROBERTSON, G.L., BHOOPALAM, N., AND ZELKOWITZ, L.J. (1973). Vincristine neurotoxicity and abnormal secretion of antidiuretic hormone. *Arch. intern. Med.* 132,717–20.

ROBINSON, R.J. (1975). Arsenical polyneuropathy due to caustic arsenical paste. *Br. med. J.* 3, 139.

ROELSEN, E. (1964). Polyneuritis after nitrofurantoin therapy. A survey and report of two new cases.*Acta med. scand.* 175, 145–54.

ROESER, H.P., HOCKER, G.A., KYNASTON, B., ROBERTS, S.J., AND WHITAKER, S.V. (1975). Advanced non-Hodgkin's lymphomas: response to treatment with combination chemotherapy and factors influencing prognosis. *Br. J. Haematol.* 30, 233–47.

ROSÉN, I. AND SÖRNÄS, R. (1982). Peripheral motor neuropathy caused by excessive intake of dapsone (*Avlosulfon*). *Arch. Psychiat. Nervenkrankheit.* 232, 63–9.

ROSENTHAL, S. AND KAUFMAN, S. (1974). Vincristine neurotoxicity. *Ann. intern. Med.* 80, 733–7.

ROTHKOFF, L., BIEDNER, B., SHOKAM, K., AND BLUMENTHAL, M. (1979). Optic atrophy after irrigation of the lacrimal ducts with chloramphenicol. *Ann. Ophthalmol.* 11, 105–6.

RUBINSTEIN, C.J. (1964). Peripheral polyneuropathy caused by nitrofurantoin. *J. Am. med. Ass.* 187, 647–9.

—— (1968). Peripheral polyneuropathy caused by nitrofurantoin. In *Drug-induced diseases*, Vol. 3. (ed. L.Meyler and H.M.Peck). Excerpta Medica, Amsterdam.

RUDMAN, D. AND WILLIAMS, P.J. (1983). Megadose vitamins: use and misuse. *New Engl. J. Med.* 309, 488–90.

RUSSELL, J.A. AND POWLES, R.L. (1974). Neuropathy due to cytosine arbinoside. *Br. med. J.* 4, 652–3.

SAAVEDRA, A., MARIATEGUI, J., AND BOGIANO, L. (1960). La imiprámima en los estados depresivos. *Rev. Neuropsiquiat.* 23, 195–228.

SABOUR, M.S., AND FADEL, H.E. (1970). The carpal-tunnel syndrome — a new complication ascribed to the pill. *Am. J. Obstet. Gynecol,* 107, 1265–7.

SAHENK, Z., MENDELL, J.R. COURT, D., AND NACHTMAN, J. (1978). Polyneuropathy from inhalation of nitrous oxide cartridges through a whipped cream dispenser. *Neurol., Minneanpolis.* 28. 485–7.

SAID, G. (1978). Perhexiline neuropathy: a clinicopathological study. *Ann. Neurol.* 3, 259–66.

——, GOASQUEU, J., AND LAVERDANT, C. (1978). Polynevrites au cours des traitements prolongés par le métronidazole. *Rev. neurol.* 134, 515–21.

SAKAMOTO, A. (1974). Physical activity: a possible determinant of vincristine (NSC-67574) neuropathy. *Cancer Chemother. Rep.* 58, 413–15.

SAMUELS, M.L., LEARY, W.B., ALEXANIAN, R., HOWE, C.D., AND FREI, E. (1967). Clinical trails with N-iso-propyl-α-(2-methylhydrazinol-p-toluamide hydrochloride in malignant lymphoma and other disseminated neoplasia. *Cancer, NY* 20, 1187–94.

SANDERSON, P.A., KUWABARA, T., AND COGAN, D.G. (1976). Optic neuropathy presumably caused by vincristine therapy. *Am. J. Ophthalmol.* 81, 146–50.

SANDLER, R.M. AND GONSALKORALE, M. (1977). Chronic lymphatic leukaemia, chlorambucil and sensorimotor peripheral neuropathy. *Br. med. J.* 2, 1265–6.

SANDLER, S.G., TOBIN, W., AND HENDERSON, E.S. (1969). Vincristine induced neuropathy. A clinical study of fifty leukaemic patients. *Neurology, Minneapolis* 19, 367–74.

SAUNDERS, M.I., DISCHE, S., ANDERSON, P., AND FLOCKHART, I.R., (1978). The neurotoxicity of misonidazole and its relationship to dose, half-life and concentration in the serum. *Br. J. Cancer* 37 (Suppl.3), 268–70.

SCHAUMBUG, H., KAPLAN, J., WINDEBANK, A., VICK, N., RASMUS, S., PLEASURE, D., AND BROWN, M.J. (1983). Sensory neuropathy from pyridoxine abuse. *New Engl. J. Med.* 309, 445–8.

—— AND SPENCER, P.S. (1979). Toxic neuropathies. *Neurology, Minneapolis.* 29, 429–31.

SCHIRREN, C.G. (1966). Schwere Allgemeinvergiftung nach ortlicher Anwendung von Podophyllin spiritus bei spitzen

Condylomen. *Hautarzt. Zeitschr. Dermatol., Venerol., verwandte Gebiete* 17, 321-2.

SCHALEPFER, W.W. (1971). Vincristine-induced axonal alterations in rat peripheral nerve. *J. Neuropathol. exp. Neurol.* 30, 488-505.

SCHLENSKA, G.K. (1977). Unusual neurological complications following tetanus toxoid administration. *J. Neurol.* 215, 299-302.

SCHWADE, J.G., MAKUCH, R.W., STRONG, J.M., AND GLATSTEIN, E. (1982). Dose-response curves for predicting misonidazole-induced peripheral neuropathy. *Cancer treat. Rep.* 66, 1743-50.

SÉBILLE. A. AND PARLIER, Y. (1981). Phase précoce de la neuropathie à la vincristine Arguments électrophysiologiques en faveur d'une axonopathie distale rétrograde. *Nouv. Presse Méd.* 10, 2417-20.

—— AND ROZENSTAJN, L. (1978). Prevalence of latent perhexiline neuropathy. *Br. Med. J.* 1, 1321-2.

SELAWRY, O.S. AND HANANIAN, J. (1963). Vincristine treatment of cancer in children. *J. Am. med. Ass.* 183, 741-6.

SELHORST, J.B., KAUFMAN, B., AND HORWITZ, S.J. (1972). Diphenyl hydantoin-induced cerebellar degeneration. *Arch. Neurol.* 27, 453-5.

SHAH, R.R., OATES, N.S., IDLE, J.R., SMITH, R.L., AND LOCKHART, J.D.F. (1982). Impaired oxidation of debrisoquine in patients with perhexiline neuropathy. *Br. med. J.* 1, 295-9.

——, ——, ——, ——, AND —— (1983). Prediction of subclinical perhexiline neuropathy in a patient with inborn error of debrisoquine hydroxylation. *Am. Heart J.* 105, 159-61.

SHARP, J.C.M. AND McDONALD, S. (1967). Effects of rabies vaccine in man. *Br. Med. J.* 3, 20-1.

SHARP, W.L. (1960). Convulsions associated with antidepressant drugs. *Am. J. Psychiat.* 117, 458-9.

SHAW, R.K. AND BRUNER, J.A. (1964). Clinical evaluation of vincristine (NSU 67574). *Cancer Chemother Rep.* 42, 45-8.

SHELANSKI, M.L., AND WISNIEWSKI, M. (1969). Neurofibrillary degeneration induced by vincristine therapy. *Arch. Neurol.* 20. 199-206.

SIDENIUS, P. AND JAKOBSEN, J. (1983). Peripheral neuropathy in rats induced by insulin treatment. *Diabetes.* 32, 383-6.

SLAMOVITS, T.L., BURDE, R.M., AND GLINGELE, T.G. (1980). Bilateral optic atrophy caused by chronic oral ingestion and topical application of hexachlorophene. *Am. J. Ophthalmol.* 89, 676-9.

SLATER, L.M. WAINER, R.A., AND SERPICK, A.A. (1969). Vincristine neurotoxicity with hyponataemia. *Cancer* 23, 122-5.

SLYTER, H., LIWINCZ, B., HERRICK, M.K., AND MASON, R. (1980). Fatal myeloencephalopathy caused by intrathecal vincristine. *Neurol., NY* 30, 867-71.

SMITH, M.S. (1979). Amitriptyline ophthalmoplegia. *Ann. intern. Med.* 91, 793.

SNIDER, D.E. (1980). Pyridoxine supplementation during isoniazid therapy. *Tubercle* 61, 191-6.

SO, E.L. AND PENRY, J.K. (1981). Adverse effects of phenytoin on peripheral nerves and neuromuscular junction: a review. *Epilepsia* 22, 467-73.

SPAULDING, M.B., BARROU, S.S., COMIS, R.L., RICHARDS, F., KENNEDY, B.J., AND PAJAK, T.F. (1982). Severe neurotoxicity with methyl G: CALGB experience. *Med. Pediat. Oncol.* 10, 521-4.

SPECTOR, R.H. AND SCHNAPPER, R. (1981). Amitriptyline-induced ophthalmoplegia. *Neurology, NY* 31, 1188-90.

SPENCER, P.S. AND SCHAUMBURG, H.H. (1980). Recent morphological studies of toxic neuropathy. *Dev. Toxicol. environ. Sci.* 8, 3-11.

STACPOOLE, P.W., MOORE, G.W., AND KORNHAUSER, D.M. (1979). Toxicity of chronic dichloroacetate. *New Engl. J. Med.* 300, 372.

STOLINSKY, D.C., BOGDON, D.L., SOLOMON, J., AND BATEMAN, J.R. (1972). Hexamethylmelamine (NSC 18375) alone and in combination with 5-(3,3-dimethyltrizeno) imidazole-4-carboxamide (NSC 45388) in the treatment of advanced cancer. *Cancer* 30, 654-9.

STUART-HARRIS, R., PONDER, B.A.J., AND WRIGLEY, P.F.M. (1980). Tetany associated with cis-platin, *Lancet* ii, 1303.

SWIFT, T.R., GROSS, J.A., WARD, C., AND CROUT, B.O. (1981). Peripheral neuropathy in epileptic patients. *Neurology, NY* 31, 826-31.

TAKEUCHI, H., TAKAHASHI, M., KANG, J., UENO, S., TARUL, S., NAKAO, Y., AND OTORI, T. (1980a). Ethambutol neuropathy: clinical and electroneuromyographic studies. *Folia Psychiat. Neurol. Jap.* 34, 45-55.

——, ——, TARUI, S., SANAGI, S., AND TAKENAKA, H. (1980b). Peripheral nerve conduction function in patients treated with antituberculotic agents, with special reference to ethambutol and isoniazid. *Folia Psychiat. Neurol. Jap.* 34, 57-64.

THANT, M., HAWLEY, R.J., SMITH, M.T., COHEN, M.H., MINNA, J.D., BUNN, P.A., IHDE, D.C., WEST, W., AND MATTHEWS, M.J. (1982). Possible enhancement of vincristine neuropathy by VP-16 *Cancer* 49, 859-64.

THYLEFORS, B. AND ROLLAND, A. (1979). The risk of optic atrophy following suramin treatment of ocular onchocerciasis. *Bull. WHO* 57, 479-80.

TINDALL, R.S. AND WILLERSON, J. (1978). Subacute phenytoin intoxication syndrome. *Arch. intern. Med.* 138, 1168-9

TOBIN, W. AND SANDLER, S.G. (1968). Neurophysiologic alterations induced by vincristine (NSC 67574). *Cancer Chemother. Rep.* 52, 519-26.

TOGHILL, P.J. AND BURKE, J.D. (1970). Death from paralytic ileus following vincristine therapy. *Postgrad. med. J.* 46, 330-1.

TOOLE, J.F. AND PARRISH, M.L. (1973). Nitrofurantoin polyneuropathy. *Neurology, Minneapolis* 23, 554-9.

TSUIJIMOTO, G., HORAI, Y., ISHIZAKI, T., AND ITOH, K. (1981). Hydralazine-induced peripheral neuropathy seen in a Japanese slow acetylator patient. *Br. J. clin. Pharmacol.* 11, 622-5.

TUBERCULOSIS CHEMOTHERARY CENTRE, MADRAS (1963). The prevention and treatment of isoniazid toxicity in the therapy of pulmonary tuberculosis. *Bull. WHO* 29, 457.

TURPIN, J.C., PLUOT, M., ALBOUZ, S., BAJOLET, A., CAULET, T., AND BAUMANN, N. (1983). Etude d'une thésaurismose induite par le maléate de perhexiline. Confirmation des données expérimentales. *Sem. Hôp. Paris.* 59, 58-61.

TZORTZATOU, F., DACOU-VOUTETAKIS, C., HAIDAS, S., PAPADELLIS, F., AND THOMAIDIS, (1979). Electrolyte abnormalities in lmyphosarcoma after chemotherapy. *Acta paediat. scand.* 68, 621-3,

UESU, C.T. (1962). Peripheral neuropathy due to nitrofurantoin. *Ohio State med. J.* 58, 53-6.

URTASUN, R.C., CHAPMAN, J.D., FELDSTEIN, M.L., BAND, R.P., RABIN, H.R., WILSON, A.F., MARYNOWSKI, B. STARREFELD, E., AND SHNITKA, T. (1978). Peripheral neuropathy related to misonidazole: incidence and pathology. *Br. J. Cancer* 37, (Suppl.3), 271-5.

UY, Q.L., MOEK, T.H., JOHNS, R.J., AND OWENS, A.H. (1967). Vincristine neurotoxicity in rodents. *Johns Hopk. med. J.* 121, 349.

VALDIVIESO, M., BEDIKIAN, A.Y. BOEDY, G.P., AND FREIREICH, E.J. (1981*b*). Broad Phase II study of vindesine. *Cancer treat. Rep.* 65, 877-9.

——, RICHMAN, S., BURGESS, A.M., BODEY, G.P., AND FREIREICH, E.J. (1981*a*). Initial clinical studies of vindesine. *Cancer treat. Rep.* 65, 873-5.

VALLAT, J.M., VITAL, L., VALLAT, M., JULIENS, J., RIEMENS, V., AND LE BLANC. M (1973). Neuropathie périphérique à la vincristine. Etude ultrastructurale d'une biopsie du muscle et du nerf périphérique. *Rev. neurol., Paris* 129, 365-8.

VAN DER GRIENT, A.J. (1968). Antibacterial drugs. In *Side-effects of drugs.* Vol. VI (ed. L. Meyler and A. Herxheimer) Excerpta Medica Foundation. Amsterdam.

VAN ECHO, D.A., CHIUTEN, D.F., GORMLEY, P.E., LICHTENFELD, J.L. SCOLTOCK, M., AND WIERNIK, P.H. (1979). Phase I clinical and pharmacological study of 4'-(9-acridinylamino)-methane-sulfon-m-anisidide using an intermittent biweekly schedule. *Cancer Res.* 39, 3881-4.

VERGARA, I., TORO, G., ROMAN, G., AND MENDOZA, G. (1979). Fatal Guillain-Barré syndrome with reduced-dose antirabies vaccination. *Arch. Neurol.* 36, 254.

VICKERS, F.J. (1965). Peripheral neuropathy due to nitrofurantoin. *J. Kv. med. Ass.* 63, 38-40.

VOLDEN, G. (1977). Peripheral neuropathy: a side-effect of sulphones. *Br. med. J.* 2, 1193.

WALKER, B.K., RAICH, P.C., FONTANA, J., SUBRAMANIAN, V.P., ROGERS, J.S. 2ND., KNOST, J.A. AND DENNING, B. (1982). Phase II study of vindesine in patients with advanced breast cancer. *Cancer treat Rep.* 66, 1729-32.

WALKER, M.D. AND STRIKE, T.A. (1980). Misonidazole peripheral neuropathy. *Cancer clin. Trials* 3, 105-9.

WALLACE, D.C. (1979). Perhexiline Maleate. *Med. J. Aust* 2, 37.

WALLS, T. PEARCE, S.J., AND VENABLES. G.S. (1980). Motor neuropathy associated with cimetidine. *Br. Med. J.* 2, 974-5.

WARD, J.W., CLIFFORD, W.S., MONACO, R., AND BICKERSTAFF, H.J. (1954). Fatal systemic posoning following podophyllin treatment of condyiomata acuminata. *Southern med. J.* 47, 1204-6.

WAROT, P., GOLDEMAND, M., AND HABAY, D. (1965). Troubles neurologiques provoqués par les alkaloides de Vinca rosea. *Rev. Neurol., Paris.* 113, 464-7.

WARREN, G. (1980). Dapsone and nerve damage. *Leprosy Rev.* 51, 94-7.

WASSERMAN, T.H., PHILLIPS, T.L., JOHNSON, R.J., GOMER, C.J., LAWRENCE, G.A., SADEE, W., MARQUES, R.A. LEVIN, V.A, AND VAN RAALTE, G. (1979). Initial United States clinical and pharmacologic evaluation of misonidazole (Ro-07-0582): a hypoxic cell radiosensitizer. *Int. J. Radiat. Oncol. biol. Phys.* 5, 775-86.

——, ——, VAN RAALTE, G., URTASUN, R., PARTINGTON, J., KOZIOL, D., SCHWADE, J.G., GANGJI, D., AND STRONG, J.M., (1980). The neurotoxicity of misonidazole: potential modifying role of phenytoin sodium and dexamethasone. *Br. J. Radiol.* 53, 172-3.

WATKINS, S.M. AND GRIFFIN, J.P. (1978). High incidence of vincristine neurotoxicity in lymphomas. *Br. med. J.* 1, 610-12.

WATSON, A.J., McCANN, S.R., AND TEMPERLEY, I.J. (1981). Tetany following aminoglycoside therapy. *Ir. J. med. Sci.* 150, 316-17.

WATSON, C.P., ASHBY, P, AND BILBAO, J.M. (1980). Disulfiram neuropathy. *Can. med. Ass. J.* 123, 123-6.

WAXMAR, H.L., GROH, W.C., MARCHLINSKI, E.E., BUXTON, A.E. SADOWSKI, L.M., HOROWITZ, L.N., JOSEPHSON, M.E., AND KASTER, J.A. (1982). Amiodarone for control of sustained ventricular tachyarrhythmias: clinical and electrophysiologic effects in 51 patients. *Am. J. Cardiol.* 50, 1066-74.

WEIDEN, P.L. AND WRIGHT, S.E. (1972). Vincristine neurotoxicity. *New Engl. J. Med.* 286, 1369-70.

WEISS, M.D., WALKER, M.D, AND WIERNIK, P.H. (1974). Neurotoxicity of commonly used antineoplastic agents (second of two parts) *New Engl. J. Med.* 291, 127-33.

WELLS, J. AND TEMPLETON, J. (1977). Femoral neuropathy associated with anticoagulant therapy. *Clin. Orthop.* 124, 155-6.

WESTBURY, G. (1962). Treatment of advanced cancer by extracorporeal perfusion and continuous intra-arterial infusion. *Proc. R. Soc. Med.* 55, 643-6.

WHARTON, J.T., RUTLEDGE, F., SMITH, J.P., HERSON, J., AND HODGE, M.P. (1979). Hexamethylmelamine: an evaluation of its role in the treatment of ovarian cancer. *Am. J. Obstet. Gynecol.* 133, 833-44.

WHISNANT, J.P. ESPINOSA, R.E., KIERLAND, R.R., AND LAMBERT, E.H. (1963). Chloroquine neuromyopathy. *Proc. Staff Meet. Mayo Clinic* 38, 501-13.

WHITELAW, D.M. COWAN, D.H., CASSIDY, F.R., AND PATTERSON, T.A. (1963). Vincristine. *Cancer Chemother. Rep.* 30, 13.

WHITTAKER, J.A. AND GRIFFITH, I.P. (1977). Recurrent laryngeal nerve paralysis in patients receiving vincristine and vinblastine. *Br. med. J.* 1, 1251-2.

WIJESEKERA, J.C., CRITCHLERY, E.M.R., FAHIM, Y., LYNCH, P.G., AND WRIGHT, J.S. (1980). Peripheral neuropathy due to perhexilene maleate. *J. neurol. Sci.* 46, 303-9.

WILLETT, R.W. (1963). Peripheral neuropathy due to nitrofurantoin, *Neurology, Minneapolis* 13, 344-5.

WILSON, R.S. TRAUNFELDER, F.T., AND LANDERS, F.H. (1979). The National Registry of drug-induced ocular side effects and toxic drug effects on the retina. *J. Arkansas med. Soc.* 76, 159-63.

WINDMILLER, J., BERRY, D.H., HADDY, T.B., VIETTI, T.J., AND SUTOW, W.W. (1966). Vincristine sulfate in the treatment of neuroblastoma in children.*Am. J. Dis. Child.* 3, 75-8.

WIRTH. G. (1965). Reversible Kochlearisschadigung nach Tetanol Injektion? *Münch. med. Wschr.* 107, 379-81.

WISNIEWSKI, H., TERRY, R.D., AND MIRANO, A. (1970). Neurofibrillary pathology, *J. Neuropath. exp. Neurol* 29, 163-76.

WOOLLING, K.R. AND RUSHTON, J.G. (1950). Serum Neuritis: report of two cases and a brief review of the syndrome *Arch. Neurol. Psychiat.* 64, 568-73.

WORNER, T.M. (1982). Peripheral neuropathy after disulfiram administration: reversibiltiy despite continued therapy. *Drug Alcohol Dependence* 10, 199-201.

WYATT, E.H. AND STEVENS, J. (1972). Dapsone-induced peripheral neuropathy. *Br. J. Dermat.* 86, 521-3.

YAP, H., Y., BLUMENSCHEIN, G.R., LASHIMA, C.K., HORTOBAGYI, G.N., BUZDAR, A.U., AND WINEMAN, C.L. (1979). Combination chemotherapy with vincristine and methotrexate for advanced breast cancer. *Cancer, NY* 44, 32-4.

YIANNIKAS, C., POLLARD, J.D., AND McLEOD, J.G. (1981). Nitrofurantoin neuropathy. *Aust. N.Z. J. Med.* 11, 400-5.

YOSHIKAWA, T.T. AND NAGAMI, P.H. (1982). Adverse drug reactions in TB therapy: risks and recommendations. *Geriatrics* 37, 61-8.

YOU, B. AND GRILLIAT, J.P. (1981). La neuropathie au metronidazole. *Nouv. Presse. Méd.* 10, 708.

27 Drug-induced disorders of central nervous function

RONALD D. MANN.

Information on the side-effects known to be caused by a particular drug is readily available. In an unquantified way it is given in the Data Sheet of prescription medicines. This information would be more useful if it were quantified, ranking the side-effects by the frequency with which they are reported. But, at all events, the available information assists the clinician who has prescribed a particular drug — indicating what adverse effects to look out for and what laboratory tests to order in an attempt to detect difficulties at an early stage.

Information of this type does not help diagnose the obscure clinical picture. There, the symptoms and signs come first and one wants to know, for each symptom, what drugs must enter the differential diagnosis, and in what order. In respect to drug-related disorders of the central nervous system the first part of this chapter presents tables for drugs-within-reactions as shown in the data reported to the Committee on Safety of Medicines between 1964 and the present date.

The master tables from which Tables 27.1 and 27.2 were derived have certain characteristics. Only the suspected drug is included; if the patient was having more than one drug, then a specialist medical assessor has decided which is the drug most likely to have caused the reported side-effect. In a similar way, only the most important adverse reaction is included. Reports are discarded if the most important reaction started before the suspect drug was commenced. Deaths considered to be due to the drug are included in both the 'reports' and 'deaths' columns. Data for 1964–78 are given in one columns then those for the separate years 1979 to 1984

Table 27.1   The 50 most frequently reported central nervous system drug reactions

Side-effect	1964–84	
	Reactions	Deaths
Headache	2 789	2
Vomiting	1 853	10
Dizziness	1 600	0
Convulsions	1 025	19
Somnolence	1 000	0
Paraesthesia	991	0

Side-effect	1964–84	
	Reactions	Deaths
Syncope	955	15
Extrapyramidal disorder	886	7
Confusion	724	4
Vertigo	684	0
Tremor	651	1
Convulsions, Grand Mal	607	20
Deafness	519	0
Neuropathy	508	11
Ataxia	446	0
Migraine	407	0
Cerebrovascular disorder	344	50
Tinnitus	301	0
Dystonia	278	0
Diplopia	239	0
Thrombosis, cerebral	231	46
Oculogyric crisis	213	1
Twitching	205	3
Sensory disturbance	188	0
Encephalopathy	187	33
Cerebral haemorrhage	183	157
Cramp, legs	159	0
Respiratory depression	130	24
Coma	129	16
Neuritis	126	1
Amnesia	118	0
Hypoaesthesia	116	0
Speech disorder	108	0
Hemiparesis	102	2
Hemiplegia	100	2
Stupor	99	1
Dyskinesia	99	0
Co-ordination, abnormal	88	0
Meningism	83	0
Cerebrovascular accident	81	16
Taste loss	76	0
Choreoathetosis	70	0
Migraine, worsened	70	0
Hyperkinesia	68	0
Photophobia	67	0
Thrombosis, cerebral, arterial	65	14
Paresis	62	0
Cardiorespiratory depression	61	41
Taste perversion	61	0
Deafness, nerve	60	0

Table 27.2   Drugs within reactions

Side-effect (n)	Drug	1964–78 Reports	1964–78 Deaths	1983 Reports	1983 Deaths	1964–84 Reports	1964–84 Deaths
Akathisia (17)	Metoclopramide	5	0	0	0	12	0
	Mianserin	0	0	1	0	10	0
	Totals for reaction	6	0	5	0	42	0
Amnesia (60)	Indomethacin	14	0	1	0	19	0
	Totals for reaction	64	0	12	0	118	0
Ataxia (147)	Flurazepam	8	0	0	0	10	0
	Phenytoin	14	0	0	0	14	0
	Diazepam	11	0	0	0	13	0
	Lorazepam	10	0	0	0	15	0
	Mianserin	5	0	1	0	15	0
	Fenfluramine	10	0	0	0	13	0
	Indomethacin	8	0	4	0	19	0
	Carbamazepine	13	0	1	0	25	0
	Perhexiline	20	0	1	0	23	0
	Cimetidine	3	0	0	0	18	0
	Totals for reaction	259	0	33	0	446	0
Blindness (32)	Practolol	15	0	0	0	15	0
	Totals for reaction	43	0	4	0	59	0
Cerebral haemorrhage (51)	Chlorpromazine	2	2	0	0	2	2
	Trifluoperazine	6	4	0	0	9	6
	Phenelzine	3	3	0	0	3	3
	Phenylbutazone	2	2	0	0	3	3
	Oxyphenbutazone	1	1	0	0	2	2
	Sodium aurothiomalate, other gold compounds	3	3	0	0	3	3
	Lignocaine	2	2	0	0	2	2
	Warfarin	20	19	0	0	25	24
	Phenindione	7	7	0	0	7	7
	Anticoagulant NOS	54	53	0	0	56	55
	Cimetidine	0	0	1	1	3	3
	Ethinyloestradiol	13	5	0	0	14	5
	Mestranol	5	3	0	0	5	3
	Corticosteroids NOS	9	9	0	0	9	9
	Totals for reaction	161	141	2	2	183	157
Cerebrovascular accident (41)	Warfarin	2	2	0	0	2	2
	Anticoagulant NOS	1	1	0	0	2	2
	Ethinyloestradiol	0	0	7	0	24	0
	Totals for reaction	5	4	22	6	81	16
Cerebrovascular disorder (65)	Phenylbutazone	2	2	0	0	2	2
	Indomethacin	5	2	0	0	5	2
	Clonidine	2	2	0	0	2	2
	Phenindione	2	2	0	0	2	2
	Anticoagulant NOS	2	2	0	0	2	2
	Ethinyloestradiol	137	11	2	0	149	11
	Mestranol	74	2	0	0	74	2

(continued overleaf)

Table 27.2 *Continued*

Side-effect (n)	Drug	1964–78		1983		1964–84	
		Reports	Deaths	Reports	Deaths	Reports	Deaths
	Oestrogens NOS	16	1	0	0	16	1
	Corticosteroids NOS	11	11	0	0	11	11
	Totals for reaction	323	50	2	0	344	50
Choreoathetosis (34)	Ethinyloestradiol	9	0	1	0	11	0
	Mestranol	10	0	0	0	10	0
	Totals for reaction	48	0	2	0	70	0
Coma (73)	Methaqualone	13	0	0	0	13	0
	Totals for reaction	111	13	3	0	129	16
Confusion (240)	Methaqualone	11	0	0	0	11	0
	Triazolam	0	0	2	0	15	0
	Mianserin	7	0	4	0	15	0
	Pentazocine	12	0	0	0	13	0
	Indomethacin	26	0	11	0	47	0
	Carbamazepine	8	0	1	0	12	0
	L-dopa	9	1	0	0	10	1
	Propranolol	11	0	1	0	12	0
	Atenolol	7	0	0	0	12	0
	Cimetidine	29	1	4	0	65	1
	Ranitidine hydrochloride	0	0	11	0	16	0
	Nalidixic acid	7	0	0	0	10	0
	Totals for reaction	417	3	93	1	724	4
Convulsions (198)	Mianserin	2	0	3	0	18	0
	Clomipramine	3	1	0	0	10	1
	Maprotiline	10	0	0	0	20	0
	Bupivacaine	1	0	6	0	13	1
	Measles vaccine (live attenuated)	304	0	6	0	362	1
	Pertussis vaccine	122	4	4	0	190	4
	Diphtheria vaccine	58	0	2	0	81	1
	Totals for reaction	660	8	71	0	1 025	19
Convulsions, grand mal (180)	Phenytoin	2	2	0	0	2	2
	Imipramine	8	2	0	0	11	2
	Amitriptyline	14	0	0	0	19	0
	Mianserin	11	0	3	0	39	0
	Dothiepin	5	0	1	0	10	0
	Maprotiline	85	0	0	0	102	0
	Zimelidine hydrochloride	0	0	5	0	16	1
	Bupivacaine	1	0	1	0	12	0
	Ethinyloestradiol	9	1	1	0	10	1
	Measles vaccine (live attenuated)	22	1	0	0	24	1
	Pertussis vaccine	12	0	0	0	15	0
	Totals for reaction	393	15	38	1	607	20
Deafness (115)	Salsalate	26	0	0	0	30	0
	Benorylate	66	0	4	0	104	0

Table 27.2   Drugs within reactions

Side-effect (n)	Drug	1964–78 Reports	Deaths	1983 Reports	Deaths	1964–84 Reports	Deaths
	Practolol	114	0	0	0	160	0
	Frusemide	13	0	0	0	13	0
	Neomycin	14	1	0	0	15	1
	Totals for reaction	344	1	21	0	519	1
Deafness, nerve (33)	Practolol	9	0	0	0	19	0
	Totals for reaction	39	0	1	0	60	0
Diplopia (119)	Lorazepam	6	0	1	0	14	0
	Carbamazepine	6	0	1	0	12	0
	Totals for reaction	147	0	16	0	239	0
Dizziness (278)	Mianserin	11	0	4	0	30	0
	Fenfluramine	24	0	2	0	29	0
	Mazindol tablets	14	0	0	0	15	0
	Buprenorphine	0	0	15	0	46	0
	Dihydrocodeine tartrate	14	0	2	0	18	0
	Pentazocine	22	0	0	0	26	0
	Dextropropoxyphene	25	0	1	0	31	0
	Piroxicam	0	0	1	0	13	0
	Diflunisal	6	0	1	0	11	0
	Paracetamol	10	0	1	0	12	0
	Fenbufen	0	0	1	0	26	0
	Benoxaprofen	0	0	0	0	12	0
	Flurbiprofen	14	0	0	0	19	0
	Ibuprofen	8	0	2	0	13	0
	Diclofenac sodium	0	0	3	0	15	0
	Naproxen	6	0	2	0	12	0
	Feprazone	8	0	0	0	13	0
	Indomethacin	41	0	32	0	106	0
	Azapropazone	5	0	0	0	10	0
	Carbamazepine	14	0	0	0	16	0
	Prazosin	69	0	0	0	72	0
	Indapamide hemihydrate	2	0	0	0	14	0
	Propranolol	6	0	1	0	14	0
	Atenolol	65	0	3	0	87	0
	Oxprenolol	11	0	1	0	15	0
	Nadolol	0	0	0	0	10	0
	Perhexilene	15	0	0	0	19	0
	Nifedipine	37	0	6	0	63	0
	Cimetidine	36	0	2	0	68	0
	Ranitidine hydrochloride	0	0	5	0	11	0
	Ketotifen	0	0	1	0	11	0
	Ethinyloestradiol	7	0	0	0	11	0
	Hydrochlorothiazide	7	0	1	0	17	0
	Bendrofluazide	3	0	2	0	11	0
	Chlorthalidone	2	0	4	0	21	0
	Clopamide	10	0	0	0	56	0
	Minocycline hydrochloride	11	0	1	0	17	0
	Co-trimoxazole	10	0	1	0	12	0
	Nalidixic acid	27	0	0	0	30	0

(*continued overleaf*)

Table 27.2   Drugs within reactions

Side-effect (n)	Drug	1964–78		1983		1964–84	
		Reports	Deaths	Reports	Deaths	Reports	Deaths
	Metronidazole	6	0	4	0	17	0
	Totals for reaction	869	0	154	0	1 600	0
Dyschromatopsia (20)	Nalidixic acid	9	0	0	0	13	0
	Totals for reaction	24	0	1	0	39	0
Dyskinesia (39)	Chlorpromazine	10	0	0	0	10	0
	Metoclopramide	2	0	1	0	15	0
	Totals for reaction	54	0	10	0	99	0
Dystonia (49)	Prochlorperazine	23	0	1	0	30	0
	Perphenazine	31	0	1	0	33	0
	Trifluoperazine	18	0	1	0	20	0
	Thiethylperazine	19	0	1	0	20	0
	Haloperidol	20	0	0	0	25	0
	Metoclopramide	37	0	15	0	80	0
	Totals for reaction	190	0	22	0	278	0
Embolism, cerebral (10)	Ethinyloestradiol	13	5	0	0	13	5
	Mestranol	6	2	0	0	6	2
	Totals for reaction	27	8	1	0	28	8
Encephalopathy (50)	Penicillin NOS	2	2	0	0	2	2
	Measles vaccine (live-attenuated)	45	8	1	0	49	8
	Pertussis vaccine	30	5	0	0	35	5
	Smallpox vaccine	17	7	0	0	17	7
	Diphtheria vaccine	11	2	0	0	11	2
	Totals for reaction	152	31	5	1	187	33
Extrapyramidal disorder (134)	Prochlorperazine	49	0	3	0	62	0
	Perphenazine	46	0	0	0	50	0
	Trifluoperazine	19	0	0	0	22	0
	Thiethylperazine	7	0	0	0	10	0
	Fluphenazine	56	1	0	0	62	2
	Flupenthixol dihydrochloride	13	0	6	0	26	0
	Thioridazine	9	0	0	0	10	0
	Haloperidol	45	1	6	2	53	3
	Metoclopramide	246	1	9	0	329	1
	Diazoxide	12	0	0	0	12	0
	Totals for reaction	669	3	37	2	886	7
Headache (479)	Chlormethiazole	15	0	3	0	23	0
	Temazepam	6	0	2	0	12	0
	Triazolam	0	0	2	0	10	0
	Trifluoperazine	9	0	0	0	10	0
	Phenelzine	9	0	0	0	11	0
	Mianserin	7	0	3	0	32	0
	Nomifensine hydrogen maleate	12	0	3	0	26	0

Table 27.2   *Continued*

Side-effect (*n*)	Drug	1964–78		1983		1964–84	
		Reports	Deaths	Reports	Deaths	Reports	Deaths
	Trazodone	0	0	7	0	19	0
	Viloxazine						
	hydrochloride	20	0	0	0	21	0
	Zimelidine						
	hydrochloride	0	0	20	0	42	0
	Fenfluramine	18	0	0	0	18	0
	Buprenorphine	0	0	7	0	20	0
	Piroxicam	0	0	4	0	19	0
	Diflunisal	11	0	0	0	21	0
	Fenbufen	0	0	3	0	24	0
	Benoxaprofen	0	0	0	0	27	0
	Flurbiprofen	10	0	1	0	19	0
	Ibuprofen	8	0	3	0	13	0
	Diclofenac sodium	0	0	0	0	34	0
	Naproxen	4	0	2	0	12	0
	Feprazone	8	0	0	0	10	0
	Indomethacin	76	0	124	0	246	0
	Sulindac	7	0	0	0	10	0
	Lignocaine	31	0	0	0	34	0
	Clonidine	8	0	0	0	15	0
	Hydrallazine	9	0	3	0	27	0
	Indapamide						
	hemihydrate	3	0	3	0	20	0
	Labetalol	6	0	1	0	11	0
	Propranolol	19	0	0	0	22	0
	Atenolol	49	0	8	0	77	0
	Oxprenolol	16	0	0	0	24	0
	Metoprolol tartrate	5	0	1	0	10	0
	Glyceryl trinitrate	3	0	18	0	30	0
	Isosorbide						
	mononitrate	0	0	11	0	15	0
	Sorbide nitrate	6	0	4	0	28	0
	Dipyridamole	2	0	5	0	19	0
	Perhexiline	13	0	0	0	15	0
	Nifedipine	59	0	24	0	137	0
	Terfenadine	0	0	14	0	19	0
	Cimetidine	70	0	9	0	116	0
	Ranitidine						
	hydrochloride	0	0	7	0	21	0
	Ketotifen	0	0	0	0	16	0
	Ethinyloestradiol	149	0	5	0	192	0
	Mestranol	95	0	0	0	95	0
	Beclomethasone	5	0	1	0	10	0
	Human antihaemo-						
	philiac fraction	0	0	0	0	11	0
	Hydrochlorothiazide	7	0	2	0	18	0
	Clopamide	1	0	0	0	20	0
	Amoxycillin	6	0	3	0	13	0
	Tetracycline	12	0	0	0	16	0
	Minocycline						
	hydrochloride	2	0	2	0	10	0
	Chlorthalidone	5	0	2	0	24	0
	Oxytetracycline	7	0	0	0	11	0
	Co-trimoxazole	27	0	3	0	35	0
	Nitrofurantoin	5	0	0	0	12	0
	Nalidixic acid	17	0	0	0	20	0
	Metronidazole	17	0	1	0	31	0

(continued overleaf)

Table 27.2   Drugs within reactions

Side-effect (n)	Drug	1964–78		1983		1964–84	
		Reports	Deaths	Reports	Deaths	Reports	Deaths
	Griseofulvin	15	0	0	0	21	0
	Ketoconazole	0	0	7	0	19	0
	Influenza vaccine (inactivated)	9	0	1	0	12	0
	Carbenoxolone	11	0	0	0	11	0
	Danazol	3	0	4	0	17	0
	Totals for reaction	1 376	2	425	0	2 789	2
Hemiparesis (48)	Ethinyloestradiol	25	0	1	0	35	0
	Totals for reaction	63	1	5	0	102	2
Hemiplegia (39)	Ethinyloestradiol	23	0	0	0	36	0
	Mestranol	15	0	0	0	15	0
	Totals for reaction	70	0	3	0	100	2
Hypertension, intracranial (24)	Nalidixic acid	8	1	1	0	10	1
	Totals for reaction	33	4	1	0	48	4
Meningism (28)	Iophendylate	33	0	0	0	33	0
	Totals for reaction	57	0	6	0	83	0
Meningitis (11)	Corticosteroids NOS	3	3	0	0	4	4
	Iophendylate	15	1	0	0	16	1
	Totals for reaction	28	4	1	0	37	6
Migraine (117)	Indomethacin	3	0	3	0	12	0
	Cimetidine	13	0	1	0	21	0
	Ethinyloestradiol	91	0	3	0	112	0
	Mestranol	56	0	1	0	58	0
	Totals for reaction	234	0	53	0	407	0
Migraine, worsened (40)	Ethinyloestradiol	12	0	0	0	15	0
	Totals for reaction	40	0	6	0	70	0
Myopathy (29)	Corticosteroids NOS	8	5	0	0	8	5
	Totals for reaction	28	6	6	1	54	7
Myositis (12)	D-penicillamine	1	1	0	0	2	2
	Totals for reaction	6	1	2	0	15	2
Neuritis (69)	Nitrofurantoin	12	0	1	0	14	0
	Totals for reaction	102	1	2	0	126	1
Neuropathy (151)	Zimelidine hydrochloride	0	0	14	0	15	0
	Indomethacin	3	0	2	0	10	0
	Amiodarone	0	0	5	0	13	0
	Perhexiline	87	3	3	0	117	4
	Cimetidine	9	0	1	0	35	1
	Nitrofurantoin	27	0	1	0	32	0

Table 27.2   Drugs within reactions

Side-effect (*n*)	Drug	1964–78		1983		1964–84	
		Reports	Deaths	Reports	Deaths	Reports	Deaths
	Metronidazole	8	0	1	0	16	0
	Vincristine	7	0	2	0	10	0
	Totals for reaction	302	6	50	2	508	11
Oculogyric crisis (38)	Prochlorperazine	11	0	1	0	22	0
	Perphenazine	16	0	0	0	22	0
	Trifluoperazine	15	0	0	0	16	0
	Metoclopramide	46	0	6	0	90	0
	Totals for reaction	121	0	16	1	213	1
Optic neuritis (23)	Ethambutol	11	0	0	0	12	0
	Totals for reaction	32	0	2	0	43	0
Paraesthesia (233)	Methaqualone	24	0	0	0	24	0
	Triazolam	0	0	2	0	18	0
	Sulthiame	8	0	0	0	10	0
	Amitriptyline	9	0	0	0	10	0
	Mianserin	4	0	5	0	31	0
	Nomifensine hydrogen maleate	6	0	0	0	11	0
	Piroxicam	0	0	2	0	12	0
	Benoxaprofen	0	0	0	0	15	0
	Diclofenac sodium	0	0	2	0	14	0
	Indomethacin	10	0	2	0	14	0
	Clonidine	9	0	1	0	16	0
	Methyldopa	7	0	0	0	11	0
	Labetalol	22	0	1	0	38	0
	Propranolol	21	0	0	0	31	0
	Atenolol	19	0	1	0	31	0
	Oxprenolol	10	0	1	0	15	0
	Perhexilene	19	0	0	0	23	0
	Nifedipine	6	0	3	0	24	0
	Cimetidine	19	0	2	0	43	0
	Ethinyloestradiol	39	0	1	0	52	0
	Mestranol	21	0	0	0	22	0
	Hydrocortisone	18	0	0	0	25	0
	Acetazolamide	6	0	0	0	10	0
	Co-trimoxazole	8	0	0	0	10	0
	Nitrofurantoin	7	0	2	0	11	0
	Metronidazole	4	0	3	0	10	0
	Totals for reaction	515	0	93	0	991	0
Photophobia (28)	Nalidixic acid	30	0	2	0	33	0
	Totals for reaction	50	0	4	0	67	0
Respiratory arrest (30)	Disopyramide	3	1	0	0	6	2
	Totals for reaction	14	4	6	1	44	12
Respiratory depression (56)	Diazepam	2	0	1	0	11	2
	Buprenorphine	6	1	8	0	24	1
	Dihydrocodeine tartrate	0	0	0	0	6	2

(continued overleaf)

Table 27.2   *Continued*

Side-effect (n)	Drug	1964–78		1983		1964–84	
		Reports	Deaths	Reports	Deaths	Reports	Deaths
	Dextromoramide	1	0	1	0	4	2
	Pentazocine	5	2	0	0	5	2
	Totals for reaction	65	14	17	3	130	24
Sensory disturbance (102)	Cimetidine	9	0	0	0	10	0
	Totals for reaction	142	0	9	0	188	0
Somnolence (232)	Lorazepam	10	0	3	0	15	0
	Metoclopramide	10	0	3	0	21	0
	Amitriptyline	10	0	0	0	10	0
	Mianserin	32	0	4	0	60	0
	Fenfluramine	19	0	1	0	23	0
	Piroxicam	0	0	4	0	22	0
	Diflunisal	6	0	0	0	10	0
	Fenbufen	0	0	7	0	13	0
	Indomethacin	8	0	9	0	22	0
	Clonidine	6	0	1	0	11	0
	Indoramin	0	0	6	0	20	0
	Atenolol	12	0	2	0	17	0
	Terfenadine	0	0	13	0	31	0
	Cimetidine	20	0	4	0	40	0
	Ketotifen	0	0	2	0	108	0
	Pizotifen hydrochloride	6	0	1	0	13	0
	Metronidazole	7	0	1	0	12	0
	Pertussis vaccine	7	0	0	0	11	0
	Totals for reaction	462	0	119	0	1 000	0
Subarachnoid haemorrhage (20)	Phenindione	2	2	0	0	2	2
	Ethinyloestradiol	9	1	0	0	13	2
	Corticosteroids NOS	3	3	0	0	3	3
	Totals for reaction	36	11	1	0	43	12
Syncope (287)	Methaqualone	13	0	0	0	13	0
	Mianserin	7	0	8	0	36	0
	Buprenorphine	0	0	5	0	13	0
	Dihydrocodeine tartrate	14	0	0	0	16	0
	Pentazocine	12	0	1	0	15	0
	Dextropropoxyphene	20	0	0	0	23	0
	Indomethacin	19	0	5	0	30	0
	Lignocaine	11	0	0	0	16	0
	Prilocaine	8	0	1	0	10	0
	Prazosin	68	0	0	0	76	0
	Atenolol	10	0	1	0	13	0
	Nifedipine	2	0	3	0	11	0
	Heparin	18	2	0	0	18	2
	Cimetidine	7	0	0	0	11	0
	Ethinyloestradiol	14	0	0	0	16	0
	Cyproterone	1	1	1	1	2	2
	Bromocriptine	4	0	1	0	11	0
	Neomycin	8	0	2	0	10	0
	Nalidixic acid	29	0	0	0	29	0
	Pertussis vaccine	11	0	0	0	14	0

Table 27.2   Drugs within reactions

Side-effect (*n*)	Drug	1964–78		1983		1964–84	
		Reports	Deaths	Reports	Deaths	Reports	Deaths
	Salmonella typhi	1	0	6	0	13	0
	Diphtheria vaccine	5	0	1	0	12	0
	Totals for reaction	647	12	86	2	955	15
Taste, abnormal (31)	Flurazepam	12	0	0	0	15	0
	Totals for reaction	27	0	13	0	58	0
Taste loss (44)	D-penicillamine	6	0	1	0	10	0
	Totals for reaction	26	0	12	0	76	0
Thrombosis, cerebral arterial (11)	Ethinyloestradiol	22	6	0	0	24	7
	Mestranol	28	4	0	0	28	4
	Oestrogens NOS	5	2	0	0	5	2
	Totals for reaction	60	13	0	0	65	14
Thrombosis, cerebral (25)	Dexamphetamine	2	2	0	0	2	2
	Ethinyloestradiol	103	15	1	1	113	17
	Mestranol	75	10	0	0	78	10
	Oestrogens NOS	12	4	0	0	13	5
	Totals for reaction	209	42	3	1	231	46
Tinnitus (128)	Salsalate	20	0	0	0	26	0
	Benorylate	27	0	0	0	36	0
	Indomethacin	10	0	4	0	18	0
	Totals for reaction	149	0	22	0	301	0
Tremor (177)	Sodium valproate	12	0	1	0	20	0
	Metoclopramide	7	0	0	0	14	0
	Mianserin	5	0	0	0	11	0
	Orciprenaline	32	0	2	0	40	0
	Fenoterol hydrobromide	6	0	1	0	16	0
	Salbutamol	62	0	4	0	81	0
	Pirbuterol hydrochloride	0	0	9	0	11	0
	Terbutaline	16	0	0	0	22	0
	Nifedipine	6	0	3	0	20	0
	Cimetidine	9	0	1	0	14	0
	Clopamide	6	0	1	0	13	0
	Co-trimoxazole	10	0	2	0	17	0
	Totals for reaction	394	0	58	1	651	1
Twitching (65)	Mianserin	1	0	3	0	13	0
	Pertussis vaccine	4	0	0	0	11	0
	Totals for reaction	58	0	13	0	113	0
Unconsciousness (76)	Prazosin	83	0	0	0	84	0
	Totals for reaction	174	3	7	0	205	3
Vertigo (213)	Mianserin	6	0	1	0	11	0
	Fenfluramine	16	0	0	0	16	0

(*continued overleaf*)

Table 27.2   Drugs within reactions

Side-effect (n)	Drug	1964–78		1983		1964–84	
		Reports	Deaths	Reports	Deaths	Reports	Deaths
	Buprenorphine	0	0	7	0	14	0
	Pentazocine	11	0	0	0	13	0
	Diflunisal	9	0	1	0	12	0
	Diclofenac sodium	0	0	1	0	13	0
	Indomethacin	30	0	2	0	45	0
	Sulindac	10	0	0	0	12	0
	Carbamazepine	7	0	0	0	10	0
	Prazosin	16	0	0	0	17	0
	Atenolol	22	0	0	0	23	0
	Perhexilene	10	0	0	0	12	0
	Nifedipine	5	0	2	0	14	0
	Cimetidine	32	0	2	0	42	0
	Clopamide	3	0	0	0	12	0
	Minocycline hydrochloride	10	0	2	0	16	0
	Nalidixic acid	9	0	0	0	10	0
	Totals for reaction	434	0	49	0	684	0
Vomiting (348)	Imipramine	11	0	0	0	12	0
	Nomifensine hydrogen maleate	4	0	1	0	10	0
	Viloxazine hydrochloride	13	0	0	0	13	0
	Caffeine	15	0	0	0	15	0
	Buprenorphine	0	0	60	0	152	0
	Dihydrocodeine tartrate	45	0	1	0	54	0
	Pentazocine	42	0	1	0	44	0
	Dextropropoxyphene	27	1	0	0	28	1
	Phenylbutazone	10	0	0	0	11	0
	Piroxicam	0	0	2	0	19	0
	Diflunisal	8	0	0	0	16	0
	Ketoprofen	13	0	2	0	17	0
	Fenbufen	0	0	2	0	28	0
	Benoxaprofen	0	0	1	0	25	0
	Indoprofen	0	0	9	0	11	0
	Flurbiprofen	12	0	2	0	26	0
	Diclofenac sodium	0	0	0	0	14	0
	Naproxen	9	0	5	0	20	0
	Mefenamic acid	8	0	1	0	14	0
	Indomethacin	34	1	22	0	65	1
	Sulindac	11	0	0	0	12	0
	Zomepirac	0	0	4	0	12	0
	L-dopa	15	0	0	0	18	0
	Aminophylline	9	0	0	0	11	0
	Propranolol	11	0	1	0	16	0
	Atenolol	12	0	0	0	18	0
	Mexiletine hydrochloride	7	0	0	0	11	0
	Nifedipine	5	0	1	0	16	1
	Oxypentifylline	10	0	2	0	12	0
	Cimetidine	23	0	2	0	39	0
	Ketotifen	0	0	1	0	11	0
	Ethinyloestradiol	7	0	2	0	11	0
	Mestranol	11	0	0	0	11	0
	Salcatonin	3	0	0	0	14	0

Table 27.2   *Continued*

Side-effect (*n*)	Drug	1964–78		1983		1964–84	
		Reports	Deaths	Reports	Deaths	Reports	Deaths
	Bromocriptine	6	0	0	0	10	0
	Clavulanic acid potassium salt	0	0	5	0	10	0
	Heroin (diamorphine)	0	0	0	0	2	2
	Minocycline hydrochloride	10	0	0	0	11	0
	Erythromycin	10	0	1	0	23	0
	Co-trimoxazole	31	0	0	0	40	0
	Nitrofurantoin	23	0	1	0	25	0
	Nalidixic acid	22	0	1	0	25	0
	Metronidazole	11	0	1	0	16	0
	Ketoconazole	0	0	3	0	10	0
	Influenza vaccine (inactivated)	16	0	1	0	18	0
	Pertussis vaccine	27	0	2	0	34	0
	BCG	21	0	0	0	21	0
	Diphtheria vaccine	14	0	1	0	21	0
	Diatrizoic acid	29	0	0	0	32	0
	Sodium iothalamate	7	0	0	0	11	0
	Totals for reaction	1 077	6	212	0	1 853	10

Drugs within reactions. Data derived from the master tables of the Committee on Safety of Medicines. Table 27.2 retains, for each of the listed drug reactions, the 'Totals for reactions' of the master tables; it also shows the number of drugs given in the master tables for each reaction as '*n*'. However, Table 27.2 omits the details of all drugs, if for the total period of data collection 1964–84 the number of reports is less than 10 or the number of deaths is less than two. See text, p. 586, and Mann (1984).

(cut-off point, 16 February 1984); then there is a total for the whole period of data collection, 1964–84.

In Table 27.2 the drug has been omitted if, for the total period 1964–84, the number of reports is less than 10 or the number of deaths is less than two. This not only greatly restricts the size of Table 27.2, it also cuts out drugs where the figures might well represent the natural history of the diseases concerned, and other such 'background' factors. It seems probable that the remaining figures show drug-related effects. In addition, the individual years between 1979 and 1984 are represented only by the data for 1983 — the last complete year and the one best showing the incidence of side-effects at present. While it is an abbreviated form of the mastertables, Table 27.2 retains the annual totals for the reaction, as in the master tables. The total of drugs in the master tables is shown as *n*.

The most frequently reported side-effects are ranked in Table 27.1, the drug reactions considered being those attributable to the central nervous system or likely to provide problems in the diagnosis of CNS lesions.

Individual side-effects are then considered in Table 27.2. Points of special interest are that dystonia has been reported as being caused in 278 patients and by 49 different drugs, metoclopramide being the most common

culprit. Both cimetidine and indomethacin are not infrequently associated with confusion. The commonest drug-related causes of convulsions are live attenuated measles vaccine and pertussis vaccine. Maprotiline is the drug most frequently noted as associated with Grand Mal convulsions. Indomethacin was the commonest among the reports of drug-related dizziness. Encephalopathy due to live attenuated measles vaccine was reported only once in 1983. When extrapyramidal disorder is the most important drug reaction, metoclopramide is by far the most common cause. Headache, when reported, is most frequently associated with indomethacin. Perhexiline has been the commonest cause of drug-attributed neuropathy — and metoclopramide of oculogyric crises. The relationship between prazosin and unconsciousness has disappeared (with dosage adjustment) just as the many grave CNS lesions caused by the female sex hormones have lessened (with the adoption of the low-dose oestrogen products). Practolol gathered 160 reports of deafness over the last 20 years — and another 19 reports of nerve deafness. Ketotifen is to be remembered as being the drug most often associated with somnolence, as is buprenorphine with vomiting. Over the years the commonest drug cause of cerebral haemorrhage has been the anticoagulants.

Faced with a presenting symptom due to a drug and involving the CNS or the differential diagnosis of the lesions of this system, Table 27.2 indicates the most likely causative agent. It is, of course, obvious that the place of a drug in any such ranking depends, among other things, on the frequency of its use, as well as its intrinsic propensity to cause the particular side-effect. To some extent it may also depend on the date of marketing. Weber (1983) showed that reporting with some classes of drugs reaches a peak in the second year after product launch; after that, interest in reporting seems to wane and the reports diminish, although the number of prescriptions for the drug continues to increase.

It is clear that there is more than one type of information about drug side-effects. The following paragraph appears in the recent paper *Iatrogenic neuromuscular disorders: a review* by Michael Swash and Martin S. Schwartz (1983).

More severe neuropathies occur with several different groups of drugs. Sensorimotor neuropathy, often with unpleasant dysaesthesiae, is an almost invariable complication of cytotoxic drug therapy. Vincristine, chlorambucil, procarbazine and cytosine arabinoside are particularly well known examples. Some of these drugs induce predominantly sensory symptoms, but in most patients there is clinical and electrophysiological evidence of involvement of both motor and sensory nerve fibres. Perhexiline and metronidazole may also cause severe neuropathy. Only one in 1000 patients treated with perhexiline develop a severe sensorimotor neuropathy, which may be accompanied by a myopathy, but as many as 65% show electrophysiological evidence of a subclinical neuropathy (Sebille 1978). Severe peripheral neuropathy also occurs in patients treated with isoniazid if they are not protected by coincidental treatment with pyridoxine. Slow acetylators of the drug are particularly vulnerable. In all these sensorimotor neuropathies, sensory symptoms precede other manifestations.

This informative summary can be compared with part of the data on neuropathy shown in Table 27.2 and given here as Table 27.3.

The close relation between the paragraph and table 27.3 is obvious. Perhexiline features in both, but Table 27.3 shows only what doctors report; there seem to be a number of reasons for not reporting side-effects. A common one is that the lesion, even if serious, is already well known.

Thus, the literature is informative in a slightly different way. It is convenient to consider it under the following headings:

1. Drug-induced convulsions;
2. Drug-induced encephalopathy;
3. Drug-induced demyelinating conditions;
4. Myeloneuropathy;
5. Extrapyramidal reactions;
6. Other drug-induced involuntary movements;
7. Tardive dyskinesia;
8. A number of other syndromes and drug effects, i.e.
   Drug-induced increased intracranial pressure;
   Dexamethasone deleterious in cerebral malaria;
   Transmission of Creutzfeldt–Jakob disease
   Hexachlorophene and new-born babies;
   Cimetidine and CNS disorders;
   Reye's syndrome and aspirin;
   Neurological effects of recombinant human interferon.

Drug-induced convulsions

Drugs can induce convulsions in patients who have neither epilepsy nor other diseases affecting the nervous system. The frequency of drug-attributed convulsions had not been reported prior to a study on 12,617 medical in-patients monitored by the Boston Collaborative Drug Surveillance Program (1972). In this series drug-attributed convulsions occurred in 17 (1.3 per 1000). The risk of this reaction was greatest in patients receiving infusions of penicillin (3.2 per 1000), insulin (3.9 per 1000), and lignocaine (5.7 per 1000).

Penicillin

In 1945 Johnson and Walker noted that, one hour after 50 000 units of penicillin were injected into the intraven-

Table 27.3

Side-effects	Drug	1983		1964–84	
		Reports	Deaths	Reports	Deaths
Neuropathy	Zimelidine hydrochloride	14	0	15	0
	Indomethacin	2	0	10	0
	Amiodarone	5	0	13	0
	Perhexiline	3	0	117	4
	Cimetidine	1	0	35	1
	Nitrofurantoin	1	0	32	0
	Metronidazole	1	0	16	0
	Vincristine	2	0	10	0

tricular system of a 22-year-old man, the patient lost consciousness and developed clonic spasms. These workers investigated this further (Johnson and Walker 1945), and showed that application of commercial penicillin to the cerebral cortex of cats, dogs, monkeys, and man gave rise to convulsive manifestations. Since these findings were reported, the association between intracisternal injection of penicillin and convulsions has been established.

Bloomer et al. (1967) described four cases in which massive doses of penicillin G were given intravenously to four patients in renal failure; this was followed by a reduced level of consciousness, myoclonic jerking, and seizures. Weinstein and colleagues (1964) had earlier described similar observations following administration of 40×10^6 units of penicillin G to a uraemic 12-year-old patient, and Oldstone and Nelson (1966) had observed convulsions of a myoclonic type in a 71-year-old woman with renal calculi who had been treated with procaine penicillin for a pyrexia of unknown origin. Bloomer and his associates (1967) also described loss of consciousness and myoclonic convulsions in a 41-year-old patient treated with sodium cephalothin, 6 g daily i.v. In this patient renal function was impaired due to glomerulonephritis.

The importance of the concentration of penicillin in the cerebrospinal fluid was shown by Smith and coworkers (1967), for they found that, even when high doses were given, there was little danger of fits, provided the concentration in the CSF did not exceed 10 units per ml.

Another aspect of this problem was reported by Seamans et al. (1968). Four patients undergoing open-heart surgery with cardiopulmonary bypass at the Royal Victoria Hospital, Montreal, developed status epilepticus and coma in the immediate post-operative period, and three of them died. Each had received large doses of sodium penicillin intravenously over a period of eight hours before, during, and after the bypass procedure, with the aim of preventing bacterial endocarditis or infection at the site of the valve prosthesis. None of these patients had azotaemia. Penicillin had not been used in this manner in the unit before January 1965 and was not given after March 1965; there were no previous or subsequent cases. The authors then carried out experiments on dogs and found that, if large amounts of penicillin were given intravenously alone, no fits or electroencephalographic changes were observed, but out of eight animals in which this was combined with a full bypass operation seven developed fits. None of the five animals having full bypass but not receiving penicillin developed fits. It was suggested that a breakdown, as yet unexplained, of the blood–brain barrier occurs after cardiopulmonary bypass. This does not appear to be related either to changes in the blood pH or to haemolysis, and there has been no convincing evidence of air embolism, particulate embolism, or thrombosis of vessels. These authors in another paper do suggest, however, that small aggregates of material such as fat globules may become temporarily trapped in the cerebral capillaries and produce local deficiency in the blood–brain barrier, which may allow passage of abnormal quantities of penicillin.

Whether this is the explanation or not, these cases are another example of the potential neurotoxic effects of penicillin when allowed to come in contact with the brain in high concentration. There is also the possibility that during cardiopulmonary bypass operations other drugs usually given with safety may more readily be able to exert direct influence on the brain and produce unexpected, unfamiliar, and perhaps undesirable effects.

Kurtzman et al. (1970) described neurotoxic reactions to the semi-synthetic penicillin carbenicillin. In both the patients described by Kurtzman et al. there were raised blood urea levels at the time of carbenicillin-induced convulsions, the maximum blood urea levels being 150 mg/100 ml and 105 mg/100 ml, respectively.

Lignocaine

Bedyneck et al. (1966) described convulsions induced by intravenous infusions of lignocaine used as an antiarrhythmic and considered that the convulsions so induced were dose-related. In both the cases of lignocaine-induced convulsions described by the Boston group there had been a sudden increase in rate of lignocaine infusion prior to the convulsion.

Hypoglycaemic agents

Grand mal convulsions occurred in three out of 763 insulin recipients (3.9 per 1000) in the Boston survey. All three of these patients had extreme hypoglycaemia, which was too low to be measured in two of them. Any drug producing marked hypoglycaemia may induce convulsions.

Convulsions precipitated by amitriptyline and zimeldine

Betts et al.(1968) described seven patients in whom amitriptyline induced grand mal seizures, and a further case was described by Houghton (1971).

In September 1983 the new antidepressant zimeldine was discontinued in all countries by Astra Pharmaceuticals because of the company's concern over reports of serious neurological side-effects, including the Guillain–Barré syndrome. Zimeldine was chemically novel and unrelated in chemical structure to the tricyclic and tetracyclic antidepressants. In a report accepted for publication in August 1983 Chapman *et al.* reported serious convulsions of the *grand mal* type in two cases of self-poisoning with zimeldine overdose.

Camphor

Camphor is obtained by distillation from the wood of *Cinnamomum camphorae*, or it may be prepared synthetically. Its use in medicine dates back to the ancient Chinese; traditionally it has been used as a circulatory and cardiac stimulant, aphrodisiac, suppressor of lactation, contraceptive, abortifacient. Currently, camphor, menthol, and eucalyptus are commonly found constituents of 'cold remedies' in the form of ointments and vapour rubs. These preparations serve as rubifacients and counter-irritants but substantial concentrations of volatile oils can be inhaled from them.

Camphor is readily absorbed from mucous membranes and the lethal dose is between 50 and 500 mg kg^{-1}. A dessertspoonful of a preparation containing 3–10 per cent camphor could cause intoxication in a child. Major manifestations of camphor intoxication are central nervous system stimulation, vertigo, confusion, delirium, convulsions, vomiting, coma, and death. In a report of the U.S Committee on Drugs (1978) entitled 'Camphor — Who needs it', the following conclusion was reached.

1. Camphor has no established role in scientific medicine.

2. Camphor has potent, serious toxicological actions. The ingestion of even small amounts has proven fatal.

3. Although accidental oral ingestion is the most common route of intoxication, significant quantities can be absorbed percutaneously or by inhalation.

4. Transplacental transfer may be toxic to the fetus.

5. Camphorated oil in particular is the worst offender since it is mistaken for a variety of over-the-counter products and accidentally ingested.

Weiss and Catalano (1973) reported the case of a pregnant woman who ingested 50 ml of camphorated oil, mistaking it for castor oil, and subsequently had three *grand mal* convulsions. She went into premature labour 20 hours later; the infant was healthy but smelt strongly of camphor. A similar case reported by Riggs *et al.* (1965) resulted in neonatal death.

Skoglund *et al.* (1977) described the case of a 7½-year-old boy who attended a neurology clinic for follow-up. He had crawled through some spilt 10 per cent camphor at the age of 15 months and had become ataxic and

developed major seizures over the ensuing 48 hours. A year later he was exposed to a camphorated vaporizer preparation to relieve the symptoms of a respiratory illness. Concurrent with exposure to this vapour, a major motor seizure occurred.

Reports of poisonings are common; in the United States, The National Clearinghouse of Poison Control Centers reported 494 cases in 1973 of which over 415 were in children aged less than five years.

Aronow and Spigiel (1976 *a–c*) reported 94 cases from the Children's Hospital of Michigan, Detroit. In 1973 Sibert commenting on 175 admissions of children aged six months to five years to Newcastle General Hospital for poisoning noted that 10 were related to camphor poisoning. Dupeyron *et al.* (1976) described a near-fatal case in a six-month-old infant after rubbing the chest and nose with an ointment containing camphor, menthol, and thymol.

Convulsions are reported to occur in 20 per cent of all reported cases of camphor intoxication according to the US National Clearinghouse for Poison Control. Rubin *et al.* (1949) reported convulsions in five of 14 cases and Craig (1953) reported convulsions in nine of 19 cases. Convulsions have been reported in different series within 4 to 120 minutes of ingestion, although in one case convulsions as late as 48 hours after ingestion were documented.

Convulsions provoked in epileptic patients by other drugs

Oral-contraceptive therapy may induce water retention which may impair the control of the epilepsy.

Previously controlled epilepsy can be exacerbated by the phenothiazines, and this group of drugs can also produce convulsions in patients who are not epileptic. In the Boston survey prochlorperazine induced convulsions in one in 1214 patients exposed and chlorpromazine in one of 453 patients treated; in both these cases there were repeated convulsions.

On rare occasions paradoxical effects occur with antiepileptic agents. An usual recent example is provided by the report of Eldridge *et al.* (1983) on 'Baltic' myoclonus epilepsy, a hereditary disorder made worse by phenytoin.

Factor VIII, aluminium, and recent reports

During 1983 seizures with thiabendazole and Factor VIII were first reported. Tchao and Templeton (1983) noted the occurrence of *grand mal* seizures in a patient with Down's syndrome treated for chronic diarrhoea caused by *Strongyloides stercoralis* with thiabendazole 1000 mg orally twice a day; the seizures recurred on challenge.

From Glasgow, Small *et al.* (1983) reported three episodes in which seizures followed the rapid self-administration of Factor VIII concentrates in a severe haemophiliac who suffered also from asthma but had no family history of epilepsy. Elliott and Fell (1983), commenting on this case, asked whether aluminium toxicity had been considered a possible cause. These workers noted that Factor VIII concentrates contained an average aluminium content of 1300 μg l^{-1}. As aluminium toxicity is now recognized as the cause of the dialysis encephalopathy syndrome, in which electroencephalographic abnormalities and seizures are prominent features, these observations may have relevance to the clinical use of other proteins and blood products.

Although the list of drugs able to cause these side-effects grows, it will be apparent that certain classes of drugs stand out as having a potential to induce convulsions, namely myelography agents, local anaesthetics, tricyclic antidepressants, and hypoglycaemic agents. There are also some drugs where, according to clinical hearsay, drug-related convulsions have been reported to have been seen clinically but where well-documented cases have not been found in the literature, e.g. digitalis, glycosides, and glucagon.

Drug-induced encephalopathy

Aluminium

In 1962 McLaughlin *et al.* described a patient who had worked for 13½ years in an aluminium powder factory. He developed left-sided focal epilepsy, hesitancy of speech, and his wife noticed that he was becoming forgetful. Over the next few months he became increasingly demented, had more frequent epilepsy, and then died of bronchopneumonia. At autopsy he had a fibrotic reaction to aluminium in his lungs.

Ten years later, Alfrey *et al.* (1972) first described a progressive cerebral disorder in renal failure. It was characterized by speech difficulty, progressive dementia, myoclonus, epilepsy which might be focal, and focal neurological signs: features similar to those in the patient described by McLaughlin *et al.* By 1976 measurements of aluminium by flameless atomic absorption spectrometry had been developed to the point at which physiological levels could be measured. Alfrey *et al.* (1976) found a higher than normal aluminium level in grey matter of patients who had died with dialysis encephalopathy. Initially this was thought to come from aluminium-containing, phosphate-binding gels commonly used in the management of patients undergoing chronic dialysis. However, the incidence of the disease corresponded poorly with aluminium consumption in this form.

Flendrig *et al.* (1976) reported a remarkable difference in the incidence of dialysis encephalopathy in two hospitals in the same town in Holland where apparently identical methods of treatment were used. In the unit with the higher incidence of encephalopathy aluminium anodes had been introduced into the boiler system of the hospital's water supply to protect against corrosion. The unit was moved to a hospital with a lower aluminium level in its water supply and the incidence of encephalopathy fell.

In Scotland, (Elliot *et al.* 1978) it was found that, in areas where the aluminium concentration of the water was high, patients on home haemodialysis were more likely to develop high blood aluminium levels and dialysis encephalopathy. Parkinson *et al.* (1979) in a survey of 1293 patients on haemodialysis in 18 dialysis centres in the United Kingdom showed a clear log/linear correlation between the percentage incidence of dialysis encephalopathy and the mean water aluminium level, and also a log/linear correlation between the percentage incidence of fractures due to dialysis osteodystrophy and the mean water aluminium levels. The mean aluminium levels in the water supply ranged from 10 to 410 μg l^{-1} in different centres. The authors recommended that the aluminium levels in the haemodialysate should be less than 50μg l^{-1}, and ideally less than 20 μg l^{-1}. In Newcastle since the dialysate water has been reduced to low levels by deionization or reverse osmosis, no cases of encephalopathy have occurred.

King *et al.* (1981) reviewed the literature on aluminium-associated encephalopathy and osteroporosis and confirmed the correlation of the incidence of dialysis encephalopathy with the level of aluminium in the tapwater.

Aluminium encephalopathy was also reported in patients on continuous ambulatory peritoneal dialysis (CAPD) and levels in these fluids should be controlled at 0.5–1.0 μmol l^{-1}.

Prevention of dialysis encephalopathy

Patients on haemodialysis or CAPD should have regular monitoring of their plasma aluminium levels, and the aluminium levels in the dialysis fluids should be monitored.

Treatment of dialysis encephalopathy

An effective treatment of dialysis encephalopathy has not been described. In most cases the syndrome deteriorates progressively until the patient dies. One case was described where a patient successfully recovered following renal transplantation. Attempts to lower plasma aluminium levels by plasmaphoresis, or dialysis with fluids which are virtually aluminium-free (less than 2 μg l^{-1}) have been made.

Ackrill et al. (1980) and Pogglitsch et al. (1981) all reported some degree of success in treating dialysis encephalopathy by removing aluminium by haemodialysis and desferrioxamine.

A growing body of knowledge suggests that the toxicity of aluminium may have wide implications. The primary lesion in Alzheimer's disease and in dialysis dementia has been postulated to be an impairment of the permeability of the blood–brain barrier, so allowing neurotoxins, such as aluminium, to reach the central nervous system. Banks and Kastin (1983) showed that aluminium itself affects the permeability of the blood–brain barrier of rats to small peptides. These workers demonstrated that intraperitoneal injection of aluminium chloride increased the permeability of the blood–brain barrier to iodinated N-Tyr-delta-sleep-inducing peptide and beta-endorphin by some 60 to 70 per cent. Thus, it seems possible that aluminium might affect the blood–brain barrier in ways that might be involved in the pathogenesis of dementia.

As interest increases in the toxicity of aluminium, knowledge of the sources of this toxic agent increase. Randall (1983), for example, described the clinical findings in an 11-month-old girl who had never been dialysed and yet died of a progressive encephalopathy associated with high serum and cerebrospinal fluid levels of aluminium. The case suggested that aluminium-containing phosphate binders can result in a significant body burden of the metal.

A number of aspects of the subject were reviewed by Wills and Savory (1983). These authors showed that the excess aluminium may be derived from the concentrate or tap water used to prepare the dialysate for use in patients with chronic renal failure on long-term intermittent haemodialysis; other sources are aluminium-containing gels taken orally to control hyperphosphataemia. The result can be not only progressive encephalopathy but also one form of osteomalacia; anaemia associated with haemodialysis and metastatic (in particular vascular) calcification can, in some subjects, be traced to the same cause. The condition can be prevented by avoiding aluminium in the sources mentioned. Once established, aluminium intoxication can be treated with the chelating agent desferrioxamine with symptomatic relief of the encephalopathy and osteomalacia.

Bismuth

In 1974 Burns et al. reported the occurrence of a chronic encephalopathy in five Australian patients who had had colostomies for several years and had been taking bismuth subgallate to control their colostomies. In each case the neurological state recovered after the bismuth had been withdrawn. In 1974, the Australian Drug Evaluation Committee reported that a further 24 patients who had been taking bismuth subgallate for ileostomy or colostomy control had been reported to them as having developed a reversible encephalopathy.

Other reports of bismuth central nervous system toxicity have appeared in the literature, mostly arising in either Australia or France. In France the disease spread from the west towards the east and south from 1974 to 1977. The high blood levels of bismuth correlated well with the severity of disease. However, the reason for the raised blood levels in some individuals remains a mystery, as does the mechanism of absorption of the insoluble bismuth salts which are ingested. The current suggestion, as yet without strong supporting evidence, is that an organism in the intestinal tract converts bismuth into a soluble form, e.g. by methylation — and that the spread of this organism determines the incidence of the disease (Martin-Bouyer 1978)

In all, about 50 cases of bismuth encephalopathy occurred in Australia and over 100 in France between the years 1973 and 1977 (Buge et al. 1981).

Buge et al. (1981) reviewed 70 patients with bismuth encephalopathy admitted to 'La Salpêtrière' between 1973 and 1978. There were 14 male patients aged 30 to 70 years and 56 females aged 24 to 84 years. All had received orally 5 to 20 g of bismuth subnitrate over periods ranging from four weeks to 30 years for gastrointestinal complaints. During the acute phase, EEG were recorded daily and every two days during the recovery phase and monthly thereafter for a period of one year. Bismuth blood levels were sampled every day and ranged from 150 to 2200 μg l^{-1} (normal less than 20 μg l^{-1}). Levels of 200–9600 μg l^{-1} were found in urine and of 10–100μg l^{-1} in CSF. Clinically bismuth encephalopathy was separated into two clinical phases.

1. A prodromal period lasting from a week to several months during which cognitive and affective disorders were predominant. The patients were asthenic, somnolent, depressed, anxious, sometimes with visual hallucinations and even delusions of persecution (three cases). Jerky movements of varying severity were seen in this phase (15 patients), and disturbances of writing and speech occurred more rarely.

2. A second phase of encephalopathy of rapid onset appeared abruptly in 24 to 48 hours. Four symptoms were constant: confusion (reaching coma or dementia), dysarthria, disturbances of walking and standing, and pseudo-tremor accompanied by myoclonic jerks. Myoclonic jerks were always present in this phase;

sometimes they predominated in the upper limbs and distally, at other times they were diffuse and involved the facial and axial muscles. They were increased by voluntary movements and by stimuli (change of position, noise). All the patients exhibited myoclonic jerks, but no paroxysmal features ever appeared on EEC. Computed tomography showed hyperdensities in basal ganglia, cerebellum, and cerebral cortex most marked when bismuth blood levels exceeded 2000 μg l^{-1}. Seizures were observed in 22 patients, but epileptic EEG patterns appeared only when the bismuth blood level was below 1500 μg/l^{-1}. It is suggested that a high cortical intracellular bismuth concentration induces a 'cortical inhibition' which causes suppression of physiological electrical brain activity, the absence of EEG paroxysmal phenomena during myoclonic jerks, and explains the rarity of epileptic seizures.

All patients recovered within 3 to 12 weeks after ceasing bismuth administration; however, some disorder of behaviour or memory persisted longer in some cases. The cerebral densities observed on computed tomography disappeared slowly and became imperceptible two to three months after withdrawal of bismuth.

One of the patients died of an accident three months after clinical recovery and at autopsy intracerebral bismuth concentration was high despite a normal blood level.

Reports of bismuth encephalopathy continued to appear and Goas et al. (1981) reported the case of a 66-year-old woman who was hospitalized for grand mal epilepsy. The patient had been taking bismuth preparations from 1935 to 1975 for 'enterocolitis', but the drug was then discontinued. In June 1980 she recommenced taking 15 sachets daily of a preparation of bismuth subnitrate 6.5 g in combination with kaolin. By December 1980 it was apparent that her memory had become impaired but this improved spontaneously during the time the patient took only one sachet daily. However, in a period of 45 days in March–May 1981 she ingested 136.5 g of bismuth — she was then admitted to hospital with grand mal epilepsy. On admission, she had myoclonic jerking, dysarthria, hallucinations, amnesia, and EEG changes. The blood level of bismuth at the time of admission was 770 μg l^{-1} and urine levels of 900μg l^{-1} were measured. Recovery occurred.

Lithium

The dosage of lithium is usually in the region of 0.25–2.0 g/day adjusted by regular monitoring to achieve a plasma concentration of 0.6–1.2 mmol l^{-1}. The signs of overdosage are blurred vision, anorexia, vomiting, diarrhoea, increasing drowsiness, and sluggishness increasing

to giddiness with ataxia, coarse tremor, and dysarthria. With severe overdosage (plasma concentrations over 2.0 mmol l^{-1}), hyper-reflexia, hyperextension of limbs, toxic psychosis, fits and coma, and occasionally death occur. These changes are reversible on lowering of the lithium plasma levels.

In 1974 Cohen and Cohen reported on four female patients aged 34–6 years treated with haloperidol and lithium in whom severe irreversible brain damage had occurred. In these patients, lethargy, fever of up to 104°F, confusion, extrapyramidal and cerebellar dysfunction had occurred with plasma lithium levels within the therapeutic range for that unit (about 1.3–1.4 mEq l^{-1}). Two patients suffered widespread and irreversible brain damage; two others were left with persistent dyskinesias. In the authors' clinic in New York, 50 patients had been treated with this particular combination of drugs but only four had developed this form of encephalopathy. Baastrup et al. (1976) reviewed the hospital records of 425 patients who had been treated simultaneously with halperidol and lithium. None of these patients had developed a syndrome resembling the encephalopathy described by Cohen and Cohen (1974). Strayhorn and Nash (1977) reviewed the literature on neurotoxicity of lithium and found nine other cases that had been treated with various neuroleptics in combination with lithium who had developed severe neurotoxicity at therapeutic levels of lithium. Destee et al. (1978) described two further cases of encephalopathy arising in patients treated with haloperidol and lithium. Boudouresques et al. (1980) described a single case of encephalopathy in a woman treated with lithium and chlorpromazine. Spring (1979) described four cases on a combination of lithium and thioridazine who developed severe neurotoxic symptoms, e.g. delirium, seizures, encephalopathy, and abnormal EEG at levels of plasma lithium within the normal therapeutic range. All the patients were female aged levels 29–61 years whose plasma lithium levels were 0.9, 0.6–1.3, 0.9, and 0.5 mEq l^{-1}1 at time of presentation.

The mechanism of this toxicity is unknown but it has been suggested that there is an interaction between flupenthixol and lithium on striatal adenyl cyclase levels (Geisler and Klysner 1977).

There are those who doubt whether the cases represent the result of a true drug interaction or whether these reports represent a rare variation on lithium intoxication (Ayd 1976). However, further evidence indicating that lithium/haloperidol combinations have a unique effect in producing permanent brain damage was produced by Thomas et al. (1982) in a study which exonerated lithium/chlorpromazine combinations from producing any similar deleterious effect.

Bromism

Bromides, formerly widely used as hypnotics, sedatives, and anticonvulsants until superseded by more modern drugs, are still available to the public without prescription in cough cures, 'nerve tonics', and other proprietary preparations. Patients may also be given repeat prescriptions for bromides (e.g. Mist.Pot.Brom.Compound Bromochloral, B.P.C.) with inadequate medical supervision. Despite sporadic case reports, bromism, the syndrome of chronic intoxication, still occurs in Britain.

The clinical picture of chronic bromide intoxication is well known. Numerous articles and case-reports on the subject have been published. In brief, the common signs and symptoms usually seen when blood bromide levels exceed 150mg/100ml consist of any combination of headache, lethargy, dizziness, ataxia, mental confusion, hallucinations, delusions, dermatitis, mydriasis, slurred speech, ptosis, and a diffuse slowing of the electroencephalogram. Treatment involves sedation when required, and various medications and procedures which will promote the excretion of bromide. These include oral ammonium chloride (6 to 12 g per day), diuretics, or even haemodialysis. Despite the warnings on the dangers of bromides for over 50 years many preparations are still available in many countries of the world, and reports of iatrogenic bromism are still appearing in the world literature; for example, Carney (1971) described five cases from Lancashire, Muller (1968) reported a series from Texas, and Martin (1967) reported a series of cases from Australia.

Benzodiazepines

Taclob and Needle (1976) described five patients who had been on chronic maintenance haemodialysis for more than eight months: a syndrome involving altered consciousness, asterixis, and abnormal electroencephalogram developed after they had been given flurazepam and diazepam. All five patients were adequately treated by haemodialysis. Hepatic, pulmonary, and cardiac decompensation were not present. The encephalopathy and other abnormalities cleared when the drugs were withdrawn. Symptoms were also produced by accidental rechallenge.

Metronidazole

The increased use of metronidazole in major gram-negative infections was reflected in the report by Bailes *et al.* (1983). Following abdominal surgery in a 12-year-old boy, peritoneal cultures grew *Bacteroides fragilis*. Metronidazole 250 mg intravenously every six hours was

begun. After receiving 4 g of metronidazole the boy had a *grand mal* seizure. He suffered a total of eight seizures over a three-hour period. This was followed by a major neurological deficit and it was 11 weeks before discharge from hospital was possible. The authors expressed the view that intravenous metronidazole should be used with caution in children. In a separate report Schentag *et al.* (1982), noted gross but reversible, apparently dose-related, mental confusion in a 65-year-old man.

Drug-induced demyelinating conditions

Hydroxyquinolines and subacute myelo-optic neuropathy (SMON)

Clioquinol (iodochlorhydroxyquin), a halogenated hydroxyquinoline, was synthesized in Germany at the turn of the century: shortly thereafter it was produced on an industrial scale and marketed by the Gesellschaft für Chemische Industrie in Basel, Switzerland, later to be known as CIBA. Within a decade, clioquinol had come into widespread use as a topical agent (*Vioform*) for use in infections of the skin and the mucous membranes: by 1913 it was marketed for this purpose as far away as Japan.

In an era when the range of anti-infective agents suitable for internal use was very limited, clioquinol was an obvious candidate for this role. Clinical findings suggested some degree of efficacy in diarrhoea and later too in amoebic dysentery, and around 1930 clioquinol was introduced in various countries for this purpose. Although patents soon expired and the drug was used under a multiplicity of names it became most widely known under the connotation bestowed upon it by the original manufacturer; *Entero-Vioform*. Doses of some 250 mg were usually administered twice daily for diarrhoea and up to six times daily for amoebic dysentery.

Although originally intended and tested as a therapeutic agent there is no doubt that it was widely regarded and used as a prophylactic drug and sometimes taken for very long periods.

In these ways the drug came to be used in medicine for many years, topically as an antibacterial in dermatology, intravaginally in the treatment of *Trichomonas vaginalis*, and orally as an amoebicide. Very little in the way of adverse reactions was reported following its use before the Second World War apart from the fact that patients taking the drug both orally and topically for prolonged periods had elevated plasma-bound iodine levels (Hodgson-Jones 1970).

Two clinical papers by Grawitz (1935) and Barros (1935) were published in *La Semana Medica* in Argentina describing two different patients in whom neurological signs and symptoms had developed during treatment with the drug: the latter author also drew the manufacturers's attention to his findings. Both of the patients concerned belonged to a series of 153 individuals who had been receiving the unusually high daily dose of 1500 mg of clioquinol for a month. In one of these cases, a woman,

. . .there was a lack of feeling in the legs several days after the start of dosage. Thereafter, there was no change in the heavy feeling in the legs during repeated stoppages and resumptions of drug administration. There was a worsening of sensory and motor disturbances in the lower limbs. She dragged her legs and had to lean against a wall for support when walking. She repeatedly fell down. These symptoms worsened day by day, but after completion of the total prescribed dose of *Vioform* the atony in both lower limbs gradually disappeared until she could walk, but with a manifestly spastic gait.

Examination of this woman revealed decreased pain sensation, exaggerated tendon reflexes in both lower limbs, clonus of both ankles and knees, disappearance of skin-abdominal reflexes, strongly positive Babinski's sign, and, after a time, contracture. The examining physician appears to have concluded that the condition was irreversible. The second patient, a male, exhibited less severe symptoms but he had similar sensory abnormalities accompanied by symmetrical paralysis and glycosuria. The latter finding does not appear to have been commented on in subsequent literature, but it is just possible that this patient was suffering diabetic neuropathy, which could have predisposed him to a drug-induced neurological syndrome.

Whether or not as a result of these reports, the Swiss manufacturer in 1939 performed studies with clioquinol in cats and found that the drug had epileptogenic potential: the animals also developed an unsteady gait and appeared to be 'dazed' (quoted decision of Tokyo District Court, page 23.) Earlier work had already pointed to a neurotoxic effect in frogs, mice, and in chicken embryo cultures; later work, in 1944 and in 1952, showed that clioquinol could produce paralysis and sensory impairment in rabbits during acute toxicity testing (quoted decision of Tokyo District Court, page 340.) None of this evidence appears to have become public knowledge at that time. In 1958, however, the independent Swiss veterinary practitioner Hangartner began to observe repeated instances of severe convulsions, epileptiform attacks, and psychic disorders in cats and dogs in which *Entero-Vioform* had been used to control gastrointestinal disease, and he published a report on his findings (Hangartner 1965). Almost at the same time, comparable findings were reported by veterinary practitioners in Sweden (Schantz and Wikström

1965) and they were later to be confirmed by other workers.

In the meantime, the human use of clioquinol apparently continued to expand, the total production at one stage far exceeding 1000 tons yearly (Gholz and Arons 1964). Reports on adverse reactions were relatively few. As late as 1971, the AMA Drug Evaluations stated:

Incidence of adverse effects is low. Those reported are flatulence, gastrointestinal distress, diarrhoea, pruritus ani and mild iodism. Administration of this agent may interfere with thyroid function tests, so these should not be performed less than one month after discontinuation of therapy. Contraindicated in patients with an idiosyncrasy to iodine or with liver disease. (AMA Drug Evaluation 1971).

Had the medical literature of the world been as readily accessible at that time as it was subsequently to become, and had medical and veterinary science not been so isolated from one another, that statement would perhaps not have been made. There were indeed isolated reports which suggested that clioquinol could on occasion prove neurotoxic in man. In 1966, Berggren and Hansson reported in Sweden on a case in which treatment with clioquinol for acrodermatitis enteropathica appeared to have induced optic atrophy in a young boy. Two years earlier, Gholz and Arons (1964) noted transient derangements of gait in 20 of 4000 patients treated with the drug over a long period. These and similar reports were however too scattered and too limited in number to generate a hypothesis, at least in the medical world at large; moreover it was still widely believed, despite evidence to the contrary, that clioquinol was scarcely absorbed from the gastrointestinal tract; the belief that it might cause neurological and optic disorders, and that the possibility of such an association should be actively and prospectively studied, was yet to be catalysed by events in Japan.

SMON in Japan

Japan has, for a number of reasons, been characterized in recent decades by a relatively heavy consumption of drugs in general, and clioquinol was no exception to the rule. The production of the compound by two companies in Japan during the period 1965-9 has been estimated at between 40 and 44 tons yearly (Takahashi 1979). It was around this time that attention was first drawn to the apparently increasing frequency of an unfamiliar neurological syndrome which had first appeared sporadically in 1955. From 1957 onwards a series of eight regional epidemics of the condition were observed. At the 61st General Meeting of the Japanese Society of Internal Medicine, held in May 1964, a symposium was devoted to the condition by then known as 'non-specific myeloencephalopathy' since it was by that time evident that the eyes could also be involved. The meeting proposed to refer to the entity henceforth as 'subacute myelo-opti-coneuropathy' which was soon abbreviated to SMON (Kono 1975).

The characteristics of the condition, as they came to be recognized by Japanese physicians at that period, have been described on many occasions. An account by Kusui is cited below *in extenso* and literally, since it is both reliable and complete. (Decision of Tokyo District Court, pages 55–7.)

Basic symptoms

(1) Abdominal symptoms
The main symptoms are diarrhoea and abdominal pain, sometimes accompanied by nausea, vomiting, constipation, a bloated feeling in the abdomen, meteorism, etc.

(2) Neural symptoms
The neural symptoms are acute or subacute in nature and sensory disturbances are characteristic. These disturbances are bilateral. They start in the bottom of the feet and gradually rise up to the thighs and reach the level of the navel. They are more severe in the periphery and are indistinct at the upper levels. There are sensations as if the limbs were in contact with some object or bound, a stinging feeling, a feeling of numbness or the sensation of walking barefoot on a gravel road. Another characteristic sensory disturbance was the feeling that nails were piercing the bottoms of the feet.

Reference items

(1) There are often deep sensory disturbances in the lower limbs. These are especially severe in the peripheries. The sense of vibration decreases and positional sensory disturbances, incoordination occurs and in a large number of cases the Romberg phenomenon became positive (loss of spinal movement coordination with body swaying becoming more pronounced when the patient is standing with the feet close together and both eyes closed).

(2) Motor disturbances: such disturbances most often appear in the lower limbs. The muscular power decreases, it becomes impossible to stand up or there are walking disturbances. There are also often pyramidal signs such as exaggeration of the leg tendon reflexes, positive Babinski phenomenon, and spastic paralytic walking.

(3) Although they are only slight, sensory and motor disturbances also appear in the upper limbs. There are decreases in sensation and abnormal sensations in the fingertips and hands, a decrease in the muscular power of the hands and wrists, exaggerated tendon reflexes. Hoffman's and Trommer's signs, and so forth.

(4) The following symptoms also occur at times;
(a) Bilateral visual disturbances in 20–40% of SMON cases. They may be mild or severe and range from blurring of vision to loss of sight due to atrophy of the optic nerve.
(b) Mental symptoms including the following: disturbance of consciousness, swallowing and speech disturbances, convulsions, vertigo, fainting and involuntary movements. Almost all cases showed mental symptoms such as nervousness, depression and insomnia.
(c) Abdominal symptoms accompanied by a green coated tongue and green faeces.
(d) Bladder and rectal disorders occurring around the time of the sensory and motor disturbances in the legs but disappearing comparatively quickly except in severe cases.

(5) The course is generally protracted and recurrent.

(6) The fact that there are no remarkable changes in blood and spinal fluid findings in SMON is extremely important in differentiating it from similar diseases such as subacute combined spinal degeneration (SCD) in pernicious anemia which resembles SMON clinically and pathologically, and the Guillain–Barré (G–B) syndrome which is similar clinically. In SCD, haematological findings are characterized by the appearance of megaloblasts, while in the G–B syndrome, there is no cell increase but a remarkable protein increase in the spinal fluid.

With descriptions as complete as this to hand it became possible to study epidemiologically a phenomenon which was gradually reaching alarming proportions. Up to and including 1961, 60 confirmed and 24 suspected cases had been reported in the entire country. By 1966 the total had risen to 1859. In 1967 alone 1452 new cases were reported, in 1968, 1770, and in 1969, 2340. Children under the age of 10 were very rarely affected, the peak incidence being in adults aged between 60 and 69; the incidence was about twice as high in women as in men (Kono 1975).

SMON was frequently severe. Of the patients who had been accorded this diagnosis up to 1968, more than 90 per cent had paraesthesia and dysthesia of the lower extremities: weakness of the legs was found in 73.8 per cent and complete paraplegia in 11.5 per cent. Blurred vision was a symptom in 18.9 per cent of cases and total blindness occurred in 2.5 per cent; 3.9 per cent of patients suffered from proctoparalysis and paralysis of the bladder (Decision of Tokyo District Court, pages 55–7.)

In the autumn of 1969, the Japanese Ministry of Health and Welfare established a special SMON Research Commission, charged with elucidating the epidemiology and aetiology of the disease and finding effective therapeutic and preventive measures.

The infection theory

From the start, a number of elements suggested that SMON was an infectious condition. The disorder appeared suddenly in particular areas where it became endemic for a period and then faded away. In some instances, familial occurrence of SMON was observed, as were hospital-centred or doctor-centred outbreaks. In Okayama it was noticed that any second case occurring in a particular family tended to appear some 10 weeks after the first. Morbidity was relatively higher in the medical and paramedical professions, but there was no other clear occupational correlation. Finally, the condition tended to be particularly prevalent during the late summer. On the other hand, the pathohistological findings suggested a metabolic disorder, a vitamin deficiency, or an intoxication rather than infection; inflammatory changes, pyrexia, rash, and changes in the blood and cerebrospinal fluid were strikingly absent.

After a number of unsuccessful attempts had been made to find a postulated association between the disorder and Echo 21 virus, a solution appeared to be in sight when in 1970 Inoue and his associates claimed to have isolated a cytopathic agent from the faeces and spinal fluid of SMON patients, detectable when this material was inoculated into a BAT-6 cell culture (derived from a hamster tumour induced by a bovine adenovirus type 3). Inoue and his staff reported positive findings in all five cases studied from one district, in eight of ten from another, and 23 of 29 from a third area of the country. Various other laboratories at once sought to repeat and confirm this work, yet none succeeded in doing so. Kono and his collaborators were among those who concluded that whilst the BAT-6 cell line grew very rapidly it was also subject to rapid spontaneous degeneration, and that the degenerative changes could easily be mistaken for cytopathic effects (Kono 1975). Later work by Nakamura and Inoue (1972), who claimed that they had succeeded in transmitting SMON to suckling mice by using material obtained from patients, could not be confirmed by other workers. The only clearly positive finding from all these microbiological studies, and one

which in retrospect is of particular interest, was that there was an intensive change in the intestinal flora of SMON patients, whilst mycoplasmata were isolated more frequently from the tongue and faeces of SMON patients than from controls (Kono 1975).

The clioquinol hypothesis

The bacteriological findings and the apparent involvement of the flora of the tongue and gastrointestinal tract clearly recall some of the clinical findings as described by Kusi, including the abdominal symptoms which head the list. The finding that in some SMON patients the tongue was coloured green was reported by Takasu et al. early in 1970. In May 1970, Igata et al. found two SMON patients who excreted greenish urine. Since a large amount of green sediment could be obtained from this urine it was analysed. Yoshioka and Tamura (1970) soon concluded that the green pigment (present in the faeces as well as the urine) was the iron chelate of clioquinol, whilst the urine of these patients contained cystals of free clioquinol in large amounts. Ideka and his collaborators soon thereafter found a high iron and zinc content in the green fur taken from the tongue, and subsequently clioquinol was identified in this material as well (Kono 1975). The hypothesis that clioquinol might be the causal agent in SMON was immediately advanced by the neurologist Tsubaki and in the early summer of 1970 he performed, with his collaborators, a rapid survey of seven hospitals to which SMON patients had been admitted. His findings, as summarized by Kono (1975) seemed to leave little room for doubt that the causal agent had been found:

1 Of 171 SMON patients entering these hospitals, 166 had been treated with clioquinol prior to the appearance of neurological symptoms.

2 Neurological symptoms appeared most often after a total dose of 10–50 g of clioquinol had been taken.

3 The mean interval between drug administration and the onset of neural symptoms was 48.8 days at a daily dose of 600 mg and 29.4 days at a daily dose of 1200 mg.

4 The larger the dose of clioquinol taken, the more severe the condition.

5 In one of the hospitals studied it was possible to check the monthly incidence of SMON and the consumption of clioquinol tablets. There was a dose parallel between the two.

6 SMON was found to be present in 29 of 263 cases with enteric disorders who had been treated with clioquinol, but no neurological complication was found in a series of 706 similar patients who had not received the drug. This latter finding was of especial significance, since one might otherwise have postulated that an unknown third factor had on the one hand induced

SMON and on the other given rise to abdominal symptoms leading the patient to take clioquinol.

With evidence such as this available as well as figures pointing to a parallel between the annual sales of clioquinol in Japan and the incidence of SMON, the Ministry of Health and Welfare took rapid action. The findings of Tsubaki's group were published in the *Asahi Daily News* on 7 August 1970; as of 8 September the sale of clioquinol was prohibited. From these moments onwards, the incidence of new cases of SMON fell dramatically (Figs. 27.1 and 27.2).

From 1971 onwards, new cases of SMON virtually ceased to occur in Japan.

Subsequent events in Japan

The SMON Research Commission continued its work after 1970 and it has produced impressive additional evidence to confirm the relationship between the consumption of clioquinol and the incidence of SMON in Japan (Decision of the Tokyo District Court, page 318). The figures produced so rapidly by Tsubaki in 1970 have

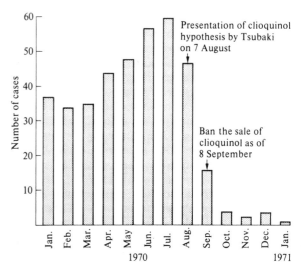

Fig. 27.2

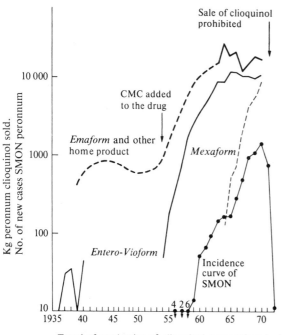

Fig. 27.1

——— Trend of production of clioquinol preparations in Japan
Broken line shows presumative estimate, since no exact information is available before 1961
——— Trend of importation of *Entero-Vioform*
– – – – Trend of importation of *Mexaform*
●—●—● Annual incidence of clinically confirmed SMON

been only slightly modified by the analysis of much larger populations. A small minority of SMON patients have always denied having taken clioquinol, but the proportion of such cases is hardly sufficient to weaken the remaining evidence; it is probably that, as it is so often the case in such studies, the evidence given by some patients is unreliable (clioquinol has in a number of cases been detected in their urine) (Shigematsu 1975), and in a few instances the diagnosis was at fault.

Typical studies of this period showed a history of chronic abdominal symptoms, consisting of pain and diarrhoea preceding the neurological illness. There was usually a prodromal exacerbation of the abdominal symptoms before the acute or subacute onset of sensory neuropathy in the lower limbs. In two-thirds of 752 patients reviewed by Sobue *et al.*(1971) there were painful dysaesthesia and ataxic gait. Half of these patients developed muscular weakness of the lower limbs and 25 per cent of them had visual impairment. Sobue *et al.* (1971) reported the follow-up on 684 patients with this condition over periods varying from 6 months to 10 years; 19 per cent had chronic relapses, 7 per cent were cured, 54 per cent improved, and 25 per cent were unchanged. Overall 93 per cent of patients were left with persisting painful dysaesthesia.

Other surveys from Japan also supported an association between SMON and clioquinol ingestion. Nakae *et al.* (1973) conducted an epidemiological survey of SMON in Toda City (population 65 000) and Warabi City (population 76 000), Japan. In Toda City, after an outbreak in 1964, the occurrence was sporadic. In Warabi City the incidence was lower and less variable. The highest incidence was in middle-aged females; no cases found in children. Out of 47 cases examined, 36 had been treated at two particular medical institutions (A and B) at the time of onset of neurological symptoms, although there were 52 medical institutions in the areas. The remaining 11 cases had been treated in other medical institutions inside and outside the areas. Review of the National Health Insurance returns from 30 medical institutions in Warabi City revealed that the highest dosage of clioquinol was used in medical institution B. No case of SMON was observed in 20 medical institutions which did not use clioquinol. In medical institution A, 17 out of 21 patients with confirmed SMON (81 per cent) had been given clioquinol before the onset of neurological symptoms. A comparative study showed that the dosage of clioquinol to SMON patients had been greater than in other patients with similar enteric symptoms.

The most extensive survey was that conducted by Kono and colleagues on behalf of the SMON Research Commission (Nakae *et al.*1973). This nationwide survey of the relation between SMON and clioquinol administration was made in Japan. Questionnaires disclosed a total of 1839 cases of SMON, of whom 1381 (75 per cent) had received clioquinol before the onset of this disease (appearance of neurological symptoms). In a subgroup of 1092 patients who had been under treatment in a given institution when neurological symptoms began, 944 (86.4 per cent) had received clioquinol before the onset of SMON. The average total dose of clioquinol in the six months before onset of SMON was 40.1 g. The dose–response pattern was similar for sex, age, and different clioquinol preparations. There was a relation between total dose and the degree of visual impairment and greenish discoloration of the tongue. The average total dose in non-fatal cases was 136 g.

Clioquinol derivatives have been identified in the green pigmentation noted on the tongue of many SMON victims. It is estimated that there are at least 10 000 cases of SMON in Japan and the overwhelming evidence points to some association with clioquinol ingestion. In March 1977 there were reported to be 3268 persons seeking compensation from the Japanese Government and 20 pharmaceutical manufacturers of clioquinol.

Particularly interesting is the close parallel which has in retrospect emerged between the consumption of the drug in various areas of the country and the incidence of the disease in those areas (Decision of Tokyo District Court, page 175). Finally, substantial additional work has been undertaken in animals to confirm the neurotoxic potential of the drug. The human neuropathology of SMON has been closely mimicked in studies carried out in dogs and cats: monkeys have proved relatively resistant, but degenerative lesions in the dorsal tract can

nevertheless be produced by high doses. (Tateishi and Otsuki 1975; Tateishi 1980).

That clioquinol is indeed absorbed in fair quantities has been amply confirmed with the development of more sensitive methods for the detection of the drug in body fluids than those originally available. Tamura (1975) found some $10\mu g/100$ ml of the compound in the serum of a patient who had ceased to take clioquinol a month earlier; a minimum toxic level of clioquinol in nerve cells from rats and mice cultivated *in vitro* appears to be 6–8 $\mu g/100$ ml (Katahira *et al.* 1980).

SMON outside Japan

Coming as it did at a time when there were extremely few reports in the Western literature on cases of SMON, the action of the Japanese authorities led to an urgent but frustrating debate as to the need for measures to be taken elsewhere. A number of countries rapidly issued warnings to practitioners; others drastically restricted the indications for the drug or imposed prescription requirements. In a few countries the drug was banned or the manufacturers withdrew it from sale after consultation with the regulatory authorities. In a large number of countries, however, including many where the drug has been widely used, it remained on free sale, a situation which has persisted up to the time of writing (Katahira *et al.* 1980).

Since the most stringent measures were, generally speaking, taken in those countries having the most advanced systems for adverse-reaction monitoring it was evident that from that time onwards it would be difficult to undertake prospective studies of the link between SMON and clioquinol. Retrospective work could, however, be performed and existing cases of neurological disorders possibly involving SMON could be studied. Between January 1970 and December 1978 94 cases of SMON were published outside Japan, and a further 17 suspected cases described in the literature. A questionnaire to a large number of national adverse-reaction monitoring centres produced evidence of some 60 cases, some of them no doubt identical with those described in the literature. Data available to the Swiss manufacturer related to 179 non-Japanese cases over the period 1935–75; a further 28 cases became known to the manufacturer during the period 1976–9 (Katahira *et al.* 1980)

These figures are not necessarily complete and they may indeed be extremely incomplete. In its mildest form, after all, SMON is a reversible condition, and it is entirely possible that in the course of the years a large number of unexplained but transient neurological disorders have in fact been due to the halogenated hydroxyquinolines. Certainly tenacious retrospective surveys in some countries have identified cases which had neither appeared in the literature nor been reported to the health authorities. In Sweden, for example, 16 cases were published in 1978 but by February 1980 a study by Hansson had brought the total number of cases detected in that country to 40 (Hansson 1980). Even taking these problems of detection into account, however, the incidence outside Japan was clearly of a different order of magnitude to that within the country.

The actual characteristics of the SMON cases reported outside Japan were in many cases entirely compatible with the descriptions given in that country. As in Japan, however, the complete syndrome did not necessarily occur. Perhaps because of different patterns of use of the drug in the Western world, two alternative pictures tend to be represented in this material. One relates to optic atrophy in young children who had been treated with high doses of clioquinol for acrodermatitis entero-pathica, prior to recognition of the fact that this condition primarily involved zinc deficiency, and that clio-quinol was effective merely because as a chelating agent it carried zinc into the body. The other group was characterized by acute but reversible encephalopathy with amnesia following the use of high doses of the drug for short periods of time — again a picture which one also finds in the Japanese material (Katahira *et al*.1980; Hansson 1980; *Lancet* 1980).

Retrospective and laboratory studies by Japanese workers and others since 1970 have thrown a great deal of light on matters which had earlier been regarded as obscure. Determination of the drug in biological material by means of gas chromatography provided a much clearer picture of its absorption and distribution (Tamura 1975). Ambitious literature studies covering even nineteenth-century journals underlined the fact, which had received too little attention, that a consi-derable number of compounds more or less structurally related to the halogenated hydroxyquinolines had at various times shown somewhat similar patterns of neuro-toxicity and oculotoxicity, either in the clinic or experi-mental animals. The retinal toxicity of chloroquine and other 4-aminoquinolines is of course well known, but it is not always realized that McKendrick and Dewar studied the toxicity of the quinoline derivatives as early as 1875; by 1926 it had been found that the toxicity of 8-hydroxyquinoline varied considerably from one species to another (Hansson 1980). This older experi-mental work was followed up energetically in the years following the SMON epidemic in Japan.

Most of the evidence relating to SMON naturally con-cerned cases where clioquinol itself had been admi-nistered. Measures taken in various parts of the world to restrict the use of this drug however as a rule related also to closely related compounds which might be expected to have similar toxic effects, in particular diiodohydroxy-quinoline, broxyquinoline, and chlorquinaldol. Cases of SMON have indeed been attributed to the first two of these compounds and in view of their closely similar structure and properties this is hardly surprising, though these compounds would certainly never have been con-demned on the basis of the limited data relating to them without extrapolation from the clioquinol material.

The paradoxes

The nature and volume of the evidence from Japan must be regarded as constituting proof that during the late-1960s there was an extraordinarily close correlation between the consumption of clioquinol in that country and the occurrence of subacute myelo–opticoneuropthy. The proof was accepted by the Tokyo District Court in an impressive judgement of 3 August 1978 in the most spec-tacular of the cases brought *en masse* by SMON victims against the manufacturers of clioquinol and the State of Japan, and this judgement has formed the basis for settlements between these patients and the companies concerned. Criticism of these positive conclusions has however naturally been forthcoming both within and outside Japan.

In part, this criticism reflects a degree of unfamiliarity with the facts. The assertion, for example, that the distri-bution of the disease correlated poorly with the overall consumption of the drug did not hold good within Japan, where the regional variation in SMON showed an asto-nishingly close parallel to the use of the drug. Such mis-understandings apart, however, the critics of the issue have correctly drawn attention to some paradoxes in the entire story and to certain questions which require an answer.

One valid point of criticism relates to the fact that from the vast amount of material available it has proved all too tempting and too simple to select the data which one requires to prove any point one may wish to make. The figures compiled by Oakley (1980) on dose–effect rela-tionships, for example, appear to be broadly reliable, but there are numerous cases which do not fit this pattern and where some individual variation in susceptibility to the disorder must be assumed to have been present.

The major paradox clearly relates to the remarkable difference in the absolute incidence of SMON as between Japan and the rest of the world. Although the production figures cited earlier (figures which themselves are very subject to manipulation, to judge by the varying esti-mates given in various places) suggest that something like one-twentieth of all the clioquinol produced yearly in the world was up to 1970 being consumed in Japan, the drug has nevertheless been used on a large scale — and chronically — elsewhere. In particular it has remained on free sale as a popular remedy for diarrhoea in large parts of the world, particularly the Far East, and some of these countries have relatively well-developed medical services which would most certainly have detected any epidemic of proportions comparable to that seen in Japan. This has not happened. There have been only nine published case reports from India since 1970, three from Indonesia, and none from the whole of South America. Indonesia, indeed, presents an especially intriguing case since the total annual consumption of clioquinol in that country in 1977 was still estimated to be some 95 tons, or twice the amount which was being sold annually in Japan in 1970 (Takahashi 1979; Katahira *et al.*1980).

The possibility that bioavailability differences are involved has been raised (Lunde 1980) but dismissed, particularly since various pharmaceutical forms, multiple brands, and probably even a number of related compounds have been involved in the causation of SMON, and precisely the same products held responsible for SMON in Japan have not been incriminated elsewhere.

A second explanation, and a much more likely one, related to differences in dosage. Oakley (1980) has, as pointed out above, compiled from the literature the information necessary to draw a dose–response curve for clioquinol and SMON, and broadly speaking it holds good. At doses of 750 mg per day given for four weeks or less there is little risk of toxic reactions. Neurological symptoms develop in some 1 per cent of patients who take doses of 750–1500 mg daily for less than two weeks. Approximately 35 per cent of those who take 750–1500 mg daily for a longer period develop symptoms. A dose of 1800 mg per day can cause the onset of symptoms as early as the fifth day. At high doses, the onset of toxic reactions may be noted within 24 hours of starting therapy. Whether clioquinol was indeed taken in high doses and for longer periods in Japan than elsewhere is not clear, but as pointed out earlier the pattern of drug use in Japan is such that this could well have been the case. Almost 40 per cent of total expenditure in medical care in Japan is devoted to drugs (Sunahara 1980), a situation which one expects to find in some developing countries but not in advanced industrialized states with ample medical care. Certainly, the actual amount of clioquinol taken by many SMON patients was far in excess of that recommended in the international standard reference works at the time; the average total dose in confirmed cases was in excess of 40 g (Toyokura and Takusu 1975). Finally, it may be noted that at the peak of its career in Japan clioquinol was present in no less than 186 preparations manufactured by 103 companies (Decision

of Tokyo District Court, page 27); it is thus entirely possible that some patients were receiving more than one clioquinol containing product at the same time.

A second explanation for differences between Japan and the remainder of the world might be genetic (Sunahara 1980). Genetic factors can result in marked differences in drug metabolism — for example isoniazid (INH) — but the hypothesis that genetic elements were indeed involved in this instance is entirely unproven. Were racial factors to have played a role one might have expected to have observed a certain incidence of SMON in genetically allied races or at least in the large Japanese communities which exist in other parts of the world.

Again, environmental or dietary factors could be involved. Japanese dietary habits differ in various respects from those of any other country, but it is not at all clear what dietary element could increase susceptibility to the injurious effects of basically neurotoxic substance; certainly there was no evidence of deficiency disease (e.g. involving vitamins of the B group) which might have adversely affected the condition of the nervous system. Nor is there any reason to give credence to the suggestions which have been made that the use of pesticides in Japan was unusually intensive; a little experimental work in animals was performed to study this question and it certainly did not point to involvement of these substances (Takeishi and Otsuki 1975).

The possibility of a drug interaction might deserve consideration. In any community in which drugs are widely used there is a strong trend towards polypharmacy. The literature on clioquinol and SMON is surprisingly silent on the question of concurrent therapy, but clioquinol was being administered in Japan for a range of conditions, some of them poorly defined, for which the doctor would have been prone to prescribe more than one drug. One of the brands most widely used was recommended for 'gastroenteritis, summer diarrhoea, amoebic dysentery, and bacillary dysentery, as well as for 'chronic diarrhoea' and 'abnormal intestinal fermentation'. Clioquinol preparations were also advertised as 'intestinal regulators' (Decision of Tokyo District Court, pages 353 and 356). What other drugs might have been given concurrently? One possibly is diphenoxylate, a known cause of CNS depression and numbness of the extremities. Another might be belladonna, again a CNS depressant and capable of affecting the gastrointestinal absorption of other substances. Neomycin or sulphonamides might have been given. A most intriguing thought is that a light metal might have been involved. Aluminium encephalopathy is a condition about which much more is known today than in 1970 (Reinicke 1980). Bismuth, now known to produce a similar encephalopathy (Reinicke 1980), is a widely used remedy in gastrointestinal disorders. Neither of these latter possibilities appears to have been examined in the Japanese studies or elsewhere. One is not suggesting that any of these agents was the true cause of SMON, but any of them, if prescribed alongside clioquinol, might have rendered the nervous system more susceptible to the drug's toxic effects; it is even conceivable that clioquinol, in view of its chemical properties, might have acted as a carrier of these substances, just as it acted as a carrier for zinc in acrodermatitis enteropathica. One recalls in this connection a description by Shiraki (1975) of two patients in whom SMON rapidly developed and proved fatal when clioquinol was given; in one there had been a preexisting neuropathy induced by isoniazid, in the other a degree of neurological impairment had been induced by dermatomyositis.

In the light of the interesting geographical distribution of what is apparently an iatrogenic disease other work should be mentioned. In this context there is the viral theory of SMON aetiology, which is supported by the work of Nakamura and Inoue (1972).

A virus isolated from SMON patients produced in mice a disease similar to human SMON. C57BL/6 mice were found to be susceptible to the virus when it was inoculated intracerebrally at birth. The diseased mice had paralysis of the legs after an incubation period of 2–3 weeks or more. The main pathological findings were symmetrical axonal degeneration and demyelination in Goll's tract and the pyramidal tract of the spinal cord, without inflammatory change.

The other experimental work that should be cited is the attempts that have been made to produce SMON-like conditions in animals by administration of clioquinol. Tateishi et al. (1971) produced neurological signs in cats and dogs at doses of 60 to 144 mg/kg/day. Histological changes in the optic nerve, posterior columns, and dorsal root ganglia consistent with demyelination were demonstrated. The same workers (Tateishi et al.1972) demonstrated a difference in sensitivity to clioquinol between mongrels and beagles. The beagles were much more resistent to clioquinol and required higher doses to produce toxicity. (The mean cumulative dose required by beagles to produce symptoms was 13 000 mg/kg compared with 4160 mg/kg for mongrels.)

Since the removal of clioquinol from the Japanese market the number of new cases of SMON has declined rapidly. Shimada and Kasaka (1973), however, pointed out that the incidence of SMON in the Ibara district of Japan started to decline one year before the ban on supply of clioquinol. They also point out that in the Okayama prefecture 20 new cases of SMON have occurred since this ban, and report two new SMON cases neither of whom had ever received clioquinol. A feature of both these cases was that a virus was isolated from

their cerebrospinal fluid as reported by Inoue and colleagues.

There is no doubt that many questions remain to be answered concerning this intriguing association between clioquinol and SMON. The current evidence is probably in favour of a cause-and-effect relationship between clioquinol and SMON but undoubtedly other factors are also involved.

Like any other retrospective study of a widespread iatrogenic accident, the SMON study has its shortcomings; the arguments and counter-arguments closely resemble those relating to analgesic nephropathy, and they will be heard again whenever a comparable problem arises. The rearguard action which has been fought to prolong the existence of clioquinol for internal use in large areas of the world has been fought largely for economic and legal reasons. Therapeutically, the drug has had its day; it causes diarrhoea rather than preventing or curing it, it has more worthy rivals as a treatment of amoebiasis, and its use in acrodermatitis enteropathica has been found to rest on a false hypothesis. The question of an efficacy/safety balance thus hardly arises, and one can only express sympathy with the view that, after a career of more than 50 years, oral clioquinol should quietly disappear, but it is unlikely to disappear without leaving a trail of unanswered questions in its wake.

Chloramphenicol

Polyneuritis and optic neuritis due to prolonged chloramphenicol therapy have been described in elderly patients receiving the drug at a dose of 1.5 g/day for several months. Peripheral neuritis and optic neuritis have also been observed in children treated with chloramphenicol because of cystic fibrosis. Biopsy material has revealed changes in the Schwann cells and the myelin sheath.

A case of cerebellar degeneration and other changes reminiscent of SMON was described in Japanese patient treated for prolonged periods with chloramphenicol (Kuroda et al. 1974)

Myeloneuropathy

Nitrous oxide

Layzer (1978) described a neurological disorder which arose after prolonged exposure to nitrous oxide in 15 patients, all but one of whom were dentists. Thirteen patients had abused nitrous oxide to some extent for periods ranging from three months to several years, but two patients were exposed to nitrous oxide only professionally, but working in poorly ventilated surgeries. Symptoms included early sensory complaints, Lhermitte's sign, loss of balance, leg weakness, gait ataxia, impotence, and sphincter disturbances, Neurological examination showed sensorimotor polyneuropathy, often combined with signs of involvement of the posterior and lateral columns of the spinal cord. Electrodiagnostic tests pointed to an axonal polyneuropathy, but other laboratory results were normal, including examination of the spinal fluid. The neurological picture is similar to that of subacute combined degeneration of the spinal cord, and it is possible that nitrous oxide interferes with the action of vitamin B_{12} in the nervous system.

Guillain–Barré syndrome

Interesting information on the background incidence of the Guillain–Barré syndrome, and an adjustment of the rates reported for this syndrome among recipients of swine flu vaccine, was reported by Greenstreet (1983). As a manifestation of presumed hypersensitivity to the now-withdrawn antidepressant, zimeldine, a flu-like clinical picture was reported. In isolated cases these symptoms were associated with reversible parasthesia and muscle weakness; in rare cases the clinical features amounted to those of the Guillain–Barré syndrome.

Extrapyramidal reactions

Phenothiazines

Since the introduction of chlorpromazine in 1951 and the large number of related phenothiazine tranquillizers in the late-1950s, reports have appeared in the literature concerning various neurological disorders resulting from their use (Schwab et al. 1956; Kinross-Wright 1959; Orland 1959; Hollister et al. 1960; Ayd 1961; McGeer et al. 1961; Schiele 1962; Moser 1966).

All members of the three groups of phenothiazine tranquillizers (dimethylaminopropyl side-chain, piperazine side-chain, piperidine side-chain) have been implicated in extrapyramidal reactions to a greater or lesser degree. It is difficult to link frequency of side-effects to chemical structure (Fig.27.3) since there is a considerable variation in patient response, and in this respect it is perhaps significant that, in spite of the large number of phenothiazines currently available, chlorpromazine, the earliest member of this family, is still widely prescribed and used with considerable benefit.

The study by Ayd (1961) is one of the most detailed and comprehensive in the literature. He surveyed 3775 patients treated with phenothiazine tranquillizers, of whom 1472 developed extrapyramidal reactions; 21.2 per cent had acathisia, 15.4 per cent had parkinsonism, and 2.3 per cent dyskinesia. There was correlation between the absolute frequency of these reactions and the chemical structure and milligram potency of the phenothiazine derivative used.

In this series the commonest extrapyramidal reaction to phenothiazines was akinesia. This condition was characterized by weakness and muscular fatigue. It caused the patient to be almost constantly aware of fatigue in a limb used for ordinary, repetitive motor acts such as walking or writing. In advanced form the patients complained of aches and pains in the musculature of the affected limb. Patients were apathetic and were disinclined to initiate activity or to expand energy to complete a task.

The acathisia (motor restlessness) of phenothiazine intoxication causes the patient to walk or pace about; when sitting he constantly taps his feet or shifts his legs, if standing he may continually rock his body backwards and forwards or from side to side. At the same time, there may be chewing movements of the jaw, rolling or smacking of the tongue, and twisting of the fingers. In Ayd's series 65 per cent of cases with acathisia were female and 35 per cent were male; the cause of the difference in sex incidence is unexplained.

Dyskinesia or dystonic reactions are characterized by abrupt onset of retrocollis, torticollis, facial grimacing and distortions, dysarthria, laboured breathing, and involuntary muscle movements. These may be accompanied by scoliosis, lordosis, opisthotonos, tortipelvis, and dystonic gait. Another form of dyskinesia associated with the phenothiazine group is the oculogyric crisis. The attack begins with a fixed stare which is momentary and is followed by rotation of the eyes upwards and to the side and the eyes are then fixed in this position. The patient is unable to move his eyes. At the same time, the head is tilted backwards and laterally, and the mouth is open. Dyskinesia was nearly twice as common in men as in women; again this sex difference is unexplained.

Children under 15 years showed the severest and most bizarre neuromuscular symptoms with phenothiazines. Usually their hyperkinesia was generalized, resulting in a clinical picture resembling advanced cases of dystonia musculorum. The older the patient the more the involvement was restricted to the muscles of the neck, face, tongue, and upper limbs. Thus the drug-induced dyskinesia closely resembles the infantile, juvenile, and late forms of dystonia. Also, the incidence of drug-induced dyskinesia diminishes with increasing age, as is the case with genuine dystonia.

According to Schwab et al.(1956) genuine parkinsonism begins with tremor (50 per cent), rigidity (30 per cent), or impairment of motor activity (10 per cent). By

1. PHENOTHIAZINES

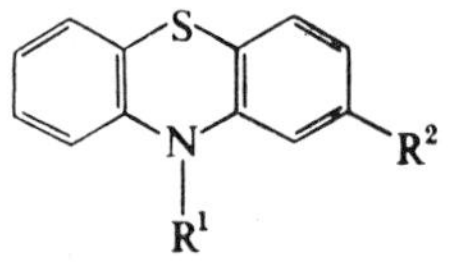

Basic structure of phenothiazine

PREPARATION	STRUCTURE
Dimethylaminopropyl side-chain	$\left(R^1 = -(CH_2)_3 - N\begin{smallmatrix}CH_3\\CH_3\end{smallmatrix} \right)$
Chlorpromazine	$R^2 = -Cl$
Promazine	$R^2 = -H$
Acetopromazine (Acetylpromazine) (Acetazine) (Acepromazine)	$R^2 = -COCH_3$
Methotrimeprazine	$R^1 = -CH_2\overset{\underset{\displaystyle CH_3}{\mid}}{C}HCH_2N(CH_3)_2$
(Levomepromazine) (Levomeprazine)	$R^2 = -OCH_3$

Piperazine side-chain

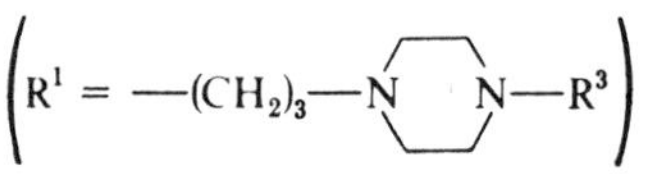

$$\left(R^1 = -(CH_2)_3 - N\diagdown\diagup N - R^3 \right)$$

Fluphenazine	$R^2 = -CF_3$ $R^3 = -CH_2CH_2OH$
Perphenazine (Chlorpiprazine) (Chlorpiprozine)	$R^2 = -Cl$ $R^3 = -CH_2CH_2OH$
Pipothiazine (Pipotiazine) (long acting)	$R^2 = -SO_2N(CH_3)_2$ $R^3 = -CH_2CH_2OH$
Prochlorperazine (Chlormeprazine) (Prochlorpemazine) (Proclorperazine)	$R^2 = -Cl$ $R^3 = -CH_3$
Thiopropazate	$R^2 = -Cl$ $R^3 = -CH_2CH_2OHOCOCH_3$
Trifluoperazine (Triftazin) (Triphthasine)	$R^2 = -CF_3$ $R^3 = -CH_3$

Piperidine side-chain

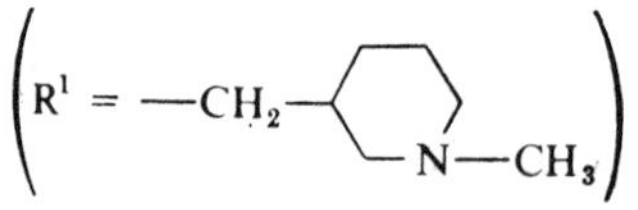

$$\left(R^1 = -CH_2 - \diagbox{N} - CH_3 \right)$$

Pecazine (Mepazine) (Mepasin)	$R^2 = -H$

Thioridazine

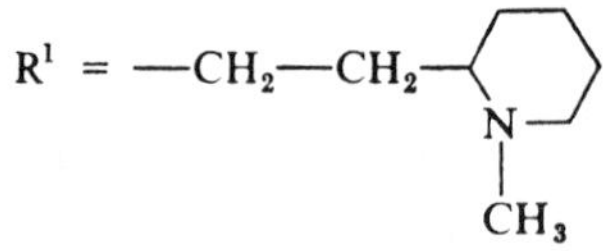

$$R^1 = -CH_2-CH_2-\text{(piperidine-N-CH}_3\text{)}$$

$$R^2 = -SCH_3$$

2. LONG-ACTING PHENOTHIAZINES

Fluphenazine decanoate
Fluphenazine enanthate
Pipothiazine (P.potiazine)
Pipothiazine palmitate
Pipothiazine undecenoate

3. PHENOTHIAZINE-LIKE COMPOUNDS

Clothiapine

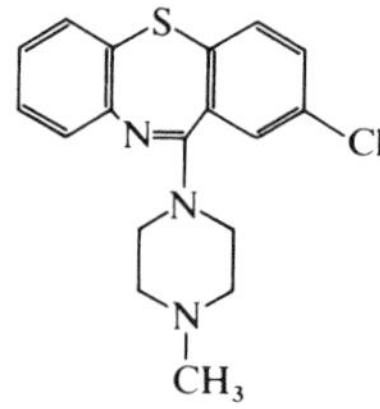

4. THIOXANTHENES

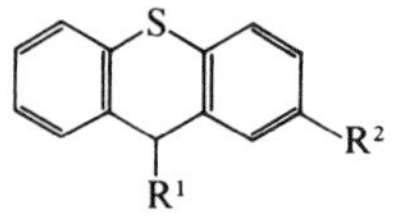

Basic structure of derivatives

Chlorprothixene	$R^1 = =CH(CH_2)_2N(CH_3)_2$
	$R^2 = -Cl$
Clopenthixol	$R^1 = =CH(CH_2)_2-N\diagdown\diagup N-CH_2CH_2OH$
	$R^2 = -Cl$
Flupenthixol	$R^1 = =CH(CH_2)_2-N\diagdown\diagup N-CH_2CH_2OH$
	$R^2 = CF_3$
Thiothixene	$R^1 = =CHCH_2CH_2$ (piperazine-N-CH_3)
	$R^2 = -SO_2N(CH_3)_2$

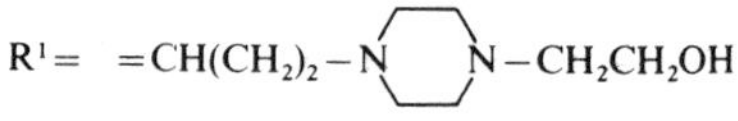

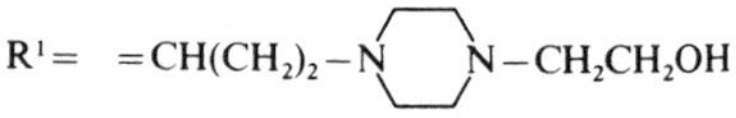

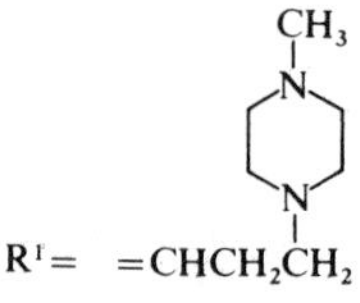

Fig. 27.3 Some drugs associated with extrapyramidal side-effects.

Trifluperidol

5. BUTYROPHENONES AND RELATED COMPOUNDS

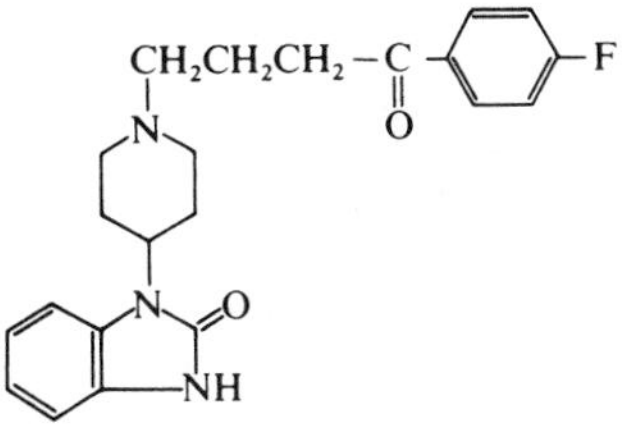

Benperidol

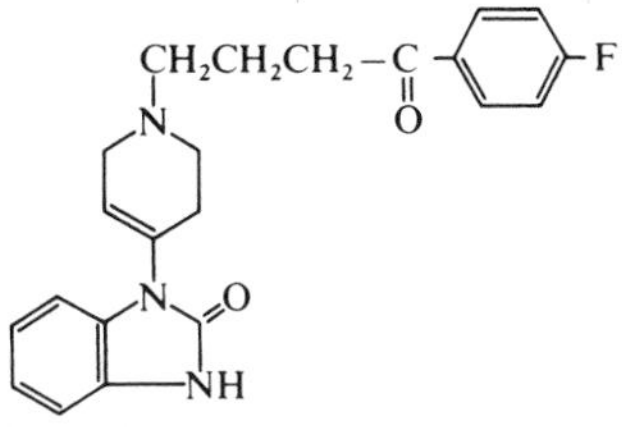

Droperidol

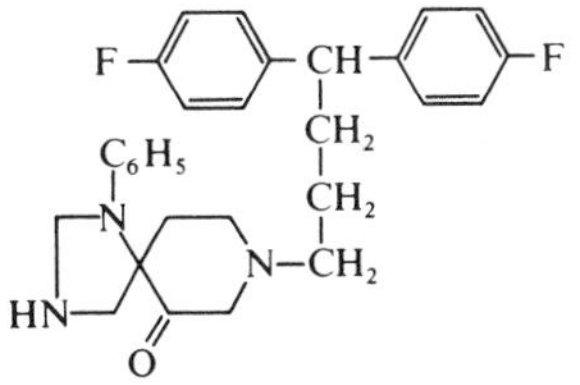

6. RAUWOLFIA ALKALOIDS

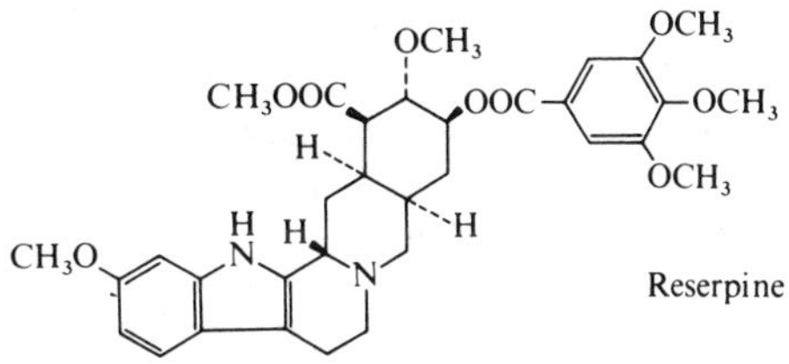

Reserpine

Fluspirilene

7. ANTIMALARIALS

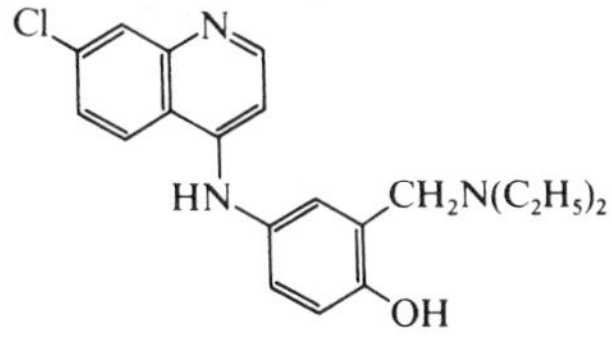

Amodiaquine

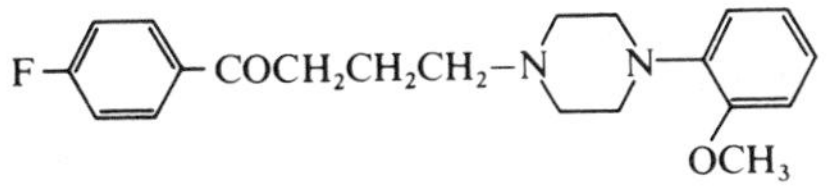

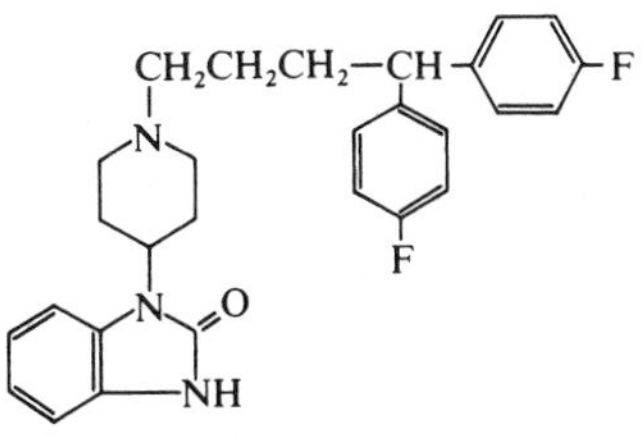

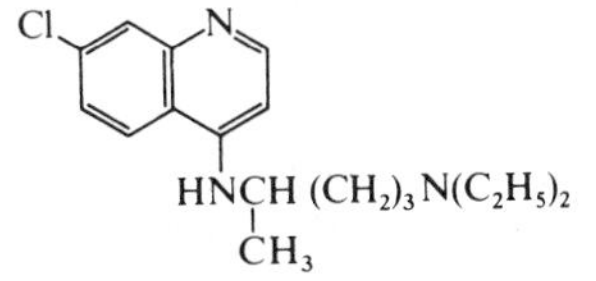

Haloanisone
(Fluanisone)

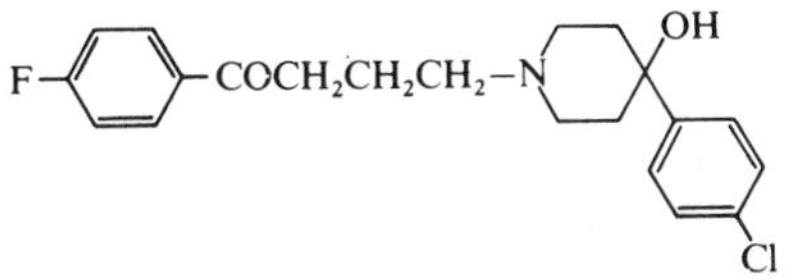

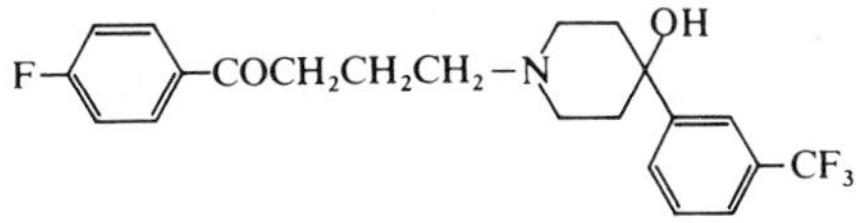

Chloroquine

Haloperidol

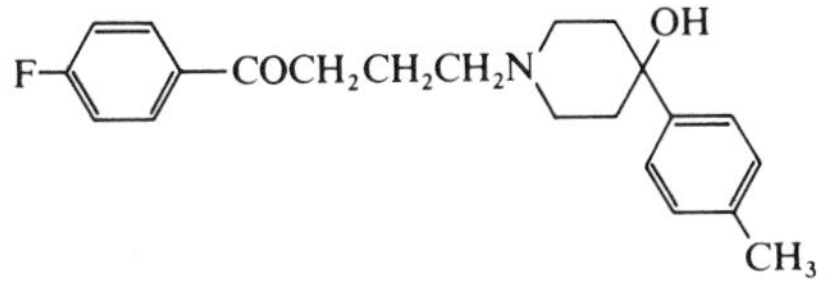

Methylperidol
(Moperone)

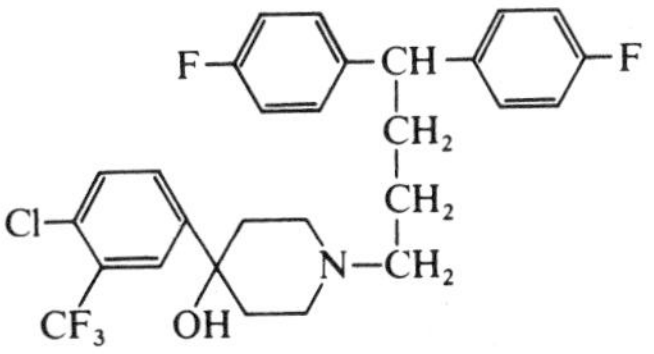

8. OTHER DRUGS

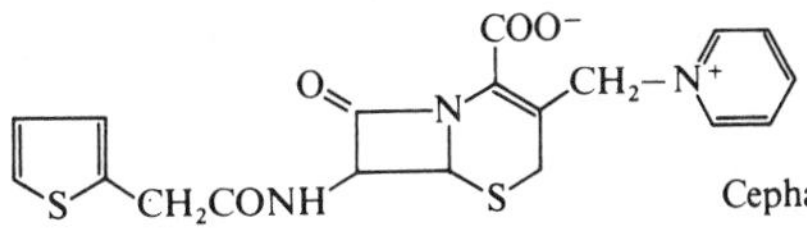

Cephaloridine

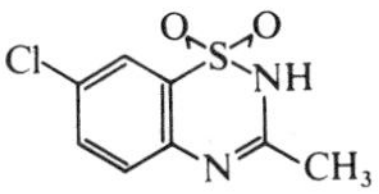

Diazoxide

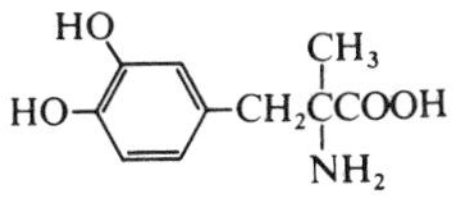

Penfluridol

Methyldopa

Pimozide

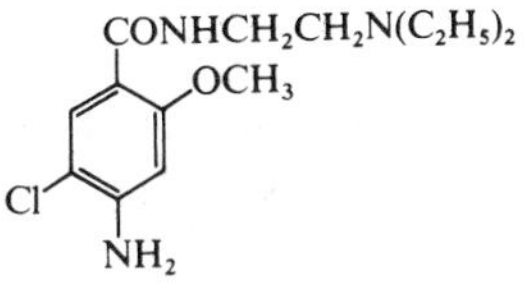

Metoclopramide

contrast, muscular rigidity, impairment of normal associated movements, and cog-wheel phenomena were the initial symptoms of 65 per cent of cases of drug-induced parkinsonism. Tremor, which ultimately appeared in 60 per cent of cases of drug-induced parkinsonism, was the initial symptom in 35 per cent. The age distribution of phenothiazine-induced parkinsonism closely paralleled that of genuine paralysis agitans.

Another difference between drug-induced parkinsonism and genuine parkinsonism was the frequent early manifestation of bulbar symptoms of dysarthria, dysphagia, salivation, and drooling.

With rare exceptions, drug-induced parkinsonism is completely reversible on removal of the offending drug. More rapid reversal may be obtained by treatment with any of a number of antiparkinsonian agents. McGeer and colleagues (1961) postulated that the syndrome was due to interference with the normal functioning of central catecholamines, particularly dopamine, and of central histamine. Extrapyramidal reactions caused by reserpine or phenothiazines were treated with 4–32 g per day of dopa (dihydroxyphenylalanine), the precursor of the catecholamines, with 0.4 to 0.6 g per day of the antihistaminic diphenhydramine or with each drug in turn. Dopa was mildly beneficial in four out of 22 patients, whereas diphenhydramine was completely effective in each of 11 cases. Ayd (1961) reported that dystonic reactions could be relieved promptly by parenteral benztropine methanesulphonate or by biperiden in doses of 1–2 mg. The latter two agents, given parenterally, exerted their effect within about 10 minutes with a maximum effect within 30 minutes. In McGeer's experience dystonic reactions were not a contraindication to further phenothiazine therapy; a recurrence could be avoided either by giving lower doses of the responsible phenothiazine or by the co-administration of an antiparkinsonian drug. Motor restlessness was also controlled by benztropine or biperiden (2 mg i.m.) or other antiparkinsonian drug by mouth; but acathisia, in Ayd's experience, was the most difficult to manage of the extrapyramidal reactions. Only partial relief was obtained with moderate doses of the antiparkinsonian drugs and larger doses were inadvisable due to their own psychotropic effects. The safest control was moderate dosage of the antiparkinsonian agent together with small doses of a barbiturate.

Long-acting phenothiazines

In a series of 140 schizophrenic patients treated with fluphenazine decanoate (depot injection) and observed for a period of 15 months it was noted that 34 patients developed side-effects (Johnson 1973). In this series 34 patients showed acathisia, and 27 developed signs and symptoms of phenothiazine-induced parkinsonism. The manifestations of parkinsonism appeared despite the fact that all patients in this study were given prophylactic benzhexol 5 mg twice daily.

Two surveys on the incidence of side-effects with pipotiazine palmitate or undecylenate in the form of a depot injection were reported. (Ahlfors and Katila 1973; Kristjansen et al 1973). In the former study 42 patients out of 93 developed the signs of parkinsonism, and in the later, 75 of 160 patients developed parkinsonism.

Flupenthixol

Dick and Saunders (1981) described the case of a 17-year-old boy treated for a relapse of his psychiatric condition with depot flupenthixol monthly. After three months he was observed walking on the outside of his feet. Within two months this symptom became more pronounced, and he developed involuntary limb jerking and dystonic hyperextension of the neck and trunk which persisted for two weeks after each injection. A symptom-free period between injections gradually shortened until the movements were continually present. A total of 360 mg flupenthixol was administered over a nine-month period. His condition persisted despite withdrawal of treatment and administration of anticholergics. He had severe, continuous movements of neck, trunk, extremities, with spasmodic myoclonic jerking of arms and legs. Limb myoclonus improved on administration of intravenous clonazepam. The patient's condition improved over the next six months sufficiently to allow him to walk unaided. Dystonic posturing of neck and trunk persisted particularly on exercise.

Butyrophenones

Extrapyramidal side-effects have been reported with haloperidol, pimozide, and penfluridol. Gessa et al. (1972) demonstrated that in 13 chronic schizophrenic patients with extrapyramidal side-effects due to haloperidol, apomorphine caused obvious reduction or complete remission of the extrapyramidal side-effects. These workers consider that these observations endorse the hypothesis that haloperidol induces parkinsonism by blockade of dopamine receptors in the brain.

Pimozide is considered less likely to produce parkinsonian side-effects than haloperidol. With penfluridol extrapyramidal side-effects tend to occur only with doses of 100 mg/week or more, and only rarely is routine antiparkinson medication required to be given simultaneously.

Other drugs causing extrapyramidal side-effects

Metoclopramide

Robinson (1973) reported on a total of 64 cases of meto-clopramide-induced parkinsonism (33 cases reported to Beecham's medical department and 31 cases reported to the Committee on Safety of Medicines). Metoclopramide (*Maxolon*) increases gut peristalsis and is used to speed gastric emptying and as an anti-emetic and is frequently administered to children. The age distribution of reports of metoclopramide-induced extrapyramidal side-effects showed that 22 reports referred to children aged 5–15 years, eight reports referred to young people of 15–19 years, and 33 reports referred to patients of 20 years and upwards. Many of the young children had received quantities in excess of the recommended dosage.

The extrapyramidal effects are transient and respond to withdrawal of the drug, and administration of a sedative or specific antiparkinson drugs. In all respects the extrapyramidal effects are similar to those caused by phenothaizines. De Silva *et al.*(1973) described a series of 10 patients with drug-induced parkinsonism and of these four were due to metoclopramide.

Amodiaquine and chloroquine

Extrapyramidal signs of a parkinsonian type have been reported to occur with both these antimalarial agents.

Akindele and Odejide (1976) desribed cases of parkinsonian side-effects occurring in three young women and a 7-year-old boy given amodiaquine prophylactic treatment in a malarial area. Umez-Eronini and Eronini (1977) described five cases of chloroquine-induced involuntary movements in five patients receiving antimalarial treatment with this agent.

It is of interest that all cases reported of extrapyramidal side-effects with these two antimalarial agents have been young adults or children, and have originated from Nigeria. All cases of amodiaquine or chloroquine-induced involuntary movements responded to antiparkinson drugs and withdrawal of the offending agent.

Cephaloridine

Mintz *et al.* (1971) reported a single case of cephalo-ridine-induced parkinsonism following intravenous administration of 4 g cephaloridine daily. In this patient a macular rash was also noted and considered to be related to the cephaloridine therapy.

Diazoxide

In a study of 100 patients with severe hypertension treated with oral diazoxide, Neary *et al.*(1973) reported that 15 developed extrapyramidal side-effects. The mean serum diazoxide level was higher in patients who developed parkinsonism than in patients who did not. Diazoxide-induced extrapyramidal symptoms responded to antiparkinson agents.

Methyldopa

Parkinsonism has been reported during the treatment of hypertension with high doses of methyldopa, e.g. 2 g daily and above (Yamadori and Albert 1972).

Rauwolfia alkaloids

Reserpine and deserpidine are also major tranquillizers and produce much the same spectrum of extrapyramidal signs and symptoms as the phenothiazines. The mechanism by which these extrapyramidal reactions are produced by the phenothiazines and the rauwolfia alkaloids is not known. Much of the available experimental evidence links the mechanism of action of these drugs to interference with the retention in brain tissue of a number of amines, such as 5-hydroxytryptamine (5-HT) and the catecholamines (adrenaline, noradrenaline, and hydroxytyramine). The rauwolfia alkaloids, such as reserpine, have been shown to deplete the brain of serotonin and catecholamines by competing for their binding sites. The phenothiazines, on the other hand, do not affect the storage of amines but block their action after release. Clinically, the observed results are very similar.

The drugs previously described as effective in the control of phenothiazine-induced extrapyramidal reactions are also effective in those due to reserpine and rauwolfia alkaloids.

Other drug-induced involuntary movements

Levodopa

The commonest dose-limiting effect of the use of this antiparkinsonian drug is its propensity to induce involuntary movements, which predominantly affect tongue, lips, jaw, neck, trunk, and limbs. The movements may be entirely oral dyskinesias, or paddling movements of the limbs, or choreoathetoid movements of limbs and trunk. These movements may necessitate reduction of dosage with the problem of weighing the balance between parkinsonian akinesia and levodopa-induced choreoathetosis.

Diphasic dyskinesia can also be induced by levodopa and may present about 3–4 hours after dosing and lasts

30–60 minutes during which time the patient demonstrates abnormal violent movements of the limbs, retroversion of the neck, open mouth and forced vocalization, and the repetitive use of meaningless words or phrases (Barbeau 1976).

Klawans *et al.*(1975) described 12 patients with Parkinson's disease who developed intermittent myoclonic body jerking due to levodopa. The movements occurred most frequently during sleep and consisted of unilateral or bilateral abrupt jerks of the extremities. Barbeau (1971) drew attention to the syndrome of 'subtle mental changes, persistence of stereotyped movements, and hypotonic-akinetic episodes' in patients on long-term levodopa therapy. These changes would appear to be much more common than the more dramatic choreoathetosis.

The use of dopa decarboxylase inhibitors in conjunction with levodopa has reduced the incidence of many of the side-effects attributed to levodopa, but these agents do not affect the incidence of involuntary movements (Lieberman *et al.*1975).

Phenytoin

McLellan and Swash (1974) described two patients with intractable epilepsy who had been treated with various combinations of anticonvulsant drugs and who developed phenytoin encephalopathy. In both patients choreoathetoid involuntary movements were prominent. Blood phenytoin concentrations were above 30 μg ml^{-1}. When phenytoin was given in smaller doses and its level in the blood fell, the involuntary movements and other clinical manifestations disappeared.

Perhexiline maleate

Gordon and Gordon (1981) reported on the case of a 78-year-old woman who developed parkinsonism while receiving 100 mg perhexiline maleate daily for severe angina. In the month after treatment commenced, the control of her angina was satisfactory but she began to experience falls, difficulty in rising from a chair, and shaking of the lower limbs. After five months of perhexiline therapy, she began to have increasing difficulty walking. A rest-and-action tremor developed in her hands with evidence of mild parkinsonism and bradykinesia, cogwheeling, and gait disturbance. Fasciculations were evident in the small muscles of her hands. Perhexiline was discontinued. During the next two months there was gradual improvement in muscle strength and the parkinsonian symptoms disappeared completely. Neurologic symptoms decreased and she became fully mobile again.

Cimetidine

Bateman *et al.* (1981) described three cases of postural and action tremor following cimetidine therapy for gastric ulcer. In a 68-year-old man postural and action tremor developed within four days of commencing cimetidine 800 mg daily; a 72-year-old woman developed postural and action tremor of the arms with increased muscle tone within 10 days of starting cimetidine 1 g daily (this patient was also receiving metoclopramide). In these two patients the symptoms subsided when cimetidine was discontinued. In these two patients tremor recurred when they were rechallenged with 200 mg cimetidine under controlled conditions. The third patient, a 46-year-old woman, developed marked tremor of all four extremities following use of cimetidine 1 g daily for peptic ulcer. Cimetidine therapy was discontinued resulting in disappearance of tremor. One week later a diagnosis of thyrotoxicosis was made and the patient was restarted on cimetidine; severe tremor reappeared. Her condition was brought under control with propranolol.

Neuroleptic malignant syndrome

This is a rare but potentially lethal syndrome characterized by muscular rigidity, hyperpyrexia, altered consciousness, and autonomic disturbance. It represents a complication of therapy with neuroleptic drugs. Numerous cases have been reported from France and the United States. A second British case was reported by Cope and Gregg (1983).

Tardive dyskinesia

Within the last two decades 'tardive dyskinesia' (Faurbye *et al.* 1964) has become generally established as the name for a syndrome (or a group of syndromes) of abnormal involuntary movements, particularly involving the tongue, mouth, and lower face, which is associated with long-term treatment with neuroleptic drugs. Unlike the acute dystonias, akathisia, and parkinsonian extrapyramidal syndromes induced by neuroleptics, tardive dyskinesia (TD) appears late in the course of treatment and rarely before three to six months: most reported cases have been on continuous treatment for more than two years (Crane 1973). It is often first evident after drug doses are reduced or abruptly withdrawn and it is unrelieved, and may be aggravated, by anticholinergic medication. Striatal dopaminergic mechanisms are believed to be implicated but the pathophysiology is uncertain and no consistently and enduringly effective treatment has yet been established.

The first detailed descriptions of the syndrome in association with phenothiazines are credited to Schönecker (1957) and Sigwald et al. (1959): Bourgeois (1977) cites still earlier accounts of persistent movements of the lower part of the face after stopping chlorpromazine. At first considered a rare development (mainly in elderly and predominantly female patients in mental hospitals with chronic schizophrenia who had received antipsychotic doses of neuroleptics for many years and often showed presumptive evidence of brain damage), TD is now recognized as a common result of prolonged neuroleptic treatment; it can occur in young patients and when these drugs are used for nonpsychotic indications (Sigwald et al. 1959; Evans 1965; Klawans et al. 1974).

High dosages for short periods of time or moderate or low dosages for less than a year may sometimes induce TD. Spontaneous cases occur, particularly in the elderly, and occasional cases have been reported in association with other drugs, but the casual role of the neuroleptics has now been generally regarded as established by epidemiological and experimental evidence to a degree which raises profound questions about the indications for use of these drugs, as well as medico-legal difficulties, especially in the USA (Ayd 1977; Sovner et al. 1978). Since many patients on prolonged neuroleptic treatment appear not to develop TD even after many years, it is not yet known whether neuroleptics alone are a sufficient cause for the syndrome to appear (Perris et al. 1979). Age and possibly female sex appear the most firmly established predisposing factors but the relationship to dosage of neuroleptics, duration of treatment, or other variables is less clear.

The prediction of Hunter et al. (1964) that a 'spate of cases' would come to light, which was based on their detailed and illustrated account of the syndrome at Friern Hospital, has been amply fulfilled. The practical and theoretical importance of the syndrome has generated a large and rapidly expanding literature. Many unsolved enigmas and areas of controversy remain, so that no short account can do justice to the wealth of experimental evidence now available. Gerlach's (1979) review was especially comprehensive and included a careful examination of potential avenues of future research. A recent monograph (Fann et al. 1980) with over 70 contributors report animal and human investigations. Other valuable reviews include those of Crane (1968a, 1973), Paulson (1975), Tarsy and Baldessarini (1976), Fann et al. (1977a), Barnes and Kidger (1979), Baldessarini and Tarsy (1979), and Klawans et al. (1980b).

An important landmark was a report of a joint task-force of the Neuropharmacology Division of the FDA Bureau of Drugs and the American College of Neuropsychopharmacology (Schiele et al. 1973); this clearly identi-fied the syndrome, differentiated it from other neurological complications of antipsychotic drug use, and detailed the precautions in prescribing which were thought necessary to minimize its occurrence. The FDA added a statement on tardive dyskinesia to the package inserts of antipsychotic drugs.

Clinical features

Involuntary movements in the region of the mouth, which have been named orofacial dyskinesia or the bucco-linguo-masticatory syndrome (Sigwald et al.1959) or BLM triad, are the most characteristic clinical features in adults, diagnosed by inspection. The onset is usually insidious. Writhing movements and protrusion of the tongue occur with exaggerated mouthing, lipsmacking, sucking, lateral chewing and pouting of the lips and cheeks. In some cases the tongue is repetitively protruded and withdrawn in 'fly-catching' movements similar to those described in encephalitis lethargica or it may repeatedly distend the cheeks (the 'bon-bon' sign). There may be accompanying grimacing and facial tics. Movements may also occur in localized or more generalized fashion almost anywhere else in the body. There may be choreiform or athetoid movements of the arms, repetitive 'piano-playing' movements of the fingers, shoulder-shrugging, or swaying or rocking movements of the head, neck, or trunk and sometimes repetitive movements of the legs, perhaps with foot-tapping or other movements of the toes. These peripheral or axial movements are not always accompanied by orofacial dyskinesia. The abdominal musculature or diaphragm may be involved. Grunting or irregularities of respiration and difficulties in talking (Portnoy 1979), drinking, and swallowing have been described (Hunter et al. 1964). In the most severely affected cases movement and co-ordination may become disorganized and even life-threatening emergencies have rarely occurred (Casey and Rabins 1978).

Though some patients are greatly distressed by the movements, it is a striking feature that many, perhaps most, affected patients do not complain at all and may even not display awareness of the disorder. It was suggested that patients may be too embarrassed to complain (Smith et al .1979a). Younger patients may be more obviously distressed than chronic psychotic in-patients, but Alexopoulos (1979) found lack of complaints and of awareness was common amongst out-patients.

The movements (and accompanying noises) are socially stigmatizing and may become intolerable to the patient's family or other close associates. Sometimes affected psychotic patients incorporate the experience into a delusional system (Asnis et al. 1979). The movements can be suppressed voluntarily, at least partially and briefly, and they disappear during sleep. Aggravating factors include distraction, voluntary movements elsewhere, or emotional stress or anxiety. The clinical picture tends to take a consistent form for each individual though its intensity may change in response to drug therapy and other factors (Gerlach 1979).

It was suggested by several observers that vermicular movements of the tongue may possibly be an early diagnostic sign, evident before other abnormalities. Early diagnosis may be helped by observing the patient unawares, perhaps when he is walking or concentrating on a mental task, such as reading. Impaired ability to maintain tongue protrusion was used as a measure of TD severity by Klawans and Rubovits (1974) but the specificity of this sign is uncertain (Gardos et al. 1977a). Repeated observation of the patient by a skilled observer is important for early diagnosis (Crane 1973).

Differentiation from other varieties of dyskinesia, akathisia, schizophrenic stereotypes, or from other varieties of facial tic or manneristic habits may occasionally be difficult and other disorders may coexist. Crane (1980a) stressed that the main features of TD are the involvement of specific areas and similarity with normal functions such as chewing and grasping; slow rhythm, and modification with changes in attention and the performance of other tasks. Hypotonia occurs in various combinations with the dyskinesias and causes an exaggeration of associated movement.

When oral dyskinesia persists there may be accompanying hypertrophy of the tongue and coarsening of the features. Intellectual deterioration has not been demonstrated as a consequence of the disorder. One study (Mehta et al. 1978) reported that geriatric mental patients with tardive dyskinesia had a significantly higher mortality rate than a matched group without dyskinesia. Kucharski et al. (1979), however, found TD was no commoner in a small number of psychiatric patients who died than in matched survivors.

Assessment and prevalence

Reported assessments of the prevalence of tardive dyskinesia in chronic mental hospital patients treated with neuroleptics have varied widely from under 0.5 per cent (Hoff and Hofmann 1967), in a survey of some 10 000 patients in 14 institutions, to 56 per cent (Jus et al. 1976a). The latter authors studied 332 adult in-patients with chronic schizophrenia (mean age 48.6 years); average duration of neuroleptic treatment had been 14.5 years. In patients aged 70 or over the prevalence was 75 per cent, while in those under 49 it was 40 per cent. These high figures may reflect detection of very early dyski-

nesia. Of those affected nearly 60 per cent had slight movements only. In psychiatric out-patients with schizophrenia treated with neuroleptics, prevalence figures as high as 31 per cent (Chouinard et al. 1979a) and 43.4 per cent (Asnis et al. 1977) have been recorded.

A review of 44 epidemiologic studies on the prevalence of tardive dyskinesia by Tepper and Haas (1979) gave close attention to the rigour of the methodology employed in identifying the syndrome and the factors which may be associated with it. Five studies which included a rating of severity and tests of inter-observer reliability showed a prevalence between 24 per cent and the 56 per cent found by Jus et al. in chronic in-patients treated with neuroleptics. All but four of the 44 studies were considered to use 'acceptable' definitions of the syndrome. Some sort of scale was used to rank severity in 22 studies and nine included measures to assure inter-observer agreement. Five investigators attempted to introduce objective measures of assessment; these included EEG and film records, a drawing test, measures of EMG and evoked potentials, and polygraph recording.

The syndrome is not rigorously delimited so that diagnostic methodology can be expected to influence estimates of prevalence, as also will sampling procedures and the type of population studied. Brandon et al. (1971), for instance, only included patients if they had orofacial dyskinesia. Increasing prevalence over time has been reported where the same population has been repeatedly studied. Crane (1973), for example, found in a survey of chronic patients that of 184 patients without neurologic symptoms attributable to drugs, 18 had very definite dyskinesia one year later. Of 221 patients studied in 1970 and again in 1974, the overall prevalence of TD nearly doubled and 25 patients were rated as 'severe' in 1974 against only five in 1970 (Crane 1977). Gibson (1978a) reported a prospective study of 374 out-patients receiving depot fluphenazine or depot flupenthixol: in three years the percentage showing the buccolinguo-masticatory syndrome rose from 8 to 22 per cent, although patients had received various neuroleptics for a mean of 13 years previously. Seventy-five per cent of affected patients had the condition in mild degree.

The difficulties of developing a reliable and valid means of assessing the presence and severity of TD, a particularly important need for studies of prevalence, natural history, and response to trials of treatment, were reviewed by Gardos et al. (1977a). The more objective methods may have less clinical validity. Several instrumental techniques have been used, such as various methods of recording finger movements. Denny and Casey (1975) used a pneumatic transducer with an inflatable balloon in the patient's mouth. Crayton et al.

(1977) found absence of the H reflex in patients with TD in recordings from the alpha motor neurones of the soleus muscle. Measurements of vocal function and computer-assisted triaxial vector accelerometry were used by Fann *et al.* (1977*b*)

Such techniques will not be readily applicable outside a few specialist research centres. Simpler methods have included frequency counts of oral movements (Kazamatsuri *et al.* 1972) or duration of tongue extension (Klawans and Rubovits 1974). Many investigators have employed some form of rating scale incorporating global or multi-item assessments or both. Four such scales were detailed by Fann *et al.* (1980). In a comparison of five scales by Chien *et al.* (1980*a*) with a piezo-electric recording technique using an intraoral rubber bulb, the Abnormal Involuntary Movement Scale (AIMS) developed by the Psychopharmacology Research Branch of NIMH (Guy 1976) appeared to correlate best with the objective method. Smith *et al.* (1979*a*) found the inter-rater and test/retest reliabilities of the scale generally reliable: they reported a TD prevalence of 30 per cent using a criterion rating of 3 (moderate symptoms) or more on the AIMS among 293 in-patients with a primary or secondary diagnosis of schizophrenia. On the basis of a criterion rating of 2.5 prevalence was similar in 213 out-patients with schizophrenia, being highest (at about 50 per cent) in the age group 50–69 (Smith *et al.* 1979*b*).

Simpson's Tardive Dyskinesia Rating Scale (Simpson *et al.* 1979) has been shown to have good reliability and validity in studies in New York and Boston, as also has the related 'abbreviated dyskinesia scale' containing 13 items. R.C. Smith (Bell and Smith 1978) developed a Tardive Dyskinesia Scale with the assistance of Crane. In a survey of 1329 patients sampled from a US state mental hospital system the prevalence of definite TD was 26 per cent; about 12 per cent had moderately severe symptoms and at least 40 per cent showed some minimal symptoms. A survey of 15 earlier published studies showed that the recorded prevalence of TD tended to be nearly three times higher when assessments were made on the basis of ordinal rating scales than when questionnaires or global clinical judgements were used (Smith *et al.* 1979*b*).

In view of the many factors which may affect the severity of TD in the short term, such as the patient's awareness of being examined, his state of concentration or motor activity, and the time since drug administration, standardization of the conditions under which observations are made is important for research purposes (Barnes and Kidger 1979). Videotape methods have been widely employed and are particularly useful for training raters and achieving reliability.

There is no generally agreed operational definition of tardive dyskinesia. Oral dyskinesia or the BLM syndrome has been regarded as the core of the syndrome (Brandon *et al.* 1971) but some authors appear readier than others to include axial and peripheral movements. It is not always certain that akathisia, which pharmacologically appears akin to the parkinsonian side-effects of neuroleptics and is accompanied by subjective restlessness, is clearly distinguished. The 'rabbit syndrome' described by Villeneuve (1972) was first regarded as a form of TD but can be distinguished by polygraph recording (Jus *et al.* 1973) and response to anticholinergic drugs (Sovner and Di Mascio 1977; Jus *et al.* 1979). The demarcation between TD and the variety of movement disorders which may occur in schizophrenia is not always clear-cut. Abnormal blinking, for instance, might be due to schizophrenia or part of TD: Stevens (1978) reported abnormalities of eye movement in 34 of 44 patients with schizophrenia who were not taking drugs.

Gibson (1979) found that six out of 167 patients developed generalized chorea within months of starting flupenthixol. A questionnaire sent to 827 UK consultant psychiatrists revealed 279 such cases reported by 112 doctors; 263 doctors said they had seen none, two said they had seen scores of cases. On the basis of estimated prescription figures, Gibson concluded the proportion developing chorea might be one in 230 patients on depot flupenthixol, one in 400 on depot fluphenazine, and one in 1800 receiving oral neuroleptics.

Chronic dystonias attributed to neuroleptics have also been described which, together with irreversible tics, ballismus, and chorea, have been called 'chemical encephalopathies' by Crane (1980*a*).

Aetiological and pathogenic factors

1. Neuroleptic drugs

Reluctance by psychiatrists to accept that neuroleptic drugs could give rise to such adverse effects and the occurrence of TD mainly among chronic institutionalized patients have been cited as reasons why recognition of the association with neuroleptics was delayed (*Lancet* 1979). That the association is causal is now unlikely to be doubted, but some observers have considered such an assumption might be premature and the frequency of occurrence exaggerated (e.g. Kline 1968; Curren 1973). Spontaneous occurrence of the syndrome in elderly patients who have never received neuroleptics is known. In 15 studies reviewed by Tepper and Haas (1979) in which the incidence of TD was recorded in patients who have not received neuroleptics, the prevalence ranged from 0 to over 30 per cent, the highest figures being recorded in geriatric populations. The majority of reported cases of TD have received prolonged treatment with neuroleptics, not necessarily for schizophrenia, and it is a fact of clinical experience that the syndrome is commonly observed as becoming worse or evident for the first time within about two weeks of doses being reduced or stopped. Crane (1968*b*) compared groups of chronic psychiatric patients in the USA and Turkey who were comparable except in relation to exposure to drugs. No cases of dyskinesia were observed in the Turkish patients, of whom only a very few had had neuroleptics, but the syndrome was common in the US patients who had been exposed. Greenblatt *et al*. (1968) surveyed geriatric patients in a nursing home and found TD in 39 per cent of those who had been treated with phenothiazines and in only 2 per cent of those not so treated.

2. *Dosage and duration of treatment*

The relationship between dose of neuroleptic and the occurrence of TD is still controversial. Total dose, exposure to high doses, and duration of treatment have each been thought to be contributory factors, and it has been suggested that all neuroleptics can cause TD (Crane 1973), but the evidence is conflicting. Crane (1970), for instance, found a higher incidence in patients receiving high doses (average daily dose 510 mg equivalents of chlorpromazine) than those on lower average treatment regimes (250 mg of chlorpromazine equivalents daily). Prevalence of TD increased abruptly in patients over 55 receiving more than 200 mg chlorpromazine equivalents daily for over six months and increased again when duration of treatment was over two years (Crane 1974). By contrast, Jus *et al*. (1976*a,b*) found the presence of TD was not related to total amounts of neuroleptics, nor to mean duration or type of drug, but prevalence appeared significantly increased when treatment begins at a higher age.

Among the studies surveyed by Tepper and Haas (1979), four studies concluded that higher cumulative doses of neuroleptics were associated with TD and seven failed to show an association with either mean daily intake or total lifetime intake. Among eight studies with details of duration of treatment four suggested higher risk with longer duration but four could not. Perris *et al*. (1979) also found that neither duration of treatment nor total amount of drugs differentiated their TD patients from controls nor did these variables significantly correlate with severity of TD as assessed by the Simpson scale. Simpson *et al*. (1978*a*) also found prevalence of TD unrelated to drug dosage in another mental hospital study and R.C. Smith *et al*. (1978) found TD significantly negatively correlated to total amount of neuroleptics. Mallya *et al*. (1979) found average total amount of neuropletics slightly (but not significantly) lower in patients with TD than in those without. Ogita *et al*. (1975)

reported there was no difference in the prevalence of TD between Japan and France, although much higher doses of neuroleptics were used in the latter country.

As Crane (1980*c*) pointed out, caution is necessary in interpreting these largely retrospective studies of drug history because the administration of drugs and their dosage are not randomly determined.

The possibility that high concentrations of neuroleptics may contribute to the pathophysiology of TD in at least a subgroup of patients was suggested by Jeste *et al*. (1979*a*). Comparisons of serum neuroleptic activity in eight elderly chronic patients with TD and eight without (five of whom were matched for daily dose of neuroleptic) showed significantly greater serum neuroleptic activity in patients with TD, in both the dose-matched subjects and the total group.

It is possible that neuroleptic dosage and method of administration at the time of assessment will influence the observed prevalence of TD. Jeste *et al*. (1977), for instance, showed that three subjects with TD given chlorpromazine in four-times-a-day dosage had less severe TD than on once-daily dosage. A negative correlation between dose level at the time of examination and prevalence of TD was reported by Crane and Paulson (1967).

3. *Drug type*

It is not yet certain whether some neuroleptics are more likely to cause TD than others. Many of the patients with chronic schizophrenia in whom TD has been described have received a variety of antipsychotic drugs over the years. Recent evidence suggests that long-acting injectable drugs may be associated with greater prevalence of TD; their increased use might contribute to the impression that TD has become more common. Treatment with fluphenazine has been shown a significantly related to prevalence and severity by multiple regression analysis in one study (Chouinard *et al*. 1979*a*). Total amount of long-acting fluphenazine was one variable, along with duration of low-potency neuroleptic therapy, previous ECT, length of neuroleptic therapy, and abnormal EEGs which differentiated patients with and without TD on discriminant function analysis in another (Gardos *et al*. 1977*b*). R.C. Smith *et al*. (1978) found only one drug, fluphenazine, had small-to-moderate positive correlations with TD scores. There was a much higher correlation when the relationship was examined between dyskinesias and the proportion of treatment with intramuscular fluphenazine the patient had received. Outpatients with schizophrenia reported by Gibson (1978*a*) developed an increasing prevalence of TD when they were given injectable fluphenazine or flupenthixol after many years of oral medicines. The development of TD was not obviously relatable to differences in total amounts of neuroleptics received, but the longer the patient had been receiving depot fluphenazine the greater the risk of developing TD. Patients receiving larger than average doses of depot injections took longer than the affected group as a whole to develop the disorder but large doses did not confer protection. Whereas only 5 per cent of the patients who had been receiving it for a year had developed some signs of TD, almost 40 per cent of patients who had been maintained on depot fluphenazine for 11 years had some signs of TD. Flupenthixol seemed to be associated with similar prevalence. Patients transferred to oral pimozide, thioridazine, and depot fluspirilene tended to be relieved of symptoms, but recurrences were later observed in some (Gibson 1980).

If depot preparations carry greater risk of TD, the reason is not yet known. More consistent plasma levels of neuroleptic from parenteral administration and better compliance, or repeated periods of relative drug 'withdrawal' associated with fluctuating plasma levels (Nasrallah 1979a), are among suggested possibilities. Gibson (1980) cited the finding of Curry et al. (1970) that low chlorpromazine levels after oral administration occur in some patients due to reduced absorption in the gut or greater first-pass metabolism: this could not occur with parenteral administration. In a study of serum levels of fluphenazine in nine patients with chronic schizophrenia maintained on depot injections of fluphenazine decanoate, Tune et al. (1980) found levels were low and they tended to be quite stable over two to three weeks for each patient.

Other antipsychotic neuroleptics have been suggested to be more or less likely to cause TD according to chemical structure, capacity to induce parkinsonism, or selectivity or potency of dopamine-blocking action, but the evidence is not strongly consistent and will not be reviewed in detail here. Impressions of lower incidence with pimozide and substituted benzamide drugs might reflect lesser duration of exposure. Clozapine (which has been associated with agranulocytosis) might be different from other antipsychotic drugs in not causing TD (Berger 1980). Metoclopramide, a dopamine antagonist reportedly without antipsychotic effect, has been associated with TD (Lavy et al. 1978). Thioridazine was recommended by Klawans et al. (1980b) when continuation of neuroleptic therapy is absolutely required after TD has developed: its relatively low tendency to induce parkinsonism and a finding that it did not produce hypersensitivity in animal studies were among reasons for the choice, but thioridazine has been associated with TD.

4. Continuity of treatment

For patients requiring long-term treatment with neuroleptics, trials of drug withdrawal or periodic brief 'drug holidays' have been proposed as a way of helping to reduce the prevalence, severity, or likelihood of irreversibility of TD, as well as serving as a useful diagnostic test by possibly unmasking latent disorder (Ayd 1970; Crane 1972b; Schiele et al. 1973). Such proposals have appeared to many to have a commonsense or logical appeal and have been reported useful in practice (Gibson 1980), but are not yet firmly established by evidence. Among the patients of Jus et al. (1976b) treated with piperazine phenothiazines or butyrophenones, mean duration of drug-free intervals was shorter in patients with TD than those without, but this was not so for other classes of neuroleptics. Jeste et al. (1979b), in a study lasting 13 months of 21 hospitalized patients over 50 with TD, found that discriminant function analysis clearly separated the group of patients (12) whose TD abated after withdrawal of neuroleptics and antidepressants for three months from the group with persistent disorders. The best discriminator, to the authors' surprise, was the number of drug-free intervals. Nine patients with persistent dyskinesia had had significantly longer neuroleptic treatment (10.8 years) and a greater number (mean 5.6) of drug interruptions of at least two months duration than did the reversible dyskinesia group. The replicability or significance of this result is not yet known. Chien et al. (1980b) found the number of drug-free intervals in 31 patients with TD was not significantly different from the number in matched controls without TD: mean numbers of drug-free periods were actually slightly higher in the dyskinetic group, who had also received greater amounts of neuroleptics. The authors noted, however, that dyskinetic patients had significantly longer periods of neuroleptic treatment and felt this might indirectly support the value of drug holidays.

5. Antiparkinson agents

It has been suggested that treatment with antiparkinson anticholinergic drugs may lower the threshold for developing TD (Klawans 1973). Three studies (Kennedy et al. 1971; Jus et al. 1976b; Bell and Smith 1978) showed use of anticholinergic drugs was not associated with either increased prevalence of TD or severity scores but a significant positive correlation between severity and amount of cholinergic drugs received was found by Perris et al. (1979). In another study, total lifetime dosage of anticholinergic drugs was higher in 32 dyskinetic patients than in matched controls. (Chien et al. 1980b).

These associations might reflect other factors such as the occurrence of extrapyramidal symptoms (Crane 1972*b*) or exposure to neuroleptics in high dosage or high potency rather than the action of anticholinergic drugs *per se*. It is now well established that the administration of antiparkinson agents can aggravate TD or may precipitate TD which was not previously evident (Gerlach 1977; Smith and Kiloh 1979; Chouinard *et al.* 1979*b*). The latter might be used as an early diagnostic test. The effect of anticholinergic drugs on TD cannot be attributed to any effect on plasma levels of neuroleptics (Gerlach 1979). There was a report (Birket-Smith 1974) of six patients who developed TD after anticholinergic drugs alone and others were cited by Smith (1979).

The possibility that prior treatment with antiparkinson agents makes TD more likely after they have been withdrawn has not been clearly demonstrated (Gerlach 1979). Such a possibility, however, strengthens the current recommendation that antiparkinson drugs should not be routinely prescribed for prophylaxis when neuroleptics are administered. Crane's (1972*a*) suggestion that TD would be more likely, on withdrawal of neuroleptics, to occur in patients who developed drug-induced pseudoparkinsonism does not appear to have been confirmed (R.C. Smith *et al.* 1978).

6. Age, sex, and other possible predisposing factors

Increasing age appears to be significantly associated with higher prevalence of TD (e.g. in nine of 15 studies surveyed by Tepper and Haas 1979). Smith *et al.* (1979*a*) found the relationship with age appeared to be linear in females but curvilinear in males. Age was the variable most related to the prevalence of TD in the study of Jus *et al.* (1976*a*) and was also found to be significantly related to severity by Chouinard *et al.* (1979*a*). The latter relationship was not observed by Perris *et al.* (1979).

Greater prevalence in females than males was reported in 12 of 17 studies (Tepper and Haas 1979). The observed sex difference has not been accounted for solely by age factors or duration of treatment. Chouinard *et al.* (1979*a*) found males tended to show more severe disorder. Smith *et al.* (1979*a*) however, observed that, only when more severe symptoms were used as criteria, did female predominance in prevalence become marked. Significantly more choreoathetoid movements in females than males were noted by Jus *et al.* (1976*a*) but overall prevalence of TD did not show a difference between the sexes.

Other factors which have been examined as possible predisposing factors for TD include organic brain disease, age at onset of neuroleptic treatment, psychiatric diagnosis, duration of hospitalization, prior ECT or leu-cotomy, insulin therapy, and alcohol or drug addiction. No causal relationship with any of these factors has been established, though organic brain disease has commonly been thought to be contributory.

Evidence of structural brain damage associated with TD might point to a consequence of the underlying disorder or a predisposing factor. Association of evidence of possible brain damage with prevalence of TD was reported in some hospital surveys (e.g. Crane and Paulson 1967; Edwards 1970; Perris *et al.* 1979) but not others (e.g. Greenblatt *et al.* 1968; Heinrich *et al.* 1968). In one study (Chouinard *et al.* 1979*a*) brain damage seemed to predispose to more severe forms of disorder but not to its initial appearance. Christensen *et al.* (1970) described atrophic changes in the substantia nigra and brainstem gliosis in 27 of their 28 cases of TD, but the findings have been questioned. The patients had started taking neuroleptics at an unusually late age. Previous small-scale studies had shown contrary results. Hunter *et al.* (1968) found no significant pathology in three patients. Three cases with atrophic caudate nuclei reported by Gross and Kaltenbach (1968) had other probable causes of damage. Two studies of CT scans in patients with tardive dyskinesia (Gelenberg 1976; Jeste *et al.* 1980) showed no significant abnormality. The latter examined 12 severely affected patients with chronic TD assessed on AIMS with matched controls. By contrast, Famuyiwa *et al.* (1979) found abnormality in CT scans in 31 of 45 long-stay patients, 17 of whom had TD. On one learning test the TD group performed significantly worse than the 33 control patients with schizophrenia, and the TD group had significantly more abnormality of Ventricular Index, but severity of cortical atrophy and increases in ventricular size showed no significant differences. The criterion for diagnosis of TD and severity are not recorded.

A high incidence of affective disorders in patients with TD was noted by Davis *et al.* (1976) and Alpert *et al.* (1976) and parallelism in response to treatment between dyskinesia and depression was reported by Rosenbaum *et al.* (1977). Thirteen of 18 consecutive patients with TD, none of whom fulfilled diagnostic criteria for schizophrenia, had elevated urinary free cortisol levels (Rosenbaum *et al.* 1979).

The edentulous state has been suggested as contributing to the occurrence of TD. Ill-fitting dentures have been said to cause an orofacial mandibular syndrome or conversely TD may make it difficult to retain dentures (Sutcher *et al.* 1971; Brandon *et al.* 1971).

7. Individual vulnerability

Present evidence suggests that there must be a variable

individual predisposition to develop tardive dyskinesia, since some patients appear much more vulnerable than others. It is therefore not possible to define risk limits for drug dosage or duration of treatment (Gerlach 1979).

A claim to the youngest recorded onset of persistent TD is made by Chouinard and Jones (1979). A 23-year-old depressed man was found to have definite dyskinesia of the tongue after one month of treatment with neuroleptics (trifluoperazine in low dosage, followed by chlorpromazine increasing to 600 mg/day); six months later, after continued treatment with a depot neuroleptic, the disorder was manifest after procyclidine challenge and was worse than before. Other reported cases of early onset of TD or occurrence on low dosages of neuroleptics (Thornton and Thornton 1973; Klawans et al. 1974) were reviewed by Fann et al. (1977a) who cite a report of a patient who developed short-lived oral dyskinesia after receiving 900 mg of mesoridazine in 72 hours. One of Evans' (1965) cases, a 50-year-old woman, developed a TD syndrome after trifluoperazine 2 mg b.i.d. for some five months.

The prevalence of such cases is not known. They may be well commoner than the occasional reports suggest, but at present they appear atypical. In a study of 39 geriatric patients not previously treated with neuroleptics, Crane and Smeets (1974) found moderate or severe TD was detected after drug withdrawal only in patients treated for at least seven months, with at least 72 mg of chlorpromazine equivalent per day and a cumulative dose of at least 14 g.

Association with other drugs

Syndromes with similarities to neuroleptic-induced TD have rarely been attributed to other drug treatments. Oral dyskinesias following amphetamine and L-dopa are well recognized. Isolated cases of transient dyskinesia were reported in association with many drugs but these usually differ from TD. Fann et al. (1976) reported two adults aged 37 and 44 who developed a brief BLM syndrome and choreoathetoid movements respectively on amitriptyline. Both had received neuroleptics but re-exposure to the antidepressant led to reappearance of the movement disorder with remission on withdrawal. A case of oral dyskinesia was attributed to imipramine (Dekret et al. 1977). Four cases of TD-like syndromes were attributed to H_1-antihistamines (Thach et al. 1975; Davis 1976; Hale and Heins 1978). All were aged over 50 and had taken treatment for chronic rhinitis or sinusitis for many years. In three, blepharospasm was the first manifestation (Smith and Domino 1980). Phenothiazine antihistamines were not particularly implicated. One case of transient oral-buccal dyskinesic symptoms attributed to low-dosage benzodiazepines occurred in a 63-year-old man who had three episodes when given benzodiazepines (Kaplan and Murkofsky 1978) and others were reported by Rosenbaum and de la Fuente (1979). Two cases of involuntary chewing movements after tranylcypromine plus lithium were described by Stancer (1979). Phenytoin intoxication can cause oral dyskinesias (Chadwick et al. 1976), which have also been reported in association with ethosuximide (Ehyai et al. 1978).

Neuroleptic-induced dyskinesias in children and young adults

Dyskinesic symptoms have been reported in association with neuroleptics in children (McAndrew et al. 1972; Polizos et al. 1973; Paulson et al. 1975). Oral dyskinesia appears to be uncommon, but repetitive choreiform movements of the extremities may often occur after neuroleptic treatment is withdrawn. They tend to remit spontaneously (e.g. within a month in nine of 18 cases cited by Schiele et al. 1973). Occurrence during drug treatment is very rare, according to Polizos and Engelhardt (1980), who found that 51 per cent of 184 children withdrawn from neuroleptics developed what they call 'withdrawal emergent symptoms.' The children, mostly aged 6–12, were out-patients in New York being treated for 'childhood schizophrenia with autistic features'; hyperactivity, mood disturbance, severe handicap in self-management, and retarded functioning are among reported features. Drugs were given for 3–24 months (average 8 months) in dosage comparable to adult doses (e.g. chlorpromazine mean 251 mg/day). Forty-one per cent had involuntary movements and ataxia, the latter accompanied by generalized hypotonia. Movements were primarily choreoathetoid, but myoclonic and hemiballismic movements, posturing, and head-rocking were occasionally also observed. Oral dyskinesia occurred in 20 per cent and in only six was it the main symptom: it was always mild. In the majority of children, active treatment had to be reinstated within a week because of behavioural relapse, but spontaneous remission was observed in 35 per cent of those affected and in 80 per cent of the children dyskinetic symptoms disappeared within 14 days of drug withdrawal. No relationship was found between the occurrence of abnormal movements and age, sex, dosage, or duration of treatment. The authors had seen only two cases of TD during (high-dosage) neuroleptic treatment. Both subsided rapidly on reduction of dose. Paulson et al. (1975), however, found that 21 of 103 mentally retarded children aged 11–16 who had recieved phenothiazines for long periods had mild-to-moderate tardive dyskinesia. Seven children were clearly worse when phenothiazines were withdrawn. Of 15 followed up four years later, after continued intermittent drug treatment, six were unchanged, four worse, five better. Further single cases were reported by Browning and Ferry (1976), Kumar (1976), McClean and Casey (1978), and Petty and Spar (1980)

The predominance of involvement of the extremities and trunk and the tendency to remission described in children have also been noted in young adult patients. To illustrate the 'several' cases they had seen of young adults developing extensive and disabling TD, Tarsy *et al*. (1977) reported two cases, aged 19 and 23, who developed severe TD during treatment with phenothiazines for psychotic illness. Exacerbation of symptoms occurred on withdrawal; haloperidol brought improvement.

Cerebrospinal fluid (CSF) accumulations of 5-hydroxy indole acetic acid (5-HIAA) and homovanillic acid (HVA) after probenecid did not differentiate five children who developed dyskinetic withdrawal symptoms after discontinuation of chronic neuroleptic treatment from six who did not. CSF 5-HIAA showed significant decrease after three to four weeks after withdrawal (Winsberg *et al*. 1978). Studies of CSF constituents in adults with TD (Nagao *et al*. 1979) have also failed to show consistent changes.

Pathophysiology

Much interest has centred on the possibility that tardive dyskinesia reflects a hyperdopaminergic state associated with hypersensitivity (Klawans 1973) or supersensitivity (Tarsy and Baldessarini 1973; Marsden 1976) of cerebral dopamine receptors. Alternative or contributory mechanisms might be cholinergic hypofunction (Fann *et al*. 1974) or imbalance in cholinergic-dopaminergic systems (Gerlach *et al*. 1974). Disturbances of other neurotransmitter systems, such as GABA, may be involved.

The antipsychotic action of neuroleptic drugs appears to be related to dopamine blockade (Van Rossum 1966). TD bears some resemblance to Huntington's chorea which appears to be associated with hyperdopaminergic activity in the nigrostriatal system. Similar kinds of dyskinesia have been produced in patients with parkinsonism by L-dopa administration. Amphetamines can cause oral dyskinesias.

According to the hypersensitivity theory, the neuroleptic drug induces incomplete blockade of the dopamine receptor, which then becomes hypersensitive as a consequence of 'chemical denervation' and liable to respond abnormally to any dopamine or dopamine-agonist which reaches it. The clinical observation that TD commonly first appears or worsens when a chronically administered neuroleptic drug is reduced or withdrawn (Degkwitz *et al*. 1967; Crane 1968*a*) fits in with the theory, as does the fact that the reinstitution or increased dosage of the original neuroleptic or administration of another dopamine-blocking drug often relieves TD, but perhaps only temporarily and at a stage when the disorder is still reversible. According to the theory, hypersensitivity will again develop so that the abnormal movements 'break through' once again. The drug that causes the disorder relieves it for a time, only, it is said, at the risk of making it worse and possibly permanent.

Biochemical and behavioural hypersensitivity to dopamine-agonists has been repeatedly induced in experimental animals after chronic neuroleptic pretreatment, and an increase in drug-specific receptor binding sites was described (Klawans *et al*. 1977; Burt *et al*. 1977). In most studies the induced hypersensitivity was short-lived, unlike the more persistent human disorder. Recently, induced supersensitivity of striatal dopamine mechanisms persisting for six months after withdrawal of continuous administration of neuroleptics for up to a year in rats was reported (Clow *et al*. 1979). The animal model most closely analogous to human TD appeared to be the Cebus monkey (Gunne and Bárány 1976). Chronic haloperidol administration induced a persistent spontaneous syndrome of oral dyskinesia closely similar to TD. Recent studies in rodents suggest induced supersensitivity may be dose-related: neither low does of chlorpromazine nor any doses of thioridazine studied produced behavioural supersensitivity (Klawans *et al*. 1980).

Two sorts of dopamine receptors. D_1 and D_2 are now identified (Kebabian and Calne, 1979). Drugs which blockade D_1 or D_2 receptors predominantly have been reported to cause TD. The finding that relatively selective D_2 blockers such as pimozide, metoclopramide, or other benzamides can inhibit TD has led to a speculation that D_2 receptor hypersensitivity might primarily be implicated (*Lancet* 1979). but this has been criticized for lack of supporting evidence and by virtue of contrary evidence from the occurrence of TD with D_1 blocking drugs such as flupenthixol (Jenner and Marsden 1979).

Prolactin secretion is under dopaminergic control. Attempts have therefore been made to confirm the hypersensitivity theory by looking for enhanced suppression of prolactin or release of growth hormone in response to dopamine-agonists (Ettigi *et al*. 1976; Smith *et al*. 1977; Tamminga *et al*: 1977*a*; Cohen *et al*. 1979). The predicted changes in endocrine response have not been observed, perhaps because secretion of prolactin does not reflect dopaminergic hypersensitivity. In animals, increase in receptor sensitivity produced by prolonged administration of neuroeptics has led to a reduction in dopamine synthesis. Measurements of CSF HVA and cyclic AMP in patients with TD after probenecid administration failed to show the decrease and rise, respectively, of these CSF consituents which might possibly have been expected if increased nigrostriatal dopamine receptor activity occurs in human TD (Bowers *et al.* 1979). In rats limbic hypersensitivity to dopamine-agonists has also been induced by neuroleptics (Davis and Rosenberg 1979). Although limbic mechanisms have been thought implicated in the pathogenesis of schizo-phrenia, the occurrence and severity of tardive dyski-nesia is not related to psychotic symptomatology.

Recent evidence suggests that antipsychotic drugs accelerate the metabolism of dopamine in the human brain in a regionally specific manner. Bacopoulos *et al.* (1980) found the concentration of homovanillic acid (HVA) was significantly increased in selected cortical areas but not in the putamen or nucleus accumbens in deceased patients with schizophrenia who had recieved antipsychotic drugs, but was normal in patients with schizophrenia who had not received antipsychotic drugs.

Another approach to possible understanding of the pathophysiology of TD is to see it as a kind of obverse to parkinsonism, reflecting cholinergic hypofunction or cholinergic-dopaminergic imbalance. The well esta-blished precipitation or aggravation of TD by antipar-kinson agents or the more questionable possibility that they lower the threshold for the later development of TD (Klawans 1973) would fit it with this view. It has been observed that when TD and parkinsonian symptoms coexist in the same patient, as is now commonly reported, treatment which relieves one condition com-monly exacerbates the other (Fann and Lake 1974; Chouinard *et al.* 1979*b*).

Although biochemical approaches to the pathophysio-logy of TD have attracted greatest interest, there exists the important possibility that tardive dyskinesia, par-ticularly when it becomes apparently irreversible, may reflect strutural changes in the brain. Mention was already made of the conflicting evidence on the possi-bility of an association with gross physical pathology of the brain, either as a cause or a consequence of TD. The possibility that TD reflects pathology in the macro- or microstructure of selective areas of the brain has not been excluded.

The significance of the unusually high incidence of B-mitten EEG patterns in patients with tardive dyskinesia reported by Wegner *et al.* (1977) and Struve *et al.* (1979) is not yet known. This finding could prove a valuable pointer to identifying the risk of TD development. The dysrhythmia has been thought possibly linked with affective disturbance.

It is possible that several pathophysiologic mecha-nisms may be involved in TD, perhaps evident in dif-ferent kinds of movement disorder, or related to

reversible or irreversible forms of the syndrome. Gerlach (1979) reviewed evidence that reversible TD may result from dopamine hypersensitivity and that irreversible dis-order, occurring mainly in elderly patients as a conti-nuation of reversible TD, reflects reduced biological 'buffer capacity', possibly due to degenerative changes in dopaminergic, cholinergic, or GABA-ergic systems and possible to cell loss related both to age and neuroleptic treatment.

The occurrence of oral dyskinesias among the motor abnormalities in chronic manganese poisoning, and evi-dence that chronic chlorpromazine administration can increase manganese concentrations in the caudate nucleus and cerebellar hemispheres in guinea-pigs raised the possibility that drug-induced alterations in caudate manganese levels could be a factor in the pathogenesis of TD (Weiner *et al.* 1980).

Trials of drug treatment for tardive dyskinesia

Many drugs believed to alter cholinergic–dopaminergic balance, as well as a great variety of other agents, have been given to patients with tardive dyskinesia in an attempt to elucidate the differential pharmacology of the disorder and to discover effective treatment. Many agents have been reported to affect its severity in the short term, but the variability of individual responses has been striking. Although several agents have shown pro-mise, the prevailing view is still that a consistently and enduringly effective form of treatment has yet to be dis-covered (Kobayashi 1977; *Drug and Therapeutics Bul-letin* 1978; *Medical Letter* 1979; *British Medical Journal* 1979; Casey 1978; *Lancet* 1979; Berger 1980). It was sug-gested that the differential pharmacology will have to be explored further on an individual basis (Casey and Denney 1977; Nasrallah 1979*b*), perhaps by exploratory drug 'challenges' (Kobayashi 1977). In this vein, Carroll *et al.* (1977) described a patient who took 19 different drugs in two years, with no lasting benefit from any. In some reports it was noted that different components of the movement disorder may appear to be differently affected by different drugs (Gerlach 1979).

Many of the reported trials of pharmacotherapy have been very small-scale non-comparative studies, usually lasting no more than a few weeks and conducted in chronic in-patients. New treatment approaches have been claimed from single-patient studies. These and other methodological shortcomings were emphasized in a review of drug treatments by Mackay and Shepherd (1979). The review of Jeste and Wyatt (1979) included tabulated summaries of some 170 trials. Reports of the studies by major research groups were included in Fann *et al.* (1980). Only a few illustrative references will be cited here. Drugs will be grouped according to notions of their rationale, as follows:

(a) Drugs expected to aggravate TD

It is now well established that anticholinergic antiparkinson drugs do not relieve TD and are likely to exacerbate it. A few patients, none the less, have been said to seem to benefit (e.g. Cole, cited by Quitkin et al. 1977; Uhrbrand and Faurbye 1960).

In line with the hypersensitivity theory, dopamine agonists would also be expected to aggravate the disorder. Movements worsened in 12 of 40 patients given L-dopa 100 mg IV (Hippius and Logemann 1970) and there were other reports of aggravation by L-dopa (e.g. Gerlach et al. 1974), but improvement in isolated cases was also described (e.g. Carroll et al. 1977). L-dopa in doses of 6 g/day was administered for two months to a small number of patients with TD by Alpert and Friedhoff (1980), in an attempt to reduce hypothesized supersensitivity at dopamine receptors. The two of seven patients who completed the treatment remained free of TD for two years. Slow build-up of dosage was found necessary because of aggravation of psychosis. Intravenous d-amphetamine was found to aggravate TD by Smith et al. (1977) but a trial of methylphenidate in 17 patients by Fann et al. (1973) showed no significant effect. The apparently paradoxical finding that the dopamine agonist apomorphine was found beneficial to some patients (Carroll et al. 1977; Smith et al. 1977; Tolosa 1978) was presumed to be due to its acting on inhibitory DA receptors at low dosage. Some drugs, such as the ergolines, may be capable of both dopamine agonist and antagonist effects. A controlled trial of bromocriptine by Chase was reported by Mackay and Shepherd (1979) to have shown no effect. Conflicting results were reported for amantadine.

(b) Dopamine-antagonists

The short-term efficacy of dopamine-antagonists in suppressing tardive dyskinesia was shown in several double-blind comparative studies. In 32 trials surveyed by Jeste and Wyatt 1979) 70.1 per cent of 259 patients benefited. Haloperidol (Kazamatsuri et al. 1973) and thiopropazate (Smith and Kiloh 1979) were particularly studied. A clinical impression of the value of haloperidol H was given by Gilbert (1979). In an open study of thiopropazate, Smith and Kiloh (1979) reported significant improvement at six months (but not earlier) in the 11 of 21 patients who completed this unusually long trial. TD deteriorated when the drug was withdrawn but was not worse than before starting.

Pimozide in doses of 6–28 mg/day was shown to be effective in controlled trial (Claveria et al. 1975) and its initial effect was reported by Gibson (1980) to have been 'almost too striking to be believed' in 16 of 17 patients given 4–8 mg/day who had developed TD on depot fluphenazine. Dyskinesic symptoms were observed 'creeping back' in 10 of 16 patients after two to three years but they were mild, and the time to reappearance seemed longer than the few weeks reported by Kazamatsuri et al. with haloperidol. The efficacy of the non-antipsychotic, dopamine-antagonist, metoclopramide has been reported as negligible but may be slight at high dosage (Bateman et al. 1979). There is recent evidence that other selective D_2 receptor antagonists such as oxiperomide (Casey and Gerlach 1980), sulpiride (Casey et al. 1979), and tiapride may be effective. Clozapine has been reported as possible 'curative' (Gerbino et al. 1980) or at least beneficial (Simpson et al. 1978b), but has been associated with agranulocytosis. The use of neuroleptics to treat TD has seemed to some illogical and potentially hazardous in the long term, since they may be aggravating the underlying pathophysiology to the extent of making a potentially reversible disorder into an irreversible one. Although eventual recurrence in some degree is likely, as reported by Gibson (1978a), the hypothesis that they will cause long-term worsening or irreversibility has not been adequately tested (Berger 1980). Chien (1980), in a review of the studies he had conducted on chronic patients at Boston State Hospital with Kazamatsuri and Cole (e.g. Kazamatsuri et al. 1973), described the results of an 18-month follow-up. During this interval, 22 patients participated in several trials of drugs affecting catecholamines, including an 18-week study of haloperidol at dosages of 8 and 16 mg/day. Dyskinetic symptoms were about the same as at the beginning.

Some authors have condemned the treatment of TD with neuroleptics as comparable to treating delirium tremens with alcohol or narcotic dependence with opiates. Such analogies are emotive and may be misleading in cases of schizophrenia where a continuing need for neuroleptics is the price for freedom from schizophrenic symptoms.

In discussing the possibility that an association between intermittent drug treatment and TD could be a causal one, Jeste et al. (1979b) mention the phenomenon of 'kindling' whereby intermittent stimulation by drugs etc. may produce sensitization. They stated there was no evidence as to whether intermittent use of neuroleptics causes kindling other than the demonstration by Weiss and Santelli (1978) that dyskinesia in primates can be induced by intermittent haloperidol administration.

(c) Dopamine-depleting drugs

Among dopamine-depleting agents there is evidence of some efficacy for tetrabenazine from several studies (e.g. Kazamatsuri *et al.* 1972). α-Methyldopa has not proved effective. Efficacy has also been claimed for oxypertine which may be a dopamine-depletor. Reserpine has only rarely been reported to cause TD (Uhrbrand and Faurbye 1960) and Klawans (1973) suggested this was consistent with the hypersensitivity theory. It was re-explored as an antipsychotic drug by Bacher and Lewis (1978) and there have been some reports of improvements in TD but it may cause depression. The synthesis-inhibitor α-methyl-para-tyrosine (AMPT) showed partial effects in studies reviewed by Gerlach (1979).

(d) Cholinergic drugs

On the basis that TD may reflect a relative hypocholinergic state, drugs which might increase cholinergic activity have been explored. Physostigmine, which has to be given parenterally, was reported to be effective in a comparative trial against placebo (Tamminga *et al.* 1977*b*) and some open studies, but others found no value or a worsening of symptoms (Tarsy *et al.* 1974). Earlier favourable reports of Deanol, which had been thought to act as a precursor of acetylcholine, were confirmed by later controlled studies (Lindeboom and Lakke 1978; Jus *et al.* 1978; de Montigny *et al.* 1979). Promising results were reported with choline (Growdon *et al.* 1977) and lecithin (Wurtman *et al.* 1977; Gelenberg *et al.* 1979). Choline has a bitter taste and causes a fishy body smell, but lecithin, the usual dietary source of choline, may offer a practicable treatment. Fifty g/day was found effective in all six patients studied in cross-over trial for 14 days by Jackson *et al.* (1979). There were no side-effects and five patients continued on neuroleptics.

(e) GABA-potentiating drugs

GABA, an inhibitory neurotransmitter, may be linked with nigrostriatal dopaminergic functions. In rats chronic blockade of dopamine receptors by antipsychotic drugs enhances GABA binding in the nigrostriatum (Gale 1980). Conflicting results in the treatment of TD were reported for sodium valproate (Linoila *et al.* 1976; Gibson 1978*b*; Casey and Hammerstad 1979) and baclofen (Gerlach *et al.* 1978; Nair *et al.* 1978) which may potentiate GABA. Fifty per cent of the elderly patients in the trial of Gerlach *et al.* experienced sedation, muscular weakness, and confusion. Benzodiazepines, too, may potentiate GABA and favourable results, possible due to sedation, have been reported in some patients treated with clonazepam. Muscimol, a psychotomimetic drug which is thought to be a GABA agonist, seemed effective in a trial by Tamminga *et al.* (1979).

(f) Miscellaneous drugs and other treatments

Other drugs tried in TD include: pyridoxine, manganese, cyproheptadine. *Hydergine*, L-tryptophan, fusaric acid, disulfiram, MSH, MSH-inhibiting factor 1, and phenytoin. Early encouraging reports on the use of lithium were not confirmed by later studies (e.g. Jus *et al.* 1978). Rosenbaum *et al.* (1977) reported good results with a combination of lithium and imipramine. Propranolol 20–40 mg/day gave good results in an open trail in 10 patients (Bacher and Lewis 1980). Papaverine, which may cause dopamine blockade, gave modest benefit in a trial by Gardos *et al.* (1979).

Isolated cases have been reported benefited by non-drug treatments including new dentures, deconditioning biofeedback, stereotactic surgery, and ECT.

Prevention and management

In 1973, the American College of Neuropsychopharmacology Task Force (Schiele *et al.* 1973) proposed three major approaches which might reduce the risk of TD, namely:

1. Minimizing the unnecessary use of neuroleptic medication (especially at high doses) in long-term patients.

2. Discontinuation, where possible, of neuroleptics at the first manifestation of TD.

3. Special precautions should be observed in patients over the age of 50, especially women.

Drug 'holidays' were urged as part of the first admonition both as a way of ascertaining whether the patient has TD and possibly to offer a small amount of protection. Their effectiveness, however, was regarded as not proven by the *Medical Letter* (1979), which urged that antipsychotic drugs should not be used for the treatment of anxiety, chronic pain, insomnia, or personality disorders and that phenothiazines should be used as little as possible for nausea and vomiting. It also said that patients and their families should be warned of the risk before long-term drug treatment is started. Where neuroleptics are needed, the drugs should be used in the lowest dosage that will control the patient's symptoms. Similar or somewhat less restrictive advice is given by other authors. The recurrent theme of the need to minimize long-term prescribing and dosage of neuroleptics was taken a step further by *The Lancet* (1979) whose editorial stated: 'We should avoid prolonged administration' but went on to say that (for patients with schizophrenia) 'in some patients the TD may be the lesser of the two evils.'

Crane (1972b, 1977) has been a determined proponent of the view that neuroleptic drugs are prescribed in gross excess, and he bemoans the lack of impact on the prescribing habits of US physicians evident since the risks of TD became publicized. He sees a need to 'dent the complacency' of many psychiatrists (Crane 1980b) and urges that doses of chlorpromazine seldom need to be more than 400 mg a day, more than 150 mg a day for patients over 55, or more than 75 mg a day in geriatrics (Crane 1977).

Knowledge of the risks of TD has highlighted the need for good prescribing practices but other more specific guidelines to treatment remain uncertain, given the experimental status of trials of drugs for the disorder. 'Drug holidays' are as yet of questionable value.

Their capacity to demonstrate 'covert' dyskinesia was shown by the finding of Carpenter et al. (1980) that 14 of 52 ambulatory patients with schizophrenia developed abnormal movements for the first time when drugs were withdrawn for four weeks as part of a research study. All eventually required restitution of drug therapy for psychotic relapse but dosages given (200–600 mg equivalents of chlorpromazine) did not achieve complete suppression of movements in 13. The average age of the affected patients was 30.9 years, with an estimated average duration of exposure to neuroleptic drugs of 4.2 years. The results might raise the question of whether the diagnosis of 'covert' TD by drug withdrawal justifies the risk in the ordinary clinical practice of treating schizophrenia. A trial of drug holidays for eight to nine weeks in 18 patients was reported by Gibson (1980). Three had an exacerbation of psychosis and TD first appeared or became worse in three others; the latter was regarded as a welcome sign, since it led to a change in management. It was decided not to restart fluphenazine until the TD symtoms disappear, unless frank psychosis develops.

Gibson (1980) provide a helpful guide to the kind of management which may be appropriate in a community-orientated psychiatric out-patient service for schizophrenia. Use of depot neuroleptics at the lowest adequately effective dose, early detection of TD, and a readiness to change drug treatment, where feasible, to oral medicines which may reduce symptoms of TD are among its practices.

The introduction of long-acting injectable neuroleptics has been widely acknowledged as the second major revolution in the pharmacotherapy of schizophrenia. Some patients requiring long-term therapy, however, are able to pursue oral treatment consistently, so that there may be commonsense grounds for recommending that depot injections, with their possibly greater risk of TD,

should be reserved for patients presenting major problems of compliance. The decision, however, must be a highly individual one, closely dependent on available facilities for continuing support and follow-up.

Reduction of dosage in patients needing long-term neuroleptics may lead to an improvement or disappearance of TD, though there may be an initial deterioration. Gibson (1980) found that nine of 31 patients with TD could be maintained on dosages as low as 12.5 mg depot fluphenazine monthly. Three years later, symptoms had reappeared even at the lower dose in two of the nine. Of a further 100 patients without TD selected for a dose-reduction trial, 61 could be maintained on doses of 12.5 mg or 18.75 mg a month without psychiatric deterioration. Such low doses may be impracticable in patients with unstable illness.

Very slow stepwise diminution of the dose of neuroleptics and of antiparkinson agents, together with treatment with small, slowly increasing doses of haloperidol (0.5–2 mg daily) or reserpine (0.25–1 mg daily) was successfully undertaken in 62 patients with TD and chronic schizophrenia in a four-year treatment programme (Jus et al 1979). TD disappeared in 23 patients, improved in 26, and was unaffected in 13. Patients whose TD disappeared had a mean age significantly lower than other groups. The rationale of the trial was to achieve slow, progressive unblocking of dopaminergic receptor sites and 'desensitization.' In 37 of the 49 patients whose TD was improved it was possible to reduce the neuroleptic dose about 35 per cent (e.g. from a mean daily dose of 461 mg of chlorpromazine equivalents to 323 mg in 19 of 23 patients whose TD disappeared). Dose reduction was possible in only two of 13 patients in the group without improvement. Nine of the 62 patients were withdrawn from 'incisive' neuroleptics and given more sedative neuroleptics such as thioridazine. Antiparkinson drugs were withdrawn in 12 patients, and reduced in another 13. The authors stressed the probable value of using very small doses of dopamine-blocking drugs to achieve 'desensitization.' After withdrawal of haloperidol or reserpine, 18 patients had a transitory reappearance of TD but improvement followed when small doses were reintroduced.

There is clearly a need to avoid prescribing neuroleptics unless they are really necessary, for some short-term indications or long-term for schizophrenia. Relapse in schizophrenia is commonly delayed weeks or months after drug withdrawal so that the continuing need for neuroleptics may not be quickly obvious to the patient or his doctor. Very often the patient's difficulties in complying with medication outside hospital will have provided abundant evidence of the risk of relapse into psychosis to those who know him well, but he may go adrift mentally, socially, and geographically as relapse sets in, so that contact is easily broken and the benefits of perhaps years of previous treatment can be quickly lost. Long-term follow-up studies by Johnston (1979) indicated that patients with definite schizophrenia are likely to require maintenance neuroleptic therapy for a minimum of four years. Even after an interval of four years off drugs, the risk of relapse was twice as high as for control patients maintained on drugs. As early as two months following discontinuation a trend was developing for patients without drugs to be at higher risk of relapse and this was significant at three months. A review by Chan et al. (1980) of 24 controlled studies of neuroleptic treatment versus placebo in schizophrenia concluded that over the period of two years the risk of relapse on placebo was almost three times greater than on neuroleptic maintenance. Up to 20 per cent of patients do not relapse on placebo maintenance for as long as two years but there as yet appears to be no sure way of identifying this group.

Prognosis

Some early reports emphasized the tendency for tardive dyskinesia to be irreversible (eg. Hunter et al. 1964), and it has sometimes been argued that the name should be reserved for persistent syndromes. A distinction has been made between 'withdrawal dyskinesias' or 'withdrawal emergent dyskinesias' which are likely to be completely reversible within six to 12 weeks (Gardos et al. 1978) and the more persistent disorders named as TD. An alternative view is that disorders nameable as tardive dyskinesia can occur along a continuum of persistence (Casey 1978) from the transient to the permanent. There seems to be no sure way of predicting the likely time course; the disorder may fluctuate in severity and extent spontaneously. Paulson (1975) stated that at least 50 per cent of patients slowly improve in the six months after drugs are eliminated. Improvement may continue well beyond this time. Younger patients or those in whom the syndrome is less well developed may remit in relatively shorter periods (Schiele et al. 1973).

A view now commonly stated is that symptoms are more likely to remit if the diagnosis is recognized and drugs withdrawn as early as possible. The study of Quitkin et al. (1977), which concluded that the length of time symptoms persist prior to drug withdrawal rather than age of onset of symptoms was the important determining factor, has been frequently cited in support. Twelve psychiatric patients with TD (including three aged over 50 years) were observed for up to five years from inception of the syndrome. In 11 patients symptoms remitted after antipsychotic drugs were reduced or discontinued. Improvement occurred within 16 weeks in seven patients. Some observations of improvement were delayed, even up to two years after discontinuation in one case. In eight patients recurrence of psychiatric disorder necessitated re-introduction of neuroleptics. In another study TD persisted in 11 of 13 elderly patients followed by Mehta et al. (1977) for five years; severity diminished in two of these and increased in two. Both of those in whom TD disappeared were receiving neuroleptics. A trend towards improvement, even in older patients, was found in 14 patients with severe TD followed for up to five years by Itoh and Yagi (1979).

It has been estimated that some 200 000 patients chronically treated with neuroleptics in the USA might have TD (Tepper and Haas 1979). The urgency of the need to clarify the orgins and natural history of the disorder and to discover antipsychotic drugs with reduced hazard from this complication is clear. It is important, however, that the benefits that neuroleptics can offer the patient with some forms of severe mental illness should not be swept aside in a rush of anti-drug sentiment, which is already propelled by many other forces, including the patient's understandable fears of losing his mind. Psychiatrists familiar with the ravages of schizophrenia may continue to believe, on present evidence, that the probability of cumulative damage to the individual from that disease, if inadequately treated, outweighs the long-term hazards of tardive dyskinesia.

Recent publications include the report of D.R. Ross et al. (1983) on the development of akathisia in a patient treated with amoxapine; these authors suggest that this extrapyramidal side-effect raises the possibility that tardive dyskinesia might develop after therapy with amoxapine.

Wihelm and colleagues (1984) provided a valuable summary of the clinical findings in cases of tardive dyskinesia associated with the use of metoclopramide and reported to the Swedish Adverse Drug Reactions Advisory Committee from 1977 to 1981. The data suggest that treatment for six months or more in patients over 70 years of age yields an incidence of more than one in a 1000 patients. Thus, elderly people given long-term treatment with metoclopramide have a substantial risk of developing this disabling condition.

Interest in tardive dyskinesia has of course been worldwide. With permission, the following note to national information officers from the World Health Organization (26 October 1983) is reproduced (in part) as this document, originating in New Zealand, provides a useful summary of the essential clinical features of the condition:

Tardive dyskinesia is a syndrome of iatrogenic involuntary

movements which complicates the long-term use of antipsychotic medication. Although it was described as a rare occurrence in the late 1950s, it has become increasingly prevalent and is now recognised 'as a public health problem of major proportion.'

Tardive dyskinesia is insidious in onset and although elderly patients are more vulnerable, it can occur in younger people and has been reported in children.

In its severe form tardive dyskinesia comprises bizarre involuntary movements of the tongue, face, jaws, and pharynx, and usually of the limbs and trunk. At present, there is no effective cure for severe tardive dyskinesia and it is therefore crucially important to recognize the earliest signs of the disorder.

The earliest signs are 'wormlike' movements in the tongue. This is not a tremor. Rhythmic lateral movements occur later, initially of the tip and then of the body of the tongue. Movements of the fingers and thumb, usually unilateral, are very common and are amongst the early signs.

The movements of tardive dyskinesia can be suspended voluntarily for short periods, so it is important to observe the patient 'indirectly'. For instance, the tongue may be best observed while ostensibly examining the teeth and gums.

The most obvious signs of tardive dyskinesia is the 'cephalic triad' or the 'Bucco-linguo-masticatory'(BLM) syndrome which comprises continual munching, sweeping movements and protrusion of the tongue, and forced swallowing, in various degrees and combinations.

Present evidence suggests that tardive dyskinesia results from hypersensitivity to dopamine of post-synaptic receptors in the corpus striatum after months or years of blockade by antipsychotic drugs. There is also growing evidence that multiple neurotransmitter systems must be involved.

Although severe and disabling tardive dyskinesia progresses from a milder form, mild signs do not inevitably progress in severity.

The potential for reversal of the symptoms seems to be inversely related to age and to the severity and duration of symptoms, but there is well documented evidence that tardive dyskinesia is irreversible in some patients.

At present, it is not possible to predict accurately the course of tardive dyskinesia for an individual patient.

It is thus evident that chronic administration of neuroleptic drugs (e.g phenothiazines, butyrophenones, diphenylbutylpiperidines, thioxanthenes, and substituted benzamides) may cause tardive dyskinesias, especially in the elderly. The most common manifestation — the orobuccolingual syndrome — is not usually a serious disability, and the patient himself may not even be aware of it. However, the more serious tardive dystonia may prevent walking, distort manual dexterity, or impair vision, speech, and swallowing, and these problems may persist when neuroleptic treatment is stopped, thus causing long-lasting disability and distress. In a series of 42 patients with tardive dystonia seen in three movement-disorder clinics in London, New York, and Houston, 19 had been treated inappropriately with neuroleptics (Burke *et al*. 1982).

Six patients had been treated for anxiety, six for depression, four for behavioural disturbances associated with mental retardation, and three for an acute single episode precipitated by emotional stress (acute psychogenic psychosis). These patients, many of whom were young adults, had received various phenothiazines or butyrophenones for an average of three years before tardive dystonia developed. Many of them continued to take the offending drug for months or years after the onset of their abnormal movements. Three patients had generalized dystonia, 12 had segmental dystonia, and four had focal dystonia; the face was affected in 10, the neck in 17, the trunk in 10, the arms in 11, and the legs in four. In only one patient, with torticollis of a two-year duration, did the tardive dystonia remit — one year after stopping neuroleptic treatment. In the remaining 18 patients, the dystonia has persisted for an average of three years (range seven months to six years). Treatment has proved unsatisfactory, despite the use of a wide range of measures.

The therapeutic difficulties, the prolonged and possibly permanent nature of the dystonia in some cases, and the disability caused by the dystonic movements, all serve to emphasize the need to avoid the unnecessary use of neuroleptic drugs to treat conditions which can be managed by alternative means.

A number of other syndromes and drug effects

Drug-induced increased intracranial pressure

Nitroprusside

Cottrell *et al*. (1978), Candia *et al*. (1978), and Griswold *et al*. (1981) all reported cases of intracranial hypertension developing in patients with metabolic encephalopaty or in post-neurosurgical patients in whom nitroprusside had been administered to control systemic blood pressure.

In the patient described by Griswold *et al*. (1981), the intracranial pressure rose from 20 to 38 mg Hg within five minutes of commencing the nitroprusside drip at 1 μg kg^{-1} min^{-1}. The blood pressure fell from 200/100 mmHg to 179/90 mmHg over the same period. In Cottrell's *et al*. report the mean rise in the intracranial pressure was from 14 to 27 mmHg during sodium nitroprusside administration. The best hypothesis to explain the phenomenon is that sodium nitroprusside by acting as a potent vasodilator causes a large increase in the volume of blood in the cerebral vessels and the increased volume of blood raises the intracranial pressure.

It is recommended that monitoring of intracranial

pressure is undertaken if sodium nitroprusside is used in comatose patients as a hypotensive agent.

Perhexilene

The UK data sheet on perhexilene draws attention to the fact that signs and symptoms thought to be manifestations of benign intracranial hypertension (pseudotumour cerebri) have been reported. These signs include papilloedema, decreased visual acuity, retinal haemorrhage, sixth-nerve palsy, optic atrophy, and loss of vision.

Tetracyclines

Benign intracranial hypertension has been repeatedly reported with tetracycline. This relatively rare complication manifests itself through non-specific signs of increased intracranial pressure such as headache, nausea, vomiting, dizziness, tinnitus, blurred vision, or double vision. Objective signs are papilloedema, venous congestion, exudates, or even haemorrhages.

The syndrome has been repeatedly associated with tetracycline therapy in infants. Maroon and Mealy (1971) described a case in a seven-year-old boy and cases in adults have been reported by Bhowhick (1972), Koch-Weser and Gilmore (1967), and Ohlrich and Ohlrich (1977). As a rule the symptoms develop 12–96 hours after the commencement of tetracycline therapy, but Giles and Soble (1971) described a case developing after five months of therapy with tetracycline for acne.

In three cases of benign intracranial hypertension (Koch-Weser and Gilmore 1967; Opfer 1963; Businco *et al.* 1968), infants were re-exposed to tetracycline with a repetition of the syndrome — a procedure which, unless inadvertent, would not now be undertaken.

Dexamethasone deleterious in cerebral malaria

About 2 per cent of patients with falciparum malaria develop cerebral symptoms such as coma, convulsions, or confusion. Mild cerebral malaria may improve after small doses of intravenous quinine or chloroquine but deep coma may proceed to death — the death rate from cerebral malaria remains at 20 per cent.

Woodruff and Dickinson (1968) reported on one patient in whom 10 mg dexamethasone given 24 hours after onset of coma had had a beneficial effect. Harding (1968) also on a single case concluded that dexamethasone was not beneficial and Reid and Nkrumah (1972) in a series of four cases could demonstrate no beneficial effect. Nevertheless, most textbooks recommend that corticosteroids have a role in the treatment of cerebral malaria.

Warrell *et al.* (1982) compared dexamethasone with placebo in a double-blind trial in 100 patients (aged 6 to 70) in Thailand. In adults the initial dose of dexamethasone was 0.5 mg kg^{-1} followed by seven doses of 10 mg each. The total duration of treatment was 48 hours. Dexamethasone increased the duration of coma (63–47 hours) and the incidence of complications including pneumonia and gastrointestinal bleeding. The authors concluded that dexamethasone is deleterious in cerebral malaria and should no longer be used. Additional consideration of this matter by Warrell *et al.* (1983) and Bent Juel-Jensen (1983) emphasized the same conclusion.

Transmission of Creutzfeldt–Jakob disease by investigative procedures

Creutzfeldt–Jakob disease is one of the spongiform encephalopathies that include kuru and scrapie. Evidence has been accumulating that these diseases are transmitted by particles which are small compared with the conventional virus and the scrapie agent is highly resistant to physical and chemical inactivation.

Creutzfeldt–Jakob disease presents a picture of subacute, rapidly progressive dementia, often accompanied by florid psychiatric symptoms including visual and auditory hallucinations, and invariably associated with myoclonus. Indeed, the combination of myoclonus with a rapidly evolving dementia strongly suggests the diagnosis of Creutzfeldt–Jakob disease. A startle reaction may often be elicited. The patient becomes apathetic and confused with a poor memory and is dysphasic. Next he develops dysarthria; bulbar or pseudobulbar palsy appears and dysphagia is common. Some patients present with cortical blindness, while others develop it in the course of the disease. Ataxia may appear early as evidence of cerebellar damage. Most patients finally become mute and progressive rigidity and spasticity make them bedbound. Repetitive jerking of the muscles is still evident late in the disease. The duration of survival depends largely on the quality of the nursing care.

Less frequently, the disease may pursue a more chronic course without myoclonus but accompanied by muscle atrophy and fasciculations, thus resembling motor neurone disease. Then form may be confused with amyotrophic lateral sclerosis, Parkinson's disease, or Alzheimer's disease.

Laboratory investigations are of little help in diagnosis. The blood and cerebrospinal fluid show no abnormalities.

Research into Creutzfeldt–Jakob disease took off with the first report of its transmission to a chimpanzee by a cerebral biopsy specimen from a patient with the disease.

The agent has also been passaged into other laboratory animals, including guinea–pigs, hamsters, and mice. Accidental transmission of Creutzfeldt–Jakob disease from man to man was reported (Will and Matthews 1982). Two young epileptics developed the disease after they had been studied with depth electrodes used previously in a patient with Creutzfeldt–Jakob disease; in another case, a neurosurgeon was affected.

Following this evidence that this disease can be transmitted as a result of investigative procedures, mandatory recommendations have been made to all those who nurse patients with the disease or handle pathological or postmortem specimens from the victims of such disease by the Advisory Group on the Management of Patients with Spongiform Encephalopathy (Creutzfeldt–Jakob Disease (DJD) (1982).

Creutzfeldt–Jakob disease has been reported following the use of human growth hormone preparations prepared from human pituitary glands (see Chapter 36).

Hexachlorophane and newborn babies

In 1971 three studies raised serious doubts regarding the saftety of the practice, which was then common, of the total body-bathing of newborn infants with hexachlorophane-containing preparations.

Curley *et al.* (1971) reported that 50 newborn infants, who had been bathed daily with 3 per cent hexachlorophane, had blood levels of 0.009 to 0.646 μg ml^{-1} of this substance on the day of hospital discharge.

Gaines and Kimbrough (1971) showed that

Rats fed hexachlorophene to achieve mean hexachlorophene blood levels of 1.21 micrograms/ml. showed brain changes characterized by cerebral edema limited to the white matter, and cystic spaces of the brain believed produced by fluid accumulation.

Finally, the Food and Drugs Administration reported (1971) that studies submitted to them by Winthrop Laboratories demonstrated that

Newborn monkeys washed daily with 3% hexachlorophene for 90 days showed mean hexachlorophene and plasma levels of 2.3 micrograms/ml. When they were sacrificed, the white matter of the brain, particularly, the cerebellum, brain stem and all parts of the cord, showed lesions consisting of cystic spaces like those described above.

Vernon Udall (1972) of the Wellcome Research Laboratories referred to the slightly earlier work of his own group reporting that in both sheep and calves hexachlorophane could produce acute toxic effects and gross visual changes. The changes suggested acute raised intracranial pressure but this was not confirmed by direct measurement. Udall commented that 'When high doses of hexachlorophane were given to young rats, exophthalmos occurred, whilst, in older rats, cerebral oedema was found.' These reports contrasted with the widespread use of this substance — uses which seemed to assume that it was innocuous.

In the United Kingdom, in February 1972, precautions regarding the use of hexachlorophane were developed by the Committee on Safety of Medicines and advised to the appropriate professions.

Cimetidine and CNS disorders

Mention was previously made of the association of cimetidine and postural and action tremor.

The first published report of an association between cimetidine and mental confusion was by Grimson (1977), who described two cases. A man aged 50, treated for nine days with 1 g/day for pain due to reflux oesophagitis, was inadvertently given cimetidine in double dosage on leaving hospital. After 48 hours on the higher dose, he felt light-headed and dizzy, experienced some pain in his neck and profuse sweating and flushing, and became mentally confused. These symptoms disappeared 24 hours after he reverted to the normal dose. The second case was a 34-year-old man, also receiving cimetidine for reflux oesophagitis, who agreed to have his dose doubled from the 1 g/day he had received for four days, to see what happened. Twenty–four hours later he became dizzy, drowsy, and confused, with flushing and sweating. When the dose was reduced to 1 g/day, he was back to normal within 24 hours and had no memory of the episode. Five more cases were soon reported (Robinson and Mulligan 1977; Delaney and Ravey 1977; Grave *et al.* 1977; Menzies-Gow 1977). The patients were males aged 55–81. One had renal failure and received 800 mg i.v./day. The rest had other medical complications but recevied normal doses of cimetidine.

After 1977, published reports of acute, reversible mental disorder associated with cimetidine began to appear more widely (Wood *et al.* 1978; Deheneffe *et al.* 1978; Beraud *et al.* 1978; Vickery 1978; Kinnell and Webb 1979; Ewers 1979). By mid-1980 at least 33 reports (involving 45 adults and three children) had been published in English-language, French, and German journals.

Although other possible causes of mental disorder are suggested in most cases by the clinical details reported, the relationship in time between the exhibition of cimetidine and the onset of disorder, together with the rapid reversal on cessation or reduction of dosage, has led authors to believe the association might be a causal one. In the few cases (referred to below) where cimetidine is reported to have been given a second time, a similar sequence has been observed.

By November 1978, the manufacturers had received 57 reports of confusional states in some kind of association with cimetidine out of a world-wide total of several million patients treeated (Flind and Rowley-Jones 1979). No relationship to dose was apparent. Of 44 reports with sufficient information, 35 concerned elderly patients or those with serious concomitant illnesses. There seemed, nevertheless, sufficient evidence that a cause-and-effect relationship existed to justify an addition to the UK data sheet for cimetidine, noting that 'Reversible confusional states have also occurred, usually in elderly or already very ill patients (e.g. those with renal failure)'

Preliminary results of an intensive post-marketing out-patient surveillance programme have now been reported (Gifford et al. 1980). During the first phase data were obtained over a three-month period for 9907 patients. Among 136 cases of CNS adverse effects, dizziness and dizzy feeling were reported most often (34 reports), followed by headache (23); there were five reports of confusion. The manufacturers' Worldwide Spontaneous Reporting System (Davis et al. 1980), identified 1.1 reports of mental confusion per 100 000 patients treated with cimetidine. Most of the patients with confusion were elderly, in intensive care, or had multi-organ disease, multi-drug therapy, or both. The survey was based on an estimated 6.2 million patients receiving cimetidine in the US and UK from November 1976 to February 1979.

CNS changes associated with cimetidine in published reports included restlessness, agitation, aggressive behaviour, confusion, disorientation, speech disturbance, focal twitching, and impaired consciousness. Depressive or suicidal features and paranoid delusions were reported to accompany these signs in some cases or were reported without evidence of organic cerebral impairment. In many reports the changes are only briefly described but it appears from those accounts which include a formal diagnosis or specific mention of objective findings, such as disorientation or impaired consciousness, that at least 29 adult cases have developed features indicative of an acute organic toxic-confusional state. Several other cases, including those of Grimson (1977), have been said to have mental confusion (not otherwise specified) or aggravation of pre-existing mental impairment. Of the 29 adult cases with reported objective signs, 21 were male. Seventeen were aged 60 or more and eight were over 75. The youngest was aged 49 (Bunodiere 1978). At least 12 had renal insufficiency. Complicated medical or surgical histories are a striking feature of most reports. Doses of cimetidine greater than 1 g/day had been given intravenously or orally in 12.

Time of onset of confusional state was most commonly one to two or a few days but ranged from four hours after the first dose of cimetidine (300 mg i.v. given to a 58-year-old man on the second day post-cholecystectomy described by Barnhart and Bowden 1979) up to two weeks (Corbeil 1979). In nearly all cases complete or near-complete recovery was observed on withdrawal of cimetidine or reduction of dose. In two cases where unspecified confusion lasted many days before cimetidine was withdrawn (24 days in Quap's (1978) case or 25 days in Spears' (1978)), some residual impairment was still evident at 72 hours after withdrawal.

Among the rarer features reported, localized twitching or generalized myoclonus were noted by Grave et al. (1977), Schentag et al. (1979), Edmonds et al. (1979), and McMillen et al. (1978). Four of five patients had renal failure. One of the three cases of Edmonds et al. (1979) had a Jacksonian fit which progressed to status epilepticus after haemodialysis was stopped; another, with meningitis, had convulsions only while he was receiving cimetidine. Central pontine myelinolysis was initially suspected in one of the two cases of McMillen et al., a 56-year-old man who received 2400 mg i.v./day post-operatively and became confused and unresponsive with divergent squint, myoclonus, and bilateral extensor plantar responses. These authors' other case, a 71-year-old man given cimetidine 1200 mg prophylactically on the second day after cholecystectomy, developed persistent post-operative delirium which progressed to coma and respiratory arrest; improvement followed within 24 hours of drug withdrawal. The occurrence of visual hallucinations seemed associated in time with cimetidine administration in a 72-year-old man who was confused following bilateral subdural haematomata and severe hyponatraemia (Agarwal 1978). Auditory illusions and hallucinations in a 49-year-old man receiving cimetidine 1200 mg/day for alcoholic gastritis were attributed to the drug rather than alcohol withdrawal by Arneson (1979). Signs suggestive of brainstem ischaemia, including transient bilateral visual impairment, left-sided deafness, dysarthria, double incontinence, pyramidal weakness in the left arm and both legs, ataxic gait, and impaired posterior column sensory modalities, occurred in a 54-year-old man receiving 1 g/day reported by Cumming and Foster (1978); there was no impairment of higher nervous functions. Depressive or suicidal features accompanying confusion were noted by Barbier and Hirsch (1978) and, in a 46-year-old man described by Petite and Bloch (1979), depression associated with cimetidine recurred on re-exposure. An unexplained illness of anxiety and depression lasting six months in a 37-year-old woman was thought possibly attributable to cimetidine by Johnson and Bailey (1979), and another case of depression in an 81-year-old man given 1200 mg/day was described by Jefferson (1979).

In contrast to earlier authors who described one or two cases, Schentag *et al.* (1979) reported that moderate-to-severe changes in mental status attributable to cimetidine occurred in six out of 36 critically ill elderly patients given the drug in dosages which were mostly 1200 mg i.v./day or 300–600 mg i.v. in those with renal failure. Of the six affected, three had both renal and hepatic failure and one had hepatic but not renal failure. Cimetidine trough concentrations over 1.2 μg ml^{-1} were recorded in five of the six patients who had levels measured. Trough levels above 2.0 μg ml^{-1} were not associated with mental status changes in two patients; one was unresponsive on admission and the other had impaired mental status on intial exposure to the drug. Two patients inadvertently given cimetidine a second time had a recurrence of mental deterioration. The average ratio of CSF to serum concentration of cimetidine was 0.241. Comparison of the percentage mental status change from baseline with serum cimetidine trough concentrations was considered by the authors to have shown a close correlation, but this was questioned by Flind and Rowley-Jones (1979).

The study demonstrated that cimetidine can enter the CSF and provided some support for the argument that mental changes associated with cimetidine may be dose-related in elderly, critically ill in-patients. Mental status scores were arrived at by scoring of nurses' comments extracted from case notes and changes of 200 per cent increase over baseline were considered significant. Two case examples given illustrate severe changes. Both became confused and hallucinated. In one, confusion recurred on re-exposure; no cimetidine levels were measured. The other had myoclonic contractions, became unresponsive, had grand mal seizures and apnoeic periods: cimetidine trough concentration was μg ml^{-1}. Mental state changes in the other patients, however, are not detailed. No changes were noted in the 11 patients with neurological disease who had severely impaired mental status before cimetidine was given.

Edmonds *et al.* (1979) also detected cimetidine in the CSF. In one patient plasma cimetidine was high when status epilepticus occurred. In another, the occurrence of twitching but not confusion appeared to corelate with plasma cimetidine only when the patient was uraemic. CSF/plasma ratios of 0.11 and 0.43 were recorded. Two other reports include plasma cimetidine concentrations. In the case described by Graham (1979), a 54-year-old woman who developed an acute confusional state post-gastrectomy, plasma cimetidine levels were at the lower end of the normal range. Kimelblatt *et al.* (1980) reported dose-related confusion in a 56-year-old woman with severe liver disease and mildly abnormal renal function. Confusion, initially with agitation and paranoid features, occurred twice when dosage was 600 mg i.v./day but rapidly receded on 300 mg i.v. daily. Cimetidine trough concentration on the higher dose was 0.25 μg ml^{-1}, 0.1 μg ml^{-1} on the lower. CSF/serum concentration during the first episode was high at 0.5.

Children

CNS reactions associated with cimetidine in children were described in three reports. An 11-year-old girl went into transient coma 15 minutes after an i.v. injection of 2 mg/kg but rapidly recovered and had no recurrence despite continuation of 6 mg/kg/day (Bacigalupo *et al.* 1978). A two-month-old boy became severely obtunded 48 hours after receiving 40 mg/kg/day: he recovered after the drug was stopped and had no recurrence on a continued dosage of 25 mg/kg/day (Thompson and Lilly 1979). A four-year-old girl developed dysarthria, hallucinations, and episodes of altered consciousness after 10 days of 15 mg/kg/day. She was also receiving hydroxyzine and pethidine. Recovery was rapid after all drugs were withdrawn (Bale *et al.* 1979).

Overdose

Evidence from reports of cimetidine taken in acute overdose does not support a simple dose–response effect. Wilson (1979) reported a man on antipsychotic medication in whom transient respiratory depression occurred after an overdose of about 60 tablets (12 g) and Nelson (1977) described a man of 27 years who became disorientated with slurred and nonsensical speech after allegedly taking 60 tablets. No neurotoxicity attributable to cimetidine could, however, be identified in three cases of overdosage reported by Illingworth and Jarvie (1979), in one of whom a plasma cimetidine level of 57 mg/l was recorded. Meredith and Volans (1979), reporting four children and 14 adults, also comment on the apparent lack of toxicity. None of the children had any symptoms. In adults, CNS symptoms and signs were attributable to alcohol or other drugs. Three of the seven adults who remained completely symptom-free had taken 20 g of cimetidine. In one, the blood cimetidine level was 45.8 mg/l three and a half hours after ingestion. The suggestion of Mogelnicki *et al.* (1979), based on experience with two cases, that response to physostigmine may be of diagnostic value in detecting confusion due to cimetidine was criticized as illogical and possibly risky by Fiore (1979).

Discussion

In many of the reported cases of mental confusion other potential causes are readily identifiable. Specialist psychiatric assessment is mentioned in only some reports. Such episodes are common in elderly patients. Many factors such as renal insufficiency, sepsis, surgical operation, cerebral hypoxia or degenerative disease, electrolyte imbalance, other drugs, or even psychosocial factors may be contributory and often no clear cause is ascertainable (Cutting 1980). Episodes are commonly transient and self-limiting so that apparent recovery after cimetidine withdrawal may be coincidental.

If cimetidine can rarely be a sufficient or contributory cause of confusion, what is the mechanism? Several authors have speculated on possible effects on putative H_2-histamine receptors in the brain. Indirect evidence of an effect on the brain was shown by Brynskov *et al.* (1980) who studied effects of cimetidine or placebo on a computer-assisted Romberg test in six healthy volunteers; sway was significantly greater after infusion of 400 mg but not 200 mg of active drug. It is known that the blood–brain barrier may be more permeable at the extremes of life. Renal failure may also facilitate entry of drugs to the brain and, since most of the disposal of cimetidine occurs via the kidneys, it will also reduce drug clearance. The possible contribution of hepatic failure, suggested by Schentag *et al.* (1979), is more questionable. Deep coma after four doses of 300 mg i.v. was reported by Levine (1978) in a 53-year-old man with alcoholic cirrhosis and raised blood ammonia levels, but others have given full doses to patients in hepatic failure without obvious adverse effect (Flind and Rowley-Jones 1979).

Present evidence suggests that CNS reactions, including severe mental disorder, are a rare association of cimetidine treatment. Caution in individualizing the dose in the elderly, small children, or the critically ill would seem appropriate. The manufacturers' advice about dosage in renal failure should be heeded. The clinical significance of recent evidence that cimetidine may delay diazepam clearance (Klotz and Reimann 1980) is not yet established, but caution may be required where benzodiazepines or other centrally-acting drugs are given concurrently.

Reye's syndrome and aspirin

In 1963 Reye *et al.* described the clinical and pathological features of a group of 21 children admitted to the Royal Alexandra Hospital for Children between March 1951 and March 1962, who appeared to have an illness which, seems to represent a clinicopathological entity. The syndrome is characterized by fatty degeneration of the viscera, of unknown cause.

The outstanding clinical features were profoundly disturbed consciousness, fever, convulsions, vomiting, disturbed respiratory rhythm, altered muscle-tone, and altered reflexes. The onset was usually associated with cough, rhinorrhoea, sore throat, or earache. There was often hypoglycaemia and a low cerebrospinal fluid (CSF) glucose and the serum glutamic–oxalacetic acid transaminase (SGOT) and serum glutamic–pyruvic acid transaminase (SGPT) levels were increased in each of the seven patients in whom they were measured.

Seventeen of the children died, and at necropsy remarkably uniform pathological changes were found. To the unaided eye, these were expressed as cerebral swelling, a slightly enlarged, firm and uniformly bright-yellow liver, and pallor and slight widening of the renal cortex.

The possibility that Reye's syndrome was associated with viral illnesses and that salicylate ingestion could be an aggravating factor was postulated by a number of works (Lyon and Nevins 1974; Sillanpää *et al.* 1975; Rosenfeld and Liebhaber 1976; Christoffersen *et al.* 1980, Mäkelä *et al.* 1980) on the basis of case reports.

Four case-control studies also indicated an association between salicylate ingestion and Reye's syndrome. In a survey by Starko *et al.* (1980) during an outbreak of influenza A, seven patients with Reye's syndrome and 16 ill classmate control subjects were evaluated for characteristics of the patients' prodromal illness and the control subjects' illness and for medication usage. Patients during the prodrome and control subjects had similar rates of sore throat, coryza, cough, headache, and gastrointestinal complaints except for documented fever which occurred significantly more often in patients than in control subjects (p = < 0.05). While medications which did not contain salicylate were taken as frequently by patients as control subjects, patients took more salicylate-containing medications than did control children ($p < 0.01$). All seven patients took salicylate whereas only eight of 16 control subjects did so ($p <$ 0.05). Patients took larger doses of salicylate than did the entire control group ($p < 0.01$). When the eight control subjects who took salicylate were compared with the patients, the patients still tended to take larger doses (p = 0.08). Patients with fever took salicylate more frequently than control subjects with fever ($p < 0.01$). In addition, salicylate consumption was correlated with severity of Reye's syndrome ($p < 0.05$). It was postulated that salicylate, operating in a dose-dependent manner, possibly potentiated by fever, represents a primary causative agent of Reye's syndrome.

These conclusions were challenged by Clark and Fitzgerald (1980) who stated

Salicylate levels have been determined in the majority of patients with Reye's syndrome at the time of admission to our institution since 1973. On reviewing the hospital records of 68 patients we found that salicylate levels were determined in 51 patients as follows: 0 to 5 mg/100 ml, 26 patients; 6 to 10 mg/100 ml, seven patients; 11 to 15 mg/100 ml, ten patient; 16 to 20 mg/100 ml, seven patients; > 20 mg/100 ml, one patient. The mean level was 8.01 mg/100 ml and the majority of values were 5 mg/100 ml or less, with a range of 0 to 36 mg/ml. Whereas it is plausible that salicylates may act synergistically with the viral illness and accompanying fever to induce the 'toxic-like tissue damage seen with Reye's syndrome,' as Starko et al. suggest, these levels hardly seem noteworthy, especially in the presence of obvious hepatic dysfunction.

The histopathology of aspirin-induced hepatotoxicity has been well described. The histologic findings of focal necrosis, hepatocyte 'ballooning' and degeneration, and mononucleur infiltration of portal zones are similar to those encountered in hepatitis. Indeed, the term 'aspirin hepatitis' is frequently used. This histologic picture is quite different from that found in Reye's syndrome which includes universal small-droplet fat-infiltration without inflammation or necrosis of hepatocytes and infrequent portal inflammation. While fat deposition has been seen with salicylate toxicity, the droplets are large and non-uniform in distribution.

Other workers were quick to critize Starko et al.'s (1980) paper (Gall et al. 1980; Pascoe 1980). Tonsgard and Huttenlocher (1981) reviewed 43 consecutive cases of Reye's syndrome seen at the University of Chicago since January 1975. Salicylate levels had been determined at the time of admission in 37 of these 43 patients. They stated that

only one of the cases of Reye's syndrome occurred on a background of chronic aspirin use for possible juvenile rheumatoid arthritis. However, 40% of the cases had not had sufficient exposure to aspirin for salicylates to be detectable in the blood or urine at the time of hospital admission. Lack of exposure to aspirin was especially common in infants with Reye's syndrome. Thus, although aspirin may play a role in the pathogenesis of Reye's syndrome in some cases, it would be difficult to implicate aspirin in more than one third of our patients.

In a futher survey by Partin et al. (1982) from New York, serum salicylate concentration was measured at admission in 130 children with liver-biopsy-confirmed Reye's disease. Mean serum salicylate was 12.3 mg dl^{-1} and mean salicylate concentrations by neurological grade (Lovejoy) were: stage 1, 12 mg dl^{-1}; stage 2, 13 mg dl^{-1}; stage 3, 11 mg dl^{-1}. However, mean serum salicylate (15 mg dl^{-1}) at admission in 21 patients who died or had serious neurological deficits was significantly higher than that in 103 patients who survived without neurological sequelae (10 mg dl^{-1}). Serum salicylate in a group of 27 age-matched, community-matched control children collected consecutively over the period 1978–80 was less than 2 mg dl^{-1}, and children with varicella or influenza had salicylate concentrations indistinguishable from apparently well classmates or siblings. Partin et al. con-

sidered it impossible to determine from these data whether salicylates are involved in the aetiology of or in determining the outcome of Reye's disease. Increased concentrations of salicylates at admission could be the result of excessive dosage because of a greater severity of the prodromal illness, or the result of diminished excretion because of impaired hepatic metabolism. To Partin et al. it seemed likely that serum salicylate concentrations entered the toxic range in many patients with Reye's disease before they presented for treatment. Most had been vomiting and had had diminished oral intake for 33 to 55 hours before hospital admission. Since the average number of hours from the beginning of vomiting to admission was no different in non-comatose and comatose cases, the time at which salicylate concentration was measured in relation to the last dose was probably similar in the two groups and therefore does not account for the higher levels in children with poor outcome. Salicylates are mitochondrial toxins and mitochondria are known to be significantly injured in Reye's disease; therefore, it seems wise to avoid the use of aspirin in children during outbreaks of Reye's disease.

Waldman et al. (1982) reported on a study of 56 cases of Reye's syndrome in school-aged children in Michigan during Winter 1979–80. The parents of 25 of these children were interviewed in Spring 1980, as were controls matched to the cases for age, race, school grade, and nature of antecedent viral illness. Children with Reye's syndrome were more likely to have received medication containing aspirin during their vital illness than controls (24/25 vs. 34/46), even when the child's highest measured fever was added as a criterion for matching (14/14 vs. 14/19). During Winter 1980–81, a second study was undertaken to examine this observed difference more carefully. All 12 school-aged children with Reye's syndrome and controls matched to the cases for race, school grade, nature of antecedent viral illness, and peak temperature, were interviewed as cases occurred. Again, the ill children had received aspirin-containing products more frequently (12/12 vs. 13/29). Although Reye's syndrome can occur in the absence of aspirin ingestion, their data lead them to conclude that aspirin taken during viral illness may contribute to the development of Reye's syndrome.

Rodgers et al. (1982) studied the salicylate kinetics in two patients with biopsy-proven Reye's syndrome. Both patients had measurable levels of salicylate (5.4 and 12.7 mg dl^{-1}, respectively) at the time of presentation, despite a typical prodromal period of emesis and a negative history for salicylate ingestion within 48 hours of admission. Both patients, a seven-year-old girl and a six-year-old-boy, presented in stage 2 and progressed to stage 3, and serum half-lives were 18.9 and 14.7 hours, respectively. Half-lives in these patients would have been expected to have been in the range 2–4 hours.

The Centre for Disease Control in the United States, Arizona Michigan using the above data and the *Morbidity and Mortality Weekly Report* data (12 February 1982) issued a warning that the use of aspirin to treat children with influenza and chickenpox has been linked with Reye's syndrome. They stated that 'salicylates may be a factor in the pathogenesis of Reye's syndrome, although the observed epidemiologic association does not prove causality', but went on to recommend 'until the nature of the association between salicylate and Reye syndrome is clarified the use of salicylates should be avoided where possible in children with varicella infection and during influenza outbreaks'.

In the United Kingdom there have been no reports associating Reye's syndrome with aspirin ingestion. In the UK Reye's syndrome is relatively rare; the National Childhood Encephalopathy study identified 37 cases of Reye's syndrome in a three-year period (Bellman *et al.* 1982). These workers' figures gave an estimated incidence of Reye's syndrome in Britain of 0.7 per 100 000 children per year. The diagnostic features were neither consistently positive in their cases, nor negative in 11 others later considered not to have Reye's syndrome. The prognosis was poor; the fatality rate was 46 per cent and 60 per cent of the survivors were handicapped. A surveillance scheme to investigate pathological, clinical, and epidemiological factors in this rare condition is in progress.

Sullivan-Bolyai and Cary (1981) in a review of the epi-

demiology of Reye's syndrome identified viral and chemical toxic associations with the development of the condition (see Table 27.4).

Recent additions to the literature included the comment by Susan Hall (1983) that the possible causative relationship between salicylate and Reye's syndrome presents an issue which remains unresolved. Mowat (1983) remarked that: 'Reye's syndrome is currently thought to be an acute, apparently self-limiting derangement of hepatic mitochondria with both structural changes and impairment of many aspects of their function. Similar mitochondrial structural changes may occur in muscle and brain.'

The subject is of national interest and a two-year epidemiological study is planned to determine the part played by salicylates, insecticides, and other risk factors in the aetiology. Hall and Bellman (1984), in a paper of considerable value, provided the first annual report of the joint British Paediatric Association and Communicable Disease Surveillance Centre scheme for the study of Reye's syndrome in the British Isles.

Neurological effects of recombinant human interferon

Smedley *et al.* (1983) reported that in 10 women given up to 12 weeks therapy with recombinant leucocyte interferon 20×10^6 U/m^2 daily or 50×10^6 U/m^2 three times weekly, influenza-like symptoms began within one hour of administration. One week later these symptoms were superseded by lethargy, anorexia, nausea, and loss of weight. Associated side-effects included profound somnolence, confusion, paraesthesia, and (in one patient) signs of an upper motor neurone lesion in the legs. All of these effects, together with increased slow-wave activity in the electroencephalograms from all patients during treatment disappeared when the interferon was withdrawn. Interestingly, the effects did not recur on reintroducing the drug at lower dosage.

The 'toxicity' of interferon has become the subject of considerable interest; an editorial (*Lancet* 1983) noted that really high doses of alpha-interferon, such as $100–200 \times 10^6$ U/m^2, given parenterally can induce serious but reversible central nervous system dysfunction which may prove to be the major dose-limiting factor for systematic interferon.

Miscellaneous subjects recently discussed

A careful study by Herbert *et al.* (1983) showed that the impairment of mental functioning following general anaesthesia is such that patients should be advised not to undertake hazardous tasks, such as driving a car, for 48 hours after such procedures.

Table 27.4 Viral and chemical factors that have been associated with the development of Reye's syndrome (after Sullivan-Bolyai and Carey 1981)

Viral associations	Chemical associations
Adenovirus	Aflatoxins
Coxsackie A	Insecticides
Coxsackie B	Isopropyl alcohol
Cytomegalovirus	Anti emetics
Dengue	Pteridine
Echo	Salicylates
Epstein–Barr	Valproic acid
Herpes simplex	Warfarin
Influenza	
Mumps	
Parainfluenza	
Poliomyelitis	
Reovirus	
Respiratory syncytial	
Rotavirus	
Rubella	
Rubeola	
Vaccinia	
Varicella-zoster	

A new warning relating to a common drug was provided by Barnes and Goodwin (1983) who reported severe narcosis caused by dihydrocodeine tartrate given to an anuric patient receiving haemodialysis — these authors urged caution in prescribing this drug even in conventional dosage in patients with severely impaired renal function.

It is established that infants born to women with epilepsy have an increased incidence of fetal abnormality. Valproic acid and carbamazepine have been advised in preference to phenytoin and phenobarbitone for use as anticonvulsants in women of child-bearing age — but two cases reported by Bailey et al. (1983) call into doubt the wisdom of this advice in respect to valproic acid.

Controversy regarding the central nervous system adverse effects of phenylpropanolamine has continued and been made difficult by the fact that the available systems for reporting side-effects for commonly used but self-administered, non-prescription medicines are inadequate, despite the exposure of massive patient populations. Discussing two case-reports of CNS effects coincident with chronic overdosage of phenylpropanolamine, Waggoner (1983) suggested 'that little useful purpose is served in reporting such cases where the propensity to overdosage and adverse effects is attributable to the patient rather than to the drug.' Refutation of this concept might seem important — it being essential that medicines taken without medical supervision and advice are safe enough for the patients who will receive them. A somewhat wider therapeutic margin may be needed for such drugs than is possessed by many of the now widely used sympathomimetic agents.

The fact that some patients show severe side-effects at low doses of the phenothiazines, whereas others tolerate high doses satisfactorily was investigated by Wright et al. (1983). These authors followed D-glucaric acid excretion in psychotic patients and control subjects and found that about half of the psychiatric patients had a phenothiazine-related rise in D-glucaric acid excretion compatible with enzyme induction. It was concluded that individual differences in metabolism of phenothiazones may in part account for the variability in clinical response to drugs of this class.

In May 1983 the Adverse Reactions to Drugs Committee of the National Board of Health and Welfare of Sweden published a notice describing an influenza-like reaction associated with the use of Zelmid (zimeldine); this same report discussed seven cases in which this flu-like reaction was associated with neurological symptoms. One of these reactions was of the Guillain–Barré type. Zimeldine was an antidepressant and, in August 1983, the Committee on Safety of Medicines in the UK noted (Current Problems, number 11) that although there had been only some 100 000 prescriptions for the product, over 300 adverse reactions associated with its use had been reported. Of these 60 were serious and included convulsions and liver damage; neuropathies and cases of the Guillain–Barré syndrome had also been notified. In a Parliamentary written answer (23 February 1984) it was shown that seven deaths had been notified to the Committee on Safety of Medicines in which zimeldine was the suspect drug. In the UK the manufacturers withdrew the product from the market in September 1983 and surrendered the product licence.

Acknowledgements

The textual sections of this chapter are heavily dependent on the work of its earlier contributors. The present section on clioquinol and SMON is that of Dr M.N.G. Dukes (Update 1981) with minor additions. The section now given on tardive dyskinesia is that of Dr P.G. Campbell (Update 1981). The basic text of this chapter was originally written by Dr. J.P. Griffin (2nd edn. 1979, Update 1982, and Update 1983). The present tables, providing this chapter with the bulk of its new material, are derived from data in the possession of the Committee on Safety of Medicines. Their generous permission in allowing access to these data is acknowledged.

REFERENCES

ACKRILL, P., RALSTON, A.J., DAY, J.P., AND HODGE, K.C. (1980). Successful removal of aluminium from a patient with dialysis encephalopathy. Lancet ii, 692–693.

ADVISORY GROUP ON THE MANAGEMENT OF PATIENTS WITH SPONGI-FORM ENCEPHALOPATHY (1982). Report of the Chief Medical Officers of the Department of Health and Social Security, the Scottish Home and Health Department, and the Welsh Office, pp. 1981–5. HMSO, London.

AGARWAL, S.K. (1978). Cimetidine and visual hallucinations. J. Am. med. Ass. 240, 214.

AHLFORS, V.G. AND KATILA, O. (1973). Clinical evaluation of a new depot neuroleptic: a pilot study with pipotiazine undecylenate. Acta psychiat. scand. (Suppl.) 241, 43–9.

AKINDELE, M.O. AND ODEJIDE, A.O. (1976). Amodiaquine-induced involuntary movements. Br. med. J. 3, 214–15.

ALEXOPOULOS, G.A. (1979). Lack of complaints in schizophrenics with tardive dyskinesia. J. nerv. ment. Dis. 167, 125–7.

ALFREY, A.C., MISHALL, J.M., AND BURKS, J. (1972). Syndrome of dyspraxia and multifocal seizures associated with chronic hemodialysis. Trans. Am. Soc. artificial intern. Organs 18, 257–67.

——, L.E. GENDRE, G.R., AND KAEHNY, W.D. (1976). The dialysis encephalopathy syndrome. Possible aluminium intoxication. New Engl. J. Med. 294, 184–8.

ALPERT, M., DIAMOND, F., AND FRIEDHOFF, A.J. (1976). Tremographic studies in tardive dyskinesia. *Psychopharmacol. Bull.* **12**, 5–7.

—— AND FRIEDHOFF, A.J. (1980). Clinical application of receptor modification treatment. In *Tardive dyskinesia* (ed. W.E. Fann, R.C. Smith, J.M. Davis, and E.F. Domino), pp. 471–3. Spectrum Publications, New York.

AMA DRUG EVALUATIONS (1971). Clioquinol, p. 463.

ARNESON, G.A. (1979). More on toxic psychosis with cimetidine. *Am. J. Psychiat.* **136**, 1348–9.

ARONOW, R. AND SPIGIEL, R.W. (1976 *a*, *b*). Camphor poisoning. *J. Am. med. Ass.* **235**, 1260, 2476.

—— AND —— (1976c). Implications of camphor poisoning: therapeutic and administrative. *Drug Intel. clin. Pharm.* **10**, 631–4.

ASNIS, G.M., LEOPOLD, M.A., DUVOISIN, R.C., AND SCHWARTZ, A.H. (1977). A survey of tardive dyskinesia in psychiatric outpatients. *Am. J. Psychiat.* **134**, 1367–70.

——, NATHAN, R.S., DAVIES, S.O., AND HALBRIECH, U. (1979). Tardive dyskinesia presenting as psychosis. *J. nerv. ment. Dis.* **167**, 762–3.

AUSTRALIAN DRUG EVALUATION COMMITTEE (1974). Adverse effects of bismuth subgallate. *Med. J. Aust.* **2**, 664

AYD, F.J. JR. (1961). A survey of drug-induced extrapyramidal reactions. *J. Am. med. Ass.* **175**, 1054–60.

—— (1970). Prevention of recurrence (maintenance therapy). In *Clinical handbook of psychopharmacology* (ed. A. Di Mascio and R.I. Shader), pp. 297–310. Science House, New York.

—— (1976). Lithium-haloperidol for mania, is it safe or hazardous? *Int. Drug Ther. Newslett.* **10**, 29–36.

—— (1977). Ethical and legal dilemmas posed by tardive dyskinesia. *Int. Drug. Ther. Newslett.* **12**, 29–36.

BAASTRUP, P.C., HOLLNAGEL, P., SORENSEN, R., AND SCHOU, M. (1976). Adverse reactions in treatment with lithium carbonate and haloperidol. *J. Am. med. Ass.* **236**, 2645–6.

BACHER, N.M. AND LEWIS, H.A. (1978). Addition of reserpine to antipsychotic medication in refractory chronic schizophrenic outpatients. *Am. J. Psychiat.* **135**, 488–9.

—— AND —— (1980). Low-dose propranolol in tardive dyskinesia. *Am J. Psychiat.* **137**, 495–7.

BACIGALUPO, A., VAN LINT, M.T., AND MARMONT, A.M. (1978). Cimetidine-induced coma. *Lancet* **ii**, 45–6.

BACOPOULOS, N.C., SPOKES, E.G., BIRD, E.D., AND ROTH, R.H. (1980). Antipsychotic drug action in schizophrenic patients: effect on cortical dopamine metabolism after long-term treatment. *Science* **205**, 1405–7.

BAILES, J. *et al.* (1983). Encephalopathy with metronidazole in a child. *Am. J. Dis. Childh.* **137**, 290–1.

BAILEY, C.J., POOL, R.W., POSKITT, EM.E., AND HARRIS, F. (1983). Valproic acid and fetal abnormality. *Br. med. J.* **286**, 190.

BALDESSARINI, R.J. AND TARSY, D. (1979). Relationship of the actions of neuroleptic drugs to the pathophysiology of tardive dyskinesia. *Int. Rev. Neurobiol.* **21**, 1–45.

BALE, J.F., ROBERTS, C., AND BOOK, L.S. (1979). Cimetidine induced cerebral toxicity in children. *Lancet* **i**, 725–6.

BANKS, W.A. AND KASTIN A.J. (1983). Aluminium increases permeability of the blood–brain barrier to labelled DSIP and β-endorphin: possible implications for senile and dialysis dementia. *Lancet* **ii**, 1227–9.

BARBEAU, A. (1971). Long-term side-effects of levodopa. *Lancet* **i**, 395.

—— (1976). Six years of high-level levodopa therapy in severely akinetic parkinsonian patients. *Arch. Neurol.* **33**, 333–8.

BARBIER, J.-PH. AND HIRSCH, J.F. (1978). Confusion mentale chez un malade traité par la cimétidine. *Nouv. Presse. Méd.* **7**, 1484.

BARNES, J.N. AND GOODWIN, F.J. (1983). Dihydrocodeine narcosis in renal failure. *Br. med. J.* **286**, 438–9.

BARNES, T. AND KIDGER, T. (1979). Tardive dyskinesia and problems of assessment. In *Current themes in Psychiatry*, **2**, (ed. R.M. Gaind and B.L. Hudson), pp. 145–162. Macmillan, London.

BARNHART, C.C. AND BOWDEN, C.L. (1979). Toxic psychosis with cimetidine. *Am. J. Psychiat.* **136**, 725–6.

BARROS, E. (1935). Amebas Y Mas Amebas. *Semana Med.* **42**, 907–8.

BATEMAN, D.N., BEVAN, P., LANGLEY, B.P., MASTAGLIA, E., AND WANDLESS, I. (1981). Cimetidine induced postural tremor. *J. Neurol. Neurosurg. Psychiat.* **44**, 94.

——, DUTTA, D.K., McCLELLAND, H.A., AND RAWLINS, M.D. (1979). Metoclopramide and haloperidol in tardive dyskinesia. *Br. J. Psychiat.* **135**, 505–8.

BEDYNECK, J.L., WEINSTEIN, K.H., KAH, R.E., AND MINTON, P.R. (1966). Ventricular tachycardia: control of intermittent, intravenous administration of lidocaine hydrochloride. *J. Am. med. Ass.* **198**, 553–5.

BEHRENS, M.M. (1974). Optic atrophy in children after diiodohydroxyquin therapy. *J. Am. med. Ass.* **228**, 693–4.

BELL, R.C.H. AND SMITH, R.C. (1978). Tardive dyskinesia: characterization and prevalence in a statewide system. *J. clin. Psychiat.* **39**, 39–47.

BELLMAN, M.H., ROSS, E.M., AND MILLER, D.L. (1982). Reye's Syndrome in children under three years old. *Arch. Dis. Childh.* **57**, 259–63.

BERAUD, J.J., MONTEIL, A.L., MUNOZ, A., AND MIROUZE, J. (1978). Confusion mentale au cours d'un traitement par la cimétidine. *Nouv. Presse Méd.* **7**, 2570.

BERGER, P.A. (1980). Tardive dyskinesia: recent developments. *Int. Drug. Ther. Newslett.* **15**, 17–20.

BERGGREN, L. AND HANSSON, O. (1966). Treating acrodermatitis enteropathica *Lancet* **i**, 52.

BETTS, T.A., KALRA, P.L., COOPER, R., AND JEAVONS, P.M. (1968). Epileptic fits as a probable side-effect of amitriptyline: report of seven cases. *Lancet* **i**, 390–2.

BHOWHICK, B.K. (1972). Benign intracranial hypertension after antibiotic therapy. *Br. med. J.* **3**, 30.

BIRKET-SMITH, E. (1974). Abnormal involuntary movements induced by anticholinergic therapy. *Acta neurol. scand.* **50**, 801–11.

BLOOMER, H.A., BARTON, L.J., AND MADDOCK, R.K.JR. (1967). Penicillin-induced encephalopathy in uremic patients. *J. Am. med. Ass.* **200**, 121–3.

BOSTON COLLABORATIVE DRUG SURVEILLANCE PROGRAM (1972). Drug-induced convulsions. *Lancet* **ii**, 677–9.

BOURDOURESQUES, C., PONCET, M., CHERIF, A.A., TAFANI, B., AND BOUDOURESQUES, J. (1980). Encephalopathie aigue, au cours d'un traitement associant phenothiazine et lithium. *Nouv. Press Méd.* **9**, 2580.

BOURGEOIS, M. (1977). Les dyskinésies tardives des neuroleptiques: enquête chez 3140 malades d'hôpital psychiatrique. *Encephale* **3**, 299–320.

BOWERS, M.B., MOORE, D., AND TARSY, D. (1979). Tardive dyskinesia: a clinical test of the supersensitivity hypothesis. *Psychopharmacology* **61**, 137–41.

BRANDON, S., MC CLELLAND, H.A., AND PROTHEROE, C. (1971). A

study of facial dyskinesia in a mental hospital population. *Br. J. Psychiat.* **118**, 171–84.

BRITISH MEDICAL JOURNAL (1979). Editorial: Tardive dyskinesia. *Br. med. J.* **2**, 1313.

BROWNING, D.H. AND FERRY, P.C. (1976). Tardive dyskinesia in a ten-year-old boy. *Clin. Paediatr., Phil.* **15**, 955–7.

BRYNSKOV, J., THYSSEN, H., JANSEN, E., AND MUNSTER-SWENDSEN, J. (1980). Cimetidine and Romberg's test. *Lancet* **i**, 1421.

BUGE, A., SUPINO-VITERBO, V., RANCUREL, G., AND PONTES, C. (1981). Epileptic phenomena in bismuth toxic encephalopathy. *J. Neurol. Neurosurg. Psychiat.* **44**, 62–7.

BUNODIERE, M. (1978). Cimétidine: complications neuropsychiatriques. *Nouv. Presse Méd.* **7**, 2870.

BURKE, R.E., FAHN, S., JANKOVIC, J., MARSDEN, C.D., LONG, A.E., GOLLOMP, S., AND ILSON, J. (1982). Tardive dyskinesia and inappropriate use of neuroleptic drugs. *Lancet* **i**, 1299.

BURNS, R., THOMAS, D.W., AND BARRON, V.J. (1974). Reversible encephalopathy possibly associated with bismuth subgallate ingestion. *Br. med. J.* **1**, 220–3.

BURT, D.R., CREESE, I., AND SNYDER, S.H. (1977). Antischizophrenic drugs: chronic treatment elevates dopamine receptor binding in brain. *Science* **196**, 326–8.

BUSINCO, L., LENDVAI, D., AND CARDI, E. (1968). Reazione allergica alla tetracicline e sindrome della 'bulging fontanel' in un bambino di quaranta giorni: Sensibilizzazione attraverso il latte materno. *Acta paediat. lat (Reggio Emilia)* **21**, 834.

CANDIA, G.J., HEROS, R.C., LAVYNE, M.H., ZERVAS, N.T., AND NELSON, C.N. (1978). Effects of intravenous sodium nitroprusside on cerebral blood flow and intracranial pressure. *Neurosurgery* **3**, 50–3.

CARNEY, M.W.P. (1971). Five cases of bromism. *Lancet* **ii**, 523–4.

CARPENTER, W.T., REY, A.C., AND STEPHENS, J.H. (1980). Covert dyskinesia in ambulatory schizophrenia. Letter. *Lancet* **ii**, 212–13.

CARROLL, B.J., CURTIS, G.C., AND KOKMEN, E. (1977). Paradoxical response to dopamine agonists in tardive dyskinesia. *Am. J. Psychiat.* **134**, 785–9.

CASEY, D.E. (1978). Managing tardive dyskinesia. *J. clin. Psychiat.* **39**, 748–53.

—— AND DENNEY, D. (1977). Pharmacological characterization of tardive dyskinesia. *Psychopharmacology* **54**, 1–8.

—— AND GERLACH, J. (1980). Oxiperomide in tardive dyskinesia. *J. Neurol. Neurosurg. Psychiat.* **43**, 264–7.

——, ——, AND SIMMELSGAARD, H. (1979). Sulpiride in tardive dyskinesia. *Psychopharmacology* **66**, 73–7.

—— AND HAMMERSTAD, J.P. (1979). Sodium valproate in tardive dyskinesia. *J. clin. Psychiat.* **40**, 483–5.

—— AND RABINS, P. (1978). Tardive dyskinesia as a life-threatening illness. *Am. J. Psychiat.* **135**, 486–8.

CASTAIGNE, R., RONDOT, P., LENOEL, Y., DUMAS, J.L.R., AND AUTRET, A. (1973). Myélopathie sévère neuropathie périphérique et névrite optique survenues au cours d'un traitement par la chloroiodoquine (clioquinol). *Thérapie* **28**, 393.

CHADWICK, D., REYNOLDS, E.H., AND MARSDEN, C.D. (1976). Anticonvulsant induced dyskinesias: a comparison with dyskinesias induced by neuroleptics. *J. Neurol. Neurosurg. Psychiat.* **39**, 1210–18.

CHAN, C.H., DAVIS, J.M., SMITH, R.C., AND REED, K. (1980). Maintenance antipsychotic therapy and the risks of tardive dyskinesia. In *Tardive dyskinesia* (ed. W.E. Fann, R.C. Smith, J.M. Davis, and E.F. Domino), pp. 511–22. Spectrum Publications, New York.

CHAPMAN, B.J., PROUDFOOT, A.T., AND DAWLING, S. (1983). Convulsions after self poisoning with zimeldine. *Br. med. J.* **287**, 1672–3.

CHIEN, C-P. (1980). Tardive dyskinesia: controlled studies of several therapeutic agents. In *Tardive dyskinesia* (ed. W.E. Fann, R.C. Smith, J.M. Davis, and E.F. Domino), pp. 429–69. Spectrum Publications, New York.

——, JUNG, K., AND ROSS-TOWNSEND, A. (1980a). Methodological approach to the measurement of tardive dyskinesia: piezoelectric recording and concurrent validity test of five clinical rating scales. In *Tardive dyskinesia,* (ed. W.E. Fann, R.C. Smith, J.M. Davis, and E.F. Domino), pp. 233–41. Spectrum Publications, New York.

——, ROSS-TOWNSEND, A., AND DONNELLY, M. (1980b). Past history of drug and somatic treatments in tardive dyskinesia. In *Tardive dyskinesia* (ed. W.E. Fann, R.C. Smith, J.M. Davis, and E.F. Domino). Spectrum Publications, New York.

CHOUINARD, G., ANNABLE, L., ROSS-CHOUINARD, A., AND NESTOROS, J.N. (1979a). Factors related to tardive dyskinesia. *Am. J. Psychiat.* **136**, 79–82.

——, DE MONTIGNY, C., AND ANNABLE, L. (1979b). Tardive dyskinesia and antiparkinsonian medication. *Am. J. Psychiat.* **136**, 228–9.

—— AND JONES, B.D. (1979). Early onset of tardive dyskinesia: case report. *Am. J. Psychiat.* **136**, 1323–4.

CHRISTENSEN, E., MØLLER, J.E., AND FAURBYE, A. (1970). Neuropathological investigation of 28 brains from patients with dyskinesia. *Acta psychiat. scand.* **46**, 14–23.

CHRISTOFFERSEN, P., FAARUP, P., GEERTINGER, P., AND KROGH, P. (1980). Reye's syndrome in a child on long-term salicylate medication. *Forensic Sci. Int.* **15**, 129–33.

CLARK, J.H. AND FITZGERALD, J.F. (1980). Doubts relationship of salicylate and Reye's syndrome. *Pediatrics* **68**, 467.

CLAVERIA, L.E., TEYCHENNE, P.F., CALNE, D.B., HASKAYNE, L., PETRIE, A., PALLIS, C.A., AND LODGE-PATCH, I.C. (1975). Tardive dyskinesia with pimozide. *J. Neurol. Sci.* **24**, 393–401.

CLOW, A., JENNER, P., AND MARSDEN, C.D. (1979). Changes in dopamine-mediated behaviour during one year's neuroleptic administration. *Eur. J. Pharmacol.* **57**, 36–7.

COHEN, K.L., COOPER, R.A., AND ALTSHUL, S. (1979). Prolactin levels in tardive dyskinesia. Letter. *New Engl. J. Med.* **300**, 46.

COHEN, W.J. AND COHEN, N.H. (1974). Lithium carbonate, haloperidol and irreversible brain damage. *J. Am. med. Ass.* **230**, 1283–7.

COMMITTEE ON DRUGS. (1978). Camphor, who needs it. *Paediatrics* **62**, 404–6.

COMMITTEE ON SAFETY OF MEDICINES (1975–83). Current problems, 1–12. HMSO, London.

COPE, R.V. AND GREGG, E.M. (1983). Neuroleptic malignant syndrome. *Br. med. J.* **286**, 1938.

CORBEIL, R. (1979). Cimétidine et état confusionnel. *Vie Med. Can. Franc.* **8**, 15–16.

COTTRELL, J.E., PATEL, K., TURNDOFF, H., AND RANSOHOFF, J. (1978). Intercranial pressure changes induced by sodium nitroprusside in patients with intracranial mass lesions. *J. Neurosurg.* **48**, 329–31.

CRAIG, J.O. (1953). Poisoning by the volatile oils in childhood. *Arch. Dis. Childh.* **28**, 475–83.

CRANE, G.E. (1968a). Tardive dyskinesia in patients treated with major neuroleptics. *Am. J. Psychiat.* **124** (Suppl.), 40–8.

—— (1968*b*). Dyskinesia and neuroleptics. *Arch. gen. Psychiat.* **19**, 700–3.

—— (1970). High doses of trifluoperazine and tardive dyskinesia. *Arch. Neurol.* **22**, 176–80.

—— (1972*a*). Pseudoparkinsonism and tardive dyskinesia. *Arch. Neurol.* **27**, 426–30.

—— (1972*b*). Prevention and management of tardive dyskinesia. *Am. J. Psychiat.* **129**, 466–7.

—— (1973). Persistent dyskinesia *Br. J. Psychiat.* **122**, 395–405.

—— (1974). Factors predisposing to drug-induced neurologic effects. In *Advances in biochemical psychopharmacology.* Vol. 9 (ed. T.S. Forrest, C.J. Carr, and E. Usdin), pp. 269–79. Raven Press, New York.

—— (1977). The prevention of tardive dyskinesia. *Am. J. Psychiat.* **134**, 756–8.

—— (1980 *a, b, c*). (*a*) A classification of the neurologic effects of neuroleptic drugs. (*b*) Preface. (*c*) Drug history and other factors associated with the prevalence of tardive dyskinesia. In *Tardive dyskinesia* (ed. W.E. Fann, R.C. Smith, J.M. Davis, and E.F. Domino. Spectrum Publications, New York.

—— AND PAULSON, G. (1967). Involuntary movements in a sample of chronic mental patients and their relation to the treatment with neuroleptics. *Int. J. Neurol. Psychiat.* **3**, 286–91.

—— AND SMEETS, R.A. (1974). Tardive dyskinesia and drug therapy in geriatric patients. *Arch. gen. Psychiat.* **30**, 341–3.

CRAYTON, J.W., SMITH, R.C., KLASS, D. *et al.* (1977). Electrophysiological (H-reflex) studies of patients with tardive dyskinesia. *Am. J. Psychiat.* **134**, 775–81.

CUMMING, W.J.K. AND FOSTER, J.B. (1978). Cimetidine-induced brainstem dysfunction. *Lancet* i, 1096.

CURLEY A., KIMBROUGH R.D., HAWK, R.E.; MATHENSON, G; AND FINBERG, L. (1971). Dermal aborption of hexachlorophene in infants. *Lancet* ii, 296–7.

CURREN, J.P. (1973). Tardive dyskinesia — side effect or not? *Am. J. Psychiat.* **130**, 406–10.

CURRY, S.H., DAVIS, J.M., JANOWSKY, D.S., AND MARSHALL, H.H.L. (1970). Factors affecting chlorpromazine plasma levels in psychiatric patients. *Arch. gen. Psychiat.* **22**, 209–14.

CUTTING, J. (1980). Physical illness and psychosis. *Br. J. Psychiat.* **136**, 109–19.

DAVIS, K.L., BERGER, P.A., AND HOLLISTER, L.E. (1976). Tardive dyskinesia and depressive illness. *Psychopharmacol. Commun.* **2**, 125–30.

—— AND ROSENBERG, G.S. (1979). Is there a limbic system equivalent of tardive dyskinesia?. *Biol. Psychiat.* **14**, 699–703.

DAVIS, T.G., PICKETT, D.L., AND SCHLOSSER, J.H. (1980). Evaluation of a worldwide spontaneous reporting system with cimetidine. *J. Am. med. Ass.* **243**, 1912–14.

DAVIS, W.A. (1976). Dyskinesia associated with chronic antihistamine use. *New Engl. J. Med.* **294**, 113.

DECISION OF THE TOKYO DISTRICT COURT, 3 August, 1978. SMON Patients vs. the State; Ciba-Geigy (Japan) Limited; Takeda Chemical Industries Limited; Tanabe Seiyaku Co. Ltd. *et al.* Organising Committee of the Kyoto International Conference Against Drug Induced Sufferings (KICADIS), Tokyo, 1979.

DEGKWITZ, R., BINSACK, K.F., AND HERKERT, H., LUXENBURGER, O., AND WENZEL, W. (1967). Zum Problem der persistierenden extrapyramidalen Hyperkinesen nach langfristiger Anwendung von Neuroleptika. *Nervenarzt* **38**, 170–7.

DEHENEFFE, Y., REYNAERT, M., AND TREMOUROUX, J. (1978). Con-fusion mentale au cours de pancréatites traitées par cimétidine. *Nouv. Presse Méd.* **7**, 4303.

DEKRET, J.J., MAANY, I., RAMSEY, T.A., AND MENDELS, J. (1977). A case of oral dyskinesia associated with imipramine treatment. *Am. J. Psychiat.* **134**, 1297–8.

DELANEY, J.C. AND RAVEY, M. (1977). Cimetidine and mental confusion. *Lancet* ii, 512.

DE MONTIGNY, C., CHOUINARD, G., AND ANNABLE, L. (1979). Ineffectiveness of deanol in tardive dyskinesia: a placebo controlled study. *Psychopharmacology* **65**, 219–23.

DENNY, D. AND CASEY, D.E. (1975). An objective method for measuring dyskinetic movements in tardive dyskinesia. *Electroencephalogr. clin. Neurophysio.* **38**, 645–6.

DE SILVA, K.L., MULLER, P.J., AND PEARCE, J. (1973). Acute drug-induced parkinsonism. *Practitioner* **211**, 316–20.

DESTEE, A., LEHEMBRE, P., PETIT, H., AND WAROT, P. (1978). Encephalopathie toxique par l'association lithium-haloperidol. *Lille Medical* **23**, 88–91.

DICK, D.J. AND SAUNDERS, M. (1981). Persistent involuntary movements after treatment with flupenthixol. *Br. med. J.* **282**, 1756.

DRUG AND THERAPEUTICS BULLETIN (1978). Treatment of tardive dyskinesia. *Drug Ther. Bull.* **16**, 55–6.

DUKES, M.N.G. (1977). The moments of truth. In *Side effects of drugs annual 1*, pp. v–ix. Excerpta Medica, Amsterdam.

DUPEYRON, J.P., QUATTROCCHI, F., CASTAING, H., AND FABIANI, D. (1976). Intoxication aigvë du nourisson par application cutanée d'une pommade révulsive locale et antiseptique pulmonaire. *Eur. J. Toxicol Environ. Hyg.* **9**, 313–20.

EDMONDS, M.E., ASHFORD, R.F.V., BRENNER, M.K., AND SAUNDERS, A. (1979). Cimetidine: does neurotoxicity occur? Report of three cases. *J. R. Soc. Med.* **72**, 172–5.

EDWARDS, H. (1970). The significance of brain damage in persistent oral dyskinesia. *Br. J. Psychiat.* **116**, 271–5.

EHYAI, A., KILROY, A.W., AND FENICHEL, G.M. (1978). Dyskinesia and akathisia induced by ethosuximide. *Am. J. Dis. Childh.* **132**, 527–8.

ELDRIDGE, R., IIVANAINEN, M., STERN, R., KOERBER, T., AND WILDER, B.J. (1983). 'Baltic' myoclonus epilepsy: hereditary disorder of childhood made worse by phenytoin. *Lancet* ii, 838–42.

ELLIOTT, H.L., DRYBURGH, F., FELL, G.S., SABET, S., AND MACDOUGALL, A.I. (1978). Aluminium toxicity during regular haemodialysis. *Br. med. J.* **1**, 1101–3.

—— AND FELL, G.S. (1983). Seizures after infusion of factor VIII. *Br. med. J.* **286**, 1900–1.

ETTIGI, P., NAIR, N.P.V., LAL, S. *et al.* (1976). Effect of apomorphine on growth hormone and prolactin secretion in schizophrenic patients, with or without oral dyskinesia, withdrawn from chronic neuroleptic therapy. *J. Neurol. Neurosurg. Psychiat.* **39**, 870–6.

EVANS, J.H. (1965). Persistent oral dyskinesias in treatment with phenothiazine derivatives. *Lancet* i, 458–60.

EWERS, H.R. (1979). Sopor unter Cimetidin-Therapie. *Dt. med. Wochenschr.* **104**, 749.

FAMUYIWA, O.O., ECCLESTON, D., DONALDSON, A.A., AND GARSIDE, R.F. (1979). Tardive dyskinesia and dementia. *Br. J. Psychiat.* **135**, 500–4.

FANN. W.E., DAVIS, J., AND WILSON, I. (1973). Methylphenidate in tardive dyskinesia. *Am. J. Psychiat.* **130**, 922–4.

—— AND LAKE, C.R. (1974). On the coexistence of parkinsonism and tardive dyskinesia. *Dis. nerv. Syst.* **35**, 324–6.

——, ——, GERBER, C.J., AND MC KENZIE, G.M. (1974). Cholinergic suppression of tardive dyskinesia. *Psychopharmacologia* **37**, 101–7.

——, SMITH, R.C., DAVIS, J.M., AND DOMINO, E.F. (Eds.)(1980). *Tardive dyskinesia*. Spectrum Publications, New York.

——, STAFFORD, J.R., MALONE, R.L., FROST, J.D., AND RICHMAN, B.W. (1977*b*). Clinical research techniques in tardive dyskinesia. *Am. J. Psychiat.* **134**, 759–62.

——, ——, AND WHELESS, J. (1977*a*). Tardive dyskinesia and antipsychotics. In *Animal models in psychiatry and neurology* (ed. I. Hanin and E. Usdin), pp. 457–67. Pergamon, Oxford.

——, SULLIVAN, J.L., AND RICHMAN, B.W. (1976). Dyskinesias associated with tricyclic antidepressants. *Br. J. Psychiat.* **128**, 490–3.

FAURBYE, A., RASCH, R.J., PETERSEN, P.B., BRANDBORG, G., AND PAKKENBERG, H. (1964). Neurological symptoms in pharmacotherapy of psychosis. *Acta psychiat. scand.* **40**, 10–27.

FIORE, J.P. (1979). Reversal of cimetidine-induced stupor by physostigmine. *J. Am. med. Ass.* **242**, 1141.

FLENDRIG, J.A., KRUIS, H., AND DAS, H.A. (1976). Aluminium and dialysis dementia. *Lancet* **i**, 1235.

FLIND, A.C. AND ROWLEY-JONES, D. (1979). Mental confusion and cimetidine. *Lancet* **i**, 379.

FOOD AND DRUGS ADMINISTRATION (1971). Hexachlorophene and newborns. *FDA Drug Bulletin*. (Studies submitted by Winthrop Laboratories to FDA on November 18, 1971).

GAINES, T.B. AND KIMBROUGH, R.D. (1971). Paper read at the 10th annual meeting of the Society of Toxicology, Washington, D.C., March 7–11, 1971. See also Kimbrough, R.D. and Gaines, T.B. (1971). *Arch. environ. Hlth* **23**, 114–18.

GALE, K. (1980). Chronic blockade of dopamine receptors by antischizophrenic drugs enhances GABA binding in substantia nigra. *Nature, London* **283**, 569–70.

GARDOS, G., COLE, J.O., AND LA BRIE, R. (1977*a*). The assessment of tardive dyskinesia. *Arch. gen. Psychiat.* **34**, 1206–12.

——, —— AND —— (1977*b*). Drug variables in the etiology of tardive dyskinesia: application of discriminant function analysis. *Prog. Neuro-Psychopharmacol.* **1**, 147–54.

——, ——, AND TARSY, D. (1978). Withdrawal syndromes associated with antipsychotic drugs. *Am. J. Psychiat.* **135**, 1321–4.

——, GRANACHER, R.P., COLE, J.O., AND SNIFFIN, C. (1979). The effect of papaverine in tardive dyskinesia. *Prog. Neuro-Psychopharmacol.* **3**, 543–50.

GEISLER, A. AND KLYSNER, R. (1977). Combined effect of lithium and flupenthixol on striatal adenyl cyclase. *Lancet* **i**, 430–1.

GELENBERG, A.J. (1976). Computerized tomography in patients with tardive dyskinesia. *Am. J. Psychiat.* **133**, 578–9.

——, DOLLER-WOJCIK, J.C., AND GROWDON, J.H. (1979). Choline and lecithin in the treatment of tardive dyskinesia: preliminary results from a pilot study. *Am. J. Psychiat.* **136**, 772–6.

GERBINO, L., SHOPSIN, B., AND COLLORA, M. (1980). Clozapine in the treatment of tardive dyskinesia: an interim report. In *Tardive dyskinesia* (eds. W.E. Fann, R.C. Smith, J.M. Davis, E.F. Domino), pp. 475–89. Spectrum Publications, New York.

GERLACH, J. (1977). The relationship between parkinsonism and tardive dyskinesia. *Am. J. Psychiat.* **134**, 781–4.

—— (1979). Tardive dyskinesia. *Dan. med. Bull.* **26**, 209–45.

——, REISBY, N., AND RANDRUP, A. (1974). Dopaminergic hypersensitiviy and cholinergic hypofunction in the pathophysiology of tardive dyskinesia. *Psychopharmacologia* **34**, 21–35.

——, RYE, T., AND KRISTJANSEN, P. (1978). Effect of baclofen on tardive dyskinesia. *Psychopharmacology* **56**, 145–51.

GESSA, R., TAGLIAMONTE, A., AND GESSA, G.L. (1972). Blockade by apomorphine of haloperidol-induced dyskinesia in schizophrenic patients. *Lancet* **ii**, 981.

GHOLZ, L.M. AND ARONS, W.L. (1964). Prophylaxis and therapy of amebiasis and shigellosis: iodochlorhydroxyquine. Am. J.trop. Med. Hyg. **13**, 396–401.

GIBSON, A.C. (1978*a*). Depot injections and tardive dyskinesia. *Br. J. Psychiat.* **132**, 361–5.

—— (1978*b*). Sodium valproate and tardive dyskinesia. *Br. J. Psychiat* **133**, 82.

—— (1979). Questionnaire on severe tardive dyskinesia. *Br. J. Psychiat.* **134**, 549–50.

—— (1980). Depot fluphenazine and tardive dyskinesia in an outpatient population. In *Tardive dyskinesia* (ed. W.E. Fann, R.C. Smith, J.M. Davis, and E.F. Domino), pp. 315–24. Spectrum Publications, New York.

GIFFORD, L.M., AEUGLE, M.E., MYERSON, R.M., AND TANNENBAUM, P.J. (1980). Cimetidine postmarket outpatient surveillance program: interim report on phase I. *J. Am. med. Ass.* **243**, 1532–5.

GILBERT, G.J. (1979). Tardive dyskinesia. Letter. *Lancet* **ii**, 798.

GILES, C.L. AND SOBLE, A.R. (1971). Intracranial hypertension and tetracycline therapy. *Am. J. Opthal.* **72**, 981–2.

GOAS, J.Y., BORSOTTI, J.P., MISSOUM, A., ALLAIN, P., AND CHALEU, D. (1981). Encephalopathie myoclonique par les sous-nitrate de bismuth. *Nouv. Presse Méd* **10**, 3855.

GORDON, M. AND GORDON, A.S. (1981). Perhexiline maleate as a cause of reversible parkinsonism and peripheral neuropathy. *J. Am. geriat. Soc.* **29**, 259–62.

GRAHAM, J.R. (1979). Psychotic reaction to cimetidine, presumably an idosyncrasy. *Med. J. Aust.* **2**, 491–2.

GRAVE, W., NADORP, J.H.S.M., AND RUTTEN, J.J.M.H. (1977). Cimetidine and renal failure. *Lancet* **ii**, 719–20.

GRAWITZ, P.B. (1935). Nuevas orientaciones en la terapeutica de la amebiasis. *Semana Med.* **42**, 525–9.

GREENBLATT, D.L., DOMINICK, J.R., STOTSKY, B.A., AND DI MASCIO, A. (1968). Phenothiazine-induced dyskinesia in nursing-home patients. *J. Am. geriatr. Soc.* **16**, 27–34.

GREENSTREET, R. (1983). Adjustment of rates of Guillain–Barré syndrome among recipients of swine flu vaccine, 1976–1977. *J. R. Soc. Med.* **76**, 620–1.

GRIMSON, T.A. (1977). Reactions to cimetidine. *Lancet* **i**, 858.

GRISWOLD, W.R., REZNIK, V., AND MENDOZA, S.A. (1981). Nitroprusside-induced intracranial hypertension. *J. Am. med. Ass.* **246**, 2679–80.

GROSS, H. AND KALTENBACH, E. (1968). Neuropathological findings in persistent hyperkinesia after neuroleptic long-term therapy. In *The present status of psychotropic drugs* (ed. A. Cerletti and F.J. Bove), pp. 474–6. Excerpta Medica, Amsterdam.

GROWDON, J.H., HIRSCH, M.J., WURTMAN, R.J., AND WEINER, W. (1977). Oral choline administration to patients with tardive dyskinesia. *New Engl. J. Med.* **297**, 524–7.

GUNNE, L.M. AND BARANY, S. (1976). Haloperidol-induced tardive dyskinesia in monkeys. *Psychopharmacology* **50**, 237–40.

GUY, W. (1976). *ECDEU assessment manual for psychopharmacology*. US Government Printing Office, Washington, DC.

HALE, C. AND HEINS, T. (1978). Tardive dyskinesia and antihistamines. *Med. J. Aust.* **1**, 112–13.

HALL, A. (1982). Editorial: Dexamethasone deleterious in cerebral malaria. *Br. med. J.* **284**, 1588.

HALL, S.M. (1983). Aspirin and Reye's syndrome. *Lancet* i, 583–4.

—— AND BELLMAN, M. (1984). Reye's syndrome in the British Isles: first annual report of the joint British Paediatric Association and Communicable Disease Surveillance Centre surveillance scheme. *Br. med. J.* **288**, 548–50.

HANGARTNER, P. (1980). Clinical study of clioquinol intoxication in dogs and cats. In *Drug-induced sufferings* (ed. T. Soda), pp. 459–63. Excerpta Medica, Amsterdam.

HANSSON, O. (1980). Oxyquinoline intoxication outside Japan, its recognition and the scope of the problem. In *Drug-induced sufferings* (ed. T. Soda), pp. 429–32. Excerpta Medica, Amsterdam.

HARDING, T. (1968). Cerebral malaria. *Br. med. J.* iii, 250.

HEINRICH, K., WEGENER, I., AND BENDER, H.J. (1968). Späte extrapyramidale Hyperkinesen bei neuroleptischer Langzeittherapie. *Pharmakopsychiatr Neuropsychopharmakologie* **1**, 169–95.

HERBERT, M., HEALY, T.E.J., BOURKE, J.B., FLETCHER, I.R., AND ROSE, J.M. (1983). Profile of recovery after general anaesthesia. *Br. med. J.* **286**, 1539–42.

HIPPIUS, H. AND LOGEMANN, G. (1970). Zur Wirkung von Dioxyphenylalanin (L-DOPA) auf extrapyramidalmotorische Hyperkinesen nach langfristiger neuroleptische Therapie. *Arzneim Forsch.* **30**, 894–6.

HODGSON-JONES, I.S. (1970). Cliniquinol and iodine metabolism. *Trans. St. John's Hosp. Dermatol. Soc.* **56**, 51–3.

HOFF, H. AND HOFMANN, G. (1967). Das persistierende extrapyramidale Syndrom bei Neuroleptikatherapie. *Wien. Med. Wochenschr.* **117**, 14–17.

HOLLISTER, L.E., CAFFEY, E.M.JR., AND KLETT, C.J. (1960). Abnormal symptoms, signs and laboratory tests during treatment with phenothiazine derivatives. *Clin. Pharmacol. Ther.* **1**, 284–93.

HOUGHTON, A.W.J. (1971). Convulsions precipitated by amitriptyline. *Lancet* i, 138.

HUNTER, R., BLACKWOOD, W., SMITH, M., AND CUMMINGS, J. (1968). Neuropathological findings in three cases of persistent dyskinesia following phenothiazine medication. *J. neurol. Sci.* **7**, 263–73.

——, EARL, C.J., AND THORNICROFT, S. (1964). An apparently irreversible syndrome of abnormal movements following phenothiazine medication *Proc. R. Soc. Med.* **57**, 758–62.

JACKSON, I.V., NUTTALL, E.A., IBE, I.O., AND PEREZ-CRUET, J. (1979). Treatment of tardive dyskinesia with lecithin. *Am. J. Psychiat.* **136**, 1458–60.

JEFFERSON, J.W. (1979). Central nervous system toxicity of cimetidine — a case of depression. *Am. J. Psychiat.* **136**, 346.

JENNER, P. AND MARSDEN, C.D. (1979). Tardive dyskinesias. Letter. *Lancet* ii, 900.

JESTE, D.V., OLGIATE, S.G., AND GHALI, A.Y. (1977). Masking of tardive dyskinesia with four times-a-day administration of chlorpromazine. *Dis. nerv. Syst.* **38**, 755–8.

——, POTKIN, S.G., SINHA, S., FEDER, S., AND WYATT, R.J. (1979b). Tardive dyskinesia-reversible and persistent. *Arch. gen. Psychiat.* **36**, 585–9.

——, ROSENBLATT, J.E., WAGNER, R.L., AND WYATT, R.J. (1979a). High serum neuroleptic levels in tardive dyskinesia. Letter. *New Engl. J. Med.* **301**, 1184.

——, WAGNER, R.L., WEINBERGER, D.R., RIETH, K.G., AND WYATT,

R.J. (1980). Evaluation of CT scans in tardive dyskinesia. *Am. J. Psychiat.* **137**, 247–8.

—— AND WYATT, R.J. (1979). In search of treatment for tardive dyskinesia: a review of the literature. *Schizophren Bull* **5**, 251–93.

JOHNSON, D.A.W. (1973). The side-effects of fluphenazine decanoate. *Br. J. Psychiat.* **123**, 519–22.

—— (1979). Further observations on the duration of depot neuroleptic maintenance therapy in schizophrenia. *Br. J. Psychiat.* **135**. 524–30.

JOHNSON, H.C. AND WALKER, A.E. (1945). Intraventricular penicillin: Note of warning. *J. Am. med. Ass.* **127**, 217–19.

JOHNSON, J. AND BAILEY, S. (1979). Cimetidine and psychiatric complications. *Br. J. Psychiat.* **134**, 315–16.

JUS, A., JUS, K., AND FONTAINE, P. (1979). Long term treatment of tardive dyskinesia. *J. clin Psychiat.* **40**, 72–7.

——, PINEAU, R., LACHANCE, R., PELCHAT, G., JUS, K., PIRES, P., AND VILLENEUVE, R. (1976a,b). Epidemiology of tardive dyskinesia: (a) part I, (b) part II. *Dis. Nerv. Syst.* **37**, (a) 210–14, (b) 257–61.

——, VILLENEUVE, A., GAUTIER, J., JUS, K., VILLENEUVE, C., PIRES, P., AND VILLENEUVE, R. (1978). Deanol, lithium and placebo in the treatment of tardive dyskinesia: a double-blind cross-over study. *Neuropsychobiology* **4**, 140–9.

JUS, K., JUS, A., AND VILLENEUVE, A. (1973). Polygraphic profile of oral tardive dyskinesia and of rabbit syndrome for quantitative and qualitative evaluation. *Dis. nerv. Syst.* **34**, 27–32.

KAESER, H.E. AND WUTHRICK, R. (1970). Zur Frage de Neurotoxizitat der oxychinoline. *Deutsch. med. Wschr.* **95**, 1685–8.

KAPLAN, S.R. AND MURKOFSKY, C. (1978). Oral-buccal dyskinesia symptoms associated with low-dose benzodiazepine treatment. *Am. J. Psychiat.* **135**, 1558–9.

KATAHIRA, K., TESHIMA, K., AND SUGISAWA, H. *et al.* (1980). An international survey on the recent reports concerning intoxication with halogenated oxyquinoline derivatives and the regulations against their use. *Drug-induced sufferings.* (ed. T. Soda), pp. 441–55. Excerpta Medica, Amsterdam.

KAZAMATSURI, H., CHIEN, C.P., AND COLE, J.O. (1972). Treatment of tardive dyskinesia I. Clinical efficacy of a dopamine-depleting agent, tetrabenazine. *Arch. gen. Psychiat.* **27**, 95–9.

——, ——, AND —— (1973). Long-term treatment of tardive dyskinesia with haloperidol and tetrabenazine. *Am. J. Psychiat.* **130**, 479–83.

KEAN, B.H. (1972). Subacute myelo-optic neuropathy. A probable case in the United States. *J. Am. med. Ass.* **220**, 243–4.

KEBABIAN, J.W. AND CALNE, D.B. (1979). Multiple receptors for dopamine. *Nature, London.* **277**, 93–6.

KENNEDY, P.F., HERSHON, H.I., AND MC GUIRE, R.J. (1971). Extrapyramidal disorders after prolonged phenothiazine therapy. *Br. J. Psychiat.* **118**, 509–18.

KIMELBLATT, B.J., CERRA, F.B., CALLERI, G., BERG, M.J., MC MILLEN, M.A., AND SCHENTAG, J.J. (1980). Dose and serum concentration relationships in cimetidine-associated mental confusion. *Gastroenterology* **78**, 791–5.

KING, S.W., SAVORY, J., AND WILLS, M.R. (1981). Aluminium toxicity in relation to kidney disorders. *Ann. clin. Lab. Sci.* **11**, 337–41.

KINNELL, H.G. AND WEBB, A. (1979). Confusion associated with cimetidine. *Br. med. J.* **2**, 1438.

KINROSS-WRIGHT, J. (1959). Newer phenothiazine drugs in

treatment of nervous disorders. *J. Am. med. Ass.* **170**, 1283–8.

KLAWANS, H.L. (1973). The pharmacology of tardive dyskinesia. *Am. J. Psychiat.* **130**, 82–6.

——, BERGEN, D., BRUYN, G.W., AND PAULSON, G.W. (1974). Neuroleptic-induced tardive dyskinesias in non-psychotic patients. *Arch. Neurol.* **30**, 338–9.

——, CARVEY, P., NAUSIEDA, P.A., GOETZ, C.G., AND WEINER, W.J. (1980*a*). Effect of dose and type of neuroleptic in an animal model of tardive dyskinesia. *Neurology* **30**, 383.

——, GOETZ, C.G., AND PERLIK, S. (1980*b*). Tardive dyskinesia: review and update. *Am. J. Psychiat.* **137**, 900–8.

——, GOETZ, E., AND BERGEN, D. (1975). Levodopa-induced myoclonus. *Arch. Neurol.* **32**, 331–4.

——, HITRI, A., NAUSIEDA, P.A., AND WEINER, W.J. (1977). Animal models of dyskinesia. In *Animal models in psychiatry and neurology* (ed. I. Hanin and E. Usdin), pp. 351–63. Pergamon, Oxford.

—— AND RUBOVITS, R. (1974). Effects of cholinergic and anticholinergic agents on tardive dyskinesia. *J. Neurol. Neurosurg. Psychiat.* **37**, 941–7.

KLINE, N.S. (1968). On the rarity of irreversible oral dyskinesias following phenothiazines. *Am. J. Psychiat.* **124**, 48–54.

KLOTZ, U. AND REIMANN, I. (1980). Delayed clearance of diazepam due to cimetidine. *New Engl. J. Med.* **302**, 1012–14.

KOBAYASHI, R.M. (1977). Drug therapy of tardive dyskinesia. *New Engl. J. Med.* **296**, 257–60.

KOCH-WESER, J. AND GILMORE, E.B. (1967). Benign intracranial hypertension in an adult after tetracycline therapy. *J. Am. med. Ass.* **200**, 345–7.

KONO, R. (1975). Introductory review of subacute myeloopticoneuropathy (SMON) and its studies done by the SMON Research Commission. *Jap. J. med. Sci. Biol.* **28** (Suppl.), 1–21.

KRISTJANSEN, P., DENCKER, S.J., ELLEY, J., HAKOLA, A., HESHE, J., MALM, V., ROBAK, O.H., SALVESEN, C.H.R., AND VAKSDAL, K. (1973). Clinical experience with 19,552 R.P. in Nordic countries. *Acta psychiat. scand.* (Suppl.) **246**, 42–7.

KUCHARSKI, L.T., SMITH, J.W., AND DUNN, D.D. (1979). Mortality and tardive dyskinesia. Letter. *Am. J. Psychiat* **136**, 1228.

KUMAR, B.B. (1976). Treatment of tardive dyskinesia with deanol. Letter. *Am. J. Psychiat.* **133**, 978.

KURODA, S., TATEISHI, J., AND YOKAYAMA, S. (1974). A case of cerebellar degeneration accompanying chloramphenicol. *Clin. Neurol., Tokyo* **14**, 315.

KURTZMAN, N.A., ROGERS, P.W., AND HARTER, H.R. (1970). Neurotoxic reaction to penicillin and carbenicillin. *J. Am. med. Ass.* **214**, 1320–1.

LANCET (1977). Comment: SMON and clioquinol. *Lancet* **i**, 534.

—— (1979). Editorial: Tardive dyskinesia. *Lancet* **ii**, 447–8.

—— (1980). Editorial: Idiosyncratic neurotoxicity: clioquinol and bismuth. *Lancet* **i**, 857–8.

—— (1983). Editorial: 'Toxicity' of interferon. *Lancet* **i**, 1256.

LAVY, S., MELAMED, E., AND PENÇHAS, S. (1978). Tardive dyskinesia associated with metoclopramide. *Br. med. J.* **1**, 77–8.

LAYZER, R.B. (1978). Myeloneuropathy after prolonged exposure to nitrous oxide. *Lancet* **ii**, 1227–30.

LE QUESNE, P.M. (1981). Toxic substances and the nervous system: the role of clinical observation. *J. Neurol. Neurosurg. Psychiat.* **44**, 1–8.

LEVINE, M.L. (1978). Cimetidine-induced coma in cirrhosis of the liver. *J. Am. med. Ass.* **240**, 1238.

LIEBERMAN, A., GOODGOLD, A., JONAS, S., AND LEIBOWITZ, M. (1975). Comparison of dopa decarboxylase inhibitor (carbidopa) combined with levodopa and levodopa alone in Parkinson's disease. *Neurology* **25**, 911–16.

LINDEBOOM, S.F. AND LAKKE, J.P.W.F. (1978). Deanol and physostigmine in the treatment of L-dopa induced dyskinesias. *Acta neurol. scand.* **58**, 134–8.

LINNOILA, M., VIUKARI, M., AND HEITALA, O. (1976). Effect of sodium valproate on tardive dyskinesia. *Br. J. Psychiat.* **129**, 114–19.

LUNDE, P.K.M. (1980). Discussion comment to Oakley, G.P., *vide infra*.

LYON, L.J. AND NEVINS, M.A. (1974). Viral hepatitis and salicylism simulating Reye's syndrome. *J. med. Soc. N. Jersey* **71**, 657–60.

MACKAY, A.V.P. AND SHEPPARD, G.P. (1979). Pharmacotherapeutic trials in tardive dyskinesia. *Br. J. Psychiat.* **135**, 489–99.

MÄKELÄ, A-C., LANG, H., AND KORPELA, P. (1980). Toxic encephalopathy with hyperammonaemia during high dose salicylate therapy. *Acta neurol. scand.* **61**, 146–51.

MALLYA, A., JOSE, C., BAIG, M., WILLIAMS, R., CHO, D., MEHTA, D., AND VOLAVKA, J. (1979). Antiparkinsonics, neuroleptics, and tardive dyskinesia. *Biol. Psychiat.* **14**, 645–9.

MANN, R.D. (1984). *Modern drug use. An enquiry on historical principles,* p. 694. MTP Press, Lancaster.

MAROON, J.C. AND MEALY, J. (1971). Benign intracranial hypertension. Sequal to tetracycline therapy in a child. *J. Am. med. Ass.* **216**, 1479.

MARSDEN, C.D. (1976). Drugs and involuntary movements: the pharmacological pathology of the dyskinesias. In *Twelfth Symposium on Advanced Medicine* (ed. P.K. Peters). Pitman Medical, Tunbridge Wells.

MARTIN, I. (1967). Bromism induced by safe medications old and new: Some psychological considerations. *Med. J. Aust.* **1**, 95–8.

MARTIN-BOUYER, G. (1978). Intoxications par les sels de bismuth administres par voie orale. *Gastroenterol clin. Biol.* **2**, 349–56.

McANDREW, J.B., CASE, Q., AND TREFFERT, D.A. (1972). Effects of prolonged phenothiazine intake on psychotic and other hospitalised children. *J. Autism Child Schizo.* **2**, 75–91.

McCLEAN, P. AND CASEY, D.E. (1978). Tardive dyskinesia in an adolescent. *Am. J. Psychiat.* **135**, 969–71.

McEWEN, L.M. (1971). Neuropathy after clioquinol. *Br. med. J.* **4**, 169–70.

McGEER, P.L., BOULDING, J.E., GIBSON, W.C., AND FOULKES, R.G. (1961). Drug-induced extrapyramidal reactions. *J. Am. med. Ass.* **177**, 665–70.

McLAUGHLIN, A.I.G., KAZANTZIS, C., KING, E., TEARE, D., PORTER, R.J., AND OWEN, R. (1962). Pulmonary fibrosis and encephalopathy associated with the inhalation of aluminium dust. *Br. J. ind. Med.* **19**, 253–63.

McLELLAN, D.L. AND SWASH, M. (1974). Choreoathetosis and encephalopathy induced by phenytoin *Br. med. J.* **2**, 204–5.

McMILLEN, M.A., AMBIS, D., AND SIEGAL, J.H. (1978). Cimetidine and mental confusion. *New Engl. J. Med.* **298**, 284–5.

MEDICAL LETTER ON DRUGS AND THERAPEUTICS (1979). Prevention and treatment of tardive dyskinesia. *Med. Lett. Drugs. Ther.* **21**, 34–5.

MEHTA, D., MALLYA, A., AND VOLAVKA, J. (1978). Mortality of patients with tardive dyskinesia. *Am. J. Psychiat.* **135**, 371–2.

——, MEHTA, S., AND MATHEW, P. (1977). Tardive dyskinesia in psychogeriatric patients: a five-year follow-up. *J. Am. geriat. Soc.* **25**, 545–7.

MENZIES-GOW, N. (1977). Cimetidine and mental confusion. *Lancet* **ii**, 928.

MEREDITH, T.J. AND VOLANS, G.N. (1979). Management of cimetidine overdose. *Lancet* **ii**, 1367.

MEYBOOM, R.H.B. (1975). Metals. In *Side-effects of drugs*, Vol. 8 (ed. M.N.G. Dukes), Excerpta Medica, Amsterdam.

MINTZ, U., LIEBERMAN, U.A., AND VRIES, A. (1971). Parkinsonism syndrome due to cephaloridine. *J. Am. med. Ass.* **216**, 1200.

MOGELNICKI, S.R., WALLER, J.L., AND FINLAYSON, D.C. (1979). Physostigmine reversal of cimetidine-induced mental confusion. *J. Am. med. Ass.* **241**, 826.

US MORBIDITY AND MORTALITY WEEKLY REPORT (1980, 1980, 1982). 11 July 1980, 7 November 1980, 12 February 1982.

MORGAN, F.P. AND BILLINGS, J.J. (1974). Is this subgallate poisoning? *Med. J. Aust.* **2**, 662–3.

MOSER, R.H. (1966). Reactions to phenothiazine and related drugs. *Clin Pharmacol. Ther.* **7**, 683–97.

MOWAT, A.P. (1983). Reye's syndrome: 20 years on. *Br. med. J.* **286**, 1999–2001.

MULLER, D.J. (1968). Bromide intoxication continues to occur. *Texas Medicine* **64**, 72–3.

NAGAO, T., OHSHIMO, T., MITSUNOBU, K., SATO, M., AND OTSUKI, S. (1979). Cerebrospinal fluid monoamine metabolites and cyclic nucleotides in chronic schizophrenic patients with tardive dyskinesia or drug-induced tremor. *Biol Psychiat.* **14**, 509–23.

NAIR, N.P.V., YASSA, R., RUIZ-NAVARRO, J., AND SCHWARTZ, G. (1978). Baclofen in the treatment of tardive dyskinesia. *Am. J. Psychiat.* **135**, 1562–3.

NAKAE, K., YAMAMOTO, S., SHIGAMATSU, I., AND KONO, R. (1973). Relation between subacute myelo-optic neuropathy (SMON) and clioquinol: nationwide survey. *Lancet* **i**, 171–3.

NAKAMURA, Y. AND INOUE, Y.K. (1972). Pathogenicity of virus associated with subacute myelo-optic neuropathy *Lancet* **i**, 223–6.

NASRALLAH, H.A. (1979*a*). Tardive dyskinesia and depot fluphenazine. Letter. *Br. J. Psychiat.* **134**, 550.

—— (1979*b*). Methodological issues in tardive dyskinesia research. *Schizophren. Bull.* **5**, 1–3.

NEARY, D., THURSTON, H., AND POHL, J.E.F. (1973). Development of extrapyramidal symptoms in hypertensive patients treated with diazoxide. *Br. med. J.* **3**, 474–5.

NELSON, P.G. (1977). Cimetidine and mental confusion. *Lancet* **ii**, 928.

OAKLEY, G.P. (1980). The neurotoxicity of the halogenated hydroxyquinolines. In *Drug-induced sufferings* (ed. T. Soda), pp. 90–6. Excerpta Medica, Amsterdam.

OGITA, K., YAGI, G., AND ITOH, H. (1975). Comparative analysis of persistent dyskinesias of long-term usage with neuroleptics in France and in Japan . *Folia Psychiatr. Neurol. Jpn* **29**, 315–20.

OHLRICH, G.D. AND OHLRICH, J.G. (1977). Papilloedema in an adolescent due to tetracycline. *Med. J. Aust.* **1**, 334.

OLDSTONE, M.B.A. AND NELSON, E. (1966). Central nervous system manifestations of penicillin toxicity in man. *Neurology* **16**, 693–700.

OPFER, K. (1963). The bulging fontanelle. *Lancet* **ii**, 116.

ORLAND, F. (1959). Use and overuse of tranquillizers. *J. Am. med. Ass.* **171**, 633–6.

PANNEKOEK, J.H. (1972). Neurotoxische verschijnselen na clioquinol (*Enterovioform*). *Ned. T. Geneesk.* **116**, 1611–15.

PARKINSON, I.S. WARD, M.K., FEEST, T.G., FAWCETT, R.W.P., AND KERR, D.N.S. (1979). Fracturing dialysis osteodystrophy and dialysis encephalopathy. *Lancet* **i**, 406–9.

PARTIN, J.S., PARTIN, J.C., SCHUBERT, W.K., AND HAMMOND, J.G. (1982). Serum salicylate concentrations in Reye's disease. A study of 130 biopsy proven cases. *Lancet* **i**, 191–4.

PASCOE, J.M. (1980). Salicylate and Reye's syndrome. *Pediatrics* **68**, 610–11.

PAULSON, G.W. (1975). Tardive dyskinesia. *Ann. Rev. Med.* **26**, 75–81.

——, RIZVI, C.A., AND CRANE, G.E. (1975). Tardive dyskinesia as a possible sequel of long-term therapy with phenothiazines. *Clin. Pediatr., Phil* **14**, 953–5.

PERRIS, C., DIMITRIJEVIC, P., JACOBSSON, L., PAULSSON, P., RAPP, W., AND FROBERG, H. (1979). Tardive dyskinesia in psychiatric patients treated with neuroleptics. *Br. J. Psychiat.* **135**, 509–14.

PETITE, J.P. AND BLOCH, F. (1979). Syndrome dépressif au cours d'un traitement par la cimétidine, *Nouv. Presse Méd.* **8**, 1260.

PETTY, L.K. AND SPAR, C.J. (1980). Haloperidol-induced tardive dyskinesia in a 10-year-old girl. *Am. J. Psychiat* **137**, 745–6.

POGGLITSCH, H., PETEK, W., WAWSCHINEK, O., AND HOLZER, W. (1981). Treatment of early stages of dialysis encephalopathy by aluminium depletion. *Lancet* **ii**, 1344–5.

POLIZOS, P. AND ENGELHARDT, D.M. (1980). Dyskinetic and neurological complications in children treated with psychotropic medication. In *Tardive dyskinesia* (ed. W.E. Fann, R.C. Smith, J.M. Davis, E.F. Domino), pp. 193–9. Spectrum Publications, New York.

——, ——, HOFFMAN, S.P., AND WAIZER, J. (1973). Neurological consequences of psychotropic drug withdrawal in schizophrenic children. *J. Autism Child Schizo.* **3**, 247–53.

PORTNOY, R.A. (1979). Hyperkinetic dysarthria as an early indicator of impending tardive dyskinesia. *J. Speech. Hear. Dis.* **44**, 214–19.

QUAP, C.W. (1978). Confusion: an adverse reaction to cimetidine therapy. *Drug Intel. clin. Pharm.* **12**, 121.

QUITKIN, F., RIFKIN, A., GOCHFELD, L., AND KLEIN, D.F. (1977). Tardive dyskinesia: are first signs reversible? *Am. J. Psychiat.* **134**, 84–7.

RANDALL, M.E. (1983). Aluminium toxicity in an infant not on dialysis. *Lancet* **i**, 1327–8.

REID, H.A. AND NKRUMAH, F.K. (1972). Fibrin degradation products in cerebral malaria. *Lancet* **i**, 218–21.

REINICKE, G. (1980). The mystery of dialysis encephalopathy. In *Side effects of drugs annual 4* (ed. M.N.G. Dukes), pp. 163–6. Excerpta Medica, Amsterdam.

REYE, R.D.K., MORGAN, G., AND BARAL, J. (1963). Encephalopathy and fatty degeneration of the viscera. A disease entity in children. *Lancet* **ii**, 749–52.

RIGGS, J., HAMILTON, R., HAMEL, S., AND McCABE, J. (1975). Camphorated oil intoxication in pregnancy. *Obstet. Gynecol.* **25**, 255.

ROBINSON, O.P.W. (1973). Metoclopramide — side effects and safety. *Postgrad. med. J.* **49**, (Suppl 4), 77–80.

ROBINSON, T.J. AND MULLIGAN, T.O. (1977). Cimetidine and mental confusion. *Lancet* **ii**, 719.

RODGERS, G.C., WEINER, L.B., AND McMILLAN, J.A. (1982). Salicylate and Reye's syndrome. *Lancet* **i**, 616.

ROSENBAUM, A.H. AND DE LA FUENTE, J.R. (1979). Benzodiazepines and tardive dyskinesia. Letter. *Lancet* ii, 900.

——, MARUTA, T., JIANG, N.S., AUGER, R.G., DE LA FUENTE, J.R., AND DUANE, D.D. (1979). Endocrine testing in tardive dyskinesias: preliminary report. *Am. J. Psychiat.* **136**, 102–3.

——, NIVEN, R.G., HANSON, N.P. *et al.* (1977). Tardive dyskinesia: relationship with a primary affective disorder. *Dis. Nerv. Syst.* **38**, 423–7.

ROSENFELD, R.G. AND LIEBHABER, M.I. (1976). Acute encephalopathy in siblings Reye syndrome vs salicylate intoxication. *Am. J. Dis. Childh.* **130**, 295–7.

ROSS, D.R. *et. al.* (1983). Akathisia induced by amoxapine. *Am. J. Psychiat.* **140**, 115–16.

RUBIN, M.B., RECINOS, A., WASHINGTON, J.A., AND KOPPANYI, T. (1949). Ingestion of poisons in children: A survey of 250 admissions to childrens hospitals. *Clin Proc. Child Hosp.,* Washington, DC.

SCHANTZ, B. AND WIKSTRÖM, B. (1965). Suspected poisoning with oxychinoline preparation in dogs. *Svensk. vet. Tidn.* **17**, 106.

SCHENTAG, J.J., CERRA, F.B., CALLERI, G., DE GLOPPER, E., ROSE, J.Q., AND BERNHARD, H. (1979). Pharmacokinetic and clinical studies in patients with cimetidine-associated mental confusion. *Lancet* i, 177–81.

—— *et al.* (1982). Mental confusion in a patient treated with metronidazole — a concentration-related effect? *Pharmacotherapy* **2**, 384–7.

SCHIELE, B.C. (1962). Newer drugs for mental illness. *J. Am. med. Ass.* **181**, 126–33.

——, GALLANT, D., SIMPSON, G., GARDNER, E.A., COLE, J., CRANE, G., CHASE, T., AYD, F., LEVINE J., AND OCHOTA, L. (1973). Neurological syndromes associated with antipsychotic drug use: a special report. *Arch gen. Psychiat.* **28**, 463–7.

SCHÖNECKER, M. (1957). Ein eigentumliches Syndrom im oralen Bereich bei Megaphenapplikation. *Nervenarzt* **28**, 35.

SCHWAB, R.S., DOSHAY, L.J., GARLAND, H., BRADSHAW, P., GARVEY, E., AND CRAWFORD, B. (1956). Shift to older age distribution in Parkinsonism: report on 1,000 patients covering past decade from three centres. *Neurology* **6**, 783–90.

SEAMANS, K.B., GLOOR, P., DOBELL, R.A.R., AND WYNANT, J.D. (1968). Penicillin-induced seizures during cardiopulmonary bypass. A clinical and electroencephalographic study. *New Engl. J. Med.* **278**, 861–8.

SELBY, G. (1972). Subacute myelo-optic neuropathy in Australia. *Lancet* i, 123–5.

SHIGEMATSU, I. (1975). Subacute myelo-optico-neuropathy (SMON) and clioquinol. *Jap. J. med. Sci. Biol.* **28**, 35–55.

SHIMADA, Y. AND KASAKA, K. (1973). New cases of SMON in Japan. *Lancet* i, 268.

SHIRAKI, H. (1975). The neuropathology of subacute myelo-optico-neuropathy, 'SMON' in the humans. *Jap. J. med. Sci. Biol.* **28**, 101–64.

SIBERT, J.R. (1973). Poisoning in children. *Br. med. J.* **1**, 803.

SIGWALD, J., BOUTTIER, D., RAYMONDEAU, C., AND PIOT, C. (1959). Quatre cas de dyskinesies facio-bucco-linguo-masticatrices à évolution prolongée secondaire à un traitement par les neuroleptiques. *Rev. Neurol.* **100**, 751–5.

SILLANPÄÄ, M., MAKELA, A-L., AND KIOVIKKI, A. (1975). Acute encephalopathy (Reye's Syndrome) during salicylate therapy. *Acta paediat. scand.* **64**, 877–80.

SIMPSON, G.M., LEE, J.H., AND SHRIVASTAVA, R.K. (1978b). Clozapine in tardive dyskinesia. *Psychopharmacology* **56**, 75–80.

——, ——, ZOUBOK, B., AND GARDOS, G. (1979). A rating scale for tardive dyskinesia. *Psychopharmacology,* **64**, 171–9.

——, VARGA. E., LEE, J.H., AND ZOUBOK, B. (1978a).Tardive dyskinesia and psychotropic drug history. *Psychopharmacology* **58**, 117–24.

SKOGLUND, R.R., WARE, L.L., AND SCHANBERGER, J.E. (1977). Prolonged seizures due to contact and inhalation exposure to camphor: A case report. *Clin. Pediat.* **16**, 901.

SMALL, M., DURWARD, W.F., AND FORBES, C.D. (1983). Seizures after infusion of factor VIII. *Br. med. J.* **286**, 1106–7.

SMEDLEY, H., KATRAK, M., SIKORA, K., AND WHEELER, T. (1983). Neurological effects of recombinant human interferon. *Br. med. J.* **286**, 262–4.

SMITH, H., LERNER, P.I., AND WEINSTEIN, L. (1967). Neurotoxicity and 'massive' intravenous therapy with penicillin. A study of possible predisposing factors. *Arch. intern. Med.* **120**, 47–53.

SMITH, J.M., KUCHARSKI, L.T., EBLEN, C., KNUTSEN, E., AND LINN, C. (1979b). An assessment of tardive dyskinesia in schizophrenic outpatients. *Psychopharmacology* **64**, 99–104.

——, ——, OSWALD, W.T., AND WATERMAN, L.J. (1979a). A systematic investigation of tardive dyskinesia in inpatients. *Am. J. Psychiat.* **136**, 918–22.

——, KUCHARSKT, T., AND WATERMAN, L.J. (1978). Tardive dyskinesia: age and sex differences in hospitalised schizophrenics. *Psychopharmacology* **58**, 207–11.

SMITH, J.S. (1979). Tardive dyskinesia and anticholinergic drugs. *Aust. Prescrib.* **3**, 74–5.

—— AND KILOH, L.G. (1979). Six month evaluation of thiopropazate hydrochloride in tardive dyskinesia. *J. Neurol. Neurosurg. Psychiat.* **42**, 576–9.

SMITH, R.C., STRIZICH, M., AND KLASS, D. (1978). Drug history and tardive dyskinesia. *Am. J. Psychiat.* **135**, 1402–3.

——, TAMMINGA, C.A., HARASZTI, J., PANDEY, G.N., AND DAVIS, J.M. (1977). Effects of dopamine agonists in tardive dyskinesia. *Am. J. Psychiat.* **134**, 763–8.

SMITH, R.E. AND DOMINO, E.F. (1980). Dystonic and dyskinetic reactions induced by H_1 antihistaminic medication. In *Tardive dyskinesia* (ed. W.E. Fann, R.C. Smith, J.M. Davis, and E.F. Domino), pp. 325–32. Spectrum Publications, New York.

SOBUE, I., AMDO, K., HDA, M., TAKAYANAGIT, T., YAMAMURA, Y., AND MATSUDA, Y. (1971). Myeloneuropathy, with abdominal disorders in Japan: a clinical study of 752 cases. *Neurology* **21**, 168–73.

SODA, T. (ed.) (1980). Drug-induced sufferings — medical, pharmaceutical and legal aspects. In *Proceedings of the Kyoto International Conference Against Drug-Induced Sufferings.* Excerpta Medica, Amsterdam.

SOVNER, R. AND DI MASCIO, A. (1977). The effects of benztropine mesylate on the rabbit syndrome and tardive dyskinesia. *Am. J. Psychiat.* **134**, 1301–2.

——, ——, BERKOWITZ, D., AND RANDOLPH, P. (1978). Tardive dyskinesia and informed consent. *Psychosomatics.* **19**, 172–7.

SPEARS, J.B. (1978). Cimetidine and mental confusion, *Am. J. hosp. Pharm.* **35**, 1035.

SPRING, G.K. (1979). Neurotoxicity with combined use of lithium and thioridazine. *J. clin. Psychiat.* **40**, 135–8.

STANCER, H.C. (1979). Tardive dyskinesia not associated with neuroleptics. Letter. *Am. J. Psychiat.* **136**, 727.

STARKO, K.M., RAY, G., DOMINGUEZ, L.B., STROMBERG, W.L., AND WOODALL, D.F. (1980). Reye's syndrome and salicylate use. *Pediatrics* **66**, 859–64.

STEVENS, J.R. (1978). Eye blink and schizophrenia: psychosis or tardive dyskinesia? *Am. J. Psychiat.* **135**, 223–6.

STRAYHORN, J. AND NASH, J. (1977). Severe neurotoxicity despite 'therapeutic' serum lithium levels. *Dis. nerv. Syst.* **38**, 107–111.

STRUVE, F.A., KANE, J.M. WEGNER, J.T., AND KANTOR, J. (1979). Relationship of mitten patterns to neuroleptic drug induced dyskinesias in psychiatric patients: early investigative findings. *Clin. Electroenceph.* **10**, 151–63.

SULLIVAN-BOLYAI, J.Z. AND CAREY, L. (1981). Epidemiology of Reye syndrome. *Epidermol. Rev.* **3**, 1–26.

SUNAHARA, S. (1980). Drugs and drug-induced sufferings. In *Drug-induced sufferings* (ed. T. Soda), pp. 5–10. Excerpta Medica, Amsterdam.

SUTCHER, H.D., UNDERWOOD, R.B., BEATTY, R.A. *et al.* (1971). Orofacial dyskinesia: a dental dimension. *J. Am. med. Ass.* **216**, 1459–63.

SWASH, M. AND SCHWARTZ, M.S. (1983). Iatrogenic neuromuscular disorders: a review. *J. R. Soc. Med.* **76**, 149–51.

SZABADI, E. (1984). Neuroleptic malignant syndrome. *Br. med. J.* **288**, 1399–400.

TACLOB, L. AND NEEDLE, M. (1976). Drug-induced encephalopathy in patients on maintenance haemodialysis. *Lancet* **ii**, 704–5.

TAKAHASHI, I. (1979). Truth of SMON lawsuit. [In Japanese.] *Hito no Nippon*, March 1979.

TAKASU, T., IGATA, A., AND TOYOKURA, Y. (1970). On the green tongue observed in SMON patients. [In Japanese.] *Igaku no Ayumi* **72**, 539–40.

TAMMINGA, C.A., CRAYTON, J.W., AND CHASE, T.N. (1979). Improvement in tardive dyskinesia after muscimol therapy. *Arch gen. Psychiat.* **36**, 595–8.

——, SMITH, R.C., ERICKSEN, S.E., CHANG, S., AND DAVIS, J.M. (1977*b*). Cholinergic influences in tardive dyskinesia. *Am. J. Psychiat.* **134**, 769–74.

——, ——, PANDEY, G., FROHMAN, L.A., AND DAVIS, J.M. (1977*a*). A neuroendocrine study of supersensitivity in tardive dyskinesia. *Arch. gen. Psychiat.* **34**, 1199–203.

TAMURA, Z. (1975). Clinical chemistry of clioquinol. *Jap. J. med. Sci. Biol.* **28**, 69–77.

TARSY, D. AND BALDESSARINI, R.J. (1973). Pharmacologically-induced behavioural supersensitivity to apomorphine. *Nature (New Biol.), London* **245**, 262–3.

—— AND —— (1976). The tardive dyskinesia syndrome. In *Clinical neuropharmacology* (ed. H.L. Klawans), Vol. I, pp. 29–61. Raven Press, New York.

——, GRANACHER, R., AND BRALOWER, M. (1977). Tardive dyskinesia in young adults. *Am. J. Psychiat.* **134**, 1032–4.

——, LEOPOLD, N., AND SAX, D. (1974). Physostigmine in choreiform movement disorders. *Neurology, Minneapolis* **24**, 28–33.

TATEISHI, J. (1980). Reproduction of experimental SMON in animals by oral administration of clioquinol. In *Drug-induced sufferings* (ed. T. Soda), pp. 464–9. Excerpta Medica, Amsterdam.

——, KURODA, S., SATTO, A., AND OTSUKI, S. (1971). Myelo-optic neuropathy induced by clioquinol in animals. *Lancet* **ii**, 1263–4.

——, ——, ——, AND —— (1972). Strain differences in dogs for neurotoxicity of clioquinol. *Lancet* **i**, 1289–90.

—— AND OTSUKI, S. (1975). Experimental reproduction of SMON in animals by prolonged administration of clioquinol:

clinico-pathological findings. *Jap. J. med. Sci. Biol.* **28**, 165–86.

TCHAO, P. AND TEMPLETON, T. (1983). Thiabendazole-associated *grand mal* seizures in a patient with Down syndrome, *J. Pediat.* **102**, 317–18.

TEPPER, S.J. AND HAAS, J.F. (1979). Prevalence of tardive dyskinesia. J. clin. Psychiat. **40**, 508–16.

THACH, B.T., CHASE, T.N., AND BOSMA, J.F. (1975). Oral-facial dyskinesia associated with prolonged use of antihistamine decongestants. *New Engl. J. Med.* **293**, 486–7.

THOMAS, C., TATHAM, A., AND JUKUBOWSKI, S. (1982). Lithium/haloperidol combinations and brain damage. *Lancet* **i**, 626.

THOMPSON, J. AND LILLY, J. (1979). Cimetidine-induced cerebral toxicity in children. *Lancet* **i**, 725.

THORNTON, W.E. AND THORNTON, B.P. (1973). Tardive dyskinesia. Letter. *J. Am. med. Ass.* **226**, 674.

TOLOSA, E.S. (1978). Modification of tardive dyskinesia and spasmodic torticollis by apomorphine. *Arch. Neurol.* **35**, 459–62.

TONSGARD, J.H. AND HUTTENLOCHER, P.R. (1981). Salicylates and Reye's syndrome. *Pediatrics* **68**, 747–8.

TOYOKURA, Y. AND TAKASU, T. (1975). Clinical features of SMON. *Jap. J. med. Sci. Biol.* **28**, 87–99.

TSUBAKI, T., HONMA, Y., AND HOSHI, M. (1971). Neurological syndrome associated with clioquinol. *Lancet* **i**, 696–7.

TUNE, L.E., CREESE, I., COYLE, J.T., PEARLSON, G., AND SNYDER, S.H. (1980). Low neuroleptic serum levels in patients receiving fluphenazine decanoate. *Am. J. Psychiat.* **137**, 80–2.

UDALL, V. (1972). Drug-induced blindness in some experimental animals and its relevance to toxicology. *Proc. R. Soc. Med.* **65**, 197–200.

UHRBRAND, L. AND FAURBYE, A. (1960). Reversible and irreversible dyskinesia after treatment with perphenazine, chlorpromazine, reserpine and electroconvulsive therapy. *Psychopharmacologia* **1**, 408–18.

UMEZ-ERONINI, E.M. AND ERONINI, E.A. (1977). Chloroquine-induced involuntary movements. *Br. med. J.* **1**, 945–6.

VAN ROSSUM, J.M. (1966). The significance of dopamine receptor blockade for the action of neuroleptic drugs. In *Neuro-psychopharmacology*, International Congress Series No. 129 (ed. H. Brill, J.O. Cole, P. Deniker, *et al.*), pp. 96–9. Excerpta Medica, Amsterdam.

VICKERY, T.R. (1978). Cimetidine reaction. *Drug Intel. clin. Pharm.* **12**, 242.

VILLENEUVE, A. (1972). The rabbit syndrome. A peculiar extrapyramidal reaction. *Can psychiat. Ass. J.* Suppl. **17**, 69–72.

WAGGONER, W.C. (1983). Phenylpropanolamine overdosage. *Lancet* **ii**, 1503–4.

WALDMAN, R.J., HALL, W.N., McGHEE, H., AND VAN AMBURG, G. (1982). Aspirin as a risk factor in Reye's syndrome. *J. Am. med. Ass.* **247**, 3089–94.

WARRELL, D., LOOAREESUWAN, S., AND WARRELL, M.J. (1982). Dexamethasone proves deleterious in cerebral malaria. A double blind trial in 100 comatose patients. *New Engl. J. Med.* **206**, 313–19.

——, WHITE, N.J., AND WARRELL, M.J. (1983). Is dexamethasone deleterious in cerebral malaria? Letter. *Br. med. J.,* **286**, 1355.

WEGNER, J.T., STRUVE, F.A., AND KANE, J.M. (1977). the B-mitten EEG pattern and tardive dyskinesia: a possible association. *Am. J. Psychiat.* **134**, 1143–5.

WEBER, J.C.P. (1984). Epidemiology of adverse reactions to nonsteroidal anti-inflammatory drug. In *Advances in inflammation research* (eds. K.D. Rainsford and G.P. Velo), Vol.6, pp.1–7. Raven Press, New York.

WEINER, W.J., NAUSIEDA, P.A., AND KLAWANS, H.L. (1980). Regional brain manganese levels in an animal model of tardive dyskinesia. In *Tardive dyskinesia* (ed. W.E. Fann, R.C. Smith, J.M. Davis, and E.F. Domino), pp. 159–63. Spectrum Publications, New York.

WEINSTEIN, L., LERNER, P.I., AND CHEW, W.R. (1964). Clinical and bacteriological studies of the effect of 'massive' doses of penicillin G on infections caused by Gram-negative bacilli. *New Engl. J. Med.* **271**, 525–33.

WEISS, B. AND SANTELLI, S. (1978). Dyskinesia evoked in monkeys by weekly administration of haloperidol. *Science* **200**, 799–801.

WEISS, J. AND CATALANO, P. (1973). Camphorated oil intoxication during pregnancy. *Pediatrics* **52**, 713.

WIHELM, B-E, MORTIMER, Ö., BOETHIUS, G., AND HÄGGSTRÖM, J.E. (1984). Tardive dyskinesia associated with metoclopramide. *Br. med. J.* **288**, 545–7.

WILL, R.G. AND MATTHEWS, W.B. (1982). Evidence for case to case transmission of Creutzfeldt-Jacob disease. *J. Neurol. Neurosurg. Psychiat.* **45**, 235–8.

WILLS, M.R. AND SAVORY, J. (1983). Aluminium poisoning: dialysis encephalopathy, osteomalacia, and anaemia. *Lancet* **ii**, 29–34.

WILSON, J.B. (1979). Cimetidine overdosage. *Br. med. J.* **1**, 955.

WINSBERG, B.G., HURWIC, M.J., SVERD, J., AND KLUTCH, A. (1978). Neurochemistry of withdrawal emergent symptoms in children. *Psychopharmacology* **56**, 157–61.

WOOD, C.A., ISAACSON, M.L., AND HIBBS, M.S. (1978). Cimetidine and mental confusion. *J. Am. med. Ass.* **239**, 2550–1.

WOODRUFF, A.W. AND DICKINSON, C.J. (1968). Use of dexamethasone in cerebral malaria. *Br. med. J.* **iii**, 31–2.

WORLD HEALTH ORGANIZATION, (1983), October 26). Antipsychotic drugs: tardive dyskinesia (New Zealand). In *PHA (DIA)* 83.10.

WRIGHT, J.H., WHITAKER, S.B., WELCH, C.B., AND TELLER, D.N. (1983). Hepatic enzyme induction patterns and phenothiazine side effects. *Clin. Pharmacol. Ther.* **34**, 533–7.

WURTMAN, R.J., HIRSCH, M.J., AND GROWDON, J.H. (1977). Lecithin consumption raises serum free choline levels. *Lancet* **ii**, 68–9.

YAMADORI, A. AND ALBERT, M.L. (1972). Involuntary movement disorder caused by methyldopa. *New Engl. J. Med.* **286**, 610.

YOSHIOKA, M. AND TAMURA, Z. (1970). On the nature of the green pigment found in SMON patients. [In Japanese.] *Igaku no Ayumi* **74**, 320–2.

28 Drug-induced psychiatric disorders

DAVID J. KING

It has been said that 'adverse reactions account for an appreciable amount of psychiatric morbidity which is likely to increase as new and more potent drugs are evolved' (Davison 1981). However, it can sometimes be too facile to attribute psychiatric disturbances to concomitant drug administration, and uncritical reporting of drug effects or interactions can have serious medico-legal repercussions. The decision in any individual case must finally depend upon a clinical judgement. If an unexpected psychiatric disturbance arises suddenly in a person of good previous personality, shortly after a drug of any sort has been taken, no matter how harmless it usually is, it is clearly wise to suspect a drug-induced reaction and, if possible, to discontinue or reduce the dose of the suspected medication. It is also good practice to avoid unnecessary polypharmacy, to attempt to treat one psychiatric condition with one drug if possible, and to remember that the use of two drugs from the same group (antidepressant, neuroleptic, minor tranquillizer, etc.) can rarely, if ever, be justified.

The central nervous system responds in a limited number of ways to injury or metabolic disturbance, although the particular content and manifestation of the disturbance is highly variable and will reflect and be coloured by the unique personality attributes of the individual concerned. The majority of drugs, whether used to produce CNS effects or not, have therefore relatively non-specific effects on CNS functions. Although advances in neuropharmacology have produced drugs with more selective effects in the CNS, the major neuro-psychiatric side-effects of drugs are clouding of consciousness, psychotic reactions, mood changes, and idiosyncratic behavioural reactions. These will be considered under separate headings below.

Psychotropic drugs are the single most widely prescribed group of drugs accounting for a fifth of all general practice prescriptions (Skegg et al. 1977). A steady increase in the prescribing of these drugs since 1966 to a peak in 1975 is shown in Fig. 28.1. These data were derived from the computerized pricing of general practice prescriptions in Northern Ireland and it can be estimated that in 1975 approximately 12.5 per cent of the total adult population were receiving these drugs, which is comparable to the findings in most other Western

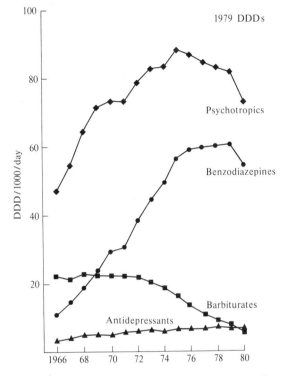

Fig. 28.1 Psychotropic drug utilization in Northern Ireland, 1966–80. Prescribing levels are expressed as defined daily doses (DDD)/1000 population/day on the basis of DDD's provided by the Nordic Council of Medicines, as described by McDevitt and McMeekin (1979). (From King and Griffiths 1984.)

European Countries (Balter et al. 1974; Böethius and Westerholm 1977). From Fig. 28.2 it can be seen that three-quarters of these psychotropic drugs were benzodiazepines (in contrast to which antidepressants constituted only about 7.7 per cent of total psychotropic drug prescribing).

Although generally very safe and free of serious adverse reactions or significant interactions, an increasing number of reports have demonstrated that both psychological and physical dependence on benzodiazepines can occur (Clare 1971; Committee on the Review of Medicine 1980; Howe 1980; Petursson and

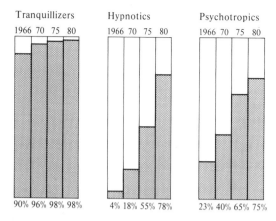

Fig. 28.2 Benzodiazepine proportion of general practice prescribing of tranquillizers, hypnotics, and total psychotropic drugs in Northern Ireland, 1966–80. (From King *et al.* 1982.)

Lader 1981*a*; Ratna 1981), sometimes after therapeutic doses for relatively short periods of time (Tyrer *et al.* 1981, 1983; Petursson and Lader 1981*b*; Schöpf 1983). There is also increasing recognition that, because of the long plasma elimination half-lives of most of these drugs and their active metabolites, they frequently have residual 'hangover' effects when used as hypnotics (Oswald 1979; *British Medical Journal* 1980*a*; Ogura *et al.* 1980; Hockings and Ballinger 1983), and can impair psychomotor performance the next day (Wittenborn 1979; File and Bond 1979; Nicholson 1980) with potentially serious implications for road traffic accidents (*British Medical Journal* 1978; Skegg *et al.* 1979; Biehl 1979; Hindmarch and Gudgeon 1980; Honkanen *et al.* 1980; Betts and Birtle 1982).

Thus a short section on drug dependency as a drug-induced psychiatric disorder is included at the end of this chapter. Apart from this and the other psychiatric symptoms listed above, the general adverse effects of psychotropic drugs will not be dealt with in this chapter. For these the reader is referred to more appropriate texts (e.g. Dukes 1975).

Delirium

A very wide range of drugs have been associated with toxic confusional reactions (acute brain syndrome) which are characterized by a fluctuating clouding of consciousness, restlessness, emotional changes (usually fear and perplexity), and paranoid delusions and/or visual hallucinations in severe cases. Detailed descriptions of delirious states were given in the classic paper by Wolfe and Curran (1935) who reviewed 106 cases asso-

ciated with 27 different precipitating noxious agents. The only drugs involved in these cases were alcohol, barbiturates, bromide, lead, and copper, the majority being associated with intracranial or systemic disease. However, a much wider range of drugs can be associated with such reactions, either during administration or in withdrawal, and examples of the main groups likely to be involved are listed in Table 28.1 and were also discussed by McClelland (1978).

The syndrome was the subject of a lengthy monograph (Lipowski 1980*a*) which deals extensively with the aetiology, pathogenesis, differential diagnosis, and management of the disorder. The various causes, such as drugs and toxins, metabolic and infective disorders, and brain lesions, and aggravating circumstances, such as recent surgery, the puerperium, and old age, are dealt with in detail. The fact that anticholinergic drugs are implicated particularly frequently and that physostigmine is effective in delirium induced not only by anticholinergics but also by alcohol and cimetidine (Lipowski 1980*b*) and in post-operative delirium (Greene 1971) has focused attention on the importance of cholinergic transmission in this syndrome, as with dementia of the Alzheimer type in which a deficiency of choline acetyltransferase has been demonstrated (Davies and Maloney 1976; Perry *et al.* 1977). Tune *et al.* (1981) recently provided direct evidence for raised serum levels of anticholinergic drugs in post-operative delirium using a radioreceptor assay technique. Thirty-four per cent of 29 patients undergoing cardiac surgery had some evidence of delirium (defined as 'an acute change in the patient's mental state characterized by alteration in the level of consciousness, by disorientation in time and place, and by cognitive impairment') during the first post-operative week. Seven of eight delirious patients had serum levels of anticholinergic drugs of greater than 1.5 pmol of atropine

Table 28.1   Examples of drugs which may be associated with delirious states

Withdrawal delirium

Drugs with central depressant actions
  Alcohol
  Bromides
  Barbiturates
  Non-barbiturate hypnotics (e.g. chlormethiazole, glutethimide, ethchlorvynol, methaqualone, chloral hydrate)
  Benzodiazepines

Delirium following overdosage or intolerance

Psychotropic drugs
  Lithium carbonate
  Disulfiram

Table 28.1 *Continued*

Any central depressant drug in elderly including both
  major and minor tranquillizers

Anticonvulsants
  Phenytoin

Drugs with anticholinergic properties
  Atropine
  Anti-parkinsonism drugs (e.g. benzhexol, orphenadrine,
    benztropine)
  Tricyclic antidepressant drugs (e.g. amitriptyline,
    imipramine, clomipramine)

Dopaminergic drugs
  Levodopa
  Amantadine
  Bromocriptine

Antihistamines
  e.g. promethazine, diphenhydramine, chlorpheniramine

Sympathomimetic drugs
  Isoprenaline
  Ephedrine

Anticholinesterase agents
  Physostigmine
  Parathion

Opiate analgesics
  Heroin, morphine, pethidine
  Pentazocrine, codeine, dextropropoxyphene

Non-narcotic analgesics
  Aspirin
  Phenylbutazone
  Indomethacin

Hormones
  ACTH
  Adrenal glucocorticoids (e.g. hydrocortisone, prednisone)

Cardiovascular drugs
  Digitalis glycosides
  Propranolol
  Tocainide

Antimicrobial drugs
  Sulphonamides
  Antituberculous drugs (e.g. PAS, isoniazid)
  Chloroquine
  Piperazine
  Cephaloridine

Inhalants
  Carbon tetrachloride
  Solvents in glues (e.g. trichloroethylene, toluene, acetone,
    benzene)

Heavy metals
  Lead, mercury

Xanthines
  Aminophylline

Diuretics
  Frusemide

equivalents/sample, compared with only four of 17 non-delirious patients with levels in this range, the differences being highly significant ($p < 0.001$).

The aetiology of delirium is, however, more complex than this, since, in spite of the beneficial effect of physostigmine in many cases, cholinergic drugs can themselves lead to delirium, particularly in overdose (Pradhan and Dutta 1971; Roberts and Breckenridge 1975; Granacher and Baldessarini 1976).

Evidently many factors are likely to be involved in post-operative states. Another recent study also reported an incidence of post-operative psychosis of 35 per cent after open-heart surgery (Naber *et al.* 1983). Here there was a significant correlation between the severity of depression and serum cortisol levels in the 3- to 5-day recovery period, but plasma drug levels were not measured and it is not clear whether the changes in serum cortisol were a cause or an effect of depression. A steady decline in the incidence of post-operative psychiatric disturbances, from estimates as high as one in 250 in 1910, to about one in 1600 operations 50 years later has been attributed largely to improved anaesthetic technique and post-operative care (Knox 1961). Occasionally newer anaesthetic agents may be associated with more psychological disturbances in the recovery period (Garfield 1974). Ketamine, in particular has been noted for its emergence delirium and unpleasant hallucinosis, but these appear to be avoidable by the judicious use of benzodiazepines towards the end of the anaesthetic period (Coppel *et al.* 1973). Non-pharmacological factors can also often play an important part in determining whether or not delirium will occur, both pre-operative anxiety and the degree of post-operative sensory deprivation having been particularly implicated in this respect (see Knox 1961 and Garfield 1974).

With the introduction of new drugs there are an increasing number which have been associated with delirium. It should be remembered that lithium does not cause drowsiness at normal therapeutic levels, and that if this does occur it should immediately be discontinued or the dose reduced before delirium develops (Shopsin and Gershon 1973; Ghose 1977) since permanent extrapyramidal and cerebellar damage can result once this happens (von Hartitzsch *et al.* 1972). Although cimetidine does not cross the blood–brain barrier to any significant extent (CSF/plasma ratio: 0.24/1), confusional states have been reported in patients with impaired hepatic and renal function (Schentag *et al.* 1979). It appears that this might be due to impaired clearance of the major metabolite, cimetidine sulphoxide (Totte *et al.* 1981). However, it is not now thought to be directly due to histamine H_2-receptor blocking effects since the reported incidence of delirium is less with its rival, ranitidine (*Lancet* 1982*a*;

Drug and Therapeutics Bulletin 1982; Silverstone and Epstein 1984) in spite of similar CSF/plasma ratios (Kagevi and Wåhlby 1985).

Solvent abuse has been a cause of increasing concern and the features of an acute encephalopathy associated with toluene intoxication were reported in detail by King *et al.* (1981). In that report on 19 children, seven presented with a history of euphoria and hallucinations and two with behaviour disturbance and diplopia. Five were described as still having psychological impairment and personality change on discharge and a sixth had persistent cerebellar ataxia after one year. There was also a report of the delayed onset of symptoms of malaise, difficulty in concentration, and deterioration in work, in a 25 year-old man 48 hours after exposure to methylene chloride (in Nitromors paint stripper) in a confined space for 3 to 4 hours (Memon and Davidson 1981). A 13-year-old boy was found to be 'barely rousable and talking gibberish' after an accidental overuse of a proprietary preparation containing 4 per cent menthol (Olbas oil) for use as an inhalant for nasal congestion (O'Mullane *et al.* 1982).

A true dementia, with irreversible cognitive impairment and a chronic course, is an extremely rare drug effect. Such changes were attributed by some to levodopa in the early years of its use in Parkinson's disease (Barbeau 1971, Wolf and Davis 1973), but are now generally thought to be part of the natural progression of the disease rather than a drug effect (Marsden 1982; Perry *et al.* 1983).

In view of the putative role of choline acetyltransferase deficiency in dementia of Alzheimer-type one might expect anticholinergic drugs, which are well known causes of delirium, to be associated with memory deficits or deterioration of dementia. Potamianos and Kellett (1982) have in fact now reported a double-blind placebo study in 13 non-demented elderly patients (aged 75–92 years) in whom a small (2 mg) dose of benzhexol produced significant impairments in a number of cognitive tests. The implications are that these drugs and others with marked anticholinergic properties, such as the tricyclic antidepressants and thioridazine, should be used sparingly, if at all, in the elderly particularly if signs of dementia are present.

Psychotic states in clear consciousness

A variety of drugs reported to cause psychotomimetic effects are shown in Table 28.2. The most worrying aspect of this type of drug-induced reaction is whether such drugs can precipitate a permanent psychosis in certain individuals. Breakey *et al.* (1974) reported an

Table 28.2   Examples of drugs which may be associated with psychotic states in clear consciousness

1. Hallucinogens
   (a) Cannabis, mescaline, bufotenine, dimethyltryptamine, LSD, psilocin

2. Central stimulants
   (a) Cocaine
   (b) Amphetamine and related drugs (e.g. phenmetrazine, diethylpropion)
   (c) Methylphenidate

3. Dopamine agonists (e.g. bromocriptine)

4. β-Adrenergic receptor blocking drugs

5. Opiate analgesics (e.g. dihydrocodeine, pentazocine)

increased use of cannabis and other hallucinogenic drugs in patients subsequently diagnosed as schizophrenic compared with the incidence of use in the general population. A six-year follow up of 51 regular drug abusers found that 45 per cent of those taking hallucinogens or psychostimulants were clinically psychotic, that 57 per cent of those on central depressants such as barbiturates or benzodiazepines had serious depression, but that none of the opiate abusers showed any increase in psychiatric symptomatology (McLellan *et al.* 1979). A past history of petrol sniffing was significantly higher in a group of schizophrenics than in matched non-psychiatric controls in the Gilbert Islands (Daniels and Latcham 1984). However, no study has been able to exclude the possibility that pre-existing personality factors might have determined the incidence, extent, and/or choice of drug abuse. If persistent effects do occur, therefore, they are generally attributed to an interaction of drug effects with psychological and constitutional factors in predisposed individuals.

Hallucinogenic drugs

The American literature on hallucinogenic drug-induced schizophrenic syndromes with particular emphasis on LSD (lysergic acid diethylamide) and cannabis was extensively reviewed by Stone (1973), and a wealth of descriptive case material was also presented. The course of these reactions is usually uncomplicated but there is a risk of sudden death if there is an acute panic reaction during a 'bad trip', with paranoid delusions, an urge to run or to fly, or suicidal or homicidal impulses (Dimijian 1976). This is even more likely if the drug was taken unwittingly and without the subject's knowledge (Stone 1973). Such reactions can also occur many months after the drug effects have worn off and occur as transient but very vivid memories known as 'flash-backs'. While the majority of these are not unpleasant, particularly the perceptual ones, and are enjoyed as a 'free LSD trip', the somatic and emotional disturbances causing a persistent depersonalization syndrome can be very distressing (Ungerleider and Frank 1976) and can lead to panic and suicidal reactions similar to those in the acute drug situation. The flash-backs may be precipitated by an acute stress or environmental upset, but the panic is often intensified by a fear that permanent brain damage has occurred or that the subject is going out of his mind. It was suggested that some of these persisting panic attacks may be related to temporal-lobe dysrhythmias (Jacobs 1979).

Both the acute and chronic effects of cannabis have been the subject of much concern and speculation (Commission of Inquiry into the Nonmedical Use of Drugs 1972; *British Medical Journal* 1976; Graham 1976; *Advisory Council on the Misuse of Drugs* 1982). The acute reactions generally last a few hours but occasionally go on for as long as seven days. These may be associated with severe panic, particularly in inexperienced users or those consuming unexpectedly potent material, and paranoid delusions, restlessness, violence, or destructiveness. Periodicity, short duration, insight between episodes, and precipitation by increased dose are the usual features of psychotic reactions in chronic users. A report of psychological testing in heavy cannabis users in India found significant deficits in memory, concentration, and psychomotor speed in 50 subjects who had taken an average daily dose of 150 mg of THC (tetrahydrocannabinol, the most active constituent of cannabis) for 4–10 years, compared with 25 controls who had been carefully matched for age, sex, education, and occupation and who had never taken cannabis (Mendhiratta *et al.* 1978). A recent study by Rottanburg *et al.* (1982) confirmed an association between high cannabis use and a rapidly resolving psychosis characterized by a schizophreniform illness with marked hypomanic features. They compared the mental state of 20 psychotic men with high urinary cannabinoid levels on admission to a psychiatric hospital with that of 20 matched cannabis-free psychotic controls, using the standardized Present State Examination diagnostic technique. Although there was no difference in amount of medication received between the two groups, the cannabis group showed marked improvement after one week, whereas the controls were unchanged.

Central stimulants

Central stimulants, such as cocaine, amphetamines, and methylphenidate, are also well known to induce temporary psychotic disturbances particularly if taken in large doses. Chronic amphetamine addiction can produce a clinical picture almost indistinguishable from clinical schizophrenia, with auditory hallucinations, paranoid delusions, and thought disorder in clear consciousness (Connell 1958; Snyder 1973). Snyder (1973) proposed that the hallmarks of the chronic amphetamine psychosis which are most useful in distinguishing it from clinical schizophrenia are stereotyped compulsive behaviour (such as repetitive searching, sorting, or grooming and preoccupation with fine details), tactile hallucinations (the 'cocaine bug'), and an increase in libido with excessive, prolonged, and frequently perverse sexual appetites. Patients with a history of methamphetamine-induced psychoses seem to have an increased susceptibility to further episodes on re-exposure to the drug even after periods of abstinence of up to five years (Sato *et al.* 1983). Khat leaves contain a potent phenylalkylamine alkaloid and the chewing of these has been associated with isolated reports of an amphetamine-like psychosis (Kalix, 1984; Gough and Cookson 1984).

Any β-adrenergic agonist can have amphetamine-like central stimulant effects. The mood-elevating properties of salbutamol were demonstrated with some success in an uncontrolled (Simon *et al.* 1978) and a controlled comparative trial with clomipramine (Lecrubier *et al.* 1980) in depressed patients. The use of high doses of salbutamol and isoxuprine has been associated with psychotic states with the development of both auditory and visual hallucinations and delusions in clear consciousness (Gluckman 1974; Feline and Jouvent 1977). Addiction to phenylephrine nasal spray was reported in a 26-year-old woman who presented with psychotic symptoms (Snow *et al.* 1980).

Sympathomimetics used as anorectics such as phenmetrazine and diethylpropion (van Praag 1968) or mazindol (Adverse Drug Reactions Advisory Committee 1980) can similarly cause psychotic states. Fenfluramine, however, which enhances central 5-hydroxytryptamine, causes drowsiness, and, on withdrawal, depression (Harding 1972; Steel and Briggs 1972).

Levodopa and dopamine agonists

Very detailed and vivid accounts of the perceptual and cognitive changes that occurred in patients with severe post-encephalitic parkinsonism when treated with levodopa were provided by Sachs (1973). Much of the phenomenology described in this classic of clinical observation is indistinguishable from other schizophreniform psychoses. The dopamine agonist bromocriptine, was also reported to cause hallucinations in patients treated for Parkinson's disease (Debono *et al.* 1976) and acromegaly (White and Murphy 1977). In addition to these there were reports of less frequent but similar psychotic reactions to lower doses of bromocriptine (Pearson 1981). One of these raised the possibility of an increased risk of such reactions if there is a family history of schizophrenia, analogous to the genetic vulnerability to affective reactions to L-dopa in those with a family history of such disorders (Le Feuvre *et al.* 1982).

Antidepressants

There was a report of paranoid symptoms following nomifensine (which enhances dopamine function in the CNS by blocking its reuptake) given to a severely depressed man subsequently found to be hyperthyroid (Katona 1982), but a casual connection between the drug and these symptoms seems doubtful and was challenged (Cohen 1982).

β-Adrenergic blocking drugs

These drugs have been reported to cause a number of different neuropsychiatric symptoms (Greenblatt *et al.* 1976) and both 'mental confusion' and 'visual hallucinations' occurred in some of the earliest reports of adverse reactions to propranolol (Stephen 1966). Both an organic syndrome with clouding of consciousness (Fraser and Carr 1976; Topliss and Bond 1977; Helson and Duque 1978) and perceptual disturbances in clear consciousness (Shopsin *et al.* 1975) were subsequently well documented.

Visual perceptual disturbances occurring in either hypnapompic or hypnagogic states were found with an incidence of 17.5 per cent in one study (Fleminger 1978) which may account for the sleep disturbances, nightmares, and bizarre dreams described in other studies (Greenblatt *et al.* 1976; Stephen 1966). Nevertheless, a recent survey of 1302 Swedish women failed to find any significant differences in the incidence of sleep disturbances or nightmares in 165 who were taking antihypertensive drugs (including 84 who were on β-blocking drugs either alone or in combination with other drugs) and in those who were not, or between the different types of antihypertensive drug treatment (Bengtsson *et al.* 1980). It appears likely that quite disturbing hallucinatory states, which are generally associated with insight (i.e. are non-psychotic 'pseudohallucinations'), can be attributed to β-blockers in certain individuals, but that the overall incidence is not high.

There have, however, been several reports of full-blown acute schizophreniform psychotic states following treatment with β-blocking drugs (Hinshelwood 1969;

Koehler and Guth 1977; Steinert and Pugh 1979). Although antipsychotic properties were previously attributed to these drugs by some (Atsmon *et al.* 1971; Yorkston *et al.* 1977; Sheppard 1979), others could not demonstrate this (Gardos *et al.* 1973; King *et al.* 1980; Peet *et al.* 1981*a*; Myers *et al.* 1981) and it appears likely that the earlier results were due to a pharmacokinetic interaction with concomitant neuroleptic drugs (Peet *et al.* 1981*b*). Increased dopamine turnover in the absence of dopamine blockade was proposed as a possible mechanism for the psychotomimetic adverse effects of these drugs (King *et al.* 1983). It seems probable that β-blocking drugs with high lipid solubility such as propranolol, particularly if also associated with low plasma protein binding (and therefore high CSF concentrations) such as pindolol, are the most likely to have these central effects (Turner 1979; Taylor *et al.* 1979).

Analgesics

Although opiates are generally associated with sedation and clouding of consciousness, hallucinations can occur particularly with pentazocine (Wood *et al.* 1974). A persistent delusional state in clear consciousness was also reported following pentazocine (Blazer and Haller 1975). In a study designed to assess the incidence of these phenomena, major psychotomimetic disturbances were found in 10 per cent of patients taking pentazocine and 4 per cent of patients taking dihydrocodeine but the difference from matched controls who had taken other drugs was only significant for pentazocine (Taylor *et al.* 1978).

The non-opiate, indomethacin, which is a methoxylated indole derivative, is known to cause severe headaches, depression, and other psychiatric disturbances (Prescott 1975) and a persistent, florid, paranoid psychotic state was reported in association with indomethacin treatment in a 65-year-old woman (Carney 1977). Salicylates were reported to be a cause of hallucinations in chronic misuse (Greer *et al.* 1965).

Proprietary preparations

Abuse of a preparation containing pseudoephedrine and triprolidine (*Actifed*) by a young man with a six-year history of manic-depressive illness was associated with a transient paranoid psychosis (Leighton 1982). A recent editorial also drew attention to psychotic reactions to phenylpropanolamine (a sympathomimetic amine found in remedies for sinusitis and the common cold such as *Contac-C* and *Ru-Tuss*) sometimes occurring in previously healthy people after ingestion of modest doses, and also its potential as a drug of abuse (*Lancet* 1982*b*).

Other drugs

Hallucinations were also reported in association with a number of other drugs, such as digoxin, methyldopa, aminophylline, some sulphonamides, tricyclic antidepressants, and some of the benzodiazepines (Boston Collaborative Drug Surveillance Program 1971; McClelland 1977). It would appear from a recent report that acyclovir given to patients with renal failure should also now be added to this list (Tomson *et al.* 1985).

Drug-induced 'steroid psychoses' are usually associated with marked mood changes and are discussed below. An acute psychosis following oral sodium and potassium replacement therapy, in association with a hypertensive crisis, was reported in a 49-year-old man with polycystic kidney disease (Peer *et al.* 1983). A series of transient paranoid psychotic reactions were reported in a 53-year-old woman with severe presenile dementia who was being treated with intranasal DDAVP (1-desamino-8-D-arginine vasopressin) (Collins *et al.* 1981).

Exacerbations of schizophrenia

Exacerbations of psychotic symptoms can occur in schizophrenic patients both with the above drugs and with a number of others which are not hallucinogenic or psychotomimetic in normals (Table 28.3). The phenomena of the amphetamine psychosis and parallel experiments in animals, together with advances in the understanding of the mode of action of neuroleptic drugs, have in recent years provided the basis for a dopaminergic hypothesis of schizophrenia (Stevens 1973; Snyder *et al.* 1974). It is therefore of significant theoretical interest that the dopamine agonist, bromocriptine, causes hallucinations in some patients (see above), and I have known a schizophrenic patient to have had an exacerbation when treated with bromocriptine for puerperal lactation.

Table 28.3 Examples of drugs reported to cause exacerbations of schizophrenia

Type of drug	Example
Central stimulants	Amphetamine Methylphenidate
Dopaminergic drugs	Levodopa Bromocriptine
GABA agonists	Sodium valproate
Antidepressants	Monoamine oxidase inhibitors Tricyclics Nomifensine
Others	Disulfiram

Although schizophrenic patients can distinguish the effects of LSD and other hallucinogens from the symptoms of their illness, amphetamines and methylphenidate produce a true exacerbation of the pre-existing psychosis (Janowsky and Davis 1974; Snyder *et al.* 1974). Indeed there appears to be a good correlation between this effect and subsequent response to neuroleptic drugs (Angrist *et al.* 1980). Nevertheless, attention was drawn to the variability in reactions of schizophrenics to amphetamine by Van Kammen *et al.* (1982) who cited cases in the literature reported to have improved after this drug. They also described a controlled study of the effect of a single (20 mg) dose of d-amphetamine given to 45 schizophrenics and transient changes (both improvements and deteriorations) in psychotic ratings over the subsequent 20 to 60 minutes. The changes were small and the 'psychosis ratings were not affected significantly by d-amphetamine'. Nevertheless, these authors appeared to think they were not observing spontaneous fluctuations and went on to speculate about an alternative state-dependent dopamine receptor sensitivity hypothesis of schizophrenia. The value of this report was to remind us that single doses of amphetamine do not invariably cause a deterioration in schizophrenic symptoms, but it said nothing about the effect of very large or chronic doses.

GABA-agonists and GABA-mimetics such as sodium valproate are also associated with worsening of psychotic symptoms in the majority of patients with acute or chronic schizophrenia (Lautin *et al.* 1980; Meldrum 1982). This opposes a GABA-deficiency hypothesis of schizophrenia, but the mechanism is not understood, although an effect on pre-synaptic GABA receptors was postulated (Meldrum 1982).

A number of other drugs can also precipitate a florid exacerbation of schizophrenia, such as levodopa, tricyclic antidepressants, MAO inhibitors, and anticholinergic drugs (Shepherd *et al.* 1968; DiMascio *et al.* 1970a; Janowsky *et al.* 1972; Goodwin and Sack 1974). The MAO inhibitors in particular, were consistently found to be associated with exacerbations of schizophrenia when given with the amino acid, methionine, in tests of the abnormal methylation hypotheses in this disorder (Wyatt *et al.* 1971). Antidepressants are particularly liable to be taken by the patients themselves seeking symptomatic relief and obtaining the tablets from friends or family doctors who are not aware of the underlying psychotic condition. Anticholinergic drugs were reported in some studies to be associated with a deterioration in therapeutic response when added to neuroleptic drug treatment (Singh and Smith 1973; Singh and Kay 1975; Johnstone *et al.* 1983). Pharmacokinetic reasons for this were proposed (Rivera-Calimlim *et al.* 1973), either as a result of effects on gut motility (Rivera-Calimlim 1976) or hepatic enzyme induction (Loga *et al.* 1975), but a recent study failed to demonstrate any significant effect of benzhexol (trihexyphenidyl) on chlorpromazine plasma levels (Simpson *et al.* 1980). Disulfiram (*Antabuse*) has been known for many years to exacerbate schizophrenic symptoms (MacDonald and Ebaugh 1954; Heath *et al.* 1965), and it was suggested that this may be related to its dopamine-β-hydroxylase (DBH)-inhibiting properties since the DBH inhibitor, fusaric acid, has been found to intensify delusions and other psychotic symptoms in manic patients (Sack and Goodwin 1974).

The possibility that the very drugs used to treat schizophrenia can themselves cause an exacerbation of the underlying psychosis by producing a supersensitivity of mesolimbic dopamine receptors, similar to that in the nigrostriatal tracts in tardive dyskinesia (Klawans 1973), was proposed by Chouinard and Jones (1980). They suggested the term 'supersensitivity psychosis' and reported 10 cases in which increasing doses of depot neuroleptics were required, with rapid relapses after dose reduction, associated with high prolactin levels, and in eight of whom there were signs of tardive dyskinesia. This concept would appear to be supported by the report of three schizophrenic patients who appeared to have been made worse by high doses of intramuscular neuroleptics (Barnes and Bridges 1980). However, the existence of such an entity as distinct from spontaneous deterioration or exacerbation of the illness is difficult to establish because:

1. Tardive dyskinesia is usually associated with an improvement and not a deterioration in mental state, while stereotyped abnormal movements can be a feature of schizophrenia itself (Marsden *et al.* 1975);

2. In animal studies supersensitivity has been found to develop in striatal but not mesolimbic dopamine areas after six months exposure to neuroleptics (Clow *et al.* 1980);

3. Two of Barnes and Bridges' (1980) cases improved with a change to oral medication, but, since steady neuroleptic plasma levels after depot drugs persist for several months after injection (Kolakowska and Wiles 1981; Wistedt *et al.* 1981), this may have increased rather than decreased the total amount of circulating neuroleptic drug.

Mood disorders

Primary affective (mood) disorders, i.e. mania or depression, can be associated with a wide range of drugs. Excited (manic) states are easily confused with schizo-

phreniform psychoses and indeed an element of elation and excitement may be part of the clinical picture in the states described above. There is therefore a certain amount of overlap between reactions discussed in the previous section and those which have been reported as 'manic' reactions. True hypomanic and manic states probably only occur in manic-depressive patients.

Elation

In dealing with bipolar patients, i.e. patients with manic-depressive illnesses who have had a history of both types of affective mood swing, there can sometimes be a dramatic 'switch' from one phase to the other. Occasionally such switches can be induced by drugs such as levodopa, amphetamines, monoamine oxidase inhibitors, tricyclic antidepressants, and corticosteroids, and the relationship between the pharmacological actions of these drugs and the pre-existing mental state of the individual in producing a rapid switch in psychopathology was discussed by Bunney and Murphy (1974). An atypical tricyclic antidepressant, dibenzepine, which lacks anticholinergic properties, was associated with a switch into a rapid 48-hour cycling disorder in a 63-year-old depressed man with no previous history of hypomania (Lerer *et al.* 1980). This was associated with parallel fluctuations in urinary cyclic AMP excretion and was presumably a variant of the previously known antidepressant-induced switches from one phase of a manic-depressive disorder to another.

Milder forms of elation or euphoria are more common and can occur in psychiatrically normal individuals. They are typically associated with high-dose regimens of systemic steroids (de Lange and Doorenbos 1975; Hall *et al.* 1979). Although steroid-containing inhalers are generally regarded as being free of this effect, there was a report of a manic reaction in a $5\frac{1}{2}$-year-old asthmatic girl who took two 50-μg puffs of budesonide twice daily (Lewis and Cochrane 1983). Since the child had no recollection of the episode, a toxic confusional state may have occurred.

A case of puerperal mania four days after starting the dopamine agonist, bromocriptine, for suppression of lactation was reported (Brook and Cookson 1978), which is similar to the psychotic reactions with bromocriptine described in the previous section.

Baclofen, a GABA-agonist, has been associated with manic reactions both in high doses given for tardive dyskinesia in a manic-depressive patient on lithium (Wolf *et al.* 1982) and in withdrawal (Arnold *et al.* 1980), but whether GABA, dopaminergic, or phenylethylaminergic mechanisms are involved is unclear.

A hypomanic reaction with delusions was described in

a 68-year-old hypertensive man who had discontinued clonidine (0.4 mg daily) seven days previously; the reaction remitted after the clonidine had been reinstated (Tollefson 1981). The possibility of a rebound central hyperadrenergic state, in addition to hypertension, occurring after clonidine withdrawal is postulated. Other instances of hypomania after clonidine withdrawal were referred to by Jimerson *et al.* (1980) and Freedman *et al.* (1982).

Two hypomanic episodes in a young woman with Hodgkin's disease, but no previous personal or family history of affective disorder were reported following the MAO inhibitor, procarbazine, which was given in conjunction with antitumour chemotherapy (Carney *et al.* 1982). The second reaction occurred in spite of lithium and phenothiazine medication, but there was no reaction when the procarbazine was omitted from the chemotherapeutic regime.

There have been recent reports of excited states following the abuse of high doses of anticholinergic drugs, such as procyclidine (Coid and Strang 1982) and trihexyphenidyl (benzhexol) (Kaminer *et al.* 1982), taken for their euphoriant effects, and attention drawn to the potential for this form of abuse in schizophrenic patients taking these drugs unnecessarily for spurious extrapyramidal side-effects.

Finally, a range of other drugs including pentazocine, cyclizine, aminophylline, isoniazid, and other antituberculous drugs, have been associated with excited or euphoric states (McClelland 1977). The anti-knock agent in petrol, tetraethyl lead, was reported to cause a short-term psychosis characterized by mania, hallucinations, and fear (Ross and Shohton 1983).

Depression

Depressed mood is a frequent and rather non-specific symptom in itself. It has been reported in association with a number of drugs, such as hypotensive agents, oral contraceptives, neuroleptics, levodopa, steroids, and anticonvulsants, and a wide range of miscellaneous drugs (*Lancet* 1977). The experience of depression probably depends on both a subjective cognitive interpretation by the subject and a physiological 'substrate'. It is the latter that can be altered by drugs, but it is generally reported that there is a higher incidence of 'depression' as a side-effect, regardless of the particular drug implicated, in those who have been treated for depression before. This does not resolve the dilemma entirely, however, because it can be argued either that:

1. Those who have previously experienced depression interpret certain autonomic effects such as lethargy and retardation as 'depression' whereas those who have not suffered from depression experience only a peripheral and 'physical' side-effect; or

2. Those who have suffered from depression before have a constitutional predisposition such that they are physiologically more vulnerable to the autonomic effects of certain drugs.

Antihypertensive drugs

Severe depression and a number of suicides were first reported in the 1950s in patients on high doses of rauwolfia alkaloids for prolonged periods (Freis 1954; Muller *et al.* 1955). Since reserpine was known to have central actions particularly on stores of serotonin (5-hydroxytryptamine) and other amines (Pletscher *et al.* 1956), this formed the basis of the first biogenic amine hypothesis of mood disorders (Pare and Sandler 1959). Subsequently practically all drugs used in the treatment of hypertension were reported to have been associated with depression, including the adrenergic neurone-blocking agents (guanethidine, debrisoquine, and bethanidine) and methyldopa, clonidine, and propranolol (McClelland 1977). The majority of these, however, were from general physicians and lacked any systematic attempt at a proper psychiatric diagnosis of depression. Furthermore a review of the literature on biogenic amine depletion and mood threw considerable doubt on the idea that depletion of brain noradrenaline, dopamine, or serotonin, even when caused by reserpine, is sufficient of itself to cause depression (Mendels and Frazer 1974).

A survey of 264 patients in general practice on a wide range of antihypertensive drugs, using a validated self-rating scale for depression (which sought to distinguish depression from non-specific drowsiness or fatigue), failed to show any relationship between methyldopa or other drugs and depressive illness (Snaith and McCoubrie 1974). Bant (1978) studied 89 hypertensive patients at intervals over one year and compared the prevalence of depression with that in 46 patients suffering from various chronic medical conditions other than hypertension. The overall prevalence was similar in the two groups (nearly 50 per cent) and comparable to the 55 per cent found by Mindham *et al.* (1976) in patients being treated for Parkinson's disease. Within the hypertensive group the severity of depression was less in those not on any drugs, and there was a slight increase in prevalence in those with a psychiatric history who were treated with methyldopa. In a major review (369 references) of the psychiatric effects of antihypertensive drugs other than reserpine (Paykel *et al.* 1982) only methyldopa was considered to be clearly associated with depression in large series of patients. Sedation and sleep disturbances, however, rather than true depression, were the most common side-effects associated with methyldopa, clonidine, and propranolol.

The position of β-adrenergic blocking drugs has been controversial (McClelland 1983). The incidence of depression and tiredness with β-blockers was not significantly greater than in the normal population in a large epidemiological Swedish survey (Bengtsson *et al.* 1980). Nevertheless a well-documented report of three female patients with depressive syndromes associated with propranolol has appeared (Petrie *et al.* 1982). These episodes were distinct from non-specific symptoms of lassitude and fatigue, occurred in a dose-dependent manner, and completely remitted when the drug was discontinued. One patient had a previous history of depression and one had a family history, but one had neither of these. Further reports implicating propranolol (Cremona-Barbaro 1983) and nadolol (Russell and Schuckit 1982) are equally convincing. Thus this adverse effect appears to be a genuine and potentially dangerous one in certain patients.

In conclusion it appears that reserpine, methyldopa, and lipophilic β-adrenergic blocking drugs can precipitate true depression, particularly in patients with a history of psychiatric illness. However, as McClelland (1983) emphasized, the incidence of drug-induced depression in hypertensive patients has probably been overestimated and, unless adequate attention is paid to the emotional responses to the social, psychological, and physical consequences of the disease, the management of these patients may be inappropriate.

Oral contraceptives

In the 1960s there were numerous reports of depression and other psychiatric symptoms in women taking oral contraceptives, and these were then the main reasons for women discontinuing this form of contraception (Herzberg *et al.* 1971). A number of ingenious theories of the possible biochemical mechanism for this were advanced but the literature was fraught with clinical and theoretical contradictions (see *Update 1981*). For example, although depression and irritability occurred in anything from 5 to 30 per cent of women on oral contraceptives in some studies, this was balanced by 10–20 per cent who have relief from premenstrual tension and depression (*British Medical Journal* 1969; Herzberg and Coppen 1970).

The problem of deciding whether progesterone and oral contraceptives have significant adverse or beneficial effects on psychiatric symptoms was reviewed by Glick and Bennett (1981). These authors showed that wide-

ranging effects have been attributed to progesterone and progestins, such as depression, loss of libido, fatigue, anxiety, irritability, and increased activation, together with a mood-stabilizing action, and concluded that they might have a causal relationship in the premenstrual syndrome and post-partum disorders. They are inconclusive about the effects of oral contraceptives but agree that the evidence from the literature is that most women using these drugs do not experience any significant change of mood.

Clinical studies face a number of methodological problems (Talwar 1979) and the difficulties of controlling for the effects of previous personality, history of depression, and psychological effects such as suggestion and fear of adverse reactions. Nevertheless, Kutner and Brown (1972) failed to find any association between oral contraceptive use and depression in their large-scale survey. Another very thorough study of 686 women, 335 of whom were taking a contraceptive pill, by Fleming and Seager (1978), found that depression was no higher in those using oral contraceptives and that a higher proportion of users experienced sexual satisfaction. The incidence of depressive symptoms was related more to age, personality, and occupational factors.

Depression is one of a number of subjective symptoms normally present in the population and which can be spontaneously elicited as a placebo response. Aznar-Ramos *et al.* (1969) studied 147 Mexican women over 424 months of observation. They had all had spontaneous abortions and were interested in becoming pregnant again but agreed to take what they believed was a contraceptive compound for a trial period. They were in fact given a placebo and the incidence of spontaneous subjective complaints and direct inquiries concerning libido were recorded. A large variety of 30 different placebo side-effects were reported in 67 per cent of the months of study: the most frequent being decreased libido (29.5 per cent months) followed by 'headache' (15.6 per cent months); 'nervousness' was sixth (6.4 per cent months).

Goldzieher (1968) found no difference in the incidence of depression between women using an intrauterine contraceptive device and those taking oral contraceptive tablets. In a large double-blind placebo-controlled cross-over study of 398 women during 1538 cycles, Goldzieher *et al.* (1971) found no difference between placebo and active treatment groups in the incidence of abdominal discomfort, breast tenderness, or depression. Nervousness and headache were present in 20 to 30 per cent of women in the pre-treatment cycle and progressively decreased during subsequent cycles regardless of treatment given.

Thus 'the burden of proof must now rest on those who

claim that a given symptom is attributable to the contra-ceptive agent rather than to coincidence or to psychic induction' (Goldzieher *et al.* 1971).

Neuroleptics

The problem of depression in schizophrenia is complex, and can be severe and of suicidal intensity. Many schizo-phrenics succeed in killing themselves, often impulsively, and suicide has been established to occur 50 times more often than in the normal population (Markowe *et al.* 1967). Accordingly it is difficult to distinguish an effect due to the drugs used to treat schizophrenia.

It was thought by some that oral phenothiazines inten-sified pre-existing depression (Cohen *et al.* 1964), while others maintained that these drugs were actually anti-depressive (see Freeman 1978). In an anecdotal report De Alarcon and Carney (1969) drew attention to a number of severe depressions and suicides which occurred in patients who happened to be receiving depot injections of fluphenazine, and it was subsequently widely believed that there was a causal relationship between the two (*Lancet* 1977). Freeman (1978), however, pointed out that whether or not the patients were receiving their neuroleptics by injection was totally irrelevant to the problem.

In subsequent double-blind placebo-controlled trials, however, no increase in depression was found in patients receiving active drug compared with those on placebo (Hirsch *et al.* 1973), and little difference in the incidence of depression between patients receiving depot fluphena-zine and flupenthixol (which had been introduced as a neuroleptic drug with a mood-elevating effect) was found by Carney and Sheffield (1976) or Knights *et al.* (1979).

This vexed question was extensively investigated in four recent studies by Johnson (1981). In a survey of the prevalence of depression in schizophrenia he found that this was present in about half of new, untreated, acute cases and about a third of relapsed chronic cases whether or not they were on depot injections, although the presence of extrapyramidal symptoms (EPS) or the use of high doses of depot neuroleptics was associated with a higher prevalence of depressive symptoms. A two-year longitudinal study of 30 schizophrenics treated with depot neuroleptics showed that the morbidity from depression was greater than that due to schizophrenic symptoms and this seemed to be independent of life events, EPS, or changes in drug treatment. Finally, two studies of the efficacy of orphenadrine and nortriptyline for depression associated with schizophrenia failed to find any significant benefit from either drug. This work

therefore confirmed that, although high doses of neuro-leptics and neuroleptic side-effects may exacerbate depression in schizophrenia, this symptom is more often a part of the illness itself and is not helped by the addition of antidepressants which tend only to increase the inci-dence of side-effects.

Indeed there are many possible causes of depression in schizophrenia, since the symptom is rather ill-defined and is common in all phases of the illness (*British Medical Journal* 1980*b*), including both fundamental aspects of the disorder itself and psychological reactions to it, drug side-effects ('akinetic' pseudo-depression), or an asso-ciation with the subsequent personality impairments of the chronic state. Thus depression in schizophrenia cannot be seen merely as a pharmacological effect of antipsychotic drugs and, indeed, it is doubtful whether the drugs alone can cause a true depression. In fact the risk of suicide in untreated schizophrenics is almost certainly greater than in those receiving neuroleptics (*Lancet* 1977).

Levodopa

A similar problem occurs with depression in Parkinson's disease and disentangling the effects of levodopa therapy. Recent studies have clearly shown that depres-sion occurs as part of the Parkinson syndrome and is not merely a reaction to the disability (Celesia and Wana-maker 1972; Robins 1976). Reports of the effects of levodopa, however, are conflicting. Celesia and Wana-maker (1972) reported a slightly lower incidence of depression including a few cases of hypomania in patients receiving levodopa than in those not treated with it, whereas Mindham *et al.* (1976), in a study designed specifically to examine this relationship in 50 patients with Parkinson's disease, found that the vast majority (91 per cent) of those who developed a depressive disorder were on levodopa or levodopa with carbidopa. They found no significant antidepressant effect of levo-dopa, and, although the physical symptoms of the disease improved with levodopa, there was on the whole a greater frequency of affective disturbance. Other studies in manic-depressive and schizophrenic patients suggest that the effect of levodopa on mood depends on the pre-existing mental state, and, while there is generally an increase in activation or arousal, this is not synony-mous with an improvement in mood and can be manifest as a decrease in psychomotor retardation in some cases, an activation of psychotic features in others, and overt hypomania in certain bipolar patients (Goodwin and Sack 1974). There is not, therefore, a simple correlation between the effects of levodopa and mood.

Corticosteroids

It has long been recognized that Addison's disease and Cushing's syndrome are frequently associated with affective disturbances (Cobb 1960; Michael and Gibbons 1963), and since the introduction of hormone replacement therapies similar changes in mental state have frequently been recorded (Beach 1948; Cobb 1960). Although euphoria is the commonest affective change with corticosteroids, depression can also occur as it does in Cushing's disease (de Lange and Doorenbos 1975). Certain patients, notably those with Addison's disease, may have quite severe mood and behavioural disturbances when put on ACTH or corticosteroid therapy, particularly at high doses. The relationship between adrenal activity and mood is an intriguing one. The administration of hydrocortisone was found to increase 'anxiety-proneness' in normal subjects (Weiner *et al.* 1963), but not all patients with Cushing's syndrome or treated with steroids have affective disturbances, nor is an increase in adrenocortical activity an invariable concomitant of primary affective illness (Michael and Gibbons 1963). Quarton *et al.* (1955), reviewing 36 different hypotheses for the association of mental disturbances with ACTH and cortisone treatment, came to the conclusion that there was no simple relationship between the two.

Anticonvulsants

Various mental symptoms are well known to develop during the course of long-standing epilepsy and, indeed, the degree of mental disturbance is often thought to be complementary to the degree of control of the epilepsy in individual cases (Trimble 1977). However, the problem of attributing the mental disturbances to the therapeutic agents involved is similar to that with the use of neuroleptics in schizophrenia. When anticonvulsants were found to cause a depletion of folic acid, evidence for a relationship between serum folate concentration and the development of mental symptoms was reported (Reynolds *et al.* 1966; Reynolds 1968). Prevention of mental deterioration (but with an increase in convulsions) by the use of adjunctive folic acid was subsequently reported by some (Neubauer 1970) but not by others (Grant and Stores 1970). Similarly, changes in calcium metabolism due to increased metabolism of calciferol were reported in patients receiving long-term anticonvulsants (Christiansen *et al.* 1972; Stamp *et al.* 1972) and might conceivably be related to the development of mental symptoms, although hypocalcaemia *per se* might be expected to cause an elevation in mood or cyclic mood changes (Carmen *et al.* 1977).

The available literature on the association between mental symptoms and anticonvulsant drugs was reviewed by Trimble and Reynolds (1976), but very little in the way of definite conclusions could be drawn. It does appear, however, that anticonvulsants, particularly phenytoin, may be associated with subtle mental changes such as impairment of drive and initiative, psychomotor slowing, and depression of mood, in the absence of physical signs of toxicity and even with serum drug levels within the generally regarded 'therapeutic' range, especially if the duration of therapy has been prolonged. Attempts to correct these mental changes with folic acid supplements should generally only be tried if a low serum folic acid level is found.

Other drugs

Two cases of 'paradoxical' depression associated with theophylline were reported (Murphy *et al.* 1980). These were both asthmatic patients on salbutamol and beclomethasone, and although the authors speculated that the depressive reaction was due to catecholamine depletion, an alternative explanation is that theophylline-induced inhibition of phosphodiesterase led to increased sensitivity of post-synaptic β-adrenergic receptors in the central nervous system. This is precisely the biochemical basis of depressive illness proposed by one of the most influential recent theories based on the observation that chronic antidepressant drug treatment and ECT cause a 'down-regulation' in the sensitivity of post-synaptic β-adrenergic receptors in the central nervous system (Sulser *et al.* 1978; Bergstrom and Kellar 1979).

There was a single case report of a severe anxiety-depressive syndrome in a 37-year-old woman taking therapeutic doses of cimetidine which recovered when the drug was discontinued (Johnson and Bailey 1979).

Attention was drawn in a review by Good and Shader (1982) to the serious toxic effects of chloroquine, which is fatal in moderately low overdose, but which can evidently cause a range of psychiatric disturbances of both neurotic and psychotic type in therapeutic doses. The authors were concerned that severe suicidal depression might have been under-reported and could be an unrecognized but preventable cause of potentially lethal chloroquine overdosage.

Depression can be the first sign of an electrolyte imbalance, particularly hypokalaemia due to diuretics, and can be present before clouding of consciousness is apparent. Antidiabetic agents and antithyroid drugs can also be indirectly responsible for depression as a result of their normal pharmacological actions (*Lancet* 1977).

Behavioural and other reactions

McClelland (1977) used the term 'minimal or borderline reactions (behavioural toxicity)' to describe these rather idiosyncratic responses to a wide range of drugs, and the Boston Collaborative Drug Surveillance Program (1971) also used the term 'borderline psychiatric reactions' to cover an number of symptoms such as insomnia, drowsiness, hyperactivity, paraesthesia, tinnitus, vertigo, confusion, and disorientation. McClelland (1977) also included nightmares, mild depression, excitement or euphoria, anxiety, irritability, increased sensitivity to noise, listlessness, and restlessness.

'Paradoxical' reactions to benzodiazepines

The existence of this entity is controversial, and appears only to have been accepted as such in the American literature. An association between chlordiazepoxide, but not oxazepam, and hostility was first reported by Gardos *et al.* (1968). Subsequently, however, it appeared that such reactions were more common in impulsive personalities with a history of aggressive and destructive behaviour and could, in fact, be regarded as quite predictable responses occurring in these patients to a non-specific disinhibiting effect of benzodiazepine tranquillizers (DiMascio *et al.* 1970*b*; Lion *et al.* 1975). A review by Hall and Zisook (1981) reported that the incidence was extremely small and the occurrence of such reactions was idiosyncratic and difficult to predict. Although depression is frequently attributed to these drugs the evidence is not wholly convincing since the patients were usually taking other additional drugs and/or were suffering from cerebrovascular or convulsive disorders. Furthermore the relief of anxiety may 'unmask' coexisting depressive features. Paradoxical 'rage reactions' are even less well substantiated. Benzodiazepines have been shown to have anti-aggressive effects in a range of clinical situations including hyperactive psychotic patients, prisoners, juvenile delinquents, epileptics, neurotic out-patients, and hyperactive children. The isolated case reports of aggressive outbursts tend to be anecdotal and appear to be limited to patients with a history of violence in settings likely to induce interpersonal frustration. Since properly controlled studies are lacking, Hall and Zissook (1981) concluded that the question of the frequency, severity, and specificity of benzodiazepine-induced hostility remains unsettled.

A similar situation applies to the '*van der Kroef syndrome*' where 34 different symptoms, which included verbal and physical aggression and a continuous fear of going insane, were reported following the benzodiazepine hypnotic, triazolam (van der Kroef 1979). This was taken up by the Dutch media and the drug branded as a cause of 'stark raving madness' which led to its suspension in the Netherlands for six months (Lasagna 1980*a*). This in turn led to further controversy in the medical press (Drost 1980; Lasagna 1980*b*) and is an object lesson in how important it is to assess carefully reports of causal relationships between subjective and behavioural symptoms and the pharmacological effects of psychotropic drugs.

There has, however, been another report documenting day-time anxiety following triazolam (Morgan and Oswald 1982). The rapid elimination of hypnotics with very short half-lives seems to lead to an increase in rebound effects the next day, which are dose-related, and are therefore a disadvantage in patients with day-time anxiety who would benefit more from a benzodiazepine with a longer plasma elimination half-life.

Other drugs

Three cases of strange perceptual reactions to the calcium antagonist, verapamil, were reported from Canada (Kumana and Mahon 1981). The symptoms included numbness, restless legs, coldness, paraesthesiae, anorexia, and insomnia. These symptoms seem to be a result of a local action on the peripheral circulation or a psychological reaction to such symptoms. Indeed, in view of the crucial role of calcium ions in neuronal transmission, one must assume that the relative lack of central effects of calcium antagonists must be due to their failure to cross the blood–brain barrier. On the other hand, an intriguing report has appeared in which a manic patient is described who refused lithium treatment but who appears to have responded to verapamil (160 mg daily) (Dubovsky *et al.* 1982). Attention was drawn in that report to certain actions of lithium which are shared with calcium antagonists, such as blockade of calcium channels, decreased rate of spontaneous depolarization of the sinoatrial node, and inhibition of adenylate cyclase.

Somnambulism was reported to occur for the first time in 10 patients who were on a lithium-neuroleptic combination (Charney *et al.* 1979). This was thought to be related to an increase in slow-wave sleep induced by the drug combination, but is evidently a rare side-effect possibly only occurring in persons with a certain neurophysiological predisposition. This was borne out by the fact that the somnambulistic episodes were transient if the EEG was normal, but persistent in association with diffuse EEG irregularities.

McClelland (1977) drew attention to the fact that the frequency of drowsiness as a reaction to many drugs such as antihistamines and antihypertensive drugs is often unrecognized. Additive effects between any drugs with central depressant effects should be expected. The possibility of a similar potentiation of haloperidol-related drowsiness by indomethacin was also reported (Bird *et al.* 1983). Fatigue without sedation, as a particular symptom of reduced exercise tolerance due to β-adrenergic blocking drugs, is also an increasingly frequent drug-induced effect as these drugs are being more widely prescribed (*Lancet* 1980).

Finally, it should be remembered that anxiety as a symptom can be either secondary to alcohol and/or drug withdrawal, or a result of excessive caffeine intake (Greden 1974; MacCallum 1979).

Dependence

There is insufficient scope in this chapter to deal at length with dependence as a drug-induced disorder. Recent theories of the pharmacological mechanisms of drug addiction are provided in the texts by Barchas *et al.* (1977) and Green and Costain (1981). Dependence is a result of an interaction between drug, environmental, and personal factors, and should be seen as a continuum from heroin, which can cause severe dependence in a matter of weeks, to alcohol in which years of abuse generally precede addiction, rather than any simple dichotomy of the addiction/habituation or 'hard' drug/'soft' drug type. Dependence may or may not be associated with physical withdrawal symptoms (an abstinence syndrome), and may or may not be associated with tolerance. Although physical dependence is easier to define, psychological dependence and associated changes in behaviour and personality are socially more important and probably more responsible for the tendency to relapse, than the tolerance/withdrawal syndrome. Table 28.4 shows the relative importance of these features for different classes of drugs. From this it can be seen that tolerance does not explain withdrawal and the two can occur independently. A withdrawal syndrome only occurs with drugs which depress CNS activity and is always associated with tolerance, but tolerance can develop without withdrawal.

Controversy surrounds cannabis dependence. There is recent good evidence, however, that delta-9-tetrahydrocannabinol (Δ-9-THC), the most active constituent of cannabis, taken in regular doses of up to 180 mg daily for 10–18 days, will cause both tolerance and a withdrawal syndrome (Jones *et al.* 1976).

Table 28.4   Relative severity of symptoms of dependence with drugs liable to misuse

	Tolerance	Dependence	
		Psychological	Physical
Controlled drugs			
Opiates	+ +	+ + +	+ +
Cocaine	–	+ + +	–
Amphetamine	+	+ +	–
LSD	+	±	–
Cannabis	+	+	+
Non-controlled drugs			
Barbiturates	+ +	+ +	+ +
Alcohol	+ +	+ +	+ +
Benzodiazepines	+	+	+
Solvents	+	+	–

The problem of dependence to the benzodiazepines was mentioned in the introduction and recent studies (Tyrer *et al.* 1981; Petursson and Lader 1981*b*; Tyrer *et al.* 1983) have been able to distinguish a true withdrawal syndrome from recurrences of pre-existing anxiety in about 45 per cent of patients who discontinue these drugs after taking therapeutic doses for six months or more. The extent of this problem remains uncertain. Marks (1978) estimated from figures based on available reports of dependence that it would be of the order of one case per 50 million 'patient months' in therapeutic use, or one per 5 million 'patient months' if both therapeutic use and drug abusers were included, but in the Tyrer *et al.* (1981) series of 40 patients withdrawal symptoms were present in 27–45 per cent depending on the criterion for withdrawal used, suggesting that the overall incidence could be much higher.

Drugs with central stimulant actions can produce an amphetamine-like dependence and this was reported with the β-agonist, salbutamol (Gaultier *et al.* 1976; Edwards and Holgate 1979) and the MAO inhibitor, tranylcypromine (Griffin *et al.* 1981). The latter may differ from amphetamine dependence in being associated with distinct withdrawal symptoms but the report concerned four patients none of whom had been on tranylcypromine alone, since they had either been taking *Parstelin* (tranylcypromine and trifluoperazine) or *Parstelin* and a benzodiazepine.

Solvent abuse is usually associated with non-specific signs of intoxication (see under 'Delirium' above). Tolerance and psychological dependence can occur but not a physical withdrawal syndrome. Subsequent encephalopathy may, however, predispose to schizophrenia later (Daniels and Latcham 1984).

Acknowledgements

Figures 28.1 and 28.2 were reproduced with the kind permission of the editors and publishers of *Acta Medica Scandinavica* and *Psychological Medicine* respectively. I am indebted to Miss Donna Crangle and Mrs Mavis Scullion for their patience and skill in typing the manuscript.

REFERENCES

ADVERSE DRUG REACTIONS ADVISORY COMMITTEE (1980). Mazindol. *Med. J. Aust.* **1**, 85.

ADVISORY COUNCIL ON THE MISUSE OF DRUGS (1982). Report of the expert group on the effects of cannabis use. *Home Office,* London.

ANGRIST, B., ROTROSEN, J., AND GERSHON, S. (1980). Responses to apomorphine, amphetamine, and neuroleptics in schizophrenic subjects. *Psychopharmacology,* **67**, 31–8.

ARNOLD, E.S., BUDD, S.M. AND KIRSHNER, H. (1980). Manic psychosis following rapid withdrawal from baclofen. *Am. J. Psychiat.* **137**, 1466–7.

ATSMON, A., BLUM, I., WIJSENBEEK, H., MAOZ, B., STEINER, M., AND ZIEGELMAN, G. (1971). The short-term effects of adrenergic-blocking agents in a small group of psychotic patients. *Psychiat. Neurol. Neurochir.* **74**, 251–8.

AZNAR-RAMOS, R., GINER-VELÁZQUEZ, J., LARA-RICALDE, R. AND MARTÍNEZ-MANAUTOU, J. (1969). Incidence of side effects with contraceptive placebo. *Amer. J. Obst. Gynec.* **105**, 1144–9.

BALTER, M.B., LEVINE, J., AND MANHEIMER, D.I. (1974). Cross-national study of the extent of anti-anxiety/sedative drug use. *New Engl. J. Med.* **290**, 769–74.

BANT, W.P. (1978). Antihypertensive drugs and depression: a reappraisal. *Psychol. Med.* **8**, 275–83.

BARBEAU, A. (1971). Long-term side-effects of levodopa. *Lancet* **i**, 395.

BARCHAS, J.D., BERGER, P.A., CIARANELLO, R.D., AND ELLIOTT, G.R. (Eds.) (1977). *Psychopharmacology. From theory to practice,* pp. 291–403. Oxford University Press, London.

BARNES, T.R.E. AND BRIDGES, P.K. (1980). Disturbed behaviour induced by high-dose antipsychotic drugs. *Br. med. J.* **281**, 274–5.

BEACH, F.A. (1948). *Hormones and behaviour,* pp. 70–2, 110–16. Paul B. Hoecher, New York.

BENGTSSON C., LENNARTSSON, J., LINDQUIST, O., NOPPA, H., AND SIGURDSSON, J. (1980). Sleep disturbances, nightmares and other possible central nervous disturbances in a population sample of women, with special reference to those on anti-hypertensive drugs. *Eur. J. clin. Pharmacol.* **17**, 173–7.

BERGSTROM, D.A. AND KELLAR, K.J. (1979). Effect of electro-convulsive shock on monoaminergic receptor binding sites in rat brain. *Nature* **278**, 464–6.

BETTS, T.A. AND BIRTLE, J. (1982). Effect of two hypnotic drugs on actual driving performance next morning. *Br. med. J.* **285**, 852.

BIEHL, B. (1979). Studies of clobazam and car-driving. *Br. J. clin. Pharmacol.* **7**, 85s–90s.

BIRD, H.A., LeGALLEZ, P., AND WRIGHT, V. (1983). Drowsiness due to haloperidol/indomethacin in combination. *Lancet* **i**, 830–1.

BLAZER, D.G. AND HALLER, L. (1975). Pentazocine psychosis. A case of persistent delusions. *Dis. nerv. Syst.* **36**, 404–5.

BÖETHIUS, G. AND WESTERHOLM, B. (1977). Purchases of hypnotics, sedatives and minor tranquillizers among 2566 individuals in the county of Jamtland, Sweden — a 6 years' follow up. *Acta psychiat. scand.* **56**, 147–59.

BOSTON COLLABORATIVE DRUG SURVEILLANCE PROGRAM (1971). Psychiatric side-effects of nonpsychiatric drugs. *Semin. Psychiat.* **3**, 406–20.

BREAKEY, W.R., GOODELL, H., LORENZ, P.C., AND McHUGH, P.R. (1974). Hallucinogenic drugs as precipitants of schizophrenia. *Psychol. Med.* **4**, 255–61.

BRITISH MEDICAL JOURNAL (1969). Leading article: Oral contraception and depression. *Br. med. J.* **4**, 380–1.

—— (1976). Leading article: Cannabis psychosis. *Br. med. J.* **2**, 1092–3.

—— (1978). Leading article: Road accidents: are drugs other than alcohol a hazard? *Br. med. J.* **2**, 1415–17.

—— (1980a). Leading article: Hypnotics and hangover. *Br. med. J.* **1**, 743.

—— (1980b). Leading article: Use of antidepressants in schizophrenia. *Br. med. J.* **280**, 1037–8.

BROOK, N.M. AND COOKSON, I.B. (1978). Bromocriptine-induced mania? *Br. med. J.* **1**, 790.

BUNNEY, W.E. AND MURPHY, D.L. (1974). Switch processes in psychiatric illnesses. In *Factors in depression* (ed. N.S. Kline), pp. 139–58. Raven Press, New York.

CARMEN, J.S., POST, R.M., GOODWIN, F.K., AND BUNNEY, W.E. (1977). Calcium and electroconvulsive therapy of severe depressive illness. *Biol. Psychiat.* **12**, 5–17.

CARNEY, M.W.P. (1977). Paranoid psychosis with indomethacin. *Br. med. J.* **2**, 994–5.

——, RAVINDRAN, A., AND LEWIS, D.S. (1982). Manic psychosis associated with procarbazine. *Br. med. J.* **284**, 82–3.

—— AND SHEFFIELD, B.F. (1976). Comparison of antipsychotic depot injections in the maintenance treatment of schizophrenia. *Br. J. Psychiat.* **129**, 476–81.

CELESIA, G.G. AND WANAMAKER, W.M. (1972). Psychiatric disturbances in Parkinson's disease. *Dis. nerv. Syst.* **33**, 577–83.

CHARNEY, D.S. KALES, A., SOLDATOS, C.R., AND NELSON, J.C. (1979). Somnambulistic-like episodes secondary to combined lithium-neuroleptic treatment. *Br. J. Psychiat.* **135**, 418–24.

CHOUINARD, G. AND JONES, B.D. (1980). Neuroleptic-induced supersensitivity psychosis: clinical and pharmacologic characteristics. *Am. J. Psychiat.* **137**, 16–21.

CHRISTIANSEN, C., KRISTENSEN, M., AND RODBRO, P. (1972). Latent osteomalacia in epileptic patients on anticonvulsants. *Br. med. J.* **3**, 738–9.

CLARE, A.W. (1971). Diazepam, alcohol and barbiturate abuse. *Br. med. J.* **4**, 340.

CLOW, A., THEODOROU, A., JENNER, P., AND MARSDEN, C.D. (1980). A comparison of striatal and mesolimbic dopamine function in the rat during 6-month trifluoperazine administration. *Psychopharmacology* **69**, 227–33.

COBB, S. (1960). Some clinical changes in behaviour accompanying endocrine disorders. *J. nerv. ment. Dis.* **130**, 97–106.

COHEN, S.I. (1982). Paranoid symptoms after nomifensine. *Lancet* **ii**, 928.

COHEN, S., LEONARD, C.V., FARBEROW, N.L., AND SCHNEIDMAR, E.S.

(1964). Tranquillizers and suicide in the schizophrenic patient. *Arch. gen. Psychiat.* **11**, 312–21.

COID, J. AND STRANG, J. (1982). Mania secondary to procyclidine ('Kemadrin') abuse. *Br. J. Psychiat.* **141**, 81–4.

COLLINS, G.B., MARZEWSKI, D.J., AND ROLLINS, M.B. (1981). Paranoid psychosis after DDAVP therapy for Alzheimer's dementia. *Lancet* **ii**, 808.

COMMISSION OF INQUIRY INTO THE NONMEDICAL USE OF DRUGS (1972). Report on cannabis. Information Canada, Ottawa.

COMMITTEE ON THE REVIEW OF MEDICINES (1980). Systematic review of benzodiazepines. *Br. med. J.* **280**, 910–12.

CONNELL, P.H. (1958). *Amphetamine psychosis.* Oxford University Press, London.

COPPEL, D.L., BOVILL, J.G., AND DUNDEE, J.W. (1973). The taming of ketamine. *Anaesthesia* **28**, 293–6.

CREMONA-BARBARO, A. (1983). Propranolol and depression. *Lancet* **i**, 185.

DANIELS, A.M. AND LATCHAM, R.W. (1984). Petrol sniffing and schizophrenia in a pacific island paradise. *Lancet* **i**, 389.

DAVIES, P. AND MALONEY, A.J.F. (1976). Selective loss of central cholinergic neurons in Alzheimer's disease. *Lancet* **ii**, 1403.

DAVISON, K. (1981). Diagnoses not to be missed: Toxic psychosis. *Br. J. hosp. Med.* **26**, 530–7.

DE ALARCON, R. AND CARNEY, M.W.P. (1969). Severe depressive mood changes following slow-release intramuscular fluphenazine injection. *Br. med. J.* **3**, 564–7.

DEBONO, A.G., MARSDEN, C.D., ASSELMAN, P., AND PARKES, J.D. (1976). Bromocriptine and dopamine receptor stimulation. *Br. J. clin. Pharmacol* **3**, 977–82.

DE LANGE, W.E. AND DOORENBOS, H. (1975). Corticotrophins and corticosteroids. In *Meyler's side effects of drugs,* 8th edn. (ed. M.N.G. Dukes), pp. 812–40. Excerpta Medica, Amsterdam.

DIMASCIO, A., SHADER, R.I., AND HARMATZ, J.S. (1970*a*). Behavioural toxicity. Part V. Gross behaviour patterns. In *Psychotropic drug side effects* (ed. R.I. Shader and A. DiMascio), pp. 142–8. Williams and Wilkins, Baltimore.

——, —— AND GILLER, D.R. (1970*b*). Behavioural toxicity. Part IV: Emotional (mood) states. In *Psychotropic drug side effects* (ed. R.I. Shader and A. DiMascio), pp. 132–41. Williams and Wilkins, Baltimore.

DIMIJIAN, G.G. (1976). Differential diagnosis of emergency drug reactions. In *Acute drug abuse emergencies* (ed. P.G. Bourne), Chapter 1. Academic Press, London.

DROST, R.A. (1980). The Halcion story. *Lancet* **i**, 1027–8.

DRUG AND THERAPEUTICS BULLETIN (1982). Rantitidine v. cimetidine in peptic ulcer. *Drug Ther. Bull.* **20**, 57–9.

DUBOVSKY, S.L., FRANKS, R.D., LIFSCHITZ, M., AND COEN, P. (1982). Effectiveness of verapamil in the treatment of a manic patient. *Am. J. Psychiat.* **139**, 502–4.

DUKES, M.N.G. (ED.) (1975). *Meyler's side effects of drugs,* Vol. VIII. Excerpta Medica, American Elsevier, New York.

EDWARDS, J.G. AND HOLGATE, S.T. (1979). Dependency upon salbutamol inhalers. *Br. J. Psychiat.* **134**, 624–6.

FELINE, A. AND JOUVENT, R. (1977). Manifestations psycho-sensorielles observées chez des psychotiques soumises à des médications bêta-mimétiques. *L'Encéphale* **3**, 149–58.

FILE, S.E. AND BOND, A.J. (1979). Impaired performance and sedation after a single dose of lorazepam. *Psychopharmacology* **66**, 309–13.

FLEMING, O. AND SEAGER, C.P. (1978). Incidence of depressive symptoms in users of the oral contraceptive. *Br. J. Psychiat.* **132**, 431–40.

FLEMINGER, R. (1978). Visual hallucinations and illusions with propranolol. *Br. med. J.* **1**, 1182.

FRASER, H.S. AND CARR, A.C. (1976). Propranolol psychosis. *Br. J. Psychiat.* **129**, 508–9.

FREEDMAN, R. KIRCH, D., BELL, J., ADLER, L.E., PECEVICH, M., PACHTMAN, E., AND DENVER, P. (1982). Clonidine treatment of schizophrenia. Double-blind comparison to placebo and neuroleptic drugs. *Acta psychiat. scand.* **65**, 35–45.

FREEMAN, H.L. (1978). Drug-induced depression. *Lancet* **i**, 274.

FREIS, E.D. (1954). Mental depression in hypertensive patients treated for long periods with large doses of reserpine. *New Engl. J. Med.* **251**, 1006–8.

GARDOS, G., COLE, J.O., VOLICER, L., ORZACK, M.H., AND OLIFF, A.C. (1973). A dose-response study of propranolol in chronic schizophrenics. *Curr. Ther. Res.* **15**, 314–23.

——, DIMASCIO, A., SALZMAN, C., AND SHADER, R.I. (1968). Differential actions of chlordiazepoxide and oxazepam on hostility. *Arch. gen. Psychiat.* **18**, 757–60.

GARFIELD, J.M. (1974). Psychologic problems in anesthesia. *Am. fam. Physn.* **10**, 60–7.

GAULTIER, M., GERVAIS, P., LAGIER, G., AND DANAN, L. (1976). Pharmacodépendance psychique au salbutamol en aérosol chez une asthmatique. *Thérapie,* **31**, 465–70.

GHOSE, K. (1977). Lithium salts: therapeutic and unwanted effects. *Br. J. hosp. Med.* **18**, 578–83.

GLICK, I.D. AND BENNETT, S.E. (1981). Psychiatric complications of progesterone and oral contraceptives. *J. clin. Psychopharmacol.* **1**, 350–67.

GLUCKMAN, L. (1974). Ventolin Psychosis. *N.Z. med. J.* **80**, 411.

GOLDZIEHER, J.W. (1968). The incidence of side-effects with oral or intrauterine contraceptives. *Am. J. Obstet. Gynec.* **102**, 91–4.

——, MOSES, L.E., AVERKIN, E., SCHEEL, C. AND TABER, B.Z. (1971). A placebo-controlled double blind cross-over investigation of the side-effects attributed to oral contraceptives. *Fert. Steril.* **22**, 609–23.

GOOD, M.I. AND SHADER, R.I. (1982). Lethality and behavioral side effects of chloroquine. *J. clin. Psychopharmacol.* **2**, 40–7.

GOODWIN, F.K. AND SACK, R.L. (1974). Central dopamine function in affective illness: evidence from precursors, enzyme inhibitors, and studies of central dopamine turnover. In *Neuropsychopharmacology of monoamines and their regulatory enzymes* (ed. E. Usdin), pp. 261–79. Raven Press, New York.

GOUGH, S.P. AND COOKSON, I.B. (1984). Khat-induced schizophreniform psychosis in U.K. *Lancet* **i**, 455.

GRAHAM, J.D.P. (1976). *Cannabis and health.* Academic Press, London.

GRANACHER, R.P. AND BALDESSARINI, R.J. (1976). The usefulness of physostigmine in neurology and psychiatry. In *Clinical neuropharmacology,* Vol I. (ed. H.L. Klawans), pp. 63–79. Raven Press, New York.

GRANT, R.H.E. AND STORES, O.P.R. (1970). Folic acid in folate-deficient patients with epilepsy. *Br. med. J.* **4**, 644–8.

GREDEN, J.F. (1974). Anxiety or caffeinism: a diagnostic dilemma. *Am. J. Psychiat.* **131**, 1089–92.

GREEN, A.R. AND COSTAIN, D.W. (1981). *Pharmacology and biochemistry of psychiatric disorders,* pp. 155–73. John Wiley and Sons.

GREENBLATT, D.J., SHADER, R.I., AND KOCH-WESER, J. (1976). The psychopharmacology of beta adrenergic blockade: pharmacokinetic and epidemiologic aspects. In *Neuropsychiatric effects of adrenergic beta-receptor blocking agents* (ed. C.

Carlsson, J. Engel, and L. Hansson). *Adv. clin. Pharmacol.* **12**, 6–12.

GREENE, L.T. (1971). Physostigmine treatment of anticholinergic delirium in postoperative patients. *Anaesth. Analg.* **50**, 222–6.

GREER, H.D., WARD, H.P., AND CORBRIN, K.B. (1965). Chronic salicylate intoxication in adults. *J. Am. med. Ass.* **193**, 555–8.

GRIFFIN, N., DRAPER, R.J. AND WEBB, M.G.T. (1981). Addiction to tranylcypromine. *Br. med. J.* **283**, 346.

HALL, R.C.W., POPKIN, M.K., STICKNEY, S.K., AND GARDNER, E.R. (1979). Presentation of steroid psychoses. *J. nerv. ment. Dis.* **167**, 229–36.

—— AND ZISOOK, S. (1981). Paradoxical reactions to benzodiazepines. *Br. J. clin. Pharmacol.* **11**, 99s–104s.

HARDING, T. (1972). Depression following fenfluramine withdrawal. *Br. J. Psychiat.* **121**, 338–9.

HEATH, R.G., NESSELHOF, W., BISHOP, M.P., AND BYERS, L.W. (1965). Behavioural and metabolic changes associated with administration of tetraethylthiuram disulfide (Antabuse). *Dis. nerv. Syst.* **26**, 99–105.

HELSON, L. AND DUQUE, L. (1978). Acute brain syndrome after propranolol. *Lancet* **i**, 98.

HERZBERG, B. AND COPPEN, A. (1970). Changes in psychological symptoms in women taking oral contraceptives. *Br. J. Psychiat.* **116**, 161–4.

——, DRAPER, K.C., JOHNSON, A.L., AND NICOL, G.C. (1971). Oral contraceptives, depression and libido. *Br. med. J.* **3**, 495–500.

HINDMARCH, I. AND GUDGEON, A.C. (1980). The effects of clobazam and lorazepam on aspects of psychomotor performance and car handling ability. *Br. J. clin. Pharmacol.* **10**, 145–50.

HINSHELWOOD, R.D. (1969). Hallucinations and propranolol. *Br. med. J.* **2**, 445.

HIRSCH, S.R., GAIND, R., ROHDE, P.D., STEVENS, B.C., AND WING, J.K. (1973). Outpatient maintenance of chronic schizophrenic patients with long-acting fluphenazine: double-blind placebo trial. *Br. med. J.* **1**, 633–7.

HOCKINGS, N. AND BALLINGER, B.R. (1983). Hypnotics and anxiolytics. *Br. med. J.* **286**, 1949–51.

HONKANEN, R., ERTAMA, L., LINNOILA, M., ALHA, A., LUKKART, I., KARLSSON, M., KIVILUOTO, O., AND PURO, M. (1980). Role of drugs in traffic accidents. *Br. med. J.* **281**, 1309–12.

HOWE, J.G. (1980). Lorazepam withdrawal seizures. *Br. med. J.* **280**, 1163–4.

JACOBS, D. (1979). Psychiatric symptoms and hallucinogenic compounds. *Br. med. J.* **2**, 49.

JANOWSKY, D.S. AND DAVIS, J.M. (1974). Dopamine, psychomotor stimulants, and schizophrenia: effects of methylphenidate and the stereoisomers of amphetamine in schizophrenics. In *Neuropsychopharmacology of monoamines and their regulatory enzymes* (ed. E. Usdin.), pp. 317–23. Raven Press, New York.

——, EL-YOUSEF, M.K., DAVIS, J.M. AND SEKERKE, H.J. (1972). A cholinergic–adrenergic hypothesis of mania and depression. *Lancet* **ii**, 632–5.

JIMERSON, D.C., POST, R.M. STODDARD, F.J., GILLIN, J.C., AND BUNNEY, W.E. (1980). Preliminary trial of the noradrenergic agonist clonidine in psychiatric patients *Biol. Psychiat.* **15**, 45–57.

JOHNSON, D.A.W. (1981). Studies of depressive symptoms in schizophrenia. *Br. J. Psychiat.* **139**, 89–101.

JOHNSON, J. AND BAILEY, S. (1979). Cimetidine and psychiatric complications. *Br. J. Psychiat.* **134**, 315–16.

JOHNSTONE, E.C., CROW, T.J., FERRIER, N., FRITH, C.D., OWENS, D.G.C., BOURNE, R.C., AND GAMBLE, S.J. (1983). Adverse effects of anticholinergic medication on positive schizophrenic symptoms. *Psychol. Med.* **13**, 513–27.

JONES, R.T., BENOWITZ, N., AND RACKMAN, J. (1976). Clinical studies of cannabis tolerance and dependence. *N.Y. Acad. Sci.* **282**, 221–39.

KAGEVI, I. AND WÄHLBY, L. (1985). CSF concentrations of ranitidine. *Lancet* **i**, 164–5.

KALIX, P. (1984). Amphetamine psychosis due to Khat leaves. *Lancet* **i**, 46.

KAMINER, Y., MUNITZ, H., AND WIJSENBEEK, H. (1982). Trihexyphenidyl (Artane) abuse: euphoriant and anxiolytic. *Br. J. Psychiat.* **140**, 473–4.

KATONA, C. (1982). Paranoid symptoms after nomifensine. *Lancet* **ii**, 384–5.

KING, D.J., COOPER, S.J., AND LIDDLE, J. (1983). The effect of propranolol on CSF amine metabolites in psychiatric patients. *Br. J. clin. Pharmacol.* **15**, 331–7.

—— AND GRIFFITHS, K. (1984). Patterns in drug utilization — national and international aspects — psychoactive drugs. In *Drug utilization studies: implications for medical care* (ed. F. Sjöqvist and I. Agenäs). *Acta med. scand.* Suppl. **683**, 71–7.

——, ——, REILLY, P.M., AND MERRETT, J.D. (1982). Psychotropic drug use in Northern Ireland 1966–80; prescribing trends, inter- and intra-regional comparisons and relationship to demographic and socioeconomic variables. *Psychol. Med.* **12**, 819–33.

——, TURKSON, S.N.A., LIDDLE, J., AND KINNEY, C.K. (1980). Some clinical and metabolic aspects of propranolol in chronic schizophrenia. *Br. J. Psychiat.* **137**, 458–68.

KING, M.D., DAY, R.E., OLIVER, J.S., LUSH, M., AND WATSON, J.M. (1981). Solvent encephalopathy. *Br. med. J.* **283**, 663–5.

KLAWANS, H.L. (1973). The pharmacology of tardive dyskinesias. *Am. J. Psychiat.* **130**, 82–6.

KNIGHTS, A., OKASHA, M.S., SALIH, M.A., AND HIRSCH, S.R. (1979). Depressive and extrapyramidal symptoms and clinical effects: a trial of fluphenazine versus flupenthixol in maintenance of schizophrenic out-patients. *Br. J. Psychiat.* **135**, 515–23.

KNOX, S.J. (1961). Severe psychiatric disturbances in the postoperative period — a five-year survey of Belfast hospitals. *J. ment. Sci.* **107**, 1078–96.

KOEHLER, K. AND GUTH, W. (1977). Schizophrenie-ähnliche Psychose nach Einnahme von propranolol. *Munch. Med. Wochenschr.* **119**, 443–4.

KOLAKOWSKA, T. AND WILES, D. (1981). Plasma drug and prolactin levels during maintenance treatment with depot neuroleptics. *Br. J. Psychiat.* **139**, 249.

KUMANA, C.R. AND MAHON, W.A. (1981). Bizarre perceptual disorder of extremities in patients taking verapamil. *Lancet* **i**, 1324–5.

KUTNER, S.J. AND BROWN, W.L. (1972). Types of oral contraceptives, depression and premenstrual symptoms. *J. nerv. ment. Dis.* **155**, 153–62.

LANCET (1977). Leading article: Drug-induced depression. *Lancet* **ii**, 1333–4.

—— (1980). Leading article: Fatigue as an unwanted effect of drugs. *Lancet* **i**, 1285–6.

—— (1982a). Leading article: Cimetidine and ranitidine. *Lancet* **i**, 601–2.

—— (1982*b*). Leading article: Phenylpropranolamine over the counter. *Lancet* **i**, 839.

LASAGNA, L. (1980*a*). The Halcion story: trial by media. *Lancet* **i**, 815–16.

—— (1980*b*). The Halcion story. *Lancet* **i**, 1304–5.

LAUTIN, A., ANGRIST, B., STANLEY, M., GERSHON, S., HECKL, K., AND KAROBATH, M. (1980). Sodium valproate in schizophrenia: some biochemical correlates. *Br. J. Psychiat.* **137**, 240–4.

LECRUBIER, Y., PUECH, A.J., JOUVENT, R., SIMON, P., AND WIDLÖCHER, D. (1980). A beta-adrenergic stimulant (salbutamol) versus clomipramine in depression: a controlled study. *Br. J. Psychiat.* **136**, 354–8.

LEFEUVRE, C.M., ISAACS, A.J., AND FRANK, O.S. (1982). Bromocriptine-induced psychosis in acromegaly. *Br. med. J.* **285**, 1315.

LEIGHTON, K.M. (1982). Paranoid psychosis after abuse of Actifed. *Br. med. J.* **284**, 789–90.

LERER, B., BIRMACHER, B., EBSTEIN, R.P., AND BELMAKER, R.H. (1980). 48-hour depressive cycling induced by antidepressant. *Br. J. Psychiat.* **137**, 183–5.

LEWIS, L.D. AND COCHRANE, G.M. (1983). Psychosis in a child inhaling budesonide. *Lancet* **ii**, 634.

LION, J.R., AZCARATE, C.L., AND KOEPKE, H.H. (1975). 'Paradoxical rage reactions', during psychotropic medication. *Dis. nerv. Syst.* **36**, 557–8.

LIPOWSKI, Z.J. (1980*a*). *Delirium*. C. Thomas, Springfield, Illinois.

—— (1980*b*). Delirium updated. *Comp. Psychiat.* **21**, 190–6.

LOGA, S., CURRY, S., AND LADER, M. (1975). Interactions of orphenadrine and phenobarbitone and chlorpromazine: plasma concentrations and effects in man. *Br. J. clin. Pharmacol.* **2**, 197–208.

MacCALLUM, W.A.G. (1979). Excess coffee and anxiety states. *Int. J. social Psychiat.*, **25**, 209–10.

MAC DONALD, J.M. AND EBAUGH, F.G. (1954). Antabuse in alcoholism. *Med. Clins. N. Am.* **38**, 515–24.

MARKOWE, M., STEINERT, J., AND HEYWORTH-DAVIS, F. (1967). Insulin and chlorpromazine in schizophrenia: a ten year comparative survey. *Br. J. Psychiat.* **113**, 1101–6.

MARKS, J. (1978). The benzodiazepines. Use, overuse, misuse, abuse. MTP Press Ltd., Lancaster, England.

MARSDEN, C.D. (1982). Basal ganglia disease. *Lancet* **ii**, 1141–7.

——, TARSY, D. AND BALDESSARINI, R.J. (1975). Spontaneous and drug-induced movement disorders in psychotic patients. In *Psychiatric aspects of neurological disease* (ed. D.F. Benson, and D. Blumer), pp. 219–66. Grune and Stratton, New York.

McCLELLAND, H.A. (1977). Psychiatric disorders. In: *Textbook of adverse drug reactions* (ed. D.M. Davies), pp. 335–53. Oxford University Press.

—— (1978). Drug-induced delirium. *Adv. Drug Reaction Bull.* **72**, 256–9.

—— (1983). Psychiatric reactions to antihypertensive drugs. *Adv. Drug Reaction Bull.* **99**, 364–7.

McDEVITT, D.G. AND McMEEKIN, C. (1979). Data collection in Northern Ireland. In *Studies in drug utilization: methods and applications* (ed. U. Bergman, A. Grimsson, A.H.W. Wahba, and B. Westerholm), pp. 103–11. WHO Regional Publications, European Series No. 8.

McLELLAN, A.T., WOODY, G.E., AND O'BRIEN, C.P. (1979). Development of psychiatric illness in drug abusers. Possible role of drug preference. *New Engl. J. Med.* **301**, 1310–14.

MELDRUM, B. (1982). GABA and acute psychoses. *Psychol. Med.* **12**, 1–5.

MEMON, N.A. AND DAVIDSON, A.R. (1981). Multisystem disorder after exposure to paint stripper (Nitromors). *Br. med. J.* **282**, 1033–4.

MENDELS, J. AND FRAZER, A. (1974). Brain biogenic amine depletion and mood. *Arch. gen. Psychiat.* **30**, 447–51.

MENDHIRATTA, S.S., WIG, N.N., AND VERMA, S.K. (1978). Some psychological correlates of long-term heavy cannabis users. *Br. J. Psychiat.* **132**, 482–6.

MICHAEL, R.P. AND GIBBONS, J.L. (1963). Interrelationships between the endocrine system and neuropsychiatry. *Int. Rev. Neurobiol.* **5**, 243–302.

MINDHAM, R.H.S., MARSDEN, C.D., AND PARKES, J.D. (1976). Psychiatric symptoms during L-dopa therapy for Parkinson's disease and their relationship to physical disability. *Psychol. Med.* **6**, 23–33.

MORGAN, K. AND OSWALD, I. (1982). Anxiety caused by a short-life hypnotic. *Br. med. J.* **284**, 942.

MULLER, J.C., PRYOR, W.W., GIBBONS, J.E., AND ORGAIN, E.S. (1955). Depression and anxiety occurring during rauwolfia therapy. *J. Am. med. Ass.* **159**, 836–9.

MURPHY, M.B., DILLON, A., AND FITZGERALD, M.X. (1980). Theophylline and depression. *Br. med. J.* **281**, 1322.

MYERS, D.H., CAMPBELL, P.L., COCKS, N.M., FLOWERDEW, J.A., AND MUIR, A. (1981). A trial of propranolol in chronic schizophrenia. *Br. J. Psychiat.* **139**, 118–21.

NABER, D., SCHMIDT-HABELMANN, P., BULLINGER, M., NEFF, A., BÜCHLER, A., AND DIETZFELBINGER, A. (1983). Serum cortisol correlates with depression score after open-heart surgery. *Lancet* **i**, 1052–3.

NEUBAUER, C. (1970). Mental deterioration in epilepsy due to folate deficiency. *Br. med. J.* **2**, 759–64.

NICHOLSON, A.N. (1980). Hypnotics and skilled work. In *Drug-induced emergencies*. (ed. P.F. D'Arcy and J.P. Griffin), pp. 183–97. Wright and Sons Ltd. Bristol.

OGURA, C., NAKAZAWA, K., MAJIMA, K., NAKAMURA, K., UEDA, H., UMEZAWA, Y., AND WARDELL, W.M. (1980). Residual effects of hypnotics: triazolam, flurazepam, and nitrazepam. *Psychopharmacology*, **68**, 61–5.

O'MULLANE, N.M., JOYCE, P., KAMATH, S.V., THAM, M.K., AND KNASS, D. (1982). Adverse CNS effects of menthol-containing Olbas oil. *Lancet* **i**, 1121.

OSWALD, I. (1979). The why and how for hypnotic drugs. *Br. med. J.* **1**, 1167–8.

PARE, C.M.B. AND SANDLER, M. (1959). A clinical and biochemical study of a trial of iproniazid in the treatment of depression. *J. Neurol. Neurosurg. Psychiat.* **22**, 247–51.

PAYKEL, E.S., FLEMINGER, R., AND WATSON, J.P. (1982). Psychiatric side effects of antihypertensive drugs other than reserpine. *J. clin. Psychopharmacol.* **2**, 14–39.

PEARSON, K.C. (1981). Mental disorders from low dose bromocriptine. *New Engl. J. Med.* **305**, 173.

PEER, G., WIGLER, I., WEINTRAUB, M., LIRON, M., AND AVIRAM, A. (1983). Acute psychosis and hypertensive crisis following oral salt replacement. *Lancet* **ii**, 634–5.

PEET, M., BETHELL, M.S., COATES, A., KHAMNEE, A.K., HALL, P., COOPER, S.J., KING, D.J., AND YATES, R.A., (1981*a*). Propranolol in schizophrenia. I. Comparison of propranolol, chlorpromazine and placebo. *Br. J. Psychiat.* **139**, 105–11.

——, MIDDLEMISS, D.N., AND YATES, R.A. (1981*b*). Propranolol in schizophrenia. II. Clinical and biochemical aspects of combining propranolol with chlorpromazine. *Br. J. Psychiat.* **139**, 112–17.

PERRY, E.K., GIBSON, P.H., BLESSED, G., PERRY, R.H., AND TOMLINSON, B.E. (1977). Neurotransmitter enzyme abnormalities in senile dementia. *J. neurol. Sci.* **34**, 247–65.

PERRY, R.H., TOMLINSON, B.E., CANDY, J.M., BLESSED, G., FOSTER, J.F., BLOXHAM, C.A., AND PERRY, E.R. (1983). Cortical cholinergic deficit in mentally impaired parkinsonian patients. *Lancet* **ii**, 789–90.

PETRIE, W.M., MAFFUCCI, R.J., AND WOOSLEY, R.L. (1982). Propranolol and depression. *Am. J. Psychiat.* **139**, 92–4.

PETURSSON, H. AND LADER, M.H. (1981*a*). Withdrawal reaction from clobazam. *Br. med. J.* **282**, 1931–2.

—— AND —— (1981*b*). Withdrawal from long-term benzodiazepine treatment. *Br. med. J.* **283**, 643–5.

PLETSCHER, A., SHORE, P.A., AND BRODIE, B.B. (1956). Serotonin as a mediator of reserpine action in brain. *J. Pharmacol. exp. Ther.* **116**, 84–9.

POTAMIANOS, G. AND KELLETT, J.M. (1982). Anticholinergic drugs and memory: the effects of benzhexol on memory in a group of geriatric patients. *Br. J. Psychiat.* **140**, 470–2.

PRADHAN, S.N. AND DUTTA, S.N. (1971). Central cholinergic mechanism and behavior. *Int. Rev. Neurobiol.* **14**, 173–231.

PRESCOTT, L.F., (1975). Anti-inflammatory analgesics and drugs used in the treatment of rheumatoid arthritis and gout. In *Meyler's side effects of drugs*, Vol. VIII (ed. M.N.G. Dukes), pp. 207–40. Excerpta Medica, Amsterdam.

QUARTON, G.C., CLARK, L.D., COBB, S., AND BAUER, W. (1955). Mental disturbances associated with ACTH and cortisone: a review of explanatory hypotheses. *Medicine, Baltimore* **34**, 13–50.

RATNA, L. (1981). Addiction to temazepam. *Br. med. J.* **282**, 1837–8.

REYNOLDS, E.H. (1968). Mental effects of anticonvulsants and folate metabolism. *Brain* **91**, 197–214.

——, CHANARIN, I., MILNER, G., AND MATTHEWS, D.M. (1966). Anticonvulsant therapy, folic acid and vitamin B_{12} metabolism and mental symptoms. *Epilepsia* **7**, 261–70.

RIVERA-CALIMLIN, L. (1976). Impaired absorption of chlorpromazine in rats given trihexyphenidyl. *Br. J. Pharmacol.* **56**, 301–5.

——, CASTANEDA, L., AND LASAGNA, L. (1973). Effects of mode of management of plasma chlopromazine in psychiatric patients. *Clin. Pharmacol. Ther.* **14**, 978–86.

ROBERTS, J.B. AND BRECKENRIDGE, A.M. (1975). Drugs affecting autonomic functions. In *Meyler's side effects of drugs*, Vol. VIII (ed. M.N.G. Dukes), pp. 304–30. Excerpta Medica and American Elsevier, New York.

ROBINS, A.H. (1976). Depression in patients with parkinsonism. *Br. J. Psychiat.* **128**, 141–5.

ROSS, W.D. AND SHOHTON, M.C. (1983). Specificity of psychiatric manifestations in relation to neurotoxic chemicals. *Acta psychiat. scand.* (Suppl. 303), 100–4.

ROTTANBURG, D., ROBINS, A.H., BEN-ARIE, O., TEGGIN, A., AND ELK, R. (1982). Cannabis-associated psychosis with hypomanic features. *Lancet* **ii**, 1364–6.

RUSSELL, J.W. AND SCHUCKIT, M.A. (1982). Anxiety and depression in patients on nadolol. *Lancet* **ii**, 1286–7.

SACHS, O. (1973). *Awakenings*. Duckworth, London.

SACK, R.L. AND GOODWIN, F.K. (1974). Inhibition of dopamine-β-hydroxylase in manic patients. *Arch. gen. Psychiat.*, **31**, 649–54.

SATO, M., CHEN, C.C., AKIYAMA, K., AND OTSUKI, S. (1983). Acute exacerbation of paranoid psychotic state after long-term abstinence in patients with previous methamphetamine psychosis. *Biol. Psychiat.* **18**, 429–40.

SCHENTAG, J.J., CALLERI, G., ROSE, J.Q., CERRA, F.B., DE GLOPPER, E., AND BERNHARD, H. (1979). Pharmacokinetic and clinical studies in patients with cimetidine-associated mental confusion. *Lancet* **i**, 177–81.

SCHÖPF, J. (1983). Withdrawal phenomena after long-term administration of benzodiazepines. A review of recent investigations. *Pharmacopsychiatry* **16**, 1–8.

SHEPHERD, M., LADER, M., AND RODNIGHT, R. (1968). Antidepressant drugs. In *Clinical psychopharmacology*. pp. 126–56. English Universities Press, London.

SHEPPARD, G.P. (1979). High dose propranolol in schizophrenia. *Br. J. Psychiat.* **134**, 470–6.

SHOPSIN, B. AND GERSHON, S. (1973). Pharmacology-toxicology of the lithium ion. In *Lithium its role in psychiatric research and treatment* (ed. S. Gershon, and B. Shopsin), pp. 107–46. Plenum Press, New York.

——, HIRSCH, J., AND GERSHON, S. (1975). Visual hallucinations and propranolol. *Biol. Psychiat.* **10**, l05–7.

SILVERSTONE, P.H. AND EPSTEIN, C.M. (1984). Ranitidine and confusion. *Lancet* **i**, 1071.

SIMON, P., LECRUBIER, Y., JOUVENT, R., PUECH, A.J., ALLILAIRE, J.F., AND WIDLÖCHER, D. (1978). Experimental and clinical evidence of the antidepressant effect of a beta-adrenergic stimulant. *Psychol. Med.* **8**, 335–8.

SIMPSON, G.M., COOPER, T.B., BARK, N., SUD, I., AND LEE, J.H. (1980). Effect of antiparkinsonian medication on plasma levels of chlorpromazine. *Arch. gen. Psychiat.* **37**, 205–8.

SINGH, M.M. AND KAY, S.R. (1975). Therapeutic reversal with benztropine in schizophrenics. *J. nerv. ment. Dis.* **160**, 258–66.

—— AND SMITH, J.M. (1973). Reversal of some therapeutic effects of an antipsychotic agent by an anti-parkinsonism drug. *J. nerv. ment. Dis.* **157**, 50–8.

SKEGG, D.C.G., DOLL, R., AND PERRY, J. (1977). Use of medicines in general practice. *Br. med. J.* **1**, 1561–3.

——, RICHARDS, S.M., AND DOLL, R. (1979). Minor tranquillisers and road accidents. *Br. med. J.* **1**, 917–19.

SNAITH, R.P. AND McCOUBRIE, M. (1974). Antihypertensive drugs and depression. *Psychol. Med.* **4**, 393–8.

SNOW, S.S., LOGAN, T.P., AND HOLLENDER, M.H. (1980). Nasal spray 'Addiction' and psychosis: a case report. *Br. J. Psychiat.* **136**, 297–9.

SNYDER, S.H. (1973). Catecholamines in the brain as mediators of amphetamine psychosis. In *Annual review of the schizophrenic syndrome*, Vol. 3 (ed. R. Cancro), pp. 137–66. Brunner/Mazal, New York.

——, BANERJEE, S.P., YAMAMURA, H.I., AND GREENBERG, D. (1974). Drugs, neurotransmitters and schizophrenia. *Science, N.Y.* **184**, 1243–53.

STAMP, T.C.B., ROUND, J.M., ROWE, D.J.F., AND HADDAD, J.G. (1972). Plasma levels and therapeutic effect of 25-hydroxycholecalciferol in epileptic patients taking anticonvulsant drugs. *Br. med. J.* **4**, 9–12.

STEEL, J.M. AND BRIGGS, M. (1972). Withdrawal depression in obese patients after fenfluramine treatment. *Br. med. J.* **3**, 26–7.

STEINERT, J. AND PUGH, C.R. (1979). Two patients with schizophrenic-like psychosis after treatment with beta-adrenergic blockers. *Br. med. J.* **1**, 790.

STEPHEN, S.A. (1966). Unwanted effects of propranolol. *Am. J. Cardiol.* **18**, 463–8.

STEVENS, J.R. (1973). An anatomy of schizophrenia? *Arch. gen. Psychiat.* **29**, 177–89.

STONE, M.H. (1973). Drug-related schizophrenic syndromes. *Int. J. Psychiat.* **11**, 391–437.

SULSER, F., VETULANT, J., AND MOBLEY, P.L. (1978). Mode of action of antidepressant drugs. *Biochem. Pharmacol.* **27**, 257–61.

TALWAR, P.P. (1979). A prospective, randomized study of oral contraceptives: the effect of study design on reported rates of symptoms. *Contraception* **20**, 329–37.

TAYLOR, E.A., CARROLL, D., AND JEFFERSON, D. (1979). CSF plasma ratios of propranolol and pindolol in man. *Br. J. clin. Pharmacol.* **8**, 381P–2P.

TAYLOR, M., GALLOWAY, D.B., PETRIE, J.C., DAVIDSON, J.F., GALLON, S.C., AND MOIR, D.C. (1978). Psychomimetic effects of pentazocine and dihydrocodeine tartrate. *Br. med. J.* **2**, 1198.

TOLLEFSON, G.D. (1981). Hyperadrenergic hypomania consequent to the abrupt cessation of clonidine. *J. clin. Psychopharmacol.* **1**, 93–5.

TOMSON, C.R., GOODSHIP, T.H.J., AND RODGER, R.S.C. (1985). Psychiatric side-effects of acyclovir in patients with chronic renal failure. *Lancet* **ii**, 385–6.

TOPLISS, D. AND BOND, R. (1977). Acute brain syndrome after propranolol treatment. *Lancet* **ii**, 1133–4.

TOTTE, J., SCHARPE, S., VERKERK, R., NEELS, H., VANHAEVERBEEK, M., SMITZ, S., AND ROUSSEAU, J.J. (1981). Neurological dysfunction associated with abnormal levels of cimetidine metabolite. *Lancet* **i**, 1047.

TRIMBLE, M. (1977). The relationship between epilepsy and schizophrenia: a biochemical hypothesis. *Biol. Psychiat.* **12**, 299–304.

—— AND REYNOLDS, E.H. (1976). Anticonvulsant drugs and mental symptoms: a review. *Psychol. Med.* **6**, 169–78.

TUNE, L.E., HOLLAND, A., FOLSTEIN, M.F., DAMLOUJI, N.F., GARDNER, T.J., AND COYLE, J.T. (1981). Association of postoperative delirium with raised serum levels of anticholinergic drugs. *Lancet* **ii**, 651–2.

TURNER, P. (1979). Central nervous actions of beta-adrenoceptor blocking drugs in man. *Trends Pharmacol. Sci.* October, 49–51.

TYRER, P., OWEN, R., AND DAWLING, S. (1983). Gradual withdrawal of diazepam after long-term therapy. *Lancet* **i**, 1402–6.

——, RUTHERFORD, D., AND HUGGETT, T. (1981). Benzodiazepine withdrawal symptoms and propranolol. *Lancet* **i**, 520–2.

UNGERLEIDER, J.T. AND FRANK, I.M. (1976). Management of acute panic reactions and flashbacks resulting from LSD ingestion. In *Acute drug abuse emergencies* (ed. P.G. Bourne), pp. 133–8. Academic Press, London

VAN DER KROEF, C. (1979). Reactions to triazolam. *Lancet* **ii**, 526.

VAN KAMMEN, D.P. BUNNEY, W.E., DOCHERTY, J.P., MARDER, S.R., EBERT, M.H., ROSENBLATT, J.E., AND RAYNER, J.N. (1982). d-Amphetamine-induced heterogeneous changes in psychotic behavior in schizophrenia. *Am. J. Psychiat.* **139**, 991–7.

VAN PRAAG, H.M. (1968). Abuse of, dependence on and psychosis from anorexigenic drugs. In *Drug-induced diseases,* Vol. 3 (ed. L. Meyler, and H.M. Peck), pp. 281. Excerpta Medica, Amsterdam.

VON HARTITZSCH, B., HOENICH, N.A., LEIGH, R.J., WILKINSON, R., FROST, T.H., WEDDEL, A., AND POSEN, G.A. (1972). Permanent neurological sequelae despite haemodialysis for lithium intoxication. *Br. med. J.* **4**, 757–9.

WEINER, S., DORMAN, D., PERSKY, H., STACH, T.W., NORTON, J., AND LEVITT, E.E. (1963). Effect on anxiety of increasing the plasma hydrocortisone level. *Psychosom. Med.* **25**, 69–77.

WHITE, A.C. AND MURPHY, T.J.C. (1977). Hallucinations caused by bromocriptine. *Br. J. Psychiat.* **130**, 104.

WISTEDT, B., WILES, D., AND KOLAKOWSKA, T. (1981). Slow decline of plasma drug and prolactin levels after discontinuation of chronic treatment with depot neuroleptics. *Lancet* **i**, 1163.

WITTENBORN, J.R. (1979). Effects of benzodiazepines on psychomotor performance. *Br. J. clin. Pharmacol.* **7**, 61S–7S.

WOLF, M.E., ALMY, G., TOLL, M., AND MOSNAIM, A.D. (1982). Mania associated with the use of baclofen. *Biol. Psychiat.* **17**, 757–9.

WOLF, S.M. AND DAVIS, R.L. (1973). Permanent dementia in idiopathic parkinsonism treated with levodopa. *Arch. Neurol.* **29**, 276–8.

WOLFE, H.G. AND CURRAN, D. (1935). Nature of delirium and allied states: the dysergasic reaction. *Arch. Neurol. Psychiat.* **33**, 1175–215.

WOOD, A.J.J., MOIR, D.C., CAMPBELL, C., DAVIDSON, J.F., GALLON, S.C., HENNEY, E., AND McALLION, S. (1974). Medicine evaluation and monitoring group: central nervous system effects of pentazocine. *Br. med. J.* **1**, 305–7.

WYATT, R.J., TERMINI, B.A., AND DAVIS, J. (1971). Biochemical and sleep studies of schizophrenia: a review of the literature — 1960-1970. *Schizophren. Bull.* **4**, 9–66.

YORKSTON, N.J., GRUZELIER, J.H., ZAKI, S.A., HOLLANDER, D., PITCHER, D.R., AND SERGEANT, H.G.S. (1977). Propranolol as an adjunct to the treatment of schizophrenia. *Lancet* **ii**, 575–8.

29 Impaired performance

ANTHONY N. NICHOLSON

Introduction

Over the past few years there has been increasing interest in the way in which drugs impair performance. This interest has arisen because some drugs may impair day-to-day skills such as car driving although for many years a degree of impairment in car drivers either through the use of alcohol or the use of drugs has been tacitly tolerated, though it is generally agreed to be unacceptable. However, there is a more serious interest in the safe use of drugs by those whose occupations demand vigilance and motor skill, who are involved in decision-making or for whom interpersonal relations are crucial. This involves a much stricter approach, and for many years, at least in certain occupations, e.g. airline pilots, where impaired performance could be a danger to others, the use of any drug has precluded employment.

Recent advances in therapeutics and a greater understanding of drug action in man has made this rather uncomplicated view of life less tenable. There is now an increasing desire that advances in therapy should, if at all possible, be available to everyone. In this way the adverse effect which a drug may have on performance is an important aspect of its clinical profile. Hypnotics for transient insomnia which may arise from the irregularity of rest inherent in some occupations need to be free of residual effects, antihistamines which are sedative must be avoided, and beta-adrenoceptor antagonists used in the management of mild hypertension, often in early middle age, must be as free as possible from central effects. It must be emphasized that these are but a few examples of drugs which are used by active, healthy or nearly healthy, individuals.

Drugs with central effects are also used in the treatment of anxiety and depression, to prevent impairment of consciousness in patients with epilepsy, and to relieve chronic pain. These patients, because of their illness, may be excluded from certain occupations, but there is still the need to ensure that treatment itself is as free as possible from adverse effects on the nervous system. It is clear that the issues involved in the safe use of a particular drug by a particular individual are complex and, as with all aspects of therapy, it is usually necessary to balance efficacy and adverse effects.

This chapter will discuss the issues involved in deciding whether a certain drug or drugs can be used safely. In this context, antihistamines and hypnotics will be considered in detail as they present examples of the two main problems which arise — whether individuals can work safely under the influence of a drug (antihistamines), and how long after ingestion of a drug which impairs performance is it safe to carry out skilled work (hypnotics). The aim is to provide a framework for discussion rather than specific recommendations of drug use, and to outline various approaches which may be adopted to ensure that a drug can be used by those involved in skilled work.

Methodology

Broadly speaking, there are two approaches to studying performance and thus to predicting the effect which a drug may have on the day-to-day life of an individual. A profile of the activity of a drug can be built up using a variety of laboratory tests directed towards assessing specific skills relevant to the individual's work; or the skill itself may be simulated with as much accuracy as possible. In the two approaches there are common considerations of methodology, and these will be considered initially.

Experimental design

It is essential in all drug studies that dose- and time-response data are obtained, and that the dose range is relevant to the projected therapeutic use. In the case of psychoactive drugs it may also be appropriate to include a dose which, though outside the anticipated range, is high enough to impair performance, as this would give an indication of the margin of safety. Dose- and time-related studies have the advantage that the analysis of the data is more likely to detect change than analysis of single assessments, and that variations in performance related to the circadian activity of the individual are taken into account. Measurements must, of course, be related to ingestion of placebo, and an active control should be used whenever possible to ensure that absence of impaired performance is not due to the relative insensitivity of the testing procedures.

A large variety of tests is used in the assessment of impaired performance. These include the deceptively simple paper-and-pencil tests, such as digit symbol substitution (Fig. 29.1), tests which assess memory and attention, and those which assess psychomotor skill. Tests also relate to specific senses such as vision. Those which measure a neurological entity such as body sway,

Fig. 29.1 The digit symbol substitution test. A series of one hundred different sheets each with two hundred randomized digits (0–9) arranged in 10 rows is presented. Under each digit there is a space where the subjects are required to write the appropriate symbol indicated by a code at the top of the page. The code is different for each of the 100 sheets. In each session subjects are given two sheets, and two minutes, timed separately, to complete as many spaces as possible for each sheet. In all tests and for all subjects errors are extremely rare, and so only the number attempted are analysed. It is important that subjects are trained on the test until they reach steady performance.

eye movement, critical flicker fusion, the electrical activity of the brain, and, more recently, drowsiness (by measuring sleep latencies during the day) are also used. These avoid the problems of measuring performance and so may be appropriate when there are difficulties in the assessment of performance, as in the elderly. However, more needs to be known about the significance of changes in such functions to the overall capability of the individual before they can be used to indicate the degree of impaired performance.

In the choice of tests there are two main issues — whether information is needed on the skill impaired and whether the persistence of effect needs to be determined. This is not the place to review the relative roles of the many tests which are available, but, if information on persistence of effect is sought, i.e. any evidence of modified central activity no matter how minimal, accurate and well designed studies with pencil and paper have proved reliable, and observations on psychomotor skills related to the peripheral nervous or oculomotor systems are useful.

Inevitably a question arises concerning the relevance of performance tests carried out in the laboratory to the day-to-day work of the individual. At first sight simulation is a more attractive approach, but there are serious doubts as to whether simulation itself is specially relevant to the real situation. Indeed, the question must be asked whether studies using simulators (including car driving tests) are as useful as laboratory tests in providing reliable information on drug effects. It is beyond discussion that laboratory studies provide accurate information on discrete skills and on persistence of effect, but the place of simulation needs to be carefully assessed.

Simulation

Simulation of the occupation or day-to-day skill of the individual may bring increased reality and motivation to the subject's participation, but simulation may nevertheless lack sensitivity in the measurement of performance itself. The sensitivity of a simulation to psychoactive drugs must always be established. An absence of an effect is of limited value unless it is known at what dose of that drug, or of a similar drug, performance would be impaired. Similarly, it is of limited usefulness to know that a particular dose leads to impaired performance as the clinician needs information on efficacy and adverse effects spread over the whole of the therapeutic dose range. Dose- and time–response data, as in all studies with drugs, are essential. Clearly, uncertain or insensitive measures obtained with simulation have no advantage over accurate measures from the laboratory. It must also be realized that simulators, including car driving tests, are often testing isolated functions in a complex, expensive, and uncontrolled way. There are many factors which influence measurement in a complex situation, and these, together with the inherent variability of the situation, can lead to difficulties in establishing a drug effect.

It is, indeed, important that a spurious confidence in the use of simulation and in car-driving tests does not arise. Car accidents are only seldom related to loss of control, and it is far more important to study the decision as to whether a specific manoeuvre is possible than the ability to carry out accurately the manoeuvre itself, and to study the ability to cope with an unexpected situation than the skill involved in negotiating stationary obstacles. It is essential that the more subtle effects of drugs on man are borne in mind. Effects on decision-making and on the behavioural integrality of man have yet to be adequately explored, and it is highly likely that the impairment of such skills has a wide and far-reaching importance to the well-being of the individual and of society.

However, it must be emphasized that there can be a measure of agreement between careful studies using simulation and similar studies with laboratory tests, and that it is the contention that laboratory tests do not provide such useful information as simulation or car driving that causes most dispute. Laboratory tests measure skills which need to be preserved and which, in the case of car accidents, may be identified from epidemiological studies, but with simulation careful assesment of its relation to the overall task in question is needed.

It would be useful if a simple test, used widely in studies on the central effects of drugs, were included in all studies. Information with a relatively simple task known to be sensitive to psychoactive drugs, such as digit symbol substitution, would allow comparative information to be built up between centres using laboratory tests and simulation. For the moment laboratory tests must be preferred in the analysis of drug effects as they are known to provide reliable information on the nature of the skill impaired and on persistence of effect, and do not require expensive equipment.

Persistence of activity and residual effects

The approach used to explore the effects of a drug on performance must relate to the way in which it is to be used. It may be relevant to establish the residual effects of a drug, or it may be important to establish whether it is safe to work while the drug is acting. If persistence of activity is the issue, then hypnotics are a useful example as they are used overnight, and the question is whether it is safe to carry out skilled work the next day. Duration of action depends on absorption, distribution into the tissues, and elimination, not only of the parent drug but also of its metabolites, and so close agreement of pharmacokinetic data and pharmacodynamic studies is necessary before recommendations which relate to safe use can be made with confidence.

Pharmacokinetics

Slow absorption leads not only to slow onset but also to a somewhat longer duration of action (Fig. 29.2), and this can be seen pharmacodynamically in studies with oxazepam and diazepam (Fig. 29.3). The profile of impaired performance with diazepam is different from that of oxazepam (Nicholson 1979a). With diazepam maximum impairment of performance is seen around half an hour after ingestion and is followed by rapid recovery, but with oxazepam, which is more slowly absorbed, performance is not impaired at that time, but is impaired around 2 to 5 hours after ingestion and recovery is somewhat slower.

However, variations in formulation may modify absorption, and such variations have been used for drugs which are slowly absorbed. A particular example is the group of 3-hydroxy 1,4-benzodiazepines. Peak plasma levels tend to be delayed until 2 to 3 hours after ingestion, but absorption can be quickened by a change in formulation and may reduce the absorption half-life by about half. The pharmacodynamic correlate of this change is seen in studies on performance. The immediate effect on performance using a soft gelatine capsule containing a

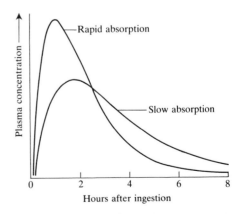

Fig. 29.2 Rate of absorption alters the profile of the plasma concentration of a drug and thus its persistence of activity.

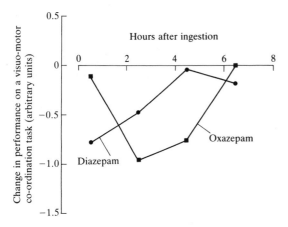

Fig. 29.3 Effect of diazepam (quickly absorbed) and oxazepam (slowly absorbed) on performance. With diazepam (10 mg) there is an immediate impairment of performance, but with oxazepam (30 mg) impairment is delayed.

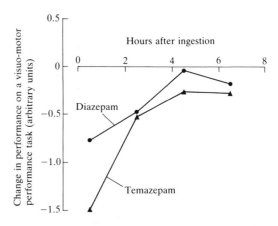

Fig. 29.5 Impaired performance and its recovery with diazepam (10 mg) and temazepam (20 mg) are similar. With both drugs impaired performance relates to the distribution phase.

solution of the drug may be much more pronounced than that with a wet granulation tablet, and so any lack of effect of the tablet on sleep onset latencies may be overcome, and undue persistence of effect avoided (Pierce *et al* 1984).

The importance of the distribution phase to duration of action is not widely appreciated. If the minimum concentration for an effect is within the distribution phase, then even a drug with a relatively long elimination half-life will have as short a duration of action as a drug which is rapidly eliminated (Fig. 29.4). This can be shown by studies on performance with diazepam which has an elimination half-life of around 40 hours and temazepam

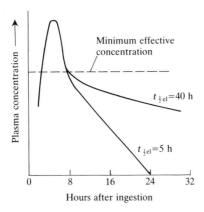

Fig. 29.4 Drugs with different rates of elimination may have the same duration of action if the minimal concentration for an effect is related to the distribution phase.

which has an elimination half-life of around 10 hours (Nicholson 1979*a*). Recovery from impaired performance with each drug is rapid as the minimum effective concentration is related to the distribution phase (Fig. 29.5). It is only when we have to consider repeated daily dosage that the difference between diazepam and temazepam is clinically relevant. Under these circumstances ingestion of diazepam will lead to accumulation of the parent drug (and of its metabolite), and so to a daytime anxiolytic effect which, however, may itself be useful.

Influence of elimination

An area of current interest is that of the rapidly eliminated hypnotics. These drugs are free of accumulation with daily ingestion, but persistence of their activity varies according to their elimination. For instance, midazolam has a marked distribution phase and an elimination half-life of around two hours with a pharmacodynamic profile very similar to that seen with temazepam (Fig. 29.6). The reason is that the rapid elimination of midazolam leads to a plasma decay similar to that seen during the distribution phase of temazepam. Clinically both are useful for sleep onset insomnia.

The fall in the plasma level of hypnotics which are rapidly eliminated and adequately absorbed relates essentially to the elimination half-life; thus with hypnotics which have mean elimination half-lives of ca. 4–5 hours, e.g. brotizolam, the fall in plasma concentration is clearly slower than that of midazolam. It is, therefore, of interest that the profile of impaired performance with brotizolam is different from that of

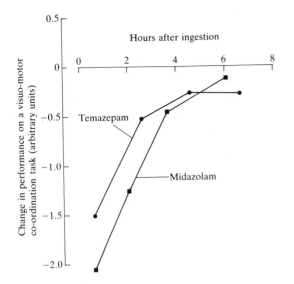

Fig. 29.6 Impaired performance and its recovery with temazepam (20 mg) and midazolam (20 mg) are similar. Impaired performance relates to the distribution phase of temazepam and the elimination of midazolam.

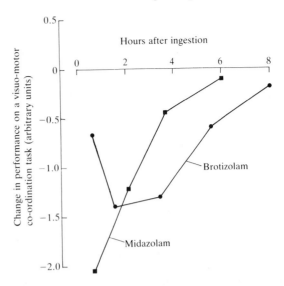

Fig. 29.7 Impaired performance with midazolam (20 mg) and brotizolam (0.4 mg). Both drugs are rapidly eliminated, but the elimination half-life of brotizolam (4.5 hours) leads to a ∪-shaped curve, whereas that of midazolam (2 hours) leads to a rapid recovery. Midazolam in a dose of 15 mg is recommended for sleep-onset insomnia, whereas brotizolam with a maximum dose of 0.25 mg is also useful for sustaining sleep and is eliminated rapidly enough to be free from residual effects.

midazolam. Brotizolam is absorbed adequately and performance is impaired around 0.5 h after ingestion, but the subsequent profile is U-shaped (Fig. 29.7). This is consistent with the use of brotizolam for frequent nocturnal awakenings; at the same time elimination is sufficiently rapid for it to be free from residual effects the next day (Nicholson et al. 1980). Slower rates of elimination would not preserve this useful profile and residual effects would be increasingly likely.

Finally, we have hypnotics with persistent activity during the day after overnight ingestion and these are useful in the management of insomnia with daytime anxiety. An example is potassium chlorazepate which is absorbed as nordiazepam. It has a useful hypnotic profile with a steady anxiolytic effect the next day related to the long elimination half-life of nordiazepam. Potassium chlorazepate has the added advantage that nordiazepam has little effect on performance, and so its prolonged anxiolytic effect is unlikely to be accompanied by significant residual sequelae (Borland and Nicholson 1976). However, other hypnotics with slow elimination may lead to residual effects.

Metabolites are also important in understanding the activity of some benzodiazepines, and flurazepam hydrochloride is of interest. The parent compound has a very short half-life and the clinical usefulness of flurazepam is due largely to its metabolites. Recovery of performance after flurazepam occurs around 16 hours

after ingestion (Borland and Nicholson 1975), and so may well be related to the activity of the hydroxyethyl metabolite which has a distribution phase with a half-life of around 2 hours and an elimination half-life of around 16–20 hours. However, though the hypnotic effect of flurazepam may well be related to the hydroxyethyl metabolite, the effect of the drug which develops with repeated ingestion is likely to be related to the desalkyl metabolite which has a long elimination half-life.

Summary

It is essential that in the consideration of the persistent effects of a drug, absorption, distribution, and elimination of both the parent drug and of its metabolites are taken into consideration. Familiarity with the pharmacokinetic profile and with the pharmacodynamic effects will lead to more skilful use of such drugs, and avoid the possibility of impaired daytime performance.

Attenuation of central effects

Hypnotics may be used safely by those involved in skilled

work when their effect does not persist beyond the sleep period. Their effect precedes work, but most drugs are used over the period of work itself, and so may adversely affect performance. An example is the antihistamines which are used for their peripheral anti-allergic properties though they often lead to drowsiness and impaired performance. This has always been a problem with the clinical use of antihistamines and various approaches, which are of general interest, have been adopted to minimize their central effects.

Drowsiness with the H_1-antihistamines has been attributed to various mechanisms such as inhibition of histamine N-methyltransferase and blockade of central histaminergic receptors, though serotonergic antagonism, anticholinergic activity, and blockade of central alpha-adrenoreceptors may also be involved. However, whatever may be the mechanism, central effects are dependent on the ability of a particular drug to cross the blood–brain barrier, and most H_1-antihistamines pass the blood–brain barrier with ease.

Nevertheless, the central effects of the antihistamines vary with respect to severity and persistence. With triprolidine there are immediate effects on performance which persist for several hours (Nicholson 1979b), and with clemastine and promethazine decrements are delayed, though equally marked. With chlorpheniramine impairments are limited and appear shortly after ingestion (Fig. 29.8). The effect of these drugs also vary from one individual to another, and so careful choice may be a useful initial approach to minimise sedation (Clarke and Nicholson 1978).

Sustained release and selective occupancy

Another approach has been the overnight ingestion of sustained release compounds. Impaired performance may appear a little later with such formulations, but recovery occurs within a few hours, and so if ingested overnight performance would not persist beyond the sleep period (Fig. 29.9). Sustained release preparations, such as brompheniramine maleate and triprolidine hydrochloride may therefore be useful if their clinical activity persisted through the next day (Nicholson 1979). At least, both wheal and flare measurements show an antihistaminic effect up to 24 h after triprolidine hydrochloride, and there is pharmacokinetic evidence of maintained blood levels with brompheniramine maleate.

Another approach which may minimize sedation — particularly with low doses — is the use of a drug with greater affinity for peripheral H_1-receptors than for central H_1-receptors. This has been suggested as an explanation for the absence of sedative effects of mequitazine at doses which are, nevertheless, optimum for its peripheral activity. At the normal therapeutic dose mequitazine would appear to be free of sedation, and there is no consistent effect of 5 mg mequitazine on performance. However, with the higher dose of 10 mg (double the recommended dose) performance is impaired (Nicholson and Stone 1983).

Blood–brain barrier

It is, however, with antihistamines which have considerable difficulty in crossing the blood–brain barrier that

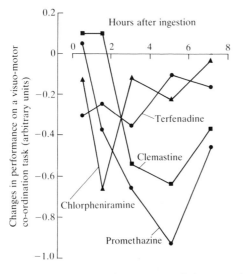

Fig. 29.8 Impaired performance with four antihistamines (chlorpheniramine 4 mg, clemastine 1 mg, promethazine 10 mg, and terfenadine 60 mg).

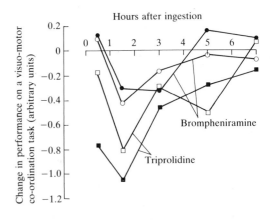

Fig. 29.9 Impaired performance with sustained release formulations of triprolidine hydrochloride (10 mg) and brompheniramine maleate (12 mg) compared with usual formulations (2.5 and 4.0 mg, respectively). Open symbols refer to sustained release formulations.

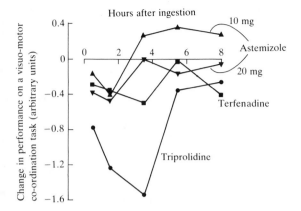

Fig. 29.10 Absence of effects of terfenadine (60 mg) and astemizole (10 and 20 mg) on performance compared with the effect of triprolidine hydrochloride (10 mg — sustained release).

current interest is centred. Such drugs may be free of impaired performance as tolerance would be able to develop gradually without any immediate central effects, and this possibility has prompted the search for such compounds as terfenadine (60–120 mg) and astemizole (10–20 mg). Studies on visual–motor co-ordination (Fig. 29.10), on skills such as arithmetical ability and digit symbol substitution, on tests of central activity such as critical flicker fusion, and on visual function have demonstrated freedom of central effects in the doses cited (Nicholson and Stone 1982): Nicholson *et al.* 1982).

It must be realized that, though recommended doses of H_1-antihistamines are used in various studies on performance and central effects, they may not necessarily produce equivalent histamine antagonism or even possess equivalent clinical efficacy. Further, tolerance of the central nervous system to some antihistamines may develop quite quickly, and that would be useful clinically. Indeed, with sustained-release preparations overnight there is impaired performance for only the initial part of its peripheral effect, and there is even evidence of some tolerance with repeated ingestion of triprolidine, which is a potent antagonist.

Though many H_1-antagonists modulate central nervous function and impair performance, there would be less likelihood that the less liposoluble H_2-antagonists would have such effects. Indeed, with healthy individuals there is no evidence of impaired performance with cimetidine (200–400 mg), though it has led to drowsiness in patients with impaired renal function and liver disease. This is of interest as cimetidine would appear to increase slow-wave sleep even in healthy individuals. There is, therefore, the possibility that even with the less lipo-

soluble H_2-antagonists there may under some circumstances be central effects.

A similar approach to that with the H_1-antagonists has been used in evaluating the H_2-antagonists. There is no consistent evidence of adverse changes in performance, central nervous function, or subjective assessments of mood (Nicholson and Stone 1984). Oral ingestion of cimetidine (200–400 mg) or ranitidine (150–300 mg) in individuals free of renal or hepatic disease is highly unlikely to lead to impaired performance, and both may be used by those involved in skilled activity.

Summary

It is clear that sedation and/or impaired performance need not be an inevitable accompaniment of the clinical use of H_1-antihistamines. Terfenadine is clearly a useful antihistamine (Nicholson 1982). In a similar manner astemizole is promising, but it is slowly eliminated with a half-life of many days. Further information on the significance of this property is needed, and there is uncertainty concerning the appropriate dose regime. Mequitazine would appear to have a therapeutic window around 5 mg with useful anti-allergic activity free of central activity. With the currently available H_2-antagonists, clinical effects other than those on performance may have to be taken into consideration, but impairment of central function is of minimal, if any, relevance.

Interpretation

When the presence or absence of impaired performance has been established, the findings need careful interpretation. Impaired performance not only implies impairment of a particular skill, but also that the central nervous system is modified so that other skills which are less obvious and less easily measured may also be affected. For instance, impaired skill on a co-ordination task may also imply an impairment of the ability to anticipate a response. The implications of such a possibility may be very relevant to the individual. Further, it must be appreciated that the inability to demonstrate impaired performance does not necessarily mean a drug is free of adverse effects, as there is no test or group of tests which indicates whether human performance *in toto* is preserved.

It must also be appreciated that much of the data on impaired performance with drugs is obtained in young healthy adults, that the effects of psychoactive drugs may vary with age and gender and the ability of the individual to metabolize and excrete the drug, and that effects may be enhanced by the concomitant use of other drugs.

There is clear evidence that the elderly are more sensitive to psychoactive drugs, and for this reason it is usually recommended that they should commence therapy with half the normal dose. Impairment of the ability of the individual to metabolize a drug, e.g. in hepatic disease, and to excrete the drug, e.g. renal failure, must also be taken into consideration.

It is unfortunate that much of the work on drugs does not provide very useful information on their relative merits. Quite frequently considerations other than those of clinical pharmacology prevail in the design, analysis, and even interpretation of the data. Studies are more often than not designed to maximize the effectiveness or minimize the adverse effects of a particular compound, and this had led to much confusion, not least amongst those who encourage and carry out such trials. It is for this reason that careful attention must be paid to the design of studies, and publications must not be accepted on their face value.

There are no simple answers to the questions which are raised by performance studies with drugs. The physician must be wary of studies which claim absence of performance deficits, and interpret studies which claim to show impairments with caution. It is important that the most sensitive techniques are used. Ideally, a drug in which impaired performance of any nature cannot be demonstrated in an adequately designed experiment is required. However, other things being equal, limited impairment of performance may have to be accepted, and the drug which is least likely to impair performance may have to be chosen.

General considerations

Though rapidly eliminated hypnotics and antihistamines which cross the blood–brain barrier with difficulty are useful, problems remain with many other drugs. Particular difficulties arise with psychoactive drugs, such as antidepressants and anxiolytics, which are used for their daytime effects. Of course, in patients with disorders of mood, treatment itself may lead to an improvement in performance, but the possibility of impairment with these drugs most certainly arises if they have sedative effects. The emergence of antidepressants free of sedation has been helpful, and some of the newer anxiolytics may have less sedative effects than others.

There is, however, increasing concern that some drugs, such as the beta-adrenoceptor antagonists, may have central effects of a subtle nature. The safe use of these drugs may well rest with compounds which cross the blood–brain barrier with difficulty, but the question arises whether tolerance to drugs of this group, even if they cross the blood–brain barrier very slowly, develops in the same way as it would appear to do so with other centrally active drugs. Much more needs to be known about the possible central effects of drugs such as the beta-adrenoceptor antagonists, and care must be taken in their use.

Most certainly a cautious approach to therapeutics is essential in the proper management of those who are engaged in occupations where impaired performance would not be acceptable. In this context the possibility that the individual may respond adversely with any drug must always be excluded, even for those drugs free of any experimental evidence of a central effect. Nevertheless, treatment has often to be decided on the basis of that which is least likely to cause harm, and it is fortunate that there are often many drugs available which, though they may have similar therapeutic efficacy, have different effects on performance. Choice is an important approach, and the solution of any one problem demands the ingenuity of the pharmaceutical industry and careful assessment by clinical pharmacologists.

RECOMMENDED FURTHER READING

NICHOLSON, A.N. (1983). Antihistamines and sedation. *Lancet* ii, 211–12.

—— AND MARKS, J. (1983). *Insomnia. A guide for medical practitioners*. MTP Press Ltd, Lancaster, England.

—— AND WARD, J. (Ed.) (1984). Psychomotor drugs and performance. Supplement to *Br. J. clin. Pharmacol.*

WILLETTE, R.E. AND WALSH, J.M. (Ed.) (1983). *Drugs, driving and traffic safety*. World Health Organization, Geneva, Switzerland.

REFERENCES

BORLAND, R.G. AND NICHOLSON, A.N. (1975). Comparison of the residual effects of two benzodiazepines (nitrazepam and flurazepam hydrochloride) and pentobarbitone sodium on human performance. *Br. J. clin. Pharmacol.* **2**, 9–17.

—— AND —— (1976). Residual effects of potassium clorazepate, a precursor of nordiazepam. *Br. J. clin. Pharmacol.* **4**, 86–9.

CLARKE, C.H. AND NICHOLSON, A.N. (1978). Performance studies with antihistamines. *Br. J. clin. Pharmacol.* **6**, 31–5.

NICHOLSON, A.N. (1979a). Performance Studies with diazepam and its hydroxylated metabolites. *Br. J. clin. Pharmacol.* **8**, 39–42S.

—— (1979b). Effect of the antihistamines brompheniramine maleate and triprolidine hydrochloride, on performance in man. *Br. J. clin. Pharmacol.* **8**, 321–24.

—— (1982). Antihistaminic activity and central effects of terfenadine. A review of European studies. *Arzneimittel-Forsch.* **32**, 1191–3.

——, SMITH, P.A., AND SPENCER, M.B. (1982). Antihistamines and visual function: Studies on dynamic acuity and the pupillary response to light. *Br. J. clin. Pharmocol.* **14**. 683–90.

—— AND STONE, B.M. (1982). Performance studies with the H₁-receptor antagonists, astemizole and terfenadine. *Br. J. clin. Pharmacol.* **13**, 199–202.

—— AND —— (1983). The H₁-antagonist mequitazine: Studies on performance and visual function. *Eur. J. clin. Pharmacol.* **25**, 563–6.

—— AND —— (1984). The H₂-antagonists, cimetidine and ranitidine: Studies on performance. *Eur. J. clin. Pharmacol.* **26**, 579–82.

——, ——, AND PASCOE, P.A. (1980). Studies on sleep and performance with a triazolo-1, 4 thienodiazepine (brotizolam). *Br. J. clin. Pharmacol.* **10**, 75–81.

PIERCE, D.M., FRANKLIN, R.A., HARRY, T.V.A., AND NICHOLSON, A.N. (1984). Pharmacodynamic correlates of modified absorption: Studies with lormetazepam. *Br. J. clin. Pharmacol.* **18**, 31–5.

30 Drug-induced disorders of the eye

P.F. D'ARCY

Ocular toxicity

Ocular structures may exhibit the earliest evidence of toxicity of drugs used in general medicine. Ocular complications of drugs and iatrogenic eye disease were extensively reviewed by Perkins (1965), Leopold (1968), Marzulli (1968), *British Medical Journal* (1969), Cant (1969), Grant (1969), and in children by Mushin (1972). Blake (1975) reviewed chemical damage to the eye; Davidson (1980) drug-induced disorders of the eye; Burstein (1980) corneal cytotoxicity of topically applied drugs, vehicles, and preservatives; Lien and Koda (1981) glaucoma and cataracts associated with drugs and chemicals; Crombie (1981) ocular complications of systemic drug therapy; Petursson *et al.* (1981) the pharmacology of ocular drugs: oral contraceptives; Olansky (1982) antimalarials and ophthalmologic safety; and Spiteri and James (1983) drug-induced ocular reactions. Common to all these reviews is the emphasis on detecting toxic changes as early as possible since at that stage the observed changes may be reversible.

All ocular structures are potentially vulnerable to drug toxicity; however, the most frequently affected structures are the conjunctiva, cornea, sclera, lens, retina, optic nerve, and extraocular muscles. Drugs can produce functional disturbance of accommodation and intraocular pressure and sometimes damage to the retina and optic nerve which is irreversible. Damage may progress even after the withdrawal of the implicated agent, as for example in chloroquine retinopathy. Local application of preparations to the eye may produce toxic symptoms and these may be due to the active principle or to a component of the formulation.

Drugs specifically implicated in the majority of reports of eye toxicity are shown in Table 30.1; these include the coronary vasodilator, amiodarone; anticholinesterase drugs; the anticonvulsants, carbamazepine and phenytoin; anti-infective agents; antimalarials; antiparkinson agents; some beta-blockers; cancer chemotherapy; the antidiabetic drug, chlorpropamide; the fertility drug, clomiphene citrate; corticosteroids; digitalis preparations; the solvent, dimethylsulphoxide; some non-steroidal anti-inflammatory drugs; oxygen; phenothiazine tranquillizers; the mydriatic, phenylephrine; psoralens; the traditional antiglaucoma agents, pilocarpine and adrenaline (epinephrine); the tricyclic antidepressants; and vitamin A deficiency. Discussion is given individually about these drugs or groups of drugs in this text.

Amiodarone

In a review of the literature between 1976 and 1981, Lubbe and Mercer (1982) assessed that 41 to 100 per cent of patients on the coronary vasodilator amiodarone (> 600 mg daily for more than three months) had corneal deposits which invariably resolved on drug discontinuation. Resolution could also be achieved by dosage at two- or three-day intervals.

In a study involving 37 patients (aged 24–76 years) treated with amiodarone, Chew *et al.* (1982) showed that 35 of these developed cornea verticillata about four weeks after starting treatment. None of the patients had impaired visual acuity, although one saw haloes around lights. One patient stopped amiodarone after 14 months and the keratopathy disappeared over nine months.

In a study on 27 patients, Kaplan and Cappaert (1982) demonstrated that the verticillate epithelial keratopathy caused by amiodarone can be classified as having three stages. Patients on low dosage (100–200 mg daily) retain clear corneas or show stage 1 changes only (coalescence of fine, punctate, greyish golden-brown opacities into a horizontal linear pattern in the inferior cornea) regardless of duration of treatment or total amount of drug ingested; patients receiving higher dosage of 400 to 1400 mg daily show stage 2 (additional arborizing and horizontal lines) and stage 3 (a verticillate, whorl-like pattern; arborizing lines may extend into the visual axis) changes, depending on the duration of treatment. This keratopathy progresses, even with reduced dosage; however, complete regression occurs once the drug is discontinued.

Ingram (1983) assessed ocular complications for three months to 10 years in 103 patients receiving long-term

Table 30.1   Drugs involved or implicated in oculotoxicity

Classification	Drug
Antibiotics and anti-infective agents	Broxyquinoline, chloramphenicol, clioquinol, (iodochlorhydroxyquin), hexachlorophene, ethambutol, rifampicin
Anticholinesterases	Demecarium bromide, dyflos, ecothipate iodide
Anticonvulsants	Carbamazepine (plus macrolide antibiotic), phenytoin sodium
Antidiabetic agents	Chlorpropamide
Antimalarials (non-malarial use)	Amodiaquine, chloroquine, hydroxychloroquine, mepacrine (quinacrine), quinine
Antileprotic drugs	Clofazimine
Antiparkinsonian drugs	Benzhexol, levodopa
Beta-blockers (systemic and topical)	Metoprolol, nadolol, oxprenolol, practolol, timolol
Cancer chemotherapy	Adriamycin, busulphan, chlorambucil, cisplatin, cyclophosphamide, fluorouracil, methotrexate, mithramycin, mitomycin C, nitrogen mustard, procarbazine, tamoxifen, vincristine
Chlolinergics (antiglaucoma agents (topical)	Pilocarpine
Coronary vasodilators	Amiodarone
Corticosteroids (systemic and topical)	Betamethasone, dexamethasone, flurandrenolone, hydrocortisone, medrysone, prednisolone, prednisone
Digitalis glycosides	Digitoxin, digoxin
Fertility drugs	Clomiphene citrate
Hypotensive agents (topical)	Guanethidine
Medicinal gases	Oxygen
Non-steroidal anti-inflammatory drugs (NSAIDS)*	Benoxaprofen (withdrawn), chloroquine, ibuprofen, indomethacin, phenylbutazone (restricted use), oxyphenbutazone (withdrawal recommended)[†], salicylates
Parasitic infections: chemotherapy	Diethylcarbamazine, levamisole, mebendazole
Phenothiazine tranquillizers	Chlorpromazine, perphenazine, piperdichlorophenothiazine, prochlorperazine, promethazine, thioridazine, trifluoperazine
Psoralens	Methoxsalen (8-MOP) with u.v.-radiation (PUVA)
Solvents	Dimethylsulphoxide (DMSO)
Sympathomimetics (topical)	Adrenaline (epinephrine) hydrochloride, phenylephrine hydrochloride
Tricyclic antidepressants	Amitriptyline, amoxaprine, nortriptyline, and possibly other tricyclics
Vitamins	Vitamin A deficiency, vitamin D.

* See text for information on current status of availability in the UK.
[†] Now withdrawn

amiodarone (200 mg t.i.d. for one week followed by 200 mg b.i.d. for one week then 200–300 mg daily thereafter). Dose- and duration-dependent keratopathy developed in 101 (98 per cent) of these patients and it was characterized by bilateral corneal microdeposits. Six per cent of this group experienced visual symptoms associated with the microdeposits. These included photophobia (3 per cent), haloes (2 per cent), and blurring of vision (1 per cent). Visual acuity was unimpaired and consequently it was unnecessary to reduce the dosage of amiodarone. No colour vision defects, dry eyes, or intraocular pressure changes were attributable to the drug.

Ingram concluded from this study that amiodarone appears to be ophthalmologically safe and that routine ophthalmological supervision may not be necessary in the majority of patients during long-term therapy. A similar conclusion was reached by Lubbe and Mercer (1982); they emphasized that despite the adverse effects of amiodarone, the relatively small risk of hazard had to be weighed against the possibility of obtaining, in the majority of patients, complete control of their incapacitating and potentially lethal arrhythmias.

Anticholinesterase drugs and cataract

In 1960, Harrison published the first documented evidence of lens opacities produced by a miotic drug. A 13-year-old girl with esotropia had been treated with 0.25 per cent dyflos (di-isopropylfluorophosphate) eye drops for three months when rosette-like anterior capsular lens opacities were noted. They did not interfere with vision and slowly disappeared after treatment was stopped.

In 1963, Axelsson and Holmberg (1966) commenced a carefully documented study of three years' duration in which they planned to compare the relative efficacy of the long-acting cholinesterase inhibitor ecothiopate iodine (phospholine iodide) with pilocarpine in the treatment of glaucoma. Within a year these workers had become impressed by the increased incidence of cataracts in the ecothiopate-treated cases, and so altered the direction of their study to compare the frequency of cataract formation in the two treatment groups. Of 103 eyes treated with pilocarpine, 10 eyes (about 10 per cent) developed, or had an increase in, lens opacities. Seventy-eight eyes were treated with ecothiopate iodide and 39 had an increase in lens opacities. Of the 87 eyes with clear lenses when therapy was started, three out of 47 were listed as positive after the pilocarpine treatment, and 16 out of 40 after ecothiopate treatment.

In later studies, Holmberg (1966) reported that similar lens changes were seen with other long-acting anticholinesterase drugs including dyflos and demecarium bromide, and de Roetth (1966) described 19 cases of anterior subcapsular lens changes in patients treated with ecothiopate iodide. Shaffer and Hetherington (1966) reviewed the literature to that date and in addition published the results of their own studies which were largely aimed at determining whether the findings of Axelsson and Holmberg (1966) in Sweden were comparable with results in California. They also found that anticholinesterase agents (dyflos, demecarium bromide, and ecothiopate iodide) could initiate cataract formation and cited a 38 per cent incidence in their anticholinesterase-treated patients. Few lens opacities were seen in patients under 60 years of age. This incidence of cataracts in the younger age groups did not conform with the experience of Mushin (1972), whose surveys suggest that ecothiopate iodide induced lens changes in children attending hospital treatment for squint.

Ecothiopate iodide is known to cause anterior subcapsular lens opacities in elderly patients with glaucoma. Cysts at the pupillary margin also occur in children as a result of treatment with strong miotic drops, especially when ecothiopate iodide is used to treat a partially accommodative convergent squint (Mushin 1972).

Cant (1969) in his comprehensive review on iatrogenic eye disease cautioned against the unnecessary use of miotics. They can interfere with vision and, although the reduction in pupil size enables the presbyope to read without spectacles, more commonly the surroundings are darkened and, if any lens opacity is present, vision can be greatly reduced with serious consequences. He especially warned of the danger of the stronger miotics such as dyflos and ecothiopate iodide which may produce cataract or serious systemic side-effects (McGavin 1965; de Roetth 1966).

Marmion (1971) warned of an interaction involving long-acting anticholinesterase drugs and suxamethonium bromide; increased sensitivity to the neuromuscular blocking drug and possibly its potentiation is occasionally seen during anaesthesia (about 1/2000 to 1/3000) in those susceptible children undergoing strabismus surgery who had pre-operative treatment with long-acting miotics. This interaction following the prolonged use of ecothiopate eye drops in the treatment of glaucoma has resulted in prolonged apnoea and death when suxamethonium was administered (*Drug and Therapeutics Bulletin* 1964; Himes *et al.* 1967; Cavallard *et al.* 1968; Lipson *et al.* 1969; Kinyon 1969). The mechanism of this interaction is that ecothiopate diminishes serum pseudocholinesterase levels and this reduces the metabolic hydrolysis of suxamethonium with a resultant prolongation of effect.

Anticonvulsants

Carbamazepine

Straughan (1982) from the Universtiy of Cape Town/Ciba-Geigy Medicines Safety Centre reported that he had received case details of two children who showed toxic symptoms to the anticonvulsant carbamazepine when it was given together with erythromycin. The clinical picture was one of diplopia, blurred vision, nystagmus, dizziness, ataxia, nausea, and vomiting. The plasma carbamazepine concentration was measured in one of the children after the toxic episode had resolved and was within the normal therapeutic range.

A true drug–drug interaction was suspected since there had also been a report of an interaction between carbamazepine and another macrolide antibiotic, troleandomycin (Dravet et al. 1977). Troleandomycin and erythromycin are so similar in chemical structure that they might well be expected to follow the same biotransformation pathways and are therefore likely to be involved in similar interactions. The mechanism of these interactions is thought to be inhibition of hepatic microsomal metabolizing enzymes by the antibiotics. The consequences of these interactions are potentially serious, and the potential for such interaction is high due to the extensive use of erythromycin in paediatric practice.

Phenytoin sodium

The drug regulatory authority in Japan (Japanese Ministry of Health and Welfare 1983) drew attention to case reports and animal experimental reports in which cataract was observed after the long-term administration of phenytoin. The regulatory authority also drew up revised precautions for use of this anticonvulsant in which the warning is given that symptoms such as double vision, visual disorders, nystagmus, and cataract may rarely occur; periodic visual activity testing is recommended. The literature reports on which this warning is based are those of Jain (1981) and Bar et al. (1983). Jain (1981) reported cataract in three patients with idiopathic grand mal epilepsy who had received phenytoin for one to four years; the progress of the cataract ceased in each case after discontinuation of the drug. Bar et al. (1983) reported two similar cases; in one of these cataract was also associated with other signs of phenytoin toxicity, namely gingival hyperplasia.

Anti-Infective agents

Chloramphenicol

In 1951, Gewin and Friou reported the history of a patient with staphylococcal endocarditis who was treated with 247.5 g of chlortetracycline and 190.5 g of chloramphenicol. Three days after medication was stopped, the patient complained of blurred vision. Ophthalmoscopy revealed blurred nerveheads. Since then there have been other reports of optic neuritis associated with chloramphenicol therapy in adults and children (for example, Wallenstein and Snyder 1952; Lasky et al. 1953; Dinning et al. 1963; Keith 1964; Chang et al. 1966; Cocke et al. 1966; Huang et al. 1966). Optic atrophy has been reported in children treated with chloramphenicol in large doses over several weeks for cystic fibrosis; this may commence with an optic neuritis and is only partially reversible (Keith 1964).

In all patients with associated optic neuritis, large total doses of chloramphenicol were given and therapy was prolonged. These changes have always been bilateral and central scotomata have been present; withdrawal of the antibiotic does not ensure a complete return of vision. The mechanism by which the optic neuritis develops is not known; it may be related to the neurological and hypersensitivity reactions occasionally reported with this agent.

Local ocular therapy should only be used according to specific indications otherwise potentially serious consequences can follow. A review in the Medical Letter (1982) warned that, although systemic reactions to ophthalmic drugs are relatively infrequent, there are several well-documented severe and potentially fatal reactions associated with some agents including chloramphenicol. Fraunfelder and Bagby (1983), from the US National Registry of Drug-Induced Ocular Side Effects, listed 10 cases of aplastic anaemia possibly associated with the ocular use of chloramphenicol. Their earlier publication (Fraunfelder et al. 1982) reported a fatal case. The only suggested risk factor in this series is a family history of a drug-related haemopoietic toxicity.

Clioquinol

Clioquinol (iodochlorhydroxyquin, chinoform) is used clinically in the treatment of intestinal amoebiasis and frequently, with less justification, by self-medication in the prophylactic treatment on intestinal infections. Its use has been casually linked with subacute myelo-optic neuropathy (SMON).

SMON has been particularly prevalent in Japan; it first presented in the mid-1950s and the incidence became so

684   **Iatrogenic Diseases**

high that in 1967 a SMON Research Commission was formed by the Japanese Government. A nation-wide survey revealed 7856 cases of which 5048 were confirmed. The majority of these patients had taken clioquinol, in high dosage, as self-medication for digestive disorders. The peak incidence of the syndrome was in 1969 and it fell rapidly when the Japanese Ministry of Health prohibited the production and sale of the drug in September 1970.

In most of the cases of clioquinol intoxication there was a history of chronic abdominal symptoms, usually of pain and diarrhoea preceding the neurological illness. There was usually a prodromal exacerbation of the abdominal symptoms before the acute or subacute onset of sensory neuropathy, usually below L_1-T_{10}. There was peripheral neuropathy, accompanied by paraesthesia in lower limbs and sometimes optic-nerve involvement. Neuropathologically, there was axonal damage and demyelination of the optic nerve, lateral and posterior columns of the spinal cord, and peripheral nerves. Other cases showed partial sensory disturbance in lower limbs, gait disturbance, complaints of visual disturbance, abnormality of deep-tendon reflexes, and psychic disorders. About a quarter of the patients had visual impairment and in some there was complete optic atrophy. In some cases, clioquinol intoxication was fatal and these cases were associated with an average intake of 141 g of the drug.

Cases of SMON have been reported in countries other than Japan, for example Australia, Germany, Norway, USA, and the United Kingdom, but only in relatively small numbers. They have also shown an association with clioquinol-containing preparations usually *Entero-Vioform* (250 mg clioquinol per tablet).

Clioquinol preparations were restricted to prescription control in Sweden in 1972 and banned in 1975 because the therapeutic benefit was not in reasonable proportion to the side-effects. Clioquinol was also banned in Norway and taken off the market in the USA by the manufacturers after the FDA demanded revised and much stricter labelling. In West Germany, after a television programme on SMON in October 1976, public opinion forced the medical authorities to re-evaluate the drug, and since January 1977 it has been available only on prescription. In Denmark, Finland, and France likewise the drug is supplied only on a prescription. Formerly, clioquinol was freely available in the United Kingdom on non-prescription sale as *Entero-Vioform* for 'diarrhoea winter and summer' and for 'the prevention and treatment of holiday diarrhoea'. In July 1977, however, the Committee on Safety of Medicines advised that oral preparations containing clioquinol should remain on the market but should be available only on prescription. In

this respect it is important to note that the scientific evidence for the value of clioquinol preparations against 'travellers' or 'holiday diarrhoea' is not only deemed to be scanty (Dunne *et al.* 1976) but also non-existent (Iwarson 1977).

Clofazimine

Ohman and Wahlberg (1975) investigated the use of the antileprotic agent, clofazimine (B663, *Lamprene*) in other dermatological disorders. They reported that 10 out of 26 patients who had received this drug orally (100–300 mg) for 1–15 months exhibited corneal changes. These were in the form of fine linear brownish subepithelial opacifications, and they diminished or disappeared within two months of stopping the treatment. One patient, with corneal pigmentation, who had normal fundi two years earlier, had speckled pigmentation in the macular area. Another patient, without corneal changes, had a clump of fine pigmentations in the macula. The retinal changes remained unchanged after 4–6 months. Ocular side-effects have not been reported with the antileprotic use of this drug (Imkamp 1968; Waters 1969; Browne 1975).

Diethylcarbamazine and mebendazole

Both these antiparasitic drugs are used in the treatment of onchocerciasis (river blindness), a disease characterized by the presence of some adult *Onchocerca volvulus* worms in subcutaneous nodules, and invasion of the skin and eyes by millions of microfilariae.

Ocular complications of these treatments are common and include restricted visual field, limbitis, bulbar conjunctival chemosis, punctate keratitis, uveitis, optic disc changes, and choriopetinitis with oral or transepidermal application (by lotion) of diethylcarbamazine (Taylor and Green 1981). Limbitis, punctate keratitis, retinal pigment, epithelial atrophy, chorioretinitis, and optic-nerve changes occur with mebendazole (Rivas-Alcala *et al.* 1981). Undoubtedly some of this eye damage is caused by or complicated by the migration of microfilariae into the eyes, and the formation of inflammatory infiltrate around microfilariae that are killed by the treatment.

Ethambutol

This antitubercular drug has been causally linked with oculotoxicity. Leibold (1966), who examined 118 patients on ethambutol, divided his cases into high-dosage (> 35 mg/kg/day) and low-dosage (< 30 mg/kg/day) groups. Eleven of the 59 cases in the high-dosage group showed ocular toxicity, while only two cases did so in the low-dosage group. This correlation between the incidence of toxicity and dosage was confirmed by others (Donomae and Yamamoto 1966; Place *et al.* 1966; Pyle 1966). Leibold found that ocular toxicity takes one of two forms: an axial type in which there is a loss of central vision and disturbance of colour perception, and a periaxial type in which there is a contraction of the peripheral field with little decrease in visual acuity. Recovery of vision is usually complete in all affected cases when the drug is stopped.

There is a problem in detecting the causative agent when oculotoxicity presents during antitubercular chemotherapy since patients are exposed to multiple drugs and may have a history of a variety of different combinations of drugs being tried. This is well illustrated by a case of toxic amblyopia described by Bowen and Vaterlaws (1971). The patient had drug-resistant pulmonary tuberculosis and had been treated for over 20 years on a variety of antitubercular regimens. She complained of symptoms of photophobia and vague ocular discomfort five months after ethambutol and rifampicin were started. She was found to have a toxic neuropathy of the axial type with severe depression of the central vision, but the field defect which developed was very unusual and took the form of an arcuate scotoma. This reaction was attributed to ethambutol; the clinicians discarded the possibility of rifampicin involvement since they were not aware of the reports of ocular toxicity associated with rifampicin.

Bartholomew (1976) described four patients who developed toxic amblyopia from ethambutol, all of whom had received the recommended dosage. On withdrawal of ethambutol varying degrees of recovery occurred over 5–12 months. No other drugs were mentioned in the report.

Kuming and Braude (1979) described a patient who showed an unusual side-effect of ethambutol, namely flame-shaped haemorrhages and retinal oedema associated with the signs and symptoms of an anterior optic neuritis, rather than the more usual retrobulbar type of neuritis. Because of the patient's renal failure (he was on haemodialysis) these symptoms and signs appeared with a dosage of ethambutol (300 mg daily, with 150 mg after dialysis) lower than that usually prescribed. The patient's anaemia was discounted as a cause of this atypical reaction to ethambutol. The oedema and haemorrhages resolved in five days after ethambutol was stopped and vision improved to 6/6 in each eye after 14 days. No residual fundal lesion was detected.

A study by Pattisson *et al.* (1978) found no evidence of oculotoxicity with ethambutol. The contrast between these findings and those of earlier positive studies may possibly be explained by the relatively low dosage (14.4 mg/kg daily) used in the latter study.

Optic neuritis is also a well-recognized complication of ethambutol therapy. A case in which a woman on ethambutol developed peripheral neuropathy several months before optic neutitis suggests that the former may serve as an early warning of development of the more serious visual toxicity. When ethambutol was stopped there was prompt improvement in peripheral neuropathy and ocular symptoms (Nair *et al.* 1980).

The mechanism of ethambutol's oculotoxicity is unknown, although experimental studies have provided some possible clues. Ethambutol may cause copper and zinc depletion in the dog, and there are several enzymes containing these elements, or requiring them as co-enzymes which may be affected by such a process (Buyske *et al.* 1966). Doses of ethambutol of the order given to man will cause decoloration of the tapetum lucidum of dogs and this might be related to the high zinc content of that ocular tissue. There is, however, no equivalent structure in man.

Hexachlorophane

Hexachlorophane is an effective antistaphylococcal germicide which has been shown to be a neurotixic agent in laboratory animals and in children. A review by Hackenberger (1980) presented many of the early investigations which established its toxicity, and restricted its use. The optic nerve and chiasm in animals (Towfighi *et al.* 1974; Rose *et al.* 1975) and in man (Martinez *et al.* 1974; Goutières and Aicardi 1977) are particularly susceptible to its toxic effects.

Slamovits *et al.* (1980) described a case of optic atrophy in a 31-year-old woman which was probably caused by chronic hexachlorophane ingestion. The patient underwent surgery to remove a growth from her right ear; after this she began taking an estimated 10–15 ml of hexachlorophane (*pHiso-hex*) orally each day during the next 10–11 months. She also applied large amounts of hexachlorophane to her face every day as self-treatment for pimples. Some 11 months after her ear operation, she noted decreasing vision, first in her right eye. She was taking no other medication than hexachlorophane and did not report any other exposure to chemical substances.

Ophthalmic examination revealed a visual acuity of R.E.: hand motions and L.E.: 6/90 (20/300). Both pupils were round and 6 mm in diameter. The left pupil reacted to light more rapidly than the right, and there was no afferent pupillary defect on the right. The near response was normal in both eyes. Slit-lamp examinations showed severe bilateral optic atrophy, which was worse in the right than in the left eye.

The patient was discharged after hospitalization with a diagnosis of hexachlorophane-induced optic-nerve atrophy and followed-up as an out-patient. During subsequent checks her visual fields improved and eventually stabilized as also did her visual acuities; 14 months after discharge visual acuity was R.E.: 6/120 (20/400) and L.E.: 6/7.5 (20/25) with no subsequent changes.

The authors of this report, in reviewing the literature, commented that it is not certain at what level oral doses of hexachlorophane are likely to cause toxicity; their case showed that anterior visual system damage can be present when no other significant signs of poisoning are evident.

Rifampicin

Loss of acuity of vision and colour-blindness were reported in a case treated with combined antitubercular drugs, rifampicin and ethambutol (Lees *et al.* 1971) and a case of reversible oculotoxicity was recorded with rifampicin alone (Cayley and Majumdar 1976). In this latter case, a 34-year-old man, suffering from pulmonary tuberculosis, received initial treatment with a triple oral regimen of isoniazid, rifampicin, and ethambutol. After a week both eyes were painful, tender, and red, and congested with thick, white secretions. All drug treatment was stopped and the eyes cleared up within 72 hours. Treatment was changed to isoniazid, PAS, and streptomycin, but this had to be stopped due to nausea and vomiting. Isoniazid alone was given for one week and there were no adverse reactions, but when rifampicin was added, the same type of ocular reactions again developed with 48 hours and cleared up within 48 hours after rifampicin was stopped. The patient was subsequently treated with isoniazid, ethambutol, and streptomycin without any untoward reactions.

Girling (1976) commented that conjunctival reactions to both daily and intermittent rifampicin are well documented and not uncommon, although the case described by Cayley and Majumdar (1976) was perhaps unusual in its severity. In support of this, he cited a study involving 396 Chinese patients treated with rifampicin plus ethambutol in Hong Kong. Cutaneous reactions consisting of flushing and/or itching, with or without a rash, involving particularly the face and scalp, and often including redness and watering of the eyes were observed. The incidence of these reactions was 4 per cent in patients on daily treatment with the drug combination, 8 per cent in those on twice-weekly dosage, 14 per cent on weekly dosage, and 5 per cent in patients on daily and then once-weekly treatment. These effects occurred during the first year of chemotherapy and, in almost all cases, were attributed to rifampicin (Aquinas *et al.* 1972; Hong Kong Tuberculosis Treatment Services 1974).

Streptomycin

Nerve-fibre-bundle defect in the retina or optic nerve with field loss was formerly thought to be pathognomic of glaucoma, but it is now well established that it can occur in a variety of disorders (Harrington 1965). Most of these defects are vascular, and a likely explanation for the defect is closure of small vessels supplying a nerve-fibre-bundle. It has been described in patients treated with streptomycin (Thomas 1950). Streptomycin has also been incriminated in causing possible paralysis of the eye muscles and nystagmus (Meyler 1966); this is not surprising since streptomycin is known to have curare-like neuromuscular-blocking properties and patients on prolonged intramuscular therapy of 1 g of streptomycin per day have developed muscular weakness, fatigability, and blurred vision. Other aminoglycoside antibiotics share in these neuromuscular effects.

Tetracyclines

Transient myopia was reported as a result of tetracycline therapy in children (Mushin 1972); this may also be complicated by papilloedema and benign intracranial hypertension (Koch-Weser and Gilmore 1967; van Dyke and Swann 1969).

Antimalarials

Apart from their obvious use in the suppression and treatment of malaria, antimalarial drugs have been used by dermatologists for over 30 years since Page (1951) established and popularized their use in the treatment of discoid lupus erythematosus. Since then they have also been used in the treatment of rheumatoid arthritis, and in polymorphous light eruptions, porphyria cutanea tarda, and sarcoidosis. Very early on in this non-malaria use, it was shown that their continued high dosage caused ocular damage of varying severity and reports in the literature have implicated all the commonly used synthetic 4-aminoquinolines: chloroquine, hydroxychloroquine, amodiaquine, and the earlier and now largely superseded aminoacridine mepacrine (quinacrine) (Kersley and Palin 1959; Hobbs *et al.* 1961; Grant 1962; Merwin and Winkelmann 1962; Scherbel 1983 (review)). It must be emphasized that ocular damage does not normally occur with the lower dosage used in the suppression and treatment of malaria.

Chloroquine and its near relative, hydroxychloroquine, are the two synthetic compounds of this group most commonly implicated in ocular damage, largely because they have greater use in their non-malaria role than any of the other synthetic antimalarials. Chloroquine and hydroxychloroquine are known to produce a number of ocular complications such as whitening of the lashes, exraocular muscle palsy, subepithelial corneal deposits, decreased corneal sensitivity, and retinal damage. Retinal damage is the most serious of these complications (see Plate 3(c) and (d)) and it may progress to a severe retinal degeneration even after cessation of therapy (*Journal of American Medical Association* 1966; Reed and Karlinsky 1967).

Henkind and Rothfield (1963) examined 56 clinic patients suffering from either rheumatoid arthritis or systemic or chronic discoid lupus erythematosus who were undergoing long-term medication with chloroquine (250–750 mg per day), hydroxychloroquine (200–400 mg per day), or mepacrine (100 mg per day). Treatment had continued for a number of years. There was a high incidence of ocular involvement with corneal deposits (68 per cent), retinopathy (18 per cent), and decreased corneal sensitivity in 50 per cent of the patients, all of which appeared to be directly related to the medication. Lens opacities in the posterior subcapsular region were present in 37 per cent of patients, but could not be attributed to the medication.

A factor of major importance revealed by this study was that retinopathy was present in asymptomatic patients and that, in the asymptomatic stage, it could be reversed by discontinuing therapy. The objective demonstration of retinopathy before symptoms were present emphasized the importance of periodic ophthalmic examination of patients on such a treatment regimen. This need for routine and regular ophthalmic monitoring was also emphasized by Percival and Meanock (1968) who assessed out-patients on long-term chloroquine therapy for rheumatoid arthritis with respect to clinical benefit and ocular toxicity.

The mechanism of retinal damage by antimalarial drugs is unknown; several workers proposed that damage is initially produced by vascular spasm, although vascular spasm has not been noted in all cases (Hobbs *et al.* 1961; Smith 1962). Others suggested that chloroquine may be a retinal enzyme inhibitor (Schmidt and Mueller-Limmroth 1962). Chloroquine has local anaesthetic properties (Mandel 1960) and the reduction in corneal sensitivity experienced by patients may be attributable to this property (Henkind and Rothfield 1963). The presence of corneal deposits appears to depend on a time–dose relationship and they disappear after cessation of therapy without causing permanent ocular damage.

Corneal opacities were reported in the literature as occurring with all the commonly used 4-aminoquinolines even at low dosage when given over prolonged periods of time. Henkind and Rothfield (1963) reviewed and summarized the progress of such corneal pigmentation.

Leopold (1968) in his comprehensive review on the ocular complications of drugs cited an incidence of about 4 per cent retinopathy in patients with collagen disease who were receiving long-term chloroquine therapy, and in common with other authors emphasized the importance of periodically examining the ocular function of such patients. Earnshaw *et al.* (1966) studied methods of screening patients at risk, to determine which test or combinations of tests were most useful. They concluded that no single test was suitable to detect early chloroquine retinopathy but that a combination of tests of colour vision, elicitation of recent symptoms, especially photophobia, and ophthalmoscopy was reliable. In their opinion the electrooculogram was not satisfactory as a screening test.

From a retrospective view, it is clear that by the mid-1960s, fear of retinal toxicity and the availability of alternative therapies severely limited the use of antimalarials in roles other than the prophylaxis and treatment of malaria. None the less, continued experience with alternative drugs such as systemic corticosteroids, nonsteroidal anti-inflammatory agents, gold, and immunosuppressive and cytotoxic drugs and their often devastating toxicities led to a renewed interest in antimalarials as is evidenced by reviews discussing these agents in dermatological journals (e.g. Logan 1980; Tanenbaum and Tuffanelli 1980; Zuehlke *et al.* 1981). Such reviews, although excellent, have, in the opinion of Olansky (1982), propagated some apparent misconceptions by disregarding or de-emphasizing data suggesting that irreversible retinal toxicity due to antimalarials can be easily avoided by judicious daily dosage and regular ophthalmic follow-up.

Okansky (1982) in his review on antimalarials and ophthalmic safety discussed the historical basis of these misconceptions and detailed the subsequent studies which suggest that antimalarial retinal toxicity can be avoided without sacrificing the therapeutic efficacy of antimalarials.

Such evidence supports the belief that antimalarial drugs can be safely and effectively used in the treatment of a number of dermatologic disorders. The retinopathy due to these agents can be easily avoided, or at least recognized at an early reversible stage, by the administration of daily doses of chloroquine, no greater than 250 mg per day, or of hydroxychloroquine, no greater than 400 mg per day (both expressed in terms of their salts), coupled with pretreatment and serial ophthalmologic examinations at four- to six-month intervals during such treatment.

The evidence implies also, contrary to previously held beliefs, that no upper limit of total dose administered nor total duration of therapy exists above which retinopathy is likely to occur. This enthusiastic view of Olansky recommends that antimalarial therapy be given careful reappraisal. The data supporting his enthusiasm and his concepts of safety in use of these antimalarials are consistently more scientific than those from which the old concepts arose.

Some practitioners, in the past, felt that the best way to avoid worry about retinal toxicity from antimalarials was to use mepacrine (quinacrine) since retinopathy had not been convincingly reported with this drug. The generalized yellow pigmentation that it produces is not, however, acceptable to most Caucasians, but there are more pressing reasons other than cutaneous pigmentation, why mepacrine should not be taken lightly since many of the non-ocular adverse effects of antimalarials are more frequent and more severe with mepacrine than with either chloroquine or hydroxychloroquine. Furthermore, as Olansky has reviewed, it would appear to be less effective than chloroquine in the treatment of both discoid and systemic lupus erythematosus.

Quinine

Many cases of partial or complete permanent blindness as a result of accidental or suicidal ingestion of quinine have been published (see review by Grant 1974). Apart from its antimalarial use (which has been rejuvenated in combined treatments), quinine is also useful in the relief of night cramps and it is in this context of general availability that Valman and White (1977) described a case of a $3\frac{1}{2}$-year-old boy who consumed a quantity of various tablets (paracetamol, quinine, phenylbutazone, aspirin, practolol, chloramphenicol, and various diuretics) which were stored in a shoe box on a window sill. Subsequent screening of the urine demonstrated the presence of quinine but no other drug. No action was taken until two hours after the ingestion, and no abnormal signs other than vomiting were evident on admission to an emergency and accident department.

Next morning (15 hours later) the boy was drowsy but responded to command. Both pupils were dilated and did not react to light and he was completely blind. The fundi appeared normal. About 28 hours after ingestion of the tablets, bilateral stellate block was carried out by paratracheal approach using bupivacaine. About 30 minutes after the bilateral block, he began moving his eyes in relation to a light and a face. Four and a half hours later a second bilateral block was carried out, and 12 hours later

central vision was normal but peripheral vision was poor.

Three days after the ingestion of the tablets, retinal oedema and blurring of the optic-disc margins were present for the first time. A peripheral vision defect was still present. A month after admission, both discs were pale and there was gross attenuation of the retinal vessels in each eye.

Apart from the oculotoxic and stellate block aspects of this case, there is an obvious moral to be drawn regarding the safe storage of drugs (any drugs!) out of the sight and touch of inquisitive young hands. This was an avoidable tragedy and it is a moral which must unceasingly be instilled into patients who have or are in contact with small children.

Antiparkinsonian agents

Benzhexol

Benzhexol was reported by Friedman and Neumann (1972) to have induced three cases of blindness in parkinsonian patients.

Benzhexol has peripheral and central atropine-like effects and thus its use in patients with angle-closure glaucoma is contraindicated. Total blindness of one eye in two patients and practical blindness with tubular vision in one eye of a third patient occurred as a result of angle-closure glaucoma due to long-term treatment with benzhexol, 15 mg daily for one or two years.

Levodopa

Spiers and his colleagues (1970) reported that the pupillary diameter of 11 patients with Parkinson's disease was significantly decreased four hours after ingestion of L-dopa. Dosage ranged from 1 to 6.8 g with a mean of 3.2 g.

The extent of the miosis in individual patients showed no correlation with the dose of L-dopa, the development of hypotension, or the response of the parkinsonism to therapy. These authors suggested that the miosis might be caused by diminished noradrenaline output at sympathetic nerve endings. Reduced sympathetic activity during therapy with a drug which is a precursor of noradrenaline may appear to be paradoxical, but the authors explained their hypothesis by commenting that L-dopa can cause release of noradrenaline from sympathetic nerve endings, and may produce depletion of noradrenaline stores. Alternatively, Spiers and his colleagues suggested that L-dopa may exert a central sympatholytic action, possibly by modulating levels of brain amines.

Miosis is not always detectable during L-dopa therapy; for example, Godwin-Austen et al. (1969) did not detect any change in resting pupillary size in a series of 18 patients.

Beta-blockers

Practolol

Practolol (*Eraldin*) was first marketed in the United Kingdom in June 1970 as a cardioselective β-adrenergic blocking agent and was subsequently widely used in the treatment of angina and hypertension. Pharmacological and toxicological studies (Shanks 1969; Barrett 1971) and early clinical trials (George *et al*. 1970; Sandler and Clayton 1970) with practolol did not reveal any adverse ocular or cutaneous reactions. However, the number of patients in these trials was small and each clinical study lasted only 12 weeks. Likewise comparative studies on practolol in angina pectoris by Balcon (1971), Pentecost *et al*. (1971), Prichard *et al*. (1971), and Sandler (1971) did not mention these side-effects. However, as the clinical use of practolol increased, reports of adverse skin reactions emerged although there was, as yet, no evidence of ocular effects (Wiseman 1971; Rowland and Stevenson 1972; Zacharias 1972; Raftery and Denman 1973; Felix and Ive 1974).

By the autumn of 1974, experience with practolol amounted to some 250 000 patient-years; it was at this stage of usage that adverse effects on the eye became evident. In June 1974, Wright reported the occurrence of eye changes and psoriasiform skin lesions with practolol (see Plate 2(a)–(f)). In October 1974, Ismail reported one case of keratoconjunctivitis sicca in Shaab Hospital, Khartoum, and in November 1974, Felix and his colleagues reported a total of 21 patients suffering from practolol-induced skin rashes seen over the preceding two years and observed that persistent ocular damage was a feature in three cases. Later in March 1975, Felix and his colleagues reported that their original observations had been extended to a total of 48 patients with cutaneous reactions to practolol. They made no comments on relative additional ocular involvement, but regarding subsequent treatment of these patients with alternative β-blockers, they demonstrated a lack of cross-reactivity between practolol and oxprenolol and/or propranolol.

From July 1974 to July 1975, doctors in the United Kingdom received a total of five letters on the side-effects of practolol. The first letter, from the manufacturers of practolol (ICI) went out on 12 July 1974 drawing attention to the probable association between practolol and eye damage; the second letter, also from ICI, went out on 9 October 1974 saying that patients developing skin or eye conditions should be withdrawn from practolol treatment. In January 1975, the Committee on Safety of Medicines sent out a letter following the report from Professor Read of Bristol about sclerosing peritonitis; this was followed on 18 April 1975 by a letter from

ICI dealing with sclerosing peritonitis and recommending that practolol be reserved for patients where there was a special benefit compared with other treatment. On 30 July 1975, ICI sent out another letter recommending that the drug be restricted to minimal indications.

In their letter of January 1975, the Committee on Safety of Medicines included the warning that by the end of 1974, 187 reports had been received of adverse effects on the eye occurring in patients who had been treated with practolol. Two-thirds of these reports described diminished tear secretion and conjunctivitis, and the remainder, corneal damage leading on occasion to impairment or loss of vision. These effects on the eye had been noted in patients who had received practolol for periods ranging from a few weeks to several years.

The letter from the Committee further warned that several hundred reports of psoriasiform or hyperkeratotic skin reactions had been received and that 25 patients had complained of deafness. Fourteen patients had developed a syndrome resembling systemic lupus erythematosus and eight had developed an unusual form of sclerosing peritonitis. Half the patients with eye changes had a rash and in others these adverse effects were multiple. The mild eye changes and the majority of skin reactions were said to usually recover when practolol was withdrawn, but the outcome of corneal involvement was said to be less certain and it might be irreversible.

During 1975 and 1976, a number of clinicians drew attention to the oculomucocutaneous syndrome associated with practolol administration (Bendtzen and Sobørg 1975; Dyer and Varley 1975; Fan *et al*. 1975; Farr *et al*. 1975; Gurry *et al*. 1975; Thompson and Jackson 1975; Windsor *et al*. 1975; Wright 1975*a, b*).

In June 1975, a leading article in the *British Medical Journal* reviewed and emphasized the ocular and other side-effects of practolol and reported that the drug had been withdrawn in some countries and its use restricted in others. Practolol tablets were withdrawn from general use in Britain on 1 October 1975, and supplies were restricted to hospitals for use for patients with certain cardiac dysrhythmias only. Finally, in September 1976, ICI withdrew the tablets from the market. The injectable form of practolol is, however, still available for short-term use in treating patients with certain cardiac dysrhythmias. The required FDA processes in the United States did not progress quickly enough to allow practolol to become available for clinical use; with hindsight of its toxicity spectrum, it is easy to appreciate the comment made by Wardell (1974) that this illustrated a serendipitous pay-off from the United States 'drug-lag'.

Reports continue to appear in the literature of oculotoxicity as part of a wider spectrum of side-effects suffered by patients who had earlier treatment with practolol (for example Bartsch and Reginster 1977; Thompson and Jackson 1977).

Mechanism of oculotoxicity

The mechanism by which practolol induces eye damage is uncertain but observations by Amos and his colleagues (1975) are of obvious interest and of probable significance in this respect. They found that an antibody, which sticks to the intercellular region of xenogenic epidermal tissue, was present in the serum of patients with practolol-induced eye damage. The fact that the intercellular region of the epidermis was this antibody's target *in vitro* suggested a possible relationship to the antibody found in pemphigus. Therefore these antibodies in practolol-treated patients and those found in patients with pemphigus were compared for their ability to bind to isolated, trypsinized epidermal cells. Binding was, however, achieved only with the pemphigus antibody, which suggested that it might have a different specificity from the antibody associated with practolol-induced eye damage.

The significance of this intercellular antibody in the patients' sera is unknown but Amos *et al.* (1975) suggested, with some evidence, that its titre might reflect the degree of eye-associated tissue damage. They postulated that this antibody titre might be valuable in monitoring practolol treatment or even predicting likely damage.

This discovery in patients with ocular damage of high titres of antinuclear antibodies and antibodies binding to the intercellular region of xenogenic epithelial tissues (Amos *et al.* 1975) and the demonstration of the deposition of IgG, IgM, and C_3 at the dermo–epidermal junction and circulating antinuclear antibodies in 24 per cent of patients with skin reactions (Felix *et al.* 1974) suggested that an abnormal immune response might be involved. Further, the possibility of a primary immunological mechanism for the oculomucocutaneous syndrome was suggested when a survey of reports on patients carried out by ICI found that the adverse reactions were not related to dosage and, therefore, were likely to be pharmacologically or toxicologically induced. Added weight to the possibility of immunological mechanisms being implicated in practolol's toxicity was given by the report of Behan *et al.* (1976) who showed that patients with and without adverse reactions to practolol therapy showed altered immune responses; it was not possible; however, to demonstrate that the several anomalies detected in the immunological status of patients, in whom the oculomucocutaneous syndrome developed in response to practolol, were of primary importance in the genesis of the condition.

Evidence in favour of a primary immunological mechanism being responsible for the oculomucocutaneous syndrome of practolol was given indirectly by the work of Raftery (1974). He reported that a prospective study of 50 patients receiving oxprenolol revealed no instance of a raised antinuclear factor (ANF) titre before or after treatment for a period of three months. The same was true for 24 patients taking propranolol. There were no instances of disseminated lupus erythematosus (DLE) syndrome in these two groups. However, in a retrospective study of 67 patients on practolol, seven had an ANF titre greater than 1/40 (11 per cent incidence), and in a prospective study of 71 patients on practolol, the ANF titre before treatment was greater than 1/40 in only one patient and after an average of six months treatment this had risen to five (11 per cent). One of these patients developed a florid DLE syndrome. Raftery suggested that practolol was unique amongst the commonly used β-blockers in producing the DLE syndrome and in producing a rising ANF titre. This finding may well have a significant bearing on the relative sparsity of reports of oculomucocutaneous reactions to other β-blocking drugs, which are discussed later in this chapter, and is counter-suggestive to the proposal that the pharmacological effects of β-blockade *per se* are responsible for the syndrome.

Gaylarde and Sarkany (1975) proposed an alternative and, in their view, a more rational explanation for the various cellular and humoral changes described as a result of practolol therapy. Whilst agreeing that certain manifestations, such as antibody formation and the presence of lupus erythematosus cells would seem to support an immune mechanism, they suggested another possible explanation which is based upon three sequelae of β-blockade.

Firstly, Benner *et al.* (1968) showed that β-blockade increases antibody production in response to antigen challenge and that adrenaline suppresses the normal response to antigenic stimulation. Secondly, β-blockers interfere with the action of anti-inflammatory agents (Riesterer and Jaques 1968) and propranolol prevents the response of lymphocytes to phytohaemagglutinin *in vitro* (Smith *et al.* 1971). Thirdly, adrenaline activates epidermal adenyl-cyclase activity and reduces cell division (Yoshikawa *et al.* 1975). β-blockers antagonize this effect, suppressing cyclic AMP formation and encouraging cell division, and probably predispose to the development of psoriasiform changes in the skin.

This explanation postulates that the side-effects are therefore the direct and specific result of pharmacologically induced β-blockade, and it is interesting to note that Brown *et al.* (1974), in discussing their report of sclerosing peritonitis and practolol, came to a similar conclusion by a similar process of pharmacological deduction. Certainly, if this postulate were true, then it might well be asked why validated reports of oculocutaneous or oculomucocutaneous reactions have not as frequently occurred with the use of β-blockers other than practolol? The reactions to oxprenolol, which are discussed later give support to this theory, but the single and controversial report of ocular symptoms to propranolol and the absence, so far, of related side-effects with other β-blockers, even when they were introduced as substitute therapy, must cast some doubt on the validity of this otherwise interesting and ingenious explanation. Marks (1975) made similar points against this explanation with the comment that the clinical features of the oculomucocutaneous syndrome, the development of a positive antinuclear-factor test, and the irregularity of the occurrence of the condition favours an immunological rather than a pharmacological mechanism. However, these mechanisms are not mutally exclusive, nor indeed do they exclude other hypotheses.

Beta-blockers other than practolol

In view of the problems that have emerged with practolol (see D'Arcy 1979, 1980; Vogel 1983) a pertinent and obvious question is whether beta-blockers other than practolol may be associated with the development of the oculomucocutaneous syndrome?

Confusion due to premature conclusions may be inadvertently introduced into the literature if it is not appreciated that beta-blockers, like so many other drugs, are associated with skin rashes of various kinds and that many patients have pre-existing skin or eye diseases; also many patients may have had earlier practolol treatment.

Timolol

Some reports suggest that there may be a causal relationship between timolol use and ocular damage. Frais and Bayley (1979) reported a case of dry eyes with timolol; a 40-year old man was treated with chlorthalidone (100 mg daily) and timolol (15–75 mg daily). After 14 months he complained of dryness of eyes and nose which had persisted since starting treatment. Symptoms improved immediately after timolol was withdrawn, although three months later tear production was still poor.

A report by van Buskirk (1979) described three glaucoma patients who developed symptomatic superficial punctate keratitis associated with complete corneal anaesthesia 1–2 months after being treated with timolol maleate eye drops (0.5 per cent) twice daily. This cleared after the timolol was withdrawn and normal corneal

sensitivity was restored. Corneal sensitivity measured in 25 other patients using timolol maleate drops, showed marked diminution in four patients, all of whom were elderly and had been using the drops for a minimum of three months.

In contrast, Saari et al. (1978) treated five patients with topical timolol (0.5 per cent) once or twice daily for secondary glaucoma in chronic uveitis and no ocular or systemic side-effects were noted.

McMahon et al. (1979) assessed the effect of adding timolol eye drops (0.25 or 0.5 per cent) to previously prescribed medication regimens in 165 glaucomatous patients. After three months treatment, 38 patients reported adverse effects and these were sufficiently severe in 15 cases for the eye drops to be withdrawn. CNS and cardiovascular effects, respiratory distress, and sexual impotence were among the adverse effects recorded but signs of the oculomucocutaneous syndrome were not included amongst these except for some complaints of deep-seated eye pain.

Much of the recent work on the ophthalmic use of timolol has evaluated its efficacy in lowering intraocular pressure often in comparison with pilocarpine. Some of the reports also indicate the acceptability to the patient of timolol eye drops in respect to adverse effects on the eye, and also to non-ocular reactions notably bronchospasm, bradycardia, and congestive heart failure.

Although in this present review the ocular side-effects of timolol must be emphasized, it must be apparent that they are far less serious than the adverse respiratory or cardiovascular effects. The latter are mentioned here so as to present a balanced account of the full toxicity spectrum of timolol ophthalmic solutions.

Willcockson and Willcockson (1980a) treated 40 patients with open-angle glaucoma in a cross-over study with timolol 1 per cent ophthalmic solution, then with pilocarpine 2 per cent, 1 drop/eye t.i.d. for two weeks. Six instances of adverse reactions occurred in four patients on timolol in contrast to 33 instances of reactions in 19 patients on pilocarpine. Blurred visions, browache, and headache occurred most frequently. No patient on timolol had to be withdrawn from the trial but three on pilocarpine withdrew, two for allergic eyelid reactions and one for chest tightness.

Willcockson and Willcockson (1980b), in an additional study, evaluated the long-term efficacy and tolerability of ophthalmic timolol in 136 patients with chronic open-angle glaucoma during treatment periods extending to 15 months. A total of 104 patients received the beta-blocker for one year or more. Symptoms of ocular irritation were reported by 11 of the 136 patients, of whom eight were receiving other antiglaucoma therapy as well. Only one patient experienced an adverse reaction, probably an allergic response attributable to timolol.

Attalla and his colleagues (1981) reported the results of a post-marketing surveillance programme on the use of timolol in hypertension carried out by 129 Canadian physicians in 509 patients. In total, less than 2.6 per cent of all the adverse reactions were classified as potentially serious, and all of these had been reported in previous studies. A total of 123 adverse reactions was reported in 90 patients; among these there was one report of blurred vision which was severe enough to cause discontinuation of medication.

A further multicentre study in Canada was carried out by Le Blanc and Krip (1981); this collaborative study involved 39 investigators and showed the effects of timolol maleate when used short-term on 418 patients. Intraocular pressure response was excellent; patient complaints were monitored during the study and very few adverse reactions were reported. In some patients blurred vision was a problem, but in all cases in this study it was found to be related to the discontinuation of miotics. Only seven patients were reported to have adverse effect necessitating discontinuation of the study and these were non-ocular effects notably dizziness, headache, and three cases of wheezing and shortness of breath. In one patient, a well-documented history of asthma had been withheld, while in another underlying asthma was unknown.

Apart from these multicentre studies on relatively large numbers of patients, there have been reports on individual, or small numbers of patients, who have experienced unwanted non-ocular side-effects during the ophthalmic use of timolol. For example, Yates (1980) reported the case of a 78-year-old woman, on no other medication, who was started on timolol eye drops for open-angle glaucoma which responded well. The following year she suffered an episode of syncope whilst swimming and nearly drowned. The following year she again experienced confusion, an inability to write or speak, and vivid visual hallucinations. The timolol was stopped and hallucinations ceased within a few hours and did not occur. She did not experience any further attacks of transient ischaemic-like episode or of syncope.

Other clinicians have associated the use of ophthalmic timolol with one case of bradycardia (Kim and Smith 1980); 14 cases of rebound tachycardia (Ros and Dake 1980); one case of bradycardia and congestive heart failure (Linkewich and Herling 1981); a similar case (Altus 1981); one case of bronchospasm in an asthmatic patient (Charan and Lakshminarayan 1980); a similar case (Laursen and Bjerrum 1982); one case of bronchorr-hoea in a patient with chronic but stable asthmatic bronchitis (Guzman 1980); five cases of patients with

reversible air-flow obstruction who showed a significant decrease in vital capacity and air-flow (Schoene et al. 1981); one case of apnoea in a neonate (Olson et al. 1979); and acute episodes of cyanosis, bradycardia, and respiratory distress in an 18-month old toddler (Williams and Ginther 1982).

Systemic reactions to ophthalmic drugs were comprehensively reviewed in the *Medical Letter* (1982); included in this is the bradycardia, exacerbation of asthma and congestive heart failure, and the induction and masking of diabetic hypoglycaemia relative to the use of timolol in the eye. Less significant systemic reaction including hypotension, syncope, hallucinations, headaches, nausea, fatigue, and depression were also reviewed in the same article.

The first report of allergic conjunctivitis after the use of ophthalmic timolol was made by Baldone et al. (1982). The patient, a 61-year-old man with a five-year history of chronic open-angle glaucoma, had his medication changed from a pilocarpine/adrenaline (epinephrine) combination to timolol (1 drop in each eye, twice daily). After about one week, he suffered red, itchy eyes. These symptoms abated six days after stopping the timolol and three days after restarting the pilocarpine/adrenaline combination drops. Two weeks later a challenge to the right eye with one drop of timolol provoked an acute conjunctivitis within two hours; in contrast one drop of a diluent formulation containing no timolol gave no adverse response when instilled into the left eye. The authors of this report concluded that the case represented a true allergy to timolol maleate.

These reports of reactions to ophthalmic timolol must, however, be seen in perspective. The general census of opinion therefore is that although timolol eye drops are a welcome supplement to the present treatment of glaucoma, care has to be exercised in their use in patients with arrhythmias and chronic obstructive lung disease. Adverse effects of timolol on the eye are minor and seem to be less evident and serious than those caused by pilocarpine or other antiglaucoma treatment (Willcockson and Willcockson 1980a, b). An excellent survey on the use of timolol eye drops over a period of one to more than three years in 130 glaucoma patients was published by Gillies et al. (1983).

Metoprolol, oxprenolol, and nadolol

Since the oculomucocutaneous syndrome of practolol use was established, a careful watch has been maintained for similar reactions associated with the non-ophthalmic use of other beta-blockers. An extensive review by Cocco et al. (1982) summarized the current status of these agents relative to the total symptoms of oculomucocutaneous toxicity. Confirmed ocular effects have only been reported with metoprolol and oxprenolol.

With metoprolol, pain and soreness of the eyes were seen in one patient; these symptoms disappeared on stopping the drug. With oxprenolol, three patients developed ocular reactions, although one patient who presented with corneal perforation had previously received practolol for 32 months. No ocular reactions were reported with acebutolol, atenolol, labetalol, pindolol, propranolol, sotalol, or timolol, although varied skin reactions, retroperitoneal fibrosis, pulmonary fibrosis, and sclerosing peritonitis were amongst

Table 30.2   Reports of oculotoxic symptoms causally linked with oxprenolol

Reference	Number of patients sex (age)	Details of therapy	Toxic outcome
Clayden (1975)	1 M (72 years)	Oxprenolol for 8 months for ischaemic heart disease	Dry and gritty eyes
Holt and Waddington (1975)	1 F (65 years)	Oxprenolol, digoxin, and methyldopa for 6 months for hypertensive heart disease	Psoriasiform lesions, eye dryness, and reduced tear secretions
Knapp and Galloway (1975)	1 F	Oxprenolol for 15 months; concomitant treatment with clonidine, bendrofluazide, frusemide, and digoxin	Red eyes, conjunctival, oedema, and congestion of conjunctival vessels; punctate epithelial opacities in both corneas
Lyall (1975)	1 M (55 years)	Oxprenolol for 15 months for angina; prior exposure to propranolol and practolol	Dry eyes and perforated left cornea; previous exposure to practolol for 32 months had produced dry eyes
Lewis *et al.* (1976)	1 F (63 years)	Oxprenolol for 2 months for angina; prior exposure to amiloride with hydrochloro-thiazide (*Moduretic*) and perhexiline	Burning eyes, photophobia, and skin rash
Pearson *et al.* (1976)	7 F 2 M (40–65 years)	Double-blind, cross-over study; placebo versus guanethidine versus oxprenolol versus oxprenolol with guanethidine for 4 weeks for essential hypertension	Blurred vision, 1 patient on oxprenolol alone, 1 on guanethidine, and 1 patient on the combination
Becker (1976)	20 F 3 M	Double-blind comparative study of oxprenolol and propranolol in anxiety states	One patient on oxprenolol experienced blurred vision
Hansen *et al.* (1977)	15 F 13 M (40–70 years)	Double-blind, cross-over study of cyclopenthiazide plus methyldopa with oxprenolol plus hydralazine over 5 weeks in essential hypertension	Blurred vision in 4 patients receiving oxprenolol plus hydralazine
Weber (1982)	1 M (52 years)	Oxprenolol prescribed for angina; patient misunderstood medication instructions and instead of replacing with propranolol took both drugs concomitantly	Diplopia one hour after taking both drugs; did not recur when on either drug alone

adverse reactions that were collectively recorded for these agents.

An earlier report on eye problems linked with the use of metoprolol was that of Scott (1977). A woman patient reported pain and soreness in both eyes after taking metoprolol, 100 mg then 200 mg daily for about four months. Symptoms abated when the drug was withdrawn but within two or three days of restarting treatment, the patient suffered from very dry and painful sore eyes and had difficulty opening her eye lids.

A survey of the literature on oxprenolol has produced eight published reports, involving 12 patients, in which oculotoxic symptoms have been causally linked with this drug. A summary of the details of these cases is given in Table 30.2; some of these cases are anecdotal.

One case of papilloedema following the use of nadolol in the treatment of hypertension was reported by Kaul *et al.* (1982). The patient, a 58-year-old man, was started on nadolol 40 mg daily. After four doses his vision blurred and he stopped taking the drug. He was found to have bilateral papilloedema and tunnel vision, with no obvious cause.

Eight months later, the papilloedema has resolved, but optic atrophy subsequently developed. His rapid onset of symptoms was suggestive of an idiosyncratic reaction to nadolol.

Cancer chemotherapy

A general review

In their review on the eye toxicity of cancer chemotherapy, Vizel and Oster (1982) emphasized that ocular side-effects of these drugs are not major problems although there are quite a few toxicity reports in the literature.

Ocular problems in cancer patients can be due to metastases to the eye or CNS, they can be the result of radiotherapy or chemotherapy, or alternatively they can be totally independent ocular illnesses.

Alkylating agents (e.g. busulphan, chlorambucil, cyclophosphamide, nitrogen mustard) are known to cause ocular side-effects varying from cataract, diplopia, bilateral papilloedema, retinal haemorrhage, to blurred vision and necrotizing vasculitis of the choroidal vessels. Vizel and Oster (1982) cited as examples of this, lenticular opacities developing in 8/19 patients treated for chronic myelogenous leukaemia with busulphan, and blurred vision in 5/59 children on high-dosage cyclophosphamide. They also cite reports of fluorouracil-associated eye symptoms developing as late as 14 months after treatment, or as early as after the first dose. Generally, however, symptoms resolved within a few weeks of chemotherapy discontinuation.

Oculomotor disturbances were reported after fluorouracil in two patients, who had weakness of convergence and divergence, and these were probably an expression of neurotoxicity affecting the brainstem. Adriamycin, methotrexate, mithramycin, mitomycin C, procarbazine, tamoxifen, and cisplatin have also been associated with ocular problems (see Vizel and Oster 1982).

Cisplatin (cis-dichlorodiammineplatinum II)

Pippitt *et al.* (1981) documented the case of a patient with cervical cancer who developed cortical blindness after single-agent therapy with cisplatin. This patient, a 36-year-old woman, received two courses of cisplatin (100 mg/m²) separated by 30 days and was discharged from hospital on prochlorperazine. Thirteen days later she became totally blind and partially deaf. Extensive audiologic and neurologic assessment demonstrated a high-frequency hearing deficit bilaterally and an EEG abnormality that suggested an encephalopathy originating in the occipital lobes with diffuse bilateral spread. She was treated with phenytoin and dexamethasone and fully anticoagulated with heparin for suspected bilateral infarcts. Within 10 hours of initiation of treatment she regained her vision; an EEG performed two days later was normal and ophthalmologic examination showed no field defects. This patient had received a total dose of 306 mg of cisplatin.

An earlier report by Berman and Mann (1980) also documented transient cortical blindness in a 30-year-old man with testicular carcinoma during treatment with cisplatin in combination with vinblastine and bleomycin.

Such reports associating cisplatin with cortical blindness, especially when used in single-agent therapy, are important since its use alone or in combination with other agents in gynaecological malignancies is expanding (Thigpen *et al.* 1980).

Vincristine

Ocular toxicity occurs more commonly with vincristine than with vinblastine. Vizel and Oster (1982) cited, for example, that 14 of 20 patients had vincristine related ptosis, 13 had oculomotor disturbances, and two suffered corneal hypoaesthesia. These abnormalities improved with drug discontinuation or dosage reduction.

Shurin *et al.* (1982) reported bilateral optic-nerve atrophy in a 15-year-old girl who had received concomitant neuraxis radiation and intravenous vincristine sulphate (2 mg weekly) subsequent to the removal of a medulloblastoma from the fourth ventricle. At the time of the tenth maintenance dose of vincristine (reduced to 1 mg i.v. because of lower extremity weakness and inco-ordination) she complained of blurred vision, diplopia, and pain in her eyes. Visual acuity had decreased from 20/20 in each eye to 20/80 in the right eye and 20/400 in the left. In both eyes there were central scotomata and peripheral constriction of the visual fields.

Vincristine was stopped and concomitant prednisone (40 mg/m^2/day) was continued for two weeks then tapered off. Visual acuity began to improve as did the patient's strength and co-ordination. One year later her reflexes had returned to normal and her visual acuity continued to improve.

The authors of this report commented that CNS toxicity of vincristine is rare except in overdosage and they suggested that this patient's blood–brain barrier could have been impaired by surgery and irradiation.

Chlorpropamide

Visual abnormality is always a matter of concern in the diabetic patient and it is therefore important to record that Wymore and Carter (1982) have reported one case of chlorpropamide-induced optic atrophy.

The patient, a 65-year-old woman, had a one-year history of adult-onset diabetes; she had been treated during that time with chlorpropamide (250 mg daily). She complained of bilateral visual loss and on examination was found to have a corrected visual acuity of 20/400 OD and OS and she could not see colour testing plates or identify bright colours; she had no foveal reflex in either eye. The patient stopped taking chlorpropamide on her own accord and four days later her visual acuity measured 20/40 OD and OS. However, visual acuity decreased to 20/70 OD and 20/60 OS when the drug was reintroduced for five days.

Chlorpropamide was subsequently withdrawn and her diabetes was controlled with diet and weight loss. The authors of this report commented that chlorpropamide-induced optic neuropathy was an unusual but reversible entity that must be considered in the differential dignosis of visual loss in diabetic subjects.

Clomiphene citrate

Clomiphene citrate stimulates the secretion of pituitary gonadotrophic hormones, particularly the luteinizing hormone, and is used in the treatment of anovular infertility and in a variety of menstrual disorders.

Roch *et al.* (1967) reported that ocular symptoms developed in four out of 58 patients during clomiphene administration; examination revealed scotomata devel-

opment which disappeared after the drug was discontinued in one patient. It recurred with readministration of the drug, but not when a placebo was given. The ocular symptoms in three patients were reversed when clomiphene was stopped, and symptoms disappeared while the drug was being administered to the fourth patient.

Other clinicians have also noted ocular symptoms associated with the use of clomiphene, although many such reports have simply stated that some patients experienced blurred vision or visual disturbances, without giving any information on the percentage incidence. These reports implied that ocular symptoms occurred rarely (Jones and de Moraes-Ruehsen 1965). Some investigators have, however, given an indication of the incidence of ocular symptoms, although they did not always define precisely the nature of these defects. For example, one of the 22 patients with endometrial hyperplasia treated by Charles *et al.* (1964) had visual disturbances, and four of the 37 cases of Naville *et al.* (1964) complained of blurred vision. Two of five men given clomiphene by Morse *et al.* (1963) noted subjective luminous sensation by pressure on the eyeball (phosphene), and two of the 34 women treated by Riley and Evans (1964) noted blurring of vision which lasted almost a week after clomiphene was stopped; reinstitution of the drug caused a prompt return of blurred vision. Thompson and Mellinger (1965) noted visual disturbances in five of 41 patients, and in two of these the symptoms were severe enough to warrant stopping the drug. Four of 36 women treated by Pildes (1965) complained of blurred vision.

Johnson *et al.* (1966) documented the total incidence of visual symptoms related to clomiphene, which were reported to them, as 2 per cent in 2616 patients.

Corticosteroids

Raised intraocular pressure

Elevation of intraocular pressure has been a well-documented side-effect of both local and systemic administration of corticosteroids for some time (e.g. Becker and Mills 1963; Perkins 1965; David and Berkowitz 1969).

The onset of raised intraocular pressure is usually a matter of weeks after local, and months after systemic, therapy. This occur in about 30 per cent of patients receiving local therapy and in a lower percentage when the steroid is given systemically. Severe increases resembling those of acute glaucoma have been reported, and cupping of the optic discs and visual-field defects produced by the raised pressure are similar to those of open-angle glaucoma. The changes are usually reversible providing treatment is withdrawn. The fact that some individuals have a rise in intraocular pressure in response to corticosteroids does not necessarily indicate that they are in a preglaucoma state, and although the tendency to this response is probably hereditary, the exact manner of inheritance is uncertain. Raised intraocular pressure after topical corticosteroids is common in diabetes without proliferative retinopathy and in patients susceptible to steroid-induced carbohydrate abnormality.

The mechanism of steroid-induced ocular hypertension may involve increased aqueous production, but an increase in resistance to the outflow tract seems to be the most important contributing factor. The trabecular meshwork, which separates the anterior chamber from the canal of Schlemm, contains collagen strands and a single layer of endothelial cells. Corticosteroids may cause swelling of the collagen strands by increasing viscosity and water-binding capacity of the mucopolysaccharides. This would block the outflow tract and increase resistance. The incidence of ocular hypertension may be correlated with the anti-inflammatory effect of the topical steroid used. In glaucomatous eyes there may be a greater accumulation of mucopolysaccharides in the degenerated trabecular meshwork, thus causing heightened susceptibility to the increased outflow resistance induced by corticosteroids. Other effects of topical corticosteroids that could raise outflow resistance include increased vasoconstriction and pupil dilatation, both being a potentiation of the normal sympathetic tone of the eye.

What is claimed to be the first case of glaucoma associated with the use of flurandrenolone is described by Brubaker and Halpin (1975). An 18-year-old man with idiopathic hypogammaglobulinaemia, who complained of intermittent blurring of vision, was found to have raised intraocular pressure in both eyes and a visual-field defect consisting of a superior arcuate scotoma on the left. As the patient had been applying a relative's skin ointment containing flurandrenolone and neomycin to his eyes for self-treatment of blepharo-conjunctivitis, the diagnosis of corticosteroid-induced scotoma was considered. After the use of 1 per cent pilocarpine, the intraocular pressure of 42 mm Hg in each eye had fallen to 10 mm Hg in the right and 24 mm Hg in the left eye. The uncorrected visual acuity improved to 6/6 in each eye, but the left arcuate scotoma persisted. The patient refused further treatment or to attend for follow-up. Four years later the right eye was normal but the left was nearly blind, the left disc being totally cupped and the intraocular pressure on that side being 43 mm Hg. The right eye had been the least exposed to the steroid, but with the left eye, the steroid-induced glaucoma progressed after the use of the steroid was terminated.

This case highlights the danger of injudicious use of corticosteroid-containing ointment in self-medication of the eye.

Posterior subcapsular cataracts

Black and his co-workers in 1960 first described posterior subcapsular cataracts (PSC) in rheumatoid arthritis patients on long-term corticosteroid therapy; this was followed by other reports giving varying incidences of the association of posterior subcapsular cataracts and long-term corticosteroid therapy in adults (for example Spencer and Andelman 1965; Fürst *et al.* 1966) and in children (for example Havre 1965; Braver *et al.* 1966, 1967).

Steroid-induced cataract is almost always bilateral; the lesion usually occupies the polar region of the posterior cortex, just within the posterior lens capsule; it extends forward into the cortex irregularly but its borders are sharply defined (Oglesby *et al.* 1961). Vision is not usually impaired early in the development of a steroid-induced cataract, and slit-lamp examination is necessary for early detection.

There is some contention as to whether steroid-induced cataracts can be distinguished from cataracts associated with intraocular disease and from cataracts which occasionally develop in adult rheumatoid-arthritic patients. There is agreement, however, that the steroid-induced cataract can usually be distinguished from the senile cataract, which tends to spread peripherally, is associated with other cortical changes, and tends to cause early loss of vision. David and Berkowitz (1969) reviewed the literature on this topic. Similar cataracts were reported following oral-contraceptive medication.

Spencer and Andelman (1965) surveyed the literature for the incidence of PSC formation in patients receiving long-term steroid therapy for various diseases (e.g. asthma, rheumatoid and non-rheumatoid arthritis, systemic lupus erythematosus, and nephrotic syndrome). This varies from 0 to 42 per cent. In the control studies reported in the literature on patients with rheumatoid arthritis who were not treated with corticosteroids, the incidence of PSC formation varied from 0 to 3 per cent.

In a specific study on 58 rheumatoid-arthritis patients who had previously received intramuscular injections of triamcinolone acetonide, Spencer and Andelman (1965) reported an overall incidence of 60 per cent PSC formation. Generally the greater the total amount of drug received, the higher the incidence of PSC formation. For example, there was PSC formation in 33 per cent of patients receiving between 500 and 1000 mg of steroid and this incidence rose to 100 per cent in patients who received more than 2000 mg of the drug.

In a related study on 113 patients comprising 56 children with Still's disease and 57 adults with classical rheumatoid arthritis, receiving maintenance therapy, Fürst *et al.* (1966) found seven adults and six children to have posterior subcapsular lens opacities. The duration of therapy with the steroid (prednisone) seemed to be less important than the average daily dose; while this tended to be higher in the patients developing cataract, particularly among the juveniles, eight of the 13 cataracts occurred in patients whose daily dose of prednisone was 10 mg or less.

In specific studies on children, Braver *et al.* (1966, 1967) confirmed that in children, as in adults, PSC is associated with prolonged corticosteroid therapy and is not associated with a particular disease. No children on therapy for less than two years developed PSC and the authors reported that when PSC did occur in three children there was no regression of the PSC three months after stopping corticosteroids, suggesting that if regression did occur it would be very slow.

There is some evidence to suggest that PSC complications of systemic corticosteroids are more common in children (Braver *et al.* 1967) with lower dosage and with shorter periods of therapy (Fürst *et al.* 1966). The opinion of these early investigators was that the incidence of cataracts with systemic corticosteroids depended predominantly upon the dose and duration of drug therapy rather than on the underlying disease (Giles *et al.* 1962; Braver *et al.* 1967). However, while the cataractogenic effects of corticosteroids are beyond dispute, disagreement exists concerning effects of total dose, intensity of dose, and duration of administration on cataract formation. It is therefore of particular interest to study the more recent observations of Skalka and Prchal (1980) on factors possibly relevant to posterior subcapsular (PSC) cataract formation and to note that some of their findings are at variance with those of the earlier studies.

These investigators studied 106 adult male patients attending a Hematology/Oncology clinic; patients were matched for age, race, and socio-economic status. Possible PSC cataract formation was compared among those with (39) and without (67) a history of systemic corticosteroid therapy. Almost all the patients were between 40 and 80 years of age; the average age of the patients who had received corticosteroid therapy was 57.3 years, while that of the control group was 57.9 years.

Difference in incidence of PSC opacities among patients with and without a history of corticosteroid therapy was statistically significant ($p < 0.001$). Twenty of the 39 patients (51.3 per cent) who had received prednisone showed evidence of some degree of PSC opacity, whereas in the control group only 10 of 67 patients (14.9

per cent) showed opacities. However, no statistically significant correlation was found between the PSC opacities and total steroid dose, intensity of dose, time course of dosage, or age of patient.

The findings of this study suggest that the most important factor in steroid-induced PSC cataract formation in adults may be variability·in individual susceptibility to the side-effects of corticosteroids and that possibly constitutional (genetic) factors may be important dominants.

Skalka and Prchal suggested abandoning the earlier concept of a 'safe' dose of corticosteroid and urged that the possibility of PSC cataract changes should neither be ignored with short-term corticosteroid therapy, nor be 'expected' when total prednisone therapy reached a certain level; it should be diligently looked for throughout the course of such therapy. They recommended that patients in whom PSC changes develop should have their corticosteroid treatment reduced to the bare minimum consistent with the control of their disease and should be considered for alternate-day therapy whenever possible.

There have been only a few reports of lens changes related to local corticosteroid therapy. It is therefore important to note the report of Yablonski and Burde (1978), who found that in 11 diabetic patients, progressive lenticular changes occurred in nine of 11 eyes treated with topical dexamethasone (0.1 per cent) compared with one of 11 untreated eyes. Eight of the nine treated eyes that developed lenticular changes demonstrated pathologic conditions of the posterior subcapsular lens region. The patients were treated for 14–36 months (mean 21.6 months), in most cases, with dexamethasone drops applied four times daily.

The reporting clinicians suggested that this high incidence of cataract formation was related to a basic diabetic metabolic defect.

Digitalis preparations

The total incidence of ocular manifestations in patients with digitalis intoxication has been estimated to be as high as 25 per cent (Robertson *et al.* 1966; Leopold 1968). In approximately 10 per cent of patients with digitalis toxicity, the ocular complications may occur prior to other symptoms of toxicity although with the majority, adverse ocular effects appear simultaneously with, or after, signs of cardiac toxicity.

The spectrum of oculotoxic effects with digitalis preparations is wide and has long been known; indeed in 1785, Withering noted that 'foxglove when given in large and quickly repeated doses occasions sickness, vomiting, giddiness, confused vision, objects appearing green or yellow'.

The most common of these side-effects are blurred and disturbed colour-vision; although yellow vision (xanthopsia) is usually described, red, green, blue, brown, and white chromatopsia have also been noted. Photophobia, 'snowy vision', flickerings, flashes, sparks, and scintillating scotomata occur. Disturbances of visual acuity are common and both transient and permanent amblyopias accompanied by central scotomata have been described. Central scotomata may be secondary to retrobulbar neuritis or to toxic effects of digitalis on the retinal receptor cell. There is evidence that the cone-dark-adaptation threshold may be elevated during digitalis therapy. Transient or permanent total blindness probably is rare.

The precise site in the visual pathway where digitalis acts to cause ocular disturbance is unclear. There have been reports suggesting toxic involvement of the retina, optic nerve, and visual cortex. The interval between the initial orally administered dose and the first manifestation of toxicity may be as short as one day. However, the symptoms more often appear within the first two weeks of therapy, although they may not occur until after several years of cardiac glycoside administration. The duration of symptoms after termination of therapy varies; most disappear within two or three weeks, and only rarely are they permanent (Leopold 1968).

Objective findings reported with some of the subjective symptoms include mild conjunctivitis, pupillary variations such as mydriasis and miosis, exophthalmos, retrobulbar neuritis, possible nystagmus, and ocular-muscle palsies. The 'possible nystagmus' is doubtful, although it has frequent mention in the literature. It is probably attributed to a case reported by Galentine in 1854, and is possibly an incorrect implication of digitalis. Likewise exophthalmos is also suspect; it has been suggested that an abrupt rise in venous pressure subsequent to cardiac dysfunction may have been the cause rather than the glycoside (Robertson *et al.* 1966).

The work by Church *et al.* (1962) indicated that there was no correlation between the specific drug preparations and the frequency of visual manifestations. However, Dubnow and Burchell (1965) found that, whereas 17 per cent of 98 patients with digitoxin toxicity experienced visual disturbances, none of the 38 patients with digoxin intoxication had such disturbances. This difference may be ascribed to the relatively slower rate of excretion of digitoxin and its cumulative effects.

Dimethyl sulphoxide (DMSO)

Oral or cutaneous administration of this solvent caused lens changes in several animals species (Rubin and Mattis 1966; Wood 1966; Denko *et al.* 1967; Kleberger 1967; Kolb *et al.* 1967; Rubin and Barnett 1967; Smith *et al.* 1967; Wood *et al.* 1967; Marzulli 1968; Mason 1971). Similar changes have not, however, been demonstrated in man and available evidence would suggest that DMSO is not oculotoxic to man (Kutschera 1966; Scherbel 1966; Gordon 1967; Jacob and Wood 1967*a, b*; Gordon and Kleberger 1968; Hull *et al.* 1969; Jacob 1971).

It is difficult, however, to reconcile the clear-cut picture of DMSO oculotoxicity in several animal species with the negative results from human studies, notwithstanding the fact that many of the human studies have included a precise objective of monitoring for such adverse effects on the eye. It must also be appreciated that, in the experimental animal studies, it was generally true that reported adverse effects on the eye had occurred only when concentrations of DMSO were employed which were far in excess of those likely to be used clinically.

Non-steroidal anti-inflammatory drugs

Ibuprofen

The identification of causative oculotoxic components of therapy is a problem with patients suffering from rheumatic disorders. Chloroquine has already been implicated but treatment may include several drugs capable of causing ocular side-effects. The situation may be further complicated by lesions due to the disease process itself, the effects of previous therapy, or coincident ocular disease in the middle-aged and elderly patient.

The publication by Collum and Bowen (1971) of a description of two cases of toxic amblyopia attributed to ibuprofen treatment was soon followed by other reports. Palmer (1972) described deterioration of vision in a 57-year-old woman who had been taking ibuprofen for fibrositis for some five months. The patient had difficulty with blue/green colour vision which was especially noticeable when she was choosing clothes. At the time of her first attendance, she had taken about 168 g of ibuprofen. Ophthalmoscopically, the eyes were normal but her comprehension was inadequate for satisfactory visual-field studies. She was found to be almost totally colour blind, and, although reliable studies were impossible, a diagnosis of toxic amblyopia was made. Her corrected visual acuity had improved from 6/24 to 6/9, four months after ibuprofen was stopped and further improvement was seen some two months later.

These three cases seemed to have certain features in common. The symptoms of toxic amblyopia appeared early in the course of treatment and after a comparatively small dosage of the drug. The symptoms were sufficiently rapid in progression and sufficiently severe, especially the deteriorating visual acuity, to compel the patient either to stop the drug or seek ophthalmic advice. The signs and symptoms of toxic amblyopia disappeared within a few days of weeks of stopping ibuprofen.

A different aspect on the oculotoxic status of ibuprofen was presented by Thompson (1972) who, in a retrospective study, observed that three patients out of a total of 38 receiving ibuprofen for rheumatic disorders had evidence of visual-field defects and pigmentary changes in the region of the macula. These changes were presumed to be degenerative; they were not associated with impairment of vision and they remained unchanged when therapy with ibuprofen was stopped for four months. However, in the same study six out of a total of 34 patients suffering from rheumatic disorders, and who had not received ibuprofen, showed ocular abnormalities, including one example of macula degeneration and one of retinal pigmentation. Thompson commented that, in view of the previously reported apparent examples of ibuprofen-induced toxic amylyopia (Collum and Bowen 1971; Palmer 1972), prospective studies involving 'base-line' ophthalmic examination might be helpful. He believed that the occurrence might be limited to a few patients who had a special liability.

A more recent report of oculotoxicity involving ibuprofen was made by Tullio (1981) who described a 35-year-old woman with degenerative joint disease of the lumbar spine of $2\frac{1}{2}$ years duration who had received non-steroidal anti-inflammatory drugs (indomethacin, naproxen, and aspirin) previously, but was changed to ibuprofen, initially 800 mg three times daily for seven days, then a decreased dosage of 400 mg two or three times a day.

About three months later at a routine follow-up appointment she complained of shooting streaks in her vision and ear fullness. The streaks were shooting from the lateral towards the central portion of vision. Questioning revealed that she had not experienced any relief from the 400 mg t.d.s. dosage of ibuprofen, and approximately one month after starting the treatment she had therefore increased her dose to 800 mg thrice daily.

Her colour vision, undilated fundus, dilated fundus, and visual fields were within normal limits and aspirin was substituted in place of the ibuprofen. However, during a return visit two weeks later the patient admitted that due to nausea with the aspirin dosage, she had started taking the ibuprofen again and as before she experienced the shooting streaks in her vision. The patient was cautioned against taking the ibuprofen and was instructed to take an antacid along with the aspirin; the streaks in her vision then stopped.

This case, although complicated by previous treatment with anti-inflammatory drugs associated with oculotoxicity suggested a strong relationship between ibuprofen and the development of visual disturbances. It also, incidentally, exemplified the dangers inherent in noncompliance with medication instructions.

Other anti-inflammatory drugs

Whereas ocular damage by chloroquine is established and ibuprofen is incriminated, the evidence for the oculotoxicity of other non-steroidal anti-inflammatory agents is less evident. However, indomethacin has been reported to have chloroquine-like oculotoxicity, causing punctate corneal opacities reversible on withdrawal of the drug, and alterations in the retinal pigment epithelium leading to disturbances of macular vision. This reaction has been reported in children (Burns 1968; Mushin 1972).

Phenylbutazone* and its cogener oxyphenbutazone* have also been associated with ocular effects, especially in children (Hertzberger 1965). These ocular effects are features of a Stevens–Johnson syndrome, and include adhesion and gross scarring of the lids, corneal ulceration and scarring, vascularization of the cornea, and loss of vision; retinal haemorrhage may occur. Phenylbutazone has also been implicated in optic neuritis (Crews 1962).

Retinal haemorrhages due to the prolonged use of salicylates are rare. However, Mortada and Abboud (1973) reported two such cases. The first patient, a 60-year-old woman had been taking 6 g of sodium salicylate daily for two months; she complained of deterioration of vision. Flame-shaped retinal haemorrhages were present and arteriosclerotic retinopathy was incorrectly diagnosed. When salicylates were stopped, the visual acuity in each eye improved from 4/60 to 6/9 and the retinal haemorrhages resolved after two months.

The second patient, a 10-year-old girl, who had been taking 4 g of sodium salicylate daily for 40 days, complained of diminution of vision and was found to have bilateral flame-shaped retinal haemorrhages. These gradually absorbed when salicylates were stopped, and visual acuity improved from 2/60 to 6/12 in the right eye, and 6/60 in the left eye.

Of the newer anti-inflammatory drugs benoxaprofen gained notoriety during the autumn of 1982 due to its association with over 3500 reports of adverse reactions in the United Kingdom; included among those reports were 61 fatal cases, predominantly in the elderly. Because of concern about serious toxic effects of the drug on various organ systems, particularly the gastrointestinal tract, the liver, and bone marrow, in addition to known effects on skin, nails, and eyes, the Licensing Authority, acting on the advice of the Committee on Safety of Medicines, suspended its product licence. Similar action was rapidly taken by drug regulatory authorities in other countries.

Dodd *et al.* (1981) associated benoxaprofen with a case of toxic optic neuropathy in a 65-year-old woman who had a previous clinical history of aortic and mitral valve replacement for valve damage secondary to previous rheumatic fever and was subsequently treated with warfarin sodium. Four years later she developed an inflammatory polyarthritis, and one year after this (May 1980) she developed a gastric ulcer and was started on cimetidine. In July 1980, benoxaprofen 600 mg nightly was started for her active rheumatoid arthritis with definite improvement in the degree of pain and synovitis. Two months later (September 1980) she complained of progressive blurring of vision over 10 days unassociated with any ocular discomfort. Acuities were reduced to 6/12 in each eye with a small hypermetropic spectacle correction. She showed a severe red–green colour defect and had bilateral central scotomata to a 15-mm red target (Bjerrum's screen). Electroretinography showed rod and cone responses and normal dark adaptation pattern. Visual evoked responses to a 50 flash/minute pattern reversal stimulus showed low-amplitude potentials that were grossly delayed (P100 = 185 ms right and left: normal 108 ± SD 2 ms). Benoxaprofen was stopped, and over the next month her vision improved.

By March 1981, visual acuities were 6/9 in each eye. The right visual field was full to both at 3-mm white and 15-mm red target, but in the left eye there was still a relative central scotoma to a 15-mm red target. An electro-retinogram remained normal and the latency of the P100 component of the visual evoked response had improved to 138 ms right and 140 ms left. The optic discs remained ophthalmoscopically normal.

Since concomitant treatment (warfarin and cimetidine) remained unchanged since before the benoxaprofen treatment, the authors concluded that the visual deterioration in this patient was the result of a toxic neuropathy due to benoxaprofen. Confirmation by reintroducing the drug was not ethically justifiable.

It may be possible in this case that the adverse effects of benoxaprofen were potentiated by the concomitant treatment with cimetidine. Cimetidine is known to inhibit the action of hepatic microsomal enzymes and benoxaprofen is metabolized by the liver. The possibility of such an interaction between benoxaprofen and cimetidine was apparently not considered by Dodd and his colleagues.

Other reports linking benoxaprofen with ocular damage were made by Halsey and Cardoe (1982) who surveyed the side-effect profile in 300 patients; (one patient had blurred vision) and Larkin *et al.* (1982), who found that one of their 24 patients taking benoxaprofen had a photosensitive rash and noticed photophobia-discomfort in and between the eyes in bright light.

Oxygen

The fact that oxygen has an extremely toxic effect on the developing retinal vessels has been known for some time. Its extensive use in small premature infants has led to cases of retinal ischaemia followed by neovascularization, haemorrhage, fibrosis, and eventual blindness in the condition of retrolental fibroplasia (RLF) (Ashton 1957; Patz 1966, 1969; Mushin 1971, 1972). It is a matter of concern that RLF still occurs in premature infants treated with oxygen (Gerhard *et al.* 1979) despite improved methods of monitoring PaO_2 and this has stimulated Bougle and his colleagues (1980) at the Neonatal Medicine and Resuscitation Service, Nancy, France, to study its pathophysiology by measuring retinal superoxide dismutase (SOD) concentration in kittens.

The retina of the newborn kitten has a similar degree of immaturity to that of a human infant at 26-week's gestation and responds to hyperoxia with changes in RLF identical to those seen in premature babies (Ashton *et al.* 1954). Superoxide dismutase is a ubiquitous enzyme which protects cells against the toxic effects of oxygen or free oxygen radicals (Fridovich 1978) and the tolerance of tissue to hyperoxia is directly related to an increased activity of this enzyme (Crapo and Tierney 1974; Fridovich 1978).

Bougle *et al.* (1980) exposed newborn kittens to ambient oxygen concentration of 80 per cent for 72 hours; this was sufficient to produce irreversible retinal lesions. Littermates kept in warm air for the same period were used as controls. Total SOD was than measured in the retina; retinal SOD activity was 525 ± 159 (SD) U g^{-1} in the seven controls and was significantly reduced to 260 ± 96 (SD) U g^{-1} in the seven kittens exposed to oxygen. The authors of the report suggested that RLF may be related to suppression, or a lack of induction, of retinal SOD in response to hyperoxia.

Phenothiazine tranquillizers

Chlorpromazine was introduced into the clinic in 1953 and since that time several ocular effects caused by it and its numerous derivatives have been reported. They include diplopia due to paresis of extraocular muscles (Crews 1962), transient myopia due to promethazine (Bard 1964), allergic conditions of conjunctiva or lids

(Dencker and Enoksson 1966; McClanahan *et al.* 1966; Mathalone 1967), optic atrophy with chlorpromazine (Rab *et al.* 1969), with thioridazine (Bonaccorsi 1967), and with perphenazine combined with large doses of thioridazine (Bonaccorsi 1967).

Oculogyric crises have occasionally been observed as a part of the parkinsonian syndrome which is sometimes produced (Ayd 1961) and blurred vision, due to an atropine-like action causing a temporary paralysis of accommodation, has been reported (Giacobini and Lassenius 1954; Sinha and Mitra 1955). Jonas (1959) reported a chlorpromazine-induced miosis and Bock and Swain (1963) observed two cases of raised intraocular pressure. There is, therefore, little doubt that chlorpromazine and its related phenothiazine tranquillizers have a multifarious range of toxic effects on the eye. It is, however, only since the report of Greiner and Berry (1964) that the long-term effects of phenothiazine therapy on the eye have been fully appreciated relative especially to pigmentation, lenticular and corneal changes, and to retinopathy.

Greiner and Berry (1964) reported lenticular opacity and cutaneous pigment in 70 women, out of a series of many thousands of patients undergoing prolonged and intensive chlorpromazine therapy. Cairns *et al.* (1965) also reported four cases of *un visage mauve* in women with chronic catatonic schizophrenia. All received chlorpromazine in large doses because of periods of acute excitement. The combination of cataract and skin changes on the exposed areas of the face suggested photosensitization with the visible light rays acting upon sensitized lenticular protein. The cases were linked with a summer period of unusually long periods of sunshine. There were similar reports by Perrott and Bourjala (1962) of one case, (due to which they proposed the term *un visage mauve*), by Zelickson and Zeller (1964) of eight cases, by Hays *et al.* (1964) of five cases, by Feldman and Frierson (1964) of one case, and by Massey (1965) of one case.

The initial report of Greiner and Berry (1964) thus alerted other clinicians to carry out ophthalmoscopic examination on patients known to have prolonged exposure to chlorpromazine or other phenothiazines. Thus Margolis and Goble (1965) found lenticular opacities and pigmentary deposits in eight of 31 schizophrenic patients who had received a succession of phenothiazine derivatives for between three and five years, with chlorpromazine having a relatively minor role and trifluoperazine a major one. None of the patients in this study showed unusual cutaneous pigmentation.

Barsa *et al.* (1965) performed visual acuity, ocular tension, slit-lamp biomicroscopy, and ophthalmoscopic examinations on 658 women under treatment for psychiatric disorders; all except seven were receiving, or had received, phenothiazine derivatives. Characteristic opacities of both lens and cornea were found in 33 patients and lens opacity alone in 145 patients. In all, 178 patients (27 per cent) were affected with lens opacities; these opacities were characteristic and could easily be distinguished from the varying degrees of senile cataract found in 175 patients. At the time of the eye examinations, none of the patients were showing evidence of skin photosensitivity. However, 103 patients had previously manifested an excessive tendency to sunburn during their phenothiazine therapy. Eighteen had developed marked suntanning and 97 had acquired a grey or violaceous pigmentation.

Mathalone (1967) commented that the majority of reports of ocular changes with chlorpromazine therapy had emanated from North America (e.g. Feldman and Frierson 1964; Greiner and Berry 1964; Greiner *et al.* 1964; Zelickson and Zeller 1964; Ban and Lehmann, 1965; Barsa *et al.* 1965; DeLong *et al.* 1965; Margolis and Goble 1965; Massey 1965; Wetterholm *et al.* 1965) where the drug was given in larger doses than in Britain. He therefore undertook a clinical survey of patients treated with chlorpromazine in England, and 462 patients in seven mental hospitals were examined for changes in the eyes and skin. Of those patients receiving more than 300 mg of chlorpromazine daily for three years or more, 36 per cent showed lens changes and 17 per cent showed corneal changes. Except for occasional complaints of slight blurring of vision no subjective ocular symptoms could be elicited. There was evidence of photosensitization in exposed areas of skin in many patients, the degree of pigmentation depending upon the extent of exposure to sunlight and varying from patient to patient. Approximately 10 per cent of patients showed some conjunctival pigmentation, but the significance of this was thought to be doubtful since many normal people show patches of similar pigmentation. No scleral pigmentation was seen.

The typical findings in the lens and cornea in patients with opacities due to phenothiazines are as follows: the sites of the lesions are the anterior part of the lens and the back of the cornea, and this leads, as Delong *et al.* (1965) suggested, to the strong implication that the drug or some metabolites accumulate in the anterior chamber and are responsible for these changes.

The corneal stroma contains a diffuse, yellow, granular pigmentation; the pigment does not accumulate on the cornea shielded by the upper eyelid. On the lens the opacities are situated to the central pupillary zone. These opacities consist of fine, yellow, granular clusters. Photosensitivity would therefore appear to play a part in the development of these opacities.

The toxic effect of phenothiazines on the eye is not only confined to the lens and cornea; indeed Cant (1969) wrote that 'prolonged administration of all phenothiazines must be considered as liable to lead to retinopathy'. Phenothiazines accumulate in the pigment cells, particularly within the melanin fraction (Potts 1962) and pigmentary retinal changes have been reported following the use of chlorpromazine and prochlorperazine (Weekley *et al.* 1960), piperdichlorphenothiazine, and thioridazine (Leopold 1968).

The retinal effects superficially resemble retinitis pigmentosa in that night-blindness occurs and the retinal arteries show diffuse narrowing. Central vision is affected early. However, unlike typical retinitis pigmentosa, the pigment usually aggregates in large clumps rather than in bone-corpuscle configurations.

Much of the investigational work into phenothiazine and ocular toxicity was done two decades or more ago, but ocular reactions continue to be reported with the phenothiazines (see reviews by Inove 1979; Bond and Yee 1980). Fortunately, most of these reactions are transient and are generally of limited clinical consequence, but a few have led to organic changes in the affected structure or tissue, and have resulted in deterioration of visual efficiency and, in some cases, in permanent visual loss.

The pathogenesis and treatment of the ocular and cutaneous effects of chronic phenothiazine therapy were reviewed by Bond and Yee (1980). Lenticular and corneal lesions, including epithelial keratopathy, were associated primarily with several years of high-dose chlorpromazine therapy; in most cases the lesions do not significantly affect the patient's vision.

Pigmentary retinopathy is the most serious of the phenothiazine-induced ocular effects, thioridazine being the most common offender. Critical factors in the development of thioridazine-induced retinopathy are the dosage and duration of therapy. High and continued dosage of thioridazine is required before ocular damage is observed (e.g. above 200 mg daily for many months), and most instances of reported complications were seen in patients who had received 600–800 mg of thioridazine hydrochloride daily. Pigmentary retinopathy has occurred when thioridazine was administered for as little as 3 weeks. Applebaum (1963) failed to find characteristic changes at smaller dosage.

If a patient develops potentially serious ocular side-effects while receiving a prolonged high-dose regimen of phenothiazine, the clinician should consider temporarily discontinuing phenothiazine therapy and prescribing other non-phenothiazine neuroleptics (e.g. haloperidol) or phenothiazines that are effective at low doses (e.g. piperazine derivatives). If the patient's overall condition warrants a continued high-dose phenothiazine regimen, then frequent and careful examination of the patient's visual status are needed.

Phenylephrine

The use of phenylephrine as a mydriatic agent for routine fundoscopic examination, refraction, and during retinal surgery is not without hazard. The National Registry of Drug-induced Ocular Side Effects (see Fraunfelder 1977) in the USA listed two cases of epithelial oedema and sloughing-off of the complete epithelium with the use of 10 per cent phenylephrine hydrochloride.

Edelhauser *et al.* (1979), on the basis of animal studies, expressed concern at the prolonged use of 10 per cent phenylephrine in the eye, especially in diabetic patients whose corneas may have epithelial ulceration as well as a slow rate of re-epithelialization. They suggested the use of ointment formulations to give safer delivery of phenylephrine.

However, a study by Olsen *et al.* (1980), from Denmark, showed that corneal thickness is higher in the diabetic patient than in non-diabetic controls in the same age range. They tentatively suggested that increased permeability of the corneal endothelium might be the cause of this increase. This may well influence the rate of delivery of phenylephrine in formulations other than simple solutions.

Psoralens

The psoralens are well-known photosensitizing compounds that are being increasingly used to treat psoriasis. One of these agents, methoxsalen (8-MOP) had been detected in human lenses as well as in ocular tissues of treated animals (Lerman and Borkman 1977; Lerman *et al.* 1977, 1980*a*).

In the lens, 8-MOP is capable of forming photoadducts with pyrimidine bases in DNA as well as with tryptophan residues in lens proteins. This is not a problem in the absence of photic stimuli since the 8-MOP will diffuse out of the lens within 24 hours. However, if there is exposure to ambient room light or to direct 360-nm ultraviolet radiation, then 8-MOP photoadducts will be formed and retained in the lens.

Since 8-MOP requires long-wave u.v. radiation (320–400 nm) for excitation, it has generally been considered that, even if this drug were present in other ocular tissues, the ocular lens would absorb almost all this radiation and would so prevent the generation of 8-MOP photoreactions in the vitreous humour and retina. Evidence by Lerman and Borkman (1978), however, indicated that the young lens transmits a considerable amount of u.v. radiation (longer than 300 nm) but that there is an increased absorption of this radiation as the lens ages. Lerman (1980) commented that by the time the person reaches 20 to 25 years of age, the ocular lens has become a highly effective filter of long-wave u.v. radiation.

However, since 8-MOP can be detected in the human lens, repeated therapy with this drug could result in the accumulation of photoproducts with the ocular lens; the retina in children and persons with aphakia could also be susceptible. Evidence supporting this view has been gained from studies in young rats and newborn dogfish which demonstrated that 8-MOP can be found in a variety of ocular tissues and that it can be photobound in the retina as well as in the lens providing that the lens transmits sufficient long-wave u.v. radiation (Lerman 1980; Lerman et al. 1980b; Gardner et al. 1980).

Gardner et al. (1980) showed that the Spectra-Shield lens or lens coating can completely prevent this type of reaction, since it reflects all u.v. radiation up to 400 nm; Lerman (1980) therefore suggested that, since hazard may result from the accumulation of photosensitizing agents in ocular tissues, all patients undergoing photo-therapy should be protected by wearing suitable glasses (including side-pieces) to prevent reflected radiation. He recommends that such glasses should be worn imme-diately after ingestion of the psoralen and for at least 24 hours afterwards.

Reassurance of the absence of hazard was given by Hammershøy and Jessen (1982) in their retrospective study of cataract formation in 96 patients treated with PUVA (8-MOP and u.v. radiation). No patient developed cataract during the PUVA treatment period; of the 96 patients, 36 (37.5 per cent) had cataract when commencing the PUVA treatment. The remaining 60 patients (62.5 per cent) were free from any sign of cataract both before and after the PUVA treatment; 28 (77.8 per cent) of the 36 patients had 'unchanged cataract'. In the remaining eight patients slit-lamp examination gave the impression of cataract growth. These findings were found to correspond to those in a 'standard population' with respect to prevalence of cataract formation.

Traditional antiglaucoma agents

Pilocarpine and adrenaline (epinephrine)

It is a well-recognized fact that side-effects of traditional antiglaucoma agents occur all too frequently. In an *aide mémoire* on systemic reactions to ophthalmic drugs, the *Medical Letter* (1982) drew attention to nausea, abdo-minal pains, and parasympathetic effects that occur with high-dosage pilocarpine. The literature on these tradi-tional agents, pilocarpine and adrenaline, suggests that roughly 20 to 30 per cent of patients experience unplea-sant symptoms, especially burning, smarting, and blurred vision (Katz 1978; Plane et al. 1978; Thomas et al. 1978; Heel et al. 1979). Moreover, allergy or intole-rance to adrenaline can be a problem for many patients (Yablonski et al. 1977; Kohn et al. 1979). In some patients side-effects are of sufficient severity to cause discontinuation of the treatment; however, because alternative therapies have not been available until recent years (i.e. beta-blocking agents applied locally), the majority of patients have learned to tolerate the side-effects of these antiglaucoma agents even though symptoms may interfere with normal daily activities.

It is therefore of significant importance that Stafford (1981) has conducted a clinical survey to assess the inci-dence and severity of 10 side-effects commonly asso-ciated with the use of traditional antiglaucoma agents. Reports from 240 patients were analysed: 167 using pilo-carpine, 43 using adrenaline (epinephrine), and 30 using fixed combinations of the two agents.

Responses indicated that at least one of the 10 specific side-effects (burning, itching, smarting, blurring of vision, difficulty with night vision, red eye, difficulty in reading, photophobia, headache, browache, nervous-ness, or spasm of accommodation) was reported by 201 (84 per cent) of the 240 patients. The average number of side-effects per patient was 3.1.

The four side-effects reported most often were burning, itching, and smarting; blurring of vision; diffi-culty with night vision; and red eye. Red eye and burning, itching, and smarting were more severe in patients using adrenaline, while blurring of vision and difficulty with night vision were more severe in patients using pilocar-pine or pilocarpine/adrenaline combinations.

Eighty-five per cent of the 197 patients who provided data on the frequency of side-effects reported that at least one side-effect occurred daily. Of the 188 patients who furnished data on the severity of the side-effects, 65 per cent reported at least one symptom of moderate-to-severe intensity. Of 229 respondents, 25 per cent reported restriction of normal activities due to the side-effects they experienced, while 12 per cent had to give up some activity entirely, particularly driving and reading.

This survey is timely and illuminating; it shows clearly that traditional antiglaucoma therapy is associated with a much higher incidence of side-effects than was formerly realized. In the past, ophthalmologists have felt that good control of glaucoma was possible in terms of intra-ocular pressure, although not necessarily in terms of freedom from annoying or even incapacitating side-effects. Alternative treatments to these traditional agents are now available (e.g. timolol maleate eye drops) and with them it may be possible to consider the quality of the patient's life as well as good control of his glaucoma.

Apart from the use of timolol, there has been some other work to examine the advantage of other antiglaucoma treatments as alternatives to the traditional agents. Hung *et al.* (1982) from Taiwan demonstrated the ocular hypotensive actions of N-demethylated carbachol hydrochloride (DMC) in a 6 or 9 per cent solution in patients with open-angle glaucoma. In comparison with a reference 1 per cent pilocarpine hydrochloride solution, DMC solutions had slightly less potent ocular hypotensive action, but they had fewer of the side-effects common to pilocarpine, such as sympathetic ganglion stimulation, exocrine secretion, miosis, and ocular irritation.

An interesting study was also reported by Hitchings and Glover (1982) from Moorfields Eye Hospital, London, in which adrenaline (1 per cent solution) with guanethidine (1 per cent solution) was compared against adrenaline solution alone in a randomized, prospective, double-blind, cross-over study in 20 patients.

The patients received treatment for eight weeks with each preparation and a significant increase in the ocular hypotensive effect was induced by the addition a significant increase in the ocular hypotensive effect was induced by the addition of the guanethidine. It was also significant, however, that the incidence of side-effects was greater with the combination formulation than with the adrenaline solution alone (Table 30.3). Approximately 25 per cent of the patients receiving the combined eye drops were not prepared to tolerate such treatment for the eight-week period of the study; this was a slightly greater number than those who were not prepared to tolerate adrenaline alone. An earlier study by Murray *et al.* (1981) suggested an advantage of combining 1 per cent

solutions of adrenaline and guanethidine, over adrenaline alone.

It is therefore suggested by these two latter studies that those patients with an inadequate ocular hypotensive response to adrenaline alone could well benefit by being changed to a combination of adrenaline 1 per cent with guanethidine 1 per cent. However, there may be a corresponding increase in the side-effects experienced.

Tricyclic antidepressants

Amitriptyline

The use of anticholinergic (parasympatholytic) drugs is well established in ocular treatments in producing mydriasis and cycloplegia; they are also well known to produce unwanted ocular effects in systemic poisoning even after topical administration (Polak 1980). It is not surprising therefore that other agents having anticholinergic properties can produce adverse effects on the eye especially when taken in overdosage.

Two cases have demonstrated that the tricyclic antidepressant, amitriptyline, which together with its active major metabolites, nortriptyline, has a potent anticholinergic action, is likely to produce adverse effects on the eye when taken in gross overdosage.

The first case reported by Spector and Schnapper (1981) was a 30-year-old pregnant woman who was admitted to hospital, unconscious, after taking 1–1.5 g of amitriptyline (normal maximum dosage range 100–150 mg daily). Her limbs were flaccid and areflexic and her eyes responded only slowly to light, and did not respond to touching the cornea. They were in midposition and did not react to aural irrigation with 200 ml cold water. Two intravenous doses of physostigmine 2 mg, one hour apart, increased the level of consciousness and motor activity, but did not alter the paralysed eye movement. Further cold caloric stimulation initially produced tonic conjugate deviation without nystagmus, then, as consciousness increased, conjugate deviation with nystagmus; voluntary eye movements were normal next day.

The second case, a 26-year-old woman took an unknown quantity of amitriptyline when she was suffering from chronic depression. Delaney and Light (1981) described how she responded incoherently to verbal commands and to pain stimulation, but there were no horizontal or vertical eye movements. Corneal reflexes were present. She became coherent and her gaze paresis resolved after 24 hours.

Amoxaprine

Amoxaprine is a tricyclic antidepressant which shares in

Table 30.3 Comparative incidence of side-effects with the use of ophthalmic solutions of adrenaline (epinephrine) plus guanethidine (1 per cent of each) versus adrenaline (1 per cent alone). The study utilized 20 patients with ocular hypertension in a double-blind, cross-over design*

Symptom	Number of patients	
	Guanethidine and adrenaline	Adrenaline alone
Headache/brow ache	4	1
Red eye	7	2
Burning	0	0
Ptosis	0	0
Ocular pain	1	0
Blurred vision	2	1

* Source: Hitchings and Glover (1982).

the anticholinergic side-effects that are seen with imipramine, amitriptyline, and nortriptyline. This is evidenced in the eye by blurred vision and impaired accommodation. Generally, however, these and other anticholinergic effects (e.g. dry mouth and constipation) are mild and seldom require withdrawal of therapy. Their incidence and severity are not greatly influenced whether the recommended daily 200–400 mg is given as a single or as a divided dose. There have been no long-term demonstrations of changes in visual acuity, visual fields, colour discrimination, or ocular motility in the use of amoxaprine (Jue *et al.* 1982).

Vitamin A deficiency

Ocular effects (e.g. night-blindness, conjunctival xerosis, Bitot's spots, xerophthalmia) from vitamin A deficiency have long been recognized; commonly they present as a symptom of nutritional problems amongst children in many developing countries. In this respect it is interesting to note that Sommer *et al.* (1980), working in Indonesia, showed that corneal xerophthalmia responds just as rapidly to double-dose vitamin A given orally as to parenteral plus oral dosage. This is important since the parenteral form is more expensive than the oral and is more complicated to give. Fears that oral therapy would delay healing, especially in children with diarrhoea or protein–energy malnutrition, appear to have been unfounded.

Vitamin A deficiency in intestinal disorders involving malabsorption is also known to affect night vision and a communication from Partamian *et al.* (1979) described iatrogenic night-blindness and keratoconjunctival xerosis as a previously undocumented and untoward manifestation of vitamin A deficiency due to a jejunoileal bypass procedure. They emphasized that small-bowel bypass patients should be monitored routinely by the dark-adaptation test and by ophthalmic examination.

The effects caused by an excessive chronic intake of vitamin A are not so familiar; indeed their origin may even go unrecognized. Vitamin A is often accompanied by adverse effects when the patient, of his own accord, increases his dosage and continues the practice for years. One such example of hypervitaminosis A, involving ocular toxicity, was described by Turtz and Turtz (1960); the patient, a 17-year-old boy, was treated for acne with 200 000 i.u. of vitamin A daily for 18 months. In addition to malaise and pains in the joints, he developed blurred vision and occasional diplopia and other neurological disturbances resembling those of a brain tumour. Slit-lamp examinations of the eyes revealed no abnormality, but fundus examination disclosed retinal haemorrhage

and oedema of the optic disc or papilloedema. There was a regression of symptoms when the vitamin A treatment was stopped.

Overdosage with vitamin D has also been implicated in ocular side-effects in children. It may lead to calcification of the cornea in the exposed area between the lids; this has been termed band-shaped degeneration of the cornea (Mushin 1972). Although this condition is rare, it has been suggested that it might occur in the presence of renal insufficiency (Duke 1967).

* *Phenylbutazone and oxyphenbutazone.* In the UK, phenylbutazone has recently (March 1984) been restricted to the treatment of ankylosing spondylitis and its supply has been limited to hospitals only; the CSM has also recommended that products containing oxyphenbutazone should have their licences revoked (Goldberg 1984). Oxyphenbutazone has now been withdrawn.

RECOMMENDED FURTHER READING

BALLANTYNE, B. AND SWANSTON, D.W. (1977). The scope and limitations of acute eye irritation tests. In *Current approaches in toxicology* (ed. B. Ballantyne), pp. 139–57. Wright, Bristol.

——, GAZZARD, M.F., AND SWANSTON, D.W. (1977). Applanation tonometry in ophthalmic toxicology. In *Current approaches in toxicology* (ed. B. Ballantyne), 158–92. Wright, Bristol.

CREWS, S.J. (1977). Ocular adverse reactions to drugs. *Practitioner* **219**, 72–7.

FRAUNFELDER, F.T. (1976). *Drug-induced ocular side-effects and drug interactions.* Lea and Febiger, Philadelphia.

LIEN, E.J., AND KODA, R.T. (1981). Structure-side effect sorting of drugs: V. Glaucoma and cataracts associated with drugs and chemicals. *Drug Intel. clin. Pharm.* **15**, 434–9.

MEDICAL LETTER (1982). Adverse systemic effects from ophthalmic drugs. *Med. Lett.* **24**, 53–4.

MORGAN, J.F. (1979). Complications associated with contact lens solutions. *Opth. A.A.O.* **86**, 1107–19.

RASMUSSEN, J.E. (1983). Antimalarials — are they safe to use in children? *Ped. Dermatol.* **1**, 89–91.

ROBERTSON, D.M. (1972). The eye and adverse drug effects. Some selected considerations. *Minn. Med.* **55**, 927–34.

REFERENCES

ALTUS, P. (1981). Timolol-induced congestive heart failure. *South. Med. J.* **74**, 88.

AMOS, H.E., BRIGDEN, W.D., AND McKERRON, R.A. (1975). Untoward effects associated with practolol: Demonstration of antibody binding to epithelial tissue. *Br. med. J.* **1**, 598–600.

APPLEBAUM, A. (1963). Ophthalmoscopic study of patients under treatment with thioridazine. *Arch. Ophthal.* **69**, 578–80.

AQUINAS, M., ALLAN, W.G.L., HORSFALL, P.A.L., JENKINS, P.K., WONG, H-Y., GIRLING, D., TALL, R., AND FOX, W. (1972). Adverse reactions to daily and intermittent rifampicin regimens for pulmonary tuberculosis in Hong Kong. *Br. med. J.* **1**, 765–71.

ASHTON, N. (1975). Experimental retrolental fibroplasia. *Ann Rev. Med.* **8**, 441–54.

——, WARD, B., AND SERPELL, G. (1954). Effect of oxygen on developing retinal vessels with particular reference to the problems of retrolental fibroplasia. *Br. J. Ophthalmol.* **38**, 397–432.

ATTALLA, F.M., SAHEB, W.H., RANDALL, R.F., AND DORIAN, W.D. (1981). Timolol (Blocadren ®) post marketing surveillance program in hypertension. *Curr. Ther. Res.* **29**, 423–37.

AXELSSON, U. AND HOLMBERG, A. (1966). The frequency of cataract after miotic therapy. *Acta ophthal. (Kbh.)* **44**, 421–9.

AYD, F. (1961). A survey of drug-induced extrapyramidal reactions. *J. Am. med. Ass.* **175**, 1054–60.

BALCON, R. (1971). Assessment of drugs in angina pectoris: 3. *Postgrad. med. J.* **47**, 53–6.

BALDONE, J.A., HANKIN, J.S., AND ZIMMERMAN, T.J. (1982). Allergic conjunctivitis associated with timolol therapy in an adult. *Ann. Opthalmol.* **14**, 364–5.

BAN, T.A. AND LEHMANN, H.E. (1965). Skin pigmentation, a rare side effect of chlorpromazine. *Can. psychiat. Ass. J.* **10**, 112–24.

BAR, S., FELLER, N., AND SAVIR, H. (1983). Presenile cataracts in phenytoin-treated epileptic patients. *Arch. Ophthalmol.* **101**, 422–5.

BARD, L.A. (1964). Transient myopia associated with promethazine (*Phenergan*) therapy. *Am. J. Ophthal.* **58**, 682–6.

BARRETT, A.M. (1971). The pharmacology of practolol. *Postgrad. med. J.* **47**, 7–12.

BARSA, J.A., NEWTON, J.C., AND SAUNDERS, J.C. (1965). Lenticular and corneal opacities during phenothiazine therapy. *J. Am. med. Ass.* **193**, 10–12.

BARTHOLOMEW, R.S. (1976). Ocular complications of ethambutol. *Br. med. J.* **1**, 1535.

BARTSCH, P. AND REGINSTER, M. (1977). Chronic obstructive lung disease after practolol. *Lancet* **i**, 908–9.

BECKER, A.L. (1976). Oxprenolol and propanolol in anxiety states. A double-blind comparative study. *S. Afr. med. J.* **50**, 627–9.

BECKER, B. AND MILLS, D.W. (1963). Elevated intraocular pressure following corticosteroid eye drops. *J. Am. med. Ass.* **185**, 884–6.

BEHAN, P.O., BEHAN, W.M.H., ZACHARIAS, F.J., AND NICHOLLS, J.T. (1976). Immunological abnormalities in patients who had the oculomucocutaneous syndrome associated with practolol therapy. *Lancet* **ii**, 984–7.

BENDTZEN, K. AND SOBØRG, M. (1975). Sclerosing peritonitis and practolol. *Lancet* **i**, 629.

BENNER, M.M., ENTNER, T., LOCKEY, S., MAKINO, S., AND REED, C.E. (1968). The immunosuppressive effect of epinephrine and the adjuvant effect of beta-adrenergic blockade. *J. Allergy* **41**, 110–11.

BERMAN, J.J. AND MANN, M.P. (1980). Seizures and transient cortical blindness associated with cis-platinum (II) diammine-dichloride (PDD) therapy in a thirty year old man. *Cancer* **45**, 764–6.

BLACK, R.L., OGLESBY, R.B., VON SALLMANN, L., AND BUNIM, J.J. (1960). Posterior subcapsular cataracts induced by corticosteroids in patients with rheumatoid arthritis. *J. Am. med. Ass.* **174**, 166–71.

BLAKE, J. (1975). Eye hazard in rural communities. *Practitioner* **214**, 641–5.

BOCK, R. AND SWAIN, J. (1963). Ophthalmological findings in patients on long-term chlorpromazine therapy. *Am. J. Ophthal.* **56**, 808–10.

BONACCORSI, M.-T. (1967). Atrophie optique et atteinte systémique: Deux cas de réactions secondaires aux phénothiazines chez des enfants. *Laval. méd.* **38**, 84–8.

BOND, W.S. AND YEE, G.C. (1980). Ocular and cutaneous effects of chronic phenothiazine therapy. *Am. J. hosp. pharm.* **37**, 74–8.

BOUGLE, D., VERT, P. REICHART, E., AND HARTEMANN, D. (1980). Retrolental fibroplasia and retinal superoxide dismutase in kittens. *Lancet* **i**, 268.

BOWEN, D.I. AND VATERLAWS, A.L. (1971). Toxic amblyopia due to ethambutol in a case of drug-resistant pulmonary tuberculosis. *Br. J. Dis. Chest* **65**, 105–10.

BRAVER, D.A., RICHARDS, R.D., AND GOOD, T.A. (1966). Posterior subcapsular cataracts in corticosteroid-treated children. *J. Pediat.* **69**, 735–8.

——, ——, AND —— (1967). Posterior subcapsular cataracts in steroid-treated children. *Arch. Ophthal,* **77**, 161–2.

BRITISH MEDICAL JOURNAL (1969). Leading article: Iatrogenic symptoms in ophthalmology. *Br. med. J.* **2**, 199–200.

—— (1975). Leading article: Side-effects of practolol. *Br. med. J.* **2**, 577–8.

BROWN, P., BADDELEY, H., READ, A.E., DAVIES, J.D., AND McGARRY, J. (1974). Sclerosing peritonitis, an unusual reaction to a β-adrenergic blocking drug (practolol). *Lancet* **ii**, 1477–81.

BROWNE, S.G. (1975). The drug treatment of leprosy. *Practitioner* **215**, 493–500.

BRUBAKER, R.F. AND HALPIN, J.A. (1975). Open-angle glaucoma associated with topical administration of fluandrenolide to the eye. *Mayo Clin. Proc.* **50**, 322–6.

BURNS, C.A. (1968). Indomethacin, reduced retinal sensitivity, and corneal deposits. *Am. J. Ophthal.* **66**, 825–35.

BURSTEIN, N.L. (1980). Corneal cytotoxicity of topically applied drugs, vehicles and preservatives. *Surv. Ophthalmol.* **25**, 15–30.

BUYSKE, D.A., PEETS, E., AND STERLING, W. (1966). Pharmacological and biochemical studies on ethambutol in laboratory animals. *Ann. NY Acad. Sci* **135**, 711–25.

CAIRNS, R.J., CAPOORE, H.S., AND GREGORY, I.D.R. (1965). Oculocutaneous changes after years on high doses of chlorpromazine. *Lancet* **i**, 239–41.

CANT, J.S. (1969). Iatrogenic eye disease. *Practitioner* **202**, 787–95.

CAVALLARD, R.J., KRUMPERMAN, L.W., AND KUGLER, F. (1968). Effect of echothiopate therapy on the metabolism of succinylcholine in man. *Anaesth. Analg. Curr. Res.* **47**, 570–4.

CAYLEY, F.E. AND MAJUMDAR, S.K. (1976). Ocular toxicity due to rifampicin. *Br. med. J.* **1**, 199–200.

CHANG, N., GILES, C.L., AND GREGG, R.H. (1966). Optic neuritis and chloramphenicol. *Am. J. Dis. Childh.* **112**, 46–8.

CHARAN, M.B. AND LAKSHMINARAYAN, S. (1980). Pulmonary effects of topical timolol. *Arch intern. Med.* **140**, 843–4.

CHARLES, D., BARR, W., AND McEWAN, H.P. (1964). The use of clomiphene in dysfunctional bleeding due to endometrial hyperplasia. *J. Obstet. Gynaec. Brit. Comm.* **71**, 66–73.

CHEW, E., GHOSH, M., AND McCULLOCH, C. (1982). Amiodarone-induced corneal verticillata. *Can. J. Opthalmol.* **17**, 96–9.

CHURCH, G., SCHAMROTH, L., SCHWARTZ, N.L., AND MARRIOTT, H.J. (1962). Deliberate digitalis intoxication: A comparison of the toxic effects of four glycoside preparations. *Ann. intern. Med.* **57**, 946-56.

CLAYDEN, J.R. (1975). Ocular reactions to beta-blockers. *Br. med. J.* **2**, 557.

COCCO, G., SANSANO, C., HAUSMAN, M., VALLINI, R., STROZZI, C., AND CHU, D. (1982). A review on the side effects of β-adrenoceptor blocking drugs on the skin, mucosae and connective tissue. *Curr. ther. Res.* **31**, 362-78.

COCKE, J.G. JR., BROWN, R.E., AND GEPPERT, L.J. (1966). Optic neuritis with prolonged use of chloramphenicol. Case report and relationship to fundus changes in cystic fibrosis. *J. Pediat.* **68**, 27-31.

COLLUM, L.M.T. AND BOWEN, D.I. (1971). Occular side-effects of ibuprofen. *Br. J. Ophthal.* **55**, 472-7.

COMMITTEE ON SAFETY OF MEDICINES (1975). Practolol and ocular damage. *Adverse Reaction Series*, No. 11.

CRAPO, J.D. AND TIERNEY, D.F. (1974). Superoxide dismutase and pulmonary oxygen toxicity. *Am. J. Physiol.* **226**, 1401-7.

CREWS, S.J. (1962). Toxic effects on the eye and visual apparatus resulting from the systemic absorption of newly introduced chemical agents. *Trans, ophthal. Soc. UK.* **82**, 387-406.

CROMBIE, A.L. (1981). Drugs causing eye problems. *Prescribers' J.* **21**, 222-7.

D'ARCY, P.F. (1979). Disorders of the eye. In: D'Arcy, P.F. and Griffin, J.P. *Iatrogenic diseases*, 2nd edn., pp. 367-96, Oxford University Press, Oxford.

—— (1980). Oculotoxicity and ototoxicity. In *Drug-induced emergencies*. P.F. D'Arcy and J.P. Griffin. pp. 304-34 Wright, Bristol.

DAVID, D.S. AND BERKOWITZ, J.S. (1969) Ocular effects of topical and systemic corticosteroids *Lancet* ii, 149-51.

DAVIDSON, S.I. (1980). Drug-induced disorders of the eye. *Br. J. hosp. Med.* **24**, 24-8.

DELANEY, P. AND LIGHT, R. (1981). Gaze paresis in amitriptyline overdose. *Ann. Neurol.* **9**, 513.

DELONG, S., POLLEY, B.J., AND McFARLANE, J.R. (1965). Ocular changes associated with long-term chlorpromazine therapy *Arch Ophthal,* **73**, 611-17.

DENCKER, S.J. AND ENOKSSON, P. (1966). Ocular changes produced by chlorpromazine. *Acta ophthal. (Kbh.)* **44**, 397-403.

DENKO, C.W., GOODMAN, R.M., MILLER, R., AND DONOVAN, T. (1967). Distribution of dimethyl sulfoxide ³⁵S in the rat. *Ann. NY Acad. Sci.* **141**, 77-84.

DE ROETTH, A. JR. (1966). Lenticular opacities in glaucoma patients receiving echothiopate iodide therapy. *J. Am. med. Ass.* **195**, 664-6.

DINNING, C.R., BRUCE, G.M., AND SPALTER, H.F. (1963). Optic neuritis in chloramphenicol-treated patients with cystic fibrosis, *J. Pediat.* **63**, 878.

DODD, M.J. GRIFFITHS, I.D., HOWE, J.W., AND MITCHELL, K.W. (1981). Toxic optic neuropathy caused by benoxaprofen. *Br. med. J.* **283**, 193-4.

DONOMAE, I. AND YAMAMOTO, K. (1966). Clinical evaluation of ethambutol in pulmonary tuberculosis. *Ann. NY Acad. Sci.* **135**, 849-81.

DRAVET, C., MESDJIAN, E., CENRAUD, B., AND ROGER, J. (1977). Interaction between carbamazepine and triacetyloleandomycin. *Lancet* i, 810-11.

DRUG AND THERAPEUTICS BULLETIN (1964). Topical treatment of chronic simple glaucoma. *Drug Ther. Bull.* **2** 18-19.

DUBNOW, M.H. AND BURCHELL, H.B. (1965). A comparison of digitalis intoxication in two separate periods. *Ann. intern. Med.* **62**, 956-65.

DUKE, P.S. (1967). Ocular side effects of drug therapy. *Med. J. Aust.* **1**, 927-9.

DUNNE, M., FLOOD, M., AND HERXHEIMER, A. (1976). Clioquinol: availability and instructions for use. *J. antimicrob. Chemother.* **2**, 21-9.

DYER, N.H. AND VARLEY, C.C. (1975). Practolol-induced pleurisy and constrictive pericarditis. *Br. med. J.* **2**, 443.

EARNSHAW, E.R., MILES, D.W., AND STEWART, T.W. (1966). Screening for chloroquine retinopathy, *Br. J. Derm.* **78**, 669-74.

EDELHAUSER, H.F., HINE, J.E., PEDERSON, H., VAN HORN, D.L., AND SCHULTZ, R.O. (1979). The effect of phenylephrine on the cornea. *Arch. Ophthalmol.* **97**, 937-47.

FAN, Y.S., CONDON, J.R., BROOKS, P.L., RHODES, J., DEXTER, H., AND ROBINSON, V. (1975). Sclerosing peritonitis and practolol. *Lancet* i, 629-30.

FARR, M.J., WINGATE, J.P., AND SHAW, J.N. (1975). Practolol and the nephrotic syndrome. *Br. med. J.* **2**, 68-9.

FELDMAN, P.E. AND FRIERSON, B.D. (1964). Dermatological and ophthalmological changes associated with prolonged chlorpromazine therapy. *Am. J. Psychiat.* **121**, 187-8.

FELIX, R.H. AND IVE, F.A. (1974). Skin reactions to practolol. *Br. med. J.* **2**, 333.

——, ——, AND DAHL, M.G.C. (1974). Cutaneous and ocular reactions to practolol. *Br. med. J.* **4**, 321-4.

——, ——, AND —— (1975). Skin reactions to beta-blockers. *Br. med. J.* **1**, 626.

FRAIS, M.A. AND BAYLEY, T.J. (1979). Ocular reactions to timolol maleate. *Postgrad, med. J.* **55**, 884-5.

FRAUNFELDER, F.T. (1977). National registry for drug-induced ocular side-effects. *J. Am. med. Ass.* **237**, 466.

—— AND BAGBY, G.C. JR. (1983). Ocular chloramphenicol and aplastic anemia. *New Engl. J. Med.* **308**, 1536.

——, ——, AND KELLY, D.J. (1982). Fatal aplastic anaemia following topical administration of ophthalmic chloramphenicol. *Am. J. Ophthalmol.* **93**, 356-60.

FRIDOVICH, I. (1978). The biology of oxygen radicals. *Science* **201**, 875-80.

FRIEDMAN, Z. AND NEUMANN, E. (1972). Benzhexol-induced blindness in Parkinson's disease. *Br. med. J.* **1**, 605.

FÜRST, C., SMILEY, W.K., AND ANSELL, B.M. (1966). Steroid cataract. *Am. rheum. Dis.* **25**, 364-7.

GALENTINE, C.B. (1854). Cumulative effects of digitalis. *Boston Med. Surg. J.* **49**, 205-6.

GARDNER, K.H., LERMAN, S., MEGAW, J.M., AND BORKMAN, R.F. (1980). The prevention of direct and photosensitized UV radiation damage to the ocular lens. *Invest. Ophthalmol. Vis. Sci.* **19** (Suppl.), 88.

GAYLARDE, P.M., AND SARKANY, I. (1975). Side-effects of practolol. *Br. med. J.* **3**, 435.

GEORGE, C.F., NAGLE, R.E., AND PENTECOST, B.L. (1970). Practolol in treatment of angina pectoris: A double-blind trial. *Br. med. J.* **1**, 402-4.

GERHARD, J.P., WILLARD, F., RISSE, J.F., KUSS, J.J., AND MESSER, J. (1979). Les lésions de la périphérie rétinienne chez le prématuré. *Arch. F. Pédiat.* **36**, 573-81.

GEWIN, H.M. AND FRIOU, G.J. (1951). Manifestations of vitamin deficiency during aureomycin and chloramphenicol therapy of endocarditis due to *Staphylococcus aureus*: Report of a case. *Yale J. Biol. Med.* **23**, 332-8.

GIACOBINI, E. AND LASSENIUS, B. (1954). Chlorpromazine therapy in psychiatric practice: Secondary effects and complications. *Nord. Med.* **52**, 1693-9.

GILES, C.L., MASON, G.L., DUFF, I.F., AND McLEAN, J.A. (1962). The association of cataract formation and systemic corticosteroid therapy. *J. Am. med. Ass.* **182**, 719-22.

GILLIES, W.E., WEST, R.H., AND CEBON, L. (1983). Timoptol — three years on. A study of timolol maleate drops over a longer period. *Aust. J. Ophthalmol.* **11**, 155-7.

GIRLING, D.J. (1976). Ocular toxicity due to rifampicin. *Br. med. J.* **1**, 585.

GODWIN-AUSTEN, R.B., LIND, N.A., AND TURNER, P. (1969). Mydriatic responses to sympathomimetic amines in patients treated with L-dopa. *Lancet* ii, 1043-4.

GOLDBERG, A. (1984). Phenylbutazone and oxyphenbutazone. Letter to doctors and pharmacists. Committee on Safety of Medicines, London, 7th March 1984.

GORDON, D.M. (1967). Dimethylsulfoxine in ophthalmology, with especial reference to possible toxic effects. *Ann. NY Acad. Sci.* **141**, 392-401.

—— AND KLEBERGER, K.E. (1968). The effect of dimethyl sulfoxide (DMSO) on animal and human eyes. *Arch. Ophthal.* **79**, 423-7.

GOUTIÈRES, F. AND AICARDI, J. (1977). Accidental percutaneous hexachlorophene intoxication in children. *Br. med. J.* **2**, 663-5.

GRANT, W.M. (1962). *Toxicology of the eye*. Thomas, Springfield, Illinois.

—— (1969). Ocular complications of drugs: Glaucoma. *J. Am. med. Ass.* **207**, 2089-91.

—— (1974). Toxicology of the eye, 2nd edn. Thomas, Springfield, Illinois.

GREINER, A.C. AND BERRY, K. (1964). Skin pigmentation and corneal and lens opacities with prolonged chlorpromazine therapy. *Can. med. Ass. J.* **90**, 663-5.

——, NICOLSON, G.A., AND BAKER, R.A. (1964). Therapy of chlorpromazine melanosis: A preliminary report. *Can. med. Ass. J.* **90**, 636-8.

GURRY, J.F., CUNNINGHAM, I.G.E., AND BROOKE, B.N. (1975). Beta-blockers and fibrinous peritonitis. *Br. med. J.* **2**, 498.

GUZMAN, C.A. (1980). Exacerbation of bronchorrhea induced by topical timolol. *Am. Rev. resp. Dis.* **121**, 899-900.

HACKENBERGER, F. (1980). Antiseptic drugs. In: *Meyler's side effects of drugs,* 9th edn. (ed. M.N.G. Dukes), pp. 391-407. Excerpta Medica, Amsterdam, Oxford, Princeton.

HALSEY, J.P. AND CARDOE, N. (1982). Benoxaprofen: side-effect profile in 300 patients. *Br. med. J.* **284**, 1365-8.

HAMMERSHØY, O., AND JESSEN, F. (1982). A retrospective study of cataract formation in 96 patients treated with PUVA. *Acta dermatovener*; *(Stockholm)* **61**, 444-6.

HANSEN, M., HANSEN, O.P., AND LINDHOLM. J. (1977). Controlled clinical study on antihypertensive treatment with a diuretic and methyldopa compared with a β-blocking agent and hydrallazine. *Acta. med. scand.* **202**, 385-8.

HARRINGTON, D.O. (1965). The Bjerrum scotoma. *Am. J. Ophthal.* **59**, 646-56.

HARRISON, R. (1960). Bilateral lens opacities associated with the use of di-isopropylfluorophosphate eyedrops. *Am. J. Ophthal.* **50**, 153-4.

HAVRE, D.C. (1965). Cataracts in children on long-term corticosteroid therapy. *Arch. Ophthal.* **73**, 818-21.

HAYS, G.B., LYLE, C.B. AND WHEELER, C.E. (1964). Slate-gray color in patients receiving chlorpromazine. *Arch. Derm.* **90**, 471-6.

HEEL, R.C., BROGDEN, R.N., SPEIGHT, T.M., AND AVERY, G.S. (1979). Timolol: A review of its therapeutic activity in the topical treatment of glaucoma. *Drugs* **17**, 38-55.

HENKIND, P. AND ROTHFIELD, N.F. (1963). Ocular abnormalities in patients treated with synthetic antimalarial drugs. *New Engl. J. Med.* **269**, 433-9.

HERTZBERGER, R. (1965). Drugs affecting vision. *Med. J. Aust.* **1**, 990-1.

HIMES, J.A., EDDS, G.T., KIRKHAM, W.W., AND NEAL, F.C. (1967). Potentiation of succinylcholine by organophosphate compounds in horses. *J. Am. vet. med. Ass.* **151**, 54-9.

HITCHINGS, R.A. AND GLOVER, D. (1982). Adrenaline 1% combined with guanethidine 1% versus adrenaline 1%: a randomised prospective double-blind cross-over study. *Br. J. Ophthalmol.* **66**, 247-9.

HOBBS, H.E., EADIE, S.P., AND SOMERVILLE, F. (1961). Ocular lesions after treatment with chloroquine. *Br. J. Ophthal.* **45**, 284-97.

HOLMBERG, A. (1966). cited by Shaffer, R.N., and Hetherington, J.Jr. (1966).

HOLT, P.J.A. AND WADDINGTON, E. (1975). Oculocutaneous reaction to oxyprenolol. *Br, med. J.* **2**, 539-40.

HONG KONG TUBERCULOSIS TREATMENT SERVICES. BROMPTON HOSPITAL (ENGLAND). BRITISH MEDICAL RESEARCH COUNCIL INVESTIGATION (1974) A controlled clinical trial of daily and intermittent regimens of rifampicin plus ethambutol in the retreatment of patients with pulmonary tuberculosis in Hong Kong *Tubercle* **55**, 1-27.

HUANG, N.N., HARLEY, R.D., PROMADHATTAVEDI, V., AND SPROUL, A. (1966). Visual disturbances in cystic fibrosis following chloramphenicol administration. *J. Pediat.* **68**, 32-44.

HUNG, P.T., HSIEH, J.W., AND CHIOU, G.C.Y. (1982). Ocular hypotensive effect of N-demethylated carbachol on open angle glaucoma. *Arch. Ophthalmol.* **100**, 262-4.

HULL, F.W., WOOD, D.C., AND BROBYN, R.D. (1969). Eye effects of DMSO. Report of negative results. *Northwest Med.* **68**, 39-41.

IMKAMP, F.M.J.H. (1968). A treatment of corticosteroid-dependent lepromatous patients in persistent erythema nodosum leprosum. A clinical evaluation of G. 30320 (B663). *Leprosy Rev.* **39**, 119-25.

INGRAM, D.V. (1983) Ocular effects in long-term amiodarone therapy. *Am. Heart J.* **106**, 902-5.

INOVE, F., (1979). Adverse reactions of antipsychotic drugs. *Drug Intel. clin. Pharm.* **13**, 198-208.

ISMAIL, S.A. (1974). Practolol and the eye. *Br. med. J.* **4**, 104.

IWARSON, S. (1977). Clioquinol. *Lancet* i, 859.

JACOB, S.W. (1971). Pharmacology of DMSO. In *Dimethyl sulfoxide. I, Basic concepts.* (ed. S.W. Jacobs, E.E. Rosenbaum, and D.C. Wood, pp. 99-112. Marcel Dekker Inc., New York.

—— AND WOOD, D.C. (1967a). Report on the Third Internation Congress on dimethyl sulphoxide, Vienna, 1966. *Arzneimittel-Forsch.* **17**, 1086, 1553.

—— AND —— (1967b). Dimethyl sulfoxide (DMSO). Toxicology, pharmacology, and clinical experience. *Am. J. Surg.* **114**, 414-26.

JAIN, I.S. (1981). Dilantin cataract. *Ann. Ophthalmol.* **13**, 1010.

JAPANESE MINISTRY OF HEALTH AND WELFARE (1983). *Information on Adverse Reactions to Drugs*, No. 64. Pharmaceuticals and Chemicals Safety Division, Pharmaceutical Affairs

Bureau, Ministry of Health and Welfare, Tokyo, Japan. December 1983.

JOHNSON, J.E., BUNDE, C.A., AND HOEKENGA, M.T.. (1966). Clinical experience with clomiphene. *Pacif. Med. Surg.* **74**, 153–8.

JONAS, S. (1959). Miosis following administration of chlorpromazine. *Am. J. Psychiat* **15**, 817–18.

JONES, G.S. AND DE MORAES-RUEHSEN, M. (1965). Induction of ovulation with human gonadotropins and with clomiphene. *Fertil. Steril.* **16**, 461–83.

JOURNAL OF THE AMERICAN MEDICAL ASSOCIATION (1966). Editorial: Chloroquine retinopathy. *J. Am. med. Ass.* **195**, 774.

JUE, S.G., DAWSON, G.W., AND BROGDEN, R.N. (1982). Amoxaprine: A review of its pharmacology in depressed states. *Drugs* **24**, 1–23.

KAPLAN, L.J. AND CAPPAERT, W.E. (1982). Amiodarone keratopathy. *Arch. Ophthalmol.* **100**, 601–2.

KATZ, I.M. (1978). Treatment of chronic open-angle glaucoma with timolol maleate ophthalmic solution. In Timolol maleate ophthalmic solution in the treatment of glaucoma. *Proceedings of the International Symposium on Glaucoma, XXIII International Congress of Ophthalmology*, Kyoto, Japan, May 1978, pp. 29–40. Merck Sharp and Dohme International, Rahaway, New Jersey, USA.

KAUL, S., WONG, M., SINGH, B.H., AND HEPLER, R.S. (1982). Nadolol and papilledema. *Ann. intern. Med.* **97**, 454.

KEITH, C.G. (1964). Optic atrophy induced by chloramphenicol. *Br. J. Ophthal.* **48**, 567–70.

KERSLEY, G.D. AND PALIN, A.G. (1959). Amodiaquine and hydroxychloroquine in rheumatoid arthritis. *Lancet* **ii**, 886–8.

KIM, J.W. AND SMITH, P.H. (1980). Timolol-induced bradycardia. *Anesth. Analg. curr. Res.* **59**, 301–3.

KINYON, G.E. (1969). Anticholinesterase eye drops — need for caution. *New Engl. J. Med.* **280**, 53.

KLEBERGER, K.E. (1967). An ophthalmological evaluation of DMSO. *Ann. N.Y. Acad. Sci.* **141**, 381–5.

KNAPP, M.S. AND GALLOWAY, N.R. (1975). Ocular reactions to beta-blockers. *Br. med. J.* **2**, 557.

KOCH-WESER, M. AND GILMORE, E.B. (1967). Benign intracranial hypertension in an adult after tetracycline therapy. *J. Am. med. Ass.* **200**, 345–7.

KOHN, A.N., MOSS, A.P., HARGETT, N.A., RITCH, R., SMITH, H. JR., AND PODOS, S.M. (1979). Clinical comparison of dipivalyl epinephrine and epinephrine in the treatment of glaucoma. *Am. J. Ophthalmol.* **87**, 196–201.

KOLB, K.H., JAENICKE, G., KRAMER, M., AND SCHULZE, P.E. (1967). Absorption, distribution and elimination of labeled dimethyl sulfoxide in man and animals. *Ann. NY Acad. Sci.* **141**, 85–95.

KUMING, B.S. AND BRAUDE, L. (1979). Anterior optic neuritis caused by ethambutol toxicity. *S. Afr. med. J.* **55**, 4.

KUTSCHERA, E. (1966). In *DMSO Symposium, Vienna* (ed. G. Laudahn), p. 117. Saladruck, Berlin.

LARKIN, J., PULLAR, T., AND MASON, D. (1982). Side effects of benoxaprofen. *Br, med. J.* **284**, 1784.

LASKY, M.A., PINCUS, M.H., AND KATLAN, N.R. (1953). Bilateral optic neuritis following chloramphenicol therapy. *J. Ann. med. Ass.* **151**, 1403–4.

LAURSEN, S.Ø. AND BJERRUM, P. (1982). Timolol eyedrop-induced severe bronchospasm. *Acta med. scand.* **211**, 505–6.

LE BLANC, R.P. AND KRIP. G. (1981). Timolol. Canadian Multicenter Study. *Ophthalmology* **88**, 244–8.

LEES, A.W., ALLAN, G.W., SMITH, J., TYRRELL, W.E. AND FALLON, R.J. (1971). Toxicity from rifampicin plus isoniazid and rifampicin plus ethambutol therapy. *Tubercle* **52**, 182–90.

LEIBOLD, J.E. (1966). The ocular toxicity of ethambutol and its relation to dose. *Ann. NY Acad. Sci.* **135**, 904–9.

LEOPOLD, I.H. (1968). Ocular complications of drugs Visual changes. *J. Am. med. Ass.* **205**, 631–3.

LERMAN, S. (1980). Ocular side effects of psoriasis therapy. *New Engl. J. Med.* **303**, 941–2.

—— AND BORKMAN, R.F. (1977). A method for detecting 8-methoxypsoralen in the ocular lens. *Science* **197**, 1287–8.

—— AND —— (1978). Photochemistry and lens aging. In Interdisciplinary topics in gerontology (Ed. O. Hockwin), Vol 13, pp. 154–82. S. Karger, Basle.

—— JOCOY, M., AND BORKMAN, R.F. (1977). Photosensitization of the lens by 8-methoxypsoralen. *Invest. Ophthalmol. Vis. Sci.* **16**, 1065–8.

——, MEGAW, J.M., TAKEI, Y., AND WILLIS, I. (1980b) Localization of 8-methoxypsoralen in ocular tissues. *Invest. Ophthalmol. Vis. Sci.* **19** (Suppl.), 37–8.

——, —— AND WILLIS, I. (1980a). The photoreaction of 8-MOP with tryptophan and lens proteins. *Photochem. Photobiol.* **31**, 235–43.

LEWIS, B.S., SETZEN, M., AND KOKORIS, N. (1976) Ocular reactions to oxprenolol. A case report. *S. Afr. med. J.* **50**, 482–3.

LIEN, E.J. AND KODA, R.T. (1981). Structure-side effect sorting of drugs V. Glaucoma and cataracts associated with drugs and chemicals. *Drug Intel. clin. Pharm.* **15**, 434–9.

LINKEWICH, J.A. AND HERLING, I.M. (1981). Bradycardia and congestive heart failure associated with ocular timolol maleate. *Am. J. hosp. Pharm.* **36**, 699–701.

LIPSON, M.L., HOLMES, J.H., AND ELLIS, P.P. (1969). Oral administration of pralidoxime chloride in echothiopate iodide therapy. *Arch. Ophthal.* **82**, 830–5.

LOGAN, W.S. (1980). Antimalarials. *Prog. Dermatol.* **141**, 1–6.

LUBBE, W.F. AND MERCER, C.J. (1982). Amiodarone its side effects, adverse reactions and dosage schedules. *NZ med. J.* **95**, 502–4.

LYALL, J.R.W. (1975). Ocular reactions to beta-blockers. *Br. med. J.* **2**, 747.

MANDEL, E.H. (1960). New local anaesthetic with anticoagulant properties, chloroquine (*Aralen*) dihydrochloride. *Arch Derm.* **81**, 260–3.

MARGOLIS, L.H. AND GOBLE, J.L. (1965). Lenticular opacities with prolonged phenothiazine therapy. *J. Am. med. Ass.* **193**, 7–9.

MARKS, R. (1975). Oculocutaneous reactions to beta-blocking drugs. *Br. med. J.* **4**, 648

MARMION, V.J. (1971). Pharmacogenetics in ophthalmology. *Proc. R. Soc. Med.* **64**, 16.

MARTINEZ, A.J., BOEHM, R., AND HADFIELD, M.G. (1974). Acute hexachlorophene encephalopathy. Clinico-neuropathological correlation. *Acta Neuropathol* **28**, 93–103.

MARZULLI, F.N. (1968). Ocular side-effects of drugs. *Food Cosmet. Toxicol.* **6**, 221–34.

MASON, M.M. (1971). Toxicology of DMSO in animals. In *Dimethyl sulfoxide, I. Basic concepts* (ed. S.W. Jacob, E.E. Rosenbaum and D.C. Wood), pp. 113–31. Marcel Dekker Inc., New York.

MASSEY, L.W.C. (1965). Skin pigmentation and lens opacities with prolonged chlorpromazine treatment. *Can. med. Ass. J.* **92**, 186–7.

MATHOLONE, M.B.R. (1967). Eye and skin changes in psychiatric patients treated with chlorpromazine. *Br. J. Ophthal.* **51**, 86–93.

McCLANAHAN, W.S. HARRIS, J.E., KNOBLOCH, W.H., TREDICI, L.M., AND UDASCO, R.L. (1966). Ocular manifestations of chronic phenothiazine derivative administration. *Arch Ophthal.* **75**, 319–25.

McGAVIN, D.D.M. (1965). Depressed levels of serum-pseudocholinesterase with echothiopate-iodide eyedrops. *Lancet* **ii**, 272–3.

McMAHON, C.D., SHAFFER, R.N., HOSKINS, H.D. JR., AND HETHERINGTON, J.JR. (1979). Adverse effects experienced by patients taking timolol. *Am. J. Ophthalmol.* **88**, 736–8.

MEDICAL LETTER (1982). Adverse systemic effects from ophthalmic drugs. *Med. Letter.* **24**, 53–4.

MERWIN, C.F. AND WINKELMANN, R.K. (1962), Antimalarial drugs in therapy of lupus erythematosus. *Proc. Mayo Clin* **37**, 253–68.

MEYLER, L. (Ed.) (1966). *Side-effects of drugs*, Vol. V. Excerpta Medica Foundation, New York.

MORSE, W.I., WARREN, W.P., PARKER, G.D., AMAD, N., AND BROWN, J.B. (1963). Effect of clomiphene on urinary oestrogens in men. *Br. med. J.* **1**, 798–9.

MORTADA, A. AND ABBOUD, I. (1973). Retinal haemorrhages after prolonged use of salicylates. *Br. J. Ophthal.* **57**, 199–200.

MURRAY, A., GLOVER, D., AND HUTCHINGS, R.A. (1981). Low-dose combined guanethidine 1% and adrenaline 0.5% in the treatment of chronic simple glaucoma. a prospective study. *Br.J. Ophthalmol.* **65**, 533–5.

MUSHIN, A.S. (1971). Ocular changes in premature babies receiving controlled oxygen therapy in the neonatal period. *Proc. Soc. Med.* **64**, 779–80.

—— (1972). Ocular damage by drugs in children. *Adverse Drug Reaction Bulletin* no. **36**, 112–15.

NAVILLE, A.H., KISTNER, R.W., AND ROCK, J. (1964). Induction of ovulation with *Clomid. Fert. Sterility* **15**, 290–309.

NAIR, V.S., LEBRUN, M., AND KASS, I. (1980). Peripheral neuropathy associated with ethambutol. *Chest* **77**, 98–100.

OGLESBY, R.B., BLACK, R.L., VON SALLMANN, L., AND BUNIM, J.J. (1961). Cataracts in rheumatoid arthritis patients treated with corticosteroids. *Arch. Ophthal.* **66**, 519–23.

OHMAN, L. AND WAHLBERG, I. (1975). Ocular side-effects of clofazimine. *Lancet* **ii**, 933–4.

OLANSKY, A.J. (1982). Antimalarials and ophthalmologic safety. *J. Am. Acad. Dermatol.* **6**, 19–23.

OLSEN, T., BUSTED, N., AND SCHMITZ, O. (1980). Corneal thickness in diabetes mellitus. *Lancet* **i**, 883.

OLSON, R.J., BROMBERG, B.B., AND ZIMMERMAN, T.J. (1979). Apneic spells associated with timolol in a neonate. *Am. J. Ophthalmol.* **88**, 120–2.

PAGE, F. (1951). Treatment of lupus erythematosus with mepacrine. *Lancet* **ii**, 755–8.

PALMER, C.A.L. (1972). Toxic amblyopia from ibuprofen *Br. med. J.* **3**, 765.

PARTAMIAN, L.G., SIDRYS, L.A., AND TRIPATHI, R.C. (1979). Iatrogenic night blindness and keratoconjunctival xerosis. *New Engl. J. Med.* **301**, 943–4.

PATTISON, P.R.M., KUN, O.J., FRIEDMAN, A.I., DARBYSHIRE, J.H., AND TALL, R. (1978). Ethambutol and intra-ocular pressure. *Tubercle* **59**, 33–40.

PATZ. A. (1966). The effect of oxygen on immature retinal vessels. In *Symposium on vascular disorders of the eye* (ed. J.W. Bettman), Mosby, St. Louis.

—— (1969). Retrolental fibroplasia. *Survey Ophthalmol.* **14**, 1–29.

PEARSON, R.M., BENDING, M.R., BULPITT, C.J., GEORGE, C.F., HOLE, D.R., WILLIAMS, F.M., AND BRECKENRIDGE, A.M. (1976). Trial of combination of guanethidine and oxprenolol in hypertension. *Br, med. J.* **1**, 933–6.

PENTECOST, B.L., GEORGE, C.F., AND NAGLE, R.E. (1971). Assessment of drugs in angina pectoris: 1. *Postgrad. med. J.* **47**, 48–50.

PERCIVAL, S.P.B. AND MEANOCK, I. (1968). Chloroquine: ophthalmological safety, and clinical assessment in rheumatoid arthritis. *Br. med. J.* **3**, 579–84.

PERKINS, E.S. (1965). Steroid-induced glaucoma. *Proc. R. Soc. Med.* **58**, 531–3.

PERROTT AND BOURJALA (1962). Langue noir pilense. Cas pour diagnostic: 'Un visage mauve'. *Bull. Soc. franç. Derm. Syph.* **69**, 631.

PETURSSON, G.J., FRAUNFELDER, F.T., AND MEYER, S.M. (1981). Pharmacology of ocular drugs. 6. Oral contraceptives. *Ophthalmology* **88**, 368–71.

PILDES, R.B. (1965). Induction of ovulation with clomiphene. *Am. J. Obstet. Gynecol*, **91**, 466–79.

PIPPITT, C.H. JR., MUSS, H.B., HOMESLEY, H.D., AND JOBSON, V.W. (1981). Case report: Cisplatin-associated cortical blindness. *Gynecol. Oncol.* **12**, 253–5.

PLACE, V.A., PEETS, E.A., BUYSKE, D.A., AND LITTLE R.R. (1966). Metabolic and special studies of ethambutol in normal volunteers and tuberculous patients. *Am. NY Acad Sci* **135**, 775–95.

PLANE, C., SOLÉ, P., OURGAUD, A.G., HAMARD, H., AND VIDAL, R. (1978). Double-observer comparison of timolol maleate and pilocarpine in open-angle glaucoma. In Timolol maleate ophthalmic solution in the treatment of glaucoma. *Proceedings of the International Symposium on Glaucoma, XXIII International Congress of Ophthalmology*, Kyoto, Japan, May 1978, pp. 41–8. Merck Sharp and Dohme International, Rahaway, New Jersey.

POLAK, B.C.P. (1980). Drugs used in ocular treatment. In *Meyler's side effects of drugs*, 9th edn. (ed. N.M.G. Dukes, pp. 772–85. Excerpta Medica, Amsterdam.

POTTS, A.M. (1962). Concentration of phenothiazines in the eye of experimental animals. *Invest. Ophthal.* **1**, 522–30.

PRICHARD, B.N.C., LIONEL, N.D., AND RICHARDSON, G.A. (1971). Comparison of practolol and propranolol in angina. *Postgrad. med. J.* **47**, 59–63.

PYLE, M.M. (1966). Ethambutol in the retreatment and primary treatment of tuberculosis; a four-year clinical investigation. *Ann. NY. Acad. Sci.* **135**, 835–45.

RAB, S.M., ALAM, M.N., AND SADEQUZZAMAN, M.D. (1969). Optic atrophy during chlorpromazine therapy. *Br. J. Ophthalmol.* **53**, 208–9.

RAFTERY, E.B. (1974). Cutaneous and ocular reactions to practolol. *Br. med. J.* **4**, 653.

—— AND DENMAN, A.M. (1973). Systemic lupus erythematosus syndrome induced by practolol. *Br. med. J.* **2**, 452–5.

REED, H., AND KARLINSKY, W. (1967). Delayed onset of chloroquine retinopathy. *Can. med. Ass. J.* **97**, 1408–11.

REISTERER, L. AND JAQUES, R. (1968). Interference by beta-adrenergic blocking agents, with the anti-inflammatory action of various drugs. *Hely. Physiol. Pharmacol. Acta.* **26**, 287–93.

RILEY, G.M. AND EVANS, T.N. (1964). Effects of clomiphene citrate on anovulatory ovarian function. *Am. J. Obstet. Gynecol.* **89**, 97–110.

RIVAS-ALCALÁ, A.R., GREENE, B.M., TAYLOR, H.R., DOMÍNGUEZ-VAŻQUEZ, A., RUVALCABA-MACÍAS, A.M., LUGO-PFEIFFER, C., MACKENZIE, C.D., AND BELTRAN, F. (1981). Chemotherapy of onchocerciasis: A controlled comparison of mebendazole, levamisole, and diethylcarbamazine. *Lancet* ii, 485–90.

ROBERTSON, D.M., HOLLENHORST, R.W., AND CALLAHAN, J.A. (1966). Ocular manifestations of digitalis toxicity. *Arch. Ophthalmol* **76**, 640–5.

ROCH, L.M.H., GORDON, D.L., BARR, A.B., AND PAULSEN, L.A. (1967). Visual changes associated with clomiphene citrate therapy. *Arch. Ophthalmol.* **77**, 14–17.

ROS, F.E. AND DAKE, C.L. (1980). Timolol eye drops: bradycardia and tachycardia. *Documenta ophthalmol.* **48**, 283–89.

ROSE, A.L., WISNIEWSKI, H.M., AND CAMMER, W. (1975). Neurotoxicity of hexachlorophene. New pathological and biochemical observations. *J. neurol. Sci.* **24**, 425–35.

ROWLAND, M.G.M. AND STEVENSON, C.J. (1972). Exfoliative dermatitis and practolol. *Lancet* i, 1130.

RUBIN, L.F. AND BARNETT, K.C. (1967). Ocular effects of oral and dermal application of dimethyl sulfoxide in animals. *Ann. NY. Acad. Sci.* **141**, 333–45.

—— AND MATTIS, P.A. (1966). Dimethyl sulfoxide: Lens changes in dogs during oral administration. *Science NY* **153**, 83–4.

SAARI, K.M., AIRAKSINEN, P.J., AND JAANIO, E-A, T.(1978). Hypotensive effect of timolol on secondary glaucoma in chronic uveitis. *Lancet* i, 442.

SANDLER, G. (1977). Assessment of drugs in angina pectoris: 2. *Postgrad. med. J.* **47**, 50–3.

—— AND CLAYTON, G.A. (1970). Clinical evaluation of practolol, a new cardioselective β-blocking agent in angina pectoris. *Br. med. J.* **2**, 399–402.

SCHERBEL, A.L. (1966). In *DMSO Symposium, Vienna*, (ed. G. Laudahn) p. 165. Saladruck, Berlin.

—— (1983). Use of synthetic antimalarial drugs and other agents for rheumatoid arthritis: historic and therapeutic perspectives. *Am. J. Med.* **75**, 1–4.

SCHMIDT, B AND MUELLER-LIMMROTH, W. (1962). Electroretinographic examinations following application of chloroquine. *Acta ophthal. (Kbh.)* **70**. (Suppl.) 245–51.

SCHOENE, R.B., MARTIN, T.R., CHARAN, N.B., AND FRENCH, C.L. (1981). Timolol-induced bronchospasm in asthmatic bronchitis. *J. Am. med. Ass.* **245**, 1460–61.

SCOTT, D. (1977). Another beta-blocker causing eye symptoms? *Br. med. J.* **2**, 1221.

SHAFFER, R.N. AND HETHERINGTON, J. JR. (1966). Anticholinesterase drugs and cataracts. *Am. J. Ophthalmol.* **62**, 613–18.

SHANKS, R.G. (1969). The properties of beta adrenergic blocking agents. *Irish J. med. Sci.* **2**, 351–67.

SHURIN, S.B., REKATE, H.L., AND ANNABLE, W. (1982). Optic atrophy induced by vincristine. *Pediatrics* **70**, 288–91.

SINHA, G.B. AND MITRA, S.K. (1955). Case of acute chlorpromazine hydrochloride poisoning. *J. Indian med. Ass.* **24**, 557–8.

SKALKA, H.W. AND PRCHAL, J.T. (1980). Effect of corticosteroids on cataract formation. *Arch. Ophthalmol.* **98**, 1773–7.

SLAMOVITS, T.L., BURDE, R.M., AND KLINGELE, T.G. (1980). Bilateral optic atrophy caused by chronic oral ingestion and topical application of hexachlorophene. *Am. J. Ophthalmol.* **89**, 676–9.

SMITH, E.R., MASON, M.M., AND EPSTEIN, E. (1967). The influence of dimethyl sulfoxide on the dog with emphasis on the ophthalmologic examination. *Ann. NY Acad. Sci.* **141**, 386–91.

SMITH, J.L. (1962). Chloroquine macular degeneration. *Arch. Ophthal.* **68**, 186–90.

SMITH, J.W., STEINER, A.L., NEWBERRY, W.M.JR., AND PARKER, C.W. (1971). Cyclic adenosine 3′, 5′ -monophosphate in human lymphocytes. Alterations after phytohemagglutinin stimulation. *J. clin. Invest.* **50**, 432–41.

SOMMER, A., DJUNAEDI, M.E., TARWOTJO, I., AND GLOVER, J. (1980). Oral versus intramuscular vitamin A in the treatment of xerophthalmia. *Lancet* i, 557–9.

SPECTOR, R.H., AND SCHNAPPER, R. (1981). Amitriptyline-induced ophthalmoplegia. *Neurology* **31**, 1188–90.

SPENCER, R.W., AND ANDELMAN, S.Y. (1965). Steroid cataracts: posterior subcapsular cataract formation in rheumatoid arthritis patients on long-term steroid therapy. *Arch. Ophthalmol* **74**, 38–41.

SPIERS, A.S.D., CALNE, D.B., AND FAYERS, P.M. (1970). Miosis during L-dopa therapy. *Br. med. J.* **2**, 639–40.

SPITERI, M.A. AND JAMES, D.G. (1983). Adverse ocular reactions to drugs. *Postgrad. med. J.* **59**, 343–9.

STAFFORD, W.R. (1981). Traditional antiglaucoma therapy: A clinical survey of side effects. *Curr. Ther. Res.* **29**, 265–74.

STRAUGHAN, J. (1982). Erythromycin-carbamazepine interaction? *S. Afr. med. J.* **61**, 420–1.

TANENBAUM, L. AND TUFFANELLI, D.L. (1980). Antimalarial agents . . . chloroquine, hydroxychloroquine, quinacrine. *Arch. Dermatol.* **116**, 587–91.

TAYLOR, H.R. AND GREENE, B.M. (1981). Ocular changes with oral and transepidermal diethylcarbamazine therapy of onchocerciasis. *Br. J. Ophthalmol.* **65**, 494–502.

THIGPEN, J.T., SHINGLETON, H., HOMESLEY, H.D., DISAIA, P., LAGASSE, L., AND BLESSING, J. (1980). Cisplatin in the treatment of advanced or recurrent cervix or uterine cancer. In *Cisplatin current status and new developments* (ed. A.W. Prestayko, S.T. Crooke, and S.K. Carter), pp. 411–21. Academic Press, New York.

THOMAS, E.B. (1950). Scotomas in conjunction with streptomycin therapy; report of 11 cases. *Arch. Ophthalmol.* **43**, 729–41.

THOMAS, J.V., GRAGOUDAS, E.S., BLAIR, N.P., AND LAPUS, J.V. (1978). Correlation of epinephrine use and macular edema in aphakic glaucomatous eyes. *Arch. Ophthalmol.* **96**, 625–8.

THOMPSON, M. (1972). Toxic amblyopia from ibuprofen. *Br. med. J.* **4**, 550.

THOMPSON, R.J. AND MELLINGER, R.C. (1965). The effects of clomiphene citrate in patients with pituitary-gonadal disorders. *Am. J. Obstet. Gynecol.* **92**, 412–20.

THOMPSON, R.P.H. AND JACKSON, B.T. (1975). Beta-blockers and fibrinous peritonitis. *Br. med. J.* **2**, 747.

—— AND —— (1977). Sclerosing peritonitis due to practolol. *Br. med. J.* **1**, 1393–4.

TOWFIGHI, J., GONATAS, N.K., AND McCREE, L. (1974). Hexachlorophene retinopathy in rats. *Lab. Invest.* **32**, 330–8.

TULLIO, C.J. (1981). Ibuprofen-induced visual disturbance. *Am. J. hosp. Pharm.* **38**, 1362.

TURIZ, C.A., AND TURTZ, A.I. (1960). Vitamin A intoxication. *Am. J. Ophthal.* **50**, 165–6.

VALMAN, H.B. AND WHITE, D.C. (1977). Stellate block for quinine blindness in a child. *Br. med. J.* **1**, 1065.

VAN BUSKIRK, E.M. (1979). Corneal anesthesia after timolol maleate therapy. *Am. J. Ophthalmol.* **88**, 739–43.

VAN DYKE, H.J.L. AND SWANN, K.C. (1969). In *Ocular therapy* (ed. I.H. Leopold), p. 71. Mosby, St. Louis.

VIZEL, M. AND OSTER, M.W. (1982). Ocular side effects of cancer chemotherapy. *Cancer* 49, 1999–2002.

VOGEL, P. (1983). Beta-adrenoceptor blocking drugs and the eye. *J. clin. hosp. Pharm.* 8, 209–18.

WALLENSTEIN, L. AND SNYDER, J. (1952). Neurotoxic reactions to chloromycetin. *Ann. intern. Med.* 36, 1526–8.

WARDELL, W.M. (1974). Therapeutic implications of the drug lag. *Clin. Pharmacol. Ther.* 15, 73–96.

WATERS, M.F.R. (1969). G 30 320 or B 663-lampren (Geigy). *Leprosy Rev.* 40, 21–47.

WEBER, J.C.P. (1982). Beta-adrenoreceptor antagonists and diplopia. *Lancet* ii, 826–7.

WEEKLEY, R.D., POTTS, A.M., REBOTON, J., AND MAY, R.H. (1960). Pigmentary retinopathy in patients receiving high doses of a new phenothiazine. *Arch. Ophthalmol.* 64, 65–76.

WETTERHOLM, D.H., SNOW, H.L., AND WINTER, F.C. (1965). A clinical study of pigmentary changes in cornea and lens in chronic chlorpromazine therapy. *Arch. Ophthalmol.* 74, 55–6.

WILLCOCKSON, J. AND WILLCOCKSON, T. (1980a). Timolol: Double-blind comparison with pilocarpine in open-angle glaucoma. *Curr. Ther. Res.* 27, 538–44.

—— AND ——. (1980b). Long-term use of timolol in open-angle glaucoma. *Curr. Ther. Res.* 27, 545–55.

WILLIAMS, T. AND GINTHER, W.H. (1982). Hazard of ophthalmic timolol. *New Engl. J. Med.* 306, 1485–6.

WINDSOR, W.O., KURREIN, F., AND DYER, N.H. (1975). Fibrinous peritonitis: a complication of practolol therapy. *Br. med. J.* 2, 68.

WISEMAN, R.A. (1971). Practolol-accumulated data on unwanted effects. *Postgrad. med. J.* 47, (Suppl.) 68–74.

WITHERING, W. (1785). cited by White, P.D. (1965). Important toxic effect of digitalis overdosage on vision. *New Engl. J.*

Med. 272, 904–5.

WOOD, D.C. (1966). In *DMSO Symposium, Vienna* (ed. G. Laudahn), p. 58, Saladruck, Berlin.

——, SWEET, D., VAN DOLAH, J., SMITH, J.C., AND CONTAXIS, I. (1967). A study of DMSO and steroids in rabbit eyes. *Ann. NY Acad. Sci.* 141, 346–80.

WRIGHT, P. (1974). Skin reactions to practolol. *Br. med. J.* 2, 560.

—— (1975a). Untoward effects associated with practolol administration: oculomucocutaneous syndrome. *Br. med. J.* 1, 595–8.

—— (1975b). Ocular reactions to beta-blocking drugs. *Br. med. J.* 4, 577.

WYMORE, J. AND CARTER, J.E. (1982). Chlorpropamide-induced optic neuropathy. *Arch. intern. Med.* 142, 381.

YABLONSKI, M.E. AND BURDE, R.M. (1978). Cataracts induced by topical dexamethasone in diabetics. *Arch. Ophthalmol.* 96, 474–6.

——, SHIN, D.H., KOLKER, E., KASS, M., AND BECKER, B. (1977). Dipivefrin use in patients with intolerance to topically applied epinephrin. *Arch. Ophthalmol.* 95, 2157–8.

YATES, D. (1980). Syncope and visual hallucinations, apparently from timolol. *J. Am. med. Ass.* 244, 768–69.

YOSHIKAWA, K., ADACHI, K., HALPRIN, K.M., AND LEVINE, V. (1975). The effects of catecholamine and related compounds on the adenyl cyclase system in the epidermis. *Br. J. Derm.* 93, 29–36.

ZACHARIAS, F.J. (1972). In *New perspectives in beta-blockade*, p. 238. International Symposium, Ciba Foundation, Denmark.

ZELICKSON, A.S. AND ZELLER, H.C. (1964). A new and unusual reaction to chlorpromazine. *J. Am. med. Ass.* 188, 394–7.

ZUEHLKE, R.L., LILLIS, P., AND TICE, A. (1981). Antimalarial therapy for lupus erythematosus. *Int. J. Dermatol.* 20, 57–61.

31 Disorders of the ear

E.S. HARPUR

The drugs involved and clinical features of ototoxicity

The administration of ototoxic substances (Table 31.1) to man may result in damage either to auditory or vestibular mechanisms or both. Most of the aminoglycoside antibiotics can damage both the cochlea and the vestibular system but a predilection for one of these organs is often characteristic of a particular drug. For example, streptomycin primarily affects the vestibular system in man whereas neomycin is potently cochleotoxic.

Vestibular toxicity can give rise to symptoms such as dizziness or vertigo and can be objectively detected as a decrease in the nystagmic response induced by caloric or rotational stimulation of the vestibular end organs. Although the lesion may be permanent, most patients can adequately compensate with time for the loss of vestibular function.

Auditory effects range from potentially transient symptoms such as diplacusis, tinnitus, or a mild reduction in auditory acuity to total and permanent deafness. Salicylates and other non-steroidal anti-inflammatory agents commonly cause tinnitus, which is reversible when

Table 31.1 Drugs involved or implicated in ototoxicity

Classification	Drugs
Antibiotics	
Aminoglycosides	Amikacin, dibekacin, dihydrostreptomycin, framycetin, gentamicin, kanamycin, neomycin, netilmicin, paromomycin, ribostamycin, sisomicin, streptomycin, tobramycin
Others	Ampicillin, capreomycin, chloramphenicol, colistin (polymyxin E), erythromycin, minocycline, polymyxin B, rifampicin, vancomycin, viomycin
Anti-inflammatory agents	Fenoprofen, ibuprofen, indomethacin, naproxen, phenylbutazone, salicylates
Antimalarials	Chloroquine, quinine
Antitumour agents	
Cytotoxics	Actinomycin, bleomycin, cis-platinum, nitrogen mustards (e.g. mustine)
Hypoxic cell radiosensitizers	Misonidazole
Beta-blockers	Practolol, propranolol
Contraceptives	Medroxyprogesterone
Loop diuretics	Bumetanide, ethacrynic acid, frusemide (furosemide), piretanide
Topical applications	
Quaternary ammonium compounds	Benzalkonium chloride, benzethonium chloride, chlorhexidine
Iodine disinfectants	Iodine–potassium iodine (Lugol's) solution in 70% alcohol, iodophor (*Iodopax*) solution in 70% alcohol, povidone–iodine solution and scrub
Others	Bonain's solution (cocaine, phenol, and thymol), formaldehyde–gelatin (absorbable gelatin sponge), lignocaine
Tricyclic antidepressants	Imipramine
Miscellaneous	Alcohol, nicotine (in tobacco), quinidine

713

the drug is stopped, and only rarely associated with permanent hearing loss (Chapman, 1982*b*). Therapeutic doses of quinine characteristically cause tinnitus but excessive doses of quinine or chloroquine may result in permanent deafness (McKenzie *et al.* 1968; Toone *et al.* 1965). Large intravenous doses of the loop diuretics, such as ethacrynic acid or frusemide, may cause a transient hearing impairment, although the deafness may be permanent (Quick and Hoppe 1975) particularly if the patient is uraemic or is also administered an aminoglycoside antibiotic (Mathog and Klein 1969).

The antitumour drug, cis-platinum, frequently causes tinnitus and permanent hearing loss which appears to be dose-related (Vermorken *et al.* 1983). Hearing loss induced by aminoglycosides may develop suddenly during therapy or, more commonly, the onset can be gradual and progress even after dosage has been discontinued. On occasions, the hearing loss may not even appear until some time after the drug administration has stopped. With the progressive, permanent lesions which are a feature of aminoglycoside and cis-platinum ototoxicity, audiometric assessment reveals a hearing loss which is initially confined to high frequencies but which eventually may involve frequencies in the speech range and result in a permanent disability. The degree of vestibular or auditory impairment with aminoglycosides may depend upon the drug itself, the unit or total dose, the route and period of administration, and the patient's age or pathological state; for example, the risk is substantially increased by poor renal function (Jackson and Arcieri 1971). It should not be forgotten that the aminoglycosides and cis-platinum are nephrotoxic.

Sites of action

Much of our understanding of drug-induced ototoxicity has been derived from histopathological studies of the lesions in the inner ear of animals. Permanent hearing loss caused by ototoxic drugs, such as the aminoglycoside antibiotics and cis-platinum, always seems to be associated with loss of hair cells in the organ of Corti. Although effects are somewhat variable between drugs and between species of animals it has generally been found that the cochlear lesion originates in the first row of outer hair cells of the basal turn of the organ of Corti (Wersäll 1981). The damage then progresses to affect the other rows of outer hair cells and to involve cells nearer the apex of the cochlea. The inner hair cells are usually damaged only in regions where there is extensive loss of outer hair cells (McDowell 1982). In agreement with clinical findings this progression of hair-cell damage is

paralleled by a hearing loss which is initially confined to high frequencies (Stebbins *et al.* 1981) but eventually may involve all frequencies.

The hair cells in the organ of Corti are generally much more sensitive to damage than are the supporting cells (Wersäll 1981). Degeneration of the afferent nerve fibres and finally the efferent nerve fibres also follows the death of the hair cells (Hawkins and Johnsson 1981; Wersäll 1981). It has been postulated that the degeneration of hair cells in the organ of Corti caused by aminoglycosides may occur secondary to drug effects on other tissues in the cochlea, notably the spiral ligament, the *stria vascularis*, Reissner's membrane, or the region of the outer sulcus (Hawkins *et al.* 1972; Johnsson and Hawkins 1972; Hawkins 1973). However, the exact time sequences of the damage to these tissues and hair-cell degeneration is not known (Hawkins 1973) so that a cause-and-effect relationship is speculative. Certainly, direct damage to hair cells by ototoxic drugs cannot be excluded.

In the vestibular system damage is seen initially in the hair cells at the tip of the *cristae ampullaris* and then progresses to affect cells on the sides of the *cristae* (Wersäll 1981). The most vulnerable cells are the Type I sensory cells although in more severe lesions the Type II cells are also destroyed. Degeneration of the cells in the *macula utriculi* and *macula sacculi* occurs at a later stage. As in the cochlea, the supporting cells and nerve fibres survive the hair cells but eventually they too may be destroyed (Wersäll 1981).

Temporal bones from patients who had suffered drug-induced cochlear or vestibular damage are available only occasionally for histopathological study. Furthermore, interpretation of any observed lesion is frequently complicated by a knowledge of prior or simultaneous exposure to other ototoxic trauma. Paradoxically, the lack of such knowledge can be just as confounding. However, some studies of human temporal bones do suggest that the pattern of damage so carefully delineated in animals is at least similar to that found in man (Johnsson *et al.* 1981; Keene *et al.* 1982; Wright and Schaefer 1982).

The *stria vascularis* is now known to be the primary site in the cochlea for the ototoxic action of the loop diuretics. Intravenous administration of ethacrynic acid or frusemide to guinea pigs and rats causes transient morphological changes in the *stria vascularis*, the time course of which parallels the loss and recovery of electrical activity in the cochlea (Brummett *et al.* 1977; Bosher 1980*a, b*; Pike and Bosher 1980). It is not known whether the hair-cell loss in the organ of Corti, produced by high doses or repetitive administration of ethacrynic acid (Mathog *et al.* 1970; Johnsson and Hawkins 1972;

Crifò 1973), results from a direct action on hair cells or is secondary to prolonged changes in the *stria vascularis*.

Mechanisms of action

Pharmacokinetics

The organ-directed toxicity of the aminoglycosides may be attributable, in part, to the pattern of their tissue distribution. They achieve particularly high concentrations in the renal cortex, both of animals (Toyoda and Tachibana 1978) and man (Schentag *et al.* 1977) and are known to cause lesions of the proximal tubules (Kaloyanides and Pastoriza-Munoz 1980). However, the concentrations of aminoglycosides in the fluids (Toyoda and Tachibana 1978; Harpur *et al.* 1981; Brummett and Fox 1982) and tissues (Desrochers and Schacht 1982) of the inner ear are not high compared with the kidney. Aminoglycosides have a long half-life in perilymph but this is probably not a unique feature of ototoxic drugs (Federspil 1981). Accumulation of aminoglycosides in cochlear perilymph and tissues is modest and variable (Brummett and Fox 1982; Desrochers and Schacht 1982) and probably also occurs in tissues which do not show toxicity, e.g. the liver (Toyoda and Tachibana 1978; Desrochers and Schacht 1982). Brummett and Fox (1982) pointed out that the concentration of aminoglycosides in perilymph associated with complete hair-cell destruction following chronic drug administration was 5/1000 the cytotoxic drug concentration *in vitro*. They concluded that the cytotoxic effect of aminoglycosides on hair cells is highly selective and not simply a result of preferential accumulation of the drug in perilymph. Penetration of aminoglycosides to endolymph is delayed (Watanabe *et al.* 1971; Tran Ba Huy *et al.* 1981) and might explain the slow development of a toxic effect on the hair cells. It may be that the interaction between loop diuretics and aminoglycosides results from a greatly increased entry of the aminoglycoside into endolymph (Tran Ba Huy *et al.* 1983), although Orsulakova and Schacht (1981) thought that it was the aminoglycoside which increased the tissue penetration of the loop diuretic. It is clear that, at present, pharmacokinetic considerations cannot fully explain the ototoxicity of aminoglycoside antibiotics.

Biochemical effects

It was orignally thought that the toxic effects of aminoglycosides on mammalian cells, like their antibacterial action, was the result of inhibition of protein synthesis. There is now considerable evidence to discount this theory (Brown and Feldman 1978; Weiner and Schacht 1981). The most favoured alternative hypotheses have

centred around inhibition of both aerobic and anaerobic untilization of glucose and effects on glucose transport (Brown and Feldman 1978). Although there is no evidence to suggest that energy generation processes are any different in the ear to other organs, it was shown that kanamycin inhibited the glycolytic enzymes hexokinase and phosphofructokinase in the kidney and the organ of Corti (Tachibana *et al.* 1976), whereas the activities of these enzymes were unaffected in the tissues of the lateral wall of the cochlea or in the liver or brain. The dependence of the hair cells on glucose as a primary source of energy, the knowledge that amino sugars compete with glucose for transport pathways, and the presence of amino sugar moieties in aminoglycoside molecules led to the hypothesis that the selective toxicity on the cochlea could result from reduced availability of glucose (Garcia-Quiroga *et al.* 1978). Evidence for this hypothesis is lacking because, although it was found that hyperglycaemia protected against kanamycin-induced ototoxicity, the most likely explanation appeared to be increased renal elimination of kanamycin (Garcia-Quiroga *et al.* 1978). Further evidence that aminoglycoside ototoxicity is not mediated by impairment of glucose uptake or metabolism was published recently (Takada *et al.* 1983).

Inhibition of polyphosphoinositide turnover in the cell membrane has been proposed as a unifying rational for the toxic effects of aminoglycosides on the ear and kidney (Weiner and Schacht 1981; Humes *et al.* 1981). There is rapid turnover of the monoesterified groups of the polyphosphoinositides in the ear and the kidney and these events are thought to be involved in the regulation of membrane permeability. This proposed mechanism of ototoxicity was considered to be compatible with the results of earlier biochemical investigations since alterations in cell metabolism, such as impaired glycolysis, might occur as a consequence of disruption of cell-membrane function.

Sokabe *et al.* (1982) suggested that the oxidized product of phosphatidylinositol diphosphate (TPI), known as lyso-TPI, may be the membrane receptor for aminoglycosides. There is evidence that the transduction process in the cochlea involves changes in the permeability of a monovalent cation channel on the hair cell membrane. In an *in-vitro* system, using an artificial bilayer, lyso-TPI has been shown to form a calcium-sensitive monovalent cation channel. Sokabe *et al.* (1982) found that aminoglycosides suppressed the potassium permeability of the lyso-TPI channel in a manner competitive with calcium and in parallel with their ototoxic actions.

It must be stressed that much of the evidence for the effects of aminoglycosides on polyphosphoinositide metabolism comes from *in-vitro* studies or acute cochlear perfusion studies. After systemic administration of aminoglycosides, decreased polyphosphoinositide turnover has been found only at times when extensive hair-cell death would already have occurred. If inhibition of phospholipid metabolism is the primary event in the development of ototoxicity then it should be evident *in vivo* in advance of other biochemical changes and certainly before extensive morphological changes occur.

Several investigators have sought evidence for a biochemical effect of the loop diuretics on the cochlea similar to the mechanisms originally postulated to be involved in their action on the kidney. It was shown (Kusakari *et al*. 1978) that ethacrynic acid moderately impairs energy generation but both ethacrynic acid and frusemide strongly inhibit energy utilization in the *stria vascularis*. At one time it was thought that the ototoxic effect of ethacrynic acid was due to inhibition of $Na^+-K^+ATPase$, the activity of which was high in the *stria vascularis*. However, it was subsequently shown that the concentrations of both ethacrynic acid and frusemide which inhibited strial $Na^+-K^+ATPase$ *in vitro* were very much higher than the concentrations which inhibited the function of the *stria vascularis* when the drugs were perfused through the perilymphatic spaces of the cochlea (Kusakari *et al*. 1978). Thus it seemed very unlikely that inhibition of $Na^+-K^+ATPase$ could be the mechanism of action *in vivo* unless the drugs were concentrated in strial tissue. Ethacrynic acid and frusemide were found to be potent inhibitors of strial adenylate cyclase *in vitro* (Paloheimo and Thalmann 1977) and the inhibitory potency of ethacrynic acid *in vitro* was comparable to the *in-vivo* potency of ethacrynic acid on strial function. Marks and Schacht (1981) confirmed that ethacrynic acid inhibited cochlear adenylate cyclase *in vitro* although only at a concentration similar to that which inhibited $Na^+-K^+ATPase$. Furthermore, frusemide did not inhibit adenylate cyclase except at very high concentrations. Therefore, Marks and Schacht (1981) rejected inhibition of adenylate cyclase as a common mechanism for the ototoxic action of the loop diuretics.

Further research is required before any definitive theory about the mechanisms of ototoxicity of aminoglycosides and loop diuretics can be formulated. Very little is known about the mechanisms of action of other ototoxic compounds.

Incidence

In 1973 the Boston Collaborative Drug Surveillance Program (BCDSP) published details of the frequency of occurrence of deafness due to drug administration in a large series of consecutively monitored patients in the medical wards of hospitals in several countries. A subsequent report was published by Porter and Jick (1977) and data from the two reports are summarized in Table 31.2. In almost 39 000 patients the overall incidence of drug-induced deafness was 1.6 per 1000 exposed. The drugs most frequently implicated were aspirin (11 per 1000 exposed) and the aminoglycoside antibiotics (collectively, 7 per 1000 exposed). Out of a total of 53 cases of deafness it was possible to follow the progress of 40 patients. Thirty-two recovered their hearing completely, in three patients there was partial recovery, and five remained deaf.

Aminoglycoside antibiotics

The aminoglycoside antibiotics undoubtedly constitute the largest class of ototoxic drugs (Table 31.1). They are also probably the most important because of their extensive use. It was reported that in 1980 and 1981 approximately 4.3 million courses of aminoglycosides were prescribed in the USA during each 12-month period. The figure for 1982 was only slightly lower at 3.9 million (Data from National Disease and Therapeutic Index, IMS America Ltd., cited by Whelton and Solez 1983). Thus, given their propensity for causing *permanent* disabilities, whether vestibular or auditory or both, even a 0.7 per cent incidence of ototoxicity (Porter and Jick 1977) represents a significant problem.

The aminoglycosides are all polybasic, hydrophilic molecules and, consequently, they are negligibly absorbed from the gastrointestinal tract and must be administered parenterally for systemic effect. There is no evidence for any metabolism and the drugs are eliminated from the body almost exclusively by renal excretion. All aminoglycosides are toxic to the inner ear, the kidney (Kaloyanides and Pastoriza-Munoz 1980), and the neuromuscular junction (Caputy *et al*. 1981) although there are important qualitative and quantitative differences between compounds. Small differences in the molecular structure of aminoglycosides can significantly alter their toxicity profile.

Streptomycin, the first aminoglycoside to be introduced (1945), is still used in the treatment of tuberculosis. Dihydrostreptomycin, the reduction product of streptomycin, was withdrawn from clinical use because it was shown (Hawkins 1959) to cause a much higher incidence of auditory toxicity than streptomycin. Soon after the introduction of neomycin (Waksman and Lechevalier 1949) it became evident that the drug was very toxic both to the kidney and the cochlea. Therefore, following the introduction in 1958 of kanamycin, a less toxic drug with similar therapeutic effects, the parenteral use of neomycin was almost entirely abandoned. The enteral and topical use of neomycin is still widespread and it should be recognized that administration even by these routes carries considerable risk of ototoxicity. Indeed, there is abundant evidence (see Harpur 1977, for review; Bamford and Jones 1978; Kavanagh and McCabe 1983) that sufficient neomycin to result in severe permanent deafness can be absorbed through the surfaces of body cavities irrigated with a solution of the drug; through the inflamed or ulcerated gastrointestinal mucosa and even through the normal gastrointestinal mucosa if neomycin is given orally in high dosage or for prolonged periods or through broken skin such as severe burns if applied topically. The use of framycetin and paromomycin, two aminoglycosides in the neomycin group, has largely been superseded by other drugs.

Jackson (1984) remarked that streptomycin, kanamycin, and gentamicin each had a key role to play in meeting a major medical problem at the time of their introduction: streptomycin for tuberculosis and some pyogenic infections; kanamycin for the inhibition of penicillinase-producing *Staphylococcus aureus* and common gram-negative bacilli; gentamicin for the treatment of the growing number of infections caused by kanamycin-resistant species of enterobacteriaceae and *Pseudomonas aeruginosa*. Gentamicin rapidly became the drug of first choice for treating life-threatening sepsis, especially in patients with nosocomical infections and in patients with a compromised immune system (Noone 1982). Time, the changing character of patients and procedures, and bacterial evolution with the selection of resistant strains led to the virtual eclipse of streptomycin and kanamycin (Jackson 1984).

For the past 15 years gentamicin has undoubtedly been the most important aminoglycoside for systemic use. The extent of use of subsequent drugs such as tobramycin, amikacin, sisomicin, ribostamycin, dibekacin, and, most recently netilmicin varies between countries and may depend on local circumstances (Noone 1984). In some countries, notably the USA tobramycin has gradually eroded the pre-eminence of gentamicin as the major aminoglycoside in clinical use because it was claimed to be less nephrotoxic and ototoxic than gentamicin in both animals and man.

The principal advantage of amikacin is its activity against gentamicin-resistant bacteria and its use is usually restricted to treatment of infections with such organisms. Sisomicin is unlikely to become widely available and the use of dibekacin and ribostamycin has mainly been confined to Japan and one or two European countries. Ribostamycin may be one of the less toxic aminoglycosides (Harpur *et al.* 1981), although its spectrum of antibacterial activity is probably no better than that of kanamycin. Netilmicin appears to be an efficacious antibiotic which may offer the advantage of reduced toxicity over its predecessors (Lane 1984; Noone 1984).

Waldvogel (1984) summarized the major indications for systemic use of the aminoglycoside antibiotics. These include treatment of serious infections due to Gram-negative aerobic bacilli, notably *Pseudomonas aeruginosa*, and, in combination with an antipseudomonal β-lactam antibiotic, treatment of leukopenic patients with pyrexia. The wide variety of third-generation cephalosporins now available encompass the same spectrum of antimicrobial activity as aminoglycosides, apparently with a lesser degree of toxicity, and in addition some agents are active against anaerobic organisms or organisms resistant to aminoglycosides. However, Waldvogel (1984) thought that, although some of the new β-lactam antibiotics will offer future alternatives to therapy with aminoglycosides, it is unlikely that they will completely supplant them.

Incidence of aminoglycoside ototoxicity

Streptomycin

Streptomycin is very rarely used other than for the treatment of tuberculosis and therefore would not have been used in the medical wards included in the Boston Collaborative Drug Surveillance Programme (Table 31.2). Hawkins (1976) stated that by 1948, i.e. three years after the first clinical use of streptomycin, vestibular disturbances had become extremely common in tuberculous patients receiving the drug. Glorig and Fowler (1947) reported that only six of 32 patients who received 1.8 to 2.0 g daily failed to show any symptoms of ototoxicity. Northington (1950) found that all patients receiving 2.5 to 3.2 g daily showed vestibular effects, usually during the second week of treatment.

There is ample evidence that the ototoxic effects of streptomycin are predominantly upon vestibular function. Indeed, there is no more convincing testimony to the selectivity of streptomycin for the vestibular system than its use to destroy this organ in Ménière's disease. The vestibular system could be consistently ablated without affecting hearing (Schuknecht 1957). Walby and Kerr (1982) reviewed the English language literature from 1945 to June 1980 and found reports of only 271 cases of hearing loss which could be attributed with certainty to streptomycin. Only 24 patients suffered deafness severe enough to be a handicap and seven of these patients had coincidental renal disease. Such a low reported incidence over a 35-year period seems to confirm that streptomycin is not markedly toxic to the cochlea. Walby and Kerr also concluded that streptomycin was cochleotoxic only when given in a daily dose of 1.0 to 2.0 g for a prolonged period.

Kanamycin

Porter and Jick (1977) found an incidence of 1.3 per cent auditory toxicity with kanamycin (Table 31.2). this is a somewhat lower figure than the early incidence of ototoxicity with kanamycin. Finegold (1966) reported data collated by Bristol Laboratories which indicated that, prior to 1963, 4.9 per cent of 1815 patients treated with kanamycin had some degree of clinically apparent ototoxicity; 1.7 per cent showed severe damage.

It is difficult, and perhaps misleading, to state an overall incidence of ototoxicity with kanamycin because the incidence varies so markedly depending on its use. It is clear that the long-term use of kanamycin in the treatment of tuberculosis greatly increases the risk of ototoxicity. Published data reviewed by the Food and Drugs Administration (cited by Finegold 1966) indicated that the incidence was about 1 per cent in patients who had therapy with kanamycin lasting less than two weeks. Long-term therapy with a daily dose of 1.0 g produced ototoxicity in 30 per cent of patients. Frost et al. (1960) found that 17 of 21 patients receiving 0.5 to 1.5 g of kanamycin daily for three months or more developed ototoxicity. Haapanen (1963) estimated that the risk of auditory damage was 23 per cent for patients treated for three months and 57 per cent for patients treated for six months.

Kreis (1966) reviewed the incidence of ototoxicity with kanamycin reported in European studies between 1959 and 1964. More than 1000 patients were included in 21 studies. The incidence of audiometric changes ranged from 0 to 82 per cent and the average incidence was 30 per cent. However, Kreis advised that the figure for average incidence had no precise meaning because it depended on the type of hearing impairment taken into account, the nature of the therapy, and patient- and disease-related variables.

Gentamicin

In one of the first clinical studies of gentamicin (Jao and Jackson 1964), five of 57 patients (9 per cent) had serious ototoxicity which was confined to the vestibular system in all cases. Subsequently, Jackson and Arcieri (1971) reviewed the case histories of 1484 patients treated with gentamicin during its 'pre-licence' clinical trials. There were 42 reports of gentamicin treatment associated with symptoms and/or signs of vestibular or auditory dysfunction, i.e. an incidence of 2.8 per cent. On critical analysis, however, they concluded that one-third of the toxic effects reported probably were unrelated to gentamicin, making the true incidence approximately 2 per cent. Their study showed further that in approximately two-thirds of the patients with ototoxic side-effects, dysfunction was limited to the vestibular system.

Other studies have found widely divergent incidences of gentamicin-induced ototoxicity. There are a number of factors which probably contribute to this apparent variability in the frequency of ototoxicity. The greatest differences between studies occur in the characteristics of the populations studied; these range from large heterogeneous groups to small numbers of patients who all have a recognized predisposing factor such as renal failure. Studies may be prospective or retrospective, and only in the former case are transient effects likely to be recorded. There are also differences in whether or not an attempt is made to regulate dosage, the basis for doing this, and the rigour with which such intentions are implemented and controlled. Not least, of course, differences exist in the methods of assessment of toxicity and the criteria used to define toxicity. Comparison of the following studies illustrates the difficulties of arriving at a true idea of the vestibular toxicity of gentamicin.

Nordström et al. (1973) investigated prospectively a small group of patients (34) receiving gentamicin therapy. Vestibular toxicity which was attributable unequivocally to gentamicin occurred in three patients (9 per cent). These patients were reported to have 'severely abnormal electronystagmograms' after treatment and all three patients exhibited symptoms of vestibular dysfunction. Dayal et al. (1974) reported that 12 of 22 patients (55 per cent) had evidence of vestibular ototoxicity after treatment with gentamicin. All of the 22 patients were in chronic renal failure and undergoing haemodialysis.

Table 31.2 Drugs implicated in the development of deafness and frequency of occurrence

Drug	Number of patients exposed	Number developing deafness	Number per 1000 developing deafness
Aspirin	2 974	33	11
Kanamycin	372	5*	13
Gentamicin	1 125	4	4
Neomycin	802	7	9
Paromomycin	75	1	13
Ethacrynic acid (i.v.)	184	2	11
Quinidine	1 024	1	1
Propranolol	853	1	1

* One patient developed deafness after treatment with ethacrynic acid and again after treatment with kanamycin.

Source: Porter and Jick (1977).

Vestibular function was assessed by hot caloric stimulation (Davey *et al.* 1982*a*) and, if abnormalities were found, full bithermal caloric tests were done. Three of four burn patients, who were given gentamicin by both topical and systemic routes, also showed vestibular toxicity.

In a prospective study Winkel *et al.* (1978) used the hot caloric test. They found that three out of 20 patients (15 per cent) had reduced vestibular function after gentamicin, although only one patient experienced symptoms. Fee *et al.* (1978) reported an incidence of 11 per cent vestibular toxicity with gentamicin although subsequently, with an increased study population, the incidence had risen to 15 per cent (Fee 1980). Caloric testing was done periodically during therapy, at the end of therapy, and at a follow-up examination.

Vestibular function testing is problematical under ideal conditions but the problems are amplified when applied to sick patients. The tests are relatively insensitive and large intertest variability has been found in normal healthy subjects (Davey *et al.* 1982*a*; Roeder *et al.* 1981). Prospective monitoring of vestibular function using caloric testing has not often been attempted during therapy with ototoxic drugs. In contrast there are sensitive reproducible tests available for testing auditory function. Pure-tone audiometry, using portable equipment, ought to be applicable to monitoring patients in a hospital ward for ototoxic hearing loss. Once again, the reported incidences for gentamicin-induced auditory toxicity vary greatly between studies (Table 31.3). It is also apparent that the incidences detected by the prospective use of sensitive tests are always higher than those found by Jackson and Arcieri (1971) in their retrospective analysis of patients who had shown symptomatic changes.

There are several important differences in methodolody between these studies (Table 31.3). Data were based on audiograms recorded in an anechoic chamber (Fee 1980) or in a ward (Bender *et al.* 1979). Unilateral changes were regarded as significant in most studies but only bilateral changes were accepted in one (Bender *et al.* 1979). The change in threshold regarded as significant varied between 10 dB (Winkel *et al.* 1978) and 40 dB (Meyers 1970). It is apparent from the data in Table 31.3 that the reported incidence of toxicity is inversely related to the change in threshold which was regarded as significant. Pure-tone audiometry is capable of measuring hearing threshold to within 10 dB (Engelberg 1978). However, this applies to relatively fit people tested in sound-proofed rooms and cannot necessarily be extended to the testing of patients with serious infections in a ward. It would seem unwise to regard as significant any differences in threshold of <15 dB, if the audio-

Table 31.3 The incidence of gentamicin auditory toxicity reported in four prospective studies using different criteria for significant hearing loss. In all studies changes at any frequency were regarded as significant

Study	Criteria for significant hearing loss	Number of patients in study	Percentage of significant hearing loss
Bender *et al.* (1979)	Bilateral 20dB	32	6.0
Meyers (1970)	Unilateral 40dB	46	7.5
Fee (1980)	Unilateral 20dB	45	16.4
Winkel *et al.* (1978)	Unilateral 10dB	20	45.0

grams are recorded outside an anechoic chamber (Davey *et al.* 1983).

In a prospective study, if observed transient changes in hearing are actually attributable to undefined variability in the test procedure, the subject's condition, or their ability to co-operate, a spurious conclusion may be drawn about the incidence of drug-induced effects. Conversely, if aminoglycosides do affect the cochlea in a reversible manner, this may be missed in a retrospective study. Davey *et al.* (1982*b*, 1983) looked prospectively at the incidence of hearing loss in hospital patients who received a course of gentamicin therapy and in a control group who had a similar spectrum of clinical conditions but did not receive gentamicin. Davey *et al.* (1982*b*, 1983) found that eight of the 27 control patients had a transient but significant ($\geqslant 20$ dB) increase in hearing threshold suggesting that when transient hearing losses are observed in aminoglycoside-treated patients they may not necessarily be attributable to the drug. Although acute reversible changes in cochlear electrophysiology caused by aminoglycosides have been reported in both animals (Logan *et al.* 1974) and man (Wilson and Ramsden 1977), it is not clear whether these changes represent a genuine ototoxic effect and, even if they do, whether or not they relate to permanent hearing impairment which is always associated with a lesion in the organ of Corti. Davey *et al.* (1983) concluded that transient changes in threshold recorded in sick patients by audiometry should be ignored.

Davey *et al.* (1982*b*, 1983) also addressed the question of how early in aminoglycoside treatment should the first audiogram be recorded. It is customary in prospective trials of aminoglycoside ototoxicity to perform an audiogram before or soon after the drug is started to obtain a 'baseline' measurement. However, this measurement will necessarily be made when the patient is suffering from a serious infection so that the improvement in hearing which may accompany the patient's recovery could mask a drug-induced hearing loss. Davey *et al.* (1983) reported sustained hearing improvement both in patients receiving gentamicin and in untreated control patients. Neu and Bendush (1976) also noted equal numbers of patients with improvement and deterioration in hearing when they analysed audiograms taken at the beginning and end of tobramycin therapy of 99 patients. Since sustained improvements in hearing, probably attributable to improvement in the patient's clinical condition, could mask a drug-induced hearing loss, Davey *et al.* (1983) concluded that an audiogram taken at the start of antibiotic therapy is not a reliable measurement of the patient's normal hearing.

One way to avoid this problem would be to assess hearing after the patient has recovered from the infection which was treated with an aminoglycoside and to compare this measurement with age- and sex-adjusted normal values. Davey *et al.* (1983) recorded pure-tone audiograms in an anechoic chamber from 21 patients who had not received aminoglycosides and 22 patients who had received gentamicin within the past six months. The incidence of hearing abnormalities was not significantly different in the two groups (9/21 and 11/22 respectively). In both groups approximately half of the patients with abnormal hearing had a history of noise exposure which might explain the hearing loss. This retrospective approach was therefore considered to be inappropriate.

Finally, Davey *et al.* (1983) found a similar incidence of sustained hearing loss in patients who were treated with gentamicin and those who were not administered a known ototoxic drug (between 7 and 9 per cent in both groups). They, therefore, recommended that future studies of aminoglycoside auditory toxicity should be prospective and should include a control group of patients who did not receive an aminoglycoside. Ideally, the initial audiogram should be recorded before therapy when the patients are clinically well. Davey *et al.* suggested that a suitable study population which could meet these criteria would be patients with cystic fibrosis or leukaemia, both of whom are likely to require recurrent courses of broad-spectrum antibiotics.

Comparative ototoxicity of newer aminoglycosides

In the early development of aminoglycosides the main objectives in introducing new drugs were to improve the antibacterial spectrum and to overcome the problems of bacterial resistance. These objectives have largely been achieved and the emphasis has switched to the reduction or elimination of toxicity — nephrotoxicity as well as ototoxicity.

Our knowledge of the comparative ototoxicity of aminoglycoside antibiotics is derived both from studies in animals and clinical studies. Gentamicin first became

available for clinical trial in the USA in 1962. It was rapidly established as the drug of first choice for treating life-threatening sepsis. The therapeutic efficacy and toxicity of new aminoglycosides, such as tobramycin, amikacin, and netilmicin, have been evaluated compared with gentamicin. Thus most controlled comparisons of the toxicities of aminoglycosides postdate gentamicin. It is therefore difficult to be certain about the relative toxicity of earlier compounds, many of which now have been almost entirely superseded by gentamicin or its successors.

Although the ototoxic effects of gentamicin in man are said to be predominantly upon the vestibular system, in animals gentamicin is both cochleotoxic and vestibulotoxic. Tobramycin was found to be less cochleotoxic than gentamicin in guinea pigs (Brummett *et al.* 1972), although Aran *et al.* (1982) thought that any advantage of tobramycin over gentamicin was confined to a reduced toxicity on the vestibular system. Brummett *et al.* (1978*a*) compared the cochleotoxicities of tobramycin, gentamicin, amikacin, and sisomicin in guinea pigs and found once again that tobramycin was less ototoxic than gentamicin. Sisomicin was comparably toxic to gentamicin whereas the ototoxic potential of amikacin, on an equal-dose basis, was less than that of gentamicin and similar to that of tobramycin. When compared to gentamicin and tobramycin at equitherapeutic doses, amikacin is probably no less toxic in guinea pigs (Brummett and Fox 1982). Parravicini *et al.* (1982) concluded on the basis of studies in guinea pigs that the ototoxic potential of gentamicin, tobramycin, and amikacin would be the same at recommended therapeutic doses.

Christensen *et al.* (1977) compared the cochlear and vestibular toxicities of amikacin with gentamicin in the cat — an animal which is particularly suited to vestibular function testing; the doses of amikacin were five times the doses of gentamicin which is in approximately the same ratio as their clinical doses. Gentamicin was primarily vestibulotoxic but amikacin did not affect vestibular function. Amikacin was selectively cochleotoxic and the damage was greater than that caused by gentamicin. Aran *et al.* (1982) found dibekacin to be much less cochleotoxic in guinea pigs than either tobramycin or gentamicin but comparably vestibulotoxic to tobramycin, which in turn was less vestibulotoxic than gentamicin. Gentamicin is a mixture of three components; gentamicin C_1, gentamicin C_{1a}, and gentamicin C_2. Gentamicin C_1 was found to be less cochleotoxic in guinea pigs than gentamicin (Fox *et al.* 1980) but any therapeutic advantage with gentamicin C_1 was likely to be small because of its lesser antibacterial potency (Brummett and Fox 1982).

One of the most recent aminoglycosides to achieve a

product licence, netilmicin (1-N-ethyl sisomicin), has been extensively studied in animals. Luft (1978) reviewed the early studies of the toxicity of netilmicin in animals. Without exception these studies found that netilmicin had a very low ototoxic potential. Netilmicin was much less vestibulotoxic (Miller *et al.* 1976; Igarishi *et al.* 1978) and cochleotoxic (Brummett *et al.* 1978b) compared to gentamicin. Subsequent studies in guinea pigs have confirmed the low ototoxic potential of netilmicin compared with gentamicin and also have shown netilmicin to be much less cochleotoxic than both amikacin (Wersäll 1980; Parravicini *et al.* 1982) and tobramycin (Parravicini *et al.* 1982).

Parravicini *et al.* (1983) found that dibekacin induced only moderate impairment of auditory and vestibular function in guinea pigs in accord with the findings of Aran *et al.* (1982) whereas netilmicin did not cause any ototoxic changes. Anniko *et al.* (1982) compared the ototoxic potentials of netilmicin and gentamicin in three models: the 'standard' *in-vivo* guinea pig model; organ culture of the embryonic inner ear of the mouse; and perilymphatic perfusions of the guinea-pig cochlea. When given by subcutaneous injections, netilmicin produced no damage to hair cells in the cochlear or the vestibular sensory epithelia at doses which with gentamicin produced extensive damage in both systems. In organ culture, concentrations of netilmicin had to be at least 10-fold higher than those of gentamicin in order to produce comparable damage. When the drugs were perfused in the cochlea, an equitoxic dose of netilmicin was three times higher than that of gentamicin. Anniko *et al.* (1982) concluded that the low ototoxicity of netilmicin observed *in vivo* was the result of lower intrinsic toxicity and lesser penetration into the inner ear.

The data from these animal studies would seem to predict that there would be little to choose between the ototoxicities of gentamicin, tobramycin, amikacin, or sisomicin at equitherapeutic doses. Tobramycin might tend to be less toxic than gentamicin for both auditory and vestibular systems and amikacin less toxic than gentamicin for the vestibular system but the differences are certainly not clear-cut. Dibekacin might be less cochleotoxic than gentamicin or tobramycin and less vestibulotoxic than gentamicin. However, the only unequivocal indicator is that netilmicin should be significantly less toxic than gentamicin, tobramycin, amikacin, and even dibekacin.

A number of comparative ototoxicological investigations have tended to confirm these findings in man, although there are insufficient data to determine the comparative ototoxic potential of sisomicin or dibekacin with other aminoglycosides in man. Neither Smith *et al.* (1980) nor Fee (1980) found any difference between the auditory toxicity of gentamicin and tobramycin in man although Fee (1980) did find a significantly lower incidence of vestibular toxicity with tobramycin (4.6 per cent) than with gentamicin (15.1 per cent). There were no differences in the auditory toxicity of gentamicin and amikacin (Smith *et al.* 1977; Matz and Lerner 1981) or in the vestibular toxicity of these drugs (Matz and Lerner 1981). Gatell *et al.* (1983) found no difference in auditory toxicity between tobramycin and amikacin.

Early clinical studies (Nordström *et al.* 1979; Tjernström 1980; Vesterhauge *et al.* 1980) seemed to confirm that the ototoxic potential of netilmicin in man is very low. Nordström *et al.* (1979) evaluated vestibular and auditory function in 52 patients treated with netilmicin and claimed that two suspected cases of ototoxicity, one auditory and one vestibular, could not reliably be attributed to the drug. Tjernström (1980) tested the vestibular function of 76 patients treated with netilmicin; the auditory function of 74 of these patients was also tested. Only one possible case of ototoxicity was detected — a vestibular lesion which was described as subclinical and reversible. Vesterhauge *et al.* (1980) reported that there were no signs or symptoms that could be ascribed to an ototoxic effect in 30 patients treated with netilmicin.

Two comparative studies found that netilmicin produced a lower incidence of ototoxicity than amikacin but the differences were not significant (Barza *et al.* 1980; Bock *et al.* 1980). Recently, two multi-centre studies have produced evidence that netilmicin is less ototoxic in man than tobramycin (Lerner *et al.* 1983; Daschner *et al.* 1984). In both studies patients with serious systemic infections were treated either with netilmicin combined with ticarcillin or with tobramycin. Auditory function was assessed by pure-tone audiometry. Lerner *et al.* (1983) found that the incidences of auditory toxicity associated with tobramycin and netilmicin were 12 and 3 per cent, respectively, a difference which was just statistically significant. In the study of Daschner *et al.* (1984) two patients treated with tobramycin–ticarcillin had drug-related decreases in auditory acuity compared with a zero incidence in the netilmicin-treated patients. When patients with transient symptoms suggestive of drug-related vestibular dysfunction were included, the comparative incidences of ototoxic reactions were 10 per cent associated with tobramycin and 3 per cent associated with netilmicin. This difference between tobramycin and netilmicin was not significant. It should be noted that a majority of the patients evaluated in the study of Daschner *et al.* (1984) had also been included in the study of Lerner *et al.* (1983).

From the results of these clinical studies it can be seen that the clear advantage displayed by netilmicin compared with other aminoglycosides in ototoxicological studies in animals is difficult to confirm in clinical use. There are undoubtedly many reasons why it is very difficult to establish unequivocally whether one aminoglycoside antibiotic is more or less ototoxic in man than another. The two main reasons are probably, firstly, that the assessment of auditory and vestibular function in sick patients is beset with numerous problems (Davey *et al.* 1982*a, b*) and, secondly, that the incidence of ototoxicity with all the drugs is quite low so that large populations would be required to detect any differences. The incidence of ototoxicity with aminoglycoside antibiotics has undoubtedly been reduced as clinicians have learned to monitor therapy and adjust dose to suit the needs of each patient, particularly where renal impairment is present. It has long been suspected that at least part of the difference in ototoxicity between aminoglycosides is actually attributable to differences in nephrotoxic potential which, if present, would lead to accumulation of the drug in the body and hence increase the risk of ototoxicity. Thus if dose is adjusted during a study both to take account of pre-existing renal impairment and to minimize the development of nephrotoxicity, this may also serve to decrease the incidence of ototoxicity with a more nephrotoxic drug.

Several reviews of the literature on comparative clinical trials of aminoglycosides have been published. Since there is considerable overlap in the literature reviewed, only two of these reviews will be mentioned here. Smith and Lietman (1982) reviewed the English language literature until the end of 1979 and highlighted the difficulties of establishing differences in ototoxicity between aminoglycosides. Before the end of 1979 Smith and Lietman could find only six studies which attempted to compare the auditory toxicity of aminoglycosides using audiometry and only two studies which used objective methods to monitor vestibular function. The mean frequencies of ototoxicity were gentamicin 8 per cent, amikacin 7 per cent, tobramycin 6 per cent, sisomicin 5 per cent, and netilmicin 3 per cent (overall mean 7 per cent). None of the studies found any difference in incidence between aminoglycosides. The mean number of patients studied in the experimental groups and control (i.e. gentamicin) groups was only 35. Thus, with an average rate of 8 per cent in the control group, Smith and Lietman stated that it would be impossible to show a statistically significant improvement in an experimental group even if the incidence in this group was zero. Smith and Lietman felt that most studies at that time could not exclude chance as a cause of even two or threefold differences in ototoxicity. They also concluded that almost nothing was known about the relative ototoxicity of aminoglycosides and called for larger, well-designed

studies to detect differences which may be clinically meaningful.

Kahlmeter and Dahlager (1984) reviewed the clinical literature relating to aminoglycoside toxicity published between 1975 and 1982. They selected, from 400 publications, 144 reports of clinical trials each of which included at least 15 patients evaluable for nephrotoxicity and/or ototoxicity provided relevant data were given on methodology, patient material, and aminoglycoside dosage. No attempt was made to separate out 'clinically significant' reactions and Kahlmeter and Dahlager accepted the definitions given by the respective authors; in some cases this was no more than a general statement that a toxic reaction had occurred. However, Kahlmeter and Dahlager analysed separately those reports of cochlear toxicity where the changes in auditory function were identified using audiometry, although even here the criteria used to define toxicity were widely divergent. Kahlmeter and Dahlager presented separately the frequencies of ototoxicity occurring in all studies (Table 31.4) and those found in comparative trials of two or more aminoglycosides. From Table 31.4 it is apparent that the incidence of both cochlear toxicity and vestibular toxicity was lowest with netilmicin although there seems little to choose between any of the drugs in terms of their vestibular toxicity. The data also indicate that amikacin was more frequently cochleotoxic than both gentamicin and tobramycin which produced very similar incidences of auditory dysfunction. If analysis is restricted to cochlear toxicity detected in comparative trials by the use of serial audiometry, the highest incidence was still seen with amikacin and the lowest with netilmicin but the

Table 31.4 Aminoglycoside nephro- and ototoxicity in prospective clinical trials 1975–82

Toxicity	No. of trials	No. of patients	Toxicity (per cent)
Renal			
Gentamicin	70	4 023	14.0
Tobramycin	38	2 130	12.9
Netilmicin	50	2 360	8.7
Amikacin	21	1 144	9.4
Cochlear			
Gentamicin	28	1 895	8.3
Tobramycin	14	572	6.1
Netilmicin	36	1 338	2.4
Amikacin	13	713	13.9
Vestibular			
Gentamicin	8	535	3.2
Tobramycin	6	289	3.5
Netilmicin	21	990	1.4
Amikacin	6	217	2.8

Source: Kahlmeter and Dahlager (1984).

orders of tobramycin and gentamicin were reversed. This probably indicates that there is little to choose between these latter two drugs in terms of their ototoxicity.

Kahlmeter and Dahlager (1984) stressed the difficulties encountered in undertaking a survey of this kind, not least because data were being pooled from many individual trials where the results may be strongly influenced by variability in therapeutic traditions, laboratory facilities, methodology, definitions of toxic reactions, the aim of the study, underlying disorders, etc. This was best illustrated by the wide variation in the frequency of nephrotoxicity in the prospective comparative trials: gentamicin 4 to 53 per cent, tobramycin 2.6 to 58 per cent, netilmicin 0 to 31 per cent, and amikacin 0 to 38 per cent. Even more striking, and more puzzling, was the pronounced difference sometimes observed in the frequency of toxicity observed with one drug depending on which drug it was being compared with. Thus, the mean cochlear toxicity of gentamicin in seven trials with netilmicin was only 2.1 per cent compared with 11.4 per cent in 10 trials with amikacin. The toxicity of netilmicin was only 1.5 per cent in seven trials with gentamicin but 6.3 per cent when compared in five trials with amikacin.

Such anomalies are totally unexplained but highlight the need for caution in the use of these data. Nevertheless, it is apparent from the survey of Kahlmeter and Dahlager, from consideration of numerous individual clinical trials, and from the results of extensive investigation in animals that netilmicin is consistently found to be the least ototoxic of the therapeutically important aminoglycosides. Any distinction between the ototoxicities of gentamicin, tobramycin, and amikacin is less clear-cut.

Risk factors associated with aminoglycoside ototoxicity

Impaired renal function

The occurrence of ototoxicity with the aminoglycosides has been linked to impaired renal function. This is not surprising since their elimination from the body is by excretion of the unchanged drug in urine. The situation is further complicated in that all aminoglycosides are, to some degree, nephrotoxic.

The earliest clinical reports of the use of neomycin revealed an association between ototoxicity and impaired renal function. Four of 63 patients who were treated with neomycin developed hearing loss and all four patients had impaired renal function (Waisbren and Spink 1950). Subsequently, severe hearing loss following therapy with neomycin was reported by several investigators. Patients with renal disorders, even of slight degree, were especially sensitive to the ototoxic effects of neomycin. It

was clear from the early literature on neomycin that not only was this drug very ototoxic but it was also very toxic to the kidneys. Thus, there is little doubt that neomycin could either exacerbate renal impairment from another cause or damage healthy kidneys to the extent that the risk of ototoxicity would be greatly increased.

As with neomycin, the earliest use of kanamycin showed that the risk of ototoxicity was increased in patients with renal dysfunction (Frost *et al.* 1959). One of these patients had severe renal dysfunction and developed total loss of hearing after a relatively small dose of 40 g. Frost *et al.* (1959) emphasized the need for caution in using kanamycin to treat patients with impaired renal function. Finegold *et al.* (1959) observed ototoxic effects in 22 of 106 patients treated with kanamycin and one of the principal factors contributing to ototoxicity was renal dysfunction, either pre-existing or as a result of kanamycin therapy. About 50 per cent of patients with ototoxicity had renal impairment compared to 20 per cent of patients without ototoxicity. The nephrotoxicity of kanamycin was thought to be important only in so far as it predisposed to ototoxicity.

Jao and Jackson (1964) reported the first cases of ototoxicity with gentamicin. They observed severe vestibular toxicity in five of 57 patients treated with gentamicin. All but one of the five patients with ototoxicity had severe renal disease. Jao and Jackson commented that ototoxicity did not occur in patients without renal insufficiency except where the dose was exceptionally high. In their review of the case histories of 1484 patients treated with gentamicin during its 'pre-licence' clinical trials, Jackson and Arcieri (1971) analysed 14 patient or treatment variables to determine whether they contributed to the development of ototoxicity. They concluded that the most dominant factor, by far, was the functional status of the kidneys. Two-thirds of all patients with ototoxicity had renal impairment.

Aspects of the dosage regimen

From the very first studies of therapy with aminoglycosides, physicians have attempted to identify any relationship which may exist between dose, or duration of therapy, and ototoxicity. Northington (1950), who described the vestibular toxicity of streptomycin in man, reported that patients receiving 1 g of streptomycin daily were almost all symptom-free, even after 75 days of treatment, whereas all patients receiving 2.5 to 3.2 g daily showed vestibular toxicity usually during the second week of administration.

Variability in patient response was clearly noted by Cawthorne and Ranger (1957) who found that very small doses of streptomycin, sometimes as little as a few grams, caused vestibular toxicity in susceptible patients. Normally, unless the daily dose exceeded 1 g, ototoxicity appeared only after prolonged therapy. With higher doses of 2 to 3 g daily, symptoms of vestibular toxicity usually appeared within three or four weeks.

Finegold *et al.* (1959) found the ototoxicity of kanamycin to be related to the amount of drug given. Patients with ototoxicity had received significantly more kanamycin, both in terms of average daily dose adjusted for body weight and average total dose, than had patients without ototoxicity. Patients with ototoxicity were also treated for a longer period, on average, but the difference was not significant. Two reveiws of kanamycin toxicity (Finegold 1966; Kreis 1966) confirmed that the risk of ototoxicity with this drug increased with increasing total dose and duration of therapy. Comparison of the incidence of ototoxicity after intravenous and intra-muscular administration of kanamycin (Kreis 1966) suggested that high blood levels did not incease the risk of ototoxicity. Kreis concluded that ototoxicity was more related to the length of time kanamycin remained in the blood than to the blood level *per se*; thus dividing a 1.0 g daily dose into two doses each of 0.5 g was disadvanta-geous. This conclusion of Kreis (1966) was echoed by the work of Line *et al.* (1970) who measured serum levels of streptomycin at intervals up to 24 h after administration. Line *et al.* found that the 24-h levels were less variable and more informative than the peak levels (1–3 h). Twenty-four hour levels greater than 3mg l⁻¹ were likely to be associated with toxicity, as manifested by dizziness.

In recent years there have been numerous attempts to correlate various aspects of aminoglycoside dosage regimens with toxicity. The general availability of reliable antibiotic assays has prompted many workers to monitor peak and trough serum levels of the drugs and to include such data in the analyses (Table 31.5). It is readily apparent that there is no consensus as to which factors are the determinants of ototoxicity. Although both high peak serum levels and high trough serum levels have been associated with ototoxicity neither of these factors *per se* appears to be the cause of ototoxicity (Davey *et al.* 1980). Thus, although it is widely believed that it is desirable to monitor the concentrations of aminoglycosides in serum in order to ensure efficacy and the absence of nephro- or ototoxicity, it remains uncertain what constitutes a safe, effective serum concentration range for aminoglycosides (Barza and Lauermann 1978). Evidence from animal studies suggests that serum levels of aminoglycosides are not a particularly useful guide to what is happening in the inner ear (Harpur 1982). Despite the uncertainty about the relationship between serum concentrations of amino-glycosides and toxicity, it is desirable that therapy should be monitored if only because measurement of the serum levels of aminoglycosides is one of the most sensitive and reliable methods of detecting a change in renal function (from whatever cause) during therapy.

It is well recognized that because aminoglycosides are eliminated by the kidney the dose should be adjusted in renal impairment. There is also considerable variability in individual pharmacokinetic handling of aminoglyco-

Table 31.5   Relationships of daily dose, total dose, duration of treatment, and serum levels to aminoglycoside ototoxicity

Reference	Drug	Correlation between ototoxicity and				
		Daily dose adjusted for body weight	Total dose	Duration of treatment	Peak serum levels	Trough serum levels
Jackson and Arcieri (1971)	Gentamicin	√	√ *	X	O	O
Nordström *et al.* (1973)	Gentamicin	X	X	√	X	√
Mawer *et al.* (1974)	Gentamicin	O	O	√	O	√
Black *et al.* (1976)	Amikacin	X	√	√	√	√
Winkel *et al.* (1978)	Gentamicin	O	X	X	X	X
Barza *et al.* (1980)	Amikacin } Netilmicin }	O	O	X	O	O
Bock *et al.* (1980)	Amikacin	O	√	√	X	X
	Netilmicin	O	X	X	√ †	√ †
Fee (1980)	Gentamicin } Tobramycin }	X	√	√	X	X
Matz and Lerner (1981)	Gentamicin	O	X	X	X	√
	Amikacin	O	√	√	X	X
Gatell *et al.* (1983)	Amikacin } Tobramycin }	O	X	X	X	X

√ Correlation present   X Correlation absent   O Correlation not attempted or data insufficient

* Only in patients with renal impairment; in patients with normal renal function there was no correlation with total dose.

† Nephrotoxicity was also present in these patients.

sides even in the presence of essentially normal renal function. A great variety of methods of varying complexity have been advocated for determining an individual patient's dosage of an aminoglycoside (see Davey *et al.* 1980 for review). It has been shown that basing a patient's dose on individual characteristics, such as weight and serum creatinine (the latter to take some account of renal function), is better than giving all patients the same dose. The advantage of more rigorous estimations and use of individual pharmacokinetic parameters remains to be proved (Davey *et al.* 1980).

Age and cochlear development

Ototoxicity in the elderly. Advanced age is often cited as a potential risk factor for the development of ototoxicity. However, close study of the literature shows that there is far from universal agreement about any increased risk among elderly patients.

Finegold *et al.* (1959), who studied the background to 22 cases of kanamycin-induced ototoxicity, concluded that age did not contribute to ototoxicity, whereas renal function did. Haapanen (1963) reported that patients over 45 years of age developed ototoxicity after kanamycin treatment more frequently, earlier in treatment, and in a more severe form than younger patients. However, pre-existing hearing loss, e.g. from previous administration of ototoxic drugs such as viomycin, predisposed to additional loss. A review of toxicity associated with kanamycin use in Europe (Kreis 1966) revealed consistent evidence that ototoxicity with kanamycin was associated with advancing age but it was also pointed out that this might merely indicate that a significant hearing loss was more likely to develop in a patient whose hearing was already impaired. Jackson and Arcieri (1971), in a review of the case histories of 1484 patients who were treated with gentamicin found that patients with normal renal function who developed ototoxicity were older, on average, than those who did not, but the difference was not significant. However, patients with impaired renal function who developed ototoxicity were significantly younger than those who did not. This finding seems to emphasize Jackson and Arcieri's main conclusion that impaired renal function was the dominant characteristic of patients who developed ototoxicity. A number of subsequent studies failed to find any association between age and the development of ototoxicity with gentamicin, tobramycin, amikacin, or netilmicin (Black *et al.* 1976; Barza *et al.* 1980; Fee 1980; Gatell *et al.* 1983).

It is likely that this lack of agreement arises from the application of statistical methods to small populations. This is illustrated by the result of a study of two groups,

each of 54 patients, who were administered gentamicin or amikacin for similar clinical indications in a randomized double-blind manner. There was no difference in the average age of the two groups of patients, nor in the incidence of ototoxicity — 11 per cent with gentamicin and 13 per cent with amikacin. In the gentamicin group, there was no difference between the average age of the patients who developed ototoxicity and those who did not, whereas the average of the patients who developed ototoxicity following amikacin therapy was significantly greater than those who did not. It seems improbable that any dependence of ototoxicity on age could be drug-specific; rather, this finding suggests an influence from other undetected differences between patients in the two groups.

The results of animal studies do not help resolve this issue. Henry *et al.* (1981) found greater cochlear damage induced by kanamycin in old mice compared to young adult mice but McDowell (1982) found minimal differences between the effect of gentamicin on the cochlea of guinea pigs aged 4 or 24 weeks at the start of his study. It remains to be established unequivocally whether or not the risk of ototoxicity in man or animals increases with advancing age.

Ototoxicity in children. The large retrospective study of gentamicin-induced ototoxicity by Jackson and Arcieri (1971) included children among the clinical material. Jackson and Arcieri reported that ototoxicity occurred infrequently in children compared to adults, but no information was provided about the actual incidence or the ages of the children.

Relatively few studies have specifically addressed the question of ototoxicity in children and they have tended to focus on special groups such as patients with cancer or cystic fibrosis. Assael *et al.* (1982) suggested that children with cancer may be at higher risk because they require large doses of aminoglycosides and prolonged treatment. Children with cystic fibrosis often required repeated treatment, which may be an additional risk factor. However, most evidence suggests that the incidence of aminoglycoside-induced ototoxicity among children is low.

Crifò *et al.* (1980) audiometrically tested the hearing of 30 young people aged between 4 and 26 years who received repeated courses of gentamicin (5 mg kg^{-1} daily) and found only one child with a hearing loss. Although it is now recognized (Kearns *et al.* 1982; Kelly *et al.* 1982) that patients with cystic fibrosis require larger doses of aminoglycosides than were given by Crifò *et al.* (1980), nevertheless Thomsen and Friis (1979) found that treatment of cystic fibrosis in patients with an average age of 11.8 years with large doses of tobramycin (10 mg kg^{-1} daily) produced no evidence of vestibular damage and a hearing loss in only one patient out of 53.

Faden *et al.* (1982) prospectively evaluated auditory toxicity associated with amikacin therapy in 12 children aged between 4 and 18 years, all of whom had cancer and were neutropenic. The therapy was monitored by measurement of peak and trough concentrations of amikacin in serum. None of the measured trough concentrations exceeded 2 mg l^{-1}. Two children (16.5 per cent) had mild, transient, high-frequency unilateral hearing loss. Faden *et al.* (1982) contrasted the low incidence and mild degree of hearing loss found in their study with reports (Black *et at.* 1976; Lau *et al* 1977) of incidences of 20 to 30 per cent amikacin-induced ototoxicity in adults, with bilateral and persistent losses occurring in half of the affected subjects.

Comparisons between studies may give rise to misleading conclusions unless the study populations are similar and the methods of assessment of, and criteria for defining, toxicity are the same. Most studies of ototoxicity in children, especially that of Faden *et al.* (1982), have used a small group of patients with a relatively homogeneous clinical background. Although Faden *et al.* (1982) demonstrated the absence of serious auditory impairment in children treated with amikacin when serum concentrations were maintained within recommended limits, they may have been unwise to contrast their results with those of other studies and suggest, as they did, that children are at less risk than adults from aminoglycoside therapy. Furthermore, the importance of the possible occurrence of asymptomatic high-frequency hearing loss, which would not be detected even by audiometry, should not be discounted.

Assael *et al.* (1982) felt that the data from studies of aminoglycoside ototoxicity in childhood were inadequate to calculate an incidence. The most that could be said was that the majority of the reported cases of ototoxicity in children were mild and, in several cases, transient. Assael *et al.* (1982) emphasized the need for further controlled, prospective investigations.

Fetal and neonatal ototoxicity. It is known that when aminoglycosides are administered during pregnancy they diffuse across the placenta and are found in both fetal serum and tissues. Rasmussen (1969) reviewed the early literature relating to clinical use of streptomycin and dihydrostreptomycin during pregnancy and reported that concentrations in fetal serum averaged about 50 per cent of that in the mother's serum. Recently, substantial transfer into human fetal serum has been reported for kanamycin (Good and Johnson 1971), gentamicin (Yoshioka *et al.* 1972), amikacin (Bernard *et al.* 1977*a*), and tobramycin (Bernard *et al.* 1977*b*). In these studies fetal serum concentrations ranged from 15 to 48 per cent of maternal serum concentrations.

Studies in guinea pigs have suggested that the concentrations of aminoglycosides in fetal serum relative to maternal serum are of the order of 9 per cent for streptomycin (Riskaer *et al.* 1952), 7 per cent for kanamycin (Anderson and Pederson 1971), and 10 per cent for sisomicin (Matsuzawa *et al.* 1981). Matsuzawa *et al.* (1981) found that, although the concentrations of sisomicin in fetal tissues were much lower than in maternal tissues, the pattern of distribution was the same in both cases with highest concentrations in the kidney and measurable concentrations in the cochlea.

The intrauterine ototoxicity of kanamycin has been demonstrated in guinea pigs using both electrophysiological and histopathological methods (Uziel *et al.* 1977, 1979; Raphael *et al.* 1983). Using behavioural methods, Kameyama *et al.* (1982) found evidence of raised auditory thresholds in rats which had been exposed *in utero* to dihydrostreptomycin, neomycin, or kanamycin. Very few clinical studies have been conducted but such evidence as there is suggests that the risk to the fetus of ototoxicity is not great when streptomycin or dihydrostreptomycin is administered during pregnancy. Only five cases of impaired hearing in children of mothers who had received streptomycin or dihydrostreptomycin during pregnancy had been described between 1950 and 1969 (Rasmussen 1969).

Rasmussen (1969) examined a further 36 children aged 2–15 years, whose mothers had received one or both of these drugs during pregnancy. All the children had normal vestibular function but one child had a slight (39 dB) unilateral, sensorineural, high-tone hearing loss. One of 33 mothers tested had reduced vestibular function and eight had loss of hearing which could be attributed to dihydrostreptomycin treatment. Whereas Crifò and Nobili-Benedetti (1978) reported that deafness in 2572 children was in 11.4 per cent of cases due to treatment of the child with an ototoxic drug, Robinson and Cambon (1964) found that, out of a total of 300 pre-school children with hearing impairment, in only two instances could the auditory defect be clearly attributed to treatment of the mothers with streptomycin during pregnancy. Two further retrospective studies of children born to mothers who had been treated with streptomycin and/or dihydrostreptomycin during pregnancy showed somewhat contrasting incidences. Conway and Birt (1965) examined 17 children (aged 6–13 years) and found six had some loss of vestibular function and four had a unilateral high-frequency hearing impairment. Varpela *et al.* (1969) found a very low incidence of inner-ear defects, comprising one case of bilateral high-frequency hearing loss out of 40 tested and two cases of vestibular dysfunction out of 34 tested. In neither of these studies were any of the disorders disabling.

Yet another retrospective study (Donald and Sellers 1981) tended to confirm the impression that fetal ototoxicity is an infrequent occurrence when streptomycin is administered during pregnancy. Donald and Sellars (1981) made no attempt to assess vestibular function but found hearing loss, of a minor degree, in only two of 33 children who had been exposed to streptomycin *in utero*. The hearing loss was in the high-frequency range and did not give rise to any handicap.

Jones (1973) described a remarkable case of transplacental ototoxicity. At 28 weeks of gestation, the mother received 4.5 g of kanamycin over five days for a urinary-tract infection. Despite a good clinical response, the patient developed oliguric renal failure. Kanamycin was discontinued and intravenous ethacrynic acid was prescribed to induce diuresis. Within two weeks the mother had a pronounced hearing impairement which progressed to total deafness. Her child was also found to have a severe sensorineural deafness when tested at three years of age. There seems little doubt that the deafness in both mother and child was attributable to kanamycin, enhanced by maternal renal failure and the addition of a second ototoxic drug, ethacrynic acid (see section on concurrent administration of other ototoxic drugs).

Assael *et al.* (1982) felt that conclusive data about the transplacental ototoxicity of aminoglycosides were lacking, particularly with regard to the risk associated with their use at different periods of gestation. Robinson and Cambon (1964) suggested that the greatest risk of ototoxicity should occur during the first half of pregnancy, because the ear reaches complete maturation by mid-term. The results of recent animal studies would tend to suggest exactly the opposite, i.e. that the risk of ototoxicity is slight before the cochlea in functionally mature. Uziel *et al.* (1977) studied the transplacental ototoxicity of kanamycin in the guinea pig and observed a trend in their results which suggested that the ototoxic effect of kanamycin was greater during the last 15 days of pregnancy — coincident with the onset of auditory function in the guinea pig as indicated by the appearance of cochlear potentials (Romand 1971). This suggestion was confirmed by the work of Raphael *et al.* (1983) who injected pregnant guinea pigs with kanamycin daily for periods of eight days during the first trimester (days 14–20), the second trimester (days 33–40), or the third trimester (days 55–62). The average duration was 67 days. Cochlear damage (hair-cell loss) in the progeny was slight when the drug was administered during the first or second trimester but substantial when it was given during the third trimester. Raphael *et al.* (1983) thought that the fetal ototoxicity of kanamycin during late pregnancy could be explained both in terms of increased transplacental transfer of the drug and increased sensitivity of the cochlea.

There is further evidence from studies in animals that the cochlea may be more vulnerable to aminoglycoside-induced ototoxicity during the final stages of its maturation. For a number of animal species this will include the neonatal period since some species have poorly developed auditory function at birth. Doses of aminoglycosides which caused little or no ototoxicity in the adult mouse (Henry *et al.* 1981), rat (Osako *et al.* 1979; Marot *et al.* 1980), or cat (Bernard 1981) were markedly ototoxic in the neonatal animal. Two of these studies suggested that there may be a short, highly critical period of ototoxic sensitivity. Marot *et al.* (1980) found that kanamycin was weakly ototoxic when given to rats from birth to the eighth postnatal day but was strongly ototoxic when given from days 8 to 16 after birth. Osako *et al.* (1979) found kanamycin to be very ototoxic to the neonatal rat when given from days 11 to 20 after birth but ineffectual when given for 7 to 10 days either preceding or following this critical period.

Although auditory maturation is not complete at birth in the human and development may continue for approximately the first year of life (Starr *et al.* 1977), the results of studies of ototoxicity in neonatal animals must be extrapolated to the human situation with caution. Bernard (1981) thought that the neonatal cat was a valid model for the *premature* human. Indeed, his study in cats (Bernard 1981) was prompted by his finding abnormalities in the development of auditory brainstem evoked responses in premature human neonates treated with gentamicin or tobramycin. These data suggested that, when sensitive electrophysiological methods were employed, it could be shown that administration of aminoglycosides to premature neonates impaired the normal development of hearing. Apart from differences in the stage of development of the cochlea between the premature and full-term neonate, differences in the disposition of aminoglycosides between these two groups (Assael *et al.* 1982) might also contribute to an increased risk of ototoxicity in the premature neonate.

Despite the findings of Bernard (1981), the results of a number of follow-up studies of children who were treated with aminoglycosides during the neonatal period suggest that the risk of a permanent auditory disability is low. For example, the incidence of auditory toxicity with kanamycin was seen to be low (High *et al.* 1958; Yow *et al.* 1962; Sanders *et al.* 1967). The findings of Yow *et al.* (1962) implied that the premature neonate might be at greater risk, but the number of patients studied was very small. Two other studies found no increased risk of aminoglycoside-induced auditory toxicity in the neonate.

These studies of kanamycin or streptomycin (Eichenwald 1966) and of kanamycin or gentamicin (Finitzo-Hieber *et al.* 1979) included large study groups and matched control groups. Thirty per cent of the patients in the study of Eichenwald (1966) were premature, suggesting that this group of neonates were not at greater risk. This would be supported by the results of another study (Davidson *et al.* 1980) where it was shown that premature infants treated with normal doses of kanamycin were at no greater risk of developing hearing deficits than untreated controls.

Although most studies of ototoxicity in neonates have concentrated on auditory function, Finitzo-Hieber *et al.* (1979) also assessed vestibular function which was unaffected by aminoglycoside therapy. However, careful vestibular function testing revealed minor abnormalities produced by administration to neonates of kanamycin alone or gentamicin in combination with either amikacin or tobramycin (Eviatar and Eviatar 1981*a*, 1982).

Assael *et al.* (1982) felt that the low reported incidence of ototoxicity in neonates, despite the presence of many risk factors (Siegel and McCracken 1981), most likely reflected the inadequacy of methods currently used to study the problem. Indeed, several authors have highlighted the many problems of assessment of auditory (Siegel and McCracken 1981; Assael *et al.* 1982) and vestibular (Eviatar and Eviatar 1981*b*) function in infants. Some of the problems of assessment of auditory function in infants may be overcome by the use of brain-stem evoked response audiometry (Bernard 1981; Finitzo-Hieber 1981). It would seem that further data are required to validate the belief that paediatric patients are less susceptible than adults to the ototoxic effects of drugs.

Concurrent administration of other ototoxic drugs

There is unequivocal evidence from animal studies that co-administration of a loop diuretic with an aminoglycoside antibiotic produces cochlear damage very much greater than that caused by either drug alone. When a large i.v. dose of ethacrynic acid or frusemide was administered shortly after a single non-ototoxic dose of kanamycin, the result was a rapid-onset, permanent depression of cochlear function associated with hair-cell destruction (West *et al.* 1973; Brummett *et al.* 1975). The interaction occurred with the newer loop diuretics, bumetanide and piretanide (Brummett 1981*a*), but not with diuretics which have a different mechanism of action, such as hydrochlorothiazide or mannitol (Brummett *et al.* 1974). All aminoglycoside antibiotics were found to interact with the loop diuretics (Brummett 1981*a*). It has also been shown that the loop diuretics

greatly augmented the ototoxic effect of non-amino-glycoside antibiotics such as viomycin and polymyxin B (Davis *et al.* 1982) and the antitumour drug, cis-platinum (Brummett 1981*b*).

There are occasional reports of permanent hearing loss caused by loop diuretics in man (Quick and Hoppe 1975). Permanent hearing loss caused by ethacrynic acid appears on occasions to have been associated with concurrent administration of an aminoglycoside antibiotic (e.g. Mathog and Klein 1969; Meriwether *et al.* 1971). However, in many studies of aminoglycoside-induced ototoxicity no association has been found between the occurrence of ototoxicity and concurrent administration of frusemide, the loop diuretic most frequently used in the past decade. Brummett (1981*a*) thought the comparative rarity of ototoxicity associated with concurrent administration of frusemide and an aminoglycoside consistent with his finding that frusemide was much less potent than ethacrynic acid in interacting with kanamycin.

Smith and Lietman (1983) retrospectively analysed data from three clinical trials of aminoglycoside toxicity and found the overall incidence of auditory toxicity was no different in patients given frusemide compared with those not given frusemide. They concluded that frusemide should not be considered a risk factor for the development of aminoglycoside-induced auditory toxicity. Nevertheless, it would seem advisable to avoid administering a loop diuretic, intravenously in high dose, to patients receiving aminoglycosides.

Previous therapy with other ototoxic drugs

There is conflicting evidence in the literature as to whether or not previous therapy with an ototoxic drug increases the risk of ototoxicity during aminoglycoside therapy. When kanamycin was first used in man, attention was drawn to the possibility that an existing hearing loss might predispose to the development of ototoxicity (Frost *et al.* 1959). However, Hawkins (1959) studied the long-term effects of kanamycin in patients who had previously received extensive treatment with strepto-mycin and/or dihydrostreptomycin. He concluded that there was no clear evidence that patients with a hearing loss due to streptomycin or dihydrostreptomycin were more susceptible to the ototoxic effect of kanamycin. This conclusion of Hawkins contrasts sharply with that of Finegold (1966) who reviewed the toxicity of kanamycin during the previous eight years and recommended, on the basis of accumulated experience, that in order to reduce the risk of ototoxicity with kanamycin the concurrent or sequential use of other ototoxic or nephrotoxic drugs should be avoided.

In an extensive review of gentamicin-induced ototoxicity, Jackson and Arcieri (1971) established clearly that, in patients with normal renal function, the risk of ototoxicity was not increased by previous treatment with an ototoxic drug, such as gentamicin, kanamycin, neomycin, streptomycin, or vancomycin. However, previous treatment with an ototoxic drug was correlated with the development of gentamicin-induced ototoxicity in patients whose renal function was impaired.

Fee (1980) found no association between the development of ototoxicity with either gentamicin or tobramycin and prior exposure to aminoglycosides. Matz and Lerner (1981), on the other hand, found a correlation between previous aminoglycoside therapy and ototoxicity caused by gentamicin, but not amikacin. This latter finding serves to emphasize that the results of such studies should be interpreted with caution, because there were no evident differences between the two groups of patients, who received gentamicin or amikacin (Matz and Lerner 1981) and there is unlikely to be an intrinsic difference between gentamicin and amikacin in this context.

Black *et al.* (1976) found a very significant correlation between amikacin-induced ototoxicity and previous therapy with either gentamicin or amikacin. The incidence of auditory toxicity was 48 per cent in those patients who had had previous aminoglycoside therapy compared with 6 per cent in those who had not. This strong correlation may reflect the fact that the clinical material of Black *et al.* (1976) included patients who were treated with amikacin for gentamicin-resistant infections. Thus, the inclusion of a higher proportion of subjects who had had prior treatment with another aminoglycoside would increase the chances of quantifying the contribution to ototoxicity made by such prior exposure. Black *et al.* (1976) provided no information on the time interval between the successive courses of aminoglycoside therapy but it may be that, because of the clinical circumstances, the interval would have been very short. This would contrast with some other studies where, for example, patients were entered into the study only if they had received no aminoglycosides during the preceding 14 days (Matz and Lerner 1981).

Miscellaneous factors

In animal studies it has been shown that factors such as noise (Brown *et al.* 1978, 1980), decreased hydration (Prazma *et al.* 1981), semistarvation (Prazma *et al.* 1983), and hyperthermia (Henry *et al.* 1983) augmented the ototoxic effects of various aminoglycosides. The augmentation of aminoglycoside ototoxicity in dehydrated (Prazma *et al.* 1981) or semistarved (Prazma *et al.* 1983) guinea pigs may, in both cases, have been attributable to a decrease in glomerular filtration rate and, therefore, reduced excretion of the drug. The clinical significance of such findings is unclear.

Noise exposure. Noise exposure has been shown to produce a pattern of damage to the cochlea similar to that caused by the aminoglycosides and greatly increased the damage to the guinea-pig cochlea resulting from administration of neomycin (Brown *et al.* 1978) and kanamycin (Brown *et al.* 1980). Brown *et al.* (1980) thought that patients undergoing therapy with aminoglycosides should be protected from sources of intense sound. Dayal *et al.* (1971) studied in guinea pigs the effects of combined exposure to low-level noise and administration of low doses of kanamycin. This study was intended to simulate the clinical situation where premature babies may be treated with low doses of kanamycin while being kept in an incubator which generates low-frequency noise at about 70 dB. Guinea pigs exposed to both low-level noise and low-dose kanamycin showed cochlear hair-cell damage whereas, acting alone, neither of these agents was harmful.

Hyperthermia. Henry *et al.* (1983) thought that the interaction between hyperthermia and aminoglycosides might involve metabolic disturbances in the cochlea or altered renal function since hyperthermia would raise the water demand of the kidney and, if this were not met, renal dysfunction could occur. Hyperthermia may well be important clinically since aminoglycosides are frequently given to patients who are febrile. Furthermore, there is evidence that changes in body temperature produce changes in serum levels of gentamicin without any change in renal function. In normal humans elevation of body temperature resulted in lower serum concentrations of gentamicin than when the subjects were afebrile (Pennington *et al.* 1975). Despite the fact that fever leads to increased renal blood flow, there was no increase in renal excretion of gentamicin and half-life was unaltered in fever. In patients, mainly children, Siber *et al.* (1975) found that elevated body temperature was associated with lower peak concentrations of gentamicin in serum and a shorter half-life. A decrease in temperature during the course of therapy was associated with an increase in peak serum concentrations and a longer half-life but renal clearance was unchanged. Siber *et al.* felt that the lower peak concentrations of gentamicin might be due to more rapid distribution into tissues or more rapid excretion by the kidneys during the 60 minutes preceding the measurement of peak concentrations.

Despite the suggestion that the tissue distribution of aminoglycosides might be altered during fever, there is very little evidence to suggest that fever increases the risk of ototoxicity. However, Fee (1980) did find a positive association between elevated body temperature and the development of cochlear toxicity with tobramycin, but not with gentamicin. When the groups of patients treated with either drug were combined there was a marginally significant association between the development of vestibular toxicity and elevated body temperature.

Haematocrit. Fee (1980) reported the somewhat paradoxical finding that ototoxicity with tobramycin was associated with a low haematocrit whereas ototoxicity with gentamicin was associated with a high haematocrit. If haematocrit were a significant factor in the development of ototoxicity, its influence would presumably be exerted through an alteration of the distribution of an aminoglycoside. Riff and Jackson (1971) reported an inverse correlation between peak concentrations of gentamicin in serum and haematocrit but there has been no consistent confirmation of this finding. Siber *et al.* (1975) did not find higher peak concentrations in patients with low haematocrits nor were changes in haematocrit during therapy associated with significant changes in peak concentrations.

Familial sensitivity. Although streptomycin is rarely incriminated as a cause of ototoxic deafness there are reports in the literature of moderate-to-severe deafness occurring in several members of the same family after treatment with streptomycin (Prazic *et al.* 1964; Johnsonbaugh *et al.* 1974; Viljoen *et al.* 1983). Familial aggregation of susceptibility to streptomycin-induced ototoxicity is rare but implies a genetic basis for the reaction in these subjects. In one study (Viljoen *et al.* 1983) eight members of a large, nonconsanguineous South African family of mixed ancestry developed deafness after streptomycin treatment. The available evidence was consistent with autosomal dominant inheritance of susceptibility to cochlear damage. Viljoen *et al.* (1983) speculated that the wide range of individual sensitivity to streptomycin ototoxicity in the general population might be attributable to some inherited characteristic — possibly related to receptors on the membrane of the hair cells in the inner ear.

Antibiotics other than aminoglycosides

Vancomycin

Although vancomycin has been clinically available for more than 25 years, its use rapidly declined after the

introduction of semisynthetic penicillins in the 1960s. In recent years there has been renewed interest in its use to treat staphylococcal infections, particularly methicillin-resistant staphylococcus aureus (Cook and Farrar 1978). Soon after its introduction into clinical practice vancomycin was reported to be ototoxic (Geraci *et al.* 1958; Dutton and Elmes 1959), but there is now less than universal agreement about the proclivity of vancomycin to cause ototoxicity. Schaad *et al.* (1981) found no otological damage in any of 33 children (aged 1 week to 16 years) treated with vancomycin for staphylococcal infections; all 29 patients with proven infections responded to therapy. Schaad *et al.* (1981) concluded that vancomycin is an effective and safe agent for the treatment of staphylococcal infections in paediatric patients. They reviewed the literature and concluded that most purported cases of ototoxicity were inadequately documented and that this effect is rare and usually reversible. When ototoxicity occurred, it was often associated with markedly elevated levels of the drug in serum or was possibly caused by synergistic toxicity with other ototoxic drugs. Fekety (1982) thought that, with the purified preparations of vancomycin now available and with better awareness of the hazards of rapid administration, adverse reactions are less frequent than when vancomycin was first introduced.

Davis *et al.* (1982) were unable to demonstrate an ototoxic interaction in guinea pigs between vancomycin and the loop diuretic, ethacrynic acid. Brummett *et al.* (1981*b*) showed that any of the aminoglycoside antibiotics will interact with any of the loop diuretics to produce an ototoxic effect which is very much greater than that seen with either drug alone. This interaction is considered to be a sensitive model for investigation of the cochlear toxicity of a drug (Brummett 1982, personal communication). Davis *et al.* (1982) extended these observations to non-aminoglycoside antibiotics. Viomycin, capreomycin, and, to a lesser extent, polymyxin B interacted with ethacrynic acid to produce cochlear destruction resembling that produced by the aminoglycoside antibiotics in combination with ethacrynic acid. Combinations of ethacrynic acid with spectinomycin, polymyxin E, or vancomycin did not produce a measurable ototoxic effect. This does not, of course exonerate vancomycin but it does at least suggest that it has a lower ototoxic potential than the aminoglycoside antibiotics or acts at a site other than the cochlea.

Erythromycin

Although erythromycin has been used for about 30 years it was not until 1973 that it was first reported to cause transient hearing impairment (Mintz *et al.* 1973). There have been a number of subsequent reports, involving more than 20 patients, in every case documenting a reversible effect, occasionally associated with tinnitus and vertigo (e.g. Quinnan and McCabe 1978). In many cases the hearing loss was characterized by sudden onset (Mèry and Kanfer 1979; Beckner *et al.* 1981; Miller 1982) and was first noted by the patient, i.e. the threshold shift was substantial and the speech frequencies were involved.

Hearing loss during erythromycin therapy does not seem to be associated with any particular salt of erythromycin since the lactobionate (Quinnan and McCabe 1978), the stearate (van Marion *et al.* 1978), the gluceptate (Beckner *et al.* 1981), the ethylsuccinate, and the propionate salts (Kanfer *et al.* 1982) have all been implicated. The effect has usually been associated with administration of large doses, e.g. 1 g every six hours, either intravenously (Beckner *et al.* 1981; Miller 1982; Kroboth *et al.* 1983) or orally (Eckman *et al.* 1975; van Marion *et al.* 1978). Kroboth *et al.* (1983) reported that 10 of 16 previous cases had received a daily dose of at least 4 g. In many cases the patients had chronic renal failure and were being dialysed (van Marion *et al.* 1978; Mèry and Kanfer 1979; Kroboth *et al.* 1983).

In the very few cases where serum levels of erythromycin were measured they were found to be exceptionally high (Taylor *et al.* 1981; Kroboth *et al.* 1983). The three patients in these two reports all had chronic renal failure and were undergoing either peritoneal dialysis (Taylor *et al.* 1981) or haemodialysis (Kroboth *et al.* 1983). Erythromycin is known to be eliminated predominantly via the hepatic route. Consequently, it is not normally considered necessary to adjust the dose in patients with renal impairment. However, Kunin and Finland (1959) provided evidence that the half-life of erythromycin is prolonged in advanced renal failure, something which was confirmed in the patient described by Kroboth *et al.* (1983). Although liver function tests were normal in both their patients, Taylor *et al.* (1981) felt that the hepatic metabolism of erythromycin could be grossly disturbed in patients with end-stage renal failure. Kanfer *et al.* (1982) showed that both the peak levels of erythromycin and the areas under the plasma concentration curves were increased in uraemic patients compared with control subjects. These findings suggest that hearing loss associated with erythromycin is not necessarily an idiosyncratic reaction but can be expected with very high serum concentrations which may be likely in patients with hepatic disease but may also be achieved unexpectedly in patients with renal failure. Kroboth *et al.* (1983) concluded that patients with decreased renal function who receive 4 g or more of erythromycin daily are at increased risk of hearing loss.

In contrast to other ototoxic antibiotics, notably the aminoglycosides, which always seems first to affect high frequencies with a gradual involvement of lower frequencies, the hearing impairment associated with erythromycin characteristically affects both low and high frequencies. This, together with the rapidity of onset and the reversibility of the hearing loss, distinguishes the ototoxic effect of erythromycin from that caused by the aminoglycoside antibiotics. It seems possible that erythromycin may act at different sites and/or by different mechanisms than the aminoglycosides. It has, to date, not proved possible to demonstrate an erythromycin-induced cochlear lesion in animals. It is tempting to speculate that the effects of erythromycin on hearing may be the result of central effects of the drug. In the two patients studied by Taylor *et al.* (1981) hearing loss was accompanied by diplopia or slurred speech and confusion. It remains to be seen whether the hearing loss associated with erythromycin is simply one manifestation of a more general toxicity to the central nervous system.

Minocycline

Soon after its introduction into clinical practice the broad-spectrum tetracycline, minocycline, was found to produce symptomatic vestibular disturbances which seemed to be more frequent in women (Nicol and Oriel 1974). In some studies there was a very high incidence of vestibular reactions (Williams *et al.* 1974; Jacobson and Daniel 1975), which in many patients were so severe as to necessitate discontinuing therapy. The onset of symptoms, including dizziness or vertigo, usually occurred soon after the start of therapy but rapidly resolved when the drug administration was stopped.

Allen (1976) reviewed five years of clinical experience with minocycline and, in six studies conducted in the USA after 1974, found that the frequency of vestibular reactions ranged from 12 to 90 per cent of patients treated with the drug; 12 to 52 per cent of patients were severely disabled with vestibular symptoms. Allen (1976) concluded that minocycline could not be recommended for general clinical use. Greco *et al.* (1979) found an 11 per cent incidence of vestibular symptoms in 112 women treated for 10 days with a dosage regimen of 50 mg four times daily instead of the recommended 100 mg twice daily. Only six patients had vestibular symptoms severe enough to discontinue treatment. Greco *et al.* (1979) attributed the low incidence of vestibular symptoms to lower peak plasma levels of the drug produced by the modified dosage regimen.

Viomycin, capreomycin, and the polymyxins

The polypeptide antibiotics, viomycin and capreomycin, are regarded as 'second-line' antitubercular agents. Viomycin is said to resemble streptomycin except that it is less active and more toxic (Garrod *et al*. 1973). There is clinical evidence to implicate viomycin in both nephrotoxicity and ototoxicity (Werner *et al*. 1951; Hackney *et al*. 1953; Edge and Weber 1960). Daly and Cohen (1965) found that viomycin was more ototoxic than streptomycin but also predominantly affected the vestibular system.

Capreomycin has been shown to be capable of causing mild vestibular toxicity but severe nephrotoxicity (Garfield *et al*. 1966). Miller *et al*. (1966) studied 294 patients treated with capreomycin for advanced tuberculosis and concluded that capreomycin infrequently caused serious damage to auditory, renal, and haematopoietic function; vestibular function was not assessed. In a study in guinea pigs Akiyoshi *et al*. (1971) concluded that capreomycin was less toxic to the cochlea than kanamycin. Garrod *et al*. (1973) stated that streptomycin-resistant bacilli may be sensitive to capreomycin, which is possibly less toxic than either viomycin or kanamycin.

Systemic polymyxin B was reported by Mittelman (1972) to be toxic to vestibular function and Walike and Snyder (1972) warned that it might cause ataxia. Polymyxin E (colistin) was shown by Meuwissen and Robinson (1967) to cause severe ataxia and partial deafness when excessively high plasma levels were attained. Kohonen and Tarkkanen (1969) demonstrated damage to the organ of Corti when either polymyxin B or polymyxin E was introduced into the middle ear of the guinea pig. Nilges and Northern (1971) reported three cases of deafness following treatment with colistin but in each case other known ototoxic drugs were also given. Walike and Snyder (1972) concluded that colistin was probably not ototoxic alone.

Ampicillin, chloramphenicol, and rifampicin

Only two reports associate ampicillin with disturbances of auditory or vestibular function (Morrison 1970; Ajodhia and Dix 1976). Morrison interpreted the effect as part of a hypersensitivity reaction.

There is evidence from studies in guinea pigs that chloramphenicol is ototoxic if applied in the middle ear either as a powder (D'Angelo *et al*. 1967) or in solution (Koide *et al*. 1966; Proud *et al*. 1968). In the light of these findings it might seem surprising that the topical application of chloramphenicol in man to treat otitis media has not been associated with ototoxicity. It may be that the drug escaped blame because hearing loss could have occurred secondary to the natural course of chronic otitis media or tympanomastoid surgery (Proud *et al*. 1968). It would seem that the only reference to the ototoxicity of chloramphenicol in man was a single case report by Gargye and Dutta (1959) who attributed complete bilateral deafness in a child to very high dosage of the drug (125 mg kg^{-1} intramuscularly) for 26 days.

No direct evidence can be found to implicate rifampicin in ototoxicity although transient, low-frequency hearing loss has been described in a review as a rare side-effect (Drugs 1971). In another extensive review, Springett (1975) did not mention ototoxicity among the side-effects of rifampicin.

Loop diuretics

Ethacrynic acid was one of a novel class of derivatives of phenoxyacetic acid described by Schultz *et al*. (1962) and which proved to be profoundly active saluretic–diuretic agents (Beyer *et al*. 1965). The clinical use of ethacrynic acid declined with the introduction of frusemide which had a lower incidence of side-effects and a less steep dose-response curve. A third loop diuretic, bumetanide, which is about 40 times more potent than frusemide has rapidly gained wide clinical acceptance. These three drugs are termed loop diuretics because their primary site of action has been localized to the ascending thick limb of the loop of Henle where they are known to inhibit the active transport of chloride ions and consequently sodium ions. Despite the similarity of their pharmacological effects the three compounds are chemically dissimilar. Although they are all carboxylic acids and frusemide and bumetanide are also sulphonamides, they share few other molecular structural features.

The development of acute transient deafness following the intravenous use of ethacrynic acid in man was first described by Maher and Schreiner (1965). Subsequently there have been numerous reports of the transient effects of ethacrynic acid on auditory function and occasional reports of permanent deafness even after oral administration (Pillay *et al*. 1969). Data from the BCDSP (Table 31.2) revealed an incidence of deafness due to ethacrynic acid of seven per 1000. Mathog and Klein (1969) observed permanent hearing loss in three uraemic patients who received combinations of ethacrynic acid and an aminoglycoside, although neither in high doses nor for prolonged periods. They postulated an additive ototoxic effect of ethacrynic acid and the aminoglycosides but also noted the implication that uraemic patients are extraordinarily sensitive to ethacrynic acid. Indeed many of the cases of reversible and permanent deafness have occurred in uraemic patients (Meriwether *et al.* 1971). Schneider and Becker (1966) thought that deafness in such patients might result from retention of the cysteine conjugate of ethacrynic acid which is a major metabolite of the drug. The cysteine conjugate of ethacrynic acid was shown to be more ototoxic in guinea pigs than the parent compound (Brown 1975*a*). Meriwether *et al.* (1971) also observed hearing loss in two non-uraemic patients who had been given standard intravenous doses of ethacrynic acid but both of whom were concurrently receiving what they considered to be non-toxic doses of aminoglycosides (Brummett 1980).

Ethacrynic acid normally affects only the cochlea and hearing loss may be preceded by tinnitus (Schneider and Becker 1966; Hanzelik and Peppercorn 1969). There have also been infrequent reports of vertigo or dizziness resulting from the use of ethacrynic acid (Maher and Schreiner 1965; Pillay *et al.* 1969).

High doses of intravenously administered frusemide can also cause vertigo, tinnitus, and transient hearing loss in patients with impaired renal function (Schwartz *et al.* 1970). Furthermore, frusemide can cause permanent hearing loss (Quick and Hoppe 1975) even after oral administration and in the absence of renal disease or concurrent aminoglycoside therapy (Gallagher and Jones 1979). Bumetanide appears to be significantly less ototoxic than either ethacrynic acid or frusemide and even was well tolerated in three patients who had experienced frusemide-induced ototoxicity (Bourke 1976).

Experimental studies in animals have supported the clinical findings relating to the ototoxicity of the loop diuretics and also provided some insight into the nature of their effects on the cochlea. Ernstson (1972) demonstrated that ethacrynic acid caused a reversible high-tone hearing loss with evidence of recruitment in guinea-pigs. Brown (1975*a*) studied the depression of the N_1 component of the compound action potential in the cochlear nerve of cats after intravenous injection of ethacrynic acid, frusemide, and bumetanide. All three drugs were thought to act on the same cochlear sites and by way of the same mechanism(s) because they produced parallel dose–response curves. However, Brown and McElwee (1972) thought that frusemide acted directly on the cochlea whereas ethacrynic acid probably acted through a metabolite. Brown (1975*a*) showed that ethacrynic acid was more potent in its effect on the cochlea than frusemide but the cysteine conjugate of ethacrynic acid was more potent than either. The absolute ototoxic potentcy of bumetanide was 6.5 times that of frusemide (Brown 1975*b*). However, when adjusted for diuretic potency, bumetanide was very much less ototoxic than frusemide.

High doses or repetetive administration of ethacrynic acid may lead to permanent hearing loss and damage to the hair cells in the organ of Corti (Mathog *et al.* 1970; Johnsson and Hawkins 1972; Crifò 1973). It is not known whether this results from a direct action on the hair cells or secondary to prolonged changes in the *stria vascularis*. When ethacrynic acid or frusemide is administered shortly after a single non-ototoxic dose of an aminoglycoside, the result is a rapid-onset, permanent depression of cochlear function associated with hair-cell destruction (West *et al.* 1973; Brummett *et al.* 1975). Little is known of the site or mechanism of this dramatic potentiation (Russell *et al.* 1979; Bosher 1980*a*) although it has been speculated both that the interaction occurs in the *stria vascularis* (Prazma *et al.* 1974) and at the hair cells (West *et al.* 1973).

Antitumour drugs

Although antitumour drugs are probably more toxic than any other widely used therapeutic agents, only a few have been specifically implicated in ototoxicity and in some cases the evidence is tenuous. There have been occasional reports of ototoxicity due to administration of nitrogen mustard (Cummings 1968; Tsunoda 1970).

Walike and Snyder (1972) reported that actinomycin was ototoxic and Nilges and Northern (1971) implicated it in a case of bilateral hearing loss following high-dosage therapy; however, unknown amounts of kanamycin and cephaloridine had also been administered and renal damage was present. Additive or synergistic damage to the cochlea was postulated.

Engström (1972) claimed that bleomycin, a mixture of antineoplastic antibiotics, may also produce ototoxic damage when it is given systemically in high dosage, or when applied topically to the middle ear. However, Dal *et al.* (1973) found no ototoxic effects when the drug was given to 21 patients, or when administered systemically to guinea pigs. Only when bleomycin was injected across the tympanic membrane into the middle ear of guinea pigs was cochlear damage seen.

Most recent interest has centred on the ototoxic side-effects of the first of a new class of antitumour drugs, cis-dichlorodiammineplatinum II (cis-platinum).

Cis-platinum

Cis-platinum (CDDP), is an antitumour drug used in the treatment of cancers of the bladder, testes, prostate, cervix, ovaries, and breast, in advanced squamous-cell carcinoma and other malignancies of the head and neck, and also in a variety of childhood tumours, including neuroblastoma and osteosarcoma. Side-effects of treatment with CDDP include nausea, vomiting, myelosuppression, nephrotoxicity, and ototoxicity. Therapy with CDDP is usually accompanied by rigorous hydration, achieved by intravenous infusion of large volumes of fluid. Diuresis is sometimes induced by administration of mannitol or, less commonly, fruse- mide. Although these measures may protect against nephrotoxicity, Helson et al. (1978) thought that they did not affect ototoxicity.

Aguilar-Markulis et al. (1981) reviewed the early litera- ture on ototoxicity of CDDP. Tinnitus, audiometrically- detected high-frequency hearing loss, symptomatic hearing loss, i.e. involving the speech frequencies, or even total deafness have been reported. Unilateral effects have been observed but bilateral effects seem to predominate (Aguilar-Markulis et al. 1981; Chapman 1982a; Strauss et al. 1983; Fausti et al. 1984). Tinnitus appears to be a common, but not an invariable, symptom of CDDP ototoxicity. For example, in one study (Reddel et al. 1982) tinnitus occurred in 14 out of 15 patients who had abnormal audiograms after CDDP therapy and in a further seven patients who did not develop any audiometric abnormalities. Reddel et al. (1982) reported that an abnormal audiogram was always preceded by one or more auditory symptoms. In contrast, Strauss et al. (1983) found that none of six patients with audiometrically-detected hearing loss experienced tinnitus or other symptoms.

It is a consistent finding that CDDP-induced auditory changes are first detected and most severe at high frequencies. Standard clinical audiometry does not assess hearing above 8000 Hz but the importance of monitoring high frequencies was emphasized by a study of Fausti et al. (1984). Specially developed instrumenta- tion and a modified technique were used to record hearing thresholds up to 20 000 Hz in 13 patients receiving CDDP. Eight of the 13 patients showed significant ($\geq$ 10 dB) changes in auditory thresholds at high frequencies, but only three of those subjects showed significant changes within the conventional test range. The observed changes progressed during the course of therapy and were irreversible.

Although CDDP-induced auditory changes are usually found to be permanent there have been reports of transient changes particularly after low doses (Aguilar- Markulis et al. 1981). One remarkable case of reversible deafness was reported by Aguilar-Markulis et al. (1981). A patient, who after 12 months of treatment with CDDP (1190-mg cumulative dose) had severe auditory impair- ment involving both speech and high-frequency ranges, was found to have completely recovered at a follow-up examination two years later.

Permanent hearing loss caused by ototoxic drugs seems always to be associated with hair-cell loss in the cochlea. Studies in guinea pigs (Fleischman et al. 1975; Estrem et al. 1981; Nakai et al. 1982) have shown that the cochlear lesion caused by CDDP is very similar to that produced by the aminoglycoside antibiotics. The outer hair cells, particularly the first row, are most affected and, consistent with the observation that the hearing loss is initially confined to high frequencies, the most severe damage is seen in the basal turns. The small amount of available evidence suggests that the lesion in man is similar. Wright and Schaefer (1982) studied temporal- bone histopathology in five patients who had each experienced a sensorineural hearing loss during the course of CDDP therapy. There were varying degrees of hair-cell loss in the organ of Corti, mainly affecting the outer hair cells but occasionally involving the inner hair cells. There was loss of myelinated nerve fibres in regions of the osseous spiral lamina adjacent to areas of exten- sively damaged organ of Corti, i.e. in the basal turns. All five patients had various degrees of hearing loss prior to therapy and the presence of age-related hair-cell degeneration made it difficult to identify unequivocally hair-cell loss due solely to CDDP ototoxicity. No such problems attended the study of temporal-bone histo- pathology in a nine-year-old boy, with a recurrent malignant astrocytoma of the frontal lobe, who died four weeks after receiving high-dose CDDP therapy, 50 mg m^{-2} on two successive days (Strauss et al. 1983). The child had developed a moderately severe bilateral hearing loss, most marked at high frequencies. The vestibular neurosensory epithelia were well preserved. In the left cochlea, there was extensive loss of outer hair cells, parti- cular in the basal coil where all three rows were absent. Inner hair-cell damage was confined to the basal and middle turns. Although the supporting cells in the organ of Corti appeared well preserved, there was degeneration of the nerve cells in the spiral ganglion. Damage was even more extensive in the right cochlea where, in the basal and middle turns, the myelinated nerve fibres and spiral ganglion cells were severely degenerated. There was also axonal degeneration in the cochlear nerve. This damage was not seen in other cranial nerves, in the spinal roots or the peripheral nerves. Strauss et al. (1983) commented that it was unusual to see degenerative changes in the spiral ganglion and cochlear nerve when the supporting cells in the organ of Corti were well preserved. Experience of mechanical injury or aminoglycoside toxicity has shown that cochlear nerve changes occur only when damage within the organ of Corti extends to the supporting cells. Strauss et al. (1983) were, therefore, uncertain whether their findings indicated a dual mode of CDDP ototoxicity with simultaneous involvement of the cochlea and its nerve or whether the neural changes were secondary to hair-cell damage.

Although vestibular function has rarely been objectively monitored during CDDP therapy, there are several reports of the absence of symptoms of vestibular toxicity (Aguilar-Makulis *et al.* 1981); Reddel *et al.* 1982; Strauss *et al.* 1983) suggesting that this is an uncommon complication of CDDP therapy. Schaefer *et al.* (1981) prospectively monitored both auditory and vestibular function in patients receiving CDDP for head and neck squamous-cell carcinoma. Vestibular function was assessed by a caloric-induced electronystagmogram (ENG) before each dose of CDDP. A 67-year-old patient who received 800 mg of the drug over a six-month period experienced only mild symptoms of vestibular dysfunction — transient dizziness — but an ENG confirmed the presence of a severe bilateral vestibular abnormality. The patient, who also developed a bilateral hearing loss, died three months after the last dose of CDDP. Wright and Schaefer (1982) histologically examined the temporal bones and found very extensive degeneration of all the vestibular neurosensory epithelia on both sides. Schaefer *et al.* (1981) commented on the need to be alert to possible symptoms of vestibular-function impairment, particularly in older patients receiving relatively high doses of CDDP. Vestibular toxicity may be a subtle complaint heralded by nausea, vomiting, and/or ataxia — symptoms not uncommon in patients with cancer receiving cytotoxic drugs. Schaefer *et al.* (1981) advocated monitoring vestibular function in patients receiving CDDP.

Black *et al.* (1982) prospectively monitored vestibular function in a small group of patients receiving CDDP therapy. They used tests thought to be more sensitive than the caloric test employed by Schaefer *et al.* (1981); these were ENG induced by rotational stimuli and automated recording and analysis of posture. Black *et al.* (1982) found that 31 per cent of their 16 patients had abnormal vestibular function before therapy but only three patients (19 per cent) showed evidence of vestibular ototoxicity. The fact that two of these three patients had abnormal vestibular function prior to therapy suggested that pre-existing vestibular dysfunction increased the risk of drug-induced vestibular ototoxicity. However, the clinical material was much too small to draw any firm conclusion. Since all patients experienced nausea and vomiting during CDDP therapy, there was no correlation between these symptoms and vestibular toxicity. Black *et al.* (1982) did not monitor auditory function but concluded on the basis of results of other studies that the incidence of auditory toxicity with CDDP was greater than that of vestibular toxicity.

Incidence of ototoxicity with cis-platinum

Patients receiving therapy with CDDP are frequently well enough to be audiometrically tested under favourable conditions in an anechoic chamber. Doses are usually administered at intervals of about four weeks so that hearing can be assessed prior to the first and each subsequent dose, i.e. at a time when the effect on hearing of any previous doses may have stabilized. Thus, in theory at least, some of the problems which frustrate the assessment of hearing in sick patients receiving aminoglycosides can be avoided. Nevertheless, as with aminoglycosides, the reported incidence of CDDP ototoxicity varies greatly between studies. Some of this variability in reported incidence is no doubt real and attributable both to differences between the clinical material of each study and to differences in the dosage protocols used. However, the care with which any monitoring procedure is implemented and the rigour of the criteria used to define ototoxicity may also contribute to apparent variability in incidence.

The reported incidence of ototoxicity with CDDP in early studies was reviewed by Helson *et al.* (1978) and varied from 9 to 20 per cent. In their own study Helson *et al.* (1978) found that 95 out of 104 patients had hearing threshold elevations ($\geq$ 5 dB) after treatment with CDDP. Aguilar-Makulis *et al.* (1981) studied a relatively uniform group of 50 male patients with genitourinary cancer who were treated with CDDP for at least 12 months. Hearing threshold elevations ($\geq$ 10 dB) were evident in 64 per cent of the patients and in two patients the effect was severe, affecting the speech frequencies. Reddel *et al.* (1982) found that, of 32 patients treated with CDDP for a variety of tumours, 47 per cent developed an audiometrically-detected hearing loss ($\geq$ 15 dB). The influence of dose and other factors on the incidence of ototoxicity is considered below.

Aspects of the dosage regimen. It is a frequent finding that CDDP-induced ototoxicity increases with increasing cumulative dose (Piel *et al.* 1974; Helson *et al.* 1978; Aguilar-Markulis *et al.* 1981). The relationship between ototoxicity and cumulative dose was clearly evident in two recent studies. Reddel *et al.* (1982) found that both the incidence and the severity of hearing loss increased with increasing cumulative dose of CDDP. Only with higher cumulative doses were lower frequencies affected. In another study (Vermorken *et al.* 1983) patients were evaluated over four treatment cycles at intervals of three to six weeks. The incidence of ototoxicity increased with each additional treatment cycle from 29 per cent after the first cycle to 65 per cent after the fourth cycle.

It is difficult to say whether there is a threshold total

dose below which toxicity will not occur but a cumulative dose of 200mg has been mentioned in several studies. Helson *et al.* (1978) thought that ototoxicity was a side-effect in virtually all patients treated with dosages of 1–3 mg kg^{-1} or a 200-mg or greater total dose. Strauss *et al.* (1983) found that all patients who developed ototoxicity had received more than 200 mg of CDDP and a minimum of three treatments. Although Fausti *et al.* (1984) did not report the doses used in their patients they found that ototoxicity began after the first or second treatment in all cases. The fact that Fausti *et al.* (1984) used audiometric equipment which was specially adapted to monitor high frequencies suggests that, under normal circumstances, undetected damage occurs early in treatment and then progresses, after subsequent doses to the extent that it can be detected by conventional means. Recent studies have confirmed that CDDP-induced ototoxicity can occur at low total doses and the appearance of ototoxicity is dependent not only on the cumulative dose but also the magnitude and manner of administration of unit doses. Reddel *et al.* (1982) studied 32 patients who received CDDP at intervals according to one of three protocols: 100 mg m^{-2} as a bolus injection; 100 mg m^{-2} as a two-hour infusion; 50 mg m^{-2}, also as a slow infusion. Those patients treated with a 100 mg m^{-2} bolus had the highest incidence of audiometric abnormality (86 per cent) at the lowest mean total dose (183 mg m^{-2}). Indeed, three patients developed an audiometrically-detected hearing loss after a single bolus dose of 100 mg m^{-2}. When this dose was given as a two-hour infusion, the incidence of ototoxicity was 43 per cent and first appeared after a mean total dose of 217 mg m^{-2}. This latter incidence was significantly less than when the drug was given as a bolus injection. The lowest incidence of ototoxicity (27 per cent) was seen in those patients receiving a slow infusion of 50 mg m^{-2} CDDP, first detected after a mean total dose of 217 mg m^{-2}. Vermorken *et al.* (1983) monitored 48 patients who received CDDP at intervals of three to six weeks either as a single low dose (25 to 50 mg m^{-2}), a single high dose (70 to 120 mg m^{-2}), or divided as five doses (20 mg m^{-2} daily). Ototoxicity was more severe and occurred at a lower cumulative dose in those patients treated with a single high dose compared with the other two regimens, which did not differ. At each dose level the mode of administration varied from rapid infusion to 24-h infusion. The latter was significantly less toxic than rapid infusion. Both Reddel *et al.* (1982) and Vermorken *et al.* (1983) recommended that administration of CDDP by rapid infusion should be avoided unless it was certain to lead to a therapeutic advantage over slow infusion.

The influence of age and pre-existing hearing loss. Few studies have examined the influence of age on CDDP-induced ototoxicity and, in some cases, the age range within the clinical material was too narrow to permit any conclusions to be drawn (e.g. Stauss *et al.* 1983). Amongst the few reports in the literature, Helson *et al.* (1978) noted that younger patients had lower increases in hearing threshold per administered dose than older patients, whereas Vermorken *et al.* (1983) found that, in one small group of patients, age did not influence the development of ototoxicity.

It has been suggested that patients with a pre-existing hearing loss are more susceptible to CDDP-induced ototoxicity (Helson *et al.* 1978; Aguilar-Markulis *et al.* 1981), but both Stauss *et al.* (1983) and Vermorken *et al.* (1983) found no significant influence of pre-existing hearing loss on ototoxicity. However, ototoxicity in patients with pre-existing hearing loss did result in clinical hearing loss, i.e. affected speech frequencies, at lower cumulative doses than in patients with normal hearing at the start of treatment (Vermorken *et al.* 1983). In this latter study, all six patients who developed clinical hearing loss had pre-existing hearing loss in one or both ears. The importance of pre-existing hearing loss was also stressed by Helson *et al.* (1978). A threshold increase of 30 dB from a baseline of 5 or 10 dB in a young adult is clinically inapparent and patients with losses of 30 dB or more in the 6000–8000 Hz frequency range are generally unaware of their condition. A rise in threshold of 30 dB in a patient with a pre-treatment level of 60 dB may prove debilitating.

Otoxicity of cis-platinum in children. Recent evidence suggests that young children might be more susceptible than adults to the ototoxic effects of CDDP. McHaney *et al.* (1983*a*) studied 24 children, aged 3.5 to 17.5 years, who were treated for a variety of solid tumours with 90 mg m^{-2} CDDP, given as a six-hour infusion every three weeks. Bilateral high-frequency hearing loss was found in 21 patients (88 per cent). In agreement with what has been found in adults, the percentage of patients with a significant hearing loss increased with increasing cumulative dose. High frequencies were first affected after a cumulative dose of 270 mg m^{-2} whereas at the speech frequencies (500 to 2000 Hz) hearing remained within normal limits until the total dose exceeded 990 mg m^{-2}. There was no recovery of hearing loss in any child retested up to 15 months after cessation of therapy. At any particular dosage or test frequency, younger children showed greater hearing loss. Since half of the patients were younger than seven years of age when first tested, McHaney *et al.* (1983*a*) thought that the results could indicate an increased susceptibility of the less mature cochlea. Mahoney *et al.* (1983) also found a high incidence of ototoxicity in children. All of 12 children with primary or recurrent brain tumours who received 60 mg m^{-2} CDDP daily for two days, repeated at three-week intervals, developed high-frequency hearing loss, first detected after a mean cumulative dose of 240 mg m^{-2}. At a mean cumulative dose of 480 mg m^{-2}, six of the 12 children experienced a progressive hearing loss at critical speech frequencies and required hearing aids. Five of eight children treated with 90 mg m^{-2} at three-week intervals developed high-frequency hearing loss after a mean cumulative dose of 300 mg m^{-2}. In one of these children, clinical hearing loss occurred after a cumulative dose of 480 mg m^{-2}. In four other children who received 90 mg m^{-2} at six-week intervals, no hearing loss occurred despite total cumulative doses of 270 to 430 mg m^{-2}. This suggested to Mahoney et al. (1983) that the frequency of drug administration, as well as the cumulative dose, might be important for ototoxicity.

Mahoney et al. (1983) commented on the severity of the ototoxicity seen in children treated with CDDP for brain tumours and wondered whether prior cranial irradiation might sensitize these patients. Support for this suggestion came from McHaney et al. (1983b) who identified rapidly progressive hearing loss after the initial dose of CDDP to five patients (aged 4 to 24 years), each of whom had a brain tumour and had received cranial irradiation. Three of these five patients had significantly elevated thresholds in the speech frequency range, and two required hearing aids. Comparison of these patients with 14 who received CDDP only suggested that radiation therapy might have potentiated the ototoxic effect of CDDP.

McHaney et al. (1983a,b) felt that the use of CDDP, where it was clearly the drug of choice, should not be discouraged but they recommended audiological monitoring of children receiving CDDP, particularly if the therapy was preceded by cranial irradiation.

Concomitant administration of frusemide. Diuretics are sometimes given concurrently with CDDP to increase the renal clearance of CDDP and thereby reduce its nephrotoxicity. There is no doubt that the loop diuretics, including frusemide, can potentiate the ototoxic effect of aminoglycoside antibiotics (Brummett et al. 1975, 1981a, b) and even interact with some non-aminoglycoside antibiotics to cause ototoxicity in animal models (Davis et al. 1982). There is now clear evidence from studies in animals that the loop diuretics can dramatically augment the ototoxic action of CDDP (Komune and Snow 1981; Brummett 1981b). This interaction could theoretically have contributed to the high incidence of ototoxicity in patients receiving CDDP by bolus injection in the study by Reddel et al. (1982) since all of the patients who

received a bolus injection of CDDP were also given an intravenous injection of 40 mg frusemide. Frusemide was also administered to the patients in the study of Vermorken et al. (1983), although it may not have contributed to the higher incidence of ototoxicity in patients receiving a single high dose of CDDP since frusemide was also given to the patients receiving lower doses of CDDP. However, it must be considered possible that the interaction between frusemide and CDDP is more likely to occur if the CDDP is given by rapid injection. Thus, in addition to avoiding, where possible, the use of rapid intravenous injection of CDDP (Reddel et al. 1982; Vermorken et al. 1983) it would seem wise to avoid concurrent administration of loop diuretics. Mannitol is frequently used clinically in combination with CDDP and Brummett (1981b) found that mannitol had no detrimental effects on the guinea-pig cochlea either alone or in combination with CDDP. Brummett concluded that, bearing in mind the limitations of animal experimentation, mannitol could be considered to be compatible with CDDP therapy in man.

Misonidazole

Misonidazole is a 2-nitroimidazole which has been tried as an adjunct to antitumour irradiation therapy. In laboratory animals misonidazole selectively sensitizes hypoxic tumour cells to the action of X- or gamma-irradiation. It became evident during early clinical evaluation of misonidazole that the most significant side-effect of the drug was the development of peripheral sensory neuropathies (Kogelnik et al. 1978), although a few patients experienced transient tinnitus or hearing impairment (Kogelnik et al. 1978; Wasserman et al. 1979). Abratt and Blackburn (1980) reported a case of severe hearing loss due to misonidazole therapy where there were no other neurological abnormalities. Over a period of one year there was only partial reversal of the hearing loss.

Waltzman and Cooper (1981) used serial audiometry to prospectively monitor the auditory toxicity of misonidazole given in conjunction with irradiation therapy to 21 patients with a variety of advanced neoplasms. Eleven of the 21 patients (52 per cent) had an increase in hearing threshold ranging from 15 to 65 dB (mean 35 dB). Only two of these patients experienced tinnitus. Seven of the 11 patients had other signs of neurotoxicity but in all cases the increase in auditory threshold was the first sign of toxicity to appear. Hearing loss first occurred after cumulative doses ranging from 6 to 12 g m^{-2}. In all cases the hearing loss was sensorineural and cochlear in origin. High-frequency losses were most common but in some patients all frequencies were affected. All patients experienced at least partial reversibility of the hearing loss. In five patients full recovery occurred within one month of completing drug therapy.

The ototoxicity induced by misonidazole appears similar in many respects to that caused by other ototoxic drugs, i.e. it is sensorineural, cochlear in origin, and predominantly high-frequency. The main difference seems to be the partial or complete reversibility of the misonidazole-induced lesion (Waltzman and Cooper 1981). Further study of the ototoxicity of misonidazole in man is unlikely as its neurotoxicity is likely to limit its clinical use (White *et al.* 1980).

Anti-inflammatory agents

Salicylates

Tinnitus is well known to be a side-effect of salicylate administration particularly when large doses are taken for conditions such as acute rheumatic fever or rheumatoid and osteoarthritis. Chronic dosage, or acute dosage in very large amounts, may also result in hearing impairment which has been reported either to affect all frequencies equally (Myers and Bernstein 1965) or mainly to involve high frequencies (Waltner 1955; McCabe and Dey 1965). The hearing loss is usually completely reversible within a few days of stopping the drug (Jager and Alway 1946; Waltner 1955). Although there have been isolated case reports of permanent deafness following aspirin administration (Kapur 1965; Jarvis 1966), the evidence is equivocal. The incidence of hearing loss can be high; for example, Jager and Alway (1946) reported both tinnitus and moderate reduction in hearing in 34 of 38 patients being treated for acute rheumatic fever with large doses of salicylates. In contrast, only 33 of 2974 patients treated with aspirin developed hearing loss in a study conducted by the BCDSP (Porter and Jick 1977).

Salicylate-induced hearing loss in man appears to be correlated with plasma salicylate levels in the range 30 to 40 mg dl^{-1}. In another BCDSP study (1973) the mean plasma salicylate level in 10 patients at or near the onset of deafness was 31 mg dl^{-1}. Myers and Bernstein (1965) found that the severity of the hearing loss in man rose with increasing plasma levels of salicylates until it reached a maximum (about 40 dB) at levels of 40 mg dl^{-1}. Behavioural studies in squirrel monkeys (Myers and Bernstein 1965) confirmed the transient nature of salicylate-induced hearing loss. The reversibility of the hearing loss was correlated with plasma salicylate levels and subsequent light- and electron-microscopic examination of the cochlea failed to show any permanent damage.

Persuasive evidence of the relationship of aspirin-induced hearing loss to dose (Table 31.6) has come from a series of studies conducted under the auspices of the BCDSP (Miller and Jick 1977; Porter and Jick 1977; Miller 1978). Hearing loss occurred in eight (0.3 per cent) of 2391 recipients of plain aspirin tablets (Miller and Jick 1977) but 33 (1.1 per cent) of 2974 recipients of plain, buffered, enteric-coated, and long-acting aspirin tablets (Porter and Jick 1977). Further analysis of these differences (Miller 1978) showed that clinically diagnosed deafness occurred in 28 per cent of 32 recipients of long-acting aspirin tablets but only 8 per cent of 105 patients who were treated with comparable doses (4 g or more daily) of plain aspirin tablets. Although the long-acting tablets were given less often they were given in higher unit doses than the plain tablets. Miller (1978) thought that the high incidence of deafness in the patients receiving long-acting aspirin tablets might have been due to higher peak salicylate concentrations in these patients. Indeed, Hollister (1972) showed that some long-acting aspirin tablets produced peak salicylate levels as high as those produced by equal doses of plain aspirin tablets. A high proportion of the patients who experienced aspirin-induced deafness in the BCDSP studies were women. This was not due to higher dosage but might have been related to the slower clearance of salicylic acid in women compared to men (Graham *et al.* 1977).

Although the mechanism of salicylate-induced ototoxicity is not known, there is evidence from clinical studies that the cochlea is the site of toxic action of aspirin (McCabe and Dey, 1965; Myers and Bernstein 1965). The results of electrophysiological studies in animals are also consistent with an intracochlear site of action (Gold and Wilpizeski 1966; Silverstein *et al.* 1967).

Other non-steroidal anti-inflammatory agents

Tinnitus, transient hearing loss, and symptoms of vestibular dysfunction such as vertigo and dizziness are occasionally reported to result from therapy with non-

Table 31.6   Unit doses of aspirin* received by patients who experienced hearing loss

Aspirin dosage (mg)	Number of patients with hearing loss	Number exposed	Deafness rate/1000
600	0	312	0
600–899	13	2 273	1
900–1 199	12	269	45
1200 +	18	120	150

* Aspirin, buffered aspirin or enteric-coated aspirin.

Source: Porter and Jick (1977).

steroid anti-inflammatory agents (NSAIs). For some NSAIs, such as phenylbutazone, the evidence is tenuous and the problem is insignificant compared with the other toxic effects of the drug. Both vestibular disturbance and deafness have been associated with indomethacin (Robinson 1965; Kinsella *et al* 1967; Pinals and Frank 1967) and deafness was an occasional side-effect in a clinical trial of ibuprofen (Dick-Smith 1969). Tinnitus and hearing loss were shown to be infrequent side-effects of therapy with fenoprofen compared with aspirin (Huskisson *et al*. 1974).

Chapman (1982*b*) described the occurrence of acute renal failure and sudden hearing loss in an elderly man who was receiving naproxen. The patient recovered from the renal failure but his hearing did not improve over a 12-month period. Chapman presented information from the Committee on Safety of Medicines (CSM) concerning reports of five patients who developed hearing loss while taking naproxen; in three cases the hearing loss was permanent. A senior official of the CSM informed Chapman that therapy with other new NSAIs had been reported with equal frequency to be associated with hearing loss. Chapman (1982*b*) observed that hearing loss, as a result of treatment with NSAIs, is rare in the usual doses prescribed but cautioned that, in contrast to salicylate ototoxicity, the hearing loss may be permanent.

Antimalarials

It is characteristic of quinine that therapeutic doses of the drug produce mild toxic symptoms such as tinnitus, visual disturbances, giddiness, nausea, and tremors. However, excessive doses may cause permanent deafness or blindness. The use of very large doses of quinine to procure abortion have resulted in deafness in the mother (McKenzie *et al*. 1968) or in the progeny where the abortion attempt was unsuccessful (McKinna 1966). When a solution of quinine was placed in the middle ear of guinea pigs, cochlear potentials were impaired and there were signs of vestibular toxicity (Hennebert and Fernàndez 1959). Histological examination showed degeneration of outer hair cells and damage to the *stria vascularis*.

Chloroquine is used not only as an antimalarial but also, in high doses, for diseases such as rheumatoid arthritis. Large doses of chloroquine very occasionally may lead to deafness (Dewar and Mann 1954) which may be permanent (Toone *et al*. 1965) or congenital following treatment of the mother with chloroquine during pregnancy (Hart and Naunton 1964).

Topical applications

In addition to the potential hazard of the application of ototoxic antibiotics in the form of otic drops (*Lancet* 1976; Ajodhia and Dix 1976), a number of other substances may prove to be ototoxic if applied to the middle ear.

Disinfectants

Aursnes (1981*a*, *b*, 1982*a*, *b*) conducted a series of studies of the ototoxic effect of disinfectants applied locally in the middle-ear space of guinea pigs. These studies were prompted by concern about the possible harmful effects of substances used prior to middle-ear surgery for skin disinfection. Bicknell (1971) thought that pre-operative disinfection of the ear with 0.5 per cent chlorhexidine in 70 per cent alcohol might have been the cause of severe sensorineural deafness in 14 patients who had myringoplasty operations. He speculated that the solution might have penetrated the round-window membrane and damaged the organ of Corti.

Aursnes studied the effect on the cochlea (1981*a*) and the vestibular system (1981*b*) of the application of chlorhexidine to the middle ear of guinea pigs. Subsequently, he studied the effect of other quaternary ammonium compounds (1982*a*) and iodine solutions (1982*b*). In ears exposed to 0.5 per cent chlorhexidine, total destruction of the entire organ of Corti (Aursnes 1981*a*) and almost total destruction of the vestibular neurosensory epithelia (Aursnes 1981*b*) was seen after three weeks, even when the exposure time was only 10 minutes. The type and degree of damage caused by 0.1 per cent benzethonium chloride or 0.1 per cent benzalkonium chloride (Aursnes 1982*a*) was very similar to that caused by chlorhexidine. In none of these studies (Aursnes 1981*a*, *b*, 1982*a*) did the use of water or 70 per cent alcohol as the vehicle influence the degree of damage caused by chlorhexidine, benzethonium chloride, or benzalkonium chloride. Furthermore, neither water nor 70 per cent alcohol caused any damage in control animals.

Iodophors, mixtures of iodine and non-ionic surface active substances, are often used for skin disinfection prior to surgery (Aursnes 1982*b*). Aursnes (1982*b*) tested the ototoxicity of iodophor (*Iodopax*) and iodine–potassium iodide (Lugol's) solution in both distilled water and 70 per cent alcohol. In each case the free iodine concentration was 0.1 per cent. There was no difference in the degree of damage caused by iodine or iodophor in 70 per cent alcohol. The pattern of damage was similar to that produced by the quaternary ammonium compounds but it was less pronounced. The same preparations in aqueous solution caused no histological damage to the inner ear. Thus the results did not indicate that iodine was ototoxic *per se*. Since 70 per cent alcohol on its own was also non-toxic, Aursnes (1982*b*) concluded that, in some unknown way, the iodine or iodophor and alcohol interacted to cause inner-ear damage.

The ototoxicity of iodine preparations applied in the middle ear was also studied by Morizono and Sikora (1982). They studied povidone–iodine solution and povidone–iodine scrub, which has detergents present. Morizono and Sikora (1982) stated that povidone–iodine solution is used to sterilize the external meatus pre-operatively and in some operating theatres is used to irrigate the middle ear, virtually ensuring contact of the antiseptic with the round-window membrane (RWM). However, it seems improbable that, in practice, the povidone-iodine scrub would come into contact with the RWM. Morizono and Sikora applied dilutions of these substances directly to the RWM of chinchillas and observed changes in the compound action potential of the auditory nerve, particularly with the povidone-iodine scrub. The value of these results is questionable because Morizono and Sikora (1982) did not examine the inner ear histologically and so could not distinguish an effect on the inner ear from an effect on the middle ear.

Aursnes (1982*a*) concluded that the extensive inner-ear destruction seen after comparatively short periods of middle-ear exposure in guinea pigs shows that quaternary ammonium compounds are among the most ototoxic pharmacological preparations that could possibly gain access to the middle-ear cavity. He felt that, because brief exposure to chlorhexidine in the concentration and solution used for pre-operative disinfection resulted in almost total destruction of the auditory (Aursnes 1981*a*) and vestibular (Aursnes 1981*b*) sensory organs, the results strongly supported the clinical observations indicating the ototoxicity of chlorhexidine when applied locally to the middle ear (Bicknell 1971). The type and degree of damage caused by benzethonium chloride and benzalkonium chloride (Aursnes 1982*a*) was almost identical with that caused by chlorhexidine. Consequently, Aursnes (1982*a*) thought it wise to keep these disinfectants out of the human middle-ear cavity and also to bear in mind that these substances are frequently used as preservatives in pharmacological preparations recommended for local treatment of middle-ear infections.

Although Aursnes (1982*b*) thought that his findings did not imply that alcoholic iodine disinfectants are ototoxic in man, he thought it reasonable to recommend that they should be prevented from entering the middle ear when used for pre-operative disinfection of skin. Morizono and Sikora (1982) concurred with the recommendations made by Aursnes (1981*a*, *b*, 1982*a*, *b*). They felt that, while the animal model might over-estimate the toxicity expected in humans, it provides a warning against avoidable iatrogenic deafness. Clearly antiseptics and detergents should be used carefully in middle-ear surgery and avoided when possible (Morizono and Sikora 1982).

Local anaesthetics

Ajodhia and Dix (1976) claimed that ototopical application of lignocaine or Bonain's solution has led to sudden deafness. Concentrations of lignocaine as low as 0.5 per cent have been shown experimentally to damage the cochlea (Rahm *et al.* 1962). In man lignocaine caused vertigo and vomiting when it entered the middle ear (Simmons *et al.* 1973).

Absorbable gelatin sponge

Bellucci and Wolff (1960) showed from experiments on cats that the application of an absorbable gelatin sponge (*Gelfoam*) to the oval window resulted in damage to the membranous labyrinth although Richards (1973) used *Gelfoam* during myringoplasty surgery in cats and found no histological damage in the labyrinth. Shenoi (1973) thought that the application of an absorbable gelatin sponge (*Sterispon*) to the oval window during stapedectomy operations might have contributed to a higher incidence of post-operative sensorineural hearing loss in these patients compared with those where fat had been used as an oval window seal. Absorbable gelatin sponge contains a small percentage of formaldehyde and Shenoi (1973) demonstrated extensive degenerative changes in the cochlea of cats after application of various quantities of formaldehyde to the oval window. These changes did not occur when small quantities of formaldehyde (< 20 μg) were used, nor when gelatin was placed in the oval window.

Miscellaneous substances

Oral contraceptives and medroxyprogesterone

Symptoms of vestibular dysfunction have occasionally been associated with the use of oral contraceptives (Peitersen 1969; Sellars 1971; Walike and Snyder 1972), although the mechanism was more likely to have been a thromboembolism in the vestibular arterial supply rather than direct effects of the drugs on the neurosensory epithelia of the vestibular end-organs. Occlusion of the cochlear artery was also thought to have been the cause of tinnitus and unilateral deafness which had a sudden onset in a woman injected with medroxyprogesterone (Sellars 1971).

Tricyclic antidepressants

Therapy with imipramine appears to have been the cause in some patients of transient deafness (Barker *et al.* 1960) or tinnitus (Racy and Ward-Racy 1980). Racy and Ward-Racy (1980) commented that tinnitus might be more commonly associated with tricyclic antidepressant therapy than suggested by the dearth of published reports. They thought that the tinnitus could be minimized by reducing the dose or substituting another tricyclic antidepressant for the offending agent. Brummett (1980) pointed out that patients with severe, chronic tinnitus often complained of depression. Treatment of the depression with a tricyclic antidepressant might aggravate the existing tinnitus.

Beta-blockers

Hearing loss (McClean *et al.* 1967; Porter and Jick 1977; Table 31.2) and dose-limiting tinnitus (Lloyd-Mostyn 1969) seem to be very rare complications of therapy with propranolol. Secretory otitis media, tinnitus, and deafness formed part of the toxic syndrome associated with practolol in some patients (Wright 1975). The other adverse effects of practolol were so severe as to lead to its withdrawal in some countries and to greatly restricted use in others (*British Medical Journal* 1975).

Alcohol and nicotine

Nicotine in tobacco and abuse of alcohol have been implicated as causes of hearing loss and the hazard may be increased in the subject exposed chronically to both agents (Tibbling 1970; Hammond 1970; Robin 1971). It has been postulated that nicotine may produce cochlear ischaemia through stimulation of the sympathetic innervation of cochlear blood vessels.

RECOMMENDED FURTHER READING

BROWN, R.D. AND DAIGNEAULT, E.A. (Eds.) (1981). *Pharmacology of hearing. Experimental and clinical bases.* Wiley, New York.

BRUMMETT, R.E. (1980). Drug-induced ototoxicity. *Drugs* 19, 412–28.

FILLASTRE, J.-P. (Ed.) (1982). *Nephrotoxicity, ototoxicity of drugs.* Inserm and University of Rouen.

HARPUR, E.S. (1981). Ototoxicological testing. In *Testing for toxicity* ed. J.W. Gorrod, pp. 219–40. Taylor and Francis, London.

—— (1982). The pharmacology of ototoxic drugs. *Br. J. Audiol* 16, 81–93.

HAWKINS, J.E. JR. (1976). Drug ototoxicity. In *Handbook of sensory physiology*, Vol. 5, Part 3, (ed. W.D. Keidel and W.D. Neff, pp. 707–48. Springer-Verlag, Berlin.

KLINKE, R., LAHN, W., QUERFURTH, H., AND SCHOLTHOLT, J. (Eds.) (1981). *Ototoxic side effects of diuretics. Scand. Audiol.* (Suppl.) 14.

LERNER, S.A., MATZ, G.J., AND HAWKINS, J.E. JR. (Eds.) (1981). *Aminoglycoside ototoxicity.* Little, Brown and Company, Boston.

WHELTON, A. AND NEU, H.C. (Eds.) (1982). *The aminoglycosides, Microbiology, clinical use and toxicology.* Marcel Dekker Inc, New York.

REFERENCES

ABRATT, R.P. AND BLACKBURN, J.D. (1980). Loss of hearing after misonidazole. *Br. J. Radiol.* 53, 1208.

AGUILAR-MARKULIS, N.V., BECKLEY, S., PRIORE, R., AND METTLIN, C. (1981). Auditory toxicity effects of long-term cis-dichloro-diammineplatinum II therapy in genitourinary cancer patients. *J. Surg. Oncol.* 16, 111–23.

AJODHIA, J.M. AND DIX, M.R. (1976). Drug-induced deafness and its treatment. *Practitioner* 216, 561–70.

AKIYOSHI, M., SATO, K., SHOJI, T., AND SUGAHIRO, K. (1971). Ototoxicity of capreomycin. *Audiol. Jap.* 14, 33–42.

ALLEN, J.C. (1976), Drugs five years later: minocycline. *Ann. intern. Med.* 85, 482–7.

ANDERSON, J.B. AND PEDERSON, C.B. (1971). A method for the determination of transplacental transmission of drugs. *Acta Pathol. Microbiol. Scand.* B 79, 204–8.

ANNIKO, M., TAKADA, A., AND SCHACHT, J. (1982). Comparative ototoxicities of gentamicin and netilmicin in three model systems. *Am. J. Otolaryngol.* 3, 422–33.

ARAN, J-M., ERRE, J-P., GUILHAUME, A., AND AUROUSSEAU, C. (1982). The comparative ototoxicities of gentamicin, tobramycin and dibekacin in the guinea-pig. A functional and morphological cochlear and vestibular study. *Acta Oto-laryngol.* (Suppl.) 390, 1–30.

ASSAEL, B.M., PARINI, R., AND RUSCONI, F. (1982). Ototoxicity of aminoglycoside antibiotics in infants and children. *Pediat. infect. Dis.* 1, 357–65.

AURSNES, J. (1981a). Cochlear damage from chlorhexidine in guinea pigs. *Acta Oto-laryngol.* 92, 259–71.

—— (1981b). Vestibular damage from chlorhexidine in guinea pigs. *Acta Oto-laryngol.* 92, 89–100.

—— (1982a). Ototoxic effect of quaternary ammonium compounds. *Acta Oto-laryngol.* 93, 421–33.

—— (1982b). Ototoxic effect of iodine disinfectants. *Acta Oto-laryngol.* 93, 219–26.

BAMFORD, M.F.M. AND JONES, L.F. (1978). Deafness and biochemical imbalance after burns treatment with topical antibiotics in young children. *Arch. Dis. Childh.* 53, 326–9.

BARKER, P.A., ASHCROFT, G.W., AND BURNS, J.K. (1960). Imipramine in chronic depression. *J. ment. Sci.* 106, 1447–51.

BARZA, M. AND LAUERMANN, M. (1978). Why monitor serum levels of gentamicin? *Clin. Pharmacokin.* 3, 202–15.

——, ——, TALY, F.P., AND GORBACH, S.L. (1980). Prospective, randomised trial of netilmicin and amikacin with emphasis on eigth-nerve toxicity. *Antimicrob. Agents. Chemother.* **17**, 707–14.

BECKNER, R.R., GANTZ, N., HUGHES, J.P., AND FARRICY, J.P. (1981). Ototoxicity of erythromycin gluceptate. *Am. J. Obstet. Gynecol.* **139**, 738–9.

BELLUCCI, R.J. AND WOLFF, D. (1960). Tissue reaction following reconstruction of the oval window in experimental animals. *Ann. Otol., St. Louis* **69**, 517–39.

BENDER, J.F., FORTNER, C.L., SCHIMPFF, S.C., GROVE, W.R., HAHN, D.M., LOVE, L.J., AND WIERNIK, P.H. (1979). Comparative auditory toxicity of aminoglycoside antibiotics in leukopenic patients. *Am. J. hosp. Pharm.* **36**, 1083–7.

BERNARD, B., ABATE, M., THIELEN, P.F., ATTAR, H., BALLARD, C.A., AND WEHRLE, P.F. (1977a). Maternal–fetal pharmacological activity of amikacin. *J. infect. Dis.* **135**, 925–31.

——, GARCIA-CÁZARES, S.J., BALLARD, C.A., THRUPP, L.D., MATHIES, A.W., AND WEHRLE, P.F. (1977b). Tobramycin: maternal–fetal pharmacology, *Antimicrob. Agents. Chemother.* **11**, 688–94.

BERNARD, P.A. (1981). Freedom from ototoxicity in aminoglycoside-treated neonates. A mistaken notion. *Laryngoscope* **91**, 1985–94.

BEYER, K.H., BAER, J.E., MICHAELSON, J.K., AND RUSSO, H.F. (1965). Renotropic characteristics of ethacrynic acid: a phenoxy-acetic saluretic diuretic agent. *J. Pharm. Exp. Ther.* **147**, 1–22.

BICKNELL, P.G. (1971). Sensorineural deafness following myringoplasty operations. *J. Laryngol. Otol.* **85**, 957–61.

BLACK, F.O., MYERS, E.N., SCHRAMM, V.L., JOHNSON, J., SIGLER, B., THEARLE, L., AND BURNS, D.S. (1982). Cisplatin vestibular ototoxicity: preliminary report. *Laryngoscope* **92**, 1363–8.

BLACK, R.E., LAU, W.K., WEINSTEIN, R.J., YOUNG, L.S., AND HEWITT, W.L. (1976). Ototoxicity of amikacin. *Antimicrob. Agents Chemother.* **9**, 956–61.

BOCK, B.V., EDELSTEIN, P.H., AND MEYER, R.D. (1980). Prospective comparative study of efficacy and toxicity of netilmicin and amikacin. *Antimicrob. Agents Chemother.* **17**, 217–25.

BOSHER, S.K. (1980a). The nature of the ototoxic actions of ethacrynic acid upon the mammalian endolymph system. I. Functional aspects. *Acta Oto-laryngol.* **89**, 407–18.

—— (1980b). The nature of the ototoxic actions of ethacrynic acid upon the mammalian endolymph system. II. Structural–functional correlates in the *stria vascularis. Acta Oto-laryngol.* **90**, 40–54.

BOSTON COLLABORATIVE DRUG SURVEILLANCE PROGRAM (1973). Drug-induced deafness: A co-operative study. *J. Am. med. Ass.* **244**, 515–16.

BOURKE, E. (1976). Frusemide, bumetanide and ototoxicity. *Lancet* **i**, 917–18.

BRITISH MEDICAL JOURNAL (1975). Leading article: Side-effects of practolol. *Br. med. J.* **2**, 577–8.

BROWN, J.J., BRUMMETT, R.E., FOX, K.E., AND BENDRICK, T.W. (1980). Combined effects of noise and kanamycin. Cochlear pathology and pharmacology. *Arch. Otolaryng.* **106**, 744–50.

——, ——, MEIKLE, M.B., AND VERNON, J. (1978). Combined effects of noise and neomycin: cochlear changes in the guinea pig. *Acta Oto-laryngol.* **86**, 394–400.

BROWN, R.D. (1975a). Comparison of the cochlear toxicity of sodium ethacrynate, furosemide, and the cysteine adduct of sodium ethacrynate in cats. *Toxicol. appl. Pharmacol.* **31**, 270–82.

—— (1975b). Cochlear N_1 depression produced by the new

'loop' diuretic, bumetanide, in cats. *Neuropharmacology* **14**, 547–53.

—— AND FELDMAN, A.M. (1978). Pharmacology of hearing and ototoxicity. *Ann. Rev. Pharmacol. Toxicol.* **18**, 233–52.

—— AND McELWEE, T.W., JR. (1972). Effects of intra-arterially and intravenously administered ethacrynic acid and furosemide on cochlear N_1 in cats. *Toxicol. appl. Pharmacol.* **22**, 589–94.

BRUMMETT, R.E. (1980). Drug-induced ototoxicity. *Drugs* **19**, 412–28.

—— (1981a). Effects of antibiotic–diuretic interactions in the guinea pig model of ototoxicity. *Rev. infect. Dis.* **3**, (Suppl.), 216–23.

—— (1981b). Ototoxicity resulting from the combined administration of potent diuretics and other agents. *Scand. Audiol.* (Suppl.) **14**, 215–24.

——, BENDRICK, T., AND HIMES, D. (1981a). Comparative ototoxicity of bumetanide and furosemide when used in combination with kanamycin. *J. clin. Pharmacol.* **21**, 628–36.

—— AND FOX, K.E. (1982). Studies of aminoglycoside ototoxicity in animal models. In *The Aminoglycosides. Microbiology, clinical use and toxicology* (ed. A. Whelton and H.C. Neu), pp. 419–51. Marcel Dekker, Inc, New York.

——, —— BENDRICK, T.W., AND HIMES, D.L. (1978a). Ototoxicity of tobramycin, gentamicin, amikacin and sisomicin in the guinea pig. *J. Antimicrob. Chemother.* **4**, (Suppl. A), 73–83.

——, ——, BROWN, R.T., AND HIMES, D.L. (1978b). Comparative ototoxic liability of netilmicin and gentamicin. *Arch. Otolaryngol.* **104**, 579–84.

——, ——, RUSSELL, N.J., AND DAVIS, R.R. (1981b). Interaction between aminoglycoside antibiotics and loop-inhibiting diuretics in the guinea pig. In *Aminoglycoside ototoxicity* (ed. S.E. Lerner, G.J. Matz, and J.E. Hawkins, Jr.), pp. 67–77. Little, Brown and Company, Boston.

——, HIMES, D., SAINE, B., AND VERNON, J. (1972). A comparative study of the ototoxicity of tobramycin and gentamicin. *Arch. Otolaryngol.* **96**, 505–12.

——, SMITH, C.A., UENO, Y., CAMERON, S., AND RICHTER, R. (1977). The delayed effects of ethacrynic acid on the stria vascularis of the guinea-pig. *Acta Oto-laryngol.* **83**, 98–112.

——, TRAYNOR, J., BROWN, R., AND HIMES, D. (1975). Cochlear damage resulting from kanamycin and furosemide. *Acta Oto-laryngol.* **80**, 86–92.

——, WEST, B.A., TRAYNOR, J., AND MANOR, N. (1974). Ototoxic interaction between aminoglycoside antibiotics and diuretics. *Toxicol. Appl. Pharmacol.* **29**, 97.

CAPUTY, A.J., KIM, Y.I., AND SANDERS, D.B. (1981). The neuromuscular blocking effects of therapeutic concentrations of various antibiotics on normal rat skeletal muscle: a quantitative comparison. *J. Pharmacol. Exp. Ther.* **217**, 369–78.

CAWTHORNE, T. AND RANGER, D. (1957). Toxic effect of streptomycin upon balance and hearing. *Br. med. J.* **1**, 1444–6.

CHAPMAN, P. (1982a). Rapid onset hearing loss after cisplatinum therapy: case reports and literature review. *J. Laryngol. Otol.* **96**, 159–62.

—— (1982b). Naproxen and sudden hearing loss. *J. Laryngol. Otol.* **96**, 163–6.

CHRISTENSEN, E.F., REIFFENSTEIN, J.C., AND MADISSOO, H. (1977). Comparative ototoxicity of amikacin and gentamicin in cats. *Antimicrob. Agents Chemother.* **12**, 178–84.

CONWAY, N. AND BIRT, B.D. (1965). Streptomycin in pregnancy:

effect on the foetal ear. *Br. med. J.* **2**, 260–3.

COOK, F.V. AND FARRAR, W.E. (1978). Vancomycin revisited. *Ann. intern. Med.* **88**, 813–18.

CRIFÒ, S. (1973). Ototoxicity of sodium ethacrynate in the guinea-pig. *Arch. Oto-rhino-laryngol.* **206**, 27–38.

——, ANTONELLI, M., GAGLIARDI, M., LUCARELLI, N., AND MARCOLINI, P. (1980). Ototoxicity of aminoglycoside antibiotics in long-term treatment for cystic fibrosis. *Int. J. Pediatr. Otorhinolaryngol.* **2**, 251–3.

—— AND NOBILI-BENEDETTI, F. (1978). Ototoxic deafness in children, *Min. Oto.* **28**, 15–20.

CUMMINGS, C.W. (1968). Experimental observations on the ototoxicity of nitrogen mustard. *Laryngoscope* **78**, 530–8.

DAL, I., EDSMYR, F., AND STAHLE, J. (1973). Bleomycin therapy and ototoxicity. *Acta Oto-laryngol.* **75**, 323–4.

DALY, J.F. AND COHEN, N.L. (1965). Viomycin ototoxicity in man. A cupulometric study. *Ann. Otol., St. Louis* **74**, 521–34.

D'ANGELO, E.P., PATTERSON, W.C., AND MORROW, R.C. (1967). Chloramphenicol: topical application in the middle ear. *Arch. Otolaryngol.* **85**, 682–4.

DASCHNER, F.D., JUST, H.-M., JANSEN, W., AND LORBER, R. (1984). Netilmicin versus tobramycin in multi-centre studies. *J. Antimicrob. Chemother.* **13** (Suppl. *A*), 37–42.

DAVEY, P.G., GONDA, I., HARPUR, E.S., AND SCOTT, D.K. (1980). Review of recent studies on control of aminoglycoside antibiotic therapy. *J. clin. hosp. Pharm.* **5**, 175–95.

——, HARPUR, E.S., JABEEN, F., SHANNON, D., AND SHENOI, P.M. (1982*a*). Variability and habituation of nystagmic responses to hot caloric stimulation of normal subjects. Evidence that this test may be inapplicable to monitoring drug-induced vestibular toxicity. *J. Laryngol. Otol.* **96**, 599–612.

——, JABEEN, F., HARPUR, E.S., SHENOI, P.M., AND GEDDES, A.M. (1982*b*). The use of pure-tone audiometry in the assessment of gentamicin auditory toxicity. *Br. J. Audiol.* **16**, 151–4.

——, ——, ——, —— AND —— (1983). A controlled study of the reliability of pure-tone audiometry for the detection of gentamicin auditory toxicity. *J. Laryngol. Otol.* **97**, 27–36.

DAVIDSON, S., BRISH, M., REIN, N., RUBINSTEIN, M., AND RUBINSTEIN, E. (1980). Ototoxicity in premature infants treated with kanamycin. *Pediatrics* **66**, 479–80.

DAVIS, R.R., BRUMMETT, R.E., BENDRICK, T.W., AND HIMES, D.L. (1982). The ototoxic interaction of viomycin, capreomycin and polymyxin B with ethacrynic acid. *Acta Oto-laryngol.* **93**, 211–17.

DAYAL, V.S., KOKSHANIAN, A., AND MITCHELL, D.P. (1971). Combined effects of noise and kanamycin. *Ann. Otol., St. Louis* **80**, 897–902.

——, SMITH, E.L., AND MCCAIN, W.G. (1974). Cochlear and vestibular gentamicin toxicity. *Arch. Otolaryngol.* **100**, 338–40.

DESROCHERS, C.S. AND SCHACHT, J. (1982). Neomycin concentrations in inner ear tissues and other organs of the guinea-pig after chronic drug administration. *Acta Oto-laryngol.* **93**, 233–6.

DEWAR, W.A. AND MANN, H.M. (1954). Chloroquine in lupus erythematosus. *Lancet* **i**, 780–1.

DICK-SMITH, J.B. (1969). Ibuprofen, aspirin and placebo in the treatment of rheumatoid arthritis — a double-blind clinical trial. *Med. J. Aust.* **2**, 853–9.

DONALD, P.R. AND SELLARS, S.L. (1981). Streptomycin ototoxicity in the unborn child. *S. Afr. med. J.* **60**, 316–18.

DRUGS (1971). Review: Rifampicin. *Drugs* **1**, 354–98.

DUTTON, A.A.C. AND ELMES, P.C. (1959). Vancomycin: report on treatment of patients with severe staphylococcal infections. *Br. med. J.* **1**, 1144–9.

ECKMAN, M.R., JOHNSON, T., AND REISS, R. (1975). Partial deafness after erythromycin. *New Engl. J. Med.* **292**, 649.

EDGE, J.R. AND WEBER, J.C.P. (1960). Ethionamide (1314TH) and viomycin in the treatment of resistant pulmonary tuberculosis. *Tubercle, London* **41**, 424–9.

EICHENWALD, H.F. (1966). Some observations on dosage and toxicity of kanamycin in premature and full-term infants. *Ann. NY. Acad. Sci.* **132**, 984–91.

ENGELBERG, M.W. (1978). Functional hearing level. *Otolaryng. Clin. N. Am.* **11**, 741–57.

ENGSTRÖM, H. (1972). Cited by Ballantyne, J. (1973). Ototoxicity: a clinical review. *Audiology* **12**, 325–36.

ERNSTSON, S. (1972). Ethacrynic acid-induced hearing loss in guinea pigs. *Acta Oto-laryngol.* **73**, 476–83.

ESTREM, S.A., BABIN, P.W., RYU, J.H., AND MOORE, K.C. (1981). Cis-diamminedichloroplatinum (II) ototoxicity in the guinea pig. *Otolaryngol. Head Neck Surg.* **89**, 638–45.

EVIATAR, L. AND EVIATAR, A. (1981*a*). Vestibular effects of aminoglycosides in infants. In *Aminoglycoside ototoxicity* (ed. S.A. Lerner, G.J. Matz, and J.E. Hawkins, Jr.), pp. 301–6. Little, Brown and Company, Boston.

—— AND —— (1981*b*). Methods of vestibular testing in infants and children exposed to ototoxic drugs. In *Aminoglycoside ototoxicity* (ed. S.A. Lerner, G.J. Matz, and J.E. Hawkins, Jr.), pp. 295–300. Little, Brown and Company, Boston.

—— AND —— (1982). Development of head control and vestibular responses in infants treated with aminoglycosides. *Develop. Med. child Neurol.* **24**, 372–9.

FADEN, H., DESHPANDE, G., AND GROSSI, M. (1982). Renal and auditory toxic effects of amikacin in children with cancer. *Am. J. Dis. Childh.* **136**, 223–5.

FAUSTI, S.A., SCHECTER, M.A., RAPPAPORT, B.Z., FREY, R.H., AND MASS, R.E. (1984). Early detection of cisplatin ototoxicity. *Cancer* **53**, 224–31.

FEDERSPIL, P. (1981). Pharmacokinetics of aminoglycoside antibiotics in the perilymph. In *Aminoglycoside ototoxicity* (eds. S.A. Lerner, G.J. Matz, and J.E. Hawkins, Jr.), pp. 99–108. Little, Brown and Company, Boston.

FEE, W.E. (1980). Aminoglycoside ototoxicity in the human. *Laryngoscope* **90**, (Suppl. 24) 1–19.

——, VIERRA, V., AND LATHROP, G.R. (1978). Clinical evaluation of aminoglycoside toxicity: tobramycin versus gentamicin, a preliminary report. *J. Antimicrob. Chemother.* **4**, (Suppl. A), 31–6.

FEKETY, R. (1982). Vancomycin. *Med. Clin. N. Am.* **66**, 175–81.

FINEGOLD, S.M. (1966). Toxicity of kanamycin in adults. *Ann. NY Acad. Sci.* **132**, 942–56.

——, WINFIELD, M.E., NISHIZAWA, A., KANTOR, E.A., KVINGE, V.E., AND HEWITT, W.L. (1959). Clinical evaluation of kanamycin. In *Antibiotics annual 1958-1959*, pp 606–22. Medical Encyclopedia, Inc, New York.

FINITZO-HIEBER, T. (1981). Auditory brainstem response in assessment of infants treated with aminoglycoside antibiotics. In *Aminoglycoside ototoxicity* (ed. S.A. Lerner, G.J. Matz, and J.E. Hawkins, Jr.), pp. 269–80. Little, Brown and Company, Boston.

——, McCRACKEN, G.H., ROESER, R.J., ALLEN, D.A., CHRANE, D.F., AND MORROW, J. (1979). Ototoxicity in neonates treated with gentamicin and kanamycin: results of a four-year controlled follow-up study. *Pediatrics* **63**, 443–50.

FLEISCHMAN, R.W., STANDNICKI, S.W., ETHIER, M.F., AND SCHAEPPI, U. (1975). Ototoxicity of cis-dichlorodiammineplatinum (II) in the guinea pig. *Toxicol. appl. Pharmacol.* **33**, 320–32.

FOX, K.E., BRUMMETT, R.E., BROWN, R., AND HIMES, D. (1980). A comparative study of the ototoxicity of gentamicin and gentamicin C_1. *Arch. Otolaryngol.* **106**, 44–9.

FROST, J.O., DALY, J.F., AND HAWKINS, J.E., JR. (1959). The ototoxicity of kanamycin in man. In *Antibiotics annual 1958–1959*, pp. 700–7. Medical Encyclopedia, Inc, New York.

——, HAWKINS, J.E., JR., AND DALY, J.F. (1960). Kanamycin II. Ototoxicity. *Am. Rev. resp. Dis.* **82**, 23–30.

GALLAGHER, K.L. AND JONES, J.K. (1979). Furosemide-induced ototoxicity. *Ann. intern. Med.* **91**, 744–5.

GARCIA-QUIROGA, J., NORRIS, C.H., GLADE, L., BRYANT, G.M., TACHIBANA, M., AND GUTH, P.S. (1978). The relationship between kanamycin ototoxicity and glucose transport. *Res. Commun. Chem. Path. Pharmacol.* **22**, 535–47.

GARFIELD, J.W., JONES, J.M., COHEN, N.L., DALY, J.F., AND McCLEMENT, J.H. (1966). The auditory, vestibular and renal effects of capreomycin in humans. *Ann. NY Acad. Sci.* **135**, 1039–46

GARGYE, A.R. AND DUTTA, D.V. (1959). Nerve deafness following chloromycetin therapy. *Ind. J. Pediat.* **26**, 265–6.

GARROD, L.P., LAMBERT, H.P., AND O'GRADY, F. (1973). Tuberculosis. In *Antibiotic and Chemotherapy*, 4th edn. pp. 432–56. Churchill Livingstone, London.

GATELL, J.M., SAN MIGUEL, J.G., ZAMORA, L., ARAUJO, V., BONET, M., BOHÉ, M., JIMENEZ DE ANTA, M.T., FARRE, M., ELENA, M., BALLESTA, A., AND MARIN, J.L. (1983). Comparison of the nephrotoxicity and auditory toxicity of tobramycin and amikacin. *Antimicrob. Agents Chemother.* **23**, 897–901.

GERACI, J.E., HEILMAN, F.R., NICHOLS, D.R., AND WELLMAN, W.E. (1958). Antibiotic therapy of bacterial endocarditis VII. Vancomycin for acute micrococcal endocarditis. *Proc. Mayo Clin.* **33**, 172–81.

GLORIG, A. AND FOWLER, E.P., JR. (1947). Tests for labyrinth function following streptomycin therapy. *Ann. Otol., St. Louis* **56**, 379–94.

GOLD, A. AND WILPIZESKI, C.R. (1966). Studies in auditory adaptation: II. Some effects of sodium salicylate on evoked auditory potentials in cats. *Laryngoscope* **76**, 674–85.

GOOD, R.G. AND JOHNSON, G.H. (1971). The placental transfer of kanamycin during late pregnancy. *Obstet. Gynecol.* **38**, 60–2.

GRAHAM, G.G., CHAMPION, G.D., DAY, R.O., AND PAULL, P.D. (1977). Patterns of plasma concentrations and urinary excretion of salicylate in rheumatoid arthritis. *Clin. Pharmacol. Ther.* **22**, 410–20.

GRECO, T.P., BONADIO, M., LEE, R.V., AND ANDRIOLE, V.T. (1979). Minocycline toxicity, experience with an altered dosage regimen. *Curr. Ther. Res.* **25**, 193–201.

HAAPANEN, J.H. (1963). Untoward phenomena during antituberculosis treatment. I. Auditory toxicity of kanamycin in tuberculous patients. *Ann. Med. Int. Finn.* **52**, (Suppl. 42), 3–38.

HACKNEY, R.L., KING, E.Q., MARSHALL, E.E., HARDEN, K.A., AND PAYNE, H.M. (1953). Clinical observations on viomycin sulphate in the treatment of tuberculosis. *Dis. Chest* **24**, 591–600.

HAMMOND, V. (1970). Perceptive deafness in adults. *Br. med. J.* **2**, 523–5.

HANZELIK, E. AND PEPPERCORN, M. (1969). Deafness after ethacrynic acid. *Lancet* i, 416.

HARPUR, E.S. (1977). Neomycin and deafness. *Pharm. J.* **218**, 494–5.

—— (1982). The pharmacology of ototoxic drugs. *Br. J. Audiol.* **16**, 81–93.

——, JABEEN, F., KINGSTON, R., GONDA, I., BRAMMER, K.W., AND GREGORY, M.H. (1981). Single and multiple dose pharmacokinetics of ribostamycin in serum and perilymph of guinea-pigs in relation to its low toxicity. In *Organ-directed toxicity, chemical indices and mechanisms* (ed. S.S. Brown and D.S. Davies), pp. 31–6. Pergamon Press, Oxford.

HART, C.W. AND NAUNTON, R.F. (1964). The ototoxicity of chloroquine phosphate. *Arch. Otolaryngol.* **80**, 407–12.

HAWKINS, J.E., JR., (1959). Antibiotics and the inner ear. *Trans. Am. Acad. Ophthal. Otolaryng.* **63**, 206–18.

—— (1973) Ototoxic mechanisms. *Audiology* **12**, 383–93.

—— (1976). Drug ototoxicity. In *Handbook of sensory physiology*, Vol. 5, Part 3, (ed. W.D. Keidel and W.D. Neff), pp. 707–48. Springer-Verlag, Berlin.

—— AND JOHNSSON, L.-G. (1981). Histopathology of cochlear and vestibular ototoxicity in laboratory animals. In *Aminoglycoside ototoxicity* (ed. S.A. Lerner, G.J. Matz, and J.E. Hawkins, Jr.), pp. 175–95. Little, Brown and Company, Boston.

——, ——, AND PRESTON, R.E. (1972). Cochlear microvasculature in normal and damaged ears. *Laryngoscope* **82**, 1091–104.

HELSON, L., OKONKWO, E., ANTON, L., AND CVITKOVIC, E. (1978). Cis-platinum ototoxicity. *Clin. Toxicol.* **13**, 469–78.

HENNEBERT, D. AND FERNÁNDEZ, C. (1959). Ototoxicity of quinine in experimental animals. *Arch. Otolaryngol.* **70**, 321–33.

HENRY, K.R., CHOLE, R.A., MCGINN, M.D., AND FRUSH, D.P. (1981). Increased ototoxicity in both young and old mice. *Arch. Otolaryng.* **107**, 92–5.

——, GUESS, M.B., AND CHOLE, R.A. (1983). Hyperthermia increases aminoglycoside ototoxicity. *Acta Oto-laryngol.* **95**, 323–7.

HIGH, R.H., SARRIA, A., AND HUANG, N.N. (1958). Kanamycin in the treatment of infections in infants and children. *Ann. NY Acad. Sci.* **76**, 289–307.

HOLLISTER, L.E. (1972). Measuring Measurin: problems of oral prolonged-action medications. *Clin. Pharmacol. Ther.* **13**, 1–5.

HUMES, H.D., WEINER, N.H., AND SCHACHT, J. (1981). The biochemical pathology of aminoglycoside-induced nephro and ototoxicity. In *Nephrotoxicity, ototoxicity of drugs* (ed. J.-P. Fillastre), pp. 333–43. Inserm and University of Rouen.

HUSKISSON, E.C., WOJTULEWSKI, J.A., BERRY, H., SCOTT. J., AND DUDLEY HART, F. (1974). Treatment of rheumatoid arthritis with fenoprofen: comparison with aspirin. *Br. med. J.* **1**, 176–80.

IGARASHI, M., LEVY, J.K., AND JERGER, J. (1978). Comparative toxicity of netilmicin and gentamicin in squirrel monkeys. *J. infect. Dis.* **137**, 476–80.

JACKSON, G.G. (1984). The key role of aminoglycosides in antibacterial therapy and prophylaxis. *J. Antimicrob. Chemother.* **13**, (Suppl. A), 1–7.

—— AND ARCIERI, G. (1971). Ototoxicity of gentamicin in man: a survey and controlled analysis of clinical experience in the United States. *J. infect. Dis.* **124**, (Suppl.), 130–7,

JACOBSON, J.A. AND DANIEL, B. (1975). Vestibular reactions associated with minocycline. *Antimicrob. Agents Chemother.* **8**, 453–6.

JAGER, B.V. AND ALWAY, R. (1946). The treatment of acute rheumatic fever with large doses of sodium salicylate with special reference to dose management and toxic manifestations. *Am. J. med. Sci.* **211**, 273–85.

JAO, R.L. AND JACKSON, G.G. (1964). Gentamicin sulphate, new antibiotic against gram-negative bacilli. Laboratory, pharmacologic and clinical evaluation. *J. Am. med. Ass.* **189**, 817–22.

JARVIS, J.F. (1966). A case of unilateral permanent deafness following acetylsalicylic acid. *J. Laryngol. Otol.* **80**, 318–20.

JOHNSONBAUGH, R.E., DREXLER, H.G., LIGHT, I.J., AND SUTHERLAND, J.M. (1974). Familial occurrence of drug-induced hearing loss. *Am. J. Dis. Childh.* **127**, 245–7.

JOHNSSON, L.-G. AND HAWKINS, J.E., JR. (1972). Strial atrophy in clinical and experimental deafness. *Laryngoscope* **82**, 1105–25.

——, ——, KINGSLEY, T.C., BLACK, F.O., AND MATZ, G.J. (1981). Aminoglycoside-induced cochlear pathology in man. *Acta Oto-laryngol.* (Suppl. 383), 1–19.

JONES, H.C. (1973). Intrauterine ototoxicity. A case report and review of the literature. *J. nat. med. Ass.* **65**, 201–3.

KAHLMETER, G. AND DAHLAGER, J.I. (1984). Aminoglycoside toxicity — a review of clinical studies published between 1975 and 1982. *J. Antimicrob. Chemother.* **13**, (Suppl. A), 9–22.

KALOYANIDES, G.J. AND PASTORIZA-MUNOZ, E. (1980). Aminoglycoside nephrotoxicity. *Kidney Int.* **8**, 571–82.

KAMEYAMA, T., NABESHIMA, T., ITOH, J., YAMAGUCHI, K., AND TAKAHASHI, K. (1982). Ototoxicity induced by pre- and postnatal exposure to aminoglycoside antibiotics. *J. Pharm. Dyn.* **5**, S-38.

KANFER, A., MERY, J.P., TORLOTIN, J.C., AND FREDJ, G. (1982). Alterations in pharmacokinetics and ototoxicity of erythromycin in renal failure. *Kidney Int.* **21**, 899.

KAPUR, Y.P. (1965). Ototoxicity of acetylsalicylic acid. *Arch. Otolaryngol.* **81**, 134–8.

KAVANAGH, K.T. AND McCABE, B.F. (1983). Ototoxicity of oral neomycin and vancomycin. *Laryngoscope* **93**, 649–53.

KEARNS, G.L., HILMAN, B.C., AND WILSON, J.T. (1982). Dosing implications of altered gentamicin disposition in patients with cystic fibrosis. *J. Pediat.* **100**, 312–18.

KEENE, M., HAWKE, M., BARBER, H.O., AND FARKASHIDY, J. (1982). Histopathological findings in clinical gentamicin ototoxicity. *Arch. Otolaryngol.* **108**, 65–70.

KELLY, H.B., MENENDEZ, R., FAN, L., AND MURPHY, S. (1982). Pharmacokinetics of tobramycin in cystic fibrosis. *J. Pediat.* **100**, 318–21.

KINSELLA, T.D., MACKENZIE, K.R., KIM, S.O., AND JOHNSSON, L.-G. (1967). Evaluation of indomethacin by a controlled cross-over technique in 30 patients with ankylosing spondylitis. *Can. med. Ass. J.* **96**, 1454–9.

KOGELNIK, H.D., MEYER, H.J., JENTZSCH, K., SZEPESI, T., KÄRCHER, K.H., MAIDA, E., MAMOLI, B. WESSELY, P., AND ZAUNBAUER, F. (1978). Further clinical experiences of a phase I study with the hypoxic cell radiosensitizer misonidazole. *Br. J. Cancer* **37**, (Suppl. III), 281–5.

KOHONEN, A. AND TARKKANEN, J. (1969). Cochlear damage from ototoxic antibiotics by intratympanic application. *Acta Oto-laryngol.* **68**, 90–7.

KOIDE, Y., HATA, A., AND HANDO, R. (1966). Vulnerability of the organ of Corti in poisoning. *Acta Oto-laryngol.* **61**, 332–44.

KOMUNE, S. AND SNOW, J.B., JR. (1981). Potentiating effects of cisplatin and ethacrynic acid in ototoxicity. *Arch. Otolaryngol.* **107**, 594–7.

KREIS, B. (1966). Kanamycin toxicity in adults. *Ann. NY Acad. Sci.* **132**, 957–67.

KROBOTH, P.D., MCNEIL, M.A., KREEGER, A., DOMINGUEZ, J., AND RAULT, R. (1983). Hearing loss and erythromycin pharmacokinetics in a patient receiving haemodialysis. *Arch. intern. Med.* **143**, 1263–5.

KUNIN, C.M. AND FINLAND, M. (1959). Persistence of antibiotics in blood of patients with acute renal failure. III: Penicillin, streptomycin, erythromycin and kanamycin. *J. clin. Invest.* **38**, 1509–19.

KUSAKARI, J., ISE, I., COMEGYS, T.H., THALMANN, I., AND THALMANN, R. (1978). Effect of ethacrynic acid, furosemide and ouabain upon the endolymphatic potential and upon high energy phosphates of the stria vascularis. *Laryngoscope* **88**, 12–37.

LANCET (1976). Editorial: Ear-drops. *Lancet* i, 896.

LANE, A.Z. (1984). Clinical experience with netilmicin. *J. Antimicrob. Chemother.* **13**, (Suppl. A.), 67–72.

LAU, W.K., YOUNG, L.S., BLACK, R.E., WINSTON, D.J., LINNE, S.R., WEINSTEIN, R.J., AND HEWITT, W.L. (1977). Comparative efficacy and toxicity of amikacin/carbenicillin versus gentamicin/carbenicillin in leukopenic patients. *Am. J. Med.* **62**, 959–66.

LERNER, A.M., REYES, M.P., CONE, L.A., BLAIR, D.C., JANSEN, W., WRIGHT, G.E., AND LORBER, R.R. (1983). Randomised, controlled trial of the comparative efficacy, auditory toxicity, and nephrotoxicity of tobramycin and netilmicin. *Lancet* i, 1123–6.

LINE, D.H., POOLE, G.W., AND WATERWORTH, P.M. (1970). Serum streptomycin levels and dizziness. *Tubercle, London* **51**, 76–81.

LLOYD-MOSTYN, R.H. (1969). Tinnitus and propranolol. *Br. med. J.* **2**, 766.

LOGAN, T.B., PRAZMA, J., THOMAS, W.G., FISCHER, H.D., AND HILL, C. (1974). Tobramycin ototoxicity. *Arch. Otolaryngol.* **99**, 190–3.

LUFT, F.C. (1978). Netilmicin: a review of toxicity in laboratory animals. *J. int. med. Res.* **6**, 286–99.

MAHER, J.E. AND SCHREINER, G.E. (1965). Studies on ethacrynic acid in patients with refractory oedema. *Ann. intern. Med.* **62**, 15–29.

MAHONEY, D.J., WEAVER, T., STEUBER, C.P., AND STARLING, K.A. (1983). Ototoxicity with cisplatin therapy. *J. Pediat.* **103**, 1006.

MARKS, S.C. AND SCHACHT, J. (1981). Effects of ototoxic diuretics on cochlear Na$^+$/K$^+$-ATPase and adenylate cyclase. *Scand. Audiol.* (Suppl. 14), 131–7.

MAROT, M., UZIEL, A., AND ROMAND, R. (1980). Ototoxicity of kanamycin in developing rats: relationship with the onset of the auditory function. *Hearing Res.* **2**, 111–13.

MATHOG, R.H. AND KLEIN, W.J., JR. (1969). Ototoxicity of ethacrynic acid and aminoglycoside antibiotics in uraemia. *New Engl. J. Med.* **280**, 1223–4.

——, THOMAS, V.G., AND HUDSON, W.R. (1970). Ototoxicity of new and potent diuretics. *Arch. Otolaryngol.* **92**, 7–13.

MATSUZAWA, T., NAKATA, M., IKEDA, C., TACHIBANA, A., AND YANO, K. (1981). Transfer of sisomicin to unborn and suckling guinea pigs. *Chemotherapy, Basel* **27**, 297–302.

MATZ, G.J. AND LERNER, S.A. (1981). Prospective studies of aminoglycoside ototoxicity in adults. In *Aminoglycoside ototoxicity* (ed. S.A. Lerner, G.J. Matz, and J.E. Hawkins, Jr.), pp.

327–36. Little, Brown and Company, Boston.

MAWER, G.E., AHMAD, R., DOBBS, S.M., MCGOUGH, J.G., LUCAS, S.B., AND TOOTH, J.A. (1974). Prescribing aids for gentamicin. *Br. J. clin. Pharmacol.* **1**, 45–50.

McCABE, P.A. AND DEY, F.L. (1965). The effect of aspirin upon auditory sensitivity. *Ann. Otol., St. Louis* **74**, 312–75.

McCLEAN, C.E., STOUGHTON, P.V., AND KAGEY K.S. (1967). Experiences with beta-adrenergic blockade. *Vasc. Surg.* **1**, 108–26.

McDOWELL, B. (1982). Patterns of cochlear degeneration following gentamicin administration in both old and young guinea pigs. *Br. J. Audiol.* **16**, 123.

McHANEY, V.A., THIBADOUX, G., HAYES, F.A., AND GREEN, A.A. (1983a). Hearing loss in children receiving cisplatin chemotherapy. *J. Pediat.* **102**, 314–17.

——, ——, ——, AND —— (1983b). Ototoxicity with cisplatin therapy. Reply. *J. Pediat.* **103**, 1006–7.

McKENZIE, I.F.C., MATHEW, T.H., AND BAILIE, M.J. (1968). Peritoneal dialysis in the treatment of quinine overdose. *Med. J. Aust.* **1**, 58–9.

McKINNA, A.J. (1966). Quinine induced hypoplasia of the optic nerve. *Can. J. Ophthalmol.* **1**, 261–6.

MERIWETHER, W.D., MANGI, R.J., AND SERPICK, A.A. (1971). Deafness following standard intravenous dose of ethacrynic acid. *J. Am. med. Ass.* **216**, 795–8.

MÉRY, J.-P. AND KANFER, A. (1979). Ototoxicity of erythromycin in patients with renal insufficiency. *New Engl. J. Med.* **301**, 944.

MEUWISSEN, H.J. AND ROBINSON, G.C. (1967). The ototoxic antibiotics. A survey of current knowledge. *Clin. Pediat.* **6**, 262–9.

MEYERS, R.M. (1970). Ototoxic effects of gentamicin. *Arch. Otolaryngol.* **92**, 160–2.

MILLER, G.H., ARCIERI, G., WEINSTEIN, M.J. AND WAITZ, J.A. (1976). Biological activity of netilmicin, a broad-spectrum semisynthetic aminoglycoside antibiotic. *Antimicrob. Agents Chemother.* **10**, 827–36.

MILLER, J.D., POPPLEWELL, A.G., LANDWEHR, A., AND GREENE, M.E. (1966). Toxicological studies in patients on prolonged therapy with capreomycin. *Ann. NY. Acad. Sci.* **135**, 1047–56.

MILLER, R.R. (1978). Deafness due to plain and long-acting aspirin tablets. *J. clin. Pharmacol.* **18**, 468–71.

—— AND JICK, H. (1977). Acute toxicity of aspirin in hospitalized medical patients. *Ann. J. med. Sci.* **274**, 271–9.

MILLER, S.M. (1982). Erythromycin ototoxicity. *Med. J. Aust.* **2**, 242–3.

MINTZ, U., AMIR, J., PINKHAS, J., AND DE VRIES, A. (1973). Transient perceptive deafness due to erythromycin lactobionate. *J. Am. med. Ass.* **225**, 1122–3.

MITTELMAN, H. (1972). Ototoxicity of 'ototopical' antibiotics: past, present and future. *Trans. Am. Acad. Ophthal. Otolaryngol.* **76**, 1432–43.

MORIZONO, T. AND SIKORA, M.A. (1982). Ototoxicity of topically applied Povidone–iodine preparations. *Arch. Oto-laryngol.* **108**, 210–13.

MORRISON, A. (1970). Cited by Ballantyne, J. (1973). Ototoxicity: a clinical review. *Audiology* **12**, 325–36.

MYERS, E.N. AND BERNSTEIN, J.M. (1965). Salicylate ototoxicity. A clinical and experimental study. *Arch. Otolaryngol.* **82**, 483–93.

NAKAI, Y., KONISHI, K., CHANG, K.C., OHASHI, K., MORISAKI, N., MINOWA, Y., AND MORIMOTO, A. (1982). Ototoxicity of the anti-cancer drug cisplatin. *Acta. Oto-laryngol.* **93**, 227–32.

NEU, H.C. AND BENDUSH, C.L. (1976). Ototoxicity of tobramycin: a clinical overview. *J. infect. Dis.* **134**, (Suppl.), 206–18.

NICOL, C.S. AND ORIEL, J.D. (1974). Minocycline: possible vestibular side effects. *Lancet* **ii**, 1260.

NILGES, T.C. AND NORTHERN, J.L. (1971). Iatrogenic hearing loss. *Ann. Surg.* **173**, 281–9.

NOONE, P. (1982). Clinical application of aminoglycosides. *Br. J. Audiol.* **16**, 141–6.

—— (1984). Netilmicin in the treatment of immuno-compromised patients. *J. Antimicrob. Chemother.* **13** (Suppl. A), 51–8.

NORDSTRÖM, L., BANCK, G., BELFRAGE, S., JUHLIN, I., TJERNSTRÖM, Ö., AND TOREMALM, N.G. (1973). Prospective study of the ototoxicity of gentamicin. *Acta path. microbiol. scand.* **B81** (Suppl. 241), 54–7.

——, CHRISTENSSON, P., HAEGER, K., JUHLIN, I., TJERNSTROM, O., AND WALLMARK, E. (1979). Netilmicin: clinical evalution of efficacy and toxicity of a new aminoglycoside. *J. int. med. Res.* **7**, 117–26.

NORTHINGTON, P. (1950). Syndrome of bilateral vestibular paralysis and its occurrence from streptomycin therapy. *Arch. Otolaryngol.* **52**, 380–96.

OSAKO, S., TOKIMOTO, T., AND MATSUURA, S. (1979). Effects of kanamycin on the auditory evoked responses during postnatal development of the hearing of the rat. *Acta Oto-laryngol.* **88**, 359–68.

ORSULAKOVA, A. AND SCHACHT, J. (1981). A biochemical mechanism of the ototoxic interaction between neomycin and ethacrynic acid. *Acta Oto-laryngol.* **93**, 43–8.

PALOHEIMO, S. AND THALMANN, R. (1977). Influence of 'loop' diuretics upon Na^+/K^+-ATPase and adenylate cyclase of the stria vascularis. *Arch. Oto-rhino-laryngol.* **217**, 347–59.

PARRAVICINI, L., ARPINI, A., BAMONTE, F., MARZANATTI, M., AND ONGINI, E. (1982). Comparative ototoxicity of amikacin., gentamicin, netilmicin and tobramycin in guinea pigs. *Toxicol. appl. Pharmacol.* **65**, 222–30.

——, FORLANI, A., MARZANATTI, M., AND ARPINI, A. (1983). Comparative ototoxicity of dibekacin and netilmicin in guinea pigs. *Acta Pharmacol. Toxicol.* **53**, 230–5.

PEITERSEN, E. (1969). Disturbances in the central vestibular system after oral contraceptives. *J. Laryngol. Otol.* **83**, 725–9.

PENNINGTON, J.E., DALE, D.C., REYNOLDS, H.Y., AND MACLOWRY, J.D. (1975). Gentamicin sulphate pharmacokinetics: lower levels of gentamicin in blood during fever. *J. infect. Dis.* **132**, 270–5.

PIEL, I.J., MEYER, D., PERLIA, C., AND WOLFE, V. (1974). Effects of cis-dichlorodiammineplatinum II (NSC-119875) on hearing function in man. *Cancer Chemother. Rep.* **58**, 871–5.

PIKE, D.A. AND BOSHER, S.K. (1980). The time course of the strial changes produced by intravenous furosemide. *Hearing Res.* **3**, 79–89.

PILLAY, V.K.G., SCHWARTZ, F.D., AIMI, K., AND KARK, R.M. (1969). Transient and permanent deafness following treatment with ethacrynic acid in renal failure. *Lancet* **i**, 77–9.

PINALS, R.S. AND FRANK, S. (1967). Relative efficacy of indomethacin and acetylsalicylic acid in rheumatoid arthritis. *New Engl. J. Med.* **276**, 512–14.

PORTER, J. AND JICK, H. (1977). Drug induced anaphylaxis, convulsions, deafness and extrapyramidal symptoms. *Lancet* **i**, 587–8.

PRAZIC, M., SALAJ, B., AND SUBOTIC, R. (1964). Familial sensitivity to streptomycin. *J.Laryngol. Otol.* **78**, 1037–43.

PRAZMA, J., BROWDER, J.P., AND FISCHER, N.D. (1974). Ethacrynic acid ototoxicity potentiation by kanamycin. *Ann. Otol., St. Louis* **83**, 111–18.

——, FERGUSON, S.D., KIDWELL, S.A., GARRISON, H.G., DRAKE, A., AND FISCHER, J. (1981). Alteration of aminoglycoside antibiotic ototoxicity by hyper- and hypohydration. *Am. J. Otolaryngol.* **2**, 299–306.

——, GARRISON, H.G., WILLIFORD S.K., FERGUSON, S.D., FISHER, J., DRAKE, A., AND KLINGLER, L.E. (1983). Alteration of aminoglycoside antibiotic ototoxicity: effect of semistarvation. *Ann. Otol., St. Louis* **92**, 178–82.

PROUD, G.O., MITTLEMAN, H., AND SEIDEN, G.D. (1968). Ototoxicity of topically applied chloramphenicol. *Arch. Otolaryngol.* **87**, 580–7.

QUICK, C.A. AND HOPPE, W. (1975). Permanent deafness associated with furosemide administration. *Ann. Otol., St. Louis* **84**, 94–101.

QUINNAN, G.V., JR. AND McCABE, W.R. (1978). Ototoxicity of erythromycin. *Lancet* **i**, 1160–1.

RACY, J. AND WARD-RACY, E.A. (1980). Tinnitus in imipramine therapy. *Am. J. Psychiat.* **137**, 854–5.

RAHM, W.E. JR., STROTHER, W.F., CRUMP, J.F., AND PARKER, D.E. (1962). The effect of anesthetics upon the ear. *Ann. Otol., St.Louis* **71**, 116–23.

RAPHAEL, Y., FEIN, A., AND NEBEL, L. (1983). Transplacental kanamycin ototoxicity in the guinea pig. *Arch. Oto-rhino-laryngol.* **238**, 45–51.

RASMUSSEN, F. (1969). The ototoxic effect of streptomycin and dihydrostreptomycin on the foetus. *Scan. J. resp. Dis.* **50**, 61–7.

REDDEL, R.R., KEFFORD, R.F., GRANT, J.M., COATES, A.S., FOX, R.M., AND TATTERSALL, M.H.N. (1982). Ototoxicity in patients receiving cisplatinum: importance of dose and method of drug administration. *Cancer treat. Rep.* **66**, 19–24.

RICHARDS, S.H. (1973). Ototoxicity of absorbable gelatin sponge. *Proc. R. Soc. Med.* **66**, 196.

RIFF, L. AND JACKSON, G. (1971). Pharmacology of gentamicin in man. *J. infect. Dis.* **124** (Suppl.), 98–105.

RISKAER, N., CHRISTENSEN, E., AND HERTZ, H. (1952). The toxic effects of streptomycin and dihydrostreptomycin in pregnancy, illustrated experimentally. *Acta tuberc. scand.* **27**, 211–16.

ROBIN, I.G. (1971). Sensorineural deafness. In *Scott-Brown's diseases of the ear, nose and throat*, 3rd edn. (ed. J. Ballantyne and J. Groves). pp. 419–74. Butterworth, London.

ROBINSON, G.C. AND CAMBON, K.G. (1964). Hearing loss in infants of tuberculous mothers treated with streptomycin during pregnancy. *New Engl. J. Med.* **271**, 949–51.

ROBINSON, R.G. (1965). Indomethacin in rheumatic disease: a clinical assessment. *Med. J. Aust.* **1**, 266–9.

ROEDER, J.W., MOWRY, H.J., MATZ, G.J., AND LERNER, S.A. (1981). Serial vestibular testing of normal subjects. In *Aminoglycoside ototoxicity* (ed. S.A. Lerner, G.J. Matz, and J.E. Hawkins, Jr.), pp. 309–19. Little, Brown and Company, Boston.

ROMAND, R. (1971). Maturation des potentiels cochleaires dans la période périnatale chez le chat et chez le cobaye. *J. Physiol., Paris* **63**, 763–82.

RUSSELL, N.J., FOX, K.E., AND BRUMMETT, R.E. (1979). Ototoxic effects of the interaction between kanamycin and ethacrynic

acid. *Acta Oto-laryngol.* **88**, 369–81.

SANDERS, D.Y., ELIOT, D.S., AND CRAMBLETT, H.B. (1967). Retrospective study for possible kanamycin ototoxicity among neonatal infants. *J. Pediat.* **70**, 960.

SCHAAD, U.B., NELSON, J.D., AND McCRACKEN, G.H. (1981). Pharmacology and efficacy of vanomycin for staphylococcal infections in children. *Rev. infect. Dis.* **3** (Suppl.), 282–8.

SCHAEFER, S.D., WRIGHT, C.G., POST, J.D., AND FRENKEL, E.P. (1981). Cis-platinum vestibular toxicity. *Cancer* **47**, 857–9.

SCHENTAG, J.J., JUSKO, W.J., VANCE, J.W., CUMBO, T.J., ABRUTYN, E., DE LATTRE, M., AND GERBRACHT, L.M. (1977). Gentamicin disposition and tissue accumulation on multiple dosing. *J. Pharmacokin. Biopharm.* **5**, 559–77.

SCHNEIDER, W.J. AND BECKER, E.L. (1966). Acute transient hearing loss after ethacrynic acid therapy. *Arch. intern. Med.* **117**, 715–17.

SCHUKNECHT, H.F. (1957). Ablation therapy in the management of Meniere's disease. *Acta Oto-laryngol.* **132** (Suppl.), 1–42.

SCHULTZ, E.M., CRAGOE, E.J., JR., BICKING, J.B., BOLHOFER, W.A., AND SPRAGUE, J.A. (1962). Alpha, beta-unsaturated ketone derivatives of aryloxyacetic acids, a new class of diuretics. *J. Med. Pharm. Chem.* **5**, 660–2.

SCHWARTZ, G.H., DAVID, D.S., RIGGIO, R.R., STENZEL, K.H., AND RUBIN, A.L. (1970). Ototoxicity induced by furosemide. *New Engl. J. Med.* **282**, 1413–14.

SELLARS, S.L. (1971). Acute deafness associated with depoprogesterone. *J. Laryngol. Otol.* **85**, 281–2.

SHENOI, P.M. (1973). Ototoxicity of absorbable gelatin sponge. *Proc. R. Soc. Med.* **66**, 193–6.

SIBER, G.R., ECHEVERRRIA P., SMITH, A.L., PAISLEY, J.W., AND SMITH, D.H. (1975). Pharmacokinetics of gentamicin in children and adults. *J. infect. Dis.* **132**, 637–51.

SIEGEL, J.D. AND McCRACKEN, G.H.JR. (1981). Aminoglycoside ototoxicity in children. In *Aminoglycoside ototoxicity* (ed. S.A. Lerner, G.J. Matz, and J.E. Hawkins, Jr.), pp. 341–53. Little, Brown and Company, Boston.

SILVERSTEIN, H., BERNSTEIN, J.M., AND DAVIES, D.G. (1967). Salicylate ototoxicity. A biochemical and electrophysiological study. *Ann. Otol., St. Louis* **76**, 118–27.

SIMMONS, F.B., GLATTKE, T.J., AND DOWNIE, D.B. (1973). Lidocaine in the middle ear: a unique cause of vertigo. *Arch. Otolaryngol.* **98**, 42–3.

SMITH, C.R., BAUGHMAN, K.L., EDWARDS, C.Q., ROGERS, J.F., AND LIETMAN, P.S., (1977). Controlled comparison of amikacin and gentamicin. *New Engl. J. Med.* **296**, 349–53.

—— AND LIETMAN, P.S. (1982). Comparative clinical trials of aminoglycosides. In *The aminoglycosides. Microbiology, clinical use and toxicology* (ed. A. Whelton, and H.C. Neu), pp. 497–509. Marcel Dekker, Inc, New York.

—— AND —— (1983). Effect of furosemide on aminoglycoside-induced nephrotoxicity and auditory toxicity in humans. *Antimicrob. Agents Chemother.* **23**, 133–7.

——, LIPSKY, J.J., LASKIN, O.L., HELLMAN, D.B., MELLITS, E.D., LONGSTRETH, J., AND LIETMAN, P.S. (1980). Double-blind comparison of the nephrotoxicity and auditory toxicity of gentamicin and tobramycin. *New Engl. J. Med.* **302**, 1106–9.

SOKABE, M., HAYASE, J., AND MIYAMOTO, K. (1982). Neomycin effect on lysotriphosphoinositide channel as a model for an acute ototoxicity. *Proc. Jap. Acad. Series B* **58**, 177–80.

SPRINGETT, V.H. (1975). The treatment of tuberculosis. *Practitioner* **215**, 480–6.

STARR, A., AMLIE, R., AND MARTIN, M. (1977). Development of

auditory function in newborn infants revealed by auditory brainstem potentials. *Pediatrics* **60**, 831–9.

STEBBINS, W.C., McGINN, C.S., FEITOSA, M.A.G., MOODY, D.B., PROSEN, C.A., AND SERAFIN, J.V. (1981). Animal models in the study of ototoxic hearing loss. In *Aminoglycoside ototoxicity* (ed. S.A. Lerner, G.J. Matz, and J.E. Hawkins, Jr.), pp. 5–25. Little, Brown and Company, Boston.

STRAUSS, M., TOWFIGHI, J., LORD, S., LIPTON, A., HARVEY, H.A., AND BROWN, B. (1983). Cis-platinum ototoxicity: clinical experience and temporal bone histopathology. *Laryngoscope* **93**, 1554–9.

TACHIBANA, M., MIZUKOSHI, O., AND KURIYAMA, K. (1976). Inhibitory effects of kanamycin on glycolysis in cochlea and kidney — possible involvement in the formation of oto- and nephrotoxicities. *Biochem. Pharmacol.* **25**, 2297–301.

TAKADA, A., CANLON, B., AND SCHACHT, J. (1983). Gentamicin ototoxicity dissociated from glucose uptake and utilization. *Res. Commun. Chem. Pathol. Pharmacol.* **42**, 203–12.

TAYLOR, R., SCHOFIELD, I.S., RAMOS, J.M., BINT, A.J., AND WARD, M.K. (1981). Ototoxicity of erythromycin in peritoneal dialysis patients. *Lancet* **ii**, 935–6.

THOMSEN, J. AND FRISS, B. (1979). High dosage tobramycin treatment of children with cystic fibrosis: bacteriological effect and clinical ototoxicity. *Int. J. Pediat. Otorhinolaryngol.* **1**, 33–40.

TIBBLING, L. (1969). The influence of tobacco smoking, nicotine, CO, and CO_2 on vestibular nystagmus. *Acta Oto-laryngol.* **68**, 118–26.

TJERNSTRÖM, Ö. (1980). Prospective evaluation of vestibular and auditory function in 76 patients treated with netilmicin. *Scand. J. infect. Dis.* **23** (Suppl.), 122–5.

TOONE, E.C., JR., HAYDEN, G.D., AND ELLMAN, H.M. (1965). Ototoxicity of chlorquine. *Arthritis Rheum.* **8**, 475–6.

TOYODA, Y. AND TACHIBANA, M. (1978). Tissue levels of kanamycin in correlation with oto- and nephrotoxicity, *Acta Oto-laryngol.* **86**, 9–14.

TRAN BA HUY, P., MANUEL, C., MEULEMANS, A., STERKERS, O., AND AMIEL, C. (1981). Pharmacokinetics of gentamicin in perilymph and endolymph of the rat as determined by radioimmunoassay. *J. infect. Dis.* **143**, 476–86.

——, ——, ——, ——, WASSEF, M., AND AMIEL, C. (1983). Ethacrynic acid facilitates gentamicin entry into endolymph of the rat. *Hearing Res.* **11**, 191–202.

TSUNODA, Y. (1970). Ototoxicity of nitrogen mustard N-oxide. *J. Otolaryngol. Jap.* **73**, 581–99.

UZIEL, A., GABRION, J., AND ROMAND, R. (1979). Hair cell degeneration in guinea pigs intoxicated with kanamycin during intrauterine life. *Arch. Otolaryngol.* **224**, 187.

——, ROMAND, R., AND GABRION, J. (1977). Intrauterine ototoxicity of kanamycin in the guinea pig. In *Inner ear biology* (ed. M. Portmann and J.M. Aran), pp. 347–58. Inserm, Paris.

VAN MARION, W.F., VAN DER MEER, J.W.M., KALFF, M.W., AND SCHICHT, S.M. (1978). Ototoxicity of erythromycin. *Lancet* **ii**, 214–15.

VARPELA, E., HIETALHATI, J., AND ARO, J.T. (1969). Streptomycin and dihydrostreptomycin medication during pregnancy and their effect on the child's inner ear. *Scand. J. resp. Dis.* **50**, 101–9.

VERMORKEN, J.B., KAPTEIJN, T.S., HART, A.A.M., AND PINEDO, H.M. (1983). Ototoxicity of cis-diamminedichloroplatinum (II): Influence of dose, schedule and mode of administration. *Eur. J. Cancer Clin. Oncol.* **19**, 53–8.

VESTERHAUGE, S., JOHNSEN, N.J., THOMSEN, J., AND SVARE, J. (1980). Netilmicin treatment followed by monitoring of vestibular and auditory function using highly sensitive methods. *Scand. J. infect. Dis.* **23** (Suppl.), 117–21.

VILJOEN, D.L., SELLARS, S.L., AND BEIGHTON, P. (1983). Familial aggregation of streptomycin ototoxicity: autosomal dominant inheritance? *J. Med. Genet.* **20**, 357–60.

WAISBREN, B.A. AND SPINK, W.W. (1950). A clinical appraisal of neomycin. *Ann. Intern. Med.* **33**, 1099–119.

WAKSMAN, S.A. AND LECHEVALIER, H.A. (1949). Neomycin, a new antibiotic against streptomycin-resistant bacteria, including tuberculosis organisms. *Science* **109** 305–7.

WALBY, A.P. AND KERR, A.G. (1982). Streptomycin sulphate and deafness: a review of the literature. *Clin. Otolaryngol.* **7**, 63–8.

WALDVOGEL, F.A. (1984). Future perspectives of aminoglycoside therapy. *J. Antimicrob. Chemother.* **13**, (Suppl. A) 73–8.

WALIKE, J.W. AND SNYDER, J.M. (1972). Recognizing and avoiding ototoxicity. *Postgrad. Med.* **52**, 141–5.

WALTNER, J. (1955). The effects of salicylates on the inner ear. *Ann. Otol., St.Louis* **64**, 617–22.

WALTZMAN, S.B. AND COOPER, J.S. (1981). Nature and incidence of misonidazole produced ototoxicity. *Arch. Otolaryngol.* **107**, 52–4.

WASSERMAN, T.H., PHILLIPS, T.L., JOHNSON, R.J., GOMER, C.J., LAWRENCE, G.A., SADEE, W., MARQUES, R.A., LEVIN, V.A., AND VAN RAALTE, G. (1979). Initial United States clinical and pharmacologic evaluation of misonidazole (RO-07-0582), an hypoxic cell radiosensitizer. *Int. J. Radiat. Oncol. Biol. Phys.* **5**, 775–86.

WATANABE, Y., NAKAJINA, R., ODA, R., UNO, M., AND NAITO, T. (1971). Experimental study on the transfer of kanamycin to inner ear fluids. *Med. J. Osaka Univ.* **21**, 257–63.

WEINER, N.D. AND SCHACHT, J. (1981). Biochemical model of aminoglycoside-induced hearing loss. In *Aminoglycoside ototoxicity* (ed. S.A. Lerner, G.J. Matz, and J.E. Hawkins, Jr.), pp. 113–21. Little, Brown and Company, Boston.

WERNER, C.A., TOMPSETT, R., MUSCHENHEIM, C., AND McDERMOTT, W. (1951). The toxicity of viomycin in humans. *Am. Rev. Tuberc.* **63**, 49–61.

WERSÄLL, J. (1980). The ototoxic potential of netilmicin compared with amikacin. An animal study in guinea pigs. *Scand. J. infect. Dis.* **23** (Suppl.), 104–13.

—— (1981). Structural damage to the organ of Corti and the vestibular epithelia caused by aminoglycoside antibiotics in the guinea pig. In *Aminoglycoside ototoxicity* (ed. S.A. Lerner, G.J. Matz, and J.E. Hawkins, Jr.), pp. 197–214. Little, Brown and Company, Boston.

WEST, B.A., BRUMMETT, R.E., AND HIMES, D.L. (1973). Interaction of kanamycin and ethacrynic acid. Severe cochlear damage in guinea pigs. *Arch. Otolaryngol.* **98**, 32–7.

WHELTON, A. AND SOLEZ, K. (1983). Pathophysiologic mechanisms in aminoglycoside nephrotoxicity. *J. clin. Pharmacol.* **23**, 453–60.

WHITE, R.A.S., WORKMAN, P., AND BROWN, J.M. (1980). The pharmacokinetics and tumour and neural tissue penetrating properties of SR-2508 and SR-2555 in the dog — hydrophilic radiosensitizers potentially less toxic than misonidazole. *Radiat. Res.* **84**, 542–61.

WILLIAMS, D.N., LAUGHLIN, L.W., AND LEE, Y.H. (1974). Minocycline: possible vestibular side effects. *Lancet* **ii**, 744–6.

WILSON, P. AND RAMSDEN, R.T., (1977). Immediate effects of

tobramycin on human cochlea and correlation with serum tobramycin levels. *Br. med. J.* **1**, 259–61.

WINKEL, O., HANSEN, M.M., KAABER, K., AND ROZARTH, K. (1978). A prospective study of gentamicin ototoxicity. *Acta Otolaryngol.* **86**, 212–16.

WRIGHT, C.G. AND SCHAEFER, S.D. (1982). Inner ear histopathology in patients treated with cis-platinum. *Laryngoscope* **92**, 1408–13.

WRIGHT, P. (1975). Untoward effects associated with practolol administration: oculomucocutaneous syndrome. *Br. med. J.* **1**, 595–8.

YOSHIOKA, H., MONMA, T., AND MATSUDA, S. (1972). Placental transfer of gentamicin. *J. Pediat.* **80**, 121–3.

YOW, M.D., TENGG, N.E., BANGS, J., BANGS, T., AND STEPHENSON, W. (1962). The ototoxic effects of kanamycin sulphate in infants and children. *J. Pediat.* **60**, 230–42.

32 Skin disease

A. McQUEEN

Introduction

It is generally agreed among dermatologists that the number of patients they are asked to see who are suffering from skin disorders of iatrogenic origin has increased dramatically in recent years. It should be remembered, however, that almost 2500 pharmaceutical preparations are advertised in MIMS in the United Kingdom, that just under half of those contain two or more pharmacologically active ingredients, and that approximately 50 new products are introduced each year.

Moreover, there are about 20 000 known additives to food, beverages, and drugs, used for the purposes of colouring, flavouring, preserving, thickening, emulsifying, and stabilizing. These substances are chemicals, and some of them are identical to or cross-react with drugs (Levantine and Almeyda 1974) so that the occurrence of a skin rash suggestive of a drug reaction in a patient not admitting to a drug history, may be more readily understood in the light of this knowledge. A leading article in the *British Medical Journal* (Miller 1982) was devoted to sensitivity to tartrazine, a dye used extensively in medicines and foods. Patients sensitive to aspirin may also be sensitive to tartrazine, and may develop allergic symptoms of urticaria, rhinitis, and asthma. This article included a discussion of possible association with a humoral immune response to part of the tartrazine molecule resulting in the production of antibodies of the IgD class.

Thus the number of drugs or related substances having the potential to produce adverse reactions, either alone, or in combination, is considerable and, while it is not possible to define the incidence accurately, a skin manifestation of an adverse drug reaction is most frequently the first indication to the patient or his physician that something is amiss. A recent study in general practice by Steele (1984) of 2409 patients seen during an eight-week period showed that 199 (8.2 per cent) presented with dermatological problems, and that nine (4.5 per cent) were due to drug reactions.

The skin in disease has a very limited range of reaction patterns although manifestations of these reactions vary considerably. This is also true of drug-induced skin changes, and the recognition of any such change today still depends on the careful clinical observation of any patient under treatment. There is no simple laboratory test yet available which will specifically confirm the cause-and-effect relationship between a drug and the adverse effect it is suspected of producing in the skin. In general the histological appearances seen in skin biopsies from lesions due to drugs tend to be rather non-specific in character, although certain patterns are now being recognized. At best it is often only possible to draw the clinician's attention to the possibility of an adverse drug reaction. Many sophisticated techniques have been developed for the study of adverse drug reactions, particularly in the field of immunology, but these remain of experimental interest and are of secondary importance to well-trained clinical experience. An immunofluorescence serological finding in patients screened for adverse drug reactions was described by McQueen and Behan (1982), but its use is limited since a 'positive' test is non-specific, and the number of drugs so far found to cause the finding is fairly small. A survey of some immunological tests including the drug-induced lymphocyte transformation test was given in the *Year book of dermatology* (Dobson 1979). Skin-patch tests have been advocated (Felix and Comaish 1974) but have not as yet found universal acceptance and in addition are completely unreliable when the question of drug interaction arises.

Similarly studies in experimental animals are of limited value since it is difficult if not impossible to produce recognizable drug-induced skin changes in animals that simulate those seen in man.

Iatrogenic dermatological reactions vary very considerably in site, extent, and severity; they may be of abrupt or gradual onset, disappear quickly or slowly after withdrawal of the drug responsible, and they may occur in patients who have previously used the drug without suffering any ill effect. An injection of insulin may produce a mild urticaria which can easily be controlled by the use of an antihistamine agent. However, what may start as an apparently trivial erythematous skin reaction to, for example, a sulphonamide, may quickly progress to widespread confluent purpuric or blistering lesions with involvement of the mucous membranes as in the Stevens–Johnson syndrome where there is real danger to the patient's life.

Any of the classical skin disorders may be mimicked by an adverse drug reaction, a fact well known to dermatologists who as a result take great care to obtain the fullest possible drug history from patients attending their clinics. General practitioners are equally aware that cutaneous eruptions are the most common indication of untoward reaction by the body to drugs, and probably are most aware of the various skin reactions which may occur in childhood as a result of administration of antibiotics. Very often drugs may produce different types of skin lesions in different people, and these lesions may be modified by genetic or environmental factors as well as by diet or current health status. Thus a certain dose of belladonna will produce a scarlatiniform type of rash on the neck and upper trunk in some infants, whereas others will tolerate as much as twice that dose before erythema appears.

The management of an adverse drug eruption may require only modification of the dose of the drug, or its withdrawal, when in most cases disappearance of the rash occurs within a few days. Rarely will it be necessary to resort to, for example, the use of corticosteroid therapy. In some cases residual staining of the skin due to melanin deposition or haemosiderin will take several weeks to fade and, in the case of 'arsenical dermatitis', there is a considerable delay before the skin resumes its normal appearance. The rash produced in some patients by certain beta-blocking agents (practolol was a notable example) may persist long after the drug has been withdrawn, or may only appear after withdrawal of the drug. Some women who develop chloasma as a result of the use of oral contraceptives find only marginal improvement in their skin colour after stopping the drug.

Skin lesions may be produced either by topical or by systemic administration of drugs, and indeed the causative agent may not, strictly speaking, be regarded as a drug. 'Enema rash' is an example of this, and was once common when soap-and-water enemas were routinely administered to patients being prepared for abdominal surgery. Today cases may still be seen occasionally in the domiciliary treatment of old people.

Precise knowledge of the mechanisms underlying adverse reactions to drugs in general is far from complete, so that in many cases where there is strong clinical evidence as to the aetiological agent involved, the pathogenesis of the presenting lesion remains obscure. At best the mechanisms involved can be classified in a somewhat broad fashion as toxicity, idiosyncrasy, allergy, or photosensitivity (D'Arcy 1965a, b; 1966). Toxic mechanisms are dependent on the pharmacological properties of the drug, and the severity of the reaction is related to the dose; all persons are potentially susceptible, provided the dose is high enough for that

person. Idiosyncrasy is an inborn predisposition to respond to a specific stimulus in a qualitatively abnormal way independent of antigen–antibody reaction where the drug or its metabolites, if not already protein, may combine with protein in the body to form an antigenic combination. The mechanism involved in photosensitivity may be similar since it has been suggested that the causative drugs, whose site of attack is the Malpighian layer of the epidermis, are metabolized or conjugated with a protein in such a way as to form a photosensitizing substance with an absorption spectrum different from that of the administered drug.

Various agencies collect reports of adverse drug reactions (see Chapter 4) and many case reports are published each year on cutaneous reactions due to drugs, some of them being hitherto unknown reactions to an established drug or interaction with another drug or drugs (see Chapter 39); other reports concern newer drugs and only after consistent reports of the same kinds of reaction occurring in many patients is it possible to accept them as recognized side-effects. To complicate matters, drug interaction may not be between only two drugs, and every large hospital can quote patients admitted who have been taking as many as 20 or even 30 different forms of medication daily. The interaction potentials in such a situation are legion, and the possible expressions of drug metabolism cannot be foreseen. Thus, any review of this nature must be limited, and in this chapter no consideration of contact reaction on the skin due to local applications or handling of drugs will be taken. Rather an outline of some of the more important adverse reactions in the skin due to drugs which have been given systemically will be attempted. For convenience and easy reference these are classified in alphabetical order, and a short list of commonly prescribed drugs together with their cutaneous side-effects is given (Table 32.1). Certain drugs which have more recently been found to have multiple cutaneous side-effects will be detailed towards the end of this chapter.

Table 32.1 Commonly prescribed drugs and their side-effects on the skin

Antibiotics

All grades of urticaria	Exanthematous eruptions
Dermatitis reactions	Toxic epidermal necrolysis
Erythema multiforme	(Lyell's disease)
Anaphylactoid reactions	Cross-reactivity with other
Fixed drug eruptions	antibiotics

Benzodiazepines

Exanthematous eruptions	Erythema multiforme
Urticaria	Erythema nodosum
Photosensitivity	Fixed drug eruptions
Purpura	

Table 32.1 *Continued*

Indomethacin

Mouth ulcers	Pruritus
Vasculitis and purpura	Cross-reaction with aspirin
All grades of urticarial reaction	Alopecia
Exanthematous eruptions	

Methyldopa

Papular eruption	Seborrhoeic dermatitis
Breast enlargement	Alopecia
Lichenoid eruption	Urticaria
Lupus erythematosus	

Oral contraceptives

All grades of urticaria (sometimes late onset)	Lupus erythematosus
	Porphyria
Chloasma	Photosensitivity
Alopecia	Erythema nodosum
Hirsutism	Purpura
Acne	Perioral dermatitis
Fungal infections	

Quinidine

Exfoliative dermatitis	Exanthematous eruptions
Lichenoid eruption	Photosensitivity
Livedo reticularis (sun-exposed sites)	Photodermatitis
	Urticaria

Salicylates

All grades of urticaria, acute and chronic,	Vasculitis
	Purpura
Erythema multiforme including Stevens–Johnson syndrome	Erythema nodosum
	Mouth ulcers
Toxic epidermal necrolysis (Lyell's disease)	Bullous eruptions
	Psoriasis

Sulphonamides

(Often seen in over-fifties age group due to sensitization years previously with sulphonamide powder wound dressings)

All grades of urticaria	Porphyria
Dermatitis reactions	Exanthematous eruptions
Fixed eruptions	Photosensitivity
Stevens–Johnson syndrome	Vasculitis
Toxic epidermal necrolysis (Lyell's disease)	Purpura
	Lupus erythematosus

Reviews in the literature

From time to time major reviews appear which by their nature are very useful sources from which to build up a knowledge of the subject. The subject of drug allergy was comprehensively detailed by Van Arsdel (1982*a*) and was based on the four classes of reactions categorized by Coombs and Gell almost 20 years previously (Coombs and Gell 1975). The value of patch and provocation tests and the ways in which drug allergies develop were discussed, and he described serological tests, concluding that objective tests are reasonably reliable in confirming or identifying allergy in a limited number of drugs. The same author (Van Arsdel 1982*b*) in a review of allergy and adverse drug reactions, summarized the features which distinguish allergic from toxic and idiosyncratic reactions. He also discussed prevention and management; simple guidelines for treatment were given to cover all from the most benign to life-threatening circumstances. Griffin (1983) gave a most lucid account of drug-induced allergy and hypersensitivity, defining the various types of reactions and their mechanisms. In five tables he listed the most common drugs reported to the CSM between 1964 and 1983 as causing anaphylaxis, bronchospasm, angioneurotic oedema, urticaria, and the lupus erythematosus syndrome. Another article on allergic reactions to drugs and biological agents reviewed the immunological aspects of adverse drug reactions (Patterson and Anderson 1982).

Allergic reactions to insulin occur in an estimated 10–15 per cent of all diabetics; the fact that these usually include the skin is pointed out in a review commentary by Jegasothy (1980). It is interesting to note that 2.5 per cent of those patients with insulin allergy also have a history of allergy to penicillin. In the same journal Chin and Fellner (1980) review allergic reactions to lidocaine hydrochloride and refer to urticarial dermographism secondary to allergic hypersensitivity.

Hardie and Savin (1979) gave a condensed account of the causes of drug eruptions and concluded that the inexcusably high incidence could be easily reduced by an avoidance of polypharmacy. Verbov (1979) in his review drew attention to the fact that an urticaria due to penicillin may present as long as four weeks after the last dose of the drug. Moreover, the presence of minute amounts of penicillin in dairy products or the production of penicillanic acid in the skin by fungi may result in persistence of a penicillin urticaria. A recent leading article in the *British Medical Journal* by Beeley (1984) discussed allergy to penicillin, with particular reference to the types of patients at risk, the types of reactions which can occur, the patterns of skin rashes, and the incidence of those side-effects. The importance of realizing the possibility of cross-reactions to cephalosporins was stressed, and the status of skin tests to establish the presence or absence of hypersensitivity was discussed.

August (1980) gave an account of common patterns of drug eruptions together with a very useful list of the main causative agents. He included a description of the medicaments which may cause allergic contact dermatitis (accounting for as much as 30 per cent of all such cases), and pointed out the possibility of sensitization, so that subsequent systemic treatment with the same drug may produce a widespread reaction. Also included was a list of potential primary irritants used in dermatology, and the side-effects of topical steroids were described. A table listing the relative potencies of commonly prescribed topical corticosteroids defined 'weak' and 'strong' preparations. Akers (1980) gave a review of data from 2849 patients who had been using eight topical steroid preparations in six known steroid-responsive dermatoses. Cutaneous side-effects noted were irritation, itching, burning, dryness, and vesicle formation. The possibility that the adverse reactions found may have been due to the drug vehicles was also considered. Miller and Levene (1982), in their review of the use of steroids in dermatology, referred to well known side-effects of topical treatment such as skin thinning, telangiectasia, striae, purpura, acneiform eruptions, and pigmentary changes. They also drew attention to less well-known effects such as localized hair growth and, more seriously, an increased risk of raised intraocular pressure, cataracts, and ocular herpes simplex if steroids repeatedly enter the conjunctival sac. When potent topical steroids are applied to widespread psoriasis on a regular basis, there is a danger of making the condition unstable, and perhaps even of precipitating pustular psoriasis. Reduction in plasma cortisol levels may occur if application is continued in sufficient amounts, especially if under occlusion, but significant adrenal suppression has been found in only a few cases. The risk is increased if liver failure is present, or if the patient is a young child with a large absorptive area in relation to body weight; young children also have increased permeability on account of their low epidermal barrier. However, plain topical hydrocortisone is known to have relatively little systemic effect in children.

Ives (1980) listed both proprietary and official names of drugs in a pragmatic review of drug eruptions, and it was clear that doctors would do well to attend a short postgraduate course in dermatology. In an illustrated guide to identifying drug eruptions, Finlay (1983) suggested clues to the diagnosis which should be of great help to the beginner and included 23 excellent photographs of the most common types of eruption which will occur in practice, whether hospital or general. Bailin and Matkaluk (1982) dealt with adverse drug eruptions occurring in the treatment of the rheumatic diseases. They pointed out that the same drug may produce different reactions in different patients, or in the same patient at different times. The drugs used in current rheumatological practice were discussed, and tables indicating both common and rare eruptions associated with these drugs were provided. ·

A comprehensive account of the cutaneous complications of chemotherapeutic agents was given by Bronner and Hood (1983). First they tabulated the classification of chemotherapeutic drugs by listing alkylating agents, antimetabolites, vinca alkaloids, antibiotics, and a miscellaneous group; then they listed the side-effects together with the drugs which produce them. A wide range of reactions may be seen especially in multiple chemotherapy, and the authors pointed out that some reactions precede and herald systemic toxicity. This is an important review and contains much information in an easily digestible form.

Holmberg et al. (1980) wrote about adverse reactions to nitrofurantoin; they analysed 921 reports in which 261 patients developed an exanthematous eruption as the initial evidence of a reaction to the drug, while 36 presented with cyanosis and 29 with jaundice. They concluded that risk of an adverse reaction to nitrofurantoin is low in childhood, but increases with age.

Harber et al. (1982) produced a comprehensive survey of the phototoxic and photoallergic mechanisms whereby adverse skin reactions occur after exposure to a drug plus light. The clinical features of both groups were described, and details of risk factors together with the drugs commonly responsible for initiating such reactions were outlined. The technique of photopatch testing was described, and the hazards and complications of the tests were listed. Reference was also made to the phenomenon of persistent light reaction in which patients known to be photosensitive to a drug continue to react adversely to sunlight after the drug has been withdrawn.

Horio (1981) gave a fascinating description of the evaluation of the phototoxicity of a number of drugs by means of long-wave ultraviolet irradiation of fungal spores in the presence of those drugs. Positive results were obtained with a number of drugs, but the point was made that an in-vitro system cannot detect phototoxicity of drug metabolites since passage through the liver or any other organ does not take place. Nevertheless, this system appears to be a useful addition to the screening profile for possibly phototoxic drugs.

The adverse reactions on the skin by the use of metoprolol and other β-adrenoreceptor-blocking agents were reviewed by Neumann and van Joost (1981), and they described psoriasiform and/or eczematous skin reactions in patients during long-term treatment with metoprolol. The use of the oral provocation test was mentioned, and the results were described.

A detailed account of drug- and heavy-metal-induced hyperpigmentation was given by Granstein and Sober (1981). They also provided a useful account of skin colour changes which may result from the use of chemotherapeutic agents in current practice, and there is an extensive list of references.

A major review of retinoids, cancer, and the skin by Elias and Williams (1981) described the chemistry, metabolism, mechanism of action, and the uses of the retinoids in specific disorders. Comparison of the mucocutaneous side-effects of natural and synthetic oral retinoids with the percentage frequency of such effects was presented in table form. Thomas and Doyle (1981) wrote of the therapeutic uses of topical vitamin A acid, referred to such side-effects as erythema, hypo- and hyperpigmentaton, and susceptibility to irritation from wind, cold, and dryness, and mentioned the potentiating effects of products with high concentrations of alcohol, including soaps and cosmetics.

Voorhees and Orfanos (1981) gave a summary of the highlights of a symposium held in Berlin in 1980 on the use of oral retinoids. The side-effects included one or more of the features of hypervitaminosis A and were seen in the skin of more than 75 per cent of all patients treated, such as cheilitis, hair loss, dry mucosae (eyes, mouth, nose), skin thinning, palmoplantar desquamation, nosebleeds, easy bruising, and paronychia. It was reported that with few exceptions these side-effects were tolerable and completely reversible. In a later publication (Cunliffe and Miller 1984), the proceedings of an International Symposium on Retinoid Therapy held in London in 1983 included detailed accounts of side-effects of the drugs involved. Of particular interest is the newer drug, isotretinoin (13-cis-retinoic acid), which can be expected to induce such side-effects as cheilitis, facial dermatitis, eczema craquelé-like dermatitis of the rest of the body, nasal dryness and nose-bleeds, scalp folliculitis, pyogenic granuloma (rare), skin thinning, and occasionally arthralgia. The important point made about this drug, the use of which is restricted, was that the acne being treated may flare during the second to sixth week of treatment to a degree worse than the pre-treatment state. The flare settles and does not affect the eventual outcome if therapy is continued, but, clearly, to obtain patient compliance a full explanation of this must be given to the patient before treatment is begun.

In a concise article Thiers (1981) discussed the hazards associated with the treatment of diseases of the skin. The possible hepatotoxicity of the methotrexate treatment of severe psoriasis was emphasized, and a number of references were given to help dermatologists determine risk factors in patients before prescribing this treatment. Photochemotherapy was likewise discussed, and adverse effects such as cutaneous neoplasia and ocular complications, especially in the long-term, together with possible mechanisms and prophylactic measures were suggested.

Gamma benzene hexachloride, used topically in the treatment of scabies, may be absorbed from the application of a 1 per cent cream, and it is claimed may produce neurological toxicity. Similarly, the use of dinitrochlorobenzene as an immune enhancer for extended periods on the skin surface calls for caution since glutathione depletion in rat skin and a strong mutagenic effect in *Salmonella typhimurium* are both known to occur. Thiers (1981) also referred to the recent reports of the association between minocycline therapy used in acne vulgaris, and cutaneous pigmentation (Simons and Morales 1980; Fenske *et al.* 1980; McGrae and Zelickson 1980). It is stated that the pigmentation is probably an idiosyncratic response to the drug and consists of localized or diffuse blue-grey, brown, or brown-black discoloration of the skin particularly in sun-exposed areas and sites of previous inflammation of the skin. A further report of discoloration of the teeth, fingernails, and skin, particularly in a scar on the knee was given by Caro (1980). The patient was a 42-year-old woman who had been using minocycline, 100 mg two or three times a day for four to five years; she also noticed that bruises which developed tended to persist for much longer than usual. After one month of stopping the drug the grey colour of her teeth started to fade.

Finally, in this list of reviews, one of considerable importance to general practitioners and dermatologists alike was concerned with nail changes secondary to systemic drugs or ingestants (Daniel and Scher 1984). The changes in the nails vary from mild pigmentation to nail shedding and matrix scarring, and are often seen in association with other skin eruptions, and in particular, with hair problems. The article provides a most useful reference list of drugs which produce nail changes, and is both comprehensive and comprehendable.

Acneiform eruptions and rosacea

Acne vulgaris is a common and often disfiguring skin disorder, typically seen between puberty and the age of about 25 years, when the condition tends to clear spontaneously. It is characterized by the presence of the comedo, or blackhead, accompanied by a varying degree of inflammatory change in the subjacent pilosebaceous follicle, producing swelling pustule formation, followed by pitting of the skin surface as scarring takes place. It is usually seen in the skin of the face, forehead, upper chest, and back above the nipple line, although atypical distributions are well known to dermatologists. A variety of drugs produce similar pustulonodular, swollen, rather umbilicated lesions, and these can often be distinguished from true acne by the absence of comedo formation in many cases, the atypical distribution, the abrupt onset, and often by their presence in older age-groups. It should be remembered, however, that exacerbation of true acne vulgaris may occur as an adverse drug reaction and, when acne vulgaris fails to respond to orthodox therapy, the drug history of the patient should be reviewed.

The following drugs have all been reported as capable of producing acneiform lesions which tend to regress on withdrawal or even adjustment of dosage: halogen compounds (expectorants, sedatives, vitamin preparations, halothane, etc.); glucocorticoids; ACTH (Sullivan and Zeligman 1956); androgens (Kennedy 1965); oral contraceptives; isoniazid (Hesse 1966); prothionamide; ethionamide (Lees 1963); and cyanocobalamin (Fellner and Baer 1965). It was traditionally held that the anticonvulsant drugs, especially phenytoin, are associated with the development or exacerbation of acne. Greenwood *et al.* (1983), however, assessed the severity of acne and rate of excretion of sebum in 243 epileptic patients who were taking various anticonvulsants, and concluded that anticonvulsant therapy does not cause acne. Lithium carbonate, used in the treatment of manic-depressive states, was reported as being capable of producing acne-like lesions (Okrasinski 1977). Heng (1982) pointed out that the various cutaneous reactions to lithium, including acneiform eruptions, are dose-dependent and should alert the clincian to a possible underlying toxic state induced by the drug. He further stated that treatment of acne with tetracycline should be avoided because of interaction with lithium which may lead to a nephrotoxic effect. This, however, was refuted in a letter by Jefferson (1982) whose clinical experience is that the drugs can be used together compatibly without a need for lower than usual doses. Finally, Pembroke *et al.* (1981) described two cases of acneiform eruptions occurring in middle-aged women under treatment with dantrolene for spasticity. In each case the eruption consisted of mainly darkly pigmented blackheads which developed on sites of chronic trauma and friction, including the face, breasts, and buttocks.

Rosacea-like lesions were reported in a 74-year-old man by Wilkin (1980) and were ascribed to the use of the vasodilator drug pentaerythritol following an episode of myocardial infarction; exacerbation of rosacea may occur during treatment with glyceryltrinitrate. Rosacea-like lesions and skin atrophy have also been seen in patients using topical applications of 1 per cent hydrocortisone for periods varying between a few weeks and over a year (Guin 1981). Case reports of six patients with ages ranging from 22 to 72 years were detailed, and attention directed to the fact that topical 1 per cent hydrocortisone had usually been regarded as a relatively safe form of treatment. A most unusual reaction to etretinate was seen in a 67-year-old man during treatment for psoriasis, namely, rosacea, by Crivellato (1982). The relationship between the drug and the appearance of the rosacea was proved by re-introducing the drug.

Bullous eruptions

A number of drugs may produce transient erythematous blistering eruptions; these bullae vary considerably in their extent, size, disposition, contents, and degree of tension.

A leading article in the *British Medical Journal* (1981) on drug-induced bullous disorders stated that such reactions had been reported with increasing frequency over the previous 10 years, and that frusemide was capable of producing the greatest variety of bullae, from erythema multiforme to the rare condition, acquired epidermolysis bullosa. It was suggested that some drugs or their metabolites may have a strong affinity for the basal-cell layer of human epidermis. Either hapten–protein carrier complexes which can behave as immunogens may be formed, or basal-cell antibodies may result from death of the cells due to an accumulation of the drug or its derivatives. Thus, drugs may well be causal in the production of many of the appearances of classical dermatoses, and the possibility that a classical dermatosis may also be modified by an adverse drug reaction has to be considered.

Characteristic large bullae are found in about 6.5 per cent of cases of acute barbiturate poisoning (see Plate 1a), and any unconscious patient displaying such bullae should be routinely screened for barbiturates (Groschel *et al.* 1970). Bullae associated with barbiturate treatment may occur early, or alternatively develop later, and have a tendency to heal with residual pigmentation. Small doses of barbiturate may evoke an impressive bullous reaction in sensitive patients, but dose-related toxic effects also occur. A menopausal woman, known to have had adverse reactions from barbiturate therapy in the past, was inadvertently given a dose of *Bellergal* which contains 25 mg of phenobarbitone. Soon after ingestion she developed widespread haemorrhagic bullae and was admitted to hospital where the eruption cleared in two weeks with only bland local applications (Morley and McQueen 1977).

In this respect it is well to recall that many intravenous anaesthetics are barbiturate derivatives, and serious reactions can occur if they are used in sensitive patients.

Bullous lesions have been reported in the skin in nitrazepam overdosage (Ridley 1971), during treatment with acetazolamide, gold salts, the thiazide diuretics, and frusemide (Fellner and Katz 1976), sulphonamides and the phenazone-like antipyretic analgesics. Salicylates are notoriously liable to lead to a number of severe skin reactions (Baker and Moore-Robinson 1970), and these include bullous urticaria, bullous erythema multiforme, and dermatitis herpetiformis-like eruptions. Cases of photosensitization and phototoxicity have been reported with the urinary antiseptic nalidixic acid; these lesions begin as erythematous blotches and progress to frank bullous eruptions (Burry and Crosby 1966; Mathew 1966). The urticarial lesions seen in some patients sensitive to the penicillins may sometimes proceed to frank bulla formation and bullous lesions resembling pemphigus, both clinically and immunologically, were reported by several authors including Tan and Rowell (1976).

Since 1969 numerous cases of penicillamine-induced pemphigus have been reported, and Yung and Hambrick reviewed 48 case reports in the literature in 1982. They made several observations; there is a close time sequence between the D-penicillamine administration and the pemphigus eruption, and there are characteristic histological findings with positive direct or indirect immunofluorescence tests as in the naturally occurring form of the disease. Both the incubation period and the drug dosage are variable (1–48 months and 250–2000 mg daily, respectively). The duration of the eruption seems to be related to the duration and dosage of administration of the drug, and there is a variable response to systemic steroids or immunosuppressive agents. The authors postulate that it is possible that epidermal-cell surface proteins may be altered to become antigenic to the host under a number of conditions, including D-penicillamine administration, thus explaining the pathogenesis of the pemphigus group of disorders.

Troy et al. (1981), however, produced immunological data to support the suggestion that a penicillamine-associated bullous eruption may not be the same disease as spontaneously occurring pemphigus. A case is described in detail in which some of the histological features of pemphigus were present, but the immunofluorescence studies suggested pemphigoid. This report also claimed that cases of penicillamine-associated bullous eruptions in the literature differ in important respects both clinically and histologically from the spontaneously occurring form of pemphigus.

Matkaluk and Bailin (1981) described a case of penicillamine-induced pemphigus foliaceus. They pointed out that of the cases reported in the literature since the first report by Degos et al. in 1969, the vast majority have been of the foliaceus type. Moreover, this form of the disease generally has a good prognosis. However, the case reported by the authors had a fatal outcome, only the second such reported.

The first report of a bullous eruption in a patient who had taken an overdose of carbamazepine was given by Godden and McPhie (1983). Sixteen hours before admission to hospital the patient, a 68-year-old woman, had taken 40 tablets of carbamazepine. On admission to hospital in coma, she was noticed to have bullae on the dorsi of both feet and on the skin over the right lower tibia. Further lesions developed on the left iliac crest area and on the left breast. Direct and indirect immunofluorescence tests were negative. Routine histology showed a subepidermal bulla with a necrotic roof. The dermis was congested and fibrinoid necrosis of occasional small vessels was noted, with a scattering of inflammatory cells. Although not stipulated by the authors, the histological picture described is that of the bullous form of erythema multiforme as caused by an adverse reaction to a drug.

Erythema multiforme (EM)

As the name suggests, this disorder may present in many forms. Usually a small ring of erythema appears in the skin, and it may become slightly raised. The target lesion is so-called since its rings of erythema of varying hues give it a target-like appearance. The sites of election include the dorsal surfaces of the hands and feet, the forearms, the legs, and the trunk, but lesions have been described in most other skin sites. Mucosal lesions may also occur. There may be trivial or severe systemic upset. EM may occur during the course of bacterial or viral infection; it may recur at intervals with no apparent cause. It may be the result of an adverse drug reaction, and in this respect, two potentially lethal forms are known — the Stevens–Johnson syndrome and toxic epidermal necrolysis (see pp. 770–1).

The number of drugs reported as being associated with EM is legion. In an extensive article about EM, Huff et al. (1983) listed 40 drugs and drug categories which have all been implicated. Among the most important are the sulphonamides including co-trimoxazole (Beck and Portnoy 1979), the penicillins, phenylbutazone, barbiturates, tetracyclines, chloramphenicol, isoniazid, codeine (Ponte 1983), sulindac, oestrogens, phenolphthalein, quinine, mianserin (Quraishy 1981), meprobamate, cimetidine, carbamazepine (M.W. Taylor et al. 1981), the hydantoins, and aspirin. The first well-documented case of histologically proven EM caused by oral frusemide was reported by Zugerman and LaVoo (1980), and the report contained a useful review of side-effects of the drug.

Erythema nodosum

These erythematous, slightly blue, subcutaneous lesions are nodular and tender, generally occur on the shins, but occasionally elsewhere, and vary in size from about 1 to 4 cm in diameter. They may be accompanied by malaise and pyrexia, and take several weeks to resolve, undergoing a variety of colour changes as they do, from red to bluish-green, to blue and purple, and eventually to a mottled brown which takes some time to disappear. The basis of the lesions is probably a state of hypersensitivity affecting the vessels in the subcutaneous fat, and the condition may occur during the course of streptococcal and viral infections, tuberculosis, certain fungal infections, and sarcoidosis.

Certain drugs are also regarded as being capable of producing similar appearances in the skin, and investigation of erythema nodosum should include a careful drug history. A number of single case reports have appeared over the years implicating such drugs as bromides, iodides, ibuprofen (Khan 1984), sulphonamides, the sulphonylurea group of oral hypoglycaemic agents, penicillins, salicylates, thiouracil, oral contraceptives (Holcomb 1965; Darlington 1974), and gold salts.

It must be emphasized that of the many causes of erythema nodosum, adverse drug reactions must be regarded as uncommon, and every effort should be made to exclude the common infective causes before drugs taken by the patient are held responsible.

Exanthematous eruptions

An exanthematous drug eruption is one which exhibits the characteristics of the rash associated with one of the eruptive fevers, for example, scarlatiniform, morbilliform, or maculopapular. It differs from the rash of measles or scarlet fever, however, in that it lacks the same degree of specificity since many drugs can produce the same appearances in different individuals. Drugs may produce these types of skin manifestation within 2–3 days of administration, or later, depending on whether the patient has previously been sensitized to the drug, and on the pharmacokinetics of the drug in any given patient. An urticarial element is often present at the onset, and the skin may feel hot, burning, or itchy. The exanthematous rash is the one which is most often seen in the skin as a result of an adverse drug reaction, and over the years innumerable case reports have appeared in the literature. From the practical point of view, the most common are:

Macular or maculo-papular: the penicillins, nitrofur-

antoin, sulphonamides, erythromycin, chloramphenicol, isoniazid, diazepam, allopurinol, penicillamine, *p*-aminosalicylic acid (PAS), phenacetin, protriptyline, chlorambucil, barbiturates, carbamazepine, trazodone (Cohen 1984), gold salts, and hydantoin derivatives. In almost all cases of infectious mononucleosis treated with ampicillin, a rash appears, and the explanation of this phenomenon remains a mystery despite extensive pharmacological and immunological investigations. A report of a similar rash occurring in a patient with a glandular fever-like illness treated with amoxycillin was published (Copeman and Scrivener 1977).

Erythema: barbiturates, captopril (Wilkin *et al.* 1980), carbamazepine, cimetidine (Merrett *et al.* 1981), frusemide (furosemide), gold salts, pancuronium bromide, phenylbutazone, phenytoin, perhexiline maleate (Tomlinson and Rosenthal 1977) and tetracycline.

Morbilliform: allopurinol, barbiturates, chloramphenicol, ampicillin and other penicillins, nalidixic acid, PAS, protriptyline, phenacetin, and sulphonamides with possible cross-sensitivity to other drugs in this group including certain related antidiabetic drugs, diuretics, and streptomycin.

Dermatitis

A dermatitis or eczematous eruption may occur during certain drug therapy, and while it may be due directly to the drug, it may also develop as a result of allergy to a topical preparation which is subsequently given systemically. The dermatitis may be mild and patchy, or it may be diffuse, and in some cases, go on to the exfoliative form (see below). Kwong *et al.* (1978) drew attention to the development of a diffuse dermatitis reaction in a 52-year-old male who was receiving warfarin for thrombophlebitis; this disappeared when the drug was withdrawn and reappeared when warfarin was recommenced.

In a review of 38 patients with the allopurinol hypersensitivity syndrome, Lupton and Odom (1979) included dermatitis varying from patchy to diffuse, and sometimes exfoliative, among its features, and pointed out that such effects may only begin a considerable time after treatment has been started.

Evans blue dermatitis was described in five patients who underwent lymphangiography (Guill and Odom 1979). In a large series of patients using sodium cromoglycate for the treatment of asthma, Settipane *et al.* (1979) found a 2 per cent overall frequency of adverse reactions, among which was dermatitis occurring in just over 1 per cent of the patients. The beta-blocking drug, metoprolol, was also incriminated as a cause of a dermatitis reaction (Neumann *et al.* 1979). Cross-sensitization may occur in a number of drugs and lead to a dermatitis eruption; such drugs include the sulphonamides and various antibiotics, tolbutamide, and chlorpropamide.

Cimetidine was reported as being responsible for the appearance of a seborrhoeic dermatitis-like eruption (Kanwar *et al.* 1981). The patient had no previous history of seborrhoeic dermatitis and, after the rash had cleared on stopping cimetidine, was given an oral challenge of the drug which led to reappearance of the rash. Another report (Greist and Epinette 1982) described the development of xerosis and asteatotic dermatitis in two patients, one of whom developed the same eruption on being given cimeditine some time after the first incident. The authors suggested that the skin changes described by them are due to the anti-androgenic properties of the drug, and to its effect on sebum secretion rate.

One of the new methods of drug delivery is by the transdermal route. Clonidine, an antihypertensive drug, was given by this route to 21 patients with hypertension (Boekhorst 1983). Three of the patients developed allergic contact dermatitis to clonidine and this was confirmed by patch tests. These reactions developed many weeks after therapy had been started, and one patient developed a generalized reaction four weeks after the dermatitis had appeared and while he was continuing to use the drug.

Other drugs which may cause a dermatitis eruption include aloes (Morrow *et al.* 1980), antibiotics, antihistamines, thiazide diuretics, chlorpromazine, carbamazepine, quinine, meprobamate, iodides, and chloral hydrate.

Exfoliative dermatitis

This is a generalized inflammatory skin disease which may be fulminating in onset and may be accompanied by pyrexia and prostration. There is redness, dryness, and peeling of hyperkeratotic scales from the skin; itching, irritation, and secondary infection due to scratching are common. The nails and hair may also be affected, with loss or damage to both. In its most severe form the patient is very ill indeed and, because of the extensive involvement of the skin, high-output cardiac failure may develop resulting in death. The condition is also known as erythroderma and may be due to a variety of underlying causes including systemic drug therapy. That it may constitute a life-threatening condition has been recognized for many years, and Rostenberg and Fagelson (1965) included this disorder in their review of life-

threatening drug eruptions. It is also recognized that other acute drug eruptions may progress to a state of generalized exfoliative dermatitis, and that it is possible for this reaction to occur in a patient given a drug systemically who has previously undergone contact sensitization by the same or related drug (Petrozzi and Shore 1976).

The number of drugs reported to be associated with the development of exfoliative dermatitis is large, but those most commonly encountered include the antibiotics, arsenicals, barbiturates, beta-blocking agents, cephalosporins (Kannangara *et al.* 1982), chloroquine, anticonvulsants, sulphonamides, griseofulvin, mercurial diuretics, gold salts, thiouracil, phenothiazines, pyrazolone derivatives, phenindione, salicylates, streptomycin, isoniazid, carbamazepine, ketoprofen (Morley 1977), and lidocaine (lignocaine) hydrochloride (Hofmann *et al.* 1975). Fisher (1976) drew attention to a case in which previous sensitization by a topical antihistamine resulted in a widespread, severe exfoliative dermatitis after ingestion of a similar type of antihistamine some time later, and reviewed 17 cases in all. Van Joost and Smitt (1982) reported a case of exfoliative dermatitis which started as an itchy eruption three days after starting propranolol. Three months after recovery, epicutaneous tests using four differing beta-blocking agents in a petroleum base gave positive reactions to propranolol, oxprenolol hydrochloride, and atenolol, but were negative for metoprolol. Twenty normal control subjects were also tested in similar fashion, and negative results were obtained in all cases. A leading article in the *Lancet* (1982) was devoted to adverse reactions to dapsone which is still the first-line treatment for leprosy. There are over twelve million sufferers in the world, and two million are under treatment. In the West the drug may be used to treat dermatitis herpetiformis; its mode of action is not clear, but possibly it acts as an immunomodulator. Most adverse reactions to the drug appear to be dose-related, but some are idiosyncratic. Two rare reactions are described. One has been long recognized in leprosy and is a hypersensitivity reaction with fever, hepatitis, and exfoliative dermatitis. It may also occur as hepatitis or exfoliative dermatitis alone. The second reaction so far has only been seen in association with the treatment of dermatitis herpetiformis, and only in three cases. The skin is not affected primarily, but a slow, relentless hypoalbuminaemia develops, which may lead to death.

Fixed drug eruptions

A fixed drug eruption is a reaction in the skin to a drug which has reached it by way of the circulation, and may be solitary or multiple. It is localized, appears rather suddenly, usually subsides quickly when the drug is withdrawn, and reappears in identical fashion at the same site or sites when the drug is taken again by the patient, although the reaction may vary in extent. It may produce itching, heat, or discomfort and, as it subsides, pigmentation occurs, which on repeated stimulus becomes darker. It may also occur on mucous surfaces producing stomatitis, vaginitis, conjunctivitis, proctitis, or urethritis, and its mode of production is unknown. It may occur as a result of food additives or drugs acting alone or in conjunction with the food or drug ingested. It is often a diagnostic puzzle, since patients may be taking drugs of which the physician is unaware, and which may indeed seem harmless to both doctor and patient. The appearance of recurrent lesions with long intervals between may easily escape the attention of both parties, and the significance of their relationship to drug therapy is often missed.

In appearance the lesion is a slightly raised reddish patch or cluster of macules, which may feel hot or itchy, or may evoke no subjective sensation at all. After each occurrence and resolution, more melanin pigment is deposited in the papillary dermis until eventually a slightly indurated brownish patch of skin results. Weiss and Kile (1935) described an unusual phenolphthalein-induced eruption, and through the years this has come to be recognized as the classical example of a fixed drug eruption (see Plate 1b). In 1964, Derbes quoted 57 drugs as being capable of producing a fixed drug eruption; now the number is almost double that figure as reports concerning newer drugs have accumulated. Pasricha (1979) investigated 40 patients who had fixed drug eruptions and subjected them to provocation tests. Twenty-eight patients completed the provocation tests and the drugs implicated were, in order of frequency of reaction: tetracyclines, metamizol (dipyrone), oxyphenbutazone, phenobarbitone, sulphadiazine, sulphaphenazole, penicillin, sulphadimethoxine, a propylphenazone–phenacetin–caffeine analgesic preparation (*Saridone*), sulphadimidine, and sulphamethoxypyridazine. Evidence of cross-sensitivity was found between tetracycline and demethylchlortetracycline (demeclocycline) and between oxyphenbutazone, but not between different sulphonamides. The author did not describe the appearance of the eruptions produced, nor was there any indication of the severity of the reactions produced by the provocation tests.

Independent lesions of fixed eruption due to two unrelated drugs in the same patient were documented by Pasricha and Shukla (1979). The drugs responsible were metamizol and tetracycline, and they were proved on provocation testing.

Described as a 'lesion in drug usage', Olumide (1979) referred to fixed drug eruption as a common finding in Nigeria where the population tend to take different kinds of drugs obtained either on prescription or directly from a pharmacy. The results encountered with a pyrazolone analgesic and a benzodiazepine were described, and the necessity for enquiry into home remedies was emphasized.

The first report of a recurrent fixed-drug eruption due to trimethoprim was by Gibson (1982). It occurred as a well- circumscribed area of swelling on the left side of the neck, and in all, eight or nine recurrences at the same site occurred over a two-year period. Despite being warned about the possible relationship between the drug and the eruption, the patient, a 49-year-old woman, continued to take the drug at intervals, usually for urinary-tract infections. On the latest occasion, flare of the rash took place within one day of taking trimethoprim. Neither the Committee on Safety of Medicines nor the manufacturers had had any report of a trimethoprim fixed drug eruption.

The occurrence of fixed genital drug eruption may lead to considerable confusion and involve the patient in full screening for venereal disease. Talbot (1980) reported two cases of penile eruptions, one of which showed erosions and the other, ulceration. Another report by Presley (1981) also described a penile eruption with blistering. In each of the three cases the drug involved was found to be co-trimoxazole in the form of *Septrin*, but it was not stated whether the plain or the soluble form of the drug was used. This may be of some interest since at least one case of morbilliform drug rash is known as a result of switching from the plain to the soluble form of the drug in a patient who had previously used only the plain form (McQueen, unpublished). Coskey (1982) described the appearance of an hourglass, 2-cm ulcer on the glans penis of a 54-year-old man, when taking chlorphenesin and salicylates, and who, as a result, was extensively investigated. On stopping the drugs the ulcer had almost completely healed within a week, but recurred at the same site some months later after he took chlorphenesin for two days. After a month's treatment on stopping the chlorphenesin, the ulcer healed with residual pigmentation. The first case report of a fixed drug eruption due to minocycline was made by LePaw (1983). The patient was a 48-year-old man who was given minocycline, 50 mg daily for rosacea. Eight weeks later he developed a pruritic, scaly eruption on the scrotum, and this cleared on stopping the drug. Nine months later he took minocycline again, and within two days had developed a deeply erythematous, oozing lesion on the glans penis and this recurred two and a half months later when he again took the same drug. On examination two years later, there was still brown pigmentation at the site of the reaction. An unusual type of fixed drug eruption occurring as pigmented lesions of the tongue was described by Westerhof and his colleagues (1983) from Amsterdam. The patient was a 32-year-old Negroid man who had dark macular pigmentation and localized ulceration on the dorsum of the tongue. He had paid no attention to the pigmentation but had noticed some link between the occurrence of ulceration of his tongue and the smoking of drugs. Eight other cases occurring in Negroid men were referred to and all followed the inhalation of heroin pyrolysate and methaqualone vapours. The authors pointed out that fixed eruptions have been reported for morphine, opium, and codeine (Derbes 1964), but not for heroin.

In a report on two patients who developed fixed drug eruptions after taking chlormezanone (Mohamed 1983), the degree of severity appears to have been considerable, and it is not altogether clear whether the Stevens–Johnson syndrome described in one of the patients is regarded as evidence of increasing severity of a fixed drug eruption. However, the author was convinced of the safety of a provocation test in the investigation of fixed drug eruptions, and pointed out that a patient may be reactive to more than one drug.

From the practitioner's point of view, the most important causative drugs include the antihistamines, barbiturates, anticonvulsants, antibiotics (Coskey and Bryan 1975), chlordiazepoxide, chlormezanone (Kader 1983), codeine, dapsone, meprobamate, phenolphthalein, salicylates, sulphonamides, quinidine, phenacetin, quinine, tetracycline, pyrazolone derivatives (e.g. phenylbutazone and oxyphenbutazone) with cross-sensitivity between members of this group (Pandhi and Bedi 1975), and paracetamol (Wilson 1975).

Hair disorders

Hair loss

Hair loss varying from thinning and glabrous patches to a total loss, usually from the scalp, but on occasion from all the normally hirsute areas of the body, has been reported in connection with a remarkable variety of drugs, many of which are currently in common use. In some instances hair loss may begin several weeks or even months after administration of the drug involved, such as with the anticoagulants and occasionally the antithyroid drugs, while with other drugs hair loss may commence within a few days of administration and may sometimes be sudden and dramatic, causing considerable distress to the patient and his relatives.

Drug-induced alopecia may present as one of two types

which may be distinguished by the appearance of shed hairs and the onset of hair shedding relative to drug exposure. Drug-induced damage to hair follicles during their growth phase (anagen) causes marked thinning of hair shafts and subsequent loss of broken hairs. Hairs with damaged, finely tapered roots are also shed. Injury is evident within a few days of drug exposure and the resulting alopecia may be profound since up to 90 per cent of scalp hairs are normally in the anagen phase.

The second pattern of increased hair fall (telogen effluvium) is similar to that sometimes occurring after childbirth or serious illness. The anagen phase of hair growth ends prematurely and the follicles enter a resting phase (telogen). Hair loss begins about three months after the initiating factor and is less severe than anagen effluvium. Shed hairs are like those from the normal scalp and have club-shaped roots.

Some drugs have been strongly implicated as causing alopecia; with others the association is less certain. Ethionamide, cimetidine (Vireburger *et al.* 1981), clofibrate, colchicine, vitamin A, PAS, chloroquine, mephenesin, troxidone, allopurinol, indomethacin, lithium, propranolol (Martin *et al.* 1973), oral contraceptives (Griffiths 1973), anticoagulants, antithyroid drugs, nicotinyl alcohol, nitrofurantoin, and sodium valproate have all been reported as causal agents of alopecia in greater or less degrees. Most have caused alopecia of the telogen effluvium type. In some instances, for example with cholesterol-lowering agents, the alopecia can be explained by interference of the drug with the process of keratinization, whereas in others, the mechanism remains unknown. The most spectacular types of drug-induced alopecia (anagen effluvium) are seen in thallium poisoning, or during treatment with the cytostatic and immunosuppressive drugs. In many cases hair loss must be expected during the use of cytostatic drugs such as adriamycin, cyclophosphamide, actinomycin D (dactinomycin), vincristine, hexamethylmelamine, and methotrexate. All types of cytostatic drugs, whether antibiotic, antimetabolite, alkaloid, or alkylating agents have this potential for producing alopecia as a direct toxic action. Total baldness may occur very rapidly and appears to be dose-related in many instances, with severe cases showing loss of all hair. It is possible, however, to reassure patients that as soon as treatment is stopped normal hair growth will begin again, though there may be some alteration in hair colour.

Although these effects are seen mainly during the treatment of neoplastic disorders, it should be remembered that other conditions, such as psoriasis, ankylosing spondylitis, pemphigus, dermatomyositis, polyarteritis nodosa, idiopathic autoimmune haemolytic anaemia, rheumatoid arthritis, systemic lupus erythematosus, iridocyclitis, ulcerative colitis, chronic active hepatitis, transplantation operations, and steroid-resistant nephrotic syndrome, may be treated with drugs such as azathioprine, cyclophosphamide, or methotexate, and that adjustment of dosage in such cases can usually mitigate the adverse effect of the treatment. The subject of skin reactions in general to the cytostatic agents is reviewed by Levantine and Almeyda (1974) and the same authors reviewed the subject of drug-induced alopecia in 1973.

Allied to the production of alopecia by cytostatic drugs is that due to the effects of radiation therapy, either alone or in combination with chemotherapy. When the effects of radiation are directly upon hair follicles, these are destroyed and no regrowth of hair can be expected after the treatment is completed.

Hirsutism

Hypertrichosis occurs frequently during the systemic administration of corticosteroids, and in a large series of patients suffering from a variety of dermatological conditions reviewed some years ago, both hair loss and hypertrichosis were attributed to the use of oral contraceptives (Merklen and Melki 1966). The same ambivalent types of reaction have also been noted in patients treated with androgens or anabolic steroids, as part of the virilizing effect associated with these hormones.

The virilizing effects seen in babies due to administration of some hormones to the mother during pregnancy may include a degree of hirsutism, as can also result from the use of the anticonvulsant agent diphenylhydantoin (phenytoin) by pregnant women. This latter drug has also been reported as leading to a generalized type of hypertrichosis in children and adults. Fortunately this regresses, albeit slowly, on cessation of treatment.

Chlorpromazine, in addition to producing amenorrhoea, weight gain, and hyperactivity of sebaceous glands in some female patients, has also been reported as leading to hypertrichosis, as also have the psoralens. The hypotensive drug diazoxide has been reported as causing hirsutism during long-term treatment of cases, the condition appearing 4–17 months after the beginning of treatment (Burton *et al*. 1975). Hirsutism has also been reported during the long-term administration of the diuretic acetazolamide (Weiss 1974). Penicillamine has also been recorded as giving rise to this increase in body hair (*Lancet* 1975), and Rampen (1983) reports the phenomenon moderate to severe in degree, in 15 out of 23 females who were receiving PUVA therapy.

Hair colour

In view of the number of drugs which are reported as being associated with some disorder of skin pigmentation it is remarkable that so few appear to have noticeable effects on hair colour. When colour change does occur it takes the form of depigmentation, resulting in a greying or a varying degree of bleaching of the hair. Haloperidol, a tranquillizing drug used both in psychiatric hospitals and in general practice, was reported as having this side-effect (Simpson *et al*. 1964), while chloroquine was reported to have caused lightening of the hair colour in blonde, brunette, and red-haired patients (Saunders *et al*. 1959). The lightening of colour begins at the hair root 2–3 months after treatment has started, and eventually the entire length of the hair all over the body becomes white. This change has been seen in the the three hair colours mentioned above, but does not occur in black-haired people; on withdrawing the drug, hair colour is restored slowly. This effect of loss of hair colour is not seen with hydroxychloroquine.

Mephenesin carbamate, a centrally acting skeletal-muscle relaxant, may also lead to bleaching of the hair 2–3 months after treatment has been started (Spillane 1963). This action may be seen in both males and females, and the effect appears to be dose-related. Colour is restored after withdrawal of the drug.

Lichenoid drug eruptions

A number of drugs are recognized as possessing as a side-effect the ability to produce a skin eruption very similar indeed to lichen planus, in appearance and in distribution, both on the skin and on mucous membranes. The condition is known as a lichenoid drug eruption, and its differentiation from true lichen planus can be difficult. However, in the drug-induced eruptions, scaling and hyperpigmentation with skin atrophy and hair loss may be seen more frequently. Anhydrosis as a result of atrophy of sweat glands in the neighbourhood of lesions may occur. The eruption may commence some weeks after the drug therapy has been instituted, and there is some evidence that it may be dose-related in many cases. On withdrawal of the offending agent, clearing of the eruption tends to be slow, and in some cases a mottled type of pigmentation may remain. It should also be noted that in some cases at least, a true lichen planus may be induced by a drug acting as a precipitating factor. Although written in 1971, the review by Almeyda and Levantine provides a concise guide to the drug-induced lichenoid states, and the reader is strongly advised to refer to this article.

The main drugs associated with lichenoid eruptions are amiphenazole, chloroquine, mepacrine, quinine, quinidine (Maltz and Becker 1980), dapsone, sulphasalazine (Dawes and Shadforth 1984), thiazide derivatives, frusemide, spironolactone (Downham 1978), gold salts, streptomycin, methyldopa, PAS, bismuth, and phenothiazines; the first report of a lichenoid reaction following the use of penicillamine was made by Staak et al. (1975), followed by Seehafer et al. (1981) who described its appearance in six patients with primary biliary cirrhosis treated with penicillamine. Other drugs have also been implicated, but many of these are as isolated case reports; beta-blocking agents have been known to cause lichenoid reactions, and Taylor et al. (1982) described a 59-year-old man who developed diffuse lichenoid plaques, papules with Wickham's striae, and areas of follicular plugging, while taking acebutolol. The eruption gradually became worse over a period of two months, and biopsy of several lesions showed the histology of a lichenoid tissue reaction in one area with features of lupus erythematosus in others; in another site features of a non-specific chronic dermatitis were seen. Acebutolol was withdrawn over 10 days and replaced by another anti-anginal agent, nifedipine. During the next three weeks, the rash diminished rapidly, then more slowly over the next two months, leaving post-inflammatory pigmentation. The interest in this case is that, when beta-blocking drugs give rise to lichenoid reactions, the onset of symptoms is usually within 10 months of starting the drug. In Taylor et al.'s (1982) case the patient had been taking acebutolol for seven years before the adverse reaction began. Another case concerned a 67-year-old man who developed bullous lichen planus after 12 weeks treatment with labetalol, 400 mg thrice daily, and clonidine, 0.3 mg per day (Grange and Wilson-Jones 1978). The eruption was itchy, starting on the penis and spreading to the trunk and limbs. After 10 weeks Wickham's striae were seen on the glans penis, with many bullous lesions on the legs. On replacement of the drugs with atenolol and *Moduretic* (hydrochlorothiazide and amiloride hydrochloride), the symptoms settled within two days, but 15 days after restarting labetalol, the rash started to develop again, and the drug was stopped.

Lastly, the first report of a lichen planus-like eruption in association with treatment of rheumatoid arthritis with the non-steroidal, anti-inflammatory drug, naproxen, was made by Heymann et al. (1984). The rash began 10 days after treatment with naproxen was started and, despite withdrawal of the drug, was still present three months later. It cleared within two weeks of treatment with systemic corticosteroids, and had not recurred one year later.

Lupus erythematosus; the 'collagen' diseases

For many years it has been recognized that a number of drugs can give rise to a syndrome in certain patients which has a very close resemblance to the systemic form of lupus erythematosus (SLE), and in some cases is virtually indistinguishable from it clinically. Most of the manifestations of SLE may be seen as adverse reactions to a number of drugs, but on the whole renal involvement is not seen, and the severity of the disease is less in the drug-induced type. Joint pain, muscle pain and tenderness, a variable degree of lymphadenophy and hepatosplenamegaly, lung involvement, pericarditis, Raynaud's phenomenon, and rash may all be seen, and pyrexial attacks may also occur. Laboratory investigations often give results typical of true SLE, including the presence of LE cells, antinuclear antibodies, and antibodies to single-strand DNA. Antibodies to native DNA are very rarely found, and the DNA-binding capacity to date has been negative in the drug-associated SLE in contradistinction to the finding in true SLE.

It is still a matter for speculation whether drug-induced SLE is an entity distinct from the spontaneously occurring type, or whether the administration of certain drugs leads to an exhibition of SLE by patients in whom the disease has been dormant, or who have a genetic predisposition to the disorder. It is known, however, that some cases may persist after discontinuation of therapy responsible, but recovery is the more likely outcome when the drug is stopped. Occasionally, treatment of drug-induced SLE by the use of corticosteroids is necessary, but this must be regarded as the exception rather than the rule. Hydrallazine was the first drug to be recognized as having a potential for leading to the appearance of SLE, and this occurred in approximately 10 per cent of patients treated with the drug for hypertension on a long-term basis. An autoimmune basis was postulated, and antibodies to hydrallazine could be demonstrated. In 1973, Harpey listed 27 drugs as being related to the appearance of SLE, the most frequently blamed being hydrallazine, isoniazid, phenytoin, and procainamide. Other important drugs include PAS, tetracyclines, streptomycin, sulphonamides, methyldopa, methylthiouracil D-penicillamine, and oral contraceptives. Reports were published of exacerbation of LE by griseofulvin (Watsky and Lynfield 1976), by rifampicin (Bagnato and Nigro 1974), and by clofibrate (Howard and Brown 1973). The following were suggested as being responsible for an LE-like syndrome : a combination of pyrithyldione and diphenhydramine (*Peroben*) (Van Neste et al. 1979); nalidixic acid

(Rubenstein 1979); labetalol (Griffiths and Richardson 1979); pindolol (Bensaid *et al*. 1979); and spironolactone (Uddin *et al*. 1979).

Hughes (1982), in a leading article, gave an outline of the present status of hypotensive agents, beta-blockers, and drug-induced lupus. He observed that, while anti-nuclear antibody is usually present in high titre in lupus associated with those drugs, anti-DNA antibodies are notably absent, and he postulated that there may well be a genetic predisposition as the basis for the emergence of clinical signs of lupus in some individuals. Taking this into account along with the other literature on the subject of drug-induced collagen disease, there is still considerable investigation needed to elucidate the pathogenesis of iatrogenic lupus erythematosus and other disorders in this group such as dermatomyositis, polyarteritis, etc. A remarkable report of dermatomyositis following BCG vaccination in a 12-year-old boy was made by Kass *et al*. (1979), and in the same article, a second boy with a similar illness was described.

Photosensitivity

The terms photosensitivity, photodermatitis, photoallergy, and phototoxicity have all gained attention and detailed clinical description in recent years. That many drugs may also lead to the appearance of the skin lesions associated with these conditions has also been confirmed by numerous reports in the literature, and a review on the subject was published by D'Arcy (1966).

Drug-induced susceptibility to adverse effects of light is of particular concern in tropical countries where climatic conditions and local customs permit large areas of the body to be exposed to strong and prolonged sunlight. They are nonetheless of concern in more temperate climates where a lesser degree of sunlight but perhaps a greater degree of drug ingestion may occur. While it is true that the most severe lesions of the skin in sufferers are seen in exposed sites, direct spread of the lesions may occur to involve non-exposed sites. The lesions vary from an exaggeration of the normal erythematous and tanning effects associated with exposure to the sun, to blistering, oedema, flaking of the skin, eczema, maculopapular eruptions, urticaria, and an appearance strikingly similar to that seen in contact dermatitis. Moreover, in certain individuals, these effects may occur without exposure to strong sunlight and may be precipitated merely by ordinary daylight. It has been suggested that the drugs responsible, the site of accumulation or reaction of which is in the Malpighian layer of the epidermis, are metabolized or conjugated with protein in such a way as to form a photosensitizing substance. In phototoxic reactions, as opposed to photoallergic reactions, however, the effects are dose-related, in terms both of the amount of drug absorbed and the degree of exposure to sunlight. One of the best known phototoxic agents is methoxsalen (8-methoxypsoralen) which dramatically increases susceptibility to the effects of sunlight when taken orally or applied topically in the form of a cream or paint. Various psoralen-containing compounds have been used for centuries in Egypt to induce repigmentation of the skin in vitiligo or leucoderma; treatment is accompanied by exposure of the depigmented skin areas to sunlight rich in ultraviolet rays. Frequently the successful course of such treatment depends entirely on the degree to which the patient can discipline himself to remain in the sun (D'Arcy 1966). In other countries such compounds have sometimes been used as suntan accelerators, often with painful and disastrous results. Their use in conjunction with exposure of the patient to UVA (photochemotherapy, PUVA) in the treatment of psoriasis, mycosis fungoides, and certain other dermatoses has become standard treatment of selected patients in centres having access to the necessary equipment. Psoralens are known mutagens and possible carcinogens, and it is still not clear just what the long-term side-effects of this therapy will be. Many reports on side-effects have already been made; freckling, pruritus, nausea, dryness of the skin, burning sensations, and bullous eruptions usually of a phototoxic nature are well known.

A great volume of clinical reports on the subject of PUVA has now accumulated. Bickers (1981) considered the risk–benefit ratios in the treatment of psoriasis by PUVA with particular reference to the three potentially severe forms of risk, namely mutagenesis, cancer, and teratogenesis. The same author in 1983 further developed the same theme and warned that PUVA should only be used under strict conditions with accurate monitoring of the specialized equipment involved.

Hofmann *et al*. (1979) reported four patients in whom skin tumours developed during long-term treatment of psoriasis with PUVA. Two of the patients had a history of arsenic intake some years previously, but the histological appearances of their tumours did not seem typical for arsenic-induced carcinoma. Reshad *et al*. (1984) published a study of 216 patients treated for psoriasis over a seven-year-period. They found a total of eight patients who had between them, developed 25 skin carcinomata while on treatment. The patients had lesions of skin cancer in covered as well as exposed sites, and the authors suggested that a history of previous skin cancer, arsenic therapy, or radiotherapy are relative contraindications to PUVA therapy for psoriasis. This report is, to date, the first to have such a long follow-up period.

Parrish (1981) listed the skin disorders which may respond to PUVA and described the mechanisms of action involved. A four-year follow-up of 631 psoriatic patients treated with PUVA by Roenigk and Caro (1981) showed that 10 patients developed skin cancer, while eight developed actinic keratoses. Those with skin cancer were on average 14 years older than the other group. Other risk factors such as previous ionizing radiation, previous arsenic therapy, and previous skin cancer were also evaluated. The authors concluded that comparison with other studies may well indicate less risk of skin cancer than previously suggested. Another important study by Henseler *et al.* (1981) reviewed the treatment of 3175 patients in a multicentre study involving 18 European cities. The patients had severe psoriasis, and the study extended over 39 months. The impressive response of psoriasis to photochemotherapy was confirmed, and all side-effects were listed. Erythema (32.38 per cent), and pruritus (25.58 per cent) were the most common side-effects, while a Koebner effect was found in 1.96 per cent of the patients. The question of long-term side-effects of prolonged PUVA was discussed and suggestions concerning the reduction in such hazards were made.

Several papers have appeared recently on the question of melanocytic lesions and PUVA, including that of Rhodes *et al.* (1983). They reported 11 PUVA-induced pigmented macules in seven psoriatic patients 4 to 6 years after starting PUVA therapy, and compared them with sun-induced pigmented macules and five specimens of light-protected skin from control subjects. A significantly increased proportion of large melanocytes, some atypical, was found in the PUVA patients.

Marx *et al.* (1983) reported two patients receiving PUVA who developed malignant melanoma *in situ* approximately four years after starting therapy. The younger patient, a 34-year-old man developed a superficial spreading melanoma *in situ* seven months after stopping therapy.

An interesting report by Sina and Adrian (1983) described phototoxicity in two patients treated with PUVA for psoriasis. Both patients had received excessive doses of PUVA, and eventually multiple keratoacanthomas developed, 14 in one case, and nine in the other. After histological confirmation of the diagnosis in the first case, the PUVA was discontinued and the lesions disappeared within three months. No new lesions appeared during the following three years. In the other case, four lesions were surgically removed and one was interpreted as a well-differentiated squamous-cell carcinoma, while the others appeared to be keratoacanthomas. In this patient, several new lesions are still under follow-up, but the interest is that the lesions developed in the sites of the most marked phototoxic reactions, and the role of PUVA in the production of the lesions was discussed.

Another aspect of carcinogenicity was discussed by MacKie and Fitzsimons (1983) when they reported the development of multiple squamous- and basal-cell carcinomas in two patients receiving concomitant PUVA and intravenous boluses of methotrexate for severe psoriasis. In their editorial, the authors reviewed the carcinogenicity of these treatments given either separately or together. They mentioned the problem of mutations occurring in the lymphocytes of patients receiving PUVA, thus posing the question of the possibility of the development of a malignant lymphoid neoplasm in such patients, and discussed the results of animal studies, reviews of case control studies, and warned about the dangers of combination therapy as a routine practice.

Quinidine is known to cause photosensitivity in some individuals, and Lang (1983) reported the case of a 73-year-old man who developed an eruption in sun-exposed areas of his skin after taking quinidine for some time. Clinically he was thought to have a photodermatitis, and the cardiologists were reluctant to substitute another drug for his quinidine. It was only after two further episodes of serious generalized erythroderma requiring hospitalization, that procainamide was begun and the quinidine was stopped. The dermatitis cleared, and subsequent phototpatch tests to quinidine sulphate were found to be positive. Eight months later there had been no recurrence of the skin lesions. Yung *et al.* (1981) recorded two cases of photosensitivity which developed immediately after administration of the drug, dacarbazine, a widely used chemotherapeutic agent in the treatment of malignant melanoma. One of the patients was phototested, and a biopsy of a lesion showed appearances of a phototoxic reaction. It is suggested that patients under treatment with dacarbazine should be warned to avoid sun-exposure after therapy.

The major photosensitizing drugs include certain tetracyclines, nalidixic acid, chlorpromazine, chlordiazepoxide, thiazide diuretics, the vasodilator amiodarone (Chalmers *et al.* 1982), sulphonylurea compounds, phenothiazines, griseofulvin, and protriptyline. The cytostatic drugs may also induce a varying degree of photosensitivity, and occasional reports have implicated many other drugs such as diphenhydramine (Horio 1976), gold salts, oral contraceptives, and quinine. A case of sulphonamide-induced photosensitivity is shown in Plate 1c.

Pigmentation

Disorders of skin pigmentation are well recognized in certain pathological conditions such as Addison's disease, haemochromatosis, Peutz–Jegher's syndrome, pellagra, jaundice, malabsorption, neurofibromatosis, carcinomatosis, vitiligo (leucoderma), pinta, chronic renal failure, Addisonian anaemia, and Cushing's syndrome. However, it is probably true to state that, in terms of numbers of patients affected, drugs are responsible for the greater proportion of pigmentation disorders of skin seen in medical practice. The present section therefore deals with some of the known causal drugs, and applies in particular to effects seen in the white-skin races. The various iatrogenic 'contact' types of pigmentation abnormalities are not included since the cause-and-effect relationships are usually self-evident, while fixed drug eruptions and the effects of photosensitivity were already discussed earlier in this chapter.

Older remedies (arsenic, gold, and silver salts)

It is amazing to recall that arsenical compounds were used as recently as 25 years ago in the treatment of a variety of diseases ranging from 'senile heart disease', psoriasis, epilepsy, syphilis, and other central nervous system disorders like disseminated sclerosis, to its use as a general 'tonic'. Fortunately the numbers of people in Britain exposed to arsenicals have fallen dramatically, and the effects of arsenic are to be expected mainly in those whose occupation exposes them to the risk of unwitting absorption. Various skin manifestations of arsenic ingestion are known, and the amount of arsenical producing the effects varies considerably from individual to individual. Moreover, these effects may be recognized in the skin many years after exposure and vary from 'raindrop' pigmentation, hyperkeratosis, and melanosis of the palms of the hands and the soles of the feet, to keratotic lesions with the occasional production of skin carcinoma.

Gold salts are still in use today for the treatment of rheumatoid disorders, and a rare complication of such treatment is chrysiasis. This is characterized by a grey-blue to brown or purple pigmentation of the skin of exposed parts of the body due to the deposition of gold granules, mainly within macrophages and in blood-vessel walls in the upper dermis. There is also an increase in the melanin content of the epidermis and dermis. The effect is dose-related, and the pigmentation tends to be permanent.

The long-term ingestion or local application of silver compounds causes a bluish-grey, slate-coloured, or almost cyanotic discoloration of the skin, most manifest in the exposed regions and in the finger-nails. Histological examination of the skin in such cases shows an increase of melanin pigment in the basal layer of the epidermis and scattered through the papillary dermis. The silver granules tend to lie freely in the dermis, rather than within cells, and are best seen around the sweat glands and the pilosebaceous follicles. They can be seen on examination of routine stained sections, but show up brilliantly by the use of dark ground illumination.

Permanent blue-grey discoloration of the face and neck was reported by Macintyre and McLay (1978) in one case associated with the use of *Respaton*, an antismoking lozenge. This preparation contains silver acetate and ammonium chloride and is said to act by precipitation of insoluble silver chloride on the buccal mucosa producing an objectionable taste with tobacco smoke.

A somewhat similar clinical picture to that seen in argyria is met with as a result of prolonged bismuth administration. The 'bismuth line' (a dark margination of the gums) is seen, and in rare cases, jaundice has been recorded.

Amiodarone

This is an anti-arrhythmic agent used in cardiology and was the subject of an investigation by Chalmers *et al.* (1982). In addition to the photosensitivity previously referred to (see p. 764), they found that some patients developed a disfiguring pigmentation, usually of a slatey-blue type. Diffey *et al.* (1984) reviewed the photobiology of amiodarone, using *in-vitro* and *in-vivo* studies, and made the point that, although the drug had been available in Europe for about 15 years, the incidence of this pigmentation was low. Zachary *et al.* (1984), however, in a study of the pathogenesis of amiodarone-induced pigmentation and photosensitivity, used sophisticated biochemical, histological, and ultrastructural techniques. They were able to detect significant amounts of amiodarone and its major metabolite, desethylamiodarone, in the light-exposed skin, and found the concentrations there to be 10 times those in non-exposed skin. They also found lipofuscin granules within dermal macrophages, thus confirming the findings of earlier workers.

Antileprosy drugs

In recent years leprosy has been diagnosed in this country more frequently than it was a generation ago. Sulphones such as dapsone are used in the treatment of this condition, and may lead to the 'fifth-week dermatitis' from which may develop post-inflammatory melanosis in the more severely affected parts of the skin. It should be remembered that dapsone is sometimes used in the treatment of dermatitis herpetiformis, and the appearance of a slightly cyanotic colour due to methaemoglobin formation is not uncommon. Clofazimine is a lipid-soluble orange dye, and may be used in the treatment of leprosy as an alternative to dapsone; it frequently leads to the appearance of a reddish tint in the skin in light-exposed areas. Rifampicin, an important antibiotic used in the treatment of both leprosy and tuberculosis, can lead to the appearance of a yellow discoloration of the skin and mucous membranes.

Antimalarials

The development of a photodermatitis in a 60-year-old female six days after taking a second dose of the drug combination pyrimethamine–sulphadoxine (*Fansidar*) for malarial prophylaxis was reported by Olsen *et al.* (1982). Unaware of any association between the drugs and her symptoms, she took a third dose. As a result she developed fever, paraesthesiae, increase in the skin lesions on exposed sites, and eosinophilia, which lasted for about two weeks.

Antimalarial drugs, although not used widely as such in the United Kingdom, may be prescribed for other conditions, and on occasion may produce colour changes in the skin. The best known is the yellowish hue, generalized in distribution, seen so frequently in troops, stationed in malarial countries during the Second World War, due to long-term prophylactic use of the drug mepacrine (quinacrine) (see Plate 1d) a feature also associated on occasion with amodiaquine to a lesser extent. The skin coloration gradually fades on withdrawal of the drug concerned. However, an appreciable number of patients receiving those drugs for prolonged periods show a greenish-grey or bluish-black discoloration of the nail beds, nose, and ears. The 4-aminoquinolines, notably chloroquine and hydroxychloroquine, have now replaced the early antimalarials, and may also be used in higher dosage in the treatment of certain collagen diseases; these drugs too may produce colour changes in the skin, ranging from grey to blue to black. These changes are seen particularly on the face, the pretibial skin, and the nail beds, but pigmentation of the buccal mucosa may also occur. Levy (1982) reported 10 patients who developed generalized hyperpigmentation and pigmentation of the gums while taking a combination drug, pyrimethamine and chloroquine, as an antimalarial preventive measure.

Bluish discoloration of the nails is also seen as an occasional effect of chloroquine therapy.

Corticosteroids and ACTH

Long-term systemic administration of corticosteroids has been known to lead to lessening of normal skin pigmentation, while ACTH may occasionally cause the development of deepening skin pigmentation similar in appearance to that seen in Addison's disease.

Hydantoin anticonvulsants

A chloasma-like picture was reported in a series of patients treated with the hydantoin type of anticonvulsant drugs by Kuske and Krebs in 1964, who recorded 13 such cases, 10 of them being women.

Universal depigmentation occurring in a 10-year-old black girl was described by Smith and Burgdorf (1984). The girl developed a severe hypersensitivity reaction three weeks after starting treatment with the anticonvulsant drug, phenytoin. She developed an acute systemic illness, and the skin showed the appearances of toxic epidermal necrolysis, with bullous lesions which sloughed off over the entire body surface. She recovered from this, but five months after her discharge from hospital she displayed lack of skin colour and absence of hair. Both routine histological and ultrastructural studies were performed on a skin biopsy, and it was apparent that no melanocytes or melanosomes remained.

Methysergide

Methysergide, a drug used in the prophylaxis of migraine, gives rise to a number of side-effects even at normal dosage, among which is the development of a dermatitis reaction and thickened 'orange peel' reddened skin (Graham 1964).

Non-barbiturate hypnotics

Carbromal sometimes produces reddish-brown patches overlying petechial haemorrhages as part of a non-thrombocytopenic purpura syndrome. Such lesions are strikingly similar in appearance to those seen in dermatitis lichenoides purpura et pigmentosa. Glutethimide may occasionally give rise to both skin rashes and brownish staining of the skin in a patchy fashion after prolonged use. Similarly bromides when used in excess may lead to the appearance of a brown macular rash, principally on light-exposed surfaces.

Oral contraceptives

It is now well documented that oral contraceptive drugs may lead to the production of chloasma (Carruthers

1966: Resnik 1967), but why it should occur in some women and not in others remains a mystery. The distribution of this distressing brown pigmentation due to increased melanin deposition in the basal layer of the epidermis and the dermis is the same as the chloasma which sometimes occurs during pregnancy. The colour deepens with exposure to sunlight, and it appears that the incidence of this complication increases markedly after five years of medication. A disturbing feature of the condition is that at best only slight regression may occur after the oral contraceptive drug is stopped. There is as yet no effective treatment for this type of chloasma, and in some instances recourse has to be made to the use of camouflage cosmetics to allay the considerable self-consciousness caused in some women.

Phenacetin and iron (haemochromatosis)

The striking slatey-bluish-grey discoloration of the skin due to iron deposition in the tissues resulting from haemolytic anaemia may be seen in the abuse of phenacetin, an analgesic which is now restricted in its use. The combination of hypotension and haemo-siderosis which results may closely mimic Addison's disease (Messens 1965).

Iron itself, if administered for long periods of time to patients with haemolytic anaemia, may give rise to cirrhosis and hyperpigmentation of the skin (Pletcher *et al.* 1963). Blood transfusions given repeatedly over a period of time have likewise been held responsible for the appearance of a variable degree of pigmentation (Oliver 1959).

Phenolphthalein

This drug is taken regularly in a chocolate-based proprietary laxative preparation by many people, and the fact is often unknown to the physician, so that the appearance of distinctly dark grey patches on the skin, quite different from the fixed eruptions so commonly associated with phenolphthalein, may be very puzzling indeed (see Plate 1b).

Phenothiazines

It is more than 25 years since the major tranquillizer group of drugs known as the phenothiazines was introduced, and they continue to be used widely in psychiatry and in general medicine. Many adverse effects have been recorded during the use of these drugs and, apart from their photosensitizing effect, colour changes in the skin may be seen, particularly when the drugs are administered for prolonged periods of time. These changes are generally confined to light-exposed skin, and the precise mechanism of their production cannot yet be clearly defined.

Chlorpromazine, still the most widely prescribed, has been held responsible for hypermelanosis, bronzing of the skin, a grey hue, and even a purplish-grey appearance reminiscent of cyanosis. Histochemical studies have shown that the differing colours seen clinically are due largely to an increase in melanin pigment deposited at various levels in the skin, and electron microscopy (Zelickson 1965) has also revealed the presence of electron-dense bodies in dermal phagocytes and blood-vessel walls. These are believed to be related to degradation products of chlorpromazine, but the questions relating to pathogenesis remain unanswered. Oestrogenic-like effects, such as intense pigmentation of the breast areolae, have been noted during treatment with phenothiazines and, incidentally, also with griseofulvin (Durand *et al.* 1964).

It should be mentioned that phenothiazine derivatives and analogues have been incriminated in the production of jaundice of the cholestatic obstructive type.

Tetracyclines

Yellow pigmentation of the fingernails and toenails developed in a 26-year-old man after a month's treatment with tetracycline for acne in a total daily dose of 2 g in four divided doses (Hendricks 1980). The lunulae were affected and gave a yellow fluorescence on examination with a Wood's lamp. Treatment was discontinued after six months, and two months later, the nails had returned to normal. Three months later the acne worsened, and tetracycline therapy was recommenced in a dose of 25 mg four times daily. Within three months the nails, although normal in colour on this occasion, again gave a yellow fluorescence under Wood's lamp examination. It is suggested that Wood's lamp examination may help to distinguish tetracycline pigmentation from the yellow-nail syndrome and other causes of yellow nails. Another use could be the monitoring of patient compliance when using doses of tetracycline in excess of 1 g per day.

Vitamins

Vitamins are responsible for some effects on skin colour. High doses of vitamin A can result in signs of hypervitaminosis A. Skin changes occur early in the course of intoxication, and may indeed be the only noticeable changes in some patients. Dryness and scaling of the lips are early symptoms and the skin becomes dry, itchy, shows flaking hyperkeratosis, and a distinct tanning effect is seen. Such effects are more readily seen in children given high daily doses of the vitamin, but they are also seen in adults, and are quickly reversible as soon as the intake of vitamin A is stopped. Attempts to exploit the tanning effect of vitamin A by using high doses of vitamin A or retinoic acid have been made by unscrupulous commercial interests who have ignored the toxic effects of such high dosage.

β-carotene is given orally as a photoprotective agent in cases of erythropoietic protoporphyria and certain other light-sensitive states. Its use is associated with the development of a bright orange pigmentation of the skin.

Nicotinic acid, in addition to the expected flushing effect, can also cause a browning of the skin after lengthy administration. In rare but disturbing instances, the picture of acanthosis nigricans may develop (Elgart 1981), presenting a difficult differential diagnostic puzzle to the physician.

In this section it is probably appropriate to include the occurrence of a vitamin-deficiency disorder, namely pellagra, which may present as scaly, erythematous, hyperpigmented areas on the hands and face, and may be induced, particularly in undernourished individuals, by the use of isoniazid in the treatment of tuberculosis. This effect was reported by several workers including Harrington (1977).

Pityriasis rosea

Pityriasis rosea is a rare side-effect of drug therapy. Cases have been described in association with treatment with only eight drugs according to Wilkin and Kirkendall (1982). They give a list of the eight drugs in a report of a further two cases which occurred during treatment with the antihypertensive drug captopril. In one of the two patients, the rash cleared on reducing the dosage of the drug, while the other patient continued to exhibit the eruption when still taking the drug in the same dosage as before the rash appeared. The other drugs quoted and with references provided are arsenicals, bismuth compounds, gold, tripelennamine hydrochloride, methoxypromazine, barbiturates, and clonidine. However, Maize and Tomecki in 1977 reported a case of pityriasis rosea in association with the treatment of trichomoniasis with metronidazole, and metoprolol has also been incriminated.

Porphyria

In this condition of abnormal porphyrin metabolism (see also Chapter 6) several types are recognized as having a genetic basis, but acquired types of the disease are known and, in general, the classification remains unsatisfactory. Several types of hepatic porphyria have been described and many drugs have been blamed for initiating attacks of the disorder. In porphyria cutanea tarda, an acquired condition, the older literature contains many reports of its occurrence in alcoholics or after ingestion of barbiturates, and modern reports have added considerably to the list of agents responsible for provoking or exacerbating attacks. Among the stigmata of the condition are blistering erythemata of light-exposed areas, hirsutism, and deepening pigmentation.

The common drugs which have been recorded as capable of affecting porphyrin metabolism in some patients include sulphonamides, oral contraceptives (Behm and Unger 1974), griseofulvin, hydroxychloroquine (Baler 1976), and, of course, barbiturates. (For a comprehensive list of drugs see pp. 69–70.)

Lastly, patients having radiotherapy for malignant disease not uncommonly develop localized areas of deepening of normal skin pigmentation.

Pruritus (see also p. 774, leucopheresis)

Pruritus, local or generalized, is more often associated with drug therapy than is generally realized. Sometimes the precursor of overt skin lesions, the symptom may exist by itself, varying hourly or daily in intensity. Allergic reactions are the probable cause of pruritus. The neomycin-induced type of dermatitis is often preceded by diffuse redness, pruritus, then scaling. With streptomycin, skin hypersensitivity is usually characterized by intense itching and the reaction may sometimes go on to the development of generalized exfoliative dermatitis. With chloramphenicol, the symptoms of local hypersensitivity are itching at the site of application of the topical preparation and the development within six hours of a typical allergic contact dermatitis. If the patient subsequently takes an oral dose of chloramphenicol, then inflammation of that part of the skin previously exposed to the topical preparation usually appears. The penicillins, lincomycin, and the antibiotics mentioned above may all give rise to pruritus in certain patients. Codeine, arsenicals, bismuth, phenolphthalein, and gold salts are among the older remedies which may give rise to pruritus; the monoamine oxidase inhibitors, anticoagulants, oral contraceptives, oral hypoglycaemics, nicotinic acid, vitamin A and its derivatives, phenothiazine drugs, methyprylone, imipramine, dichloralphenazone and other phenazone derivatives, pentazocine, and indomethacin are some of the more modern drugs which may induce varying degrees of itch in patients taking them.

Pruritus is a toxic reaction commonly seen in Africans using oral chloroquine for the prevention and treatment of malaria. Spencer *et al.* (1982) made what appears to be the first report of this occurring in a European. The patient, a 23-year-old nurse, worked in a malarial part of Kenya and had used *Maloprim* as a prophylactic, albeit on an irregular basis. She developed what appeared to be an attack of malaria and was given oral chloroquine, the first dose being 600 mg, then 300 mg at 8, 24, and 48 hours. She awoke the morning after the first two doses with severe itch affecting her whole body, and most severe on the palms and the soles. The itching did not increase in severity after the 24 hour dose and lessened before the last dose. The total duration of the itch was about 55 hours. She had taken 600 mg of the drug several years previously for malaria and, on that occasion, suffered no itch. According to the authors, the reported incidence in Nigerians varies from 8 to 28 per cent of all those using the drug.

An interesting cause of fluoroderma resulting in a pruritic eruption in two patients was reported by Blask and Spencer (1979). The causative agent in both cases was an acidulated phosphate–fluoride gel used as a cariostatic agent during the course of radiotherapy for malignant disease.

Pruritus ani is a distressing and socially embarrasing condition which can be severe enough to interfere with sleep. Psychological factors undoubtedly play a great part in the initiation or continuance of the itch, and when the itch cycle is established, by whatever means, it tends to be perpetuated by the emotional attitude of the patient. Previous generations were well aware that drugs such as the opiates, belladonna, and the arsenicals could produce pruritus ani in some patients. Today, antibiotic-induced pruritus ani is probably the most common type associated with drugs, and can often be ascribed to alteration of the status of the normal bowel flora, either alone or in combined form.

The tetracyclines are still the prime offenders in this respect, and pruritic lesions of the vulva, vagina, and perianal regions are usually due to overgrowth of *Candida* species following suppression of the normal bowel flora by the drug. Often itch persists long after the drug is withdrawn, and the treatment given may further aggravate the condition, e.g., sensitivity to the bases of local preparations, local anaesthetics, or antihistamines used to relieve the itch.

Psoriasis

Psoriasis is not one of the commoner iatrogenic skin disorders and taken in isolation such an eruption is best described as 'psoriasiform'. In patients with pure psoriasis, however, attacks may be initiated or aggravated by a number of drugs including the salicylates, gold salts, antimalarial drugs, and iodides (Shelley 1967). An interesting report by Vickers and Sneddon (1963) described two patients who developed severe psoriasis following hypoparathyroidism produced by surgery, and who showed improvement after adjustment of the serum calcium in each case. Lowe and Ridgeway (1978) reported a previously stable psoriasis which was converted into generalized pustular psoriasis after three weeks of treatment with lithium carbonate for a manic-depressive illness. The pustular psoriasis resolved on cessation of the lithium but four years later when the drug was readministered, pustular psoriasis again developed. Palmoplantar pustular psoriasis occurring in a 49-year-old woman after six months of lithium therapy was reported by White (1982). Topical medications did not control the condition well, and remission only occurred after the lithium was stopped.

David *et al.* (1981) referred to the rarity of adverse skin reactions in relation to digoxin therapy. They reported a psoriasiform eruption in a 77-year-old male occurring during treatment with digoxin, furosemide (frusemide), chlorothiazide, methyldopa, and allopurinol. A biopsy showed a psoriasiform tissue reaction with a perivascular infiltrate of mononuclear cells; direct IF tests were negative. Macrophage migratory inhibition factor (MIF) assay was carried out, using all the drugs mentioned as antigens. Positive results were obtained with methyldopa and digoxin, and three patients who were taking these drugs acted as controls. In all of them the MIF tests were negative. On stopping these two drugs the eruption gradually disappeared; a month later he developed the same type of eruption on being given digoxin. He was not at the time receiving methyldopa, and the eruption again cleared when the digoxin was withdrawn. This appears to be the first report of a psoriasiform eruption due to digoxin.

Reshad *et al.* (1983) described two patients who developed generalized pustular psoriasis, one after taking phenylbutazone, and the other after taking oxyphenbutazone. One of the patients had been given three doses of oxyphenbutazone and had to be admitted to hospital with generalized pustular psoriasis; it was discovered that, when she had been given phenylbutazone seven years previously, a similar eruption had occurred. She had no further episodes of skin disease until she took the oxyphenbutazone. The authors discuss the role of reduction of prostaglandin synthesis and suggest that there is a pathophysiological mechanism to explain the worsening of psoriasis by such drug-induced reduction.

Among drugs which have been reputed to cause a worsening of psoriasis or the development of a psoriasiform rash apart from those already mentioned, are corticosteroids, gold, indomethacin (Katayama and Kawada 1981), practolol (see Plates 2a and 2b) (Cochrane *et al.* 1975), propranolol (Cochrane *et al.* 1976), and metoprolol (Neumann and van Joost 1981).

Purpura

The appearance of haemorrhage in the skin, varying from scattered, patchy, or generalized petechiae, to areas of bruising, constitutes the condition known as purpura. Many causes are known, but all grades of purpura may derive from the action of drugs commonly prescribed in both hospital and general practice. From a practical point of view in this connection, two types of purpura are recognized, thrombocytopenic, and non-thrombocytopenic (vascular) purpura. In both types drug-induced hypersensitivity reactions may be operative. Thrombocytopenic purpura may occur as a result of destruction of megakaryocytes, either directly and selectively, or as part of marrow hypoplasia. Chloramphenicol is well known in this respect, but similar reactions have been seen occasionally with the tetracyclines, sulphonamides, phenylbutazone, gold salts, immunosuppressant drugs, cytostatic drugs, D-penicillamine, and meprobamate among others. Thrombocytopeneic purpura may also be due to drug effects on the circulating platelets, and the classical example of this was with apronal (*Sedormid*), where the drug was shown to act as a hapten. Sensitization occurred in certain individuals when the drug combined with platelets to form the antigen, and an antigen–antibody reaction took place. The platelet antibodies formed produced rapid thrombocytopenia whenever the drug was readministered, and 'Sedormid-purpura' resulted. A very large number of drugs have now been reported as leading to thrombocytopenic purpura, and it seems likely that a similar type of mechanism is involved. Among the more important are quinidine, quinine, methyldopa, chloroquine, PAS, amitriptyline, diazoxide, digitoxin, frusemide, paracetamol, tolbutamide, barbiturates, and sulphonamides. It should be remembered that, while the purpura, usually of sudden onset, is seen in the skin of patients who have suffered the adverse reaction to the drug concerned, there is every likelihood that other organs are likewise affected and that mucous membranes will also be involved.

The vascular type of purpura produced by drugs initiating an allergic reaction in the vessel walls or as a toxic action may be seen in patients using barbiturates, ethambutol, indomethacin, salicylates, PAS, some tranquillizers, including meprobamate, sulphonamides, thiouracil, carbromal and bromvaletone, iodides, and occasionally the penicillins.

It is true to state that in any purpuric eruption, no matter which type, the possibility of a drug aetiology should be considered, and the list of drugs mentioned here can in no way be regarded as complete. Of particular interest to the general practitioner is the possibility of purpuric eruptions occurring in newly-born babies as a result of use by the mother of drugs such as quinidine and thiouracil. Lastly, the possibility of purpuric eruptions arising in patients due to the transfusion of stored blood should be borne in mind.

Stevens–Johnson syndrome

An erythemato-bullous eruption of the erythema multiforme type which may occasionally lead to a fatal outcome has been described by a number of workers in association with a number of drugs. The onset is acute with fever which persists for 10–14 days, myalgia and arthralgia, and a rash which varies considerably in appearance in the same patient, but which usually displays a few target lesions. There is oedema, confluent erosive lesions, and haemorrhagic bullae which lead to ulceration of the mucosal surfaces of the mouth, genitalia, and anus. The eyes may be affected and, in extreme cases, panophthalmitis can occur. The exact cause of this condition is unknown, and many cases are idiopathic, but an association with infections, pregnancy, food allergies, deep X-ray therapy, and certain types of drug therapy has long been suspected (Rostenberg and Fagelson 1965; Bianchine *et al.* 1968). The drugs especially implicated in this syndrome are sulphonamides (Bowell 1965), barbiturates, benoxaprofen (A.E.M. Taylor *et al.* 1981; Morgan and Behn 1981), chloramphenicol and other antibiotics, chlorpropamide (Tullet 1966), diphenylhydantoin (Bray 1959; Watts 1962), meprobamate, carbamazepine (Coombes 1965), pyrazolone derivatives (e.g. oxyphenbutazone, phenylbutazone), thiacetazone, sulindac (Levitt and Pearson 1980), and the antimalarial combination drug *Fansidar* (Olsen *et al.* 1982).

It should be emphasized that, while there may be considerable suspicion by the physician that a particular drug is responsible for the appearance of this syndrome, it is not possible in scientific terms to prove the relationship.

Toxic epidermal necrolysis (Lyell's disease)

Toxic epidermal necrolysis (TEN) is a serious and some-times fatal disorder which was given its name by Lyell in 1956. It has a striking clinical picture, initially resembling scalded skin which is exquisitely tender, and which subsequently undergoes desquamation of large areas, leaving a red, inflamed dermis (see Plate 1e). There is severe constitutional upset with pyrexia and electrolyte imbalance, presumably through loss of body fluids through the damaged skin. Unlike the lesions seen in scalding due to exogenous wet heat, healing occurs without residual scarring. It is now generally accepted that at least two different mechanisms are involved in the production of an almost identical cutaneous syndrome. One is predominantly seen in infants, and young children when it is almost invariably found in association with toxins derived from *Staphylococci* of phage type 71. The other is seen in adults and has been reported as an adverse effect of a great variety of drugs by a large number of authors over the years. The staphylococcal type is characterized by high epidermal cleavage; the adult drug type shows the full histological spectrum of the erythema multiforme tissue reaction. This is not surprising in that many cases show mucosal involvement reminiscent of that seen in the Stevens–Johnson syndrome and, of course, the same drugs may be implicated in both syndromes. Indeed, it may well be that these two syndromes are really variants of each other. Levitt and Pearson (1980) illustrated this very point in an account of a patient who exhibited signs of the Stevens–Johnson syndrome with features of TEN. The drug involved was sulindac, and the point was made that no details of the type or severity of possible cutaneous reactions were given in the manufacturer's advertising literature although it did indicate an incidence of 3 to 9 per cent of cutaneous reactions. Of four cases reviewed by Park *et al.* (1982), one died as a result of sulindac-induced TEN. They also reviewed eight other cases reported in the literature and showed clearly that the majority of cases displayed both skin and liver involvement. Kvasnicka *et al.* (1979) gave details of eight patients in whom evidence of disseminated intravascular coagulation was recorded during the course of TEN following administration of various drugs. One death occurred in the series, and it was suggested that alteration of haemostasis and interrelated biological systems such as activation of components of complement, kinins, and immunoglobulins may affect the outcome of TEN.

An authoritative account of TEN was given by Lyell (1979). He described in detail the current information on this dangerous and complex disorder, and gave a lucid guide to its diagnosis. Drug-induced TEN was given a detailed review, and the interesting thought that some cases could in fact be severe fixed drug eruptions was considered. Certain drugs have been more frequently held responsible for TEN than others, and prominent in the literature are reports on hydantoins, phenylbutazone, sulphonamides, antibiotics including penicillins, barbiturates, phenolphthalein, and cytostatic drugs; other drugs reported from time to time in association with TEN include allopurinol, acetylsalicylic acid, benoxaprofen, thiazide diuretics, nitrofuran derivatives, pentazocine, and dapsone (Lyell 1956; Jarkowski and Martmer 1962; Bailey *et al.* 1965; Carpentier 1965; Srivastava and Gour 1966).

Tuberose lesions

These can present great difficulty in diagnosis, although fortunately they tend to occur as multiple rather than as solitary lesions. The classical type is that produced by bromides in some patients, where lesions simulating tumours appear as warty growths, mostly located on the extremities. Their tumour-like appearance is due to a combination of papillomatosis and pseudo-epitheliomatous hyperplasia of the epidermis, often in association with some degree of secondary infection. Similar granulomatous lesions are seen occasionally as a result of an adverse reaction to iodides. The gingival hyperplasia (see Plate 1f) which occurs following the long-term administration of hydantoin derivatives in 40–60 per cent of patients at risk varies in degree and appears to be dose-related. Various studies have reported higher percentages of patients taking the drug succumbing to this effect on the gums, and there appear to be differences in population terms in the incidence of the condition. A report of gingival hypertrophy following the use of oral contraceptives was published by El Ashiry *et al.* in 1971. Also described in association with the use of oral contraceptives was the condition of acanthosis nigricans (Curth 1975). This latter condition has also been seen in relation to treatment with nicotinic acid as a rare adverse reaction (Elgart 1981).

Arsenic, notorious for its production of arsenical keratosis, can and does on occasion lead to the appearance of squamous-cell carcinoma of the epidermis. Such tumours usually appear many years after ingestion of the arsenical, and it is of interest that in such cases the arsenical content of the tumour can often be demonstrated.

What is fortunately a rare occurrence nowadays is the type of skin cancer occurring perhaps many years after radiotherapy. Keloid formation is seen from time to time after radiotherapy, but in this country is probably seen more in children as a late and rare effect of some vaccination procedures.

Within the past year, reports have appeared in which descriptions of excessive granulation tissue production have been made, and the phenomenon linked to the use of retinoid therapy for cystic acne and psoriasis. Campbell *et al.* (1983) gave details of eight such patients, and concluded that the response appears to be an idiosyncratic one which is not related to the daily dose or the total cumulative dose of the retinoid in question.

Finally, a condition known as pseudolymphoma where there is generalized lymphadenopathy, hepatosplenomegaly, and a widespread erythematous skin eruption simulating malignant lymphoid neoplasia both clinically and pathologically, has been known to occur in association with administration of phenytoin. A recorded case developed the syndrome after only six weeks of medication, and immunological studies suggested a state of immunologic hyper-responsiveness (Charlesworth 1977).

Urticaria

This is one of the most commonly seen adverse drug reactions in the skin, and consists of localized patchy, or generalized single or cropping, erythematous lesions which vary in appearance from the 'heat spot' type to large wheals. Sometimes the lesions are confluent, and on occasion they may proceed to actual blister formation. The lesions are usually intensely itchy, though the main complaint from some patients may be that of heat. In severe cases the patient may feel that his skin is about to burst. Taken together, the picture is essentially that of an acute hypersensitivity reaction which, particularly in the case of injected drugs, occurs very rapidly after the administration of the offending agent.

Some cases, mild-to-moderate in nature, respond well and quickly to oral antihistamines, but the more serious types which may produce dangerous angioneurotic oedema require immediate injection of adrenaline, antihistamine, or corticosteroid, followed by oral therapy as necessary. The anaphylactic nature of some of these reactions to drugs is such that life is threatened, and they must therefore be treated as emergencies. McCall and Cooper (1980) described an unusual anaphylactic reaction to tolmetin sodium. The patient took the drug because of an injured muscle and stopped therapy after two weeks. When his symptoms recurred he took another tablet, and within 30 minutes a severe anaphylactic reaction occurred necessitating admission to hospital. The

patient ascribed the reaction to aged cheese he had consumed just before taking the tolmetin, and two weeks later he again took tolmetin with the same result, this time within 15 minutes. On a lesser scale, Fallah-Sony and Figueredo (1979) described a generalized urticarial reaction to doxorubicin. The first report of an anaphylactoid reaction to N-acetylcysteine was by Walton *et al.* (1979), and is important as this drug is the treatment of choice in the early stage of paracetamol poisoning.

In 1968, Montgomery and Jackson listed 100 consecutive patients with drug reactions in their dermatological practice over a period of seven months, and urticaria headed that list with 28 patients, exceeding by 12 its nearest competitor.

The drugs in common use which lead to urticarial reactions are numerous, and include allopurinol, the barbiturates, salicylates, penicillin and many other antibiotics, sulphonamides, insulin, nalidixic acid, thyroid hormones, captopril, (Wilkin *et al.* 1980), oral contraceptives, indomethacin, some of the tranquillizing drugs, and dextran (Fothergill and Heaney 1976). Urticaria is also caused by food and drug additives, and cross-sensitization between drugs as well as additives may occur.

It should also be remembered that any vaccination procedure or desensitizing therapy for allergic disorders may bring about an allergic type of reaction, the common skin manifestation of which is a varying degree of urticaria. For this reason, suitable precautions should be observed and the availability of emergency treatment checked before the injection is given.

Vasculitis

In 1981, Peacock and Weatherall drew attention to the development of necrotizing vasculitis after long-term treatment with hydrallazine. The patient had been taking the drug in a dose of 75 mg per day for seven years and developed severe, widespread lesions of cutaneous vasculitis. The following year, Howitt *et al.* (1982) reported a case of warfarin-induced vasculitis in a 52-year-old female undergoing treatment for right external jugular-vein thrombosis. The patient developed tenderness and swelling of the right breast, with smaller lesions of the left flank and right elbow. These lesions gradually resolved over a period of some weeks but, five days after the warfarin was discontinued, thrombosis of the left external jugular vein occurred. Heparin was administered over the next four weeks, then warfarin was re-introduced in a daily dose of 3 mg. At the time of the report the patient had been receiving warfarin for five months without recurrence of the vasculitis. It was suggested that the reaction may be due to a direct toxic effect of the initial high dose through damage to the capillaries.

Spironolactone-induced vasculitis was reported in 1984 by Phillips and Williams. The patient, an 80-year-old man, developed a symmetrical purpuric rash on three occasions, 1–4 days after taking spironolactone; on each occasion the rash faded after the drug was withdrawn. The possibility that the reaction occurs due to the raised concentration of circulating immune complexes in the presence of normal complement concentrations was discussed, and this is believed to be the first report of vasculitis found in association with spironolactone treatment.

The development of necrotizing vasculitis together with nephritis in two patients after the intravascular injection of radiocontrast media, and the possible mechanisms involved were recently recorded by Kerdel *et al.* (1984). This appears to be the first report of such an incident.

In addition to the above drugs, reports have been made on a large number of other drugs; these include the sulphonamides, hydantoins, indomethacin, the thiazides, the pyrazolones, cimetidine, naproxen, corticosteroids, allopurinol, guanethidine, bromides, and amitriptyline.

Miscellaneous adverse effects

Allergic reactions and human insulin

With the introduction recently of human insulin it was hoped that the side-effects of porcine insulin, in particular, allergy, would be minimal. There have been several reports, however, of adverse reactions to human insulin, including that by Altman *et al.* (1983). They reported three patients who showed varying degrees of allergic-type responses during the use of human insulin, including swelling at the injection site, dyspnoea, swelling of hands, feet, and ears, with a persistent rash. The authors concluded that desensitization protocols with highly purified porcine insulin have generally proved effective and that, in patients with insulin allergy, change to human insulin is not likely to be the answer.

Anticoagulants and skin necrosis

Several reports of skin necrosis occurring during the course of anticoagulant therapy have been recorded. Scandling and Walker (1980) described a 66-year-old woman who was given a single dose of warfarin sodium (20 mg) for suspected pulmonary embolism. About three days later she developed painful purpuric lesions on the nose and extremities, leading eventually to debridement of the nose and to amputation of the right leg, the right arm, and later still, the left leg because of gangrene. The patient died on the 62nd day of her illness.

Schleicher and Fricker (1980) described a 75-year-old female who developed necrosis of the skin soon after injection of coumarin, with similar lesions resulting from inadvertent rechallenging. They pointed out that the incidence of such effects is rare considering the widespread use of this anticoagulant.

The eighth case in the literature of the 'purple toe syndrome' resulting from oral anticoagulants was reported by Akle and Joiner in 1981. A 66-year-old woman developed pain and discoloration of her toes within a month of starting treatment. The symptoms improved markedly on stopping warfarin and being given heparin instead. Four years later asymptomatic residual discoloration of the toes was still present. This rare effect of oral anticoagulant therapy was the first such to be reported in the UK.

Yet another report of warfarin-induced skin necrosis was by Slutzki *et al.* (1984). They described necrosis of the leg occurring in a 16-year-old girl after a single dose of the drug, and a 59-year-old woman also developed gangrene of most of her left breast, four days after anticoagulant therapy had been started. The authors agreed that the basic pathogenesis of the necrosis remains obscure, and made the point that once initiated, the clinical course of the necrosis seems to be unaffected by withdrawal or continuation of the drug, or indeed by administration of treatment such as hypothermia, vasodilators, nerve blocks, or vitamin K.

Jackson and Pollock (1981) described a 79-year-old man admitted for surgical treatment of an ischaemic foot. He was given 5000 units of porcine sodium heparin subcutaneously into the abdominal wall before femoral arteriography, 10 000 units intravenously five days later during exploration of the left femoral artery, and three further doses of 5000 units subcutaneously on the second and third days after operation. Areas of skin necrosis were noted at the sites of the three post-operative heparin injections by the fourth day. Histological examination showed acute haemorrhagic and necrotizing vasulitis affecting the small dermal blood vessels and extending into the subcutaneous fat. The findings suggested a hypersensitivity reaction to heparin, or possibly to the preservative, and the presence of fibrinogen, IgG, and complement (C3) in the walls of affected blood vessels could be demonstrated by immunoperoxidase techniques. The subsequent clinical course of the patient was not recorded.

Penicillamine reactions (see also p. 756)

Penicillamine was regarded as inducing a 'dermolytic dermatosis' in a patient with Wilson's disease by Bardach and Gebhart (1981). A 29-year-old female developed the dermatosis after two years of treatment with penicillamine. The lesions occurred on skin areas exposed to trauma, and the report contains light- and electron-microscope studies which show alteration to connective tissue.

A comprehensive study of the ultrastructure of penicillamine-induced skin lesions in three patients was provided by Hashimoto *et al.* (1981). Two cases of Wilson's disease and one of cystinuria were treated with penicillamine for 13 years, 10 years, and 20 years, respectively. Two patients developed haemorrhagic skin lesions, while the third developed elastosis perforans serpiginosa-like lesions. In all three the most striking finding was that of marked variation in the thickness of individual collagen fibres, but with normal banding patterns and periodicity. This report was well illustrated and the discussion reviewed the present state of knowledge of elastin biochemistry and the ultrastructure of the lesions with suggestions for the mechanisms whereby these are produced.

Leucopheresis

A wholly unexpected and unpleasant result of volunteering to submit to leucopheresis was reported by Parker *et al.* (1982) in four subjects, three of whom were authors of the communication. Hetastarch (2-hydroxyethyl starch) is widely used as a sedimenting agent to increase the yield of granulocytes during leucopheresis. The four healthy volunteers on separate occasions, and in different establishments, suffered intractable pruritus after the procedure, causing considerable discomfort and lasting for three to six months. None of the four had any history of allergy, and the itching developed about two weeks after the leucopheresis, with no visible skin changes. On investigation, it appears that this undesirable effect only occurs in those donors who have received more than 1 litre of hetastarch within a short time.

The development of graft-versus-host disease (GVHD) due to transfusion of leucocytes was described by Tolbert *et al.* (1983). The patient was a 30-year-old female who was immunosuppressed because of a renal transplant. Two weeks after transfusion of one unit of leucocytes per day for five days she developed a maculopapular rash over most of the body which, during the first week, became intensely itchy. A biopsy showed the features of GVHD, and she was given oral corticosteroids with improvement in the rash and the gastrointestinal symp-

toms which had also developed. However, she developed neutropenia, then pancytopenia, and the patient died of sepsis, despite intense therapy. The authors comment that while treatment of this condition in such circumstances is unsatisfactory, it can be prevented by irradiation of blood products prior to administration.

Retinoids and oedema

An unusual adverse reaction to the retinoid Ro 10-9359 (*Tigason*) was reported from Greece by Moulopoulou-Karakitsou *et al.* (1981). It occurred in a female under treatment for severe psoriasis on a dose of 70 mg daily for 20 days on three different occasions during a six-month period. On each occasion, after 14 days of treatment, the patient developed generalized oedema which started on the face, then spread rapidly to the rest of the body. All laboratory tests were normal and, when the drug was stopped on each occasion, the oedema disappeared within seven days. This appears to be the first report of oedema in association with this drug.

Panniculitis and procaine povidone

Kossard *et al.* (1980) described an unusual reaction in a 60-year-old female who had been given intramuscular injections of procaine povidone for seven years at weekly intervals on account of severe angina pectoris. The dose of 8–10 ml was given into the arms or buttocks, and, in all, it was estimated that more than 3 l of the drug had been administered during the course of the seven years. The injections were discontinued when symptoms and signs of panniculitis developed at injection sites, during which episodes the patient became pyrexial. Full laboratory findings were given with the histological and immunofluorescence examinations reported in detail.

To end this section on miscellaneous reactions it seems appropriate to refer the reader to the paper by Schmoeckel *et al.* (1979), in which the treatment of alopecia areata by anthralin (dithranol)-induced dermatitis was described — surely a virtue of expediency!

The therapeutic rose

'There is no therapeutic rose which does not have its thorn' is a quotation familiar to the students of Sir Derrick Dunlop. The skin changes resulting from the administration of drugs offer visual confirmation of this remark and present frequent difficulties in diagnostic decision-making. The history of drug ingestion should also be a routine part of the medical consultation, bearing in mind that what are regarded as drugs by the doctor may not be so regarded by the patient. Very few skin patterns induced by drugs can be accepted as diagnostic, and the absence of reliable laboratory tests adds to the difficulties, while the provocation test may carry grave danger to the patient. Drug interactions must be taken into consideration before a conclusion is reached, and it should also be borne in mind that many of the bizarre eruptions caused by drugs may themselves be mimicked by some of the skin markers of malignancy. Thus the diagnosis of 'drug rash' is not to be made lightly and care must be taken to exclude other possible causes. As knowledge of drug metabolism and its relationship with immunological and genetic influences grows, hopefully more precise methods of identifying or even predicting drug eruptions will emerge. When the thalidomide tragedy became understood, many lessons were learned about methods of anticipating possible side-effects of drugs in terms of teratogenicity.

The practolol story (*British Medical Journal* 1975) has thrown more light on certain aspects of immunological findings in drug-induced disorders (Amos *et al.* 1975; Behan *et al.* 1976); this is discussed on p. 690. In May 1974, Felix *et al.* reported rashes resembling those of eczema, lupus erythematosus, lichen planus, and a highly characteristic toxic erythematous psoriasiform eruption in a total of 21 patients taking practolol. In June 1974, and in subsequent reports, Wright (1974, 1975) reported the occurrence of eye changes (see also pp. 689–91) and psoriasiform skin lesions with practolol. In December 1974, Brown *et al.* published reports on three patients who developed sclerosing peritonitis while on long-term practolol therapy. Of considerable concern with this latter manifestation is that it was found to develop in some patients several weeks after the drug had been stopped (Marshall *et al.* 1977).

Since these initial reports, many others have confirmed the oculomucocutaneous manifestations of the 'practolol syndrome' (see Plates 2a–f, 3a, b). This unusual series of iatrogenic disorders attributable to practolol could not have been predicted; indeed it took several years before the seriousness of these reactions were realized. Nevertheless, the results of continuing investigation of patients still affected by the adverse effects of this drug, which has been withdrawn in some countries with limited use only permitted in others, suggest new ways in which new drugs can be monitored. In doing so it is hoped that adverse effects of drugs, not only on the skin but on other body organs, will in the future be held at an acceptable level.

The benoxaprofen affair was another example of a drug producing multisystem effects. This non-steroidal, anti-inflammatory drug used in the treatment of rheumatoid arthritis and osteoarthritis was hailed as a 'break-through' when it was launched in the UK in 1980

under the trade name *Opren*. It was the subject of more than 3000 reports of adverse reactions to the Committee on Safety of Medicines before suspension of the product licences took place at the beginning of August 1982. The product had been marketed for more than a year in Europe, and had only recently been introduced in the USA. Many of the reports of adverse reactions referred to the skin, and an early report was contained in a letter to the *British Medical Journal* by Fenton *et al.* (1981). They had noted the development of photosensitization which was not due to UVA alone, but occurred in the UVB range as well. They established this by performing provocation tests in patients with known photosensitivity to benoxaprofen. The manufacturer's literature referred to the possibility of mild photosensitive skin reactions as well as mild-to-moderate areas of separation of the nail from the nail bed, but indicated that such reactions would not normally interfere with benoxaprofen therapy. They warned against exposure to bright sunlight unless using ultraviolet screens, and warned against the use of ultraviolet and infra-red lamps during treatment with the drug. Nevertheless, a flood of reports began to appear, indicating that a wide variety of skin changes could derive from treatment with benoxaprofen. These included the Stevens–Johnson syndrome (A.E.M. Taylor *et al.* 1981), erythroderma, bulla formation, and conjunctivitis (Morgan and Behn 1981), severe bullous dermatitis as a phototoxic reaction on exposed sites (Vivier 1982), eruptive tumours on sun-exposed skin (Finlay and Hull 1982), toxic epidermal necrolysis (Fenton and English 1982), jaundice (Goudie *et al.* 1982), and, later in the same year, Taggart and Alderdice reported fatal cholestatic jaundice in elderly patients. Hindson *et al.* (1982a), in a survey of cutaneous side-effects they had encountered, listed photosensitivity, erythema multiforme, Stevens–Johnson syndrome, milia, onycholysis, and toxic epidermal necrolysis. In the same journal, Halsey and Cardoe (1982) described the side-effects of benoxaprofen they had observed in 300 patients attending a rheumatology clinic; cutaneous side-effects accounted for 69.5 per cent of all 259 side-effects reported. The most common side-effect was photosensitivity, followed by onycholysis, an effect which increased in incidence in patients over the age of 70 years. They also found pruritus, hypertrichosis, milia, and a variety of mild non-specific rashes. Once again the unpredictability of adverse effects of a new drug had been demonstrated, but at the same time, unexpected therapeutic effects also came to light. A number of reports appeared which claimed a beneficial effect in the treatment of psoriasis. Two letters to the *British Medical Journal* (Gordon 1982; Kingston and Marks 1982) referred to improvement in psoriasis during treatment with the drug. The letter by Kingston and Marks gave details of an open trial on 19 patients which continued for eight weeks; 12 patients, with biopsy evidence before and after treatment, showed improvement. Another short report by Allen and Littlewood (1982) was of a small pilot study on 13 patients, which continued for 4–8 weeks with a dose of 600 mg benoxaprofen daily, the ages of the patients varying from 21 to 71 years. Four patients showed complete clearance of their psoriasis, four were greatly improved, and five were unchanged. The authors suggested that possibly the drug acts by inhibiting the migration of leucocytes into the skin.

Hindson *et al.* (1982*a*) reported a 60-year-old female under their care for nodular prurigo who suddenly improved when she was given benoxaprofen for osteoarthritis by her family doctor. Hindson *et al.* then treated a 55-year-old man who had nodular prurigo affecting both legs for seven years, with 600 mg of benoxaprofen daily for two months. The lesions became flat, and healing took place with some atrophy and scarring. A biopsy was obtained before treatment, and another after one month's treatment, showing marked improvement in the second biopsy.

Yet another therapeutic use of benoxaprofen in skin disorders was reported by Hindson *et al.* (1982*b*) in the treatment of nodular acne, a form of acne much less responsive to conventional topical and systemic therapy. It was suggested to the authors by Dr A. Kligman that, because of the macrophage-inhibiting action of benoxaprofen, it might possibly play a useful role in the treatment of nodular acne. In all, 17 males were treated in a dose of 600 mg of benoxaprofen daily for one month as sole therapy. This was then replaced by tetracyclines or erythromycin, and, after a further month, any patient with new or still active nodules was given another course of 600 mg of benoxaprofen daily for a month. Marked flattening of nodules occurred in all cases after one month, and 10 cases showed regression of nodules to atrophic, depressed areas after the second course of benoxaprofen. Photomicrographs of a nodule biopsied before treatment, and of a similar lesion from the same patient after one month's treatment with benoxaprofen show impressive differences. A detailed review of benoxaprofen and the skin was given by Allen (1983) in which he discussed the role of prostaglandins, the effect on lipoxygenase activity, and other properties of the drug, including its possible use in Behçet's syndrome.

It would seem, therefore, that the last chapter in the benoxaprofen affair remains to be written, and clearly there are many lessons still to be learned from it.

RECOMMENDED FURTHER READING

AMOS, H.E. (1976). *Allergic drug reactions*. Arnold, London.
ARNDT, K.A. AND JICK, H. (1976). Rates of custaneous reactions to drugs. A report from the Boston Collaborative Drug Surveillance Program. *J.Am. med. Ass.* **235**, 918–23.
BRUINSMA, W. (1982). *A guide to drug eruptions*. Excerpta Medica, Amsterdam.
DUKES, M.N.G. (Ed.) (1984). *Meyler's side-effects of drugs*, 10th edn., Elsevier Amsterdam-New York-Oxford.
JACKSON, R. (1976). Systemic drug rashes; pitfalls in diagnosis and treatment. *Cutis* **17**, 386–8.
MOORE, D.E. (1977). Photosensitization by drugs. *J. Pharm. Sci.* **66**, 1282–4.
WINTROUB, B. U. AND STERN, R.S. (1984). Cutaneous drug reactions. In *Current perspectives in immunodermatology* (ed. R.M. Mackie), pp. 75–88. Churchill Livingstone, Edinburgh.

REFERENCES

AKERS, W.A. (1980). Risks of unoccluded topical steroids in clinical trials. *Arch. Dermatol.* **116**, 786–8.
AKLE, C.A. AND JOINER, C.L. (1981). Purple toe syndrome. *J.R. Soc. Med.* **74**, 219.
ALLEN, B.R. (1983). Benoxaprofen and the skin. *Br.J. Dermatol.* **109**, 361–4.
—— AND LITTLEWOOD, S.M. (1982). Benoxaprofen: effect on cutaneous lesions in psoriasis. *Br. med. J.* **285**, 1241.
ALMEYDA, J. AND LEVANTINE, A., (1971). Lichenoid drug eruptions. *Br. J. Dermatol.* **85**, 604–7.
ALTMAN, J.J., PEHUET, M., SLAMA, G., AND TCHOBROUTSKY, C. (1983). Three cases of allergic reaction to human insulin. *Lancet* ii, 524.
AMOS, H.H., BRIDGEN, W.D., AND McKERRON, R.A. (1975). Untoward effects assoicated with practolol: Demonstration of antibody binding to epithelial tissue. *Br. med. J.* **1**, 598–600.
AUGUST, P.J. (1980). Iatrogenic skin disease. *Practitioner* **224**, 471–8.
BAGNATO, A. AND NIGRO, M. (1974). Iatrogenic lupus erythematosus in the course of treatment with rifampicin and isoniazid. *G. Ital. Mal. Torace* **28**, 183–6.
BAILEY, G., ROSENBAUM, J.M., AND ANDERSON, B. (1965). Toxic epidermal necrolysis. *J. Am. med. Ass.* **191**, 979–82.
BAILIN, P.L. AND MATKALUK, R.M. (1982). Cutaneous reactions to rheumatological drugs. *Clin. rheum. Dis.* **8**, 493–516.
BAKER, H. AND MOORE-ROBINSON, M. (1970). Cutaneous responses to aspirin and its derivatives. *Br. J. Dermatol,* **82**, 319–20.
BALER, G.R. (1976). Porphyria preciptated by hydroxychloroquine treatment of systemic lupus erythematosus. *Cutis* **17**, 96–8.
BARDACH, H. AND GEBHART, W. (1981). Penicillamine-induced dermolytic dermatosis in a patient with Wilson's disease. *Dermatologica, Basle* **142**, 473–83.
BECK, M.H. AND PORTNOY, B. (1979). Severe erythema multiforme complicated by fatal gastro-intestinal involvement following co-trimoxazole therapy. *J. clin. exp. Dermatol.* **4**, 201–4.
BEELEY. L. (1984). Allergy to penicillin. *Br. med. J.* **288**, 511–12.
BEHAN, P.O., BEHAN, W.M.H., ZACHARIAS, F.J., AND NICHOLLS, J.T. (1976). Immunological abnormalities in patients who had the

oculomucocutaneous syndrome associated with practolol therapy. *Lancet* **ii**, 984–7.

BEHM, A.R. AND UNGER, W.P. (1974). Oral contraceptives and porphyria cutanea tarda. *Can. med. Ass. J.* **110**, 1052–4.

BENSAID, J., ALDIGIER, J.C., AND GUALDE, N. (1979) Systemic lupus erythematosus syndrome induced by pindolol. *Br. med. J.* **1**, 1603–4.

BIANCHINE, J.R., MACARAEG, P.V.J., JR., LASAGNA, L., AZARNOFF, D.L., BRUNK, S.F., HVIDBERG, E.G., AND OWEN, J.A., JR. (1968). Drugs as etiologic factors in the Stevens–Johnson syndrome. *Am. J. Med.* **44**, 390–405.

BICKERS, D.R. (1981). Photochemotherapy of psoriasis; risk: benefit ratios. *J. Am. Acad. Dermatol.* **4**, 90–3.

—— (1983). Position paper — PUVA therapy. *J. Am. Acad. Dermatol.* **8**, 265–70.

BLASK, L.G. AND SPENCER, S.K. (1979). Fluoderma. *Arch. Dermatol.* **115**, 1334–5.

BOEKHORST, J. C. (1983). Allergic contact dermatitis with transdermal clonidine. *Lancet* **ii**, 103–12.

BOWELL, G.R. (1965). Stevens–Johnson syndrome and long-acting sulphonamides. *Aust. dent. J.* **10**, 85.

BRAY, P.F. (1959). Diphenylhydantoin (dilantin) after 20 years, a review with re-emphasis by treatment of 84 patients. *Pediatrics* **23**, 151–61.

BRITISH MEDICAL JOURNAL (1975). Leading article: Side effects of practolol. *Br. med. J.* **2**, 577–8.

—— (1981). Leading article: Drug-induced bullous eruptions. *Br. med. J.* **282**, 421–2.

BRONNER, A.K. AND HOOD A.F. (1983). Cutaneous complications of chemotherapeutic agents. *J. Am. Acad. Dermatol.* **9**, 645–63.

BROWN, P., BADDELEY, H., READ, A.E., DAVIES, J.D., AND McGARRY, J. (1974). Sclerosing peritonitis, and unusual reaction to a beta-adrenergic blocking drug (Practolol). *Lancet* **ii**, 1477–81.

BURRY, J.N. AND CROSBY, R.W.L. (1966). A case of phototoxicity to nalidixic acid. *Med. J. Aust.* **2**, 698–700.

BURTON, J.L. SCHUTT, W.M., AND CALDWELL, I.W. (1975). Hypertrichosis due to diazoxide. *Br. J. Dermatol.* **93**, 707–11.

CAMPBELL, J.P., GREKIN, R.C., ELLIS, C.N., MATSUDA-JOHN, S.S., SWANSON, N.A., AND VOORHEES, J.J. (1983). Retinoid therapy is associated with excess granulation tissue responses. *J. Am. Acad. Dermatol.* **9**, 708–13.

CARO, I. (1980). Discoloration of the teeth related to minocycline therapy for acne. *J. Am. Acad. Dermatol.* **3**, 317–18.

CARPENTIER, E. (1965). A case of Lyell's syndrome. *Arch. belges Derm.* **21**, 363–5.

CARRUTHERS, R. (1966). Chloasma and oral contraceptives. *Med. J. Aust.* **2**, 17–20.

CHALMERS, R.J.G., HAYDN, L.M., SRINIVAS, V., AND BENNETT, D.H. (1982). High incidence of amiodarone induced photosensitivity in North-West England. *Br. med. J.* **285**, 341.

CHARLESWORTH, E.N. (1977). Phenytoin induced pseudolymphoma syndrome. *Arch. Dermatol.* **113**, 477–80.

CHIN, T.M. AND FELLNER, M.J. (1980). Allergic hypersensitivity to lidocaine. *Int. J. Dermatol.* **19**, 147–8.

COCHRANE, R.E.L., THOMSON, J. BEAVERS, D.G., AND McQUEEN, A. (1976). Skin reactions associated with propranolol. *Arch. Dermatol.* **112**, 1173–4.

——, ——, FLEMING, K., AND McQUEEN, A. (1975). The psoriasiform eruption induced by practolol. *J. cutan. Pathol.* **2**, 314–19.

COHEN, L. (1984). Drug eruption secondary to trazodone: A recently introduced antidepressant. *J. Am. Acad. Dermatol.* **10**, 303–5.

COOMBES, B.W. (1965). Stevens–Johnson syndrome associated with carbamazepine *(Tegretol)*. *Med. J. Aust.* **1**, 895–6.

COOMBS, R.R.A. AND GELL, P.G.H. (1975). Classification of allergic reactions responsible for clinical hypersensitivity and disease. In *Clinical aspects of immunology*, 3rd edn. (ed. P.G.H. Gell, R.R.A. Coombs, and P.J. Lachmann), pp. 761–81. Blackwell Scientific Publications, Oxford.

COPEMAN, P.W.M. AND SCRIVENER, R. (1977). Amoxycillin rash. *Br. med. J.* **1**, 1354.

COSKEY, R.J. (1982). Fixed drug eruption from chlorphenesin carbamate. *J. Am. med. Ass.* **248**, 30–1.

—— AND BRYAN, H.G. (1975). Fixed drug eruption due to penicillin. *Arch. Dermatol.* **111**, 791–2.

CRIVELLATO, E. (1982). A rosacea-like eruption induced by Tigason (Ro 10-9359). *Acta Dermatol. Venereol.* **62**, 450–2.

CUNLIFFE, W.J. AND MILLER, A.J. (Eds.) (1984). *Retinoid therapy*. MTP Press Limited, Lancaster, Boston, The Hague, Dordrecht.

CURTH, H.O. (1975). Acanthosis nigricans following use of oral contraceptives. *Arch. Dermatol.* **111**, 1069.

DANIEL, C.R. AND SCHER, R.K. (1984). Nail changes secondary to systemic drugs or ingestants. *J. Am. Acad. Dermatol.* **10**, 250–8.

D'ARCY, P.F. (1965*a*). The pharmacological basis of drug treatment in dermatology. *Pharm. J.* **194**, 637–43.

—— (1965*b*). How drugs act. *J. mond. Pharm., La Haye* **2–3**, 79–92.

—— (1966). The sun and the skin. *Pharm. J.* **196**, 477–81.

DARLINGTON, L.G. (1974). Erythema nodosum and oral contraceptives. *Br. J. Dermatol.* **90**, 209–12.

DAVID, M., LIVNI, E., STERN, E., FEUERMAN, E.J., AND GRINBLATT, J. (1981). Psoriasiform eruption induced by digoxin: confirmed by re-exposure. *J. Am. Acad. Dermatol.* **5**, 702–3.

DAWES, P.T. AND SHADFORTH, M.F. (1984). Sulphasalazine induced oral lichen planus. *Br. med. J.* **288**, 194.

DEGOS, R., TOURAINE, R.M., BELAICH, S., AND RUVUZ, J. (1969). Pemphigus chez un malade traite par penicillamine pour maladie de Wilson. *Bull. Soc. Fr. Dermatol.* **76**, 751–3.

DERBES, V.J. (1964). The fixed eruption. *J. Am. med. Ass.* **190**, 765–6.

DIFFEY, B.L., CHALMERS, R.J.G., AND MUSTON, H.L. (1984). Photobiology of amiodarone: preliminary in vitro and in vivo studies. *Clin. exp. Dermatol.* **9**, 248–55.

DOBSON, R.L. (1979). *Year book of dermatology*. Year Book Medical Publications, Chicago.

DOWNHAM, T.F. (1978). Spironolactone induced lichen planus. *J. Am. med. Ass.* **240**, 1138.

DURAND, P., BORRONE, C., SCARABICCHI, S., AND RASSI, A. (1964). Hyperpigmentation of breast areolae and external genitals with gynaecomastia following griseofulvin treatment. *Minerva med.* **55**, 2422–5.

EL ASHIRY, G.M., AL KAFRAWY, A.H., NASR, M.F., AND YOUNIS, N. (1971). Effects of oral contraceptives on the gingiva. *J. Periodont.* **42**, 273–5.

ELGART, M.L. (1981). Acanthosis nigricans and nicotinic acid. *J. Am. Acad. Dermatol.* **5**, 709–10.

ELIAS, P.M. AND WILLIAMS, M.L. (1981). Retinoids, cancer and the skin. *Arch. Dermatol.* **117**, 160–80.

FALLAH-SONY, E. AND FIGUEREDO, A.T. (1979). Generalized urticarial reaction to doxorubicin. *J. Am. med. Ass.* **241**, 1108–9.

FELIX, R.H. AND COMAISH, J.S. (1974). The value of patch and other skin tests in drug eruptions. *Lancet* i, 1017–19.

——, IVE, F.A., AND DAHL, M.G.C. (1974). Cutaneous and ocular reactions to practolol. Br. med. J. **4**, 321–4.

FELLNER, M.J. AND BAER, R.L. (1965). Cutaneous reactions to drugs: With particular reference to penicillin sensitivity. *Med. Clins. N. Am.* **49**, 709–24.

—— AND KATZ, J.M. (1976). Occurrence of bullous pemphigoid after furosemide therapy. *Arch. Dermatol.* **112**, 75–7.

FENSKE, N., MILNS J., AND GREER, K. (1980). Minocycline-induced pigmentation at sites of cutaneous inflammation. *J. Am. med. Ass.* **244**, 1103–4.

FENTON, D.A., AND ENGLISH, J.S. (1982). Toxic epidermal necrolysis: a further complication of benoxaprofen therapy. *J. exp. Dermatol.* **7**, 277–80.

——, WILKINSON, J.D., AND ENGLISH, J.S. (1981). Photosensitisation to benoxaprofen not due to ultraviolet A alone. *Lancet* ii, 1230–1.

FINDLAY, G.H. AND HULL, P.R. (1982). Eruptive tumours on sun-exposed skin after benoxaprofen. *Lancet* ii, 95.

FINLAY, A. (1983). An illustrated guide to identifying drug eruptions. *Modern med.* **18**, 66–75.

FISHER, A.A. (1976). Antihistamine dermatitis. *Cutis* 18, 329–36.

FOTHERGILL, R. AND HEANEY, G.A. (1976). Reactions to dextran. *Br. med. J.* **2**, 1502.

GIBSON, J.R. (1982). Recurrent trimethoprim-associated fixed skin eruption. *Br. med. J.* **284**, 1529–30.

GODDEN, D.J. AND McPHIE, J.L. (1983). Bullous skin eruption associated with carbamazepine overdosage. *Postgrad. med. J.* **59**, 336–7.

GORDON, M.J. (1982). Benoxaprofen: effect on cutaneous lesions in psoriasis. *Br. med. J.* **285**, 1741.

GOUDIE, B.M. BIRNIE, G.F., WATKINSON, G., McSWEEN, R.N.M., KISSEN, L.H. AND CUNNINGHAM, N.E. (1982). Jaundice associated with the use of benoxaprofen. *Lancet* i, 959.

GRAHAM, J.R.S. (1964). Methysergide for prevention of headache. *New Engl. J. Med.* **270**, 67–72.

GRANGE, R.W. AND WILSON-JONES, E. (1978). Bullous lichen planus caused by labetalol. *Br. med. J.* **1**, 816.

GRANSTEIN, R.D. AND SOBER, A.J. (1981). Drug and heavy metal-induced hyperpigmentation. *J.Am. Acad. Dermatol.* **5**, 1–18.

GREENWOOD, R., FENWICK, P.B.C., AND CUNLIFFE, W.J. (1983). Acne and anticonvulsants. *Br. med. J.* **287**, 1669–70.

GREIST, M.C. AND EPINETTE, W.W. (1982). Cimetidine-induced xerosis and asteatotic dermatitis. *Arch. Dermatol.* **118**, 253–4.

GRIFFIN, J.P. (1983). Drug-induced allergic and hypersensitivity reactions. *Practitioner* **227**, 1283–97.

GRIFFITHS, J.D. AND RICHARDSON, J. (1979). Lupus-type illness associated with labetalol. *Br. med. J.* **2**, 496–7.

GRIFFITHS, W.A.D. (1973). Diffuse hair loss and oral contraceptives. *Br. J. Dermatol.* **88**, 31–6.

GROSCHEL, D., GERSTEIN, A.R., AND ROSENBAUM, J.M. (1970). Skin lesions as a diagnostic aid in barbiturate poisoning. *New Engl. J. Med.* **283**, 409–10.

GUILL, M.A. AND ODOM, R.B. (1979). Evans blue dermatitis. *Arch. Dermatol.* **115**, 1071–3.

GUIN, J.D. (1981). Complications of topical hydrocortisone. *J. Am. Acad. Dermatol.* **4**, 417–22.

HARLSEY, J.P. AND CARDOE, N. (1982). Benoxaprofen: side-effect profile in 300 patients. *Br. med. J.* **284**, 1365–8.

HARBER, L.C., BICKERS, D.R., ARMSTRONG, R.B., AND KOCHEVAR, I.E.

(1982). Drug photosensitivity: phototoxic and photoallergic mechanisms. *Sem. Dermatol.* **1**, 183–95.

HARDIE, R.A. AND SAVIN, J.A. (1979). Drug induced skin diseases. *Br. med. J.* **1**, 935–7.

HARPEY, J.P. (1973). Drugs and disseminated lupus erythematosus. *Adverse Drug React. Bull.* **43**, 140–3.

HARRINGTON, C.I. (1977). A case of pellagra induced by isoniazid. *Practitioner* **218**, 716–17.

HASHIMOTO, K., McEVOY, B., AND BELCHER, R. (1981). Ultrastructure of penicillamine induced skin lesions. *J. Am. Acad. Dermatol.* **4**, 300–15.

HENDRICKS, A.A. (1980). Yellow lunulae with fluorescence after tetracycline therapy. *Arch. Dermatol.* **116**, 438–40.

HENG, M.C. (1982). Cutaneous manifestations of lithium toxicity. *Br. J. Dermatol.* **106**, 107–9.

HENSELER, T., WOLFF, K., HONIGSMANN, H., AND CHRISTOPHERS, E. (1981). Oral 8-methoxypsoralen photochemotherapy of psoriasis. *Lancet* i, 853–7.

HESSE, P.G. (1966). Die antituberkulose Therapie und des akneiform Exanthem. *Derm. Wschr.* **152**, 305–12.

HEYMANN, W.R., LERMAN, J.S., AND LUFTSCHEIN, S. (1984). Naproxen-induced lichen planus. *J. Am. Acad. Dermatol.* **10**, 299–301.

HINDSON, C., DAYMOND, T., DIFFEY, B., AND LAWLOR, F. (1982a). Side effects of benoxaprofen. *Br. med. J.* **284**, 1368–9.

——, LAWLOR, F., AND WACKS, H. (1982b). Benoxaprofen for nodular acne. *Lancet* i, 1415.

HOFMANN, C., PLEWIG, G., AND BRAUN-FALCO, O. (1979). Bowenoid lesions, Bowen's disease and keratoacanthoma in long term PUVA-treated patients. *Br. J. Dermatol.* **101**, 685–92.

HOFMANN, H., MAIBACH, H.I., AND PROUT, E. (1975). Presumed generalized exfoliative dermatitis to lidocaine. *Arch. Dermatol.* **111**, 266.

HOLCOMB, F.D. (1965). Erythema nodosum associated with the use of oral contraceptive. Report of a case. *Obstet. gynecol.* **25**, 156–7.

HOLMBERG, L., BOMAN, G., BÖTTIGER, L.E., ERIKSSON, B., SPROSS, R., AND WESSLING, A. (1980). Adverse reactions to nitrofurantoin. *Am. J. Dermatol.* **69**, 733–8.

HORIO, T. (1976). Allergic and photoallergic dermatitis from diphenylhydramine. *Arch. Dermatol.* **112**, 1124–6.

—— (1981). Evaluation of drug phototoxicity by photosensitization of Trichophyton mentagrophytes. *Br. J. Dermatol.* **105**, 365–70.

HOWARD, E.J. AND BROWN, S.M. (1973). Clofibrate induced antinuclear factor and lupus-like syndrome. *J. Am. med. Ass.* **226**, 1358–9.

HOWITT, A.J., WILLIAMS, A.J., AND SKINNER, C. (1982). Warfarin induced vasculitis: a dose related phenomenon in susceptible individuals? *Postgrad. med. J.* **58**, 233–4.

HUFF, J.C., WESTON, W.L., AND TONNESEN, M.G. (1983). Erythema multiforme: a critical review of characteristics, diagnostic criteria, and causes. *J. Am. Acad. Dermatol.* **88**, 763–75.

HUGHES, G.R.V. (1982). Leading article: Hypotensive agents, beta-blockers, and drug induced lupus. *Br. med. J.* **284**, 1358–9.

IVES, A. (1980). Drug eruptions. *Medicine 30*, 1572–6.

JACKSON, A.M. AND POLLOCK, A.V. (1981). Skin necrosis after heparin injections. *Br. med. J.* **283**, 1087–8.

JARKOWSKI, T.L. AND MARTMER, E.E. (1962). Fatal reaction to sulphadimethoxine (Madribon). A case showing toxic epidermal necrosis and leukopenia. *Am. J. Dis Childh.* **106**, 669–74.

JEFFERSON, J. W. (1982). Lithium and tetracycline. *Br. J. Dermatol.* **107**, 370.

JEGASOTHY, B.J. (1980). Allergic reactions to insulin. *Int. J. Dermatol.* **19**, 139–41.

KADER, N.M. (1983). Fixed drug eruption caused by chloromezanone. *Int. J. Dermatol.* **22**, 548.

KANNANGARA, D.W., SMITH, B., AND COHEN, K. (1982). Exfoliative dermatitis during cefoxitin therapy. *Arch. intern. Med.* **142**, 1031–2.

KANWAR, A.J., MAJD, A., GARG, M.P., AND SINGH, G. (1981). Seborrhoeic dermatitis-like eruption caused by cimetidine. *Arch. Dermatol.* **117**, 65–6.

KASS, E., STRAUME, S., MELLBYE, O.J., MUNTHE, E., AND SOCHEIM, B.G. (1979). Dermatomyositis associated with BCG vaccination. *Scand. J. Rheum.* **8**, 187–91.

KATAYAMA, H. AND KAWADA, A. (1981). Exacerbation of psoriasis induced by indomethacin. *J. Dermatol., (Tokyo)* **8**, 323.

KENNEDY, B.J. (1965). Systemic effects of estrogenic hormones in advanced breast cancer. *J. Am. Geriat. Soc.* **13**, 230–5.

KERDEL, F. A., FRAKER, D.L., AND HAYNES, H.A. (1984). Necrotizing vasculitis from radiographic contrast media. *J. Am. Acad. Dermatol.* **10**, 25–9.

KHAN, A.R. (1984). Erythema nodosum after ibuprofen. *Br. med. J.* **288**, 1048.

KINGSTON, T. AND MARKS, R. (1982). Benoxaprofen and psoriasis. *Br. med. J.* **285**, 1741.

KOSSARD, S., ECKER, R.L., AND DICKEN, C.H. (1980). Povidone panniculitis. *Arch. Dermatol.* **116**, 704–6.

KUSKE, H. AND KREBS, A. (1964). Chloasma-type hyperpigmentation after treatment with hydantoin preparations. *Dermatologica, Basle* **129**, 121–39.

KVASNICKA, J., REZAC, J., SVEJDA, J., DUCHKOVA, H., KAZE, F., ZALUD, P., AND RICHTER, J. (1979). Disseminated intravascular coagulation associated with toxic epidermal necrolysis (Lyell's syndrome). *Br. J. Dermatol.* **100**, 551–8.

KWONG, P., ROBERTS, P., PRESCOTT, S.M., AND TIKOFF, G. (1978). Dermatitis induced by warfarin. *J. Am. med. Ass.* **239**, 1884–5.

LANCET (1975). Leading article: D-penicillamine in rheumatoid arthritis. *Lancet* **i**, 1123.

—— (1982). Leading article: Adverse reactions to dapsone. *Lancet* **ii**, 184–5.

LANG, P.G. (1983). Quinidine-induced photodermatitis confirmed by photopatch testing. *J. Am. Acad. Dermatol.* **9**, 124–8.

LEES, A.W. (1963). Toxicity in newly diagnosed cases of pulmonary tuberculosis treated with ethionamide. *Am. Rev. resp. Dis.* **88**, 347–54.

LEPAW, M. I. (1983). Fixed drug eruption due to minocycline — report of one case. *J. Am. Acad. Dermatol.* **8**, 263–4.

LEVANTINE, A. AND ALMEYDA, J. (1973). Drug induced alopecia. *Br. J. Dermatol.* **89**, 549–53.

—— AND —— (1974). Cutaneous reactions to cytostatic agents. *Br. J. Dermatol.* **90**, 239–42.

LEVITT, L. AND PEARSON, R.W. (1980). Sulindac-induced Stevens–Johnson syndrome. *J. Am. med. Ass.* **243**, 1262–3.

LEVY, H. (1982). Chloroquine-induced pigmentation. *S. Afr. med. J.* **62**, 735–7.

LOWE, N.J. AND RIDGEWAY, H.B., (1978). Generalized pustular

psoriasis precipitated by lithium carbonate. *Arch. Dermatol.* **114**, 1788–9.

LUPTON, G.P. AND ODOM, R.B. (1979). The allopurinol hypersensitivity syndrome. *J. Am. Acad. Dermatol.* **1**, 365–74.

LYELL, A. (1956). Toxic epidermal necrolysis: Eruption resembling scalding of skin. *Br. J. Dermatol.* **68**, 355–61.

—— (1979). Toxic epidermal necrolysis (the scalded skin syndrome): a reappraisal. *Br. J. Dermatol.* **100**, 69–86.

MACINTYRE, D. AND McLAY, A.L.C. (1978). Silver poisoning associated with an antismoking lozenge. *Br. med. J.* **2**, 1749–50.

MacKIE, R.M. AND FITZSIMONS, C.P. (1983). Risk of carcinogenicity in patients with psoriasis treated with methotrexate or PUVA singly or in combination. *J. Am. Acad. Dematol.* **9**, 467–9.

MAIZE, J.C. AND TOMECKI, K.J. (1977). Pityriasis rosea-like drug eruption secondary to metronidazole. *Arch. Dermatol.* **113**, 1457–8.

MALTZ, B.L. AND BECKER, L.E. (1980). Quinidine induced lichen planus. *Int. J. Dermatol.* **19**, 96–7.

MARSHALL, A.J. BADDELEY, H., BARRITT, D.W., DAVIES, J.D., LEE, R.E.J., LOW-BEER, T.B., AND READ A.E. (1977). Practolol peritonitis. A study of 16 cases and survey of small bowel function in patients taking beta-adrenergic blockers. *Quart. J. Med.* **46**, 135–49.

MARTIN, C.M., SOUTHWICK, E.G., AND MAIBACH, H.I. (1973). Propranolol-induced alopecia. *Am. Heart. J.* **86**, 236–7.

MARX, J.L., AUERBACH, R., POSSICK, P., MYROW, R., GLADSTEIN, A.H., AND KOPF, A.W. (1983). Malignant melanoma in situ in two patients treated with psoralens and ultraviolet-A. *J. Am. Acad. Dermatol.* **9**, 904–11.

MATHEW, T.H. (1966). Nalidixic acid. *Med. J. Aust.* **2**, 243–4.

MATKALUK, R.M. AND BAILIN, P.L. (1981). Penicillamine-induced pemphigus foliaceus. A fatal outcome. *Arch. Dermatol.* **117**, 156–7.

McCALL, C.Y. AND COOPER, J.W. (1980). Tolmetin anaphylactoid reaction. *J. Am. med. Ass.* **243**, 1263.

McGRAE, J. AND ZELICKSON, A. (1980). Skin pigmentation secondary to minocycline therapy. *Arch. Dermatol.* **116**, 1262–5.

McQUEEN, A. AND BEHAN, W.N.H. (1982). The 'string of pearls' phenomenon — An immunofluorescent serological finding in patients screened for adverse drug reactions. *Am. J. Dermatopathol.* **4**, 155–9.

MERKLEN, F.P. AND MELKI, J.R.(1966). Incidences dermatologiques de contraceptifs oraux. *Bull. Acad. nat. Med., Paris* **150**, 624–32.

MERRETT, A.C., MARKS, R., AND DUDLEY, F.J. (1981). Cimetidine induced erythema annulare centrifugum: no cross sensitivity with ranitidine. *Br. med. J.* **282**, 698.

MESSENS, Y. (1965). A propos d'un cas d'intoxiation chronique par la phenacetine. *Rev. med., Liege* **20**, 516–22.

MILLER, J.A. AND LEVENE, G.M. (1982). Steroids in dermatology. *Br. J. hosp. Med.* **28**, 331–8.

MILLER, K. (1982). Leading article: Sensitivity to tartrazine. *Br. med. J.* **285**, 1597–8.

MOHAMED, K.N. (1983). Fixed drug eruption caused by chloromezanone. *Int. J. Dermatol.* **22**, 548.

MONTGOMERY, D.C. AND JACKSON, R. (1968). One hundred consecutive patients with drug reactions. *Can. med. Ass. J.* **99**, 712–14.

MORGAN, S.H. AND BEHN, A.R. (1981). Association between Stevens–Johnson syndrome and benoxaprofen. *Br. med. J.* **283**, 144.

MORLEY, W.N. (1977). Ketoprofen dermatitis: personal communication.

—— AND McQUEEN, A. (1977). Bullous lesions from phenobarbitone sensitivity, unpublished data.

MORROW, D.M., RAPAPORT, M.J., AND STRICK, R.A. (1980). Hypersensitivity to aloe. *Arch. Dermatol.* **116**, 1064–5.

MOULOPOULOU-KARAKITSOU, K., MAVRIKAKIS, M., AND ANASTASIOUNANA, M. (1981). An unusual adverse reaction to Ro. 10–9359. *Br. J. Dermatol.* **104**, 709.

NEUMANN, H.A.M. AND VAN JOOST, R. (1981). Adverse reations of the skin to metoprolol and other beta-adrenoreceptor-blocking agents. *Dermatologica, Basle* **161**, 330–5.

——, VAN JOOST, T.H., AND WESTERHOF, W. (1979). Dermatitis as side-effect of long term metoprolol. *Lancet* **ii**, 745.

OKRASINSKI, H. (1977). Lithium acne. *Dermatologica, Basle* **154**, 251–3.

OLIVER, R.A.M. (1959). Siderosis following transfusions of blood. *J. Path. Bact.* **77**, 171–94.

OLSEN, V.V., LOFTS, S., AND CHRISTENSEN, K.D. (1982). Serious reactions during malaria prophylaxis with pyrimethamine–sulfadoxine. *Lancet* **ii**, 994.

OLUMIDE, Y. (1979). Fixed drug eruption; a lesson in drug usage. *Int. J. Dermatol* **18**, 818–21.

PANDHI, R.K. AND BEDI, T.R. (1975). Fixed skin eruption caused by oxyphenbutazone with cross-reactivity to phenylbutazone. *Arch. Dermatol.* **111**, 1331.

PARK, G.D., SPECTOR, R., HEADSTREAM, T., AND GOLDBERG, M. (1982). Serious adverse reactions associated with sulindac. *Arch. intern. Med.* **142**, 1292–4.

PARKER, N.E., PORTER, J.B., WILLIAMS, H.J.M., AND LEFTLEY, N. (1982). Pruritus after administration of hetastarch. *Br. med. J.* **284**, 385.

PARRISH, J.A. (1981). Phototherapy and photochemotherapy of skin diseases. *J. Invest. Dermatol.* **77**, 167–71.

PASRICHA, J.S. (1979). Drugs causing fixed eruptions. *Br. J. Dermatol.* **100**, 183–5.

—— AND SHUKLA, S.R. (1979). Independent lesions of fixed eruption due to two unrelated drugs in the same patient. *Br. J. Dermatol.* **101**, 361–2.

PATTERSON, R. AND ANDERSON, J. (1982). Allergic reactions to drugs and biologic agents. *J. Am. med Ass.* **248**, 2637–45.

PEACOCK, A. AND WEATHERALL, D. (1981). Hydralazine-induced vasculitis. *Br. med. J.* **282**, 1121–2.

PEMBROKE, A.C., SAXENA, S.R., KATARIA, M., AND ZILKHA, K.D. (1981). Acne induced by dantrolene. *Br. J. Dermatol.* **104**, 465–8.

PETROZZI, J.W. AND SHORE, R.N. (1976). Generalized exfoliative dermatitis from ethylenediamine. Arch. Dermatol. **112**, 525–6.

PHILLIPS, G.W.L. AND WILLIAMS, A.J. (1984). Spironolactone induced vasculitis. *Br. med. J.* **288**, 368.

PLETCHER, W.D., BRODY, G.L., AND MEYERS, M.C. (1963). Hemochromatosis following prolonged iron therapy in a patient with hereditary nonspherocytic hemolytic anemia. *Am. J. med. Sci.* **246**, 27–34.

PONTE, C.D. (1983). A suspected case of codeine-induced erythema multiforme. *Drug Intel. clin. Pharm.* **17**, 128–30.

PRESLEY, P. (1981). Fixed genital drug eruption. *Mod. Med.* **26**, 64.

QURAISHY, E. (1981). Erythema multiforme during treatment with mianserin. *Br. J. Dermatol.* **104**, 481.

RAMPEN, F.H.J. (1983). Hypertrichosis in PUVA-treated patients. *Br. J. Dermatol.* **109**, 657–60.

RESHAD, H., CHALLONER, F., POLLOCK, D.J., AND BAKER, H. (1984). Cutaneous carcinoma in psoriatic patients treated with PUVA. *Br. J. Dermatol.* **110**, 299–305.

——, HARGREAVES, G.K., AND VICKERS, C.F.H. (1983). Generalized pustular psoriasis precipitated by phenylbutazone and oxyphenbutazone. *Br. J. Dermatol.* **108**, 111–13.

RESNIK, S. (1967). Melasma induced by oral contraceptive drugs. *J. Am. med. Ass.* **199**, 601–5.

RHODES, A.R., HARRIST, T.J., AND MONTAZ, T. K. (1983). The PUVA-induced pigmented macule: a lentiginous proliferation of large, sometimes cytologically atypical, melanocytes. *J. Am. Acad. Dermatol.* **9**, 47–58.

RIDLEY, C.M. (1971). Bullous lesions in nitrazepam overdosage. *Br. med. J.* **3**, 28.

ROENIGK, H.H. AND CARO, W.A. (1981). Skin cancer in the Puva-48 cooperative study. *J. Am. Acad. Dermatol.* **4**, 319–24.

ROSTENBERG, A., JR. AND FAGELSON, H.J. (1965). Life threatening drug eruptions. *J. Am. med. Ass.* **194**, 660–2.

RUBENSTEIN, A. (1979). Lupus erythematosus-like disease caused by nalidixic acid. *New Engl. J. Med.* **301**, 1288.

SAUNDERS, T.S., FITZPATRICK, T.B., SEIJI, M., BRUNET, P., AND ROSENBAUM, E.E. (1959). Decrease in human hair colour and feather pigment of fowl following choroquine diphosphate. *J. Invest. Dermatol.* **33**, 87–90.

SCANDLING, J. AND WALKER, B.K. (1980). Extensive tissue necrosis associated with warfarin sodium therapy. *South. Med. J.* **73**, 1470–2.

SCHLEICHER, S.M. AND FRICKER, M.P. (1980). Coumarin necrosis. *Arch. Dermatol.* **116**, 444–5.

SCHMOECKEL, C., WEISSMAN, I., PLEWIG, G., AND BRAUN-FALCO, O. (1979). Treatment of alopecia-areata by anthralin-induced dermatitis. *Arch. Dermatol.* **115**, 1254–5.

SEEHAFER, J.R., ROGERS, R.S. FLEMING, C.R., AND DICKSON. E.R. (1981). Lichen planus-like lesions caused by penicillamine in primary biliary cirrhosis. *Arch. Dermatol.* **117**, 140–2.

SETTIPANE, G.A., KLEIN, D.E. BOYD, G.K., STURAM, J.H., FREYE, H.B., AND WELLMAN, J.K. (1979). Adverse reactions to cromolyn. *J. Am. med. Ass.* **241**, 811–13.

SHELLEY, W.R. (1967). Generalized pustular psoriasis induced by potassium iodide. *J. Am. med. Ass.* **201**, 1009–14.

SIMONS, J.J. AND MORALES, A. (1980). Minocycline and generalized cutaneous pigmentation. *J. Am. Acad. Dermatol.* **3**, 244–7.

SIMPSON, G.M., BLAIR, J.H., AND CRANSWICK, E.H. (1964). Cutaneous effects of a new butyrophenone drug. *Clin. Pharmacol. Ther.* **5**, 310–21.

SINA, B. AND ADRIAN, R.M. (1983). Multiple keratoacanthomas possibly induced by psoralens and ultraviolet A photochemotherapy. *J. Am. Acad. Dermatol.* **9**, 686–8.

SLUTZKI, S., BOGOKOWSKY, H., GILBOA, Y., AND HALPERN, Z. (1984). Coumarin-induced skin necrosis. *Int. J. Dermatol.* **23**, 117–19.

SMITH, D.A. AND BURGDORF, M.D. (1984). Universal cutaneous depigmentation following phenytoin-induced toxic epidermal necrolysis. *J. Am. Acad. Dermatol.* **10**, 106–9.

SPENCER, H.C., POULTER, N.R., LURY, J.D., AND POULTER, C.J. (1982). Chloroquine associated pruritus in a European. *Br. med. J.* **285**, 1703–4.

SPILLANE, J.D. (1963). Brunette to blonde: Depigmentation of human hair during oral treatment with mephenesin. *Br. med. J.* 1, 997–1000.

SRIVASTAVA, B.N. AND GOUR, K.A., (1966). Cutaneous manifestations of drug toxicity. *Indian Practit.* 19, 465–73.

STAAK, W.J.B.M. VAN DER, COTTON, D.W.K., JONCKHEER-VENNESTE, M.M.H., AND BOERBOOMS, A.M.T.H. (1975). Lichenoid eruption following penicillamine. *Dermatologica, Basle* 150, 372–4.

STEELE, K. (1984). Primary dermatological care in general practice. *J. R. Coll. Gen. Pract.* 34, 22–3.

SULLIVAN, M. AND ZELIGMAN, I. (1956). Acneiform eruption due to corticotropin. *Arch. Dermatol.* 73, 133–41.

TAGGART, H.McA. AND ALDERDICE, J.M. (1982). Fatal cholestatic jaundice in elderly patients taking benoxaprofen. *Br. med. J.* 284, 1372.

TALBOT, M.D. (1980). Fixed genital drug eruption. *Practitioner* 224, 823–4.

TAN, S. AND ROWELL, N.R. (1976). Pemphigus-like syndrome induced by D-penicillamine. *Br. J. Dermatol.* 95, 99–100.

TAYLOR, A.E.M., GOFF, G., AND HINDSON, T.C. (1981). Association between Stevens–Johnson syndrome and benoxaprofen. *Br. med. J.* 282, 1433.

——, HINDSON, C., AND WACKS, H., (1982). A drug eruption due to acebutolol with combined lichenoid and lupus erythematosus features. *J. clin. exp. Dermatol.* 7, 219–21.

TAYLOR, M.W., SMITH, C.C., AND HERN, J.E.C. (1981). An unexpected reaction to carbamazepine. *Practitioner* 225, 219–20.

THIERS, B.J. (1981). Hazards of therapy. *J. Am. Acad. Dermatol.* 4, 495–9.

THOMAS, J.R. AND DOYLE, J.A. (1981). The therapeutic uses of topical vitamin A acid. *J. Am. Acad. Dermatol.* 4, 505–13.

TOLBERT, B., KAUFMAN, C.E., JR., BURGDORF, W.H.C., AND BRUBAKER, D. (1983). Graft-versus-host disease from leukocyte transfusions. *J. Am. Acad. Dermatol.* 9, 416–19.

TOMLINSON, I.E. AND ROSENTHAL, F.D. (1977). Proximal myopathy after perhexiline maleate treatment. *Br. med. J.* 1, 1319–20.

TROY, J.L., SILVERS, D.N., GROSSMAN, M.E., AND JAFFE, I.A. (1981). Penicillamine-associated pemphigus: Is it really pemphigus? *J. Am. Acad. Dermatol.* 4, 547–55.

TULLET, G.L. (1966). Fatal case of toxic erythema after chlorpropamide (Diabinese). *Br. med. J.* 1, 148.

UDDIN, M.S., LYNFIELD, Y.L., GROSBERG, S.J., AND STIEFLER, R. (1979). Cutaneous reaction to spironolactone resembling lupus erythematosus. *Cutis* 24, 198–200.

VAN ARSDEL, P.P. (1982a). Diagnosing drug allergy. *J. Am. med. Ass.* 247, 2576–81.

—— (1982b). Allergy and adverse drug reactions. *J. Am. Acad. Dermatol.* 6, 833–45.

VAN JOOST, T. AND SMITT, J.H.S. (1982). Skin reactions to propranolol and sensitivity to beta-adrenoceptor blocking agents. *Arch. Dermatol.* 117, 600–1.

VAN NESTE, D., PIERARD, G.E., AND HERMANNS, J.F. (1979). Immune complex disease associated with Peroben intake. *Dermatologica, Basle* 158, 417–26.

VERBOV, J. (1979). Drug eruptions. *Practitioner* 222, 400–9.

VICKERS, H.R. AND SNEDDON, I.B. (1963). Psoriasis and hypoparathyroidism. *Br. J. Dermatol.* 35, 419–21.

VIREBURGER, M.I., PRELEVIC, G.M., BRKIC, S., ANDREJEVIC, M.M., AND PERIC, L.A. (1981). Transitory alopecia and hypergonadotrophic hypogonadism during cimetidine treatment. *Lancet* i, 1160–1.

VIVIER, A. DU (1982). Bullous dermatitis associated with benoxaprofen. *Lancet* i, 47.

VOORHEES, J.J. AND ORFANOS, C.E. (1981). Oral retinoids, broad spectrum dermatologic therapy for the 1980s. *Arch. Dermatol.* 117, 418–21.

WALTON, N.G., NANN, T.A.N. AND SHAW, K.M. (1979). Anaphylactic reaction to N-acetylcysteine. *Lancet* ii, 1298.

WATSKY, M.S. AND LYNFIELD, Y.L. (1976). Lupus erythematosus exacerbated by griseofulvin. *Cutis* 17, 316–3.

WATTS, J.C. (1962). Fatal case of erythema multiforme exudativum (Stevens–Johnson syndrome) following therapy with Dilantin. *Pediatrics* 30, 592–4.

WEISS, I.S. (1974). Hirsutism after chronic administration of acetazolamide. *Am. J. Ophthal.* 78, 327–8.

WEISS, R.S. AND KILE, R.L. (1935). Unusual phenolphthalein eruption. Report of a case. *Arch. Derm. Syph., Berlin.* 32, 915–21.

WESTERHOF, W., WOLTERS, E.CH., BROOKBAKKER, J.T.W., BOELEN, R.E., AND SCHIPPER, M.E.I. (1983). Pigmented lesions of the tongue in heroin addicts — fixed drug eruption. *Br. J. Dermatol.* 109, 605–10.

WHITE, S. W. (1982). Palmoplantar pustular psoriasis provoked by lithium therapy. *J. Am. Acad. Dermatol.* 7, 660–2.

WILKIN, J.K. (1980). Vasodilator rosacea. *Arch. Dermatol.* 116, 598.

——, HAMMON, J.J., AND KIRKENDALL, W.M. (1980). The captopril induced eruption. *Arch. Dermatol.* 116, 902–5.

—— AND KIRKENDALL, W.M. (1982). Pityriasis rosea like rash from captopril. *Arch. Dermatol.* 118, 186–7.

WILSON, H.T.H. (1975). A fixed drug eruption due to paracetamol. *Br. J. Dermatol.* 92, 213–14.

WRIGHT, P. (1974). Skin reactions to practolol. *Br. med. J.* 2, 560.

—— (1975). Untoward effects associated with practolol administration: Oculomucocutaneous syndrome. *Br. med. J.* 1, 595–8.

YUNG, C.W. AND HAMBRICK, G.W. (1982). D-Penicillamine-induced pemphigus syndrome. *J. Am. Acad. Dermatol.* 6, 317–24.

——, WINTON, E.M., AND LORINEZ, A.L. (1981). Dacarbazine induced photosensitivity reaction. *J. Am. Acad. Dermatol.* 4, 541–3.

ZACHARY, C.B., SLATER, D.N., HOLT, D.W., STOREY, G.C.A., AND MAC DONALD, D.M. (1984). The pathogenesis of amiodarone-induced pigmentation and photosensitivity. *Br. J. Dermatol.* 110, 451–6.

ZELICKSON, A.S. (1965). Skin pigmentation and chlorpromazine. *J. Am. med. Ass.* 194, 670–2.

ZUGERMAN, C. AND LA VOO, E.J. (1980). Erythema multiforme caused by oral furosemide. *Arch. Dermatol.* 116, 518–19.

33 Drug-induced disorders of temperature regulation

JENS SCHOU

The body temperature results from a balance between heat production by the energy production of basal metabolic processes and muscle activity and heat loss by radiation, conduction, and convection. The thermoregulatory centre in the hypothalamus functions as a thermostat to keep the body temperature at the normal level using vasodilatation and evaporation (sweat) to get rid of overproduction.

Elevation of body temperature caused by therapeutic agents administered to patients is seen rather often. Adverse reactions to drugs occur in about 10 per cent of hospitalized patients and about 2.5 per cent of outpatients receiving any form of drug therapy. Elevated body temperature as the sole or predominant clinical feature constitutes 3–5 per cent of all adverse reactions to drugs (Burnum 1976; Caldwell and Cluff 1974; Cluff and Johnson 1964). Fever, which in this context means any elevation of body temperature above normal, should therefore occur in one of every 200–400 patients as a result of the drug therapy they are receiving. Drug fever shall here be used for any elevation in body temperature caused by drug therapy. The description of the types of drug fever follows the general lines for the mechanisms of fever induction as reviewed by Bernheim *et al.* (1979). At the end of the chapter the effect of fever on drug metabolism is mentioned. This could mean that the drug fever elicited by a certain medicament might even influence the metabolism and thereby the efficacy of the therapy.

Types of drug fever

Drug hypersensitivity of allergy

This type of drug fever is by far the most commonly encountered condition, and to most clinicians drug fever simply means fever resulting from drug hypersensitivity. Fever may be the solitary symptom or it may occur as part of a complex of reactions to the drug often simulating the serum sickness syndrome with urticarial rash, myalgias, and arthralgias.

Typically, it occurs on the seventh to tenth day of medication. Factors increasing the likelihood of this drug reaction, as well as other drug allergies, are serious medical problems, administration of multiple pharmacological agents, and previous adverse reactions; therefore, women and the elderly are at increased risk (Lipsky and Hirschmann 1981). Fever from drug hypersensitivity occurs more commonly in patients with atopy, severe infections, and systemic lupus erythematosus (Cluff and Johnson 1964). Drug-induced lupus erythematosus always shows hyperpyrexia (Stahl *et al.* 1979).

Almost any drug is able to elicit drug fever, as is the case for all other reaction types of drug allergy. A number of typical examples are given in Table 33.1. Usually the low-molecular-weight drug acts as a hapten that binds to an endogenous protein compound to form a functional allergen, but larger molecules like interferons may themselves be allergens (Bocci 1980). The exact mechanism of fever production is not known, but one possibility is that the antigen–antibody complex elicits the release of endogenous pyrogens, from granulocytes which act on the thermoregulatory centre (Snell 1969; Chusid and Atkins 1972; Root and Wolf 1968). As a result the thermostat is set at a higher level.

Monosymptomatic drug fever constitutes less than 50 per cent of the total number of drug fevers. Usually fever is accompanied by rash or slight arthralgia; very occasionally severe reactions are seen such as systemic lupus erythematosus (Stahl *et al* 1979). The monosymptomatic drug fever should always be kept in mind when the fever of a patient continues longer than should be expected. As for other allergic drug reactions, antibiotics are the commonest cause. If a patient with tonsillitis or pneumonia continues to show fever after effective antibiotic therapy, the drug itself might be the reason. When the administration of the responsible agent is discontinued, the body temperature returns to the normal level. Drugs are said to be the cause of 1–2 per cent of prolonged cases of fever in both children and adults (Petersdorf and Beeson 1961; Howard *et al.* 1977; Lohr and Hendley 1977).

Drug fever most often occurs on the seventh to tenth day of medication, but may be seen even on the first day. Early onset may result from either previous exposure to the same drug or may have the character of an intolerance reaction. A diagnostic aid is that about a fourth of all cases show a discrepancy between the temperature curve and the pulse rate, as the tachycardia

Table 33.1 Drugs reported to cause fever and their considered mechanism of action. H—hypersensitivity, P—pyrogenic effect, T—thermoregulatory effect, U—unknown (Sources of information: Lipsky and Hirschmann 1981; *CSM Adverse Reactions Register*)

Drug	Cause
Allopurinol	H
Amphetamines	T
Amphotericin B	P
Antihistamines	U
Asparginase	U (?P)
Antiparkinson agents, for example, benztropine, L-dopa	U (?T)
Atropine	T
Azathioprine	H
Barbiturates	U
Benzodiazepines	U
Bleomycin sulphate	P
Bretylium tosylate	U (?T)
Cephalosporins	H
Cimetidine	T
Cis-diammine dichloroplatinum	P
Cocaine derivatives	T
Colaspase	P
Guanethidine	U (?T)
Hydrallazine derivatives	H
Hydroxyurea	P or U
Iodines	H
Iron dextran	U
Isoniazid	H
Isoprenaline	T
Methyldopa	H
Monamine oxidase inhibitors	T
Nitrofurantoin	H
Oxyphenbutazone	H
Para amino salicylic acid	H
Penicillins	H
Phenylbutazone	H
Phenytoin	H
Procainamide	H
Propylthiouracil	U
Quinidine	H
Rifampicin	H
Salicylates*	U (?H)
Streptokinase	P
Streptomycin	H
Sulphonamides	H
Tricyclic antidepressants	T
Vaccines, for example, Diphtheria, DTP, influenza, TAB, typhoid vaccines, typhus, tetanus, rubella, polio, plague, measles	H
Vancomycin	U
Warfarin	U (?P)

Other drugs which rarely have been reported to cause fever are chloramphenicol, digitalis, oxprenolol. Drugs associated with malignant hyperpyrexia have been given the test and are not included in the table. Drugs causing haemolytic anaemia and inducing hepatitis are listed in the appropriate sections of *Iatrogenic diseases*, 2nd edition, pp. 100 and 180, respectively.
* The normal action of salicylates is antipyretic.

typically seen with fevers caused by infections diseases is not found in a number of drug fevers (Schou 1956). Apart from these signs the only certain sign is that the drug fever disappears within the first day to two after discontinuation of the responsible therapy, and positive provocation by rechallenge, meaning a fever spike after readministration of a single dose, is diagnostic. Rechallenge is not without risk, and sulindac rechallenge has led to serious reactions (Levites *et al.* 1981; Park *et al.* 1982). The duration of fever after discontinuation is related to the half-life of the responsible agent in the body. Usually, allergic drug fever is a rather harmless condition, but severe reactions simulating septic shock may be seen, e.g. with methyldopa (De Bard 1979). Interestingly, drug fever due to oxacillin has been observed in the treatment of severe combined immune deficiency (Peerless 1981).

Drugs affecting temperature regulation

A direct pharmacological effect on the centre for temperature regulation is believed to be the cause of some drug fevers (see Table 33.1). A typical example is amphetamine which stimulates the central nervous system and is believed also to stimulate the thermoregulatory centre to increase body temperature. However, like other symphomimetic drugs, amphetamine also constricts peripheral vessels and in this way decreases heat emission, which again results in body temperature increase (Tierney 1978). In atropine poisoning hyperpyrexia is a dominating symptom due to prevention of sweating and probably also central-nervous-system stimulation. Increased energy production is a direct effect of thyroid hormone, and any drug with this effect, e.g. levothyroxine has a tendency to increase body temperature. This means that the pharmacological targets for effects resulting in elevated body temperature can be the thermoregulatory centre, the processes of energy metabolism, and the regulatory functions for peripheral heat dissipation.

Fever secondary to pharmacological action

The pharmacological actions of certain drugs may lead to release of endogenous pyrogens and pyrogen formation may be due to endotoxins released from microorganisms. The first-mentioned mechanism causes fever in connection with chemotherapy for malignant diseases. Numerous reports describe bleomycin-induced fever (Carter *et al.* 1983; Levy and Chiarillo 1980) which is thought to be due to granulocytes releasing endogenous pyrogen (Dinarello *et al.* 1973). This condition has even led to death.

Drug-induced hepatitis may also lead to tissue destruction and fever, as is also seen with haemolytic reactions such as may be induced by antimalarial drugs in patients with glucose-6-phosphate dehydrogenase deficiency, or after haemolysis following blood transfusion with mismatched or incompletely matched blood. Rifampicin has been reported to give mild intravascular haemolysis associated with a flu-like syndrome, and fever may be a warning sign for haemolysis and hepatic and renal complications caused by drug treatment (Mattson and Jänne 1982).

The Jarish–Herxheimer reaction reviewed by Gelford *et al.* (1976) is a condition with pyrexia due to endotoxin release from dying spirochaetes when syphilis is treated with penicillins or other spirochaeticide drugs. Similar reactions are seen after antibiotic therapy for leptospirosis and Borrelia infections.

Drug fever and biochemical defects

Biochemical defects that cause drug reactions with fever are most often of hereditary (genetic) character. Mentioned above is the glucose-6-phosphate deficiency which leads to haemolysis after a number of drugs, as also does the sickle-cell anaemia. The most dramatic of these febrile reactions, however, is malignant hyperpyrexia.

Malignant hyperpyrexia

Malignant hyperpyrexia (hyperthermia) is a rare pharmacogenetic disease occurring both in man and in a strain of pig, the Landrace strain. The mode of inheritance is autosomal dominant.

Malignant hyperpyrexia occurs as a complication of anaesthesia in individuals who have an underlying disorder of muscle. In these individuals skeletal muscle shows increased contractility in response to caffeine, halothane, and other compounds, indicating a rise in calcium-ion concentration in the myoplasm.

In susceptible individuals any potent inhalational anaesthetic agent or any skeletal-muscle relaxant causes fever, rigidity, hyperventilation, cyanosis, hypoxia, respiratory and metabolic acidosis, and hyperphosphataemia with a raised glucose. An initial hyperkalaemia and hypercalcaemia are followed by hypokalaemia and hypocalcaemia. Later there is elevation in the serum of enzymes of cardiac and skeletal muscle origin.

In the series reported by Britt and Kalow (1968) the mean maximum temperature was 108°F (42.2°C) and accompanying this there were signs of increased muscle metabolism, in particular tachycardia, tachypnoea, sweating, and blotchy cyanosis. The mortality of the condition is high (60–70 per cent).

Malignant hyperpyrexia has been divided into two clinical categories, a rigid and non-rigid type. Furniss (1970) showed that the age of the patient is significant when attempting to differentiate between the rigid and non-rigid conditions. Those reacting with hypertonus were usually under 20 years of age and the abnormality was probably hereditary, whilst most of the nonrigid cases occurred in patients over 20 years old and appeared to be sporadic.

The susceptibility to develop malignant hyperpyrexia is due to a genetic anomaly, and in the first family in which this was recognized the abnormality was clearly inherited as a dominant characteristic. Subsequently, further examples of the familial occurrence of malignant hyperpyrexia have been described, and in a review of 115 cases in 1969, 43 of these had other members in the family who had also been affected. The inherited defect is associated with a mild or even subclinical myopathy, which can be detected in affected individuals and their relatives by finding raised serum creatine-phosphokinase levels (Britt *et al.* 1969). Isaacs and Barlow (1970) investigated 99 members of a single family with a history of malignant hyperpyrexia. They found that the resting creatine-phosphokinase and aldolase levels were high in many of these patients, and concluded that the high level of muscle enzyme in serum was evidence of a subclinical myopathy. They also suggested that anaesthetic agents, such as halothane, and neuromuscular blockers like suxamethonium damaged the abnormal muscle mechanisms and triggered off a fulminating hyperpyrexia, the rigidity following immediately on the administration of suxamethonium or alternatively developing progressively during the halothane anaesthetic. Suxamethonium and halothane are most frequently described as the causative agents, but two cases precipitated by enflurane were described by Sutherland and Carter (1975) and Caropreso *et al.* (1975).

The observation by Denborough and his colleagues that the father of a child who had died from sudden infant death syndrome (SIDS) had a myopathy which predisposes to malignant hyperpyrexia prompted them to investigate further. Denborough *et al.* (1982) found that the muscle disorder which predisposes to malignant hyperpyrexia was present in five of 15 parents of 13 children who had died of SIDS. (Three of these five parents who had this muscle abnormality described a family history of events in first-degree relatives that could have been malignant pyrexia including two deaths under anaesthesia.) These observations raise the possibility that an acute disorder of myoplasmic calcium metabolism may be implicated is some sudden infant deaths.

Denborough *et al.* (1970) described a patient who had

survived malignant hyperpyrexia, and three of his close relatives were found to have very high levels of serum creatine phosphokinase. Although the patient's muscles seemed normal, two of the three relatives had a mild but definite myopathy, affecting predominantly the lower muscles of the thigh. It seems that malignant hyperpyrexia develops in individuals with a myopathy which is inherited as an autosomal dominant and which may be subclinical.

King *et al.* (1972) described the results of a family study which was carried out on all the known cases of malignant hyperpyrexia in Australia and New Zealand. Serum creatine-phosphokinase levels, clinical examinations, and retrospective information on 18 propositi and their families have identified at least three groups of individuals who are susceptible to malignant hyperpyrexia. All six survivors tested had high serum creatine-phosphokinase levels, and the results of this study suggest that all individuals who develop malignant hyperpyrexia have one of a number of specific susceptible myopathies. The first susceptible group of patients have a dominantly inherited myopathy. This was found in nine of the 18 families, but in six of these the myopathy was not clinically detectable in affected individuals. In three of the nine families an overt myopathy was present, and in the most severely affected of these a characteristic clinical picture was found. All nine families may have had the same myopathy, with considerable heterogeneity of clinical expression. Three propositi from normal families were probably mutants for the dominant myopathy (or myopathies). The second group of susceptible individuals consists of some patients who have myotonia congenita, another example of which was found in this series. The third group consists of patients with physical abnormalities and a progressive congenital myopathy. This was found in five young males with short stature, cryptorchidism, pectus carinatum, lumbar lordosis, and thoracic kyphosis. Three had a similar facies.

In an extensive study on metabolic rate and the usual clinical laboratory tests for blood hormones and metabolite levels only very small variations from normal levels wer found (Campbell *et al.* 1981). This investigation was carried out in individuals susceptible to malignant hyperpyrexia both at rest and in response to food and mild exercise.

Denborough *et al.* (1973) described the electron-microscopic structure of a biopsy of the rectus abdominus muscle of a 71-year-old aunt of the propositus in a family where there had been deaths from malignant hyperpyrexia. This elderly lady had shown the features of a myopathy since childhood. The striking histological abnormality in this patient was the presence of cores in 55 per cent of her Type I muscle fibres. This histological appearance in a myopathy has been called central-core disease (Shy and Magee 1956; Dubowitz and Pearse 1960). (Core-like structures have been noted in the muscle fibres of some Landrace pigs which show a susceptibility to develop malignant hyperpyrexia.) An *in-vitro* pharmacological study on muscle from this patient showed a markedly abnormal response to halothane. The correlation between the histological appearance of cores in muscle and the pharmacologically abnormal behaviour is still to be defined.

Kalow and his colleagues (1970) investigated three patients who had suffered malignant hyperpyrexia and found changes in metabolism in muscle biopsy specimens from those who had experienced rigidity. The calcium uptake in the sarcoplasmic reticulum was low after exposure to halothane, and they concluded that the lesion was an inability of the sarcoplasmic reticulum to store calcium. This meant that, as the concentration of calcium in the cytoplasm remained high, the enzyme myosin adenosinetriphosphate was not activated and thus the myofibrils remained locked together as in contraction. Moulds and Denborough (1974a) also produced data on raised levels of calcium concentration in the myoplasm and showed that there was a defect in calcium binding to the muscle-cell membrane and reticulum.

Harrison (1971), whilst describing anaesthetic-induced malignant hyperpyrexia, pointed out that caffeine-induced rigor in muscle was due to an enhanced release and a depressed rebinding of Ca^{2+} by the sarcoplasmic reticulum (Weber 1968; Weber and Herz 1968) and has long been known to be blocked by procaine (Feinstein 1963) which acts directly on the muscle.

This discussion therefore raised the interesting possibility that, if halothane and caffeine produced similar changes in sarcoplasmic reticular function, and, if in fact they enhanced each other's action, would not procaine block the action of halothane in the same way that it did for the caffeine rigor? The other question was of course, what effect would procaine have on a suxamethonium-induced syndrome?

Harrison (1971) conducted experiments in Landrace pigs to find answers to these questions. It was known from pre-screening with halothane that these pigs were susceptible to malignant hyperpyrexia. He showed that procaine blocked anaesthetic-induced hyperpyrexia induced by both halothane and suxamethonium. Pretreatment with tubocurarine prevented only the trigger action of suxamethonium.

Procaine was then used in a preliminary study to treat the established syndrome in five pigs, two of which survived. On the basis of these findings, a treatment for hyperpyrexia was suggested:

1. Discontinuation of anaesthetic agents;
2. Rapid correction of acidosis;
3. Rapid whole-body cooling;
4. Administration of procaine intravenously in a loading dose of 30–40 mg/kg, followed by a slow infusion of 0.2 mg/kg/min;
5. Support of circulation by isoprenaline infusion;
6. Correction of hyperkalaemia.

Strunin (1975) recommended that the treatment of malignant hyperpyrexia should consist of stopping the anaesthesia, active cooling, sodium-bicarbonate intravenous infusion, measures to reduce the hyperkalaemia, and artificial ventilation. Strunin did not consider that steroids were of any value and thought that procaine may be detrimental.

Investigations then turned towards the development of a diagnostic test, when Ellis (1973) realized the necessity of screening relatives of patients who had developed malignant hyperpyrexia, so that appropriate precautions could be taken if these people ever required an anaesthetic. They described a technique in which a biopsy of the vastus medialis muscle was taken under *local anaesthesia*. The motor point was identified by electrical stimulation and the specimens were taken to include the motor innervation. The muscle specimens were then set up in an isolated organ bath and exposed first to halothane and then to halothane plus suxamethonium. Four out of seven specimens taken developed contracture to halothane alone and a further specimen developed a contracture to halothane in the presence of suxamethonium.

The value of *in-vitro* testing of muscle biopsy specimens from relatives of patients who had suffered from malignant hyperpyrexia was confirmed by Moulds and Denborough (1974*b*).

The diagnosis of patients prone to the non-rigid type of hyperpyrexia was somewhat simpler. These patients can be distinguished from normals, not on the basis of abnormal serum enzymes, as with the rigid type, but oddly enough on the failure of intra-arterial suxamethonium to produce a neuromuscular block.

Considerable attention has been paid to the possibility of detecting sensitive individuals and of making anaesthesia possible in those patients. The family anamnesis is important in detection. Gronert (1981) recommended that susceptible patients who need anaesthesia should be given dantrolene 4 mg/kg orally in four divided doses in a 24-hour period prior to anaesthesia. The pre-anaesthetic administration of tranquillizers, such as benzo-diazapines, barbiturates, and opiates, is emphasized, and adequate monitoring is necessary. Also clodanolone, a muscle relaxant closely related to dantrolene, may be used (Sullivan *et al.* 1982). Monitoring the end-tidal carbon dioxide concentration is recommended as a rapid and reliable method for detecting elevated carbon dioxide production, one of the early clinical signs of impending malignant hyperthermia (Triner and Sherman 1981).

A related syndrome? In the present context, it is interesting to note that Pollock and Watson (1971) described a syndrome similar to anaesthetic-induced malignant hyperpyrexia after giving a combination of tricyclic antidepressants and MAO-inhibitors. There were a number of common factors in the two syndromes; the patients were usually young (20–30 years) and they experienced sweating, hyperpyrexia, and rigidity within a few hours of starting therapy. The mortality was also about 60–70 per cent. The authors suggested that this drug combination might induce a hypermetabolic state by increasing the intracellular concentration of cyclic adenosine monophosphate.

Combined therapy with these two types of antidepressants, as is now well known, is potentially hazardous and earlier reports of such interaction episodes have described sweating, hyperpyrexia, and rigidity amongst the sequelae.

Isoflurane (Forane). Two reports of malignant hyperthermia associated with isoflurane anaesthesia in a 20-year-old man and a 17-year-old man were made by Joseph *et al.* (1982) and Boheler *et al.* (1982). The hyperthermia occurred suddenly after 2½ hours and 1 hour of uneventful anaesthesia, respectively; in one case the temperature rose to over 100°F (38°C) and in the other to 105.1°F (40.6°C) and in both cases there was marked metabolic and respiratory acidosis.

Neutroleptic malignant syndrome

A condition similar to malignant hyperpyrexia is the neuroleptic malignant syndrome. It was first described as a combination of hyperpyrexia (drug fever) associated with neurologic and autonomic abnormalities developed during treatment with phenothiazines (Delay and Deniker 1968).

The syndrome comprises hyperpyrexia, altered consciousness, muscular rigidity, and autonomic dysfunction. It is a rare idiosyncratic reaction to major tranquillizers, including phenothiazines, butyrophenones, and thioxanthines; haloperidol and fluphenazine are the drugs most commonly incriminated (Smego and Durack 1982).

The pathogenesis of this disorder is obscure. In contrast to malignant hyperpyrexia there is no clear genetic basis demonstrated. The syndrome occurs after therapy rather than after toxic doses of neuroleptic drugs, and is unrelated to the duration of therapy. The syndrome is attributed to a disturbance in the central dopaminergic systems within the basal ganglia and hypothalamus that probably participates in the thermoregulation. Dopamine depletion probably plays a pathogenetic role (Burke *et al*. 1981; Henderson and Wooten 1981).

The neuroleptic malignant syndrome has a mortality of about 20 per cent, and early diagnosis and withdrawal of the neuroleptic drug is crucial. Therefore, it should be taken into consideration as a differential diagnosis in any febrile patient with a history of neuroleptic treatment. It has also occurred with a combined lithium and haloperidol therapy (Baastrup *et al*. 1976), in which case severe rigidity, mutism, and development of irreversible dyskinesia were reported as well as hyperpyrexia (Spring and Frankel 1981). However, with the same drug combination, lithium intoxication symptoms may also occur with delirium, seizures, and abnormal EEG. The possibility of the neuroleptic malignant syndrome is an argument against the use of lithium with long-acting parenteral therapy with neuroleptics in which cases the drug may remain in the body for weeks after occurrence of the syndrome (Grünhaus *et al*. 1979; Allen and White 1972; Meltzer 1973).

Drug fever related to parenteral administration

The classical febrile reaction elicited by intravenous infusion of fluids containing pyrogens was ended many years ago when the pharmacopoeias demanded that pyrogen-free water be used. These pyrogens were typically polysaccharides or lipopolysaccharides from microorganisms.

Fever can result from intravenous infusions if unauthorized dilution (Kantor *et al*. 1983) or failing hygiene with needles and catheters lead to bacterial contamination or if there is thrombophlebitis from physical stimuli. These stimuli can be mechanical lesion by the needle or damage to the vessel walls by irritating infusion fluids, resulting in an aseptic inflammation and coagulation of the blood.

Another related cause of fever is sterile abscess formation after multiple intramuscular injections or after injection of oil-based medicaments where non-absorbable oil may exert a prolonged irritation. A similar reaction was reported in connection with an artificial oleothorax, used earlier in treatment of pulmonary tuberculosis (Schou 1956). The patient had recurrent episodes of fever lasting for about one week with afebrile intervals of a little longer duration. After surgical removal of the remnants of the oil with granulated masses of sterile material the febrile episodes ceased.

The withdrawal of a therapeutic agent may also lead to fever. A baby born to a methadone-addicted mother developed hyperpyrexia accompanied by vomiting, diarrhoea, irritability, tremor, and convulsions (Amato *et al*. 1982).

This shows the multifariousness of drug-related febrile reactions.

The effect of fever on drug metabolism

Studies have shown that in adults fever may impair drug metabolism. Trenholme *et al*. (1976) showed that, in experimentally-induced fever, the plasma level of quinine to quinine metabolites increased. Elin *et al*. (1975) found that the plasma antipyrine clearance in adults may be decreased by as much as 16 per cent during etiocholanolone-induced fever.

Forsyth *et al*. (1982) demonstrated that the elimination of antipyrine from saliva measured in six children aged between five months and five years during fever caused by respiratory-tract infections, was less during the febrile period than during a subsequent afebrile period. The mean (± S.D.) saliva antipyrine clearance during fever, 32 ± 13 ml kg^{-1} h^{-1}, was nearly 50 per cent less than that when the children were afebrile, 59 ± 22 ml kg^{-1} h^{-1} ($p < 0.02$). The mean saliva antipyrine half-life during fever, 15.20 ± 5.40 h, was almost twice as long as that found when body temperature was normal, 9.18 ± 2.49 h ($p < 0.01$). The apparent volume of distribution of antipyrine was not significantly affected by fever.

The mechanism underlying the reduction of drug metabolism during fever is unknown. Body temperature itself may have a direct effect but the elevation which occurs in fever is relatively small. In addition, there was no apparent correlation between the magnitude of the temperature change and the reduction of antipyrine elimination in either the study in adults by Elin *et al*. (1975) or in children by Forsyth *et al*. (1982). Liver blood flow is increased during fever (Bradley 1949) and this could affect the rate of drug metabolism but would tend to increase the rate of drug extraction from the plasma rather than decrease it (Shand *et al*. 1971). Other factors which could change drug metabolism during fever are altered co-factor availability, circulating endogenous pyrogens, and secondary effects such as altered catecholamine and prostaglandin metabolism.

Fever occurs frequently in children and often requires the administration of drugs including antipyretics, antibiotics, and anticonvulsants. These studies indicate that in fever the metabolism of these and other drugs in children may be retarded and some modification of dosage and dose frequency may be required. This may be a relevant factor in the apparent association between the febrile prodromal illness and aspirin ingestion with Reye's syndrome (see Chapter 17, this volume).

Drug-induced fever as a potential mechanism of drug interaction

The number of patients in whom drug-induced fever arises as an adverse reaction is quite appreciable. Adverse reactions occur in about 10 per cent of all hospitalized patients and 2.5 per cent of all out-patients receiving drugs, and fever, as the sole or most prominent clinical feature, constitutes 3 to 5 per cent of all adverse reactions to drugs. It is, therefore, apparent that between one in 200 and one in 400 of all patients receiving medication will develop a drug-induced fever.

Since fever is an important cause of reduced elimination of a number of drugs, it also must be accepted that drug-induced fever will be a factor in reducing the elimination of concurrently administered medicaments. The extent and frequency of drug-induced fever as a mechanism of drug interaction has yet to be determined but it is clearly a factor which cannot be ignored.

REFERENCES

ALLEN, R.C. AND WHITE, H.C. (1972). Side-effects of parenteral long-acting phenothiazines. *Br. med. J.* **1**, 221.

AMATO, M., FREI, H., AND BÜHLMANN, U. (1982). Drogen-entzugs-syndrom beim Neugeborenen. *Schweiz. med. W.* **112**, 442–3.

BAASTRUP, P.G., HOLLNAGEL, P., AND SØRENSEN, R. (1976). Adverse reactions in treatment with lithium carbonate and halo-peridol. *J. Am. med. Ass.* **236**, 2645–6.

BERNHEIM, H.A., BLOCK, L.H., AND ATKINS, E. (1979). Fever patho-genesis, pathophysiology and purpose. *Ann. intern. Med.* **91**, 261–70.

BOCCI, V. (1980). Possible causes of fever after interferon administration. *Biomedicine* **32**, 159–62.

BOHELER, J., HAMRICK, J.C., MCKNIGHT, R.L., AND EGER, E.I. (1982). Isoflurane and malignant hyperthermia. *Anesthes. Analges.* **61**, 712–13.

BRADLEY, S.E. (1949). Variations in hepatic blood flow in man during health and disease. *New Engl. J. Med.* **240**, 456–61.

BRANCH, R.A., SHAND, D.G. WILKINSON, G.R., AND NIES, A.S. (1974). Increased clearance of antipyrine and d-propranolol after phenobarbitol treatment in the monkey: Relative contributions of enzyme induction and increased hepatic blood flow. *J. clin. Invest.* **53**, 1101–7.

BRITT, B.A. AND KALOW, W. (1968). Hyperrigidity and hyperthermia associated with anaesthesia *Ann. NY. Acad. Sci.* **151**, 947–58.

——, LOCHER, W.G., AND KALOW, W. (1969). Hereditary aspects of malignant hyperthermia. *Can. anaesth. Soc. J.* **16**, 89–98.

BURKE, R.E., FAHN, S., MAYEUX, R., WEINBERG, H., LOUIS, K., AND

WILLNER, J.H. (1981). Neuroleptic malignant syndrome caused by dopamine-depleting drugs in a patient with Huntington's disease. *Neurology* **31**, 1022–6.

BURNUM, J.F. (1976). Preventability of adverse drug reactions. *Ann intern. Med.* **85**, 80–1.

CALDWELL, J.R. AND CLUFF, L.E. (1974). Adverse reactions to anti-microbial agents. *J. Am. med. Ass.* **230**, 77–80.

CAMPBELL, I.T., ELLIS, F.R., AND EVANS R.T. (1981). Metabolic rate and blood hormone and metabolic levels of individuals susceptible to malignant hyperpyrexia at rest and in response to food and mild exercise. *Anesthesiology* **55**, 46–52.

CAROPRESO, P.R. GITTLEMAN, M.A., REILLY, D.J., AND PATERSON, L.J. (1975). Malignant hyperthermia associated with enflurance anaesthesia. *Arch. Surg.* **110**, 1491–3.

CARTER, J.J., MCLAUGHLIN, M.L., AND BERN, M.M. (1983). Bocomycin induced fatal hyperpyrexia. *Am. J. Med.* **74**, 523–8.

CHUSID, M.J. AND ATINS, E. (1972). Studies on the mechanism of penicillin-induced fever. *J. exp. Med.* **136**, 227–40.

CLUFF, L. AND JOHNSON, J.E., III (1964). Drug fever. *Prog. Allergy* **8**, 149–94.

DE BARD, M.L. (1979). Methyldopa reaction simulating septic shock. *Arch. intern. Med.* **139**, 196–7.

DELAY, J. AND DENIKER, P. (1968). Drug-induced extrapyramidal syndrome. In *Handbook of clinical neurology: diseases of the basal ganglia*, (ed. P.J. Vincken and G.W. Bruyn), Vol.6, pp. 248–66. Elsevier North Holland, New York.

DENBOROUGH, M.A., DENNETT, X., AND ANDERSON, R.McD. (1973). Central core disease and malignant hyperpyrexia. *Br. med. J.* **1**, 272–3.

——, EBELING, P., KING, J.O., AND ZAPF, P. (1970). Myopathy and malignant pyrexia. *Lancet* **1**, 1138–40.

——, GALLOWAY, G.J., AND HOPKINSON, K.C. (1982). Malignant hyperpyrexia and sudden infant death. *Lancet* **ii**, 1068–9.

DINARELLO, C.A., WARD, S.B., AND WOLFF, S.M. (1973). Pyrogenic properties of bleomycin. *Cancer Chemother. Rep.* **57**, 393–8.

DUBOWITZ, V. AND PEARSE, A.G.E. (1960). Oxidative enzymes and phosphorylase in central-core disease of muscle. *Lancet* **ii**, 23–4.

ELIN, R.J., VESELL, E.S., AND WOLFF, S.M. (1975). Effects of etiocholanolone-induced fever on plasma antipyrine half-lives and metabolic clearance. *Clin. Pharm. Ther.* **17**, 447–57.

ELLIS, F.G. (1973). Malignant hyperpyrexia. *Anaesthesia* **28**, 245–52.

FEINSTEIN, M.B. (1963). Inhibition of caffeine rigor and radio-active calcium movements by local anaesthetics in frog sar-torius muscle. *J. gen. Physiol.* **47**, 151–72.

FORSYTH, J.S. MORELAND, T.A., AND RYLANCE, G.W. (1982). The effects of fever on antipyrine metabolism in children. *Br. J. clin. Pharmacol* **13**, 811–15.

FURNISS, P. (1970). Hyperpyrexia during anaesthesia. *Br. med. J.* **4**, 745.

GELFORD, J.A., ELIN, R.J., AND BERRY, F.W. (1976). Endotoxemia associated with Jarish–Herxheimer reaction *New Engl. J. Med.* **295**, 211–13.

GRONERT, G.A. (1981). Puzzles in malignant hyperthermia. *Anesthesiology* **54**, 1–2.

GRÜNHAUS, L. SANCOVICI, S., AND RIMON, R. (1979). Neuroleptic malignant syndrome due to depot fluphanazine. *J. clin. Psychia.* **40**, 99–100.

HARRISON, G.G. (1971). Anaesthetic-induced hyperpyrexia. A

suggested method of treatment. *Br. med. J.* **3**, 454–6.

HENDERSON, V.W. AND WOOTEN, G.F. (1981). Neuroleptic malignant syndrome: Apathogenetic role for dopamine receptor blockade? *Neurology* **31**, 132–7.

HOWARD, P.H., HAHN, H.H., AND PAHNER, R.L. (1977). Fever of unknown origin: A prospective study of 100 patients. *Tex. Med.* **73**, 56–9.

ISAACS, H. AND BARLOW, M.B. (1970). Malignant hyperpyrexia during anaesthesia: possible association with subclinical myopathy. *Br. med. J.* **1**, 275–7.

JOSEPH, M.M., SHAH, K., AND VILHOEN, J.F., (1982). Malignant hyperthermia associated with isoflurane anesthesia. *Anesthes. Analges.* **61**, 711–12.

KALOW, W., BRITT, B.A., TERREAU, M.E., AND HAIST, C. (1970). Metabolic error of muscle metabolism after recovery from malignant hyperthermia. *Lancet* **ii**, 895–8.

KANTOR, R.J., CARSON, L.A., GRAHAM, D.R., PETERSEN, N.J., AND FAVERO, M.S. (1983). Outbreak of pyrogenic reactions at a dialysis center. *Am. J. Med.* **73**, 449–56.

KING, J.O., DENBOROUGH, M.A., AND ZAPF, P.W. (1972). Inheritance of malignant hyperpyrexia. *Lancet* **i**, 365–70.

LEVITES, R., HAFITZ, G., AND KIRUBAKARAN, M. (1981). Febrile reaction to sulindac. *J. Am. med. Ass.* **246**, 213–14.

LEVY, R.L. AND CHIARILLO, S. (1980). Hyperpyrexia, allergic-type response and death occurring with low-dose bleomycin administration. *Oncology* **37**, 316–17.

LIPSKY, B.A. AND HIRSCHMANN, J.V. (1981). Drug induced fever. *J. Am. med. Ass.* **245**, 851–4.

LOHR, J.A. AND HENDLEY, J.O. (1977). Prolonged fever of unknown origin: A record of experiences with 54 childhood patients. *Clin. Pediat.* **16**, 768–73.

MATTSON, K AND JÄNNE, J. (1982). Mild intravasal haemolysis associated with flu-syndrome during intermittent rifampicin treatment. *Eur. J. resp. Dis.* **63**, 68–72.

MELTZER, H.Y. (1973). Rigidity, hyperpyrexia, and coma following fluphenazine enanthate. *Psychopharmacologia* **29**, 337–46.

MOULDS, R.F.W. AND DENBOROUGH, M.A. (1947a). Identification of susceptibility to malignant hyperpyrexia. *Br. med. J.* **2**, 245–7.

—— AND —— (1974b). Biochemical basis of malignant hyperpyrexia. *Br. med. J.* **2**, 241–4.

PARK, G.D., SPECTOR, R., HEADSTREAM, T., AND GOLDBERG, M. (1982). Serious adverse reactions associated with sulindac. *Arch. intern. Med.* **142**, 1292–4.

PEERLESS, A.G. (1981). Drug fever in severe combined immuno-

deficiency, a dilemma letter. *Lancet* **ii**, 638.

PETERSDORF, R.G. AND BEESON, P.B. (1961). Fever of unexpected origin: Report on 100 cases. *Medicine* **40**, 1–30.

POLLOCK, R.A. AND WATSON, R.L. (1971). Malignant hyperthermia associated with hypokalcaemia. *Anaesthesiology* **34**, 188–94.

ROOT, R.K. AND WOLFF, S.M. (1968). Pathogenic mechanisims in experimental immune fever. *J. exp. Med.* **128**, 309–23.

SCHOU, J. (1956). Farmakon-feber [Drug fever]. *Ugeskr. f. Laeger* **118**, 1231–6.

SHAND, D.G., EVANS, G.H., AND NIES, A.S. (1971). The almost complete hepatic extraction of propranolol during intravenous administration in the dog. Part 1. *Life Sci.* **10**, 1417–21.

SHY, G.M. AND MAGEE, K.R. (1956). A new congenital non-progressive myopathy. *Brain* **79**, 610–21.

SMEGO, R.A. AND DURACK, D.T. (1982). The neuroleptic malignant syndrome. *Arch. intern. Med.* **142**, 1183–5.

SNELL, E. (1969). Hypersensitivity fever. *Br. J. clin. Pract.* **23**, 73–7.

SPRING, G. AND FRANKEL, M. (1981). New data on lithium and haloperidol incompatibility. *Am. J. Psychiat.* **138**, 818–21.

STAHL, N.I., KLIPPEL, J.H., AND DECKER, J.L. (1979). Fever in systemic lupus erythematosus. *Am. J. Med.* **67**, 935–40.

STRUNIN, L. (1975). General anaesthetics and therapeutic gases. In *Side-effects of drugs*, Vol. VIII, (ed. M.N.G. Dukes), pp. 241–55. Excerpta Foundation, Amsterdam.

SULLIVAN, J.S., GALLOWAY, G.J., AND DENBOROUGH, M.A. (1982). Clodanolene sodium and malignant hyperthermia. *Br. J. Anaesthiol.* **54**, 1237.

SUTHERLAND, F.S., AND CARTER, J.R. (1975). Malignant hyperpyrexia during enflurane anaesthesia. *J. Tenn. med. Ass.* **68**, 785–6.

TIERNEY, L.M. (1978). Drug fever. Medical staff conference. University of California, San Francisco. *Western med. J.* **129**, 321–6.

TRENHOLME, G.M. WILLIAMS, R.L., RIECKMANN, K.H., FRISCHER, H., AND CARSON, P.E. (1976). Quinine disposition during malaria and during induced fever. *Clin Pharmacol. Ther.* **19**, 459–67.

TRINER, L. AND SHERMAN, J. (1981). Potential value of expiratory carbon dioxide measurement in patients considered to be susceptible to malignant hyperthermia. *Anesthesiology* **55**, 482.

WEBER, A. (1968). The mechanism of action of caffeine on sarcoplasmic reticulum. *J. gen. Physiol.* **52**, 760–72.

—— AND HERZ, R. (1968). The relationship between caffeine contracture of intact muscle and effect of caffeine on reticulum. *J. gen. Physiol.* **52**, 750–9.

34 Renal disease associated with drugs

MARY G. McGEOWN

Nephrotoxic drugs

A great many drugs and therapeutic agents may cause damage to the kidneys and urinary tracts. Drugs may damage the kidneys directly or indirectly and often the mechanism is not fully understood. Some drugs are known to have more than one side-effect. Drug reactions may be considered in the following categories.

1. General hypersensitivity and idosyncratic reactions;
2. Direct toxicity to the kidney;
3. Indirect toxicity to the kidney;
4. Drug-induced abnormalities of the lower urinary tract.

Drugs not harmful to patients with normal kidney function may be injurious in patients with already impaired renal function; and in elderly people. Drugs harmless when used alone may become nephrotoxic when given along with other drugs. Hypotension, which may of itself cause renal damage, potentiates the toxic effects of many drugs. Great care should be taken when selecting drugs for treatment of patients with impaired renal function, and combinations of drugs should be avoided when possible and, if necessary, should be carefully considered.

General hypersensitivity and idiosyncratic reactions

These reactions are neither dose- nor dose-duration-related. They occur without warning, after varying periods of exposure, and are often part of a generalized reaction involving skin, blood vessels, and other systems. Involvement of the kidneys in hypersensitivity reactions is fairly uncommon but tends to be severe. Damage to the kidneys may be severe and progressive, but is more often reversible after withdrawal of the drug; corticosteroid therapy may be beneficial.

Hypersensitivity reactions cause damage to glomeruli, which may be slight or severe, and may be accompanied by interstitial oedema, infiltration with lymphocytes and plasma cells and sometimes large numbers of eosinophils. When the renal damage does not appear as part of a generalized hypersensitivity reaction, as occasionally happens, the glomerular lesions may be difficult to distinguish from other forms of glomerulonephritis, but the presence of interstitial oedema and a cellular infiltrate, especially if eosinophils are prominent, may suggest a drug-induced hypersensitivity reaction.

Routine toxicological testing of new drugs does not reveal the potential to produce hypersensitivity reactions. Hypersensitivity is an abnormal response on the part of the patient, and this hazard is revealed only by clinical experience.

Direct toxicity

Some drugs may be toxic in normal dosage, while others are toxic only when taken in high dosage or when excretion is impaired due to already reduced renal function. Damage may affect any part of the kidneys or renal tracts so that many different clinical pictures may be seen.

Direct toxic effects on the kidney are sometimes predictable from testing in animals. However, some direct toxic effects occur only when renal function is already reduced or when other drugs are administered at the same time, and may not be predicted from animal experiments.

Indirect toxicity

Drugs may have an indirect effect on the kidney as a consequence of a primary extrarenal effect. The extrarenal effect may be pre-renal or post-renal. Examples of pre-renal drug-induced renal disease are acute haemolysis leading to acute renal failure; severe potassium deficiency due to over-use of purgatives leading to acute or chronic impairment of renal function. Retroperitoneal fibrosis is an example of the post-renal effect of certain drugs. For practical purposes post-renal effects of drugs are usually considered separately.

Indirect toxicity of drugs is not likely to be predicted from animal experiments and is usually discovered through association of a disease pattern with a particular drug.

Drug-induced abnormalities of the lower urinary tract

Drug-induced abnormalities of the lower urinary tract include obstructive renal failure, lesions of the bladder, and drug-induced retention or incontinence of urine. Obstructive renal failure can easily be missed and may masquerade as acute or chronic renal failure; *it is a sound clinical rule that an obstructive element should be sought in every patient with renal failure*. The obstruction may be within the renal pelvis, or ureters, or the ureters may be obstructed from without, as in retroperitoneal fibrosis. Lesions of the trigone of the bladder may involve both ureters; lesions below the bladder neck obstruct both upper renal tracts. To lead to renal failure, obstruction which arises above bladder level must involve both ureters, or the ureter or pelvis of a solitary functioning kidney. Blockage within the lumen may be due to clumps of crystal, stone, blood clot, sloughed renal pyramid, or even thick pus. More rarely obstruction may be due to thickening or oedema of the ureteric wall, or retroperitoneal fibrosis.

An important clue to the possibility of obstruction as the cause of renal failure is great variation in urine volume from day to day, from near or even complete anuria to large volumes, later followed by low volumes. This cycle of events may be repeated many times and is probably due to a variable amount of oedema around the original obstruction.

Ureteric obstruction may remain undetected by animal studies unless they are very prolonged, but drug-induced urinary retention should be discovered. Haemorrhagic cystitis is most likely to be a hypersensitivity reaction and is therefore not predictable from animal studies.

Urinary retention due to drugs usually subsides when the drug is withdrawn although a short period of catheter drainage may be necessary. Cystitis also recovers after the treatment is discontinued. Retroperitoneal fibrosis is treated by freeing the ureters from the dense mass of new fibrosis tissue and by wrapping them in omentum to prevent recurrence of fibrosis. Corticosteroids have been used in the hope of preventing recurrence of the condition but without convincing benefit. Stenosis and thickening of the ureteric wall is treated by removal of the affected length of ureter and anastomosis of the proximal end to the bladder and, if necessary, by inserting a piece of ileum to lengthen the remaining ureter.

Drug effects on renal function

Drug effects may be classified according to the pharmacological use of the drugs or according to the nature of the toxic effects. The most logical classification of drugs affecting the kidney seems to be according to the part of the renal tract involved, commencing at the glomerulus and proceeding downwards. An indication is given, when possible, as to whether the effect is thought to be due to hypersensitivity, direct toxicity, or indirect toxicity. The drugs are considered under each heading in alphabetical order.

1. Drugs which affect the glomerulus, causing proteinuria or the nephrotic syndrome.
2. Drugs causing acute renal failure thought to be due to acute tubular necrosis, haemolysis, or myolysis.
3. Drugs known or thought to impair renal function by causing interstitial nephritis. The more frequent use of renal biopsy in patients with impaired renal function has shown that acute interstitial nephritis is a very common cause of drug-induced renal disease.
4. Drugs causing impairment of distal tubular function.
5. Drugs causing analgesic nephropathy and papillary necrosis.
6. Drugs causing obstructive renal disease.
7. Drugs causing disturbance of micturition.
8. Cosmetic agents known to affect renal function have been included although they are not strictly therapeutic agents.

Drugs affecting the glomerulus

Damage to glomeruli produced by drugs causes proteinuria and is probably the commonest toxic effect on the kidney. The urinary loss of protein may be comparatively small and detected only if the urine is tested. A large number of drugs may cause urinary protein losses large enough to lead to hypoproteinaemia and oedema — the nephrotic syndrome. Much less commonly , drugs may cause a syndrome resembling acute glomerulonephritis: the patient passes very small amounts of urine loaded with albumin, red cell, and granular casts, 'the face and feet become oedematous', and the blood pressure rises. More serious damage to the glomeruli by necrotizing vasculitis occasionally occurs. In most cases the proteinuria subsides after the drug is discontinued though it sometimes persists for several months. Serious permanent damage to kidney function follows necrotizing glomerular disease.

A list of drugs known to cause glomerular damage and proteinuria is shown in Table 34.1. Some of the drugs

Table 34.1 Drugs causing glomerular damage, proteinuria with/without nephrotic syndrome

Drug	Use	Nephrotic syndrome	Other known renal effects	Contributing factor	Recovery of renal function
Aloes	purgative	+	acute renal failure via K^+ loss	–	±
Amphotericin B	antifungal, antibiotic	–	acute renal failure	–	±
Bacitracin	antibiotic	–	tubular dysfunction	–	+
Bismuth salts	antacid	+	acute renal failure	–	+
Captopril	hypotensive	+	acute renal failure	renal artery stenosis	+
Cephaloridine	antibiotic	–	–	high doses	+
Dapsone	antileprotic, dermatitis herpetiformis	+	haematuria	deficiency of G6PD	+
Diloxanide	amoebiasis	–	–	–	+
Disodium edetate	chelating agent	–	–	–	+
Ethosuximide	anticonvulsant	–	–	–	+
Fenclofenac	antirheumatic	–	acute interstitial nephritis	–	+
Fenoprofen	antirheumatic	+	acute interstitial nephritis	–	+
Glue-sniffing	addiction	–	acute renal failure	–	±
Gold salts	antirheumatic	+	–	–	±
Halothane	anaesthetic	+	–	–	+
Hexamine	antibacterial	–	haematuria	–	+
Hydrallazine	hypotensive	–	–	–	±
Indomethacin	antirheumatic	–	–	pre-existing glomerular lesion	–
Ipecacuanha	emetic, expectorant	–	–	–	+
Lithium	antidepressant	+	acute interstitial nephritis	–	+
Male fern	antihelmintic	–	–	impaired renal function	+
Mersalyl and other mercurials	diuretic	+	–	–	±
Methsuximide	anticonvulsant	+	–	–	+
Nafcillin	antibiotic	–	–	–	+
Naproxen	antirheumatic	–	–	–	+
Paramethadione	anticonvulsant	+	–	–	+
Penicillin	antibiotic	–	–	–	+
Penicillamine	antirheumatic, chelating agent	+	–	–	±
Pheneturide	anticonvulsant	+	–	–	+
Phenindione	anticoagulant	+	–	acute renal failure	+
Phenolphthalein	laxative	–	–	–	+
Phensuximide	anticonvulsant	–	+	–	+
Phenylbutazone	antirheumatic	–	–	–	±
Piroxicam	antirheumatic	–	–	–	+
Polymyxins	antibiotics	–	tubular dysfunction	–	+

Table 34.1 *Continued*

Drug	Use	Nephrotic syndrome	Other known renal effects	Contributing factor	Recovery of renal function
Probenecid	uricosuric	+	tubular dysfunction	–	?
Sodium aminosalicylate	antitubercular agent	–	acute renal failure	–	+
Sodium cacodylate	antisyphilitic	–	–	–	±
Sodium calcium edetate	chelating agent	–	–	–	+
Suramin	trypanocide	–	acute renal failure	–	+
Theophylline monoethanolamine	bronchodilator	–	–	–	+
Tolbutamide	hypoglycaemic agent	+	–	–	+
Troxidone	anticonvulsant	+	–	–	+

cited may also cause other types of renal damage, which may or may not be reversible.

Captopril

There have been numerous accounts of proteinuria associated with the use of the potent hypotensive agent captopril, which inhibits conversion of angiotensin I to angiotensin II. Although it may be severe enough to cause the nephrotic syndrome, the proteinuria may subside despite continued therapy. It also seems to occur less frequently than originally appeared. Case *et al.* (1980) reported that six of 81 patients treated with captopril developed proteinuria, although two had previously had proteinuria. The proteinuria varied from 976 mg to 13.8 g/24 h, and two patients developed the nephrotic syndrome. Two patients had a renal biopsy and both showed membranous glomerulonephritis as was also found by Hoorntje *et al.* (1980). The proteinuria regressed spontaneously in five of the six patients despite continued therapy with captopril in the same dosage.

There have been several reports of acute renal failure developing in patients with severe hypertension associated with bilateral renal artery stenosis when treated with captopril. Hricik *et al.* (1983) reported 11 patients who developed acute renal failure when on captopril, all of whom had renal artery stenosis bilaterally or in a solitary kidney. The renal failure resolved promptly after the drug was discontinued. They suggested that the renal failure may be due to blockage of the renin–angiotension system in the presence of reduced renal artery perfusion pressure. Silas *et al.* (1983) suggested that captopril-induced reversible renal failure is a marker for renal artery stenosis affecting a solitary kidney. Curtis *et al.* (1983) reported that captopril caused acute loss of renal transplant function in four patients with graft artery stenosis but not in five patients known not to have stenosis. Pirson *et al.* (1983) reported six similar cases. The occurrence of acute renal failure with captopril is not always a marker for graft artery stenosis as Nath *et al.* (1983) reported acute renal failure in several patients without stenosis. Moreover, renal function deteriorated again on rechallenge with captopril. See 'Drugs causing interstitial nephritis'.

Fenclofenac

Proteinuria and the nephrotic syndrome has been reported in patients given this anti-inflammatory drug for rheumatoid arthritis. A further case of severe nephrotic syndrome which recovered after the drug was discontinued was reported (Hamilton *et al.* 1979). A renal biopsy showed mild focal proliferative glomerulonephritis and mild intestinal fibrosis but arteritis and amyloid were not present. Eosinophilia was present and the patient recovered after treatment with prednisolone.

Glue-sniffing

Venkataraman (1981) reported a patient who developed progressive renal failure after five years of glue sniffing (*Evostik*) He had a heavy proteinuria (12.5 g/24 h). Renal biopsy showed mesangiocapillary glomerulonephritis with immune complex depositions. Renal function continued to deteriorate despite discontinuance of glue sniffing. *Evostik* contains organic hydrocarbon solvents which are thought to be the cause of the nephritis.

Hexamine

The *British National Formulary* (1981) advised against the use of hexamine, because of the frequency of side effects such as haematuria and proteinuria. It is bacteriostatic rather than bactericidal.

Hydrallazine

Rapidly progressive glomerulonephritis associated with hydrallazine therapy was reported by Bjorck *et al.* (1983) and Kincaid-Smith and Whitworth (1983). Biopsies were carried out on Kincaid-Smith's four cases and showed segmental proliferative and necrotizing glomerular lesions with immune deposits in glomerular basement membranes. All four patients were slow acetylators. The drug was withdrawn and they were given prednisolone and cyclophosphamide. Six months to two years later one patient had normal renal function while the other three had mild-to-moderate impairment.

Hydrallazine has been implicated in the production of necrotizing vasculitis with acute renal failure (Finlay 1981). It is well known that hydrallazine may produce a systemic lupus erythematosis-like syndrome.

Indomethacin

Although indomethacin has been advocated for the treatment of glomerulonephritis, it can also cause renal disease. Marsh *et al.* (1971) reported a patient who developed non-thrombocytopenic purpura and acute glomerulonephritis after indomethacin therapy. Tan *et al.* (1979) reported reversible acute renal failure in a patient given indomethacin for only five days. Indomethacin administration in our experience is a very effective method of carrying out medical bilateral nephrectomy. Indeed, Dr C.C. Doherty used it for this purpose in a patient with end-stage renal failure due to amyloid disease who continued to excrete large amounts of protein in his urine while requiring haemodialysis therapy. After three days on indomethacin 50 mg b.d. he became totally anuric (Unpublished observation 1982). Dr L. Baker used indomethacin therapy for the same purpose with similar results (Personal communication 1982). It would appear from these experiences that great caution is necessary if indomethacin is given to a patient whose renal function is already impaired.

Lithium

Lithium is a drug widely used for treatment of manic-depressive disorders which can cause sodium diuresis and polyuria and it has also been shown to be associated with interstitial renal disease and glomerulonephritis. Richman *et al.* (1980) reported a patient who developed heavy proteinuria (21.5 g per day), severe oedema, and thrombocytopenia after 10 month's treatment with lithium. The proteinuria and thrombocytopenia disappeared within two weeks of withdrawing the drug. After two years without proteinuria, lithium was given again and five weeks later the nephrotic syndrome recurred but resolved again when it was withdrawn. Renal biopsies were done during both episodes of the nephrotic syndrome. Both biopsies appeared normal on light microscopy, immunofluorescence was absent, and on electron microscopy there was the loss of foot processes of the endothelial cells of the glomeruli typical of minimal-change glomerulonephritis. Moskovitz and Miller (1981) reported a patient who developed the nephrotic syndrome with 22 g proteinuria when taking lithium and haloperidol. The proteinuria rapidly subsided when both drugs were stopped and did not recur when haloperidol was again given (see also under 'Drugs causing interstitial nephritis').

Penicillamine

Proteinuria seems to be considerably more frequent when penicillamine is given for rheumatoid arthritis or cystinuria (30 per cent) than for Wilson's disease (4 per cent). The occurrence of proteinuria seems to be related both to the maintenance dose and the rate at which the dosage is increased (Panayi *et al.* 1978).

Renal biopsy of patients with proteinuria associated with penicillamine therapy has shown a variety of lesions including minimal change, mesangio-proliferative, membranous, and crescentric glomerulonephritis. A few patients have developed clinical features of Goodpasture's syndrome and have had linear immunofluorescence on biopsy. A fatal renal vasculitis has also been reported (Falck *et al.* 1979). The commonest lesion is membranous nephropathy (Bacon *et al.* 1976).

Before commencing penicillamine therapy, renal function should be assessed, including measurements of 24-hour excretion of protein. It seems to be important to exclude urinary infection in patients with cystine calculi (*British Medical Journal* 1981). Dodd *et al.* (1980) suggest that adverse reaction to penicillamine is more likely if the patient has previously had toxic side-effects to gold therapy. The starting dose should not exceed 250 mg per day and should be increased very gradually, with frequent measurement of urinary excretion of protein. Proteinuria of less than 2 g/24 h daily is probably not a contraindication to continued treatment, and may subside. Some clinicians are prepared to continue treatment when proteinuria exceeds 2 g/24 h, even up to 5–6 g/24 h. Heavy proteinuria leading to the nephrotic syndrome may develop rapidly but usually subsides when the drug is stopped, and steroid therapy may hasten recovery. It is essential to continue to monitor urinary protein excretion as long as penicillamine is continued.

Piroxicam

Piroxicam is a new anti-arthritic agent which has been shown to be a potent inhibitor of prostaglandin synthesis. Goebel and Mueller-Brodmann (1982) reported that two patients receiving piroxicam developed acute nephropathy with the characteristic features of Henoch–Schönlein purpura. Both recovered on stopping the drug. One patient was given a second dose and again developed signs of Henoch–Schönlein purpura.

Drugs causing acute renal failure due to acute tubular necrosis, haemolysis, or myolysis

Acute renal failure may be caused by direct toxicity leading to acute tubular necrosis or by the products of haemolysis or myolysis leading to cast formation in the renal tubules. Table 34.2 lists drugs causing acute renal failure of this type.

Table 34.2   Drugs causing acute renal failure

Drug	Use	Probable mechanism	Haemolysis	Other known renal effects	Contributing factors	Recovery of renal function
Acyclovir	antiviral	direct toxicity	–	–	–	±
Amidopyrine (aminopyrine)	analgesic, antipyretic	?hypersensitivity	–	–	other drugs	±
Aminocaproic acid		indirect toxicity	myolysis	–	–	+
Amphotericin B	antifungal, antibiotic	hypersensitivity	–	proteinuria, urinary retention	–	+
Aspirin	analgesic, antipyretic, antirheumatic	direct toxicity	–	proteinuria, haematuria, casts	impaired renal function	+
Barbiturates	sedatives	indirect toxicity	myolysis	–	overdose	+
Bismuth salts	antacid, antisyphilitic	direct toxicity	–	Fanconi's syndrome, proteinuria	–	incomplete
Blood (incompatible)	transfusion	indirect toxicity	+	–	–	+
Boric acid (borax)	disinfectant lavage	direct toxicity	–	–	lavage body cavities or large denuded areas	+
Camphor	expectorant, rubifacient	direct toxicity	–	–	–	+
Carbenoxolone	gastric and duodenal ulcers	indirect toxicity	myolysis	K$^+$ loss, Na$^+$ and water retention	thiazide diuretic	+
Carbon tetrachloride	anthelmintic	direct toxicity	–	–	–	+
Cephalexin	antibiotic	direct toxicity	–	proteinuria	impaired renal function with amino-glycosides	+
Cephaloridine	antibiotic	direct toxicity	–	–	large dose	+
Chlorprothixene	sedative, anti-emetic	direct toxicity	–	–	impaired renal function	+
Cisplatin	cyctotoxic	direct toxicity	–	–	–	±
Colistin (polymyxin E)	antibiotic	direct toxicity	–	hypokalaemia	impaired renal function	+
Contraceptive pill	contraceptive		+	interstitial nephritis		+

(continued overleaf)

Table 34.2 *Continued*

Drug	Use	Probable mechanism	Haemolysis	Other known renal effects	Contributing factors	Recovery of renal function
Contrast media	radiology	direct toxicity	–	–	dehydration, myeloma, jaundice	+
Co-trimoxazole	antibacterial	?hypersensitivity	–	proteinuria, red cell and granular casts	impaired renal function	±
Dextran	plasma expander	?	–	–	impaired renal function, vascular disease	
5-Fluorouracil	cytotoxic	indirect toxicity	+	–	–	–
Glycerol	cerebral oedema	indirect toxicity	+	oxalosis	–	–
Gold salts	antirheumatic	direct toxicity	–	proteinuria	–	±
Ibuprofen	antirheumatic		–	–	–	+
Iodine	pre-operative for thyrotoxicosis, contrast media	direct toxicity	–	–	–	+
Mephenesin	muscle relaxant	indirect toxicity	+	–	–	+
Mersalyl and other mercurials	diuretic, skin lightener	direct toxicity	–	proteinuria	–	±
Methoxyflurane	anaesthetic	indirect and direct toxicity, deposition of calcium oxalate	–	failure of concentration	prolonged anaesthesia, other drugs	–
Niridazole	schistosomicide	indirect toxicity	+	–	only when deficiency of G6PD	+
Paracetamol	analgesic	direct toxicity	–	–	–	+
Phencyclidine	addiction	indirect toxicity	myolysis	–	–	+
Phenindione	anticoagulant	?hypersensitivity	+	proteinuria	–	high mortality
Phenol	disinfectant, skin sclerosing, dental use	indirect toxicity	+	–	–	+
Phenylbutazone	antirheumatic	?hypersensitivity	–	–	known allergy	±
Phenylhydrazine	polycythaemia vera	indirect toxicity	+	–	–	+
Potassium chlorate	astringent, mouth wash	indirect toxicity	+	–	–	+
Quinine salts	antimalarial, abortifacient	indirect toxicity (shock), hyper-sensitivity	+	–	malaria	+
Rifampicin	antitubercular antibiotic	?hypersensitivity	–	–	?intermittent use	+
Sodium aminosalicylate	antitubercular agent	hypersensitivity	+	SLE	impaired renal function	+
Sodium chlorate	mouth wash (weed killer)	indirect toxicity	+	–	–	+
Sodium chloride (hypertonic solution)	electrolyte imbalance	indirect toxicity	+	–	–	+ or death
Sulphonamides	antibacterials	hypersensitivity	–	crystalluria, activate SLE	–	+
Teniposide	cytotoxic		+	–	–	+
Tetrachloro-ethylene	antithelmintic	direct toxicity	–	–	–	+
Tetracycline	antibiotic	direct toxicity	–	elevation of urea, tubular dysfunction	large doses IV (impaired renal function)	+
Trichlorethylene	anaesthetic	direct toxicity	–	–	–	+

Acyclovir

The antiviral agent acyclovir given intravenously may cause elevation of serum urea and serum creatinine. This was reported by Chapman and Brigden (1981). On the basis of animal work, Weller *et al.* (1982) suggested that the renal impairment may be due to transient crystal nephropathy when the concentration of acyclovir in the collecting ducts exceeds drug solubility. However, the author observed acute renal failure due to acute tubular necrosis in a renal transplant. The patient had a renal biopsy for a suspected episode of rejection which was unremarkable but later, after the acyclovir was given intravenously, the patient became oliguric and a second renal biopsy showed severe acute tubular necrosis; recovery of renal function was slow and incomplete (Unpublished observation).

Aminocaproic acid

Acute renal failure due to myoglobinuria has been reported during therapy with aminocaproic acid to inhibit fibrinolysis after subarachnoid haemorrhage. Muscle biopsy has shown massive necrosis. The renal failure is severe, requiring haemodialysis, though recovery occurs. Brodkin (1980) reviewed the literature and suggested that the dose should not exceed 24 g per day and should not be given for longer than 28 days. Severe muscle pain is a warning sign of this serious complication.

Barbiturates

Several cases of acute tubular necrosis have been reported after over-dosage with barbiturates. The patients were found in deep coma and necrosis of muscle masses on which they were lying caused myoglobinuria and acute renal failure.

Carbenoxolone

Carbenoxolone is used for the treatment of gastric ulcer. Well-recognized side-effects are retention of sodium and water for which thiazide diuretics have been prescribed. There have been two reports of rhabdomyolysis leading to acute renal failure (Descamps *et al.* 1977; Hurley 1977). Both patients were very ill and required dialysis but ultimately recovered. Both carbenoxolone and thiazide diuretics cause potassium deficiency and it was suggested that this was severe enough to account for the muscle necrosis.

Cisplatin

This cytotoxic drug seems to be very nephrotoxic and necrosis of proximal tubules occurs in about 50 per cent of cases (Maher 1981). The nephrotoxicity may be lethal (Levi *et al.* 1983).

Contraceptive pill

Whicher *et al.* (1981) reported haemolytic uraemic syndrome in association with oestrogen-containing oral contraceptive pills. See 'Drugs causing interstitial nephritis'.

Contrast media

Acute renal failure was reported more frequently with older urographic contrast media such as idiopyracet *(Diodrast)* and sodium acetrizoate *(Urokon)*, but has also occurred with meglumine *(Renografin)* and sodium diatrizoate *(Hypaque)*. There is special risk to the kidneys when high concentrations are delivered into the aorta or renal arteries for arteriography. Patients with myelomatosis are considered to be at special risk; Marshall (1971) reviewed the evidence and concluded that excretory urography carries a small but definite risk for patients with myelomatosis. Jaundiced patients also seem to be at special risk when given contrast media (McEvoy *et al.* 1970). The risk of contrast media is increased by deprivation of fluids and purgation and these should be avoided in patients known to have impaired renal function. A new series of 'low-osmolality' intravascular contrast media have reduced cardiotoxic and haemodynamic effects and this may minimize any pre-renal element in acute renal failure. It is claimed that they have lower intrinsic nephrotoxicity than the older media (Dawson *et al.* 1983). However, in the majority of patients with undiagnosed renal failure contrast-medium studies are no longer necessary. Dilatation of the collecting system can almost always be diagnosed by ultrasound and kidney size can be defined by computed tomography. Webb *et al.* (1983) recommended a plain film of abdomen to check for calculi, with renal tomography and ultrasound to define renal size and exclude a dilated collecting system.

Cyclosporin

This drug was introduced as an immunosuppressive agent for organ transplantation. It was first used by Calne *et al.* (1979) as the sole immunosuppressive drug for kidney transplantation. They reported acute reduction in graft function within the first few days after transplantation although biopsy of the graft looked almost normal. The reduction in graft function occurred even when the kidney was taken from a living related donor (when acute tubular necrosis is extremely rare). The toxicity is dose-related and renal function improves again when the dose is reduced or when azathioprine and steroid is substituted. All transplant centres using cyclosporin for immunosuppression for renal transplantation have repeated the Cambridge experience. Prolonged oliguric renal failure has been reported (Bear *et al.* 1983). Nephrotoxicity occurs in the previously normal kidneys of patients receiving cyclosporin for bone-marrow, liver, and heart transplantation and haemodialysis is sometimes required. In renal transplantation it is very difficult to distinguish cyclosporin toxicity from acute rejection, as the usual clinical signs of rejection are suppressed. It has been observed that eventual renal function in stable grafts treated with cyclosporin is less good than in those treated with azathioprine and steroid. Keown *et al.* (1981) reported interstitial fibrosis in graft biopsies after three months and this suggests that decline in graft function may occur later.

5-Fluorouracil and Mitomycin C

Jones *et al.* (1980) and Lempert (1980) reported progressive renal failure in patients receiving longer-term therapy with these drugs. The patients died.

Glue-sniffing

Will and McLaren (1981) reported reversible acute renal failure after glue-sniffing. The author has seen a similar case.

Ibuprofen

Fong and Cohen (1982) reported acute reversible renal failure in two patients with systemic lupus erythematosis (SLE) who received ibuprofen, a phenylproprionic anti-inflammatory agent, for painful swollen joints. Before receiving ibuprofen one patient had minimal proteinuria, the other was otherwise normal. Both developed impaired renal function shortly after treatment was commenced, and one required peritoneal dialysis. Both patients had renal biopsies which showed acute tubular necrosis as well as features of SLE. Ibuprofen was discontinued and both recovered normal renal function.

Incompatible blood

Transfusion of incompatible blood is rapidly followed by haemolysis, loin pain, and oliguria to complete anuria. It is now seldom due to mismatch of blood in the laboratory and is usually the result of giving the matched blood to the wrong patient in theatre or ward. It is essential to follow strictly the checking procedure with patient's personal particulars and hospital and laboratory numbers before giving blood.

Paracetamol

Gabriel *et al.* (1982) reported acute reversible renal failure without other known cause in two men taking paracetamol in normal doses. Renal biopsies in both patients showed healing acute tubular necrosis. They suggested that, as paracetamol is not regarded as nephrotoxic in standard dosage and is widely used in the treatment of trivial illness, acute renal failure due to it may not be recognized. Direct enquiry as to paracetamol ingestion should be made in patients with renal failure of unknown cause. Acute renal failure often accompanies hepatic failure in paracetamol poisoning but it is less well recognized that renal failure may occur without hepatic failure. Cobden *et al.* (1982) reported 10 patients in whom acute renal failure occurred in the absence of severe liver damage; 10 required peritoneal dialysis, all recovered.

Phencyclidine

Akmal *et al.* (1981) reported rhabdomyolysis in patients intoxicated with phencyclidine, a drug of addiction. Ten out of 1000 cases treated developed acute renal failure and seven had mild impairment of renal function.

Phenylbutazone and oxyphenbutazone

Acute renal failure, with anuria, is a rare complication associated with the use of these analgesic and anti-inflammatory drugs. The rarity suggests that these reactions may be due to hypersensitivity. Acute tubular necrosis was reported by a number of authors, and a review by Bloch *et al.* (1966) revealed 50 cases of renal complications due to phenylbutazone and reported one personal case. Doyle *et al.* (1983) reported reversible acute tubular necrosis in their cases.

Quinine

The effects of quinine on the kidney are indirect and usually result from haemolysis or hypotension. Some patients are hypersensitive to quinine and haemolysis can occur. Quinine given to a patient with inadequately controlled malarial infection may precipitate an attack of blackwater fever with severe haemolysis.

Teniposide

Renal failure due to acute intravascular haemolysis following the use of teniposide, a cytotoxic agent used mainly for treatment of lymphomas and brain tumours, was reported by Habibi *et al.* (1981). Immediately after the 34th infusion of the drug (total dose 3000 mg) in a patient with a brain tumour, malaise, chills, fever, myalgia, and loin pain developed. Haemoglobinuria and jaundice appeared a few hours later. The urinary output fell to 200 ml, the haematocrit from 30 per cent to 19 per cent. An antiteniposide antibody in high titre was demonstrated which was still present 17 months later. Renal function recovered completely.

Tetracyclines

Severe renal failure has occurred when large doses of tetracyclines are given intravenously. Most but not all patients reported were pregnant women being treated for acute pyelonephritis, and the mortality rate was high. The kidneys showed areas of cortical necrosis. Tetracyclines are known to depress glomerular filtration rate in patients with already impaired renal function (Phillips *et al.* 1974); this may be sufficient to precipitate severe and even irreversible renal failure. Tetracyclines also increase catabolism which may contribute to the rise in blood urea. Doxycycline is said to be safe in patients with impaired renal function, but all other tetracyclines should be avoided. See also 'Drugs causing interstitial nephritis'.

Drugs causing interstitial nephritis

The increasing use of renal biopsy has shown that many drugs cause acute interstitial nephritis rather than glomerular damage or acute tubular necrosis. It is important to recognize acute interstitial nephritis as complete or almost complete recovery occurs after stopping the drug. If unrecognized the patient may die. The presence of skin rash may be a clue, but is often absent. The biopsy shows oedema, interstitial infiltration with lymphocytes, plasma cells, and sometimes numerous eosinophils. Eosinophils in peripheral blood are sometimes present. Steroid therapy is believed to be helpful though many patients recover without it. Linton *et al.* (1980) reviewed the literature and reported nine cases of their own presenting as acute renal failure and stated that drug-induced acute interstitial nephritis may account for 8 per cent of all cases of acute renal failure, not counting very ill patients in whom biopsy was not possible. Drugs most frequently reported included methicillin, ampicillin, cep-

Table 34.3 Drugs associated with acute interstitial nephritis

Drug	Use	Recovery of renal function
Amoxicillin	antibiotic	+
Ampicillin	antibiotic	+
Bendrofluazide	diuretic	+
Captopril	hypotensive	+
Cephalexin	antibiotic	+
Cimetidine	H_2-receptor blocker	+
Clafenin	analgesic	+
Clofibrate	anticholesterol formation	+
Contraceptive pill	contraceptive	+
Co-trimoxazole	antibiotic	+
Diflunisal	analgesic	+
Diphosphonate	anti-Paget's disease	+
Erythromycin	antibiotic	+
Fenoprofen	antirheumatic	+
Frusemide	diuretic	+
Gentamicin	antibiotic	+
Hydrochlorothiazide	diuretic	+
Interferon	antifungal	+
Lithium	treatment of mania	+
Magnapen	antibiotic	+
Mefenamic acid	analgesic	+
Methicillin	antibiotic	+
Minocycline	antibiotic	+
Naproxen	antirheumatic	+
Penicillin	antibiotic	+
Rifampicin	antibiotic	±
Triamterene	diuretic	+
Tienilic acid		

halosporin, sulphonamides, phenindione, allopurinol, non-steroidal anti-inflammatory agents, thiazides, frusemide, and phenylbutazone.

Transient minor elevation of the blood urea, which subsides fairly quickly, occurs commonly in ill patients in hospital. This is often unexplained or attributed to dehydration and/or hypotension which in themselves, especially in elderly patients, may be sufficient to cause short periods of oliguria and rise in blood urea, and may potentiate drug toxicity. Very ill patients commonly receive multiple drugs and it is often not possible to do a renal biopsy or, if one is done, to decide which drug is the offending one.

Table 34.3 lists drugs associated with biopsy-proven acute interstitial nephritis.

Antibiotics

Nephrotoxicity due to antibiotics continues to be reported frequently. Four of Linton *et al.*'s (1980) nine patients with acute interstitial nephritis had received an antibiotic, and two had received a sulphonamide. Pusey *et al.* (1983) reported nine episodes of acute renal failure in seven patients, four of which were due to antibiotics and three to co-trimoxazole. The antibiotics most frequently involved are ampicillin, methicillin, cephalosporins, and rifampin (Linton *et al.* 1980).

Captopril

A renal biopsy in Farrow and Wilkinson's (1979) patient with acute renal failure during treatment with captopril showed patchy tubular atrophy and inflammation. Hooke *et al.* (1982) reported interstitial infiltration in which eosinophils were prominent, which appeared to be due to captopril in a renal transplant.

Cimetidine

There have been few reports of acute renal failure associated with cimetidine. Richman *et al.* (1981) reported acute renal failure due to acute interstitial nephritis associated with cimetidine therapy. Payne *et al.* (1982) reported a patient who developed acute renal failure on two occasions after ingestion of cimetidine. On the first occasion the serum creatinine rose to 500 μmol/l^{-1}, but renal function recovered after cimetidine was discontinued. A renal biopsy was carried out during the second episode and showed focal interstitial inflammatory-cell infiltrates. On both occasions the serum alkaline phosphatase rose considerably.

Clafenin

Proesmans *et al.* (1981) reported recurrent attacks of acute renal failure which occurred over 18 months in a 7-year-old boy. Renal biopsy showed acute tubulo-interstitial nephritis. Each episode subsided spontaneously in hospital. Examination of the urine showed clafenin in high concentration. This is an analgesic popular in Europe, particularly in France and South Africa. The mother eventually confessed to poisoning her child with this drug. Reversible acute renal failure has been reported in adults who had taken supra-therapeutic doses (Rosen *et al.* 1978).

Clofibrate

Cumming (1980) reported an insulin-treated diabetic patient who developed a skin rash and renal failure after treatment with clofibrate for three weeks. The renal biopsy showed a heavy interstitial infiltrate with lymphocytes, plasma cells, polymorphonuclear leucocytes, and occasional eosinophils. The patient was never oliguric although the serum urea rose to 56 mmol l^{-1}, after which renal function recovered spontaneously without dialysis. Cumming pointed out that their case differs from previous cases of clofibrate toxicity which were associated

with myalgia and muscle damage or water and electrolyte depletion. The risk of nephrotoxicity seems to be greater when pre-existing renal disease is present.

Contraceptive pill

The author has seen a young woman who developed progressive renal failure after about one year on the pill. No other drugs had been taken. She was normotensive and had little proteinuria despite a blood urea of over 40 mmol l^{-1} and a serum creatinine of 600 μmol l^{-1}. Renal biopsy showed interstitial oedema and heavy inflammatory infiltration, although eosinophils were not prominent and the glomeruli were normal. Rapid improvement in renal function followed cessation of the drug and it returned to normal after about a month.

Diflunisal

This is a new analgesic drug related to acetylsalicylic acid. Chan *et al.* (1980) reported the development of acute anuric renal failure in a young woman three weeks after she had taken diflunisal for two weeks. The acute renal failure was preceded by a rash for which she had already been given prednisolone. The renal biopsy showed intense inflammation although eosinophils were not prominent; the glomeruli were normal. Rapid improvement in renal function followed cessation of the drug and it returned to normal after about a month.

Diphosphonate

Bounameaux *et al.* (1983) reported three patients treated with intravenous ethylidene diphosphonate for hypercalcaemia due to malignancy. At necropsy severe tubulo-interstitial disease was found. One patient had received only EHDP before the onset of renal failure though other drugs had been given to the other patients.

Erythromycin

Biopsy-proven acute interstitial nephritis after erythromycin was reported by Rosenfeld *et al.* (1983).

Fenoprofen

Several cases of acute renal failure after fenoprofen therapy have been reported. Brezin *et al.* (1979) reported two cases, one of whom required peritoneal dialysis and the renal biopsy showed interstitial nephritis. Renal function returned to normal in both patients. The biopsy in Curtis *et al.*'s (1980) patient showed tubular necrosis, severe interstitial inflammation, and oedema but also had fusion of glomerular epithelial cell foot processes on electron microscopy. She required haemodialysis and was given 1-g boluses of methylprednisolone intravenously and, after 10 days of oliguria, renal function began to return and eventually became normal. Abraham and Keane (1984) in a report on 36 cases of nephropathy associated with non-sterodial anti-inflammatory drugs incriminated fenoprofen in 22 (61 per cent). They point out that a heavy proteinuria is usual in these cases, although the glomeruli appear normal and the lesion is intestinal fibrosis.

Frusemide

Frusemide may cause interstitial nephritis and this should be suspected when renal function suddenly deteriorates in patients with nephrotic syndrome who are not volume depleted. Renal function often improves again when frusemide is stopped.

Interferon

Averbuch (1984) reported acute interstitial nephritis and also nephrotic syndrome following recombinant leucocyte A interferon for mycosis fungoides.

Mefenamic acid

Following the report of Robertson et al. (1980) of non-oliguric renal failure in elderly patients treated with mefenamic acid (*Ponstan*) other reports have appeared. Venning et al. (1980) reported a previously fit 60-year-old patient who developed acute renal failure requiring haemodialysis after taking this drug for sciatic pain. Renal biopsy showed a heavy interstitial inflammatory infiltrate containing many eosinophils. Renal function improved with steroid therapy but was still impaired three weeks later at the time of reporting. Malik et al. (1980) reported a patient with Still's disease who had remained controlled on D-penicillamine and a small dose of prednisolone for two years. One week after commencing mefenamic acid she developed a macular rash. Despite discontinuation of the drug a haemorrhagic bullous eruption affecting skin and mucous membranes, digital ischaemia, and disseminated intravascular coagulation developed. There was evidence of widespread vasculitis with peripheral gangrene and at post mortem there were areas of infarction in all major organs.

Renal impairment occurring in an otherwise fit patient with leg ulceration was reported by Drury et al. (1981). After the serum creatinine and urea rose, the drug was accidentally discontinued with immediate improvement in renal function. Five days later it was given again and renal function again deteriorated and improved again when withdrawn.

Minocycline

Walker et al. (1979) reported acute interstitial nephritis with eosinophilia and severe renal failure 10 days after minocycline was given for upper respiratory infection. Complete recovery followed withdrawal of the tetracycline.

Tienilic acid

Walker et al. (1980) reported acute interstitial nephritis in a patient who had received tienilic acid.

Triamterene

Bailey et al. (1982) reported mild acute renal failure, with the renal biopsy characteristic of interstitial nephritis, in a 52-year-old woman treated with triamterene; this was first noted after 25 days. The biopsy was performed after 36 days and after 40 days the drug was discontinued and renal function gradually improved. Magil et al. (1980) reported biopsy-proven interstitial nephritis in three elderly patients after treatment with *Dyazide* (a combination of hydrochlorothiazide 25 mg with triamterene 50 mg).

Drugs causing disturbance of renal tubular function

A large number of drugs cause disturbances of renal tubular function. Iatrogenic calculi may follow prolonged use of some drugs. Table 34.4 lists renal tubular abnormalities associated with drug administration.

Acetazolamide

Acetazolamide is a diuretic used mainly for treatment of glaucoma and was formerly used for epilepsy. It reduces excretion of citrate and produces systemic acidosis. The acidosis increases urinary excretion of calcium. Citrate may be important as a chelator of calcium. Calcium oxalate stones and nephrolithiasis have been reported following prolonged used of acetazolamide for treatment of epilepsy and the author has seen a patient who had a stone following acetazolamide for treatment of glaucoma.

Allopurinol

Allopurinol is a synthetic isomer of hypoxanthine which inhibits the conversion of hypoxanthine to xanthine by

Table 34.4　Drugs causing disturbance of renal tubular function

Abnormality	Drugs
Acidaemia	acetazolamide ammonium chloride amphotericin B aspirin methyl alcohol (methanol)
Fanconi's syndrome (loss of potassium, glucose, phosphate aminoacids)	bismuth salts outdated tetracyclines
Nephrocalcinosis	amphotericin B calciferol (vitamin D) calcium salts methoxyflurane xylitol
Polyuria	caffeine salts calciferol (vitamin D) chlorpromazine diuretics fluopromazine lithium lobelia methoxyflurane primidone triclofenol piperazine
Potassium loss	amphotericin B capreomycin sulphate carbenoxolone colistin (polymyxin E) diuretics liquorice polymyxin B
Sodium loss	diuretics
Sodium and water retention	aldosterone carbenoxolone corticosteroids guanethidine liquorice phenylbutazone probenecid reserpine

xanthine oxidase. As xanthine oxidase also converts xanthine to uric acid, allopurinol increases the concentration of both hypoxanthine and xanthine while decreasing the production of uric acid. It is used for control of hyperuricaemia and the treatment of gout.

Xanthine calculi are occasionally formed when allopurinol is used for treatment of conditions in which there

is overproduction of purines. Such conditions include malignant disease treated with cytotoxic drugs, chronic myeloproliferative disease, and the very rare Lesch–Nyhan syndrome (due to deficiency of the enzyme hypoxanthine-guanine-phosphoribosyltranferase).

Ascorbic acid (vitamin C)

Hyperoxaluria appears to occur in some individuals after ingestion of very large doses of vitamin C, for example 1–3 g daily as a prophylaxis against the common cold.

Contrast media

Ramsay *et al.* (1982) reported crystalluria following excretory urography.

Lithium

Lithium commonly causes a sodium diuresis lasting for a few days which may be followed by mild thirst and polyuria for a few weeks. In about 10 per cent of patients impairment of renal tubular concentrating ability persists but usually disappears when the drug is withdrawn. Occasional patients develop irreversible nephrogenic diabetes insipidus (Hestbech *et al.* 1977).

Wallin *et al.* (1982) reported renal function studies on 278 patients receiving long-term lithium treatment in whom the plasma lithium was controlled between 0.7 and 1.2 mmol l^{-1}. The mean time of treatment was 6.5 years and the longest was 15 years. Forty-nine per cent of the patients were unable to concentrate their urine above 800 mosmol kg^{-1} and the decrease in urinary concentration capacity was related to duration of treatment. Seventeen per cent had reduced creatinine clearance and in the whole group of patients it tended to be around the lower end of the normal range. The decrease in filtration rate was also related to duration of treatment.

Walker *et al.* (1982) reported a study of both renal function and renal histology in 47 patients who had received maintenance lithium therapy. They compared the results with 32 psychiatric patients who had not taken lithium. The renal biopsies of all patients currently receiving lithium showed vacuolation of the cells of the collecting ducts, together with strongly PAS-positive granules and strands. This lesion is considered to be specific and was not found in the biopsies of the psychiatric patients not receiving lithium at the time of the biopsy. Interstitial fibrosis did not differ in degree between treated and non-treated patients when matched for age. The lithium-treated patients showed marked loss of urinary concentrating ability and urinary acidification. They suggested that the marked distal-nephron dysfunction may lead to pre-renal impairment of glomerular filtration, as the interstitial fibrosis did not differ between treated and non-treated patients.

It is still not clear whether long-term treatment with lithium is likely to lead to deterioration in renal function in a significant number of patients. It is widely used for the treatment of patients with affective disorders and such patients may be given other drugs which may potentiate the nephrotoxicity. In patients requiring treatment with lithium who also have renal disease, renal function ought to be assessed before and from time to time during treatment. (See also under 'Drugs affecting the glomerulus' and 'Drugs causing interstitial nephritis'.)

Magnesium trisilicate

Magnesium trisilicate is used as an antacid for peptic ulceration. Silica stones have been reported, some patients apparently not having exceeded the usual therapeutic amount. Joekes et al. (1973) reported such a case and commented that it is surprising that silica calculi are not more common — possibly because calculi are often not analysed.

Triamterene

Stones formed of triamterene have been reported in patients given this diuretic for control of hypertension (Ettinger et al. 1979). Jick, et al. (1982) in a group of patients reported that the risk is small. This diuretic is also known to cause a fall in creatinine clearance and a rise in blood urea when given to patients already seriously ill. Fairley et al. (1983) reported casts in the urine of normal volunteers taking this drug.

Magil et al. (1980) reported renal insufficiency during treatment with *Dyazide* (hydrochorthiazide 25 mg and triamterene 50 mg) in three elderly patients. All had previously normal renal function and renal biopsies showed acute interstitial nephritis. Nephrogenic diabetes insipidus was reported in a patient treated with *Dyazide* by Macleod et al. (1981).

Uric acid stones

Uric acid crystals and stones may result from use of probenecid for treatment of gout, especially if a high fluid intake is not maintained and the urine is not made alkaline.

Patients with leukaemia or malignant neoplasms treated with cytotoxic drugs have increased excretion of uric acid, and urate stones may be formed.

Vitamin D

Calciferol and the potent metabolites of vitamin D, 1 α-hydroxycholecalciferol (One-alpha), 1 α, 25-dihydroxycholecalciferol (calcitriol), and the analogue dihydrotachysterol are used for treatment of hypoparathyroidism (usually post-operative), renal osteodystrophy, and certain forms of osteomalacia. The dose of vitamin D needs to be controlled by regular estimations of urinary and serum calcium. The author has seen several patients with multiple renal calculi as a result of being prescribed vitamin D for post thyroidectomy tetany without adequate follow-up.

Drugs causing analgesic nephropathy and papillary necrosis

The cause of analgesic nephropathy is still debated. Earlier animal experiments suggested that aspirin is more toxic than phenacetin but recently Burry and Dieppe (1982) concluded that the original experiments in rodents were misleading because the papillae of the rat kidney is attenuated and has a precarious blood supply. Analgesic nephropathy continues to occur in Australia after banning of phenacetin (Nanra et al. 1978) unlike experience in Finland (Sillanpää, et al. 1982).

The presence of persistent pyuria is suggestive of analgesic nephropathy. It is useful to test the urine for analgesic metabolites when this condition may be present.

Dubach et al. (1983) reported a long-term follow-up of 623 women with objective evidence of intake of phenacetin-containing analgesics and matched controls over a period of up to 11 years. There was no significant difference in mortality due to urinary or cardiovascular disease in the two groups, although the incidence was low in both groups. Akyol et al. (1982) studied 16 patients who had taken large amounts of aspirin (5 to 37 kg) for rheumatoid arthritis. None had clinical features of analgesic nephropathy; all had normal plasma urea and creatinine concentrations; none had impaired urinary concentrating ability (the most sensitive measure of medullary function). One patient died and at autopsy evidence of analgesic nephropathy was not found. This, and an earlier study by the same authors, strongly suggests that aspirin taken alone, even in large doses over long periods, is not nephrotoxic. Most large series of cases seem to implicate mixtures of drugs containing aspirin and phenacetin (Gault and Wilson 1978; Murray 1978).

Analgesic nephropathy may well be under-diagnosed as patients often underestimate or even conceal their ingestion of analgesic drugs. Murray (1978) commented on the wide variation in the frequency of analgesic nephropathy in the United Kingdom. It seems to be particularly common in Glasgow where three-quarters of the cases resulted from ingestion of *Askit* powder, which contained aspirin, caffeine, and phenacetin, until phenacetin was withdrawn. He suggested that the mixture may be more habit-forming than either aspirin or paracetamol alone, which tend to be taken relatively infrequently and for appropriate reasons such as relief of pain. *Askit* is more likely to be taken with excessive frequency for its supposed mood-elevating properties. Working class women with psychiatric problems are especially prone to take *Askit* daily in large and increasing doses. Murray suggests that the dose is increased because of tolerance and also for symptoms caused by the powder itself as well as for the effect on mood. The classical radiological appearances of papillary necrosis may be absent in many cases — Cove-Smith and Knapp (1978) found papillary necrosis in 26 per cent of their patients on intravenous pyelography but it was present in all cases examined by autopsy.

The author's view is that caffeine may be important in the apparently greater toxicity of mixtures both by contributing to habituation and by causing diuresis followed by dehydration, thus increasing the concentration of the other analgesics in the renal medulla. The concentrations of salicylate occurring in the medulla are sufficient to result in uncoupling of oxidative phosphorylation. The hexose phosphate shunt is inhibited and this may lead to diminished ability to generate reducing agents and so to oxidative damage (Eknoyan 1984). It is known that, in paracetamol-treated rats, papillary necrosis can be completely prevented by diuresis combined with free access to water containing glucose. Figure 34.1 suggests how analgesic mixtures might lead to papillary necrosis.

The pathological hallmark of analgesic nephropathy is papillary necrosis and radiological evidence of papillary necrosis raises suspicion of analgesic abuse. However, papillary necrosis may sometimes occur without abuse of analgesics, and Friedreich postulated as long ago in 1877 that sloughing of papillae may be due to vascular changes. Papillary necrosis occurs in such diverse conditions as diabetes mellitus, severe urinary infection, analgesic nephropathy, and sickle-cell disease. In sickle-cell disease the vasa recta are blocked by sickling erythrocytes. The effect of aspirin and other non-steriodal anti-inflammatory agents may be to diminish blood flow in the papillae by inhibiting vasodilator prostaglandins. Their effect may be more important in patients with already mildly impaired renal function in whom there is evidence that renal blood flow is critically dependent on prostacyclin production.

Ciabattoni *et al.* (1984) reported that, in patients with impaired renal function compared to controls, ibuprofen reduces the urinary excretion of urinary 6-keto-prosta-

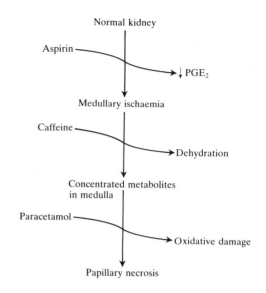

Fig. 34.1. Pathogenesis of papillary necrosis by ingestion of analgesic mixtures.

glandins F1a and E$_2$ with increase in serum creatinine and fall in creatinine and para-aminohippurate clearance whereas sulindac has no such effect. Sulindac may therefore be a safe non-steroidal anti-inflammatory drug.

Prescott (1982) reported 151 cases of papillary necrosis in patients taking aspirin or aspirin plus non-steroidal anti-inflammatory agents. Phenacetin alone did not cause papillary necrosis whereas aspirin alone and many non-steroidal anti-inflammatory agents have done so. Renal papillary necrosis has been found in between 20 and 60 per cent of patients reaching necropsy with rheumatoid arthritis (Clausen and Pedensen 1961; Lawson and Maclean 1966) although screening for nephrotoxicity among patients attending arthritis clinics revealed little evidence of nephrotoxicity (New Zealand Rheumatism Association 1974; Emkey and Mills 1982).

Erwin and Boulton-Jones (1982) reported a man who developed renal papillary necrosis (which was fresh at the time of autopsy) after taking benoxaprofen for 10 weeks. He had received no other drugs for four months although he had previously taken indomethacin and flurbiprofen. The ingestion of benoxaprofen was considered to be the most likely cause of the papillary necrosis.

Sillanpää *et al.* (1982) reported on the statistics of patients dying of renal disease between 1957 and 1977. In Finland phenacetin consumption increased after the Second World War; its use was restricted in 1962 and banned in 1965. Both the epidemiological study and the frequency of papillary necrosis at autopsy showed strong correlation with increased phenacetin consumption and, within four years of phenacetin being banned, there was a decline in mortality from nephropathy. Robertson *et al.* (1980) reported non-oliguric renal failure in six elderly women treated for two to six weeks with mefenamic acid, all of whom recovered after the drug was discontinued, though renal function did not become normal for their age. One of them had intravenous pyelography which showed the calyceal deformities of papillary necrosis. Hoffrand (1978) reported papillary necrosis in a patient treated with dapsone.

The risk of carcinoma developing in patients with analgesic nephropathy has recently been emphasized. Handa and Tewari (1981) reported cancer of the urinary tract in two of 19 patients followed up for an average of 5.1 years. Aurell *et al.* (1981) found carcinoma of the renal pelvis in four of 88 patients with end-stage renal failure. The epithelium had undergone metaplasia in 27 of 56 patients in which this was examined. They regarded these patients as being at high risk of developing neoplasms during regular dialysis therapy or after transplantation, and recommended bilateral nephroureterectomy.

Bengtsson *et al.* (1978) believed that the tumours are due to the metabolism of phenacetin to an orthoaminophenol, chemically similar to known carcinogens known from the dye and rubber industries.

Drugs causing obstructive nephropathy

Obstructive nephropathy can be classified according to the level at which the obstruction occurs, and also according to the cause of obstruction (Table 34.5).

Drug-induced stone disease has been discussed already under 'Drugs causing impairment of tubular function'.

Table 34.5   Obstructive renal disease

Stone	
Crystals	
Blood clot	Pelvis and ureters
Papilla	
Tumour	
Stenosis of wall	
Oedema	
Retroperitoneal fibrosis	Ureters
Ligation during surgery	
Stone	
Blood clot	
Bladder neck retention	Bladder
Tumour	

Ureteric stenosis and analgesic abuse

Fibrous thickening of the ureteric wall was reported in

association with analgesic abuse by MacGregor *et al.* (1973) and by Lewis *et al.* (1975). There is a concentric thickening which progressively occludes the lumen. Like other forms of ureteric obstruction, one ureter may become occluded without producing symptoms and only when the second ureter becomes obstructed does the patient see a doctor. Surgery is required to remove the obstructed length of ureter, replacing this by an isolated piece of ileum.

Retroperitoneal fibrosis due to drugs

In 1964 Graham reported that two out of 500 patients treated with methysergide for migraine developed retroperitoneal fibrosis. Other reports followed and by 1968 Graham had collected 55 patients with retroperitoneal fibrosis who had received methysergide. The lowest dosage was 2 mg daily for 21 months, and the shortest period of treatment was seven months.

The symptoms of retroperitoneal fibrosis are often vague and easily dismissed. They range from backache to vague abdominal pain, tenderness, and mild fever; the diagnosis may remain unsuspected until the patient becomes uraemic or even completely anuric. If anuric, as in other forms of ureteric obstruction, the anuria may resolve spontaneously after a few days and the blood urea fall towards normal. However, the urine volume soon falls again. The great vessels, lymphatics, or even bowel may also be involved in the fibrotic process so that a wide variety of symptoms apart from those of ureteric obstruction may occur.

In the typical case the intravenous pyelogram will show the ureters drawn medially from their normal lateral positions, with distensions of the upper ureters, renal pelvis, and calyces. However, the author has seen two patients, in whom the retroperitoneal fibrosis was far advanced at operation, yet there was no dilatation of the urinary tracts above the area of fibrosis on the pyelogram. The obstruction of the ureters was demonstrated by renography and by failed attempts to perform a Braasch catheterization of the ureters. Retroperitoneal fibrosis is a rare condition and may occur without any history of drug ingestion.

The author has encountered retroperitoneal fibrosis in a 68-year-old woman who had received atenolol for nine months (Doherty *et al.* 1978). A similar case was reported by Johnson and McFarland (1980). The manufacturers of the drug contend that these cases of retroperitoneal fibrosis cannot be due to the beta-blocker because the appearance of the fibrous tissue differs from that associated with practolol sclerosing peritonitis. Nevertheless, the retroperitoneal fibrosis associated with methysergide does not differ histologically from that of idiopathic retroperitoneal fibrosis and it seems likely that the fibroblastic reaction will be similar in the same region of the body whatever the aetiological agent.

We also found retroperitoneal fibrosis in a patient treated with oxprenolol (McCluskey *et al.* 1980). Laakso *et al.* (1982) reported retroperitoneal fibrosis in a patient taking sotalol, Rimmer *et al.* (1983) in a patient receiving timolol. A further case associated with metoprolol therapy was reported by Thompson and Julian (1982). By February 1982 there were 22 reports on the Adverse Drug Reactions Register of the Committee on Safety of Medicines of retroperitoneal fibrosis associated with beta-blocker therapy, including acebutolol, atenolol, metoprolol, oxprenolol, pindolol, propranolol, and sotalol. Pierce *et al.* (1981) reported retroperitoneal fibrosis in a patient taking propanolol for angina rather than for hypertension. In their patient it therefore seems unlikely that hypertension and retroperitoneal fibrosis could have antedated the drug therapy.

Table 34.6 Drugs causing urinary retention

Drug	Use	Other effects on renal function
Amphotericin B	antifungal	acute renal failure
Chlormezanone	tranquillizer	—
Chlorphentermine	obesity, symphathomimetic	—
Dipipanone	narcotic analgesic	—
Ephedrine	bronchial spasm, sympathomimetic	—
Fentanyl citrate	narcotic analgesic	—
Hydrallazine	hypotensive agent	SLE syndrome
Imipramine	antidepressant, tricyclic	—
Morphine and derivatives	narcotic analgesic	—
Pentazocine	narcotic analgesic	—
Pethidine	narcotic analgesic	—
Phenazocine	narcotic analgesic	—
Phendimetrazine	obesity	—
Phenelzine	antidepressant MAO-inhibitor	—
Vinblastine	cytotoxic agent	—
Vincristine	cytotoxic agent	—

Drugs causing incontinence of urine

The bladder and urethra have both parasympathetic and sympathetic innervation. Parasympathetic stimulation via acetycholine causes the emptying phase of the bladder by contraction of smooth muscle. The sympathetic nervous system controls filling of the bladder, alpha-adrenergic stimulation via norepinephrine causing contraction of the smooth muscle in the bladder neck, and proximal urethra and beta-adrenergic stimulation causing relaxation of the smooth muscle in the body of the bladder. Most drugs which cause incontinence of urine probably act on the alpha-adrenergic fibres in the proximal urethra (Kiruluta and Andrews 1983).

Neuroleptic drugs used in the treatment of psychoses, including thiohexane, chloropromazine, and haloperidol, all may cause incontinence which clears up immediately after the drug is discontinued. Phentolamine and phenoxybenzamine are used for neurogenic bladder retention. In patients on long-term treatment stress urinary incontinence may occur.

Drugs used for treatment of hypertension may occasionally cause urinary incontinence. Methyldopa reduces peripheral sympathetic innervation and Raz (1974) reported urinary incontinence in a patient receiving it alone. Prazosin produces its hypotensive effect by blocking alpha-adrenergic receptors. Both Kiruluta and Andrews (1983) and Thien *et al.* (1978) reported incontinence in patients taking prazosin, which cleared up when it was discontinued.

Drugs causing difficulty in micturition and urinary retention

A large number of drugs have been reported to cause difficulty in micturition and urinary retention (Table 34.6). Urinary retention is easily missed and it is not uncommon to be called to see a patient with apparent anuria which turns out to be simple urinary retention. It is particularly easy to miss a distended bladder when the abdominal wall is obese. Although urinary retention may be due to drug effect alone, in the elderly male, who has mild prostatic hypertrophy, confinement in bed may precipitate retention. Diuretic therapy with resultant high urinary volume may lead to retention of urine. Elderly females may also develop retention. Indeed an elderly confused patient who is restless, or who has incontinence, often has urinary retention.

Drugs affecting bladder directly

Cyclophosphamide

Haemorrhagic cystitis can complicate cyclophosphamide therapy. Droller *et al* (1982) reported that eight of 97 patients treated with cyclophosphamide for malignancy developed massive clot haemorrhage. They found that this may be prevented by a regime of diuresis and catheter drainage or very frequent voiding.

Rowland and Eble (1983) reported the development of a leiomyosarcoma of the bladder following cyclophosphamide therapy.

Cosmetic agents known to affect renal function

The association between the use of mercury-containing skin creams and proteinuria is well known and is common in Africa. One of the author's patients with the nephrotic syndrome due to membranous glomerulonephritis also suffers from psoriasis. Over a period of six months the proteinuria diminished and finally disappeared. The patient admitted to using an ointment containing mercury for a long period prior to the appearance of the nephrotic syndrome. The nephrotic syndrome and psoriasis occur together not uncommonly. Many, if not most, of the standard applications used for psoriasis contain crude coal tar or other hydrocarbon, which may be absorbed and are perhaps nephrotoxic.

Paraphenylenediamine ('Para') is used as a hair dye and for dyeing furs. In susceptible persons contact with the dye may cause severe local and systemic reactions, including oedema and severe dermatitis of the scalp, cardiovascular damage, vertigo, gastritis, asthma, diplopia, and exfoliative dermatitis. The author had a patient who developed severe necrotizing arteritis with necrotizing skin lesions and permanent kidney damage. There had been repeated use of 'para' for over two years. Chugh *et al.* (1982) reported two patients who developed intravascular haemolysis and acute tubular necrosis after ingestion of 'para'; one recovered, the other died.

Danger of drugs in patients with renal failure

It should be remembered that the toxicity of many drugs is greater and the side-effects more numerous in patients whose kidney function is already impaired. Very high blood levels of drugs and their metabolites may result from failure of excretion, and may cause further renal damage. Moreover, many drugs contain sodium as part of their formulation and the sodium load may be sufficient to cause worsening of hypertension, congestive heart failure, or hypertensive encephalopathy.

Drugs should NOT be given to patients with impaired renal function unless they are ABSOLUTELY necessary. When they are needed, then consideration must be given to the choice of the least toxic drug available for the required purpose, and care must also be taken to avoid combinations of drugs which enhance each other's toxicity.

Acknowledgements

I am grateful to Dr Dermot O'Reilly for reading the typescript and for Fig. 34.1 and several useful suggestions and to Mrs Caroline Weir for her patience and careful typing from manuscript. I gratefully acknowledge the support of the Northern Ireland Kidney Research Fund.

RECOMMENDED FURTHER READING

ABRAHAM, P.A. AND KEANE, W.F. (1984). Glomerular and interstitial disease induced by non-steroidal anti-inflammatory drugs. *Amer. J. Nephrol.* **4**, 1–6.

APPEL, G.B. AND NEU, H.C. (1977). The nephrotoxicity of anti-microbial agents. *New Engl. J. Med.* **296**, 663–70; 722–8; 784–7.

LINTON, A.L., CLARK, W.F., DRIEDGER, A.A., TURNBULL, D.I., AND LINDSAY, R.M. (1980). Acute interstitial nephritis due to drugs. *Ann. intern. Med.* **93**, 735–41.

McGEOWN, M.G. (Ed.) (1983). Drug-induced electrolyte disturbances. In *Clinical management of electrolyte disorders* pp. 28–31, 96–7. Martinus Nijhoff, The Hague.

REFERENCES

ABRAHAM, P.A. AND KEANE, W.E. (1984). Glomerular and interstitial disease induced by non-steroidal anti-inflammatory drugs. *Am, J. Nephrol.* **4**, 1–6.

AKMAL, M., VALDIN, J.R., McCARRON, M.M., AND MASSRY, S.G. (1981). Rhabdomyolysis with and without acute renal failure in patients with phencyclidine intoxication. *Am. J. Nephrol.* **1**, 91–6.

AKYOL, S.M., THOMPSON, M., AND KERR, D.N.S. (1982). Renal function after prolonged consumption of aspirin. *Br. med. J.* **284**, 631–2.

AURELL, M., SVALANDER, C., WALLIN, L., AND ALLING, C. (1981). Renal pelvic neoplasms and atypical urothelium in patients with end stage analgesic nephropathy. *Kidney Int.* **20**, 671–5.

AVERBUCH, S.D. (1984). Acute interstitial nephritis with the nephrotic syndrome following recombinant leucocyte A interferon therapy for mycosis fungoides. *New Engl. J. Med.* **310**, 32–5.

BACON, P.A., TRIBE, C.R., MACKENZIE, J.C., JONES, J.V., CUMMING, R.H.,A ND AMER, B. (1976). Penicillamine nephropathy in rheumatoid arthritis — a clinical, pathological and immunological study. *Quart. J. Med.* **55**, 661–84.

BAILEY, R.R., LYNN, K.L., DRENNAN, C.J., AND TURNER, G.A.L. (1982). Triamterene-induced acute interstitial nephritis. *Lancet* i, 22.

BEAR, R., WALKER, F., AND LANG, A. (1983). Prolonged oliguric renal failure related to cyclosporin A in a renal transplant recipient. *Am. J. Nephrol.* **3**, 293–4.

BENGTSSON, U., JOHANNSON, S., AND ANGERVALL, L. (1978). Malignancies of the urinary tract and their relation to analgesic abuse. *Kidney Int.* **13**, 107–13.

BJORCK, S., WESTBERG, G., SVALANDER, C., AND MULEC, H. (1983). Rapidly progressive glomerulonephritis after hydrallazine. *Lancet* ii, 42.

BLOCH, M.H., GORINS, A., AND MEYEROVITCH, A. (1966). Les accidents de la phénylbutazone. *Presse Méd* **74**, 2671–4.

BOUNAMEAUX, H.M., SCHIFFERLI, J., MONTANI, J-P., JUNG, A., AND CHATELAANT, E. (1983). Renal failure associated with intravenous diphosphonates. *Lancet* **i**, 471.

BREZIN, J.H., KATZ, S.M., SCHWARTZ, A.B., AND CHINITZ, J.L. (1979). Reversible renal failure and nephrotic syndrome associated with non-steroidal anti-inflammatory drugs. *New Engl. J. Med.* **301**, 1271–3.

BRITISH MEDICAL JOURNAL (1981). Penicillamine nephropathy. *Br. med. J.* **282**, 761–2.

BRITISH NATIONAL FORMULARY (1981). Urinary antimicrobial drugs. In British national formulary, pp. 172–3. British Medical Association and Pharmaceutical Society of Great Britain, London.

BRODKIN, H.M. (1980). Myoglobinuria following epsilon-aminocaproic acid (EACA) therapy. *J. Neurosurg* **53**, 690–2.

BURRY, H.C. AND DIEPPE, P.A. (1982). Renal function after prolonged consumption of aspirin. *Br. med. J.* **284**, 1117–8.

CALNE, R.Y., ROLLES, K., WHITE, D.J.G., THIRU, S., EVANS, D.B., McMASTER, P., DUNN, D.C., CRADDOCK, G.N., HENDERSON, R.G., AZIZ, S., AND LEWIS, P. (1979). Cyclosporin A initially as the only immunosuppressant in 34 recipients of cadaveric organs. *Lancet* **ii**, 1033–6.

CASE, D.B., ATLAS, S.A., MOURADIAN, J.A., FISHMAN, R.A., SHERMAN, R.L., AND LARAGH, J.H. (1980). Proteinuria during long-term captopril therapy. *J. Am. med. Ass.* **244**, 346–9.

CHAN, L.K., WINEARLS, C.C., OLIVER, D.O., AND DUNNELL, M.S. (1980). Acute interstitial nephritis and erythema associated with diflunisal. *Br. med. J.* **1**, 84–5.

CHAPMAN, J.R. AND BRIGDEN, D. (1981). Transient renal impairment during intravenous acyclovir therapy. *Lancet* **ii**, 1103.

CHUGH, K.S., MALIK, G.H., AND SINGHAL, P.C. (1982). Acute renal failure following paraphenylene diamine (hair dye) poisoning. *J. Med.* **13**, 131–7.

CIABATTONI, G., CINOTTI, G.A., PIERUCCI, A., SIMONETTI, B.M., MANZI, M., PUGLIESE, F., BARSOTTI, P., PECCI, G., TAGGI, F., AND PATRONO, C. (1984). Effects of sulindac and ibuprofen in patients with chronic glomerular disease. *New Engl. J. Med.* **310**, 279–83.

CLAUSEN, E. AND PEDERSEN, J. (1961). Necrosis of the renal papillae in rheumatoid arthritis. *Acta. med. scand.* **170**, 631–3.

COBDEN, I., RECORD, C.O., WARD, M.K., AND KERR, D.N.S. (1982). Paracetamol-induced acute renal failure in the absence of fulminant liver damage. *Br. med. J.* **284**, 21–2.

COVE-SMITH, J.R. AND KNAPP, M.S. (1978). Analgesic nephropathy, an important cause of chronic renal failure. *Quart. J. Med.* **185**, 49–69.

CUMMING, A. (1980). Acute renal failure and interstitial nephritis after clofibrate treatment. *Br. med. J.* **281**, 1529–30.

CURTIS, G.A., KALDANY, A., WHITELEY, L.G., CROSSON, A.W., ROLLA, A., AND MERINO, M.J. (1980). Reversible rapidly progressive renal failure with nephrotic syndrome due to fenoprofen calcium. *Ann. intern. Med.* **92**, 72–3.

CURTIS, J.J., LUKE, R.G., WHELCHEL, J.D., DIETHELM, A.G., JONES, P., AND DUSTAN, J.P. (1983). Inhibition of angiotensin converting enzyme in renal transplant recipients with hypertension. *New Engl. J. Med.* **308**, 377–81.

DAWSON, P., HEMINGWAY, A., AND ALLISON, D.J. (1983). Renal failure after contrast radiology. *Br. med. J.* **287**, 691.

DESCAMPS, C., VANDENBROUCKE, J.M., AND VAN YPERSELE De STRIHOU, C. (1977). Rhabdomyolysis and acute tubular necrosis associated with carbenoxolone and diuretic treatment. *Br. med. J.* **1**, 272.

DODD, M.J., GRIFFITHS, I.D., AND THOMPSON, M. (1980). Adverse reactions to D-penicillamine after gold toxicity. *Br. med. J.* **1**, 1498–500.

DOHERTY, C.C., McGEOWN, M.G., AND DONALDSON, R.A. (1978). Retroperitoneal fibrosis after treatment with atenolol. *Br. med. J.* **2**, 1786.

DOYLE, C.D., CAMPBELL, E., GAVIN, N., GARRETT, P., AND CARMODY, M. (1983). Phenylbutazone nephrotoxicity: a light and electron microscope study. *Ir. J. med. Sci.* **152**, 435–9.

DROLLER, M.J., SARAL, R., AND SANTOS, G. (1982). Prevention of cyclophosphamide-induced haemorrhagic cystitis. *Urology* **20**, 256–8.

DRURY, P.L., ASIRDAS, L.G., AND BULGER, G.V. (1981). Mefenamic acid nephropathy: further evidence. *Br. med. J.* **282**, 865–6.

DUBACH, U.C., ROSNER, B., AND PFISTER, E. (1983). Epidemiologic study of abuse of analgesics containing phenacetin. *New Engl. J. Med.* **308**, 357–62.

EKNOYAN, G. (1984). Analgesic nephropathy and renal papillary necrosis. *Sem. Nephrol.* **4**, 65–76.

EMKEY, R.D. AND MILLS, J.A. (1982). Aspirin and analgesic nephropathy. *J. Am. med. Ass.* **247**, 55–7.

ERWIN, L. AND BOULTON-JONES, J.M. (1982). Benoxaprofen and papillary necrosis. *Br. med. J.* **285**, 694.

ETTINGER, M.D., WEIL, E., MANDEL, N.S., AND DARLING, S. (1979). Triamterene-induced nephrolithiasis. *Ann. intern. Med.* **91**, 745.

FAIRLEY, K.F., BIRCH, D.F., AND HAINES, I. (1983). Abnormal urinary sediment in patients on triamterene. *Lancet* **i**, 421–2.

FALCK, H.M., TORNROTH, T., KOCK, B., AND WEGELIUS, O. (1979). Fatal renal vasculitis and minimal change glomerulonephritis complicating treatment with D-penicillamine. *Acta. med. scand.* **205**, 133–8.

FARROW, P.R. AND WILKINSON, R. (1979). Reversible renal failure during treatment with captopril. *Br. med. J.* **1**, 1680.

FINLAY, A.Y. (1981). Hydrallazine-induced necrotising vasculitis. *Br. med. J.* **282**, 1703.

FONG, H.J. AND COHEN, A.H. (1982). Ibuprofen-induced acute renal failure with acute tubular necrosis. *Am. J. Nephrol.* **2**, 28–31.

FRIEDREICH, N. (1877). Ueber der Nierenpapillen bei Hydronephrose. *Virchow's Arch. Path. Anat.* **69**, 308.

GABRIEL, R., CALDWELL, J., AND HARTLEY, R.B. (1982). Acute tubular necrosis caused by therapeutic doses of paracetamol. *Clin. Nephro.* **18**, 269–71.

GAULT, M.H. AND WILSON, D.R. (1978). Analgesic nephropathy in Canada: clinical syndromes, management and outcome. *Kidney Int.* **13**, 58–63.

GOEBEL, K.M. AND MUELLER-BRODMANN, W. (1982). Reversible overt nephropathy with Henoch–Schönlein purpura due to piroxicam. *Br. med. J.* **284**, 311–12.

GRAHAM, J.R. (1964). Methysergide for prevention of headaches: experience of 500 patients over 3 years. *New Engl. J. Med.* **270**, 67–72.

—— (1968). Fibrosis associated with methysergide therapy. In *Drug-induced diseases*, Vol 3 (ed. L. Meyler, and H.M. Peck), pp. 249–69. Excerpta Medica Foundation, Amsterdam.

HABIBI, B., BAUMELOU, A., AND SERDAU, M. (1981). Acute intravascular haemolysis and renal failure due to teniposide related antibody *Lancet* **i**, 1423–4.

HAMILTON, D.V., PRYOR, J.S., AND CARDOE, N. (1979). Fenclofenac-induced nephrotic syndrome. *Br. med. J.* **2**, 391.

HANDA, S.P. AND TEWARI, H.D. (1981). Urinary tract carcinoma in patients with analgesic nephropathy. *Nephron* **28**, 62–4.

HESTBECH, J., HANSEN, H.E., AMDISEN, A., AND OLSEN, S. (1977). Chronic renal lesions following long-term treatment with lithium. *Kidney Int.* **12**, 205–13.

HOFFRAND, B.I. (1978). Dapsone and renal papillary necrosis. *Br. med. J.* **1**, 78.

HOOKE, D., WALKER, R.G., WALTER, N.M.A., D'APICE, A.J.F., WHITWORTH, J.A., AND KINCAID-SMITH, P. (1982). Repeated renal failure with use of captopril in cystinotic renal allograft recipient. *Br. med. J.* **285**, 1538.

HOORNTJE, S.J., KALLENBERG, C.G.M., WEENING, J.J., DONKER, Ab J.M., AND HOEDEMAEKER, P.J. (1980). Immune complex glomerulonephritis in patients treated with captopril. *Lancet* **i**, 1212–14.

HRICIK, D.E., BROWNING, P.J., KOPELMAN, R., GOORNO, W.E., MADIAS, N.E., AND DZAU, V.J. (1983). Captopril-induced functional insufficiency in patients with bilateral renal artery stenosis or renal artery stenosis in a single kidney. *New Engl. J. Med.* **308**, 373–81.

HURLEY, B. (1977). Acute renal failure associated with carbenoxolone treatment. *Br. med. J.* **1**, 1472.

JICK, H., DINAN, B.J., AND HUNTER, J.R. (1982). Triamterene and renal stones. *J. Urol.* **127**, 224–5.

JOEKES, A.M., ROSE, G.A., AND SUTOR, J. (1973). Multiple renal silica calculi. *Br. med. J.* **1**, 146–7.

JOHNSON, J.N. AND McFARLAND, J.B. (1980). Retroperitoneal fibrosis associated with atenolol. *Br. med. J.* **1**, 864.

JONES, B.G., FIELDING, J.W., NEWMAN, C.E., HOWELL, A., AND BROOKS, V.S. (1980). Intravascular haemolysis and renal impairment after blood transfusion in 2 patients on long-term fluorouracil and mitomycin C therapy. *Lancet* **i**, 1275–7.

KEOWN, P.A., STILLER, C.R., ULAN, R.A., SINCLAIR, N.R., WALL, W.J., CARRUTHERS, G., AND HOWSON, W. (1981). Immunological and pharmacological monitoring in the clinical use of cyclosporin A. *Lancet* **i**, 686–9.

KINCAID-SMITH, P. AND WHITWORTH, J.A. (1983). Hydrallazine-associated glomerulonephritis. *Lancet* **ii**, 348.

KIRULUTA, H.G. AND ANDREWS, K. (1983). Urinary incontinence secondary to drugs. *Urology* **22**, 88–90.

LAAKSO, M., ARVALA, I., TERVONEN, S., AND SOTARAUTA, M. (1982). Retroperitoneal fibrosis associated with sotalol. *Br. med. J.* **285**, 1085–6.

LAWSON, A.A.H. AND MACLEAN, N. (1966). Renal disease and drug therapy in rheumatoid arthritis. *Ann. rheum. Dis.* **25**, 441–9.

LEMPERT, K.D. (1980). Haemolysis and renal impairment syndrome in patients on 5-fluorourcil and mitomycin C. *Lancet* **ii**, 369.

LEVI, F.A., HRUSHESKY, W.J.M., HALBERG, F., LANGEVIN, T.R., HAUS, E., AND KENNEDY, B.J. (1983). Lethal nephrotoxicity and hematologic toxicity of cis-diamminedichloroplatinum ameliorated by optimal circadian timing and hydration. *J. Urol.* **129**, 446.

LEWIS, C.T., MOLLAND, E.A., MARSHALL, V.R., TRESSIDER, G.C., AND BLANDY, J.P. (1975). Analgesic abuse, ureteric obstruction and retroperitoneal fibrosis. *Br. med. J.* **2**, 76–8.

LINTON, A.L., CLARK, W.F., DRIEDGER, A.A., TURNBULL, D.I., AND LINDSAY, R.M. (1980). Acute interstitial nephritis due to drugs. Review of the literature with report of 9 cases. *Ann. intern. Med.* **93**, 735–41.

MACGREGOR, G.A., JONES, N.F., BARRACLOUGH, M.A., WING, A.J., AND CRANSTON, W.I. (1973). Ureteric stricture and analgesic nephropathy. *Br. med. J.* **2**, 271–2.

MACLEOD, M.D., BELL, G.M., AND IRVINE, W.J. (1981). Nephrogenic diabetes insipidus associated with Dyazide (triamterene–hydrochlorothiazide). *Br. med. J.* **283**, 1155–6.

MAGIL, A.B., BALLON, H.S., CAMERON, E.C., AND RAE, A. (1980). Acute interstitial nephritis associated with thiazide diuretics: clinical and pathological observations in 3 cases. *Am. J. Med.* **69**, 939–43.

MAHER, J.F. (1981). Nephrotoxicity due to drugs. *Sem. Nephrol,* **1**, 27–35.

MALIK, S., ARTHURTON, I., AND GRIFFITHS, I.D. (1980). Mefenamic acid nephropathy. *Lancet* **ii**, 746.

MARSH, F.P., ALMEYDA, J.R., AND LEVY, I.S. (1971). Non-thrombocytopenic purpura and acute glomerulonephritis after indomethacin therapy. *Ann. rheum. Dis.* **30**, 501–5.

MARSHALL, V.F. (1971). Methods of urographic diagnosis. In *Clinical urography*, Vol 1, (ed. J.L. Emmett, and D.M. Written), Saunders, Philadelphia.

McCLUSKEY, D., DONALDSON, R.A., AND McGEOWN, M.G. (1980). Retroperitoneal fibrosis in a patient receiving oxprenolol. *Br. med. J.* **281**, 1459.

McEVOY, J., McGEOWN, M.G., AND KUMAR, R. (1970). Renal failure after radiological contrast media. *Br. med. J.* **4**, 717–18.

MOSKOVITZ, R. AND MILLER, J.H. (1981). Lithium induced nephrotic syndrome. *Am. J. Psychiat.* **138**, 382–383.

MURRAY, R.M. (1978). Genesis of analgesic nephropathy in the United Kingdom. *Kidney Int.* **13**, 50–7.

NANRA, R.S., STUART-TAYLOR, I., DE LEON, A.H., AND WHITE, K.H. (1978). Analgesic nephropathy: etiology, clinical syndrome and clinicopathologic correlations in Australia. *Kidney Int.* **13**, 79–92.

NATH, K.A., CRUMBLEY, A.J., MURRAY, B.M., AND SIBLEY, R.K. (1983). Captopril and renal insufficiency. *New Engl. J. Med.* **309**, 665.

NEW ZEALAND RHEUMATISM ASSOCIATION. (1974). Aspirin and the kidney. *Br. med. J.* **1**, 593–6.

PANAYI, G.S., WOOLEY, P., AND BATCHELOR, J.R. (1978). Genetic basis of rheumatoid disease; HLA antigens, disease manifestations and toxic reactions to drugs. *Br. med. J.* **2**, 1326–8.

PAYNE, C.R., ACKRILL, P., AND RALSTON, A.J. (1982). Acute renal failure and rise in alkaline phosphate activity caused by cimetidine. *Br. med. J.* **285**, 100.

PHILLIPS, M.E., EASTWOOD, J.E., CURTIS, J.R., GOWER, P.E., AND DE WARDENER, H.E. (1974). Tetracycline poisoning in renal failure. *Br. med. J.* **2**, 149–51.

PIERCE, J.R., TROSTLE, D.C., AND WARNER, J.J. (1981). Propanolol and retroperitoneal fibrosis. *Ann. intern. Med.* **95**, 244.

PIRSON, Y., MEYER, M., PLAEN, J-F., SQUIFFLET, J-P., ALEXANDRE, G.P., AND VAN YPERSELE DE STRIHOU, C. (1983). Captopril and renal insufficiency. *New Engl. J. Med.* **309**, 667.

PRESCOTT, L.F. (1982). Analgesic nephropathy. A reassessment of the role of phenacetin and other analgesics. *Drugs,* **23**, 75–149.

PROESMANS, W., KYELE AKUMOLA SINA, J., DEBUCQUOY, P., RENOIRTE, A-M., AND EECKELS, R. (1981). Recurrent acute renal failure due to non-accidental poisoning with clafenin in a child. *Clin. Nephrol.* **16**, 207–10.

PUSEY, C.D., SALTISSI, D., BLOODWORTH, L., RAINFORD, D.J., AND CHRISTIE, J.L. (1983). Drug associated acute interstitial nephritis: clinical and pathological features and the response to high dose steroid therapy. *Quart. J. Med.* **52**, 194–211.

RAMSAY, A.W., SPECTOR, M., RODGERS, A.L., MILLER, R.L., AND KNAPP, D.R. (1982). Crystalluria following excretory urography. *Br. J. Urol.* **54**, 341–5.

RAZ, S. (1974). Adrenergic influence on the internal urinary sphincter *Isr. J. med. Sci.* **10**, 608.

RICHMAN, A.V., MASCO, H.L., RIFKIN, S.I., AND ACHARYA, M.K. (1980). Minimal change disease and the nephrotic syndrome associated with lithium therapy. *Ann. intern. Med.* **92**, 70–1.

——, NARAYAN, J.L., AND HIRSCHFIELD, J.S. (1981). Acute interstitial nephritis and acute renal failure associated with cimetidine therapy. *Am. J. Med.* **70**, 1272–4.

RIMMER, E., RICHENS, A., FORSTER, M.E., AND RESS, R.W.M. (1983). Retroperitoneal fibrosis associated with timolol. *Lancet* **i**, 300.

ROBERTSON, C.E., FORD, M.J., VAN SOMERSEN, V., DLUGOLECKA, M., AND PRESCOTT, L.F. (1980). Mefenamic nephropathy. *Lancet* **ii**, 232–3.

ROSEN, J., HOMBERG, J.C., OFFENSTADT, G., HÉRICORD, P., DAMECOUR, C., AND DURON, F. (1978). Insufficiance rénale aigue à la clafénine. *Nouv. Presse. Méd.* **7**, 3255.

ROSENFELD, J., GURA, V., BONER, G., BEN-BASSAT, M., AND LIVNI, E. (1983). Interstitial nephritis with acute renal failure after erythromycin. *Br. med. J.* **286**, 938–9.

ROWLAND, R.G. AND EBLE, J.N. (1983). Bladder leiomyosarcoma and pelvic fibroblastic tumour following cyclophosphamide therapy. *J. Urol.* **130**, 344–6.

SILAS, J.H., KLENKA, Z., SOLOMON, S.A., AND BONE, J.M. (1983). Captopril-induced reversible renal failure: a marker of renal artery stenosis affecting a solitary kidney. *Br. med. J.* **286**, 1702–3.

SILLANPÄÄ, M., KASANEN, A., AND ELONEN, A. (1982). Changes of panorama in renal disease mortality in Finland after phenacetin restriction. *Acta. med. scand.* **212**, 313–17.

TAN, S.Y., SHAPIRO, R., AND KISH, M.A. (1979). Reversible acute renal failure induced by indomethacin. *J. Am. med. Ass.* **241**, 2732.

THIEN, T.H., DELAERE, K.P.J., DEBRUYNE, F.M.T., AND KOENE, R.A.P. (1978). Urinary incontinence caused by prazosin. *Br. med. J.* **1**, 622–3.

THOMPSON, J. AND JULIAN, D.G. (1982). Retroperitoneal fibrosis associated with metoprolol. *Br. med. J.* **284**, 83–4.

VENKATARAMAN, G. (1981). Renal change and glue sniffing. *Br. med. J.* **283**, 1467.

VENNING, V., DIXON, A.J., AND OLIVER, D.O. (1980). Mefenamic acid nephropathy. *Lancet* **ii**, 745–6.

WALKER, R.G., BENNETT, W.M., DAVIES, B.M., AND KINCAID-SMITH, P. (1982). Structural and functional effects of long-term lithium therapy, *Kidney Int.* **21**, 511, 513–519.

——, THOMSON, N.M., DOWLING, J.P., AND OGG, C.S. (1979). Minocycline-induced interstitial nephritis. *Br. med. J.* **1**, 524.

——, WHITWORTH, J.A., AND KINCAID-SMITH, P. (1980). Acute interstitial nephritis in a patient taking tienilic acid. *Br. med. J.* **1**, 1212.

WALLIN, L., ALLING, G., AND AURELL, M. (1982). Impairment of renal function in patients on long-term lithium treatment. *Clin. Nephrol.* **18**, 23–8.

WEBB, J.A., FRY, I.K., AND CATTELL, W.R. (1983). Renal failure after contrast radiography. *Br. med. J.* **287**, 423.

WELLER, I.V.D., CARREND, V., FOWLER, M.J.F, MONJARDINO, J., MAKINEN, D., THOMAS, H.C., AND SHERLOCK, S. (1982). Acyclovir inhibits hepatitis B replication in man. *Lancet* **i**, 273.

WHICHER, J.T., MARTIN, M.F.R., AND DIEPPE, P.A. (1981). Oestrogen-containing oral contraceptives, decreased prostacyclin production and haemolytic uraemic syndrome *Lancet* **i**, 328–9.

WILL, A.M. AND McLAREN, E.A. (1981). Reversible renal damage due to glue sniffing. *Br. med. J.* **283**, 525.

35 Iatrogenic neoplasia

G.E. DIGGLE

Effective forms of therapy, by their nature, embody risks. The hazards include, in some cases, neoplastic disease. These generalizations are not, of course, restricted to the administration of drugs, and they certainly apply to other forms of treatment such as surgery and radiotherapy. The apparent prevalence of iatrogenic neoplasia, as judged from adverse reaction reports, surveys, and mortality statistics, is very low. The true prevalence, however, is almost certainly higher. Aside from the other difficulties of attribution, the major reason for the discrepancy is the length of the latency period between the administration of therapy and the appearance of a neoplasm. The impact of improved information technology and the increasing co-ordination of clinical records upon these problems in the future is likely to be very considerable (Diggle 1973). Occasionally, as with immunosuppressant and cytotoxic drugs, ionizing radiation, and some endocrine treatments, a carcinogenic risk is to be anticipated from the inherent action of the therapy. In other cases conditions contributing to carcinogenesis may be suspected and new forms of therapy which might create these circumstances are subjected to special scrutiny: this has occurred with the advent of H_2-receptor antagonists, and with speculation that they could produce conditions similar to those which lead to gastric carcinoma in pernicious anaemia and following surgery for peptic ulcer, after many years. Well-recognized examples of long latent periods are the 20 or more years which elapse before the carcinogenic effects of *Thorotrast* (^{232}thorium dioxide) are seen, and periods up to 30 years before the neoplastic manifestations of arsenic occur. On average 20–5 years elapse before the appearance of surgery-induced gastric carcinoma (Bushkin 1976).

It is essential that the risks associated with treatments should be seen in perspective, particularly in relation to environmental causes of cancer. Ames (1983) referred, for example, to the very wide range of mutagenic and carcinogenic substances which have been detected in common foods, sometimes in substantial quantities, following the development of cheap, short-term mutagenicity assays. Some 80 per cent of all cancers are thought to be environmental and 30 per cent of cancer deaths in the UK and USA are ascribed to the effects of tobacco (American Cancer Society 1983). The prevalence of iatrogenic neoplasia is very small indeed when considered in this context. By design, drugs are substances which are biologically active, since they are intended to change disordered body function, and favourably to modify the course of disease. Similar considerations apply to radiotherapy and surgical procedures. It is to be expected, therefore, that on occasions treatment will produce toxic, and sometimes carcinogenic, effects. The low prevalence may perhaps owe as much to the relative long-term safety of most forms of treatment, as it does to the difficulties involved in establishing causality. Information about iatrogenic neoplasia is sometimes presented inaccurately, sensationally, and irresponsibly by the media, as the examples in Chapter 1 illustrate. Fortunately many of those who are responsible for treatment decisions are aware of the character of such reportage and make their own assessment of risk and benefit in spite of it; patients may not be able to be so objective, however, and may be distressed by what they learn from the media.

The neoplastic effects described in this chapter are mediated through a variety of mechanisms, many of which are poorly understood. A final common mechanism may eventually be demonstrated, but this is not yet established, although there are many publications in the field. An exhaustive review cannot be given here, but attention is drawn to several relevant principles. In this chapter, 'neoplasm' signifies an abnormal tissue whose growth exceeds (and is not co-ordinated with) that of the corresponding normal tissue, and which continues in the same manner after the inducing stimuli have ceased. The term therefore includes malignant and benign tumours, and neoplasms which are not solid tumours (such as leukaemic cells).

'Carcinogen' is used in the literature in a number of senses. While some drugs may act directly as carcinogens, others may act as co-carcinogens and enhance or accelerate response to a carcinogen. A drug may also act as an 'initiator' producing a permanent change in tissue, which may not however be detectable, but may be seen to develop when the initiator is followed by a 'promoter'. Such distinctions may be of purely academic interest since, in most cases where drugs are incriminated in neoplastic transformation, there is usually little information available regarding the precise mechanism of action.

Traditionally, genotoxic carcinogenesis involves effects from radiation, chemical, or viral agents. In some circumstances, the genome may not be altered, and the neoplastic change is referred to as 'epigenetic', with changes of translation or transcription within the cell producing transformation. Both genotoxic and epigenetic mechanisms may be involved in iatrogenic neoplasia. Radiation-induced carcinogenesis can result from X-rays used in radiotherapy and diagnostic radiology, as discussed in this chapter. The other ionizing radiations are of course potent inducers. *Thorotrast*, for example, contains the radionucleide Th^{232} and is responsible for the tumours described later in this chapter. Th^{232} emits alpha particles during the first stage of decay and other emissions during subsequent stages.

When the development of neoplasia is associated with administration of drugs, the mechanism may relate to some form of chemical carcinogenesis, sometimes involving the formation of a reactive intermediate, or the process may be the outcome of a different type of activity, for example, viral or hormonal carcinogenesis. Some antineoplastic drugs are directly-acting, although others such as chlornaphazine are pre-carcinogens and act indirectly following activation. Chlornaphazine is metabolized to produce beta-napthylamine, which is excreted in the urine (Videbaeck 1964).

Viral induction may be related to therapy when, for example, treatments which reduce immunity are employed. The use of cyclosporin-A, which is considered below, depresses T-lymphocytes and may be followed by neoplasia because of supposed viral activity. Treatment which alters immunity may of course facilitate the ability of other carcinogens, as well as oncogenic viruses, to exert their effects.

'Traumatic' carcinogenesis in the original sense has for many years been regarded as an outmoded concept (Boyd 1953). Neoplasia following surgical operations are discussed later in this chapter. Hormonal carcinogenesis is exemplified by the development of tumours in the liver after the administration of certain steroidal agents and by the development of thyroid carcinoma in response to goitrogens. Malignant and benign tumours have been ascribed to androgenic/anabolic compounds. Hormonal induction is sometimes claimed to be epigenetic, usually on the basis of negative results in mutagenicity testing, but it is not in fact known whether transformation is preceded by a change in the genome.

Some chemial carcinogens, e.g. the nitrosamines which particularly affect the liver and oesophagus, are markedly organotropic. Similarly, with drugs, neoplasms tend to develop at sites where there is maximum concentration of the drug. Thus, carcinogenic drugs which are excreted via the kidneys reach their maximum concentration in the genitourinary tract, and it is at this site that neoplasms are likely to develop; for example, *Thorotrast* and renal carcinoma, chlornaphazine and bladder cancer. Bengtsson *et al.* (1968) also described transitional-cell tumours of the renal pelvis and bladder in patients taking excessive doses of phenacetin-containing analgesics. Similarly, carcinogenic drugs that are metabolized and excreted by the liver may induce liver neoplasia, for example *Thorotrast-induced angiosarcoma* (da Silva Horta *et al.* 1965). Drugs may also be deposited in specific organs producing high local concentrations. Thus, arsenic deposited in the skin may ultimately lead to the development of basal-cell carcinoma (Fierz 1965). When a drug is a carcinogen there is likely to be a dose–effect relationship between the total dose and the incidence of neoplasia. This has been shown, for example, in chlornaphazine-induced bladder tumours (Laursen 1970).

The difficulties involved in adducing clinical inferences from the results of animal carcinogenicity studies are well known, one reason being kinetic and metabolic differences between man and laboratory animal species. Zbinden (1983), referring to the carcinogenic properties of clofibrate in rodents, pointed out that, whereas the site of fatty acid oxidation is the peroxisome in rodents, it is the mitochondrion in primates. Clofibrate induces peroxisome proliferation and liver nodules and tumours in rats, but has not been shown to do so in man. Several widely marketed drugs are known to induce tumours in animals, but the significance for man is unknown. Examples quoted by Zbinden (1983) are given here. The antiprotozoan agent metronidazole has induced lung tumours in mice and other tumours in various species, and mutagenicity tests have been positive in bacterial systems. The antifibrinolytic agent tranexamic acid (*Cyclokapron*) has induced cholangiomas and cholangiocarcinomas in one strain of rats. Some contraceptive progestogens (as well as progesterone itself) readily produce mammary tumours in the beagle bitch, and the other progestogens are known to do so at the higher doses at which they are hormonally active. The progestogen medroxyprogesterone acetate (Department of Health and Social Security 1984) is of particular interest. Oestrogens induce renal carcinomata in male hamsters, leiomyomas in guinea pigs, lymphoid tumours in mice, ovarian carcinomata in dogs, and adrenal cortical carcinomata in rats. The beta-blocker pamatolol induces hepatic nodules and hepatocellular carcinomata in rats. In the United States, beta-blockers are regarded as potential carcinogens, and this accounts for the relatively late introduction of these drugs into clinical use in that country. The reasons, according to the *Food and Drug Administration Drug Bulletin* (1978), are that:

two early beta-blockers available in Europe, pronethalol and a closely related derivative, proved to be carcinogenic and two others, practolol and alprenolol, gave some indication of tumourigenicity in rodents. In addition, a variety of other beta-blockers have proved to be tumourigenic or otherwise hazardous. Two cardioselective agents, pamatolol and tolamolol, have produced carcinomas of the liver in rats and tolamolol has produced mammary carcinomas in mice as well. Other drugs of this general class appear to have produced benign tumours although some of the results are equivocal, necessitiating repeat studies. .... in view of these findings two expert FDA advisory committees in late 1972 concluded that all members of this class of drugs posed serious potential problems and advised that clinical studies with them should be limited to 30 days and approval withheld until the tumour-inducing potential of each agent had been carefully evaluated in animals. FDA accepted this advice and so informed the manufacturers. The carcinogenicity study requirements have now been met for many of these drugs and the 30-day limitation removed for six: acebutolol, metoprolol, nadolol, oxprenolol, pindolol, and timolol.

When carcinogenicity studies have not been entirely reassuring, this information should be readily available to clinicians, in order to assist them in weighing the balance of risk and benefit when prescribing for individual patients, although such judgements are, of course, extremely difficult in many cases. Unfortunately, clinicians do not always find it easy to discover the results of the carcinogenicity testing of medicines which they are invited to prescribe. Studies having positive or equivocal results are rarely published in the open scientific literature or revealed at open scientific gatherings. In the UK, the data sheet is one of the media which responsible pharmaceutical companies employ to convey such information to doctors. It was pointed out (Melrose 1982) that, while safety problems may be dealt with properly in the product literature of developed countries, they may be omitted completely in other countries.

Survey of carcinogens

In 1979 the International Agency for Research on Cancer published a list of chemical substances, including drugs, which are known with specified levels of certainty to be carcinogenic in man. The list is divided into four groups according to the adequacy of the available evidence in support of causality.

Group 1 contains the substances (and industrial processes) known to be human carcinogens; this category is used only where there is sufficient evidence to support a causal association between the exposure and human cancer.

4-Aminobiphenyl	Diethylstilboestrol
Arsenic and certain arsenic compounds	Underground haematite mining
Asbestos	Manufacture of isopropyl alcohol by the strong acid process
Manufacture of auramine	Melphalan
Benzene	Mustard gas
Benzidine	2-Naphthylamine
N, N-bis (2-chloroethyl)-2-naphthylamine (chlornaphazine)	Nickel refining
Bis (chloromethyl) ether and technical grade chloromethyl methyl ether	Soots, tars, and mineral oils
Chromium and certain chromium compounds	Vinyl chloride

Group 2 contains the substances which are probably carcinogenic for humans. This category includes chemicals for which the evidence of human carcinogenicity is almost 'sufficient' as well as chemicals for which it is only suggestive. To reflect this range this category has been divided into higher (subgroup A) or lower (subgroup B) degrees of evidence. The data from experimental animal studies played an important role in assigning chemicals to group 2, and particularly to subgroup 2B.

Subgroup 2A

Aflatoxins	Cyclophosphamide
Cadmium and certain cadmium compounds	Nickel and certain nickel compounds
Chlorambucil	Tris (1-aziridinyl) phosphine sulphide (thiotepa)

Subgroup 2B

Acrylonitrile	Dimethylsulphate
Amitrole (aminotriazole)	Ethylene oxide
Auramine	Iron dextran
Beryllium and certain beryllium compounds	Oxymetholone
Carbon tetrachloride	Phenacetin
Dimethylcarbamoyl chloride	Polychlorinated biphenyls

Group 3 contains the substances which it was not considered possible to classify as to their human carcinogenicity.

Chloramphenicol	Isopropyl oils
Chlordane/heptachlor	Lead and certain lead compounds
Chloroprene	Phenobarbitone

Dichlorodiphenyltri-
chloroethane (DDT)
Dieldrin
Epichlorohydrin
Haematite
Hexachlorocyclohexane
(technical grade
HCH/lindane)
Isoniazid

N-Phenyl-2-naphthyl-
amine
Phenytoin
Reserpine
Styrene
Trichloroethylene

Tris (aziridinyl) -para-
benzoquinone (triazi-
quone)

The best evidence of carcinogenicity, in the sense of scientific certainty, is that which rests on human, rather than animal, data. There are three major sources of such data. Firstly, the spontaneous case reports of individual cancer patients with a history of exposure to the putative neoplastic agent which may provide the first intimations of an association. Secondly, non-controlled epidemiological studies, which seek a temporal or geographical variation in incidence, associated with exposure. Thirdly, as the foregoing methods generally provide only suggestive evidence, there is usually a need for epidemiological methods of an analytic nature, such as cohort or case-control studies (Diggle 1973) in which individual exposure may be found to be associated with increased relative risk. Greater confidence in the evidence for an association between the agent and a neoplasm is inspired if it can be shown that positive bias and positive confounding have been sought ruthlessly and eliminated as far as is practicable. Confidence is also increased by a demonstration that the association is most unlikely to arise by chance alone, when a dose–response relationship can be established, and, of course, when the association is strong. Many possible associations between therapies and neoplasms are discussed in this book in addition to those given in the lists above; it will be noted that the confidence deserved by the evidence varies widely. The greatest confidence for a claimed causal relationship arises when positive findings are confirmed by independent studies, conducted according to the principles mentioned above and under different circumstances.

Negative findings are interpreted according to similar rules. In particular, negative bias and negative confounding should be considered and shown, as thoroughly as is practicable, not to be present. The degree of exposure to a putative neoplastic agent must, of course, be classified accurately and the possible effects of misclassification weighed. The same applies to the classification of outcome. In order to be convincing, a negative study must possess a large population: in a small study, where the upper confidence limit for relative risk may be well over unity, the study cannot exclude those (increased)

risks which lie between 1.0 and this limit. As with positive findings, confidence in negative results is increased when they are confirmed by several independent, well-conducted studies under different circumstances. A valid negative conclusion must be restricted to exposure levels at (or below) those actually found. In view of the long latency period which applies in many cases, a negative conclusion can only be accepted if sufficient time has elapsed.

A large amount of information about the carcinogenicity of drugs has, of necessity, been omitted from this book. This comprises most of the data obtained by pharmaceutical companies about their products. Although these data are submitted confidentially to regulatory authorities in countries where marketing authorization is sought, they are rarely released publicly, for legal reasons. Such findings are therefore not accessible to the normal scientific processes of publication, open discussion, and further work to confirm or refute findings. Moreover there is evidence (Berry 1983) that the testing requirements of regulatory authorities are often carried out with an insufficient understanding of the biological principles involved. This factor sometimes combines with a desire to cut costs and leads to unsatisfactory studies as, for example, when poor animal husbandry prevents the survival of adequate numbers of even the undosed control animals in a carcinogenicity study.

Arsenical carcinogenesis

The first suggestion that the therapeutic administration of arsenic could lead to cancer was made by Hutchinson in 1887 who described five cases of skin cancer following the medical use of arsenic. It is now generally accepted that arsenic given by mouth can cause cancer of the skin, which is usually preceded by arsenical pigmentation, keratosis, and dermatosis. Fierz (1965) found that, in a series of 262 patients treated with Fowler's solution for from 6 to 26 years, carcinoma of the skin developed in 21 patients (8 per cent). This type of cancer most frequently encountered was a basal-cell carcinoma, and in 16 of these patients there were multiple basal-cell carcinomata.

Sommers and Mc Manus (1953) described a series of 27 cases of arsenical skin cancer and drew attention to the fact that 10 of these patients also had other primary sites of neoplastic change. Two cases had bronchial carcinoma and three had primary carcinomata arising in the genito-urinary tract. This association of arsenic with carcinogenesis was further elaborated by Robson and Jelliffe (1963) who described six cases (four female and 2 male) who all developed bronchial carcinoma with an average latent period of 32 years after receiving arsenic therapy. Arsenic had been prescribed in these patients for psoriasis, rheumatic fever, 'convulsions', or as a 'tonic'. Each of these patients had the dermatological stigmata of arsenic ingestion.

The major importance of discussing arsenic-induced cancers at this current time is that, owing to the long latent period of onset of clinical neoplasia, such patients may still present themselves for treatment. It should also be borne in mind that although there are no indications today for using arsenic, the organo-arsenicals have been used in chemotherapy, and tryparsamide, which contains about 25 per cent of pentavalent arsenic in organic combination, is still used in the treatment of African trypanosomiasis.

Quinacrine carcinogenesis

Quinacrine, like arsenic, forms positive ions and is deposited in sweat glands. Like arsenic, it also induces squamous-cell carcinoma up to 30 years after ingestion. Bauer (1978) reported that the palms were involved in nine of 10 malignancies developed by three patients, two of whom developed axillary metastases.

These findings were reported in Australian servicemen, after prophylactic quinacrine received during the Second World War. The tumours arose only after the development of quinacrine-induced lesions resembling lichen planus and Calloway (1979) pointed out that carcinomata are described not infrequently in various atrophic lesions of the palmar surface including radiation dermatitis and acrokeratosis verruciformis, as well as in arsenical keratoses.

Chlornaphazine-induced bladder carcinoma

Three reports of individual cases of chlornaphazine-induced bladder cancer were cited by Meyler (1966). The first case was a woman of 68 being treated for polycythaemia vera; she had received a total of 156.6 grams of chlornaphazine over a period of 10 years. The second patient, a woman of 45, had received 149.5 grams of the drug over a period of five years for the treatment of Hodgkin's disease. The third patient, a man of 30, had been treated for Hodgkin's disease with chlornaphazine for six years.

It would seem from these cases that a high dose of chlornaphazine is associated with neoplastic changes in the bladder epithelium. This is in keeping with the findings of Thiede et al. (1964) who treated 60 patients suffering from polycythaemia rubra vera with chlornaphazine. Only 20 of these patients received over 100 g of chlornaphazine, and, of these, seven patients developed bladder carcinoma. Laursen (1970) reported two patients treated for Hodgkin's disease with chlornap-

hazine who developed cancer of the bladder five and six years after the drug had been stopped. In these cases the total doses were only 78 and 85 g respectively.

Isoniazid: is it a carcinogen?

The observation that isoniazid caused pulmonary tumours in experimental animals (Biancifiori and Severi 1966) suggested the need to re-examine the lung-cancer mortality of tuberculosis patients and to determine whether this was increased by isoniazid therapy. Two earlier investigations, one from Australia (Campbell 1961) and the other from Israel (Steinitz 1965) had already demonstrated that patients with tuberculosis had an excess mortality from lung cancer.

Campbell and Guilfoyle (1970) examined a group of 3064 tuberculous ex-servicemen and compared them with a control population of 14241 ex-servicemen who had been prisoners of war. All these men had served with the Australian Armed Forces during the Second World War. It was found that the tuberculous patients experienced a significant excess mortality from lung cancer (33 observed, 16 expected) and carcinoma of the upper respiratory and digestive tracts (12 observed, 5 expected). Nevertheless, the use of isoniazid was not shown conclusively in this small series to cause the higher incidence of lung cancer in patients who had been treated for tuberculosis.

Phenacetin-induced genito-urinary tumours

Phenacetin-induced tumours of the renal pelvis and bladder were described by Bengtsson et al. (1968) in patients taking excessive amounts of phenacetin-containing analgesics. This was fully discussed in Chapter 34.

Sarcoma after intramuscular iron

Animal studies have indicated that the intramuscular injection of iron dextran preparations may be associated with malignant change; usually these have been fibrosarcomas at the site of injection. Studies of human tissue after iron dextran failed to show malignant change in several investigations. Robinson et al. (1960) described a single case of fibrosarcoma, and Mac Kinnon and Bancewicz (1973) described a reticulum-cell sarcoma of the buttock in a 56-year-old woman and a pleomorphic sarcoma in the buttock of a 22-year-old woman. In all three of these cases, the patients had received iron sorbitol or iron dextran into their buttocks. Greenberg (1976) described a further three cases of sarcoma of the buttock associated with previous iron dextran injections.

Weinbren et al. (1978) evaluated the evidence for iron

compounds as local carcinogens in man. Histological material and clinical reports were reviewed in seven of the above eight published cases of tumours developing at the site of intramuscular injections. The microscopical appearances suggested benign lesions in two cases and a variety of tumours in the other five. In only two cases (a rhabdomyosarcoma and a fibrosarcoma) was the interval between injections and tumour development longer than six years. Of the remaining three tumours, one was considered to be a rather slowly growing haemangiopericytoma (with an interval of two years), one appeared to be a subcutaneous lymphoma with no evidence of having arisen in the gluteal muscles, and one was a pleomorphic sarcoma with a possible five-year interval. Sarcomas induced experimentally by iron compounds in rodents and rabbits differ in being less variable in type and in containing abundant iron-containing macrophages, which were negligible in these human tumours.

Although the total number of patients who have received intramuscular injections of iron compounds is not known, the present findings, in contrast to experimental work in animals, does not support the view that such treatment carries a strong risk of tumour development.

Is phenytoin-induced lymphadenopathy a premalignant state?

A benign lymphadenopathy resembling malignant lymphoma, and accompanied by fever, an exanthematous eruption, eosinophilia, often hepatosplenomegaly, and sometimes arthralgia, was first reported in association with hydantion therapy when *Nirvanol* (5-ethyl-5-phenylhydantoin) was introduced in the 1920s for the treatment of Sydenham's chorea. When phenytoin (5,5-diphenylhydantoin) became popular as an anticonvulsant in the late-1930s it was not long before pseudolymphoma was reported in epileptic patients treated with this drug (Coope and Burrow 1940). By 1959, Saltzstein and Ackerman had collected seven patients with this condition, and reviewed 75 further cases which had been reported over the preceding 20 years. Hydantoin-induced pseudolymphoma has the features of a delayed drug-hypersensitivity reaction, and resembles the lymphadenopathy which occurs with a variety of drugs, including *p*-aminosalicylic acid, iron dextran, sulphonamides, phenylbutazone, and meprobamate. Most reported cases haver arisen within four weeks of starting hydantoin treatment, although sometimes the duration of treatment was much longer. The syndrome rapidly clears after withdrawal of the drug, and recurs if it is given again (Oates and Tonge 1971). Several different hydantoin drugs have been incri-

minated including phenytoin, ethotoin, and methoin.

The characteristic triad of the syndrome is fever, rash, and lymphadenopathy. The cervical glands are most commonly involved, but a more generalized enlargement is not rare. Histologically the normal architecture of the lymph-nodes is distorted, with reticulum-cell hyperplasia, frequent mitoses, infiltration with eosinophils, plasma cells, and polymorphs, and focal necrosis, often haemorrhagic. But the features which characterize malignant lymphoma — namely, the Reed–Sternberg cells of Hodgkin's disease and the monotonous cellular pattern and capsular involvement of lymphosarcoma — are absent.

Differential diagnosis can usually be made on the clinical course after withdrawal of the drug, and lymph-node biopsy examination should, therefore, be unnecessary in most cases.

Sorre *et al.* (1971) demonstrated that in patients on long-term phenytoin therapy there were laboratory data to indicate a depression of immunological function. Sixty-three patients on long-term oral therapy with phenytoin sodium (sodium diphenylhydantoin) were screened for abnormalities of immunological function. They were compared with 92 controls and 28 patients with lymphoma. Depression of cellular or humoral immunity, or both, was found in a significant number of phenytoin-treated and lymphoma subjects. Phenytoin therapy was associated with low immunoglobulin A (21 per cent), failure of antibody response to *Salmonella typhi* antigen (9 per cent), absence of delayed hypersensitivity to three common skin-test antigens (22 per cent), and depression of *in-vitro* lymphocyte transformation by phytohaemagglutinin (27 per cent). Lymphoma patients manifested low IgM (22 per cent), and inability to make antibody to *S. typhi* (11 per cent) and to tetanus toxoid (21 per cent): delayed hypersensitivity was absent in 36 per cent; lymphocyte transformation was depressed in 17 per cent. Abnormal lymphocyte transformation did not correlate with depression of cellular or humoral immunity in either group.

Nitrofurantoin and nodular hyperplasia of liver

The case of a 6-year-old girl given prophylactic nitrofurantoin 20 mg daily for two months for recurrent acute urinary-tract infections was reported by Anttinen *et al.* (1982) who recorded that approximately four months later, when she again had an acute urinary-tract infection, nitrofurantoin therapy was reinstituted. In another two months the patient's liver was palpable 3 cm below the costal margin. A large tumour, 5 × 5 × 8 cm, was seen in the right lobe of liver on compound ultrasonogram, aortic abdominal angiography, and ^{99m}Tc colloid liver scan. At operation the right lobe and medical segment of the left lobe were resected. Gross and microscopic pathologic findings were consistent with focal nodular hyperplasia of the liver. The authors attributed this growth to the nitrofurantoin therapy.

Isotretinoin (Roaccutane)

Drugs which show teratogenic effects in a particular species should (Schottenfeld 1982) be suspected of carcinogenicity in that species. *Roaccutane*, a Vitamin-A derivative and a known animal teratogen, was licensed in 1982 for limited use in intractable acne in the UK. However, it was in use more widely and at higher doses in the USA.

In August 1983 it was reported that 13 cases of adverse pregnancy outcome had occurred in the USA (Rosa 1983). Birth defects occurred in five of these cases: two had hydrocephaly alone, two had hydrocephaly with a cardiovascular defect and microtia, one had microtia alone. The outcome in the remaining eight of these cases was spontaneous abortion. In addition, it is known that a further 13 inadvertent pregnancy exposures have occurred, the outcomes of which will occur between September 1983 and January 1984, if they continue to term.

Although it is known that *Roaccutane* does not interfere with the action of oral contraceptives (*Current Problems* 1983), it is clearly a major human teratogen. It appears to constitute the most clear-cut example of drug-induced birth defects since thalidomide.

Neovascular inflammatory nodules have been reported following the treatment of acne with oral isotretinoin (Valentic *et al.* 1983). Two of 12 patients treated with 1–2 mg/kg/day oral isotretinoin developed inflammatory neovascular nodules during the third month of treatment. The nodules were confined to the back and upper part of the arms. The lesions resolved spontaneously, without scarring, within several weeks of stopping isotretinoin therapy. Reports of granulomatous dermal lesions have also appeared in the recent literature (Spear and Muller *et al.* 1983; Exner 1983).

Aristolochic acid

This substance occurs in plants of the Aristolochiaceae family, such as *A. serpentaria* (serpentary) and *A. clematitis* (birthwort). Structually the molecule is a phenanthene derivative, but contains a nitro group and two epoxide groups. Plant preparations containing aristolochic acid in relatively high concentrations have been used as anthroposophic remedies and, in low concentrations, in homoeopathy.

Anthroposophy was the term used by the Austrian polymath Rudolph Steiner, 1861–1925, for a system of philosophical ideas which he obtained from an undisclosed source. Much secrecy and speculation surrounds the origins of Steiner's ideas, which probably lie in esoteric Middle-eastern traditions (Scott 1983). In August 1981 the World Health Organization issued a notice (World Health Organization 1981) stating that the Federal German Health Office had banned the use of all but highly diluted preparations of aristolochic acid because of striking carcinogenicity findings in a rodent toxicity study. These findings have now been published (Mengs *et al.* 1982) and confirmed (Mengs 1983). In the Wistar rat, doses of 0.1, 1.0, and 10.0 mg/kg/day induce a variety of tumours in a time- and dose-related manner. Doses of 1.0 and 10.0 induce severe papillomatosis of the forestomach, with occasional malignant change by three months. By 3–6 months, without further dosing, metastasizing squamous-cell carcinomata of the forestomach appeared. Carcinomata, papillomata, and hyperplasia of the renal pelves and bladder appeared, together with adenomata of the renal cortices and anaplastic changes to the tubular epithelium. The lowest dose, 0.1 mg/kg/day, was administered to various animals for 3–12 months. Although no tumours appeared during the first six months, squamous-cell carcinomata and papillomata appeared in the forestomach after 12–16 months, together with hyperplastic change of the urethelium. The consequences of human exposure, especially to those products containing the higher concentrations of aristolochic acid, are not yet known (Penn 1983).

Sassafras

The roots and bark of Sassafras spp. have been used in herbal medicine for a variety of indications. Extracts containing the major ingredient safrole have been used as a flavouring and aromatic agent. In 1961 it was shown that sassafras produced hepatic tumours in rats in a dose-related manner (Homburger *et al.* 1961) at relatively high dose levels. These findings were confirmed in 1963 (Long *et al.* 1963), when the dose-related induction of benign and malignant liver tumours was demonstrated in the rat. In subsequent studies (Borchert *et al.* 1973) it was shown that the rat and mouse metabolite hydroxysafrole acted as a proximate carcinogen in those species.

Phenothiazines and pituitary tumours

Asplund *et al.* (1982) reported two cases of pituitary tumours associated with chronic exposure to phenothiazines. A 65-year-old man, who had received chlorpromazine for 14 years for affective psychosis, had been taking 75 mg of the drug daily in recent years. He developed impotence, symptoms suggestive of hypothyroid-ism, and later, bitemporal hemianopia. A pituitary tumour with suprasellar extension was diagnosed. Endocrine evaluation showed partial pituitary insufficiency and the tumour was removed. Microscopic examination of the tumour disclosed a chromophobic adenoma.

The second patient, a 56-year-old woman with an affective psychosis, was treated for 10 years with perphenazine. She developed deteriorating memory, widebase gait, dizziness, and galactorrhea. Computed tomography showed a pituitary tumour with suprasellar extension which was removed. Microscopic examination showed a chromophobe adenoma. It was postulated that long-standing pituitary stimulation by chronic phenothiazine therapy may cause enlargement of the pituitary gland and development of pituitary macroadenomata. This class of drugs has been known to induce pituitary adenomata in rodent toxicity studies for a number of years. This aspect of rodent toxicity is common to most phenothiazines and prolactin-releasing drugs. It is not surprising therefore to see this toxic effect being reported in patients receiving long-term phenothiazine therapy.

Benoxaprofen (*Opren*)

The non-steroidal anti-inflammatory agent benoxaprofen was marketed in the UK in July 1980. It was withdrawn in September 1982 as a result of major adverse reactions affecting the liver, kidney, and other organs, although trials in severe psoriasis were permitted in the UK. Limited investigational use in arthritis was also authorized in the US, on 'compassionate' grounds (SCRIP 1984; *Washington Post* 1983).

In a report from South Africa (Findlay and Hull 1982), tumours of the skin were reported in four subjects who had taken benoxaprofen during the preceding months and who had suffered from excessive sun-exposure. Two of these patients suffered onycholysis and two had photodermatitis (both reactions being recognized, non-fatal, adverse effects of benoxaprofen). When benoxaprofen was withdrawn, no new tumours appeared. The tumours varied from milium-like lesions or closed comedones in appearance, to prominent, smooth convex papules and nodules. Sometimes these were grouped in clusters. Histologically the lesions from all cases consisted of single or lobulated epidermoid cysts opening on to normal epidermis. They could be regarded as a tumour-forming transformation of the hair follicle infundibulum.

Animal carcinogenicity studies completed in 1983 indicated (SCRIP 1984) that there was an increased incidence of liver cancers in mice receiving doses between three and 10 times the recommended human dose of 600 mg daily.

All use of benoxaprofen in man has now ceased as a result of these findings. A circular letter to the doctors who were still prescribing the drug, from the manufacturers, explained that: '. . . it is not possible directly to correlate these findings in mice to possible long-term adverse effects in humans. Nevertheless, we have decided that the prudent course of action is to discontinue' (SCRIP 1984).

Ethylene oxide

Ethylene oxide is used in the preparation of medicinal products in the pharmaceutical industry. It is employed as a sterilizing agent for crude drugs, powders, and herbs in bulk, containers, and in situations where other methods are inappropriate. Residues of ethylene oxide are therefore present in some medicines. Workers exposed to ethylene oxide were shown (Hogstedt *et al.* 1978, 1979) to suffer from increased rates of leukaemia. An increased rate of gastric carcinoma was also shown (Hogstedt *et al.* 1978) in one of these studies.

These findings do not, of course, conclusively incriminate ethylene oxide as a carcinogen, as the workers were concurrently exposed to other substances. Moreover, no results of adequate animal studies are available (International Agency for Research on Cancer 1979). Nevertheless, it is essential that residues in medicinal products should be limited to the lowest possible levels.

Carcinogenic effects of plasticizers

Plastic tubing and bags used in renal dialysis, blood transfusions, and the infusion of intraveous medicines are constructed from materials such as polyvinyl chloride (PVC) which require the addition of a 'plasticizer' to improve flexibility. The most commonly used plasticizer is di(2-ethyl-hexyl) phthalate (DEHP).

The International Agency for Research on Cancer (1982) concluded that there is sufficient evidence to establish that DEHP is carcinogenic in mice and rats.

When administered orally, the compound has been found significantly to increase the incidence of benign and malignant liver-cell tumours in animals of both species and a dose–response relationship was seen. DEHP has caused dominant lethal mutations in mice, following parenteral administration. The compound induces testicular damage in rats, and there is evidence that it, and one of its metabolites, are teratogenic and embryo-lethal in rodents.

These findings do not, of course, establish that DEHP is carcinogenic in man. Moreover, no adequate epidemiological study has been conducted (International Agency on Cancer Research 1982) and there are no data relating to mutagenic or teratogenic effects in man. However, in a small prospective cohort study, eight deaths were observed among 221 workers exposed to DEHP for periods between 3 months and 24 years. One carcinoma of the pancreas and one bladder papilloma were reported (Thiess *et al*. 1978).

In a study of patients who had received blood stored in PVC products, or undergone haemodialysis, the levels of DEHP, expressed in μg per g of wet tissue, were found to be: brain 1.9; heart 0.5; kidney 1.2–2.2; liver 1.5–4.6; lung 1.4–2.2; and spleen 2.2–4.7 (Chen *et al*. 1979). The levels of DEHP in neonatal heart tissue from infants who had undergone umbilical catheterization, with or without the infusion of blood products, were reported to be higher than those in untreated controls (Hillman *et al*. 1975). Whole blood or plasma stored in the PVC bags has been found to contain DEHP in concentrations up to 250 mg/l (Pik *et al*. 1979; National Toxicology Program 1982). Blood stored at 4°C in PVC bags leached DEHP from the plastic at a linear rate of 2.5 ± 0.3 mg/l daily (Jaeger and Rubin 1972).

There is a clear need for epidemiological work to establish whether DEHP in medicinal products and equipment poses a hazard to man. Occupational exposure limits exist in many developed countries, including the US; if it is deemed that industrial workers are exposed to some degree of risk, such a possibility cannot be ignored in patients receiving long-term infusions of medicines known to be capable of leaching DEHP from plastics.

Psoralens and ultraviolet irradiation and cancer

As a cause of skin damage, especially skin cancer, ultraviolet (UV) radiation is arousing serious concern. Sun-induced skin injury is caused by the UV part of the solar emission spectrum. Solar UV radiation can conveniently, but arbitrarily, be divided into three zones — UV-A (long wavelength UV, 320–400 nm); UV-B (sunburn UV, 290–320 nm); and UV-C (short-wavelength 'germicidal' UV, below 290 nm) (Bridges 1978).

Notwithstanding the potentially harmful actions of high-dosage UV, phototherapy has been used for at least half a century in the treatment of inflammatory, pigmentary, and proliferative disorders of skin. There is intense interest, in particular, in long-term treatment of psoriasis by a combination of a systemically administered photosensitizing psoralen and high-intensity longwave-length (UV-A) irradiation (photochemotherapy, PUVA) (see also Chapter 32).

In psoriasis the rate of epidermal cellular proliferation is increased, and the prevailing view is that resolution of psoriatic lesions in response to PUVA is due to production of DNA damage which in turn leads to inhibition of epidermal DNA synthesis. Psoralens form photo-adducts and interstrand cross-links with DNA in the presence of UV-A and mounting evidence indicates that these events are mutagenic. It has been known for some time that, in *Escherichia coli* cultures, 8-methoxy-psoralen produces mutations in the presence of UV-A. More recently extensive chromosome damage has been produced by long-wavelength UV in Chinese hamster cell cultures containing 8-methoxypsoralen. In man, PUVA may well give rise to similar chromosomal damage. Although sister chromatid exchange studies after PUVA treatment *in vivo* have revealed no evidence of chromosome lesions, *in vitro* irradiation of human lymphocytes by UV-A in the presence of 8-methoxypsoralen significantly increased the frequency of chromosomal aberrations. Evidence of chromosomal damage does not by itself prove that PUVA treatment is carcinogenic in the skin, but there is certainly cause for disquiet in view of the ability of parenteral 8-methoxypsoralen combined with UV-A to promote skin cancer in mice and to increase the rate of squamous epithelioma in patients with xeroderma pigmentosum.

Stern *et al*. (1979) conducted a prospective study of 1373 patients given oral 8-methoxypsoralen photo-chemotherapy for psoriasis which revealed 30 patients with a total of 48 basal-cell and squamous-cell carcinoma. The observed incidence of cutaneous carcinoma was 2.63 times greater than that expected for an age, sex, and geographically-matched population. Relative risk to patients with a history of ionizing radiation was 3.68. Patients with a previous cutaneous carcinoma had a relative risk of 10.22. A higher than expected proportion of squamous-cell carcinoma and an excess of squamous cell carcinoma in skin areas not exposed to sun were seen.

Several workers have reported effects on DNA synthesis in circulating lymphocytes of patients treated with PUVA (Kraemer and Weinstein 1977; Lischka *et al*. 1977). This raises the possibility of hazard from DNA damage to cells circulating in the blood as well as to those in the skin. Presumed gene mutations (HGPRT variants) have also been detected in lymphocytes from PUVA-treated psoriatics (Strauss and Albertini 1978). Two reports of leukaemic conditions arising in psoratic patients treated with PUVA were reported in the Scandinavian literature (Wagner *et al*. 1978; Hansen 1979).

The presence of oil of bergamot in various sun-tan preparations has recently been raised in this context since oil of bergamot contains bergapten (5-methoxypsoralen). It has been inferred that the use of these sun-tan preparations may be associated with an increased risk of skin neoplasia including malignant melanoma (Ashwood-Smith 1979; Kersey 1980; Epenetos 1980). This suggestion was the subject of an alarmist article appearing in the *Sunday Times* (18 May 1980). Oil of bergamot is also present in Earl Grey tea.

There is at present no positive evidence to incriminate oil of bergamot in either sun-tan preparations or Earl Grey tea as a safety hazard: nevertheless, the situation merits scrutiny.

L-dopa and malignant melanoma

The formation of melanin depends upon a metabolic pathway which is impaired in Parkinson's disease. It is therefore to be expected that melanin synthesis might be decreased in this condition and that the prevalence of melanin-forming tumours might possibly be lowered (Van Rens *et al.* 1982).

In 1972 a case of Parkinson's disease with malignant melanoma was reported (Skibba *et al.* 1972), in which recurrence with multiple skin melanomata occurred two months after commencing L-dopa therapy. It must at the same time be remembered that 3.9 per cent of patients with cutaneous melanomata have been reported (Beardmore *et al.* 1975) to possess multiple primary lesions. This was followed by further reports suggesting a relation between cutaneous malignant melanoma, Parkinson's disease, and the use of L-dopa (Robinson *et al.* 1973; Lieberman and Shupack 1974; Fermaglich and Delaney 1977; Bernstein *et al.* 1980). In 1982, three cases of uveal-tract melanoma were reported in patients suffering from Parkinson's disease and treated with L-dopa (Van Rens *et al.* 1982).

It was suggested (Lieberman and Shupack 1974) that L-dopa might possess inducing or promoting properties, although the compound was found to inhibit melanoma cells *in vitro* (Wick 1977) and *in vitro* in the mouse (Wick *et al.* 1979).

An increased prevalence of cutaneous melanomata in Parkinson's disease was reported (Fermaglich and Delaney 1977) but this was not statistically significant. In another survey in 1099 patients with the tumour, only one patient with Parkinson's disease who had used L-dopa was found (Sober and Wick 1978, 1979). A possible association between malignancy and L-dopa was suggested by Barbeau (1972) but malignancies were not mentioned in a recent review of the adverse effects of antiparkinsonian drugs (Parkes 1981). The case for an association must be regarded as unproven until further evidence is adduced.

Cancer and chlorinated water

Studies of a rather preliminary nature have indicated that the continued consumption of chlorinated water might increase the risk of rectal, colonic, and bladder cancer.

Five such studies have been conducted over the past few years and cover some 11 500 individuals in various areas of the USA. The relative increased risks that have been propounded vary from 1.13 to 1.93 and, although the results are suggestive of a causal association, relative risks of less than 2.0 are always equivocal. For what they are worth, the studies so far seem to show, in those drinking chlorinated water, a slightly increased risk of rectal cancer and a lesser one of colonic and bladder cancer. But, even if taken at face value, the risks seem negligible in the face of the catastrophic problems that would arise were chlorination abandoned: a slightly increased risk of cancer would be a small price to pay for protection against other water-borne diseases (National Academy of Sciences 1980; Council on Environmental Quality 1980; Maugh 1981; *Lancet* 1981*a*).

Is alcohol a carcinogen?

In interview data from the USA's Third National Cancer Survey, alcohol ingestion was associated with a higher occurrence of cancers of the breast, thyroid, and malignant melanoma. Williams (1976) suggests that a unifying hypothesis to explain all these apparently diverse associations might be advanced on the basis that alcohol might stimulate anterior pituitary secretion of prolactin, thyroid-stimulating hormone, and melanocyte-stimulating hormone. Under the trophic effects of these pituitary hormones the target organs exhibit increased mitotic activity and an increased susceptibility to malignancy.

The mechanism advanced by Williams is a hypothesis but the epidemiological association is a fact that could have far-reaching consequences (Breslow and Engstrom 1974).

Neoplasia following the use of cytotoxic and immunosuppressant agents

Many antineoplastic drugs are themselves carcinogenic. Some of these compounds have come to be employed in non-neoplastic diseases; for example, the alkylating agent cyclophosphamide has been used in psoriasis, rheumatoid arthritis, systemic lupus erythematosus, Sjögren's syndrome, and other conditions. Some anticancer agents have also been used for immunosuppression in organ transplantation; again, cyclophosphamide is an example. Finally, several compounds, other than antineoplastic agents and steroids, have been found recently to possess significant immunosuppressant actions in their own right; the T-lymphocyte suppressor cyclosporin-A is the best-known example. Drugs which fall into one or more of the above categories, and which are associated with new neoplasia, are dealt with in this section.

Among the cytotoxic agents, it appears to be the antimetabolites (and particularly methotrexate) which carry the least carcinogenic risk. However it is now clear (International Agency for Research on Cancer 1979, 1982) that the alkylating agents (e.g. cyclophosphamide, carmustine, and chlorambucil), some antitumour antibiotics (e.g. doxorubicin, bleomycin, and mitomycin-C), and individual compounds (e.g. cisplatin and procarbazine) are known or suspected carcinogens.

Treatment for Hodgkin's disease followed by new neoplasia

A survey conducted in Italy by Valagussa *et al.* (1980) investigated a total of 764 patients with Hodgkin's disease treated with radiotherapy, chemotherapy, or both; they were reviewed 3–186 months (median 43 months) after initial treatment to assess the incidence of second malignancies. Incidences of solid tumours and acute non-lymphoblastic leukaemia (ANLL) were calculated by a life-table method and percentages of patients affected derived from life-table plots.

Within 10 years after initial treatment the overall incidence of second solid tumours was 7.3 per cent and over a comparable period 2.4 per cent of patients developed ANLL. Solid tumours occurred only in patients given radiotherapy with or without adjuvant chemotherapy, and ANLL occurred only after treatment with MOPP (mustine, vincristine, procarbazine, and prednisolone) or modified MOPP regimes. Neither solid tumours nor ANLL occurred in patients given ABVD (adriamycin, bleomycin, vinblastine, and dacarbazine).

The highest incidence of leukaemia (5.4 per cent) occurred after treatment with excessive radiotherapy plus MOPP; hence the benefits of this approach in Hodgkin's disease must be weighed against its carcinogenic potential.

Leukaemia following treatment of ovarian cancer with alkylating agents

Greene *et al.* (1982) assessed the occurrence of acute non-lymphocytic leukaemia (ANL) among 1399 women with ovarian cancer, who were treated in five randomized clinical trials. Of the 1399 women, 998 had been treated with alkylating agents and, among these, 12 cases of ANL were observed; the expected number was 0.11. Ten patients with ANL had received melphalan, and two chlorambucil. ANL was not observed in 401 women who had been treated with surgery or radiation or both, without alkylating agents. The excess risk of ANL that was associated with alkylating-agent therapy was 5.8 cases per 1000 women per year, and the cumulative seven-year risk of ANL was 9.6 ± 3.3 per cent ($\pm$ S.E.). The risk of ANL among patients who were treated with chemotherapy alone was indistinguishable from that observed in patients receiving both radiation and chemotherapy. A positive correlation between initial drug dose and the risk of ANL was suggested. These data underscore the need to assess other cytotoxic agents and regimens of drug administration to identify those that do not have harmful late effects.

It is now acknowledged that alkylating agents such as chlorambucil are carcinogenic at therapeutic doses. The current datasheet (Association of the British Pharmaceutical Industry 1983) for *Leukeran* contains the following statements:

Leukeran has been shown to cause chromatid or chromosome damage in man. Development of acute leukaemia after Leukeran therapy for chronic lymphocytic leukaemia has been reported. However, it was not clear whether the acute leukaemia was part of the natural history of the disease or if the chemotherapy was the cause.

A comparison of patients with ovarian cancer who received alkylating agents with those who did not, showed that the use of alkylating agents, including Leukeran, significantly increased the incidence of acute leukaemia.

Acute myelogenous leukaemia has been reported in a small proportion of patients receiving Leukeran as long-term adjuvant therapy for breast cancer.

The leukaemogenic risk must be balanced against the potential therapeutic benefit when considering the use of Leukeran.

Cyclophosphamide and bladder cancer

Cyclophosphamide is an alkylating agent and has been widely used in cancer chemotherapy since 1958 and more recently as an immunosuppressant in systemic lupus erythematosus.

Haemorrhagic cystitis was recognized as a complication of cyclophosphamide therapy as early as 1959 by Coggins *et al.*

Johnson and Meadows 1971 analysed the frequency of bladder fibrosis in relation to the dose and length of treatment with cyclophosphamide. Necropsy in 40 children treated with cyclophosphamide for various cancers showed that 10 had bladder fibrosis and three had telangiectasia of the mucosa. This fibrosis could affect the whole thickness of the bladder wall and all the changes were irreversible. Though nine out of the 10 patients with bladder fibrosis had been treated for 20 weeks, the duration of therapy did not seem as important as total dosage. The fibrosis occurred when the total dosage exceeded 6 gm^{-2} of body surface area, and the more severe changes were associated with the highest doses. Only five patients had any symptoms and all of these had macroscopic haematuria. There was no evidence of actual tumours in the bladder, and up to this point in time no cyclophosphamide-induced tumours of the bladder have been reported in the literature.

However, later in 1971, Worth reported one patient with Hodgkin's disease and another with lymphosarcoma, in each of whom transitional-cell carcinomas of the bladder developed after treatment with cyclophosphamide for two and four years, respectively.

Dale and Smith (1974) described two patients receiving cyclophosphamide for prolonged periods (one with myeloma, the other with Waldenstrom's macroglobulinaemia) in whom localized transitional-cell carcinoma of the urinary bladder developed. Wall and Clausen (1975) documented five patients, four with myeloma and one with Hodgkin's disease, who received large cumulative doses of cyclophosphamide over a prolonged period with the development of squamous-cell carcinoma of the urinary bladder.

In 1982 Elliott *et al.* reported two cases of women who developed bladder cancer after prolonged treatment with cyclophosphamide. The first report was of a 28-year-old woman who in 1969 developed lupus erythematosus, and was treated with 50–100 mg cyclophosphamide daily for intermittent periods over the next six years. In 1981 following investigation for urinary-tract infection, a bladder tumour was diagnosed. At biopsy a poorly differentiated keratinizing squamous-cell carcinoma of the bladder with extensive haemorrhage and necrosis was found. Treatment was a short course of deep X-ray therapy followed by total cystectomy and transplantation of the ureters into an ileal conduit.

The second case reported by Elliot *et al.* was a 45-year-old woman who also presented in 1969 with lupus erythematosus. In 1969 treatment was commenced with prednisone 50 mg and cyclophosphamide 100 mg daily; both were reduced to a lower maintenance dosage. In 1976 while taking 7.5 mg prednisone and 20 mg cyclophosphamide daily, investigation for left loin pain disclosed a bladder tumour. Biopsy showed a poorly differentiated transitional-cell carcinoma infiltrating muscle. Deep X-ray treatment was followed by total cystectomy and transplantation of the ureters into an ileal conduit.

Carney *et al.* (1982) described a further case of cyclophosphamide-induced, low-grade, non-classified sarcoma of the bladder in a 14-year-old boy who had commenced treatment for stage-1 Hodgkin's disease with a regime including cyclophosphamide at age 11 years.

Donadia *et al.* (1981) doubted whether there was any hard evidence that indicated any beneficial effect of cyclophosphamide in the treatment of lupus erythe-

matosus nephritis; therefore, in view of the risks, it would seem totally reprehensible to continue to use the drug for this purpose.

Lomustine (CCNU) and leukaemia

Cohen (1980) described the case of a 60-year-old woman given methyl CCNU (200 mg/m^2 of body surface 6 weekly) after a subtotal removal of a left frontal oligodendroglioma. Bone marrow was normal at the start of CCNU therapy, but after 11 months there were megaloblastic and hypoplastic changes. Dosage was reduced and after 20 months dosage was stopped due to the development of pancytopaenia. The patient remained pancytopaenic for about a further year when she was diagnosed as having developed myelomonoblastic leukaemia.

Chlornaphazine-induced bladder carcinoma

Three reports of individual cases of chlornaphazine-induced bladder cancer were cited by Meyler (1966). The first case was a woman of 68 being treated for polycythaemia vera; she had received a total of 156.6 grams of chlornaphazine over a period of 10 years. The second patient, a woman of 45, had received 149.5 grams of the drug over a period of five years for the treatment of Hodgkin's disease. The third patient, a man of 30, had been treated for Hodgkin's disease with chlornaphazine for six years.

It would seem from these cases that a high dose of chlornaphazine is associated with neoplastic changes in the bladder epithelium. This is in keeping with the findings of Thiede *et al.* (1964) who treated 60 patients suffering from polycythaemia rubra vera with chlornaphazine. Only 20 of these patients received over 100 g of chlornaphazine, and, of these, seven patients developed bladder carcinoma. Laursen (1970) reported two patients treated for Hodgkin's disease with chlornaphazine who developed cancer of the bladder five and six years after the drug had been stopped. In these cases the total doses were only 78 and 85 g, respectively.

Carcinogenesis associated with immunosuppressant therapy

Burnet (1959) postulated that the immune system plays an important role in preventing and restricting the growth of neoplastic cells, and Swanson and Schwartz (1967) carried this concept a stage further when they suggested that neoplasia might occur as a complication of treatment with immunosuppressive agents.

Unfortunately this concept has been shown to be true

and Doak and colleagues (1968) reported two such cases in which prednisone and azathioprine had been used to suppress the immunological mechanisms after renal transplantation. Both of these patients developed reticulum-cell sarcoma; the first patient was a 34-year-old man with chronic renal failure due to glomerulonephritis who had had a cadaver kidney transplanted. The other patient, a 46-year-old woman, had been similarly treated by renal transplant because she had renal failure due to pyelonephritis. Both patients developed infections with *Candida albicans*, and the woman also developed labial herpes simplex and a systemic staphylococcal infection.

In this context it is of interest to note the observations of Schneck and Penn (1971) who reviewed the reports of 24 mesenchymal neoplasms that had arisen *de novo* in patients given immunosuppressant therapy following renal transplantation. Eleven of 24 mesenchymal neoplasms with organ homografts throughout the world had involved the brain. In eight cases the tumours were present only in this organ. Nine neoplasms were reticulum-cell sarcomata and two were unclassified lymphomata. These lesions appeared sooner after transplantation than did non-cerebral mesenchymal or epithelial tumours. The brain has no lymphatic system, and little proliferation of reactive cells takes place in response to foreign antigens unless prior immunization has occurred. It is suggested that tumour cells could arise in the brain itself or be carried there from other sites. The poor immunological reactions of the unimmunized and immunosuppressed brain suggest the possibility that neoplastic cells would grow more readily in this relatively immunologically privileged environment than in other tissues.

In a report from Australia, Walder *et al.* (1971) described seven patients who developed malignant skin tumours following immunosuppressive treatment following kidney allograft out of a total series of 51 cases. The malignant skin tumours appeared 4–45 months after transplantation and included squamous-cell carcinoma, basal-cell carcinoma, and kerato-acanthoma and Bowen's disease. All but two patients had multiple primaries; one case had seven primary skin tumours excised. The primary tumours all arose in areas exposed to sunlight, and sunlight must be regarded as an important initiating factor. However, in Australia basal-cell carcinoma is 11 times commoner than squamous-cell carcinoma, but in this series 16 primary squamous-cell carcinomata were diagnosed compared to one basal-cell carcinoma.

More recently, a fatal myeloproliferative syndrome diagnosed as subacute myelomonocytic leukaemia was reported by Hochberg and Schulman (1978) as developing in patient with Sjögren's syndrome following treatment with cyclophosphamide. Although patients with Sjögren's syndrome have a recognized increased risk of developing lymphoproliferative disorders, particularly reticulum-cell sarcoma, there was, hitherto, no evidence to support a similar predisposition to leukaemia. These rheumatologists also reviewed recent reports of patients developing fatal myeloproliferative disorders after receiving immunosuppressive therapy and emphasized that great caution should be exercised in the use of these agents in patients with chronic, non-fatal disorders such as rheumatoid arthritis and Sjögren's disease.

Proposed reasons for the increased incidence of malignant neoplasms in patients on immunosuppressive treatment include facilitation of infection by oncogenic viruses, direct effect of the immunosuppressive drug inducing neoplastic transformation, or suppression of an immunosurveillance mechanism normally responsible for destruction of mutant cells with malignant potential.

Immunosuppressant therapy using cytotoxic agents for non-malignant conditions was suspected as a cause of neoplasia after it had been established that such treatments, used for malignant disease, can be responsible for second tumours. In 1971, attention was first drawn to a number of reports of acute myeloblastic leukaemia arising in patients whose neoplastic disease had been treated with cytotoxic drugs. Most of the early cases arose after treatment of myeloma or chronic lymphatic leukaemia but there are now many reports of acute myeloid leukaemia as a late sequel to treatment of Hodgkin's disease now that more of these patients are having their survival prolonged. Acute myeloid leukaemia has also been recorded after chemotherapy for cancer of breast, lung, and ovary. Moreover, there are reports linking acute myeloid leukaemia with cytotoxic immunosuppressant therapy in patients with non-neoplastic disorders such as renal transplantation and rheumatoid arthritis.

Tchernia and his colleagues (1976) reviewed 11 cases, including three of their own, in which acute myeloblastic leukaemia arose after immunosuppressive therapy for 'primary non-malignant' disease. Apart from one patient who had received intra-articular injections of radioactive gold and yttrium, none had been irradiated. Most had been given alkylating agents, such as cyclophosphamide, chlorambucil, or melphalan. Two, however, had received only azathioprine, in one after renal transplantation and in the other for chronic hepatitis. Most had received therapy for two years or longer before the onset of leukaemia, but one patient with psoriasis had been on busulphan and methotrexate for only six months. Several patients were persistently cytopaenic before the diagnosis of acute myeloblastic leukaemia, but in at least three cases there was no recognizable preleukaemic syndrome.

One had an episode of marrow aplasia, seemed to recover from it, and then became leukaemic. Roberts and Bell (1976) described three patients with chronic renal disease who acquired acute myeloblastic leukaemia after receiving cyclophosphamide for two to seven years.

Several large epidemiological surveys have been conducted into this problem and have indicated that the use of these agents is associated with an increased risk of malignant change.

Kinlen *et al.* (1979, 1981) published the results of a collaborative study including centres in the United Kingdom, Australia, and New Zealand which was instituted in 1970 to determine the incidence of cancer in patients treated for at least three months with azathioprine, cyclophosphamide, or chlorambucil. Follow-up of 3823 renal transplant recipients showed an almost 60-fold increase of non-Hodgkin's lymphoma together with an excess of squamous-cell skin cancer (34 cases observed against 0.58 expected), and mesenchymal tumours. Fifteen of the 34 non-Hodgkin's lymphomas affected the brain and presented typically as space-occupying lesions. A series of 1349 patients without transplants showed an excess of the same tumours, though to a less extent (four cases as against 0.34 expected).

Kinlen *et al.* (1979) suggested that in view of the short induction period for the neoplasia to occur, sometimes within months of transplantation of the kidney, and the fact that the risk was greater if the cytotoxic drugs were given together with renal transplantation than when given alone, a viral origin for tumourogenesis was likely.

Immunosuppression and malignant melanoma

Immune function seems to play a prominent part in the biology of human cutaneous malignant melanoma (CMM), and yet it is not generally appreciated that the risk of CMM is significantly raised in patients with immune dysfunction. In particular, the striking excess of non-Hodgkin's lymphoma in renal-transplant recipients has overshadowed the observation that squamous-cell carcinoma and melanoma of the skin are also excessively common in this population.

Greene *et al.* (1981) reviewed 14 patients who had developed CMM after renal transplantation and subsequently all patients received corticosteroids and azathioprine to prevent rejection; in 10 patients these were the only immunosuppressant drugs used. The remaining four patients received varying combinations of antilymphocyte serum, cyclophosphamide, actinomycin D, allograft irradiation, and splenectomy. Histological review of the primary melanomas revealed that all 10 evaluable patients had a precursor naevus from which the CMM arose, seven of which were dysplastic, and that 13 of the

14 patients had an abnormal host response to the tumour, characterized by absence of the usual lymphocyte/macrophage infiltrate. These observations suggest that CMM in renal-transplant recipients evolves from precursor naevi in immunosuppressed patients who are unable to mount an appropriate cellular immune response to neoplastic cells. The clinical detection of dysplastic naevi offers an opportunity to identify in advance those immunosuppressed patients who are prone to melanoma and to modify their medical management accordingly.

Similar increased risk of developing malignant melanoma in immunosuppressed renal transplant patients has been found (Birkeland *et al.* 1975; Hoover 1977; Kinlen *et al.* 1979) and increased risk of malignant melanoma has been described in immunosuppressed chronic lymphatic leukaemia patients (Greene *et al.* 1978) and in nontransplant immunosuppressed patients (Kinlen *et al.* 1979).

Cyclosporin-A and lymphoma

Cyclosporin-A was first marketed in the UK in 1983, as an immunosuppressant to facilitate bone marrow transplantation. It has subsequently been marketed in the UK for use in the transplantation of several specific organs. It possesses unique advantage (Kostakis *et al.* 1977; Bordes-Asnar and Tilney 1982) and is used very widely. However, there is now evidence that the use of cyclosporin-A is linked with the production of malignant lymphomata, probably in association with viral agents of the herpes group, and particularly the Epstein–Barr virus, in man and in animals.

Of the 34 patients who initially received cyclosporin-A for organ allografts, three subsequently developed lymphomata (Calne *et al.* 1979). In a more recent study (Penn 1982*a*), 15 cases of malignant neoplasm in transplant patients were analysed. The size of the patient-population in which these 15 cases were found is not stated. Lymphomata were diagnosed in 13 of these cases. Although all these patients received cyclosporin-A, all but one received other immunosuppressants concomitantly.

Various non-primate mammals have been studied and negative oncogenicity findings reported (Thiru *et al.* 1981). The species examined included rat, mouse, rabbit, pig, and dog. In the rhesus monkey, following renal allograft, no lymphomas were found (Borleffs 1982). These macaques received cyclosporin-A alone, or in combination with prednisolone and azathioprine, and were followed for more than one year. Similar results were obtained in animals which received the immunosuppressants, but without grafting. However, in striking contrast to all the above results, a study at Stanford (Reitz and Bieber 1982) demonstrated the induction of lymphomata in macaques. Of 97 macaques at risk two weeks or longer after transplantation, 12 developed lymphomata. All were among the 55 which had received cyclosporin-A, either alone or with other agents. Most of the animals not receiving cyclosporin-A died (because of rejection problems) at a relatively early stage and proper statistical comparison between the two groups is therefore not possible. The neoplastic cells showed the electron-microscopical appearances of B-lymphocytes and herpes-like viral particles were seen in most cases studied by electron microscope. Epidemics of lymphoma among macaques are not unknown (Terrell *et al.* 1980) and this, together with the isolation of a virus having antigens cross-reactive with Epstein–Barr virus, strongly supports a viral aetiology. The precise role of cyclosporin-A, and whether reactivation of latent virus or new infection have occurred, remain open questions. There are, however, several clear, if circumstantial, indications as to possible aetiology. In infectious mononucleosis, the Epstein–Barr virus (which is implicated in the genesis of lymphoma as well as nasopharyngeal carcinoma in man) induces B-cell proliferation. The attack is normally terminated when this proliferation is brought under control by T-cell action. If T-cell function is deficient, however, unchecked growth can lead to neoplasia. Cyclosporin-A works primarily though its potent anti-T-cell action.

Risks to those handling anticancer agents

In view of the fact that many anticancer agents are known to be both carcinogenic and mutagenic a report from Finland (Falck *et al.* 1979) raised concern about the safety of handling cytotoxic agents by nurses and other personnel. These workers tested the urine of patients receiving cytotoxic agents and the nurses administering the drugs and a control group. All the patients and most of the nurses showed the presence of mutagenic substances in their urine. The level of mutagenic substances was very much higher in the patients than the nurses. In the nurses' urine the level of mutagenic activity was much less after a nurses' duty-free weekend. The studies by Falck *et al.* indicated that precautions should be taken by medical and nursing staff handling these substances; both gloves and masks should be used. These problems are considered in detail in Chapter 41.

Razoxane-associated malignant disease

Razoxane is a novel anticancer drug derived from a series of diimidine analogues of ethylenediamine tetraacetic acid (EDTA). It exhibits antitumour activity in several tumour models. Though its precise mechanism of action is unknown, it does inhibit the progression of cells through the post-DNA synthesis G_2-M phase of the life-cycle. Its action therefore is both cell-cycle dependent and phase-specific. Razoxane is used in the management of certain sarcomata, lymphomata, and leukaemias (Bakowski 1976). It has been undergoing evaluation in a range of other neoplastic conditions, the largest study being its assessment as adjuvant therapy following resection of colorectal cancer (Dukes stages B and C). As a result of some encouraging findings by Atherton *et al.* (1980) it has also been evaluated in severe psoriasis.

Razoxane is generally well tolerated and it has there-

Table 35.1 Leukaemias associated with razoxane treatment

Case no.	Date	Disease	Leukaemia	Razoxane treatment		
				Cumulative dose (g) (approx.)	Duration dosing (MOS)	Time to diagnosis (months)
1	Dec. 81	Colonic* cancer	AMML (M4)	43	17	21
2	Dec. 81	Pancreatic* cancer	AMML (M4)	155	31	49
3	Oct. 83	Psoriasis*	AML (M2)	43	9	21
4	Oct. 83	Psoriasis*	AML (M2)	450	80	80
5	Oct. 83	Psoriasis*	AML (–)	42	14	14
6	Dec. 83	Colorectal cancer	AML (M2)	420	84	84
7	Jan. 84	Psoriasis	Acute monocytic leukaemia	162	54	54
8	Jan. 84	Colorectal	Pre-leukaemia	?	?	?

* Dead.

fore been used in prolonged low-dose therapeutic regimes, which is in contrast to the normal high-dose cyclical therapy favoured for classical antimitotic agents.

In December 1981 Joshi *et al.* reported two cases of acute myelomonocytic leukaemia (AMML). One patient received therapy for 17 months (cumulative dose 43 g) and the other for 31 months (cumulative dose 155 g). As a result of this observation, all patients in clinical trials were carefully monitored for abnormalities of circulating white cells. In October, 1983, three further cases of AMML in patients receiving razoxane were reported (personal communication to ICI Pharmaceuticals Division). In the following two months, two further cases of AMML were identified, as well as a case of prolonged marrow suppression, which is suspected as being a preleukaemic condition. In summary, there have been a total of seven cases of AMML and one suspected case associated with prolonged razoxane therapy (see Table 35.1). In addition, six cases of skin epitheliomata (three basal-cell; three squamous cell) were reported.

As a consequence, the Licensing Authority suspended further trials of razoxane, both in adjuvant therapy for colorectal cancer, and in non-malignant conditions, specifically severe psoriasis, though such an action does not imply that a causal relationship has been firmly established. At present it is not possible to assess the relative risk of developing AMML during razoxane treatment, because the total number of patients exposed is not known and neither is the dose/duration data available for a cohort of razoxane-treated patients.

There is certain circumstantial evidence to suggest that there may be a causal connection. Prolonged administration of razaxane intraperitoneally to mice (40 and 80 mg/kg) and rats (48 and 96 mg/kg) induced adenocarcinomata in female rats and lymphoreticular tumours in female mice. The design and execution of this study have been criticized. Razoxane is not mutagenic in bacteria (Ames test) but there is preliminary evidence of chromosomal alteration of bone marrow cells of the mouse and Chinese hamster.

In the mouse micronuclues test, razoxane in a single dose of 200 mg/kg i.p. causes a fivefold increase in micronuclei (chromatids) in polychromatic red cells within 24 hours of dosing. Cytogenic studies in the Chinese hamster show dose-dependent (20–500 mg/kg) changes in the chromosomes comprising chromosome gaps, breaks, and translocations (Watkins and Albanese, personal communication). Similar findings were reported previously in cultured human lymphocyte preparations (Sharpe *et al.* 1970). Preliminary studies in the patients who have developed AMML show hyperdiploidy with a ring chromosome in one case, and an abnormality of chromosome-7 in the other (Joshi, per-

sonal communication). The peripheral blood films of patients treated with razoxane consistently show macrocytosis of the red blood cells and polyploidy of the white cells.

In some patients the abnormal white cell appearances persist for at least six weeks after discontinuing therapy. The effects of razoxane on cultured mouse L cells are morphologically indistinguishable from the appearances due to irradiation or the effect of alkylating agents on similar cells (Stephens and Creighton 1974).

Clearly, it is now important to examine in detail the effects of razoxane on chromosome morphology in human haemopoietic neoplasms to determine whether there are specific rearrangements of chromosomes associated with razoxane therapy. If changes are present, their relationship to those observed in patients exposed to carcinogens who have a high incidence of abnormalities affecting chromosomes 5 and 7 could be determined.

These unanticipated associated effects of razoxane have arisen at a particularly unfortunate time, since there is now clear evidence of its efficacy in the control of the very severe forms of psoriasis, as well as there being suggestive evidence of it improving the prognosis in some forms of colorectal cancer following resection.

Hormonal substances and carcinogenesis

Diethylstilboestrol and adenocarcinoma of the vagina

The first confirmed reports of the transplacental transmission of cancer in man by means of a hormone, diethylstilboestrol (stilboestrol), have recently been published. The evidence provided by this extremely important research, and its significance, need immediate and careful assessment (see also p. 400).

Herbst and Scully (1970) reported seven cases of adenocarcinoma of the vagina in adolescent girls in the New England area during a period of four years. The patients' ages ranged from 15 to 22 years. They had symptoms of irregular vaginal bleeding for up to one year. Five were treated by radical surgery and one by wide excision. All were alive one to four years after operation. The seventh in whom the disease was too advanced at surgical exploration, died within six months. The authors were puzzled about the causation of this apparent clustering of cases, as carcinoma of the vagina is uncommon and usually occurs at a much older age. An eighth case was added in a retrospective study of factors that might have been associated with the appearance of these tumours (Herbst *et al.* 1971). Herbst and colleagues (1972) noticed that maternal bleeding when the girl's mother was pregnant with the patient and in previous pregnancies was more common than in a control group. But of greater significance than that was the finding that seven of the eight mothers had been treated with diethylstilboestrol during the first trimester of the maternal pregnancy, while none of the control group was so treated. A separate study by Greenwald and colleagues (1971) confirmed this association, adding five more cases in which the actual dosage of synthetic oestrogen used has been obtained. All 13 patients were born between 1946 and 1953, a period when diethylstilboestrol was being given for repeated or threatened abortion. All the mothers who took diethylstilboestrol began treatment in the first two months of pregnancy. They received either a constant dose administered throughout pregnancy or a continually increasing dose given almost to term. The actual dose varied but followed roughly that suggested by Smith (1948) beginning at 5 mg by mouth during the sixth or seventh week of pregnancy and increasing by 5 mg at 2-weekly intervals to the 15th week when 25 mg daily was being given. The dose was then increased by 5 mg at weekly intervals until the 35th week, at which time as much as 125 mg of diethylstilboestrol was being taken by mouth daily.

The original series of seven cases exceeded the number of cases in the entire world literature for a tumour of this type in adolescent girls born before 1945. Indeed, adenocarcinoma of the vagina was thought to have some relationship to vaginal adenosis or to originate from Mullerian-duct or mesonephric remnants. Moreover, if these neoplasms were the result simply of high-risk pregnancies, this should have become apparent before 1945. It was therefore suspected that exposure to diethylstilboestrol and vaginal carcinoma in the offspring might have a cause-and-effect relationship. The suggestion is reinforced by the fact that diethylstilboestrol was used only infrequently in general obstetric practice. Even at the Boston Hospital for Women, where a special high-risk pregnancy clinic was being conducted, only about one in 21 patients delivered in the wards had received diethylstilboestrol during the five-year period 1946 to 1951. Thus when the expectancy of a chance association is less than 5 per cent, the occurrence of maternal diethylstilboestrol therapy in 12 out of 13 cases of vaginal adenocarcinoma in young women cannot be considered coincidental.

The teratogenic effects of drugs in pregnancy have been appreciated since the thalidomide disaster. These studies add another dimension, carcinogenesis. The administration of diethylstilboestrol to the mother in early pregnancy must now be considered as a probable cause of adenocarcinoma developing in their daughters. In a Senate debate reported in the *Washington Post* (Rich 1975) on 10 September 1975, Senator Kennedy claimed that 220 cases of vaginal carcinoma were now known to

have appeared in the daughters of women treated with the drug. A more worrying problem is that of diethylstilboestrol residues in meat; apparently 75 per cent of cattle in America have been fed diethylstilboestrol to increase weight. In Britain it has been used on a small scale in veal production and also in the poultry industry. The United States, Britain, and Sweden have now banned the use of this hormone for growth-promotion in cattle. The results of this action on grounds of safety of course has economic consequences in terms of cost of beef production.

In the 1977 Newsletter from the Registry for Research on Hormonal Transplacental Carcinogenesis, 333 cases of clear-cell adenocarcinoma (213 vaginal and 120 cervical) had been established as having been definitely caused by intrauterine exposure to diethylstilboestrol. Recently new cases are being documented at the rate of 50 per year. The youngest patient with adenocarcinoma due to maternal ingestion of diethylstilboestrol was 7 years old at diagnosis, and the oldest was 27 years; most girls present with vaginal adenocarcinoma between the ages of 14 and 19 years.

An important finding of the Registry is that three cases of adenocarcinoma of the vagina have been documented following intrauterine exposure to progestogens and another has occurred in a girl whose mother had taken an oestrogen-progestogen mixture during pregnancy. The first and so far the only case of vaginal clear-cell adenocarcinoma associated with maternal ingestion of stilboestrol was reported by Monaghan and Sirisena (1978).

Benign liver tumours in patients on oral contraceptives

Primary benign hepatoma of the liver is rare and in recent years a number of vascular haematomata of the liver have been described in young women taking oral contraceptives.

The association was first made between this rare tumour and oral-contraceptive usage by Baum *et al.* (1973) who described seven cases. Other similar cases were described by Contastovalos (1973), Knapp and Reubner (1974), Kelso (1974), and Horvath *et al.* (1972), who each documented a single case, and O'Sullivan and Wilding (1974) who described a further three cases. These cases presented with pain or a mass in the abdomen, rather than haemorrhage into the peritoneal cavity. In all these reported cases the tumour was solitary, superficially situated in the liver in close relation to the liver capsule, and the tumour was usually partially or almost totally encapsulated. In two cases the tumour was pedunculated. The sizes of the tumours have varied from 5 to 18 cm in diameter.

The histological appearance showed normal or almost

typical appearance of the hepatocyte, the cells arranged in cords or forming pseudo-acini with retained bile. The predominant feature of the tumours was in each case the grossly dilated endothelium-lined vascular channels.

The course of these tumours was benign; only four of the above 14 patients died, the others all doing well following resection of the tumour. Tigano *et al.* (1976) described one fatal case of malignant hepatoma in a 22-year-old woman who had been on oral-contraceptive therapy for two years. This raised the question of whether some of these hepatic tumours might be malignant.

Oral contraceptive therapy and hepatocellular carcinoma

This is discussed in the section on hepatic dysfunction. (See Chapter 17.)

The use of oestrogens in the menopausal period and increased risk of endometrial carcinoma

Two epidemiological studies (Smith *et al.* 1975; Ziel and Finkle 1975) showed an increased incidence of endometrial carcinoma associated with the use of oestrogens in the menopausal and post-menopausal period (Smith *et al.* (1975) showed that the incidence was 4.5 times greater in women treated with oestrogens than in control women who were not so exposed. Ziel and Finkle (1975) investigated the possibility that the use of conjugated oestrogens increases the risk of endometrial carcinoma. Conjugated oestrogens containing principally sodium oestrone sulphate (*Premarin*) were found to have been taken by 57 per cent of 94 patients with endometrial carcinoma but by only 15 per cent of matched controls without. These workers showed that the increased risk compared to controls of developing endometrial carcinoma was 5.6 in women taking conjugated oestrogens for one to five years and, with greater than a seven-year exposure to conjugated oestrogens, the increased risk was 13.9 compared with matched controls, i.e. the longer the duration of therapy the greater the risk. These findings of increased risk of developing endometrial carcinoma in women taking exogenous oestrogens coincide well with the findings of MacDonald and Siiteri (1974) and Siiteri *et al.* (1974) who demonstrated evidence for a causal relationship between endogenous oestrogens production and endometrial carcinoma. These workers demonstrated rates of conversion of androstenedione to oestrone two to three times higher in women with endometrial cancer than in controls, and Schindler *et al.* (1972) showed that adipose tissue of patients with endometrial carcinoma converted androstenedione to oestrone at four times the rate of controls.

McDonald *et al.* (1977) demonstrated that there is an estimated relative risk of endometrial cancer associated with conjugated oestrogen treatment of six months or longer of 4.9 times that seen in untreated women, and this increased relative risk was 7.9 for women taking conjugated oestrogens for three years.

Gray *et al.* (1977) also demonstrated an increased relative risk of endometrial cancer developing in women taking hormone-replacement oestrogen therapy. These workers gave an increased risk factor of 11.5 in women taking oestrogen-replacement therapy for 10 years or more.

Oestrogen therapy and endometrial cancer

Gynaecologists have long suspected that oestrogen therapy may induce endometrial cancer. During the 1940s case-reports linking endometrial cancer to previous oestrogen medication were not infrequent, but it was not until 1954 that evidence strongly suggestive of an association was first published. Curiously this detailed investigation by Danish gynaecologists attracted little attention and the subject was seldom discussed in the ensuing 20 years.

A large case-control study by Antunes *et al.* (1979) produced evidence that the association between hormone replacement (unopposed) oestrogen therapy and endometrial cancer is real. This case-control study of the relation between oestrogen use and endometrial cancer involved 451 cases and 888 controls. The overall risk of endometrial carcinoma was sixfold for oestrogen users as compared with non-users; long-term users (five years) had a 15-fold risk. Excess risk was present for both diethylstilbestrol and conjugated oestrogens. The risk associated with cyclic use was as great as that for continuous use. Increased risk was associated with oestrogen use for all histologic grades of the tumour. The risk of advanced-stage carcinoma was fourfold for oestrogen users, but the confidence interval was wide, and this question requires further study.

Finally, this investigation contradicts the speculation that this association between cancer and oestrogen use can be explained by swifter diagnosis for oestrogen users, misclassification of oestrogen-related hyperplasia, or treatment of early symptoms of the tumour with oestrogen.

The reports of two case-control studies by Horwitz and Feinstein (1979) into the use of intravaginal oestrogen creams and endometrial cancer revealed that no increased risk of endometrial cancer in oestrogen cream users was proven. The small size of these studies detracts from their value and cannot be used to give a clean bill of safety to these products.

Cyclical low-dose oestrogen therapy with 7–13 days of progestogen has been claimed not to be associated with increased risk of endometrial hyperplasia or carcinoma (Thom *et al*. 1979) but such claims should be regarded as premature and not enough epidemiological evidence has been accumulated to exonerate this regime from this complication.

Several studies have shown that among young women with endometrial cancer who had used oral contraceptives a greater proportion had taken sequential preparations than might have been expected in view of relative usage of the various types of oral contraceptives by women in the general population (Silverberg and Makowski 1975; Cohen and Deppe 1977; Silverberg *et al*. 1977).

In a study by Weiss and Sayvetz (1980) residents of King and Pierce Counties in the state of Washington in whom endometrial cancer was diagnosed duirng 1975–77 were interviewed concerning prior use of oral contraceptives. Their responses were compared with those of a random sample of women from the same population. Women who had taken *Oracon* (0.1 mg of ethinyl oestradiol and 25 mg of dimethisterone) were estimated to have a risk of endometrial cancer 7.3 times that of other women ($p = 0.007$). This evaluation in risk was not seen in users of other sequential preparations. Women who had used combined oral contraceptives had only 50 per cent of the incidence of endometrial cancer of non-users ($p = 0.05$), or those who subsequently took menopausal oestrogens for more than two years. This study by Weiss and Sayvetz (1980) indicated that oestrogen/progestogen combined preparations might have a protective effect against endometrial cancer and is therefore compatible with the finding of Thom *et al*. (1979).

Since 1975, there have been reports of many case-control studies, mostly in North America, of oestrogen replacement therapy and endometrial cancer covering several thousand patients with the disease. Almost all studies have indicated a positive association; the average estimate of relative risk for women who had taken oestrogens at any time was about 6.0. In those studies in which it was sought, a relationship between the extent of the relative risk and the dose and duration of use of oestrogens was generally found. There was little information distinguishing the effects of continuous from cyclic administration of oestrogen, but such as there was suggested that both regimens are associated with similar risks. In most studies, conjugated equine oestrogens (*Premarin*) were used, usually in a dose of 0.625–1.25 mg per day.

Endometrial cancers occurring in women using oestrogens tend to be less advanced at detection and of a more favourable grade than those in other women; thus it may be that tumours resulting from oestrogen replacement therapy are of low malignancy. The survival of patients who developed endometrial cancer and had taken oestrogens has been compared with that of patients who had developed the disease in the absence of such exposure. The 5- and 10-year survival rates of the oestrogen users were higher than those of the non-users, but the difference disappeared when the grade of the neoplasm was taken into account. For example, in a study of 860 women with endometrial cancer registered at a regional cancer treatment centre in Ontario, Canada (Collins *et al*. 1980), 259 (30 per cent) gave a history of oestrogen use for six months or more at some time before diagnosis, and 568 (66 per cent) were non-users. Oestrogen use was associated with younger age, earlier stage, lower grade of tumour, less common myometrial invasion. All patients' treatment included radiation and hysterectomy. The five-year survival rate of oestrogen users was 92 ± 2 per cent compared with 68 ± 2 per cent for non-users. Further analysis controlling for differences in the distribution of the variables associated with survival showed that for women who have endometrial cancer, the risk of any death for non-users was approximately 2.7 times greater than for oestrogen users, and the risk of death from endometrial cancer was 5.4 times greater for non-users. A history of oestrogen use was not associated with superior survival of women who have poorly differentiated tumours or myometrial invasion.

The British Gynaecological Cancer Group have examined the relationship between oestrogen replacement in post-menopausal women and the occurrence of endometrial cancer. Their assessment and conclusions were published in 1981. Data from a number of studies in which women with climacteric symptoms had been treated with oestrogens plus progestogens and in whom endometrial biopsies were obtained were reviewed. Ingestion of cyclical unopposed oestrogens carries a risk of about 13 per cent for the development of hyperplasia. This risk seems to reduce to about 3 per cent when seven days of progestogen are given per cycle. The ingestion of progestogens for 13 days in each month apparently protects the endometrium completely from the development of hyperplasia regardless of whether the ostrogen is taken orally or given by subcutaneous implant.

As a result of their examination of the data, the British Gynaecological Cancer Group were able to make the following four clear-cut points:

1. There is an increased risk of endometrial cancer when oestrogen replacement is given after the climacteric; the precise risks with various regimens are uncertain and it seems unlikely that these will be defined in the near future. In the meantime there will be a con-

tinuing demand for oestrogen prescription for post-climacteric women. In the circumstances all that can be offered are suggestions as to the regimens which in the light of the information currently available seem likely to carry the lowest risk of causing cancer.

2. If it is decided to administer oestrogen and the patient has not had a hysterectomy, this is best done cyclically and combined with a progestogen in the last 7–13 days of the cycle.

3. Endometrial out-patient sampling by curettage, Vabra aspiration, etc. for histological and/or cytological examination should be performed if unopposed oestrogens have been prescribed for over six months. If cyclic administration with added progestogen is used, sampling of the endometrium is recommended at two-yearly intervals if treatment is continuous.

4. Any irregular bleeding should be regarded as an indication for immediate curettage.

Now that it is generally accepted that there is a cause-and-effect relationship between oestrogen use and the incidence of endometrial cancer it is useful to have available the results of the first two large detailed studies from two Canadian centres, on the prognosis of this cancer in oestrogen users and non-users (Elwood 1981). These studies concern 827 patients seen at the Ontario Cancer Treatment and Research Foundation clinic in London, Ontario from 1967 to 1976, and 494 patients with newly diagnosed endometrial cancer seen at the A. Maxwell Evans Clinic in Vancouver from 1968 to 1972. The London series presents information on survival up to 10 years, and the Vancouver series up to eight years. Thirty-one per cent of the London patients and 27 per cent of the Vancouver series were defined as oestrogen users. Both studies showed oestrogen users to be younger than non-users, with lesions that were better staged. Information on menstrual history and menopausal symptoms in the Vancouver study showed no difference between users and non-users. Women who had used oestrogens had a considerably better overall survival in both studies, with a five-year survival rate of 92 per cent for oestrogen users and 68 per cent for non-users in the London series, and 89 and 74 per cent, respectively, in the Vancouver study.

Disagreement between the two studies on a major point however reopens the argument about pathological interpretation. The London study found that after controlling for differences in age, pathological tumour grading, and presence of myometrial invasion, the difference in survival persisted (oestrogen non-users having an overall risk of death 2.7 times that of users, and a risk of death from endometrial cancer 5.4 times that of the users). This suggests that tumours in women who have used oestrogens are considerably less aggressive than those in women who have not used oestrogens. The Vancouver results, on the other hand, show no difference in survival between oestrogen users and non-users when differences in staging and invasion are considered, suggesting that the overall difference in survival is due to earlier diagnosis in oestrogen users. It can be inferred from both studies that the improved survival may be great enough to outweigh the increased incidence of endometrial cancer in oestrogen users. The authors emphasize, however, that this should not be seen as evidence of the safety of these drugs — endometrial cancer is far from trivial even when it is not fatal.

Hypophyseal tumours associated with oestrogen therapy

Peillon et al. (1970) described 13 cases of hypophyseal tumours associated with oestrogen therapy. Twelve of these occurred in amenorrhoeic women treated with oestrogens and progestogens and one affected a man with cancer of the prostate treated with oestrogens. Five of these tumours were associated with acromegaly and eight were chromophobe adenomas.

Oral contraceptives and carcinoma of the uterine cervix

The incidence of cervical neoplasia in 6838 women who entered the Oxford/Family Planning Association contraception study was reported by Vessey et al. (1983). The study showed a small number of invasive carcinomas, 13 in Pill users and none in IUD users. This work does not establish a causal link, however, between the use of the Pill and cervical carcinoma. In studies of this type, despite the efforts of the investigators, some unidentified factor may always influence the results. Sexual behaviour, for example, was only investigated in a small sample of the women studied because of the need to maintain continuing co-operation, and this is likely to be an important factor. All the cases of carcinoma in the study, which were detected by cervical cytology screening, were treated effectively. As the authors say, long-term Pill users should undergo regular cytological examination.

Oral contraceptives and endometrial cancer

The possibility that oral contraceptives (OCs) might reduce the risk of endometrial cancer was investigated by the Centres for Disease Control in the Cancer and Steroid Hormone (CASH) Study. Data from a multicentre, population-based, case-control study were analysed (Centres for Disease Control CASH Study 1983 a, b).

Cases were all women 20 to 54 years old with a first diagnosis of endometrial cancer ascertained through eight population-based cancer registries. The controls were women selected at random from the population of these eight areas. Analysis of the first 187 cases and 1320 controls showed that women who had used combination OCs at some time in their lives had a relative risk of endometrial cancer developing of 0.5 (95 per cent confidence interval, 0.4 to 0.8) compared with never-users. The protective effect occurred in women who had used combination OCs for at least 12 months, and it persisted for at least 10 years after the cessation of OC use. The protective effect was most notable for nulliparous women. Nulliparous combination OC users had a risk 0.4 times (95 per cent confidence interval, 0.2 to 0.9) that of nulliparous never-users. These results were not accounted for by differences between cases and controls in health status, parity, infertility, or other potentially confounding variables. It was estimated that approximately 2000 cases of endometrial cancer are averted each year by past and current OC use among women in the United States.

Oral contraceptives and malignant melanoma

A positive association between the use of oral contraceptives (OCs) and the subsequent development of malignant melanoma was first suggested by a single case report (Ellerbroek 1968). The possibility was then investigated in a small prospective study, which provided some support for the association (Beral et al. 1977). It is known that OCs alter skin pigmentation in some subjects (Jelinek 1970) and oestrogen receptors have been identified in melanomata (Fisher et al. 1976; Chaudhuri et al. 1980). In guinea-pigs, skin melanin content and melanocyte numbers are increased in response to oestrogen and progesterone (Snell and Bischitz 1960).

Stevens et al. (1980) sought an association by comparing rates in females with those in males during and following the introduction of OCs in several countries. The results indicated a relative risk close to unity.

In a case-control study (Adam et al. 1981), 169 women aged 15-49 years with malignant melanoma notified to the Oxford and South-western cancer registries during the years 1971-76, together with 507 matched controls, were investigated. Data about medical, reproductive, drug, and smoking histories were obtained both by reviewing general practitioner (GP) records and from the women themselves by postal questionnaires. There was no significant evidence of any overall increase in the risk of melanoma in OC users (data from GP records — ever use vs. never use, relative risk 1.34, 95 per cent confidence limits 0.92-1.96; corresponding data from postal questionnaires — relative risk 1.13, limits 0.73-1.75). However, although not significant, the risk estimated from data in the postal questionnaire was higher in women who had used OCs for five years or more (use for five years or more than five years vs. never-use, relative risk 1.57, limits 0.83-3.03). Previously demonstrated risk factors for melanoma, such as fair skin, blonde or red hair, and Celtic origin were found to be commoner in cases that in the controls. Data from the Oxford/Family Planning Association contraceptive study were also examined. Unexpectedly there was a strong suggestion of a negative association between OC use and melanoma risk, but analysis was based on only 12 women with the disease.

In a further case-control study by Bain et al. (1982), there was no overall relationship between a prior history of OC use and the development of melanoma among 141 cases of non-fatal malignant melanoma and 2820 age-matched controls drawn from respondents to a large postal survey of registered US nurses; crude relative risk was 0.93 and 95 per cent confidence limits were between 0.64 and 1.36. Adjustment for a number of additional variables did not alter this estimate materially. Duration of OC use and interval since first use were similarly unrelated to the occurrence of melanoma. For women diagnosed before age 40, there was a crude positive association of 'ever' use of OC and melanoma (relative risk 1.78; 95 per cent confidence limit, 1.11-2.86). However, adjustment for geography and other variables diminished this association and rendered it statistically not significant (relative risk 1.43; 95 per cent confidence limit, 0.83-2.46). These data do not support the hypothesis that OC use is an independent risk factor for melanoma.

In conclusion, it appears from the later studies that the association suggested by Ellerbroek (1968) and supported by Beral et al. (1977) does not exist.

Prolactinomas and oral contraceptives

A recent epidemiological survey (Shy et al. 1983) examined the question of whether the prior use of oral contraceptives (OCs) was a risk factor for pituitary prolactinoma. A total of 72 women having such a tumour, and diagnosed between 1976 and 1980, were identified in Washington State, USA. A control group of 303 women was selected on a random basis using telephone directories from the same area. The histories of the control subjects were taken by telephone. Taking the risk for women who had never used OCs as 1.0, the relative risk of acquiring this tumour was 1.3 for women who had used OC for birth control (95 per cent confidence interval 0.7-2.6). However, the relative risk was increased to 7.7 for women who had used OCs for menstrual regulation

(95 per cent confidence interval 3.7–17.0). It is suggested, from the high risk found in the latter group, that the use of OC to treat menstrual irregularity resulting from undiagnosed prolactinoma may explain the associations between prolactinomata and OCs seen in earlier studies (Vaisrub 1979; Annegers *et al.* 1978; Coulman *et al.* 1979; Wingrave *et al.* 1980).

Oral contraceptives and ovarian cancer

This chapter concentrates upon positive associations between therapies and carcinogenesis. It is not inappropriate, however, to refer briefly to a very widely-used group of drugs which appear to exert a protective effect against a neoplasm which ranks as the fourth leading cause of cancer mortality among women in the United States.

The possibility that the use of oral contraceptives (OCs) might reduce the prevalence of ovarian cancer was suggested by the observation that a history of full-term pregnancy reduces the frequency of this tumour (Stewart *et al.* 1966; Joly *et al.* 1974; Annegers *et al.* 1979). It was argued that, as OCs mimic some physiological aspects of pregnancy (e.g. suppression of gonadotrophins and ovulation), they might also inhibit the development of ovarian cancer. The truth of this hypothesis was supported by the work of Willett *et al.* (1981) and Weiss *et al.* (1981). This result was confirmed by case-control studies in 1982 (Rosenburg *et al.* 1982; Cramer *et al.* 1982) and in 1983 was again confirmed by the Cancer and Steroid Hormone (CASH) study (Centres for Disease Control 1983*a,b*).

Androgenic anabolic steroids and hepatocellular carcinoma

In 1971 Bernstein *et al.* reported a case of hepatocellular carcinoma in a patient with Fanconi's anaemia treated with oxymetholone, and considered that the tumour had been oxymetholone-induced.

Johnson *et al.* (1972) reported four patients treated with androgenic anabolic steroids for long periods for aplastic anaemia who developed hepatocellular carcinoma. Two of Johnson *et al.*'s cases, a $2\frac{1}{2}$-year-old girl and a 17-year-old girl, had been treated with oxymetholone; the two others, a 27-year-old man and a 21-year-old man had been treated with methyltestosterone and methandienone respectively. Henderson *et al.* (1973) reported a case of a 16-year-old boy with hypoplastic anaemia who had been treated from the age of 8 years with methyltestosterone for 2 years, norethandrolone for 3 months, then stanozol for 18 months, and finally after a period in which no anabolic steroids were given he

recommended treatment with oxymetholone. He too developed and died from hepatocellular carcinoma.

The discussion was then raised that there might be an association between Fanconi's anaemia and hepatocellular carcinoma (Guy and Auslander 1973). However, this hypothesis was refuted firstly on the basis that no cases of hepatocellular carcinoma had been reported in patients who had Fanconi's anaemia but had not been treated with C-17-alkylated androgenic–anabolic steroids. Secondly, cases were reported where hepatocellular carcinoma was associated with treatment with androgenic anabolic steroids for other reasons. Zeigenfuss and Carabasi (1973) reported a case of hepatocellular carcinoma in a 68-year-old man who had been receiving methyltestosterone for 30 years because of impotence. Farrell *et al.* (1975) reported three young men with hepatocellular carcinoma, none of whom had Fanconi's anaemia. Two of Farrell *et al.*'s patients had been treated for hypogonadism, due in one case to hypopituitarism in a 40-year-old man and in the other due to cryptorchism in a 33-year-old man. In the former case methyltestosterone 25 mg twice daily had been given for six years, in the other methyltestosterone 50 mg/day and testosterone propionate 50 mg/month had been given for eight years. The third patient had been treated with oxymetholone 100 mg/day increasing to 150 mg/day for five years for paroxysmal nocturnal haemoglobinuria.

In view of all the available evidence it would appear that long-term therapy with C-17-alkylated androgenic-anabolic steroids carries a risk of induction of liver tumours.

Regression of the liver tumours on stopping androgen therapy was noted in one of the four cases reported by Johnson *et al.* (1972) and two of the three cases reported by Farrell *et al.* (1975). The use of anabolic steroids by athletes for muscle-building would appear to be totally unjustified in view of the carcinogenic risk.

Synthetic progestogens

A number of synthetic progestogens (as well as progesterone itself) have been associated with tumours of the mammary glands when tested in the beagle bitch. However many of these compounds have proved not to be tumourigenic in other species and marketing authorisation has been granted in some of these cases. It is considered that there is a high probability of false positive results when compounds of this kind are tested in the beagle model. A depot formulation of medroxyprogesterone acetate (*Depo-Provera*) has now been approved in the UK for long-term use as an injectable contraceptive, in specific and limited clinical circumstances. This product is unusual (if not unique) in that it is the subject of an official publication which deals with questions of benefits and possible risks, including carcinogenicity testing (Department of Health and Social Security 1984).

Nitrosamines

The nitrosamines are a class of N-nitroso compounds many of which exhibit organotropic carcinogenicity in a wide range of animal models. There is increasing evidence (Craddock 1983) that certain cancers in man are associated with human exposure to nitrosamines. Generalized molecular structures are:

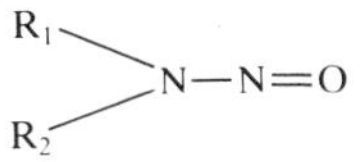

Acyclic nitrosamine 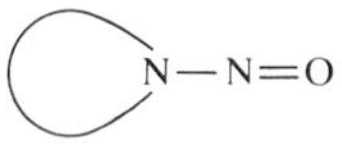   Cyclic nitrosamine

The carcinogenic potential of nitrosamines operates after metabolic activation, and characteristic sites are upper gastrointestinal tract, liver, nasal cavities, lung, and oesophagus. Many nitrosamides are also carcinogenic, but prior metabolic activation is not required, so that there is greater localization of lesions: sites depend upon the route of administration. Nitrosamines, which are generally more stable than nitrosamides, occur widely in the environment, at low concentrations. They are also formed in reactions between nitrate (which, although it may occur naturally, is also added in appreciable amounts to certain foods as preservatives) and amino groups such as occur in many drugs and foods. To place the problem in perspective, it has been shown that nitrosamines are formed *in vivo* following the ingestion of meals of luncheon meat and milk (Walters 1977), and of bacon and spinach (Fine *et al.* 1977). The sequence of events required for nitrosamine-induced carcinogenesis following ingestion is: formation of N-nitroso compounds *in vivo*, metabolic activation when required, reaction (probably alkylation) with DNA, and DNA replication before repair occurs.

One of the mechanisms by which nitrite could react with tertiary amines, to form nitrosomines is:

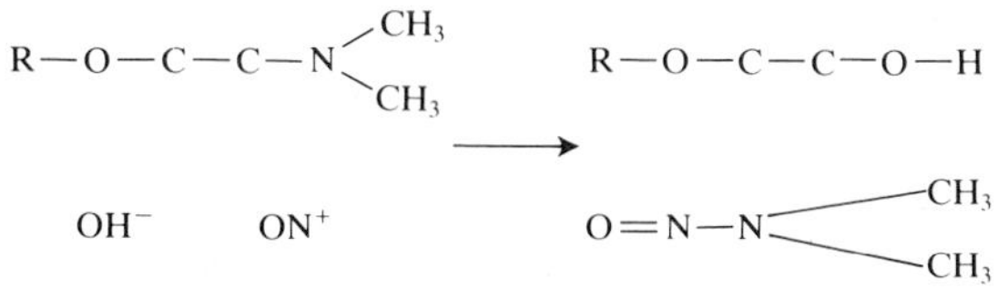

Nitrite and hydrogen ion concentrations in fasting gastric juice are inversely related. This leads to the production of N-nitroso compounds in the hypochlorhydric and achlorhydric stomach and could explain the association between the achlorhydria of pernicious anaemia and increased gastric carcinoma in these patients. Ruddell *et al.* (1978; Ruddell 1979) found that the mean nitrite concentration in 13 fasting patients with pernicious anaemia was nearly 50 times greater than age-matched controls. The number of bacteria in the gastric juice of these patients was also greatly increased. Small amounts of volatile nitrosamines were detected in simulated gastric juice *in vitro* after addition of nitrite to achieve concentrations similar to those found *in vivo*.

Aminopyrine

Aminopyrine (aminophenazone, amidopyrine) is an analgesic and antipyretic having therapeutic actions similar to those of phenazone. Although its toxicity, especially its ability to induce agranulocytosis, has been known for some years, the readiness with which it forms dimethyl nitrosamine (Lijinsky and Greenbalt 1972; Taylor and Lijinsky 1975) is a more recent finding. There are no licensed medicines containing aminopyrine on the British market.

Methapyrilene

Methapyrilene is an antihistamine containing, like aminopyrine, a tertiary amino group. *In vitro*, the drug reacts with nitrite to form dimethylnitrosamine. In rodent studies it is carcinogenic when administered with nitrite, but has subsequently been found to be a rat liver carcinogen in its own right. Methapyrilene is not mutagenic in mammalian and bacterial test systems, even after metabolic activation. In Britain the sale of all medicines containing methapyrilene was discontinued in 1979.

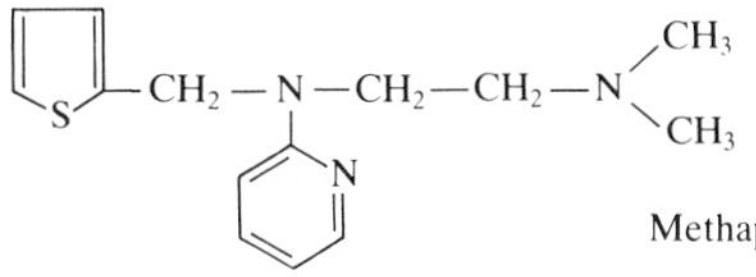

Methapyrilene

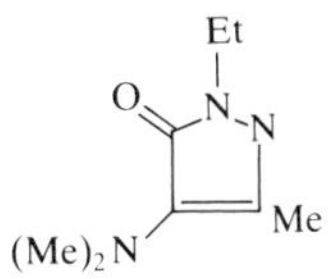

Aminopyrine

Me methyl group
Et ethyl group

Disulfiram

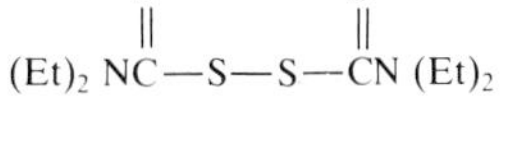

Disulfiram

Disulfiram, the aldehyde dehydrogenase inhibitor used in alcohol aversion therapy, contains two tertiary amino groups and reacts with nitrite at low pH *in vivo* to form diethylnitrosamine. In combination with nitrite it has been shown to be carcinogenic in the rat, and it is conceivable that it might make a small contribution to the risk of neoplasia in man, if used for long periods.

Piperazine

Piperazine, the anthelminthic, is readily nitrosated under *in vitro* conditions which resemble the intragastric environment (Coulston and Dume 1978). Under these conditions conversion to mono-nitroso-piperazine (MNPZ) and, to a lesser extent, di-nitroso-piperazine (DNPZ) occurs. These nitrosation products have been shown to be carcinogenic in rats (Garcia *et al.* 1970; Love *et al.* 1977) and mice (Schmähl and Thomas 1965), and to be mutagenic in various test systems (Zeiger and Sheldon 1978; Larciner *et al.* 1980).

In 1981 it was demonstrated (Bellander *et al.* 1981) that high levels of MNPZ are formed in the human stomach following a normal oral dose of piperazine. The intragastric concentration of MNPZ ranged from 0.08 to 0.59 μg ml^{-1}, although the peak was probably not reached within the 30-min study period. MNPZ was also excreted in the urine, the peak being reached within two hours: in one subject the amount recovered was as high as 300 μg.

In Sweden, piperazine products have been withdrawn from the market, as a result of these findings (Swedish Department of Drugs 1982).

Amyl nitrite and Kaposi's sarcoma in male homosexuals

Amyl nitrite and isobutyl nitrite are readily available and widely used by male homosexuals — evidence indicates that amyl nitrite is preferred. Both drugs cause immediate vasodilatation of short duration, and relaxation of the anal sphincter. In the United States there has been an outbreak of Kaposi's sarcoma in New York, Florida, and California among male homosexuals.

It has been pointed out (Jørgensen 1982) that volatile nitrites can act as nitrosating agents and that they are mutagenic in bacterial test systems (Quinot 1980). The suggestion (Jørgensen 1982) that amyl nitrite might cause Kaposi's sarcoma in male homosexuals through the formation of nitrosamines is, however, unlikely to prove correct. The amount of nitrite administered is small in comparison with dietary nitrite and nitrate (Mirvish *et al.* 1980) and no association between the sarcoma and nitrites used for ischaemic heart disease has been reported. If the incrimination of amyl nitrite is not artefactual, the compound is more likely to act through effects on the immune system.

The data generated by Goedert *et al.* (1982) suggest that nitrites may be immunosuppressive in the setting of repeated viral antigenic stimulation and may thus contribute to the high frequency of Kaposi's sarcoma and opportunist infections in homosexual males. Altered T-cell function due to nitrites is not likely to be the whole explanation of this outbreak; otherwise it would be expected that similar outbreaks would have been seen in patients with angina who take large quantities of nitrites. It is more likely that what is being seen are manifestations of AIDS.

Propranolol and N-nitrosopranolol

Propranolol, in common with other beta-adrenergic receptor blockers, contains an amino group. The possibility that N-nitrosopropranolol (NNP) might form under simulated intragastric conditions was investigated by Chen and Raisfeld-Danse (1983). In solutions of HCl within the pH range found in the stomach, the optimum pH for the formation of NNP was 3. The yield of NNP increased linearly as incubation time and concentration of propranolol increased and exponentially as the concentration of nitrite was raised. Under optimal conditions in hydrochloric acid, the minimum concentration of nitrite required for the production of detectable amounts of NNP was 10^{-5}M.

In an extension of this work (Raisfeld-Danse and Chen 1983), NNP formation was examined in human gastric juice and in the presence of organic nitrate ester vasodilator drugs. In comparison to HCl solutions, equivalent concentrations of propranolol and nitrite produced similar amounts of NNP in gastric juice; however, the yield increased as the pH was lowered and the kinetics of nitrosamine formation were different. Endogenous nitrite concentrations in 22 samples of human gastric juice were below the minimum concentration (10^{-5}M) required for production of detectable levels of NNP. Maximal therapeutic dosages of propranolol (10^{-2}M) incubated with isosorbide dinitrate (3.4–6.8 $\times$ 10^{-3}M) or nitroglycerin (8.6 $\times$ 10^{-4}M) also failed to produce NNP. However, NNP formed adventitiously during the concentration of aqueous and methylene chloride solutions that contained propranolol and organic nitrates, underscoring the importance of avoiding artefactual formation of nitrosamines. Furthermore, synthetic NNP was not mutagenic in either a bacterial test of mutagenicity using several strains of *E. coli*, or in a hepatocyte-mediated mammalian cell mutagenesis assay.

It was concluded that NNP is unlikely to form in the stomach under conditions normally present in patients. It was also concluded that, even if NNP was formed under exceptional circumstances, this compound is unlikely to be a carcinogen. With respect to the potential formation of nitrosamines during drug dissolution in the stomach, long-term therapy with propranolol hydrochloride appears to be safe.

Cimetidine

Cimetidine is an H_2 antagonist and is now widely used in a variety of patients with gastric hyperacidity and gastric and duodenal ulceration.

On theoretical grounds, an association between cimetidine therapy and gastric cancer might be suspected for four reasons. Firstly, cimetidine induces hypochlorhydria and it is known that in pernicious anaemia, achlorhydria is present and the frequency of gastric cancer is 2–4 times normal. Secondly, it is known that cimetidine treatment creates an intragastric environment which favours the formation, by the mechanisms illustrated in the flow-diagram, below, of nitrosamines. Thirdly, cimetidine can itself be nitrosated to form N-nitrosocimetidine, a known mutagen which is closely related to methylnitro-nitrosoguanidine (MNNG), one of the most potent gastric carcinogens known. Fourthly, the molecular structure of cimetidine closely resembles that of tiotidine, which induces gastric carcinoma in rats.

Concern arose in 1979 and 1980 with the appearance of reports about several patients in whom gastric cancer had been diagnosed following cimetidine treatment (Elder *et al.* 1979; Reed *et al.* 1979; Taylor *et al.* 1979; Hawker *et al.* 1980). There was no firm link between cimetidine and tumour formation in these cases, although it was clear that the possibility of a carcinogenic action which had been raised justified serious examination.

The molecular structure of cimetidine closely resembles that of tiotidine (ICI 125211), an experimental H_2 antagonist (Domschke and Domschke 1980). It will be noted that both compounds contain the same N-methyl-methylene-diamine moiety.

Histamine

Cimetidine

Tiotidine

According to information provided to the US Gastrointestinal Advisory Committee of the Food and Drugs Administration on 12 January 1981, and made available under the US Freedom of Information Act, clinical trials of tiotidine in the United States were terminated in summer 1980 following interim results of rat carcinogenicity studies (Food and Drug Administration 1981; Scrip 1981).

In rats treated for 12 months at 150 mg/kg/day, and above, there was a dose-related incidence of so-called 'intestinal metaplasia' of the gastric epithelium. No pathological changes occurred at 30 mg/kg/day. Intramucosal carcinomas were found in the pyloric region in five out of 55 rats treated for 11 months at 1500 mg/kg/day. The lesions may resemble those induced by N-methyl-N'-nitro-N-nitroso-guanidine (MNNG) Morson *et al.* 1980; Saito *et al.* 1980; Kunze *et al.* 1979).

Notwithstanding these findings, however, safety tests in small animals including conventional carcinogenicity tests, have failed to establish any link between cimetidine and the development of stomach cancer (*Current Problems* No. 6, July 1981). During the six years ending in January 1981, 21 reports of stomach cancer were known to the Committee on Safety of Medicines from 'yellow cards' submitted and from the literature. These do not, of course, provide evidence of any causal relationship since, in the majority of cases, it is likely that the cancers existed before the start of cimetidine treatment. There is, therefore, no reason why cimetidine should not continue to be used prudently, for the licensed indications, although there is good evidence (Schade and Donaldson 1981) that, in the US at least, less than 10 per cent of in-patients receiving cimetidine were being treated for the indications approved by the Food and Drug Administration. Moreover, the compound was being administered to between 2.5 and 5.6 per cent *(sic)* of in-patients for a variety of reasons. It is quite clear that cimetidine is being used widely and indiscriminately in the treatment of inadequately investigated cases of dyspepsia and it is inevitable that some of these patients will eventually be found to have had an undiagnosed malignancy. The current indiscriminate use of cimetidine without proper investigation is to be deplored.

Further research is required to clarify the question of any 'nitrosamine hazard' arising from cimetidine therapy. In particular, it would be helpful to know the levels of nitrosamines appearing in subjects taking ordinary diets, either alone, or in conjunction with cimetidine therapy or therapy with amino compounds. A considerable step forward in providing this knowledge has come from the work of Reed and Walters and their colleagues (Reed *et al.* 1981*a*,*b*).

Reed *et al.* (1981*a*) measured the concentrations of total extractable N-nitroso compounds, pH, and nitrite levels in samples of fasting gastric juice, which were also cultured for bacteria from 50 healthy volunteers and 217 patients with upper gastrointestinal conditions (duodenal ulcer 70 patients, vagotomy 24 patients, partial gastrectomy 14 patients, atrophic gastritis 13 patients, gastric ulcer 33 patients, pernicious anaemia 16 patients, carcinoma of the stomach 23 patients, other conditions 24 patients).

The concentration of N-nitroso compounds and pH levels rose significantly with age. Sex and smoking had no significant effect. There was a positive correlation between pH and N-nitroso concentration, and between pH and nitrite and N-nitroso levels and growth of nitrate reductase-positive microorganisms was demonstrated. Reed *et al.* in these studies demonstrated for the first time in man the interrelationship between N-nitrosamine concentration, pH, gastric juice nitrite, and nitrite-reducing bacteria.

Reed *et al.* (1981*b*) in an extension of the study considered above measured total extractable N-nitroso compounds, pH, nitrite levels, and microorganisms cultured from fasting gastric juice from 140 patients receiving cimetidine and 267 subjects not taking cimetidine. Significantly higher mean N-nitrosamine concentrations and pH levels were demonstrated in the cimetidine-treated patients, N-nitrosamine concentration increasing with pH. In 30 patients studied cimetidine significantly increased gastric pH and N-nitrosamine concentrations, while in 23 patients withdrawal of cimetidine treatment resulted in significant falls in pH but not of N-nitrosamine concentration. In the cimetidine-treated patients the gastric juice nitrite levels were often raised and the nitrate-reducing bacteria cultured were similar to those associated with other causes of hypochlorhydria. Reed and his colleagues discussed their results in terms of their implication for gastric cancer induction and produced a flow diagram of events which is reproduced in Fig. 35.1.

It should be recognized that conventional regimens of dosing with cimetidine do not produce continuous achlorhydria or even hypochlorhydria. The pH of the gastric juice fluctuates markedly over a 24-hour period presumably in contrast to the situation in post-peptic-ulcer surgery patients or in patients with pernicious anaemia. Although colonization with bacteria may occur in cimetidine-treated patients, it seems unlikely that permanent

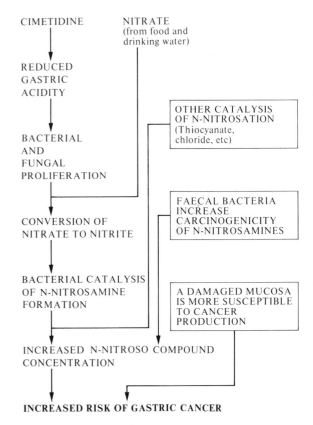

Fig. 35.1 Flow diagram linking intragastric nitrosation with possible induction of gastric cancer.

bacterial overgrowth will be present, unless treatment at high dosage is continued over many months.

Further work towards completing the 24-hour picture of gastric conditions (in terms of pH, levels of nitrite and nitrate, total N-nitroso compounds, and bacterial counts) was undertaken by Milton-Thompson *et al.* (1982). Eight healthy subjects were studied half-hourly or hourly for 24-hour periods before, during the first and second weeks treatment with cimetidine (200 mg t.d.s. and 400 mg nocte), and five weeks after ceasing to take the drug. No significant differences in intragastric bacterial counts or bacterial species or in intragastric nitrite or N-nitroso-compound concentrations were found as a result of cimetidine treatment. Bacterial counts and nitrite concentrations tended to increase with pH, but N-nitroso-compound concentrations did not. This study provides no evidence that cimetidine treatment may increase the risk of gastric carcinoma by raising N-nitroso-compound concentrations.

This study, however, was conducted in young healthy, non-fasting subjects with no gastrointestinal disease over a two-week period and, for this and several other reasons, cannot be regarded as entirely a realistic picture of what happens or might happen in patients with gastric hyperacidity, possibly with some element of pyloric stenosis who are treated chronically for many months. The assay methods used were not the same as those used by other workers (Smith *et al.* 1983). Nevertheless, the results are, as far as they go, reassuring.

Preliminary data are available from a post-marketing surveillance study in which 9940 patients taking cimetidine were entered (Collin-Jones *et al.* 1982). A total of 9504 were observed for at least a year. Seventy-four cases of gastric cancer were identified in those taking cimetidine, but 23 of these were diagnosed before the use of the drug and 29 others with advanced malignancy had received cimetidine within the previous six months only. Ten of the remaining 22 had gastric cancer diagnosed within a year of starting treatment, and 12 after more than a year; only four of the total group had histologically 'early' cancer. the occurrence of gastric cancer a long time after starting cimetidine treatment cannot be explained in every case, but it is noteworthy that in the control group (which was not directly comparable) gastric cancer was observed in eight patients. The hypothesis that cimetidine treatment predisposes to gastric cancer cannot be excluded by these preliminary findings, but the data do not appear to support such an association.

Further reassurance is obtained from the work of Habs *et al.* (1982*a,b*) who studied the carcinogenicity of N-nitrosocimetidine (NC), which is known to be mutagenic, in rats. Two groups of 40 Sprague–Dawley rats each received 500 and 50 mg/kg NC, respectively, by gavage twice weekly for one year with subsequent lifetime follow-up. An additional 40 rats were given the structurally-related carcinogen methylnitro-nitrosoguanidine (MNNG) orally at weekly doses of 80 mg/kg for 3 months. 100 animals served as untreated controls. MNNG-treated rats died with carcinomas of the forestomach (median induction time: 226 days). No evidently treatment-related tumours were seen in animals receiving NC. The median survival time in both NC-treated groups was significantly reduced (400 and 393 vs. 630 days in the control). No evidence of carcinogenicity of NC was seen in this bioassay. These results suggest that NC is not carcinogenic in rats, or that it is much less so than MNNG. In another study, however, one of 16 rats receiving NC for six months developed a gastric carcinoma after 15 months (Elder *et al.* 1982).

There is no hard evidence that cimetidine causes gastric cancer. The speculation which arose in 1979 still continues, but it has now been appreciated that cimetidine is often given, inadvertently, to patients with undiagnosed gastric carcinoma (Schotcher *et al.* 1981; Collin-Jones *et al.* 1982). Furthermore some 30 per cent of patients dying of gastric carcinoma have evidence of a previous duodenal ulcer (Ellis *et al.* 1979). It would be facile to assume that all diagnoses of gastric carcinoma made after one or two years of cimetidine therapy must incriminate the drug. However it is possible that over longer periods cimetidine treatment, like partial gastrectomy and other conditions characterized by achlorhydria, could predispose to the induction of stomach cancer. The unifying hypothesis is therefore that the chronic elevation of intragastric pH which occurs in each of these instances, might well be responsible for the higher prevalences of carcinoma of the stomach seen in these conditions. In the flow diagram in Fig. 35.1, the main pathway starts with cimetidine; it could well commence with any other cause of elevated pH. It is known, for example, that after partial gastrectomy nitrate-reducing bacteria and level of nitrite and nitroso compounds all increase in the gastric remnant (Reed *et al.* 1982).

In-vitro screening procedures

The screening of drug substances in order to detect those which react readily to form nitrosamines should become routine. There is a place for an easily performed and inexpensive chemical test of nitrosatability, or nitrosation assay procedure (NAP); however, a satisfactory NAP has yet to be devised. In 1978 an international group of experts, working under the informal auspices of the WHO recommended a NAP in which the test drug and nitrite at concentrations of 10 and 40 mmol^{-1} respectively, are allowed to react at 37°C for 1 and 4 hours at pH 3–4 (WHO 1978; Coulston and Dume 1978).

Estimates of the *in-vitro* nitrosatability of various drugs (Lijinksy 1972, 1974; Andrews *et al.* 1978) have been made. In descending order of nitrosamine yield these are: aminophenazone, phencetin, lucanthone, tolazamide, oxytetracycline, chlorpheniramine, quinacrine, disulfiram, methapyrilene, chlorpromazine, methadone, and *d*-propoxyphene. Clearly these results should be interpreted in the light of those obtained when nitrate and drug have been studied in appropriate animal models (Taylor and Lijinsky 1975; Lijinsky and Taylor 1977*a,b*) and in conjunction with other relevant investigations, such as epidemiological surveys.

In the latter context the most interesting factor is the high *in-vitro* nitrosatability of phenacetin which is the only one of these agents which to date has also been proved to be a carcinogen in man inducing tumours both of the renal pelvis and bladder (Bengtsson and Angervall 1970).

Gillat *et al.* (1984, 1985) examined the *in-vitro* nitrosatability of a range of drugs under simulated intragastric conditions. The conditions used were based upon the NAP test mentioned above (WHO 1978; Coulston and Dume 1978), but possessed the added refinement that nitrite level was stabilized over a 3-hour incubation period in order to reproduce the replenishment which occurs as a result of the swallowing of saliva. This work concentrated upon drugs which are in common use, are prescribed for long periods, and which possess relevant molecular structures. Of 22 drugs containing either a NN-dimethylamino, NN-diethylamino or N-morpholino group, volatile N-nitrosamines were obtained from eight. A molar nitrite/drug ratio of 4/1 was used, at a pH of 3.0. The greatest yields were obtained from aminopyrine (which is discussed above) and minocycline (*Minocin*). Much smaller yields were obtained from oxytetracycline, chlortetracycline, tetracycline, promethazine, chlorpromazine, imipramine, and disulfiram. No yield of volatile N-nitrosamines was obtained following reaction with amitriptyline, clomiphene, clomipramine, propoxyphene, diphenhydramine, disopyramide, erythomycin, mepyramine, methapyriline, procaine, tamoxifen, trimeprazine, tripelennamine, and penicillin-G procaine salt.

In addition, these workers examined the products of nitrosation from a selection of 57 drugs by means of a group selective procedure, in order to include both volatile and non-volatile N-nitroso compounds. The responses from aminopyrine and minocycline could be accounted for by the yield of N-nitrosodimethylamine (NDMA). However, compounds responding as N-nitrosamines in excess of NDMA were obtained from the other three tetracyclines employed. In general, drugs giving the greatest yields contained secondary, rather than tertiary, amines. A considerable range of susceptibilities was seen; 10 drugs containing a secondary or tertiary amino or an amido or a hydrazido group did not give rise to N-nitroso compounds under the conditions used.

The results obtained with minocycline are of particular interest. Other workers, at the Laboratory of the Government Chemist (Webb *et al.* 1983) have noted an increase in the concentration of N-nitrosodimethylamine in human urine following the administration of oxytetracycline.

Further research is now required to establish whether the effects observed take place under *in-vivo* conditions in the normal and achlorhydric stomachs.

Cancer following surgery

Although the majority of iatrogenic neoplasia result from drug or radiation therapy, it is a well-documented fact that some surgical procedures also can lead to malignancies. This chapter would be incomplete without a survey of this field. The author is particularly indebted to the exhaustive and detailed review by Dr Israel Penn (Penn 1982*b*).

Surgically-induced neoplasms may be classified into the rare and the very rare. The first group consists of gastric carcinoma as a late complication of stomach surgery for benign disease, carcinoma of the colon many years after cholecystectomy, carcinoma of the colon after prolonged exposure of the gut to urine, lymphangiosarcoma of the upper limb after the development of post-mastectomy lymphoedema, and, finally, sarcomata induced by foreign bodies. These five subclasses of rare tumours following surgery receive further attention below. The 'very rare' class is based upon occasional reports of late malignancies at the site of surgery for benign disease. These include tracheal cancer (at least five cases) following tracheostomy and prolonged use of a tracheostomy tube (Weisman and Konrad 1979). The pathogenesis may resemble foreign-body reactions, they may be unrelated to the surgery, or they may result from processes of repair which extend to 'overhealing' (Haddow 1974) and then neoplasia. Other isolated reports include an intrathoracic desmoid at a 3-year-old thoracotomy site (Giustra *et al.* 1979), and an adenocarcinoma at the choledochoenteric anastomosis 14 years after pancreaticoduodenectomy for chronic pancreatitis (Shields 1977).

Gastric surgery for benign disease, followed by cancer of the stomach

Gastric cancer may develop many years after surgery for benign gastric disease. Operations for peptic ulcer are thought by some authors (Helsingen and Hillestad 1956) to be three times more likely to be followed by gastric cancer when the ulcer site was gastric rather than duodenal. Other workers, however, found no such differences (Bushkin 1976; Dougherty *et al* 1982). Billroth I and II procedures and gastrojejunostomies may all be followed by gastric cancer (Bushkin 1976; Domellof *et al.* 1976; Dougherty *et al.* 1982) so that the type of ulcer operation does not itself appear to be critical. This is to be expected, since it has long been believed that hypochlorhydria and reduced gastric mobility, whether brought about by pharmacological or surgical means, may create carcinogenic conditions. The effects of known carcinogens have been studied in rats following vagotomy and pyloroplasty (Junghanns *et al.* 1979; Morgenstern 1979).

Procedures which permit duodenal reflux constitute

probable risk factors, and carcinoma of the gastric remnant arises in the area of the stoma in 85–90 per cent of cases (Bushkin 1976; Kobayashi *et al.* 1970; Morgenstern *et al.* 1973). Dougherty *et al.* (1982) proposed that the permeability of cell membranes to carcinogens (including oncogenic viruses) may be increased by duodenal contents, and it is known (Du Plessis 1962) that persistent reflux can lead to chronic atrophic gastritis with the 'intestinal' type of metaplasia. Moreover, it is believed (Morson 1955) that 30 per cent of stomach cancers develop in regions of 'intestinal' metaplasia. Further, these metaplastic cells contain the enzyme aminopeptidase, as do the cells of gastric carcinoma, unlike normal and inflamed gastric mucosal cells (Wattenburg 1959; Willighagen and Planteydt 1959). The use of procedures associated with reflux has decreased with the widespread use of histamine H_2-antagonists and with the advent of parietal-cell vagotomy which preserves the pylorus and avoids the creation of a stoma.

The incidence of stomach cancer following gastric surgery for benign disease is controversial. Various estimates suggest that it is higher or lower than, or equal to, that of stomach cancer in normal populations (Dougherty *et al.* 1982) The same author studied seven large series of stomach cancer cases and found that there had been prior gastric surgery for benign disease in 1.9 to 5.4 per cent of patients. Using other data and an alternative method, he found that stomach cancer followed surgery for benign gastric disease in 0.5 to 5.4 per cent of cases. These studies made use of the large series which have been published; altogether more than 1200 cases of stomach cancer following gastric surgery for benign disease are available in the literature (Bushkin 1976; Schmähl *et al.* 1977).

A number of authors have commented upon the age at which surgically-induced gastric cancer is likely to arise, Dougherty *et al.* (1982) pointed out that these tumours tend to occur in the sixth decade of life, irrespective of the date of the original operation. The prognosis is generally poor in this age-group, since total gastric resection is usually unavoidable (Bushkin 1976; Dougherty *et al.* 1982). Whereas a monitoring period of 5–10 years after surgery for a supposedly-benign condition is adequate to eliminate the possibility of an undetected pre-existing cancer, some 20–25 years is required for the emergence of a surgically-induced malignancy (Bushkin 1976). Taksdal and Stalsberg (1973) estimated that, at 35 years after gastric surgery, the risk of therapeutically-induced stomach cancer is eight times greater than normal. The conclusion to be drawn from this review must be that prolonged follow-up is essential if prompt diagnosis is to be ensured.

Cholecystectomy and colorectal cancer

Various investigators have established that an association exists between colorectal cancer and prior cholecystectomy (Capron *et al.* 1978; Peters and Keimes 1979; Vernick and Kuller 1981; Vernick *et al.* 1980), although Castleden *et al.* (1978) did not find an association in their earlier work. A careful study by Linos *et al.* (1981) eliminated most of the usual sources of bias that may be present in case-control designs, and confirmed the existence of the association. However, the risk level is not thought to be great (*Lancet* 1981b).

Final agreement on the carcinogenic mechanisms has not been reached. One hypothesis worthy of consideration is that cholelithiasis is a cause common both to colorectal cancer and gall bladder disease requiring cholecystectomy. This was not supported by the findings of Doouss and Castleden (1973) and Linos *et al.* (1981) who sought an association between cholelithiasis and colorectal cancer. It is known that, in animals, increased bile-acid secretion leads to a higher frequency of colonic carcinomata (Nigro *et al.* 1973; Reddy *et al.* 1974; Williamson *et al.* 1979). One of the companies which markets chenodeoxycholic acid states in its datasheet that this bile acid induces malignant liver-cell tumours in female rats and benign liver-cell tumours in female rats and male mice. There is evidence (Hill *et al.* 1975) that bile acids may also be relevant to colonic carcinogenesis in humans. Gonadal steroids may also be involved, and this might explain the higher incidence known to occur in women (Romsdahl 1980). Cholecystectomy is known to increase the enterohepatic recirculation, and thence the bacterial exposure, of bile. The increased production, by enterobacteria, of secondary bile acids and deoxycholic acid may account for the changes in the bile-acid pool which are known (Pomare and Heaton 1973) to follow cholecystectomy, the overall result being a greater concentration of carcinogenic substances in the colon. It is known that, in the mouse, the frequency of colonic neoplasia induced by dimethylhydrazine is much higher after cholecystectomy (Werner *et al.* 1977). High-fibre diets might have prophylactic value, through reduction of deoxycholic acid levels in bile. This occurs either by reducing reabsorption of the acid in the small intestine, or through reducing production of the acid by enterobacteria (*Lancet* 1981b).

Tumours resulting from prolonged intestinal exposure to urine

Carcinomas are known to arise in portions of gut which are exposed to urine in the long term. This occurs following surgical procedures used in the treatment of urinary-tract diseases and anomalies, in which urine is diverted into the colon, or into an isolated loop of colon or ileum (Leadbetter *et al.* 1979; Shapiro *et al.* 1974; Sooriyaarachchi *et al.* 1977). Segments of ileum may also be used to replace portions of ureter or bladder.

The most common tumour to occur is carcinoma of the colon following ureterosigmoidostomy, and at least 45 cases have occurred (Leadbetter *et al.* 1979). Some 20 per cent of the tumours reported are adenomata or inflammatory polyps. The incidence of colonic cancer, following this operation, is estimated to be 500 times the normal incidence at age 45 (Leadbetter *et al.* 1979; Urdaneta *et al.* 1966). Chrissey *et al.* (1980) calculated that the incidence in children is 7000 times that in all individuals under 25. According to Leadbetter *et al.* (1979) ureterosigmoidostomy confers a 5 per cent risk of colonic cancer occurring within 6–50 years. The tumours are usually adenocarcinomas of polypoid or exophytic morphology, located near a ureteral orifice (Leadbetter *et al.* 1979; Shapiro *et al.* 1974; Sooriyaarachchi *et al.* 1977). Some tumours were located at the ureteral stump in cases where the uretersoigmoidostomy had been disconnected years previously because of ascending infections (Chrissey *et al.* 1980; Leadbetter *et al.* 1979; Shapiro *et al.* 1974).

In the 82 per cent of patients who underwent ureterosigmoidostomy because of non-malignant indications, the median age at diagnosis of the colonic tumour was only 33 (range 16–75) years, and the average time from the diversion operation to the diagnosis of the tumours was 22.5 (range 7–46) years. In the 18 per cent of patients who underwent ureterosigmoidostomy because of bladder carcinoma, the median age at diagnosis of the colonic tumour was 63 and the average time from the diversion to the diagnosis of the lesion in the colon was 5.6 (range 2–9) years (Shapiro *et al.* 1979). In some patients there is delay because the level of suspicion is too low (Shapiro *et al.* 1974). In some patients a hydronephrotic kidney has been removed and the obstruction only discovered later at the anastamosis (Sooriyaarachchi *et al.* 1977). Local excision of the colonic tumour may be feasible in a minority of cases, but resection of a portion of colon is more usual (Sooriyaarachchi *et al.* 1977).

The mechanism by which colonic tumours are induced by prolonged exposure to urine is unknown. It was shown (Chrissey *et al.* 1980) that, in rat, both urine and the faecal stream are required for carcinogenesis. This suggests several possibilities: urinary enzymes may deconjugate faecal carcinogens; the urethelium may produce a local activator. Faecal bacteria may convert urinary nitrates to N-nitroso compounds (cf. Chrissey *et al.* 1981). When ureterosigmoid anastomoses are left *in situ* following a later nephrectomy or ileal loop diversion, tumours still arise at the anastomoses. This suggests (Aldis 1961) that the projecting lower end of the ureter may be subjected to repeated trauma and irritation by the faecal stream (Shapiro *et al.* 1974; Sooriyaarachchi *et al.* 1977). Bristol and Williamson (1981) suggested that surgical trauma or suture material provoke hyperplasia leading to neoplasia. This is unlikely since malignancies very rarely arise in other colonic anastomoses. The tendency for tumours to be located in the suture line may be explained by local sensitization to unknown carcinogens, however (Chrissey *et al.* 1981).

Tumours of the ileum are rare following procedures causing ileal exposure to urine. Cases following cystectomy for cancer are likely to be attributable to metastasis (Banigo *et al.* 1975; Grabstald 1974; Soloway *et al.* 1972; Wajsman *et al.* 1975) although isolated cases have occurred in the absence of prior malignant disease (Shousha *et al.* 1978; Egbert *et al.* 1980). These findings may result from some difference between ileal and colonic mucosa, in their sensitization to carcinogens following surgery (see above) or they may be a consequence of the villous atrophy which is known to occur within two years of the diversion procedure (Goldstein *et al.* 1067; Sherlock 1976).

Five cases of carcinoma of the urethra following urethroplasty have been reported (Williams and Ashken 1980), although these probably result from chronic irritation before operation.

The early detection of a colonic tumour is facilitated by the recognition of urinary obstruction, by pyelography (which may reveal hydronephrosis), by endoscopy and biopsy of the tumour, and by lavage and brush cytology. Regular testing of faeces and urine for occult blood, endoscopy, and radiological studies of the colon using water-soluble media (to avoid damage to the urinary tract) are also of value (Leadbetter *et al.* 1979; Shapiro *et al.* 1979; Sooriyaarachchi *et al.* 1977; Sherlock 1976). The anastamosis should always be excised completely when a ureterosigmoidostomy is discontinued for any reason. The early detection of a tumour arising in ileal tissue may be achieved by periodic cystoscopic examination of conduits and by urine cytology (Sherlock 1976; Soloway *et al.* 1972).

The Stewart–Treves syndrome (lymphangiosarcoma following mastectomy)

This aggressive tumour arises in the arm as a rare, late complication of mastectomy (Stewart and Treves 1948). Less than 200 cases have been reported (Unruh *et al.* 1979) and the prevalence is less than 1 per cent of mastectomy patients (Schreiber *et al.* 1979). The average age at diagnosis is 64 (Woodward *et al.* 1972). Lymphoedema usually precedes the tumour by some years (Unruh *et al.* 1979) although lymphoedema may not appear at all in some cases (Eby *et al.* 1967).

The clinical and pathological features of the syndrome are well described in the literature (McConnell and Halsam 1959; Woodward *et al.* 1972; Danese *et al.* 1967; Schreiber *et al.* 1979; Unruh *et al.* 1979). It is clear that the prognosis is very poor: according to Woodward *et al.* (1972) the five-year survival rate is only 8.5 per cent (129 cases) and the median survival time is 19 months.

The mechanism of carcinogenesis is obscure, and it must be borne in mind that few cases unassociated with radiotherapy have been reported. It is also relevant that, although such a tumour may occur in the arm when congenital lymphoedema exists (Schmähl *et al.* 1977), it does not occur in other conditions involving chronic lymphoedema such as filariasis (Danese *et al.* 1967). It is therefore unlikely that lymphoedema alone is the aetiological factor. Various hypotheses have been advanced, and these are summarized briefly. A common carcinogen acts on the breast and an area of low resistance in the arm (Stewart and Treves 1948). The protein of the oedema fluid undergoes alteration (Danese *et al.* 1967; Schreiber *et al.* 1979). The fluid becomes carcinogenic (McConnell and Halsam 1959). Lymphostasis stimulates the growth of new vessels; lymphangiomatosis occurs and leads to malignant change (Danese *et al.* 1967; Taswell *et al.* 1962). The atrophy of connective tissue which occurs in chronic lymphoedema leads to sarcomatous change (McConnell and Halsam 1959). The dearth of lymphatic tissue impairs the transport of tumour antigen to sites of immunocyte production, and dilution of antigen results from lymphoedema (Schreiber *et al.* 1979).

The prevention of lymphangiosarcoma following breast surgery rests on the reduction of lymphoedema and, although radical surgery is now less popular, unnecessary interruption of channels, especially around the axillary vein, should be avoided (Schreiber *et al.* 1979). Cellulitis requires prompt treatment; radiotherapy should be used only when appropriate; elasticated sleeves and pressure-gradient therapy should be used to treat lymphoedema (Eby *et al.* 1967; Schreiber *et al.* 1979).

Sarcomata induced by foreign bodies (FBs)

This subject was dealt with exhaustively by Brand (1975; Brand *et al.* 1973) and readers requiring a highly detailed account will wish to consult this source. Unlike rat and mouse, man is relatively resistant to these tumours, in common with guineapig and chicken (Schmähl *et al.* 1977; Stinson 1964; Ott 1970; Brand 1975; Brand *et al.*

1973). The period of latency is, of course, much longer in man than in small laboratory animals.

It is the physical, rather than chemical, natures of FBs which determine their induction potential (Bischoff 1972; Furst and Haro 1969; Stinson 1964) and chemically inert substances such as plastics and glass can induce sarcomata. Hard, smooth surfaces which are hydrophobic and which posses a large area with concavities or perforations are particularly active (Carter *et al.* 1971; Bates and Klein 1966; Bischoff and Bryson 1964; Ott 1970). These properties produce a fibrous, rather than cellular, reaction and this more readily leads to sarcoma formation. Species, such as the male rat and mouse, which possess a more vigorous fibrous reaction, show the greatest prevalence of sarcomata (Bashey *et al.* 1964; Murphy 1971; Brand 1975; Brand *et al.* 1973; Stinson 1964; Ott 1970). Latency in rat and mouse is increased by inflammatory complications which delay fibrosis. In man, subcutaneous FBs are surrounded initially by vascular, cellular tissue which is replaced gradually by a fibrous capsule over one to two years (Ott 1970).

Foreign-body sarcomata are invasive, rapidly-growing, malignant tumours showing varying degrees of anaplasia which rarely metastasize. After serial transplantations in isogenic individuals they may become transplantable to allogenic recipients (Brand 1975; Brand *et al.* 1973; Ott 1970). The cells of origin include fibroblasts, macrophages, endothelial cells, pericytes, and smooth muscle cells (Johnson *et al.* 1973). It is not necessary that the cells of origin be in contact with the FB in order that neoplastic transformation occurs, although neoplastic maturation does depend on such contact (Brand 1975; Brand *et al.* 1973).

The frequency of associated tumours following plastic surgery for breast enlargement is very low (de Cholnoky 1970) although Lewis (1980) reported four cases of inflammatory breast carcinoma in women aged 31–3 years, who had received silicone implants into both breasts for cosmetic reasons 6–8 years earlier. No causal link has been established between silastic prostheses inserted in mastectomy procedures and subsequent tumours (Bowers and Radlauer 1969; Johnson and Lloyd 1974). Breast cancer has been reported in a very small proportion of patients in whom cardiac pacemakers have been implanted, but there is no good evidence of causality (Zafiracopoulos and Rouskas 1974; Dalal *et al.* 1980; Magilligan and Isshak 1980; Hamaker al. 1976).

The majority of reports of FB sarcomata relate to isolated cases, in which metal or silicone implants were used for othopaedic purposes (Arden and Bywaters 1978; Delgado 1958; Dube and Fisher 1972; McDonald 1981; McDougall 1956; Digby and Wells 1981). An interesting case of primary extra-skeletal chondroscarcoma was reported following extrapleural *Lucite* ball plombage (Thomson and Entin 1969). Arterial grafts of *Dacron* and *Teflon–Dacron* have also been implicated (Burns *et al.* 1972; O'Connell *et al.* 1976; Weinberg and Maini 1980).

The aetiology of FB sarcomata has not been elucidated, although various factors have been suggested, including: the release of chemicals from some FBs, interruption of intercellular communication; anoxia and insufficient exchange of metabolities; disturbed regulation of cell growth; and the liberation of oncogenic viruses (Alexander and Horning 1959; Bischoff and Brysen 1964; Brand 1975; Brand *et al.* 1973; Carter *et al.* 1971; Johnson *et al.* 1973; Oppenheimer *et al.* 1959).

Brand (1975) suggested the following preventative principles: unnecessary cosmetic procedures should be avoided; implants should be as small as possible; regular follow-up should occur; a centralized registry should be established; further research on safety and aetiology is required.

Neoplasia induced by radiation

Neoplastic effects of radiation

In human experience the best-documented carcinogen is ionizing radiation which for practical purposes means X-rays as applied in the course of diagnosis and therapy. In the early days of the use of X-rays it was commonplace to find that radiologists suffered from mutiple skin carcinomata of the hands. One hundred radiologists were said to have died of malignant disease induced by radiation prior to 1922.

Fortunately in recent years much more attention has been paid to the increasing incidence of leukaemia attributable to ionizing radiation from the medical use of X-rays. Data concerning this incidence can be summarized as follow:

1. Patients heavily irradiated for the treatment of ankylosing spondylitis showed a significant incidence of myeloid leukaemia (Court-Brown and Doll 1957).
2. Children irradiated *in utero* have a greater risk of developing leukaemia in the first decade of life (Stewart *et al.* 1956).
3. There is a considerably higher death rate from leukaemia amongst radiologists than among other members of the medical profession (Hunter 1978).

In all cases there is evidence that there is a quantitative relationship between dose of irradiation and the risk of developing leukaemia. A latent period of 3–4 years after a single heavy dose of radiation is common.

Other neoplastic conditions have also been reported following X-irradiation; these were classified by Bailey and Love (1959) as follows:

1. Skin: rodent ulcers and squamous-cell carcinoma.
2. Soft-tissue and bone sarcomata: sarcomata may arise in the mesenchyme.
3. Thyroid carcinoma: this may follow irradiation of the thymus gland in infancy, or repeated irradiation of tuberculous lymph-nodes sufficient to produce skin damage in adolescents.
4. Myelogenous leukaemia.

Neoplasia following therapeutic irradiation

Attempts have been made a predict the risks associated with therapeutic irradiation, using data on leukaemia and other neoplasia in Japanese survivors of the nuclear fission exposures at Hiroshima and Nagasaki (UNSCEAR 1977; BEIR III 1980). The induction of leukaemia in the survivors is dose-related and the limited data available (UNSCEAR 1977) are not inconsistent with a liner relationship. The latent period is unusually short (2–3 years) and the type of leukaemia acute. The risk for leukaemia is about 1 in 500 per Sievert. The treatment of ankylosing spondylitis by irradiation, while involving a different radiation spectrum and not being confined to a single dose, produced comparable effects (BEIR III 1980). The induction rates for both forms of radiation lie within the range 50 to 300 cases per million per year per Gray. Irradiation for ankylosing spondylitis is discussed elsewhere in this chapter.

Similar data on the induction rates of carcinoma of the breast, thyroid, and lung in atomic bomb survivors has been obtained (UNSCEAR 1977; BEIR III 1980). These data, also, are consistent with the carcinogenic effects of irradiation administered for non-malignant conditions. Patients treated for ankylosing spondylitis, for example, with an average dose of 4 Gray had a twice-normal incidence of lung cancer; a similar increase was seen in those survivors from Hiroshima who received comparable exposure (BEIR III 1980). In view of these findings, it might be expected that the consistent dose–response findings following atomic explosions and radiotherapy for benign conditions would also apply to patients irradiated for malignant tumours.

For reasons which are not understood, the risk of a second neoplasm (especially leukaemia) following radiotherapy for some common malignancies (particularly cancer of the ovary, cervix, testis, and breast) is claimed to be considerably lower than that predicted from the findings discussed in the previous paragraph. Harwood and Yaffe (1982) reviewed data from several souces on large numbers of patients treated with irradiation for cancer of the uterine cervix. These data suggest strongly that there is no increase in the incidence of leukaemia in these patients; indeed decreases in the incidence of carcinoma of the breast were noted, possibly resulting from the effects of radiation in the ovaries. In a study of 6596 patients with carcinoma of the ovary (Reimer *et al.* 1977) no excess risk of leukaemia was found following irradiation, whereas 13 cases were found (0.62 expected) in 5455 patients treated with alkylating agents. Similar findings were reported in a review (Harwood and Yaffe 1982) of unpublished data in which radiotherapy, chemotherapy, and a combination of the two treatments were compared in ovarian cancer. In a review (Harwood and Yaffe 1982) of 652 patients irradiated for seminoma of the testis, no cases of leukaemia were seen in 5900 patient-years at risk, although the treatment is similar in dose and volume to that used in ankylosing spondylitis. In a review of ongoing work in carcinoma of the breast treated by irradiation, only one case of leukaemia was seen among 526 patients, 104 of whom have been followed for 10 years or more (Harwood and Yaffe 1982).

Various explanations have been proposed to explain the alleged discrepancies between expected and observed frequency of oncogenesis following therapeutic irradiation. It has been suggested that patients with particular tumours are less susceptible to radiogenic leukaemia than others, or that exposure of large volumes of the body such as occurs in radiotherapy for ankylosing spondylitis and in atomic bomb victims is the determining factor.

It seems likely that relevant factors are the types, i.e. wavelengths, of ionizing radiations involved, and the manner in which exposure periods are timed. With modern (megavoltage) irradiation, the proportion of unwanted absorption is considerably less than that which occurred with the earlier (orthovoltage) treatments which were used, for example, in ankylosing spondylitis.

To put the problem in perspective, the incidence of bone sarcomata following modern radiotherapy is less than 0.2 per cent (Harwood and Yaffe 1982) and this is similar to the mortality associated with the administration of general anaesthesia (Dripps *et al.* 1961).

Modan *et al.* (1974) reported on the results of a survey of 11 000 children irradiated for ringworm of the scalp, and two matched control groups who were followed up for 12 to 23 years. The irradiated group had a significantly higher risk of both malignant and benign head and neck tumours especially those in the brain (meningioma), parotid gland, and thyroid. The overall increase in malignancies in the irradiated group was 2.5–4.0 times greater than the incidence in the two control groups. The association between thyroid tumours and irradiation has been known for many years

and the association of parotid tumours was greatly increased in survivors of atomic-bomb explosions (Belsky *et al.* 1972) but Modan *et al.* (1974) made the first association between meningioma and irradiation.

It is interesting that Sypkens-Smith and Meyler (1970) described a case of myeloid leukaemia induced by cyclophosphamide which had been used for the treatment of bilateral papilliferous ovarian cyst adenomas in a 44-year-old woman. They drew attention to the fact that the activity of alkylating agents such as thiotepa (triethylene thiophosphoramide) and cyclophosphamide in many respects resembled that of ionizing radiation.

Mammography

Simon (1977) and Bailar (1977) drew attention to the fact that the investigative procedure of mammography may of itself cause breast cancer. Bailar cautioned that in women under 60–65 years mammography should only be undertaken if there is specific reason for concern or where mammography may give results which will affect the management of the case. Whereas early diagnosis of breast cancer is desirable, we must be sure that efforts to stop the disease do not lead to greater problems than they are intended to prevent.

Ahmed and Steel (1972) described mammary carcinoma occurring 30 years after *Thorotrast* mammography conducted in 1939.

Radiotherapy for ankylosing spondylitis

Court-Brown and Doll (1957) identified over 14 000 patients with ankylosing spondylitis who had been treated with one or more courses of X-irradiation from 1935 to 1954 at one of 87 radiotherapy centres in Great Britain and Northern Ireland. The early reports of Court-Brown and Doll (1957, 1965) analysed the mortality from leukaemia and other causes, particularly cancer, but these analyses included many patients who had been irradiated more than once. This complicated the interpretation and quantification of risk and in a paper by Smith and Doll (1982) they avoided this difficulty by examining the death rate from leukaemia and other radiation-induced cancers at different times after a single course of irradiation.

Mortality was studied in 14 111 patients with ankylosing spondylitis given a single course of X-ray treatment during the period 1935–54. Mortality from all causes combined was 66 per cent greater than that of members of the general population of England and Wales. There were substantial excesses of deaths from non-neoplastic conditions, but these appeared to be associated with the disease itself rather than its treatment. A nearly fivefold excess of deaths from leukaemia and a 62 per cent excess of deaths from cancers of sites that would have been in the radiation fields ('heavily irradiated sites') were likely to have been a direct consequence of the radiation treatment itself. The excess death rate from leukaemia was greatest three to five years after treatment and was close to zero after 18 years. In contrast, the excess of cancers of heavily irradiated sites did not become apparent until nine or more years after irradiation and continued for a further 11 years. More than 20 years after irradiation the excess risk declined, but the fall was not statistically significant. The number of cancers of sites not considered to be in the radiation fields was 20 per cent greater than expected. This excess, although not statistically significant, may also have been due to radiation scattered from beams directed at other parts of the body.

The risk of a radiation-induced leukaemia or other cancer was related to the age of the patient at the time of treatment. Those irradiated when aged 55 years or more had an excess death rate from leukaemia more than 15 times that of those treated under 25 years of age, and a similar difference was apparent for cancers of heavily irradiated sites. The radiation dose to the bone marrow was estimated for the patients who died with leukaemia and for a 1 in 15 sample of the total study population. A clear dose–response relationship between radiation dosage and leukaemia incidence and latent period to development of leukaemia was not established.

Thorotrast (232-thorium dioxide)

All ionizing radiations, and not merely X-rays, are capable of incuding neoplasia; these include gamma rays, as well as the electromagnetic energy associated with emissions of neutrons, protons, alpha and beta particles (Upton 1975). *Thorotrast* is a rare and interesting example of a radionucleide (^{232}Th), incorporated into a medicine which produces radiation-induced neoplasia after a single exposure.

It was used in the first half of the present century as a radiological contrast medium because of its radio-opacity. The safety implications of the radioactivity of ^{232}Th, as with Thorium-X (224-radium), were not appreciated at that time. ^{232}Th is an alpha-emitter with a half-life of 1.4×10^{10} years. Alpha particles are emitted with an energy of four million electron-volts. Subsequent stages in the decay of the nucleide yield further ionizing radiations. *Thorotrast* consists of a colloidal suspension of 232-thorium dioxide. This agent has been used extensively since 1928 as a radiographic contrast medium, either to outline body cavities such as the renal pelvis or the cerebral ventricles, or for visualization of blood vessels.

In 1961 Suckow and co-workers drew attention to the association between *Thorotrast* administration and haemangio-endothelioma of the liver. Sherlock (1968) described this tumour as very rare and also described its association with *Thorotrast*. The clinical course of hepatic haemangio-endothelioma is rapid deterioration with cachexia, blood-stained ascites; bruit may be heard over the liver.

In Portugal, Da Silva Horta *et al.* (1965) checked records of 2377 individuals who had received injections of *Thorotrast* between 1930 and 1952. A total of 1107 cases were traced; of these 699 had died and 408 were still living; certified cause of death was obtained for the former group.

Twenty-two cases of haemangio-endothelioma had occurred; this tumour is virtually thorium-dioxide specific in the liver. The latent period between *Thorotrast* injection and onset of this fatal condition was, with a single exception, 20 years or more. In this investigation, liver cirrhosis was recorded in 42 cases, and 17 of these were fatal, and this incidence of cirrhosis was significantly higher than in the general population. There were also 16 cases of fatal blood dyscrasia: six acute and two chronic myeloid leukaemia, six aplastic anaemia (pancytopenia), and two recorded as fatal purpuras. Eighty-one local granulomata occurred and eight of these caused death. Blood dyscrasias and leukaemias were significantly higher than in the general population.

As a result of this earlier survey Boyd *et al.* (1968) undertook a survey on 137 patients who had received *Thorotrast* by intra-arterial injection between the years 1933–1955. Although mortality amongst the patients reflected the disease (mainly cerebrovascular) that brought them to original investigation, it was noted that there were 12 deaths from neoplasia among 109 patients who survived more than one year after the *Thorotrast* administration. Four cases of primary liver tumours were found in this group, three were tumours of the intrahepatic bile ducts, and one case was a haemangio-endothelioma. A further case with oat-cell carcinoma also showed, at laparotomy, liver changes suggestive of haemangio-endothelioma.

Thorium dioxide (*Thorotrast*) is no longer used; however, patients with tumours caused by its use in the past are still presenting themselves and the possibility of prior use of this agent in cases of primary liver tumour should be investigated. *Thorotrast* has also been reported to cause renal carcinoma (Kruckemeyer *et al.* 1960). Meyler (1966) reviewed the incidence of tumours in other organs caused by *Thorotrast*.

Benjamin and Albukerk (1982) reported the case of a 76-year-old woman who had undergone carotid angiography at the age of 44 years and who developed a 5-cm large neck mass at the site of the angiography. An excisional biopsy showed changes suggestive of thorium dioxide-induced local reaction (thorotrastoma). On final hospital admission she was unconscious and anaemic (haemoglobin 5.1 g). She developed hypotension, bradycardia, and respiratory arrest. An abdominal flat plate showed foci of radio density suggestive of radiographic contrast material. Despite supportive therapy she died soon after admission. At autopsy, thorium dioxide crystals were noted in the liver, spleen, and brain. Microscopic examination of the liver showed a wide spectrum of changes indicative of angiosarcoma. Although thorium dioxide is no longer being utilized, the authors stress that new cases may continue to arise because of the long latency period and long biologic half-life (about 400 years), of this contrast medium.

Radium-induced carcinoma

The isolation and purification of radium was achieved during the period 1898–1902 by the Curies and, by the 1920s, the element was in widespread use by the painters of watch dials. These young women tipped their brushes with the lips and the resulting radium poisoning induced osteosarcomata, aplastic anaemia, and severe oral lesions. The cause was recognized and few workers were exposed after 1925 (Martland 1929, 1931). Unfortunately, however, the therapeutic use of intravenous and oral radium choride continued (Brues and Kirsh 1976) for hypertension, arthritis, gout, poliomyelitis, veneral disease, mental illness, and as a tonic (sic).

In 1929 the dangers and ineffectiveness of therapeutic radium emerged in the medical press (Flinn 1929; *Journal of the American Medical Association* 1929) and physicians began to abandon the systemic administration of the element and its salts. Radium localizes in osteon formations of resorbed Haversian systems in bone (Rowland and Marshal 1959). During its radioactive decay, radium and the isotopes derived from it emit gamma rays and alpha and beta particles. The alpha particles travel 30–50 μm in tissue and are therefore able to disrupt chromosomal material. In the USA alone, records have been brought together showing 82 verified cases of osteosarcoma and other sarcomata involving bone, and 33 cases of nasopharyngeal, paranasal sinus and mastoid carcinoma (Brues and Kirsh 1976). The mastoids and paranasal sinuses contain large surface areas of mucous membrane applied closely to alpha-emitting bone and this results in the predilection of radium-induced malignant tumours for these sites.

The number of patients treated with intravenous and oral radium is unknown, and many of those who received this therapy were undoubtedly unaware of its nature or subsequently forgot the details. Some of these patients however are still living and are, from time to time, referred to otolaryngologists with mastoid and paranasal sinus carcinoma (Beal *et al.* 1965; Applebaum 1979), presenting up to 40 years after the initial treatment. Others may demonstrate the typical radiological changes of radium retention in the skeleton, but are frequently misdiagnosed as cases of osteoporosis, Paget's disease, multiple myeloma, and metastatic carcinoma (Brues and Kirsh 1976).

Cancer following diagnostic irradiation

In diagnostic radiological procedures, the total dose administered rarely exceeds 0.1 Gray. To permit statistical verification of neoplastic effects in the range 0.01–0.1 Gray would require an epidemiological study containing 10^7 people in each group (Webster 1981). Until a mechanism for the radiation-induction of neoplasia has been established, uncertainty about the precise degree of risk to the radiosensitive tissues (marrow, thyroid, lung, breast) will persist.

A great deal of work has been done to establish the shape of the dose–effect curve at doses in the diagnostic range. Various mathematical functions have been proposed, based on animal studies (Webster 1981), radiobiological models of induction (Rossi 1977; Brown 1977; Upton 1977), radiotherapy for benign disease, and on data relating to atomic bomb survivors (BEIR III 1980). The major functions which have been suggested to describe the relation between dose and neoplastic effect (Harwood and Yaffe 1982) are

$$E = aD \qquad (35.1)$$
$$E = bD + cD^2 \qquad (35.2)$$
$$E = dD^2 \qquad (35.3)$$

where E is neoplastic effect in arbitrary units, D is dose in Grays, and a, b, c, and d, are constants. The curves of these functions are shown in Fig. 35.2. In the diagnostic (0.001–0.1 Gray) range, the pure quadratic curve, eqn. (35.3), predicts the lowest risk, but the linear function, eqn. (35.1), appears to be more consistent with the radiation-induction of breast cancer (Brown 1977; Upton 1977; BEIR III 1980). In survivors of the Nagasaki fission bomb exposed to 0.01 Gray, the predicted risk of breast cancer is 115 times greater using the linear rather than the quadratic function; for leukaemia the risk is 156 times greater (Webster 1981).

It must be emphasized that insufficient data are available to establish which of these functions is the most accurate representation for low-dose induction in general. The linear function is the simplest and most conservative of the three, and it does not presuppose any no-

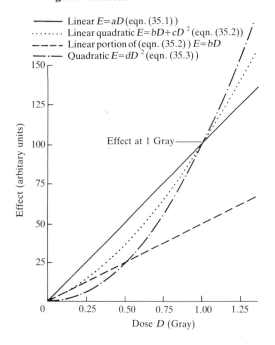

Legend for figure:
- Linear $E=aD$ (eqn. (35.1))
- .... Linear quadratic $E=bD+cD^2$ (eqn. (35.2))
- – – – Linear portion of (eqn. (35.2)) $E=bD$
- –·–·– Quadratic $E=dD^2$ (eqn. (35.3))

Effect (arbitary units)

Effect at 1 Gray

Dose D (Gray)

effect threshold, for which there is insufficient evidence. Using this approach and other data Harwood and Yaffe (1982) obtained the estimates in Table 35.2 for the risk of cancer induced by diagnostic radiological procedures.

If every care is taken to minimize exposure, these results show that '. . . with the exception of the barium enema and positive-mode mammogram, the estimates of total risk of cancer-induction from a diagnostic examination are about 1.0 in 10^6 per year. This corresponds approximately to the risk of smoking 0.75 cigarette per year . . .' The risk is, of course, considerably higher when exposure is not rigorously minimized, especially in certain procedures such as barium swallows and enemas (Harwood and Yaffe 1982).

On average in each year in the UK individuals receive an effective dose equivalent of approximately 500 micro-Sieverts (ie about half the dose imparted in a chest X-ray) as a result of medical procedures. This constitutes by far the greater part of all ionising radiation exposure of artificial origin (including weapons fallout, discharges to environment, occupational exposure, luminous watches, television receivers, visual display units and increased cosmic ray exposure during air travel). Approximate values for the effective dose equivalents arising from artificial sources are given in Table 35.3. The average risk of fatal cancer induced by radiation is about 1 in 80 per Sv. per year (BEIR III 1980; UNSCEAR 1977). In the UK, this would correspond to roughly 300 additional cancer deaths per year, from all medical uses of radiation.

Table 35.3 Approximate effective dose-equivalents (EDE) in Sieverts [*]

Examples of therapeutic procedures	
i) Radiotherapy for carcinoma of larynx. Small area (5 × 5 cm). 15 doses of 3.3 Sv given over 18 days	50.0
ii) Radiotheapy for basal cell carcinoma. Single dose to skin, 2 cm circle.	18.0
iii) Superficial X-ray therapy for eczema of hand. 3 doses of 2 Sv at weekly intervals.	6.0
Human LD_{50} (single dose, whole body)	5.0
Serious non-stochastic effects (radiation sickness) gut, marrow, CNS, skin.	1.0
Early non-stochastic effects (lymphocyte death, cataracts)	0.5
Major diagnostic radiology procedure.	0.1
Chest X-ray (EDE to lung)	0.001
Artificial Sources (whole body) average EDE annually	
i) Medical procedures.	0.0005
ii) Weapons Fallout.	0.00001
iii) Occupational Exposure.	0.000009
iv) Miscellaneous (including television receivers, VDUs, luminized watches, air travel, etc).	0.000008
v) Industrial discharges to environment.	0.000003

[*] The average risk for radiation-induced cancer is approximately 1 in 80 per Sievert.

Table 35.2 Risk of cancer induced by radiological procedures (Harwood and Yaffe 1982)

Procedure	Procedures per person year	Possible risk of cancer per 10^6 procedures/year				
		Thyroid	Breast	Lung	Active marrow	Total risk
Chest X-ray	0.24	0.007–0.1	0.09–0.6	0.07–0.05	0.007–0.04	0.2–1.2
IVP	0.02	0.0009–0.002	—	0.16–2.3	0.32–4.4	0.5–9
Barium swallow	0.03	1–150	0.14–8	0.9–58	0.3–9	2–225
Barium enema	0.018	0.007–0.15	—	1.3–11	4.7–37	6–48
Gall bladder	0.02	0.003–0.08	—	0.9–12	0.2–2.5	1–15
Xeromammogram positive mode			5.5			5.5
Film screen (Gd_2O_2S)			0.46			0.5

Acknowledgements

The assistance of the following individuals is acknowledged with gratitude: Professor H.K.Weinbren, who provided much helpful advice on the first part of the chapter; Dr D Fitzgerald, medical director of Imperial Chemical Industries, who contributed the item on razoxane and leukaemia; Dr C.L.Walters, who gave permission for the use of material (Gillatt *et al.* 1984) before its publication; Dr P.I. Reed for advice on the item dealing with cimetidine.

REFERENCES

ADAM, S.A., SHEAVES, J.K., WRIGHT N.H., MOSSER, G., HARRIS, R.W., AND VESSEY, M.P. (1981). A case-control study of the possible association between oral contraceptives and malignant melanoma. *Br. J. Cancer* **44**, 45–50.

AHMED, M.Y. AND STEELE, H.D. (1972). Breast carcinoma 30 years after *Thorotrast* mammography. *Can. J. Surg.* **15**, 45–9.

ALDIS, A.S. (1961). Carcinoma of colon following transplantation of the ureters, and at the site of transplantation. *Proc. R. Soc. Med.* **54**, 159–60.

ALEXANDER, P. AND HORNING, E.S. (1959). Observations on the Oppenheimer methods of inducing tumours by subcutaneous implantation of plastic films. In *CIBA Foundation Symposium on Carcinogenesis. Mechanisms of action* (ed. G.E.W. Wolstenholme and M. O'Connor), pp. 12–25. Little, Brown and Co, Boston.

AMERICAN CANCER SOCIETY (1983). *Cancer facts and figures 1983*. American Cancer Society, New York.

AMES B.N. (1983). Dietary carcinogens and anticarcinogens. *Science* **221**, 1256–64.

ANDREWS, A.W., THIBAULT, L.H., AND LIJINSKY, W. (1978). The relationship between mutagenicity and carcinogenicity of some nitrosamines. *Mutat. Res.* **51**, 319–26.

ANNEGERS, J.F., COULAM, C.B., ABDOUD, C.F., *et al.* (1978). Pituitary adenoma in Olmstead Country, Minnesota, 1935–1977. *Mayo Clin. Proc.* **53**, 641–3.

——, STROM, H., DECKER, D.G., DOCKERTY, M.B., AND O'FALLON, M. (1979). Ovarian cancer: incidence and case control study. *Cancer* **43**, 723–9.

ANTTINEN, H., AHONEN, A., LEINONEN, A., KALLIOINEN, M., AND HEIKKINEN, E.S. (1982). Diagnostic imaging of focal nodular hyperplasia of the liver developing during nitrofurantoin therapy. *Acta med. scand.* **211**, 227–32.

ANTUNES, C.M.F., STOLLEY, P.D., ROSENSHEIN, N.B., DAVIES, J.L., TONASCIA, J.A., BROWN, C., BURNETT, L., RUTLEDGE, A., POLEMPNER, M., AND GARCIA, R. (1979). Endometrial cancer and oestrogen use. Report of a large case-control study. *New Engl. J. Med.* **300**, 9–13.

APPLEBAUM, E.L. (1979). Radiation-induced carcinoma of the temporal bone, *Otolaryngol. Head Neck Surg.* **87**, 604–9.

ARDEN, G.P. AND BYWATERS, E.G.L. (1978). Tissue reaction. In *Surgical management of juvenile chronic polyarthritis* (ed. G.P. Arden and B.M. Ansell), pp. 263–75. Academic Press, London.

ASHWOOD-SMITH, M.J. (1979). Possible cancer hazard associated with 5-methoxypsoralen in sun tan preparations. *Br. med. J.* **2**, 1144.

ASPLUND, K., HAGG, E., LINDGVIST, M., AND RAPP, W. (1982). Phenothizine drugs and pituitary tumours. *Ann. intern. Med.* **96**, 533.

ASSOCIATION OF THE BRITISH PHARMACEUTICAL INDUSTRY (1983). *Datasheet compendium 1983–84*, p. 1392. Association of the British Pharmaceutical Industry, London.

ATHERTON, D.J., WELLS, R.S., LAURENT, M.R., AND WILLIAMS, Y.F. (1980). Razoxane (ICRF 159) in the treatment of psoriasis. *Br. J. Dermatol.* **102**, 307–17.

BAILAR, J.C. (1977). Mammography — a time for caution. *J. Am. med. Ass.* **237**, 997–8.

BAILEY, H. AND LOVE, M.C.N. (1959). *A short practice of surgery*. H.K. Lewis, London.

BAIN, C., HENNETEUS, C.H., SPEIZER, F.E., ROSNER, B., WILLETT, W., AND BELANGER, C. (1982). Oral contraceptive use and malignant melanoma. *J. Nato. Cancer Inst.*, **68**, 537–9.

BAKOWSKI, M.T. (1976). ICRF 159, ($\pm$) 1,2-di (3,5-dioxo-piperazin-1-yl) propane, NSC-129, 943; Roxane. *Cancer treat. Rev.* **3**, 95–107.

BANIGO, O.G., WAISMAN, J., AND KAUFMAN, J.J. (1975). Papillary (transitional) carcinoma in an ileal conduit. *J. Urol.* **114**, 626–7.

BARBEAU, A. (1972). Long-term appraisal of Levodopa therapy. *Neurology* **22** (suppl.), 22–4.

BASHEY, R.I., WOESSNER, J.F., JR., AND BOUCEK, R.J. (1964). Connective tissue development in subcutaneously implanted polyvinyl sponge. III. Ribonucleic acid changes during development. *Arch. Biochem. Biophys.* **104**, 32–8.

BATES, R.R. AND KLEIN, M. (1966). Importance of a smooth surface in carcinogenesis by plastic film. *J. Nat. Cancer Inst.* **37**, 145–51.

BAUER, F. (1978). Late sequelae of atabrine dermatitis — a new pre-malignant entity. *Aust. J. Dermatol.* **19**, 9–12.

BAUM, J.K., HOLTZ, F., BOOKSTEIN, J.J., AND KLEIN, E.W. (1973). Possible association between benign hepatomas and oral contraceptives. *Lancet* ii, 926–9.

BEAL, D.D., LINDSAY, J.R., AND WARD, P.H. (1965). Radiation-induced carcinoma of the mastoid. *Arch. Otolaryngol.* **81**, 9–16.

BEARDMORE, G.L. AND DAVIS, N.C. (1975). Multiple primary cutaneous melanomas. *Arch. Dermatol.* **111**, 603–9.

BEIR III (1980). *The effects on populations of exposure to low levels of ionizing radiation*. National Academy of Sciences, Washington, DC.

BELLANDER, B.T.D., HAGMAR, L.E. AND OSTERWALL, B.G. (1981). Letter: Nitrosation of piperazine in the stomach. *Lancet* i, 372.

BELSKY, J.L., TACHIKAWA, K.K., CIHAK, R.W., AND YAMAMOTO, T. (1972). Salivary gland tumours in atomic bomb survivors, Hiroshima–Nagasaki, 1957 to 1970 *J. Am. med. Ass.* **219**,864–8.

BENGTSSON, U. AND ANGERWALL, L. (1970). Analgesic abuse and tumours of the renal pelvis. *Lancet* i, 306.

——, ——, EKMAN, H., AND LEHMANN, L. (1968). Transitional cell tumours of the renal pelvis in analgesic abusers. *Scand. J. Urol. Nephrol.* **2**, 145–50.

BENJAMIN, A.G. AND ALBUKERK, J.N. (1982). Thorotrast-induced angiosarcoma of liver. *NY State J. Med.* **82**, 751-3.

BERAL, V., RAMCHARAN, S., AND FARIS, R. (1977). Malignant melanoma and oral contraceptive use among women in California. *Br. J. Cancer* **36**, 804-9.

BERNSTEIN, J.E., MEDENICA, M., SOLTANI, K., *et al.* (1980). Levodopa administration and multiple primary cutaneous melanomas. *Arch. Dermatol.* **116**, 1041-4.

BERNSTEIN, M.S., HUNTER, R.L., AND RACHNIN, S. (1971). Hepatoma and peliosis hepatis developing in a patient with Franconi's anemia. *New Engl. J. Med.* **284**, 1135-6.

BERRY, C.L. (1983). Reproductive toxicity. In *Animals and alternatives in toxicity testing* (ed. M. Balls, R.J. Riddell, and A.N. Worden). Academic Press, London.

BIANCIFIORI, C. AND SEVERI, L. (1966). The relation of isoniazid (INH) and allied compounds to carcinogenesis in some species of small laboratory animals. *Br. J. Cancer* **20**, 528-38.

BIRKELAND, S.A., KEMP, E., AND HAUGE, M. (1975). Renal transplantation and cancer: The Scandia transplant material. *Riss. Antigens* **6**, 28-36.

BISCHOFF, F., (1972). Organic polymer biocompatibility and toxicology. *Clin. Chem.* **18**, 869-94.

—— AND BRYSON, G. (1964). Carcinogenesis through solid state surfaces. *Prog. Exp. Tumour Res.* **5**, 85-133.

BORCHERT, P., MILLER, J.A., MILLER, E.C., AND SHIRES, T.K. (1973). 1-hydroxy-safrole, a proximate carcinogenic metabolite of safrole in the rat and mouse. The metabolism of the naturally-occurring hepato-carcinogen safrole to 1-hydroxy-safrole and the electrophilic reactivity of 1-acetoxy-safrole. *Cancer Res.* **33**, 575-600.

BORDES-ASNAR, J. AND TILNEY, N.L. (1982). Cyclosporin A: A more specific immuno-suppressive agent? *J. clin. Surg.* **1**, 53-61.

BORLEFFS, J.C.C., (1982). Kidney transplantation in rhesus monkeys, pp. 102-4. PhD thesis, Primate Center, Rijswik, The Netherlands.

BOWERS, D.G., JR. AND RADLAUER, C.B. (1969). Breast cancer after prophlactic subcutaneous mastectomies and reconstruction with silastic prostheses. *Plastic reconstruct. Surg.* **44**, 541-4.

BOYD, J.T., LONGLANDS, A.O., AND MACCABE, J.J. (1968). Long-Term hazards of Thorotrast. *Br. med. J.* **2**, 517-21.

BOYD, W. (1953). *Textbook of pathology*, 6th edn. Henry Kimpton, London.

BRAND, K.G. (1975). Foreign body induced sarcomas. In *Cancer a comprehensive treatise. 1. Etiology: chemical and physical carcinogenesis* (ed. F. F. Becker), pp. 485-511. Plenum Press, New York.

——, BUOEN, L.C., AND BRAND, I. (1973). Foreign-body tumorigenesis in mice: most probable number of originator cells. *J. Nat. Cancer Inst.* **51**, 1071-4.

BRESLOW, N.E. AND ENSTROM, J.E. (1974). Geographic correlations between mortality rates and alcohol tobacco consumption in the United States. *J. nat. Cancer Inst.* **53**, 631-4.

BRIDGES, B.A. (1978). Possible long-term hazards of photochemotherapy with psora-lens and near ultra violet light. *Clin. exp. Dermatol.* **3**, 349-53.

BRISTOL, J.B. AND WILLIAMSON, R.C.N. (1981). Ureterosignoidostomy and colon carcinogenesis. *Science* **214**, 351.

BROWN, J.M. (1977). The shape of the dose–response curve for radiation carcinogenesis extrapolation to low doses. *Radiat. Res.* **71**, 34-50.

BRUES, A.M. AND KIRSH, I.E. (1976). The fate of individuals containing radium. *Trans. Am. clin. Climatol. Ass.* **88**, 211-18.

BURNET, F.M. (1959). *The clonal selection theory of acquired immunity.* Cambridge University Press, Cambridge.

BURNS, W.A., KANHOUWA, S., TILLMAN, L., SAINI, N., AND HERRMANN, J.B. (1972). Fibrosarcoma occurring at the site of a plastic vascular graft. *Cancer* **29**, 66-72.

BUSHKIN, F.L. (1976). Gastric remnant carcinoma. In *Postgastrectomy syndromes* (ed. F.L. Bushkin and E.R Woodward), pp. 106-13. W.B. Saunders Co, Philadelphia.

CALLOWAY, J.L. (1979). Late sequelae of quinacrine dermatitis, a new premalignant entity. *Am. Acad. Dermatol.* **1** (5), 456.

CALNE, R.Y., WHITE, D.J.G., ROLLES, K., THIRU, S., EVANS, D.B., MCMASTER, P., DUNN, D.C., CRADDOCK, G.H., HENDERSON, R.G., AZIZ, S., AND LEWIS, P. (1979). Cyclosporin A initially as the only immunosuppressant in 34 recipients of cadaveric organs: 32 kidneys, 2 pancreases, and 2 livers. *Lancet* **ii**, 1022-6.

CAMPBELL, A.H., (1961). The association of lung cancer and tuberculosis. *Aust. Ann. Med.* **10**, 129-36.

—— AND GUILFOYLE, P. (1970). Pulmonary tuberculosis, isoniazid, and cancer. *Br. J. Dis. Chest* **64**, 141-9.

CAPRON, J.P., DELAMARRE, J., CANARELLI, J.P., BROUSSE, N., AND DUPAS, J.L. (1978). La cholecystectomie favorise-t-elle l'apparition du cancer rectocolique? *Gastroenterol. Clin. Biolog.* **2**, 383-9.

CARNEY, C.N., STEVENS, P.S., FRIED, F.A., AND MANDELL, J. (1982). Fibroblastic tumour of the urinary bladder after cyclophosphamide therapy. *Arch. Pathol. Lab. Med.* **106**, 247-9.

CARTER, R.L., ROE, F.J.C., AND PETO, R. (1971). Tumour induction by plastic films: Attempt to correlate carcinogenic activity with certain physicochemical properties of the implant. *J. Nat. Cancer Ins.* **46**, 1277-89.

CASTLEDEN, W.M., DOOUSS, T.W., JENNINGS, K.P., AND LEIGHTON, M. (1978). Gallstones, carcinoma of the colon and diverticular disease. *Clin. Oncol.* **4**, 139-44.

CENTRES FOR DISEASE CONTROL: CANCER AND STEROID HORMONE STUDY (CASH) (1983*a*). *J. Am. med. Ass.* **249**, 1596-9.

—— (1983*b*). *J. Am. med. Ass.* **249**, 1600-4.

CHAUDHURI, P.K., WALKER, M.J., BRIELE, H.A., BEATTIE, C.W., AND DAS GUPTA, T.K. (1980). Incidence of estrogen receptor in benign nevi and human malignant melnoma. *J. Am. med. Ass.* **244**, 791-3.

CHEN, J. AND RAISFELD-DANSE, I.H. (1983). Drug interactions II: Formation of Nitrosamines from therapeutic drugs. Properties and kinetics of the formation of N-nitrosopropanolol from nitrite and the secondary amine propranolol hydrochloride. *J. Pharmacol. Exp. Ther.* **225**, 705-12.

CHEN, W.S., KERKAY, J., PEARSON, K.H., PAGANINI, E.P., AND NAKAMOTO, S. (1979). Tissue bis(2-ethylhexyl) phthalate levels in uremic subjects. *Anal. Lett.* **12**, 1517-35.

CHRISSEY, M.D., STEELE, G.D., AND GITTES, R.F. (1980). Rat model for carcinogenesis in ureterosigmoidostomy. *Science* **207**, 1079-80.

——, ——, AND —— (1981). Ureterosigmoidostomy and colon carcinogenesis. *Science* **214**, 351.

COGGINS, P.R., RAVDIN, R.G., AND EISMAN, S.H. (1959). Clinical pharmacology and preliminary evaluation of cytoxan (cyclophosphamide). *Cancer Chemother. Rep.* **1**, 9-11.

COHEN, C.J. AND DEPPE, G. (1977). Endometrial carcinoma and oral contraceptive agents. *Obstet. Gynecol.* **49**, 390-2.

COHEN, R.J. (1980). Leukemia after therapy with methyl CCNU. *New Engl. J. Med.* **302**, 120.

COLLIN-JONES, D.G., LANGMAN, M.J.S., LAWSON, D.H., AND VESSEY, M.P., (1982). Cimetidine and gastric cancer preliminary report from post-marketing surveillance study. *Br. med.* **285**, 1311–13.

COLLINS, J., DONNER, A., ALLEN, L.H., AND ADAMS, O. (1980). Oestrogen use and survival in endometrial cancer. *Lancet* ii, 961–4.

CONTASTOVALOS, P.L. (1973). Benign hepatomas and oral contraceptives. *Lancet* ii, 1200.

COOPE, R. AND BURROW, R.G.R. (1940). Treatment of epilepsy with sodium diphenylhydantoinate. *Lancet* i, 490–2.

COULMAN, C.B., ANNEGERS, J.F., ABDOUD, C.F., *et al.* (1979). Pituitary adenoma and oral contraceptives: A case-control study. *Fertil. Steril.* **31**, 25–8.

COULSTON, F. AND DUME, I.F.(Eds) (1978). *The potential carcinogenicity of nitrosatable drugs.* WHO symposium, Geneva, June 1978. Ablex Publishing, Norwood.

COUNCIL OF ENVIRONMENTAL QUALITY (1980). *Drinking water and cancer: review of recent findings and assessment of risk.* Council of Environmental Quality, Washington, DC.

COURT-BROWN W.M. AND DOLL, R. (1957). *Leukaemia and aplastic anaemia in patients irradiated for ankylosing spondylitis.* Medical Research Council Special Report Series, no. 295. HMSO, London.

—— AND —— (1965). Mortality from cancer and other causes after radiotherapy for ankylosing spondylitis. *Br. med. J.* **2**, 1327–32.

CRADDOCK, V.M. (1983). Nitrosamines and human cancer: proof of an association? Nature **306**, 638.

CRAMER, D.W., HUTCHINSON, G.B., WELCH, W.R., SCULLY, R.E., AND KNAPP, R.C. (1982). Factors affecting the Association of oral contraceptives and ovarian cancer. *New Engl. J. Med.* **307**, 1047–51.

CURRENT PROBLEMS (1983). Human teratogenicity and a treatment for acne. *Current Problems*, no. 12.

DALAL, J.J., WINTERBOTTAM, T., WEST, R.R., AND HENDERSON, A.H. (1980). Letteri Implanted pacemakers and breast cancer. *Lancet* ii, 311.

DALE, G.A. AND SMITH, R.B. (1974). Transitional cell carcinoma of the bladder associated with cyclophosphamide. *J. Urol.* **112**, 603–4

DANESE, C.A., GRISHMAN, E.O.C., AND DREILING, D.A. (1967). Malignant vascular tumours of the lymphedematous extremity. Ann. Surg. **166**, 245–53.

DA SILVA HORTA, J., DA MOTTA, L.C., ABBATT, J.D., AND RORIZ, M.L. (1965). Malignancy and other latent effects following administration of Thorotrast. *Lancet* ii, 201–5.

DE CHOLNOKY, T. (1970). Augmentation mammaplasty. Survey of complications in 10, 941 patients by 265 surgeons. *Plastic reconstruct. surg.* **45**, 573–7.

DELGADO, E.R. (1958). Sarcoma following a surgically treated fractured tibia. A case report. *Clin. Orthopaed. rel. res.* **12**, 315–18.

DEPARTMENT OF HEALTH AND SOCIAL SECURITY (1984). Report of the panel of persons appointed by the licensing authority to hear the application by Upjohn Ltd for a product licence to market the drug Depo-Provera as a long-term contraceptive. ISBN-O-946539-32-4. HMSO, London.

DIGBY, J.M. AND WELLS, A.L. (1981). Malignant lymphoma with intranodal refractile particles after insertion of silicone prostheses. *Lancet* ii, 580.

DIGGLE, G.E. (1973). The processing of medical information. *Br. clin. J.* **I**(6), 30–3.

DOAK, P.B., MONTGOMERIE, J.Z., NORTH, J.D.K., AND SMITH, F. (1968). Reticulum cell sarcoma after renal homotransplantation azathioprine and prednisone therapy. *Br. med. J.* **4**, 746–8.

DOMELLOF, L., ERIKSSON, S., AND JANUNGER, K.G.(1976). Late precancerous changes and carcinoma of the gastric stump after Billroth I resection. *Am. J. Surg.* **132**, 26–31.

DOMSCHKE, S. AND DOMSCHKE, W. (1980). New histamine H_2-receptor antagonists. *Hepato-Gastroenterol.* **27**, 163–8.

DONADIA, J.V., HOLLEY, K.E., AND ILSTRUP, D.M. (1981). Adrenocorticoid and cytotoxic drug treatment of lupus nephropathy. *Proc. 8th Int. Cong. Nephrology*, Athens, pp. 642–8. Karger, Basle.

DOOUSS, T.W. AND CASTLEDEN, W.M. (1973). Gallstones and carcinoma of the large bowel. *NZ Med. J.* **77**, 162–5.

DOUGHERTY, S.H., FOSTER, C.A., AND EISENBERG, M.M. (1982). Stomach cancer following gastric surgery for benign disease. *Arch. Surg.* **117**, 294–7.

DRIPPS, R.D., LANORT, A., AND ECKENHOFF, J.E. (1961). The role of anaethesia in surgical mortality. *J. Am. med. Ass.* **178**, 261–6.

DUBE, V.E. AND FISHER, D.E. (1972). Hemangioendothelioma of the leg following metallic fixation of the tibia. *Cancer* **30**, 1260–6.

DU PLESSIS, D.J. (1962). Gastric mucosal changes after operations on the stomach. *S. Afr. med. J.* **36**, 471–8.

EBY, C.S., BRENNAN, M.J., AND FINE, G. (1967). Lymphangiosarcoma: a lethal complication of chronic lymphedema: report of two cases and review of the literature. *Arch. Surg.* **94**, 223–30.

EGBERT, B.M., KRAFT, J.K., AND PERKASH, I. (1980). Undifferentiated sarcoma arising in an augmented ileocystoplast patch. *J. Urol.* **123**, 272–4.

ELDER, J.B., GANGULI, P.C., AND GILLESPIE, L.E. (1979). Cimetidine and gastric cancer. *Lancet* i, 1005–6.

——, WELLS, S., KOFFMAN, C.G., GANULI, P.C., AND WILLIAMS G. (1982). Cimetidine and human gastric carcinoma. In *Nitrosamines and human cancer*, Banbury Report no. 12 (ed. P. N. Magee), pp. 335–49. Cold Spring Harbor, Cold Spring Harbor Laboratory, New York.

ELLERBROEK, W.C. (1968). Oral contraceptives and malignant melanoma. *J. Am. med. Ass.* **206**, 649–50.

ELLIOTT, R.W., ESSENHIGH, D.M., AND MORLEY, A.R. (1982). Cyclophosphamide treatment of systemic lupus erythematosis: risk of bladder cancer exceeds benefits. *Br. med. J.* **284**, 1160–1.

ELLIS, D.J., KINGSTON, R.D., BROOKES, V.S., AND WATERHOUSE, J.A.H. (1979). Gastric carcinoma and previous peptic ulceration. *Br. J. Surg.* **66**, 117–19.

ELWOOD, J.M. (1981). Estrogens and endometrial cancer: some answers and further questions. *Can. med. Ass. J.* **124**, 1129–31.

EPENETOS, A.A. (1980). Sunshine and malignant melanoma. *Br. med. J.* **280**, 112.

EXNER, J.H., DAHOD, S., AND POCHI, P.E. (1983). Pyogenic granuloma-like acne lesions during isotretinoin therapy. *Arch. Dermatol.* **119**, 808–11.

FALCK, K., GROHN, P., SORSA, M., VAIRUO, H., NEINONEN, E., AND HOLSTI, L.R. (1979). Mutagenicity in urine of nurses handling cytostatic drugs. *Lancet* i, 1250–1.

FARRELL, G.C., JOSHUA, D.E., UREN, R.F., BAIRD, P.J., PERKINS, K.W.,

AND KRONEBERG, H. (1975). Androgen induced hepatoma. *Lancet* i, 430–1.

FERMAGLICH, J. AND DELANEY, P. (1977). Parkinson's disease, melanoma and levodopa. *J. Neurol.* **215**, 221–4.

FIERZ, U. (1965). Katamnestiche Untersuchungen uber die Nebenwirkungen der Therapie von HautKrankeiten mit an organischem Arsen bei Hautkrankheit. *Dermatologica, Basel* **131**, 41–58.

FINDLAY, G.H. AND HULL, P.R. (1982). Letter: Eruptive tumours on sun-exposed skin after benoxaprofen. *Lancet* i, 95.

FINE, D.H., ROSS, R., ROUNBEHLER, D.P., SILVERGLEID, A., AND SONG, L. (1977). Formation in vivo of volatile N-nitrosamines in man after ingestion of cooked bacon and spinach. *Nature, London* **265**, 753–5.

FISHER, R.I., NEIFELD, J.P., AND LIPPMAN, M.E. (1976). Oestrogen receptors in human malignant melanoma. *Lancet* ii, 337–9.

FLINN, F.B. (1929). Dangers of radium taken internally. *J. Am. med. Ass.* **93**, 1581.

FOOD AND DRUG ADMINISTRATION (1981). Transcript of proceedings of Gastrointestinal Drugs Advisory Committee, 12 January 1981 (made available under US Freedom of Information Act.)

FOOD AND DRUG ADMINISTRATION DRUG BULLETIN (1978). Status report on beta-blockers. *Food Drug Admin. Drug Bull.* **8**, 13.

FURST, A. AND HARO, R.T. (1969). A survey of metal carcinogenesis. *Prog. exp. Tumor Res.* **12**, 102–33.

GARCIA, H., KEEFER, L., LIPINSKY, W., AND WENYON, C.E.M. (1970). Carcinogenicity of nitrosothiomorpholine and 1-nitrosopiperazine in rats. *Z. Krebsforsch.* **74**, 179–84.

GILLATT, P.N., HART, R.J., AND WALTERS, C.L. (1984). The susceptibilities of drugs to nitrosation under standardized chemical conditions. *Food Chem. Toxicol.* **22**(4), 269–74.

GILLAT, P.N., PALMER, R.C., SMITH, P.L.R., WALTERS, C.L. AND REED, P.I. (1985). *Food Chem. Toxicol.* (in press).

GIUSTRA, P.E., WHITE, H.O., AND KILLORAN, P.J. (1979). Intrathoracic desmoid at previous thoracotomy site. *J. Can. Ass. Radiol.* **30**, 122–3.

——, PALMER, R.C., SMITH, P.L.R., WALTERS, C.L., AND REED, P.I. (1985). *Food Chem. Toxicol.* [in Press].

GOEDERT, J.J., NEULAND, C.Y., WALLEN, W.C., GREEN, M.H., MANN, D.L., MURRAY, C., STRONG, D.M., FRAUMENI, J.F., AND BLATHNER, W.A. (1982). Amylnitrate may alter T-lymphocytes in homosexual men. *Lancet.* i, 412–15.

GOLDSTEIN, M.J., MELAMED, M.R., GRABSTALD, H., AND SHERLOCK, P. (1967). Progressive villous atrophy of the ileum used as a urinary conduit. *Gastroenterology* **52**, 859–64.

GRABSTALD, H. (1974). Carcinoma of ileal bladder stoma. *J. Urol.* **112**, 332–4.

GRAY, L.A., CHRISTOPHERSON, W.M., AND HOOVER, R.N. (1977). Estrogens and endometrial cancer. *J. obstet. Gynecol.* **49**, 385.

GREENBERG, G. (1976). Sarcoma after intramuscular iron injection. *Br. med. J.* **2**, 1508–9.

GREENE, M.H., BOICE, J.D., GREER, B.E., BLESSING, J.A., AND DEMBO, A.J. (1982). Acute non-lymphocytic leukaemia after therapy with alkylating agents for ovarian cancer. *New Engl. J. Med.* **307**, 1416–22.

——, HOOVER, R.N., AND FRAUMENT, J.F. (1978). Subsequent cancer in patients with chronic lymphocytic leukaemia, a possible immunological mechanism. *J. nat. Cancer Inst.* **61**, 337–40.

——, YOUNG, T.I., AND CLARK, W.H. (1981). Malignant melanoma in renal transplant recipients. *Lancet* i, 1196–8.

GREENWALD, P., BARLOW, J.J., NASCA, P.C., AND BURNETT, W.S. (1971). Vaginal cancer after maternal treatment with synthetic estrogens. *New Engl. J. Med.* **285**, 390–2.

GUY, J.T. AND AUSLANDE, M.O. (1973). Androgenic steroids and hepatocellular carcinoma. *Lancet* i, 148.

HABS, M., SCHMÄHL, D., EISENBRAND, G., AND PREUSSMANN, R. (1982a). Carcinogenesis studies with N-nitrosocimetidine (part 2). Oral adminitration to Sprague–Dawley rats. *Banbury Report* **12**, 403–5.

——, EISENBRAND, G., AND SCHMÄHL, D. (1982b). No evidence of carcinogenicity of N-nitrosocimetidine in rats, Hepatogastroenterol. **29**, 265–6.

HADDOW, A. (1974). Addendum to 'molecular repair, would healing and carcinogenesis: Tumor production a possible overhealing?' In *Advances in cancer research*, Vol. 20 (ed. G. Klein and S. Weinhouse), pp. 343–66. Academic Press, New York.

HAMAKER, W.R., LINDELL, M.E., AND GOMEZ, A.C. (1976). Plasmacytoma arising in a pacemaker pocket. *An. thorac. Surg.* **21**, 354–6.

HANSEN, N.E. (1979). Development of acute myeloid leukaemia in a patient with psoriasis treated with oral 8-methoxypsoralen and long-wave ultra violet light. *Scand. J. Haematol.* **22**, 57–60.

HARWOOD, A.R. AND YAFFE, M. (1982). Cancer in man after diagnostic or therapeutic irradiation. *Cancer Surv.* **1**(4), 703–31.

HAWKER, P.C., MUSCROFT, T.J., AND KEIGHLEY, M.R.B. (1980). Gastric cancer after cimetidine in patient with two negative pretreatment biopsies. *Lancet* i, 709–10.

HELSINGEN, N. AND HILLESTAD, L. (1956). Cancer development in the gastric stump after partial gastrectomy for ulcer. *Ann. Surg.* **143**, 173–9.

HENDERSON, J.T., RICHMOND, J., AND SUMERLING, M.D. (1973). Androgenic–anabolic steroid therapy and hepatocellular carcinoma. *Lancet* i, 934.

HERBST, A.L., KURMAN, R.J., SCULLY, R.E., AND POSKANZER, D.C. (1972). Clear-cell adenocarcinoma of the genital tract in young females. *New Engl. J. Med.* **287**, 1259–67.

—— AND SCULLY, R.E. (1970). Adenocarcinoma of the vagina in adolescence: a report of 7 cases including 6 clear-cell carcinomas (so-called mesonephromas). *Cancer* **25**, 745–7.

——, ULFELOER, H., AND POSKANZER, D.C. (1971). Adenocarcinoma of the vagina: association of maternal stilboestrol therapy with tumour appearance in young women. *New Engl. J. Med.* **284**, 878–81.

HILL, M.J., DRASAR, B.S., WILLIAMS, R.E.O., MEADE, T.W., COX, A.G., SIMPSON, J.E.P., AND MORSON, B.C. (1975). Faecal bile-acids and clostridia in patients with cancer of the large bowel. *Lancel* i, 535–9.

HILLMAN, L.S., GOODWIN, S.L., AND SHERMAN, W.R. (1975). Identification and measurement of plasticizer in neonatal tissue after umbilical catheters and blood products. *New Engl. J. Med.* **292**, 381–6.

HOCHBERG, M.C. AND SHULMAN, L.E. (1978). Acute leukemia following cyclophosphamide therapy for Sjogren's syndrome. *Johns Hopk. med. J.* **142**, 211–14.

HOGSTEDT, C., MALAQVIST, N, AND WADMAN, B. (1979). Leukemia in workers exposed to ethylene oxide. *J. Am. med. Ass.* **241**, 1132–3.

——, RÖHLÉN, O., BERNDTSSON, B.S., AXELSON, O., AND EHREN-BERG, L. (1978). Kohortstudie av dödsorsaker hos anställda i etylenoxidframställning. *Läkartidningen* **75**, 3285–7.

HOMBURGER, F., KELLY, T., FRIEDLER, G., AND RUSSFIELD, A.B. (1961). Toxic and possible carcinogenic effects of 4-allyl-1-2-methylene-dioxy benzene (Safrole) in rats on deficient diets. *Medicina Experimentalis* **4**(1), 1–11.

HOOVER, R.N. (1977). Effects of drugs: immunosuppression. In *Origins of human cancer* (ed. H.H. Hiatt, J.O. Watson, and J.A. Winston), pp. 369–79. Cold Spring Harbour Laboratory, New York.

HARVATH, E., KOVACS, K., AND ROSS, R.C. (1972). Ultrastructural findings in a well-differentiated hepatoma. *Digestion* **7**, 74–82.

HORWITZ, R.I. AND FEINSTEIN, A.R. (1979). Intravaginal oestrogen creams and endometrial carcinoma. *J. Am. med. Ass.* **241**, 1266–7.

HUNTER, D. (1978). *The diseases of occupations*, 6th edn. pp. 911–13. Hodder and Stoughton, London.

HUTCHINSON, J. (1887). Arsenic cancer. *Br. med. J.* **2**, 1280–1.

INTERNATIONAL AGENCY FOR RESEARCH ON CANCER (1979). *Evaluation of the carcinogenic risk of chemicals to humans; Monographs, Supplement I.* IARC, Lyon, France.

—— (1982). *Some industrial chemicals and dyestuffs.* IARC monographs on evaluation of the carcinogenic risk of chemicals to humans, no. 29. IARC, Lyon, France.

JAEGER, R.J. AND RUBIN, R.J. (1972). Migration of a phthalate ester plasticizer from polyvinyl chloride blood bags into stored human blood and its localization in human tissues. *New Engl. J. Med.* **287**, 1114–18.

JELINEK, J.E. (1970). Cutaneous side effects of oral contraceptives. *Arch. Dermatol.* **101**, 131–86.

JOHNSON, F.L., FEAGLER, J.R., LERNER, K.G., MAJERUS, P.W., SIEGEL, M., HARTMANN, J.R., AND THOMAS, E.D. (1972). Association of androgenic-anabolic steroid therapy with development of hepatocellular carcinoma. *Lancet* **ii**, 1273–6.

JOHNSON, K.H., GHOBRIAL, H.K.G., BUOEN, L.C., BRAND, I., AND BRAND, K.G. (1973). Nonfibroblastic origin of foreign body sarcomas implicated by histological and electron microscope studies. *Cancer Res.* **33**, 3139–54.

JOHNSON, M. AND LLOYD, H.E.D. (1974). Bilaterial breast cancer 10 years after an augmentation mammaplasty. *Plastic* reconstruct. Surg. **53**, 88–90.

JOHNSON, W.M. AND MEADOWS, D.C. (1971). Urinary bladder fibrosis and telangectasia associated with long-term cyclophosphamide therapy. *New. Engl. J. Med.* **284**, 390–4.

JOLY, D.J. LILIENFELD, A.M., DIAMOND, E.L., AND BROSS, I.D.J. (1974). An epidemiological study of the relationship of reproductive experience to cancer and the ovary. *Am. J. Epidemiol.* **99**, 190–209.

JØRGENSEN, K.A. (1982). Amyl nitrite and kaposi's sarcoma in homosexual men. *New Engl. J. Med.* **307**(14), 893.

JOSHI, R., SMITH, B., PHILLIPS, R.H., AND BARRETT, A.J. (1981). Acute myelomonocytic leukaemia after razoxane therapy. *Lancet* **ii**, 1343–4.

JOURNAL OF THE AMERICAN MEDICAL ASSOCIATION (1929). Editorial: Radioactive waters and solutions. *J. Am. med. Ass.* **93**, 771–2.

JUNGHANNS, K., SEUFERT, R., VON GERSTENBERGK, L., AND IVANKOVIC, S. (1979). Does vagotomy and pyloroplasty change the location of gastrointestinal tumours? *World J. Surg.* **3**, 497–500.

KELSO, D.R. (1974). Benign hepatomas and oral contraceptives. *Lancet* **i**, 315–16.

KERSEY, P. (1980). Possible cancer hazard associated with 5-methoxypsoralen in sun-tan preparations. *Br. med. J.* **280**, 940.

KINLEN, L.J., PETO, J., AND DOLL, R. (1981). Letter: Cancer in patients treated with immunosuppressive drugs. *Br. med. J.* **282**, 474.

——, SHEIL, A.G.R., PETO, J., AND DOLL, R. (1979). Collaborative United Kingdom–Australasian study of cancer in patients treated with immunosuppressive drugs. *Br. med. J.* **2**, 1461–6.

KNAPP, W.A. AND RUEBNER, B.H. (1974). Hepatomas and oral contraceptives. *Lancet* **i**, 270–1.

KOBAYASHI, S., PROLLA, J.C., AND KIRSNER, J.B. (1970). Later gastric carcinoma developing after surgery for benign conditions. Endoscopic and histologic studies of the anastomosis and diagnostic problems. *Am. J. dig. Dis.* **15**, 905–12.

KOSTAKIS, A.J., WHITE D. J.G., AND CALNE, R.Y. (1977). Prolongation of rat heart allograft survival by cyclosporin A. *Int. Res. Commun. System J. Med. Sci.* **5**, 280.

KRAEMER, K.H. AND WEINSTEIN, G.D. (1977). Decreased thymidine incorporation in circulating leukocytes after treatment of psoriasis with psoralen and long-wave ultraviolet light. *J. invest. Dermatol.* **69**, 211–14.

KRUCKEMEYER, K., LESSMANN, H.D., AND PUDWITZ, K.R. (1950). Nierenkarzinom als Thorotranstschaden. *Fortsch. Rontgenstr.* **93**, 313–21.

KÜNZE, E., SCHAUER, A., EDER, M., AND SEEFELDT, C. (1979). Early sequential lesions during development of experimental gastric cancer with special reference to dysplasias. *J. Cancer Res. clin. Oncol.* **95**, 247–64.

LANCET (1982a) Editorial: Cancer and chlorinated water. *Lancet* **i**, 1142.

—— (1982b) Editorial: Large-bowel cancer after cholecystectomy. *Lancet* **ii**, 562–3.

LARCINER, F.W., HARDIGREE, A.P., LIPINSKY, W., AND EPLER, I.L. (1980). Mutagenicity of N-nitrosospiperaine dirivatives in Saccharomyces cerevisiae *Mutation Res.* **77**, 143–8.

LAURSEN, B. (1970). Cancer of the bladder in patients treated with chlornaphazine. *Br. med. J.* **3**, 684–5.

LEADBETTER, G. W., JR., ZICKERMAN, P., AND PIERCE, E. (1979). Ureterosigmoidostomy and carcinoma of the colon. *J. Urol.* **121**, 732–5.

LEWIS, C.M. (1980). Inflammatory carcinoma of breast following silicone injections. *Plastic reconstruct. Surg.* **66**, 134–6.

LIEBERMAN, A.N. AND SHUPACK, J.L. (1974). Levodopa and melanoma. *Neurology* **24**, 340–3.

LIJINSKY, W. (1972). Carcinogenic nitrosamines formed from drug-nitrate interactions. *Br. J. Cancer* **10**, 114–22.

—— (1964). Reaction of drugs with nitrous acid as a source of carcinogenic nitrosamines. *Cancer Res.* **34**, 255–8.

—— AND GREENBLATT, M. (1972). Carcinogenic dimethylnitrosamine produced in vivo from nitrate and aminopyrine. *Nature New Biol.* **236**, 177–8.

—— AND TAYLOR, H.W. (1977a). *Cold Spring Habour symposium on the origins of human cancer*, Book C. Cold Spring Harbour Laboratory, New York.

—— AND —— (1977b). Feeding tests in rats on mixtures of nitrate with secondary and tertiary amines of environmental importance. *Food Cosmet. Toxicol.* **15**, 269–74.

LINOS, D.A., BEARD, C.M., O'FALLON, W.M., DOCKERTY, M.B., BEART

R.W., JR. AND KURLAND, L.T. (1981). Cholecystectomy and carcinoma of the colon. *Lancet* ii, 379–81.

LISCHKA, G., BOHNERT, E., BACHTOLD, G., AND JUNG, E.G. (1977). Effects of 8-methoxypsoralen (8-MOP) and UVA on human lymphocytes. *Arch. derm. Res.* **259**, 293–8.

LONG, E.L., NELSON, A.A., FITZHUGH, O.G., AND HANSEN, W.H. (1963). Liver tumours produced in rats by feeding safrole. *Arch. Pathol.* **75**, 33–42.

LOVE, L.A., LIPINSKY, W., KEEFER, L., AND GARCIA, H. (1977). Chronic oral administration of 1-nitrosopiperazin at high doses to MRC rats. *Z. Krebforsch.* **89**, 69–73.

MacDONALD, P.C. AND SIITERI, P.K. (1974). The relationship between the extraglandular production of estrone and the occurrence of endometrial neoplasia. *Gynecol. Oncol.* **2**, 259–63.

MACKINNON, A.E. AND BANCEWICZ, J. (1973). Sarcoma after injection of intramuscular iron. *Br. med. J.* **2**, 277–9.

MAGILLIGAN, H.S. (1929). Occupational poisoning in manufacture of luminous watch dials. *J. Am. med. Ass.* **92**, 466–73, 552–9.

—— (1931). The occurrence of malignancy in radioactive person. *Am. J. Cancer* **15**, 2435–516.

MAUGH, T.H. (1981). New study links chlorination and cancer. *Science* **211**, 694.

McCONNELL, E.M. HALSAM, P. (1959). Angiosarcoma in post-mastectomy lymphedema: A report of 5 cases and a review of the literature. *Br. J. Surg.* **46**, 322–32.

McDONALD, I. (1981). Malignant lymphoma associated with internal fixation of a fractured tibia. *Cancer* **48**, 1009–11.

McDONALD, T.W., ANNEGERS, J.F., O'FALLON, W.M., DOCKERTY, M.B., MALKASIAN, G.D., AND KURLORD, L.T. (1977). Exogenous estrogen and endometrial carcinoma — a case-controlled study. *Am. J. Obstet. Gynecol.* **127**, 572.

McDOUGALL, A. (1956). Malignant tumour at the site of bone plating. *J. Bone Joint Surg.*, British Volume, **38**, 709–13.

MELROSE, D. (1982). Bitter pills: medicines and the third world poor. Oxfam, Oxford.

MENGS, U. (1983). On the histopathogenesis of rat forestomach carcinoma caused by aristolochic acid. *Arch. Toxicol.* **51**, 209–20.

——, LANG, W., AND POCH, J.A. (1982). The carcinogenic action of aristolochic acid in rats. *Arch. Toxicol.* **51**, 107–19.

MEYLER, L. (1966). Radioactive isotopes. In *Side-effects of drugs*, Vol. V (ed. L. Meyler, and A. Herxheimer). Excerpta Medica Foundation, Amsterdam.

MILTON-THOMPSON, G.J., LIGHTFOOT, N.E., AHMET, Z., HUNT, R.H., BARNARD, J., BAVIN, P.M.G., BRIMBLECOMBE, R.W., DARKIN, D.W., MOORE, P.J., AND VINEY, H. (1982). Intragastric acidity, bacteria, nitrite and N-nitroso compounds before during and after cimetidine treatment. *Lancet* i, 1091–5.

MIRVISH, S.S., BULAY, O., RUNGE, R.G., AND PATIL, K. (1980). Study of the carcinogenicity of large doses of dimethylnitramine, N-nitroso-L-proline, and sodium nitrite administered in drinking water to rats. *Jinat. Cancer Inst.* **64**, 1435–42.

MODAN, B., BAIDATZ, D., MART, H., STEINITZ, R., AND LEVIN, S.G. (1974). Radiation-induced head and neck tumours. *Lancet* i, 277–9.

MONAGHAN, J.M. AND SIRISENA, L.A.W. (1978). Stilboestrol and vaginal clear-cell adenocarcinoma syndrome. *Br. med. J.* **1**, 1588–90.

MORGENSTERN, L. (1979). Invited commentary. *World J. Surg.* **3**, 499.

——, YAMAKAWA, T., AND SELTZER, D. (1973). Carcinoma of the gastric stump. *Am. J. Surg.* **125**, 29–38.

MORSON, B.C. (1955). Intestinal metaplasia of the gastric mucosa. *Br. J. Cancer* **9**, 365–85.

——, SOBIN, L.H., GRUNDMANN, E., JOHANSEN, A., NAGAYO, T., AND SERCK-HARSSEN, A. (1980). Recancerous conditions and epithelial displasia in the stomach. *J. clin. Pathol.* **33**, 711–21.

MURPHY, W.M. (1971). Tissue reaction of rats and guinea-pigs to Co–Cr implants with different surface finishes. *Br. J. Exp. Pathol.* **52**, 353–9.

NATIONAL ACADEMY OF SCIENCES (1980). *Drinking water and health*, Vol. 3 National Academy of Sciences, Washington DC.

NATIONAL TOXICOLOGY PROGRAM (1982). *Carcinogenesis bioassay of di(2-ethylhexyl) phthalate (CAS No. 1117–81–7) in F334 Rats and B6C3F1 mice (feed study)*. Technical Report Series, No. 217 (NH Publ. No. 82–1773). Research Triangle Park, North Carolina.

NIGRO, N.D., BHADRACHARI, N., AND CHOMCHAI, C. (1973). A rat model for studying colonic cancer: effect of cholestyramine on induced tumours. *Dis. Colon Rectum* **16**, 438–43.

OATES, R.K. AND TONGE, R.E. (1971). Phenytoin and pseudolymphoma syndrome. *Med. J. Aust.* **2**, 371–3.

O'CONNELL, T.X., FEE, H.J., AND GOLDING, A. (1976). Sarcoma associated with dacron prosthetic material. Case report and review of the literature. *J. thorac. cardiovasc. Surg.* **72**, 94–6.

OPPENHEIMER, B.S., OPPENHEIMER, E.T., STOUT, A.P., DANISHEFSKY, J., AND WILLHITE, M. (1959). Studies of the mechanism of carcinogenesis by plastic films. *Acta: Unio Int. contra Cancrum* **15**, 659–83.

O'SULLIVAN, J.P. AND WILDING, R.P. (1974). Liver harmartomas in patients on oral contraceptives. *Br. med. J.* **3**, 7–10.

OTT, G. (1970). Fremdkorpersarkome. *Exp. Med. Pathol. Klin.* **32**, 1–118.

PARKES, J.D. (1981). Adverse effects of antiparkinsonian drugs. *Drugs* **21**, 341–53.

PEILLON, E., VILA-PORCILE, E., OILVIER, L., AND RACADOT, J. (1970). L'action des oestrogens sur les adenomas hypophysaires chez l'homme. *Soc. d'Endocrinol.* **31**, 259–70.

PENN, I. (1982a). Malignancies following the use of cyclosporin-A in man. *Cancer Surv.* **1**(4), 621–4.

—— (1982b). Surgically induced malignancies. *Cancer Surv.* **1**(4), 746–61.

PENN, R.G. (1983). Adverse reactions to herbal medicines. *Adv. Drug Reaction Bull.*, No. 102.

PETERS, H. AND KEIMES, A.M. (1979). Die cholezystektomie als pradisponierender faktor in der genese des kolorektalen karzinoms? *Deutsche med. Wochenschr.* **104**, 1581–3.

PIK, J., BRAIER, J., KALMAN, Z., AND JOZEF, M. (1979). Determination of bis(2-ethylhexyl) phthalate (DEHP) in blood preparations stored in poly(vinyl chloride) bags (Rum.). *Probl. Gematol. Per. Krovi.* **24**, 49–51. [Chem. Abstr. **91**, 27213n].

POMARE, E.W. AND HEATON, K.W. (1973). The effect of cholecystectomy on bile salt metabolism. *Gut* **14**, 753–62.

QUINOT I. (1980). Mutagenicity of alkylnitrites in the Salmonella test. *Boll. Soc. Ital. Biol. Sper.* **56**(3), 816–20.

RAISFELD-DANSE, I.H. AND CHEN, J. (1983). Drug interactions III: formation of nitrosamines from therapeutic drugs. Formation, mutagenic properties and safety formation of N-nitrosopropranolol under conditions found in patients. *J. Pharmacol. Exp. Ther.* **225**, 713–19.

REDDY, B.S., WEISBURGER, J.H., AND WYNDER, E.L. (1974). Effects of dietary fat level and dimethylhydrazine on fecal acid and neutral sterold excretion and colon carcinogenesis in rats. *J. nat. Cancer Inst.* **52**, 507–11.

REED, P.I., CASSELL, P.G., AND WALTERS, C.L. (1979). Gastric cancer in patients who have taken cimetidine. *Lancet* i, 1234–5.

——, SMITH, P.L.R., HAINES, K., HOUSE, F.R., AND WALTERS, C.L. (1981*a*). Gastric juice and N-nitrosamines in health and gastro-intestinal disease. *Lancet* ii, 550–2.

——, ——, ——, ——, AND —— (1981*b*). Effect of cimetidine on gastric juice N-nitrosamine concentration. *Lancet* ii, 553–6.

——, SUMMERS, K., SMITH, P.L.R., WALTERS, C.L., AND HOUSE, F.R. (1982). The effect of Vitamin C on gastric juice nitrite and nitroso compound levels in achlorhydric subjects. *Scand. J. Gastroenterol.* **17**, (Suppl. 78), 239.

REGISTRY FOR RESEARCH ON HORMONAL TRANSPLACENTAL CARCINOGENESIS (1977). *Newsletter.* Chicago.

REIMER, R.R., HOOVER, R., FRAUMERI, J.F., AND YOUNG, R.C. (1977). Acute leukaemia after alkylating-agent therapy of ovarian cancer. *New Engl. J. Med.* **297**, 177–81.

REITZ, B.A. AND BIEBER, C.P. (1982). Cancer after use of cyclosporin-A in animals. *Cancer Surv.* **1**(4), 613–19.

RICH, S. (1975). Senate votes to ban D.E.S. *Washington Post*, September 10th.

ROBERTS, M.M. AND BELL, R. (1976). Acute leukaemia after immunosuppressive therapy. *Lancet* ii, 768–70.

ROBINSON, C.E.G., BELL, D.N., AND STURDY, J.H. (1960). Possible association of malignant neoplasm with iron-dextran injection. *Br. med. J.* **2**, 648–50.

ROBINSON, E., WAJSBORT, J., AND HIRSHOWITZ, B. (1973). Levodopa and malignant melanoma. *Arch. Pathol.* **95**, 213.

ROBSON, A.O. AND JELLIFFE, A.M. (1963). Medicinal arsenic poisoning and lung cancer. *Br. med. J.* **2**, 207–9.

ROMSDAHL, M.D. (1980). The national large bowel cancer project. Summary of the 1979 workshop. Approaches to prevention and treatment of large bowel cancer. *Cancer* **45**, 1264–71.

ROSA, F.W. (1983). Vitamin-A cogener teratology: isotretinoin (*Accutane*). *Adv. Drug Reactions Highlights*, No. 83–5.

ROSENBERG, L., SHAPIRO, S., SLONE, D., KAUFMAN, D.W., IDELMRICH, S.P., MIETHINEN, O.S., STOLLEY, P.D., ROSENSHEIN, N.B., SCHOTTENFELD, D., AND ENGLE, R.L. (1982). Epithelial ovarian cancer and combination oral contraceptives. *J. Am. med. Ass.* **247** 2310–12.

ROSSI, H.H. (1977). The effects of small doses of ionizing radiation: fundamental biophysical characteristics. *Radiat. Res.* **71**, 1–8.

ROWLAND, R.E. AND MARSHAL, J.H. (1959). Radium in human bone: The dose in microscopic volumes of bone. *Radiat. Res.* **11**, 299–313.

RUDDELL, W.S.J. (1979). Gastric cancer in patients who have taken cimetidine. *Lancet* i, 1234.

——, BONE, E.S., HILL, M.J., AND WALTERS, C.L. (1978). Pathogenesis of gastric cancer in pernicious anaemia. *Lancet* i, 521–3.

SAITO, T., SASAKI, O., IWAMATSU, M., TAMADA, R., AND INOKUCHI, K. (1980). Experimental gastric carcinoma induced by N-methyl-N-nitro-N-nitrosoguanidine. *J. Cancer Res. clin. Oncol.* **97**, 51–62.

SALTZSTEIN, S.L. AND ACKERMAN, L.V. (1959). Lymphadenopathy induced by anticonvulsant drugs and mimicing clinically and pathologically malignant lymphomas. *Cancer* **12**, 164.

SHADE, R.R. AND DONALDSON, R.M. (1981). How physicians use cimetidine. *New Engl. J. Med.* **304**, 1281–4.

SCHINDLER, A.E., EBERT, A., AND PRIEDRICH, E. (1972). Conversion of androstenedione to estrone by human fat tissue. *J. clin. Endocrinol. Metab.* **35**, 627–30.

SCHMÄHL, D. AND THOMAS, C. (1965). Erzeugung von Lungen-und Lebertumoren bei Mausen mit N, N-dinitrosopiperazine. *Z. Krebsforsch.* **67**, 11–15.

——, ——, AND AUER, R. (1977). *Iatrogenic carcinogenesis*, pp. 63–100. Springer-Verlag, Berlin.

SCHNECK, S.A. AND PENN, I. (1971). De novo brain tumours in renal transplant recipients. *Lancet* i, 913–16.

SCHOTCHER, S., SLKORA, K., AND FREEDMAN, L. (1981). Gastric cancer and cimetidine: Does delay in diagnosis matter? *Lancet* ii, 630–1.

SCHOTTENFELD, (1982). Cancer risks of medical treatment. *Cancer J. Clinicians* **32**(V), 258–79.

SCHREIBER, H., BARRY, F.M., RUSSELL, W.C., MACON, W.L., IV, PONSKY, J.L., AND PORIES, W.J. (1979). Stewart-Treves syndrome. A lethal complication of postmastectomy lymphedema and regional immune deficiency. *Arch. Surg.* **114**, 82–5.

SCOTT, E. (1983). *The people of the secret.* Octagon Press, London.

SCRIP (1981). ICI details tiotidine animal findings to US FDA Committee *SCRIP*, no. 557 p.15.

—— (1984). Lilly abandons Benoxaprofen. *SCRIP* no. 859.

SHAPIRO, A., BERLATZKY, Y., PFEFFERMAN, R., LIJOVETZKY, G., AND CAINE, M. (1979). Carcinoma of the colon after ureterocolic anastomosis. Implantation on calyceal mucosa. *Urology* **13**, 617–20.

SHAPIRO, S.R., BAEZ, A., COLODNY, A.H., AND FOLKMAN, J. (1974). Adenocarcinoma of colon at ureterosigmoidostomy site 14 years after conversion to ileal loop. *Urology* **3**, 229–31.

SHARPE, H.B.A., FIELD, E.O., AND HELLMANN, K. (1970). Mode of action of the cytostatic agent ICRF 159. *Nature* **226**, 524–6.

SHERLOCK, P. (1976). Colonic neoplasms complicating ureterosigmoidostomy. *Gastroenterology* **70**, 459–60.

SHERLOCK, S. (1968). *Diseases of the liver and biliary system*, 4th edn. Blackwell, Oxford.

SHIELDS, H.M. (1977). Occurrence of an adenocarcinoma at the choledochoenteric anastomosis 14 years after pancreatoduodenectomy for benign disease. *Gastroenterology* **72**, 322–4.

SHOUSHA, S., SCOTT, J., AND POLAK, J. (1978). Ileal loop carcinoma after cystectomy for bladder exstrophy. *Br. med. J.* **2**, 397–398.

SHY, K.K., MCTIERNAN, A.M., DALING, J.R., AND WEISS, N.S. (1983). Oral contraceptive use and the occurrence of pituitary prolactinoma. *J. Am. med. Ass.* **249**, 2204–7.

SIITERI, P.K., SCHWARTZ, B.E., AND MACDONALD, P.C. (1974). Estrogen receptors and the oestrone hypothesis in relation to endometrial and breast cancer. *Gynaecol. Oncol.* **2**, 228–38.

SILVERBERG, S.G. AND MAKOWSKI, E.L. (1975). Endometrial carcinoma in young women taking oral contraceptive agents. *Obstet. Gynecol.* **46**, 503–6.

——, ——, AND ROCHE, W.D. (1977). Endometrial carcinoma in

women under 40 years of age: comparison of cases in oral contraceptive users and non-users. *Cancer, NY* **39**, 592–8.

SIMON, N. (1977). Breast cancer induced by radiation. *J. Am. med. Ass.* **237**, 789–90.

SKIBBA, J.L., PINCKLEY, J., GIBLERT, E.F., AND JOHNSON, R.O. (1972). Multiple primary melanoma following administration of levodopa. *Arch. Pathol.* **93**, 556–61.

SMITH, D.C., ROSS, P., THOMPSON, D.J., AND HERRMANN, W.L. (1975). Association of exogenous estrogen and endometrial carcinoma. *New Engl. J. Med.* **293**, 1164–7.

SMITH, O.W. (1948). Diethylstilboestrol in the prevention and treatment of complications of pregnancy. *Am. J. Obstet. Gynecol.* **56**, 821–34.

SMITH, P.G. AND DOLL, R. (1982). Mortality among patients with ankylosing spondylitis after a single treatment course with x-rays. *Br. med. J.* **284**, 449–60.

SMITH, P.L.R, WALTERS, C.L., AND REED, P.I. (1983). Importance of selectivity in the determination of N-nitroso compounds as a group. *The Analyst* **108**, 896–8.

SNELL, R.S. AND BISCHITZ, P.G. (1960). The effect of large doses of estrogen and estrogen and progesterone on melanin pigmentation. *J. invest. Dermatol.* **35**, 73–82.

SOBER, A.J. AND WICK, M.M. (1978). Levodopa therapy and malignant melanoma. *J. Am. med. Ass.* **240**, 554–5.

—— AND —— (1979). Reply to letter: Levodopa and melanoma. *J. Am. med. Ass.* **241**, 883–4.

SOLOWAY, M.S., MYERS, G.H., JR., BURDICK, J.F., AND MALMGREN, R.A. (1972). Ileal conduit exfoliative cytology in the diagnosis of recurrent cancer. *J. Urol.* **107**, 835–9.

SOMMERS, S.C. AND MCMANUS, R.G. (1953). Arsenical tumours of the skin and viscera. *Cancer, Philadelphia* **6**, 547–9.

SOORIYAARACHCHI, G.S., JOHNSON, R.O., AND CARBONE, P.P. (1977). Neoplasms of the large bowel following ureterosigmoidostomy. *Arch. Surg.* **112**, 1174–7.

SORRELL, T.C., FORBES, I.J., BURNES, F.R., AND RISCHBIETH, R.H.C. (1971). Depression of immunological function in patients treated with phenytoin sodium. *Lancet* **ii**, 1233–5.

SPEAR, K.L AND MULLER, S.A.(1983). Treatment of cystic acne with 13-cis-retinoic acid. *Mayo Clin. Proc.* **58**, 509–14.

STEINITZ, R. (1965). Pulmonary tuberculosis and carcinoma of the lung. *Am. Rev. resp. Dis.* **92**, 758–66.

STEPHENS, T.C. AND CREIGHTON, A.M. (1974). Mechanism of action studies with ICRF 159: effects on the growth and morphology of BHK 21S cells. *Br. J. Cancer* **29**, 99–102.

STERN, R.S., THIBODEAU, L.A., KLEINERMAN, R.A., PARRISH, J.A., FITZPATRICK, T.B., AND 22 PARTICIPATING INVESTIGATORS (1979). Risk of cutaneous carcinoma in patients treatment with oral methoxsalen photochemotherapy for psoriasis. *New Engl. J. Med.* **300**, 809–13.

STEVENS, R.G., LEE, J.A., AND MOOLGAVKAR, S.H. (1980). Letter: No association between Oral contraceptives and malignant melanomas. *New Engl. J. Med.* **302**, 966.

STEWART, A., WEBB, J., GILES, D., AND HEWITT, D. (1956). Malignant disease in childhood and diagnostic irradiation in utero. *Lancet* **ii**, 447.

STEWART, F.W. AND TREVES, N. (1948). Lymphangiosarcoma in post-mastectomy lymphedema. A report of six cases in elephantiasis chirurgica. *Cancer* **1**, 64–81.

STEWART, H.L., DUNHAM, L.J., CASPER, J., DORN, H.F., THOMAS, L.B., EDGCOMB, J.H., AND SYMEONIDIS, A. (1966). Epidemiology of cancers of the uterine cervix and corpus, breast and ovary in Israel and New York City. *J. nat. Cancer Inst.* **37**, 195.

STINSON, N.E. (1964). The tissue reaction in rats and guinea-pigs by polymethylmetharcrylate (acrylic) and stainless steel (18/8/mo). *Br. J. Exp. Pathol.* **45**, 21–9.

STRAUSS, G.H. AND ALBERTINI, R.J. (1978). 6-Thioguanine-resistant peripheral blood lymphocytes in patients with psoriasis receiving 8-methoxypsoralen and long-wave ultraviolet light treatment (PUVA). Abstracts of Enivronmental Mutagen Society Meeting, San Francisco.

SUCKOW, E.E., HENEGAR, G.C., AND BASERGA, R. (1961). Tumours of the liver following administration of Thorotrast. *Am. J. Pathol.* **38**, 663–77.

SWANSON, H.A. AND SCHWARTZ, R.S. (1967). Immunosuppressive therapy. *New Engl. J. Med.* **277**, 163–70.

SWEDISH DEPARTMENT OF DRUGS (1982). Information fran socialstyrelsens lakemedelsavdelning: Vol. 7. Nr. 7. p.5. P.O. Box 607, 751 25 Uppsala, Sweden. (11 November 1982).

SYPKENS-SMIT, G.C. AND MEYLER, L. (1970). Acute myeloid leukaemia after treatment with cytostatic agents. *Lancet* **ii**, 671–2.

TAKSDAL, S. AND STALSBERG, H. (1973). Histology of gastric carcinoma occurring after gastric surgery for benign conditions. *Cancer* **32**, 162–6.

TASWELL, H.F., SOULE, H.E., AND COVENTRY, M.B. (1962). Lymphangiosarcoma arising in chronic lymphedematous extremities. Report of thirteen cases and review of literature. *J. Bone Joint Surg.* **44A**, 277–94.

TAYLOR, H.W. AND LIJINSKY, W. (1975). Tumour induction in rats by feeding aminopyrine or oxytetracycline with nitrite. *Int. J. Cancer* **16**, 211–15.

TAYLOR, T.V., LEE, D., HOWATSON, A.G., ANDERSON, J., AND MACLEOD, I.B. (1979). Gastric cancer in patients who have taken cimetidine. *Lancet* **i**, 1235–6.

TCHERNIA, G., MIELOT, F., SUBTIL, E., AND PARMENTIER, K. (1976). Acute myeloblastic leukaemia after immunodepressive therapy for primary non-malignant disease. *Nouv. Rev. franc. Hematol. Blood Cells* **17**, 67–80.

TERRELL, T., GRIBBLE, D., AND OSBURN, B. (1980). Malignant lymphoma in macaques: a clinicopathologic study of 45 cases. *J. nat. Cancer Inst.* **63**, 561–6.

THIEDE, T., CHIEVITZ, E., AND CHRISTENSEN, B.C.H. (1964). Chlornaphazine as a bladder carcinogen. *Acta med. scand.* **175**, 721–5.

THIESS, A.M., FRENTZEL-BEYME, R., AND WIELAND, R. (1978). Mortality study in workers exposed to di-2-ethylhexylphthalate (DOP) (Ger.). In *Möglichkerten and Grenzen des Biological Monitoring. Arbeitsmedizinische Probleme des Dienstleistungsgewerbes. Arbeitsmedizinisches Kolloquium* [Possibilities and limits of biological monitoring. Problems of occupational medicine in small industries. Colloquium in Occupational Medicine], pp. 155–64. A.W. Gentner, Stuttgart.

THIRU, S., CALNE, R.Y., AND NAGINGTON, J. (1981). Lymphoma in renal allograft patients treated with cyclosporin-A as one of the immunosuppressive agents. *Trans. Proc.* **13**, 359–63.

THOM, M.H., WHITE, P.J., WILLIAMS, R.M., STURDEE, D.W. PATTERSON, M.E.L., WADE-EVANS, T., AND STUDD, J.W.W. (1979). Prevention and treatment of endometrial disease in climacteric women receiving oestrogen therapy. *Lancet* **ii**, 455–7.

THOMPSON, R.J. AND ENTIN, S.D. (1969). Primary extraskeletal chondrosarcoma. Report of a case arising in conjunction with extrapleural Lucite ball plombage. *Cancer* **23**, 936–9.

TIGANO, F., FERLAZZO, B., AND BARRILE, A., (1976). Oral contraceptives and malignant hepatoma. *Lancet* ii, 196.

UNRUH, H., ROBERTSON, D.J., AND KARESEWICH, E. (1979). Postmastectomy lymphangiosarcoma: Experience with three patients and electron microscopic observations in one. *Can. J. Surg.* 22, 586-90.

UNSCEAR (1977). *Ionizing radiation levels and effects*, Vol. 11. Effects. Annex G, 361-423. (A report of the United Nations Scientific Committee on the effects of atomic radiation to the General Assembly, with annexes). United Nations, New York.

UPTON, A.C. (1975). Physical carcinogenesis: radiation — history and sources. In *Cancer, a comprehensive treatise* (ed. F.F. Becker), Vol. 1, pp. 387-403. Plenum Press, New York.

—— (1977). Radiological effects of low doses. *Radiat. Res.* 71, 51-74.

URDANETA, L.F., DUFFELL, D., CREEVY, C.D., AND AUST, J.B. (1966). Late development of primary carcinoma of the colon following ureterosigmoidostomy: Report of three cases and literature review. *Ann. Surg.* 164, 503-13.

VAISRUB, S. (1979). Pituitary prolactinoma and estrogen contraceptives. *J. Am. med. Ass.* 242, 177-8.

VALAGUSSA, P., SANTORO, A., KENDA, P., BELLANI, F.F., FRANCHI, F., BANFI, A., RILKE, E., AND BONADONNA, G. (1980). Second malignancies in Hodgkin's disease: a complication of certain forms of treatment. *Br. med. J.* 280, 216-19.

VALENTIC, J.P, BARR, R.J, AND WEINSTEIN, G.D. (1983). Inflammatory neovascular nodules associated with oral isotretinoin treatment of severe acne. *Arch. Dermatol.* 119, 871-2.

VAN RENS, G.H., DE JONG, P.T.V.M., DE MOLS, E.E.J.L.R., GEERTRUYDEN, M.F.B. (1982). Uveal malignant melanoma and L-dopa therapy in Parkinson's disease *Opthalmology* 89(12), 1464-6.

VERNICK, L.J. AND KULLER, L.H. (1981). Cholecystectomy and right-sided colon cancer: an epidemiological study. *Lancet* ii, 381-3.

——, ——, LOHSOONTHORN, P., RYCHECK, R.R., AND REDMOND, C.K. (1980). Relationship between cholecystectomy and ascending colon cancer. *Cancer* 45, 392-5.

VESSEY, M.P., LAWLESS, M., MCPHERSON, K., AND YEATES, D. (1983). Neoplasia of the cervix uteri and contraception: a possible adverse effect of the Pill. *Lancet* ii, 930-4.

VIDEBAECK, A. (1964). Chlornaphazine (Erysan-R) may induce cancer of the urinary bladder. *Acta. med. scand.* 176, 45-50.

WAGNER, J., MANTHORPE, R., PHILIP, P., AND FROST, F. (1978). Preluekaemia developing in a patient with psoriasis treated with 8-methoxypsoralen and ultra violet light (PUVA treatment). *Scand. J. Haematol.* 21, 299-304.

WAJSMAN, Z., BAUMGARTNER, G., AND MERRIN, C. (1975). Transitional cell carcinoma of ileal loop following cystectomy. *Urology* 5, 255-6.

WALDER, B.K., ROBERTSON, M.R., AND JEREMY, D. (1971). Skin cancer and immunosuppression. *Lancet* ii, 1282-3.

WALL, R.L. AND CLAUSEN, K.P. (1975). Carcinoma of the urinary bladder in patients receiving cyclophosphamide. *New Engl. J. Med.* 293, 271-3.

WALTERS, C.L. (1977). *Proceedings of Fourth Meeting on analysis and formation of N-nitroso compounds*. International Agency for Research on Cancer, Lyons.

WASHINGTON POST (1983). Eli Lilly Notifies FDA and Doctors of Cancer Incidence in Oraflex Study. *Washington Post*, pp. Al, A8, 24 December 1983.

WATTENBURG, L.W. (1959). Histochemical study of aminopeptidase in metaplasia and carcinoma of the stomach. *Arch. Pathol.* 67, 281-6.

WEBB, K.S., WOOD, B.J., AND GOUGH, T.A. (1983). The effect of the intake of a nitrosatable drug on the nitrosamine levels in human urine. *J. anal. Toxicol.* 7, 181-4.

WEBSTER, E.W. (1981). On the question of cancer induction by small x-ray doses. *Am. J. Roentgenol.* 137, 647-66.

WEINBERG, D.S. AND MAINI, B.S. (1980). Primary sarcoma of the aorta associated with a vascular prosthesis: A case report. *Cancer* 46, 398-402.

WEINBREN, K., SALM, R., AND GREENBERG, G. (1978). Intramuscular injections of iron compounds and oncogenesis in man. *Br. med. J.* 1, 683-5.

WEISMAN, R.A. AND KONRAD, H.R. (1979). Tracheal carcinoma after tracheostomy. *Arch. Otolaryngol.* 105, 364-6.

WEBSTER, E.W. (1980). Estimates of cancer risks from low-level exposure to ionising radiation. In *Biological risks of medical irradiations* (ed. G.D. Fullerton). American Institute of Physics.

WEISS, N.S., LYON, J.L., LIFF, J.M., VOLLMER, W.M., AND DALING, J.L. (1981). Incidence of ovarian cancer in relation to the use of oral contraceptives. *Int. J. Cancer* 28, 669-71.

—— AND SAYVETZ, T.A. (1980). Incidence of endometrial cancer in relation to the use of oral contraceptives. *New Engl. J. Med.* 302, 551-4.

WERNER, B., DE HEER, K., AND MITSCHKE, H. (1977). Cholecystectomy and carcinoma of the colon. An experiment study. *Z. Krebsforsch.* 88. 223-30.

WORLD HEALTH ORGANISATION (1978). Informal consultation on the potential carcinogenicity of nitrostable drugs. Report of a WHO meeting, Geneva 12-16 June 1978.

—— (1981). *Aristolochic acid*, Report no. PHAM/10/832 (19 August 1981). World Health Organisation, Geneva.

WICK, M.M. (1977). L-Dopa methyl ester as a new antitumour agent. *Nature* 269, 512-13.

——, BYERS, L., AND RATCLIFF, J. (1979). Selective toxicity of 6-hydroxydopa for melanoma cells. *J. invest. Dermatol.* 72, 67-9.

WILLETT, W.C., BAIN, C., HENNEKEUS, C.H., ROSNER, B., AND SPEIZER, F.E. (1981). Oral contraceptives and risk of ovarian cancer. *Cancer* 48, 1684-7.

WILLIAMS, G. AND ASKHEN, M.H. (1980). Urethral carcinoma following urethroplasty. *J. R. Soc. Med.* 73, 370-1.

WILLIAMS, R.R. (1976). Breast and thyroid cancer and malignant melanoma promoted by alcohol-induced pituitary secretion of prolactin, T.S.H. and M.S.H. *Lancet* ii, 996-9.

WILLIAMSON, R.C.N., BAUER, F.L.R., ROSSE, J.S., WATKINS, J.B., AND MALT, R.A. (1979). Enhanced colonic carcinogenesis with Azyoxymethane in rats after pancreaticobiliary diversion to mid-small bowel. *Gastroenterology* 76, 1386-92.

WILLIGHAGEN, R.G.J. AND PLANTEYDT, H.T. (1959). Aminopeptidase activity in cancer cells. *Nature* 183, 263-4.

WINGRAVE, S.J. KAY, C.R., AND VESSERY, M.P. (1980). Oral contraceptives and pituitary adenomas. *Br. med. J.* 1, 685-6.

WOODWARD, A.H., IVINS, J.C., AND SOULE, E.H. (1972). Lymphangiosarcoma arising in chronic lymphedematous extremities. *Cancer* 30, 562-72.

WORTH, P.H.L. (1971). Cyclophosphamide and the bladder. *Br. med. J.* 3, 182.

ZAFIRACOPOULOS, P. AND ROUSKAS, A. (1974). Breast cancer at site of implantation of pacemaker generator. *Lancet* i, 1114.

ZBINDEN, G. (1983). Risk assessment of drug carcinogenicity. *Trends pharmacol. Sci.* **4**(4), 146–7.

ZEIGENFUSS, J. AND CARBABSI, R. (1973). Androgens and hepatocellular carcinoma. *Lancet* **i**, 262.

ZEIGER, E. AND SHELDON, A.T. (1978). The mutagenicity of hetero-cyclic N-nitrosamines for Salmonella typhimurium. *Mutation Res.* **57**, 1–10.

ZIEL, H.K. AND FINKLE, W.D. (1975). Increased risk of endometrial carcinoma among users of conjugated estogens. *New Engl. J. Med.* **293**, 1167–70.

36 Drug-, device-, or procedure-induced or aggravated infective conditions

P.F. D'ARCY

The potential hazard from drug-induced or aggravated infective conditions is largely confined to four classes of drugs: antibiotics; corticosteroids; immunosuppressive and antineoplastic agents; and oral contraceptives. A search of the recent literature has, however, suggested that the emphasis on causative agents for induced or aggravated infection has moved away from these traditional areas towards, for example, infection and toxic shock with intrauterine contraceptive devices and vaginal tampons, and the more frequent problem of iatrogenic septicaemia from the use of intravenous fluid infusions and total parenteral nutrition solutions. Co-trimoxazole-induced complications in the diagnosis of malaria, Reye's syndrome and aspirin, and the complication of travellers' diarrhoea by drugs are all relatively new facets of drug-aggravated infection and some description of them is included in this review.

Nasotracheal intubation, bladder catheterization, and colonic irrigation have all been incriminated in infective conditions and this reminder that devices and clinical procedures can also induce or aggravate infective conditions, in the absence of drugs, has determined the change in the title of this present chapter from that of the previous edition; it now embraces devices and procedures as well as drugs.

Antibiotics

Three problems, concerning the role of antibiotics in induced or aggravated infective conditions, are highlighted in this chapter. The first is the continuing saga of superinfection, the second is the related problem of antibiotics and colitis, while the third, which is far more serious and far-reaching, is the current continuing upsurge in the development of antibiotic-resistant strains of bacteria. This latter problem is little helped by poor antibiotic-control policies in many countries and the spread of antibiotic-resistant organisms is facilitated by the relative ease of world travel.

Superinfection

Even in the early days of antibiotic therapy, reports (e.g. Weinstein 1947) appeared on the development of new infections due to organisms that were not susceptible to the agent being used to treat the primary disease. It appeared that, while one group of microorganisms were suppressed, others that were present multiplied more rapidly, invaded tissue, and caused a superinfection. For example, Harris (1950), reported on 91 female and 44 male patients treated with *Aureomycin* (chlortetracycline) and/or chloramphenicol for chronic brucellosis. These antibiotics, although effective in curing the brucellosis, produced in a large percentage of the female patients (60.7 per cent) and in a smaller percentage of the males (19.3 per cent) side-effects of a type and severity not previously reported. These included lesions of the mucous membranes of the mouth, pharynx, cheeks, lips, and tongue. In all of the 24 female patients with oral lesions, there was an accompanying vaginitis and vulval and anal irritation. *Candida albicans* was isolated in culture in all of the patients treated with *Aureomycin*.

Woods and his co-workers (1951) observed 25 cases of moniliasis developing as a result of administration of penicillin, chloramphenicol, or *Aureomycin*. Tomaszewski (1951) studied 126 patients treated with chloramphenicol and *Aureomycin*; side-effects were more pronounced in women than in men and were attributable to *Candida albicans* infestation.

It was observations of this type that caused the Council on Pharmacy and Chemistry of the American Medical Association, in April 1951, to request that a warning statement be added to the labelling of *Aureomycin*, chloramphenicol, and *Terramycin* (oxytetracycline). The statement was to the effect that, while these antibiotics were highly bacteriostatic for many bacteria, they could cause suppression of susceptible bacteria and thus encourage the replacement by *Monilia* and other yeast-like organisms of the normal or abnormal bacterial flora.

Throughout the intervening years such side-effects of antibiotic therapy have become well recognized and, although superinfection associated with antibiotic usage still presents some hazard, recent evidence suggests that it is an infrequent and mostly only a minor problem.

In a series of 14 077 hospitalized medical patients receiving antibiotics, superinfection developed in 95 (0.7 per cent) during drug therapy; these figures drawn from the records of the Boston Collaborative Drug Surveillance Program (Walker *et al.* 1979) are one of the few indications of epidemiologic data available concerning endemic hospital-acquired infection. This survey provides data on host characteristics, infecting agents, and severity of infection. Patients 60 years or older had a risk of superinfection slightly higher than that of younger patients (0.8 vs. 0.6 per cent, respectively), and women were at a slightly higher risk than men. Among 12 362 patients in whom admission BUN level was recorded, the superinfection rate increased progressively with higher BUN levels - 0.6 per cent at levels less than 25 mg dl^{-1} to 1.3 per cent for levels at 50 mg dl^{-1} or greater. Immunosuppressed patients on antimetabolites or corticosteroids tended to have infections more commonly than others.

Among the 95 patients in whom a superinfection developed, 101 presumed infecting organisms were identified; the overwhelming majority (71 per cent) were yeasts and fungi, principally *Candida* species. Gram-positive organisms accounted for 8 per cent of the infections and gram-negative made up the remainder; none of the infecting organisms appeared to be endemic in any of the hospitals studied.

Thirteen per cent of the infections recorded were considered to be major by the attending physicians and about one-third were thought to be of only minor importance. Three patients died of their superinfections; the causative organisms and types of infection were: *C. albicans* pneumonia, gram-negative septicaemia, and chloramphenicol-resistant *Escherichia coli* sepsis.

The survey also provided some information on the site of infection; the great majority of fungal and yeast infections were located in the oropharynx, vagina, or urinary tract and bore no regular association to the site of initial infection. This was not so, however, for bacterial infections, where generally the site of the superinfection occurred at the site of the original infections for which antibiotic treatment had been instituted.

Antibiotic combination with penicillin did not lead to a greater incidence of superinfection than did penicillin alone.

Walker and his colleagues, in concluding their report urged caution in interpreting their results since they reflected endemic, not epidemic, infection rates. It was not possible, they warned, to infer from their data information on superinfection rates during hospital epidemics with virulent or highly resistant organisms.

Colitis

Colitis associated with antibiotics is a continuing conundrum; as they suppress normal gut flora so may they also predispose some patients to colitis caused by *Clostridium difficile*. Normally *Cl. difficile* is found in the stools of less than 2 per cent of healthy adults.

Clindamycin and lincomycin accounted for 80 per cent of all antibiotic-associated colitis cases reported to the Committee on Safety of Medicines in Britain between 1964 and 1978, although ampicillin, tetracycline, chloramphenicol, co-trimoxazole, cephalosporins, penicillin, and metronidazole have been associated with other cases.

Clostridium difficile and its toxin are apparently closely involved not only in pseudomembranous colitis but also in antibiotic-associated colitis and in diarrhoea without pseudomembranous changes. There is a complication, however, in that it has also been suggested that *Cl. difficile* toxin may contribute to relapse in some cases of inflammatory bowel disease even when no antibiotic has recently been taken. Former treatment with corticosteroids may also complicate the outcome. For example, Bolton *et al.* (1980) showed that nine of 56 patients with diarrhoea had a positive test for *Cl. difficile*; five of these patients had severe inflammatory bowel disease and were on systemic corticosteroids, two had been on steroids for other conditions, one had been on antibiotics, and one had no apparent predisposing factor. In each case clearance of the toxin was associated with clinical improvement. La Mont and Trnka (1980) reported the presence of *Cl. difficile* toxin in the stools of six patients with chronic inflammatory disease during symptomatic relapse. Only two of these patients had taken antibiotics known to cause pseudomembranous colitis and none had pseudomembranes; all patients improved when the toxin disappeared.

These two latter papers raise the question of whether inflammation of the bowel, caused by some other factors, favours the colonization by *Cl. difficile* or whether multiplication of the organism caused by some other factor is responsible for the inflammation (*Lancet* 1980). It is also questioned whether factors are involved in the pathogenesis of the disorder other than the simple presence of a certain concentration of toxin in the colon. Lishman *et al.* (1981) found that none of 26 patients with ulcerative colitis, none of eight patients with non-specific diarrhoea, and none of the 27 control subjects had *Cl. difficile* or its toxin in their stools. However, stool samples from 10 out of 53 patients who had mild-to-severe diarrhoea during or following a course of antibiotics contained *Cl. difficile* toxin with titres from 10^{-2} to 10^{-5}. Antibiotics or antibacterials taken by these previously fit patients were clindamycin, amoxycillin, flucloxacillin, cephradine, or co-trimoxazole. Only one patient had histologically proven pseudomembranous colitis, while two had severe colitis without pseudomembranes.

Also of significance in this latter study, the stools of four of the 53 control subjects, who had taken antibiotics for at least seven days but did not develop diarrhoea, also contained *Cl. difficile* toxin in a concentration of 10^{-4} to 10^{-5}. The authors of the report suggested that perhaps some adults retained the resistance to the toxin seen in infants, or perhaps they subsequently acquired an immunity.

A further question was asked by Greenfield *et al.* (1981): is pseudomembraneous colitis infectious? A cluster of four male and four female patients in two adjacent 28-bed medical wards developed acute diarrhoea within an 11-day period. All their stool samples contained *Cl. difficile* toxin and the organism was isolated from each one; tests for *Shigella*, *Campylobacter*, *Salmonella*, pathogenic *E. coli*, and ova cysts and parasites were negative. The three patients given rectal biopsies had findings consistent with pseudomembranous colitis. Four of the patients had neither received antibiotics nor had they undergone surgery but two of them had been given corticosteroids. The two patients with no predisposing cause for *Cl. difficile* infection both had bloody diarrhoea, abnormal sigmoidoscopy findings, and rectal biopsy findings compatible with pseudomembranous colitis. All patients responded to oral vancomycin (250 or 500 mg) and none relapsed. All patients were potentially in contact and were attended by the same doctors and students.

The treatment of pseudomembranous colitis has also been under fire by Ahmad (1980); the use of the antiperistaltic agent *Lomotil* (diphenoxylate with atropine) may prolong and worsen the condition by retarding clearance of clostridial toxin in the stools and reports in the literature cited by this author indicate a worse prognosis in pseudomembranous colitis patients given *Lomotil*. He warns that *Lomotil*, corticosteroids, and corticotrophin have no place in the treatment of antibiotic-associated colitis.

A similar stricture was advocated by Morrison and Little (1981) following the death of a severely dehydrated 22-month-old child suffering from vomiting and diar-

rhoea; he had been treated with diphenoxylate plus atropine and had been ill for three days. It was suggested that antidiarrhoeal preparations were ineffective and even harmful. Kaolin may facilitate viral penetration into the gastric mucosa, while opiates and diphenoxylate plus atropine may encourage pathogen proliferation by delaying the passage of liquid stools.

One may justifiably conclude that there is a surfeit of unanswered questions on the relationship between some antibiotics, *Cl. difficile* and its toxin, and colitis. It is evident that there is such a tripartite relationship although it is clearly evident that other factors, as yet undetermined, are also involved in the great colitis mystery. In particular, there is a history of corticosteroid treatment, in a number of cases where tests for *Cl. difficile* are positive (Bolton *et al.* 1980; Greenfield *et al.* 1981).

Currently vancomycin seems to be the best treatment for antibiotic-induced pseudomembranous colitis (Batts *et al.* 1980), although relapse may be a problem after stopping this antibiotic (Bowman and Riley 1980; Rampling *et al.* 1980; Roberts and Seneviratne 1980). Oral metronidazole may also be a suitable and less expensive alternative (Bolton 1980).

Antibiotic resistance

A leading article in the *Lancet* (1981), 'Gentamicin and staphylococci', is a timely warning of the way in which antibiotics themselves, by inducing bacterial resistance, can worsen the whole spectrum of the treatment of infective disease.

In the 20 years which followed the introduction of antibiotics, hospital-acquired infections due to *Staphylococcus aureus* became a curse, especially since the organism was often resistant to penicillin and other antibiotics. Then in the late-1960s this trend went into reverse probably because of the composite influence of control of infection procedures, restraint in the use of antibiotics, availability of beta-lactamase stable penicillins, and perhaps also due to a natural reduction in the virulence and transmissibility of *Staph. aureus*.

Multiple antibiotic resistance means an enhanced risk not only to the individual patient but also to the other patients; it is not surprising therefore that the leading article expressed concern that the cycle may once again be repeated. First reports of infection due to gentamicin-resistant *Staph. aureus* emerged some five years ago and more recently a report from Dublin (Hone *et al.* 1981) revealed that between 1976 and 1979, 55 patients in Dublin hospitals had a bacteraemia due to gentamicin-resistant strains of *Staph. aureus* that were resistant to penicillin, methicillin, kanamycin, tobramycin, and erythromycin as well. Seventy-six per cent of these strains were resistant to fusidic acid, and 94 per cent were sensitive to amikacin and chloramphenicol.

The Dublin patients were in 12 hospitals with 3047 beds and all the major specialties were represented in the series; 19 of the 55 patients died and in 14 of these *Staph. aureus* septicaemia was deemed to be the cause of death. Seven of the fatal infections began at intravenous infusion or shunt sites and of the non-fatal infections, two began in subclavian lines and eight in i.v. infusion or shunt sites.

This epidemic, according to Hone and his colleagues, resulted from a multiplicity of factors but it was clear that a lack of isolation facilities and difficulty in operating a uniform and effective antibiotic policy throughout that group of Irish hospitals bore much of the blame (McCormack and Hone 1981).

Unfortunately, however, this Irish epidemic is by no means an isolated occurrence; antibiotic resistance is being reported around the world and inadequate hospital and general-practice antibiotic-prescribing policies are undoubtedly part of the cause for this (Mabeck 1982; Grimwood *et al.* 1983; Jewesson *et al.* 1983; Swindell *et al.* 1983).

In the midwestern United States, an outbreak of infection was caused by strains of *Staph. aureus* resistant to methicillin and aminoglycoside antibiotics (Crossley *et al.* 1979). In America also, Nelson (1980) commented with concern that the frequency of beta-lactamase-producing strains of type β *H. influenzae* isolated from children with meningitis or arthritis has increased steadily from 5 per cent in 1975 to 22 per cent in 1979. Five of the 25 isolates of *H. influenzae* from patients with aryepiglottis in the previous two years were beta-lactamase positive; he emphasized that ampicillin-resistant *H. influenzae* occurred in all areas of the USA.

In New Mexico and California, beta-lactamase-producing gonococci have complicated the treatment of gonococcal urethritis and endocervical infection (Borchers *et al.* 1980), as also with gonococcal infection of the eyes in newborn children in Britain (Dunlop *et al.* 1980). Penicillin-resistant pneumococci are on the increase in New Guinea and serotyping of 57 strains of pneumococci from patients with severe pneumococcal infections identified 21 types, of which 10 were penicillin-resistant (Gratten *et al.* 1980).

In Britain, Gross *et al.* (1981) reported that 2370 strains of *Shigella dysenteriae, Sh. flexneri*, and *Sh. boydii* isolated in England and Wales from 1974 to 1978 were tested for resistance to 12 antimicrobial drugs. Eighty per cent of the strains were resistant to one or more drugs with sulphonamide resistance occurring most frequently. Resistance to streptomycin, tetracyclines,

ampicillin, and chloramphenicol increased during this period, as did the incidence of multiple resistance. The proportion of infection due to these subgroups increased from 2.4 per cent of all shigella infection in 1965 to 16 per cent in 1978.

Most infections due to *Sh. dysenteriae, Sh. flexneri*, and *Sh. boydii* are acquired abroad (Gross *et al.* 1979) notably on the Indian subcontinent and in the Mediterranean countries of North Africa, and the increasing incidence of drug resistance amongst these organisms contrasts with the decreasing incidence of resistance among the indigenous *Sh. sonnei*. The lesson to be drawn from this is fairly obvious — it indicates the need for better control of antibiotic usage particularly in developing countries where antibiotic and antibacterial use in hospital or the community is often subject to little or no restriction.

The eternal battle with drug-resistant tuberculosis continues and an editorial in the *British Medical Journal* (1981) emphasized that keeping up the pressure with the proper multiple-drug regimens is absolutely vital. Drug resistance to tuberculosis across the world varies from about 2 to 52 per cent with averages in the UK and USA of around 3–7 per cent, though it must be recognized that in some areas there may be confusion between true primary resistance and acquired resistance from previous, undisclosed treatment. It is likely in developing countries, where drugs are usually in short supply (or not available) and money for health care is even shorter, that inadequate regimes of single or two-drug regimes will be tried. This is indeed the formula for the development of resistant strains.

It is thus very apparent that the seeds of the future diminishing prospects for many of the current antibiotics lies in their present usage and overusage. Of importance in the development of resistant bacterial clones is the strong selective pressure exerted by actual exposure to antibiotics, often when used inappropriately in topical formulations (Lacy 1975). One only has to look at the tetracyclines as an example of resistance being coupled so closely to usage, to see the likely fate of many of the presently useful antibiotics. Indeed with the tetracyclines the problem of resistance has in some places reached a level such that it interferes significantly with the treatment for customary indications. Governmental legislation by many countries to attempt to reduce the selective impact of tetracyclines, in the hope of diminishing the incidence of resistant organisms, has generally had little effect — it was probably too late in any case, especially since the use (legal and illegal) of tetracyclines for therapeutic and prophylactic purpose in poultry and other farm animals is continually tending to increase the incidence of resistant bacteria where there are a few resistant cells within a predominantly sensitive population. The evolution of resistant populations under these conditions is consequently due to elimination of sensitive competitors by the antibiotic, thus opening up a new ecological niche which can be exploited by resistant bacteria. The prospect for the tetracyclines in the 1980s are rapidly diminishing as the review by Chopra *et al.* (1981) so clearly indicates. It is likely that other antibiotics will share the same fate and the treatment of infective disease will be all the poorer because of this.

Corticosteroids

It is salutary to consider that, although the corticosteroids have been in major clinical use for over 30 years, it is still necessary on occasion to remind practitioners of their inherent dangers. Thorn (1981) writing in the *Prescribers' Journal* reminds us that the corticosteroids are responsible for more deaths per year than any other class of drugs and that the morbidity following their use may be extensive. Osteoporosis, diabetes, hypertension, growth retardation, peptic ulcer, susceptibility to infection, and cosmetic effects such as moon face and hirsutism are facets of their iatrogenic potential. Use must, he emphasizes, be contemplated only when it is likely to do more good than harm; as length of use increases so does the risk of complications, which may last for long periods after treatment is stopped.

It is by now so well established that corticosteroids can interfere with the host defence mechanisms against bacterial, fungal, viral, and parasitic infections that there is little new that can be added to the warnings that have been expressed so frequently. It is true that one no longer reads the types of papers that were so common in the early days of corticosteroid therapy in which investigators showed that cortisone and its analogues would lower the resistance of man and laboratory animals to a whole host of infections; however, odd snippets do still appear in papers which indicate that the problem although dormant is still viable.

Bacterial infections

Tuberculosis

Tuberculosis is a major bacterial infection of concern associated with corticosteroid treatment. D'Arcy Hart and Rees (1950) were the first to show that the severity of a tuberculous infection in mice could be increased by the concurrent administration of massive doses of cortisone. Other investigators found similar results in other animal species. It is now universally recognized that in man

manifestations of tuberculous infection, or aggravation of existing tuberculosis as well as reactivation of completely quiet disease, can occur during treatment with corticosteroids (Espersen 1963). The course is often severe owing to a decreased reaction around the lesion and can continue undiagnosed due to the masking effect of the corticosteroid (Veuthey 1962).

Mayfield (1962) discussed tuberculosis occurring in association with steroid treatment and assembled some revealing statistics on the extent of the danger. In 1959, of a total of 7785 patients treated with corticosteroids for asthma, rheumatoid arthritis, eczema, and leukaemia, 13 developed tuberculosis during or within six months of the end of therapy. In 1960, there were 21 such patients out of a total of 6584. In an earlier study, Shubin *et al.* (1959) reported that during 1958 they had seen 58 patients in whom active tuberculosis developed during or after steroid therapy without antitubercular drugs.

Patients with active or doubtfully quiescent tuberculosis should not be given corticosteroids except as adjuncts to treatment with tuberculostatic drugs. The addition of corticosteroids or corticotrophin to antitubercular chemotherapy may lead to earlier improvement in the patient's condition and in radiographic clearing of shadows, but generally the slight long-term benefit is not considered to justify their routine use. However, they may be of benefit in seriously ill and toxic patients, in case of massive effusions, in tuberculous meningitis (see p. 862), or in the treatment of severe drug hypersensitivity reactions (Stevenson 1974).

Bacterial meningitis

In addition to the hope of controlling oedema in meningitis, corticosteroids have been recommended for their anti-inflammatory effects in the treatment of bacterial meningitis. Again, however, the evidence underlying this recommendation is uncontrolled and largely anecdotal. For example, in animal studies Nolan *et al.* (1978) and Sheld and Brodeur (1981) showed that corticosteroids would suppress meningeal inflammation in pneumococcal meningitis in rabbits; Brady *et al.* (1981) reported the cases of two children with pneumococcal meningitis who had been treated with both penicillin and dexamethasone. Both children improved initially but later relapsed on therapy, one at five days and the other at 10 days. In both these cases, chloramphenicol prescribed at the time of clinical diagnosis was withdrawn on the second day when culture results became available.

Penicillin crosses the blood–brain barrier well only when the meninges are inflamed (Hieber and Nelson 1977) and since the anti-inflammatory effect of steroids is known to lessen cerebrospinal levels of ampicillin in animals (Sheld and Brodeur 1981), Brady and his colleagues (1981) thought that the dexamethasone might have critically reduced the CSF penicillin concentrations in their two patients and thus caused their relapse.

There have been few controlled trials of corticosteroids in meningitis, however, although one study of prednisone (60 mg daily) in pneumococcal meningitis (Bademosi and Osuntokun 1979) and another of low-dose dexamethasone (4.8 mg/m² daily) in meningitis of mixed aetiologies in children showed no difference in the outcome of the treated and control groups (Belsey *et al.* 1969).

An editorial in the *Lancet* (1982) reviewed the use of corticosteroids in bacterial meningitis and concluded that there is no unequivocal evidence of benefit from their use in pyogenic meningitis. Indeed it was suggested that harm might possibly be done by lessening CSF penetration by penicillin. The latter effect, it suggested, could, however, be easily overcome by addition of daily intrathecal penicillin if the anti-inflammatory effect of steroids were deemed necessary in a severe case. The editorial emphasized that the theoretical advantages of steroids remained and that the failure to control brain oedema in cerebral falciparum malaria, where the aetiology was ischaemic rather than vasogenic, was not in itself an argument against the possible value of corticosteroids in acute bacterial meningitis. However, the onus of proof for this usage rests with the steroid protagonists and clearly well-controlled trials with steroids at high dosage are required to determine whether practice confirms theory.

The editorial also commented that the use of corticosteroids in tuberculous meningitis was even more controversial than their use in pneumococcal infection since all the evidence in favour of their use was anecdotal. This uncertainty will probably remain since it is unlikely that controlled trials will ever be feasible in this condition. Gordon and Parsons (1972), some years earlier, commented that, apart from their possible value when spinal block is threatened or in sustaining life in a moribund patient while the antitubercular drugs are taking effect, corticosteroids induce serious risks of complications and may give an artificial sense of security. In their view corticosteroids should be avoided in the treatment of tuberculous meningitis. Current opinion is in accord with this earlier recommendation.

Opportunistic infection

A typical report of opportunistic infection was made by Krylov and Murzin (1979) from Leningrad. They described two cases of severe purulent disease in patients

with rheumatoid arthritis. Empyema of the appendix with diffuse peritonitis and bilateral extensive paranephritis were described. These fatal conditions were characterized by an atypical course and were viewed against a background of prolonged therapy with corticosteroids.

Prevention of respiratory distress syndrome by corticoids has also been associated with opportunistic infection. Pre-natal glucocorticoid therapy may be accompanied by a significantly higher risk of maternal infections, particularly when the membrane has been ruptured for more than 48 hours (Taeusch *et al.* 1979).

Although, to date, the majority of the reports that have emerged on corticosteroids and infection relate to systemic administration, it is worth bearing in mind that the local use of corticosteroids is extensive and this may also present an infection hazard. For example, inhalation therapy with beclomethasone diproprionate has proven its value in the treatment of bronchial asthma; also, given intranasally, it has value in the treatment of allergic rhinitis. Oral candidiasis is a frequent side-effect of such aerosol therapy; fortunately, however, it is not usually associated with clinical risk since the infection does not extend to the lungs (Sahay *et al.* 1979).

Topically hydrocortisone and hydrocortisone acetate have been judged to be safe and effective for over-the-counter sale by the FDA Advisory Review Panel in the United States and these formulations can now be sold in 0.25 and 0.5 per cent concentrations without a prescription. It is not certain, however, whether the liberal use of hydrocortisone could aggravate skin infections and there has been some correspondence (*Medical Letter* 1980) expressing concern that use in such conditions may allow infections to spread and cause ulceration and scarring.

Viral infections

The defences against viral infections include the natural resistance of cells at the primary site of infection, the activity of macrophages, and the effects of circulating antibody and cell-mediated immunity (Allison 1972). Drugs may interfere with one or all of these protective mechanisms.

Increased susceptibility of all kinds of viral infection has been noted and viral infections may run a very severe course in patients on corticosteroid treatment. Even in the early days of cortisone-availability it was known from animal studies that corticosteroids and ACTH permit a more rapid multiplication of such viruses as poliomyelitis, Rift Valley fever, Coxsackie, and encephalomyocarditis in host tissues (Findlay and Howard 1952).

Infections with herpes virus seem especially dangerous; dissemination of herpes zoster has been reported

(Irons 1964; Merselis *et al.* 1964) and ocular herpes simplex may be exacerbated by corticosteroids (Crompton 1965). Duckworth (1973) also warned that corticosteroids should not be used in the treatment of herpes infection of the mouth as their use could lead to the generalized spread of herpes infection. Petrie *et al.* (1974) contraindicated oral corticosteroids with measles and smallpox vaccines. Rosenbaum *et al.* (1966) reported that generalized vaccinia developed after a smallpox vaccination in a 66-year-old woman who had lymphosarcoma and hypogammaglobulinaemia and who was being treated with 15 mg of prednisone daily. Levy (1969) also warned that vaccination against smallpox may result in severe gangrenous vaccinia in patients taking corticosteroids.

Children would seem to be at a special risk from exacerbation of viral infection whilst on corticosteroids. For example, Roschlau (1967) reported that generalized varicella occurred in two children given large doses of corticosteroids; both died. Five children developed varicella during corticosteroid therapy and were managed by gradual withdrawal of the steroid and by transfusion of fresh leucocytes and plasma from adults recently convalescent from varicella; no child developed progressive varicella (Spirer 1975).

Meyler (1963) reviewed some of the early literature and commented that chickenpox is not uncommonly a fatal disease in children on steroid treatment. There is also a greater susceptibility to poliomyelitis virus and simple respiratory infections may progress to fatal illness in these children, especially if resistant staphylococci are involved. Thirteen cases, nine of which were fatal, of interstitial plasma-cell pneumonia were cited by Meyler (1963); most of the cases were receiving corticosteroids.

Gerbeaux *et al.* (1963) produced some revealing statistical data which emphasize the true nature of this problem; 194 cases of infectious disease were reported in 154 children on long-term corticosteroid treatment for various complications of tuberculosis. Viral, fungal, and bacterial diseases are included in their list which is summarized in Table 36.1.

Protozoal infections

There are relatively few reports of parasitic protozoal infection in man being enhanced by corticosteroids, although early work in experimental animals showed that this was likely to occur (Kass and Finland 1953). Meyler (1963) reviewed the literature on amoebiasis and corticosteroids and commented that colitis with ulceration due to amoebiasis under the influence of steroid therapy had been reported although aggravation of a latent amoebiasis is rather a rare complication. A case described was

Table 36.1 Cases of infectious disease reported in 154 children on long-term corticosteroid treatment for various complications of tuberculosis

Disease	Number of cases
Chickenpox	102
Pyogenic infections	17
Poliomyelitis	3
Measles	16
Ocular herpes	2
Whooping cough	11
Mumps	28
Lymphocytic meningitis	1
Moniliasis (candidiasis)	3
German measles	6
Herpes zoster	4
Undetermined (fatal)	1

Source: Gerbeaux *et al.* (1963).

observed in a young soldier treated with corticosteroids for severe arthritic complaints associated with Reiter's disease.

Mody (1959) reported two cases of corticosteroid treatment unmasking latent infection of the bowel with *Entamoeba histolytica*. The first case was diagnosed as non-specific ulcerative colitis and treatment by retention enemas of hydrocortisone hemisuccinate was given. Subsequent examination of stools and scrapings from the ulcers revealed swarms of *E. histolytica* trophozoites and cysts to an abnormal degree. In the second case, initial diagnosis was non-specific protocolitis; amoebiasis had been excluded by careful examination. Prednisolone suppositories (5 mg) were used twice daily for over a month. On examination a most profuse infection with *E. histolytica* trophozoites was found in the stools. The ulcers on scraping yielded copious evidence of amoebic infection. The author suggested that corticosteroids might alter the environmental or nutritive condition of *E. histolytica* in a like manner to that known to occur with cholesterol which acts as a growth factor for the organism (Sharma 1959).

Corticosteroid treatment has also been associated with infection by *Pneumocystis carinii* (Woodward and Sheldon 1961). This organism inhabits the lungs of many animal species without normally causing disease; it has also been associated with the use of immunosuppressant drugs combined with corticosteroids (Ruskin and Remington 1967; Montgomerie *et al.* 1969; Lessof 1972); *Pneumocystis carinii* has not yet been classified as a protozoon or a fungus.

Corticosteroid treatment has also been associated with toxoplasmosis (Cohen 1970) and also with strongyloidiasis. Four patients, aged 5 to 32 years, taking prednisone 40–60 mg daily for the nephrotic syndrome, and a

23-year-old woman taking dexamethasone, 20 mg daily for eczema, developed fatal strongyloidiasis during or shortly after treatment. *Strongyloides stercoralis* was responsible in each instance (Cruz *et al.* 1966).

Malaria

Gilles (1981) vividly emphasized the danger of malaria which is still spreading feverishly around the world; protection with chemoprophylaxis is only relative not absolute. Falciparum malaria presents the greatest problem both in number and sequelae although in semi-immune people it produces few clinical problems — that is except in two conditions. The first of these is pregnancy and the second is in patients immunosuppressed with corticosteroids.

Corticosteroids, usually dexamethasone, are frequently prescribed in various forms of brain oedema (Fishman 1982) although convincing evidence of their effect is confined to cases where oedema is associated with mass lesions (Katzman *et al.* 1977). The use of steroids in cerebral malaria and in bacterial meningitis, although formerly recommended, is now controversial and the weight of current opinion is that their use should be avoided. An account of the studies underlying this opinion is given in the following paragraphs.

Cerebral malaria

Glucocorticoids, such as dexamethasone, can reduce the vasogenic oedema associated with some types of cerebral disease, and because of this they have been used extensively since the 1960s in the treatment of cerebral malaria (*Plasmodium* falciparum), on the assumption that cerebral oedema is a consistent feature of that disease. Claims that this treatment reduced mortality and speeded the recovery of consciousness were, however, based largely on uncontrolled studies in small groups of patients who were also receiving antimalarial agents and other drugs. For example, Daroff *et al.* (1967) described 19 patients with cerebral malaria, not all of whom were in coma; besides antimalarials those more severely affected were given dexamethasone, dextran, and diphenylhydantoin (phenytoin). They reported that this polypharmacy was effective. Woodruff and Dickinson (1968) described one patient with falciparum coma; although treated with intravenous chloroquine his coma persisted and 24 hours after its onset, 10 mg of dexamethasone was given intravenously. Four hours later the patient woke and it was concluded that dexamethasone had a dramatic and life-saving effect; it was therefore recommended that dexamethasone should be given routinely in cases of cerebral malaria. Other reports drew similar comments and

conclusions (Oriscello 1968; Blount 1969; Smitskamp and Wolthuis 1971).

Only a few investigators argued against this uncritical acceptance of the use of corticosteroids and foremost among these was Harding (1968) who called for a controlled study. In addition, another single case was reported by Smith and Harper (1968) in which the use of a corticosteroid had not been helpful and this added weight to the lobby which urged definitive and controlled trials. However, De Swiet (1968) considered that a controlled trial was not indicated and Woodruff (1969) considered such a trial unethical.

As was clearly related in an almost blow-by-blow account by Hall (1982), the bandwagon continued to roll and since 1968 all but a few books and review articles have recommended treatment with corticosteroids for cerebral malaria. In the intervening years, however, several authoritative studies and reviews have shown that dexamethasone has no use in cerebral falciparum malaria; for example, Reid and Nkrumah (1972) found no benefit from dexamethasone in a controlled study in four patients. A review in the *British Medical Journal* by Hall in 1976 again concluded that a controlled trial of corticosteroids in falciparum malaria was long overdue because their value had not been established.

The latest evidence against the use of dexamethasone was produced by Warrell *et al.* (1982) who compared the steroid with placebo in a double-blind trial in 100 comatose patients with strictly defined cerebral malaria in Thailand; the ages of these patients ranged from 6 to 70 years. The two treatment groups (steroid or placebo) proved comparable on admission to the trial and there were eight deaths in the dexamethasone group and nine in the placebo group (no significant difference: $p = 0.8$). At post-mortem examination the brain showed features diagnostic of cerebral malaria in all but one of the patients who died.

In adults the initial dose of dexamethasone was 0.5 mg/kg by slow intravenous injection followed by seven doses each of 10 mg. For children, the dose regimen was 0.6 mg/kg given immediately followed by seven doses of 0.2 mg/kg, all intravenously, at six-hour intervals. The total duration of treatment in each regimen was 48 hours.

Dexamethasone significantly increased the duration of coma (63 versus 47 hours: $p = 0.02$) and the incidence of complications including pneumonia and gastrointestinal bleeding (26 versus 11: $p = 0.004$). The authors of this latter report therefore concluded that dexamethasone was deleterious in cerebral malaria and recommended that it should no longer be used. They specifically requested (as has been done here) that those writing textbooks and review articles should note these conclusions.

Fungal infections

Fungal infections may run a more severe course in corticosteroid-treated patients and such patients frequently have a greater susceptibility to them. Nabarro (1960) reviewed much of the earlier use of ACTH and adrenal steroids in general medicine and mentioned under the dangers of steroid therapy that death due to pulmonary aspergillosis occurred in a patient on prolonged steroid therapy for pemphigus. Boyd and Chappell (1961) reported a case of a 17-year old girl who had recurrent acute rheumatic fever; a fatal mycetosis occurred due to *Candida albicans* which was associated with long-continued corticosteroid and intensive antibiotic therapy.

Jacobs (1963) reported a case of a 64-year old woman with post-necrotic liver cirrhosis treated with prednisone 20 mg daily. The woman died three years later of sepsis caused by *Cryptococcus neoformans*. Bennington *et al.* (1964) reported a case of fatal cryptococcal meningitis in an 84-year old man who had received steroid treatment for over a year for his asthma. Other cases of fungal infection after steroid therapy were reviewed by Meyler (1966) and include cryptococcal osteitis in one scapula followed by fatal cryptococcal meningitis in a 33-year old man given prednisone for Beock's sarcoid, and a primary cutaneous cryptococcosis which developed after long-term corticosteroid treatment for arthropathy associated with psoriasis in a 41-year old man.

Apart from systemic fungal infections, the topical infections due to yeasts are also a hazard during long-term treatment. Hayes (1965) reported two cases of oesophageal moniliasis while Lehner (1964) found oral thrush in a group of 44 patients who had been given corticosteroids in association with antibiotics or cytotoxic agents. Dennis and Itkin (1964) reported monilial infection after the use of an aerosol containing dexamethasone in asthmatic patients. Five of the 25 patients developed infection of the oropharynx, and two an infection of the larynx with *Candida albicans*.

The topical application of corticosteroids may also promote fungal infection especially if occlusive dressings are used (Gill *et al.* 1963). Grant Peterkin and Khan (1969), in discussing iatrogenic skin disease, remarked that not only did topical corticosteroids increase the incidence of fungal infections but they also rendered them atypical.

Immunosuppressive and antineoplastic agents

Azathioprine and corticosteroids

Opportunistic infection is a common complication of renal transplantation and an important factor in most deaths. This complication is largely due to the use of immunosuppressive drugs, often combined with corticosteroids, which reduce the rejection process; immunosuppression and bone-marrow toxicity increase the risk of infection and reduce the patient's ability to cope with it. The combination of drugs most frequently used is azathioprine and prednisone or prednisolone, although other combinations are common.

The toxic effects of azathioprine include bone-marrow depression and infections frequently reported as the cause of death following renal transplantation have usually been bacterial or fungal. However, infection with viruses of the herpes group (herpes zoster, herpes simplex, varicella, and cytomegalic virus) have also been noted and there is evidence that esoteric infections such as fungal, viral, and *Pneumocystis* involvement are less frequent in patients maintained on long-term intermittent chemotherapy than in those on continuous medication (Frei 1972).

Mackowiak (1979), in his review on microbial synergism in human infection, mentioned that, although polymicrobial infections occur with regularity in immunosuppressed patients, the combination of *Pneumonocystis carinii* and cytomegalovirus infection occurs much more frequently in these patients than would be expected from mere chance alone. Of the many hypotheses that have been proposed to explain this association, that of Wang *et al.* (1970) is especially intriguing. Cytomegalovirus-like bodies have been observed with *P. carinii* in patients with combined pneumocystis and cytomegalovirus pneumonia and this has led to speculation that, in patients with dual infection, the parasitized pneumocystis might act as a vector for cytomegalovirus. Alternatively, it has been suggested that the patient and *P. carinii* might be infected simultaneously by cytomegalovirus.

Montgomerie *et al.* (1969) described four cases in which normally benign infections of herpes simplex were fatal. In one patient, the herpes infection involved the face, mouth, oesophagus, ileum, and anogenital areas. All four of Montgomerie's patients also showed evidence of infection with cytomegalic virus. There were two cases of candidiasis, and two patients with cysts in the lungs typical of *Pneumocystis carinii* infection.

Pneumonia due to *Pneumocystis carinii* also developed in five patients being treated with corticosteroids and other immunosuppressant drugs for lymphoma and systemic lupus erythematosus (Ruskin and Remington 1967). Three patients treated by Park *et al.* (1967) developed persistent varicella, severe herpes simplex infection, and fatal disseminated histoplasmosis, respectively. These unusual skin infections followed treatment with immunosuppressive drugs such as azathioprine, actinomycin D, and prednisone.

Evans *et al*. (1975) reported congenital cytomegalovirus infection in an infant whose mother had received immunosuppression with azathioprine and prednisone (for renal transplantation) before and during pregnancy.

The effects of azathioprine are enhanced by allopurinol and the dose of azathioprine should be reduced to about one-quarter when allopurinol is given concomitantly. The bone-marrow depressant effects of azathioprine may be enhanced if the drug is given with sulphamethoxazole and trimethoprim (i.e. as *Bactrim* or *Septrin*). This interaction is important since the enhanced effects of azathioprine may favour the emergence of opportunistic infection, an infection that, in the case of the latter combination, may not be controlled by the antibacterials.

Azathioprine is, in the main, slowly metabolized to mercaptopurine; it is not surprising therefore that the potential for drug-induced or aggravated infection of the two drugs is similar. The effects of mercaptopurine are also enhanced by allopurinol.

Cyclophosphamide

Cyclophosphamide produces leucopenia which mainly effects the neutrophils and is usually reversible. It should not be given to patients with infections and, if patients develop an infection during treatment, it is advisable to withdraw the cyclophosphamide.

Experimental studies in animals (Allison 1972) showed that dosage with cyclophosphamide potentiates the infectivity of a number of arboviruses, enteroviruses, and herpes viruses. Much the same picture is evident in man and infection with herpes zoster, herpes simplex, varicella, and cytomegalic virus has complicated treatment with immunosuppressive drugs especially when combined with corticosteroids (Meadow *et al*. 1969; Montgomerie *et al*. 1969; Lessof 1972).

Meadow *et al*. (1969) described the case of a 4-year-old boy who had a relapsing nephrotic syndrome treated with prednisone and cyclophosphamide. He developed measles and died of a giant-cell pneumonia. Before treatment he had been immunologically competent, but in his terminal illness measles antibody failed to develop.

Cyclophosphamide should not be used in patients who have not experienced varicella. Drummond (1969) reported fatalities from varicella in children following the use of the drug. All patients about to receive cytotoxic therapy should also be screened for hepatitis B antigen, and those at risk should be closely observed for alterations in liver function. These were the conclusions reached by Galbraith *et al*. (1975) who observed fulmi-

nant and fatal acute hepatitis B in three patients after the withdrawal of antineoplastic therapy. HB_s Ag was detected in each patient at least six months before the hepatitis appeared and it was considered that immunosuppression permitted diffuse infection of the hepatocytes without the normal response; withdrawal of the immunosuppressants instigated the rapid destruction of the infected hepatocytes with consequent liver damage. Wands (1975) reported two similar cases.

An indication of the extent to which induced infection occurs during cyclophosphamide treatment was given by Feng *et al*. (1973) In their series 42 patients received the drug for treatment of systemic lupus erythematosus. Prednisolone was given initially during treatment and was continued in some patients. Infective side-effects included herpes zoster in five patients, tuberculous infection in two, and fungal infestation in two others.

Medved and Maxwell (1974) also reported that, of two patients with pemphigus successfully treated with intermittent cyclophosphamide dosage, one developed disseminated candidiasis.

Methotrexate

An unusual pulmonary toxicity of uncertain relation to methotrexate was described in children and young adults receiving remission-maintenance therapy for acute leukaemia (Chabner *et al*. 1975). The illness consisted of a clinical syndrome of cough, fever, and shortness of breath, accompanied by radiologic findings of patchy bibasilar infiltrates that closely resembled the picture of *Pneumocystis carinii* or other interstitial pneumonia.

Procedures

Although the potential hazard from opportunistic infection associated with the use of immunosuppressant therapy is well appreciated, perhaps not so well established is that some of the procedures by which cancer chemotherapy is administered may also present an infection hazard. For example, D'Orsi *et al*. (1979) reported three cases of intrahepatic gas-forming abscess as a complication of intrahepatic arterial drug infusion. Maki *et al*. (1979) described nine cases of staphylococcal septic endarteritis originating from percutaneously inserted brachial artery catheters for regional cancer chemotherapy. Discontinuation of hexachlorophane for scrub of the extremity prior to cannulation was incriminated. These infections produced a distinctive clinical syndrome which facilitated implicating the catheter in the resulting fevers; early localized pain, haemorrhage, and Osler's nodes distally were later followed by local inflammation, purulence, and signs of systemic sepsis. The duration of cannulation did not influence susceptibility to infection but susceptibility was related to difficulty in cannulation or the need for repositioning the catheter. Prior radiation therapy, leukopenia, and hypoalbuminaemia were all associated with septicaemia. Implementation of specific control measures prevented catheter-related septicaemia and these were described by Maki *et al.* (1979).

Oral contraceptives

Bacteriuria

Bacterial infections of the urinary tract are the second most common type of infection in the general population after respiratory infections (Freeman 1979). There is some evidence to suggest an association between the use of oral contraceptives and bacteriuria (Takahashi and Loveland 1974; Evans *et al.* 1978). Although both oestrogens and progestogens alter urinary-tract motility and oestrogen administration appears to increase susceptibility to urinary-tract infections in animals, the biological basis of the association remains unclear.

Bacteriuria in adult women may clear spontaneously or with the aid of antimicrobial treatment; however, the factors influencing its clearance are not well understood. It is, therefore, of interest and importance that work by Evans *et al.* (1980) from Boston, USA suggested that, among bacteriuric women who take oral contraceptives, clearance is related to discontinuation of the contraceptive.

In a community-based study, Evans and his associates showed that 289 of 8352 women were bacteriuric; of these 12 were pre-menopausal oral-contraceptive users. Clearance of bacteriuria was again causally related to discontinuation of oral contraceptive use. No other differences could be found between the women who stopped taking oral contraceptives and those who continued which might account for clearance of bacteriuria.

Intractable vaginal candidiasis

Several reports of severe candidiasis in women taking oral contraceptives have appeared. In many of these women, local antifungal therapy was ineffective until oral contraceptives were stopped (Yaffee and Grots 1965). Pregnant women are at a higher risk of contracting vaginal candidiasis than non-pregnant women; diabetics also have a higher risk of vaginal candidiasis than normal women. Since oral contraceptives induce a pseudopregnancy state and have a diabetogenic action it

is not surprising that they increase the risk of vaginal candidiasis. With all these conditions, pregnancy, diabetes, and oral-contraceptive usage, there is an increase in epithelial-cell glycogen, a rise of pH, and a fall in the Doberlein's bacillus population.

In a comparative study of 275 pregnant women, 77 women taking oral contraceptives and 100 control women, the overall incidence of candidiasis during pregnancy was 15.3 per cent compared with 20 per cent in women on the Pill for 3 to 9 months, and 3 per cent in the controls (Gruber *et al.* 1970). In similar study, Oriel *et al.* (1972) showed that the incidence of vaginal yeast infections (mainly *Candida albicans*) in 241 women taking oral contraceptives was 32 per cent compared with 18 per cent in women not taking oral contraceptives. Ridley (1972) published a brief review, mainly of a negative association between candidiasis and oral contraceptives.

Intrauterine contraceptive devices

Risk factors

Pelvic inflammatory disease is one of the most important diseases occurring among young women; in America, for example, more than 850 000 cases occur annually, resulting in more than 200 000 hospital admissions and 2 500 000 physician visits (Layde 1983). It is of significance therefore that a major study by Vessey *et al.* (1981) has helped to quantify the risk factors in the use of intrauterine contraceptive devices (IUDs).

They carried out a large cohort investigation among parous women enrolled in the Oxford Family Planning Association contraceptive study (17 032 women attending 17 family planning clinics in England and Wales). Hospital admission rates for 'acute definite disease' were 1.5 per 1000 woman-years among those currently using an IUD and 0.14 per 1000 among those using other methods of birth control (age standardized relative risk 10.5 to 1). There was little evidence of an increased risk of such disease in ex-users of an IUD. Hospital admission for a 'chronic definite disease' was, however, commoner in ex-users than in current users.

Acute definite disease was more common during the early months of IUD use, and while the rate of such disease was increased in users of each type of device, the highest rate (8.1 per 1000 woman-years) was observed in users of the *Dalkon* shield. This rate, however, was based only on three different women. This and other suggestions (Kaufman *et al.* 1983; Layde 1983) that acute definite pelvic disease might occur more frequently in association with a *Dalkon* shield than any other device is interesting in view of the evidence that this IUD has been associated with an increased risk of septic abortion in the second trimester (Cates *et al.* 1976).

Unfortunately, Vessey and his colleagues did not provide any information on the bacteria or other organisms associated with pelvic inflammatory disease. However, isolated reports continue to appear on staphylococcal septicaemia associated with IUDs. For example, Geddes (1980) described how, three days after having a *Lippes Loop* device fitted, a 31-year-old woman developed vomiting, sweating, cough, and confusion. During the next 48 hours she became disorientated and hallucinated. On admission to hospital she had a temperature of 38.8°C (101.8°F) and a blood-pressure reading of 95/6 mm Hg. Chest X-ray examination showed patchy consolidation of the upper lobes of both lungs. The blood cultures drawn on admission grew *Staphylococcus aureus* resistant to penicillin. The patient was treated with high-dose i.v. cloxacillin and 24 hours later her condition had markedly improved. The *Lippes loop* was removed and cultures from the coil and from a high vaginal swab yielded *Staph. aureus* similar in antibiogram to that cultured from the blood.

Geddes commented that the patient was both toxic and shocked and that the case provided another example of the importance of *Staph. aureus* as a cause of infection associated with the insertion of foreign bodies into and through the vagina.

This comment is especially valid and timely in view of the recent spate of reports on toxic shock and tampons, and the considerable publicity that has been given to this in America (US Department of Health and Human Services/Public Health Services 1980*a, b, c*).

Toxic shock syndrome

Toxic shock syndrome (TSS) is an uncommon condition characterized by rash, fever, hypotension, desquamation of skin, and multisystem involvement. It occurs most commonly in menstruating women and is estimated to afflict nine per 100 000 per year (Hulka 1982), although it has also been reported in non-menstruating women and in men and in association with surgical wound infections (Bartlett *et al.* 1982). Toxin(s)-producing strains of *Staphylococcus aureus* have been cultured from some body sites in almost all cases. The *Staphylococcus* is presumed to be aetiologic, although identical strains have been cultured from normal persons (Hulka 1982).

Intrauterine contraceptive devices

In letters to the *New England Journal of Medicine*, Harvey (1981), Hymowitz (1981), Loomis and Feder (1981), and Jaffe (1981) discussed toxic shock syndrome

associated with diaphragm use. The case reported by Loomis and Feder (1981) will serve to illustrate the course of the syndrome which was similar to the others. The patient, a 23-year-old woman, presented with an eight-hour history of fever, vomiting, and diarrhoea. Her most recent menstrual period, during which she used tampons, had ended two weeks earlier. Two days before presentation, she had inserted a diaphragm, which she was unable to remove. Physical examination did not show any focus of infection; pelvic examination gave normal results, and the diaphragm was easily removed. The patient was discharged to be followed as an out-patient.

Thirty-six hours later she returned to hospital reporting fever, arthralgias, myalgias, and pharyngitis. Her temperature was raised (38.9°C) and her blood pressure was low (84/50). She had a fine erythematous rash which was generalized, conjunctivitis, pharyngeal oedema without exudate, and a strawberry tongue. A serosanguineous cervical discharge was noted on pelvic examination.

She was treated as an out-patient and given intravenous fluids over four hours; she was also treated with intramuscular procaine penicillin and placed on oral penicillin for presumed scarlet fever. When examined two days later she was afebrile and clinically well. A throat culture was negative but a cervical culture was positive for *Staphylococcus aureus* resistant to penicillin and ampicillin.

Because of the possibility of toxic shock syndrome, the penicillin was discontinued and she was treated for 10 days with dicloxacillin. Desquamation of her palms and soles occurred 11 days after the onset of her illness. Transverse furrows were present on all her nails three months after the rash desquamated. On follow-up for nine months, repeat cervical cultures were negative for *Staph. aureus*. She continued to use her diaphragm without recurrence of toxic-shock syndrome.

Vaginal tampons

The toxic shock syndrome (TSS) has been associated with the presence of *Staph. aureus* (Todd *et al.* 1978; US Department of Health and Human Services/Public Health Services 1980*a, b*) and Fuller *et al.* (1980) suggested that the vaginal tampon acts as a blocking agent resulting in the accumulation and absorption of staphylococcal toxins. Siegel and Gleicher (1981) believe that the circulatory failure in patients with TSS, which may lead to a case fatality rate of 10–15 per cent in serious cases, occurs because of the effect of *Staph. aureus* toxin on human platelets. This toxin degranulates platelets and induces membrane changes which, in turn, allow the free diffusion of platelet products into the surrounding medium. Amongst released products is platelet factor 3, which enhances coagulation, and platelet serotonin as well as histamine, which may induce vasoconstriction. Intravascular coagulation and peripheral vasoconstriction may then lead to the clinical picture of TSS.

The first case of toxic shock syndrome associated with the use of tampons reported in Britain was by Holt (1980); this was followed by much correspondence in the columns of the *British Medical Journal* (*British Medical Journal* 1980; Goulding 1980; Lea and Ellis-Pegler 1980; McCormick and Allwright 1980; Smith 1980).

Two articles published in the same issue of the *Journal of the American Medical Association* raised the question of whether tampon use enhances the risk of TSS amongst menstruating women who are vaginal carriers of *Staphylococcus aureus*, and, if so, whether *Rely* brand tampons in particular further enhance the risk? (*Rely* brand tampons were voluntarily removed from the market in September 1980 by the manufacturer, Procter and Gamble.)

The first of these articles by Schlech *et al.* (1982), representing the US Centers for Disease Control, related how 50 patients with menstrually-associated TSS were interviewed and asked to provide information about the type of menstrual sanitary products used during the menstrual period associated with their illness. The same questions were also asked to 150 age-matched control subjects.

All 50 cases (with TSS) and 125 of the 150 controls used tampons, and among women using tampons, TSS cases were more likely to have used *Rely* brand tampons when compared with controls. No differences were found between cases and controls in the absorbency of tampon products used. No other factors were found to be significantly associated with the development of menstrually associated toxic shock syndrome.

The second article by Harvey *et al.* (1982) reviewed the events that transpired between the first report of TSS (Todd *et al.* 1978) and its later epidemiologic association with tampons. They refuted the validity of the epidemiologic association of TSS with tampons on the grounds of bias in study design and data collection.

The main evidence that tampons are an aetiologic co-factor in the development of TSS comes from epidemiologic case-control studies, and according to Harvey and her colleagues the patients chosen as cases in those studies were assembled from reports submitted to health agencies in response to publicity that may have influenced physicians to diagnose TSS and to submit reports particularly in situations where the patient was a menstruating tampon user. When submitted reports were checked for fulfilment of TSS diagnostic criteria, and when cases or controls were asked about antecedent tampon use, suitable scientific precautions were not used to achieve 'blinded' objective decisions. Harvey and her colleagues are of the opinion that since these biases would have distorted the statistical relationships, the aetiologic role of tampons in TSS has not been scientifically proven.

Hulka (1982) in an editorial to the two conflicting articles has with skill and patience attempted to evaluate the findings of each of the five published case-controlled studies of TSS (Davies *et al.* 1980; Shands *et al.* 1980; Kehrberg *et al.* 1981; Osterholm *et al.* 1982; Schlech *et al.* 1982). She found that each of these studies was deficient in some aspect of their design or methods and she posed two critical and related questions: Were biases introduced by these methodological deficiencies? and, if so, were they of sufficient degree to produce the tampon–TSS association, if in fact such association existed?

One of the major problems facing these studies, for example, was the publicity concerning the possible association between tampon use and TSS to which study participants could have been exposed by the media. To what extent such information might distort a case or control subject's memory of whether or not she used tampons is speculative.

Hulka emphasized that such potential distortions introduced by external, uncontrollable events as well as by investigative strategies must be recognized. She emphasized also that the TSS–tampon association should also be viewed in the context of existing data on clinical presentation and recurrences, pathogenesis, and microbiology with particular emphasis on *Staph. aureus*-TSS associated strains, colonization patterns, and toxin production. In a 'Portia-like' judgement, she concluded that only substantial new research evidence evoking alternative explanations for the existing observations would be sufficient to negate the association between TSS in menstruating women and tampon use.

In the UK, the general view, as expressed by Eykyn (1982) from St. Thomas's Hospital, London, is that there is no justification at present for women to avoid using tampons, since the risk of developing toxic shock syndrome is extremely small. Up to the end of April 1982, 25 'probable' cases of TSS had been reported in the UK to the Central Public Health Laboratory at Colindale, as compared with 941 cases reported in the USA by January 1981. The low figure of cases reported in the UK is thought to be unlikely to represent underdiagnosis because of the widespread publicity given to the condition. Details of cases in Britain were published by de Saxe *et al.* (1982); 24 cases were associated with menstruation and one occurred post-partum. Tampons of various kinds had been used by all the menstrually-associated patients.

Eykyn (1982) emphasized that TSS is not specific, however, to menstruating women, nor indeed is it solely a 'tampon' disease since similar symptoms have been reported in children. She believes that it is emerging as a staphylococcal disease of multifactorial aetiology. This viewpoint is well confirmed by a report from Thomas *et al.* (1982) who described the case of a previously healthy 25-year-old woman hospitalized for submucous resection and rhinoplasty. The patient's nose was packed with gauze post-operatively and she developed TSS. Her last menstrual period had been three weeks earlier and was normal. She had never used vaginal tampons. Random culture of 10 specimens of the prepacked sterile gauze that had been used for nasal packing failed to show bacterial contamination. Reingold *et al.* (1982) also described a number of patients with TSS that was not associated with menses and tampon use.

Actinomyces infection

An editorial in the *Journal of the American Medical Association* (Gupta and Woodruff 1982) suggested that women who had *Actinomyces-positive* vaginal smears should use a contraceptive method other than the intrauterine device. This stimulated Duguid and her colleagues (1982) from Dundee, Scotland, to report in the correspondence columns of that journal that they had examined cervical material from 197 plastic-device users and 209 copper-users, and found a significant difference in the prevalence of actinomycete infestation between the two groups (42 per cent and 2 per cent, respectively). In every case, the type of device and duration of usage were noted, the comparison being almost exclusively between *Saf-T-Coil* and *Gravigard*.

Duguid *et al.* suggested that this difference in the actinomycete isolation rate between copper-device and plastic-device users might be due to several factors. It is possible that will prolonged use of plastic devices (more than two years), calcium carbonate deposits flake from a device and form a nidus for *Actinomyces* growth, or that copper acts as an inhibitor of their growth. Some preliminary studies reported by this group suggested that when, the IUD is removed from *Actinomyces*-positive cases or replaced by a copper device, the bacterial flora revert to normal and that this occurs more rapidly after a copper device has been inserted.

Dugid *et al.* suggest that, in the absence of clinical evidence of infection, alternative means of contraception need only be considered in those uncommon cases where *Actinomyces* colonization occurs in women using copper devices, and it may well be that even in those cases, a fresh copper device could be inserted with safety even after a lapse of one to two months.

Co-trimoxazole

Delayed diagnosis of malaria

Delayed diagnosis and therefore delayed institution of corrective treatment is a special facet of aggravation of infective diseases. That such a delay is caused by drugs is relatively rare and it is important to record this when it occurs. In this respect, Williams *et al.* (1982) from the Departments of Infectious Diseases and Laboratory Medicine, Ruchill Hospital, Glasgow reported on two patients in whom the diagnosis of falciparum malaria was delayed; both had taken co-trimoxazole shortly before admission to hospital. A description of one of these cases will suffice to illustrate the nature of the problem.

A 56-year-old construction engineer was admitted two weeks after returning from a 10-day stay in Madagascar. He had worked in the bush and had frequently been bitten by mosquitoes. For antimalarial prophylaxis he had taken only three tablets of an unknown type given to him irregularly by a workmate. For three days he had been shivering and sweating and had increased lassitude, headache, and cough. He had vomited twice and he complained of dysuria and frequency. Co-trimoxazole at the standard dosage was started on the day before admission, on diagnosis of possible enteric fever.

On admission the marks of insect bites were visible on his legs but no malaria parasites were seen on a thin blood film. Co-trimoxazole treatment was continued since enteric fever still remained a possible diagnosis. Thin blood-films on the third and fifth days after admission still failed to show malaria parasites, but on re-examination of the third-day film for over two hours, a few altered gametocytes of *P. falciparum* were finally seen. No parasites were seen on a retrospective search of the blood film taken on admission although pigment in macrophages was noted. Serum from day 6 gave a strong reaction by indirect fluorescence to *P. falciparum* antigens. A standard course of treatment with chloroquine was then given and the patient made an uneventful recovery.

Williams *et al.* believe that the earlier treatment with co-trimoxazole delayed the diagnosis of the falciparum malaria in both their cases. This is not entirely surprising since co-trimoxazole is known to modify and even cure falciparum malaria whether caused by chloroquine-sensitive or -resistant strains (Benjapongs *et al.* 1970; Fasan 1971; Hansford and Hoyland 1982). It acts against the asexual forms in the red blood cell with little effect on other stages of the parasite. Its role as an antimalarial drug, however, remains controversial and it is not routinely recommended for prophylaxis.

Williams and his colleagues warn that doctors should be aware of the possibility of modifying a large range of bacterial and protozoal illnesses when prescribing co-trimoxazole for an undiagnosed fever. To this may also be added the warning of the dangers of self-medication with co-trimoxazole in tropical or Third World areas where frequently such medication is available without undue restriction. The failure of current methods of communicating to travellers the need for regualr malaria prophylaxis is also well shown by these two cases. An article by Logan *et al.* (1982) described well the advice that should be given to such travellers.

Intravenous infusion fluids and total parenteral nutrition

Intravenous infusion fluids

Intravenous infusion fluids have been in clinical use for nearly a century and a half, but it is only in relatively recent years that the full spectrum of iatrogenic hazards, including septicaemia, associated with their use has been fully recognized and documented (Maki 1976; Lawson and Henry 1977; Holmes and Allwood 1979; D'Arcy 1980).

Fortunately, with careful quality assurance during manufacturing and with government-enforced good manufacturing procedures, septicaemia due to infected fluid supplies is now very unusual. Much more commonly sepsis arises in the hospital which is a less easily controllable *milieu* for the development and spread of infection. Thus most present contamination of infusion or total parenteral nutrition (TPN) fluids is probably extrinsic; that is it occurs when microorganisms are introduced into the infusion fluid during the manipulation of the infusion apparatus to prepare it for use, or during the actual administrative procedure, and is especially related to the catheter, its insertion technique, and the care of its entry site into the skin. It cannot be overemphasized that hospital infection surveillance together with strict guide-lines for the preparation and administration of intravenous fluids, especially parenteral nutrition, is the only way of ensuring the continued safety of this therapy. If the system breaks down at any stage, then opportunistic infection will follow.

Problems with total parenteral nutrition

It is only in the past 10 years or so that the intravenous administration of hypertonic glucose solutions with protein hydrolysates has been developed and expanded into the present highly effective system of total parenteral nutrition. However, during the early years of intravenous hyperalimentation and in particular during 1969–73, many hospitals in the USA reported extraordinary rates of associated septicaemia reaching 27 per cent in several hospitals and averaging 7 per cent in a nation-wide survey of major centres (Goldman and Maki 1973). Inexplicably *Candida* species were implicated in over half of these incidents, many of which had a fatal outcome. This problem still exists and, as far as current reports testify, it does not seem to be diminishing.

Endogenous *Candida* endophthalmitis has been the subject of several reports; these were reviewed by Cantrill *et al.* (1980) and they stressed the association of endogenous infection with certain predisposing factors including the use of intravenous catheters especially for hyperalimentation. Other predisposing factors are the use of broad-spectrum antibiotics, extensive abdominal surgery, immunosuppressive therapy, and disease associated with impaired host defenses.

Montgomerie and Edwards (1978) reported the results of a prospective study of 23 patients receiving hyperalimentation; five of these experienced asymptomatic fundus lesions compatible with the clinical diagnosis of fungal endophthalmitis. Candidaemia was verified in three of these patients. It is also of interest to note that *Candida* endophthalmitis can occur in drug addicts in whom the organisms are directly introduced via 'mainline' injections (Aguilar *et al.* 1979).

From France, Trunet *et al.* (1980) surveyed the role of iatrogenic disease in admissions to intensive care. Of 325 hospitalized patients who were admitted to a multidisciplinary intensive care unit over a one-year period, 41 were hospitalized because of iatrogenic disease. Of these, five patients suffered septicaemia due to central venous catheterization (two were fatal and one was life-threatening), two patients developed septicaemia after peripheral venous catheterization (one case was fatal), and two patients developed bacterial arthritis after intra-articular corticosteroid administration which was given in their homes.

Steel *et al.* (1981) from Boston, USA also surveyed iatrogenic illness on a general medical service at their University hospital; of 815 consecutive patients, 290 (36 per cent) had one or more adverse reactions. Intravenous therapy accounted for 34 complications (whether infective or not was not stated) out of a total of 497, and aspiration pneumonia, nosocomial infection, and problems that were secondary to invasive procedures accounted for the majority of the major complications.

Much has been written in recent years in American literature about the team concept in the TPN support of the hospitalized patient (Ferguson 1980); such teams are generally composed of an attending physician, a pharmacist, a nurse, clinical dieticians, and an infection-control co-ordinator. Nehme (1980) provided some interesting findings from a two-year prospective study of this team concept in the nutritional support of the hospitalized patient.

This study well confirmed earlier reports for the need of a protocol for the administration of TPN to reduce complications. It also showed that, where a protocol is closely adhered to, there are few complications. Total parenteral nutrition has an enormous therapeutic potential when properly administered under the close supervision of persons with expertise who can manage its potentially hazardous complications.

These concepts are especially relevant in view of the current interest in the USA for home parenteral nutrition. This relatively new innovation has, it is claimed, revolutionized the treatment of formerly fatal intestinal problems, in much the same way that renal dialysis has affected the outlook for patients with chronic renal failure. Catheter care for such patients is of the prime importance. Grundfest and Steiger (1980) have done much work in this area and in their paper on home parenteral nutrition they related how 43 patients have been treated at home; catheter care is a problem and infection, when it has occurred, has usually been due to fungal infestation by *Candida* or *Torulopsis*, although on occasion *Staph. aureus* and a variety of gram-negative bacteria have been cultured from catheter tips. The general ground-rules for preventing catheter-induced infection may have to be more strictly applied in the home environment simply because the catheter is most likely to be indwelling for longer periods of time than would be normal in hospitalized cases.

A final point of relevance in this particular discussion is the disquieting report from Fischer *et al.* (1980) who showed, in animal studies, that *Intralipid*, a lipid emulsion used in parenteral nutrition, impaired bacterial clearance and enhanced bacterial virulence in mice. In addition, it inhibited the chemotaxis of human neutrophils *in vitro*. Since certain patients, notably premature infants and debilitated subjects who are malnourished or have malignant disease, are at high risk from bacterial sepsis, this study strongly suggests that *Intralipid* may compromise human host defence mechanisms and put patients at risk from invasive bacterial disease. This is worrying since the use of lipid emulsion in some special care babies has become standard practice (Coran 1972; Gustafson *et al.* 1972; Bryan *et al.* 1976). It is thought that *Intralipid*, whilst improving caloric intake, may diminish neutrophil and macrophage function and so allow microbial invasion.

Reye's syndrome and aspirin

A warning issued by the United States Department of Health and Human Services on September 20th 1982 linked Reye's syndrome, a rare but life-threatening disease, with the use of aspirin and other salicylate-containing drugs by children under 16 years of age who have influenza or chickenpox (varicella). The announcement of this hazard was made by Richard S. Schweiker, the Secretary of Health and Human Services, who also outlined his plans for radio and public service announcements aimed at reaching parents and physicians before the next influenza season.

Currently in America, the Reye's syndrome awareness campaign is enlisting the help of consumer and health professional groups to circulate information to the public. One of the problems inherent in such a campaign of public education is the general acceptance that aspirin or other salicylates are safe, effective, well-accepted, and well-tried analgesic/antipyretic agents which are well suited for self-medication. Directly linked with this is the second problem, which is the myriad formulations of aspirin or other salicylate-containing medicines which are available both on and off prescription.

The Americans have drawn up appropriate warnings on medicine labels and for many parents and physicians these warnings will come very much as an incredulous surprise; aspirin has been and is still commonly given to children to combat the symptoms of many ailments and few appear to suffer from this. Not surprisingly therefore the informed paediatric community is divided on the issue of the hazards of aspirin and on the necessity for warning labels (Bianchine *et al.* 1982; Mortimer 1982; Brown *et al.* 1983; Daugherty *et al.* 1983).

Reye's syndrome was well described by Lichtenstein *et al.* (1983); it was recognized in 1963, and is an acute life-threatening condition characterized by vomiting and lethargy that may progress to delirium and coma. The disease commonly occurs in children between the ages of 5 and 16 years who are recovering rom viral infections especially chickenpox and influenza. Some 600–1200 cases are known to occur each year in America and the estimated mortality rate is some 20–30 per cent. Some survivors suffer permanent brain damage. Antipyretics and anti-emetics administered during the antecedal viral illness have been suspected of having some role in the development of the syndrome. Other accusations have centred on multifarious factors ranging from extrinsic toxins such as afatoxin and pesticide emulsifiers, to genetic predisposition, or metabolic abnormalities.

The basis for the incrimination of salicylates in this syndrome largely stems from the case-control epidemiological studies initiated in three American centres (Arizona, Michigan, and Ohio). The Arizona study by Starko *et al.* (1980) reported an investigation of seven children with the disease and compared them with 16 controls who had similar prodromal illness but who did not develop Reye's syndrome. Compared with controls, nearly every child who subsequently developed the disease had been given aspirin during the antecedent illness. The Michigan study of Waldman *et al.* (1982) showed that 24 out of 25 children with Reye's syndrome received salicylates compared with 34 out of 46 controls matched for age, race, school grade, and nature of antecedal viral illness. A second study by the same clinicians on 12 children of school age with Reye's syndrome following an epidemic of influenza A and 29 matched controls, confirmed the association of salicylates with the development of the syndrome. The Ohio study reported by Halpin *et al.* (1982) gave similar findings; 94 out of 97 cases exhibiting the syndrome had been given salicylates or salicylates-containing medicines compared with 110 of 156 controls.

There three studies were the basis upon which subsequent authoritative reviews have been written and upon which authoritative recommendations have been issued. The US Centers for Disease Control (1982), commenting on these studies, warned physicians and parents of the possible increased risk of Reye's syndrome associated with the use of salicylates in children with chickenpox or an influenza-like illness. A Food and Drugs Administration working group subsequently audited the raw data from the three studies as also did the American Academy of Pediatrics Committee on Infectious Diseases. The conclusions of both these expert bodies was that the use of salicylates should be avoided in children suffering from influenza or chickenpox and that a warning to physicians and parents was warranted.

These viewpoints are currently controversial and there are counter-claims that the official warnings are not substantiated by the quality of the evidence available. This was illustrated by an article in the *Medical World News* of September 13th 1982 (Eichenwald 1982) entitled 'Aspirin's bum rap' in studies linking it to Reye's. The author of that article is the professor of paediatrics at the University of Texas Southwesten Medical School in Dallas, and he writes also on behalf of five other US paediatric professors. They urged the Secretary of the United States Department of Health and Human Services not to relabel aspirin with warnings until he has had the raw data from the studies done to date evaluated by a panel of scientists who have not previously taken a position on the issue.

The basis of Eichenwald's argument is that many methological flaws invalidate the Arizona, Michigan,

and Ohio case-control studies that suggested the aspirin–Reye's syndrome link. He questioned, for example, whether all children counted as disease cases actually had the syndrome, whether some received aspirin only after they had developed the syndrome, and whether all children counted as taking aspirin really got it. Moreover, he claimed that the controls were not properly matched as to the severity of the preceding viral infection. Of concern also, he suggested that the drug data gathering underlying the studies did not determine whether a child received a liquid acetaminophen (paracetamol)-containing form of an antipyretic or tablets with the same brand name that contained aspirin. When parents named such a product, he commented that it was counted as aspirin.

An editorial in *Reactions* of October 29th 1982, produced an excellent review of other critical views of the interpretations made on the evidence so far available, as also did the article 'Take two aspirins and call me in the morning' by Hoekelman (1982). The Reye's syndrome (RS) Working Group of the Aspirin Foundation of America (1982) were also especially critical and questioned the biological plausibility of the claimed association. This group concluded that 'the only responsible statements that can be made in relation to RS at this time are those that pertain to early recognition of RS and to prudent use of all medications in children'.

There is little need for futher comment at this time; enough has already been written about the limited data that are currently available. More evidence is badly needed to confirm the association between salicylates and the syndrome or the suggest that Reye's syndrome and salicylate toxicity are separate entities. In the interim period, it is necessary to emphasize that the association has been suggested and that the use of aspirin or other salicylates cannot be advised with safety in children suffering from influenza or chickenpox. Until such time as other studies clarify the position, the use of alternative analgesics and antipyretics would seem to be a sensible precaution.

Travellers' diarrhoea

A review on travellers' diarrhoea by Gorbach (1982) starts appropriately with the view that 'Travel expands the mind and loosens the bowels', and ends somewhat realistically with the conclusions that 'if you can't completely stop the runs, you can at least cut your losses'. Concern over diarrhoeal illness has led the travellers to a variety of nostrums for treatment — some salutary, others ineffective, and most potentially harmful as well as ineffective. It is this latter iatrogenic component that justifies an account of such treatment in this present context.

For centuries, a virtual pharmacopoeia of starches, talcs, chalks, gums, astringents, adsorbents, herbal preparations, reptile parts, opiates, and heavy metal salts have been used to relieve the coprologic agonies of travel. In more recent times, prevention of travellers' diarrhoea has been attempted with more sophisticated measures, and the broad application of antibiotics or antibacterial agents to prevent diarrhoea has led to the timely concern that these will induce resistance in bacteria, particularly the toxin-producing coliforms. Certainly there is ample evidence for such concern, for such studies that have been done, and they are relatively few, have shown that certain areas of the world already have a high incidence of resistance among enterotoxigenic *E. coli* strains. For example the Philippines, Korea, Indonesia, and Honduras (and probably many other regions as yet undetermined) have at least 60 per cent of toxin-producing coliform isolates resistant to tetracyclines or other antibiotics (Santosham *et al*. 1981; Gorbach 1982). Such a high incidence of resistance therefore renders the utility of such antimicrobials highly questionable as anti-diarrhoeals.

A recent example of this antimicrobial-induced resistance was presented by Murray *et al*. (1982) who showed emergence of high-level trimethoprim resistance in faecal *E. coli* during oral dosage of trimethoprim alone or a trimethoprim–sulphamethoxazole combination (co-trimoxazole) in 136 American students during a two-week diarrhoea-prevention study in Mexico.

Ninety-six per cent of the 165 trimethoprim-resistant *E. coli* isolates were shown to be resistant to at least four other antimicrobial agents and 25 per cent were resistant to seven. Trimethoprim resistance was transferrable to the laboratory strain of *E. coli* J53 at a high frequency in 40 of the 100 strains tested.

The major benefit of prophylactic trimethoprim alone or its combination in co-trimoxazole has been the suppression of faecal or introital *Enterobacteriaceae* and, as the recent study in Mexico by DuPont *et al*. (1982) showed, these agents have marked benefit in the early symptomatic treatment of *E. coli*-induced diarrhoea, shigellosis, and diarrhoea not associated with an enteropathogen. Thus the emergence of resistance and the subsequent failure to suppress such growth are clearly of major importance.

Agents other than trimethoprim and co-trimoxazole that are currently used to treat travellers' diarrhoea, appart from creating secondary problems of side-effects and toxicity, can also worsen the diarrhoea. For example, DuPont and Hornick (1973) reported that diphenoxylate, a synthetic opiod analogue, enhanced

diarrhoeal symptoms and prolonged the excretion of *Shigella* organisms. A closely related compound, difenoxine, was shown by Steffen and Gsell (1981) to be disadvantageous for prophylaxis in travellers' diarrhoea. In their study involving 653 travellers and several prophylactic agents, difenoxine showed a tendency (*p* = 0.07) to increase the proportion of travellers with diarrhoea (53 per cent vs. 38 per cent on placebo). It was suggested that this untoward effect might be explained by closer contact between microorganism and mucosal cells (DuPont and Hornic 1973).

It must be emphasized, however, that current authoritative views are that neither antibiotics nor sulphonamides should be taken to treat diarrhoea even when a bacterial cause is suspected because they may prolong rather than shorten to time taken to control diarrhoea and carrier states. Bacterial resistance to antibiotics may develop and they may cause a bacterial diarrhoea and pseudomembranous colitis. The first line of treatment in acute diarrhoea, as in gastroenteritis, is prevention or treatment of fluid and electrolyte depletion with oral electrolyte–carbohydrate solutions. As far as prevention of travellers' diarrhoea is concerned, the best advice is that already given by Kuehnle (1983): 'boil it, cook it, peel it, or forget it' sums up what the traveller should or should not eat or drink.

Nasotracheal intubation, bladder catheterization, and colonic irrigation

Three papers published during the second half of 1982 are a reminder that clinical procedures in which drugs are not involved can also induce or aggravate infective conditions.

The first paper by Knodel and Beekman (1982) described how acute maxillary sinusitis developed in three critically ill patients whose conditions required mechanical ventilation. In all three fever was the initial clinical manifestation. An obtunded state and an inability to speak delayed localization of the site of infection.

Nasotracheal tubes had been in place for 4, 8, and 11 days, respectively; results of routine evaluations including examination, cultures, and chest roentgenograms were unrewarding. There was no response to empirical antibiotic therapy. None of the patients indicated pain in the sinuses but they were sedated or comatose for many of the days during intubation. It was not until purulent drainage was seen in the intubated nares that the diagnosis of sinusitis was diagnosed expectantly by portable sinus X-ray films and sinus aspiration. Measures to promote sinus drainage and administration of antibiotics resulted in rapid improvement in all three cases; a beta-haemolytic non-group A, B, or D *Streptococcus* was isolated from sinus aspirates taken from the third patient in this series.

Knodel and Beekman concluded that prolonged intubation, simultaneous use of nasal gastric tubes, poor healing, and large-diameter endotracheal tubes most likely contributed to the disease pathogenesis in these cases. They cautioned that acute maxillary sinusitis may be the cause of unexplained fevers in patients with nasotracheal intubation. Earlier reports by Arens *et al.* (1974) and Stauffer *et al.* (1981) also described sinusitis after nasotracheal intubation; the incidence was, respectively, four of 200, and 3 of 16 patients.

The second paper described a prospective study by Platt *et al.* (1982) in which 131 of 1458 patients acquired 136 urinary-tract infections (defined as $\geqslant 10^5$ colony-forming units/ml) during 1474 indwelling bladder catheterizations. Seventy-six patients (25 infected and 51 non-infected) died during hospitalization; death rates were 19 per cent in infected patients and 4 per cent in non-infected patients. Multiple logistic regression analysis demonstrated that seven of 21 prospectively monitored variables were associated with mortality among the catheterized patients. The adjusted odds ratio for mortality between those who acquired infection and those who did not was 2.8 (95 per cent confidence limits 1.5 to 5.1).

The acquisition of infection was not associated with the severity of the underlying disease; among patients who died, infections occurred in 38 per cent of those classified as having non-fatal underlying disease (15 of 39) and in 27 per cent of those classified as having fatal disease (10 of 37).

Platt and his colleagues thought that 12 deaths may have been caused by acquired urinary-tract infections. Two patients had urinary-tract pathogens in pre-mortem blood cultures. Another 10 died with clinical pictures compatible with serious infection, but no diagnostic cultures were performed. They concluded that the acquisition of urinary-tract infection during indwelling bladder catheterization was associated with nearly a threefold increase in mortality among hospitalized patients. They further emphasized this finding by extrapolating the death rate (5.2 per cent) among all catheterized patients in their hospital (New England Deaconess Hospital) to the approximately 7.5 million persons who are catheterized each year in acute-care hospitals in the USA. Their calculations suggested that there are approximately 400 000 deaths per year in that population. Their survey data indicated that 56 000 (14 per cent) of these may represent the excess mortality associated with catheter-related infection.

The third paper by Istre *et al.* (1982) related how an outbreak of amoebiasis was spread by colonic irrigation at a chiropractice clinic in Western Colorado. During a period of 2½ years at least 36 cases of amoebiasis (*Entamoeba histolytica*) occurred in persons who had received colonic-irrigation therapy. Of 10 persons who required colectomy, principally for widespread colonic necrosis with perforations, six died. Of 176 persons who had been in the clinic during the last four months of the 2½ year period, 80 had received colonic-irrigation therapy and 96 had received other forms of treatment (e.g. spinal and manipulation). Twenty-one per cent of the colonic-irrigation group had bloody diarrhoea, as compared with 1 per cent of the non-irrigation group ($p = 0.0001$). Thirty-seven per cent of the colonic-irrigation group who submitted specimens had evidence of amoebic infection on either stool examination or serum titre, as compared with 2.4 per cent in the non-irrigation group ($p = 0.0001$).

Persons who were given colonic irrigation immediately after a person who subsequently developed bloody diarrhoea were at the highest risk of developing amoebiasis. Tests of the colonic-irrigation machine after routine cleaning showed heavy contamination with faecal coliform bacteria. The severity of the disease in this outbreak may have been related to the route of inoculation (the normal route of contamination is by oral ingestion of amoebic cysts) since such irrigation may well have injured the intestinal mucosa or stripped it of protective mucus; alternatively, the procedure may have altered the bacterial flora. Any of these events could have facilitated amoebic invasion.

Not surprisingly Istre *et al.* urge that more aggressive action should be taken in the USA to license, limit, control, or prohibit the use of colonic irrigation. They warn that the present report probably deals with only a small fraction of a more widespread problem of bacterial or viral infections spread by colonic-irrigation practices. Corticosteroid usage could also contribute greatly to the fulminant nature of such infections, although no data were given in the present report on whether patients were taking such medication.

ADDENDUM

Acquired immune deficiency syndrome (AIDS)

Whilst this text was in the proof stage, there was increased concern about AIDS (acquired immune deficiency syndrome) and it is appropriate to give some mention to this disease as an addendum to the chapter. Currently, the incidence of AIDS is continually increasing in the United States and some cases have been reported in the United Kingdom, Canada, Europe, Japan, and Australia. Most of the patients were active homosexuals and intravenous drug addicts.

It is now evident that AIDS is caused by a transmissable agent, presumably a virus. The two main contenders for this agent are the human T-cell leukaemia viruses and an unrelated lymphotropic retrovirus. In his 1985 review on this topic, Williams emphasized that anxiety about possible transfer of AIDS in blood products (e.g. Factor VIII) has concerned haemophiliacs and blood transfusion recipients even more during the past year. He reviewed the cases so far reported.

Growth hormone and Creutzfeldt–Jakob disease

Another item of concern comes from a press report in the *Guardian* newspaper of 2 August 1985. This alleges that 500 people given growth hormone as children may unknowingly be at risk from a lethal, slow-acting virus which attacks brain cells. This allegation is based on recent reports in the *Lancet* (Anon 1985; Powell-Jackson *et al.* 1985; Taylor *et al.* 1985) and it suggests that the effects of infection in the early 1970s are beginning to show only now. The growth hormone was extracted from the pituitary glands of cadavers and it is suggested that Creutzfeldt–Jakob disease is an outcome of viral contamination of the extracted hormone; the virus has been found in pituitary glands and contaminated equipment used to extract the hormone from the gland might be to blame for spreading the infection. The hormone was banned by the UK Department of Health and the US Food and Drug Administration in May 1985 after three patients died in the USA.

Whether these fears are exaggerated or not, it is obvious that growth hormone produced as a genetically engineered product would not suffer from such possible contamination. Such products are now available.

RECOMMENDED FURTHER READING

ANON (1985) Ban on growth hormone. *Lancet* i, 1172.

POWELL-JACKSON, J., KENNEDY, P., WHITCOMBE, E.M., WELLER, R.O., PREECE, M.A., AND NEWSOM-DAVIS, J. (1985) Creutzfeldt–Jakob disease after administration of human growth hormone. *Lancet* ii, 244–246.

TAYLOR, D.M., FRASER, H., SALACINSKI, P.R., DICKINSON, A.G., ROBERTSON, P.A. AND LOWRY, P.J. (1985) Preparation of growth hormone free from contamination with unconventional slow viruses. *Lancet* ii, 260–262.

WILLIAMS, J.R.B. (1985) Blood and blood products. In: *Side Effects of Drugs Annual* 9, M.N.G. Dukes ed. Elsevier, Amsterdam – New York – Oxford. pp. 286–295.

REFERENCES

AGUILAR, G.L., BLUMENKRANTZ, M.S., EGBERT, P.R., AND MCCULLEY, J.P. (1979). *Candida* endophthalmitis after intravenous drug abuse. Arch. Ophthalmol. **97**, 96–100.

AHMAD, S. (1980). Antiperistaltic agents contraindicated in pseudomembranous colitis. J. Am. med. Ass. **243**, 1036.

ALLISON, A.C. (1972). Immunity against viruses. In *The scientific basis of medicine*, (ed. I. Gilliland and J. Francis). p.51. Annual Reviews, London.

ARENS, J.F., LEJEUNE, F.E.JR., AND WEBRE, D.R. (1974). Maxillary sinusitis: A complication of nasotracheal intubation. *Anesthesiology* **40**, 415–16.

BADEMOSI, O. AND OSUNTOKUN, B.O. (1979). Prednisolone in the treatment of pneumococcal meningitis. *Trop. Geog. Med.* **31**, 53–6.

BARTLETT, P., REINGOLD, A.L., GRAHAM, D.R., DAN, B.B., SELINGER, D.S., TANK, G.W., AND WICHTERMAN, K.A. (1982). Toxic shock syndrome with surgical wound infections. J. Am. med. Ass. **247**, 1448–50.

BATTS, D.H., MARTIN, D., HOLMES, R., SILVA, J., AND FEKETY, F.R. (1980). Treatment of antibiotic-associated Clostridium difficile diarrhea with oral vancomycin. J. Pediat. **97**, 151–3.

BELSEY, M.A., HOFPAUIR, C.W., AND SMITH, M.H. (1969). Dexamethasone in the treatment of acute bacterial meningitis: the effect of study design on the interpretation of results. *Pediatrics* **44**, 503–15.

BENJAPONGS, W., SADUI, N., AND NOEYPATIMANOND, S. (1970). Trimethoprim/sulphamethoxazole combination in the treatment of falciparum malaria. J. med. Ass. Thai. **53**, 849–56.

BENNINGTON, J.L., HABER, S.L., AND MORGENSTERN, N.L. (1964). Increased susceptibility to cryptococcosis following steroid therapy. Dis. Chest **45**, 262–3.

BIANCHINE, J.R., ALEXANDER, M.S., ANDRESEN, B.D., AND NG, K.J. (1982). The aspirin / Reye's syndrome link. *Lancet* ii, 1333.

BLOUNT, R.E., JR. (1969). Acute falciparum malaria: field experience with quinine/pyrimethamine combined therapy. *Ann. intern. Med.* **70**, 142–7.

BOLTON, R.P. (1980). Vancomycin dose for pseudomembraneous colitis. *Lancet* ii, 428.

——, SHERRIFF, R.J., AND READ, A.E. (1980). Clostridium difficile associated diarrhoea: a role in inflamatory bowel disease. *Lancet* i, 383–4.

BORCHERS, S.L., SKEELS, M.R., AND GEROW, P. (1980). Penicillin-producing gonococcal urethritis. *Morbid. Mortal. Weekly Rep.* **29**, 381–2.

BOWMAN, R.A. AND RILEY, T.V. (1980). Vancomycin therapy. Clostridium difficile. *Med. J. Australia* 2, 98.

BOYD, J.F. AND CHAPPELL, A.G. (1961). Fatal mycetosis due to *Candida albicans* after combined steroid and antibiotic therapy. *Lancet* ii, 19–22.

BRADY, M.T., SHELDON, L.K., AND TABER, L.H. (1981). Association between persistence of pneumococcal meningitis and dexamethasone administration. J. Pediat. **99**, 924–6.

BRITISH MEDICAL JOURNAL (1980). Leading article: Toxic shock and tampons. Br. med. J. **281**, 1426.

—— (1981). Editorial: Drug-resistant tuberculosis. Br. med. J. **283**, 336–7.

BROWN, A.K., FIKRIG, S., AND FINBERG, L. (1983). Aspirin and Reye syndrome. J. Pediat. **102**, 157–8.

BRYAN, H., SHENNAN, A., GRIFFIN, E., AND ANGEL, A. (1976). Intralipid — its rational use in parenteral nutrition of the newborn. *Pediatrics* **58**, 787–90.

CANTRILL, H.L., RODMAN, W.P., RAMSAY, R.C., AND KNOBLOCH, W.H. (1980). Postpartum *Candida* endophthalmitis. *J. Am. med. Ass.* **243**, 1163–5.

CATES, W., ORY, H.W., ROCHAT, R.W., AND TYLER, C.W. (1976). The intrauterine device and deaths from spontaneous abortion. *New Engl. J. Med.* **295**, 1155–9.

CHABNER, B.A., MYERS, C.E., COLEMAN, C.N., AND JOHNS, D.G. (1975). The clinical pharmacology of antineoplastic agents, I. *New Engl. J. Med.* **292**, 1107–13.

CHOPRA, I., HOWE, T.G.B., LINTON, A.H., LINTON, K.B., RICHMOND, M.H., AND SPELLER, D.C.E. (1981). The tetracyclines: prospects at the beginning of the 1980s. *J. Antimicrob. Chemother.* **8**, 5–21.

COHEN, S.N. (1970). Toxoplasmosis in patients receiving immunosuppressive therapy. *J. Am. med. Ass.* **211**, 657–60.

CORAN, A.G. (1972). The intravenous use of fat for the total parenteral nutrition of the infant. *Lipids* **7**, 455–8.

COUNCIL ON PHARMACY AND CHEMISTRY (1951). Warning statement to be included in *Aureomycin* hydrochloride, chloramphenicol and *Terramycin* hydrochloride labelling. *J. Am. med. Ass.* **143**, 1267.

CROMPTON, D.O. (1965). Corticosteroids exacerbate ocular herpes simplex. *Med. J. Australia* **1**, 487.

CROSSLEY, K., LOESCH, D., LANDESMAN, B., MEAD, K., CHERN, M., AND STRATE, R. (1979). An outbreak of infection caused by strains of *Staphylococcus aureus* resistant to methicillin and aminoglycosides. I. Clinical studies. *J. infect. Dis.* **139**, 273–9.

CRUZ, T., REBOUCAS, G., AND ROCHA, H. (1966). Fatal strongyloidiasis in patients receiving corticosteroids. *New Engl. J. Med.* **275**, 1093–6.

D'ARCY, P.F. (1980). Acute problems due to drug-induced or exacerbated infections. In *Drug-induced emergencies* (ed. P.F. D'Arcy and J.P. Griffin), pp. 271–84. Wright, Bristol.

D'ARCY HART, P. AND REES, R.J.W. (1950). Enhancing effect of cortisone on tuberculosis in the mouse. *Lancet* **ii**, 391–5.

DAROFF, R.B., DELLER, J.J. JR., KASTL, A.J. JR., AND BLOCKER, W.W. (1967). Cerebral malaria. *J. Am. med. Ass.* **202**, 679–82.

DAUGHERTY, C.C., MCADAMS, A.J., AND PARTIN, J.S. (1983). Aspirin and Reye's syndrome. *Lancet* **ii**, 104.

DAVIS, J.P., CHESNEY, P.J., WAND, P.J., LAVENTURE, M., AND THE INVESTIGATION AND LABORATORY TEAM (1980). Toxic-shock syndrome. Epidemiologic features, recurrence, risk factors and prevention. *New Engl. J. Med.* **303**, 1429–35.

DENNIS, M. AND ITKIN, I.H. (1964). Effectiveness and complications of aerosol dexamethasone phosphate in severe asthma. *J. Allergy* **35**, 70–6.

DE SAXE, M.J., WIENEKE, A., DE AZAVEDO, J., AND ARBUTHNOTT, J.P. (1982). Toxic shock syndrome in Britain. *Br. med. J.* **284**, 1641–2.

DE SWIET, J. (1968). Cerebral malaria. *Br. med. J.* **3**, 377.

D'ORSI, C.J. ENSMINGER, W., SMITH, E.H., AND LEW, M. (1979). Gasforming intrahepatic abscess: A possible complication of arterial infusion chemotherapy. *Gastrointest. Radiol.* **4**, 157–61.

DRUMMOND, K.N. (1969). Cyclophosphamide and the nephrotic syndrome. *Br. med. J.* **2**, 576–7.

DUCKWORTH, R. (1973). Acute periodontal conditions — their pathogenesis and management. *Br. dent. J.* **135**, 168–9.

DUGUID, H., DUNCAN, I., PARRATT, D., AND TRAYNOR, R. (1982). Actinomyces and intrauterine devices. *J. Am. med. Ass.* **248**, 1579.

DUNLOP, E.M.C., RODIN, P., SETH, A.D., AND KOLATOR, B. (1980). Ophthalmia neonatorum due to beta-lactamase-producing gonococci. *Br. med. J.* **281**, 483.

DuPONT, H.L. AND HORNICK, R.B. (1973). Adverse effects of Lomotil therapy in shigellosis. *J. Am. med. Ass.* **226**, 1525–8.

——, REVES, R.R., GALINDO, E., SULLIVAN, P.S., WOOD, L.V., AND MENDIOLA, J.G. (1982). Treatment of traveler's diarrhea with trimethoprim/sulfamethoxazole and with trimethoprim alone. *New Engl. J. Med.* **307**, 841–4.

EICHENWALD, H.F. (1982). Aspirin's bum rap in studies linking it to Reye's. *Medical World News*, 13 September 1982.

ESPERSEN, E. (1963). Corticosteroids and pulmonary tuberculosis. Activation of four cases. *Acta tuberc. pneumol. belg.* **43**, 1–8.

EVANS, D.A., HENNEKENS, C.H., MIAO, L., LAUGHLIN, L.W., CHAPMAN, W.G., ROSNER, B., TAYLOR, J.O., AND KASS, E.H. (1978). Oral contraceptive use and bacteriuria in a community-based study. *New Engl. J. Med.* **299**, 536–7.

——, MIAO, L., HENNEKENS, C.H., AND KASS, E.H. (1980). Clearance of bacteriuria on discontinuing oral contraception. *Br. med. J.* **280**, 152.

EVANS, T.J., MC COLLUM, J.P.K., AND VALDIMARSSON, H. (1975). Congenital cytomegalovirus infection after maternal renal transplantation. *Lancet* **i**, 1359–60.

EYKYN, S.J. (1982). Toxic shock syndrome: Some answers but questions remain. *Br. med. J.* **284**, 1585–6.

FASAN, P.O.(1971). Trimethoprim plus sulphamethoxazole compared with chloroquine in the treatment and suppression of malaria in African school-children. *Ann. trop. Med. Parasitol.* **65**, 117–21.

FENG, P.H., JAYARATNAM, F.J., TOCK, E.P.C., AND SEAH, C.S. (1973). Cyclophosphamide in treatment of systemic lupus erythematosus: 7 years' experience. *Br. med. J.* **2**, 450–2.

FERGUSON, D.J. (1980). Editorial: Total parenteral nutrition and the team. *J. Am. med. Ass.* **243**, 1931.

FINDLAY, G.M. AND HOWARD, E.M. (1952). The effects of cortisone and adrenocorticotrophic hormone on poliomyelitis and on other virus infections. *J. Pharm. Pharmacol.* **4**, 37–42.

FISCHER, G.W., HUNTER, K.W., WILSON, S.R., AND MEASE, A.D. (1980). Diminished-bacterial defences with intralipid. *Lancet* **ii**, 819–20.

FISHMAN, R.A. (1982). Steroids in the treatment of brain edema. *New Engl. J. Med.* **306**, 359–60.

FREEMAN, R.B. (1979). Urinary tract infections. *Medical Times* **107**, 40–5.

FREI, E., III (1972). Combination cancer therapy. *Cancer Res.* **32**, 2593–607.

FULLER, A.F. JR., SWARTZ, M.N., WOLFSON, J.S., AND SALZMAN, R. (1980). Toxic-shock syndrome. *New Engl. J. Med.* **303**, 881.

GALBRAITH, R.M., EDDLESTON, A.L.W.F., WILLIAMS, R., ZUCKERMAN, A.J., AND BAGSHAWE, K.D. (1975). Fulminant hepatic failure in leukaemia and choriocarcinoma related to withdrawal of cytotoxic drug therapy. *Lancet* **ii**, 528–30.

GEDDES, A.M. (1980). Staphylococcal septicaemia after insertion of an intrauterine device. *Br. med. J.* **281**, 1639.

GERBEAUX, J., COUVREUR, R., BACULARD-BEAUCHEF, A., AND JOLY, J.B. (1963). Maladies infectieuses sont corticothérapie au long cours (194 observations dont 102 varicelles). *Sem. Hôp. Paris* **39**, 61–75.

GILL, K.A., JR., KATZ, H.I., AND BAXTER, D.L. (1963). Fungus infections occurring under occlusive dressings. *Arch. Dermatol.* **88**, 348–9.

GILLES, H.M. (1981). Malaria. *Br. med. J.* **283**, 1382–5.

GOLDMAN, D.A. AND MAKI, D.G. (1973). Infection control in total parenteral nutrition. *J. Am. med. Ass.* **233**, 1360–4.

GORBACH, S.L. (1982). Editorial: Travelers' diarrhea. *New Engl. J. Med.* **307**, 881–3.

GORDON, A. AND PARSONS, M. (1972). The place of corticosteroids in the management of tuberculous meningitis. *Br. J. hosp. Med.* **7**, 651–5.

GOULDING, R. (1980). Toxic shock syndrome. *Br. med. J.* **281**, 1570.

GRANT PETERKIN, G.A. AND KHAN, S.A. (1969). Iatrogenic skin disease. *Practitioner* **202**, 117–25.

GRATTEN, M., NARAQI, S., AND HANSMAN, D. (1980). High prevalance of penicillin-insensitive pneumococci in Port Moresby, Papua New Guinea. *Lancet* **ii**, 192–5.

GREENFIELD, C., BURROUGHS, A., SZAWATHOWSKI, M., BASS, N., NOONE, P., AND POUNDER, R. (1981). Is pseudomembraneous colitis infectious? *Lancet* **i**, 371–2.

GRIMWOOD, K., COOK, J.J., AND ABBOTT, G.D. (1983). Antimicrobial prescribing errors in children. *NZ med. J.* **96**, 785–7.

GROSS, R.J., ROWE, B., CHEASTY, T., AND THOMAS, L.V. (1981). Increase in drug resistance among Shigella dysenteriae, Sh. flexneri, and Sh. boydii. *Br. med. J.* **283**, 575–6.

——, THOMAS, L.V., AND ROWE, B. (1979). Shigella dysenteriae, Sh. flexneri, and Sh. boydii infection in England and Wales: the importance of foreign travel. *Br. med. J.* **2**, 744.

GRUBER, W., ZEIBEKIS, N., AND GOLOB, E. (1970). Über den Soorbefall der Scheide während der Schwangerschaft und unter Einnahme oraler Kontrazeptiva. *Wien. med. Wschr.* **120**, 898–900.

GRUNDFEST, S. AND STEIGER, E. (1980). Home parenteral nutrition. *J. Am. med. Ass.* **244**, 1701–3.

GUPTA, P.K. AND WOODRUFF, J.D. (1982). Editorial: Actinomyces in vaginal smears. *J. Am. med. Ass.* **247**, 1175–6.

GUSTAFSON, A., KJELLMER, I., OLEGARD, R., AND VICTORINI, L. (1972). Nutrition in low-birth-weight infants. I. Intravenous injection of fat emulsion. *Acta pediat. scand.* **61**, 149–58.

HALL, A.P. (1976). The treatment of malaria. *Br. med. J.* **1**, 323–8.

—— (1982). Dexamethasone deleterious in cerebral malaria. *Br. med. J.* **284**, 1588.

HALPIN, T.J., HOLTZHAUER, F.J., CAMPBELL, R.J., HALL, L.J., CORREA-VILLASEÑOR, A., LANESE, R., RICE, J., AND HURWITZ, E.S. (1982). Reye's syndrome and medication use. *J. Am. med. Ass.* **248**, 687–91.

HANSFORD, C.F. AND HOYLAND, J. (1982). An evaluation of co-trimoxazole in the treatment of *Plasmodium falciparum* malaria. *S. Afr. med. J.* **61**, 512–14.

HARDING, T. (1968). Cerebral malaria. *Br. med. J.* **3**, 250.

HARRIS, H.J. (1950). Aureomycin and chloramphenicol in brucellosis with special reference to side-effects. *J. Am. med. Ass.* **142**, 161–5.

HARVEY, M. (1981). Absorption of staphylococcl toxin in toxic-shock syndrome. *New Engl. J. Med.* **305**, 1652–3.

——, HORWITZ, R.I., AND FEINSTEIN, A.R. (1982). Toxic shock and tampons. Evaluation of the epidemiologic evidence. *J. Am. med. Ass.* **248**, 840–6.

HAYES, M. (1965). Esophageal moniliasis. *Am. J. Gastroenterol.* **43**, 143–9.

HIEBER, J.P. AND NELSON, J.D. (1977). A pharmacologic evaluation of penicillin in children with purulent meningitis. *New Engl. J. Med.* **297**, 410.

HOEKELMAN, R.A. (1982). Take two aspirins and call me in the morning. Salicylate use and Reye's syndrome. *Am. J. Dis. Childh.* **136**, 973–4.

HOLMES, C.J. AND ALLWOOD, M.C. (1979). The microbial contamination of intravenous infusions during clinical use. *J. appl. Bacteriol.* **46**, 247–67.

HOLT, P. (1980). Tampon-associated toxic shock syndrome. *Br. med. J.* **281**, 1321–2.

HONE, R., CAFFERKEY, M., KEANE, C.T., HARTE-BARRY, M., MOORHOUSE, E., CARROLL, R., MARTIN, F., AND RUDDY, R. (1981). Bacteraemia in Dublin due to gentamicin resistant *Staphylococcus aureus*. *J. hosp. Inf.* **2**, 119–26.

HULKA, B. S., (1982). Editorial: Tampons and toxic shock syndrome. *J. Am. med. Ass.* **248**, 872–4.

HYMOWITZ, E.E. (1981). Toxic-shock syndrome and the diaphragm. *New Engl. J. Med.* **305**, 834.

IRONS, G.V. (1964). Steroids and herpes zoster. *J. Am. med. Ass.* **189**, 649.

ISTRE, G.R., KREISS, K., HOPKINS, R.S., HEALEY, G.R., BENZIGER, M., CANFIELD, T.M., DICKINSON, P., ENGLERT, T.R., COMPTON, R.C., MATTHEWS, H.M., AND SIMMONS, R.A. (1982). An outbreak of amebiasis spread by colonic irrigation at a chiropractice clinic. *New Engl. J. Med.* **307**, 339–42.

JACOBS, H.W. (1963). Unusual fatal infectious complications of steroid-treated liver disease. *Gastroenterology* **44**, 519–26.

JAFFE, R. (1981). Toxic shock syndrome associated with diaphragm use. *New Engl. J. Med.* **305**, 1585–6.

JEWESSON, P.J., HO, R., JANG, Q., WATTS, G., AND CHOW, A.W. (1983). Auditing antibiotic use in a teaching hospital: focus on cefoxitin. *Can. med. Ass. J.* **128**, 1075–8.

KASS, E.H. AND FINLAND, M. (1953). Adrenocortical hormones in infection and immunity. *Ann. Rev. Microbiol.* **7**, 361–88.

KATZMAN, R., CLASEN, R., KLATZO, I., MEYER, J.S., PAPPIUS, H.M., AND WALTER, A.G. (1977). Brain edema in stroke. *Stroke* **8**, 512–40.

KAUFMAN, D.W., WATSON, J., ROSENBERG, L., HELMRICH, S.P., MILLER, D.R., MIETTINEN, O.S., STOLLEY, P.D., AND SHAPIRO, S. (1983). The effect of different types of intrauterine devices on the risk of pelvic inflammatory disease. *J. Am. med. Ass.* **250**, 759–62.

KEHRBERG, M.W., LATHAM, R.H., HASLAM, B.T., HIGHTOWER, A., TANNER, M., JACOBSON, J.A., BARBOUR, A.G., NOBLE, V., AND SMITH, C.B. (1981). Risk factors for staphylococcal toxic-shock syndrome. *Am. J. Epidemiol.* **114**, 873–9.

KNODEL, A.R. AND BEEKMAN, J.F. (1982). Unexplained fevers in patients with nasotracheal intubation. *J. Am. med. Ass.* **248**, 868–70.

KRYLOV, A.A. AND MURZIN, A.S. (1979). On severe purulent disease in patients with rheumatoid polyarthritis during prolonged treatment with steroid hormones. *Vopr. Revm. (USSR)* **19**, 57–9.

KUEHNLE, J.C. (1983). Traveler's diarrhea. *New Engl. J. Med.* **308**, 464.

LACY, R.W. (1975). Antibiotic resistant plasmids of *Staphylococcus aureus* and their clinical importance. *Bact. Rev.* **39**, 1–32.

LA MONT, J.T. AND TRNKA, Y.M. (1980). Therapeutic implications of Clostridium difficile toxin during relapse of chronic inflammatory bowel disease. *Lancet* **i**, 381–3.

LANCET (1980). Editorial: Clostridium difficile and chronic bowel disease. *Lancet* i, 402–3.

—— (1981). Leading article: Gentamicin and staphylococci. *Lancet* ii, 127–8.

—— (1982). Editorial: Steroids in bacterial meningitis — helpful or harmful? *Lancet* i, 1164.

LAWSON, D.H. AND HENRY, D.A. (1977). Drug therapy reviews: Intravenous fluid therapy. *Am. J. hosp. Pharm.* **34**, 1332–8.

LAYDE, P.M. (1983). Editorial: Pelvic inflammatory disease and the Dalkon shield. *J. Am. med. Ass.* **250**, 796–7.

LEA, S. AND ELLIS-PEGLER, R.B. (1980). Toxic shock and tampons. *Br. med. J.* **281**, 1639.

LEHNER, T. (1964). Oral thrush, or acute pseudomembranous candidiasis. A clinicopathologic study of forty-four cases. *Oral. Surg.* **18**, 27–37.

LESSOF, M.H. (1972). The current status of transplantation immunology. In *The scientific basis of medicine* (ed. I. Gilliland and J. Francis), p.22. Annual Reviews, London.

LEVY, J.S. (1969). Vaccinia gangrenosum; rare complication of smallpox vaccination. *South. med. J.* **62**, 1408–11.

LICHTENSTEIN, P.K., HEUBI, J.E., DAUGHERTY, C.C., FARELL, M.K., SOKOL, R.J., ROTHBAUM, R.J., SUCHY, F.J., AND BALISTRERI, W.F. (1983). Grade I Reye's syndrome. A frequent cause of vomiting and liver dysfunction after varicella and upper-respiratory-tract infection. *New Engl. J. Med.* **309**, 133–9.

LISHMAN, A.H., AL JUMAILI, I.J., AND RECORD, C.O. (1981). Spectrum of antibiotic-associated diarrhoea. *Gut* **22**, 34–7.

LOGAN, R., HARRON, D.W.G., AND D'ARCY, P.F. (1982). Tropical diseases: The pharmacist's role. *Pharmacy Int* **3**, 297–300.

LOOMIS, L. AND FEDER, H.M. JR. (1981). Toxic-shock syndrome associated with diaphragm use. *New Engl. J. Med.* **305**, 1585.

MABECK, C.E. (1982). Prescription of antibacterial drugs for the treatment of otitis media and upper respiratory tract infections in general practice in Denmark. *Acta Otolaryngol.* **93**, 69–72.

MACKOWIAK, P.A. (1979). Microbial synergism in human infections, Part 2. *New Engl. J. Med.* **298**, 83–7.

MAKI, D.G. (1976). Sepsis arising from extrinsic contamination of the infusion and measures for control. In *Microbial hazards of infusion therapy.* (ed. I. Phillips, P.D. Meers, and P.F. D'Arcy), pp. 99–143. MTP Press, Lancaster.

——, McCORMICK, R.D., UMAN, S.J., AND WIRTANEN, G.W. (1979). Septic endarteritis due to intra-arterial catheters for cancer chemotherapy. I Evaluation of an outbreak, II Risk factors, clinical features and management, III Guidelines for prevention. *Cancer* **44**, 1228–40.

MAYFIELD, R.B. (1962). Tuberculosis occurring in association with corticosteroid treatment. *Tubercle, Edinburgh* **43**, 55–60.

McCORMACK, T.T. AND HONE, R. (1981). Antibiotic usage in a surgical unit. *Irish med. J.* **74**, 284–6.

McCORMICK, J. AND ALLWRIGHT, S. (1980). Toxic shock and tampons. *Br. med. J.* **281**, 1639.

MEADOW, S.R., WELLER, R.O., AND ARCHIBALD, R.W.R. (1969). Fatal systemic measles in a child receiving cyclophosphamide for nephrotic syndrome. *Lancet* ii, 876–8.

MEDICAL LETTER (1980). Medical letter on drugs and therapeutics: topical hydrocortisone without a prescription. *Med. Lett.* **22**, 38–9.

MEDVED, A. AND MAXWELL, I. (1974). Intermittent cyclophosphamide in pemphigus vulgaris and bullous pemphigoid. *Can. med. Ass. J.* **111**, 245–50.

MERSELIS, J.G., KAYE, D., AND HOOK, E.W. (1964). Disseminated herpes zoster. A report of 17 cases. *Arch. intern. Med.* **113**, 679–86.

MEYLER, L. (Ed.) (1963). *Side-effects of drugs*, 4th edn., p. 236. Excerpta Medica, Amsterdam.

—— (Ed.) (1966). *Side-effects of drugs*, 5th edn., pp. 415–16. Excerpta Medica, Amsterdam.

MODY, V.R. (1959). Corticosteroids in latent amoebiasis. *Br. med. J.* **2**, 1399.

MONTGOMERIE, J.Z., BECROFT, D.M.O., CROXSON, M.C., DOAK, P.B., AND NORTH, J.D.K. (1969). Herpes simplex infection after renal transplantation. *Lancet* ii, 867–71.

—— AND EDWARDS, J.E. (1978). Association of infection due to *Candida albicans* with intravenous hyperalimentation. *J. infect. Dis.* **137**, 197–201.

MORRISON, P.S. AND LITTLE, T.M. (1981). How is gastroenteritis treated? *Br. med. J.* **283**, 1300.

MORTIMER, E.A. JR. (1982). Reye syndrome and salicylates. *Int. J. Epidemiol.* **11**, 314–15.

MURRAY, B.E., RENSIMER, E.R., AND DUPONT, H.L. (1982). Emergence of high-level trimethoprim resistance in fecal *Escherichia coli* during oral administration of trimethoprim or trimethoprim-sulfamethoxazole. *New Engl. J. Med.* **306**, 130–5.

NABARRO, J.D.N. (1960). The pituitary and adrenal cortex in general medicine. *Br. med. J.* **2**, 533–8, 625–33.

NEHME, A.E. (1980). Nutritional support of the hospitalized patient. The team concept. *J. Am. med. Ass.* **243**, 1906–8.

NELSON, J.D. (1980). The increasing frequency of β-lactamase producing Haemophilus influenzae B. *J. Am. med. Ass.* **244**, 2390.

NOLAN, C.M., McALLISTER, C.K., WALTERS, E., AND BEATY, H.N. (1978). Experimental pneumococcal meningitis IV. The effect of methyl prednisolone on meningeal inflammation. *J. Lab. clin. Med.* **91**, 979–88.

ORIEL, J.D., PARTRIDGE, B.M., DENNY, M.J., AND COLEMAN, J.C. (1972). Genital yeast infections. *Br. med. J.* **4**, 761–4.

ORISCELLO, R.G. (1968). Cerebral malaria. *Br. med. J.* **3**, 617–18.

OSTERHOLM, M.T., DAVIS, J.P., GIBSON, R.W., MANDEL, J.S., WINTERMEYER, L.A., HELMS, C.M., FORFANG, J.C., RONDEAU, J., VERGERONT, J.M., AND THE INVESTIGATION TEAM (1982). Tri-state toxic-shock syndrome: I. Epidemiologic findings. *J. infect. Dis.* **145**, 431–40.

PARK, R.Z., GOLTZ, R.W., AND CAREY, T.B. (1967). Unusual cutaneous infections associated with immunosuppressive therapy. *Arch. Dermatol.* **95**, 345–50.

PETRIE, J.C., DURNO, D., AND HOWIE, J.G.R. (1974). Drug interaction in general practice. In *Clinical effects of interaction between drugs* (ed. L.E. Cluff and J.C. Petrie), pp. 235–53. Excerpta Medica, Amsterdam.

PLATT, R., POLK, B.F., MURDOCH, B., AND ROSNER, B. (1982). Mortality associated with nosocomial urinary-tract infection. *New Engl. J. Med.* **307**, 637–41.

RAMPLING, A., WARREN, R.E., AND SYKES, H.V. (1980). Relapse of Clostridium colitis after vancomycin therapy. *J. Antimicrob. Chemother.* **6**, 551–2.

REACTIONS (1982). Editorial: Reye's syndrome and aspirin. *Reactions*, no. 63, pp. 8–9. Adis Press, New York.

REID, H.A., AND NKRUMAH, F.K. (1972). Fibrin-degradation products in cerebral malaria. *Lancet* i, 218–21.

REINGOLD, A.L., DANN, B.B., SHANDS, K.N., AND BROOME, C.V.

(1982). Toxic shock syndrome not associated with menstruation. *Lancet* i, 1–4.

RIDLEY, C.M. (1972). A review of the recent literature on diseases of the vulva. II Vulvitis infections. *Br. J. Dermatol.* 87, 58–69.

REYE SYNDROME WORKING GROUP, ASPIRIN FOUNDATION OF AMERICA INC. (1982). Reye syndrome and salicylates: spurious association. *Pediatrics* 70, 158–60.

ROBERTS, R.K. AND SENEVIRATNE, E. (1980). Vancomycin therapy. Clostridium difficile. *Med. J. Australia* 2, 98.

ROSCHLAU, G. (1967). Two cases of generalized fatal varicella during treatment with large doses of corticosteroids. *Münch. med. Wschr.* 109, 1889–92.

ROSENBAUM, E.H., COHEN, R.A., AND GLATSTEIN, H.R. (1966). Vaccination of a patient receiving immunosuppressive therapy for lymphosarcoma. *J. Am. med. Ass.* 198, 737–40.

RUSKIN, J. AND REMINGTON, J.S. (1967). The compromised host and infection. I. *Pneumocystis carinii* pneumonia. *J. Am. med. Ass.* 202, 1070–4.

SAHAY, J.N., CHATTERJEE, S.S., AND STANBRIDGE, T.N. (1979). Inhaled corticosteroid aerosols and candidiasis. *Br. J. Dis. Chest.* 73, 164–8.

SANTOSHAM, M., SACK, R.B., FROEHLICH, J., GREENBERG, H., YOLKEN, R., KAPIKAIN, A., JAVIER, C., MEDINA, C., ORSKOV, F., AND ORSKOV, I. (1981). Biweekly prophylactic doxycycline for travelers' diarrhea. *J. infect. Dis.* 143, 598–602.

SCHLECH, W.F., SHANDS, K.N., REINGOLD, A.L., DANN, B.B., SCHMID, G.P., HARGRETT, N.T., HIGHTOWER, A., HERWALDT, L.A., NEILL, M.A., BRAND, J.D., AND BENNETT, J.V. (1982). Risk factors for development of toxic shock syndrome. Association with a tampon brand. *J. Am. med. Ass.* 248, 835–9.

SCHWEIKER, R.C. (1982). Secretary, US Department of Health and Human Services. Press release. P 82–40, dated September 20th 1982.

SHANDS, K.N., SCHMID, G.P., DAN, B.B., BLUM, D., GUIDOTTI, R.J., HARGRET, N.T., ANDERSON, R.L., HILL, D.L., BROOME, C.V., BAND, J.D., AND FRASER, D.W. (1980). Toxic-shock syndrome in menstruating women: Association with tampon use and *Staphylococcal aureus* and clinical features in 52 cases. *New Engl. J. Med.* 303, 1436–42.

SHARMA, R. (1959). Effect of cholesterol on the growth and virulence of *Entamoeba histolytica*. *Trans. R. Soc. trop. Med. Hyg.* 53, 278–81.

SHELD, W.M. AND BRODEUR, J.P. (1981). Effects of methyl prednisolone on antibiotic entry into the cerebrospinal fluid. *Clin. Res.* 29, 369A.

SHUBIN, H., LAMBERT, R.E., HEIKIN, C.A., SOKMENSUER, A., AND GLASKIN, A. (1959). Steroid therapy and tuberculosis. *J. Am. med. Ass.* 170, 1885–90.

SIEGEL, I. AND GLEICHER, N. (1981). Toxic shock syndrome and human platelets. *Lancet* i, 493.

SMITH, A.J. (1980). Toxic shock syndrome. *Br. med. J.* 281, 1570.

SMITH, D.H. AND HARPER, P.S. (1968). Cerebral malaria. *Br. med. J.* 3, 377.

SMITSKAMP, H. AND WOLTHUIS, F.H. (1971). New concepts in treatment of malignant tertian malaria with cerebral involvement. *Br. med. J.* 1, 714–16.

SPIRER, Z. (1975). Prevention and treatment of varicella during steroid therapy. *Lancet*, i, 635.

STARKO, K.M., RAY, C.G., DOMINGUEZ, L.B., STROMBERG, K.L., AND WOODALL, D.F. (1980). Reye's syndrome and salicylate use. *Pediatrics* 66, 859–64.

STAUFFER, J.L., OLSON, D.E., AND PETTY, T.L. (1981). Complications and consequences of endotracheal intubation and tracheostomy. *Am. J. Med.* 70, 65–76.

STEFFEN, R. AND GSELL, O. (1981). Prophylaxis of traveller's diarrhoea. *J. trop. Med. Hyg.* 84, 239–42.

STEEL, K., GERTMAN, P.M., CRESCENZI, C., AND ANDERSON, J. (1981). Iatrogenic illness on a general medical service at a university hospital. *New Engl. J. Med.* 304, 638–42.

STEVENSON, D.K. (1974). Pulmonary tuberculosis. *Practitioner* 212, 320–6.

SWINDELL, P.J., REEVES, D.S., BULLOCK, D.W., DAVIES, A.J., AND SPENCE, C.E. (1983). Audits of antibiotic prescribing in a Bristol hospital. *Br. med. J.* 286, 118–22.

TAEUSCH, H.W., FRIGOLETTO, F., KITZMILLER, J., AVERY, M.E., HEHRE, A., FROMM, B., LAWSON, E., AND NEFF, R.K. (1979). Risk of respiratory distress syndrome after prenatal dexamethasone treatment. *Pediatrics* 63, 64–72.

TAKAHASHI, M. AND LOVELAND, D.B. (1974). Bacteriuria and oral contraceptives. Routine health examinations of 12076 middle-class women. *J. Am. med. Ass.* 227, 762–5.

THOMAS, S.W., BAIRD, I.M., AND FRAZIER, R.D. (1982). Toxic shock syndrome following submucous resection and rhinoplasty. *J. Am. med. Ass.* 247, 2402–3.

THORN, P.A. (1981). Misuse of corticosteroids. *Prescribers' J.* 21, 191–6.

TODD, J., FISHAUT, M., KAPRAL, F., AND WELCH, T. (1978). Toxic-shock syndrome associated with phage-group-I staphylococci. *Lancet* ii, 1116–18.

TOMASZEWSKI, T. (1951). Side-effects of chloramphenicol and aureomycin with special reference to oral lesions. *Br. med. J.* 1, 388–92.

TRUNET, P., LE GALL, J.-R., L'HOSTE, F., REGNIER, B., SAILLARD, Y., CARLET, J., AND RAPIN, M. (1980). The role of iatrogenic disease in admission to intensive car. *J. Am. med. Ass.* 244, 2617–20.

UNITED STATES CENTERS FOR DISEASE CONTROL (1982). *Morbid. Mortal. Weekly Rep.* 31, 53.

UNITED STATES DEPARTMENT OF HEALTH AND HUMAN SERVICES/PUBLIC HEALTH SERVICES (1980a). Toxic-shock syndrome. *Morbid. Mortal. Weekly Rep.* 29, 229–30.

—— (1980). Follow-up on toxic-shock syndrome. *Morbid. Mortal. Weekly Rep.* 29, 441–5.

—— Toxic-shock syndrome. *Morbid. Mortal. Weekly Rep.* 29, 495–6.

VESSEY, M.P., YEATES, D., FLAVEL, R., AND MCPHERSON, K. (1981). Pelvic inflammatory disease and the intrauterine device: findings in a large cohort study. *Br. med. J.* 282, 855–7.

VEUTHEY, J. (1962). Tuberculose aigüe au cours de traitements par la cortisone. A propos de cinq observations. *Praxis* 51, 403–6.

WALDMAN, R.J., HALL, W.N., MCGEE, H., AND VAN AMBURG, G. (1982). Aspirin as a risk factor in Reye's syndrome. *J. Am. med. Ass.* 247, 3089–94.

WALKER, A.M., JICK, H., AND PORTER, J. (1979). Drug-related superinfection in hospitalized patients. *J. Am. med. Ass.* 242, 1273–5.

WANDS, J.R. (1975). Subacute and chronic active hepatitis after withdrawal of chemotherapy. *Lancet* ii, 979.

WANG, N.-S., HUANG, N.-S., AND THURLBECK, W.M. (1970). Combined *Pneumocystis carinii* and cytomegalovirus infection. *Arch. Pathol.* 90, 529–35.

WARRELL, D.A., LOOAREESUWAN, S., WARRELL, M.J., KASEMSARN, P., INTARAPRASERT, R., BUNNAG, D., AND HARINASUTA, T. (1982).

Dexamethasone proves deleterious in cerebral malaria. A double-blind trial in 100 comatose patients. *New Engl. J. Med.* **306**, 313–19.

WEINSTEIN, L. (1947). The spontaneous occurrence of new bacterial infections during the course of treatment with streptomycin or penicillin. *Am. J. med. Sci.* **214**, 56–63.

WILLIAMS, G.R., TAW, T.L., KENNEDY, D.H., AND LOVE, W.C. (1982). Delayed diagnosis of malaria. *Br. med. J.* **284**, 1616–17.

WOODRUFF, A.W. (1969). Current practice in tropical medicine. *Trans. med. Soc., London* **85**, 111–25.

—— AND DICKINSON, C.J. (1968). Use of dexamethasone in cerebral malaria. *Br. med. J.* **3**, 31–2.

WOODS, J.W., MANNING, I.H., JR, AND PATTERSON, C.N. (1951). Monilial infections complicating the therapeutic use of antibiotics. *J. Am. med. Ass.* **143**, 207–11.

WOODWARD, S.C. AND SHELDON, W.H. (1961). Subclinical *Pneumocystis carinii* pneumonitis in adults. *Bull. Johns Hopk. Hosp.* **109**, 148–59.

YAFFEE, H.S. AND GROTS, I. (1965). Moniliasis due to *Norethynodrel* with *Mestranol. New Engl. J. Med.* **272**, 647.

37 Adverse effects associated with vaccines

J.A. HOLGATE

Introduction

Vaccines from their first recorded use have been associated with unwanted effects ranging from no reaction at all to such severe brain damage that the patient becomes decerebrate or dies. These unwanted effects may be more capable of interpretation if some attempt is made at classification.

Some of the more serious and severe unwanted effects which occurred in the past were not in the nature of 'adverse effects' as normally defined but were due to product defects. Two which fell into this category were the inclusion in a BCG vaccine of a fully virulent strain of Mycobacterium tuberculosis and the inclusion in a supposedly inactivated poliomyelitis vaccine of living poliovirus. Such episodes are sufficiently rare to become historical landmarks and their rarity may well be accounted for by a combination of highly responsible manufacturers and systems of check scrutiny imposed by certain National Authorities. In the United Kingdom inspection of manufacture of vaccines together with scrutiny of protocols and testing of samples of production batches has been in operation for some 60 years — initially under the Therapeutic Substances Acts of 1925 and 1956 and latterly under the Medicines Act (1968). The stimulus to such control came originally from an observation by the British Pharmacopoeia Committee in 1909 to the effect that it was impossible to verify the quality of certain recent products such as immune sera by means of the tests then at their disposal. This observation was taken up by a Departmental Committee (Ministry of Health) appointed to consider and advise upon the legislative and administrative measures to be taken for the effective control of the quality and authenticity of such therapeutic substances offered for sale to the public as cannot be tested adequately by direct chemical means. Their report published by HMSO (Ministry of Health 1921) and containing much still very valid today, led to the passing of the Therapeutic Substances Act 1925. This was replaced in 1956 by a further Therapeutic Substances Act which contained a Part II dealing with control of sale and supply, thereby being totally unrelated to Part I which, like its predecessor in 1925, was concerned with quality control. The first schedule to the 1956 Act included the substances commonly known as vaccines, sera, toxins, antitoxins and antigens. Regulations made under the Act in the following years detailed minimal requirements of manufacture and quality control of vaccines as they appeared on the market — the last, in 1971, covering poliomyelitis and rubella vaccines when prepared in monkey kidney or human diploid cells, or rabbit kidney or human diploid cells, respectively. The eventual replacing of the Therapeutic Substances Acts by the implementation of the Medicines Act 1968 did not result in the loss of these minimal requirements which reappeared as the *Compendium of licensing requirements for the manufacture of (certain) biological medicinal products* — the first edition (without certain in its title) in 1977 and the second edition (including certain) in 1982. These requirements take the form of Standard Provisions for Licences and as such, are more flexible than the older Regulations but they, nevertheless, form the same safeguards when combined with laboratory scrutiny as did the former texts.

The laboratory control in early years was vested in the Medical Research Council — Dr (later Sir) Henry Dale as director of the laboratories having been a member of the 1920 Advisory Committee. In the mid-1950s with the advent of poliomyelitis vaccines the Ministry of Health in conjunction with the Medical Research Council additionally set up the Division of Immunological Products Control at Holly Hill, Hampstead. In the early-1970s the Division of Biological Standards and that of Immunological Products Control were brought together in Holly Hill as the National Institute for Biological Standards and Control eventually under the Biological Standards Board set up by the Biological Standards Act 1975.

The most common of the unwanted effects associated with vaccines are more correctly classified as adverse reactions although they also may be regarded as partly associated with quality of product. These consist of the local responses at the injection site — redness, swelling, pain, tenderness — and the assorted general responses of headache, fever, malaise, etc. which may be associated. These reactions are most probably associated with the injection of foreign protein. This foreign protein may be either specific or non-specific and arises from several sources, including the medium in which or on which the vaccine was prepared, additives to the medium, e.g. antibiotics, additives to the final preparation of the vaccine, and, finally, from the vaccine itself. Thus both local and general reactions are commonly seen in connection with bacterial vaccines containing intact dead organisms, such as pertussis, typhoid, and paratyphoid bacteria. A similar situation occurs with whole-virus vaccines even when inactivated. Thus the original influenza vaccines (inactivated) frequently give rise to reactions but these have been reduced by successive purification through the 'split' virus vaccine to the purified surface antigens of haemagglutinin and neuraminidase. Some reactions still occur even when the latter purified vaccine is given possibly indicating the 'foreign' nature of the ultimate antigens.

The more 'reactive' antibiotics such as penicillin are no longer used in the preparation of vaccines and label statements that poliomyelitis vaccine may contain penicillin merely reflect the fact that the original seed strain was grown in the presence of this antibiotic and trace carry-over cannot be guaranteed not to occur. Reactions in vaccinees will only occur in the most hypersensitive individuals. Antibiotics permitted in production at present are less reactogenic and may be present in larger traces than penicillin.

The case of vaccines where the virus has been grown in eggs is rather similar. Modern methods of purification have reduced the amount of egg protein left in the final preparation to a considerable degree. Recent trials of measles vaccine have shown it capable of being without untoward effect to children known to be sensitive to egg, but the label and data sheet warnings should still be maintained as such.

Although the impurities of antibiotics and egg proteins have been included in this section of local and general adverse effects, the nature of the reactions in these cases are somewhat different from those occurring to non-specific impurities, being anaphylactic in character and only in vaccinees already primed. Similar reactions can occur when an individual has been over-exposed to the active antigen of a vaccine even when present in a highly purified form.

Modern methods of purification combined with production involving biotechnology may well result in future vaccines which consist essentially of the immunizing antigen. Only when this is available in any given case will a study be possible to determine the extent to which the 'extraneous' elements have contributed to these common reactions. The speed with which such developments occur will be determined by the cost/benefit equation and will obviously take second place to those developments where known toxic factors are present in the vaccine which may be separable from an immunizing antigen, for example, pertussis vaccine.

Having dealt with unwanted effects arising from faulty manufacture or control of quality and the adverse reactions possibly due to impurities in the product, such reactions as are left must be due to the vaccine itself assuming such a separation can be made. These will be dealt with in this chapter under the separate heads of the vaccines. There are, however, some general observations applying to many or all vaccines.

The most obvious unwanted effects which were mentioned at the start of this introduction are those of no response at all or an excessive response often presenting as the disease the vaccine is being given to prevent. When not due to errors of product, such extremes are caused by 'errors' in the vaccinee. Such errors may be no more than the fact that the vaccinee is an individual at one or other extreme of the frequency distribution curve representing response. Such biological variability alone will result in unwanted effects, even death, resulting from vaccination and illustrates the statement made by Sir Graham Wilson (1967) that the use of any vaccine or antiserum for the purpose of immunization carries some risk of causing an adverse reaction. Such risks are increased if the position of the individual on the response curve is shifted by other factors, for example, immunosuppression, and consideration of these provide many of the items on lists of contraindications to vaccination. It thus becomes of extreme importance to study and comply with advice given in connection with vaccination.

The whole question of risk–benefit ratio present with virtually all drugs is brought to the highest pitch in the case of vaccination since the prophylactic agent is usually being given to an otherwise entirely healthy individual and the benefits of proper use are gained not only by that individual but also by the community. The acceptability of cetain vaccines has recently been questioned both by the medical profession and also the mass media as a result of publicity about adverse reactions. This has produced two results — the good being a realization that vaccination produces great benefits but needs care and attention in administration, the bad being the decreased use of several good vaccines which have been shown to provide protection.

Adverse-reaction reporting is an essential part of mass prophylaxis by vaccination and the framework for full study of such reactions has now been further developed. Such study is essential to provide information on which to base any policy of use and any list of contraindications such as are published in the vaccine manufacturer's data sheet and package inserts and also in the Department of Health and Social Security's (1983) publication, *Immunization against infectious disease.* Policy of use must be formulated on hard scientific evidence which, if not already available, must be actively sought. It must recognize the nature of the disease against which protection is being required; often as a result of many years of virtual eradication of a disease a previously valid risk–benefit ratio is attacked without justification as the nature of the disease has been forgotten while the risk has been magnified by media publicity. Such a shift should not be possible if both benefit and risk are closely monitored and all the evidence is constantly updated.

Vaccines of bacterial origin

These range from highly purified preparations of toxoids (tetanus, diphtheria) through simple suspensions of bacterial cells (typhoid, paratyphoid, cholera, pertussis, autogenous vaccine) to live attenuated organisms. While the latter are used in veterinary medicine, the only example at present used in the human field is BCG. Autogenous vaccines are very rarely used at the present time and can be particularly dangerous, especially if containing living organisms or, occasionally, if used on individuals not providing the original strains present in the killed suspension. Addition of adjuvants, superficial injection, or freezing can also lead to moderate-to-severe local or general reactions.

BCG vaccine

The bacillus of Calmette and Guerin used in this live attenuated bacterial vaccine (BCG) was accidentally changed at an early stage in its development on one occasion for a fully virulent strain of Mycobacterium tuberculosis due to incompetent manufacture in the laboratory. This brought about the Lubeck disaster in which many babies and young children contracted tuberculosis, and many of them died; as a result the use of this vaccine fell into disrepute after 1930.

After more than 20 years during which time improvements had been introduced in the manufacture of vaccines such as better organization, documentation, and quality-control backed up by legislation, interest was again aroused in the use of this vaccine. Freeze-drying techniques had been developed for penicillin production during the war and these were used to make a stable vaccine which could be transported and used safely in hot

countries, thereby overcoming loss of potency and failure to immunize.

The most common adverse reactions in connection with BCG vaccine involve the skin and underlying tissues at the site of injection. When the injection is correctly given intradermally, occasionally a discharging ulcer may result or an abscess with enlarged regional lymph glands. Such severe reactions are probably more frequently due to errors of technique in vaccination including subcutaneous instead of intradermal injection, giving in an inappropriate site, and using an inappropriate needle or errors in reading of the previous tuberculin response. (*British Medical Journal* 1983). In some instances hypertrophic or keloid scars can result, the frequency not being known but depending upon the site of injection, the presence of infection, and the pigmentation of the skin (Sanders and Dickson 1982).

An exaggerated response and a more widespread infection may result in patients who have generalized eczema, infective skin conditions, or a history of deficient immunity. Similarly patients with a depressed immune response due to treatment with systemic corticosteroids, irradiation, or antimetabolites may respond in like fashion and all of these conditions are contraindicated when the use of BCG vaccine is being considered. Vaccines prepared from bacteria or toxoids should not be administered during the period seven days before to 10 days after BCG vaccine is given though diphtheria/tetanus vaccine has been successfully given in an opposite limb at the same time as BCG vaccine.

Diphtheria vaccine

This vaccine takes the form of a toxoid which is the cell-free toxin treated so as to remove its toxic properties while retaining antigenicity. The plain toxoid is usually adsorbed on to an adjuvant (aluminium phosphate or hydroxide) to improve the immune response. It is more commonly given for primary immunization in combination with tetanus or tetanus and pertussis vaccines. An alternative method of delayed release has been the combination of the toxoid with antitoxin prepared in horses, the combination taking the form of a floccular suspension.

The main indication for use of diphtheria vaccine alone is for specific immunization of persons at risk over the age of 10 years. If the normal vaccine is to be used, it is necessary to carry out a Schick test in order to determine whether the individual is susceptible to the disease or hypersensitive to the vaccine but, if low-dose vaccine prepared for use in adults is available, this may be safely used, it is claimed, without the need for a preliminary Schick test. Formerly the toxoid–antitoxin floccules would have been used in these circumstances assuming the Schick test indicated it was necessary and safe to immunize but there was then introduced the risk of sensitization or reactions to horse serum protein.

The most common adverse effects of diphtheria vaccine are those non-specific local and general reactions described in the introduction. Allergic reactions are rare if the precautions and contraindications are fully observed. Occasionally, a painless nodule may be formed at the site of injection which may last for some weeks. Any more serious reactions should be very fully investigated before ascribing to the vaccine.

Tetanus vaccine

Tetanus vaccine is available as toxoid in simple solution or as toxoid adsorbed on aluminium hydroxide. The adsorbed vaccine, which must not be injected intradermally if superficial nodules are to be avoided, is preferred for primary immunization but subsequent doses may be of either vaccine. A complete course of injections produces a very durable immunity and it is not uncommon for adults receiving booster doses to have severe local reactions due to over–immunization especially with adsorbed vaccine.

These reactions which are almost restricted to adults but can also occur rarely in children, consist of pain, swelling, and redness and usually appear within a few hours of injection. In some cases these reactions are delayed for a period of up to 10 days following administration. Urticaria and other signs of serum sickness have been reported following the use of these two tetanus vaccines even though neither contains horse serum. All of these reactions usually resolve quickly in a few days.

Since it is clear that local reactions increase in both incidence and severity with repetition of booster doses recommendations have been made concerning the frequency with which such reinforcing injections are given.

Inadvertent freezing of the vaccines is likely to spoil them and thereby reduce their efficacy as well as increase the chances of an adverse reaction.

Diphtheria and tetanus vaccines

The combination of diphtheria and tetanus vaccines has been available for some time mainly for use in reinforcement in the absence of the pertussis component. With the reduction in acceptance of this latter vaccine the dual combination has been used much more frequently for primary immunization than was the case. As a result there has been an extensive increase in the frequency and range of adverse effects claimed to be due to the combined vaccines. While the vast majority of such reactions are minor in character and as described above (and increase in frequency with second and third doses), others are both different in character and more serious including simple or febrile convulsions, encephalitis or encephalopathy, and infantile spasms. These mirror the reactions claimed to be causally related to pertussis vaccine and each needs to be investigated with great care before accepting such a causal relationship. Such an investigation was included in the National Childhood Encephalopathy Study and failed to show a statistically significant difference in the incidence of convulsions and encephalitis in previously normal children in each seven-day period up to 28 days after immunization with diphtheria/tetanus vaccine compared with controls except for an increase in the first week followed by a deficit in second and third weeks suggesting the vaccine might be a precipitating factor in those children prone to convulsions. There was no statistically significant difference between previously normal children vaccinated up to 28 days before developing infantile spasms and the controls but here again, when each seven-day period was separately examined, it appeared that vaccination might have been a precipitating factor.

Unexpected death in infants (cot deaths) has also been claimed on occasions to be related to vaccination but a case-control study from April 1979 to March 1982 (Taylor and Emery 1982) showed babies dying unexpectedly were less likely to have received any form of immunization (including diphtheria/tetanus) than the matched controls.

Pertussis vaccine

Although pertussis vaccine either as a simple solution (plain) or adsorbed on aluminium hydroxide (adsorbed) has been regularly available, its use has been small as a monocomponent vaccine. The original investigations carried out by the Medical Research Council were, of course, with the pertussis vaccine alone and there has been more extensive use in recent years in immunizing children in whom the primary course had consisted of diphtheria and tetanus vaccines. The bulk of the literature dealing with adverse effects has concerned the triple vaccine of diphtheria, tetanus, and pertussis.

Diphtheria, tetanus, and pertussis vaccine

By far the most common adverse effects are the local reactions at the site of injection already described in the introduction. Occasionally a nodule may appear at the site of injection particularly with adsorbed vaccine.

More rarely fever and restlessness together with irritability may occur a few hours after inoculation and there may be crying and loss of appetite. These systemic reactions are less common after the use of adsorbed vaccine. Allergic reactions with pallor and dyspnoea have been reported. An increased incidence of all such reactions has been noted as being associated with failing to resuspend the vaccine before injection or to too rapid injection.

In the past when wild polio virus was present in the community, provocation paralysis — a paralysis of the muscle groups which move the affected limb above the site of inoculation — was seen especially after the use of adsorbed triple vaccine. This has not now been reported for many years.

Of more serious import than the above are the numerous adverse effects, mainly connected with the central nervous system, which have been claimed at one time or another, singly or in combination, to be causally related to the pertussis component of the triple vaccine. These effects include unexplained cot deaths, infantile spasms, convulsions both febrile and afebrile, and encephalopathy/encephalitis. A brief account of the recent developments relating to these effects is given in an attempt to put them into context.

Prior to 1974 the DTP vaccine acceptance rate in the UK population had risen to a level of 80 per cent. In that year considerable adverse publicity appeared in the daily press and on the radio following critical comment in medical journals that the risk of brain damage outweighed any advantage that the vaccine might offer. As a consequence public confidence was undermined and the acceptance rate fell to a level of 31 per cent in 1978. Vaccination is usually carried out in infants from 3 months to 3 years of age and it is of considerable scientific interest that an epidemic of whooping cough began at the end of 1977 and lasted until the middle of 1979. The number of whooping cough cases notified was in the region of 102 500, a number far greater than in any other epidemic since whooping cough vaccination was introduced and promoted on a national scale in 1957 by the then Ministry of Health. Twenty-seven children died in this epidemic and there were 17 cases of severe neurological illness associated with whooping cough itself reported to the National Childhood Encephalopathy Study. This epidemic subsided to an inter-epidemic level more than 10 times greater than the levels which occurred between epidemics before 1974.

A further similar epidemic took place in 1982–83, the number of cases notified up to the end of the first quarter of 1983 being 80 000 with 21 deaths. In this second epidemic the attack rate in children aged five to nine years was higher than in the 1977–79 event.

Studies during the epidemics clearly showed that the vaccine resulted in protective efficacy in vaccinated children and that vaccinated children who contracted whooping cough had a milder clinical attack than unvaccinated children.

As a result of the adverse publicity mentioned above, information began to accumulate on the adverse reactions suffered by vaccinees. The information in the possession of the Association of Parents of Vaccine-Damaged Children (APVDC) was examined by two committees specially set up by the Secretary of State for this purpose. They were

1. The Advisory panel on Serious Reactions to Vaccines;
2. The Advisory panel on the Collection of Data Relating to Adverse Reactions to Pertussis Vaccine.

The APVDC data consisted of information in documents of 555 children born between 1940 and 1975 who were thought to have suffered damage after receiving a pertussis-containing vaccine. This data was somewhat variable in scientific quality and ranged from accounts of incidents by parents, nurses, or doctors which occurred soon after to a long time after the suspected event. Further information was also provided by Professor Stewart of Glasgow University. The work of these panels is briefly summarized as follows.

Advisory panel on Serious Reactions to Vaccines

50 summaries of case histories of children reported to have suffered serious adverse reactions following injection of pertussis antigen were examined. These 50 cases were selected on the basis that their documentation was relatively good and the reactions were reported between 1956 and 1976. The panel divided the cases into four main groups:

1. Insufficient information   5 cases
2. Unlikely   7 cases
3. Possible   34 cases
4. Exacerbation of pre-existing neurological
   disease   4 cases

Careful scrutiny of the information failed to show any specific syndrome which could be associated with the use of pertussis antigens but it was possible to see three separate clinical patterns which were:

1. Chronic epilepsy;
2. Acute encephalopathy;
3. Infantile spasms (West's syndrome).

Mental retardation followed in all but three of the 50 cases.

The timing of reactions in relation to vaccination in children with chronic epilepsy suggested that an association seemed possible and to a lesser degree this was also the case with acute encephalopathy. The evidence was more convincing in some cases than in others and was most suggestive when convulsions occurred shortly after each of two or three injections.

Association with infantile spasms appeared weaker since there was often a long delay between vaccination and spasms which in this group occurred at the age at which infantile spasms not related to vaccination are most likely to occur.

It was noted in the reports of the 50 cases examined that several patients were vaccinated depite contraindications listed in the manufacturer's data sheet. In a few cases a second injection was given even though a major neurological reaction had occurred after a previous dose.

Advisory panel on the Collection of Data Relating to Adverse Reactions to Pertussis Vaccine

This panel examined all the data previously mentioned, including the 50 cases seen by the above committee. They found the case records unsuitable for epidemiological analysis because the methods for the detection of adverse events after the use of vaccines were inadequate. This could have led to underreporting which may have varied in extent because of the non-specific nature of some of the clinical events and the differing degrees of severity of the different reactions reported. Control information on the incidence of neurological events in non-immunized children was not available. A total of 229 reports from the UK for the period 1970–4 were examined and the results are summarized in Table 37.1a taken from the panel's report. The most common neurological events from the 305 events reported for the 229 children were

grand mal convulsions (38 per cent), infantile spasm (15 per cent), and screaming (8 per cent). Pyrexia was the commonest non-neurological event and was often associated with grand mal convulsions, which were however not always associated with fever.

The appearance of infantile spasms was preceded by other neurological signs such as eye-rolling or altered responsiveness. Contraindications had been present in at least 110 of these 229 cases, 50 of whom had a personal or family history of neurological disorder or defect. Another 30 children had had a possible reaction to an earlier injection of vaccine. The outcome of these reported adverse reactions ranged from full recovery to varying severity of brain damage. There were 14 deaths, seven in the acute epsiode and seven delayed by months or even years. The resulting clinical state of these children is given in Table 37.1b

The prime concern of the panel was with the occurrence of brain damage which, because of the manner of reporting, often followed severe reactions. An attempt was made to calculate the frequency of such brain damage or death for the period 1970–4 when uptake of vaccine was about 80 per cent. Since this date uptake has fallen to 30–40 per cent and as a result of publicity concerning contraindications, users of vaccine have taken greater note of these conditions. Since 1974 they have probably been more selective when carrying out whooping cough immunization. The attempted estimates of frequencies were:

1. Brain damage after any neurological event (68) 1 in 155 000 injections

Table 37.1a Major initial neurological events reported among 229 children having serious adverse reactions to pertussis vaccine in the UK (1970–4)*

Reported events	No.	Per cent
Grand-mal convulsions	115	38
Infantile spasms	47	15
Screaming	25	8
Twitching	15	5
Altered response	14	5
Halted progress	13	4
Paralysis	12	4
Coma	10	3
Others	54	18
Total	305	100

* Taken from the report of the Advisory Panel on the Collection of Data Relating to Adverse Reactions to Pertussis Vaccine (Committee on the Safety of Medicines 1981).

† More than one reaction per case may be shown.

Table 37.1b Final clinical state of the children in Table 37.1a

Present condition	No. (deaths)	Per cent†
Severly subnormal	90 ‡(3)	39
Educationally subnormal	65**	28
Epilepsy	11 (1)	5
Febrile convulsions	12	5
Other handicap	23 (10)	10
Normal	23	10
Not known	5	2
Total	229 (14)	99

Each child appears once only in Table 37.1b listed according to its most severe handicap. Thus a child who was severely subnormal and also had febrile convulsions will appear in the first line only.

* Taken from the report of the Advisory Panel on the Collection of Data Relating to Adverse Reactions to Pertussis Vaccine (Committee on the Safety of Medicines 1981).

† Because of rounding, numbers do not add up to 100.

‡ 49 with other handicaps as well.

** 43 with other handicaps as well.

2. Brain damage after a
   convulsive episode (57)   1 in 180 000 injections
3. Brain damage after a grand
   mal convulsion within 24
   hours of immunization (42
   convulsions, 15 with brain
   damage)   1 in 695 000 injections.
4. Deaths (8)   1 in 1 300 000
                              injections

The conclusions reached by these two committees examining the same data were fundamentally different. The committee on Serious Reactions to Vaccines recognized that there was no specific syndrome which could be associated with the use of pertussis antigens and, hence, the matter could be taken little further with regard to those reactions which appeared time-associated. In the case of infantile spasms no clear time relationship existed and thus the causal relationship seemed dubious. The panel on Collection of Data Relating to Adverse Reactions to Pertussis Vaccine by publishing estimates of frequencies pressed the concept of cause rather than coincidence further than the data could permit (Fenichel 1983).

However, both panels did comment on one feature of the data — the high frequency of second and even third inoculations taking place despite adverse effects following the first. Not only did such cases suggest a causal relationship (even if in the nature of a precipitating factor) more than others but also demonstrated the need for stress being laid on the main contraindication to use of the vaccine.

Even before study of retrospective data had been started it was recognized that a prospective study would be the only way to give more definitive answers. Ideally this should take the form of a control trial of use of the vaccine with the controls remaining unvaccinated. Such a trial with current background knowledge on efficacy was clearly unethical as well as being extremely expensive in the light of the necessary scale and follow up. However, since a number of children were being given diptheria and tetanus vaccine as a preliminary course, it was possible to compare such children with those receiving the triple vaccine in a certain area, the North West Thames Region, over a period of seven years (Pollock and Morris 1983). While it was the case that more reactions were reported following triple than diphtheria and tetanus vaccines it was felt this could have been due to bias since no such differences were found when the immunization histories of children admitted to hospital with such conditions were compared. Furthermore, no convincing evidence that triple vaccine caused major neurological damage

emerged from the large and lengthy study.

The other form of prospective investigation was set up as the National Childhood Encephalopathy study following advice from the Joint Committee on Vaccination and Immunisation (Miller *et al.* 1981). This investigation, as the title suggests, was not restricted to any one vaccine. The aim was to identify acute neurological illnesses leading to hospital admission of children aged 2 to 36 months during the period July 1976 to June 1979. Each child had two matched controls and a full history including pertussis immunization obtained for both cases and controls. The course of each child's illness was followed up and other risk factors and family history noted.

Serious neurological illnesses as studied in these 1000 cases, are very rare events and a total of 35 children were found whose neurological illness started within seven days of DTP vaccination. Three of these were known to have had a neurological problem before immunization. The 32 previously normal children were examined a year later and 21 were growing normally and appeared to have recovered. Three children had minor defects and six had major defects which could not be accounted for by any other explanation than DTP vaccination. The attributable risk (which is that part of the incidence of serious neurological illness in young children attributable to DTP vaccine) was estimated as one in 110 000 immunizations for previously normal children irrespective of eventual clinical outcome and one in 310 000 for those who later had neurological damage. The risk to a child receiving three injections on this basis could be about one in 100 000.

This risk is the difference between the incidence in the immunized children and that in the non-immunized children. The authors point out that due to the various assumptions made and to the wide confidence limits which apply, these estimates of risk should be accepted with caution.

No significant relationship was found between DT-vaccinated children and control children unlike children receiving measles vaccination (Schwarz strain) who showed an onset of neurological illness 7–14 days after vaccination compared with controls. Two of 16 children so affected had an abnormal history and the remaining 14 made a complete recovery except two who were left with mild defects. The roughly estimated attributable risk was one in 87 000 immunizations. It was concluded that most cases of acute neurological illness in early childhood are attributable to causes other than immunization.

Such illnesses occur more frequently within seven days and particularly within 72 hours after DTP vaccine and within 7 to 14 days after measles vaccine than would be expected by chance. Most affected children made a complete recovery.

Taking account of possible alternative explanations of the clinical findings in cases associated with DTP and of the fact that similar cases occur after DT vaccine, it seems likely that permanent damage as a result of pertussis immunization is a very rare event and attribution of a cause in individual cases is precarious.

With regard to infantile spasms and their possible association with pertussis immunization Bellman *et al.* (1983) concluded from an analysis of 269 cases reported to the National Childhood Encephalopathy study that neither triple nor diphtheria and tetanus vaccines cause infantile spasms but may trigger their onset in those children in whom the disorder is destined to develop.

In a similar manner a prospective study by the National Institute of Child Health and Development, Bethesda, Maryland clearly excluded triple vaccine as a causal factor in sudden infant death syndrome (cot death) and this has been confirmed more recently by work done in Sheffield and referred to above. Again the confusion of causal and coincidal relationships arising from retrospective studies required prospective investigation.

In summary it may be said that the present pertussis component of triple vaccine is far less pure and homogeneous than the two toxoids which accompany it. Hopefully work going on in many places might remedy this situation. In the meantime the few reactions which may occur and the very few serious consequences following vaccination in the absence of contraindications do not outweigh the benefits gained from the prevention of whooping cough and its sequelae.

Typhoid vaccine

This vaccine consists of not less than 1000 million dead organisms per ml and is well recognized as a reactive vaccine especially in people over 35 years old or in those who have been vaccinated against typhoid previously. Its use is justified because of its efficacy (70–90 per cent recipients protected) against such a serious disease.

The local reactions consist of swelling, pain, and tenderness and occur two to three hours or later. The systemic reactions can be severe and appear as nausea, malaise, headache, and fever. When these are severe the patient should go to bed and the symptoms usually disappear within 36–48 hours. Systemic reactions may be reduced by intradermal administration when given as a second dose after primary immunization which should be subcutaneous or intramuscular but this does not prevent local reactions from appearing.

Reactions can be avoided by not vaccinating children

up to 12 months of age and people with acute infections or chronic illness.

Cholera vaccine

Cholera vaccine is a heat-killed suspension of the Inaba and Agawa serotypes of *Vibrio cholerae* and is not very reactive when compared with typhoid vaccine. The El Tor biotype may also be included. The vaccine is liable to settle out as a gelatinous precipitate which must be vigorously shaken before adminstration to avoid erratic dosing.

Reactions are usually mild and occur within a few hours of administration. They consist of pain, redness, and swelling and such subjects can be expected to suffer similar reactions should they receive a further dose of vaccine. It is rare for reactions to be severe and this hazard can be reduced by not vaccinating people suffering acute infections or chronic illness.

Vaccines of viral origin

While most modern vaccines concern viruses it is often forgotten that the earliest two vaccines to be used were viral although not recognized at the time. Pustular material from smallpox cases used to infect others at risk was introduced into England in 1718 by Lady Mary Wortley Montague from Turkey. This process of variolation possibly derived from the Far East. Possibly as a result of serious adverse effects the method was not widely adopted but gave way to Jennerian vaccine prepared from cowpox lesions passed from person to person. Again the results while frequently dramatic were not susceptible to proper quality control and the method was not universally adopted. Nevertheless, the principle involved of attenuating a variola virus isolated in Cologne in the 1890s by passage in animals was established and such a vaccine grown in the skin of sheep, calves, buffaloes, or other convenient domestic animals was used on a world basis. Under the supervision of the World Health Organization use of this vaccine has enabled the world to be officially declared free from the disease and interest in adverse effects, which were, not surprisingly, numerous in such a crude preparation, is now purely academic. It is of interest to speculate whether such adverse effects would have been so frequent and diverse if the use of smallpox vaccine had persisted long enough for clean and controlled manufacturing methods to have been developed.

The other vaccine of viral origin in use for many years was rabies vaccine prepared in the brains of rabbits. Again such uncontrolled production gave rise to very many severe reactions including demyelinating encephalomyelitis but, despite this, the vaccine was widely used for many years because of the ability to prevent or even abort a much more serious disease. Rabies not having been eliminated still calls for prophylaxis and more modern and much safer vaccines are now used.

The possibility of more scientific production of viral vaccines stemmed from the discovery by Smith *et al.* in 1933 that human influenza virus could be transmitted to ferrets (thus providing an indicator) and could be cultivated in hen's eggs. The resulting harvest could be further purified and, if necessary, adsorbed to reduce reactions and improve antibody production. This established a method which permitted production of other viral vaccines such as measles, rubella, mumps, and yellow fever.

The portal of entry of poliovirus for many years was considered to be the nasopharynx but it was later shown to be an enterovirus which infected people through the intestine. Enders was able to show that poliovirus could replicate well in clean monkey kidney tissue and so paved the way to the development by Salk of inactivated poliomyelitis vaccine.

The formaldehyde inactivation curve flattens out long before total inactivation of the virus has been achieved and it is a matter of fine judgement where this point is placed. Although this is still a safe and acceptable vaccine, its history was marred by the Cutter incident in 1955 when insufficient time was allowed for the complete inactivation of the virus harvest by formaldehyde and the few remaining infectious virus particles brought about deaths and paralysis in many recipients of the vaccine. This was complicated by the protection afforded to the live virus particles during inactivation by particulate protein, a problem later solved by filtration.

This event gave a considerable boost to the use of live attenuated vaccine developed by Sabin, which has the advantages of cheapness and ease of adminstration. Such vaccine can also be used for viral interference and monotypic vaccine has been used to stop epidemics caused by heterologous types.

Modern attenuated-virus vaccines are only developed as the result of very extensive trials for adverse effects as also are inactivated virus vaccines. One major difficulty is the possibility of contamination of the tissue cultures by natural wild viruses. These can only be excluded by a very extensive and expensive series of tests on the parent cells and virus suspensions carried out both in other tissue cultures and in a range of laboratory animals. Established cell-lines offer hope for greater freedom from adventitious agents.

This short history of the development of viral vaccines again serves to illustrate the general principles described in the introduction. Control of manufacture can eliminate the more obvious examples of product defect (the Cutter incident) and improved methods of manufacture and purification can reduce those effects due to extraneous material (rabies vaccine, influenza vaccine). However with the live attenuated vaccines a further problem arises — the degree of attenuation in relation to the host response. Some perfectly safe vaccines have proved unacceptable in some communities by virtue of the extent of symptoms produced in comparison with the incidence and severity of the natural disease.

Influenza vaccine

There are three main types of inactivated influenza vaccines all of which are produced in eggs. Some work has been done on a live attenuated virus vaccine for nasal administration aimed at producing a local as well as a general immunity but difficulties in marketing combined with inconclusive trial results have led to its being dropped. The inactivated vaccines range from simple suspensions of influenza virus obtained from allantoic fluid to suspensions of disrupted and partially purified virus particles adsorbed on aluminium hydroxide and, more recently, to a very pure suspension of haemagglutinin and neuraminidase surface antigens also adsorbed on aluminium hydroxide.

These vaccines consist of the current A and B serotypes and it is worth noting that the B component, which only changes about once in every 10 years is considered to produce an undue proportion of the reactions. Sensitivity to eggs or other chicken antigens may be a contraindication to the use of these vaccines but the degree varies with the purity of the vaccine itself.

The most common adverse reactions are the local and general responses previously described. Encephalopathy has on rare occasions been seen but no causal relationship has been clearly demonstrated. However, this together with other neurological complications calls for further surveillance.

A number of cases of Guillain–Barré syndrome were reported in the USA in 1976 after extensive use of monocomponent swine-influenza vaccine (581 cases in 4 months). This rare syndrome has been associated with other active immunizing procedures and particularly with the older rabies vaccines. Although the swine-influenza component (A/New Jersey/8/76/Hsw_1N_1) was included in the UK influenza vaccine for a short period, there were no cases of Guillain–Barré syndrome reported in the UK (*Br. med. J.* 1977). Clinically this syndrome is an acute motor and sensory peripheral polyneuritis and paralysis may ascend or descend to the medulla but the prognosis is good if intensive-care nursing is available. The neurological symptoms occur within three weeks of an upper respiratory or gastrointestinal illness and chickenpox, influenza A, mumps, herpes, glandular fever (Epstein–Barr virus), parainfluenza 3, and tickborne encephalitis have been regarded as being associated with the syndrome. The later appearance of the syndrome suggests that it may be due to an allergic reaction following the original vaccination or infection. Studies in the USA in later years following the use of vaccines not containing the swine-influenza component did not reveal a significant risk of the syndrome. The vaccine may cause a depression of liver enzymes concerned with metabolism of some drugs, e.g. warfarin. This possibility is still under investigation.

Contraindications to the use of influenza vaccines include concurrent acute febrile illness or infections of the upper respiratory tract, and those conditions or therapies which may alter the immune state of the prospective recipient. The vaccine should also be restricted to the age group indicated in the accompanying literature since reactions tend to be more severe in children but less frequent in the case of highly purified surface antigen vaccines than the less pure earlier forms.

Measles vaccine

The vaccine was introduced in the United Kingdom on a national scale in 1968 at which time there were two products available each from a different strain of attenuated virus. One was soon withdrawn as it appeared to cause a greater frequency of adverse reactions than the other which was based on the Schwarz strain and this vaccine has been almost the only one used to the present time. Adverse effects fall into clear-cut categories although these may be linked.

Most common are measles-like illnesses with or without convulsions. A mild cough and coryza may occur some five to 10 days after injection and may further develop into rash, pyrexia, and pharyngitis. Convulsions may develop with a peak incidence at about seven and 14 days of between 0.2 and 0.9 per 1000 vaccinations. Measles vaccination being carried out in the second year of life tends not to coincide with the highest frequency of natural fits and thus, unlike the case of pertussis vaccine, there is less dispute about causal relationship in the case of measles vaccine, the background incidence in this age group being, probably, about 0.1 per 1000 per week.

Thus the Medical Research Council (1966) trials involving the Schwarz strain vaccine showed nine convulsions occurring between seven and 14 days after vaccination, an incidence of 0.9 per 1000 while the North West Thames study reported a figure of 0.2 per 1000. Of 132 instances of convulsions reported since 1975 90 occurred between 5 and 10 days and 19 between 11 and 15 days after immunization. The National Childhood Encephalopathy study showed of 1000 children admitted to hospital 14 apparently normal before onset included nine with febrile convulsions and five with encephalitis or encephalopathy occurring within the period 7 to 14 days — an estimated rate of one in 87 000 injections. A rate of one in 83 000 was estimated from the adverse reaction reporting system between 1970 and 1979.

Taking the serious involvement of the central nervous system to include convulsions and encephalitis together the outcome of the 14 cases in the NCES study was favourable but for two which had further febrile convulsions, while the outcome of the 55 cases reported revealed 22 with sequelae and eight deaths.

However, while there is little doubt about this effect of vaccination occasionally occurring, the incidence of neurological complications following the natural disease is of the order of 10 times greater. Thus Miller (1964) found an incidence of one per 8000–9000 and Miller (1978) reported convulsions at a rate of one in 200 cases and encephalitis at one in 4500 cases.

Less common reactions include anaphylaxis which occur within a few hours of vaccination. Some of these may be due to egg protein or antibiotic sensitivity and as such are included in the contraindications together with the usual causes of immune suppression. An immunoglobulin preparation is available for use in cases where it is felt essential to immunize against the disease but wise to avoid potential excessive response.

Rubella vaccine

There are three live attenuated rubella vaccines available which are prepared from tissue cultures made from eggs, primary rabbit kidney cells, or human diploid cells respectively. Each type of cell culture is used to produce virus after inoculation with a special strain of attenuated virus specially developed for each particular tissue. Although it can be used at any age it is not recommended during the first year of life because of maternal antibody.

Adverse reactions are uncommon and resemble the symptoms of rubella when they do occur which is around the ninth day following inoculation. Usually these are mild and transient and consist of malaise, mild fever, headache, sore throat, rash, and lymphadenopathy. Older women in particular (men are rarely immunized against rubella) may develop arthralgia with signs of inflammation. Such arthritis may be short-lived but on rare occasions may last for several weeks in adult women.

This reactivity would seem to vary inversely with the attenuation of the virus strain which tends to relate to the antibody response. The egg-adapted strain is least attenuated in this respect.

This tendency for the vaccine (like the disease) to give arthralgia has given rise to a suggestion that the virus might be a possible aetiological factor in juvenile rheumatoid arthritis. However, such claims have not received support and the current view with regard to the immunizing against rubella of a patient suffering from juvenile rheumatoid arthritis is to first check the rubella antibody status and then, if necessary, defer immunization until the arthritis is in remission.

On very rare occasions, serious disorders of the central nervous system (encephalitis, transverse myelitis) have been claimed as resulting from rubella vaccination as has, equally rarely, thrombocytopaenia.

The major contraindication to rubella vaccination remains that of pregnancy which should not only not be present at the time but avoided for at least two months after vaccination.

Poliomyelitis vaccine

At present there are two types of vaccine in use in the UK. The one (IPV) containing inactivated virus produced in monkey kidney cells, the other (OPV) containing a live attenuated virus produced in monkey kidney or human diploid cells. Both vaccines contain the three types of virus.

The inactivated vaccine was the first to be used on a national scale in 1956. Surveys by Geffen and Spicer (1960), Geffen (1960), and Roden (1964) indicated that the vaccine was both efficacious and safe. No instances of poliomyelitis arising from contamination with active virus were reported and such adverse effects as were seen were not readily separable from those possibly due to triple vaccine given simultaneously (quadruple vaccine) or closely associated in time. At the present time only 1000–2000 doses are used each year — insufficient to give any meaningful information on adverse effects, such use being for patients in whom OPV is contraindicated (e.g. pregnancy). IPV suffers from three disadvantages over OPV: it has to be given by injection; it only immunizes the recipient; and it cannot block the spread of wild virus during an epidemic. However, it cannot be charged with causing the disease.

The live attenuated virus vaccine given by mouth (OPV) became the major factor controlling poliomyelitis following 1962 either given as a primary three-dose course followed by reinforcement or to control outbreaks. Adverse reactions consist almost entirely of the occurrence of recipient vaccine-associated cases or contact vaccine-associated cases of paralytic or non-paralytic poliomyelitis.

The diagnosis and attribution of such cases has never been easy although, with the decline in the natural disease, it is becoming easier. Only when reliable marker tests are available can a case be said to be caused by the vaccine as opposed to associated with its administration, though if the disease is not very obviously poliomyelitis the possibility of other virus causation is always present. The decline from 20 recipient and 11 contact cases (1.6 and 0.9 per million doses of vaccine given) in 1962–64 to, at present, about three recipient and four contact (0.3 and 0.4 per million doses given) presumably reflects the reduction in wild virus in the population. The principle of recommending non-immunized parents of children being given OPV to be immunized at the same time is an attempt to reduce the incidence of contact vaccine-associated disease still further.

Although other conditions have been reported as having been caused by OPV including convulsions, encephalomyelitis, and polyneuritis, these have been few in number and proof of causation is lacking (Miller and Galbraith 1965; Miller *et al.* 1970; Smith and Wherry 1978; Collingham *et al.* 1978). In many cases the picture is confused by the giving of other vaccines at the same time.

Correctly used with observation of all contraindications there would seem little doubt that this vaccine is one of the safest and most efficacious.

Rabies vaccine

The older inactivated vaccines of the Semple type prepared from the brains of rabbits are now considered too toxic to use, as demyelinating encephalomyelitis was well known as an adverse reaction.

The new human diploid-cell vaccine has been shown to be efficacious in post-exposure clinical trials in Iran and has not been associated with anything more than very mild local reactions. These consist of mild erythema, swelling, and pain and are of infrequent occurrence in people undergoing active protective immunization. There are no contraindications.

Yellow-fever vaccine

The old yellow-fever vaccine which has not been made in the UK for many years was prepared from what would now be regarded as an insufficiently attenuated strain of virus. It was too reactogenic and severe reactions including fever and encephalitis were reported after its use. This led to the development of the modern yellow-fever vaccine which is prepared from a special attenuated strain of yellow-fever virus known to be avirulent in man and free from leucosis virus. The virus is cultivated in leucosis-free chick embryos and harvested in the presence of polymyxin and neomycin. People who are hypersensitive to these antibiotics or to chick proteins should not be inoculated with this vaccine.

Adverse reactions after its use are very rare and consist of redness and swelling at the site of injection. Headache may also occur and encephalitis has been reported in infants under nine months of age. Children under this age should not be vaccinated unless there is a serious risk of exposure to yellow fever.

The contraindications are those normally indicated for live-virus vaccines and are listed as impaired immune responsiveness whether it be idiopathic or caused by disease or by treatment with cytotoxic drugs, steroids, or radiation.

Hepatitis B vaccine

This vaccine, prepared from blood containing the hepatitis B surface antigen, has been introduced too recently and used on too limited a scale to reveal, as yet, any clear pattern of adverse effects. Clear evidence has been obtained as to its efficacy and, by use of three separate steps of inactivation during manufacture each of which appears capable of virtual elimination of known contamination combined with rigorous testing, the vaccine would appear to be safe. However, since the first use of the vaccine following its development, the condition of acquired immune deficiency syndrome (AIDS) has been recognized. This syndrome, although occurring in differing groups of world population, has a high incidence in the homosexual communities in large cities, particularly in the USA. These communities also have a high incidence of hepatitis B and thus are excellent providers of source materials for the production of this particular form of vaccine. AIDS has also been reported in a small number of haemophiliacs treated with large amounts of Factor VIII Concentrate. AIDS has been shown by both French and American workers to be a retro-virus and has been classified as HTL VIII. (Human T-cell lymphotropic virus); hence the possibility of AIDS being regarded as an adverse effect of Hepatitis B vaccine cannot be dismissed. Careful surveying of the earliest cases of use of the vaccine have not revealed any instance of such adverse effect except in two individuals amongst many hundreds. These two individuals both belonged to that social group most at risk of hepatitis B and AIDS and it has been generally accepted that the cause of the syndrome was other than the vaccine. Current advice remains that the vaccine continues to be promoted for

use in those groups at high risk of contracting hepatitis B where the benefit outweighs the potential hazard (Furesz and Boucher 1983).

Miscellaneous vaccines

Small quantities of special vaccines are occasionally made for protection against rare conditions such as anthrax and for people working with dangerous pathogens such a Clostridium botulinum. The main difficulties are in the manufacture of these vaccines but they do not appear to have produced any serious unwanted effects in the relatively few recipients. A few thousand doses of anthrax vaccine are distributed every year for administration to workers handling hides and animal products who are at particular risk.

Autogenous vaccines are used less now than hitherto and not uncommonly cause local reactions which soon regress. Their main danger is the variable quality of manufacture when prepared privately. This rarely matches that of normal commercial vaccines which have to be made under carefully licensed and inspected conditions in accordance with the Medicines Act 1968.

Therapeutic allergens

Therapeutic allergens come within the definition of an antigen for the purposes of the Medicines Act as set out in Statutory Instrument 1971, No. 1200, being substances which on administration to a human being or animal are capable of eliciting a specific immunological response. As therapeutic allergens are prepared from sensitizing substances, which bring about allergic reactions, it is not surprising that this class of medicinal products are prone to produce adverse reactions. It is necessary that initial injections contain extremely small quantities of allergens and, as sensitive subjects may react to tiny amounts, considerable skill and experience are needed to safely administer these products.

Most allergens would appear to be proteins, some may contain carbohydrate and are of biological origin though there are natural or synthetic chemicals which function as allergens probably because they bind to body protein. In allergic subjects they stimulate the formation of antibody which can bind to certain cells, for example, mast cells, thereby sensitizing them to the offending allergen. This is considered to be the mechanism in Type 1 immediate allergy which is believed to be mediated by reagin (Coca and Grove 1925), and is now recognized as IgE antibody (Pepys 1974).

Therapeutic allergens have been used for many years to desensitize people with an immediate allergy such as hay-fever after Noon first introduced de-sensitization to pollen in 1911. The allergic mechanism in these cases is thought to be due to the degranulation of mast cells which have been previously sensitized by the incorporation of IgE on to their surface membranes, in response to the patient being exposed to the offending allergen. A subsequent re-exposure leads to the new allergen binding to the bivalent IgE determinants on the cell surface. This triggers off the release from the granules of a number of pharmacologically active substances including histamine, which produce a clinical effect in a tissue or target organ. This triggering action could theoretically be prevented if sufficient IgG were to be present as a blocking antibody which could combine with the allergen antigen and so prevent it combining with the IgE determinants on the mast cells.

Such blocking antibody was discovered by Cooke *et al.* in 1935 and there is some evidence that its presence correlates with a reduction in symptom scores (Lichtenstein *et al.* 1968; Starr and Weinstock 1970) in allergic subjects. The general view, however, is against a direct relationship of this type though blocking antibody is believed to have some part in hyposensitization.

Desensitizing procedures have other effects which may be useful such as the reduction of IgE levels which occur after continued therapy (Sherman *et al.* 1940). The normal seasonal rise in IgE antibody may also be lowered in some patients (Levy *et al.* 1971). The release of histamine by basophils in sensitive subjects in the presence of specific allergen may also be reduced as the result of hyposensitization (Evans *et al.* 1976; Lichtenstein *et al.* 1973). Immunotherapy may also reduce the increase in the lymphocyte count which occurs in sensitive patients on exposure to allergen (Nagaya *et al.* 1977).

Therapeutic allergens are used to stimulate the formation of specific IgG in the same way as conventional vaccines but suffer from the defect of being likely to bring about an allergic response, especially when first used. It is very important, therefore, to control the manufacture of therapeutic allergens so that the allergens are constant from batch to batch, and that the quantities of each allergen present are also standardized. This is not easy to carry out and, if there is too much quantitative or qualitative variation, allergic reactions could occur or an insufficient IgG response take place. The control of quality is beset with the multiplicity of allergens (antigens) and, although therapeutic allergens are widely used to de-sensitize people with single or multiple allergies, difficulties do arise in relation to safety and efficacy. Until quality, especially in relation to consistency of manufacture, has been put on a firm scientific basis, safety and efficacy are unlikely to be brought under useful clinical control.

Such good evidence as there is consists of trials involving rather small numbers of people and the clinical benefits described are of necessity to a large extent subjective. Reproducibility of such findings is dependent on uniform quality of the allergen preparation from batch to batch. This is not easy to achieve for the following reasons:

1. Allergic subjects are commonly sensitive to a number of different allergens.
2. An individual extract of grass pollen, for example, may contain many allergic proteins. It is believed that harvests of an individual grass pollen may vary in composition from year to year possibly due to differing climatic factors in the growing season.
3. Even if the particular allergic proteins are known and hopefully the subject did not change his pattern of allergies, it is only recently that it has been found possible to estimate the potency of a particular allergen.

The old system of estimating the total protein nitrogen of allergens and expressing the answer as protein nitrogen units (PNU) is not sufficiently specific to be more than a guide to the total protein present.

Clinicians have tested for allergenicity by intradermal skin tests of allergen in human volunteers and measuring the response. Apart from lack of good proof that this relates to potency, this is not a practical way as there are insufficient volunteers and in addition there are ethical and legal difficulties. At present allergenicity is best measured by an *in-vitro* test called the radioallergosorbent test (RAST) (Wide *et al.* 1967). This consists of covalently binding the proteins of a relevant allergen to filter paper discs. This bound allergen is then allowed to react with allergic serum (from a pool) containing IgE antibodies against the specific allergen under test and the disc well washed to remove other IgE antibodies and other substances present in the serum. The solid-phase-bound IgE antibody is then reacted with affinity chromatographically purified antibody to human IgE which has been radiolabelled and washed free of soluble impurities. The radioactivity of the paper disc can be measured in a gamma counter and the results compared with a similar estimation done on a standard preparation of the allergen.

Similarly, improvements were shown to be possible in determining the range of potentially allergic proteins present in a bulk extract of grass pollen. This was achieved by iso-electric focusing using a method based upon polyacrylamide-gel electrophoresis. Samples of the allergen extract at the correct concentration are placed in holes in a flat plate of polyacrylamide gel. An electric current is run through the gel which has a pH gradient of 3.5 to 9.5 whereupon the proteins after some 2–5 hours form bands at their isoelectric points. After staining with a dye such as Coomassie blue it is possible to see the protein bands. The presence of major allergens in an extract of grass pollen can then be compared with a standard extract by examination of the protein bands obtained after similar testing. Variations can thereby be detected.

The antigenic composition of the vaccine can then be more accurately specified and hopefully will not from time to time contain new major allergens to which the subject is sensitive. This should improve safety in terms of avoidance of adverse reactions. Similarly any absence or low potency of allergens can be detected during bulk manufacture and corrected and thereby improve efficacy. This system of quality control should result in more uniform products being made. It will require many new standard preparations to be prepared from among the several hundreds of allergens available. Those in greatest demand are already being developed.

A course of injections for an allergic subject will need to be carried out very carefully to avoid a serious allergic response and usually consists of a long series of injections (12) of increasing potency starting with a very highly diluted extract. This type of treatment is commonly used for specific desensitizing vaccines made especially for individual patients. Another approach to this problem is to adsorb the allergen on an adjuvant which releases the antigen very slowly.

Adverse reactions such as urticaria, local pain, and swelling have been observed with both types of preparation though more rarely with the adsorbed type. On examination one such adverse reaction (severe anaphylactic shock) which occurred many years ago was shown to be due to a change in pH on storage which allowed the allergen to elute into the supernatant. As a result, this aspect of quality control has for long been very carefully examined before release. Unlike vaccines (except influenza) some of these allergens are seasonal in character. Use of grass pollen extracts is contraindicated during the grass pollinating season as the patient may receive too much antigen when the air pollen count is high and severe adverse reactions result. The course of subcutaneous injections should be started early in the year and no later than the middle of April. Care must be taken not to give the injection into a blood vessel as bronchospasm or anaphylactic shock may ensue. Every effort should be taken to follow the warnings and contra-indications listed in the data sheet. These include fever and pregnancy, especially in the first three months, and no patient with an acute attack of asthma undergoing a course of desensitizing injections should be injected with therapeutic allergens until 24 hours after the condition has returned to normal.

Adverse reactions

Adverse reactions are largely those due to release of histamine and similar pharmacologically active drugs into the tissues or target organ. They occur at the site where the offending allergen, which can be one of several hundreds possible, reacts with the cells sensitized by IgE. Local reactions consist of itching and swelling at the site of injection and there may be pain lasting for several hours. If these persist for more than a few hours treatment with antihistamines may be necessary. A reaction to a slow-release adsorbed vaccine can last for several days. Severe local reactions may require the use of adrenaline and a tourniquet placed above the site of injection.

Systemic reactions usually consist of rhinitis, urticaria, or bronchospasm. These can be severe and anaphylactic shock with severe bronchospasm, laryngeal oedema, and collapse is a life-threatening emergency requiring immediate treatment with adrenaline and oxygen. Severe asthma or generalized urticaria suggest that the vaccine being used is unsuitable for the patient who should be referred back to an allergy clinic. Severe bronchospasm may require treatment with a sympathomimetic bronchodilator and or an anti-inflammatory steroid. Therapeutic allergens are unusual vaccines insofar as they are made from a selection of hundreds of natural substances which cause allergic disorders. Each may consist of many antigenic determinants and in the case of grass pollens a single pollen may posses up to 20–30 different proteins though these need not all be allergenic. By the very nature of the allergic response in sensitized subjects they are able to produce severe reactions unless the vaccine is tailor-made to the recipient's requirements. This is done as the result of careful skin testing which requires expert interpretation as the relationship of reactions to provocation tests with subsequent effective desensitization is not simple.

Advances in this field are promising and a good example is the recent introduction of desensitizing vaccines to such potentially lethal conditions as allergy to bee and wasp venom. These allergens are obtained relatively pure (each consists of a mixture of toxins) from individual insects stings. Adverse reactions to these preparations are however likely to be similar to other therapeutic allergens the basic cause being the individual idiosyncrasy of each patient in relation to relevant allergens which bring about the characteristic response.

Future development

This chapter has dealt with adverse effects of agents used to prevent a disease or to lessen its worst manifestations.

Such adverse effects form the numerator in the risk–benefit ratio and, especially in the instances where the disease may not be prevalent at the time and its severity forgotten, any adverse effects must, clearly, be reduced to an absolute minimum. This can be achieved by a combination of the following factors.

1. Continued improvement in methods of manufacture and quality control of the product attempting to produce the pure, single efficacious antigen. The new development of biotechnology re-opens this whole area.

2. The acceptance of the necessity for carefully controlled large-scale clinical trials to evaluate the efficacy and safety of a procedure before marketing.

2. The careful observance of all precautions and contra-indications connected with the particular therapy in wide-scale use.

4. The accurate reporting, full recording, and positive follow-up of claimed adverse affects, with instant readiness to conduct high-quality research into any areas where doubt has arisen with regard to either safety or efficacy.

It may have been noticed that no section has been allocated in this chapter to smallpox vaccine. This is a result of the dramatic decline in its use following the success of the eradication programme mounted by the WHO. If any doubt remains regarding the value and importance of the area of preventive medicine covered by this chapter, it should be dispelled by consideration of this simple fact.

REFERENCES

BELLMAN, M.H., ROSS, E.M., AND MILLER, D.L. (1983). Infantile spasms and pertussis immunisation. *Lancet* i, 1031–3.

BIOLOGICAL STANDARDS ACT (1975). Ch.4. HMSO, London.

BRITISH MEDICAL JOURNAL (1977). Editorial: Guillain–Barré syndrome and influenza vaccine. *Br. med. J.* i, 1373–4.

—— (1983). Communicable diseases — BCG vaccination. *Br. med. J.* **286**, 876–7.

COCA, A.F. AND GROVE, E. (1925). Studies in hypersensitiveness: a study of the atopic reagins. *J. Immunol.* **10**, 445–64.

COMMITTEE ON THE SAFETY OF MEDICINES (1981). *Whooping cough*. HMSO, London.

COLLINGHAM, K.E., POLLOCK, T.M., AND ROEBUCK, M.O. (1978). Paralytic poliomyelitis in England and Wales 1976–7. *Lancet* i, 976–7.

COOKE, R.A., BARNARD, J.H., HEBALD, S., AND STULL, A. (1935). Serological evidence of immunity with co-existing sensitization in a type of human allergen (hayfever). *J. exp. Med.* **62**, 733–50.

COMPENDIUM OF LICENSING REQUIREMENTS FOR THE MANUFACTURE OF CERTAIN BIOLOGICAL MEDICINAL PRODUCTS Issue 2. HMSO (1982). London.

DEPARTMENT OF HEALTH AND SOCIAL SECURITY (1983). *Immunisation against infectious disease*. HMSO, London.

EVANS, R., PENCE, H., KAPLAN, H., AND ROCKLIN, R.E. (1976). The effect of immunotherapy on humoral and cellular responses in ragweed hayfever. *J. clin. Invest.* **57**, 1378–85.

FENICHEL, G.M. (1983). The Pertussis controversy — the danger of case reports. *Arch. Neurol.* **40**, 193–4.

FURESZ, J. AND BOUCHER, D.W. (1983). Safety of hepatitis B vaccine. *Can. med. Ass. J.* **129**, 17–18.

GEFFEN, T.J. (1960). Poliomyelitis in vaccinated subjects: England and Wales 1959. *Monthly Bull. Min. Hlth Publ. Health Lab. Serv.* **19**, 196–9.

—— AND SPICER, C.C. (1960). Poliomyelitis in the vaccinated: England and Wales, 1958. *Lancet* ii, 87–9.

LEVY, D.A., LICHTENSTEIN, L.M., GOLDSTEIN, E.O., AND ISHIZAKA, K. (1971). Immunologic and cellular changes accompanying the therapy of pollen allergy. *J. clin. Invest.* **50**, 472–82.

LICHTENSTEIN, L.M., ISHIZAKA, K., NORMAN, P.S., SOBTOKA, A.K., AND HILL, B.M. (1973). IgE antibody measurements in ragweed hayfever: relationship to clinical severity and the results of the immunotherapy. *J. clin. Invest.* **52**, 472–82.

——, NORMAN, P., AND WINKENWERDER, W. (1968). Clinical and in-vitro studies on the role of immunotherapy in ragweed hayfever. *Am. J. Med.* **44**, 515–24.

MEDICAL RESEARCH COUNCIL (1966). Vaccination against measles: a clincial trial of live measles vaccine given alone and live vaccine preceded by killed vaccine. *Br. med. J.* i, 441–6.

MEDICINES ACT (1968). Ch. 67 HMSO, London.

MILLER, C.L. (1978). Severity of notified measles. *Br. med. J.* i, 1253.

MILLER, D.L., (1964). Frequency of complications of measles 1963: report on a national inquiry by the Public Health Laboratory Service in collaboration with the Society of Medical Officers of Health. *Br. med. J.* ii, 75–78.

—— AND GALBRAITH, N.S. (1965). Surveillance of the safety of oral poliomyelitis vaccine in England and Wales 1962–4. *Br. med. J.* ii, 504–9.

——, REID, D. AND DIAMOND, J.R. (1970). Poliomyelitis surveillance in England and Wales 1965–8. *Pub. Hlth, London* **84**, 265–85.

——, ROSS, E.M., ALDERSLADE, R., BELLMAN, M.H., AND RAWSON, N.S.B. (1981). Pertussis immunisation and serious acute neurological illness in children. *Br. med. J.* **282**, 1595–9.

MINISTRY OF HEALTH (1921). *Report of the departmental committee appointed to consider and advise upon the legislative and administrative measures to be taken for the effective control of the quality and authenticity of such therapeutic substances offered for sale to the public as cannot be tested adequately by direct chemical means*. HMSO, London.

NAGAYA, H., LEE, S.K., REDDY, P.M., PASCUAL, H., JEROME, D., SADAI, J., GUPTA, S., AND LAURIDSEN, J. (1977). Lymphocyte response to grass pollen antigens: a correlation with radioallergosorbent test and effect of immunotherapy. *Ann. Allergy* **39**, 246–52.

PEPYS, J. (1974). *Clinical immunology — allergy in paediatric medicine*. Blackwell, Oxford.

POLLOCK, T.M. AND MORRIS, J. (1983). A 7-year survey of disorders attributed to vaccination in North West Thames Region. *Lancet* i, 753–7.

RODEN, A.T. (1964). Surveillance of Salk Vaccination in England and Wales 1956–61. *Proc. R. Soc. Med.* **57**, 464–6.

SANDERS, R. AND DICKSON, M.G. (1982). BCG vaccination scars: an avoidable problem? *Br. med. J.* **285**, 1679–80.

SHERMAN, W.B., STULL, A., AND COOKE, R.A. (1940). Serologic

changes in hayfever cases treated over a period of years. *J. Allergy* **11**, 225–44.

SMITH, J.W.G. AND WHERRY, P.J. (1978). Poliomyelitis surveillance is England and Wales 1969–1975. *J. Hygiene, Cambridge* **80**, 155–67.

SMITH, W., ANDREWES, C.H., AND LAIDLAW, P.P. (1933). A virus obtained from influenza patients. *Lancet* **ii**, 66–8.

STARR, M.S. AND WEINSTOCK, M. (1970). The relationship between blocking antibody levels and symptomatic relief following hypo-sensitization with Allpyral in hayfever subjects. *Int.*

Arch. Allergy. appl. Immunol. **38**, 514–21.

TAYLOR, E.M. AND EMERY, J.L. (1982). Immunisation and cot deaths. *Lancet* **ii**, 721.

THERAPEUTIC SUBSTANCES ACT (1956). 4 and 5 Eliz.2, Ch.25. HMSO, London.

WIDE, L., BENNICK, H., AND JOHANNSON, S.G.O. (1967). Diagnosis of allergy by an in-vitro test for allergen antibodies. *Lancet* **ii**, 1105–7.

WILSON, G.S. (1967). *The hazards of immunisation*. Athlone Press, London.

38 Adverse reactions to herbal and other unorthodox medicines

R.G. PENN

Introduction

Medicines derived from plants formed the majority of the early materia medica. Many of these herbs stood the test of critical assessment and found their way into the therapeutic armamentarium of the qualified physician often latterly as the isolated and chemically standardized active ingredient. The roll-call is long and drugs derived from plants are still being added. Included are morphine, digitalis, cocaine, ephedrine, coumarins, curare, quinine, reserpine, senna, colchicine, and the ergot and vinca alkaloids. Orthodox medicine does not reject medicines derived from herbs but it does reject quackery.

In the Western world there are still many other herbal substances which have survived perhaps hundreds of years of use. These herbal medicines are usually prescribed by non-medically qualified herbalists or are self-prescribed under the firm belief that such medicines are both effective and harmless. It is the intention of this chapter to show that such a belief may be badly mistaken. A brief glance through any of the more scientific herbals will show the presence of toxic substances in a large proportion of herbs but it should be emphasized that this review deals with herbal medicines when used therapeutically and not where there is accidental or deliberate ingestion of poisonous plants.

The paucity of adverse reaction reports about herbal medicines in the literature is probably due to the lesser side-effects not being seen by medical practitioners whose advice would be called upon only in serious cases. Certainly there is scant evidence of controlled studies of efficacy and toxicity as are available for orthodox medicines.

As investigations into herbal medicines proceed, more and more are being shown to have serious potential hazards raising doubts on the claims by herbalists that these medicines be judged on their long history of use rather than by modern standards of safety, quality, and efficacy.

Herbalists attribute too great a reliability to the discriminative powers of patient and doctors of bygone days, and it is a philosophic question whether serious toxic effects would have been recognized other perhaps than the acute toxicities of the immediate post-ingestion phase. The more subtle toxicities such as carcinogenicity, mutagenicity, and hepatotoxicity are undoubtedly very recent concepts and surely would have been missed. Even in orthodox medicine today any event related to exposure to a medicine, but with a long latent period before manifestation, is likely to be overlooked.

The philosophies underlying the use of herbal medicines are not easy to understand for the non-herbalist and perhaps have only minor relevance in a discussion on the adverse effects of herbs. Whilst a definition of herbal medicine will not be attempted, it must be realized how varied it is, ranging across what might be called the survival of the old Western folk medicine; fashionable crazes such as ginseng, laetrile, and 'health foods'; and the influence of non-Western herbal knowledge and traditions in the immigrant community.

Herbal medicines exist in a variety of forms which serve to confuse the issues of determining their efficacy and toxicity. The professional herbalist prescribes for his patient medicines derived from single plant extracts, though even these are a mixture of substances of which a large part seem merely to be bitters and astringents. It is an important part of the herbalist's canon that the activity of such a product must be viewed as a whole and not as the pharmacological properties of the individual components.

Unfortunately, most herbal preparations, especially those self-prescribed, are a mixture of extracts; these are often large in number and are seemingly illogically incorporated with herbs of contrary properties in the same nostrum. These herbal mixtures may be even more insidiously dangerous when they contain, usually unknown to the taker, ingredients of non-herbal origin.

The quality of herbal preparations is also often suspect. Chemical standardization of such heterogeneous mixtures is obviously not possible and they are rarely assayed in any biological test system thus casting doubts on both stability and efficacy. It is also not unknown for incorrect and perhaps toxic plants to be mistakenly (or illegally) used in the preparation of herbal medicines.

Toxicity from herbal medicines may therefore be considered under several headings:

1. Herbal extracts of proven or suspected toxicity;
2. Herbal extracts which have been adulterated either by the use of the wrong plant or contain toxic or potentially hazardous non-herbal substances;
3. Interactions between the effects of herbal medicines and drugs used in orthodox therapy.

These divisions are somewhat artificial and many instances of toxicity are not easy to classify under one heading. Most literature of herbal toxicity in man is of the case-report type and, whilst many of these would appear not to be equivocal, some have not had chance association ruled out.

Much of the available evidence is also of the more dubious kind due to the noncritical discussions of such medicines in newspapers and magazines and on television. It is also important for the non-botanist to realize the pitfalls in identifying plant species and, when, for example, comfrey is named, this may be one of several species whose toxicities may not be identical. Even a plant of one properly identified species may have differing toxicities in its various parts, e.g. root, stem, leaves, and fruits; and even at separate times of the year.

Toxic herbal extracts

Under this heading are considered those herbs where there is a proven or *prima facie* case for toxicity due to a normal constituent of the herbal extract itself. Classification is not easy as knowledge of toxicities is still in a very elementary state. Whereas a classification either by botanical group or by specific adverse reaction would be more scientific, it is more practical for the present purpose generally to list the herbs by their common names and to discuss the adverse reactions occurring with each.

Herbal teas

It is not possible here even to being to list the vast number of herbal species that may be used in herbal teas (tisanes). Siegel (1976) commented that there are some 396 distinct herbs and spices commercially available and used either singly or in blended mixtures as herbal teas. These may be defined as drinks containing a small concentration of a herbal substance obtained by maceration (steeping in cold water), decoction (boiling in water for a short time), or infusion (boiling water is poured over the herb and the mixture is left to stand for a while before straining).

Their purpose is also varied and includes general systemic indications such as 'tonic', hypnotic, analgesic, diuretic, and local uses such as foments, compresses, and would dressings.

That some herbal teas are toxic is beyond doubt as is illustrated by the following examples.

Teas containing pyrrolizidine alkaloids

These alkaloids are typically found in plants of *Senecio, Crotalaria,* and *Heliotropium* species and cause characteristic liver pathology. Hill *et al.* (1951) described a new type of progressive liver disease occurring in West Indian children which they called 'serous hepatitis'. Stuart and Bras (1957) described in Jamaicans three over-lapping stages of liver damage: an acute stage especially in the younger age group (1–6 years old) with hepato-megaly and ascites; a subacute stage characterized by a persistent and often symptomless hepatomegaly; and a chronic cirrhotic stage. Histologically the picture was of a widespread occlusion of the small hepatic veins with sinusoidal congestion and pressure necrosis of parenchymal cells around the central vein with a non-portal type of cirrhosis. They suggested that the aetiology was from plant toxins derived from *Senecio* and *Crotalaria* species and these were later identified as pyrrolizidine alkaloids (Mclean 1970). Huxtable (1980) in a review of pyrrolizidine alkaloid poisoning in the USA further emphasized the importance of this group and its toxicity.

Other workers showed that such teas were used world wide. Datta *et al.* (1978) described six cases of hepatic veno-occlusive disease from India of whom three died. These patients had been taking Ayurvedic medicines derived from *Heliotropium* species for such indications as fever and inflammation and as diuretics. *H. eichwaldii* was shown to contain 1–2 per cent of the toxic pyrrolizi-dine alkaloid, heliotrine. Two of the patients died of a fulminant hepatic failure of sudden onset with a notice-able delay after ceasing to take the herb; the delay to symptoms in the first fatal case was 45 days after 20 days of herbal intake, and 90 days and 50 days, respectively, in the second case.

Teas responsible for similar pictures of liver damage were also reported from South Africa (Mokhobo 1976), from Hong Kong (Kumana *et al.* 1983), and from the USA. For example, Stillman *et al.* (1977) described how in Arizona a six-month-old girl had developed the picture of hepatic veno-occlusive disease after being given a tea made from *Senecio longilobus*. Fox *et al.* (1978) described a similar case in which a two-month-old American boy was given a herb tea ('gordolobo') made from *S. longilobus* to treat nasal congestion. The herb contained 1.5 per cent of pyrrolidizine alkaloids and it was estimated that he had consumed 66 mg of the alkaloid over four days. He died six days after admission to hospital; liver histology showed a severe centrilobular necrosis. Fox commented that the LD_{50} (rat, intraperitoneal, 72h) for retrosine, the major alkaloid found in S. *longilobus* was 35 mg kg^{-1}.

This syndrome is also known in Britain; McGee *et al.* (1976) described a 26-year-old woman, presenting with ascites and hepatomegaly, who had been drinking large quantities of herbal tea obtained from a local health food store. This tea (maté, Paraguay tea) was made from a mixture of the leaves of various Ilex plants. Although analysis showed only trace amounts of pyrrolizidine alkaloids, the patient had been taking large quantities of the tea for two years.

Herbs used for their psychoactive effects. Siegel (1976) commented that 43 of the 396 herbs commercially available in herbal teas contained psychoactive agents, although some of these were present only in trace amounts and unlikely to have much effect. This group included such herbs as yohimbe, catnip, hops, kavakava, kola, lobelia, mandrake, maté, nutmeg, passionflower, periwinkle, snakeroot, thornapple, valerian, and wormwood. Nevertheless, intoxications have been reported and a few illustrative examples are given below. A case of intoxication with maté tea is described under the previous heading 'Teas containing pyrrolizidine alkaloids'.

Kavakava: the rhizome of *Piper methysticum* comes from Polynesia and is a primary stimulant of the central nervous system followed by sedation. It is used for its alleged euphoric and relaxing effect. Long-term use of this tea results in a yellowing of the skin, probably due to kava pyrones depositied in keratin. The central relaxant effects may be due to dihydromethysticin and cause ataxia, somnolence and difficulty in hearing and vision (Siegel 1976).

Datura: *Datura stramonium*, also known as Jimson weed, Jamestown weed, thornapple, devil's apple, stinkweek, 'loco' weed, is indigenous to Europe and N. America. It has a thorny fruit with blackish-brown seeds containing hyoscyamine. The plant itself contains atropine, scopolamine, hyoscine, and hyoscyamine. Extracts have been used for some time for their anti-asthma properties but use has increased recently for its mind-altering effects.

Mahler (1976) in discussing the acute toxicity of stramonium said that primary signs were of atropine poisoning with disturbances of thought and hallucinations. The differential diagnosis included LSD intoxication and acute schizophrenia, but physostigmine will reverse the central and peripheral manifestations.

Belton and Gibbons (1979) commented on the presence in west Cornwall of certain trees of the *Datura* genus (*D. sanguina, D. cornigera, D. aurea*). They reported five cases admitted to hospital who had either brewed tea from the leaves of this type of shrub or had eaten them. Symptoms were visual hallucinations and atropine-like effects. Treatment consisted of gastric lavage and sedation with intramuscular or intravenous diazepam. All recovered within 36 hours.

Khat, Catha (*Catha edulis*): leaves of the khat shrub are widely used as a stimulant in E. Africa and the Arab states. The leaves contain a phenylalkylamine, (-)-cathinone, which has similar propertites to(+)-amphetamine causing euphoria, hyperactivity, and sympathomimetic effects (Kalix 1984). Khat leaves may also cause a toxic psychosis of the amphetamine type and this has occurred in the UK (Gough and Cookson 1984). The leaves must be fresh to be active but this is possible world-wide with international air travel and refrigeration.

Herbal cigarettes

The problem of 'tobacco', i.e. *Nicotiana spp.*, and carcinogenicity is well known and is not further discussed here. It is, however, of interest that the carcinogenic potential of *Nicotiana tabacum* was recognized as long ago as 1761 by Sir John Hill who associated the excessive use of snuff with the appearance of fatal malignant polyps ('polypusses') of the nose (Redmond 1970).

According to Siegel (1976) there are presently 192 distinct herbs commercially available and used as smoking substances. Forty-four per cent of these contain substances with known psychoactive effects, and the group includes yohimbe, broom, Californian poppy, catnip, cinnamon, damiana, hops, hydrangea, juniper, kavakava, kola, lobelia, passion flower, periwinkle, prickly poppy, snakeroot, thorn apple, tobacco, wild lettuce, wormwood. The use of these smoking substances has given rise to a number of adverse effects, especially of a stimulant, euphoriant, or hallucinogenic nature.

Other herbal teas

Casterline (1980) reported a case of a 54-year-old woman who drank a cup of camomile tea and developed severe anaphylactic reaction within 20 min with upper-airway obstruction and pharyngeal oedema. She was treated symptomatically and recovered. Camomile tea is made from camomile flower heads which are known to cross-react with ragweed, chrysanthemums, and other members of the Compositae family. This patient had a strongly positive reaction with specific radio-allergosorbent tests against ragweed pollen. Skin testing with camomile was thought inadvisable.

Even the more familiar coffee and tea are not without their detractors but recent reports (MacMahon et al. 1981) that coffee drinking may cause pancreatic cancer were summarily dismissed in a critical review by Feinstein et al. (1981) of the case-control technique that was used. This view was supported by Benarde and Weiss (1982) who were more impressed with an association between cigarette-smoking and pancreatic cancer.

Eisele and Reay (1980) reported on two deaths which they said were related to coffee given by enema. Naturopathic therapies relying mainly on changes in diet are increasing in popularity in the USA and one regimen used for patients with cancer or chronic degenerative disease also includes coffee enemas. The first of the two cases quoted was a 46-year-old woman with chronic cholecystitis and degenerative arthritis who was receiving coffee enemas as well as a complex diet. Enemas of coffee were at times being given as frequently as three or four per hour. Death was attributed to bronchopneumonia and hypokalaemia. The second case was of a 37-year-old woman with carcinomatosis following carcinoma of the breast. She was similarly treated with special diets and coffee enemas and death was attributed to fluid and electrolyte imbalance.

It seems doubtful that in these cases enough caffeine was absorbed to produce a substantial toxic effect but that the enemas contribute to death by electrolyte depletion. These case reports seem more of a commentary on the credulity of man rather than on the toxicity of coffee.

Miscellaneous herbs and other treatments

Alfalfa

Alfalfa seeds are said to lower plasma cholesterol and to prevent atherosclerosis. Malinow et al. (1981) observed splenomegaly and pancytopenia in a 59-year-old man who, in a dietary study, ingested 80–160 g of ground alfalfa seeds daily on eight occasions for periods of up to six weeks. His plasma cholesterol fell from 218 ± 17 to 130–60 mg dl^{-1}. When alfalfa was removed from his diet, the haematological values and spleen size slowly reverted to normal. The authors agreed that the evidence for alfalfa being the cause of the pancytopenia is circumstantial but pointed out that they had observed a similar syndrome in monkeys fed on alfalfa. They postulated

that the toxic substance could be canavanine, an arginine analogue present in high concentrations in alfalfa seeds.

Roberts and Hayashi (1983) commented that systemic lupus erythematosus (SLE) can be induced in monkeys by alfalfa and that L-canavine has been implicated. They reported on two patients whose clinical and serologically quiescent SLE was reactivated in association with the ingestion of alfalfa.

Alfalfa also has allergenic properties and exposure to it may intensify and exacerbate symptoms caused by existing sensitivities (Polk 1982).

Aristolochic acid

Aristolochic acid is a phenanthrene derivative found in Aristolochia spp., such as A. serpentaria (Virginian snakeroot) and A. reticulata (Texas snakeroot, Red River snakeroot, Serpentary), and is of some interest being one of the few naturally occurring organic compounds to contain a nitro-group. Serpentary (B.P.C. 1949) is the dried rhizome and roots of A. reticulata or A. serpentaria and is used either as an infusion or as a tincture.

Extracts of Aristolochia spp. are used for a variety of herbal purposes world-wide such as the treatment of dermatitis, rheumatism, and gout. The root of A. indica L. is reputed to have emmenagogic and abortifacient properties and this was confirmed experimentally in the mouse (Pakrashi and Shaha 1977).

Aristolochic acid was shown to have a significant activity against some experimental neoplasms in the mouse (Kupchan and Doskotch 1962) although human clinical trials were abandoned because of its nephrotoxic effect (Jackson et al. 1964) and it has been implicated as a probable causative agent in Balkan endemic nephropathy (Dammin 1972).

Aristolochic acid recently became of concern when a long-term toxicity test in rats showed a dose and time-dependent carcinogenic effect (Mengs 1983; Mengs et al. 1982). In this study aristolochic acid was given to three groups of rats daily by mouth at 0.1, 1, and 10 mg/kg, respectively. Rats in the 10-mg group examined at three months had developed malignant papillomata of the stomach. Other animals in this dosage group were allowed a further 3–6 months drug-free period and were then found to have squamous-cell carcinomata of the stomach and multiple carcinomatous lesions especially in the kidney and bladder. Similar but less severe changes were found in the 1-mg/kg group. Even in the 0.1-mg/kg group, although no tumours were detected in animals sacrificed at the end of a 6-month dosage period, papillomatous changes and squamous-cell carcinoma of the stomach occurred in animals given a further 12–

16-month drug-free period. Further details are given in Chapter 35.

The carcinogenic potential of aristolochic acid in this study is emphasized by the short period of exposure necessary to induce malignant change and by the fact that malignant changes occur in many tissues and continue to develop after withdrawal of the drug.

It is difficult to quantify the human carcinogenic risk as the dose of aristolochic acid used has a wide range. One preparation of the pure acid (Martindale 1977) is used in doses of 150–300 μg three times daily but other dosage received would depend on the amount of herb used and the method of preparation. It is difficult to justify any use of aristolochic acid having regard to its therapeutic value and to the carcinogenicity in the rat although it may perhaps be permissible in the higher homoeopathic dilutions.

Castor beans and ricin

Castor beans (sometimes known as mole beans as they are used for killing moles in residential lawns) are the seed of the castor oil plant (*Ricinus communis*) which is cultivated commercially especially in the southern USA. The castor oil plant and more especially the seed, contains ricin, a member of the class of proteins called phytotoxins and one of the most toxic substances known. The lethal intravenous dose in animals is 0.3 μg/kg (Martindale 1977) and, even though ingestion lessens its toxicity, one castor seed contains enough ricin to kill a child.

After a latent period of some hours to days, ricin causes a marked gastrointestinal irritation with subsequent widespread organ toxicity and central nervous disturbances. A recent victim of ricin poisoning was the Bulgarian exile Georgi Markov who in September 1978 was killed by a sphere 1.5 mm in diameter containing ricin which was implanted in his thigh by an assassin's umbrella (Knight 1979).

Allergic reactions such as urticaria, asthma, and anaphylaxis can by precipitated by direct contact or inhalation such as by agricultural workers handling the castor oil plant or by women wearing necklaces made of the rather ornamental beans (Henry *et al.* 1981).

Cremophor EL, a polyethoxylated derivative of castor oil and used as a non-ionic emulsifying agent, has been implicated in 'histaminoid' type of reactions. Polyethoxylated castor oils are used in the formulation of such products as propanidid and *Althesin* (alphaxalone/alphadione) to solubilize their active ingredients. The problem is well reviewed by Watkins (1982).

Comfrey (Symphytum spp)

Comfrey *(Symphytum officinale)* is indicated in the British Herbal Pharmacopoeia (1976) for gastric and duodenal ulcer, haematemesis, and colitis. It is also recommended externally for ulcers, wounds, fractures, and hernias by application of the fresh root preparation. Attention has focused on the possible carcinogenic and hepatoxic effects of this commonly used herb which contains up to eight pyrrolizidine alkaloids such as symphytine, lycopsamine, and echimidine (Culvenor *et al.* 1980; Furuya and Araki 1968). The liver toxicity of the pyrrolizidine alkaloids was discussed above.

A paper by Taylor and Taylor (1963) reported a protective effect of comfrey-leaf extract in mice bearing spontaneous and transplanted tumours. The tumours were smaller than those in the untreated animals and tumour-bearing treated mice survived longer when given the extract than did the tumour-bearing controls. Nevertheless, the study was conducted for only 40 days and the feeding schedule for only 5 days in each week. Also, even assuming that the study showed an antitumour effect of comfrey, this did not indicate that it was free of carcinogenic activity since many cytotoxic agents are known carcinogens.

Hirono *et al.* (1978) reported a study in rats fed either comfrey leaves or comfrey roots. A high incidence of tumours of the liver was reported in all treated groups compared with their controls. The tumours were diagnosed mainly as hepatocellular adenomas with a few haemangiosarcomata and an increased incidence of bladder tumours. The evidence suggested that comfrey is a carcinogen for rats and may thus have a carcinogenic hazard in man.

The carcinogenic action of comfrey in rats was demonstrated after continuous high dosage for long periods and the consumption by man is usually at a much lower level. Nevertheless, unless the findings of Hirono and his co-workers can be disproved, it is unlikely that comfrey can ever again be regarded as completely safe in man.

Mattocks (1980) found that the highest levels of total alkaloids in the Russian comfrey S. × *uplandicum* were in the young, small leaves (0.049 per cent) especially early in the season and that the large, mature leaves contained the least (0.003 per cent).

He futher commented that the external use of comfrey preparations should not be hazardous as the alkaloids are not in themselves toxic but are converted to toxic pyrroles in the liver after absorption following oral administration (Mattocks 1968). While this may be so, it is not clear whether the alkaloids themselves are absorbed when used locally which would make this extrapolation by Mattocks an unwarranted one.

Estimates of how much total pyrrolizidine alkaloid is ingested during the use of comfrey as a herbal tea were made by Roitman (1981) who found this could be as much as 26 mg/cup when an infusion of comfrey root was made a according to the package directions.

Ginseng

It is not intended in this review to discuss ginseng in detail but merely to point out some hazards of this modern Western craze. The powdered roots *Panax ginseng* (*panax* = all heal) has a millenia-old tradition of use in China and Korea for its antifatigue and antistress properties and to give health, longevity, and virility. It has also been known in the Western world for some time but it is only recently that ginseng has become something of a cult object.

Ginseng is an interesting example of a herbal product with a long pedigree and some demonstrable pharmacological effects but where the truth and attitudes to the drug are clouded by the aura of nonrespectability and the implicit and sometimes explicit association with sexual potency. It is to be regretted that the claim that ginseng is an adaptogen (i.e. increases the general capacity of the organism to overcome stess, fatigue, and ageing) is not more universally and dispassionately investigated.

Previously found only in the wild, commercial ginseng is now mainly cultivated and there are several varieties of the Asian species. *P. notoginseng* in found in south China, *P. pseudoginseng* in India and south China, and *P. japonicum* in Japan. A separate and native species is found in North America (*P. quinquefolium*). There are many other ginseng-type plants distributed world-wide such as *Eleutherococcus senticosus* (Russian or Siberian ginseng). The whole topic is a confusion of names and minor variations in spelling.

The herbal extract of ginseng contains a complex mixture of sugars, steroids, and saponin glycosides (Popov and Goldwag 1973). Some of these glycosides have been further characterized but the terminology varies according to the different schools of workers (primarily Russian or Japanese). The Russians have isolated panaxosides A-F whose aglycones, oleanolic acid, (20S)-protopanaxadiol, and (20S)-protopanaxatriol are attached to sugar components differing in position, number, or type (glucose, rhamnose, xylose, and arabinose). The Japanese workers have characterized these glycosides as 'ginsenosides' but the relationship between them and the panaxosides is not always unequivocal.

Ginseng as commercially available is of substantial economic importance and Siegel (1977) commented on the widespread mislabelling of this expensive herbal. He said that some products packaged as ginseng contained adulterants such as *Mandragora officinarum* (scopolamine), *Rauwolfia serpentina* (reserpine), and *Cola* species, all of which are both cheaper and have more noticeable pyschoactive effects than ginseng. The picture is greatly confused by the dubious nature of some commercial preparations of ginseng. Corrigan and Connolly (1981) tested 38 preparations of ginseng on the Irish market and commented very adversely on their quality. Many of the products contained only very small amounts of ginsenosides or contained such plants as *Eleutherococcus* rather than *Panax*. Many of the Irish products were also very poorly formulated.

Reports of side-effects to ginseng have appeared but without some degree of standardization of the commercial products, the picture is most unclear for one cannot be certain that ginseng is actually involved. Palmer *et al*. (1978) described a case of a 70-year-old woman who developed swollen, tender breasts after taking ginseng for three weeks. Symptoms settled after she stopped taking ginseng but they readily recurred after further challenges. There is little doubt that this was due to an oestrogenic-like substance present in ginseng. Five more cases of breast enlargement in women taking ginseng were reported briefly by Koreich (1978); the herb was also associated with claims of an increase in sexual responsiveness.

Punnonen and Lukola (1980) confirmed the oestrogenic-like effect of ginseng by its effect on the vaginal epithelium of a 62-year-old woman who had been taking ginseng and multivitamins for a year. Gas chromatography of the tablets taken by this patient showed no oestrogens, but a crude methanolic extract competed very strongly with 17-oestradiol and R5020 for the oestrogen-and progesterone-binding sites in human myometrial cytosol. The oestrogenic activity of ginseng was thought to be the reason for vaginal bleeding in a 72-year-old woman (Greenspan 1983).

Siegel (1979) described an interesting study in which 133 ginseng users were followed over a period of two years. He labelled as the 'ginseng abuse syndrome' (GAS) the reported effects from the long-term use of ginseng in this group. These were mainly symptoms of central nervous excitation and arousal, nervousness, and tremor. Twenty-two of the group became hypertensive (not defined but said to be 'confirmed by examination'). Siegel likened the effects experienced in GAS to corticosteroid poisoning. Whilst this study is very suggestive, it is unfortunate that no control group was included and that the route and dosage of ginseng was variable. The only objective measurement was the blood pressure and details of this are not given.

Hammond and Whitworth (1981) reported the case of a 39-year-old man with hypertension (140/100 mmHg) who had been taking a variety of different ginseng products orally for three years. He was advised to stop taking them and five days later was normotensive (140/85 mmHg) and remained so for an observation period of three months. Whether ginseng can now be definitely added to the list of drugs causing clinically significant hypertension needs confirmation.

Gossypol

Gossypol is the active principle in an alcoholic extract of cotton root bark, a Chinese folk medicine for treating bronchitis. Animal studies revealed its antifertility action in males by depression of spermatogenesis and it has been tried successfully clinically (Qian *et al.* 1981).

Although highly effective and apparently associated with few serious side-effects, the use of gossypol has led to the development of hypokalaemic paralysis (Qian *et al.* 1980). In the study by Qian and his colleagues, 148 men in Nanjing took gossypol as a contraceptive during 1972–77. Seven of these men developed a hypokalaemic paralysis with serum potassium values of 2.00–2.73 mEq l[-1]. This hypokalaemia seems to be due to increased renal potassium loss not brought about by hyperaldosteronism. The Nanjing diet was found by Qian and his colleagues to be low in potassium, and this may have been a contributory factor as the incidence of hypokalaemia varied from region to region.

Although gossypol appears to be effective as a male contraceptive, Qian feels that more toxicological studies are necessary. Recovery of fertility in the majority of men appears complete within 3–6 months of stopping the drug but there may be some who take up to 3 to 4 years or even become permanently sterile.

Green-lipped mussel

A recent natural product promoted for the treatment of chronic inflammatory joint disease is 'Seatone', an extract of the New Zealand green-lipped mussel (*Perna canaliculus*). A study by Gibson *et al.* (1980) suggested that *Seatone* might be of value in rheumatoid arthritis although a further study by Huskisson *et al.* (1981) found no useful effect. It may be, as Gibson and Gibson (1981a, b) suggested that *Seatone* requires some time to have an overt effect, say 3–6 months rather like penicillamine.

Ahern *et al.* (1980) reported the case of a 64-year-old woman presenting with a three-week history of epigastric pain, anorexia, nausea, and jaundice. She had been taking *Seatone* for non-erosive, seronegative, sym-

metrical arthritis of the wrists and metacarpophalangeal joints.

Biochemical investigation and liver biopsy confirmed the clinical diagnosis of granulomatous hepatitis which resolved on ceasing *Seatone* medication. These workers suggested that the temporal relationship of the onset and resolution of the disorder with the ingestion of *Seatone* supports the view that it was the cause of the hepatitis. Rechallenge was not attempted as this was thought to be ethically improper.

Hallucinogenic fungi

Young *et al.* (1982) commented on the use of indigenous hallucinogenic fungi in the UK and reported a series of 49 patients admitted to Glasgow hospitals after the deliberate ingestion of *Psilocybe semilanceata* ('liberty cap', 'magic mushrooms'). This gill fungus is found particularly in western Britain and contains psychoactive substances such as psilocybin and psilocin which can act on indoles in the central nervous system to produce symptoms which can mimic acute toxic and schizophrenic states. The freshly picked mushrooms contain 3–12 mg of psilocybin in 20 to 30 g (equivalent to 5 g of the dried mushrooms or 30 fruits), 3–6 mg of which would be sufficient to produce hallucinations. The effect is potentiated by alcohol and there is obviously a serious risk of interaction with other psychoactive drugs which may be taken concurrently. The authors of the study comment on the easy availability of *P. semilanceata* and the encouragement given to its ingestion by some subcultures of society. Possession of the raw mushrooms is not an offence in law though processing them to extract the drugs may be.

Karela (Mormodica charantia)

The fruit of *Mormodica charantia* (karela, bitter gourd) is indigenous to South America and Asia. The use of karela, a traditional antidiabetic agent used either in meals or traditional Indian medicines, may interfere with the control of diabetes mellitus in patients on orthodox hypoglycaemics (Aslam and Stockley 1979). Akhtar *et al.* (1981) confirmed the hypoglycaemic effect of karela in both normal and alloxan-diabetic rabbits and this was confirmed clinically in diabetics by Leatherdale *et al.* (1981).

Kelp

Kelp, a seaweed, is a staple diet in Japan and is also used in the fashionable 'eat kelp and grow thin' diets. The weed contains large amounts of potassium iodide and a somewhat speculative editorial in the *Journal of the American Medical Association* (1975) commented that 1–3 per cent of the American population has an underlying sensitivity to chronic iodide exposure, and further, that a small percentage might get myxoedema especially if a latent Hashimoto thyroiditis was present.

Laetrile (amygdalin) and other cyanogenic glycosides

Laetrile is a phenomenon of this century especially in the USA as an alleged treatment for cancer. It is to be hoped that the long saga of laetrile is coming to an end. Over the years it achieved a status in the alternative-medicine field of the USA as being a potent and safe remedy for cancer which, the pro-laetrile lobby maintained, the orthodox physician, the pharmaceutical manufacturers, and the government were determined to suppress for underhand reasons of their own. The Food and Drugs Administration (FDA) certainly consistently refused to approve the use of laetrile in the absence of acceptable laboratory or clinical evidence that it was safe and effective. Nevertheless, public pressure forced the legislatures in over half of the individual states to pass laws legalizing the use of laetrile within their own boundaries.

Chemistry and pharmacology. There is considerable confusion in the terminology of laetrile in both lay and medical literature. The terms amygdalin, Laetrile, laetrile (no initial capital), vitamin B–17, nitrilosides, and others have been used interchangeably though the chemical identity of the substance referred to is not constant. These substances are also cyanogenic glycosides (glucosides), i.e. they break down to yield cyanide and glucose. The interested enquirer is referred to the *Federal Register* (1977) where the Food and Drug Administration (FDA) give a fascinating account and to a review by Herbert (1979).

Amygdalin is α-cyanobenzyl-6-0-β-D-glucopyranosyl-β-D-glucopyranoside (mandelonitrile-β-D-glucosido-6-β-D-glucoside). It is found in vetches, clover, sorghum, cassava, lima beans, acacia, but especially in the kernels of apricots, almonds, peaches, and plums. Amygdalin is hydrolysed by β-glucosidases to glucose and mandelonitrile which is further hydrolysed to benzaldehyde and hydrogen cyanide.

The kernels of apricots, almonds, and similar fruits contain not only amygdalin but also a β-glucosidase which readily hydrolyses the amygdalin to produce cyanide whenever the kernel is crushed or disrupted. Other fruits or vegetables such as celery, peaches, mushrooms, and beansprouts also contain similar β-glucosidases but no amygdalin. They will, however, obviously facilitate the breakdown of amygdalin in the gut if taken simultaneously. The importance of this is that these substances are often taken in 'metabolic diets' associated with laetrile regimes for the treatment of cancer.

The word Laetrile (initial capital) has been used to refer to a specific compound prepared in 1952 by Krebs (cyanobenzyl-β-D-glucopyranosiduronic acid or mandelonitrile-β-D-glucuronic acid). Laetrile has also been used as a synonym for amygdalin and also for mixtures often of a non-specific composition. It is not possible in discussing cases of toxicity of 'laetrile' to be always sure what are the components of the product.

The original theory to explain the alleged antitumour effect of laetrile was that the cancerous cells were said to contain an increased amount of β-glucosidase and laetrile administration would therefore produce a higher local concentration of cyanide than in normal cells. This differential toxicity was compounded by cancer cells having a low rhodanese (thiosulfurtansferase) content which was plentiful in normal cells and served to transform the toxic cyanide to thiocyanate (in passing it should be noted that thiocyanate is goitrogenic and a potential carcinogen).

No acceptable evidence is available to support this theory and later revised versions have proved no more attractive. One recent proposed mechanism is that laetrile is a vitamin (B_{17}) and that cancer is the result of a deficiency which may be cured or prevented by the administration of laetrile. This theory is no more persuasive than the others. The dispute continues as to the mode of action of laetrile but it would be profitless to consider it further at this time but instead we will concentrate on the toxicity.

Toxicity of laetrile and other cyanogenic glycosides. There is no argument that amygdalin is a cyanogenic glucoside and that there is a β-glucosidase present in the microflora of the gut lumen which could release free cyanide from it or any other cyanogenic herbal extract. Sayre and Kaymakcalan (1964) discussed the fashion in Anatolia of eating apricot and peach seeds which are a plentiful source of amygdalin; the cyanide content of these is not small, for example, peaches contain 88 mg, cultivated apricots 8.9 mg, and wild apricots 217 mg per 100 g moist seed in each case. Children have died in Anatolia from eating just a few wild apricot seeds.

Some Nigerians exist on a diet of cassava which is rich in cyanogenic glycosides such as linamrin. Chronic cyanide poisoning is common, presenting as an ataxic-neuropathy with lesions of the skin, mucous membranes, optic and auditory nerves, spinal cord, and peripheral nerves.

Laetrile has been shown to be toxic in animals. Schmidt *et al.* (1978) fed dogs on laetrile in doses 1–5 times the human dose. Six out of 10 dogs died with symptoms characteristic of cyanide poisoning. Willhite (1982) showed in hamsters that laetrile was teratogenic by the oral route but not by the intravenous route, causing skeletal deformities. As the effect was antagonized by thiosulphate administration, it was suggested that the teratogenicity was due to cyanide released by glucosidase activity in the gut.

Human toxicity is of course far more difficult to document and the picture is usually complicated by the presence of terminal carcinoma, these being the type of patients attracted to such a therapy. Nonetheless there are a number of disturbing reports. For example, Braico et al. (1979) reported a case of an 11-month-old girl, previously healthy, who accidentally swallowed 1–5 tablets (500 mg) of laetrile. Within half-an-hour she had become lethargic and rapidly went into coma, shock and died. Her serum cyanide measured 3 hours after ingestion of the laetrile was 0.029 mg dl^{-1} and it was estimated that each 500 mg tablet of laetrile could have given 12–26 mg of cyanide. Graham et al. (1977) estimated that the lethal dose of potassium cyanide was 50–200 mg in an adult.

Sadoff et al. (1978) described a case of a 17-year-old girl who had a right frontal astrocytoma removed surgically. She had been self-administering laetrile intravenously for four weeks (12 g daily in divided doses) but, unable to give the dose intravenously one morning, swallowed about 10 g. She collapsed in 10 minutes with headache, dizziness, and tetanic convulsions. She went into coma and died in 24 hours. Unfortunately samples of blood, urine, or gastric contents were not taken until 36 hours after ingestion of the laetrile and no cyanide was present. Sadoff et al. commented that although the evidence was admittedly circumstantial, the symptoms were characteristic of cyanide poisoning and demonstrated the increased toxicity of laetrile taken orally and subjected to rapid intestinal hydrolysis.

Increased blood cyanide was shown in other cases though these were not fatal. Maxwell (1978) related a case of a 69-year-old man with sarcoma of the left humerus and pulmonary metastases. He had been taking laetrile for 12 months and was admitted with episodes of syncope 1 hour after taking laetrile; blood cyanide was 0.6 mg dl^{-1}. Maxwell gave normal values of blood cyanide as 0.01–0.02 mg dl^{-1}, acute intoxication and coma with recovery 0.02–0.75 mg dl^{-1}, and fatal cases 0.26–3.1 mg dl^{-1}.

Morse et al. (1979) reported a case of a 48-year-old woman with lung cancer who for nine days had been taking laetrile, intravenously, intramuscularly, orally, and rectally together with a 'detoxification' diet of raw fruits and vegetables and coffee enemas. She was admitted in shock but responded to symptomatic treatment. Blood cyanide was 0.116 mg dl^{-1}.

Smith et al. (1977) reported two cases where toxicity was attributed to laetrile. The first was a 48-year-old woman who had been taking laetrile intravenously and orally. Two months later she was admitted to hospital with fever, malaise, headache, and abdominal cramps. Laetrile was discontinued and the symptoms cleared in two days. Two months later she resumed taking laetrile and was admitted 15 days later with the same symptoms as before. Blood cyanide was given as 1 mg dl^{-1} though this seems somewhat higher than other workers have found.

The second case was that of a 46-year-old man with lung cancer who had been taking 500 mg laetrile daily for six months when he complained of neuromuscular weakness of the extremities with bilateral ptosis. This resolved within 48 hours of stopping laetrile and has not recurred. The patient did not take any more laetrile.

In 1978 the National Cancer Institute (NCI) yielded to public pressure and carried out a retrospective analysis (Ellison et al. 1978). A mail request was sent to 385 000 physicians and 70 000 health professionals and to pro-laetrile groups for details of the estimated 70 000 patients who had used laetrile. Only 93 cases were submitted of which 26 were eliminated due to bad documentation. The remaining 67 case records were assessed as alleged examples of the beneficial effect of laetrile but the NCI could find evidence of a possible complete response in only two cases and a possible partial response in four. It was concluded that these results allowed no definite conclusions supporting the anticancer action of laetrile. Under public and political pressure the National Cancer Institute (NCI) requested clinical trials with laetrile and this was agreed by the FDA.

The previous knowledge of the toxicity and pharmacology of laetrile was very fragmentary and Moertel et al. (1981) reported on the first stage of the NCI investigation. This was a pharmacological and toxicological study in which six patients with advanced cancer were treated with amygdalin in doses similar to those used by laetrile practitioners and a 'metabolic therapy' programme. Amygdalin was given intravenously at doses of 4.5 g m^{-2} daily and was largely excreted unchanged in the urine and produced no clinical or laboratory evidence of toxic reaction. Amygdalin given orally at 0.5 g three times daily produced significant blood cyanide levels of 0.2 mg dl^{-1} thus confirming the importance of oral administration although there were no clinical or laboratory evidence of toxic reaction. The importance of the accompanying diet was confirmed in this study when raw almonds were also given to two patients to produce a large β-glucosidase load. One of these patients had transient signs of cyanide intoxication. The identity of the amygdalin in the preparations used was established by a variety of analytical techniques and was found to be pure amygdalin at 100 per cent of the labelled strength. The small number of patients in the study by Moertel and his colleagues precluded any conclusions about therapeutic efficacy.

Further public pressure forced a prospective study sponsored by the NCI at four treatment centres: the Department of Oncology at the Mayo Clinic; the UCLA Johnson Comprehensive Cancer Center in Los Angeles; the Memorial Sloan-Kettering Cancer Center in New York; and the University of Arizona Cancer Centrer. These centres studied 178 patients to whom they had given laetrile and concluded that it is non-effective as a cancer treatment and that it is toxic and can produce cyanide poisoning after oral administration (Moertel *et al.* 1982).

It is unfortunate that the laetrile lobby has countered these negative findings of the NCI with allegations of impropriety and bias by the Institute. Moertel and his colleagues admit that their study was not designed in the classic mould of being controlled, randomized, and blind but they nevertheless did try to ensure that their therapeutic regime was consistent with that of laetrile practitioners. It was perhaps inevitable that their regime failed to meet the approval of the pro-laetrile lobby.

Laetrile may or may not have received its death blow in the USA but there are regrettable indications that this regime is being promoted in other countries including the UK. There are no products containing laetrile on the UK market which may legitimately be commercially promoted for the treatment of cancer or, indeed, for the treatment of any other medical condition. However, in the UK a doctor may prescribe for a particular patient any drug, licensed or not, which he may consider necessary to the specific case. With regard to laetrile the doctor must balance in his own mind the various factors such as the long history of emotional publicity, the known toxicity, the lack of effectiveness in the NCI studies, and the absence of peer review. He has to remember that a drug may be unsafe to use for several reasons. It may be inherently toxic or it may be non-effective in a serious condition for which other and effective treatment is available. Laetrile falls under both of these headings. It should also not be thought that, even if a substance is not itself toxic, that a product containing it is necessarily safe to administer. Whereas the pharmaceutical quality of licensed products in the UK is legally controlled, that of unlicensed products like laetrile is not. Samples of amygdalin, for example, have been found to have a high frequency of microbial and endotoxic contamination (Davignon *et al.* 1978).

Such contaminants were implicated by Liegner *et al.* (1981) when describing the case of a 61-year-old woman who had been treating herself with laetrile (among other things) for five years. She developed acute agranulocytosis but her blood picture returned to normal when she stopped all medication. The agranulocytosis recurred acutely within a few days of restarting laetrile. The patient stopped her medication again; the blood picture reverted to normal and remained so for the 18-month laetrile-free follow-up period. These workers suggested that, while the agranulocytosis could have been due to the cyanide produced by the laetrile, it was also possibly due to contaminants and impurities in the product.

While the involvement of laetrile seems reasonable in this case, it is puzzling that the patient took it for five years with no problem. The authors regrettably do not say if she changed the source of her laetrile in the latter period.

Cassileth (1982) in an interesting commentary entitled 'After laetrile, what?' suggested that laetrile is being replaced in unorthodox cancer therapy in the USA by the 'natural' approach to malignant disease which has eclectic roots in homoeopathy, naturopathy, Asian medicine, and philosophy, together with somewhat more than variant notions of human physiology and pathology. This approach rejects medicines as such and emphasizes dietary regimes and cures through 'purification' of the body together with the capacity of the body to heal itself.

Liquorice

Liquorice (Licorice) (*Glycyrrihiza glabra*) is one of the most widely used medicinal plants. The herbalists' indication for liquorice root for gastric ulcer received some credence when the saponin-like compounds glycyrrhizin and glycyrrhizinic acid were isolated and found to be effective in healing duodenal and gastric ulceration. Their mode of action is unknown and is probably due to an increased mucous secretion which protects the gastric cells and an increased cell turnover which promotes epithelialization of damaged areas. Carbenoxolone is a derivative of enoxolone, a complex triterpene prepared by the hydrolysis of glycyrrhizinic acid and has anti-ulcer properties.

Liquorice has considerable and popular use in confectionary; excessive use is not, however, without hazard as several reports have shown. Koster and David (1968) described a 20-year-old girl with a severe hypertension (220/140) with water and sodium retention. Her diet included 100 g of liquorice daily and her blood pressure soon returned to normal on stopping this. The aetiology was confirmed by several rechallenges with liquorice.

Bannister *et al.* (1977) reported a case of a 58-year-old woman with typical signs of hypokalaemia (1.8 mmol l⁻¹) who went into ventricular fibrillation. She had been eating 1.8 kg of liquorice sweets a week, but on stopping these her condition rapidly returned to normal and remained so. The hypokalaemia and sodium retention caused by the mineralocorticoid action of liquorice was the cause of a report by Cumming *et al.* (1980) who described a case of a 70-year-old woman with a flaccid quadriplegia due to profound hypokalaemia caused by the ingestion of small amounts of liquorice in a laxative preparation. The diagnosis was confirmed by the controlled administration of the preparation.

Maklua

Maklua, a local Thai fruit (*Diospyros mollis Griff* of the family *Ebenaceae*) is one of the oldest herbal remedies in Thailand and neighbouring countries with a useful activity against hookworm and roundworm and has been used in recent mass chemotherapy campaigns. Some cases of blindness have been reported (Sadavongvivad 1980; Chotibutr *et al.* 1981) either with the herb or one of its active principles, diospyrol, (2,3-bis 3 methyl-1, 8-dihydroxynaphthalene).

The pathology is an optic neuritis but the underlying cause is not known. It seems likely that children under 10 are the most susceptible and that an individual hypersensitivity to some of the constituents of the Maklua extract plays a major part.

Margosa oil (the neem tree, Azadirachta indica A. Juss)

Margosa oil is another traditional treatment from South-east Asia and is an extract of the seeds of the neem tree (*Azadirachta indica A. Juss*). Sinniah and Baskaran (1981) reported on 13 infants who had been given the oil for minor ailments such as constipation and respiratory-tract infection and who were admitted to hospital with vomiting, drowsiness, metabolic acidosis, polymorphonuclear leucocytosis, and encephalopathy. Liver biopsy showed pronounced fatty infiltration. Treatment was symptomatic and 11 of the infants survived. Commercial Margosa oil is of variable constitution and toxicity but there was no suggestion in these cases of contamination. The actual toxic consituents have not been identified.

Mistletoe (Viscum album)

'Mistletoe' is the popular name of some 1300 species of perennial evergreen parasites occurring world-wide of which *Viscum album* is the European variety (Anderson and Phillipson 1982).

Mistletoe has been shown to have many pharmacological properties including being hypotensive, diuretic, and antispasmodic. It is also claimed by some unorthodox practitioners to have an anticancer action.

Mistletoe extract is a complex mixture containing at least three types of potentially toxic compounds; alkaloids, some of which may be cytotoxic; proteins of small molecular weight called viscotoxins; and lectins which have haemagglutinin and mitogenic actions (Franz *et al.* 1981).

Harvey and Colin-Jones (1981) reported a case of a 49-year-old woman who presented with nausea and general malaise. Liver function test suggested hepatitis and liver biopsy showed light inflammatory-cell infiltration of the portal tracts. Hepatitis B surface antigen was not detected and a cholecystogram was normal and the symptoms settled with conservative treatment.

Two years later she presented with similar symptoms and findings and it was discovered that both illnesses had coincided with the ingestion of a herbal remedy for 'nervous tension' which contained motherwort, kelp, wild lettuce, skullcap, and mistletoe. Once again the condition settled after symptomatic treatment but recurred after a formal challenge test with the herbal tablets. Symptoms returned after 10 days of ingestion of the tablets and, after 14 days, liver biopsy showed severe changes with heavy infiltration of lymphocytes and plasma cells and considerable focal necrosis.

Harvey and Colin-Jones (1981) stated that they could find no reports of toxic reactions to kelp, motherwort, and skullcap, although wild lettuce has given the occasional case of toxicity. These workers also pointed out that the herbal tablets used by their patient contained 90 mg of mistletoe extract which is not a large dose as the recommended dose of mistletoe in some herbals is 2–8 g. This point was taken up on behalf of the herbalists by Fletcher Hyde (1981) who said that experienced herbal practitioners would be incredulous that a dose of mistletoe of less than one-twentieth of the pharmacopoeial dose could cause hepatitis in a few days.

This may be so but, nevertheless, the evidence provided by Harvey and Colin-Jones is persuasive that the herbal tablets were involved although, admittedly, mistletoe *per se* is incriminated solely on the basis of its potential toxic components. The time sequence of the illness in relation to the taking of the tablets, its resolution when the tablets were stopped, and the response of the patient to the formal challenge lead to a high of probability of causation in such a case report.

Pennyroyal

Pennyroyal oil, derived from the leaves of *Mentha pulegium* or *Hedeoma pulegioides* is a folklore abortifacient and contains 85 per cent of pulegone, a cyclohexanone. Sullivan *et al.* (1979) described two cases where the oil was taken as an abortifacient and gave rise to severe hepatotoxicity. One of the cases was fatal, the liver showing a massive centrilobular necrosis.

Pulegone may possibly be metabolized to an epoxide or furan and cause liver damage in a similar way to paracetamol.

Sassafras

Sassafras is the dried inner bark of the root of *Sassafras albidum (S. variifolium)* and contains about 3 per cent of volatile oil and is used as a carminative and antirheumatic and also to flavour soft drinks. The volatile oil contains 80–95 per cent of safrole (isosafrole, 1-ally-3,4,-methylenedioxybenzene). Safrole and its metabolite 1-hydroxysafrole were shown in rats and mice to be both hepatoxic and a hepatocarcinogen (Borchert *et al.* 1973; Segelman *et al.* 1976, Kapadia *et al.* 1978). Its use as a flavouring agent in foods has been prohibited (Food Standards Committee 1965).

Safrole also inhibits liver microsomal enzymes and so may prolong the action of many drugs which are inactivated by this system (Jaffe *et al.* 1968). It is a minor component of many other essential oils including those derived from star anise, camphor, mace, nutmeg, bay laurel, and cinnamon.

Starch blockers

'Starch blocker' is the generic name for a group of products manufactured from such legumes as the red kidney bean (*Phaseolus vulgaris*). It is alleged that these products contain an alpha-amylase inhibitor which, on ingestion, prevents pancreatic amylase from hydrolysing starch to dextrins and maltose. The starch is therefore presumed to pass through the gut unchanged and thereby to facilitate dieting aimed at losing weight.

Bo-Linn and his colleagues (1982) using a one-day calorie balance technique and a high-starch (100 g) diet found no difference in the excretion of faecal calories after normal subjects had taken either placebo or starch-blocker tablets. Each starch-blocker tablet is alleged to produce enough anti-amylase activity to block the digestion and absorption of 100 g of starch (400 kcal). If the starch-blocker tablets had prevented the digestion of starch, the faecal calorie excretion should have increased by 400 kcal. The starch blockers were shown by *in-vitro*

testing to have the stated amount of anti-amylase activity and Bo-Linn and his group postulated that the most likely explanation of their results is the capacity of the human pancreas to secrete amylase in sufficient amounts to override the effect of the inhibited enzyme. They did not comment on any adverse reactions in their subjects and their study may be criticized for being somewhat unphysiological.

Li (1983) postulated that the lack of increase of faecal calorie excretion in the study by Bo-Linn *et al.* could also be due to the hydrolysis of excess starch by colonic bacteria with subsequent absorption of short-chain fatty acids. They agree, however, that starch blockers are clinically not effective.

Another study by Garrow and his colleagues (1983), while using a completely different technique, had essentially similar results. By using ^{13}C-labelled starch volunteers, they showed that there was no significant change in the absorption of ^{13}C-label (measured by excretion and expired air) before and after the administration of starch blockers. This showed that the starch blockers did not affect the digestion and metabolism of the labelled starch to any measurable degree. Yet another study in volunteers by Carlson *et al.* (1983) showed that a commercial formulation of starch blockers had no effect on the response of blood glucose, insulin, or breath hydrogen to a standardized starch meal.

While the studies by these various groups are open to criticism, nevertheless, it is compelling that three studies using different techniques and marketed formulations of starch blockers have not confirmed the pharmacological mechanisms by which they are said to work.

Other information on efficacy or adverse reactions to starch blockers is sadly lacking although complaints have been made of nausea, vomiting, diarrhoea, and stomach pains probably caused by fermentation of undigested starch in the lower bowel. Other potential adverse effects were said by some commentators to be imbalance in the nutritional state due to relative increases of fat and protein absorption and the inhibition of pancreatic enzymes causing enlargement of the organ which might be tumorigenic.

Macri *et al.* (1977) found that chickens whose diet included 3 per cent of amylase inhibitor derived from wheat flour had a 41 per cent increase in pancreatic weight after only four weeks on the diet. Histological examination showed that the pancreatic enlargement was due to an increase of cell size rather than cell numbers. There were also degenerative phenomena and an increase in connective tissue. The inhibition of alpha amylase by plant proteins has been known for some 50 years (Lindner 1982) especially that ('phaseolamin') derived from the red kidney bean by Marshall and Lauda in 1975, and by Powers and Whitaker in 1977. The purity of such extracts gives grounds for concern and Lindner (1982) commented that they can contain phytates which inhibit mineral absorption, haemaglutinins, and protease inhibitors (especially trypsin inhibitors).

That the problem is a real one was supported by Kilpatrick *et al*. (1983) who found that the phytohaemagglutinin content of a commercially available 'starch blocker' was appreciable. Puztai *et al*. (1979) confirmed the enteral toxicity of phytohaemagglutinins in rats by studies of the damage caused to the microvilli of the small intestine. It is too soon to be certain about starch blockers; whether they are effective or whether they possess a real toxicity sufficient to advise caution in human usage. They certainly form yet another milestone in the commercial exploitation of the credulity of *Homo sapiens* facilitated by the sensation-seeking of the media.

Tonka bean

The tonka (tonquin) bean obtained from *Coumarouna odorata* (*Dipteryx odorta*) contains 1–10 per cent of coumarin and is used as a flavouring agent and as a fixative in perfumes.

A 25-year-old woman with menometrorrhagia was found to have clotting functions typical of those after coumarin-type drugs. She had been taking large quantities of a home-made 'seasonal tonic' containing not only tonka bean but other natural coumarin-containing herbs such as melilot and sweet woodruff. She had also been taking paracetamol, propoxyphene, Vitamin A, and bromelain which may have enhanced the actions of the coumarins. Cessation of medication brought about a return to normal (Hogan 1983).

Sensitivity reactions

The possibility of sensitivity reactions to camomile tea and alfalfa was discussed above. Reactions to other herbal products have also been reported. Mitchell (1980) described the case of a 41-year-old Chinese with contact sensitivity to garlic. Morrow *et al*. (1980) described a case of hypersenstitivity to aloe in a 47-year-old man manifested by generalized nummular and eczematous and papular dermatitis and confirmed by patch test. The patient had been taking aloe by mouth and also applying it to his face as part of a nutritional regime.

Adulterated herbal mixtures

The use of the phrase 'adulterated herbal mixtures' is not to prejudge any product in an emotive fashion, but to label a large group of mixtures of herbal extracts and non-herbal substances. Sometimes these non-herbal substances will be derived from the cultural origin of the herbal mixture itself, e.g. the use of heavy metals in some Asian medicines. At other times the mixtures are of herbal extracts with potent non-herbal orthodox (Western) drugs. Also in the adulterated herbal group it is pertinent to include those mixtures where some or all of the herbal extracts are mislabelled. The 'adulterated herbal mixtures' will therefore be considered under the following subheadings; Pure herbal extracts and mixtures derived from the wrong plants and herbal mixtures containing synthetic or non-herbal substances.

Herbal mixtures derived from the wrong plants

The quality control of manufacture of many herbal products leaves much to be desired. Even the basic materials, i.e. the plant itself, may be wrongly identified especially by amateurs. This can lead to severe toxicity and even fatality.

Bryson *et al*. (1978) described a 26-year-old woman who was admitted to hospital with typical signs of atropine poisoning which responded to physostigmine. She had brewed a tea made from burdock root by steeping it in water. The roots and seeds of burdock (*Arctium minus, A. lappa*) have long been used as a tea whose constituents include volatile oils, inulin, tannin, and bitter glycoside called arctiin. Normally burdock root does not contain atropine-like alkaloids but the offending dried teas was found to contain 30 mg/g of such alkaloids presumably present as a contaminant.

An urgent warning had to be given in Britain (*CMO Letter* 1983) about one brand of imported comfrey leaves which was found to be contaminated with up to 20 per cent belladonna. Some 20 cases of typical belladonna poisoning were reported following consumption of a tea made from from the product.

Amateur herbalists are especially at risk of misidentifying plants and Dickstein and Kunkel (1980) reported a case of an 85-year-old man who picked some leaves from an unfamiliar plant in his back garden to make a herbal tea. Unfortunately the plant he chose was *Digitalis purpurea* and he became typically toxic from cardiac glycoside overdosage, but survived after treatment.

Herbal mixtures containing synthetic or non-herbal substances

General

Many Western herbal mixtures contain substances which are non-herbal, the logic of including which would seem to be at variance with the herbalists' expounded principles, and may owe more to commercialization than these principles.

The growing interest in the Western world in medicines derived from other, i.e. non-Western or Asian cultures, is a potential fertile source of toxicity and worthy of some detailed comment. Certainly recent mass movements of refugees or other immigrants have posed interesting questions of cultural clashes. An example of this is the massive influx of immigrants into the UK from its former dependent territories such as Pakistan, India, Bangladesh, and East Africa. The various cultural and religious groups concerned have sought to retain their identity and may therefore form separate communities which do not integrate well with each other, let alone their indigenous neighbours. These groups tend to follow and retain their old traditions and standards and may, indeed, be stricter than their country of origin has become in the interim.

Many of these Asian immigrants have also retained their own native system of medicine. There are two basic systems of medicine in India and Pakistan — the western and the traditional — (Bannerman *et al*. 1983). The traditional type of healer although superficially similar may also be subdivided into the Tabib, following the Unani–Tibb or Graeco-Arab sytem of medicine, and the Vaid, following the Ayurvedic system. The Tabib or Vaid spends much of his time treating conditions for which Western medicine has no effective answer, especially those of a psychological and sexual nature. Unfortunately some of the medicines used give rise to current concern. Both types of healers rely on mixtures of crude herbals, minerals, and metals.

Preparations called 'Kushtay' used especially as tonics and aphrodisiacs contain oxidized heavy metals such as arsenic, mercury, tin, zinc, and lead, and a typical kushtay may contain 10–12 per cent of each of several or more of these metals. The efficacy of these mixtures has not yet been demonstrated epidemiologically and few, if any, have been subjected to detailed scrutiny. The toxicity of ingredients such as arsenic, mercury, and lead has, of course, been adequately demonstrated.

Lead-containing herbal mixtures

Lead is a major constituent of herbal products worldwide, and has been incriminated in toxicity. Chan *et al*. (1977) described how a 4-month-old child in Hongkong was shown to have acute plumbism after ingesting certain Chinese medicines. The analysis of 11 related brands showed a mean lead content of up to 7.5 mg per dose.

Chinese herbal medicines were also implicated in the lead intoxication suffered by a 59-year-old Californian women who, while under treatment from a herbalist-acupuncturist for post-traumatic arthralgia, was given two types of herbal pills. Four months later she was admitted to hospital with severe lead intoxication (her 24-hour urinary lead was 1044 μg 24 hours compared with normal 0–80 μg). The pills which had been obtained by the acupuncturist by mail order from Hong Kong contained 0.5 mg lead each. The patient was therefore ingesting up to 15 mg lead daily (Lightfoote *et al*. 1977). Imported Chinese medicines were also thought to be responsible for 24 cases of lead poisoning in Laotian refugees in Minnesota, USA (*Journal of the American Medical Association* 1983).

Several instances of lead-containing Asian medicines were also reported from Britain. Ali *et al*. (1978) and Aslam *et al*. (1980) reported on the cosmetic use of surmas in Asian children, which are painted around the eyes supposedly to prevent eye disease and vision defects. Some varieties contain up to 86 per cent lead and were held by Ali and his co-workers to be potential sources of plumbism. Mean lead concentrations in those who had not used surma was 0.98 ± 0.42 mmol l^{-1} compared with 1.65 ± 0.68 mmol l^{-1} for those who had used surma.

However, Attenburrow *et al*. (1980) in Glasgow measured the blood levels of 217 Asian children aged 4 months to 18 years. In this series there was no essential difference in the blood levels of lead in the 17 (8 per cent) children who used surmas regularly (0.799 ± 0.285 mmol l^{-1} compared with the 178 (82 per cent) who never used surmas (0.760 ± 0.302 mmol l^{-1}).

Arsenic-containing herbal mixtures

Like lead, arsenic also occurs in herbal mixtures worldwide. Tay and Seah (1975) found arsenical poisoning in 74 patients in Singapore occurring over a 15-month period; 64 of these cases were caused by an anti-asthmatic herbal preparation containing 12 000 p.p.m. of inorganic sulphide of arsenic. Six other brands were responsible for the other patients.

Investigation showed a total of 29 brands of Chinese herbal mixtures containing up to 107 000 p.p.m. arsenic. Nearly 40 per cent of the patients investigated had taken the medicines for less than six months; the others had a longer history of exposure ranging from 1–15 years. The toxicity was confined mainly to the skin in 91 per cent, nervous system in 51 per cent, gastrointestinal tract in 23 per cent, and blood in 23 per cent of patients. Malignancy of the skin was present in six patients and of the visceral organs in four patients. Using hair and urine samples, toxicological confirmation of arsenic poisoning was found in half the cases investigated, but there seemed no correlation between the clinical state of the patient and their tissue content of arsenic.

Herbal mixtures containing undeclared 'Western' drugs

A disturbing recent trend has been the incorporation of orthodox Western drugs in herbal mixtures, undeclared on the label and without regard to the known toxicity of these drugs.

Ries and Sahud (1975) gave details of four non-Chinese patients who presented with agranulocytosis after taking Chinese herbal medicines for arthritis and back pain. Six preparations were involved, containing in each tablet 23–43 mg aminopyrine, 17–34 mg phenylbutazone, with a total dose of 12–18 tablets each day. Neither aminopyrine or phenylbutazone were declared on the label. The authors of this report commented that

these substances were a recent addition to a traditional Chinese product whose packaging was such as to appeal to Western tastes. It is possible that the Chinese themselves were not taking this type of adulterated tablet, which was illegally imported into the United States from Taiwan and Hong Kong.

A more recent case was described by Forster *et al.* (1979) of a 44-year-old Dutch woman with rheumatoid arthritis who took a Chinese medicine which on analysis was found to contain dexamethasone and indomethacin though the authors comment that other contaminants have been found elsewhere in different batches of the medicine including aminopyrine and phenylbutazone.

Herbal preparation —— drug interactions

In this section a classification is given which is designed to identify those herbal extracts which have or may have actions which could interact with conventional drugs in the therapeutic situation. It is modified from Griffin and D'Arcy (1984) to which reference should be made for further details. Some of the interactions are also discussed above.

1. Herbs reputed to have action on the heart and known to contain cardiac glycosides

Adonis vernalis (Adonis, False Hellebore, Pheasant's Eye)
Convallaria majalis (Lily of the Valley)
Cytisus scoparius (Broom)

Digitalis lanata (Yellow Foxglove)
Digitalis purpurea (Purple Foxglove)
Scilla maritima (White Squill)
Strophanthus kombé (Strophanthus)

Combination	Interaction	Management
/digoxin, digitoxin	Potentiation may produce digitalis toxicity	Advise patient to refrain from self-medication
/other drugs	Varied dependent on particular drug	

2. Herbs reputed to have diuretic actions

Adonis vernalis (Adonis, False Hellebore, Pheasant's Eye)
Alchemilla arvensis (Lady's Mantle, Parsley Piert)
Alisma plantago (Water Plantain)
Agrimonia eupatoria (Agrimony)
Agropyrum repens (Couch Grass)
Anacyclus pyrethrum (Pellitory)
Arctostaphylos uva-ursi (Bearberry)
Barosma betulina (Buchu)
Capsella bursa-pastoris (Shepherd's Purse)
Carum petroselinium (Parsley)
Chelidonium majus (Celandine)
Chimaphila umbellata (Ground Holly, Pipsesswa, Prince's Pine)

Chondodendron tomentosum (Pareira Brava Root)
Collinsonia canadensis (Stone Root)
Cytisus scoparius (Broom)
Daucus carota (Wild Carrot)
Galium aparine (Cleavers, Goosegrass)
Herniaria glabra (Rupture-wort)
Hydrangea aborescens (Hydrangea)
Hypericum perforatum (St. John's Wort)
Larix americana (Larch, Tamarac)
Oxalis acetosella (Wood Sorrel)
Sassafras variifolium (Sassafras)
Senecio species (Ragwort)
Taraxacum officinale (Dandelion)
Viola tricolor (Heartsease, Pansy)

Combination	Interaction	Management
/antihypertensive therapy	Hypertension may be difficult to control and hypotensive episodes may occur	Advise patient to avoid self-medication
/digitalis alkaloids	The effect and toxicity of cardiac glycosides is enhanced by K^+ loss associated with diuresis	Advise patient to refrain from self-medication
muscle relaxants	Increased responsiveness to tubocurarine and gallamine may occur if there is associated K^+ depletion due to the diuretic action	Advise patient to avoid self-medication

3. Herbs reputed have sedative actions

Aconitum napellus (Aconite, Monkshood, Wolfsbane)
Chelidonium majus (Celandine)
Conium maculatum (Hemlock)
Humulus lupulus (Common Hop)
Lactuca virosa (Wild Lettuce)
Papaver somniferum (Opium Poppy)
Passiflora incarnata (Passion Flower)

Scopolia carniolica (Scopulia)
Scutellaria laterifolia (Skullcap)
Valeriana officinalis (Valerian)
Plants containing tropane alkaloids, e.g. *Atropa belladona* (Deadly Nightshade, *Datura stramonium* (Jimson Weed, Thorn Apple), *Hyoscyamus niger* (Henbane)

Combination	Interaction	Management
/alcohol /antihistamines /hypnotics	Potentiation of sedative effects with deterioration of performance in driving cars or operating machinery	Warn the patient

4. Herbs reputed to lower blood pressure

Crataegus oxyacantha (English Hawthorn)
Rauwolfia serpentina (Chotachand, Rauwolfia)

Veratrum viride (Hellebore, Green Hellebore)
Viscum album (Mistletoe)

Combination	Interaction	Management
/antihypertensive drugs /diuretics	Hypotension may ensue, or due to the variable potency of these herbal preparations, the hypertension may be difficult to stabilize	Advise patients to refrain from self-medication

5. Herbal preparations with actions which raise blood pressure

Glycyrrihiza glabra (Liquorice root or extracts)
Liquorice can exert a DOCA-like action causing oedema and hypertension (see above)

Combination	Interaction	Management
Liquorice/antihypertensive drugs	Retention of Na + and water and development of oedema and hypertension	Patients receiving therapy for their hypertension should be warned not to take liquorice derivatives

Note: There are several preparations for the treatment of gastric conditions containing liquorice. Carbenoxolone is a liquorice derivative.

6. Herbs containing substances with anticoagulant actions

Aesculus hippocastanum (Horse-chestnut)
The esculin (esculoside) in leaves and bark is a coumarin-like substance.

Coumarouna odorata (Tonka Bean)
Contains 1–10 per cent coumarin (see above)

Combination	Interaction	Management
/other drugs	normally to potentiate any anti-coagulant action of coumarin	Advise patient to refrain from self-medication

7. Herbs containing substances causing gynaecomastia

Panax ginseng and other *Panax* species (Ginseng, Ninjin, Pannag)
Rauwolfia serpentina (Chotachand, Rauwolfia)

Ginseng contains small quantities of oestrone, oestradiol, and oestriol in the root which may induce swelling of the breasts (see above).
Rauwolfia alkaloids have been known to produce gynaecomastia and galactorrhoea.

Combination	Interaction	Management
Rauwolfia alkaloids or Ginseng /other drugs causing gynaecomastia, e.g. digitalis alkaloids, ethionamide, griseofulvin, methyldopa, phenothiazine derivatives, spironolactone	Any drug causing gynaecomastia or galactorrhoea will, when given in combination with another drug with the same side-effect, potentiate these effects	Awareness of the interaction and avoidance of unnecessary medication is all that can be advised

8. Herbs producing hallucinations

Catharanthus roseus (Periwinkle)
Cinnamomum camphora (Cinnamon)*
Corynanthe yohimbi (Yohimbe)
Datura stramonium (Thorn Apple)
Eschscholtzia californica (California Poppy)*
Humulus lupulus (Common Hop)*
Hydrangea paniculata (Hydrangea)*

Lobelia inflata (Lobelia)*
Mandragora officinarum (Mandrake)
Myristica fragrans (Nutmeg)
Piper methysticum (Kava, Kava-Kava)*
Passiflora incarnata (Passion Flower)*
Herbs marked with an asterisk have been used as marihuana (cannabis) substitute for smoking or used in combination with marihuana.

Combinations	Interactions	Management
Herbal hallucinogens/ propranolol	Visual hallucinations may occur with propranolol. Combination of herb and drug would increase the risk of hallucination	Advise patient to refrain from self-medication

9. Miscellaneous herbs

Combination	Interaction	Management
(a) Ink cap/alcohol	*Coprinus atramentarius* (Ink cap) contains bis (diethylthiocarbamoyl) disulphide which is disulfiram, the active component of *Antabuse*. It reacts with alcohol to give a typical unpleasant syndrome of systemic reactions	Health food addicts should be warned of this interaction since Coprinus is a component in some of these foods
(b) karela/chlorpropamide other antidiabetics	Karela (*Momordica charantia*) and is used in Indian curries. It has a hypoglycaemic action and may upset the control of diabetes mellitus (Aslam and Stockley 1979; Leatherdale *et al.* 1981). Garlic (*Allium sativum*) has a similar effect	Warn patient

REFERENCES

AHERN, M.J., MILAZZO, S.C., AND DYMOCK, R. (1980). Granulomatous hepatitis and seatone. *Med. J. Aust.* 2, 151–2.

AKHTAR, M.S., ATHAR, M.A., AND YAQUB, M.(1981). Effect of *Momordica charantia* on blood glucose level of normal and alloxan-diabetic rabbits. *Planta Med.* 42, 205–12.

ALI, A.R., SMALES, O.R.C., AND ASLAM, M. (1978). Surma and lead poisoning. *Br. med. J.* 2, 915–16.

ANDERSON, L.A. AND PHILLIPSON, J.D. (1982). Mistletoe: the magic herb. *Pharm. J.* 229, 437–9.

ASLAM, M. AND STOCKLEY, I.H. (1979). Interaction between curry ingredient (karela) and drug (chlorpropamide). *Lancet* i, 607.

——, HEALY, M.A., DAVIS, S.S., AND ALI, A.R. (1980). Surma and blood lead in children. *Lancet* i, 658–9.

ATTENBURROW, A.A., CAMPBELL, S., LOGAN, R.W., AND GOEL, K.M. (1980). Surma and blood levels in Asian children in Glasgow. *Lancet* i, 323.

BANNERMAN, R.H., BURTON, J., AND CH'EN, W.C., (1983). *Traditional medicine and health care coverage.* World Health Organisation, Geneva.

BANNISTER, B., GINSBURG, R., AND SHNEERSON, J. (1977). Cardiac arrest due to liquorice-induced hypokalaemia. *Br. med. J.* 2, 738–9.

BELTON, P.A. AND GIBBONS, D.O. (1979). *Datura* intoxication in West Cornwall. *Br. med. J.* 1, 585–6.

BENARDE, M.A. AND WEISS, W. (1982). Coffee consumption and pancreatic cancer; temporal and spatial correlation. *Br. med. J.* 284, 400–2.

BO-LINN, G.W., SANTA ANA, C.A., MORAWSKI, S.G., AND FORDTRAN, J.S. (1982). Starch blockers — their effect on calorie absorption from a high starch meal. *New Engl. J. Med.* 307, 1413–16.

BORCHERT, P., MILLER, J.A., MILLER, E.C., AND SHIRES, T.K. (1973). l'Hydroxysafrole, a proximate carcinogenic metabolite of safrole in the rat and the mouse. *Cancer Res.* 33, 590–600.

BRAICO, K.T., HUMBERT, J.R., TERPLAN, K.L., AND LEHOTAY, J.M. (1979). Laetrile intoxication: report of a fatal case. *New Engl. J. Med.* 300, 238–40.

BRYSON, P.D., WATANABE, A.S., RUMACK, B.H., AND MURPHY, R.C. (1978). Burdock root tea poisoning. *J. Am. med. Ass.* 239, 2157.

CARLSON, G.L., LI, B. U. K., BASS, P., AND OLSEN, W.A. (1983). A bean alpha-amylase inhibitor formulation (starch blocker) is ineffective in man. *Science* 219, 393–4.

CASSILETH, B.R. (1982). After Laetrile, What? *New Engl. J. Med.* 306, 1482–4.

CASTERLINE, C.L. (1980). Allergy to chamomile tea. *J. Am. med. Ass.* 244, 330–1.

CHAN, H., BILLMEIER, C.J. JR, AND EVANS, W.E. (1977). Lead poisoning from ingestion of Chinese herbal medicines. *Clin. Toxicol.* 10, 273–8.

CHOTIBUTR, S., RASMIDATTA, S., LAWTIANTONG, T., AND KANCHANARANYA, C. (1981). Toxic effect of local Thai anthelminthic (Maklua) cases report. *J. med. Ass. Thailand* **64**, 574–9.

CMO LETTER (1983). Comfrey leaves: contamination with belladonna, *CMO Lett.* no. 83, p.3.

CORRIGAN, D. AND CONNOLLY, P.J. (1981). A survey of ginseng preparations on the Irish market. *Irish pharm. J.* **59**, 303–4.

CULVENOR, C.C.J., CLARKE, M., EDGAR, J.A., FRAHN, J.L., JAGO, M.V., PETERSON, J.E., AND SMITH, L.W. (1980). Structure and toxicity of the alkaloids of Russian comfrey *(Symphytum X Uplandicum Nyman)* a medicinal herb and item of human diet. *Experientia* **36**, 377–9.

CUMMING, A.M.M., BODDY, K., BROWN, J.J., FRASER, R., LEVER, A.F., PADFIELD, P.L., AND ROBERTSON, J.I.S. (1980). Severe hypokalaemia with paralysis induced by small doses of liquorice. *Postgrad. med. J.* **56**, 526–9.

DAMMIN, G.J. (1972). Endemic nephropathy in Yugoslavia. *Arch. Pathol.* **93**, 372–4.

DATTA, D.V., KHUROO, M.S., MATTOCKS, A.R., AIKAT, B.K., AND CHHUTTANI, P.N. (1978). Herbal medicines and veno-occlusive disease in India. *Postgrad. med. J.* **54**, 511–15.

DAVIGNON, J.P., TRISSEL, L.A., AND KLEINMAN, L.M. (1978). Pharmaceutical assessment of amygdalin (Laetrile) products. *Cancer treat. Rep.* **62**, 99–104.

DICKSTEIN, E.S. AND KUNKEL, F.W. (1980). Foxglove tea poisoning. *Am. J. Med.* **69**, 167–9.

EISELE, J.W. AND REAY, D.T. (1980). Deaths related to coffee enemas. *J. Am. med. Ass.* **244**, 1608–9.

ELLISON, N.M., BYAR, D.P., AND NEWELL, G.R. (1978). Special report on laetrile: the NCI laetrile review. *New Engl. J. Med.* **299**, 549–52.

FEDERAL REGISTER (1977). Laetrile. *Fed. Reg.* **42**, 39768–806.

FEINSTEIN, A.R., HORWITZ, R.I., SPITZER, W.O., AND BATTISTA, R.N. (1981). Coffee and pancreatic cancer. *J. Am. med. Ass.* **246**, 957–61.

FLETCHER HYDE, F. (1981). Mistletoe hepatitis. *Br. med. J.* **282**, 739.

FOOD STANDARDS COMMITTEE (1965). *Report on flavouring agents*. HMSO, London.

FORSTER, P.J.G., CALVERLEY, M., HUBBALL, S., AND McCONKEY, B. (1979). Chuei-Fong-Tou-Geu-Wan in rheumatoid arthritis. *Br. med. J.* **2**, 308.

FOX, D.W., HART, M.C., BERGESON, P.S., JARRETT, P.B., STILLMAN, A.E., AND HUXTABLE, R.J. (1978). Pyrrolizidine (Senecio) intoxication mimicking Reye syndrome. *J. Pediat.* **93**, 980–2.

FRANZ, H., ZISKA, P., AND KINDT, A. (1981). Isolation and properties of three lectins from mistletoe *(Viscum album* L). *Biochem. J.* **195**, 481–4.

FURUYA, T. AND ARAKI, K. (1968). Studies on consituents of crude drugs — alkaloids of *Symphytum officinale*. L. *Chem. Pharm. Bull* **16**, 2512–16.

GARROW, J.J., SCOTT, P.F., HEELS, S., NAIR, K.S., AND HALLIDAY, D. (1983). Starch blockers are ineffective in man. *Lancet* **i**, 60–1.

GIBSON, R.G. AND GIBSON, S.L.M. (1981*a*). Green-lipped mussel extract in arthritis. *Lancet* **i**, 439.

—— AND —— (1981*b*). Seatone in arthritis. *Br. med. J.* **282**, 1795.

——, —— CONWAY, V., AND CHAPPELL, D. (1980). *Perna canaliculus* in the treatment of arthritis. *Practitioner* **224**, 955–60.

GOUGH, S.P. AND COOKSON, I.B. (1984). Khat-induced schizophreniform pyschosis in UK. *Lancet* **i**, 455.

GRAHAM, D.L., LAMAN, D., THEODORE, J., AND ROBIN, E.D. (1977). Acute cyanide poisoning complicated by lactic acidosis and pulmonary edema. *Arch int. Med.* **137**, 1051–5.

GREENSPAN, E.M. (1983). Ginseng and vaginal bleeding. *J. Am. med. Ass.* **249**, 2018.

GRIFFIN, J.P. AND D'ARCY, P.F. (1984). *A manual of adverse drug interactions*, 3rd edn., pp. 363–8. Wright, Bristol.

HAMMOND, T.G. AND WHITWORTH, J.A. (1981). Adverse reactions to ginseng. *Med. J. Aust.* **i**, 492.

HARVEY, J. AND COLIN-JONES, D.G. (1981). Mistletoe hepatitis. *Br. med. J.* **282**, 186–7.

HENRY, G.W., SCHWENK, G.R., AND BONHERT, P.A. (1981). Umbrellas and mole beans: a warning about acute ricin poisoning. *J. Indiana State med. Ass.* **74**, 572–3.

HERBERT, V. (1979). Laetrile: the cult of cyanide: promoting poison for profit. *Am. J. clin. Nutr.* **32**, 1121–58.

HILL, K.R., RHODES, K.R., STAFFORD, J.L., AND AUB, B. (1951). Liver disease in Jamaican children. *West Indian med. J.* **1**, 49–63.

HIRONO, I., MORI, H., AND HAGA, M. (1978). Carcinogenic activity of *Symphytum officinale*. *J. nat. Cancer Inst.* **61**, 865–9.

HOGAN, R.P. (1983). Hemorrhagic diathesis caused by drinking an herbal tea. *J. Am. med. Ass.* **249**, 2679–80.

HUSKISSON, E.C., SCOTT, J., AND BRYANS, R. (1981). Seatone is ineffective in rheumatic arthritis. *Br. med. J.* **282**, 1358–9.

HUXTABLE, R.J. (1980). Herbal teas and toxins: novel aspects of pyrrolizidine poisoning in the United States. *Perspectives Biol. Med.* **24**, 1–14.

JACKSON, L., KOFMAN, S., WEISS, A., AND BRODOVSKY, H. (1964). Aristolochic acid(NSC-50413) phase 1 clinical study. *Cancer Chemotherapy Rep.* **42**, 35–7.

JAFFE, H., FUJII, K., SENGUPTA, M., GUERIN, H., AND EPSTEIN, S.S. (1968). In vivo inhibition of mouse liver microsomal hydroxylation systems by methylenedioxyphenyl insecticidal synergists and related compounds. *Life Sci.* **7**, 1051–62.

JOURNAL OF THE AMERICAN MEDICAL ASSOCIATION (1975). Kelp diets can produce myxedema in iodide-sensitive individuals. *J. Am. med. Ass.* **233**, 9–10.

—— (1983). Folk remedy associated lead poisoning in Hmong children. *J. Am. med. Ass.* **250**, 3149–50.

KALIX, P. (1984). Amphetamine psychosis due to Khat leaves. *Lancet* **i**, 46.

KAPADIA, G.J., CHUNG, E.B., GHOSH, B., SHUKLA, Y.N., BASAK, S.P., MORTON, J.F., AND PRADHAN, S.N. (1978). Carcinogenicity of some folk medicinal herbs in rats. *J. nat. Cancer Inst.* **60**, 683–6.

KILPATRICK, D.C., GREEN, C., AND YAP, P.L. (1983). Lectin content of slimming pills. *Br. med. J.* **286**, 305.

KNIGHT, B. (1979). Ricin — a potent homicidal poison. *Br. med. J.* **1**, 350–1.

KOREICH, O.M. (1978). Ginseng and mastalgia. *Br. med. J.* **1**, 1556.

KOSTER, M. AND DAVID, G.K. (1968). Reversible severe hypertension due to licorice ingestion. *New Engl. J. Med.* **278**, 1381–3.

KUMANA, C.R., NG, M., LIN, H.J., KO, W., WU, P.C., AND TODD, D. (1983). Hepatic veno-occlusive disease due to toxic alkaloid in herbal tea. *Lancet* **ii**, 1360–1.

KUPCHAN, S.M. AND DOSKOTCH, R.W. (1962). Tumor inhibitors I aristolochic acid, the active principle of *Aristolochia indica*. *J. med. pharm. Chem.* **5**, 657–9.

LEATHERDALE, B.A., PANESAR, R.K., SINGH, G., ATKINS, T.W., BAKEY C.J., AND BIGNELL, A.H.C. (1981). Improvement in glucose tolerance due to *Momordica charantia* (karela). *Br. med. J.* **282**, 1823–4.

LI, B.U.K. (1983). Starch blockers. New Engl. J. Med. **308**, 902.

LIEGNER, K.B., BECK, E.M., AND ROSENBERG, A. (1981). Laetrile induced agranulocytosis. *J. Am. med. Ass.* **246**, 2841–2.

LIGHTFOOTE, J., BLAIR, J., AND COHEN, J.R. (1977). Lead intoxication in an adult caused by Chinese herbal medication. *J. Am. med. Ass.* **238**, 1539.

LINDNER, P.G. (1982). "Starch-blocker" for obesity: breakthrough or hazard? *Obesity/Bariatric Med.* **11**, 65–6.

MacMAHON, B., YEN, S., TRICHOPOULOS, D., WARREN, K., AND NARDI, G. (1981). Coffee and cancer of the pancreas. *New Engl. J. Med.* **304**, 603–33.

MACRI, A., PARLAMENTI, R., SILANO, V., AND VALFRE, F. (1977). Adaptation of the domestic chicken, *Gallus domesticus*, to continuous feeding of albumin amylase inhibitors from wheat flour as gastroresistant microgranules. *Poultry Sci.* **56**, 434–41.

MAHLER, D.A. (1976). Anticholinergic poisoning from Jimson Weed. *J. Am. Coll. emerg. Physns* **5**, 440–2.

MALINOW, M.R., BARDANA, E.J., JR., AND GOODNIGHT, S.H., JR. (1981). Pancytopenia during ingestion of alfalfa seeds. *Lancet* **i**, 615.

MARSHALL, J.J. AND LAUDA, C.M. (1975). Purification and properties of phaseolamin, an inhibitor of alpha amylase from the kidney bean, *Phaseolus vulgaris. J. Biol. Chem.* **250**, 8030–7.

MARTINDALE: THE EXTRA PHARMACOPOEIA (27th edn) (1977). The Pharmaceutical Press, London.

MATTOCKS, A.R. (1968). Toxicity of pyrrolizidine alkaloids. *Nature* **217**, 723–8.

—— (1980). Toxic pyrrolizidine alkaloids in comfrey. *Lancet* **ii**, 1136–7.

MAXWELL, D.M. (1978). Increased cyanide values in a laetrile user. *Can. med. Ass. J.* **119**, 18.

McGEE, J.O., PATRICK, R.S., WOOD, C.B., AND BLUMGART, L.H. (1976). A case of veno-occlusive disease of the liver in Britain associated with herbal tea consumption. *J. clin. Pathol.* **29**, 788–94.

McLEAN, E.K. (1970). The toxic actions of pyrrolizidine (Senecio) alkaloids. *Pharmac. Rev.* **22**, 429.

MENGS, U. (1983). On the histopathogenesis of rat forestomach carcinoma caused by aristolochic acid. *Arch. Toxicol.* **52**, 209–20.

——, LANG, W., AND POCH, J.A. (1982). The carcinogenic action of aristolochic acid in rats. *Arch. Toxicol.* **51**, 107–19.

MITCHELL, J.C. (1980). Contact sensitivity to garlic (Allium). *Contact Derm.* **6**, 356–7.

MOERTEL, C.G., AMES, M.M., KOVACH, J.S., MOYER, T.P., RUBIN, J.R., AND TINKER, J.H. (1981). A pharmacologic and toxicological study of amygdalin. *J. Am. med. Ass.* **245**, 591–4.

——, FLEMING, T.R., RUBIN, J., KVOLS, L.K., SARNA, G., KOCH, R., CURRIES, V.E., YOUNG, C.W., JONES, S.E., AND DAVIGNON, J.P. (1982). A clinical trial of amygdalin (Laetrile) in the treatment of human cancer. *New Engl. J. Med.* **306**, 201–6.

MOKHOBO, K.P. (1976). Herb use and necrodegenerative hepatitis. *S. Afr. med. J.* **50**, 1096–9.

MORROW, D.M., RAPAPORT, M.J., AND STRICK, R.A. (1980). Hypersensitivity to aloe. *Arch. Dermatol.* **116**, 1064–5.

MORSE, D.L., BOROS, L., AND FINDLEY, P.A. (1979). More on cyanide poisoning from laetrile. *New Engl. J. Med.* **301**, 892.

PAKRASHI, A. AND SHAHA, C. (1977). Effect of a sesquiterpene form *Aristolochia indica Linn* on fertility in female mice. *Experientia* **33**, 1498–9.

PALMER, B.V., MONTGOMERY, A.C.V., AND MONTEIRO, J.C.M.P. (1978). Ginseng and mastalgia. *Br. med. J.* **1**, 1284.

POLK, I.J. (1982). Alfalfa pill treatment of allergy may be hazardous. *J. Am. med. Ass.* **247**(10), 1493.

POPOV, I.M. AND GOLDWAG, W.J. (1973). A review of the properties and clinical effects of ginseng. *Am. J. chinese Med.* **1**, 263–70.

POWERS, J.R., AND WHITAKER, J.R. (1977). Purification and some physical and chemical properties of red kidney bean (*Phaseolus vulgaris*) alpha amylase inhibitor. *J. food Biochem.* **1**, 217–38.

PUNNONEN, R. AND LUKOLA, A. (1980). Oestrogen like effect of ginseng. *Br. med. J.* **281**, 1110.

PUSZTAI, A., CLARKE, E.M.W., AND KING, T.P. (1979). The nutritional toxicity of *Phaseolus vulgaris* lectins. *Proc. nutr. Soc.* **38**, 115–20.

QIAN, S.Z., JING, G.W., WU, X.Y., XU, Y., LI, Y.Q., AND ZHOU, S.Z. (1980). Gossypol related hypokalemia: Clinico-Pharmacologic studies. *Chinese med. J.* **93**, 477–82.

——, HU, J.H., HO, L.X., SUN, M.X., HUANG, Y.Z., AND FANG, J.H. (1981). The first clinical trial of gossypol on male antifertility. *Clinical pharmacology and therapeutics: Proceedings of the First World Conference* (ed. P. Turner), pp. 489–92.

REDMOND, D.E. (1970). Tobacco and cancer, the first clinical report 1761. *New Engl. J. Med.* **282**, 18–23.

RIES, C.A. AND SAHUD, M.A. (1975). Agranulocytosis caused by Chinese herbal medicines. *J. Am. med. Ass.* **231**, 352–5.

ROBERTS, J.L. AND HAYASHI, J.A. (1983). Exacerbation of S.L.E. associated with alfalfa ingestion. *New Engl. J. Med.* **308**, 1361.

ROITMAN, J.N. (1981). Comfrey and liver damage. *Lancet* **i, 944.**

SADAVONGVIVAD, C. (1980). An ideal herbal drug faces ordeal. *Trends pharmacol. Sci* **1**(12), VII–VIII.

SADOFF, L., FUCHS, K., AND HOLLANDER, J. (1978). Rapid death associated with laetrile ingestion. *J. Am. med. Ass.* **239**, 1532.

SAYRE, J.W. AND KAYMAKCALAN, S. (1964). Cyanide poisoning from apricot seeds among children in central Turkey. *New Engl. J. Med.* **270**, 1113–15.

SCHMIDT, E.S., NEWTON, G.W., SANDERS, S.M., LEWIS, J.P., AND CONN, E.E. (1978). Laetrile toxicity studies in dogs. *J. Am. med. Ass.* **239**, 943–7.

SEGELMAN, A.B., SEGELMAN, F.P., KARLINER, J., AND SOFIA, R.D. (1976). Sassafras and herb tea: potential health hazards. *J. Am. med. Ass.* **236**, 477.

SIEGEL, R.K. (1976). Herbal intoxication: psychoactive effects from herbal cigarettes, tea and capsules. *J. Am. med. Ass.* **236**, 473–6.

—— (1977). Kola, ginseng and mislabelled drugs. *J. Am. med. Ass.* **237**, 24–5.

—— (1979). Ginseng abuse syndrome. *J. Am. med. Ass.* **241**, 1614–15.

SINNIAH, D., AND BASKARAN, G. (1981). Margosa oil poisoning as a cause of Reye's syndrome. *Lancet* **ii**, 487–9.

SMITH, F.P., BUTLER, T.P., COHAN, S., AND SCHEIN, P.S. (1977). Laetrile toxicity: a report of two cases. *J. Am. med. Ass.* **238**, 1351.

STILLMAN, A.E., HUXTABLE, R., CONSROE, P., KOHNEN, P., AND SMITH, S. (1977). Hepatic veno-occlusive disease due to pyrrolizidine (Senecio) poisoning in Arizona. *Gastroenterology* **73**, 349–52.

STUART, K.L., AND BRAS, G. (1957). Veno-occlusive disease of the liver. *Quart. J. Med.* NS **26**, 291–315.

SULLIVAN, J.B. RUMACK, B.H., THOMAS, H., PETERSON, R.G., AND BRYSON, P. (1979). Pennyroyal oil poisoning and hepatotoxicity. *J. Am. med. Ass.* **242**, 2873–4.

TAY, C.H. AND SEAH, C.S. (1975). Arsenic poisoning from anti-asthmatic herbal preparations. *Med. J. Aust.* **2**, 424–8.

TAYLOR, A. AND TAYLOR, N.C. (1963). Protective effect of *Symphytum officiale* on mice bearing spontaneous and transplant tumours. *Proc. Soc. exp. Biol. Med.* **114**, 772–4.

WATKINS, J. (1982). Hypersensitivity response to drugs and plasma substitutes used in anaesthesia and surgery. In *Trauma, stress and immunity in anaesthesia and surgery* (ed. J. Watkins and M. Salo), pp. 254–91. Butterworth Scientific, London.

WILLHITE, C.C. (1982). Congenital malformations induced by laetrile. *Science* **215**, 1513–15.

YOUNG, R.E., MILROY, R., HUTCHINSON, S., AND KESSON, C.M. (1982). The rising price of mushrooms. *Lancet* **i**, 213–15.

Part III

39 Drug interactions

P.F. D'ARCY

Format of the chapter

The corresponding chapters to this one in the two previous editions of *Iatrogenic diseases* and its *Updates 1981, 1982, and 1983*, gave information on drug interactions. in a tabular form and each entry was referenced to the relevant literature. However, the whole field of drug interaction has grown so rapidly that it is no longer possible within the confines of a single chapter to continue with this format; it would indeed take a whole book to do full justice to the subject. A compromise is therefore made in this present chapter; some preliminary discussion is given to specific drug interactions that present a distinct clinical hazard and this is followed by 24 tables which give relevant but summarized information on drug interactions involving major drug groupings. References are not, however, given to individual table entries; these entries are in the main based on an enlarged text recently published by Griffin and D'Arcy (1984) in the 3rd edition of their *Manual of adverse drug interactions*. References to the present tables are therefore given in that book and it is suggested that the reader refers to it if more comprehensive information is required on the interactions that are summarized here. The only exception to this is new information or where a review article is thought to be helpful; in such cases references are given in the present tables.

As far as is possible, indications of the established or likely mechanisms of the interactions are given in the present tables; more detail on the nature of the various mechanisms by which drug–drug interactions can occur is given in the other book (Griffin and D'Arcy 1984).

Not all drug interactions are serious

Much attention has been focused on adverse drug interactions during the last decade or so, and as a result many drug–drug interactions are now predictable and many of the unwanted consequences of using drug combinations can be avoided by simply adjusting the dosage of one or more of the interactants. As a result of this there has been a considerable improvement in the safety and efficacy of therapy with drug combinations.

Unfortunately, however, because much has been written with a lack of clinical perspective, the literature has become clogged with a sticky morass of irrelevant information much of which has been generated in animal studies or in single-dose pharmacokinetic studies in healthy human volunteers. Such studies can be of predictive value, but only if they mimic the clinical situation and if they relate to drug combinations and dosage regimens that are normally used in the clinic. Likewise a large number of reports of uncorroborated observations on individual patients have appeared; some of these are largely anecdotal. Such reports can be useful if they stimulate other practitioners to write of experiences in their own patients; indeed they often serve as an 'early-warning' system (see review: D'Arcy 1983*a*). However, if uncorroborated by subsequent findings, then the original report should be accepted with extreme caution.

Fortunately, only a relatively small number of drugs enter into clinically important drug interactions, and fewer still enter into those interactions which present life-threatening clinical emergencies. Discussion is given later in this text to these two categories which include: anti-arrhythmic agents (notably quinidine) (see Table 39.10), anticoagulants (see Table 39.3), anticonvulsants (phenytoin) (see Table 39.4), beta-blockers (see Tables 39.8 and 39.10), H$_2$-receptor blockers (particularly cimetidine) (see Table 39.7), cardiac glycosides (see Table 39.10), lithium salts (see Table 39.14), antidiabetic agents (see Table 39.6), psychotrophic agents (notably anti-depressants and neuroleptics) (see Tables 39.5 and 39.16), and theophylline (Table 39.21).

It should be noted from this list that most of the drugs involved are those on which patients are carefully stabilized for relatively long periods. Past experience has shown that it is these drug-stabilized patients who are at special risk from any changes in therapy or environment which will influence the potency or availability of their normal medication. It should also be recognized that removal of a drug from a stabilized regimen of therapy may also evoke an interaction sequel.

Oral contraceptives (see Table 39.18) are another group of drugs on which patients are carefully stabilized, albeit this time — healthy women. These drugs can also enter into interactions and these interactions must be considered as important, not because they present life-threatening emergencies, but because quite the opposite — they lead to unwanted pregnancies.

Clinically important drug interactions

Anti-arrhythmic agents and cardiac glycosides

Anti-arrhythmic agents may be broadly divided into those agents which act on ventricular and supraventricular arrhythmias (e.g. quinidine), those which act mainly on ventricular arrhythmias (e.g. lignocaine), and those that act upon supraventricular arrhythmias (e.g. verapamil). It is only too clear, however, that most of the problems with interactions have occurred with quinidine; not surprisingly, since they are frequently used together, digoxin is the other component in most of the interactions.

The quinidine–digoxin interaction has received considerable attention over the last few years (see review articles by Bigger 1979; Hooymans and Merkus 1979; Small and Marshall 1979; Bigger and Leahey 1982). Quinidine causes an increase in serum digoxin concentration in about 90 per cent of cases. The increase is variable but it averages twofold and remains as long as quinidine is given. The mechanism of this interaction has been suggested to be a quinidine-induced reduction of both renal and non-renal clearance of digoxin from tissue binding sites, notably a reduction in the ratio of skeletal muscle to serum digoxin levels. This is suggestive of reduced binding of digoxin in muscle. Serum protein sites are unlikely to be involved as quinidine does not affect serum protein binding of digoxin *in vitro* or *in vivo*. It is also clear that, while extrarenal digoxin excretion is of little or no importance in normal situations, it becomes the main mechanism of elimination in patients with impaired renal function (Doering 1979; Hager *et al.* 1979; Leahey *et al.* 1979; Aronson and Carver 1981; Fichtl and Doering 1981; Schenck-Gustafsson 1981; Schenck-Gustafsson and Dahlqvist 1981; Schenck-Gustafsson *et al.* 1981; Walker *et al.* 1983).

It has been suggested that the adverse effects of this interaction can be avoided if the digoxin dosage is halved before starting quinidine (Doering 1979); however, due to the variable nature of the interaction, adjustment of digoxin dosage is better made 4–5 days after starting quinidine, provided that serum digoxin concentrations indicate that a reduction in dosage is required. Patients should be monitored for signs of too much or too little digoxin (Bigger 1979). It must also be emphasized that the serum concentration and clinical effect of digoxin

will be reduced if quinidine is *withdrawn* from a patient previously stabilized on both drugs (Leahey *et al.* 1979; Moench 1980); it will then be necessary to raise the dose of digoxin until the patient is re-stabilized.

Although the digoxin–quinidine interaction is by now well recognized and its mechanisms at least partially explained, there is a corresponding interaction between digitoxin and quinidine which is still somewhat controversial (Ochs *et al.* 1980; Rollins and Garty 1980; Fenster *et al.* 1981; Garty *et al.* 1981; Melvin and Kates 1981). Current evidence would suggest that the interaction only becomes evident at relatively high steady-state quinidine serum concentrations and that it has less clinical significance than the digoxin–quinidine interaction.

Other interactions involving digoxin which are of proven clinical importance or which are suspected as likely to cause problems are shown in summary in Table 39.10.

Anticoagulants

Of all drug interactions, those involving the oral anticoagulants (coumarins and indanediones) exhibit the widest spectrum of mechanisms which operate singly or simultaneously. Although anticoagulant interactions reported in the very extensive world literature are rare, this in no way diminishes their seriousness when they occur. It is useful, however, to have some guidance as to which interactions are the most common, and it was with this objective in view that a study was set up by Williams *et al.* (1976) in a group of 277 anticoagulated patients, mostly receiving warfarin, over a period of six months. Prothrombin times were determined at each visit to the clinic and patients were questioned about side-effects of treatment and drugs taken since their last visit to the doctor.

The most obvious conclusion drawn from the study was that the more drugs a patient received, the more unstable was his anticoagulant control. When drugs, which were theoretically supposed to interact with anticoagulants, were taken at a steady constant daily dosage throughout the six months, they did not seem to introduce any major problems with anticoagulant control. Examples of this were phenobarbitone and/or phenytoin taken by epileptic patients, or oral antidiabetic drugs, or even alcohol taken at a steady level. However, in the case of drugs known to interact but taken for short periods, for example antibiotics, or of drugs taken spasmodically or irregularly, for example barbiturate hypnotics, anticoagulant control proved to be difficult. Fluctuating cardiac status in patients requiring digoxin and diuretics was also shown to be a risk factor in controlling anticoagulation.

Drug interactions involving anticoagulants are summarized in Table 39.3.

Anticonvulsants

Fluid retention caused by oral contraceptives may precipitate epileptic seizures in patients on anticonvulsant therapy (McArthur 1967; *Oral Contraceptives* 1974); this is due to an increase in volume of distribution of anticonvulsants which thereby lowers their plasma and CSF concentrations. Nine out of 20 patients receiving chloramphenicol and phenytoin simultaneously exhibited signs of CNS toxicity (Christensen and Skovsted 1969; Ballek *et al*. 1973; Koup *et al*. 1978; Harper *et al*. 1979). The mechanism of the interaction is blockade of the liver enzyme metabolism of phenytoin. This latter interaction could be of particular concern in Third World countries where the use of chloramphenicol is widespread and often indiscriminate.

These examples are but two of the many interactions (see Table 39.4) that can upset the stabilized control of the epileptic patient who is receiving phenytoin, sodium valproate, or other anticonvulsant. Potentiation of anticonvulsant action, especially that of phenytoin, can result in frank toxicity since the drug has a narrow therapeutic index. Antagonism of anticonvulsant action is likely to result in an increased frequency and severity of convulsions, and drugs like the oral contraceptives or the tricyclic antidepressants can cause convulsions in the stabilized epileptic patient simply because they themselves are capable of producing seizures even in the non-epileptic patient (McArthur 1967; *Oral Contraceptives* 1974; Betts *et al*. 1968; Dallos and Heathfield 1969; Houghton 1971).

Antidiabetic agents and beta-blockers

Drug-induced hypoglycaemia can be a medical emergency and many drug interactions with insulin or with the oral hypoglycaemia drugs have this life-threatening sequel (see Table 39.6). In particular, the beta-blockers can present a number of problems for the controlled diabetic patient. Propranolol, for example, will precipitate hypoglycaemia in the insulin-dependent diabetic; it increases sensitivity to insulin by damping down the rebound of blood glucose level after its initial fall, and it may act as a stimulator of insulin secretion (Abramson *et al*. 1966; Kotler *et al*. 1966; Sussman *et al*. 1967; Divitiis *et al*. 1968). Beta-blockers reduce splanchnic blood flow and reduce the access of sulphonylureas to pancreatic tissue; this greatly restricts their ability to stimulate the secretion of insulin (Divitiis *et al*. 1968; see review by Hansten 1980). However, the most important action of beta-blocking drugs with respect to the control of the diabetic patient on insulin or oral hypoglycaemic drugs is undoubtedly that the block of beta-adrenergic receptors abolishes those signs of hypoglycaemia that are mediated by adrenergic mechanisms, in particular, shaking and sweating (Griffin and D'Arcy 1974). It is by these signs that the experienced diabetic or his relatives can recognize the impending hypoglycaemia and can take appropriate preventive measures (e.g. glucose sweet, hot sweet drink, etc). The loss of these prodromal warning signs can result in the patient passing into a life-threatening hypoglycaemic coma. Interestingly, more recent work has shown that cardioselective beta-blockade with acebutolol or metoprolol does not influence these parameters that reflect diabetic control (Zaman *et al*. 1982; Kølendorf *et al*. 1982). This field was reviewed by Hansten (1980).

Histamine H_2-receptor blockers

The H_2-receptor blocking agent, cimetidine, has been generally available in Britain and in the United States for just over six years. During this time there has been a growing list of reports of the extent to which cimetidine may participate in drug interactions.

The earliest reports of such participation came from Fluid (1978), Silver and Bell (1979), and from Serlin *et al*. (1979); these investigators reported that cimetidine potentiated the anticoagulant effects of warfarin and suggested that it did so by inhibiting the hepatic microsomal enzyme oxidase activity. It was suggested, as a result of this knowledge, that cimetidine might also interact with other drugs that were metabolized by liver microsomal enzymes (Serlin *et al*. 1979).

That this prediction was justified has been well shown by subsequent reports in the literature, and it is now certain that cimetidine has the potential to interact with a wide range of drugs including some benzodiazepines (diazepam, chlordiazepoxide, prazepam), carbamazepine, chlormethiazole, methadone, morphine, phenytoin, theophylline, vitamin B_{12} (uptake of food-bound vitamin), frusemide, and digitoxin/quinidine (a double interaction due to inhibition of liver enzymes). It has also become evident that cimetidine will potentiate the action of the beta-blockers, atenolol, labetalol, metoprolol, and propranolol, and the anti-arrhythmic activity of lignocaine (lidocaine), probably by reducing liver blood flow, although there is some evidence to suggest that this effect is only transient and that chronic treatment with cimetidine does not reduce apparent liver blood flow (Daneshmend *et al*. 1984). The literature on drug interactions involving cimetidine has become almost voluminous and to save time in the present text the reader is referred to the reference sources cited in the following reviews: Bauman and Kimelblatt (1982), D'Arcy (1982*a*), Somogyi and Gugler (1983), and Sorkin and Darvey (1983).

Lithium and other psychotropic drugs

A knowledge of the pharmacokinetic drug interactions involving lithium salts is of considerable practical importance; firstly because of the growing and widespread use of lithium in the treatment of mania, and secondly, because the therapeutic range of the lithium ion is extraordinarily narrow (plasma lithium ion levels: 0.8 to 1.4 mEq l^{-1}) and any excess in this may lead to lithium intoxication. The early effects of lithium toxicity are ataxia, coarse tremor, confusion, vomiting, diarrhoea, thirst and dryness of mouth, drowsiness, and slurred speech. Severe intoxication results in coma with hyperreflexia, muscle tremors, and occasional epileptiform seizures.

An example of such an interaction is that between indomethacin and lithium (Frölish *et al.* 1979); studies in psychiatric patients and in normal subjects have shown that indomethacin decreases renal lithium ion elimination by 23–30 per cent, causing plasma lithium ion concentrations to increase by 30–60 per cent. It has been suggested that this renal clearance of lithium may be affected by a prostaglandin-dependent mechanism, possibly in the distal renal tubule. If this suggestion is correct, then it might well be expected that all non-steroidal anti-inflammatory drugs (NSAIDs) will cause elevations in plasma lithium levels and precipitate lithium toxicity. So far, this interaction has also been shown to occur with diclofenac (Reimann and Frölich, 1981) but apparently not with aspirin although both drugs suppressed renal prostaglandin E_2 excretion by over 50 per cent. This field was reviewed by Salem (1982).

In the past, interactions with other psychoactive drugs have largely centred on the interaction between MAOIs and tyramine-containing foods or indirectly acting sympathomimetic amines (see review: D'Arcy and Griffin 1980). Such agents are now less frequently prescribed and, except in specific treatments, have been largely superseded by the tricyclic and tetracyclic compounds which seem to be the drugs of choice in the more typical endogenous types of depression. Interactions with the tricyclic compounds can, however, be dangerous (see Table 39.5 and, apart from those which are well recognized (e.g. MAOI-tricyclic combinations, tricyclic–sympathominetic amines, tricyclic–anticholinergic drugs, tricyclic–adrenergic blocking hypotensives, etc.), there are later reports of interactions between some neuroleptic agents (perphenazine, haloperidol,

fluphenamine, thiothixene) and imipramine or desipramine which have resulted in severe CNS side-effects and urinary retention due to raised plasma concentrations of the tricyclic or of its major active metabolite (Nelson and Jatlow 1980; Siris *et al.* 1982).

The mechanism of this type of interaction is almost certainly due to inhibition of liver microsomal metabolism of the tricyclic antidepressant. Adjustment of tricyclic dosage when concomitant antipsychotic treatment is given should prevent major problems arising from this interaction.

Theophylline

Theophylline, at serum concentrations of 10–20 μg ml^{-1}, is an effective and relatively safe drug for controlling the symptoms of chronic asthma. When elimination is rapid, however, as is typical among children, excessive fluctuations in serum concentrations occur during chronic therapy. Since bronchodilatation and blocking of exercise-induced bronchospasm becomes less effective as serum concentrations of theophylline decrease, and the risk of toxicity increases as serum concentrations exceed 20 μg ml^{-1}, it follows that any drug interaction which upsets the stabilized balance of the theophylline-treated asthmatic is a potential hazard.

When considering the long-established use of theophylline, it is surprising that it is only relatively recently that reports of interactions involving theophylline have emerged (see review by McElnay *et al.* 1982). In all probability this emergence correlates with the more enthusiastic and vigorous use of theophylline in recent years as an acceptable alternative to the sympathomimetic bronchodilators and the corticosteroids in the treatment of asthma.

With the sole exception of smoking (enzyme induction), all the important interactions so far reported (see Table 39.21) have a common sequel of inhibiting the metabolism of theophylline, increasing its elimination half-life, and elevating the steady-state serum theophylline concentration. Such interactions would be especially dangerous if intravenous theophylline were involved since it is clear from literature reports (Zwillich *et al.* 1975; Hendeles *et al.* 1977; Kordash *et al.* 1977; Culberson 1979; Hendeles and Weinberger 1980; Hughey *et al.* 1982) that dosage is critical with seizures and fatalities accompanying regimens where dosage has been excessive.

Oral contraceptives

Interactions involving the combined type (oestrogen plus progestogen) oral contraceptives are often unsuspected and even unestablished; the sequel, an unplanned pregnancy, is often mistakenly blamed by the consulting physician on poor patient compliance with medication instructions (see review by D'Arcy and Griffin 1976). Interactions involving oral contraceptives are summarized in Table 39.18; it should not be overlooked that oral contraceptives themselves may modify the activity of co-administered drugs; a list of these latter interactions is also given in Table 39.18.

The latest interaction involving oral contraceptives is that with griseofulvin (Van Dijke and Weber 1984); the Committee on Safety of Medicines in Britain and the Netherlands Centre for Monitoring of Adverse Reactions to Drugs have received a total of 22 reports of a possible interaction. Twenty women receiving long-term oral contraception were reported to have experienced transient intermenstrual bleeding or amenorrhoea in the first or second cycle after starting treatment with griseofulvin (0.5–1.0 g daily). Two others had unintended pregnancies. There was a possible association of sulphonamides with griseofulvin in the latter two cases which is interesting since sulphonamides or sulphonamide-containing preparations (e.g. co-trimoxazole) are not an established cause of failure or oral contraceptives.

Griseofulvin therefore enters a list of drugs which may reduce the efficacy of oral contraceptives; the other drugs on the list are: tetracycline (Bacon and Shenfield 1980); rifampicin (Nocke-Finck et al. 1973; Skolnick et al. 1976); phenytoin (Janz and Schmidt 1974; Coulam and Annegers 1979); barbiturates (Back et al. 1980); and other hypnotics, sedatives, and CNS depressants (e.g. dichloralphenazone, glutethimide, meprobamate, etc.) (Janz and Schmidt 1974; Azarnoff and Hurwitz 1970).

Ampicillin tentatively held a place on that list for some time (Dossetor 1975), but later evidence from controlled studies failed to demonstrate any effect of ampicillin on various parameters of the contraceptive cycle (Friedman et al. 1980; Back et al. 1982).

Medical plastics and vaccines

The reader is also specifically referred to interactions involving medical plastics (Table 39.24) and vaccines (Table 39.23). Both these fields have developed within the last few years.

Sorption of drugs on to medical plastics is the major hazard since drugs can be lost from solution when they are in contact with intravenous fluid containers, delivery sets, syringes, terminal filters, or other plastic materials used in infusion systems. Polyvinyl chloride (PVC) is the major offender in this respect. Other interactions with plastics range from the solvent action of paraldehyde on polystyrene, styrene–acrylonitrile co-polymer, and rubber to drug interactions causing colouration of hydrophilic contact lenses. These interactions were reviewed by D'Arcy (1983b).

Clinically important interactions between a vaccine and a drug have so far only been reported with influenza vaccine and three drugs: phenytoin, theophylline, and warfarin, and with BCG vaccine and theophylline. Some of these interactions are as yet unconfirmed but they are worrying because the underlying mechanism is non-specific and is thought to be due to the vaccine (as an interferon inducer) inactivating the hepatic cytochrome P-450 system. This results in depressed drug metabolism and reduced clearance. Because this mechanism is non-specific it could well occur with other vaccines (e.g. yellow fever, cholera, tetanus, etc.) and it is possible that other vaccine–drug interactions may not yet have been recognized. It is known, for example, that immunologic stimulation by endotoxin (Gorodischer et al. 1976), by BCG, Corynebacterium parvum, and Bordetella pertussis (Farquhar et al. 1976; Soyka et al. 1976; Renton 1979), and by interferon (Sonnenfeld et al. 1980) depress hepatic biotransformation in animals.

Vaccine–drug interactions were recently reviewed by D'Arcy (1984a).

Table 39.1 Alcohol (ethanol)*

Combination	Interaction
Acetaminophen	Enhanced hepatotoxicity; can be fatal.
Anticoagulants	Acute: coumarins potentiated; chronic: coumarin half-life reduced.
Anticonvulsants	Phenytoin metabolism increased; failure to control epilepsy.
Antidiabetic agents	Risk of severe hypoglycaemia; reduced half-life of sulphonylureas; increased risk of lactic acidosis with biguanides.
Aspirin	Increased magnitude and duration of bleeding time.

(continued overleaf)

Table 39.1   *Continued*

Combination	Interaction
Benzodiazepines	Impaired psychomotor and driving skills with lorazepam.
Beta-lactam antibiotics	'*Antabuse*-like' effects.
Bromocriptine	Enhanced nausea and abdominal symptoms.
Calcium carbimide	Potentially dangerous cardiovascular changes.
Chlormethiazole	Reduced lethal dose; fatal cases of self-poisoning.
Chlorpropamide	Facial flush.
Cimetidine	Increased alcohol intoxication.
Clorazepate	Euphoric effects of alcohol increased.
CNS depressants e.g. antihistamines, barbiturates, benzodiazepines, chloral hydrate and derivatives, dextropropoxypene, hypnotics, glutethimide, minor and major tranquillizers, sedatives, narcotic analgesics	Enhanced CNS depression, can be fatal; impaired driving skills.
Naloxone	Reversal of ethanol-induced coma.
Procainamide	Reduced half-life, increased total body clearance, and acetylation rate of procainamide.
Propranolol	Increased absorption of propranolol.
Tolbutamide	Facial flush.
Tricyclic antidepressant	Additive sedative effect; impaired driving skills.

* See also review by D'Arcy and Merkus (1981).

Table 39.2   Antibiotics and antibacterial agents

I.   Aminoglycosides (amikacin, gentamicin, kanamycin, neomycin, streptomycin, tobramycin, viomycin, etc.)

Combination	Interaction
Anticoagulant	Decreased vitamin K production may potentiate anticoagulant action; depends on duration of aminoglycoside treatment.
Diuretics (ethacrynic acid and frusemide)	Combinations may potentiate ototoxicity especially in patients with diminished renal function.
Penicillins (azlocillin, carbenicillin, mezlocillin)	Gentamicin, tobramycin, and netilmicin are inactivated *in vitro* by azlocillin, carbenicillin, and mezlocillin; extent of inactivation depends upon penicillin

Table 39.2 *Continued*

Combination	Interaction
	concentration, contact time, and temperature. Interaction of likely clinical importance if aminoglycosides were in contact with high concentrations of these penicillins for lengthy periods (e.g. in renal failure or when aminoglycoside serum levels are determined).
Skeletal muscle relaxant	Kanamycin, neomycin, viomycin, and streptomycin all produce neuromuscular blockade and potentiate skeletal muscle-relaxant drugs. Neomycin, streptomycin, and viomycin produce a curare-like, and kanamycin a depolarizing neuromuscular block.
Ticarcillin	Gentamicin and tobramycin are both antagonized by ticarcillin *in vivo*; decreased half-life of both aminoglycosides. Mechanism thought to be physicochemical.

II. Antifungal antibiotics (amphotericin, griseofulvin, nystatin)

Combination	Interaction
Amphotericin/digitalis glycosides	Severe hypokalaemia may follow amphotericin therapy and this effect may potentiate the toxicity of digitalis glycosides.
Amphotericin/miconazole	Drugs antagonize each other in their antifungal activity.
Amphotericin/skeletal muscle relaxant	Hypokalaemia following amphotericin therapy may enhance curariform actions of skeletal muscle relaxants.
Griseofulvin/alcohol	Metabolism of alcohol affected; activity enhanced; impaired psychomotor and driving skills.
Griseofulvin/ anticoagulant	Increased rate of coumarin metabolism due to liver enzyme induction; antagonism of anticoagulant effects.
Griseofulvin/barbiturate	Combination results in an increased inactivation of griseofulvin by liver microsomal enzymes; phenobarbitone also interferes with intestinal absorption of griseofulvin; antifungal activity almost nullified.
Griseofulvin/ bromocriptine	Response to bromocriptine blocked in one case.
Griseofulvin/oral contraceptive	See Table 39.18(I)
Nystatin/riboflavine	Activity of nystatin against *Candida albicans* almost completely inhibited by presence of riboflavine.

III. Beta-lactam antibiotics (cephalosporin and related antibiotics)

Combination	Interaction
Alcohol	'Antabuse-like' effects with cefoperazone, cephamandole, latamoxef, and possibly other cephalosporins.
Anticoagulants	Anticoagulant efffects potentiated, especially with latamoxef; this coagulopathy is reversible with vitamin K. Mechanism suggested to be platelet inhibition mediated by antibiotic binding to the platelet surface. Time schedule of interaction too short for effect to be due to suppression of vitamin K-producing gut microflora.
Creatinine estimation	Cefoxitin sodium acts like creatinine in assays for creatinine determination and causes an apparent rise in the serum creatinine level. Cephalothin has been reported to act in the same way, but not cephapirin, cefazolin, or cefamandole.

(continued overleaf)

Table 39.2   *Continued*

IV.   Chloramphenicol

Combination	Interaction
Anticoagulant	Marked potentiation of anticoagulant action due to inhibition of liver metabolism and also to decreased vitamin K production by intestinal bacteria.
Antidiabetic agent	Plasma half-life of tolbutamide and chlorpropamide increased at least threefold due to inhibition of microsomal enzyme activity; danger of hypoglycaemic episodes.
Paracetamol	Half-life of chloramphenicol prolonged fivefold; mechanism suggested to be decreased chloramphenicol elimination due to paracetamol competition; risk of chloramphenicol-induced agranulocytosis increased.
Penicillins	A bacteriostatic antibiotic such as chloramphenicol can inhibit the bactericidal action of penicillin; reduced penicillin effect.
Phenytoin	Rise in serum concentrations of phenytoin and prolonged half-life due to inhibition of liver enzyme metabolism by chloramphenicol; danger of phenytoin toxicity.

V.   Macrolide antibiotics (erythromycin, troleandomycin (triacetyloleandomycin))

Combination	Interaction
Erythromycin/ carbamazepine	Carbamazepine intoxication when used in combination.
Erythromycin/theophylline	See Table 39.21.
Erythromycin/warfarin	Warfarin anticoagulation increased in one case.
Troleandomycin/ carbamazepine	Carbamazepine intoxication when used in combination.
Troleandomycin/oral contraceptive	See Table 39.18(II).

VI.   Nalidixic acid

Combination	Interaction
Nitrofurantoin	Antibacterial action of nalidixic acid *in vitro* inhibited. Combination contraindicated in urinary-tract infections.
Anticoagulants	Anticoagulant effects of nicoumalone and warfarin increased; mechanism probably due to displacement of coumarins from binding sites.
Laboratory diagnostic tests	Nalidixic acid may cause false positive reactions in urine tests for glucose using copper reduction methods. Nalidixic acid also reported to give false elevations of urinary 17-ketosteroids and to interfere with estimation of plasma 11-hydroxycorticosteroids.

VII.   Nitrofuran derivatives (furazolidone, nitrofurantoin)

Combination	Interaction
Furazolidone/alcohol	Intolerance to alcohol.
Furazolidone/ amitriptyline	Toxic psychosis in one case after commencement of furazolidone treatment; mechanism likely to be due to MAOI properties of furazolidone.

Table 39.2 *Continued*

Combination	Interaction
Furazolidone/amphetamine and tyramine-containing foods	Hypersensitivity to amphetamine and tyramine in presence of furazolidone; mechanism due to inhibition of intestinal monoamine oxidase; risk of hypertensive crisis.
Nitrofurantoin/alcohol (absence of interaction)	An error has crept into the literature regarding an alleged '*Antabuse*-like' interaction between alcohol and nitrofurantoin; this was perpetuated in recent time by the American FDA including nitrofurantoin in a list of 48 drugs submitted to the US Department of Health, Education, and Welfare, as requiring appropriate warnings or precautionary statements on their labels regarding an interaction with alcohol. The source of this misinformation has been indentified (Rowles and Worthen 1982) and there are studies in animals and clinical experiences which prove that this interaction does not occur (see also review: D'Arcy 1982*b*).
Nitrofurantoin/antacids (absence of general interaction)	Nitrofurantoin is adsorbed *in vitro* on to magnesium trisilicate and to a lesser extent on to bismuth oxycarbonate, talc, kaolin, and magnesium oxide; aluminium hydroxide and calcium carbonate exhibited no or low adsorption properties. *In vivo* administration of magnesium trisilicate with nitrofurantoin reduced the rate and extent of its absorption (Naggar and Khalil 1979). This is a specific adsorption interaction and it does not occur *in vivo* when aluminium hydroxide gel is given concomitantly with nitrofurantoin (Jaffe *et al*. 1976). The widely reported and generalized nitrofurantoin/antacid interaction is therefore false; it has been continually and wrongly reported based solely on secondary references.
Nitrofurantoin/ nalidixic acid	See Table 39.2(VI).

VIII. Penicillins

Combination	Interaction
Ampicillin or Amoxycillin/allopurinol	Administration of allopurinol in presence of ampicillin or amoxycillin doubled or quadrupled, respectively, the incidence of skin rashes.
Ampicillin/oral contraceptives	See Table 39.18(I).
Ampicillin/ sulphasalazine	Only a small fraction of orally administered sulphasalazine is absorbed; remainder reaches the coion intact where it is metabolized by azoreduction to yield active metabolites, sulphapyridine and 5-aminosalicylate. Ampicillin reduces plasma concentrations of active metabolites and may reduce activity of sulphasalazine. Mechanism thought to be ampicillin effect on gut microflora responsible for sulphasalazine disposition.
Carbenicillin/gentamicin	See Table 39.2(I).
Penicillin/bacteriostatic antibiotic (e.g. chloramphenicol)	See Table 39.2(IV).
Phenoxymethyl-penicillin/neomycin	Oral neomycin decreases absorption of phenoxymethylpenicillin (penicillin V) presumably due to production of a malabsorption syndrome.
Ticarcillin/gentamicin or tobramycin	See Table 39.2(I).

(*continued overleaf*)

Table 39.2 *Continued*

IX. Polymyxins (colistin sulphate and sulphomethate sodium, polymyxin B sulphate)

Combination	Interaction
Colistin/cephalosporins	Increased hazard of renal damage.
Polymyxins/skeletal muscle relaxants	Respiratory paralysis has been reported extensively with the polymyxins both when taken alone and in combination with skeletal muscle relaxants.
Polymyxin B/calcium and magnesium ions	Bactericidal action of polymyxin B antagonized by Mg^{2+} and Ca^{2+}; it is suggested that interaction of divalent cations with cell wall prevents antibiotic access to bacterial cytoplasmic membrane.

X. Sulphonamides

Combination	Interaction
Co-trimoxazole/immuno-suppressive agents	Haematological toxicity (bone-marrow depression) of co-trimoxazole increased by accompanying immunosuppression with azathioprine. Use of methotrexate with co-trimoxazole may lead to pancytopenia with megaloblastosis. *In-vitro* evidence of 6-mercaptopurine suppression potentiated.
Co-trimoxazole/phenytoin	Effects of phenytoin potentiated; increased risk of intoxication.
Co-trimoxazole/warfarin	Anticoagulant effect of warfarin potentiated; interaction stereoselective with S-isomer of warfarin. Mechanism uncertain but possible displacement of warfarin from plasma-binding sites, or decreased synthesis of vitamin K due to antibacterial effect on gut flora.
Sulphadiazine/phenytoin	Effects of phenytoin potentiated; increased risk of intoxication.
Sulphamethizole/phenytoin, tolbutamide, and warfarin	Hepatic metabolism of all three drugs inhibited; clinical effects enhanced with dangers of overactivity.
Sulphamethoxazole/theophylline determinations	Sulphamethoxazole in large quantities could interfere with HPLC determination of theophylline concentrations in body fluids.
Sulphathiazole/hexamine mandelate	Sulphathiazole forms an insoluble compound with formaldehyde in urine; administration with hexamine (which liberates formaldehyde in urine) is contraindicated.
Sulphaphenazole/phenytoin	Effects of phenytoin potentiated; increased risk of intoxication.
Sulphonamides/anticoagulants	Inhibition of vitamin K synthesis by intestinal bacteria may potentiate effects of anticoagulants; sulphonamides also displace coumarin from protein-binding which also potentiates their anticoagulant effect.
Sulphonamides/antidiabetic drugs	Enhanced hypoglycaemic action of tolbutamide when concomitant treatment with sulphaphenazole or sulphafurazole (sulfisoxazole) and similar interaction with chlorpropamide and sulphadimidine (sulphamethazine). Mechanism due to displacement of plasma-protein binding and also (with tolbutamide) inhibition of decarboxylation metabolism.
Sulphonamides/methotrexate	Methotrexate displaced from plasma binding and its toxicity increased.
Sulphonamides/non-steroidal anti-inflammatory drugs	Sulphonamide effects may be enhanced due to displacement from plasma-binding sites. Duration of long-acting sulphonamides may be reduced.
Sulphonamides/paraldehyde	Acetylation of sulphonamides increased by paraldehyde; increased risk of crystalluria due to formation of less soluble sulphonamide derivatives.
Sulphonamides/procaine and similar local anaesthetics	Local anaesthetics which are derivatives of *p*-aminobenzoic acid (PABA) are hydrolysed to that compound and this may reduce their antibacterial action which depends on inhibition of PABA in the microorganism.

Table 39.2 *Continued*

XI.   Tetracyclines

Combination	Interaction
Demeclocycline/diuretic	Diuresis induced by amiloride and metolazone enhanced when demeclocycline was given concomitantly in one case.
Doxycycline/anticonvulsant	Plasma half-life of doxycycline reduced significantly when combined with carbamazepine, phenytoin, or combination of the two, or with barbiturate therapy due to induced metabolism; reduced antibacterial activity.
Doxycycline/ferrous sulphate	Serum concentration lowered when oral iron given concomitantly but not if antibiotic and iron given several hours apart.
Tetracyclines/antacids and milk	Diminished absorption of tetracyclines from intestine due to formation of unabsorbed, stable complexes with soluble salts of divalent and trivalent metals.
Tetracyclines/anticoagulants	Tetracyclines reduce plasma prothrombin activity by impairing prothrombin utilization. They also decrease vitamin K production by intestinal bacteria; anticoagulant activity increased.
Tetracycline/cimetidine	See Table 39.7(II).
Tetracycline/digoxin	See Table 39.10(I).
Tetracyclines/iron salts	Absorption of tetracyclines from intestine diminished by iron salts due to formation of a non-absorbable complex. Tetracyclines also impair iron absorption.
Tetracycline/lithium salts	See Table 39.14.
Tetracycline/methoxyflurane	Fatal renal failure due to potentiation of toxic effects of tetracycline on the kidney; may occur with other tetracyclines.
Tetracycline/oral contraceptives	See Table 39.18(I).
Tetracyclines/penicillins	Combination of bacteriostatic agent (tetracyclines) with a bactericidal agent (penicillin) inhibits the action of the latter.
Tetracycline/phenformin	Lactic acidosis precipitated.

Table 39.3   Anticoagulants (coumarins)

I.   Drugs reported to potentiate anticoagulants

Alcohol
Allopurinol
Amiodarone
Anabolic steroids
Antibiotics and antibacterials
Antidiabetic agents (biguanides)
Antihyperlipidaemic agents (e.g. clofibrate, cholestyramine)
Cimetidine
Co-trimoxazole
Danazol
Disulfiram
Erythromycin
Influenza vaccine
Latamoxef (and cephalosporins)
Non-steroidal anti-inflammatory drugs
Paracetamol (acetaminophen)
Phenothiazine tranquillizers
Sulphafurazole (sulfisoxazole)
Sulphinpyrazone
Thyroid hormones
Tricyclic antidepressants

II.   Drugs reported to reduce the effects of anticoagulants

Anti-epileptic drugs (e.g. phenobarbitone, phenytoin)
Barbiturate hypnotics
Carbamazepine
Diflunisal
Vitamin K (in weight-reducing diets)

Table 39.4 Anticonvulsants

I. Drugs reported to increase the effects of phenytoin sodium sometimes in single or isolated cases (*)

Aspirin	Oestrogens*
Azapropazone	Oral contraceptives
Chloramphenicol*	Pheneturide
Chlorpheniramine*	Phenyramidol
Cimetidine	Prochlorperazine*
Dexamethasone	Propranolol*
Dextropropoxyphene	Sulphadiazine
(propoxyphene)	Sulphamethizole
Diazepam	Sulphaphenizole
Dicoumarol	Sulthiame
Disulfiram	Thioridazine
Frusemide*	Thyroid hormones
Halothane*	Tolbutamide
Isoniazid (in slow	Viloxazine
inactivators)	Warfarin*
Lithium salts	
Methylphenidate*	

II. Drugs reported to decrease the effects of phenytoin sodium

Alcohol
Antacids
Cancer chemotherapy
(e.g. bleomycin,
cisplatin,
vinblastine)
Dichloralphenazone
Phenobarbitone
Sodium valproate

III. Drugs reported to increase the risk of epileptiform seizures

Oral contraceptives
Tricyclic anti-depressants

IV. Drug actions modified by phenytoin sodium

Combination	Interaction
Corticosteroids	Metabolism increased; efficacy decreased.
Disopyramide	Metabolism increased; efficacy decreased.
Methadone	Accelerated metabolism; efficacy decreased.
Mexiletine	Enhanced metabolic clearance; efficacy decreased.
Mianserin	Enhanced metabolic clearance; efficacy decreased.
Oestrogens	Menopausal replacement effect of conjugated oestrogens antagonized.
Pethidine (meperidine)	Increased metabolism; reduced analgesia.
Sodium valproate	Drowsiness or other alterations in consciousness.
Theophylline	Enhanced metabolic clearance; efficacy decreased.

Table 39.4   *Continued*

V.   Drug interactions with other anticonvulsants (carbamazepine, sodium valproate (valproic acid))

Combination	Interaction
Antacids	Increased bioavailability of valproic acid.
Aspirin	Raised serum concentrations of valproic acid; possible cumulation to toxic levels.
Cimetidine	Controversial interaction; raised serum concentration of carbamazepine with neurological toxicity in one case but in another report (Sonne *et al.* 1983) there were no significant alterations in steady-state plasma carbamazepine concentrations in seven epileptics receiving both drugs.
Dextropropoxyphene (propoxyphene)	Raised serum concentrations of carbamazepine; toxic symptoms.
Isoniazid	Raise serum concentrations of carbamazepine; toxic symptoms.
Macroline antibiotics (e.g. erythromycin, troleandomycin)	Raised serum concentrations of carbamazepine; toxic symptoms.

Table 39.5   Antidepressants

I.   MAOIs*

Combination	Interaction
Alcohol	See Table 39.1.
Antihypertensive agents (e.g guanethidine, methyldopa)	Hypotensive action reduced.
(e.g. propranolol)	Hypertensive episodes.
Appetite suppressants (e.g. amphetamine-like compounds, fenfluramine, and mazindol)	Hypertensive episodes.
Sympathomimetic amines	Adrenergic effects potentiated; hypertensive crisis.
Tricyclic anti-depressants	Increased toxicity of tricyclic drug.
Tyramine-containing foods	Hypertensive episodes.

* See also review by D'Arcy and Griffin (1980).

II.   Tricyclic agents

Combination	Interaction
Alcohol	Enhanced sedation; impairment of driving ability.
Anticholinergic drugs (e.g. anti-emetics, antihistaminics, atropine-like drugs, antiparkinsonian agents, orphenadrine (mephenamine))	Anticholinergic effects enhanced.

(*continued overleaf*)

Table 39.5 *Continued*

Combination	Interaction
Anticoagulants	See Table 39.3(I).
Anticonvulsants	See Table 39.4(III).
Antihypertensive agents (e.g. bethanidine, clonidine, debrisoquine, guanethidine)	Reduced hypotensive effects.
Baclofen	See Table 39.19.
Enflurane	Seizures after induction anaesthesia.
Phenelzine	Hypothermia; mental impairment.
Phenothiazine tranquillizers (also haloperidol and thiothixene)	Plasma concentration of tricyclic drugs increased; CNS toxicity and urinary retention.
Physostigmine	Bradycardia and asystole; fatal in one case.
Smoking	Reduced plasma tricyclic concentrations.
Sympathomimetic amines	Enhanced cardiovascular effects; hypertensive crisis.

Table 39.6 **Antidiabetic agents**

Combination	Interaction
Alcohol	See Table 39.1
Beta-blockers	Increased duration and magnitude of insulin-induced hypoglycaemia; inhibition of prodromal signs of hypoglycaemia.
Guanethidine	Reduced insulin requirements.
Nifedipine	Impairment of glucose tolerance.
Non-steroidal anti-inflammatory drugs (e.g. azapropazone, fenclofenac)	Severe hypoglycaemia.
Ritodrine	Hyperglycaemia and metabolic acidosis.
Smoking	Increased insulin requirements; not confirmed in a later study (Mühlhauser *et al.* 1984).
Sulphinpyrazone	Effect of tolbutamide potentiated.

Table 39.7   Antihistamines (H$_1$ and H$_2$ receptor blockers)

I.   H$_1$-receptor blockers

Combination	Interaction
CNS depressants (e.g. alcohol, barbiturates, hypnotics, sedatives, and tranquillizers)	Enhanced sedative or hypnotic effects some fatal; impaired driving ability.

II.   H$_2$-receptor blockers (cimetidine)*

Combination	Interaction
Antacids	Reduced intestinal absorption of cimetidine.
Anticholinergic drugs (e.g. metoclopramide, propantheline)	Reduced intestinal absorption of cimetidine.
Anticoagulants (e.g. warfarin)	Excess anticoagulation (but not with phenprocoumon).
Benzodiazepines	Elimination of chlordiazepoxide, diazepam, and lorazepam impaired; 'hang-over' and impaired performance of skilled work (including driving).
Beta-blockers	Increased plasma levels (and increased bioavailability) of propranolol, labetalol, and metoprolol, but not of atenolol.
Carbamazepine	See Table 39.4(V).
Chlormethiazole	Increased plasma concentration of chlormethiazole increased sedation.
Digitoxin and quinidine	Cardiotoxicity.
Insulin	See Table 39.6.
Lignocaine	Reduced systemic clearance; lignocaine toxicity.
Morphine	CNS-depressant effects enhanced.
Phenytoin	See Table 39.4(I).
Procainamide	Renal clearance of procainamide reduced; possible toxicity.
Tetracycline	Absorption of tetracycline from capsules reduced, but not from tablet formulations.
Theophylline (aminophylline)	See Table 39.21.

* See also review articles: Bauman and Kimelblatt 1982; D'Arcy 1982a; Somogyi and Gugler 1983; Sorkin and Darvey 1983.

Table 39.8 Antihypertensive agents

I. Adrenergic neurone-blocking agents

Combination	Interaction
Clonidine/desipramine	See Table 39.5(II).
Clonidine/naloxone	Naloxone successfully reversed coma and apnoea in young child who swallowed unknown quantity of clonidine tablets.
Debrisoquine/phenylephrine	Combination contraindicated since debrisoquine has properties of a MAO-inhibitor; danger of hypertensive crisis.
Debrisoquine/PUVA	Hydroxylation of debrisoquine inhibited by PUVA therapy in psoriatic patients; adverse effect on patients' ability to handle the drug.
Debrisoquine/tyramine-containing foods	Debrisoquine has MAO-inhibitor properties; danger of hypertensive crisis with ingestion of cheese or other tyramine-rich foods or beverages.
Guanethidine/antidepressants MAOIs	See Table 39.5(I).
Tricyclics	See Table 39.5(II).
Guanethidine/antidiabetic agents	See Table 39.6.
Guanethidine/drugs interfering with its uptake into adrenergic neurons (e.g amphetamines, diethylpropion, ephedrine, mephentermine, methylphenidate, tricyclic antidepressants)	Hypotensive actions of guanethidine antagonized; marked hypotensive episode may occur if one of these drugs is withdrawn in an otherwise stabilized patient.
Guanethidine/mazindol	Pressor action of guanethidine potentiated.
Guanethidine/oral contraceptives	See Table 39.18(II).
Guanethidine/phenothiazine tranquillizers	Conflicting reports on nature of interaction; some suggest inhibition of uptake of guanethidine into adrenergic neurones with antagonism of hypotensive action, others suggest phenothiazines potentiate action of guanethidine due to their alpha-adrenergic blocking action.
Guanethidine/sympathomimetic agents (e.g. amphetamines, ephedrine, noradrenaline, phenylephrine, phenylprop-anolamine*)	Adrenergic receptor is excessively sensitive to direct-acting sympathomimetic amines and there is increased tendency for cardiac arrhythmias. Pupillary response of phenylephrine eye drops enhanced; amphetamines and ephedrine antagonize hypotensive effects of guanethidine.

* See footnote on p. 949

II. Beta-blocking drugs (See also Table 39.10(II))

Combination	Interaction
Oxprenolol/indomethacin	Antihypertensive effects reduced; mechanism may be due to inhibition of prostaglandin synthesis which plays some role in antihypertensive action of oxprenolol.
Propranolol/cimetidine	See Table 39.7(II).
Propranolol/insulin	See Table 39.6.

Table 39.8 *Continued*

III. Enzyme-inhibitors (captopril, methyldopa)

Combination	Interaction
Captopril/indomethacin	See Table 39.17(II).
Captopril/potassium-sparing diuretics or potassium supplements	Captopril decreases aldosterone production and serum potassium levels may rise especially in patients with renal failure. Potassium-sparing diuretics or potassium supplements should be used with caution since they may lead to a significant increase in serum potassium concentrations.
Methyldopa/antidepressant MAOI	Hypotensive effects diminished; interaction may cause hypertension and central excitation.
Tricyclics	Most reports suggest methyldopa exerts full hypotensive effects in presence of tricyclics; one report, however, of amitriptyline apparently reducing expected hypotensive effects of methyldopa.
Methyldopa/ephedrine	Ephedrine has both direct and indirect effect on adrenergic receptors; indirect effect is due to release of noradrenaline from storage sites in adrenergic nerve endings. Methyldopa reduces amount of noradrenaline available for release thus ephedrine is less effective in this combination.
Methyldopa/levodopa	See Table 39.13.
Methyldopa/lithium	See Table 39.14.
Methyldopa/mazindol	Mazindol potentiates the pressor action of catecholamines; drugs should not be given together.
Methyldopa/propranolol and oxprenolol	Severe hypertension on occasion; also reported with oxprenolol and phenylpropanolamine.
Methyldopa/sympathomimetic amine	Hypotensive action diminished by amphetamine and other sympathomimetic drugs.

IV. Rauwolfia alkaloids

Combination	Interaction
Anticonvulsant	Reserpine lowers the convulsive threshold; anticonvulsant dosage may have to be adjusted to control epilepsy.
Antidepressant MAOI	MAOIs cause accumulation of noradrenaline in storage sites within the adrenergic neurone. If reserpine is given in this situation, *theoretically* it might be expected to evoke exaggerated responses (hypertension and central excitation). However, alternatively if patient already on reserpine is given a MAOI, then *theoretically* there should be no interaction since noradrenaline stores would have been depleted.
Tricyclic	Reserpine is contraindicated in depression; also *theoretically* antagonism of hypotensive effect would be expected.
Digitalis glycosides	Reserpine and related compounds may enhance bradycardia produced by digitalis; mechanism probably release of catecholamines by reserpine.
Levodopa	Hypotensive effects and CNS-depressant effects of reserpine potentiated; levodopa effects may be diminished due to dopamine depletion in brain.

Table 39.10 Cardiac drugs

I. **Cardiac glycosides** (digoxin, digitoxin)

Combination	Interaction
Amiodarone	Raised plasma digoxin concentration toxicity.
Antibiotics (e.g. erythromycin, tetracycline)	Raised plasma digoxin concentration; toxicity.
Cancer chemotherapy (e.g. COPP, COP)	Reduced plasma digoxin concentration; cardiac decompensation.
Chloroquine and hydroxychloroquine	Raised plasma digoxin concentration.
Diuretics K^+-losing (e.g. thiazides, frusemide, ethacrynic acid, etc.)	Effects of digoxin increased; toxicity.
K^+-losing (e.g. spironolactone)	Increased digitoxin half-life; possible toxicity.
Ion-exchange resins (e.g. cholestyramine, colestipol)	Reduced gastrointestinal absorption of digoxin; reduced half-life of digitoxin.
Nifedipine	Raised plasma digoxin concentration.
Quinidine	Raised plasma digoxin concentration.
Rifampicin	Reduced plasma digoxin concentration.
Suxamethonium	See Table 39.19.
Verapamil	Raised plasma digoxin concentration; toxicity.

II. **Beta-blockers**

Combination	Interaction
Antidiabetic agents	See Table 39.6
Antidepressants	See Table 39.5(I).
Antithyroid drugs and Radioiodine	Plasma propranolol steady-state concentration increased significantly after correction of thyroid disorders.
Food/Alcohol	First-pass hepatic clearance of propranolol and labetalol increased by food; increased amount of drug enters circulation. Alcohol increases gastrointestinal absorption of propranolol.
Indomethacin	See Table 39.8.
Lignocaine (lidocaine)	Propranolol reduces metabolic clearance of lignocaine; toxicity.
Nifedipine	Negative inotropic effect of nifedipine worsened by beta-blockage; severe hypotension and cardiac failure.
Procainamide	Elimination half-life increased and plasma clearance of procainamide reduced; clinical significance uncertain.

(continued overleaf)

Table 39.10   *Continued*

Combination	Interaction
Rifampicin	Metoprolol metabolism increased; some loss of beta-blocking activity.
Smoking	Effects on heart negate beneficial effects of propranolol.

III.   Other agents (amiodarone, aprindine, disopyramide, lignocaine, nitrates, nitroglycerin, prenylamine, procainamide, quinidine) (see also Tables 39.10 (I and II))

Combination	Interaction
Amiodarone	Combination with aprindine raises serum concentration of aprindine to toxic levels.
Amiodarone	Combination with quinidine gives dangerous atypical ventricular tachycardia (torsades des pointes).
Calcium adipinate with calciferol	Atrial fibrillation when given with verapamil due to calcium reversing its negative inotropic effect.
Nitrates (sublingual)	Combination with prenylamine and lignocaine caused A–V block.
Plastic (PVC) containers, administration sets, filters, and infusion pump tubing	Decreased nitroglycerin concentrations in intravenous infusions due to sorption onto plastic surfaces (see also Table 39.24).
Propranolol	Reduced metabolic clearance of lignocaine; toxicity.
Rifampicin	Disopyramide metabolism increased; reduced anti-arrhythmic effect.
Rifampicin	Quinidine metabolism increased; reduced anti-arrhythmic effect.
Smoking	Plasma free-fraction of lignocaine reduced; reduced therapeutic effect.

Table 39.11   Diuretics

Combination	Interaction
Aminoglycoside antibiotics	See Table 39.2(I).
Captopril	See Table 39.8(III).
Fenfluramine	Hypotensive action of hydrochlorothiazide antagonized.
Flurbiprofen	Diuretic activity of frusemide reduced.
Indomethacin	Combination is nephrotoxic.
Skeletal muscle relaxants	See Table 39.19.

Table 39.12 Hypnotics and sedatives

I. Barbiturates

Combination	Interaction
Alcohol	See Table 39.1.
Amidopyrine	Acute toxic actions of amidopyrine inhibited by barbiturates but not chronic toxic effects; danger that immediate effects of overdosage are masked. Combination results in increased hypnosis.
Anticholinesterase agent (e.g. ambenonium, neostigmine, pyridostigmine, demecarium, dyflos, ecothiopate)	Increased muscle depolarization; toxic effects (muscle twitching) may be controlled by small dose of barbiturate.
Anticoagulants	See Table 39.3(II).
Antidepressant MAOI	MAOIs slow the metabolism of barbiturates and prolong their duration of action.
Tricyclics	Metabolism of tricyclics stimulated and efficacy reduced. Barbiturates enhance respiratory depression of toxic doses of tricyclics.
Antihistamines (H_1-blockers)	See Table 39.7(I).
Chloramphenicol	Serum chloramphenicol concentrations reduced during phenobarbitone treatment; reduced antibacterial activity.
Corticosteroids	Hydroxylation of endogenous cortisol and exogenous corticosteroids by liver microsomal enzymes is enhanced; possible reduced efficacy of corticosteroids.
Dextropropoxyphene (propoxyphene)	Serum concentrations of phenobarbitone increased by mean of 20 per cent after one week's treatment with dextropropoxyphene; danger of enhanced barbiturate activity.
Griseofulvin	Increased hepatic inactivation of griseofulvin and intestinal absorption of griseofulvin reduced; antifungal activity seriously impaired.
Metronidazole	One case of therapeutic failure in treatment of vaginal trichomoniasis due to increased metabolism of metronidazole. Increased dosage of metronidazole was effective.
Oestrogens or low-oestrogen oral contraceptives	Metabolism of oestrogens increased; danger of reduced oral contraceptive efficacy.
Pethidine (meperidine)	Enhanced sedation in patients previously tolerant to pethidine. Combination should be avoided since phenobarbitone enhances production of toxic metabolite norpethidine by increasing N-demethylation.
Phenylbutazone	Increased hypnosis; pretreatment with barbiturates increases liver microsomal activity and reduces plasma half-life of phenylbutazone.
Sodium valproate	Metabolism of phenobarbitone inhibited; danger of drowsiness, stupor, or coma with combination.
Tetracycline (doxycycline)	Plasma-half life of doxycycline reduced due to barbiturate-induced hepatic microsomal enzyme induction; reduced antibacterial effect.

Table 39.12 *Continued*

II. Benzodiazepines

Combination	Interaction
Aminophylline	Diazepam sedation antagonized.
Anticoagulants	Antagonism of anticoagulants by benzodiazepines is controversial but chlordiazepoxide, diazepam, and nitrazepam do not seem to interfere with warfarin anticoagulation, nor nitrazepam with phenprocoumon. However, in presence of digoxin and diuretics, anticoagulant control becomes unstable when benzodiazepines were given.
Antidepressant (tricyclic)	Enhanced sedation or atropine-like effects with concomitant use of chlordiazepoxide and a tricyclic antidepressant.
Cimetidine	Metabolism of chlordiazepoxide and diazepam inhibited by cimetidine; no interaction with lorazepam and oxazepam due to different metabolism but intravenous cimetidine variably increased absorption of oral diazepam and lorazepam; plasma concentrations of latter two agents were raised compared with non-cimetidine treated controls.
Doxapram	Reverses sedation after intravenous diazepam.
Levodopa	Marked reduction in effect of levodopa after chlordiazepoxide; mechanism not established.
Oral contraceptive	Total metabolic clearance of diazepam impaired; possible increased effects of diazepam.
Penicillamine	Single case report of exacerbation of intravenous diazepam-induced phlebitis after oral penicillamine. Effect of penicillamine on collagen synthesis may have prevented subclinical healing of phlebitis.
Phenytoin sodium	Chlordiazepoxide reported in rare clinical cases to inhibit phenytoin metabolism.
Skeletal muscle relaxants	Preliminary reports of diazepam increasing duration of neuromuscular block produced by gallamine, and reduces that produced by suxamethonium. Interaction thought to be at presynaptic site.
Thyroid hormones	Thyroxine and tri-iodothyronine displaced by diazepam from plasma protein-binding. May interfere with thyroid function tests.
Sodium valproate	Diazepam displaced from protein binding sites by sodium valproate; unbound fraction in serum increased twofold.

III. Chloral hydrate and derivatives

Combination	Interaction
Alcohol	See Table 39.1.
Anticoagulants	Two-phase interaction: firstly, displacement of warfarin and other coumarins from plasma protein-binding sites with potentiation of anticoagulant effect; secondly, metabolism of coumarins increased by induction of hepatic microsomal enzymes with reduced anticoagulant effect.
Phenytoin	Clearance of phenytoin significantly increased by nightly administration of dichloralphenazone due to induced microsomal enzyme activity; reduced efficacy of phenytoin.

IV. Other hypnotics and sedatives (chlormethiazole, ethchlorvynol, glutethimide, metaqualone, paraldehyde)

Combination	Interaction
Chlormethiazole/alcohol	See Table 39.1.

(continued overleaf)

Table 39.12   *Continued*

Combination	Interaction
Ethchlorvynol/alcohol or other CNS depressant	Effects of ethchlorvynol may be enhanced by alcohol or barbiturates; fatalities have been reported.
Ethchlorvynol/anticoagulant	Metabolism of warfarin and other coumarins increased due to enzyme induction; reduced effect of anticoagulant. Removal of ethchlorvynol from stable anticoagulated patient may cause haemorrhagic episodes.
Ethchlorvynol/tricyclic antidepressant	Transient delirium reported with concomitant treatments; mechanism unknown.
Glutethimide/alcohol	See Table 39.1.
Glutethimide/anticoagulant	Metabolism of warfarin and other coumarins increased due to enzyme induction; reduced effect of anticoagulant. Removal of glutethimide from stable anticoagulated patient may cause haemorrhagic episodes.
Glutethimide/tricyclic antidepressant	Increased anticholinergic effects (dry mouth, constipation, blurred vision, urinary retention, sweating, etc.) and possibly also ileus and onset of glaucoma in elderly patients.
Methaqualone/alcohol or other CNS depressant	Effects are enhanced by alcohol or other CNS depressant.
Methaqualone/anticoagulant	Instability of anticoagulant control due to induction of metabolizing enzymes.
Methaqualone/diphenhydramine	Effects of methaqualone enhanced when given concurrently with diphenhydramine; abuse of combination (*Mandrax*) occurs and may lead to dependence of the barbiturate type. Cases of fatal overdosage with combination reported.
Paraldehyde/alcohol	Fatalities have been reported following the administration of paraldehyde to drunks.
Paraldehyde/phenytoin	Hypnotic effects of paraldehyde may be grossly extended by previous and concomitant phenytoin treatment; single case report.
Paraldehyde/plastics and rubber	Paraldehyde has a solvent action upon rubber, polystyrene, styrene–acrylonitrile co-polymer, and should not be administered using a plastic syringe (see also Table 39.24).

Table 39.13   *Continued*

Combination	Interaction
Antidepressants (MAOIs)	Hypertension and hypertensive crisis.
Antihypertensive	Hypotensive effects of guanethidine, methyldopa, and reserpine potentiated.
Beta-blockers	Combined levodopa and propranolol reduces tremor in parkinsonism.
Chlordiazepoxide	Reduced effect of levodopa in treatment of parkinsonism.
Pyridoxine (vitamin B_6)	Reduced effect of levodopa in treatment of parkinsonism.

Table 39.14 Lithium salts*

Combination	Interaction
Carbamazepine	Enhanced CNS toxicity.
Diclofenac	Reduced renal clearance; increased steady-state plasma lithium concentration; danger of lithium toxicity.
Diuretics	Acetazolamide increases lithium excretion; reduced antipsychotic effects. Hydrochlorothiazide increases serum lithium concentration; danger of lithium toxicity.
Indomethacin	Reduced renal clearance; increased plasma lithium concentration; danger of lithium toxicity.
Mazindol	Enhanced lithium toxicity.
Methyldopa	Reduced renal excretion of lithium; danger of lithium toxicity.
Phenothiazine tranquillizers	Reduced response to therapeutic doses of chlorpromazine.
Phenytoin	Increased risk of lithium toxicity.
Sodium bicarbonate (in antacid formulations)	Renal excretion of lithium increased.
Theophylline (aminophylline)	Renal excretion of lithium increased.
Tetracycline	Increased plasma lithium concentration and toxicity due to nephrotoxic effect of long-acting tetracycline.

* See also review by Salem (1982).

Table 39.15 Local anaesthetics

Combination	Interaction
Antidepressants (MAOIs and tricyclics)	Vasopressor drugs in local anaesthetic injections can induce hypertensive episodes.
Cimetidine	See Table 39.7(II).
Prenylamine/nitrates	See Table 39.10(III).
Propranolol	See Table 39.10(III).
Smoking	See Table 39.10(III).

Table 39.16   Neuroleptics (major tranquillizers)

Combination	Interaction
Chlorpromazine/antacids	Antacids such as aluminium hydroxide decrease phenothiazine absorption by forming an adsorption complex with them; space dosage of individual drugs by at least two hours.
Chlorpromazine/anticoagulant	Reduced metabolism and increased activity of anticoagulant due to liver microsomal enzyme inhibition by chlorpromazine; monitoring of patient essential.
Chlorpromazine/corticosteroids	Reduced gut motility may enhance corticosteroid absorption; space out dosage of two drugs as far as possible.
Chlorpromazine/digoxin	Reduced gut motility may enhance absorption of digoxin; space out dosage of two drugs as far as possible.
Chlorpromazine/lithium carbonate	Peak plasma chlorpromazine concentrations reduced to about 60 per cent of levels with chlorpromazine alone in normal subjects. Area under chlorpromazine concentration/time curve reduced by 26.6 per cent when lithium was taken as well. Interaction may explain inadequate responses to chlorpromazine in combined treatment and sudden onset of chlorpromazine toxicity when lithium is withdrawn.
Chlorpromazine/pethidine (meperidine)	Potentially toxic combination which is not advised; debilitating lethargy and fall in systolic and diastolic blood pressures. Interaction may be due to alteration of pethidine's metabolism.
Chlorpromazine plus amitriptyline/prazocin	Acute agitation in a single case; episodes of sudden loss of consciousness in others.
Chlorpromazine/sympathomimetic amine	Chlorpromazine is an alpha-adrenergic blocker and might therefore be expected to block or reverse a variety of actions of adrenaline and antagonize adrenaline-induced hypertension.
Neuroleptic/alcohol and other CNS-depressant drugs	Alcohol potentiates the effect of neuroleptics; effect of narcotic analgesics and phenothiazines are additive; increased CNS depression with other combinations.
Neuroleptic/levodopa	See Table 39.13.
Neuroleptic/lithium salt	Lithium appears to potentiate the neurological complications produced by haloperidol or *vice versa*; lithium has been reported to be excreted more rapidly during concomitant treatment with chlorpromazine (see also Table 39.14).
Neuroleptic/other drug affecting ADH secretion	A drug-induced syndrome of inappropriate secretion of antidiuretic hormone (ADH) has been described secondarily to cyclophosphamide or vincristine therapy and to chlorpropamide. The syndrome has also been described with a number of other drugs including: amitriptyline, carbamazepine, clofibrate, diuretics, fluphenazine, haloperidol, thiothixene, thioridazine, and vinblastine.

Combination of these may increase the likelihood of the syndrome occurring with resulting water retention and natriuresis. |
Neuroleptic/tricyclic antidepressant	Neuroleptic drugs inhibit the metabolism of tricyclic antidepressants. Combined treatments are often prescribed and they do not present interaction problems but if prescribed together beyond these doses they may evoke troublesome CNS depression of anticholinergic effects.
Perphenazine/disulfiram	One case of disulfiram inactivating oral perphenazine dosage due to its activation of liver enzymes; parenteral dosage of perphenazine recommended in such cases to avoid 'first pass' effect in liver.
Phenothiazine/antidiabetic agent	Phenothiazines have been reported to cause hyperglycaemia; mechanism is not known. They should be used with care in the presence of stabilized diabetic treatment.

(continued overleaf)

Table 39.16 *Continued*

Combination	Interaction
Phenothiazine/guanethidine and haloperidol/guanethidine and thiothixene/guanethidine	Conflicting reports as to the nature of this interaction; some report antagonism of antihypertensive action; others report potentiation of the effects of guanethidine and other hypotensive drugs. Patients should be monitored for inadequate or excessive response to antihypertensive treatments.
Phenothiazine/MAOI	Combination may lead to hypertension and increased extrapyramidal reactions. Mechanism uncertain but it is possible that MAOIs inhibit the metabolism of phenothiazines.
Phenothiazine/oral contraceptives	Oestrogen-containing oral contraceptives may potentiate phenothiazine-stimulated prolactin secretion resulting in mammary hypertrophy and galactorrhoea.
Phenothiazine/phenytoin sodium	In rare instances chlorpromazine and prochlorperazine impair the metabolism of phenytoin and increase the risk of phenytoin intoxication. Other phenothiazines might be expected to interact in the same way.
Phenothiazines/plastic intravenous delivery systems	Loss of some phenothiazines (including chlorpromazine, thioridazine, and trifluoperazine) has been attributed to sorption processes during simulated infusions through plastic infusion sets. Loss can be reduced by using short lengths of small-diameter tubing made of high-density polyethylene (see review: D'Arcy 1983*b*).
Phenothiazine/skeletal muscle relaxant	There is some evidence that phenothiazines lower serum and erythrocyte cholinesterase levels. Methotrimeprazine has been reported to prolong tubocurarine-induced muscle relaxation; possibility of interaction with suxamethonium has been suggested. Promazine has been shown to produce prolonged apnoea with suxamethonium.
Thioridazine/phenytoin sodium and other phenothiazines/phenytoin	Phenytoin intoxication in two cases when drugs used together. Drugs compete for cytochrome P-450 hydroxylation and metabolism of phenytoin is inhibited. Similar interactions have been reported with chlorpromazine and prochlorperazine (see earlier entry in this section).

Table 39.17 Non-steroidal anti-inflammatory drugs, corticosteroids, gold salts, and other antirheumatic drugs

I. Aspirin and salicylates

Combination	Interaction
Alcohol	Aspirin damages gastric mucosa and can cause bleeding; alcohol can increase this blood loss.
Anticoagulants	Aspirin displaces coumarins from plasma protein-binding sites and potentiates their anticoagulant action. Aspirin also tends to reduce plasma prothrombin if taken in large doses; it also reduces platelet adhesiveness, both of which potentiate anticoagulation.
Corticosteroids	Corticosteroids decrease the blood salicylate concentration by increasing glomerular filtration rate; possibility of salicylate toxicity. Both drugs are ulcerogenic.
Fenoprofen calcium	Concomitant dosage reduces plasma concentrations of fenoprofen due to impaired intestinal absorption; reduced fenoprofen effect.
Insulin and oral hypoglycaemic agents	Aspirin and salicylates displace tolbutamide and chlorpropamide *in vitro* from plasma protein binding sites thus increasing unbound active sulphonylurea; possibility of enhanced hypoglycaemia.
	Insulin requirements can be reduced by aspirin when used in high dosage.
Methotrexate	Salicylates displace methotrexate from its plasma binding; renal tubular excretion of methotrexate reduced by 35 per cent; possible methotrexate toxicity.

(continued overleaf)

Table 39.17   *Continued*

Combination	Interaction
Phenytoin sodium	Aspirin displaces phenytoin from binding sites but acute interaction unlikely because increased free phenytoin level compensated by increased availability of drug for metabolic pathways and more rapid clearance.
Probenecid	Uricosuric action of probenecid antagonized by salicylates; concomitant administration should be avoided.
Sodium valproate	Steady-state serum free fractions of valproate increased by antipyretic doses of aspirin; danger of cumulation of valproate to toxic levels.
Sulphinpyrazone	Uricosuric action of sulphinpyrazone antagonized by salicylates; concomitant administration should be avoided.

II.   Other NSAIDs (azapropazone, diclofenac, diflunisal, fenclofenac, flurbiprofen, indomethacin, naproxen, sulindac)

Combination	Interaction
Azapropazone/phenytoin	Drugs compete for hepatic hydroxylation systems; increased plasma phenytoin concentrations; danger of phenytoin toxicity.
Axapropazone/tolbutamide	Plasma clearance of tolbutamide inhibited due to azapropazone inhibited hepatic metabolism; danger of severe hypoglycaemia.
Diclofenac/lithium salts	Diclofenac inhibits prostaglandin synthesis; reduced renal clearance of lithium; increased steady-state plasma lithium concentration; danger of lithium toxicity.
Diflunisal/antacids	Aluminium hydroxide gel decreased peak plasma concentration of diflunisal in fasting state by 46 per cent, in contrast magnesium hydroxide increased peak levels by 130 per cent at 5 hours and 64 per cent at 1 hour in fasting state.
Diflunisal/warfarin	Total warfarin serum concentrations decreased; reduced anticoagulant effect.
Fenclofenac/chlorpropamide and metformin	Hypoglycaemic action of chlorpropamide–metformin combination potentiated by fenclofenac probably due to displacement from protein binding sites; danger of severe hypoglycaemia extending to coma.
Flurbiprofen/frusemide	Diuretic effect antagonized.
Flurbiprofen/nicoumalone	Anticoagulant action increased.
Indomethacin/appetite suppressant	See Table 39.20, also footnote on p. 949.
Indomethacin/bumetanide	Reduction of natriuretic effect of bumetanide.
Indomethacin/captopril	Blood pressure lowering and renin-stimulating effect of captopril reduced.
Indomethacin or naproxen/ corticosteroids	Increased free prednisolone serum levels after concurrent administration of indomethacin or naproxen with corticosteroids; steroid-sparing effect.
Indomethacin/lithium salts	Indomethacin effect on prostaglandin-dependent mechanism in distal renal tubules caused reduced renal clearance and raised plasma lithium concentrations; danger of lithium toxicity.
Indomethacin/oxprenolol	Antihypertensive effect of oxprenolol reduced possibly due to inhibition of prostaglandin synthesis; loss of control of hypertension.
Indomethacin/triamterene	Nephrotoxicity of triamterene enhanced; reversible acute renal failure.
Indomethacin/warfarin	Possible displacement of warfarin from plasma binding sites; danger of excess anticoagulation.
Sulindac/warfarin	Anticoagulant effect potentiated (conflicting evidence).

Table 39.17 *Continued*

III. Corticosteroids

Combination	Interaction
Antacids	Aluminium and magnesium hydroxide-containing antacids reduce relative bioavailability of prednisolone by 26–43 per cent; reduced efficacy of steroid.
Indomethacin or naproxen	See Table 39.17(II).
Salicylates	See Table 39.17(I).
Troleandomycin	Immediate and continued inhibition of methylprednisolone metabolism; steroid-sparing effect.

IV. Gold salts and other antirheumatic drugs

Combination	Interaction
Penicillamine/gold salt	Conflicting reports on effects of gold therapy (sodium aurothiomalate) on development of penicillamine toxicity and vice versa. But patients who had proteinuria or bone-marrow depression with gold treatment had increased likelihood of similar side-effects with penicillamine.

Table 39.18 Oral contraceptives*

I. Drugs implicated (or suspected of implication) in oral contraceptive failure

Antibiotics (ampicillin[1], griseofulvin[2], rifampicin, tetracycline).

Anticonvulsants (phenobarbitone, phenytoin, primidone).

Cholesterol-lowering agents (clofibrate).

Non-steroidal anti-inflammatory agents (phenylbutazone).

Hypnotics and sedatives (barbiturates, chloral hydrate and derivatives, ethchlorvynol, glutethimide, meprobamate).

Other drugs (largely circumstantial evidence) (amidopyrine, aspirin, chloramphenicol, chlorpromazine, chlordiazepoxide, diazepam, dihydroergotamine, ethosuximide, isoniazid, neomycin, nitrofurantoin, phenazone, phenacetin, phenoxymethylpenicillin, promethazine, oxyphenbutazone, sulphamethoxypyridazine).

[1] Evidence against interaction given by Back *et al*. (1982).
[2] Recent evidence by Van Dijke and Weber (1984).

II. Drug activities modified by oral contraceptives

Combination	Interaction
Aminocaproic acid	Oestrogen component augments blood levels of clotting factors, VII, VIII, IX, and X; possible hypercoagulable state.
Anticoagulants	Oestrogen component increases plasma concentration of clotting factors; reduced anticoagulant efficacy.
Antidiabetic agents	Impairment of glucose tolerance; development of overt diabetes mellitus; increased requirement for insulin and oral hypoglycaemic drugs.
Antihypertensives	Na^+ and fluid retention reduces efficacy of guanethidine, cyclopenthiazide, and methyldopa.

(continued overleaf)

Table 39.18 *Continued*

Combination	Interaction
Folic acid and vitamin B$_{12}$ (cyanocobalamin)	Impaired folate metabolism; degree of folate depletion.
Pethidine (meperidine)	Pethidine metabolism inhibited; possible increased analgesia and CNS depression.
Phenothiazine tranquillizers	Phenothiazine-stimulated prolactin secretion potentiated; mammary hypertrophy and galactorrhoea.
Troleandomycin	Pruritus and jaundice following combined use; both components cause jaundice.

III.   Drugs potentiating side-effects of oral contraceptives

Combination	Interaction
Drugs causing liver enzyme inhibition (e.g allopurinol, aspirin, chloramphenicol, disulfiram, hydrocortisone (cortisol), isoniazid, methandienone, methylphenidate, MAOI-antidepressants, PAS, phenothiazines, phenyramidol, prednisone, sulphaphenazole, triparanol, etc.)	*Theoretically* concomitant administration could potentiate the actions of oral contraceptives by delaying hepatic metabolism of oestrogenic and progestogen components. May present as increased side-effects (e.g. fluid retention, diabetogenic action, hypertension, increased risk of thromboembolic disorders, etc.).

* See review by D'Arcy and Griffin (1976).

Table 39.19   Skeletal muscle relaxants

Combination	Interaction
Antibiotics, e.g.	
Aminoglycosides	See Table 39.2(I).
Amphotericin	See Table 39.2(VI).
Polymyxins	See Table 39.2 (VII).
Anticholinesterases (ecothiopate)	Prolonged apnoea and death after suxamethonium.
Antidepressants (tricyclics)	Antispastic effect of baclofen potentiated by nortriptyline or imipramine.
Cancer chemotherapy	Cyclophosphamide and thiotepa both lower serum pseudocholinesterase level and may induce apnoea with depolarizing drugs (e.g. suxamethonium).
Cardiac glycosides	Conduction and ventricular irritability effects of digitalis potentiated by suxamethonium.
Diuretics	Thiazides increase responsiveness to tubocurarine and gallamine due to K$^+$ depletion. Frusemide may potentiate tubocurarine and other non-depolarizers due to K$^+$ depletion.

Table 39.20 Sympathomimetic amines

Combination	Interaction
Antidepressants MAOIs	See Table 39.5(I).
Tricyclics	See Table 39.5(II).
Antihypertensive agents	See Table 39.8.
Indomethacin	Severe systemic hypertension with appetite suppressants containing phenylpropanolamine (see footnote to this section*). Mechanism due to indomethacin inhibition of prostaglandin synthesis evoking enhanced sympathomimetic effects of phenylpropanolamine.
Insulin	Ritodrine given by intravenous infusion to an insulin-dependent diabetic to delay contractions in preterm labour caused hyperglycaemia and ketoacidosis; baby stillborn.
Mazindol	Pressor actions of catecholamines potentiated; danger of cough or cold remedies or local anaesthetics (all may contain sympathomimetic amines) during or for one month after mazindol treatment.
Phenothiazine tranquillizer	Chlorpromazine blocks or reverses a variety of actions of adrenaline and antagonizes adrenaline-induced hypertension by its alpha-adrenergic blocking action; other phenothiazines might be expected to produce the same type of interaction.

* The United Kingdom Government has recently proposed that all medicinal products containing phenylpropanolamine will be available only *on prescription* except where products are indicated for the relief of cough, colds, and hay fever with a recommended maximum daily dose of not more than 75 mg. The maximum permitted strength of nasal sprays and nasal drops will remain at 2.0 per cent. Phenylpropanolamine-containing slimming aids are still available, however, with little restriction in America and Australia (see review, D'Arcy 1984*b*).

Table 39.21 Theophylline (aminophylline)*

Combination	Interaction
Allopurinol	Theophylline metabolism inhibited; danger of accumulation and toxicity.
Cimetidine	Theophylline metabolism inhibited; toxicity.
Diazepam	See Table 39.12(II).
Erythromycin	Effects controversial but up to 50 per cent total body clearance of bronchodilator reported; danger of toxicity.
Frusemide (furosemide)	Reduced volume of distribution of theophylline and increased steady-state plasma concentration; danger of toxicity.
Hydrocortisone (cortisol)	Rapid increase in plasma theophylline concentration; toxicity.
Influenza vaccine	See Table 39.23.
Ketamine	Extensor-type seizures after induction of anaesthesia with ketamine.
Phenytoin	See Table 39.4(IV).
Probenecid	Clearance of dyphylline (dihydroxypropyl theophylline) slowed; increased half-life.
Smoking	Theophylline metabolism induced, clearance increased and effect reduced.
Thiabendazole	Increased plasma concentration of theophylline; danger of toxicity.
Troleandomycin	Decreased theophylline clearance; serum concentrations doubled; danger of toxicity.

* See also review by McElnay *et al.* (1982).

Table 39.22   Tuberculostatic agents

I.   Aminoglycoside antibiotics

Combination	Interaction
Diuretics (e.g. ethacrynic acid, frusemide (furosemide))	See Table 39.2(I).
Skeletal muscle relaxants	See Table 39.2(I).

II.   Rifampicin

Combination	Interaction
Corticosteroids	Metabolism of cortisol, prednisolone (possibly other corticosteroids) increased; reduced activity.
Digoxin	See Table 39.10(I).
Disopyramide	See Table 39.10(III).
Metoprolol	See Table 39.10(II).
Oral contraceptives	See Table 39.18(I).
Quinidine	See Table 39.10(III).
Vitamin D metabolism	Consistent fall in plasma calcifediol of 70 per cent during rifampicin treatment; danger of metabolic bone disease when nutrition is suboptimal.

III.   Other agents (isoniazid)

Combination	Interaction
Carbamazepine	See Table 39.4(V).
Food (cheese, fish, red wine)	Monoamine and diamine oxidase inhibitory activity of isoniazid potentiates effects of tyramine or histamine present in these foods. Danger of hypertensive episodes and symptoms of histamine intoxication.

Table 39.23   Vaccines (influenza vaccines)*

Combination	Interaction
Phenytoin sodium	Epileptic patients who are immunized against influenza or who develop a febrile illness may experience an increase in seizure frequency. Serum phenytoin concentrations may be reduced below effective levels.
Theophylline and aminophylline	Theophylline toxicity has been reported in stabilized asthmatic children during an influenza B outbreak. Decreased elimination of theophylline has also been reported after influenza immunization.
Warfarin	One case report of patient stabilized on warfarin for 12 years who suffered massive upper gastrointestinal tract haemorrhage and had prothrombin time of 48 seconds 10 days after influenza immunization. Mechanism thought to be decreased inactivation of warfarin due to depression of hepatic metabolism.

* See also review article (D'Arcy 1984a)

Table 39.24 Medical plastics*

Combination	Interaction
Chlormethiazole/ plastic materials	Chlormethiazole lost from aqueous solutions when stored in plastic infusion bags and run through plastic infusion sets. No problems with glass containers.
Diazepam/plastic materials	Diazepam is substantially 'sorbed' by the plastics in flexible containers, volume-control-set chambers, and tubing of i.v. administration sets. No problems with glass containers.
Glyceryl trinitrate plastic materials	Concentrations of glyceryl trinitrate (nitroglycerin) can be reduced by up to 80 per cent due to 'sorption' onto plastic surfaces of i.v. bags and admixture sets made from polyvinyl chloride (PVC). No problems with glass containers.
Insulin/plastic materials	Insulin in common with many polypeptides may be strongly 'sorbed' onto glassware, onto polyethylene, and onto PVC.
Isosorbide nitrate/ plastic materials	Appreciable loss (up to 80 per cent) of drug from aqueous solution when stored in plastic infusion bags or in the isolated burettes of giving sets. No problems with glass containers.
Paraldehyde/plastic materials	Paraldehyde has a solvent action upon polystyrene, styrene–acrylonitrile co-polymer, and upon rubber.
Various drugs/plastic contact lenses	Some drugs can enter into surface interactions with plastic contact lenses. For example, an orange colouration of contact lenses has been associated with the use of rifampicin, and adrenochrome pigmentation of hydrophilic lenses has been reported in three elderly patients within two months of starting topical treatment with adrenaline hydrochloride or epinephryl borate eye drops. Dense brown pigmentation was frightening to these patients and the lenses had to be discarded. There are also reports of deliberate adrenochrome pigmentation of lenses for cosmetic purposes.
Vitamin A/plastic materials	Reports suggest that vitamin A acetate is strongly bound to plastic bags (up to 80 per cent), especially when dextrose or sodium chloride injections are the diluent. Vitamin A palmitate does not bind to PVC but it is unstable to light.

* See also review by D'Arcy (1983b).

RECOMMENDED FURTHER READING

D'ARCY, P.F. AND GRIFFIN, J.P. (1984). *A manual of adverse drug interactions*, 3rd ed. John Wright & Sons, Bristol.

NOACK, E., LEDWOCH, W., AND SCHREY, A. (Eds.) (1983). *Arzneimittel-interaktionen* [Adverse drug interactions]. Wolf und Sohn, Munich [German text.]

SODA, T. (Ed.) (1980). *Drug-induced sufferings. Medical, pharmaceutical and legal aspects*, International Congress Series 513. Excerpta Medica, Amsterdam-Oxford-Princeton.

WINICK, M. (Ed) (1983). *Nutrition and drugs*. Series on Current concepts in nutrition, Vol 12. John Wiley & Sons, New York, Chichester, Brisbane, Toronto, Singapore.

RECOMMENDED FURTHER READING

D'ARCY, P.F. AND GRIFFIN, J.P. (1984). *A manual of adverse drug interactions*, 3rd ed. John Wright & Sons, Bristol.

NOACK, E., LEDWOCH, W., AND SCHREY, A. (Eds.) (1983). *Arzneimittel-interaktionen* [Adverse drug interactions]. Wolf und Sohn, Munich [German text.]

SODA, T. (Ed.) (1980). *Drug-induced sufferings. Medical, pharmaceutical and legal aspects*, International Congress Series 513. Excerpta Medica, Amsterdam-Oxford-Princeton.

WINICK, M. (Ed) (1983). *Nutrition and drugs*. Series on Currents concepts in nutrition, Vol 12. John Wiley & Sons, New York, Chichester, Brisbane, Toronto, Singapore.

REFERENCES

ABRAMSON, E.A., ARKY, R.A., AND WOEBER, K.A. (1966). Effects of propranolol on the hormonal and metabolic responses to insulin-induced hypoglycaemia. *Lancet* **i**, 1386-9.

ARONSON, J.K. AND CARVER, J.G. (1981). Interaction of digoxin with quinine. *Lancet* **i**, 1418.

AZARNOFF, D.L. AND HURWITZ, A. (1970). Drug interactions. *Pharmacol. Physcns* **4**, 1-7.

BACK, D.J., BATES, M., BOWDEN, A., BRECKENRIDGE, A.M., HALL, M.J., JONES, H., MACIVER, M., ORME, M., PERUCCA, E., RICHENS, A., ROWE, P.H., AND SMITH, E. (1980). The interaction of pheno-barbital and other anticonvulsants with oral contraceptive steroid therapy. *Contraception* **2**, 495-503.

——, BRECKENRIDGE, A.M., MACIVER, M., ORME, M., ROWE, P.H., STAIGER, CH., THOMAS, E., AND TJIA, J. (1982). The effects of ampicillin on oral contraceptive steroids in women. *Br. J. clin. Pharmacol.* **14**, 43–8.

BACON, J.F. AND SHENFIELD, G.M. (1980). Pregnancy attributable to interaction between tetracycline and oral contraceptives. *Br. med. J.* **i**, 293.

BALLEK, R.E., REIDENBERG, M.M., AND ORR, L. (1973). Inhibition of diphenylhydantoin metabolism by chloramphenicol. *Lancet* **i**, 150.

BAUMAN, J.H. AND KIMELBLATT, B.J. (1982). Cimetidine as an inhibitor of drug metabolism. *Drug Intel. clin. Pharm.* **16**, 380–6.

BETTS, T.A., KALRA, P.L., COOPER, R., AND JEAVONS, P. (1968). Epileptic fits as a probable side effect of amitriptyline. Report of seven cases. *Lancet* **i**, 390–2.

BIGGER, J.T., JR. (1979). The quinidine–digoxin interaction. What do we know about it? *New Engl. J. Med.* **301**, 779–81.

—— AND LEAHEY, E.B., JR. (1982). Quinidine and digoxin: An important interaction. *Drugs* **24**, 229–39.

CHRISTENSEN, L.K. AND SKOVSTED, L. (1969). Inhibition of drug metabolism by chloramphenicol. *Lancet* **ii**, 1397–9.

COULAM, C.B. AND ANNEGERS, J.F. (1979). Do anticonvulsants reduce the efficacy of oral contraceptives? *Epilepsia* **20**, 519–25.

CULBERSON, G. (1979). Paper presented at the October 1979 meeting of the American Neurological Association, St. Louis.

DALLOS, V. AND HEATHFIELD, K. (1969). Iatrogenic epilepsy due to anticonvulsant drugs. *Br. med. J.* **4**, 80–2.

DANESHMEND, T.K., ENE, M.D., PARKER, G., AND ROBERTS, C.J.C. (1984). Effects of chronic cimetidine on apparent liver blood flow and hepatic microsomal enzyme activity in man. *Gut* **25**, 125–8.

D'ARCY, P.F. (1982a). Interactions with H_2-receptor antagonists. *Drug Intel. clin. Pharm.* **16**, 669–70.

—— (1982b). Nitrofurantoin does not require an alcohol warning label. *Pharm. Int.* **3**, 120–1.

—— (1983a). Early reports on drug interactions. *Drug Intel. clin. Pharm.* **17**, 105–9.

—— (1983b). Drug interactions with medical plastics. *Drug Intel. clin. Pharm.* **17**, 726–31.

—— (1983c). Handling anticancer drugs. *Drug Intel. clin. Pharm.* **17**, 532–8.

—— (1984a). Vaccine-drug interactions. *Drug Intel. clin. Pharm.* **18**, 697–700.

—— (1984b). Phenylpropanolamine to go on prescription in the UK. *Pharm. Int.* **5**, 86–7.

—— AND GRIFFIN, J.P. (1976). Drug interactions with oral contraceptives. *J. Fam. Plann. Doctors* **2**, 48–51.

—— AND —— (1980). *Drug-induced emergencies*, pp. 44–51. John Wright & Sons, Bristol.

—— AND —— (1984). *A manual of adverse drug interactions*, 3rd edn. John Wright & Sons, Bristol.

—— AND MERKUS, F.W.H.M. (1981). Alcohol and drug interactions. *Pharm. Int.* **2**, 273–80.

DIVITIIS, O. DE, GIORDANO, F., GALLO, B., AND JACONO, A. (1968). Tolbutamide and propranolol. *Lancet* **i**, 749.

DOERING, W. (1979). Quinidine–digoxin interaction: Pharmacokinetics, underlying mechanism and clinical implications. *New Engl. J. Med.* **301**, 400–5.

DOSSETOR, J. (1975). Drug interactions with oral contraceptives. *Br. med. J.* **4**, 467–8.

FARQUHAR, D., LOO, T.L., GUTTERMAN, J.U., HERSH, E.M., AND LUNA, M.A. (1976). Inhibition of drug metabolizing enzymes in the rat after Bacillus Calmette–Guérin treatment. *Biochem. Pharmacol.* **25**, 1529–35.

FENSTER, P.E., MARCUS, F.I., CONRAD, K., GRAVES, P.E., HAGER, W.D., AND POWELL, J.R. (1981). Noninteraction of digitoxin and quinidine. *New Engl. J. Med.* **304**, 118.

FICHTL, B. AND DOERING, W. (1981). The digoxin–quinidine interaction: serum protein binding unlikely to be involved. *Br. J. clin. Pharmacol.* **11**, 94–6.

FLIND, A.C. (1978). Cimetidine and oral anticoagulants. *Br. med. J.* **2**, 1367.

FRIEDMAN, C.I., HUNEKE, A.L., KIM, M.H., AND POWELL, J. (1980). The effect of ampicillin on oral contraceptive effectiveness. *Obstet. Gynecol.* **55**, 33–6.

FRÖLICH, J.C., LEFTWICH, R., RAGHEB, M., OATES, J.A., REIMANN, I., AND BUCHANAN, D. (1979). Indomethacin increases plasma lithium. *Br. med. J.* **1**, 1115–16.

GARTY, M., SOOD, P., AND ROLLINS, D.E. (1981). Digitoxin elimination reduced during quinidine therapy. *Ann. intern. Med.* **94**, 35–7.

GORODISCHER, R., KRASNER, J., MCDEVITT, J.J., NOLAN, J.P., AND YAFFE, S.J. (1976). Hepatic microsomal drug metabolism after administration of endotoxin in rats. *Biochem. Pharmacol.* **25**, 351–3.

GRIFFIN, J.P. AND D'ARCY, P.F. (1974). Drug interactions: 4. With hypoglycaemic agents. *Prescribers' J.* **14**, 103–6.

—— AND —— (1984). *A manual of adverse drug interactions*, 3rd edn. John Wright & Sons, Bristol.

HAGER, W.D., FENSTER, P., MAYERSOHN, M., PERRIER, D., GRAVES, P., MARCUS, F.I., AND GOLDMAN, S. (1979). Digoxin–quinidine interaction. Pharmacokinetic evaluation. *New Engl. J. Med.* **300**, 1238–41.

HANSTEN, P.D. (1980). Beta-blocking agents and antidiabetic drugs. *Drug Intel. clin. Pharm.* **14**, 46–50.

HARPER, J.M., YOST, R.L., STEWART, R.B., AND CIEZKOWSKI, J. (1979). Phenytoin–chloramphenicol interaction; retrospective study. *Drug Intel. clin. Pharm.* **13**, 425–9.

HENDELES, L., BIGHLEY, L., RICHARDSON, R.H., HEPLER, C.D., AND CARMICHAEL, J. (1977). Frequent toxicity from IV aminophylline infusions in critically ill patients. *Drug Intel. clin. Pharm.* **11**, 12–18.

—— AND WEINBERGER, M. (1980). Editorial: Poisoning patients with intravenous theophylline. *Am. J. hosp. Pharm.* **37**, 49–50.

HOOYMANS, P.M. AND MERKUS, F.W.H.M. (1979). The mechanism of the interaction between digoxin and quinidine. *Pharmaceutisch Weekblad (Sci. Edit.)* **1**, 36–40.

HOUGHTON, A.W.J. (1971). Convulsions precipitated by amitriptyline. *Lancet* **i**, 138.

HUGHEY, M.C. JR, YOST, R.L., ROBINSON, J.D., AND HARMAN, E.M. (1982). Intravenous theophylline regimen. *Drug Intel. clin. Pharm.* **16**, 301–5.

JAFFE, J.M., HAMILTION, B., AND JEFFERS, S. (1976). Nitrofurantoin–antacid interaction. *Drug Intel. clin. Pharm.* **10**, 419–20.

JANZ, D. AND SCHMIDT, D. (1974). Anti-epileptic drugs and failure of oral contraceptives. *Lancet* **i**, 1113.

KØLENDORF, K., BONNEVIE-NIELSEN, V., AND BROCH-MØLLER, B. (1982). A trial of metoprolol in hypertensive insulin-dependent diabetic patients. *Acta med. scand.* **211**, 175–8.

KORDASH, T.R., VAN DELLEN, R.G., AND MCCALL, J.T. (1977). Theophylline concentrations in asthmatic patients after administration of aminophylline. *J. Am. med. Ass.* **238**, 139–41.

KOTLER, M.N., BERMAN, L., AND RUBENSTEIN, H. (1966). Hypoglycaemia precipitated by propranolol. *Lancet* **ii**, 1389–90.

KOUP, J.R., GIBALDI, M., MCNAMARA, P., HILLIGOSS, D.M., COLBURN, W.A., AND BRUCK, E. (1978). Interaction of choramphenicol with phenytoin and phenobarbital. Case report. *Clin. Pharmacol. Ther.* **24**, 571–5.

LEAHEY, E.B.JR, REIFFEL, J.A., HEISSENBUTTEL, R.H., DRUSIN, R.E., LOVEJOY, W.P., AND BIGGER, J.T.JR. (1979). Enhanced cardiac effect of digoxin during quinidine treatment. *Arch. intern. Med.* **139**, 519–21.

McARTHUR, J. (1967). Notes and comments: Oral contraceptives and epilepsy. *Br. med. J.* **3**, 162.

McELNAY, J.C., SIMITH, G.D., AND HELLING D.K. (1982). Guide to interactions involving theophylline kinetics. *Drug Intel. clin. Pharm.* **16**, 533–42.

MELVIN, K.R. AND KATES, R.E. (1981). Noninteraction of digitoxin and quinidine. *New Engl. J. Med.* **304**, 118.

MOENCH, T.R. (1980). The quinidine–digoxin interaction. *New Engl. J. Med.* **302**, 864.

MÜHLHAUSER, I., CUPPERS, H.J., AND BERGER, M. (1984). Smoking and insulin absorption from subcutaneous tissue. *Br. med. J.* **288**, 1875–6.

NAGGAR, V.F. AND KHALIL, S.A. (1979). Effect of magnesium trisilicate on nitrofurantoin absorption. *Clin. Pharmacol. Ther.* **25**, 857–63.

NELSON, J.C. AND JATLOW, P.I. (1980). Neuroleptic effect on desipramine steady-state plasma concentrations. *Am. J. Psychiat.* **137**, 1232–4.

NOCKE-FINCK, L., BREUER, H., AND REIMERS, D. (1973). Effects of rifampicin on the menstrual cycle and on estrogen excretion in patients taking oral contraceptives. *Dtsch. Med. Wochenschr.* **98**, 1521–3.

OCHS, H.R., PABST, J., GREENBLATT, D.J., AND DENGLER, H.J. (1980). Noninteraction of digitoxin and quinidine. *New Engl. J. Med.* **303**, 672–4.

ORAL CONTRACEPTIVES (1974). *Oral contraceptives*, Bulletin Vol.5, No.3. Health Protection Branch, Department of Health and Welfare, Canada.

REIMANN, I.W. AND FRÖLICH, J.C. (1981). Effects of diclofenac on lithium kinetics. *Clin. Pharmacol. Ther.* **30**, 348–52.

RENTON, K.W. (1979). The deleterious effects of *Bordetella pertussis* vaccine and poly (rI:rC) on the metabolism and disposition of phenytoin. *J. Pharmacol. exp. Ther.* **208**, 267–70.

ROLLINS, D.E. AND GARTY, M. (1980). Noninteraction of digitoxin and quinidine. *New Engl. J. Med.* **304**, 118.

ROWLES, B. AND WORTHEN, D.B. (1982). Clinical drug information: A case of misinformation. *New Engl. J. Med.* **306**, 113–14.

SALEM, R.B. (1982). A pharmacist's guide to monitoring lithium drug–drug interactions. *Drug Intel. clin. Pharm.* **16**, 745–7.

SCHENCK-GUSTAFSSON, K. (1981). Digitalis and quinidine. *Lancet* **i**, 105.

—— AND DAHLQVIST, R. (1981). Pharmacokinetics of digoxin in patients subjected to the quinidine–digoxin interaction. *Br. J. clin. Parmacol.* **11**, 181–6.

——, JOGESTRAND, T., NORLANDER, R., AND DAHLQVIST, R. (1981). Effects of quinidine on digoxin concentration in skeletal muscle and serum in patients with atrial fibrillation. Evidence for reduced binding of digoxin in muscle. *New Engl. J. Med.* **305**, 209–11.

SERLIN, M.J., SIBEON, R.G., MOSSMAN, S., BRECKENRIDGE, A.M., WILLIAMS, J.R.B., ATWOOD, J.L., AND WILLOUGHBY, J.M.T. (1979). Cimetidine interaction with oral anticoagulants in man. *Lancet* **ii**, 317–19.

SILVER, B.A. AND BELL, W.R. (1979). Cimetidine potentiation of the hypoprothombinemic effect of warfarin. *Ann. Intern. Med.* **90**, 348–9.

SIRIS, S.G., COOPER, T.B., RIFKIN, A.E., BRENNER, R., AND LIEBERMAN, J.A. (1982). Plasma imipramine concentrations in patients receiving concomitant fluphenazine decanoate. *Am. J. Psychiat.* **139**, 104–6.

SKOLNICK, J.L., STOLER, B.S., KATZ, D.B., AND ANDERSON, W.H. (1976). Rifampicin, oral contraceptives and pregnancy. *J. Am. med. Ass.* **236**, 1382.

SMALL, R.E. AND MARSHALL, J.H. (1979). Quinidine–digoxin interaction. A discussion of a case report and review of the literature. *Drug Intel.clin. Pharm.* **13**, 286–9.

SOMOGYI, A. AND GUGLER, R. (1983). Clinical pharmacokinetics of cimetidine. *Clin. Pharmacokinetics* **8**, 463–95.

SONNE, J., LÜHDORF, K., LARSEN, N.E., AND ANDREASEN, P.B. (1983). Lack of interaction between cimetidine and carbamazepine. *Acta neurol. scand.* **68**, 253–6.

SONNENFELD, G., HARNED, C.L., THANIYAVARN, S., HUFF, T., MANDEL, A.D., AND NERLAND, D.E. (1980). Type II interferon inducing and passive-transfer depress the murine cytochrome P-450 drug metabolism system. *Antimicrob. Agents. Chemother.* **17**, 969–72.

SORKIN, E.M. AND DARVEY, D.L. (1983). Review of cimetidine drug interactions. *Drug Intel. clin. Pharm.* **17**, 110–20.

SOYKA, L.E., HUNT, W.G., KNIGHT, S.E., AND FOSTER, R.S., JR. (1976). Decreased liver and lung drug metabolizing activity in mice treated with Corynebacterium parvum. *Cancer Res.* **36**, 4425–8.

SUSSMAN, K.E., STJERNHOLM, M.R., AND VAUGHAN, G.D. (1967). Propranolol and hypoglycaemia. *Lancet* **i**, 626.

VAN DIJKE, C.P.H. AND WEBER, J.C.P. (1984). Interaction between oral contraceptives and griseofulvin. *Br. med. J.* **288**, 1125–6.

WALKER, A.M., CODY, R.J.JR, GREENBLATT, D.J., AND JICK, H. (1983). Drug toxicity in patients receiving digoxin and quinidine. *Am. Heart J.* **105**, 1025–8.

WILLIAMS, J.R.B., GRIFFIN, J.P., AND PARKINS, A. (1976). Effect of concomitantly administered drugs on the control of long term anticoagulant therapy. *Quart. J. Med.* **45**, 63–73.

ZAMAN, R., KENDALL, M.J., AND BIGGS, P.I. (1982). The effect of acebutol and propranolol on the hypoglycaemic action of glibencamide. *Br. J. clin. Pharmacol.* **13**, 507–12.

ZWILLICH, C.W., SUTTON, F.D., NEFF, T.A., CORN, W.M., MATTHAY, R.A., AND WEINBERGER, M.M. (1975). Theophylline induced seizures in adults: correlation with serum concentrations. *Ann. intern. Med.* **82**, 784–7.

40 Drugs causing interference with laboratory tests

SHIMONA YOSSELSON-SUPERSTINE

Laboratory tests are an essential tool for diagnosis as well as for the follow-up of therapy. This is why it is crucial that the technology of the tests is reliable, specific, and free of the influence of foreign substances such as foods, chemicals, or drugs in the patient's biological fluids.

Drugs can affect the results of laboratory tests either by their pharmacological actions or by their chemical or physical interaction with the test procedure. It is this last form of interference which will be discussed in this chapter. Many of the drug interferences which have been repeatedly cited in the literature have been omitted from this review because of lack of supporting original studies or because they are not applicable to modern methods of analysis or therapy.

Drug interference with routine blood clinical chemistry tests

The routine determination of blood constituents is done nowadays, in most hospitals and clinics by multi-channel automatic analysers. This makes it possible to perform a large number of tests on a large number of samples in a short period of time. However, in some cases specificity is compromised (Yosselson-Superstine et al. 1980a).

Acid phosphatase

The Technicon acid phosphatase method is based on an enzymatic hydrolysis, using sodium thymolphthalein monophosphate (STMP). When phenyl phosphate was used, heparin in a low dose inhibited the enzyme activity (De Chatelet et al. 1972).

Albumin

The SMA 12/60 method for measuring serum albumin is based on the quantitative binding of an anionic dye known as HABA. The Technicon SMAC high-speed computer-controlled biochemical analyser utilizes a different reagent, bromocresol greeen (BCG). Drug interferences with both methods are summarized in Table 40.1. The interferences of ampicillin (given i.v.), salicylates, sulphonamides, paramethadione, and disulphine blue seem to be clinically significant.

Alkaline phosphatase

The test is based on the enzymatic hydrolysis of p-nitrophenyl phosphate. An infusion of albumin derived from placental sources can cause sudden and prolonged elevations in serum alkaline phosphatase level (Bark 1969a, b; Mackie et al. 1971; Wilkins 1971). Young et al. (1975) list many chemicals as having an effect on enzyme activity; whether or not this is relevant to medications containing such chemicals is not known. Sulphobromophthalein and anticoagulants are also potential interferents (Technicon SMAC, Product Labelling 1976). Theophylline can cause a decrease of 10–15 per cent in alkaline phosphatase activity at the $20~\mu g~ml^{-1}$ level (Young and Panek 1976; Vinet and Letellier 1977; Panek et al. 1978; Wright and Foster 1980). The postulated mechanism is enzyme inhibition. This mechanism plus the yellow colour of the drug was also suggested for the interference caused by therapeutic concentrations of nitrofurantoin (Panek et al. 1978; Wright and Foster 1980).

Bilirubin (total and direct)

The test is based on the diazo reaction. Al-Damluji and Meek (1980) showed that a propranolol metabolite, a conjugate of 4-hydroxypropranolol which accumulates in the plasma of undialysed patients with chronic renal failure, interferes with the diazo reaction to give falsely raised bilirubin concentrations. Phenazopyridine is postulated to increase colour with diazotization (Young 1975).

Blood urea nitrogen (BUN)

The SMA 12/60 and the SMAC methods are based

Table 40.1   Drug interference with serum albumin determination

Drug	Method	Reference	Result of interference*	Medium studied	Dose or therapeutic range	
					Studied	Known
Ampicillin	HABA	Beng and Lim (1973)	–	Pooled sera	1 mg/ml	1–100 μg/ml
	BCG	Beng and Lim (1973)	+ (0.1 g/100 ml)	Pooled sera	1 mg/ml	
Aspirin	HABA	Notrica *et al.* (1972)	– (10 per cent)	Serum	40 mg/dl (salicylic acid)	3–6 mg/dl (analgesic) 20–30 mg/dl (rheumatic diseases)
		Routh and Paul (1976)	–		3 250 mg/day	
Carbenicillin	BCG	Panek *et al.* (1978)	– (5 per cent)	Serum	2 mg/ml	10–300 μg/ml
Disulphine blue	BCG	Halloran and Torrens (1983)	+	Patient's serum	recommended	
Heparin	HABA	Niall and Owen (1962)	+	Plasma of patients Plasma of 25 volunteers	640 u/dl 1 800–3 600 u/dl	1–3 u/ml
Paramethadione	HABA	Rosenberg and Tobey (1971)	+ (1.4 g/100 ml)	Patient's serum		
Penicillin	HABA	Arvan and Ritz (1969)	–	Pooled sera	10 000 u/ml (potassium penicillin G)	20 u/ml (after 1 000 000 u given i.m.)
Phenazopyridine	HABA	Caraway and Kammeyer (1972)	+			
Sulphonamides	HABA	Arvan and Ritz (1969)	– (2–13 per cent)	Pooled sera	10–50 mg/dl (sulfisoxazole)	6–17 mg/dl

* + = False positive; – = false negative

on a reaction with diacetyl-monoxime. Acetohexamide, guanethidine, hydantoin derivatives, and sulphonylurea were listed by Young *et al.* (1975) as potential interferents. However, no clinical studies were cited. Sulphamethoxazole in therapeutic doses can cause overestimation of urea when a reagent strip method is used (Gregory and Lester 1982). This method is based on the reaction of amides with *o*-phthalaldehyde.

Calcium

The test is based on the formation of a coloured complex. Fluorides can precipitate calcium to form insoluble salts leading to decreased calcium values (Henry 1964). Sulphobromophthalein produces a violet colour which contributes to absorbance or obscure end-points in titrimetric methods (Caraway and Kammeyer 1972).

Carbon dioxide

The SMA 12/60 and the SMAC carbon dioxide methods are a modification of the Skeggs and Hochstrasser (1964) method. Anticoagulants such as EDTA, potassium oxalate, and sodium fluoride when added *in vitro* to the blood sample have been reported to alter the pH of the

sample resulting in invalid results for CO_2 (Zaroda 1964; Henry 1964; Young *et al.* 1975). Aspirin and ascorbic acid reduced the coefficient of variation of CO_2 levels by 2.0 per cent when added *in vitro* in therapeutic concentrations (Wright and Foster 1980).

Chloride

The SMA 12/60 chloride colorimetric method is that of Skeggs and Hochstrasser (1964). This method was modified by Morgenstern *et al.* (1973) for the SMAC system. A study (Fingerhut 1972) indicated that, for each mEq l^{-1} of bromide present in the serum sample, an absorbance equal to approximately 1.0 mEq l^{-1} of chloride occurs. This interference might be significant when some sedative hypnotic agents which release bromide ion in the body are administered (Ricci *et al.* 1982). Medications which contain iodide can lead to the same interference as do the bromides (Baker *et al.* 1980).

Cholesterol

Most automated analysers measure cholesterol with methods based on a modification of the Lieberman–Burchard reaction (Huang *et al.* 1961; Levine *et al.*

1968). Amphotericin B in a very high and unrealistic concentration interfered with the assay *in vitro* (Singh *et al.* 1972). Gamma globulin is capable of increasing serum viscosity causing less sample to be aspirated. On the other hand, it can also introduce a photometric interference, leading to the elevation of serum cholesterol (Holub and Galli 1972).

Creatine phosphokinase (CPK)

A number of substances have been reported to cause physiological changes in serum CPK activity. However, interferences with the method of analysis have not been reported to date.

Creatinine

The assay is based on the Jaffé reaction in which saturated picric acid reacts with creatinine in an alkaline medium (Chasson *et al.* 1961). It has been suggested (Young *et al.* 1975) that reducing agents such as ascorbic acid and levodopa might falsely elevate creatinine levels. However, only ascorbic acid was studied and was found to interfere only when given in a large amount (Siest *et al.* 1976; Vinet and Letellier 1977). Two other drugs which interfere with serum creatinine estimation in concentrations higher than those achieved clinically are methyldopa (Maddocks *et al.* 1973) and nitrofurantoin (Haeckel 1980, 1981).

The interference of cephalosporins with the creatinine assay has been studied very extensively. Only therapeutic concentrations of cephalothin (Watkins *et al.* 1976; Swain and Briggs 1977; Rankin *et al.* 1979; Wright and Foster 1980; Allen *et al.* 1982) and of cefoxitin (Swain and Briggs 1977; Durham *et al.* 1979; Hyneck *et al.* 1981; Saah *et al.* 1982; Allen *et al.* 1982; Guay *et al.* 1983) were found to significantly elevate creatinine levels. Other rarely used agents which were found to interfere in very high concentration and in *in-vitro* solution only were cephaloglycin, cephaloridine, and cephacetrile (Swain and Briggs 1977). No effect or a negligible one was found with cefamandole, cefazolin, cefotaxime, cephalexin, and other rarely used cephalosporins (Swain and Briggs 1977; Polk and Stephens 1981; Hyneck *et al.* 1981; Allen *et al.* 1982; Guay *et al.* 1983; Souney *et al.* 1983).

Glucose

The SMA 12/60 glucose method is based on the reduction of cupricneocuproine chelate by glucose in an alkaline medium resulting in highly-coloured complex (Brown 1961; Bittner and McCleary 1963). Several drugs were investigated *in vitro* and found to interfere with the assay in very high concentrations (Singh *et al.* 1972; Pennock *et al.* 1973; Young *et al.* 1975; Siest *et al.* 1976). The only clinically significant interferences were demonstrated with an overdosage of acetaminophen and with therapeutic concentration of aminosalicylic acid (Singh *et al.* 1972).

The more recommended assays for the estimation of serum glucose today are those based on the glucose-oxidase or the glucose-hexokinase methods. As can be seen from Table 40.2 the most significant interferences with the various glucose-oxidase methods are those with acetaminophen (paracetamol) and with ascorbic acid. Tetracycline, in higher concentrations than usually found in sera, could interfere with the hexokinase method as well (Neeley 1972).

Inorganic phosphorus

Inorganic phosphate, in a protein-free filtrate of serum, reacts with molybdate to form phosphomolybdate which is reduced to a blue complex (Amador and Urban 1972). The following compounds, if present, complex with molybdate and prevent full colour development: citrate, mannitol in levels higher than those achieved in plasma (Cook and Simmons 1962); oxalates; tartrates (Caraway and Kammeyer 1972); and promethazine (El-Dorry *et al.* 1972). The Technicon SMAC Product Labelling (1976) lists other interfering substances such as adrenaline, antacids, general anaesthetics, heparin, insulin, methicillin, tetracyclines, and vitamin D; however, this information is not based on experimental data and it is not clear whether the changes in the phosphate values are caused by the pharmacological or toxic effects of the drugs or by their interference with the procedure of the inorganic phosphorus assay. Methicillin preparations contain phosphorus salt so this can explain the increase in phosphorus levels (Panek *et al.* 1978).

Iron

The serum iron method is based on the interaction of the chromagen ferrozine with protein-free iron (Stookey 1970). Disulphine-blue can increase iron levels, as measured by this method, for up to 2½ days after its administration (Halloran and Torrens 1983). When tripyridl-s-triazine is used as a complexing agent, EDTA and deferoxamine might decrease (Young *et al.* 1975; Schenken and Gross 1971) and dextran might increase serum iron values (Young *et al.* 1975).

Lactate dehydrogenase (LDH)

Theophylline has a minor effect on the LDH assay which

Table 40.2 Drug interference with serum glucose, glucose oxidase methods

Drug	Reference	Method	Result of interference*	Medium studied	Dose or therapeutic range Studied	Known
Acetaminophen	Kaufmann-Raab et al. (1976)	GOD-Perid (ABTS)	–	Serum of 3 volunteers	1.5–2.0 g/day	up to 4.0 g/d
	(1976)		– (8 μmol/1)	Solution of drug	2 μmol/1	
	Wright and Foster (1980)	Trinder	– (1.4 per cent)	Solution of drug	6 μg/ml	5–20 μg/ml
	Farrance and Aldons (1981)	YSI	+ (1.4 mmol/1)	Solution of drug	151 μg/ml	
	Fleetwood and Robinson (1981)	YSI	+ (0.5 mmol/1)	Solution of drug	23 μg/ml	
	Roddis (1981)	YSI	+ (20 mmol/1)	Solution of drug	400 μg/ml	
	Farah et al. (1982)	YSI	+ (40 mmol/1)	Patient's serum	680 μg/ml	
p-Aminophenol	Kaufmann-Raab et al. (1976)	GOD-Perid (ABTS)	–	Solution of drug	50–200 mg/1	
Ampicillin	Neeley (1972)	o-toluidine	+ (3 per cent)	Solution of drug	100 μg/ml	1–100 μg/ml
Ascorbic acid	Neeley (1972)	o-toluidine	+ (5 per cent)	Solution of drug	100 μg/ml	9.7–27.5 μg/ml (after 2 g of vitamin C)
		Hall and Tucker (Ferrocyanide)	– (4 per cent)	Solution of drug	100 μg/ml	
	Pennock et al. (1973)	God-Perid (ABTS)	– (25 per cent)	Solution of drug	1 000 μg/ml	
	Romano (1973)	o-dianisidine	– (17 per cent)	Solution of drug	100 μg/ml	
	Wright and Foster (1980)	Trinder	– (4 per cent)	Solution of drug	60 μg/ml	
Chlorpropamide	Pennock et al. (1973)	God-Perid (ABTS)	–	Solution of drug	1 000 μg/ml	26 μg/ml
	Sharp (1972)	God-Perid (ABTS)	no effect	Solution of drug		
Dextran	Neeley (1972)	o-toluidine	turbidity	Solution of drug	10 mg/ml	
Hydrallazine	Sharp (1972)	God-Perid (ABTS)	–	Solution of drug	0.8–4.0 mmole/1	3–5.1 μg/ml
Iproniazid	Sharp (1972)	God-Perid (ABTS)	–	Solution of drug	0.8–4.0 mmole/1	
Isoniazid	Sharp (1972)	God-Perid (ABTS)	–	Solution of drug	0.8–4.0 mmole/1	1–4.5 μg/ml
Levodopa	Neeley (1972)	Hall and Tucker (Ferrocyanide)	–	Solution of drug	100 μg/ml	1 μg/ml
	Romano (1973)	o-dianisidine	– (51 per cent)	Solution of drug	100 μg/ml	
Methyldopa	Wright and Foster (1980)	Trinder	– (1.3 per cent)	Solution of drug	7 μg/ml	4 μg/ml
Oxyphenbutazone	Kaufmann-Raab et al. (1976)	God-Perid	–	Solution of drug	50–200 mg/1	5.4 μg/ml
Phenformin	Pennock et al. (1973)	God-Perid (ABTS)	–	Solution of drug	1 000 μg/ml	
Tetracycline	Neeley (1972)	o-toluidine	+ (4 per cent)	Solution of drug	100 μg/ml	1–5 μg/ml
	Wright and Foster (1980)	Trinder	– (2 per cent)	Solution of drug	4 μg/ml	
Tolazamide	Sharp et al. (1972)	God-Perid (ABTS)	–	Solution of drug	50 μg/ml	50 μg/ml
		o-dianisidine	–	Solution of drug		
	Neeley (1972)	Hall and Tucker (Ferrocyanide)		Solution of drug	100 μg/ml	
Tolbutamide	Sharp et al. (1972)	God-Perid (ABTS)	+ (2 per cent)	Solution of drug	50 μg/ml	80–240 μg/ml
		o-dianisidine	+ (2 per cent)	Solution of drug	50 μg/ml	
	Pennock et al. (1973)	God-Perid (ABTS)	–	Solution of drug	1 000 μg/ml	

* + = False positive; – = false negative

could be ignored at therapeutic levels of the drug (Vinet and Letellier 1977). *p*-Aminosalicylate at therapeutic concentration decreased LDH activity by as much as 15 per cent (Panek *et al.* 1978).

Potassium and sodium

No drug interferences with the flame photometer or the potentiometric procedures for the estimation of potassium or sodium in serum have been reported to date.

Serum glutamic-oxaloacetic transaminase (SGOT) and serum glutamic-pyruvic transaminase (SGPT)

Therapeutic doses of erythromycin estolate (Sabath *et al.* 1968) and of para-aminosalicylic acid (Glynn *et al.* 1970) have been found to affect the SGOT values as determined by the diazonium method (Morgenstern *et al.* 1966). A therapeutic concentration of isoniazid, tested *in vitro*, read as 10 units of SGOT (Singh *et al.* 1972).

The Technicon SMAC, SGOT, and SGPT methods are enzymatic assays based on transamination and dehydrogenase reactions. The activities of the enzymes are determined by a three-point rate reaction (Kessler *et al.* 1975). The Technicon SMAC product labelling (1976) lists acetaminophen, fluorides, and isoniazid as interfering substances; however this is not substantiated in the literature. Nitrofurantoin in five times therapeutic levels caused a 7 per cent decrease in SGOT concentrations as measured by an enzymatic assay (Wright and Foster 1980).

Total protein

Most laboratories measure serum total protein by a procedure based on a modified biuret reaction (Skeggs and Hochstrasser 1964). The usage of alkaline serum blank in this procedure eliminates error caused by substances such as bromosulphophthalein (Caraway and Kammeyer 1972) and phenazopyridine (Naumann 1967), but will not correct for interference caused by dextran (Crowley 1969). Total protein serum levels as measured by the biuret method can be increased by 5 per cent by therapeutic concentration of carbenicillin (Panek *et al.* 1978) and for up to 2½ days by disulphine-blue (Halloran and Torrens 1983).

Table 40.3 Drug interference with the phosphotungstate uric acid test

Drug	Reference	Result of interference*	Medium studied	Dose or therapeutic range Studied	Known
Acetaminophen	Singh *et al.* (1972)	+ (17 per cent)	Sera of 11 volunteers	0.2 g	up to 4.0 g/day
	Young and Panek (1976)	+	Solution of drug	Therapeutic	
	Panek *et al.* (1978)	+ (50 per cent)	Solution of drug	Toxic	
Ascorbic acid	Caraway (1969)	+ (1.34 mg/dl)	Pooled sera and Sera of 21 patients	10 μg/ml	9.7–27.5 μg/ml (after 2 g of vitamin C)
	Singh *et al.* (1972)	+ (2 mg/dl)	Solution of drug	250 μg/ml	
	Siest *et al.* (1976)	+	Solution of drug	>125 μg/ml	
	Young and Panek (1976)	+ (>5 per cent)	Solution of drug	>125 μg/ml	
	Vinet and Letellier (1977)	+ (0.25 mg/dl)	Pooled sera	50 μg/ml	
	Wright and Foster (1980)	+ (2.2 per cent)	Pooled sera	60 μg/ml	
Aspirin	Grayzel *et al.* (1961)	+ (7–119 per cent)	Sera of 12 patients	at least 13 mg/dl	3–30 mg/dl
Caffeine	Young *et al.* (1975)	+			
Chloral hydrate	Young *et al.* (1975)	+			
Dextran	Young *et al.* (1975)	+			
Epinephrine	Singh *et al.* (1972)	+	Solution of drug	183 μg/ml	
Gentamicin	Panek *et al.* (1978)	− (10 per cent)	Solution of drug	400 μg/ml	4–11 μg/ml
Hydrallazine	Singh *et al.* (1972)	+ (3.0 mg/dl)	Solution of drug	16 μg/ml	3–5.1 μg/ml
Iproniazid	Panek *et al.* (1978)	+	Solution of drug	>30 μg/ml	3 μg/ml
Isoniazid	Panek *et al.* (1978)	+ (25 per cent)	Solution of drug	135 μg/ml	1–4.5 μg/ml
Levodopa	Cawein and Hewins (1969)	+ (20 per cent)	Sera of 18 patients	3–7 g/day	3–6 g/day
Methyldopa	Small *et al.* (1976)	No change	Sera of 17 patients	500–1 500 mg/day	500–3 000 mg/day
		+ (>0.1 mg/dl)	Solution of drug	>60 μg/ml	4 μg/ml
	Wright and Foster (1980)	No change	Pooled sera	7 μg/ml	
6-Mercaptopurine	Singh *et al.* (1972)	+ (8.7 mg/dl)	Solution of drug	15.2 mg/dl	
Phenacetin	Singh *et al.* (1972)	+ (20 per cent)	Sera of 10 patients	2.0 g	up to 3.0 g/day
Phenalzine	Singh *et al.* (1972)	+ (0.3 mg/dl)	Solution of drug	136 μg/ml	
Propylthiouracil	Singh *et al.* (1972)	+ (0.5 mg/dl)	Solution of drug	170 μg/ml	
Theophylline	Yosselson-Superstine *et al.* (1980c)	No change	Solution of drug	5–40 μg/ml	10–20 μg/ml
		No change	Sera of 3 volunteers	2.4–13.0 μg/ml	
		No change	Sera of 15 patients	7.4–18.0 μg/ml	

* + = False positive; − = false negative.

Triglycerides

A number of substances have been reported to cause physiological changes in serum triglycerides; however, interferences of drugs with the method of analysis have not been reported to date.

Uric acid

The phosphotungstate uric acid test (Musser and Ortigoza 1966) is subject to many drug interferences (Yosselson-Superstine *et al.* 1980*b*). These interferences are summarized in Table 40.3. Analysis of this table demonstrates that the most significant interferences are those with acetaminophen, ascorbic acid, aspirin, phenacetin, and levodopa.

The uricase method (Sigma Chemical Company 1977), although specific for uric acid, is subject to interference due to inhibition of the enzyme activity by inhibitors such as 6-mercaptopurine and its metabolites, 6-thiouric acid and various other purine analogues (Baum *et al.* 1956; Bergmann *et al.* 1963). Ascorbic acid in high concentration (>125 μg ml⁻¹) (Young and Panek 1976) and disulfiram (Panek *et al.* 1978) were found to reduce serum uric acid levels as measured by the uricase with indamine dye method.

Drug interference with routine urine clinical chemistry tests

Bilirubin

The tests are based on a diazo reaction with various dizaonium salts used as reagents. High concentrations of ascorbic acid can inhibit the reaction. Chlorpromazine, salicylates (Bryant and Flynn 1955), and metabolites of mefenamic acid (Kater 1968) and of phenazopyridine (Naumann 1967) can cause false elevations of urine bilirubin.

Glucose

Two types of methods for testing urine glucose are currently used. The first type comprises the copper reduction methods (e.g *Clinitest*) which are semi-quantitative and are recommended for insulin-dependent diabetics who spill large amounts of glucose in the urine. The second type comprises the glucose oxidase methods (e.g. *Diastix*, *Tes-Tape*) which are only qualitative but on the other hand are more specific. However, these last methods are also prone to drug interferences since many drug metabolites have reducing properties whereby they can affect the results of the tests. Drug interferences with the various urine glucose methods are summarized in Table 40.4. Many other drugs have been incriminated in the literature, however their interferences are unsubstantiated.

Table 40.4   Drug interference with urine glucose determination

Drug	Dose/concentration	Method	Result of interference*	Reference
Aminoglycosides (amikacin, gentamicin, tobramycin, streptomycin)	Therapeutic concentration	*Clinitest, Diastix, Tes-Tape*	0	MacCara and Parker (1981)
Ampicillin	4 mg/ml; 10 mg/ml 0.5 mg/ml 10 mg/ml 0.5–10 mg/ml	*Clinitest* *Diastix, Tes-Tape*	+ (0.25 per cent instead of 0 per cent) − (0.25 per cent instead of 0.5 per cent) − (1 per cent instead of 2 per cent) 0	MacCara and Parker (1981)
Ascorbic acid	>0.09 mg/ml 1–3 g/day 3 g/day, 9 g/day	*Clinistix, Diastix, Labstix, Tes-Tape,* 2-drop *Clinitest*	− − 0	Feldman and Lebovitz (1973) Brandt *et al.* (1977) Smith and Young (1977)
Aspirin	>2.4 g/daily 0.05 mg/dl gentisic acid	*Clinistix, Clinitest Clinistix, Diastix*	− + −	Feldman *et al.* (1970) Feldman and Lebovitz (1973)
Carbenicillin	10 mg/ml; 20 mg/ml 0.5 mg/ml 10 mg/ml; 20 mg/ml 0.5–20 mg/ml	*Clinitest* *Diastix, Tes-Tape*	+ (0.25 per cent instead of 0 per cent) − (0.25 per cent instead of 0.5 per cent) − (1 per cent instead of 2 per cnet) 0	MacCara and Parker (1981)

Table 40.4   *Continued*

Drug	Dose/concentration	Method	Result of interference*	Reference
Cefamandole	2.5 mg/ml; 5 mg/ml	*Clinitest*	+ (0.25 per cent instead of 0 per cent)	Kowalsky and Wishnoff (1982)
	5 mg/ml	*Clinitest*	+ (0.75 per cent instead of 0.5 per cent)	McManus and Barriere (1983)
		Diastix, Tes-tape	0	
Cefazolin	3.5 mg/ml; 7 mg/ml	*Clinitest*	+ (0.75 per cent instead of 0.5 per cent)	McCara and Angaran (1978)
	1.75–7 mg/ml	*Diastix, Tes-Tape*	0	
Cefotaxime	2.5 mg/ml; 5 mg/ml	*Clinitest*	+ (0.25 per cent instead of 0 per cent)	Kowalsky and Wishnoff (1982)
	0.1 mg/ml; 1 mg/ml	*Clinitest*	− (0.25 per cent instead of 0.5 per cent)	McManus and Barriere (1983)
	2 mg/ml		+ (0.75 per cent instead of 0.5 per cent)	
	5 mg/ml		+ (1 per cent instead of 0.5 per cent)	
		Diastix, Tel-Tape	0	
Cefoxitin	2.5 mg/ml; 5 mg/ml	*Clinitest*	+ (0.25 per cent instead of 0 per cent)	Kowalsky and Wishnoff (1982)
	2 mg/ml; 5 mg/ml	*Clinitest*	+ (0.75 per cent instead of 0.5 per cent)	McManus and Barriere (1983)
		Diastix, Tes-Tape	0	
Cephalexin	2 g/day	*Clinitest*	+	Morrill *et al.* (1974)
		Tes-Tape	0	
Cephalothin	2 g/day	*Clinitest*	+ (0.75 per cent instead of 0.25 per cent)	Morrill *et al.* (1974)
	7 mg/ml	*Clinitest*	+ (0.75 per cent instead of 0.5 per cent)	MacCara and Angaran (1978)
		Diastix, Tes-Tape	0	
Cephradine	1.75 mg/ml; 3.5 mg/ml	*Clinitest*	+ (0.25 per cent instead of 0 per cent)	MacCara and Angaran (1978)
	7 mg/ml		+ (0.75 per cent instead of 0 per cent)	
	1.75–7 mg/ml		+ (0.75 per cent instead of 0.5 per cent)	
		Diastix, Tes-Tape	0	
Levodopa	0.75–5 g/day	*Clinitest, Clinistix*	−	Feldman *et al.* (1970)
Moxalactam	1.25–5 mg/ml	*Clinitest, Tes-Tape*	0	Kowalsky and Wishnoff (1982)
	0.1–5 mg/ml	*Clinitest, Diastix, Tes-Tape*	0	McManus and Barriere (1983)
Nalidixic acid	Therapeutic	*Clinitest,*	+ (0.5 per cent)	Islam and Sreedharan (1965)
		Clinistix, Tes-Tape	0	Klumpp (1965)
Penicillin G	16.2 mg/ml	*Clinitest*	+ (0.25 per cent instead of 0 per cent)	MacCara and Parker (1981)
	0.48 mg/ml		− (0.25 per cent instead of 0.5 per cent)	
	9 mg/ml; 16.2 mg/ml		− (1 per cent instead of 2 per cent)	
		Diastix, Tes-Tape	0	
Phenazopyridine	600 mg/day	*Clinistix,*	−	Naumann (1967)
		Combstix,		
		Tes-Tape	+	

* + = False positive; − = false negative; 0 = no change.

Ketones

The urine ketones assay is based on the reaction with sodium nitroprusside in an alkaline medium and the formation of a violet dye. The drugs which have been reported to cause false positive results with this test when administered to patients in therapeutic doses are levodopa (Pocelinko *et al.* 1969; Dawson 1970; Cawein *et al.* 1970), methyldopa (Cawein *et al.* 1970), phthaleins (Caraway and Kammeyer 1972), phenazopyridine (Naumann 1967), and paraldehyde (Hadden and Metzner 1969).

Protein

Various methods are used in clinical practice for the estimation of protein in urine. They are discussed in the previous edition of this book (Yosselson-Superstine 1983). Drug interferences with these methods are summarized in Table 40.5. The most significant interferences are those of agents that render the urine very alkaline and of phenylazopyridine with reagents strip, of X-ray contrast media, tolbutamide, cephaloridine, cephalothin, and tolmetin with the sulphosalicylic acid method, and of massive doses of penicillin, oxacillin, and nafcillin with this method as well as with the trichloroacetic and turbidity test.

Urobilinogen

The test is based on the Ehrlich aldehyde reaction or the formation of a red azo dye. Several drugs have been reported to react with p-dimethylaminobenzaldehyde (Ehrlich reagent) or with the diazonium salt, produce a colour reaction, and thus interfere with the test. The list of such drugs include phenazopyridine and tetracycline (McEwen and Paterson 1972; Naumann 1967). Radiographic contrast media (Hurt 1960; Koneman and Schessler 1965), para-aminosalicylic acid, procaine, antipyrine, sulphonamides (Bauer et al. 1962), phenothiazines (Reio and Wetterberg 1969), cascara sagrada (Carmichael and Neill 1961), and methyldopa (Pierach et al. 1977). Many of these reports are not detailed enough and as a result it is hard to draw conclusions about their clinical significance.

Drug interference with plasma drug assays

The estimation of plasma drug concentration is becoming a routine procedure during the period of therapy with various classes of drugs. A detailed review of drug interferences with commonly used drug assays was published elsewhere (Yosselson-Superstine 1984). The most significant interferences are as follows:

Digoxin

Digoxin is measured in most laboratories by radioimmunoassay methods. Canrenone, a metabolite of spironolactone (Lichey et al. 1977; Silber et al. 1979), digitoxin (Múller et al. 1978; Scherrmann and Bourdon 1980), and prednisolone (Phillips 1973) could cross-react with these methods. The interference of spironolactone depends on the antiserum used (Bergdähl and Molin 1981).

Gentamicin

Various antimicrobial agents, including cephalosporins (Shanson and Hince, 1977; Habbal, 1979), co-trimoxazole (Shanson and Hince 1977), and tetracyclines (Daigneault et al. 1974; Manos and Jacobs 1979), can interfere with microbiological methods for the analysis of gentamicin, as well as with spectrofluorometric, enzymatic, and radiometric methods. Many of the aminoglycosides can interfere or cross-react as well. Gallium-67, used in radioactive scanning, can cause a false elevation of serum gentamicin when the radioenzyme method is used (Shannon et al. 1980). Heparin in high concentrations can interfere with gentamicin measurement by biological, radioenzymatic, and homogeneous enzyme immunoassays (Nilsson 1980; Krogstad et al. 1982).

Procainamide

Chlordiazepoxide in therapeutic doses was found to interfere with a colorimetric method (Sterling et al. 1974). There is also a suggestion, but no documentation, that sulphonamides can react with a similar assay (Sitar et al. 1976). Drug interferences with HPLC procainamide assays were reported with acetaminophen, aminophylline, and codeine in toxic concentration and with caffeine (Stearns 1981) and metronidazole (Gannon and Phillips 1982) in therapeutic concentrations. Sulphathiazole (Shukur et al. 1977) and a presumed metabolite of quinidine (Dutcher and Strong 1977) could affect the estimation of N-acetyl-procainamide (NAPA) when measured by an HPLC method. Butacaine and procaine were most cross-reactive towards the EMIT procainamide assay (Pape 1982).

Quinidine

Several drugs have been found to interfere with the HPLC assays of quinidine. Chlordiazepoxide was only partly separated from the quinidine peak in the Powers and Sadee (1978) method, contributing as much as 1 μg ml^{-1} to quinidine blood level. A metabolite of disopyramide (Ahokas et al. 1980), procainamide (Achari et al. 1978; Peat and Jennison 1978), and propranolol (Achari et al. 1978; Flood et al. 1980) were shown to interfere as well. Since all of these three last mentioned drugs are used for treating arrhythmia, these interferences can be clinically significant. The diuretic triamterene could also contribute to the fluorescence reading of quinidine (Osinga and de Wolff 1976).

Table 40.5   Drug interference with urine protein determination

Drug	Dose/concentration	Method	Result of interference
Alkaline agents		RS	+
(acetazolamide,		SSA	–
bicarbonates)		HA	–
Cephaloridine	6.0 g daily i.v.	SSA	+
	40 mg/ml (*in vitro*)		+ (1 +)
Cephalothin	10 g daily i.v.	SSA	+
	10–40 mg/ml (*in vitro*)		+ (1 +)
Chlorpromazine			
Cloxacillin	200 mg/dl (*in vitro*)	TCA	+ (200 mg/dl)
Nafcillin	12 α 16 g daily i.v.	SSA	+ (3 + – 4 +)
	100 mg/dl (*in vitro*)		+ (3 +)
	200 mg/dl (*in vitro*)		+ (4 +)
	12 g daily i.v.	TCA	uninterpretable
	200 mg/dl (*in vitro*)		+ (225 mg/dl)
Oxacillin	250 mg/dl (*in vitro*)	SSA	+ (1 +)
	500 mg/dl (*in vitro*)		+ (4 +)
	500 mg/dl (*in vitro*)	TCA	+ (30 mg/dl)
PAS	12 g daily orally	SSA	+ (5–100 mg/dl)
Penicillin G	20 million u.i.v.	SSA	+
	4 000 u/ml (*in vitro*)		
	500 mg/dl (*in vitro*)	TCA	+ (40 mg/dl)
	100 mg/dl (*in vitro*)		+ (30 mg/dl)
	> 5 g daily	Biuret	+
	massive dose	HA	+
Phenformin	high concentration (*in vitro*)	RS	+ (trace)
Phenylazopyridine	600 mg daily orally	RS	discoloration
Promazine			
Quaternary			
Ammonium compounds		RS	+
Radio-opaque X-ray	normal dose	SSA	+ (1 +)
Contrast media			
	normal dose	HA	+ (1 +)
Rifampicin	12–36 mg/dl (*in vitro*)	Urine-Pak	+ (4–17 mg/dl)
Sulphonylureas			
Tolbutamide	high concentration (*in vitro*)	RS	+ (trace)
	therapeutic dose	SSA	+ (strongly positive)
Other sulphonylureas	high concentration (*in vitro*)		+ (1 + – 2 +)
	high concentration	HA	+ (1 + – 2 +)
	(*in vitro*, animal urine)		
Sulphisoxazole		SSA	+
Tolmetin	therapeutic dose	SSA	+ (strongly positive)

* + = False-positive; – = false-negative.
RS = reagent strip; SSA = sulphosalicylic acid; HA = heat and acetic acid; TCA = trichloroacetic acid.

Table 40.5 *Continued*

Comment	Reference
When urine pH > 10.0	Gyure (1977)
When urine pH > 10.0	Gyure (1977)
No evidence in the literature	
Clinically significant	Levy and Eliakim (1972)
Clinically significant	Levy and Eliakim (1972)
Confusing results with tubidity tests	Free *et al.* (1957)
Needs to be verified *in vivo*	Muir and Hensley (1979)
Clinically significant	Line *et al.* (1976)
	Felice-Johnson and Nappi (1981)
Clinically significant	Felice-Johnson and Nappi (1981)
	Line *et al.* (1976)
Clinically significant	Line *et al.* (1976)
Significance questionable due to high drug concentration	Line *et al.* (1976)
Only with formaldehyde urine preservative	Opstad (1958); Gyure (1977)
Clinically significant	Whipple and Bloom (1950); Lippman (1952)
	Line *et al.* (1976)
	Line *et al.* (1976)
	Muir and Hensley (1979)
Could also reduce protein levels at high urinary protein concentrations	Andrassy *et al.* (1978)
No evidence in the literature	Bradley *et al.* (1979)
Significance questionable due to very high drug concentration	Wachter *et al.* (1960)
Could be prevented by the addition of dithionite	Naumann (1967)
Confusing results with turbidity tests	Free *et al.* (1957)
No studies are cited	Bradley *et al.* (1979)
Clinically significant	Seedrof *et al.* (1952, 1953); Free *et al.* (1957)
	Giordano *et al.* (1957); Hurt (1960)
Clinically significant	Aguzzi *et al.* (1980)
Significance questionable due to very high drug concentration	Wachter *et al.* (1960)
Clinically significant	Free *et al.* (1957); Free and Fancher (1958)
Significance questionable due to very high drug concentrations	Wachter *et al.* (1960)
Significance questionable due to very high drug concentrations	Wachter *et al.* (1960)
No evidence in the literature	Bauer *et al.* (1974); Bradley *et al.* (1979)
Clinically significant	Ehrlich and Worthman (1975)

Salicylates

Theophylline was shown to interfere with salicylic acid determination in the HPLC method of Miceli *et al.* (1979) which also measures paracetamol simultaneously. In therapeutic concentration it gave an apparent salicylic acid concentration of 80 μg ml^{-1}.

Phenobarbitone and phenytoin

The most significant interferences with these assays are by other anticonvulsant agents. Other barbiturates could interfere with spectrophotometric methods (Jeremić 1976), GLC (St Onge *et al.* 1979), HPLC (Soldin and Hill 1977*a*; Kinberger and Holmen 1982), as well as with radioimmunoassays (Kawashima *et al.* 1980), substrate-labelled fluorescent immunoassay (Krausz *et al.* 1980), and EMIT (Oellerich *et al.*, 1977). Carbamazepine could interfere with a GLC method (Latham and Varlow 1976). Ethosuximide, mephenytoin, and primidone could increase phenytoin levels when the drug is measured by a spectrophotometric method (Svensmark and Kristensen 1963). The last two agents can affect phenobarbitone determination in a similar manner as well and ethotoin could decrease its levels. A false increase caused by ethotoin and mephenytoin was also detected when radio-immunoassay techniques were employed (Stanley and Peikert 1978). Caffeine and theophylline can co-elute with the internal standards used in the GLC assays causing an apparent increase (St Onge *et al.* 1979) or decrease (Schier and Gan 1979) in phenobarbitone levels, depending on their concentrations in the plasma. Other interferences which could be of clinical significance are caused by propoxyphene, which emerges with phenytoin in one of the HPLC methods (Adams and Vandemark 1976), and probenecid which, in therapeutic doses, elevates phenytoin levels as measured by an on-column alkylation GLC procedure (Steyn and Hundt 1977).

Theophylline

Close to 40 different drugs were found to interfere with the spectrophotometric procedure for the estimation of plasma theophylline. They are summarized in the previous edition of this book (Yosselson-Superstine 1983). As a result of the impressive non-specificity of these procedures, many high-pressure liquid-chromatographic theophylline assays have been developed during the last few years. However, it did not take long to find out that these newer methods were also subject to potential drug interferences. Thus, drugs such as acetaminophen in therapeutic doses (Quattrone and Putnam 1981), acetazolamide (Robinson and Dobbs 1978), aspirin (Leslie and Miller 1982), paraxanthine which is a metabolite of caffeine (Jonkman *et al.* 1982), dimenhydrinate (Adams and Vandemark 1976), procainamide (Marion *et al.* 1981), and various anti-microbial agents, such as ampicillin and methicillin (Soldin and Hill 1977*b*), cephalexin, cephalothin, and cefazolin (Kelly *et al.* 1978; Frutkoff *et al.* 1982), chloramphenicol (Weidner *et al.* 1980), and sulphonamides (Bowman *et al.* 1980; Weidner *et al.* 1980), were found to cause an apparent increase in theophylline plasma concentrations when the liquid chromatographic techniques were employed. Caffeine and dimenhydrinate can cross-react with the theophylline radioimmunoassay (Cook *et al.* 1976; Hahn 1980). As a result of all these reported interferences, laboratories which can afford the cost nowadays use homogeneous enzyme immunoassays for the quantitation of theophylline in patient's sera.

REFERENCES

ACHARI. R.G., BALDRIDGE, J.L., KOZIOL, T.R., AND YU, L. (1978). Rapid determination of quinidine in human plasma by high-performance liquid chromatography. *J. Chromatogr. Sci.* **16**, 271–3.

ADAMS, R.F. AND VANDEMARK, F.L. (1976). Simultaneous high-pressure liquid-chromatographic determination of some anti-convulsants in serum. *Clin Chem.* **22**, 25–31.

AGUZZI, F., CHIARA, T., AND MAGGI, M. (1980). Valutazione di un nuovo metodo per il dosaggio della proteinuria. *J. Res. Lab. Med.* **7**, 189–92.

AHOKAS, J.T., DAVIES, C., AND RAVENSCROFT, P.J. (1980). Simultaneous analysis of disopyramide and quinidine in plasma by high-performance liquid chromatography. *J. Chromatogr.* **183**, 65–71.

AL-DAMLUJI, S. AND MEEK, J.H. (1980). Interference of a pro-pranolol metabolite with serum bilirubin estimation in chronic renal failure. *Br. med. J.* **280**, 1414.

ALLEN, L.C., MICHALKO, K., AND COONS, C. (1982). More on cephalosporin interference with creatinine determinations. *Clin. Chem.* **28**, 555–6.

AMADOR, E. AND URBAN, J. (1972). Simplified serum phosphorus analyses by continuous-flow spectrophotometry. *Clin. Chem.* **18**, 601–4.

ANDRASSY, K., RITZ, E., KODEVISCH, J., SALZMANN, W., AND BOMMER, J. (1978). Pseudoproteinuria in patients taking penicillin. *Lancet* **ii**, 154.

ARVAN, D.A. AND RITZ, A. (1969). Measurement of serum albumin by the HABA-dye technique: A study of the effect of free and conjugated bilirubin, of bile acids and of certain drugs. *Clin. Chem. Acta* **26**, 505–16.

BAKER, C., KAHN, S.E., AND BERMES, E.W. (1980). Effect of bromide and iodide on chloride methodologies in plasma or serum. *Ann. Clin. Lab. Sci.* **10**, 523–8.

BARK, C.J. (1969*a*). Artifactual serum alkaline phosphatase from placental albumin. *J. Am. med. Ass.* **207**, 953.

—— (1969*b*). Artifactual elevations of serum alkaline

phosphatase following albumin infusions. *Am. J. clin. Pathol.* **52**, 466–7.

BAUER, J.D., ACKERMANN, G.P., AND TORO, G. (1974). *Clinical laboratory methods*, 8th ed., pp. 410–11. C.V. Mosby Company, Saint Louis, Missouri.

——, TORO, G., AND ACKERMANN, P.G. (1962). *Bray's clinical laboratory methods*. C.V. Mosby Company, Saint Louis, Missouri.

BAUM, H., HUBSCHER, G., AND MAHLER, H.R. (1956). Studies on uricase. II. The enzyme-substrate complex. *Biochem. Biophys. Acta* **22**, 514–27.

BENG, C.G. AND LIM. K.L. (1973). An improved automated method for determination of serum albumin using bromcresol green. *Am. J. clin. Pathol.* **59**, 14–21.

BERGDÄHL, B. AND MOLIN, L. (1981). Precision of digoxin radio-immunoassays and matrix effects. Four kits compared. *Clin. Biochem.* **14**, 67–71.

BERGMANN, F., KWIETNY-GORRIN, H., URGAR-WARON, H., KALMUS, A., AND TAMARI, M. (1963). Relation of structure to the inhibitory activity of purines against urate oxidase. *Biochem. J.* **86**, 567–74.

BITTNER, D.L. AND MCCLEARY, M.L. (1963). The cupric-phenanthroline chelate in the determination of monosaccharides in whole blood. *Am. J. clin. Pathol.* **1 40**, 423.

BOWMAN, D.B., ARAVIND, M.K., KAUFMAN, R.E., AND MICELI, J.N. (1980). Sulfamethoxazole interferes with liquid-chromatographic analysis for theophylline in serum. *Clin. Chem.* **26**, 1622.

BRADLEY, M., SCHUMANN, G.B., AND WARD, C.J. (1979). Examination of urine, proteinuria. In *Clinical diagnosis and management by laboratory methods*, 16th ed. (ed. J.B. Henry), pp. 604–5. W.B. Saunders Company, Philadelphia, Pennsylvania.

BRANDT, R., GUYER, K.E., AND BANKS, W.L. (1977). Urinary glucose and vitamin C. *Am. J. clin. Pathol.* **68**, 592–4.

BROWN, M.E. (1961). Ultra-micro sugar determination using 2, 9-dimethyl-1, 10-phenanthroline hydrochloride (neocuproine). *Diabetes* **10**, 60–2.

BRYANT, D. AND FLYNN, F.V. (1955). An assessment of new tests for detecting bilirubin in urine. *J. clin. Pathol.* **8**, 163–5.

CARAWAY, W.T. (1969). Non-urate chromagens in body fluids. *Clin. Chem.* **15**, 720–6.

—— AND KAMMEYER, C.W. (1972). Chemical interference by drugs and other substances with clinical laboratory test procedures. *Clin. Chim. Acta* **41**, 395–434.

CARMICHAEL, R. AND NEILL, D.W. (1961). A possible source of error in the detection of urobilinogen. *Clin. Chim. Acta* **6**, 590–1.

CAWEIN, M.J. AND HEWINS, J. (1969). False rise in serum uric acid after L-dopa. *New Engl. J. Med.* **281**, 1489–90.

——, WILLIAMSON, M.A., EBENEZER, C., AND HEWINS, J.P. (1970). Levodopa and tests for ketonuria (contd.). *New Engl. J. Med.* **283**, 659.

CHASSON, A.L., GRADY, H.T., AND STANLEY, M.A. (1961). Determination of creatinine by means of automatic chemical analysis. *Am. J. clin. Pathol.* **35**, 83–8.

COOK, B.S. AND SIMMONS, D.H. (1962). Mannitol interference in phosphate determination: Method of correction. *J. Lab. clin. Med.* **60**, 160–3.

COOK, C.E., TWINE, M.E., MYERS, M., AMERSON, E., KEPLER, J.A., AND TAYLOR, G.F. (1976). Theophylline radioimmunoassay.

Synthesis of antigen and characterization of antiserum. *Res. Commun. Chem. Pathol. Pharmacol.* **13**, 497–505.

CROWLEY, L.V. (1969). Interference with certain chemical analyses caused by dextran. *Am. J. clin. Pathol.* **51**, 425–6.

DAIGNEAULT, R., CAGNE, M., AND BRAZEUAU, M. (1974). A comparison of the methods of gentamicin assay: An enzymatic procedure and an agar diffusion technique. *J. infect. Dis.* **130**, 642–5.

DAWSON, W.L. (1970). Levodopa and tests for ketonuria. *New Engl. J. Med.* **283**, 264.

DE CHATELET, L.R., McCALL, C.E., COOPER M.R., AND SHIRLEY, P.S. (1972). Inhibition of leukocyte acid phosphatase by heparin. *Clin. Chem.* **18**, 1532–4.

DURHAM, S.R., BEGNELL, A.H.C., AND WISE, R. (1979). Interference of cefoxitin in the creatinine estimation and its clinical relevance. *J. clin. Pathol.* **32**, 1148–51.

DUTCHER, J.S. AND STRONG, J.M. (1977). Determination of plasma procainamide and N-acetylprocainamide concentration by high-pressure liquid chromatography. *Clin. Chem.* **23**, 1318–20.

EHRLICH, G.E. AND WORTHMAN, G.F. (1975). Pseudoproteinuria in tolmetin-treated patients. *Clin. Pharmacol. Ther.* **17**, 467–8.

EL-DORRY, H.F.A., MEDINA, H., AND BACILA, M. (1972). Interference of phenothiazine compounds in the colorimetric determination of inorganic phosphate. *Anal. Biochem.* **47**, 329–36.

FARAH, D.A., BOAG, D., MORAN, F., AND MCINTOSH, S. (1982). Paracetamol interference with blood glucose analysis: A potentially fatal phenomenon. *Br. med. J.* **285**, 172.

FARRANCE, I. AND ALDONS, J. (1981). Paracetamol interference with YSI glucose analyzer. *Clin. Chem.* **27**, 782–3.

FELDMAN, J.M., KELLEY, W.N., AND LEBOVITZ, H.E. (1970). Inhibition of glucose oxidase paper tests by reducing metabolites. *Diabetes* **19**, 337–43.

—— AND LEBOVITZ, F.L. (1973). Test for glucosuria. An analysis of factors that cause misleading results. *Diabetes* **22**, 115–21.

FELICE-JOHNSON, J. AND NAPPI, J.M. (1981). Nafcillin interference with quantitative protein urinalysis. *Am. J. hosp. Pharm.* **38**, 1360–1.

FINGERHUT, B. (1972). A non-mercurimetric automated method for serum chloride. *Clin. Chim. Acta* **41**, 247–53.

FLEETWOOD, J.A., AND ROBINSON, S.M.A. (1981). Paracetamol interference with glucose analyzer. *Clin. Chem.* **27**, 1945.

FLOOD, J.G., BOWERS, G.N., AND McCOMB, R.B. (1980). Simultaneous liquid-chromatographic determination of three antiarrhythmic drugs: disopyramide, lidocaine, and quinidine. *Clin. Chem.* **26**, 197–200.

FREE, A.H. AND FANCHER, D.E. (1958). Urine protein tests in presence of tolbutamide metabolite. *Am J. med. Technol.* **24**, 64–5.

——, RUPE, C.O., AND METZLER, I. (1957). Studies with a new colimetric test for proteinuria. *Clin. Chem.* **3**, 716–27.

FRUTKOFF, I.W., MENCZEL, J., AND KIDRONI, G. (1982). Monitoring of serum theophylline by high performance liquid chromatography without interferences by various coadministered drugs. *Isr. J. med. Sci.* **18**, 639–41.

GANNON, R.H. AND PHILLIPS, L.R. (1982). Metronidazole interferences with procainamide HPLC assay. *Am. J. hosp. Pharm.* **39**, 1966–7.

GIORDANO, A.S., ALLEN, N., WINSTEAD, M., AND PAYTON, M.A. (1957). A new colorimetric test for albuminuria. *Am. J. med. Technol.* **23**, 216–19.

GLYNN, K.P., CAFARO, A.F., FOWLER, C.W., AND STEAD, W.W. (1970). False elevations of serum glutamic-oxalacetic transaminase due to para-aminosalicylic acid. *Ann. intern. Med.* **72**, 525–7.

GRAYZEL, A.L., LIDDLE, L., AND SEEGMILLER, J.E. (1961). Diagnostic significance of hyperuricemia in arthritis. *New Engl. J. Med.* **265**, 763–8.

GREGORY, J. AND LESTER, E. (1982). Drug interference with reagent strip method for measuring urea. *Lancet* **ii**, 443.

GUAY, D.R.P., MEATHERALL, R.C., AND MACAULAY, P.A. (1983). Interference of selected second and third generation cephalosporins with creatinine determination. *Am. J. hosp. Pharm.* **40**, 435–8.

GYURE, W.L. (1977). Comparison of several methods for semi-quantitative determination of urinary protein. *Clin. Chem.* **23**, 876–9.

HABBAL, Z.M. (1979). Spectrofluorometric assay of gentamicin in serum. *Clin. Chim. Acta* **95**, 301–9

HADDEN, J.W. AND METZNER, R.J. (1969). Pseudoketosis and hyperacetaldehydemia in paraldehyde acidosis. *Am. J. Med.* **47**, 642–7.

HAECKEL, R. (1980). Simplified determinations of the "true" creatinine concentration is serum and urine. *J. clin. Biochem.* **18**, 385–94.

—— (1981). Assay of creatinine in serum, with use of Fuller's earth to remove interferents. *Clin. Chem.* **27**, 179–83.

HAHN, E. (1980). Dimenhydrinate interferes with radioimmunoassay of theophylline. *Clin. Chem.* **26**, 1759–60.

HALLORAN, S.P. AND TORRENS, D.J. (1983). Disulphine-blue interferes with laboratory measurements of protein and iron. *Lancet* **i**, 188.

HENRY, R.J. (1964). *Clinical chemistry principles and techniques.* Harper and Row, New York.

HOLUB, W.R. AND GALLI, F.A. (1972). Automated direct method for measurement of serum cholesterol, with use of primary standards and a stable reagent. *Clin. Chem.* **18**, 239–43.

HUANG, C., CHEN, C.P., WEFLER, V., AND RAFTERY, A. (1961). A stable reagent for the Liebermann-Burchard reaction application to rapid serum cholesterol determinations. *Anal. Chem.* **33**, 1405–7.

HURT, R. (1960). The effect of radiographic contrast media on urinalysis. *Am. J. med. Technol.* **26**, 122–4.

HYNECK, M.L., BERARDI, R.R., AND JOHNSON, R.M. (1981). Interference of cephalosporins and cefoxitin with serum creatinine determination. *Am. J. hosp. Pharm.* **38**, 1348–52.

ISLAM, M.A. AND SREEDHARAN, T. (1965). Convulsions, hyperglycemia and glycosuria from overdose of nalidixic acid. *J. Am. med. Ass.* **192**, 158–9.

JEREMIĈ, V. (1976). An improved UV-spectrophotometric method for routine barbiturate monitoring. *J. clin. Chem. clin. Biochem.* **14**, 479–83.

JONKMAN, J.H.G., DE ZEEUW, R.A., AND SCHONMAKER, R. (1982). Interference in the proposed selected method for determination of theophylline by liquid chromatography. *Clin. Chem.* **28**, 1987–8.

KATER, R.M.H. (1968). Double blind evaluation of the analgesic and toxic effects of flufenamic acid and mefenamic acid in patients with chronic pain. *Med. J. Aust.* **1**, 848–5.

KAUFMANN-RAAB, I., JONEN, H.G., JAHNCHEN, E., KALL, G.F., AND GROTH, U. (1976). Interference by acetaminophen in the glucose oxidase–peroxidase method for blood glucose determination. *Clin. Chem.* **22**, 1729–31.

KAWASHIMA, K., ISHIJIMA, B., YOSHIMIZU, N., AND SATO, F. (1980). Determination of dose-plasma concentration relationship of phenobarbitol in epileptic patients by a new specific radioimmunoassay. *Arch. Int. Pharmacodyn.* **244**, 166–76.

KELLY, R.C., PRENTICE, D.E., AND HEARNE, G.M. (1978). Cephalosporin antibiotics interfere with the analysis for theophylline by high-performance liquid chromatography. *Clin. Chem.* **24**, 838–9.

KESSLER, G., MORGENSTERN, S., SNYDER, L., AND VARADY, R. (1975). Improved point assays for ALT and AST in serum using the Technicon SMAC high-speed, computor-controlled biochemical analyzer to eliminate the common errors found. In *Enzyme analysis*, Ninth International Congress on Clinical Chemistry, Toronto, Canada.

KINBERGER, B. AND HOLMEN, A. (1982). Analysis for carbamazepine and phenytoin in serum with a high-speed liquid chromatography system (Perkin–Elmer). *Clin. Chem.* **28**, 718–19.

KLUMPP, T.G. (1965). Nalidixic acid — false positive glycosuria and hyperglycemia. *J. Am. med. Ass.* **193**, 122.

KONEMAN, E.W. AND SCHESSLER, J. (1965). Unusual urinary crystals. *Am. J. Clin. Pathol.* **44**, 358.

KOWALSKY, S.F. AND WISHNOFF, F.G. (1982). Evaluation of potential interaction of new cephalosporins with clinitest. *Am. J. hosp. Pharm.* **39**, 1499–501.

KRAUSZ, L.M., HITZ, J.B., BUCKLER, R.T., AND BURD, J.F. (1980). Substrate-labelled fluorescent immunoassay for phenobarbital. *Ther. Drug Monit.* **2**, 261–72.

KROGSTAD, D.J., GRANICH, G.G., MURRAY, P.R., PFALLER, M.A., AND VALDES, R. (1982). Heparin interferes with the radioenzymatic and homogeneous enzyme immunoassays for aminoglycosides. *Clin. Chem.* **28**, 1517–21.

LATHAM, A.N. AND VARLOW, G. (1976). Simultaneous quantitative gas-chromatographic analysis of ethosuximide, phenobarbitone, primidone and diphenylhydantoin. *Br. J. clin. Pharmacol.* **3**, 145–50.

LESLIE, J. AND MILLER, A.K. (1982). Interferences in a high pressure liquid chromatographic assay of theophylline. *Ther. Drug Monit.* **4**, 323–4.

LEVINE, J., MORGENSTERN, S., AND VLASTELICA, D. (1968). A direct Liebermann–Burchard method for serum cholesterol. In *Automation in analytical chemistry*, Technicon symposia, 1967. Medical Inc, White Plains, New York.

LEVY, M. AND ELIAKIM, M. (1972). Urinary precipitate during cephalothin–cephaloridine treatment. *J. Am. med. Ass.* **219**, 908.

LICHEY, J., SCHRODER, R., AND RIETBROCK, N. (1977). The effect of oral spironolactone and intravenous canrenoate-K on the digoxin radioimmunoassay. *Int. J. clin. Pharmacol.* **15**, 557–9.

LINE, D.E., ALDER, S., FRALEY, D.S., AND BURNS F.J. (1976). Massive pseudoproteinuria caused by nafcillin. *J. Am. med. Ass.* **235**, 1259.

LIPPMAN, R.W. (1952). Effect of antibiotic agents on the tests for protein and reducing sugar in uring. *Am. J. clin. Pathol.* **22**, 1186–8.

MACCARA, M.E. AND ANGARAN, D.M. (1978). Cephalosporin–Clinitest interaction: Comparison of cephalothin, cefazolin and cephradine. *Am. J. hosp. Pharm.* **35**, 1064–7.

—— AND PARKER, W.A. (1981). In vitro effect of penicillin and aminoglycosides on commonly used tests for glycosuria. *Am. J. hosp. Pharm.* **38**, 1340–5.

MACKIE, J.A., ARVAN, D.A., MULLEN J.L., AND RAWNSLEY, H.M. (1971). Elevated serum alkaline phosphatase levels after the administration of certain preparations of human albumin. *Am. J. Surg.* **121**, 57–61.

MADDOCKS, J., HANN, S., HOPKINS, M., AND COLES, G.A. (1973). Effect of methyldopa on creatinine estimation. *Lancet* i, 157.

MANOS, J.P. AND JACOBS, P.F. (1979). Evaluation of the Bactec-serum gentamicin assay. *Antimicrob. Agents Chemother.* **16**, 631–4.

MARION, A., LESKO, L.J., AND OLIVER, C. (1981). Procainamide interference with liquid chromatography of theophylline in serum. *Ther. Drug Monit.* **3**, 107–8.

McEWEN, J. AND PATERSON, C. (1972). Drugs and false-positive screening tests for porphyria. *Br. med. J.* **1**, 421.

McMANUS, M.C. AND BARRIERE, S.L. (1983). Interaction between new cephalosporins and Clinitest, Diastix, and Tes-Tape. *Am. J. hosp. Pharm.* **40**, 1544–5.

MICELI, J.N., ARAVIND, M.K., COHEN, S.N., AND DONE, A.K. (1979). Simultaneous measurement of acetaminophen and salicylate in plasma by liquid chromatography. *Clin. Chem.* **25**, 1002–4.

MORGENSTERN, S., OKLANDER, M., AUERBACH, I., KAUFMAN, J., AND KLEIN, B. (1966). Automated determination of serum glutamic oxaloacetic transaminase. *Clin. Chem.* **12**, 95–111.

——, RUSH, R., AND LEHMAN, D. (1973). Chemical methods for increased specificity. In *Advances in automated analysis*, Vol. 1, pp. 27–31. Technicon International Congress, 1972. Medical Inc, White Plains, New York.

MORRILL, J., DAVIS, L.J., AND BURRIS, D.M. (1974). Interference with urinary glucose determination by cephalothin. *J. Am. med. Ass.* **230**, 822–3.

MUIR, A. AND HENSLEY, W.J. (1979). Pseudoproteinuria due to penicillins, in the turbidometric measurement of proteins with trichloroacetic acid. *Clin. Chem.* **25**, 1662–3.

MÜLLER, H., BRAUER, H., AND RESCH. B. (1978). Cross reactivity of digitoxin and spironolactone in two radioimmunoassays for serum digoxin. *Clin. Chem.* **24**, 706–9.

MUSSER, W.A. AND ORTIGOZA, C. (1966). Automated determination of uric acid by the hydroxylamine method. *Tech. Bul. Reg. Med. Tech.* **30**, 21–5.

NAUMANN, H.N. (1967). Prevention of pyridium interference in urinalysis by dithionite reduction or butanol extraction. *Am. J. cline. Pathol.* **48**, 337–41.

NEELEY, W.E. (1972). Simple automated determination of serum or plasma glucose by a hexokinase/glucose-6-phosphate dehydrogenase method. *Clin. Chem.* **18**, 509–15.

NIALL, M.M. AND OWEN, J.A. (1962). Heparin as a source of error in the determination of plasma albumin on a basis of selective dye-binding. *Clin. Chim. Acta* **7**, 155–8.

NILSSON, L. (1980). Factors affecting gentamicin assay. *Antimicrob. Agents Chemother.* **17**, 918–21.

NOTRICA, S., MIYADA, D.S., BAYSINGER, V., AND NAKAMURA, R.M. (1972). Effects of various medications on values from the HABA and BCG methods for determining albumin. *Clin. Chem.* **18**, 1537–8.

OELLERICH, M., KÜLPMANN, W.R., HAECKEL, R., AND HEYER, R. (1977). Determination of phenobarbital and phenytoin in serum by a mechanized enzyme immunoassay (EMIT) in comparison with a gas-liquid chromatographic method. *J. clin. Chem. clin. Biochem.* **15**, 353–8.

OPSTAD, E.T. (1958). False albuminuria due to para-amino-salicylic acid. *Minn. Med.* **41**, 111–20.

OSINGA, A. AND DE WOLFF, F.A. (1976). Determination of quinidine in human serum in the presence of diuretics. *Clin. Chim. Acta* **73**, 505–12.

PANEK, E., YOUNG, D.S., AND BENTE, J. (1978). Analytical interferences of drugs in clinical chemistry. *Am. J. med. Technol.* **44**, 217–24.

PAPE, B.E. (1982). Enzyme immunoassay, liquid chromatography, and spectrofluorometry compared for the determination of procainamide and N-acetylprocainamide in serum. *J. Anal. Toxicol.* **6**, 44–8.

PEAT, M.A. AND JENNISON, T.A. (1978). High-performance liquid chromatography of quinidine in plasma with use of a microparticulate silica column. *Clin. Chem.* **24**, 2166–8.

PENNOCK, C.A., MURPHY, D., SELLERS, J., AND LONGDON, K.J. (1973). A comparison of autoanalyser methods for the estimation of glucose in blood. *Clin. Chim. Acta* **48**, 193–201.

PHILLIPS, A.P. (1973). The improvement of specificity in radio-immunoassays. *Clin. Chim. Acta* **44**, 333–40.

PIERACH, C.A., CARDINAL, R.A., PETRYKA, Z.J., AND WATSON, C.J. (1977). Unusual Watson–Schwartz test from methyldopa. *New Engl. J. Med.* **296**, 577.

POCELINKO, R., SOLOMON, H.M., AND GAUT, Z.N. (1969). Doped dipstick. *New Engl. J. Med.* **281**, 1075.

POLK, R.E. AND STEPHENS, G.H. (1981). Effect of cefazolin and moxalactam on serum chemistry values determined by autoanalyzer. *Am. J. hosp. Pharm* **38**, 866–8.

POWERS, J.L. AND SADEE, W. (1978). Determination of quinidine by high-performance liquid chromatrography. *Clin. Chem.* **24**, 299–302.

QUATTRONE, A.J. AND PUTNAM, R.S. (1981). A single liquid-chromatographic procedure for therapeutic monitoring of theophylline, acetaminophen, or ethosuximide. *Clin. Chem.* **27**, 129–32.

RANKIN, L.I., SWAIN, R.R., AND LUFT, F.C. (1979). Effect of cephalothin on measurement of creatinine concentration. *Antimicrob. Agents Chemother.* **15**, 666–9.

REIO, L. AND WETTERBERG, L. (1969). False porphobilinogen reactions in the urine of mental patients. *J. Am. med. Ass.* **207**, 148–50.

RICCI, N., TOMA, P.M., PAZZI, P., STABELLIN, G., AND VANARA, F. (1982). Bromide can interfere with chloride estimation. *Lancet* ii, 100–1.

ROBINSON, C.A. AND DOBBS, J. (1978). Acetazolamide interference with theophylline analysis by high-performance liquid chromatography. *Clin. Chem.* **24**, 2208–9.

RODDIS, M.J. (1981). Paracetamol interference with glucose analysis. *Lancet* ii, 634–5.

RAMANO, A.T. (1973). Automated glucose methods: Evaluation of a glucose oxidase peroxidase procedure. *Clin. Chem.* **9**, 1152–7.

ROSENBERG, J.M. AND TOBEY, H.C. (1971). A false elevation of serum albumin in a patient receiving paramethadione. *Am. J. hosp. Pharm.* **281**, 493.

ROUTH, J.I. AND PAUL, W.D. (1976). Assessment of interference by aspirin with some assays commonly done in the clinical laboratory. *Clin. Chem.* **22**, 837–42.

SAAH, A.J., KOCH, T.R., AND DRUSANO, G.L. (1982). Cefoxitin falsely elevates creatinine levels. *J. Am. med. Ass.* **247**, 205–6.

SABATH, I.D., GERSTEIN, D.A., AND FINLAND, M. (1968). Serum glutamic oxalacetic transaminase, false elevation during administration of erythromycin. *New Engl. J. Med.* **279**, 1137–9.

SCHENKEN, J.R. AND GROSS, I. (1971). Deferoxamine and auto-assay of serum iron. *Clin. Toxicol.* **4**, 641–2.

SCHERRMANN, J.M. AND BOURDON, R. (1980). Cross reactivity of digitoxin in radioimmunoassay and enzyme-linked immunoassay for digoxin in plasma. *Clin. Chem.* **26**, 670–1.

SCHIER, G.M. AND GAN, I.E.T. (1979). Interference of theophylline and caffeine in the gas-chromatographic estimation of phenobarbital. *Clin. Chem.* **25**, 1191.

SEEDORF, E.E., POWELL, W.N., AND GREENLEE, R.G. (1953). Telepaque and pseudoalbuminaria. *J. Am. med. Ass.* **152**, 1332–3.

——, ——, ——, AND DYSART, D.N. (1952). Pseudo-albuminaria following Monophen and Priodax. *Radiology* **59**, 422–3.

SHANNON, K., WARREN C., AND PHILLIPS, I. (1980). Interference with gentamicin assays by gallium-67. *J. Antimicrob. Chem.* **6**, 285.

SHANSON, D.C. AND HINCE, C.J. (1977). Factors affecting plate assay of gentamicin. III Klebsiella assay strains. *J. Antimicrob. Chem.* **3**, 563–70.

SHARP, P. (1972). Interferences in glucose oxidase–peroxidase blood glucose methods. *Clin. Chim. Acta* **40**, 115–20.

——, RILEY, C., COOK, J.G.H., AND PINK, P.J.F. (1972). Effect of two sulphonylureas on glucose determinations by enzymic methods. *Clin. Chim. Acta* **36**, 93–8.

SHUKUR, L.R., POWERS, J.L., MARQUES, R.A., WINTER, M.E., AND SADEE, W. (1977). Measurement of procainamide and N-acetylprocainamide in serum by high-performance liquid chromatography. *Clin. Chem.* **23**, 636–8.

SIEST, G., APPEL, W., BLIJENBERG, G.B., CAPOLAGHI, B., GALTEAU, M.M., HEUSGHEM, C., HJELM, M., LAUER, K.L., LE PERRON, B., LOPPINET, V., LOVE, C., ROYER, R.J., TOGNONI, G., AND WILDING, P. (1976). Drug interference in clinical chemistry. In *Drug interference and drug measurement in clinical chemistry* (ed. G. Siest and D.S. Young), pp. 1–9. Karger, Basle.

SIGMA CHEMICAL COMPANY (1977). *The enzymatic–calorimetric determination of uric acid in serum or urine.* Technical Bulletin No. 680. St Louis, Missouri.

SILBER, B., SHEINER, L.B., POWERS, J.L., WINTER, M.E., AND SADEE, W. (1979). Spironolactone-associated digoxin radioimmunoassay interference. *Clin. Chem.* **25**, 48–50.

SINGH, H.P., HEBERT, M.A., AND GAULT, M.H. (1972). Effect of some drugs on clinical laboratory values as determined by the Technicon SMA 12/60. *Clin. Chem.* **18**, 137–44.

SITAR, D.S., GRAHAM, D.N., RANGNO, R.E., DUSFRESNE, L.R., AND OGILVIE, R.I. (1976). Modified colorimetric method for procainamide in plasma. *Clin. Chem.* **22**, 379–80.

SKEGGS, L.T. AND HOCHSTRASSER, H. (1964). Multiple automated sequential analysis. *Clin. Chem.* **10**, 918–36.

SMALL, R.E., FREEDY, H.R., AND SMALL, B.J. (1976). Alpha-methyldopa interference with the phosphotungstate uric acid test. *Am. J. hosp. Pharm.* **33**, 556–60.

SMITH, D. AND YOUNG, W.W. (1977). Effect of large-dose ascorbic acid on the two-drop Clinitest determination. *Am. J. hosp. Pharm.* **34**, 1347–9.

SOLDIN, S.J. AND HILL, J.G. (1977a). Routine dual-wavelength analysis of anticonvulsant drugs by high-performance liquid chromatography. *Clin. Chem.* **23**, 2352–3.

—— AND —— (1977b). A rapid micromethod for measuring theophylline serum by reverse-phase high-performance liquid chromatography. *Clin. Biochem.* **10**, 74–7.

SOUNEY, P.F., MENARD, C., CHANG, J.T., AND CHURCHILL, W.W.

(1983). Effect of cephem antibiotics on creatinine assay. *Am. J. hosp. Pharm.* **40**, 1152–3.

STANLEY, P.E. AND PEIKERT, M.R. (1978). Determination of phenytoin in plasma: A comparison of procedures using a new radioimmunoassay, gas chromatography, and enzyme immunoassay. *Epilepsia* **19**, 265–72.

STEARNS, F.M. (1981). Determination of procainamide and N-acetylprocainamide by high performance liquid chromatography. *Clin. Chem.* **27**, 2064–7.

STERLING, J., COX, S., AND HANEY, W.G. (1974). Comparison of procainamide analyses in plasma by spectrophotofluorometry, colorimetry and GLC. *J. Pharm. Sci.* **63**, 1744–7.

STEYN, J.M. AND HUNDT, H.K.L. (1977). Probenecid, a possible interferent in the gas chromatographic determination of diphenylhydantoin. *J. Chromatogr.* **143**, 207–9.

ST ONGE, L.M., DOLAR, E., ANGLIM, M.A., AND LEAST, C.J. (1979). Improved determination of phenobarbital, primidone, and phenytoin by use of a preparative instrument for extraction, followed by gas chromatography. *Clin. Chem.* **25**, 1373–6.

STOOKEY, L.L. (1970). Ferro zinc — A new spectrophotometric reagent for iron. *Anal. Chem.* **42**, 779–81.

SVENSMARK, O. AND KRISTENSEN, P. (1963). Determination of diphenylhydantoin and phenobarbital in small amounts of serum. *J. Lab. clin. Med.* **61**, 501–7.

SWAIN, R.R. AND BRIGGS, S.L. (1977). Positive interference with the Jaffe reaction by cephalosporin antibiotics. *Clin. Chem.* **23**, 1340–2.

TECHNICON SMAC HIGH-SPEED COMPUTER-CONTROLLED BIO-CHEMICAL ANALYSER, PRODUCT LABELLING (1976). Technicon Instrument Corporation, Terrytown, New York.

VINET, B. AND LETELLIER, G. (1977). The in vitro effect of drugs on biochemical parameters determined by a SMAC system. *Clin. Biochem.* **1**, 47–51.

WACHTER, J.P., SMEBY, R.R., AND FREE, A.H. (1960). Urinalysis and oral hypoglycemic agents. *Am. J. med. Technol.* **26**, 125–30.

WATKINS, R.E., FELDKAMP, C.S., THIBERT, R.J., AND ZAK, B. (1976). Interesting interferences in a direct serum creatinine reaction. *Microchem. J.* **21**, 370–84.

WEIDNER, N., DIETZLER, D.N., LADENSON, J.H., KESSLER, G., LARSON, L.. SMITH, C.H., JAMES, T., AND McDONALD, J.M. (1980). A clinically applicable high-pressure liquid chromatographic method for measurement of serum theophylline with detailed evaluation of interferences. *Am. J. clin. Pathol.* **73**, 79–86.

WHIPPLE, R.L. AND BLOOM, W.L. (1950). The occurrence of false positive tests for albumin and glucose in the urine during the course of massive penicillin therapy. *J. Lab. clin. Med.* **36**, 635–9.

WILKINS, D.T. (1971). Albumin-induced elevation of alkaline phosphate. *J. Am. med. Ass.* **215**, 486.

WRIGHT, L.A. AND FOSTER, M.G. (1980). Effect of some commonly prescribed drugs on certain chemistry test. *Clin. Biochem.* **6**, 249–52.

YOSSELSON-SUPERSTINE, S. (1983). Drugs causing interference with laboratory tests. In D'Arcy, P.F., and Griffin, J.P. (eds.) *Iatrogenic diseases, Update 1983*, pp. 214–17. Oxford University Press, Oxford.

—— (1984). Drug interferences with plasma assays in therapeutic drug monitoring. *Clin. Pharmacokinet.* **9**, 67–87.

——, GRANIT, D., AND SUPERSTINE, E. (1980a). Drug interferences with tests performed by a 12-channel autoanalyzer. *Am. J. hosp. Pharm.* **37**, 1333–8.

——, ——, AND —— (1980*b*). Drug interference with the phosphotungstate uric acid test. *Am. J. hosp. Pharm.* **37**, 1458–62.

——, ——, AND —— (1980*c*). Theophylline interference with the phosphotungstate uric acid test. *Am. J. hosp. Pharm.* **37**, 1522–4.

YOUNG, D.S. AND PANEK, E. (1976). Effects of drugs on the anlytical procedures of a multitest analyzer. *In Drug interference and drug measurement in clinical chemistry* (ed. G. Siest and D.S. Young), pp. 10–20. Karger, Basle.

——, PESTANER, L.C., AND GIBBERMAN, V. (1975). Effects of drugs on clinical laboratory tests. *Clin. Chem.* **21**, 1D–432D.

ZARODA, R.A. (1964). Effect of various anticoagulants on carbon dioxide-combining power of blood. *Am. J. clin. Pathol.* **41**, 377–80.

41 Handling anticancer drugs

P.F. D'ARCY

The past 20 years or so have witnessed impressive changes in the drug treatment of cancers. Not only has there been a substantial increase in the range of effective drugs but there has also been a corresponding increase in the number of practitioners prescribing and health-care personnel handling or administering these potent and inherently toxic agents.

All of the cancer chemotherapeutic agents kill or impair susceptible tumour cells by blocking a drug-sensitive biochemical or metabolic pathway. Even in therapeutic doses, cytotoxic drugs produce toxic side-effects due to poor selectivity between target and normal cells. Rapidly dividing cells are particularly damaged, for example those of bone marrow, gut mucosa, hair follicles, testes, thymus, and the fetus. These toxic effects of antineoplastic drugs are medicated directly by their antimitotic or cytotoxic properties and indirectly by their myelosuppressive or cytotoxic immunosuppressive action or both.

This picture of clinical toxicity has been well documented over the years (see for example *Lancet* 1982; Penn 1982) and is largely well recognized relative to adverse effects of treatment in patients. However, many of these drugs have a direct irritant effect on the skin, eyes, mucous membranes, and other tissues and some come into contact with these chemicals. Handled without due care, especially when powdered drugs are being prepared for injection, most cytotoxic drugs can cause local toxic or allergic reactions or both; furthermore they present hazards of carcinogenicity and mutagenicity and this spectrum of potential risk should always be kept in mind by personnel administering or handling these drugs. This is especially important in oncology departments where just a few individuals may routinely and frequently reconsitute many doses of cytotoxic agents. Danger can arise from contact with skin or through environmental contamination following, for example, spillage or aspiration of charged syringes.

It is therefore appropriate in this context to review problems in handling anticancer drugs, to warn practitioners about the inherent dangers, and to suggest ways in which these hazards can be reduced or prevented. The hazards stem indeed from reactions to drugs; however, unlike the normal *milieu* of adverse drug reactions, the practitioner is also at hazard as well as the treated patient. It is a *milieu* of drug therapy that has very few counterparts in the whole field of drug reactions; chronic exposure of anaesthetists to environmental concentrations of volatile anaesthetists, exposure of radiologists and radiographers to radiation hazards, the one-time exposure of dental surgeons and their assistants to the peculiar hazards of mercury vapour, and the exposure of workers in the pharmaceutical industry to skin-sensitizing substances, e.g. phenothiazones, are the only other examples of such 'practitioner–drug' interactions that come readily to mind.

The nature of the problem

The first report in the literature of any potential danger to personnel admixing chemotherapeutic agents appeared in 1979. Falck *et al.* (1979) from the Institute of Occupational Health in Helsinki found mutagenic activity in the urine of oncology nurses who mixed and administered anticancer drugs. The significance of this discovery can only be a matter of speculation, but the potential hazard is obvious when cytotoxic drugs or their metabolities are excreted in urine. Falck and his colleagues commented in this respect that it is known that patients on cyclophosphamide may be at risk from urinary-tract tumours and this might correlate with the excretion of mutagenic urinary metabolites of that drug. These authors inferred that personnel handling cyclophosphamide might also share in that hazard.

Walsvik *et al.* (1981) from the Norsk Hydro's Institute for Cancer Research and The Norwegian Radium Hospital were also concerned with possible health hazards in handling cytotoxic drugs and therefore studied the chromosomes of two groups of nurses who had worked daily with such chemicals. There were 10 nurses in the first group (group 1) and they were selected for this cytogenetic study because of their long-term experience (average 2150 hours) of preparing solutions of cytotoxic drugs and drawing them into syringes ready for infusion. All the nurses claimed that they had used hand gloves and surgical masks all the time, whereas a fume cabinet had only been used by seven of them during the previous 10 months. Each nurse was asked to rank the 10 cytostatic agents most frequently handled and pooling information gave the following ranking list: cyclo-phosphamide, vincristine, doxorubicin, methotrexate, vinblastine, mechlorethamine, 5-FU, dactinomycin, dacarbazine, bleomycin, cisplatin, lomustine (capsules), ifosfamide, abrin, VP-16-213, thiotepa, 9-hydroxyellip-ticine, mithramycin, procarbazine (capsules), cytara-bine, and AMSA. Ten women working as clerks in the same building were selected as controls (group 2); a second group of 11 nurses (group 3) were also studied — they had handled cytotoxic agents for a shorter period of time (average 1078 hours). A fourth group of nurses were engaged in therapeutic and diagnostic radiology, a fifth group of nine nurses were exposed to anaesthetic gases in their work in operating theatres, and a sixth group of seven nurses were working in a post-operative ward in which exposure to mutagenic agents was almost absent.

Chromosome damage was assessed by three indices: the number of gaps per 100 cells, the number of breaks per 100 cells, and the number of sister chromatic exchanges (SCEs) per 30 cells from each subject.

Analyses showed that compared with controls there was an increased frequency of chromosome gaps and a slight increase in SCE frequency among the group-1 nurses who had long-term exposure to anticancer drugs. These indices did not differ from controls in other groups of nurses except for an increased frequency of chromo-some gaps in the radiology group (group 4).

Waksvik et al. concluded that the exposure of group-1 nurses to mutagenic agents must have been fairly low; their findings, however, pointed to the handling of cytostatic drugs as a possible health hazard and indicated that protective measures (gloves, masks, and fume cabinets) should be used.

Evidence from other centres is largely anecdotal but none the less disturbing. For example, both Crudi (1980) from Hartford Hospital and Ladik et al. (1980) from Magee Women's Hospital in Pittsburg reported unpleasant effects occurring in their staff who prepared antineoplastic drugs. These effects included light-headedness, dizziness, headache, facial flushing, nausea, and nasal sores.

It was also disturbing to note from Tortorici's (1980) survey of 21 comprehensive cancer centres in the USA, that only four were using vertical laminar flow hoods to prepare chemotherapy, that only four required staff to wear masks, and that only eight insisted that gloves be worn. Of the respondents to his questionnaire, 86 per cent said that they were concerned with the possible adverse effects of handling cytotoxic drugs, but it was Tortorici's conclusion that most institutions did

Table 41.1 Methods of preparation, administration, and disposal of injectable antineoplastic medications at comprehensive cancer centres (n = 21) (from LeRoy et al. 1983)

Variable	No.
Responsibility for preparing chemotherapy medication	
Pharmacist or pharmacy technician	9
Pharmacist or nurse	5
Pharmacist	3
Nurse, pharmacist, or pharmacy technician	2
Nurse	1
Not answered	1
Responsibility for administering chemotherapy medication	
Nurse	9
Nurse or physician	9
Pharmacist, nurse, or physician	1
Pharmacist, nurse, or physician assistant	1
Medical technician, nurse, or physician	1
Location of preparing injectable antineoplastics	
Vertical laminar-flow hood	10
Clean countertop	4
Horizontal laminar-flow hood	4
Clean countertop and horizontal laminar-flow hood	3
Method of handling waste if pharmacy disposes of it separately[a]	
Incinerated	5
Same as other hazardous waste	3
Same as radioactive waste	2
Autoclaved	1
Returned to manufacturer or incinerated	1
Same as other hazardous waste or incinerated	1

[a] Thirteen respondents noted that antineoplastic drug waste is disposed of separately.

absolutely nothing to protect their personnel from such hazards.

A later survey by LeRoy et al. (1983) of procedures for handling injectable antineoplastic drugs in compre-hensive cancer centres in the United States is somewhat more reassuring. In May 1982, these investigators sent a survey questionnaire to directors of pharmacy at 27 institutions designated as comprehensive cancer centres. Some questions duplicated the 1979 survey of Tortorici, while others addressed points in guidelines on handling antineoplastic medications that had been published since that first survey (e.g. Davis 1981; Harrison 1981; Wilson and Solimando 1981a,b; Zimmerman et al. 1981; Ander-son et al. 1982).

Table 41.2 Precautions for handling injectable antineoplastic medications at comprehensive cancer centres (n = 21) (from LeRoy et al. 1983)

Variable	No. responses	
	Yes	No
Training programme		
Preparation of antineoplastic medications	9	12
Administration of antineoplastic medication	10	11
Written policies and procedures		
Preparation and handling of antineoplastics	17	4
Administration of antineoplastics	17	4
Disposal of antineoplastics	11	10
Tests performed on personnel handling antineoplastics		
Urine testing (Ames test)	2	19
Complete blood count	2	19
Physical examination	5	16
Health assessment of some kind	6	15
Precautions used for bulk or unit dose repackaging of oral antineoplastics[a]		
Separate counting trays	2	18
Routine cleaning of trays	9	11
Separate unit dose packaging	0	20
Precautions used for preparing antineoplastics[b]		
Gloves	20	1
Hand washing	16	5
Masks	13	8
Gowns	11	10
Goggles	7	14
Laboratory coats	6	15
Dispose of antineoplastic drug waste separately	13	8
One cabinet used exclusively for preparation of antineoplastics[c]	4	15

[a] One respondent did not answer this question.
[b] The yes responses indicate institutions that currently use or plan to use the precautions within the next six months.
[c] Two respondents did not answer this question.

Representatives of 21 out of 27 institutions contacted responded to the questionnaire (Tables 41.1 and 41.2). These 21 institutions represented 13 638 beds of which 1848 were for oncology patients. Seventeen institutions had written policies for the preparation of antineoplastic drugs, but only nine had a training programme. The pharmacist or pharmacy technician prepared these medications in 12 institutions. Ten institutions prepared antineoplastics in a vertical laminar-flow hood. Gloves and masks were worn by employees in 20 and 13 of the institutions, respectively. Six institutions in some way assessed the health of the employees handling antineoplastics. Eleven instititions had written policies on disposal of antineoplastics and 13 institutions disposed of their waste separately. Ten institutions had a training programme for administration of antineoplastics.

Compared with the data from the 1979 survey (Tortorici 1980). the trend shown by this later survey is towards increased protection of persons handling injectable antineoplastic agents. Institutions may also be more aware of the problems associated with the handling of antineoplastic waste than they were in 1979. However, lack of standardized policies for all these procedures can be a problem. Centres do not agree on policies concerning the safe handling of antineoplastic drugs and waste, and community pharmacies and hospitals have a difficult time in interpreting published guidelines. If voluntary attempts to standardize these policies are not successful, then LeRoy and her colleagues (1983) rightly fear that potentially expensive and cumbersome governmental legislation may be introduced in the near future in the USA and that this will probably be mirrored in legislation in other countries.

Until recently, there was no information in the literature on ambient concentrations of antineoplastics in drug-preparation areas. Such data have now become available from the work of Neal et al. (1983) at the School of Public Health, University of Illinois. They surveyed practices for handling antineoplastic drugs in 10 hospitals with out-patient oncology clinics in the Chicago area and also conducted ambient air-sampling for cyclophosphamide, doxorubicin, fluorouracil, and methotrexate at three drug-preparation sites.

Nine clinics had no ventilation hoods, and drugs were prepared by nurses in eight clinics. Routine use of gloves (three clinics) and masks (one clinic) was uncommon. Wastes were disposed of in uncovered receptacles in four of the clinics. Eating and drinking occurred in seven of the preparation rooms.

At the main sampling site, fluorouracil (0.12–82.26 ng m^{-3}) was detected in air during 200 of the 320 hours monitored; cyclophosphamide (370 ng m^{-3}) was present during 80 hours. In the two other sites, fluorouracil was found in minimal detectable levels, but no detectable amounts of methotrexate or doxorubicin were present.

The authors commented that personnel handling antineoplastic drugs are subject to potential systemic absorption of these agents by inhalation. Although fluorouracil is not considered to be as carcinogenic in man as the alkylating agents (e.g. cyclophosphamide), the long-term health effects of daily low-level exposure to fluorouracil and other antineoplastics are unknown.

Hospital pharmacists take precautions

Three reports have shown that hospital pharmacists avoid hazards from handling anticancer drugs simply because they take appropriate precautions — precautions incidentally that are almost identical with those involved in standard aseptic procedures for preparing pharmaceutical products.

The first of these reports was from the pharmacists at the US National Institutes of Health (Staiano et al. 1981) who spent most of their time reconstituting and admixing antineoplastic drugs for intravenous administration. Eight subjects took part in this study and no mutagenic activity was detected in the urine of any of these in spite of the fact that they had commonly handled bleomycin, 5-FU, vincristine, methotrexate, cyclophosphamide, daunomycin, doxorubicin, and actinomycin D. They took the precaution, however, of handling these agents in vertical or horizontal laminar-air flow hoods and they wore protecting gloves during these operations.

The second report came from pharmacists at the Walter Reed Army Medical Center. Pharmacy personnel there have mixed and administered an average of over 100 doses of chemotherapy per month for a number of years. There were no complaints of any unusual symptoms from these personnel and urine examinations for mutagenicity were negative (Wilson and Solimando 1981a,b).

The third report from the Anderson Hospital and Tumor Institute at Houston, showed that over 200 doses of chemotherapy were prepared daily and that no reports of any unusual symptoms had been made by the pharmacy personnel (Ballentine 1982). Mutagenicity studies had, however, been recently instituted at that hospital as a precaution for those personnel who had the highest exposure to such drugs.

The value of urinary mutagenicity assays in monitoring exposure of nursing and pharmacy personnel to cytotoxic drugs was, however, questioned by Venitt et al. (1984) from the Institute of Cancer Research, Pollards Wood Research Station, Buckinghamshire, UK. These assays gave positive results in a high proportion of non-exposed control subjects as well as in 'exposed' hospital staff and were therefore considered unsuitable for routine monitoring.

This conclusion is not altogether surprising since mutation assays of biological material are beset with technical problems. Some arise from the use of assays employing reverse mutation of bacteria from aminoacid auxotrophy to prototrophy (Venitt and Bosworth 1983). Others are peculiar to urinary mutation assays: for example, the urine of smokers is mutagenic (Yamasaki

and Ames 1977; Dolara et al. 1981; Kriebel et al. 1983) and the urine of non-smokers subjected to passive smoking has been claimed to be weakly mutagenic (Bos et al. 1983) and to contain nicotine (Feyerbrand et al. 1982). Diet may be another confounding factor; for example, Baker et al. (1982) detected mutagenic acitvity in human urine after meals of fried pork or bacon.

Since platinum-containing drugs have a widespread use in oncology clinics (e.g. cisplatin and the newer cis-diammine-1, 1-cyclobutanedicarboxylate Pt II (CBDCA)), Venitt et al. (1984) thought that determination of urinary platinum levels with atomic-absorption spectrophotometry might be a useful alternative marker to urinary mutagenic assays for monitoring drug absorption by hospital personnel. They examined untreated urine from two pharmacists and eight nurses who routinely handled cytotoxic drugs and detected platinum levels (mean 10.2 ng Pt/ ml urine: range 0.6 to 23.1 ng Pt/ ml) although these were generally below the reliable limit of detection. In contrast, the urine of cisplatin-treated patients contained an average level of 7 μg Pt/ml (range 4.1 to 11.3 μg Pt/ ml). This latter result also suggested that contact with urine from patients undergoing intensive chemotherapy might be hazardous.

Venitt et al. (1984) concluded from their study that the universal adoption of safe methods for handling cytotoxic antitumour drugs is likely to be the best method of eliminating possible risks to hospital staff. However, should it be necessary to monitor drug absorption, their studies suggested that direct chemical analysis of urine would be a more reliable guide to drug absorption than the use of urinary mutagenicity assays.

Safe handling

A number of review articles have made practical recommendations on the way in which cytotoxic drugs should be prepared and handled to minimize the risk to operatives. Foremost amongst these articles are those by Davis (1980) from Woden Valley Hospital, A.C.T. Australia, Hoffman (1980) from the Sloan-Kettering Cancer Center, New York, Knowles and Virden (1980) from the Western General Hospital, Edinburgh, the Derbyshire Drug Information Service, UK (Trent Regional Health Authority 1982), and an editorial by Zellmer (1981) in the *American Journal of Hospital Pharmacy*.

More recently, the American Society of Hospital Pharmacy selected the topic 'Safe handling of cytotoxic drugs' as the 1983–84 Practice Spotlight Program (Stolar and Power 1983). Pharmacists were urged to review their respective institutions's policies and procedures for preparing, administering, and disposing of antineoplastic drugs and other cytotoxic agents. This has been the stimulus for a number of useful and explicit publications, in particular: *Reccommendations for handling cytotoxic drugs in hospitals* (Stolar *et al.* 1983), and *Pharmacy program for improved handling of antineoplastic agents* (Scott *et al.* 1983). correspondence columns of journals have also contributed useful advice (e.g. Crudi 1982; Hoffman 1982; Spross 1982; Cooke 1983; Hirst *et al.* 1983; Kennelly, 1983; Stolar 1983; Warrington and Howden 1983) not only on handling cytotoxic drugs but also on associated problems of compatibility of drugs with infusion soft-ware (e.g. Forrest 1984).

In addition, in the UK, the official Health and Safety Executive (1983) have published a pamphlet *Precautions for the safe handling of cytotoxic drugs* under the provisions of the Health and Safety at Work Act 1974. This gives advice on precautions aimed at reducing the exposure of health-care personnel (pharmacists, nurses, and doctors) to cytotoxic drugs during reconstitution, preparation, and administration. It is intended for use as a basis for the preparation of detailed local rules at places where cytotoxic drugs are used. A similar document *Recommendations for handling cytotoxic agents* was widely distributed in America in August 1983 by the independent National Study Commission on cytotoxic exposure (see Gallina 1983).

The message contined in these reviews is relatively simple and largely uniform in its context; it is that the pharmacy department should take the leadership in making hospital staff aware of the potential risks in handling anticancer drugs. Furthermore, it is clear that pharmacy departments in hospitals and institutes should spearhead a multidisciplinary assessment of the desirability of producing local guidelines for working with these drugs. Such guidelines are largely a matter of informed common sense and are based on the simple principle of designing and using equipment and work techniques to ensure that personnel have minimal contact with these toxic chemicals. Protective gloves are an important part of these pecautionary measures and it is important to note that Thomsen and Mikkelsen (1975) showed that only gloves made from polyvinylchloride (PVC) protected against the penetration of nitrogen mustard; rubber gloves afforded no protection nor did gloves made from polyethylene.

The recommendations of Zimmerman *et al.* (1981) from the US National Institutes of Health for the safe handling of antineoplastic drug products outline clearly the containment equipment and the work techniques that reduce contact with anticancer drugs to a minimum. This paper is also a useful *aide mémoire* of the way in which personnel may be exposed to antineoplastics. While exposure via skin contact with drug solution is conspicuous, inhalation of the aerosolized drug product is not. Such an aerosol of an injection solution may be formed by withdrawing a needle from a vial, opening an ampoule, or by expelling air from a syringe. In this respect, it is also important to note the findings of Kleinberg and Quinn (1981) who showed that there were detectable levels of drug in the air after manipulation of injectable drug solutions in a horizontal laminar air-flow hood. Indeed the National Institutes of Health (Zimmerman *et al.* 1981) recommended that Class II biological safety cabinets be used for preparing injectable doses of anticancer drugs and this is prudent advice. Horizontal laminar air-flow hoods seem to predominate in most pharmacy-based intravenous fluid admixture services, and although this type of equipment is quite adequate for handling most injectables in terms of maintaining sterility, it is becoming quite clear that greater protection is needed for the operator when antineoplastic agents are involved.

Knowles and Virden (1980) gave sound and practical advice on the handling of injectable antineoplastics and in particular they presented their recommendations in tabular form for 21 drugs; a similar list was also prepared by the Derbyshire Drug Information Service (Trent Regional Health Authority 1982) and recommended handling precautions are given for the same 21 injectable antineoplastic agents. These two lists have been collated together and other information has been added from manufacturers' literature in Table 41.3 in this chapter.

Frequently, solutions of antineoplastic drugs are incorporated as admixtures into other injection solutions; it is therefore appropriate in this present context to give some information about major *in-vitro* incompatibilities; these are summarized in Table 41.4. Table 41.5 summarizes the action that should be taken on spillage of anticancer drugs during the preparation of injection solutions from vials of powdered drug.

Important observations, not only on the utility and safety but also on the economy of a pharmacy-based anticancer drug reconstitution service, were given by Anderson *et al.* in their paper 'Development and operation of a pharmacy-based intravenous cytotoxic reconstitution service' which appeared in the *British Medical Journal* in January 1983. These pharmacists and their clinical oncology colleague at the London Hospital described how the pharmacy department took over the cytotoxic reconstitution service and the responsibility for presenting these drugs in a readily usable form. In doing so, they claimed to have saved much time and also to have eliminated the potential hazards that would otherwise have faced medical and nursing staff who were not experienced in handling these toxic chemicals.

Table 41.3 Handling anticancer drugs: recommended precautions in preparing injection solutions (based on data from Knowles and Virden (1980); Trent Regional Health Authority (1982); Anderson *et al.* (1983); Association of the British Pharmaceutical Industry *Data Sheet Compendium 1983–84* (1983); and other manufacturers' literature)

Drug/route*	Expiry time after reconstitution (store in dark)	Effect on skin, eyes and mucous membranes	Handling precautions (manufacturer's advice)	Action on contamination
Actinomycin D (Dactinomycin, *Cosmegen*, *Lyovac*) i.v.	Discard unused solution	Corrosive	Gloves and eye shield	Rinse in running water for 10 min; finally rinse with buffered phosphate solution
Amsacrine (Acridinyl anisidine, *m*-AMSA) i.v.	Two sterile liquids to be mixed aseptically before use; stable at room temperature for 48 h. Use only with glass syringes.	One of the admixture liquids (lactic acid) is irritant	Gloves and eye shield	Wash thoroughly with water
Azathioprine (*Imuran*) i.v.	Discard unused solution	Alkaline and irritant	Gloves and eye shield	Wash off with water quickly
Bleomycin (*Bleomycin Lundbeck*) i.m., i.v., i.a.	Use freshly prepared	Locally toxic, allergenic	Gloves and mask	Rinse thoroughly with water; then wash with soap and water
Cisplatin (*Neoplatin*) i.v.	20 h at room temperature. Refrigeration causes precipitation	Potentially allergenic	Gloves and mask necessary only if spillage	Wash thoroughly with water
Colaspase (*Crasnitin*) i.v.	14 days at 5°C	Not irritant	No special precautions	Wash with water
Cyclophosphamide (*Endoxana*) i.v.	6 days at 4°C	Skin irritation is rare	No special precautions	Wash thoroughly with water
Cytarabine (*Cytosar*) i.v., s.c., i.t.	Store at room temperature; use within 48 h	Not absorbed through intact skin	No special precautions	Wash thoroughly with water
Dacarbazine (*DTIC*) i.v.	72 h at 4°C	Irritant	Gloves	Wash with soap and water immediately; irrigate eyes with water
Daunorubicin (*Cerubidin*) i.v.	48 h at room temperature	Irritant	Gloves	Immediate washing with water or isotonic saline
Doxorubicin (*Adriamycin*) i.v.	48 h at 4°C	Irritant	Gloves	Copious washing with soap and water
Estraumustine sodium phosphate (*Estracyt, Emyct*) Oral (rarely by injection)	Store capsules at 2–8°C. If solutions are prepared, use immediately	Irritant, drug must not come into contact with skin or mucous membranes	Special precautions (e.g. gloves, eye shield, face mask) must be taken if the capsule has to be opened.	Rinse copiously with water; if eyes are contaminated, rinse quickly with isotonic saline followed by an isotonic solution (2.98 per cent) of sodium thiosulphate. Seek advice from an ophthalmologist if irritation continues. Wipe up spillage with a mixture of acetone and concentrated solution of

(continued overleaf)

Table 41.3   *Continued*

Drug/route*	Expiry time after reconstitution (store in dark)	Effect on skin, eyes and mucous membranes	Handling precautions (manufacturer's advice)	Action on contamination
				ammonia followed by through washing with water
Ethoglucid (*Epodyl*) i.v., i.a.	Prepare immediately before use. Use only with glass syringes	Irritant in undiluted form	Gloves during preparation	Wash immediately with large amounts of water
Fluorouracil (*Fluoro-uracil*) i.v. infusion, i.a.	Dilute ampouled solution immediately before parenteral use	Minor local inflammation if skin is broken	No special precautions but avoid contact with skin and mucous membranes	Flush affected parts with copious amounts of water
Ifosfamide (*Mitoxana*) i.v.	Use within 2 h of preparation	Irritation rare	No special precautions	Wash thoroughly with water
Lomustine (*CCNU Lundbeck*, *CeeNU*) oral, i.v.	Crystalline form stable for 2 years after manufacture provided package unopened and protected from light	Highly toxic, drug must not come into contact with skin or mucous membranes	Speical precautions (e.g. gloves, eye shield, face mask) must be taken if in exceptional circumstances capsule has to be opened	Skin and mucous membranes should be washed copiously with water. Rinse all utensils which have been in contact with powdered drug with 10 per cent solution of concentrated (25 per cent) aqueous ammonia in acetone or alcohol. Utensils which have been in contact with solutions of the drug should be rinsed with dilute aqueous ammonia
Melphalan (*Alkeran*) i.v., i.a.	Use within 15–30 min	Not irritant	Gloves and eye shield	Solution of 3 per cent sodium carbonate should be used if spillage
Methotrexate i.m., i.v., i.t., i.a.	2 weeks at room temperature	Irritant	Gloves	Wash with water; apply a bland cream for transient stinging. For systemic absorption of significant quantities, give calcium folinate (*Leucovorin*) cover
Mithramycin (*Mithracin*) i.v.	Prepare immediately before use	Irritation rare	No special precautions	Wash with water
Mitomycin (*Mitomycin C*, *Mutamycin*) i.v., i.p., i.a.	7 days at 4°C	Irritant	Gloves	Thorough and immediate washing with water; irrigate eyes with water
Mustine (*Mustargen*) i.v., i.p., i.a.	Prepare immediately before use	Vesicant and strong nasal irritant	Gloves and eye shield	Wash with copious amounts of water or 3 per cent sodium carbonate solution or isotonic solution of sodium thiosulphate (2.98 per cent). Wash eyes with large amounts of water or isotonic

Table 41.3 *Continued*

Drug/route*	Expiry time after reconstitution (store in dark)	Effect on skin, eyes and mucous membranes	Handling precautions (manufacturer's advice)	Action on contamination
				solution of sodium thiosulphate
Vinblastine (*Velbe*) i.v.	30 days at 4°C	Irritant	Gloves	Thorough and immediate washing with large amounts of water. If accidental injection into subcutaneous tissues. Apply heparin (*Hirudoid*) cream to affected area
Vincristine (*Oncovin*) i.v.	14 days at 4°C	Irritant	Gloves	As for vinblastine
Vindesine (*Eldisine*) i.v.	30 days at 4°C	Irritant; contact with eye may cause corneal ulceration	Gloves and eye shield	As for vinblastine; wash eyes thoroughly and immediately with water

* i.v. = intravenous; i.m. = intramuscular; i.a. = intra-arterial; s.c. = subcutaneous; i.t. = intrathecal; i.p. = intraperitoneal.

Table 41.4 **Anticancer drugs:** *In-vitro* **incompatibilities/interactions***

Drug	Nature of incompatibility/interaction
Actinomycin D	Drug binds to cellulose filters; avoid such filtration
Amsacrine	Solutions react with plastics; use glass syringes. Combined solution (N, N-dimethylacetamide: lactic acid) unstable in solutions containing Cl^-; danger of precipitation
Azathioprine	Admixture to any alkaline solution should be avoided since drug will be metabolized to mercaptopurine
Bleomycin	Drug chelates divalent and trivalent cations; do not admix with solutions containing these ions (especially copper). Drug may be precipitated by hydrophobic anions; do not mix drug solutions with solutions of essential amino acids, riboflavine, ascorbic acid, dexamethasone, aminophylline, or frusemide. Drug is inactivated by compounds containing sulphydryl groups (e.g. glutathione)
Cisplatin	Chemical reaction with sodium bisulphite; ciplatin may be inactivated if added to solutions containing this anti-oxidant
Cyclophosphamide	Decomposition is faster in water preserved with benzyl alcohol than in water for injection
Cytarabine	Incompatible with solutions of fluorouracil or methotrexate
Dacarbazine	Incompatible with hydrocortisone sodium succinate
Daunorubicin	Incompatible with heparin sodium
Doxorubicin	Incompatible with heparin sodium; may be incompatible with aluminium (hub of syringe needle), aminophylline, cephalothin sodium, dexamethasone, fluorouracil, and hydrocortisone
Ethoglucid	Concentrated solutions react with plastics; use glass syringes
Etoposide	Injection formulation (*Vepesid*) dissolves cellulose, single-use filter (*Millex*) within seconds due to polyethylene glycol content. Nylon (*Cuno Zetapor*) or PTFE filters (*Fluripore*) are more suitable (Forrest 1984).

(continued overleaf)

Table 41.4 *Continued*

Drug	Nature of incompatibility/interaction
Fluorouracil	Incompatible with cytarabine, diazepam, doxorubicin, and methotrexate. Drug solution is alkaline so avoid admixture with any acidic agents (e.g. amino acids, insulin, multi-vitamins, penicillin, or tetracyclines)
Methotrexate	Incompatible with cytarabine, fluorouracil, and prednisolone sodium phosphate
Mithramycin	Readily chelates with divalent cations (especially Fe^{2+}); avoid admixture with trace element solutions

* Sources: Association of the British Pharmaceutical Industry *Data Sheet Compendium* 1983–84 (1983); Dorr and Fritz (1980); McRae and King (1976); and as cited in the text of the table.

Table 41.5 Precautions to be taken on spillage of anticancer drugs

1. Put on protective gloves
2. Wear a face mask and eye protection if there is a powder spill
3. Place spilled materials in a polythene bag
4. Wipe up remains with a damp cloth or cotton waste and place in bag
5. Seal bag, place in second bag, and seal
6. Label bag stating contents and mark *DANGER*
7. Wash contaminated surfaces with copious amounts of water; wash exposed skin areas with soap and water
8. Wash eyes copiously with water or isotonic saline; seek ophthalmological advice if eye irritation continues
9. Dispose of washing materials used as in 3–6 above; send waste for disposal by incineration
10. Take precautions to prevent further spillage

The basis of this pharmaceutical service, which has now been operating for almost 2½ years, is to provide the appropriate patient dose in a labelled disposable syringe, sealed with a sterile blind hub. The only manoeuvre then required by nursing and medical staff is to remove the blind hub and replace it with a needle; the reconstitution of cytotoxic drugs and the syringe-filling procedures are carried out by gowned, masked, and gloved personnel in a vertical downflow, laminar safety cabinet (Howie Class 2 type) supplied by Microflow Pathfinder Ltd. This type of cabinet gives a high degree of protection for the operator as well as providing a Class 1 working environment suitable for preparing intravenous doses. It avoids the hazards to the operator which are inherent in working with a horizontal laminar flow cabinet (see Donner 1978).

The pharmacy service at the London Hospital is currently reconstituting some 98 per cent of all the intravenous cytotoxic agents prescribed in that hospital. Not only is this a safe and effective procedure; it has also cut the annual cost of injectable cytotoxic drugs by over 10 per cent.

A similar economy was reported by Bennett (1984) from the Christie Hospital, Manchester, UK. Introduction of a pharmacy-based cytotoxic reconstitution service for out-patients resulted in annual savings of over £16 000 on the drug bill. It is predicted that when the service is extended to include in-patients, the overall saving will be over £60 000 per year. Before introduction of the present pharmacy-based service for out-patients, drug wastage accounted for 16 per cent (£64 000 per year) of the cytotoxic drug budget.

RECOMMENDED FURTHER READING

D'ARCY, P.F. (1983). Reactions and interactions in handling anti-cancer drugs. *Drug Intel. clin. Pharm.* 17, 532–8.
—— AND GRIFFIN, J.P. (1982). Drug-induced cancer in animals and man. *Cancer Surv.* 1, 653–80.
DORR, R.T. AND FRITZ, W.L. (1980). *Cancer chemotherapy handbook*. Elsevier, New York.
KINLEN, L.J. (1982). Immunosuppressive therapy and cancer. *Cancer Surv.* 1, 565–83.
ROSNER, F., GRÜNWALD, H.W., AND ZARRABI, H.M. (1982). Cancer after the use of alkylating and non-alkylating cytotoxic agents in man. *Cancer Surv.* 1, 599–612.

REFERENCES

ANDERSON, M., BRASSINGTON, D., AND BOLGER, J. (1983). Development and operation of a pharmacy-based intravenous cytotoxic reconstitution service. *Br. med. J.* 286, 32–6.
ANDERSON, R.W., PUCKETT, W.H., JR, DANA, W.J., NGUYEN, T.V., THEISS, J.C., AND MATNEY, T.S. (1982). Risk of handling injectable antineoplastic agents. *Am. J. hosp. Pharm.* 39, 1881–7.
ASSOCIATION OF THE BRITISH PHARMACEUTICAL INDUSTRY (1983). *Data sheet compendium 1983–84.* Datapharm Publications Ltd, London.
BAKER, R., ARLAUSKAS, A., BONIN, A., AND ANGUS, D. (1982). Detection of mutagenic activity in human urine following fried pork or bacon meals. *Cancer Lett.* 16, 81–9.
BALLENTINE, R. (1982). Editorial: Cancerphobia — or whatever happened to red M & Ms ? *Drug Intel. clin. Pharm.* 16, 60–1.
BENNETT, S. (1984). Reducing cytotoxic drug wastage. *Br. med. J.* 288, 194.

BOS, R.P. THEUWS, J.L.G., AND HENDERSON, P.TH. (1983). Excretion of mutagens in urine after passive smoking. *Cancer Lett.* **19**, 85-90.

COOKE, J. (1983). Handling of cytotoxic drugs. *Pharm. J.* **230**, 274.

CRUDI, C.B. (1980). A compounding dilema: I've kept the drug sterile but have I contaminated myself? *Nat. IV Ther. Ass.* **3**, 77-8.

—— (1982). More on hazards of working with cytotoxic agents. *Oncol. Nurs. Forum.* **9**, 9.

DAVIS, M.R. (1980). Handling and preparation of cytotoxic drugs — minimising the risk. *Aust. J. hosp. Pharm.* **10**, 127-30.

—— (1981). Guidelines for safe handling of cytotoxic drugs in pharmacy departments and hospital wards. *Hosp. Pharm.* **16**, 17-20.

DOLARA, P., MAZZOLI, S., ROSI, D., BUIATTI, S., TURCHI, A., AND VANUCCI, V. (1981). Exposure to carcinogenic chemicals and smoking increases urinary excretion of mutagens in humans. *J. Toxicol. environ. Hlth.* **8**, 95-103.

DONNER, A.L. (1978). Possible risks of working with antineoplastic drugs in horizontal laminar flow hoods. *Am. J. hosp. Pharm.* **35**, 900.

DORR, R.T. AND FRITZ, W.L. (1980). *Cancer chemotherapy handbook* Elsevier, New York.

FALCK, K., GRÖHN, P., SORSA, M., VAINO, H., HEINONEN, E., AND HOLSTI, L.R. (1979). Mutagenicity in urine of nurses handling cytostatic drugs. *Lancet* i, 1250-1.

FEYERBRAND, C., HIGENBOTTAM, T., AND RUSSEL, M.A. (1982). Nicotine concentrations in urine and saliva of smokers and non-smokers. *Br. med. J.* **284**, 1002-4.

FORREST, S.C. (1984). Vepesid injection. *Pharm. J.* **232**, 88.

GALLINA, J.N. (1983). Recommendations for handling cytotoxic drugs in hospitals. *Am. J. hosp. Pharm.* **40**, 2133-4.

HARRISON, B.R. (1981). Developing guidelines for working with antineoplastic drugs. *Am. J. hosp. Pharm.* **38**, 1686-93.

HEALTH AND SAFETY EXECUTIVE (1983). *Precautions for the safe handling of cytotoxic drugs.* Guidance note from the Health and Safety Executive, *Medical Series 21*, pp. 1-3 HMSO, London.

HIRST, M., TSE, S., MILLS, D.G., LEVINE, L., AND WHITE, D.F. (1983). Caution on handling antineoplastic drugs. *New Engl. J. Med.* **309**, 188-9.

HOFFMAN, D.M. (1980). The handling of antineoplastic drugs in a major cancer center. *Hosp. Pharm.* **15**, 302-4.

—— (1982). More on hazards of working with cytotoxic agents. *Oncol. Nurs. Forum.* **9**, 9.

KENNELLY, B. (1983). Recommendations for handling cytotoxic drugs in hospitals. *Am. J. hosp. Pharm.* **40**, 2134.

KLEINBERG, M.L. AND QUINN, M.J. (1981). Airborne drug levels in a laminar-flow hood. *Am. J. hosp. Pharm.* **38**, 1301-3.

KNOWLES, R.S. AND VIRDEN, J.E. (1980). Handling of injectable antineoplastic agents. *Br. med. J.* **281**, 589-91.

KRIEBEL, D., COMMONER, B., BOLLINGER, D., BRONSDON, A., GOLD, J., AND HENRY, J. (1983). Detection of occupational exposure to genotoxic agents with a urinary mutagen assay. *Mutat. Res.* **108**, 67-79.

LADIK, C.F., STOEHR, G.P., AND MAURER, M.A. (1980). Precautionary measures in the preparation of antineoplastics. *Am. J. hosp. Pharm.* **37**, 1184, 1186.

LANCET (1982). Editorial: Hazards of cancer chemotherapy. *Lancet* ii, 1317-18.

LeROY, M.L., ROBERTS, M.J., AND THEISEN, J.A. (1983). Procedures for handling antineoplastic injections in comprehensive cancer centers. *Am. J. hosp. Pharm.* **40**, 601-3.

McRAE, M.P. AND KING, J.C. (1976). Compatibility of antineoplastic, antibiotic and corticosteroid drugs in intravenous admixtures. *Am. J. hosp. Pharm.* **33**, 1010-13.

NEAL, A. DEW., WADDEN, R.A., AND CHIOU, W.L. (1983). Exposure of hospital workers to airborne antineoplastic agents. *Am. J. hosp. Pharm.* **40**, 597-601.

PENN, I. (1982). Mechanisms of therapy-induced malignancies — editorial review. *Cancer Surv.* **1**, 763-82.

SCOTT, S.A., SCHROTT, D.B., AND LOESCH, G.A. (1983). Pharmacy program for improved handling of antineoplastic agents. *Am. J. hosp. Pharm.* **40**, 1179-82.

SPROSS, J. (1982). More on hazards of working with cytotoxic agents. *Oncol. Nurs. Forum.* **9**, 9-10.

STAIANO, N., GALLELLI, J.F., ADAMSON, R.H., AND THORGEIRSSON, S.S. (1981). Lack of mutagenic activity in urine from hospital pharmacists admixing antitumour drugs. *Lancet* i, 615-16.

STOLAR, M.H. (1983). Recommendations for handling cytotoxic drugs in hospitals. *Am. J. hosp. Pharm.* **40**, 2134.

—— AND POWER, L.A. (1983). The 1983-84 ASHP Practice Spotlight: Safe handling of cytotoxic drugs. *Am. J. hosp. Pharm.* **40**, 1161.

——, ——, AND VIELE, C.S. (1983). Recommendations for handling cytotoxic drugs in hospitals. *Am. J. hosp. Pharm.* **40**, 1163-71.

THOMSEN, K. AND MIKKELSEN, H.I. (1975). Protective capacity of gloves used for handling of nitrogen mustard. *Contact Dermatitis* **1**, 268-9.

TORTORICI, M.P. (1980). Precautions followed by personnel involved with the preparation of parenteral antineoplastic medications. *Hosp. Pharm.* **15**, 293-301.

TRENT REGIONAL HEALTH AUTHORITY (Derbyshire drug information service) (1982). Hazards of antineoplastic agents. *Pharmascan* June 1982, 94-8.

VENITT, S. AND BOSWORTH, D. (1983). The development of anaerobic methods for bacterial mutation assays: Aerobic and anaerobic fluctuation tests of human faecal extracts and reference mutagens. *Carcinogenesis* **4**, 339-45.

——, CROFTON-SLEIGH, C., HUNT, J., SPEECHLEY, V., AND BRIGGS, K. (1984). Monitoring exposure of nursing and pharmacy personnel to cytotoxic drugs: Urinary mutation assays and urinary platinum as markers of absorption. *Lancet* i, 74-7.

WAKSVIK, H., KLEPP, O., AND BRØGGER, A. (1981). Chromosome analyses of nurses handling cytotoxic agents. *Cancer treat. Rep.* **65**, 607-10.

WARRINGTON, P. AND HOWDEN, L. (1983). Handling of cytotoxic drugs. *Pharm. J.* **230**, 274.

WILSON, J.P. AND SOLIMANDO, D.A. JR. (1981a). Antineoplastics: a safety hazard? *Am. J. hosp. Pharm.* **38**, 624.

—— AND —— (1981b). Aseptic techniques as a safety precaution in the preparation of antineoplastic agents. *Hosp. Pharm.* **16**, 575-81.

YAMASAKI, E. AND AMES, B.N. (1977). Concentration of mutagens from urine by adsorption with the nonpolar resins XAD-2: Cigarette smokers have mutagenic urine. *Proc. nat. Acad. Sci. USA* **74**, 3555-9.

ZELLMER, W.A. (1981). Editorial: Reducing occupational exposure to potential carcinogens in hospitals. *Am. J. hosp. Pharm.* **38**, 1679.

ZIMMERMAN, P.F., LARSEN, R.K., BARKLEY, E.W., AND GALLELLI, J.F. (1981). Recommendations for the safe handling of injectable antineoplastic drug products. *Am. J. hosp. Pharm.* **38**, 1693-5.

Part IV

Cross-index of official and proprietary drug names

Where possible the British Pharmacopeia Commission approved name is given as the official name for each drug.

The proprietary names listed are those in most common use in the United Kingdom (bold type), the United States, Europe and Commonwealth Countries; the list is not intended to be comprehensive.

The index covers only those drugs mentioned in the text of the book.

PART 1: OFFICIAL NAME–PROPRIETARY NAME

Official name	Proprietary name
Absorbable gelatin sponge	*Sterispon; Gelfoam*
Acebutolol	*in Secadrex; Neptall Prent*
Acecainide	*not registered ASL–601 (American critical care–USA)*
Acenocoumarol	see *nicoumalone*
Acepromazine	*Plégicil*
Acetaminophen	see *paracetamol*
Acetanilide	*(obsolete)*
Acetazine	see *acepromazine*
Acetazolamide	*Diamox; Defiltran, Hydrazol*
Acetohexamide	*Dimelor; Dymelor*
Acetopromazine	see *acepromazine*
N–acetylcysteine	*Airbron, Parvolex*
β-acetyldigoxin	*Novodigal*
Acetyl-β-Methylcholine	see *methacholine*
N-Acetylprocainamide	see *acecainide*
Acetylpromazine	see *acepromazine*
Actinomycin	see *actinomycin D*
Actinomycin C	*Sanamycin (withdrawn)*
Actinomycin D	*Cosmegan Lyovac*
Acycloguanosine	see *acyclovir*
Acyclovir	*Zovirax*
Adicillin	*Penicillin N*
Adriamycin	see *doxorubicin hydrochloride*
Ajmaline	*Cardiorythmine, Gilurytmal*
Albuterol sulphate	see *salbutamol*
Alclofenac	*Prinalgin (withdrawn); Allopydin, Argun, Mervan, Mirvan, Zumaril*
Alcuronium	*Alloferin*
Aldosterone	*Aldocorten*
Alfacalcidol	*One Alpha; Delakmin, Eins–Alpha, Etalpha, Un–Alfa*
Allopurinol	*Zyloric; Bloxanth, Epidropal, Foligan, Urosin, Zyloprim*
Alpha-methyldopa	see *methyldopa*
Alphadolone acetate } Alphaxalone	*Althesin; Alfethesin*

Official name	Proprietary name
Alprazolam	*Xanax*
Alprenolol	*Aptin, Betacard, Gubernal*
Alprostadil	*Prostin VR Sterile Solution*
Aluminium hydroxide gel	*Aludrox; Amphogel*
Aluminium hydroxide } Magnesium hydroxide	*Aludrox; Melox*
Amantadine	*Symmetrel; Contention, Mantadix, PK–Merz*
Ambenonium	*Mytelase*
Amethocaine	*in Eludril Spray, Locan Norgotin, Noxyflex with Amethocaine Hydrochloride; Contralgin, Decicain, Pantocain, Pantocaine*
Amethopterin	see *methotrexate*
Amidopyrine	*Gentamidon, Pyramidon*
Amidotrizoic acid	see *sodium Diatrizoate*
Amikacin	*Amikin*
Amiloride hydro- } chlorothiazide	*Moduretic*
Amineptine	*Survector*
Aminocaproic Acid	*Epsikapron; Afibrin, Amicar, Capracid, Capralense, Capramol, Caproamin, Eaca, Ecapron, Ekaprol, Epsamon, Hemocaprol, etc.*
Aminoglutethimide	*Orimeten*
Aminophenazone	see *amidopyrine*
Aminophylline	*Cardophylin, Phyldrox Suppositories, Phyllocontin Tablets, Riddovydrin Capsules; Aminodur, Caréna, Corophyllin, Euphyllin, Lixaminol*
Aminophylline } Ephedrine	*Amesec*
Aminopromazine	see *proquamezine*
Aminopterin	*Aminopterin (withdrawn)*
Aminopyrine	see *amidopyrine*
Aminosidin	see *paromomycin*

Official name	Proprietary name
Aminotriazole	*Amizol, Cytrol, Weedazol (all Herbicides)*
Amiodarone	*Cordarone* X
Amiphenazole	*Daptazole*
Amitriptyline	*Amizol, Domical, Lentizol, Saroten, Tryptizol; Deprex, Elatrol, Elavil, Endep, Levate, Mareline Novotriptyn, Tryptanol*
Amitrole	see *aminotriazole*
Amobarbital	see *amylobarbitone sodium*
Amodiaquine	*Basoquin; Flavoquine*
Amoxapine	*Demelox*
Amoxicillin	see *amoxycillin*
Amoxycillin	*Amoxil; Clamoxyl, Imacillin, Larotid, Polymox*
Amphotericin	*Fungilin; Fungizone Intravenous; Ampho-Moronal*
Amphotericin B	see *amphotericin*
Ampicillin	*Amfipen, Penbritin, Pentrexyl; Vidopen; Alpen, Amblosin, Amperil, Bintotal, Deripen, Omnipen, Penbrock, Péniciline, Principen, Suractin, Totacillin, Totapen*
Ampicillin } Cloxacillin }	*Ampiclox*
αMPT	see *metirosine*
Amrinone	*Inocor*
Amsacrine	not registered NSC 156303 NSC 249992
Amygdalin	see *laetrile*
Amyl nitrite	*Nitrit*
Amylobarbitone	*Amytal, Eunoctal, Isonal, Mylodorm, Mylosed, Schiwanox, Stadadorm*
Amylobarbitone sodium	*Sodium Amytal; Amal Sodium, Amsal, Amylbarb Sodium, Amylobeta, Amylosol, Neur-Amyl Sodium, Novamobarb, Sedal Sodium*
Ancrod	*Arvin; Venacil*
Androstenedione	*Androtex*
Antazoline	*Antistin-Privine: Vasocon-A*
Anthralin	*Dithrocream, Dithrolan, Psoradrate, Stie-Lasan*
Antidiuretic hormone	see *vasopressin*
Antipyrine	see *phenazone*
Aprindine	*Amidonal, Fibocil, Fiboran*
Apronal	*Sedormid*
Apronalide	*Obsolete*
Aprotinin	*Trasylol; Antagosan, Gordox, Iniprol, Midran Tzalol, Zymofren*
Arginine	*Arginine-Sorbitol (withdrawn); Argamin, R-Gene, Spermargin*
Aristolochia sodium	*Tardolyt*
Aristolochic acid	see *aristolochia sodium*
Arsphenamine	*obsolete*

Official name	Proprietary name
Ascorbic Acid	*Redoxon; Abriscor, Adenex, Cemil, Cenolate, Cetane, Cevalin, C-Vita, Lauroscorbine, Scarbid, Solucap C, Visascorbol, etc.*
L-Asparaginase	see *colaspase*
Aspirin } Caffeine } Codeine phosphate }	*Hypon*
Astemizole	*Hismanal*
Atenolol	*Tentormin*
Atracurium	*Tracrium*
Azacytidine	*Azacytidine, NSC-102816*
Azapropazone	*Rheumox; Prolixan*
Azaribine	*Triazure*
Azathioprine	*Imuran; Imurek, Imurel*
Azatidine	*Optimine Syrup; Idulian, Zadine*
Azauridine	*Azauridine*
Azepinamide	*Parinase*
Azlocillin sodium	*Securopen*
B663	see *clofazimine*
Bacampicillin	*Ambaxin; Bacacil, Penglobe, Spectrobid*
Bacitracin	in *Baciguent*
Baclofen	*Lioresal*
Barbital	see *barbitone*
Barbitone	*Somnytic, Neuronidia*
B-carotene	see *betacarotene*
BCNU	see *carmustine*
Beclomethasone dipropionate	*Beconase Nasal Spray, Becotide, Propaderm: Aldecin*
Belladonna alkaloids } Ergotamine } Phenobarbitone }	*Bellergal Bellergal Retard*
Bemegride	*Eukraton, Megimide*
Bendrofluazide	*Aprinox, Berkozide, Centyl, Centyl K, Neo-Naclex, Neo-Naclex-K; Aprinox-M, Bristuric, Naturetin, Naturine, Pluryl*
Bendroflumethiazide	see *bendrofluazide*
Benorylate	*Benoral (range); Benortan, Benotamol, Salipran, Winolate*
Benoxaprofen	*Opren; Oraflex (both withdrawn)*
Benperidol	*Anquil; Frénactil, Glianimon,*
Benzalkonium chloride	*Capital, Empigen Bac, Hyamine 3500, Morpan BC, Roccal, Silquat, Vantoc CL,* In *Cetanorm, Drapolene, Polycide, Strombar,* etc.
Benzathine penicillin	*Penidural; LPG, Permapen Tardocillin Dulcepen-G, Extencilline,*
Benzathine penicillin G	see *benzathine Penicillin*
Benzethonium chloride	*Benzalcan, Desamon, Hyamine 1622, Phemerol Chloride*

Official name	Proprietary name
Deoxycorticosterone	see *deoxycortone*
Deoxycortone	*Deoxycortone Acetate Implants; Cortiron, Doca, Percorten, Syncortyl*
Deptropine	*Brontina; Brontine*
Deserpidine	*Harmonyl*
Desipramine	*Pertofran; Norpramin, Pertofrane*
Deslanoside	*Cedilanid; Cedilanid-D, Cedilanid Desacetyl, Cedlanide, Desace, Desaci, Verdiana*
Desmethyldiazepam	*Madar*
Desmethylimipramine	see *desipramine*
Dexamethasone	*Decadron Dexacortisyl, Oradexon; Aeroseb-D, Cortisumman, Deronil, Dexameth, Dexamethadrone, Dexa-sine, Dexmethsone Dexone, Fortecortin, Gammacorten, Hexadrol, Miral, Predni-F, etc.*
Dexamphetamine	*Dexamed, Dexedrine; Curban, Dexamine, Dexamphate, Dexaspan, Ferndex, Phetadex*
Dextran	*Dextraven, Gentran, Hyskon, Lomodex, Macrodex, Perfudex, Polidexide, Rheomacrodex, LMD*
Dextroamphetamine	see *dexamphetamine*
Dextromoramide	*Palfium (range); Jetrium, Narcolo*
Dextropropoxyphene	*Depronal SA, Dolasan, Doloxene, SK-65; Algaphan, Antalvic, Darvon, Darvon-N, Develin, Dolene, Dolocap, Erantin, Mardon, Pro-65, Progesic, Propoxychel, Proxagesic*
Dextropropoxyphene } Aspirin }	*Napsalgesic*
Di-iodohydroxyquinoline	*Diodoquin, Embequin, Floraquin; Direxiode, Gynovules, Ioquin, Moebiquin, Vaam-DHQ*
Di-isopropylfluorophosphonate	see *dyflos*
Di-isopropylfluorophosphate	see *dyflos*
Diacetylcholine	see *suxamethonium (obsolete)*
Diallylbarbituric acid	
Diaminodiphenyl sulphone	see *dapsone*
Diamorphine	*Diamorphine Hydrochloride*
Diatrizoic acid	see *sodium diatrizoate*
Diazepam	*Atensine, Diazemuls, Evacalm, Sedapam, Solis, Valium; E-Pam, Paxel, Serenack, Vivol*
Diazoxide	*Eudemine; Hyperstat*
Dibekacin	*Decabicin, Klobamicina,*

Official name	Proprietary name
	Orbicin, Panimycin
Dibenzepin	*Noveril*
Dicarnitine chloride	see *carnitine*
Dichloracetate	*not registered*
Dichloralphenazone	*Dormwell, Welldorm; Bonadorm, Chloralol*
Diclofenac	*Voltarol*
Diclofensine	*not registered, RO-8-4650 (Roche)*
Dicloxacillin sodium	*Constaphyl, Dichlor-Stapenor, Diclocil, Dycill, Dynapen, Pathocil, Veracillin*
Dicoumarol	*Dufalone*
Dicumarol	see *dicoumarol*
Dicyclomine	*Debendox, Merbentyl; in Diarrest, Kolanticon, Kolantyl, Oval; Atumin, Bendectin, Bentyl, Bentyol, Diclomyl, Dicycol, Dyspas, Procyclomin, Wyovin*
Diethazine	*Diparcol*
Diethylaminoethanol	*Cérébrol*
Diethyldithiocarbamate sodium	see *dithiocarb*
Diethylpropion	*Apisate, Tenuate Dospan; Prefamone, Regenon retard, etc.*
Diethylstilboestrol	see *stilboestrol*
Diflunisal	*Dolobid*
Digitoxin	*Digitaline Nativelle; Crystodigin, Digilong, Digimed, Digimerck, Digitox, Ditaven, Purodigin*
Digoxin	*Diganox Nativelle, Lanoxin; Cardiox, Coragoxine, Davoxin, Dialoxin, Digacin, Digolan, Digoxine Nativelle, Fibroxin, Lanatoxin, Lanicor, Natigoxin, Natigoxine, Nativelle, Novodigal Ampulle, Prodigox, Rougoxin, Winoxin, etc.*
Dihydralazine	*Nepresol, Népressol*
Dihydroergocornine } Dihydroergocristine } Dihydroergocryptine }	*Hydergine*
Dihydroergotamine	*Dihydergot (range); Det MS, DH-Ergotamin-Retard, Diidergot, Ergotonin, Ikaran, Orstanorm*
Dihydroflumethiazide	see *hydroflumethiazide*
Dihydrotachysterol	*AT-10, Tachyrol; Atecen, Calcamine, Dygratyl, Hytakerol*
1α 25-Dihydroxy-cholecalciferol	see *calcitriol*
Diloxanide	*Furamide*
Diltiazem	*Herbesser*
Dimenhydrinate	*Dramamine, Gravol; Andrumin, Dramavol, Dymenol, Epha-Retard,*

Official name	Proprietary name
	Neo-Metic, Novodimenate, Novomina, Prevenause, Travamine, Vomex A, Vomital
Dimethindene maleate	**Fenostil-Retard**; *Fenistil, Forhistal, Triten*
Dinoprost trometamol	**Prostin F₂ Alpha**; *Amoglandin, Minoprostin F$_{2\alpha}$, Prostalmon F, Prostamodin-F*
Dinoprostone	**Prostin E₂**; *Minprostin, Minprostin E₂, Prostarmon-E*
Dioctyl sodium sulfosuccinate	see *docusate sodium*
Diodone	**Umbradil**; *Diodrast*
Diphenhydramine	**Benadryl, Histergan**; *Alergicap, Benhydramil, Bidramine, Dabylen, Lensen*
Diphenoxylate } Atropine }	**Lomotil**
Diphenylhydantoin sodium	see *phenytoin sodium*
Dipipanone	**Diconal, Wellconal**
Diprophylline	**Silbephylline**; *Aerophylline, Airet, Asthmolysin, Dilin, Dilor, Droxine, Dyflex, Emfabid, Glyfyllin, Lufyllin, Neothylline, Neutraphylline, Protophylline*
Dipyrone	*Lagalgin, Minalgin, Novalgin, Novalgine, Novaminosulfon, Sulfonovin, Tapal*
Disodium edetate	**Nervanaid BA2, Sequestrene NA2**; *Disotate, Endrate Disodium, Sodium Versenate*
Disodium etidronate	**Didronel Tablets**; *Etidron*
Disopyramide	**Norpace, Rythmodan**
Disulfiram	**Antabuse 200**; *Espéral*
Dithiazanine iodide	**Telmid**
Dithiocarb	*Cupral*
Dithranol	see *anthralin*
Dobutamine	**Dobutrex**
Docusate sodium	**Dioctyl, Dioctyl Forte, Molcer, Soliwax, Waxsol**; in **Klyx, Migraleve Normax**: *Afko-Lube, Bu-Lax, Colace, Coloxyl, Comfolax, Constiban, Dioctylal Forte, Dilax, Disonate, Doxinate, Rapidax, Regulex, Mollax,* etc.
Domperidone	**Motilium**
Dopamine	**Intropin**
Dothiepin	**Prothiaden**
Doxacillin	*not registered*
Doxapram	**Dopram**
Doxepin	**Sinequan**; *Adaptin, Apanal, Quitaxon, Sinquan*
Doxorubicin hydrochloride	**Adriamycin**; *Adriblastin, Adriblastina, Farmablastina*

Official name	Proprietary name
Doxycycline hyclate	see *doxycycline hydrochloride*
Doxycycline hydrochloride	**Doxatet, Vibramycin**; *Doxin, Doxitard, Doxychel, Doxylin, Dumoxin, Idocyklin, Vibra-Tabs, Vibravenös*
Droperidol	**Droleptan**; *Inapsine*
DTIC	see *dacarbazine*
Dyflos	*DFP-Oel, Diflupyl, Floropryl*
Dyphylline (Dihydroxypropyl-theophylline)	see *diprophylline*
Echothiopate	see *ecothiopate*
Ecothiopate Iodide	**Phospholine Iodide**
Ectylurea	*Cronil, Disteol, Ektyl, Levanil, Neuroprocin, Nostyn*
Edetic acid	*Sequestrene AA, Versene Acid*
Edrophonium	**Tensilon**
EDTA	see *edetic acid*
EHDP	see *disodium etidronate*
Emepronium	**Cetiprin**
Enalapril maleate	*not registered, MK-421*
Endralazine	*not registered, BQ-22-708 (Sandoz)*
Enflurane	**Ethrane**
Enoxolone	*PO12 Pommade*
Ephedrine } Trifluoperazine } Diphenylpyraline }	**Expansyl Spansule**
Epimestrol	*Stimovul*
Epipodophyllotoxin	see *etoposide*
Epsilon-aminocaproic acid	see *aminocaproic acid*
Ergocalciferol	see *calciferol*
Ergotamine	**Femergin, Lingraine, Medihaler Ergotamine**; *Ergomar, Ergotart, Gynergen*
Ergotamine tartrate } Caffeine }	**Cafergot**
Erythromycin	**Erycen, Erythrocin, Erythromid, Erythroped, Ilosone, Ilotycin, Betcin**; *Abboticine, Bristamycin, Chemthromycin, Emcinka, E-Mycin, Eratrex, Eromycin, Erostin, Erycinum, Erythromyctine, Mycin, Neo-Erycinium, Novorythro, Propiocine, Robimycin,* etc.
Erythromycin estolate	**Ilosone (range)**; *Chemthromycin, Cimetrin, Eromycin, Neo-Erycinum, Neo-Ilotycin, Ritromin,* etc.
Erythromycin ethyl succinate	**Erythrocim IM. Erythroped** *(range)*
Erythromycin stearate	**Erythrocin**
Eserine	see *physostigmine*

Official name	Proprietary name
Ibufenac	*Dytransin (withdrawn)*
Ibuprofen	*Apsifen, Brufen, Ebufac, Ibu-Slo; Algofen, Amersol, Focus, Inflam, Motrin, Rebugen, etc.*
Idoxuridine	*Dendrid, Kerecid, Ophthalmadine; Herplex, Iduviran, Stoxil, Synmiol, Virunguent*
Idoxuridine Dimethylsulphoxide }	*Herpid, Iduridin*
Ifosfamide	*Mitoxana; Holoxan*
Imipramine	*Berkomine, Co-Caps Imipramine, Dimipressin, Norpramine, Oppanyl, Praminil, Tofranil; Censtim, Chemipramine, Iramil, Imavate, Impranil, Impril, Imiprin, Janimine, Melipramine, Novopramine, Presamine, Prodepress, SK-Pramine, Somipra, etc.*
Indapamide	*Natrilix; Fludex, Ipamix*
Indomethacin	*Artracin, Imbrilon, Indocid, Indoflex, Mobilan, Osmosin (withdrawn); Amuno, Confortid, Indosmos (withdrawn), Osmogits (withdrawn), Rheumacin*
Indoprofen	*Flosint (withdrawn)*
Iocarmate	see *iocarmic acid*
Iocarmic acid	*Dimerx*
Iodochlorhydro-xyquinoline	see *clioquinol*
Iodohydroxyquin	see *clioquinol*
Iodopyracet	see *diodone*
Iofendylate	see *iophendylate injection*
Iopanoic acid	*Telepaque; Bilijodon-Natrium, Biliopaco, Colegraf, Jopanonsyre, Neocontrast, Nigrantil, Panjopaque, Teletrast*
Iophendylate injection	*Myodil; Ethiodan, Pantopaque*
Iothalamate	see *iothalamic acid*
Iothalamic acid	*Cardio-Conray, Conray; Angio-Conray, Angio-Contrix, Contrix, Medio-Contrix, Vascoray*
Ipratropium bromide	*Atrovent; In Duovent*
Iprindole	*Prondol*
Iproniazid	*Marsilid*
Iron dextran	*Direx, Imferon, Ironorm Injection, Niferex: Inferdex*
Isocarboxazide	*Marplan*
Isoflurane	*Forane*
Isoflurophate	see *dyflos*
Isoniazid	*Rimifon; Cedin, INH, Isotamine, Isotinyl, Isozid, Neoteben, Niconyl, Nydrazid, Panazid, Tb-Phlogin*

Official name	Proprietary name
Isoprenaline	*Aleudrin, Iso-Autohaler, Lomupren, Medihaler Iso, Medihaler Iso Forte, Prenomiser, Saventrine, Suscardia; Aludrin, Ingelan, Iso-Intranefrin, Isolin, Isovon, Isuprel, Luf-Iso, Neo-Epinine, Norisodrine, Proternol, Vapo-N-Iso.*
Isopropylantipyrine	see *propyphenazone*
Isopropylmeprobamate	see *carisoprodol*
Isoproterenol	see *isoprenaline*
Isosorbide dinitrate	*Cedocard, Isoket, Isordil, Soni-Slo, Sorbid, Sorbitrate, Vascardin; Carvasin, Coronex, Isotrate, Iso-Bid, Maycor, Risordan, Sorbide TD, Sorquad, etc.*
Isotretinoin	*Roaccutane*
Isoxepac	*HP-549*
Isoxsuprine	*Defencin CP, Duvadilan; Cardilan, Fenam, Isolait, Suprilent, Vasodilan, Vasoplex*
Kanamycin	*Kannasyn, Kantrex; Kamycine, Kanabristol, Kanasig, Ophtalmokalixan, Optokalixan*
Ketamine	*Ketalar; Ketaject, Ketanest, Nixoral*
Ketoconazole	
Ketoprofen	*Alrheumat, Orudis; Profenid*
Ketotifen	*Zaditen (range)*
Labetalol	*Trandate*
Laetrile (vitamin B$_{17}$)	not licensed in UK or USA
Lanatoside C	*Cedilanid; Allocor, Ceglunat, Celadigal, Celanat, Ceto Sanol, Isolanid, Lanimerck, Lanoside, etc.*
Latamoxef sodium	*Moxalactam; Moxam*
Latiazem	see *diltiazem*
Lenperone	*Elanone-V, Lenperol*
Leptazol	*Cardiazol, Metalex-P, Metrazol*
Lergotrile mesylate	not registered (withdrawn)
Levallorphan	*Lorfan*
Levamisole	*Ketrax; Decaris, Ergamisol, Solaskil*
Levodopa	*Berkdopa, Brocadopa, Larodopa, Veldopa; Bendopa, Dopar, Helfodopa, Ledopa, Sobiodopa, Speciadopa, Syndopa*
Levodopa Benserazide }	*Madopar*
Levodopa Carbidopa }	*Sinemet*
Levomeprazine	see *methotrimeprazine*
Levomepromazine	see *methotrimeprazine*
Levonorgestrel	see *d-norgestrel*
Levothyroxine	see *thyroxine sodium*
Lidocaine	see *lignocaine*

Official name	Proprietary name
Metahexamide	*Euglycin, Isodiane, Melanex*
Metamizol	see *dipyrone*
Metaproterenol	see *orciprenaline*
Metenolone	see *methenolone*
Metformin	*Glucophage, Metiguanide;* *Diabex SR, Diabexyl,* *Haurymellin,*
Methacholine chloride	*Amechol*
Methacyline	*Rondomycin; Megamycine*
Methadone	*Physeptone; Dolorphine* *Hydrochloride,* *L-Polamidon, Westadone*
Methandienone	*Dianabol; Danabol*
Methandrostenolone	see *methandienone*
Methapyrilene	*Allergin, Dozar, Driliton,* *Dylhista, Histadyl, M-P,* *Pyrteen, Semikon,* *Thenylene*
Methaqualone	*Revonal; Dormir, Hyptor,* *Mequelon, Methalone,* *Normi-Nox, Oblioser,* *Optinoxan, Parest,* *Pexaqualone, Quaalude,* *Rouqualone, Sleepinal* *Somnafac, Sopor,* *Thendorm, Tiqualoine,* *Triador, Tualone*
Methaqualone ⎫ Diphenhydramine ⎬	*Mandrax*
Methazolamide	*Neptazane*
Methenamine	see *hexamine*
Methenolone	*Primobolan, Primobolan S,* *Primobolan Depot*
Methicillin	*Celbenin; Azapen,* *Cinopenil, Flabelline, Metin,* *Penistaph, Staphcillin*
Methimazole	*Favistan, Tapazole,* *Thacapzol*
Methionine	*Litrison (withdrawn),* *Meonine, Unihepa* *(withdrawn); Antamon,* *Lobamine, Methine, Monile,* *Ninol, Pedameth, Uracid,* *Uranap*
Methohexital	see *methohexitone*
Methohexitone	*Brietal Sodium; Brevimytal* *Natrium, Brevital Sodium*
Methoin	*Mesontoin (withdrawn);* *Mesantoin, Sedantoinal*
Methopromaxine	see *methoxypromazine*
Methotrexate	*Methotrexate; Ledertrexate*
Methotrimeprazine	*Nozinan, Veractil;* *Levoprome, Minozinan,* *Neurocil*
Methoxsalen	*Meladinine; Oxsoralen,* *Soloxsalen*
Methoxyflurane	*Penthrane*
Methoxypromazine	*Mopazine; Tentone,* *Vetomazin*
8-Methoxypsoralen	see *methoxsalen*
Methoxypyrimal	see *sulphamethoxydiazine*
Methsuximide	*Celontin; Petinutin*

Official name	Proprietary name
Methyclothiazine	*Enduron, Enduronyl;* *Aquatensen, Diuretic*
Methylamphetamine	*Desoxyn, Fetamin,* *Neodrine, Pervitin, Syndrox*
Methylandrostanalone	see *mestanolone*
Methyl-CCNU	see *semustine*
Methyldopa	*Aldomet, Co-Caps* *Methyldopa, Dopamet,* *Medomet; Aldometil,* *Presinol, Sembrina*
Methyldopate	*Aldomet Injection*
Methylestrenolone	see *normethandrone*
Methyl lomustine	see *semustine*
Methyl methacrylate	*In CMW Orthopaedic Bone* *Cement, Garamycin Chains,* *Palacos R, Septopal Chains,* *Surgical Simplex Plain Bone* *Cement*
Methylnortestosterone	see *normethandrone*
Methylperidol	see *moperone*
Methylphenidate	*Ritalin; Methidate*
Methylphenobarbitone	*Prominal; Mebaral,* *Promitone*
Methylphenylethylhydantoin	see *methoin*
Methylprednisolone	*Depo-Medrone, Medrone* *Veriderm; Depo-Medrate,* *Depo-Medral, Medrate,* *Medrol, Urbason* *(obsolete)*
Methylsulphonal	
Methyltestosterone	*Perandren, Virormone Oral;* *Android, Dumogran,* *Glosso-Sténrandryl,* *Metandren, Metestine,* *Neohombreol M, Oreton* *Methyl, Testomet,* *Testostelets, Testoviron,* *Testred*
Methylthiouracil	*Thyreostat*
Methyltryptamine	*DMT, not registered*
Methyl-*p*-tyrosine	see *metirosine*
Methyprednisolone	see *methylprednisolone*
Methyprylone	*Noludar*
Methysergide	*Deseril; Désernil, Sansert*
Metiamide	*SK 92058 (not marketed)*
Metirosine	*Demser*
Metoclopramide	*Maxolon, Primperan; In* *Migravess, Paramax;* *Cerucal, Donopon-GP,* *Maxeran, Metoclol,* *Moriperan, Pasperin, Plasil,* *Regulan*
Metolazone	*Zaroxolyn*
Metoprolol	*Betaloc, Lopresor; In* *Co-Betaloc, Lopresoretic;* *Beloc, Lopressor, Selokeen,* *Seloken*
Metrizamide	*Amipaque*
Metronidazole	*Flagyl; Clont, Neo-Tric,* *Novonidazol, Sanatrichom,* *Trichazol, Trikacide,* *Trikamon*
Metronidazole ⎫ Nystatin ⎬	*Flagyl Compak*

Official name	Proprietary name
Metyrapone	*Metopirone*
Mexiletine	*Mexitil*
Mezlocillin	*Baypen*
Mianserin	*Bolvidon, Norval; Tolvin, Tolvon*
Miconazole	*Daktarin, Dermonistat, Gyno-Daktarin, Monistat; Albistat, Brentan, Conofite, Daktar, Micatin*
Midazolam	*Hypnovel*
Minocycline	*Minocin; Klinomycin, Minomycin, Mynocine, Ultramycin, Vectrin*
Minoxidil	*Loniten*
Misonidazole	not registered, RO-07-0582
Mithramycin	*Mithracin*
Mitoguazone	*Methyl GAG*
Mitomycin	*Mitomycin C; Amétycine, Mitomycine, Mutamycin*
Mitomycin C	see *mitomycin*
Mitotane	*Lysodren*
Mofeburazone	*Arcomonol, Mobutazon, Monazan*
Monophenylbutazone	see *mofeburazone*
Moperone	*Luvatren, Luvatrena*
Moxalactam disodium	see *latamoxef sodium*
Methylglyoxal	see *mitoguazone*
Mustine hydrochloride	*Mustine Hydrochloride; Caryolysine, Cloramin, Dichloren, Dimitan, Erasol, Mustargen, Onco-Imine*
Nadolol	*Corgard; in Corgaretic; Solgol*
Nafcillin	*Unipen*
Naftidrofuryl oxalate	*Praxilene, Praxilene Forte, Citoxid, Dusodril*
Nalidixic acid	*Negram; Neggram, Nogram*
Naloxone	*Narcan; Nalonee, Narcanti*
Naphazoline	*Privine*
Naproxen	*Naprosyn, Proxen*
Neoarsphenamine	*Novarsenobillon (withdrawn)*
Neocarzinostatin	not registered, NSC-69856
Neomycin	*Minima Neomycin Sulphate, Mycifradin, Neomin, Nivemycin, Tampovagen N; Bykomycin, Herisan Antibiotic, Myacyne, Neobiotic, Neobram, Neocin, Neomate, Neo-Morrhuol, Neopt, Otobiotic*
Neostigmine	*Prostigmin; Juvastigmin*
Netilmicin	*Netillin; Netilyn, Netromicin, Netromicina*
Nialamide	*Niamid*
Nicotinamide	*Nicamid, Nicobion, Nicotamide*
Nicotinic acid	*Diacin, Efacin, Niac, Nicobid, Nicolar, Niconacid, Nico-Span, Nicotinex, Nisco-400, Nicyl, Vasotherm, Wamocap*
Nicotinyl alcohol	see *nicotinyl tartrate*
Nicotinyl tartrate	*Ronicol, Timespan; Roniacol*

Official name	Proprietary name
Nicoumalone	*Sinthrome; Sintrom*
Nifedipine	*Adalat*
Niflumic acid	*Actol, Flaminon, Forenol, Inflaryl, Landruma, Nifluril, Niflux, Noflame*
Nikethamide	*Coramine; Cormed, Juvacor, Kardonyl*
Niridazone	*Ambilhar*
Nitrazepam	*Mogadon, Remnos; Mogadan*
Nitrofural	see *nitrofurazone*
Nitrofurantoin	*Berkfurin, Co-Caps Nitrofurantoin, Furadantin, Furan, Macrodantin, Urodantin; Cyantin, Furadöine, Fura-Med, Furanex, Furanite, Furatine, Ituran, Nephronex, Nifuran, Novofuran, Parfuran, Trantoin, Urolong, Uro-tablinen*
Nitrofurazone	*Furacin; Acutol, Furacine, Furesol*
Nitrogen mustard	see *mustine hydrochloride*
Nitroglycerin	see *glyceryl trinitrate*
Nitroprusside	see *sodium nitroprusside*
Nomifensine	*Merital; Alival, Anametrin, Merival, Psicronizer*
Nonoxynol-9	*C-Film, Delfen, Double Check, Duracreme, Duragel, Ortho-Creme, Ortho-Forms, Staycept, in Emko*
Noramidopyrine	see *dipyrone*
Nordiazepam	*Madar*
Norethandrolone	*Nilevar*
Norethindrone	see *norethisterone*
Norethisterone	*Micronar, Noriday, Primolut N; Micronovum, Norfor, Norluten, Norlutin, Nor-od.*
Norethisterone acetate	*SH 420; Norlutate, Primolut-Nor*
Norethisterone acetate Ethinyloestradiol }	*Anovlar 21, Controlvar, Gynovlar 21, Loestrin 20, Minovlar, Norlestrin, Orlest 21, Primodos*
Norethisterone enanthate	*Noristerat, Nur-Isterate*
Norethisterone oenanthate	see *norethisterone enanthate*
Norethisterone Ethinyloestradiol }	*Ovysmen, Brevinor*
Norethisterone Mestranol }	*Menophase, Norinyl-1, Norinyl-2, Ortho-Novin 1/50, Ortho-Novin 1/80, Ortho-Novin 2 MG*
Norethynodrel Mestranol }	*Conovid, Conovid E, Enavid, Enavid-E; Enovid*
Norgestrel	*Neogest*
Norgestrel Ethinyloestradiol }	*Eugynon 30, Eugynon 50, Microgynon 30, Ovran, Ovran 30, Ovranette*
Norgestrienone	*Orgyline, Planor*
Normethandrone	*Orgasteron*

Official name	Proprietary name	Official name	Proprietary name
Nortriptyline	*Allegron, Aventyl; Acetexa, Nortrilen, Nortab, Psychostyl*		*Papalease, Pap-Kaps, Pavabid, Pavacaps, Pavacen, Pavadel, Pavagrant, Vasal, Vasopan, etc.*
Novaminsulfon	see *dipyrone*	Para-aminosalicylic acid	*Pamisyl, Parasal, Rezipas, Teebacin-Acid*
Novobiocin	*Albamycin; Cathomycine, Inamycin*	Paracetamol	*Calpol, Pamol, Panadol, Panasorb, Salzone, Ticelgesic: Anuphen,*
Nystatin	*Nystan, Nystatin-Dome, Nystavescent; Candio-Hermal, Diastatin, Korostatin, Moronal, Mycostatin, Mycostatine, Nilstat*		*Apamide, Atasol, Campain, Capital, Cen-App, Ceetamol, Chemcetaphen, Dolamin, Dolanex, Doliprane, Dymadon, Korum, Paracet, Paralgin, Parmol, Phendex, Proval,*
Oestradiol	*Oestradiol Implants; Aquadiol, Estrace, Gynoestryl, Progynon*		*Tapar, Temlo, Tylenol, etc*
Oleandomycin	*Oleandocyn*	Paracetamol	
Opipramol	*Insidon; Ensidon*	Codeine	
Orciprenaline	*Alupent (range); Metaprel*	Docusate sodium	*Migraleve*
Orciprenaline } Bromhexine }	*Alupent Expectorant*	with or without buclizine }	
Orphenadrine	*Disipal, Norflex, Mephenamin*	Paramethadione	*Paradione*
Oxacillin	*Bactocill, Bristopen, Cryptocillin, Prostaphlin, Stapenor*	Paramethasone	*Haldrate, Metilar; Dilar, Haldrone, Stemex, Monocortin*
Oxamniquine	*Mansil, Vancil, Vansil*	Pargyline	*Eutonyl*
Oxazepam	*Serenid-D, Serenid Forte; Adumbran, Praxiten, Serax, Serepax, Seresta*	Paromomycin	*Humatin (withdrawn); Gabbromycin, Humycin*
Oxtoxynol	*Staycept Jelly*	Pecazine	*Lacumin, Pacatal, Ravenil*
Oxymetholone	*Anapolon; Anadroi, Nastenon*	Pempidine	*Perolysen*
Oxypentifylline	*Trental; Elorgan, Tarontal, Torental*	Penfluridol	*Semap*
Oxypertine	*Integrin*	Penicillamine	*Cuprimine, Depamine, Distamine; Metalcaptase, D-Penamine, Trolovol*
Oxyphenbutazone	*Tandacote, Tanderil (both withdrawn); Oxalid, Phlogase, Tandearil (withdrawn)*	D-Penicillamine	see *penicillamine*
		Penicillin G	see *benzylpenicillin*
		Penicillin V	see *phenoxymethylpenicillin*
Oxyphenisatin	*Varipaque; Bisco-Zitron, Laxatan, Obstilax, Schokolax, Vinco-Abführperlen, Darmoletten*	Pentachlorophenol	*Sanobrite*
		Pentamethonium	*Penthonium*
		Pentamidine isethionate	*Pentamidine Isethionate; Lomidine, M&B 800*
Oxyprenolol	*Trasicor*	Pentaquine	*obsolete*
Oxytetracycline	*Berkmycen, Chemocycline, Clinimycin, Galenomycin, Imperacin, Oppamycin, Oxydon, Oxymycin, Stecsolin, Terramycin, Unimycin; Bobbamycin, Clinmycin, Macocyn, Oxlopar, Oxycycline, Oxy-Kesso-Tetra, Terravenös, Tetra-Tablinen, Uritet, Vendarcin*	Pentazocine	*Fortral; in Fortagesic; Fortal, Fortralin, Liticon, Peltazon, Sosegon, Sosenyl*
		Pentobarbital	see *pentobarbitone*
		Pentobarbitone	*Nembutal; Barbopent, Hypnol, Ibatal, Nabralin, Neodorm, Penbarb, Penbon, Pental, Pentanca, Pentobeta, Pentogen, Pentone, Petab, Sodepent*
Oxytocin	*Pitocin, Syntocinon, Orasthin, Partocon,*	Pentolinium	*Ansolysen*
Pamaquin	*Aminoquin; Plasmochin, Plasmoquine (all withdrawn)*	Pentoxifylline	see *oxypentifylline*
		Pentylenetetrazol	see *leptazol*
Pancuronium	*Pavulon*	Perhexiline maleate	*Pexid*
Papaverine	*Cerebid, Cerespan, Dilaspan, Dylate, Kavrin, Lempav, Myobid, Panergon,*	Pericyazine	*Neulactil; Aolept, Nemactil, Neuleptil*
		Perphenazine	*Fentazin; Decentan, Trilafon, Trilifan*
		Pethidine	*Pethidine Roche; Dermol, Dolantin, Dolosal, Pethoid, Phytadon*

Official name	Proprietary name
Pirbuterol hydrochloride	*not registered, CP-24314-1 (Pfrizer)*
Piretanide	*Arelix*
Piroxicam	**Feldene**
Pizotifen	**Sanomigran;** *Mosegor, Sandomigran, Sandomigrin, Sanmigran (obsolete)*
Plasmocid	
Polidexide	**Secholex** *(withdrawn)*
Polymyxin B sulphate	**Aerosporin**
Polymyxin E sulphate	see *colistin sulphate*
Polythiazide	**Nephril;** *Drenusil, Renese*
Potassium clorazepate	see *clorazepate potassium*
Potassium chloride	**Kay-Cee-L, K Contin Continus, Leo-K, Nu-K, Sando-K, Selora, Slow-K,** etc
Potassium menaphthosulphate	*Vikastab*
Potassium perchlorate	**Peroidin;** *Thyronorman*
Povidone-iodine	**Betadine (range), Disadine (range) Pevidine (range);** *Bridine, Efodine, Isodine, Providine K, Proviodine, Ultradine*
Practolol	**Eraldin;** *Dalzic (withdrawn)*
Prazepam	**Centrax**
Praziquantel	*Biltricide, Cesol, Droncit*
Prazosin	**Hypovase, Sinetens, Minipress**
Prednisolone	**Cordelcortone, Delta-Cortef, Deltacortril Enteric, Deltalone, Delta-Phoricol, Deltastab, Di-Adreson-F, Marsolone, Precortisyl;** *Adnisolone, Deltasolone, Decortin H, Hostacortin H, Hydrocortancyl, Keteocort H, Meticortelone, Optocort, Panafcortelone, Paracortol, Prednicoelin, Predni H, Predniretard, Prednis, Prelone, Ropredlone, Scherisolon, Solone, Ulacort,* etc.
Prednisone	**Decortisyl, Deltacortone, Di-Adreson, Marsone;** *Adasone, Colisone, Constancyl, Dabroson, Decortin, Delta-Dome, Erftopred, Hostacortin, Inocortyl, Keteocort, Lisacort, Orasone, Paracort, Prednilonga, Presone, Propred, Sone, Ultracorten, Urtilone, Wescopred, Winpred,* etc.
Prenylamine	**Synadrin;** *Bismetin, Crepasin, Sedolatam, Segontin,* etc.
Prilocaine	**Citanest (range), Xylonest**
Primaquine	**Primaquine Phosphate**
Primidone	**Mysoline;** *Cyral, Elmidone, Lepsiral, Liskantin, Midone, Mylepsin, Primoline*

Official name	Proprietary name
Probenecid	**Benemid;** *Benacen, Proben, Probexin, Urecid*
Probucol	**Lurselle;** *Lorelco*
Procainamide	**Procainamide Durules, Pronestyl;** *Novocamid, Procapan*
Procaine	**Novutox (withdrawn);** *Novocaine, Westocaine,*
Procaine penicillin	**Depocillin;** *Aquacaine G, Aquacillin, Ayercillin, Cilicaine, Crysticillin A-S, Duracillin AS, Evacillin, Flocilline, in Megacillin; Megapen, Pfizerpen-AS, Procillin, Therapen IM, Wycillin,* etc.
Procarbazine	**Natulan;** *Matulane, Natulanar,*
Prochlorpemazine	see *prochlorperazine*
Prochlorperazine	**Stemetil, Vertigon Spansule;** *Anti-Naus. Compazine, Témentil*
Procyclidine	**Kemadrin;** *Osnervan*
Proguanil	**Paludrine;** *Paludrinol*
Promazine	**Sparine;** *Liranol*
Promethazine	**Phenergan;** *Atosil, Ganphen, Histantil, Lemprometh, Meth-Zine, Progan, Promethapor, Prothazine, Quadnite, Remsed, Zipan obsolete*
Pronethalol	*Rytmonorm*
Propafenone	
Propanidid	**Epontol**
Propantheline	**Pro-Banthine;** *Banlin, Neo-Banex, Pantheline*
Propicillin potassium	*Baycillin, Bayercillin, Bayercilline, Delprosyn, Oricillin*
Propoxyphene	see *dextropropoxyphene*
Propranolol	**Inderal;** *Avlocardyl, Dolciton, Inderalici, Sumial*
Propylthiouracil	*Propycil, Propyl-Thyracil, Thyreostat II*
Propyphenazone	**Saridone (withdrawn);** *in Baukal*
Proquamezine	*Myspamol*
Prostaglandin E_1	see *alprostadil*
Prostaglandin E_2	see *dinoprostone*
Prostaglandin $F_{2\alpha}$	see *dinoprost trometamol*
Prothionamide	**Trevintix;** *Ektebin, Peteha*
Protriptyline	**Concordin;** *Maximed, Triptil, Vivactil*
Proxicromil	*withdrawn, (Fisons UK)*
Proxyphylline	*Brontyl 300, Thean, Purophyllin, Spasmolysin, Theon*
Pseudoephedrine	**Sudafed;** *Bensan, D-Feda, Novafed, Sudabid, Tussaphed*
Puromycin	*not registered, CL 16536, CL 13900, L3123*
Pyramidon	see *amidopyrine*
Pyrazinamide	**Zinamide;** *Pyrafat, Tebrazid*

Official name	Proprietary name	Official name	Proprietary name
Stibophen	*Fantorin, Fuadin*		*S-Guanidine, Shigatox*
Stilbamidine	*M&B 744 (obsolete)*	Sulphamerazine	*Solumédine*
Stilboestrol	***Parestrol;*** *Cyren-A, Desma, Dicorvin, Distilbene, Oekolp, Oestromon, Stilbetin, Stibilium*	Sulphamethazine	see *sulphadimidine*
		Sulphamethizole	***Methisul, Urolucosil;*** *Famet, Proklar-M, Rufol, S-Methizole, Thiosulfil, Urolex, Uroz, Utrasul*
Streptokinase	***Kabinkinase, Streptase***		
Streptomycin	***Orastrep, Streptomycin Sulphate;*** *Strepolin, Strept-Evanules, Streptosol*	Sulphamethoxazole	***Gantanol;*** in ***Bactrim, Septrin***
Streptonigrin	see *rufocromomycin*	Sulphamethoxydiazine	***Durenate;*** *Bayrena, Durenat, Kirocid, Kiron, Sulla*
Streptozocin	*Zanosar, NSC-85998, U-9889*	Sulphamethoxypy-ridazine	***Lederkyn, Midicel;*** *Davosin, Kynex, SDM, Sultirène*
Streptozotocin	see *streptozocin*	Sulphan Blue	*Disulphine Blue Intravenous Injection*
Succinylcholine	see *suxamethonium*		
Sucralfate	***Antepsin;*** *Sulcrate, Ulcerban, Ulcerlimin, Ulsanic*	Sulphanilamide	*Amindan, Exoseptoplix, Pyodental, Tablamide*
		Sulphaphenazole	***Orisulf;*** *Orisul, Sulfabid*
Sulbenicillin sodium	*Lilacillin, Sulbenil*	Sulphaproxyline	*In Dosulfin (withdrawn)*
Sulfacetamide	see *sulphacetamide sodium*	Sulphapyridine	*M&B 693; Dagenan, Eubasinum*
Sulfadimethoxine	see *sulphadimethoxine*		
Sulfadoxine ⎫ Pyrimethamine ⎭	***Fansidar***	Sulphapyrimidine	see *sulphadiazine*
		Sulphasalazine	***Salazopyrin;*** *Azulfidine, Colo-Pleon, Rorasul, Salazopyrine, Salisulf, SAS-500, Sulcolon,* etc.
Sulfafurazole	see *sulphafurazole*		
Sulfamerazine	see *sulphamerazine*		
Sulfamethizole	see *sulphamethoxazole*		
Sulfamethoxazole	see *sulphamethoxazole*	Sulphasomidine	*Aristamid, Elcosine, Elkosin, Elkosina, Elkosine, Isosulf, Pepsilphen*
Sulfamethoxypyridazine	see *sulphamethoxpyridazine*		
Sulfamethyldiazine	see *sulphamerazine*		
Sulfamethylthiadiazole	see *sulphamethizole*	Sulphathiazole	***Thiazamide;*** *Cibazol, Sulfamul, Thiazomide*
Sulfametin	see *sulphamethoxydiazine*		
Sulfametopyrazine	***Kelfizine W;*** *Kelfizina, Longum, Policydal*	Sulphinpyrazone	***Anturan***
		Sulphobromo-phthalein sodium	*Bromophthalein, Bromsulfan, Bromosulfophthalein, Bromsulphalein, Bromthalein, Hepartest (obsolete)*
Sulfamonomethoxine	see *sulphamethoxydiazine*		
Sulfanilamide	see *sulphanilamide*		
Sulfanilylguanidine	see *sulphaguanidine*		
Sulfapyridine	see *sulphapyridine*		
Sulfasalazine	see *sulphasalazine*	Sulphonal	
Sulfinpyrazone	see *sulphinpyrazone*	Sulpiride	***Dolmatil;*** *Co-Sulpir, Dogmatil, Drominetas, Equilid, Miradol, Mirbanil,* etc.
Sulfisomezole	see *sulphamethoxazole*		
Sulfisomidine	see *sulphasomidine*		
Sulfisoxazole	see *sulphafurazole*		
Sulfoxone sodium	*Diasone Sodium*	Sulthiame	***Ospolot;*** *Elisal*
Sulindac	*Clinoril*	Suxamethonium	***Anectine, Brevidil M. Suxamethonium Chloride;*** *Lysthenon, Pantolax, Quelicin, Scoline, Succinyl-Asta, Succinyl HaF, Succinyl Vitrium, Sucostrin*
Sulphacetamide sodium	***Albucid, Bleph-10 Liquifilm, Minims Sulphacetamide Sodium;*** *Sulphacalyre;* in ***Cortucid, Isopto Cetamide, Ocusol, Sulfapred,*** etc		
		Tamoxifen	***Nolvadex;*** *Novaldex*
Sulphadiazine	*Adiazine, Diazyl, Sulfadets*	Temazepam	***Euhypnos, Euhypnos Forte, Normison;*** *Levanxol, Maeva, Restoril*
Sulphadimethoxine	***Madribon***		
Sulphadimidine	***Sulphamezathine;*** *Dimethazine, Diminsul, S-Dimidine, Sulfadine, Mezin*	Teniposide	*Vehem, Vumon, NSC 122819*
		Terbutaline	***Bricanyl, Filair;*** *Brethine, Feevone*
Sulphafurazole	***Gantrisin;*** *Chemovag, Gantrisine, Novosoxazole, Sk-Soxazole, Sosol, Soxomide, Sulfagan, Sulfagen, Sulfalar, Sulfazole, Sulfisin, Sulfizole, Urogan, US-67*		
		Terfenadine	***Triludan;*** *Teldane*
		Testosterone	*Testoral Sublings, Testosterone Implants; Andronaq, Malogen, Neo-Hombreol, Neo-Hombreol F, Oreton, Testoviron T*
Sulphaguanidine	*Diacta, Guannicil, Guanidan, Resulfon,*		

Official name	Proprietary name
Testosterone enanthate	*Primoteston-Depot; Androtardyl, Delatestryl, Malogen La, Malogex, Testate, Testostroval-Pa, Testoviron-Depot*
Testosterone propionate	*Tes PP (withdrawn)*
Tetracaine	see *amethocaine*
Tetracosactrin	*Cortrosyn Depot, Synacthen; Actholain, Cortrophin-S, Cortrosinta*
Tetracycline	*Achromycin, Co-Caps Tetracycline, Economycin, Oppacyn, Steclin, Sustamycin Capsules, Tetrabid, Tetrachel, Tetracyn, Tetrex, Totomycin; Austramycin, Bristacycline, Centet, Chemcycline, Cyclopar, Decabiotic, Decycline, Dema, Florocycline, Hydracycline, Hexacycline, Hostacyclin, Kesso-Tetra, Lexacycline, Miramycine, Panmycin, Polycycline, QIDtet, Quatrax, Robitet, SK-Tetracycline, Tetracap, Tetralution, U-Tet, Withtracin, etc.*
Tetraethylammonium bromide	*obsolete*
Tetramisole	*Appercol, Athelvet, Concurat, Nilverm, Ripercol*
Thalidomide	*Distaval (withdrawn)*
Thenaldine	*Sandosten*
Thenyldiamine	*In Bronchilator, Franol Plus, Hayphryn*
Theophylline	*Neulin, Slo-Phyllin, Theo-Dur, Theograd, Theosol, Uniphyllin, Unicontin; in Asmapax, Franol, Labophylline, Taumasthman, Tedral; Accurbron, Aerolate, Bronkodyl, Elixicon, Elixomin, Labid, Physpan, etc.*
Theophylline monoethanolamine	*Monotheamin; Inophyline*
Theophylline olamine	see *theophylline monoethanolamine*
Theophylline Ethylenediamine }	see *aminophylline*
Thiabendazole	*Mintezol; Minzolum*
Thiacetazone	*Tebewas, Thetazone, Thioparamizone (withdrawn)*
Thiamazole	see *methimazole*
Thiambutosine	*Ciba 1906*
Thiamine hydrochloride	*Benerva; Berin, Betabion, Betalin S, Betamin, Beta-Sol, Beta-Tabs, Betaxim, Betaxin, Invite B₁, Vibex, Vitobun.*

Official name	Proprietary name
Thiamphenicol	*Descocin, Efnicol, Fricol, Glitisol, Hyrazin, Igralin, Racenicol, Rigelon, Rincrol, Thiamcol, Thiofact, etc.*
Thiazosulfone	*obsolete*
Thiethylperazine	*Torecan (range)*
Thioguanine	*Lanvis*
Thiopentone sodium	*Intraval Sodium, Pentothal; Nesdonal, Trapanal*
Thiopropazate	*Dartalan; Dartal*
Thioridazine	*Melleril (range); Malloral, Mellaril, Melleretten, Mellerettes, Novoridazine, Thioril*
Thiotepa	*Thio-Tepa*
Thiothixene	*Navane (withdrawn); Orbinamon*
Thiouracil	*obsolete*
Thonzylamine	*Tonamil, in Biomydrin*
Thorium dioxide	*Thorotrast (withdrawn)*
Thyroid extract	*Thyroid Dellipsoids D12*
Thyroid stimulating hormone (TSH)	see *thyrotrophin*
Thyrotrophin	*Thytropar; Actyron, Thyratrop, Thyreostimulin, Thyréostimuline, Thyrotron*
Thyroxine sodium	*Eltrexin; Cytolen, Euthyrox, Letter, Levoid, Oroxine, Percutacrine, Synthroid, Thyrine, Thyroxevan, Thyroxinal, Thyroxinique*
Ticarcillin sodium	*Ticar; Aerugipen, Tarcil, Ticarpen*
Ticrynafen	*Diflurex, Selcryn (both withdrawn)*
Tienilic acid	see *ticrynafen (withdrawn)*
Timolol	*Blocadren*
Timolol maleate eye drops	*Timoptol*
Tiopronin	*Epatiol, Mucolysin, Sutilan, Thiola, Thiosol, Vincol*
Tiotidine	*ICI-125211 (withdrawn)*
Tiotixene	see *thiothixene*
Tobramycin	*Nebcin*
Tocainide	*Tonocard*
Tocopheryl acetate	*Ephynal, Vita-E; Aquasol E, Daltose, E-Ferol, E Mulsin, Eprolin, Invite E, Lan-E, Lethoperol, Phytoferol, Solucap E, Tocerol, Tocomine, Tocovite*
Tofenacin	*Elamol*
Tolamolol	*not registered, UK 6558-01 (Pfizer)*
Tolazamide	*Tolanase; Tolinase, Tolisan, Norglyncin*
Tolbutamide	*Pramidex, Rastinon; Artosin, Chembutamide, Dolipol, Mellitol, Mobenol, Neo-Dibetic, Novobutamide, Oramide, Oribetic, Orinase, Tolbutol, Tobutone, Wescotol*

Official name	Proprietary name	Official name	Proprietary name
Tolfenamic acid	*Clotam*	Trinitroglycerine	see *glyceryl trinitrate*
Tolmetin sodium	**Tolectin; Tolmex**	Trional	see *methylsulphonal*
Tranexamic acid	**Cyclokapron;** *Anvitoff, Exacyl, Frenolyse, Ugorol, etc.*	Trioxsalen	*Trisoralen*
		Triparanol	*Clotrox, Hipocolestina, Trianel, Triparin (all withdrawn)*
Tranylcypromine	**Parnate;** *Tylciprine*		
Tranylcypromine	**Parnate;** *Tylciprine*	Tripelennamine	*Pyribenzamine*
Trifluoperazine		Triphthasine	see *trifluoperazine*
Trifluoperazine }	**Parstelin**	Triprolidine }	**Actifed Syrup**
Trazodone	**Molipaxin;** *Thombran, Trittico*	Pseudoephedrine }	
		Trisodium edetate	**Linclair, Sequestrene Na3**
Tretamine	*not registered, TEM, NSC, 9706*	Trithiozine	see *tritiozine*
		Tritiozine	*Tresanil*
Tretinoin	**Retin A (range);** *Aberal, Aberala, Acid a Vit, Acnavit, Acretin, Airol, A-Vitaminsyre, Cordes Vas, Effederm, Epi-Aberel, Eudyna, etc.*	Troleandomycin	see *triacetyloleandomycin*
		Troxidone	**Tridione;** *Épidone, Trimedone*
		L-Tryptophan	**Pacitron; in Optimax;** *Trofan, Tryptacin*
Triacetyloleandomycin	*Aovine, Cetilmin, Oleandocyn, Tao, Triacet, Wytrion*	Tubocurarine	**Tubarine, Tubocurarine Chloride**
		Uracil Mustard	see *uramustine*
Triamcinolone acetonide	**Adcortyl, Kenalog, Ledercort Acetonide;** *Aristocort, Kenacort, Kenalone, Tédarol, Triamalone, Triamcin*	Uramustine	*Uracil Mustard*
		Urethane	*Pracarbamin*
		Valproic acid	see *sodium valproate*
		Vancomycin	**Vancocin**
Triamterene	**Dytac;** *Dyrenium, Jatropur, Teriam*	Vasopressin	**Pitressin;** *Prostacton*
		Verapamil	**Cordilox;** *Isoptin*
Triaziquone	**Trenimon (withdrawn)**	Vercuronium bromide	**Norcuron**
Triazolam	**Halcion**	Vidarabine	*Vira-A, CI-673*
Trichlorethylene	**Trilene;** *Anamenth*	Viloxazine	**Vivalan;** *Vicilan*
Trichlormethiazide	*Esmarin, Flutra, Metahydrin, Naqua*	Vinblastine	**Velbe;** *Velban*
		Vincristine	**Oncovin**
Triclofenol piperazine	*CI-416 (obsolete)*	Vindesine	**Eldisine**
Triclofos sodium	**Tricloryl;** *Triclos*	Viomycin	**Viomycin Sulphate** *(withdrawn); Viocin*
Triethelenemelamine	see *tretamine*	Vitamin A	**Ro-A-Vit;** *Alphalin, A-Mulsin, Aquasol-A, Arovit, Atamin Forte, Avibon, A-Vicotrat, Carotin, Halivite, Solu-A, Vi-Dom-A, Vogan*
Triethylene thiophosphoramide	see *thiotepa*		
Trifluoperazine	**Stelazine;** *Calmazine, Chemflurazine, Eskazine, Fluazine, JatroNeural, Novoflurazine, Pentazine, Solazine, Terfluzin, Terfluzine, Trifluoper-Ez-Ets, Triflurin*		
		Vitamin B$_1$	see *thiamine hydrochloride*
		Vitamin B$_{12}$	see *cyanocobalamin*
		Vitamin B$_6$	see *pyridoxine*
		Vitamin C	see *ascorbic acid*
		Vitamin D$_2$	see *calciferol*
Trifluoperidol	**Triperidol;** *Psicoperidol*	Vitamin D$_3$	see *cholecalciferol*
Trifluopromazine	see *fluopromazine*	Vitamin E	see *tocopheryl acetate*
Trifluridine	*Bephen, TFT, Triherpine, Viroptic*	Vitamin K$_1$	see *phytomenadione*
		Vp 16-213	see *etoposide*
Triftazin	see *trifluoperazine*	Warfarin sodium	**Marevan, Warfarin Sodium;** *Coumadin, Coumadine, Panwarfin, Waran, Warfilone, Warnerin Diurexan; Aquafor, Aquaphor*
Trihexylphenidyl	see *benzhexol*		
Triiodothyronine	see *liothyronine sodium*		
Trimeprazine	**Vallergan;** *Panectyl, Repeltin, Temaril, Theralene*		
		Xipamide	
Trimetaphan camsylate	*Arfonad*	Xylitol	*Xylit*
Trimethadione	see *troxidone*	Zimelidine	**Zelmid (withdrawn)**
Trimethoprim	**Co-fram, Coptin, Ipral, Monotrim, Syraprim, Trimopan; in Bactrim, Septrin**	Zinc pyrithione	**Head and Shoulders Shampoo**
		Zinostatin	*not registered, NSC 157365*
Trimethoprim }	see *co-trimoxazole*	Zolamine	*not registered, 194-B, WI 291*
Sulphamethoxazole }			
Trimethylpsoralen	see *trioxsalen*	Zomepirac sodium	*Zomax (withdrawn)*
Trimipramine acid maleate	**Surmontil**	Zoxazolamine	*Flexin (withdrawn)*

PART 2: PROPRIETARY NAME–OFFICIAL NAME

Proprietary name	Official name	Proprietary name	Official name
Amizol	Amitriptyline	*Antamon*	Methionine
Amoglandin	Dinoprost trometamol	*Antegan*	Cyproheptadine
Amotril	Clofibrate	*Antelmina*	Piperazine
Amoxil	Amoxycillin	*Antepar (range)*	Piperazine
Amperil	Ampicillin	*Antepsin*	Sucralfate
Amphicol	Chloramphenicol	*Anthisan*	Mepyramine
Ampho-Moronal	Amphotericin	*Anti-Naus*	Prochlorperazine
Amphogel	Aluminium hydroxide gel	*Anti-Spas*	Benzhexol
Ampiclox	{ Ampicillin Cloxacillin	*Antial*	Brompheniramine
		Antibiopto	Chloramphenicol
Amplivix	Benziodarone	*Antiderm*	Pyridoxine
Amsal	Amylobarbitone sodium	*Antihydral*	Hexamine
A-Mulsin	Vitamin A	*Antilirum*	Physostigmine
Amuno	Indomethacin	In *Antistin-Privine*	Antazoline
Amylobarb Sodium	Amylobarbitone sodium	*Antitrem*	Benzhexol
Amylobeta	Amylobarbitone sodium	*Antivert*	Meclozine
Amylosol	Amylobarbitone sodium	*Antuitrin*	Human chorionic gonadotro-
Amytal	Amylobarbitone		phin
Anabactyl	Carbenicillin	*Antuitrin-T*	Growth hormone
Anacobin	Cyanocobalamin	*Anturan*	Sulphinpyrazone
Anadroi	Oxymetholone	In *Anugesic-HC Supposi-*	
Anaesthol	Lignocaine	*tories*	Bismuth subgallate
Anafranil	Clomipramine	*Anuphen*	Paracetamol
Anamenth	Trichlorethylene	In *Anusol*	Bismuth subgallate
Anametrin	Nomifensine	In *Anusol HC*	Bismuth subgallate
Anapolon	Oxymetholone	*Anvitoff*	Tranexamic acid
Anarexol	Cyproheptadine	*Aolept*	Pericyazine
Anatensol	Fluphenazine	*Aovine*	Triacetyloleandomycin
Ancazine	Piperazine	*Apamide*	Paracetamol
Ancef	Cephazolin sodium	*Aparkane*	Benzhexol
Ancoban	Flucytosine	*Apedine*	Phenmetrazine
Ancoloxin	Meclozine-pyridoxine	In *Aperient Dellipsoids D9*	Phenolphthalein
Ancotil	Flucytosine	*Apisate*	Diethylpropion
Ancovil	Pheniramine	*Apl*	Human chorionic gona-
Androcur	Cyproterone acetate		dotrophin
Android	Methyltestosterone	*Aponal*	Doxepin
Andronaq	Testosterone	*Appercol*	Tetramisole
Androstalone (withdrawn)	Mestanolone	*Apresoline*	Hydrallazine
Androtardyl	Testosterone enanthate	*Aprinox*	Bendrofluazide
Androtex	Androstenedione	*Aprinox-M*	Bendrofluazide
Andrumin	Dimenhydrinate	*Apsedon*	Chlorphentermine
Anectine	Suxamethonium	*Apsifen*	Ibuprofen
Anergomycil	Rolitetracycline	*Aspin VK*	Phenoxymethylpenicillin
Anestacon	Lignocaine	*Aptin*	Alprenolol
Aneural	Meprobamate	*Aquacaine G*	Procaine penicillin
Anginin	Pyridinolcarbamate	*Aquachloral*	Chloral hydrate
Angio-Conray	{ Iothalamic acid Sodium iothalamate	*Aquacillin*	Procaine penicillin
		Aquadiol	Oestradiol
Angio-Contrix	{ Iothalamic acid Sodium iothalamate	*Aquafor*	Xipamide
		Aquamephyton	Phytomenadione
Angioxine	Pyridinolcarbamate	*Aquamycetin*	Chloramphenicol
Anhydron	Cyclothiazide	*Aquaphor*	Xipamide
Anorex	Phendimetrazine	*Aquasol-A*	Vitamin A
Anorex	Phenmetrazine	*Aquatensen*	Methyclothiazide
Anovlar 21	{ Norethisterone acetate- Ethinyloestradiol	*Aracytine*	Cytarabine
		Aralen	Chloroquine
Anquil	Benperidol	*Arcasin*	Phenoxymethylpenicillin
Ansolysen	Pentolinium	*Arcomonol*	Mofebutazone
Anspor	Cephradine	*Arcotrol*	Phendimetrazine
Antabuse 200	Disulfiram	*Arcylate*	Salsalate
Antagosan	Aprotinin	*Arelix*	Piretanide
Antalvic	Dextropropoxyphene	*Arfonad*	Trimetaphan camsylate

Proprietary name	Official name	Proprietary name	Official name
Argamin	Arginine	*Azolid*	Phenylbutazone
Arginine-Sorbitol	Arginine	*Azulfidine*	Sulphasalazine
(withdrawn)		*B6-Vicotrat*	Pyridoxine
Argun	Alclofenac	*Bacacil*	Bacampicillin
Aristamid	Sulphasomidine	*Bacarate*	Phendimetrazine
Aristocort	Triamcinolone acetonide	In *Bacinguent*	Bacitracin
Arlef	Flufenamic acid	*Bactocill*	Oxacillin
Arlibide	Buphenine	**Bactrim**	Co-trimoxazole
Arlidin	Buphenine	In **Bactrim**	{ Sulphamethoxazole
Arovit	Vitamin A		Trimethoprim
Artane	Benzhexol	*Bafucin*	Gramicidin
Arthri-Sel	Lithium citrate	*Banlin*	Propantheline
Artosin	Tolbutamide	*Barbopent*	Pentobarbitone
Artracin	Indomethacin	In **Barquinol HC**	Clioquinol
Arvin	Ancrod	**Basoquin**	Amodiaguine
Ascalix	Piperazine	In **Baukal**	Propyphenazone
Asellacrin	Growth hormone	**Baxan**	Cefadroxil
In **Asmapax**	Theophylline	*Baycaron*	Mefruside
Asthmolysin	Diprophylline	*Baycillin*	Propicillin potassium
Astonin	Cyproheptadine	*Bayercillin*	Propicillin potassium
Astonin-H	Fludrocortisone	*Bayercilline*	Propicillin potassium
At-10	Dihydrotachysterol	*Baymicin*	Sisomicin
Atabrine	Mepacrine	*Baymicine*	Sisomicin
Atamin Forte	Vitamin A	**Baypen**	Mezlocillin
Atan	Gamma benzene hexa-	*Bayrena*	Sulphamethoxydiazine
	chloride	*Becaptan*	Mercaptamine
Atarax	Hydroxyzine	*Bécilan*	Pyridoxine
Atasol	Paracetamol	**Beconase Nasal Spray**	Beclomethasone dipropio-
Atecen	Dihydrotachysterol		nate
Atensine	Diazepam	**Becotide**	Beclomethasone dipro-
Athelvet	Tetramisole		pionate
Atheropront 500	Clofibrate	*Bedoz*	Cyanocobalamin
Ativan	Lorazepam	*Beflavin*	Riboflavine
Atosil	Promethazine	*Beflavina*	Riboflavine
Atromid S 500	Clofibrate	*Beflavine*	Riboflavine
Atrophan	Cinchophen	**Beflavit** (withdrawn)	Riboflavine
Atrovent	Ipratropium bromide		{ Belladonna alkaloids
Atumin	Dicyclomine	**Bellergal**	{ Ergotamine
In **Augmentin**	Clavulanic acid potassium		{ Phenobarbitone
	salt		{ Belladonna alkaloids
Aureomycin	Chlortetracycline	**Bellergal Retard**	{ Ergotamine
Auréomycine	Chlortetracycline		{ Phenobarbitone
Austramycin	Tetracycline	*Beloc*	Metoprolol
Austrastaph	Cloxacillin	*Belustine*	Lomustine
Austrawolf	Rauwolfia serpentina	*Benacen*	Probenecid
Aventyl	Nortriptyline	**Benadon**	Pyrodoxine
Avibon	Vitamin A	**Benadryl**	Diphenhydramine
Avil	Pheniramine	*Bendectin*	Dicyclomine
A-Vicotrat	Vitamin A	*Bendopa*	Levodopa
Aviletten	Pheniramine	*Bendralan*	Phenethicillin potassium
Avilettes	Pheniramine	**Benemid**	Probenecid
A-Vitaminsyre	Tretinoin	*Benhydramil*	Diphenhydramine
Avlocardyl	Propranolol	*Benol*	Cyanocobalamin
Avloclor	Chloroquine	**Benoral** (range)	Benorylate
Avlosulfon	Dapsone	*Benortan*	Benorylate
Ayercillin	Procaine penicillin	*Benotamol*	Benorylate
Azacytidine	Azacytidine	*Bensan*	Pseudoephedrine
Azapen	Methicillin	*Bentyl*	Dicyclomine
Azauridine	Azauridine	*Bentylol*	Dicyclomine
Azionyl	Clofibrate	*Bentyol*	Dicyclomine
Azo-Stat	Phenazopyridine	**Benuride**	Pheneturide
Azodine	Phenazopyridine	*Benzalcan*	Benzethonium chloride

Proprietary name	Official name
Bephen	Trifluridine
Berivine	Riboflavine
Berkdopa	Levodopa
Berkfurin	Nitrofurantoin
Berkmycen	Oxytetracycline
Berkomine	Imipramine
Berkozide	Bendrofluazide
Berubi	Cyanocobalamin
Berubigen	Cyanocobalamin
Beta-Cardone	Sotalol
Beta-Chlor	Chloral betaine
Betacard	Alprenalol
Betadine (range)	Povidone-iodine
Betalin 12	Cyanocobalamin
Betaloc	Metoprolol
Betnelan	Betamethasone
Betnesol	Betamethasone
Betnesol V	Betamethasone valerate
Betnovat	Betamethasone valerate
Betnovate (range)	Betamethasone valerate
In *Betnovate C*	Clioquinol
Bextasol Inhaler	Betamethasone
Bextasol Inhaler	Betamethasone valerate
Bi-Quinate	Quinine bisulphate or sulphate
Bicarnésine	Carnitine
Bidramine	Diphenhydramine
Bilijodon-Natrium	Iopanoic acid
Biliopaco	Iopanoic acid
Biltricide	Praziquantel
Bintotal	Ampicillin
Biogastrone	Carbenoxolone
In *Biomyorin*	Thonzylamine
Bioral	Carbenoxolone
Biquin	Quinidine bisulphate or sulphate
Bisco-Zitron	Oxyphenisatin
Bismetin	Prenylamine
In *Bismodyne*	Bismuth subgallate
Blenoxane	Bleomycin
Bleomycin	Bleomycin
Bleph-10 Liquifilm	Sulphacetamide sodium
Blocadren	Timolol
Bloxanth	Allopurinol
Bobbamycin	Oxytetracycline
Bolvidon	Mianserin
Bonadorm	Dichloralphenazone
Bonamine	Meclozine
Bonine	Meclozine
Bontril PDM	Phendimetrazine
BQ 22708	Endralazine
Bramcillin	Phenoxymethylpenicillin
Brentan	Miconazole
Brethine	Terbutaline
Bretylate	Bretylium
Brevidil M	Suxamethonium
Brevimytal Natrium	Methohexitone
Brevital Sodium	Methohexitone
Bricanyl	Terbutaline
Bridine	Povidone-iodine
Brietal Sodium	Methohexitone
Brinaldix	Clopamide

Proprietary name	Official name
Brinaldix K	Clopamide
Bristacin	Rolitetracycline
Bristacin-A	Rolitetracycline
Bristacycline	Tetracycline
Bristamycin	Erythromycin
Bristopen	Oxacillin
Bristuric	Bendrofluazide
Brocadopa	Levodopa
Bromophthalein	Sulphobromophthalein sodium
Bromosulfophthalein	Sulphobromophthalein sodium
Bromsulfan	Sulphobromophthalein sodium
Bromsulphalein	Sulphobromophthalein sodium
Bromthalein	Sulphobromophthalein sodium
Bromural	Bromvaletone
Bronalin	Hexoprenaline
In *Bronchilator*	Thenyldiamine
Bronkodyl	Theophylline
Brontina	Deptropine
Brontine	Deptropine
Brontyl 300	Proxyphylline
Brotazona	Feprazone
Broxil	Phenethicillin potassium
Brufen	Ibuprofen
Bu-Lax	Docusate sodium
Bucrol	Carbutamide
Budoform	Clioquinol
Bufedon	Buphenine
Bufemid	Fenbufen
Buprex	Buprenorphine
Burinex	Bumetanide
Buscopan	Hyoscine N-butyl bromide
Butacal	Phenylbutazone
Butacote	Phenylbutazone
Butagesic	Phenylbutazone
Butalan	Phenylbutazone
Butaphen	Phenylbutazone
Butazolidin	Phenylbutazone
Butazone	Phenylbutazone
Butoz	Phenylbutazone
Bykomycin	Neomycin
Cabral	Phenyramidol
Cafegot	Ergotamine tartrate / Caffeine
Calcamine	Dihydrotachysterol
Calcitare	Calcitonin (pork)
Calcium Leucovorin	Calcium folinate
Calmazine	Trifluoperazine
Calmoden Capsules	Chlordiazepoxide
Calmonal	Meclozine
Calpol	Paracetamol
Calsynar	Salcatonin
Calthor	Ciclacillin
Camcolit	Lithium carbonate
Canpain	Paracetamol
Candio-Hermal	Nystatin
Capastat	Capreomycin
Capital	Paracetamol

Proprietary name	Official name
Capitol	Benzalkonium chloride
Capoten	Captopril
Capracid	Aminocaproic acid
Capralense	Aminocaproic acid
Capramol	Aminocaproic acid
Caproamin	Aminocaproic acid
Caprocin	Capreomycin
Carbachol	Carbachol
Carbapen	Carbenicillin
Carbazole	Carbenoxolone
Carbocaine	Mepivacaine
Carbolith	Lithium carbonate
Carbostesin	Bupivacaine
Carbrital	{ Carbromal
(withdrawn)	{ Phenobarbitone
Cardabid	Glyceryl trinitrate
Cardiazol	Leptazol
Cardilan	Isoxsuprine
Cardio-Conray	Iothalamic acid
Cardiografin	Sodium diatrizoate
Cardiorythmine	Ajmaline
Cardiox	Digoxin
Cardivix	Benziodarone
Cardophylin	Aminophylline
Caréna	Aminophylline
Carindapen	Carindacillin
Carisoma	Carisoprodol
Carmustine	Carmustine
Carotin	Vitamin A
Carotin	Betacarotene
Carvasin	Isosorbide Dinitrate
Caryolysine	Mustine hydrochloride
In *Castel-Minus*	Resorcinol
Castilium	Clobazam
Catapres	Clonidine
Catapresan	Clonidine
Catapressan	Clonidine
Cathomycine	Novobiocin
Catilan	Chloramphenicol
Catron	
(withdrawn)	Pheniprazine
Ccnu Lunbeck	Lomustine
Ce-Cobalin Syrup	Cyanocobalamin
Cebutid	Flurbiprofen
Cecenu	Lomustine
Ceclor	Cefaclor
Cedilanid	Lanatoside C
Cedilanid Inject-ion	Deslanoside
Cedilanid Desa-cetyl	Deslanoside
Cedilanid-D	Deslanoside
Cedin	Isoniazid
Cedlanide	Deslanoside
Cedocard	Isosorbide Dinitrate
Ceenu	Lomustine
Ceenu	Lomustine
Ceetamol	Paracetamol
Cefacidal	Cephazolin sodium

Proprietary name	Official name
Cefadyl	Cefapirin
Céfalotine	Cephalothin
Cefamid	Cephradine
Cefobid	Cefoperazone sodium
Cefobine	Cefoperazone sodium
Cefobis	Cefoperazone sodium
Cefradex	Cephradine
Cefril	Cephradine
Ceglunat	Lanatoside C
Celadigal	Lanatoside C
Celanat	Lanatoside C
Celbenin	Methicillin
Celestan	Betamethasone
Celestan V	Betamethasone valerate
Celestene	Betamethasone
Celestona Valerat	Betamethasone valerate
Celestone	Betamethasone
Celestone V	Betamethasone valerate
Celmetin	Cephazolin sodium
Celontin	Methsuximide
Célospor	Cephacetrile
Cen-App	Paracetamol
Censtim	Imipramine
Centet	Tetracycline
Centrax	Prazepam
Centyl	Bendrofluazide
Centyl K	Bendrofluazide
Ceporan	Cephaloridine
Ceporex	Cephalexin
Ceporexine	Cephalexin
Ceporin	Cephaloridine
Cepovenin	Cephalothin
Cerebid	Papaverine
Cérébrol	Diethylaminoethanol
Cerespan	Papaverine
Cerubidin	Daunorubicin
Cerucal	Metoclopramide
Cesol	Praziquantel
Cetal	Chlorhexidine
In *Cetanorm*	Benzalkonium chloride
Cetilmin	Triacetyloleandomycin
Cetiprin	Emepronium
Ceto Sanol	Lanatoside C
C-Film	Nonoxynol-9
Chembutamide	Tolbutamide
Chembutazone	Phenylbutazone
Chemceptaphen	Paracetamol
Chemcycline	Tetracyline
Chemdipoxide	Chlordiazepoxide
Chemflurazine	Trifluoperazine
Chemhydrazide	Hydrochlorothiazide
Chemipramine	Imipramine
Chemocycline	Oxytetracycline
Chemovag	Sulphafurazole
Chemthromycin	Erythromycin
Chemthromycin	Erythromycin estolate
Chendal	Chenodeoxycholic acid
Cheno-Caps	Chenodeoxycholic acid
Chenoacid	Chenodeoxycholic acid
Chenodol	Chenodeoxycholic acid
Chenofalk	Chenodeoxycholic acid
Chenofalk	Chenodeoxycholic acid

Proprietary name	Official name
DBI	Phenformin
Debendox	Dicyclomine
Deca-Nephrine	Phenylephrine
Decabicin	Dibekacin
Decabiotic	Tetracycline
Decadron	Dexamethasone
Decaris	Levamisole
Decentan	Perphenazine
Decicain	Amethocaine
Declimone	{ Ethinyloestradiol / Methyltestosterone }
Declinax	Debrisoquine
Declomycin	Demeclocycline
Declostatin	Demeclocycline
Decontractyl	Mephenesin
Decortin	Prednisone
Decortin H	Prednisolone
Decortisyl	Prednisone
Decycline	Tetracycline
Defencin CP	Isoxsuprine
Defiltran	Acetazolamide
Degest	Phenylephrine
Dehydrometine Roche	Dehydroemetine
Delakmin	Alfacalcidol
Delalutin	Hydroxyprogesterone hexanoate
Delatestryl	Testosterone enanthate
Delfen	Nonoxynol-9
Delprosyn	Propicillin potassium
Delta-Cortef	Prednisolone
Delta-Dome	Prednisone
Delta-Phoricol	Prednisolone
Deltacortone	Prednisone
Deltacortril Enteric	Prednisolone
Deltalin	Calciferol
Deltalone	Prednisolone
Deltasolone	Prednisolone
Deltastab	Prednisolone
Dema	Tetracycline
Demelox	Amoxapine
Demser	Metirosine
Demulen	{ Mestranol / Ethynodiol }
Demulen 50	{ Ethinyloestradiol / Ethylnodiol }
Dendrid	Idoxuridine
Deoxycortone Acetate Implants	Deoxycortone
Dep-75	Diethylpropion
Depakene	Sodium valproate
Depakine	Sodium valproate
Depamine	Penicillamine
Depixol Injection	Flupenthixol decanoate
Depixol-Conc. Injection	Flupenthixol decanoate
Depo-Clinovir	Medroxyprogesterone
Depo-Heparin	Heparin
Depo-Medrate	Methylprednisolone
Depo-Medrol	Methylprednisolone
Depo-Medrone	Methylprednisolone
Depo-Provera	Medroxyprogesterone
Depocillin	Procaine penicillin
Deprakine	Sodium valproate

Proprietary name	Official name
Deprex	Amitriptyline
Depronal SA	Dextropropoxyphene
Derbac (range)	Malathion
Derbac Soap	Gamma benzene hexachloride
Dercusan	Chloramine
Deripen	Amipicillin
Dermatol	Bismuth subgallate
Dermo-6	Pyridoxine
Dermohex	Hexachlorophane
In **Dermovate-NN**	Clobetasol propionate
Dermol	Pethidine
Dermonistat	Miconazole
Dermovate	Clobetasol propionate
Deronil	Dexamethasone
Desace	Deslanoside
Desaci	Deslanoside
Desamon	Benzethonium chloride
Descocin	Thiamphenicol
Deseril	Methysergide
Désernil	Methysergide
Desma	Stilboestrol
Desopimon	Chlorphentermine
Desoxyn	Methylamphetamine
Det MS	Dihydroergotamine
Develin	Dextropropoxyphene
Dexa-Sine	Dexamethasone
Dexacortisyl	Dexamethasone
Dexambutol	Ethambutol
Dexamed	Dexamphetamine
Dexameth	Dexamethasone
Dexamethadrone	Dexamethasone
Dexamine	Dexaphetamine
Dexamphate	Dexamphetamine
Dexaspan	Dexamphetamine
Dexedrine	Dexamphetamine
Dexmethsone	Dexamethasone
Dexone	Dexamethasone
Dextraven	Dextran
D-Feda	Pseudoephedrine
DFP-OEL	Dyflos
DH-Ergotamin-Retard	Dihydroergotamine
Di-Ademil	Hydroflumethiazide
Di-Adreson	Prednisone
Di-Adreson-F	Prednisolone
Di-Hydan	Phenytoin sodium
Di-Paraline	Chlorcyclizine
Dia-Tablinen	Carbutamide
Diabeta	Glibenclamide
Diabetoral	Chlorpropamide
Diabex SR	Metformin
Diabexyl	Metformin
Diabinese	Chlorpropamide
Diacin	Nicotinic acid
Diacta	Sulphaguanidine
Diaginol Viscous	Sodium acetrizoate
Dialoxin	Digoxin
Diamicron	Gliclazide
Diamorphine Hydrochloride	Diamorphine
Diamox	Acetazolamide
Dianabol	Methandienone
Diapax	Chlordiazepoxide

Proprietary name	Official name
In *Diarrest*	Dicyclomine
Diasone Sodium	Sulfoxone sodium
Diastatin	Nystatin
Diazemuls	Diazepam
Diazyl	Sulphadiazine
Dibenyline	Phenoxybenzamine
Dibenzyline	Phenoxybenzamine
Dibotin	Phenformin
Dichlor-Stapenor	Dicloxacillin sodium
Dichloren	Mustine hydrochloride
Dichlotride	Hydrocholorothiazide
Diclocil	Dicloxacillin sodium
Diclomyl	Dicyclomine
Diconal	Dipipanone
Dicorvin	Stilboestrol
Dicycol	Dicyclomine
Didrex (withdrawn)	Benzphetamine
Didronel Tablets	Disodium etidronate
Diételmin	Piperazine
Diflupyl	Dyflos
Diflurex	Ticrynafen
Digacin	Digoxin
Diganox Nativelle	Digoxin
Digilong	Digitoxin
Digimed	Digitoxin
Digimerck	Digitoxin
Digitaline Nativelle	Digitoxin
Digitox	Digitoxin
Digolan	Digoxin
Digoxine Nativelle	Digoxin
Dihydergot (range)	Dihydroergotamine
Diidergot	Dihydroergotamine
Dilabid	Phenytoin sodium
Dilafurane	Benziodarone
Dilantin	Phenytoin sodium
Dilar	Paramethasone
Dilaspan	Papaverine
Dilatol	Buphenine
Dilax	Docusate sodium
Dilhydrin	Buphenine
Dilin	Diprophylline
Dilor	Diprophylline
Dimegan	Brompheniramine
Dimelor	Acetohexamide
Dimerx	Iocarmic acid
Dimetane	Brompheniramine
Dimethazine	Sulphadimidine
Diminsul	Sulphadimidine
Dimipressin	Imipramine
Dimitan	Mustine hydrochloride
Dimotane	Brompheniramine
Dimotane (range)	Brompheniramine
Dimotapp	Brompheniramine
Dindevan	Phenindione
Dioctyl	Docusate sodium
Dioctyl Forte	Docusate sodium
Dioctylal Forte	Docusate sodium
Diodoquin	Di-iodohydroxyquinoline
Diodrast	Diodone
Dipar	Phenformin
Diparcol	Diethazine
Diphentyn	Phenytoin sodium
Diprosone	Betamethasone
Direma	Hydrochlorothiazide
Direx	Iron dextran
Direxiode	Di-iodohydroxyquinoline
Disadine (range)	Povidone-iodine
Disalcid	Salsalate
Disipal	Orphenadrine
Disonate	Docusate sodium
Disotate	Disodium edetate
Distaclor	Cefaclor
Distamine	Penicillamine
Distaquaine V-K	Phenoxymethylpenicillin
Distaval (withdrawn)	Thalidomide
Disteol	Ectylurea
Distilbene	Stilboestrol
Distivit	Cyanocobalamin
Distraneurin	Chlormethiazole
Distraneurine	Chlormethiazole
Disulphine Blue Intravenous Injection	Sulphan Blue
Ditaven	Digitoxin
Dithrocream	Anthralin
Dithrolan	Anthralin
Ditoin	Phenytoin sodium
Diucardin	Hydroflumethiazide
Diucen-H	Hydrochlorothiazide
Diuchlor H	Hydrochlorothiazide
Diuretic	Methyclothiazide
Diurexan	Xipamide
Diuril	Chlorothiazide
Diurilix	Chlorothiazide
Divermex	Piperazine
Divulsan	Phenytoin sodium
Dixarit	Clonidine
D-Mulsin	Cholecalciferol
Doburil	Cyclothiazide
Dobutrex	Dobutamine
Doca	Deoxycortone
Docémine	Cyanocobalamin
Dodex	Cyanocobalamin
Dogmatil	Sulpiride
Dolamin	Paracetamol
Dolanex	Paracetamol
Dolantin	Pethidine
Dolasan	Dextropropoxyphene
Dolciton	Propranolol
Dolene	Dextropropoxyphene
Dolipol	Tolbutamine
Doliprane	Paracetamol
Dolmatil	Sulpiride
Dolobid	Diflunisal
Dolocap	Dextropropoxyphene
Dolorphine Hydrochloride	Methadone
Dolosal	Pethidine
Doloxene	Dextropropoxyphene
Domical	Amitriptyline
Domistan	Histapyrrodine
Donopon-GP	Metoclopramide
Dopamet	Methyldopa
Dopar	Levodopa
Dopram	Doxapram
In **Dorbanex**	Danthron

Proprietary name	Official name
Enavid E	{ Norethynodrel { Mestranol
Endep	Amitriptyline
Endoxan	Cyclophosphamide
Endoxana	Cyclophosphamide
Endrate Disodium	Disodium edetate
Enduron	Methylclothiazide
Enduronyl	Methylclothiazide
Enovid	{ Mestranol { Norethynodrel
Ensidon	Opipramol
Ensobarb	Phenobarbitone
Entacyl	Piperazine
Enterfram	Framycetin
Entero-Valodon	Clioquinol
Entero-Vioform	Clioquinol
Enteroquin	Clioquinol
Envacar (withdrawn)	Guanoxan sulphate
E-Pam	Diazepam
Epanutin	Phenytoin sodium
Epatiol	Tiopronin
Epha-Retard	Dimenhydrinate
Epi-Aberel	Tretinoin
Épidone	Troxidone
Epidropal	Allopurinol
Epilim	Sodium valproate
Epilol	Phenobarbitone
Epodyl	Ethoglucid
Epontol	Propanidid
Epsamon	Aminocaproic acid
Epsikapron	Aminocaproic acid
Epsylone	Phenobarbitone
Equanil	Meprobamate
Equilid	Sulpiride
Equipose	Hydroxyzine
Eraldin	Practolol
Erantin	Dextropropoxyphene
Erasol	Mustine hydrochloride
Eratrex	Erythromycin
Eraverm	Piperazine
Erftopred	Prednisone
Ergamisol	Levamisole
Ergenyl	Sodium valproate
Ergomar	Ergotamine
In *Ergosol R*	Resorcinol
Ergotart	Ergotamine
Ergotonin	Dihydroergotamine
Eributazone	Phenylbutazone
Eromycin	Erythromycin estolate
Erostin	Erythromycin
Ertonyl	Ethinyloestradiol
Erycen	Erythromycin
Erycinum	Erythromycin
Erysan	Chlornaphazine
Erythrocin	Erythromycin stearate
Erythrocin IM	Erythromycin ethyl succinate
Erythromid	Erythromycin
Erythromyctine	Erythromycin
Erythroped (range)	Erythromycin ethyl succinate
Erythroped	Erythromycin

Proprietary name	Official name
Esbaloid	Bethanidine
Esbatal	Bethanidine
Esidrex	Hydrochlorothiazide
Esidrix	Hydrochlorothiazide
Eskabarb	Phenobarbitone
Eskacef	Cephradine
Eskalith	Lithium carbonate
In *Eskamel*	Resorcinol
Eskaserp	Reserpine
Eskazine	Trifluoperazine
Esmarin	Trichlormethiazide
Espéral	Disulfiram
Estigyn	Ethinyloestradiol
Estinyl	Ethinyloestradiol
Estrace	Oestradiol
Estracyt	Estramustine sodium phosphate
Etalpha	Alfacalcidol
Etherone	Ethisterone
Ethibute	Phenylbutazone
Ethiodan	Iophendylate injection
Ethrane	Enflurane
Ethyl II	Ethinyloestradiol
Ethymal	Ethosuximide
Etibi	Ethambutol
Etidron	Disodium etidronate,
Etoscol	Hexoprenaline
Etrenol	Hycanthone
Etumine	Clothiapine
Eubasinum	Sulphapyridine
Endemine	Diazoxide
Eudorm	Chloral hydrate
Eudyna	Tretinoin
Euglucon	Glibenclamide
Euglycin	Metahexamide
Eugynon 30	{ Norgestrel { Ethinyloestradiol
Eugynon 50	{ Ethinyloestradiol { Norgestrel
Euhypnos	Temazepam
Euhypnos Forte	Temazepam
Eukraton	Bemegride
Eunoctal	Amylobarbitone
Euphyllin	Aminophylline
Eusaprim	Co-trimoxazole
Euthyrox	Thyroxine sodium
Eutonyl	Pargyline
Evac-U-Gen	Phenolphthalein
Evacalm	Diazepam
Evacilin	Procaine penicillin
Evadyne	Butriptyline
Exacyl	Tranexamic acid
Exoseptoplix	Sulphanilamide
Espansyl Spansule	{ Ephedrine { Trifluoperazine { Diphenylpyraline
In *Expulin*	Chlorpheniramine maleate
Extencilline	Benzathine penicillin
Extramycin	Sisomicin
Exyphen	Brompheniramine
Fahahistin	Mebhydrolin
Famet	Sulphamethizole

Proprietary name	Official name
Fansidar	{ Pyrimethamine / Sulfadoxine
Fantorin	Stibophen
Farlutal	Medroxyprogesterone
Farmablastina	Doxorubicin hydrochloride
Fastin	Phentermine
Favistan	Methimazole
Fe-Cap	Ferrous glycine sulphate
Fe-Cap C	{ Ferrous glycine sulphate / Ascorbic acid
Fe-Cap Folic	{ Ferrous glycine sulphate / Folic acid
Feclan	Fenclofenac
Feevone	Terbutaline
Feldene	Piroxicam
Felsol	Phenazone
Felsules	Chloral hydrate
Femergin	Ergotamine
Feminone	Ethinyloestradiol
Fenam	Isoxsuprine
Fenamine	Pheniramine
Fenilor	Broxyquinoline
Fenistil	Dimethindene maleate
Fenopron Progesic	Fenoprofen calcium
Fenostil-Retard	Dimethindene maleate
Fenox	Phenylephrine
Fentazin	Perphenazine
Fepron	Fenoprofen calcium
Feprona	Fenoprofen calcium
Ferndex	Dexamphetamine
Ferro-Gradumet	Ferrous sulphate
Fertodur	Cyclofenil
Fetamin	Methylamphetamine
Fibocil	Aprindine
Fiboran	Aprindine
Fibroxin	Digoxin
Filair	Terbutaline
Fiocortril	Hydrocortisone
Fivepen	Benzylpenicillin
FL 6654	Caroxazone
Flabelline	Methicillin
Flagyl	Metronidazole
Flagyl Compak	{ Metronidazole / Nystatin
Flaminon	Niflumic acid
Flatax	Phenolphthalein
In *Flavelix*	Mepyramine
Flavoquine	Amodiaquine
Flaxedil	Gallamine
Flenac	Fenclofenac
Flexartal	Carisoprodol
Flexazone	Phenylbutazone
Flexin (withdrawn)	Zoxazolamine
Flocilline	Procaine penicillin
Flonatril	Clorexolone
Floraquin	Di-iodohydroxquinoline
Florinef	Fludrocortisone
Florinef Acetate	Fludrocortisone
Flurocycline	Tetracycline
Floropryl	Dyflos
Flosint (withdrawn)	Indoprofen
Fluanoxal	Flupenthixol decanoate

Proprietary name	Official name
Fluanoxal Retard	Flupenthixol decanoate .
Fluazine	Trifluoperazine
Fludex	Indapamide
Fluibil	Chenodeoxycholic acid
Fluoro-uracil	Fluorouracil
Fluoroplex	Fluorouracil
Fluothane	Halothane
Flutra	Trichlormethiazide
Focus	Ibuprofen
Folasic	Folic acid
Foldine	Folic acid
Folettes	Folic acid
Folic Acid	Folic acid
Foligan	Allopurinol
Folsan	Folic acid
Folvite	Folic acid
Forane	Isoflurane
Forenol	Niflumic acid
Forhistal	Dimethindene maleate
Forpen	Benzylpenicillin
In *Fortagesic*	Pentazocine
Fortal	Pentazocine
Fortecortin	Dexamethasone
Fortral	Pentazocine
Fortralin	Pentazocine
Framygen	Framycetin
In *Franol*	Theophylline
In *Franol Plus*	Thenyldiamine
Frénactil	Benperidol
Frenal	Sodium cromoglycate
Frenolyse	Tranexamic acid
Fricol	Thiamphenicol
Frisium	Clobazam
Froben	Flurbiprofen
Fructines-Vichy	Phenolphthalein
Frusid	Frusemide
Fuadin	Stibophen
Fucidin	Sodium fusidate
Fucidin H	Sodium fusidate
Fudr	Floxuridine
Fulcin 125	Griseofulvin
Fulcin-S	Griseofulvin
Fulvicin-U/F	Griseofulvin
Fungilin	Amphotericin
Fungizone Intravenous	Amphotericin
Fura-Med	Nitrofurantoin
Furacin	Nitrofurazone
Furacine	Nitrofurazone
Furadantin	Nitrofurantoin
Furadöine	Nitrofurantoin
Furamide	Diloxanide
Furan	Nitrofurantoin
Furanex	Nitrofurantoin
Furanite	Nitrofurantoin
Furatine	Nitrofurantoin
Furesol	Nitrofurazone
Furoxane	Furazolidone
Furoxone	Furazolidone
Gabbromycin	Paromomycin
Galactoquin	Quinidine polygalacturonate
Galenomycin	Oxytetracycline

Proprietary name	Official name
Gameme	Gamma benzene hexa-chloride
Gammacorten	Dexamethasone
Ganphen	Promethazine
Gantanol	Sulphamethoxazole
Gastrisin	Sulphafurazole
Gantrisine	Sulphafurazole
Garamycin	Gentamicin
In **Garamycin Chains**	Methyl methacrylate
Gardenal	Phenobarbitone
Gastro-Conray	Sodium iothalamate
Gastrografin	Sodium diatrizoate
GBH	Gamma benzene hexachloride
Gelfoam	Absorbable gelatin sponge
Gentalline	Gentamicin
Gentamidon	Amidopyrine
Genticin	Gentamicin
Gentran	Dextran
Geocillin	Carindacillin
Geopen	Carbenicillin
Gesinal	Medroxyprogesterone
Gestapuran	Medroxyprogesterone
Gestone-Oral	Ethisterone
Gidalon	Fenclofenac
Gilurytmal	Ajmaline
Giracid	Phenazopyridine
Glianimon	Benperidol
Glibenese	Glipizide
Glifan	Glafenine
Glifanan	Glafenine
Glitisol	Thiamphenicol
Glosso-Stérandryl	Methyltestosterone
Glucidoral	Carbutamide
Glucophage	Metformin
Gludorm	Glutethimide
Glyconormal	Glymidine
Glyfyllin	Diprophylline
Godafon	Glymidine
Gonadex	Human chorionic gona-dotrophin
Gonadotraphon Lh	Human chorionic gona-dotrophin
Gordox	Aprotinin
In *Gorun*	Cinchophen
Gotimycin	Chloramphenicol
GPV	Phenoxymethylpenicillin
Gramaxin	Cephazolin sodium
In **Graneodin Ointment**	Gramicidin
Gravol	Dimenhydrinate
Grifulvin V	Griseofulvin
Gris-Peg	Griseofulvin
Grisactin	Griseofulvin
Griséfuline	Griseofulvin
Grisona	Feprazone
Grisovin	Griseofulvin
Guanicil	Sulphaguanidine
Guanidan	Sulphaguanidine
Guanutil	Guanoxan sulphate
Gubernal	Alprenalol
Gynergen	Ergotamine
Gyno-Daktarin	Miconazole

Proprietary name	Official name
Gynoestryl	Oestradiol
Gynolett	Ethinyloestradiol
Gynovlar 21	Ethinyloestradiol / Norethisterone acetate
Gynovules	Di-iodohydroxyquinoline
Haelan	Flurandrenolone
In **Haelan C**	Clioquinol
Haemopan	Phenindione
Halcion	Triazolam
Haldol	Haloperidol
Haldrate	Paramethasone
Haldrone	Paramethasone
Halivite	Vitamin A
Holotestin	Fluoxymesterone
Halothane	Halothane
Hamocura	Heparin
Harmonyl	Deserpidine
Harvatropin	Human chorionic gonadotrophin
Haurymellin	Metformin
Haymine	Chlorpheniramine maleate
Haynon	Chlorpheniramine
In **Hayphryn**	Thenyldiamine
Head and shoulders shampoo	Zinc pyrithione
Hedulin	Phenindione
Hekbilin	Chenodeoxycholic acid
Helfodopa	Levodopa
Helmezine Elixir	Piperazine
Hemineurin	Chlormethiazole
Hémineurine	Chlormethiazole
Heminevrin	Chlormethiazole
Hemocaprol	Aminocaproic acid
Hepacon-B₁₂	Cyanocobalamin
Hepalean	Heparin
Heparin Retard Injection	Heparin
Hepartest	Sulphobromophthalein sodium
Hepathrom	Heparin
Heprinar	Heparin
Herbesser	Diltiazem
Herisan Antibiotic	Neomycin
Herpid	Idoxuridine / Dimethysulphoxide
Herplex	Idoxuridine
Hexa-Betalin	Pyridoxine
Hexachlorone	Hexachlorophane
Hexacycline	Tetracycline
Hexadrol	Dexamethasone
Hexastat	Hexamethylmelamine
Hexavibex	Pyridoxine
Hexicid	Gamma benzene hexa-chloride
Hexobion	Pyridoxine
Hexol	Chlorhexidine
Hibitan	Chlorhexidine
Hibitane	Chlorhexidine
Hip-Réx	Hexamine hippurate
Hipcolestina	Triparanol
Hiprex	Hexamine hippurate
Hirudoid Gel	Heparin

Proprietary name	Official name
Hismanal	Astemizole
Histadyl	Methapyrilene
Histaids	Chlorpheniramine
In *Histamed*	Mepyramine
Histantil	Promethazine
Histaspan	Chlorpheniramine
Histergan	Diphenhydramine
HMS Liquifilm	Medrysone
Holoxan	Ifosfamide
Hostacortin	Prednisone
Hostacortin H	Prednisolone
Hostacyclin	Tetracycline
HP-549	Isoxepac
Humatin *(withdrawn)*	Paromomycin
Humorsol	Demecarium
Humycin	Paromomycin
Hyamine 1622	Benzethonium chloride
Hyamine 3500	Benzalkonium chloride
Hyasorb	Benzylpenicillin
Hydergine	Dihydroergocornine / Dihydroergocristine / Dihydroergocryptine
Hydergine	Co-dergocrine mesylate
Hydracycline	Tetracycline
Hydrazide	Hydrochlorothiazide
Hydrazol	Acetazolamide
Hydrea	Hydroxyurea
Hydrenox	Hydroflumethiazide
Hydrid	Hydrochlorothiazide
Hydro-Aquil	Hydrochlorothiazide
Hydro-Long	Chlorthalidone
Hydroclonazone	Chloramine
Hydrocortancyl	Prednisolone
Hydrocortistab	Hydrocortisone
Hydrocortone	Hydrocortisone
Hydrodiuretex	Hydrochlorothiazide
Hydrodiuril	Hydrochlorothiazide
Hydromedin	Ethacrynic acid
Hydrosaluret	Hydrochlorothiazide
Hydrosaluric	Hydrochlorothiazide
Hydrosone	Hydrocortisone
In *Hygroton K*	Chlorthalidone
Hyeloril	Hydrochlorothiazide
Hygroton	Chlorthalidone
Hylenta	Benzylpenicillin
Hypaque	Sodium diatrizoate
Hypercal	Rauwolfia serpentina
Hyperstat	Diazoxide
Hipertane	Rauwolfia serpentina
Hypertensan	Rauwalfia serpentina
Hypnol	Pentobarbitone
Hypnolone	Phenobarbitone
Hypnomidate	Etomidate
Hypnorex	Lithium carbonate
Hypnovel	Midazolam
Hypon	Aspirin / Caffeine / Codeine phosphate
Hypoten	Sodium nitroprusside
Hypovase	Prazosin
Hyproval PA	Hydroxyprogesterone Hexanoate

Proprietary name	Official name
Hyptor	Methaqualone
Hyrazin	Thiamphenicol
Hyskon	Dextran
Hytakerol	Dihydrotachysterol
Ibatal	Pentobarbitone
Ibu-Slo	Ibuprofen
I-Care	Phenylephrine
ICI-54450 *(withdrawn)*	Fenclozic acid
ICI 59118	Razoxane
ICI-125211 *(withdrawn)*	Tiotidine
Icipen	Phenoxymethylpenicillin
Idalon	Floctafenine
Idarac	Floctafenine
Idocyklin	Doxycycline hydrochloride
Idrogestene	Hydroxyprogesterone Hexanoate
Idulian	Azatidine
Iduridin	Idoxuridine / Dimethylsulphoxide
Iduviran	Idoxuridine
Igralin	Thiamphenicol
Ikaran	Dihydroergotamine
Iktorivil	Clonazepam
Ilosone *(range)*	Erythromycin estolate
Ilotycin	Erythromycin
Ilvin	Brompheniramine
Imacillin	Amoxycillin
Imadyl	Carprofen
Imap	Fluspirilene
Imavate	Imipramine
Imbrilon	Indomethacin
Imferon	Iron dextran
Imiprin	Imipramine
Imménoctal	Quinalbarbitone
Imodium	Loperamide
Imperacin	Oxytetracycline
Impranil	Imipramine
Impril	Imipramine
Imuran	Azathioprine
Imurek	Azathioprine
Imurel	Azathioprine
Inamycin	Novobiocin
Inapetyl	Benzphetamine
Inapsine	Droperidol
Incidal	Mebhydrolin
Incidaletten	Mebhydrolin
Inderal	Propranolol
Inderalici	Propranolol
Indocid	Indomethacin
Indoflex	Indomethacin
Indosmos *(withdrawn)*	Indomethacin
Inferdex	Iron dextran
Inflam	Ibuprofen
Inflaryl	Niflumic acid
Ingelan	Isoprenaline
INH	Isoniazid
Inhiston	Pheniramine
Iniprol	Aprotinin
Inocor	Amrinone
Inocortyl	Prednisone
Inofal	Mesoridazine
Inophyline	Theophylline mono-ethanolamine

Proprietary name	Official name
Insidon	Opipramol
Insoral	Phenformin
Intal	Sodium cromogylcate
Integrin	Oxypertine
Intermedine	Melanocyte-stimulating hormone (MSH)
Intestopan	Broxyquinoline
Intraval Sodium	Thiopentone sodium
Intropin	Dopamine
Invenol	Carbutamide
Inversine	Mecamylamine
Ionamin	Phentermine
Ioquin	Di-iodohydroxyquinoline
Ipamix	Indapamide
Ipradol	Hexoprenaline
Ipral	Trimethoprim
Iramil	Imipramine
Ironorm Injection	Iron dextran
Ismelin	Guanethidine
Iso-Autohaler	Isoprenaline
Iso-Bid	Isosorbide dinitrate
Iso-Intranefrin	Isoprenaline
Isocillin	Phenoxymethylpenicillin
Isodiane	Metahexamide
Isodine	Povidone-iodine
Isoglaucon	Clonidine
Isoket	Isosorbide dinitrate
Isolait	Isoxsuprine
Isolanid	Lanatoside C
Isolin	Isoprenaline
Isonal	Amylobarbitone
Isophrim	Phenylephrine
Isoptin	Verapamil
Isopto-Carpine	Pilocarpine
Isopto-Cetamide	Sulphacetamide sodium
Isopto-Phenylephrine	Phenylephrine
Isopto-Eserine	Physostigmine
Isopto-Pilocarpine	Pilocarpine
Isordil	Isosorbide dinitrate
Isosulf	Suphasomidine
Isotamine	Isoniazid
Isotinyl	Isoniazid
Isotrate	Isosorbide dinitrate
Isovon	Isoprenaline
Isozid	Isoniazid
Ispenoral	Phenoxymethylpenicillin
Isuprel	Isoprenaline
Itorex	Cefuroxime
Ituran	Nitrofurantoin
Janimine	Imipramine
Jatroneural	Trifluoperazine
Jatropur	Triamterene
Jetrium	Dextromoramide
Jopanonsyre	Iopanoic acid
Juvacor	Nikethamide
Juvastigmin	Neostigmine
Kabikinase	Streptokinase
Kabipenin	Phenoxymethypenicillin
Kalspare	Chlorthalidone
Kamycine	Kanamycin
Kanabristol	Kanamycin

Proprietary name	Official name
Kanasig	Kanamycin
Kannasyn	Kanamycin
Kantrex	Kanamycin
Ka-Pen	Benzylpenicillin
Kappadione	Menadiol sodium diphosphate
Kardonyl	Nikethamide
Kavrin	Papaverine
Kay-Cee-L	Potassium chloride
K Contin Continus	Potassium chloride
Kefadol	Cephamandole
Keflex	Cephalexin
Keflin	Cephalothin
Keflodin	Cephaloridine
Kefloral	Cephalexin monohydrate
Kefocin	Cephaloglycin
Keforal	Cephalexin
Kefspor	Cephaloridine
Kefzol	Cephazolin sodium
Kelfizina	Sulfametopyrazine
Kelfizine W	Sulfametopyrazine
Kelfolate	{ Ferrous glycine sulphate { Folic acid
Kemadrin	Procyclidine
Kemicetine	Chloramphenicol
Kenacort	Triamcinolone acetonide
Kenalog	Triamcinolone acetonide
Kenalone	Triamcinolone acetonide
Kerecid	Idoxuridine
Kesso-Pen	Benzylpenicillin
Kesso-Pen-K	Phenoxymethylpenicillin
Kesso-Tetra	Tetracycline
Kessodanten	Phenytoin sodium
Kessodrate	Chloral hydrate
Ketaject	Ketamine
Ketalar	Ketamine
Ketanest	Ketamine
Keteocort	Prednisone
Keteocort H	Prednisolone
Ketrax	Levamisole
Kidrolase	Colaspase
Kinidin Durules	Quinidine bisulphate or sulphate
Kirocid	Sulphamethoxydiazine
Kiron	Sulphamethoxydiazine
Klinomycin	Minocycline
Klobamicina	Dibekacin
Klorokin	Chloroquine
Klortee	Chloramine
In *KLYX*	Dicusate sodium
Kolanticon	Dicyclomine
Kolantyl	Dicyclomine
Konakion	Phytomenadione
Korostatin	Nystatin
Korum	Paracetamol
Kwell	Gamma benzene hexachloride
Kwellada	Gamma benzene hexachloride
Kynex	Sulphamethoxypyridazine
L 3123	Puromycin
Labid	Theophylline

Proprietary name	Official name
Lorfan	Levallorphan
Loridine	Cephaloridine
Lorphen	Chlorpheniramine
Loxitane	Loxapine
LPG	Benzathine penicillin
L-Polamidon	Methadone
Lucofen SA (withdrawn)	Chlorphentermine
Lucostine	Lomustine
Ludiomil	Maprotiline
LUF-ISO	Isoprenaline
Lufyllin	Diprophylline
Lumbrioxyl	Piperazine
Luminal	Phenobarbitone
Lundbeck	Bleomycin
Lurselle	Probucol
Luvatren	Moperone
Luvatrena	Moperone
Lycanol	Glymidine
Lycoral	Chloral hydrate
Lyndiol 2.5	{ Lynoestrenol / Mestranol
Lynoral	Ethinyloestradiol
Lyogen	Fluphenazine
Lysodren	Mitotane
Lysthenon	Suxamethonium
M&B 693	Sulphapyridine
M&B 744 (obsolete)	Stilbamidine
M&B 800	Pentamidine isethionate
Macocyn	Oxytetracycline
Macrodantin	Nitrofurantoin
Macrodex	Dextran
Madar	{ Desmethyldiazepam / Nordiazepam
Madopar	{ Levodopa / Benserazide
Madribon	Sulphadimethoxine
Maeva	Temazepam
Malaquin	Chloroquine
Malarivan	Chloroquine
Malloral	Thioridazine
Malocide	Pyrimethamine
Malogen	Testosterone
Malogen LA	Testosterone enanthate
Malogex	Testosterone enanthate
Maloprim	{ Dapsone / Pyrimethamine
Mandalay	Hexamine mandelate
Mandaze	Hexamine mandelate
Mandelamine	Hexamine mandelate
Mandelurine	Hexamine mandelate
Mandokef	Cephamandole
Mandol	Cephamandole
Mandrax	{ Methaqualone / Diphenhydramine
Mansil	Oxamniquine
Mantadix	Amantadine
Maolate	Chlorphenesin carbamate
Marcain	Bupivacaine
Marcaina	Bupivacaine
Marcaine	Bupivacaine
Marcoumar	Phenprocoumon
Marcumar	Phenprocoumon

Proprietary name	Official name
Mardon	Dextropropoxyphene
Mareline	Amitriptyline
Marevan	Warfarin sodium
Marezine	Cyclizine
Marplan	Isocarboxazid
Marsilid	Iproniazid
Marsolone	Prednisolone
Marsone	Predisone
Marzine	Cyclizine
Masmoran	Hydroxyzine
Matulane	Procarbazine
Maxeran	Metoclopramide
Maximed	Protriptyline
Maxisporin	Cephradine
Maxolon	Metoclopramide
Maycor	Isosorbide dinitrate
Meanverin Ultra	Bupivacaine
Mebadin	Dehydroemetine
Mebaral	Methylphenobarbitone
Mebendacin	Mebendazole
Mecazine	Meclozine
Medihaler Ergotamine	Ergotamine
Medihaler Iso	Isoprenaline
Medihaler Iso Forte	Isoprenaline
Médio-Contrix	Sodium iothalamate
Medio-Contrix	Iothalamic acid
Medomet	Methyldopa
Medomin	Heptabarbitone
Médomine	Heptabarbitone
Medrate	Methylprednisolone
Medrocort	Medrysone
Medrol	Methylprednisolone
Medrone Veriderm	Methylprednisolone
Mefoxin	Cefoxitin sodium
Mefoxitin	Cefoxitin sodium
Megace	Megestrol
Megacef	Cephradine
Megacillin	{ Benzylpenicillin / Procaine penicillin
Megaclor	Clomocycline
Megamycine	Methacycline
Megapen	Procaine penicillin
Megaphen	Chlorpromazine
Megimide	Bemegride
Meladinine	Methoxsalen
Melanex	Metahexamide
Melfiat	Phendimetrazine
Melipramine	Imipramine
Melitase	Chlorpropamide
Mellaril	Thioridazine
Mellereten	Thioridazine
Melleretten	Thioridazine
Mellerettes	Thioridazine
Melleril (range)	Thioridazine
Mellitol	Tolbutamide
Melox	{ Aluminium hydroxide / Magnesium hydroxide
Meltrol	Phenformin
Menolet	{ Ethinyloestradiol / Methyltestosterone
Menolyn	Ethinyloestradiol
Meonine	Methionine

Proprietary name	Official name
Mepavlon	Meprobamate
Mephenamin	Orphenadrine
Mephine	Mephentermine
Mephyton	Phytomenadione
Mepilin	{ Ethinyloestradiol / Methyltestosterone }
Meprate	Meprobamate
Meprocompren	Meprobamate
Meprosa	Meprobamate
Mequelon	Methaqualone
Merbentul	Dicyclomine
Merbentyl	Chlorotrianisene
Merbentyl	Dicyclomine
Merital	Nomifensine
Merival	Nomifensine
Mersalyl	Mersalyl
Mervan	Alclofenac
Mesantoin	Methoin
Mesontoin (withdrawn)	Methoin
Mestinon	Pyridostigmine bromide
Metahydrin	Trichlormethiazide
Metalcaptase	Penicillamine
Metalex-P	Leptazol
Metandren	Methyltestosterone
Metaplexan	Mequitazine
Metaprel	Orciprenaline
Metestine	Methyltestosterone
Meth-Zine	Promethazine
Methalone	Methaqualone
Methandine	Hexamine mandelate
Methendalate	Hexamine mandelate
Methidate	Methylphenidate
Methisul	Sulphamethizole
Methnine	Methionine
Methotrexate	Methotrexate
Methrazone	Feprazone
Méthyl-Gag	Mitoguazone
Meticortelone	Prednisolone
Metiguanide	Metformin
Metilar	Paramethasone
Metilpen	Phenethicillin potassium
Metin	Methicillin
Metoclol	Metoclopramide
Metopirone	Metyrapone
Metrazol	Leptazol
Metrospan	Meprobamate
Metrulen	Mestranol-ethynodiol
Metrulen 50	{ Ethinyloestradiol / Ethynodiol }
Metrulen M	{ Mestranol / Ethynodiol }
Mevasine	Mecamylamine
Mexaform	Clioquinol
Mexitil	Mexiletine
Mexocine	Demeclocycline
Mezin	Sulphadimidine
MI 1727	Calvacin
In *Migravess*	Metoclopramide
Mi-Pilo	Pilocarpine
Mialex	Fenclozic acid
Micatin	Miconazole
Microcillin	Carbenicillin

Proprietary name	Official name
Microgynon 30	{ Ethinyloestradiol / Norgestrel }
Micronor	Norethisterone
Micronovum	Norethisterone
Mictrin	Hydrochlorothiazide
Midicel	Sulphamethoxypyridazine
Midone	Primidone
Midran	Aprotinin
Migraleve	{ Paracetamol } with or / Codeine } without / Docusate sodium } buclizine
Milafol	Folic acid
Milonorm	Meprobamate
Milontin (withdrawn)	Phensuximide
Miltaun	Meprobamate
Miltown	Meprobamate
Minalgin	Dipyrone
Mindiab	Glipizide
Minidiab	Glipizide
Minilyn	{ Ethinyloestradiol / Lynoestrenol }
Minims Neomycin Sulphate	Neomycin
Minims Phenylephrine Hydrochloride	Phenylephrine
Minima Sulphacetamide Sodium	Sulphacetamide sodium
Minipress	Prazosin
Minocin	Minocycline
Minodiab	Glipizide
Minomycin	Minocycline
Minoprostin $F_{2\alpha}$	Dinoprost trometamol
Minovlar	{ Ethinyloestradiol / Norethisterone- / acetate }
Minozinan	Methotrimeprazine
Minprostin	Dinoprostone
Minprostin E_2	Dinoprostone
Mintezol	Thiabendazole
Minzolum	Thiabendazole
Miostat	Carbachol
Miracil D	Lucanthone
Miradol	Sulpiride
Miral	Dexamethasone
Miramycine	Tetracycline
Mirapront	Phentermine
Mirbanil	Sulpiride
Mircol	Mequitazine
Mirfudorm	Carbromal
Mirvan	Alclofenac
Mistura C	Carbachol
Misulban	Busulphan
Mithracin	Mithramycin
Mitomycin C	Mitomycin
Mitomycine	Mitomycin
Mitoxanna	Ifosfamide
Mixogen	{ Ethinyloestradiol / Methyltestosterone }
MK 421	Enalapril maleate
Mobenol	Tolbutamide
Mobilan	Indomethacin
Mobutazon	Mofebutazone

Proprietary name	Official name
Modecate	Fluphenazine
Moditen	Fluphenazine
Moduretic	{ Amiloride { Hydrochlorothiazide
Moebiquin	Di-iodohydroxyquinoline
Mogadan	Nitrazepam
Mogadon	Nitrazepam
Molcer	Docusate sodium
Molipaxin	Trazodone
Mollax	Docusate sodium
Monazan	Mofebutazone
Monile	Methionine
Monistat	Miconazole
Monocortin	Paramethasone
Monotheamin	Theophylline monoethanolamine
Monotrim	Trimethoprim
Mopazine	Methoxypromazine
Moriperan	Metoclopramide
Moronal	Nystatin
Morpan BC	Benzalkonium chloride
Mosegor	Pizotifen
Motilium	Domperidone
Motrin	Ibuprofen
Movecil	Pyridinolcarbamate
Moxalactam	Latamoxef sodium
Moxam	Latamoxef sodium
M-P	Methapyrilene
Mucolysin	Tiopronin
Mustargen	Mustine hydrochloride
Mustine Hydrochloride	Mustine hydrochloride
Mutamycin	Mitomycin
Myacyne	Neomycin
Myambutol	Ethambutol
Myanesin	Mephenesin
Mycifradin	Neomycin
Mycin	Erythromycin
Mycivin	Lincomycin
Mycostatin	Nystatin
Mycostatine	Nystatin
Mylepsin	Primidone
Myleran	Busulphan
Mylodorm	Amylobarbitone
Mylosed	Amylobarbitone
Mynocine	Minocycline
Myobid	Papaverine
Myochrysine	Sodium aurothiomalate
Myocrisin	Sodium aurothiomalate
Myodil	Iophendylate injection
Mysoline	Primidone
Myspamol	Proquamezine
Mytelase	Ambenonium
Nabralin	Pentobarbitone
NaClex	Hydroflumethiazide
Nadisan (*obsolete*)	Carbutamide
Nagemid	Brompheniramine
Nalcrom	Sodium cromoglycate
Nalfon	Fenoprofen calcium
Nalgésic	Fenoprofen calcium
Nalonee	Naloxone
Naprosyn	Naproxen
Napsalgesic	{ Dextropropoxyphene { Aspirin

Proprietary name	Official name
Naqua	Trichlormethiazide
Narcan	Naloxone
Narcanti	Naloxone
Narcolo	Dextromoramide
Narcozep	Flunitrazepam
Nardelzine	Phenelzine
Nardil	Phenelzine
Nastenon	Oxymetholone
Natigoxin	Digoxin
Natigoxine	Digoxin
Natisédine	Quinidine phenylethylbarbiturate
Nativelle	Digoxin
Natrilix	Indapamide
Natulan	Procarbazine
Natulanar	Procarbazine
Naturetin	Bendrofluazide
Naturine	Bendrofluazide
Navane (withdrawn)	Thiothixene
Navidrex	Cyclopenthiazide
Navidrex-K	Cyclopenthiazide
Navidrix	Cyclopenthiazide
Nebcin	Tobramycin
Nece-Pen	Benzylpenicillin
Nefrolan	Clorexolone
Neggram	Nalidixic acid
Negram	Nalidixic acid
Nemactil	Pericyazine
Nembutal	Pentobarbitone
Neo-Atromid	Clofibrate
Neo-Banex	Propantheline
Neo-Dibetic	Tolbutamide
Neo-Epinine	Isoprenaline
Neo-Erycinium	Erythromycin
Neo-Erycinum	Erythromycin estolate
Neo-Flumen	Hydrochlorothiazide
Neo-Flucin	Griseofulvin
Neo-Hombreol	Testosterone
Neo-Hombreol F	Testosterone
Neo-Ilotycin	Erythromycin estolate
Neo-Mercazole	Carbenoxolone
Neo-Metic	Dimenhydrinate
Neo-Morphazole	Carbenoxolone
Neo-Morrhuol	Neomycin
Neo-NaClex	Bendrofluazide
Neo-NaClex-K	Bendrofluazide
Neo-Serp	Reserpine
Neo-Thyreostat	Carbenoxolone
Neo-Tran	Meprobamate
Neo-Tric	Metronidazole
Neobiotic	Neomycin
Neobram	Neomycin
Neocin	Neomycin
Neocontrast	Iopanoic acid
Neodorm	Pentobarbitone
Neodrine	Methylamphetamine
Neogest	Norgestrel
Neohombreol M	Methyltestosterone
Neomate	Neomycin
Neomin	Neomycin
Neophryn	Phenylephrine
Neoplatin	Cisplatin

Proprietary name	Official name
Ortho-Forms	Nonoxynol-9
Ortho-Novin 1/50	{ Mestranol { Norethisterone
Ortho-Novin 1/80	{ Mestranol { Norethisterone
Ortho-Novin 2 MG	{ Mestranol { Norethisterone
Orudis	Ketoprofen
Osmogits (withdrawn)	Indomethacin
Osmosin (withdrawn)	Indomethacin
Osnervan	Procyclidine
Ospen	Phenoxymethylpenicillin
Ospolot	Sulthiame
Ostelin	Calciferol
Ostoforte	Calciferol
Otobiotic	Neomycin
Ovanon	{ Lynoestrenol { Mestranol
Ovol	Dicyclomine
Ovran	{ Ethinyloestradiol { Norgestrel
Ovran 30	{ Ethinyloestradiol { Norgestrel
Ovranette	{ Ethinyloestradiol { Norgestrel
Ovulen IMG	{ Mestranol { Ethynodiol
Ovulen 50	{ Ethinyloestradiol { Ethynodiol
Ovysmen	{ Norethisterone { Ethinyloestradiol
Oxalid	Oxyphenbutazone
Oxamycin	Cycloserine
Oxlopar	Oxytetracycline
Oxsoralen	Methoxalen
Oxy-Kesso-Tetra	Oxytetracycline
Oxycycline	Oxytetracycline
Oxydon	Oxytetracycline
Oxymycin	Oxytetracycline
Oxypel	Piperazine
Pabestrol	Stilboestrol
Pabracort	Hydrocortisone
Pacatal	Pecazine
Pacitron	L-Tryptophan
Paclin G	Benzylpenicillin
Paclin VK	Phenoxymethylpenicillin
In *Palacos R*	Methyl methacrylate
Palfium (range)	Dextromoramide
Paludrine	Proguanil
Paludrinol	Proguanil
Pamisyl	Para-aminosalicylic acid
Pamol	Paracetamol
Panadol	Paracetamol
Panafcortelone	Prednisolone
Panasorb	Paracetamol
Panazid	Isoniazid
Panectyl	Trimeprazine
Panergon	Papaverine
Panheprin	Heparin
Panimycin	Dibekacin
Panjopaque	Iopanoic acid
Panmycin	Tetracycline

Proprietary name	Official name
Panoral	Cefaclor
Panta	Chloropyrilene citrate
Pantheline	Propantheline
Pantocain	Amethocaine
Pantocaine	Amethocaine
Pantolax	Suxamethonium
Pantopaque	Iophendylate injection
Panwarfin	Warfarin sodium
Pap-Kaps	Papaverine
Papalease	Papaverine
Paracet	Paracetamol
Paracort	Prednisone
Paracortol	Prednisolone
Paradione	Paramethadione
Paralgin	Paracetamol
Paralut	{ Ethinyloestradiol { Ethisterone
In *Paramax*	Metoclopamide
In *Paramethyl Methacrylate*	Methyl methacrylate
Parasal	Para-aminosalicylic acid
Paravermin	Piperazine
Parest	Methaqualone
Parfuran	Nitrofurantoin
Pargitan	Benzhexol
Parinase	Azepinamide
Parkemed	Mefenamic acid
Parlodel	Bromocriptine
Parmol	Paracetamol
Parnate	Tranlcypromine
Parstelin	{ Tranylcypromine { Trifluoperazine
Partocon	Oxytocin
Parvolex	N-Acetylcysteine
Pas Depress	Hydroxyzine
Paspertin	Metoclopramide
Pathocil	Dicloxacillin sodium
Pathomycin	Sisomicin
Pavabid	Papaverine
Pavacaps	Papaverine
Pavacen	Papaverine
Pavadel	Papaverine
Pavagrant	Papaverine
Pavulon	Pancuronium
Paxel	Diazepam
Pedameth	Methionine
Peganone	Ethotoin
Peltazon	Pentazocine
Penapar VK	Phenoxymethylpenicillin
Penbard	Pentobarbitone
Penbon	Pentobarbitone
Penbritin	Ampicillin
Penbrock	Ampicillin
Penevan	Benzylpenicillin
Penglobe	Bacampicillin
Péniciline	Ampicillin
Penicillin N	Adicillin
Penicillin VK	Phenoxymethylpenicillin
PenicilliN-V-Lilly	Phenoxymethylpenicillin
Penicillin-V-Potassium	Phenoxymethylpenicillin
Penidural	Benzathine penicillin
Peniset	Benzylpenicillin
Penistaph	Methicillin

Proprietary name	Official name
Penitardon	Buphenine
Pensig	Phenethicillin potassium
Pensol	Benzylpenicillin
Pental	Pentobarbitone
Pentamidine Isethionate	Pentamide isethionate
Pentanca	Pentobarbitone
Pentazine	Trifluoperazine
Penthonium	Pentamethonium
Penthrane	Methoxyflurane
Pentids	Benzylpenicillin
Pentobeta	Pentobarbitone
Pentogen	Pentobarbitone
Pentone	Pentobarbitone
Pentothal	Thiopentone sodium
Pentrexyl	Ampicillin
Peperzinal	Piperazine
Pepsilphen	Sulphasomidine
Perandren	Methyltestosterone
Perazil	Chlorcyclizine
Percorten	Deoxycortone
Percutacrine	Thyroxine sodium
Percutol	Glyceryl trinitrate
Perfudex	Dextran
Pergestron	Hydroxyprogesterone hexanoate
In *Pergoids*	Phenolphthalein
Periactin	Cyproheptadine
Periactinol	Cyproheptadine
Perideca	Cyproheptadine
Permapen	Benzathine penicillin
Permitil	Fluphenazine
Peroidin	Potassium perchlorate
Perolysen	Pempidine
Pertofran	Desipramine
Pertofrane	Desipramine
Pervadil	Buphenine
Pervitin	Methylamphetamine
Petab	Pentobarbitone
Peteha	Prothionamide
Pethidine Roche	Pethidine
Pethoid	Pethidine
Petinimid	Ethosuximide
Petinutin	Methsuximide
Petnidan	Ethosuximide
Pevidine (range)	Povidone-iodine
Pexaqualone	Methaqualone
Pexid	Perhexiline maleate
Pfizerpen G	Benzylpenicillin
Pfizerpen VK	Phenoxymethylpenicillin
Pfizerpen-AS	Procaine penicillin
Pharlon	Hydroxyprogesterone hexandate
Pharmacillin	Benzylpenicillin
Phasal	Lithium carbonate
Phemerol Chloride	Benzethonium chloride
Phenaemal	Phenobarbitone
Phenaemaletten	Phenobarbitone
Phenased	Phenobarbitone
Phenbutazol	Phenylbutazone
Phendex	Paracetamol
Phenephrin	Phenylephrine
Phenergan	Promethazine

Proprietary name	Official name
Phenhydan	Phenytoin sodium
Phenobarbitone Spansule	Phenobarbitone
Phentoin	Phenytoin sodium
Phenybute	Phenylbutazone
Phenyl Idium	Phenazopyridine
Phetadex	Dexamphetamine
Phiso-Hex	Hexachlorophane
Phiso-Med	Hexachlorophane
Phlogase	Oxyphenbutazone
Phospholine Iodine	Ecothiopate iodide
Phthazol	Phthalysulphathiazole
Phyldrox Suppositories	Aminophylline
Phyllocontin Tablets	Aminophylline
Phyone	Growth hormone
Physeptone	Methadone
Physpan	Theophylline
Phytadon	Pethidine
Pilocar	Pilocarpine
Pilopt	Pilocarpine
Pimal	Meprobamate
Pindione	Phenindione
Piperol Forte	Piperazine
Piportil L4	Pipothiazine
Pipracil	Piperacillin sodium
Pipril	Piperacillin sodium
Piprosan	Piperazine
Piranex	Chlorpheniramine
Piriton	Chlorpheniramine maleate
Pitocin	Oxytocin
Pitressin	Vasopressin
PK-Merz	Amantadine
Placidyl	Ethchlorvynol
Planor	Norgestrienone
Planovin	{ Megestrol / Ethinyloestradiol
Plaquenil	Hydroxychloroquine
Plasil	Metoclopramide
Plasmochin	Pamaquin
Plasmoquine (withdrawn)	Pamaquin
Platinex	Cisplatin
Platinol	Cisplatin
Plégicil	Acepromazine
Plegine	Phendimetrazine
Pluryl	Bendrofluazide
PO12 Pommade	Enoxolone
Policydal	Sulfametopyrazine
Polidexide	Dextran
In *Polycide*	Benzalkonium chloride
Polycycline	Tetracycline
Polymox	Amoxycillin
Ponderal	Fenfluramine
Pondéral	Fenfluramine
Ponderax	Fenfluramine
Pondimin	Fenfluramine
Ponstan	Mefenamic acid
Ponstel	Mefenamic acid
Ponstyl	Mefenamic acid
Postafen	Meclozine
Pracarbamin	Urethane
Praecirheumin	Phenylbutazone
Pramidex	Tolbutamide
Praminil	Imipramine

Proprietary name	Official name
Psychostyl	Nortriptyline
Psyquil	Fluopromazine
Pularin	Heparin
Pulmadil	Rimiterol
Pulmicort	Budesonide
Puri-Nethol	Mercaptopurine
Purodigin	Digitoxin
Purophyllin	Proxyphylline
PV Carpine	Pilocarpine
Pydox	Pyridoxine
Pyelokon-R	Sodium acetrizoate
Pyknolepsinum	Ethosuximide
Pyodental	Sulphanilamide
Pyopen	Carbenicillin
Pyoredol	Phenytoin sodium
Pyrafat	Pyrazinamide
Pyramidon	Amidopyrine
Pyrazodine	Phenazopyridine
Pyribenzamine	Tripelennamine
Pyridacil	Phenazopyridine
Pyridiate	Phenazopyridine
Pyridium	Phenazopyridine
Pyroxin	Pyridoxine
Pyrteen	Methapyrilene
Qid Bamate	Meprobamate
Qidpen G	Benzylpenicillin
Qidtet	Tetracycline
Quaalude	Methaqualone
Quadnite	Promethazine
Quantalan	Cholestyramine
Quatrax	Tetracycline
Quelicin	Suxamethonium
Quellada Application	Gamma benzene hexachloride
Quellada Lotion	Gamma benzene hexachloride
Quensyl	Hydroxychloroquine
Questran	Cholestyramine
Quietal	Meprobamate
Quilonum	Lithium carbonate
Quinacrine	Mepacrine
Quinaglute	Quinidine gluconate
Quinaltone	Quinalbarbitone
Quinate	Quinine bisulphate or sulphate
Quinbar	Quinalbarbitone
Quinbisan	Quinine bisulphate or sulphate
Quinclor	Halquinol
Quinicardine	Quinidine bisulphate or sulphate
Quinidex	Quinidine bisulphate or sulphate
Quinora	Quinidine bisulphate or sulphate
Quinsan	Quinine bisulphate or sulphate
Quitaxon	Doxepin
Quixalin	Halquinol
Quixalud	Halquinol
R 818	Flecainide acetate
Racenicol	Thiamphenicol

Proprietary name	Official name
Radiosélectan	Sodium diatrizoate
Radiostol	Calciferol
Rapidax	Docusate sodium
Rastinon	Tolbutamide
Rau-Sed	Reserpine
Raudixin	Rauwolfia serpentina
Raufonol	Rauwolfia serpentina
Raupina	Rauwolfia serpentina
Raupinetten	Rauwolfia serpentina
Raurine	Reserpine
Rautabs	Rauwolfia serpentina
Rautensin	Rauwolfia serpentina
Rauval	Rauwolfia serpentina
Ravenil	Pecazine
Rawlina	Rauwolfia serpentina
Razoxin	Razoxane
Rebugen	Ibuprofen
Rectules	Chloral hydrate
Redeptin	Fluspirilene
Redul	Glymidine
Refobacin	Gentamicin
Regelan	Clofibrate
Regelan N	Clofibrate
Regenon Retard	Diethylpropion
Regitin	Phentolamine
Regitine	Phentolamine
Reglan	Metoclopramide
Regonol	Pyridostigmine bromide
Regulex	Docusate sodium
Rehibin	Cyclofenil
Rela	Carisoprodol
Relaxil	Chlordiazepoxide
Relutin	Hydroxyprogesterone hexanoate
Remnos	Nitrazepam
Remsed	Promethazine
Renese	Polythiazide
Reno-M	Sodium diatrizoate
Renografin	Sodium diatrizoate
Renovist	Sodium diatrizoate
Repeltin	Trimeprazine
Resedrex	Reserpine
Resercrine	Reserpine
Reserpanca	Reserpine
Reserpine Dellipsoids D29	Reserpine
Reserpoid	Reserpine
Resochin	Chloroquine
Restenil	Meprobamate
Restoril	Temazepam
Resulfon	Sulphaguanidine
Retcin	Erythromycin
Retin A (range)	Tretinoin
Reverin	Rolitetracycline
Revonal	Methaqualone
Rezipas	Para-aminosalicylic acid
R-Gene	Arginine
Rheomacrodex	Dextran
Rheumacin	Indomethacin
Rheumox	Azapropazone
Rhex	Mephenesin
Riddovydrin Capsules	Aminophylline
Rifa	Rifampicin

Proprietary name	Official name	Proprietary name	Official name
Rifadin	Rifampicin	*Sanomigran*	Pizotifen
Rigelon	Thiamphenicol	*Sanorex*	Mazindol
Rimactane	Rifampicin	*Sansert*	Methysergide
Rimifon	Isoniazid	*Sarcomycin*	Sarkomycin
Rincrol	Thiamphenicol	*Saridone* (withdrawn)	Prophenazone
Rinlaxer	Chlorphenesin carbamate	*Sarnacaine*	Lignocaine
Ripercol	Tetramisole	*Saroten*	Amitriptyline
Risordan	Isosorbide dinitrate	*Sarpagan*	Rauwolfia serpentina
Riston	Ristocetin	*Sas-500*	Sulphasalazine
Ritalin	Methylphenidate	*Saventrine*	Isoprenaline
Ritmos Elle	Lorajmine hydrochloride	*Savitol*	Calciferol
Ritromin	Erythromycin estolate	*Scandicain*	Mepivacaine
Rivadescin	Rauwolfia serpentina	*SCB*	Quinalbarbitone
Rivotril	Clonazepam	*SCH-13475*	Sisomycin
RO 070582	Misonidazole	*Scherisolon*	Prednisolone
RO 84650	Diclofensine	*Scherofluron*	Fludrocortisone
RO-A-Vit	Vitamin A	*Scheroson F*	Hydrocortisone
Roaccutane	Isotretinoin	*Schiwanox*	Amylobarbitone
Robimycin	Erythromycin	*Schokolax*	Oxyphenisatin
Robitet	Tetracycline	*Scoline*	Suxamethonium
Rocaltrol	Calcitriol	*S-Dimidine*	Sulphadimidine
Roccal	Benzalkonium chloride	*SDM*	Sulphamethoxypyridazine
Rogitine	Phentolamine	*Seatone*	Green-lipped muscle extract
Rohipnol	Flunitrazepam	In *Secadrex*	Acebutolol
Rohypnol	Flunitrazepam	*Secaps*	Quinalbarbitone
Roipnol	Flunitrazepam	*Secholex (withdrawn)*	Polidexide
Rondomycin	Methacycline	*Secocaps*	Quinalbarbitone
Roniacol	Nicotinyl tartrate	*Secogen*	Quinalbarbitone
Ronicol	Nicotinyl tartrate	*Seconal Sodium*	Quinalbarbitone
Ropredlone	Prednisolone	*Securopen*	Azlocillin sodium
Rorasul	Sulphasalazine	*Seda-Tablinen*	Phenobarbitone
Rougaxin	Digoxin	*Sedal Sodium*	Amylobarbitone sodium
Rouqualone	Methaqualone	*Sedalande*	Fluanisone
Rovamycin	Spiramycin	*Sedantoinal*	Methoin
Rubrine	Cyanocobalamin	*Sedapam*	Diazepam
Rufocromomycine	Rufocromomycin	*Sedaraupin*	Reserpine
Rufol	Sulphamethizole	*Sedolatam*	Prenylamine
Ruvite	Cyanocobalamin	*Sedonan*	Phenazone
Rynacrom	Sodium cromoglycate	*Sedormid*	Apronal
Ryser	Reserpine	*Sefril*	Cephradine
Rythmodan	Disopyramide	*Segontin*	Prenylamine
Rytmonorm	Propafenone	*Selcryn* (withdrawn)	Ticrynafen
Salazopyrin	Sulphasalazine	*Selectomycin*	Spiramycin
Salazopyrine	Sulphasalazine	*Selokeen*	Metoprolol
Salipran	Benorylate	*Seloken*	Metoprolol
Salisulf	Sulphasalazine	*Selora*	Potassium chloride
Salpix	Sodium acetrizoate	*Semap*	Penfluridol
Saluric	Chlorothiazide	*Sembrina*	Methyldopa
Saluron	Hydroflumethiazide	*Semikon*	Methapyrilene
Salzone	Paracetamol	In *Septopal Chains*	Methyl methacrylate
Sanamycin (withdrawn)	Actinomycin C	*Septra*	Co-trimoxazole
Sanatrichom	Metronidazole	*Septrin*	{ Sulphamethoxazole
Sancos Co	Chlorpheniramine maleate		{ Trimethoprim
Sandimmun	Cyclosporin A	*Septrin*	Co-triamoxazole
Sando-K	Potassium chloride	*Sequestrene AA*	Edetic acid
Sandomigran	Pizotifen	*Sequestrene Na2*	Disodium edetate
Sandomigrin	Pizotifen	*Sequestrene Na2Ca*	Sodium calcium edetate
Sandosten	Thenalidine	*Sequestrene Na3*	Trisodium edetate
Sandril	Reserpine	*Serax*	Oxazepam
Sanmigran	Pizotifen	*Serazone*	Chlorpromazine
Sanobrite	Pentachlorophenol	*Serenace*	Haloperidol
Sanoma	Carisoprodol	*Serenack*	Diazepam

Proprietary name	Official name
Tosmilen	Demecarium
Totacillin	Ampicillin
Totapen	Ampicillin
Totomycin	Tetracycline
Toxichlor	Chlordane
Trabest	Clemastine
Tracrium	Atracurium
Trancopal	Chlormezanone
Trandate	Labetalol
Transcycline	Rolitetracycline
Transiderm-Nitro	Glyceryl trinitrate
Trantoin	Nitrofurantoin
Tranxene	Clorazepate
Tranxilen	Clorazepate
Tranxilium	Clorazepate
Trapanal	Thiopentone sodium
Trasicor	Oxprenolol
Trasylol	Aprotinin
Travamine	Dimenhydrinate
Trecator	Ethionamide
Trelmar	Meprobamate
Tremin	Benzhexol
Trental	Oxypentifylline
Tresanil	Tritiozine
Trescatyl	Ethionamide
Trevintix	Prothionamide
Triacet	Triacetyloleandomycin
Triador	Methaqualone
Triamalone	Triamcinolone acetonide
Triamcin	Triamcinolone acetonide
Trianel	Triparanol
Triazure	Azaribine
Trib	Co-trimoxazole
Trichazol	Metronidazole
Tricloryl	Triclofos sodium
Triclos	Triclofos sodium
Tricofuron	Furazolidone
Tridil	Glyceryl trinitrate
Tridione	Troxidone
Trifluoper-Ez-Ets	Trifluoperazine
Triflurin	Trifluoperazine
Triherpine	Trifluridine
Trihexy	Benzhexol
Triiodothyronine Injection	Liothyronine sodium
Trikacide	Metronidazole
Trikamon	Metronidazole
Trilafon	Perphenazine
Trilene	Trichlorethylene
Trilifan	Perphenazine
Triludan	Terfenadine
Trimedone	Troxidone
Trimopan	Trimethoprim
Trimstat	Phendimetrazine
Trinuride	Pheneturide / Phenytoin / Phenobarbitone
Triocos	Chlorpheniramine maleate
In *Triominic*	Mepyramine
In *Triotussic*	Mepyramine
Triparin (withdrawn)	Triparanol
Triperidol	Trifluperidol
Triptil	Protriptyline

Proprietary name	Official name
Trisoralen	Trioxsalen
Triten	Dimethindene maleate
Trithyrone	Liothyronine sodium
Trittico	Trazodone
Trixyl	Benzhexol
Trofan	L-Tryprophan
Trolovol	Penicillamine
Tromexan	Ethylbiscoumacetate
Tropium	Chlordiazepoxide
Tropium	Chlordiazepoxide
Truxal	Chlorprothixene
Truxaletten	Chlorprothixene
Tryptacin	L-Tryptophan
Tryptanol	Amitriptyline
Tryptizol	Amitriptyline
Tualone	Methaqualone
Tubarine	Tubocurarine
Tubocurarine Chloride	Tubocurarine
Tuinal	Quinalbarbitone
Tussaphed	Pseudoephedrine
Tylciprine	Tranylcypromine
Tylenol	Paracetamol
Tzalol	Aprotinin
U-9889	Streptozocin
Ugurol	Tranexamic acid
UK-655801	Tolamolol
Ulacort	Prednisolone
Ulcerban	Sucralfate
Ulcerlimin	Sucralfate
Ulmenid	Chenodeoxycholic acid
Ulsanic	Sucralfate
Ultracef	Cefadroxil
Ultracillin	Ciclacillin
Ultracorten	Prednisone
Ultradine	Povidone-iodine
Ultramycin	Minocycline
Ultranden	Fluoxymesterone
Ultromix	Cefuroxime
Umbradil	Diodone
Un-Alfa	Alfacalcidol
Unicontin	Theophylline
Unihepa (withdrawn)	Methionine
Unimycin	Oxytetracycline
Unipen	Nafcillin
Uniphyllin	Theophylline
Uracid	Methionine
Uracil Mustard	Uramustine
Uranap	Methionine
Urbanol	Clobazam
Urbanyl	Clobazam
Urbason	Methylprednisolone
Urbilat	Meprobamate
Urecid	Probenecid
Urekene	Sodium valproate
Urex	Hexamine hippurate
Urid	Chlorthalidone
Uridon	Chlorthalidone
Uritet	Oxytetracycline
Uritone	Hexamine
Uro-Tablinen	Nitrofurantoin
Urodantin	Nitrofurantoin
Urogan	Sulphafurazole

Proprietary name	Official name
Urografin	Sodium diatrizoate
Urolex	Sulphamethizole
Urolin	Chlorthalidone
Urolong	Nitrofurantoin
Urolucosil	Sulphamethizole
In Uromide	Phenazopyridine hydrochloride
Uropyridin	Phenazopyridine
Urosin	Allopurinol
Urovison	{ Sodium diatrizoate / Meglumine diatrizoate
Urovist	Sodium diatrizoate
Uroz	Sulphamethizole
Urtilone	Prednisone
US-67	Sulphafurazole
U-Tet	Tetracycline
Utopar	Ritodrine
Utrasul	Sulphamethizole
Uvilon	Piperazine
VAAM-DHQ	Di-iodohydroxyquinoline
Valamin	Ethinamate
Valisone	Betamethasone valerate
Valium	Diazepam
Vallergan	Trimeprazine
Valmid	Ethinamate
Valoid	Cyclizine
Valsera	Flunitrazepam
Vancil	Oxamniquine
Vancocin	Vancomycin
Vansil	Oxamniquine
Vantoc CL	Benzalkonium chloride
Vapo-N-ISO	Isoprenaline
Varipaque	Oxyphenisatin
Vasal	Papaverine
Vasapril	Pyridinolcarbamate
Vascardin	Isosorbide dinitrate
Vascoray	Iothalamic acid
Vasocil	Pyridinolcarbamate
Vasocon-A	Antazoline
Vasodilan	Isoxsuprine
Vasopan	Papaverine
Vasoplex	Isoxsuprine
Vasotherm	Nicotinic acid
Vatensol	Guanoclor sulphate
V-CIL-K	Phenoxymethylpenicillin
Vectrin	Minocycline
Vegolysen (withdrawn)	Hexamethonium
Vehem	Teniposide
Velacycline	Rolitetracycline
Velban	Vinblastine
Velbe	Vinblastine
Veldopa	Levodopa
Velocef	Cephradine
Velosef	Cephradine
Veltane	Brompheniramine
Venacil	Ancrod
Vendarcin	Oxytetracycline
Ventolin	Salbutamol
Vepesid	Etoposide
Veracillin	Dicloxacillin sodium
Veractil	Methotrimeprazine

Proprietary name	Official name
Veranterol	Pyridinolcarbamate
Verdiana	Deslanoside
Veritab	Meclozine
Vermicompren	Piperazine
Vermolina	Piperazine
Vermox	Mebendazole
Versene Acid	Edetic acid
Vertigon Spansule	Prochlorperazine
Vertizine	Meclozine
Vesprin	Fluopromazine
Vetomazin	Methoxypromazine
Vetren	Heparin
Viaductor	Lorajmine hydrochloride
Vibra-Tabs	Doxycycline hydrochloride
Vibramycin	Doxycycline hydrochloride
Vibravenös	Doxycycline hydrochloride
Vicilan	Viloxazine
Vicin	Phenoxymethylpenicillin
Vi-dom-A	Vitamin A
Vidopen	Ampixillin
Vigigan	Mequitazine
Vikastab	Potassium menaphthosulphate
Vilexin	Phenyramidol
Vimicon	Cyprohepradine
Vinco-Abführperlen	Oxyphenisatin
Vincol	Tiopronin
Vio-Bamate	Meprobamate
Vio-Serpine	Reserpine
Viocin	Viomycin
Vioform	Clioquinol
Viomycin Sulphate (withdrawn)	Viomycin
Vira-A	Vidarabine
Viroptin	Trifluridine
Virormone Oral	Methyltestosterone
Virunguent	Idoxuridine
Visken	Pindolol
Visorsan D3	Cholecalciferol
Vistamycin	Ribostamycin
Vistaril	Hydroxyzine
Vivactil	Protriptyline
Vivalan	Viloxazine
Vivol	Diazepam
Vogan	Vitamin A
Volidan 21	{ Ethinyloestradiol / Megestrol
Voltarol	Diclofenac
Vomex A	Dimenhydrinate
Vomital	Dimenhydrinate
Vumon	Teniposide
Wamocap	Nicotinic Acid
Waran	Warfarin sodium
Warfarin Sodium	Warfarin sodium
Warfilone	Warfarin sodium
Warnerin	Warfarin sodium
Waxsol	Docusate sodium
We-941-BS	Brotizolam
Weedazol (herbicide)	Aminotriazole
Wellconal	Dipipanone
Welldorm	Dichloralphenazone
Wescomep	Meprobamate

Index

reserpine 385
thioridazine 386
thioxanthenes 386
disruption of hypothalamo-pituitary
    link 382
suppression by ergot preparations
    507
oral contraceptives (breast
    milk) 500
Lactation suppressors, inhibition of
    prolactin secretion 392
Lactic acidosis
associated with
    biguanides 336–8
    diazepam 339
    dithiazanine 339
    fructose 339
    metformin 337–8
    methanol 339
    phenformin 336–8
    terbutaline 339
    xylitol 338–9
implication of tetracyclines 338
Lactotropes, drug effects 389
    stimulation of oestrogen 390
Lactulose, causing hypernatraemia 351
Laetrile
causing
    agranulocytosis 907
    animal teratogenicity 905–6
    blood dyscrasias 148
    cyanide poisoning 4, 905–6
    neuromyopathy 566
    prussic-acid poisoning 4
    non-efficacy in cancer 907
Law commissions 118–19
Laxatives, effect on intestinal absorp-
    tion 243
Lead poisoning
    associated with porphyria 75
    herbal remedies 911
Lecithin, effect on tardive dyskinesia
    629
Legislation, teratogenic potential 402
Lenticular opacities
associated with
    anticholinesterase drugs 682
    antimalarial drugs 687
    busulphan 694
    dimethylsulphoxide 5, 698
    phenothiazines 701–2
    see also Corneal . . .; Ocular . . .
Lergotrile, induction of lowered serum
    prolactin levels 392
Lethargy, associated with
    benzodiazepines (breast milk) 503
    interferon 639
    phenothiazines (breast milk) 503–4
    reserpine (breast milk) 497
Leucocyte transfusion, causing ARDS
    213
Leucopheresis, associated with skin
    disease 774
Leucotomy, tardive dyskinesia 624
Leukaemia
associated with
    alkylating agents 374, 821

ethylene oxide 818
Hodgkin's disease therapy 821
immunosuppressant therapy 823–4
lomustine 822
razoxane treatment 825–6
    (table) 825
therapeutic radiation 842–4
childhood
    associated with combination
        chemotherapy 374
Levamisole, causing
    blood dyscrasias 141–2
    lupus erythematosus syndrome 88
Levodopa
affecting male libido 520
alteration of taste sensation 224
associated with
    dementia 654
    depression 661
    elation (mania) 658
    exacerbation of schizophrenia 657
    haemolytic anaemia 85, 155
    hypokalaemia 352
    involuntary movements 617–18
    lupus erythematosus
        syndrome 535–5
    malignant melanoma 820
    oral dyskinesia 625
    post-menopausal vaginal
        bleeding 372
    psychotic states 654
    pyridoxine deficiency 343
interactions with other drugs
    (table) 942
interference with uric acid test 959
    creatine test 956
    ketone urine test 960
oculotoxicity 688
treatment for tardive dyskinesia 628
Levothyroxine affecting temperature
    regulation 783
Leydig-cell dysfunction, association
    with
    combination chemotherapy 374, 375
    seminiferous tubule dysfunction 388
Liability
    civil, Royal Commission 118, 119,
        120
    legal 117–23
    medical practitioners 118
    strict 118–19, 120–1, 122
Liberty cap mushroom, adverse effects
    904
Libido, female
    drugs affecting 522–4
    effect of cytotoxic-induced ovarian
        failure 522
Libido, male
    drugs affecting 519–20
    loss in glaucoma 517
Lichenoid drug eruptions, drugs
    causing 761–2
Lidocaine, see Lignocaine
Lignocaine
    affecting sperm motility 520–1

allergic reactions to 752
causing
    adult respiratory distress
        syndrome 212
    anaphylaxis 83
    convulsions 599
    dermatitis 758
    lupus erythematosus syndrome 88
    interaction with amiodarone 190
    ototoxicity 740, (table) 713
Limb defects caused by maternal
    antithyroid agents 463
Limb reduction deformities
    associated with maternal contra-
        ceptive use 461
    vaginal spermicides 464–7
Limbs, symptoms caused by
    amitriptyline 569
Lincomycin
associated with
    colitis 269, 858–9
    pruritis 768
    breast milk, infant toxicity 486
    use in pregnancy 431
Lipid metabolism disorders, caused by
    β-adrenergic blocking drugs 339–40
    clofibrate 340
    combination oestrogen–progestogen
        products 340–1
    medroxyprogesterone 341
    metenolone 341
    norethisterone 341–2
Lipidosis, drug-induced
    amiodarone 531, 537
    carnitine 531
    chloroquine 532–3
    emetine 531
Lipids
    effect of β-adrenergic blockers on
        plasma levels 186
    malabsorption 243, 244, 246, 259
Lipoatrophy, associated with insulin
    342
Lipogranuloma, associated with
    chlorpromazine 385
Liquid paraffin, causing mal-
    absorption of lipids 243
Liquorice, causing
    hypokalaemia 907
    impotence 516
    salt and water retention 907
Literature reports, monitoring ADRs
    93
Lithium
    aggravation of myasthenia
        gravis 546
    alteration of taste sensation 224
associated with
    acne 755
    alopecia 760
    breast milk, infant toxicity 504
    diabetes insipidus 350, 370
    dyscrasias 143
    encephalopathy 603
    glomerulonephritis 794